DATE DUE

Handbook of Electrical Tables and Design Criteria

Handbook of Electrical Tables and Design Criteria

V. F. Christoffer

McGraw-Hill
New York San Francisco Washington, D.C. Auckland Bogotá
Caracas Lisbon London Madrid Mexico City Milan
Montreal New Delhi San Juan Singapore
Sydney Tokyo Toronto

Library of Congress Cataloging-in-Publication Data

Christoffer, V. F.
Handbook of electrical tables and design criteria / V.F. Christoffer.
p. cm.
ISBN 0-07-913722-9
1. Electric engineering—Tables. I. Title.
TK151.C468 1998
621.3′02′1—dc21 98-10517
CIP

McGraw-Hill

A Division of The **McGraw-Hill** *Companies*

1 2 3 4 5 6 7 8 9 0 KGP/KGP 9 0 3 2 1 0 9 8

P/N 012050-1
PART OF
ISBN 0-07-913722-9

The sponsoring editor for this book was Larry Hager, the editing supervisor was Ruth Mannino, and the production supervisor was Tina Cameron. The book designer was Nicholas A. Bernini.

Printed and bound by Quebecor/Kingsport.

This book is printed on recycled, acid-free paper containing a minimum of 50% recycled, de-inked fiber.

McGraw-Hill books are available at special quantity discounts to use as premiums and sales promotions, or for use in corporate training programs. For more information, please write to the Director of Special Sales, McGraw-Hill, 11 West 19th Street, New York, N.Y. 10011. Or contact your local bookstore.

Information contained in this work has been obtained by The McGraw-Hill Companies, Inc. ("McGraw-Hill") from sources believed to be reliable. However, neither McGraw-Hill nor its authors guarantee the accuracy or completeness of any information published herein and neither McGraw-Hill nor its authors shall be responsible for any errors, omissions, or damages arising out of use of this information. This work is published with the understanding that McGraw-Hill and its authors are supplying information but are not attempting to render engineering or other professional services. If such services are required, the assistance of an appropriate professional should be sought.

Contents

Preface

The purpose of this book is to provide the electrical professional a quick and easy way to design electrical systems with reasonable accuracy. It was not intended as a learning tool for the novice electrical designer or engineer. Only persons experienced in the design of electrical systems who can make sound judgments of its calculations and results should use this text.

Some of the information contained in tables is extrapolated to give the user data on which to base calculations when the data is unavailable in other texts. The 700, 800, and 900 MCM conductor data, for example, has been extrapolated because the code does not give the impedance values for these conductors. Even though these wire sizes are given throughout the code, as in the ampacity tables, they are inconsistently left out in the impedance tables.

The information in this book leans toward a "worse case" scenario, and is meant to give a "conservative" calculation. As an example, the conductors shown in all the tables are for stranded conductors only, and in the case of DC voltage drop the results will be slightly higher for stranded conductors than for solid conductors.

Copyright dates shown on the following pages are for the original work as submitted to the Copyright Office, which also applies to updates to the original work. The information contained herein is based on information contained in the latest edition of the NEC ®.

The fault-current tables and calculations were based on data from Bussman Division, Cooper Industries and their "Electrical Protection Handbook" or "SPD" as it has been called in the past. They can be reached by mail: Bussman Division, P.O. Box 14460, St. Louis, MO 63178-4460; Phone: (314) 527-3877; EMAIL; http://www.bussman.com with questions about their products and information bulletins.

Handbook of Electrical Tables and Design Criteria

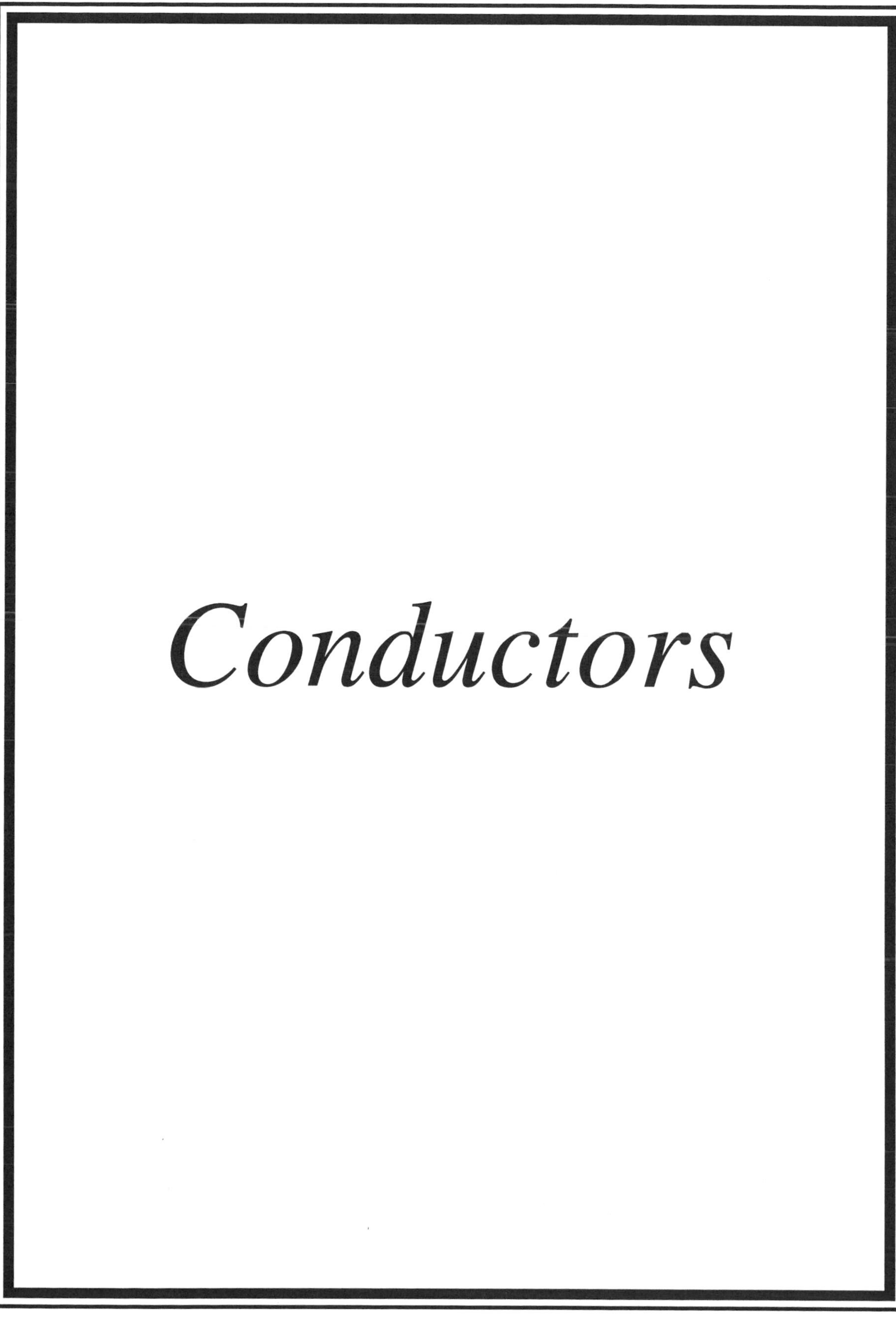

Conductors

Conductor Tables
and
Voltage-Drop Data

Example Voltage-Drop Calculations

The Christoffer Tables (tm) were designed in a unique fashion to give you a quick and easy way to do your calculations simply by reviewing the tables and using a four function calculator. In some cases you don't even need to use a calculator, all you have to do is look at the appropriate table and copy the answer. To get a precise answer or get the actual volts drop, for example, you will only have to do simple multiplication and division. That's it! Remember, there are **two** types of tables used for conductor calculations that are explained below. The following examples will give you enough direction in these calculations to do almost any voltage-drop study.

1 - "Voltage Drop Factors per 1000 Ampere Feet." This first type of table, shown in example #1, gives voltage drop **factors** for three-phase, single-phase line to line and single-phase line to neutral circuits, each on separate pages. These tables are extremely flexible and allow you to calculate the voltage drop for virtually any circuit through 600 volts. They are also categorized by the type of conduit and whether the conductors used are copper or aluminum.

Example #1: A three phase, 240 volt, 40 amp load at 80% power factor; #6 copper installed in 180 feet of galvanized rigid conduit. What is the actual drop in volts? What is the actual drop in percent? If the distance increases to 350 feet, what size wire must be used to maintain a maximum 3% drop?

Step One: Look in the table of contents for "Voltage Drop Factors - Steel Conduit - Copper - Three Phase", page 10.

Step Two: Find the voltage drop factor in the table where #6 in the left-hand column intersects the .80 power factor column.

Step Three: Multiply the actual load amps (40) by the distance (180) and then by the voltage drop factor (.745) to get the ampere feet. Divide the result by the number of runs and then divide by 1000 to get the actual volts drop.

Step Four: To get the actual percent drop, divide the actual volts drop by the L-L volts and multiply by 100.

Step Five: To calculate the minimum size wire required for a maximum 3% percent drop (max. drop factor), multiply the L-L volts times the maximum % drop. Then multiply the distance by the load amps and divide by 1000 for the ampere feet. Next, divide the maximum volts drop by the ampere-feet and then multiply by the number of runs. Choose a number in the .80 column that is less than or equal to the result. This corresponds to the minimum wire size in the left column.

Actual Volts Drop = Distance x Load Amps x Factor / Runs / 1000 =
180 x 40 x .745 / 1 / 1000 = **5.364** volts drop

Actual Percent Drop = Volts Drop / Voltage L-L x 100 =
5.364 / 240 x 100 = **2.235** % drop

Max. Drop Factor = Max. Volts Drop / (Distance x Load Amps / 1000) x Runs =
(.03 x 240) / (350 x 40 / 1000) x 1 = **.5143 = # 4 wire (.492)**

2 -"Circuit Design Data for Copper (or AL) Conductors." The second type of table, shown in example #2, includes voltage drop data for three-phase, single-phase line to line and single-phase line to neutral circuits also, but only shows data for circuits at 85% power factor and one run. This is the power factor chosen by the NEC ® as typical for circuits in the "AC Resistance and Reactance" table and one that seems reasonable for most circuit calculations. This is especially important when doing circuit design estimates where the actual power factor information is unavailable. This second type of table also has other useful information for designers like the number of THWN conductors in a conduit, breaker and fuse information, and the area in square inches for each size of conductor. Each table, like the first type, is distinguished by the type of conduit and whether the conductors used are copper or aluminum.

Example #2: A three phase, 480 volt, 52 amp load at approximately 85% power factor with #6 copper installed in 180 feet of galvanized rigid conduit. What is the actual drop in percent? What is the actual drop in volts?

Step One: Since the power factor is 85%, look in the table of contents for "Conductor Design-Data for Copper Conductors (Galv. Rigid)", page 6.

Step Two: Find where #6 in the left-hand column intersects the 480 volts three phase "Maximum Distance.." column on the right.

Step Three: To get the actual percent drop, divide the actual distance (180 ft.) by the maximum distance in the table (118 ft.).

Step Four: To get the actual volts drop, divide the result in step three by 100 and multiply by the L-L volts.

Actual Percent Drop = Actual Distance / Maximum Distance =
180 / 118 = **1.53** % drop

Actual Volts Drop = (Actual Distance / Maximum Distance) / 100 x L-L Volts =
1.53 / 100 x 480 = **7.34** volts drop

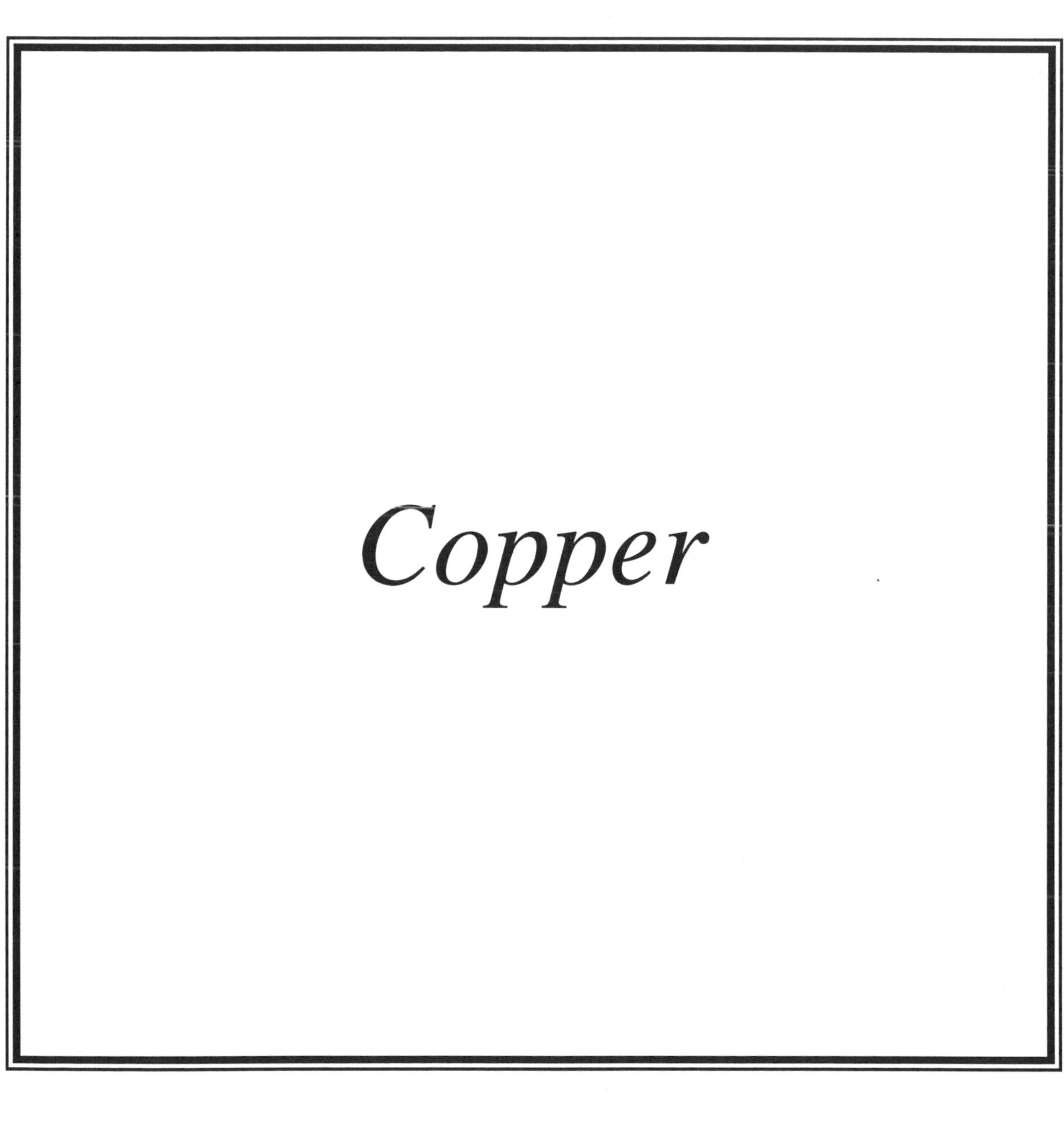
Copper

Circuit Design-Data for Copper Conductors

60 Hz - Galv. Rigid Conduit - 75° THWN - 30° C Ambient
Not More Than Three Current-Carrying Conductors

									Maximum Distance - 1 Phase for 1% Drop @ .85 P.F.*				Maximum Distance - 3 Phase for 1% Drop @ .85 P.F.*				
Wire AWG MCM	Rated Amps 125%	**Cont. Amps 80%**	Area Square Inches	Conduit Size 3-Each	Conduit Size 4-Each	Ins. Eq. Ground Size	Conduit 3 Ea. With Eq. Grd.	Conduit 4 Ea. With Eq. Grd.	120 Volts	208 Volts	240 Volts	277 Volts	208 Volts	240 Volts	480 Volts	600 Volts	NEC Fuse 125%	NEC C.B. 125%
14	15	**12**	.0097	1/2	1/2	14	1/2	1/2	37	32	37	86	37	43	86	108	15	15
12	20	**16**	.0133	1/2	1/2	12	1/2	1/2	43	35	40	100	43	50	100	125	20	20
10	30	**24**	.0211	1/2	1/2	10	1/2	1/2	47	41	47	110	48	55	110	137	30	30
8	50	**40**	.0366	1/2	3/4	10	3/4	3/4	43	37	43	99	43	50	99	124	50	50
6	65	**52**	.0507	3/4	3/4	8	3/4	1	51	44	51	118	51	59	118	148	70	70
4	85	**68**	.0824	1	1	8	1	1-1/4	60	52	60	138	60	69	138	173	90	90
3	100	**80**	.0973	1	1-1/4	8	1	1-1/4	61	53	62	142	62	71	142	178	100	100
2	115	**92**	.1158	1	1-1/4	6	1-1/4	1-1/4	65	57	65	151	65	75	151	188	125	125
1	130	**104**	.1562	1-1/4	1-1/4	6	1-1/4	1-1/2	70	60	70	160	69	80	160	200	150	150
1/0	150	**120**	.1855	1-1/4	1-1/2	6	1-1/4	1-1/2	76	66	76	176	76	88	176	220	150	150
2/0	175	**140**	.2223	1-1/2	2	6	1-1/2	2	76	65	76	175	76	87	175	219	175	175
3/0	200	**160**	.2679	1-1/2	2	6	2	2	79	69	79	182	79	91	183	229	200	200
4/0	230	**184**	.3237	2	2	4	2	2-1/2	82	70	81	188	81	94	188	235	250	250
250	255	**204**	.3970	2	2-1/2	4	2	2-1/2	81	69	80	186	80	93	185	232	300	300
300	285	**228**	.4608	2	2-1/2	4	2-1/2	2-1/2	81	70	81	187	81	93	186	233	300	300
350	310	**248**	.5242	2-1/2	3	4	2-1/2	3	82	70	81	189	81	94	188	235	350	350
400	335	**268**	.5863	2-1/2	3	4	2-1/2	3	80	70	81	185	81	93	187	233	350	350
500	380	**304**	.7073	3	3	3	3	3	79	68	79	182	80	92	184	229	400	400
600	420	**336**	.8676	3	3-1/2	3	3	3-1/2	76	67	77	175	76	88	176	220	450	450
700	460	**368**	.9887	3	3-1/2	2	3-1/2	4	72	63	72	167	72	84	167	209	500	500
750	475	**380**	1.0496	3	3-1/2	2	3-1/2	4	73	64	73	170	73	84	168	211	500	500
800	490	**392**	1.1085	3-1/2	4	2	3-1/2	4	73	63	73	168	74	85	170	213	500	500
900	520	**416**	1.2311	3-1/2	4	1	3-1/2	4	70	61	70	162	70	81	163	203	600	600
1000	545	**436**	1.3478	3-1/2	4	1	4	5	69	60	70	159	70	81	162	202	600	600

* Lagging

A. To determine actual percent voltage drop:
1. Divide the actual distance by the max distance in the table for 1% drop.

B. To determine actual voltage drop:
1. Divide the actual distance by the max distance in the table for 1% drop.
2. Divide the result by 100.
3. Multiply the result by the L-L voltage.

Formulas:
A. Percent Voltage Drop = Actual Distance / Max Distance
B. Actual Volts Drop = (Actual Dist. / Max Dist.) / 100 x L-L Volts
C. New Max Dist. = (Cont. Amps / Actual FLA) x Max Distance

Temperature Correction Factors for Other then 30° C Ambient	
21-25° C (70-77° F)	1.05
26-30° C (78-86° F)	1.00
31-35° C (87-95° F)	.94
36-40° C (96-104° F)	.88
41-45° C (105-113° F)	.82
46-50° C (114-122° F)	.75
51-55° C (123-131° F)	.67
56-60° C (132-140° F)	.58
61-70° C (141-158° F)	.33

Conductor Derating Factors for More Than Three Current-Carrying Conductors	
4 Through 6	.80
7 Through 9	.70
10 Through 20	.50
21 Through 30	.45
31 Through 40	.40
41 and Above	.35

Circuit Design-Data for Copper Conductors

60 Hz - Intermediate Metallic Conduit - 75° THWN - 30° C Ambient
Not More Than Three Current-Carrying Conductors

Wire AWG MCM	Rated Amps 125%	**Cont. Amps 80%**	Area Square Inches	Conduit Size 3-Each	Conduit Size 4-Each	Ins. Eq. Ground Size	Conduit 3 Ea. With Eq. Grd.	Conduit 4 Ea. With Eq. Grd.	Maximum Distance - 1 Phase for 1% Drop @ .85 P.F.* 120 Volts	208 Volts	240 Volts	277 Volts	Maximum Distance 3 Phase for 1% Drop @ .85 P.F.* 208 Volts	240 Volts	480 Volts	600 Volts	NEC Fuse 125%	NEC C.B. 125%
14	15	**12**	.0097	1/2	1/2	14	1/2	1/2	37	32	37	86	37	43	83	108	15	15
12	20	**16**	.0133	1/2	1/2	12	1/2	1/2	43	35	40	100	43	50	100	125	20	20
10	30	**24**	.0211	1/2	1/2	10	1/2	1/2	47	41	47	110	48	55	110	137	30	30
8	50	**40**	.0366	1/2	3/4	10	1/2	3/4	43	37	43	99	43	50	99	124	50	50
6	65	**52**	.0507	3/4	3/4	8	3/4	1	51	44	51	118	51	59	118	148	70	70
4	85	**68**	.0824	3/4	1	8	1	1	60	52	60	138	60	69	138	173	90	90
3	100	**80**	.0973	1	1	8	1	1-1/4	61	53	62	142	62	71	142	178	100	100
2	115	**92**	.1158	1	1-1/4	6	1-1/4	1-1/4	65	57	65	151	65	75	151	188	125	125
1	130	**104**	.1562	1-1/4	1-1/4	6	1-1/4	1-1/2	70	60	70	160	69	80	160	200	150	150
1/0	150	**120**	.1855	1-1/4	1-1/2	6	1-1/4	1-1/2	76	66	76	176	76	88	176	220	150	150
2/0	175	**140**	.2223	1-1/4	1-1/2	6	1-1/2	2	76	65	76	175	76	87	175	219	175	175
3/0	200	**160**	.2679	1-1/2	2	6	1-1/2	2	79	69	79	182	79	91	183	229	200	200
4/0	230	**184**	.3237	2	2	4	2	2	82	70	81	188	81	94	138	235	250	250
250	255	**204**	.3970	2	2-1/2	4	2	2-1/2	81	69	80	186	80	93	135	232	300	300
300	285	**228**	.4608	2	2-1/2	4	2-1/2	2-1/2	81	70	81	187	81	93	136	233	300	300
350	310	**248**	.5242	2-1/2	2-1/2	4	2-1/2	3	82	70	81	189	81	94	138	235	350	350
400	335	**268**	.5863	2-1/2	3	4	2-1/2	3	80	70	81	185	81	93	137	233	350	350
500	380	**304**	.7073	2-1/2	3	3	3	3	79	68	79	182	80	92	134	229	400	400
600	420	**336**	.8676	3	3-1/2	3	3	3-1/2	76	67	77	175	76	88	176	220	450	450
700	460	**368**	.9887	3	3-1/2	2	3	3-1/2	72	63	72	167	72	84	137	209	500	500
750	475	**380**	1.0496	3	3-1/2	2	3-1/2	4	73	64	73	170	73	84	138	211	500	500
800	490	**392**	1.1085	3	3-1/2	2	3-1/2	4	73	63	73	168	74	85	170	213	500	500
900	520	**416**	1.2311	3-1/2	4	1	3-1/2	4	70	61	70	162	70	81	163	203	600	600
1000	545	**436**	1.3478	3-1/2	4	1	3-1/2	4.0 **	69	60	70	159	70	81	162	202	600	600

* Lagging

A. To determine actual percent voltage drop:
1. Divide the actual distance by the max distance in the table for 1% drop.

B. To determine actual voltage drop:
1. Divide the actual distance by the max distance in the table for 1% drop.
2. Divide the result by 100.
3. Multiply the result by the L-L voltage.

Formulas:
A. Percent Voltage Drop = Actual Distance / Max Distance
B. Actual Volts Drop = (Actual Dist. / Max Dist.) / 100 x L-L Volts
C. New Max Dist. = (Cont. Amps / Actual FLA) x Max Distance

Temperature Correction Factors for Other then 30° C Ambient	
21-25° C (70-77° F)	1.05
26-30° C (78-86° F)	1.00
31-35° C (87-95° F)	.94
36-40° C (96-104° F)	.88
41-45° C (105-113° F)	.82
46-50° C (114-122° F)	.75
51-55° C (123-131° F)	.67
56-60° C (132-140° F)	.58
61-70° C (141-153° F)	.33

Conductor Derating Factors for More Than Three Current-Carrying Conductors	
4 Through 6	.80
7 Through 9	.70
10 Through 20	.50
21 Through 30	.45
31 Through 40	.40
41 and Above	.35

** .0954 sq. in.too small, use this size rigid.

Circuit Design-Data for Copper Conductors

60 Hz - EMT Conduit - 75° THWN - 30° C Ambient
Not More Than Three Current-Carrying Conductors

Wire AWG MCM	Rated Amps 125%	**Cont. Amps 80%**	Area Square Inches	Conduit Size 3-Each	Conduit Size 4-Each	Ins. Eq. Ground Size	Conduit 3 Ea. With Eq. Grd.	Conduit 4 Ea. With Eq. Grd.	Maximum Distance - 1 Phase for 1% Drop @ .85 P.F.* 120 Volts	208 Volts	240 Volts	277 Volts	Maximum Distance - 3 Phase for 1% Drop @ .85 P.F.* 208 Volts	240 Volts	480 Volts	600 Volts	NEC Fuse 125%	NEC C.B. 125%
14	15	**12**	.0097	1/2	1/2	14	1/2	1/2	37	32	37	86	37	43	86	108	15	15
12	20	**16**	.0133	1/2	1/2	12	1/2	1/2	43	35	40	100	43	50	100	125	20	20
10	30	**24**	.0211	1/2	1/2	10	1/2	1/2	47	41	47	110	48	55	110	137	30	30
8	50	**40**	.0366	1/2	3/4	10	3/4	3/4	43	37	43	99	43	50	99	124	50	50
6	65	**52**	.0507	3/4	3/4	8	3/4	1	51	44	51	118	51	59	118	148	70	70
4	85	**68**	.0824	1	1	8	1	1-1/4	60	52	60	138	60	69	138	173	90	90
3	100	**80**	.0973	1	1-1/4	8	1	1-1/4	61	53	62	142	62	71	142	178	100	100
2	115	**92**	.1158	1	1-1/4	6	1-1/4	1-1/4	65	57	65	151	65	75	151	188	125	125
1	130	**104**	.1562	1-1/4	1-1/4	6	1-1/4	1-1/2	70	60	70	160	69	80	160	200	150	150
1/0	150	**120**	.1855	1-1/4	1-1/2	6	1-1/2	1-1/2	76	66	76	176	76	88	176	220	150	150
2/0	175	**140**	.2223	1-1/2	2	6	1-1/2	2	76	65	76	175	76	87	175	219	175	175
3/0	200	**160**	.2679	1-1/2	2	6	2	2	79	69	79	182	79	91	183	229	200	200
4/0	230	**184**	.3237	2	2	4	2	2-1/2	82	70	81	188	81	94	188	235	250	250
250	255	**204**	.3970	2	2-1/2	4	2	2-1/2	81	69	80	186	80	93	185	232	300	300
300	285	**228**	.4608	2	2-1/2	4	2-1/2	2-1/2	81	70	81	187	81	93	186	233	300	300
350	310	**248**	.5242	2-1/2	2-1/2	4	2-1/2	2-1/2	82	70	81	189	81	94	188	235	350	350
400	335	**268**	.5863	2-1/2	2-1/2	4	2-1/2	3	80	70	81	185	81	93	187	233	350	350
500	380	**304**	.7073	2-1/2	3	3	2-1/2	3	79	68	79	182	80	92	184	229	400	400
600	420	**336**	.8676	3	3	3	3	3-1/2	76	67	77	175	76	88	176	220	450	450
700	460	**368**	.9887	3	3-1/2	2	3	3-1/2	72	63	72	167	72	84	167	209	500	500
750	475	**380**	1.0496	3	3-1/2	2	3	3-1/2	73	64	73	170	73	84	168	211	500	500
800	490	**392**	1.1085	3	3-1/2	2	3	3-1/2	73	63	73	168	74	85	170	213	500	500
900	520	**416**	1.2311	3	4	1	3-1/2	4	70	61	70	162	70	81	163	203	600	600
1000	545	**436**	1.3478	3-1/2	4	1	3-1/2	4	69	60	70	159	70	81	162	202	600	600

* Lagging

A. To determine actual percent voltage drop:
1. Divide the actual distance by the max distance in the table for 1% drop.

B. To determine actual voltage drop:
1. Divide the actual distance by the max distance in the table for 1% drop.
2. Divide the result by 100.
3. Multiply the result by the L-L voltage.

Formulas:
A. Percent Voltage Drop = Actual Distance / Max Distance
B. Actual Volts Drop = (Actual Dist. / Max Dist.) / 100 x L-L Volts
C. New Max Dist. = (Cont. Amps / Actual FLA) x Max Distance

Temperature Correction Factors for Other then 30° C Ambient	
21-25° C (70-77° F)	1.05
26-30° C (78-86° F)	1.00
31-35° C (87-95° F)	.94
36-40° C (96-104° F)	.88
41-45° C (105-113° F)	.82
46-50° C (114-122° F)	.75
51-55° C (123-131° F)	.67
56-60° C (132-140° F)	.58
61-70° C (141-158° F)	.33

Conductor Derating Factors for More Than Three Current-Carrying Conductors	
4 Through 6	.80
7 Through 9	.70
10 Through 20	.50
21 Through 30	.45
31 Through 40	.40
41 and Above	.35

Circuit Design-Data for Copper Conductors

60 Hz - Liquidtight Flex Metal - 75° THWN - 30° C Ambient
Not More Than Three Current-Carrying Conductors

Wire AWG MCM	Rated Amps 125%	**Cont. Amps 80%**	Area Square Inches	Conduit Size 3-Each	Conduit Size 4-Each	Ins. Eq. Ground Size	Conduit 3 Ea. With Eq. Grd.	Conduit 4 Ea. With Eq. Grd.	Maximum Distance - 1 Phase for 1% Drop @ .85 P.F.*				Maximum Distance - 3 Phase for 1% Drop @ .85 P.F.*				NEC Fuse 125%	NEC C.B. 125%
									120 Volts	208 Volts	240 Volts	277 Volts	208 Volts	240 Volts	480 Volts	600 Volts		
14	15	**12**	.0097	1/2	1/2	14	1/2	1/2	37	32	37	86	37	43	86	108	15	15
12	20	**16**	.0133	1/2	1/2	12	1/2	1/2	43	35	40	100	43	50	100	125	20	20
10	30	**24**	.0211	1/2	1/2	10	1/2	1/2	47	41	47	110	48	55	110	137	30	30
8	50	**40**	.0366	1/2	3/4	10	3/4	3/4	43	37	43	99	43	50	99	124	50	50
6	65	**52**	.0507	3/4	3/4	8	3/4	1	51	44	51	118	51	59	118	148	70	70
4	85	**68**	.0824	1	1	8	1	1-1/4	60	52	60	138	60	69	138	173	90	90
3	100	**80**	.0973	1	1-1/4	8	1	1-1/4	61	53	62	142	62	71	142	178	100	100
2	115	**92**	.1158	1	1-1/4	6	1-1/4	1-1/4	65	57	65	151	65	75	151	188	125	125
1	130	**104**	.1562	1-1/4	1-1/4	6	1-1/4	1-1/2	70	60	70	160	69	80	130	200	150	150
1/0	150	**120**	.1855	1-1/4	1-1/2	6	1-1/4	1-1/2	76	66	76	176	76	88	176	220	150	150
2/0	175	**140**	.2223	1-1/2	2	6	1-1/2	2	76	65	76	175	76	87	175	219	175	175
3/0	200	**160**	.2679	1-1/2	2	6	2	2	79	69	79	182	79	91	133	229	200	200
4/0	230	**184**	.3237	2	2	4	2	2-1/2	82	70	81	188	81	94	138	235	250	250
250	255	**204**	.3970	2	2-1/2	4	2	2-1/2	81	69	80	186	80	93	185	232	300	300
300	285	**228**	.4608	2	2-1/2	4	2-1/2	2-1/2	81	70	81	187	81	93	186	233	300	300
350	310	**248**	.5242	2-1/2	3	4	2-1/2	3	82	70	81	189	81	94	188	235	350	350
400	335	**268**	.5863	2-1/2	3	4	2-1/2	3	80	70	81	185	81	93	187	233	350	350
500	380	**304**	.7073	3	3	3	3	3	79	68	79	182	80	92	184	229	400	400
600	420	**336**	.8676	3	3-1/2	3	3	3-1/2	76	67	77	175	76	88	176	220	450	450
700	460	**368**	.9887	3	3-1/2	2	3-1/2	4	72	63	72	167	72	84	167	209	500	500
750	475	**380**	1.0496	3	4	2	3-1/2	4	73	64	73	170	73	84	168	211	500	500
800	490	**392**	1.1085	3-1/2	4	2	3-1/2	4	73	63	73	168	74	85	170	213	500	500
900	520	**416**	1.2311	3-1/2	4	1	3-1/2	4 **	70	61	70	162	70	81	163	203	600	600
1000	545	**436**	1.3478	3-1/2	4 **	1	4	5 **	69	60	70	159	70	81	162	202	600	600

* Lagging

A. To determine actual percent voltage drop:
1. Divide the actual distance by the max distance in the table for 1% drop.

B. To determine actual voltage drop:
1. Divide the actual distance by the max distance in the table for 1% drop.
2. Divide the result by 100.
3. Multiply the result by the L-L voltage.

Formulas:
A. Percent Voltage Drop = Actual Distance / Max Distance
B. Actual Volts Drop = (Actual Dist. / Max Dist.) / 100 x L-L Volts
C. New Max Dist. = (Cont. Amps / Actual FLA) x Max Distance

Temperature Correction Factors for Other then 30° C Ambient	
21-25° C (70-77° F)	1.05
26-30° C (78-86° F)	1.00
31-35° C (87-95° F)	.94
36-40° C (96-104° F)	.88
41-45° C (105-113° F)	.82
46-50° C (114-122° F)	.75
51-55° C (123-131° F)	.67
56-60° C (132-140° F)	.58
61-70° C (141-158° F)	.33

Conductor Derating Factors for More Than Three Current-Carrying Conductors	
4 Through 6	.80
7 Through 9	.70
10 Through 20	.50
21 Through 30	.45
31 Through 40	.40
41 and Above	.35

** Liquidtight conduit too small, use this size rigid.

Voltage Drop Factors per 1000 Ampere Feet

60 Hz - Steel Conduit - Copper Conductors - Three Phase

*Load Power Factor **

Conductor AWG	DC	.45	.50	.55	.60	.65	.70	.75	.80	.85	.90	.95	1.00
# 14	6.280	2.529	2.794	3.059	3.323	3.586	3.849	4.111	4.371	4.630	4.887	5.140	5.369
# 12	3.960	1.664	1.834	2.004	2.173	2.341	2.509	2.676	2.842	3.006	3.169	3.328	3.464
# 10	2.480	1.033	1.134	1.234	1.334	1.434	1.533	1.631	1.728	1.824	1.918	2.009	2.078
# 8	1.556	.708	.773	.837	.901	.964	1.026	1.088	1.148	1.208	1.265	1.319	1.351
# 6	.982	.481	.520	.559	.598	.636	.673	.710	.745	.780	.812	.841	.849
# 4	.616	.334	.358	.382	.405	.428	.450	.471	.492	.511	.529	.543	.537
# 3	.490	.286	.305	.323	.342	.359	.376	.392	.408	.422	.434	.443	.433
# 2	.388	.244	.259	.273	.287	.300	.313	.325	.336	.346	.355	.360	.346
# 1	.308	.213	.224	.235	.245	.255	.264	.273	.281	.288	.292	.294	.277
# 1/0	.244	.179	.186	.194	.201	.207	.214	.219	.223	.227	.229	.227	.208
# 2/0	.193	.161	.168	.173	.179	.184	.188	.192	.195	.196	.197	.194	.173
# 3/0	.153	.142	.146	.150	.154	.157	.160	.162	.164	.164	.162	.158	.137
# 4/0	.122	.128	.131	.134	.136	.138	.139	.140	.140	.139	.137	.131	.109
# 250	.103	.123	.125	.127	.128	.129	.130	.130	.129	.127	.123	.117	.094
# 300	.086	.114	.115	.117	.117	.118	.118	.117	.115	.113	.109	.102	.078
# 350	.073	.108	.109	.109	.110	.110	.109	.108	.106	.103	.099	.091	.068
# 400	.064	.103	.104	.104	.104	.104	.103	.102	.099	.096	.092	.084	.061
# 500	.052	.097	.097	.097	.097	.096	.095	.093	.090	.086	.081	.074	.050
# 600	.043	.094	.094	.093	.092	.091	.090	.087	.085	.081	.075	.067	.043
# 700	.037	.092	.092	.091	.090	.089	.087	.085	.082	.078	.072	.064	.040
# 750	.034	.091	.090	.089	.088	.087	.085	.082	.079	.075	.069	.061	.036
# 800	.032	.088	.088	.087	.086	.084	.082	.080	.077	.072	.067	.058	.035
# 900	.029	.088	.087	.086	.085	.083	.081	.079	.075	.071	.065	.057	.033
# 1000	.026	.085	.085	.084	.082	.081	.079	.076	.073	.068	.063	.054	.031

* Lagging

A. To determine voltage drop:
1. Determine the load power factor and conductor size to be used.
2. Select the corresponding factor from the table.
3. Multiply the circuit distance, load amps and the table factor together, divide by the number of runs and then by 1000.

Formulas:

A. Voltage Drop = Distance x Load Amps x Factor / Runs / 1000

B. Max Drop Factor = Max Volts Drop / (Distance x Load Amps / 1000) x Runs

B. To determine minimum wire size:
1. Multiply the circuit distance by the load amps and divide by 1000 for the ampere-feet.
2. Divide the maximum volts drop allowed by the ampere-feet and then multiply by the number of runs to determine the maximum table factor.
3. Select the wire size from the column having the appropriate load power factor and a factor less than or equal to the maximum table factor.

Voltage Drop Factors per 1000 Ampere Feet

60 Hz - Steel Conduit - Copper Conductors - Single Phase L-L

*Load Power Factor **

Conductor AWG	DC	.45	.50	.55	.60	.65	.70	.75	.80	.85	.90	.95	1.00
# 14	6.280	2.920	3.226	3.532	3.837	4.141	4.444	4.747	5.048	5.347	5.644	5.936	6.200
# 12	3.960	1.921	2.118	2.314	2.509	2.703	2.897	3.090	3.282	3.472	3.659	3.842	4.000
# 10	2.480	1.193	1.309	1.425	1.541	1.656	1.770	1.883	1.996	2.106	2.215	2.319	2.400
# 8	1.556	.818	.893	.967	1.040	1.113	1.185	1.256	1.326	1.394	1.461	1.523	1.560
# 6	.982	.555	.601	.646	.690	.734	.777	.820	.861	.900	.938	.971	.980
# 4	.616	.386	.414	.441	.468	.494	.520	.544	.568	.590	.610	.626	.620
# 3	.490	.330	.352	.374	.394	.415	.434	.453	.471	.487	.501	.512	.500
# 2	.388	.282	.299	.315	.331	.347	.361	.375	.388	.400	.410	.416	.400
# 1	.308	.246	.259	.271	.283	.295	.305	.315	.324	.332	.338	.340	.320
# 1/0	.244	.206	.215	.224	.232	.240	.247	.253	.258	.262	.264	.262	.240
# 2/0	.193	.186	.194	.200	.206	.212	.217	.221	.225	.227	.227	.224	.200
# 3/0	.153	.164	.169	.174	.178	.182	.185	.187	.189	.189	.188	.183	.158
# 4/0	.122	.148	.151	.154	.157	.159	.161	.162	.162	.161	.158	.152	.126
# 250	.103	.141	.144	.146	.148	.149	.150	.150	.149	.147	.143	.135	.108
# 300	.086	.132	.133	.135	.136	.136	.136	.135	.133	.130	.125	.117	.090
# 350	.073	.124	.126	.126	.127	.127	.126	.125	.122	.119	.114	.105	.078
# 400	.064	.119	.120	.120	.120	.120	.119	.117	.115	.111	.106	.097	.070
# 500	.052	.112	.112	.112	.112	.111	.109	.107	.104	.100	.094	.085	.058
# 600	.043	.108	.108	.108	.107	.105	.104	.101	.098	.093	.087	.077	.050
# 700	.037	.106	.106	.105	.104	.103	.101	.098	.094	.090	.083	.074	.046
# 750	.034	.105	.104	.103	.102	.100	.098	.095	.091	.086	.080	.070	.042
# 800	.032	.102	.101	.101	.099	.097	.095	.092	.088	.084	.077	.067	.040
# 900	.029	.101	.100	.099	.098	.096	.094	.091	.087	.082	.075	.065	.038
# 1000	.026	.098	.098	.097	.095	.093	.091	.088	.084	.079	.073	.063	.036

* Lagging

A. To determine voltage drop:

1. Determine the load power factor and conductor size to be used.
2. Select the corresponding factor from the table.
3. Multiply the circuit distance, load amps and the table factor together, divide by the number of runs and then by 1000.

Formulas:

A. Voltage Drop = Distance x Load Amps x Factor / Runs / 1000

B. Max Drop Factor = Max Volts Drop / (Distance x Load Amps / 1000) x Runs

B. To determine minimum wire size:

1. Multiply the circuit distance by the load amps and divide by 1000 for the ampere-feet.
2. Divide the maximum volts drop allowed by the ampere-feet and then multiply by the number of runs to determine the maximum table factor.
3. Select the wire size from the column having the appropriate load power factor and a factor less than or equal to the maximum table factor.

Voltage Drop Factors per 1000 Ampere Feet

60 Hz - Steel Conduit - Copper Conductors - Single Phase L-N

*Load Power Factor **

Conductor AWG	DC	.45	.50	.55	.60	.65	.70	.75	.80	.85	.90	.95	1.00
# 14	6.280	1.460	1.613	1.766	1.918	2.070	2.222	2.373	2.524	2.673	2.822	2.968	3.100
# 12	3.960	.961	1.059	1.157	1.254	1.352	1.449	1.545	1.641	1.736	1.830	1.921	2.000
# 10	2.480	.596	.655	.713	.770	.828	.885	.942	.998	1.053	1.107	1.160	1.200
# 8	1.556	.409	.446	.483	.520	.556	.592	.628	.663	.697	.730	.761	.780
# 6	.982	.278	.300	.323	.345	.367	.389	.410	.430	.450	.469	.485	.490
# 4	.616	.193	.207	.221	.234	.247	.260	.272	.284	.295	.305	.313	.310
# 3	.490	.165	.176	.187	.197	.207	.217	.227	.235	.244	.251	.256	.250
# 2	.388	.141	.149	.158	.166	.173	.181	.188	.194	.200	.205	.208	.200
# 1	.308	.123	.129	.136	.142	.147	.153	.158	.162	.166	.169	.170	.160
# 1/0	.244	.103	.108	.112	.116	.120	.123	.126	.129	.131	.132	.131	.120
# 2/0	.193	.093	.097	.100	.103	.106	.109	.111	.112	.113	.114	.112	.100
# 3/0	.153	.082	.085	.087	.089	.091	.092	.094	.094	.095	.094	.091	.079
# 4/0	.122	.074	.076	.077	.079	.080	.081	.081	.081	.080	.079	.076	.063
# 250	.103	.071	.072	.073	.074	.075	.075	.075	.074	.073	.071	.068	.054
# 300	.086	.066	.067	.067	.068	.068	.068	.067	.067	.065	.063	.059	.045
# 350	.073	.062	.063	.063	.063	.063	.063	.062	.061	.059	.057	.053	.039
# 400	.064	.060	.060	.060	.060	.060	.059	.059	.057	.056	.053	.049	.035
# 500	.052	.056	.056	.056	.056	.055	.055	.053	.052	.050	.047	.043	.029
# 600	.043	.054	.054	.054	.053	.053	.052	.050	.049	.047	.043	.039	.025
# 700	.037	.053	.053	.053	.052	.051	.050	.049	.047	.045	.042	.037	.023
# 750	.034	.052	.052	.052	.051	.050	.049	.047	.046	.043	.040	.035	.021
# 800	.032	.051	.051	.050	.050	.049	.048	.046	.044	.042	.038	.034	.020
# 900	.029	.051	.050	.050	.049	.048	.047	.045	.043	.041	.038	.033	.019
# 1000	.026	.049	.049	.048	.048	.047	.045	.044	.042	.040	.036	.031	.018

* Lagging

A. To determine voltage drop:
1. Determine the load power factor and conductor size to be used.
2. Select the corresponding factor from the table.
3. Multiply the circuit distance, load amps and the table factor together, divide by the number of runs and then by 1000.

Formulas:

A. Voltage Drop = Distance x Load Amps x Factor / Runs / 1000
(For 3-Phase multiply by 1.732, Single Phase multiply by 2)

B. Max Drop Factor = Max Volts Drop / (Distance x Load Amps / 1000) x Runs
(For 3-Phase divide by 1.732, Single Phase divide by 2)

B. To determine minimum wire size:
1. Multiply the circuit distance by the load amps and divide by 1000 for the ampere-feet.
2. Divide the maximum volts drop allowed by the ampere-feet, then multiply by the number of runs to determine the maximum table factor.
3. Select the wire size from the column having the appropriate load power factor and a factor less than or equal to the maximum table factor.

Circuit Design-Data for Copper Conductors

60 Hz - PVC Sch 40 Conduit - 75° THWN - 30° C Ambient
Not More Than Three Current-Carrying Conductors

Wire AWG MCM	Rated Amps 125%	**Cont. Amps 80%**	Area Square Inches	Conduit Size 3-Each	Conduit Size 4-Each	Ins. Eq. Ground Size	Conduit 3 Ea. With Eq. Grd.	Conduit 4 Ea. With Eq. Grd.	Maximum Distance - 1 Phase for 1% Drop @ .85 P.F.*				Maximum Distance - 3 Phase for 1% Drop @ .85 P.F.*				NEC Fuse 125%	NEC C.B. 125%
									120 Volts	208 Volts	240 Volts	277 Volts	208 Volts	240 Volts	480 Volts	600 Volts		
14	15	**12**	.0097	1/2	1/2	14	1/2	1/2	38	33	38	87	38	43	87	108	15	15
12	20	**16**	.0133	1/2	1/2	12	1/2	1/2	43	38	43	100	43	50	100	125	20	20
10	30	**24**	.0211	1/2	1/2	10	1/2	1/2	48	41	48	110	48	55	110	138	30	30
8	50	**40**	.0366	1/2	3/4	10	3/4	3/4	43	38	43	100	43	50	100	125	50	50
6	65	**52**	.0507	3/4	3/4	8	3/4	1	52	45	52	120	52	60	120	150	70	70
4	85	**68**	.0824	1	1	8	1	1-1/4	61	53	61	141	61	71	141	176	90	90
3	100	**80**	.0973	1	1-1/4	8	1	1-1/4	63	55	63	146	63	73	146	182	100	100
2	115	**92**	.1158	1	1-1/4	6	1-1/4	1-1/4	71	61	71	163	70	81	163	203	125	125
1	130	**104**	.1562	1-1/4	1-1/2	6	1-1/4	1-1/2	76	66	76	175	76	88	175	219	150	150
1/0	150	**120**	.1855	1-1/4	1-1/2	6	1-1/2	1-1/2	80	69	80	185	80	92	134	230	150	150
2/0	175	**140**	.2223	1-1/2	2	6	1-1/2	2	79	69	80	183	80	92	134	230	175	175
3/0	200	**160**	.2679	1-1/2	2	6	2	2	85	74	86	197	86	99	137	247	200	200
4/0	230	**184**	.3237	2	2	4	2	2-1/2	88	76	88	203	88	101	202	253	250	250
250	255	**204**	.3970	2	2-1/2	4	2	2-1/2	89	77	89	206	89	103	206	258	300	300
300	285	**228**	.4608	2	2-1/2	4	2-1/2	3	89	77	89	206	89	103	206	258	300	300
350	310	**248**	.5242	2-1/2	3	4	2-1/2	3	91	78	90	211	91	105	210	263	350	350
400	335	**268**	.5863	2-1/2	3	4	2-1/2	3	91	79	91	211	91	105	211	263	350	350
500	380	**304**	.7073	3	3	3	3	3-1/2	92	79	91	212	91	105	211	263	400	400
600	420	**336**	.8676	3	3-1/2	3	3	3-1/2	89	77	89	206	90	104	207	259	450	450
700	460	**368**	.9887	3	3-1/2	2	3-1/2	4	86	73	85	198	86	99	198	247	500	500
750	475	**380**	1.0496	3-1/2	4	2	3-1/2	4	88	76	88	202	87	100	201	251	500	500
800	490	**392**	1.1085	3-1/2	4	2	3-1/2	4	87	75	86	202	87	100	201	251	500	500
900	520	**416**	1.2311	3-1/2	4	1	3-1/2	5	85	74	85	196	85	98	196	244	600	600
1000	545	**436**	1.3478	3-1/2	5	1	4	5	86	75	86	199	85	98	197	246	600	600

* Lagging

A. To determine actual percent voltage drop:
1. Divide the actual distance by the max distance in the table for 1% drop.

B. To determine actual voltage drop:
1. Divide the actual distance by the max distance in the table for 1% drop.
2. Divide the result by 100.
3. Multiply the result by the L-L voltage.

Formulas:
A. Percent Voltage Drop = Actual Distance / Max Distance
B. Actual Volts Drop = (Actual Dist. / Max Dist.) / 100 x L-L Volts
C. New Max Dist. = (Cont. Amps / Actual FLA) x Max Distance

Temperature Correction Factors for Other then 30° C Ambient	
21-25° C (70-77° F)	1.05
26-30° C (78-86° F)	1.00
31-35° C (87-95° F)	.94
36-40° C (96-104° F)	.88
41-45° C (105-113° F)	.82
46-50° C (114-122° F)	.75
51-55° C (123-131° F)	.67
56-60° C (132-140° F)	.58
61-70° C (141-158° F)	.33

Conductor Derating Factors for More Than Three Current-Carrying Conductors	
4 Through 6	.80
7 Through 9	.70
10 Through 20	.50
21 Through 30	.45
31 Through 40	.40
41 and Above	.35

Voltage Drop Factors per 1000 Ampere Feet

60 Hz - PVC Conduit - Copper Conductors - Three Phase

*Load Power Factor **

Conductor AWG	DC	.45	.50	.55	.60	.65	.70	.75	.80	.85	.90	.95	1.00
# 14	6.280	2.506	2.772	3.037	3.302	3.566	3.830	4.093	4.356	4.617	4.876	5.132	5.369
# 12	3.960	1.642	1.813	1.983	2.153	2.323	2.492	2.660	2.827	2.994	3.158	3.320	3.464
# 10	2.480	1.013	1.114	1.215	1.316	1.417	1.517	1.616	1.715	1.812	1.908	2.002	2.078
# 8	1.556	.688	.753	.818	.883	.947	1.010	1.073	1.135	1.196	1.255	1.312	1.351
# 6	.982	.461	.501	.541	.580	.619	.657	.695	.732	.768	.802	.834	.849
# 4	.616	.316	.340	.365	.389	.412	.435	.458	.479	.500	.519	.536	.537
# 3	.490	.268	.287	.306	.325	.343	.361	.379	.395	.411	.425	.437	.433
# 2	.388	.218	.232	.246	.260	.273	.286	.298	.310	.321	.330	.337	.329
# 1	.308	.188	.199	.209	.220	.229	.239	.248	.256	.263	.269	.272	.260
# 1/0	.244	.162	.170	.178	.186	.193	.200	.206	.212	.217	.220	.221	.208
# 2/0	.193	.144	.151	.157	.164	.169	.174	.179	.183	.186	.188	.188	.173
# 3/0	.153	.125	.130	.134	.138	.142	.145	.148	.150	.152	.152	.149	.133
# 4/0	.122	.112	.115	.118	.121	.124	.126	.128	.129	.129	.128	.124	.107
# 250	.103	.104	.107	.109	.111	.113	.114	.115	.115	.114	.112	.108	.090
# 300	.086	.098	.100	.101	.103	.103	.104	.104	.104	.102	.100	.095	.076
# 350	.073	.091	.093	.094	.095	.095	.096	.095	.094	.092	.089	.084	.066
# 400	.064	.088	.089	.089	.090	.090	.089	.089	.087	.085	.082	.076	.057
# 500	.052	.081	.082	.082	.082	.082	.081	.080	.078	.075	.072	.066	.047
# 600	.043	.078	.078	.078	.078	.077	.076	.075	.072	.069	.065	.059	.040
# 700	.037	.077	.077	.076	.076	.075	.074	.072	.070	.066	.062	.056	.036
# 750	.034	.074	.073	.073	.072	.071	.070	.068	.066	.063	.058	.052	.033
# 800	.032	.073	.073	.072	.071	.070	.069	.067	.064	.061	.057	.050	.031
# 900	.029	.070	.070	.070	.069	.068	.066	.064	.062	.059	.054	.048	.029
# 1000	.026	.069	.068	.068	.067	.066	.064	.062	.059	.056	.051	.045	.026

* Lagging

A. To determine voltage drop:

1. Determine the load power factor and conductor size to be used.
2. Select the corresponding factor from the table.
3. Multiply the circuit distance, load amps and the table factor together, divide by the number of runs and then by 1000.

Formulas:

A. Voltage Drop = Distance x Load Amps x Factor / Runs / 1000

B. Max Drop Factor = Max Volts Drop / (Distance x Load Amps / 1000) x Runs

B. To determine minimum wire size:

1. Multiply the circuit distance by the load amps and divide by 1000 for the ampere-feet.
2. Divide the maximum volts drop allowed by the ampere-feet and then multiply by the number of runs to determine the maximum table factor.
3. Select the wire size from the column having the appropriate load power factor and a factor less than or equal to the maximum table factor.

Voltage Drop Factors per 1000 Ampere Feet

60 Hz - PVC Conduit - Copper Conductors - Single Phase L-L

*Load Power Factor **

Conductor AWG	DC	.45	.50	.55	.60	.65	.70	.75	.80	.85	.90	.95	1.00
# 14	6.280	2.894	3.200	3.507	3.813	4.118	4.423	4.727	5.030	5.331	5.631	5.926	6.200
# 12	3.960	1.896	2.094	2.290	2.486	2.682	2.877	3.071	3.265	3.457	3.647	3.834	4.000
# 10	2.480	1.169	1.287	1.404	1.520	1.636	1.751	1.866	1.980	2.093	2.204	2.311	2.400
# 8	1.556	.795	.870	.945	1.019	1.093	1.166	1.239	1.310	1.381	1.449	1.514	1.560
# 6	.982	.532	.578	.624	.670	.715	.759	.802	.845	.887	.926	.963	.980
# 4	.616	.365	.393	.421	.449	.476	.503	.528	.554	.578	.600	.619	.620
# 3	.490	.309	.331	.354	.375	.396	.417	.437	.456	.475	.491	.504	.500
# 2	.388	.251	.268	.284	.300	.315	.330	.345	.358	.370	.381	.389	.380
# 1	.308	.217	.230	.242	.254	.265	.276	.286	.295	.303	.310	.314	.300
# 1/0	.244	.187	.196	.205	.214	.223	.231	.238	.245	.250	.254	.255	.240
# 2/0	.193	.167	.174	.182	.189	.195	.201	.207	.212	.215	.217	.217	.200
# 3/0	.153	.144	.150	.155	.160	.164	.168	.171	.174	.175	.175	.173	.154
# 4/0	.122	.129	.133	.137	.140	.143	.145	.147	.148	.149	.147	.143	.124
# 250	.103	.120	.123	.126	.128	.130	.131	.132	.132	.132	.129	.124	.104
# 300	.086	.113	.115	.117	.118	.120	.120	.120	.120	.118	.115	.109	.088
# 350	.073	.106	.107	.109	.110	.110	.110	.110	.109	.107	.103	.097	.076
# 400	.064	.101	.102	.103	.104	.104	.103	.102	.101	.098	.094	.088	.066
# 500	.052	.094	.095	.095	.095	.094	.094	.092	.090	.087	.083	.076	.054
# 600	.043	.090	.091	.090	.090	.089	.088	.086	.084	.080	.075	.068	.046
# 700	.037	.089	.089	.088	.088	.087	.085	.083	.080	.077	.072	.064	.042
# 750	.034	.085	.085	.084	.084	.082	.081	.079	.076	.072	.067	.060	.038
# 800	.032	.084	.084	.083	.082	.081	.079	.077	.074	.071	.066	.058	.036
# 900	.029	.081	.081	.081	.080	.078	.077	.074	.072	.068	.063	.055	.034
# 1000	.026	.080	.079	.078	.077	.076	.074	.071	.068	.064	.059	.052	.030

* Lagging

A. To determine voltage drop:

1. Determine the load power factor and conductor size to be used.
2. Select the corresponding factor from the table.
3. Multiply the circuit distance, load amps and the table factor together, divide by the number of runs and then by 1000.

Formulas:

A. Voltage Drop = Distance x Load Amps x Factor / Runs / 1000

B. Max Drop Factor = Max Volts Drop / (Distance x Load Amps / 1000) x Runs

B. To determine minimum wire size:

1. Multiply the circuit distance by the load amps and divide by 1000 for the ampere-feet.
2. Divide the maximum volts drop allowed by the ampere-feet and then multiply by the number of runs to determine the maximum table factor.
3. Select the wire size from the column having the appropriate load power factor and a factor less than or equal to the maximum table factor.

Voltage Drop Factors per 1000 Ampere Feet

60 Hz - PVC Conduit - Copper Conductors - Single Phase L-N

*Load Power Factor **

Conductor AWG	DC	.45	.50	.55	.60	.65	.70	.75	.80	.85	.90	.95	1.00
# 14	6.280	1.447	1.600	1.753	1.906	2.059	2.211	2.363	2.515	2.666	2.815	2.963	3.100
# 12	3.960	.948	1.047	1.145	1.243	1.341	1.439	1.536	1.632	1.728	1.824	1.917	2.000
# 10	2.480	.585	.643	.702	.760	.818	.876	.933	.990	1.046	1.102	1.156	1.200
# 8	1.556	.397	.435	.472	.510	.547	.583	.619	.655	.690	.725	.757	.780
# 6	.982	.266	.289	.312	.335	.357	.379	.401	.423	.443	.463	.481	.490
# 4	.616	.182	.197	.211	.224	.238	.251	.264	.277	.289	.300	.309	.310
# 3	.490	.154	.166	.177	.188	.198	.209	.219	.228	.237	.245	.252	.250
# 2	.388	.126	.134	.142	.150	.158	.165	.172	.179	.185	.191	.195	.190
# 1	.308	.109	.115	.121	.127	.132	.138	.143	.148	.152	.155	.157	.150
# 1/0	.244	.093	.098	.103	.107	.111	.115	.119	.122	.125	.127	.128	.120
# 2/0	.193	.083	.087	.091	.094	.098	.101	.103	.106	.108	.109	.108	.100
# 3/0	.153	.072	.075	.077	.080	.082	.084	.086	.087	.088	.088	.086	.077
# 4/0	.122	.065	.067	.068	.070	.071	.073	.074	.074	.074	.074	.072	.062
# 250	.103	.060	.062	.063	.064	.065	.066	.066	.066	.066	.065	.062	.052
# 300	.086	.056	.058	.058	.059	.060	.060	.060	.060	.059	.057	.055	.044
# 350	.073	.053	.054	.054	.055	.055	.055	.055	.054	.053	.052	.049	.038
# 400	.064	.051	.051	.052	.052	.052	.052	.051	.050	.049	.047	.044	.033
# 500	.052	.047	.047	.047	.047	.047	.047	.046	.045	.043	.041	.038	.027
# 600	.043	.045	.045	.045	.045	.045	.044	.043	.042	.040	.038	.034	.023
# 700	.037	.044	.044	.044	.044	.043	.043	.042	.040	.038	.036	.032	.021
# 750	.034	.042	.042	.042	.042	.041	.040	.039	.038	.036	.034	.030	.019
# 800	.032	.042	.042	.042	.041	.041	.040	.039	.037	.035	.033	.029	.018
# 900	.029	.041	.041	.040	.040	.039	.038	.037	.036	.034	.031	.028	.017
# 1000	.026	.040	.040	.039	.039	.038	.037	.036	.034	.032	.030	.026	.015

* Lagging

A. To determine voltage drop:
1. Determine the load power factor and conductor size to be used.
2. Select the corresponding factor from the table.
3. Multiply the circuit distance, load amps and the table factor together, divide by the number of runs and then by 1000.

B. To determine minimum wire size:
1. Multiply the circuit distance by the load amps and divide by 1000 for the ampere-feet.
2. Divide the maximum volts drop allowed by the ampere-feet, then multiply by the number of runs to determine the maximum table factor.
3. Select the wire size from the column having the appropriate load power factor and a factor less than or equal to the maximum table factor.

Formulas:

A. Voltage Drop = Distance x Load Amps x Factor / Runs / 1000
(For 3-Phase multiply by 1.732, Single Phase multiply by 2)

B. Max Drop Factor = Max Volts Drop / (Distance x Load Amps / 1000) x Runs
(For 3-Phase divide by 1.732, Single Phase divide by 2)

Circuit Design-Data for Copper Conductors

60 Hz - Aluminum Rigid - 75° THWN - 30° C Ambient
Not More Than Three Current-Carrying Conductors

Wire AWG MCM	Rated Amps 125%	**Cont. Amps 80%**	Area Square Inches	Conduit Size 3-Each	Conduit Size 4-Each	Ins. Eq. Ground Size	Conduit 3 Ea. With Eq. Grd.	Conduit 4 Ea. With Eq. Grd.	Maximum Distance - 1 Phase for 1% Drop @ .85 P.F.* 120 Volts	208 Volts	240 Volts	277 Volts	Maximum Distance - 3 Phase for 1% Drop @ .85 P.F.* 208 Volts	240 Volts	480 Volts	600 Volts	NEC Fuse 125%	NEC C.B. 125%
14	15	**12**	.0097	1/2	1/2	14	1/2	1/2	38	33	38	87	38	43	87	108	15	15
12	20	**16**	.0133	1/2	1/2	12	1/2	1/2	43	38	43	100	43	50	100	125	20	20
10	30	**24**	.0211	1/2	1/2	10	1/2	1/2	48	41	48	110	48	55	110	138	30	30
8	50	**40**	.0366	1/2	3/4	10	3/4	3/4	43	38	43	100	43	50	100	125	50	50
6	65	**52**	.0507	3/4	3/4	8	3/4	1	52	45	52	120	52	60	120	150	70	70
4	85	**68**	.0824	1	1	8	1	1-1/4	61	53	61	141	61	71	141	176	90	90
3	100	**80**	.0973	1	1-1/4	8	1	1-1/4	63	55	63	146	63	73	146	182	100	100
2	115	**92**	.1158	1	1-1/4	6	1-1/4	1-1/4	67	58	67	155	67	78	156	195	125	125
1	130	**104**	.1562	1-1/4	1-1/4	6	1-1/4	1-1/2	72	63	72	166	72	83	166	208	150	150
1/0	150	**120**	.1855	1-1/4	1-1/2	6	1-1/4	1-1/2	75	65	75	172	75	86	172	216	150	150
2/0	175	**140**	.2223	1-1/2	2	6	1-1/2	2	79	69	80	183	80	92	134	230	175	175
3/0	200	**160**	.2679	1-1/2	2	6	2	2	82	71	82	188	82	94	139	236	200	200
4/0	230	**184**	.3237	2	2	4	2	2-1/2	83	72	83	191	83	96	132	240	250	250
250	255	**204**	.3970	2	2-1/2	4	2	2-1/2	84	73	84	194	84	97	134	243	300	300
300	285	**228**	.4608	2	2-1/2	4	2-1/2	2-1/2	84	72	84	193	83	96	191	239	300	300
350	310	**248**	.5242	2-1/2	3	4	2-1/2	3	83	73	84	193	84	97	194	242	350	350
400	335	**268**	.5863	2-1/2	3	4	2-1/2	3	84	73	84	195	84	97	195	243	350	350
500	380	**304**	.7073	3	3	3	3	3	82	72	83	190	82	95	190	238	400	400
600	420	**336**	.8676	3	3-1/2	3	3	3-1/2	81	70	80	187	80	93	186	232	450	450
700	460	**368**	.9887	3	3-1/2	2	3-1/2	4	76	66	77	175	76	88	176	220	500	500
750	475	**380**	1.0496	3	3-1/2	2	3-1/2	4	79	68	78	182	78	90	180	226	500	500
800	490	**392**	1.1085	3-1/2	4	2	3-1/2	4	77	67	77	177	77	89	177	222	500	500
900	520	**416**	1.2311	3-1/2	4	1	3-1/2	4	76	66	76	175	76	87	175	219	600	600
1000	545	**436**	1.3478	3-1/2	4	1	4	5	76	67	78	176	77	89	178	222	600	600

* Lagging

A. To determine actual percent voltage drop:
1. Divide the actual distance by the max distance in the table for 1% drop.

B. To determine actual voltage drop:
1. Divide the actual distance by the max distance in the table for 1% drop.
2. Divide the result by 100.
3. Multiply the result by the L-L voltage.

Formulas:
A. Percent Voltage Drop = Actual Distance / Max Distance
B. Actual Volts Drop = (Actual Dist. / Max Dist.) / 100 x L-L Volts
C. New Max Dist. = (Cont. Amps / Actual FLA) x Max Distance

Temperature Correction Factors for Other then 30° C Ambient	
21-25° C (70-77° F)	1.05
26-30° C (78-86° F)	1.00
31-35° C (87-95° F)	.94
36-40° C (96-104° F)	.88
41-45° C (105-113° F)	.82
46-50° C (114-122° F)	.75
51-55° C (123-131° F)	.67
56-60° C (132-140° F)	.58
61-70° C (141-158° F)	.33

Conductor Derating Factors for More Than Three Current-Carrying Conductors	
4 Through 6	.80
7 Through 9	.70
10 Through 20	.50
21 Through 30	.45
31 Through 40	.40
41 and Above	.35

Voltage Drop Factors per 1000 Ampere Feet

60 Hz - Aluminum Conduit - Copper Conductors - Three Phase

*Load Power Factor **

Conductor AWG	DC	.45	.50	.55	.60	.65	.70	.75	.80	.85	.90	.95	1.00
# 14	6.280	2.506	2.772	3.037	3.302	3.566	3.830	4.093	4.356	4.617	4.876	5.132	5.369
# 12	3.960	1.642	1.813	1.983	2.153	2.323	2.492	2.660	2.827	2.994	3.158	3.320	3.464
# 10	2.480	1.013	1.114	1.215	1.316	1.417	1.517	1.616	1.715	1.812	1.908	2.002	2.078
# 8	1.556	.688	.753	.818	.883	.947	1.010	1.073	1.135	1.196	1.255	1.312	1.351
# 6	.982	.461	.501	.541	.580	.619	.657	.695	.732	.768	.802	.834	.849
# 4	.616	.316	.340	.365	.389	.412	.435	.458	.479	.500	.519	.536	.537
# 3	.490	.268	.287	.306	.325	.343	.361	.379	.395	.411	.425	.437	.433
# 2	.388	.225	.241	.256	.270	.284	.298	.311	.324	.335	.346	.353	.346
# 1	.308	.196	.208	.219	.230	.241	.251	.261	.269	.278	.284	.288	.277
# 1/0	.244	.169	.179	.187	.196	.204	.212	.219	.226	.232	.236	.238	.225
# 2/0	.193	.144	.151	.157	.164	.169	.174	.179	.183	.186	.188	.188	.173
# 3/0	.153	.129	.134	.139	.143	.148	.151	.155	.157	.159	.160	.158	.142
# 4/0	.122	.116	.120	.123	.126	.129	.132	.134	.135	.136	.135	.132	.116
# 250	.103	.108	.111	.114	.116	.118	.120	.121	.122	.121	.120	.116	.099
# 300	.086	.102	.104	.106	.108	.109	.110	.111	.111	.110	.107	.103	.085
# 350	.073	.095	.097	.099	.100	.101	.102	.102	.101	.100	.097	.092	.074
# 400	.064	.091	.093	.094	.095	.095	.096	.095	.094	.092	.089	.084	.066
# 500	.052	.085	.086	.087	.087	.087	.087	.086	.085	.083	.079	.074	.055
# 600	.043	.082	.083	.083	.083	.083	.082	.081	.079	.077	.073	.067	.048
# 700	.037	.081	.081	.081	.081	.081	.080	.078	.077	.074	.070	.064	.045
# 750	.034	.077	.078	.078	.078	.077	.076	.075	.073	.070	.066	.060	.042
# 800	.032	.077	.077	.077	.077	.076	.075	.073	.071	.069	.065	.058	.040
# 900	.029	.074	.075	.074	.074	.073	.072	.071	.069	.066	.062	.056	.038
# 1000	.026	.072	.072	.072	.071	.070	.069	.067	.065	.062	.058	.051	.033

* Lagging

A. To determine voltage drop:

1. Determine the load power factor and conductor size to be used.
2. Select the corresponding factor from the table.
3. Multiply the circuit distance, load amps and the table factor together, divide by the number of runs and then by 1000.

Formulas:

A. Voltage Drop = Distance x Load Amps x Factor / Runs / 1000

B. Max Drop Factor = Max Volts Drop / (Distance x Load Amps / 1000) x Runs

B. To determine minimum wire size:

1. Multiply the circuit distance by the load amps and divide by 1000 for the ampere-feet.
2. Divide the maximum volts drop allowed by the ampere-feet and then multiply by the number of runs to determine the maximum table factor.
3. Select the wire size from the column having the appropriate load power factor and a factor less than or equal to the maximum table factor.

Voltage Drop Factors per 1000 Ampere Feet

60 Hz - Aluminum Conduit - Copper Conductors - Single Phase L-L

*Load Power Factor **

Conductor AWG	DC	.45	.50	.55	.60	.65	.70	.75	.80	.85	.90	.95	1.00
# 14	6.280	2.894	3.200	3.507	3.813	4.118	4.423	4.727	5.030	5.331	5.631	5.926	6.200
# 12	3.960	1.896	2.094	2.290	2.486	2.682	2.877	3.071	3.265	3.457	3.647	3.834	4.000
# 10	2.480	1.169	1.287	1.404	1.520	1.636	1.751	1.866	1.980	2.093	2.204	2.311	2.400
# 8	1.556	.795	.870	.945	1.019	1.093	1.166	1.239	1.310	1.381	1.449	1.514	1.560
# 6	.982	.532	.578	.624	.670	.715	.759	.802	.845	.887	.926	.963	.980
# 4	.616	.365	.393	.421	.449	.476	.503	.528	.554	.578	.600	.619	.620
# 3	.490	.309	.331	.354	.375	.396	.417	.437	.456	.475	.491	.504	.500
# 2	.388	.260	.278	.295	.312	.328	.344	.360	.374	.387	.399	.408	.400
# 1	.308	.226	.240	.253	.266	.278	.290	.301	.311	.320	.328	.333	.320
# 1/0	.244	.196	.206	.216	.226	.236	.245	.253	.261	.267	.272	.274	.260
# 2/0	.193	.167	.174	.182	.189	.195	.201	.207	.212	.215	.217	.217	.200
# 3/0	.153	.149	.155	.160	.166	.170	.175	.179	.182	.184	.184	.182	.164
# 4/0	.122	.134	.138	.142	.146	.149	.152	.155	.156	.157	.156	.153	.134
# 250	.103	.125	.128	.131	.134	.136	.138	.140	.140	.140	.138	.134	.114
# 300	.086	.117	.120	.122	.124	.126	.127	.128	.128	.126	.124	.119	.098
# 350	.073	.110	.112	.114	.116	.117	.117	.117	.117	.115	.112	.107	.086
# 400	.064	.106	.107	.109	.110	.110	.110	.110	.109	.107	.103	.097	.076
# 500	.052	.098	.100	.100	.101	.101	.101	.100	.098	.095	.092	.085	.064
# 600	.043	.095	.096	.096	.096	.096	.095	.094	.092	.089	.084	.078	.056
# 700	.037	.093	.094	.094	.094	.093	.092	.091	.088	.085	.081	.074	.052
# 750	.034	.089	.090	.090	.090	.089	.088	.086	.084	.081	.076	.069	.048
# 800	.032	.089	.089	.089	.088	.088	.086	.085	.082	.079	.075	.067	.046
# 900	.029	.086	.086	.086	.086	.085	.084	.082	.080	.076	.072	.065	.044
# 1000	.026	.083	.083	.083	.082	.081	.079	.077	.075	.071	.066	.059	.038

* Lagging

A. To determine voltage drop:

1. Determine the load power factor and conductor size to be used.
2. Select the corresponding factor from the table.
3. Multiply the circuit distance, load amps and the table factor together, divide by the number of runs and then by 1000.

Formulas:

A. Voltage Drop = Distance x Load Amps x Factor / Runs / 1000

B. Max Drop Factor = Max Volts Drop / (Distance x Load Amps / 1000) x Runs

B. To determine minimum wire size:

1. Multiply the circuit distance by the load amps and divide by 1000 for the ampere-feet.
2. Divide the maximum volts drop allowed by the ampere-feet and then multiply by the number of runs to determine the maximum table factor.
3. Select the wire size from the column having the appropriate load power factor and a factor less than or equal to the maximum table factor.

Voltage Drop Factors per 1000 Ampere Feet

60 Hz - Aluminum Conduit - Copper Conductors - Single Phase L-N

*Load Power Factor **

Conductor AWG	DC	.45	.50	.55	.60	.65	.70	.75	.80	.85	.90	.95	1.00
# 14	6.280	1.447	1.600	1.753	1.906	2.059	2.211	2.363	2.515	2.666	2.815	2.963	3.100
# 12	3.960	.948	1.047	1.145	1.243	1.341	1.439	1.536	1.632	1.728	1.824	1.917	2.000
# 10	2.480	.585	.643	.702	.760	.818	.876	.933	.990	1.046	1.102	1.156	1.200
# 8	1.556	.397	.435	.472	.510	.547	.583	.619	.655	.690	.725	.757	.780
# 6	.982	.266	.289	.312	.335	.357	.379	.401	.423	.443	.463	.481	.490
# 4	.616	.182	.197	.211	.224	.238	.251	.264	.277	.289	.300	.309	.310
# 3	.490	.154	.166	.177	.188	.198	.209	.219	.228	.237	.245	.252	.250
# 2	.388	.130	.139	.148	.156	.164	.172	.180	.187	.194	.200	.204	.200
# 1	.308	.113	.120	.126	.133	.139	.145	.150	.156	.160	.164	.166	.160
# 1/0	.244	.098	.103	.108	.113	.118	.122	.127	.130	.134	.136	.137	.130
# 2/0	.193	.083	.087	.091	.094	.098	.101	.103	.106	.108	.109	.108	.100
# 3/0	.153	.074	.077	.080	.083	.085	.087	.089	.091	.092	.092	.091	.082
# 4/0	.122	.067	.069	.071	.073	.075	.076	.077	.078	.079	.078	.076	.067
# 250	.103	.062	.064	.066	.067	.068	.069	.070	.070	.070	.069	.067	.057
# 300	.086	.059	.060	.061	.062	.063	.064	.064	.064	.063	.062	.059	.049
# 350	.073	.055	.056	.057	.058	.058	.059	.059	.058	.058	.056	.053	.043
# 400	.064	.053	.054	.054	.055	.055	.055	.055	.054	.053	.052	.049	.038
# 500	.052	.049	.050	.050	.050	.050	.050	.050	.049	.048	.046	.043	.032
# 600	.043	.047	.048	.048	.048	.048	.047	.047	.046	.044	.042	.039	.028
# 700	.037	.047	.047	.047	.047	.047	.046	.045	.044	.043	.040	.037	.026
# 750	.034	.045	.045	.045	.045	.044	.044	.043	.042	.040	.038	.035	.024
# 800	.032	.044	.044	.044	.044	.044	.043	.042	.041	.040	.037	.034	.023
# 900	.029	.043	.043	.043	.043	.042	.042	.041	.040	.038	.036	.032	.022
# 1000	.026	.042	.042	.041	.041	.040	.040	.039	.037	.036	.033	.030	.019

* Lagging

A. To determine voltage drop:
1. Determine the load power factor and conductor size to be used.
2. Select the corresponding factor from the table.
3. Multiply the circuit distance, load amps and the table factor together, divide by the number of runs and then by 1000.

Formulas:

A. Voltage Drop = Distance x Load Amps x Factor / Runs / 1000
(For 3-Phase multiply by 1.732, Single Phase multiply by 2)

B. Max Drop Factor = Max Volts Drop / (Distance x Load Amps / 1000) x Runs
(For 3-Phase divide by 1.732, Single Phase divide by 2)

B. To determine minimum wire size:
1. Multiply the circuit distance by the load amps and divide by 1000 for the ampere-feet.
2. Divide the maximum volts drop allowed by the ampere-feet, then multiply by the number of runs to determine the maximum table factor.
3. Select the wire size from the column having the appropriate load power factor and a factor less than or equal to the maximum table factor.

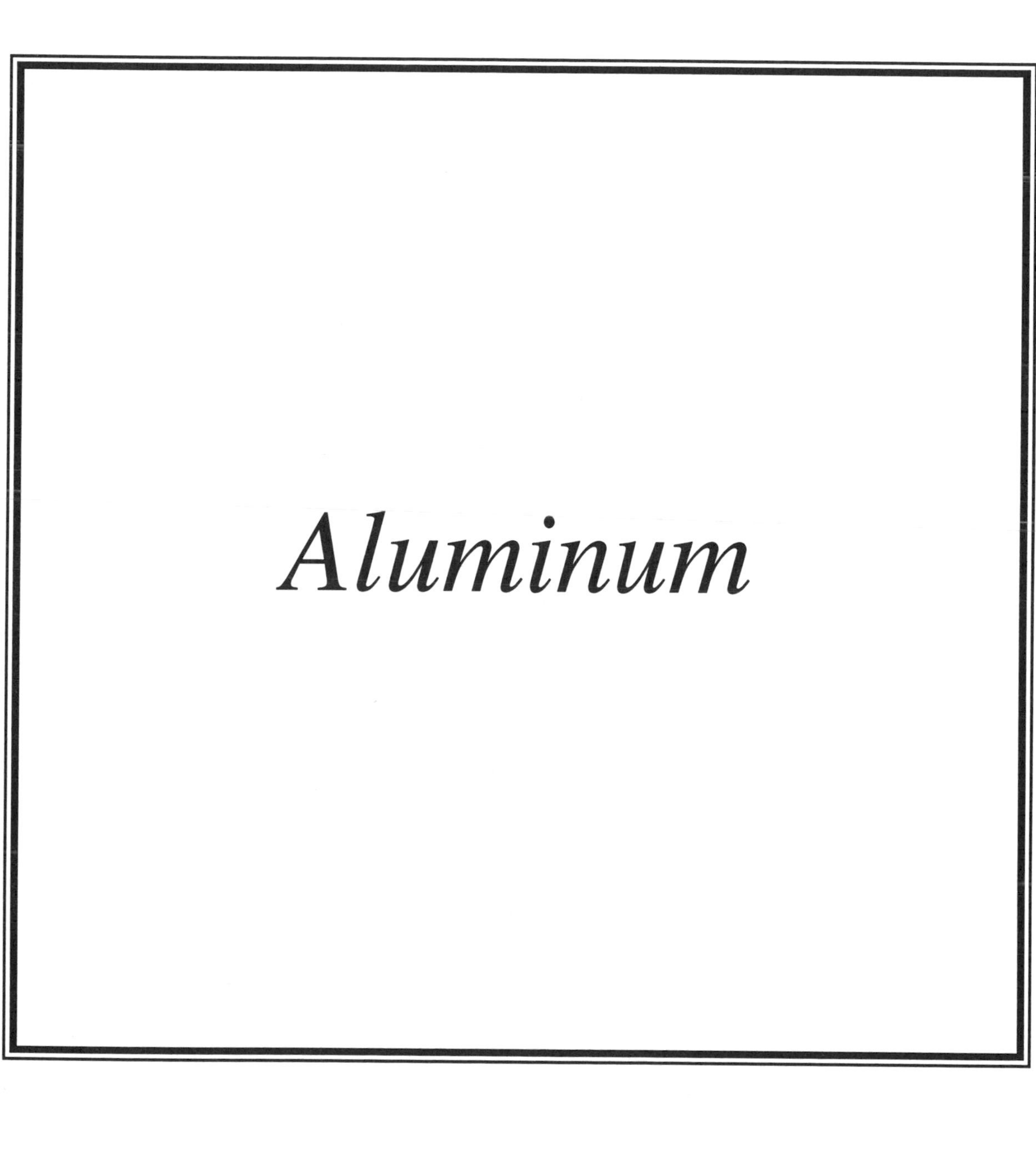

Aluminum

Circuit Design-Data for Aluminum Conductors

60 Hz - Galv. Rigid Conduit - 75° THWN - 30° C Ambient
Not More Than Three Current-Carrying Conductors

Wire AWG MCM	Rated Amps 125%	**Cont. Amps 80%**	Area Square Inches	Conduit Size 3-Each	Conduit Size 4-Each	Ins. Eq. Ground Size	Conduit 3 Ea. With Eq. Grd.	Conduit 4 Ea. With Eq. Grd.	Maximum Distance - 1 Phase for 1% Drop @ .85 P.F.* 120 Volts	1 Phase 208 Volts	1 Phase 240 Volts	1 Phase 277 Volts	Maximum Distance - 3 Phase for 1% Drop @ .85 P.F.* 208 Volts	3 Phase 240 Volts	3 Phase 480 Volts	3 Phase 600 Volts	NEC Fuse 125%	NEC C.B. 125%
14	N/A	**N/A**	N/A	N/A	N/A	N/A	N/A	N/A	N/A	N/A	N/A	N/A	N/A	N/A	N/A	N/A	N/A	N/A
12	15	**12**	.0133	1/2	1/2	12	1/2	1/2	36	31	36	84	36	42	84	105	15	15
10	25	**20**	.0211	1/2	1/2	10	1/2	1/2	35	30	35	80	35	40	80	100	25	25
8	40	**32**	.0366	1/2	3/4	8	3/4	3/4	33	29	33	76	33	38	76	95	40	40
6	50	**40**	.0507	3/4	3/4	8	3/4	1	42	36	42	96	42	48	96	120	50	50
4	65	**52**	.0824	1	1	6	1	1-1/4	50	43	50	115	50	57	115	143	70	70
3	75	**60**	.0973	1	1-1/4	6	1	1-1/4	54	47	54	124	54	62	124	156	80	80
2	90	**72**	.1158	1	1-1/4	6	1-1/4	1-1/4	55	48	55	127	55	64	127	159	90	90
1	100	**80**	.1562	1-1/4	1-1/4	6	1-1/4	1-1/2	62	54	62	142	62	71	143	179	100	100
1/0	120	**96**	.1855	1-1/4	1-1/2	4	1-1/4	1-1/2	63	54	63	145	63	72	145	181	125	125
2/0	135	**108**	.2223	1-1/2	2	4	1-1/2	2	68	59	68	156	68	78	156	195	150	150
3/0	155	**124**	.2679	1-1/2	2	4	2	2	70	61	70	162	70	81	162	202	175	175
4/0	180	**144**	.3237	2	2	4	2	2-1/2	74	64	74	172	74	85	171	214	200	200
250	205	**164**	.3970	2	2-1/2	2	2	2-1/2	73	63	73	169	73	84	168	210	225	225
300	230	**184**	.4608	2	2-1/2	2	2-1/2	3	74	64	74	171	74	85	171	213	250	250
350	250	**200**	.5242	2-1/2	3	2	2-1/2	3	75	65	75	173	75	87	174	217	250	250
400	270	**216**	.5863	2-1/2	3	2	2-1/2	3	76	66	77	176	76	88	176	220	300	300
500	310	**248**	.7073	3	3	1	3	3	76	66	76	175	76	88	176	220	350	350
600	340	**272**	.8676	3	3-1/2	1	3	3-1/2	76	66	77	176	76	88	176	221	350	350
700	375	**300**	.9887	3	3-1/2	1	3-1/2	4	73	63	73	168	73	84	168	211	400	400
750	385	**308**	1.0496	3	3-1/2	1	3-1/2	4	75	66	76	173	76	88	175	219	400	400
800	395	**316**	1.1085	3-1/2	4	1	3-1/2	4	77	68	78	179	78	90	181	226	400	400
900	425	**340**	1.2311	3-1/2	4	1/0	3-1/2	4	75	65	75	173	76	87	174	218	450	450
1000	445	**356**	1.3478	3-1/2	4	1/0	4	5	75	64	74	173	74	85	171	213	450	450

* Lagging

A. To determine actual percent voltage drop:
1. Divide the actual distance by the max distance in the table for 1% drop.

B. To determine actual voltage drop:
1. Divide the actual distance by the max distance in the table for 1% drop.
2. Divide the result by 100.
3. Multiply the result by the L-L voltage.

Formulas:
A. Percent Voltage Drop = Actual Distance / Max Distance
B. Actual Volts Drop = (Actual Dist. / Max Dist.) / 100 x L-L Volts
C. New Max Dist. = (Cont. Amps / Actual FLA) x Max Distance

Temperature Correction Factors for Other then 30° C Ambient	
21-25° C (70-77° F)	1.05
26-30° C (78-86° F)	1.00
31-35° C (87-95° F)	.94
36-40° C (96-104° F)	.88
41-45° C (105-113° F)	.82
46-50° C (114-122° F)	.75
51-55° C (123-131° F)	.67
56-60° C (132-140° F)	.58
61-70° C (141-158° F)	.33

Conductor Derating Factors for More Than Three Current-Carrying Conductors	
4 Through 6	.80
7 Through 9	.70
10 Through 20	.50
21 Through 30	.45
31 Through 40	.40
41 and Above	.35

Circuit Design-Data for Aluminum Conductors

60 Hz - Intermediate Metallic Conduit - 75° THWN - 30° C Ambient
Not More Than Three Current-Carrying Conductors

Wire AWG MCM	Rated Amps 125%	**Cont. Amps 80%**	Area Square Inches	Conduit Size 3-Each	Conduit Size 4-Each	Ins. Eq. Ground Size	Conduit 3 Ea. With Eq. Grd.	Conduit 4 Ea. With Eq. Grd.	Maximum Distance - 1 Phase for 1% Drop @ .85 P.F.* 120 Volts	208 Volts	240 Volts	277 Volts	Maximum Distance - 3 Phase for 1% Drop @ .85 P.F.* 208 Volts	240 Volts	480 Volts	600 Volts	NEC Fuse 125%	NEC C.B. 125%
14	N/A	**N/A**	N/A	N/A	N/A	N/A	N/A	N/A	N/A	N/A	N/A	N/A	N/A	N/A	N/A	N/A	N/A	N/A
12	15	**12**	.0133	1/2	1/2	12	1/2	1/2	36	31	36	84	36	42	84	105	15	15
10	25	**20**	.0211	1/2	1/2	10	1/2	1/2	35	30	35	80	35	40	80	100	25	25
8	40	**32**	.0366	1/2	3/4	8	3/4	3/4	33	29	33	76	33	38	76	95	40	40
6	50	**40**	.0507	3/4	3/4	8	3/4	1	42	36	42	96	42	48	96	120	50	50
4	65	**52**	.0824	3/4	1	6	1	1	50	43	50	115	50	57	115	143	70	70
3	75	**60**	.0973	1	1	6	1	1-1/4	54	47	54	124	54	62	124	156	80	80
2	90	**72**	.1158	1	1-1/4	6	1-1/4	1-1/4	55	48	55	127	55	64	127	159	90	90
1	100	**80**	.1562	1-1/4	1-1/4	6	1-1/4	1-1/2	62	54	62	142	62	71	143	179	100	100
1/0	120	**96**	.1855	1-1/4	1-1/2	4	1-1/4	1-1/2	63	54	63	145	63	72	145	181	125	125
2/0	135	**108**	.2223	1-1/4	1-1/2	4	1-1/2	2	68	59	68	156	68	78	153	195	150	150
3/0	155	**124**	.2679	1-1/2	2	4	1-1/2	2	70	61	70	162	70	81	162	202	175	175
4/0	180	**144**	.3237	2	2	4	2	2	74	64	74	172	74	85	171	214	200	200
250	205	**164**	.3970	2	2-1/2	2	2	2-1/2	73	63	73	169	73	84	168	210	225	225
300	230	**184**	.4608	2	2-1/2	2	2-1/2	2-1/2	74	64	74	171	74	85	171	213	250	250
350	250	**200**	.5242	2-1/2	2-1/2	2	2-1/2	3	75	65	75	173	75	87	174	217	250	250
400	270	**216**	.5863	2-1/2	3	2	2-1/2	3	76	66	77	176	76	88	176	220	300	300
500	310	**248**	.7073	2-1/2	3	1	3	3	76	66	76	175	76	88	176	220	350	350
600	340	**272**	.8676	3	3-1/2	1	3	3-1/2	76	66	77	176	76	88	176	221	350	350
700	375	**300**	.9887	3	3-1/2	1	3	3-1/2	73	63	73	168	73	84	168	211	400	400
750	385	**308**	1.0496	3	3-1/2	1	3-1/2	4	75	66	76	173	76	88	175	219	400	400
800	395	**316**	1.1085	3	3-1/2	1	3-1/2	4	77	68	78	179	78	90	181	226	400	400
900	425	**340**	1.2311	3-1/2	4	1/0	3-1/2	4	75	65	75	173	76	87	174	218	450	450
1000	445	**356**	1.3478	3-1/2	4	1/0	3-1/2	4.0 **	75	64	74	173	74	85	171	213	450	450

* Lagging

A. To determine actual percent voltage drop:
 1. Divide the actual distance by the max distance in the table for 1% drop.

B. To determine actual voltage drop:
 1. Divide the actual distance by the max distance in the table for 1% drop.
 2. Divide the result by 100.
 3. Multiply the result by the L-L voltage.

Formulas:
A. Percent Voltage Drop = Actual Distance / Max Distance
B. Actual Volts Drop = (Actual Dist. / Max Dist.) / 100 x L-L Volts
C. New Max Dist. = (Cont. Amps / Actual FLA) x Max Distance

Temperature Correction Factors for Other then 30° C Ambient	
21-25° C (70-77° F)	1.05
26-30° C (78-86° F)	1.00
31-35° C (87-95° F)	.94
36-40° C (96-104° F)	.88
41-45° C (105-113° F)	.82
46-50° C (114-122° F)	.75
51-55° C (123-131° F)	.67
56-60° C (132-140° F)	.58
61-70° C (141-158° F)	.33

Conductor Derating Factors for More Than Three Current-Carrying Conductors	
4 Through 6	.80
7 Through 9	.70
10 Through 20	.50
21 Through 30	.45
31 Through 40	.40
41 and Above	.35

** .1247 sq. in.too small, use use this size rigid.

Circuit Design-Data for Aluminum Conductors

60 Hz - EMT Conduit - 75° THWN - 30° C Ambient
Not More Than Three Current-Carrying Conductors

Wire AWG MCM	Rated Amps 125%	**Cont. Amps 80%**	Area Square Inches	Conduit Size 3-Each	Conduit Size 4-Each	Ins. Eq. Ground Size	Conduit 3 Ea. With Eq. Grd.	Conduit 4 Ea. With Eq. Grd.	Maximum Distance - 1 Phase for 1% Drop @ .85 P.F.* 120 Volts	208 Volts	240 Volts	277 Volts	Maximum Distance - 3 Phase for 1% Drop @ .85 P.F.* 208 Volts	240 Volts	480 Volts	600 Volts	NEC Fuse 125%	NEC C.B. 125%
14	N/A	**N/A**	N/A	N/A	N/A	N/A	N/A	N/A	N/A	N/A	N/A	N/A	N/A	N/A	N/A	N/A	N/A	N/A
12	15	**12**	.0133	1/2	1/2	12	1/2	1/2	36	31	36	84	36	42	84	105	15	15
10	25	**20**	.0211	1/2	1/2	10	1/2	1/2	35	30	35	80	35	40	80	100	25	25
8	40	**32**	.0366	1/2	3/4	8	3/4	3/4	33	29	33	76	33	38	76	95	40	40
6	50	**40**	.0507	3/4	3/4	8	3/4	1	42	36	42	96	42	48	96	120	50	50
4	65	**52**	.0824	1	1	6	1	1-1/4	50	43	50	115	50	57	115	143	70	70
3	75	**60**	.0973	1	1-1/4	6	1	1-1/4	54	47	54	124	54	62	124	156	80	80
2	90	**72**	.1158	1	1-1/4	6	1-1/4	1-1/4	55	48	55	127	55	64	127	159	90	90
1	100	**80**	.1562	1-1/4	1-1/4	6	1-1/4	1-1/2	62	54	62	142	62	71	143	179	100	100
1/0	120	**96**	.1855	1-1/4	1-1/2	4	1-1/2	2	63	54	63	145	63	72	145	181	125	125
2/0	135	**108**	.2223	1-1/2	2	4	1-1/2	2	68	59	68	156	68	78	156	195	150	150
3/0	155	**124**	.2679	1-1/2	2	4	2	2	70	61	70	162	70	81	162	202	175	175
4/0	180	**144**	.3237	2	2	4	2	2-1/2	74	64	74	172	74	85	171	214	200	200
250	205	**164**	.3970	2	2-1/2	2	2	2-1/2	73	63	73	169	73	84	168	210	225	225
300	230	**184**	.4608	2	2-1/2	2	2-1/2	2-1/2	74	64	74	171	74	85	171	213	250	250
350	250	**200**	.5242	2-1/2	2-1/2	2	2-1/2	2-1/2	75	65	75	173	75	87	174	217	250	250
400	270	**216**	.5863	2-1/2	2-1/2	2	2-1/2	3	76	66	77	176	76	88	176	220	300	300
500	310	**248**	.7073	2-1/2	3	1	2-1/2	3	76	66	76	175	76	88	176	220	350	350
600	340	**272**	.8676	3	3	1	3	3-1/2	76	66	77	176	76	88	176	221	350	350
700	375	**300**	.9887	3	3-1/2	1	3	3-1/2	73	63	73	168	73	84	168	211	400	400
750	385	**308**	1.0496	3	3-1/2	1	3	3-1/2	75	66	76	173	76	88	175	219	400	400
800	395	**316**	1.1085	3	3-1/2	1	3	3-1/2	77	68	78	179	78	90	181	226	400	400
900	425	**340**	1.2311	3	4	1/0	3-1/2	4	75	65	75	173	76	87	174	218	450	450
1000	445	**356**	1.3478	3-1/2	4	1/0	3-1/2	4	75	64	74	173	74	85	171	213	450	450

* Lagging

A. To determine actual percent voltage drop:
1. Divide the actual distance by the max distance in the table for 1% drop.

B. To determine actual voltage drop:
1. Divide the actual distance by the max distance in the table for 1% drop.
2. Divide the result by 100.
3. Multiply the result by the L-L voltage.

Formulas:
A. Percent Voltage Drop = Actual Distance / Max Distance
B. Actual Volts Drop = (Actual Dist. / Max Dist.) / 100 x L-L Volts
C. New Max Dist. = (Cont. Amps / Actual FLA) x Max Distance

Temperature Correction Factors for Other then 30° C Ambient	
21-25° C (70-77° F)	1.05
26-30° C (78-86° F)	1.00
31-35° C (87-95° F)	.94
36-40° C (96-104° F)	.88
41-45° C (105-113° F)	.82
46-50° C (114-122° F)	.75
51-55° C (123-131° F)	.67
56-60° C (132-140° F)	.58
61-70° C (141-158° F)	.33

Conductor Derating Factors for More Than Three Current-Carrying Conductors	
4 Through 6	.80
7 Through 9	.70
10 Through 20	.50
21 Through 30	.45
31 Through 40	.40
41 and Above	.35

Circuit Design-Data for Aluminum Conductors

60 Hz - Liquidtight Flex Metal - 75° THWN - 30° C Ambient
Not More Than Three Current-Carrying Conductors

Wire AWG MCM	Rated Amps 125%	**Cont. Amps 80%**	Area Square Inches	Conduit Size 3-Each	Conduit Size 4-Each	Ins. Eq. Ground Size	Conduit 3 Ea. With Eq. Grd.	Conduit 4 Ea. With Eq. Grd.	Maximum Distance - 1 Phase for 1% Drop @ .85 P.F.* 120 Volts	208 Volts	240 Volts	277 Volts	Maximum Distance - 3 Phase for 1% Drop @ .85 P.F.* 208 Volts	240 Volts	480 Volts	600 Volts	NEC Fuse 125%	NEC C.B. 125%
14	N/A	**N/A**	N/A	N/A	N/A	N/A	N/A	N/A	N/A	N/A	N/A	N/A	N/A	N/A	N/A	N/A	N/A	N/A
12	15	**12**	.0133	1/2	1/2	12	1/2	1/2	36	31	36	84	36	42	84	105	15	15
10	25	**20**	.0211	1/2	1/2	10	1/2	1/2	35	30	35	80	35	40	80	100	25	25
8	40	**32**	.0366	1/2	3/4	8	3/4	3/4	33	29	33	76	33	38	76	95	40	40
6	50	**40**	.0507	3/4	3/4	8	3/4	1	42	36	42	96	42	48	96	120	50	50
4	65	**52**	.0824	1	1	6	1	1-1/4	50	43	50	115	50	57	115	143	70	70
3	75	**60**	.0973	1	1-1/4	6	1	1-1/4	54	47	54	124	54	62	124	156	80	80
2	90	**72**	.1158	1	1-1/4	6	1-1/4	1-1/4	55	48	55	127	55	64	127	159	90	90
1	100	**80**	.1562	1-1/4	1-1/4	6	1-1/4	1-1/2	62	54	62	142	62	71	143	179	100	100
1/0	120	**96**	.1855	1-1/4	1-1/2	4	1-1/2	2	63	54	63	145	63	72	145	181	125	125
2/0	135	**108**	.2223	1-1/2	2	4	1-1/2	2	68	59	68	156	68	78	156	195	150	150
3/0	155	**124**	.2679	1-1/2	2	4	2	2	70	61	70	162	70	81	162	202	175	175
4/0	180	**144**	.3237	2	2	4	2	2-1/2	74	64	74	172	74	85	171	214	200	200
250	205	**164**	.3970	2	2-1/2	2	2-1/2	2-1/2	73	63	73	169	73	84	168	210	225	225
300	230	**184**	.4608	2	2-1/2	2	2-1/2	3	74	64	74	171	74	85	171	213	250	250
350	250	**200**	.5242	2-1/2	3	2	2-1/2	3	75	65	75	173	75	87	174	217	250	250
400	270	**216**	.5863	2-1/2	3	2	2-1/2	3	76	66	77	176	76	88	176	220	300	300
500	310	**248**	.7073	3	3	1	3	3	76	66	76	175	76	88	176	220	350	350
600	340	**272**	.8676	3	3-1/2	1	3	3-1/2	76	66	77	176	76	88	176	221	350	350
700	375	**300**	.9887	3	3-1/2	1	3-1/2	4	73	63	73	168	73	84	138	211	400	400
750	385	**308**	1.0496	3	4	1	3-1/2	4	75	66	76	173	76	88	175	219	400	400
800	395	**316**	1.1085	3-1/2	4	1	3-1/2	4	77	68	78	179	78	90	131	226	400	400
900	425	**340**	1.2311	3-1/2	4	1/0	3-1/2	4 **	75	65	75	173	76	87	174	218	450	450
1000	445	**356**	1.3478	3-1/2	4 **	1/0	4	5 **	75	64	74	173	74	85	171	213	450	450

* Lagging

A. To determine actual percent voltage drop:
1. Divide the actual distance by the max distance in the table for 1% drop.

B. To determine actual voltage drop:
1. Divide the actual distance by the max distance in the table for 1% drop.
2. Divide the result by 100.
3. Multiply the result by the L-L voltage.

Formulas:
A. Percent Voltage Drop = Actual Distance / Max Distance
B. Actual Volts Drop = (Actual Dist. / Max Dist.) / 100 x L-L Volts
C. New Max Dist. = (Cont. Amps / Actual FLA) x Max Distance

Temperature Correction Factors for Other then 30° C Ambient	
21-25° C (70-77° F)	1.05
26-30° C (78-86° F)	1.00
31-35° C (87-95° F)	.94
36-40° C (96-104° F)	.88
41-45° C (105-113° F)	.82
46-50° C (114-122° F)	.75
51-55° C (123-131° F)	.67
56-60° C (132-140° F)	.58
61-70° C (141-158° F)	.33

Conductor Derating Factors for More Than Three Current-Carrying Conductors	
4 Through 6	.80
7 Through 9	.70
10 Through 20	.50
21 Through 30	.45
31 Through 40	.40
41 and Above	.35

** Liquidtight conduit too small, use this size rigid.

Voltage Drop Factors per 1000 Ampere Feet

60 Hz - Steel Conduit - Aluminum Conductors - Three Phase

*Load Power Factor **

Conductor AWG	DC	.45	.50	.55	.60	.65	.70	.75	.80	.85	.90	.95	1.00
# 14	N/A	N/A	N/A	N/A	N/A	N/A	N/A	N/A	N/A	N/A	N/A	N/A	N/A
# 12	6.500	2.599	2.873	3.147	3.420	3.692	3.964	4.235	4.505	4.773	5.039	5.302	5.542
# 10	4.080	1.656	1.826	1.996	2.166	2.335	2.503	2.670	2.837	3.002	3.165	3.325	3.464
# 8	2.560	1.114	1.223	1.332	1.441	1.549	1.657	1.763	1.869	1.973	2.076	2.174	2.252
# 6	1.616	.730	.797	.864	.930	.996	1.061	1.126	1.189	1.251	1.311	1.367	1.403
# 4	1.016	.490	.532	.573	.613	.653	.693	.731	.769	.806	.840	.872	.883
# 3	.806	.403	.435	.466	.497	.528	.558	.587	.616	.643	.668	.690	.693
# 2	.638	.338	.363	.387	.412	.435	.458	.481	.503	.523	.542	.557	.554
# 1	.506	.283	.302	.321	.339	.356	.374	.390	.406	.420	.433	.442	.433
# 1/0	.402	.241	.256	.270	.284	.298	.311	.323	.334	.345	.353	.359	.346
# 2/0	.318	.208	.220	.231	.241	.251	.261	.270	.278	.285	.290	.292	.277
# 3/0	.252	.182	.191	.199	.207	.215	.222	.228	.234	.239	.242	.242	.225
# 4/0	.200	.158	.165	.170	.176	.181	.186	.189	.193	.195	.195	.193	.173
# 250	.169	.147	.152	.157	.161	.165	.169	.171	.173	.174	.173	.170	.149
# 300	.141	.135	.139	.142	.145	.148	.150	.152	.153	.153	.151	.146	.125
# 350	.121	.126	.130	.132	.135	.137	.138	.139	.139	.138	.136	.131	.109
# 400	.106	.119	.121	.123	.125	.126	.127	.128	.127	.126	.123	.117	.095
# 500	.085	.109	.111	.112	.113	.114	.114	.113	.112	.110	.106	.100	.078
# 600	.071	.104	.105	.106	.106	.106	.105	.104	.103	.100	.095	.088	.066
# 700	.061	.102	.102	.103	.103	.103	.102	.100	.098	.095	.091	.084	.061
# 750	.056	.098	.099	.099	.099	.098	.097	.095	.093	.089	.085	.077	.054
# 800	.053	.095	.095	.095	.094	.093	.092	.090	.088	.084	.079	.071	.048
# 900	.047	.093	.093	.093	.092	.091	.090	.088	.085	.081	.076	.068	.045
# 1000	.042	.091	.091	.090	.090	.089	.087	.085	.082	.079	.074	.066	.043

* Lagging

A. To determine voltage drop:

1. Determine the load power factor and conductor size to be used.
2. Select the corresponding factor from the table.
3. Multiply the circuit distance, load amps and the table factor together, divide by the number of runs and then by 1000.

Formulas:

A. Voltage Drop = Distance x Load Amps x Factor / Runs / 1000

B. Max Drop Factor = Max Volts Drop / (Distance x Load Amps / 1000) x Runs

B. To determine minimum wire size:

1. Multiply the circuit distance by the load amps and divide by 1000 for the ampere-feet.
2. Divide the maximum volts drop allowed by the ampere-feet and then multiply by the number of runs to determine the maximum table factor.
3. Select the wire size from the column having the appropriate load power factor and a factor less than or equal to the maximum table factor.

Voltage Drop Factors per 1000 Ampere Feet

60 Hz - Steel Conduit - Aluminum Conductors - Single Phase L-L

*Load Power Factor **

Conductor AWG	DC	.45	.50	.55	.60	.65	.70	.75	.80	.85	.90	.95	1.00
# 14	N/A	N/A	N/A	N/A	N/A	N/A	N/A	N/A	N/A	N/A	N/A	N/A	N/A
# 12	6.500	3.001	3.318	3.634	3.949	4.263	4.577	4.890	5.202	5.512	5.819	6.122	6.400
# 10	4.080	1.913	2.109	2.305	2.501	2.696	2.890	3.083	3.276	3.466	3.655	3.839	4.000
# 8	2.560	1.286	1.413	1.539	1.664	1.789	1.913	2.036	2.158	2.278	2.397	2.511	2.600
# 6	1.616	.843	.921	.998	1.074	1.150	1.225	1.300	1.373	1.444	1.514	1.579	1.620
# 4	1.016	.566	.614	.661	.708	.754	.800	.844	.888	.930	.970	1.006	1.020
# 3	.806	.465	.502	.539	.574	.610	.644	.678	.711	.742	.771	.797	.800
# 2	.638	.390	.419	.447	.475	.503	.529	.555	.580	.604	.626	.644	.640
# 1	.506	.327	.349	.370	.391	.412	.431	.450	.468	.485	.500	.511	.500
# 1/0	.402	.278	.295	.312	.328	.344	.359	.373	.386	.398	.408	.414	.400
# 2/0	.318	.240	.254	.266	.278	.290	.301	.311	.321	.329	.335	.338	.320
# 3/0	.252	.210	.220	.230	.239	.248	.256	.264	.270	.276	.279	.279	.260
# 4/0	.200	.183	.190	.197	.203	.209	.214	.219	.222	.225	.225	.222	.200
# 250	.169	.170	.176	.181	.186	.191	.195	.198	.200	.201	.200	.196	.172
# 300	.141	.156	.160	.164	.168	.171	.174	.175	.176	.176	.174	.169	.144
# 350	.121	.146	.150	.153	.156	.158	.160	.161	.161	.160	.157	.151	.126
# 400	.106	.137	.140	.142	.144	.146	.147	.147	.147	.145	.142	.135	.110
# 500	.085	.126	.128	.130	.131	.131	.132	.131	.130	.127	.123	.115	.090
# 600	.071	.120	.121	.122	.122	.122	.122	.120	.118	.115	.110	.102	.076
# 700	.061	.117	.118	.119	.119	.118	.118	.116	.114	.110	.105	.096	.070
# 750	.056	.114	.114	.114	.114	.113	.112	.110	.107	.103	.098	.089	.062
# 800	.053	.109	.109	.109	.109	.108	.106	.104	.101	.097	.091	.083	.056
# 900	.047	.107	.107	.107	.106	.105	.104	.101	.098	.094	.088	.079	.052
# 1000	.042	.105	.105	.104	.104	.102	.101	.098	.095	.091	.085	.076	.050

* Lagging

A. To determine voltage drop:

1. Determine the load power factor and conductor size to be used.
2. Select the corresponding factor from the table.
3. Multiply the circuit distance, load amps and the table factor together, divide by the number of runs and then by 1000.

B. To determine minimum wire size:

1. Multiply the circuit distance by the load amps and divide by 1000 for the ampere-feet.
2. Divide the maximum volts drop allowed by the ampere-feet and then multiply by the number of runs to determine the maximum table factor.
3. Select the wire size from the column having the appropriate load power factor and a factor less than or equal to the maximum table factor.

Formulas:

A. Voltage Drop = Distance x Load Amps x Factor / Runs / 1000

B. Max Drop Factor = Max Volts Drop / (Distance x Load Amps / 1000) x Runs

Voltage Drop Factors per 1000 Ampere Feet

60 Hz - Steel Conduit - Aluminum Conductors - Single Phase L-N

*Load Power Factor **

Conductor AWG	DC	.45	.50	.55	.60	.65	.70	.75	.80	.85	.90	.95	1.00
# 14	N/A	N/A	N/A	N/A	N/A	N/A	N/A	N/A	N/A	N/A	N/A	N/A	N/A
# 12	6.500	1.501	1.659	1.817	1.974	2.132	2.289	2.445	2.601	2.756	2.910	3.061	3.200
# 10	4.080	.956	1.055	1.153	1.250	1.348	1.445	1.542	1.638	1.733	1.827	1.920	2.000
# 8	2.560	.643	.706	.769	.832	.894	.956	1.018	1.079	1.139	1.198	1.255	1.300
# 6	1.616	.422	.460	.499	.537	.575	.613	.650	.686	.722	.757	.789	.810
# 4	1.016	.283	.307	.331	.354	.377	.400	.422	.444	.465	.485	.503	.510
# 3	.806	.233	.251	.269	.287	.305	.322	.339	.355	.371	.386	.398	.400
# 2	.638	.195	.209	.224	.238	.251	.265	.278	.290	.302	.313	.322	.320
# 1	.506	.163	.174	.185	.196	.206	.216	.225	.234	.243	.250	.255	.250
# 1/0	.402	.139	.148	.156	.164	.172	.179	.186	.193	.199	.204	.207	.200
# 2/0	.318	.120	.127	.133	.139	.145	.151	.156	.160	.164	.168	.169	.160
# 3/0	.252	.105	.110	.115	.120	.124	.128	.132	.135	.138	.140	.140	.130
# 4/0	.200	.091	.095	.098	.102	.105	.107	.109	.111	.112	.113	.111	.100
# 250	.169	.085	.088	.091	.093	.095	.097	.099	.100	.100	.100	.098	.086
# 300	.141	.078	.080	.082	.084	.086	.087	.088	.088	.088	.087	.084	.072
# 350	.121	.073	.075	.076	.078	.079	.080	.080	.080	.080	.078	.075	.063
# 400	.106	.069	.070	.071	.072	.073	.073	.074	.073	.073	.071	.068	.055
# 500	.085	.063	.064	.065	.065	.066	.066	.065	.065	.064	.061	.058	.045
# 600	.071	.060	.061	.061	.061	.061	.061	.060	.059	.058	.055	.051	.038
# 700	.061	.059	.059	.059	.059	.059	.059	.058	.057	.055	.052	.048	.035
# 750	.056	.057	.057	.057	.057	.057	.056	.055	.054	.052	.049	.044	.031
# 800	.053	.055	.055	.055	.054	.054	.053	.052	.051	.049	.046	.041	.028
# 900	.047	.054	.054	.054	.053	.053	.052	.051	.049	.047	.044	.039	.026
# 1000	.042	.052	.052	.052	.052	.051	.050	.049	.048	.045	.043	.038	.025

* Lagging

A. To determine voltage drop:
1. Determine the load power factor and conductor size to be used.
2. Select the corresponding factor from the table.
3. Multiply the circuit distance, load amps and the table factor together, divide by the number of runs and then by 1000.

Formulas:

A. Voltage Drop = Distance x Load Amps x Factor / Runs / 1000
(For 3-Phase multiply by 1.732, Single Phase multiply by 2)

B. Max Drop Factor = Max Volts Drop / (Distance x Load Amps / 1000) x Runs
(For 3-Phase divide by 1.732, Single Phase divide by 2)

B. To determine minimum wire size:
1. Multiply the circuit distance by the load amps and divide by 1000 for the ampere-feet.
2. Divide the maximum volts drop allowed by the ampere-feet, then multiply by the number of runs to determine the maximum table factor.
3. Select the wire size from the column having the appropriate load power factor and a factor less than or equal to the maximum table factor.

Circuit Design-Data for Aluminum Conductors

60 Hz - PVC Sch 40 Conduit - 75° THWN - 30° C Ambient
Not More Than Three Current-Carrying Conductors

Wire AWG MCM	Rated Amps 125%	**Cont. Amps 80%**	Area Square Inches	Conduit Size 3-Each	Conduit Size 4-Each	Ins. Eq. Ground Size	Conduit 3 Ea. With Eq. Grd.	Conduit 4 Ea. With Eq. Grd.	Maximum Distance - 1 Phase for 1% Drop @ .85 P.F.* 120 Volts	208 Volts	240 Volts	277 Volts	Maximum Distance - 3 Phase for 1% Drop @ .85 P.F.* 208 Volts	240 Volts	480 Volts	600 Volts	NEC Fuse 125%	NEC C.B. 125%
14	N/A	**N/A**	N/A	N/A	N/A	N/A	N/A	N/A	N/A	N/A	N/A	N/A	N/A	N/A	N/A	N/A	N/A	N/A
12	15	**12**	.0133	1/2	1/2	12	1/2	1/2	36	32	36	84	36	42	84	105	15	15
10	25	**20**	.0211	1/2	1/2	10	1/2	1/2	35	30	35	80	35	40	80	100	25	25
8	40	**32**	.0366	1/2	3/4	8	3/4	3/4	33	29	33	76	33	38	76	96	40	40
6	50	**40**	.0507	3/4	3/4	8	3/4	1	42	36	42	97	42	48	97	121	50	50
4	65	**52**	.0824	1	1	6	1	1-1/4	50	44	50	116	50	58	116	145	70	70
3	75	**60**	.0973	1	1-1/4	6	1-1/4	1-1/4	55	47	55	126	55	63	127	158	80	80
2	90	**72**	.1158	1	1-1/4	6	1-1/4	1-1/4	56	49	56	130	56	65	130	163	90	90
1	100	**80**	.1562	1-1/4	1-1/2	6	1-1/4	1-1/2	63	55	63	146	63	73	146	183	100	100
1/0	120	**96**	.1855	1-1/4	1-1/2	4	1-1/2	2	65	56	65	150	65	75	149	187	125	125
2/0	135	**108**	.2223	1-1/2	2	4	1-1/2	2	70	61	70	161	70	81	162	202	150	150
3/0	155	**124**	.2679	1-1/2	2	4	2	2	73	63	73	168	73	84	168	210	175	175
4/0	180	**144**	.3237	2	2	4	2	2-1/2	78	68	78	180	78	90	180	225	200	200
250	205	**164**	.3970	2	2-1/2	2	2	2-1/2	78	67	78	180	78	90	180	224	225	225
300	230	**184**	.4608	2	2-1/2	2	2-1/2	3	80	69	80	184	80	92	184	230	250	250
350	250	**200**	.5242	2-1/2	3	2	2-1/2	3	82	71	82	190	83	95	190	238	250	250
400	270	**216**	.5863	2-1/2	3	2	2-1/2	3	83	72	83	191	83	96	192	239	300	300
500	310	**248**	.7073	3	3	1	3	3-1/2	85	74	85	196	85	98	196	244	350	350
600	340	**272**	.8676	3	3-1/2	1	3	3-1/2	87	75	87	200	86	99	198	248	350	350
700	375	**300**	.9887	3	3-1/2	1	3-1/2	4	85	74	85	196	86	99	198	247	400	400
750	385	**308**	1.0496	3-1/2	4	1	3-1/2	4	87	76	88	200	88	101	202	253	400	400
800	395	**316**	1.1085	3-1/2	4	1	3-1/2	4	88	77	88	204	89	103	205	257	400	400
900	425	**340**	1.2311	3-1/2	4	1/0	3-1/2	5	84	73	84	194	84	97	193	242	450	450
1000	445	**356**	1.3478	3-1/2	5	1/0	4	5	86	75	86	200	86	99	198	248	450	450

* Lagging

A. To determine actual percent voltage drop:
1. Divide the actual distance by the max distance in the table for 1% drop.

B. To determine actual voltage drop:
1. Divide the actual distance by the max distance in the table for 1% drop.
2. Divide the result by 100.
3. Multiply the result by the L-L voltage.

Formulas:
A. Percent Voltage Drop = Actual Distance / Max Distance
B. Actual Volts Drop = (Actual Dist. / Max Dist.) / 100 x L-L Volts
C. New Max Dist. = (Cont. Amps / Actual FLA) x Max Distance

Temperature Correction Factors for Other then 30° C Ambient	
21-25° C (70-77° F)	1.05
26-30° C (78-86° F)	1.00
31-35° C (87-95° F)	.94
36-40° C (96-104° F)	.88
41-45° C (105-113° F)	.82
46-50° C (114-122° F)	.75
51-55° C (123-131° F)	.67
56-60° C (132-140° F)	.58
61-70° C (141-158° F)	.33

Conductor Derating Factors for More Than Three Current-Carrying Conductors	
4 Through 6	.80
7 Through 9	.70
10 Through 20	.50
21 Through 30	.45
31 Through 40	.40
41 and Above	.35

Voltage Drop Factors per 1000 Ampere Feet

60 Hz - PVC Conduit - Aluminum Conductors - Three Phase

*Load Power Factor **

Conductor AWG	DC	.45	.50	.55	.60	.65	.70	.75	.80	.85	.90	.95	1.00
# 14	N/A	N/A	N/A	N/A	N/A	N/A	N/A	N/A	N/A	N/A	N/A	N/A	N/A
# 12	6.500	2.578	2.852	3.126	3.400	3.674	3.946	4.219	4.490	4.760	5.029	5.294	5.542
# 10	4.080	1.636	1.807	1.978	2.148	2.317	2.487	2.655	2.823	2.990	3.155	3.318	3.464
# 8	2.560	1.094	1.204	1.314	1.423	1.532	1.640	1.748	1.855	1.961	2.066	2.167	2.252
# 6	1.616	.710	.778	.845	.912	.979	1.045	1.111	1.175	1.239	1.301	1.360	1.403
# 4	1.016	.472	.514	.555	.597	.637	.678	.717	.757	.795	.831	.865	.883
# 3	.806	.384	.417	.449	.481	.512	.543	.573	.603	.632	.659	.684	.693
# 2	.638	.319	.345	.370	.395	.419	.444	.467	.490	.512	.533	.551	.554
# 1	.506	.266	.285	.305	.324	.342	.360	.377	.394	.410	.424	.436	.433
# 1/0	.402	.224	.239	.254	.269	.283	.297	.310	.323	.335	.345	.353	.346
# 2/0	.318	.191	.203	.215	.226	.237	.247	.257	.266	.275	.282	.287	.277
# 3/0	.252	.166	.176	.185	.193	.202	.210	.217	.224	230	.234	.237	.225
# 4/0	.200	.141	.148	.155	.161	.167	.172	.177	.181	.185	.187	.187	.173
# 250	.169	.130	.135	.140	.145	.150	.154	.157	.160	.163	.163	.162	.147
# 300	.141	.119	.123	.127	.131	.134	.137	.139	.141	.142	.142	.139	.123
# 350	.121	.109	.113	.116	.119	.121	.123	.125	.126	.126	.125	.122	.106
# 400	.106	.104	.107	.109	.112	.113	.115	.116	.116	.116	.114	.110	.094
# 500	.085	.094	.096	.097	.099	.100	.100	.101	.100	.099	.096	.092	.074
# 600	.071	.088	.090	.091	.091	.092	.092	.091	.090	.089	.086	.080	.062
# 700	.061	.084	.085	.086	.086	.086	.086	.085	.083	.081	.078	.072	.054
# 750	.056	.081	.082	.083	.083	.083	.082	.081	.080	.077	.074	.068	.050
# 800	.053	.080	.080	.081	.081	.080	.080	.079	.077	.074	.071	.065	.047
# 900	.047	.079	.080	.080	.080	.079	.079	.077	.076	.073	.069	.063	.045
# 1000	.042	.075	.075	.075	.075	.075	.074	.072	.070	.068	.064	.058	.040

* Lagging

A. To determine voltage drop:

1. Determine the load power factor and conductor size to be used.
2. Select the corresponding factor from the table.
3. Multiply the circuit distance, load amps and the table factor together, divide by the number of runs and then by 1000.

Formulas:

A. Voltage Drop = Distance x Load Amps x Factor / Runs / 1000

B. Max Drop Factor = Max Volts Drop / (Distance x Load Amps / 1000) x Runs

B. To determine minimum wire size:

1. Multiply the circuit distance by the load amps and divide by 1000 for the ampere-feet.
2. Divide the maximum volts drop allowed by the ampere-feet and then multiply by the number of runs to determine the maximum table factor.
3. Select the wire size from the column having the appropriate load power factor and a factor less than or equal to the maximum table factor.

Voltage Drop Factors per 1000 Ampere Feet

60 Hz - PVC Conduit - Aluminum Conductors - Single Phase L-L

*Load Power Factor **

Conductor AWG	DC	.45	.50	.55	.60	.65	.70	.75	.80	.85	.90	.95	1.00
# 14	N/A	N/A	N/A	N/A	N/A	N/A	N/A	N/A	N/A	N/A	N/A	N/A	N/A
# 12	6.500	2.976	3.294	3.610	3.926	4.242	4.557	4.871	5.185	5.497	5.807	6.114	6.400
# 10	4.080	1.889	2.087	2.284	2.480	2.676	2.871	3.066	3.260	3.453	3.644	3.831	4.000
# 8	2.560	1.263	1.390	1.517	1.643	1.769	1.894	2.019	2.142	2.265	2.385	2.502	2.600
# 6	1.616	.820	.898	.976	1.054	1.131	1.207	1.282	1.357	1.431	1.502	1.571	1.620
# 4	1.016	.545	.593	.641	.689	.736	.783	.828	.874	.918	.960	.999	1.020
# 3	.806	.444	.481	.519	.555	.591	.627	.662	.696	.730	.761	.789	.800
# 2	.638	.368	.398	.427	.456	.484	.512	.540	.566	.591	.615	.636	.640
# 1	.506	.307	.330	.352	.374	.395	.416	.436	.455	.473	.490	.504	.500
# 1/0	.402	.259	.276	.293	.310	.327	.343	.358	.373	.386	.398	.407	.400
# 2/0	.318	.221	.234	.248	.261	.273	.285	.297	.308	.317	.325	.331	.320
# 3/0	.252	.192	.203	.213	.223	.233	.242	.251	.258	.265	.271	.273	.260
# 4/0	.200	.163	.171	.178	.186	.192	.199	.204	.209	.213	.216	.216	.200
# 250	.169	.150	.156	.162	.168	.173	.178	.182	.185	.188	.189	.187	.170
# 300	.141	.137	.142	.147	.151	.155	.158	.161	.163	.164	.164	.161	.142
# 350	.121	.126	.130	.134	.137	.140	.143	.144	.146	.146	.145	.141	.122
# 400	.106	.120	.123	.126	.129	.131	.133	.134	.134	.134	.132	.128	.108
# 500	.085	.108	.111	.112	.114	.115	.116	.116	.116	.114	.111	.106	.086
# 600	.071	.102	.104	.105	.106	.106	.106	.106	.104	.102	.099	.093	.072
# 700	.061	.098	.099	.099	.100	.100	.099	.098	.096	.094	.090	.083	.062
# 750	.056	.094	.095	.095	.096	.095	.095	.094	.092	.089	.085	.079	.058
# 800	.053	.092	.093	.093	.093	.093	.092	.091	.089	.086	.082	.075	.054
# 900	.047	.091	.092	.092	.092	.092	.091	.089	.087	.084	.080	.073	.052
# 1000	.042	.087	.087	.087	.087	.086	.085	.083	.081	.078	.074	.067	.046

* Lagging

A. To determine voltage drop:

1. Determine the load power factor and conductor size to be used.
2. Select the corresponding factor from the table.
3. Multiply the circuit distance, load amps and the table factor together, divide by the number of runs and then by 1000.

B. To determine minimum wire size:

1. Multiply the circuit distance by the load amps and divide by 1000 for the ampere-feet.
2. Divide the maximum volts drop allowed by the ampere-feet and then multiply by the number of runs to determine the maximum table factor.
3. Select the wire size from the column having the appropriate load power factor and a factor less than or equal to the maximum table factor.

Formulas:

A. Voltage Drop = Distance x Load Amps x Factor / Runs / 1000

B. Max Drop Factor = Max Volts Drop / (Distance x Load Amps / 1000) x Runs

Voltage Drop Factors per 1000 Ampere Feet

60 Hz - PVC Conduit - Aluminum Conductors - Single Phase L-N

*Load Power Factor **

Conductor AWG	DC	.45	.50	.55	.60	.65	.70	.75	.80	.85	.90	.95	1.00
# 14	N/A	N/A	N/A	N/A	N/A	N/A	N/A	N/A	N/A	N/A	N/A	N/A	N/A
# 12	6.500	1.488	1.647	1.805	1.963	2.121	2.279	2.436	2.592	2.748	2.904	3.057	3.200
# 10	4.080	.945	1.043	1.142	1.240	1.338	1.436	1.533	1.630	1.726	1.822	1.916	2.000
# 8	2.560	.631	.695	.758	.822	.885	.947	1.009	1.071	1.132	1.193	1.251	1.300
# 6	1.616	.410	.449	.488	.527	.565	.603	.641	.679	.715	.751	.785	.810
# 4	1.016	.272	.297	.321	.344	.368	.391	.414	.437	.459	.480	.499	.510
# 3	.806	.222	.241	.259	.278	.296	.314	.331	.348	.365	.380	.395	.400
# 2	.638	.184	.199	.214	.228	.242	.256	.270	.283	.296	.308	.318	.320
# 1	.506	.154	.165	.176	.187	.197	.208	.218	.228	.237	.245	.252	.250
# 1/0	.402	.129	.138	.147	.155	.163	.171	.179	.186	.193	.199	.204	.200
# 2/0	.318	.110	.117	.124	.130	.137	.143	.148	.154	.159	.163	.165	.160
# 3/0	.252	.096	.101	.107	.112	.116	.121	.125	.129	.133	.135	.137	.130
# 4/0	.200	.082	.086	.089	.093	.096	.099	.102	.105	.107	.108	.108	.100
# 250	.169	.075	.078	.081	.084	.086	.089	.091	.093	.094	.094	.094	.085
# 300	.141	.069	.071	.073	.075	.077	.079	.080	.081	.082	.082	.080	.071
# 350	.121	.063	.065	.067	.069	.070	.071	.072	.073	.073	.072	.070	.061
# 400	.106	.060	.062	.063	.064	.065	.066	.067	.067	.067	.066	.064	.054
# 500	.085	.054	.055	.056	.057	.058	.058	.058	.058	.057	.056	.053	.043
# 600	.071	.051	.052	.052	.053	.053	.053	.053	.052	.051	.049	.046	.036
# 700	.061	.049	.049	.050	.050	.050	.050	.049	.048	.047	.045	.042	.031
# 750	.056	.047	.047	.048	.048	.048	.047	.047	.046	.045	.043	.039	.029
# 800	.053	.046	.046	.047	.047	.046	.046	.045	.044	.043	.041	.038	.027
# 900	.047	.046	.046	.046	.046	.046	.045	.045	.044	.042	.040	.037	.026
# 1000	.042	.043	.044	.044	.043	.043	.043	.042	.041	.039	.037	.033	.023

* Lagging

A. To determine voltage drop:
1. Determine the load power factor and conductor size to be used.
2. Select the corresponding factor from the table.
3. Multiply the circuit distance, load amps and the table factor together, divide by the number of runs and then by 1000.

Formulas:

A. Voltage Drop = Distance x Load Amps x Factor / Runs / 1000
(For 3-Phase multiply by 1.732, Single Phase multiply by 2)

B. Max Drop Factor = Max Volts Drop / (Distance x Load Amps / 1000) x Runs
(For 3-Phase divide by 1.732, Single Phase divide by 2)

B. To determine minimum wire size:
1. Multiply the circuit distance by the load amps and divide by 1000 for the ampere-feet.
2. Divide the maximum volts drop allowed by the ampere-feet, then multiply by the number of runs to determine the maximum table factor.
3. Select the wire size from the column having the appropriate load power factor and a factor less than or equal to the maximum table factor.

Circuit Design-Data for Aluminum Conductors

60 Hz - Aluminum Rigid - 75° THWN - 30° C Ambient
Not More Than Three Current-Carrying Conductors

Wire AWG MCM	Rated Amps 125%	**Cont. Amps 80%**	Area Square Inches	Conduit Size 3-Each	Conduit Size 4-Each	Ins. Eq. Ground Size	Conduit 3 Ea. With Eq. Grd.	Conduit 4 Ea. With Eq. Grd.	Maximum Distance - 1 Phase for 1% Drop @ .85 P.F.* 120 Volts	208 Volts	240 Volts	277 Volts	Maximum Distance 3 Phase for 1% Drop @ .85 P.F.* 208 Volts	240 Volts	480 Volts	600 Volts	NEC Fuse 125%	NEC C.B. 125%
14	N/A	**N/A**	N/A	N/A	N/A	N/A	N/A	N/A	N/A	N/A	N/A	N/A	N/A	N/A	N/A	N/A	N/A	N/A
12	15	**12**	.0133	1/2	1/2	12	1/2	1/2	36	32	36	84	36	42	84	105	15	15
10	25	**20**	.0211	1/2	1/2	10	1/2	1/2	35	30	35	80	35	40	80	100	25	25
8	40	**32**	.0366	1/2	3/4	8	3/4	3/4	33	29	33	76	33	38	76	96	40	40
6	50	**40**	.0507	3/4	3/4	8	3/4	1	42	36	42	97	42	48	97	121	50	50
4	65	**52**	.0824	1	1	6	1	1-1/4	50	44	50	116	50	58	116	145	70	70
3	75	**60**	.0973	1	1-1/4	6	1	1-1/4	54	46	54	124	54	62	124	155	80	80
2	90	**72**	.1158	1	1-1/4	6	1-1/4	1-1/4	56	49	56	130	56	65	130	163	90	90
1	100	**80**	.1562	1-1/4	1-1/4	6	1-1/4	1-1/2	61	53	61	141	61	71	141	176	100	100
1/0	120	**96**	.1855	1-1/4	1-1/2	4	1-1/4	1-1/2	62	54	62	143	62	72	143	179	125	125
2/0	135	**108**	.2223	1-1/2	2	4	1-1/2	2	70	61	70	161	70	81	162	202	150	150
3/0	155	**124**	.2679	1-1/2	2	4	2	2	73	63	73	168	73	84	168	210	175	175
4/0	180	**144**	.3237	2	2	4	2	2-1/2	72	63	72	167	73	84	168	209	200	200
250	205	**164**	.3970	2	2-1/2	2	2	2-1/2	75	65	75	172	75	86	172	215	225	225
300	230	**184**	.4608	2	2-1/2	2	2-1/2	3	76	66	76	175	76	88	175	219	250	250
350	250	**200**	.5242	2-1/2	3	2	2-1/2	3	78	68	78	180	78	90	179	224	250	250
400	270	**216**	.5863	2-1/2	3	2	2-1/2	3	78	68	78	181	78	90	181	226	300	300
500	310	**248**	.7073	3	3	1	3	3	79	68	79	183	79	91	183	228	350	350
600	340	**272**	.8676	3	3-1/2	1	3	3-1/2	80	69	79	185	80	92	184	230	350	350
700	375	**300**	.9887	3	3-1/2	1	3-1/2	4	75	65	75	174	75	87	174	217	400	400
750	385	**308**	1.0496	3	3-1/2	1	3-1/2	4	80	69	80	184	79	92	183	229	400	400
800	395	**316**	1.1085	3-1/2	4	1	3-1/2	4	83	71	82	191	82	95	190	237	400	400
900	425	**340**	1.2311	3-1/2	4	1/0	3-1/2	4	78	69	79	181	79	92	183	229	450	450
1000	445	**356**	1.3478	3-1/2	4	1/0	4	5	80	69	79	185	79	91	182	228	450	450

* Lagging

A. To determine actual percent voltage drop:
1. **Divide the actual distance by the max distance in the table for 1% drop.**

B. To determine actual voltage drop:
1. **Divide the actual distance by the max distance in the table for 1% drop.**
2. **Divide the result by 100.**
3. **Multiply the result by the L-L voltage.**

Formulas:
A. Percent Voltage Drop = Actual Distance / Max Distance
B. Actual Volts Drop = (Actual Dist. / Max Dist.) / 100 x L-L Volts
C. New Max Dist. = (Cont. Amps / Actual FLA) x Max Distance

Temperature Correction Factors for Other then 30° C Ambient	
21-25° C (70-77° F)	1.05
26-30° C (78-86° F)	1.00
31-35° C (87-95° F)	.94
36-40° C (96-104° F)	.88
41-45° C (105-113° F)	.82
46-50° C (114-122° F)	.75
51-55° C (123-131° F)	.67
56-60° C (132-140° F)	.58
61-70° C (141-158° F)	.33

Conductor Derating Factors for More Than Three Current-Carrying Conductors	
4 Through 6	.80
7 Through 9	.70
10 Through 20	.50
21 Through 30	.45
31 Through 40	.40
41 and Above	.35

Voltage Drop Factors per 1000 Ampere Feet

60 Hz - Aluminum Conduit - Aluminum Conductors - Three Phase

*Load Power Factor **

Conductor AWG	DC	.45	.50	.55	.60	.65	.70	.75	.80	.85	.90	.95	1.00
# 14	N/A	N/A	N/A	N/A	N/A	N/A	N/A	N/A	N/A	N/A	N/A	N/A	N/A
# 12	6.500	2.578	2.852	3.126	3.400	3.674	3.946	4.219	4.490	4.760	5.029	5.294	5.542
# 10	4.080	1.636	1.807	1.978	2.148	2.317	2.487	2.655	2.823	2.990	3.155	3.318	3.464
# 8	2.560	1.094	1.204	1.314	1.423	1.532	1.640	1.748	1.855	1.961	2.066	2.167	2.252
# 6	1.616	.710	.778	.845	.912	.979	1.045	1.111	1.175	1.239	1.301	1.360	1.403
# 4	1.016	.472	.514	.555	.597	.637	.678	.717	.757	.795	.831	.865	.883
# 3	.806	.392	.426	.459	.491	.523	.555	.586	.617	.646	.675	.700	.710
# 2	.638	.319	.345	.370	.395	.419	.444	.467	.490	.512	.533	.551	.554
# 1	.506	.274	.294	.314	.334	.353	.372	.390	.408	.425	.440	.453	.450
# 1/0	.402	.232	.248	.264	.279	.294	.309	.323	.337	.349	.361	.369	.364
# 2/0	.318	.191	.203	.215	.226	.237	.247	.257	.266	.275	.282	.287	.277
# 3/0	.252	.166	.176	.185	.193	.202	.210	.217	.224	.230	.234	.237	.225
# 4/0	.200	.149	.157	.164	.171	.178	.184	.190	.195	.199	.202	.203	.191
# 250	.169	.134	.139	.145	.150	.155	.160	.164	.167	.170	.171	.170	.156
# 300	.141	.123	.127	.132	.136	.140	.143	.146	.148	.149	.149	.147	.132
# 350	.121	.113	.117	.121	.124	.127	.129	.132	.133	.134	.133	.130	.114
# 400	.106	.108	.111	.114	.117	.119	.121	.122	.123	.123	.122	.119	.102
# 500	.085	.098	.100	.102	.104	.105	.106	.107	.107	.106	.104	.100	.083
# 600	.071	.092	.094	.095	.097	.097	.098	.098	.097	.096	.093	.089	.071
# 700	.061	.090	.091	.093	.094	.094	.094	.094	.093	.092	.089	.084	.066
# 750	.056	.085	.086	.087	.088	.088	.088	.088	.087	.085	.082	.076	.059
# 800	.053	.083	.084	.084	.085	.085	.085	.084	.082	.080	.077	.072	.054
# 900	.047	.081	.082	.083	.083	.083	.082	.081	.080	.077	.074	.068	.050
# 1000	.042	.078	.079	.079	.079	.079	.078	.077	.076	.074	.070	.064	.047

* Lagging

A. To determine voltage drop:

1. Determine the load power factor and conductor size to be used.
2. Select the corresponding factor from the table.
3. Multiply the circuit distance, load amps and the table factor together, divide by the number of runs and then by 1000.

Formulas:

A. Voltage Drop = Distance x Load Amps x Factor / Runs / 1000

B. Max Drop Factor = Max Volts Drop / (Distance x Load Amps / 1000) x Runs

B. To determine minimum wire size:

1. Multiply the circuit distance by the load amps and divide by 1000 for the ampere-feet.
2. Divide the maximum volts drop allowed by the ampere-feet and then multiply by the number of runs to determine the maximum table factor.
3. Select the wire size from the column having the appropriate load power factor and a factor less than or equal to the maximum table factor.

Voltage Drop Factors per 1000 Ampere Feet

60 Hz - Aluminum Conduit - Aluminum Conductors - Single Phase L-L

*Load Power Factor **

Conductor AWG	DC	.45	.50	.55	.60	.65	.70	.75	.80	.85	.90	.95	1.00
# 14	N/A	N/A	N/A	N/A	N/A	N/A	N/A	N/A	N/A	N/A	N/A	N/A	N/A
# 12	6.500	2.976	3.294	3.610	3.926	4.242	4.557	4.871	5.185	5.497	5.807	6.114	6.400
# 10	4.080	1.889	2.087	2.284	2.480	2.676	2.871	3.066	3.260	3.453	3.644	3.831	4.000
# 8	2.560	1.263	1.390	1.517	1.643	1.769	1.894	2.019	2.142	2.265	2.385	2.502	2.600
# 6	1.616	.820	.898	.976	1.054	1.131	1.207	1.282	1.357	1.431	1.502	1.571	1.620
# 4	1.016	.545	.593	.641	.689	.736	.783	.828	.874	.918	.960	.999	1.020
# 3	.806	.453	.491	.530	.567	.604	.641	.677	.712	.747	.779	.808	.820
# 2	.638	.368	.398	.427	.456	.484	.512	.540	.566	.591	.615	.636	.640
# 1	.506	.316	.340	.363	.386	.408	.430	.451	.471	.490	.508	.523	.520
# 1/0	.402	.268	.286	.304	.322	.340	.357	.373	.389	.403	.416	.426	.420
# 2/0	.318	.221	.234	.248	.261	.273	.285	.297	.308	.317	.325	.331	.320
# 3/0	.252	.192	.203	.213	.223	.233	.242	.251	.258	.265	.271	.273	.260
# 4/0	.200	.172	.181	.189	.198	.205	.213	.219	.225	.230	.234	.235	.220
# 250	.169	.154	.161	.167	.174	.179	.185	.189	.193	.196	.198	.197	.180
# 300	.141	.142	.147	.152	.157	.161	.165	.168	.171	.172	.173	.170	.152
# 350	.121	.131	.135	.139	.143	.147	.150	.152	.154	.154	.154	.150	.132
# 400	.106	.125	.128	.132	.135	.137	.140	.141	.142	.142	.141	.137	.118
# 500	.085	.113	.116	.118	.120	.122	.123	.124	.124	.123	.120	.116	.096
# 600	.071	.107	.109	.110	.112	.113	.113	.113	.112	.111	.108	.102	.082
# 700	.061	.104	.106	.107	.108	.109	.109	.109	.108	.106	.102	.097	.076
# 750	.056	.098	.100	.101	.102	.102	.102	.101	.100	.098	.094	.088	.068
# 800	.053	.096	.097	098	.098	.098	.098	.097	.095	.093	.089	.083	.062
# 900	.047	.094	.095	095	.096	.095	.095	.094	.092	.089	.085	.079	.058
# 1000	.042	.090	.091	.092	.092	.091	.091	.089	.088	.085	.081	.074	.054

* Lagging

A. To determine voltage drop:

1. Determine the load power factor and conductor size to be used.
2. Select the corresponding factor from the table.
3. Multiply the circuit distance, load amps and the table factor together, divide by the number of runs and then by 1000.

Formulas:

A. Voltage Drop = Distance x Load Amps x Factor / Runs / 1000

B. Max Drop Factor = Max Volts Drop / (Distance x Load Amps / 1000) x Runs

B. To determine minimum wire size:

1. Multiply the circuit distance by the load amps and divide by 1000 for the ampere-feet.
2. Divide the maximum volts drop allowed by the ampere-feet and then multiply by the number of runs to determine the maximum table factor.
3. Select the wire size from the column having the appropriate load power factor and a factor less than or equal to the maximum table factor.

Voltage Drop Factors per 1000 Ampere Feet

60 Hz - Aluminum Conduit - Aluminum Conductors - Single Phase L-N

*Load Power Factor **

Conductor AWG	DC	.45	.50	.55	.60	.65	.70	.75	.80	.85	.90	.95	1.00
# 14	N/A	N/A	N/A	N/A	N/A	N/A	N/A	N/A	N/A	N/A	N/A	N/A	N/A
# 12	6.500	1.488	1.647	1.805	1.963	2.121	2.279	2.436	2.592	2.748	2.904	3.057	3.200
# 10	4.080	.945	1.043	1.142	1.240	1.338	1.436	1.533	1.630	1.726	1.822	1.916	2.000
# 8	2.560	.631	.695	.758	.822	.885	.947	1.009	1.071	1.132	1.193	1.251	1.300
# 6	1.616	.410	.449	.488	.527	.565	.603	.641	.679	.715	.751	.785	.810
# 4	1.016	.272	.297	.321	.344	.368	.391	.414	.437	.459	.480	.499	.510
# 3	.806	.226	.246	.265	.284	.302	.321	.339	.356	.373	.389	.404	.410
# 2	.638	.184	.199	.214	.228	.242	.256	.270	.283	.296	.308	.318	.320
# 1	.506	.158	.170	.181	.193	.204	.215	.225	.236	.245	.254	.261	.260
# 1/0	.402	.134	.143	.152	.161	.170	.178	.187	.194	.202	.208	.213	.210
# 2/0	.318	.110	.117	.124	.130	.137	.143	.148	.154	.159	.163	.165	.160
# 3/0	.252	.096	.101	.107	.112	.116	.121	.125	.129	.133	.135	.137	.130
# 4/0	.200	.086	.091	.095	.099	.103	.106	.110	.113	.115	.117	.117	.110
# 250	.169	.077	.081	.084	.087	.090	.092	.095	.097	.098	.099	.098	.090
# 300	.141	.071	.074	.076	.078	.081	.082	.084	.085	.086	.086	.085	.076
# 350	.121	.065	.068	.070	.072	.073	.075	.076	.077	.077	.077	.075	.066
# 400	.106	.062	.064	.066	.067	.069	.070	.071	.071	.071	.071	.069	.059
# 500	.085	.056	.058	.059	.060	.061	.061	.062	.062	.061	.060	.058	.048
# 600	.071	.053	.054	.055	.056	.056	.057	.057	.056	.055	.054	.051	.041
# 700	.061	.052	.053	.053	.054	.054	.054	.054	.054	.053	.051	.048	.038
# 750	.056	.049	.050	.050	.051	.051	.051	.051	.050	.049	.047	.044	.034
# 800	.053	.048	.048	.049	.049	.049	.049	.048	.048	.046	.044	.041	.031
# 900	.047	.047	.047	.048	.048	.048	.047	.047	.046	.045	.043	.039	.029
# 1000	.042	.045	.046	.046	.046	.046	.045	.045	.044	.042	.040	.037	.027

* Lagging

A. To determine voltage drop:
1. Determine the load power factor and conductor size to be used.
2. Select the corresponding factor from the table.
3. Multiply the circuit distance, load amps and the table factor together, divide by the number of runs and then by 1000.

Formulas:

A. Voltage Drop = Distance x Load Amps x Factor / Runs / 1000
(For 3-Phase multiply by 1.732, Single Phase multiply by 2)

B. Max Drop Factor = Max Volts Drop / (Distance x Load Amps / 1000) x Runs
(For 3-Phase divide by 1.732, Single Phase divide by 2)

B. To determine minimum wire size:
1. Multiply the circuit distance by the load amps and divide by 1000 for the ampere-feet.
2. Divide the maximum volts drop allowed by the ampere-feet, then multiply by the number of runs to determine the maximum table factor.
3. Select the wire size from the column having the appropriate load power factor and a factor less than or equal to the maximum table factor.

Conductor Calculation Forms and Examples

Conductor Voltage-Drop Calculations

Project: **Project No.:** **By:** **Page** **of**

Cir. No.	Name	Volts	Ph.	% Drop Max.	Volts Drop Max.	Cont. Amps	Non-Cont. Amps	Total Dsgn Amps	Wire Type	Wire AWG	No. of Runs	Conduit Type	Dist. in Feet	Load P.F. %	Table Factor	Volts Drop	Percent Drop	Percent Above Max.	Max. Drop Factor	New Wire AWG	New Table Factor	New Volts Drop	New Percent Drop

Example Voltage-Drop Form Calculations

We have provided an input form that makes it easy to enter and calculate more than one circuit at a time. If you decide to use the spreadsheet you can save even more time and let the computer do calculations the automatically. Note that while some items are optional they should be entered for reference purposes.

This form takes you through voltage drop calculations in two steps. The first step calculates the voltage drop for the initial conductors selected. The second step allows you to recalculate the voltage drop if the initial calculation shows too much loss in the conductors.

The header row at the top shows the names that identify each of the column entries as follows:

1. **Cir. No.** *(optional)* is the circuit number or letter that identifies each circuit (A, 1, X1, etc.).

2. **Name** *(optional)* is the circuit name that identifies each circuit (Recpt, Fan, etc.).

3. **Volts** is the voltage for each circuit (120, 480, etc.).

4. **Ph.** is the phases for each circuit (1 or 3).

5. **% Drop Max.** is the maximum percent drop for each circuit (1, 3, 5, etc.).

6. **Volts Drop Max.** is the maximum volts drop, automatically calculated for each circuit (col. 5 x col. 3).

7. **Cont. Amps** is the continuous amps for each circuit (10, 3.2, 100, etc.).

8. **Non-Cont. Amps** is the non-continuous amps for each circuit (10, 3.2, 100, etc.).

9. **Total Dsgn Amps** is the sum of the continuous (125%) and non-continuous amps (100%), automatically calculated for each circuit.

10. **Wire Type** *(optional)* is the wire type for each circuit (CU or AL), if left blank assume copper.

11. **Wire AWG** is the wire size for each circuit (12, 1/0, etc.).

12. **No. of Runs** is the number of conductors per phase for each circuit (1, 2, 6, etc.).

13. **Conduit Type** *(optional)* is the conduit type for each circuit , AL=Aluminum, STL=Steel or PVC).

14. **Dist. in Feet** is the one-way distance for each circuit (100, 266, etc.).

15. **Load P.F. %** is the power factor in percent for each circuit (100, 85, etc.).

16. **Table Factor** is the factor from the appropriate table for each circuit (1.728, .745, etc.).

17. **Volts Drop** is the volts drop, automatically calculated for each circuit.

18. **Percent Drop** is the percent drop, automatically calculated for each circuit.

19. **Percent Above Max.** is the percent drop above the desired drop, automatically calculated for each circuit. If there is a number in this cell then you will have to recalculate with larger conductors.

20. **Max. Drop Factor** is the maximum drop factor, automatically calculated for each circuit. If you have to recalculate the circuit this number tells you what factor choose to select the larger conductors. The number chosen must be less than or equal to the maximum drop factor.

21. **New Wire AWG** is the new wire size based on the maximum drop factor in column twenty for each circuit if you have to recalculate the circuit and select larger conductors.

22. **New Table Factor** is the table factor you enter based on the new wire size if you have to recalculate the circuit and select larger conductors.

23. **New Volts Drop** is the volts drop, automatically calculated based on the new table factor if you have to recalculate the circuit and select larger conductors.

24. **New Percent Drop** is the percent drop, automatically calculated based on the new table factor if you have to recalculate the circuit and select larger conductors.

The following page gives several examples to show you how to go about using the form. If your using the spreadsheet, it's best to copy the form first and rename it for a particular job and keep the original in tact.

Conductor Voltage-Drop Calculations

Project: Example **Project No.:** 0001 **By:** V.F. Christoffer **Page** 1 **of** 1

Cir. No.	Name	Volts	Ph.	% Drop Max.	Volts Drop Max.	Cont. Amps	Non-Cont. Amps	Total Dsgn Amps	Wire Type	Wire AWG	No. of Runs	Conduit Type	Dist. in Feet	Load P.F. %	Table Factor	Volts Drop	Percent Drop	Percent Above Max.	Max. Drop Factor	New Wire AWG	New Table Factor	New Volts Drop	New Percent Drop
1	40 HP Fan	480	3	3	14.40	52.0		65.0		6	1	STL	350	95	.841	15.31	3.19	**0.19**	.791	**4**	**.543**	**9.88**	**2.06**
2	Feeder	208	3	3	6.24	100.0		125.0		1	1	PVC	275	85	.263	7.23	3.48	**0.48**	.227	**1/0**	**.217**	**5.97**	**2.87**
3	Lighting	277	1	3	8.31	16.0		20.0		12	1	STL	375	90	1.830	10.98	3.96	**0.96**	1.385	**10**	**1.107**	**6.64**	**2.40**
4	Oven	208	1	3	6.24	61.0		76.3	AL	2	1	AL	535	100	.640	20.89	10.04	**7.04**	.191	**250**	**.180**	**5.87**	**2.82**
5	Panel LA	208	3	3	6.24		600.0	600.0		350	3	PVC	350	85	.092	6.44	3.10	**0.10**	.089	**400**	**.085**	**5.95**	**2.86**

Conductor Combination and Derating Calculations

Project: **Project No.:** **By:** **Page** **of**

Cir. No.	Name	Volts	Ph.	Wire	Cont. Amps	Non-Cont. Amps	Total Dsgn Amps	No. of CCC	No. of CCC Used	Derating Factor	Minimum Ampacity	75° C AWG CU	Ampacity	Ambient Temp. ° C	Temp. Derating Factor	New Ampacity Required	**New 75° Wire**	Dist. in Feet	Percent Drop	**Neutral**	**Eq. Ground**

Example Conductor Combination and Derating Form Calculations

We have provided an input form that makes it easy to enter and calculate combinations and derating for all the circuits in a conduit. If you decide to use the spreadsheet you can save even more time and let the computer do calculations the automatically. Note that while some items are optional they should be entered for reference purposes. Ampacities and wire sizes automatically calculated are for copper conductors only.

The header row at the top shows the names that identify each of the column entries as follows:

1. **Cir. No.** (*optional)* is the circuit number or letter that identifies each circuit (A, 1, X1, etc.).

2. **Name** (*optional)* is the circuit name that identifies each circuit (Recpt, Fan, etc.).

3. **Volts** is the voltage for each circuit (120, 480, etc.).

4. **Ph.** is the phase for each circuit (1 or 3).

5. **Wire** is the number of wires based on the phase and voltage of the circuit (2, 3, or 4).

6. **Cont. Amps** is the continuous amps for each circuit (10, 3.2, 100, etc.).

7. **Non-Cont. Amps** is the non-continuous amps for each circuit (10, 3.2, 100, etc.).

8. **Total Dsgn Amps** is the sum of the continuous (125%) and non-continuous amps (100%), automatically calculated for each circuit.

9. **No. of CCC** is the number of current carrying conductors (CCC) automatically calculated for each circuit that will be added to the *total* number and used to determine the final derating factor.

10. **No. of CCC Used** is the *total* number of current carrying conductors, automatically calculated, that are in the raceway and is the sum of all CCC the circuits on the form.

11. **Derating Factor** is the code derating factor automatically calculated for conductors based on the total number of CCC used on the form (col. 10).

12. **Minimum Ampacity** is the *required* amp rating automatically calculated for the circuit conductor based on the total number of CCC along with the sum of the continuous and non-continuous amps of the circuit.

13. **75° AWG CU** is the minimum copper wire size automatically calculated for each circuit based on the minimum ampacity (col. 12).

14. **Ampacity** is the copper amp rating automatically calculated for the 75° conductor that was calculated in col. 13.

15. **Ambient Temp. °C** is the ambient temperature in Celsius that will effect the conductors (21-80 °C).

16. **Temp. Derating Factor** is the code derating factor automatically calculated for conductors based on the ambient temp. that was entered in col. 15.

17. **New Ampacity Required** is the amp rating *required* for the circuit, automatically calculated based on the temp. derating factor, number of conductors used, and the total design amps of the circuit.

18. **New 75° Wire** is the new 75° copper wire size automatically calculated based on the new required ampacity.

19. **Dist. in Feet** is the one-way distance for each circuit (100, 266, etc.). This step was added to show voltage drop calculations if desired and may be omitted.

20. **Percent Drop** is the percent drop you calculated for the circuit. This step was added to show voltage drop calculations if desired and may be omitted.

21. **Neutral** is the AWG of the neutral conductor required, automatically calculated for each circuit based on all conductors being the same size. If the circuit does not require a neutral a size will not appear in this cell.

22. **Eq. Ground** is the AWG of the equipment ground conductor required that you calculated for the circuit based on the type, size and application of the overcurrent device being used. The final equipment ground used should be the largest one required for any of the circuits.

The following page gives several examples to show you how to go about using the form. If your using the spreadsheet, it's best to copy the form first and rename it for a particular job and keep the original in tact. Once all the calculations are completed, you must then add all the areas of the conductors required to get the size of the required conduit.

Conductor Combination and Derating Calculations

Project: Example **Project No.:** 0001 **By:** V.F.Christoffer **Page** 1 **of** 1

Cir. No.	Name	Volts	Ph.	Wire	Cont. Amps	Non-Cont. Amps	Total Dsgn Amps	No. of CCC	No. of CCC Used	Derating Factor	Minimum Ampacity	75° C AWG CU	Ampacity	Ambient Temp. ° C	Temp. Derating Factor	New Ampacity Required	**New 75° Wire**	Dist. in Feet	Percent Drop	**Neutral**	**Eq. Ground**
1	40 HP Fan	480	3	3	52.0		65.0	3	10	0.50	130.0	1	130	30	1.00	130.0	**1**				**8**
2	Feeder	208	3	4	100.0		125.0	3	10	0.50	250.0	250	255	30	1.00	255.0	**250**			**250**	**6**
3	Lighting	277	1	2	16.0		20.0	1	10	0.50	40.0	8	50	30	1.00	50.0	**8**			**8**	**12**
4	Oven	208	1	2	61.0		76.3	2	10	0.50	152.5	2/0	175	30	1.00	175.0	**2/0**				**8**
5	Receptacles	120	1	2	12.0		15.0	1	10	0.50	30.0	10	30	30	1.00	30.0	**10**			**10**	**14**

Motors

Circuit Design-Data for Three-Phase Motors

60 Hz - Rigid Conduit - Copper THWN Conductors - 30° C

Calculations based on Premium Efficiency Squirrel Cage Motors

Horse Power	Runs	75° Wire	Volts L-L	Conduit Size In.	Ins. Eq. Ground Size *	Conduit With Equip. Ground	Full Load Amps 100%	Design Amps 125%	Maximum Distance For 1% Drop	Approx. Percent Efficiency	Approx. Percent P.F.	Load KVA	NEC Switch Size 115%	NEMA Starter Size	Inst. C.B. HMCP Amps / 800%	Dual Ele. T.D. Fuse 175%	Inv. Time C.B. Amps 250%
1/2	1	14	480	1/2	14	1/2	1.1	1.4	998	.77	.80	0.91	30	00	3 / 9	3	5
	1	14	240	1/2	14	1/2	2.2	2.8	250	.77	.80	0.91	30	00	3 / 18	6	10
	1	14	208	1/2	14	1/2	2.4	3.0	198	.77	.80	0.86	30	00	3 / 19	6	10
3/4	1	14	480	1/2	14	1/2	1.6	2.0	686	.77	.80	1.32	30	00	3 / 13	3	5
	1	14	240	1/2	14	1/2	3.2	4.0	172	.77	.80	1.32	30	00	7 / 26	6	10
	1	14	208	1/2	14	1/2	3.5	4.4	136	.77	.80	1.25	30	00	7 / 28	10	10
1	1	14	480	1/2	14	1/2	2.1	2.6	523	.80	.80	1.74	30	00	3 / 17	6	10
	1	14	240	1/2	14	1/2	4.2	5.3	131	.80	.80	1.74	30	00	7 / 34	10	15
	1	14	208	1/2	14	1/2	4.6	5.8	103	.80	.80	1.65	30	00	7 / 37	10	15
1 1/2	1	14	480	1/2	14	1/2	3.0	3.8	346	.84	.83	2.48	30	00	7 / 24	6	10
	1	14	240	1/2	14	1/2	6.0	7.5	86	.84	.83	2.48	30	00	7 / 48	15	15
	1	14	208	1/2	14	1/2	6.6	8.3	68	.84	.83	2.37	30	00	15 / 53	15	20
2	1	14	480	1/2	14	1/2	3.4	4.3	305	.85	.85	2.81	30	00	7 / 27	6	10
	1	14	240	1/2	14	1/2	6.8	8.5	76	.85	.85	2.81	30	0	15 / 54	15	20
	1	14	208	1/2	14	1/2	7.5	9.4	60	.85	.85	2.69	30	0	15 / 60	15	20
3	1	14	480	1/2	14	1/2	4.8	6.0	216	.86	.86	3.97	30	0	7 / 38	10	15
	1	14	240	1/2	14	1/2	9.6	12.0	54	.86	.86	3.97	30	0	15 / 77	20	25
	1	14	208	1/2	14	1/2	10.6	13.3	42	.86	.86	3.80	30	0	15 / 85	20	30
5	1	14	480	1/2	14	1/2	7.6	9.5	136	.87	.87	6.29	30	0	15 / 61	15	20
	1	12	240	1/2	12	1/2	15.2	19.0	53	.87	.87	6.29	30	1	25 / 122	30	40
	1	10	208	1/2	10	1/2	16.7	20.9	68	.87	.87	5.99	30	1	25 / 133	30	45
7 1/2	1	14	480	1/2	14	1/2	11.0	13.8	89	.89	.89	9.10	30	1	15 / 88	20	30
	1	10	240	1/2	10	1/2	22.0	27.5	57	.89	.89	9.10	30	1	30 / 176	40	60
	1	10	208	1/2	10	1/2	24.2	30.3	45	.89	.89	8.67	30	1	30 / 193	45	70
10	1	12	480	1/2	12	1/2	14.0	17.5	108	.90	.92	11.58	30	1	25 / 112	25	35
	1	8	240	1/2	8	3/4	28.0	35.0	68	.90	.92	11.58	60	2	50 / 224	50	70
	1	8	208	1/2	8	3/4	30.8	38.5	53	.90	.92	11.04	60	2	50 / 246	60	80
15	1	10	480	1/2	10	1/2	21.0	26.3	114	.91	.93	17.37	30	2	25 / 168	40	60
	1	6	240	3/4	6	3/4	42.0	52.5	68	.91	.93	17.37	60	2	50 / 330	80	110
	1	6	208	3/4	6	3/4	46.2	57.8	54	.91	.93	16.56	60	3	70 / 369	90	125
20	1	8	480	1/2	8	3/4	27.0	33.8	135	.91	.93	22.33	60	2	50 / 216	50	70
	1	4	240	1	6	1	54.0	67.5	82	.91	.93	22.33	100	3	70 / 432	100	150
	1	4	208	1	6	1	59.4	74.3	64	.91	.93	21.29	100	3	70 / 457	110	150
25	1	8	480	1/2	8	3/4	34.0	42.5	107	.91	.93	28.12	60	2	50 / 272	60	90

Horse Power	Runs	75° Wire	Volts L-L	Conduit Size In.	Ins. Eq. Ground Size *	Conduit With Equip. Ground	Full Load Amps 100%	Design Amps 125%	Maximum Distance For 1% Drop	Approx. Percent Efficiency	Approx. Percent P.F.	Load KVA	NEC Switch Size 115%	NEMA Starter Size	Inst. C.B. HMCP Amps / 800%	Dual Ele. T.D. Fuse 175%	Inv. Time C.B. Amps 250%
	1	4	240	1	6	1	68.0	85.0	65	.91	.93	28.12	100	3	100 / 544	125	175
	1	3	208	1	6	1	74.8	93.5	63	.91	.93	26.81	100	3	100 / 598	150	200
30	1	8	480	1/2	8	3/4	40	50	91	.92	.93	33.08	60	3	50 / 320	70	100
	1	3	240	1	6	1	80	100	68	.92	.93	33.08	200	3	100 / 640	150	200
	1	2	208	1	4	1-1/4	88	110	66	.92	.93	31.54	200	4	150 / 704	175	225
40	1	6	480	3/4	6	3/4	52	65	110	.93	.93	43.01	60	3	70 / 416	100	150
	1	1	240	1-1/4	4	1-1/4	104	130	78	.93	.93	43.01	200	4	150 / 832	200	300
	1	1/0	208	1-1/4	4	1-1/2	114	143	80	.93	.93	40.86	200	4	150 / 912	200	300
50	1	4	480	1	6	1	65	81	136	.93	.93	53.76	100	3	100 / 520	125	175
	1	2/0	240	1-1/2	3	1-1/2	130	163	95	.93	.93	53.76	200	4	150 / 1040	250	350
	1	3/0	208	1-1/2	3	2	143	179	92	.93	.93	51.25	200	5	250 / 1144	300	400
60	1	3	480	1	6	1	77	96	141	.94	.94	63.68	100	4	100 / 616	150	200
	1	3/0	240	1-1/2	3	2	154	193	99	.94	.94	63.68	200	5	250 / 1232	300	400
	1	4/0	208	2	2	2	169	211	94	.94	.94	60.57	200	5	250 / 1352	300	500
75	1	1	480	1-1/4	4	1-1/4	96	120	170	.93	.94	79.40	200	4	150 / 768	175	250
	1	250	240	2	2	2	192	240	107	.93	.94	79.40	400	5	250 / 1536	350	500
	1	300	208	2	1	2-1/2	211	264	97	.93	.94	75.62	400	5	250 / 1688	400	600
100	1	2/0	480	1-1/2	3	1-1/2	124	155	200	.94	.94	102 55	200	4	150 / 992	225	350
	1	350	240	2-1/2	1/0	2-1/2	248	310	106	.94	.94	102.55	400	5	400 / 1984	500	700
	1	500	208	3	1/0	3	273	341	103	.94	.94	97.84	400	6	400 / 2184	500	700
125	1	3/0	480	1-1/2	3	2	156	195	195	.94	.96	129.02	200	5	250 / 1248	300	400
	1	600	240	3	1/0	3	312	390	115	.94	.96	129.02	400	6	400 / 2496	600	800
	1	700	208	3	2/0	3-1/2	343	429	95	.94	.96	122.93	400	6	400 / 2744	700	1,000
150	1	4/0	480	2	2	2	180	225	204	.94	.96	148.87	400	5	250 / 1440	350	500
	2	4/0	240	2	2/0	2	360	450	102	.94	.96	148.87	600	6	600 / 2880	700	1,000
	2	250	208	2	2/0	2-1/2	396	495	90	.94	.96	141.92	600	6	600 / 3168	700	1,000
200	1	350	480	2-1/2	1	2-1/2	240	300	220	.95	.96	198.49	400	5	400 / 1920	500	600
	2	350	240	2-1/2	4/0	2-1/2	480	600	110	.95	.96	198.49	600	6	600 / 3840	1,000	1,200
	2	400	208	2-1/2	4/0	3	528	660	94	.95	.96	189.23	800	7	800 / 4224	1,000	1,600

* Based on Inverse Time C.B.

A. To determine actual percent voltage drop:

1. Divide the actual distance by the max distance in the table for 1% drop .

Formulas:

A. Percent Voltage Drop = Actual Distance / Max Distance

B. Actual Volts Drop = (Actual Dist. / Max Dist.) / 100 x L-L Volts

B. To determine actual voltage drop:

1. Divide the actual distance by the max distance in the table for 1% drop.
2. Divide the result by 100.
3. Multiply the result by the L-L voltage.

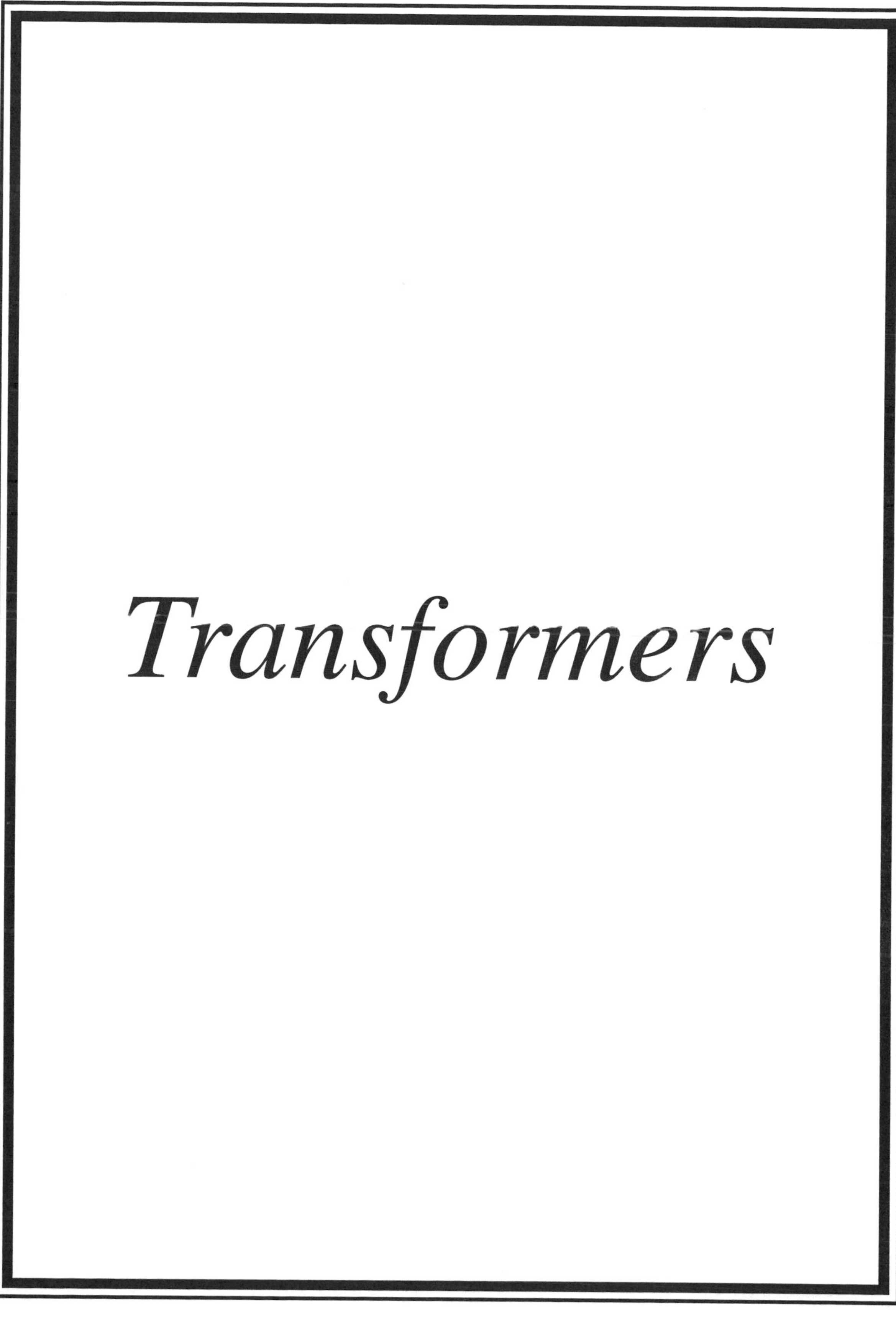

Transformers

Three-Phase Transformer Circuit Data

60 Hz - Rigid Conduit - Copper THWN Conductors - 30° C

XFMR KVA	Runs	75° Wire	Volts L-L	Conduit Size In. 3-Each	Conduit With Full Size Neutral	Ins. Eq. Ground Size	Conduit With Neutral & Eq. Ground	Grounding Electrode Cond.	FLA 3ph 100%	Design Amps 125%	Maximum Distance 1% Drop .85 P.F.	% Imp.	S/C Amps With Infinite Primary	S/C Amps With 50% Motor Load	S/C Amps With 100% Motor Load	NEC Fuse Amps 125%	NEC C.B. Amps 125%
3	1	14	480	1/2	1/2	14	1/2	8	3.6	4.5	287	8.0	45	52	70	6	5
	1	14	240	1/2	1/2	14	1/2	8	7.2	9.0	72	8.0	90	105	141	10	10
	1	14	208	1/2	1/2	14	1/2	8	8.3	10.4	54	8.0	104	121	162	15	15
6	1	14	480	1/2	1/2	14	1/2	8	7.2	9.0	144	3.5	206	221	257	10	10
	1	12	240	1/2	1/2	12	1/2	8	14.4	18.0	55	3.5	412	441	513	20	20
	1	10	208	1/2	1/2	10	1/2	8	16.7	20.8	68	3.5	476	509	592	25	25
9	1	14	480	1/2	1/2	14	1/2	8	10.8	13.5	96	3.0	361	383	437	15	15
	1	10	240	1/2	1/2	10	1/2	8	21.7	27.1	61	3.0	722	765	873	30	30
	1	8	208	1/2	3/4	10	3/4	8	25.0	31.2	69	3.0	833	883	1,008	35	35
15	1	10	480	1/2	1/2	10	1/2	8	18.0	22.6	146	2.3	784	821	911	25	25
	1	8	240	1/2	3/4	10	3/4	8	36.1	45.1	55	2.3	1,569	1,641	1,822	50	50
	1	6	208	3/4	3/4	10	1	8	41.6	52.0	64	2.3	1,810	1,894	2,102	60	60
25	1	8	480	1/2	3/4	10	3/4	8	30.1	37.6	132	2.5	1,203	1,263	1,413	40	40
	1	4	240	1	1	8	1-1/4	8	60.1	75.2	78	2.5	2,406	2,526	2,827	80	80
	1	3	208	1	1-1/4	8	1-1/4	8	69.4	86.7	71	2.5	2,776	2,915	3,262	90	90
30	1	8	480	1/2	3/4	10	3/4	8	36.1	45.1	110	3.0	1,203	1,275	1,455	50	50
	1	3	240	1	1-1/4	8	1-1/4	8	72.2	90.2	79	3.0	2,406	2,550	2,911	100	100
	1	2	208	1	1-1/4	6	1-1/2	8	83.3	104.1	72	3.0	2,776	2,942	3,359	110	110
45	1	4	480	1	1	8	1-1/4	8	54.1	68	174	3.0	1,804	1,913	2,183	70	70
	1	1/0	240	1-1/4	1-1/2	6	1-1/2	6	108.3	135	98	3.0	3,609	3,825	4,366	150	150
	1	2/0	208	1-1/2	2	6	2	4	124.9	156	85	3.0	4,164	4,414	5,038	175	175
75	1	2	480	1	1-1/4	6	1-1/2	8	90	113	154	3.5	2,578	2,758	3,209	125	125
	1	4/0	240	2	2	4	2-1/2	2	180	226	96	3.5	5,155	5,516	6,418	250	250
	1	300	208	2	2-1/2	4	2-1/2	2	208	260	88	3.5	5,948	6,365	7,405	300	300
112.5	1	2/0	480	1-1/2	2	6	2	4	135	169	181	2.0	6,766	7,037	7,713	175	175
	1	500	240	3	3	3	3	1/0	271	338	103	2.0	13,532	14,073	15,427	350	350
	1	600	208	3	3-1/2	3	3-1/2	1/0	312	390	82	2.0	15,614	16,238	17,800	400	400
150	1	4/0	480	2	2	4	2-1/2	2	180	226	191	2.0	9,021	9,382	10,284	250	250
	2	4/0	240	2	2	2	2-1/2	1/0	361	451	96	2.0	18,043	18,764	20,569	500	500
	2	300	208	2	2-1/2	1	3	1/0	416	520	88	2.0	20,819	21,651	23,733	600	600
225	1	500	480	3	3	3	3	1/0	271	338	206	2.0	13,532	14,073	15,427	350	350

XFMR KVA	Runs	75° Wire	Volts L-L	Conduit Size In. 3-Each	Conduit With Full Size Neutral	Ins. Eq. Ground Size	Conduit With Neutral & Eq. Ground	Grounding Electrode Cond.	FLA 3ph 100%	Design Amps 125%	Maximum Distance 1% Drop .85 P.F.	% Imp.	S/C Amps With Infinite Primary	S/C Amps With 50% Motor Load	S/C Amps With 100% Motor Load	NEC Fuse Amps 125%	NEC C.B. Amps 125%
	2	500	240	3	3	1/0	3-1/2	2/0	541	677	103	2.0	27,064	28,147	30,853	700	700
	2	600	208	3	3-1/2	1/0	3-1/2	3/0	625	781	82	2.0	31,228	32,477	35,600	800	800
300	1	500	600	3	3	3	3	1/0	289	361	266	4.5	6,415	6,993	8,436	400	400
	1	700	480	3	3-1/2	2	4	2/0	361	451	171	4.5	8,019	8,741	10,545	500	500
	3	350	240	2-1/2	3	2/0	3	2/0	722	902	97	4.5	16,038	17,481	21,090	1,000	1,000
	3	500	208	3	3	3/0	3-1/2	3/0	833	1,041	87	4.5	18,505	20,171	24,335	1,200	1,200
400	2	250	600	2	2-1/2	2	2-1/2	1/0	385	481	303	4.5	8,554	9,323	11,248	500	500
	2	350	480	2-1/2	3	1/0	3	2/0	481	601	194	4.5	10,692	11,654	14,060	601	700
	3	600	240	3	3-1/2	4/0	3-1/2	3/0	962	1,203	92	4.5	21,384	23,309	28,120	1,600	1,600
	3	750	208	3	3-1/2	4/0	4	3/0	1,110	1,388	75	4.5	24,674	26,894	32,446	1,600	1,600
500	2	350	600	2-1/2	3	1/0	3	2/0	481	601	290	4.5	10,692	11,654	14,060	601	700
	2	500	480	3	3	1/0	3-1/2	2/0	601	752	186	4.5	13,365	14,568	17,575	800	800
	4	500	240	3	3	4/0	3-1/2	3/0	1,203	1,504	93	4.5	26,730	29,136	35,150	1,600	1,600
	4	700	208	3	3-1/2	250	4	3/0	1,388	1,735	77	4.5	30,842	33,618	40,558	2,000	2,000
750	2	700	600	3	3-1/2	2/0	4	3/0	722	902	193	5.75	12,551	13,995	17,603	1,000	1,000
	3	500	480	3	3	3/0	3-1/2	3/0	902	1,128	186	5.75	15,689	17,494	22,004	1,200	1,200
	5	700	240	3	3-1/2	350	4	3/0	1,804	2,255	85	5.75	31,379	34,987	44,009	2,500	2,500
	6	700	208	3	3-1/2	400	4	3/0	2,082	2,602	77	5.75	36,206	40,370	50,779	3,000	3,000
1000	3	600	600	3	3-1/2	4/0	3-1/2	3/0	962	1,203	218	5.75	16,735	18,660	23,471	1,600	1,600
	4	500	480	3	3	4/0	3-1/2	3/0	1,203	1,504	186	5.75	20,919	23,325	29,339	1,600	1,600
	7	700	240	3	3-1/2	500	4	3/0	2,406	3,007	90	5.75	41,838	46,650	58,678	4,000	4,000
	8	700	208	3	3-1/2	500	4	3/0	2,776	3,470	77	5.75	48,275	53,826	67,705	4,000	4,000
1500	4	700	600	3	3-1/2	250	4	3/0	1,443	1,804	213	5.75	25,103	27,990	35,207	2,000	2,000
	5	700	480	3	3-1/2	350	4	3/0	1,804	2,255	171	5.75	31,379	34,987	44,009	2,500	2,500
	10	700	240	3	3-1/2	700	4	3/0	3,609	4,511	85	5.75	62,757	69,974	88,017	5,000	5,000
	11	750	208	3	3-1/2	800	4	3/0	4,164	5,205	73	5.75	72,412	80,740	101,558	6,000	6,000
2000	6	600	600	3	3-1/2	350	3-1/2	3/0	1,925	2,406	231	5.75	33,471	37,320	46,942	2,500	2,500
	7	700	480	3	3-1/2	500	4	3/0	2,406	3,007	179	5.75	41,838	46,650	58,678	4,000	4,000
2500	7	700	600	3	3-1/2	500	4	3/0	2,406	3,007	224	5.75	41,838	46,650	58,678	4,000	4,000
	8	750	480	3	3-1/2	500	4	3/0	3,007	3,759	170	5.75	52,298	58,312	73,348	4,000	4,000

A. To determine actual percent voltage drop:

1. Divide the actual distance by the max distance in the table for 1% drop .

Formulas:

A. Percent Voltage Drop = Actual Distance / Max Distance

B. Actual Volts Drop = (Actual Dist. / Max Dist.) / 100 x L-L Volts

B. To determine actual voltage drop:

1. Divide the actual distance by the max distance in the table for 1% drop.
2. Divide the result by 100.
3. Multiply the result by the L-L voltage.

Power Factor
Correction

Factors for Determining Capacitor Kilovars

Desired Power Factor

Original Power Factor	.80	.81	.82	.83	.84	.85	.86	.87	.88	.89	.90	.91	.92	.93	.94	.95	.96	.97	.98	.99	1.00
.05	19.225	19.251	19.277	19.303	19.329	19.355	19.382	19.408	19.435	19.463	19.491	19.519	19.549	19.580	19.612	19.646	19.683	19.724	19.772	19.832	19.975
.10	9.200	9.226	9.252	9.278	9.304	9.330	9.357	9.383	9.410	9.438	9.466	9.494	9.524	9.555	9.587	9.621	9.658	9.699	9.747	9.807	9.950
.20	4.149	4.175	4.201	4.227	4.253	4.279	4.306	4.332	4.359	4.387	4.415	4.443	4.473	4.504	4.536	4.570	4.607	4.648	4.696	4.756	4.899
.30	2.430	2.456	2.482	2.508	2.534	2.560	2.586	2.613	2.640	2.667	2.695	2.724	2.754	2.785	2.817	2.851	2.888	2.929	2.977	3.037	3.180
.40	1.541	1.567	1.593	1.619	1.645	1.672	1.698	1.725	1.752	1.779	1.807	1.836	1.865	1.896	1.928	1.963	2.000	2.041	2.088	2.149	2.291
.45	1.235	1.261	1.287	1.313	1.339	1.365	1.391	1.418	1.445	1.472	1.500	1.529	1.559	1.589	1.622	1.656	1.693	1.734	1.781	1.842	1.985
.46	1.180	1.206	1.232	1.258	1.284	1.311	1.337	1.364	1.391	1.418	1.446	1.475	1.504	1.535	1.567	1.602	1.639	1.680	1.727	1.788	1.930
.47	1.128	1.154	1.180	1.206	1.232	1.258	1.285	1.311	1.338	1.366	1.394	1.422	1.452	1.483	1.515	1.549	1.586	1.627	1.675	1.736	1.878
.48	1.078	1.104	1.130	1.156	1.182	1.208	1.234	1.261	1.288	1.315	1.343	1.372	1.402	1.432	1.465	1.499	1.536	1.577	1.625	1.685	1.828
.49	1.029	1.055	1.081	1.107	1.133	1.159	1.186	1.212	1.239	1.267	1.295	1.323	1.353	1.384	1.416	1.450	1.487	1.528	1.576	1.637	1.779
.50	.982	1.008	1.034	1.060	1.086	1.112	1.139	1.165	1.192	1.220	1.248	1.276	1.306	1.337	1.369	1.403	1.440	1.481	1.529	1.590	1.732
.51	.937	.963	.989	1.015	1.041	1.067	1.093	1.120	1.147	1.174	1.202	1.231	1.261	1.291	1.324	1.358	1.395	1.436	1.484	1.544	1.687
.52	.893	.919	.945	.971	.997	1.023	1.049	1.076	1.103	1.130	1.158	1.187	1.217	1.247	1.280	1.314	1.351	1.392	1.440	1.500	1.643
.53	.850	.876	.902	.928	.954	.980	1.007	1.033	1.060	1.088	1.116	1.144	1.174	1.205	1.237	1.271	1.308	1.349	1.397	1.458	1.600
.54	.809	.835	.861	.887	.913	.939	.965	.992	1.019	1.046	1.074	1.103	1.133	1.163	1.196	1.230	1.267	1.308	1.356	1.416	1.559
.55	.768	.794	.820	.846	.873	.899	.925	.952	.979	1.006	1.034	1.063	1.092	1.123	1.156	1.190	1.227	1.268	1.315	1.376	1.518
.56	.729	.755	.781	.807	.834	.860	.886	.913	.940	.967	.995	1.024	1.053	1.084	1.116	1.151	1.188	1.229	1.276	1.337	1.479
.57	.691	.717	.743	.769	.796	.822	.848	.875	.902	.929	.957	.986	1.015	1.046	1.079	1.113	1.150	1.191	1.238	1.299	1.441
.58	.655	.681	.707	.733	.759	.785	.811	.838	.865	.892	.920	.949	.979	1.009	1.042	1.076	1.113	1.154	1.201	1.262	1.405
.59	.618	.644	.670	.696	.723	.749	.775	.802	.829	.856	.884	.913	.942	.973	1.006	1.040	1.077	1.118	1.165	1.226	1.368
.60	.583	.609	.635	.661	.687	.714	.740	.767	.794	.821	.849	.878	.907	.938	.970	1.005	1.042	1.083	1.130	1.191	1.333
.61	.549	.575	.601	.627	.653	.679	.706	.732	.759	.787	.815	.843	.873	.904	.936	.970	1.007	1.048	1.096	1.157	1.299
.62	.515	.541	.567	.593	.620	.646	.672	.699	.726	.753	.781	.810	.839	.870	.903	.937	.974	1.015	1.062	1.123	1.265
.63	.483	.509	.535	.561	.587	.613	.639	.666	.693	.720	.748	.777	.807	.837	.870	.904	.941	.982	1.030	1.090	1.233
.64	.451	.477	.503	.529	.555	.581	.607	.634	.661	.688	.716	.745	.775	.805	.838	.872	.909	.950	.998	1.058	1.201
.65	.419	.445	.471	.497	.523	.549	.576	.602	.629	.657	.685	.714	.743	.774	.806	.840	.877	.919	.966	1.027	1.169
.66	.388	.414	.440	.466	.492	.519	.545	.572	.599	.626	.654	.683	.712	.743	.775	.810	.847	.888	.935	.996	1.138
.67	.358	.384	.410	.436	.462	.488	.515	.541	.568	.596	.624	.652	.682	.713	.745	.779	.816	.857	.905	.966	1.108
.68	.328	.354	.380	.406	.432	.459	.485	.512	.539	.566	.594	.623	.652	.683	.715	.750	.787	.828	.875	.936	1.078
.69	.299	.325	.351	.377	.403	.429	.456	.482	.509	.537	.565	.593	.623	.654	.686	.720	.757	.798	.846	.907	1.049
.70	.270	.296	.322	.348	.374	.400	.427	.453	.480	.508	.536	.565	.594	.625	.657	.692	.729	.770	.817	.878	1.020
.71	.242	.268	.294	.320	.346	.372	.398	.425	.452	.480	.508	.536	.566	.597	.629	.663	.700	.741	.789	.849	.992
.72	.214	.240	.266	.292	.318	.344	.370	.397	.424	.452	.480	.508	.538	.569	.601	.635	.672	.713	.761	.821	.964
.73	.186	.212	.238	.264	.290	.316	.343	.370	.396	.424	.452	.481	.510	.541	.573	.608	.645	.686	.733	.794	.936

Desired Power Factor

Original Power Factor	.80	.81	.82	.83	.84	.85	.86	.87	.88	.89	.90	.91	.92	.93	.94	.95	.96	.97	.98	.99	1.00
.74	.159	.185	.211	.237	.263	.289	.316	.342	.369	.397	.425	.453	.483	.514	.546	.580	.617	.658	.706	.766	.909
.75	.132	.158	.184	.210	.236	.262	.289	.315	.342	.370	.398	.426	.456	.487	.519	.553	.590	.631	.679	.739	.882
.76	.105	.131	.157	.183	.209	.235	.262	.288	.315	.343	.371	.400	.429	.460	.492	.526	.563	.605	.652	.713	.855
.77	.079	.105	.131	.157	.183	.209	.235	.262	.289	.316	.344	.373	.403	.433	.466	.500	.537	.578	.626	.686	.829
.78	.052	.078	.104	.130	.156	.183	.209	.236	.263	.290	.318	.347	.376	.407	.439	.474	.511	.552	.599	.660	.802
.79	.026	.052	.078	.104	.130	.156	.183	.209	.236	.264	.292	.320	.350	.381	.413	.447	.484	.525	.573	.634	.776
.80	0	.026	.052	.078	.104	.130	.157	.183	.210	.238	.266	.294	.324	.355	.387	.421	.458	.499	.547	.608	.750
.81	0	0	.026	.052	.078	.104	.131	.157	.184	.212	.240	.268	.298	.329	.361	.395	.432	.473	.521	.581	.724
.82	0	0	0	.026	.052	.078	.105	.131	.158	.186	.214	.242	.272	.303	.335	.369	.406	.447	.495	.556	.698
.83	0	0	0	0	.026	.052	.079	.105	.132	.160	.188	.216	.246	.277	309	.343	.380	.421	.469	.530	.672
.84	0	0	0	0	0	.026	.053	.079	.106	.134	.162	.190	.220	.251	283	.317	.354	.395	.443	.503	.646
.85	0	0	0	0	0	0	.026	.053	.080	.107	.135	.164	.194	.225	.257	.291	.328	.369	.417	.477	.620
.86	0	0	0	0	0	0	0	.027	.054	.081	.109	.138	.167	.198	.230	.265	.302	.343	.390	.451	.593
.87	0	0	0	0	0	0	0	0	.027	.054	.082	.111	.141	.172	.204	.238	.275	.316	.364	.424	.567
.88	0	0	0	0	0	0	0	0	0	.027	.055	.084	.114	.145	.177	.211	.248	.289	.337	.397	.540
.89	0	0	0	0	0	0	0	0	0	0	.028	.057	.086	.117	.149	.184	.221	.262	.309	.370	.512
.90	0	0	0	0	0	0	0	0	0	0	0	.029	.058	.089	.121	.156	.193	.234	.281	.342	.484
.91	0	0	0	0	0	0	0	0	0	0	0	0	.030	.060	.093	.127	.164	.205	.253	.313	.456
.92	0	0	0	0	0	0	0	0	0	0	0	0	0	.031	.063	.097	.134	175	.223	.284	.426
.93	0	0	0	0	0	0	0	0	0	0	0	0	0	0	.032	.067	.104	145	.192	.253	.395
.94	0	0	0	0	0	0	0	0	0	0	0	0	0	0	0	.034	.071	112	.160	.220	.363
.95	0	0	0	0	0	0	0	0	0	0	0	0	0	0	0	0	.037	.078	.126	.186	.329
.96	0	0	0	0	0	0	0	0	0	0	0	0	0	0	0	0	0	041	.089	.149	.292
.97	0	0	0	0	0	0	0	0	0	0	0	0	0	0	0	0	0	0	.048	.108	.251
.98	0	0	0	0	0	0	0	0	0	0	0	0	0	0	0	0	0	0	0	.061	.203
.99	0	0	0	0	0	0	0	0	0	0	0	0	0	0	0	0	0	0	0	0	.142

A. To determine kilovars needed:

1. Determine the original load KW and power factor to be used.
2. Locate the original power factor from the column on the far left side of the table.
3. Across that row find where the corresponding desired power factor from the heading intersects the original power factor.
4. Multiply the corresponding factor with the load KW to detemine the KVARS needed.

KVAR = KW [TAN(ACOS(pf1 - TAN(ACOS(pf2)]

Formula:

A. Kilovars = Factor x KW

NEMA Ratings for Enclosures

NEMA Ratings for Enclosures

Exposure Description	*Enclosure Designation* 1	2	3	3R	3S	4	4X	5	6	6P	7	8	9	12	12K	13
Outdoor Locations			X	X	X	X	X		X	X						
Hazardous Locations											X	X	X			
Indoor Locations	X	X				X	X	X	X	X				X	X	X
Incidental contact with the enclosed equipment	X	X	X	X	X	X	X	X	X	X				X	X	X
Falling Dirt	X	X				X	X	X	X	X				X	X	X
Falling liquids and light splashing		X				X	X	X	X	X				X	X	X
Circulating dust, lint, fibers and flyings						X	X		X	X				X	X	X
Settling airborne dust, lint, fibers and flyings						X	X	X	X	X				X	X	X
Hosedown and splashing water						X	X		X	X						
Oil and coolant seepage														X	X	X
Oil or coolant spraying and splashing																X
Corrosive agents							X			X						
Occasional temporary submersion									X	X						
Occasional prolonged submersion										X						
Rain, snow, and sleet			X	X	X	X	X		X	X						
Sleet *					X											
Windblown dust			X		X	X	X		X	X						
Hosedown						X	X		X	X						
Class I											X	X				
Class II													X			
Class III													X			
Group A											X	X				
Group B											X	X				
Group C											X	X				
Group D											X	X				
Group E													X			
Group F													X			
Group G													X			
Group 10													X			

* Mechanism shall be operable when ice covered.

Fault-Current Calculations with Short-Circuit Tables and Data

Fault-Current Calculations

Secondary Fault-Current for Three-Phase Transformers

Project: ______ **Project No.:** ______ **Date:** ______ **Designer:** ______ **Page** ___ **of** ___

Transformer Data

XFMR KVA	% Imp.	FLA Amps @ 208 Volts	S/C Amps @ 208 Volts	208 V. w/50% Motor Load	FLA Amps @ 240 Volts	S/C Amps @ 240 Volts	240 V. w/50% Motor Load	FLA Amps @ 480 Volts	S/C Amps @ 480 Volts	480 V. w/50% Motor Load
3	8.0	8	104	121	7	90	105	4	45	52
6	3.5	17	476	509	14	412	441	7	206	221
9	3.0	25	833	883	22	722	765	11	361	383
15	2.3	42	1,810	1,894	36	1,569	1,641	18	784	821
25	2.5	69	2,776	2,915	60	2,406	2,526	30	1,203	1,263
30	3.0	83	2,776	2,942	72	2,406	2,550	36	1,203	1,275
45	3.0	125	4,164	4,414	108	3,609	3,825	54	1,804	1,913
75	3.5	208	5,948	6,365	180	5,155	5,516	90	2,578	2,758
112.5	2.0	312	15,614	16,238	271	13,532	14,073	135	6,766	7,037
150	2.0	416	20,819	21,651	361	18,043	18,764	180	9,021	9,382
225	2.0	625	31,228	32,477	541	27,064	28,147	271	13,532	14,073
300	4.5	833	18,505	20,171	722	16,038	17,481	361	8,019	8,741
400	4.5	1,110	24,674	26,894	962	21,384	23,309	481	10,692	11,654
500	4.5	1,388	30,842	33,618	1,203	26,730	29,136	601	13,365	14,568
750	5.75	2,082	36,206	40,370	1,804	31,379	34,987	902	15,689	17,494
1000	5.75	2,776	48,275	53,826	2,406	41,838	46,650	1,203	20,919	23,325
1500	5.75	4,164	72,412	80,740	3,609	62,757	69,974	1,804	31,379	34,987
2000	5.75	5,552	96,550	107,653	4,811	83,676	93,299	2,406	41,838	46,650
2500	5.75	6,940	120,687	134,566	6,014	104,596	116,624	3,007	52,298	58,312

Circuit Breaker AIC Ratings

AIC in Thousands

Volts	10	14	18	22	25	35	42	50	65	100	200
120	X			X					X		
120/240	X			X			X		X		
240	X		X	X			X		X	X	X
277		X		X	X				X		
480		X		X	X	X		X	X	X	
600		X	X	X	X	X		X	X		

Note: All results are given in symmetrical amperes

Transformer Calculations

Infinite Primary Isc:

1.) ______ (KVA) / ______ (KV x 1.732) = ______ (XFMR FLA)

2.) 100 / ______ (%Z) x ______ (XFMR FLA) = ______ (Isca)

Known Primary Isc:

1.) ______ (Iscp) x ______ (E L-Lp) x 1.732 x ______ (%Z) = ______ (a.)

2.) 100,000 x ______ (KVA) = ______ (b.)

3.) f. = ______ (a.) / ______ (b.) = ______

4.) M = 1 / ______ (f + 1) = ______

5.) ______ (E L-Lp) / ______ (E L-Ls) x ______ (M) x ______ (Iscp) = ______ (Isca)

Legend:

Isc = Short Circuit or Fault Current
Iscp = Available Fault-Current at Primary
E L-Lp = Primary Line to Line Volts
E L-Ls = Secondary Line to Line Volts
KV = Line to Line Kilovolts
Isca = Available Fault-Current

______ Fuse Let-Thru

______ Isc Used

Three-Phase Bolted Fault Calculations

Motor Contribution: (FLA) X 4 = Proj. No.: By: Page of

Node	From Node	Name	Wire Size	(Isca	x	1.732	x	Length)	/	(Runs	x	C-Value	x	L-L Volts)	+	1	(1/x)	x	Isca	=	Isc	+	Motor Contribution	=	Isc @ Node
					x	1.732	x		/		x		x		+	1	(1/x)	x		=		+		=	
					x	1.732	x		/		x		x		+	1	(1/x)	x		=		+		=	
					x	1.732	x		/		x		x		+	1	(1/x)	x		=		+		=	
					x	1.732	x		/		x		x		+	1	(1/x)	x		=		+		=	
					x	1.732	x		/		x		x		+	1	(1/x)	x		=		+		=	
					x	1.732	x		/		x		x		+	1	(1/x)	x		=		+		=	
					x	1.732	x		/		x		x		+	1	(1/x)	x		=		+		=	
					x	1.732	x		/		x		x		+	1	(1/x)	x		=		+		=	
					x	1.732	x		/		x		x		+	1	(1/x)	x		=		+		=	
					x	1.732	x		/		x		x		+	1	(1/x)	x		=		+		=	
					x	1.732	x		/		x		x		+	1	(1/x)	x		=		+		=	
					x	1.732	x		/		x		x		+	1	(1/x)	x		=		+		=	
					x	1.732	x		/		x		x		+	1	(1/x)	x		=		+		=	
					x	1.732	x		/		x		x		+	1	(1/x)	x		=		+		=	
					x	1.732	x		/		x		x		+	1	(1/x)	x		=		+		=	
					x	1.732	x		/		x		x		+	1	(1/x)	x		=		+		=	
					x	1.732	x		/		x		x		+	1	(1/x)	x		=		+		=	
					x	1.732	x		/		x		x		+	1	(1/x)	x		=		+		=	
					x	1.732	x		/		x		x		+	1	(1/x)	x		=		+		=	
					x	1.732	x		/		x		x		+	1	(1/x)	x		=		+		=	
					x	1.732	x		/		x		x		+	1	(1/x)	x		=		+		=	
					x	1.732	x		/		x		x		+	1	(1/x)	x		=		+		=	
					x	1.732	x		/		x		x		+	1	(1/x)	x		=		+		=	
					x	1.732	x		/		x		x		+	1	(1/x)	x		=		+		=	
					x	1.732	x		/		x		x		+	1	(1/x)	x		=		+		=	
					x	1.732	x		/		x		x		+	1	(1/x)	x		=		+		=	
					x	1.732	x		/		x		x		+	1	(1/x)	x		=		+		=	
					x	1.732	x		/		x		x		+	1	(1/x)	x		=		+		=	
					x	1.732	x		/		x		x		+	1	(1/x)	x		=		+		=	
					x	1.732	x		/		x		x		+	1	(1/x)	x		=		+		=	
					x	1.732	x		/		x		x		+	1	(1/x)	x		=		+		=	
					x	1.732	x		/		x		x		+	1	(1/x)	x		=		+		=	
					x	1.732	x		/		x		x		+	1	(1/x)	x		=		+		=	

Example Fault-Current Form Calculations

We have provided an input form that makes it easy to enter and calculate secondary fault-currents for several nodes at once. If you decide to use the spreadsheet you can save even more time and let the computer do calculations the automatically. Note that while some items are optional they should be entered for reference purposes.

Before starting the fault current calculations you must determine any significant motor contribution. This is done by adding the full-load-amps for all the motors. This number is entered at the top of the form after "**Motor Contribution: (FLA)**." If you are using the spreadsheet, it will automatically multiply this number by four and add it to the final fault current for each circuit.

The header row at the top shows the names that identify each of the column entries as follows:

1. **Node** is the node or bus indicated by a letter that identifies each circuit being calculated (A, A1, D6, etc.).

2. **From Node** is the node or bus indicated by a letter that identifies the source node upstream from the circuit being calculated (A, A1, D6, etc.).

3. **Name** *(optional)* is the circuit name that identifies each circuit (MSB, Fan, etc.).

4. **Wire Size** is the wire size entered for the circuit (12, 1/0, etc.).

5. **Isca** is the fault-current available from the circuit source node that is automatically entered for each circuit. The first node *must* be entered manually, then all succeeding nodes will be entered automatically by the computer.

6. **Length** is the distance in feet for each circuit (100, 266, etc.).

7. **Runs** is the number of conductors per phase for each circuit (1, 2, 6, etc.).

8. **C-Value** is the number taken from the c-value tables for the particular conductor being used.

9. **L-L Volts** is the voltage for each circuit (208, 480, etc.).

10. **Isca** is the fault-current available from the circuit source node that is automatically entered for each circuit.

11. **Isc** is the calculated fault-current available for the circuit, without motor contribution, that is automatically calculated.

12. **Motor Contribution** is the contribution in amps that was entered at the top of the form. It is automatically entered and added to the fault-current calculations for each circuit.

13. **Isc @ Node** is the calculated fault-current available for the circuit, with motor contribution, that is automatically calculated.

The following page gives several examples to show you how to go about using the form. If your using the spreadsheet, it's best to copy the form first and rename it for a particular job and keep the original in tact. *Note that the example form is indicative of all the calculation forms whether they are single or three phase calculations.*

Three-Phase Bolted Fault Calculations

Motor Contribution: (FLA) 1804 X 4 = 7,216 Proj. No.: 0001 By: V.F. Christoffer Page 1 of 1

Node	From Node	Name	Wire Size	(Isca	x	1.732	x	Length)	/	(Runs	x	C-Value	x	L-L Volts)	+	1	(1/x)	x	Isca	=	Isc	+	Motor Contribution	=	Isc @ Node
A		Panel MSB	500	51,540	x	1.732	x	25	/	6	x	22,185	x	480	+	1	(1/x)	x	51,540	=	49,801	+	7,216	=	57,017
B	A	Panel L1	500	49,801	x	1.732	x	50	/	1	x	22,185	x	480	+	1	(1/x)	x	49,801	=	35,445	+	7,216	=	42,661
C	B	Panel L2	250	35,445	x	1.732	x	125	/	2	x	16,483	x	480	+	1	(1/x)	x	35,445	=	23,869	+	7,216	=	31,085
D	C	40 HP Motor	6	23,869	x	1.732	x	225	/	1	x	2,425	x	480	+	1	(1/x)	x	23,869	=	2,655	+	7,216	=	9,871
					x	1.732	x		/		x		x		+	1	(1/x)	x		=		+		=	
					x	1.732	x		/		x		x		+	1	(1/x)	x		=		+		=	
					x	1.732	x		/		x		x		+	1	(1/x)	x		=		+		=	
					x	1.732	x		/		x		x		+	1	(1/x)	x		=		+		=	
					x	1.732	x		/		x		x		+	1	(1/x)	x		=		+		=	
					x	1.732	x		/		x		x		+	1	(1/x)	x		=		+		=	
					x	1.732	x		/		x		x		+	1	(1/x)	x		=		+		=	
					x	1.732	x		/		x		x		+	1	(1/x)	x		=		+		=	
					x	1.732	x		/		x		x		+	1	(1/x)	x		=		+		=	
					x	1.732	x		/		x		x		+	1	(1/x)	x		=		+		=	
					x	1.732	x		/		x		x		+	1	(1/x)	x		=		+		=	
					x	1.732	x		/		x		x		+	1	(1/x)	x		=		+		=	
					x	1.732	x		/		x		x		+	1	(1/x)	x		=		+		=	
					x	1.732	x		/		x		x		+	1	(1/x)	x		=		+		=	
					x	1.732	x		/		x		x		+	1	(1/x)	x		=		+		=	
					x	1.732	x		/		x		x		+	1	(1/x)	x		=		+		=	
					x	1.732	x		/		x		x		+	1	(1/x)	x		=		+		=	
					x	1.732	x		/		x		x		+	1	(1/x)	x		=		+		=	
					x	1.732	x		/		x		x		+	1	(1/x)	x		=		+		=	
					x	1.732	x		/		x		x		+	1	(1/x)	x		=		+		=	
					x	1.732	x		/		x		x		+	1	(1/x)	x		=		+		=	
					x	1.732	x		/		x		x		+	1	(1/x)	x		=		+		=	
					x	1.732	x		/		x		x		+	1	(1/x)	x		=		+		=	
					x	1.732	x		/		x		x		+	1	(1/x)	x		=		+		=	
					x	1.732	x		/		x		x		+	1	(1/x)	x		=		+		=	
					x	1.732	x		/		x		x		+	1	(1/x)	x		=		+		=	
					x	1.732	x		/		x		x		+	1	(1/x)	x		=		+		=	
					x	1.732	x		/		x		x		+	1	(1/x)	x		=		+		=	
					x	1.732	x		/		x		x		+	1	(1/x)	x		=		+		=	

"C"-Values for Copper Conductors

Conductor AWG	Three Single Conductors						Three-Conductor Cable					
	Magnetic Duct			Non-Magnetic Duct			Magnetic Duct			Non-Magnetic Duct		
	600V	5KV	15KV	**600V**	5KV	15KV	**600V**	5KV	15KV	**600V**	5KV	15KV
# 14	**389**	389	389	**389**	389	389	**389**	389	389	**389**	389	389
# 12	**617**	617	617	**617**	617	617	**617**	617	617	**617**	617	617
# 10	**981**	981	981	**981**	981	981	**981**	981	981	**981**	981	981
# 8	**1,557**	1,551	1,557	**1,558**	1,555	1,558	**1,559**	1,557	1,559	**1,559**	1,558	1,559
# 6	**2,425**	2,406	2,389	**2,430**	2,417	2,406	**2,431**	2,424	2,414	**2,433**	2,428	2,420
# 4	**3,806**	3,750	3,695	**3,825**	3,789	3,752	**3,830**	3,811	3,778	**3,837**	3,823	3,798
# 3	**4,760**	4,760	4,760	**4,802**	4,802	4,802	**4,760**	4,790	4,760	**4,802**	4,802	4,802
# 2	**5,906**	5,736	5,574	**6,044**	5,926	5,809	**5,989**	5,929	5,827	**6,087**	6,022	5,957
# 1	**7,292**	7,029	6,758	**7,493**	7,306	7,108	**7,454**	7,364	7,188	**7,579**	7,507	7,364
# 1/0	**8,924**	8,543	7,973	**9,317**	9,033	8,590	**9,209**	9,086	8,707	**9,472**	9,372	9,052
# 2/0	**10,755**	10,061	9,389	**11,423**	10,877	10,318	**11,244**	11,045	10,500	**11,703**	11,528	11,052
# 3/0	**12,843**	11,804	11,021	**13,923**	13,048	12,360	**13,656**	13,333	12,613	**14,410**	14,118	13,461
# 4/0	**15,082**	13,605	12,542	**16,673**	15,351	14,347	**16,391**	15,890	14,813	**17,482**	17,019	16,012
# 250	**16,483**	14,924	13,643	**18,593**	17,120	15,865	**18,310**	17,850	16,465	**19,779**	19,352	18,001
# 300	**18,176**	16,292	14,768	**20,867**	18,975	17,408	**20,617**	20,051	18,318	**22,524**	21,938	20,163
# 350	**19,703**	17,385	15,678	**22,736**	20,526	18,672	**22,646**	21,914	19,821	**24,904**	24,126	21,982
# 400	**20,565**	18,235	16,365	**24,296**	21,786	19,731	**24,253**	23,371	21,042	**26,915**	26,044	23,517
# 500	**22,185**	19,172	17,492	**26,706**	23,277	21,329	**26,980**	25,449	23,125	**30,028**	28,712	25,916
# 600	**22,965**	20,567	17,962	**28,033**	25,203	22,097	**28,752**	27,974	24,896	**32,236**	31,258	27,766
# 750	**24,136**	21,386	18,888	**28,303**	25,430	22,690	**31,050**	30,024	26,932	**32,404**	31,338	28,303
# 1000	**25,278**	22,539	19,923	**31,490**	28,083	24,887	**33,864**	32,688	29,320	**37,197**	35,748	31,959

Three Phase Calculations:

1. $f = \dfrac{1.732 \times \text{Length} \times I_{sca}}{\text{Runs} \times \text{C-Value} \times E_{L\text{-}L}}$

2. $M = \dfrac{1}{1 + f}$

3. Isc = Isca x M

4. Add Motor Contribution = Motor FLA x 4

"C"-Values for Aluminum Conductors

Conductor AWG	Three Single Conductors						Three-Conductor Cable					
	Magnetic Duct			Non-Magnetic Duct			Magnetic Duct			Non-Magnetic Duct		
	600V	5KV	15KV	**600V**	5KV	15KV	**600V**	5KV	15KV	**600V**	5KV	15KV
# 12	**375**	375	375	**375**	375	375	**375**	375	375	**375**	375	375
# 10	**598**	598	598	**598**	598	598	**598**	598	598	**598**	598	598
# 8	**951**	950	951	**951**	950	951	**951**	951	951	**951**	951	951
# 6	**1,480**	1,476	1,472	**1,481**	1,478	1,476	**1,481**	1,480	1,478	**1,482**	1,481	1,479
# 4	**2,345**	2,332	2,319	**2,350**	2,341	2,333	**2,351**	2,347	2,339	**2,353**	2,349	2,344
# 3	**2,948**	2,948	2,948	**2,958**	2,958	2,958	**2,948**	2,956	2,948	**2,958**	2,958	2,958
# 2	**3,713**	3,669	3,626	**3,729**	3,701	3,672	**3,733**	3,719	3,693	**3,739**	3,724	3,709
# 1	**4,645**	4,574	4,497	**4,678**	4,631	4,580	**4,686**	4,663	4,617	**4,699**	4,681	4,646
# 1/0	**5,777**	5,669	5,493	**5,838**	5,766	5,645	**5,852**	5,820	5,717	**5,875**	5,851	5,771
# 2/0	**7,186**	6,968	6,733	**7,301**	7,152	6,986	**7,327**	7,271	7,109	**7,372**	7,328	7,201
# 3/0	**8,826**	8,466	8,163	**9,110**	8,851	8,627	**9,077**	8,980	8,750	**9,242**	9,164	8,977
# 4/0	**10,740**	10,167	9,700	**11,174**	10,749	10,386	**11,184**	11,021	10,642	**11,408**	11,277	10,968
# 250	**12,122**	11,460	10,848	**12,862**	12,343	11,847	**12,796**	12,636	12,115	**13,236**	13,105	12,661
# 300	**13,909**	13,009	12,192	**14,922**	14,182	13,491	**14,916**	14,698	13,973	**15,494**	15,299	14,658
# 350	**15,484**	14,280	13,288	**16,812**	15,857	14,954	**15,413**	16,490	15,540	**17,635**	17,351	16,500
# 400	**16,670**	15,355	14,188	**18,505**	17,321	16,233	**18,461**	18,063	16,921	**19,587**	19,243	18,154
# 500	**18,755**	16,827	15,657	**21,390**	19,503	18,314	**21,394**	20,606	19,314	**22,987**	22,381	20,978
# 600	**20,093**	18,427	16,484	**23,451**	21,718	19,635	**23,633**	23,195	21,348	**25,750**	25,243	23,294
# 750	**21,766**	19,685	17,686	**23,491**	21,769	19,976	**26,431**	25,789	23,750	**25,682**	25,141	23,491
# 1000	**23,477**	21,235	19,005	**28,778**	26,109	23,482	**29,864**	29,049	26,608	**32,938**	31,919	29,135

Three Phase Calculations:

1. $f = \dfrac{1.732 \times \text{Length} \times \text{Isca}}{\text{Runs} \times \text{C-Value} \times \text{E L-L}}$

2. $M = \dfrac{1}{1 + f}$

3. Isc = Isca x M

4. Add Motor Contribution = Motor FLA x 4

Three-Phase Fault-Current Tables

Short-Circuit Tables and Data: 600 Volts Three-Phase

Three-Phase Bolted Fault-Current* Table

One-Way Distance in Feet

Available Isc in Thousands (Isca)	5	10	15	20	25	30	40	50	60	70	80	90	100	125	150	175	200	225	250	300	350	400	500	600
1	964	931	900	871	844	818	771	729	692	658	627	600	574	519	473	435	403	375	350	310	278	252	212	183
3	2,700	2,454	2,249	2,076	1,927	1,799	1,587	1,420	1,284	1,173	1,079	999	930	793	691	613	550	499	457	391	341	303	247	209
5	4,218	3,647	3,212	2,870	2,594	2,366	2,013	1,751	1,550	1,390	1,260	1,152	1,061	887	762	667	594	535	487	412	357	316	256	215
10	7,294	5,740	4,732	4,026	3,502	3,100	2,520	2,123	1,834	1,614	1,442	1,302	1,188	973	824	715	631	565	511	430	371	326	262	220
15	9,637	7,099	5,619	4,649	3,965	3,457	2,751	2,285	1,953	1,706	1,514	1,361	1,236	1,006	848	732	645	576	520	436	375	329	265	221
20	11,481	8,051	6,199	5,040	4,246	3,668	2,883	2,375	2,019	1,756	1,554	1,393	1,263	1,023	860	741	652	582	525	439	378	331	266	222
25	12,970	8,756	6,609	5,307	4,434	3,808	2,969	2,433	2,061	1,787	1,578	1,413	1,279	1,033	867	747	656	585	528	441	379	332	267	223
30	14,197	9,299	6,914	5,502	4,569	3,907	3,029	2,473	2,090	1,809	1,595	1,426	1,290	1,041	872	751	659	587	530	443	380	333	267	223
35	15,226	9,730	7,149	5,650	4,671	3,981	3,073	2,502	2,111	1,825	1,607	1,436	1,298	1,046	876	753	661	589	531	443	381	334	267	223
40	16,102	10,080	7,336	5,767	4,750	4,038	3,107	2,525	2,127	1,837	1,616	1,443	1,304	1,050	879	755	663	590	532	444	381	334	268	223
45	16,856	10,370	7,489	5,860	4,814	4,084	3,134	2,543	2,139	1,846	1,624	1,449	1,308	1,053	881	757	664	591	533	445	382	334	268	223
50	17,512	10,615	7,616	5,938	4,866	4,122	3,156	2,557	2,149	1,854	1,630	1,454	1,312	1,055	883	758	665	592	533	445	382	335	268	224
55	18,088	10,824	7,722	6,003	4,909	4,153	3,174	2,569	2,158	1,860	1,634	1,458	1,315	1,057	884	759	666	592	534	446	382	335	268	224
60	18,598	11,004	7,814	6,058	4,946	4,179	3,190	2,579	2,165	1,865	1,638	1,461	1,318	1,059	885	760	666	593	534	446	383	335	268	224
65	19,052	11,162	7,893	6,105	4,978	4,202	3,203	2,588	2,171	1,870	1,642	1,464	1,320	1,060	886	761	667	593	535	446	383	335	268	224
70	19,459	11,300	7,962	6,146	5,005	4,221	3,214	2,595	2,176	1,874	1,645	1,466	1,322	1,062	887	762	667	594	535	446	383	335	268	224
75	19,827	11,423	8,023	6,182	5,029	4,238	3,224	2,602	2,181	1,877	1,647	1,468	1,324	1,063	888	762	668	594	535	447	383	335	269	224
80	20,160	11,533	8,077	6,214	5,050	4,253	3,233	2,607	2,185	1,880	1,650	1,470	1,325	1,064	888	763	668	594	535	447	383	335	269	224
85	20,463	11,632	8,125	6,243	5,069	4,266	3,241	2,612	2,188	1,882	1,652	1,471	1,327	1,065	889	763	668	595	536	447	383	336	269	224
90	20,741	11,721	8,168	6,269	5,086	4,278	3,247	2,617	2,191	1,885	1,654	1,473	1,328	1,065	890	764	669	595	536	447	383	336	269	224
95	20,995	11,802	8,208	6,292	5,101	4,289	3,254	2,621	2,194	1,887	1,655	1,474	1,329	1,066	890	764	669	595	536	447	383	336	269	224
100	21,230	11,875	8,243	6,313	5,115	4,299	3,259	2,624	2,197	1,889	1,657	1,475	1,330	1,067	890	764	669	595	536	447	384	336	269	224
105	21,447	11,943	8,276	6,332	5,127	4,308	3,264	2,628	2,199	1,890	1,658	1,476	1,330	1,067	891	764	669	596	536	447	384	336	269	224
110	21,648	12,005	8,306	6,349	5,138	4,316	3,269	2,631	2,201	1,892	1,659	1,477	1,331	1,068	891	765	670	596	536	447	384	336	269	224
115	21,834	12,062	8,333	6,365	5,149	4,323	3,273	2,633	2,203	1,893	1,660	1,478	1,332	1,068	891	765	670	596	537	447	384	336	269	224
120	22,008	12,115	8,358	6,380	5,159	4,330	3,277	2,636	2,205	1,895	1,661	1,479	1,333	1,068	892	765	670	596	537	448	384	336	269	224
125	22,171	12,164	8,381	6,393	5,167	4,336	3,281	2,638	2,206	1,896	1,662	1,480	1,333	1,069	892	765	670	596	537	448	384	336	269	224
130	22,323	12,210	8,403	6,406	5,176	4,342	3,284	2,640	2,208	1,897	1,663	1,480	1,334	1,069	892	766	670	596	537	448	384	336	269	224
135	22,466	12,253	8,423	6,418	5,183	4,347	3,287	2,642	2,209	1,898	1,664	1,481	1,334	1,070	892	766	670	596	537	448	384	336	269	224
140	22,601	12,293	8,442	6,428	5,190	4,352	3,290	2,644	2,210	1,899	1,664	1,481	1,335	1,070	893	766	671	596	537	448	384	336	269	224
145	22,727	12,330	8,460	6,439	5,197	4,357	3,292	2,646	2,212	1,900	1,665	1,482	1,335	1,070	893	766	671	596	537	448	384	336	269	224
150	22,847	12,365	8,476	6,448	5,203	4,361	3,295	2,648	2,213	1,901	1,666	1,483	1,336	1,070	893	766	671	597	537	448	384	336	269	224
155	22,959	12,398	8,492	6,457	5,209	4,365	3,297	2,649	2,214	1,901	1,666	1,483	1,336	1,071	893	766	671	597	537	448	384	336	269	224
160	23,066	12,429	8,506	6,466	5,215	4,369	3,299	2,651	2,215	1,902	1,667	1,483	1,336	1,071	893	766	671	597	537	448	384	336	269	224
165	23,167	12,458	8,520	6,474	5,220	4,373	3,302	2,652	2,216	1,903	1,667	1,484	1,337	1,071	894	766	671	597	537	448	384	336	269	224
170	23,263	12,486	8,533	6,481	5,225	4,376	3,303	2,653	2,217	1,904	1,668	1,484	1,337	1,071	894	767	671	597	537	448	384	336	269	224
175	23,355	12,512	8,545	6,488	5,229	4,380	3,305	2,654	2,217	1,904	1,668	1,485	1,337	1,071	894	767	671	597	537	448	384	336	269	224
180	23,442	12,537	8,557	6,495	5,234	4,383	3,307	2,655	2,218	1,905	1,669	1,485	1,338	1,072	894	767	671	597	537	448	384	336	269	224
185	23,524	12,561	8,568	6,501	5,238	4,385	3,309	2,656	2,219	1,905	1,669	1,485	1,338	1,072	894	767	671	597	537	448	384	336	269	224
190	23,603	12,583	8,578	6,507	5,242	4,388	3,310	2,657	2,220	1,906	1,670	1,486	1,338	1,072	894	767	671	597	538	448	384	336	269	224
195	23,679	12,605	8,588	6,513	5,245	4,391	3,312	2,658	2,220	1,906	1,670	1,486	1,338	1,072	894	767	671	597	538	448	384	336	269	224
200	23,751	12,625	8,598	6,518	5,249	4,393	3,313	2,659	2,221	1,907	1,670	1,486	1,339	1,072	894	767	672	597	538	448	384	336	269	224

# 14 AWG	1 Run(s)	600 Volts	C-Value = 389	Magnetic Duct	Copper

Calculations:

1. $f = \dfrac{1.732 \times \text{Length} \times \text{Isca}}{\text{Runs} \times \text{C-Value} \times E_{L\text{-}L}}$

2. $M = \dfrac{1}{1 + f}$

3. Isc = Isca x M

4. Add Motor Contribution = Motor FLA x 4

* All results are given in symmetrical amperes

Three-Phase Bolted Fault-Current* Table

One-Way Distance in Feet

Available Isc in Thousands (Isca)

Isca	5	10	15	20	25	30	40	50	60	70	80	90	100	125	150	175	200	225	250	300	350	400	500	600
1	977	955	934	914	895	877	842	810	781	753	728	704	681	631	588	550	517	487	461	41[illegible]	379	348	299	263
3	2,803	2,631	2,478	2,342	2,221	2,111	1,921	1,763	1,629	1,513	1,413	1,326	1,248	1,089	966	868	788	721	665	57[illegible]	507	454	374	318
5	4,476	4,052	3,701	3,406	3,155	2,938	2,583	2,305	2,080	1,896	1,741	1,610	1,497	1,274	1,109	982	881	798	730	62[illegible]	544	483	394	333
10	8,104	6,813	5,876	5,166	4,609	4,160	3,483	2,995	2,627	2,339	2,108	1,919	1,761	1,460	1,247	1,088	966	868	788	66[illegible]	576	507	410	344
15	11,104	8,814	7,308	6,241	5,446	4,830	3,940	3,327	2,879	2,537	2,268	2,050	1,871	1,535	1,301	1,129	998	893	809	68[illegible]	587	516	416	348
20	13,625	10,332	8,321	6,965	5,989	5,253	4,217	3,522	3,024	2,649	2,357	2,123	1,931	1,575	1,330	1,151	1,014	907	820	688	593	520	419	350
25	15,775	11,523	9,076	7,487	6,371	5,545	4,403	3,651	3,118	2,721	2,414	2,169	1,969	1,600	1,348	1,164	1,025	915	827	693	596	523	420	351
30	17,629	12,481	9,661	7,880	6,653	5,757	4,536	3,742	3,184	2,771	2,453	2,201	1,995	1,618	1,360	1,174	1,032	921	831	695	599	525	421	352
35	19,244	13,270	10,127	8,187	6,871	5,920	4,636	3,810	3,233	2,808	2,482	2,224	2,014	1,630	1,369	1,180	1,037	925	835	698	600	526	422	353
40	20,664	13,930	10,507	8,434	7,044	6,048	4,714	3,862	3,271	2,837	2,504	2,242	2,029	1,640	1,376	1,185	1,041	928	837	700	602	527	423	353
45	21,923	14,491	10,822	8,636	7,185	6,151	4,776	3,904	3,301	2,859	2,522	2,256	2,040	1,647	1,381	1,189	1,044	930	839	701	603	528	423	353
50	23,045	14,973	11,089	8,805	7,301	6,236	4,828	3,938	3,325	2,878	2,536	2,267	2,050	1,653	1,385	1,192	1,046	932	841	702	603	529	424	354
55	24,053	15,392	11,317	8,948	7,399	6,308	4,870	3,967	3,346	2,893	2,548	2,277	2,057	1,658	1,389	1,195	1,048	934	842	703	604	529	424	354
60	24,963	15,760	11,515	9,071	7,483	6,368	4,907	3,991	3,363	2,906	2,558	2,284	2,064	1,663	1,392	1,197	1,050	935	843	704	605	530	424	354
65	25,788	16,085	11,687	9,178	7,556	6,421	4,938	4,011	3,377	2,916	2,566	2,291	2,069	1,666	1,394	1,199	1,051	936	844	705	605	530	425	354
70	26,540	16,374	11,839	9,272	7,619	6,467	4,965	4,029	3,390	2,926	2,574	2,297	2,074	1,669	1,397	1,200	1,053	937	845	705	605	530	425	354
75	27,229	16,634	11,974	9,354	7,675	6,507	4,988	4,044	3,401	2,934	2,580	2,302	2,078	1,672	1,398	1,202	1,054	938	845	706	606	531	425	355
80	27,861	16,868	12,095	9,428	7,724	6,542	5,009	4,058	3,410	2,941	2,585	2,306	2,082	1,674	1,400	1,203	1,055	939	846	706	606	531	425	355
85	28,443	17,079	12,204	9,493	7,768	6,574	5,027	4,070	3,419	2,948	2,590	2,310	2,085	1,676	1,401	1,204	1,055	939	846	707	606	531	425	355
90	28,982	17,272	12,302	9,553	7,808	6,602	5,044	4,081	3,427	2,953	2,595	2,314	2,088	1,678	1,403	1,205	1,056	940	847	707	607	531	425	355
95	29,482	17,448	12,391	9,606	7,844	6,628	5,059	4,091	3,434	2,958	2,599	2,317	2,090	1,680	1,404	1,206	1,057	941	847	707	607	531	426	355
100	29,947	17,610	12,472	9,655	7,876	6,651	5,072	4,100	3,440	2,963	2,602	2,320	2,093	1,681	1,405	1,207	1,057	941	848	707	607	532	426	355
105	30,380	17,759	12,547	9,700	7,906	6,672	5,085	4,108	3,445	2,967	2,605	2,322	2,095	1,683	1,406	1,207	1,058	941	848	708	607	532	426	355
110	30,785	17,897	12,615	9,741	7,933	6,691	5,096	4,115	3,451	2,971	2,608	2,325	2,097	1,684	1,407	1,208	1,058	942	848	708	607	532	426	355
115	31,164	18,024	12,678	9,778	7,958	6,709	5,106	4,122	3,455	2,974	2,611	2,327	2,098	1,685	1,408	1,209	1,059	942	849	708	607	532	426	355
120	31,520	18,143	12,737	9,813	7,981	6,725	5,116	4,128	3,460	2,978	2,614	2,329	2,100	1,686	1,408	1,209	1,059	943	849	708	608	532	426	355
125	31,854	18,253	12,791	9,845	8,002	6,741	5,124	4,133	3,464	2,981	2,616	2,331	2,101	1,687	1,409	1,210	1,060	943	849	708	608	532	426	355
130	32,170	18,356	12,842	9,875	8,022	6,755	5,133	4,139	3,467	2,983	2,618	2,332	2,103	1,688	1,409	1,210	1,060	943	849	709	608	532	426	355
135	32,467	18,453	12,889	9,903	8,040	6,768	5,140	4,144	3,471	2,986	2,620	2,334	2,104	1,689	1,410	1,210	1,060	943	850	709	608	532	426	355
140	32,749	18,543	12,933	9,929	8,058	6,780	5,147	4,148	3,474	2,988	2,622	2,335	2,105	1,689	1,411	1,211	1,061	944	850	709	608	532	426	355
145	33,015	18,628	12,974	9,953	8,074	6,791	5,154	4,152	3,477	2,990	2,623	2,337	2,106	1,690	1,411	1,211	1,061	944	850	709	608	532	426	355
150	33,267	18,708	13,013	9,976	8,089	6,802	5,160	4,156	3,480	2,993	2,625	2,338	2,107	1,691	1,412	1,212	1,061	944	850	709	608	532	426	355
155	33,507	18,784	13,050	9,998	8,103	6,812	5,165	4,160	3,482	2,994	2,626	2,339	2,108	1,691	1,412	1,212	1,061	944	850	709	608	533	426	355
160	33,735	18,855	13,084	10,018	8,116	6,821	5,171	4,164	3,485	2,996	2,628	2,340	2,109	1,692	1,412	1,212	1,062	944	850	709	608	533	426	355
165	33,952	18,923	13,117	10,037	8,128	6,830	5,176	4,167	3,487	2,998	2,629	2,341	2,110	1,692	1,413	1,212	1,062	945	851	709	608	533	426	355
170	34,159	18,987	13,147	10,055	8,140	6,838	5,181	4,170	3,489	3,000	2,630	2,342	2,111	1,693	1,413	1,213	1,062	945	851	709	609	533	426	355
175	34,356	19,048	13,177	10,072	8,151	6,846	5,185	4,173	3,491	3,001	2,632	2,343	2,112	1,693	1,413	1,213	1,062	945	851	710	609	533	426	356
180	34,544	19,105	13,204	10,088	8,162	6,853	5,189	4,176	3,493	3,003	2,633	2,344	2,112	1,694	1,414	1,213	1,062	945	851	710	609	533	426	356
185	34,724	19,160	13,230	10,103	8,172	6,861	5,194	4,178	3,495	3,004	2,634	2,345	2,113	1,694	1,414	1,213	1,063	945	851	710	609	533	426	356
190	34,897	19,213	13,255	10,118	8,182	6,867	5,197	4,181	3,497	3,005	2,635	2,346	2,114	1,695	1,414	1,214	1,063	945	851	710	609	533	427	356
195	35,062	19,263	13,279	10,132	8,191	6,874	5,201	4,183	3,498	3,006	2,636	2,346	2,114	1,695	1,415	1,214	1,063	945	851	710	609	533	427	356
200	35,220	19,310	13,302	10,145	8,199	6,880	5,204	4,185	3,500	3,008	2,637	2,347	2,115	1,695	1,415	1,214	1,063	945	851	710	609	533	427	356

12 AWG | 1 Run(s) | 600 Volts | C-Value = 617 | Magnetic Duct | Copper

Calculations:

1. $f = \dfrac{1.732 \times \text{Length} \times \text{Isca}}{\text{Runs} \times \text{C-Value} \times E_{L\text{-}L}}$

2. $M = \dfrac{1}{1 + f}$

3. Isc = Isca x M

4. Add Motor Contribution = Motor FLA x 4

* All results are given in symmetrical amperes

Three-Phase Bolted Fault-Current* Table

One-Way Distance in Feet

Available Isc in Thousands (Isca)	5	10	15	20	25	30	40	50	60	70	80	90	100	125	150	175	200	225	250	300	350	400	500	600
1	986	971	958	944	931	919	895	872	850	829	809	791	773	731	694	660	630	602	576	531	493	459	405	362
3	2,873	2,757	2,649	2,550	2,458	2,372	2,217	2,081	1,961	1,854	1,758	1,672	1,593	1,426	1,291	1,179	1,085	1,005	935	822	734	662	554	476
5	4,657	4,359	4,096	3,863	3,655	3,469	3,148	2,881	2,656	2,463	2,297	2,151	2,023	1,761	1,559	1,399	1,268	1,160	1,069	924	813	726	598	509
10	8,717	7,726	6,938	6,295	5,762	5,311	4,593	4,046	3,616	3,268	2,981	2,741	2,536	2,138	1,847	1,626	1,452	1,312	1,197	1,018	885	783	636	536
15	12,288	10,407	9,025	7,967	7,131	6,454	5,424	4,677	4,111	3,668	3,310	3,017	2,771	2,302	1,968	1,719	1,526	1,372	1,246	1,053	912	804	650	546
20	15,453	12,590	10,623	9,187	8,093	7,232	5,963	5,073	4,414	3,907	3,504	3,176	2,905	2,393	2,035	1,770	1,566	1,404	1,273	1,072	926	815	657	551
25	18,277	14,404	11,885	10,116	8,806	7,796	6,341	5,344	4,618	4,065	3,631	3,280	2,992	2,452	2,077	1,802	1,591	1,424	1,289	1,084	935	822	662	554
30	20,813	15,934	12,908	10,848	9,355	8,223	6,621	5,541	4,764	4,179	3,721	3,354	3,053	2,493	2,107	1,824	1,608	1,438	1,300	1,092	941	826	665	556
35	23,103	17,242	13,753	11,439	9,791	8,558	6,836	5,692	4,875	4,263	3,788	3,408	3,098	2,523	2,128	1,840	1,621	1,448	1,309	1,097	945	829	667	557
40	25,181	18,374	14,464	11,926	10,146	8,828	7,008	5,810	4,961	4,329	3,840	3,450	3,132	2,546	2,144	1,852	1,630	1,455	1,315	1,102	948	832	668	558
45	27,075	19,362	15,069	12,334	10,440	9,050	7,147	5,905	5,031	4,382	3,882	3,484	3,160	2,564	2,157	1,862	1,637	1,461	1,319	1,105	950	834	670	559
50	28,808	20,232	15,591	12,682	10,688	9,236	7,262	5,983	5,088	4,425	3,915	3,511	3,182	2,579	2,167	1,869	1,643	1,466	1,323	1,108	952	835	671	560
55	30,400	21,005	16,046	12,981	10,900	9,393	7,359	6,049	5,135	4,461	3,943	3,533	3,201	2,591	2,176	1,876	1,648	1,470	1,327	1,110	954	837	671	561
60	31,868	21,696	16,446	13,242	11,083	9,529	7,442	6,105	5,175	4,491	3,967	3,552	3,216	2,601	2,183	1,881	1,652	1,473	1,329	1,112	956	838	672	561
65	33,225	22,316	16,800	13,471	11,242	9,647	7,514	6,153	5,210	4,517	3,987	3,569	3,230	2,610	2,189	1,886	1,656	1,476	1,332	1,113	957	839	673	562
70	34,484	22,877	17,116	13,673	11,383	9,750	7,576	6,195	5,240	4,540	4,005	3,583	3,241	2,617	2,195	1,890	1,659	1,478	1,333	1,115	958	839	673	562
75	35,655	23,387	17,400	13,853	11,508	9,841	7,631	6,232	5,266	4,560	4,020	3,595	3,251	2,624	2,199	1,893	1,662	1,481	1,335	1,116	959	840	674	562
80	36,747	23,852	17,656	14,015	11,619	9,923	7,680	6,265	5,289	4,577	4,034	3,606	3,260	2,629	2,203	1,896	1,664	1,482	1,337	1,117	959	841	674	562
85	37,768	24,277	17,888	14,161	11,719	9,996	7,724	6,294	5,310	4,593	4,046	3,615	3,268	2,634	2,207	1,899	1,666	1,484	1,338	1,118	960	841	674	563
90	38,724	24,669	18,100	14,293	11,810	10,062	7,763	6,320	5,329	4,606	4,057	3,624	3,275	2,639	2,210	1,901	1,668	1,485	1,339	1,119	961	842	675	563
95	39,621	25,030	18,293	14,414	11,892	10,121	7,799	6,343	5,345	4,619	4,066	3,632	3,281	2,643	2,213	1,903	1,669	1,487	1,340	1,119	961	842	675	563
100	40,465	25,364	18,471	14,524	11,967	10,175	7,831	6,364	5,360	4,630	4,075	3,639	3,287	2,647	2,215	1,905	1,671	1,488	1,341	1,120	962	842	675	563
105	41,260	25,674	18,635	14,625	12,035	10,225	7,860	6,384	5,374	4,640	4,083	3,645	3,292	2,650	2,218	1,907	1,672	1,489	1,342	1,121	962	843	675	563
110	42,010	25,963	18,787	14,718	12,098	10,270	7,887	6,401	5,387	4,650	4,090	3,651	3,297	2,653	2,220	1,908	1,673	1,490	1,343	1,121	962	843	676	563
115	42,719	26,232	18,927	14,804	12,157	10,312	7,911	6,417	5,398	4,658	4,097	3,656	3,301	2,656	2,222	1,910	1,674	1,491	1,343	1,122	963	843	676	564
120	43,391	26,484	19,058	14,884	12,210	10,351	7,934	6,432	5,409	4,666	4,103	3,661	3,305	2,658	2,224	1,911	1,675	1,492	1,344	1,122	963	844	676	564
125	44,028	26,720	19,180	14,959	12,260	10,387	7,955	6,446	5,418	4,673	4,108	3,665	3,308	2,661	2,225	1,912	1,676	1,492	1,345	1,123	963	844	676	564
130	44,633	26,941	19,293	15,028	12,307	10,420	7,975	6,459	5,428	4,680	4,114	3,669	3,312	2,663	2,227	1,913	1,677	1,493	1,345	1,123	964	844	676	564
135	45,207	27,149	19,400	15,092	12,350	10,451	7,993	6,471	5,436	4,686	4,118	3,673	3,315	2,665	2,228	1,914	1,678	1,494	1,346	1,123	964	844	676	564
140	45,755	27,346	19,500	15,153	12,390	10,480	8,010	6,482	5,444	4,692	4,123	3,677	3,318	2,667	2,230	1,915	1,679	1,494	1,346	1,124	964	844	676	564
145	46,276	27,531	19,594	15,210	12,428	10,507	8,026	6,492	5,451	4,698	4,127	3,680	3,321	2,669	2,231	1,916	1,680	1,495	1,347	1,124	965	845	677	564
150	46,774	27,707	19,683	15,263	12,464	10,533	8,041	6,502	5,458	4,703	4,131	3,683	3,323	2,670	2,232	1,917	1,680	1,495	1,347	1,124	965	845	677	564
155	47,249	27,873	19,767	15,313	12,498	10,556	8,054	6,511	5,464	4,707	4,135	3,686	3,325	2,672	2,233	1,918	1,681	1,496	1,348	1,125	965	845	677	564
160	47,703	28,030	19,846	15,361	12,529	10,579	8,068	6,520	5,470	4,712	4,138	3,689	3,328	2,673	2,234	1,919	1,681	1,496	1,348	1,125	965	845	677	564
165	48,138	28,180	19,921	15,405	12,559	10,600	8,080	6,528	5,476	4,716	4,141	3,692	3,330	2,675	2,235	1,919	1,682	1,497	1,348	1,125	965	845	677	564
170	48,555	28,322	19,992	15,448	12,587	10,620	8,092	6,535	5,481	4,720	4,144	3,694	3,332	2,676	2,236	1,920	1,682	1,497	1,349	1,125	965	845	677	565
175	48,954	28,458	20,059	15,488	12,614	10,639	8,103	6,543	5,486	4,724	4,147	3,696	3,334	2,677	2,237	1,921	1,683	1,497	1,349	1,126	966	845	677	565
180	49,338	28,587	20,123	15,526	12,639	10,657	8,113	6,549	5,491	4,727	4,150	3,698	3,335	2,678	2,237	1,921	1,683	1,498	1,349	1,126	966	846	677	565
185	49,706	28,710	20,184	15,563	12,663	10,674	8,123	6,556	5,496	4,731	4,153	3,700	3,337	2,679	2,238	1,922	1,684	1,498	1,349	1,126	966	846	677	565
190	50,060	28,828	20,242	15,597	12,686	10,691	8,132	6,562	5,500	4,734	4,155	3,702	3,339	2,680	2,239	1,922	1,684	1,498	1,350	1,126	966	846	677	565
195	50,400	28,940	20,298	15,630	12,708	10,706	8,141	6,568	5,504	4,737	4,157	3,704	3,340	2,681	2,240	1,923	1,685	1,499	1,350	1,126	966	846	677	565
200	50,728	29,048	20,351	15,661	12,728	10,721	8,150	6,573	5,508	4,740	4,160	3,706	3,342	2,682	2,240	1,923	1,685	1,499	1,350	1,126	966	846	677	565

10 AWG | 1 Run(s) | 600 Volts | C-Value = 981 | Magnetic Duct | Copper

Calculations:

1. $f = \dfrac{1.732 \times \text{Length} \times \text{Isca}}{\text{Runs} \times \text{C-Value} \times E_{L\text{-}L}}$

2. $M = \dfrac{1}{1 + f}$

3. Isc = Isca x M

4. Add Motor Contribution = Motor FLA x 4

* All results are given in symmetrical amperes

Three-Phase Bolted Fault-Current* Table

One-Way Distance in Feet

Available Isc in Thousands (Isca)

Isca	5	10	15	20	25	30	40	50	60	70	80	90	100	125	150	175	200	225	250	300	350	400	500	600
3	2,919	2,842	2,769	2,700	2,634	2,571	2,454	2,347	2,249	2,159	2,076	1,999	1,928	1,770	1,636	1,520	1,420	1,332	1,255	1,124	1,018	930	793	692
5	4,779	4,576	4,390	4,218	4,059	3,912	3,648	3,416	3,213	3,032	2,871	2,726	2,595	2,316	2,092	1,907	1,752	1,620	1,507	1,322	1,178	1,062	887	762
7	6,573	6,196	5,859	5,558	5,285	5,038	4,608	4,245	3,936	3,668	3,434	3,229	3,046	2,669	2,376	2,140	1,947	1,786	1,649	1,431	1,263	1,131	935	797
10	9,152	8,436	7,824	7,295	6,833	6,426	5,742	5,189	4,734	4,352	4,027	3,747	3,504	3,014	2,645	2,356	2,124	1,934	1,775	1,524	1,335	1,188	974	825
15	13,169	11,736	10,585	9,639	8,848	8,178	7,101	6,275	5,621	5,090	4,651	4,282	3,967	3,351	2,901	2,557	2,286	2,067	1,886	1,605	1,397	1,237	1,006	848
20	16,872	14,590	12,852	11,484	10,379	9,468	8,054	7,008	6,202	5,562	5,042	4,611	4,248	3,549	3,048	2,671	2,376	2,141	1,947	1,650	1,431	1,263	1,024	860
25	20,296	17,082	14,747	12,974	11,581	10,458	8,760	7,536	6,612	5,890	5,310	4,834	4,437	3,680	3,144	2,744	2,434	2,187	1,986	1,677	1,452	1,279	1,034	868
30	23,472	19,278	16,355	14,202	12,550	11,242	9,303	7,934	6,917	6,131	5,505	4,995	4,572	3,772	3,211	2,795	2,474	2,220	2,013	1,696	1,466	1,290	1,041	873
35	26,426	21,226	17,736	15,232	13,347	11,878	9,734	8,246	7,153	6,315	5,653	5,117	4,674	3,841	3,261	2,833	2,504	2,244	2,032	1,710	1,476	1,298	1,046	876
40	29,180	22,967	18,936	16,108	14,015	12,404	10,085	8,496	7,340	6,461	5,770	5,212	4,753	3,895	3,299	2,862	2,527	2,262	2,047	1,721	1,484	1,304	1,050	879
45	31,754	24,533	19,987	16,863	14,583	12,847	10,375	8,702	7,493	6,579	5,864	5,289	4,816	3,937	3,330	2,885	2,544	2,276	2,059	1,729	1,490	1,309	1,053	881
50	34,165	25,947	20,916	17,519	15,072	13,224	10,620	8,873	7,620	6,676	5,941	5,352	4,869	3,972	3,355	2,903	2,559	2,288	2,068	1,736	1,495	1,313	1,056	883
55	36,428	27,232	21,743	18,096	15,496	13,550	10,829	9,019	7,727	6,759	6,006	5,404	4,912	4,001	3,375	2,919	2,571	2,297	2,076	1,741	1,499	1,316	1,058	885
60	38,556	28,404	22,484	18,606	15,869	13,834	11,010	9,144	7,818	6,828	6,061	5,449	4,949	4,026	3,393	2,932	2,581	2,305	2,083	1,746	1,502	1,319	1,060	886
65	40,560	29,477	23,151	19,061	16,198	14,084	11,168	9,252	7,897	6,889	6,109	5,487	4,980	4,046	3,407	2,943	2,589	2,312	2,088	1,750	1,505	1,321	1,061	887
70	42,453	30,464	23,755	19,468	16,492	14,305	11,336	9,347	7,967	6,941	6,150	5,520	5,008	4,064	3,420	2,952	2,597	2,318	2,093	1,753	1,508	1,323	1,062	888
75	44,241	31,374	24,305	19,836	16,755	14,503	11,429	9,431	8,027	6,987	6,186	5,550	5,032	4,080	3,431	2,960	2,603	2,323	2,097	1,756	1,510	1,325	1,063	888
80	45,935	32,217	24,808	20,169	16,992	14,680	11,539	9,506	8,081	7,028	6,218	5,575	5,053	4,094	3,441	2,968	2,609	2,327	2,101	1,758	1,512	1,326	1,064	889
85	47,541	32,998	25,269	20,473	17,207	14,840	11,638	9,573	8,130	7,065	6,247	5,598	5,072	4,107	3,450	2,974	2,614	2,331	2,104	1,761	1,514	1,327	1,065	890
90	49,065	33,726	25,693	20,751	17,403	14,986	11,727	9,633	8,173	7,098	6,272	5,619	5,089	4,118	3,458	2,980	2,618	2,335	2,107	1,763	1,515	1,329	1,066	890
95	50,515	34,404	26,085	21,006	17,582	15,118	11,808	9,687	8,212	7,127	6,295	5,637	5,104	4,128	3,465	2,985	2,622	2,338	2,110	1,765	1,516	1,330	1,067	891
100	51,894	35,039	26,448	21,241	17,746	15,239	11,882	9,737	8,248	7,154	6,316	5,654	5,118	4,137	3,471	2,990	2,626	2,341	2,112	1,766	1,518	1,331	1,067	891
105	53,209	35,633	26,785	21,458	17,898	15,351	11,950	9,782	8,281	7,179	6,335	5,669	5,130	4,145	3,477	2,994	2,629	2,344	2,114	1,76[illegible]	1,519	1,331	1,068	891
110	54,464	36,191	27,100	21,659	18,037	15,453	12,012	9,824	8,310	7,201	6,353	5,683	5,142	4,152	3,482	2,998	2,632	2,346	2,116	1,76[illegible]	1,520	1,332	1,068	892
115	55,662	36,717	27,393	21,846	18,167	15,548	12,069	9,862	8,338	7,222	6,369	5,696	5,152	4,159	3,487	3,002	2,635	2,348	2,118	1,770	1,521	1,333	1,069	892
120	56,808	37,212	27,668	22,020	18,287	15,636	12,122	9,898	8,363	7,240	6,384	5,708	5,162	4,165	3,491	3,005	2,638	2,350	2,119	1,771	1,522	1,333	1,069	892
125	57,904	37,679	27,925	22,183	18,399	15,718	12,171	9,931	8,386	7,258	6,397	5,719	5,171	4,171	3,495	3,008	2,640	2,352	2,121	1,772	1,522	1,334	1,070	893
130	58,954	38,121	28,167	22,335	18,504	15,795	12,217	9,961	8,408	7,274	6,410	5,729	5,179	4,176	3,499	3,011	2,642	2,354	2,122	1,773	1,523	1,335	1,070	893
135	59,961	38,540	28,395	22,478	18,602	15,866	12,260	9,989	8,428	7,289	6,422	5,738	5,187	4,181	3,503	3,013	2,644	2,355	2,124	1,774	1,524	1,335	1,070	893
140	60,928	38,937	28,610	22,613	18,694	15,933	12,300	10,016	8,447	7,303	6,432	5,747	5,194	4,186	3,506	3,016	2,646	2,357	2,125	1,775	1,524	1,336	1,071	893
145	61,856	39,314	28,813	22,739	18,781	15,996	12,337	10,041	8,465	7,317	6,443	5,755	5,200	4,190	3,509	3,018	2,648	2,358	2,126	1,776	1,525	1,336	1,071	893
150	62,749	39,672	29,005	22,859	18,862	16,055	12,372	10,064	8,481	7,329	6,452	5,763	5,207	4,194	3,512	3,020	2,649	2,360	2,127	1,777	1,525	1,336	1,071	894
155	63,607	40,014	29,187	22,972	18,939	16,110	12,405	10,086	8,497	7,340	6,461	5,770	5,212	4,198	3,514	3,022	2,651	2,361	2,128	1,777	1,526	1,337	1,071	894
160	64,433	40,339	29,360	23,079	19,011	16,163	12,436	10,106	8,511	7,351	6,470	5,777	5,218	4,202	3,517	3,024	2,652	2,362	2,129	1,778	1,526	1,337	1,072	894
165	65,229	40,650	29,524	23,180	19,080	16,213	12,466	10,126	8,525	7,362	6,478	5,783	5,223	4,205	3,519	3,026	2,654	2,363	2,130	1,779	1,527	1,338	1,072	894
170	65,997	40,946	29,680	23,276	19,145	16,260	12,493	10,144	8,538	7,371	6,485	5,789	5,228	4,208	3,521	3,027	2,655	2,364	2,130	1,779	1,527	1,338	1,072	894
175	66,737	41,230	29,829	23,368	19,207	16,304	12,520	10,161	8,550	7,380	6,492	5,795	5,232	4,211	3,523	3,029	2,656	2,365	2,131	1,730	1,528	1,338	1,072	894
180	67,451	41,502	29,971	23,455	19,266	16,346	12,545	10,178	8,562	7,389	6,499	5,800	5,237	4,214	3,525	3,030	2,657	2,366	2,132	1,730	1,528	1,338	1,072	894
185	68,141	41,762	30,107	23,538	19,322	16,387	12,568	10,193	8,573	7,397	6,505	5,805	5,241	4,217	3,527	3,032	2,658	2,367	2,133	1,781	1,528	1,339	1,072	895
190	68,808	42,011	30,236	23,617	19,375	16,425	12,591	10,208	8,583	7,405	6,511	5,810	5,245	4,219	3,529	3,033	2,659	2,367	2,133	1,781	1,529	1,339	1,072	895
195	69,453	42,251	30,360	23,692	19,426	16,461	12,612	10,222	8,593	7,412	6,517	5,814	5,249	4,222	3,531	3,034	2,660	2,368	2,134	1,781	1,529	1,339	1,073	895
200	70,077	42,481	30,479	23,764	19,474	16,496	12,633	10,235	8,603	7,420	6,522	5,819	5,252	4,224	3,532	3,035	2,661	2,369	2,134	1,782	1,529	1,339	1,073	895

# 8 AWG	1 Run(s)	600 Volts	C-Value = 1,557	Magnetic Duct	Copper

Calculations:

1. $f = \dfrac{1.732 \times \text{Length} \times \text{Isca}}{\text{Runs} \times \text{C-Value} \times E_{L\text{-}L}}$

2. $M = \dfrac{1}{1 + f}$

3. Isc = Isca x M

4. Add Motor Contribution = Motor FLA x 4

* All results are given in symmetrical amperes

Three-Phase Bolted Fault-Current* Table

One-Way Distance in Feet

Available Isc in Thousands (Isca)	5	10	15	20	25	30	40	50	60	70	80	90	100	125	150	175	200	225	250	300	350	400	500	600
3	2,947	2,897	2,847	2,800	2,754	2,710	2,625	2,545	2,471	2,400	2,333	2,270	2,211	2,074	1,954	1,846	1,750	1,663	1,585	1,448	1,333	1,235	1,077	955
5	4,856	4,719	4,590	4,468	4,352	4,242	4,039	3,853	3,684	3,529	3,387	3,256	3,134	2,867	2,642	2,449	2,283	2,138	2,010	1,795	1,622	1,479	1,258	1,094
7	6,720	6,462	6,222	6,000	5,793	5,600	5,250	4,941	4,667	4,421	4,200	4,000	3,818	3,429	3,111	2,848	2,625	2,435	2,270	2,000	1,787	1,615	1,355	1,167
10	9,438	8,936	8,485	8,077	7,707	7,369	6,774	6,269	5,834	5,455	5,122	4,828	4,565	4,019	3,590	3,243	2,958	2,719	2,515	2,188	1,936	1,736	1,438	1,228
15	13,771	12,727	11,831	11,053	10,371	9,768	8,750	7,925	7,242	6,667	6,177	5,754	5,385	4,641	4,078	3,637	3,281	2,990	2,745	2,360	2,069	1,842	1,511	1,281
20	17,872	16,154	14,737	13,549	12,538	11,667	10,244	9,131	8,236	7,500	6,886	6,364	5,916	5,030	4,375	3,871	3,471	3,146	2,877	2,456	2,143	1,901	1,550	1,309
25	21,762	19,266	17,284	15,672	14,335	13,208	11,414	10,048	8,975	8,109	7,395	6,797	6,288	5,297	4,575	4,027	3,596	3,248	2,962	2,518	2,190	1,937	1,574	1,326
30	25,455	22,106	19,535	17,501	15,850	14,483	12,354	10,770	9,546	8,572	7,778	7,119	6,563	5,491	4,719	4,138	3,684	3,320	3,022	2,561	2,222	1,963	1,591	1,338
35	28,966	24,706	21,539	19,092	17,144	15,556	13,126	11,352	10,001	8,937	8,077	7,369	6,775	5,638	4,828	4,221	3,750	3,374	3,066	2,593	2,246	1,981	1,603	1,346
40	32,308	27,097	23,334	20,489	18,262	16,471	13,771	11,832	10,371	9,231	8,317	7,568	6,943	5,754	4,913	4,286	3,801	3,415	3,100	2,617	2,264	1,995	1,612	1,353
45	35,494	29,303	24,951	21,725	19,238	17,261	14,319	12,234	10,679	9,474	8,514	7,731	7,079	5,847	4,981	4,338	3,842	3,448	3,127	2,636	2,279	2,007	1,620	1,358
50	38,533	31,344	26,416	22,827	20,097	17,950	14,790	12,576	10,938	9,678	8,678	7,866	7,192	5,924	5,036	4,380	3,875	3,474	3,149	2,652	2,290	2,016	1,626	1,362
55	41,436	33,238	27,749	23,816	20,859	18,555	15,198	12,870	11,160	9,851	8,817	7,980	7,288	5,989	5,083	4,415	3,902	3,496	3,167	2,665	2,300	2,023	1,630	1,365
60	44,211	35,001	28,967	24,707	21,540	19,092	15,557	13,126	11,352	10,001	8,937	8,077	7,369	6,044	5,122	4,445	3,926	3,515	3,182	2,675	2,308	2,029	1,634	1,368
65	46,868	36,646	30,084	25,515	22,151	19,571	15,873	13,351	11,520	10,131	9,040	8,162	7,439	6,091	5,156	4,470	3,945	3,531	3,195	2,685	2,315	2,034	1,638	1,371
70	49,413	38,183	31,113	26,251	22,704	20,001	16,155	13,549	11,667	10,245	9,131	8,236	7,501	6,132	5,186	4,492	3,963	3,545	3,206	2,693	2,321	2,039	1,641	1,373
75	51,853	39,624	32,063	26,925	23,206	20,390	16,407	13,726	11,799	10,346	9,211	8,301	7,555	6,168	5,211	4,512	3,978	3,557	3,216	2,699	2,326	2,043	1,643	1,374
80	54,195	40,977	32,943	27,542	23,663	20,742	16,635	13,885	11,916	10,436	9,282	8,359	7,602	6,200	5,234	4,529	3,991	3,567	3,225	2,706	2,330	2,046	1,646	1,376
85	56,444	42,250	33,761	28,112	24,082	21,063	16,841	14,028	12,021	10,516	9,346	8,411	7,645	6,228	5,254	4,544	4,003	3,577	3,232	2,711	2,334	2,050	1,648	1,377
90	58,606	43,450	34,522	28,638	24,467	21,357	17,028	14,158	12,116	10,589	9,404	8,457	7,684	6,254	5,272	4,557	4,013	3,585	3,239	2,716	2,338	2,052	1,649	1,379
95	60,686	44,583	35,234	29,126	24,823	21,627	17,199	14,276	12,203	10,655	9,456	8,499	7,718	6,277	5,289	4,569	4,022	3,592	3,245	2,720	2,341	2,055	1,651	1,380
100	62,688	45,654	35,899	29,579	25,151	21,876	17,357	14,385	12,282	10,715	9,503	8,537	7,750	6,297	5,303	4,581	4,031	3,599	3,251	2,724	2,344	2,057	1,652	1,381
105	64,617	46,669	36,524	30,002	25,456	22,107	17,501	14,484	12,354	10,770	9,546	8,572	7,778	6,316	5,317	4,591	4,039	3,605	3,256	2,727	2,347	2,059	1,654	1,382
110	66,477	47,631	37,110	30,397	25,740	22,320	17,635	14,575	12,420	10,820	9,586	8,604	7,805	6,334	5,329	4,600	4,046	3,611	3,261	2,731	2,349	2,061	1,655	1,383
115	68,271	48,545	37,663	30,766	26,004	22,519	17,759	14,660	12,482	10,867	9,622	8,633	7,829	6,349	5,340	4,608	4,052	3,616	3,265	2,734	2,351	2,063	1,656	1,383
120	70,002	49,414	38,184	31,113	26,252	22,704	17,874	14,738	12,538	10,910	9,656	8,660	7,851	6,364	5,351	4,616	4,058	3,621	3,269	2,736	2,353	2,064	1,657	1,384
125	71,675	50,242	38,676	31,439	26,483	22,877	17,981	14,811	12,591	10,950	9,687	8,686	7,872	6,378	5,360	4,623	4,064	3,625	3,272	2,739	2,355	2,065	1,658	1,385
130	73,291	51,031	39,142	31,746	26,701	23,040	18,081	14,878	12,640	10,987	9,716	8,709	7,891	6,390	5,369	4,629	4,069	3,629	3,276	2,741	2,357	2,067	1,659	1,385
135	74,854	51,783	39,583	32,036	26,906	23,192	18,174	14,942	12,686	11,021	9,743	8,730	7,909	6,402	5,377	4,636	4,074	3,633	3,279	2,743	2,358	2,068	1,659	1,386
140	76,366	52,503	40,002	32,310	27,099	23,335	18,262	15,001	12,728	11,053	9,768	8,751	7,925	6,413	5,385	4,641	4,078	3,637	3,282	2,745	2,360	2,069	1,660	1,386
145	77,830	53,191	40,400	32,569	27,281	23,470	18,345	15,057	12,768	11,084	9,792	8,770	7,941	6,423	5,392	4,647	4,082	3,640	3,284	2,747	2,361	2,070	1,661	1,387
150	79,248	53,849	40,779	32,815	27,453	23,597	18,422	15,109	12,806	11,112	9,814	8,787	7,955	6,432	5,399	4,652	4,086	3,643	3,287	2,749	2,362	2,071	1,662	1,387
155	80,622	54,480	41,140	33,048	27,616	23,718	18,496	15,158	12,841	11,139	9,835	8,804	7,969	6,441	5,405	4,656	4,090	3,646	3,289	2,751	2,364	2,072	1,662	1,388
160	81,955	55,085	41,484	33,270	27,770	23,831	18,565	15,205	12,875	11,164	9,854	8,820	7,982	6,450	5,411	4,661	4,093	3,649	3,291	2,752	2,365	2,073	1,663	1,388
165	83,247	55,666	41,813	33,480	27,917	23,940	18,630	15,249	12,906	11,187	9,873	8,834	7,994	6,458	5,417	4,665	4,096	3,651	3,293	2,754	2,366	2,074	1,663	1,388
170	84,501	56,224	42,127	33,681	28,057	24,042	18,692	15,290	12,936	11,210	9,890	8,848	8,005	6,465	5,422	4,669	4,099	3,653	3,295	2,755	2,367	2,075	1,664	1,389
175	85,718	56,760	42,427	33,873	28,190	24,140	18,751	15,330	12,964	11,231	9,906	8,861	8,016	6,472	5,427	4,672	4,102	3,656	3,297	2,756	2,368	2,075	1,664	1,389
180	86,900	57,276	42,715	34,056	28,317	24,232	18,807	15,367	12,991	11,251	9,922	8,874	8,026	6,479	5,431	4,676	4,105	3,658	3,299	2,757	2,369	2,076	1,665	1,389
185	88,049	57,773	42,990	34,231	28,437	24,321	18,861	15,403	13,016	11,270	9,937	8,886	8,036	6,485	5,436	4,679	4,107	3,660	3,300	2,758	2,369	2,077	1,665	1,390
190	89,166	58,252	43,255	34,399	28,553	24,405	18,911	15,436	13,040	11,288	9,951	8,897	8,045	6,491	5,440	4,682	4,109	3,662	3,302	2,760	2,370	2,077	1,665	1,390
195	90,252	58,713	43,509	34,559	28,663	24,486	18,960	15,469	13,063	11,305	9,964	8,908	8,054	6,497	5,444	4,685	4,112	3,663	3,303	2,761	2,371	2,078	1,666	1,390
200	91,308	59,158	43,753	34,713	28,769	24,563	19,006	15,499	13,085	11,322	9,977	8,918	8,062	6,502	5,448	4,688	4,114	3,665	3,305	2,762	2,372	2,078	1,666	1,390

# 6 AWG	1 Run(s)	600 Volts	C-Value = 2,425	Magnetic Duct	Copper

Calculations:

1. $f = \dfrac{1.732 \times \text{Length} \times \text{Isca}}{\text{Runs} \times \text{C-Value} \times E_{L\text{-}L}}$

2. $M = \dfrac{1}{1 + f}$

3. Isc = Isca x M

4. Add Motor Contribution = Motor FLA x 4

* All results are given in symmetrical amperes

Three-Phase Bolted Fault-Current* Table

One-Way Distance in Feet

Available Isc in Thousands (Isca)	5	10	15	20	25	30	40	50	60	70	80	90	100	125	150	175	200	225	250	300	350	400	500	600
3	2,966	2,933	2,901	2,869	2,839	2,808	2,750	2,694	2,640	2,588	2,538	2,490	2,444	2,336	2,237	2,146	2,062	1,984	1,912	1,78[illegible]	1,670	1,571	1,403	1,268
5	4,907	4,817	4,731	4,648	4,567	4,489	4,341	4,203	4,073	3,951	3,836	3,728	3,625	3,392	3,187	3,005	2,843	2,698	2,567	2,33[illegible]	2,148	1,987	1,726	1,527
7	6,819	6,647	6,484	6,328	6,180	6,038	5,774	5,532	5,309	5,103	4,913	4,737	4,572	4,208	3,897	3,629	3,395	3,190	3,008	2,700	2,449	2,241	1,915	1,672
10	9,635	9,295	8,979	8,683	8,406	8,146	7,672	7,250	6,873	6,532	6,224	5,943	5,687	5,133	4,678	4,297	3,973	3,695	3,453	3,053	2,736	2,479	2,087	1,802
15	14,193	13,468	12,813	12,220	11,678	11,183	10,309	9,561	8,915	8,350	7,853	7,411	7,017	6,193	5,542	5,015	4,580	4,214	3,902	3,39[illegible]	3,011	2,702	2,243	1,917
20	18,590	17,366	16,293	15,345	14,501	13,745	12,447	11,374	10,470	9,700	9,035	8,456	7,946	6,906	6,106	5,473	4,958	4,532	4,173	3,603	3,170	2,830	2,330	1,980
25	22,835	21,015	19,464	18,126	16,960	15,935	14,217	12,833	11,695	10,742	9,933	9,237	8,632	7,418	6,503	5,789	5,217	4,747	4,355	3,738	3,274	2,912	2,385	2,020
30	26,936	24,439	22,366	20,618	19,122	17,829	15,706	14,034	12,684	11,571	10,637	9,843	9,159	7,804	6,798	6,022	5,405	4,902	4,485	3,833	3,347	2,970	2,424	2,047
35	30,899	27,658	25,032	22,862	21,038	19,484	16,975	15,039	13,499	12,245	11,205	10,327	9,577	8,105	7,025	6,200	5,547	5,019	4,583	3,905	3,401	3,012	2,452	2,068
40	34,732	30,689	27,490	24,895	22,747	20,941	18,071	15,893	14,183	12,805	11,672	10,723	9,916	8,347	7,206	6,340	5,660	5,111	4,660	3,9[illegible]0	3,443	3,045	2,474	2,083
45	38,440	33,549	29,763	26,744	24,282	22,234	19,026	16,627	14,765	13,278	12,063	11,052	10,197	8,545	7,353	6,454	5,750	5,185	4,721	4,0[illegible]4	3,476	3,071	2,491	2,095
50	42,030	36,252	31,871	28,434	25,667	23,390	19,866	17,264	15,266	13,681	12,395	11,330	10,433	8,710	7,476	6,548	5,824	5,245	4,771	4,0[illegible]0	3,503	3,092	2,505	2,105
55	45,508	38,810	33,831	29,984	26,923	24,429	20,610	17,824	15,701	14,030	12,681	11,568	10,635	8,850	7,579	6,626	5,887	5,296	4,812	4,0[illegible]0	3,526	3,110	2,516	2,113
60	48,878	41,235	35,659	31,411	28,068	25,368	21,274	18,319	16,084	14,335	12,929	11,775	10,809	8,971	7,667	6,694	5,940	5,339	4,848	4,0[illegible]5	3,545	3,125	2,526	2,120
65	52,146	43,537	37,367	32,729	29,116	26,220	21,871	18,759	16,423	14,604	13,147	11,955	10,961	9,075	7,743	6,752	5,985	5,375	4,878	4,117	3,561	3,137	2,534	2,126
70	55,316	45,724	38,967	33,950	30,078	26,998	22,410	19,154	16,724	14,842	13,340	12,114	11,095	9,167	7,809	6,802	6,025	5,407	4,904	4,135	3,575	3,148	2,541	2,131
75	58,392	47,806	40,469	35,085	30,965	27,711	22,898	19,510	16,995	15,055	13,512	12,256	11,213	9,247	7,868	6,846	6,060	5,435	4,927	4,152	3,587	3,157	2,547	2,135
80	61,379	49,790	41,882	36,141	31,785	28,366	23,344	19,832	17,239	15,246	13,666	12,382	11,319	9,319	7,920	6,886	6,090	5,460	4,948	4,1[illegible]6	3,598	3,166	2,553	2,139
85	64,280	51,682	43,212	37,128	32,546	28,970	23,751	20,126	17,461	15,419	13,804	12,496	11,414	9,383	7,966	6,921	6,118	5,482	4,966	4,179	3,607	3,173	2,558	2,142
90	67,099	53,488	44,468	38,052	33,253	29,529	24,126	20,394	17,662	15,576	13,930	12,599	11,500	9,441	8,008	6,952	6,142	5,502	4,982	4,190	3,616	3,180	2,562	2,145
95	69,839	55,216	45,656	38,918	33,913	30,048	24,471	20,640	17,846	15,719	14,044	12,692	11,578	9,494	8,045	6,981	6,165	5,519	4,997	4,201	3,623	3,186	2,566	2,148
100	72,504	56,868	46,780	39,731	34,529	30,531	24,790	20,867	18,016	15,850	14,149	12,778	11,649	9,541	8,080	7,006	6,185	5,536	5,010	4,210	3,630	3,191	2,569	2,150
105	75,097	58,451	47,846	40,498	35,106	30,981	25,087	21,076	18,172	15,971	14,245	12,856	11,714	9,585	8,111	7,030	6,203	5,550	5,022	4,218	3,637	3,196	2,572	2,152
110	77,621	59,968	48,858	41,220	35,648	31,403	25,362	21,270	18,316	16,082	14,333	12,928	11,774	9,625	8,139	7,051	6,220	5,564	5,033	4,226	3,642	3,200	2,575	2,154
115	80,077	61,424	49,820	41,903	36,157	31,797	25,619	21,451	18,449	16,185	14,415	12,994	11,829	9,662	8,166	7,071	6,235	5,576	5,043	4,233	3,648	3,204	2,578	2,156
120	82,470	62,823	50,735	42,549	36,637	32,168	25,859	21,619	18,573	16,280	14,491	13,056	11,880	9,696	8,190	7,089	6,249	5,587	5,052	4,240	3,652	3,208	2,580	2,158
125	84,801	64,166	51,608	43,161	37,090	32,517	26,084	21,776	18,689	16,369	14,561	13,113	11,927	9,727	8,212	7,106	6,262	5,597	5,060	4,246	3,657	3,212	2,582	2,159
130	87,073	65,459	52,441	43,742	37,518	32,845	26,295	21,923	18,797	16,452	14,627	13,166	11,971	9,756	8,233	7,121	6,274	5,607	5,068	4,251	3,661	3,215	2,585	2,161
135	89,288	66,703	53,236	44,294	37,924	33,155	26,493	22,060	18,898	16,529	14,688	13,216	12,012	9,783	8,253	7,136	6,285	5,616	5,076	4,256	3,665	3,218	2,586	2,162
140	91,449	67,901	53,997	44,819	38,308	33,449	26,680	22,190	18,993	16,602	14,745	13,262	12,050	9,809	8,271	7,149	6,296	5,624	5,082	4,261	3,668	3,220	2,588	2,164
145	93,556	69,056	54,725	45,319	38,673	33,727	26,857	22,312	19,083	16,670	14,799	13,305	12,086	9,833	8,287	7,162	6,306	5,632	5,089	4,266	3,672	3,223	2,590	2,165
150	95,612	70,170	55,422	45,797	39,020	33,990	27,024	22,427	19,167	16,734	14,849	13,346	12,119	9,855	8,303	7,174	6,315	5,640	5,095	4,270	3,675	3,225	2,591	2,166
155	97,619	71,245	56,090	46,252	39,350	34,241	27,182	22,536	19,246	16,795	14,897	13,385	12,151	9,876	8,318	7,185	6,323	5,646	5,100	4,274	3,678	3,228	2,593	2,167
160	99,579	72,283	56,732	46,687	39,665	34,479	27,331	22,638	19,321	16,852	14,942	13,421	12,181	9,895	8,332	7,195	6,332	5,653	5,106	4,277	3,680	3,230	2,594	2,168
165	101,493	73,286	57,348	47,104	39,965	34,705	27,474	22,736	19,392	16,906	14,984	13,455	12,209	9,914	8,345	7,205	6,339	5,659	5,111	4,281	3,683	3,232	2,595	2,169
170	103,363	74,256	57,940	47,503	40,252	34,921	27,609	22,828	19,459	16,957	15,024	13,487	12,236	9,932	8,358	7,214	6,346	5,665	5,115	4,284	3,685	3,233	2,597	2,169
175	105,191	75,195	58,510	47,885	40,526	35,127	27,737	22,916	19,523	17,005	15,062	13,518	12,261	9,948	8,369	7,223	6,353	5,670	5,120	4,287	3,688	3,235	2,598	2,170
180	106,977	76,103	59,059	48,252	40,788	35,324	27,860	23,000	19,584	17,051	15,099	13,547	12,285	9,964	8,381	7,231	6,359	5,675	5,124	4,290	3,690	3,237	2,599	2,171
185	108,723	76,983	59,587	48,604	41,040	35,513	27,977	23,080	19,642	17,095	15,133	13,575	12,308	9,979	8,391	7,239	6,366	5,680	5,128	4,293	3,692	3,238	2,600	2,172
190	110,431	77,835	60,096	48,942	41,281	35,693	28,089	23,156	19,697	17,137	15,165	13,601	12,329	9,993	8,401	7,247	6,371	5,685	5,131	4,296	3,694	3,240	2,601	2,172
195	112,102	78,661	60,588	49,268	41,512	35,866	28,196	23,228	19,749	17,176	15,197	13,626	12,350	10,007	8,411	7,254	6,377	5,689	5,135	4,298	3,696	3,241	2,602	2,173
200	113,736	79,463	61,062	49,581	41,734	36,031	28,298	23,298	19,799	17,214	15,226	13,650	12,369	10,019	8,420	7,261	6,382	5,693	5,138	4,300	3,697	3,243	2,603	2,174

4 AWG | 1 Run(s) | 600 Volts | C-Value = 3,806 | Magnetic Duct | Copper

Calculations:

1. $f = \dfrac{1.732 \times \text{Length} \times \text{Isca}}{\text{Runs} \times \text{C-Value} \times E_{L\text{-}L}}$
2. $M = \dfrac{1}{1 + f}$
3. $Isc = Isca \times M$
4. Add Motor Contribution = Motor FLA x 4

* All results are given in symmetrical amperes

Three-Phase Bolted Fault-Current* Table

One-Way Distance in Feet

Available Isc in Thousands (Isca)	5	10	15	20	25	30	40	50	60	70	80	90	100	125	150	175	200	225	250	300	350	400	500	600
3	2,973	2,946	2,920	2,895	2,869	2,845	2,796	2,750	2,705	2,661	2,619	2,578	2,538	2,444	2,357	2,276	2,200	2,129	2,062	1,941	1,833	1,736	1,571	1,434
5	4,925	4,853	4,782	4,714	4,648	4,583	4,459	4,342	4,230	4,125	4,024	3,928	3,837	3,626	3,437	3,267	3,112	2,972	2,844	2,618	2,426	2,259	1,987	1,773
7	6,855	6,715	6,581	6,452	6,328	6,209	5,984	5,774	5,579	5,396	5,225	5,065	4,914	4,573	4,277	4,016	3,786	3,580	3,396	3,079	2,816	2,594	2,242	1,973
10	9,706	9,428	9,166	8,918	8,683	8,461	8,048	7,673	7,332	7,020	6,733	6,469	6,225	5,688	5,237	4,851	4,519	4,229	3,974	3,547	3,203	2,919	2,480	2,156
15	14,347	13,749	13,199	12,691	12,221	11,784	10,998	10,310	9,704	9,164	8,682	8,248	7,855	7,019	6,344	5,787	5,320	4,923	4,581	4,023	3,585	3,234	2,704	2,323
20	18,856	17,837	16,921	16,096	15,347	14,664	13,467	12,450	11,576	10,817	10,151	9,562	9,038	7,949	7,094	6,405	5,838	5,363	4,960	4,312	3,813	3,418	2,831	2,416
25	23,238	21,709	20,368	19,183	18,129	17,184	15,562	14,220	13,091	12,128	11,297	10,573	9,936	8,635	7,636	6,843	6,200	5,667	5,219	4,506	3,964	3,539	2,914	2,476
30	27,499	25,382	23,568	21,996	20,621	19,407	17,364	15,710	14,343	13,195	12,218	11,375	10,641	9,163	8,045	7,170	6,467	5,890	5,407	4,645	4,072	3,624	2,971	2,518
35	31,642	28,872	26,548	24,570	22,866	21,384	18,929	16,980	15,395	14,080	12,972	12,026	11,209	9,581	8,366	7,424	6,673	6,060	5,550	4,750	4,152	3,688	3,014	2,548
40	35,673	32,191	29,328	26,933	24,900	23,152	20,301	18,076	16,290	14,826	13,603	12,566	11,676	9,920	8,623	7,626	6,836	6,194	5,662	4,832	4,215	3,737	3,047	2,572
45	39,597	35,352	31,930	29,111	26,750	24,743	21,515	19,032	17,062	15,462	14,137	13,020	12,068	10,201	8,835	7,791	6,968	6,302	5,753	4,898	4,265	3,776	3,073	2,590
50	43,417	38,366	34,368	31,125	28,441	26,183	22,595	19,872	17,735	16,013	14,595	13,408	12,400	10,438	9,012	7,928	7,078	6,392	5,827	4,952	4,306	3,808	3,094	2,605
55	47,139	41,244	36,659	32,992	29,991	27,491	23,563	20,617	18,326	16,493	14,993	13,744	12,686	10,640	9,162	8,044	7,170	6,467	5,890	4,997	4,340	3,835	3,111	2,617
60	50,764	43,993	38,815	34,728	31,419	28,686	24,435	21,282	18,849	16,915	15,342	14,036	12,935	10,814	9,291	8,144	7,249	6,531	5,943	5,035	4,368	3,857	3,126	2,628
65	54,298	46,622	40,848	36,346	32,738	29,781	25,226	21,879	19,316	17,290	15,649	14,293	13,153	10,966	9,403	8,230	7,317	6,586	5,988	5,068	4,393	3,877	3,139	2,637
70	57,744	49,140	42,767	37,858	33,960	30,789	25,945	22,418	19,735	17,625	15,923	14,521	13,346	11,100	9,501	8,305	7,376	6,634	6,028	5,096	4,414	3,893	3,150	2,644
75	61,104	51,552	44,583	39,274	35,095	31,719	26,602	22,907	20,113	17,926	16,168	14,725	13,518	11,218	9,588	8,371	7,428	6,676	6,063	5,121	4,433	3,908	3,159	2,651
80	64,382	53,866	46,304	40,603	36,152	32,580	27,205	23,352	20,456	18,198	16,389	14,908	13,672	11,324	9,665	8,430	7,474	6,714	6,093	5,143	4,449	3,920	3,167	2,657
85	67,582	56,088	47,936	41,852	37,139	33,380	27,760	23,760	20,768	18,445	16,589	15,073	13,810	11,419	9,734	8,482	7,516	6,747	6,121	5,163	4,464	3,932	3,175	2,662
90	70,705	58,222	49,486	43,029	38,063	34,125	28,273	24,135	21,054	18,670	16,771	15,223	13,936	11,505	9,796	8,530	7,553	6,777	6,145	5,180	4,477	3,942	3,181	2,667
95	73,754	60,275	50,961	44,140	38,930	34,819	28,749	24,481	21,316	18,876	16,937	15,360	14,051	11,583	9,853	8,572	7,586	6,804	6,168	5,196	4,489	3,951	3,187	2,671
100	76,733	62,249	52,365	45,190	39,744	35,469	29,191	24,800	21,558	19,065	17,090	15,485	14,155	11,654	9,904	8,611	7,617	6,828	6,188	5,210	4,499	3,959	3,193	2,675
105	79,643	64,151	53,704	46,184	40,511	36,079	29,602	25,097	21,782	19,240	17,230	15,600	14,251	11,719	9,951	8,647	7,645	6,851	6,206	5,223	4,509	3,967	3,197	2,678
110	82,487	65,983	54,983	47,126	41,234	36,651	29,986	25,372	21,989	19,402	17,359	15,706	14,340	11,779	9,994	8,679	7,670	6,871	6,223	5,235	4,518	3,973	3,202	2,681
115	85,267	67,750	56,204	48,020	41,917	37,190	30,346	25,629	22,182	19,552	17,479	15,804	14,422	11,834	10,034	8,709	7,693	6,890	6,238	5,246	4,526	3,980	3,206	2,684
120	87,985	69,455	57,372	48,871	42,563	37,698	30,683	25,870	22,361	19,691	17,591	15,895	14,497	11,885	10,071	8,737	7,715	6,907	6,252	5,256	4,533	3,985	3,210	2,687
125	90,644	71,101	58,491	49,680	43,176	38,178	31,000	26,095	22,529	19,821	17,694	15,980	14,568	11,932	10,104	8,762	7,735	6,923	6,265	5,265	4,540	3,991	3,213	2,689
130	93,244	72,692	59,563	50,451	43,757	38,632	31,299	26,306	22,687	19,943	17,791	16,059	14,633	11,976	10,136	8,786	7,753	6,938	6,277	5,274	4,547	3,996	3,216	2,691
135	95,789	74,229	60,591	51,187	44,310	39,061	31,580	26,504	22,834	20,057	17,882	16,132	14,695	12,017	10,165	8,808	7,770	6,951	6,289	5,281	4,552	4,000	3,219	2,693
140	98,279	75,716	61,578	51,890	44,835	39,469	31,847	26,692	22,973	20,164	17,967	16,201	14,752	12,056	10,193	8,828	7,786	6,964	6,299	5,289	4,558	4,004	3,222	2,695
145	100,717	77,155	62,527	52,561	45,336	39,857	32,098	26,868	23,104	20,264	18,047	16,266	14,806	12,092	10,218	8,848	7,801	6,976	6,309	5,296	4,563	4,008	3,225	2,697
150	103,105	78,548	63,438	53,204	45,813	40,225	32,337	27,035	23,227	20,359	18,122	16,327	14,856	12,125	10,242	8,866	7,815	6,987	6,318	5,302	4,568	4,012	3,227	2,699
155	105,443	79,897	64,316	53,820	46,269	40,576	32,563	27,193	23,344	20,449	18,193	16,385	14,904	12,157	10,265	8,883	7,828	6,998	6,327	5,308	4,572	4,016	3,229	2,700
160	107,733	81,206	65,161	54,410	46,705	40,911	32,779	27,343	23,454	20,533	18,260	16,439	14,949	12,187	10,286	8,899	7,841	7,008	6,335	5,314	4,577	4,019	3,231	2,702
165	109,977	82,474	65,975	54,977	47,122	41,231	32,983	27,486	23,559	20,614	18,323	16,491	14,991	12,215	10,306	8,914	7,852	7,017	6,342	5,319	4,581	4,022	3,233	2,703
170	112,176	83,705	66,760	55,521	47,521	41,536	33,178	27,621	23,658	20,690	18,383	16,539	15,032	12,242	10,325	8,928	7,863	7,026	6,349	5,324	4,584	4,025	3,235	2,705
175	114,331	84,899	67,518	56,044	47,903	41,828	33,364	27,750	23,752	20,762	18,440	16,585	15,070	12,267	10,343	8,941	7,874	7,034	6,356	5,329	4,588	4,028	3,237	2,706
180	116,445	86,059	68,249	56,547	48,270	42,107	33,542	27,872	23,842	20,830	18,494	16,629	15,106	12,291	10,360	8,954	7,884	7,042	6,363	5,334	4,591	4,030	3,239	2,707
185	118,517	87,185	68,956	57,031	48,623	42,375	33,712	27,990	23,928	20,896	18,546	16,671	15,140	12,314	10,376	8,966	7,893	7,049	6,369	5,338	4,594	4,033	3,240	2,708
190	120,549	88,280	69,639	57,498	48,961	42,632	33,874	28,102	24,010	20,958	18,595	16,710	15,173	12,335	10,392	8,977	7,902	7,057	6,375	5,342	4,597	4,035	3,242	2,709
195	122,543	89,344	70,300	57,947	49,287	42,879	34,030	28,208	24,088	21,018	18,642	16,748	15,204	12,356	10,406	8,988	7,910	7,063	6,380	5,346	4,600	4,037	3,243	2,710
200	124,499	90,380	70,939	58,381	49,601	43,116	34,179	28,311	24,162	21,074	18,686	16,784	15,234	12,375	10,420	8,999	7,918	7,070	6,385	5,350	4,603	4,039	3,244	2,711

# 3 AWG	1 Run(s)	600 Volts		C-Value = 4,760	Magnetic Duct	Copper

Calculations:

1. $f = \dfrac{1.732 \times \text{Length} \times \text{Isca}}{\text{Runs} \times \text{C-Value} \times E_{L\text{-}L}}$

2. $M = \dfrac{1}{1 + f}$

3. Isc = Isca x M

4. Add Motor Contribution = Motor FLA x 4

* All results are given in symmetrical amperes

Three-Phase Bolted Fault-Current* Table

One-Way Distance in Feet

Available Isc in Thousands (Isca)	5	10	15	20	25	30	40	50	60	70	80	90	100	125	150	175	200	225	250	300	350	400	500	600
3	2,978	2,957	2,935	2,915	2,894	2,874	2,834	2,795	2,757	2,721	2,685	2,650	2,616	2,535	2,459	2,387	2,320	2,256	2,195	2,083	1,983	1,891	1,731	1,596
5	4,940	4,881	4,823	4,767	4,712	4,658	4,555	4,456	4,361	4,270	4,182	4,099	4,018	3,830	3,659	3,502	3,358	3,226	3,104	2,885	2,695	2,528	2,250	2,027
7	6,882	6,768	6,658	6,552	6,448	6,348	6,157	5,977	5,808	5,647	5,496	5,352	5,216	4,903	4,626	4,378	4,156	3,955	3,773	3,454	3,185	2,955	2,582	2,293
10	9,761	9,534	9,317	9,110	8,911	8,721	8,365	8,036	7,732	7,451	7,189	6,945	6,717	6,207	5,770	5,390	5,057	4,763	4,501	4,055	3,689	3,384	2,904	2,543
15	14,470	13,975	13,514	13,082	12,677	12,296	11,599	10,976	10,417	9,913	9,455	9,037	8,655	7,827	7,144	6,570	6,082	5,661	5,295	4,638	4,206	3,814	3,215	2,778
20	19,068	18,219	17,442	16,729	16,072	15,465	14,378	13,434	12,606	11,875	11,223	10,640	10,114	9,001	8,109	7,378	6,768	6,251	5,807	5,036	4,523	4,073	3,397	2,913
25	23,561	22,278	21,128	20,090	19,150	18,294	16,792	15,519	14,425	13,475	12,642	11,906	11,252	9,892	8,825	7,966	7,259	6,668	6,166	5,358	4,738	4,246	3,516	3,001
30	27,951	26,164	24,591	23,197	21,953	20,835	18,909	17,309	15,959	14,804	13,806	12,933	12,164	10,590	9,377	8,413	7,629	6,978	6,430	5,557	4,892	4,370	3,601	3,062
35	32,242	29,887	27,853	26,078	24,515	23,130	20,780	18,864	17,272	15,927	14,777	13,782	12,912	11,152	9,815	8,764	7,916	7,218	6,633	5,708	5,009	4,463	3,664	3,107
40	36,438	33,459	30,930	28,756	26,868	25,212	22,446	20,227	18,407	16,888	15,600	14,495	13,536	11,615	10,171	9,047	8,146	7,409	6,794	5,826	5,100	4,535	3,712	3,142
45	40,542	36,887	33,837	31,252	29,035	27,111	23,939	21,431	19,399	17,719	16,307	15,103	14,065	12,002	10,467	9,280	8,335	7,565	6,925	5,922	5,174	4,593	3,751	3,170
50	44,556	40,181	36,588	33,585	31,037	28,849	25,284	22,503	20,273	18,445	16,920	15,628	14,519	12,331	10,716	9,476	8,492	7,694	7,033	6,001	5,234	4,640	3,782	3,192
55	48,483	43,347	39,195	35,769	32,894	30,446	26,502	23,463	21,049	19,086	17,457	16,085	14,912	12,614	10,929	9,642	8,625	7,803	7,124	6,068	5,284	4,680	3,809	3,211
60	52,327	46,394	41,670	37,819	34,619	31,919	27,611	24,328	21,743	19,654	17,931	16,486	15,257	12,860	11,113	9,785	8,740	7,896	7,202	6,124	5,327	4,713	3,831	3,227
65	56,090	49,328	44,022	39,746	36,227	33,280	28,624	25,111	22,366	20,162	18,353	16,842	15,561	13,075	11,274	9,909	8,839	7,977	7,269	6,172	5,363	4,742	3,850	3,240
70	59,774	52,156	46,259	41,561	37,729	34,544	29,554	25,824	22,930	20,619	18,731	17,160	15,832	13,266	11,415	10,018	8,925	8,048	7,327	6,214	5,395	4,767	3,866	3,252
75	63,383	54,882	48,391	43,274	39,135	35,719	30,410	26,475	23,441	21,032	19,071	17,445	16,075	13,436	11,541	10,115	9,002	8,110	7,379	6,251	5,423	4,788	3,880	3,262
80	66,917	57,512	50,425	44,893	40,454	36,815	31,201	27,072	23,909	21,407	19,379	17,703	16,293	13,588	11,653	10,200	9,070	8,165	7,424	6,284	5,448	4,808	3,893	3,271
85	70,380	60,051	52,366	46,425	41,695	37,839	31,933	27,622	24,336	21,749	19,659	17,936	16,490	13,725	11,754	10,278	9,131	8,214	7,465	6,313	5,469	4,825	3,904	3,278
90	73,774	62,505	54,222	47,878	42,863	38,799	32,614	28,130	24,730	22,063	19,915	18,149	16,670	13,849	11,845	10,347	9,186	8,259	7,502	6,339	5,489	4,840	3,914	3,285
95	77,100	64,876	55,998	49,257	43,965	39,699	33,248	28,600	25,093	22,351	20,150	18,343	16,834	13,962	11,927	10,410	9,235	8,299	7,535	6,363	5,507	4,854	3,923	3,292
100	80,361	67,170	57,698	50,568	45,006	40,546	33,840	29,037	25,428	22,617	20,366	18,522	16,985	14,065	12,003	10,467	9,280	8,335	7,565	6,384	5,523	4,866	3,931	3,297
105	83,559	69,389	59,328	51,816	45,992	41,345	34,394	29,444	25,740	22,864	20,565	18,687	17,123	14,160	12,072	10,520	9,322	8,368	7,592	6,404	5,537	4,877	3,938	3,303
110	86,695	71,538	60,892	53,005	46,926	42,098	34,914	29,825	26,030	23,092	20,750	18,839	17,251	14,248	12,135	10,568	9,359	8,399	7,617	6,422	5,551	4,888	3,945	3,307
115	89,771	73,620	62,394	54,139	47,813	42,811	35,403	30,180	26,301	23,305	20,922	18,981	17,369	14,328	12,193	10,612	9,394	8,427	7,640	6,438	5,563	4,897	3,951	3,312
120	92,789	75,637	63,837	55,222	48,656	43,485	35,863	30,514	26,554	23,503	21,082	19,112	17,479	14,403	12,248	10,653	9,426	8,453	7,661	6,453	5,574	4,906	3,957	3,316
125	95,750	77,593	65,225	56,258	49,458	44,125	36,297	30,828	26,791	23,689	21,231	19,235	17,582	14,473	12,298	10,691	9,456	8,477	7,681	6,467	5,584	4,914	3,962	3,319
130	98,657	79,491	66,561	57,249	50,222	44,732	36,707	31,123	27,014	23,863	21,370	19,349	17,677	14,537	12,345	10,727	9,484	8,499	7,699	6,480	5,594	4,921	3,967	3,323
135	101,510	81,333	67,848	58,198	50,951	45,309	37,095	31,401	27,223	24,026	21,501	19,457	17,767	14,598	12,388	10,759	9,509	8,519	7,716	6,492	5,603	4,928	3,972	3,326
140	104,311	83,122	69,088	59,108	51,647	45,859	37,462	31,664	27,421	24,180	21,624	19,557	17,851	14,654	12,429	10,790	9,533	8,539	7,732	6,503	5,611	4,935	3,976	3,329
145	107,062	84,859	70,284	59,981	52,313	46,383	37,811	31,913	27,607	24,325	21,740	19,652	17,930	14,707	12,467	10,819	9,556	8,557	7,747	6,514	5,619	4,941	3,980	3,332
150	109,763	86,547	71,438	60,820	52,950	46,883	38,143	32,149	27,783	24,462	21,849	19,741	18,004	14,757	12,503	10,846	9,577	8,573	7,760	6,523	5,626	4,946	3,983	3,334
155	112,417	88,189	72,552	61,626	53,559	47,360	38,458	32,373	27,950	24,591	21,952	19,825	18,074	14,804	12,537	10,871	9,596	8,589	7,773	6,532	5,633	4,951	3,987	3,337
160	115,024	89,785	73,629	62,401	54,144	47,817	38,759	32,586	28,109	24,713	22,050	19,905	18,140	14,849	12,568	10,895	9,615	8,604	7,786	6,541	5,640	4,956	3,990	3,339
165	117,586	91,338	74,671	63,147	54,705	48,254	39,045	32,788	28,259	24,830	22,142	19,980	18,203	14,891	12,598	10,918	9,633	8,618	7,797	6,549	5,646	4,961	3,993	3,341
170	120,103	92,850	75,678	63,866	55,244	48,673	39,319	32,981	28,402	24,940	22,230	20,052	18,262	14,930	12,627	10,939	9,649	8,631	7,808	6,557	5,651	4,965	3,996	3,343
175	122,577	94,322	76,653	64,559	55,762	49,074	39,580	33,165	28,538	25,045	22,314	20,119	18,318	14,968	12,653	10,959	9,665	8,644	7,818	6,564	5,657	4,970	3,998	3,345
180	125,009	95,756	77,597	65,228	56,260	49,459	39,831	33,340	28,668	25,145	22,393	20,184	18,371	15,003	12,679	10,978	9,680	8,656	7,828	6,571	5,662	4,974	4,001	3,347
185	127,401	97,153	78,512	65,873	56,739	49,829	40,070	33,508	28,792	25,240	22,468	20,245	18,422	15,037	12,703	10,996	9,694	8,667	7,837	6,577	5,667	4,977	4,003	3,348
190	129,752	98,514	79,399	66,496	57,200	50,185	40,300	33,668	28,911	25,331	22,540	20,304	18,471	15,069	12,726	11,014	9,707	8,678	7,846	6,584	5,671	4,981	4,006	3,350
195	132,065	99,841	80,259	67,098	57,645	50,527	40,520	33,822	29,024	25,418	22,609	20,359	18,517	15,100	12,748	11,030	9,720	8,688	7,854	6,589	5,675	4,984	4,008	3,351
200	134,339	101,136	81,093	67,680	58,075	50,857	40,732	33,969	29,132	25,501	22,675	20,413	18,561	15,129	12,769	11,046	9,732	8,698	7,862	6,595	5,680	4,987	4,010	3,353

2 AWG | 1 Run(s) | 600 Volts | | C-Value = 5,906 | Magnetic Duct | Copper

Calculations:

1. $f = \dfrac{1.732 \times \text{Length} \times \text{Isca}}{\text{Runs} \times \text{C-Value} \times E_{L-L}}$

2. $M = \dfrac{1}{1 + f}$

3. $\text{Isc} = \text{Isca} \times M$

4. Add Motor Contribution = Motor FLA x 4

* All results are given in symmetrical amperes

Three-Phase Bolted Fault-Current* Table

One-Way Distance in Feet

Available Isc in Thousands (Isca)	5	10	15	20	25	30	40	50	60	70	80	90	100	125	150	175	200	225	250	300	350	400	500	600
3	2,982	2,965	2,947	2,930	2,913	2,897	2,864	2,832	2,800	2,770	2,740	2,710	2,682	2,612	2,546	2,484	2,424	2,367	2,313	2,212	2,119	2,034	1,882	1,752
5	4,951	4,903	4,856	4,810	4,764	4,720	4,633	4,550	4,469	4,392	4,316	4,244	4,174	4,008	3,855	3,714	3,582	3,459	3,345	3,137	2,954	2,791	2,513	2,286
7	6,904	6,811	6,721	6,632	6,546	6,463	6,302	6,148	6,002	5,863	5,730	5,603	5,481	5,199	4,945	4,714	4,504	4,312	4,135	3,822	3,554	3,320	2,934	2,629
10	9,806	9,619	9,439	9,266	9,099	8,938	8,633	8,348	8,081	7,830	7,595	7,373	7,164	6,690	6,274	5,907	5,581	5,289	5,026	4,571	4,192	3,871	3,356	2,963
15	14,567	14,159	13,773	13,408	13,061	12,732	12,121	11,566	11,060	10,596	10,169	9,776	9,411	8,610	7,934	7,356	6,857	6,421	6,037	5,393	4,873	4,444	3,779	3,287
20	19,238	18,533	17,877	17,266	16,695	16,161	15,190	14,328	13,559	12,868	12,244	11,678	11,162	10,052	9,142	8,384	7,742	7,191	6,713	5,926	5,304	4,800	4,033	3,478
25	23,821	22,749	21,768	20,869	20,041	19,277	17,910	16,724	15,686	14,769	13,953	13,223	12,565	11,175	10,062	9,151	8 391	7,748	7,196	6,299	5,601	5,042	4,203	3,603
30	28,318	26,815	25,464	24,242	23,132	22,119	20,338	18,823	17,518	16,382	15,384	14,501	13,714	12,075	10,786	9,746	8 888	8,170	7,559	6,575	5,818	5,217	4,324	3,692
35	32,732	30,741	28,978	27,406	25,996	24,723	22,519	20,676	19,112	17,768	16,600	15,576	14,672	12,811	11,370	10,220	9 281	8,500	7,841	6,787	5,984	5,350	4,415	3,758
40	37,065	34,532	32,323	30,379	28,656	27,118	24,489	22,325	20,512	18,971	17,646	16,494	15,483	13,426	11,851	10,607	9 599	8,767	8,067	6,956	6,114	5,454	4,486	3,809
45	41,320	38,196	35,511	33,179	31,134	29,327	26,276	23,801	21,751	20,027	18,556	17,286	16,179	13,946	12,255	10,929	9,862	8,985	8,252	7,093	6,220	5,538	4,542	3,850
50	45,497	41,739	38,553	35,820	33,449	31,372	27,906	25,130	22,856	20,960	19,354	17,977	16,782	14,392	12,598	11,201	10,083	9,168	8,406	7,207	6,307	5,607	4,589	3,883
55	49,600	45,166	41,460	38,315	35,614	33,269	29,397	26,333	23,847	21,790	20,060	18,584	17,310	14,779	12,893	11,434	10,272	9,324	8,536	7,302	6,380	5,665	4,627	3,911
60	53,631	48,484	44,239	40,677	37,646	35,035	30,768	27,427	24,741	22,534	20,688	19,122	17,777	15,117	13,150	11,636	10,434	9,457	8,648	7,384	6,442	5,714	4,660	3,934
65	57,591	51,698	46,899	42,915	39,555	36,683	32,031	28,427	25,551	23,204	21,252	19,603	18,191	15,416	13,375	11,812	10,576	9,574	8,745	7,455	6,496	5,756	4,688	3,954
70	61,482	54,811	49,447	45,039	41,352	38,224	33,200	29,344	26,290	23,812	21,760	20,035	18,562	15,682	13,575	11,967	10,700	9,675	8,830	7,516	6,543	5,793	4,712	3,971
75	65,305	57,830	51,891	47,057	43,048	39,668	34,284	30,187	26,965	24,364	22,221	20,424	18,896	15,919	13,753	12,105	10 810	9,765	8,905	7,570	6,584	5,825	4,733	3,986
80	69,064	60,758	54,236	48,978	44,649	41,024	35,292	30,966	27,585	24,869	22,640	20,778	19,199	16,133	13,912	12,228	10 908	9,845	8,971	7,618	6,620	5,853	4,752	4,000
85	72,759	63,600	56,488	50,808	46,165	42,300	36,233	31,688	28,156	25,332	23,023	21,100	19,474	16,327	14,056	12,339	10 996	9,917	9,031	7,661	6,653	5,878	4,769	4,011
90	76,392	66,358	58,654	52,553	47,601	43,503	37,112	32,358	28,684	25,759	23,375	21,395	19,725	16,503	14,186	12,440	11,076	9,982	9,084	7,700	6,682	5,901	4,784	4,022
95	79,964	69,037	60,737	54,219	48,964	44,638	37,935	32,982	29,173	26,153	23,699	21,666	19,955	16,664	14,305	12,531	11,148	10,041	9,133	7,735	6,708	5,922	4,797	4,031
100	83,477	71,640	62,743	55,812	50,260	45,712	38,708	33,565	29,628	26,518	23,998	21,916	20,167	16,811	14,413	12,614	11,214	10,094	9,177	7,766	6,732	5,940	4,809	4,040
105	86,933	74,170	64,675	57,336	51,492	46,729	39,434	34,110	30,052	26,857	24,276	22,147	20,362	16,947	14,513	12,690	11,274	10,143	9,217	7,795	6,753	5,957	4,820	4,048
110	90,332	76,631	66,538	58,795	52,666	47,694	40,119	34,621	30,448	27,173	24,534	22,362	20,543	17,072	14,605	12,760	11,330	10,187	9,254	7,822	6,773	5,972	4,830	4,055
115	93,677	79,024	68,336	60,194	53,786	48,611	40,766	35,101	30,819	27,468	24,774	22,561	20,711	17,188	14,690	12,825	11,381	10,229	9,288	7,846	6,791	5,986	4,840	4,061
120	96,968	81,354	70,070	61,536	54,855	49,482	41,377	35,553	31,167	27,744	24,998	22,747	20,868	17,296	14,768	12,885	11,428	10,267	9,320	7,868	6,808	6,000	4,848	4,067
125	100,207	83,621	71,746	62,825	55,876	50,312	41,956	35,980	31,494	28,003	25,208	22,921	21,014	17,396	14,841	12,940	11,471	10,302	9,349	7,889	6,823	6,012	4,856	4,073
130	103,395	85,830	73,366	64,063	56,854	51,103	42,504	36,383	31,802	28,246	25,405	23,084	21,151	17,490	14,909	12,992	11,512	10,335	9,376	7,908	6,838	6,023	4,863	4,078
135	106,533	87,981	74,932	65,254	57,790	51,858	43,025	36,764	32,093	28,475	25,591	23,237	21,279	17,578	14,973	13,040	11,550	10,365	9,401	7,926	6,851	6,033	4,870	4,083
140	109,623	90,078	76,448	66,400	58,687	52,579	43,521	37,125	32,368	28,691	25,765	23,380	21,400	17,660	15,032	13,086	11,585	10,394	9,424	7,943	6,864	6,043	4,876	4,087
145	112,665	92,122	77,915	67,504	59,548	53,269	43,992	37,467	32,628	28,896	25,930	23,516	21,513	17,737	15,088	13,128	11,618	10,420	9,446	7,958	6,875	6,052	4,882	4,091
150	115,660	94,115	79,336	68,568	60,374	53,930	44,442	37,793	32,874	29,089	26,085	23,644	21,620	17,809	15,141	13,168	11,650	10,445	9,467	7,973	6,886	6,060	4,888	4,095
155	118,611	96,059	80,713	69,594	61,168	54,562	44,871	38,103	33,109	29,272	26,232	23,764	21,721	17,878	15,190	13,205	11,679	10,469	9,486	7,986	6,896	6,068	4,893	4,099
160	121,516	97,956	82,048	70,585	61,932	55,169	45,280	38,397	33,331	29,446	26,372	23,879	21,817	17,943	15,237	13,240	11,706	10,491	9,504	7,999	6,906	6,075	4,898	4,102
165	124,379	99,808	83,343	71,541	62,667	55,752	45,672	38,679	33,543	29,611	26,504	23,987	21,907	18,004	15,281	13,274	11,732	10,512	9,521	8,011	6,915	6,082	4,902	4,105
170	127,199	101,615	84,600	72,465	63,375	56,311	46,047	38,947	33,745	29,768	26,630	24,090	21,993	18,062	15,323	13,305	11,757	10,532	9,538	8,023	6,923	6,089	4,906	4,108
175	129,978	103,381	85,820	73,359	64,057	56,849	46,406	39,204	33,937	29,918	26,750	24,188	22,075	18,117	15,362	13,335	11,780	10,550	9,553	8,034	6,932	6,095	4,910	4,111
180	132,716	105,106	87,005	74,223	64,716	57,367	46,750	39,449	34,121	30,060	26,864	24,281	22,152	18,169	15,400	13,363	11,802	10,568	9,567	8,044	6,939	6,101	4,914	4,114
185	135,414	106,791	88,157	75,060	65,351	57,866	47,081	39,684	34,297	30,197	26,972	24,370	22,226	18,219	15,436	13,390	11,823	10,585	9,581	8,054	6,946	6,107	4,918	4,116
190	138,074	108,438	89,276	75,870	65,964	58,346	47,398	39,910	34,465	30,327	27,076	24,455	22,297	18,266	15,470	13,416	11,843	10,601	9,594	8,063	6,953	6,112	4,921	4,119
195	140,696	110,049	90,365	76,654	66,556	58,809	47,703	40,126	34,626	30,452	27,176	24,536	22,364	18,311	15,502	13,440	11,862	10,616	9,607	8,072	6,960	6,117	4,925	4,121
200	143,280	111,624	91,424	77,415	67,129	59,256	47,997	40,333	34,780	30,571	27,271	24,614	22,428	18,354	15,533	13,463	11,880	10,630	9,618	8,080	6,966	6,122	4,928	4,123

# 1 AWG	1 Run(s)	600 Volts	C-Value = 7,292	Magnetic Duct	Copper

Calculations:

1. $f = \dfrac{1.732 \times \text{Length} \times \text{Isca}}{\text{Runs} \times \text{C-Value} \times E_{L\text{-}L}}$

2. $M = \dfrac{1}{1 + f}$

3. $\text{Isc} = \text{Isca} \times M$

4. Add Motor Contribution = Motor FLA x 4

* All results are given in symmetrical amperes

Three-Phase Bolted Fault-Current* Table

One-Way Distance in Feet

Available Isc in Thousands (Isca)	5	10	15	20	25	30	40	50	60	70	80	90	100	125	150	175	200	225	250	300	350	400	500	600
3	2,986	2,971	2,957	2,943	2,929	2,915	2,888	2,861	2,835	2,809	2,784	2,759	2,735	2,675	2,619	2,564	2,512	2,462	2,414	2,324	2,239	2,161	2,020	1,896
5	4,960	4,920	4,882	4,843	4,806	4,769	4,696	4,626	4,558	4,491	4,427	4,365	4,304	4,159	4,024	3,897	3,778	3,666	3,560	3,367	3,193	3,036	2,764	2,538
7	6,922	6,845	6,770	6,697	6,625	6,555	6,419	6,288	6,163	6,042	5,926	5,815	5,708	5,456	5,225	5,013	4,818	4,637	4,470	4,[illegible]68	3,905	3,673	3,283	2,968
10	9,841	9,687	9,537	9,392	9,252	9,115	8,854	8,608	8,375	8,154	7,944	7,745	7,556	7,121	6,733	6,385	6,072	5,788	5,529	5,[illegible]75	4,690	4,359	3,821	3,400
15	14,645	14,306	13,982	13,673	13,377	13,094	12,562	12,071	11,618	11,197	10,806	10,441	10,100	9,337	8,681	8,112	7,613	7,171	6,778	6,[illegible]08	5,559	5,101	4,378	3,835
20	19,373	18,785	18,231	17,709	17,216	16,749	15,888	15,112	14,407	13,766	13,179	12,640	12,144	11,058	10,150	9,380	8,719	8,145	7,641	6,[illegible]01	6,127	5,574	4,723	4,097
25	24,028	23,130	22,296	21,520	20,796	20,119	18,890	17,802	16,833	15,963	15,180	14,469	13,822	12,433	11,297	10,351	9,552	8,867	8,273	7,297	6,527	5,904	4,957	4,272
30	28,612	27,346	26,188	25,124	24,143	23,236	21,611	20,199	18,960	17,865	16,889	16,014	15,225	13,556	12,217	11,118	10,201	9,424	8,756	7,570	6,824	6,145	5,126	4,397
35	33,125	31,440	29,919	28,538	27,279	26,126	24,090	22,349	20,842	19,526	18,366	17,336	16,415	14,492	12,971	11,740	10,722	9,867	9,137	7,[illegible]61	7,053	6,331	5,255	4,491
40	37,569	35,417	33,498	31,777	30,224	28,815	26,358	24,287	22,518	20,989	19,655	18,480	17,438	15,283	13,602	12,254	11,149	10,227	9,446	8,[illegible]94	7,235	6,477	5,355	4,564
45	41,947	39,282	36,935	34,853	32,993	31,322	28,441	26,044	24,021	22,289	20,790	19,480	18,325	15,960	14,136	12,686	11,505	10,526	9,700	8,385	7,383	6,596	5,436	4,623
50	46,259	43,039	40,238	37,779	35,604	33,665	30,359	27,644	25,375	23,450	21,797	20,361	19,103	16,547	14,594	13,054	11,807	10,778	9,914	8,544	7,507	6,694	5,502	4,671
55	50,507	46,693	43,414	40,566	38,068	35,860	32,133	29,107	26,603	24,495	22,697	21,144	19,791	17,060	14,992	13,371	12,066	10,993	10,096	8,679	7,611	6,776	5,558	4,711
60	54,693	50,248	46,471	43,222	40,398	37,921	33,777	30,450	27,720	25,439	23,505	21,844	20,402	17,513	15,340	13,647	12,291	11,180	10,253	8,794	7,699	6,847	5,605	4,745
65	58,817	53,708	49,415	45,758	42,605	39,858	35,306	31,687	28,741	26,297	24,235	22,473	20,950	17,915	15,648	13,890	12,488	11,342	10,389	8,895	7,776	6,907	5,646	4,774
70	62,881	57,076	52,253	48,181	44,698	41,684	36,731	32,831	29,679	27,079	24,898	23,042	21,444	18,275	15,922	14,106	12,661	11,485	10,509	8,983	7,843	6,960	5,681	4,799
75	66,887	60,357	54,989	50,498	46,685	43,407	38,063	33,890	30,542	27,796	25,503	23,559	21,891	18,599	16,167	14,298	12,816	11,612	10,616	9,060	7,902	7,007	5,712	4,821
80	70,835	63,554	57,630	52,716	48,575	45,037	39,310	34,875	31,340	28,455	26,057	24,031	22,298	18,891	16,388	14,470	12,954	11,726	10,710	9,129	7,954	7,048	5,739	4,841
85	74,727	66,669	60,180	54,842	50,374	46,579	40,480	35,793	32,079	29,063	26,566	24,464	22,670	19,158	16,588	14,626	13,079	11,828	10,795	9,191	8,001	7,084	5,764	4,858
90	78,564	69,707	62,644	56,881	52,089	48,042	41,580	36,651	32,766	29,626	27,035	24,861	23,011	19,400	16,770	14,767	13,192	11,920	10,872	9,246	8,043	7,117	5,785	4,873
95	82,347	72,669	65,026	58,838	53,726	49,430	42,616	37,453	33,406	30,148	27,469	25,228	23,324	19,623	16,936	14,896	13,294	12,004	10,942	9,296	8,081	7,147	5,805	4,887
100	86,078	75,559	67,331	60,718	55,289	50,751	43,594	38,206	34,004	30,634	27,872	25,567	23,614	19,828	17,088	15,013	13,388	12,080	11,005	9,342	8,116	7,174	5,823	4,900
105	89,757	78,379	69,561	62,526	56,784	52,008	44,518	38,914	34,564	31,088	28,247	25,882	23,883	20,017	17,228	15,121	13,474	12,150	11,063	9,384	8,147	7,199	5,839	4,911
110	93,386	81,132	71,721	64,266	58,215	53,205	45,393	39,581	35,089	31,512	28,597	26,176	24,132	20,192	17,358	15,221	13,553	12,214	11,116	9,422	8,176	7,221	5,854	4,922
115	96,965	83,820	73,813	65,941	59,586	54,348	46,222	40,210	35,582	31,909	28,924	26,449	24,365	20,354	17,477	15,313	13,626	12,273	11,165	9,457	8,203	7,242	5,867	4,931
120	100,495	86,445	75,841	67,555	60,901	55,440	47,010	40,805	36,047	32,283	29,230	26,705	24,582	20,506	17,589	15,399	13,693	12,328	11,211	9,490	8,227	7,261	5,880	4,940
125	103,979	89,010	77,808	69,111	62,163	56,484	47,758	41,367	36,485	32,634	29,518	26,945	24,785	20,647	17,693	15,478	13,756	12,379	11,253	9,520	8,250	7,279	5,891	4,948
130	107,415	91,516	79,717	70,613	63,375	57,483	48,470	41,901	36,899	32,965	29,788	27,170	24,975	20,779	17,789	15,552	13,815	12,426	11,292	9,548	8,271	7,295	5,902	4,956
135	110,806	93,966	81,569	72,062	64,540	58,440	49,149	42,407	37,292	33,277	30,043	27,382	25,154	20,902	17,880	15,621	13,869	12,471	11,328	9,574	8,290	7,310	5,912	4,963
140	114,152	96,362	83,368	73,463	65,661	59,358	49,796	42,888	37,663	33,573	30,284	27,582	25,323	21,019	17,965	15,686	13,920	12,512	11,362	9,598	8,309	7,324	5,921	4,970
145	117,455	98,704	85,116	74,817	66,741	60,238	50,415	43,346	38,016	33,853	30,512	27,771	25,482	21,128	18,045	15,747	13,968	12,551	11,394	9,621	8,326	7,338	5,930	4,976
150	120,714	100,996	86,815	76,126	67,781	61,084	51,006	43,782	38,351	34,118	30,727	27,949	25,632	21,231	18,120	15,804	14,013	12,587	11,424	9,642	8,342	7,350	5,938	4,981
155	123,931	103,238	88,467	77,393	68,783	61,897	51,572	44,198	38,670	34,371	30,932	28,118	25,774	21,328	18,191	15,858	14,056	12,621	11,452	9,662	8,357	7,362	5,946	4,987
160	127,107	105,433	90,073	78,620	69,751	62,680	52,113	44,596	38,974	34,610	31,126	28,279	25,909	21,421	18,258	15,909	14,096	12,653	11,479	9,681	8,371	7,373	5,953	4,992
165	130,243	107,581	91,636	79,808	70,684	63,433	52,633	44,976	39,264	34,839	31,310	28,431	26,036	21,508	18,321	15,957	14,133	12,684	11,504	9,699	8,384	7,383	5,960	4,996
170	133,338	109,684	93,158	80,960	71,586	64,158	53,131	45,339	39,540	35,056	31,486	28,576	26,158	21,591	18,381	16,003	14,169	12,712	11,527	9,716	8,396	7,393	5,966	5,001
175	136,395	111,744	94,640	82,077	72,458	64,857	53,610	45,687	39,805	35,264	31,654	28,714	26,273	21,669	18,438	16,046	14,203	12,740	11,550	9,732	8,408	7,402	5,972	5,005
180	139,413	113,762	96,083	83,160	73,301	65,532	54,070	46,021	40,058	35,463	31,813	28,845	26,383	21,744	18,492	16,087	14,235	12,765	11,571	9,747	8,420	7,410	5,978	5,009
185	142,394	115,739	97,490	84,212	74,117	66,183	54,513	46,341	40,300	35,653	31,966	28,970	26,488	21,815	18,544	16,126	14,265	12,790	11,591	9,761	8,430	7,419	5,983	5,013
190	145,338	117,676	98,861	85,233	74,907	66,812	54,939	46,649	40,533	35,834	32,112	29,090	26,588	21,883	18,593	16,163	14,294	12,813	11,610	9,775	8,440	7,427	5,988	5,016
195	148,246	119,575	100,197	86,224	75,672	67,420	55,349	46,944	40,756	36,008	32,252	29,205	26,684	21,948	18,640	16,198	14,322	12,835	11,628	9,788	8,450	7,434	5,993	5,020
200	151,118	121,437	101,501	87,188	76,413	68,008	55,745	47,229	40,970	36,175	32,386	29,315	26,776	22,010	18,684	16,232	14,348	12,857	11,646	9,800	8,459	7,441	5,997	5,023

 # 1/0 AWG | **1 Run(s)** | **600 Volts**

C-Value = 8,924 | **Magnetic Duct** | **Copper**

Calculations:

1. $f = \dfrac{1.732 \times \text{Length} \times \text{Isca}}{\text{Runs} \times \text{C-Value} \times E_{L\text{-}L}}$

2. $M = \dfrac{1}{1 + f}$

3. $\text{Isc} = \text{Isca} \times M$

4. Add Motor Contribution = Motor FLA x 4

* All results are given in symmetrical amperes

Three-Phase Bolted Fault-Current* Table

One-Way Distance in Feet

Available Isc in Thousands (Isca)	5	10	15	20	25	30	40	50	60	70	80	90	100	125	150	175	200	225	250	300	350	400	500	600
3	2,993	2,986	2,978	2,971	2,964	2,957	2,943	2,929	2,915	2,901	2,888	2,874	2,861	2,828	2,796	2,765	2,735	2,705	2,675	2,619	2,564	2,512	2,414	2,324
5	4,980	4,960	4,940	4,920	4,901	4,882	4,843	4,806	4,769	4,732	4,696	4,661	4,626	4,541	4,459	4,380	4,304	4,230	4,159	4,024	3,897	3,778	3,560	3,367
7	6,961	6,922	6,883	6,845	6,807	6,770	6,697	6,625	6,555	6,486	6,419	6,353	6,288	6,132	5,984	5,842	5,708	5,579	5,456	5,225	5,013	4,818	4,470	4,168
10	9,920	9,841	9,763	9,687	9,611	9,537	9,392	9,252	9,115	8,983	8,854	8,729	8,608	8,318	8,048	7,794	7,556	7,332	7,121	6,733	6,385	6,072	5,529	5,075
15	14,820	14,645	14,473	14,306	14,142	13,982	13,673	13,377	13,094	12,822	12,562	12,312	12,071	11,510	10,998	10,530	10,100	9,703	9,337	8,681	8,112	7,613	6,778	6,108
20	19,682	19,373	19,074	18,785	18,504	18,231	17,709	17,216	16,749	16,307	15,888	15,490	15,112	14,242	13,466	12,771	12,144	11,575	11,058	10,150	9,380	8,719	7,641	6,801
25	24,505	24,028	23,570	23,130	22,705	22,296	21,520	20,796	20,119	19,485	18,890	18,330	17,802	16,607	15,562	14,640	13,822	13,091	12,433	11,297	10,351	9,552	8,273	7,297
30	29,289	28,612	27,965	27,346	26,755	26,188	25,124	24,143	23,236	22,394	21,611	20,881	20,199	18,674	17,363	16,224	15,225	14,342	13,556	12,217	11,118	10,201	8,756	7,670
35	34,037	33,125	32,261	31,440	30,661	29,919	28,538	27,279	26,126	25,067	24,090	23,187	22,349	20,497	18,928	17,582	16,415	15,394	14,492	12,971	11,740	10,722	9,137	7,961
40	38,747	37,569	36,462	35,417	34,431	33,498	31,777	30,224	28,815	27,532	26,358	25,280	24,287	22,116	20,300	18,760	17,438	16,289	15,283	13,602	12,254	11,149	9,446	8,194
45	43,420	41,947	40,571	39,282	38,073	36,935	34,853	32,993	31,322	29,812	28,441	27,190	26,044	23,563	21,513	19,792	18,325	17,061	15,960	14,136	12,686	11,505	9,700	8,385
50	48,057	46,259	44,591	43,039	41,591	40,238	37,779	35,604	33,665	31,927	30,359	28,938	27,644	24,865	22,594	20,702	19,103	17,733	16,547	14,594	13,054	11,807	9,914	8,544
55	52,658	50,507	48,525	46,693	44,994	43,414	40,566	38,068	35,860	33,894	32,133	30,545	29,107	26,042	23,561	21,512	19,791	18,324	17,060	14,992	13,371	12,066	10,096	8,679
60	57,223	54,693	52,376	50,248	48,286	46,471	43,222	40,398	37,921	35,729	33,777	32,028	30,450	27,112	24,434	22,237	20,402	18,848	17,513	15,340	13,647	12,291	10,253	8,794
65	61,754	58,817	56,146	53,708	51,472	49,415	45,758	42,605	39,858	37,445	35,306	33,399	31,687	28,089	25,224	22,889	20,950	19,314	17,915	15,648	13,890	12,488	10,389	8,895
70	66,250	62,881	59,838	57,076	54,558	52,253	48,181	44,698	41,684	39,051	36,731	34,672	32,831	28,983	25,943	23,480	21,444	19,733	18,275	15,922	14,106	12,661	10,509	8,983
75	70,711	66,887	63,454	60,357	57,548	54,989	50,498	46,685	43,407	40,560	38,063	35,856	33,890	29,806	26,600	24,017	21,891	20,111	18,599	16,167	14,298	12,816	10,616	9,060
80	75,139	70,835	66,997	63,554	60,447	57,630	52,716	48,575	45,037	41,979	39,310	36,960	34,875	30,565	27,203	24,508	22,298	20,454	18,891	16,388	14,470	12,954	10,710	9,129
85	79,533	74,727	70,468	66,669	63,259	60,180	54,842	50,374	46,579	43,316	40,480	37,993	35,793	31,268	27,758	24,957	22,670	20,766	19,158	16,588	14,626	13,079	10,795	9,191
90	83,894	78,564	73,871	69,707	65,987	62,644	56,881	52,089	48,042	44,578	41,580	38,960	36,651	31,920	28,271	25,371	23,011	21,052	19,400	16,770	14,767	13,192	10,872	9,246
95	88,222	82,347	77,206	72,669	68,636	65,026	58,838	53,726	49,430	45,771	42,616	39,868	37,453	32,527	28,747	25,753	23,324	21,314	19,623	16,936	14,896	13,294	10,942	9,296
100	92,518	86,078	80,476	75,559	71,208	67,331	60,718	55,289	50,751	46,901	43,594	40,723	38,206	33,094	29,188	26,107	23,614	21,556	19,828	17,088	15,013	13,388	11,005	9,342
105	96,782	89,757	83,683	78,379	73,707	69,561	62,526	56,784	52,008	47,972	44,518	41,528	38,914	33,624	29,600	26,436	23,883	21,780	20,017	17,228	15,121	13,474	11,063	9,384
110	101,014	93,386	86,829	81,132	76,136	71,721	64,266	58,215	53,205	48,990	45,393	42,288	39,581	34,120	29,984	26,742	24,132	21,987	20,192	17,358	15,221	13,553	11,116	9,422
115	105,215	96,965	89,914	83,820	78,499	73,813	65,941	59,586	54,348	49,957	46,222	43,007	40,210	34,587	30,343	27,027	24,365	22,180	20,354	17,477	15,313	13,626	11,165	9,457
120	109,385	100,495	92,942	86,445	80,797	75,841	67,555	60,901	55,440	50,878	47,010	43,688	40,805	35,026	30,681	27,295	24,582	22,359	20,506	17,589	15,399	13,693	11,211	9,490
125	113,524	103,979	95,914	89,010	83,033	77,808	69,111	62,163	56,484	51,756	47,758	44,334	41,367	35,440	30,998	27,545	24,785	22,527	20,647	17,693	15,478	13,756	11,253	9,520
130	117,633	107,415	98,830	91,516	85,210	79,717	70,613	63,375	57,483	52,593	48,470	44,947	41,901	35,830	31,296	27,781	24,975	22,685	20,779	17,789	15,552	13,815	11,292	9,548
135	121,712	110,806	101,694	93,966	87,330	81,569	72,062	64,540	58,440	53,393	49,149	45,530	42,407	36,200	31,578	28,002	25,154	22,832	20,902	17,880	15,621	13,869	11,328	9,574
140	125,762	114,152	104,505	96,362	89,395	83,368	73,463	65,661	59,358	54,158	49,796	46,085	42,888	36,550	31,844	28,211	25,323	22,971	21,019	17,965	15,686	13,920	11,362	9,598
145	129,782	117,455	107,266	98,704	91,408	85,116	74,817	66,741	60,238	54,891	50,415	46,614	43,346	36,882	32,096	28,409	25,482	23,102	21,128	18,045	15,747	13,968	11,394	9,621
150	133,773	120,714	109,978	100,996	93,370	86,815	76,126	67,781	61,084	55,592	51,006	47,119	43,782	37,197	32,334	28,596	25,632	23,225	21,231	18,120	15,804	14,013	11,424	9,642
155	137,735	123,931	112,642	103,238	95,283	88,467	77,393	68,783	61,897	56,265	51,572	47,601	44,198	37,497	32,561	28,772	25,774	23,341	21,328	18,191	15,858	14,056	11,452	9,662
160	141,670	127,107	115,260	105,433	97,150	90,073	78,620	69,751	62,680	56,910	52,113	48,062	44,596	37,783	32,776	28,940	25,909	23,452	21,421	18,258	15,909	14,096	11,479	9,681
165	145,576	130,243	117,832	107,581	98,971	91,636	79,808	70,684	63,433	57,530	52,633	48,504	44,976	38,055	32,980	29,100	26,036	23,556	21,508	18,321	15,957	14,133	11,504	9,699
170	149,454	133,338	120,360	109,684	100,748	93,158	80,960	71,586	64,158	58,126	53,131	48,927	45,339	38,315	33,175	29,252	26,158	23,656	21,591	18,381	16,003	14,169	11,527	9,716
175	153,304	136,395	122,845	111,744	102,483	94,640	82,077	72,458	64,857	58,700	53,610	49,333	45,687	38,563	33,361	29,396	26,273	23,750	21,669	18,438	16,046	14,203	11,550	9,732
180	157,128	139,413	125,288	113,762	104,178	96,083	83,160	73,301	65,532	59,252	54,070	49,722	46,021	38,801	33,539	29,534	26,383	23,840	21,744	18,492	16,087	14,235	11,571	9,747
185	160,925	142,394	127,690	115,739	105,833	97,490	84,212	74,117	66,183	59,784	54,513	50,096	46,341	39,028	33,709	29,665	26,488	23,926	21,815	18,544	16,126	14,265	11,591	9,761
190	164,695	145,338	130,053	117,676	107,451	98,861	85,233	74,907	66,812	60,297	54,939	50,456	46,649	39,246	33,871	29,791	26,588	24,007	21,883	18,593	16,163	14,294	11,610	9,775
195	168,438	148,246	132,376	119,575	109,032	100,197	86,224	75,672	67,420	60,791	55,349	50,802	46,944	39,455	34,027	29,911	26,684	24,085	21,948	18,640	16,198	14,322	11,628	9,788
200	172,156	151,118	134,661	121,437	110,578	101,501	87,188	76,413	68,008	61,269	55,745	51,135	47,229	39,656	34,176	30,027	26,776	24,160	22,010	18,684	16,232	14,348	11,646	9,800

# 1/0 AWG	2 Run(s)	600 Volts

C-Value = 8,924	Magnetic Duct	Copper

Calculations:

1. $f = \dfrac{1.732 \times \text{Length} \times \text{Isca}}{\text{Runs} \times \text{C-Value} \times \text{E L-L}}$

2. $M = \dfrac{1}{1 + f}$

3. $\text{Isc} = \text{Isca} \times M$

4. Add Motor Contribution = Motor FLA x 4

* All results are given in symmetrical amperes

Three-Phase Bolted Fault-Current* Table

One-Way Distance in Feet

Available Isc in Thousands (Isca)

	5	10	15	20	25	30	40	50	60	70	80	90	100	125	150	175	200	225	250	300	350	400	500	600
3	2,988	2,976	2,964	2,952	2,941	2,929	2,906	2,884	2,862	2,840	2,818	2,797	2,776	2,726	2,677	2,629	2,584	2,540	2,497	2,416	2,340	2,269	2,139	2,023
5	4,967	4,934	4,901	4,869	4,838	4,806	4,745	4,686	4,627	4,571	4,515	4,461	4,408	4,282	4,162	4,049	3,942	3,840	3,744	3,555	3,402	3,254	2,992	2,770
7	6,935	6,871	6,808	6,746	6,686	6,627	6,511	6,399	6,291	6,186	6,085	5,988	5,893	5,669	5,461	5,268	5,088	4,920	4,763	4,477	4,223	3,997	3,609	3,291
10	9,868	9,739	9,613	9,491	9,371	9,255	9,030	8,817	8,613	8,418	8,232	8,054	7,884	7,488	7,130	6,804	6,507	6,235	5,984	5,540	5,156	4,823	4,270	3,831
15	14,704	14,419	14,146	13,882	13,628	13,384	12,919	12,486	12,082	11,702	11,346	11,010	10,694	9,978	9,352	8,800	8,309	7,870	7,476	6,794	6,226	5,746	4,978	4,392
20	19,477	18,981	18,510	18,061	17,634	17,226	16,465	15,768	15,128	14,537	13,991	13,485	13,014	11,969	11,079	10,312	9,645	9,059	8,540	7,662	6,947	6,355	5,429	4,738
25	24,188	23,428	22,714	22,042	21,409	20,811	19,710	18,720	17,824	17,010	16,268	15,587	14,961	13,596	12,459	11,498	10,675	9,961	9,337	8,297	7,466	6,786	5,740	4,974
30	28,839	27,764	26,767	25,839	24,973	24,163	22,691	21,389	20,228	19,186	18,246	17,394	16,619	14,951	13,588	12,453	11,492	10,670	9,957	8,783	7,857	7,108	5,969	5,145
35	33,430	31,994	30,677	29,464	28,343	27,305	25,440	23,814	22,384	21,115	19,983	18,965	18,047	16,097	14,528	13,238	12,158	11,241	10,452	9,167	8,162	7,357	6,144	5,274
40	37,962	36,122	34,452	32,929	31,536	30,255	27,983	26,028	24,328	22,837	21,518	20,343	19,290	17,079	15,323	13,895	12,710	11,711	10,858	9,477	8,408	7,555	6,281	5,375
45	42,437	40,151	38,098	36,245	34,563	33,031	30,341	28,057	26,092	24,384	22,886	21,562	20,382	17,930	16,004	14,452	13,175	12,105	11,195	9,733	8,609	7,717	6,393	5,457
50	46,856	44,084	41,622	39,420	37,439	35,648	32,535	29,922	27,698	25,781	24,113	22,647	21,349	18,674	16,595	14,932	13,572	12,439	11,481	9,948	8,776	7,852	6,485	5,524
55	51,219	47,925	45,029	42,463	40,174	38,119	34,581	31,644	29,166	27,049	25,218	23,619	22,211	19,330	17,111	15,349	13,916	12,727	11,726	10,131	8,919	7,965	6,562	5,580
60	55,529	51,678	48,326	45,383	42,778	40,455	36,493	33,237	30,515	28,205	26,220	24,496	22,985	19,914	17,566	15,714	14,215	12,977	11,938	10,289	9,041	8,063	6,628	5,627
65	59,785	55,345	51,518	48,187	45,260	42,668	38,284	34,717	31,757	29,263	27,132	25,290	23,683	20,435	17,971	16,037	14,479	13,197	12,123	10,427	9,147	8,147	6,685	5,668
70	63,989	58,928	54,610	50,881	47,629	44,767	39,965	36,093	32,906	30,235	27,966	26,013	24,316	20,905	18,333	16,325	14,713	13,391	12,287	10,548	9,240	8,221	6,735	5,704
75	68,141	62,432	57,606	53,472	49,892	46,761	41,546	37,378	33,970	31,132	28,731	26,674	24,892	21,329	18,659	16,583	14,922	13,564	12,433	10,655	9,322	8,285	6,778	5,735
80	72,244	65,859	60,511	55,966	52,056	48,657	43,036	38,580	34,960	31,961	29,436	27,281	25,419	21,715	18,954	16,815	15,110	13,719	12,563	10,750	9,395	8,343	6,817	5,762
85	76,297	69,210	63,328	58,368	54,128	50,462	44,443	39,706	35,882	32,730	30,087	27,839	25,903	22,068	19,222	17,026	15,280	13,859	12,680	10,836	9,460	8,394	6,851	5,787
90	80,301	72,489	66,063	60,683	56,113	52,183	45,772	40,764	36,744	33,446	30,691	28,355	26,350	22,391	19,466	17,217	15,434	13,986	12,786	10,913	9,519	8,441	6,882	5,809
95	84,258	75,698	68,717	62,915	58,017	53,826	47,031	41,760	37,551	34,113	31,251	28,833	26,762	22,688	19,690	17,392	15,575	14,101	12,882	10,983	9,572	8,483	6,910	5,829
100	88,168	78,839	71,296	65,070	59,844	55,395	48,225	42,698	38,308	34,737	31,774	29,277	27,144	22,962	19,896	17,553	15,703	14,206	12,970	11,047	9,621	8,521	6,935	5,847
105	92,032	81,915	73,802	67,151	61,600	56,896	49,359	43,585	39,020	35,321	32,262	29,691	27,500	23,216	20,087	17,701	15,822	14,303	13,051	11,106	9,665	8,555	6,958	5,863
110	95,850	84,926	76,237	69,161	63,287	58,333	50,436	44,423	39,690	35,869	32,719	30,078	27,831	23,451	20,263	17,838	15,931	14,392	13,125	11,159	9,706	8,587	6,979	5,878
115	99,625	87,876	78,606	71,105	64,911	59,710	51,462	45,217	40,323	36,385	33,148	30,440	28,141	23,671	20,427	17,964	16,032	14,475	13,193	11,209	9,743	8,616	6,998	5,891
120	103,355	90,766	80,910	72,985	66,474	61,030	52,440	45,970	40,921	36,871	33,551	30,779	28,430	23,876	20,579	18,082	16,125	14,551	13,257	11,254	9,778	8,643	7,016	5,904
125	107,043	93,598	83,153	74,805	67,981	62,297	53,373	46,685	41,487	37,330	33,930	31,098	28,702	24,067	20,721	18,192	16,213	14,622	13,315	11,297	9,810	8,668	7,032	5,916
130	110,689	96,373	85,336	76,568	69,433	63,515	54,264	47,365	42,023	37,764	34,288	31,399	28,958	24,247	20,854	18,294	16,294	14,688	13,370	11,336	9,839	8,692	7,048	5,926
135	114,293	99,094	87,463	78,275	70,834	64,685	55,116	48,013	42,532	38,174	34,627	31,682	29,199	24,415	20,979	18,390	16,370	14,750	13,421	11,373	9,867	8,713	7,062	5,937
140	117,857	101,762	89,534	79,930	72,187	65,811	55,932	48,631	43,016	38,564	34,947	31,950	29,426	24,574	21,096	18,480	16,441	14,807	13,469	11,407	9,893	8,733	7,075	5,946
145	121,380	104,378	91,553	81,535	73,494	66,896	56,713	49,221	43,477	38,934	35,250	32,203	29,641	24,724	21,206	18,564	16,508	14,862	13,514	11,439	9,917	8,752	7,087	5,955
150	124,865	106,944	93,522	83,093	74,757	67,941	57,462	49,784	43,916	39,285	35,538	32,443	29,845	24,865	21,310	18,644	16,571	14,913	13,556	11,470	9,940	8,770	7,099	5,963
155	128,310	109,461	95,441	84,605	75,978	68,948	58,181	50,323	44,335	39,620	35,812	32,671	30,037	24,999	21,408	18,719	16,630	14,961	13,596	11,498	9,961	8,786	7,110	5,970
160	131,717	111,932	97,314	86,073	77,160	69,920	58,872	50,839	44,734	39,939	36,072	32,888	30,220	25,125	21,501	18,790	16,686	15,006	13,633	11,525	9,981	8,802	7,120	5,978
165	135,087	114,356	99,141	87,499	78,304	70,858	59,536	51,333	45,117	40,243	36,320	33,094	30,394	25,246	21,589	18,857	16,739	15,049	13,668	11,550	10,000	8,817	7,130	5,984
170	138,420	116,735	100,925	88,886	79,413	71,765	60,174	51,807	45,482	40,534	36,557	33,291	30,560	25,360	21,672	18,920	16,789	15,089	13,702	11,574	10,018	8,831	7,139	5,991
175	141,717	119,072	102,666	90,234	80,487	72,641	60,789	52,262	45,833	40,812	36,783	33,478	30,718	25,468	21,751	18,981	16,837	15,127	13,733	11,596	10,035	8,844	7,147	5,997
180	144,979	121,365	104,367	91,545	81,529	73,488	61,381	52,699	46,169	41,078	36,999	33,657	30,868	25,572	21,826	19,038	16,882	15,164	13,763	11,618	10,051	8,856	7,155	6,003
185	148,205	123,618	106,028	92,821	82,539	74,308	61,952	53,119	46,491	41,333	37,206	33,828	31,012	25,670	21,898	19,093	16,925	15,199	13,792	11,638	10,066	8,868	7,163	6,008
190	151,397	125,831	107,652	94,063	83,520	75,102	62,503	53,524	46,800	41,578	37,404	33,991	31,149	25,764	21,967	19,145	16,965	15,231	13,819	11,657	10,080	8,879	7,170	6,013
195	154,554	128,004	109,239	95,272	84,472	75,871	63,035	53,913	47,098	41,812	37,593	34,148	31,281	25,854	22,032	19,194	17,004	15,263	13,845	11,676	10,094	8,890	7,177	6,018
200	157,679	130,140	110,791	96,450	85,397	76,616	63,548	54,288	47,384	42,038	37,776	34,298	31,407	25,940	22,094	19,242	17,041	15,293	13,870	11,693	10,107	8,900	7,184	6,023

# 2/0 AWG	1 Run(s)	600 Volts	C-Value = 10,755	Magnetic Duct	Copper

Calculations:

1. $f = \dfrac{1.732 \times \text{Length} \times \text{Isca}}{\text{Runs} \times \text{C-Value} \times E_{L\text{-}L}}$

2. $M = \dfrac{1}{1 + f}$

3. $\text{Isc} = \text{Isca} \times M$

4. Add Motor Contribution = Motor FLA x 4

* All results are given in symmetrical amperes

Three-Phase Bolted Fault-Current* Table

One-Way Distance in Feet

Available Isc in Thousands (Isca)	5	10	15	20	25	30	40	50	60	70	80	90	100	125	150	175	200	225	250	300	350	400	500	600
3	2,994	2,988	2,982	2,976	2,970	2,964	2,952	2,941	2,929	2,918	2,906	2,895	2,884	2,856	2,829	2,803	2 776	2,751	2,726	2,677	2,629	2,584	2,497	2,416
5	4,983	4,967	4,950	4,934	4,918	4,901	4,869	4,838	4,806	4,776	4,745	4,715	4,686	4,613	4,543	4,475	4 408	4,344	4,282	4,162	4,049	3,942	3,744	3,565
7	6,967	6,935	6,903	6,871	6,839	6,808	6,746	6,686	6,627	6,568	6,511	6,454	6,399	6,264	6,135	6,012	5,893	5,779	5,669	5,461	5,268	5,088	4,763	4,477
10	9,933	9,868	9,803	9,739	9,675	9,613	9,491	9,371	9,255	9,141	9,030	8,922	8,817	8,563	8,324	8,098	7,884	7,681	7,488	7,130	6,804	6,507	5,984	5,540
15	14,851	14,704	14,560	14,419	14,281	14,146	13,882	13,628	13,384	13,147	12,919	12,699	12,486	11,984	11,521	11,092	10,694	10,324	9,978	9,352	8,800	8,309	7,476	6,794
20	19,735	19,477	19,226	18,981	18,742	18,510	18,061	17,634	17,226	16,837	16,465	16,109	15,768	14,976	14,259	13,608	13,014	12,470	11,969	11,079	10,312	9,645	8,540	7,662
25	24,588	24,188	23,802	23,428	23,065	22,714	22,042	21,409	20,811	20,245	19,710	19,202	18,720	17,613	16,631	15,752	14,961	14,246	13,596	12,459	11,498	10,675	9,337	8,297
30	29,408	28,839	28,291	27,764	27,257	26,767	25,839	24,973	24,163	23,404	22,691	22,021	21,389	19,957	18,704	17,600	16,619	15,741	14,951	13,588	12,453	11,492	9,957	8,783
35	34,197	33,430	32,696	31,994	31,322	30,677	29,464	28,343	27,305	26,340	25,440	24,601	23,814	22,052	20,533	19,210	18,047	17,016	16,097	14,528	13,238	12,158	10,452	9,167
40	38,954	37,962	37,019	36,122	35,267	34,452	32,929	31,536	30,255	29,075	27,983	26,970	26,028	23,938	22,158	20,625	19,290	18,117	17,079	15,323	13,895	12,710	10,858	9,477
45	43,681	42,437	41,262	40,151	39,097	38,098	36,245	34,563	33,031	31,629	30,341	29,154	28,057	25,643	23,611	21,878	20,382	19,078	17,930	16,004	14,452	13,175	11,195	9,733
50	48,377	46,856	45,428	44,084	42,817	41,622	39,420	37,439	35,648	34,020	32,535	31,174	29,922	27,192	24,919	22,996	21,349	19,922	18,674	16,595	14,932	13,572	11,481	9,948
55	53,042	51,219	49,518	47,925	46,432	45,029	42,463	40,174	38,119	36,264	34,581	33,047	31,644	28,607	26,101	24,000	22,211	20,671	19,330	17,111	15,349	13,916	11,726	10,131
60	57,678	55,529	53,534	51,678	49,946	48,326	45,383	42,778	40,455	38,372	36,493	34,789	33,237	29,903	27,176	24,905	22 985	21,339	19,914	17,566	15,714	14,215	11,938	10,289
65	62,283	59,785	57,479	55,345	53,363	51,518	48,187	45,260	42,668	40,357	38,284	36,413	34,717	31,095	28,157	25,727	23 683	21,940	20,435	17,971	16,037	14,479	12,123	10,427
70	66,860	63,989	61,354	58,928	56,687	54,610	50,881	47,629	44,767	42,230	39,965	37,931	36,093	32,195	29,056	26,475	24 316	22,482	20,905	18,333	16,325	14,713	12,287	10,548
75	71,406	68,141	65,162	62,432	59,922	57,606	53,472	49,892	46,761	44,000	41,546	39,352	37,378	33,213	29,883	27,160	24 892	22,973	21,329	18,659	16,583	14,922	12,433	10,655
80	75,924	72,244	68,904	65,859	63,071	60,511	55,966	52,056	48,657	45,674	43,036	40,687	38,580	34,159	30,646	27,789	25,419	23,422	21,715	18,954	16,815	15,110	12,563	10,750
85	80,414	76,297	72,581	69,210	66,139	63,328	58,368	54,128	50,462	47,262	44,443	41,941	39,706	35,039	31,353	28,369	25,903	23,832	22,068	19,222	17,026	15,280	12,680	10,836
90	84,874	80,301	76,196	72,489	69,127	66,063	60,683	56,113	52,183	48,768	45,772	43,123	40,764	35,860	32,009	28,905	26,350	24,209	22,391	19,466	17,217	15,434	12,786	10,913
95	89,307	84,258	79,749	75,698	72,039	68,717	62,915	58,017	53,826	50,200	47,031	44,239	41,760	36,628	32,620	29,402	26,762	24,557	22,688	19,690	17,392	15,575	12,882	10,983
100	93,712	88,168	83,243	78,839	74,878	71,296	65,070	59,844	55,395	51,562	48,225	45,294	42,698	37,348	33,189	29,864	27,144	24,879	22,962	19,896	17,553	15,703	12,970	11,047
105	98,089	92,032	86,679	81,915	77,647	73,802	67,151	61,600	56,896	52,860	49,359	46,292	43,585	38,024	33,722	30,295	27,500	25,177	23,216	20,087	17,701	15,822	13,051	11,106
110	102,439	95,850	90,058	84,926	80,348	76,237	69,161	63,287	58,333	54,098	50,436	47,239	44,423	38,661	34,222	30,697	27,831	25,454	23,451	20,263	17,838	15,931	13,125	11,159
115	106,762	99,625	93,382	87,876	82,983	78,606	71,105	64,911	59,710	55,280	51,462	48,138	45,217	39,261	34,691	31,074	28,141	25,713	23,671	20,427	17,964	16,032	13,193	11,209
120	111,058	103,355	96,652	90,766	85,555	80,910	72,985	66,474	61,030	56,410	52,440	48,992	45,970	39,827	35,133	31,428	28,430	25,955	23,876	20,579	18,082	16,125	13,257	11,254
125	115,327	107,043	99,870	93,598	88,067	83,153	74,805	67,981	62,297	57,491	53,373	49,806	46,685	40,363	35,549	31,761	28,702	26,181	24,067	20,721	18,192	16,213	13,315	11,297
130	119,570	110,689	103,036	96,373	90,520	85,336	76,568	69,433	63,515	58,526	54,264	50,581	47,365	40,871	35,942	32,074	28,958	26,394	24,247	20,854	18,294	16,294	13,370	11,336
135	123,787	114,293	106,152	99,094	92,916	87,463	78,275	70,834	64,685	59,519	55,116	51,320	48,013	41,352	36,314	32,370	29,199	26,594	24,415	20,979	18,390	16,370	13,421	11,373
140	127,978	117,857	109,219	101,762	95,257	89,534	79,930	72,187	65,811	60,471	55,932	52,027	48,631	41,810	36,666	32,650	29,426	26,782	24,574	21,096	18,480	16,441	13,469	11,407
145	132,143	121,380	112,239	104,378	97,546	91,553	81,535	73,494	66,896	61,385	56,713	52,702	49,221	42,245	37,000	32,914	29,641	26,960	24,724	21,206	18,564	16,508	13,514	11,439
150	136,283	124,865	115,212	106,944	99,784	93,522	83,093	74,757	67,941	62,264	57,462	53,348	49,784	42,659	37,318	33,165	29,845	27,128	24,865	21,310	18,644	16,571	13,556	11,470
155	140,398	128,310	118,139	109,461	101,972	95,441	84,605	75,978	68,948	63,109	58,181	53,967	50,323	43,054	37,620	33,404	30,037	27,287	24,999	21,408	18,719	16,630	13,596	11,498
160	144,488	131,717	121,021	111,932	104,112	97,314	86,073	77,160	69,920	63,922	58,872	54,561	50,839	43,431	37,907	33,630	30,220	27,438	25,125	21,501	18,790	16,686	13,633	11,525
165	148,553	135,087	123,860	114,356	106,206	99,141	87,499	78,304	70,858	64,705	59,536	55,131	51,333	43,791	38,181	33,846	30,394	27,582	25,246	21,589	18,857	16,739	13,668	11,550
170	152,594	138,420	126,657	116,735	108,256	100,925	88,886	79,413	71,765	65,460	60,174	55,678	51,807	44,136	38,443	34,051	30,560	27,718	25,360	21,672	18,920	16,789	13,702	11,574
175	156,610	141,717	129,411	119,072	110,262	102,666	90,234	80,487	72,641	66,188	60,789	56,204	52,262	44,465	38,693	34,247	30,718	27,848	25,468	21,751	18,981	16,837	13,733	11,596
180	160,602	144,979	132,125	121,365	112,226	104,367	91,545	81,529	73,488	66,891	61,381	56,710	52,699	44,781	38,932	34,434	30,868	27,971	25,572	21,826	19,038	16,882	13,763	11,618
185	164,571	148,205	134,800	123,618	114,150	106,028	92,821	82,539	74,308	67,570	61,952	57,197	53,119	45,085	39,161	34,613	31,012	28,089	25,670	21,898	19,093	16,925	13,792	11,638
190	168,516	151,397	137,435	125,831	116,034	107,652	94,063	83,520	75,102	68,226	62,503	57,666	53,524	45,376	39,380	34,785	31,149	28,202	25,764	21,967	19,145	16,965	13,819	11,657
195	172,437	154,554	140,032	128,004	117,880	109,239	95,272	84,472	75,871	68,860	63,035	58,118	53,913	45,655	39,591	34,949	31,281	28,310	25,854	22,032	19,194	17,004	13,845	11,676
200	176,336	157,679	142,592	130,140	119,688	110,791	96,450	85,397	76,616	69,473	63,548	58,555	54,288	45,924	39,793	35,106	31,407	28,413	25,940	22,094	19,242	17,041	13,870	11,693

 # 2/0 AWG | **2 Run(s)** | **600 Volts** | **C-Value = 10,755** | **Magnetic Duct** | **Copper**

Calculations:

1. $f = \dfrac{1.732 \times \text{Length} \times \text{Isca}}{\text{Runs} \times \text{C-Value} \times E_{L\text{-}L}}$
2. $M = \dfrac{1}{1 + f}$
3. $\text{Isc} = \text{Isca} \times M$
4. Add Motor Contribution = Motor FLA x 4

* All results are given in symmetrical amperes

Three-Phase Bolted Fault-Current* Table

One-Way Distance in Feet

Available Isc in Thousands (Isca)

Isca	5	10	15	20	25	30	40	50	60	70	80	90	100	125	150	175	200	225	250	300	350	400	500	600
3	2,990	2,980	2,970	2,960	2,950	2,941	2,921	2,902	2,883	2,865	2,846	2,828	2,810	2,767	2,724	2,683	2,643	2,605	2,567	2,495	2,427	2,363	2,244	2,136
5	4,972	4,944	4,917	4,890	4,863	4,837	4,785	4,734	4,684	4,635	4,588	4,541	4,495	4,384	4,279	4,178	4,082	3,991	3,903	3,739	3,588	3,449	3,201	2,986
7	6,945	6,892	6,839	6,786	6,735	6,684	6,586	6,489	6,396	6,306	6,217	6,132	6,048	5,850	5,663	5,489	5,325	5,170	5,024	4,755	4,514	4,296	3,918	3,601
10	9,889	9,780	9,674	9,570	9,468	9,368	9,175	8,990	8,812	8,641	8,476	8,317	8,165	7,807	7,479	7,177	6,899	6,641	6,402	5,973	5,597	5,266	4,708	4,258
15	14,751	14,511	14,278	14,052	13,834	13,622	13,217	12,836	12,476	12,136	11,814	11,508	11,218	10,553	9,962	9,434	8,959	8,530	8,139	7,457	6,881	6,387	5,585	4,962
20	19,560	19,140	18,737	18,350	17,979	17,623	16,952	16,330	15,752	15,213	14,710	14,239	13,798	12,805	11,945	11,194	10,532	9,943	9,417	8,516	7,772	7,148	6,158	5,410
25	24,317	23,670	23,057	22,474	21,921	21,394	20,412	19,517	18,697	17,942	17,247	16,603	16,006	14,685	13,556	12,605	11,771	11,041	10,396	9,308	8,427	7,698	6,562	5,719
30	29,022	28,105	27,244	26,435	25,672	24,952	23,627	22,436	21,359	20,380	19,488	18,670	17,918	16,279	14,915	13,761	12,774	11,918	11,170	9,924	8,928	8,114	6,863	5,946
35	33,675	32,447	31,306	30,242	29,248	28,317	26,623	25,119	23,777	22,571	21,481	20,492	19,589	17,647	16,055	14,726	13,601	12,635	11,798	10,417	9,325	8,440	7,094	6,119
40	38,279	36,700	35,247	33,904	32,659	31,503	29,420	27,595	25,984	24,550	23,266	22,110	21,063	18,834	17,031	15,544	14,295	13,232	12,317	10,819	9,646	8,703	7,279	6,255
45	42,834	40,867	39,072	37,429	35,918	34,524	32,038	29,886	28,005	26,346	24,873	23,556	22,372	19,874	17,877	16,245	14,886	13,737	12,753	11,154	9,912	8,918	7,429	6,366
50	47,340	44,949	42,787	40,824	39,033	37,393	34,494	32,012	29,863	27,985	26,329	24,858	23,542	20,792	18,617	16,854	15,396	14,170	13,125	11,438	10,135	9,099	7,554	6,457
55	51,798	48,949	46,397	44,097	42,015	40,121	36,802	33,990	31,578	29,485	27,653	26,034	24,595	21,609	19,269	17,387	15,839	14,545	13,446	11,68[illegible]	10,325	9,252	7,659	6,534
60	56,210	52,870	49,905	47,255	44,872	42,717	38,975	35,836	33,165	30,864	28,862	27,104	25,547	22,340	19,849	17,857	16,229	14,872	13,725	11,89[illegible]	10,489	9,383	7,749	6,600
65	60,575	56,714	53,316	50,302	47,611	45,192	41,025	37,562	34,637	32,136	29,971	28,079	26,412	22,999	20,367	18,275	16,573	15,161	13,971	12,075	10,632	9,497	7,827	6,656
70	64,895	60,484	56,634	53,245	50,239	47,554	42,962	39,179	36,008	33,312	30,991	28,973	27,202	23,595	20,833	18,650	16,881	15,418	14,189	12,233	10,758	9,598	7,895	6,705
75	69,170	64,181	59,863	56,090	52,764	49,810	44,795	40,697	37,287	34,403	31,934	29,795	27,925	24,138	21,255	18,987	17,157	15,648	14,383	12,382	10,869	9,686	7,954	6,748
80	73,401	67,807	63,006	58,840	55,190	51,967	46,532	42,126	38,482	35,419	32,807	30,554	28,591	24,633	21,638	19,292	17,405	15,855	14,558	12,571	10,969	9,765	8,008	6,786
85	77,588	71,366	66,067	61,501	57,525	54,032	48,180	43,473	39,603	36,366	33,618	31,256	29,205	25,088	21,988	19,570	17,631	16,042	14,715	12,627	11,058	9,836	8,055	6,820
90	81,733	74,857	69,048	64,076	59,772	56,010	49,747	44,744	40,655	37,251	34,373	31,908	29,773	25,506	22,308	19,824	17,837	16,212	14,858	12,732	11,138	9,899	8,098	6,851
95	85,836	78,284	71,954	66,571	61,937	57,906	51,238	45,946	41,645	38,081	35,078	32,515	30,300	25,892	22,603	20,056	18,025	16,367	14,989	12,838	11,211	9,957	8,136	6,878
100	89,897	81,648	74,786	68,988	64,024	59,727	52,558	47,085	42,579	38,860	35,738	33,081	30,791	26,250	22,876	20,270	18,197	16,509	15,108	12,9[illegible]5	11,278	10,009	8,171	6,903
105	93,918	84,951	77,548	71,331	66,037	61,475	54,012	48,165	43,460	39,592	36,357	33,610	31,250	26,582	23,127	20,468	18,356	16,640	15,217	12,9[illegible]5	11,339	10,057	8,203	6,926
110	97,898	88,195	80,241	73,604	67,981	63,156	55,305	49,190	44,293	40,283	36,938	34,107	31,678	26,891	23,361	20,651	18,503	16,761	15,318	13,068	11,395	10,101	8,232	6,947
115	101,838	91,380	82,870	75,809	69,858	64,773	56,541	50,166	45,082	40,935	37,486	34,573	32,080	27,180	23,579	20,820	18,640	16,873	15,411	13,136	11,446	10,142	8,259	6,966
120	105,740	94,509	85,435	77,951	71,672	66,329	57,724	51,094	45,831	41,551	38,002	35,011	32,457	27,451	23,782	20,979	18,766	16,976	15,498	13,199	11,494	10,179	8,284	6,984
125	109,603	97,583	87,939	80,030	73,426	67,829	58,856	51,980	46,542	42,134	38,489	35,425	32,812	27,704	23,972	21,126	18,885	17,073	15,578	13,257	11,538	10,214	8,307	7,000
130	113,428	100,604	90,385	82,050	75,123	69,275	59,941	52,825	47,218	42,688	38,951	35,815	33,147	27,942	24,150	21,265	18,995	17,163	15,653	13,312	11,579	10,246	8,328	7,015
135	117,216	103,573	92,774	84,014	76,766	70,669	60,983	53,632	47,862	43,213	39,388	36,184	33,463	28,167	24,318	21,394	19,098	17,247	15,724	13,362	11,618	10,276	8,348	7,029
140	120,967	106,490	95,108	85,924	78,358	72,016	61,983	54,404	48,476	43,713	39,802	36,534	33,762	28,378	24,475	21,516	19,195	17,326	15,789	13,410	11,654	10,304	8,366	7,042
145	124,682	109,359	97,390	87,782	79,900	73,316	62,944	55,142	49,062	44,189	40,196	36,866	34,045	28,578	24,624	21,631	19,287	17,401	15,851	13,454	11,687	10,330	8,384	7,054
150	128,362	112,179	99,620	89,590	81,395	74,573	63,868	55,850	49,621	44,642	40,571	37,181	34,313	28,767	24,764	21,739	19,372	17,471	15,909	13,496	11,719	10,355	8,400	7,066
155	132,006	114,952	101,801	91,350	82,845	75,789	64,757	56,529	50,157	45,075	40,928	37,481	34,568	28,946	24,896	21,841	19,453	17,537	15,963	13,535	11,748	10,378	8,415	7,077
160	135,615	117,679	103,934	93,064	84,252	76,965	65,614	57,181	50,669	45,488	41,269	37,766	34,811	29,116	25,022	21,938	19,530	17,599	16,015	13,572	11,776	10,400	8,429	7,087
165	139,190	120,362	106,021	94,734	85,618	78,103	66,440	57,807	51,160	45,884	41,594	38,038	35,042	29,277	25,141	22,029	19,603	17,658	16,064	13,607	11,802	10,420	8,443	7,096
170	142,731	123,001	108,063	96,361	86,945	79,206	67,236	58,409	51,631	46,262	41,905	38,298	35,262	29,431	25,254	22,116	19,671	17,713	16,110	13,640	11,827	10,440	8,456	7,105
175	146,239	125,597	110,062	97,947	88,235	80,274	68,004	58,988	52,083	46,625	42,202	38,546	35,472	29,577	25,362	22,198	19,737	17,766	16,154	13,672	11,851	10,458	8,468	7,114
180	149,714	128,152	112,019	99,494	89,488	81,311	68,747	59,546	52,517	46,972	42,487	38,783	35,673	29,717	25,464	22,277	19,799	17,816	16,195	13,701	11,873	10,475	8,479	7,122
185	153,157	130,667	113,936	101,003	90,707	82,316	69,464	60,083	52,934	47,306	42,759	39,010	35,865	29,850	25,562	22,352	19,858	17,864	16,235	13,730	11,894	10,492	8,490	7,129
190	156,568	133,141	115,813	102,475	91,892	83,291	70,157	60,601	53,336	47,626	43,021	39,228	36,049	29,977	25,655	22,423	19,914	17,910	16,272	13,757	11,915	10,508	8,500	7,137
195	159,948	135,577	117,651	103,912	93,046	84,238	70,827	61,100	53,723	47,934	43,272	39,437	36,226	30,099	25,745	22,491	19,968	17,953	16,308	13,782	11,934	10,522	8,510	7,143
200	163,297	137,976	119,453	105,315	94,170	85,157	71,476	61,583	54,095	48,231	43,514	39,637	36,395	30,215	25,830	22,556	20,019	17,995	16,342	13,806	11,952	10,537	8,519	7,150

# 3/0 AWG	1 Run(s)	600 Volts	C-Value = 12,843	Magnetic Duct	Copper

Calculations:

1. $f = \dfrac{1.732 \times \text{Length} \times \text{Isca}}{\text{Runs} \times \text{C-Value} \times E_{L\text{-}L}}$

2. $M = \dfrac{1}{1 + f}$

3. Isc = Isca x M

4. Add Motor Contribution = Motor FLA x 4

* All results are given in symmetrical amperes

Three-Phase Bolted Fault-Current* Table

One-Way Distance in Feet

Available Isc in Thousands (Isca)	5	10	15	20	25	30	40	50	60	70	80	90	100	125	150	175	200	225	250	300	350	400	500	600
3	2,995	2,990	2,985	2,980	2,975	2,970	2,960	2,950	2,941	2,931	2,921	2,912	2,902	2,879	2,856	2,833	2,810	2,788	2,767	2,724	2,683	2,643	2,567	2,495
5	4,986	4,972	4,958	4,944	4,931	4,917	4,890	4,863	4,837	4,811	4,785	4,759	4,734	4,672	4,611	4,552	4,495	4,439	4,384	4,279	4,178	4,082	3,903	3,739
7	6,973	6,945	6,918	6,892	6,865	6,839	6,786	6,735	6,684	6,635	6,586	6,537	6,489	6,373	6,261	6,153	6,048	5,947	5,850	5,663	5,489	5,325	5,024	4,755
10	9,944	9,889	9,834	9,780	9,727	9,674	9,570	9,468	9,368	9,271	9,175	9,081	8,990	8,768	8,557	8,357	8,165	7,982	7,807	7,479	7,177	6,899	6,402	5,973
15	14,875	14,751	14,630	14,511	14,393	14,278	14,052	13,834	13,622	13,417	13,217	13,024	12,836	12,389	11,973	11,583	11,218	10,875	10,553	9,962	9,434	8,959	8,139	7,457
20	19,778	19,560	19,348	19,140	18,936	18,737	18,350	17,979	17,623	17,281	16,952	16,635	16,330	15,613	14,957	14,354	13,798	13,283	12,805	11,945	11,194	10,532	9,417	8,516
25	24,654	24,317	23,989	23,670	23,359	23,057	22,474	21,921	21,394	20,891	20,412	19,954	19,517	18,502	17,588	16,760	16,006	15,317	14,685	13,566	12,605	11,771	10,396	9,308
30	29,503	29,022	28,556	28,105	27,668	27,244	26,435	25,672	24,952	24,272	23,627	23,016	22,436	21,105	19,924	18,868	17,918	17,059	16,279	14,915	13,761	12,774	11,170	9,924
35	34,325	33,675	33,050	32,447	31,866	31,306	30,242	29,248	28,317	27,444	26,623	25,849	25,119	23,464	22,012	20,730	19,589	18,567	17,647	16,055	14,726	13,601	11,798	10,417
40	39,121	38,279	37,473	36,700	35,959	35,247	33,904	32,659	31,503	30,426	29,420	28,478	27,595	25,610	23,891	22,388	21,063	19,886	18,834	17,031	15,544	14,295	12,317	10,819
45	43,890	42,834	41,827	40,867	39,949	39,072	37,429	35,918	34,524	33,235	32,038	30,925	29,886	27,571	25,589	23,872	22,372	21,049	19,874	17,877	16,245	14,886	12,753	11,154
50	48,634	47,340	46,113	44,949	43,841	42,787	40,824	39,033	37,393	35,885	34,494	33,207	32,012	29,370	27,132	25,210	23,542	22,082	20,792	18,617	16,854	15,396	13,125	11,438
55	53,351	51,798	50,333	48,949	47,639	46,397	44,097	42,015	40,121	38,390	36,802	35,340	33,990	31,027	28,539	26,421	24,595	23,005	21,609	19,269	17,387	15,839	13,446	11,681
60	58,043	56,210	54,489	52,870	51,345	49,905	47,255	44,872	42,717	40,761	38,975	37,340	35,836	32,558	29,829	27,523	25,547	23,836	22,340	19,849	17,857	16,229	13,725	11,891
65	62,710	60,575	58,581	56,714	54,963	53,316	50,302	47,611	45,192	43,008	41,025	39,217	37,562	33,976	31,015	28,529	26,412	24,588	22,999	20,367	18,275	16,573	13,971	12,075
70	67,351	64,895	62,612	60,484	58,496	56,634	53,245	50,239	47,554	45,142	42,962	40,983	39,179	35,294	32,110	29,453	27,202	25,270	23,595	20,833	18,650	16,881	14,189	12,238
75	71,967	69,170	66,582	64,181	61,947	59,863	56,090	52,764	49,810	47,170	44,795	42,648	40,697	36,521	33,123	30,303	27,925	25,894	24,138	21,255	18,987	17,157	14,383	12,382
80	76,558	73,401	70,493	67,807	65,319	63,006	58,840	55,190	51,967	49,100	46,532	44,219	42,126	37,668	34,063	31,088	28,591	26,465	24,633	21,638	19,292	17,405	14,558	12,511
85	81,125	77,588	74,347	71,366	68,614	66,067	61,501	57,525	54,032	50,939	48,180	45,706	43,473	38,741	34,938	31,815	29,235	26,990	25,088	21,988	19,570	17,631	14,715	12,627
90	85,668	81,733	78,144	74,857	71,836	69,048	64,076	59,772	56,010	52,693	49,747	47,113	44,744	39,747	35,754	32,491	29,773	27,475	25,506	22,308	19,824	17,837	14,858	12,732
95	90,186	85,836	81,886	78,284	74,986	71,954	66,571	61,937	57,906	54,368	51,238	48,448	45,946	40,693	36,518	33,120	30,300	27,923	25,892	22,603	20,056	18,025	14,989	12,828
100	94,680	89,897	85,574	81,648	78,067	74,786	68,988	64,024	59,727	55,970	52,658	49,715	47,085	41,584	37,234	33,707	30,791	28,340	26,250	22,876	20,270	18,197	15,108	12,915
105	99,150	93,918	89,210	84,951	81,081	77,548	71,331	66,037	61,475	57,502	54,012	50,921	48,165	42,424	37,906	34,257	31,250	28,727	26,582	23,127	20,468	18,356	15,217	12,995
110	103,597	97,898	92,793	88,195	84,030	80,241	73,604	67,981	63,156	58,970	55,305	52,069	49,190	43,218	38,538	34,773	31,678	29,089	26,891	23,361	20,651	18,503	15,318	13,068
115	108,020	101,838	96,326	91,380	86,917	82,870	75,809	69,858	64,773	60,378	56,541	53,163	50,166	43,969	39,134	35,258	32,080	29,428	27,180	23,579	20,820	18,640	15,411	13,136
120	112,420	105,740	99,810	94,509	89,743	85,435	77,951	71,672	66,329	61,728	57,724	54,207	51,094	44,680	39,697	35,714	32,457	29,745	27,451	23,782	20,979	18,766	15,498	13,199
125	116,796	109,603	103,245	97,583	92,511	87,939	80,030	73,426	67,829	63,025	58,856	55,205	51,980	45,356	40,229	36,144	32,812	30,043	27,704	23,972	21,126	18,885	15,578	13,257
130	121,150	113,428	106,632	100,604	95,221	90,385	82,050	75,123	69,275	64,271	59,941	56,158	52,825	45,998	40,734	36,551	33,147	30,323	27,942	24,150	21,265	18,995	15,653	13,312
135	125,481	117,216	109,973	103,573	97,876	92,774	84,014	76,766	70,669	65,470	60,983	57,072	53,632	46,609	41,212	36,935	33,463	30,587	28,167	24,318	21,394	19,098	15,724	13,362
140	129,790	120,967	113,268	106,490	100,478	95,108	85,924	78,358	72,016	66,624	61,983	57,946	54,404	47,190	41,666	37,300	33,762	30,837	28,378	24,475	21,516	19,195	15,789	13,410
145	134,076	124,682	116,519	109,359	103,028	97,390	87,782	79,900	73,316	67,735	62,944	58,785	55,142	47,745	42,098	37,646	34,045	31,073	28,578	24,624	21,631	19,287	15,851	13,454
150	138,340	128,362	119,726	112,179	105,527	99,620	89,590	81,395	74,573	68,807	63,868	59,591	55,850	48,275	42,510	37,974	34,313	31,296	28,767	24,764	21,739	19,372	15,909	13,496
155	142,582	132,006	122,890	114,952	107,978	101,801	91,350	82,845	75,789	69,840	64,757	60,364	56,529	48,782	42,902	38,287	34,563	31,508	28,946	24,896	21,841	19,453	15,963	13,535
160	146,802	135,615	126,012	117,679	110,380	103,934	93,064	84,252	76,965	70,838	65,614	61,108	57,181	49,266	43,276	38,585	34,811	31,710	29,116	25,022	21,938	19,530	16,015	13,572
165	151,000	139,190	129,093	120,362	112,737	106,021	94,734	85,618	78,103	71,801	66,440	61,824	57,807	49,730	43,634	38,869	35,042	31,901	29,277	25,141	22,029	19,603	16,064	13,607
170	155,177	142,731	132,134	123,001	115,049	108,063	96,361	86,945	79,206	72,732	67,236	62,512	58,409	50,175	43,976	39,140	35,262	32,084	29,431	25,254	22,116	19,671	16,110	13,640
175	159,332	146,239	135,135	125,597	117,318	110,062	97,947	88,235	80,274	73,632	68,004	63,176	58,988	50,602	44,303	39,399	35,472	32,258	29,577	25,362	22,198	19,737	16,154	13,672
180	163,466	149,714	138,097	128,152	119,544	112,019	99,494	89,488	81,311	74,503	68,747	63,816	59,546	51,012	44,617	39,647	35,673	32,424	29,717	25,464	22,277	19,799	16,195	13,701
185	167,579	153,157	141,021	130,667	121,729	113,936	101,003	90,707	82,316	75,345	69,464	64,434	60,083	51,405	44,918	39,884	35,865	32,582	29,850	25,562	22,352	19,858	16,235	13,730
190	171,672	156,568	143,908	133,141	123,874	115,813	102,475	91,892	83,291	76,162	70,157	65,030	60,601	51,784	45,207	40,112	36,049	32,734	29,977	25,655	22,423	19,914	16,272	13,757
195	175,743	159,948	146,758	135,577	125,980	117,651	103,912	93,046	84,238	76,953	70,827	65,605	61,100	52,148	45,484	40,330	36,226	32,879	30,099	25,745	22,491	19,968	16,308	13,782
200	179,794	163,297	149,572	137,976	128,048	119,453	105,315	94,170	85,157	77,719	71,476	66,162	61,583	52,499	45,751	40,540	36,395	33,018	30,215	25,830	22,556	20,019	16,342	13,806

3/0 AWG | 2 Run(s) | 600 Volts

C-Value = 12,843 | Magnetic Duct | Copper

Calculations:

1. $f = \dfrac{1.732 \times \text{Length} \times \text{Isca}}{\text{Runs} \times \text{C-Value} \times E_{L\text{-}L}}$

2. $M = \dfrac{1}{1 + f}$

3. Isc = Isca x M

4. Add Motor Contribution = Motor FLA x 4

* All results are given in symmetrical amperes

Three-Phase Bolted Fault-Current* Table

One-Way Distance in Feet

Available Isc in Thousands (Isca)

	5	10	15	20	25	30	40	50	60	70	80	90	100	125	150	175	200	225	250	300	350	400	500	600
3	2,991	2,983	2,974	2,966	2,958	2,949	2,933	2,916	2,900	2,884	2,868	2,853	2,837	2,799	2,762	2,726	2,691	2,657	2,623	2,559	2,498	2,440	2,331	2,231
5	4,976	4,953	4,929	4,906	4,883	4,860	4,816	4,772	4,728	4,686	4,644	4,604	4,563	4,466	4,372	4,283	4,197	4,114	4,035	3,885	3,745	3,616	3,382	3,176
7	6,953	6,907	6,862	6,817	6,773	6,730	6,644	6,561	6,479	6,400	6,322	6,247	6,173	5,996	5,829	5,670	5,521	5,379	5,244	4,993	4,765	4,558	4,192	3,881
10	9,905	9,812	9,721	9,631	9,543	9,457	9,289	9,127	8,970	8,819	8,672	8,531	8,393	8,069	7,769	7,491	7,232	6,990	6,764	6,352	5,988	5,664	5,110	4,655
15	14,788	14,581	14,381	14,185	13,995	13,811	13,455	13,117	12,796	12,490	12,198	11,920	11,654	11,039	10,485	9,984	9,529	9,113	8,732	8,059	7,482	6,982	6,159	5,509
20	19,624	19,263	18,914	18,578	18,253	17,940	17,344	16,787	16,264	15,773	15,311	14,875	14,463	13,527	12,705	11,977	11,328	10,745	10,220	9,309	8,548	7,901	6,863	6,067
25	24,416	23,858	23,326	22,816	22,329	21,862	20,984	20,174	19,424	18,727	18,079	17,475	16,909	15,643	14,554	13,606	12,775	12,039	11,383	10,265	9,347	8,579	7,369	6,458
30	29,163	28,371	27,621	26,910	26,234	25,592	24,397	23,308	22,313	21,399	20,557	19,779	19,057	17,465	16,118	14,964	13,964	13,089	12,318	11,019	9,968	9,100	7,750	6,749
35	33,866	32,803	31,804	30,865	29,979	29,143	27,603	26,218	24,965	23,827	22,788	21,835	20,959	19,049	17,458	16,112	14,959	13,959	13,085	11,623	10,465	9,512	8,047	6,973
40	38,525	37,155	35,880	34,689	33,574	32,529	30,622	28,927	27,409	26,043	24,807	23,682	22,655	20,440	18,619	17,096	15,803	14,692	13,727	12,133	10,871	9,846	8,285	7,151
45	43,142	41,432	39,851	38,387	37,027	35,760	33,469	31,454	29,668	28,074	26,642	25,350	24,177	21,670	19,634	17,948	16,528	15,317	14,271	12,556	11,209	10,123	8,480	7,296
50	47,717	45,633	43,724	41,967	40,347	38,847	36,159	33,818	31,762	29,942	28,319	26,863	25,549	22,766	20,530	18,693	17,159	15,857	14,738	12,917	11,496	10,356	8,643	7,416
55	52,250	49,762	47,500	45,434	43,541	41,799	38,703	36,034	33,709	31,666	29,856	28,242	26,794	23,749	21,326	19,351	17,711	16,327	15,144	13,227	11,741	10,555	8,781	7,518
60	56,742	53,819	51,183	48,793	46,617	44,626	41,114	38,115	35,523	33,262	31,271	29,505	27,928	24,636	22,038	19,936	18,200	16,742	15,500	13,458	11,954	10,727	8,900	7,604
65	61,193	57,808	54,778	52,049	49,580	47,334	43,402	40,073	37,218	34,743	32,577	30,665	28,965	25,439	22,679	20,459	18,634	17,109	15,814	13,735	12,140	10,876	9,002	7,679
70	65,605	61,730	58,286	55,207	52,437	49,931	45,575	41,919	38,805	36,122	33,787	31,734	29,917	26,171	23,258	20,929	19,024	17,437	16,094	13,946	12,304	11,008	9,092	7,744
75	69,977	65,585	61,712	58,271	55,193	52,424	47,643	43,662	40,295	37,409	34,910	32,723	30,795	26,840	23,785	21,355	19,375	17,731	16,344	14,134	12,450	11,124	9,172	7,802
80	74,311	69,377	65,058	61,245	57,854	54,819	49,613	45,311	41,695	38,613	35,956	33,641	31,606	27,454	24,266	21,742	19,693	17,997	16,570	14,302	12,580	11,228	9,242	7,853
85	78,606	73,106	68,326	64,133	60,424	57,121	51,492	46,872	43,013	39,742	36,932	34,494	32,358	28,019	24,707	22,095	19,982	18,238	16,775	14,454	12,698	11,322	9,305	7,899
90	82,863	76,775	71,520	66,939	62,909	59,336	53,285	48,354	44,258	40,801	37,846	35,290	33,057	28,542	25,112	22,419	20,247	18,458	16,960	14,592	12,804	11,406	9,362	7,940
95	87,083	80,384	74,642	69,666	65,311	61,469	54,999	49,761	45,433	41,799	38,702	36,033	33,708	29,027	25,487	22,716	20,489	18,660	17,130	14,718	12,901	11,483	9,414	7,977
100	91,266	83,935	77,694	72,317	67,636	63,525	56,638	51,099	46,547	42,739	39,507	36,730	34,317	29,477	25,833	22,991	20,713	18,845	17,286	14,833	12,989	11,553	9,461	8,010
105	95,413	87,429	80,679	74,896	69,887	65,506	58,208	52,373	47,602	43,627	40,265	37,384	34,887	29,897	26,155	23,246	20,919	19,016	17,430	14,938	13,070	11,617	9,504	8,041
110	99,523	90,869	83,599	77,406	72,068	67,418	59,713	53,588	48,603	44,467	40,979	37,999	35,422	30,289	26,455	23,482	21,110	19,173	17,562	15,035	13,144	11,675	9,543	8,069
115	103,599	94,254	86,456	79,849	74,181	69,264	61,156	54,748	49,555	45,262	41,654	38,578	35,925	30,656	26,734	23,702	21,288	19,320	17,685	15,125	13,213	11,730	9,579	8,095
120	107,639	97,587	89,251	82,228	76,229	71,047	62,542	55,856	50,461	46,017	42,292	39,125	36,399	31,000	26,996	23,907	21,453	19,456	17,799	15,208	13,276	11,780	9,612	8,119
125	111,645	100,868	91,988	84,545	78,217	72,770	63,874	56,915	51,324	46,734	42,897	39,642	36,846	31,324	27,241	24,099	21,608	19,583	17,905	15,286	13,335	11,826	9,643	8,141
130	115,616	104,098	94,668	86,804	80,146	74,437	65,154	57,930	52,148	47,415	43,470	40,131	37,269	31,628	27,471	24,280	21,752	19,702	18,004	15,358	13,390	11,869	9,672	8,161
135	119,554	107,280	97,292	89,005	82,019	76,049	66,386	58,902	52,934	48,065	44,016	40,596	37,669	31,916	27,688	24,449	21,888	19,813	18,097	15,426	13,441	11,909	9,699	8,180
140	123,459	110,414	99,862	91,151	83,838	77,611	67,573	59,835	53,686	48,684	44,534	41,036	38,048	32,188	27,892	24,608	22,016	19,917	18,184	15,489	13,489	11,947	9,724	8,198
145	127,331	113,501	102,380	93,244	85,605	79,123	68,717	60,730	54,406	49,275	45,028	41,455	38,408	32,445	28,085	24,758	22,136	20,016	18,266	15,548	13,534	11,982	9,747	8,215
150	131,171	116,541	104,848	95,287	87,324	80,589	69,820	61,589	55,095	49,839	45,499	41,854	38,750	32,689	28,267	24,900	22,249	20,108	18,343	15,604	13,577	12,015	9,769	8,230
155	134,978	119,537	107,266	97,280	88,995	82,011	70,884	62,416	55,755	50,379	45,949	42,234	39,076	32,920	28,440	25,034	22,356	20,195	18,416	15,657	13,616	12,047	9,789	8,245
160	138,754	122,489	109,637	99,226	90,621	83,389	71,912	63,211	56,389	50,896	46,378	42,597	39,386	33,140	28,604	25,161	22,457	20,278	18,484	15,706	13,654	12,076	9,809	8,258
165	142,499	125,398	111,962	101,127	92,204	84,727	72,905	63,977	56,998	51,391	46,789	42,943	39,682	33,350	28,760	25,281	22,553	20,356	18,549	15,753	13,689	12,104	9,827	8,271
170	146,213	128,265	114,242	102,983	93,744	86,027	73,865	64,715	57,583	51,867	47,183	43,275	39,965	33,549	28,908	25,396	22,644	20,430	18,611	15,797	13,723	12,130	9,844	8,284
175	149,896	131,091	116,479	104,797	95,245	87,289	74,793	65,427	58,146	52,323	47,560	43,592	40,235	33,739	29,049	25,504	22,730	20,501	18,669	15,839	13,754	12,155	9,861	8,295
180	153,550	133,877	118,673	106,570	96,707	88,515	75,692	66,114	58,687	52,761	47,922	43,895	40,493	33,921	29,184	25,608	22,813	20,568	18,725	15,879	13,785	12,178	9,876	8,306
185	157,173	136,623	120,826	108,303	98,132	89,707	76,562	66,777	59,209	53,182	48,269	44,187	40,741	34,095	29,313	25,707	22,891	20,631	18,778	15,917	13,813	12,200	9,891	8,316
190	160,768	139,331	122,939	109,997	99,521	90,867	77,405	67,417	59,712	53,588	48,603	44,466	40,979	34,261	29,435	25,801	22,966	20,692	18,828	15,953	13,840	12,222	9,905	8,326
195	164,333	142,001	125,013	111,655	100,876	91,995	78,222	68,036	60,197	53,978	48,924	44,735	41,206	34,420	29,553	25,891	23,037	20,750	18,876	15,988	13,866	12,242	9,918	8,336
200	167,870	144,634	127,049	113,276	102,198	93,093	79,014	68,635	60,665	54,354	49,232	44,993	41,425	34,572	29,665	25,978	23,106	20,805	18,922	16,021	13,891	12,261	9,931	8,345

# 4/0 AWG	1 Run(s)	600 Volts	C-Value = 15,082	Magnetic Duct	Copper

Calculations:

1. $f = \dfrac{1.732 \times \text{Length} \times \text{Isca}}{\text{Runs} \times \text{C-Value} \times \text{E L-L}}$

2. $M = \dfrac{1}{1 + f}$

3. $\text{Isc} = \text{Isca} \times M$

4. Add Motor Contribution = Motor FLA x 4

* All results are given in symmetrical amperes

Three-Phase Bolted Fault-Current* Table

One-Way Distance in Feet

Available Isc in Thousands (Isca)

	5	10	15	20	25	30	40	50	60	70	80	90	100	125	150	175	200	225	250	300	350	400	500	600
3	2,996	2,991	2,987	2,983	2,979	2,974	2,966	2,958	2,949	2,941	2,933	2,924	2,916	2,896	2,876	2,856	2,837	2,818	2,799	2,762	2,726	2,691	2,623	2,559
5	4,988	4,976	4,964	4,953	4,941	4,929	4,906	4,883	4,860	4,838	4,816	4,794	4,772	4,718	4,665	4,614	4,563	4,514	4,466	4,372	4,283	4,197	4,035	3,885
7	6,977	6,953	6,930	6,907	6,885	6,862	6,817	6,773	6,730	6,686	6,644	6,602	6,561	6,459	6,361	6,265	6,173	6,083	5,996	5,829	5,670	5,521	5,244	4,993
10	9,952	9,905	9,858	9,812	9,766	9,721	9,631	9,543	9,457	9,372	9,289	9,207	9,127	8,932	8,745	8,566	8,393	8,228	8,069	7,769	7,491	7,232	6,764	6,352
15	14,893	14,788	14,684	14,581	14,480	14,381	14,185	13,995	13,811	13,630	13,455	13,284	13,117	12,718	12,342	11,988	11,654	11,338	11,039	10,485	9,984	9,529	8,732	8,059
20	19,810	19,624	19,442	19,263	19,087	18,914	18,578	18,253	17,940	17,637	17,344	17,061	16,787	16,139	15,539	14,982	14,463	13,980	13,527	12,705	11,977	11,328	10,220	9,309
25	24,704	24,416	24,134	23,858	23,589	23,326	22,816	22,329	21,862	21,414	20,984	20,571	20,174	19,245	18,398	17,622	16,909	16,252	15,643	14,554	13,606	12,775	11,383	10,265
30	29,575	29,163	28,761	28,371	27,991	27,621	26,910	26,234	25,592	24,980	24,397	23,840	23,308	22,077	20,970	19,968	19,057	18,226	17,465	16,118	14,964	13,964	12,318	11,019
35	34,423	33,866	33,326	32,803	32,296	31,804	30,865	29,979	29,143	28,352	27,603	26,893	26,218	24,671	23,296	22,066	20,959	19,959	19,049	17,458	16,112	14,959	13,085	11,629
40	39,249	38,525	37,828	37,155	36,506	35,880	34,689	33,574	32,529	31,547	30,622	29,750	28,927	27,055	25,410	23,954	22,655	21,490	20,440	18,619	17,096	15,803	13,727	12,133
45	44,051	43,142	42,270	41,432	40,626	39,851	38,387	37,027	35,760	34,577	33,469	32,431	31,454	29,253	27,340	25,661	24,177	22,855	21,670	19,634	17,948	16,528	14,271	12,556
50	48,832	47,717	46,652	45,633	44,658	43,724	41,967	40,347	38,847	37,455	36,159	34,949	33,818	31,287	29,108	27,213	25,549	24,078	22,766	20,530	18,693	17,159	14,738	12,917
55	53,590	52,250	50,975	49,762	48,604	47,500	45,434	43,541	41,799	40,192	38,703	37,321	36,034	33,174	30,735	28,629	26,794	25,180	23,749	21,326	19,351	17,711	15,144	13,227
60	58,325	56,742	55,242	53,819	52,468	51,183	48,793	46,617	44,626	42,798	41,114	39,558	38,115	34,930	32,236	29,928	27,928	26,179	24,636	22,038	19,936	18,200	15,500	13,498
65	63,039	61,193	59,453	57,808	56,252	54,778	52,049	49,580	47,334	45,283	43,402	41,671	40,073	36,567	33,625	31,122	28,965	27,088	25,439	22,679	20,459	18,634	15,814	13,735
70	67,731	65,605	63,608	61,730	59,959	58,286	55,207	52,437	49,931	47,654	45,575	43,671	41,919	38,098	34,916	32,224	29,917	27,919	26,171	23,258	20,929	19,024	16,094	13,946
75	72,402	69,977	67,710	65,585	63,590	61,712	58,271	55,193	52,424	49,919	47,643	45,566	43,662	39,532	36,116	33,244	30,795	28,682	26,840	23,785	21,355	19,375	16,344	14,134
80	77,051	74,311	71,759	69,377	67,148	65,058	61,245	57,854	54,819	52,086	49,613	47,364	45,311	40,879	37,237	34,191	31,606	29,384	27,454	24,266	21,742	19,693	16,570	14,302
85	81,678	78,606	75,756	73,106	70,636	68,326	64,133	60,424	57,121	54,160	51,492	49,073	46,872	42,146	38,285	35,073	32,358	30,033	28,019	24,707	22,095	19,982	16,775	14,454
90	86,284	82,863	79,703	76,775	74,054	71,520	66,939	62,909	59,336	56,148	53,285	50,700	48,354	43,340	39,268	35,896	33,057	30,634	28,542	25,112	22,419	20,247	16,960	14,592
95	90,869	87,083	83,599	80,384	77,407	74,642	69,666	65,311	61,469	58,054	54,999	52,249	49,761	44,467	40,191	36,665	33,708	31,193	29,027	25,487	22,716	20,489	17,130	14,718
100	95,434	91,266	87,447	83,935	80,694	77,694	72,317	67,636	63,525	59,884	56,638	53,726	51,099	45,532	41,060	37,387	34,317	31,714	29,477	25,833	22,991	20,713	17,286	14,833
105	99,977	95,413	91,247	87,429	83,919	80,679	74,896	69,887	65,506	61,642	58,208	55,137	52,373	46,542	41,878	38,065	34,837	32,200	29,897	26,155	23,246	20,919	17,430	14,938
110	104,500	99,523	94,999	90,869	87,082	83,599	77,406	72,068	67,418	63,332	59,713	56,485	53,588	47,499	42,652	38,702	35,422	32,655	30,289	26,455	23,482	21,110	17,562	15,035
115	109,002	103,599	98,706	94,254	90,187	86,456	79,849	74,181	69,264	64,958	61,156	57,775	54,748	48,407	43,383	39,304	35,925	33,082	30,656	26,734	23,702	21,288	17,685	15,125
120	113,484	107,639	102,367	97,587	93,233	89,251	82,228	76,229	71,047	66,524	62,542	59,010	55,856	49,271	44,076	39,871	36,399	33,483	31,000	26,996	23,907	21,453	17,799	15,208
125	117,945	111,645	105,983	100,868	96,223	91,988	84,545	78,217	72,770	68,032	63,874	60,194	56,915	50,094	44,733	40,408	36,846	33,861	31,324	27,241	24,099	21,608	17,905	15,286
130	122,387	115,616	109,555	104,098	99,159	94,668	86,804	80,146	74,437	69,487	65,154	61,330	57,930	50,878	45,357	40,917	37,269	34 218	31,628	27,471	24,280	21,752	18,004	15,358
135	126,809	119,554	113,085	107,280	102,042	97,292	89,005	82,019	76,049	70,890	66,386	62,421	58,902	51,627	45,951	41,400	37,669	34,555	31,916	27,688	24,449	21,888	18,097	15,426
140	131,210	123,459	116,573	110,414	104,873	99,862	91,151	83,838	77,611	72,245	67,573	63,469	59,835	52,342	46,517	41,858	38,048	34,873	32,188	27,892	24,608	22,016	18,184	15,489
145	135,592	127,331	120,019	113,501	107,654	102,380	93,244	85,605	79,123	73,554	68,717	64,477	60,730	53,025	47,056	42,294	38,408	35,176	32,445	28,085	24,758	22,136	18,266	15,548
150	139,955	131,171	123,424	116,541	110,386	104,848	95,287	87,324	80,589	74,819	69,820	65,447	61,589	53,680	47,570	42,709	38,750	35,462	32,689	28,267	24,900	22,249	18,343	15,604
155	144,298	134,978	126,789	119,537	113,070	107,266	97,280	88,995	82,011	76,042	70,884	66,381	62,416	54,307	48,062	43,105	39,076	35,735	32,920	28,440	25,034	22,356	18,416	15,657
160	148,622	138,754	130,115	122,489	115,708	109,637	99,226	90,621	83,389	77,226	71,912	67,282	63,211	54,908	48,532	43,483	39,386	35,994	33,140	28,604	25,161	22,457	18,484	15,706
165	152,926	142,499	133,403	125,398	118,300	111,962	101,127	92,204	84,727	78,373	72,905	68,150	63,977	55,485	48,982	43,844	39,682	36,241	33,350	28,760	25,281	22,553	18,549	15,753
170	157,212	146,213	136,652	128,265	120,848	114,242	102,983	93,744	86,027	79,483	73,865	68,988	64,715	56,039	49,414	44,190	39,965	36,477	33,549	28,908	25,396	22,644	18,611	15,797
175	161,478	149,896	139,865	131,091	123,354	116,479	104,797	95,245	87,289	80,559	74,793	69,797	65,427	56,572	49,828	44,520	40,235	36,702	33,739	29,049	25,504	22,730	18,669	15,839
180	165,726	153,550	143,040	133,877	125,817	118,673	106,570	96,707	88,515	81,603	75,692	70,579	66,114	57,084	50,225	44,837	40,493	36,917	33,921	29,184	25,608	22,813	18,725	15,879
185	169,955	157,173	146,180	136,623	128,240	120,826	108,303	98,132	89,707	82,615	76,562	71,335	66,777	57,578	50,607	45,141	40,741	37,123	34,095	29,313	25,707	22,891	18,778	15,917
190	174,166	160,768	149,284	139,331	130,623	122,939	109,997	99,521	90,867	83,597	77,405	72,066	67,417	58,053	50,974	45,433	40,979	37,320	34,261	29,435	25,801	22,966	18,828	15,953
195	178,358	164,333	152,353	142,001	132,967	125,013	111,655	100,876	91,995	84,551	78,222	72,774	68,036	58,512	51,327	45,713	41,206	37,509	34,420	29,553	25,891	23,037	18,876	15,988
200	182,532	167,870	155,388	144,634	135,273	127,049	113,276	102,198	93,093	85,478	79,014	73,460	68,635	58,954	51,667	45,983	41,425	37,690	34,572	29,665	25,978	23,106	18,922	16,021

 # 4/0 AWG | **2 Run(s)** | **600 Volts**

C-Value = 15,082 | **Magnetic Duct** | **Copper**

Calculations:

1. $f = \dfrac{1.732 \times Length \times Isca}{Runs \times C\text{-}Value \times E_{L\text{-}L}}$

2. $M = \dfrac{1}{1 + f}$

3. $Isc = Isca \times M$

4. Add Motor Contribution = Motor FLA x 4

* All results are given in symmetrical amperes

Three-Phase Bolted Fault-Current* Table

One-Way Distance in Feet

Available Isc in Thousands (Isca)	5	10	15	20	25	30	40	50	60	70	80	90	100	125	150	175	200	225	250	300	350	400	500	600
3	2,992	2,984	2,977	2,969	2,961	2,953	2,933	2,923	2,908	2,894	2,879	2,865	2,850	2,815	2,781	2,747	2,715	2,683	2,652	2,592	2,534	2,479	2,376	2,281
5	4,978	4,957	4,935	4,914	4,893	4,872	4,831	4,790	4,750	4,711	4,673	4,635	4,597	4,507	4,420	4,336	4,255	4,177	4,102	3,960	3,827	3,703	3,477	3,278
7	6,957	6,915	6,874	6,832	6,792	6,752	6,673	6,596	6,520	6,447	6,375	6,304	6,236	6,070	5,913	5,764	5,622	5,487	5,358	5,118	4,898	4,697	4,340	4,033
10	9,913	9,828	9,744	9,662	9,581	9,501	9,345	9,195	9,049	8,908	8,771	8,638	8,510	8,204	7,920	7,654	7,406	7,173	6,955	6,556	6,200	5,881	5,331	4,876
15	14,806	14,616	14,431	14,251	14,076	13,904	13,574	13,259	12,958	12,670	12,395	12,132	11,879	11,292	10,760	10,276	9,834	9,428	9,054	8,389	7,815	7,314	6,484	5,823
20	19,656	19,323	19,002	18,691	18,390	18,098	17,542	17,019	16,527	16,062	15,622	15,206	14,812	13,910	13,111	12,400	11,761	11,185	10,663	9,752	8,985	8,330	7,269	6,448
25	24,464	23,951	23,459	22,987	22,534	22,098	21,274	20,510	19,799	19,135	18,515	17,933	17,387	16,157	15,090	14,155	13,329	12,594	11,936	10,80[illegible]	9,872	9,087	7,839	6,893
30	29,232	28,503	27,808	27,147	26,517	25,915	24,790	23,759	22,810	21,933	21,122	20,369	19,667	18,108	16,778	15,630	14,629	13,748	12,968	11,64[illegible]	10,568	9,673	8,271	7,225
35	33,959	32,979	32,053	31,178	30,349	29,564	28,108	26,790	25,589	24,491	23,484	22,556	21,699	19,817	18,235	16,886	15,724	14,711	13,821	12,329	11,128	10,140	8,611	7,482
40	38,646	37,381	36,197	35,085	34,039	33,054	31,245	29,624	28,163	26,839	25,634	24,533	23,522	21,326	19,505	17,970	16,659	15,527	14,539	12,89[illegible]	11,588	10,521	8,884	7,688
45	43,294	41,713	40,243	38,873	37,593	36,395	34,214	32,280	30,553	29,001	27,599	26,327	25,167	22,669	20,622	18,914	17,468	16,227	15,150	13,375	11,974	10,837	9,109	7,855
50	47,903	45,974	44,195	42,548	41,020	39,598	37,030	34,775	32,779	30,999	29,403	27,963	26,657	23,871	21,613	19,744	18,173	16,834	15,678	13,7[illegible]	12,301	11,105	9,297	7,995
55	52,473	50,168	48,057	46,116	44,326	42,670	39,703	37,122	34,856	32,851	31,063	29,461	28,015	24,954	22,497	20,479	18,794	17,365	16,138	14,1[illegible]	12,582	11,334	9,457	8,113
60	57,005	54,295	51,831	49,580	47,517	45,619	42,244	39,334	36,799	34,571	32,598	30,837	29,257	25,935	23,290	21,135	19,345	17,835	16,543	14,450	12,827	11,532	9,594	8,214
65	61,500	58,357	55,520	52,946	50,600	48,453	44,663	41,423	38,621	36,175	34,019	32,107	30,397	26,827	24,007	21,724	19,837	18,252	16,901	14,722	13,041	11,705	9,713	8,301
70	65,957	62,356	59,127	56,217	53,579	51,178	46,968	43,399	40,333	37,672	35,341	33,281	31,448	27,642	24,658	22,255	20,279	18,625	17,221	14,9[illegible]5	13,231	11,857	9,818	8,378
75	70,378	66,293	62,656	59,397	56,460	53,800	49,168	45,270	41,944	39,074	36,571	34,370	32,419	28,389	25,251	22,737	20,679	18,962	17,508	15,1[illegible]1	13,400	11,993	9,911	8,445
80	74,763	70,169	66,107	62,490	59,248	56,326	51,268	47,044	43,464	40,389	37,721	35,384	33,319	29,077	25,793	23,176	21,041	19,266	17,768	15,3[illegible]5	13,551	12,114	9,994	8,505
85	79,112	73,986	69,485	65,499	61,947	58,759	53,277	48,730	44,898	41,625	38,797	36,329	34,156	29,712	26,292	23,578	21,372	19,543	18,003	15,5[illegible]1	13,687	12,222	10,067	8,559
90	83,425	77,746	72,790	68,429	64,560	61,106	55,199	50,333	46,256	42,790	39,807	37,212	34,936	30,301	26,752	23,947	21,675	19,796	18,217	15,7[illegible]1	13,811	12,321	10,134	8,607
95	87,704	81,449	76,027	71,281	67,094	63,370	57,040	51,860	47,542	43,888	40,755	38,040	35,664	30,847	27,177	24,287	21,953	20,028	18,413	15,857	13,923	12,410	10,195	8,650
100	91,949	85,097	79,196	74,060	69,550	65,557	58,806	53,315	48,762	44,925	41,649	38,817	36,346	31,357	27,571	24,602	22,209	20,241	18,593	15,990	14,026	12,492	10,250	8,690
105	96,159	88,691	82,299	76,767	71,932	67,669	60,500	54,704	49,921	45,908	42,491	39,548	36,987	31,832	27,938	24,893	22,447	20,438	18,760	16,1[illegible]3	14,120	12,567	10,300	8,726
110	100,336	92,232	85,340	79,406	74,244	69,712	62,127	56,031	51,024	46,838	43,288	40,237	37,588	32,277	28,280	25,164	22,667	20,621	18,913	16,226	14,207	12,635	10,346	8,759
115	104,479	95,722	88,319	81,979	76,488	71,687	63,691	57,300	52,074	47,722	44,041	40,888	38,155	32,694	28,600	25,417	22,872	20,790	19,056	16,331	14,288	12,699	10,388	8,789
120	108,590	99,161	91,239	84,488	78,668	73,599	65,195	58,514	53,075	48,562	44,755	41,502	38,690	33,086	28,899	25,653	23,063	20,948	19,188	16,428	14,362	12,757	10,428	8,817
125	112,668	102,550	94,100	86,937	80,787	75,450	66,644	59,678	54,031	49,361	45,433	42,085	39,196	33,455	29,180	25,875	23,242	21,095	19,312	16,518	14,431	12,812	10,464	8,843
130	116,714	105,892	96,906	89,326	82,846	77,243	68,039	60,795	54,945	50,122	46,077	42,637	39,674	33,803	29,445	26,082	23,409	21,233	19,427	16,603	14,495	12,863	10,498	8,868
135	120,728	109,186	99,658	91,659	84,849	78,981	69,384	61,866	55,819	50,848	46,690	43,161	40,128	34,131	29,694	26,278	23,566	21,362	19,535	16,682	14,555	12,910	10,529	8,890
140	124,711	112,433	102,356	93,937	86,797	80,666	70,681	62,896	56,655	51,541	47,274	43,659	40,558	34,442	29,929	26,462	23,714	21,484	19,637	16,[illegible]56	14,612	12,954	10,559	8,911
145	128,664	115,636	105,003	96,162	88,693	82,301	71,933	63,885	57,457	52,204	47,831	44,134	40,968	34,737	30,151	26,635	23,854	21,598	19,732	16,[illegible]25	14,664	12,996	10,586	8,931
150	132,585	118,794	107,601	98,336	90,539	83,889	73,143	64,838	58,226	52,838	48,363	44,586	41,357	35,017	30,362	26,799	23,985	21,706	19,822	16,[illegible]90	14,714	13,035	10,612	8,949
155	136,477	121,908	110,150	100,460	92,337	85,430	74,312	65,754	58,964	53,445	48,871	45,018	41,728	35,282	30,561	26,955	24,109	21,807	19,907	16,[illegible]52	14,761	13,071	10,636	8,966
160	140,338	124,980	112,651	102,537	94,089	86,927	75,442	66,638	59,674	54,027	49,357	45,430	42,082	35,535	30,751	27,102	24,227	21,904	19,987	17,[illegible]10	14,805	13,106	10,659	8,982
165	144,170	128,010	115,107	104,567	95,796	88,382	76,536	67,490	60,356	54,586	49,823	45,825	42,420	35,776	30,931	27,242	24,339	21,995	20,063	17,[illegible]65	14,846	13,138	10,681	8,998
170	147,973	130,999	117,518	106,554	97,460	89,797	77,594	68,311	61,012	55,122	50,270	46,202	42,744	36,005	31,102	27,375	24,445	22,082	20,135	17,[illegible]17	14,886	13,169	10,701	9,012
175	151,747	133,948	119,886	108,497	99,083	91,173	78,620	69,105	61,644	55,638	50,698	46,564	43,053	36,225	31,266	27,501	24,546	22,164	20,203	17,166	14,923	13,198	10,720	9,026
180	155,492	136,858	122,212	110,398	100,666	92,512	79,613	69,871	62,254	56,134	51,109	46,910	43,349	36,434	31,422	27,622	24,642	22,242	20,268	17,213	14,959	13,226	10,739	9,039
185	159,209	139,729	124,496	112,259	102,211	93,815	80,576	70,612	62,841	56,611	51,504	47,243	43,633	36,635	31,571	27,737	24,733	22,317	20,330	17,258	14,992	13,253	10,756	9,051
190	162,898	142,563	126,741	114,080	103,719	95,084	81,511	71,328	63,408	57,070	51,885	47,563	43,906	36,826	31,713	27,847	24,821	22,388	20,389	17,300	15,024	13,278	10,773	9,063
195	166,560	145,359	128,947	115,864	105,192	96,320	82,417	72,022	63,955	57,513	52,250	47,870	44,167	37,010	31,849	27,952	24,904	22,456	20,445	17,341	15,055	13,301	10,788	9,074
200	170,194	148,120	131,114	117,611	106,630	97,524	83,297	72,693	64,484	57,940	52,603	48,166	44,419	37,187	31,980	28,052	24,984	22,520	20,499	17,380	15,084	13,324	10,803	9,084

250 MCM	1 Run(s)	600 Volts	C-Value = 16,483	Magnetic Duct	Copper

Calculations:

1. $f = \dfrac{1.732 \times \text{Length} \times \text{Isca}}{\text{Runs} \times \text{C-Value} \times E_{L\text{-}L}}$

2. $M = \dfrac{1}{1 + f}$

3. $\text{Isc} = \text{Isca} \times M$

4. Add Motor Contribution = Motor FLA x 4

* All results are given in symmetrical amperes

Three-Phase Bolted Fault-Current* Table

One-Way Distance in Feet

Available Isc in Thousands (Isca)

	5	10	15	20	25	30	40	50	60	70	80	90	100	125	150	175	200	225	250	300	350	400	500	600
3	2,996	2,992	2,988	2,984	2,980	2,977	2,969	2,961	2,953	2,946	2,938	2,931	2,923	2,905	2,886	2,868	2,850	2,833	2,815	2,781	2,747	2,715	2,652	2,592
5	4,989	4,978	4,967	4,957	4,946	4,935	4,914	4,893	4,872	4,851	4,831	4,810	4,790	4,741	4,692	4,644	4,597	4,552	4,507	4,420	4,336	4,255	4,102	3,960
7	6,979	6,957	6,936	6,915	6,894	6,874	6,832	6,792	6,752	6,712	6,673	6,634	6,596	6,502	6,411	6,322	6,236	6,152	6,070	5,913	5,764	5,622	5,358	5,118
10	9,956	9,913	9,870	9,828	9,786	9,744	9,662	9,581	9,501	9,422	9,345	9,269	9,195	9,013	8,839	8,671	8,510	8,354	8,204	7,920	7,654	7,406	6,955	6,556
15	14,902	14,806	14,710	14,616	14,523	14,431	14,251	14,076	13,904	13,737	13,574	13,414	13,259	12,885	12,531	12,197	11,879	11,578	11,292	10,760	10,276	9,834	9,054	8,389
20	19,826	19,656	19,488	19,323	19,161	19,002	18,691	18,390	18,098	17,816	17,542	17,277	17,019	16,408	15,839	15,308	14,812	14,347	13,910	13,111	12,400	11,761	10,663	9,752
25	24,729	24,464	24,205	23,951	23,703	23,459	22,987	22,534	22,098	21,678	21,274	20,885	20,510	19,629	18,820	18,075	17,387	16,750	16,157	15,090	14,155	13,329	11,936	10,806
30	29,611	29,232	28,863	28,503	28,151	27,808	27,147	26,517	25,915	25,340	24,790	24,263	23,759	22,584	21,520	20,552	19,667	18,855	18,108	16,778	15,630	14,629	12,968	11,645
35	34,472	33,959	33,462	32,979	32,509	32,053	31,178	30,349	29,564	28,818	28,108	27,433	26,790	25,306	23,977	22,781	21,699	20,715	19,817	18,235	16,886	15,724	13,821	12,329
40	39,312	38,646	38,003	37,381	36,779	36,197	35,085	34,039	33,054	32,124	31,245	30,413	29,624	27,820	26,223	24,799	23,522	22,370	21,326	19,505	17,970	16,659	14,539	12,897
45	44,131	43,294	42,489	41,713	40,965	40,243	38,873	37,593	36,395	35,271	34,214	33,219	32,280	30,150	28,283	26,634	25,167	23,852	22,669	20,622	18,914	17,468	15,150	13,376
50	48,929	47,903	46,919	45,974	45,067	44,195	42,548	41,020	39,598	38,271	37,030	35,867	34,775	32,315	30,180	28,309	26,357	25,188	23,871	21,613	19,744	18,173	15,678	13,786
55	53,707	52,473	51,294	50,168	49,090	48,057	46,116	44,326	42,670	41,133	39,703	38,369	37,122	34,332	31,932	29,846	28,315	26,396	24,954	22,497	20,479	18,794	16,138	14,140
60	58,464	57,005	55,617	54,295	53,034	51,831	49,580	47,517	45,619	43,867	42,244	40,737	39,334	36,216	33,555	31,259	29,257	27,496	25,935	23,290	21,135	19,345	16,543	14,450
65	63,201	61,500	59,887	58,357	56,903	55,520	52,946	50,600	48,453	46,481	44,663	42,982	41,423	37,979	35,064	32,564	30,397	28,501	26,827	24,007	21,724	19,837	16,901	14,722
70	67,918	65,957	64,106	62,356	60,699	59,127	56,217	53,579	51,178	48,983	46,968	45,113	43,399	39,633	36,469	33,773	31,448	29,422	27,642	24,658	22,255	20,279	17,221	14,965
75	72,616	70,378	68,274	66,293	64,423	62,656	59,397	56,460	53,800	51,380	49,168	47,138	45,270	41,188	37,781	34,895	32,419	30,271	28,389	25,251	22,737	20,679	17,508	15,181
80	77,293	74,763	72,393	70,169	68,078	66,107	62,490	59,248	56,326	53,678	51,268	49,066	47,044	42,652	39,010	35,940	33,319	31,054	29,077	25,793	23,176	21,041	17,768	15,375
85	81,950	79,112	76,463	73,986	71,665	69,485	65,499	61,947	58,759	55,884	53,277	50,902	48,730	44,033	40,162	36,916	34,156	31,780	29,712	26,292	23,578	21,372	18,003	15,551
90	86,588	83,425	80,486	77,746	75,187	72,790	68,429	64,560	61,106	58,002	55,199	52,654	50,333	45,338	41,244	37,829	34,536	32,454	30,301	26,752	23,947	21,675	18,217	15,711
95	91,206	87,704	84,461	81,449	78,645	76,027	71,281	67,094	63,370	60,039	57,040	54,327	51,860	46,572	42,264	38,684	35,664	33,081	30,847	27,177	24,287	21,953	18,413	15,857
100	95,805	91,949	88,390	85,097	82,040	79,196	74,060	69,550	65,557	61,998	58,806	55,926	53,315	47,743	43,225	39,488	36,346	33,668	31,357	27,571	24,602	22,209	18,593	15,990
105	100,385	96,159	92,274	88,691	85,376	82,299	76,767	71,932	67,669	63,884	60,500	57,456	54,704	48,853	44,133	40,245	36,987	34,216	31,832	27,938	24,893	22,447	18,760	16,113
110	104,946	100,336	96,113	92,232	88,652	85,340	79,406	74,244	69,712	65,701	62,127	58,921	56,031	49,909	44,993	40,959	37,588	34,731	32,277	28,280	25,164	22,667	18,913	16,226
115	109,487	104,479	99,909	95,722	91,871	88,319	81,979	76,488	71,687	67,453	63,691	60,326	57,300	50,913	45,808	41,633	38,155	35,214	32,694	28,600	25,417	22,872	19,056	16,331
120	114,010	108,590	103,661	99,161	95,035	91,239	84,488	78,668	73,599	69,143	65,195	61,674	58,514	51,870	46,581	42,270	38,690	35,669	33,086	28,899	25,653	23,063	19,188	16,428
125	118,514	112,668	107,371	102,550	98,144	94,100	86,937	80,787	75,450	70,774	66,644	62,969	59,678	52,783	47,315	42,875	39,196	36,098	33,455	29,180	25,875	23,242	19,312	16,518
130	122,999	116,714	111,040	105,892	101,200	96,906	89,326	82,846	77,243	72,349	68,039	64,213	60,795	53,654	48,014	43,448	39,674	36,504	33,803	29,445	26,082	23,409	19,427	16,603
135	127,466	120,728	114,667	109,186	104,204	99,658	91,659	84,849	78,981	73,872	69,384	65,410	61,866	54,487	48,680	43,992	40,128	36,887	34,131	29,694	26,278	23,566	19,535	16,682
140	131,914	124,711	118,255	112,433	107,158	102,356	93,937	86,797	80,666	75,344	70,681	66,561	62,896	55,284	49,315	44,510	40,558	37,251	34,442	29,929	26,462	23,714	19,637	16,756
145	136,344	128,664	121,802	115,636	110,063	105,003	96,162	88,693	82,301	76,769	71,933	67,671	63,885	56,047	49,922	45,004	40,968	37,596	34,737	30,151	26,635	23,854	19,732	16,825
150	140,756	132,585	125,311	118,794	112,920	107,601	98,336	90,539	83,889	78,148	73,143	68,740	64,838	56,779	50,501	45,474	41,357	37,924	35,017	30,362	26,799	23,985	19,822	16,890
155	145,150	136,477	128,782	121,908	115,731	110,150	100,460	92,337	85,430	79,484	74,312	69,772	65,754	57,480	51,056	45,923	41,728	38,235	35,282	30,561	26,955	24,109	19,907	16,952
160	149,525	140,338	132,214	124,980	118,496	112,651	102,537	94,089	86,927	80,778	75,442	70,767	66,638	58,154	51,587	46,352	42,082	38,533	35,535	30,751	27,102	24,227	19,987	17,010
165	153,883	144,170	135,610	128,010	121,216	115,107	104,567	95,796	88,382	82,033	76,536	71,729	67,490	58,802	52,096	46,763	42,420	38,816	35,776	30,931	27,242	24,339	20,063	17,065
170	158,223	147,973	138,969	130,999	123,893	117,518	106,554	97,460	89,797	83,251	77,594	72,658	68,311	59,425	52,584	47,156	42,744	39,086	36,005	31,102	27,375	24,445	20,135	17,117
175	162,546	151,747	142,293	133,948	126,528	119,886	108,497	99,083	91,173	84,432	78,620	73,556	69,105	60,024	53,053	47,533	43,053	39,345	36,225	31,266	27,501	24,546	20,203	17,166
180	166,851	155,492	145,581	136,858	129,121	122,212	110,398	100,666	92,512	85,579	79,613	74,425	69,871	60,602	53,504	47,894	43,349	39,592	36,434	31,422	27,622	24,642	20,268	17,213
185	171,138	159,209	148,834	139,729	131,674	124,496	112,259	102,211	93,815	86,693	80,576	75,266	70,612	61,158	53,937	48,241	43,633	39,829	36,635	31,571	27,737	24,733	20,330	17,258
190	175,408	162,898	152,054	142,563	134,187	126,741	114,080	103,719	95,084	87,775	81,511	76,080	71,328	61,695	54,354	48,574	43,906	40,056	36,826	31,713	27,847	24,821	20,389	17,300
195	179,661	166,560	155,239	145,359	136,662	128,947	115,864	105,192	96,320	88,828	82,417	76,870	72,022	62,213	54,756	48,895	44,167	40,273	37,010	31,849	27,952	24,904	20,445	17,341
200	183,897	170,194	158,391	148,120	139,099	131,114	117,611	106,630	97,524	89,851	83,297	77,635	72,693	62,713	55,143	49,203	44,419	40,482	37,187	31,980	28,052	24,984	20,499	17,380

250 MCM	2 Run(s)	600 Volts	C-Value = 16,483	Magnetic Duct	Copper

Calculations:

1. $f = \dfrac{1.732 \times \text{Length} \times \text{Isca}}{\text{Runs} \times \text{C-Value} \times E_{L-L}}$
2. $M = \dfrac{1}{1 + f}$
3. $\text{Isc} = \text{Isca} \times M$
4. Add Motor Contribution = Motor FLA x 4

* All results are given in symmetrical amperes

Three-Phase Bolted Fault-Current* Table

One-Way Distance in Feet

Available Isc in Thousands (Isca)	5	10	15	20	25	30	40	50	60	70	80	90	100	125	150	175	200	225	250	300	350	400	500	600
3	2,993	2,986	2,979	2,972	2,965	2,958	2,944	2,930	2,917	2,903	2,890	2,877	2,864	2,831	2,800	2,769	2,739	2,710	2,681	2,62[illegible]	2,571	2,520	2,423	2,333
5	4,980	4,961	4,941	4,922	4,903	4,884	4,846	4,809	4,773	4,737	4,701	4,666	4,632	4,549	4,468	4,390	4,315	4,242	4,172	4,03[illegible]	3,913	3,795	3,579	3,386
7	6,961	6,923	6,885	6,848	6,811	6,774	6,732	6,631	6,562	6,495	6,428	6,363	6,300	6,146	6,000	5,860	5,727	5,599	5,478	5,24[illegible]	5,039	4,845	4,499	4,199
10	9,921	9,844	9,767	9,692	9,618	9,545	9,403	9,264	9,130	9,000	8,873	8,749	8,629	8,344	8,076	7,825	7,589	7,367	7,158	6,773	6,427	6,115	5,574	5,121
15	14,823	14,651	14,482	14,318	14,157	13,999	13,695	13,403	13,124	12,856	12,599	12,352	12,114	11,558	11,051	10,587	10,159	9,766	9,401	8,748	8,180	7,681	6,846	6,174
20	19,687	19,384	19,090	18,805	18,529	18,260	17,745	17,259	16,799	16,362	15,948	15,554	15,179	14,316	13,546	12,855	12,230	11,664	11,148	10,2[illegible]1	9,471	8,808	7,727	6,883
25	24,513	24,045	23,595	23,161	22,743	22,339	21,574	20,859	20,190	19,563	18,973	18,418	17,895	16,708	15,568	14,751	13,935	13,204	12,546	11,4[illegible]0	10,462	9,659	8,375	7,392
30	29,302	28,636	27,999	27,390	26,807	26,248	25,198	24,228	23,330	22,497	21,721	20,997	20,319	18,802	17,496	16,360	15,362	14,479	13,692	12,3[illegible]9	11,246	10,324	8,870	7,775
35	34,054	33,157	32,306	31,498	30,730	29,998	28,333	27,388	26,246	25,196	24,227	23,329	22,496	20,651	19,086	17,742	16,574	15,551	14,646	13,1[illegible]0	11,882	10,858	9,261	8,074
40	38,769	37,611	36,520	35,491	34,518	33,597	31,395	30,357	28,961	27,688	26,521	25,449	24,461	22,295	20,482	18,942	17,617	16,465	15,455	13,7[illegible]6	12,409	11,296	9,578	8,313
45	43,447	41,998	40,643	39,372	38,179	37,055	34,396	33,153	31,495	29,994	28,631	27,385	26,244	23,767	21,718	19,994	18,523	17,254	16,148	14,3[illegible]3	12,852	11,662	9,840	8,510
50	48,091	46,322	44,678	43,147	41,718	40,380	37,947	35,790	33,865	32,137	30,576	29,160	27,869	25,093	22,819	20,924	19,319	17,942	16,749	14,733	13,230	11,972	10,059	8,674
55	52,698	50,582	48,628	46,820	45,142	43,580	40,759	38,281	36,087	34,131	32,376	30,793	29,357	26,292	23,807	21,751	20,022	18,547	17,275	15,1[illegible]1	13,556	12,239	10,247	8,813
60	57,271	54,780	52,496	50,396	48,456	46,661	43,442	40,638	38,174	35,992	34,046	32,299	30,723	27,383	24,698	22,492	20,648	19,084	17,740	15,549	13,840	12,470	10,408	8,932
65	61,810	58,918	56,285	53,877	51,666	49,630	46,004	42,872	40,139	37,733	35,600	33,695	31,983	28,379	25,505	23,160	21,210	19,562	18,152	15,855	14,090	12,672	10,549	9,035
70	66,314	62,997	59,995	57,267	54,776	52,493	48,453	44,991	41,991	39,366	37,049	34,990	33,148	29,293	26,241	23,765	21,716	19,992	18,522	16,147	14,312	12,851	10,673	9,126
75	70,784	67,017	63,631	60,571	57,791	55,255	50,797	47,005	43,740	40,899	38,404	36,197	34,229	30,134	26,914	24,315	22,175	20,380	18,855	16,[illegible]9	14,510	13,011	10,783	9,206
80	75,221	70,982	67,194	63,790	60,715	57,922	53,043	48,922	45,395	42,342	39,674	37,322	35,234	30,910	27,531	24,818	22,592	20,732	19,155	16,626	14,687	13,153	10,880	9,277
85	79,625	74,890	70,687	66,930	63,552	60,499	55,196	50,747	46,962	43,703	40,866	38,376	36,171	31,629	28,100	25,279	22,974	21,053	19,429	16,832	14,848	13,282	10,968	9,341
90	83,997	78,745	74,110	69,991	66,306	62,990	57,261	52,488	48,449	44,988	41,988	39,363	37,047	32,296	28,626	25,704	23,324	21,347	19,679	17,019	14,993	13,398	11,047	9,398
95	88,336	82,546	77,468	72,979	68,981	65,399	59,245	54,150	49,862	46,203	43,045	40,290	37,867	32,918	29,113	26,096	23,646	21,617	19,908	17,[illegible]91	15,126	13,504	11,119	9,450
100	92,643	86,295	80,761	75,894	71,580	67,730	61,152	55,739	51,206	47,355	44,042	41,163	38,637	33,498	29,566	26,460	23,944	21,866	20,119	17,[illegible]47	15,247	13,600	11,185	9,498
105	96,919	89,993	83,991	78,739	74,106	69,987	62,986	57,258	52,486	48,447	44,986	41,986	39,361	34,041	29,988	26,798	24,221	22,096	20,314	17,[illegible]92	15,359	13,689	11,244	9,541
110	101,163	93,641	87,160	81,518	76,562	72,174	64,752	58,714	53,706	49,485	45,879	42,763	40,044	34,551	30,383	27,112	24,477	22,309	20,494	17,[illegible]25	15,461	13,771	11,299	9,580
115	105,377	97,240	90,270	84,232	78,951	74,293	66,452	60,109	54,871	50,472	46,727	43,499	40,668	35,029	30,752	27,406	24,716	22,508	20,661	17,[illegible]49	15,557	13,846	11,350	9,617
120	109,560	100,791	93,322	86,883	81,276	76,348	68,092	61,447	55,984	51,412	47,531	44,195	41,297	35,479	31,098	27,681	24,940	22,693	20,817	17,[illegible]64	15,645	13,916	11,397	9,650
125	113,713	104,295	96,318	89,475	83,539	78,342	69,673	62,732	57,048	52,309	48,297	44,856	41,873	35,904	31,424	27,938	25,149	22,866	20,962	17,[illegible]71	15,727	13,981	11,441	9,681
130	117,836	107,753	99,260	92,008	85,743	80,277	71,200	63,967	58,067	53,165	49,025	45,484	42,420	36,305	31,731	28,181	25,345	23,028	21,099	18,071	15,803	14,041	11,481	9,710
135	121,929	111,166	102,148	94,484	87,890	82,156	72,674	65,154	59,044	53,982	49,720	46,081	42,938	36,684	32,020	28,409	25,529	23,180	21,226	18,[illegible]64	15,875	14,098	11,519	9,737
140	125,993	114,534	104,986	96,907	89,982	83,982	74,098	66,297	59,981	54,764	50,382	46,650	43,432	37,044	32,294	28,624	25,703	23,323	21,346	18,252	15,942	14,150	11,554	9,762
145	130,028	117,859	107,772	99,276	92,022	85,755	75,476	67,397	60,881	55,513	51,015	47,192	43,901	37,385	32,553	28,827	25,866	23,457	21,459	18,335	16,004	14,200	11,587	9,786
150	134,035	121,141	110,510	101,595	94,011	87,480	76,809	68,458	61,745	56,231	51,621	47,709	44,349	37,709	32,798	29,019	26,021	23,585	21,565	18,412	16,064	14,246	11,618	9,808
155	138,013	124,381	113,201	103,864	95,950	89,157	78,099	69,481	62,576	56,919	52,200	48,204	44,776	38,017	33,031	29,202	26,168	23,705	21,666	18,485	16,119	14,290	11,647	9,829
160	141,963	127,581	115,844	106,086	97,843	90,789	79,348	70,468	63,375	57,580	52,755	48,677	45,184	38,311	33,253	29,375	26,306	23,819	21,761	18,555	16,172	14,331	11,674	9,848
165	145,885	130,740	118,443	108,261	99,691	92,378	80,559	71,421	64,145	58,215	53,288	49,130	45,574	38,591	33,464	29,539	26,438	23,927	21,851	18,620	16,221	14,370	11,700	9,867
170	149,780	133,859	120,998	110,391	101,494	93,924	81,732	72,342	64,887	58,825	53,799	49,564	45,947	38,858	33,664	29,695	26,563	24,029	21,936	18,682	16,269	14,407	11,725	9,884
175	153,648	136,940	123,509	112,478	103,255	95,431	82,871	73,232	65,602	59,412	54,290	49,980	46,305	39,114	33,856	29,844	26,682	24,126	22,017	18,741	16,313	14,442	11,748	9,901
180	157,489	139,983	125,979	114,523	104,976	96,898	83,975	74,094	66,293	59,978	54,762	50,380	46,648	39,358	34,039	29,986	26,796	24,219	22,095	18,797	16,355	14,475	11,770	9,916
185	161,304	142,988	128,408	116,526	106,657	98,329	85,048	74,927	66,959	60,523	55,216	50,764	46,977	39,592	34,214	30,122	26,904	24,308	22,168	18,850	16,396	14,507	11,790	9,931
190	165,092	145,957	130,797	118,490	108,300	99,724	86,089	75,734	67,603	61,049	55,653	51,133	47,293	39,816	34,381	30,251	27,008	24,392	22,238	18,901	16,434	14,537	11,810	9,945
195	168,854	148,890	133,148	120,416	109,907	101,084	87,101	76,516	68,226	61,556	56,074	51,489	47,596	40,031	34,541	30,376	27,106	24,473	22,305	18,949	16,471	14,566	11,829	9,958
200	172,590	151,787	135,460	122,304	111,477	102,412	88,085	77,275	68,828	62,045	56,480	51,831	47,889	40,238	34,695	30,494	27,201	24,550	22,369	18,995	16,505	14,593	11,847	9,971

300 MCM	1 Run(s)	600 Volts	C-Value = 18,176	Magnetic Duct	Copper

Calculations:

1. $f = \dfrac{1.732 \times \text{Length} \times \text{Isca}}{\text{Runs} \times \text{C-Value} \times \text{E L-L}}$

2. $M = \dfrac{1}{1 + f}$

3. Isc = Isca x M

4. Add Motor Contribution = Motor FLA x 4

* All results are given in symmetrical amperes

Three-Phase Bolted Fault-Current* Table

One-Way Distance in Feet

Available Isc in Thousands (Isca)	5	10	15	20	25	30	40	50	60	70	80	90	100	125	150	175	200	225	250	300	350	400	500	600
3	2,996	2,993	2,989	2,986	2,982	2,979	2,972	2,965	2,958	2,951	2,944	2,937	2,930	2,913	2,896	2,880	2,864	2,847	2,831	2,800	2,769	2,739	2,681	2,625
5	4,990	4,980	4,970	4,961	4,951	4,941	4,922	4,903	4,884	4,865	4,846	4,827	4,809	4,764	4,719	4,675	4,632	4,590	4,549	4,468	4,390	4,315	4,172	4,038
7	6,981	6,961	6,942	6,923	6,904	6,885	6,848	6,811	6,774	6,738	6,702	6,666	6,631	6,545	6,461	6,379	6,300	6,222	6,146	6,000	5,860	5,727	5,478	5,249
10	9,960	9,921	9,882	9,844	9,805	9,767	9,692	9,618	9,545	9,473	9,403	9,333	9,264	9,097	8,936	8,780	8,629	8,484	8,344	8,076	7,825	7,589	7,158	6,773
15	14,911	14,823	14,737	14,651	14,566	14,482	14,318	14,157	13,999	13,846	13,695	13,548	13,403	13,056	12,726	12,413	12,114	11,830	11,558	11,051	10,587	10,159	9,401	8,748
20	19,842	19,687	19,535	19,384	19,236	19,090	18,805	18,529	18,260	17,999	17,745	17,499	17,259	16,687	16,152	15,650	15,179	14,735	14,316	13,546	12,855	12,230	11,148	10,241
25	24,754	24,513	24,277	24,045	23,818	23,595	23,161	22,743	22,339	21,950	21,574	21,210	20,859	20,030	19,264	18,554	17,895	17,281	16,708	15,668	14,751	13,935	12,546	11,410
30	29,647	29,302	28,965	28,636	28,314	27,999	27,390	26,807	26,248	25,712	25,198	24,703	24,228	23,116	22,102	21,173	20,319	19,531	18,802	17,496	16,360	15,362	13,692	12,349
35	34,520	34,054	33,599	33,157	32,726	32,306	31,498	30,730	29,998	29,300	28,633	27,997	27,388	25,976	24,702	23,547	22,496	21,534	20,651	19,086	17,742	16,574	14,646	13,120
40	39,375	38,769	38,181	37,611	37,057	36,520	35,491	34,518	33,597	32,724	31,895	31,107	30,357	28,632	27,092	25,709	24,461	23,328	22,295	20,482	18,942	17,617	15,455	13,766
45	44,210	43,447	42,711	41,998	41,310	40,643	39,372	38,179	37,055	35,996	34,996	34,049	33,153	31,106	29,297	27,686	26,244	24,944	23,767	21,718	19,994	18,523	16,148	14,313
50	49,027	48,091	47,190	46,322	45,485	44,678	43,147	41,718	40,380	39,126	37,947	36,837	35,790	33,416	31,337	29,502	27,869	26,408	25,093	22,819	20,924	19,319	16,749	14,783
55	53,825	52,698	51,618	50,582	49,586	48,628	46,820	45,142	43,580	42,122	40,759	39,481	38,281	35,577	33,230	31,174	29,357	27,740	26,292	23,807	21,751	20,022	17,275	15,191
60	58,604	57,271	55,998	54,780	53,614	52,496	50,396	48,456	46,661	44,994	43,442	41,993	40,638	37,604	34,992	32,719	30,723	28,957	27,383	24,698	22,492	20,648	17,740	15,549
65	63,365	61,810	60,329	58,918	57,571	56,285	53,877	51,666	49,630	47,748	46,004	44,383	42,872	39,509	36,636	34,152	31,983	30,074	28,379	25,505	23,160	21,210	18,152	15,865
70	68,107	66,314	64,613	62,997	61,459	59,995	57,267	54,776	52,493	50,392	48,453	46,658	44,991	41,302	38,172	35,483	33,148	31,102	29,293	26,241	23,765	21,716	18,522	16,147
75	72,831	70,784	68,849	67,017	65,280	63,631	60,571	57,791	55,255	52,933	50,797	48,828	47,005	42,993	39,612	36,724	34,229	32,051	30,134	26,914	24,315	22,175	18,855	16,399
80	77,537	75,221	73,040	70,982	69,036	67,194	63,790	60,715	57,922	55,375	53,043	50,899	48,922	44,591	40,965	37,884	35,234	32,931	30,910	27,531	24,818	22,592	19,155	16,626
85	82,225	79,625	77,185	74,890	72,728	70,687	66,930	63,552	60,499	57,726	55,196	52,878	50,747	46,102	42,237	38,969	36,171	33,748	31,629	28,100	25,279	22,974	19,429	16,832
90	86,895	83,997	81,286	78,745	76,357	74,110	69,991	66,306	62,990	59,989	57,261	54,771	52,488	47,535	43,436	39,988	37,047	34,509	32,296	28,626	25,704	23,324	19,679	17,019
95	91,547	88,336	85,343	82,546	79,926	77,468	72,979	68,981	65,399	62,170	59,245	56,583	54,150	48,894	44,568	40,945	37,867	35,220	32,918	29,113	26,096	23,646	19,908	17,191
100	96,181	92,643	89,356	86,295	83,436	80,761	75,894	71,580	67,730	64,273	61,152	58,320	55,739	50,185	45,639	41,847	38,637	35,885	33,498	29,566	26,460	23,944	20,119	17,347
105	100,798	96,919	93,328	89,993	86,888	83,991	78,739	74,106	69,987	66,302	62,986	59,986	57,258	51,414	46,652	42,698	39,361	36,509	34,041	29,988	26,798	24,221	20,314	17,492
110	105,397	101,163	97,257	93,641	90,284	87,160	81,518	76,562	72,174	68,262	64,752	61,585	58,714	52,585	47,614	43,502	40,044	37,095	34,551	30,383	27,112	24,477	20,494	17,625
115	109,978	105,377	101,145	97,240	93,625	90,270	84,232	78,951	74,293	70,154	66,452	63,122	60,109	53,701	48,527	44,263	40,688	37,647	35,029	30,752	27,406	24,716	20,661	17,749
120	114,543	109,560	104,993	100,791	96,913	93,322	86,883	81,276	76,348	71,984	68,092	64,599	61,447	54,766	49,396	44,985	41,297	38,168	35,479	31,098	27,681	24,940	20,817	17,864
125	119,090	113,713	108,801	104,295	100,148	96,318	89,475	83,539	78,342	73,754	69,673	66,021	62,732	55,785	50,223	45,669	41,873	38,659	35,904	31,424	27,938	25,149	20,962	17,971
130	123,619	117,836	112,569	107,753	103,332	99,260	92,008	85,743	80,277	75,466	71,200	67,390	63,967	56,759	51,011	46,320	42,420	39,125	36,305	31,731	28,181	25,345	21,099	18,071
135	128,132	121,929	116,299	111,166	106,466	102,148	94,484	87,890	82,156	77,125	72,674	68,709	65,154	57,692	51,763	46,940	42,938	39,566	36,684	32,020	28,409	25,529	21,226	18,164
140	132,628	125,993	119,991	114,534	109,552	104,986	96,907	89,982	83,982	78,731	74,098	69,981	66,297	58,586	52,482	47,530	43,432	39,984	37,044	32,294	28,624	25,703	21,346	18,252
145	137,107	130,028	123,645	117,859	112,590	107,772	99,276	92,022	85,755	80,288	75,476	71,208	67,397	59,444	53,169	48,093	43,921	40,382	37,385	32,553	28,827	25,866	21,459	18,335
150	141,569	134,035	127,262	121,141	115,582	110,510	101,595	94,011	87,480	81,798	76,809	72,393	68,458	60,267	53,827	48,631	44,349	40,760	37,709	32,798	29,019	26,021	21,565	18,412
155	146,014	138,013	130,843	124,381	118,528	113,201	103,864	95,950	89,157	83,262	78,099	73,538	69,481	61,059	54,458	49,145	44,776	41,121	38,017	33,031	29,202	26,168	21,666	18,485
160	150,443	141,963	134,388	127,581	121,430	115,844	106,086	97,843	90,789	84,684	79,348	74,645	70,468	61,820	55,062	49,636	45,184	41,465	38,311	33,253	29,375	26,306	21,761	18,555
165	154,855	145,885	137,898	130,740	124,288	118,443	108,261	99,691	92,378	86,064	80,559	75,715	71,421	62,552	55,642	50,107	45,574	41,793	38,591	33,464	29,539	26,438	21,851	18,620
170	159,251	149,780	141,373	133,859	127,104	120,998	110,391	101,494	93,924	87,405	81,732	76,751	72,342	63,257	56,200	50,559	45,947	42,106	38,858	33,664	29,695	26,563	21,936	18,682
175	163,631	153,648	144,814	136,940	129,878	123,509	112,478	103,255	95,431	88,708	82,871	77,754	73,232	63,937	56,736	50,992	46,305	42,407	39,114	33,856	29,844	26,682	22,017	18,741
180	167,994	157,489	148,221	139,983	132,612	125,979	114,523	104,976	96,898	89,975	83,975	78,726	74,094	64,593	57,251	51,408	46,643	42,694	39,358	34,039	29,986	26,796	22,095	18,797
185	172,341	161,304	151,595	142,988	135,307	128,408	116,526	106,657	98,329	91,207	85,048	79,667	74,927	65,225	57,748	51,808	46,977	42,969	39,592	34,214	30,122	26,904	22,168	18,850
190	176,672	165,092	154,936	145,957	137,962	130,797	118,490	108,300	99,724	92,406	86,089	80,581	75,734	65,836	58,226	52,193	47,293	43,234	39,816	34,381	30,251	27,008	22,238	18,901
195	180,987	168,854	158,244	148,890	140,579	133,148	120,416	109,907	101,084	93,573	87,101	81,466	76,516	66,426	58,687	52,563	47,596	43,487	40,031	34,541	30,376	27,106	22,305	18,949
200	185,287	172,590	161,521	151,787	143,159	135,460	122,304	111,477	102,412	94,709	88,085	82,326	77,275	66,997	59,132	52,920	47,889	43,731	40,238	34,695	30,494	27,201	22,369	18,995

300 MCM	2 Run(s)	600 Volts	C-Value = 18,176	Magnetic Duct	Copper

Calculations:

1. $f = \dfrac{1.732 \times \text{Length} \times \text{Isca}}{\text{Runs} \times \text{C-Value} \times \text{E L-L}}$

2. $M = \dfrac{1}{1 + f}$

3. $\text{Isc} = \text{Isca} \times M$

4. Add Motor Contribution = Motor FLA x 4

* All results are given in symmetrical amperes

Three-Phase Bolted Fault-Current* Table

One-Way Distance in Feet

Available Isc in Thousands (Isca)	5	10	15	20	25	30	40	50	60	70	80	90	100	125	150	175	200	225	250	300	350	400	500	600
3	2,993	2,987	2,980	2,974	2,967	2,961	2,948	2,935	2,923	2,910	2,898	2,886	2,874	2,844	2,814	2,786	2,758	2,730	2,703	2,6[illegible]1	2,600	2,551	2,459	2,374
5	4,982	4,964	4,946	4,928	4,910	4,892	4,858	4,823	4,789	4,756	4,723	4,691	4,659	4,581	4,505	4,432	4,361	4,292	4,226	4,0[illegible]9	3,980	3,867	3,660	3,473
7	6,964	6,929	6,894	6,859	6,825	6,791	6,724	6,659	6,594	6,531	6,469	6,408	6,349	6,205	6,067	5,935	5,809	5,688	5,572	5,3[illegible]3	5,151	4,964	4,627	4,333
10	9,927	9,856	9,785	9,715	9,647	9,579	9,446	9,317	9,192	9,070	8,951	8,835	8,722	8,452	8,198	7,959	7,734	7,521	7,319	6,9[illegible]7	6,610	6,305	5,772	5,322
15	14,837	14,677	14,521	14,368	14,219	14,072	13,788	13,515	13,253	13,000	12,757	12,523	12,297	11,767	11,281	10,834	10,420	10,037	9,681	9,0[illegible]0	8,479	7,983	7,147	6,469
20	19,711	19,431	19,158	18,893	18,635	18,384	17,902	17,444	17,010	16,596	16,202	15,826	15,463	14,638	13,893	13,221	12,610	12,053	11,544	10,644	9,874	9,208	8,113	7,251
25	24,550	24,117	23,698	23,294	22,903	22,525	21,805	21,130	20,496	19,898	19,335	18,802	18,298	17,149	16,135	15,235	14,430	13,705	13,050	11,[illegible]11	10,956	10,142	8,830	7,818
30	29,355	28,737	28,144	27,576	27,030	26,505	25,514	24,595	23,740	22,942	22,196	21,497	20,840	19,362	18,080	16,957	15,965	15,083	14,294	12,539	11,819	10,877	9,382	8,248
35	34,125	33,293	32,500	31,744	31,023	30,334	29,043	27,858	26,765	25,755	24,819	23,948	23,136	21,329	19,783	18,447	17,279	16,251	15,338	13,[illegible]89	12,524	11,471	9,821	8,585
40	38,861	37,786	36,768	35,804	34,889	34,019	32,404	30,935	29,594	28,364	27,233	26,188	25,220	23,087	21,287	19,748	18,416	17,252	16,227	14,503	13,110	11,961	10,178	8,857
45	43,564	42,217	40,950	39,758	38,632	37,569	35,609	33,844	32,245	30,790	29,461	28,242	27,120	24,670	22,625	20,894	19,408	18,120	16,993	15,[illegible]11	13,605	12,372	10,474	9,080
50	48,233	46,587	45,050	43,611	42,261	40,992	38,669	36,596	34,734	33,052	31,525	30,133	28,859	26,100	23,823	21,911	20,283	18,881	17,659	15,[illegible]37	14,030	12,722	10,723	9,267
55	52,870	50,899	49,069	47,366	45,778	44,293	41,594	39,205	37,075	35,165	33,442	31,880	30,457	27,401	24,902	22,820	21,060	19,552	18,245	16,[illegible]94	14,397	13,023	10,937	9,426
60	57,474	55,152	53,010	51,029	49,190	47,479	44,391	41,680	39,282	37,144	35,227	33,498	31,931	28,588	25,878	23,637	21,754	20,149	18,764	16,[illegible]96	14,718	13,285	11,121	9,563
65	62,046	59,348	56,876	54,601	52,501	50,556	47,070	44,033	41,365	39,001	36,893	35,001	33,294	29,675	26,766	24,376	22,378	20,683	19,226	16,[illegible]53	15,001	13,516	11,282	9,681
70	66,586	63,489	60,667	58,086	55,715	53,530	49,637	46,272	43,335	40,748	38,452	36,401	34,558	30,675	27,577	25,047	22,942	21,164	19,641	17,[illegible]71	15,252	13,719	11,423	9,786
75	71,094	67,575	64,387	61,487	58,837	56,406	52,100	48,406	45,200	42,393	39,914	37,709	35,734	31,599	28,321	25,659	23,455	21,599	20,016	17,[illegible]56	15,477	13,901	11,549	9,878
80	75,571	71,607	68,038	64,808	61,871	59,188	54,465	50,440	46,969	43,945	41,287	38,932	36,831	32,453	29,005	26,220	23,922	21,995	20,355	17,[illegible]14	15,679	14,064	11,661	9,960
85	80,018	75,587	71,621	68,051	64,820	61,881	56,737	52,383	48,649	45,413	42,580	40,079	37,856	33,247	29,637	26,735	24,351	22,357	20,665	17,[illegible]48	15,862	14,211	11,762	10,033
90	84,433	79,515	75,139	71,219	67,687	64,490	58,922	54,240	50,247	46,802	43,799	41,157	38,817	33,985	30,223	27,211	24,745	22,688	20,948	18,[illegible]61	16,028	14,344	11,853	10,099
95	88,819	83,393	78,592	74,314	70,477	67,017	61,025	56,017	51,768	48,119	44,950	42,173	39,718	34,674	30,767	27,651	25,108	22,993	21,207	18,356	16,180	14,466	11,936	10,159
100	93,175	87,221	81,983	77,338	73,192	69,467	63,050	57,719	53,218	49,369	46,039	43,130	40,566	35,319	31,273	28,059	25,444	23,275	21,447	18,535	16,319	14,576	12,011	10,214
105	97,501	91,001	85,314	80,296	75,835	71,844	65,002	59,350	54,602	50,558	47,071	44,034	41,366	35,923	31,746	28,439	25,756	23,536	21,668	18,700	16,447	14,678	12,080	10,264
110	101,797	94,733	88,585	83,187	78,409	74,150	66,884	60,915	55,924	51,689	48,050	44,890	42,120	36,490	32,188	28,794	26,047	23,778	21,873	18,[illegible]52	16,565	14,772	12,144	10,310
115	106,065	98,418	91,800	86,015	80,917	76,389	68,700	62,418	57,188	52,767	48,980	45,701	42,833	37,024	32,603	29,125	26,318	24,004	22,064	18,[illegible]94	16,674	14,859	12,203	10,352
120	110,304	102,057	94,958	88,782	83,361	78,563	70,454	63,862	58,398	53,795	49,865	46,470	43,508	37,528	32,993	29,436	26,571	24,214	22,242	19,126	16,775	14,939	12,257	10,391
125	114,514	105,651	98,062	91,490	85,743	80,676	72,148	65,251	59,557	54,778	50,708	47,201	44,148	38,003	33,360	29,727	26,808	24,411	22,408	19,248	16,870	15,014	12,307	10,427
130	118,696	109,201	101,113	94,140	88,067	82,730	73,786	66,588	60,669	55,717	51,512	47,897	44,756	38,453	33,706	30,002	27,031	24,596	22,563	19,363	16,958	15,084	12,354	10,461
135	122,851	112,708	104,112	96,734	90,333	84,727	75,371	67,875	61,736	56,615	52,279	48,560	45,334	38,879	34,032	30,260	27,241	24,770	22,709	19,470	17,040	15,149	12,397	10,492
140	126,978	116,172	107,061	99,275	92,545	86,669	76,904	69,117	62,761	57,476	53,012	49,192	45,885	39,283	34,342	30,505	27,439	24,933	22,847	19,571	17,117	15,210	12,438	10,521
145	131,077	119,594	109,960	101,763	94,703	88,560	78,389	70,314	63,747	58,302	53,714	49,795	46,409	39,667	34,635	30,736	27,626	25,087	22,976	19,666	17,190	15,267	12,476	10,548
150	135,150	122,975	112,812	104,201	96,811	90,400	79,827	71,469	64,695	59,094	54,385	50,372	46,910	40,032	34,913	30,954	27,802	25,233	23,098	19,755	17,258	15,321	12,512	10,574
155	139,195	126,315	115,617	106,589	98,870	92,192	81,222	72,584	65,608	59,854	55,029	50,923	47,388	40,379	35,177	31,162	27,969	25,370	23,213	19,840	17,322	15,372	12,546	10,598
160	143,214	129,616	118,376	108,930	100,880	93,938	82,574	73,662	66,487	60,585	55,646	51,451	47,845	40,711	35,428	31,359	28,128	25,501	23,322	19,919	17,383	15,419	12,578	10,621
165	147,207	132,878	121,091	111,225	102,845	95,640	83,886	74,705	67,335	61,289	56,239	51,958	48,282	41,027	35,667	31,546	28,279	25,625	23,426	19,995	17,440	15,465	12,608	10,642
170	151,174	136,102	123,763	113,475	104,766	97,299	85,159	75,713	68,153	61,966	56,808	52,443	48,701	41,329	35,895	31,724	28,422	25,742	23,524	20,066	17,495	15,507	12,636	10,662
175	155,115	139,288	126,392	115,681	106,644	98,916	86,396	76,689	68,943	62,618	57,356	52,910	49,103	41,618	36,113	31,895	28,558	25,854	23,617	20,134	17,546	15,548	12,663	10,682
180	159,031	142,437	128,979	117,845	108,480	100,494	87,597	77,634	69,705	63,246	57,883	53,358	49,489	41,895	36,321	32,057	28,688	25,960	23,706	20,199	17,595	15,586	12,689	10,700
185	162,921	145,550	131,526	119,968	110,276	102,034	88,764	78,549	70,443	63,853	58,390	53,789	49,860	42,160	36,521	32,212	28,812	26,062	23,791	20,260	17,642	15,623	12,713	10,717
190	166,786	148,627	134,034	122,050	112,034	103,537	89,900	79,437	71,156	64,438	58,879	54,204	50,216	42,415	36,711	32,360	28,931	26,159	23,872	20,319	17,686	15,658	12,736	10,733
195	170,627	151,669	136,503	124,094	113,754	105,004	91,004	80,298	71,846	65,003	59,351	54,603	50,558	42,659	36,894	32,502	29,044	26,252	23,949	20,375	17,728	15,691	12,758	10,749
200	174,443	154,677	138,935	126,101	115,437	106,437	92,078	81,133	72,513	65,550	59,806	54,988	50,888	42,893	37,070	32,638	29,153	26,340	24,023	20,428	17,769	15,722	12,779	10,764

350 MCM | 1 Run(s) | 600 Volts | C-Value = 19,703 | Magnetic Duct | Copper

Calculations:

1. $f = \dfrac{1.732 \times \text{Length} \times \text{Isca}}{\text{Runs} \times \text{C-Value} \times E_{L\text{-}L}}$

2. $M = \dfrac{1}{1 + f}$

3. Isc = Isca x M

4. Add Motor Contribution = Motor FLA x 4

* All results are given in symmetrical amperes

Three-Phase Bolted Fault-Current* Table

One-Way Distance in Feet

Available Isc in Thousands (Isca)	5	10	15	20	25	30	40	50	60	70	80	90	100	125	150	175	200	225	250	300	350	400	500	600
3	2,997	2,993	2,990	2,987	2,984	2,980	2,974	2,967	2,961	2,955	2,948	2,942	2,935	2,920	2,904	2,889	2,874	2,859	2,844	2,814	2,786	2,758	2,703	2,651
5	4,991	4,982	4,973	4,964	4,955	4,946	4,928	4,910	4,892	4,875	4,858	4,840	4,823	4,781	4,740	4,699	4,659	4,619	4,581	4,505	4,432	4,361	4,226	4,099
7	6,982	6,964	6,947	6,929	6,911	6,894	6,859	6,825	6,791	6,757	6,724	6,691	6,659	6,578	6,500	6,424	6,349	6,276	6,205	6,067	5,935	5,809	5,572	5,353
10	9,964	9,927	9,891	9,856	9,820	9,785	9,715	9,647	9,579	9,512	9,446	9,381	9,317	9,161	9,010	8,864	8,722	8,585	8,452	8,198	7,959	7,734	7,319	6,947
15	14,918	14,837	14,757	14,677	14,599	14,521	14,368	14,219	14,072	13,929	13,788	13,650	13,515	13,189	12,877	12,581	12,297	12,027	11,767	11,281	10,834	10,420	9,681	9,040
20	19,855	19,711	19,570	19,431	19,293	19,158	18,893	18,635	18,384	18,140	17,902	17,670	17,444	16,904	16,397	15,919	15,468	15,042	14,638	13,893	13,221	12,610	11,544	10,644
25	24,773	24,550	24,332	24,117	23,906	23,698	23,294	22,903	22,525	22,159	21,805	21,462	21,130	20,343	19,612	18,932	18,298	17,705	17,149	16,135	15,235	14,430	13,050	11,911
30	29,674	29,355	29,043	28,737	28,438	28,144	27,576	27,030	26,505	26,000	25,514	25,046	24,595	23,535	22,562	21,667	20,840	20,074	19,362	18,080	16,957	15,965	14,294	12,939
35	34,557	34,125	33,704	33,293	32,892	32,500	31,744	31,023	30,334	29,674	29,043	28,438	27,858	26,505	25,278	24,160	23,136	22,196	21,329	19,783	18,447	17,279	15,338	13,789
40	39,422	38,861	38,316	37,786	37,270	36,768	35,804	34,889	34,019	33,192	32,404	31,653	30,935	29,277	27,787	26,441	25,220	24,107	23,087	21,287	19,748	18,416	16,227	14,503
45	44,270	43,564	42,880	42,217	41,574	40,950	39,758	38,632	37,569	36,563	35,609	34,704	33,844	31,868	30,111	28,537	27,120	25,837	24,670	22,625	20,894	19,408	16,993	15,111
50	49,101	48,233	47,396	46,587	45,806	45,050	43,611	42,261	40,992	39,797	38,669	37,604	36,596	34,297	32,270	30,470	28,859	27,411	26,100	23,823	21,911	20,283	17,659	15,637
55	53,914	52,870	51,866	50,899	49,967	49,069	47,366	45,778	44,293	42,901	41,594	40,364	39,205	36,578	34,282	32,257	30,457	28,848	27,401	24,902	22,820	21,060	18,245	16,094
60	58,710	57,474	56,289	55,152	54,060	53,010	51,029	49,190	47,479	45,883	44,391	42,993	41,680	38,724	36,160	33,914	31,931	30,167	28,588	25,878	23,637	21,754	18,764	16,496
65	63,488	62,046	60,667	59,348	58,086	56,876	54,601	52,501	50,556	48,751	47,070	45,501	44,033	40,747	37,918	35,456	33,294	31,381	29,675	26,766	24,376	22,378	19,226	16,853
70	68,250	66,586	65,000	63,489	62,046	60,667	58,086	55,715	53,530	51,510	49,637	47,896	46,272	42,658	39,567	36,893	34,558	32,501	30,675	27,577	25,047	22,942	19,641	17,171
75	72,995	71,094	69,290	67,575	65,943	64,387	61,487	58,837	56,406	54,168	52,100	50,185	48,406	44,464	41,116	38,237	35,734	33,539	31,599	28,321	25,659	23,455	20,016	17,456
80	77,723	75,571	73,536	71,607	69,777	68,038	64,808	61,871	59,188	56,729	54,465	52,375	50,440	46,175	42,575	39,495	36,831	34,504	32,453	29,005	26,220	23,922	20,355	17,714
85	82,434	80,018	77,739	75,587	73,551	71,621	68,051	64,820	61,881	59,198	56,737	54,473	52,383	47,798	43,950	40,676	37,856	35,402	33,247	29,637	26,735	24,351	20,665	17,948
90	87,128	84,433	81,901	79,515	77,265	75,139	71,219	67,687	64,490	61,580	58,922	56,484	54,240	49,339	45,250	41,787	38,817	36,241	33,985	30,223	27,211	24,745	20,948	18,161
95	91,806	88,819	86,021	83,393	80,921	78,592	74,314	70,477	67,017	63,881	61,025	58,414	56,017	50,805	46,480	42,834	39,718	37,025	34,674	30,767	27,651	25,108	21,207	18,356
100	96,467	93,175	90,100	87,221	84,521	81,983	77,338	73,192	69,467	66,103	63,050	60,267	57,719	52,201	47,646	43,822	40,566	37,761	35,319	31,273	28,059	25,444	21,447	18,535
105	101,111	97,501	94,139	91,001	88,066	85,314	80,296	75,835	71,844	68,252	65,002	62,047	59,350	53,531	48,752	44,756	41,366	38,453	35,923	31,746	28,439	25,756	21,668	18,700
110	105,740	101,797	98,138	94,733	91,556	88,585	83,187	78,409	74,150	70,330	66,884	63,760	60,915	54,801	49,803	45,640	42,120	39,103	36,490	32,188	28,794	26,047	21,873	18,852
115	110,352	106,065	102,098	98,418	94,994	91,800	86,015	80,917	76,389	72,341	68,700	65,408	62,418	56,015	50,803	46,479	42,833	39,717	37,024	32,603	29,125	26,318	22,064	18,994
120	114,948	110,304	106,020	102,057	98,380	94,958	88,782	83,361	78,563	74,288	70,454	66,996	63,862	57,175	51,756	47,275	43,508	40,297	37,528	32,993	29,436	26,571	22,242	19,126
125	119,528	114,514	109,904	105,651	101,715	98,062	91,490	85,743	80,676	76,174	72,148	68,526	65,251	58,286	52,664	48,032	44,148	40,846	38,003	33,360	29,727	26,808	22,408	19,248
130	124,091	118,696	113,751	109,201	105,002	101,113	94,140	88,067	82,730	78,002	73,786	70,002	66,588	59,350	53,532	48,752	44,756	41,366	38,453	33,706	30,002	27,031	22,563	19,363
135	128,639	122,851	117,561	112,708	108,240	104,112	96,734	90,333	84,727	79,775	75,371	71,427	67,875	60,371	54,361	49,439	45,334	41,859	38,879	34,032	30,260	27,241	22,709	19,470
140	133,171	126,978	121,335	116,172	111,430	107,061	99,275	92,545	86,669	81,495	76,904	72,803	69,117	61,351	55,154	50,094	45,885	42,328	39,283	34,342	30,505	27,439	22,847	19,571
145	137,687	131,077	125,072	119,594	114,575	109,960	101,763	94,703	88,560	83,164	78,389	74,132	70,314	62,292	55,914	50,720	46,409	42,774	39,667	34,635	30,736	27,626	22,976	19,666
150	142,188	135,150	128,775	122,975	117,674	112,812	104,201	96,811	90,400	84,785	79,827	75,417	71,469	63,197	56,642	51,318	46,910	43,199	40,032	34,913	30,954	27,802	23,098	19,755
155	146,673	139,195	132,443	126,315	120,730	115,617	106,589	98,870	92,192	86,360	81,222	76,661	72,584	64,068	57,340	51,891	47,388	43,604	40,379	35,177	31,162	27,969	23,213	19,840
160	151,143	143,214	136,076	129,616	123,742	118,376	108,930	100,880	93,938	87,890	82,574	77,864	73,662	64,906	58,011	52,440	47,845	43,990	40,711	35,428	31,359	28,128	23,322	19,919
165	155,597	147,207	139,676	132,878	126,711	121,091	111,225	102,845	95,640	89,378	83,886	79,029	74,705	65,714	58,655	52,966	48,282	44,360	41,027	35,667	31,546	28,279	23,426	19,995
170	160,035	151,174	143,242	136,102	129,639	123,763	113,475	104,766	97,299	90,825	85,159	80,159	75,713	66,493	59,275	53,471	48,701	44,713	41,329	35,895	31,724	28,422	23,524	20,066
175	164,459	155,115	146,776	139,288	132,527	126,392	115,681	106,644	98,916	92,233	86,396	81,253	76,689	67,245	59,871	53,955	49,103	45,052	41,618	36,113	31,895	28,558	23,617	20,134
180	168,867	159,031	150,277	142,437	135,374	128,979	117,845	108,480	100,494	93,603	87,597	82,315	77,634	67,970	60,446	54,421	49,489	45,377	41,895	36,321	32,057	28,688	23,706	20,199
185	173,260	162,921	153,746	145,550	138,183	131,526	119,968	110,276	102,034	94,938	88,764	83,345	78,549	68,671	60,999	54,870	49,860	45,688	42,160	36,521	32,212	28,812	23,791	20,260
190	177,638	166,786	157,184	148,627	140,954	134,034	122,050	112,034	103,537	96,237	89,900	84,345	79,437	69,348	61,533	55,301	50,216	45,987	42,415	36,711	32,360	28,931	23,872	20,319
195	182,001	170,627	160,590	151,669	143,687	136,503	124,094	113,754	105,004	97,504	91,004	85,316	80,298	70,003	62,049	55,717	50,558	46,274	42,659	36,894	32,502	29,044	23,949	20,375
200	186,349	174,443	163,966	154,677	146,384	138,935	126,101	115,437	106,437	98,738	92,078	86,260	81,133	70,637	62,546	56,118	50,888	46,550	42,893	37,070	32,638	29,153	24,023	20,428

350 MCM | 2 Run(s) | 600 Volts | C-Value = 19,703 | Magnetic Duct | Copper

Calculations:

1. $f = \dfrac{1.732 \times \text{Length} \times \text{Isca}}{\text{Runs} \times \text{C-Value} \times \text{E L-L}}$

2. $M = \dfrac{1}{1 + f}$

3. $\text{Isc} = \text{Isca} \times M$

4. Add Motor Contribution = Motor FLA x 4

* All results are given in symmetrical amperes

Three-Phase Bolted Fault-Current* Table

One-Way Distance in Feet

Available Isc in Thousands (Isca)	5	10	15	20	25	30	40	50	60	70	80	90	100	125	150	175	200	225	250	300	350	400	500	600
3	2,994	2,987	2,981	2,975	2,969	2,963	2,950	2,938	2,926	2,914	2,902	2,890	2,879	2,850	2,822	2,794	2,767	2,740	2,714	2,664	2,615	2,568	2,478	2,395
5	4,983	4,965	4,948	4,931	4,914	4,897	4,863	4,830	4,798	4,766	4,734	4,703	4,672	4,597	4,524	4,453	4,385	4,318	4,254	4,130	4,014	3,904	3,701	3,518
7	6,966	6,932	6,898	6,865	6,832	6,800	6,735	6,672	6,610	6,550	6,490	6,431	6,374	6,234	6,101	5,973	5,850	5,733	5,620	5,406	5,209	5,025	4,694	4,404
10	9,930	9,862	9,794	9,727	9,661	9,596	9,468	9,344	9,223	9,105	8,990	8,878	8,769	8,507	8,261	8,028	7,808	7,600	7,402	7,037	6,706	6,404	5,876	5,428
15	14,844	14,691	14,541	14,394	14,250	14,109	13,835	13,571	13,318	13,073	12,838	12,610	12,391	11,875	11,400	10,961	10,555	10,178	9,827	9,153	8,636	8,142	7,307	6,627
20	19,723	19,454	19,192	18,937	18,688	18,446	17,981	17,538	17,117	16,715	16,332	15,966	15,616	14,805	14,074	13,411	12,808	12,257	11,752	10,857	10,088	9,421	8,321	7,450
25	24,569	24,152	23,750	23,360	22,984	22,619	21,923	21,268	20,652	20,070	19,520	18,999	18,506	17,377	16,379	15,488	14,690	13,970	13,317	12,179	11,220	10,401	9,076	8,050
30	29,381	28,788	28,218	27,670	27,143	26,635	25,675	24,782	23,949	23,170	22,440	21,755	21,110	19,654	18,386	17,272	16,285	15,404	14,614	13,255	12,127	11,176	9,660	8,507
35	34,161	33,361	32,598	31,869	31,171	30,504	29,252	28,098	27,032	26,044	25,125	24,269	23,470	21,684	20,150	18,820	17,654	16,624	15,708	14,148	12,870	11,804	10,126	8,866
40	38,908	37,874	36,893	35,962	35,076	34,234	32,664	31,232	29,920	28,714	27,602	26,572	25,617	23,504	21,713	20,176	18,842	17,673	16,641	14,901	13,490	12,323	10,506	9,156
45	43,622	42,326	41,105	39,953	38,863	37,831	35,923	34,199	32,632	31,203	29,894	28,690	27,579	25,146	23,107	21,374	19,882	18,586	17,448	15,544	14,015	12,760	10,822	9,395
50	48,305	46,721	45,238	43,845	42,537	41,303	39,040	37,012	35,184	33,528	32,021	30,644	29,380	26,634	24,357	22,439	20,801	19,386	18,151	16,100	14,466	13,132	11,088	9,595
55	52,956	51,058	49,292	47,644	46,102	44,657	42,023	39,682	37,588	35,705	34,001	32,452	31,038	27,989	25,486	23,394	21,619	20,095	18,771	16,586	14,857	13,454	11,317	9,765
60	57,575	55,339	53,270	51,350	49,564	47,898	44,881	42,221	39,859	37,747	35,847	34,130	32,570	29,229	26,510	24,254	22,351	20,726	19,320	17,013	15,199	13,734	11,514	9,912
65	62,164	59,565	57,175	54,969	52,927	51,032	47,621	44,637	42,005	39,666	37,574	35,692	33,989	30,367	27,443	25,032	23,011	21,291	19,811	17,393	15,501	13,980	11,687	10,040
70	66,722	63,737	61,008	58,503	56,196	54,064	50,250	46,939	44,038	41,474	39,192	37,149	35,308	31,415	28,296	25,740	23,608	21,801	20,252	17,732	15,769	14,198	11,839	10,152
75	71,250	67,856	64,772	61,955	59,373	56,998	52,776	49,136	45,966	43,180	40,712	38,511	36,536	32,384	29,079	26,387	24,151	22,264	20,650	18,036	16,010	14,393	11,974	10,251
80	75,747	71,923	68,467	65,328	62,464	59,841	55,204	51,234	47,796	44,791	42,142	39,788	37,684	33,282	29,802	26,980	24,647	22,685	21,012	18,312	16,226	14,567	12,094	10,339
85	80,215	75,939	72,097	68,624	65,471	62,595	57,539	53,239	49,537	46,317	43,489	40,987	38,757	34,117	30,469	27,526	25,101	23,069	21,342	18,561	16,422	14,725	12,203	10,418
90	84,653	79,905	75,662	71,847	68,398	65,265	59,788	55,159	51,195	47,763	44,762	42,116	39,765	34,895	31,088	28,030	25,520	23,423	21,644	18,789	16,600	14,868	12,301	10,490
95	89,062	83,822	79,165	74,998	71,248	67,855	61,954	56,997	52,775	49,135	45,965	43,179	40,711	35,622	31,664	28,498	25,907	23,748	21,921	18,998	16,763	14,998	12,390	10,554
100	93,442	87,691	82,607	78,080	74,024	70,368	64,042	58,760	54,283	50,439	47,104	44,183	41,603	36,303	32,201	28,932	26,265	24,048	22,177	19,190	16,912	15,118	12,471	10,613
105	97,793	91,512	85,989	81,095	76,728	72,808	66,057	60,451	55,723	51,681	48,185	45,133	42,444	36,942	32,702	29,336	26,598	24,327	22,414	19,367	17,050	15,227	12,546	10,667
110	102,116	95,287	89,314	84,046	79,364	75,177	68,001	62,076	57,100	52,863	49,212	46,032	43,238	37,542	33,172	29,713	26,907	24,586	22,633	19,531	17,176	15,328	12,614	10,717
115	106,411	99,016	92,583	86,934	81,935	77,479	69,879	63,637	58,419	53,992	50,188	46,885	43,990	38,107	33,612	30,066	27,197	24,827	22,837	19,683	17,294	15,422	12,678	10,762
120	110,679	102,701	95,796	89,761	84,441	79,717	71,695	65,139	59,682	55,069	51,118	47,695	44,702	38,641	34,027	30,397	27,467	25,053	23,028	19,824	17,403	15,509	12,736	10,804
125	114,918	106,341	98,956	92,530	86,887	81,893	73,450	66,585	60,894	56,099	52,004	48,466	45,379	39,145	34,417	30,708	27,721	25,264	23,206	19,956	17,504	15,589	12,790	10,844
130	119,131	109,939	102,063	95,241	89,274	84,010	75,148	67,978	62,056	57,084	52,849	49,199	46,021	39,622	34,786	31,001	27,960	25,461	23,373	20,079	17,599	15,664	12,841	10,880
135	123,316	113,493	105,120	97,897	91,604	86,070	76,792	69,320	63,173	58,028	53,657	49,899	46,633	40,075	35,134	31,278	28,184	25,647	23,530	20,195	17,688	15,735	12,888	10,914
140	127,475	117,006	108,127	100,500	93,879	88,075	78,385	70,615	64,247	58,932	54,430	50,566	47,215	40,504	35,463	31,539	28,396	25,823	23,677	20,303	17,771	15,800	12,932	10,945
145	131,607	120,479	111,086	103,051	96,101	90,029	79,928	71,865	65,280	59,800	55,169	51,204	47,771	40,912	35,776	31,785	28,596	25,988	23,816	20,405	17,849	15,862	12,973	10,975
150	135,713	123,910	113,997	105,552	98,272	91,931	81,424	73,072	66,275	60,634	55,878	51,814	48,301	41,301	36,073	32,019	28,785	26,144	23,947	20,501	17,923	15,920	13,012	11,003
155	139,793	127,303	116,862	108,003	100,393	93,785	82,875	74,239	67,233	61,435	56,558	52,398	48,808	41,671	36,355	32,241	28,964	26,292	24,071	20,592	17,992	15,975	13,049	11,029
160	143,847	130,656	119,681	110,407	102,467	95,593	84,284	75,367	68,157	62,205	57,210	52,957	49,293	42,024	36,623	32,452	29,134	26,432	24,188	20,678	18,057	16,026	13,083	11,053
165	147,875	133,971	122,457	112,765	104,495	97,355	85,651	76,458	69,048	62,947	57,837	53,494	49,758	42,361	36,879	32,653	29,296	26,565	24,300	20,759	18,119	16,075	13,116	11,076
170	151,879	137,249	125,190	115,079	106,479	99,075	86,979	77,515	69,908	63,661	58,439	54,009	50,203	42,683	37,123	32,844	29,450	26,691	24,406	20,836	18,178	16,121	13,146	11,098
175	155,857	140,490	127,880	117,348	108,419	100,752	88,269	78,538	70,740	64,350	59,019	54,504	50,630	42,992	37,356	33,027	29,596	26,812	24,506	20,910	18,234	16,165	13,176	11,119
180	159,811	143,694	130,530	119,576	110,317	102,390	89,523	79,529	71,543	65,014	59,577	54,979	51,040	43,287	37,579	33,201	29,736	26,926	24,602	20,979	18,287	16,207	13,203	11,139
185	163,740	146,863	133,139	121,762	112,175	103,989	90,743	80,491	72,320	65,655	60,115	55,437	51,435	43,570	37,792	33,367	29,869	27,036	24,693	21,046	18,337	16,246	13,229	11,157
190	167,645	149,996	135,710	123,908	113,994	105,550	91,930	81,423	73,071	66,274	60,633	55,878	51,814	43,842	37,996	33,526	29,997	27,140	24,780	21,109	18,385	16,284	13,254	11,175
195	171,525	153,095	138,241	126,015	115,776	107,075	93,085	82,328	73,799	66,872	61,134	56,302	52,178	44,103	38,192	33,678	30,119	27,240	24,863	21,169	18,431	16,320	13,278	11,192
200	175,382	156,160	140,736	128,084	117,520	108,565	94,209	83,206	74,504	67,450	61,616	56,711	52,530	44,354	38,380	33,824	30,236	27,335	24,943	21,227	18,474	16,354	13,301	11,208

400 MCM | 1 Run(s) | 600 Volts | C-Value = 20,565 | Magnetic Duct | Copper

Calculations:

1. $f = \dfrac{1.732 \times \text{Length} \times \text{Isca}}{\text{Runs} \times \text{C-Value} \times \text{E L-L}}$

2. $M = \dfrac{1}{1 + f}$

3. $\text{Isc} = \text{Isca} \times M$

4. Add Motor Contribution = Motor FLA x 4

* All results are given in symmetrical amperes

Three-Phase Bolted Fault-Current* Table

One-Way Distance in Feet

Available Isc in Thousands (Isca)

Isca	5	10	15	20	25	30	40	50	60	70	80	90	100	125	150	175	200	225	250	300	350	400	500	600
3	2,997	2,994	2,991	2,987	2,984	2,981	2,975	2,969	2,963	2,956	2,950	2,944	2,938	2,923	2,908	2,893	2,879	2,864	2,850	2,822	2,794	2,767	2,714	2,664
5	4,991	4,983	4,974	4,965	4,957	4,948	4,931	4,914	4,897	4,880	4,863	4,847	4,830	4,790	4,750	4,711	4,672	4,634	4,597	4,524	4,453	4,385	4,254	4,130
7	6,983	6,966	6,949	6,932	6,915	6,898	6,865	6,832	6,800	6,767	6,735	6,704	6,672	6,595	6,520	6,446	6,374	6,303	6,234	6,101	5,973	5,850	5,620	5,406
10	9,965	9,930	9,896	9,862	9,828	9,794	9,727	9,661	9,596	9,532	9,468	9,406	9,344	9,193	9,048	8,906	8,769	8,636	8,507	8,261	8,028	7,808	7,402	7,037
15	14,921	14,844	14,767	14,691	14,615	14,541	14,394	14,250	14,109	13,970	13,835	13,702	13,571	13,256	12,954	12,666	12,391	12,127	11,875	11,400	10,961	10,555	9,827	9,193
20	19,861	19,723	19,588	19,454	19,322	19,192	18,937	18,688	18,446	18,211	17,981	17,757	17,538	17,015	16,521	16,056	15,616	15,200	14,805	14,074	13,411	12,808	11,752	10,857
25	24,783	24,569	24,359	24,152	23,949	23,750	23,360	22,984	22,619	22,265	21,923	21,591	21,268	20,503	19,791	19,127	18,506	17,924	17,377	16,379	15,488	14,690	13,317	12,179
30	29,687	29,381	29,082	28,788	28,500	28,218	27,670	27,143	26,635	26,146	25,675	25,221	24,782	23,749	22,799	21,922	21,110	20,356	19,654	18,386	17,272	16,285	14,614	13,255
35	34,575	34,161	33,756	33,361	32,975	32,598	31,869	31,171	30,504	29,865	29,252	28,663	28,098	26,778	25,576	24,478	23,470	22,541	21,684	20,150	18,820	17,654	15,708	14,148
40	39,446	38,908	38,384	37,874	37,377	36,893	35,962	35,076	34,234	33,430	32,664	31,932	31,232	29,609	28,147	26,822	25,617	24,515	23,504	21,713	20,176	18,842	16,641	14,901
45	44,300	43,622	42,965	42,326	41,707	41,105	39,953	38,863	37,831	36,853	35,923	35,040	34,199	32,263	30,535	28,982	27,579	26,306	25,146	23,107	21,374	19,882	17,448	15,544
50	49,138	48,305	47,500	46,721	45,967	45,238	43,845	42,537	41,303	40,140	39,040	37,999	37,012	34,755	32,757	30,977	29,380	27,940	26,634	24,357	22,439	20,801	18,151	16,100
55	53,959	52,956	51,990	51,058	50,159	49,292	47,644	46,102	44,657	43,300	42,023	40,819	39,682	37,099	34,832	32,826	31,038	29,435	27,989	25,486	23,394	21,619	18,771	16,586
60	58,763	57,575	56,435	55,339	54,285	53,270	51,350	49,564	47,898	46,340	44,881	43,510	42,221	39,309	36,772	34,544	32,570	30,809	29,229	26,510	24,254	22,351	19,320	17,013
65	63,550	62,164	60,837	59,565	58,346	57,175	54,969	52,927	51,032	49,267	47,621	46,080	44,637	41,395	38,592	36,144	33,989	32,076	30,367	27,443	25,032	23,011	19,811	17,393
70	68,322	66,722	65,196	63,737	62,343	61,008	58,503	56,196	54,064	52,087	50,250	48,538	46,939	43,368	40,301	37,639	35,308	33,248	31,415	28,296	25,740	23,608	20,252	17,732
75	73,077	71,250	69,512	67,856	66,278	64,772	61,955	59,373	56,998	54,806	52,776	50,891	49,136	45,236	41,910	39,039	36,536	34,335	32,384	29,079	26,387	24,151	20,650	18,036
80	77,815	75,747	73,786	71,923	70,153	68,467	65,328	62,464	59,841	57,429	55,204	53,145	51,234	47,008	43,426	40,352	37,684	35,346	33,282	29,802	26,980	24,647	21,012	18,312
85	82,538	80,215	78,019	75,939	73,968	72,097	68,624	65,471	62,595	59,961	57,539	55,306	53,239	48,691	44,859	41,585	38,757	36,290	34,117	30,469	27,526	25,101	21,342	18,561
90	87,245	84,653	82,211	79,905	77,726	75,662	71,847	68,398	65,265	62,406	59,788	57,380	55,159	50,291	46,213	42,747	39,765	37,171	34,895	31,088	28,030	25,520	21,644	18,789
95	91,935	89,062	86,363	83,822	81,427	79,165	74,998	71,248	67,855	64,770	61,954	59,372	56,997	51,815	47,497	43,843	40,711	37,997	35,622	31,664	28,498	25,907	21,921	18,998
100	96,610	93,442	90,475	87,691	85,073	82,607	78,080	74,024	70,368	67,056	64,042	61,287	58,760	53,268	48,715	44,879	41,603	38,773	36,303	32,201	28,932	26,265	22,177	19,190
105	101,269	97,793	94,549	91,512	88,665	85,989	81,095	76,728	72,808	69,268	66,057	63,130	60,451	54,654	49,872	45,859	42,444	39,502	36,942	32,702	29,336	26,598	22,414	19,367
110	105,912	102,116	98,584	95,287	92,204	89,314	84,046	79,364	75,177	71,409	68,001	64,904	62,076	55,979	50,972	46,788	43,238	40,189	37,542	33,172	29,713	26,907	22,633	19,531
115	110,539	106,411	102,581	99,016	95,691	92,583	86,934	81,935	77,479	73,483	69,879	66,612	63,637	57,245	52,020	47,669	43,990	40,838	38,107	33,612	30,066	27,197	22,837	19,683
120	115,151	110,679	106,541	102,701	99,128	95,796	89,761	84,441	79,717	75,493	71,695	68,260	65,139	58,458	53,020	48,507	44,702	41,451	38,641	34,027	30,397	27,467	23,028	19,824
125	119,747	114,918	110,464	106,341	102,516	98,956	92,530	86,887	81,893	77,442	73,450	69,849	66,585	59,620	53,974	49,304	45,379	42,032	39,145	34,417	30,708	27,721	23,206	19,956
130	124,328	119,131	114,350	109,939	105,855	102,063	95,241	89,274	84,010	79,332	75,148	71,383	67,978	60,734	54,885	50,064	46,021	42,583	39,622	34,786	31,001	27,960	23,373	20,079
135	128,894	123,316	118,201	113,493	109,146	105,120	97,897	91,604	86,070	81,167	76,792	72,865	69,320	61,803	55,757	50,788	46,633	43,106	40,075	35,134	31,278	28,184	23,530	20,195
140	133,444	127,475	122,016	117,006	112,392	108,127	100,500	93,879	88,075	82,948	78,385	74,297	70,615	62,830	56,592	51,480	47,215	43,603	40,504	35,463	31,539	28,396	23,677	20,303
145	137,979	131,607	125,797	120,479	115,592	111,086	103,051	96,101	90,029	84,678	79,928	75,682	71,865	63,818	57,392	52,141	47,771	44,076	40,912	35,776	31,785	28,596	23,816	20,405
150	142,499	135,713	129,543	123,910	118,747	113,997	105,552	98,272	91,931	86,359	81,424	77,022	73,072	64,768	58,159	52,774	48,301	44,527	41,301	36,073	32,019	28,785	23,947	20,501
155	147,004	139,793	133,256	127,303	121,859	116,862	108,003	100,393	93,785	87,993	82,875	78,320	74,239	65,683	58,896	53,379	48,808	44,958	41,671	36,355	32,241	28,964	24,071	20,592
160	151,494	143,847	136,935	130,656	124,928	119,681	110,407	102,467	95,593	89,583	84,284	79,576	75,367	66,565	59,603	53,960	49,293	45,369	42,024	36,623	32,452	29,134	24,188	20,678
165	155,969	147,875	140,580	133,971	127,956	122,457	112,765	104,495	97,355	91,129	85,651	80,794	76,458	67,415	60,284	54,517	49,758	45,762	42,361	36,879	32,653	29,296	24,300	20,759
170	160,429	151,879	144,194	137,249	130,942	125,190	115,079	106,479	99,075	92,633	86,979	81,975	77,515	68,234	60,939	55,052	50,203	46,139	42,683	37,123	32,844	29,450	24,406	20,836
175	164,875	155,857	147,775	140,490	133,889	127,880	117,348	108,419	100,752	94,098	88,269	83,120	78,538	69,026	61,569	55,566	50,630	46,499	42,992	37,356	33,027	29,596	24,506	20,910
180	169,306	159,811	151,325	143,694	136,796	130,530	119,576	110,317	102,390	95,525	89,523	84,231	79,529	69,791	62,177	56,061	51,040	46,845	43,287	37,579	33,201	29,736	24,602	20,979
185	173,722	163,740	154,843	146,863	139,665	133,139	121,762	112,175	103,989	96,915	90,743	85,310	80,491	70,530	62,763	56,537	51,435	47,177	43,570	37,792	33,367	29,869	24,693	21,046
190	178,124	167,645	158,330	149,996	142,496	135,710	123,908	113,994	105,550	98,270	91,930	86,358	81,423	71,245	63,328	56,995	51,814	47,496	43,842	37,996	33,526	29,997	24,780	21,109
195	182,511	171,525	161,787	153,095	145,290	138,241	126,015	115,776	107,075	99,591	93,085	87,376	82,328	71,936	63,874	57,437	52,178	47,802	44,103	38,192	33,678	30,119	24,863	21,169
200	186,884	175,382	165,214	156,160	148,047	140,736	128,084	117,520	108,565	100,879	94,209	88,366	83,206	72,606	64,401	57,863	52,530	48,097	44,354	38,380	33,824	30,236	24,943	21,227

400 MCM	2 Run(s)	600 Volts	C-Value = 20,565	Magnetic Duct	Copper

Calculations:

1. $f = \dfrac{1.732 \times \text{Length} \times \text{Isca}}{\text{Runs} \times \text{C-Value} \times E_{L\text{-}L}}$

2. $M = \dfrac{1}{1 + f}$

3. $\text{Isc} = \text{Isca} \times M$

4. Add Motor Contribution = Motor FLA x 4

* All results are given in symmetrical amperes

Three-Phase Bolted Fault-Current* Table

One-Way Distance in Feet

Available Isc in Thousands (Isca)	5	10	15	20	25	30	40	50	60	70	80	90	100	125	150	175	200	225	250	300	350	400	500	600
3	2,994	2,988	2,983	2,977	2,971	2,965	2,954	2,943	2,931	2,920	2,909	2,898	2,887	2,860	2,834	2,808	2,783	2,758	2,733	2,68[illegible]	2,639	2,595	2,510	2,431
5	4,984	4,968	4,952	4,936	4,920	4,904	4,873	4,842	4,812	4,782	4,753	4,723	4,695	4,624	4,555	4,489	4,424	4,362	4,301	4,18[illegible]	4,073	3,968	3,773	3,596
7	6,968	6,937	6,906	6,875	6,844	6,814	6,754	6,695	6,637	6,580	6,525	6,470	6,416	6,284	6,159	6,038	5,921	5,809	5,702	5,49[illegible]	5,308	5,131	4,810	4,526
10	9,935	9,872	9,809	9,746	9,685	9,624	9,505	9,389	9,276	9,165	9,057	8,952	8,849	8,601	8,367	8,145	7,935	7,735	7,545	7,19[illegible]	6,871	6,577	6,058	5,616
15	14,855	14,713	14,573	14,436	14,302	14,170	13,914	13,666	13,428	13,197	12,974	12,759	12,550	12,058	11,603	11,181	10,789	10,423	10,081	9,46[illegible]	8,912	8,424	7,592	6,909
20	19,743	19,493	19,249	19,011	18,778	18,552	18,114	17,697	17,299	16,918	16,554	16,205	15,870	15,091	14,385	13,742	13,154	12,614	12,117	11,2[illegible]1	10,467	9,799	8,691	7,808
25	24,600	24,212	23,837	23,473	23,120	22,777	22,122	21,503	20,917	20,363	19,838	19,338	18,864	17,773	16,802	15,931	15,146	14,435	13,787	12,6[illegible]3	11,690	10,864	9,518	8,469
30	29,426	28,873	28,341	27,827	27,333	26,855	25,948	25,101	24,307	23,562	22,861	22,201	21,577	20,162	18,921	17,824	16,847	15,972	15,183	13,8[illegible]8	12,678	11,712	10,163	8,976
35	34,221	33,475	32,762	32,078	31,422	30,793	29,607	28,508	27,489	26,540	25,654	24,825	24,048	22,303	20,795	19,477	18,317	17,287	16,366	14,7[illegible]1	13,493	12,404	10,680	9,377
40	38,985	38,021	37,103	36,229	35,395	34,598	33,107	31,740	30,481	29,318	28,241	27,240	26,308	24,234	22,463	20,933	19,599	18,424	17,382	15,6[illegible]6	14,176	12,979	11,104	9,702
45	43,720	42,511	41,367	40,283	39,254	38,276	36,461	34,809	33,301	31,918	30,645	29,470	28,382	25,983	23,958	22,226	20,727	19,418	18,264	16,325	14,757	13,464	11,457	9,971
50	48,425	46,946	45,554	44,243	43,005	41,835	39,675	37,727	35,962	34,355	32,885	31,535	30,292	27,575	25,305	23,381	21,728	20,294	19,037	16,939	15,258	13,880	11,757	10,197
55	53,100	51,327	49,668	48,114	46,653	45,279	42,760	40,506	38,478	36,643	34,976	33,453	32,058	29,030	26,526	24,419	22,622	21,071	19,719	17,477	15,693	14,239	12,013	10,389
60	57,746	55,655	53,710	51,897	50,202	48,614	45,722	43,154	40,860	38,797	36,933	35,239	33,694	30,366	27,636	25,357	23,425	21,766	20,327	17,553	16,075	14,553	12,236	10,555
65	62,363	59,931	57,682	55,596	53,655	51,845	48,569	45,682	43,119	40,828	38,769	36,907	35,216	31,596	28,651	26,209	24,150	22,391	20,871	18,[illegible]76	16,413	14,830	12,431	10,700
70	66,951	64,156	61,586	59,213	57,017	54,978	51,307	48,096	45,264	42,746	40,494	38,467	36,633	32,733	29,583	26,986	24,808	22,956	21,361	18,[illegible]54	16,715	15,075	12,603	10,828
75	71,511	68,332	65,423	62,752	60,291	58,015	53,943	50,405	47,303	44,560	42,118	39,930	37,958	33,786	30,440	27,698	25,408	23,469	21,804	19,[illegible]95	16,985	15,295	12,756	10,940
80	76,042	72,458	69,196	66,215	63,480	60,962	56,482	52,615	49,244	46,279	43,650	41,304	39,198	34,765	31,233	28,352	25,958	23,937	22,208	19,[illegible]04	17,229	15,493	12,893	11,041
85	80,546	76,535	72,905	69,604	66,588	63,823	58,930	54,733	51,094	47,909	45,098	42,598	40,361	35,677	31,967	28,956	26,463	24,366	22,576	19,[illegible]85	17,450	15,671	13,017	11,131
90	85,022	80,565	76,553	72,921	69,618	66,602	6[illegible],290	56,763	52,859	49,458	46,467	43,818	41,454	36,529	32,649	29,514	26,929	24,760	22,914	19,[illegible]42	17,652	15,833	13,129	11,213
95	89,470	84,549	80,141	76,169	72,573	69,301	63,569	58,712	54,545	50,931	47,765	44,970	42,484	37,326	33,285	30,033	27,360	25,124	23,226	20,[illegible]77	17,836	15,981	13,230	11,287
100	93,892	88,486	83,670	79,350	75,455	71,924	65,769	60,584	56,157	52,333	48,997	46,060	43,456	38,074	33,878	30,515	27,760	25,460	23,513	20,[illegible]93	18,005	16,117	13,323	11,355
105	98,286	92,379	87,142	82,466	78,267	74,475	67,895	62,384	57,700	53,671	50,167	47,093	44,374	38,777	34,433	30,965	28,131	25,773	23,779	20,593	18,160	16,241	13,408	11,416
110	102,654	96,227	90,558	85,519	81,012	76,956	69,951	64,116	59,179	54,948	51,281	48,073	45,243	39,439	34,955	31,386	28,478	26,064	24,027	20,[illegible]79	18,304	16,356	13,486	11,473
115	106,995	100,032	93,919	88,511	83,692	79,370	71,941	65,783	60,596	56,167	52,342	49,005	46,067	40,063	35,444	31,780	28,802	26,335	24,257	20,[illegible]51	18,438	16,463	13,558	11,525
120	111,310	103,794	97,228	91,444	86,309	81,720	73,866	67,389	61,956	57,334	53,354	49,890	46,849	40,654	35,905	32,150	29,106	26,589	24,472	21,111	18,562	16,562	13,625	11,574
125	115,599	107,513	100,485	94,319	88,866	84,009	75,731	68,937	63,263	58,451	54,320	50,734	47,592	41,212	36,340	32,499	29,391	26,827	24,673	21,261	18,677	16,654	13,688	11,618
130	119,862	111,192	103,691	97,138	91,364	86,238	77,537	70,431	64,519	59,522	55,243	51,539	48,300	41,741	36,751	32,827	29,660	27,050	24,862	21,401	18,785	16,739	13,745	11,660
135	124,100	114,829	106,847	99,902	93,805	88,410	79,289	71,874	65,727	60,549	56,127	52,307	48,974	42,244	37,140	33,137	29,912	27,260	25,040	21,532	18,886	16,820	13,800	11,699
140	128,313	118,427	109,955	102,614	96,193	90,527	80,988	73,267	66,890	61,534	56,973	53,041	49,616	42,721	37,509	33,430	30,151	27,458	25,206	21,655	18,981	16,895	13,850	11,735
145	132,500	121,985	113,016	105,275	98,527	92,592	82,636	74,613	68,010	62,481	57,783	53,743	50,230	43,175	37,858	33,707	30,377	27,645	25,364	21,771	19,070	16,965	13,897	11,769
150	136,663	125,504	116,030	107,886	100,810	94,606	84,236	75,915	69,091	63,392	58,561	54,415	50,817	43,608	38,191	33,970	30,590	27,822	25,513	21,881	19,154	17,032	13,942	11,801
155	140,801	128,986	119,000	110,449	103,044	96,570	85,790	77,175	70,133	64,268	59,308	55,059	51,378	44,021	38,507	34,220	30,793	27,989	25,653	21,984	19,233	17,094	13,984	11,831
160	144,915	132,430	121,925	112,964	105,230	98,488	87,300	78,395	71,139	65,112	60,026	55,677	51,916	44,415	38,808	34,458	30,985	28,148	25,787	22,082	19,308	17,153	14,023	11,859
165	149,005	135,837	124,807	115,434	107,370	100,360	88,768	79,577	72,110	65,925	60,716	56,271	52,432	44,792	39,096	34,685	31,168	28,299	25,913	22,175	19,379	17,209	14,061	11,886
170	153,070	139,207	127,647	117,859	109,465	102,188	90,195	80,722	73,049	66,708	61,381	56,841	52,926	45,153	39,370	34,900	31,342	28,442	26,034	22,263	19,446	17,262	14,096	11,911
175	157,112	142,542	130,445	120,241	111,517	103,974	91,583	81,832	73,957	67,465	62,020	57,389	53,401	45,498	39,632	35,106	31,508	28,579	26,148	2[illegible],347	19,510	17,313	14,130	11,935
180	161,131	145,842	133,203	122,580	113,527	105,718	92,934	82,909	74,836	68,195	62,637	57,917	53,858	45,829	39,883	35,303	31,666	28,709	26,257	2[illegible],426	19,571	17,360	14,161	11,958
185	165,126	149,107	135,922	124,879	115,495	107,424	94,250	83,954	75,686	68,901	63,232	58,425	54,297	46,146	40,123	35,491	31,818	28,833	26,361	2[illegible],502	19,628	17,406	14,192	11,979
190	169,098	152,338	138,602	127,137	117,425	109,091	95,530	84,969	76,510	69,583	63,806	58,914	54,720	46,451	40,354	35,67[illegible]	31,962	28,952	26,460	2[illegible],574	19,683	17,449	14,220	12,000
195	173,046	155,536	141,243	129,357	119,315	110,721	96,778	85,954	77,308	70,242	64,360	59,387	55,127	46,744	40,575	35,844	32,101	29,066	26,555	2[illegible],643	19,736	17,490	14,248	12,019
200	176,973	158,700	143,848	131,538	121,169	112,315	97,994	86,912	78,082	70,880	64,895	59,842	55,519	47,026	40,787	36,009	32,234	29,174	26,646	2[illegible],709	19,786	17,529	14,274	12,038

500 MCM | **1 Run(s)** | **600 Volts**

C-Value = 22,185 | **Magnetic Duct** | **Copper**

Calculations:

1. $f = \dfrac{1.732 \times \text{Length} \times \text{Isca}}{\text{Runs} \times \text{C-Value} \times E_{L-L}}$
2. $M = \dfrac{1}{1 + f}$
3. $\text{Isc} = \text{Isca} \times M$
4. Add Motor Contribution = Motor FLA x 4

* All results are given in symmetrical amperes

Three-Phase Bolted Fault-Current* Table

One-Way Distance in Feet

Available Isc in Thousands (Isca)	5	10	15	20	25	30	40	50	60	70	80	90	100	125	150	175	200	225	250	300	350	400	500	600
3	2,997	2,994	2,991	2,988	2,985	2,983	2,977	2,971	2,965	2,960	2,954	2,948	2,943	2,929	2,915	2,901	2,887	2,874	2,860	2,834	2,808	2,783	2,733	2,686
5	4,992	4,984	4,976	4,968	4,960	4,952	4,936	4,920	4,904	4,889	4,873	4,858	4,842	4,805	4,767	4,731	4,695	4,659	4,624	4,555	4,489	4,424	4,301	4,183
7	6,984	6,968	6,953	6,937	6,921	6,906	6,875	6,844	6,814	6,784	6,754	6,724	6,695	6,623	6,552	6,483	6,416	6,349	6,284	6,159	6,038	5,921	5,702	5,498
10	9,968	9,935	9,903	9,872	9,840	9,809	9,746	9,685	9,624	9,564	9,505	9,447	9,389	9,248	9,111	8,978	8,349	8,723	8,601	8,367	8,145	7,935	7,545	7,192
15	14,927	14,855	14,784	14,713	14,643	14,573	14,436	14,302	14,170	14,041	13,914	13,789	13,666	13,369	13,085	12,812	12 550	12,299	12,058	11,603	11,181	10,789	10,081	9,461
20	19,871	19,743	19,617	19,493	19,370	19,249	19,011	18,778	18,552	18,330	18,114	17,903	17,697	17,202	16,734	16,291	15,870	15,471	15,091	14,385	13,742	13,154	12,117	11,231
25	24,798	24,600	24,405	24,212	24,023	23,837	23,473	23,120	22,777	22,445	22,122	21,808	21,503	20,776	20,097	19,461	18,864	18,302	17,773	16,802	15,931	15,146	13,787	12,653
30	29,710	29,426	29,147	28,873	28,604	28,341	27,827	27,333	26,855	26,394	25,948	25,518	25,101	24,116	23,206	22,362	21,577	20,846	20,162	18,921	17,824	16,847	15,183	13,818
35	34,606	34,221	33,844	33,475	33,115	32,762	32,078	31,422	30,793	30,188	29,607	29,047	28,508	27,245	26,089	25,027	24,048	23,143	22,303	20,795	19,477	18,317	16,366	14,791
40	39,486	38,985	38,497	38,021	37,557	37,103	36,229	35,395	34,598	33,836	33,107	32,409	31,740	30,182	28,770	27,484	26,308	25,228	24,234	22,463	20,933	19,599	17,382	15,616
45	44,351	43,720	43,107	42,511	41,931	41,367	40,283	39,254	38,276	37,346	36,461	35,616	34,809	32,944	31,268	29,755	28,382	27,129	25,983	23,958	22,226	20,727	18,264	16,325
50	49,200	48,425	47,674	46,946	46,240	45,554	44,243	43,005	41,835	40,726	39,675	38,677	37,727	35,546	33,603	31,862	30,292	28,870	27,575	25,305	23,381	21,728	19,037	16,939
55	54,033	53,100	52,198	51,327	50,484	49,668	48,114	46,653	45,279	43,983	42,760	41,602	40,506	38,002	35,790	33,821	32,058	30,469	29,030	26,526	24,419	22,622	19,719	17,477
60	58,851	57,746	56,681	55,655	54,665	53,710	51,897	50,202	48,614	47,124	45,722	44,401	43,154	40,324	37,842	35,648	33,694	31,944	30,366	27,636	25,357	23,425	20,327	17,953
65	63,654	62,363	61,123	59,931	58,785	57,682	55,596	53,655	51,845	50,154	48,569	47,081	45,682	42,522	39,772	37,355	35,216	33,308	31,596	28,651	26,209	24,150	20,871	18,376
70	68,442	66,951	65,524	64,156	62,845	61,586	59,213	57,017	54,978	53,079	51,307	49,650	48,096	44,607	41,589	38,954	36,633	34,573	32,733	29,583	26,986	24,808	21,361	18,754
75	73,214	71,511	69,885	68,332	66,846	65,423	62,752	60,291	58,015	55,905	53,943	52,114	50,405	46,586	43,305	40,455	37,958	35,751	33,786	30,440	27,698	25,408	21,804	19,095
80	77,971	76,042	74,207	72,458	70,789	69,196	66,215	63,480	60,962	58,637	56,482	54,480	52,615	48,468	44,926	41,867	39,198	36,848	34,765	31,233	28,352	25,958	22,208	19,404
85	82,713	80,546	78,489	76,535	74,676	72,905	69,604	66,588	63,823	61,279	58,930	56,754	54,733	50,259	46,461	43,196	40,361	37,875	35,677	31,967	28,956	26,463	22,576	19,685
90	87,440	85,022	82,734	80,565	78,508	76,553	72,921	69,618	66,602	63,836	61,290	58,940	56,763	51,966	47,916	44,451	41,454	38,836	36,529	32,649	29,514	26,929	22,914	19,942
95	92,152	89,470	86,940	84,549	82,286	80,141	76,169	72,573	69,301	66,311	63,569	61,044	58,712	53,594	49,297	45,638	42,484	39,738	37,326	33,285	30,033	27,360	23,226	20,177
100	96,850	93,892	91,109	88,486	86,011	83 670	79,350	75,455	71,924	68,709	65,769	63,070	60,584	55,150	50,610	46,761	43,456	40,587	38,074	33,878	30,515	27,760	23,513	20,393
105	101,532	98,286	95,241	92,379	89,684	87,142	82,466	78,267	74,475	71,033	67,895	65,023	62,384	56,637	51,860	47,826	44,374	41,387	38,777	34,433	30,965	28,131	23,779	20,593
110	106,200	102,654	99,336	96,227	93,306	90,558	85,519	81,012	76,956	73,287	69,951	66,907	64,116	58,061	53,051	48,837	45,243	42,142	39,439	34,955	31,386	28,478	24,027	20,779
115	110,853	106,995	103,396	100,032	96,879	93,919	88,511	83,692	79,370	75,473	71,941	68,724	65,783	59,425	54,187	49,798	46,067	42 856	40,063	35,444	31,780	28,802	24,257	20,951
120	115,492	111,310	107,420	103,794	100,404	97,228	91,444	86,309	81,720	77,595	73,866	70,479	67,389	60,732	55,273	50,713	46,849	43,532	40,654	35,905	32,150	29,106	24,472	21,111
125	120,116	115,599	111,410	107,513	103,880	100,485	94,319	88,866	84,009	79,655	75,731	72,175	68,937	61,987	56,310	51,585	47,592	44,173	41,212	36,340	32,499	29,391	24,673	21,261
130	124,726	119,862	115,364	111,192	107,310	103 691	97,138	91,364	86,238	81,656	77,537	73,814	70,431	63,192	57,303	52,417	48,300	44,782	41,741	36,751	32,827	29,660	24,862	21,401
135	129,321	124,100	119,285	114,829	110,694	106,847	99,902	93,805	88,410	83,601	79,289	75,399	71,874	64,351	58,254	53,212	48,974	45,360	42,244	37,140	33,137	29,912	25,040	21,532
140	133,902	128,313	123,172	118,427	114,034	109,955	102,614	96,193	90,527	85,492	80,988	76,934	73,267	65,465	59,166	53,972	49,616	45,911	42,721	37,509	33,430	30,151	25,206	21,655
145	138,469	132,500	127,025	121,985	117,329	113,016	105,275	98,527	92,592	87,331	82,636	78,420	74,613	66,538	60,041	54,699	50,230	46,436	43,175	37,858	33,707	30,377	25,364	21,771
150	143,021	136,663	130,846	125,504	120,582	116,030	107,886	100,810	94,606	89,120	84,236	79,860	75,915	67,572	60,881	55,396	50,817	46,937	43,608	38,191	33,970	30,590	25,513	21,881
155	147,560	140,801	134,635	128,986	123,792	119,000	110,449	103,044	96,570	90,862	85,790	81,255	77,175	68,568	61,689	56,063	51,373	47,416	44,021	38,507	34,220	30,793	25,653	21,984
160	152,084	144,915	138,391	132,430	126,960	121,925	112,964	105,230	98,488	92,557	87,300	82,608	78,395	69,530	62,465	56,704	51,913	47,874	44,415	38,808	34,458	30,985	25,787	22,082
165	156,595	149,005	142,116	135,837	130,088	124,807	115,434	107,370	100,360	94,209	88,768	83,921	79,577	70,457	63,213	57,320	52,432	48,312	44,792	39,096	34,685	31,168	25,913	22,175
170	161,092	153,070	145,810	139,207	133,177	127,647	117,859	109,465	102,188	95,818	90,195	85,196	80,722	71,354	63,934	57,912	52,926	48,731	45,153	39,370	34,900	31,342	26,034	22,263
175	165,574	157,112	149,473	142,542	136,226	130,445	120,241	111,517	103,974	97,386	91,583	86,433	81,832	72,220	64,628	58,481	53,401	49,134	45,498	39,632	35,106	31,508	26,148	22,347
180	170,043	161,131	153,106	145,842	139,236	133,203	122,580	113,527	105,718	98,915	92,934	87,636	82,909	73,057	65,298	59,029	53,858	49,520	45,829	39,883	35,303	31,666	26,257	22,426
185	174,499	165,126	156,708	149,107	142,209	135,922	124,879	115,495	107,424	100,406	94,250	88,804	83,954	73,867	65,945	59,557	54,297	49,891	46,146	40,123	35,491	31,818	26,361	22,502
190	178,940	169,098	160,281	152,338	145,146	138,602	127,137	117,425	109,091	101,861	95,530	89,940	84,969	74,652	66,569	60,066	54,720	50,248	46,451	40,354	35,671	31,962	26,460	22,574
195	183,368	173,046	163,825	155,536	148,046	141,243	129,357	119,315	110,721	103,281	96,778	91,046	85,954	75,412	67,172	60,556	55,127	50,591	46,744	40,575	35,844	32,101	26,555	22,643
200	187,783	176,973	167,339	158,700	150,910	143,848	131,538	121,169	112,315	104,667	97,994	92,121	86,912	76,148	67,756	61,030	55,519	50,921	47,026	40,787	36,009	32,234	26,646	22,709

500 MCM	2 Run(s)	600 Volts	C-Value = 22,185	Magnetic Duct	Copper

Calculations:

1. $f = \dfrac{1.732 \times \text{Length} \times \text{Isca}}{\text{Runs} \times \text{C-Value} \times E_{L\text{-}L}}$
2. $M = \dfrac{1}{1 + f}$
3. Isc = Isca x M
4. Add Motor Contribution = Motor FLA x 4

* All results are given in symmetrical amperes

Three-Phase Bolted Fault-Current* Table

One-Way Distance in Feet

Available Isc in Thousands (Isca)	5	10	15	20	25	30	40	50	60	70	80	90	100	125	150	175	200	225	250	300	350	400	500	600
3	2,994	2,989	2,983	2,978	2,972	2,966	2,955	2,944	2,934	2,923	2,912	2,902	2,891	2,865	2,839	2,814	2,790	2,765	2,742	2,695	2,650	2,607	2,524	2,446
5	4,984	4,969	4,953	4,938	4,923	4,907	4,877	4,848	4,818	4,789	4,761	4,732	4,704	4,636	4,569	4,505	4,442	4,381	4,321	4,207	4,098	3,996	3,804	3,631
7	6,969	6,939	6,909	6,879	6,849	6,820	6,762	6,705	6,649	6,594	6,540	6,486	6,434	6,306	6,184	6,066	5,952	5,843	5,738	5,538	5,352	5,178	4,861	4,581
10	9,938	9,876	9,815	9,755	9,695	9,637	9,521	9,409	9,299	9,191	9,086	8,984	8,883	8,642	8,414	8,197	7,991	7,795	7,609	7,262	6,945	6,654	6,141	5,701
15	14,860	14,722	14,587	14,455	14,325	14,197	13,948	13,708	13,476	13,251	13,034	12,824	12,620	12,139	11,693	11,279	10,892	10,532	10,195	9,581	9,037	8,551	7,721	7,038
20	19,752	19,510	19,273	19,043	18,817	18,597	18,173	17,767	17,379	17,007	16,651	16,310	15,982	15,218	14,523	13,889	13,309	12,774	12,281	11,401	10,639	9,972	8,861	7,973
25	24,613	24,238	23,875	23,522	23,179	22,846	22,208	21,605	21,034	20,492	19,978	19,488	19,022	17,949	16,991	16,130	15,352	14,645	14,001	12,868	11,906	11,077	9,723	8,664
30	29,445	28,910	28,394	27,896	27,415	26,951	26,068	25,241	24,465	23,735	23,047	22,398	21,785	20,389	19,161	18,073	17,102	16,230	15,442	14,076	12,932	11,960	10,397	9,195
35	34,247	33,525	32,833	32,169	31,532	30,919	29,762	28,689	27,691	26,759	25,888	25,073	24,306	22,582	21,085	19,775	18,618	17,589	16,668	15,087	13,781	12,682	10,938	9,616
40	39,019	38,085	37,195	36,345	35,533	34,757	33,302	31,964	30,730	29,587	28,526	27,538	26,617	24,563	22,803	21,278	19,944	18,768	17,723	15,947	14,494	13,284	11,383	9,958
45	43,762	42,591	41,481	40,427	39,425	38,472	36,697	35,079	33,597	32,236	30,981	29,820	28,742	26,361	24,345	22,614	21,114	19,800	18,640	16,686	15,102	13,793	11,755	10,242
50	48,477	47,043	45,692	44,417	43,211	42,068	39,955	38,045	36,308	34,724	33,271	31,936	30,703	28,002	25,737	23,811	22,153	20,712	19,446	17,328	15,626	14,229	12,070	10,480
55	53,162	51,443	49,832	48,319	46,895	45,552	43,085	40,872	38,875	37,064	35,414	33,904	32,519	29,504	27,000	24,889	23,083	21,522	20,159	17,892	16,083	14,607	12,341	10,684
60	57,820	55,792	53,902	52,136	50,482	48,929	46,094	43,570	41,308	39,269	37,422	35,740	34,204	30,884	28,152	25,864	23,920	22,248	20,794	18,390	16,485	14,937	12,576	10,859
65	62,449	60,090	57,904	55,870	53,975	52,204	48,989	46,148	43,618	41,350	39,307	37,457	35,772	32,158	29,206	26,751	24,677	22,901	21,363	18,834	16,841	15,229	12,782	11,013
70	67,050	64,339	61,838	59,525	57,378	55,381	51,777	48,613	45,814	43,319	41,082	39,065	37,236	33,336	30,175	27,561	25,364	23,492	21,877	19,233	17,158	15,488	12,964	11,148
75	71,624	68,539	65,708	63,102	60,695	58,465	54,462	50,973	47,904	45,183	42,755	40,574	38,605	34,429	31,067	28,304	25,992	24,030	22,342	19,591	17,444	15,720	13,126	11,267
80	76,170	72,690	69,515	66,605	63,929	61,459	57,052	53,234	49,895	46,951	44,334	41,994	39,889	35,446	31,893	28,988	26,568	24,521	22,766	19,917	17,701	15,929	13,272	11,374
85	80,689	76,795	73,259	70,035	67,082	64,368	59,550	55,403	51,796	48,630	45,828	43,332	41,094	36,394	32,659	29,619	27,097	24,971	23,154	20,213	17,934	16,118	13,402	11,470
90	85,182	80,853	76,943	73,394	70,158	67,195	61,962	57,484	53,611	50,226	47,243	44,595	42,228	37,281	33,371	30,204	27,586	25,385	23,510	20,483	18,147	16,289	13,521	11,557
95	89,647	84,866	80,569	76,685	73,159	69,943	64,291	59,484	55,346	51,746	48,586	45,789	43,297	38,112	34,035	30,747	28,038	25,768	23,837	20,731	18,342	16,446	13,629	11,635
100	94,087	88,834	84,136	79,911	76,089	72,617	66,543	61,406	57,006	53,195	49,861	46,920	44,307	38,892	34,656	31,253	28,458	26,122	24,140	20,960	18,520	16,589	13,727	11,707
105	98,500	92,758	87,648	83,072	78,950	75,218	68,720	63,256	58,597	54,577	51,073	47,992	45,262	39,626	35,238	31,725	28,849	26,451	24,421	21,171	18,685	16,722	13,817	11,773
110	102,887	96,638	91,105	86,171	81,744	77,749	70,827	65,037	60,122	55,898	52,228	49,011	46,166	40,317	35,784	32,167	29,214	26,757	24,682	21,367	18,838	16,843	13,900	11,833
115	107,248	100,476	94,508	89,209	84,473	80,214	72,867	66,753	61,586	57,161	53,329	49,979	47,025	40,970	36,297	32,581	29,555	27,043	24,925	21,549	18,979	16,956	13,977	11,889
120	111,584	104,272	97,859	92,189	87,140	82,615	74,843	68,408	62,991	58,370	54,380	50,900	47,840	41,588	36,781	32,970	29,875	27,311	25,152	21,719	19,110	17,061	14,048	11,940
125	115,895	108,027	101,158	95,112	89,747	84,955	76,758	70,004	64,342	59,528	55,384	51,779	48,615	42,172	37,237	33,336	30,175	27,562	25,365	21,877	19,233	17,159	14,114	11,988
130	120,181	111,741	104,408	97,979	92,295	87,235	78,615	71,545	65,642	60,638	56,344	52,617	49,353	42,727	37,669	33,682	30,458	27,797	25,564	22,026	19,347	17,250	14,176	12,032
135	124,442	115,415	107,609	100,792	94,788	89,459	80,416	73,034	66,893	61,704	57,263	53,418	50,057	43,253	38,078	34,008	30,725	28,019	25,752	22,165	19,455	17,335	14,234	12,073
140	128,678	119,050	110,762	103,554	97,226	91,627	82,164	74,472	68,098	62,728	58,144	54,184	50,729	43,754	38,465	34,317	30,976	28,229	25,929	22,295	19,555	17,415	14,287	12,112
145	132,890	122,646	113,869	106,264	99,611	93,743	83,861	75,864	69,259	63,713	58,989	54,917	51,371	44,230	38,833	34,610	31,215	28,426	26,095	22,418	19,650	17,490	14,338	12,148
150	137,077	126,204	116,930	108,925	101,946	95,807	85,509	77,211	70,380	64,660	59,799	55,619	51,984	44,685	39,183	34,887	31,440	28,613	26,253	22,535	19,739	17,560	14,385	12,182
155	141,241	129,725	119,946	111,538	104,231	97,823	87,111	78,514	71,462	65,572	60,579	56,292	52,572	45,118	39,516	35,151	31,654	28,790	26,402	22,644	19,823	17,627	14,430	12,214
160	145,381	133,209	122,918	114,104	106,468	99,791	88,669	79,777	72,506	66,450	61,328	56,938	53,135	45,533	39,833	35,402	31,858	28,959	26,543	22,748	19,903	17,690	14,472	12,245
165	149,497	136,657	125,848	116,624	108,659	101,713	90,183	81,001	73,516	67,297	62,048	57,559	53,676	45,929	40,136	35,641	32,051	29,118	26,677	22,847	19,978	17,749	14,512	12,273
170	153,590	140,069	128,736	119,100	110,806	103,591	91,657	82,188	74,492	68,114	62,742	58,156	54,194	46,308	40,425	35,869	32,235	29,270	26,805	22,940	20,049	17,806	14,549	12,300
175	157,660	143,446	131,583	121,532	112,908	105,427	93,091	83,339	75,436	68,903	63,411	58,730	54,692	46,671	40,702	36,083	32,411	29,415	26,926	23,029	20,117	17,859	14,585	12,325
180	161,706	146,788	134,390	123,923	114,969	107,221	94,487	84,456	76,351	69,665	64,056	59,282	55,171	47,019	40,966	36,294	32,578	29,553	27,042	23,113	20,182	17,910	14,619	12,350
185	165,730	150,096	137,158	126,273	116,988	108,976	95,847	85,541	77,236	70,401	64,678	59,815	55,632	47,354	41,220	36,493	32,738	29,685	27,152	23,194	20,243	17,958	14,651	12,372
190	169,732	153,371	139,887	128,582	118,968	110,692	97,171	86,594	78,094	71,113	65,278	60,328	56,076	47,675	41,463	36,683	32,892	29,810	27,257	23,271	20,301	18,004	14,682	12,394
195	173,711	156,612	142,578	130,853	120,909	112,370	98,463	87,618	78,926	71,802	65,858	60,823	56,503	47,983	41,696	36,865	33,038	29,931	27,358	23,344	20,357	18,048	14,711	12,415
200	177,667	159,821	145,233	133,085	122,813	114,013	99,721	88,614	79,733	72,470	66,419	61,301	56,916	48,280	41,920	37,041	33,179	30,046	27,454	23,414	20,410	18,090	14,739	12,435

600 MCM	1 Run(s)	600 Volts	C-Value = 22,965	Magnetic Duct	Copper

Calculations:

1. $f = \dfrac{1.732 \times \text{Length} \times \text{Isca}}{\text{Runs} \times \text{C-Value} \times \text{E L-L}}$

2. $M = \dfrac{1}{1 + f}$

3. Isc = Isca x M

4. Add Motor Contribution = Motor FLA x 4

* All results are given in symmetrical amperes

Three-Phase Bolted Fault-Current* Table

One-Way Distance in Feet

Available Isc in Thousands (Isca)	5	10	15	20	25	30	40	50	60	70	80	90	100	125	150	175	200	225	250	300	350	400	500	600
3	2,997	2,994	2,992	2,989	2,986	2,983	2,978	2,972	2,966	2,961	2,955	2,950	2,944	2,931	2,917	2,904	2,891	2,878	2,865	2,839	2,814	2,790	2,742	2,695
5	4,992	4,984	4,977	4,969	4,961	4,953	4,938	4,923	4,907	4,892	4,877	4,862	4,848	4,811	4,775	4,739	4,704	4,670	4,636	4,569	4,505	4,442	4,321	4,207
7	6,985	6,969	6,954	6,939	6,924	6,909	6,879	6,849	6,820	6,791	6,762	6,733	6,705	6,635	6,567	6,500	6,434	6,369	6,306	6,184	6,066	5,952	5,738	5,538
10	9,969	9,938	9,907	9,876	9,845	9,815	9,755	9,695	9,637	9,579	9,521	9,465	9,409	9,272	9,138	9,009	8,883	8,761	8,642	8,414	8,197	7,991	7,609	7,262
15	14,930	14,860	14,791	14,722	14,655	14,587	14,455	14,325	14,197	14,071	13,948	13,827	13,708	13,419	13,142	12,876	12,620	12,375	12,139	11,693	11,279	10,892	10,195	9,581
20	19,875	19,752	19,630	19,510	19,391	19,273	19,043	18,817	18,597	18,383	18,173	17,967	17,767	17,284	16,827	16,394	15,982	15,591	15,218	14,523	13,889	13,309	12,281	11,401
25	24,805	24,613	24,424	24,238	24,055	23,875	23,522	23,179	22,846	22,523	22,208	21,903	21,605	20,896	20,232	19,608	19,022	18,470	17,949	16,991	16,130	15,352	14,001	12,868
30	29,720	29,445	29,175	28,910	28,650	28,394	27,896	27,415	26,951	26,502	26,068	25,648	25,241	24,278	23,386	22,557	21,785	21,064	20,389	19,161	18,073	17,102	15,442	14,076
35	34,619	34,247	33,882	33,525	33,176	32,833	32,169	31,532	30,919	30,330	29,762	29,216	28,689	27,452	26,317	25,272	24,306	23,412	22,582	21,085	19,775	18,618	16,668	15,087
40	39,503	39,019	38,546	38,085	37,635	37,195	36,345	35,533	34,757	34,014	33,302	32,620	31,964	30,436	29,047	27,779	26,617	25,549	24,563	22,803	21,278	19,944	17,723	15,947
45	44,373	43,762	43,169	42,591	42,028	41,481	40,427	39,425	38,472	37,563	36,697	35,870	35,079	33,246	31,596	30,102	28,742	27,500	26,361	24,345	22,614	21,114	18,640	16,686
50	49,227	48,477	47,749	47,043	46,358	45,692	44,417	43,211	42,068	40,985	39,955	38,977	38,045	35,899	33,982	32,259	30,703	29,290	28,002	25,737	23,811	22,153	19,446	17,328
55	54,066	53,162	52,289	51,443	50,625	49,832	48,319	46,895	45,552	44,284	43,085	41,949	40,872	38,405	36,220	34,270	32,519	30,938	29,504	27,000	24,889	23,083	20,159	17,892
60	58,890	57,820	56,788	55,792	54,831	53,902	52,136	50,482	48,929	47,470	46,094	44,797	43,570	40,778	38,323	36,146	34,204	32,459	30,884	28,152	25,864	23,920	20,794	18,390
65	63,699	62,449	61,247	60,090	58,977	57,904	55,870	53,975	52,204	50,546	48,989	47,526	46,148	43,028	40,303	37,903	35,772	33,869	32,158	29,206	26,751	24,677	21,363	18,834
70	68,493	67,050	65,667	64,339	63,064	61,838	59,525	57,378	55,381	53,518	51,777	50,145	48,613	45,163	42,171	39,550	37,236	35,178	33,336	30,175	27,561	25,364	21,877	19,233
75	73,273	71,624	70,047	68,539	67,094	65,708	63,102	60,695	58,465	56,393	54,462	52,660	50,973	47,193	43,935	41,098	38,605	36,397	34,429	31,067	28,304	25,992	22,342	19,591
80	78,038	76,170	74,390	72,690	71,067	69,515	66,605	63,929	61,459	59,174	57,052	55,077	53,234	49,125	45,605	42,556	39,889	37,536	35,446	31,893	28,988	26,568	22,766	19,917
85	82,789	80,689	78,694	76,795	74,985	73,259	70,035	67,082	64,368	61,865	59,550	57,402	55,403	50,966	47,187	43,930	41,094	38,601	36,394	32,659	29,619	27,097	23,154	20,213
90	87,525	85,182	82,961	80,853	78,850	76,943	73,394	70,158	67,195	64,472	61,962	59,639	57,484	52,722	48,689	45,229	42,228	39,601	37,281	33,371	30,204	27,586	23,510	20,483
95	92,246	89,647	87,191	84,866	82,661	80,569	76,685	73,159	69,943	66,998	64,291	61,794	59,484	54,400	50,116	46,458	43,297	40,539	38,112	34,035	30,747	28,038	23,837	20,731
100	96,953	94,087	91,385	88,834	86,421	84,136	79,911	76,089	72,617	69,447	66,543	63,872	61,406	56,003	51,474	47,622	44,307	41,423	38,892	34,656	31,253	28,458	24,140	20,960
105	101,646	98,500	95,542	92,758	90,130	87,648	83,072	78,950	75,218	71,822	68,720	65,875	63,256	57,538	52,767	48,727	45,262	42,257	39,626	35,238	31,725	28,849	24,421	21,171
110	106,325	102,887	99,665	96,638	93,790	91,105	86,171	81,744	77,749	74,127	70,827	67,809	65,037	59,007	54,001	49,777	46,166	43,044	40,317	35,784	32,167	29,214	24,682	21,367
115	110,989	107,248	103,752	100,476	97,401	94,508	89,209	84,473	80,214	76,364	72,867	69,676	66,753	60,416	55,178	50,776	47,025	43,789	40,970	36,297	32,581	29,555	24,925	21,549
120	115,639	111,584	107,804	104,272	100,964	97,859	92,189	87,140	82,615	78,537	74,843	71,481	68,408	61,768	56,304	51,728	47,840	44,495	41,588	36,781	32,970	29,875	25,152	21,719
125	120,275	115,895	111,823	108,027	104,480	101,158	95,112	89,747	84,955	80,649	76,758	73,226	70,004	63,067	57,381	52,635	48,315	45,165	42,172	37,237	33,336	30,175	25,365	21,877
130	124,898	120,181	115,807	111,741	107,950	104,408	97,979	92,295	87,235	82,701	78,615	74,913	71,545	64,315	58,412	53,502	49,353	45,801	42,727	37,669	33,682	30,458	25,564	22,026
135	129,506	124,442	119,758	115,415	111,375	107,609	100,792	94,788	89,459	84,697	80,416	76,547	73,034	65,515	59,401	54,330	50,057	46,407	43,253	38,078	34,008	30,725	25,752	22,165
140	134,100	128,678	123,677	119,050	114,757	110,762	103,554	97,226	91,627	86,638	82,164	78,129	74,472	66,671	60,349	55,122	50,729	46,984	43,754	38,465	34,317	30,976	25,929	22,295
145	138,681	132,890	127,563	122,646	118,095	113,869	106,264	99,611	93,743	88,527	83,861	79,662	75,864	67,784	61,260	55,881	51,371	47,534	44,230	38,833	34,610	31,215	26,095	22,418
150	143,248	137,077	131,416	126,204	121,390	116,930	108,925	101,946	95,807	90,366	85,509	81,148	77,211	68,857	62,135	56,608	51,984	48,059	44,685	39,183	34,887	31,440	26,253	22,535
155	147,801	141,241	135,238	129,725	124,644	119,946	111,538	104,231	97,823	92,157	87,111	82,590	78,514	69,892	62,976	57,306	52,572	48,561	45,118	39,516	35,151	31,654	26,402	22,644
160	152,340	145,381	139,029	133,209	127,857	122,918	114,104	106,468	99,791	93,902	88,669	83,988	79,777	70,891	63,786	57,976	53,135	49,041	45,533	39,833	35,402	31,858	26,543	22,748
165	156,866	149,497	142,789	136,657	131,030	125,848	116,624	108,659	101,713	95,602	90,183	85,346	81,001	71,856	64,566	58,619	53,676	49,501	45,929	40,136	35,641	32,051	26,677	22,847
170	161,379	153,590	146,518	140,069	134,164	128,736	119,100	110,806	103,591	97,259	91,657	86,664	82,188	72,788	65,318	59,238	54,194	49,941	46,308	40,425	35,869	32,235	26,805	22,940
175	165,878	157,660	150,217	143,446	137,259	131,583	121,532	112,908	105,427	98,875	93,091	87,945	83,339	73,690	66,043	59,834	54,692	50,364	46,671	40,702	36,086	32,411	26,926	23,029
180	170,364	161,706	153,887	146,788	140,316	134,390	123,923	114,969	107,221	100,452	94,487	89,190	84,456	74,562	66,743	60,408	55,171	50,770	47,019	40,966	36,294	32,578	27,042	23,113
185	174,836	165,730	157,526	150,096	143,336	137,158	126,273	116,988	108,976	101,990	95,847	90,401	85,541	75,406	67,418	60,961	55,632	51,160	47,354	41,220	36,493	32,738	27,152	23,194
190	179,295	169,732	161,137	153,371	146,319	139,887	128,582	118,968	110,692	103,492	97,171	91,579	86,594	76,223	68,071	61,494	56,076	51,535	47,675	41,463	36,683	32,892	27,257	23,271
195	183,741	173,711	164,719	156,612	149,266	142,578	130,853	120,909	112,370	104,958	98,463	92,724	87,618	77,016	68,702	62,008	56,503	51,896	47,983	41,696	36,866	33,038	27,358	23,344
200	188,173	177,667	168,273	159,821	152,178	145,233	133,085	122,813	114,013	106,389	99,721	93,840	88,614	77,784	69,313	62,505	56,916	52,244	48,280	41,920	37,041	33,179	27,454	23,414

600 MCM	2 Run(s)	600 Volts	C-Value = 22,965	Magnetic Duct	Copper

Calculations:

1. $f = \dfrac{1.732 \times \text{Length} \times \text{Isca}}{\text{Runs} \times \text{C-Value} \times E_{L\text{-}L}}$

2. $M = \dfrac{1}{1 + f}$

3. $\text{Isc} = \text{Isca} \times M$

4. Add Motor Contribution = Motor FLA x 4

* All results are given in symmetrical amperes

Three-Phase Bolted Fault-Current* Table

One-Way Distance in Feet

Available Isc in Thousands (Isca)	5	10	15	20	25	30	40	50	60	70	80	90	100	125	150	175	200	225	250	300	350	400	500	600
3	2,995	2,989	2,984	2,979	2,973	2,968	2,958	2,947	2,937	2,926	2,916	2,906	2,896	2,871	2,847	2,823	2,799	2,776	2,753	2,708	2,665	2,623	2,544	2,469
5	4,985	4,970	4,956	4,941	4,926	4,912	4,883	4,855	4,827	4,799	4,772	4,745	4,718	4,652	4,588	4,526	4,466	4,407	4,350	4,239	4,135	4,035	3,849	3,680
7	6,971	6,942	6,913	6,885	6,856	6,828	6,773	6,719	6,665	6,612	6,561	6,510	6,459	6,337	6,219	6,105	5,996	5,890	5,788	5,595	5,414	5,244	4,934	4,659
10	9,941	9,882	9,824	9,766	9,710	9,654	9,543	9,436	9,330	9,227	9,127	9,028	8,932	8,699	8,479	8,269	8,070	7,880	7,698	7,359	7,049	6,764	6,258	5,822
15	14,867	14,736	14,607	14,480	14,356	14,234	13,996	13,765	13,542	13,326	13,117	12,915	12,718	12,252	11,819	11,416	11,039	10,686	10,356	9,752	9,214	8,733	7,907	7,224
20	19,764	19,533	19,307	19,087	18,871	18,661	18,254	17,864	17,490	17,132	16,788	16,457	16,139	15,396	14,719	14,098	13,528	13,002	12,516	11,344	10,886	10,221	9,107	8,213
25	24,632	24,274	23,927	23,589	23,261	22,942	22,329	21,749	21,197	20,673	20,174	19,699	19,246	18,198	17,259	16,412	15,645	14,945	14,306	13,179	12,216	11,384	10,020	8,948
30	29,471	28,961	28,468	27,991	27,531	27,085	26,235	25,437	24,686	23,978	23,309	22,677	22,078	20,711	19,503	18,429	17,466	16,599	15,814	14,448	13,299	12,319	10,737	9,515
35	34,282	33,594	32,932	32,296	31,684	31,095	29,980	28,942	27,974	27,068	26,220	25,422	24,672	22,977	21,500	20,201	19,051	18,024	17,102	15,516	14,198	13,087	11,316	9,967
40	39,066	38,174	37,322	36,507	35,727	34,980	33,575	32,279	31,079	29,965	28,928	27,961	27,056	25,031	23,288	21,772	20,442	19,264	18,215	16,426	14,957	13,729	11,792	10,335
45	43,821	42,702	41,639	40,627	39,663	38,744	37,029	35,458	34,016	32,686	31,456	30,316	29,255	26,902	24,899	23,174	21,672	20,353	19,186	17,211	15,605	14,273	12,192	10,640
50	48,548	47,179	45,884	44,659	43,497	42,394	40,349	38,491	36,797	35,246	33,820	32,506	31,289	28,612	26,357	24,432	22,769	21,317	20,040	17,895	16,166	14,741	12,531	10,898
55	53,249	51,605	50,061	48,605	47,233	45,935	43,543	41,388	39,436	37,659	36,036	34,547	33,176	30,182	27,684	25,568	23,752	22,177	20,798	18,497	16,655	15,147	12,823	11,118
60	57,922	55,983	54,169	52,470	50,873	49,371	46,619	44,157	41,942	39,938	38,117	36,456	34,932	31,629	28,896	26,598	24,639	22,948	21,475	19,031	17,086	15,502	13,078	11,309
65	62,568	60,311	58,212	56,254	54,423	52,708	49,582	46,806	44,325	42,094	40,076	38,243	36,570	32,966	30,008	27,537	25,442	23,644	22,083	19,507	17,469	15,817	13,301	11,475
70	67,188	64,592	62,190	59,960	57,885	55,948	52,439	49,344	46,595	44,135	41,922	39,921	38,101	34,205	31,031	28,396	26,174	24,274	22,632	19,934	17,811	16,096	13,498	11,622
75	71,781	68,826	66,106	63,592	61,262	59,097	55,196	51,778	48,758	46,072	43,666	41,498	39,536	35,357	31,976	29,186	26,843	24,849	23,130	20,320	18,118	16,347	13,674	11,752
80	76,348	73,014	69,959	67,150	64,558	62,158	57,857	54,113	50,823	47,911	45,314	42,985	40,883	36,430	32,852	29,913	27,457	25,374	23,585	20,670	18,396	16,573	13,831	11,868
85	80,888	77,156	73,753	70,638	67,775	65,135	60,428	56,355	52,796	49,661	46,876	44,388	42,150	37,433	33,665	30,586	28,023	25,857	24,001	20,989	18,648	16,777	13,973	11,973
90	85,404	81,254	77,489	74,057	70,916	68,031	62,912	58,510	54,683	51,326	48,358	45,714	43,344	38,371	34,422	31,210	28,546	26,301	24,384	21,281	18,878	16,963	14,102	12,067
95	89,893	85,307	81,167	77,409	73,985	70,850	65,315	60,583	56,490	52,915	49,765	46,970	44,471	39,252	35,129	31,790	29,031	26,712	24,736	21,549	19,089	17,133	14,219	12,153
100	94,357	89,318	84,789	80,697	76,982	73,594	67,641	62,578	58,221	54,431	51,104	48,160	45,537	40,080	35,791	32,331	29,481	27,093	25,063	21,796	19,283	17,289	14,327	12,231
105	98,797	93,285	88,356	83,922	79,912	76,267	69,892	64,500	59,881	55,879	52,378	49,291	46,547	40,860	36,412	32,837	29,901	27,447	25,365	22,025	19,461	17,433	14,425	12,303
110	103,211	97,211	91,870	87,086	82,775	78,871	72,073	66,353	61,474	57,264	53,594	50,365	47,504	41,596	36,995	33,310	30,293	27,777	25,647	22,237	19,627	17,565	14,516	12,368
115	107,600	101,095	95,332	90,190	85,575	81,409	74,186	68,140	63,005	58,590	54,754	51,388	48,413	42,291	37,544	33,755	30,660	28,085	25,910	22,434	19,780	17,688	14,599	12,429
120	111,965	104,939	98,743	93,237	88,313	83,883	76,235	69,865	64,477	59,861	55,862	52,363	49,277	42,949	38,061	34,172	31,005	28,374	26,155	22,618	19,923	17,802	14,677	12,485
125	116,306	108,743	102,103	96,228	90,992	86,296	78,223	71,531	65,894	61,080	56,922	53,294	50,100	43,573	38,550	34,566	31,328	28,645	26,385	22,789	20,056	17,908	14,749	12,538
130	120,623	112,507	105,415	99,164	93,613	88,650	80,152	73,141	67,257	62,250	57,936	54,182	50,885	44,165	39,013	34,938	31,633	28,900	26,601	22,950	20,181	18,008	14,816	12,586
135	124,916	116,233	108,679	102,047	96,178	90,947	82,025	74,697	68,571	63,374	58,909	55,031	51,633	44,728	39,452	35,289	31,921	29,140	26,804	23,101	20,297	18,100	14,879	12,631
140	129,185	119,921	111,896	104,878	98,689	93,189	83,844	76,203	69,838	64,454	59,841	55,845	52,348	45,263	39,868	35,622	32,193	29,366	26,996	23,243	20,407	18,187	14,938	12,674
145	133,430	123,570	115,067	107,659	101,147	95,378	85,612	77,661	71,060	65,494	60,737	56,623	53,032	45,774	40,263	35,937	32,450	29,580	27,176	23,377	20,510	18,269	14,993	12,713
150	137,653	127,183	118,194	110,392	103,555	97,517	87,331	79,072	72,240	66,495	61,597	57,370	53,686	46,261	40,639	36,236	32,694	29,783	27,347	23,504	20,607	18,346	15,045	12,751
155	141,852	130,760	121,277	113,076	105,914	99,605	89,003	80,440	73,380	67,460	62,423	58,087	54,314	46,725	40,998	36,521	32,925	29,975	27,509	23,623	20,699	18,419	15,094	12,786
160	146,028	134,300	124,316	115,714	108,225	101,647	90,629	81,766	74,482	68,390	63,219	58,775	54,915	47,170	41,339	36,792	33,145	30,157	27,663	23,736	20,786	18,488	15,140	12,819
165	150,182	137,805	127,314	118,307	110,490	103,642	92,212	83,052	75,548	69,288	63,985	59,437	55,492	47,595	41,666	37,053	33,355	30,330	27,808	23,843	20,868	18,553	15,184	12,850
170	154,313	141,276	130,270	120,855	112,710	105,593	93,753	84,300	76,579	70,154	64,723	60,073	56,046	48,002	41,977	37,296	33,554	30,495	27,947	23,945	20,946	18,614	15,225	12,880
175	158,421	144,712	133,186	123,361	114,886	107,500	95,254	85,512	77,578	70,991	65,435	60,686	56,579	48,393	42,276	37,531	33,745	30,652	28,079	24,042	21,020	18,673	15,264	12,908
180	162,508	148,114	136,063	125,825	117,020	109,367	96,716	86,688	78,545	71,800	66,122	61,276	57,092	48,767	42,561	37,756	33,926	30,802	28,204	24,134	21,090	18,728	15,301	12,934
185	166,572	151,483	138,900	128,248	119,113	111,193	98,141	87,832	79,482	72,583	66,785	61,845	57,586	49,127	42,835	37,972	34,100	30,945	28,324	24,222	21,157	18,781	15,336	12,959
190	170,615	154,819	141,700	130,631	121,166	112,980	99,531	88,943	80,391	73,340	67,426	62,394	58,061	49,473	43,098	38,178	34,266	31,082	28,439	24,305	21,221	18,831	15,370	12,983
195	174,636	158,123	144,463	132,975	123,180	114,729	100,886	90,024	81,273	74,073	68,045	62,924	58,520	49,805	43,350	38,376	34,426	31,213	28,548	24,385	21,282	18,879	15,402	13,006
200	178,635	161,394	147,189	135,281	125,156	116,442	102,208	91,075	82,129	74,783	68,644	63,436	58,962	50,125	43,592	38,565	34,578	31,338	28,653	24,462	21,340	18,925	15,432	13,028

750 MCM	1 Run(s)	600 Volts		C-Value = 24,136	Magnetic Duct	Copper

Calculations:

1. $f = \dfrac{1.732 \times \text{Length} \times \text{Isca}}{\text{Runs} \times \text{C-Value} \times E_{L\text{-}L}}$

2. $M = \dfrac{1}{1 + f}$

3. $\text{Isc} = \text{Isca} \times M$

4. Add Motor Contribution = Motor FLA x 4

* All results are given in symmetrical amperes

Three-Phase Bolted Fault-Current* Table

One-Way Distance in Feet

Available Isc in Thousands (Isca)	5	10	15	20	25	30	40	50	60	70	80	90	100	125	150	175	200	225	250	300	350	400	500	600
3	2,997	2,995	2,992	2,989	2,987	2,984	2,979	2,973	2,968	2,963	2,958	2,952	2,947	2,934	2,921	2,909	2,896	2,884	2,871	2,847	2,823	2,799	2,753	2,708
5	4,993	4,985	4,978	4,970	4,963	4,956	4,941	4,926	4,912	4,897	4,883	4,869	4,855	4,820	4,785	4,751	4,718	4,685	4,652	4,588	4,526	4,466	4,350	4,239
7	6,985	6,971	6,956	6,942	6,928	6,913	6,885	6,856	6,828	6,801	6,773	6,746	6,719	6,652	6,586	6,522	6,459	6,397	6,337	6,219	6,105	5,996	5,788	5,595
10	9,970	9,941	9,911	9,882	9,853	9,824	9,766	9,710	9,654	9,598	9,543	9,489	9,436	9,304	9,177	9,053	8,932	8,814	8,699	8,479	8,269	8,070	7,698	7,359
15	14,933	14,867	14,801	14,736	14,671	14,607	14,480	14,356	14,234	14,114	13,996	13,880	13,765	13,488	13,221	12,965	12,718	12,481	12,252	11,819	11,416	11,039	10,356	9,752
20	19,881	19,764	19,648	19,533	19,419	19,307	19,087	18,871	18,661	18,455	18,254	18,056	17,864	17,399	16,958	16,538	16,139	15,759	15,396	14,719	14,098	13,528	12,516	11,644
25	24,815	24,632	24,452	24,274	24,099	23,927	23,589	23,261	22,942	22,632	22,329	22,035	21,749	21,064	20,421	19,816	19,246	18,707	18,198	17,259	16,412	15,645	14,306	13,179
30	29,733	29,471	29,214	28,961	28,712	28,468	27,991	27,531	27,085	26,653	26,235	25,830	25,437	24,505	23,639	22,832	22,078	21,373	20,711	19,503	18,429	17,466	15,814	14,448
35	34,638	34,282	33,935	33,594	33,260	32,932	32,296	31,684	31,095	30,527	29,980	29,452	28,942	27,742	26,637	25,617	24,672	23,795	22,977	21,500	20,201	19,051	17,102	15,516
40	39,527	39,066	38,615	38,174	37,743	37,322	36,507	35,727	34,980	34,263	33,575	32,914	32,279	30,793	29,438	28,197	27,056	26,004	25,031	23,288	21,772	20,442	18,215	16,426
45	44,403	43,821	43,254	42,702	42,163	41,639	40,627	39,663	38,744	37,867	37,029	36,226	35,458	33,673	32,059	30,593	29,255	28,029	26,902	24,899	23,174	21,672	19,186	17,211
50	49,264	48,548	47,854	47,179	46,522	45,884	44,659	43,497	42,394	41,346	40,349	39,398	38,491	36,397	34,518	32,825	31,289	29,891	28,612	26,357	24,432	22,769	20,040	17,895
55	54,110	53,249	52,414	51,605	50,821	50,061	48,605	47,233	45,935	44,707	43,543	42,438	41,388	38,976	36,830	34,908	33,176	31,609	30,182	27,684	25,568	23,752	20,798	18,497
60	58,943	57,922	56,936	55,983	55,061	54,169	52,470	50,873	49,371	47,955	46,619	45,354	44,157	41,422	39,007	36,857	34,932	33,199	31,629	28,896	26,598	24,639	21,475	19,031
65	63,761	62,568	61,419	60,311	59,243	58,212	56,254	54,423	52,708	51,097	49,582	48,154	46,806	43,745	41,060	38,685	36,570	34,675	32,966	30,008	27,537	25,442	22,083	19,507
70	68,565	67,188	65,864	64,592	63,368	62,190	59,960	57,885	55,948	54,137	52,439	50,845	49,344	45,954	43,000	40,403	38,101	36,048	34,205	31,031	28,396	26,174	22,632	19,934
75	73,355	71,781	70,272	68,826	67,438	66,106	63,592	61,262	59,097	57,080	55,196	53,432	51,778	48,058	44,836	42,020	39,536	37,330	35,357	31,976	29,186	26,843	23,130	20,320
80	78,131	76,348	74,644	73,014	71,454	69,959	67,150	64,558	62,158	59,930	57,857	55,922	54,113	50,063	46,577	43,545	40,883	38,528	36,430	32,852	29,913	27,457	23,585	20,670
85	82,893	80,888	78,978	77,156	75,416	73,753	70,638	67,775	65,135	62,693	60,428	58,320	56,355	51,976	48,228	44,985	42,150	39,652	37,433	33,665	30,586	28,023	24,001	20,989
90	87,642	85,404	83,277	81,254	79,327	77,489	74,057	70,916	68,031	65,372	62,912	60,631	58,510	53,804	49,798	46,348	43,344	40,706	38,371	34,422	31,210	28,546	24,384	21,281
95	92,376	89,893	87,540	85,307	83,186	81,167	77,409	73,985	70,850	67,970	65,315	62,860	60,583	55,551	51,292	47,639	44,471	41,699	39,252	35,129	31,790	29,031	24,736	21,549
100	97,097	94,357	91,768	89,318	86,994	84,789	80,697	76,982	73,594	70,492	67,641	65,011	62,578	57,225	52,715	48,864	45,537	42,635	40,080	35,791	32,331	29,481	25,063	21,796
105	101,804	98,797	95,962	93,285	90,754	88,356	83,922	79,912	76,267	72,940	69,892	67,088	64,500	58,828	54,072	50,028	46,547	43,518	40,860	36,412	32,837	29,901	25,365	22,025
110	106,497	103,211	100,121	97,211	94,465	91,870	87,086	82,775	78,871	75,319	72,073	69,095	66,353	60,365	55,368	51,135	47,504	44,354	41,596	36,995	33,310	30,293	25,647	22,237
115	111,177	107,600	104,246	101,095	98,129	95,332	90,190	85,575	81,409	77,630	74,186	71,035	68,140	61,840	56,607	52,190	48,413	45,145	42,291	37,544	33,755	30,660	25,910	22,434
120	115,844	111,965	108,338	104,939	101,747	98,743	93,237	88,313	83,883	79,876	76,235	72,911	69,865	63,258	57,792	53,196	49,277	45,896	42,949	38,061	34,172	31,005	26,155	22,618
125	120,496	116,306	112,397	108,743	105,319	102,103	96,228	90,992	86,296	82,061	78,223	74,727	71,531	64,620	58,928	54,157	50,100	46,609	43,573	38,550	34,566	31,328	26,385	22,789
130	125,136	120,623	116,424	112,507	108,846	105,415	99,164	93,613	88,650	84,187	80,152	76,486	73,141	65,931	60,016	55,074	50,885	47,287	44,165	39,013	34,938	31,633	26,601	22,950
135	129,762	124,916	120,418	116,233	112,329	108,679	102,047	96,178	90,947	86,256	82,025	78,190	74,697	67,193	61,060	55,952	51,633	47,933	44,728	39,452	35,289	31,921	26,804	23,101
140	134,375	129,185	124,380	119,921	115,769	111,896	104,878	98,689	93,189	88,270	83,844	79,841	76,203	68,409	62,062	56,793	52,348	48,549	45,263	39,868	35,622	32,193	26,996	23,243
145	138,975	133,430	128,311	123,570	119,167	115,067	107,659	101,147	95,378	90,232	85,612	81,443	77,661	69,582	63,026	57,599	53,032	49,136	45,774	40,263	35,937	32,450	27,176	23,377
150	143,561	137,653	132,211	127,183	122,524	118,194	110,392	103,555	97,517	92,143	87,331	82,997	79,072	70,713	63,952	58,371	53,686	49,698	46,261	40,639	36,236	32,694	27,347	23,504
155	148,135	141,852	136,080	130,760	125,840	121,277	113,076	105,914	99,605	94,006	89,003	84,505	80,440	71,805	64,844	59,113	54,314	50,235	46,725	40,998	36,521	32,925	27,509	23,623
160	152,695	146,028	139,919	134,300	129,116	124,316	115,714	108,225	101,647	95,822	90,629	85,970	81,766	72,860	65,703	59,826	54,915	50,749	47,170	41,339	36,792	33,145	27,663	23,736
165	157,242	150,182	143,728	137,805	132,352	127,314	118,307	110,490	103,642	97,593	92,212	87,393	83,052	73,879	66,531	60,512	55,492	51,241	47,595	41,666	37,050	33,355	27,808	23,843
170	161,777	154,313	147,507	141,276	135,550	130,270	120,855	112,710	105,593	99,321	93,753	88,776	84,300	74,865	67,329	61,172	56,046	51,713	48,002	41,977	37,296	33,554	27,947	23,945
175	166,298	158,421	151,257	144,712	138,710	133,186	123,361	114,886	107,500	101,007	95,254	90,120	85,512	75,819	68,100	61,807	56,579	52,167	48,393	42,276	37,531	33,745	28,079	24,042
180	170,807	162,508	154,977	148,114	141,833	136,063	125,825	117,020	109,367	102,653	96,716	91,428	86,688	76,743	68,844	62,420	57,092	52,602	48,767	42,561	37,756	33,926	28,204	24,134
185	175,303	166,572	158,670	151,483	144,919	138,900	128,248	119,113	111,193	104,260	98,141	92,701	87,832	77,637	69,563	63,010	57,586	53,021	49,127	42,835	37,972	34,100	28,324	24,222
190	179,786	170,615	162,333	154,819	147,969	141,700	130,631	121,166	112,980	105,830	99,531	93,939	88,943	78,504	70,258	63,580	58,061	53,424	49,473	43,098	38,178	34,266	28,439	24,305
195	184,257	174,636	165,969	158,123	150,984	144,463	132,975	123,180	114,729	107,363	100,886	95,146	90,024	79,345	70,931	64,130	58,520	53,812	49,805	43,350	38,376	34,426	28,548	24,385
200	188,715	178,635	169,578	161,394	153,965	147,189	135,281	125,156	116,442	108,861	102,208	96,321	91,075	80,160	71,582	64,662	58,962	54,186	50,125	43,592	38,565	34,578	28,653	24,462

750 MCM | **2 Run(s)** | **600 Volts** | **C-Value = 24,136** | **Magnetic Duct** | **Copper**

Calculations:

1. $f = \dfrac{1.732 \times \text{Length} \times \text{Isca}}{\text{Runs} \times \text{C-Value} \times E_{L\text{-}L}}$

2. $M = \dfrac{1}{1 + f}$

3. $\text{Isc} = \text{Isca} \times M$

4. Add Motor Contribution = Motor FLA x 4

* All results are given in symmetrical amperes

Three-Phase Bolted Fault-Current* Table

One-Way Distance in Feet

Available Isc in Thousands (Isca)

	5	10	15	20	25	30	40	50	60	70	80	90	100	125	150	175	200	225	250	300	350	400	500	600
3	2,995	2,990	2,985	2,980	2,975	2,969	2,959	2,949	2,940	2,930	2,920	2,910	2,901	2,877	2,853	2,830	2,808	2,785	2,763	2,72[illegible]	2,679	2,638	2,561	2,488
5	4,986	4,972	4,958	4,944	4,930	4,916	4,888	4,861	4,834	4,808	4,782	4,756	4,730	4,667	4,606	4,546	4,488	4,431	4,375	4,26[illegible]	4,167	4,070	3,890	3,724
7	6,972	6,944	6,917	6,890	6,863	6,836	6,783	6,731	6,680	6,629	6,579	6,530	6,482	6,364	6,251	6,141	6,035	5,933	5,834	5,646	5,470	5,304	5,001	4,731
10	9,943	9,887	9,832	9,777	9,722	9,669	9,563	9,460	9,359	9,260	9,163	9,068	8,975	8,751	8,538	8,334	8,141	7,956	7,779	7,448	7,144	6,864	6,365	5,934
15	14,873	14,747	14,624	14,503	14,384	14,267	14,038	13,817	13,602	13,394	13,192	12,996	12,806	12,355	11,934	11,541	11,172	10,827	10,502	9,908	9,378	8,901	8,080	7,397
20	19,774	19,553	19,338	19,126	18,920	18,718	18,326	17,950	17,590	17,243	16,910	16,590	16,281	15,558	14,897	14,289	13,729	13,211	12,731	11,858	11,115	10,452	9,337	8,438
25	24,648	24,306	23,973	23,650	23,335	23,028	22,438	21,877	21,344	20,836	20,352	19,890	19,448	18,425	17,504	16,671	15,914	15,222	14,588	13,456	12,505	11,672	10,299	9,215
30	29,495	29,006	28,534	28,076	27,633	27,204	26,384	25,613	24,885	24,197	23,547	22,930	22,345	21,005	19,817	18,755	17,802	16,941	16,160	14,795	13,642	12,656	11,058	9,818
35	34,314	33,655	33,020	32,409	31,820	31,253	30,176	29,170	28,230	27,348	26,520	25,741	25,006	23,339	21,881	20,595	19,451	18,428	17,507	15,916	14,590	13,468	11,673	10,300
40	39,107	38,253	37,435	36,652	35,900	35,179	33,820	32,563	31,395	30,309	29,295	28,347	27,458	25,462	23,736	22,230	20,903	19,726	18,674	16,875	15,392	14,149	12,181	10,693
45	43,873	42,801	41,780	40,806	39,877	38,989	37,327	35,801	34,395	33,095	31,890	30,769	29,725	27,400	25,412	23,693	22,192	20,870	19,696	17,705	16,079	14,727	12,607	11,020
50	48,612	47,299	46,055	44,875	43,754	42,688	40,704	38,896	37,241	35,722	34,322	33,028	31,827	29,176	26,933	25,010	23,343	21,885	20,598	18,430	16,675	15,226	12,970	11,297
55	53,325	51,750	50,264	48,862	47,536	46,280	43,957	41,856	39,946	38,204	36,606	35,138	33,782	30,811	28,320	26,201	24,378	22,791	21,399	19,[illegible]69	17,197	15,659	13,284	11,534
60	58,013	56,153	54,408	52,769	51,225	49,770	47,093	44,690	42,520	40,551	38,756	37,113	35,604	32,319	29,589	27,284	25,313	23,607	22,116	19,[illegible]36	17,657	16,040	13,557	11,739
65	62,674	60,509	58,488	56,598	54,826	53,162	50,119	47,406	44,971	42,775	40,782	38,968	37,307	33,716	30,756	28,273	26,162	24,343	22,762	20,[illegible]44	18,066	16,376	13,796	11,919
70	67,310	64,819	62,505	60,351	58,341	56,460	53,040	50,011	47,309	44,884	42,696	40,711	38,902	35,014	31,832	29,180	26,936	25,013	23,345	20,[illegible]00	18,432	16,677	14,009	12,077
75	71,920	69,083	66,462	64,032	61,773	59,669	55,862	52,512	49,541	46,889	44,506	42,353	40,399	36,221	32,827	30,014	27,645	25,623	23,876	21,[illegible]12	18,761	16,946	14,198	12,217
80	76,505	73,303	70,358	67,641	65,126	62,791	58,590	54,915	51,675	48,795	46,220	43,903	41,807	37,349	33,750	30,784	28,297	26,182	24,361	21,386	19,059	17,188	14,368	12,343
85	81,066	77,479	74,197	71,181	68,401	65,830	61,227	57,226	53,716	50,611	47,846	45,367	43,132	38,403	34,609	31,497	28,898	26,696	24,805	21,728	19,330	17,408	14,522	12,456
90	85,601	81,612	77,978	74,654	71,602	68,790	63,780	59,450	55,670	52,343	49,390	46,753	44,384	39,392	35,410	32,159	29,455	27,170	25,214	22,[illegible]41	19,577	17,609	14,661	12,558
95	90,112	85,702	81,704	78,062	74,732	71,673	66,251	61,591	57,544	53,995	50,859	48,068	45,566	40,321	36,159	32,775	29,971	27,609	25,591	22,329	19,804	17,792	14,787	12,651
100	94,599	89,751	85,376	81,407	77,791	74,483	68,644	63,654	59,341	55,575	52,258	49,315	46,686	41,195	36,860	33,351	30,451	28,016	25,941	22,594	20,012	17,960	14,903	12,736
105	99,061	93,758	88,994	84,690	80,784	77,222	70,964	65,644	61,066	57,085	53,592	50,501	47,748	42,020	37,519	33,889	30,899	28,394	26,265	22,840	20,205	18,115	15,010	12,814
110	103,499	97,724	92,560	87,913	83,711	79,893	73,213	67,564	62,725	58,532	54,865	51,630	48,755	42,798	38,138	34,393	31,318	28,748	26,567	23,068	20,383	18,258	15,108	12,885
115	107,914	101,651	96,074	91,078	86,576	82,498	75,395	69,418	64,319	59,918	56,081	52,705	49,713	43,535	38,722	34,867	31,711	29,078	26,849	23,280	20,549	18,391	15,199	12,951
120	112,305	105,538	99,539	94,186	89,379	85,040	77,512	71,209	65,854	61,248	57,244	53,732	50,625	44,232	39,273	35,314	32,079	29,388	27,113	23,478	20,703	18,514	15,283	13,012
125	116,673	109,386	102,955	97,239	92,124	87,520	79,568	72,940	67,332	62,524	58,358	54,711	51,494	44,894	39,794	35,734	32,426	29,679	27,360	23,664	20,847	18,629	15,361	13,069
130	121,017	113,195	106,323	100,238	94,812	89,943	81,565	74,615	68,756	63,751	59,425	55,648	52,323	45,523	40,287	36,131	32,753	29,952	27,593	23,837	20,981	18,737	15,434	13,122
135	125,339	116,968	109,645	103,185	97,444	92,308	83,505	76,236	70,130	64,930	60,448	56,545	53,115	46,121	40,755	36,507	33,061	30,210	27,811	24,000	21,108	18,837	15,502	13,171
140	129,637	120,703	112,920	106,081	100,022	94,618	85,392	77,805	71,456	66,065	61,430	57,403	53,872	46,691	41,199	36,863	33,353	30,453	28,017	24,153	21,226	18,932	15,566	13,217
145	133,913	124,401	116,151	108,927	102,549	96,876	87,226	79,325	72,735	67,158	62,374	58,227	54,596	47,234	41,621	37,201	33,629	30,683	28,212	24,298	21,338	19,020	15,626	13,260
150	138,166	128,063	119,337	111,724	105,025	99,083	89,011	80,798	73,973	68,211	63,281	59,017	55,290	47,753	42,024	37,522	33,891	30,901	28,396	24,435	21,443	19,104	15,683	13,301
155	142,397	131,690	122,481	114,475	107,451	101,240	90,748	82,227	75,169	69,226	64,155	59,775	55,956	48,248	42,407	37,827	34,140	31,108	28,571	24,564	21,542	19,183	15,736	13,339
160	146,606	135,282	125,582	117,179	109,831	103,350	92,440	83,613	76,325	70,206	64,995	60,504	56,594	48,722	42,772	38,118	34,377	31,305	28,736	24,686	21,636	19,257	15,786	13,375
165	150,793	138,839	128,641	119,839	112,164	105,413	94,087	84,959	77,445	71,152	65,805	61,206	57,207	49,176	43,122	38,395	34,602	31,491	28,894	24,802	21,725	19,328	15,833	13,409
170	154,959	142,363	131,660	122,455	114,452	107,432	95,692	86,265	78,529	72,066	66,586	61,881	57,797	49,611	43,456	38,660	34,817	31,669	29,043	24,912	21,810	19,394	15,878	13,441
175	159,102	145,852	134,640	125,028	116,697	109,407	97,256	87,534	79,579	72,950	67,340	62,531	58,364	50,028	43,776	38,912	35,022	31,838	29,186	25,017	21,890	19,458	15,920	13,471
180	163,224	149,309	137,580	127,559	118,899	111,340	98,781	88,767	80,597	73,804	68,067	63,158	58,909	50,428	44,082	39,154	35,218	32,000	29,321	25,116	21,966	19,518	15,961	13,500
185	167,325	152,733	140,482	130,050	121,061	113,233	100,268	89,967	81,585	74,631	68,770	63,763	59,435	50,813	44,376	39,383	35,405	32,155	29,451	25,211	22,039	19,576	15,999	13,528
190	171,405	156,125	143,346	132,501	123,182	115,087	101,719	91,133	82,543	75,432	69,450	64,346	59,942	51,183	44,657	39,608	35,584	32,302	29,575	25,302	22,108	19,630	16,036	13,554
195	175,464	159,485	146,174	134,914	125,264	116,903	103,134	92,268	83,472	76,208	70,107	64,910	60,431	51,539	44,928	39,821	35,756	32,444	29,693	25,389	22,174	19,682	16,070	13,578
200	179,502	162,814	148,966	137,288	127,309	118,682	104,516	93,372	84,375	76,960	70,743	65,455	60,903	51,882	45,189	40,025	35,920	32,579	29,807	25,472	22,238	19,732	16,103	13,602

1000 MCM | 1 Run(s) | 600 Volts | C-Value = 25,278 | Magnetic Duct | Copper

Calculations:

1. $f = \dfrac{1.732 \times \text{Length} \times \text{Isca}}{\text{Runs} \times \text{C-Value} \times \text{E L-L}}$

2. $M = \dfrac{1}{1 + f}$

3. Isc = Isca x M

4. Add Motor Contribution = Motor FLA x 4

* All results are given in symmetrical amperes

Three-Phase Bolted Fault-Current* Table

One-Way Distance in Feet

Available Isc in Thousands (Isca)	5	10	15	20	25	30	40	50	60	70	80	90	100	125	150	175	200	225	250	300	350	400	500	600
3	2,997	2,995	2,992	2,990	2,987	2,985	2,980	2,975	2,969	2,964	2,959	2,954	2,949	2,937	2,925	2,913	2,901	2,889	2,877	2,853	2,830	2,808	2,763	2,720
5	4,993	4,986	4,979	4,972	4,965	4,958	4,944	4,930	4,916	4,902	4,888	4,875	4,861	4,828	4,795	4,762	4,730	4,698	4,667	4,606	4,546	4,488	4,375	4,269
7	6,986	6,972	6,958	6,944	6,931	6,917	6,890	6,863	6,836	6,809	6,783	6,757	6,731	6,667	6,604	6,542	6,482	6,422	6,364	6,251	6,141	6,035	5,834	5,646
10	9,972	9,943	9,915	9,887	9,859	9,832	9,777	9,722	9,669	9,616	9,563	9,511	9,460	9,334	9,211	9,092	8,975	8,862	8,751	8,538	8,334	8,141	7,779	7,448
15	14,936	14,873	14,810	14,747	14,686	14,624	14,503	14,384	14,267	14,152	14,038	13,927	13,817	13,549	13,292	13,045	12,806	12,576	12,355	11,934	11,541	11,172	10,502	9,908
20	19,886	19,774	19,663	19,553	19,445	19,338	19,126	18,920	18,718	18,520	18,326	18,136	17,950	17,502	17,075	16,669	16,281	15,912	15,558	14,897	14,289	13,729	12,731	11,868
25	24,823	24,648	24,476	24,306	24,139	23,973	23,650	23,335	23,028	22,729	22,438	22,154	21,877	21,215	20,591	20,003	19,448	18,922	18,425	17,504	16,671	15,914	14,588	13,466
30	29,745	29,495	29,248	29,006	28,768	28,534	28,076	27,633	27,204	26,788	26,384	25,993	25,613	24,709	23,867	23,081	22,345	21,654	21,005	19,817	18,755	17,802	16,160	14,795
35	34,654	34,314	33,981	33,655	33,335	33,020	32,409	31,820	31,253	30,705	30,176	29,665	29,170	28,004	26,928	25,931	25,006	24,144	23,339	21,881	20,595	19,451	17,507	15,916
40	39,548	39,107	38,675	38,253	37,839	37,435	36,652	35,900	35,179	34,486	33,820	33,180	32,563	31,116	29,793	28,578	27,458	26,422	25,462	23,736	22,230	20,903	18,674	16,875
45	44,429	43,873	43,330	42,801	42,284	41,780	40,806	39,877	38,989	38,140	37,327	36,548	35,801	34,060	32,481	31,042	29,725	28,515	27,400	25,412	23,693	22,192	19,696	17,705
50	49,296	48,612	47,947	47,299	46,669	46,055	44,875	43,754	42,688	41,672	40,704	39,779	38,896	36,850	35,008	33,342	31,827	30,444	29,176	26,933	25,010	23,343	20,598	18,430
55	54,150	53,325	52,526	51,750	50,996	50,264	48,862	47,536	46,280	45,088	43,957	42,880	41,856	39,496	37,388	35,494	33,782	32,228	30,811	28,320	26,201	24,378	21,399	19,069
60	58,990	58,013	57,067	56,153	55,267	54,408	52,769	51,225	49,770	48,394	47,093	45,860	44,690	42,010	39,633	37,511	35,604	33,882	32,319	29,589	27,284	25,313	22,116	19,636
65	63,816	62,674	61,572	60,509	59,481	58,488	56,598	54,826	53,162	51,596	50,119	48,725	47,406	44,401	41,755	39,406	37,307	35,421	33,716	30,756	28,273	26,162	22,762	20,144
70	68,628	67,310	66,041	64,819	63,641	62,505	60,351	58,341	56,460	54,697	53,040	51,481	50,011	46,679	43,763	41,190	38,902	36,856	35,014	31,832	29,180	26,936	23,345	20,600
75	73,428	71,920	70,473	69,083	67,747	65,462	64,032	61,773	59,669	57,703	55,862	54,135	52,512	48,850	45,666	42,871	40,399	38,196	36,221	32,827	30,014	27,645	23,876	21,012
80	78,214	76,505	74,870	73,303	71,801	70,358	67,641	65,126	62,791	60,617	58,590	56,693	54,915	50,923	47,473	44,460	41,807	39,452	37,349	33,750	30,784	28,297	24,361	21,386
85	82,986	81,066	79,232	77,479	75,803	74,197	71,181	68,401	65,830	63,445	61,227	59,159	57,226	52,904	49,190	45,962	43,132	40,631	38,403	34,609	31,497	28,898	24,805	21,728
90	87,745	85,601	83,559	81,612	79,754	77,978	74,654	71,602	68,790	66,190	63,780	61,539	59,450	54,799	50,824	47,386	44,384	41,739	39,392	35,410	32,159	29,455	25,214	22,041
95	92,491	90,112	87,852	85,702	83,656	81,704	78,062	74,732	71,673	68,855	66,251	63,836	61,591	56,614	52,380	48,736	45,566	42,784	40,321	36,159	32,775	29,971	25,591	22,329
100	97,224	94,599	92,111	89,751	87,509	85,376	81,407	77,791	74,483	71,444	68,644	66,055	63,654	58,352	53,865	50,019	46,686	43,769	41,195	36,860	33,351	30,451	25,941	22,594
105	101,944	99,061	96,336	93,758	91,314	88,994	84,690	80,784	77,222	73,961	70,964	68,200	65,644	60,020	55,284	51,240	47,748	44,701	42,020	37,519	33,889	30,899	26,265	22,840
110	106,651	103,499	100,529	97,724	95,072	92,560	87,913	83,711	79,893	76,407	73,213	70,275	67,564	61,621	56,639	52,402	48,755	45,583	42,798	38,138	34,393	31,318	26,567	23,068
115	111,344	107,914	104,689	101,651	98,784	96,074	91,078	86,576	82,498	78,786	75,395	72,283	69,418	63,159	57,936	53,511	49,713	46,419	43,535	38,722	34,867	31,711	26,849	23,280
120	116,025	112,305	108,816	105,538	102,451	99,539	94,186	89,379	85,040	81,102	77,512	74,227	71,209	64,639	59,178	54,569	50,625	47,213	44,232	39,273	35,314	32,079	27,113	23,478
125	120,693	116,673	112,912	109,386	106,073	102,955	97,239	92,124	87,520	83,355	79,568	76,110	72,940	66,062	60,369	55,580	51,494	47,968	44,894	39,794	35,734	32,426	27,360	23,664
130	125,348	121,017	116,976	113,195	109,652	106,323	100,238	94,812	89,943	85,549	81,565	77,935	74,615	67,433	61,512	56,547	52,323	48,687	45,523	40,287	36,131	32,753	27,593	23,837
135	129,990	125,339	121,008	116,968	113,188	109,645	103,185	97,444	92,308	87,686	83,505	79,705	76,236	68,754	62,609	57,472	53,115	49,372	46,121	40,755	36,507	33,061	27,811	24,000
140	134,619	129,637	125,010	120,703	116,682	112,920	106,081	100,022	94,618	89,769	85,392	81,422	77,805	70,027	63,663	58,360	53,872	50,025	46,691	41,199	36,863	33,353	28,017	24,153
145	139,236	133,913	128,982	124,401	120,134	116,151	108,927	102,549	96,876	91,798	87,226	83,088	79,325	71,256	64,678	59,211	54,536	50,649	47,234	41,621	37,201	33,629	28,212	24,298
150	143,840	138,166	132,923	128,063	123,546	119,337	111,724	105,025	99,083	93,777	89,011	84,706	80,798	72,443	65,654	60,028	55,290	51,246	47,753	42,024	37,522	33,891	28,396	24,435
155	148,432	142,397	136,835	131,690	126,918	122,481	114,475	107,451	101,240	95,708	90,748	86,278	82,227	73,589	66,594	60,813	55,956	51,817	48,248	42,407	37,827	34,140	28,571	24,564
160	153,011	146,606	140,717	135,282	130,251	125,582	117,179	109,831	103,350	97,591	92,440	87,805	83,613	74,698	67,500	61,568	56,594	52,364	48,722	42,772	38,118	34,377	28,736	24,686
165	157,577	150,793	144,570	138,839	133,546	128,641	119,839	112,164	105,413	99,428	94,087	89,290	84,959	75,770	68,374	62,294	57,207	52,888	49,176	43,122	38,395	34,602	28,894	24,802
170	162,131	154,959	148,394	142,363	136,802	131,660	122,455	114,452	107,432	101,222	95,692	90,734	86,265	76,807	69,218	62,994	57,797	53,392	49,611	43,456	38,660	34,817	29,043	24,912
175	166,673	159,102	152,189	145,852	140,022	134 640	125,028	116,697	109,407	102,974	97,256	92,139	87,534	77,811	70,033	63,668	58,364	53,875	50,028	43,776	38,912	35,022	29,186	25,017
180	171,202	163,224	155,957	149,309	143,205	137,580	127,559	118,899	111,340	104,685	98,781	93,507	88,767	78,784	70,820	64,318	58,909	54,340	50,428	44,082	39,154	35,218	29,321	25,116
185	175,719	167,325	159,696	152,733	146,351	140,482	130,050	121,061	113,233	106,357	100,268	94,838	89,967	79,728	71,581	64,945	59,435	54,787	50,813	44,376	39,386	35,405	29,451	25,211
190	180,224	171,405	163,408	156,125	149,463	143,346	132,501	123,182	115,087	107,991	101,719	96,135	91,133	80,642	72,317	65,551	59,942	55,217	51,183	44,657	39,608	35,584	29,575	25,302
195	184,717	175,464	167,093	159,485	152,540	146,174	134,914	125,264	116,903	109,588	103,134	97,399	92,268	81,529	73,030	66,136	60,431	55,632	51,539	44,928	39,821	35,756	29,693	25,389
200	189,197	179,502	170,751	162,814	155,582	148,966	137,288	127,309	118,682	111,150	104,516	98,630	93,372	82,391	73,720	66,701	60,903	56,031	51,882	45,189	40,025	35,920	29,807	25,472

1000 MCM	2 Run(s)	600 Volts	C-Value = 25,278	Magnetic Duct	Copper

Calculations:

1. $f = \dfrac{1.732 \times \text{Length} \times \text{Isca}}{\text{Runs} \times \text{C-Value} \times E_{L\text{-}L}}$

2. $M = \dfrac{1}{1 + f}$

3. Isc = Isca x M

4. Add Motor Contribution = Motor FLA x 4

* All results are given in symmetrical amperes

Short-Circuit Tables and Data: 480 Volts Three-Phase

Three-Phase Bolted Fault-Current* Table

One-Way Distance in Feet

Available Isc in Thousands (Isca)

Isca	5	10	15	20	25	30	40	50	60	70	80	90	100	125	150	175	200	225	250	300	350	400	500	600
1	956	915	878	844	812	782	729	683	642	606	574	545	519	463	418	381	350	324	301	264	235	212	177	152
3	2,634	2,347	2,117	1,927	1,769	1,635	1,420	1,255	1,124	1,018	930	856	793	670	580	511	457	413	377	321	279	247	201	170
5	4,059	3,416	2,949	2,594	2,315	2,091	1,751	1,506	1,322	1,177	1,061	966	887	736	628	548	487	437	397	335	290	256	207	173
10	6,832	5,188	4,182	3,502	3,013	2,644	2,123	1,774	1,523	1,335	1,188	1,070	973	794	671	580	511	457	413	347	299	262	211	177
15	8,846	6,273	4,859	3,965	3,349	2,899	2,285	1,885	1,605	1,397	1,236	1,109	1,006	816	686	592	520	464	419	351	302	265	213	178
20	10,376	7,005	5,287	4,246	3,547	3,046	2,375	1,946	1,649	1,430	1,263	1,130	1,023	827	694	598	525	468	422	353	303	266	213	178
25	11,577	7,532	5,582	4,434	3,678	3,142	2,433	1,985	1,676	1,451	1,279	1,143	1,033	834	699	601	528	470	424	354	304	267	214	178
30	12,545	7,931	5,798	4,569	3,770	3,209	2,473	2,012	1,695	1,465	1,290	1,152	1,041	838	702	604	530	472	425	355	305	267	214	179
35	13,342	8,242	5,963	4,671	3,839	3,259	2,502	2,031	1,709	1,475	1,298	1,158	1,046	842	704	605	531	473	426	356	305	267	214	179
40	14,010	8,492	6,092	4,750	3,893	3,297	2,525	2,046	1,720	1,483	1,304	1,163	1,050	844	706	607	532	473	427	356	306	268	214	179
45	14,577	8,697	6,197	4,814	3,935	3,328	2,543	2,058	1,728	1,489	1,308	1,167	1,053	846	707	608	533	474	427	357	306	268	215	179
50	15,065	8,868	6,284	4,866	3,970	3,353	2,557	2,067	1,734	1,494	1,312	1,170	1,055	848	709	609	533	475	428	357	306	268	215	179
55	15,489	9,014	6,356	4,909	3,999	3,373	2,569	2,075	1,740	1,498	1,315	1,172	1,057	849	709	609	534	475	428	357	306	268	215	179
60	15,861	9,139	6,418	4,946	4,023	3,390	2,579	2,081	1,745	1,502	1,318	1,174	1,059	850	710	610	534	475	428	357	306	268	215	179
65	16,191	9,247	6,472	4,978	4,044	3,405	2,588	2,087	1,748	1,504	1,320	1,176	1,060	851	711	610	535	476	428	357	307	268	215	179
70	16,484	9,342	6,518	5,005	4,062	3,418	2,595	2,092	1,752	1,507	1,322	1,178	1,062	852	711	611	535	476	429	358	307	268	215	179
75	16,747	9,426	6,559	5,029	4,078	3,429	2,602	2,096	1,755	1,509	1,324	1,179	1,063	853	712	611	535	476	429	358	307	269	215	179
80	16,984	9,500	6,595	5,050	4,092	3,439	2,607	2,100	1,757	1,511	1,325	1,180	1,064	853	712	611	535	476	429	358	307	269	215	179
85	17,199	9,567	6,627	5,069	4,104	3,448	2,612	2,103	1,760	1,513	1,327	1,181	1,065	854	713	612	536	476	429	358	307	269	215	179
90	17,394	9,627	6,656	5,086	4,115	3,456	2,617	2,106	1,762	1,514	1,328	1,182	1,065	854	713	612	536	477	429	358	307	269	215	179
95	17,573	9,682	6,682	5,101	4,125	3,463	2,621	2,108	1,763	1,516	1,329	1,183	1,066	855	713	612	536	477	429	358	307	269	215	179
100	17,737	9,731	6,705	5,115	4,134	3,469	2,624	2,111	1,765	1,517	1,330	1,184	1,067	855	714	612	536	477	429	358	307	269	215	179
105	17,888	9,777	6,727	5,127	4,142	3,475	2,628	2,113	1,767	1,518	1,330	1,184	1,067	855	714	612	536	477	429	358	307	269	215	179
110	18,028	9,818	6,746	5,138	4,150	3,480	2,631	2,115	1,768	1,519	1,331	1,185	1,068	856	714	613	536	477	430	358	307	269	215	179
115	18,157	9,857	6,764	5,149	4,156	3,485	2,633	2,116	1,769	1,520	1,332	1,185	1,068	856	714	613	537	477	430	358	307	269	215	179
120	18,277	9,892	6,781	5,159	4,163	3,489	2,636	2,118	1,770	1,521	1,333	1,186	1,068	856	714	613	537	477	430	358	307	269	215	179
125	18,389	9,925	6,796	5,167	4,168	3,493	2,638	2,120	1,771	1,521	1,333	1,186	1,069	857	715	613	537	477	430	358	307	269	215	179
130	18,494	9,955	6,811	5,176	4,174	3,497	2,640	2,121	1,772	1,522	1,334	1,187	1,069	857	715	613	537	477	430	358	307	269	215	179
135	18,592	9,983	6,824	5,183	4,179	3,500	2,642	2,122	1,773	1,523	1,334	1,187	1,070	857	715	613	537	477	430	358	307	269	215	179
140	18,684	10,010	6,836	5,190	4,183	3,504	2,644	2,123	1,774	1,523	1,335	1,188	1,070	857	715	613	537	478	430	358	307	269	215	179
145	18,770	10,035	6,848	5,197	4,188	3,507	2,646	2,125	1,775	1,524	1,335	1,188	1,070	857	715	613	537	478	430	358	307	269	215	179
150	18,851	10,058	6,858	5,203	4,192	3,509	2,648	2,126	1,775	1,524	1,336	1,188	1,070	858	715	614	537	478	430	358	307	269	215	179
155	18,928	10,080	6,869	5,209	4,196	3,512	2,649	2,127	1,776	1,525	1,336	1,189	1,071	858	715	614	537	478	430	359	307	269	215	179
160	19,001	10,100	6,878	5,215	4,199	3,515	2,651	2,127	1,777	1,525	1,336	1,189	1,071	858	715	614	537	478	430	359	307	269	215	179
165	19,069	10,119	6,887	5,220	4,202	3,517	2,652	2,128	1,777	1,526	1,337	1,189	1,071	858	716	614	537	478	430	359	307	269	215	179
170	19,134	10,138	6,896	5,225	4,206	3,519	2,653	2,129	1,778	1,526	1,337	1,189	1,071	858	716	614	537	478	430	359	307	269	215	179
175	19,196	10,155	6,904	5,229	4,209	3,521	2,654	2,130	1,779	1,527	1,337	1,190	1,071	858	716	614	537	478	430	359	307	269	215	179
180	19,255	10,171	6,911	5,234	4,211	3,523	2,655	2,131	1,779	1,527	1,338	1,190	1,072	858	716	614	537	478	430	359	307	269	215	179
185	19,311	10,187	6,918	5,238	4,214	3,525	2,656	2,131	1,779	1,527	1,338	1,190	1,072	858	716	614	537	478	430	359	308	269	215	180
190	19,364	10,202	6,925	5,242	4,217	3,527	2,657	2,132	1,780	1,528	1,338	1,190	1,072	859	716	614	533	478	430	359	308	269	215	180
195	19,415	10,216	6,932	5,245	4,219	3,529	2,658	2,133	1,780	1,528	1,338	1,191	1,072	859	716	614	533	478	430	359	308	269	215	180
200	19,463	10,229	6,938	5,249	4,221	3,530	2,659	2,133	1,781	1,528	1,339	1,191	1,072	859	716	614	538	478	430	359	308	269	215	180

# 14 AWG	1 Run(s)	480 Volts	C-Value = 389	Magnetic Duct	Copper

Calculations:

1. $f = \dfrac{1.732 \times \text{Length} \times \text{Isca}}{\text{Runs} \times \text{C-Value} \times \text{E L-L}}$

2. $M = \dfrac{1}{1 + f}$

3. Isc = Isca x M

4. Add Motor Contribution = Motor FLA x 4

* All results are given in symmetrical amperes

Three-Phase Bolted Fault-Current* Table

One-Way Distance in Feet

Available Isc in Thousands (Isca)

	5	10	15	20	25	30	40	50	60	70	80	90	100	125	150	175	200	225	250	300	350	400	500	600
1	972	945	919	895	872	851	810	774	740	710	681	655	631	578	533	494	461	432	406	368	328	299	255	222
3	2,758	2,552	2,375	2,221	2,085	1,965	1,763	1,598	1,462	1,346	1,248	1,163	1,089	940	826	737	665	606	557	479	420	374	307	260
5	4,362	3,869	3,476	3,155	2,888	2,664	2,305	2,031	1,815	1,641	1,497	1,377	1,274	1,074	928	817	730	660	602	512	445	394	320	270
10	7,737	6,310	5,327	4,609	4,062	3,630	2,995	2,548	2,218	1,963	1,761	1,597	1,460	1,203	1,023	890	788	706	640	539	466	410	331	277
15	10,427	7,991	6,477	5,446	4,698	4,130	3,327	2,785	2,395	2,101	1,871	1,686	1,535	1,254	1,059	917	809	723	654	549	473	416	334	280
20	12,620	9,218	7,261	5,989	5,097	4,436	3,522	2,920	2,494	2,177	1,931	1,735	1,575	1,280	1,078	932	820	732	661	554	477	419	336	281
25	14,442	10,154	7,829	6,371	5,370	4,642	3,651	3,008	2,558	2,225	1,969	1,766	1,600	1,297	1,090	940	827	738	666	557	479	420	337	282
30	15,981	10,891	8,261	6,653	5,570	4,790	3,742	3,070	2,603	2,259	1,995	1,787	1,618	1,308	1,098	946	831	741	669	559	481	421	338	282
35	17,297	11,487	8,599	6,871	5,722	4,902	3,810	3,115	2,635	2,283	2,014	1,802	1,630	1,316	1,104	951	835	744	671	561	482	422	339	283
40	18,436	11,979	8,871	7,044	5,841	4,989	3,862	3,151	2,660	2,302	2,029	1,814	1,640	1,323	1,108	954	837	746	672	562	483	423	339	283
45	19,431	12,391	9,095	7,185	5,937	5,059	3,904	3,178	2,680	2,317	2,040	1,823	1,647	1,328	1,112	956	839	747	674	563	483	423	339	283
50	20,308	12,742	9,283	7,301	6,017	5,117	3,938	3,201	2,696	2,329	2,050	1,830	1,653	1,332	1,115	958	841	749	675	564	484	424	340	283
55	21,087	13,044	9,442	7,399	6,083	5,165	3,967	3,220	2,709	2,339	2,057	1,836	1,658	1,335	1,117	960	842	750	676	564	484	424	340	284
60	21,783	13,307	9,580	7,483	6,140	5,205	3,991	3,235	2,721	2,347	2,064	1,842	1,663	1,337	1,119	961	843	750	676	565	485	424	340	284
65	22,409	13,538	9,699	7,556	6,189	5,240	4,011	3,249	2,730	2,354	2,069	1,846	1,666	1,340	1,120	963	844	751	677	565	485	425	340	284
70	22,974	13,742	9,803	7,619	6,231	5,271	4,029	3,261	2,738	2,360	2,074	1,850	1,669	1,342	1,122	964	845	752	677	565	485	425	340	284
75	23,488	13,925	9,895	7,675	6,268	5,297	4,044	3,271	2,746	2,366	2,078	1,853	1,672	1,343	1,123	965	845	752	678	566	485	425	340	284
80	23,957	14,088	9,978	7,724	6,301	5,321	4,058	3,280	2,752	2,370	2,082	1,856	1,674	1,345	1,124	965	846	753	678	566	486	425	341	284
85	24,387	14,236	10,052	7,768	6,330	5,342	4,070	3,288	2,757	2,375	2,085	1,858	1,676	1,346	1,125	966	846	753	679	566	486	425	341	284
90	24,782	14,369	10,118	7,808	6,357	5,360	4,081	3,295	2,762	2,378	2,088	1,861	1,678	1,347	1,126	967	847	754	679	566	486	425	341	284
95	25,146	14,491	10,178	7,844	6,380	5,377	4,091	3,301	2,767	2,382	2,090	1,863	1,680	1,349	1,126	967	847	754	679	567	486	426	341	284
100	25,484	14,602	10,233	7,876	6,402	5,392	4,100	3,307	2,771	2,385	2,093	1,864	1,681	1,349	1,127	968	848	754	679	567	486	426	341	284
105	25,797	14,705	10,283	7,906	6,421	5,406	4,108	3,312	2,775	2,387	2,095	1,866	1,683	1,350	1,128	968	848	755	680	567	486	426	341	284
110	26,088	14,799	10,329	7,933	6,439	5,419	4,115	3,317	2,778	2,390	2,097	1,868	1,684	1,351	1,128	969	848	755	680	567	486	426	341	284
115	26,360	14,886	10,371	7,958	6,456	5,431	4,122	3,321	2,781	2,392	2,098	1,869	1,685	1,352	1,129	969	849	755	680	567	486	426	341	284
120	26,614	14,967	10,411	7,981	6,471	5,441	4,128	3,325	2,784	2,394	2,100	1,870	1,686	1,353	1,129	969	849	755	680	567	487	426	341	284
125	26,852	15,042	10,447	8,002	6,485	5,451	4,133	3,329	2,786	2,396	2,101	1,871	1,687	1,353	1,130	970	849	755	680	567	487	426	341	284
130	27,076	15,112	10,481	8,022	6,498	5,460	4,139	3,332	2,789	2,398	2,103	1,873	1,688	1,354	1,130	970	849	756	680	567	487	426	341	284
135	27,286	15,177	10,512	8,040	6,510	5,469	4,144	3,335	2,791	2,399	2,104	1,874	1,689	1,354	1,130	970	850	756	681	568	487	426	341	284
140	27,485	15,238	10,541	8,058	6,521	5,477	4,148	3,338	2,793	2,401	2,105	1,874	1,689	1,355	1,131	970	850	756	681	568	487	426	341	284
145	27,672	15,296	10,569	8,074	6,532	5,484	4,152	3,341	2,795	2,402	2,106	1,875	1,690	1,355	1,131	971	850	756	681	568	487	426	341	284
150	27,849	15,350	10,594	8,089	6,541	5,491	4,156	3,344	2,797	2,404	2,107	1,876	1,691	1,356	1,131	971	850	756	681	568	487	426	341	284
155	28,017	15,400	10,619	8,103	6,551	5,498	4,160	3,346	2,798	2,405	2,108	1,877	1,691	1,356	1,132	971	850	756	681	568	487	426	341	284
160	28,176	15,448	10,641	8,116	6,559	5,504	4,164	3,348	2,800	2,406	2,109	1,878	1,692	1,356	1,132	971	850	756	681	568	487	426	341	284
165	28,327	15,494	10,663	8,128	6,567	5,509	4,167	3,350	2,801	2,407	2,110	1,878	1,692	1,357	1,132	971	851	756	681	568	487	426	341	284
170	28,471	15,537	10,683	8,140	6,575	5,515	4,170	3,352	2,803	2,408	2,111	1,879	1,693	1,357	1,132	972	851	757	681	568	487	426	341	285
175	28,608	15,577	10,702	8,151	6,582	5,520	4,173	3,354	2,804	2,409	2,112	1,880	1,693	1,357	1,133	972	851	757	681	568	487	426	341	285
180	28,739	15,616	10,721	8,162	6,589	5,525	4,176	3,356	2,805	2,410	2,112	1,880	1,694	1,358	1,133	972	851	757	681	568	487	426	341	285
185	28,863	15,653	10,738	8,172	6,596	5,529	4,178	3,358	2,807	2,411	2,113	1,881	1,694	1,358	1,133	972	851	757	681	568	487	426	341	285
190	28,982	15,687	10,754	8,182	6,602	5,534	4,181	3,359	2,808	2,412	2,114	1,881	1,695	1,358	1,133	972	851	757	682	568	487	427	341	285
195	29,096	15,721	10,770	8,191	6,608	5,538	4,183	3,361	2,809	2,413	2,114	1,882	1,695	1,358	1,133	972	851	757	682	568	487	427	341	285
200	29,205	15,753	10,785	8,199	6,614	5,542	4,185	3,362	2,810	2,413	2,115	1,882	1,695	1,359	1,133	972	851	757	682	568	487	427	341	285

# 12 AWG	1 Run(s)	480 Volts	C-Value = 617	Magnetic Duct	Copper

Calculations:

1. $f = \dfrac{1.732 \times \text{Length} \times \text{Isca}}{\text{Runs} \times \text{C-Value} \times \text{E L-L}}$

2. $M = \dfrac{1}{1 + f}$

3. Isc = Isca x M

4. Add Motor Contribution = Motor FLA x 4

* All results are given in symmetrical amperes

Three-Phase Bolted Fault-Current* Table

One-Way Distance in Feet

Available Isc in Thousands (Isca)	5	10	15	20	25	30	40	50	60	70	80	90	100	125	150	175	200	225	250	300	350	400	500	600
1	982	965	948	931	916	901	872	845	819	795	773	751	731	685	644	608	576	547	521	475	437	405	352	312
3	2,843	2,702	2,574	2,458	2,351	2,254	2,081	1,933	1,805	1,693	1,593	1,505	1,426	1,261	1,130	1,024	935	861	798	696	617	554	460	394
5	4,579	4,223	3,919	3,655	3,425	3,222	2,881	2,605	2,377	2,186	2,023	1,883	1,761	1,516	1,330	1,185	1,069	973	893	767	672	598	490	415
10	8,447	7,311	6,444	5,762	5,210	4,754	4,046	3,522	3,118	2,797	2,536	2,320	2,138	1,786	1,534	1,345	1,197	1,078	981	831	721	636	516	433
15	11,757	9,667	8,207	7,131	6,304	5,649	4,677	3,991	3,480	3,085	2,771	2,514	2,302	1,900	1,617	1,408	1,246	1,118	1,014	855	739	650	525	440
20	14,622	11,523	9,508	8,093	7,044	6,236	5,073	4,275	3,694	3,252	2,905	2,624	2,393	1,962	1,662	1,442	1,273	1,139	1,031	867	748	657	529	443
25	17,126	13,024	10,507	8,806	7,578	6,651	5,344	4,466	3,836	3,362	2,992	2,695	2,452	2,001	1,690	1,463	1,289	1,153	1,042	875	753	662	532	445
30	19,333	14,262	11,299	9,355	7,982	6,960	5,541	4,603	3,937	3,439	3,053	2,744	2,493	2,028	1,709	1,477	1,300	1,162	1,049	880	757	665	534	446
35	21,294	15,301	11,941	9,791	8,297	7,198	5,692	4,706	4,012	3,496	3,098	2,781	2,523	2,048	1,723	1,488	1,309	1,168	1,055	883	760	667	535	447
40	23,046	16,186	12,473	10,146	8,550	7,388	5,810	4,787	4,070	3,540	3,132	2,809	2,546	2,063	1,734	1,495	1,315	1,173	1,059	886	762	668	536	448
45	24,622	16,948	12,921	10,440	8,758	7,543	5,905	4,851	4,117	3,575	3,160	2,831	2,564	2,075	1,742	1,502	1,319	1,177	1,062	888	764	670	537	449
50	26,048	17,611	13,303	10,688	8,932	7,672	5,983	4,904	4,155	3,604	3,182	2,849	2,579	2,084	1,749	1,507	1,323	1,180	1,064	890	765	671	538	449
55	27,343	18,194	13,632	10,900	9,080	7,780	6,049	4,948	4,186	3,628	3,201	2,864	2,591	2,092	1,755	1,511	1,327	1,182	1,066	892	766	671	538	449
60	28,524	18,709	13,920	11,083	9,206	7,873	6,105	4,986	4,213	3,648	3,216	2,876	2,601	2,099	1,759	1,514	1,329	1,184	1,068	893	767	672	539	450
65	29,607	19,169	14,173	11,242	9,316	7,953	6,153	5,018	4,236	3,665	3,230	2,887	2,610	2,105	1,763	1,517	1,332	1,186	1,070	894	768	673	539	450
70	30,603	19,582	14,397	11,383	9,413	8,024	6,195	5,045	4,256	3,680	3,241	2,896	2,617	2,109	1,767	1,520	1,333	1,188	1,071	895	768	673	540	450
75	31,521	19,954	14,597	11,508	9,498	8,085	6,232	5,070	4,273	3,693	3,251	2,904	2,624	2,114	1,770	1,522	1,335	1,189	1,072	895	769	674	540	450
80	32,372	20,291	14,777	11,619	9,573	8,140	6,265	5,091	4,288	3,704	3,260	2,911	2,629	2,117	1,772	1,524	1,337	1,190	1,073	896	769	674	540	451
85	33,161	20,599	14,939	11,719	9,641	8,189	6,294	5,110	4,302	3,714	3,268	2,917	2,634	2,121	1,775	1,526	1,338	1,191	1,074	897	770	674	540	451
90	33,896	20,880	15,087	11,810	9,702	8,233	6,320	5,128	4,314	3,723	3,275	2,923	2,639	2,124	1,777	1,527	1,339	1,192	1,074	897	770	675	540	451
95	34,581	21,138	15,221	11,892	9,758	8,273	6,343	5,143	4,325	3,731	3,281	2,928	2,643	2,126	1,779	1,529	1,340	1,193	1,075	898	770	675	541	451
100	35,222	21,376	15,344	11,967	9,808	8,309	6,364	5,157	4,335	3,739	3,287	2,932	2,647	2,129	1,780	1,530	1,341	1,194	1,076	898	771	675	541	451
105	35,823	21,595	15,457	12,035	9,854	8,342	6,384	5,170	4,344	3,745	3,292	2,936	2,650	2,131	1,782	1,531	1,342	1,195	1,076	898	771	675	541	451
110	36,387	21,799	15,561	12,098	9,896	8,373	6,401	5,181	4,352	3,751	3,297	2,940	2,653	2,133	1,783	1,532	1,343	1,195	1,077	899	771	676	541	451
115	36,918	21,989	15,657	12,157	9,935	8,400	6,417	5,192	4,359	3,757	3,301	2,943	2,656	2,135	1,784	1,533	1,343	1,196	1,077	899	772	676	541	451
120	37,419	22,165	15,746	12,210	9,971	8,426	6,432	5,202	4,366	3,762	3,305	2,947	2,658	2,136	1,786	1,534	1,344	1,196	1,078	899	772	676	541	451
125	37,892	22,330	15,829	12,260	10,004	8,450	6,446	5,211	4,373	3,767	3,308	2,950	2,661	2,138	1,787	1,534	1,345	1,197	1,078	900	772	676	541	451
130	38,339	22,485	15,907	12,307	10,035	8,472	6,459	5,219	4,379	3,771	3,312	2,952	2,663	2,139	1,788	1,535	1,345	1,197	1,078	900	772	676	541	452
135	38,762	22,630	15,979	12,350	10,064	8,492	6,471	5,227	4,384	3,775	3,315	2,955	2,665	2,140	1,788	1,536	1,346	1,198	1,079	900	772	676	542	452
140	39,164	22,766	16,047	12,390	10,091	8,511	6,482	5,234	4,389	3,779	3,318	2,957	2,667	2,142	1,789	1,536	1,346	1,198	1,079	900	772	676	542	452
145	39,545	22,894	16,111	12,428	10,116	8,529	6,492	5,241	4,394	3,783	3,321	2,959	2,669	2,143	1,790	1,537	1,347	1,198	1,079	901	773	677	542	452
150	39,908	23,016	16,171	12,464	10,140	8,546	6,502	5,247	4,398	3,786	3,323	2,961	2,670	2,144	1,791	1,538	1,347	1,199	1,080	901	773	677	542	452
155	40,253	23,130	16,227	12,498	10,162	8,562	6,511	5,253	4,402	3,789	3,325	2,963	2,672	2,145	1,792	1,538	1,348	1,199	1,080	901	773	677	542	452
160	40,583	23,238	16,280	12,529	10,183	8,577	6,520	5,259	4,406	3,792	3,328	2,965	2,673	2,146	1,792	1,539	1,348	1,199	1,080	901	773	677	542	452
165	40,897	23,341	16,331	12,559	10,202	8,591	6,528	5,264	4,410	3,795	3,330	2,966	2,675	2,147	1,793	1,539	1,348	1,200	1,080	901	773	677	542	452
170	41,197	23,439	16,379	12,587	10,221	8,604	6,535	5,269	4,414	3,797	3,332	2,968	2,676	2,147	1,793	1,539	1,349	1,200	1,081	901	773	677	542	452
175	41,485	23,531	16,424	12,614	10,239	8,616	6,543	5,274	4,417	3,800	3,334	2,970	2,677	2,148	1,794	1,540	1,349	1,200	1,081	902	773	677	542	452
180	41,759	23,620	16,467	12,639	10,255	8,628	6,549	5,278	4,420	3,802	3,335	2,971	2,678	2,149	1,794	1,540	1,349	1,200	1,081	902	773	677	542	452
185	42,023	23,704	16,507	12,663	10,271	8,639	6,556	5,282	4,423	3,804	3,337	2,972	2,679	2,150	1,795	1,541	1,349	1,200	1,081	902	774	677	542	452
190	42,276	23,784	16,546	12,686	10,286	8,650	6,562	5,286	4,426	3,806	3,339	2,974	2,680	2,150	1,795	1,541	1,350	1,201	1,081	902	774	677	542	452
195	42,518	23,860	16,583	12,708	10,300	8,660	6,568	5,290	4,428	3,808	3,340	2,975	2,681	2,151	1,796	1,541	1,350	1,201	1,081	902	774	677	542	452
200	42,751	23,934	16,619	12,728	10,314	8,670	6,573	5,293	4,431	3,810	3,342	2,976	2,682	2,152	1,796	1,542	1,350	1,201	1,082	902	774	677	542	452

10 AWG | 1 Run(s) | 480 Volts

C-Value = 981 | Magnetic Duct | Copper

Calculations:

1. $f = \dfrac{1.732 \times \text{Length} \times \text{Isca}}{\text{Runs} \times \text{C-Value} \times E_{L\text{-}L}}$

2. $M = \dfrac{1}{1 + f}$

3. $\text{Isc} = \text{Isca} \times M$

4. Add Motor Contribution = Motor FLA x 4

* All results are given in symmetrical amperes

Three-Phase Bolted Fault-Current* Table

One-Way Distance in Feet

Available Isc in Thousands (Isca)	5	10	15	20	25	30	40	50	60	70	80	90	100	125	150	175	200	225	250	300	350	400	500	600
3	2,899	2,805	2,717	2,634	2,556	2,482	2,347	2,226	2,117	2,018	1,928	1,845	1,770	1,605	1,469	1,353	1,255	1,170	1,096	972	874	793	670	580
5	4,726	4,481	4,260	4,059	3,877	3,710	3,416	3,166	2,949	2,761	2,595	2,448	2,316	2,042	1,826	1,651	1,507	1,386	1,283	1,117	989	887	736	629
7	6,475	6,023	5,630	5,285	4,980	4,708	4,245	3,865	3,547	3,278	3,046	2,846	2,669	2,312	2,039	1,823	1,649	1,505	1,385	1,193	1,048	935	768	652
10	8,962	8,119	7,420	6,833	6,332	5,899	5,189	4,632	4,183	3,814	3,504	3,241	3,014	2,566	2,234	1,978	1,775	1,609	1,472	1,257	1,098	974	794	671
15	12,779	11,131	9,859	8,848	8,025	7,343	6,275	5,478	4,861	4,369	3,967	3,633	3,351	2,806	2,414	2,118	1,886	1,700	1,548	1,312	1,139	1,006	816	686
20	16,237	13,666	11,798	10,379	9,265	8,366	7,008	6,029	5,290	4,712	4,248	3,867	3,549	2,944	2,515	2,195	1,947	1,750	1,589	1,342	1,161	1,024	827	694
25	19,385	15,829	13,376	11,581	10,211	9,130	7,536	6,415	5,585	4,945	4,437	4,023	3,680	3,033	2,580	2,244	1,986	1,781	1,615	1,360	1,175	1,034	834	699
30	22,261	17,697	14,685	12,550	10,956	9,722	7,934	6,702	5,801	5,114	4,572	4,134	3,772	3,096	2,625	2,278	2,013	1,803	1,632	1,373	1,184	1,041	839	702
35	24,901	19,325	15,789	13,347	11,560	10,194	8,246	6,923	5,966	5,241	4,674	4,217	3,841	3,142	2,658	2,303	2,032	1,818	1,645	1,382	1,191	1,046	842	705
40	27,332	20,758	16,733	14,015	12,057	10,579	8,496	7,099	6,096	5,341	4,753	4,281	3,895	3,178	2,684	2,323	2,047	1,830	1,655	1,388	1,196	1,050	845	706
45	29,577	22,028	17,549	14,583	12,475	10,900	8,702	7,241	6,201	5,422	4,816	4,333	3,937	3,206	2,704	2,338	2,059	1,839	1,662	1,394	1,200	1,053	847	708
50	31,658	23,162	18,261	15,072	12,831	11,170	8,873	7,360	6,287	5,488	4,869	4,375	3,972	3,229	2,720	2,350	2,068	1,847	1,668	1,398	1,203	1,056	848	709
55	33,592	24,180	18,888	15,496	13,137	11,402	9,019	7,460	6,360	5,543	4,912	4,410	4,001	3,248	2,734	2,360	2,076	1,853	1,673	1,402	1,206	1,058	850	710
60	35,393	25,099	19,444	15,869	13,404	11,602	9,144	7,545	6,422	5,590	4,949	4,440	4,026	3,264	2,745	2,368	2,083	1,858	1,678	1,405	1,208	1,060	851	711
65	37,075	25,934	19,941	16,198	13,638	11,777	9,252	7,619	6,475	5,630	4,980	4,465	4,046	3,278	2,755	2,376	2,088	1,863	1,681	1,407	1,210	1,061	852	711
70	38,650	26,695	20,388	16,492	13,846	11,932	9,347	7,683	6,522	5,665	5,008	4,487	4,064	3,290	2,763	2,382	2,093	1,867	1,684	1,409	1,212	1,062	852	712
75	40,127	27,391	20,792	16,755	14,031	12,069	9,431	7,739	6,562	5,696	5,032	4,506	4,080	3,300	2,770	2,387	2,097	1,870	1,687	1,411	1,213	1,063	853	712
80	41,515	28,031	21,158	16,992	14,197	12,191	9,506	7,790	6,599	5,723	5,053	4,523	4,094	3,309	2,777	2,392	2,101	1,873	1,690	1,413	1,214	1,064	854	713
85	42,823	28,621	21,493	17,207	14,347	12,302	9,573	7,835	6,631	5,747	5,072	4,538	4,107	3,317	2,783	2,396	2,104	1,875	1,692	1,414	1,215	1,065	854	713
90	44,056	29,166	21,799	17,403	14,483	12,401	9,633	7,875	6,660	5,769	5,089	4,552	4,118	3,324	2,788	2,400	2,107	1,878	1,694	1,416	1,216	1,066	855	713
95	45,221	29,673	22,081	17,582	14,606	12,492	9,687	7,911	6,686	5,789	5,104	4,564	4,128	3,331	2,792	2,403	2,110	1,880	1,695	1,417	1,217	1,067	855	714
100	46,323	30,143	22,340	17,746	14,719	12,575	9,737	7,944	6,709	5,806	5,118	4,575	4,137	3,337	2,796	2,406	2,112	1,882	1,697	1,418	1,218	1,067	856	714
105	47,368	30,582	22,580	17,898	14,823	12,650	9,782	7,975	6,731	5,822	5,130	4,585	4,145	3,342	2,800	2,409	2,114	1,883	1,698	1,419	1,219	1,068	856	714
110	48,360	30,993	22,803	18,037	14,919	12,720	9,824	8,002	6,750	5,837	5,142	4,594	4,152	3,347	2,803	2,412	2,116	1,885	1,699	1,420	1,219	1,068	856	714
115	49,302	31,377	23,011	18,167	15,008	12,784	9,862	8,028	6,768	5,851	5,152	4,603	4,159	3,351	2,806	2,414	2,118	1,886	1,700	1,421	1,220	1,069	857	715
120	50,199	31,738	23,204	18,287	15,090	12,844	9,898	8,051	6,785	5,863	5,162	4,610	4,165	3,355	2,809	2,416	2,119	1,888	1,702	1,421	1,220	1,069	857	715
125	51,053	32,077	23,385	18,399	15,166	12,899	9,931	8,073	6,800	5,875	5,171	4,617	4,171	3,359	2,812	2,418	2,121	1,889	1,702	1,422	1,221	1,070	857	715
130	51,868	32,397	23,555	18,504	15,237	12,951	9,961	8,093	6,815	5,885	5,179	4,624	4,176	3,363	2,814	2,420	2,122	1,890	1,703	1,423	1,221	1,070	857	715
135	52,646	32,699	23,714	18,602	15,303	12,998	9,989	8,111	6,828	5,895	5,187	4,630	4,181	3,366	2,817	2,421	2,124	1,891	1,704	1,423	1,222	1,070	858	715
140	53,389	32,984	23,863	18,694	15,366	13,043	10,016	8,129	6,840	5,904	5,194	4,636	4,186	3,369	2,819	2,423	2,125	1,892	1,705	1,424	1,222	1,071	858	715
145	54,101	33,254	24,004	18,781	15,424	13,085	10,041	8,145	6,852	5,913	5,200	4,641	4,190	3,372	2,821	2,424	2,126	1,893	1,706	1,424	1,222	1,071	858	716
150	54,782	33,510	24,138	18,862	15,479	13,125	10,064	8,161	6,863	5,921	5,207	4,646	4,194	3,374	2,823	2,425	2,127	1,894	1,706	1,425	1,223	1,071	858	716
155	55,435	33,754	24,264	18,939	15,531	13,162	10,086	8,175	6,873	5,929	5,212	4,651	4,198	3,377	2,824	2,427	2,128	1,894	1,707	1,425	1,223	1,071	858	716
160	56,062	33,985	24,383	19,011	15,579	13,197	10,106	8,188	6,882	5,936	5,218	4,655	4,202	3,379	2,826	2,423	2,129	1,895	1,708	1,426	1,223	1,072	858	716
165	56,663	34,205	24,496	19,080	15,626	13,230	10,126	8,201	6,891	5,942	5,223	4,659	4,205	3,381	2,827	2,429	2,130	1,896	1,708	1,426	1,224	1,072	859	716
170	57,242	34,415	24,603	19,145	15,669	13,261	10,144	8,213	6,900	5,949	5,228	4,663	4,208	3,383	2,829	2,430	2,130	1,896	1,709	1,426	1,224	1,072	859	716
175	57,798	34,615	24,706	19,207	15,711	13,291	10,161	8,224	6,908	5,955	5,232	4,667	4,211	3,385	2,830	2,431	2,131	1,897	1,709	1,427	1,224	1,072	859	716
180	58,333	34,806	24,803	19,266	15,750	13,319	10,178	8,235	6,915	5,960	5,237	4,670	4,214	3,387	2,831	2,432	2,132	1,898	1,710	1,427	1,224	1,072	859	716
185	58,848	34,989	24,896	19,322	15,787	13,346	10,193	8,245	6,923	5,966	5,241	4,673	4,217	3,389	2,833	2,433	2,133	1,898	1,710	1,427	1,225	1,072	859	716
190	59,345	35,164	24,984	19,375	15,823	13,371	10,208	8,255	6,929	5,971	5,245	4,676	4,219	3,390	2,834	2,434	2,133	1,899	1,710	1,428	1,225	1,073	859	716
195	59,824	35,332	25,069	19,426	15,857	13,395	10,222	8,264	6,936	5,975	5,249	4,679	4,222	3,392	2,835	2,435	2,134	1,899	1,711	1,428	1,225	1,073	859	717
200	60,287	35,493	25,149	19,474	15,889	13,418	10,235	8,273	6,942	5,980	5,252	4,682	4,224	3,393	2,836	2,436	2,134	1,900	1,711	1,428	1,225	1,073	859	717

8 AWG | 1 Run(s) | 480 Volts | C-Value = 1,557 | Magnetic Duct | Copper

Calculations:

1. $f = \dfrac{1.732 \times \text{Length} \times \text{Isca}}{\text{Runs} \times \text{C-Value} \times E_{L\text{-}L}}$

2. $M = \dfrac{1}{1 + f}$

3. Isc = Isca x M

4. Add Motor Contribution = Motor FLA x 4

* All results are given in symmetrical amperes

Three-Phase Bolted Fault-Current* Table

One-Way Distance in Feet

Available Isc in Thousands (Isca)	5	10	15	20	25	30	40	50	60	70	80	90	100	125	150	175	200	225	250	300	350	400	500	600
3	2,935	2,872	2,812	2,754	2,699	2,646	2,545	2,453	2,366	2,286	2,211	2,140	2,074	1,926	1,797	1,684	1,585	1,497	1,418	1,283	1,171	1,077	928	816
5	4,821	4,654	4,498	4,352	4,216	4,088	3,853	3,644	3,457	3,288	3,134	2,995	2,867	2,591	2,363	2,172	2,010	1,870	1,748	1,547	1,387	1,258	1,059	915
7	6,653	6,340	6,054	5,793	5,554	5,333	4,941	4,603	4,308	4,048	3,818	3,613	3,429	3,041	2,732	2,480	2,270	2,094	1,942	1,697	1,507	1,355	1,128	966
10	9,308	8,705	8,175	7,707	7,289	6,914	6,269	5,734	5,283	4,898	4,565	4,275	4,019	3,497	3,094	2,775	2,515	2,300	2,119	1,830	1,611	1,438	1,185	1,007
15	13,494	12,263	11,238	10,371	9,628	8,984	7,925	7,089	6,413	5,854	5,385	4,985	4,641	3,958	3,450	3,058	2,745	2,491	2,280	1,949	1,702	1,511	1,234	1,042
20	17,410	15,413	13,828	12,538	11,468	10,566	9,131	8,039	7,180	6,487	5,916	5,437	5,030	4,237	3,660	3,222	2,877	2,599	2,370	2,015	1,752	1,550	1,259	1,061
25	21,079	18,222	16,046	14,335	12,953	11,815	10,048	8,741	7,735	6,937	6,288	5,750	5,297	4,425	3,799	3,329	2,962	2,668	2,427	2,056	1,783	1,574	1,276	1,072
30	24,526	20,741	17,969	15,850	14,178	12,825	10,770	9,282	8,156	7,273	6,563	5,979	5,491	4,559	3,898	3,405	3,022	2,716	2,467	2,085	1,805	1,591	1,286	1,080
35	27,769	23,014	19,650	17,144	15,204	13,659	11,352	9,712	8,485	7,534	6,775	6,154	5,638	4,661	3,972	3,461	3,066	2,752	2,496	2,105	1,820	1,603	1,294	1,085
40	30,826	25,075	21,133	18,262	16,077	14,360	11,832	10,061	8,751	7,742	6,943	6,293	5,754	4,739	4,029	3,504	3,100	2,779	2,519	2,121	1,832	1,612	1,300	1,090
45	33,713	26,953	22,451	19,238	16,829	14,956	12,234	10,350	8,969	7,913	7,079	6,405	5,847	4,803	4,075	3,538	3,127	2,801	2,537	2,134	1,842	1,620	1,305	1,093
50	36,443	28,670	23,630	20,097	17,483	15,470	12,576	10,593	9,151	8,054	7,192	6,497	5,924	4,854	4,112	3,566	3,149	2,819	2,551	2,144	1,849	1,626	1,309	1,096
55	39,029	30,247	24,691	20,859	18,057	15,918	12,870	10,801	9,306	8,174	7,288	6,575	5,989	4,898	4,143	3,590	3,167	2,833	2,563	2,153	1,855	1,630	1,312	1,098
60	41,483	31,699	25,650	21,540	18,565	16,312	13,126	10,981	9,439	8,276	7,369	6,641	6,044	4,934	4,169	3,609	3,182	2,845	2,573	2,160	1,861	1,634	1,315	1,100
65	43,813	33,042	26,522	22,151	19,017	16,660	13,351	11,138	9,554	8,365	7,439	6,698	6,091	4,966	4,191	3,626	3,195	2,856	2,581	2,166	1,865	1,638	1,317	1,101
70	46,029	34,287	27,318	22,704	19,423	16,971	13,549	11,276	9,656	8,443	7,501	6,747	6,132	4,993	4,211	3,641	3,206	2,865	2,589	2,171	1,869	1,641	1,319	1,102
75	48,139	35,445	28,048	23,206	19,789	17,250	13,726	11,398	9,745	8,511	7,555	6,791	6,168	5,017	4,228	3,653	3,216	2,873	2,595	2,175	1,872	1,643	1,320	1,104
80	50,151	36,523	28,719	23,663	20,121	17,501	13,885	11,508	9,825	8,572	7,602	6,830	6,200	5,038	4,243	3,664	3,225	2,879	2,601	2,179	1,875	1,646	1,322	1,105
85	52,071	37,531	29,339	24,082	20,423	17,729	14,028	11,606	9,897	8,626	7,645	6,864	6,228	5,057	4,256	3,674	3,232	2,886	2,606	2,183	1,878	1,648	1,323	1,106
90	53,906	38,475	29,913	24,467	20,699	17,937	14,158	11,695	9,961	8,675	7,684	6,895	6,254	5,073	4,268	3,683	3,239	2,891	2,610	2,186	1,880	1,649	1,324	1,106
95	55,660	39,361	30,445	24,823	20,953	18,127	14,276	11,775	10,020	8,720	7,718	6,923	6,277	5,088	4,279	3,691	3,245	2,896	2,614	2,189	1,882	1,651	1,325	1,107
100	57,340	40,193	30,941	25,151	21,187	18,302	14,385	11,849	10,073	8,760	7,750	6,948	6,297	5,102	4,288	3,698	3,251	2,900	2,618	2,191	1,884	1,652	1,326	1,108
105	58,950	40,978	31,404	25,456	21,403	18,463	14,484	11,916	10,121	8,796	7,778	6,971	6,316	5,115	4,297	3,705	3,256	2,904	2,621	2,193	1,886	1,654	1,327	1,108
110	60,493	41,718	31,836	25,740	21,603	18,612	14,575	11,978	10,166	8,830	7,805	6,993	6,334	5,126	4,305	3,711	3,261	2,908	2,624	2,195	1,887	1,655	1,328	1,109
115	61,975	42,417	32,242	26,004	21,789	18,749	14,660	12,035	10,207	8,861	7,829	7,012	6,349	5,136	4,312	3,716	3,265	2,911	2,627	2,197	1,889	1,656	1,329	1,109
120	63,399	43,079	32,623	26,252	21,962	18,878	14,738	12,087	10,245	8,890	7,851	7,030	6,364	5,146	4,319	3,721	3,269	2,914	2,629	2,199	1,890	1,657	1,329	1,110
125	64,767	43,707	32,982	26,483	22,124	18,997	14,811	12,136	10,280	8,916	7,872	7,046	6,378	5,155	4,325	3,726	3,272	2,917	2,632	2,201	1,891	1,658	1,330	1,110
130	66,084	44,303	33,320	26,701	22,276	19,109	14,878	12,182	10,312	8,941	7,891	7,062	6,390	5,163	4,331	3,730	3 276	2,920	2,634	2,202	1,892	1,659	1,330	1,111
135	67,352	44,869	33,639	26,906	22,418	19,214	14,942	12,224	10,343	8,963	7,909	7,076	6,402	5,171	4,336	3,734	3 279	2,922	2,636	2,204	1,893	1,659	1,331	1,111
140	68,574	45,408	33,942	27,099	22,552	19,312	15,001	12,264	10,371	8,985	7,925	7,089	6,413	5,178	4,341	3,738	3 282	2,925	2,638	2,205	1,894	1,660	1,331	1,111
145	69,752	45,922	34,228	27,281	22,678	19,404	15,057	12,301	10,398	9,005	7,941	7,102	6,423	5,184	4,346	3,741	3 284	2,927	2,639	2,206	1,895	1,661	1,332	1,112
150	70,889	46,411	34,499	27,453	22,797	19,491	15,109	12,336	10,423	9,023	7,955	7,113	6,432	5,190	4,350	3,744	3,287	2,929	2,641	2,207	1,896	1,662	1,332	1,112
155	71,987	46,879	34,757	27,616	22,909	19,573	15,158	12,369	10,446	9,041	7,969	7,124	6,441	5,196	4,355	3,747	3,289	2,930	2,642	2,208	1,897	1,662	1,333	1,112
160	73,047	47,327	35,002	27,770	23,015	19,651	15,205	12,399	10,468	9,057	7,982	7,134	6,450	5,202	4,358	3,750	3,291	2,932	2,644	2,209	1,897	1,663	1,333	1,112
165	74,072	47,755	35,236	27,917	23,116	19,724	15,249	12,429	10,489	9,073	7,994	7,144	6,458	5,207	4,362	3,753	3,293	2,934	2,645	2,210	1,898	1,663	1,333	1,113
170	75,063	48,165	35,459	28,057	23,212	19,794	15,290	12,456	10,509	9,088	8,005	7,153	6,465	5,212	4,365	3,755	3,295	2,935	2,646	2,211	1,899	1,664	1,334	1,113
175	76,022	48,558	35,671	28,190	23,303	19,860	15,330	12,482	10,527	9,101	8,016	7,162	6,472	5,216	4,369	3,758	3,297	2,937	2,648	2,212	1,899	1,664	1,334	1,113
180	76,950	48,935	35,874	28,317	23,389	19,922	15,367	12,507	10,545	9,115	8,026	7,170	6,479	5,221	4,372	3,760	3,299	2,938	2,649	2,213	1,900	1,665	1,334	1,113
185	77,850	49,297	36,069	28,437	23,472	19,982	15,403	12,531	10,561	9,127	8,036	7,178	6,485	5,225	4,374	3,762	3,300	2,939	2,650	2,213	1,900	1,665	1,334	1,113
190	78,721	49,645	36,255	28,553	23,550	20,039	15,436	12,553	10,577	9,139	8,045	7,185	6,491	5,228	4,377	3,764	3,302	2,941	2,651	2,214	1,901	1,665	1,335	1,114
195	79,567	49,980	36,433	28,663	23,625	20,093	15,469	12,574	10,592	9,150	8,054	7,192	6,497	5,232	4,380	3,766	3,303	2,942	2,652	2,215	1,901	1,666	1,335	1,114
200	80,387	50,303	36,604	28,769	23,697	20,145	15,499	12,595	10,607	9,161	8,062	7,199	6,502	5,236	4,382	3,768	3,305	2,943	2,653	2,215	1,902	1,666	1,335	1,114

# 6 AWG	1 Run(s)	480 Volts	Copyright © 1994 - V.F. Christoffer - All Rights Reserved	C-Value = 2,425	Magnetic Duct	Copper

Calculations:

1. $f = \dfrac{1.732 \times \text{Length} \times \text{Isca}}{\text{Runs} \times \text{C-Value} \times \text{E L-L}}$

2. $M = \dfrac{1}{1 + f}$

3. $\text{Isc} = \text{Isca} \times M$

4. Add Motor Contribution = Motor FLA x 4

* All results are given in symmetrical amperes

Three-Phase Bolted Fault-Current* Table

One-Way Distance in Feet

Available Isc in Thousands (Isca)	5	10	15	20	25	30	40	50	60	70	80	90	100	125	150	175	200	225	250	300	350	400	500	600
3	2,958	2,917	2,877	2,839	2,801	2,764	2,694	2,626	2,563	2,502	2,444	2,389	2,336	2,213	2,103	2,003	1,912	1,829	1,753	1,619	1,503	1,403	1,239	1,108
5	4,884	4,774	4,668	4,567	4,470	4,377	4,203	4,042	3,893	3,754	3,625	3,505	3,392	3,140	2,922	2,733	2,567	2,419	2,288	2,064	1,880	1,726	1,484	1,301
7	6,775	6,564	6,366	6,180	6,004	5,838	5,532	5,256	5,006	4,780	4,572	4,382	4,208	3,826	3,508	3,239	3,008	2,808	2,632	2,340	2,107	1,915	1,621	1,405
10	9,547	9,134	8,755	8,406	8,084	7,786	7,250	6,784	6,374	6,011	5,687	5,396	5,133	4,576	4,129	3,761	3,453	3,192	2,967	2,601	2,316	2,087	1,742	1,495
15	14,004	13,132	12,363	11,678	11,066	10,514	9,561	8,767	8,094	7,517	7,017	6,579	6,193	5,400	4,788	4,300	3,902	3,572	3,293	2,848	2,509	2,243	1,849	1,574
20	18,268	16,812	15,571	14,501	13,568	12,748	11,374	10,267	9,356	8,594	7,946	7,390	6,906	5,934	5,203	4,632	4,173	3,798	3,484	2,990	2,619	2,330	1,908	1,616
25	22,351	20,210	18,443	16,960	15,698	14,611	12,833	11,441	10,322	9,402	8,632	7,979	7,418	6,309	5,488	4,856	4,355	3,948	3,610	3,082	2,689	2,385	1,945	1,642
30	26,265	23,357	21,029	19,122	17,533	16,188	14,034	12,386	11,084	10,030	9,159	8,428	7,804	6,586	5,697	5,019	4,485	4,054	3,699	3,147	2,739	2,424	1,971	1,661
35	30,019	26,280	23,369	21,038	19,130	17,540	15,039	13,162	11,702	10,533	9,577	8,780	8,105	6,799	5,855	5,142	4,583	4,134	3,765	3,195	2,775	2,452	1,990	1,674
40	33,624	29,002	25,497	22,747	20,533	18,712	15,893	13,812	12,212	10,945	9,916	9,064	8,347	6,968	5,981	5,238	4,660	4,196	3,817	3,232	2,803	2,474	2,004	1,684
45	37,088	31,543	27,440	24,282	21,775	19,738	16,627	14,363	12,641	11,288	10,197	9,298	8,545	7,106	6,082	5,315	4,721	4,246	3,857	3,261	2,825	2,491	2,015	1,692
50	40,420	33,921	29,222	25,667	22,882	20,643	17,264	14,836	13,007	11,579	10,433	9,494	8,710	7,220	6,165	5,379	4,771	4,286	3,891	3,285	2,842	2,505	2,024	1,698
55	43,626	36,150	30,862	26,923	23,876	21,448	17,824	15,247	13,322	11,828	10,635	9,661	8,850	7,316	6,235	5,432	4,812	4,320	3,919	3,305	2,857	2,516	2,032	1,704
60	46,714	38,245	32,375	28,068	24,772	22,169	18,319	15,608	13,596	12,044	10,809	9,805	8,971	7,398	6,294	5,477	4,848	4,348	3,942	3,321	2,870	2,526	2,038	1,708
65	49,690	40,217	33,777	29,116	25,584	22,817	18,759	15,927	13,837	12,233	10,961	9,929	9,075	7,469	6,345	5,516	4,878	4,373	3,962	3,336	2,880	2,534	2,043	1,712
70	52,560	42,076	35,079	30,078	26,325	23,404	19,154	16,210	14,051	12,399	11,095	10,039	9,167	7,530	6,390	5,549	4,904	4,394	3,979	3,348	2,889	2,541	2,048	1,715
75	55,329	43,833	36,292	30,965	27,002	23,938	19,510	16,465	14,242	12,547	11,213	10,136	9,247	7,585	6,429	5,579	4,927	4,412	3,994	3,358	2,897	2,547	2,052	1,718
80	58,004	45,495	37,424	31,785	27,623	24,425	19,832	16,694	14,413	12,680	11,319	10,222	9,319	7,633	6,464	5,605	4,948	4,428	4,008	3,368	2,904	2,553	2,055	1,720
85	60,588	47,069	38,483	32,546	28,196	24,872	20,126	16,901	14,567	12,799	11,414	10,300	9,383	7,676	6,495	5,628	4,966	4,443	4,020	3,376	2,910	2,558	2,058	1,722
90	63,086	48,563	39,476	33,253	28,725	25,283	20,394	17,090	14,707	12,907	11,500	10,369	9,441	7,715	6,522	5,649	4,982	4,456	4,030	3,384	2,916	2,562	2,061	1,724
95	65,502	49,983	40,408	33,913	29,216	25,662	20,640	17,262	14,835	13,005	11,578	10,433	9,494	7,750	6,547	5,668	4,997	4,467	4,040	3,390	2,921	2,566	2,064	1,726
100	67,841	51,333	41,287	34,529	29,672	26,013	20,867	17,421	14,951	13,095	11,649	10,490	9,541	7,782	6,570	5,685	5,010	4,478	4,048	3,397	2,925	2,569	2,066	1,728
105	70,106	52,619	42,115	35,106	30,097	26,340	21,076	17,566	15,058	13,177	11,714	10,543	9,585	7,811	6,591	5,700	5,022	4,488	4,056	3,402	2,930	2,572	2,068	1,729
110	72,300	53,846	42,897	35,648	30,495	26,643	21,270	17,701	15,157	13,253	11,774	10,591	9,625	7,837	6,609	5,714	5,033	4,496	4,063	3,407	2,933	2,575	2,070	1,730
115	74,427	55,017	43,636	36,157	30,867	26,927	21,451	17,826	15,249	13,323	11,829	10,636	9,662	7,861	6,627	5,727	5,043	4,504	4,070	3,412	2,937	2,578	2,072	1,731
120	76,490	56,136	44,337	36,637	31,216	27,192	21,619	17,942	15,333	13,387	11,880	10,677	9,696	7,884	6,643	5,739	5,052	4,512	4,076	3,416	2,940	2,580	2,073	1,733
125	78,491	57,206	45,003	37,090	31,544	27,441	21,776	18,049	15,412	13,447	11,927	10,715	9,727	7,905	6,657	5,750	5,060	4,518	4,081	3,420	2,943	2,582	2,075	1,734
130	80,434	58,231	45,634	37,518	31,853	27,675	21,923	18,150	15,486	13,503	11,971	10,751	9,756	7,924	6,671	5,760	5,068	4,525	4,086	3,423	2,945	2,585	2,076	1,735
135	82,320	59,213	46,236	37,924	32,145	27,895	22,060	18,245	15,554	13,555	12,012	10,784	9,783	7,942	6,684	5,770	5,076	4,531	4,091	3,427	2,948	2,586	2,077	1,735
140	84,153	60,156	46,808	38,308	32,421	28,102	22,190	18,333	15,618	13,604	12,050	10,814	9,809	7,959	6,696	5,779	5,082	4,536	4,096	3,430	2,950	2,588	2,078	1,736
145	85,934	61,061	47,354	38,673	32,682	28,298	22,312	18,416	15,679	13,650	12,086	10,843	9,833	7,974	6,707	5,787	5,089	4,541	4,100	3,433	2,952	2,590	2,079	1,737
150	87,666	61,930	47,875	39,020	32,929	28,483	22,427	18,495	15,736	13,693	12,119	10,870	9,855	7,989	6,717	5,794	5,095	4,546	4,104	3,435	2,954	2,591	2,080	1,738
155	89,350	62,766	48,373	39,350	33,164	28,659	22,536	18,568	15,789	13,733	12,151	10,896	9,876	8,003	6,727	5,802	5,100	4,550	4,107	3,438	2,956	2,593	2,081	1,738
160	90,989	63,570	48,850	39,665	33,387	28,825	22,638	18,638	15,839	13,771	12,181	10,920	9,895	8,016	6,736	5,809	5,106	4,554	4,111	3,440	2,958	2,594	2,082	1,739
165	92,585	64,345	49,306	39,965	33,600	28,983	22,736	18,704	15,887	13,807	12,209	10,943	9,914	8,028	6,744	5,815	5,111	4,558	4,114	3,443	2,960	2,595	2,083	1,739
170	94,138	65,091	49,743	40,252	33,802	29,134	22,828	18,767	15,932	13,841	12,236	10,964	9,932	8,039	6,753	5,821	5,115	4,562	4,117	3,445	2,961	2,597	2,084	1,740
175	95,652	65,811	50,162	40,526	33,995	29,277	22,916	18,826	15,975	13,874	12,261	10,984	9,948	8,050	6,760	5,827	5,120	4,566	4,120	3,447	2,963	2,598	2,084	1,740
180	97,126	66,506	50,565	40,788	34,180	29,414	23,000	18,883	16,016	13,904	12,285	11,003	9,964	8,060	6,767	5,832	5,124	4,569	4,122	3,449	2,964	2,599	2,085	1,741
185	98,564	67,177	50,952	41,040	34,356	29,544	23,080	18,936	16,054	13,933	12,308	11,022	9,979	8,070	6,774	5,837	5,128	4,572	4,125	3,450	2,965	2,600	2,086	1,741
190	99,965	67,825	51,324	41,281	34,525	29,669	23,156	18,987	16,091	13,961	12,329	11,039	9,993	8,079	6,781	5,842	5,131	4,575	4,127	3,452	2,967	2,601	2,086	1,742
195	101,332	68,452	51,682	41,512	34,686	29,788	23,228	19,036	16,126	13,987	12,350	11,055	10,007	8,088	6,787	5,847	5,135	4,578	4,130	3,454	2,968	2,602	2,087	1,742
200	102,666	69,058	52,027	41,734	34,841	29,903	23,298	19,083	16,159	14,013	12,369	11,071	10,019	8,097	6,793	5,851	5,138	4,581	4,132	3,455	2,969	2,603	2,088	1,743

4 AWG | 1 Run(s) | 480 Volts | | C-Value = 3,806 | Magnetic Duct | Copper

Calculations:

1. $f = \dfrac{1.732 \times \text{Length} \times \text{Isca}}{\text{Runs} \times \text{C-Value} \times E_{L\text{-}L}}$

2. $M = \dfrac{1}{1 + f}$

3. Isc = Isca x M

4. Add Motor Contribution = Motor FLA x 4

* All results are given in symmetrical amperes

Three-Phase Bolted Fault-Current* Table

One-Way Distance in Feet

Available Isc in Thousands (Isca)

Isca	5	10	15	20	25	30	40	50	60	70	80	90	100	125	150	175	200	225	250	300	350	400	500	600
3	2,966	2,933	2,901	2,869	2,839	2,808	2,750	2,694	2,640	2,588	2,538	2,490	2,444	2,336	2,237	2,146	2,062	1,985	1,913	1,783	1,670	1,571	1,404	1,269
5	4,907	4,817	4,731	4,648	4,567	4,490	4,342	4,203	4,074	3,952	3,837	3,728	3,626	3,393	3,188	3,006	2,844	2,699	2,567	2,340	2,149	1,987	1,727	1,527
7	6,819	6,647	6,484	6,328	6,180	6,039	5,774	5,532	5,310	5,104	4,914	4,737	4,573	4,209	3,898	3,630	3,396	3,191	3,009	2,701	2,450	2,242	1,916	1,673
10	9,635	9,295	8,979	8,683	8,407	8,147	7,673	7,251	6,874	6,533	6,225	5,944	5,688	5,135	4,679	4,298	3,974	3,696	3,454	3,054	2,737	2,480	2,088	1,802
15	14,193	13,469	12,814	12,221	11,680	11,185	10,310	9,563	8,917	8,352	7,855	7,413	7,019	6,195	5,544	5,017	4,581	4,215	3,904	3,400	3,012	2,704	2,244	1,918
20	18,591	17,367	16,294	15,347	14,503	13,747	12,450	11,376	10,473	9,703	9,038	8,458	7,949	6,908	6,108	5,475	4,960	4,534	4,175	3,605	3,171	2,831	2,331	1,981
25	22,836	21,017	19,466	18,129	16,963	15,938	14,220	12,837	11,698	10,745	9,936	9,240	8,635	7,421	6,506	5,792	5,219	4,749	4,357	3,739	3,275	2,914	2,386	2,021
30	26,937	24,442	22,369	20,621	19,126	17,833	15,710	14,038	12,688	11,574	10,641	9,847	9,163	7,807	6,801	6,024	5,407	4,904	4,487	3,835	3,348	2,971	2,425	2,048
35	30,901	27,661	25,036	22,866	21,043	19,488	16,980	15,043	13,504	12,250	11,209	10,331	9,581	8,108	7,028	6,202	5,550	5,022	4,585	3,906	3,403	3,014	2,453	2,069
40	34,734	30,693	27,495	24,900	22,752	20,946	18,076	15,898	14,188	12,810	11,676	10,727	9,920	8,350	7,209	6,343	5,662	5,113	4,662	3,962	3,444	3,047	2,475	2,084
45	38,443	33,554	29,768	26,750	24,287	22,240	19,032	16,632	14,770	13,283	12,068	11,056	10,201	8,549	7,357	6,457	5,753	5,187	4,723	4,006	3,478	3,073	2,492	2,096
50	42,034	36,257	31,877	28,441	25,673	23,396	19,872	17,270	15,271	13,687	12,400	11,335	10,438	8,714	7,479	6,551	5,827	5,248	4,773	4,042	3,505	3,094	2,506	2,106
55	45,512	38,816	33,838	29,991	26,930	24,436	20,617	17,830	15,707	14,036	12,686	11,573	10,640	8,854	7,582	6,629	5,890	5,298	4,815	4,072	3,527	3,111	2,518	2,114
60	48,883	41,242	35,667	31,419	28,076	25,375	21,282	18,325	16,090	14,341	12,935	11,780	10,814	8,975	7,670	6,697	5,943	5,341	4,850	4,097	3,546	3,126	2,527	2,121
65	52,152	43,544	37,376	32,738	29,124	26,229	21,879	18,766	16,429	14,610	13,153	11,960	10,966	9,079	7,746	6,755	5,988	5,378	4,880	4,119	3,562	3,139	2,535	2,127
70	55,322	45,733	38,976	33,960	30,087	27,007	22,418	19,161	16,731	14,848	13,346	12,120	11,100	9,171	7,813	6,805	6,028	5,410	4,907	4,137	3,576	3,150	2,543	2,132
75	58,399	47,815	40,479	35,095	30,974	27,720	22,907	19,518	17,002	15,061	13,518	12,261	11,218	9,252	7,871	6,850	6,063	5,438	4,930	4,154	3,589	3,159	2,549	2,136
80	61,386	49,799	41,892	36,152	31,795	28,376	23,352	19,840	17,246	15,252	13,672	12,388	11,324	9,323	7,923	6,889	6,093	5,463	4,950	4,168	3,599	3,167	2,554	2,140
85	64,288	51,692	43,224	37,139	32,556	28,980	23,760	20,134	17,468	15,425	13,810	12,502	11,419	9,388	7,970	6,924	6,121	5,485	4,968	4,181	3,609	3,175	2,559	2,143
90	67,108	53,500	44,480	38,063	33,264	29,540	24,135	20,402	17,670	15,582	13,936	12,605	11,505	9,446	8,012	6,956	6,145	5,504	4,984	4,192	3,618	3,181	2,563	2,146
95	69,849	55,228	45,668	38,930	33,924	30,059	24,481	20,649	17,854	15,726	14,051	12,698	11,583	9,498	8,049	6,984	6,168	5,522	4,999	4,203	3,625	3,187	2,567	2,149
100	72,515	56,881	46,793	39,744	34,541	30,542	24,800	20,876	18,023	15,857	14,155	12,784	11,654	9,546	8,084	7,010	6,188	5,538	5,012	4,212	3,632	3,193	2,571	2,151
105	75,108	58,465	47,859	40,511	35,118	30,993	25,097	21,085	18,180	15,978	14,251	12,862	11,719	9,590	8,115	7,033	6,206	5,553	5,024	4,220	3,638	3,197	2,574	2,154
110	77,633	59,983	48,872	41,234	35,660	31,414	25,372	21,280	18,324	16,089	14,340	12,934	11,779	9,629	8,143	7,055	6,223	5,566	5,035	4,228	3,644	3,202	2,577	2,156
115	80,090	61,439	49,834	41,917	36,170	31,809	25,629	21,460	18,457	16,192	14,422	13,000	11,834	9,666	8,170	7,074	6,238	5,579	5,045	4,235	3,649	3,206	2,579	2,157
120	82,484	62,838	50,751	42,563	36,651	32,180	25,870	21,628	18,582	16,287	14,497	13,062	11,885	9,700	8,194	7,093	6,252	5,590	5,054	4,242	3,654	3,210	2,582	2,159
125	84,816	64,183	51,624	43,176	37,104	32,529	26,095	21,785	18,697	16,376	14,568	13,119	11,932	9,732	8,216	7,109	6,265	5,600	5,063	4,248	3,659	3,213	2,584	2,161
130	87,088	65,476	52,457	43,757	37,532	32,858	26,306	21,932	18,806	16,459	14,633	13,172	11,976	9,761	8,237	7,125	6,277	5,610	5,071	4,253	3,663	3,216	2,586	2,162
135	89,304	66,720	53,253	44,310	37,938	33,169	26,504	22,070	18,907	16,537	14,695	13,222	12,017	9,788	8,257	7,139	6,289	5,619	5,078	4,259	3,667	3,219	2,588	2,163
140	91,465	67,919	54,014	44,835	38,323	33,462	26,692	22,200	19,002	16,609	14,752	13,268	12,056	9,814	8,275	7,153	6,299	5,627	5,085	4,263	3,670	3,222	2,590	2,165
145	93,573	69,075	54,742	45,336	38,688	33,740	26,868	22,322	19,091	16,678	14,806	13,312	12,092	9,837	8,292	7,166	6,309	5,635	5,091	4,268	3,674	3,225	2,591	2,166
150	95,630	70,189	55,440	45,813	39,035	34,004	27,035	22,437	19,176	16,742	14,856	13,353	12,125	9,860	8,307	7,177	6,318	5,642	5,097	4,272	3,677	3,227	2,593	2,167
155	97,638	71,265	56,109	46,269	39,366	34,255	27,193	22,546	19,255	16,802	14,904	13,391	12,157	9,881	8,322	7,189	6,327	5,649	5,103	4,276	3,680	3,229	2,594	2,168
160	99,599	72,304	56,751	46,705	39,680	34,493	27,343	22,649	19,330	16,860	14,949	13,427	12,187	9,900	8,336	7,199	6,335	5,656	5,108	4,280	3,682	3,231	2,596	2,169
165	101,514	73,308	57,368	47,122	39,981	34,720	27,486	22,746	19,401	16,914	14,991	13,462	12,215	9,919	8,349	7,209	6,342	5,662	5,113	4,283	3,685	3,233	2,597	2,170
170	103,385	74,278	57,960	47,521	40,268	34,936	27,621	22,839	19,468	16,965	15,032	13,494	12,242	9,937	8,362	7,218	6,349	5,668	5,118	4,286	3,687	3,235	2,598	2,171
175	105,213	75,217	58,531	47,903	40,542	35,142	27,750	22,927	19,532	17,013	15,070	13,525	12,267	9,953	8,374	7,227	6,356	5,673	5,122	4,289	3,690	3,237	2,599	2,171
180	107,000	76,126	59,079	48,270	40,805	35,339	27,872	23,011	19,593	17,059	15,106	13,554	12,291	9,969	8,385	7,235	6,363	5,678	5,126	4,292	3,692	3,239	2,600	2,172
185	108,747	77,006	59,608	48,623	41,056	35,528	27,990	23,090	19,651	17,103	15,140	13,581	12,314	9,984	8,395	7,243	6,369	5,683	5,130	4,295	3,694	3,240	2,601	2,173
190	110,455	77,859	60,118	48,961	41,298	35,708	28,102	23,166	19,706	17,145	15,173	13,608	12,335	9,998	8,405	7,250	6,375	5,687	5,134	4,298	3,696	3,242	2,602	2,173
195	112,127	78,686	60,610	49,287	41,529	35,881	28,208	23,239	19,758	17,185	15,204	13,633	12,356	10,012	8,415	7,258	6,380	5,692	5,138	4,300	3,698	3,243	2,603	2,174
200	113,762	79,488	61,084	49,601	41,751	36,047	28,311	23,309	19,809	17,222	15,234	13,657	12,375	10,024	8,424	7,264	6,385	5,696	5,141	4,303	3,699	3,244	2,604	2,175

# 3 AWG	1 Run(s)	480 Volts	C-Value = 4,760	Magnetic Duct	Copper

Calculations:

1. $f = \frac{1.732 \times \text{Length} \times \text{Isca}}{\text{Runs} \times \text{C-Value} \times E_{L-L}}$

2. $M = \frac{1}{1 + f}$

3. Isc = Isca x M

4. Add Motor Contribution = Motor FLA x 4

* All results are given in symmetrical amperes

Three-Phase Bolted Fault-Current* Table

One-Way Distance in Feet

Available Isc in Thousands (Isca)	5	10	15	20	25	30	40	50	60	70	80	90	100	125	150	175	200	225	250	300	350	400	500	600
3	2,973	2,946	2,920	2,894	2,869	2,844	2,795	2,748	2,703	2,659	2,616	2,575	2,535	2,441	2,353	2,271	2,195	2,124	2,057	1,9[illegible]6	1,828	1,731	1,565	1,429
5	4,925	4,852	4,781	4,712	4,645	4,580	4,456	4,337	4,226	4,119	4,018	3,922	3,830	3,618	3,429	3,258	3,104	2,963	2,835	2,6[illegible]9	2,416	2,250	1,978	1,765
7	6,853	6,713	6,578	6,448	6,324	6,204	5,977	5,767	5,571	5,387	5,216	5,054	4,903	4,561	4,264	4,004	3,773	3,567	3,383	3,0[illegible]6	2,804	2,582	2,230	1,963
10	9,704	9,424	9,160	8,911	8,675	8,451	8,036	7,660	7,318	7,004	6,717	6,452	6,207	5,670	5,218	4,833	4,501	4,211	3,957	3,5[illegible]0	3,186	2,904	2,466	2,143
15	14,343	13,741	13,187	12,677	12,204	11,765	10,976	10,287	9,678	9,138	8,655	8,220	7,827	6,991	6,317	5,761	5,295	4,899	4,558	4,001	3,565	3,215	2,687	2,308
20	18,848	17,822	16,902	16,072	15,320	14,635	13,434	12,415	11,540	10,780	10,114	9,525	9,001	7,913	7,060	6,373	5,807	5,334	4,932	4,237	3,790	3,397	2,813	2,401
25	23,226	21,687	20,340	19,150	18,092	17,144	15,519	14,175	13,045	12,082	11,252	10,528	9,892	8,593	7,596	6,807	6,166	5,635	5,188	4,479	3,940	3,516	2,895	2,460
30	27,481	25,353	23,531	21,953	20,573	19,357	17,309	15,654	14,288	13,141	12,164	11,322	10,590	9,115	8,001	7,130	6,430	5,855	5,374	4,616	4,046	3,601	2,951	2,501
35	31,619	28,834	26,500	24,515	22,807	21,322	18,864	16,915	15,331	14,018	12,912	11,968	11,152	9,529	8,318	7,381	6,633	6,023	5,515	4,720	4,125	3,664	2,994	2,531
40	35,645	32,144	29,270	26,868	24,830	23,079	20,227	18,002	16,219	14,756	13,536	12,502	11,615	9,865	8,573	7,580	6,794	6,155	5,626	4,801	4,187	3,712	3,026	2,554
45	39,562	35,296	31,861	29,035	26,669	24,660	21,431	18,950	16,984	15,387	14,065	12,952	12,002	10,143	8,782	7,744	6,925	6,262	5,716	4,856	4,236	3,751	3,052	2,572
50	43,375	38,300	34,288	31,037	28,349	26,090	22,503	19,783	17,650	15,932	14,519	13,336	12,331	10,377	8,957	7,879	7,033	6,351	5,789	4,919	4,276	3,782	3,072	2,587
55	47,088	41,167	36,568	32,894	29,890	27,389	23,463	20,521	18,235	16,407	14,912	13,667	12,614	10,576	9,105	7,994	7,124	6,425	5,851	4,964	4,310	3,809	3,090	2,599
60	50,706	43,905	38,713	34,619	31,308	28,575	24,328	21,180	18,753	16,825	15,257	13,956	12,860	10,748	9,233	8,092	7,202	6,488	5,903	5,001	4,338	3,831	3,104	2,609
65	54,232	46,524	40,735	36,227	32,617	29,662	25,111	21,771	19,215	17,196	15,561	14,210	13,075	10,899	9,343	8,176	7,269	6,542	5,948	5,033	4,363	3,850	3,117	2,618
70	57,668	49,031	42,644	37,729	33,830	30,661	25,824	22,305	19,630	17,528	15,832	14,436	13,266	11,031	9,440	8,251	7,327	6,590	5,987	5,061	4,384	3,866	3,127	2,626
75	61,020	51,433	44,449	39,135	34,956	31,583	26,475	22,789	20,004	17,825	16,075	14,637	13,436	11,148	9,526	8,316	7,379	6,631	6,021	5,086	4,402	3,880	3,137	2,632
80	64,289	53,736	46,159	40,454	36,005	32,437	27,072	23,230	20,343	18,094	16,293	14,818	13,588	11,252	9,602	8,374	7,424	6,668	6,052	5,108	4,418	3,893	3,145	2,638
85	67,479	55,946	47,780	41,695	36,984	33,230	27,622	23,634	20,652	18,338	16,490	14,981	13,725	11,346	9,670	8,426	7,465	6,701	6,079	5,127	4,433	3,904	3,152	2,643
90	70,592	58,070	49,321	42,863	37,900	33,967	28,130	24,004	20,934	18,560	16,670	15,129	13,849	11,431	9,732	8,472	7,502	6,731	6,103	5,144	4,445	3,914	3,159	2,648
95	73,632	60,111	50,785	43,965	38,759	34,656	28,600	24,346	21,194	18,764	16,834	15,264	13,962	11,508	9,788	8,515	7,535	6,757	6,125	5,160	4,457	3,923	3,164	2,652
100	76,600	62,075	52,180	45,006	39,566	35,300	29,037	24,662	21,433	18,951	16,985	15,388	14,065	11,578	9,838	8,553	7,565	6,781	6,145	5,174	4,468	3,931	3,170	2,656
105	79,500	63,966	53,510	45,992	40,326	35,903	29,444	24,955	21,654	19,124	17,123	15,501	14,160	11,642	9,885	8,588	7,592	6,803	6,163	5,186	4,477	3,938	3,175	2,659
110	82,334	65,787	54,779	46,926	41,043	36,470	29,825	25,228	21,859	19,283	17,251	15,606	14,248	11,701	9,927	8,620	7,617	6,823	6,179	5,198	4,486	3,945	3,179	2,662
115	85,103	67,544	55,991	47,813	41,719	37,004	30,180	25,482	22,049	19,431	17,369	15,703	14,328	11,756	9,966	8,649	7,640	6,842	6,194	5,209	4,494	3,951	3,183	2,665
120	87,811	69,238	57,150	48,656	42,360	37,506	30,514	25,719	22,227	19,569	17,479	15,793	14,403	11,806	10,002	8,677	7,661	6,859	6,208	5,219	4,501	3,957	3,187	2,667
125	90,458	70,874	58,260	49,458	42,966	37,981	30,828	25,942	22,393	19,698	17,582	15,876	14,473	11,853	10,036	8,702	7,681	6,874	6,221	5,228	4,508	3,962	3,190	2,670
130	93,048	72,454	59,324	50,222	43,542	38,430	31,123	26,150	22,548	19,818	17,677	15,954	14,537	11,896	10,067	8,725	7,699	6,889	6,233	5,236	4,514	3,967	3,193	2,672
135	95,582	73,981	60,343	50,951	44,089	38,856	31,401	26,347	22,694	19,930	17,767	16,027	14,598	11,936	10,096	8,747	7,716	6,903	6,244	5,244	4,520	3,972	3,196	2,674
140	98,062	75,458	61,322	51,647	44,609	39,259	31,664	26,532	22,831	20,036	17,851	16,095	14,654	11,974	10,123	8,767	7,732	6,915	6,255	5,251	4,525	3,976	3,199	2,676
145	100,489	76,887	62,263	52,313	45,105	39,643	31,913	26,706	22,960	20,135	17,930	16,160	14,707	12,010	10,148	8,786	7,747	6,927	6,264	5,258	4,530	3,980	3,201	2,678
150	102,865	78,270	63,167	52,950	45,577	40,007	32,149	26,871	23,082	20,229	18,004	16,220	14,757	12,043	10,172	8,804	7,760	6,938	6,273	5,264	4,535	3,983	3,204	2,679
155	105,192	79,610	64,037	53,559	46,029	40,354	32,373	27,027	23,197	20,317	18,074	16,277	14,804	12,074	10,194	8,821	7,773	6,948	6,282	5,270	4,540	3,987	3,206	2,681
160	107,471	80,909	64,874	54,144	46,460	40,685	32,586	27,175	23,306	20,401	18,140	16,330	14,849	12,104	10,215	8,836	7,786	6,958	6,290	5,276	4,544	3,990	3,208	2,682
165	109,704	82,168	65,681	54,705	46,872	41,001	32,788	27,316	23,409	20,480	18,203	16,381	14,891	12,131	10,235	8,851	7,797	6,967	6,297	5,281	4,548	3,993	3,210	2,684
170	111,892	83,389	66,459	55,244	47,267	41,303	32,981	27,450	23,507	20,555	18,262	16,429	14,930	12,158	10,254	8,865	7,808	6,976	6,304	5,286	4,551	3,996	3,212	2,685
175	114,037	84,575	67,210	55,762	47,646	41,592	33,165	27,577	23,601	20,626	18,318	16,474	14,968	12,183	10,271	8,878	7,818	6,984	6,311	5,291	4,555	3,998	3,213	2,686
180	116,139	85,725	67,935	56,260	48,009	41,868	33,340	27,698	23,689	20,694	18,371	16,517	15,003	12,206	10,288	8,891	7,828	6,992	6,317	5,295	4,558	4,001	3,215	2,687
185	118,200	86,843	68,635	56,739	48,357	42,133	33,508	27,814	23,774	20,759	18,422	16,559	15,037	12,229	10,304	8,903	7,837	6,999	6,323	5,300	4,561	4,003	3,217	2,688
190	120,222	87,929	69,312	57,200	48,692	42,387	33,668	27,924	23,855	20,820	18,471	16,598	15,069	12,250	10,319	8,914	7,846	7,006	6,329	5,304	4,564	4,006	3,218	2,689
195	122,204	88,985	69,966	57,645	49,014	42,631	33,822	28,030	23,932	20,879	18,517	16,635	15,100	12,270	10,334	8,925	7,854	7,013	6,334	5,307	4,567	4,008	3,219	2,690
200	124,150	90,012	70,599	58,075	49,324	42,865	33,969	28,131	24,005	20,935	18,561	16,670	15,129	12,290	10,347	8,935	7,862	7,019	6,340	5,311	4,570	4,010	3,221	2,691

# 2 AWG	1 Run(s)	480 Volts	C-Value = 5,906	Magnetic Duct	Copper

Calculations:

1. $f = \frac{1.732 \times \text{Length} \times \text{Isca}}{\text{Runs} \times \text{C-Value} \times \text{E L-L}}$

2. $M = \frac{1}{1 + f}$

3. $\text{Isc} = \text{Isca} \times M$

4. Add Motor Contribution = Motor FLA x 4

* All results are given in symmetrical amperes

Three-Phase Bolted Fault-Current* Table

One-Way Distance in Feet

Available Isc in Thousands (Isca)	5	10	15	20	25	30	40	50	60	70	80	90	100	125	150	175	200	225	250	300	350	400	500	600
3	2,978	2,956	2,935	2,913	2,893	2,872	2,832	2,793	2,755	2,718	2,682	2,646	2,612	2,530	2,454	2,381	2,313	2,249	2,188	2,076	1,974	1,882	1,722	1,587
5	4,939	4,879	4,821	4,764	4,709	4,655	4,550	4,450	4,354	4,262	4,174	4,089	4,008	3,819	3,647	3,489	3,345	3,212	3,089	2,870	2,680	2,513	2,235	2,012
7	6,881	6,766	6,654	6,546	6,442	6,341	6,148	5,967	5,796	5,634	5,481	5,336	5,199	4,885	4,607	4,358	4,135	3,934	3,751	3,433	3,164	2,934	2,562	2,274
10	9,759	9,528	9,309	9,099	8,899	8,707	8,348	8,017	7,711	7,427	7,164	6,919	6,690	6,178	5,740	5,359	5,026	4,732	4,470	4,025	3,660	3,356	2,878	2,520
15	14,463	13,964	13,497	13,061	12,652	12,268	11,566	10,940	10,378	9,871	9,411	8,993	8,610	7,781	7,098	6,525	6,037	5,618	5,253	4,649	4,169	3,779	3,184	2,751
20	19,057	18,199	17,415	16,695	16,033	15,421	14,328	13,379	12,549	11,815	11,162	10,578	10,052	8,940	8,050	7,321	6,713	6,198	5,757	5,039	4,480	4,033	3,362	2,883
25	23,544	22,248	21,087	20,041	19,095	18,233	16,724	15,446	14,349	13,398	12,565	11,829	11,175	9,818	8,755	7,899	7,196	6,608	6,108	5,306	4,691	4,203	3,479	2,968
30	27,927	26,122	24,536	23,132	21,880	20,756	18,823	17,219	15,867	14,712	13,714	12,842	12,075	10,506	9,297	8,338	7,559	6,912	6,368	5,501	4,842	4,324	3,562	3,028
35	32,211	29,833	27,782	25,996	24,425	23,033	20,676	18,757	17,164	15,820	14,672	13,679	12,811	11,059	9,728	8,683	7,841	7,147	6,567	5,649	4,956	4,415	3,623	3,072
40	36,398	33,391	30,843	28,656	26,759	25,097	22,325	20,104	18,285	16,768	15,483	14,381	13,426	11,514	10,078	8,961	8,067	7,335	6,725	5,765	5,046	4,486	3,671	3,107
45	40,492	36,805	33,733	31,134	28,908	26,978	23,801	21,293	19,263	17,587	16,179	14,980	13,946	11,894	10,368	9,190	8,252	7,487	6,853	5,859	5,117	4,542	3,709	3,134
50	44,496	40,083	36,466	33,449	30,892	28,698	25,130	22,351	20,125	18,302	16,782	15,495	14,392	12,217	10,613	9,381	8,406	7,614	6,959	5,936	5,176	4,589	3,739	3,156
55	48,412	43,234	39,056	35,614	32,730	30,278	26,333	23,297	20,889	18,932	17,310	15,945	14,779	12,494	10,822	9,544	8,536	7,721	7,048	6,001	5,225	4,627	3,765	3,174
60	52,244	46,264	41,512	37,646	34,438	31,734	27,427	24,150	21,572	19,491	17,777	16,339	15,117	12,735	11,002	9,684	8,648	7,812	7,124	6,056	5,267	4,660	3,787	3,189
65	55,995	49,181	43,846	39,555	36,029	33,080	28,427	24,921	22,185	19,991	18,191	16,689	15,416	12,947	11,159	9,806	8,745	7,891	7,189	6,104	5,303	4,688	3,805	3,202
70	59,666	51,991	46,065	41,352	37,514	34,328	29,344	25,623	22,740	20,440	18,562	17,001	15,682	13,134	11,298	9,913	8,830	7,960	7,247	6,145	5,334	4,712	3,821	3,214
75	63,261	54,700	48,179	43,048	38,904	35,488	30,187	26,264	23,243	20,846	18,896	17,281	15,919	13,300	11,421	10,007	8,905	8,021	7,297	6,181	5,361	4,733	3,835	3,223
80	66,782	57,312	50,194	44,649	40,208	36,570	30,966	26,852	23,702	21,214	19,199	17,533	16,133	13,449	11,531	10,091	8,971	8,075	7,342	6,213	5,385	4,752	3,847	3,232
85	70,230	59,833	52,118	46,165	41,433	37,580	31,688	27,392	24,123	21,550	19,474	17,762	16,327	13,583	11,629	10,167	9,031	8,123	7,382	6,242	5,407	4,769	3,858	3,240
90	73,609	62,269	53,956	47,601	42,586	38,527	32,358	27,892	24,509	21,858	19,725	17,971	16,503	13,705	11,718	10,235	9,084	8,167	7,417	6,267	5,426	4,784	3,868	3,247
95	76,920	64,622	55,714	48,964	43,674	39,415	32,982	28,354	24,865	22,141	19,955	18,162	16,664	13,816	11,799	10,296	9,133	8,206	7,450	6,290	5,443	4,797	3,877	3,253
100	80,166	66,897	57,397	50,260	44,701	40,249	33,565	28,784	25,195	22,402	20,167	18,337	16,811	13,917	11,873	10,352	9,177	8,241	7,479	6,311	5,459	4,809	3,885	3,258
105	83,347	69,098	59,010	51,492	45,673	41,036	34,110	29,184	25,501	22,644	20,362	18,498	16,947	14,010	11,940	10,404	9,217	8,274	7,506	6,330	5,473	4,820	3,892	3,263
110	86,467	71,229	60,557	52,666	46,594	41,778	34,621	29,557	25,786	22,868	20,543	18,648	17,072	14,095	12,002	10,451	9,254	8,304	7,530	6,348	5,486	4,830	3,899	3,268
115	89,527	73,292	62,042	53,786	47,469	42,480	35,101	29,907	26,051	23,077	20,711	18,786	17,188	14,174	12,060	10,494	9,288	8,331	7,553	6,364	5,498	4,840	3,905	3,272
120	92,528	75,292	63,468	54,855	48,299	43,144	35,553	30,234	26,300	23,271	20,868	18,915	17,296	14,248	12,113	10,534	9,320	8,356	7,573	6,378	5,509	4,848	3,910	3,276
125	95,473	77,230	64,840	55,876	49,090	43,773	35,980	30,542	26,532	23,453	21,014	19,035	17,396	14,316	12,162	10,571	9,349	8,380	7,593	6,392	5,519	4,856	3,915	3,280
130	98,362	79,110	66,160	56,854	49,843	44,371	36,383	30,832	26,751	23,624	21,151	19,147	17,490	14,379	12,207	10,606	9,376	8,401	7,610	6,404	5,528	4,863	3,920	3,283
135	101,198	80,934	67,431	57,790	50,561	44,939	36,764	31,105	26,956	23,784	21,279	19,252	17,578	14,438	12,250	10,638	9,401	8,421	7,627	6,416	5,537	4,870	3,924	3,286
140	103,982	82,705	68,656	58,687	51,246	45,480	37,125	31,363	27,150	23,934	21,400	19,351	17,660	14,493	12,290	10,668	9,424	8,440	7,642	6,427	5,545	4,876	3,928	3,289
145	106,715	84,425	69,837	59,548	51,901	45,995	37,467	31,607	27,332	24,076	21,513	19,443	17,737	14,545	12,327	10,696	9,446	8,458	7,657	6,437	5,553	4,882	3,932	3,292
150	109,399	86,095	70,976	60,374	52,528	46,486	37,793	31,839	27,505	24,210	21,620	19,531	17,809	14,594	12,362	10,722	9,467	8,474	7,670	6,447	5,560	4,888	3,936	3,294
155	112,035	87,720	72,077	61,168	53,128	46,956	38,103	32,058	27,669	24,337	21,721	19,613	17,878	14,640	12,395	10,747	9,486	8,490	7,683	6,456	5,567	4,893	3,939	3,296
160	114,624	89,299	73,139	61,932	53,703	47,405	38,397	32,267	27,824	24,457	21,817	19,691	17,943	14,683	12,426	10,771	9,504	8,504	7,695	6,464	5,573	4,898	3,942	3,299
165	117,168	90,835	74,167	62,667	54,255	47,834	38,679	32,465	27,971	24,571	21,907	19,765	18,004	14,724	12,456	10,793	9,521	8,518	7,706	6,472	5,579	4,902	3,945	3,301
170	119,667	92,330	75,160	63,375	54,785	48,245	38,947	32,654	28,112	24,679	21,993	19,834	18,062	14,763	12,483	10,813	9,538	8,531	7,717	6,480	5,584	4,906	3,948	3,303
175	122,123	93,785	76,122	64,057	55,294	48,640	39,204	32,834	28,245	24,781	22,075	19,901	18,117	14,800	12,509	10,833	9,553	8,543	7,727	6,487	5,590	4,910	3,951	3,305
180	124,537	95,203	77,053	64,716	55,784	49,018	39,449	33,006	28,372	24,879	22,152	19,964	18,169	14,835	12,534	10,852	9,567	8,555	7,736	6,493	5,594	4,914	3,953	3,306
185	126,910	96,583	77,955	65,351	56,255	49,382	39,684	33,171	28,494	24,973	22,226	20,024	18,219	14,868	12,558	10,869	9,581	8,566	7,745	6,500	5,599	4,918	3,955	3,308
190	129,244	97,929	78,829	65,964	56,709	49,731	39,910	33,328	28,610	25,062	22,297	20,081	18,266	14,899	12,580	10,886	9,594	8,576	7,754	6,506	5,604	4,921	3,958	3,309
195	131,538	99,240	79,677	66,556	57,146	50,067	40,126	33,478	28,721	25,147	22,364	20,136	18,311	14,929	12,602	10,902	9,607	8,586	7,762	6,511	5,608	4,925	3,960	3,311
200	133,794	100,519	80,499	67,129	57,568	50,390	40,333	33,623	28,827	25,228	22,428	20,188	18,354	14,958	12,622	10,918	9,618	8,596	7,769	6,517	5,612	4,928	3,962	3,312

1 AWG | 1 Run(s) | 480 Volts | C-Value = 7,292 | Magnetic Duct | Copper

Calculations:

1. $f = \dfrac{1.732 \times \text{Length} \times \text{Isca}}{\text{Runs} \times \text{C-Value} \times \text{E L-L}}$

2. $M = \dfrac{1}{1 + f}$

3. $\text{Isc} = \text{Isca} \times M$

4. Add Motor Contribution = Motor FLA x 4

* All results are given in symmetrical amperes

Three-Phase Bolted Fault-Current* Table

One-Way Distance in Feet

Available Isc in Thousands (Isca)

	5	10	15	20	25	30	40	50	60	70	80	90	100	125	150	175	200	225	250	300	350	400	500	600
3	2,982	2,964	2,946	2,929	2,912	2,895	2,861	2,828	2,796	2,765	2,735	2,705	2,675	2,605	2,538	2,475	2,414	2,357	2,302	2,200	2,106	2,020	1,867	1,736
5	4,950	4,901	4,853	4,806	4,759	4,714	4,626	4,541	4,459	4,380	4,304	4,230	4,159	3,991	3,837	3,693	3,560	3,437	3,321	3,[illegible]12	2,928	2,764	2,487	2,259
7	6,902	6,807	6,715	6,625	6,537	6,452	6,288	6,132	5,984	5,842	5,708	5,579	5,456	5,171	4,914	4,681	4,470	4,277	4,099	3,786	3,516	3,283	2,898	2,594
10	9,802	9,611	9,428	9,252	9,082	8,918	8,608	8,318	8,048	7,794	7,556	7,332	7,121	6,643	6,225	5,856	5,529	5,236	4,973	4,519	4,140	3,821	3,309	2,919
15	14,559	14,142	13,749	13,377	13,025	12,691	12,071	11,510	10,998	10,530	10,100	9,703	9,337	8,532	7,854	7,277	6,778	6,343	5,961	5,320	4,803	4,378	3,720	3,233
20	19,223	18,504	17,836	17,216	16,637	16,095	15,112	14,242	13,466	12,771	12,144	11,575	11,058	9,946	9,037	8,281	7,641	7,093	6,619	5,838	5,221	4,723	3,966	3,418
25	23,797	22,705	21,708	20,796	19,957	19,183	17,802	16,607	15,562	14,640	13,822	13,091	12,433	11,045	9,935	9,029	8,273	7,635	7,088	6,200	5,509	4,957	4,129	3,539
30	28,285	26,755	25,382	24,143	23,019	21,996	20,199	18,674	17,363	16,224	15,225	14,342	13,556	11,922	10,640	9,607	8,756	8,044	7,439	6,467	5,719	5,126	4,246	3,624
35	32,687	30,661	28,871	27,279	25,853	24,569	22,349	20,497	18,928	17,582	16,415	15,394	14,492	12,640	11,208	10,067	9,137	8,365	7,713	6,572	5,879	5,255	4,334	3,688
40	37,007	34,431	32,190	30,224	28,483	26,932	24,287	22,116	20,300	18,760	17,438	16,289	15,283	13,238	11,675	10,443	9,446	8,622	7,931	6,835	6,005	5,355	4,402	3,737
45	41,247	38,073	35,352	32,993	30,930	29,110	26,044	23,563	21,513	19,792	18,325	17,061	15,960	13,743	12,067	10,755	9,700	8,834	8,110	6,967	6,107	5,436	4,456	3,776
50	45,410	41,591	38,365	35,604	33,213	31,123	27,644	24,865	22,594	20,702	19,103	17,733	16,547	14,176	12,399	11,018	9,914	9,011	8,259	7,077	6,191	5,502	4,501	3,808
55	49,496	44,994	41,242	38,068	35,348	32,990	29,107	26,042	23,561	21,512	19,791	18,324	17,060	14,551	12,685	11,243	10,096	9,161	8,385	7,169	6,262	5,558	4,538	3,835
60	53,509	48,286	43,991	40,398	37,348	34,726	30,450	27,112	24,434	22,237	20,402	18,848	17,513	14,879	12,934	11,438	10,253	9,290	8,492	7,248	6,322	5,605	4,570	3,857
65	57,450	51,472	46,621	42,605	39,226	36,344	31,687	28,089	25,224	22,889	20,950	19,314	17,915	15,168	13,152	11,608	10,389	9,402	8,586	7,316	6,373	5,646	4,597	3,876
70	61,322	54,558	49,138	44,698	40,993	37,856	32,831	28,983	25,943	23,480	21,444	19,733	18,275	15,425	13,345	11,758	10,509	9,500	8,668	7,375	6,418	5,681	4,620	3,893
75	65,125	57,548	51,551	46,685	42,659	39,272	33,890	29,806	26,600	24,017	21,891	20,111	18,599	15,655	13,516	11,892	10,616	9,587	8,740	7,427	6,458	5,712	4,640	3,907
80	68,862	60,447	53,864	48,575	44,231	40,601	34,875	30,565	27,203	24,508	22,298	20,454	18,891	15,862	13,670	12,011	10,710	9,664	8,804	7,474	6,493	5,739	4,658	3,920
85	72,535	63,259	56,086	50,374	45,718	41,850	35,793	31,268	27,758	24,957	22,670	20,766	19,158	16,049	13,809	12,118	10,795	9,733	8,861	7,515	6,524	5,764	4,674	3,931
90	76,145	65,987	58,220	52,089	47,126	43,027	36,651	31,920	28,271	25,371	23,011	21,052	19,400	16,220	13,935	12,214	10,872	9,796	8,913	7,552	6,552	5,785	4,689	3,941
95	79,694	68,636	60,272	53,726	48,462	44,137	37,453	32,527	28,747	25,753	23,324	21,314	19,623	16,375	14,049	12,302	10,942	9,852	8,960	7,586	6,577	5,805	4,702	3,951
100	83,183	71,208	62,247	55,289	49,730	45,187	38,206	33,094	29,188	26,107	23,614	21,556	19,828	16,517	14,154	12,382	11,005	9,903	9,002	7,616	6,600	5,823	4,713	3,959
105	86,614	73,707	64,148	56,784	50,936	46,181	38,914	33,624	29,600	26,436	23,883	21,780	20,017	16,648	14,250	12,456	11,063	9,950	9,041	7,644	6,621	5,839	4,724	3,966
110	89,988	76,136	65,980	58,215	52,085	47,123	39,581	34,120	29,984	26,742	24,132	21,987	20,192	16,769	14,339	12,523	11,116	9,993	9,076	7,669	6,640	5,854	4,733	3,973
115	93,307	78,499	67,747	59,586	53,180	48,017	40,210	34,587	30,343	27,027	24,365	22,180	20,354	16,881	14,420	12,586	11,165	10,033	9,109	7,692	6,657	5,867	4,742	3,979
120	96,571	80,797	69,452	60,901	54,225	48,867	40,805	35,026	30,681	27,295	24,582	22,359	20,506	16,985	14,496	12,643	11,211	10,069	9,139	7,714	6,673	5,880	4,751	3,985
125	99,783	83,033	71,098	62,163	55,223	49,677	41,367	35,440	30,998	27,545	24,785	22,527	20,647	17,082	14,566	12,697	11,253	10,103	9,167	7,734	6,688	5,891	4,758	3,990
130	102,944	85,210	72,688	63,375	56,177	50,448	41,901	35,830	31,296	27,781	24,975	22,685	20,779	17,172	14,632	12,747	11,292	10,135	9,193	7,752	6,702	5,902	4,765	3,995
135	106,055	87,330	74,225	64,540	57,091	51,183	42,407	36,200	31,578	28,002	25,154	22,832	20,902	17,256	14,693	12,793	11,328	10,164	9,217	7,769	6,715	5,912	4,772	4,000
140	109,116	89,395	75,712	65,661	57,966	51,886	42,888	36,550	31,844	28,211	25,323	22,971	21,019	17,335	14,751	12,837	11,362	10,192	9,240	7,785	6,727	5,921	4,778	4,004
145	112,130	91,408	77,151	66,741	58,806	52,558	43,346	36,882	32,096	28,409	25,482	23,102	21,128	17,410	14,804	12,877	11,394	10,217	9,261	7,800	6,738	5,930	4,783	4,008
150	115,096	93,370	78,544	67,781	59,612	53,200	43,782	37,197	32,334	28,596	25,632	23,225	21,231	17,480	14,855	12,913	11,424	10,241	9,281	7,814	6,748	5,938	4,788	4,012
155	118,018	95,283	79,893	68,783	60,386	53,816	44,198	37,497	32,561	28,772	25,774	23,341	21,328	17,546	14,903	12,951	11,452	10,264	9,299	7,828	6,758	5,946	4,793	4,015
160	120,894	97,150	81,201	69,751	61,130	54,406	44,596	37,783	32,776	28,940	25,909	23,452	21,421	17,608	14,947	12,985	11,479	10,285	9,317	7,840	6,767	5,953	4,798	4,018
165	123,727	98,971	82,469	70,684	61,846	54,973	44,976	38,055	32,980	29,100	26,036	23,556	21,508	17,667	14,990	13,017	11,504	10,305	9,333	7,852	6,776	5,960	4,802	4,021
170	126,517	100,748	83,700	71,586	62,536	55,517	45,339	38,315	33,175	29,252	26,158	23,656	21,591	17,723	15,030	13,048	11,527	10,324	9,349	7,863	6,784	5,966	4,806	4,024
175	129,266	102,483	84,894	72,458	63,200	56,040	45,687	38,563	33,361	29,396	26,273	23,750	21,669	17,776	15,068	13,076	11,550	10,342	9,363	7,873	6,792	5,972	4,810	4,027
180	131,974	104,178	86,054	73,301	63,840	56,543	46,021	38,801	33,539	29,534	26,383	23,840	21,744	17,826	15,104	13,104	11,571	10,359	9,377	7,883	6,799	5,978	4,814	4,030
185	134,642	105,833	87,180	74,117	64,458	57,027	46,341	39,028	33,709	29,665	26,488	23,926	21,815	17,874	15,139	13,129	11,591	10,375	9,391	7,892	6,806	5,983	4,818	4,032
190	137,271	107,451	88,275	74,907	65,055	57,493	46,649	39,246	33,871	29,791	26,588	24,007	21,883	17,919	15,171	13,154	11,610	10,391	9,403	7,901	6,813	5,988	4,821	4,034
195	139,862	109,032	89,339	75,672	65,631	57,943	46,944	39,455	34,027	29,911	26,684	24,085	21,948	17,963	15,202	13,177	11,628	10,405	9,415	7,909	6,819	5,993	4,824	4,037
200	142,416	110,578	90,374	76,413	66,188	58,376	47,229	39,656	34,176	30,027	26,776	24,160	22,010	18,004	15,232	13,200	11,646	10,419	9,426	7,918	6,825	5,997	4,827	4,039

# 1/0 AWG	1 Run(s)	480 Volts

C-Value = 8,924	Magnetic Duct	Copper

Calculations:

1. $f = \dfrac{1.732 \times \text{Length} \times \text{Isca}}{\text{Runs} \times \text{C-Value} \times E_{L-L}}$

2. $M = \dfrac{1}{1 + f}$

3. Isc = Isca x M

4. Add Motor Contribution = Motor FLA x 4

* All results are given in symmetrical amperes

Three-Phase Bolted Fault-Current* Table

Available Isc in Thousands (Isca)	5	10	15	20	25	30	40	50	60	70	80	90	100	125	150	175	200	225	250	300	350	400	500	600
	One-Way Distance in Feet																							
3	2,991	2,982	2,973	2,964	2,955	2,946	2,929	2,912	2,895	2,878	2,861	2,845	2,828	2,789	2,750	2,712	2,675	2,640	2,605	2,538	2,475	2,414	2,302	2,200
5	4,975	4,950	4,925	4,901	4,877	4,853	4,806	4,759	4,714	4,670	4,626	4,583	4,541	4,439	4,342	4,248	4,159	4,074	3,991	3,837	3,693	3,560	3,321	3,112
7	6,951	6,902	6,854	6,807	6,761	6,715	6,625	6,537	6,452	6,369	6,288	6,209	6,132	5,948	5,774	5,611	5,456	5,309	5,171	4,914	4,681	4,470	4,099	3,786
10	9,900	9,802	9,706	9,611	9,519	9,428	9,252	9,082	8,918	8,760	8,608	8,461	8,318	7,983	7,673	7,387	7,121	6,873	6,643	6,225	5,856	5,529	4,973	4,519
15	14,776	14,559	14,347	14,142	13,943	13,749	13,377	13,025	12,691	12,373	12,071	11,784	11,510	10,877	10,310	9,799	9,337	8,916	8,532	7,854	7,277	6,778	5,961	5,320
20	19,604	19,223	18,856	18,504	18,164	17,836	17,216	16,637	16,095	15,588	15,112	14,664	14,242	13,285	12,449	11,712	11,058	10,472	9,946	9,037	8,281	7,641	6,619	5,838
25	24,384	23,797	23,238	22,705	22,195	21,708	20,796	19,957	19,183	18,467	17,802	17,184	16,607	15,321	14,220	13,266	12,433	11,698	11,045	9,935	9,029	8,273	7,088	6,200
30	29,117	28,285	27,498	26,755	26,050	25,382	24,143	23,019	21,996	21,059	20,199	19,407	18,674	17,064	15,709	14,553	13,556	12,687	11,922	10,640	9,607	8,756	7,439	6,467
35	33,804	32,687	31,642	30,661	29,739	28,871	27,279	25,853	24,569	23,406	22,349	21,383	20,497	18,573	16,979	15,637	14,492	13,503	12,640	11,208	10,067	9,137	7,713	6,672
40	38,445	37,007	35,673	34,431	33,273	32,190	30,224	28,483	26,932	25,542	24,287	23,151	22,116	19,892	18,075	16,562	15,283	14,187	13,238	11,675	10,443	9,446	7,931	6,835
45	43,042	41,247	39,596	38,073	36,662	35,352	32,993	30,930	29,110	27,492	26,044	24,742	23,563	21,056	19,030	17,361	15,960	14,769	13,743	12,067	10,755	9,700	8,110	6,967
50	47,594	45,410	43,417	41,591	39,913	38,365	35,604	33,213	31,123	29,281	27,644	26,181	24,865	22,089	19,871	18,057	16,547	15,270	14,176	12,399	11,018	9,914	8,259	7,077
55	52,103	49,496	47,138	44,994	43,037	41,242	38,068	35,348	32,990	30,927	29,107	27,490	26,042	23,013	20,615	18,670	17,060	15,706	14,551	12,685	11,243	10,096	8,385	7,169
60	56,569	53,509	50,763	48,286	46,039	43,991	40,398	37,348	34,726	32,448	30,450	28,685	27,112	23,845	21,280	19,214	17,513	16,089	14,879	12,934	11,438	10,253	8,492	7,248
65	60,992	57,450	54,297	51,472	48,926	46,621	42,605	39,226	36,344	33,856	31,687	29,780	28,089	24,597	21,877	19,699	17,915	16,428	15,168	13,152	11,608	10,389	8,586	7,316
70	65,374	61,322	57,742	54,558	51,706	49,138	44,698	40,993	37,856	35,165	32,831	30,787	28,983	25,280	22,416	20,135	18,275	16,730	15,425	13,345	11,758	10,509	8,668	7,375
75	69,715	65,125	61,103	57,548	54,385	51,551	46,685	42,659	39,272	36,383	33,890	31,717	29,806	25,904	22,905	20,528	18,599	17,001	15,655	13,516	11,892	10,616	8,740	7,427
80	74,015	68,862	64,381	60,447	56,966	53,864	48,575	44,231	40,601	37,521	34,875	32,578	30,565	26,475	23,351	20,886	18,891	17,245	15,862	13,670	12,011	10,710	8,804	7,474
85	78,274	72,535	67,580	63,259	59,457	56,086	50,374	45,718	41,850	38,585	35,793	33,378	31,268	27,001	23,758	21,211	19,158	17,466	16,049	13,809	12,118	10,795	8,861	7,515
90	82,495	76,145	70,703	65,987	61,861	58,220	52,089	47,126	43,027	39,584	36,651	34,122	31,920	27,486	24,133	21,510	19,400	17,668	16,220	13,935	12,214	10,872	8,913	7,552
95	86,676	79,694	73,752	68,636	64,183	60,272	53,726	48,462	44,137	40,522	37,453	34,817	32,527	27,935	24,479	21,784	19,623	17,852	16,375	14,049	12,302	10,942	8,960	7,586
100	90,820	83,183	76,731	71,208	66,426	62,247	55,289	49,730	45,187	41,405	38,206	35,467	33,094	28,352	24,798	22,036	19,828	18,022	16,517	14,154	12,382	11,005	9,002	7,616
105	94,925	86,614	79,641	73,707	68,596	64,148	56,784	50,936	46,181	42,237	38,914	36,076	33,624	28,740	25,095	22,270	20,017	18,178	16,648	14,250	12,456	11,063	9,041	7,644
110	98,993	89,988	82,485	76,136	70,696	65,980	58,215	52,085	47,123	43,024	39,581	36,649	34,120	29,102	25,370	22,487	20,192	18,322	16,769	14,339	12,523	11,116	9,076	7,669
115	103,024	93,307	85,265	78,499	72,728	67,747	59,586	53,180	48,017	43,768	40,210	37,187	34,587	29,440	25,627	22,688	20,354	18,456	16,881	14,420	12,586	11,165	9,109	7,692
120	107,018	96,571	87,983	80,797	74,696	69,452	60,901	54,225	48,867	44,474	40,805	37,695	35,026	29,758	25,867	22,876	20,506	18,580	16,985	14,496	12,643	11,211	9,139	7,714
125	110,977	99,783	90,641	83,033	76,603	71,098	62,163	55,223	49,677	45,143	41,367	38,175	35,440	30,056	26,092	23,052	20,647	18,696	17,082	14,566	12,697	11,253	9,167	7,734
130	114,901	102,944	93,241	85,210	78,453	72,688	63,375	56,177	50,448	45,779	41,901	38,628	35,830	30,337	26,303	23,217	20,779	18,804	17,172	14,632	12,747	11,292	9,193	7,752
135	118,789	106,055	95,786	87,330	80,246	74,225	64,540	57,091	51,183	46,384	42,407	39,058	36,200	30,601	26,502	23,371	20,902	18,905	17,256	14,693	12,793	11,328	9,217	7,769
140	122,644	109,116	98,276	89,395	81,987	75,712	65,661	57,966	51,886	46,960	42,888	39,466	36,550	30,851	26,689	23,517	21,019	19,000	17,335	14,751	12,837	11,362	9,240	7,785
145	126,464	112,130	100,714	91,408	83,676	77,151	66,741	58,806	52,558	47,509	43,346	39,854	36,882	31,087	26,866	23,654	21,128	19,089	17,410	14,804	12,877	11,394	9,261	7,800
150	130,250	115,096	103,101	93,370	85,318	78,544	67,781	59,612	53,200	48,034	43,782	40,222	37,197	31,311	27,033	23,783	21,231	19,174	17,480	14,855	12,916	11,424	9,281	7,814
155	134,004	118,018	105,439	95,283	86,912	79,893	68,783	60,386	53,816	48,535	44,198	40,573	37,497	31,523	27,191	23,905	21,328	19,253	17,546	14,903	12,951	11,452	9,299	7,828
160	137,725	120,894	107,729	97,150	88,462	81,201	69,751	61,130	54,406	49,015	44,596	40,908	37,783	31,725	27,341	24,021	21,421	19,328	17,608	14,947	12,985	11,479	9,317	7,840
165	141,414	123,727	109,973	98,971	89,970	82,469	70,684	61,846	54,973	49,474	44,976	41,227	38,055	31,916	27,483	24,131	21,508	19,399	17,667	14,990	13,017	11,504	9,333	7,852
170	145,070	126,517	112,172	100,748	91,436	83,700	71,586	62,536	55,517	49,914	45,339	41,532	38,315	32,099	27,618	24,235	21,591	19,466	17,723	15,030	13,048	11,527	9,349	7,863
175	148,696	129,266	114,327	102,483	92,863	84,894	72,458	63,200	56,040	50,337	45,687	41,824	38,563	32,273	27,747	24,334	21,669	19,530	17,776	15,068	13,076	11,550	9,363	7,873
180	152,290	131,974	116,440	104,178	94,252	86,054	73,301	63,840	56,543	50,742	46,021	42,104	38,801	32,439	27,870	24,429	21,744	19,591	17,826	15,104	13,104	11,571	9,377	7,883
185	155,854	134,642	118,512	105,833	95,605	87,180	74,117	64,458	57,027	51,132	46,341	42,372	39,028	32,598	27,987	24,519	21,815	19,649	17,874	15,139	13,129	11,591	9,391	7,892
190	159,388	137,271	120,544	107,451	96,924	88,275	74,907	65,055	57,493	51,506	46,649	42,629	39,246	32,750	28,099	24,605	21,883	19,704	17,919	15,171	13,154	11,610	9,403	7,901
195	162,891	139,862	122,538	109,032	98,208	89,339	75,672	65,631	57,943	51,867	46,944	42,875	39,455	32,895	28,206	24,686	21,948	19,756	17,963	15,202	13,177	11,628	9,415	7,909
200	166,366	142,416	124,493	110,578	99,460	90,374	76,413	66,188	58,376	52,214	47,229	43,112	39,656	33,035	28,308	24,765	22,010	19,807	18,004	15,232	13,200	11,646	9,426	7,918

# 1/0	AWG	2	Run(s)	480	Volts	C-Value =	8,924	Magnetic Duct	Copper

Calculations:

1. $f = \dfrac{1.732 \times \text{Length} \times \text{Isca}}{\text{Runs} \times \text{C-Value} \times \text{E L-L}}$

2. $M = \dfrac{1}{1 + f}$

3. Isc = Isca x M

4. Add Motor Contribution = Motor FLA x 4

* All results are given in symmetrical amperes

Three-Phase Bolted Fault-Current* Table

One-Way Distance in Feet

Available Isc in Thousands (Isca)	5	10	15	20	25	30	40	50	60	70	80	90	100	125	150	175	200	225	250	300	350	400	500	600
3	2,985	2,970	2,955	2,941	2,926	2,912	2,884	2,856	2,829	2,803	2,776	2,751	2,726	2,665	2,606	2,551	2,497	2,446	2,397	2,304	2,218	2,139	1,996	1,870
5	4,958	4,918	4,877	4,838	4,799	4,760	4,686	4,613	4,543	4,475	4,408	4,344	4,282	4,133	3,995	3,865	3,744	3,630	3,523	3,326	3,150	2,992	2,719	2,492
7	6,919	6,839	6,762	6,686	6,612	6,539	6,399	6,264	6,135	6,012	5,893	5,779	5,669	5,411	5,176	4,961	4,763	4,580	4,410	4,107	3,842	3,609	3,219	2,906
10	9,835	9,675	9,521	9,371	9,226	9,086	8,817	8,563	8,324	8,098	7,884	7,681	7,483	7,045	6,652	6,301	5,984	5,698	5,438	4,984	4,599	4,270	3,735	3,319
15	14,632	14,281	13,947	13,628	13,324	13,032	12,486	11,984	11,521	11,092	10,694	10,324	9,973	9,208	8,548	7,976	7,476	7,035	6,643	5,977	5,432	4,978	4,266	3,732
20	19,351	18,742	18,171	17,634	17,127	16,649	15,768	14,976	14,259	13,608	13,014	12,470	11,969	10,877	9,968	9,199	8,540	7,969	7,470	6,638	5,973	5,429	4,592	3,979
25	23,994	23,065	22,206	21,409	20,666	19,974	18,720	17,613	16,631	15,752	14,961	14,246	13,596	12,204	11,071	10,130	9,337	8,659	8,073	7,110	6,352	5,740	4,813	4,144
30	28,563	27,257	26,065	24,973	23,969	23,042	21,389	19,957	18,704	17,600	16,619	15,741	14,951	13,285	11,953	10,864	9,957	9,189	8,532	7,464	6,633	5,969	4,973	4,262
35	33,059	31,322	29,758	28,343	27,057	25,882	23,814	22,052	20,533	19,210	18,047	17,016	16,097	14,183	12,675	11,457	10,452	9,610	8,893	7,739	6,849	6,144	5,094	4,350
40	37,485	35,267	33,297	31,536	29,951	28,518	26,028	23,938	22,158	20,625	19,290	18,117	17,079	14,939	13,276	11,946	10,858	9,951	9,185	7,959	7,021	6,281	5,188	4,419
45	41,841	39,097	36,691	34,563	32,669	30,972	28,057	25,643	23,611	21,878	20,382	19,078	17,930	15,586	13,784	12,356	11,195	10,234	9,425	8,138	7,161	6,393	5,264	4,474
50	46,131	42,817	39,948	37,439	35,227	33,261	29,922	27,192	24,919	22,996	21,349	19,922	18,674	16,145	14,220	12,704	11,481	10,473	9,627	8,288	7,277	6,485	5,326	4,519
55	50,354	46,432	43,077	40,174	37,637	35,402	31,644	28,607	26,101	24,000	22,211	20,671	19,330	16,633	14,597	13,005	11,726	10,676	9,798	8,415	7,374	6,562	5,378	4,556
60	54,513	49,946	46,085	42,778	39,913	37,409	33,237	29,903	27,176	24,905	22,985	21,339	19,914	17,064	14,927	13,266	11,938	10,851	9,946	8,524	7,458	6,628	5,422	4,588
65	58,609	53,363	48,978	45,260	42,066	39,293	34,717	31,095	28,157	25,727	23,683	21,940	20,435	17,445	15,218	13,496	12,123	11,004	10,075	8,618	7,530	6,685	5,460	4,615
70	62,644	56,687	51,765	47,629	44,105	41,066	36,093	32,195	29,056	26,475	24,316	22,482	20,905	17,786	15,477	13,693	12,287	11,139	10,187	8,700	7,592	6,735	5,493	4,638
75	66,618	59,922	54,449	49,892	46,039	42,738	37,378	33,213	29,883	27,160	24,892	22,973	21,329	18,093	15,709	13,883	12,433	11,259	10,287	8,773	7,648	6,778	5,522	4,659
80	70,534	63,071	57,037	52,056	47,875	44,316	38,580	34,159	30,646	27,789	25,419	23,422	21,715	18,370	15,917	14,042	12,563	11,365	10,376	8,838	7,697	6,817	5,548	4,677
85	74,392	66,139	59,534	54,128	49,622	45,809	39,706	35,039	31,353	28,369	25,903	23,832	22,068	18,621	16,106	14,183	12,680	11,461	10,456	8,896	7,740	6,851	5,571	4,693
90	78,194	69,127	61,944	56,113	51,286	47,223	40,764	35,860	32,009	28,905	26,350	24,209	22,391	18,851	16,277	14,322	12,786	11,547	10,528	8,948	7,780	6,882	5,591	4,708
95	81,941	72,039	64,272	58,017	52,871	48,564	41,760	36,628	32,620	29,402	26,762	24,557	22,688	19,061	16,433	14,443	12,882	11,626	10,593	8,995	7,815	6,910	5,609	4,721
100	85,635	74,878	66,522	59,844	54,385	49,838	42,698	37,348	33,189	29,864	27,144	24,879	22,962	19,254	16,577	14,553	12,970	11,698	10,652	9,037	7,848	6,935	5,626	4,733
105	89,275	77,647	68,699	61,600	55,830	51,049	43,585	38,024	33,722	30,295	27,500	25,177	23,216	19,432	16,709	14,655	13,051	11,763	10,707	9,076	7,877	6,958	5,641	4,743
110	92,864	80,348	70,804	63,287	57,213	52,203	44,423	38,661	34,222	30,697	27,831	25,454	23,451	19,597	16,830	14,748	13,125	11,823	10,757	9,112	7,904	6,979	5,655	4,753
115	96,403	82,983	72,843	64,911	58,537	53,303	45,217	39,261	34,691	31,074	28,141	25,713	23,671	19,750	16,943	14,835	13,193	11,879	10,802	9,145	7,929	6,998	5,667	4,762
120	99,892	85,555	74,817	66,474	59,805	54,352	45,970	39,827	35,133	31,428	28,430	25,955	23,876	19,892	17,048	14,915	13,257	11,930	10,845	9,176	7,952	7,016	5,679	4,770
125	103,332	88,067	76,731	67,981	61,022	55,355	46,685	40,363	35,549	31,761	28,702	26,181	24,067	20,025	17,145	14,990	13,315	11,978	10,884	9,204	7,973	7,032	5,690	4,778
130	106,726	90,520	78,586	69,433	62,190	56,315	47,365	40,871	35,942	32,074	28,958	26,394	24,247	20,149	17,236	15,059	13,370	12,022	10,921	9,230	7,992	7,048	5,700	4,785
135	110,072	92,916	80,386	70,834	63,311	57,233	48,013	41,352	36,314	32,370	29,199	26,594	24,415	20,265	17,321	15,124	13,421	12,063	10,955	9,254	8,011	7,062	5,709	4,791
140	113,374	95,257	82,133	72,187	64,390	58,113	48,631	41,810	36,656	32,650	29,426	26,782	24,574	20,375	17,401	15,185	13,469	12,102	10,987	9,277	8,028	7,075	5,718	4,797
145	116,631	97,546	83,829	73,494	65,427	58,957	49,221	42,245	37,030	32,914	29,641	26,960	24,724	20,477	17,476	15,242	13,514	12,138	11,017	9,298	8,044	7,087	5,726	4,803
150	119,844	99,784	85,476	74,757	66,426	59,767	49,784	42,659	37,318	33,165	29,845	27,128	24,865	20,574	17,546	15,295	13,556	12,172	11,045	9,318	8,058	7,099	5,733	4,808
155	123,014	101,972	87,076	75,978	67,389	60,545	50,323	43,054	37,620	33,404	30,037	27,287	24,999	20,666	17,613	15,346	13,596	12,204	11,071	9,337	8,072	7,110	5,740	4,813
160	126,143	104,112	88,633	77,160	68,317	61,293	50,839	43,431	37,907	33,630	30,220	27,438	25,125	20,752	17,676	15,393	13,633	12,234	11,096	9,354	8,086	7,120	5,747	4,818
165	129,230	106,206	90,146	78,304	69,213	62,013	51,333	43,791	38,181	33,846	30,394	27,582	25,246	20,834	17,735	15,438	13,668	12,263	11,119	9,371	8,098	7,130	5,753	4,822
170	132,277	108,256	91,618	79,413	70,077	62,706	51,807	44,136	38,443	34,051	30,560	27,718	25,360	20,912	17,791	15,481	13,702	12,289	11,141	9,387	8,110	7,139	5,759	4,827
175	135,285	110,262	93,051	80,487	70,913	63,374	52,262	44,465	38,693	34,247	30,718	27,848	25,468	20,985	17,844	15,521	13,733	12,315	11,162	9,402	8,121	7,147	5,765	4,831
180	138,254	112,226	94,446	81,529	71,720	64,018	52,699	44,781	38,932	34,434	30,868	27,971	25,572	21,056	17,895	15,560	13,763	12,339	11,182	9,416	8,131	7,155	5,770	4,834
185	141,185	114,150	95,804	82,539	72,501	64,639	53,119	45,085	39,161	34,613	31,012	28,089	25,670	21,122	17,943	15,596	13,792	12,362	11,201	9,429	8,141	7,163	5,775	4,838
190	144,078	116,034	97,128	83,520	73,256	65,239	53,524	45,376	39,380	34,785	31,149	28,202	25,764	21,186	17,989	15,631	13,819	12,384	11,218	9,442	8,151	7,170	5,780	4,841
195	146,935	117,880	98,418	84,472	73,988	65,819	53,913	45,655	39,591	34,949	31,281	28,310	25,854	21,247	18,033	15,664	13,845	12,404	11,235	9,454	8,160	7,177	5,784	4,844
200	149,756	119,688	99,676	85,397	74,696	66,379	54,288	45,924	39,793	35,106	31,407	28,413	25,940	21,305	18,075	15,695	13,870	12,424	11,252	9,465	8,168	7,184	5,789	4,847

2/0 AWG | 1 Run(s) | 480 Volts | C-Value = 10,755 | Magnetic Duct | Copper

Calculations:

1. $f = \dfrac{1.732 \times \text{Length} \times \text{Isca}}{\text{Runs} \times \text{C-Value} \times E_{L\text{-}L}}$

2. $M = \dfrac{1}{1 + f}$

3. Isc = Isca x M

4. Add Motor Contribution = Motor FLA x 4

* All results are given in symmetrical amperes

Three-Phase Bolted Fault-Current* Table

One-Way Distance in Feet

Available Isc in Thousands (Isca)	5	10	15	20	25	30	40	50	60	70	80	90	100	125	150	175	200	225	250	300	350	400	500	600
3	2,992	2,985	2,978	2,970	2,963	2,955	2,941	2,926	2,912	2,898	2,884	2,870	2,856	2,822	2,789	2,757	2,726	2,695	2,665	2,606	2,551	2,497	2,397	2,304
5	4,979	4,958	4,938	4,918	4,897	4,877	4,838	4,799	4,760	4,723	4,686	4,649	4,613	4,526	4,441	4,360	4,282	4,206	4,133	3,995	3,865	3,744	3,523	3,326
7	6,959	6,919	6,879	6,839	6,800	6,762	6,686	6,612	6,539	6,468	6,399	6,331	6,264	6,104	5,952	5,807	5,669	5,537	5,411	5,176	4,961	4,763	4,410	4,107
10	9,917	9,835	9,755	9,675	9,598	9,521	9,371	9,226	9,086	8,949	8,817	8,688	8,563	8,267	7,990	7,731	7,488	7,260	7,045	6,652	6,301	5,984	5,438	4,984
15	14,814	14,632	14,454	14,281	14,112	13,947	13,628	13,324	13,032	12,754	12,486	12,230	11,984	11,411	10,890	10,414	9,978	9,578	9,208	8,548	7,976	7,476	6,643	5,977
20	19,670	19,351	19,042	18,742	18,452	18,171	17,634	17,127	16,649	16,196	15,768	15,362	14,976	14,091	13,304	12,601	11,969	11,397	10,877	9,968	9,199	8,540	7,470	6,638
25	24,487	23,994	23,520	23,065	22,628	22,206	21,409	20,666	19,974	19,326	18,720	18,150	17,613	16,402	15,346	14,418	13,596	12,863	12,204	11,071	10,130	9,337	8,073	7,110
30	29,264	28,563	27,894	27,257	26,647	26,065	24,973	23,969	23,042	22,185	21,389	20,648	19,957	18,415	17,095	15,952	14,951	14,069	13,285	11,953	10,864	9,957	8,532	7,464
35	34,002	33,059	32,167	31,322	30,520	29,758	28,343	27,057	25,882	24,805	23,814	22,900	22,052	20,186	18,610	17,263	15,097	15,079	14,183	12,675	11,457	10,452	8,893	7,739
40	38,702	37,485	36,342	35,267	34,254	33,297	31,536	29,951	28,518	27,216	26,028	24,939	23,938	21,754	19,935	18,397	17,079	15,938	14,939	13,276	11,946	10,858	9,185	7,959
45	43,363	41,841	40,423	39,097	37,856	36,691	34,563	32,669	30,972	29,442	28,057	26,795	25,643	23,153	21,104	19,388	17,930	16,676	15,586	13,784	12,356	11,195	9,425	8,138
50	47,988	46,131	44,412	42,817	41,333	39,948	37,439	35,227	33,261	31,503	29,922	28,492	27,192	24,409	22,142	20,261	13,674	17,318	16,145	14,220	12,704	11,481	9,627	8,288
55	52,575	50,354	48,314	46,432	44,692	43,077	40,174	37,637	35,402	33,418	31,644	30,049	28,607	25,542	23,071	21,036	19,330	17,881	16,633	14,597	13,005	11,726	9,798	8,415
60	57,125	54,513	52,130	49,946	47,938	46,085	42,778	39,913	37,409	35,200	33,237	31,482	29,903	26,571	23,907	21,728	19,914	18,379	17,064	14,927	13,266	11,938	9,946	8,524
65	61,639	58,609	55,863	53,363	51,077	48,978	45,260	42,066	39,293	36,863	34,717	32,806	31,095	27,508	24,663	22,351	20,435	18,822	17,445	15,218	13,496	12,123	10,075	8,618
70	66,118	62,644	59,517	56,687	54,114	51,765	47,629	44,105	41,066	38,420	36,093	34,033	32,195	28,365	25,350	22,914	20,905	19,220	17,786	15,477	13,699	12,287	10,187	8,700
75	70,561	66,618	63,093	59,922	57,054	54,449	49,892	46,039	42,738	39,879	37,378	35,173	33,213	29,153	25,977	23,425	21,329	19,578	18,093	15,709	13,880	12,433	10,287	8,773
80	74,970	70,534	66,594	63,071	59,903	57,037	52,056	47,875	44,316	41,250	38,580	36,235	34,159	29,878	26,551	23,891	21,715	19,903	18,370	15,917	14,042	12,563	10,376	8,838
85	79,343	74,392	70,023	66,139	62,663	59,534	54,128	49,622	45,809	42,540	39,706	37,227	35,039	30,550	27,080	24,318	22,068	20,198	18,621	16,106	14,189	12,680	10,456	8,896
90	83,683	78,194	73,382	69,127	65,339	61,944	56,113	51,286	47,223	43,757	40,764	38,155	35,860	31,172	27,568	24,711	22,391	20,469	18,851	16,277	14,322	12,786	10,528	8,948
95	87,989	81,941	76,672	72,039	67,934	64,272	58,017	52,871	48,564	44,906	41,760	39,026	36,628	31,751	28,020	25,073	22,688	20,717	19,061	16,433	14,443	12,882	10,593	8,995
100	92,262	85,635	79,896	74,878	70,453	66,522	59,844	54,385	49,838	45,993	42,698	39,844	37,348	32,290	28,439	25,409	22,962	20,945	19,254	16,577	14,553	12,970	10,652	9,037
105	96,501	89,275	83,056	77,647	72,899	68,699	61,600	55,830	51,049	47,023	43,585	40,615	38,024	32,795	28,830	25,720	23,216	21,156	19,432	16,709	14,655	13,051	10,707	9,076
110	100,708	92,864	86,154	80,348	75,275	70,804	63,287	57,213	52,203	48,000	44,423	41,342	38,661	33,267	29,194	26,010	23,451	21,352	19,597	16,830	14,748	13,125	10,757	9,112
115	104,883	96,403	89,191	82,983	77,583	72,843	64,911	58,537	53,303	48,928	45,217	42,029	39,261	33,710	29,535	26,280	23,671	21,533	19,750	16,943	14,835	13,193	10,802	9,145
120	109,026	99,892	92,169	85,555	79,827	74,817	66,474	59,805	54,352	49,811	45,970	42,679	39,827	34,127	29,854	26,532	23,876	21,703	19,892	17,048	14,915	13,257	10,845	9,176
125	113,138	103,332	95,091	88,067	82,009	76,731	67,981	61,022	55,355	50,652	46,685	43,294	40,363	34,520	30,154	26,769	24,067	21,861	20,025	17,145	14,990	13,315	10,884	9,204
130	117,219	106,726	97,957	90,520	84,132	78,586	69,433	62,190	56,315	51,454	47,365	43,879	40,871	34,890	30,437	26,991	24,247	22,009	20,149	17,236	15,059	13,370	10,921	9,230
135	121,268	110,072	100,769	92,916	86,198	80,386	70,834	63,311	57,233	52,219	48,013	44,434	41,352	35,241	30,703	27,201	24,415	22,148	20,265	17,321	15,124	13,421	10,955	9,254
140	125,288	113,374	103,529	95,257	88,210	82,133	72,187	64,390	58,113	52,951	48,631	44,963	41,810	35,572	30,954	27,398	24,574	22,278	20,375	17,401	15,185	13,469	10,987	9,277
145	129,277	116,631	106,238	97,546	90,169	83,829	73,494	65,427	58,957	53,651	49,221	45,467	42,245	35,887	31,192	27,584	24,724	22,401	20,477	17,476	15,242	13,514	11,017	9,298
150	133,237	119,844	108,898	99,784	92,077	85,476	74,757	66,426	59,767	54,320	49,784	45,947	42,659	36,185	31,418	27,760	24,865	22,517	20,574	17,546	15,295	13,556	11,045	9,318
155	137,167	123,014	111,509	101,972	93,937	87,076	75,978	67,389	60,545	54,963	50,323	46,405	43,054	36,469	31,631	27,927	24,999	22,627	20,666	17,613	15,346	13,596	11,071	9,337
160	141,068	126,143	114,074	104,112	95,751	88,633	77,160	68,317	61,293	55,578	50,839	46,844	43,431	36,739	31,834	28,085	25,125	22,730	20,752	17,676	15,393	13,633	11,096	9,354
165	144,941	129,230	116,593	106,206	97,519	90,146	78,304	69,213	62,013	56,170	51,333	47,263	43,791	36,997	32,027	28,235	25,246	22,829	20,834	17,735	15,438	13,668	11,119	9,371
170	148,785	132,277	119,067	108,256	99,244	91,618	79,413	70,077	62,706	56,738	51,807	47,664	44,136	37,242	32,211	28,378	25,360	22,922	20,912	17,791	15,481	13,702	11,141	9,387
175	152,601	135,285	121,498	110,262	100,928	93,051	80,487	70,913	63,374	57,284	52,262	48,049	44,465	37,477	32,387	28,514	25,468	23,011	20,985	17,844	15,521	13,733	11,162	9,402
180	156,389	138,254	123,888	112,226	102,571	94,446	81,529	71,720	64,018	57,810	52,699	48,419	44,781	37,701	32,554	28,643	25,572	23,095	21,056	17,895	15,560	13,763	11,182	9,416
185	160,150	141,185	126,236	114,150	104,175	95,804	82,539	72,501	64,639	58,316	53,119	48,773	45,085	37,916	32,714	28,767	25,670	23,175	21,122	17,943	15,596	13,792	11,201	9,429
190	163,883	144,078	128,544	116,034	105,742	97,128	83,520	73,256	65,239	58,804	53,524	49,114	45,376	38,121	32,867	28,885	25,764	23,252	21,186	17,989	15,631	13,819	11,218	9,442
195	167,589	146,935	130,813	117,880	107,273	98,418	84,472	73,988	65,819	59,274	53,913	49,442	45,655	38,318	33,013	28,998	25,854	23,325	21,247	18,033	15,664	13,845	11,235	9,454
200	171,269	149,756	133,045	119,688	108,769	99,676	85,397	74,696	66,379	59,728	54,288	49,757	45,924	38,508	33,154	29,107	25,940	23,395	21,305	18,075	15,695	13,870	11,252	9,465

# 2/0 AWG	2 Run(s)	480 Volts	C-Value = 10,755	Magnetic Duct	Copper

Calculations:

1. $f = \dfrac{1.732 \times \text{Length} \times \text{Isca}}{\text{Runs} \times \text{C-Value} \times E_{L\text{-}L}}$

2. $M = \dfrac{1}{1 + f}$

3. $\text{Isc} = \text{Isca} \times M$

4. Add Motor Contribution = Motor FLA x 4

* All results are given in symmetrical amperes

Three-Phase Bolted Fault-Current* Table

One-Way Distance in Feet

Available Isc in Thousands (Isca)	5	10	15	20	25	30	40	50	60	70	80	90	100	125	150	175	200	225	250	300	350	400	500	600
3	2,987	2,975	2,963	2,950	2,938	2,926	2,902	2,879	2,856	2,833	2,810	2,788	2,767	2,714	2,663	2,614	2,567	2,522	2,478	2,395	2,317	2,244	2,111	1,992
5	4,965	4,931	4,897	4,863	4,830	4,798	4,734	4,672	4,611	4,552	4,495	4,439	4,384	4,253	4,130	4,013	3,903	3,799	3,700	3,518	3,352	3,201	2,937	2,713
7	6,932	6,865	6,799	6,735	6,672	6,610	6,489	6,373	6,261	6,153	6,048	5,947	5,850	5,619	5,405	5,208	5,024	4,853	4,693	4,402	4,146	3,918	3,529	3,211
10	9,861	9,727	9,596	9,468	9,344	9,223	8,990	8,768	8,557	8,357	8,165	7,982	7,807	7,401	7,035	6,704	6,402	6,127	5,874	5,426	5,042	4,708	4,158	3,723
15	14,690	14,393	14,108	13,834	13,570	13,316	12,836	12,389	11,973	11,583	11,218	10,875	10,553	9,825	9,190	8,633	8,139	7,699	7,304	6,625	6,061	5,585	4,828	4,251
20	19,453	18,936	18,445	17,979	17,536	17,115	16,330	15,613	14,957	14,354	13,798	13,283	12,805	11,748	10,853	10,084	9,417	8,833	8,317	7,447	6,741	6,158	5,250	4,575
25	24,152	23,359	22,617	21,921	21,266	20,649	19,517	18,502	17,588	16,760	16,006	15,317	14,685	13,312	12,174	11,215	10,396	9,688	9,071	8,046	7,229	6,562	5,541	4,794
30	28,787	27,668	26,633	25,672	24,779	23,945	22,436	21,105	19,924	18,868	17,918	17,059	16,279	14,609	13,249	12,121	11,170	10,357	9,655	8,502	7,595	6,863	5,753	4,953
35	33,360	31,866	30,501	29,248	28,094	27,027	25,119	23,464	22,012	20,730	19,589	18,567	17,647	15,701	14,141	12,864	11,798	10,895	10,120	8,861	7,880	7,094	5,915	5,072
40	37,872	35,959	34,230	32,659	31,227	29,914	27,595	25,610	23,891	22,388	21,063	19,886	18,834	16,633	14,893	13,483	12,317	11,336	10,500	9,150	8,108	7,279	6,043	5,166
45	42,324	39,949	37,826	35,918	34,193	32,625	29,886	27,571	25,589	23,872	22,372	21,049	19,874	17,439	15,536	14,008	12,753	11,704	10,815	9,389	8,295	7,429	6,146	5,241
50	46,719	43,841	41,298	39,033	37,004	35,176	32,012	29,370	27,132	25,210	23,542	22,082	20,792	18,142	16,092	14,458	13,125	12,017	11,082	9,589	8,451	7,554	6,231	5,303
55	51,055	47,639	44,650	42,015	39,673	37,579	33,990	31,027	28,539	26,421	24,595	23,005	21,609	18,761	16,577	14,848	13,446	12,285	11,310	9,759	8,582	7,659	6,303	5,355
60	55,336	51,345	47,890	44,872	42,211	39,848	35,836	32,558	29,829	27,523	25,547	23,836	22,340	19,310	17,004	15,190	13,725	12,518	11,507	9,906	8,696	7,749	6,364	5,398
65	59,561	54,963	51,023	47,611	44,626	41,993	37,562	33,976	31,015	28,529	26,412	24,588	22,999	19,800	17,383	15,491	13,971	12,723	11,679	10,033	8,794	7,827	6,416	5,436
70	63,733	58,496	54,054	50,239	46,927	44,025	39,179	35,294	32,110	29,453	27,202	25,270	23,595	20,241	17,721	15,760	14,189	12,903	11,831	10,145	8,879	7,895	6,461	5,469
75	67,851	61,947	56,988	52,764	49,123	45,952	40,697	36,521	33,123	30,303	27,925	25,894	24,138	20,639	18,026	16,000	14,383	13,064	11,966	10,244	8,955	7,954	6,501	5,497
80	71,918	65,319	59,829	55,190	51,219	47,781	42,126	37,668	34,063	31,088	28,591	26,465	24,633	21,000	18,300	16,216	14,558	13,207	12,086	10,332	9,022	8,008	6,537	5,523
85	75,933	68,614	62,582	57,525	53,224	49,521	43,473	38,741	34,938	31,815	29,205	26,990	25,088	21,329	18,550	16,412	14,715	13,337	12,195	10,411	9,083	8,055	6,568	5,545
90	79,898	71,836	65,251	59,772	55,142	51,178	44,744	39,747	35,754	32,491	29,773	27,475	25,506	21,631	18,778	16,590	14,858	13,454	12,293	10,482	9,137	8,098	6,597	5,565
95	83,815	74,986	67,840	61,937	56,979	52,756	45,946	40,693	36,518	33,120	30,300	27,923	25,892	21,908	18,986	16,752	14,989	13,561	12,382	10,547	9,186	8,136	6,622	5,583
100	87,682	78,067	70,351	64,024	58,741	54,263	47,085	41,584	37,234	33,707	30,791	28,340	26,250	22,163	19,178	16,901	15,108	13,658	12,463	10,606	9,231	8,171	6,645	5,600
105	91,503	81,081	72,790	66,037	60,431	55,702	48,165	42,424	37,906	34,257	31,250	28,727	26,582	22,400	19,355	17,038	15,217	13,748	12,537	10,660	9,271	8,203	6,667	5,615
110	95,277	84,030	75,158	67,981	62,055	57,079	49,190	43,218	38,538	34,773	31,678	29,089	26,891	22,619	19,518	17,165	15,318	13,830	12,606	10,709	9,309	8,232	6,686	5,629
115	99,006	86,917	77,459	69,858	63,615	58,396	50,166	43,969	39,134	35,258	32,080	29,428	27,180	22,823	19,670	17,282	15,411	13,906	12,669	10,755	9,343	8,259	6,704	5,641
120	102,689	89,743	79,696	71,672	65,116	59,659	51,094	44,680	39,697	35,714	32,457	29,745	27,451	23,013	19,811	17,391	15,498	13,976	12,727	10,797	9,375	8,284	6,720	5,653
125	106,329	92,511	81,871	73,426	66,560	60,869	51,980	45,356	40,229	36,144	32,812	30,043	27,704	23,191	19,943	17,492	15,578	14,042	12,781	10,836	9,404	8,307	6,735	5,663
130	109,925	95,221	83,987	75,123	67,952	62,031	52,825	45,998	40,734	36,551	33,147	30,323	27,942	23,358	20,066	17,587	15,653	14,103	12,832	10,872	9,432	8,328	6,749	5,673
135	113,479	97,876	86,045	76,766	69,294	63,147	53,632	46,609	41,212	36,935	33,463	30,587	28,167	23,514	20,181	17,676	15,724	14,160	12,879	10,906	9,457	8,348	6,762	5,682
140	116,991	100,478	88,050	78,358	70,588	64,220	54,404	47,190	41,666	37,300	33,762	30,837	28,378	23,662	20,290	17,759	15,789	14,213	12,923	10,937	9,481	8,366	6,774	5,691
145	120,463	103,028	90,002	79,900	71,837	65,252	55,142	47,745	42,098	37,646	34,045	31,073	28,578	23,800	20,391	17,837	15,851	14,263	12,964	10,967	9,503	8,384	6,785	5,699
150	123,893	105,527	91,903	81,395	73,043	66,245	55,850	48,275	42,510	37,974	34,313	31,296	28,767	23,931	20,488	17,910	15,909	14,310	13,003	10,995	9,524	8,400	6,796	5,706
155	127,285	107,978	93,756	82,845	74,209	67,203	56,529	48,782	42,902	38,287	34,568	31,508	28,946	24,055	20,578	17,979	15,963	14,354	13,039	11,021	9,543	8,415	6,806	5,713
160	130,637	110,380	95,562	84,252	75,336	68,126	57,181	49,266	43,276	38,585	34,811	31,710	29,116	24,172	20,664	18,045	16,015	14,396	13,074	11,045	9,562	8,429	6,815	5,720
165	133,951	112,737	97,324	85,618	76,426	69,016	57,807	49,730	43,634	38,869	35,042	31,901	29,277	24,283	20,745	18,107	16,064	14,435	13,106	11,068	9,579	8,443	6,824	5,726
170	137,228	115,049	99,042	86,945	77,482	69,876	58,409	50,175	43,976	39,140	35,262	32,084	29,431	24,389	20,822	18,165	16,110	14,472	13,137	11,090	9,595	8,456	6,832	5,732
175	140,468	117,318	100,719	88,235	78,504	70,706	58,988	50,602	44,303	39,399	35,472	32,258	29,577	24,489	20,895	18,221	16,154	14,508	13,166	11,111	9,611	8,468	6,840	5,738
180	143,671	119,544	102,355	89,488	79,495	71,509	59,546	51,012	44,617	39,647	35,673	32,424	29,717	24,585	20,965	18,274	16,195	14,541	13,194	11,131	9,626	8,479	6,848	5,743
185	146,839	121,729	103,953	90,707	80,455	72,285	60,083	51,405	44,918	39,884	35,865	32,582	29,850	24,676	21,031	18,324	16,235	14,573	13,220	11,149	9,639	8,490	6,855	5,748
190	149,971	123,874	105,513	91,892	81,386	73,036	60,601	51,784	45,207	40,112	36,049	32,734	29,977	24,763	21,094	18,372	16,272	14,603	13,245	11,167	9,653	8,500	6,861	5,752
195	153,069	125,980	107,037	93,046	82,290	73,763	61,100	52,148	45,484	40,330	36,226	32,879	30,099	24,846	21,154	18,418	16,308	14,632	13,268	11,184	9,665	8,510	6,868	5,757
200	156,133	128,048	108,526	94,170	83,167	74,467	61,583	52,499	45,751	40,540	36,395	33,018	30,215	24,925	21,212	18,461	16,342	14,659	13,291	11,200	9,677	8,519	6,874	5,761

 # 3/0 AWG | **1 Run(s)** | **480 Volts** | **C-Value = 12,843** | **Magnetic Duct** | **Copper**

Calculations:

1. $f = \dfrac{1.732 \times \text{Length} \times \text{Isca}}{\text{Runs} \times \text{C-Value} \times \text{E L-L}}$

2. $M = \dfrac{1}{1 + f}$

3. $\text{Isc} = \text{Isca} \times M$

4. Add Motor Contribution = Motor FLA x 4

* All results are given in symmetrical amperes

Three-Phase Bolted Fault-Current* Table

One-Way Distance in Feet

Available Isc in Thousands (Isca)	5	10	15	20	25	30	40	50	60	70	80	90	100	125	150	175	200	225	250	300	350	400	500	600
3	2,994	2,987	2,981	2,975	2,969	2,963	2,950	2,938	2,926	2,914	2,902	2,890	2,879	2,850	2,822	2,794	2,767	2,740	2,714	2,663	2,614	2,567	2,478	2,395
5	4,983	4,965	4,948	4,931	4,914	4,897	4,863	4,830	4,798	4,766	4,734	4,703	4,672	4,596	4,523	4,453	4,384	4,318	4,253	4,130	4,013	3,903	3,700	3,518
7	6,966	6,932	6,898	6,865	6,832	6,799	6,735	6,672	6,610	6,549	6,489	6,431	6,373	6,234	6,100	5,972	5,850	5,732	5,619	5,405	5,208	5,024	4,693	4,402
10	9,930	9,861	9,794	9,727	9,661	9,596	9,468	9,344	9,223	9,105	8,990	8,878	8,768	8,506	8,260	8,027	7,807	7,598	7,401	7,035	6,704	6,402	5,874	5,426
15	14,844	14,690	14,540	14,393	14,249	14,108	13,834	13,570	13,316	13,072	12,836	12,609	12,389	11,873	11,398	10,959	10,553	10,176	9,825	9,190	8,633	8,139	7,304	6,625
20	19,723	19,453	19,191	18,936	18,687	18,445	17,979	17,536	17,115	16,713	16,330	15,963	15,613	14,802	14,070	13,408	12,805	12,254	11,748	10,853	10,084	9,417	8,317	7,447
25	24,569	24,152	23,749	23,359	22,982	22,617	21,921	21,266	20,649	20,067	19,517	18,996	18,502	17,373	16,374	15,484	14,685	13,965	13,312	12,174	11,215	10,396	9,071	8,046
30	29,381	28,787	28,216	27,668	27,141	26,633	25,672	24,779	23,945	23,166	22,436	21,750	21,105	19,649	18,381	17,266	16,279	15,399	14,609	13,249	12,121	11,170	9,655	8,502
35	34,160	33,360	32,596	31,866	31,169	30,501	29,248	28,094	27,027	26,038	25,119	24,263	23,464	21,677	20,144	18,813	17,647	16,617	15,701	14,141	12,864	11,798	10,120	8,861
40	38,907	37,872	36,891	35,959	35,073	34,230	32,659	31,227	29,914	28,708	27,595	26,565	25,610	23,496	21,705	20,168	18,834	17,665	16,633	14,893	13,483	12,317	10,500	9,150
45	43,621	42,324	41,103	39,949	38,859	37,826	35,918	34,193	32,625	31,196	29,886	28,682	27,571	25,137	23,098	21,365	19,874	18,577	17,439	15,536	14,008	12,753	10,815	9,389
50	48,304	46,719	45,234	43,841	42,532	41,298	39,033	37,004	35,176	33,519	32,012	30,634	29,370	26,624	24,348	22,430	20,792	19,377	18,142	16,092	14,458	13,125	11,082	9,589
55	52,954	51,055	49,288	47,639	46,096	44,650	42,015	39,673	37,579	35,695	33,990	32,441	31,027	27,979	25,475	23,383	21,609	20,085	18,761	16,577	14,848	13,446	11,310	9,759
60	57,574	55,336	53,266	51,345	49,557	47,890	44,872	42,211	39,848	37,736	35,836	34,118	32,558	29,217	26,498	24,242	22,340	20,715	19,310	17,004	15,190	13,725	11,507	9,906
65	62,162	59,561	57,170	54,963	52,920	51,023	47,611	44,626	41,993	39,654	37,562	35,679	33,976	30,354	27,430	25,020	22,999	21,280	19,800	17,383	15,491	13,971	11,679	10,033
70	66,720	63,733	61,002	58,496	56,187	54,054	50,239	46,927	44,025	41,461	39,179	37,135	35,294	31,402	28,283	25,727	23,595	21,790	20,241	17,721	15,760	14,189	11,831	10,145
75	71,247	67,851	64,765	61,947	59,364	56,988	52,764	49,123	45,952	43,165	40,697	38,496	36,521	32,370	29,065	26,373	24,138	22,251	20,639	18,026	16,000	14,383	11,966	10,244
80	75,744	71,918	68,459	65,319	62,453	59,829	55,190	51,219	47,781	44,776	42,126	39,772	37,668	33,267	29,787	26,966	24,633	22,672	21,000	18,300	16,216	14,558	12,086	10,332
85	80,211	75,933	72,088	68,614	65,459	62,582	57,525	53,224	49,521	46,300	43,473	40,971	38,741	34,101	30,454	27,511	25,088	23,056	21,329	18,550	16,412	14,715	12,195	10,411
90	84,649	79,898	75,653	71,836	68,385	65,251	59,772	55,142	51,178	47,745	44,744	42,098	39,747	34,879	31,072	28,015	25,506	23,409	21,631	18,778	16,590	14,858	12,293	10,482
95	89,057	83,815	79,155	74,986	71,234	67,840	61,937	56,979	52,756	49,116	45,946	43,160	40,693	35,605	31,647	28,482	25,892	23,734	21,908	18,986	16,752	14,989	12,382	10,547
100	93,437	87,682	82,596	78,067	74,008	70,351	64,024	58,741	54,263	50,420	47,085	44,164	41,584	36,285	32,184	28,915	26,250	24,034	22,163	19,178	16,901	15,108	12,463	10,606
105	97,788	91,503	85,977	81,081	76,712	72,790	66,037	60,431	55,702	51,660	48,165	45,112	42,424	36,923	32,684	29,319	26,582	24,312	22,400	19,355	17,038	15,217	12,537	10,660
110	102,111	95,277	89,301	84,030	79,347	75,158	67,981	62,055	57,079	52,842	49,190	46,011	43,218	37,522	33,154	29,696	26,891	24,571	22,619	19,518	17,165	15,318	12,606	10,709
115	106,405	99,006	92,568	86,917	81,916	77,459	69,858	63,615	58,396	53,969	50,166	46,863	43,969	38,087	33,594	30,049	27,180	24,812	22,823	19,670	17,282	15,411	12,669	10,755
120	110,672	102,689	95,781	89,743	84,422	79,696	71,672	65,116	59,659	55,045	51,094	47,673	44,680	38,620	34,008	30,379	27,451	25,037	23,013	19,811	17,391	15,498	12,727	10,797
125	114,911	106,329	98,940	92,511	86,866	81,871	73,426	66,560	60,869	56,074	51,980	48,442	45,356	39,124	34,398	30,690	27,704	25,248	23,191	19,943	17,492	15,578	12,781	10,836
130	119,123	109,925	102,046	95,221	89,252	83,987	75,123	67,952	62,031	57,059	52,825	49,175	45,998	39,601	34,766	30,983	27,942	25,445	23,358	20,066	17,587	15,653	12,832	10,872
135	123,308	113,479	105,102	97,876	91,580	86,045	76,766	69,294	63,147	58,002	53,632	49,874	46,609	40,052	35,113	31,259	28,167	25,631	23,514	20,181	17,676	15,724	12,879	10,906
140	127,466	116,991	108,108	100,478	93,854	88,050	78,358	70,588	64,220	58,905	54,404	50,541	47,190	40,481	35,443	31,519	28,378	25,806	23,662	20,290	17,759	15,789	12,923	10,937
145	131,597	120,463	111,065	103,028	96,075	90,002	79,900	71,837	65,252	59,773	55,142	51,178	47,745	40,889	35,755	31,766	28,578	25,971	23,800	20,391	17,837	15,851	12,964	10,967
150	135,703	123,893	113,975	105,527	98,245	91,903	81,395	73,043	66,245	60,605	55,850	51,787	48,275	41,277	36,051	32,000	28,767	26,127	23,931	20,488	17,910	15,909	13,003	10,995
155	139,782	127,285	116,839	107,978	100,366	93,756	82,845	74,209	67,203	61,406	56,529	52,371	48,782	41,647	36,333	32,221	28,946	26,275	24,055	20,578	17,979	15,963	13,039	11,021
160	143,835	130,637	119,658	110,380	102,438	95,562	84,252	75,336	68,126	62,176	57,181	52,929	49,266	41,999	36,601	32,432	29,116	26,415	24,172	20,664	18,045	16,015	13,074	11,045
165	147,863	133,951	122,432	112,737	104,465	97,324	85,618	76,426	69,016	62,916	57,807	53,465	49,730	42,336	36,856	32,632	29,277	26,548	24,283	20,745	18,107	16,064	13,106	11,068
170	151,866	137,228	125,164	115,049	106,447	99,042	86,945	77,482	69,876	63,630	58,409	53,980	50,175	42,658	37,100	32,823	29,431	26,674	24,389	20,822	18,165	16,110	13,137	11,090
175	155,844	140,468	127,853	117,318	108,386	100,719	88,235	78,504	70,706	64,318	58,988	54,474	50,602	42,966	37,333	33,005	29,577	26,794	24,489	20,895	18,221	16,154	13,166	11,111
180	159,797	143,671	130,502	119,544	110,284	102,355	89,488	79,495	71,509	64,981	59,546	54,949	51,012	43,261	37,555	33,179	29,717	26,908	24,585	20,965	18,274	16,195	13,194	11,131
185	163,725	146,839	133,110	121,729	112,141	103,953	90,707	80,455	72,285	65,622	60,083	55,406	51,405	43,544	37,768	33,345	29,850	27,017	24,676	21,031	18,324	16,235	13,220	11,149
190	167,629	149,971	135,679	123,874	113,959	105,513	91,892	81,386	73,036	66,240	60,601	55,846	51,784	43,815	37,972	33,504	29,977	27,122	24,763	21,094	18,372	16,272	13,245	11,167
195	171,509	153,069	138,210	125,980	115,738	107,037	93,046	82,290	73,763	66,837	61,100	56,271	52,148	44,076	38,168	33,656	30,099	27,221	24,846	21,154	18,418	16,308	13,268	11,184
200	175,365	156,133	140,703	128,048	117,482	108,526	94,170	83,167	74,467	67,415	61,583	56,679	52,499	44,327	38,356	33,802	30,215	27,317	24,925	21,212	18,461	16,342	13,291	11,200

3/0 AWG | 2 Run(s) | 480 Volts

C-Value = 12,843 | Magnetic Duct | Copper

Calculations:

1. $f = \dfrac{1.732 \times \text{Length} \times \text{Isca}}{\text{Runs} \times \text{C-Value} \times E_{L\text{-}L}}$

2. $M = \dfrac{1}{1 + f}$

3. Isc = Isca x M

4. Add Motor Contribution = Motor FLA x 4

* All results are given in symmetrical amperes

Three-Phase Bolted Fault-Current* Table

One-Way Distance in Feet

Available Isc in Thousands (Isca)

	5	10	15	20	25	30	40	50	60	70	80	90	100	125	150	175	200	225	250	300	350	400	500	600
3	2,989	2,979	2,968	2,958	2,947	2,937	2,916	2,896	2,876	2,856	2,837	2,818	2,799	2,753	2,708	2,665	2,623	2,583	2,544	2,468	2,398	2,331	2,208	2,097
5	4,970	4,941	4,912	4,883	4,855	4,827	4,772	4,718	4,665	4,614	4,563	4,514	4,466	4,350	4,239	4,134	4,035	3,940	3,849	3,680	3,524	3,382	3,129	2,911
7	6,942	6,885	6,828	6,773	6,719	6,665	6,561	6,459	6,361	6,265	6,173	6,083	5,996	5,788	5,595	5,413	5,244	5,084	4,934	4,659	4,413	4,192	3,810	3,492
10	9,882	9,766	9,654	9,543	9,436	9,330	9,127	8,932	8,745	8,566	8,393	8,228	8,069	7,698	7,359	7,049	6,764	6,501	6,257	5,822	5,443	5,110	4,553	4,106
15	14,736	14,480	14,234	13,995	13,765	13,542	13,117	12,718	12,342	11,988	11,654	11,338	11,039	10,355	9,751	9,214	8,732	8,299	7,906	7,223	6,649	6,159	5,368	4,757
20	19,533	19,087	18,661	18,253	17,863	17,489	16,787	16,139	15,539	14,982	14,463	13,980	13,527	12,515	11,643	10,885	10,220	9,631	9,106	8,212	7,477	6,863	5,895	5,167
25	24,274	23,589	22,942	22,329	21,748	21,197	20,174	19,245	18,398	17,622	16,909	16,252	15,643	14,305	13,177	12,215	11,383	10,657	10,019	8,947	8,082	7,369	6,265	5,448
30	28,961	27,991	27,084	26,234	25,436	24,685	23,308	22,077	20,970	19,968	19,057	18,226	17,465	15,813	14,447	13,298	12,318	11,473	10,736	9,514	8,542	7,750	6,538	5,653
35	33,593	32,296	31,094	29,979	28,941	27,973	26,218	24,671	23,296	22,066	20,959	19,959	19,049	17,101	15,514	14,197	13,085	12,136	11,314	9,966	8,904	8,047	6,748	5,810
40	38,173	36,506	34,979	33,574	32,278	31,078	23,927	27,055	25,410	23,954	22,655	21,490	20,440	18,213	16,424	14,955	13,727	12,685	11,791	10,333	9,197	8,285	6,914	5,933
45	42,701	40,626	38,743	37,027	35,457	34,014	31,454	29,253	27,340	25,661	24,177	22,855	21,670	19,183	17,209	15,603	14,271	13,149	12,190	10,539	9,438	8,480	7,050	6,032
50	47,178	44,658	42,393	40,347	38,489	36,795	33,818	31,287	29,108	27,213	25,549	24,078	22,766	20,038	17,893	16,163	14,738	13,544	12,529	10,896	9,640	8,643	7,162	6,114
55	51,605	48,604	45,934	43,541	41,386	39,433	36,034	33,174	30,735	28,629	26,794	25,180	23,749	20,795	18,495	16,653	15,144	13,886	12,822	11,117	9,812	8,781	7,257	6,183
60	55,982	52,468	49,370	46,617	44,154	41,939	38,115	34,930	32,236	29,928	27,928	26,179	24,636	21,472	19,028	17,084	15,500	14,185	13,076	11,307	9,960	8,900	7,337	6,242
65	60,311	56,252	52,706	49,580	46,804	44,322	40,073	36,567	33,625	31,122	28,965	27,088	25,439	22,080	19,504	17,466	15,814	14,448	13,298	11,473	10,089	9,002	7,407	6,292
70	64,591	59,959	55,946	52,437	49,342	46,591	41,919	38,098	34,916	32,224	29,917	27,919	26,171	22,629	19,931	17,808	16,094	14,681	13,496	11,620	10,202	9,092	7,468	6,336
75	68,825	63,590	59,095	55,193	51,775	48,755	43,662	39,532	36,116	33,244	30,795	28,682	26,840	23,127	20,317	18,115	16,344	14,889	13,671	11,750	10,302	9,172	7,521	6,374
80	73,013	67,148	62,155	57,854	54,109	50,820	45,311	40,879	37,237	34,191	31,606	29,384	27,454	23,582	20,667	18,393	16,570	15,076	13,829	11,866	10,391	9,242	7,569	6,408
85	77,155	70,636	65,132	60,424	56,351	52,792	46,872	42,146	38,285	35,073	32,358	30,033	28,019	23,998	20,986	18,645	16,775	15,245	13,971	11,970	10,471	9,305	7,611	6,439
90	81,252	74,054	68,028	62,909	58,506	54,679	48,354	43,340	39,268	35,896	33,057	30,634	28,542	24,380	21,277	18,875	16,960	15,398	14,100	12,065	10,543	9,362	7,649	6,466
95	85,306	77,407	70,846	65,311	60,578	56,485	49,761	44,467	40,191	36,665	33,708	31,193	29,027	24,733	21,545	19,086	17,130	15,538	14,217	12,151	10,609	9,414	7,683	6,490
100	89,316	80,694	73,590	67,636	62,574	58,216	51,099	45,532	41,060	37,387	34,317	31,714	29,477	25,059	21,793	19,280	17,286	15,666	14,324	12,229	10,668	9,461	7,715	6,513
105	93,283	83,919	76,263	69,887	64,495	59,876	52,373	46,542	41,878	38,065	34,887	32,200	29,897	25,362	22,021	19,458	17,430	15,784	14,423	12,300	10,723	9,504	7,743	6,533
110	97,209	87,082	78,867	72,068	66,348	61,469	53,588	47,499	42,652	38,702	35,422	32,655	30,289	25,643	22,233	19,624	17,562	15,893	14,513	12,366	10,773	9,543	7,769	6,551
115	101,093	90,187	81,404	74,181	68,135	63,000	54,748	48,407	43,383	39,304	35,925	33,082	30,656	25,906	22,430	19,777	17,685	15,993	14,597	12,427	10,819	9,579	7,793	6,568
120	104,937	93,233	83,878	76,229	69,859	64,471	55,856	49,271	44,076	39,871	36,399	33,483	31,000	26,151	22,614	19,920	17,799	16,086	14,675	12,483	10,861	9,612	7,815	6,584
125	108,740	96,223	86,291	78,217	71,525	65,887	56,915	50,094	44,733	40,408	36,846	33,861	31,324	26,381	22,786	20,053	17,905	16,173	14,747	12,535	10,901	9,643	7,836	6,599
130	112,504	99,159	88,644	80,146	73,134	67,251	57,930	50,878	45,357	40,917	37,269	34,218	31,628	26,597	22,947	20,177	18,004	16,254	14,814	12,584	10,937	9,672	7,854	6,612
135	116,230	102,042	90,941	82,019	74,690	68,564	58,902	51,627	45,951	41,400	37,669	34,555	31,916	26,800	23,098	20,294	18,097	16,330	14,877	12,629	10,972	9,699	7,872	6,624
140	119,917	104,873	93,183	83,838	76,196	69,831	59,835	52,342	46,517	41,858	38,048	34,873	32,188	26,991	23,240	20,403	18,184	16,401	14,935	12,672	11,004	9,724	7,889	6,636
145	123,567	107,654	95,372	85,605	77,653	71,053	60,730	53,025	47,056	42,294	38,408	35,176	32,445	27,172	23,373	20,507	18,266	16,467	14,991	12,711	11,033	9,747	7,904	6,647
150	127,179	110,386	97,510	87,324	79,065	72,233	61,589	53,680	47,570	42,709	38,750	35,462	32,689	27,343	23,500	20,604	18,343	16,530	15,042	12,748	11,062	9,769	7,918	6,657
155	130,756	113,070	99,598	88,995	80,432	73,373	62,416	54,307	48,062	43,105	39,076	35,735	32,920	27,505	23,619	20,695	18,416	16,589	15,091	12,783	11,088	9,789	7,932	6,667
160	134,296	115,708	101,639	90,621	81,758	74,474	63,211	54,908	48,532	43,483	39,386	35,994	33,140	27,658	23,732	20,782	18,484	16,644	15,137	12,817	11,113	9,809	7,944	6,676
165	137,801	118,300	103,634	92,204	83,044	75,540	63,977	55,485	48,982	43,844	39,682	36,241	33,350	27,804	23,839	20,864	18,549	16,697	15,181	12,848	11,136	9,827	7,956	6,684
170	141,271	120,848	105,585	93,744	84,292	76,571	64,715	56,039	49,414	44,190	39,965	36,477	33,549	27,942	23,941	20,942	18,611	16,747	15,222	12,877	11,158	9,844	7,968	6,692
175	144,707	123,354	107,492	95,245	85,503	77,569	65,427	56,572	49,828	44,520	40,235	36,702	33,739	28,074	24,038	21,016	18,669	16,794	15,261	12,905	11,179	9,861	7,978	6,700
180	148,109	125,817	109,358	96,707	86,680	78,536	66,114	57,084	50,225	44,837	40,493	36,917	33,921	28,200	24,130	21,085	18,725	16,839	15,298	12,932	11,199	9,876	7,989	6,707
185	151,477	128,240	111,184	98,132	87,823	79,473	66,777	57,578	50,607	45,141	40,741	37,123	34,095	28,319	24,217	21,153	18,778	16,882	15,333	12,957	11,218	9,891	7,998	6,713
190	154,813	130,623	112,970	99,521	88,934	80,382	67,417	58,053	50,974	45,433	40,979	37,320	34,261	28,434	24,301	21,217	18,828	16,922	15,367	12,981	11,236	9,905	8,007	6,720
195	158,117	132,967	114,719	100,876	90,014	81,264	68,036	58,512	51,327	45,713	41,206	37,509	34,420	28,544	24,381	21,273	18,876	16,961	15,399	13,003	11,253	9,918	8,016	6,726
200	161,388	135,273	116,432	102,198	91,065	82,119	68,635	58,954	51,667	45,983	41,425	37,690	34,572	28,648	24,458	21,336	18,922	16,998	15,429	13,025	11,269	9,931	8,024	6,732

 # 4/0 AWG | **1 Run(s)** | **480 Volts** | **C-Value = 15,082** | **Magnetic Duct** | **Copper**

Calculations:

1. $f = \dfrac{1.732 \times \text{Length} \times \text{Isca}}{\text{Runs} \times \text{C-Value} \times \text{E L-L}}$

2. $M = \dfrac{1}{1 + f}$

3. Isc = Isca x M

4. Add Motor Contribution = Motor FLA x 4

* All results are given in symmetrical amperes

Three-Phase Bolted Fault-Current* Table

One-Way Distance in Feet

Available Isc in Thousands (Isca)

Isca	5	10	15	20	25	30	40	50	60	70	80	90	100	125	150	175	200	225	250	300	350	400	500	600
3	2,995	2,989	2,984	2,979	2,973	2,968	2,958	2,947	2,937	2,926	2,916	2,906	2,896	2,871	2,847	2,823	2,799	2,776	2,753	2,708	2,665	2,623	2,544	2,468
5	4,985	4,970	4,956	4,941	4,926	4,912	4,883	4,855	4,827	4,799	4,772	4,745	4,718	4,652	4,588	4,526	4,466	4,407	4,350	4,239	4,134	4,035	3,849	3,680
7	6,971	6,942	6,913	6,885	6,856	6,828	6,773	6,719	6,665	6,612	6,561	6,509	6,459	6,337	6,219	6,105	5,996	5,890	5,788	5,595	5,413	5,244	4,934	4,659
10	9,941	9,882	9,824	9,766	9,710	9,654	9,543	9,436	9,330	9,227	9,127	9,028	8,932	8,699	8,479	8,269	8,069	7,879	7,698	7,359	7,049	6,764	6,257	5,822
15	14,867	14,736	14,607	14,480	14,356	14,234	13,995	13,765	13,542	13,326	13,117	12,914	12,718	12,252	11,819	11,415	11 039	10,686	10,355	9,751	9,214	8,732	7,906	7,223
20	19,764	19,533	19,307	19,087	18,871	18,661	18,253	17,863	17,489	17,131	16,787	16,457	16,139	15,396	14,718	14,098	13 527	13,001	12,515	11,643	10,885	10,220	9,106	8,212
25	24,632	24,274	23,927	23,589	23,261	22,942	22,329	21,748	21,197	20,672	20,174	19,698	19,245	18,197	17,258	16,411	15,643	14,944	14,305	13,177	12,215	11,383	10,019	8,947
30	29,471	28,961	28,468	27,991	27,530	27,084	26,234	25,436	24,685	23,977	23,308	22,676	22,077	20,710	19,502	18,427	17,465	16,598	15,813	14,447	13,298	12,318	10,736	9,514
35	34,282	33,593	32,932	32,296	31,684	31,094	29,979	28,941	27,973	27,067	26,218	25,421	24,671	22,976	21,498	20,200	19,049	18,022	17,101	15,514	14,197	13,085	11,314	9,966
40	39,065	38,173	37,321	36,506	35,726	34,979	33,574	32,278	31,078	29,964	28,927	27,959	27,055	25,029	23,286	21,770	20,440	19,262	18,213	16,424	14,955	13,727	11,791	10,333
45	43,821	42,701	41,638	40,626	39,662	38,743	37,027	35,457	34,014	32,684	31,454	30,314	29,253	26,900	24,897	23,172	21,670	20,351	19,183	17,209	15,603	14,271	12,190	10,639
50	48,548	47,178	45,883	44,658	43,496	42,393	40,347	38,489	36,795	35,244	33,818	32,503	31,287	28,610	26,355	24,429	22,766	21,315	20,038	17,893	16,163	14,738	12,529	10,896
55	53,248	51,605	50,060	48,604	47,231	45,934	43,541	41,386	39,433	37,657	36,034	34,545	33,174	30,180	27,681	25,565	23,749	22,174	20,795	18,495	16,653	15,144	12,822	11,117
60	57,921	55,982	54,168	52,468	50,872	49,370	46,617	44,154	41,939	39,936	38,115	36,453	34,930	31,626	28,893	26,595	24,636	22,945	21,472	19,028	17,084	15,500	13,076	11,307
65	62,568	60,311	58,211	56,252	54,421	52,706	49,580	46,804	44,322	42,091	40,073	38,240	36,567	32,962	30,005	27,534	25,439	23,641	22,080	19,504	17,466	15,814	13,298	11,473
70	67,187	64,591	62,189	59,959	57,883	55,946	52,437	49,342	46,591	44,132	41,919	39,917	38,098	34,201	31,028	28,393	26,171	24,271	22,629	19,931	17,808	16,094	13,496	11,620
75	71,780	68,825	66,104	63,590	61,260	59,095	55,193	51,775	48,755	46,068	43,662	41,495	39,532	35,353	31,972	29,182	26,840	24,846	23,127	20,317	18,115	16,344	13,671	11,750
80	76,347	73,013	69,958	67,148	64,555	62,155	57,854	54,109	50,820	47,907	45,311	42,981	40,879	36,426	32,848	29,910	27,454	25,371	23,582	20,667	18,393	16,570	13,829	11,866
85	80,888	77,155	73,751	70,636	67,772	65,132	60,424	56,351	52,792	49,656	46,872	44,384	42,146	37,428	33,661	30,582	28,019	25,853	23,998	20,986	18,645	16,775	13,971	11,970
90	85,403	81,252	77,487	74,054	70,913	68,028	62,909	58,506	54,679	51,322	48,354	45,710	43,340	38,367	34,418	31,206	28,542	26,297	24,380	21,277	18,875	16,960	14,100	12,065
95	89,892	85,306	81,164	77,407	73,981	70,846	65,311	60,578	56,485	52,910	49,761	46,965	44,467	39,248	35,125	31,786	29,027	26,708	24,733	21,545	19,086	17,130	14,217	12,151
100	94,356	89,316	84,786	80,694	76,979	73,590	67,636	62,574	58,216	54,426	51,099	48,155	45,532	40,075	35,786	32,327	29,477	27,089	25,059	21,793	19,280	17,286	14,324	12,229
105	98,795	93,283	88,354	83,919	79,908	76,263	69,887	64,495	59,876	55,874	52,373	49,285	46,542	40,855	36,407	32,832	29,897	27,443	25,362	22,021	19,458	17,430	14,423	12,300
110	103,210	97,209	91,867	87,082	82,771	78,867	72,068	66,348	61,469	57,259	53,588	50,360	47,499	41,591	36,990	33,305	30,289	27,773	25,643	22,233	19,624	17,562	14,513	12,366
115	107,599	101,093	95,329	90,187	85,571	81,404	74,181	68,135	63,000	58,585	54,748	51,383	48,407	42,286	37,539	33,750	30,656	28,081	25,906	22,430	19,777	17,685	14,597	12,427
120	111,964	104,937	98,739	93,233	88,309	83,878	76,229	69,859	64,471	59,855	55,856	52,357	49,271	42,944	38,056	34,168	31,000	28,370	26,151	22,614	19,920	17,799	14,675	12,483
125	116,305	108,740	102,100	96,223	90,987	86,291	78,217	71,525	65,887	61,074	56,915	53,287	50,094	43,567	38,545	34,561	31,324	28,641	26,381	22,786	20,053	17,905	14,747	12,535
130	120,621	112,504	105,411	99,159	93,608	88,644	80,146	73,134	67,251	62,243	57,930	54,176	50,878	44,159	39,008	34,933	31,628	28,895	26,597	22,947	20,177	18,004	14,814	12,584
135	124,914	116,230	108,675	102,042	96,172	90,941	82,019	74,690	68,564	63,367	58,902	55,025	51,627	44,722	39,446	35,284	31,916	29,135	26,800	23,098	20,294	18,097	14,877	12,629
140	129,183	119,917	111,892	104,873	98,683	93,183	83,838	76,196	69,831	64,447	59,835	55,838	52,342	45,257	39,862	35,616	32,188	29,361	26,991	23,240	20,403	18,184	14,935	12,672
145	133,428	123,567	115,063	107,654	101,141	95,372	85,605	77,653	71,053	65,487	60,730	56,616	53,025	45,768	40,257	35,932	32,445	29,575	27,172	23,373	20,507	18,266	14,991	12,711
150	137,650	127,179	118,189	110,386	103,549	97,510	87,324	79,065	72,233	66,488	61,589	57,363	53,680	46,254	40,633	36,231	32,639	29,778	27,343	23,500	20,604	18,343	15,042	12,748
155	141,849	130,756	121,271	113,070	105,907	99,598	88,995	80,432	73,373	67,452	62,416	58,080	54,307	46,719	40,992	36,515	32,920	29,970	27,505	23,619	20,695	18,416	15,091	12,783
160	146,025	134,296	124,311	115,708	108,218	101,639	90,621	81,758	74,474	68,382	63,211	58,768	54,908	47,163	41,333	36,786	33,140	30,152	27,658	23,732	20,782	18,484	15,137	12,817
165	150,179	137,801	127,308	118,300	110,483	103,634	92,204	83,044	75,540	69,280	63,977	59,429	55,485	47,588	41,659	37,044	33,350	30,325	27,804	23,839	20,864	18,549	15,181	12,848
170	154,310	141,271	130,264	120,848	112,702	105,585	93,744	84,292	76,571	70,146	64,715	60,065	56,039	47,995	41,971	37,290	33,549	30,490	27,942	23,941	20,942	18,611	15,222	12,877
175	158,418	144,707	133,180	123,354	114,878	107,492	95,245	85,503	77,569	70,983	65,427	60,678	56,572	48,386	42,269	37,526	33,739	30,647	28,074	24,038	21,016	18,669	15,261	12,905
180	162,505	148,109	136,056	125,817	117,012	109,358	96,707	86,680	78,536	71,792	66,114	61,268	57,084	48,760	42,555	37,750	33,921	30,797	28,200	24,130	21,086	18,725	15,298	12,932
185	166,569	151,477	138,893	128,240	119,104	111,184	98,132	87,823	79,473	72,574	66,777	61,837	57,578	49,120	42,828	37,966	34,095	30,940	28,319	24,217	21,153	18,778	15,333	12,957
190	170,611	154,813	141,693	130,623	121,157	112,970	99,521	88,934	80,382	73,331	67,417	62,386	58,053	49,465	43,091	38,172	34,261	31,077	28,434	24,301	21,217	18,828	15,367	12,981
195	174,632	158,117	144,455	132,967	123,171	114,719	100,876	90,014	81,264	74,064	68,036	62,915	58,512	49,798	43,343	38,369	34,420	31,208	28,544	24,381	21,278	18,876	15,399	13,003
200	178,631	161,388	147,181	135,273	125,147	116,432	102,198	91,065	82,119	74,774	68,635	63,427	58,954	50,118	43,585	38,559	34,572	31,333	28,648	24,458	21,336	18,922	15,429	13,025

4/0 AWG | 2 Run(s) | 480 Volts | C-Value = 15,082 | Magnetic Duct | Copper

Calculations:

1. $f = \dfrac{1.732 \times \text{Length} \times \text{Isca}}{\text{Runs} \times \text{C-Value} \times E_{L\text{-}L}}$

2. $M = \dfrac{1}{1 + f}$

3. Isc = Isca x M

4. Add Motor Contribution = Motor FLA x 4

* All results are given in symmetrical amperes

Three-Phase Bolted Fault-Current* Table

One-Way Distance in Feet

Available Isc in Thousands (Isca)

	5	10	15	20	25	30	40	50	60	70	80	90	100	125	150	175	200	225	250	300	350	400	500	600
3	2,990	2,980	2,971	2,961	2,952	2,942	2,923	2,905	2,886	2,868	2,850	2,833	2,815	2,772	2,731	2,691	2,652	2,614	2,577	2,506	2,439	2,376	2,258	2,152
5	4,973	4,946	4,919	4,893	4,867	4,841	4,790	4,741	4,692	4,644	4,597	4,552	4,507	4,398	4,295	4,196	4,102	4,012	3,926	3,764	3,615	3,477	3,231	3,018
7	6,947	6,894	6,843	6,792	6,742	6,692	6,596	6,502	6,411	6,322	6,236	6,152	6,070	5,875	5,692	5,520	5,358	5,205	5,061	4,795	4,556	4,340	3,963	3,647
10	9,892	9,786	9,682	9,581	9,481	9,384	9,195	9,013	8,839	8,671	8,510	8,354	8,204	7,852	7,528	7,230	6,955	6,700	6,463	6,036	5,662	5,331	4,774	4,323
15	14,758	14,523	14,296	14,076	13,862	13,655	13,259	12,885	12,531	12,197	11,879	11,578	11,292	10,635	10,050	9,526	9,054	8,626	8,238	7,556	6,979	6,484	5,678	5,050
20	19,572	19,161	18,767	18,390	18,027	17,678	17,019	16,408	15,839	15,308	14,812	14,347	13,910	12,926	12,072	11,324	10,663	10,075	9,549	8,645	7,898	7,269	6,271	5,514
25	24,334	23,703	23,103	22,534	21,991	21,474	20,510	19,629	18,820	18,075	17,387	16,750	16,157	14,845	13,729	12,770	11,936	11,204	10,557	9,453	8,575	7,839	6,691	5,836
30	29,046	28,151	27,310	26,517	25,769	25,062	23,759	22,584	21,520	20,552	19,667	18,855	18,108	16,475	15,113	13,958	12,968	12,108	11,356	10,700	9,095	8,271	7,003	6,072
35	33,709	32,509	31,392	30,349	29,374	28,459	26,790	25,306	23,977	22,781	21,699	20,715	19,817	17,878	16,284	14,952	13,821	12,849	12,005	10,911	9,507	8,611	7,245	6,253
40	38,322	36,779	35,356	34,039	32,816	31,678	29,624	27,820	26,223	24,799	23,522	22,370	21,326	19,097	17,290	15,795	14,539	13,467	12,543	11,129	9,841	8,884	7,437	6,396
45	42,888	40,965	39,207	37,593	36,108	34,735	32,280	30,150	28,283	26,634	25,167	23,852	22,669	20,167	18,162	16,520	15,150	13,990	12,995	11,377	10,117	9,109	7,594	6,512
50	47,406	45,067	42,949	41,020	39,258	37,640	34,775	32,315	30,180	28,309	26,657	25,188	23,871	21,113	18,926	17,150	15,678	14,439	13,382	11,672	10,350	9,297	7,725	6,607
55	51,877	49,090	46,586	44,326	42,275	40,405	37,122	34,332	31,932	29,846	28,015	26,396	24,954	21,956	19,601	17,702	16,138	14,829	13,716	11,925	10,548	9,457	7,835	6,688
60	56,302	53,034	50,124	47,517	45,168	43,040	39,334	36,216	33,555	31,259	29,257	27,496	25,935	22,711	20,201	18,190	16,543	15,169	14,007	12,145	10,720	9,594	7,929	6,756
65	60,683	56,903	53,567	50,600	47,945	45,554	41,423	37,979	35,064	32,564	30,397	28,501	26,827	23,393	20,738	18,624	16,901	15,470	14,263	12,337	10,869	9,713	8,010	6,815
70	65,018	60,699	56,917	53,579	50,611	47,955	43,399	39,633	36,469	33,773	31,448	29,422	27,642	24,010	21,221	19,013	17,221	15,738	14,490	12,506	11,000	9,818	8,081	6,867
75	69,310	64,423	60,179	56,460	53,174	50,249	45,270	41,188	37,781	34,895	32,419	30,271	28,389	24,572	21,659	19,364	17,508	15,977	14,693	12,657	11,117	9,911	8,144	6,912
80	73,559	68,078	63,357	59,248	55,640	52,446	47,044	42,652	39,010	35,940	33,319	31,054	29,077	25,085	22,057	19,681	17,768	16,193	14,875	12,792	11,221	9,994	8,200	6,952
85	77,765	71,665	66,452	61,947	58,013	54,549	48,730	44,033	40,162	36,916	34,156	31,780	29,712	25,557	22,421	19,970	18,003	16,388	15,039	12,913	11,314	10,067	8,249	6,988
90	81,929	75,187	69,470	64,560	60,299	56,566	50,333	45,338	41,244	37,829	34,936	32,454	30,301	25,991	22,754	20,234	18,217	16,566	15,189	13,023	11,399	10,134	8,294	7,020
95	86,052	78,645	72,411	67,094	62,503	58,501	51,860	46,572	42,264	38,684	35,664	33,081	30,847	26,392	23,061	20,477	18,413	16,728	15,325	13,123	11,475	10,195	8,335	7,049
100	90,134	82,040	75,280	69,550	64,630	60,360	53,315	47,743	43,225	39,488	36,346	33,668	31,357	26,764	23,344	20,700	18,593	16,876	15,449	13,215	11,545	10,250	8,371	7,075
105	94,176	85,376	78,079	71,932	66,682	62,146	54,704	48,853	44,133	40,245	36,987	34,216	31,832	27,109	23,607	20,906	18,760	17,013	15,564	13,298	11,609	10,300	8,405	7,099
110	98,179	88,652	80,811	74,244	68,664	63,864	56,031	49,909	44,993	40,959	37,588	34,731	32,277	27,431	23,851	21,097	18,913	17,139	15,669	13,375	11,667	10,346	8,435	7,121
115	102,143	91,871	83,477	76,488	70,579	65,518	57,300	50,913	45,808	41,633	38,155	35,214	32,694	27,732	24,078	21,274	19,056	17,256	15,767	13,446	11,721	10,388	8,464	7,141
120	106,068	95,035	86,081	78,668	72,432	67,111	58,514	51,870	46,581	42,270	38,690	35,669	33,086	28,013	24,289	21,439	19,188	17,365	15,858	13,512	11,771	10,428	8,490	7,159
125	109,956	98,144	88,624	80,787	74,224	68,647	59,678	52,783	47,315	42,875	39,196	36,098	33,455	28,277	24,488	21,594	19,312	17,466	15,942	13,573	11,818	10,464	8,514	7,176
130	113,806	101,200	91,108	82,846	75,958	70,128	60,795	53,654	48,014	43,448	39,674	36,504	33,803	28,525	24,674	21,738	19,427	17,560	16,020	13,630	11,861	10,498	8,536	7,192
135	117,620	104,204	93,536	84,849	77,638	71,557	61,866	54,487	48,680	43,992	40,128	36,887	34,131	28,759	24,848	21,874	19,535	17,648	16,094	13,683	11,901	10,529	8,557	7,207
140	121,397	107,158	95,909	86,797	79,267	72,938	62,896	55,284	49,315	44,510	40,558	37,251	34,442	28,980	25,013	22,001	19,637	17,731	16,163	13,733	11,939	10,559	8,576	7,221
145	125,139	110,063	98,230	88,693	80,845	74,273	63,885	56,047	49,922	45,004	40,968	37,596	34,737	29,188	25,168	22,121	19,732	17,809	16,227	13,780	11,974	10,586	8,595	7,234
150	128,846	112,920	100,499	90,539	82,376	75,563	64,838	56,779	50,501	45,474	41,357	37,924	35,017	29,385	25,314	22,234	19,822	17,882	16,288	13,824	12,007	10,612	8,612	7,246
155	132,517	115,731	102,719	92,337	83,861	76,811	65,754	57,480	51,056	45,923	41,728	38,235	35,282	29,572	25,453	22,341	19,907	17,951	16,345	13,865	12,038	10,636	8,628	7,257
160	136,155	118,496	104,891	94,089	85,304	78,019	66,638	58,154	51,587	46,352	42,082	38,533	35,535	29,749	25,584	22,442	19,987	18,016	16,399	13,904	12,067	10,659	8,643	7,268
165	139,759	121,216	107,017	95,796	86,705	79,189	67,490	58,802	52,096	46,763	42,420	38,816	35,776	29,918	25,709	22,538	20,063	18,078	16,450	13,940	12,095	10,681	8,657	7,278
170	143,330	123,893	109,098	97,460	88,066	80,323	68,311	59,425	52,584	47,156	42,744	39,086	36,005	30,078	25,827	22,629	20,135	18,136	16,499	13,975	12,121	10,701	8,670	7,287
175	146,868	126,528	111,136	99,083	89,389	81,422	69,105	60,024	53,053	47,533	43,053	39,345	36,225	30,231	25,940	22,715	20,203	18,192	16,545	14,008	12,146	10,720	8,683	7,296
180	150,373	129,121	113,132	100,666	90,675	82,488	69,871	60,602	53,504	47,894	43,349	39,592	36,434	30,377	26,047	22,797	20,268	18,245	16,588	14,039	12,169	10,739	8,695	7,304
185	153,847	131,674	115,087	102,211	91,927	83,523	70,612	61,158	53,937	48,241	43,633	39,829	36,635	30,516	26,149	22,875	20,330	18,295	16,630	14,069	12,191	10,756	8,706	7,312
190	157,289	134,187	117,002	103,719	93,145	84,527	71,328	61,695	54,354	48,574	43,906	40,056	36,826	30,649	26,247	22,950	20,389	18,342	16,669	14,097	12,213	10,773	8,717	7,320
195	160,700	136,662	118,879	105,192	94,331	85,502	72,022	62,213	54,756	48,895	44,167	40,273	37,010	30,777	26,340	23,021	20,445	18,388	16,707	14,124	12,233	10,788	8,727	7,327
200	164,081	139,099	120,719	106,630	95,485	86,450	72,693	62,713	55,143	49,203	44,419	40,482	37,187	30,898	26,429	23,090	20,499	18,431	16,743	14,150	12,252	10,803	8,737	7,334

250 MCM | **1 Run(s)** | **480 Volts** | **C-Value = 16,483** | **Magnetic Duct** | **Copper**

Calculations:

1. $f = \dfrac{1.732 \times \text{Length} \times \text{Isca}}{\text{Runs} \times \text{C-Value} \times \text{E L-L}}$
2. $M = \dfrac{1}{1 + f}$
3. $\text{Isc} = \text{Isca} \times M$
4. Add Motor Contribution = Motor FLA x 4

* All results are given in symmetrical amperes

Three-Phase Bolted Fault-Current* Table

One-Way Distance in Feet

Available Isc in Thousands (Isca)	5	10	15	20	25	30	40	50	60	70	80	90	100	125	150	175	200	225	250	300	350	400	500	600
3	2,995	2,990	2,985	2,980	2,976	2,971	2,961	2,952	2,942	2,933	2,923	2,914	2,905	2,882	2,859	2,837	2,815	2,794	2,772	2,731	2,691	2,652	2,577	2,506
5	4,986	4,973	4,959	4,946	4,933	4,919	4,893	4,867	4,841	4,816	4,790	4,765	4,741	4,680	4,621	4,563	4,507	4,452	4,398	4,295	4,196	4,102	3,926	3,764
7	6,973	6,947	6,920	6,894	6,868	6,843	6,792	6,742	6,692	6,644	6,596	6,548	6,502	6,388	6,278	6,172	6,070	5,971	5,875	5,692	5,520	5,358	5,061	4,795
10	9,946	9,892	9,838	9,786	9,734	9,682	9,581	9,481	9,384	9,288	9,195	9,103	9,013	8,796	8,590	8,392	8,204	8,024	7,852	7,528	7,230	6,955	6,463	6,036
15	14,878	14,758	14,639	14,523	14,409	14,296	14,076	13,862	13,655	13,454	13,259	13,069	12,885	12,446	12,036	11,652	11,292	10,954	10,635	10,050	9,526	9,054	8,238	7,556
20	19,783	19,572	19,364	19,161	18,962	18,767	18,390	18,027	17,678	17,342	17,019	16,708	16,408	15,703	15,056	14,460	13,910	13,400	12,926	12,072	11,324	10,663	9,549	8,645
25	24,663	24,334	24,014	23,703	23,399	23,103	22,534	21,991	21,474	20,981	20,510	20,060	19,629	18,628	17,725	16,905	16,157	15,473	14,845	13,729	12,770	11,936	10,557	9,463
30	29,515	29,046	28,592	28,151	27,724	27,310	26,517	25,769	25,062	24,393	23,759	23,157	22,584	21,270	20,100	19,052	18,108	17,253	16,475	15,113	13,958	12,968	11,356	10,100
35	34,342	33,709	33,098	32,509	31,941	31,392	30,349	29,374	28,459	27,599	26,790	26,026	25,306	23,667	22,227	20,953	19,817	18,797	17,878	16,284	14,952	13,821	12,005	10,611
40	39,143	38,322	37,535	36,779	36,054	35,356	34,039	32,816	31,678	30,617	29,624	28,694	27,820	25,852	24,144	22,648	21,326	20,150	19,097	17,290	15,795	14,539	12,543	11,029
45	43,918	42,888	41,904	40,965	40,066	39,207	37,593	36,108	34,735	33,463	32,280	31,179	30,150	27,852	25,879	24,168	22,669	21,345	20,167	18,162	16,520	15,150	12,995	11,377
50	48,668	47,406	46,207	45,067	43,982	42,949	41,020	39,258	37,640	36,151	34,775	33,500	32,315	29,689	27,459	25,540	23,871	22,408	21,113	18,926	17,150	15,678	13,382	11,672
55	53,393	51,877	50,445	49,090	47,805	46,586	44,326	42,275	40,405	38,694	37,122	35,672	34,332	31,384	28,902	26,783	24,954	23,359	21,956	19,601	17,702	16,138	13,716	11,925
60	58,092	56,302	54,619	53,034	51,538	50,124	47,517	45,168	43,040	41,104	39,334	37,711	36,216	32,950	30,225	27,916	25,935	24,216	22,711	20,201	18,190	16,543	14,007	12,145
65	62,767	60,683	58,732	56,903	55,185	53,567	50,600	47,945	45,554	43,390	41,423	39,626	37,979	34,404	31,444	28,952	26,827	24,992	23,393	20,738	18,624	16,901	14,263	12,337
70	67,417	65,018	62,784	60,699	58,747	56,917	53,579	50,611	47,955	45,563	43,399	41,431	39,633	35,755	32,569	29,904	27,642	25,698	24,010	21,221	19,013	17,221	14,490	12,506
75	72,043	69,310	66,777	64,423	62,229	60,179	56,460	53,174	50,249	47,630	45,270	43,132	41,188	37,016	33,611	30,780	28,389	26,343	24,572	21,659	19,364	17,508	14,693	12,657
80	76,644	73,559	70,712	68,078	65,632	63,357	59,248	55,640	52,446	49,598	47,044	44,741	42,652	38,194	34,580	31,591	29,077	25,934	25,085	22,057	19,681	17,768	14,875	12,792
85	81,222	77,765	74,590	71,665	68,960	66,452	61,947	58,013	54,549	51,476	48,730	46,263	44,033	39,298	35,482	32,342	29,712	27,478	25,557	22,421	19,970	18,003	15,039	12,913
90	85,775	81,929	78,413	75,187	72,215	69,470	64,560	60,299	56,566	53,268	50,333	47,705	45,338	40,334	36,325	33,040	30,301	27,981	25,991	22,754	20,234	18,217	15,189	13,023
95	90,305	86,052	82,182	78,645	75,399	72,411	67,094	62,503	58,501	54,981	51,860	49,074	46,572	41,308	37,113	33,691	30,847	28,446	26,392	23,061	20,477	18,413	15,325	13,123
100	94,811	90,134	85,897	82,040	78,515	75,280	69,550	64,630	60,360	56,619	53,315	50,375	47,743	42,226	37,852	34,300	31,357	28,879	26,764	23,344	20,700	18,593	15,449	13,215
105	99,294	94,176	89,560	85,376	81,565	78,079	71,932	66,682	62,146	58,188	54,704	51,613	48,853	43,093	38,547	34,869	31,832	29,281	27,109	23,607	20,906	18,760	15,564	13,298
110	103,754	98,179	93,173	88,652	84,550	80,811	74,244	68,664	63,864	59,691	56,031	52,793	49,909	43,912	39,201	35,404	32,277	29,657	27,431	23,851	21,097	18,913	15,669	13,375
115	108,191	102,143	96,735	91,871	87,473	83,477	76,488	70,579	65,518	61,134	57,300	53,918	50,913	44,687	39,818	35,906	32,694	30,009	27,732	24,078	21,274	19,056	15,767	13,446
120	112,605	106,068	100,249	95,035	90,336	86,081	78,668	72,432	67,111	62,518	58,514	54,992	51,870	45,423	40,401	36,379	33,086	30,339	28,013	24,289	21,439	19,188	15,858	13,512
125	116,996	109,956	103,715	98,144	93,141	88,624	80,787	74,224	68,647	63,849	59,678	56,019	52,783	46,121	40,953	36,826	33,455	30,649	28,277	24,488	21,594	19,312	15,942	13,573
130	121,365	113,806	107,133	101,200	95,889	91,108	82,846	75,958	70,128	65,129	60,795	57,002	53,654	46,785	41,475	37,248	33,803	30,941	28,525	24,674	21,738	19,427	16,020	13,630
135	125,712	117,620	110,506	104,204	98,582	93,536	84,849	77,638	71,557	66,360	61,866	57,943	54,487	47,417	41,971	37,647	34,131	31,216	28,759	24,848	21,874	19,535	16,094	13,683
140	130,037	121,397	113,834	107,158	101,222	95,909	86,797	79,267	72,938	67,546	62,896	58,845	55,284	48,019	42,443	38,026	34,442	31,476	28,980	25,013	22,001	19,637	16,163	13,733
145	134,339	125,139	117,118	110,063	103,810	98,230	88,693	80,845	74,273	68,688	63,885	59,710	56,047	48,594	42,891	38,386	34,737	31,722	29,188	25,168	22,121	19,732	16,227	13,780
150	138,620	128,846	120,359	112,920	106,348	100,499	90,539	82,376	75,563	69,790	64,838	60,541	56,779	49,143	43,318	38,727	35,017	31,955	29,385	25,314	22,234	19,822	16,288	13,824
155	142,880	132,517	123,557	115,731	108,837	102,719	92,337	83,861	76,811	70,854	65,754	61,340	57,480	49,668	43,725	39,053	35,282	32,176	29,572	25,453	22,341	19,907	16,345	13,865
160	147,118	136,155	126,713	118,496	111,279	104,891	94,089	85,304	78,019	71,881	66,638	62,108	58,154	50,171	44,114	39,363	35,535	32 386	29,749	25,584	22,442	19,987	16,399	13,904
165	151,334	139,759	129,829	121,216	113,675	107,017	95,796	86,705	79,189	72,873	67,490	62,847	58,802	50,652	44,486	39,658	35,776	32 586	29,918	25,709	22,538	20,063	16,450	13,940
170	155,530	143,330	132,905	123,893	116,026	109,098	97,460	88,066	80,323	73,832	68,311	63,559	59,425	51,113	44,841	39,941	36,005	32,776	30,078	25,827	22,629	20,135	16,499	13,975
175	159,704	146,868	135,941	126,528	118,334	111,136	99,083	89,389	81,422	74,760	69,105	64,245	60,024	51,556	45,182	40,211	36,225	32,958	30,231	25,940	22,715	20,203	16,545	14,008
180	163,858	150,373	138,939	129,121	120,599	113,132	100,666	90,675	82,488	75,657	69,871	64,907	60,602	51,982	45,508	40,469	36,434	33,131	30,377	26,047	22,797	20,268	16,588	14,039
185	167,991	153,847	141,899	131,674	122,823	115,087	102,211	91,927	83,523	76,527	70,612	65,546	61,158	52,390	45,821	40,716	36,635	33,297	30,516	26,149	22,875	20,330	16,630	14,069
190	172,104	157,289	144,823	134,187	125,007	117,002	103,719	93,145	84,527	77,369	71,328	66,163	61,695	52,784	46,122	40,953	36,826	33,455	30,649	26,247	22,950	20,389	16,669	14,097
195	176,196	160,700	147,709	136,662	127,152	118,879	105,192	94,331	85,502	78,185	72,022	66,759	62,213	53,163	46,411	41,181	37,010	33,607	30,777	26,340	23,021	20,445	16,707	14,124
200	180,268	164,081	150,561	139,099	129,259	120,719	106,630	95,485	86,450	78,977	72,693	67,335	62,713	53,527	46,689	41,400	37,187	33,752	30,898	26,429	23,090	20,499	16,743	14,150

250 MCM | **2 Run(s)** | **480 Volts** | **C-Value = 16,483** | **Magnetic Duct** | **Copper**

Calculations:

1. $f = \dfrac{1.732 \times \text{Length} \times \text{Isca}}{\text{Runs} \times \text{C-Value} \times E_{L\text{-}L}}$
2. $M = \dfrac{1}{1 + f}$
3. $\text{Isc} = \text{Isca} \times M$
4. Add Motor Contribution = Motor FLA x 4

* All results are given in symmetrical amperes

Three-Phase Bolted Fault-Current* Table

One-Way Distance in Feet

Available Isc in Thousands (Isca)	5	10	15	20	25	30	40	50	60	70	80	90	100	125	150	175	200	225	250	300	350	400	500	600
3	2,991	2,982	2,973	2,965	2,956	2,947	2,330	2,913	2,896	2,880	2,864	2,847	2,831	2,792	2,754	2,717	2,681	2,645	2,611	2,545	2,483	2,423	2,312	2,210
5	4,975	4,951	4,927	4,903	4,879	4,855	4,309	4,764	4,719	4,675	4,632	4,590	4,549	4,448	4,352	4,260	4,172	4,087	4,006	3,853	3,711	3,579	3,342	3,134
7	6,952	6,904	6,857	6,811	6,765	6,720	6,531	6,545	6,461	6,379	6,300	6,222	6,146	5,964	5,793	5,631	5,478	5,333	5,195	4,940	4,709	4,499	4,130	3,817
10	9,902	9,805	9,711	9,618	9,527	9,438	9,264	9,097	8,936	8,780	8,629	8,484	8,344	8,012	7,705	7,422	7,158	6,912	6,683	6,257	5,900	5,574	5,019	4,564
15	14,780	14,566	14,359	14,157	13,961	13,770	13,403	13,056	12,726	12,413	12,114	11,830	11,558	10,931	10,369	9,861	9,401	8,982	8,599	7,922	7,345	6,846	6,027	5,383
20	19,611	19,236	18,876	18,529	18,194	17,871	17,259	16,687	16,152	15,650	15,179	14,735	14,316	13,366	12,535	11,801	11,148	10,563	10,037	9,128	8,369	7,727	6,700	5,913
25	24,395	23,818	23,268	22,743	22,240	21,760	20,859	20,030	19,264	18,554	17,895	17,281	16,708	15,428	14,331	13,379	12,546	11,811	11,157	10,645	9,134	8,375	7,181	6,285
30	29,132	28,314	27,540	26,807	26,112	25,452	24,228	23,116	22,102	21,173	20,319	19,531	18,802	17,197	15,845	14,690	13,692	12,820	12,053	10,765	9,726	8,870	7,542	6,560
35	33,825	32,726	31,696	30,730	29,820	28,963	27,388	25,976	24,702	23,547	22,496	21,534	20,651	18,731	17,138	15,795	14,646	13,654	12,787	11,347	10,198	9,261	7,823	6,771
40	38,472	37,057	35,743	34,518	33,374	32,304	30,357	28,632	27,092	25,709	24,461	23,328	22,295	20,074	18,255	16,739	15,455	14,354	13,399	11,526	10,584	9,578	8,048	6,939
45	43,076	41,310	39,682	38,179	36,785	35,489	33,153	31,106	29,297	27,686	26,244	24,944	23,767	21,260	19,231	17,555	16,148	14,950	13,917	12,228	10,905	9,840	8,232	7,075
50	47,636	45,485	43,520	41,718	40,059	38,527	35,790	33,416	31,337	29,502	27,869	26,408	25,093	22,314	20,089	18,268	16,749	15,464	14,362	12,570	11,175	10,059	8,385	7,188
55	52,153	49,586	47,260	45,142	43,206	41,429	38,281	35,577	33,230	31,174	29,357	27,740	26,292	23,257	20,851	18,895	17,275	15,911	14,747	12,364	11,407	10,247	8,515	7,284
60	56,627	53,614	50,905	48,456	46,233	44,204	40,638	37,604	34,992	32,719	30,723	28,957	27,383	24,107	21,531	19,452	17,740	16,304	15,084	13,119	11,608	10,408	8,626	7,365
65	61,060	57,571	54,459	51,666	49,146	46,860	42,872	39,509	36,636	34,152	31,983	30,074	28,379	24,876	22,142	19,950	18,152	16,652	15,381	13,344	11,783	10,549	8,723	7,435
70	65,452	61,459	57,926	54,776	51,951	49,404	44,991	41,302	38,172	35,483	33,148	31,102	29,293	25,575	22,694	20,397	18,522	16,963	15,645	13,542	11,938	10,673	8,807	7,496
75	69,803	65,280	61,308	57,791	54,656	51,843	47,005	42,993	39,612	36,724	34,229	32,051	30,134	26,213	23,196	20,801	18,855	17,241	15,882	13,719	12,075	10,783	8,881	7,550
80	74,115	69,036	64,609	60,715	57,264	54,184	48,922	44,591	40,965	37,884	35,234	32,931	30,910	26,799	23,653	21,168	19,155	17,492	16,095	13,878	12,198	10,880	8,948	7,598
85	78,386	72,728	67,831	63,552	59,781	56,432	50,747	46,102	42,237	38,969	36,171	33,748	31,629	27,337	24,071	21,503	19,429	17,720	16,288	14,021	12,308	10,968	9,007	7,641
90	82,619	76,357	70,978	66,306	62,212	58,593	52,488	47,535	43,436	39,988	37,047	34,509	32,296	27,835	24,456	21,809	19,679	17,928	16,463	14,151	12,408	11,047	9,060	7,679
95	86,814	79,926	74,051	68,981	64,560	60,672	54,150	48,894	44,568	40,945	37,867	35,220	32,918	28,295	24,811	22,091	19,908	18,118	16,623	14,269	12,499	11,119	9,109	7,714
100	90,970	83,436	77,054	71,580	66,831	62,674	55,739	50,185	45,639	41,847	38,637	35,885	33,498	28,723	25,139	22,351	20,119	18,292	16,770	14,377	12,581	11,185	9,152	7,745
105	95,089	86,888	79,990	74,106	69,028	64,602	57,258	51,414	46,652	42,698	39,361	36,509	34,041	29,121	25,444	22,591	20,314	18,453	16,905	14,476	12,657	11,244	9,192	7,774
110	99,172	90,284	82,859	76,562	71,154	66,460	58,714	52,585	47,614	43,502	40,044	37,095	34,551	29,493	25,727	22,814	20,494	18,602	17,030	14,567	12,727	11,299	9,229	7,800
115	103,218	93,625	85,664	78,951	73,213	68,253	60,109	53,701	48,527	44,263	40,688	37,647	35,029	29,841	25,992	23,022	20,661	18,740	17,145	14,652	12,791	11,350	9,263	7,824
120	107,228	96,913	88,408	81,276	75,208	69,984	61,447	54,766	49,396	44,935	41,297	38,168	35,479	30,167	26,239	23,216	20,817	18,868	17,252	14,730	12,851	11,397	9,294	7,846
125	111,202	100,148	91,093	83,539	77,142	71,656	62,732	55,785	50,223	45,669	41,873	38,659	35,904	30,474	26,470	23,397	20,962	18,987	17,352	14,802	12,906	11,441	9,323	7,867
130	115,142	103,332	93,720	85,743	79,018	73,271	63,967	56,759	51,011	46,320	42,420	39,125	36,305	30,762	26,688	23,566	21,099	19,099	17,445	14,870	12,958	11,481	9,350	7,886
135	119,047	106,466	96,291	87,890	80,838	74,833	65,154	57,692	51,763	46,940	42,938	39,566	36,684	31,034	26,892	23,726	21,226	19,203	17,532	14,933	13,006	11,519	9,375	7,904
140	122,919	109,552	98,808	89,982	82,604	76,345	66,297	58,586	52,482	47,530	43,432	39,984	37,044	31,291	27,085	23,875	21,346	19,301	17,614	14,993	13,050	11,554	9,398	7,920
145	126,756	112,590	101,272	92,022	84,320	77,808	67,397	59,444	53,169	48,093	43,901	40,382	37,385	31,534	27,267	24,017	21,459	19,393	17,691	15,048	13,093	11,587	9,420	7,936
150	130,561	115,582	103,686	94,011	85,987	79,225	68,458	60,267	53,827	48,631	44,349	40,760	37,709	31,764	27,439	24,150	21,565	19,480	17,763	15,100	13,132	11,618	9,440	7,950
155	134,332	118,528	106,051	95,950	87,607	80,598	69,481	61,059	54,458	49,145	44,776	41,121	38,017	31,983	27,602	24,276	21,666	19,562	17,831	15,150	13,169	11,647	9,460	7,964
160	138,072	121,430	108,368	97,843	89,182	81,929	70,468	61,820	55,062	49,636	45,184	41,465	38,311	32,190	27,756	24,395	21,761	19,640	17,895	15,196	13,204	11,674	9,478	7,977
165	141,779	124,288	110,639	99,691	90,714	83,221	71,421	62,552	55,642	50,107	45,574	41,793	38,591	32,388	27,903	24,509	21,851	19,713	17,956	15,240	13,237	11,700	9,495	7,989
170	145,455	127,104	112,865	101,494	92,205	84,474	72,342	63,257	56,200	50,559	45,947	42,106	38,858	32,576	28,042	24,616	21,936	19,782	18,014	15,281	13,269	11,725	9,511	8,000
175	149,100	129,878	115,047	103,255	93,656	85,690	73,232	63,937	56,736	50,992	46,305	42,407	39,114	32,755	28,175	24,718	22,017	19,848	18,069	15,321	13,298	11,748	9,526	8,011
180	152,715	132,612	117,187	104,976	95,070	86,872	74,094	64,593	57,251	51,408	46,648	42,694	39,358	32,926	28,301	24,816	22,095	19,911	18,121	15,358	13,327	11,770	9,540	8,021
185	156,298	135,307	119,286	106,657	96,446	88,020	74,927	65,225	57,748	51,808	46,977	42,969	39,592	33,090	28,422	24,903	22,168	19,971	18,170	15,394	13,353	11,790	9,554	8,031
190	159,852	137,962	121,345	108,300	97,788	89,136	75,734	65,836	58,226	52,193	47,293	43,234	39,816	33,246	28,538	24,997	22,238	20,028	18,217	15,427	13,379	11,810	9,567	8,040
195	163,377	140,579	123,365	109,907	99,096	90,221	76,516	66,426	58,687	52,563	47,596	43,487	40,031	33,396	28,648	25,082	22,305	20,082	18,262	15,460	13,403	11,829	9,580	8,049
200	166,872	143,159	125,347	111,477	100,371	91,277	77,275	66,997	59,132	52,920	47,889	43,731	40,238	33,540	28,754	25,163	22,369	20,134	18,305	15,490	13,426	11,847	9,591	8,057

300 MCM	1 Run(s)	480 Volts	C-Value = 18,176	Magnetic Duct	Copper

Calculations:

1. $f = \dfrac{1.732 \times \text{Length} \times \text{Isca}}{\text{Runs} \times \text{C-Value} \times E_{L\text{-}L}}$

2. $M = \dfrac{1}{1 + f}$

3. Isc = Isca x M

4. Add Motor Contribution = Motor FLA x 4

* All results are given in symmetrical amperes

Three-Phase Bolted Fault-Current* Table

One-Way Distance in Feet

Available Isc in Thousands (Isca)	5	10	15	20	25	30	40	50	60	70	80	90	100	125	150	175	200	225	250	300	350	400	500	600
3	2,996	2,991	2,987	2,982	2,978	2,973	2,965	2,956	2,947	2,939	2,930	2,922	2,913	2,892	2,872	2,851	2,831	2,812	2,792	2,754	2,717	2,681	2,611	2,545
5	4,988	4,975	4,963	4,951	4,939	4,927	4,903	4,879	4,855	4,832	4,809	4,786	4,764	4,708	4,654	4,600	4,549	4,498	4,448	4,352	4,260	4,172	4,006	3,853
7	6,976	6,952	6,928	6,904	6,880	6,857	6,811	6,765	6,720	6,675	6,631	6,588	6,545	6,441	6,339	6,241	6,146	6,054	5,964	5,793	5,631	5,478	5,195	4,940
10	9,951	9,902	9,853	9,805	9,758	9,711	9,618	9,527	9,438	9,350	9,264	9,180	9,097	8,896	8,704	8,520	8,344	8,174	8,012	7,705	7,422	7,158	6,683	6,267
15	14,889	14,780	14,672	14,566	14,462	14,359	14,157	13,961	13,770	13,584	13,403	13,227	13,056	12,646	12,262	11,899	11,558	11,236	10,931	10,369	9,861	9,401	8,599	7,922
20	19,803	19,611	19,422	19,236	19,054	18,876	18,529	18,194	17,871	17,560	17,259	16,968	16,687	16,024	15,411	14,843	14,316	13,825	13,366	12,535	11,801	11,148	10,037	9,128
25	24,694	24,395	24,103	23,818	23,540	23,268	22,743	22,240	21,760	21,300	20,859	20,436	20,030	19,081	18,219	17,431	16,708	16,043	15,428	14,331	13,379	12,546	11,157	10,045
30	29,560	29,132	28,717	28,314	27,921	27,540	26,807	26,112	25,452	24,825	24,228	23,659	23,116	21,862	20,737	19,722	18,802	17,964	17,197	15,845	14,690	13,692	12,053	10,765
35	34,402	33,825	33,266	32,726	32,203	31,696	30,730	29,820	28,963	28,153	27,388	26,663	25,976	24,403	23,009	21,767	20,651	19,644	18,731	17,138	15,795	14,646	12,787	11,347
40	39,221	38,472	37,752	37,057	36,388	35,743	34,518	33,374	32,304	31,301	30,357	29,469	28,632	26,733	25,069	23,601	22,295	21,127	20,074	18,255	16,739	15,455	13,399	11,826
45	44,017	43,076	42,174	41,310	40,480	39,682	38,179	36,785	35,489	34,281	33,153	32,097	31,106	28,877	26,946	25,257	23,767	22,444	21,260	19,231	17,555	16,148	13,917	12,228
50	48,789	47,636	46,536	45,485	44,481	43,520	41,718	40,059	38,527	37,108	35,790	34,562	33,416	30,857	28,662	26,759	25,093	23,622	22,314	20,089	18,268	16,749	14,362	12,570
55	53,539	52,153	50,837	49,586	48,395	47,260	45,142	43,206	41,429	39,793	38,281	36,880	35,577	32,691	30,238	28,127	26,292	24,682	23,257	20,851	18,895	17,275	14,747	12,864
60	58,265	56,627	55,079	53,614	52,224	50,905	48,456	46,233	44,204	42,346	40,638	39,062	37,604	34,395	31,690	29,380	27,383	25,641	24,107	21,531	19,452	17,740	15,084	13,119
65	62,969	61,060	59,264	57,571	55,972	54,459	51,666	49,146	46,860	44,777	42,872	41,122	39,509	35,981	33,032	30,529	28,379	26,512	24,876	22,142	19,950	18,152	15,381	13,344
70	67,650	65,452	63,393	61,459	59,640	57,926	54,776	51,951	49,404	47,094	44,991	43,068	41,302	37,463	34,276	31,589	29,293	27,308	25,575	22,694	20,397	18,522	15,645	13,542
75	72,308	69,803	67,466	65,280	63,232	61,308	57,791	54,656	51,843	49,306	47,005	44,910	42,993	38,849	35,433	32,569	30,134	28,037	26,213	23,196	20,801	18,855	15,882	13,719
80	76,945	74,115	71,485	69,036	66,749	64,609	60,715	57,264	54,184	51,418	48,922	46,656	44,591	40,148	36,511	33,478	30,910	28,708	26,799	23,653	21,168	19,155	16,095	13,878
85	81,559	78,386	75,451	72,728	70,194	67,831	63,552	59,781	56,432	53,439	50,747	48,313	46,102	41,370	37,518	34,323	31 629	29,327	27,337	24,071	21,503	19,429	16,288	14,021
90	86,152	82,619	79,365	76,357	73,569	70,978	66,306	62,212	58,593	55,373	52,488	49,889	47,535	42,519	38,461	35,110	32 296	29,900	27,835	24,456	21,809	19,679	16,463	14,151
95	90,723	86,814	83,228	79,926	76,877	74,051	68,981	64,560	60,672	57,226	54,150	51,388	48,894	43,604	39,346	35,846	32,918	30,432	28,295	24,811	22,091	19,908	16,623	14,269
100	95,272	90,970	87,040	83,436	80,118	77,054	71,580	66,831	62,674	59,003	55,739	52,816	50,185	44,628	40,178	36,535	33,498	30,927	28,723	25,139	22,351	20,119	16,770	14,377
105	99,799	95,089	90,804	86,888	83,296	79,990	74,106	69,028	64,602	60,709	57,258	54,179	51,414	45,597	40,962	37,182	34,041	31,390	29,121	25,444	22,591	20,314	16,905	14,476
110	104,306	99,172	94,520	90,284	86,412	82,859	76,562	71,154	66,460	62,347	58,714	55,480	52,585	46,515	41,701	37,791	34,551	31,822	29,493	25,727	22,814	20,494	17,030	14,567
115	108,791	103,218	98,188	93,625	89,468	85,664	78,951	73,213	68,253	63,923	60,109	56,724	53,701	47,386	42,400	38,364	35,329	32,228	29,841	25,992	23,022	20,661	17,145	14,652
120	113,255	107,228	101,810	96,913	92,465	88,408	81,276	75,208	69,984	65,438	61,447	57,915	54,766	48,214	43,062	38,904	35,479	32,608	30,167	26,239	23,216	20,817	17,252	14,730
125	117,698	111,202	105,386	100,148	95,406	91,093	83,539	77,142	71,656	66,897	62,732	59,055	55,785	49,001	43,689	39,416	35,904	32,967	30,474	26,470	23,397	20,962	17,352	14,802
130	122,121	115,142	108,918	103,332	98,291	93,720	85,743	79,018	73,271	68,303	63,967	60,148	56,759	49,751	44,284	39,900	36,305	33,304	30,762	26,688	23,566	21,099	17,445	14,870
135	126,523	119,047	112,406	106,466	101,123	96,291	87,890	80,838	74,833	69,659	65,154	61,196	57,692	50,467	44,850	40,358	36,684	33,623	31,034	26,892	23,726	21,226	17,532	14,933
140	130,904	122,919	115,851	109,552	103,903	98,808	89,982	82,604	76,345	70,967	66,297	62,203	58,586	51,150	45,389	40,794	37,044	33,925	31,291	27,085	23,875	21,346	17,614	14,993
145	135,266	126,756	119,254	112,590	106,632	101,272	92,022	84,320	77,808	72,229	67,397	63,171	59,444	51,802	45,902	41,208	37,385	34,211	31,534	27,267	24,017	21,459	17,691	15,048
150	139,607	130,561	122,615	115,582	109,311	103,686	94,011	85,987	79,225	73,449	68,458	64,102	60,267	52,427	46,391	41,602	37,709	34,482	31,764	27,439	24,150	21,565	17,763	15,100
155	143,928	134,332	125,936	118,528	111,943	106,051	95,950	87,607	80,598	74,628	69,481	64,998	61,059	53,024	46,859	41,978	38,017	34,740	31,983	27,602	24,276	21,666	17,831	15,150
160	148,229	138,072	129,217	121,430	114,528	108,368	97,843	89,182	81,929	75,768	70,468	65,861	61,820	53,597	47,306	42,336	38,311	34,985	32,190	27,756	24,395	21,761	17,895	15,196
165	152,511	141,779	132,459	124,288	117,067	110,639	99,691	90,714	83,221	76,871	71,421	66,693	62,552	54,147	47,733	42,678	38,591	35,218	32,388	27,903	24,509	21,851	17,956	15,240
170	156,773	145,455	135,662	127,104	119,562	112,865	101,494	92,205	84,474	77,939	72,342	67,495	63,257	54,675	48,143	43,005	38,858	35,441	32,576	28,042	24,616	21,936	18,014	15,281
175	161,015	149,100	138,827	129,878	122,014	115,047	103,255	93,656	85,690	78,973	73,232	68,270	63,937	55,182	48,536	43,318	39,114	35,653	32,755	28,175	24,718	22,017	18,069	15,321
180	165,238	152,715	141,955	132,612	124,423	117,187	104,976	95,070	86,872	79,976	74,094	69,018	64,593	55,669	48,912	43,618	39,358	35,856	32,926	28,301	24,816	22,095	18,121	15,358
185	169,442	156,298	145,047	135,307	126,792	119,286	106,657	96,446	88,020	80,948	74,927	69,740	65,225	56,139	49,274	43,906	39,532	36,050	33,090	28,422	24,909	22,168	18,170	15,394
190	173,627	159,852	148,103	137,962	129,121	121,345	108,300	97,788	89,136	81,891	75,734	70,439	65,836	56,591	49,622	44,182	39,816	36,236	33,246	28,538	24,997	22,238	18,217	15,427
195	177,793	163,377	151,123	140,579	131,411	123,365	109,907	99,096	90,221	82,806	76,516	71,115	66,426	57,026	49,957	44,447	40,031	36,414	33,396	28,648	25,082	22,305	18,262	15,460
200	181,940	166,872	154,109	143,159	133,663	125,347	111,477	100,371	91,277	83,694	77,275	71,770	66,997	57,446	50,279	44,701	40,238	36,585	33,540	28,754	25,163	22,369	18,305	15,490

300 MCM	2 Run(s)	480 Volts	C-Value = 18,176	Magnetic Duct	Copper

Calculations:

1. $f = \dfrac{1.732 \times \text{Length} \times \text{Isca}}{\text{Runs} \times \text{C-Value} \times E_{L-L}}$

2. $M = \dfrac{1}{1 + f}$

3. Isc = Isca x M

4. Add Motor Contribution = Motor FLA x 4

* All results are given in symmetrical amperes

Three-Phase Bolted Fault-Current* Table

One-Way Distance in Feet

Available Isc in Thousands (Isca)	5	10	15	20	25	30	40	50	60	70	80	90	100	125	150	175	200	225	250	300	350	400	500	600
3	2,992	2,984	2,975	2,967	2,959	2,951	2,935	2,920	2,904	2,889	2,874	2,859	2,844	2,807	2,772	2,737	2,703	2,670	2,638	2,575	2,516	2,459	2,353	2,256
5	4,977	4,955	4,932	4,910	4,888	4,866	4,823	4,781	4,740	4,699	4,659	4,619	4,581	4,486	4,396	4,309	4,226	4,146	4,069	3,922	3,786	3,660	3,430	3,227
7	6,955	6,911	6,868	6,825	6,783	6,741	6,659	6,578	6,500	6,424	6,349	6,276	6,205	6,033	5,871	5,717	5,572	5,433	5,301	5,056	4,832	4,627	4,266	3,957
10	9,909	9,820	9,733	9,647	9,562	9,479	9,317	9,161	9,010	8,864	8,722	8,585	8,452	8,137	7,845	7,573	7,319	7,082	6,859	6,454	6,094	5,772	5,220	4,765
15	14,797	14,599	14,406	14,219	14,036	13,858	13,515	13,189	12,877	12,581	12,297	12,027	11,767	11,166	10,623	10,130	9,681	9,270	8,893	8,213	7,647	7,147	6,320	5,664
20	19,640	19,293	18,958	18,635	18,322	18,020	17,444	16,904	16,397	15,919	15,468	15,042	14,638	13,719	12,908	12,188	11,544	10,964	10,440	9,519	8,764	8,113	7,064	6,255
25	24,441	23,906	23,393	22,903	22,432	21,981	21,130	20,343	19,612	18,932	18,298	17,705	17,149	15,900	14,821	13,879	13,050	12,314	11,657	10,533	9,606	8,830	7,601	6,672
30	29,198	28,438	27,716	27,030	26,377	25,755	24,595	23,535	22,562	21,667	20,840	20,074	19,362	17,786	16,446	15,295	14,294	13,416	12,639	11,028	10,264	9,382	8,006	6,983
35	33,913	32,892	31,930	31,023	30,166	29,355	27,858	26,505	25,278	24,160	23,136	22,196	21,329	19,431	17,844	16,496	15,338	14,331	13,449	11,574	10,791	9,821	8,324	7,223
40	38,587	37,270	36,040	34,889	33,808	32,793	30,935	29,277	27,787	26,441	25,220	24,107	23,087	20,880	19,058	17,529	16,227	15,104	14,127	12,509	11,224	10,178	8,579	7,414
45	43,219	41,574	40,049	38,632	37,313	36,080	33,844	31,868	30,111	28,537	27,120	25,837	24,670	22,166	20,124	18,426	16,993	15,766	14,705	12,960	11,585	10,474	8,788	7,570
50	47,811	45,806	43,962	42,261	40,686	39,225	36,596	34,297	32,270	30,470	28,859	27,411	26,100	23,314	21,066	19,213	17,659	16,338	15,201	13,244	11,891	10,723	8,963	7,699
55	52,363	49,967	47,781	45,778	43,936	42,237	39,205	36,578	34,282	32,257	30,457	28,848	27,401	24,346	21,905	19,908	18,245	16,839	15,633	13,676	12,154	10,937	9,112	7,809
60	56,875	54,060	51,510	49,190	47,070	45,125	41,680	38,724	36,160	33,914	31,931	30,167	28,588	25,279	22,657	20,527	18,764	17,279	16,013	13,965	12,382	11,121	9,239	7,902
65	61,349	58,086	55,152	52,501	50,093	47,896	44,033	40,747	37,918	35,456	33,294	31,381	29,675	26,126	23,334	21,082	19,226	17,671	16,348	14,220	12,581	11,282	9,350	7,983
70	65,783	62,046	58,710	55,715	53,011	50,557	46,272	42,658	39,567	36,893	34,558	32,501	30,675	26,898	23,949	21,582	19,641	18,021	16,647	14,445	12,758	11,423	9,447	8,054
75	70,180	65,943	62,188	58,837	55,829	53,114	48,406	44,464	41,116	38,237	35,734	33,539	31,599	27,605	24,508	22,035	20,016	18,336	16,915	14,647	12,915	11,549	9,533	8,116
80	74,540	69,777	65,586	61,871	58,553	55,574	50,440	46,175	42,575	39,495	36,831	34,504	32,453	28,255	25,018	22,447	20,355	18,620	17,157	14,828	13,055	11,661	9,609	8,171
85	78,862	73,551	68,910	64,820	61,188	57,941	52,383	47,798	43,950	40,676	37,856	35,402	33,247	28,854	25,487	22,824	20,665	18,878	17,377	14,991	13,182	11,762	9,677	8,221
90	83,148	77,265	72,160	67,687	63,737	60,222	54,240	49,339	45,250	41,787	38,817	36,241	33,985	29,409	25,919	23,170	20,948	19,114	17,576	15,140	13,296	11,853	9,739	8,265
95	87,397	80,921	75,339	70,477	66,204	62,420	56,017	50,805	46,480	42,834	39,718	37,025	34,674	29,924	26,318	23,488	21,207	19,330	17,759	15,275	13,401	11,936	9,795	8,305
100	91,611	84,521	78,450	73,192	68,595	64,541	57,719	52,201	47,646	43,822	40,566	37,761	35,319	30,403	26,688	23,782	21,447	19,529	17,926	15,399	13,496	12,011	9,846	8,342
105	95,790	88,066	81,494	75,835	70,911	66,587	59,350	53,531	48,752	44,756	41,366	38,453	35,923	30,849	27,031	24,054	21,668	19,712	18,081	15,512	13,583	12,080	9,892	8,375
110	99,934	91,556	84,474	78,409	73,157	68,564	60,915	54,801	49,803	45,640	42,120	39,103	36,490	31,267	27,351	24,307	21,873	19,882	18,223	15,617	13,663	12,144	9,935	8,405
115	104,044	94,994	87,392	80,917	75,335	70,473	62,418	56,015	50,803	46,479	42,833	39,717	37,024	31,658	27,650	24,543	22,064	20,040	18,355	15,714	13,738	12,203	9,974	8,433
120	108,120	98,380	90,250	83,361	77,449	72,320	63,862	57,175	51,756	47,275	43,508	40,297	37,528	32,025	27,930	24,763	22,242	20,186	18,478	15,804	13,806	12,257	10,010	8,459
125	112,162	101,715	93,049	85,743	79,501	74,107	65,251	58,286	52,664	48,032	44,148	40,846	38,003	32,371	28,192	24,970	22,408	20,323	18,593	15,888	13,870	12,307	10,043	8,483
130	116,171	105,002	95,791	88,067	81,495	75,836	66,588	59,350	53,532	48,752	44,756	41,366	38,453	32,696	28,439	25,163	22,563	20,451	18,700	15,966	13,930	12,354	10,075	8,505
135	120,148	108,240	98,479	90,333	83,432	77,510	67,875	60,371	54,361	49,439	45,334	41,859	38,879	33,004	28,671	25,345	22,709	20,571	18,800	16,039	13,985	12,397	10,104	8,526
140	124,092	111,430	101,113	92,545	85,315	79,133	69,117	61,351	55,154	50,094	45,885	42,328	39,283	33,295	28,891	25,516	22,847	20,683	18,894	16,107	14,037	12,438	10,131	8,545
145	128,004	114,575	103,696	94,703	87,146	80,706	70,314	62,292	55,914	50,720	46,409	42,774	39,667	33,570	29,098	25,677	22,976	20,789	18,982	16,171	14,086	12,476	10,156	8,563
150	131,885	117,674	106,228	96,811	88,928	82,232	71,469	63,197	56,642	51,318	46,910	43,199	40,032	33,831	29,294	25,829	23,098	20,889	19,066	16,232	14,131	12,512	10,180	8,580
155	135,735	120,730	108,712	98,870	90,662	83,712	72,584	64,068	57,340	51,891	47,388	43,604	40,379	34,079	29,479	25,974	23,213	20,983	19,144	16,289	14,174	12,546	10,202	8,596
160	139,554	123,742	111,148	100,880	92,350	85,149	73,662	64,906	58,011	52,440	47,845	43,990	40,711	34,315	29,656	26,110	23,322	21,072	19,218	16,342	14,215	12,578	10,223	8,611
165	143,343	126,711	113,538	102,845	93,994	86,545	74,705	65,714	58,655	52,966	48,282	44,360	41,027	34,539	29,823	26,240	23,426	21,157	19,288	16,393	14,253	12,608	10,243	8,625
170	147,101	129,639	115,883	104,766	95,595	87,901	75,713	66,493	59,275	53,471	48,701	44,713	41,329	34,753	29,983	26,364	23,524	21,237	19,355	16,441	14,290	12,636	10,262	8,638
175	150,830	132,527	118,185	106,644	97,156	89,219	76,689	67,245	59,871	53,955	49,103	45,052	41,618	34,957	30,134	26,481	23,617	21,313	19,418	16,487	14,324	12,663	10,279	8,651
180	154,530	135,374	120,444	108,480	98,678	90,501	77,634	67,970	60,446	54,421	49,489	45,377	41,895	35,152	30,279	26,593	23,706	21,385	19,478	16,530	14,357	12,689	10,296	8,663
185	158,201	138,183	122,663	110,276	100,162	91,747	78,549	68,671	60,999	54,870	49,860	45,688	42,160	35,339	30,417	26,699	23,791	21,454	19,535	16,571	14,388	12,713	10,312	8,674
190	161,843	140,954	124,841	112,034	101,610	92,961	79,437	69,348	61,533	55,301	50,216	45,987	42,415	35,517	30,550	26,801	23,872	21,520	19,590	16,610	14,417	12,736	10,327	8,685
195	165,456	143,687	126,980	113,754	103,023	94,142	80,298	70,003	62,049	55,717	50,558	46,274	42,659	35,688	30,676	26,898	23,949	21,582	19,642	16,648	14,445	12,758	10,342	8,695
200	169,042	146,384	129,081	115,437	104,402	95,292	81,133	70,637	62,546	56,118	50,888	46,550	42,893	35,853	30,797	26,991	24,023	21,642	19,691	16,683	14,472	12,779	10,355	8,705

350 MCM	1 Run(s)	480 Volts

C-Value = 19,703	Magnetic Duct	Copper

Calculations:

1. $f = \dfrac{1.732 \times \text{Length} \times \text{Isca}}{\text{Runs} \times \text{C-Value} \times E_{L\text{-}L}}$

2. $M = \dfrac{1}{1 + f}$

3. $\text{Isc} = \text{Isca} \times M$

4. Add Motor Contribution = Motor FLA x 4

* All results are given in symmetrical amperes

Three-Phase Bolted Fault-Current* Table

One-Way Distance in Feet

Available Isc in Thousands (Isca)	5	10	15	20	25	30	40	50	60	70	80	90	100	125	150	175	200	225	250	300	350	400	500	600
3	2,996	2,992	2,988	2,984	2,980	2,975	2,967	2,959	2,951	2,943	2,935	2,928	2,920	2,900	2,881	2,862	2,844	2,825	2,807	2,772	2,737	2,703	2,638	2,575
5	4,989	4,977	4,966	4,955	4,943	4,932	4,910	4,888	4,866	4,845	4,823	4,802	4,781	4,729	4,679	4,629	4,581	4,533	4,486	4,396	4,309	4,226	4,069	3,922
7	6,978	6,955	6,933	6,911	6,890	6,868	6,825	6,783	6,741	6,699	6,659	6,618	6,578	6,481	6,386	6,294	6,205	6,118	6,033	5,871	5,717	5,572	5,301	5,056
10	9,954	9,909	9,865	9,820	9,776	9,733	9,647	9,562	9,479	9,398	9,317	9,239	9,161	8,973	8,792	8,619	8,452	8,292	8,137	7,845	7,573	7,319	6,859	6,454
15	14,898	14,797	14,697	14,599	14,502	14,406	14,219	14,036	13,858	13,684	13,515	13,350	13,189	12,802	12,438	12,093	11,767	11,459	11,166	10,623	10,130	9,681	8,893	8,223
20	19,819	19,640	19,465	19,293	19,124	18,958	18,635	18,322	18,020	17,727	17,444	17,170	16,904	16,274	15,690	15,146	14,638	14,164	13,719	12,908	12,188	11,544	10,440	9,529
25	24,717	24,441	24,170	23,906	23,647	23,393	22,903	22,432	21,981	21,547	21,130	20,729	20,343	19,438	18,610	17,849	17,149	16,501	15,900	14,821	13,879	13,050	11,657	10,533
30	29,594	29,198	28,813	28,438	28,072	27,716	27,030	26,377	25,755	25,162	24,595	24,053	23,535	22,332	21,246	20,260	19,362	18,540	17,786	16,446	15,295	14,294	12,639	11,328
35	34,448	33,913	33,395	32,892	32,404	31,930	31,023	30,166	29,355	28,587	27,858	27,165	26,505	24,989	23,637	22,424	21,329	20,336	19,431	17,844	16,496	15,338	13,449	11,974
40	39,281	38,587	37,917	37,270	36,645	36,040	34,889	33,808	32,793	31,837	30,935	30,083	29,277	27,438	25,816	24,376	23,087	21,928	20,880	19,058	17,529	16,227	14,127	12,509
45	44,092	43,219	42,381	41,574	40,797	40,049	38,632	37,313	36,080	34,926	33,844	32,826	31,868	29,702	27,811	26,146	24,670	23,351	22,166	20,124	18,426	16,993	14,705	12,960
50	48,881	47,811	46,787	45,806	44,865	43,962	42,261	40,686	39,225	37,865	36,596	35,409	34,297	31,801	29,643	27,759	26,100	24,629	23,314	21,066	19,213	17,659	15,201	13,344
55	53,649	52,363	51,137	49,967	48,850	47,781	45,778	43,936	42,237	40,664	39,205	37,846	36,578	33,752	31,331	29,234	27,401	25,783	24,346	21,905	19,908	18,245	15,633	13,676
60	58,396	56,875	55,432	54,060	52,754	51,510	49,190	47,070	45,125	43,334	41,680	40,148	38,724	35,571	32,893	30,589	28,588	26,832	25,279	22,657	20,527	18,764	16,013	13,965
65	63,122	61,349	59,673	58,086	56,581	55,152	52,501	50,093	47,896	45,883	44,033	42,327	40,747	37,271	34,341	31,838	29,675	27,787	26,126	23,334	21,082	19,226	16,348	14,220
70	67,826	65,783	63,860	62,046	60,332	58,710	55,715	53,011	50,557	48,320	46,272	44,391	42,658	38,863	35,688	32,992	30,675	28,663	26,898	23,949	21,582	19,641	16,647	14,445
75	72,510	70,180	67,996	65,943	64,010	62,188	58,837	55,829	53,114	50,651	48,406	46,351	44,464	40,356	36,943	34,063	31,599	29,467	27,605	24,508	22,035	20,016	16,915	14,647
80	77,173	74,540	72,080	69,777	67,617	65,586	61,871	58,553	55,574	52,883	50,440	48,213	46,175	41,761	38,117	35,058	32,453	30,209	28,255	25,018	22,447	20,355	17,157	14,828
85	81,816	78,862	76,114	73,551	71,155	68,910	64,820	61,188	57,941	55,022	52,383	49,985	47,798	43,084	39,216	35,985	33,247	30,895	28,854	25,487	22,824	20,665	17,377	14,991
90	86,438	83,148	80,098	77,265	74,625	72,160	67,687	63,737	60,222	57,075	54,240	51,674	49,339	44,332	40,247	36,852	33,985	31,532	29,409	25,919	23,170	20,948	17,576	15,140
95	91,040	87,397	84,035	80,921	78,030	75,339	70,477	66,204	62,420	59,046	56,017	53,284	50,805	45,512	41,218	37,664	34,674	32,124	29,924	26,318	23,488	21,207	17,759	15,275
100	95,622	91,611	87,924	84,521	81,372	78,450	73,192	68,595	64,541	60,939	57,719	54,821	52,201	46,629	42,131	38,425	35,319	32,677	30,403	26,688	23,782	21,447	17,926	15,399
105	100,184	95,790	91,766	88,066	84,652	81,494	75,835	70,911	66,587	62,761	59,350	56,291	53,531	47,688	42,994	39,142	35,923	33,193	30,849	27,031	24,054	21,668	18,081	15,512
110	104,726	99,934	95,562	91,556	87,873	84,474	78,409	73,157	68,564	64,513	60,915	57,697	54,801	48,693	43,809	39,816	36,490	33,677	31,267	27,351	24,307	21,873	18,223	15,617
115	109,248	104,044	99,313	94,994	91,034	87,392	80,917	75,335	70,473	66,201	62,418	59,043	56,015	49,648	44,581	40,453	37,024	34,131	31,658	27,650	24,543	22,064	18,355	15,714
120	113,750	108,120	103,020	98,380	94,139	90,250	83,361	77,449	72,320	67,828	63,862	60,334	57,175	50,558	45,313	41,055	37,528	34,559	32,025	27,930	24,763	22,242	18,478	15,804
125	118,233	112,162	106,683	101,715	97,189	93,049	85,743	79,501	74,107	69,397	65,251	61,572	58,286	51,424	46,008	41,624	38,003	34,962	32,371	28,192	24,970	22,408	18,593	15,888
130	122,697	116,171	110,304	105,002	100,185	95,791	88,067	81,495	75,836	70,912	66,588	62,761	59,350	52,251	46,669	42,164	38,453	35,342	32,696	28,439	25,163	22,563	18,700	15,966
135	127,142	120,148	113,883	108,240	103,129	98,479	90,333	83,432	77,510	72,374	67,875	63,904	60,371	53,041	47,298	42,677	38,879	35,701	33,004	28,671	25,345	22,709	18,800	16,039
140	131,567	124,092	117,421	111,430	106,021	101,113	92,545	85,315	79,133	73,786	69,117	65,003	61,351	53,796	47,897	43,164	39,283	36,042	33,295	28,891	25,516	22,847	18,894	16,107
145	135,973	128,004	120,918	114,575	108,864	103,696	94,703	87,146	80,706	75,152	70,314	66,060	62,292	54,518	48,469	43,628	39,667	36,364	33,570	29,098	25,677	22,976	18,982	16,171
150	140,361	131,885	124,375	117,674	111,659	106,228	96,811	88,928	82,232	76,473	71,469	67,079	63,197	55,210	49,015	44,070	40,032	36,671	33,831	29,294	25,829	23,098	19,066	16,232
155	144,729	135,735	127,793	120,730	114,406	108,712	98,870	90,662	83,712	77,752	72,584	68,061	64,068	55,873	49,537	44,492	40,379	36,963	34,079	29,479	25,974	23,213	19,144	16,289
160	149,079	139,554	131,173	123,742	117,107	111,148	100,880	92,350	85,149	78,990	73,662	69,008	64,906	56,510	50,037	44,895	40,711	37,240	34,315	29,656	26,110	23,322	19,218	16,342
165	153,411	143,343	134,515	126,711	119,763	113,538	102,845	93,994	86,545	80,190	74,705	69,922	65,714	57,121	50,516	45,280	41,027	37,505	34,539	29,823	26,240	23,426	19,288	16,393
170	157,724	147,101	137,819	129,639	122,376	115,883	104,766	95,595	87,901	81,353	75,713	70,804	66,493	57,709	50,975	45,648	41,329	37,757	34,753	29,983	26,364	23,524	19,355	16,441
175	162,019	150,830	141,087	132,527	124,946	118,185	106,644	97,156	89,219	82,481	76,689	71,657	67,245	58,274	51,415	46,001	41,618	37,998	34,957	30,134	26,481	23,617	19,418	16,487
180	166,295	154,530	144,319	135,374	127,474	120,444	108,480	98,678	90,501	83,575	77,634	72,481	67,970	58,818	51,838	46,339	41,895	38,229	35,152	30,279	26,593	23,706	19,478	16,530
185	170,554	158,201	147,516	138,183	129,961	122,663	110,276	100,162	91,747	84,637	78,549	73,279	68,671	59,342	52,245	46,664	42,160	38,449	35,339	30,417	26,699	23,791	19,535	16,571
190	174,795	161,843	150,678	140,954	132,409	124,841	112,034	101,610	92,961	85,668	79,437	74,051	69,348	59,847	52,636	46,976	42,415	38,661	35,517	30,550	26,801	23,872	19,590	16,610
195	179,018	165,456	153,805	143,687	134,818	126,980	113,754	103,023	94,142	86,670	80,298	74,798	70,003	60,335	53,013	47,275	42,659	38,864	35,688	30,676	26,898	23,949	19,642	16,648
200	183,223	169,042	156,899	146,384	137,189	129,081	115,437	104,402	95,292	87,644	81,133	75,522	70,637	60,805	53,375	47,564	42,853	39,058	35,853	30,797	26,991	24,023	19,691	16,683

350 MCM | 2 Run(s) | 480 Volts

C-Value = 19,703 | Magnetic Duct | Copper

Calculations:

1. $f = \dfrac{1.732 \times \text{Length} \times \text{Isca}}{\text{Runs} \times \text{C-Value} \times E_{L\text{-}L}}$

2. $M = \dfrac{1}{1 + f}$

3. Isc = Isca x M

4. Add Motor Contribution = Motor FLA x 4

* All results are given in symmetrical amperes

Three-Phase Bolted Fault-Current* Table

One-Way Distance in Feet

Available Isc in Thousands (Isca)	5	10	15	20	25	30	40	50	60	70	80	90	100	125	150	175	200	225	250	300	350	400	500	600
3	2,992	2,984	2,976	2,969	2,961	2,953	2,938	2,923	2,908	2,893	2,879	2,864	2,850	2,815	2,780	2,747	2,714	2,682	2,651	2,591	2,533	2,478	2,375	2,280
5	4,978	4,957	4,935	4,914	4,893	4,872	4,830	4,790	4,750	4,711	4,672	4,634	4,597	4,506	4,419	4,335	4,254	4,176	4,101	3,958	3,825	3,701	3,475	3,276
7	6,957	6,915	6,873	6,832	6,791	6,751	6,672	6,595	6,520	6,446	6,374	6,303	6,234	6,068	5,911	5,762	5,620	5,484	5,356	5,[illegible]15	4,896	4,694	4,337	4,030
10	9,913	9,828	9,744	9,661	9,580	9,500	9,344	9,193	9,048	8,906	8,769	8,636	8,507	8,201	7,916	7,651	7,402	7,170	6,951	6,551	6,195	5,876	5,327	4,871
15	14,805	14,615	14,430	14,250	14,074	13,902	13,571	13,256	12,954	12,666	12,391	12,127	11,875	11,287	10,754	10,270	9,827	9,421	9,047	8,382	7,808	7,307	6,477	5,816
20	19,655	19,322	19,000	18,688	18,387	18,095	17,538	17,015	16,521	16,056	15,616	15,200	14,805	13,902	13,103	12,391	11,752	11,176	10,654	9,743	8,976	8,321	7,261	6,440
25	24,463	23,949	23,457	22,984	22,529	22,093	21,268	20,503	19,791	19,127	18,506	17,924	17,377	16,147	15,079	14,143	13,317	12,582	11,924	10,795	9,861	9,076	7,829	6,883
30	29,231	28,500	27,805	27,143	26,511	25,909	24,782	23,749	22,799	21,922	21,110	20,356	19,654	18,094	16,764	15,616	14,614	13,734	12,954	11,632	10,555	9,660	8,260	7,215
35	33,957	32,975	32,048	31,171	30,342	29,555	28,098	26,778	25,576	24,478	23,470	22,541	21,684	19,800	18,218	16,870	15,708	14,695	13,805	12,314	11,113	10,126	8,598	7,471
40	38,644	37,377	36,190	35,076	34,029	33,043	31,232	29,609	28,147	26,822	25,617	24,515	23,504	21,307	19,486	17,952	16,641	15,509	14,521	12,880	11,573	10,506	8,871	7,676
45	43,291	41,707	40,235	38,863	37,582	36,382	34,199	32,263	30,535	28,982	27,579	26,306	25,146	22,648	20,601	18,894	17,448	16,207	15,132	13,358	11,957	10,822	9,095	7,843
50	47,899	45,967	44,185	42,537	41,006	39,582	37,012	34,755	32,757	30,977	29,380	27,940	26,634	23,848	21,589	19,722	18,151	16,813	15,658	13,767	12,283	11,088	9,282	7,982
55	52,468	50,159	48,045	46,102	44,310	42,652	39,682	37,099	34,832	32,826	31,038	29,435	27,989	24,929	22,472	20,455	18,771	17,343	16,117	14,120	12,564	11,317	9,442	8,100
60	57,000	54,285	51,817	49,564	47,499	45,599	42,221	39,309	36,772	34,544	32,570	30,809	29,229	25,907	23,264	21,109	19,320	17,811	16,520	14,429	12,808	11,514	9,579	8,201
65	61,493	58,346	55,505	52,927	50,579	48,430	44,637	41,395	38,592	36,144	33,989	32,076	30,367	26,797	23,979	21,697	19,811	18,227	16,878	14,701	13,022	11,687	9,698	8,288
70	65,950	62,343	59,110	56,196	53,556	51,152	46,939	43,368	40,301	37,639	35,308	33,248	31,415	27,610	24,628	22,227	20,252	18,600	17,197	14,942	13,211	11,839	9,802	8,364
75	70,370	66,278	62,636	59,373	56,434	53,772	49,136	45,236	41,910	39,039	36,536	34,335	32,384	28,356	25,219	22,707	20,650	18,935	17,483	15,158	13,379	11,974	9,895	8,431
80	74,754	70,153	66,086	62,464	59,219	56,294	51,234	47,008	43,426	40,352	37,684	35,346	33,282	29,042	25,761	23,145	21,012	19,239	17,742	15,352	13,530	12,094	9,977	8,491
85	79,101	73,968	69,461	65,471	61,915	58,725	53,239	48,691	44,859	41,585	38,757	36,290	34,117	29,676	26,258	23,546	21,342	19,515	17,976	15,527	13,666	12,203	10,051	8,544
90	83,414	77,726	72,764	68,398	64,526	61,069	55,159	50,291	46,213	42,747	39,765	37,171	34,895	30,263	26,716	23,914	21,644	19,767	18,190	15,686	13,789	12,301	10,117	8,592
95	87,691	81,427	75,998	71,248	67,056	63,331	56,997	51,815	47,497	43,843	40,711	37,997	35,622	30,808	27,141	24,253	21,921	19,998	18,385	15,832	13,901	12,390	10,177	8,635
100	91,935	85,073	79,165	74,024	69,510	65,514	58,760	53,268	48,715	44,879	41,603	38,773	36,303	31,316	27,534	24,567	22,177	20,211	18,565	15,965	14,003	12,471	10,232	8,675
105	96,144	88,665	82,266	76,728	71,889	67,624	60,451	54,654	49,872	45,859	42,444	39,502	36,942	31,790	27,900	24,858	22,414	20,407	18,731	16,087	14,097	12,546	10,282	8,711
110	100,319	92,204	85,304	79,364	74,198	69,664	62,076	55,979	50,972	46,788	43,238	40,189	37,542	32,234	28,241	25,128	22,633	20,589	18,884	16,200	14,184	12,614	10,328	8,744
115	104,461	95,691	88,280	81,935	76,440	71,636	63,637	57,245	52,020	47,669	43,990	40,838	38,107	32,650	28,559	25,380	22,837	20,758	19,026	16,304	14,264	12,678	10,371	8,774
120	108,570	99,128	91,197	84,441	78,617	73,545	65,139	58,458	53,020	48,507	44,702	41,451	38,641	33,041	28,858	25,616	23,028	20,915	19,158	16,401	14,338	12,736	10,410	8,802
125	112,647	102,516	94,057	86,887	80,733	75,393	66,585	59,620	53,974	49,304	45,379	42,032	39,145	33,409	29,138	25,836	23,206	21,062	19,281	16,491	14,407	12,790	10,446	8,828
130	116,691	105,855	96,860	89,274	82,790	77,184	67,978	60,734	54,885	50,064	46,021	42,583	39,622	33,756	29,402	26,043	23,373	21,200	19,396	16,575	14,471	12,841	10,480	8,852
135	120,704	109,146	99,608	91,604	84,790	78,919	69,320	61,803	55,757	50,788	46,633	43,106	40,075	34,083	29,650	26,233	23,530	21,328	19,504	16,654	14,531	12,888	10,511	8,874
140	124,686	112,392	102,304	93,879	86,735	80,602	70,615	62,830	56,592	51,480	47,215	43,603	40,504	34,393	29,885	26,421	23,677	21,449	19,605	16,728	14,587	12,932	10,540	8,895
145	128,636	115,592	104,949	96,101	88,629	82,234	71,865	63,818	57,392	52,141	47,771	44,076	40,912	34,687	30,106	26,594	23,816	21,563	19,700	16,797	14,640	12,973	10,568	8,915
150	132,556	118,747	107,543	98,272	90,472	83,819	73,072	64,768	58,159	52,774	48,301	44,527	41,301	34,966	30,316	26,758	23,947	21,671	19,790	16,862	14,689	13,012	10,594	8,933
155	136,446	121,859	110,090	100,393	92,267	85,358	74,239	65,683	58,896	53,379	48,808	44,958	41,671	35,231	30,515	26,913	24,071	21,772	19,874	16,923	14,736	13,049	10,618	8,950
160	140,306	124,928	112,589	102,467	94,016	86,852	75,367	66,565	59,603	53,960	49,293	45,369	42,024	35,483	30,704	27,060	24,188	21,868	19,954	16,981	14,780	13,083	10,641	8,967
165	144,136	127,956	115,042	104,495	95,720	88,305	76,458	67,415	60,284	54,517	49,758	45,762	42,361	35,723	30,884	27,199	24,300	21,959	20,030	17,036	14,821	13,116	10,662	8,982
170	147,937	130,942	117,450	106,479	97,382	89,717	77,515	68,234	60,939	55,052	50,203	46,139	42,683	35,952	31,055	27,331	24,406	22,045	20,102	17,088	14,860	13,146	10,682	8,996
175	151,709	133,889	119,815	108,419	99,002	91,091	78,538	69,026	61,569	55,566	50,630	46,499	42,992	36,171	31,218	27,458	24,506	22,127	20,170	17,137	14,898	13,176	10,702	9,010
180	155,452	136,796	122,138	110,317	100,583	92,427	79,529	69,791	62,177	56,061	51,040	46,845	43,287	36,379	31,373	27,578	24,602	22,205	20,235	7,184	14,933	13,203	10,720	9,023
185	159,167	139,665	124,420	112,175	102,125	93,728	80,491	70,530	62,763	56,537	51,435	47,177	43,570	36,579	31,521	27,692	24,693	22,280	20,296	7,228	14,966	13,229	10,737	9,035
190	162,854	142,496	126,662	113,994	103,631	94,994	81,423	71,245	63,328	56,995	51,814	47,496	43,842	36,771	31,663	27,802	24,780	22,351	20,355	7,271	14,998	13,254	10,753	9,047
195	166,514	145,290	128,864	115,776	105,100	96,228	82,328	71,936	63,874	57,437	52,178	47,802	44,103	36,954	31,799	27,907	24,863	22,418	20,411	7,311	15,029	13,278	10,769	9,058
200	170,146	148,047	131,029	117,520	106,536	97,430	83,206	72,606	64,401	57,863	52,530	48,097	44,354	37,130	31,930	28,007	24,943	22,483	20,465	7,350	15,058	13,301	10,784	9,068

400 MCM	1 Run(s)	480 Volts	C-Value = 20,565	Magnetic Duct	Copper

Calculations:

1. $f = \dfrac{1.732 \times \text{Length} \times \text{Isca}}{\text{Runs} \times \text{C-Value} \times \text{E L-L}}$

2. $M = \dfrac{1}{1 + f}$

3. Isc = Isca x M

4. Add Motor Contribution = Motor FLA x 4

* All results are given in symmetrical amperes

Three-Phase Bolted Fault-Current* Table

One-Way Distance in Feet

Available Isc in Thousands (Isca)	5	10	15	20	25	30	40	50	60	70	80	90	100	125	150	175	200	225	250	300	350	400	500	600
3	2,996	2,992	2,988	2,984	2,980	2,976	2,969	2,961	2,953	2,946	2,938	2,931	2,923	2,904	2,886	2,868	2,850	2,832	2,815	2,780	2,747	2,714	2,651	2,591
5	4,989	4,978	4,967	4,957	4,946	4,935	4,914	4,893	4,872	4,851	4,830	4,810	4,790	4,740	4,691	4,644	4,597	4,551	4,506	4,419	4,335	4,254	4,101	3,958
7	6,979	6,957	6,936	6,915	6,894	6,873	6,832	6,791	6,751	6,711	6,672	6,633	6,595	6,501	6,410	6,321	6,234	6,150	6,068	5,911	5,762	5,620	5,356	5,115
10	9,956	9,913	9,870	9,828	9,785	9,744	9,661	9,580	9,500	9,421	9,344	9,268	9,193	9,012	8,837	8,669	8,507	8,351	8,201	7,916	7,651	7,402	6,951	6,551
15	14,902	14,805	14,710	14,615	14,522	14,430	14,250	14,074	13,902	13,735	13,571	13,412	13,256	12,881	12,527	12,192	11,875	11,573	11,287	10,754	10,270	9,827	9,047	8,382
20	19,826	19,655	19,487	19,322	19,160	19,000	18,688	18,387	18,095	17,812	17,538	17,272	17,015	16,403	15,833	15,302	14,805	14,339	13,902	13,103	12,391	11,752	10,654	9,743
25	24,729	24,463	24,204	23,949	23,700	23,457	22,984	22,529	22,093	21,673	21,268	20,879	20,503	19,621	18,811	18,066	17,377	16,739	16,147	15,079	14,143	13,317	11,924	10,795
30	29,610	29,231	28,861	28,500	28,148	27,805	27,143	26,511	25,909	25,333	24,782	24,255	23,749	22,574	21,509	20,540	19,654	18,842	18,094	16,764	15,616	14,614	12,954	11,632
35	34,471	33,957	33,459	32,975	32,505	32,048	31,171	30,342	29,555	28,808	28,098	27,422	26,778	25,292	23,963	22,767	21,684	20,699	19,800	18,218	16,870	15,708	13,805	12,314
40	39,310	38,644	38,000	37,377	36,774	36,190	35,076	34,029	33,043	32,112	31,232	30,399	29,609	27,804	26,206	24,781	23,504	22,352	21,307	19,486	17,952	16,641	14,521	12,880
45	44,129	43,291	42,484	41,707	40,958	40,235	38,863	37,582	36,382	35,257	34,199	33,203	32,263	30,131	28,263	26,613	25,146	23,831	22,648	20,601	18,894	17,448	15,132	13,358
50	48,927	47,899	46,913	45,967	45,059	44,185	42,537	41,006	39,582	38,254	37,012	35,848	34,755	32,293	30,157	28,286	26,634	25,164	23,848	21,589	19,722	18,151	15,658	13,767
55	53,704	52,468	51,288	50,159	49,080	48,045	46,102	44,310	42,652	41,113	39,682	38,347	37,099	34,308	31,907	29,820	27,989	26,371	24,929	22,472	20,455	18,771	16,117	14,120
60	58,461	57,000	55,609	54,285	53,023	51,817	49,564	47,499	45,599	43,845	42,221	40,713	39,309	36,189	33,528	31,231	29,229	27,468	25,907	23,264	21,109	19,320	16,520	14,429
65	63,198	61,493	59,878	58,346	56,890	55,505	52,927	50,579	48,430	46,456	44,637	42,955	41,395	37,949	35,033	32,534	30,367	28,471	26,797	23,979	21,697	19,811	16,878	14,701
70	67,915	65,950	64,096	62,343	60,683	59,110	56,196	53,556	51,152	48,955	46,939	45,083	43,368	39,601	36,436	33,740	31,415	29,390	27,610	24,628	22,227	20,252	17,197	14,942
75	72,611	70,370	68,263	66,278	64,406	62,636	59,373	56,434	53,772	51,349	49,136	47,105	45,236	41,153	37,746	34,860	32,384	30,237	28,356	25,219	22,707	20,650	17,483	15,158
80	77,288	74,754	72,380	70,153	68,058	66,086	62,464	59,219	56,294	53,645	51,234	49,030	47,008	42,614	38,972	35,903	33,282	31,018	29,042	25,761	23,145	21,012	17,742	15,352
85	81,945	79,101	76,449	73,968	71,644	69,461	65,471	61,915	58,725	55,848	53,239	50,864	48,691	43,993	40,122	36,877	34,117	31,742	29,676	26,258	23,546	21,342	17,976	15,527
90	86,582	83,414	80,470	77,726	75,163	72,764	68,398	64,526	61,069	57,964	55,159	52,613	50,291	45,295	41,202	37,787	34,895	32,415	30,263	26,716	23,914	21,644	18,190	15,686
95	91,200	87,691	84,443	81,427	78,619	75,998	71,248	67,056	63,331	59,997	56,997	54,283	51,815	46,528	42,219	38,641	35,622	33,041	30,808	27,141	24,253	21,921	18,385	15,832
100	95,798	91,935	88,371	85,073	82,013	79,165	74,024	69,510	65,514	61,954	58,760	55,879	53,268	47,696	43,179	39,443	36,303	33,626	31,316	27,534	24,567	22,177	18,565	15,965
105	100,377	96,144	92,253	88,665	85,346	82,266	76,728	71,889	67,624	63,837	60,451	57,407	54,654	48,804	44,085	40,198	36,942	34,173	31,790	27,900	24,858	22,414	18,731	16,087
110	104,937	100,319	96,090	92,204	88,620	85,304	79,364	74,198	69,664	65,651	62,076	58,870	55,979	49,858	44,943	40,910	37,542	34,686	32,234	28,241	25,128	22,633	18,884	16,200
115	109,477	104,461	99,884	95,691	91,837	88,280	81,935	76,440	71,636	67,400	63,637	60,272	57,245	50,860	45,756	41,583	38,107	35,168	32,650	28,559	25,380	22,837	19,026	16,304
120	113,999	108,570	103,635	99,128	94,998	91,197	84,441	78,617	73,545	69,087	65,139	61,618	58,458	51,815	46,527	42,219	38,641	35,622	33,041	28,858	25,616	23,028	19,158	16,401
125	118,502	112,647	107,343	102,516	98,104	94,057	86,887	80,733	75,393	70,716	66,585	62,910	59,620	52,725	47,260	42,822	39,145	36,050	33,409	29,138	25,836	23,206	19,281	16,491
130	122,987	116,691	111,009	105,855	101,158	96,860	89,274	82,790	77,184	72,289	67,978	64,152	60,734	53,595	47,957	43,393	39,622	36,454	33,756	29,402	26,043	23,373	19,396	16,575
135	127,453	120,704	114,635	109,146	104,160	99,608	91,604	84,790	78,919	73,809	69,320	65,346	61,803	54,426	48,622	43,936	40,075	36,837	34,083	29,650	26,238	23,530	19,504	16,654
140	131,900	124,686	118,220	112,392	107,111	102,304	93,879	86,735	80,602	75,279	70,615	66,496	62,830	55,221	49,255	44,453	40,504	37,199	34,393	29,885	26,421	23,677	19,605	16,728
145	136,329	128,636	121,766	115,592	110,013	104,949	96,101	88,629	82,234	76,701	71,865	67,603	63,818	55,982	49,860	44,945	40,912	37,543	34,687	30,106	26,594	23,816	19,700	16,797
150	140,740	132,556	125,272	118,747	112,868	107,543	98,272	90,472	83,819	78,078	73,072	68,670	64,768	56,712	50,438	45,414	41,301	37,870	34,966	30,316	26,758	23,947	19,790	16,862
155	145,132	136,446	128,741	121,859	115,676	110,090	100,393	92,267	85,358	79,411	74,239	69,699	65,683	57,412	50,992	45,862	41,671	38,181	35,231	30,515	26,913	24,071	19,874	16,923
160	149,507	140,306	132,171	124,928	118,438	112,589	102,467	94,016	86,852	80,703	75,367	70,693	66,565	58,085	51,521	46,290	42,024	38,477	35,483	30,704	27,060	24,188	19,954	16,981
165	153,864	144,136	135,565	127,956	121,156	115,042	104,495	95,720	88,305	81,956	76,458	71,652	67,415	58,731	52,029	46,700	42,361	38,760	35,723	30,884	27,199	24,300	20,030	17,036
170	158,203	147,937	138,922	130,942	123,830	117,450	106,479	97,382	89,717	83,171	77,515	72,579	68,234	59,352	52,516	47,092	42,683	39,030	35,952	31,055	27,331	24,406	20,102	17,088
175	162,524	151,709	142,243	133,889	126,462	119,815	108,419	99,002	91,091	84,350	78,538	73,475	69,026	59,950	52,984	47,468	42,992	39,287	36,171	31,218	27,458	24,506	20,170	17,137
180	166,828	155,452	145,529	136,796	129,052	122,138	110,317	100,583	92,427	85,495	79,529	74,343	69,791	60,526	53,433	47,828	43,287	39,534	36,379	31,373	27,578	24,602	20,235	17,184
185	171,114	159,167	148,780	139,665	131,602	124,420	112,175	102,125	93,728	86,606	80,491	75,182	70,530	61,081	53,865	48,174	43,570	39,770	36,579	31,521	27,692	24,693	20,296	17,228
190	175,383	162,854	151,996	142,496	134,113	126,662	113,994	103,631	94,994	87,687	81,423	75,994	71,245	61,617	54,281	48,506	43,842	39,996	36,771	31,663	27,802	24,780	20,355	17,271
195	179,635	166,514	155,179	145,290	136,585	128,864	115,776	105,100	96,228	88,737	82,328	76,782	71,936	62,133	54,682	48,826	44,103	40,213	36,954	31,799	27,907	24,863	20,411	17,311
200	183,869	170,146	158,329	148,047	139,019	131,029	117,520	106,536	97,430	89,758	83,206	77,545	72,606	62,632	55,068	49,133	44,354	40,422	37,130	31,930	28,007	24,943	20,465	17,350

400 MCM	2 Run(s)	480 Volts	C-Value = 20,565	Magnetic Duct	Copper

Calculations:

1. $f = \dfrac{1.732 \times \text{Length} \times \text{Isca}}{\text{Runs} \times \text{C-Value} \times E_{L\text{-}L}}$
2. $M = \dfrac{1}{1 + f}$
3. $\text{Isc} = \text{Isca} \times M$
4. Add Motor Contribution = Motor FLA x 4

* All results are given in symmetrical amperes

Three-Phase Bolted Fault-Current* Table

One-Way Distance in Feet

Available Isc in Thousands (Isca)	5	10	15	20	25	30	40	50	60	70	80	90	100	125	150	175	200	225	250	300	350	400	500	600
3	2,993	2,985	2,978	2,971	2,964	2,957	2,943	2,929	2,915	2,901	2,887	2,874	2,860	2,828	2,795	2,764	2,733	2,703	2,674	2,617	2,562	2,510	2,412	2,321
5	4,980	4,960	4,940	4,920	4,900	4,881	4,842	4,805	4,767	4,731	4,695	4,659	4,624	4,539	4,456	4,377	4,301	4,227	4,155	4,019	3,892	3,773	3,555	3,360
7	6,960	6,921	6,882	6,844	6,806	6,769	6,695	6,623	6,552	6,483	6,416	6,349	6,284	6,128	5,979	5,837	5,702	5,572	5,449	5,218	5,005	4,810	4,461	4,159
10	9,919	9,840	9,762	9,685	9,609	9,535	9,389	9,248	9,111	8,978	8,849	8,723	8,601	8,310	8,039	7,784	7,545	7,321	7,109	6,721	6,372	6,058	5,515	5,061
15	14,819	14,643	14,470	14,302	14,138	13,977	13,666	13,369	13,085	12,812	12,550	12,299	12,058	11,495	10,981	10,512	10,081	9,684	9,317	8,661	8,091	7,592	6,757	6,088
20	19,680	19,370	19,070	18,778	18,496	18,222	17,697	17,202	16,734	16,291	15,870	15,471	15,091	14,218	13,441	12,745	12,117	11,548	11,030	10,122	9,352	8,691	7,615	6,776
25	24,502	24,023	23,563	23,120	22,693	22,282	21,503	20,776	20,097	19,461	18,864	18,302	17,773	16,575	15,529	14,606	13,787	13,056	12,397	11,262	10,317	9,518	8,242	7,268
30	29,286	28,604	27,954	27,333	26,738	26,169	25,101	24,116	23,206	22,362	21,577	20,846	20,162	18,634	17,322	16,182	15,183	14,300	13,514	12,176	11,079	10,163	8,722	7,638
35	34,031	33,115	32,246	31,422	30,639	29,895	28,508	27,245	26,089	25,027	24,048	23,143	22,303	20,449	18,879	17,533	16,366	15,345	14,444	12,926	11,696	10,680	9,100	7,926
40	38,740	37,557	36,444	35,395	34,404	33,468	31,740	30,182	28,770	27,484	26,308	25,228	24,234	22,060	20,244	18,704	17,382	16,235	15,230	13,551	12,206	11,104	9,405	8,157
45	43,411	41,931	40,548	39,254	38,040	36,898	34,809	32,944	31,268	29,755	28,382	27,129	25,983	23,500	21,450	19,729	18,264	17,002	15,902	14,081	12,634	11,457	9,658	8,347
50	48,046	46,240	44,564	43,005	41,552	40,194	37,727	35,546	33,603	31,862	30,292	28,870	27,575	24,795	22,524	20,634	19,037	17,669	16,485	14,536	12,999	11,757	9,869	8,504
55	52,645	50,484	48,493	46,653	44,948	43,363	40,506	38,002	35,790	33,821	32,058	30,469	29,030	25,965	23,486	21,438	19,719	18,256	16,994	14,931	13,314	12,013	10,050	8,638
60	57,209	54,665	52,339	50,202	48,233	46,412	43,154	40,324	37,842	35,648	33,694	31,944	30,366	27,029	24,352	22,158	20,327	18,775	17,443	15,276	13,588	12,236	10,205	8,752
65	61,737	58,785	56,103	53,655	51,412	49,349	45,682	42,522	39,772	37,355	35,216	33,308	31,596	27,999	25,137	22,806	20,871	19,238	17,842	15,581	13,829	12,431	10,340	8,852
70	66,230	62,845	59,789	57,017	54,490	52,178	48,096	44,607	41,589	38,954	36,633	34,573	32,733	28,888	25,851	23,392	21,361	19,654	18,199	15,853	14,043	12,603	10,459	8,939
75	70,689	66,846	63,399	60,291	57,473	54,907	50,405	46,586	43,305	40,455	37,958	35,751	33,786	29,705	26,504	23,925	21,804	20,028	18,520	16,096	14,233	12,756	10,564	9,015
80	75,113	70,789	66,936	63,480	60,364	57,539	52,615	48,468	44,926	41,867	39,198	36,848	34,765	30,459	27,102	24,412	22,208	20,368	18,810	16,315	14,404	12,893	10,658	9,084
85	79,504	74,676	70,401	66,588	63,168	60,081	54,733	50,259	46,461	43,196	40,361	37,875	35,677	31,157	27,653	24,858	22,576	20,678	19,074	16,513	14,558	13,017	10,742	9,145
90	83,862	78,508	73,796	69,618	65,888	62,537	56,763	51,966	47,916	44,451	41,454	38,836	36,529	31,805	28,162	25,269	22,914	20,961	19,315	16,693	14,698	13,129	10,818	9,200
95	88,187	82,286	77,125	72,573	68,528	64,911	58,712	53,594	49,297	45,638	42,484	39,738	37,326	32,407	28,634	25,648	23,226	21,222	19,536	16,858	14,825	13,230	10,887	9,249
100	92,479	86,011	80,388	75,455	71,092	67,207	60,584	55,150	50,610	46,761	43,456	40,587	38,074	32,970	29,072	25,999	23,513	21,461	19,739	17,008	14,942	13,323	10,950	9,295
105	96,739	89,684	83,587	78,267	73,584	69,429	62,384	56,637	51,860	47,826	44,374	41,387	38,777	33,496	29,480	26,325	23,779	21,683	19,926	17,147	15,049	13,408	11,007	9,336
110	100,968	93,306	86,726	81,012	76,005	71,580	64,116	58,061	53,051	48,837	45,243	42,142	39,439	33,988	29,861	26,628	24,027	21,888	20,099	17,276	15,147	13,486	11,060	9,374
115	105,165	96,879	89,804	83,692	78,359	73,664	65,783	59,425	54,187	49,798	46,067	42,856	40,063	34,451	30,218	26,911	24,257	22,079	20,260	17,394	15,239	13,558	11,109	9,409
120	109,331	100,404	92,824	86,309	80,648	75,684	67,389	60,732	55,273	50,713	46,849	43,532	40,654	34,887	30,553	27,176	24,472	22,257	20,410	17,505	15,323	13,625	11,154	9,441
125	113,466	103,880	95,788	88,866	82,876	77,643	68,937	61,987	56,310	51,585	47,592	44,173	41,212	35,297	30,867	27,425	24,673	22,424	20,550	17,607	15,402	13,688	11,195	9,471
130	117,570	107,310	98,697	91,364	85,045	79,544	70,431	63,192	57,303	52,417	48,300	44,782	41,741	35,685	31,163	27,653	24,862	22,579	20,681	17,703	15,475	13,745	11,234	9,498
135	121,645	110,694	101,553	93,805	87,157	81,388	71,874	64,351	58,254	53,212	48,974	45,360	42,244	36,051	31,442	27,873	25,040	22,726	20,803	17,793	15,544	13,800	11,270	9,524
140	125,690	114,034	104,356	96,193	89,214	83,179	73,267	65,465	59,166	53,972	49,616	45,911	42,721	36,398	31,706	28,085	25,206	22,863	20,918	17,877	15,608	13,850	11,304	9,548
145	129,705	117,329	107,109	98,527	91,218	84,919	74,613	66,538	60,041	54,699	50,230	46,436	43,175	36,728	31,955	28,281	25,364	22,993	21,027	17,956	15,668	13,897	11,335	9,571
150	133,692	120,582	109,813	100,810	93,172	86,609	75,915	67,572	60,881	55,396	50,817	46,937	43,608	37,040	32,192	28,463	25,513	23,115	21,129	18,031	15,725	13,942	11,365	9,592
155	137,649	123,792	112,469	103,044	95,077	88,253	77,175	68,568	61,689	56,063	51,378	47,416	44,021	37,338	32,416	28,641	25,653	23,230	21,225	18,101	15,778	13,984	11,393	9,612
160	141,578	126,960	115,079	105,230	96,935	89,852	78,395	69,530	62,465	56,704	51,916	47,874	44,415	37,621	32,629	28,807	25,787	23,340	21,317	18,167	15,829	14,023	11,419	9,630
165	145,479	130,088	117,643	107,370	98,748	91,407	79,577	70,457	63,213	57,320	52,432	48,312	44,792	37,891	32,832	28,965	25,913	23,443	21,403	18,230	15,876	14,061	11,444	9,648
170	149,352	133,177	120,162	109,465	100,517	92,921	80,722	71,354	63,934	57,912	52,926	48,731	45,153	38,149	33,026	29,116	26,034	23,542	21,485	18,289	15,921	14,096	11,467	9,665
175	153,197	136,226	122,639	111,517	102,245	94,396	81,832	72,220	64,628	58,481	53,401	49,134	45,498	38,395	33,210	29,259	26,148	23,635	21,563	18,346	15,964	14,130	11,489	9,680
180	157,016	139,236	125,074	113,527	103,931	95,832	82,909	73,057	65,298	59,029	53,858	49,520	45,829	38,630	33,386	29,395	26,257	23,724	21,637	18,399	16,005	14,161	11,510	9,695
185	160,807	142,209	127,468	115,495	105,579	97,231	83,954	73,867	65,945	59,557	54,297	49,891	46,146	38,856	33,554	29,526	26,361	23,809	21,707	18,450	16,043	14,192	11,530	9,709
190	164,571	145,146	129,822	117,425	107,189	98,594	84,969	74,652	66,569	60,066	54,720	50,248	46,451	39,072	33,715	29,650	26,460	23,890	21,775	18,499	16,080	14,220	11,549	9,723
195	168,309	148,046	132,137	119,315	108,762	99,924	85,954	75,412	67,172	60,556	55,127	50,591	46,744	39,279	33,869	29,769	26,555	23,967	21,839	18,545	16,115	14,248	11,567	9,736
200	172,021	150,910	134,414	121,169	110,300	101,220	86,912	76,148	67,756	61,030	55,519	50,921	47,026	39,477	34,017	29,883	26,646	24,041	21,900	18,589	16,148	14,274	11,584	9,748

500 MCM	1 Run(s)	480 Volts	C-Value = 22,185	Magnetic Duct	Copper

Calculations:

1. $f = \frac{1.732 \times \text{Length} \times \text{Isca}}{\text{Runs} \times \text{C-Value} \times \text{E L-L}}$

2. $M = \frac{1}{1 + f}$

3. Isc = Isca x M

4. Add Motor Contribution = Motor FLA x 4

* All results are given in symmetrical amperes

Three-Phase Bolted Fault-Current* Table

One-Way Distance in Feet

Available Isc in Thousands (Isca)	5	10	15	20	25	30	40	50	60	70	80	90	100	125	150	175	200	225	250	300	350	400	500	600
3	2,996	2,993	2,989	2,985	2,982	2,978	2,971	2,964	2,957	2,950	2,943	2,936	2,929	2,911	2,894	2,877	2,860	2,844	2,828	2,795	2,764	2,733	2,674	2,617
5	4,990	4,980	4,970	4,960	4,950	4,940	4,920	4,900	4,881	4,862	4,842	4,823	4,805	4,758	4,713	4,668	4,624	4,581	4,539	4,456	4,377	4,301	4,155	4,019
7	6,980	6,960	6,941	6,921	6,902	6,882	6,844	6,806	6,769	6,732	6,695	6,659	6,623	6,535	6,449	6,366	6,284	6,205	6,128	5,979	5,837	5,702	5,449	5,218
10	9,960	9,919	9,879	9,840	9,801	9,762	9,685	9,609	9,535	9,461	9,389	9,318	9,248	9,077	8,913	8,754	8,601	8,453	8,310	8,039	7,784	7,545	7,109	6,721
15	14,909	14,819	14,730	14,643	14,556	14,470	14,302	14,138	13,977	13,820	13,666	13,516	13,369	13,015	12,680	12,361	12,058	11,770	11,495	10,981	10,512	10,081	9,317	8,661
20	19,839	19,680	19,524	19,370	19,219	19,070	18,778	18,496	18,222	17,956	17,697	17,446	17,202	16,621	16,078	15,569	15,091	14,642	14,218	13,441	12,745	12,117	11,030	10,122
25	24,748	24,502	24,260	24,023	23,791	23,563	23,120	22,693	22,282	21,885	21,503	21,133	20,776	19,934	19,158	18,439	17,773	17,153	16,575	15,529	14,606	13,787	12,397	11,262
30	29,638	29,286	28,941	28,604	28,275	27,954	27,333	26,738	26,169	25,624	25,101	24,599	24,116	22,989	21,963	21,024	20,162	19,368	18,634	17,322	16,182	15,183	13,514	12,176
35	34,509	34,031	33,567	33,115	32,675	32,246	31,422	30,639	29,895	29,185	28,508	27,862	27,245	25,815	24,528	23,363	22,303	21,336	20,449	18,879	17,533	16,366	14,444	12,926
40	39,360	38,740	38,139	37,557	36,992	36,444	35,395	34,404	33,468	32,581	31,740	30,941	30,182	28,437	26,883	25,490	24,234	23,096	22,060	20,244	18,704	17,382	15,230	13,551
45	44,191	43,411	42,658	41,931	41,228	40,548	39,254	38,040	36,898	35,823	34,809	33,851	32,944	30,876	29,052	27,432	25,983	24,679	23,500	21,450	19,729	18,264	15,902	14,081
50	49,004	48,046	47,126	46,240	45,386	44,564	43,005	41,552	40,194	38,922	37,727	36,604	35,546	33,150	31,057	29,213	27,575	26,111	24,795	22,524	20,634	19,037	16,485	14,536
55	53,797	52,645	51,542	50,484	49,468	48,493	46,653	44,948	43,363	41,886	40,506	39,214	38,002	35,277	32,916	30,851	29,030	27,413	25,965	23,486	21,438	19,719	16,994	14,931
60	58,571	57,209	55,908	54,665	53,477	52,339	50,202	48,233	46,412	44,724	43,154	41,691	40,324	37,269	34,644	32,364	30,366	28,600	27,029	24,352	22,158	20,327	17,443	15,276
65	63,326	61,737	60,225	58,785	57,413	56,103	53,655	51,412	49,349	47,444	45,682	44,046	42,522	39,139	36,254	33,765	31,596	29,689	27,999	25,137	22,806	20,871	17,842	15,581
70	68,063	66,230	64,493	62,845	61,279	59,789	57,017	54,490	52,178	50,054	48,096	46,286	44,607	40,898	37,758	35,066	32,733	30,690	28,888	25,851	23,392	21,361	18,199	15,853
75	72,780	70,689	68,713	66,846	65,077	63,399	60,291	57,473	54,907	52,560	50,405	48,420	46,586	42,555	39,167	36,278	33,786	31,614	29,705	26,504	23,925	21,804	18,520	16,096
80	77,480	75,113	72,887	70,789	68,808	66,936	63,480	60,364	57,539	54,967	52,615	50,456	48,468	44,120	40,488	37,409	34,765	32,470	30,459	27,102	24,412	22,208	18,810	16,315
85	82,160	79,504	77,015	74,676	72,475	70,401	66,588	63,168	60,081	57,282	54,733	52,400	50,259	45,599	41,731	38,467	35,677	33,264	31,157	27,653	24,858	22,576	19,074	16,513
90	86,823	83,862	81,097	78,508	76,079	73,796	69,618	65,888	62,537	59,510	56,763	54,259	51,966	47,000	42,901	39,459	36,529	34,003	31,805	28,162	25,269	22,914	19,315	16,693
95	91,467	88,187	85,134	82,286	79,622	77,125	72,573	68,528	64,911	61,656	58,712	56,037	53,594	48,328	44,005	40,391	37,326	34,693	32,407	28,634	25,648	23,226	19,536	16,858
100	96,093	92,479	89,128	86,011	83,104	80,388	75,455	71,092	67,207	63,724	60,584	57,740	55,150	49,590	45,048	41,268	38,074	35,338	32,970	29,072	25,999	23,513	19,739	17,008
105	100,701	96,739	93,078	89,684	86,528	83,587	78,267	73,584	69,429	65,718	62,384	59,372	56,637	50,789	46,035	42,096	38,777	35,943	33,496	29,480	26,325	23,779	19,926	17,147
110	105,291	100,968	96,986	93,306	89,896	86,726	81,012	76,005	71,580	67,643	64,116	60,938	58,061	51,931	46,972	42,877	39,439	36,511	33,988	29,861	26,628	24,027	20,099	17,276
115	109,863	105,165	100,852	96,879	93,208	89,804	83,692	78,359	73,664	69,501	65,783	62,442	59,425	53,019	47,860	43,616	40,063	37,046	34,451	30,218	26,911	24,257	20,260	17,394
120	114,417	109,331	104,677	100,404	96,465	92,824	86,309	80,648	75,684	71,296	67,389	63,888	60,732	54,058	48,705	44,316	40,654	37,550	34,887	30,553	27,176	24,472	20,410	17,505
125	118,954	113,466	108,462	103,880	99,670	95,788	88,866	82,876	77,643	73,032	68,937	65,278	61,987	55,050	49,508	44,981	41,212	38,026	35,297	30,867	27,425	24,673	20,550	17,607
130	123,473	117,570	112,206	107,310	102,824	98,697	91,364	85,045	79,544	74,711	70,431	66,616	63,192	55,998	50,274	45,612	41,741	38,476	35,685	31,163	27,658	24,862	20,681	17,703
135	127,975	121,645	115,912	110,694	105,927	101,553	93,805	87,157	81,388	76,335	71,874	67,905	64,351	56,906	51,005	46,213	42,244	38,903	36,051	31,442	27,878	25,040	20,803	17,793
140	132,460	125,690	119,578	114,034	108,981	104,356	96,193	89,214	83,179	77,909	73,267	69,147	65,465	57,776	51,703	46,785	42,721	39,307	36,398	31,706	28,085	25,206	20,918	17,877
145	136,927	129,705	123,207	117,329	111,987	107,109	98,527	91,218	84,919	79,433	74,613	70,345	66,538	58,610	52,369	47,330	43,175	39,691	36,728	31,955	28,281	25,364	21,027	17,956
150	141,377	133,692	126,799	120,582	114,946	109,813	100,810	93,172	86,609	80,911	75,915	71,501	67,572	59,410	53,008	47,851	43,608	40,057	37,040	32,192	28,466	25,513	21,129	18,031
155	145,810	137,649	130,353	123,792	117,859	112,469	103,044	95,077	88,253	82,343	77,175	72,618	68,568	60,179	53,619	48,348	44,021	40,405	37,338	32,416	28,641	25,653	21,225	18,101
160	150,226	141,578	133,871	126,960	120,728	115,079	105,230	96,935	89,852	83,733	78,395	73,697	69,530	60,918	54,205	48,824	44,415	40,737	37,621	32,629	28,807	25,787	21,317	18,167
165	154,626	145,479	137,354	130,088	123,553	117,643	107,370	98,748	91,407	85,083	79,577	74,740	70,457	61,629	54,767	49,280	44,792	41,054	37,891	32,832	28,965	25,913	21,403	18,230
170	159,009	149,352	140,801	133,177	126,335	120,162	109,465	100,517	92,921	86,393	80,722	75,749	71,354	62,314	55,307	49,717	45,153	41,356	38,149	33,026	29,116	26,034	21,485	18,289
175	163,375	153,197	144,214	136,226	129,076	122,639	111,517	102,245	94,396	87,666	81,832	76,726	72,220	62,973	55,826	50,136	45,498	41,646	38,395	33,210	29,259	26,148	21,563	18,346
180	167,724	157,016	147,593	139,236	131,776	125,074	113,527	103,931	95,832	88,903	82,909	77,672	73,057	63,609	56,325	50,538	45,829	41,923	38,630	33,386	29,395	26,257	21,637	18,399
185	172,057	160,807	150,937	142,209	134,436	127,468	115,495	105,579	97,231	90,106	83,954	78,588	73,867	64,222	56,805	50,924	46,146	42,188	38,856	33,554	29,526	26,361	21,707	18,450
190	176,374	164,571	154,249	145,146	137,057	129,822	117,425	107,189	98,594	91,276	84,969	79,477	74,652	64,815	57,268	51,296	46,451	42,443	39,072	33,715	29,650	26,460	21,775	18,499
195	180,674	168,309	157,528	148,046	139,639	132,137	119,315	108,762	99,924	92,414	85,954	80,338	75,412	65,387	57,714	51,653	46,744	42,688	39,279	33,869	29,769	26,555	21,839	18,545
200	184,958	172,021	160,775	150,910	142,185	134,414	121,169	110,300	101,220	93,522	86,912	81,175	76,148	65,939	58,144	51,998	47,026	42,922	39,477	34,017	29,883	26,646	21,900	18,589

500 MCM	2 Run(s)	480 Volts	C-Value = 22,185	Magnetic Duct	Copper

Calculations:

1. $f = \dfrac{1.732 \times \text{Length} \times \text{Isca}}{\text{Runs} \times \text{C-Value} \times \text{E L-L}}$

2. $M = \dfrac{1}{1 + f}$

3. Isc = Isca x M

4. Add Motor Contribution = Motor FLA x 4

* All results are given in symmetrical amperes

Three-Phase Bolted Fault-Current* Table

One-Way Distance in Feet

Available Isc in Thousands (Isca)	5	10	15	20	25	30	40	50	60	70	80	90	100	125	150	175	200	225	250	300	350	400	500	600
3	2,993	2,986	2,979	2,972	2,965	2,958	2,944	2,931	2,917	2,904	2,891	2,878	2,865	2,833	2,802	2,771	2,742	2,712	2,684	2,628	2,575	2,524	2,428	2,339
5	4,980	4,961	4,942	4,923	4,904	4,885	4,848	4,811	4,775	4,739	4,704	4,670	4,636	4,553	4,473	4,396	4,321	4,249	4,179	4,346	3,922	3,804	3,590	3,398
7	6,962	6,924	6,886	6,849	6,813	6,776	6,705	6,635	6,567	6,500	6,434	6,369	6,306	6,154	6,009	5,870	5,738	5,611	5,490	5,263	5,054	4,861	4,516	4,217
10	9,922	9,845	9,770	9,695	9,622	9,550	9,409	9,272	9,138	9,009	8,883	8,761	8,642	8,358	8,093	7,843	7,609	7,388	7,180	6,796	6,452	6,141	5,600	5,147
15	14,825	14,655	14,488	14,325	14,165	14,009	13,708	13,419	13,142	12,876	12,620	12,375	12,139	11,587	11,082	10,620	10,195	9,802	9,439	8,787	8,220	7,721	6,886	6,213
20	19,691	19,391	19,100	18,817	18,543	18,277	17,767	17,284	16,827	16,394	15,982	15,591	15,218	14,359	13,593	12,904	12,281	11,716	11,201	10,295	9,524	8,861	7,778	6,931
25	24,518	24,055	23,609	23,179	22,764	22,365	21,605	20,896	20,232	19,608	19,022	18,470	17,949	16,767	15,731	14,816	14,001	13,271	12,613	11,476	10,527	9,723	8,434	7,447
30	29,309	28,650	28,019	27,415	26,837	26,283	25,241	24,278	23,386	22,557	21,785	21,064	20,389	18,877	17,574	16,439	15,442	14,559	13,771	12,427	11,322	10,397	8,937	7,837
35	34,063	33,176	32,333	31,532	30,770	30,043	28,689	27,452	26,317	25,272	24,306	23,412	22,582	20,742	19,179	17,835	16,668	15,644	14,738	13,209	11,967	10,938	9,334	8,140
40	38,781	37,635	36,554	35,533	34,568	33,655	31,964	30,436	29,047	27,779	26,617	25,549	24,563	22,401	20,589	19,049	17,723	16,569	15,557	13,863	12,501	11,383	9,656	8,384
45	43,463	42,028	40,685	39,425	38,240	37,125	35,079	33,246	31,596	30,102	28,742	27,500	26,361	23,888	21,838	20,113	18,640	17,369	16,259	14,418	12,951	11,755	9,922	8,584
50	48,110	46,358	44,729	43,211	41,792	40,463	38,045	35,899	33,982	32,259	30,703	29,290	28,002	25,227	22,952	21,054	19,446	18,066	16,869	14,895	13,335	12,070	10,146	8,751
55	52,722	50,625	48,689	46,895	45,229	43,677	40,872	38,405	36,220	34,270	32,519	30,938	29,504	26,439	23,952	21,892	20,159	18,680	17,403	15,310	13,666	12,341	10,337	8,892
60	57,299	54,831	52,567	50,482	48,556	46,772	43,570	40,778	38,323	36,146	34,204	32,459	30,884	27,543	24,854	22,643	20,794	19,224	17,874	15,673	13,955	12,576	10,501	9,014
65	61,842	58,977	56,365	53,975	51,779	49,755	46,148	43,028	40,303	37,903	35,772	33,869	32,158	28,551	25,672	23,320	21,363	19,709	18,293	15,994	14,209	12,782	10,644	9,119
70	66,351	63,064	60,087	57,378	54,903	52,633	48,613	45,163	42,171	39,550	37,236	35,178	33,336	29,476	26,417	23,934	21,877	20,146	18,668	16,281	14,434	12,964	10,770	9,212
75	70,827	67,094	63,734	60,695	57,933	55,411	50,973	47,193	43,935	41,098	38,605	36,397	34,429	30,327	27,099	24,492	22,342	20,540	19,006	16,537	14,636	13,126	10,882	9,293
80	75,269	71,067	67,309	63,929	60,871	58,093	53,234	49,125	45,605	42,556	39,889	37,536	35,446	31,113	27,725	25,002	22,766	20,897	19,312	16,768	14,816	13,272	10,982	9,366
85	79,679	74,985	70,814	67,082	63,724	60,686	55,403	50,966	47,187	43,930	41,094	38,601	36,394	31,842	28,302	25,470	23,154	21,224	19,590	16,977	14,980	13,402	11,071	9,431
90	84,057	78,850	74,250	70,158	66,493	63,192	57,484	52,722	48,689	45,229	42,228	39,601	37,281	32,519	28,835	25,902	23,510	21,522	19,844	17,168	15,128	13,521	11,152	9,489
95	88,402	82,661	77,621	73,159	69,183	65,617	59,484	54,400	50,116	46,458	43,297	40,539	38,112	33,149	29,330	26,300	23,837	21,796	20,077	17,342	15,263	13,629	11,225	9,542
100	92,716	86,421	80,927	76,089	71,797	67,964	61,406	56,003	51,474	47,622	44,307	41,423	38,892	33,738	29,790	26,669	24,140	22,049	20,292	17,502	15,386	13,727	11,292	9,590
105	96,999	90,130	84,170	78,950	74,339	70,237	63,256	57,538	52,767	48,727	45,262	42,257	39,626	34,289	30,219	27,012	24,421	22,283	20,490	17,649	15,500	13,817	11,353	9,634
110	101,250	93,790	87,353	81,744	76,811	72,440	65,037	59,007	54,001	49,777	46,166	43,044	40,317	34,805	30,619	27,332	24,682	22,500	20,673	17,785	15,605	13,900	11,409	9,674
115	105,471	97,401	90,477	84,473	79,216	74,575	66,753	60,416	55,178	50,776	47,025	43,789	40,970	35,291	30,994	27,638	24,925	22,702	20,844	17,911	15,701	13,977	11,460	9,712
120	109,662	100,964	93,544	87,140	81,557	76,646	68,408	61,768	56,304	51,728	47,840	44,495	41,588	35,748	31,346	27,910	25,152	22,891	21,002	18,028	15,791	14,048	11,508	9,746
125	113,822	104,480	96,554	89,747	83,836	78,655	70,004	63,067	57,381	52,635	48,615	45,165	42,172	36,179	31,677	28,172	25,365	23,067	21,150	18,137	15,875	14,114	11,552	9,778
130	117,953	107,950	99,511	92,295	86,056	80,606	71,545	64,315	58,412	53,502	49,353	45,801	42,727	36,586	31,989	28,418	25,564	23,231	21,289	18,238	15,953	14,176	11,594	9,807
135	122,055	111,375	102,414	94,788	88,219	82,501	73,034	65,515	59,401	54,330	50,057	46,407	43,253	36,972	32,283	28,650	25,752	23,386	21,419	18,334	16,026	14,234	11,632	9,835
140	126,128	114,757	105,266	97,226	90,327	84,342	74,472	66,671	60,349	55,122	50,729	46,984	43,754	37,337	32,561	28,869	25,929	23,532	21,541	18,423	16,094	14,287	11,668	9,860
145	130,172	118,095	108,068	99,611	92,382	86,131	75,864	67,784	61,260	55,881	51,371	47,534	44,230	37,683	32,825	29,076	26,095	23,669	21,656	18,507	16,158	14,338	11,702	9,884
150	134,187	121,390	110,822	101,946	94,386	87,871	77,211	68,857	62,135	56,608	51,984	48,059	44,685	38,013	33,074	29,271	26,253	23,799	21,764	18,586	16,218	14,385	11,733	9,907
155	138,174	124,644	113,527	104,231	96,342	89,563	78,514	69,892	62,976	57,306	52,572	48,561	45,118	38,326	33,311	29,457	26,402	23,921	21,866	18,661	16,275	14,430	11,763	9,928
160	142,134	127,857	116,187	106,468	98,250	91,210	79,777	70,891	63,786	57,976	53,135	49,041	45,533	38,624	33,536	29,633	26,543	24,037	21,963	18,731	16,328	14,472	11,791	9,948
165	146,066	131,030	118,801	108,659	100,113	92,813	81,001	71,856	64,566	58,619	53,676	49,501	45,929	38,909	33,751	29,800	26,677	24,147	22,055	18,798	16,379	14,512	11,817	9,967
170	149,971	134,164	121,371	110,806	101,932	94,375	82,188	72,788	65,318	59,238	54,194	49,941	46,308	39,181	33,955	29,959	26,805	24,251	22,142	18,861	16,427	14,549	11,842	9,984
175	153,848	137,259	123,898	112,908	103,709	95,896	83,339	73,690	66,043	59,834	54,692	50,364	46,671	39,440	34,150	30,111	26,926	24,350	22,225	18,921	16,472	14,585	11,866	10,001
180	157,700	140,316	126,384	114,969	105,445	97,378	84,456	74,562	66,743	60,408	55,171	50,770	47,019	39,689	34,336	30,255	27,042	24,445	22,303	18,978	16,516	14,619	11,888	10,017
185	161,524	143,336	128,829	116,988	107,141	98,823	85,541	75,406	67,418	60,961	55,632	51,160	47,354	39,927	34,514	30,393	27,152	24,535	22,378	19,032	16,557	14,651	11,909	10,032
190	165,323	146,319	131,234	118,968	108,799	100,232	86,594	76,223	68,071	61,494	56,076	51,535	47,675	40,155	34,684	30,525	27,257	24,621	22,450	19,084	16,596	14,682	11,930	10,047
195	169,095	149,266	133,600	120,909	110,421	101,606	87,618	77,016	68,702	62,008	56,503	51,896	47,983	40,374	34,847	30,652	27,358	24,703	22,518	19,133	16,633	14,711	11,949	10,060
200	172,842	152,178	135,928	122,813	112,006	102,947	88,614	77,784	69,313	62,505	56,916	52,244	48,280	40,584	35,004	30,772	27,454	24,781	22,583	19,180	16,669	14,739	11,967	10,073

600 MCM	1 Run(s)	480 Volts	C-Value = 22,965	Magnetic Duct	Copper

Calculations:

1. $f = \dfrac{1.732 \times \text{Length} \times \text{Isca}}{\text{Runs} \times \text{C-Value} \times \text{E L-L}}$

2. $M = \dfrac{1}{1 + f}$

3. Isc = Isca x M

4. Add Motor Contribution = Motor FLA x 4

* All results are given in symmetrical amperes

Three-Phase Bolted Fault-Current* Table

One-Way Distance in Feet

Available Isc in Thousands (Isca)	5	10	15	20	25	30	40	50	60	70	80	90	100	125	150	175	200	225	250	300	350	400	500	600
3	2,996	2,993	2,989	2,986	2,982	2,979	2,972	2,965	2,958	2,951	2,944	2,938	2,931	2,914	2,898	2,881	2,865	2,849	2,833	2,802	2,771	2,742	2,684	2,628
5	4,990	4,980	4,971	4,961	4,951	4,942	4,923	4,904	4,885	4,866	4,848	4,829	4,811	4,766	4,722	4,678	4,636	4,594	4,553	4,473	4,396	4,321	4,179	4,046
7	6,981	6,962	6,943	6,924	6,905	6,886	6,849	6,813	6,776	6,741	6,705	6,670	6,635	6,550	6,467	6,385	6,306	6,229	6,154	6,009	5,870	5,738	5,490	5,263
10	9,961	9,922	9,884	9,845	9,807	9,770	9,695	9,622	9,550	9,479	9,409	9,340	9,272	9,106	8,946	8,791	8,642	8,498	8,358	8,093	7,843	7,609	7,180	6,796
15	14,912	14,825	14,739	14,655	14,571	14,488	14,325	14,165	14,009	13,857	13,708	13,562	13,419	13,074	12,747	12,435	12,139	11,856	11,587	11,082	10,620	10,195	9,439	8,787
20	19,844	19,691	19,539	19,391	19,244	19,100	18,817	18,543	18,277	18,018	17,767	17,522	17,284	16,717	16,185	15,687	15,218	14,776	14,359	13,593	12,904	12,281	11,201	10,295
25	24,757	24,518	24,285	24,055	23,830	23,609	23,179	22,764	22,365	21,978	21,605	21,245	20,896	20,072	19,311	18,605	17,949	17,338	16,767	15,731	14,816	14,001	12,613	11,476
30	29,651	29,309	28,976	28,650	28,331	28,019	27,415	26,837	26,283	25,752	25,241	24,750	24,278	23,173	22,164	21,240	20,389	19,604	18,877	17,574	16,439	15,442	13,771	12,427
35	34,525	34,063	33,614	33,176	32,749	32,333	31,532	30,770	30,043	29,351	28,689	28,057	27,452	26,047	24,780	23,630	22,582	21,623	20,742	19,179	17,835	16,668	14,738	13,209
40	39,381	38,781	38,199	37,635	37,086	36,554	35,533	34,568	33,655	32,788	31,964	31,181	30,436	28,719	27,186	25,808	24,563	23,432	22,401	20,589	19,049	17,723	15,557	13,863
45	44,218	43,463	42,734	42,028	41,346	40,685	39,425	38,240	37,125	36,073	35,079	34,138	33,246	31,209	29,406	27,801	26,361	25,064	23,888	21,838	20,113	18,640	16,259	14,418
50	49,037	48,110	47,218	46,358	45,529	44,729	43,211	41,792	40,463	39,217	38,045	36,941	35,899	33,534	31,462	29,631	28,002	26,542	25,227	22,952	21,054	19,446	16,869	14,895
55	53,837	52,722	51,652	50,625	49,638	48,689	46,895	45,229	43,677	42,228	40,872	39,600	38,405	35,712	33,371	31,318	29,504	27,888	26,439	23,952	21,892	20,159	17,403	15,310
60	58,618	57,299	56,038	54,831	53,675	52,567	50,482	48,556	46,772	45,114	43,570	42,128	40,778	37,755	35,148	32,879	30,884	29,118	27,543	24,854	22,643	20,794	17,874	15,673
65	63,382	61,842	60,375	58,977	57,641	56,365	53,975	51,779	49,755	47,884	46,148	44,533	43,028	39,675	36,807	34,325	32,158	30,247	28,551	25,672	23,320	21,363	18,293	15,994
70	68,127	66,351	64,666	63,064	61,539	60,087	57,378	54,903	52,633	50,543	48,613	46,825	45,163	41,484	38,358	35,671	33,336	31,287	29,476	26,417	23,934	21,877	18,668	16,281
75	72,854	70,827	68,910	67,094	65,371	63,734	60,695	57,933	55,411	53,099	50,973	49,010	47,193	43,190	39,813	36,925	34,429	32,248	30,327	27,099	24,492	22,342	19,006	16,537
80	77,563	75,269	73,108	71,067	69,137	67,309	63,929	60,871	58,093	55,558	53,234	51,097	49,125	44,802	41,179	38,098	35,446	33,139	31,113	27,725	25,002	22,766	19,312	16,768
85	82,254	79,679	77,261	74,985	72,840	70,814	67,082	63,724	60,686	57,924	55,403	53,092	50,966	46,329	42,465	39,196	36,394	33,966	31,842	28,302	25,470	23,154	19,590	16,977
90	86,927	84,057	81,370	78,850	76,481	74,250	70,158	66,493	63,192	60,203	57,484	55,000	52,722	47,775	43,677	40,226	37,281	34,737	32,519	28,835	25,902	23,510	19,844	17,168
95	91,582	88,402	85,435	82,661	80,062	77,621	73,159	69,183	65,617	62,400	59,484	56,828	54,400	49,148	44,822	41,195	38,112	35,458	33,149	29,330	26,300	23,837	20,077	17,342
100	96,220	92,716	89,458	86,421	83,584	80,927	76,089	71,797	67,964	64,519	61,406	58,580	56,003	50,454	45,905	42,108	38,892	36,132	33,738	29,790	26,669	24,140	20,292	17,502
105	100,841	96,999	93,438	90,130	87,048	84,170	78,950	74,339	70,237	66,564	63,256	60,261	57,538	51,696	46,931	42,970	39,626	36,764	34,289	30,219	27,012	24,421	20,490	17,649
110	105,444	101,250	97,377	93,790	90,457	87,353	81,744	76,811	72,440	68,539	65,037	61,876	59,007	52,879	47,904	43,784	40,317	37,359	34,805	30,619	27,332	24,682	20,673	17,785
115	110,030	105,471	101,275	97,401	93,811	90,477	84,473	79,216	74,575	70,448	66,753	63,427	60,416	54,008	48,828	44,555	40,970	37,919	35,291	30,994	27,630	24,925	20,844	17,911
120	114,598	109,662	105,133	100,964	97,112	93,544	87,140	81,557	76,646	72,293	68,408	64,919	61,768	55,086	49,708	45,287	41,588	38,447	35,748	31,346	27,910	25,152	21,002	18,028
125	119,150	113,822	108,951	104,480	100,361	96,554	89,747	83,836	78,655	74,078	70,004	66,355	63,067	56,116	50,545	45,981	42,172	38,946	36,179	31,677	28,172	25,365	21,150	18,137
130	123,684	117,953	112,730	107,950	103,559	99,511	92,295	86,056	80,606	75,806	71,545	67,738	64,315	57,102	51,344	46,641	42,727	39,419	36,586	31,989	28,418	25,564	21,289	18,238
135	128,202	122,055	116,471	111,375	106,707	102,414	94,788	88,219	82,501	77,479	73,034	69,071	65,515	58,046	52,106	47,269	43,253	39,866	36,972	32,283	28,650	25,752	21,419	18,334
140	132,702	126,128	120,174	114,757	109,807	105,266	97,226	90,327	84,342	79,100	74,472	70,356	66,671	58,952	52,834	47,867	43,754	40,291	37,337	32,561	28,869	25,929	21,541	18,423
145	137,186	130,172	123,839	118,095	112,859	108,068	99,611	92,382	86,131	80,672	75,864	71,597	67,784	59,820	53,531	48,438	44,230	40,695	37,683	32,825	29,076	26,095	21,656	18,507
150	141,654	134,187	127,468	121,390	115,865	110,822	101,946	94,386	87,871	82,196	77,211	72,795	68,857	60,654	54,198	48,984	44,585	41,080	38,013	33,074	29,271	26,253	21,764	18,586
155	146,104	138,174	131,061	124,644	118,826	113,527	104,231	96,342	89,563	83,676	78,514	73,953	69,892	61,456	54,837	49,505	45,118	41,446	38,326	33,311	29,457	26,402	21,866	18,661
160	150,539	142,134	134,618	127,857	121,743	116,187	106,468	98,250	91,210	85,111	79,777	75,072	70,891	62,227	55,450	50,004	45,533	41,795	38,624	33,536	29,633	26,543	21,963	18,731
165	154,957	146,066	138,140	131,030	124,616	118,801	108,659	100,113	92,813	86,506	81,001	76,155	71,856	62,969	56,039	50,482	45,929	42,128	38,909	33,751	29,800	26,677	22,055	18,798
170	159,358	149,971	141,627	134,164	127,447	121,371	110,806	101,932	94,375	87,861	82,188	77,203	72,788	63,684	56,604	50,941	46,308	42,447	39,181	33,955	29,959	26,805	22,142	18,861
175	163,744	153,848	145,081	137,259	130,237	123,898	112,908	103,709	95,896	89,177	83,339	78,218	73,690	64,373	57,148	51,381	46,671	42,752	39,440	34,150	30,111	26,926	22,225	18,921
180	168,113	157,700	148,501	140,316	132,986	126,384	114,969	105,445	97,378	90,458	84,456	79,201	74,562	65,037	57,671	51,803	47,019	43,044	39,689	34,336	30,255	27,042	22,303	18,978
185	172,467	161,524	151,887	143,336	135,695	128,829	116,988	107,141	98,823	91,703	85,541	80,154	75,406	65,679	58,175	52,209	47,354	43,324	39,927	34,514	30,393	27,152	22,378	19,032
190	176,804	165,323	155,241	146,319	138,366	131,234	118,968	108,799	100,232	92,915	86,594	81,079	76,223	66,298	58,660	52,600	47,675	43,593	40,155	34,684	30,525	27,257	22,450	19,084
195	181,126	169,095	158,563	149,266	140,999	133,600	120,909	110,421	101,606	94,095	87,618	81,976	77,016	66,897	59,128	52,976	47,983	43,851	40,374	34,847	30,652	27,358	22,518	19,133
200	185,432	172,842	161,854	152,178	143,595	135,928	122,813	112,006	102,947	95,244	88,614	82,846	77,784	67,475	59,580	53,338	48,280	44,099	40,584	35,004	30,772	27,454	22,583	19,180

600 MCM | 2 Run(s) | 480 Volts | C-Value = 22,965 | Magnetic Duct | Copper

Calculations:

1. $f = \dfrac{1.732 \times \text{Length} \times \text{Isca}}{\text{Runs} \times \text{C-Value} \times E_{L\text{-}L}}$
2. $M = \dfrac{1}{1 + f}$
3. $\text{Isc} = \text{Isca} \times M$
4. Add Motor Contribution = Motor FLA x 4

* All results are given in symmetrical amperes

Three-Phase Bolted Fault-Current* Table

One-Way Distance in Feet

Available Isc in Thousands (Isca)	5	10	15	20	25	30	40	50	60	70	80	90	100	125	150	175	200	225	250	300	350	400	500	600
3	2,993	2,987	2,980	2,973	2,967	2,960	2,947	2,934	2,921	2,909	2,896	2,884	2,871	2,841	2,811	2,782	2,753	2,725	2,698	2,644	2,593	2,544	2,450	2,364
5	4,981	4,963	4,945	4,926	4,908	4,890	4,855	4,820	4,785	4,751	4,718	4,685	4,652	4,573	4,496	4,422	4,350	4,280	4,213	4,034	3,963	3,849	3,640	3,452
7	6,964	6,928	6,892	6,856	6,822	6,787	6,719	6,652	6,586	6,522	6,459	6,397	6,337	6,190	6,050	5,916	5,788	5,666	5,548	5,327	5,123	4,934	4,595	4,300
10	9,926	9,853	9,781	9,710	9,640	9,571	9,436	9,304	9,177	9,053	8,932	8,814	8,699	8,425	8,168	7,926	7,698	7,483	7,279	6,904	6,565	6,258	5,722	5,271
15	14,834	14,671	14,512	14,356	14,204	14,054	13,765	13,488	13,221	12,965	12,718	12,481	12,252	11,716	11,224	10,772	10,356	9,970	9,612	8,967	8,404	7,907	7,071	6,395
20	19,705	19,419	19,142	18,871	18,609	18,354	17,864	17,399	16,958	16,538	16,139	15,759	15,396	14,559	13,807	13,130	12,516	11,956	11,445	10,543	9,773	9,107	8,016	7,158
25	24,541	24,099	23,673	23,261	22,864	22,479	21,749	21,064	20,421	19,816	19,246	18,707	18,198	17,039	16,019	15,114	14,306	13,580	12,924	11,786	10,831	10,020	8,715	7,710
30	29,342	28,712	28,109	27,531	26,975	26,442	25,437	24,505	23,639	22,832	22,078	21,373	20,711	19,223	17,935	16,808	15,814	14,932	14,143	12,790	11,674	10,737	9,252	8,128
35	34,108	33,260	32,453	31,684	30,951	30,251	28,942	27,742	26,637	25,617	24,672	23,795	22,977	21,160	19,609	18,270	17,102	16,075	15,164	13,520	12,361	11,316	9,679	8,455
40	38,839	37,743	36,707	35,727	34,798	33,916	32,279	30,793	29,438	28,197	27,056	26,004	25,031	22,890	21,086	19,546	18,215	17,054	16,032	14,316	12,932	11,792	10,025	8,718
45	43,536	42,163	40,875	39,663	38,521	37,443	35,458	33,673	32,059	30,593	29,255	28,029	26,902	24,444	22,398	20,668	19,186	17,902	16,779	14,909	13,414	12,192	10,312	8,935
50	48,199	46,522	44,959	43,497	42,127	40,841	38,491	36,397	34,518	32,825	31,289	29,891	28,612	25,848	23,571	21,663	20,040	18,644	17,429	15,420	13,826	12,531	10,554	9,116
55	52,828	50,821	48,961	47,233	45,622	44,117	41,388	38,976	36,830	34,908	33,176	31,609	30,182	27,123	24,626	22,551	20,798	19,298	18,000	15,865	14,183	12,823	10,761	9,269
60	57,425	55,061	52,884	50,873	49,010	47,278	44,157	41,422	39,007	36,857	34,932	33,199	31,629	28,285	25,581	23,349	21,475	19,879	18,504	16,256	14,495	13,078	10,939	9,401
65	61,988	59,243	56,731	54,423	52,295	50,328	46,806	43,745	41,060	38,685	36,570	34,675	32,966	29,350	26,448	24,069	22,083	20,399	18,954	16,602	14,769	13,301	11,095	9,516
70	66,519	63,368	60,503	57,885	55,484	53,274	49,344	45,954	43,000	40,403	38,101	36,048	34,205	30,328	27,240	24,723	22,632	20,867	19,357	16,910	15,013	13,498	11,231	9,617
75	71,019	67,438	64,202	61,262	58,579	56,122	51,778	48,058	44,836	42,020	39,536	37,330	35,357	31,230	27,966	25,319	23,130	21,290	19,721	17,187	15,230	13,674	11,353	9,706
80	75,486	71,454	67,831	64,558	61,586	58,875	54,113	50,063	46,577	43,545	40,883	38,528	36,430	32,064	28,633	25,865	23,585	21,674	20,050	17,437	15,426	13,831	11,461	9,785
85	79,922	75,416	71,392	67,775	64,507	61,540	56,355	51,976	48,228	44,985	42,150	39,652	37,433	32,838	29,249	26,366	24,001	22,025	20,350	17,663	15,603	13,973	11,559	9,856
90	84,327	79,327	74,886	70,916	67,346	64,119	58,510	53,804	49,798	46,348	43,344	40,706	38,371	33,559	29,819	26,829	24,384	22,347	20,624	17,870	15,764	14,102	11,647	9,920
95	88,701	83,186	78,316	73,985	70,107	66,616	60,583	55,551	51,292	47,639	44,471	41,699	39,252	34,230	30,348	27,256	24,736	22,643	20,876	18,058	15,911	14,219	11,727	9,977
100	93,045	86,994	81,683	76,982	72,793	69,037	62,578	57,225	52,715	48,864	45,537	42,635	40,080	34,858	30,840	27,653	25,063	22,916	21,108	18,232	16,045	14,327	11,799	10,030
105	97,359	90,754	84,988	79,912	75,407	71,384	64,500	58,828	54,072	50,028	46,547	43,518	40,860	35,447	31,300	28,022	25,365	23,169	21,322	18,391	16,168	14,425	11,866	10,078
110	101,642	94,465	88,235	82,775	77,952	73,660	66,353	60,365	55,368	51,135	47,504	44,354	41,596	35,999	31,730	28,366	25,647	23,404	21,521	18,539	16,282	14,516	11,927	10,122
115	105,897	98,129	91,423	85,575	80,430	75,869	68,140	61,840	56,607	52,190	48,413	45,145	42,291	36,519	32,133	28,688	25,910	23,622	21,706	18,676	16,388	14,599	11,984	10,163
120	110,122	101,747	94,555	88,313	82,844	78,013	69,865	63,258	57,792	53,196	49,277	45,896	42,949	37,008	32,512	28,989	26,155	23,826	21,878	18,803	16,486	14,677	12,036	10,201
125	114,318	105,319	97,632	90,992	85,197	80,096	71,531	64,620	58,928	54,157	50,100	46,609	43,573	37,471	32,868	29,272	26,385	24,017	22,039	18,921	16,577	14,749	12,085	10,235
130	118,486	108,846	100,656	93,613	87,491	82,120	73,141	65,931	60,016	55,074	50,885	47,287	44,165	37,908	33,203	29,538	26,601	24,196	22,189	19,032	16,662	14,816	12,130	10,268
135	122,626	112,329	103,628	96,178	89,727	84,087	74,697	67,193	61,060	55,952	51,633	47,933	44,728	38,322	33,521	29,789	26,804	24,364	22,330	19,136	16,741	14,879	12,172	10,298
140	126,737	115,769	106,549	98,689	91,909	86,000	76,203	68,409	62,062	56,793	52,348	48,549	45,263	38,714	33,821	30,025	26,996	24,522	22,463	19,233	16,816	14,938	12,211	10,326
145	130,821	119,167	109,421	101,147	94,038	87,861	77,661	69,582	63,026	57,599	53,032	49,136	45,774	39,087	34,105	30,249	27,176	24,671	22,588	19,325	16,886	14,993	12,248	10,352
150	134,877	122,524	112,244	103,555	96,115	89,673	79,072	70,713	63,952	58,371	53,686	49,698	46,261	39,441	34,374	30,461	27,347	24,811	22,706	19,411	16,952	15,045	12,282	10,377
155	138,906	125,840	115,020	105,914	98,144	91,436	80,440	71,805	64,844	59,113	54,314	50,235	46,725	39,779	34,630	30,662	27,509	24,944	22,817	19,493	17,014	15,094	12,315	10,400
160	142,908	129,116	117,751	108,225	100,125	93,153	81,766	72,860	65,703	59,826	54,915	50,749	47,170	40,100	34,874	30,852	27,663	25,071	22,923	19,569	17,072	15,140	12,346	10,422
165	146,884	132,352	120,437	110,490	102,061	94,826	83,052	73,879	66,531	60,512	55,492	51,241	47,595	40,407	35,105	31,034	27,808	25,190	23,023	19,642	17,128	15,184	12,375	10,443
170	150,833	135,550	123,079	112,710	103,952	96,457	84,300	74,865	67,329	61,172	56,046	51,713	48,002	40,700	35,327	31,206	27,947	25,304	23,117	19,711	17,180	15,225	12,402	10,462
175	154,756	138,710	125,679	114,886	105,800	98,046	85,512	75,819	68,100	61,807	56,579	52,167	48,393	40,981	35,537	31,371	28,079	25,412	23,208	19,777	17,230	15,264	12,428	10,481
180	158,653	141,833	128,237	117,020	107,607	99,596	86,688	76,743	68,844	62,420	57,092	52,602	48,767	41,249	35,739	31,528	28,204	25,515	23,293	19,839	17,277	15,301	12,452	10,498
185	162,525	144,919	130,755	119,113	109,374	101,108	87,832	77,637	69,563	63,010	57,586	53,021	49,127	41,506	35,932	31,678	28,324	25,613	23,375	19,898	17,322	15,336	12,476	10,515
190	166,371	147,969	133,233	121,166	111,103	102,583	88,943	78,504	70,258	63,580	58,061	53,424	49,473	41,752	36,117	31,821	28,439	25,706	23,453	19,955	17,365	15,370	12,498	10,530
195	170,192	150,984	135,672	123,180	112,794	104,024	90,024	79,345	70,931	64,130	58,520	53,812	49,805	41,989	36,293	31,953	28,548	25,796	23,528	20,009	17,405	15,402	12,519	10,545
200	173,989	153,965	138,074	125,156	114,449	105,430	91,075	80,160	71,582	64,662	58,962	54,186	50,125	42,216	36,463	32,090	28,653	25,882	23,599	20,060	17,444	15,432	12,539	10,560

750 MCM | **1 Run(s)** | **480 Volts**

C-Value = 24,136 | **Magnetic Duct** | **Copper**

Calculations:

1. $f = \dfrac{1.732 \times \text{Length} \times \text{Isca}}{\text{Runs} \times \text{C-Value} \times \text{E L-L}}$
2. $M = \dfrac{1}{1 + f}$
3. $\text{Isc} = \text{Isca} \times M$
4. Add Motor Contribution = Motor FLA x 4

* All results are given in symmetrical amperes

Three-Phase Bolted Fault-Current* Table

One-Way Distance in Feet

Available Isc in Thousands (Isca)

Isca	5	10	15	20	25	30	40	50	60	70	80	90	100	125	150	175	200	225	250	300	350	400	500	600
3	2,997	2,993	2,990	2,987	2,983	2,980	2,973	2,967	2,960	2,954	2,947	2,941	2,934	2,918	2,902	2,887	2,871	2,856	2,841	2,811	2,782	2,753	2,698	2,644
5	4,991	4,981	4,972	4,963	4,954	4,945	4,926	4,908	4,890	4,873	4,855	4,837	4,820	4,777	4,735	4,693	4,652	4,612	4,573	4,496	4,422	4,350	4,213	4,084
7	6,982	6,964	6,945	6,928	6,910	6,892	6,856	6,822	6,787	6,753	6,719	6,685	6,652	6,570	6,491	6,413	6,337	6,263	6,190	6,050	5,916	5,788	5,548	5,327
10	9,963	9,926	9,889	9,853	9,817	9,781	9,710	9,640	9,571	9,503	9,436	9,370	9,304	9,145	8,992	8,843	8,699	8,560	8,425	8,168	7,926	7,698	7,279	6,904
15	14,916	14,834	14,752	14,671	14,591	14,512	14,356	14,204	14,054	13,908	13,765	13,625	13,488	13,156	12,840	12,540	12,252	11,978	11,716	11,224	10,772	10,356	9,612	8,967
20	19,852	19,705	19,561	19,419	19,279	19,142	18,871	18,609	18,354	18,105	17,864	17,628	17,399	16,851	16,337	15,853	15,396	14,966	14,559	13,807	13,130	12,516	11,445	10,543
25	24,769	24,541	24,318	24,099	23,884	23,673	23,261	22,864	22,479	22,108	21,749	21,401	21,064	20,266	19,526	18,839	18,198	17,600	17,039	16,019	15,114	14,306	12,924	11,786
30	29,667	29,342	29,024	28,712	28,407	28,109	27,531	26,975	26,442	25,930	25,437	24,962	24,505	23,432	22,449	21,545	20,711	19,939	19,223	17,935	16,808	15,814	14,143	12,790
35	34,548	34,108	33,678	33,260	32,851	32,453	31,684	30,951	30,251	29,582	28,942	28,329	27,742	26,375	25,136	24,008	22,977	22,031	21,160	19,609	18,270	17,102	15,164	13,620
40	39,411	38,839	38,283	37,743	37,218	36,707	35,727	34,798	33,916	33,077	32,279	31,518	30,793	29,117	27,615	26,260	25,031	23,913	22,890	21,086	19,546	18,215	16,032	14,316
45	44,256	43,536	42,839	42,163	41,509	40,875	39,663	38,521	37,443	36,424	35,458	34,543	33,673	31,680	29,909	28,326	26,902	25,614	24,444	22,398	20,668	19,186	16,779	14,909
50	49,083	48,199	47,346	46,522	45,727	44,959	43,497	42,127	40,841	39,631	38,491	37,415	36,397	34,079	32,038	30,229	28,612	27,160	25,848	23,571	21,663	20,040	17,429	15,420
55	53,892	52,828	51,805	50,821	49,874	48,961	47,233	45,622	44,117	42,709	41,388	40,146	38,976	36,330	34,020	31,987	30,182	28,571	27,123	24,626	22,551	20,798	18,000	15,865
60	58,684	57,425	56,218	55,061	53,951	52,884	50,873	49,010	47,278	45,664	44,157	42,746	41,422	38,446	35,869	33,616	31,629	29,864	28,285	25,581	23,349	21,475	18,504	16,256
65	63,458	61,988	60,585	59,243	57,960	56,731	54,423	52,295	50,328	48,503	46,806	45,224	43,745	40,439	37,598	35,130	32,966	31,053	29,350	26,448	24,069	22,083	18,954	16,602
70	68,215	66,519	64,906	63,368	61,902	60,503	57,885	55,484	53,274	51,234	49,344	47,589	45,954	42,320	39,218	36,540	34,205	32,150	30,328	27,240	24,723	22,632	19,357	16,910
75	72,955	71,019	69,182	67,438	65,780	64,202	61,262	58,579	56,122	53,862	51,778	49,848	48,058	44,097	40,740	37,858	35,357	33,165	31,230	27,966	25,319	23,130	19,721	17,187
80	77,677	75,486	73,415	71,454	69,595	67,831	64,558	61,586	58,875	56,394	54,113	52,009	50,063	45,780	42,172	39,091	36,430	34,108	32,064	28,633	25,865	23,585	20,050	17,437
85	82,383	79,922	77,604	75,416	73,349	71,392	67,775	64,507	61,540	58,833	56,355	54,077	51,976	47,374	43,521	40,248	37,433	34,985	32,838	29,249	26,366	24,001	20,350	17,663
90	87,071	84,327	81,750	79,327	77,042	74,886	70,916	67,346	64,119	61,186	58,510	56,058	53,804	48,888	44,796	41,335	38,371	35,804	33,559	29,819	26,829	24,384	20,624	17,870
95	91,743	88,701	85,855	83,186	80,677	78,316	73,985	70,107	66,616	63,457	60,583	57,958	55,551	50,327	46,001	42,359	39,252	36,570	34,230	30,348	27,256	24,736	20,876	18,058
100	96,397	93,045	89,918	86,994	84,255	81,683	76,982	72,793	69,037	65,649	62,578	59,782	57,225	51,696	47,142	43,325	40,080	37,287	34,858	30,840	27,653	25,063	21,108	18,232
105	101,035	97,359	93,940	90,754	87,777	84,988	79,912	75,407	71,384	67,768	64,500	61,533	58,828	53,001	48,225	44,238	40,860	37,961	35,447	31,300	28,022	25,365	21,322	18,391
110	105,656	101,642	97,922	94,465	91,244	88,235	82,775	77,952	73,660	69,816	66,353	63,217	60,365	54,246	49,253	45,102	41,596	38,596	35,999	31,730	28,366	25,647	21,521	18,539
115	110,261	105,897	101,865	98,129	94,658	91,423	85,575	80,430	75,869	71,797	68,140	64,838	61,840	55,434	50,231	45,920	42,291	39,194	36,519	32,133	28,688	25,910	21,706	18,676
120	114,849	110,122	105,769	101,747	98,019	94,555	88,313	82,844	78,013	73,715	69,865	66,397	63,258	56,570	51,162	46,697	42,949	39,758	37,008	32,512	28,989	26,155	21,878	18,803
125	119,421	114,318	109,634	105,319	101,330	97,632	90,992	85,197	80,096	75,571	71,531	67,900	64,620	57,658	52,049	47,436	43,573	40,292	37,471	32,868	29,272	26,385	22,039	18,921
130	123,976	118,486	113,462	108,846	104,591	100,656	93,613	87,491	82,120	77,371	73,141	69,349	65,931	58,699	52,897	48,138	44,165	40,798	37,908	33,203	29,538	26,601	22,189	19,032
135	128,516	122,626	117,252	112,329	107,803	103,628	96,178	89,727	84,087	79,114	74,697	70,747	67,193	59,697	53,706	48,807	44,728	41,278	38,322	33,521	29,789	26,804	22,330	19,136
140	133,039	126,737	121,005	115,769	110,968	106,549	98,689	91,909	86,000	80,806	76,203	72,096	68,409	60,655	54,480	49,446	45,263	41,733	38,714	33,821	30,025	26,996	22,463	19,233
145	137,546	130,821	124,722	119,167	114,086	109,421	101,147	94,038	87,861	82,447	77,661	73,400	69,582	61,575	55,221	50,056	45,774	42,167	39,087	34,105	30,249	27,176	22,588	19,325
150	142,037	134,877	128,404	122,524	117,159	112,244	103,555	96,115	89,673	84,039	79,072	74,659	70,713	62,459	55,931	50,638	46,261	42,580	39,441	34,374	30,461	27,347	22,706	19,411
155	146,512	138,906	132,050	125,840	120,187	115,020	105,914	98,144	91,436	85,586	80,440	75,878	71,805	63,310	56,612	51,196	46,725	42,973	39,779	34,630	30,662	27,509	22,817	19,493
160	150,972	142,908	135,662	129,116	123,172	117,751	108,225	100,125	93,153	87,089	81,766	77,056	72,860	64,128	57,266	51,730	47,170	43,349	40,100	34,874	30,852	27,663	22,923	19,569
165	155,416	146,884	139,240	132,352	126,114	120,437	110,490	102,061	94,826	88,550	83,052	78,198	73,879	64,917	57,893	52,242	47,595	43,708	40,407	35,105	31,034	27,808	23,023	19,642
170	159,844	150,833	142,784	135,550	129,014	123,079	112,710	103,952	96,457	89,970	84,300	79,303	74,865	65,677	58,497	52,733	48,002	44,051	40,700	35,327	31,206	27,947	23,117	19,711
175	164,257	154,756	146,294	138,710	131,873	125,679	114,886	105,800	98,046	91,351	85,512	80,374	75,819	66,410	59,078	53,204	48,393	44,379	40,981	35,537	31,371	28,079	23,208	19,777
180	168,654	158,653	149,772	141,833	134,693	128,237	117,020	107,607	99,596	92,695	86,688	81,413	76,743	67,117	59,637	53,657	48,767	44,694	41,249	35,739	31,528	28,204	23,293	19,839
185	173,036	162,525	153,218	144,919	137,473	130,755	119,113	109,374	101,108	94,003	87,832	82,420	77,637	67,800	60,176	54,093	49,127	44,996	41,506	35,932	31,678	28,324	23,375	19,898
190	177,402	166,371	156,632	147,969	140,215	133,233	121,166	111,103	102,583	95,277	88,943	83,398	78,504	68,461	60,696	54,513	49,473	45,286	41,752	36,117	31,821	28,439	23,453	19,955
195	181,754	170,192	160,014	150,984	142,919	135,672	123,180	112,794	104,024	96,519	90,024	84,348	79,345	69,099	61,197	54,917	49,805	45,564	41,989	36,293	31,958	28,548	23,528	20,009
200	186,090	173,989	163,365	153,965	145,587	138,074	125,156	114,449	105,430	97,728	91,075	85,270	80,160	69,717	61,681	55,306	50,125	45,832	42,216	36,463	32,090	28,653	23,599	20,060

750 MCM | 2 Run(s) | 480 Volts | C-Value = 24,136 | Magnetic Duct | Copper

Calculations:

1. $f = \dfrac{1.732 \times \text{Length} \times \text{Isca}}{\text{Runs} \times \text{C-Value} \times E_{L-L}}$

2. $M = \dfrac{1}{1 + f}$

3. Isc = Isca x M

4. Add Motor Contribution = Motor FLA x 4

* All results are given in symmetrical amperes

Three-Phase Bolted Fault-Current* Table

One-Way Distance in Feet

Available Isc in Thousands (Isca)

Isca	5	10	15	20	25	30	40	50	60	70	80	90	100	125	150	175	200	225	250	300	350	400	500	600
3	2,994	2,987	2,981	2,975	2,968	2,962	2,949	2,937	2,925	2,913	2,901	2,889	2,877	2,848	2,819	2,791	2,763	2,736	2,710	2,658	2,609	2,561	2,471	2,387
5	4,982	4,965	4,947	4,930	4,912	4,895	4,861	4,828	4,795	4,762	4,730	4,698	4,667	4,590	4,516	4,445	4,375	4,308	4,243	4,118	4,001	3,890	3,685	3,501
7	6,965	6,931	6,897	6,863	6,829	6,796	6,731	6,667	6,604	6,542	6,482	6,422	6,364	6,223	6,088	5,958	5,834	5,715	5,601	5,386	5,186	5,001	4,668	4,376
10	9,929	9,859	9,790	9,722	9,655	9,589	9,460	9,334	9,211	9,092	8,975	8,862	8,751	8,486	8,236	8,001	7,779	7,569	7,370	7,002	6,668	6,365	5,835	5,387
15	14,841	14,686	14,533	14,384	14,238	14,095	13,817	13,549	13,292	13,045	12,806	12,576	12,355	11,833	11,353	10,911	10,502	10,123	9,770	9,133	8,574	8,080	7,244	6,565
20	19,719	19,445	19,179	18,920	18,668	18,422	17,950	17,502	17,075	16,669	16,281	15,912	15,558	14,740	14,003	13,337	12,731	12,178	11,670	10,773	10,004	9,337	8,239	7,372
25	24,562	24,139	23,730	23,335	22,952	22,582	21,877	21,215	20,591	20,003	19,448	18,922	18,425	17,288	16,283	15,389	14,588	13,866	13,212	12,074	11,116	10,299	8,979	7,959
30	29,371	28,768	28,189	27,633	27,099	26,585	25,613	24,709	23,867	23,081	22,345	21,654	21,005	19,540	18,266	17,149	16,160	15,279	14,489	13,131	12,006	11,058	9,551	8,405
35	34,147	33,335	32,560	31,820	31,114	30,438	29,170	28,004	26,928	25,931	25,006	24,144	23,339	21,545	20,007	18,673	17,507	16,477	15,562	14,007	12,734	11,673	10,006	8,755
40	38,890	37,839	36,844	35,900	35,003	34,150	32,563	31,116	29,793	28,578	27,458	26,422	25,462	23,341	21,546	20,008	18,674	17,508	16,478	14,744	13,340	12,181	10,376	9,038
45	43,600	42,284	41,045	39,877	38,773	37,729	35,801	34,060	32,481	31,042	29,725	28,515	27,400	24,959	22,918	21,185	19,696	18,403	17,269	15,374	13,854	12,607	10,684	9,270
50	48,277	46,669	45,165	43,754	42,429	41,182	38,896	36,850	35,008	33,342	31,827	30,444	29,176	26,425	24,148	22,232	20,598	19,187	17,958	15,918	14,294	12,970	10,944	9,465
55	52,923	50,996	49,205	47,536	45,976	44,515	41,856	39,496	37,388	35,494	33,782	32,228	30,811	27,758	25,257	23,168	21,399	19,881	18,564	16,392	14,675	13,284	11,166	9,631
60	57,536	55,267	53,169	51,225	49,419	47,735	44,690	42,010	39,633	37,511	35,604	33,882	32,319	28,977	26,261	24,011	22,116	20,498	19,101	16,809	15,009	13,557	11,359	9,774
65	62,118	59,481	57,059	54,826	52,761	50,847	47,406	44,401	41,755	39,406	37,307	35,421	33,716	30,095	27,176	24,774	22,762	21,052	19,581	17,180	15,303	13,796	11,526	9,898
70	66,669	63,641	60,876	58,341	56,009	53,856	50,011	46,679	43,763	41,190	38,902	36,856	35,014	31,125	28,013	25,467	23,345	21,550	20,011	17,510	15,565	14,009	11,674	10,007
75	71,189	67,747	64,622	61,773	59,165	56,767	52,512	48,850	45,666	42,871	40,399	38,196	36,221	32,075	28,781	26,100	23,876	22,002	20,400	17,807	15,799	14,198	11,805	10,103
80	75,679	71,801	68,300	65,126	62,233	59,586	54,915	50,923	47,473	44,460	41,807	39,452	37,349	32,956	29,488	26,681	24,361	22,413	20,753	18,075	16,010	14,368	11,923	10,189
85	80,138	75,803	71,912	68,401	65,217	62,317	57,226	52,904	49,190	45,962	43,132	40,631	38,403	33,775	30,142	27,214	24,805	22,788	21,074	18,319	16,201	14,522	12,028	10,266
90	84,568	79,754	75,459	71,602	68,121	64,963	59,450	54,799	50,824	47,386	44,384	41,739	39,392	34,537	30,747	27,707	25,214	23,133	21,369	18,541	16,374	14,661	12,124	10,335
95	88,968	83,656	78,942	74,732	70,947	67,528	61,591	56,614	52,380	48,736	45,566	42,784	40,321	35,249	31,310	28,164	25,591	23,450	21,639	18,744	16,532	14,787	12,210	10,398
100	93,338	87,509	82,364	77,791	73,699	70,016	63,654	58,352	53,865	50,019	46,686	43,769	41,195	35,915	31,835	28,587	25,941	23,743	21,888	18,931	16,677	14,903	12,289	10,455
105	97,680	91,314	85,727	80,784	76,380	72,431	65,644	60,020	55,284	51,240	47,748	44,701	42,020	36,540	32,325	28,982	26,265	24,014	22,119	19,103	16,811	15,010	12,361	10,507
110	101,993	95,072	89,031	83,711	78,992	74,776	67,564	61,621	56,639	52,402	48,755	45,583	42,798	37,128	32,784	29,350	26,567	24,267	22,333	19,262	16,934	15,108	12,428	10,555
115	106,277	98,784	92,278	86,576	81,537	77,053	69,418	63,159	57,936	53,511	49,713	46,419	43,535	37,681	33,214	29,695	26,849	24,502	22,532	19,410	17,048	15,199	12,489	10,600
120	110,533	102,451	95,470	89,379	84,020	79,266	71,209	64,639	59,178	54,569	50,625	47,213	44,232	38,202	33,619	30,018	27,113	24,721	22,717	19,548	17,154	15,283	12,546	10,640
125	114,761	106,073	98,608	92,124	86,441	81,417	72,940	66,062	60,369	55,580	51,494	47,968	44,894	38,695	34,000	30,321	27,360	24,927	22,890	19,676	17,253	15,361	12,599	10,678
130	118,962	109,652	101,693	94,812	88,802	83,509	74,615	67,433	61,512	56,547	52,323	48,687	45,523	39,161	34,359	30,606	27,593	25,119	23,053	19,796	17,345	15,434	12,648	10,714
135	123,135	113,188	104,727	97,444	91,107	85,545	76,236	68,754	62,609	57,472	53,115	49,372	46,121	39,603	34,699	30,876	27,811	25,300	23,205	19,908	17,431	15,502	12,694	10,746
140	127,282	116,682	107,712	100,022	93,358	87,526	77,805	70,027	63,663	58,360	53,872	50,025	46,691	40,022	35,020	31,130	28,017	25,471	23,348	20,013	17,512	15,566	12,736	10,777
145	131,401	120,134	110,647	102,549	95,555	89,454	79,325	71,256	64,678	59,211	54,596	50,649	47,234	40,421	35,325	31,370	28,212	25,632	23,484	20,112	17,588	15,626	12,776	10,806
150	135,494	123,546	113,535	105,025	97,701	91,332	80,798	72,443	65,654	60,028	55,290	51,246	47,753	40,800	35,614	31,598	28,396	25,783	23,611	20,206	17,659	15,683	12,814	10,833
155	139,561	126,918	116,376	107,451	99,798	93,162	82,227	73,589	66,594	60,813	55,956	51,817	48,248	41,161	35,889	31,815	28,571	25,927	23,731	20,294	17,727	15,736	12,849	10,858
160	143,601	130,251	119,173	109,831	101,847	94,945	83,613	74,698	67,500	61,568	56,594	52,364	48,722	41,505	36,151	32,020	28,736	26,063	23,846	20,377	17,790	15,786	12,883	10,882
165	147,616	133,546	121,925	112,164	103,850	96,684	84,959	75,770	68,374	62,294	57,207	52,888	49,176	41,834	36,400	32,215	28,894	26,193	23,954	20,456	17,850	15,833	12,914	10,904
170	151,605	136,802	124,633	114,452	105,809	98,379	86,265	76,807	69,218	62,994	57,797	53,392	49,611	42,149	36,638	32,401	29,043	26,316	24,056	20,531	17,907	15,878	12,944	10,925
175	155,569	140,022	127,300	116,697	107,725	100,033	87,534	77,811	70,033	63,668	58,364	53,875	50,028	42,449	36,865	32,579	29,186	26,433	24,154	20,602	17,961	15,920	12,972	10,945
180	159,508	143,205	129,925	118,899	109,599	101,647	88,767	78,784	70,820	64,318	58,909	54,340	50,428	42,737	37,082	32,748	29,321	26,544	24,247	20,670	18,013	15,961	12,999	10,965
185	163,422	146,351	132,510	121,061	111,432	103,223	89,967	79,728	71,581	64,945	59,435	54,787	50,813	43,013	37,289	32,910	29,451	26,650	24,336	20,734	18,061	15,999	13,024	10,983
190	167,311	149,463	135,056	123,182	113,227	104,761	91,133	80,642	72,317	65,551	59,942	55,217	51,183	43,278	37,488	33,065	29,575	26,752	24,420	20,796	18,108	16,036	13,049	11,000
195	171,176	152,540	137,563	125,264	114,984	106,263	92,268	81,529	73,030	66,136	60,431	55,632	51,539	43,532	37,679	33,213	29,693	26,848	24,501	20,854	18,152	16,070	13,072	11,016
200	175,017	155,582	140,033	127,309	116,704	107,731	93,372	82,391	73,720	66,701	60,903	56,031	51,882	43,777	37,862	33,355	29,807	26,941	24,578	20,910	18,195	16,103	13,094	11,032

1000 MCM | 1 Run(s) | 480 Volts | C-Value = 25,278 | Magnetic Duct | Copper

Calculations:

1. $f = \dfrac{1.732 \times \text{Length} \times \text{Isca}}{\text{Runs} \times \text{C-Value} \times \text{E L-L}}$

2. $M = \dfrac{1}{1 + f}$

3. Isc = Isca x M

4. Add Motor Contribution = Motor FLA x 4

* All results are given in symmetrical amperes

Three-Phase Bolted Fault-Current* Table

One-Way Distance in Feet

Available Isc in Thousands (Isca)	5	10	15	20	25	30	40	50	60	70	80	90	100	125	150	175	200	225	250	300	350	400	500	600
3	2,997	2,994	2,990	2,987	2,984	2,981	2,975	2,968	2,962	2,956	2,949	2,943	2,937	2,922	2,907	2,892	2,877	2,862	2,848	2,819	2,791	2,763	2,710	2,658
5	4,991	4,982	4,973	4,965	4,956	4,947	4,930	4,912	4,895	4,878	4,861	4,844	4,828	4,786	4,746	4,706	4,667	4,628	4,590	4,516	4,445	4,375	4,243	4,118
7	6,983	6,965	6,948	6,931	6,914	6,897	6,863	6,829	6,796	6,763	6,731	6,699	6,667	6,589	6,512	6,437	6,364	6,293	6,223	6,088	5,958	5,834	5,601	5,386
10	9,964	9,929	9,894	9,859	9,825	9,790	9,722	9,655	9,589	9,524	9,460	9,396	9,334	9,181	9,033	8,890	8,751	8,616	8,486	8,236	8,001	7,779	7,370	7,002
15	14,920	14,841	14,763	14,686	14,609	14,533	14,384	14,238	14,095	13,954	13,817	13,682	13,549	13,230	12,924	12,633	12,355	12,088	11,833	11,353	10,911	10,502	9,770	9,133
20	19,858	19,719	19,581	19,445	19,311	19,179	18,920	18,668	18,422	18,183	17,950	17,723	17,502	16,972	16,473	16,002	15,558	15,138	14,740	14,003	13,337	12,731	11,670	10,773
25	24,779	24,562	24,348	24,139	23,932	23,730	23,335	22,952	22,582	22,224	21,877	21,541	21,215	20,441	19,722	19,051	18,425	17,838	17,288	16,283	15,389	14,588	13,212	12,074
30	29,682	29,371	29,066	28,768	28,476	28,189	27,633	27,099	26,585	26,090	25,613	25,153	24,709	23,666	22,707	21,823	21,005	20,246	19,540	18,266	17,149	16,160	14,489	13,131
35	34,568	34,147	33,736	33,335	32,943	32,560	31,820	31,114	30,438	29,791	29,170	28,576	28,004	26,672	25,460	24,354	23,339	22,406	21,545	20,007	18,673	17,507	15,562	14,007
40	39,437	38,890	38,357	37,839	37,335	36,844	35,900	35,003	34,150	33,338	32,563	31,823	31,116	29,480	28,007	26,674	25,462	24,355	23,341	21,546	20,008	18,674	16,478	14,744
45	44,289	43,600	42,932	42,284	41,655	41,045	39,877	38,773	37,729	36,740	35,801	34,909	34,060	32,109	30,369	28,808	27,400	26,123	24,959	22,918	21,185	19,696	17,269	15,374
50	49,123	48,277	47,460	46,669	45,905	45,165	43,754	42,429	41,182	40,006	38,896	37,845	36,850	34,576	32,567	30,778	29,176	27,732	26,425	24,148	22,232	20,598	17,958	15,918
55	53,941	52,923	51,942	50,996	50,085	49,205	47,536	45,976	44,515	43,144	41,856	40,642	39,496	36,896	34,617	32,603	30,811	29,205	27,758	25,257	23,168	21,399	18,564	16,392
60	58,742	57,536	56,378	55,267	54,198	53,169	51,225	49,419	47,735	46,162	44,690	43,308	42,010	39,080	36,533	34,297	32,319	30,557	28,977	26,261	24,011	22,116	19,101	16,809
65	63,526	62,118	60,771	59,481	58,245	57,059	54,826	52,761	50,847	49,066	47,406	45,854	44,401	41,142	38,328	35,875	33,716	31,803	30,095	27,176	24,774	22,762	19,581	17,180
70	68,294	66,669	65,120	63,641	62,228	60,876	58,341	56,009	53,856	51,862	50,011	48,288	46,679	43,090	40,013	37,347	35,014	32,955	31,125	28,013	25,467	23,345	20,011	17,510
75	73,045	71,189	69,426	67,747	66,148	64,622	61,773	59,165	56,767	54,557	52,512	50,615	48,850	44,934	41,599	38,724	36,221	34,023	32,075	28,781	26,100	23,876	20,400	17,807
80	77,779	75,679	73,689	71,801	70,007	68,300	65,126	62,233	59,586	57,156	54,915	52,844	50,923	46,682	43,092	40,016	37,349	35,015	32,956	29,488	26,681	24,361	20,753	18,075
85	82,498	80,138	77,910	75,803	73,806	71,912	68,401	65,217	62,317	59,663	57,226	54,980	52,904	48,341	44,502	41,229	38,403	35,941	33,775	30,142	27,214	24,805	21,074	18,319
90	87,199	84,568	82,090	79,754	77,547	75,459	71,602	68,121	64,963	62,084	59,450	57,030	54,799	49,918	45,836	42,370	39,392	36,805	34,537	30,747	27,707	25,214	21,369	18,541
95	91,885	88,968	86,230	83,656	81,231	78,942	74,732	70,947	67,528	64,423	61,591	58,997	56,614	51,419	47,098	43,447	40,321	37,615	35,249	31,310	28,164	25,591	21,639	18,744
100	96,554	93,338	90,329	87,509	84,858	82,364	77,791	73,699	70,016	66,684	63,654	60,888	58,352	52,850	48,295	44,464	41,195	38,375	35,915	31,835	28,587	25,941	21,888	18,931
105	101,208	97,680	94,389	91,314	88,432	85,727	80,784	76,380	72,431	68,871	65,644	62,706	60,020	54,214	49,432	45,425	42,020	39,089	36,540	32,325	28,982	26,265	22,119	19,103
110	105,845	101,993	98,411	95,072	91,952	89,031	83,711	78,992	74,776	70,987	67,564	64,456	61,621	55,517	50,513	46,337	42,798	39,762	37,128	32,784	29,350	26,567	22,333	19,262
115	110,467	106,277	102,393	98,784	95,420	92,278	86,576	81,537	77,053	73,037	69,418	66,141	63,159	56,762	51,542	47,201	43,535	40,397	37,681	33,214	29,695	26,849	22,532	19,410
120	115,072	110,533	106,339	102,451	98,837	95,470	89,379	84,020	79,266	75,022	71,209	67,765	64,639	57,954	52,523	48,022	44,232	40,997	38,202	33,619	30,018	27,113	22,717	19,548
125	119,662	114,761	110,246	106,073	102,204	98,608	92,124	86,441	81,417	76,946	72,940	69,331	66,062	59,096	53,459	48,804	44,894	41,565	38,695	34,000	30,321	27,360	22,890	19,676
130	124,236	118,962	114,117	109,652	105,523	101,693	94,812	88,802	83,509	78,812	74,615	70,842	67,433	60,190	54,353	49,548	45,523	42,103	39,161	34,359	30,606	27,593	23,053	19,796
135	128,795	123,135	117,952	113,188	108,793	104,727	97,444	91,107	85,545	80,622	76,236	72,301	68,754	61,241	55,208	50,257	46,121	42,614	39,603	34,699	30,876	27,811	23,205	19,908
140	133,338	127,282	121,751	116,682	112,017	107,712	100,022	93,358	87,526	82,379	77,805	73,711	70,027	62,249	56,026	50,934	46,691	43,100	40,022	35,020	31,130	28,017	23,348	20,013
145	137,866	131,401	125,515	120,134	115,196	110,647	102,549	95,555	89,454	84,085	79,325	75,074	71,256	63,218	56,810	51,581	47,234	43,563	40,421	35,325	31,370	28,212	23,484	20,112
150	142,379	135,494	129,245	123,546	118,329	113,535	105,025	97,701	91,332	85,743	80,798	76,393	72,443	64,151	57,562	52,200	47,753	44,003	40,800	35,614	31,598	28,396	23,611	20,206
155	146,876	139,561	132,940	126,918	121,419	116,376	107,451	99,798	93,162	87,354	82,227	77,669	73,589	65,048	58,283	52,793	48,248	44,424	41,161	35,889	31,815	28,571	23,731	20,294
160	151,358	143,601	136,601	130,251	124,466	119,173	109,831	101,847	94,945	88,920	83,613	78,904	74,698	65,913	58,976	53,361	48,722	44,825	41,505	36,151	32,020	28,736	23,846	20,377
165	155,825	147,616	140,229	133,546	127,471	121,925	112,164	103,850	96,684	90,443	84,959	80,101	75,770	66,746	59,643	53,906	49,176	45,209	41,834	36,400	32,215	28,894	23,954	20,456
170	160,276	151,605	143,824	136,802	130,435	124,633	114,452	105,809	98,379	91,925	86,265	81,262	76,807	67,549	60,283	54,429	49,611	45,576	42,149	36,638	32,401	29,043	24,056	20,531
175	164,713	155,569	147,387	140,022	133,358	127,300	116,697	107,725	100,033	93,367	87,534	82,387	77,811	68,325	60,900	54,931	50,028	45,928	42,449	36,865	32,579	29,186	24,154	20,602
180	169,135	159,508	150,917	143,205	136,242	129,925	118,899	109,599	101,647	94,772	88,767	83,479	78,784	69,074	61,495	55,414	50,428	46,265	42,737	37,082	32,748	29,321	24,247	20,670
185	173,543	163,422	154,416	146,351	139,087	132,510	121,061	111,432	103,223	96,140	89,967	84,538	79,728	69,798	62,068	55,879	50,813	46,589	43,013	37,289	32,910	29,451	24,336	20,734
190	177,935	167,311	157,884	149,463	141,895	135,056	123,182	113,227	104,761	97,473	91,133	85,567	80,642	70,498	62,621	56,327	51,183	46,900	43,278	37,488	33,065	29,575	24,420	20,796
195	182,313	171,176	161,322	152,540	144,665	137,563	125,264	114,984	106,263	98,772	92,268	86,567	81,529	71,175	63,155	56,759	51,539	47,198	43,532	37,679	33,213	29,693	24,501	20,854
200	186,676	175,017	164,728	155,582	147,399	140,033	127,309	116,704	107,731	100,039	93,372	87,538	82,391	71,831	63,670	57,175	51,882	47,486	43,777	37,862	33,355	29,807	24,578	20,910

1000 MCM	2 Run(s)	480 Volts	C-Value = 25,278	Magnetic Duct	Copper

Calculations:

1. $f = \dfrac{1.732 \times \text{Length} \times \text{Isca}}{\text{Runs} \times \text{C-Value} \times \text{E L-L}}$

2. $M = \dfrac{1}{1 + f}$

3. $\text{Isc} = \text{Isca} \times M$

4. Add Motor Contribution = Motor FLA x 4

* All results are given in symmetrical amperes

Short-Circuit Tables and Data: 240 Volts Three-Phase

Three-Phase Bolted Fault-Current* Table

One-Way Distance in Feet

Available Isc in Thousands (Isca)	5	10	15	20	25	30	40	50	60	70	80	90	100	125	150	175	200	225	250	300	350	400	500	600
1	915	844	782	729	683	642	574	519	473	435	403	375	350	301	264	235	212	193	177	152	133	119	97	82
3	2,347	1,927	1,635	1,420	1,255	1,124	930	793	691	613	550	499	457	377	321	279	247	222	201	170	146	129	104	87
5	3,416	2,594	2,091	1,751	1,506	1,322	1,061	887	762	667	594	535	487	397	335	290	256	229	207	173	149	131	106	88
10	5,188	3,502	2,644	2,123	1,774	1,523	1,188	973	824	715	631	565	511	413	347	299	262	234	211	177	152	133	107	89
15	6,273	3,965	2,899	2,285	1,885	1,605	1,236	1,006	848	732	645	576	520	419	351	302	265	236	213	178	152	134	107	89
20	7,005	4,246	3,046	2,375	1,946	1,649	1,263	1,023	860	741	652	582	525	422	353	303	266	237	213	178	153	134	107	89
25	7,532	4,434	3,142	2,433	1,985	1,676	1,279	1,033	867	747	656	585	528	424	354	304	267	237	214	178	153	134	107	90
30	7,931	4,569	3,209	2,473	2,012	1,695	1,290	1,041	872	751	659	587	530	425	355	305	267	238	214	179	153	134	107	90
35	8,242	4,671	3,259	2,502	2,031	1,709	1,298	1,046	876	753	661	589	531	426	356	305	267	238	214	179	153	134	107	90
40	8,492	4,750	3,297	2,525	2,046	1,720	1,304	1,050	879	755	663	590	532	427	356	306	268	238	214	179	153	134	108	90
45	8,697	4,814	3,328	2,543	2,058	1,728	1,308	1,053	881	757	664	591	533	427	357	306	268	238	215	179	153	134	108	90
50	8,868	4,866	3,353	2,557	2,067	1,734	1,312	1,055	883	758	665	592	533	428	357	306	268	238	215	179	154	134	108	90
55	9,014	4,909	3,373	2,569	2,075	1,740	1,315	1,057	884	759	666	592	534	428	357	306	268	239	215	179	154	134	108	90
60	9,139	4,946	3,390	2,579	2,081	1,745	1,318	1,059	885	760	666	593	534	428	357	306	268	239	215	179	154	134	108	90
65	9,247	4,978	3,405	2,588	2,087	1,748	1,320	1,060	886	761	667	593	535	428	357	307	268	239	215	179	154	134	108	90
70	9,342	5,005	3,418	2,595	2,092	1,752	1,322	1,062	887	762	667	594	535	429	358	307	268	239	215	179	154	134	108	90
75	9,426	5,029	3,429	2,602	2,096	1,755	1,324	1,063	888	762	668	594	535	429	358	307	269	239	215	179	154	135	108	90
80	9,500	5,050	3,439	2,607	2,100	1,757	1,325	1,064	888	763	668	594	535	429	358	307	269	239	215	179	154	135	108	90
85	9,567	5,069	3,448	2,612	2,103	1,760	1,327	1,065	889	763	668	595	536	429	358	307	269	239	215	179	154	135	108	90
90	9,627	5,086	3,456	2,617	2,106	1,762	1,328	1,065	890	764	669	595	536	429	358	307	269	239	215	179	154	135	108	90
95	9,682	5,101	3,463	2,621	2,108	1,763	1,329	1,066	890	764	669	595	536	429	358	307	269	239	215	179	154	135	108	90
100	9,731	5,115	3,469	2,624	2,111	1,765	1,330	1,067	890	764	669	595	536	429	358	307	269	239	215	179	154	135	108	90
105	9,777	5,127	3,475	2,628	2,113	1,767	1,330	1,067	891	764	669	596	536	429	358	307	269	239	215	179	154	135	108	90
110	9,818	5,138	3,480	2,631	2,115	1,768	1,331	1,068	891	765	670	596	536	430	358	307	269	239	215	179	154	135	108	90
115	9,857	5,149	3,485	2,633	2,116	1,769	1,332	1,068	891	765	670	596	537	430	358	307	269	239	215	179	154	135	108	90
120	9,892	5,159	3,489	2,636	2,118	1,770	1,333	1,068	892	765	670	596	537	430	358	307	269	239	215	179	154	135	108	90
125	9,925	5,167	3,493	2,638	2,120	1,771	1,333	1,069	892	765	670	596	537	430	358	307	269	239	215	179	154	135	108	90
130	9,955	5,176	3,497	2,640	2,121	1,772	1,334	1,069	892	766	670	596	537	430	358	307	269	239	215	179	154	135	108	90
135	9,983	5,183	3,500	2,642	2,122	1,773	1,334	1,070	892	766	670	596	537	430	358	307	269	239	215	179	154	135	108	90
140	10,010	5,190	3,504	2,644	2,123	1,774	1,335	1,070	893	766	671	596	537	430	358	307	269	239	215	179	154	135	108	90
145	10,035	5,197	3,507	2,646	2,125	1,775	1,335	1,070	893	766	671	596	537	430	358	307	269	239	215	179	154	135	108	90
150	10,058	5,203	3,509	2,648	2,126	1,775	1,336	1,070	893	766	671	597	537	430	358	307	269	239	215	179	154	135	108	90
155	10,080	5,209	3,512	2,649	2,127	1,776	1,336	1,071	893	766	671	597	537	430	359	307	269	239	215	179	154	135	108	90
160	10,100	5,215	3,515	2,651	2,127	1,777	1,336	1,071	893	766	671	597	537	430	359	307	269	239	215	179	154	135	108	90
165	10,119	5,220	3,517	2,652	2,128	1,777	1,337	1,071	894	766	671	597	537	430	359	307	269	239	215	179	154	135	108	90
170	10,138	5,225	3,519	2,653	2,129	1,778	1,337	1,071	894	767	671	597	537	430	359	307	269	239	215	179	154	135	108	90
175	10,155	5,229	3,521	2,654	2,130	1,779	1,337	1,071	894	767	671	597	537	430	359	307	269	239	215	179	154	135	108	90
180	10,171	5,234	3,523	2,655	2,131	1,779	1,338	1,072	894	767	671	597	537	430	359	307	269	239	215	179	154	135	108	90
185	10,187	5,238	3,525	2,656	2,131	1,779	1,338	1,072	894	767	671	597	537	430	359	308	269	239	215	180	154	135	108	90
190	10,202	5,242	3,527	2,657	2,132	1,780	1,338	1,072	894	767	671	597	538	430	359	308	269	239	215	180	154	135	108	90
195	10,216	5,245	3,529	2,658	2,133	1,780	1,338	1,072	894	767	671	597	538	430	359	308	269	239	215	180	154	135	108	90
200	10,229	5,249	3,530	2,659	2,133	1,781	1,339	1,072	894	767	672	597	538	430	359	308	269	239	215	180	154	135	108	90

# 14 AWG	1 Run(s)	240 Volts	C-Value = 389	Magnetic Duct	Copper

Calculations:

1. $f = \dfrac{1.732 \times \text{Length} \times \text{Isca}}{\text{Runs} \times \text{C-Value} \times E_{L\text{-}L}}$
2. $M = \dfrac{1}{1 + f}$
3. $\text{Isc} = \text{Isca} \times M$
4. Add Motor Contribution = Motor FLA x 4

* All results are given in symmetrical amperes

Three-Phase Bolted Fault-Current* Table

Available Isc in Thousands (Isca)	5	10	15	20	25	30	40	50	60	70	80	90	100	125	150	175	200	225	250	300	350	400	500	600
1	945	895	851	810	774	740	681	631	588	550	517	487	461	406	363	328	299	275	255	222	196	176	146	125
3	2,552	2,221	1,965	1,763	1,598	1,462	1,248	1,089	966	868	788	721	665	557	479	420	374	337	307	260	226	200	162	136
5	3,869	3,155	2,664	2,305	2,031	1,815	1,497	1,274	1,109	982	881	798	730	602	512	445	394	353	320	270	233	205	165	139
10	6,310	4,609	3,630	2,995	2,548	2,218	1,761	1,460	1,247	1,088	966	868	788	640	539	466	410	366	331	277	238	209	168	140
15	7,991	5,446	4,130	3,327	2,785	2,395	1,871	1,535	1,301	1,129	998	893	809	654	549	473	416	371	334	280	240	211	169	141
20	9,218	5,989	4,436	3,522	2,920	2,494	1,931	1,575	1,330	1,151	1,014	907	820	661	554	477	419	373	336	281	241	211	170	141
25	10,154	6,371	4,642	3,651	3,008	2,558	1,969	1,600	1,348	1,164	1,025	915	827	666	557	479	420	374	337	282	242	212	170	142
30	10,891	6,653	4,790	3,742	3,070	2,603	1,995	1,618	1,360	1,174	1,032	921	831	669	559	481	421	375	338	282	242	212	170	142
35	11,487	6,871	4,902	3,810	3,115	2,635	2,014	1,630	1,369	1,180	1,037	925	835	671	561	482	422	376	339	283	243	212	170	142
40	11,979	7,044	4,989	3,862	3,151	2,660	2,029	1,640	1,376	1,185	1,041	928	837	672	562	483	423	376	339	283	243	213	170	142
45	12,391	7,185	5,059	3,904	3,178	2,680	2,040	1,647	1,381	1,189	1,044	930	839	674	563	483	423	377	339	283	243	213	170	142
50	12,742	7,301	5,117	3,938	3,201	2,696	2,050	1,653	1,385	1,192	1,046	932	841	675	564	484	424	377	340	283	243	213	170	142
55	13,044	7,399	5,165	3,967	3,220	2,709	2,057	1,658	1,389	1,195	1,048	934	842	676	564	484	424	377	340	284	243	213	170	142
60	13,307	7,483	5,205	3,991	3,235	2,721	2,064	1,663	1,392	1,197	1,050	935	843	676	565	485	424	378	340	284	243	213	171	142
65	13,538	7,556	5,240	4,011	3,249	2,730	2,069	1,666	1,394	1,199	1,051	936	844	677	565	485	425	378	340	284	243	213	171	142
70	13,742	7,619	5,271	4,029	3,261	2,738	2,074	1,669	1,397	1,200	1,053	937	845	677	565	485	425	378	340	284	243	213	171	142
75	13,925	7,675	5,297	4,044	3,271	2,746	2,078	1,672	1,398	1,202	1,054	938	845	678	566	485	425	378	340	284	243	213	171	142
80	14,088	7,724	5,321	4,058	3,280	2,752	2,082	1,674	1,400	1,203	1,055	939	846	678	566	486	425	378	341	284	244	213	171	142
85	14,236	7,768	5,342	4,070	3,288	2,757	2,085	1,676	1,401	1,204	1,055	939	846	679	566	486	425	378	341	284	244	213	171	142
90	14,369	7,808	5,360	4,081	3,295	2,762	2,088	1,678	1,403	1,205	1,056	940	847	679	566	486	425	378	341	284	244	213	171	142
95	14,491	7,844	5,377	4,091	3,301	2,767	2,090	1,680	1,404	1,206	1,057	941	847	679	567	486	426	378	341	284	244	213	171	142
100	14,602	7,876	5,392	4,100	3,307	2,771	2,093	1,681	1,405	1,207	1,057	941	848	679	567	486	426	379	341	284	244	213	171	142
105	14,705	7,906	5,406	4,108	3,312	2,775	2,095	1,683	1,406	1,207	1,058	941	848	680	567	486	426	379	341	284	244	213	171	142
110	14,799	7,933	5,419	4,115	3,317	2,778	2,097	1,684	1,407	1,208	1,058	942	848	680	567	486	426	379	341	284	244	213	171	142
115	14,886	7,958	5,431	4,122	3,321	2,781	2,098	1,685	1,408	1,209	1,059	942	849	680	567	486	426	379	341	284	244	213	171	142
120	14,967	7,981	5,441	4,128	3,325	2,784	2,100	1,686	1,408	1,209	1,059	943	849	680	567	487	426	379	341	284	244	213	171	142
125	15,042	8,002	5,451	4,133	3,329	2,786	2,101	1,687	1,409	1,210	1,060	943	849	680	567	487	426	379	341	284	244	213	171	142
130	15,112	8,022	5,460	4,139	3,332	2,789	2,103	1,688	1,409	1,210	1,060	943	849	680	567	487	426	379	341	284	244	213	171	142
135	15,177	8,040	5,469	4,144	3,335	2,791	2,104	1,689	1,410	1,210	1,060	943	850	681	568	487	426	379	341	284	244	213	171	142
140	15,238	8,058	5,477	4,148	3,338	2,793	2,105	1,689	1,411	1,211	1,061	944	850	681	568	487	426	379	341	284	244	213	171	142
145	15,296	8,074	5,484	4,152	3,341	2,795	2,106	1,690	1,411	1,211	1,061	944	850	681	568	487	426	379	341	284	244	213	171	142
150	15,350	8,089	5,491	4,156	3,344	2,797	2,107	1,691	1,412	1,212	1,061	944	850	681	568	487	426	379	341	284	244	213	171	142
155	15,400	8,103	5,498	4,160	3,346	2,798	2,108	1,691	1,412	1,212	1,061	944	850	681	568	487	426	379	341	284	244	213	171	142
160	15,448	8,116	5,504	4,164	3,348	2,800	2,109	1,692	1,412	1,212	1,062	944	850	681	568	487	426	379	341	284	244	213	171	142
165	15,494	8,128	5,509	4,167	3,350	2,801	2,110	1,692	1,413	1,212	1,062	945	851	681	568	487	426	379	341	284	244	213	171	142
170	15,537	8,140	5,515	4,170	3,352	2,803	2,111	1,693	1,413	1,213	1,062	945	851	681	568	487	426	379	341	285	244	213	171	142
175	15,577	8,151	5,520	4,173	3,354	2,804	2,112	1,693	1,413	1,213	1,062	945	851	681	568	487	426	379	341	285	244	213	171	142
180	15,616	8,162	5,525	4,176	3,356	2,805	2,112	1,694	1,414	1,213	1,062	945	851	681	568	487	426	379	341	285	244	213	171	142
185	15,653	8,172	5,529	4,178	3,358	2,807	2,113	1,694	1,414	1,213	1,063	945	851	681	568	487	426	379	341	285	244	213	171	142
190	15,687	8,182	5,534	4,181	3,359	2,808	2,114	1,695	1,414	1,214	1,063	945	851	682	568	487	427	379	341	285	244	214	171	142
195	15,721	8,191	5,538	4,183	3,361	2,809	2,114	1,695	1,415	1,214	1,063	945	851	682	568	487	427	379	341	285	244	214	171	142
200	15,753	8,199	5,542	4,185	3,362	2,810	2,115	1,695	1,415	1,214	1,063	945	851	682	568	487	427	379	341	285	244	214	171	142

One-Way Distance in Feet

\# 12 AWG | 1 Run(s) | 240 Volts | C-Value = 617 | Magnetic Duct | Copper

Calculations:

1. $f = \dfrac{1.732 \times \text{Length} \times \text{Isca}}{\text{Runs} \times \text{C-Value} \times E_{L\text{-}L}}$

2. $M = \dfrac{1}{1 + f}$

3. Isc = Isca x M

4. Add Motor Contribution = Motor FLA x 4

* All results are given in symmetrical amperes

Three-Phase Bolted Fault-Current* Table

One-Way Distance in Feet

Available Isc in Thousands (Isca)	5	10	15	20	25	30	40	50	60	70	80	90	100	125	150	175	200	225	250	300	350	400	500	600
1	965	931	901	872	845	819	773	731	694	660	630	602	576	521	475	437	405	377	352	312	280	254	214	185
3	2,702	2,458	2,254	2,081	1,933	1,805	1,593	1,426	1,291	1,179	1,085	1,005	935	798	696	617	554	503	460	394	344	305	249	211
5	4,223	3,655	3,222	2,881	2,605	2,377	2,023	1,761	1,559	1,399	1,268	1,160	1,069	893	767	672	598	539	490	415	360	318	258	217
10	7,311	5,762	4,754	4,046	3,522	3,118	2,536	2,138	1,847	1,626	1,452	1,312	1,197	981	831	721	636	570	516	433	374	329	265	222
15	9,667	7,131	5,649	4,677	3,991	3,480	2,771	2,302	1,968	1,719	1,526	1,372	1,246	1,014	855	739	650	581	525	440	379	332	267	223
20	11,523	8,093	6,236	5,073	4,275	3,694	2,905	2,393	2,035	1,770	1,566	1,404	1,273	1,031	867	748	657	586	529	443	381	334	268	224
25	13,024	8,806	6,651	5,344	4,466	3,836	2,992	2,452	2,077	1,802	1,591	1,424	1,289	1,042	875	753	662	590	532	445	382	335	269	225
30	14,262	9,355	6,960	5,541	4,603	3,937	3,053	2,493	2,107	1,824	1,608	1,438	1,300	1,049	880	757	665	592	534	446	383	336	269	225
35	15,301	9,791	7,198	5,692	4,706	4,012	3,098	2,523	2,128	1,840	1,621	1,448	1,309	1,055	883	760	667	594	535	447	384	337	270	225
40	16,186	10,146	7,388	5,810	4,787	4,070	3,132	2,546	2,144	1,852	1,630	1,455	1,315	1,059	886	762	668	595	536	448	385	337	270	225
45	16,948	10,440	7,543	5,905	4,851	4,117	3,160	2,564	2,157	1,862	1,637	1,461	1,319	1,062	888	764	670	596	537	449	385	337	270	225
50	17,611	10,688	7,672	5,983	4,904	4,155	3,182	2,579	2,167	1,869	1,643	1,466	1,323	1,064	890	765	671	597	538	449	385	338	270	226
55	18,194	10,900	7,780	6,049	4,948	4,186	3,201	2,591	2,176	1,876	1,648	1,470	1,327	1,066	892	766	671	598	538	449	386	338	271	226
60	18,709	11,083	7,873	6,105	4,986	4,213	3,216	2,601	2,183	1,881	1,652	1,473	1,329	1,068	893	767	672	598	539	450	386	338	271	226
65	19,169	11,242	7,953	6,153	5,018	4,236	3,230	2,610	2,189	1,886	1,656	1,476	1,332	1,070	894	768	673	599	539	450	386	338	271	226
70	19,582	11,383	8,024	6,195	5,045	4,256	3,241	2,617	2,195	1,890	1,659	1,478	1,333	1,071	895	768	673	599	540	450	386	338	271	226
75	19,954	11,508	8,085	6,232	5,070	4,273	3,251	2,624	2,199	1,893	1,662	1,481	1,335	1,072	895	769	674	599	540	450	386	338	271	226
80	20,291	11,619	8,140	6,265	5,091	4,288	3,260	2,629	2,203	1,896	1,664	1,482	1,337	1,073	896	769	674	600	540	451	387	338	271	226
85	20,599	11,719	8,189	6,294	5,110	4,302	3,268	2,634	2,207	1,899	1,666	1,484	1,338	1,074	897	770	674	600	540	451	387	338	271	226
90	20,880	11,810	8,233	6,320	5,128	4,314	3,275	2,639	2,210	1,901	1,668	1,485	1,339	1,074	897	770	675	600	540	451	387	339	271	226
95	21,138	11,892	8,273	6,343	5,143	4,325	3,281	2,643	2,213	1,903	1,669	1,487	1,340	1,075	898	770	675	600	541	451	387	339	271	226
100	21,376	11,967	8,309	6,364	5,157	4,335	3,287	2,647	2,215	1,905	1,671	1,488	1,341	1,076	898	771	675	601	541	451	387	339	271	226
105	21,595	12,035	8,342	6,384	5,170	4,344	3,292	2,650	2,218	1,907	1,672	1,489	1,342	1,076	898	771	675	601	541	451	387	339	271	226
110	21,799	12,098	8,373	6,401	5,181	4,352	3,297	2,653	2,220	1,908	1,673	1,490	1,343	1,077	899	771	676	601	541	451	387	339	271	226
115	21,989	12,157	8,400	6,417	5,192	4,359	3,301	2,656	2,222	1,910	1,674	1,491	1,343	1,077	899	772	676	601	541	451	387	339	271	226
120	22,165	12,210	8,426	6,432	5,202	4,366	3,305	2,658	2,224	1,911	1,675	1,492	1,344	1,078	899	772	676	601	541	451	387	339	271	226
125	22,330	12,260	8,450	6,446	5,211	4,373	3,308	2,661	2,225	1,912	1,676	1,492	1,345	1,078	900	772	676	601	541	451	387	339	271	226
130	22,485	12,307	8,472	6,459	5,219	4,379	3,312	2,663	2,227	1,913	1,677	1,493	1,345	1,078	900	772	676	601	541	452	387	339	271	226
135	22,630	12,350	8,492	6,471	5,227	4,384	3,315	2,665	2,228	1,914	1,678	1,494	1,346	1,079	900	772	676	601	542	452	387	339	271	226
140	22,766	12,390	8,511	6,482	5,234	4,389	3,318	2,667	2,230	1,915	1,679	1,494	1,346	1,079	900	772	676	602	542	452	387	339	271	226
145	22,894	12,428	8,529	6,492	5,241	4,394	3,321	2,669	2,231	1,916	1,680	1,495	1,347	1,079	901	773	677	602	542	452	387	339	271	226
150	23,016	12,464	8,546	6,502	5,247	4,398	3,323	2,670	2,232	1,917	1,680	1,495	1,347	1,080	901	773	677	602	542	452	387	339	271	226
155	23,130	12,498	8,562	6,511	5,253	4,402	3,325	2,672	2,233	1,918	1,681	1,496	1,348	1,080	901	773	677	602	542	452	387	339	271	226
160	23,238	12,529	8,577	6,520	5,259	4,406	3,328	2,673	2,234	1,919	1,681	1,496	1,348	1,080	901	773	677	602	542	452	387	339	271	226
165	23,341	12,559	8,591	6,528	5,264	4,410	3,330	2,675	2,235	1,919	1,682	1,497	1,348	1,080	901	773	677	602	542	452	387	339	271	226
170	23,439	12,587	8,604	6,535	5,269	4,414	3,332	2,676	2,236	1,920	1,682	1,497	1,349	1,081	901	773	677	602	542	452	388	339	271	226
175	23,531	12,614	8,616	6,543	5,274	4,417	3,334	2,677	2,237	1,921	1,683	1,497	1,349	1,081	902	773	677	602	542	452	388	339	271	226
180	23,620	12,639	8,628	6,549	5,278	4,420	3,335	2,678	2,237	1,921	1,683	1,498	1,349	1,081	902	773	677	602	542	452	388	339	271	226
185	23,704	12,663	8,639	6,556	5,282	4,423	3,337	2,679	2,238	1,922	1,684	1,498	1,349	1,081	902	774	677	602	542	452	388	339	271	226
190	23,784	12,686	8,650	6,562	5,286	4,426	3,339	2,680	2,239	1,922	1,684	1,498	1,350	1,081	902	774	677	602	542	452	388	339	271	226
195	23,860	12,708	8,660	6,568	5,290	4,428	3,340	2,681	2,240	1,923	1,685	1,499	1,350	1,081	902	774	677	602	542	452	388	339	271	226
200	23,934	12,728	8,670	6,573	5,293	4,431	3,342	2,682	2,240	1,923	1,685	1,499	1,350	1,082	902	774	677	602	542	452	388	339	272	226

10 AWG | 1 Run(s) | 240 Volts | C-Value = 981 | Magnetic Duct | Copper

Calculations:

1. $f = \dfrac{1.732 \times \text{Length} \times \text{Isca}}{\text{Runs} \times \text{C-Value} \times \text{E L-L}}$

2. $M = \dfrac{1}{1 + f}$

3. $\text{Isc} = \text{Isca} \times M$

4. Add Motor Contribution = Motor FLA x 4

* All results are given in symmetrical amperes

Three-Phase Bolted Fault-Current* Table

One-Way Distance in Feet

Available Isc in Thousands (Isca)	5	10	15	20	25	30	40	50	60	70	80	90	100	125	150	175	200	225	250	300	350	400	500	600
3	2,805	2,634	2,482	2,347	2,226	2,117	1,928	1,770	1,636	1,520	1,420	1,332	1,255	1,096	972	874	793	727	670	580	511	457	377	321
5	4,481	4,059	3,710	3,416	3,166	2,949	2,595	2,316	2,092	1,907	1,752	1,620	1,507	1,263	1,117	989	887	805	736	629	549	487	397	335
7	6,023	5,285	4,708	4,245	3,865	3,547	3,046	2,669	2,376	2,140	1,947	1,786	1,649	1,385	1,193	1,048	935	843	768	652	567	501	406	342
10	8,119	6,833	5,899	5,189	4,632	4,183	3,504	3,014	2,645	2,356	2,124	1,934	1,775	1,472	1,257	1,098	974	875	794	671	581	512	414	347
15	11,131	8,848	7,343	6,275	5,478	4,861	3,967	3,351	2,901	2,557	2,286	2,067	1,836	1,548	1,312	1,139	1,006	901	816	686	592	521	419	351
20	13,666	10,379	8,366	7,008	6,029	5,290	4,248	3,549	3,048	2,671	2,376	2,141	1,947	1,589	1,342	1,161	1,024	915	827	694	598	525	422	353
25	15,829	11,581	9,130	7,536	6,415	5,585	4,437	3,680	3,144	2,744	2,434	2,187	1,986	1,615	1,360	1,175	1,034	923	834	699	602	528	424	354
30	17,697	12,550	9,722	7,934	6,702	5,801	4,572	3,772	3,211	2,795	2,474	2,220	2,013	1,632	1,373	1,184	1,041	929	839	702	604	530	425	355
35	19,325	13,347	10,194	8,246	6,923	5,966	4,674	3,841	3,261	2,833	2,504	2,244	2,032	1,645	1,382	1,191	1,046	933	842	705	606	531	426	356
40	20,758	14,015	10,579	8,496	7,099	6,096	4,753	3,895	3,299	2,862	2,527	2,262	2,047	1,655	1,388	1,196	1,050	936	845	706	607	532	427	356
45	22,028	14,583	10,900	8,702	7,241	6,201	4,816	3,937	3,330	2,885	2,544	2,276	2,059	1,662	1,394	1,200	1,053	939	847	708	608	533	427	357
50	23,162	15,072	11,170	8,873	7,360	6,287	4,869	3,972	3,355	2,903	2,559	2,288	2,068	1,668	1,398	1,203	1,056	941	848	709	609	534	428	357
55	24,180	15,496	11,402	9,019	7,460	6,360	4,912	4,001	3,375	2,919	2,571	2,297	2,076	1,673	1,402	1,206	1,058	942	850	710	610	534	428	357
60	25,099	15,869	11,602	9,144	7,545	6,422	4,949	4,026	3,393	2,932	2,581	2,305	2,083	1,678	1,405	1,208	1,060	944	851	711	610	535	428	357
65	25,934	16,198	11,777	9,252	7,619	6,475	4,980	4,046	3,407	2,943	2,589	2,312	2,088	1,681	1,407	1,210	1,061	945	852	711	611	535	429	358
70	26,695	16,492	11,932	9,347	7,683	6,522	5,008	4,064	3,420	2,952	2,597	2,318	2,093	1,684	1,409	1,212	1,062	946	852	712	611	535	429	358
75	27,391	16,755	12,069	9,431	7,739	6,562	5,032	4,080	3,431	2,960	2,603	2,323	2,097	1,687	1,411	1,213	1,063	947	853	712	611	536	429	358
80	28,031	16,992	12,191	9,506	7,790	6,599	5,053	4,094	3,441	2,968	2,609	2,327	2,101	1,690	1,413	1,214	1,064	948	854	713	612	536	429	358
85	28,621	17,207	12,302	9,573	7,835	6,631	5,072	4,107	3,450	2,974	2,614	2,331	2,104	1,692	1,414	1,215	1,065	948	854	713	612	536	429	358
90	29,166	17,403	12,401	9,633	7,875	6,660	5,089	4,118	3,458	2,980	2,618	2,335	2,107	1,694	1,416	1,216	1,066	949	855	713	612	536	429	358
95	29,673	17,582	12,492	9,687	7,911	6,686	5,104	4,128	3,465	2,985	2,622	2,338	2,110	1,695	1,417	1,217	1,067	949	855	714	612	536	430	358
100	30,143	17,746	12,575	9,737	7,944	6,709	5,118	4,137	3,471	2,990	2,626	2,341	2,112	1,697	1,418	1,218	1,067	950	856	714	613	536	430	358
105	30,582	17,898	12,650	9,782	7,975	6,731	5,130	4,145	3,477	2,994	2,629	2,344	2,114	1,698	1,419	1,219	1,068	950	856	714	613	537	430	358
110	30,993	18,037	12,720	9,824	8,002	6,750	5,142	4,152	3,482	2,998	2,632	2,346	2,116	1,699	1,420	1,219	1,068	951	856	714	613	537	430	358
115	31,377	18,167	12,784	9,862	8,028	6,768	5,152	4,159	3,487	3,002	2,635	2,348	2,118	1,700	1,421	1,220	1,069	951	857	715	613	537	430	358
120	31,738	18,287	12,844	9,898	8,051	6,785	5,162	4,165	3,491	3,005	2,638	2,350	2,119	1,702	1,421	1,220	1,069	951	857	715	613	537	430	359
125	32,077	18,399	12,899	9,931	8,073	6,800	5,171	4,171	3,495	3,008	2,640	2,352	2,121	1,702	1,422	1,221	1,070	952	857	715	613	537	430	359
130	32,397	18,504	12,951	9,961	8,093	6,815	5,179	4,176	3,499	3,011	2,642	2,354	2,122	1,703	1,423	1,221	1,070	952	857	715	614	537	430	359
135	32,699	18,602	12,998	9,989	8,111	6,828	5,187	4,181	3,503	3,013	2,644	2,355	2,124	1,704	1,423	1,222	1,070	952	858	715	614	537	430	359
140	32,984	18,694	13,043	10,016	8,129	6,840	5,194	4,186	3,506	3,016	2,646	2,357	2,125	1,705	1,424	1,222	1,071	952	858	715	614	537	430	359
145	33,254	18,781	13,085	10,041	8,145	6,852	5,200	4,190	3,509	3,018	2,648	2,358	2,126	1,706	1,424	1,222	1,071	953	858	716	614	537	430	359
150	33,510	18,862	13,125	10,064	8,161	6,863	5,207	4,194	3,512	3,020	2,649	2,360	2,127	1,706	1,425	1,223	1,071	953	858	716	614	537	430	359
155	33,754	18,939	13,162	10,086	8,175	6,873	5,212	4,198	3,514	3,022	2,651	2,361	2,128	1,707	1,425	1,223	1,071	953	858	716	614	538	430	359
160	33,985	19,011	13,197	10,106	8,188	6,882	5,218	4,202	3,517	3,024	2,652	2,362	2,129	1,708	1,426	1,223	1,072	953	858	716	614	538	430	359
165	34,205	19,080	13,230	10,126	8,201	6,891	5,223	4,205	3,519	3,026	2,654	2,363	2,130	1,708	1,426	1,224	1,072	953	859	716	614	538	430	359
170	34,415	19,145	13,261	10,144	8,213	6,900	5,228	4,208	3,521	3,027	2,655	2,364	2,130	1,709	1,426	1,224	1,072	954	859	716	614	538	430	359
175	34,615	19,207	13,291	10,161	8,224	6,908	5,232	4,211	3,523	3,029	2,656	2,365	2,131	1,709	1,427	1,224	1,072	954	859	716	614	538	430	359
180	34,806	19,266	13,319	10,178	8,235	6,915	5,237	4,214	3,525	3,030	2,657	2,366	2,132	1,710	1,427	1,224	1,072	954	859	716	614	538	430	359
185	34,989	19,322	13,346	10,193	8,245	6,923	5,241	4,217	3,527	3,032	2,658	2,367	2 133	1,710	1,427	1,225	1,072	954	859	716	614	538	430	359
190	35,164	19,375	13,371	10,208	8,255	6,929	5,245	4,219	3,529	3,033	2,659	2,367	2 133	1,710	1,428	1,225	1,073	954	859	716	614	538	431	359
195	35,332	19,426	13,395	10,222	8,264	6,936	5,249	4,222	3,531	3,034	2,660	2,368	2 134	1,711	1,428	1,225	1,073	954	859	717	614	538	431	359
200	35,493	19,474	13,418	10,235	8,273	6,942	5,252	4,224	3,532	3,035	2,661	2,369	2,134	1,711	1,428	1,225	1,073	954	859	717	615	538	431	359

# 8 AWG	1 Run(s)	240 Volts	C-Value = 1,557	Magnetic Duct	Copper

Calculations:

1. $f = \dfrac{1.732 \times \text{Length} \times \text{Isca}}{\text{Runs} \times \text{C-Value} \times E_{L-L}}$

2. $M = \dfrac{1}{1 + f}$

3. $\text{Isc} = \text{Isca} \times M$

4. Add Motor Contribution = Motor FLA x 4

* All results are given in symmetrical amperes

Three-Phase Bolted Fault-Current* Table

One-Way Distance in Feet

Available Isc in Thousands (Isca)	5	10	15	20	25	30	40	50	60	70	80	90	100	125	150	175	200	225	250	300	350	400	500	600
3	2,872	2,754	2,646	2,545	2,453	2,366	2,211	2,074	1,954	1,846	1,750	1,663	1,585	1,418	1,283	1,171	1,077	997	928	816	727	656	549	472
5	4,654	4,352	4,088	3,853	3,644	3,457	3,134	2,867	2,642	2,449	2,283	2,138	2,010	1,748	1,547	1,387	1,258	1,150	1,059	915	805	719	592	504
7	6,340	5,793	5,333	4,941	4,603	4,308	3,818	3,429	3,111	2,848	2,625	2,435	2,270	1,942	1,697	1,507	1,355	1,231	1,128	966	844	750	613	519
10	8,705	7,707	6,914	6,269	5,734	5,283	4,565	4,019	3,590	3,243	2,958	2,719	2,515	2,119	1,830	1,611	1,438	1,299	1,185	1,007	876	775	630	530
15	12,263	10,371	8,984	7,925	7,089	6,413	5,385	4,641	4,078	3,637	3,281	2,990	2,745	2,280	1,949	1,702	1,511	1,358	1,234	1,042	902	796	643	540
20	15,413	12,538	10,566	9,131	8,039	7,180	5,916	5,030	4,375	3,871	3,471	3,146	2,877	2,370	2,015	1,752	1,550	1,390	1,259	1,061	916	806	650	545
25	18,222	14,335	11,815	10,048	8,741	7,735	6,288	5,297	4,575	4,027	3,596	3,248	2,962	2,427	2,056	1,783	1,574	1,409	1,276	1,072	925	813	654	548
30	20,741	15,850	12,825	10,770	9,282	8,156	6,563	5,491	4,719	4,138	3,684	3,320	3,022	2,467	2,085	1,805	1,591	1,423	1,286	1,080	930	817	657	550
35	23,014	17,144	13,659	11,352	9,712	8,485	6,775	5,638	4,828	4,221	3,750	3,374	3,066	2,496	2,105	1,820	1,603	1,432	1,294	1,085	934	820	659	551
40	25,075	18,262	14,360	11,832	10,061	8,751	6,943	5,754	4,913	4,286	3,801	3,415	3,100	2,519	2,121	1,832	1,612	1,440	1,300	1,090	938	823	661	552
45	26,953	19,238	14,956	12,234	10,350	8,969	7,079	5,847	4,981	4,338	3,842	3,448	3,127	2,537	2,134	1,842	1,620	1,445	1,305	1,093	940	825	662	553
50	28,670	20,097	15,470	12,576	10,593	9,151	7,192	5,924	5,036	4,380	3,875	3,474	3,149	2,551	2,144	1,849	1,626	1,450	1,309	1,096	942	826	663	554
55	30,247	20,859	15,918	12,870	10,801	9,306	7,288	5,989	5,083	4,415	3,902	3,496	3,167	2,563	2,153	1,855	1,630	1,454	1,312	1,098	944	827	664	554
60	31,699	21,540	16,312	13,126	10,981	9,439	7,369	6,044	5,122	4,445	3,926	3,515	3,182	2,573	2,160	1,861	1,634	1,457	1,315	1,100	945	828	665	555
65	33,042	22,151	16,660	13,351	11,138	9,554	7,439	6,091	5,156	4,470	3,945	3,531	3,195	2,581	2,166	1,865	1,638	1,460	1,317	1,101	946	829	665	555
70	34,287	22,704	16,971	13,549	11,276	9,656	7,501	6,132	5,186	4,492	3,963	3,545	3,206	2,589	2,171	1,869	1,641	1,462	1,319	1,102	947	830	666	556
75	35,445	23,206	17,250	13,726	11,398	9,745	7,555	6,168	5,211	4,512	3,978	3,557	3,216	2,595	2,175	1,872	1,643	1,464	1,320	1,104	948	831	666	556
80	36,523	23,663	17,501	13,885	11,508	9,825	7,602	6,200	5,234	4,529	3,991	3,567	3,225	2,601	2,179	1,875	1,646	1,466	1,322	1,105	949	831	666	556
85	37,531	24,082	17,729	14,028	11,606	9,897	7,645	6,228	5,254	4,544	4,003	3,577	3,232	2,606	2,183	1,878	1,648	1,468	1,323	1,106	949	832	667	556
90	38,475	24,467	17,937	14,158	11,695	9,961	7,684	6,254	5,272	4,557	4,013	3,585	3,239	2,610	2,186	1,880	1,649	1,469	1,324	1,106	950	832	667	557
95	39,361	24,823	18,127	14,276	11,775	10,020	7,718	6,277	5,289	4,569	4,022	3,592	3,245	2,614	2,189	1,882	1,651	1,470	1,325	1,107	950	833	667	557
100	40,193	25,151	18,302	14,385	11,849	10,073	7,750	6,297	5,303	4,581	4,031	3,599	3,251	2,618	2,191	1,884	1,652	1,471	1,326	1,108	951	833	668	557
105	40,978	25,456	18,463	14,484	11,916	10,121	7,778	6,316	5,317	4,591	4,039	3,605	3,256	2,621	2,193	1,886	1,654	1,473	1,327	1,108	951	833	668	557
110	41,718	25,740	18,612	14,575	11,978	10,166	7,805	6,334	5,329	4,600	4,046	3,611	3,261	2,624	2,195	1,887	1,655	1,473	1,328	1,109	952	834	668	557
115	42,417	26,004	18,749	14,660	12,035	10,207	7,829	6,349	5,340	4,608	4,052	3,616	3,265	2,627	2,197	1,889	1,656	1,474	1,329	1,109	952	834	668	557
120	43,079	26,252	18,878	14,738	12,087	10,245	7,851	6,364	5,351	4,616	4,058	3,621	3,269	2,629	2,199	1,890	1,657	1,475	1,329	1,110	952	834	668	557
125	43,707	26,483	18,997	14,811	12,136	10,280	7,872	6,378	5,360	4,623	4,064	3,625	3,272	2,632	2,201	1,891	1,658	1,476	1,330	1,110	953	834	668	558
130	44,303	26,701	19,109	14,878	12,182	10,312	7,891	6,390	5,369	4,629	4,069	3,629	3,276	2,634	2,202	1,892	1,659	1,476	1,330	1,111	953	835	669	558
135	44,869	26,906	19,214	14,942	12,224	10,343	7,909	6,402	5,377	4,636	4,074	3,633	3,279	2,636	2,204	1,893	1,659	1,477	1,331	1,111	953	835	669	558
140	45,408	27,099	19,312	15,001	12,264	10,371	7,925	6,413	5,385	4,641	4,078	3,637	3,282	2,638	2,205	1,894	1,660	1,478	1,331	1,111	954	835	669	558
145	45,922	27,281	19,404	15,057	12,301	10,398	7,941	6,423	5,392	4,647	4,082	3,640	3,284	2,639	2,206	1,895	1,661	1,478	1,332	1,112	954	835	669	558
150	46,411	27,453	19,491	15,109	12,336	10,423	7,955	6,432	5,399	4,652	4,086	3,643	3,287	2,641	2,207	1,896	1,662	1,479	1,332	1,112	954	835	669	558
155	46,879	27,616	19,573	15,158	12,369	10,446	7,969	6,441	5,405	4,656	4,090	3,646	3,289	2,642	2,208	1,897	1,662	1,479	1,333	1,112	954	836	669	558
160	47,327	27,770	19,651	15,205	12,399	10,468	7,982	6,450	5,411	4,661	4,093	3,649	3,291	2,644	2,209	1,897	1,663	1,480	1,333	1,112	954	836	669	558
165	47,755	27,917	19,724	15,249	12,429	10,489	7,994	6,458	5,417	4,665	4,096	3,651	3,293	2,645	2,210	1,898	1,663	1,480	1,333	1,113	955	836	669	558
170	48,165	28,057	19,794	15,290	12,456	10,509	8,005	6,465	5,422	4,669	4,099	3,653	3,295	2,646	2,211	1,899	1,664	1,480	1,334	1,113	955	836	669	558
175	48,558	28,190	19,860	15,330	12,482	10,527	8,016	6,472	5,427	4,672	4,102	3,656	3,297	2,648	2,212	1,899	1,664	1,481	1,334	1,113	955	836	669	558
180	48,935	28,317	19,922	15,367	12,507	10,545	8,026	6,479	5,431	4,676	4,105	3,658	3,299	2,649	2,213	1,900	1,665	1,481	1,334	1,113	955	836	670	558
185	49,297	28,437	19,982	15,403	12,531	10,561	8,036	6,485	5,436	4,679	4,107	3,660	3,300	2,650	2,213	1,900	1,665	1,481	1,334	1,113	955	836	670	558
190	49,645	28,553	20,039	15,436	12,553	10,577	8,045	6,491	5,440	4,682	4,109	3,662	3,302	2,651	2,214	1,901	1,665	1,482	1,335	1,114	955	836	670	558
195	49,980	28,663	20,093	15,469	12,574	10,592	8,054	6,497	5,444	4,685	4,112	3,663	3,303	2,652	2,215	1,901	1,666	1,482	1,335	1,114	955	836	670	558
200	50,303	28,769	20,145	15,499	12,595	10,607	8,062	6,502	5,448	4,688	4,114	3,665	3,305	2,653	2,215	1,902	1,666	1,482	1,335	1,114	955	837	670	558

# 6 AWG	1 Run(s)	240 Volts	C-Value = 2,425	Magnetic Duct	Copper

Calculations:

1. $f = \dfrac{1.732 \times \text{Length} \times \text{Isca}}{\text{Runs} \times \text{C-Value} \times E_{L-L}}$

2. $M = \dfrac{1}{1 + f}$

3. Isc = Isca x M

4. Add Motor Contribution = Motor FLA x 4

* All results are given in symmetrical amperes

Three-Phase Bolted Fault-Current* Table

One-Way Distance in Feet

Available Isc in Thousands (Isca)	5	10	15	20	25	30	40	50	60	70	80	90	100	125	150	175	200	225	250	300	350	400	500	600
3	2,917	2,839	2,764	2,694	2,626	2,563	2,444	2,336	2,237	2,146	2,062	1,984	1,912	1,753	1,619	1,503	1,403	1,316	1,239	1,108	1,003	916	780	680
5	4,774	4,567	4,377	4,203	4,042	3,893	3,625	3,392	3,187	3,005	2,843	2,698	2,567	2,288	2,064	1,880	1,726	1,596	1,484	1,301	1,158	1,043	871	748
7	6,564	6,180	5,838	5,532	5,256	5,006	4,572	4,208	3,897	3,629	3,395	3,190	3,008	2,632	2,340	2,107	1,915	1,756	1,621	1,405	1,240	1,109	917	781
10	9,134	8,406	7,786	7,250	6,784	6,374	5,687	5,133	4,678	4,297	3,973	3,695	3,453	2,967	2,601	2,316	2,087	1,899	1,742	1,495	1,310	1,165	954	808
15	13,132	11,678	10,514	9,561	8,767	8,094	7,017	6,193	5,542	5,015	4,580	4,214	3,902	3,293	2,848	2,509	2,243	2,027	1,849	1,574	1,369	1,212	985	830
20	16,812	14,501	12,748	11,374	10,267	9,356	7,946	6,906	6,106	5,473	4,958	4,532	4,173	3,484	2,990	2,619	2,330	2,098	1,908	1,616	1,401	1,237	1,002	842
25	20,210	16,960	14,611	12,833	11,441	10,322	8,632	7,418	6,503	5,789	5,217	4,747	4,355	3,610	3,082	2,689	2,385	2,143	1,945	1,642	1,421	1,252	1,012	849
30	23,357	19,122	16,188	14,034	12,386	11,084	9,159	7,804	6,798	6,022	5,405	4,902	4,485	3,699	3,147	2,739	2,424	2,174	1,971	1,661	1,435	1,263	1,019	854
35	26,280	21,038	17,540	15,039	13,162	11,702	9,577	8,105	7,025	6,200	5,547	5,019	4,583	3,765	3,195	2,775	2,452	2,197	1,990	1,674	1,445	1,271	1,024	857
40	29,002	22,747	18,712	15,893	13,812	12,212	9,916	8,347	7,206	6,340	5,660	5,111	4,660	3,817	3,232	2,803	2,474	2,214	2,004	1,684	1,452	1,276	1,028	860
45	31,543	24,282	19,738	16,627	14,363	12,641	10,197	8,545	7,353	6,454	5,750	5,185	4,721	3,857	3,261	2,825	2,491	2,228	2,015	1,692	1,458	1,281	1,031	862
50	33,921	25,667	20,643	17,264	14,836	13,007	10,433	8,710	7,476	6,548	5,824	5,245	4,771	3,891	3,285	2,842	2,505	2,239	2,024	1,698	1,463	1,285	1,033	864
55	36,150	26,923	21,448	17,824	15,247	13,322	10,635	8,850	7,579	6,626	5,887	5,296	4,812	3,919	3,305	2,857	2,516	2,248	2,032	1,704	1,467	1,288	1,035	865
60	38,245	28,068	22,169	18,319	15,608	13,596	10,809	8,971	7,667	6,694	5,940	5,339	4,848	3,942	3,321	2,870	2,526	2,256	2,038	1,708	1,470	1,290	1,037	866
65	40,217	29,116	22,817	18,759	15,927	13,837	10,961	9,075	7,743	6,752	5,985	5,375	4,878	3,962	3,336	2,880	2,534	2,262	2,043	1,712	1,473	1,292	1,038	867
70	42,076	30,078	23,404	19,154	16,210	14,051	11,095	9,167	7,809	6,802	6,025	5,407	4,904	3,979	3,348	2,889	2,541	2,268	2,048	1,715	1,475	1,294	1,039	868
75	43,833	30,965	23,938	19,510	16,465	14,242	11,213	9,247	7,868	6,846	6,060	5,435	4,927	3,994	3,358	2,897	2,547	2,273	2,052	1,718	1,477	1,296	1,040	869
80	45,495	31,785	24,425	19,832	16,694	14,413	11,319	9,319	7,920	6,886	6,090	5,460	4,948	4,008	3,368	2,904	2,553	2,277	2,055	1,720	1,479	1,297	1,041	869
85	47,069	32,546	24,872	20,126	16,901	14,567	11,414	9,383	7,966	6,921	6,118	5,482	4,966	4,020	3,376	2,910	2,558	2,281	2,058	1,722	1,481	1,298	1,042	870
90	48,563	33,253	25,283	20,394	17,090	14,707	11,500	9,441	8,008	6,952	6,142	5,502	4,982	4,030	3,384	2,916	2,562	2,284	2,061	1,724	1,482	1,299	1,043	870
95	49,983	33,913	25,662	20,640	17,262	14,835	11,578	9,494	8,045	6,981	6,165	5,519	4,997	4,040	3,390	2,921	2,566	2,288	2,064	1,726	1,483	1,300	1,043	871
100	51,333	34,529	26,013	20,867	17,421	14,951	11,649	9,541	8,080	7,006	6,185	5,536	5,010	4,048	3,397	2,925	2,569	2,290	2,066	1,728	1,484	1,301	1,044	871
105	52,619	35,106	26,340	21,076	17,566	15,058	11,714	9,585	8,111	7,030	6,203	5,550	5,022	4,056	3,402	2,930	2,572	2,293	2,068	1,729	1,486	1,302	1,044	872
110	53,846	35,648	26,643	21,270	17,701	15,157	11,774	9,625	8,139	7,051	6,220	5,564	5,033	4,063	3,407	2,933	2,575	2,295	2,070	1,730	1,486	1,303	1,045	872
115	55,017	36,157	26,927	21,451	17,826	15,249	11,829	9,662	8,166	7,071	6,235	5,576	5,043	4,070	3,412	2,937	2,578	2,297	2,072	1,731	1,487	1,304	1,045	872
120	56,136	36,637	27,192	21,619	17,942	15,333	11,880	9,696	8,190	7,089	6,249	5,587	5,052	4,076	3,416	2,940	2,580	2,299	2,073	1,733	1,488	1,304	1,046	873
125	57,206	37,090	27,441	21,776	18,049	15,412	11,927	9,727	8,212	7,106	6,262	5,597	5,060	4,081	3,420	2,943	2,582	2,301	2,075	1,734	1,489	1,305	1,046	873
130	58,231	37,518	27,675	21,923	18,150	15,486	11,971	9,756	8,233	7,121	6,274	5,607	5,068	4,086	3,423	2,945	2,585	2,302	2,076	1,735	1,490	1,305	1,046	873
135	59,213	37,924	27,895	22,060	18,245	15,554	12,012	9,783	8,253	7,136	6,285	5,616	5,076	4,091	3,427	2,948	2,586	2,304	2,077	1,735	1,490	1,306	1,047	873
140	60,156	38,308	28,102	22,190	18,333	15,618	12,050	9,809	8,271	7,149	6,296	5,624	5,082	4,096	3,430	2,950	2,588	2,305	2,078	1,736	1,491	1,306	1,047	873
145	61,061	38,673	28,298	22,312	18,416	15,679	12,086	9,833	8,287	7,162	6,306	5,632	5,089	4,100	3,433	2,952	2,590	2,307	2,079	1,737	1,491	1,307	1,047	874
150	61,930	39,020	28,483	22,427	18,495	15,736	12,119	9,855	8,303	7,174	6,315	5,640	5,095	4,104	3,435	2,954	2,591	2,308	2,080	1,738	1,492	1,307	1,047	874
155	62,766	39,350	28,659	22,536	18,568	15,789	12,151	9,876	8,318	7,185	6,323	5,646	5,100	4,107	3,438	2,956	2,593	2,309	2,081	1,738	1,492	1,307	1,048	874
160	63,570	39,665	28,825	22,638	18,638	15,839	12,181	9,895	8,332	7,195	6,332	5,653	5,106	4,111	3,440	2,958	2,594	2,310	2,082	1,739	1,493	1,308	1,048	874
165	64,345	39,965	28,983	22,736	18,704	15,887	12,209	9,914	8,345	7,205	6,339	5,659	5,111	4,114	3,443	2,960	2,595	2,311	2,083	1,739	1,493	1,308	1,048	874
170	65,091	40,252	29,134	22,828	18,767	15,932	12,236	9,932	8,358	7,214	6,346	5,665	5,115	4,117	3,445	2,961	2,597	2,312	2,084	1,740	1,494	1,308	1,048	874
175	65,811	40,526	29,277	22,916	18,826	15,975	12,261	9,948	8,369	7,223	6,353	5,670	5,120	4,120	3,447	2,963	2,598	2,313	2,084	1,740	1,494	1,309	1,048	875
180	66,506	40,788	29,414	23,000	18,883	16,016	12,285	9,964	8,381	7,231	6,359	5,675	5,124	4,122	3,449	2,964	2,599	2,314	2,085	1,741	1,494	1,309	1,049	875
185	67,177	41,040	29,544	23,080	18,936	16,054	12,308	9,979	8,391	7,239	6,366	5,680	5,128	4,125	3,450	2,965	2,600	2,315	2,086	1,741	1,495	1,309	1,049	875
190	67,825	41,281	29,669	23,156	18,987	16,091	12,329	9,993	8,401	7,247	6,371	5,685	5,131	4,127	3,452	2,967	2,601	2,315	2,086	1,742	1,495	1,309	1,049	875
195	68,452	41,512	29,788	23,228	19,036	16,126	12,350	10,007	8,411	7,254	6,377	5,689	5,135	4,130	3,454	2,968	2,602	2,316	2,087	1,742	1,495	1,310	1,049	875
200	69,058	41,734	29,903	23,298	19,083	16,159	12,369	10,019	8,420	7,261	6,382	5,693	5,138	4,132	3,455	2,969	2,603	2,317	2,088	1,743	1,496	1,310	1,049	875

#4 AWG	1 Run(s)	240 Volts	C-Value = 3,806	Magnetic Duct	Copper

Calculations:

1. $f = \dfrac{1.732 \times \text{Length} \times \text{Isca}}{\text{Runs} \times \text{C-Value} \times E_{L-L}}$

2. $M = \dfrac{1}{1 + f}$

3. Isc = Isca x M

4. Add Motor Contribution = Motor FLA x 4

* All results are given in symmetrical amperes

Three-Phase Bolted Fault-Current* Table

One-Way Distance in Feet

Available Isc in Thousands (Isca)	5	10	15	20	25	30	40	50	60	70	80	90	100	125	150	175	200	225	250	300	350	400	500	600
3	2,933	2,869	2,808	2,750	2,694	2,640	2,538	2,444	2,357	2,276	2,200	2,129	2,062	1,913	1,783	1,670	1,571	1,483	1,404	1,269	1,157	1,064	916	805
5	4,817	4,648	4,490	4,342	4,203	4,074	3,837	3,626	3,437	3,267	3,112	2,972	2,844	2,567	2,340	2,149	1,987	1,848	1,727	1,527	1,369	1,240	1,044	901
7	6,647	6,328	6,039	5,774	5,532	5,310	4,914	4,573	4,277	4,016	3,786	3,580	3,396	3,009	2,701	2,450	2,242	2,066	1,916	1,673	1,485	1,335	1,110	950
10	9,295	8,683	8,147	7,673	7,251	6,874	6,225	5,688	5,237	4,851	4,519	4,229	3,974	3,454	3,054	2,737	2,480	2,267	2,088	1,802	1,586	1,416	1,165	990
15	13,469	12,221	11,185	10,310	9,563	8,917	7,855	7,019	6,344	5,787	5,320	4,923	4,581	3,904	3,400	3,012	2,704	2,452	2,244	1,918	1,674	1,486	1,213	1,024
20	17,367	15,347	13,747	12,450	11,376	10,473	9,038	7,949	7,094	6,405	5,838	5,363	4,960	4,175	3,605	3,171	2,831	2,557	2,331	1,981	1,722	1,523	1,238	1,042
25	21,017	18,129	15,938	14,220	12,837	11,698	9,936	8,635	7,636	6,843	6,200	5,667	5,219	4,357	3,739	3,275	2,914	2,624	2,386	2,021	1,752	1,547	1,253	1,053
30	24,442	20,621	17,833	15,710	14,038	12,688	10,641	9,163	8,045	7,170	6,467	5,890	5,407	4,487	3,835	3,348	2,971	2,671	2,425	2,048	1,773	1,563	1,264	1,060
35	27,661	22,866	19,488	16,980	15,043	13,504	11,209	9,581	8,366	7,424	6,673	6,060	5,550	4,585	3,906	3,403	3,014	2,705	2,453	2,069	1,788	1,575	1,271	1,066
40	30,693	24,900	20,946	18,076	15,898	14,188	11,676	9,920	8,623	7,626	6,836	6,194	5,662	4,662	3,962	3,444	3,047	2,731	2,475	2,084	1,800	1,584	1,277	1,070
45	33,554	26,750	22,240	19,032	16,632	14,770	12,068	10,201	8,835	7,791	6,968	6,302	5,753	4,723	4,006	3,478	3,073	2,752	2,492	2,096	1,809	1,591	1,282	1,073
50	36,257	28,441	23,396	19,872	17,270	15,271	12,400	10,438	9,012	7,928	7,078	6,392	5,827	4,773	4,042	3,505	3,094	2,769	2,506	2,106	1,816	1,596	1,285	1,076
55	38,816	29,991	24,436	20,617	17,830	15,707	12,686	10,640	9,162	8,044	7,170	6,467	5,890	4,815	4,072	3,527	3 111	2,783	2,518	2,114	1,822	1,601	1,288	1,078
60	41,242	31,419	25,375	21,282	18,325	16,090	12,935	10,814	9,291	8,144	7,249	6,531	5,943	4,850	4,097	3,546	3 126	2,795	2,527	2,121	1,827	1,605	1,291	1,080
65	43,544	32,738	26,229	21,879	18,766	16,429	13,153	10,966	9,403	8,230	7,317	6,586	5,988	4,880	4,119	3,562	3 139	2,805	2,535	2,127	1,831	1,608	1,293	1,081
70	45,733	33,960	27,007	22,418	19,161	16,731	13,346	11,100	9,501	8,305	7,376	6,634	6,028	4,907	4,137	3,576	3,150	2,814	2,543	2,132	1,835	1,611	1,295	1,082
75	47,815	35,095	27,720	22,907	19,518	17,002	13,518	11,218	9,588	8,371	7,428	6,676	6,063	4,930	4,154	3,589	3,159	2,821	2,549	2,136	1,838	1,613	1,296	1,083
80	49,799	36,152	28,376	23,352	19,840	17,246	13,672	11,324	9,665	8,430	7,474	6,714	6,093	4,950	4,168	3,599	3,167	2,828	2,554	2,140	1,841	1,616	1,298	1,084
85	51,692	37,139	28,980	23,760	20,134	17,468	13,810	11,419	9,734	8,482	7,516	6,747	6,121	4,968	4,181	3,609	3,175	2,834	2,559	2,143	1,844	1,618	1,299	1,085
90	53,500	38,063	29,540	24,135	20,402	17,670	13,936	11,505	9,796	8,530	7,553	6,777	6,145	4,984	4,192	3,618	3,181	2,839	2,563	2,146	1,846	1,619	1,300	1,086
95	55,228	38,930	30,059	24,481	20,649	17,854	14,051	11,583	9,853	8,572	7,586	6,804	6,168	4,999	4,203	3,625	3,187	2,844	2,567	2,149	1,848	1,621	1,301	1,087
100	56,881	39,744	30,542	24,800	20,876	18,023	14,155	11,654	9,904	8,611	7,617	6,828	6,188	5,012	4,212	3,632	3,193	2,848	2,571	2,151	1,850	1,622	1,302	1,087
105	58,465	40,511	30,993	25,097	21,085	18,180	14,251	11,719	9,951	8,647	7,645	6,851	6,206	5,024	4,220	3,638	3,197	2,852	2,574	2,154	1,851	1,623	1,303	1,088
110	59,983	41,234	31,414	25,372	21,280	18,324	14,340	11,779	9,994	8,679	7,670	6,871	6,223	5,035	4,228	3,644	3,202	2,855	2,577	2,156	1,853	1,625	1,304	1,088
115	61,439	41,917	31,809	25,629	21,460	18,457	14,422	11,834	10,034	8,709	7,693	6,890	6,238	5,045	4,235	3,649	3,206	2,859	2,579	2,157	1,854	1,626	1,304	1,089
120	62,838	42,563	32,180	25,870	21,628	18,582	14,497	11,885	10,071	8,737	7,715	6,907	6,252	5,054	4,242	3,654	3,210	2,862	2,582	2,159	1,855	1,627	1,305	1,089
125	64,183	43,176	32,529	26,095	21,785	18,697	14,568	11,932	10,104	8,762	7,735	6,923	6,265	5,063	4,248	3,659	3,213	2,864	2,584	2,161	1,857	1,627	1,305	1,090
130	65,476	43,757	32,858	26,306	21,932	18,806	14,633	11,976	10,136	8,786	7,753	6,938	6,277	5,071	4,253	3,663	3,216	2,867	2,586	2,162	1,858	1,628	1,306	1,090
135	66,720	44,310	33,169	26,504	22,070	18,907	14,695	12,017	10,165	8,808	7,770	6,951	6,289	5,078	4,259	3,667	3,219	2,869	2,588	2,163	1,859	1,629	1,306	1,090
140	67,919	44,835	33,462	26,692	22,200	19,002	14,752	12,056	10,193	8,828	7,786	6,964	6,299	5,085	4,263	3,670	3,222	2,871	2,590	2,165	1,859	1,630	1,307	1,091
145	69,075	45,336	33,740	26,868	22,322	19,091	14,806	12,092	10,218	8,848	7,801	6,976	6,309	5,091	4,268	3,674	3,225	2,873	2,591	2,166	1,860	1,630	1,307	1,091
150	70,189	45,813	34,004	27,035	22,437	19,176	14,856	12,125	10,242	8,866	7,815	6,987	6,318	5,097	4,272	3,677	3,227	2,875	2,593	2,167	1,861	1,631	1,308	1,091
155	71,265	46,269	34,255	27,193	22,546	19,255	14,904	12,157	10,265	8,883	7,828	6,998	6,327	5,103	4,276	3,680	3,229	2,877	2,594	2,168	1,862	1,632	1,308	1,092
160	72,304	46,705	34,493	27,343	22,649	19,330	14,949	12,187	10,286	8,899	7,841	7,008	6,335	5,108	4,280	3,682	3,231	2,879	2,596	2,169	1,863	1,632	1,308	1,092
165	73,308	47,122	34,720	27,486	22,746	19,401	14,991	12,215	10,306	8,914	7,852	7,017	6,342	5,113	4,283	3,685	3,233	2,880	2,597	2,170	1,863	1,633	1,309	1,092
170	74,278	47,521	34,936	27,621	22,839	19,468	15,032	12,242	10,325	8,928	7,863	7,026	6,349	5,118	4,286	3,687	3,235	2,882	2,598	2,171	1,864	1,633	1,309	1,092
175	75,217	47,903	35,142	27,750	22,927	19,532	15,070	12,267	10,343	8,941	7,874	7,034	6,356	5,122	4,289	3,690	3,237	2,883	2,599	2,171	1,864	1,634	1,309	1,092
180	76,126	48,270	35,339	27,872	23,011	19,593	15,106	12,291	10,360	8,954	7,884	7,042	6,363	5,126	4,292	3,692	3,239	2,885	2,600	2,172	1,865	1,634	1,310	1,093
185	77,006	48,623	35,528	27,990	23,090	19,651	15,140	12,314	10,376	8,966	7,893	7,049	6,369	5,130	4,295	3,694	3,240	2,886	2,601	2,173	1,866	1,634	1,310	1,093
190	77,859	48,961	35,708	28,102	23,166	19,706	15,173	12,335	10,392	8,977	7,902	7,057	6,375	5,134	4,298	3,696	3,242	2,887	2,602	2,173	1,866	1,635	1,310	1,093
195	78,686	49,287	35,881	28,208	23,239	19,758	15,204	12,356	10,406	8,988	7,910	7,063	6,380	5,138	4,300	3,698	3,243	2,888	2,603	2,174	1,866	1,635	1,310	1,093
200	79,488	49,601	36,047	28,311	23,309	19,809	15,234	12,375	10,420	8,999	7,918	7,070	6,385	5,141	4,303	3,699	3,244	2,889	2,604	2,175	1,867	1,635	1,311	1,093

# 3 AWG	1 Run(s)	240 Volts

C-Value = 4,760	Magnetic Duct	Copper

Calculations:

1. $f = \dfrac{1.732 \times \text{Length} \times \text{Isca}}{\text{Runs} \times \text{C-Value} \times \text{E L-L}}$

2. $M = \dfrac{1}{1 + f}$

3. Isc = Isca x M

4. Add Motor Contribution = Motor FLA x 4

* All results are given in symmetrical amperes

Three-Phase Bolted Fault-Current* Table

One-Way Distance in Feet

Available Isc in Thousands (Isca)	5	10	15	20	25	30	40	50	60	70	80	90	100	125	150	175	200	225	250	300	350	400	500	600
3	2,946	2,894	2,844	2,795	2,748	2,703	2,616	2,535	2,459	2,387	2,320	2,256	2,195	2,057	1,936	1,828	1,731	1,644	1,565	1,429	1,314	1,216	1,059	938
5	4,852	4,712	4,580	4,456	4,337	4,226	4,018	3,830	3,659	3,502	3,358	3,226	3,104	2,835	2,609	2,416	2,250	2,106	1,978	1,765	1,593	1,452	1,233	1,072
7	6,713	6,448	6,204	5,977	5,767	5,571	5,216	4,903	4,626	4,378	4,156	3,955	3,773	3,383	3,066	2,804	2,582	2,394	2,230	1,963	1,753	1,583	1,327	1,142
10	9,424	8,911	8,451	8,036	7,660	7,318	6,717	6,207	5,770	5,390	5,057	4,763	4,501	3,957	3,530	3,186	2,904	2,667	2,466	2,143	1,895	1,698	1,407	1,200
15	13,741	12,677	11,765	10,976	10,287	9,678	8,655	7,827	7,144	6,570	6,082	5,661	5,295	4,558	4,001	3,565	3,215	2,927	2,687	2,308	2,023	1,800	1,476	1,250
20	17,822	16,072	14,635	13,434	12,415	11,540	10,114	9,001	8,109	7,378	6,768	6,251	5,807	4,932	4,287	3,790	3,397	3,078	2,813	2,401	2,093	1,856	1,513	1,277
25	21,687	19,150	17,144	15,519	14,175	13,045	11,252	9,892	8,825	7,966	7,259	6,668	6,166	5,188	4,479	3,940	3,516	3,175	2,895	2,460	2,138	1,891	1,536	1,293
30	25,353	21,953	19,357	17,309	15,654	14,288	12,164	10,590	9,377	8,413	7,629	6,978	6,430	5,374	4,616	4,046	3,601	3,244	2,951	2,501	2,169	1,915	1,552	1,305
35	28,834	24,515	21,322	18,864	16,915	15,331	12,912	11,152	9,815	8,764	7,916	7,218	6,633	5,515	4,720	4,125	3,664	3,295	2,994	2,531	2,192	1,933	1,564	1,313
40	32,144	26,868	23,079	20,227	18,002	16,219	13,536	11,615	10,171	9,047	8,146	7,409	6,794	5,626	4,801	4,187	3,712	3,334	3,026	2,554	2,209	1,946	1,572	1,319
45	35,296	29,035	24,660	21,431	18,950	16,984	14,065	12,002	10,467	9,280	8,335	7,565	6,925	5,716	4,866	4,236	3,751	3,365	3,052	2,572	2,223	1,957	1,579	1,324
50	38,300	31,037	26,090	22,503	19,783	17,650	14,519	12,331	10,716	9,476	8,492	7,694	7,033	5,789	4,919	4,276	3,782	3,391	3,072	2,587	2,234	1,966	1,585	1,328
55	41,167	32,894	27,389	23,463	20,521	18,235	14,912	12,614	10,929	9,642	8,625	7,803	7,124	5,851	4,964	4,310	3,809	3,412	3,090	2,599	2,243	1,973	1,589	1,331
60	43,905	34,619	28,575	24,328	21,180	18,753	15,257	12,860	11,113	9,785	8,740	7,896	7,202	5,903	5,001	4,338	3,831	3,429	3,104	2,609	2,251	1,978	1,593	1,334
65	46,524	36,227	29,662	25,111	21,771	19,215	15,561	13,075	11,274	9,909	8,839	7,977	7,269	5,948	5,033	4,363	3,850	3,445	3,117	2,618	2,257	1,984	1,597	1,336
70	49,031	37,729	30,661	25,824	22,305	19,630	15,832	13,266	11,415	10,018	8,925	8,048	7,327	5,987	5,061	4,384	3,866	3,458	3,127	2,626	2,263	1,988	1,599	1,338
75	51,433	39,135	31,583	26,475	22,789	20,004	16,075	13,436	11,541	10,115	9,002	8,110	7,379	6,021	5,086	4,402	3,880	3,469	3,137	2,632	2,268	1,992	1,602	1,340
80	53,736	40,454	32,437	27,072	23,230	20,343	16,293	13,588	11,653	10,200	9,070	8,165	7,424	6,052	5,108	4,418	3,893	3,479	3,145	2,638	2,272	1,995	1,604	1,341
85	55,946	41,695	33,230	27,622	23,634	20,652	16,490	13,725	11,754	10,278	9,131	8,214	7,465	6,079	5,127	4,433	3,904	3,488	3,152	2,643	2,276	1,998	1,606	1,342
90	58,070	42,863	33,967	28,130	24,004	20,934	16,670	13,849	11,845	10,347	9,186	8,259	7,502	6,103	5,144	4,445	3,914	3,496	3,159	2,648	2,279	2,000	1,608	1,344
95	60,111	43,965	34,656	28,600	24,346	21,194	16,834	13,962	11,927	10,410	9,235	8,299	7,535	6,125	5,160	4,457	3,923	3,503	3,164	2,652	2,282	2,003	1,609	1,345
100	62,075	45,006	35,300	29,037	24,662	21,433	16,985	14,065	12,003	10,467	9,280	8,335	7,565	6,145	5,174	4,468	3,931	3,510	3,170	2,656	2,285	2,005	1,610	1,346
105	63,966	45,992	35,903	29,444	24,955	21,654	17,123	14,160	12,072	10,520	9,322	8,368	7,592	6,163	5,186	4,477	3,938	3,515	3,175	2,659	2,287	2,007	1,612	1,346
110	65,787	46,926	36,470	29,825	25,228	21,859	17,251	14,248	12,135	10,568	9,359	8,399	7,617	6,179	5,198	4,486	3,945	3,521	3,179	2,662	2,290	2,009	1,613	1,347
115	67,544	47,813	37,004	30,180	25,482	22,049	17,369	14,328	12,193	10,612	9,394	8,427	7,640	6,194	5,209	4,494	3,951	3,526	3,183	2,665	2,292	2,010	1,614	1,348
120	69,238	48,656	37,506	30,514	25,719	22,227	17,479	14,403	12,248	10,653	9,426	8,453	7,661	6,208	5,219	4,501	3,957	3,530	3,187	2,667	2,294	2,012	1,615	1,349
125	70,874	49,458	37,981	30,828	25,942	22,393	17,582	14,473	12,298	10,691	9,456	8,477	7,681	6,221	5,228	4,508	3,962	3,534	3,190	2,670	2,295	2,013	1,616	1,349
130	72,454	50,222	38,430	31,123	26,150	22,548	17,677	14,537	12,345	10,727	9,484	8,499	7,699	6,233	5,236	4,514	3,967	3,538	3,193	2,672	2,297	2,014	1,616	1,350
135	73,981	50,951	38,856	31,401	26,347	22,694	17,767	14,598	12,388	10,759	9,509	8,519	7,716	6,244	5,244	4,520	3,972	3,542	3,196	2,674	2,298	2,015	1,617	1,350
140	75,458	51,647	39,259	31,664	26,532	22,831	17,851	14,654	12,429	10,790	9,533	8,539	7,732	6,255	5,251	4,525	3,976	3,545	3,199	2,676	2,300	2,016	1,618	1,351
145	76,887	52,313	39,643	31,913	26,706	22,960	17,930	14,707	12,467	10,819	9,556	8,557	7,747	6,264	5,258	4,530	3,980	3,548	3,201	2,678	2,301	2,017	1,618	1,351
150	78,270	52,950	40,007	32,149	26,871	23,082	18,004	14,757	12,503	10,846	9,577	8,573	7,760	6,273	5,264	4,535	3,983	3,551	3,204	2,679	2,302	2,018	1,619	1,352
155	79,610	53,559	40,354	32,373	27,027	23,197	18,074	14,804	12,537	10,871	9,596	8,589	7,773	6,282	5,270	4,540	3,987	3,554	3,206	2,681	2,303	2,019	1,620	1,352
160	80,909	54,144	40,685	32,586	27,175	23,306	18,140	14,849	12,568	10,895	9,615	8,604	7,786	6,290	5,276	4,544	3,990	3,556	3,208	2,682	2,305	2,020	1,620	1,352
165	82,168	54,705	41,001	32,788	27,316	23,409	18,203	14,891	12,598	10,918	9,633	8,618	7,797	6,297	5,281	4,548	3,993	3,559	3,210	2,684	2,306	2,021	1,621	1,353
170	83,389	55,244	41,303	32,981	27,450	23,507	18,262	14,930	12,627	10,939	9,649	8,631	7,808	6,304	5,286	4,551	3,996	3,561	3,212	2,685	2,307	2,022	1,621	1,353
175	84,575	55,762	41,592	33,165	27,577	23,601	18,318	14,968	12,653	10,959	9,665	8,644	7,818	6,311	5,291	4,555	3,998	3,563	3,213	2,686	2,307	2,022	1,622	1,353
180	85,725	56,260	41,868	33,340	27,698	23,689	18,371	15,003	12,679	10,978	9,680	8,656	7,828	6,317	5,295	4,558	4,001	3,565	3,215	2,687	2,308	2,023	1,622	1,354
185	86,843	56,739	42,133	33,508	27,814	23,774	18,422	15,037	12,703	10,996	9,694	8,667	7,837	6,323	5,300	4,561	4,003	3,567	3,217	2,688	2,309	2,024	1,622	1,354
190	87,929	57,200	42,387	33,668	27,924	23,855	18,471	15,069	12,726	11,014	9,707	8,678	7,846	6,329	5,304	4,564	4,006	3,569	3,218	2,689	2,310	2,024	1,623	1,354
195	88,985	57,645	42,631	33,822	28,030	23,932	18,517	15,100	12,748	11,030	9,720	8,688	7,854	6,334	5,307	4,567	4,008	3,571	3,219	2,690	2,311	2,025	1,623	1,354
200	90,012	58,075	42,865	33,969	28,131	24,005	18,561	15,129	12,769	11,046	9,732	8,698	7,862	6,340	5,311	4,570	4,010	3,572	3,221	2,691	2,311	2,025	1,623	1,355

# 2 AWG	1 Run(s)	240 Volts	C-Value = 5,906	Magnetic Duct	Copper

Calculations:

1. $f = \dfrac{1.732 \times \text{Length} \times \text{Isca}}{\text{Runs} \times \text{C-Value} \times E_{L\text{-}L}}$

2. $M = \dfrac{1}{1 + f}$

3. Isc = Isca x M

4. Add Motor Contribution = Motor FLA x 4

* All results are given in symmetrical amperes

Three-Phase Bolted Fault-Current* Table

One-Way Distance in Feet

Available Isc in Thousands (Isca)	5	10	15	20	25	30	40	50	60	70	80	90	100	125	150	175	200	225	250	300	350	400	500	600
3	2,956	2,913	2,872	2,832	2,793	2,755	2,682	2,612	2,546	2,484	2,424	2,367	2,313	2,188	2,076	1,974	1,882	1,799	1,722	1,587	1,471	1,371	1,207	1,079
5	4,879	4,764	4,655	4,550	4,450	4,354	4,174	4,008	3,855	3,714	3,582	3,459	3,345	3,089	2,870	2,680	2,513	2,366	2,235	2,012	1,830	1,678	1,439	1,260
7	6,766	6,546	6,341	6,148	5,967	5,796	5,481	5,199	4,945	4,714	4,504	4,312	4,135	3,751	3,433	3,164	2,934	2,736	2,562	2,274	2,044	1,856	1,568	1,357
10	9,528	9,099	8,707	8,348	8,017	7,711	7,164	6,690	6,274	5,907	5,581	5,289	5,026	4,470	4,025	3,660	3,356	3,099	2,878	2,520	2,240	2,017	1,681	1,441
15	13,964	13,061	12,268	11,566	10,940	10,378	9,411	8,610	7,934	7,356	6,857	6,421	6,037	5,253	4,649	4,169	3,779	3,456	3,184	2,751	2,421	2,162	1,781	1,514
20	18,199	16,695	15,421	14,328	13,379	12,549	11,162	10,052	9,142	8,384	7,742	7,191	6,713	5,757	5,039	4,480	4,033	3,667	3,362	2,883	2,523	2,243	1,835	1,553
25	22,248	20,041	18,233	16,724	15,446	14,349	12,565	11,175	10,062	9,151	8,391	7,748	7,196	6,108	5,306	4,691	4,203	3,807	3,479	2,968	2,588	2,294	1,870	1,578
30	26,122	23,132	20,756	18,823	17,219	15,867	13,714	12,075	10,786	9,746	8,888	8,170	7,559	6,368	5,501	4,842	4,324	3,906	3,562	3,028	2,634	2,330	1,893	1,595
35	29,833	25,996	23,033	20,676	18,757	17,164	14,672	12,811	11,370	10,220	9,281	8,500	7,841	6,567	5,649	4,956	4,415	3,980	3,623	3,072	2,667	2,356	1,911	1,607
40	33,391	28,656	25,097	22,325	20,104	18,285	15,483	13,426	11,851	10,607	9,599	8,767	8,067	6,725	5,765	5,046	4,486	4,038	3,671	3,107	2,693	2,376	1,924	1,616
45	36,805	31,134	26,978	23,801	21,293	19,263	16,179	13,946	12,255	10,929	9,862	8,985	8,252	6,853	5,859	5,117	4,542	4,083	3,709	3,134	2,713	2,392	1,934	1,623
50	40,083	33,449	28,698	25,130	22,351	20,125	16,782	14,392	12,598	11,201	10,083	9,168	8,406	6,959	5,936	5,176	4,589	4,121	3,739	3,156	2,729	2,405	1,942	1,629
55	43,234	35,614	30,278	26,333	23,297	20,889	17,310	14,779	12,893	11,434	10,272	9,324	8,536	7,048	6,001	5,225	4,627	4,152	3,765	3,174	2,743	2,415	1,949	1,634
60	46,264	37,646	31,734	27,427	24,150	21,572	17,777	15,117	13,150	11,636	10,434	9,457	8,648	7,124	6,056	5,267	4,660	4,178	3,787	3,189	2,754	2,424	1,955	1,638
65	49,181	39,555	33,080	28,427	24,921	22,185	18,191	15,416	13,375	11,812	10,576	9,574	8,745	7,189	6,104	5,303	4,688	4,201	3,805	3,202	2,764	2,432	1,960	1,642
70	51,991	41,352	34,328	29,344	25,623	22,740	18,562	15,682	13,575	11,967	10,700	9,675	8,830	7,247	6,145	5,334	4,712	4,220	3,821	3,214	2,773	2,438	1,964	1,645
75	54,700	43,048	35,488	30,187	26,264	23,243	18,896	15,919	13,753	12,105	10,810	9,765	8,905	7,297	6,181	5,361	4,733	4,237	3,835	3,223	2,780	2,444	1,968	1,647
80	57,312	44,649	36,570	30,966	26,852	23,702	19,199	16,133	13,912	12,228	10,908	9,845	8,971	7,342	6,213	5,385	4,752	4,252	3,847	3,232	2,786	2,449	1,971	1,649
85	59,833	46,165	37,580	31,688	27,392	24,123	19,474	16,327	14,056	12,339	10,996	9,917	9,031	7,382	6,242	5,407	4,769	4,265	3,858	3,240	2,792	2,453	1,974	1,651
90	62,269	47,601	38,527	32,358	27,892	24,509	19,725	16,503	14,186	12,440	11,076	9,982	9,084	7,417	6,267	5,426	4,784	4,277	3,868	3,247	2,797	2,457	1,976	1,653
95	64,622	48,964	39,415	32,982	28,354	24,865	19,955	16,664	14,305	12,531	11,148	10,041	9,133	7,450	6,290	5,443	4,797	4,288	3,877	3,253	2,802	2,461	1,979	1,655
100	66,897	50,260	40,249	33,565	28,784	25,195	20,167	16,811	14,413	12,614	11,214	10,094	9,177	7,479	6,311	5,459	4,809	4,298	3,885	3,258	2,806	2,464	1,981	1,656
105	69,098	51,492	41,036	34,110	29,184	25,501	20,362	16,947	14,513	12,690	11,274	10,143	9,217	7,506	6,330	5,473	4,820	4,307	3,892	3,263	2,810	2,467	1,983	1,657
110	71,229	52,666	41,778	34,621	29,557	25,786	20,543	17,072	14,605	12,760	11,330	10,187	9,254	7,530	6,348	5,486	4,830	4,315	3,899	3,268	2,813	2,469	1,984	1,659
115	73,292	53,786	42,480	35,101	29,907	26,051	20,711	17,188	14,690	12,825	11,381	10,229	9,288	7,553	6,364	5,498	4,840	4,322	3,905	3,272	2,816	2,472	1,986	1,660
120	75,292	54,855	43,144	35,553	30,234	26,300	20,868	17,296	14,768	12,885	11,428	10,267	9,320	7,573	6,378	5,509	4,848	4,329	3,910	3,276	2,819	2,474	1,987	1,661
125	77,230	55,876	43,773	35,980	30,542	26,532	21,014	17,396	14,841	12,940	11,471	10,302	9,349	7,593	6,392	5,519	4,856	4,335	3,915	3,280	2,822	2,476	1,989	1,662
130	79,110	56,854	44,371	36,383	30,832	26,751	21,151	17,490	14,909	12,992	11,512	10,335	9,376	7,610	6,404	5,528	4,863	4,341	3,920	3,283	2,824	2,478	1,990	1,663
135	80,934	57,790	44,939	36,764	31,105	26,956	21,279	17,578	14,973	13,040	11,550	10,365	9,401	7,627	6,416	5,537	4,870	4,346	3,924	3,286	2,827	2,480	1,991	1,663
140	82,705	58,687	45,480	37,125	31,363	27,150	21,400	17,660	15,032	13,086	11,585	10,394	9,424	7,642	6,427	5,545	4,876	4,351	3,928	3,289	2,829	2,481	1,992	1,664
145	84,425	59,548	45,995	37,467	31,607	27,332	21,513	17,737	15,088	13,128	11,618	10,420	9,446	7,657	6,437	5,553	4,882	4,356	3,932	3,292	2,831	2,483	1,993	1,665
150	86,095	60,374	46,486	37,793	31,839	27,505	21,620	17,809	15,141	13,168	11,650	10,445	9,467	7,670	6,447	5,560	4,888	4,360	3,936	3,294	2,832	2,484	1,994	1,665
155	87,720	61,168	46,956	38,103	32,058	27,669	21,721	17,878	15,190	13,205	11,679	10,469	9,486	7,683	6,456	5,567	4,893	4,364	3,939	3,296	2,834	2,486	1,995	1,666
160	89,299	61,932	47,405	38,397	32,267	27,824	21,817	17,943	15,237	13,240	11,706	10,491	9,504	7,695	6,464	5,573	4,898	4,368	3,942	3,299	2,836	2,487	1,996	1,667
165	90,835	62,667	47,834	38,679	32,465	27,971	21,907	18,004	15,281	13,274	11,732	10,512	9,521	7,706	6,472	5,579	4,902	4,372	3,945	3,301	2,837	2,488	1,996	1,667
170	92,330	63,375	48,245	38,947	32,654	28,112	21,993	18,062	15,323	13,305	11,757	10,532	9,538	7,717	6,480	5,584	4,906	4,375	3,948	3,303	2,839	2,489	1,997	1,668
175	93,785	64,057	48,640	39,204	32,834	28,245	22,075	18,117	15,362	13,335	11,780	10,550	9,553	7,727	6,487	5,590	4,910	4,378	3,951	3,305	2,840	2,490	1,998	1,668
180	95,203	64,716	49,018	39,449	33,006	28,372	22,152	18,169	15,400	13,363	11,802	10,568	9,567	7,736	6,493	5,594	4,914	4,382	3,953	3,306	2,841	2,491	1,998	1,668
185	96,583	65,351	49,382	39,684	33,171	28,494	22,226	18,219	15,436	13,390	11,823	10,585	9,581	7,745	6,500	5,599	4,918	4,384	3,955	3,308	2,843	2,492	1,999	1,669
190	97,929	65,964	49,731	39,910	33,328	28,610	22,297	18,266	15,470	13,416	11,843	10,601	9,594	7,754	6,506	5,604	4,921	4,387	3,958	3,309	2,844	2,493	2,000	1,669
195	99,240	66,556	50,067	40,126	33,478	28,721	22,364	18,311	15,502	13,440	11,862	10,616	9,607	7,762	6,511	5,608	4,925	4,390	3,960	3,311	2,845	2,494	2,000	1,670
200	100,519	67,129	50,390	40,333	33,623	28,827	22,428	18,354	15,533	13,463	11,880	10,630	9,618	7,769	6,517	5,612	4,928	4,392	3,962	3,312	2,846	2,495	2,001	1,670

# 1 AWG	1 Run(s)	240 Volts	C-Value = 7,292	Magnetic Duct	Copper

Calculations:

1. $f = \dfrac{1.732 \times \text{Length} \times \text{Isca}}{\text{Runs} \times \text{C-Value} \times \text{E L-L}}$
2. $M = \dfrac{1}{1 + f}$
3. Isc = Isca x M
4. Add Motor Contribution = Motor FLA x 4

* All results are given in symmetrical amperes

Three-Phase Bolted Fault-Current* Table

One-Way Distance in Feet

Available Isc in Thousands (Isca)

	5	10	15	20	25	30	40	50	60	70	80	90	100	125	150	175	200	225	250	300	350	400	500	600
3	2,964	2,929	2,895	2,861	2,828	2,796	2,735	2,675	2,619	2,564	2,512	2,462	2,414	2,302	2,200	2,106	2,020	1,941	1,867	1,736	1,622	1,523	1,356	1,222
5	4,901	4,806	4,714	4,626	4,541	4,459	4,304	4,159	4,024	3,897	3,778	3,666	3,560	3,321	3,112	2,928	2,764	2,618	2,487	2,259	2,070	1,910	1,655	1,459
7	6,807	6,625	6,452	6,288	6,132	5,984	5,708	5,456	5,225	5,013	4,818	4,637	4,470	4,099	3,786	3,516	3,283	3,079	2,898	2,594	2,348	2,144	1,827	1,592
10	9,611	9,252	8,918	8,608	8,318	8,048	7,556	7,121	6,733	6,385	6,072	5,788	5,529	4,973	4,519	4,140	3,821	3,547	3,309	2,919	2,611	2,361	1,983	1,709
15	14,142	13,377	12,691	12,071	11,510	10,998	10,100	9,337	8,681	8,112	7,613	7,171	6,778	5,961	5,320	4,803	4,378	4,022	3,720	3,233	2,860	2,563	2,123	1,812
20	18,504	17,216	16,095	15,112	14,242	13,466	12,144	11,058	10,150	9,380	8,719	8,145	7,641	6,619	5,838	5,221	4,723	4,311	3,966	3,418	3,003	2,678	2,201	1,868
25	22,705	20,796	19,183	17,802	16,607	15,562	13,822	12,433	11,297	10,351	9,552	8,867	8,273	7,088	6,200	5,509	4,957	4,505	4,129	3,539	3,096	2,751	2,251	1,904
30	26,755	24,143	21,996	20,199	18,674	17,363	15,225	13,556	12,217	11,118	10,201	9,424	8,756	7,439	6,467	5,719	5,126	4,645	4,246	3,624	3,161	2,803	2,285	1,928
35	30,661	27,279	24,569	22,349	20,497	18,928	16,415	14,492	12,971	11,740	10,722	9,867	9,137	7,713	6,672	5,879	5,255	4,750	4,334	3,688	3,209	2,841	2,310	1,946
40	34,431	30,224	26,932	24,287	22,116	20,300	17,438	15,283	13,602	12,254	11,149	10,227	9,446	7,931	6,835	6,005	5,355	4,832	4,402	3,737	3,246	2,870	2,329	1,960
45	38,073	32,993	29,110	26,044	23,563	21,513	18,325	15,960	14,136	12,686	11,505	10,526	9,700	8,110	6,967	6,107	5,436	4,898	4,456	3,776	3,276	2,893	2,344	1,971
50	41,591	35,604	31,123	27,644	24,865	22,594	19,103	16,547	14,594	13,054	11,807	10,778	9,914	8,259	7,077	6,191	5,502	4,952	4,501	3,808	3,300	2,911	2,357	1,979
55	44,994	38,068	32,990	29,107	26,042	23,561	19,791	17,060	14,992	13,371	12,066	10,993	10,096	8,385	7,169	6,262	5,558	4,997	4,538	3,835	3,320	2,927	2,367	1,987
60	48,286	40,398	34,726	30,450	27,112	24,434	20,402	17,513	15,340	13,647	12,291	11,180	10,253	8,492	7,248	6,322	5,605	5,035	4,570	3,857	3,337	2,940	2,375	1,993
65	51,472	42,605	36,344	31,687	28,089	25,224	20,950	17,915	15,648	13,890	12,488	11,342	10,389	8,586	7,316	6,373	5,646	5,067	4,597	3,876	3,351	2,951	2,383	1,998
70	54,558	44,698	37,856	32,831	28,983	25,943	21,444	18,275	15,922	14,106	12,661	11,485	10,509	8,668	7,375	6,418	5,681	5,096	4,620	3,893	3,363	2,961	2,389	2,002
75	57,548	46,685	39,272	33,890	29,806	26,600	21,891	18,599	16,167	14,298	12,816	11,612	10,616	8,740	7,427	6,458	5,712	5,121	4,640	3,907	3,374	2,969	2,394	2,006
80	60,447	48,575	40,601	34,875	30,565	27,203	22,298	18,891	16,388	14,470	12,954	11,726	10,710	8,804	7,474	6,493	5,739	5,143	4,658	3,920	3,384	2,976	2,399	2,009
85	63,259	50,374	41,850	35,793	31,268	27,758	22,670	19,158	16,588	14,626	13,079	11,828	10,795	8,861	7,515	6,524	5,764	5,162	4,674	3,931	3,392	2,983	2,403	2,012
90	65,987	52,089	43,027	36,651	31,920	28,271	23,011	19,400	16,770	14,767	13,192	11,920	10,872	8,913	7,552	6,552	5,785	5,180	4,689	3,941	3,400	2,989	2,407	2,015
95	68,636	53,726	44,137	37,453	32,527	28,747	23,324	19,623	16,936	14,896	13,294	12,004	10,942	8,960	7,586	6,577	5,805	5,195	4,702	3,951	3,406	2,994	2,410	2,017
100	71,208	55,289	45,187	38,206	33,094	29,188	23,614	19,828	17,088	15,013	13,388	12,080	11,005	9,002	7,616	6,600	5,823	5,210	4,713	3,959	3,413	2,999	2,413	2,019
105	73,707	56,784	46,181	38,914	33,624	29,600	23,883	20,017	17,228	15,121	13,474	12,150	11,063	9,041	7,644	6,621	5,839	5,223	4,724	3,966	3,418	3,003	2,416	2,021
110	76,136	58,215	47,123	39,581	34,120	29,984	24,132	20,192	17,358	15,221	13,553	12,214	11,116	9,076	7,669	6,640	5,854	5,234	4,733	3,973	3,423	3,007	2,419	2,023
115	78,499	59,586	48,017	40,210	34,587	30,343	24,365	20,354	17,477	15,313	13,626	12,273	11,165	9,109	7,692	6,657	5,867	5,245	4,742	3,979	3,428	3,011	2,421	2,025
120	80,797	60,901	48,867	40,805	35,026	30,681	24,582	20,506	17,589	15,399	13,693	12,328	11,211	9,139	7,714	6,673	5,880	5,255	4,751	3,985	3,432	3,014	2,423	2,026
125	83,033	62,163	49,677	41,367	35,440	30,998	24,785	20,647	17,693	15,478	13,756	12,379	11,253	9,167	7,734	6,688	5,891	5,264	4,758	3,990	3,436	3,017	2,425	2,028
130	85,210	63,375	50,448	41,901	35,830	31,296	24,975	20,779	17,789	15,552	13,815	12,426	11,292	9,193	7,752	6,702	5,902	5,273	4,765	3,995	3,440	3,020	2,427	2,029
135	87,330	64,540	51,183	42,407	36,200	31,578	25,154	20,902	17,880	15,621	13,869	12,471	11,328	9,217	7,769	6,715	5,912	5,281	4,772	4,000	3,443	3,022	2,429	2,030
140	89,395	65,661	51,886	42,888	36,550	31,844	25,323	21,019	17,965	15,686	13,920	12,512	11,362	9,240	7,785	6,727	5,921	5,288	4,778	4,004	3,446	3,025	2,430	2,031
145	91,408	66,741	52,553	43,346	36,882	32,096	25,482	21,128	18,045	15,747	13,968	12,551	11,394	9,261	7,800	6,738	5,930	5,295	4,783	4,008	3,449	3,027	2,432	2,032
150	93,370	67,781	53,200	43,782	37,197	32,334	25,632	21,231	18,120	15,804	14,013	12,587	11,424	9,281	7,814	6,748	5,938	5,302	4,788	4,012	3,452	3,029	2,433	2,033
155	95,283	68,783	53,816	44,198	37,497	32,561	25,774	21,328	18,191	15,858	14,056	12,621	11,452	9,299	7,828	6,758	5,946	5,308	4,793	4,015	3,454	3,031	2,434	2,034
160	97,150	69,751	54,405	44,596	37,783	32,776	25,909	21,421	18,258	15,909	14,096	12,653	11,479	9,317	7,840	6,767	5,953	5,313	4,798	4,018	3,457	3,033	2,436	2,035
165	98,971	70,684	54,973	44,976	38,055	32,980	26,036	21,508	18,321	15,957	14,133	12,684	11,504	9,333	7,852	6,776	5,960	5,319	4,802	4,021	3,459	3,035	2,437	2,036
170	100,748	71,586	55,517	45,339	38,315	33,175	26,158	21,591	18,381	16,003	14,169	12,712	11,527	9,349	7,863	6,784	5,966	5,324	4,806	4,024	3,461	3,036	2,438	2,036
175	102,483	72,458	56,040	45,687	38,563	33,361	26,273	21,669	18,438	16,046	14,203	12,740	11,550	9,363	7,873	6,792	5,972	5,329	4,810	4,027	3,463	3,038	2,439	2,037
180	104,178	73,301	56,543	46,021	38,801	33,539	26,383	21,744	18,492	16,087	14,235	12,765	11,571	9,377	7,883	6,799	5,978	5,333	4,814	4,030	3,465	3,039	2,440	2,038
185	105,833	74,117	57,027	46,341	39,028	33,709	26,488	21,815	18,544	16,126	14,265	12,790	11,591	9,391	7,892	6,806	5,983	5,337	4,818	4,032	3,467	3,041	2,441	2,038
190	107,451	74,907	57,493	46,649	39,246	33,871	26,588	21,883	18,593	16,163	14,294	12,813	11,610	9,403	7,901	6,813	5,988	5,341	4,821	4,034	3,469	3,042	2,441	2,039
195	109,032	75,672	57,943	46,944	39,455	34,027	26,684	21,948	18,640	16,198	14,322	12,835	11,628	9,415	7,909	6,819	5,993	5,345	4,824	4,037	3,470	3,043	2,442	2,039
200	110,578	76,413	58,376	47,229	39,656	34,176	26,776	22,010	18,684	16,232	14,348	12,857	11,646	9,426	7,918	6,825	5,997	5,349	4,827	4,039	3,472	3,044	2,443	2,040

1/0 AWG | 1 Run(s) | 240 Volts | C-Value = 8,924 | Magnetic Duct | Copper

Calculations:

1. $f = \dfrac{1.732 \times \text{Length} \times \text{Isca}}{\text{Runs} \times \text{C-Value} \times E_{L\text{-}L}}$

2. $M = \dfrac{1}{1 + f}$

3. Isc = Isca x M

4. Add Motor Contribution = Motor FLA x 4

* All results are given in symmetrical amperes

Three-Phase Bolted Fault-Current* Table

One-Way Distance in Feet

Available Isc in Thousands (Isca)

Isca	5	10	15	20	25	30	40	50	60	70	80	90	100	125	150	175	200	225	250	300	350	400	500	600
3	2,982	2,964	2,946	2,929	2,912	2,895	2,861	2,828	2,796	2,765	2,735	2,705	2,675	2,605	2,538	2,475	2,414	2,357	2,302	2,200	2,106	2,020	1,867	1,736
5	4,950	4,901	4,853	4,806	4,759	4,714	4,626	4,541	4,459	4,380	4,304	4,230	4,159	3,991	3,837	3,693	3,560	3,437	3,321	3,112	2,928	2,764	2,487	2,259
7	6,902	6,807	6,715	6,625	6,537	6,452	6,288	6,132	5,984	5,842	5,708	5,579	5,456	5,171	4,914	4,681	4,470	4,277	4,099	3,786	3,516	3,283	2,898	2,594
10	9,802	9,611	9,428	9,252	9,082	8,918	8,608	8,318	8,048	7,794	7,556	7,332	7,121	6,643	6,225	5,856	5,529	5,236	4,973	4,519	4,140	3,821	3,309	2,919
15	14,559	14,142	13,749	13,377	13,025	12,691	12,071	11,510	10,998	10,530	10,100	9,703	9,337	8,532	7,854	7,277	6,778	6,343	5,961	5,320	4,803	4,378	3,720	3,233
20	19,223	18,504	17,836	17,216	16,637	16,095	15,112	14,242	13,466	12,771	12,144	11,575	11,058	9,946	9,037	8,281	7,641	7,093	6,619	5,838	5,221	4,723	3,966	3,418
25	23,797	22,705	21,708	20,796	19,957	19,183	17,802	16,607	15,562	14,640	13,822	13,091	12,433	11,045	9,935	9,029	8,273	7,635	7,088	6,200	5,509	4,957	4,129	3,539
30	28,285	26,755	25,382	24,143	23,019	21,996	20,199	18,674	17,363	16,224	15,225	14,342	13,556	11,922	10,640	9,607	8,756	8,044	7,439	6,467	5,719	5,126	4,246	3,624
35	32,687	30,661	28,871	27,279	25,853	24,569	22,349	20,497	18,928	17,582	16,415	15,394	14,492	12,640	11,208	10,067	9,137	8,365	7,713	6,672	5,879	5,255	4,334	3,688
40	37,007	34,431	32,190	30,224	28,483	26,932	24,287	22,116	20,300	18,760	17,438	16,289	15,283	13,238	11,675	10,443	9,446	8,622	7,931	6,835	6,005	5,355	4,402	3,737
45	41,247	38,073	35,352	32,993	30,930	29,110	26,044	23,563	21,513	19,792	18,325	17,061	15,960	13,743	12,067	10,755	9,700	8,834	8,110	6,967	6,107	5,436	4,456	3,776
50	45,410	41,591	38,365	35,604	33,213	31,123	27,644	24,865	22,594	20,702	19,103	17,733	16,547	14,176	12,399	11,018	9,914	9,011	8,259	7,077	6,191	5,502	4,501	3,808
55	49,496	44,994	41,242	38,068	35,348	32,990	29,107	26,042	23,561	21,512	19,791	18,324	17,060	14,551	12,685	11,243	10,096	9,161	8,385	7,169	6,262	5,558	4,538	3,835
60	53,509	48,286	43,991	40,398	37,348	34,726	30,450	27,112	24,434	22,237	20,402	18,848	17,513	14,879	12,934	11,438	10,253	9,290	8,492	7,248	6,322	5,605	4,570	3,857
65	57,450	51,472	46,621	42,605	39,226	36,344	31,687	28,089	25,224	22,889	20,950	19,314	17,915	15,168	13,152	11,608	10,389	9,402	8,586	7,316	6,373	5,646	4,597	3,876
70	61,322	54,558	49,138	44,698	40,993	37,856	32,831	28,983	25,943	23,480	21,444	19,733	18,275	15,425	13,345	11,758	10,509	9,500	8,668	7,375	6,418	5,681	4,620	3,893
75	65,125	57,548	51,551	46,685	42,659	39,272	33,890	29,806	26,600	24,017	21,891	20,111	18,599	15,655	13,516	11,892	10,616	9,587	8,740	7,427	6,458	5,712	4,640	3,907
80	68,862	60,447	53,864	48,575	44,231	40,601	34,875	30,565	27,203	24,508	22,298	20,454	18,891	15,862	13,670	12,011	10,710	9,664	8,804	7,474	6,493	5,739	4,658	3,920
85	72,535	63,259	56,086	50,374	45,718	41,850	35,793	31,268	27,758	24,957	22,670	20,766	19,158	16,049	13,809	12,118	10,795	9,733	8,861	7,515	6,524	5,764	4,674	3,931
90	76,145	65,987	58,220	52,089	47,126	43,027	36,651	31,920	28,271	25,371	23,011	21,052	19,400	16,220	13,935	12,214	10,872	9,796	8,913	7,552	6,552	5,785	4,689	3,941
95	79,694	68,636	60,272	53,726	48,462	44,137	37,453	32,527	28,747	25,753	23,324	21,314	19,623	16,375	14,049	12,302	10,942	9,852	8,960	7,586	6,577	5,805	4,702	3,951
100	83,183	71,208	62,247	55,289	49,730	45,187	38,206	33,094	29,188	26,107	23,614	21,556	19,828	16,517	14,154	12,382	11,005	9,903	9,002	7,616	6,600	5,823	4,713	3,959
105	86,614	73,707	64,148	56,784	50,936	46,181	38,914	33,624	29,600	26,436	23,883	21,780	20,017	16,648	14,250	12,456	11,063	9,950	9,041	7,644	6,621	5,839	4,724	3,966
110	89,988	76,136	65,980	58,215	52,085	47,123	39,581	34,120	29,984	26,742	24,132	21,987	20,192	16,769	14,339	12,523	11,116	9,993	9,076	7,669	6,640	5,854	4,733	3,973
115	93,307	78,499	67,747	59,586	53,180	48,017	40,210	34,587	30,343	27,027	24,365	22,180	20,354	16,881	14,420	12,586	11,165	10,033	9,109	7,692	6,657	5,867	4,742	3,979
120	96,571	80,797	69,452	60,901	54,225	48,867	40,805	35,026	30,681	27,295	24,582	22,359	20,506	16,985	14,496	12,643	11,211	10,069	9,139	7,714	6,673	5,880	4,751	3,985
125	99,783	83,033	71,098	62,163	55,223	49,677	41,367	35,440	30,998	27,545	24,785	22,527	20,647	17,082	14,566	12,697	11,253	10,103	9,167	7,734	6,688	5,891	4,758	3,990
130	102,944	85,210	72,688	63,375	56,177	50,448	41,901	35,830	31,296	27,781	24,975	22,685	20,779	17,172	14,632	12,747	11,292	10,135	9,193	7,752	6,702	5,902	4,765	3,995
135	106,055	87,330	74,225	64,540	57,091	51,183	42,407	36,200	31,578	28,002	25,154	22,832	20,902	17,256	14,693	12,793	11,328	10,164	9,217	7,769	6,715	5,912	4,772	4,000
140	109,116	89,395	75,712	65,661	57,966	51,886	42,888	36,550	31,844	28,211	25,323	22,971	21,019	17,335	14,751	12,837	11,362	10,192	9,240	7,785	6,727	5,921	4,778	4,004
145	112,130	91,408	77,151	66,741	58,806	52,558	43,346	36,882	32,096	28,409	25,482	23,102	21,128	17,410	14,804	12,877	11,394	10,217	9,261	7,800	6,738	5,930	4,783	4,008
150	115,096	93,370	78,544	67,781	59,612	53,200	43,782	37,197	32,334	28,596	25,632	23,225	21,231	17,480	14,855	12,916	11,424	10,241	9,281	7,814	6,748	5,938	4,788	4,012
155	118,018	95,283	79,893	68,783	60,386	53,816	44,198	37,497	32,561	28,772	25,774	23,341	21,328	17,546	14,903	12,951	11,452	10,264	9,299	7,828	6,758	5,946	4,793	4,015
160	120,894	97,150	81,201	69,751	61,130	54,406	44,596	37,783	32,776	28,940	25,909	23,452	21,421	17,608	14,947	12,985	11,479	10,285	9,317	7,840	6,767	5,953	4,798	4,018
165	123,727	98,971	82,469	70,684	61,846	54,973	44,976	38,055	32,980	29,100	26,036	23,556	21,508	17,667	14,990	13,017	11,504	10,305	9,333	7,852	6,776	5,960	4,802	4,021
170	126,517	100,748	83,700	71,586	62,536	55,517	45,339	38,315	33,175	29,252	26,158	23,656	21,591	17,723	15,030	13,048	11,527	10,324	9,349	7,863	6,784	5,966	4,806	4,024
175	129,266	102,483	84,894	72,458	63,200	56,040	45,687	38,563	33,361	29,396	26,273	23,750	21,669	17,776	15,068	13,076	11,550	10,342	9,363	7,873	6,792	5,972	4,810	4,027
180	131,974	104,178	86,054	73,301	63,840	56,543	46,021	38,801	33,539	29,534	26,383	23,840	21,744	17,826	15,104	13,104	11,571	10,359	9,377	7,883	6,799	5,978	4,814	4,030
185	134,642	105,833	87,180	74,117	64,458	57,027	46,341	39,028	33,709	29,665	26,488	23,926	21,815	17,874	15,139	13,129	11,591	10,375	9,391	7,892	6,806	5,983	4,818	4,032
190	137,271	107,451	88,275	74,907	65,055	57,493	46,649	39,246	33,871	29,791	26,588	24,007	21,883	17,919	15,171	13,154	11,610	10,391	9,403	7,901	6,813	5,988	4,821	4,034
195	139,862	109,032	89,339	75,672	65,631	57,943	46,944	39,455	34,027	29,911	26,684	24,085	21,948	17,963	15,202	13,177	11,628	10,405	9,415	7,909	6,819	5,993	4,824	4,037
200	142,416	110,578	90,374	76,413	66,188	58,376	47,229	39,656	34,176	30,027	26,776	24,160	22,010	18,004	15,232	13,200	11,646	10,419	9,426	7,918	6,825	5,997	4,827	4,039

# 1/0 AWG	2 Run(s)	240 Volts	C-Value = 8,924	Magnetic Duct	Copper

Calculations:

1. $f = \dfrac{1.732 \times \text{Length} \times \text{Isca}}{\text{Runs} \times \text{C-Value} \times E_{L\text{-}L}}$

2. $M = \dfrac{1}{1 + f}$

3. Isc = Isca x M

4. Add Motor Contribution = Motor FLA x 4

* All results are given in symmetrical amperes

Three-Phase Bolted Fault-Current* Table

One-Way Distance in Feet

Available Isc in Thousands (Isca)	5	10	15	20	25	30	40	50	60	70	80	90	100	125	150	175	200	225	250	300	350	400	500	600
3	2,970	2,941	2,912	2,884	2,856	2,829	2,776	2,726	2,677	2,629	2,584	2,540	2,497	2,397	2,304	2,218	2,139	2,065	1,996	1,870	1,760	1,662	1,495	1,359
5	4,918	4,838	4,760	4,686	4,613	4,543	4,408	4,282	4,162	4,049	3,942	3,840	3,744	3,523	3,326	3,150	2,992	2,849	2,719	2,492	2,300	2,135	1,867	1,659
7	6,839	6,686	6,539	6,399	6,264	6,135	5,893	5,669	5,461	5,268	5,088	4,920	4,763	4,410	4,107	3,842	3,609	3,403	3,219	2,906	2,648	2,432	2,090	1,833
10	9,675	9,371	9,086	8,817	8,563	8,324	7,884	7,488	7,130	6,804	6,507	6,235	5,984	5,438	4,984	4,599	4,270	3,984	3,735	3,319	2,986	2,714	2,296	1,990
15	14,281	13,628	13,032	12,486	11,984	11,521	10,694	9,978	9,352	8,800	8,309	7,870	7,476	6,643	5,977	5,432	4,978	4,595	4,266	3,732	3,317	2,984	2,487	2,131
20	18,742	17,634	16,649	15,768	14,976	14,259	13,014	11,969	11,079	10,312	9,645	9,059	8,540	7,470	6,638	5,973	5,429	4,976	4,592	3,979	3,511	3,141	2,594	2,209
25	23,065	21,409	19,974	18,720	17,613	16,631	14,961	13,596	12,459	11,498	10,675	9,961	9,337	8,073	7,110	6,352	5,740	5,236	4,813	4,144	3,638	3,243	2,663	2,259
30	27,257	24,973	23,042	21,389	19,957	18,704	16,619	14,951	13,588	12,453	11,492	10,670	9,957	8,532	7,464	6,633	5,969	5,426	4,973	4,262	3,729	3,314	2,711	2,294
35	31,322	28,343	25,882	23,814	22,052	20,533	18,047	16,097	14,528	13,238	12,158	11,241	10,452	8,893	7,739	6,849	6,144	5,570	5,094	4,350	3,796	3,367	2,747	2,319
40	35,267	31,536	28,518	26,028	23,938	22,158	19,290	17,079	15,323	13,895	12,710	11,711	10,858	9,185	7,959	7,021	6,281	5,683	5,188	4,419	3,848	3,408	2,774	2,339
45	39,097	34,563	30,972	28,057	25,643	23,611	20,382	17,930	16,004	14,452	13,175	12,105	11,195	9,425	8,138	7,161	6,393	5,774	5,264	4,474	3,890	3,441	2,795	2,354
50	42,817	37,439	33,261	29,922	27,192	24,919	21,349	18,674	16,595	14,932	13,572	12,439	11,481	9,627	8,288	7,277	6,485	5,849	5,326	4,519	3,924	3,467	2,813	2,366
55	46,432	40,174	35,402	31,644	28,607	26,101	22,211	19,330	17,111	15,349	13,916	12,727	11,726	9,798	8,415	7,374	6,562	5,912	5,378	4,556	3,952	3,489	2,827	2,377
60	49,946	42,778	37,409	33,237	29,903	27,176	22,985	19,914	17,566	15,714	14,215	12,977	11,938	9,946	8,524	7,458	6,628	5,965	5,422	4,588	3,976	3,508	2,840	2,385
65	53,363	45,260	39,293	34,717	31,095	28,157	23,683	20,435	17,971	16,037	14,479	13,197	12,123	10,075	8,618	7,530	6,685	6,011	5,460	4,615	3,996	3,524	2,850	2,392
70	56,687	47,629	41,066	36,093	32,195	29,056	24,316	20,905	18,333	16,325	14,713	13,391	12,287	10,187	8,700	7,592	6,735	6,051	5,493	4,638	4,014	3,537	2,859	2,399
75	59,922	49,892	42,738	37,378	33,213	29,883	24,892	21,329	18,659	16,583	14,922	13,564	12,433	10,287	8,773	7,648	6,778	6,086	5,522	4,659	4,029	3,549	2,867	2,404
80	63,071	52,056	44,316	38,580	34,159	30,646	25,419	21,715	18,954	16,815	15,110	13,719	12,563	10,376	8,838	7,697	6,817	6,117	5,548	4,677	4,043	3,560	2,874	2,409
85	66,139	54,128	45,809	39,706	35,039	31,353	25,903	22,068	19,222	17,026	15,280	13,859	12,680	10,456	8,896	7,740	6,851	6,145	5,571	4,693	4,055	3,569	2,880	2,413
90	69,127	56,113	47,223	40,764	35,860	32,009	26,350	22,391	19,466	17,217	15,434	13,986	12,786	10,528	8,948	7,780	6,882	6,170	5,591	4,708	4,066	3,578	2,885	2,417
95	72,039	58,017	48,564	41,760	36,628	32,620	26,762	22,688	19,690	17,392	15,575	14,101	12,882	10,593	8,995	7,815	6,910	6,192	5,609	4,721	4,075	3,585	2,890	2,421
100	74,878	59,844	49,838	42,698	37,348	33,189	27,144	22,962	19,896	17,553	15,703	14,206	12,970	10,652	9,037	7,848	6,935	6,212	5,626	4,733	4,084	3,592	2,894	2,424
105	77,647	61,600	51,049	43,585	38,024	33,722	27,500	23,216	20,087	17,701	15,822	14,303	13,051	10,707	9,076	7,877	6,958	6,231	5,641	4,743	4,092	3,598	2,898	2,426
110	80,348	63,287	52,203	44,423	38,661	34,222	27,831	23,451	20,263	17,838	15,931	14,392	13,125	10,757	9,112	7,904	6,979	6,247	5,655	4,753	4,099	3,604	2,902	2,429
115	82,983	64,911	53,303	45,217	39,261	34,691	28,141	23,671	20,427	17,964	16,032	14,475	13,193	10,802	9,145	7,929	6,998	6,263	5,667	4,762	4,106	3,609	2,905	2,431
120	85,555	66,474	54,352	45,970	39,827	35,133	28,430	23,876	20,579	18,082	16,125	14,551	13,257	10,845	9,176	7,952	7,016	6,277	5,679	4,770	4,112	3,614	2,908	2,433
125	88,067	67,981	55,355	46,685	40,363	35,549	28,702	24,067	20,721	18,192	16,213	14,622	13,315	10,884	9,204	7,973	7,032	6,290	5,690	4,778	4,118	3,618	2,911	2,435
130	90,520	69,433	56,315	47,365	40,871	35,942	28,958	24,247	20,854	18,294	16,294	14,688	13,370	10,921	9,230	7,992	7,048	6,302	5,700	4,785	4,123	3,622	2,914	2,437
135	92,916	70,834	57,233	48,013	41,352	36,314	29,199	24,415	20,979	18,390	16,370	14,750	13,421	10,955	9,254	8,011	7,062	6,314	5,709	4,791	4,128	3,626	2,916	2,439
140	95,257	72,187	58,113	48,631	41,810	36,666	29,426	24,574	21,096	18,480	16,441	14,807	13,469	10,987	9,277	8,028	7,075	6,324	5,718	4,797	4,132	3,629	2,918	2,441
145	97,546	73,494	58,957	49,221	42,245	37,000	29,641	24,724	21,206	18,564	16,508	14,862	13,514	11,017	9,298	8,044	7,087	6,334	5,726	4,803	4,137	3,632	2,921	2,442
150	99,784	74,757	59,767	49,784	42,659	37,318	29,845	24,865	21,310	18,644	16,571	14,913	13,556	11,045	9,318	8,058	7,099	6,343	5,733	4,808	4,140	3,635	2,923	2,443
155	101,972	75,978	60,545	50,323	43,054	37,620	30,037	24,999	21,408	18,719	16,630	14,961	13,596	11,071	9,337	8,072	7,110	6,352	5,740	4,813	4,144	3,638	2,924	2,445
160	104,112	77,160	61,293	50,839	43,431	37,907	30,220	25,125	21,501	18,790	16,686	15,006	13,633	11,096	9,354	8,086	7,120	6,360	5,747	4,818	4,148	3,641	2,926	2,446
165	106,206	78,304	62,013	51,333	43,791	38,181	30,394	25,246	21,589	18,857	16,739	15,049	13,668	11,119	9,371	8,098	7,130	6,368	5,753	4,822	4,151	3,643	2,928	2,447
170	108,256	79,413	62,706	51,807	44,136	38,443	30,560	25,360	21,672	18,920	16,789	15,089	13,702	11,141	9,387	8,110	7,139	6,375	5,759	4,827	4,154	3,646	2,929	2,448
175	110,262	80,487	63,374	52,262	44,465	38,693	30,718	25,468	21,751	18,981	16,837	15,127	13,733	11,162	9,402	8,121	7,147	6,382	5,765	4,831	4,157	3,648	2,931	2,449
180	112,226	81,529	64,018	52,699	44,781	38,932	30,868	25,572	21,826	19,038	16,882	15,164	13,763	11,182	9,416	8,131	7,155	6,388	5,770	4,834	4,160	3,650	2,932	2,450
185	114,150	82,539	64,639	53,119	45,085	39,161	31,012	25,670	21,898	19,093	16,925	15,199	13,792	11,201	9,429	8,141	7,163	6,395	5,775	4,838	4,162	3,652	2,933	2,451
190	116,034	83,520	65,239	53,524	45,376	39,380	31,149	25,764	21,967	19,145	16,965	15,231	13,819	11,218	9,442	8,151	7,170	6,400	5,780	4,841	4,165	3,654	2,935	2,452
195	117,880	84,472	65,819	53,913	45,655	39,591	31,281	25,854	22,032	19,194	17,004	15,263	13,845	11,235	9,454	8,160	7,177	6,406	5,784	4,844	4,167	3,656	2,936	2,453
200	119,688	85,397	66,379	54,288	45,924	39,793	31,407	25,940	22,094	19,242	17,041	15,293	13,870	11,252	9,465	8,168	7,184	6,411	5,789	4,847	4,169	3,658	2,937	2,453

2/0 AWG | 1 Run(s) | 240 Volts | C-Value = 10,755 | Magnetic Duct | Copper

Calculations:

1. $f = \dfrac{1.732 \times \text{Length} \times \text{Isca}}{\text{Runs} \times \text{C-Value} \times \text{E L-L}}$

2. $M = \dfrac{1}{1 + f}$

3. Isc = Isca x M

4. Add Motor Contribution = Motor FLA x 4

* All results are given in symmetrical amperes

Three-Phase Bolted Fault-Current* Table

One-Way Distance in Feet

Available Isc in Thousands (Isca)	5	10	15	20	25	30	40	50	60	70	80	90	100	125	150	175	200	225	250	300	350	400	500	600
3	2,985	2,970	2,955	2,941	2,926	2,912	2,884	2,856	2,829	2,803	2,776	2,751	2,726	2,665	2,606	2,551	2,497	2,446	2,397	2,304	2,218	2,139	1,996	1,870
5	4,958	4,918	4,877	4,838	4,799	4,760	4,686	4,613	4,543	4,475	4,408	4,344	4,282	4,133	3,995	3,865	3,744	3,630	3,523	3,326	3,150	2,992	2,719	2,492
7	6,919	6,839	6,762	6,686	6,612	6,539	6,399	6,264	6,135	6,012	5,893	5,779	5,669	5,411	5,176	4,961	4,763	4,580	4,410	4,107	3,842	3,609	3,219	2,906
10	9,835	9,675	9,521	9,371	9,226	9,086	8,817	8,563	8,324	8,098	7,884	7,681	7,488	7,045	6,652	6,301	5,984	5,698	5,438	4,984	4,599	4,270	3,735	3,319
15	14,632	14,281	13,947	13,628	13,324	13,032	12,486	11,984	11,521	11,092	10,694	10,324	9,978	9,208	8,548	7,976	7,476	7,035	6,643	5,977	5,432	4,978	4,266	3,732
20	19,351	18,742	18,171	17,634	17,127	16,649	15,768	14,976	14,259	13,608	13,014	12,470	11,969	10,877	9,968	9,199	8,540	7,969	7,470	6,638	5,973	5,429	4,592	3,979
25	23,994	23,065	22,206	21,409	20,666	19,974	18,720	17,613	16,631	15,752	14,961	14,246	13,596	12,204	11,071	10,130	9,337	8,659	8,073	7,110	6,352	5,740	4,813	4,144
30	28,563	27,257	26,065	24,973	23,969	23,042	21,389	19,957	18,704	17,600	16,619	15,741	14,951	13,285	11,953	10,864	9,957	9,189	8,532	7,464	6,633	5,969	4,973	4,262
35	33,059	31,322	29,758	28,343	27,057	25,882	23,814	22,052	20,533	19,210	18,047	17,016	16,097	14,183	12,675	11,457	10,452	9,610	8,893	7,739	6,849	6,144	5,094	4,350
40	37,485	35,267	33,297	31,536	29,951	28,518	26,028	23,938	22,158	20,625	19,290	18,117	17,079	14,939	13,276	11,946	10,858	9,951	9,185	7,959	7,021	6,281	5,188	4,419
45	41,841	39,097	36,691	34,563	32,669	30,972	28,057	25,643	23,611	21,878	20,382	19,078	17,930	15,586	13,784	12,356	11,195	10,234	9,425	8,138	7,161	6,393	5,264	4,474
50	46,131	42,817	39,948	37,439	35,227	33,261	29,922	27,192	24,919	22,996	21,349	19,922	18,674	16,145	14,220	12,704	11,481	10,473	9,627	8,288	7,277	6,485	5,326	4,519
55	50,354	46,432	43,077	40,174	37,637	35,402	31,644	28,607	26,101	24,000	22,211	20,671	19,330	16,633	14,597	13,005	11,726	10,676	9,798	8,415	7,374	6,562	5,378	4,556
60	54,513	49,946	46,085	42,778	39,913	37,409	33,237	29,903	27,176	24,905	22,985	21,339	19,914	17,064	14,927	13,266	11,938	10,851	9,946	8,524	7,458	6,628	5,422	4,588
65	58,609	53,363	48,978	45,260	42,066	39,293	34,717	31,095	28,157	25,727	23,683	21,940	20,435	17,445	15,218	13,496	12,123	11,004	10,075	8,618	7,530	6,685	5,460	4,615
70	62,644	56,687	51,765	47,629	44,105	41,066	36,093	32,195	29,056	26,475	24,316	22,482	20,905	17,786	15,477	13,699	12,287	11,139	10,187	8,700	7,592	6,735	5,493	4,638
75	66,618	59,922	54,449	49,892	46,039	42,738	37,378	33,213	29,883	27,160	24,892	22,973	21,329	18,093	15,709	13,880	12,433	11,259	10,287	8,773	7,648	6,778	5,522	4,659
80	70,534	63,071	57,037	52,056	47,875	44,316	38,580	34,159	30,646	27,789	25,419	23,422	21,715	18,370	15,917	14,042	12,563	11,365	10,376	8,838	7,697	6,817	5,548	4,677
85	74,392	66,139	59,534	54,128	49,622	45,809	39,706	35,039	31,353	28,369	25,903	23,832	22,068	18,621	16,106	14,189	12,680	11,461	10,456	8,896	7,740	6,851	5,571	4,693
90	78,194	69,127	61,944	56,113	51,286	47,223	40,764	35,860	32,009	28,905	26,350	24,209	22,391	18,851	16,277	14,322	12,786	11,547	10,528	8,948	7,780	6,882	5,591	4,708
95	81,941	72,039	64,272	58,017	52,871	48,564	41,760	36,628	32,620	29,402	26,762	24,557	22,688	19,061	16,433	14,443	12,882	11,626	10,593	8,995	7,815	6,910	5,609	4,721
100	85,635	74,878	66,522	59,844	54,385	49,838	42,698	37,348	33,189	29,864	27,144	24,879	22,962	19,254	16,577	14,553	12,970	11,698	10,652	9,037	7,848	6,935	5,626	4,733
105	89,275	77,647	68,699	61,600	55,830	51,049	43,585	38,024	33,722	30,295	27,500	25,177	23,216	19,432	16,709	14,655	13,051	11,763	10,707	9,076	7,877	6,958	5,641	4,743
110	92,864	80,348	70,804	63,287	57,213	52,203	44,423	38,661	34,222	30,697	27,831	25,454	23,451	19,597	16,830	14,748	13,125	11,823	10,757	9,112	7,904	6,979	5,655	4,753
115	96,403	82,983	72,843	64,911	58,537	53,303	45,217	39,261	34,691	31,074	28,141	25,713	23,671	19,750	16,943	14,835	13,193	11,879	10,802	9,145	7,929	6,998	5,667	4,762
120	99,892	85,555	74,817	66,474	59,805	54,352	45,970	39,827	35,133	31,428	28,430	25,955	23,876	19,892	17,048	14,915	13,257	11,930	10,845	9,176	7,952	7,016	5,679	4,770
125	103,332	88,067	76,731	67,981	61,022	55,355	46,685	40,363	35,549	31,761	28,702	26,181	24,067	20,025	17,145	14,990	13,315	11,978	10,884	9,204	7,973	7,032	5,690	4,778
130	106,726	90,520	78,586	69,433	62,190	56,315	47,365	40,871	35,942	32,074	28,958	26,394	24,247	20,149	17,236	15,059	13,370	12,022	10,921	9,230	7,992	7,048	5,700	4,785
135	110,072	92,916	80,386	70,834	63,311	57,233	48,013	41,352	36,314	32,370	29,199	26,594	24,415	20,265	17,321	15,124	13,421	12,063	10,955	9,254	8,011	7,062	5,709	4,791
140	113,374	95,257	82,133	72,187	64,390	58,113	48,631	41,810	36,666	32,650	29,426	26,782	24,574	20,375	17,401	15,185	13,469	12,102	10,987	9,277	8,028	7,075	5,718	4,797
145	116,631	97,546	83,829	73,494	65,427	58,957	49,221	42,245	37,000	32,914	29,641	26,960	24,724	20,477	17,476	15,242	13,514	12,138	11,017	9,298	8,044	7,087	5,726	4,803
150	119,844	99,784	85,476	74,757	66,426	59,767	49,784	42,659	37,318	33,165	29,845	27,128	24,865	20,574	17,546	15,295	13,556	12,172	11,045	9,318	8,058	7,099	5,733	4,808
155	123,014	101,972	87,076	75,978	67,389	60,545	50,323	43,054	37,620	33,404	30,037	27,287	24,999	20,666	17,613	15,346	13,596	12,204	11,071	9,337	8,072	7,110	5,740	4,813
160	126,143	104,112	88,633	77,160	68,317	61,293	50,839	43,431	37,907	33,630	30,220	27,438	25,125	20,752	17,676	15,393	13,533	12,234	11,096	9,354	8,086	7,120	5,747	4,818
165	129,230	106,206	90,146	78,304	69,213	62,013	51,333	43,791	38,181	33,846	30,394	27,582	25,246	20,834	17,735	15,438	13,568	12,263	11,119	9,371	8,098	7,130	5,753	4,822
170	132,277	108,256	91,618	79,413	70,077	62,706	51,807	44,136	38,443	34,051	30,560	27,718	25,360	20,912	17,791	15,481	13,702	12,289	11,141	9,387	8,110	7,139	5,759	4,827
175	135,285	110,262	93,051	80,487	70,913	63,374	52,262	44,465	38,693	34,247	30,718	27,848	25,468	20,985	17,844	15,521	13,733	12,315	11,162	9,402	8,121	7,147	5,765	4,831
180	138,254	112,226	94,446	81,529	71,720	64,018	52,699	44,781	38,932	34,434	30,868	27,971	25,572	21,056	17,895	15,560	13,763	12,339	11,182	9,416	8,131	7,155	5,770	4,834
185	141,185	114,150	95,804	82,539	72,501	64,639	53,119	45,085	39,161	34,613	31,012	28,089	25,670	21,122	17,943	15,596	13,792	12,362	11,201	9,429	8,141	7,163	5,775	4,838
190	144,078	116,034	97,128	83,520	73,256	65,239	53,524	45,376	39,380	34,785	31,149	28,202	25,764	21,186	17,989	15,631	13,819	12,384	11,218	9,442	8,151	7,170	5,780	4,841
195	146,935	117,880	98,418	84,472	73,988	65,819	53,913	45,655	39,591	34,949	31,281	28,310	25,854	21,247	18,033	15,664	13,845	12,404	11,235	9,454	8,160	7,177	5,784	4,844
200	149,756	119,688	99,676	85,397	74,696	66,379	54,288	45,924	39,793	35,106	31,407	28,413	25,940	21,305	18,075	15,695	13,870	12,424	11,252	9,465	8,168	7,184	5,789	4,847

# 2/0 AWG	2 Run(s)	240 Volts	C-Value = 10,755	Magnetic Duct	Copper

Calculations:

1. $f = \dfrac{1.732 \times \text{Length} \times \text{Isca}}{\text{Runs} \times \text{C-Value} \times E_{L\text{-}L}}$

2. $M = \dfrac{1}{1 + f}$

3. Isc = Isca x M

4. Add Motor Contribution = Motor FLA x 4

* All results are given in symmetrical amperes

Three-Phase Bolted Fault-Current* Table

One-Way Distance in Feet

Available Isc in Thousands (Isca)

Isca	5	10	15	20	25	30	40	50	60	70	80	90	100	125	150	175	200	225	250	300	350	400	500	600
3	2,975	2,950	2,926	2,902	2,879	2,856	2,810	2,767	2,724	2,683	2,643	2,605	2,567	2,478	2,395	2,317	2,244	2,175	2,111	1,992	1,887	1,792	1,628	1,491
5	4,931	4,863	4,798	4,734	4,672	4,611	4,495	4,384	4,279	4,178	4,082	3,991	3,903	3,700	3,518	3,352	3,201	3,063	2,937	2,713	2,521	2,354	2,079	1,862
7	6,865	6,735	6,610	6,489	6,373	6,261	6,048	5,850	5,663	5,489	5,325	5,170	5,024	4,693	4,402	4,146	3,918	3,713	3,529	3,211	2,945	2,720	2,360	2,083
10	9,727	9,468	9,223	8,990	8,768	8,557	8,165	7,807	7,479	7,177	6,899	6,641	6,402	5,874	5,426	5,042	4,708	4,416	4,158	3,723	3,371	3,079	2,625	2,288
15	14,393	13,834	13,316	12,836	12,389	11,973	11,218	10,553	9,962	9,434	8,959	8,530	8,139	7,304	6,625	6,061	5,585	5,179	4,828	4,251	3,797	3,431	2,877	2,476
20	18,936	17,979	17,115	16,330	15,613	14,957	13,798	12,805	11,945	11,194	10,532	9,943	9,417	8,317	7,447	6,741	6,158	5,668	5,250	4,575	4,054	3,639	3,022	2,583
25	23,359	21,921	20,649	19,517	18,502	17,588	16,006	14,685	13,566	12,605	11,771	11,041	10,396	9,071	8,046	7,229	6,562	6,009	5,541	4,794	4,225	3,777	3,116	2,651
30	27,668	25,672	23,945	22,436	21,105	19,924	17,918	16,279	14,915	13,761	12,774	11,918	11,170	9,655	8,502	7,595	6,863	6,259	5,753	4,953	4,348	3,874	3,182	2,699
35	31,866	29,248	27,027	25,119	23,464	22,012	19,589	17,647	16,055	14,726	13,601	12,635	11,798	10,120	8,861	7,880	7,094	6,452	5,915	5,072	4,440	3,947	3,231	2,734
40	35,959	32,659	29,914	27,595	25,610	23,891	21,063	18,834	17,031	15,544	14,295	13,232	12,317	10,500	9,150	8,108	7,279	6,604	6,043	5,166	4,511	4,004	3,268	2,761
45	39,949	35,918	32,625	29,886	27,571	25,589	22,372	19,874	17,877	16,245	14,886	13,737	12,753	10,815	9,389	8,295	7,429	6,727	6,146	5,241	4,568	4,049	3,298	2,783
50	43,841	39,033	35,176	32,012	29,370	27,132	23,542	20,792	18,617	16,854	15,396	14,170	13,125	11,082	9,589	8,451	7,554	6,829	6,231	5,303	4,615	4,086	3,323	2,800
55	47,639	42,015	37,579	33,990	31,027	28,539	24,595	21,609	19,269	17,387	15,839	14,545	13,446	11,310	9,759	8,582	7,659	6,915	6,303	5,355	4,654	4,116	3,343	2,814
60	51,345	44,872	39,848	35,836	32,558	29,829	25,547	22,340	19,849	17,857	16,229	14,872	13,725	11,507	9,906	8,696	7,749	6,988	6,364	5,398	4,687	4,142	3,360	2,826
65	54,963	47,611	41,993	37,562	33,976	31,015	26,412	22,999	20,367	18,275	16,573	15,161	13,971	11,679	10,033	8,794	7,827	7,051	6,416	5,436	4,716	4,164	3,374	2,837
70	58,496	50,239	44,025	39,179	35,294	32,110	27,202	23,595	20,833	18,650	16,881	15,418	14,189	11,831	10,145	8,879	7,895	7,106	6,461	5,469	4,740	4,183	3,387	2,845
75	61,947	52,764	45,952	40,697	36,521	33,123	27,925	24,138	21,255	18,987	17,157	15,648	14,383	11,966	10,244	8,955	7,954	7,155	6,501	5,497	4,762	4,200	3,398	2,853
80	65,319	55,190	47,781	42,126	37,668	34,063	28,591	24,633	21,638	19,292	17,405	15,855	14,558	12,086	10,332	9,022	8,008	7,198	6,537	5,523	4,781	4,215	3,408	2,860
85	68,614	57,525	49,521	43,473	38,741	34,938	29,205	25,088	21,988	19,570	17,631	16,042	14,715	12,195	10,411	9,083	8,055	7,236	6,568	5,545	4,798	4,228	3,416	2,866
90	71,836	59,772	51,178	44,744	39,747	35,754	29,773	25,506	22,308	19,824	17,837	16,212	14,858	12,293	10,482	9,137	8,098	7,271	6,597	5,565	4,813	4,240	3,424	2,871
95	74,986	61,937	52,756	45,946	40,693	36,518	30,300	25,892	22,603	20,056	18,025	16,367	14,989	12,382	10,547	9,186	8,136	7,302	6,622	5,583	4,826	4,250	3,431	2,876
100	78,067	64,024	54,263	47,085	41,584	37,234	30,791	26,250	22,876	20,270	18,197	16,509	15,108	12,463	10,606	9,231	8,171	7,330	6,645	5,600	4,839	4,260	3,437	2,881
105	81,081	66,037	55,702	48,165	42,424	37,906	31,250	26,582	23,127	20,468	18,356	16,640	15,217	12,537	10,660	9,271	8,203	7,355	6,667	5,615	4,850	4,268	3,443	2,885
110	84,030	67,981	57,079	49,190	43,218	38,538	31,678	26,891	23,361	20,651	18,503	16,761	15,318	12,606	10,709	9,309	8,232	7,379	6,686	5,629	4,860	4,276	3,448	2,888
115	86,917	69,858	58,396	50,166	43,969	39,134	32,080	27,180	23,579	20,820	18,640	16,873	15,411	12,669	10,755	9,343	8,259	7,400	6,704	5,641	4,869	4,283	3,452	2,891
120	89,743	71,672	59,659	51,094	44,680	39,697	32,457	27,451	23,782	20,979	18,766	16,976	15,498	12,727	10,797	9,375	8,284	7,420	6,720	5,653	4,878	4,290	3,457	2,895
125	92,511	73,426	60,869	51,980	45,356	40,229	32,812	27,704	23,972	21,126	18,885	17,073	15,578	12,781	10,836	9,404	8,307	7,439	6,735	5,663	4,886	4,296	3,461	2,897
130	95,221	75,123	62,031	52,825	45,998	40,734	33,147	27,942	24,150	21,265	18,995	17,163	15,653	12,832	10,872	9,432	8,328	7,456	6,749	5,673	4,893	4,302	3,464	2,900
135	97,876	76,766	63,147	53,632	46,609	41,212	33,463	28,167	24,318	21,394	19,098	17,247	15,724	12,879	10,906	9,457	8,348	7,472	6,762	5,682	4,900	4,307	3,468	2,902
140	100,478	78,358	64,220	54,404	47,190	41,666	33,762	28,378	24,475	21,516	19,195	17,326	15,789	12,923	10,937	9,481	8,366	7,487	6,774	5,691	4,906	4,312	3,471	2,905
145	103,028	79,900	65,252	55,142	47,745	42,098	34,045	28,578	24,624	21,631	19,287	17,401	15,851	12,964	10,967	9,503	8,384	7,500	6,785	5,699	4,912	4,317	3,474	2,907
150	105,527	81,395	66,245	55,850	48,275	42,510	34,313	28,767	24,764	21,739	19,372	17,471	15,909	13,003	10,995	9,524	8,400	7,513	6,796	5,706	4,918	4,321	3,477	2,909
155	107,978	82,845	67,203	56,529	48,782	42,902	34,568	28,946	24,896	21,841	19,453	17,537	15,963	13,039	11,021	9,543	8,415	7,525	6,806	5,713	4,923	4,325	3,479	2,910
160	110,380	84,252	68,126	57,181	49,266	43,276	34,811	29,116	25,022	21,938	19,530	17,599	16,015	13,074	11,045	9,562	8,429	7,537	6,815	5,720	4,928	4,329	3,482	2,912
165	112,737	85,618	69,016	57,807	49,730	43,634	35,042	29,277	25,141	22,029	19,603	17,658	16,064	13,106	11,068	9,579	8,443	7,548	6,824	5,726	4,933	4,332	3,484	2,914
170	115,049	86,945	69,876	58,409	50,175	43,976	35,262	29,431	25,254	22,116	19,671	17,713	16,110	13,137	11,090	9,595	8,456	7,558	6,832	5,732	4,937	4,336	3,486	2,915
175	117,318	88,235	70,706	58,988	50,602	44,303	35,472	29,577	25,362	22,198	19,737	17,766	16,154	13,166	11,111	9,611	8,468	7,567	6,840	5,738	4,941	4,339	3,488	2,917
180	119,544	89,488	71,509	59,546	51,012	44,617	35,673	29,717	25,464	22,277	19,799	17,816	16,195	13,194	11,131	9,626	8,479	7,577	6,848	5,743	4,945	4,342	3,490	2,918
185	121,729	90,707	72,285	60,083	51,405	44,918	35,865	29,850	25,562	22,352	19,858	17,864	16,235	13,220	11,149	9,639	8,490	7,585	6,855	5,748	4,949	4,345	3,492	2,919
190	123,874	91,892	73,036	60,601	51,784	45,207	36,049	29,977	25,655	22,423	19,914	17,910	16,272	13,245	11,167	9,653	8,500	7,593	6,861	5,752	4,952	4,347	3,494	2,920
195	125,980	93,046	73,763	61,100	52,148	45,484	36,226	30,099	25,745	22,491	19,968	17,953	16,308	13,268	11,184	9,665	8,510	7,601	6,868	5,757	4,955	4,350	3,495	2,922
200	128,048	94,170	74,467	61,583	52,499	45,751	36,395	30,215	25,830	22,556	20,019	17,995	16,342	13,291	11,200	9,677	8,519	7,609	6,874	5,761	4,959	4,352	3,497	2,923

3/0 AWG | 1 Run(s) | 240 Volts

C-Value = 12,843 | Magnetic Duct | Copper

Calculations:

1. $f = \dfrac{1.732 \times \text{Length} \times \text{Isca}}{\text{Runs} \times \text{C-Value} \times E_{L\text{-}L}}$

2. $M = \dfrac{1}{1 + f}$

3. Isc = Isca x M

4. Add Motor Contribution = Motor FLA x 4

* All results are given in symmetrical amperes

Three-Phase Bolted Fault-Current* Table

One-Way Distance in Feet

Available Isc in Thousands (Isca)	5	10	15	20	25	30	40	50	60	70	80	90	100	125	150	175	200	225	250	300	350	400	500	600
3	2,987	2,975	2,963	2,950	2,938	2,926	2,902	2,879	2,856	2,833	2,810	2,788	2,767	2,714	2,663	2,614	2,567	2,522	2,478	2,395	2,317	2,244	2,111	1,992
5	4,965	4,931	4,897	4,863	4,830	4,798	4,734	4,672	4,611	4,552	4,495	4,439	4,384	4,253	4,130	4,013	3,903	3,799	3,700	3,518	3,352	3,201	2,937	2,713
7	6,932	6,865	6,799	6,735	6,672	6,610	6,489	6,373	6,261	6,153	6,048	5,947	5,850	5,619	5,405	5,208	5,024	4,853	4,693	4,402	4,146	3,918	3,529	3,211
10	9,861	9,727	9,596	9,468	9,344	9,223	8,990	8,768	8,557	8,357	8,165	7,982	7,807	7,401	7,035	6,704	6,402	6,127	5,874	5,426	5,042	4,708	4,158	3,723
15	14,690	14,393	14,108	13,834	13,570	13,316	12,836	12,389	11,973	11,583	11,218	10,875	10,553	9,825	9,190	8,633	8,139	7,699	7,304	6,625	6,061	5,585	4,828	4,251
20	19,453	18,936	18,445	17,979	17,536	17,115	16,330	15,613	14,957	14,354	13,798	13,283	12,805	11,748	10,853	10,084	9,417	8,833	8,317	7,447	6,741	6,158	5,250	4,575
25	24,152	23,359	22,617	21,921	21,266	20,649	19,517	18,502	17,588	16,760	16,006	15,317	14,685	13,312	12,174	11,215	10,396	9,688	9,071	8,046	7,229	6,562	5,541	4,794
30	28,787	27,668	26,633	25,672	24,779	23,945	22,436	21,105	19,924	18,868	17,918	17,059	16,279	14,609	13,249	12,121	11,170	10,357	9,655	8,502	7,595	6,863	5,753	4,953
35	33,360	31,866	30,501	29,248	28,094	27,027	25,119	23,464	22,012	20,730	19,589	18,567	17,647	15,701	14,141	12,864	11,798	10,895	10,120	8,861	7,880	7,094	5,915	5,072
40	37,872	35,959	34,230	32,659	31,227	29,914	27,595	25,610	23,891	22,388	21,063	19,886	18,834	16,633	14,893	13,483	12,317	11,336	10,500	9,150	8,108	7,279	6,043	5,166
45	42,324	39,949	37,826	35,918	34,193	32,625	29,886	27,571	25,589	23,872	22,372	21,049	19,874	17,439	15,536	14,008	12,753	11,704	10,815	9,389	8,295	7,429	6,146	5,241
50	46,719	43,841	41,298	39,033	37,004	35,176	32,012	29,370	27,132	25,210	23,542	22,082	20,792	18,142	16,092	14,458	13,125	12,017	11,082	9,589	8,451	7,554	6,231	5,303
55	51,055	47,639	44,650	42,015	39,673	37,579	33,990	31,027	28,539	26,421	24,595	23,005	21,609	18,761	16,577	14,848	13,446	12,285	11,310	9,759	8,582	7,659	6,303	5,355
60	55,336	51,345	47,890	44,872	42,211	39,848	35,836	32,558	29,829	27,523	25,547	23,836	22,340	19,310	17,004	15,190	13,725	12,518	11,507	9,906	8,696	7,749	6,364	5,398
65	59,561	54,963	51,023	47,611	44,626	41,993	37,562	33,976	31,015	28,529	26,412	24,588	22,999	19,800	17,383	15,491	13,971	12,723	11,679	10,033	8,794	7,827	6,416	5,436
70	63,733	58,496	54,054	50,239	46,927	44,025	39,179	35,294	32,110	29,453	27,202	25,270	23,595	20,241	17,721	15,760	14,189	12,903	11,831	10,145	8,879	7,895	6,461	5,469
75	67,851	61,947	56,988	52,764	49,123	45,952	40,697	36,521	33,123	30,303	27,925	25,894	24,138	20,639	18,026	16,000	14,383	13,064	11,966	10,244	8,955	7,954	6,501	5,497
80	71,918	65,319	59,829	55,190	51,219	47,781	42,126	37,668	34,063	31,088	28,591	26,465	24,633	21,000	18,300	16,216	14,558	13,207	12,086	10,332	9,022	8,008	6,537	5,523
85	75,933	68,614	62,582	57,525	53,224	49,521	43,473	38,741	34,938	31,815	29,205	26,990	25,088	21,329	18,550	16,412	14,715	13,337	12,195	10,411	9,083	8,055	6,568	5,545
90	79,898	71,836	65,251	59,772	55,142	51,178	44,744	39,747	35,754	32,491	29,773	27,475	25,506	21,631	18,778	16,590	14,858	13,454	12,293	10,482	9,137	8,098	6,597	5,565
95	83,815	74,986	67,840	61,937	56,979	52,756	45,946	40,693	36,518	33,120	30,300	27,923	25,892	21,908	18,986	16,752	14,989	13,561	12,382	10,547	9,186	8,136	6,622	5,583
100	87,682	78,067	70,351	64,024	58,741	54,263	47,085	41,584	37,234	33,707	30,791	28,340	26,250	22,163	19,178	16,901	15,108	13,658	12,463	10,606	9,231	8,171	6,645	5,600
105	91,503	81,081	72,790	66,037	60,431	55,702	48,165	42,424	37,906	34,257	31,250	28,727	26,582	22,400	19,355	17,038	15,217	13,748	12,537	10,660	9,271	8,203	6,667	5,615
110	95,277	84,030	75,158	67,981	62,055	57,079	49,190	43,218	38,538	34,773	31,678	29,089	26,891	22,619	19,518	17,165	15,318	13,830	12,606	10,709	9,309	8,232	6,686	5,629
115	99,006	86,917	77,459	69,858	63,615	58,396	50,166	43,969	39,134	35,258	32,080	29,428	27,180	22,823	19,670	17,282	15,411	13,906	12,669	10,755	9,343	8,259	6,704	5,641
120	102,689	89,743	79,696	71,672	65,116	59,659	51,094	44,680	39,697	35,714	32,457	29,745	27,451	23,013	19,811	17,391	15,498	13,976	12,727	10,797	9,375	8,284	6,720	5,653
125	106,329	92,511	81,871	73,426	66,560	60,869	51,980	45,356	40,229	36,144	32,812	30,043	27,704	23,191	19,943	17,492	15,578	14,042	12,781	10,836	9,404	8,307	6,735	5,663
130	109,925	95,221	83,987	75,123	67,952	62,031	52,825	45,998	40,734	36,551	33,147	30,323	27,942	23,358	20,066	17,587	15,653	14,103	12,832	10,872	9,432	8,328	6,749	5,673
135	113,479	97,876	86,045	76,766	69,294	63,147	53,632	46,609	41,212	36,935	33,463	30,587	28,167	23,514	20,181	17,676	15,724	14,160	12,879	10,906	9,457	8,348	6,762	5,682
140	116,991	100,478	88,050	78,358	70,588	64,220	54,404	47,190	41,666	37,300	33,762	30,837	28,378	23,662	20,290	17,759	15,789	14,213	12,923	10,937	9,481	8,366	6,774	5,691
145	120,463	103,028	90,002	79,900	71,837	65,252	55,142	47,745	42,098	37,646	34,045	31,073	28,578	23,800	20,391	17,837	15,851	14,263	12,964	10,967	9,503	8,384	6,785	5,699
150	123,893	105,527	91,903	81,395	73,043	66,245	55,850	48,275	42,510	37,974	34,313	31,296	28,767	23,931	20,488	17,910	15 909	14,310	13,003	10,995	9,524	8,400	6,796	5,706
155	127,285	107,978	93,756	82,845	74,209	67,203	56,529	48,782	42,902	38,287	34,568	31,508	28,946	24,055	20,578	17,979	15 963	14,354	13,039	11,021	9,543	8,415	6,806	5,713
160	130,637	110,380	95,562	84,252	75,336	68,126	57,181	49,266	43,276	38,585	34,811	31,710	29,116	24,172	20,664	18,045	16 015	14,396	13,074	11,045	9,562	8,429	6,815	5,720
165	133,951	112,737	97,324	85,618	76,426	69,016	57,807	49,730	43,634	38,869	35,042	31,901	29,277	24,283	20,745	18,107	16,064	14,435	13,106	11,068	9,579	8,443	6,824	5,726
170	137,228	115,049	99,042	86,945	77,482	69,876	58,409	50,175	43,976	39,140	35,262	32,084	29,431	24,389	20,822	18,165	16,110	14,472	13,137	11,090	9,595	8,456	6,832	5,732
175	140,468	117,318	100,719	88,235	78,504	70,706	58,988	50,602	44,303	39,399	35,472	32,258	29,577	24,489	20,895	18,221	16,154	14,508	13,166	11,111	9,611	8,468	6,840	5,738
180	143,671	119,544	102,355	89,488	79,495	71,509	59,546	51,012	44,617	39,647	35,673	32,424	29,717	24,585	20,965	18,274	16,195	14,541	13,194	11,131	9,626	8,479	6,848	5,743
185	146,839	121,729	103,953	90,707	80,455	72,285	60,083	51,405	44,918	39,884	35,865	32,582	29,850	24,676	21,031	18,324	16,235	14,573	13,220	11,149	9,639	8,490	6,855	5,748
190	149,971	123,874	105,513	91,892	81,386	73,036	60,601	51,784	45,207	40,112	36,049	32,734	29,977	24,763	21,094	18,372	16,272	14,603	13,245	11,167	9,653	8,500	6,861	5,752
195	153,069	125,980	107,037	93,046	82,290	73,763	61,100	52,148	45,484	40,330	36,226	32,879	30,099	24,846	21,154	18,418	16,308	14,632	13,268	11,184	9,665	8,510	6,868	5,757
200	156,133	128,048	108,526	94,170	83,167	74,467	61,583	52,499	45,751	40,540	36,395	33,018	30,215	24,925	21,212	18,461	16,342	14,659	13,291	11,200	9,677	8,519	6,874	5,761

# 3/0 AWG	2 Run(s)	240 Volts	C-Value = 12,843	Magnetic Duct	Copper

Calculations:

1. $f = \dfrac{1.732 \times \text{Length} \times \text{Isca}}{\text{Runs} \times \text{C-Value} \times \text{E L-L}}$

2. $M = \dfrac{1}{1 + f}$

3. $\text{Isc} = \text{Isca} \times M$

4. Add Motor Contribution = Motor FLA x 4

* All results are given in symmetrical amperes

Three-Phase Bolted Fault-Current* Table

One-Way Distance in Feet

Available Isc in Thousands (Isca)	5	10	15	20	25	30	40	50	60	70	80	90	100	125	150	175	200	225	250	300	350	400	500	600
3	2,979	2,958	2,937	2,916	2,896	2,876	2,837	2,799	2,762	2,726	2,691	2,657	2,623	2,544	2,468	2,398	2,331	2,268	2,208	2,097	1,997	1,906	1,746	1,612
5	4,941	4,883	4,827	4,772	4,718	4,665	4,563	4,466	4,372	4,283	4,197	4,114	4,035	3,849	3,680	3,524	3,382	3,250	3,129	2,911	2,721	2,555	2,277	2,053
7	6,885	6,773	6,665	6,561	6,459	6,361	6,173	5,996	5,829	5,670	5,521	5,379	5,244	4,934	4,659	4,413	4,192	3,992	3,810	3,492	3,222	2,992	2,617	2,326
10	9,766	9,543	9,330	9,127	8,932	8,745	8,393	8,069	7,769	7,491	7,232	6,990	6,764	6,257	5,822	5,443	5,110	4,816	4,553	4,106	3,739	3,432	2,948	2,583
15	14,480	13,995	13,542	13,117	12,718	12,342	11,654	11,039	10,485	9,984	9,529	9,113	8,732	7,906	7,223	6,649	6,159	5,736	5,368	4,757	4,271	3,875	3,269	2,827
20	19,087	18,253	17,489	16,787	16,139	15,539	14,463	13,527	12,705	11,977	11,328	10,745	10,220	9,106	8,212	7,477	6,863	6,343	5,895	5,167	4,598	4,143	3,457	2,967
25	23,589	22,329	21,197	20,174	19,245	18,398	16,909	15,643	14,554	13,606	12,775	12,039	11,383	10,019	8,947	8,082	7,369	6,772	6,265	5,448	4,820	4,322	3,581	3,057
30	27,991	26,234	24,685	23,308	22,077	20,970	19,057	17,465	16,118	14,964	13,964	13,089	12,318	10,736	9,514	8,542	7,750	7,092	6,538	5,653	4,980	4,450	3,669	3,121
35	32,296	29,979	27,973	26,218	24,671	23,296	20,959	19,049	17,458	16,112	14,959	13,959	13,085	11,314	9,966	8,904	8,047	7,340	6,748	5,810	5,101	4,546	3,734	3,168
40	36,506	33,574	31,078	28,927	27,055	25,410	22,655	20,440	18,619	17,096	15,803	14,692	13,727	11,791	10,333	9,197	8,285	7,538	6,914	5,933	5,196	4,621	3,784	3,204
45	40,626	37,027	34,014	31,454	29,253	27,340	24,177	21,670	19,634	17,948	16,528	15,317	14,271	12,190	10,639	9,438	8,480	7,699	7,050	6,032	5,272	4,681	3,825	3,233
50	44,658	40,347	36,795	33,818	31,287	29,108	25,549	22,766	20,530	18,693	17,159	15,857	14,738	12,529	10,896	9,640	8,643	7,833	7,162	6,114	5,334	4,730	3,857	3,256
55	48,604	43,541	39,433	36,034	33,174	30,735	26,794	23,749	21,326	19,351	17,711	16,327	15,144	12,822	11,117	9,812	8,781	7,946	7,257	6,183	5,386	4,771	3,885	3,276
60	52,468	46,617	41,939	38,115	34,930	32,236	27,928	24,636	22,038	19,936	18,200	16,742	15,500	13,076	11,307	9,960	8,900	8,043	7,337	6,242	5,431	4,806	3,908	3,292
65	56,252	49,580	44,322	40,073	36,567	33,625	28,965	25,439	22,679	20,459	18,634	17,109	15,814	13,298	11,473	10,089	9,002	8,127	7,407	6,292	5,469	4,836	3,927	3,306
70	59,959	52,437	46,591	41,919	38,098	34,916	29,917	26,171	23,258	20,929	19,024	17,437	16,094	13,496	11,620	10,202	9,092	8,200	7,468	6,336	5,502	4,862	3,944	3,318
75	63,590	55,193	48,755	43,662	39,532	36,116	30,795	26,840	23,785	21,355	19,375	17,731	16,344	13,671	11,750	10,302	9,172	8,265	7,521	6,374	5,531	4,884	3,959	3,329
80	67,148	57,854	50,820	45,311	40,879	37,237	31,606	27,454	24,266	21,742	19,693	17,997	16,570	13,829	11,866	10,391	9,242	8,322	7,569	6,408	5,556	4,904	3,972	3,338
85	70,636	60,424	52,792	46,872	42,146	38,285	32,358	28,019	24,707	22,095	19,982	18,238	16,775	13,971	11,970	10,471	9,305	8,373	7,611	6,439	5,579	4,922	3,984	3,346
90	74,054	62,909	54,679	48,354	43,340	39,268	33,057	28,542	25,112	22,419	20,247	18,458	16,960	14,100	12,065	10,543	9,362	8,419	7,649	6,466	5,600	4,938	3,994	3,353
95	77,407	65,311	56,485	49,761	44,467	40,191	33,708	29,027	25,487	22,716	20,489	18,660	17,130	14,217	12,151	10,609	9,414	8,461	7,683	6,490	5,618	4,952	4,004	3,360
100	80,694	67,636	58,216	51,099	45,532	41,060	34,317	29,477	25,833	22,991	20,713	18,845	17,286	14,324	12,229	10,668	9,461	8,499	7,715	6,513	5,635	4,965	4,012	3,366
105	83,919	69,887	59,876	52,373	46,542	41,878	34,887	29,897	26,155	23,246	20,919	19,016	17,430	14,423	12,300	10,723	9,504	8,533	7,743	6,533	5,650	4,977	4,020	3,371
110	87,082	72,068	61,469	53,588	47,499	42,652	35,422	30,289	26,455	23,482	21,110	19,173	17,562	14,513	12,366	10,773	9,543	8,565	7,769	6,551	5,664	4,988	4,027	3,376
115	90,187	74,181	63,000	54,748	48,407	43,383	35,925	30,656	26,734	23,702	21,288	19,320	17,685	14,597	12,427	10,819	9,579	8,594	7,793	6,568	5,676	4,998	4,033	3,381
120	93,233	76,229	64,471	55,856	49,271	44,076	36,399	31,000	26,996	23,907	21,453	19,456	17,799	14,675	12,483	10,861	9,612	8,621	7,815	6,584	5,688	5,007	4,039	3,385
125	96,223	78,217	65,887	56,915	50,094	44,733	36,846	31,324	27,241	24,099	21,608	19,583	17,905	14,747	12,535	10,901	9,643	8,646	7,836	6,599	5,699	5,015	4,045	3,389
130	99,159	80,146	67,251	57,930	50,878	45,357	37,269	31,628	27,471	24,280	21,752	19,702	18,004	14,814	12,584	10,937	9,672	8,669	7,854	6,612	5,709	5,023	4,050	3,392
135	102,042	82,019	68,564	58,902	51,627	45,951	37,669	31,916	27,688	24,449	21,888	19,813	18,097	14,877	12,629	10,972	9,699	8,690	7,872	6,624	5,718	5,030	4,054	3,396
140	104,873	83,838	69,831	59,835	52,342	46,517	38,048	32,188	27,892	24,608	22,016	19,917	18,184	14,935	12,672	11,004	9,724	8,710	7,889	6,636	5,727	5,037	4,059	3,399
145	107,654	85,605	71,053	60,730	53,025	47,056	38,408	32,445	28,085	24,758	22,136	20,016	18,266	14,991	12,711	11,033	9,747	8,729	7,904	6,647	5,735	5,043	4,063	3,401
150	110,386	87,324	72,233	61,589	53,680	47,570	38,750	32,689	28,267	24,900	22,249	20,108	18,343	15,042	12,748	11,062	9,769	8,747	7,918	6,657	5,743	5,049	4,066	3,404
155	113,070	88,995	73,373	62,416	54,307	48,062	39,076	32,920	28,440	25,034	22,356	20,195	18,416	15,091	12,783	11,088	9,789	8,763	7,932	6,667	5,750	5,054	4,070	3,407
160	115,708	90,621	74,474	63,211	54,908	48,532	39,386	33,140	28,604	25,161	22,457	20,278	18,484	15,137	12,817	11,113	9,809	8,779	7,944	6,676	5,756	5,059	4,073	3,409
165	118,300	92,204	75,540	63,977	55,485	48,982	39,682	33,350	28,760	25,281	22,553	20,356	18,549	15,181	12,848	11,136	9,827	8,793	7,956	6,684	5,763	5,064	4,077	3,411
170	120,848	93,744	76,571	64,715	56,039	49,414	39,965	33,549	28,908	25,396	22,644	20,430	18,611	15,222	12,877	11,158	9,844	8,807	7,968	6,692	5,768	5,069	4,079	3,413
175	123,354	95,245	77,569	65,427	56,572	49,828	40,235	33,739	29,049	25,504	22,730	20,501	18,669	15,261	12,905	11,179	9,861	8,820	7,978	6,700	5,774	5,073	4,082	3,415
180	125,817	96,707	78,536	66,114	57,084	50,225	40,493	33,921	29,184	25,608	22,813	20,568	18,725	15,298	12,932	11,199	9,876	8,833	7,989	6,707	5,779	5,077	4,085	3,417
185	128,240	98,132	79,473	66,777	57,578	50,607	40,741	34,095	29,313	25,707	22,891	20,631	18,778	15,333	12,957	11,218	9,891	8,844	7,998	6,713	5,784	5,081	4,087	3,419
190	130,623	99,521	80,382	67,417	58,053	50,974	40,979	34,261	29,435	25,801	22,966	20,692	18,828	15,367	12,981	11,236	9,905	8,855	8,007	6,720	5,789	5,085	4,090	3,420
195	132,967	100,876	81,264	68,036	58,512	51,327	41,206	34,420	29,553	25,891	23,037	20,750	18,876	15,399	13,003	11,253	9,918	8,866	8,016	6,726	5,794	5,088	4,092	3,422
200	135,273	102,198	82,119	68,635	58,954	51,667	41,425	34,572	29,665	25,978	23,106	20,805	18,922	15,429	13,025	11,269	9,931	8,876	8,024	6,732	5,798	5,092	4,094	3,424

# 4/0 AWG	1 Run(s)	240 Volts	C-Value = 15,082	Magnetic Duct	Copper

Calculations:

1. $f = \dfrac{1.732 \times \text{Length} \times \text{Isca}}{\text{Runs} \times \text{C-Value} \times E_{L-L}}$

2. $M = \dfrac{1}{1 + f}$

3. Isc = Isca x M

4. Add Motor Contribution = Motor FLA x 4

* All results are given in symmetrical amperes

Three-Phase Bolted Fault-Current* Table

One-Way Distance in Feet

Available Isc in Thousands (Isca)	5	10	15	20	25	30	40	50	60	70	80	90	100	125	150	175	200	225	250	300	350	400	500	600
3	2,989	2,979	2,968	2,958	2,947	2,937	2,916	2,896	2,876	2,856	2,837	2,818	2,799	2,753	2,708	2,665	2,623	2,583	2,544	2,468	2,398	2,331	2,208	2,097
5	4,970	4,941	4,912	4,883	4,855	4,827	4,772	4,718	4,665	4,614	4,563	4,514	4,466	4,350	4,239	4,134	4,035	3,940	3,849	3,680	3,524	3,382	3,129	2,911
7	6,942	6,885	6,828	6,773	6,719	6,665	6,561	6,459	6,361	6,265	6,173	6,083	5,996	5,788	5,595	5,413	5,244	5,084	4,934	4,659	4,413	4,192	3,810	3,492
10	9,882	9,766	9,654	9,543	9,436	9,330	9,127	8,932	8,745	8,566	8,393	8,228	8,069	7,698	7,359	7,049	6,764	6,501	6,257	5,822	5,443	5,110	4,553	4,106
15	14,736	14,480	14,234	13,995	13,765	13,542	13,117	12,718	12,342	11,988	11,654	11,338	11,039	10,355	9,751	9,214	8,732	8,299	7,906	7,223	6,649	6,159	5,368	4,757
20	19,533	19,087	18,661	18,253	17,863	17,489	16,787	16,139	15,539	14,982	14,463	13,980	13,527	12,515	11,643	10,885	10,220	9,631	9,106	8,212	7,477	6,863	5,895	5,167
25	24,274	23,589	22,942	22,329	21,748	21,197	20,174	19,245	18,398	17,622	16,909	16,252	15,643	14,305	13,177	12,215	11,383	10,657	10,019	8,947	8,082	7,369	6,265	5,448
30	28,961	27,991	27,084	26,234	25,436	24,685	23,308	22,077	20,970	19,968	19,057	18,226	17,465	15,813	14,447	13,298	12,318	11,473	10,736	9,514	8,542	7,750	6,538	5,653
35	33,593	32,296	31,094	29,979	28,941	27,973	26,218	24,671	23,296	22,066	20,959	19,959	19,049	17,101	15,514	14,197	13,085	12,136	11,314	9,966	8,904	8,047	6,748	5,810
40	38,173	36,506	34,979	33,574	32,278	31,078	28,927	27,055	25,410	23,954	22,655	21,490	20,440	18,213	16,424	14,955	13,727	12,685	11,791	10,333	9,197	8,285	6,914	5,933
45	42,701	40,626	38,743	37,027	35,457	34,014	31,454	29,253	27,340	25,661	24,177	22,855	21,670	19,183	17,209	15,603	14,271	13,149	12,190	10,639	9,438	8,480	7,050	6,032
50	47,178	44,658	42,393	40,347	38,489	36,795	33,818	31,287	29,108	27,213	25,549	24,078	22,766	20,038	17,893	16,163	14,738	13,544	12,529	10,896	9,640	8,643	7,162	6,114
55	51,605	48,604	45,934	43,541	41,386	39,433	36,034	33,174	30,735	28,629	26,794	25,180	23,749	20,795	18,495	16,653	15,144	13,886	12,822	11,117	9,812	8,781	7,257	6,183
60	55,982	52,468	49,370	46,617	44,154	41,939	38,115	34,930	32,236	29,928	27,928	26,179	24,636	21,472	19,028	17,084	15,500	14,185	13,076	11,307	9,960	8,900	7,337	6,242
65	60,311	56,252	52,706	49,580	46,804	44,322	40,073	36,567	33,625	31,122	28,965	27,088	25,439	22,080	19,504	17,466	15,814	14,448	13,298	11,473	10,089	9,002	7,407	6,292
70	64,591	59,959	55,946	52,437	49,342	46,591	41,919	38,098	34,916	32,224	29,917	27,919	26,171	22,629	19,931	17,808	16,094	14,681	13,496	11,620	10,202	9,092	7,468	6,336
75	68,825	63,590	59,095	55,193	51,775	48,755	43,662	39,532	36,116	33,244	30,795	28,682	26,840	23,127	20,317	18,115	16,344	14,889	13,671	11,750	10,302	9,172	7,521	6,374
80	73,013	67,148	62,155	57,854	54,109	50,820	45,311	40,879	37,237	34,191	31,606	29,384	27,454	23,582	20,667	18,393	16,570	15,076	13,829	11,866	10,391	9,242	7,569	6,408
85	77,155	70,636	65,132	60,424	56,351	52,792	46,872	42,146	38,285	35,073	32,358	30,033	28,019	23,998	20,986	18,645	16,775	15,245	13,971	11,970	10,471	9,305	7,611	6,439
90	81,252	74,054	68,028	62,909	58,506	54,679	48,354	43,340	39,268	35,896	33,057	30,634	28,542	24,380	21,277	18,875	16,960	15,398	14,100	12,065	10,543	9,362	7,649	6,466
95	85,306	77,407	70,846	65,311	60,578	56,485	49,761	44,467	40,191	36,665	33,708	31,193	29,027	24,733	21,545	19,086	17,130	15,538	14,217	12,151	10,609	9,414	7,683	6,490
100	89,316	80,694	73,590	67,636	62,574	58,216	51,099	45,532	41,060	37,387	34,317	31,714	29,477	25,059	21,793	19,280	17,286	15,666	14,324	12,229	10,668	9,461	7,715	6,513
105	93,283	83,919	76,263	69,887	64,495	59,876	52,373	46,542	41,878	38,065	34,887	32,200	29,897	25,362	22,021	19,458	17,430	15,784	14,423	12,300	10,723	9,504	7,743	6,533
110	97,209	87,082	78,867	72,068	66,348	61,469	53,588	47,499	42,652	38,702	35,422	32,655	30,289	25,643	22,233	19,624	17,562	15,893	14,513	12,366	10,773	9,543	7,769	6,551
115	101,093	90,187	81,404	74,181	68,135	63,000	54,748	48,407	43,383	39,304	35,925	33,082	30,656	25,906	22,430	19,777	17,685	15,993	14,597	12,427	10,819	9,579	7,793	6,568
120	104,937	93,233	83,878	76,229	69,859	64,471	55,856	49,271	44,076	39,871	36,399	33,483	31,000	26,151	22,614	19,920	17,799	16,086	14,675	12,483	10,861	9,612	7,815	6,584
125	108,740	96,223	86,291	78,217	71,525	65,887	56,915	50,094	44,733	40,408	36,846	33,861	31,324	26,381	22,786	20,053	17,905	16,173	14,747	12,535	10,901	9,643	7,836	6,599
130	112,504	99,159	88,644	80,146	73,134	67,251	57,930	50,878	45,357	40,917	37,269	34,218	31,628	26,597	22,947	20,177	18,004	16,254	14,814	12,584	10,937	9,672	7,854	6,612
135	116,230	102,042	90,941	82,019	74,690	68,564	58,902	51,627	45,951	41,400	37,669	34,555	31,916	26,800	23,098	20,294	18,097	16,330	14,877	12,629	10,972	9,699	7,872	6,624
140	119,917	104,873	93,183	83,838	76,196	69,831	59,835	52,342	46,517	41,858	38,048	34,873	32,188	26,991	23,240	20,403	18,184	16,401	14,935	12,672	11,004	9,724	7,889	6,636
145	123,567	107,654	95,372	85,605	77,653	71,053	60,730	53,025	47,056	42,294	38,408	35,176	32,445	27,172	23,373	20,507	18,266	16,467	14,991	12,711	11,033	9,747	7,904	6,647
150	127,179	110,386	97,510	87,324	79,065	72,233	61,589	53,680	47,570	42,709	38,750	35,462	32,689	27,343	23,500	20,604	18,343	16,530	15,042	12,748	11,062	9,769	7,918	6,657
155	130,756	113,070	99,598	88,995	80,432	73,373	62,416	54,307	48,062	43,105	39,076	35,735	32,920	27,505	23,619	20,695	18,416	16,589	15,091	12,783	11,088	9,789	7,932	6,667
160	134,296	115,708	101,639	90,621	81,758	74,474	63,211	54,908	48,532	43,483	39,386	35,994	33,140	27,658	23,732	20,782	18,484	16,644	15,137	12,817	11,113	9,809	7,944	6,676
165	137,801	118,300	103,634	92,204	83,044	75,540	63,977	55,485	48,982	43,844	39,682	36,241	33,350	27,804	23,839	20,864	18,549	16,697	15,181	12,848	11,136	9,827	7,956	6,684
170	141,271	120,848	105,585	93,744	84,292	76,571	64,715	56,039	49,414	44,190	39,965	36,477	33,549	27,942	23,941	20,942	18,611	16,747	15,222	12,877	11,158	9,844	7,968	6,692
175	144,707	123,354	107,492	95,245	85,503	77,569	65,427	56,572	49,828	44,520	40,235	36,702	33,739	28,074	24,038	21,016	18,669	16,794	15,261	12,905	11,179	9,861	7,978	6,700
180	148,109	125,817	109,358	96,707	86,680	78,536	66,114	57,084	50,225	44,837	40,493	36,917	33,921	28,200	24,130	21,086	18,725	16,839	15,298	12,932	11,199	9,876	7,989	6,707
185	151,477	128,240	111,184	98,132	87,823	79,473	66,777	57,578	50,607	45,141	40,741	37,123	34,095	28,319	24,217	21,153	18,778	16,882	15,333	12,957	11,218	9,891	7,998	6,713
190	154,813	130,623	112,970	99,521	88,934	80,382	67,417	58,053	50,974	45,433	40,979	37,320	34,261	28,434	24,301	21,217	18,828	16,922	15,367	12,981	11,236	9,905	8,007	6,720
195	158,117	132,967	114,719	100,876	90,014	81,264	68,036	58,512	51,327	45,713	41,206	37,509	34,420	28,544	24,381	21,278	18,876	16,961	15,399	13,003	11,253	9,918	8,016	6,726
200	161,388	135,273	116,432	102,198	91,065	82,119	68,635	58,954	51,667	45,983	41,425	37,690	34,572	28,648	24,458	21,336	18,922	16,998	15,429	13,025	11,269	9,931	8,024	6,732

# 4/0 AWG	2 Run(s)	240 Volts	C-Value = 15,082	Magnetic Duct	Copper

Calculations:

1. $f = \dfrac{1.732 \times \text{Length} \times \text{Isca}}{\text{Runs} \times \text{C-Value} \times E_{L-L}}$
2. $M = \dfrac{1}{1 + f}$
3. $\text{Isc} = \text{Isca} \times M$
4. Add Motor Contribution = Motor FLA x 4

* All results are given in symmetrical amperes

Three-Phase Bolted Fault-Current* Table

One-Way Distance in Feet

Available Isc in Thousands (Isca)	5	10	15	20	25	30	40	50	60	70	80	90	100	125	150	175	200	225	250	300	350	400	500	600
3	2,980	2,961	2,942	2,923	2,905	2,886	2,850	2,815	2,781	2,747	2,715	2,683	2,652	2,577	2,506	2,439	2,376	2,316	2,258	2,152	2,055	1,967	1,811	1,678
5	4,946	4,893	4,841	4,790	4,741	4,692	4,597	4,507	4,420	4,336	4,255	4,177	4,102	3,926	3,764	3,615	3,477	3,350	3,231	3,018	2,831	2,666	2,387	2,161
7	6,894	6,792	6,692	6,596	6,502	6,411	6,236	6,070	5,913	5,764	5,622	5,487	5,358	5,061	4,795	4,556	4,340	4,143	3,963	3,647	3,377	3,145	2,764	2,466
10	9,786	9,581	9,384	9,195	9,013	8,839	8,510	8,204	7,920	7,654	7,406	7,173	6,955	6,463	6,036	5,662	5,331	5,038	4,774	4,323	3,949	3,635	3,136	2,757
15	14,523	14,076	13,655	13,259	12,885	12,531	11,879	11,292	10,760	10,276	9,834	9,428	9,054	8,238	7,556	6,979	6,484	6,054	5,678	5,050	4,547	4,136	3,502	3,036
20	19,161	18,390	17,678	17,019	16,408	15,839	14,812	13,910	13,111	12,400	11,761	11,185	10,663	9,549	8,645	7,898	7,269	6,734	6,271	5,514	4,920	4,442	3,719	3,198
25	23,703	22,534	21,474	20,510	19,629	18,820	17,387	16,157	15,090	14,155	13,329	12,594	11,936	10,557	9,463	8,575	7,839	7,220	6,691	5,836	5,175	4,648	3,862	3,304
30	28,151	26,517	25,062	23,759	22,584	21,520	19,667	18,108	16,778	15,630	14,629	13,748	12,968	11,356	10,100	9,095	8,271	7,585	7,003	6,072	5,360	4,797	3,964	3,378
35	32,509	30,349	28,459	26,790	25,306	23,977	21,699	19,817	18,235	16,886	15,724	14,711	13,821	12,005	10,611	9,507	8,611	7,869	7,245	6,253	5,500	4,909	4,041	3,433
40	36,779	34,039	31,678	29,624	27,820	26,223	23,522	21,326	19,505	17,970	16,659	15,527	14,539	12,543	11,029	9,841	8,884	8,096	7,437	6,396	5,610	4,997	4,100	3,476
45	40,965	37,593	34,735	32,280	30,150	28,283	25,167	22,669	20,622	18,914	17,468	16,227	15,150	12,995	11,377	10,117	9,109	8,283	7,594	6,512	5,699	5,067	4,147	3,510
50	45,067	41,020	37,640	34,775	32,315	30,180	26,657	23,871	21,613	19,744	18,173	16,834	15,678	13,382	11,672	10,350	9,297	8,438	7,725	6,607	5,772	5,125	4,186	3,537
55	49,090	44,326	40,405	37,122	34,332	31,932	28,015	24,954	22,497	20,479	18,794	17,365	16,138	13,716	11,925	10,543	9,457	8,570	7,835	6,688	5,834	5,173	4,218	3,560
60	53,034	47,517	43,040	39,334	36,216	33,555	29,257	25,935	23,290	21,135	19,345	17,835	16,543	14,007	12,145	10,723	9,594	8,682	7,929	6,756	5,886	5,214	4,245	3,580
65	56,903	50,600	45,554	41,423	37,979	35,064	30,397	26,827	24,007	21,724	19,837	18,252	16,901	14,263	12,337	10,869	9,713	8,780	8,010	6,815	5,930	5,249	4,268	3,596
70	60,699	53,579	47,955	43,399	39,633	36,469	31,448	27,642	24,658	22,255	20,279	18,625	17,221	14,490	12,506	11,000	9,818	8,866	8,081	6,867	5,969	5,279	4,288	3,610
75	64,423	56,460	50,249	45,270	41,188	37,781	32,419	28,389	25,251	22,737	20,679	18,962	17,508	14,693	12,657	11,117	9,911	8,941	8,144	6,912	6,003	5,306	4,306	3,623
80	68,078	59,248	52,446	47,044	42,652	39,010	33,319	29,077	25,793	23,176	21,041	19,266	17,768	14,875	12,792	11,221	9,994	9,008	8,200	6,952	6,034	5,330	4,321	3,634
85	71,665	61,947	54,549	48,730	44,033	40,162	34,156	29,712	26,292	23,578	21,372	19,543	18,003	15,039	12,913	11,314	10,067	9,068	8,249	6,988	6,060	5,351	4,335	3,644
90	75,187	64,560	56,566	50,333	45,338	41,244	34,936	30,301	26,752	23,947	21,675	19,796	18,217	15,189	13,023	11,399	10,134	9,122	8,294	7,020	6,085	5,369	4,347	3,652
95	78,645	67,094	58,501	51,860	46,572	42,264	35,664	30,847	27,177	24,287	21,953	20,028	18,413	15,325	13,123	11,475	10,195	9,171	8,335	7,049	6,106	5,386	4,358	3,660
100	82,040	69,550	60,360	53,315	47,743	43,225	36,346	31,357	27,571	24,602	22,209	20,241	18,593	15,449	13,215	11,545	10,250	9,216	8,371	7,075	6,126	5,402	4,368	3,667
105	85,376	71,932	62,146	54,704	48,853	44,133	36,987	31,832	27,938	24,893	22,447	20,438	18,760	15,564	13,298	11,609	10,300	9,256	8,405	7,099	6,144	5,416	4,378	3,674
110	88,652	74,244	63,864	56,031	49,909	44,993	37,588	32,277	28,280	25,164	22,667	20,621	18,913	15,669	13,375	11,667	10,346	9,294	8,435	7,121	6,160	5,428	4,386	3,679
115	91,871	76,488	65,518	57,300	50,913	45,808	38,155	32,694	28,600	25,417	22,872	20,790	19,056	15,767	13,446	11,721	10,388	9,328	8,464	7,141	6,175	5,440	4,394	3,685
120	95,035	78,668	67,111	58,514	51,870	46,581	38,690	33,086	28,899	25,653	23,063	20,948	19,188	15,858	13,512	11,771	10,428	9,359	8,490	7,159	6,189	5,451	4,401	3,690
125	98,144	80,787	68,647	59,678	52,783	47,315	39,196	33,455	29,180	25,875	23,242	21,095	19,312	15,942	13,573	11,818	10,464	9,389	8,514	7,176	6,202	5,461	4,407	3,694
130	101,200	82,846	70,128	60,795	53,654	48,014	39,674	33,803	29,445	26,082	23,409	21,233	19,427	16,020	13,630	11,861	10,498	9,416	8,536	7,192	6,214	5,470	4,413	3,698
135	104,204	84,849	71,557	61,866	54,487	48,680	40,128	34,131	29,694	26,278	23,566	21,362	19,535	16,094	13,683	11,901	10,529	9,441	8,557	7,207	6,225	5,478	4,419	3,702
140	107,158	86,797	72,938	62,896	55,284	49,315	40,558	34,442	29,929	26,462	23,714	21,484	19,637	16,163	13,733	11,939	10,559	9,465	8,576	7,221	6,235	5,486	4,424	3,706
145	110,063	88,693	74,273	63,885	56,047	49,922	40,968	34,737	30,151	26,635	23,854	21,598	19,732	16,227	13,780	11,974	10,586	9,487	8,595	7,234	6,245	5,494	4,429	3,709
150	112,920	90,539	75,563	64,838	56,779	50,501	41,357	35,017	30,362	26,799	23,985	21,706	19,822	16,288	13,824	12,007	10,612	9,508	8,612	7,246	6,254	5,501	4,433	3,712
155	115,731	92,337	76,811	65,754	57,480	51,056	41,728	35,282	30,561	26,955	24,109	21,807	19,907	16,345	13,865	12,038	10,636	9,527	8,628	7,257	6,262	5,507	4,437	3,715
160	118,496	94,089	78,019	66,638	58,154	51,587	42,082	35,535	30,751	27,102	24,227	21,904	19,987	16,399	13,904	12,067	10,659	9,546	8,643	7,268	6,270	5,513	4,441	3,718
165	121,216	95,796	79,189	67,490	58,802	52,096	42,420	35,776	30,931	27,242	24,339	21,995	20,063	16,450	13,940	12,095	10,681	9,563	8,657	7,278	6,277	5,519	4,445	3,721
170	123,893	97,460	80,323	68,311	59,425	52,584	42,744	36,005	31,102	27,375	24,445	22,082	20,135	16,499	13,975	12,121	10,701	9,579	8,670	7,287	6,285	5,524	4,449	3,723
175	126,528	99,083	81,422	69,105	60,024	53,053	43,053	36,225	31,266	27,501	24,546	22,164	20,203	16,545	14,008	12,146	10,720	9,595	8,683	7,296	6,291	5,530	4,452	3,726
180	129,121	100,666	82,488	69,871	60,602	53,504	43,349	36,434	31,422	27,622	24,642	22,242	20,268	16,588	14,039	12,169	10,739	9,609	8,695	7,304	6,297	5,534	4,455	3,728
185	131,674	102,211	83,523	70,612	61,158	53,937	43,633	36,635	31,571	27,737	24,733	22,317	20,330	16,630	14,069	12,191	10,756	9,623	8,706	7,312	6,303	5,539	4,458	3,730
190	134,187	103,719	84,527	71,328	61,695	54,354	43,906	36,826	31,713	27,847	24,821	22,388	20,389	16,669	14,097	12,213	10,773	9,636	8,717	7,320	6,309	5,543	4,461	3,732
195	136,662	105,192	85,502	72,022	62,213	54,756	44,167	37,010	31,849	27,952	24,904	22,456	20,445	16,707	14,124	12,233	10,788	9,649	8,727	7,327	6,314	5,548	4,463	3,734
200	139,099	106,630	86,450	72,693	62,713	55,143	44,419	37,187	31,980	28,052	24,984	22,520	20,499	16,743	14,150	12,232	10,803	9,661	8,737	7,334	6,320	5,552	4,466	3,736

250 MCM | **1 Run(s)** | **240 Volts** | **C-Value = 16,483** | **Magnetic Duct** | **Copper**

Calculations:

1. $f = \dfrac{1.732 \times \text{Length} \times \text{Isca}}{\text{Runs} \times \text{C-Value} \times \text{E L-L}}$

2. $M = \dfrac{1}{1 + f}$

3. Isc = Isca x M

4. Add Motor Contribution = Motor FLA x 4

* All results are given in symmetrical amperes

Three-Phase Bolted Fault-Current* Table

One-Way Distance in Feet

Available Isc in Thousands (Isca)	5	10	15	20	25	30	40	50	60	70	80	90	100	125	150	175	200	225	250	300	350	400	500	600
3	2,990	2,980	2,971	2,961	2,952	2,942	2,923	2,905	2,886	2,868	2,850	2,833	2,815	2,772	2,731	2,691	2,652	2,614	2,577	2,506	2,439	2,376	2,258	2,152
5	4,973	4,946	4,919	4,893	4,867	4,841	4,790	4,741	4,692	4,644	4,597	4,552	4,507	4,398	4,295	4,196	4,102	4,012	3,926	3,764	3,615	3,477	3,231	3,018
7	6,947	6,894	6,843	6,792	6,742	6,692	6,596	6,502	6,411	6,322	6,236	6,152	6,070	5,875	5,692	5,520	5,358	5,205	5,061	4,795	4,556	4,340	3,963	3,647
10	9,892	9,786	9,682	9,581	9,481	9,384	9,195	9,013	8,839	8,671	8,510	8,354	8,204	7,852	7,528	7,230	6,955	6,700	6,463	6,036	5,662	5,331	4,774	4,323
15	14,758	14,523	14,296	14,076	13,862	13,655	13,259	12,885	12,531	12,197	11,879	11,578	11,292	10,635	10,050	9,526	9,054	8,626	8,238	7,556	6,979	6,484	5,678	5,050
20	19,572	19,161	18,767	18,390	18,027	17,678	17,019	16,408	15,839	15,308	14,812	14,347	13,910	12,926	12,072	11,324	10,663	10,075	9,549	8,645	7,898	7,269	6,271	5,514
25	24,334	23,703	23,103	22,534	21,991	21,474	20,510	19,629	18,820	18,075	17,387	16,750	16,157	14,845	13,729	12,770	11,936	11,204	10,557	9,463	8,575	7,839	6,691	5,836
30	29,046	28,151	27,310	26,517	25,769	25,062	23,759	22,584	21,520	20,552	19,667	18,855	18,108	16,475	15,113	13,958	12,968	12,108	11,356	10,100	9,095	8,271	7,003	6,072
35	33,709	32,509	31,392	30,349	29,374	28,459	26,790	25,306	23,977	22,781	21,699	20,715	19,817	17,878	16,284	14,952	13,821	12,849	12,005	10,611	9,507	8,611	7,245	6,253
40	38,322	36,779	35,356	34,039	32,816	31,678	29,624	27,820	26,223	24,799	23,522	22,370	21,326	19,097	17,290	15,795	14,539	13,467	12,543	11,029	9,841	8,884	7,437	6,396
45	42,888	40,965	39,207	37,593	36,108	34,735	32,280	30,150	28,283	26,634	25,167	23,852	22,669	20,167	18,162	16,520	15,150	13,990	12,995	11,377	10,117	9,109	7,594	6,512
50	47,406	45,067	42,949	41,020	39,258	37,640	34,775	32,315	30,180	28,309	26,657	25,188	23,871	21,113	18,926	17,150	15,678	14,439	13,382	11,672	10,350	9,297	7,725	6,607
55	51,877	49,090	46,586	44,326	42,275	40,405	37,122	34,332	31,932	29,846	28,015	26,396	24,954	21,956	19,601	17,702	16,138	14,829	13,716	11,925	10,548	9,457	7,835	6,688
60	56,302	53,034	50,124	47,517	45,168	43,040	39,334	36,216	33,555	31,259	29,257	27,496	25,935	22,711	20,201	18,190	16,543	15,169	14,007	12,145	10,720	9,594	7,929	6,756
65	60,683	56,903	53,567	50,600	47,945	45,554	41,423	37,979	35,064	32,564	30,397	28,501	26,827	23,393	20,738	18,624	16,901	15,470	14,263	12,337	10,869	9,713	8,010	6,815
70	65,018	60,699	56,917	53,579	50,611	47,955	43,399	39,633	36,469	33,773	31,448	29,422	27,642	24,010	21,221	19,013	17,221	15,738	14,490	12,506	11,000	9,818	8,081	6,867
75	69,310	64,423	60,179	56,460	53,174	50,249	45,270	41,188	37,781	34,895	32,419	30,271	28,389	24,572	21,659	19,364	17,508	15,977	14,693	12,657	11,117	9,911	8,144	6,912
80	73,559	68,078	63,357	59,248	55,640	52,446	47,044	42,652	39,010	35,940	33,319	31,054	29,077	25,085	22,057	19,681	17,768	16,193	14,875	12,792	11,221	9,994	8,200	6,952
85	77,765	71,665	66,452	61,947	58,013	54,549	48,730	44,033	40,162	36,916	34,156	31,780	29,712	25,557	22,421	19,970	18,003	16,388	15,039	12,913	11,314	10,067	8,249	6,988
90	81,929	75,187	69,470	64,560	60,299	56,566	50,333	45,338	41,244	37,829	34,936	32,454	30,301	25,991	22,754	20,234	18,217	16,566	15,189	13,023	11,399	10,134	8,294	7,020
95	86,052	78,645	72,411	67,094	62,503	58,501	51,860	46,572	42,264	38,684	35,664	33,081	30,847	26,392	23,061	20,477	18,413	16,728	15,325	13,123	11,475	10,195	8,335	7,049
100	90,134	82,040	75,280	69,550	64,630	60,360	53,315	47,743	43,225	39,488	36,346	33,668	31,357	26,764	23,344	20,700	18,593	16,876	15,449	13,215	11,545	10,250	8,371	7,075
105	94,176	85,376	78,079	71,932	66,682	62,146	54,704	48,853	44,133	40,245	36,987	34,216	31,832	27,109	23,607	20,906	18,760	17,013	15,564	13,298	11,609	10,300	8,405	7,099
110	98,179	88,652	80,811	74,244	68,664	63,864	56,031	49,909	44,993	40,959	37,588	34,731	32,277	27,431	23,851	21,097	18,913	17,139	15,669	13,375	11,667	10,346	8,435	7,121
115	102,143	91,871	83,477	76,488	70,579	65,518	57,300	50,913	45,808	41,633	38,155	35,214	32,694	27,732	24,078	21,274	19,056	17,256	15,767	13,446	11,721	10,388	8,464	7,141
120	106,068	95,035	86,081	78,668	72,432	67,111	58,514	51,870	46,581	42,270	38,690	35,669	33,086	28,013	24,289	21,439	19,188	17,365	15,858	13,512	11,771	10,428	8,490	7,159
125	109,956	98,144	88,624	80,787	74,224	68,647	59,678	52,783	47,315	42,875	39,196	36,098	33,455	28,277	24,488	21,594	19,312	17,466	15,942	13,573	11,818	10,464	8,514	7,176
130	113,806	101,200	91,108	82,846	75,958	70,128	60,795	53,654	48,014	43,448	39,674	36,504	33,803	28,525	24,674	21,738	19,427	17,560	16,020	13,630	11,861	10,498	8,536	7,192
135	117,620	104,204	93,536	84,849	77,638	71,557	61,866	54,487	48,680	43,992	40,128	36,887	34,131	28,759	24,848	21,874	19,535	17,648	16,094	13,683	11,901	10,529	8,557	7,207
140	121,397	107,158	95,909	86,797	79,267	72,938	62,896	55,284	49,315	44,510	40,558	37,251	34,442	28,980	25,013	22,001	19,637	17,731	16,163	13,733	11,939	10,559	8,576	7,221
145	125,139	110,063	98,230	88,693	80,845	74,273	63,885	56,047	49,922	45,004	40,968	37,596	34,737	29,188	25,168	22,121	19,732	17,809	16,227	13,780	11,974	10,586	8,595	7,234
150	128,846	112,920	100,499	90,539	82,376	75,563	64,838	56,779	50,501	45,474	41,357	37,924	35,017	29,385	25,314	22,234	19,822	17,882	16,288	13,824	12,007	10,612	8,612	7,246
155	132,517	115,731	102,719	92,337	83,861	76,811	65,754	57,480	51,056	45,923	41,728	38,235	35,282	29,572	25,453	22,341	19,907	17,951	16,345	13,865	12,038	10,636	8,628	7,257
160	136,155	118,496	104,891	94,089	85,304	78,019	66,638	58,154	51,587	46,352	42,082	38,533	35,535	29,749	25,584	22,442	19,987	18,016	16,399	13,904	12,067	10,659	8,643	7,268
165	139,759	121,216	107,017	95,796	86,705	79,189	67,490	58,802	52,096	46,763	42,420	38,816	35,776	29,918	25,709	22,538	20,063	18,078	16,450	13,940	12,095	10,681	8,657	7,278
170	143,330	123,893	109,098	97,460	88,066	80,323	68,311	59,425	52,584	47,156	42,744	39,086	36,005	30,078	25,827	22,629	20,135	18,136	16,499	13,975	12,121	10,701	8,670	7,287
175	146,868	126,528	111,136	99,083	89,389	81,422	69,105	60,024	53,053	47,533	43,053	39,345	36,225	30,231	25,940	22,715	20,203	18,192	16,545	14,008	12,146	10,720	8,683	7,296
180	150,373	129,121	113,132	100,666	90,675	82,488	69,871	60,602	53,504	47,894	43,349	39,592	36,434	30,377	26,047	22,797	20,268	18,245	16,588	14,039	12,169	10,739	8,695	7,304
185	153,847	131,674	115,087	102,211	91,927	83,523	70,612	61,158	53,937	48,241	43,633	39,829	36,635	30,516	26,149	22,875	20,330	18,295	16,630	14,069	12,191	10,756	8,706	7,312
190	157,289	134,187	117,002	103,719	93,145	84,527	71,328	61,695	54,354	48,574	43,906	40,056	36,826	30,649	26,247	22,950	20,389	18,342	16,669	14,097	12,213	10,773	8,717	7,320
195	160,700	136,662	118,879	105,192	94,331	85,502	72,022	62,213	54,756	48,895	44,167	40,273	37,010	30,777	26,340	23,021	20,445	18,388	16,707	14,124	12,233	10,788	8,727	7,327
200	164,081	139,099	120,719	106,630	95,485	86,450	72,693	62,713	55,143	49,203	44,419	40,482	37,187	30,898	26,429	23,090	20,499	18,431	16,743	14,150	12,252	10,803	8,737	7,334

250 MCM | 2 Run(s) | 240 Volts

C-Value = 16,483 | Magnetic Duct | Copper

Calculations:

1. $f = \dfrac{1.732 \times \text{Length} \times \text{Isca}}{\text{Runs} \times \text{C-Value} \times \text{E L-L}}$

2. $M = \dfrac{1}{1 + f}$

3. $\text{Isc} = \text{Isca} \times M$

4. Add Motor Contribution = Motor FLA x 4

* All results are given in symmetrical amperes

Three-Phase Bolted Fault-Current* Table

One-Way Distance in Feet

Available Isc in Thousands (Isca)

Isca	5	10	15	20	25	30	40	50	60	70	80	90	100	125	150	175	200	225	250	300	350	400	500	600
3	2,982	2,965	2,947	2,930	2,913	2,896	2,864	2,831	2,800	2,769	2,739	2,710	2,681	2,611	2,545	2,483	2,423	2,366	2,312	2,210	2,117	2,032	1,880	1,750
5	4,951	4,903	4,855	4,809	4,764	4,719	4,632	4,549	4,468	4,390	4,315	4,242	4,172	4,006	3,853	3,71[illegible]	3,579	3,456	3,342	3,134	2,950	2,787	2,509	2,282
7	6,904	6,811	6,720	6,631	6,545	6,461	6,300	6,146	6,000	5,860	5,727	5,599	5,478	5,195	4,940	4,709	4,499	4,307	4,130	3,817	3,548	3,315	2,929	2,624
10	9,805	9,618	9,438	9,264	9,097	8,936	8,629	8,344	8,076	7,825	7,589	7,367	7,158	6,683	6,267	5,900	5,574	5,282	5,019	4,564	4,185	3,864	3,350	2,957
15	14,566	14,157	13,770	13,403	13,056	12,726	12,114	11,558	11,051	10,587	10,159	9,766	9,401	8,599	7,922	7,345	6,846	6,410	6,027	5,383	4,863	4,435	3,771	3,280
20	19,236	18,529	17,871	17,259	16,687	16,152	15,179	14,316	13,546	12,855	12,230	11,664	11,148	10,037	9,128	8,369	7,727	7,177	6,700	5,913	5,292	4,789	4,024	3,469
25	23,818	22,743	21,760	20,859	20,030	19,264	17,895	16,708	15,668	14,751	13,935	13,204	12,546	11,157	10,045	9,134	8,375	7,732	7,181	6,285	5,588	5,030	4,192	3,594
30	28,314	26,807	25,452	24,228	23,116	22,102	20,319	18,802	17,496	16,360	15,362	14,479	13,692	12,053	10,765	9,726	8,870	8,152	7,542	6,560	5,804	5,204	4,313	3,682
35	32,726	30,730	28,963	27,388	25,976	24,702	22,496	20,651	19,086	17,742	16,574	15,551	14,646	12,787	11,347	10,158	9,261	8,481	7,823	6,771	5,969	5,336	4,403	3,748
40	37,057	34,518	32,304	30,357	28,632	27,092	24,461	22,295	20,482	18,942	17,617	16,465	15,455	13,399	11,826	10,584	9,578	8,746	8,048	6,939	6,099	5,440	4,474	3,799
45	41,310	38,179	35,489	33,153	31,106	29,297	26,244	23,767	21,718	19,994	18,523	17,254	16,148	13,917	12,228	10,905	9,840	8,964	8,232	7,075	6,204	5,524	4,530	3,840
50	45,485	41,718	38,527	35,790	33,416	31,337	27,869	25,093	22,819	20,924	19,319	17,942	16,749	14,362	12,570	11,175	10,059	9,146	8,385	7,188	6,291	5,592	4,576	3,873
55	49,586	45,142	41,429	38,281	35,577	33,230	29,357	26,292	23,807	21,751	20,022	18,547	17,275	14,747	12,864	11,407	10,247	9,301	8,515	7,284	6,363	5,650	4,615	3,900
60	53,614	48,456	44,204	40,638	37,604	34,992	30,723	27,383	24,698	22,492	20,648	19,084	17,740	15,084	13,119	11,608	10,408	9,434	8,626	7,365	6,425	5,699	4,647	3,923
65	57,571	51,666	46,860	42,872	39,509	36,636	31,983	28,379	25,505	23,160	21,210	19,562	18,152	15,381	13,344	11,783	10,549	9,549	8,723	7,435	6,479	5,740	4,675	3,943
70	61,459	54,776	49,404	44,991	41,302	38,172	33,148	29,293	26,241	23,765	21,716	19,992	18,522	15,645	13,542	11,938	10,673	9,651	8,807	7,496	6,525	5,777	4,699	3,960
75	65,280	57,791	51,843	47,005	42,993	39,612	34,229	30,134	26,914	24,315	22,175	20,380	18,855	15,882	13,719	12,075	10,783	9,740	8,881	7,550	6,566	5,809	4,720	3,975
80	69,036	60,715	54,184	48,922	44,591	40,965	35,234	30,910	27,531	24,818	22,592	20,732	19,155	16,095	13,878	12,198	10,880	9,820	8,948	7,598	6,602	5,837	4,739	3,988
85	72,728	63,552	56,432	50,747	46,102	42,237	36,171	31,629	28,100	25,279	22,974	21,053	19,429	16,288	14,021	12,308	10,968	9,891	9,007	7,641	6,634	5,862	4,755	4,000
90	76,357	66,306	58,593	52,488	47,535	43,436	37,047	32,296	28,626	25,704	23,324	21,347	19,679	16,463	14,151	12,408	11,047	9,956	9,060	7,679	6,663	5,885	4,770	4,011
95	79,926	68,981	60,672	54,150	48,894	44,568	37,867	32,918	29,113	26,096	23,646	21,617	19,908	16,623	14,269	12,499	11,119	10,014	9,109	7,714	6,689	5,905	4,784	4,020
100	83,436	71,580	62,674	55,739	50,185	45,639	38,637	33,498	29,566	26,460	23,944	21,866	20,119	16,770	14,377	12,581	11,185	10,067	9,152	7,745	6,713	5,924	4,796	4,029
105	86,888	74,106	64,602	57,258	51,414	46,652	39,361	34,041	29,988	26,798	24,221	22,096	20,314	16,905	14,476	12,657	11,244	10,115	9,192	7,774	6,735	5,940	4,807	4,036
110	90,284	76,562	66,460	58,714	52,585	47,614	40,044	34,551	30,383	27,112	24,477	22,309	20,494	17,030	14,567	12,727	11,299	10,160	9,229	7,800	6,754	5,956	4,817	4,043
115	93,625	78,951	68,253	60,109	53,701	48,527	40,688	35,029	30,752	27,406	24,716	22,508	20,661	17,145	14,652	12,791	11,350	10,201	9,263	7,824	6,772	5,970	4,826	4,050
120	96,913	81,276	69,984	61,447	54,766	49,396	41,297	35,479	31,098	27,681	24,940	22,693	20,817	17,252	14,730	12,851	11,397	10,239	9,294	7,846	6,789	5,983	4,834	4,056
125	100,148	83,539	71,656	62,732	55,785	50,223	41,873	35,904	31,424	27,938	25,149	22,866	20,962	17,352	14,802	12,906	11,441	10,274	9,323	7,867	6,804	5,995	4,842	4,061
130	103,332	85,743	73,271	63,967	56,759	51,011	42,420	36,305	31,731	28,181	25,345	23,028	21,099	17,445	14,870	12,958	11,481	10,306	9,350	7,886	6,819	6,006	4,849	4,066
135	106,466	87,890	74,833	65,154	57,692	51,763	42,938	36,684	32,020	28,409	25,529	23,180	21,226	17,532	14,933	13,006	11,519	10,337	9,375	7,904	6,832	6,016	4,856	4,071
140	109,552	89,982	76,345	66,297	58,586	52,482	43,432	37,044	32,294	28,624	25,703	23,323	21,346	17,614	14,993	13,050	11,554	10,365	9,398	7,920	6,844	6,026	4,862	4,075
145	112,590	92,022	77,808	67,397	59,444	53,169	43,901	37,385	32,553	28,827	25,866	23,457	21,459	17,691	15,048	13,093	11,587	10,392	9,420	7,936	6,856	6,034	4,868	4,080
150	115,582	94,011	79,225	68,458	60,267	53,827	44,349	37,709	32,798	29,019	26,021	23,585	21,565	17,763	15,100	13,132	11,618	10,417	9,440	7,950	6,867	6,043	4,874	4,083
155	118,528	95,950	80,598	69,481	61,059	54,458	44,776	38,017	33,031	29,202	26,168	23,705	21,666	17,831	15,150	13,169	11,647	10,440	9,460	7,964	6,877	6,051	4,879	4,087
160	121,430	97,843	81,929	70,468	61,820	55,062	45,184	38,311	33,253	29,375	26,306	23,819	21,761	17,895	15,196	13,204	11,674	10,462	9,478	7,977	6,886	6,058	4,883	4,090
165	124,288	99,691	83,221	71,421	62,552	55,642	45,574	38,591	33,464	29,539	26,438	23,927	21,851	17,956	15,240	13,237	11,700	10,483	9,495	7,989	6,895	6,065	4,888	4,094
170	127,104	101,494	84,474	72,342	63,257	56,200	45,947	38,858	33,664	29,695	26,563	24,029	21,936	18,014	15,281	13,269	11,725	10,502	9,511	8,000	6,904	6,072	4,892	4,097
175	129,878	103,255	85,690	73,232	63,937	56,736	46,305	39,114	33,856	29,844	26,682	24,126	22,017	18,069	15,321	13,298	11,748	10,521	9,526	8,011	6,912	6,078	4,896	4,099
180	132,612	104,976	86,872	74,094	64,593	57,251	46,648	39,358	34,039	29,986	26,796	24,219	22,095	18,121	15,358	13,327	11,770	10,538	9,540	8,021	6,919	6,084	4,900	4,102
185	135,307	106,657	88,020	74,927	65,225	57,748	46,977	39,592	34,214	30,122	26,904	24,308	22,168	18,170	15,394	13,353	11,790	10,555	9,554	8,031	6,927	6,089	4,904	4,105
190	137,962	108,300	89,136	75,734	65,836	58,226	47,293	39,816	34,381	30,251	27,008	24,392	22,238	18,217	15,427	13,379	11,810	10,571	9,567	8,040	6,933	6,095	4,907	4,107
195	140,579	109,907	90,221	76,516	66,426	58,687	47,596	40,031	34,541	30,376	27,106	24,473	22,305	18,262	15,460	13,403	11,829	10,586	9,580	8,049	6,940	6,100	4,910	4,109
200	143,159	111,477	91,277	77,275	66,997	59,132	47,889	40,238	34,695	30,494	27,201	24,550	22,369	18,305	15,490	13,426	11,847	10,601	9,591	8,057	6,946	6,104	4,913	4,111

300 MCM	1 Run(s)	240 Volts	C-Value = 18,176	Magnetic Duct	Copper

Calculations:

1. $f = \dfrac{1.732 \times \text{Length} \times \text{Isca}}{\text{Runs} \times \text{C-Value} \times E_{L\text{-}L}}$

2. $M = \dfrac{1}{1 + f}$

3. Isc = Isca x M

4. Add Motor Contribution = Motor FLA x 4

* All results are given in symmetrical amperes

Three-Phase Bolted Fault-Current* Table

One-Way Distance in Feet

Available Isc in Thousands (Isca)	5	10	15	20	25	30	40	50	60	70	80	90	100	125	150	175	200	225	250	300	350	400	500	600
3	2,991	2,982	2,973	2,965	2,956	2,947	2,930	2,913	2,896	2,880	2,864	2,847	2,831	2,792	2,754	2,717	2,681	2,645	2,611	2,545	2,483	2,423	2,312	2,210
5	4,975	4,951	4,927	4,903	4,879	4,855	4,809	4,764	4,719	4,675	4,632	4,590	4,549	4,448	4,352	4,260	4,172	4,087	4,006	3,853	3,711	3,579	3,342	3,134
7	6,952	6,904	6,857	6,811	6,765	6,720	6,631	6,545	6,461	6,379	6,300	6,222	6,146	5,964	5,793	5,631	5,478	5,333	5,195	4,940	4,709	4,499	4,130	3,817
10	9,902	9,805	9,711	9,618	9,527	9,438	9,264	9,097	8,936	8,780	8,629	8,484	8,344	8,012	7,705	7,422	7,158	6,912	6,683	6,267	5,900	5,574	5,019	4,564
15	14,780	14,566	14,359	14,157	13,961	13,770	13,403	13,056	12,726	12,413	12,114	11,830	11,558	10,931	10,369	9,861	9,401	8,982	8,599	7,922	7,345	6,846	6,027	5,383
20	19,611	19,236	18,876	18,529	18,194	17,871	17,259	16,687	16,152	15,650	15,179	14,735	14,316	13,366	12,535	11,801	11,148	10,563	10,037	9,128	8,369	7,727	6,700	5,913
25	24,395	23,818	23,268	22,743	22,240	21,760	20,859	20,030	19,264	18,554	17,895	17,281	16,708	15,428	14,331	13,379	12,546	11,811	11,157	10,045	9,134	8,375	7,181	6,285
30	29,132	28,314	27,540	26,807	26,112	25,452	24,228	23,116	22,102	21,173	20,319	19,531	18,802	17,197	15,845	14,690	13,692	12,820	12,053	10,765	9,726	8,870	7,542	6,560
35	33,825	32,726	31,696	30,730	29,820	28,963	27,388	25,976	24,702	23,547	22,496	21,534	20,651	18,731	17,138	15,795	14,646	13,654	12,787	11,347	10,198	9,261	7,823	6,771
40	38,472	37,057	35,743	34,518	33,374	32,304	30,357	28,632	27,092	25,709	24,461	23,328	22,295	20,074	18,255	16,739	15,455	14,354	13,399	11,826	10,584	9,578	8,048	6,939
45	43,076	41,310	39,682	38,179	36,785	35,489	33,153	31,106	29,297	27,686	26,244	24,944	23,767	21,260	19,231	17,555	16,148	14,950	13,917	12,228	10,905	9,840	8,232	7,075
50	47,636	45,485	43,520	41,718	40,059	38,527	35,790	33,416	31,337	29,502	27,869	26,408	25,093	22,314	20,089	18,268	16,749	15,464	14,362	12,570	11,175	10,059	8,385	7,188
55	52,153	49,586	47,260	45,142	43,206	41,429	38,281	35,577	33,230	31,174	29,357	27,740	26,292	23,257	20,851	18,895	17,275	15,911	14,747	12,864	11,407	10,247	8,515	7,284
60	56,627	53,614	50,905	48,456	46,233	44,204	40,638	37,604	34,992	32,719	30,723	28,957	27,383	24,107	21,531	19,452	17,740	16,304	15,084	13,119	11,608	10,408	8,626	7,365
65	61,060	57,571	54,459	51,666	49,146	46,860	42,872	39,509	36,636	34,152	31,983	30,074	28,379	24,876	22,142	19,950	18,152	16,652	15,381	13,344	11,783	10,549	8,723	7,435
70	65,452	61,459	57,926	54,776	51,951	49,404	44,991	41,302	38,172	35,483	33,148	31,102	29,293	25,575	22,694	20,397	18,522	16,963	15,645	13,542	11,938	10,673	8,807	7,496
75	69,803	65,280	61,308	57,791	54,656	51,843	47,005	42,993	39,612	36,724	34,229	32,051	30,134	26,213	23,196	20,801	18,855	17,241	15,882	13,719	12,075	10,783	8,881	7,550
80	74,115	69,036	64,609	60,715	57,264	54,184	48,922	44,591	40,965	37,884	35,234	32,931	30,910	26,799	23,653	21,168	19,155	17,492	16,095	13,878	12,198	10,880	8,948	7,598
85	78,386	72,728	67,831	63,552	59,781	56,432	50,747	46,102	42,237	38,969	36,171	33,748	31,629	27,337	24,071	21,503	19,429	17,720	16,288	14,021	12,308	10,968	9,007	7,641
90	82,619	76,357	70,978	66,306	62,212	58,593	52,488	47,535	43,436	39,988	37,047	34,509	32,296	27,835	24,456	21,809	19,679	17,928	16,463	14,151	12,408	11,047	9,060	7,679
95	86,814	79,926	74,051	68,981	64,560	60,672	54,150	48,894	44,568	40,945	37,867	35,220	32,918	28,295	24,811	22,091	19,908	18,118	16,623	14,269	12,499	11,119	9,109	7,714
100	90,970	83,436	77,054	71,580	66,831	62,674	55,739	50,185	45,639	41,847	38,637	35,885	33,498	28,723	25,139	22,351	20,119	18,292	16,770	14,377	12,581	11,185	9,152	7,745
105	95,089	86,888	79,990	74,106	69,028	64,602	57,258	51,414	46,652	42,698	39,361	36,509	34,041	29,121	25,444	22,591	20,314	18,453	16,905	14,476	12,657	11,244	9,192	7,774
110	99,172	90,284	82,859	76,562	71,154	66,460	58,714	52,585	47,614	43,502	40,044	37,095	34,551	29,493	25,727	22,814	20,494	18,602	17,030	14,567	12,727	11,299	9,229	7,800
115	103,218	93,625	85,664	78,951	73,213	68,253	60,109	53,701	48,527	44,263	40,688	37,647	35,029	29,841	25,992	23,022	20,661	18,740	17,145	14,652	12,791	11,350	9,263	7,824
120	107,228	96,913	88,408	81,276	75,208	69,984	61,447	54,766	49,396	44,985	41,297	38,168	35,479	30,167	26,239	23,216	20,817	18,868	17,252	14,730	12,851	11,397	9,294	7,846
125	111,202	100,148	91,093	83,539	77,142	71,656	62,732	55,785	50,223	45,669	41,873	38,659	35,904	30,474	26,470	23,397	20,962	18,987	17,352	14,802	12,906	11,441	9,323	7,867
130	115,142	103,332	93,720	85,743	79,018	73,271	63,967	56,759	51,011	46,320	42,420	39,125	36,305	30,762	26,688	23,566	21,099	19,099	17,445	14,870	12,958	11,481	9,350	7,886
135	119,047	106,466	96,291	87,890	80,838	74,833	65,154	57,692	51,763	46,940	42,938	39,566	36,684	31,034	26,892	23,726	21,226	19,203	17,532	14,933	13,006	11,519	9,375	7,904
140	122,919	109,552	98,808	89,982	82,604	76,345	66,297	58,586	52,482	47,530	43,432	39,984	37,044	31,291	27,085	23,875	21,346	19,301	17,614	14,993	13,050	11,554	9,398	7,920
145	126,756	112,590	101,272	92,022	84,320	77,808	67,397	59,444	53,169	48,093	43,901	40,382	37,385	31,534	27,267	24,017	21,459	19,393	17,691	15,048	13,093	11,587	9,420	7,936
150	130,561	115,582	103,686	94,011	85,987	79,225	68,458	60,267	53,827	48,631	44,349	40,760	37,709	31,764	27,439	24,150	21,565	19,480	17,763	15,100	13,132	11,618	9,440	7,950
155	134,332	118,528	106,051	95,950	87,607	80,598	69,481	61,059	54,458	49,145	44,776	41,121	38,017	31,983	27,602	24,276	21,666	19,562	17,831	15,150	13,169	11,647	9,460	7,964
160	138,072	121,430	108,368	97,843	89,182	81,929	70,468	61,820	55,062	49,636	45,184	41,465	38,311	32,190	27,756	24,395	21,761	19,640	17,895	15,196	13,204	11,674	9,478	7,977
165	141,779	124,288	110,639	99,691	90,714	83,221	71,421	62,552	55,642	50,107	45,574	41,793	38,591	32,388	27,903	24,509	21,851	19,713	17,956	15,240	13,237	11,700	9,495	7,989
170	145,455	127,104	112,865	101,494	92,205	84,474	72,342	63,257	56,200	50,559	45,947	42,106	38,858	32,576	28,042	24,616	21,936	19,782	18,014	15,281	13,269	11,725	9,511	8,000
175	149,100	129,878	115,047	103,255	93,656	85,690	73,232	63,937	56,736	50,992	46,305	42,407	39,114	32,755	28,175	24,718	22,017	19,848	18,069	15,321	13,298	11,748	9,526	8,011
180	152,715	132,612	117,187	104,976	95,070	86,872	74,094	64,593	57,251	51,408	46,648	42,694	39,358	32,926	28,301	24,816	22,095	19,911	18,121	15,358	13,327	11,770	9,540	8,021
185	156,298	135,307	119,286	106,657	96,446	88,020	74,927	65,225	57,748	51,808	46,977	42,969	39,592	33,090	28,422	24,909	22,168	19,971	18,170	15,394	13,353	11,790	9,554	8,031
190	159,852	137,962	121,345	108,300	97,788	89,136	75,734	65,836	58,226	52,193	47,293	43,234	39,816	33,246	28,538	24,997	22,238	20,028	18,217	15,427	13,379	11,810	9,567	8,040
195	163,377	140,579	123,365	109,907	99,096	90,221	76,516	66,426	58,687	52,563	47,596	43,487	40,031	33,396	28,648	25,082	22,305	20,082	18,262	15,460	13,403	11,829	9,580	8,049
200	166,872	143,159	125,347	111,477	100,371	91,277	77,275	66,997	59,132	52,920	47,889	43,731	40,238	33,540	28,754	25,163	22,369	20,134	18,305	15,490	13,426	11,847	9,591	8,057

300 MCM	2 Run(s)	240 Volts	C-Value = 18,176	Magnetic Duct	Copper

Calculations:

1. $f = \dfrac{1.732 \times \text{Length} \times \text{Isca}}{\text{Runs} \times \text{C-Value} \times E_{L\text{-}L}}$
2. $M = \dfrac{1}{1 + f}$
3. $\text{Isc} = \text{Isca} \times M$
4. Add Motor Contribution = Motor FLA x 4

* All results are given in symmetrical amperes

Three-Phase Bolted Fault-Current* Table

One-Way Distance in Feet

Available Isc in Thousands (Isca)

Isca	5	10	15	20	25	30	40	50	60	70	80	90	100	125	150	175	200	225	250	300	350	400	500	600
3	2,984	2,967	2,951	2,935	2,920	2,904	2,874	2,844	2,814	2,786	2,758	2,730	2,703	2,638	2,575	2,516	2,459	2,405	2,353	2,256	2,167	2,084	1,936	1,808
5	4,955	4,910	4,866	4,823	4,781	4,740	4,659	4,581	4,505	4,432	4,361	4,292	4,226	4,069	3,922	3,786	3,660	3,541	3,430	3,227	3,047	2,886	2,610	2,382
7	6,911	6,825	6,741	6,659	6,578	6,500	6,349	6,205	6,067	5,935	5,809	5,688	5,572	5,301	5,056	4,832	4,627	4,439	4,266	3,957	3,689	3,456	3,068	2,758
10	9,820	9,647	9,479	9,317	9,161	9,010	8,722	8,452	8,198	7,959	7,734	7,521	7,319	6,859	6,454	6,094	5,772	5,482	5,220	4,765	4,382	4,057	3,532	3,127
15	14,599	14,219	13,858	13,515	13,189	12,877	12,297	11,767	11,281	10,834	10,420	10,037	9,681	8,893	8,223	7,647	7,147	6,708	6,320	5,664	5,132	4,691	4,003	3,491
20	19,293	18,635	18,020	17,444	16,904	16,397	15,468	14,638	13,893	13,221	12,610	12,053	11,544	10,440	9,529	8,764	8,113	7,552	7,064	6,255	5,612	5,089	4,289	3,707
25	23,906	22,903	21,981	21,130	20,343	19,612	18,298	17,149	16,135	15,235	14,430	13,705	13,050	11,657	10,533	9,606	8,830	8,169	7,601	6,672	5,945	5,362	4,482	3,850
30	28,438	27,030	25,755	24,595	23,535	22,562	20,840	19,362	18,080	16,957	15,965	15,083	14,294	12,639	11,328	10,264	9,382	8,640	8,006	6,983	6,191	5,560	4,620	3,951
35	32,892	31,023	29,355	27,858	26,505	25,278	23,136	21,329	19,783	18,447	17,279	16,251	15,338	13,449	11,974	10,791	9,821	9,010	8,324	7,223	6,379	5,712	4,723	4,027
40	37,270	34,889	32,793	30,935	29,277	27,787	25,220	23,087	21,287	19,748	18,416	17,252	16,227	14,127	12,509	11,224	10,178	9,310	8,579	7,414	6,528	5,831	4,805	4,086
45	41,574	38,632	36,080	33,844	31,868	30,111	27,120	24,670	22,625	20,894	19,408	18,120	16,993	14,705	12,960	11,585	10,474	9,557	8,788	7,570	6,648	5,927	4,870	4,132
50	45,806	42,261	39,225	36,596	34,297	32,270	28,859	26,100	23,823	21,911	20,283	18,881	17,659	15,201	13,344	11,891	10,723	9,765	8,963	7,699	6,748	6,006	4,923	4,171
55	49,967	45,778	42,237	39,205	36,578	34,282	30,457	27,401	24,902	22,820	21,060	19,552	18,245	15,633	13,676	12,154	10,937	9,941	9,112	7,809	6,832	6,072	4,967	4,203
60	54,060	49,190	45,125	41,680	38,724	36,160	31,931	28,588	25,878	23,637	21,754	20,149	18,764	16,013	13,965	12,382	11,121	10,093	9,239	7,902	6,903	6,128	5,005	4,230
65	58,086	52,501	47,896	44,033	40,747	37,918	33,294	29,675	26,766	24,376	22,378	20,683	19,226	16,348	14,220	12,581	11,282	10,225	9,350	7,983	6,965	6,177	5,037	4,253
70	62,046	55,715	50,557	46,272	42,658	39,567	34,558	30,675	27,577	25,047	22,942	21,164	19,641	16,647	14,445	12,758	11,423	10,342	9,447	8,054	7,018	6,219	5,065	4,273
75	65,943	58,837	53,114	48,406	44,464	41,116	35,734	31,599	28,321	25,659	23,455	21,599	20,016	16,915	14,647	12,915	11,549	10,444	9,533	8,116	7,066	6,256	5,090	4,290
80	69,777	61,871	55,574	50,440	46,175	42,575	36,831	32,453	29,005	26,220	23,922	21,995	20,355	17,157	14,828	13,055	11,661	10,536	9,609	8,171	7,108	6,289	5,112	4,305
85	73,551	64,820	57,941	52,383	47,798	43,950	37,856	33,247	29,637	26,735	24,351	22,357	20,665	17,377	14,991	13,182	11,762	10,618	9,677	8,221	7,145	6,318	5,131	4,319
90	77,265	67,687	60,222	54,240	49,339	45,250	38,817	33,985	30,223	27,211	24,745	22,688	20,948	17,576	15,140	13,296	11,853	10,693	9,739	8,265	7,178	6,344	5,148	4,331
95	80,921	70,477	62,420	56,017	50,805	46,480	39,718	34,674	30,767	27,651	25,108	22,993	21,207	17,759	15,275	13,401	11,936	10,760	9,795	8,305	7,209	6,368	5,164	4,342
100	84,521	73,192	64,541	57,719	52,201	47,646	40,566	35,319	31,273	28,059	25,444	23,275	21,447	17,926	15,399	13,496	12,011	10,821	9,846	8,342	7,236	6,389	5,178	4,352
105	88,066	75,835	66,587	59,350	53,531	48,752	41,366	35,923	31,746	28,439	25,756	23,536	21,668	18,081	15,512	13,583	12,080	10,877	9,892	8,375	7,261	6,409	5,190	4,361
110	91,556	78,409	68,564	60,915	54,801	49,803	42,120	36,490	32,188	28,794	26,047	23,778	21,873	18,223	15,617	13,663	12,144	10,929	9,935	8,405	7,284	6,427	5,202	4,370
115	94,994	80,917	70,473	62,418	56,015	50,803	42,833	37,024	32,603	29,125	26,318	24,004	22,064	18,355	15,714	13,738	12,203	10,976	9,974	8,433	7,305	6,443	5,213	4,377
120	98,380	83,361	72,320	63,862	57,175	51,756	43,508	37,528	32,993	29,436	26,571	24,214	22,242	18,478	15,804	13,806	12,257	11,020	10,010	8,459	7,324	6,458	5,223	4,384
125	101,715	85,743	74,107	65,251	58,286	52,664	44,148	38,003	33,360	29,727	26,808	24,411	22,408	18,593	15,888	13,870	12,307	11,061	10,043	8,483	7,342	6,472	5,232	4,391
130	105,002	88,067	75,836	66,588	59,350	53,532	44,756	38,453	33,706	30,002	27,031	24,596	22,563	18,700	15,966	13,930	12,354	11,098	10,075	8,505	7,359	6,485	5,240	4,396
135	108,240	90,333	77,510	67,875	60,371	54,361	45,334	38,879	34,032	30,260	27,241	24,770	22,709	18,800	16,039	13,985	12,397	11,134	10,104	8,526	7,374	6,497	5,248	4,402
140	111,430	92,545	79,133	69,117	61,351	55,154	45,885	39,283	34,342	30,505	27,439	24,933	22,847	18,894	16,107	14,037	12,438	11,166	10,131	8,545	7,389	6,508	5,255	4,407
145	114,575	94,703	80,706	70,314	62,292	55,914	46,409	39,667	34,635	30,736	27,626	25,087	22,976	18,982	16,171	14,086	12,476	11,197	10,156	8,563	7,402	6,519	5,262	4,412
150	117,674	96,811	82,232	71,469	63,197	56,642	46,910	40,032	34,913	30,954	27,802	25,233	23,098	19,066	16,232	14,131	12,512	11,226	10,180	8,580	7,415	6,528	5,269	4,416
155	120,730	98,870	83,712	72,584	64,068	57,340	47,388	40,379	35,177	31,162	27,969	25,370	23,213	19,144	16,289	14,174	12,546	11,253	10,202	8,596	7,427	6,538	5,275	4,421
160	123,742	100,880	85,149	73,662	64,906	58,011	47,845	40,711	35,428	31,359	28,128	25,501	23,322	19,218	16,342	14,215	12,578	11,279	10,223	8,611	7,438	6,546	5,280	4,425
165	126,711	102,845	86,545	74,705	65,714	58,655	48,282	41,027	35,667	31,546	28,279	25,625	23,426	19,288	16,393	14,253	12,608	11,303	10,243	8,625	7,448	6,554	5,286	4,428
170	129,639	104,766	87,901	75,713	66,493	59,275	48,701	41,329	35,895	31,724	28,422	25,742	23,524	19,355	16,441	14,290	12,636	11,326	10,262	8,638	7,458	6,562	5,290	4,432
175	132,527	106,644	89,219	76,689	67,245	59,871	49,103	41,618	36,113	31,895	28,558	25,854	23,617	19,418	16,487	14,324	12,663	11,347	10,279	8,651	7,468	6,569	5,295	4,435
180	135,374	108,480	90,501	77,634	67,970	60,446	49,489	41,895	36,321	32,057	28,688	25,960	23,706	19,478	16,530	14,357	12,689	11,368	10,296	8,663	7,477	6,576	5,300	4,438
185	138,183	110,276	91,747	78,549	68,671	60,999	49,860	42,160	36,521	32,212	28,812	26,062	23,791	19,535	16,571	14,388	12,713	11,387	10,312	8,674	7,485	6,583	5,304	4,441
190	140,954	112,034	92,961	79,437	69,348	61,533	50,216	42,415	36,711	32,360	28,931	26,159	23,872	19,590	16,610	14,417	12,736	11,406	10,327	8,685	7,493	6,589	5,308	4,444
195	143,687	113,754	94,142	80,298	70,003	62,049	50,558	42,659	36,894	32,502	29,044	26,252	23,949	19,642	16,648	14,445	12,758	11,423	10,342	8,695	7,501	6,595	5,312	4,447
200	146,384	115,437	95,292	81,133	70,637	62,546	50,888	42,893	37,070	32,638	29,153	26,340	24,023	19,691	16,683	14,472	12,779	11,440	10,355	8,705	7,508	6,600	5,315	4,449
	350 MCM		1 Run(s)		240 Volts												C-Value = 19,703			Magnetic Duct			Copper	

Calculations:

1. $f = \frac{1.732 \times \text{Length} \times \text{Isca}}{\text{Runs} \times \text{C-Value} \times E_{L-L}}$

2. $M = \frac{1}{1 + f}$

3. Isc = Isca x M

4. Add Motor Contribution = Motor FLA x 4

* All results are given in symmetrical amperes

Three-Phase Bolted Fault-Current* Table

One-Way Distance in Feet

Available Isc in Thousands (Isca)	5	10	15	20	25	30	40	50	60	70	80	90	100	125	150	175	200	225	250	300	350	400	500	600
3	2,992	2,984	2,975	2,967	2,959	2,951	2,935	2,920	2,904	2,889	2,874	2,859	2,844	2,807	2,772	2,737	2,703	2,670	2,638	2,575	2,516	2,459	2,353	2,256
5	4,977	4,955	4,932	4,910	4,888	4,866	4,823	4,781	4,740	4,699	4,659	4,619	4,581	4,486	4,396	4,309	4,226	4,146	4,069	3,922	3,786	3,660	3,430	3,227
7	6,955	6,911	6,868	6,825	6,783	6,741	6,659	6,578	6,500	6,424	6,349	6,276	6,205	6,033	5,871	5,717	5,572	5,433	5,301	5,056	4,832	4,627	4,266	3,957
10	9,909	9,820	9,733	9,647	9,562	9,479	9,317	9,161	9,010	8,864	8,722	8,585	8,452	8,137	7,845	7,573	7,319	7,082	6,859	6,454	6,094	5,772	5,220	4,765
15	14,797	14,599	14,406	14,219	14,036	13,858	13,515	13,189	12,877	12,581	12,297	12,027	11,767	11,166	10,623	10,130	9,681	9,270	8,893	8,223	7,647	7,147	6,320	5,664
20	19,640	19,293	18,958	18,635	18,322	18,020	17,444	16,904	16,397	15,919	15,468	15,042	14,638	13,719	12,908	12,188	11,544	10,964	10,440	9,529	8,764	8,113	7,064	6,255
25	24,441	23,906	23,393	22,903	22,432	21,981	21,130	20,343	19,612	18,932	18,298	17,705	17,149	15,900	14,821	13,879	13,050	12,314	11,657	10,533	9,606	8,830	7,601	6,672
30	29,198	28,438	27,716	27,030	26,377	25,755	24,595	23,535	22,562	21,667	20,840	20,074	19,362	17,786	16,446	15,295	14,294	13,416	12,639	11,328	10,264	9,382	8,006	6,983
35	33,913	32,892	31,930	31,023	30,166	29,355	27,858	26,505	25,278	24,160	23,136	22,196	21,329	19,431	17,844	16,496	15,338	14,331	13,449	11,974	10,791	9,821	8,324	7,223
40	38,587	37,270	36,040	34,889	33,808	32,793	30,935	29,277	27,787	26,441	25,220	24,107	23,087	20,880	19,058	17,529	16,227	15,104	14,127	12,509	11,224	10,178	8,579	7,414
45	43,219	41,574	40,049	38,632	37,313	36,080	33,844	31,868	30,111	28,537	27,120	25,837	24,670	22,166	20,124	18,426	16,993	15,766	14,705	12,960	11,585	10,474	8,788	7,570
50	47,811	45,806	43,962	42,261	40,686	39,225	36,596	34,297	32,270	30,470	28,859	27,411	26,100	23,314	21,066	19,213	17,659	16,338	15,201	13,344	11,891	10,723	8,963	7,699
55	52,363	49,967	47,781	45,778	43,936	42,237	39,205	36,578	34,282	32,257	30,457	28,848	27,401	24,346	21,905	19,908	18,245	16,839	15,633	13,676	12,154	10,937	9,112	7,809
60	56,875	54,060	51,510	49,190	47,070	45,125	41,680	38,724	36,160	33,914	31,931	30,167	28,588	25,279	22,657	20,527	18,764	17,279	16,013	13,965	12,382	11,121	9,239	7,902
65	61,349	58,086	55,152	52,501	50,093	47,896	44,033	40,747	37,918	35,456	33,294	31,381	29,675	26,126	23,334	21,082	19,226	17,671	16,348	14,220	12,581	11,282	9,350	7,983
70	65,783	62,046	58,710	55,715	53,011	50,557	46,272	42,658	39,567	36,893	34,558	32,501	30,675	26,898	23,949	21,582	19,641	18,021	16,647	14,445	12,758	11,423	9,447	8,054
75	70,180	65,943	62,188	58,837	55,829	53,114	48,406	44,464	41,116	38,237	35,734	33,539	31,599	27,605	24,508	22,035	20,016	18,336	16,915	14,647	12,915	11,549	9,533	8,116
80	74,540	69,777	65,586	61,871	58,553	55,574	50,440	46,175	42,575	39,495	36,831	34,504	32,453	28,255	25,018	22,447	20,355	18,620	17,157	14,828	13,055	11,661	9,609	8,171
85	78,862	73,551	68,910	64,820	61,188	57,941	52,383	47,798	43,950	40,676	37,856	35,402	33,247	28,854	25,487	22,824	20,665	18,878	17,377	14,991	13,182	11,762	9,677	8,221
90	83,148	77,265	72,160	67,687	63,737	60,222	54,240	49,339	45,250	41,787	38,817	36,241	33,985	29,409	25,919	23,170	20,948	19,114	17,576	15,140	13,296	11,853	9,739	8,265
95	87,397	80,921	75,339	70,477	66,204	62,420	56,017	50,805	46,480	42,834	39,718	37,025	34,674	29,924	26,318	23,488	21,207	19,330	17,759	15,275	13,401	11,936	9,795	8,305
100	91,611	84,521	78,450	73,192	68,595	64,541	57,719	52,201	47,646	43,822	40,566	37,761	35,319	30,403	26,688	23,782	21,447	19,529	17,926	15,399	13,496	12,011	9,846	8,342
105	95,790	88,066	81,494	75,835	70,911	66,587	59,350	53,531	48,752	44,756	41,366	38,453	35,923	30,849	27,031	24,054	21,668	19,712	18,081	15,512	13,583	12,080	9,892	8,375
110	99,934	91,556	84,474	78,409	73,157	68,564	60,915	54,801	49,803	45,640	42,120	39,103	36,490	31,267	27,351	24,307	21,873	19,882	18,223	15,617	13,663	12,144	9,935	8,405
115	104,044	94,994	87,392	80,917	75,335	70,473	62,418	56,015	50,803	46,479	42,833	39,717	37,024	31,658	27,650	24,543	22,064	20,040	18,355	15,714	13,738	12,203	9,974	8,433
120	108,120	98,380	90,250	83,361	77,449	72,320	63,862	57,175	51,756	47,275	43,508	40,297	37,528	32,025	27,930	24,763	22,242	20,186	18,478	15,804	13,806	12,257	10,010	8,459
125	112,162	101,715	93,049	85,743	79,501	74,107	65,251	58,286	52,664	48,032	44,148	40,846	38,003	32,371	28,192	24,970	22,408	20,323	18,593	15,888	13,870	12,307	10,043	8,483
130	116,171	105,002	95,791	88,067	81,495	75,836	66,588	59,350	53,532	48,752	44,756	41,366	38,453	32,696	28,439	25,163	22,563	20,451	18,700	15,966	13,930	12,354	10,075	8,505
135	120,148	108,240	98,479	90,333	83,432	77,510	67,875	60,371	54,361	49,439	45,334	41,859	38,879	33,004	28,671	25,345	22,709	20,571	18,800	16,039	13,985	12,397	10,104	8,526
140	124,092	111,430	101,113	92,545	85,315	79,133	69,117	61,351	55,154	50,094	45,885	42,328	39,283	33,295	28,891	25,516	22,847	20,683	18,894	16,107	14,037	12,438	10,131	8,545
145	128,004	114,575	103,696	94,703	87,146	80,706	70,314	62,292	55,914	50,720	46,409	42,774	39,667	33,570	29,098	25,677	22,976	20,789	18,982	16,171	14,086	12,476	10,156	8,563
150	131,885	117,674	106,228	96,811	88,928	82,232	71,469	63,197	56,642	51,318	46,910	43,199	40,032	33,831	29,294	25,829	23,098	20,889	19,066	16,232	14,131	12,512	10,180	8,580
155	135,735	120,730	108,712	98,870	90,662	83,712	72,584	64,068	57,340	51,891	47,388	43,604	40,379	34,079	29,479	25,974	23,213	20,983	19,144	16,289	14,174	12,546	10,202	8,596
160	139,554	123,742	111,148	100,880	92,350	85,149	73,662	64,906	58,011	52,440	47,845	43,990	40,711	34,315	29,656	26,110	23,322	21,072	19,218	16,342	14,215	12,578	10,223	8,611
165	143,343	126,711	113,538	102,845	93,994	86,545	74,705	65,714	58,655	52,966	48,282	44,360	41,027	34,539	29,823	26,240	23,426	21,157	19,288	16,393	14,253	12,608	10,243	8,625
170	147,101	129,639	115,883	104,766	95,595	87,901	75,713	66,493	59,275	53,471	48,701	44,713	41,329	34,753	29,983	26,364	23,524	21,237	19,355	16,441	14,290	12,636	10,262	8,638
175	150,830	132,527	118,185	106,644	97,156	89,219	76,689	67,245	59,871	53,955	49,103	45,052	41,618	34,957	30,134	26,481	23,617	21,313	19,418	16,487	14,324	12,663	10,279	8,651
180	154,530	135,374	120,444	108,480	98,678	90,501	77,634	67,970	60,446	54,421	49,489	45,377	41,895	35,152	30,279	26,593	23,706	21,385	19,478	16,530	14,357	12,689	10,296	8,663
185	158,201	138,183	122,663	110,276	100,162	91,747	78,549	68,671	60,999	54,870	49,860	45,688	42,160	35,339	30,417	26,699	23,791	21,454	19,535	16,571	14,388	12,713	10,312	8,674
190	161,843	140,954	124,841	112,034	101,610	92,961	79,437	69,348	61,533	55,301	50,216	45,987	42,415	35,517	30,550	26,801	23,872	21,520	19,590	16,610	14,417	12,736	10,327	8,685
195	165,456	143,687	126,980	113,754	103,023	94,142	80,298	70,003	62,049	55,717	50,558	46,274	42,659	35,688	30,676	26,898	23,949	21,582	19,642	16,648	14,445	12,758	10,342	8,695
200	169,042	146,384	129,081	115,437	104,402	95,292	81,133	70,637	62,546	56,118	50,888	46,550	42,893	35,853	30,797	26,991	24,023	21,642	19,691	16,683	14,472	12,779	10,355	8,705

350 MCM	2 Run(s)	240 Volts	C-Value = 19,703	Magnetic Duct	Copper

Calculations:

1. $f = \dfrac{1.732 \times \text{Length} \times \text{Isca}}{\text{Runs} \times \text{C-Value} \times E_{L-L}}$
2. $M = \dfrac{1}{1 + f}$
3. Isc = Isca x M
4. Add Motor Contribution = Motor FLA x 4

* All results are given in symmetrical amperes

Three-Phase Bolted Fault-Current* Table

One-Way Distance in Feet

Available Isc in Thousands (Isca)

	5	10	15	20	25	30	40	50	60	70	80	90	100	125	150	175	200	225	250	300	350	400	500	600
3	2,984	2,969	2,953	2,938	2,923	2,908	2,879	2,850	2,822	2,794	2,767	2,740	2,714	2,651	2,591	2,533	2,478	2,425	2,375	2,280	2,192	2,111	1,965	1,839
5	4,957	4,914	4,872	4,830	4,790	4,750	4,672	4,597	4,524	4,453	4,385	4,318	4,254	4,101	3,958	3,825	3,701	3,585	3,475	3,276	3,098	2,938	2,663	2,436
7	6,915	6,832	6,751	6,672	6,595	6,520	6,374	6,234	6,101	5,973	5,850	5,733	5,620	5,356	5,115	4,896	4,694	4,508	4,337	4,030	3,764	3,531	3,142	2,830
10	9,828	9,661	9,500	9,344	9,193	9,048	8,769	8,507	8,261	8,028	7,808	7,600	7,402	6,951	6,551	6,195	5,876	5,588	5,327	4,871	4,488	4,160	3,630	3,220
15	14,615	14,250	13,902	13,571	13,256	12,954	12,391	11,875	11,400	10,961	10,555	10,178	9,827	9,047	8,382	7,808	7,307	6,867	6,477	5,816	5,277	4,830	4,130	3,607
20	19,322	18,688	18,095	17,538	17,015	16,521	15,616	14,805	14,074	13,411	12,808	12,257	11,752	10,654	9,743	8,976	8,321	7,755	7,261	6,440	5,786	5,253	4,435	3,838
25	23,949	22,984	22,093	21,268	20,503	19,791	18,506	17,377	16,379	15,488	14,690	13,970	13,317	11,924	10,795	9,861	9,076	8,406	7,829	6,883	6,142	5,544	4,641	3,991
30	28,500	27,143	25,909	24,782	23,749	22,799	21,110	19,654	18,386	17,272	16,285	15,404	14,614	12,954	11,632	10,555	9,660	8,905	8,260	7,215	6,404	5,757	4,789	4,100
35	32,975	31,171	29,555	28,098	26,778	25,576	23,470	21,684	20,150	18,820	17,654	16,624	15,708	13,805	12,314	11,113	10,126	9,300	8,598	7,471	6,605	5,919	4,901	4,182
40	37,377	35,076	33,043	31,232	29,609	28,147	25,617	23,504	21,713	20,176	18,842	17,673	16,641	14,521	12,880	11,573	10,506	9,619	8,871	7,676	6,765	6,047	4,989	4,245
45	41,707	38,863	36,382	34,199	32,263	30,535	27,579	25,146	23,107	21,374	19,882	18,586	17,448	15,132	13,358	11,957	10,822	9,883	9,095	7,843	6,894	6,150	5,059	4,296
50	45,967	42,537	39,582	37,012	34,755	32,757	29,380	26,634	24,357	22,439	20,801	19,386	18,151	15,658	13,767	12,283	11,088	10,105	9,282	7,982	7,002	6,236	5,116	4,337
55	50,159	46,102	42,652	39,682	37,099	34,832	31,038	27,989	25,486	23,394	21,619	20,095	18,771	16,117	14,120	12,564	11,317	10,295	9,442	8,100	7,092	6,307	5,164	4,372
60	54,285	49,564	45,599	42,221	39,309	36,772	32,570	29,229	26,510	24,254	22,351	20,726	19,320	16,520	14,429	12,808	11,514	10,458	9,579	8,201	7,169	6,368	5,205	4,401
65	58,346	52,927	48,430	44,637	41,395	38,592	33,989	30,367	27,443	25,032	23,011	21,291	19,811	16,878	14,701	13,022	11,687	10,600	9,698	8,288	7,236	6,420	5,240	4,426
70	62,343	56,196	51,152	46,939	43,368	40,301	35,308	31,415	28,296	25,740	23,608	21,801	20,252	17,197	14,942	13,211	11,839	10,725	9,802	8,364	7,294	6,466	5,270	4,448
75	66,278	59,373	53,772	49,136	45,236	41,910	36,536	32,384	29,079	26,387	24,151	22,264	20,650	17,483	15,158	13,379	11,974	10,835	9,895	8,431	7,345	6,506	5,297	4,467
80	70,153	62,464	56,294	51,234	47,008	43,426	37,684	33,282	29,802	26,980	24,647	22,685	21,012	17,742	15,352	13,530	12,094	10,934	9,977	8,491	7,390	6,542	5,320	4,483
85	73,968	65,471	58,725	53,239	48,691	44,859	38,757	34,117	30,469	27,526	25,101	23,069	21,342	17,976	15,527	13,666	12,203	11,023	10,051	8,544	7,430	6,573	5,341	4,498
90	77,726	68,398	61,069	55,159	50,291	46,213	39,765	34,895	31,088	28,030	25,520	23,423	21,644	18,190	15,686	13,789	12,301	11,103	10,117	8,592	7,466	6,602	5,360	4,511
95	81,427	71,248	63,331	56,997	51,815	47,497	40,711	35,622	31,664	28,498	25,907	23,748	21,921	18,385	15,832	13,901	12,390	11,175	10,177	8,635	7,499	6,627	5,377	4,523
100	85,073	74,024	65,514	58,760	53,268	48,715	41,603	36,303	32,201	28,932	26,265	24,048	22,177	18,565	15,965	14,003	12,471	11,241	10,232	8,675	7,529	6,650	5,392	4,534
105	88,665	76,728	67,624	60,451	54,654	49,872	42,444	36,942	32,702	29,336	26,598	24,327	22,414	18,731	16,087	14,097	12,546	11,302	10,282	8,711	7,556	6,671	5,406	4,544
110	92,204	79,364	69,664	62,076	55,979	50,972	43,238	37,542	33,172	29,713	26,907	24,586	22,633	18,884	16,200	14,184	12,614	11,357	10,328	8,744	7,581	6,691	5,419	4,553
115	95,691	81,935	71,636	63,637	57,245	52,020	43,990	38,107	33,612	30,066	27,197	24,827	22,837	19,026	16,304	14,264	12,678	11,409	10,371	8,774	7,604	6,709	5,430	4,561
120	99,128	84,441	73,545	65,139	58,458	53,020	44,702	38,641	34,027	30,397	27,467	25,053	23,028	19,158	16,401	14,338	12,736	11,456	10,410	8,802	7,625	6,725	5,441	4,569
125	102,516	86,887	75,393	66,585	59,620	53,974	45,379	39,145	34,417	30,708	27,721	25,264	23,206	19,281	16,491	14,407	12,790	11,500	10,446	8,828	7,644	6,740	5,451	4,576
130	105,855	89,274	77,184	67,978	60,734	54,885	46,021	39,622	34,786	31,001	27,960	25,461	23,373	19,396	16,575	14,471	12,841	11,541	10,480	8,852	7,662	6,754	5,460	4,582
135	109,146	91,604	78,919	69,320	61,803	55,757	46,633	40,075	35,134	31,278	28,184	25,647	23,530	19,504	16,654	14,531	12,888	11,579	10,511	8,874	7,679	6,767	5,468	4,588
140	112,392	93,879	80,602	70,615	62,830	56,592	47,215	40,504	35,463	31,539	28,396	25,823	23,677	19,605	16,728	14,587	12,932	11,614	10,540	8,895	7,694	6,779	5,476	4,594
145	115,592	96,101	82,234	71,865	63,818	57,392	47,771	40,912	35,776	31,785	28,596	25,988	23,816	19,700	16,797	14,640	12,973	11,648	10,568	8,915	7,709	6,791	5,484	4,599
150	118,747	98,272	83,819	73,072	64,768	58,159	48,301	41,301	36,073	32,019	28,785	26,144	23,947	19,790	16,862	14,689	13,012	11,679	10,594	8,933	7,723	6,801	5,491	4,604
155	121,859	100,393	85,358	74,239	65,683	58,896	48,808	41,671	36,355	32,241	28,964	26,292	24,071	19,874	16,923	14,735	13,049	11,708	10,618	8,950	7,736	6,811	5,497	4,608
160	124,928	102,467	86,852	75,367	66,565	59,603	49,293	42,024	36,623	32,452	29,134	26,432	24,188	19,954	16,981	14,780	13,083	11,736	10,641	8,967	7,748	6,820	5,503	4,613
165	127,956	104,495	88,305	76,458	67,415	60,284	49,758	42,361	36,879	32,653	29,296	26,565	24,300	20,030	17,036	14,821	13,116	11,762	10,662	8,982	7,759	6,829	5,509	4,617
170	130,942	106,479	89,717	77,515	68,234	60,939	50,203	42,683	37,123	32,844	29,450	26,691	24,406	20,102	17,088	14,860	13,146	11,787	10,682	8,996	7,770	6,838	5,514	4,620
175	133,889	108,419	91,091	78,538	69,026	61,569	50,630	42,992	37,356	33,027	29,596	26,812	24,506	20,170	17,137	14,898	13,176	11,810	10,702	9,010	7,780	6,845	5,520	4,624
180	136,796	110,317	92,427	79,529	69,791	62,177	51,040	43,287	37,579	33,201	29,736	26,926	24,602	20,235	17,184	14,933	13,203	11,833	10,720	9,023	7,790	6,853	5,524	4,627
185	139,665	112,175	93,728	80,491	70,530	62,763	51,435	43,570	37,792	33,367	29,869	27,036	24,693	20,296	17,228	14,966	13,229	11,854	10,737	9,035	7,799	6,860	5,529	4,631
190	142,496	113,994	94,994	81,423	71,245	63,328	51,814	43,842	37,996	33,526	29,997	27,140	24,780	20,355	17,271	14,998	13,254	11,874	10,753	9,047	7,807	6,867	5,533	4,634
195	145,290	115,776	96,228	82,328	71,936	63,874	52,178	44,103	38,192	33,678	30,119	27,240	24,863	20,411	17,311	15,029	13,278	11,893	10,769	9,058	7,816	6,873	5,537	4,636
200	148,047	117,520	97,430	83,206	72,606	64,401	52,530	44,354	38,380	33,824	30,236	27,335	24,943	20,465	17,350	15,058	13,301	11,911	10,784	9,068	7,823	6,879	5,541	4,639

400 MCM	1 Run(s)	240 Volts	C-Value = 20,565	Magnetic Duct	Copper

Calculations:

1. $f = \dfrac{1.732 \times \text{Length} \times \text{Isca}}{\text{Runs} \times \text{C-Value} \times E_{L\text{-}L}}$

2. $M = \dfrac{1}{1 + f}$

3. Isc = Isca x M

4. Add Motor Contribution = Motor FLA x 4

* All results are given in symmetrical amperes

Three-Phase Bolted Fault-Current* Table

One-Way Distance in Feet

Available Isc in Thousands (Isca)	5	10	15	20	25	30	40	50	60	70	80	90	100	125	150	175	200	225	250	300	350	400	500	600
3	2,992	2,984	2,976	2,969	2,961	2,953	2,938	2,923	2,908	2,893	2,879	2,864	2,850	2,815	2,780	2,747	2,714	2,682	2,651	2,591	2,533	2,478	2,375	2,280
5	4,978	4,957	4,935	4,914	4,893	4,872	4,830	4,790	4,750	4,711	4,672	4,634	4,597	4,506	4,419	4,335	4,254	4,176	4,101	3,958	3,825	3,701	3,475	3,276
7	6,957	6,915	6,873	6,832	6,791	6,751	6,672	6,595	6,520	6,446	6,374	6,303	6,234	6,068	5,911	5,762	5,620	5,484	5,356	5,115	4,896	4,694	4,337	4,030
10	9,913	9,828	9,744	9,661	9,580	9,500	9,344	9,193	9,048	8,906	8,769	8,636	8,507	8,201	7,916	7,651	7,402	7,170	6,951	6,551	6,195	5,876	5,327	4,871
15	14,805	14,615	14,430	14,250	14,074	13,902	13,571	13,256	12,954	12,666	12,391	12,127	11,875	11,287	10,754	10,270	9,827	9,421	9,047	8,382	7,808	7,307	6,477	5,816
20	19,655	19,322	19,000	18,688	18,387	18,095	17,538	17,015	16,521	16,056	15,616	15,200	14,805	13,902	13,103	12,391	11,752	11,176	10,654	9,743	8,976	8,321	7,261	6,440
25	24,463	23,949	23,457	22,984	22,529	22,093	21,268	20,503	19,791	19,127	18,506	17,924	17,377	16,147	15,079	14,143	13,317	12,582	11,924	10,795	9,861	9,076	7,829	6,883
30	29,231	28,500	27,805	27,143	26,511	25,909	24,782	23,749	22,799	21,922	21,110	20,356	19,654	18,094	16,764	15,616	14,614	13,734	12,954	11,632	10,555	9,660	8,260	7,215
35	33,957	32,975	32,048	31,171	30,342	29,555	28,098	26,778	25,576	24,478	23,470	22,541	21,684	19,800	18,218	16,870	15,708	14,695	13,805	12,314	11,113	10,126	8,598	7,471
40	38,644	37,377	36,190	35,076	34,029	33,043	31,232	29,609	28,147	26,822	25,617	24,515	23,504	21,307	19,486	17,952	16,641	15,509	14,521	12,880	11,573	10,506	8,871	7,676
45	43,291	41,707	40,235	38,863	37,582	36,382	34,199	32,263	30,535	28,982	27,579	26,306	25,146	22,648	20,601	18,894	17,448	16,207	15,132	13,358	11,957	10,822	9,095	7,843
50	47,899	45,967	44,185	42,537	41,006	39,582	37,012	34,755	32,757	30,977	29,380	27,940	26,634	23,848	21,589	19,722	18,151	16,813	15,658	13,767	12,283	11,088	9,282	7,982
55	52,468	50,159	48,045	46,102	44,310	42,652	39,682	37,099	34,832	32,826	31,038	29,435	27,989	24,929	22,472	20,455	18,771	17,343	16,117	14,120	12,564	11,317	9,442	8,100
60	57,000	54,285	51,817	49,564	47,499	45,599	42,221	39,309	36,772	34,544	32,570	30,809	29,229	25,907	23,264	21,109	19,320	17,811	16,520	14,429	12,808	11,514	9,579	8,201
65	61,493	58,346	55,505	52,927	50,579	48,430	44,637	41,395	38,592	36,144	33,989	32,076	30,367	26,797	23,979	21,697	19,811	18,227	16,878	14,701	13,022	11,687	9,698	8,288
70	65,950	62,343	59,110	56,196	53,556	51,152	46,939	43,368	40,301	37,639	35,308	33,248	31,415	27,610	24,628	22,227	20,252	18,600	17,197	14,942	13,211	11,839	9,802	8,364
75	70,370	66,278	62,636	59,373	56,434	53,772	49,136	45,236	41,910	39,039	36,536	34,335	32,384	28,356	25,219	22,707	20,650	18,935	17,483	15,158	13,379	11,974	9,895	8,431
80	74,754	70,153	66,086	62,464	59,219	56,294	51,234	47,008	43,426	40,352	37,684	35,346	33,282	29,042	25,761	23,145	21,012	19,239	17,742	15,352	13,530	12,094	9,977	8,491
85	79,101	73,968	69,461	65,471	61,915	58,725	53,239	48,691	44,859	41,585	38,757	36,290	34,117	29,676	26,258	23,546	21,342	19,515	17,976	15,527	13,666	12,203	10,051	8,544
90	83,414	77,726	72,764	68,398	64,526	61,069	55,159	50,291	46,213	42,747	39,765	37,171	34,895	30,263	26,716	23,914	21,644	19,767	18,190	15,686	13,789	12,301	10,117	8,592
95	87,691	81,427	75,998	71,248	67,056	63,331	56,997	51,815	47,497	43,843	40,711	37,997	35,622	30,808	27,141	24,253	21,921	19,998	18,385	15,832	13,901	12,390	10,177	8,635
100	91,935	85,073	79,165	74,024	69,510	65,514	58,760	53,268	48,715	44,879	41,603	38,773	36,303	31,316	27,534	24,567	22,177	20,211	18,565	15,965	14,003	12,471	10,232	8,675
105	96,144	88,665	82,266	76,728	71,889	67,624	60,451	54,654	49,872	45,859	42,444	39,502	36,942	31,790	27,900	24,858	22,414	20,407	18,731	16,087	14,097	12,546	10,282	8,711
110	100,319	92,204	85,304	79,364	74,198	69,664	62,076	55,979	50,972	46,788	43,238	40,189	37,542	32,234	28,241	25,128	22,633	20,589	18,884	16,200	14,184	12,614	10,328	8,744
115	104,461	95,691	88,280	81,935	76,440	71,636	63,637	57,245	52,020	47,669	43,990	40,838	38,107	32,650	28,559	25,380	22,837	20,758	19,026	16,304	14,264	12,678	10,371	8,774
120	108,570	99,128	91,197	84,441	78,617	73,545	65,139	58,458	53,020	48,507	44,702	41,451	38,641	33,041	28,858	25,616	23,028	20,915	19,158	16,401	14,338	12,736	10,410	8,802
125	112,647	102,516	94,057	86,887	80,733	75,393	66,585	59,620	53,974	49,304	45,379	42,032	39,145	33,409	29,138	25,836	23,206	21,062	19,281	16,491	14,407	12,790	10,446	8,828
130	116,691	105,855	96,860	89,274	82,790	77,184	67,978	60,734	54,885	50,064	46,021	42,583	39,622	33,756	29,402	26,043	23,373	21,200	19,396	16,575	14,471	12,841	10,480	8,852
135	120,704	109,146	99,608	91,604	84,790	78,919	69,320	61,803	55,757	50,788	46,633	43,106	40,075	34,083	29,650	26,238	23,530	21,328	19,504	16,654	14,531	12,888	10,511	8,874
140	124,686	112,392	102,304	93,879	86,735	80,602	70,615	62,830	56,592	51,480	47,215	43,603	40,504	34,393	29,885	26,421	23,677	21,449	19,605	16,728	14,587	12,932	10,540	8,895
145	128,636	115,592	104,949	96,101	88,629	82,234	71,865	63,818	57,392	52,141	47,771	44,076	40,912	34,687	30,106	26,594	23,816	21,563	19,700	16,797	14,640	12,973	10,568	8,915
150	132,556	118,747	107,543	98,272	90,472	83,819	73,072	64,768	58,159	52,774	48,301	44,527	41,301	34,966	30,316	26,758	23,947	21,671	19,790	16,862	14,689	13,012	10,594	8,933
155	136,446	121,859	110,090	100,393	92,267	85,358	74,239	65,683	58,896	53,379	48,808	44,958	41,671	35,231	30,515	26,913	24,071	21,772	19,874	16,923	14,736	13,049	10,618	8,950
160	140,306	124,928	112,589	102,467	94,016	86,852	75,367	66,565	59,603	53,960	49,293	45,369	42,024	35,483	30,704	27,060	24,188	21,868	19,954	16,981	14,780	13,083	10,641	8,967
165	144,136	127,956	115,042	104,495	95,720	88,305	76,458	67,415	60,284	54,517	49,758	45,762	42,361	35,723	30,884	27,199	24,300	21,959	20,030	17,036	14,821	13,116	10,662	8,982
170	147,937	130,942	117,450	106,479	97,382	89,717	77,515	68,234	60,939	55,052	50,203	46,139	42,683	35,952	31,055	27,331	24,406	22,045	20,102	17,088	14,860	13,146	10,682	8,996
175	151,709	133,889	119,815	108,419	99,002	91,091	78,538	69,026	61,569	55,566	50,630	46,499	42,992	36,171	31,218	27,458	24,506	22,127	20,170	17,137	14,898	13,176	10,702	9,010
180	155,452	136,796	122,138	110,317	100,583	92,427	79,529	69,791	62,177	56,061	51,040	46,845	43,287	36,379	31,373	27,578	24,602	22,205	20,235	17,184	14,933	13,203	10,720	9,023
185	159,167	139,665	124,420	112,175	102,125	93,728	80,491	70,530	62,763	56,537	51,435	47,177	43,570	36,579	31,521	27,692	24,693	22,280	20,296	17,228	14,966	13,229	10,737	9,035
190	162,854	142,496	126,662	113,994	103,631	94,994	81,423	71,245	63,328	56,995	51,814	47,496	43,842	36,771	31,663	27,802	24,780	22,351	20,355	17,271	14,998	13,254	10,753	9,047
195	166,514	145,290	128,864	115,776	105,100	96,228	82,328	71,936	63,874	57,437	52,178	47,802	44,103	36,954	31,799	27,907	24,863	22,418	20,411	17,311	15,029	13,278	10,769	9,058
200	170,146	148,047	131,029	117,520	106,536	97,430	83,206	72,606	64,401	57,863	52,530	48,097	44,354	37,130	31,930	28,007	24,943	22,483	20,465	17,350	15,058	13,301	10,784	9,068

400 MCM	2 Run(s)	240 Volts	C-Value = 20,565	Magnetic Duct	Copper

Calculations:

1. $f = \dfrac{1.732 \times \text{Length} \times \text{Isca}}{\text{Runs} \times \text{C-Value} \times E_{L\text{-}L}}$

2. $M = \dfrac{1}{1 + f}$

3. Isc = Isca x M

4. Add Motor Contribution = Motor FLA x 4

* All results are given in symmetrical amperes

Three-Phase Bolted Fault-Current* Table

One-Way Distance in Feet

Available Isc in Thousands (Isca)	5	10	15	20	25	30	40	50	60	70	80	90	100	125	150	175	200	225	250	300	350	400	500	600
3	2,985	2,971	2,957	2,943	2,929	2,915	2,887	2,860	2,834	2,808	2,783	2,758	2,733	2,674	2,617	2,562	2,510	2,460	2,412	2,321	2,236	2,158	2,016	1,892
5	4,960	4,920	4,881	4,842	4,805	4,767	4,695	4,624	4,555	4,489	4,424	4,362	4,301	4,155	4,019	3,892	3,773	3,660	3,555	3,360	3,186	3,029	2,757	2,531
7	6,921	6,844	6,769	6,695	6,623	6,552	6,416	6,284	6,159	6,038	5,921	5,809	5,702	5,449	5,218	5,005	4,810	4,629	4,461	4,159	3,895	3,663	3,273	2,958
10	9,840	9,685	9,535	9,389	9,248	9,111	8,849	8,601	8,367	8,145	7,935	7,735	7,545	7,109	6,721	6,372	6,058	5,774	5,515	5,061	4,676	4,346	3,807	3,388
15	14,643	14,302	13,977	13,666	13,369	13,085	12,550	12,058	11,603	11,181	10,789	10,423	10,081	9,317	8,661	8,091	7,592	7,150	6,757	6,088	5,540	5,082	4,361	3,819
20	19,370	18,778	18,222	17,697	17,202	16,734	15,870	15,091	14,385	13,742	13,154	12,614	12,117	11,030	10,122	9,352	8,691	8,117	7,615	6,776	6,103	5,552	4,703	4,079
25	24,023	23,120	22,282	21,503	20,776	20,097	18,864	17,773	16,802	15,931	15,146	14,435	13,787	12,397	11,262	10,317	9,518	8,835	8,242	7,268	6,500	5,878	4,935	4,252
30	28,604	27,333	26,169	25,101	24,116	23,206	21,577	20,162	18,921	17,824	16,847	15,972	15,183	13,514	12,176	11,079	10,163	9,387	8,722	7,638	6,794	6,118	5,103	4,376
35	33,115	31,422	29,895	28,508	27,245	26,089	24,048	22,303	20,795	19,477	18,317	17,287	16,366	14,444	12,926	11,696	10,680	9,827	9,100	7,926	7,021	6,302	5,230	4,469
40	37,557	35,395	33,468	31,740	30,182	28,770	26,308	24,234	22,463	20,933	19,599	18,424	17,382	15,230	13,551	12,206	11,104	10,184	9,405	8,157	7,202	6,447	5,329	4,542
45	41,931	39,254	36,898	34,809	32,944	31,268	28,382	25,983	23,958	22,226	20,727	19,418	18,264	15,902	14,081	12,634	11,457	10,481	9,658	8,347	7,349	6,564	5,409	4,600
50	46,240	43,005	40,194	37,727	35,546	33,603	30,292	27,575	25,305	23,381	21,728	20,294	19,037	16,485	14,536	12,999	11,757	10,731	9,869	8,504	7,471	6,661	5,475	4,647
55	50,484	46,653	43,363	40,506	38,002	35,790	32,058	29,030	26,526	24,419	22,622	21,071	19,719	16,994	14,931	13,314	12,013	10,944	10,050	8,638	7,574	6,743	5,530	4,687
60	54,665	50,202	46,412	43,154	40,324	37,842	33,694	30,366	27,636	25,357	23,425	21,766	20,327	17,443	15,276	13,583	12,236	11,129	10,205	8,752	7,662	6,813	5,577	4,720
65	58,785	53,655	49,349	45,682	42,522	39,772	35,216	31,596	28,651	26,209	24,150	22,391	20,871	17,842	15,581	13,823	12,431	11,290	10,340	8,852	7,738	6,873	5,617	4,749
70	62,845	57,017	52,178	48,096	44,607	41,589	36,633	32,733	29,583	26,986	24,808	22,956	21,361	18,199	15,853	14,043	12,603	11,432	10,459	8,939	7,804	6,925	5,652	4,774
75	66,846	60,291	54,907	50,405	46,586	43,305	37,958	33,786	30,440	27,698	25,408	23,469	21,804	18,520	16,096	14,233	12,756	11,557	10,564	9,015	7,862	6,971	5,682	4,796
80	70,789	63,480	57,539	52,615	48,468	44,926	39,198	34,765	31,233	28,352	25,958	23,937	22,208	18,810	16,315	14,404	12,893	11,670	10,658	9,084	7,914	7,012	5,709	4,815
85	74,676	66,588	60,081	54,733	50,259	46,461	40,361	35,677	31,967	28,956	26,463	24,366	22,576	19,074	16,513	14,558	13,017	11,771	10,742	9,145	7,961	7,048	5,734	4,832
90	78,508	69,618	62,537	56,763	51,966	47,916	41,454	36,529	32,649	29,514	26,929	24,760	22,914	19,315	16,693	14,698	13,129	11,862	10,818	9,200	8,002	7,081	5,755	4,848
95	82,286	72,573	64,911	58,712	53,594	49,297	42,484	37,326	33,285	30,033	27,360	25,124	23,226	19,536	16,858	14,825	13,230	11,945	10,887	9,249	8,040	7,110	5,775	4,861
100	86,011	75,455	67,207	60,584	55,150	50,610	43,456	38,074	33,878	30,515	27,760	25,460	23,513	19,739	17,008	14,942	13,323	12,020	10,950	9,295	8,074	7,137	5,792	4,874
105	89,684	78,267	69,429	62,384	56,637	51,860	44,374	38,777	34,433	30,965	28,131	25,773	23,779	19,926	17,147	15,049	13,408	12,090	11,007	9,336	8,105	7,161	5,808	4,885
110	93,306	81,012	71,580	64,116	58,061	53,051	45,243	39,439	34,955	31,386	28,478	26,064	24,027	20,099	17,276	15,147	13,486	12,153	11,060	9,374	8,134	7,183	5,823	4,896
115	96,879	83,692	73,664	65,783	59,425	54,187	46,067	40,063	35,444	31,780	28,802	26,335	24,257	20,260	17,394	15,239	13,558	12,212	11,109	9,409	8,160	7,204	5,836	4,905
120	100,404	86,309	75,684	67,389	60,732	55,273	46,849	40,654	35,905	32,150	29,106	26,589	24,472	20,410	17,505	15,323	13,625	12,266	11,154	9,441	8,184	7,223	5,849	4,914
125	103,880	88,866	77,643	68,937	61,987	56,310	47,592	41,212	36,340	32,499	29,391	26,827	24,673	20,550	17,607	15,402	13,688	12,317	11,195	9,471	8,207	7,240	5,860	4,922
130	107,310	91,364	79,544	70,431	63,192	57,303	48,300	41,741	36,751	32,827	29,660	27,050	24,862	20,681	17,703	15,475	13,745	12,363	11,234	9,498	8,227	7,256	5,871	4,929
135	110,694	93,805	81,388	71,874	64,351	58,254	48,974	42,244	37,140	33,137	29,912	27,260	25,040	20,803	17,793	15,544	13,800	12,407	11,270	9,524	8,247	7,271	5,880	4,936
140	114,034	96,193	83,179	73,267	65,465	59,166	49,616	42,721	37,509	33,430	30,151	27,458	25,206	20,918	17,877	15,608	13,850	12,448	11,304	9,548	8,265	7,285	5,890	4,943
145	117,329	98,527	84,919	74,613	66,538	60,041	50,230	43,175	37,858	33,707	30,377	27,645	25,364	21,027	17,956	15,668	13,897	12,486	11,335	9,571	8,282	7,298	5,898	4,949
150	120,582	100,810	86,609	75,915	67,572	60,881	50,817	43,608	38,191	33,970	30,590	27,822	25,513	21,129	18,031	15,725	13,942	12,522	11,365	9,592	8,297	7,311	5,906	4,954
155	123,792	103,044	88,253	77,175	68,568	61,689	51,378	44,021	38,507	34,220	30,793	27,989	25,653	21,225	18,101	15,778	13,984	12,556	11,393	9,612	8,312	7,322	5,914	4,960
160	126,960	105,230	89,852	78,395	69,530	62,465	51,916	44,415	38,808	34,458	30,985	28,148	25,787	21,317	18,167	15,829	14,023	12,588	11,419	9,630	8,326	7,333	5,921	4,965
165	130,088	107,370	91,407	79,577	70,457	63,213	52,432	44,792	39,096	34,685	31,168	28,299	25,913	21,403	18,230	15,876	14,061	12,618	11,444	9,648	8,339	7,343	5,927	4,969
170	133,177	109,465	92,921	80,722	71,354	63,934	52,926	45,153	39,370	34,900	31,342	28,442	26,034	21,485	18,289	15,921	14,096	12,646	11,467	9,665	8,352	7,353	5,934	4,974
175	136,226	111,517	94,396	81,832	72,220	64,628	53,401	45,498	39,632	35,106	31,508	28,579	26,148	21,563	18,346	15,964	14,130	12,673	11,489	9,680	8,363	7,362	5,940	4,978
180	139,236	113,527	95,832	82,909	73,057	65,298	53,858	45,829	39,883	35,303	31,666	28,709	26,257	21,637	18,399	16,005	14,161	12,699	11,510	9,695	8,375	7,371	5,945	4,982
185	142,209	115,495	97,231	83,954	73,867	65,945	54,297	46,146	40,123	35,491	31,818	28,833	26,361	21,707	18,450	16,043	14,192	12,723	11,530	9,709	8,385	7,379	5,951	4,985
190	145,146	117,425	98,594	84,969	74,652	66,569	54,720	46,451	40,354	35,671	31,962	28,952	26,460	21,775	18,499	16,080	14,220	12,746	11,549	9,723	8,395	7,387	5,956	4,989
195	148,046	119,315	99,924	85,954	75,412	67,172	55,127	46,744	40,575	35,844	32,101	29,066	26,555	21,839	18,545	16,115	14,248	12,768	11,567	9,736	8,405	7,394	5,960	4,992
200	150,910	121,169	101,220	86,912	76,148	67,756	55,519	47,026	40,787	36,009	32,234	29,174	26,646	21,900	18,589	16,148	14,274	12,789	11,584	9,748	8,414	7,401	5,965	4,996

500 MCM	1 Run(s)	240 Volts	C-Value = 22,185	Magnetic Duct	Copper

Calculations:

1. $f = \dfrac{1.732 \times \text{Length} \times \text{Isca}}{\text{Runs} \times \text{C-Value} \times E_{L-L}}$

2. $M = \dfrac{1}{1 + f}$

3. $\text{Isc} = \text{Isca} \times M$

4. Add Motor Contribution = Motor FLA x 4

* All results are given in symmetrical amperes

Three-Phase Bolted Fault-Current* Table

One-Way Distance in Feet

Available Isc in Thousands (Isca)	5	10	15	20	25	30	40	50	60	70	80	90	100	125	150	175	200	225	250	300	350	400	500	600
3	2,993	2,985	2,978	2,971	2,964	2,957	2,943	2,929	2,915	2,901	2,887	2,874	2,860	2,828	2,795	2,764	2,733	2,703	2,674	2,617	2,562	2,510	2,412	2,321
5	4,980	4,960	4,940	4,920	4,900	4,881	4,842	4,805	4,767	4,731	4,695	4,659	4,624	4,539	4,456	4,377	4,301	4,227	4,155	4,019	3,892	3,773	3,555	3,360
7	6,960	6,921	6,882	6,844	6,806	6,769	6,695	6,623	6,552	6,483	6,416	6,349	6,284	6,128	5,979	5,837	5,702	5,572	5,449	5,218	5,005	4,810	4,461	4,159
10	9,919	9,840	9,762	9,685	9,609	9,535	9,389	9,248	9,111	8,978	8,849	8,723	8,601	8,310	8,039	7,784	7,545	7,321	7,109	6,721	6,372	6,058	5,515	5,061
15	14,819	14,643	14,470	14,302	14,138	13,977	13,666	13,369	13,085	12,812	12,550	12,299	12,058	11,495	10,981	10,512	10,081	9,684	9,317	8,661	8,091	7,592	6,757	6,088
20	19,680	19,370	19,070	18,778	18,496	18,222	17,697	17,202	16,734	16,291	15,870	15,471	15,091	14,218	13,441	12,745	12,117	11,548	11,030	10,122	9,352	8,691	7,615	6,776
25	24,502	24,023	23,563	23,120	22,693	22,282	21,503	20,776	20,097	19,461	18,864	18,302	17,773	16,575	15,529	14,606	13,787	13,056	12,397	11,262	10,317	9,518	8,242	7,268
30	29,286	28,604	27,954	27,333	26,738	26,169	25,101	24,116	23,206	22,362	21,577	20,846	20,162	18,634	17,322	16,182	15,183	14,300	13,514	12,176	11,079	10,163	8,722	7,638
35	34,031	33,115	32,246	31,422	30,639	29,895	28,508	27,245	26,089	25,027	24,048	23,143	22,303	20,449	18,879	17,533	16,366	15,345	14,444	12,926	11,696	10,680	9,100	7,926
40	38,740	37,557	36,444	35,395	34,404	33,468	31,740	30,182	28,770	27,484	26,308	25,228	24,234	22,060	20,244	18,704	17,382	16,235	15,230	13,551	12,206	11,104	9,405	8,157
45	43,411	41,931	40,548	39,254	38,040	36,898	34,809	32,944	31,268	29,755	28,382	27,129	25,983	23,500	21,450	19,729	18,264	17,002	15,902	14,081	12,634	11,457	9,658	8,347
50	48,046	46,240	44,564	43,005	41,552	40,194	37,727	35,546	33,603	31,862	30,292	28,870	27,575	24,795	22,524	20,634	19,037	17,669	16,485	14,536	12,999	11,757	9,869	8,504
55	52,645	50,484	48,493	46,653	44,948	43,363	40,506	38,002	35,790	33,821	32,058	30,469	29,030	25,965	23,486	21,438	19,719	18,256	16,994	14,931	13,314	12,013	10,050	8,638
60	57,209	54,665	52,339	50,202	48,233	46,412	43,154	40,324	37,842	35,648	33,694	31,944	30,366	27,029	24,352	22,158	20,327	18,775	17,443	15,276	13,588	12,236	10,205	8,752
65	61,737	58,785	56,103	53,655	51,412	49,349	45,682	42,522	39,772	37,355	35,216	33,308	31,596	27,999	25,137	22,806	20,871	19,238	17,842	15,581	13,829	12,431	10,340	8,852
70	66,230	62,845	59,789	57,017	54,490	52,178	48,096	44,607	41,589	38,954	36,633	34,573	32,733	28,888	25,851	23,392	21,361	19,654	18,199	15,853	14,043	12,603	10,459	8,939
75	70,689	66,846	63,399	60,291	57,473	54,907	50,405	46,586	43,305	40,455	37,958	35,751	33,786	29,705	26,504	23,925	21,804	20,028	18,520	16,096	14,233	12,756	10,564	9,015
80	75,113	70,789	66,936	63,480	60,364	57,539	52,615	48,468	44,926	41,867	39,198	36,848	34,765	30,459	27,102	24,412	22,208	20,368	18,810	16,315	14,404	12,893	10,658	9,084
85	79,504	74,676	70,401	66,588	63,168	60,081	54,733	50,259	46,461	43,196	40,361	37,875	35,677	31,157	27,653	24,858	22,576	20,678	19,074	16,513	14,558	13,017	10,742	9,145
90	83,862	78,508	73,796	69,618	65,888	62,537	56,763	51,966	47,916	44,451	41,454	38,836	36,529	31,805	28,162	25,269	22,914	20,961	19,315	16,693	14,698	13,129	10,818	9,200
95	88,187	82,286	77,125	72,573	68,528	64,911	58,712	53,594	49,297	45,638	42,484	39,738	37,326	32,407	28,634	25,648	23,226	21,222	19,536	16,858	14,825	13,230	10,887	9,249
100	92,479	86,011	80,388	75,455	71,092	67,207	60,584	55,150	50,610	46,761	43,456	40,587	38,074	32,970	29,072	25,999	23,513	21,461	19,739	17,008	14,942	13,323	10,950	9,295
105	96,739	89,684	83,587	78,267	73,584	69,429	62,384	56,637	51,860	47,826	44,374	41,387	38,777	33,496	29,480	26,325	23,779	21,683	19,926	17,147	15,049	13,408	11,007	9,336
110	100,968	93,306	86,726	81,012	76,005	71,580	64,116	58,061	53,051	48,837	45,243	42,142	39,439	33,988	29,861	26,628	24,027	21,888	20,099	17,276	15,147	13,486	11,060	9,374
115	105,165	96,879	89,804	83,692	78,359	73,664	65,783	59,425	54,187	49,798	46,067	42,856	40,063	34,451	30,218	26,911	24,257	22,079	20,260	17,394	15,239	13,558	11,109	9,409
120	109,331	100,404	92,824	86,309	80,648	75,684	67,389	60,732	55,273	50,713	46,849	43,532	40,654	34,887	30,553	27,176	24,472	22,257	20,410	17,505	15,323	13,625	11,154	9,441
125	113,466	103,880	95,788	88,866	82,876	77,643	68,937	61,987	56,310	51,585	47,592	44,173	41,212	35,297	30,867	27,425	24,673	22,424	20,550	17,607	15,402	13,688	11,195	9,471
130	117,570	107,310	98,697	91,364	85,045	79,544	70,431	63,192	57,303	52,417	48,300	44,782	41,741	35,685	31,163	27,658	24,862	22,579	20,681	17,703	15,475	13,745	11,234	9,498
135	121,645	110,694	101,553	93,805	87,157	81,388	71,874	64,351	58,254	53,212	48,974	45,360	42,244	36,051	31,442	27,878	25,040	22,726	20,803	17,793	15,544	13,800	11,270	9,524
140	125,690	114,034	104,356	96,193	89,214	83,179	73,267	65,465	59,166	53,972	49,616	45,911	42,721	36,398	31,706	28,085	25,206	22,863	20,918	17,877	15,608	13,850	11,304	9,548
145	129,705	117,329	107,109	98,527	91,218	84,919	74,613	66,538	60,041	54,699	50,230	46,436	43,175	36,728	31,955	28,281	25,364	22,993	21,027	17,956	15,668	13,897	11,335	9,571
150	133,692	120,582	109,813	100,810	93,172	86,609	75,915	67,572	60,881	55,396	50,817	46,937	43,608	37,040	32,192	28,466	25,513	23,115	21,129	18,031	15,725	13,942	11,365	9,592
155	137,649	123,792	112,469	103,044	95,077	88,253	77,175	68,568	61,689	56,063	51,378	47,416	44,021	37,338	32,416	28,641	25,653	23,230	21,225	18,101	15,778	13,984	11,393	9,612
160	141,578	126,960	115,079	105,230	96,935	89,852	78,395	69,530	62,465	56,704	51,916	47,874	44,415	37,621	32,629	28,807	25,787	23,340	21,317	18,167	15,829	14,023	11,419	9,630
165	145,479	130,088	117,643	107,370	98,748	91,407	79,577	70,457	63,213	57,320	52,432	48,312	44,792	37,891	32,832	28,965	25,913	23,443	21,403	18,230	15,876	14,061	11,444	9,648
170	149,352	133,177	120,162	109,465	100,517	92,921	80,722	71,354	63,934	57,912	52,926	48,731	45,153	38,149	33,026	29,116	26,034	23,542	21,485	18,289	15,921	14,096	11,467	9,665
175	153,197	136,226	122,639	111,517	102,245	94,396	81,832	72,220	64,628	58,481	53,401	49,134	45,498	38,395	33,210	29,259	26,148	23,635	21,563	18,346	15,964	14,130	11,489	9,680
180	157,016	139,236	125,074	113,527	103,931	95,832	82,909	73,057	65,298	59,029	53,858	49,520	45,829	38,630	33,386	29,395	26,257	23,724	21,637	18,399	16,005	14,161	11,510	9,695
185	160,807	142,209	127,468	115,495	105,579	97,231	83,954	73,867	65,945	59,557	54,297	49,891	46,146	38,856	33,554	29,526	26,361	23,809	21,707	18,450	16,043	14,192	11,530	9,709
190	164,571	145,146	129,822	117,425	107,189	98,594	84,969	74,652	66,569	60,066	54,720	50,248	46,451	39,072	33,715	29,650	26,460	23,890	21,775	18,499	16,080	14,220	11,549	9,723
195	168,309	148,046	132,137	119,315	108,762	99,924	85,954	75,412	67,172	60,556	55,127	50,591	46,744	39,279	33,869	29,769	26,555	23,967	21,839	18,545	16,115	14,248	11,567	9,736
200	172,021	150,910	134,414	121,169	110,300	101,220	86,912	76,148	67,756	61,030	55,519	50,921	47,026	39,477	34,017	29,883	26,646	24,041	21,900	18,589	16,148	14,274	11,584	9,748

500 MCM	2 Run(s)	240 Volts	Copyright © 1994 - V.F. Christoffer - All Rights Reserved	C-Value = 22,185	Magnetic Duct	Copper

Calculations:

1. $f = \dfrac{1.732 \times \text{Length} \times \text{Isca}}{\text{Runs} \times \text{C-Value} \times \text{E L-L}}$
2. $M = \dfrac{1}{1 + f}$
3. $\text{Isc} = \text{Isca} \times M$
4. Add Motor Contribution = Motor FLA x 4

* All results are given in symmetrical amperes

Three-Phase Bolted Fault-Current* Table

One-Way Distance in Feet

Available Isc in Thousands (Isca)	5	10	15	20	25	30	40	50	60	70	80	90	100	125	150	175	200	225	250	300	350	400	500	600
3	2,986	2,972	2,958	2,944	2,931	2,917	2,891	2,865	2,839	2,814	2,790	2,765	2,742	2,684	2,628	2,575	2,524	2,475	2,428	2,339	2,256	2,178	2,039	1,916
5	4,961	4,923	4,885	4,848	4,811	4,775	4,704	4,636	4,569	4,505	4,442	4,381	4,321	4,179	4,046	3,922	3,804	3,694	3,590	3,398	3,226	3,070	2,800	2,574
7	6,924	6,849	6,776	6,705	6,635	6,567	6,434	6,306	6,184	6,066	5,952	5,843	5,738	5,490	5,263	5,054	4,861	4,682	4,516	4,217	3,955	3,724	3,334	3,017
10	9,845	9,695	9,550	9,409	9,272	9,138	8,883	8,642	8,414	8,197	7,991	7,795	7,609	7,180	6,796	6,452	6,141	5,858	5,600	5,147	4,762	4,431	3,889	3,466
15	14,655	14,325	14,009	13,708	13,419	13,142	12,620	12,139	11,693	11,279	10,892	10,532	10,195	9,439	8,787	8,220	7,721	7,279	6,886	6,213	5,661	5,198	4,468	3,918
20	19,391	18,817	18,277	17,767	17,284	16,827	15,982	15,218	14,523	13,889	13,309	12,774	12,281	11,201	10,295	9,524	8,861	8,285	7,778	6,531	6,251	5,692	4,828	4,192
25	24,055	23,179	22,365	21,605	20,896	20,232	19,022	17,949	16,991	16,130	15,352	14,645	14,001	12,613	11,476	10,527	9,723	9,033	8,434	7,447	6,667	6,035	5,073	4,375
30	28,650	27,415	26,283	25,241	24,278	23,386	21,785	20,389	19,161	18,073	17,102	16,230	15,442	13,771	12,427	11,322	10,397	9,612	8,937	7,837	6,977	6,288	5,251	4,507
35	33,176	31,532	30,043	28,689	27,452	26,317	24,306	22,582	21,085	19,775	18,618	17,589	16,668	14,738	13,209	11,967	10,938	10,073	9,334	8,140	7,217	6,482	5,385	4,606
40	37,635	35,533	33,655	31,964	30,436	29,047	26,617	24,563	22,803	21,278	19,944	18,768	17,723	15,557	13,863	12,501	11,383	10,449	9,656	8,384	7,408	6,636	5,491	4,683
45	42,028	39,425	37,125	35,079	33,246	31,596	28,742	26,361	24,345	22,614	21,114	19,800	18,640	16,259	14,418	12,951	11,755	10,761	9,922	8,584	7,564	6,760	5,576	4,745
50	46,358	43,211	40,463	38,045	35,899	33,982	30,703	28,002	25,737	23,811	22,153	20,712	19,446	16,869	14,895	13,335	12,070	11,025	10,146	8,751	7,693	6,863	5,646	4,795
55	50,625	46,895	43,677	40,872	38,405	36,220	32,519	29,504	27,000	24,889	23,083	21,522	20,159	17,403	15,310	13,666	12,341	11,250	10,337	8,892	7,802	6,950	5,704	4,837
60	54,831	50,482	46,772	43,570	40,778	38,323	34,204	30,884	28,152	25,864	23,920	22,248	20,794	17,874	15,673	13,955	12,576	11,445	10,501	9,014	7,896	7,024	5,754	4,873
65	58,977	53,975	49,755	46,148	43,028	40,303	35,772	32,158	29,206	26,751	24,677	22,901	21,363	18,293	15,994	14,209	12,782	11,616	10,644	9,119	7,976	7,088	5,797	4,904
70	63,064	57,378	52,633	48,613	45,163	42,171	37,236	33,336	30,175	27,561	25,364	23,492	21,877	18,668	16,281	14,434	12,964	11,766	10,770	9,212	8,047	7,144	5,834	4,930
75	67,094	60,695	55,411	50,973	47,193	43,935	38,605	34,429	31,067	28,304	25,992	24,030	22,342	19,006	16,537	14,636	13,126	11,899	10,882	9,293	8,109	7,193	5,867	4,953
80	71,067	63,929	58,093	53,234	49,125	45,605	39,889	35,446	31,893	28,988	26,568	24,521	22,766	19,312	16,768	14,816	13,272	12,018	10,982	9,366	8,164	7,236	5,895	4,974
85	74,985	67,082	60,686	55,403	50,966	47,187	41,094	36,394	32,659	29,619	27,097	24,971	23,154	19,590	16,977	14,980	13,402	12,126	11,071	9,431	8,213	7,275	5,921	4,992
90	78,850	70,158	63,192	57,484	52,722	48,689	42,228	37,281	33,371	30,204	27,586	25,385	23,510	19,844	17,168	15,128	13,521	12,222	11,152	9,489	8,258	7,309	5,944	5,009
95	82,661	73,159	65,617	59,484	54,400	50,116	43,297	38,112	34,035	30,747	28,038	25,768	23,837	20,077	17,342	15,263	13,629	12,310	11,225	9,542	8,298	7,341	5,965	5,023
100	86,421	76,089	67,964	61,406	56,003	51,474	44,307	38,892	34,656	31,253	28,458	26,122	24,140	20,292	17,502	15,386	13,727	12,391	11,292	9,590	8,334	7,369	5,984	5,037
105	90,130	78,950	70,237	63,256	57,538	52,767	45,262	39,626	35,238	31,725	28,849	26,451	24,421	20,490	17,649	15,500	13,817	12,464	11,353	9,634	8,367	7,395	6,001	5,049
110	93,790	81,744	72,440	65,037	59,007	54,001	46,166	40,317	35,784	32,167	29,214	26,757	24,682	20,673	17,785	15,605	13,900	12,532	11,409	9,674	8,398	7,419	6,016	5,060
115	97,401	84,473	74,575	66,753	60,416	55,178	47,025	40,970	36,297	32,581	29,555	27,043	24,925	20,844	17,911	15,701	13,977	12,594	11,460	9,712	8,426	7,441	6,031	5,070
120	100,964	87,140	76,646	68,408	61,768	56,304	47,840	41,588	36,781	32,970	29,875	27,311	25,152	21,002	18,028	15,791	14,048	12,652	11,508	9,746	8,452	7,461	6,044	5,079
125	104,480	89,747	78,655	70,004	63,067	57,381	48,615	42,172	37,237	33,336	30,175	27,562	25,365	21,150	18,137	15,875	14,114	12,706	11,552	9,778	8,476	7,480	6,056	5,088
130	107,950	92,295	80,606	71,545	64,315	58,412	49,353	42,727	37,669	33,682	30,458	27,797	25,564	21,289	18,238	15,953	14,176	12,755	11,594	9,807	8,498	7,497	6,067	5,096
135	111,375	94,788	82,501	73,034	65,515	59,401	50,057	43,253	38,078	34,008	30,725	28,019	25,752	21,419	18,334	16,026	14,234	12,802	11,632	9,835	8,518	7,513	6,078	5,103
140	114,757	97,226	84,342	74,472	66,671	60,349	50,729	43,754	38,465	34,317	30,976	28,229	25,929	21,541	18,423	16,094	14,287	12,845	11,668	9,860	8,538	7,528	6,088	5,110
145	118,095	99,611	86,131	75,864	67,784	61,260	51,371	44,230	38,833	34,610	31,215	28,426	26,095	21,656	18,507	16,158	14,338	12,886	11,702	9,884	8,556	7,542	6,097	5,117
150	121,390	101,946	87,871	77,211	68,857	62,135	51,984	44,685	39,183	34,887	31,440	28,613	26,253	21,764	18,586	16,218	14,385	12,925	11,733	9,907	8,572	7,555	6,105	5,123
155	124,644	104,231	89,563	78,514	69,892	62,976	52,572	45,118	39,516	35,151	31,654	28,790	26,402	21,866	18,661	16,275	14,430	12,961	11,763	9,928	8,588	7,567	6,113	5,128
160	127,857	106,468	91,210	79,777	70,891	63,786	53,135	45,533	39,833	35,402	31,858	28,959	26,543	21,963	18,731	16,328	14,472	12,995	11,791	9,948	8,603	7,579	6,121	5,134
165	131,030	108,659	92,813	81,001	71,856	64,566	53,676	45,929	40,136	35,641	32,051	29,118	26,677	22,055	18,798	16,379	14,512	13,027	11,817	9,967	8,617	7,590	6,128	5,139
170	134,164	110,806	94,375	82,188	72,788	65,318	54,194	46,308	40,425	35,869	32,235	29,270	26,805	22,142	18,861	16,427	14,549	13,057	11,842	9,984	8,630	7,600	6,135	5,143
175	137,259	112,908	95,896	83,339	73,690	66,043	54,692	46,671	40,702	36,086	32,411	29,415	26,926	22,225	18,921	16,472	14,585	13,086	11,866	10,001	8,643	7,610	6,141	5,148
180	140,316	114,969	97,378	84,456	74,562	66,743	55,171	47,019	40,966	36,294	32,578	29,553	27,042	22,303	18,978	16,515	14,619	13,113	11,888	10,017	8,655	7,619	6,147	5,152
185	143,336	116,988	98,823	85,541	75,406	67,418	55,632	47,354	41,220	36,493	32,738	29,685	27,152	22,378	19,032	16,557	14,651	13,139	11,909	10,032	8,666	7,628	6,153	5,156
190	146,319	118,968	100,232	86,594	76,223	68,071	56,076	47,675	41,463	36,683	32,892	29,810	27,257	22,450	19,084	16,596	14,682	13,163	11,930	10,047	8,677	7,636	6,158	5,160
195	149,266	120,909	101,606	87,618	77,016	68,702	56,503	47,983	41,696	36,866	33,038	29,931	27,358	22,518	19,133	16,633	14,711	13,187	11,949	10,060	8,687	7,644	6,163	5,163
200	152,178	122,813	102,947	88,614	77,784	69,313	56,916	48,280	41,920	37,041	33,179	30,046	27,454	22,583	19,180	16,669	14,739	13,209	11,967	10,073	8,697	7,651	6,168	5,167

600 MCM	1 Run(s)	240 Volts	C-Value = 22,965	Magnetic Duct	Copper

Calculations:

1. $f = \dfrac{1.732 \times \text{Length} \times \text{Isca}}{\text{Runs} \times \text{C-Value} \times E_{L\text{-}L}}$

2. $M = \dfrac{1}{1 + f}$

3. Isc = Isca x M

4. Add Motor Contribution = Motor FLA x 4

* All results are given in symmetrical amperes

Three-Phase Bolted Fault-Current* Table

One-Way Distance in Feet

Available Isc in Thousands (Isca)	5	10	15	20	25	30	40	50	60	70	80	90	100	125	150	175	200	225	250	300	350	400	500	600
3	2,993	2,986	2,979	2,972	2,965	2,958	2,944	2,931	2,917	2,904	2,891	2,878	2,865	2,833	2,802	2,771	2,742	2,712	2,684	2,628	2,575	2,524	2,428	2,339
5	4,980	4,961	4,942	4,923	4,904	4,885	4,848	4,811	4,775	4,739	4,704	4,670	4,636	4,553	4,473	4,396	4,321	4,249	4,179	4,046	3,922	3,804	3,590	3,398
7	6,962	6,924	6,886	6,849	6,813	6,776	6,705	6,635	6,567	6,500	6,434	6,369	6,306	6,154	6,009	5,870	5,738	5,611	5,490	5,263	5,054	4,861	4,516	4,217
10	9,922	9,845	9,770	9,695	9,622	9,550	9,409	9,272	9,138	9,009	8,883	8,761	8,642	8,358	8,093	7,843	7,609	7,388	7,180	6,796	6,452	6,141	5,600	5,147
15	14,825	14,655	14,488	14,325	14,165	14,009	13,708	13,419	13,142	12,876	12,620	12,375	12,139	11,587	11,082	10,620	10,195	9,802	9,439	8,787	8,220	7,721	6,886	6,213
20	19,691	19,391	19,100	18,817	18,543	18,277	17,767	17,284	16,827	16,394	15,982	15,591	15,218	14,359	13,593	12,904	12,281	11,716	11,201	10,295	9,524	8,861	7,778	6,931
25	24,518	24,055	23,609	23,179	22,764	22,365	21,605	20,896	20,232	19,608	19,022	18,470	17,949	16,767	15,731	14,816	14,001	13,271	12,613	11,476	10,527	9,723	8,434	7,447
30	29,309	28,650	28,019	27,415	26,837	26,283	25,241	24,278	23,386	22,557	21,785	21,064	20,389	18,877	17,574	16,439	15,442	14,559	13,771	12,427	11,322	10,397	8,937	7,837
35	34,063	33,176	32,333	31,532	30,770	30,043	28,689	27,452	26,317	25,272	24,306	23,412	22,582	20,742	19,179	17,835	16,668	15,644	14,738	13,209	11,967	10,938	9,334	8,140
40	38,781	37,635	36,554	35,533	34,568	33,655	31,964	30,436	29,047	27,779	26,617	25,549	24,563	22,401	20,589	19,049	17,723	16,569	15,557	13,863	12,501	11,383	9,656	8,384
45	43,463	42,028	40,685	39,425	38,240	37,125	35,079	33,246	31,596	30,102	28,742	27,500	26,361	23,888	21,838	20,113	18,640	17,369	16,259	14,418	12,951	11,755	9,922	8,584
50	48,110	46,358	44,729	43,211	41,792	40,463	38,045	35,899	33,982	32,259	30,703	29,290	28,002	25,227	22,952	21,054	19,446	18,066	16,869	14,895	13,335	12,070	10,146	8,751
55	52,722	50,625	48,689	46,895	45,229	43,677	40,872	38,405	36,220	34,270	32,519	30,938	29,504	26,439	23,952	21,892	20,159	18,680	17,403	15,310	13,666	12,341	10,337	8,892
60	57,299	54,831	52,567	50,482	48,556	46,772	43,570	40,778	38,323	36,146	34,204	32,459	30,884	27,543	24,854	22,643	20,794	19,224	17,874	15,673	13,955	12,576	10,501	9,014
65	61,842	58,977	56,365	53,975	51,779	49,755	46,148	43,028	40,303	37,903	35,772	33,869	32,158	28,551	25,672	23,320	21,363	19,709	18,293	15,994	14,209	12,782	10,644	9,119
70	66,351	63,064	60,087	57,378	54,903	52,633	48,613	45,163	42,171	39,550	37,236	35,178	33,336	29,476	26,417	23,934	21,877	20,146	18,668	16,281	14,434	12,964	10,770	9,212
75	70,827	67,094	63,734	60,695	57,933	55,411	50,973	47,193	43,935	41,098	38,605	36,397	34,429	30,327	27,099	24,492	22,342	20,540	19,006	16,537	14,636	13,126	10,882	9,293
80	75,269	71,067	67,309	63,929	60,871	58,093	53,234	49,125	45,605	42,556	39,889	37,536	35,446	31,113	27,725	25,002	22,766	20,897	19,312	16,768	14,816	13,272	10,982	9,366
85	79,679	74,985	70,814	67,082	63,724	60,686	55,403	50,966	47,187	43,930	41,094	38,601	36,394	31,842	28,302	25,470	23,154	21,224	19,590	16,977	14,980	13,402	11,071	9,431
90	84,057	78,850	74,250	70,158	66,493	63,192	57,484	52,722	48,689	45,229	42,228	39,601	37,281	32,519	28,835	25,902	23,510	21,522	19,844	17,168	15,128	13,521	11,152	9,489
95	88,402	82,661	77,621	73,159	69,183	65,617	59,484	54,400	50,116	46,458	43,297	40,539	38,112	33,149	29,330	26,300	23,837	21,796	20,077	17,342	15,263	13,629	11,225	9,542
100	92,716	86,421	80,927	76,089	71,797	67,964	61,406	56,003	51,474	47,622	44,307	41,423	38,892	33,738	29,790	26,669	24,140	22,049	20,292	17,502	15,386	13,727	11,292	9,590
105	96,999	90,130	84,170	78,950	74,339	70,237	63,256	57,538	52,767	48,727	45,262	42,257	39,626	34,289	30,219	27,012	24,421	22,283	20,490	17,649	15,500	13,817	11,353	9,634
110	101,250	93,790	87,353	81,744	76,811	72,440	65,037	59,007	54,001	49,777	46,166	43,044	40,317	34,805	30,619	27,332	24,682	22,500	20,673	17,785	15,605	13,900	11,409	9,674
115	105,471	97,401	90,477	84,473	79,216	74,575	66,753	60,416	55,178	50,776	47,025	43,789	40,970	35,291	30,994	27,630	24,925	22,702	20,844	17,911	15,701	13,977	11,460	9,712
120	109,662	100,964	93,544	87,140	81,557	76,646	68,408	61,768	56,304	51,728	47,840	44,495	41,588	35,748	31,346	27,910	25,152	22,891	21,002	18,028	15,791	14,048	11,508	9,746
125	113,822	104,480	96,554	89,747	83,836	78,655	70,004	63,067	57,381	52,635	48,615	45,165	42,172	36,179	31,677	28,172	25,365	23,067	21,150	18,137	15,875	14,114	11,552	9,778
130	117,953	107,950	99,511	92,295	86,056	80,606	71,545	64,315	58,412	53,502	49,353	45,801	42,727	36,586	31,989	28,418	25,564	23,231	21,289	18,238	15,953	14,176	11,594	9,807
135	122,055	111,375	102,414	94,788	88,219	82,501	73,034	65,515	59,401	54,330	50,057	46,407	43,253	36,972	32,283	28,650	25,752	23,386	21,419	18,334	16,026	14,234	11,632	9,835
140	126,128	114,757	105,266	97,226	90,327	84,342	74,472	66,671	60,349	55,122	50,729	46,984	43,754	37,337	32,561	28,869	25,929	23,532	21,541	18,423	16,094	14,287	11,668	9,860
145	130,172	118,095	108,068	99,611	92,382	86,131	75,864	67,784	61,260	55,881	51,371	47,534	44,230	37,683	32,825	29,076	26,095	23,669	21,656	18,507	16,158	14,338	11,702	9,884
150	134,187	121,390	110,822	101,946	94,386	87,871	77,211	68,857	62,135	56,608	51,984	48,059	44,685	38,013	33,074	29,271	26,253	23,799	21,764	18,586	16,218	14,385	11,733	9,907
155	138,174	124,644	113,527	104,231	96,342	89,563	78,514	69,892	62,976	57,306	52,572	48,561	45,118	38,326	33,311	29,457	26,402	23,921	21,866	18,661	16,275	14,430	11,763	9,928
160	142,134	127,857	116,187	106,468	98,250	91,210	79,777	70,891	63,786	57,976	53,135	49,041	45,533	38,624	33,536	29,633	26,543	24,037	21,963	18,731	16,328	14,472	11,791	9,948
165	146,066	131,030	118,801	108,659	100,113	92,813	81,001	71,856	64,566	58,619	53,676	49,501	45,929	38,909	33,751	29,800	26,677	24,147	22,055	18,798	16,379	14,512	11,817	9,967
170	149,971	134,164	121,371	110,806	101,932	94,375	82,188	72,788	65,318	59,238	54,194	49,941	46,308	39,181	33,955	29,959	26,805	24,251	22,142	18,861	16,427	14,549	11,842	9,984
175	153,848	137,259	123,898	112,908	103,709	95,896	83,339	73,690	66,043	59,834	54,692	50,364	46,671	39,440	34,150	30,111	26,926	24,350	22,225	18,921	16,472	14,585	11,866	10,001
180	157,700	140,316	126,384	114,969	105,445	97,378	84,456	74,562	66,743	60,408	55,171	50,770	47,019	39,689	34,336	30,255	27,042	24,445	22,303	18,978	16,516	14,619	11,888	10,017
185	161,524	143,336	128,829	116,988	107,141	98,823	85,541	75,406	67,418	60,961	55,632	51,160	47,354	39,927	34,514	30,393	27,152	24,535	22,378	19,032	16,557	14,651	11,909	10,032
190	165,323	146,319	131,234	118,968	108,799	100,232	86,594	76,223	68,071	61,494	56,076	51,535	47,675	40,155	34,684	30,525	27,257	24,621	22,450	19,084	16,596	14,682	11,930	10,047
195	169,095	149,266	133,600	120,909	110,421	101,606	87,618	77,016	68,702	62,008	56,503	51,896	47,983	40,374	34,847	30,652	27,358	24,703	22,518	19,133	16,633	14,711	11,949	10,060
200	172,842	152,178	135,928	122,813	112,006	102,947	88,614	77,784	69,313	62,505	56,916	52,244	48,280	40,584	35,004	30,772	27,454	24,781	22,583	19,180	16,669	14,739	11,967	10,073

600 MCM | 2 Run(s) | 240 Volts | | C-Value = 22,965 | Magnetic Duct | Copper

Calculations:

1. $f = \dfrac{1.732 \times \text{Length} \times \text{Isca}}{\text{Runs} \times \text{C-Value} \times E_{L\text{-}L}}$

2. $M = \dfrac{1}{1 + f}$

3. $\text{Isc} = \text{Isca} \times M$

4. Add Motor Contribution = Motor FLA x 4

* All results are given in symmetrical amperes

Three-Phase Bolted Fault-Current* Table

One-Way Distance in Feet

Available Isc in Thousands (Isca)	5	10	15	20	25	30	40	50	60	70	80	90	100	125	150	175	200	225	250	300	350	400	500	600
3	2,987	2,973	2,960	2,947	2,934	2,921	2,896	2,871	2,847	2,823	2,799	2,776	2,753	2,698	2,644	2,593	2,544	2,496	2,450	2,364	2,283	2,208	2,071	1,950
5	4,963	4,926	4,890	4,855	4,820	4,785	4,718	4,652	4,588	4,526	4,466	4,407	4,350	4,213	4,084	3,963	3,849	3,741	3,640	3,452	3,282	3,129	2,861	2,636
7	6,928	6,856	6,787	6,719	6,652	6,586	6,459	6,337	6,219	6,105	5,996	5,890	5,788	5,548	5,327	5,123	4,934	4,759	4,595	4,300	4,040	3,810	3,420	3,103
10	9,853	9,710	9,571	9,436	9,304	9,177	8,932	8,699	8,479	8,269	8,070	7,880	7,698	7,279	6,904	6,565	6,258	5,978	5,722	5,271	4,886	4,554	4,008	3,579
15	14,671	14,356	14,054	13,765	13,488	13,221	12,718	12,252	11,819	11,416	11,039	10,686	10,356	9,612	8,967	8,404	7,907	7,466	7,071	6,395	5,837	5,369	4,626	4,064
20	19,419	18,871	18,354	17,864	17,399	16,958	16,139	15,396	14,719	14,098	13,528	13,002	12,516	11,445	10,543	9,773	9,107	8,527	8,016	7,158	6,466	5,896	5,013	4,359
25	24,099	23,261	22,479	21,749	21,064	20,421	19,246	18,198	17,259	16,412	15,645	14,945	14,306	12,924	11,786	10,831	10,020	9,322	8,715	7,710	6,913	6,266	5,277	4,558
30	28,712	27,531	26,442	25,437	24,505	23,639	22,078	20,711	19,503	18,429	17,466	16,599	15,814	14,143	12,790	11,674	10,737	9,940	9,252	8,128	7,247	6,539	5,469	4,701
35	33,260	31,684	30,251	28,942	27,742	26,637	24,672	22,977	21,500	20,201	19,051	18,024	17,102	15,164	13,620	12,361	11,316	10,433	9,679	8,455	7,506	6,749	5,616	4,808
40	37,743	35,727	33,916	32,279	30,793	29,438	27,056	25,031	23,288	21,772	20,442	19,264	18,215	16,032	14,316	12,932	11,792	10,837	10,025	8,718	7,713	6,916	5,731	4,892
45	42,163	39,663	37,443	35,458	33,673	32,059	29,255	26,902	24,899	23,174	21,672	20,353	19,186	16,779	14,909	13,414	12,192	11,174	10,312	8,935	7,882	7,051	5,823	4,960
50	46,522	43,497	40,841	38,491	36,397	34,518	31,289	28,612	26,357	24,432	22,769	21,317	20,040	17,429	15,420	13,826	12,531	11,458	10,554	9,116	8,022	7,163	5,900	5,015
55	50,821	47,233	44,117	41,388	38,976	36,830	33,176	30,182	27,684	25,568	23,752	22,177	20,798	18,000	15,865	14,183	12,823	11,702	10,761	9,269	8,141	7,258	5,964	5,061
60	55,061	50,873	47,278	44,157	41,422	39,007	34,932	31,629	28,896	26,598	24,639	22,948	21,475	18,504	16,256	14,495	13,078	11,913	10,939	9,401	8,243	7,339	6,018	5,100
65	59,243	54,423	50,328	46,806	43,745	41,060	36,570	32,966	30,008	27,537	25,442	23,644	22,083	18,954	16,602	14,769	13,301	12,098	11,095	9,516	8,331	7,408	6,065	5,134
70	63,368	57,885	53,274	49,344	45,954	43,000	38,101	34,205	31,031	28,396	26,174	24,274	22,632	19,357	16,910	15,013	13,498	12,261	11,231	9,617	8,408	7,469	6,106	5,163
75	67,438	61,262	56,122	51,778	48,058	44,836	39,536	35,357	31,976	29,186	26,843	24,849	23,130	19,721	17,187	15,230	13,674	12,406	11,353	9,706	8,476	7,523	6,141	5,189
80	71,454	64,558	58,875	54,113	50,063	46,577	40,883	36,430	32,852	29,913	27,457	25,374	23,585	20,050	17,437	15,426	13,831	12,535	11,461	9,785	8,536	7,570	6,173	5,211
85	75,416	67,775	61,540	56,355	51,976	48,228	42,150	37,433	33,665	30,586	28,023	25,857	24,001	20,350	17,663	15,603	13,973	12,652	11,559	9,856	8,590	7,612	6,201	5,231
90	79,327	70,916	64,119	58,510	53,804	49,798	43,344	38,371	34,422	31,210	28,546	26,301	24,384	20,624	17,870	15,764	14,102	12,757	11,647	9,920	8,638	7,650	6,226	5,249
95	83,186	73,985	66,616	60,583	55,551	51,292	44,471	39,252	35,129	31,790	29,031	26,712	24,736	20,876	18,058	15,911	14,219	12,853	11,727	9,977	8,682	7,685	6,249	5,265
100	86,994	76,982	69,037	62,578	57,225	52,715	45,537	40,080	35,791	32,331	29,481	27,093	25,063	21,108	18,232	16,045	14,327	12,941	11,799	10,030	8,722	7,716	6,270	5,280
105	90,754	79,912	71,384	64,500	58,828	54,072	46,547	40,860	36,412	32,837	29,901	27,447	25,365	21,322	18,391	16,168	14,425	13,021	11,866	10,078	8,759	7,745	6,288	5,293
110	94,465	82,775	73,660	66,353	60,365	55,368	47,504	41,596	36,995	33,310	30,293	27,777	25,647	21,521	18,539	16,282	14,516	13,095	11,927	10,122	8,792	7,771	6,306	5,305
115	98,129	85,575	75,869	68,140	61,840	56,607	48,413	42,291	37,544	33,755	30,660	28,085	25,910	21,706	18,676	16,383	14,599	13,163	11,984	10,163	8,823	7,794	6,321	5,316
120	101,747	88,313	78,013	69,865	63,258	57,792	49,277	42,949	38,061	34,172	31,005	28,374	26,155	21,878	18,803	16,485	14,677	13,226	12,036	10,201	8,851	7,817	6,336	5,327
125	105,319	90,992	80,096	71,531	64,620	58,928	50,100	43,573	38,550	34,566	31,328	28,645	26,385	22,039	18,921	16,577	14,749	13,285	12,085	10,235	8,877	7,837	6,349	5,336
130	108,846	93,613	82,120	73,141	65,931	60,016	50,885	44,165	39,013	34,938	31,633	28,900	26,601	22,189	19,032	16,662	14,816	13,339	12,130	10,268	8,901	7,856	6,362	5,345
135	112,329	96,178	84,087	74,697	67,193	61,060	51,633	44,728	39,452	35,289	31,921	29,140	26,804	22,330	19,136	16,741	14,879	13,390	12,172	10,298	8,924	7,874	6,373	5,353
140	115,769	98,689	86,000	76,203	68,409	62,062	52,348	45,263	39,868	35,622	32,193	29,366	26,996	22,463	19,233	16,815	14,938	13,438	12,211	10,326	8,945	7,890	6,384	5,361
145	119,167	101,147	87,861	77,661	69,582	63,026	53,032	45,774	40,263	35,937	32,450	29,580	27,176	22,588	19,325	16,886	14,993	13,482	12,248	10,352	8,965	7,905	6,394	5,368
150	122,524	103,555	89,673	79,072	70,713	63,952	53,686	46,261	40,639	36,236	32,694	29,783	27,347	22,706	19,411	16,952	15,045	13,524	12,282	10,377	8,983	7,920	6,403	5,374
155	125,840	105,914	91,436	80,440	71,805	64,844	54,314	46,725	40,998	36,521	32,925	29,975	27,509	22,817	19,493	17,014	15,094	13,564	12,315	10,400	9,001	7,933	6,412	5,381
160	129,116	108,225	93,153	81,766	72,860	65,703	54,915	47,170	41,339	36,792	33,145	30,157	27,663	22,923	19,569	17,072	15,140	13,601	12,346	10,422	9,017	7,946	6,421	5,386
165	132,352	110,490	94,826	83,052	73,879	66,531	55,492	47,595	41,666	37,050	33,355	30,330	27,808	23,023	19,642	17,128	15,184	13,636	12,375	10,443	9,033	7,958	6,428	5,392
170	135,550	112,710	96,457	84,300	74,865	67,329	56,046	48,002	41,977	37,296	33,554	30,495	27,947	23,117	19,711	17,180	15,225	13,669	12,402	10,462	9,047	7,969	6,436	5,397
175	138,710	114,886	98,046	85,512	75,819	68,100	56,579	48,393	42,276	37,531	33,745	30,652	28,079	23,208	19,777	17,230	15,264	13,701	12,428	0,481	9,061	7,980	6,443	5,402
180	141,833	117,020	99,596	86,688	76,743	68,844	57,092	48,767	42,561	37,756	33,926	30,802	28,204	23,293	19,839	17,277	15,301	13,730	12,452	0,498	9,074	7,990	6,449	5,407
185	144,919	119,113	101,108	87,832	77,637	69,563	57,586	49,127	42,835	37,972	34,100	30,945	28,324	23,375	19,898	17,322	15,336	13,759	12,476	0,515	9,086	8,000	6,456	5,411
190	147,969	121,166	102,583	88,943	78,504	70,258	58,061	49,473	43,098	38,178	34,266	31,082	28,439	23,453	19,955	17,365	15,370	13,786	12,498	0,530	9,098	8,009	6,461	5,415
195	150,984	123,180	104,024	90,024	79,345	70,931	58,520	49,805	43,350	38,376	34,426	31,213	28,548	23,528	20,009	17,405	15,402	13,812	12,519	0,545	9,109	8,017	6,467	5,419
200	153,965	125,156	105,430	91,075	80,160	71,582	58,962	50,125	43,592	38,565	34,578	31,338	28,653	23,599	20,060	17,444	15,432	13,836	12,539	0,560	9,120	8,026	6,472	5,423

750 MCM	1 Run(s)	240 Volts	C-Value = 24,136	Magnetic Duct	Copper

Calculations:

1. $f = \dfrac{1.732 \times \text{Length} \times \text{Isca}}{\text{Runs} \times \text{C-Value} \times E_{L-L}}$

2. $M = \dfrac{1}{1 + f}$

3. Isc = Isca x M

4. Add Motor Contribution = Motor FLA x 4

* All results are given in symmetrical amperes

Three-Phase Bolted Fault-Current* Table

One-Way Distance in Feet

Available Isc in Thousands (Isca)

	5	10	15	20	25	30	40	50	60	70	80	90	100	125	150	175	200	225	250	300	350	400	500	600
3	2,993	2,987	2,980	2,973	2,967	2,960	2,947	2,934	2,921	2,909	2,896	2,884	2,871	2,841	2,811	2,782	2,753	2,725	2,698	2,644	2,593	2,544	2,450	2,364
5	4,981	4,963	4,945	4,926	4,908	4,890	4,855	4,820	4,785	4,751	4,718	4,685	4,652	4,573	4,496	4,422	4,350	4,280	4,213	4,084	3,963	3,849	3,640	3,452
7	6,964	6,928	6,892	6,856	6,822	6,787	6,719	6,652	6,586	6,522	6,459	6,397	6,337	6,190	6,050	5,916	5,788	5,666	5,548	5,327	5,123	4,934	4,595	4,300
10	9,926	9,853	9,781	9,710	9,640	9,571	9,436	9,304	9,177	9,053	8,932	8,814	8,699	8,425	8,168	7,926	7,698	7,483	7,279	6,904	6,565	6,258	5,722	5,271
15	14,834	14,671	14,512	14,356	14,204	14,054	13,765	13,488	13,221	12,965	12,718	12,481	12,252	11,716	11,224	10,772	10,356	9,970	9,612	8,967	8,404	7,907	7,071	6,395
20	19,705	19,419	19,142	18,871	18,609	18,354	17,864	17,399	16,958	16,538	16,139	15,759	15,396	14,559	13,807	13,130	12,516	11,956	11,445	10,543	9,773	9,107	8,016	7,158
25	24,541	24,099	23,673	23,261	22,864	22,479	21,749	21,064	20,421	19,816	19,246	18,707	18,198	17,039	16,019	15,114	14,306	13,580	12,924	11,786	10,831	10,020	8,715	7,710
30	29,342	28,712	28,109	27,531	26,975	26,442	25,437	24,505	23,639	22,832	22,078	21,373	20,711	19,223	17,935	16,808	15,814	14,932	14,143	12,790	11,674	10,737	9,252	8,128
35	34,108	33,260	32,453	31,684	30,951	30,251	28,942	27,742	26,637	25,617	24,672	23,795	22,977	21,160	19,609	18,270	17,102	16,075	15,164	13,620	12,361	11,316	9,679	8,455
40	38,839	37,743	36,707	35,727	34,798	33,916	32,279	30,793	29,438	28,197	27,056	26,004	25,031	22,890	21,086	19,546	18,215	17,054	16,032	14,316	12,932	11,792	10,025	8,718
45	43,536	42,163	40,875	39,663	38,521	37,443	35,458	33,673	32,059	30,593	29,255	28,029	26,902	24,444	22,398	20,668	19,186	17,902	16,779	14,909	13,414	12,192	10,312	8,935
50	48,199	46,522	44,959	43,497	42,127	40,841	38,491	36,397	34,518	32,825	31,289	29,891	28,612	25,848	23,571	21,663	20,040	18,644	17,429	15,420	13,826	12,531	10,554	9,116
55	52,828	50,821	48,961	47,233	45,622	44,117	41,388	38,976	36,830	34,908	33,176	31,609	30,182	27,123	24,626	22,551	20,798	19,298	18,000	15,865	14,183	12,823	10,761	9,269
60	57,425	55,061	52,884	50,873	49,010	47,278	44,157	41,422	39,007	36,857	34,932	33,199	31,629	28,285	25,581	23,349	21,475	19,879	18,504	16,256	14,495	13,078	10,939	9,401
65	61,988	59,243	56,731	54,423	52,295	50,328	46,806	43,745	41,060	38,685	36,570	34,675	32,966	29,350	26,448	24,069	22,083	20,399	18,954	16,602	14,769	13,301	11,095	9,516
70	66,519	63,368	60,503	57,885	55,484	53,274	49,344	45,954	43,000	40,403	38,101	36,048	34,205	30,328	27,240	24,723	22,632	20,867	19,357	16,910	15,013	13,498	11,231	9,617
75	71,019	67,438	64,202	61,262	58,579	56,122	51,778	48,058	44,836	42,020	39,536	37,330	35,357	31,230	27,966	25,319	23,130	21,290	19,721	17,187	15,230	13,674	11,353	9,706
80	75,486	71,454	67,831	64,558	61,586	58,875	54,113	50,063	46,577	43,545	40,883	38,528	36,430	32,064	28,633	25,865	23,585	21,674	20,050	17,437	15,426	13,831	11,461	9,785
85	79,922	75,416	71,392	67,775	64,507	61,540	56,355	51,976	48,228	44,985	42,150	39,652	37,433	32,838	29,249	26,366	24,001	22,025	20,350	17,663	15,603	13,973	11,559	9,856
90	84,327	79,327	74,886	70,916	67,346	64,119	58,510	53,804	49,798	46,348	43,344	40,706	38,371	33,559	29,819	26,829	24,384	22,347	20,624	17,870	15,764	14,102	11,647	9,920
95	88,701	83,186	78,316	73,985	70,107	66,616	60,583	55,551	51,292	47,639	44,471	41,699	39,252	34,230	30,348	27,256	24,736	22,643	20,876	18,058	15,911	14,219	11,727	9,977
100	93,045	86,994	81,683	76,982	72,793	69,037	62,578	57,225	52,715	48,864	45,537	42,635	40,080	34,858	30,840	27,653	25,063	22,916	21,108	18,232	16,045	14,327	11,799	10,030
105	97,359	90,754	84,988	79,912	75,407	71,384	64,500	58,828	54,072	50,028	46,547	43,518	40,860	35,447	31,300	28,022	25,365	23,169	21,322	18,391	16,168	14,425	11,866	10,078
110	101,642	94,465	88,235	82,775	77,952	73,660	66,353	60,365	55,368	51,135	47,504	44,354	41,596	35,999	31,730	28,366	25,647	23,404	21,521	18,539	16,282	14,516	11,927	10,122
115	105,897	98,129	91,423	85,575	80,430	75,869	68,140	61,840	56,607	52,190	48,413	45,145	42,291	36,519	32,133	28,688	25,910	23,622	21,706	18,676	16,388	14,599	11,984	10,163
120	110,122	101,747	94,555	88,313	82,844	78,013	69,865	63,258	57,792	53,196	49,277	45,896	42,949	37,008	32,512	28,989	26,155	23,826	21,878	18,803	16,486	14,677	12,036	10,201
125	114,318	105,319	97,632	90,992	85,197	80,096	71,531	64,620	58,928	54,157	50,100	46,609	43,573	37,471	32,868	29,272	26,385	24,017	22,039	18,921	16,577	14,749	12,085	10,235
130	118,486	108,846	100,656	93,613	87,491	82,120	73,141	65,931	60,016	55,074	50,885	47,287	44,165	37,908	33,203	29,538	26,601	24,196	22,189	19,032	16,662	14,816	12,130	10,268
135	122,626	112,329	103,628	96,178	89,727	84,087	74,697	67,193	61,060	55,952	51,633	47,933	44,728	38,322	33,521	29,789	26,804	24,364	22,330	19,136	16,741	14,879	12,172	10,298
140	126,737	115,769	106,549	98,689	91,909	86,000	76,203	68,409	62,062	56,793	52,348	48,549	45,263	38,714	33,821	30,025	26,996	24,522	22,463	19,233	16,816	14,938	12,211	10,326
145	130,821	119,167	109,421	101,147	94,038	87,861	77,661	69,582	63,026	57,599	53,032	49,136	45,774	39,087	34,105	30,249	27,176	24,671	22,588	19,325	16,886	14,993	12,248	10,352
150	134,877	122,524	112,244	103,555	96,115	89,673	79,072	70,713	63,952	58,371	53,686	49,698	46,261	39,441	34,374	30,461	27,347	24,811	22,706	19,411	16,952	15,045	12,282	10,377
155	138,906	125,840	115,020	105,914	98,144	91,436	80,440	71,805	64,844	59,113	54,314	50,235	46,725	39,779	34,630	30,662	27,509	24,944	22,817	19,493	17,014	15,094	12,315	10,400
160	142,908	129,116	117,751	108,225	100,125	93,153	81,766	72,860	65,703	59,826	54,915	50,749	47,170	40,100	34,874	30,852	27,663	25,071	22,923	19,569	17,072	15,140	12,346	10,422
165	146,884	132,352	120,437	110,490	102,061	94,826	83,052	73,879	66,531	60,512	55,492	51,241	47,595	40,407	35,105	31,034	27,808	25,190	23,023	19,642	17,128	15,184	12,375	10,443
170	150,833	135,550	123,079	112,710	103,952	96,457	84,300	74,865	67,329	61,172	56,046	51,713	48,002	40,700	35,327	31,206	27,947	25,304	23,117	19,711	17,180	15,225	12,402	10,462
175	154,756	138,710	125,679	114,886	105,800	98,046	85,512	75,819	68,100	61,807	56,579	52,167	48,393	40,981	35,537	31,371	28,079	25,412	23,208	19,777	17,230	15,264	12,428	10,481
180	158,653	141,833	128,237	117,020	107,607	99,596	86,688	76,743	68,844	62,420	57,092	52,602	48,767	41,249	35,739	31,528	28,204	25,515	23,293	19,839	17,277	15,301	12,452	10,498
185	162,525	144,919	130,755	119,113	109,374	101,108	87,832	77,637	69,563	63,010	57,586	53,021	49,127	41,506	35,932	31,678	28,324	25,613	23,375	19,898	17,322	15,336	12,476	10,515
190	166,371	147,969	133,233	121,166	111,103	102,583	88,943	78,504	70,258	63,580	58,061	53,424	49,473	41,752	36,117	31,821	28,439	25,706	23,453	19,955	17,365	15,370	12,498	10,530
195	170,192	150,984	135,672	123,180	112,794	104,024	90,024	79,345	70,931	64,130	58,520	53,812	49,805	41,989	36,293	31,958	28,548	25,796	23,528	20,009	17,405	15,402	12,519	10,545
200	173,989	153,965	138,074	125,156	114,449	105,430	91,075	80,160	71,582	64,662	58,962	54,186	50,125	42,216	36,463	32,090	28,653	25,882	23,599	20,060	17,444	15,432	12,539	10,560

750 MCM	2 Run(s)	240 Volts	C-Value = 24,136	Magnetic Duct	Copper

Calculations:

1. $f = \dfrac{1.732 \times \text{Length} \times \text{Isca}}{\text{Runs} \times \text{C-Value} \times \text{E L-L}}$
2. $M = \dfrac{1}{1 + f}$
3. $\text{Isc} = \text{Isca} \times M$
4. Add Motor Contribution = Motor FLA x 4

* All results are given in symmetrical amperes

Three-Phase Bolted Fault-Current* Table

One-Way Distance in Feet

Available Isc in Thousands (Isca)

Isca	5	10	15	20	25	30	40	50	60	70	80	90	100	125	150	175	200	225	250	300	350	400	500	600
3	2,987	2,975	2,962	2,949	2,937	2,925	2,901	2,877	2,853	2,830	2,808	2,785	2,763	2,710	2,658	2,609	2,561	2,515	2,471	2,387	2,308	2,234	2,100	1,982
5	4,965	4,930	4,895	4,861	4,828	4,795	4,730	4,667	4,606	4,546	4,488	4,431	4,375	4,243	4,118	4,001	3,890	3,784	3,685	3,501	3,334	3,183	2,918	2,693
7	6,931	6,863	6,796	6,731	6,667	6,604	6,482	6,364	6,251	6,141	6,035	5,933	5,834	5,601	5,386	5,186	5,001	4,829	4,668	4,376	4,119	3,890	3,501	3,183
10	9,859	9,722	9,589	9,460	9,334	9,211	8,975	8,751	8,538	8,334	8,141	7,956	7,779	7,370	7,002	6,668	6,365	6,089	5,835	5,337	5,002	4,669	4,120	3,686
15	14,686	14,384	14,095	13,817	13,549	13,292	12,806	12,355	11,934	11,541	11,172	10,827	10,502	9,770	9,133	8,574	8,080	7,639	7,244	6,555	6,003	5,529	4,775	4,202
20	19,445	18,920	18,422	17,950	17,502	17,075	16,281	15,558	14,897	14,289	13,729	13,211	12,731	11,670	10,773	10,004	9,337	8,754	8,239	7,372	6,670	6,090	5,188	4,519
25	24,139	23,335	22,582	21,877	21,215	20,591	19,448	18,425	17,504	16,671	15,914	15,222	14,588	13,212	12,074	11,116	10,299	9,594	8,979	7,559	7,147	6,485	5,472	4,733
30	28,768	27,633	26,585	25,613	24,709	23,867	22,345	21,005	19,817	18,755	17,802	16,941	16,160	14,489	13,131	12,006	11,058	10,249	9,551	8,405	7,504	6,778	5,679	4,887
35	33,335	31,820	30,438	29,170	28,004	26,928	25,006	23,339	21,881	20,595	19,451	18,428	17,507	15,562	14,007	12,734	11,673	10,775	10,006	8,755	7,782	7,004	5,837	5,003
40	37,839	35,900	34,150	32,563	31,116	29,793	27,458	25,462	23,736	22,230	20,903	19,726	18,674	16,478	14,744	13,340	12,181	11,206	10,376	9,038	8,005	7,184	5,961	5,094
45	42,284	39,877	37,729	35,801	34,060	32,481	29,725	27,400	25,412	23,693	22,192	20,870	19,696	17,269	15,374	13,854	12,607	11,566	10,684	9,270	8,187	7,330	6,062	5,167
50	46,669	43,754	41,182	38,896	36,850	35,008	31,827	29,176	26,933	25,010	23,343	21,885	20,598	17,958	15,918	14,294	12,970	11,871	10,944	9,465	8,339	7,452	6,145	5,228
55	50,996	47,536	44,515	41,856	39,496	37,388	33,782	30,811	28,320	26,201	24,378	22,791	21,399	18,564	16,392	14,675	13,284	12,133	11,166	9,631	8,467	7,554	6,214	5,278
60	55,267	51,225	47,735	44,690	42,010	39,633	35,604	32,319	29,589	27,284	25,313	23,607	22,116	19,101	16,809	15,009	13,557	12,361	11,359	9,774	8,577	7,642	6,273	5,320
65	59,481	54,826	50,847	47,406	44,401	41,755	37,307	33,716	30,756	28,273	26,162	24,343	22,762	19,581	17,180	15,303	13,796	12,560	11,526	9,898	8,673	7,717	6,324	5,357
70	63,641	58,341	53,856	50,011	46,679	43,763	38,902	35,014	31,832	29,180	26,936	25,013	23,345	20,011	17,510	15,565	14,009	12,735	11,674	10,007	8,756	7,783	6,368	5,388
75	67,747	61,773	56,767	52,512	48,850	45,666	40,399	36,221	32,827	30,014	27,645	25,623	23,876	20,400	17,807	15,799	14,198	12,892	11,805	10,103	8,830	7,841	6,407	5,416
80	71,801	65,126	59,586	54,915	50,923	47,473	41,807	37,349	33,750	30,784	28,297	26,182	24,361	20,753	18,075	16,010	14,368	13,032	11,923	10,189	8,895	7,893	6,441	5,441
85	75,803	68,401	62,317	57,226	52,904	49,190	43,132	38,403	34,609	31,497	28,898	26,696	24,805	21,074	18,319	16,201	14,522	13,158	12,028	10,266	8,954	7,939	6,472	5,463
90	79,754	71,602	64,963	59,450	54,799	50,824	44,384	39,392	35,410	32,159	29,455	27,170	25,214	21,369	18,541	16,374	14,661	13,272	12,124	10,335	9,006	7,980	6,500	5,482
95	83,656	74,732	67,528	61,591	56,614	52,380	45,566	40,321	36,159	32,775	29,971	27,609	25,591	21,639	18,744	16,532	14,787	13,376	12,210	10,398	9,054	8,018	6,524	5,500
100	87,509	77,791	70,016	63,654	58,352	53,865	46,686	41,195	36,860	33,351	30,451	28,016	25,941	21,888	18,931	16,677	14,903	13,471	12,289	10,455	9,097	8,052	6,547	5,516
105	91,314	80,784	72,431	65,644	60,020	55,284	47,748	42,020	37,519	33,889	30,899	28,394	26,265	22,119	19,103	16,811	15,010	13,558	12,361	10,507	9,137	8,083	6,567	5,530
110	95,072	83,711	74,776	67,564	61,621	56,639	48,755	42,798	38,138	34,393	31,318	28,748	26,567	22,333	19,262	16,934	15,108	13,638	12,428	10,555	9,173	8,111	6,586	5,544
115	98,784	86,576	77,053	69,418	63,159	57,936	49,713	43,535	38,722	34,867	31,711	29,078	26,849	22,532	19,410	17,048	15,199	13,712	12,489	10,600	9,207	8,137	6,603	5,556
120	102,451	89,379	79,266	71,209	64,639	59,178	50,625	44,232	39,273	35,314	32,079	29,388	27,113	22,717	19,548	17,154	15,283	13,780	12,546	10,640	9,237	8,161	6,619	5,567
125	106,073	92,124	81,417	72,940	66,062	60,369	51,494	44,894	39,794	35,734	32,426	29,679	27,360	22,890	19,676	17,253	15,361	13,844	12,599	10,678	9,266	8,184	6,634	5,577
130	109,652	94,812	83,509	74,615	67,433	61,512	52,323	45,523	40,287	36,131	32,753	29,952	27,593	23,053	19,796	17,345	15,434	13,903	12,648	10,714	9,292	8,204	6,647	5,587
135	113,188	97,444	85,545	76,236	68,754	62,609	53,115	46,121	40,755	36,507	33,061	30,210	27,811	23,205	19,908	17,431	15,502	13,958	12,694	10,746	9,317	8,223	6,660	5,596
140	116,682	100,022	87,526	77,805	70,027	63,663	53,872	46,691	41,199	36,863	33,353	30,453	28,017	23,348	20,013	17,512	15,566	14,010	12,736	10,777	9,340	8,241	6,672	5,604
145	120,134	102,549	89,454	79,325	71,256	64,678	54,596	47,234	41,621	37,201	33,629	30,683	28,212	23,484	20,112	17,588	15,626	14,058	12,776	10,806	9,362	8,258	6,683	5,612
150	123,546	105,025	91,332	80,798	72,443	65,654	55,290	47,753	42,024	37,522	33,891	30,901	28,396	23,611	20,206	17,659	15,683	14,104	12,814	10,833	9,382	8,274	6,693	5,619
155	126,918	107,451	93,162	82,227	73,589	66,594	55,956	48,248	42,407	37,827	34,140	31,108	28,571	23,731	20,294	17,727	15,736	14,147	12,849	10,858	9,401	8,289	6,703	5,626
160	130,251	109,831	94,945	83,613	74,698	67,500	56,594	48,722	42,772	38,118	34,377	31,305	28,736	23,846	20,377	17,790	15,786	14,187	12,883	10,882	9,419	8,302	6,712	5,632
165	133,546	112,164	96,684	84,959	75,770	68,374	57,207	49,176	43,122	38,395	34,602	31,491	28,894	23,954	20,456	17,850	15,833	14,225	12,914	10,904	9,435	8,315	6,720	5,638
170	136,802	114,452	98,379	86,265	76,807	69,218	57,797	49,611	43,456	38,660	34,817	31,669	29,043	24,056	20,531	17,907	15,878	14,262	12,944	10,925	9,451	8,328	6,728	5,644
175	140,022	116,697	100,033	87,534	77,811	70,033	58,364	50,028	43,776	38,912	35,022	31,838	29,186	24,154	20,602	17,961	15,920	14,296	12,972	10,945	9,466	8,340	6,736	5,649
180	143,205	118,899	101,647	88,767	78,784	70,820	58,909	50,428	44,082	39,154	35,218	32,000	29,321	24,247	20,670	18,013	15,961	14,328	12,999	10,965	9,481	8,351	6,743	5,654
185	146,351	121,061	103,223	89,967	79,728	71,581	59,435	50,813	44,376	39,386	35,405	32,155	29,451	24,336	20,734	18,061	15,999	14,359	13,024	10,983	9,494	8,361	6,750	5,659
190	149,463	123,182	104,761	91,133	80,642	72,317	59,942	51,183	44,657	39,608	35,584	32,302	29,575	24,420	20,796	18,103	16,036	14,389	13,049	11,000	9,507	8,371	6,756	5,664
195	152,540	125,264	106,263	92,268	81,529	73,030	60,431	51,539	44,928	39,821	35,756	32,444	29,693	24,501	20,854	18,152	16,070	14,417	13,072	11,016	9,519	8,380	6,763	5,668
200	155,582	127,309	107,731	93,372	82,391	73,720	60,903	51,882	45,189	40,025	35,920	32,579	29,807	24,578	20,910	18,195	16,103	14,443	13,094	11,032	9,531	8,389	6,768	5,672

1000 MCM | **1 Run(s)** | **240 Volts** | **C-Value = 25,278** | **Magnetic Duct** | **Copper**

Calculations:

1. $f = \dfrac{1.732 \times \text{Length} \times \text{Isca}}{\text{Runs} \times \text{C-Value} \times E_{L\text{-}L}}$

2. $M = \dfrac{1}{1 + f}$

3. Isc = Isca x M

4. Add Motor Contribution = Motor FLA x 4

* All results are given in symmetrical amperes

Three-Phase Bolted Fault-Current* Table

One-Way Distance in Feet

Available Isc in Thousands (Isca)	5	10	15	20	25	30	40	50	60	70	80	90	100	125	150	175	200	225	250	300	350	400	500	600
3	2,994	2,987	2,981	2,975	2,968	2,962	2,949	2,937	2,925	2,913	2,901	2,889	2,877	2,848	2,819	2,791	2,763	2,736	2,710	2,658	2,609	2,561	2,471	2,387
5	4,982	4,965	4,947	4,930	4,912	4,895	4,861	4,828	4,795	4,762	4,730	4,698	4,667	4,590	4,516	4,445	4,375	4,308	4,243	4,118	4,001	3,890	3,685	3,501
7	6,965	6,931	6,897	6,863	6,829	6,796	6,731	6,667	6,604	6,542	6,482	6,422	6,364	6,223	6,088	5,958	5,834	5,715	5,601	5,386	5,186	5,001	4,668	4,376
10	9,929	9,859	9,790	9,722	9,655	9,589	9,460	9,334	9,211	9,092	8,975	8,862	8,751	8,486	8,236	8,001	7,779	7,569	7,370	7,002	6,668	6,365	5,835	5,387
15	14,841	14,686	14,533	14,384	14,238	14,095	13,817	13,549	13,292	13,045	12,806	12,576	12,355	11,833	11,353	10,911	10,502	10,123	9,770	9,133	8,574	8,080	7,244	6,565
20	19,719	19,445	19,179	18,920	18,668	18,422	17,950	17,502	17,075	16,669	16,281	15,912	15,558	14,740	14,003	13,337	12,731	12,178	11,670	10,773	10,004	9,337	8,239	7,372
25	24,562	24,139	23,730	23,335	22,952	22,582	21,877	21,215	20,591	20,003	19,448	18,922	18,425	17,288	16,283	15,389	14,588	13,866	13,212	12,074	11,116	10,299	8,979	7,959
30	29,371	28,768	28,189	27,633	27,099	26,585	25,613	24,709	23,867	23,081	22,345	21,654	21,005	19,540	18,266	17,149	16,160	15,279	14,489	13,131	12,006	11,058	9,551	8,405
35	34,147	33,335	32,560	31,820	31,114	30,438	29,170	28,004	26,928	25,931	25,006	24,144	23,339	21,545	20,007	18,673	17,507	16,477	15,562	14,007	12,734	11,673	10,006	8,755
40	38,890	37,839	36,844	35,900	35,003	34,150	32,563	31,116	29,793	28,578	27,458	26,422	25,462	23,341	21,546	20,008	18,674	17,508	16,478	14,744	13,340	12,181	10,376	9,038
45	43,600	42,284	41,045	39,877	38,773	37,729	35,801	34,060	32,481	31,042	29,725	28,515	27,400	24,959	22,918	21,185	19,696	18,403	17,269	15,374	13,854	12,607	10,684	9,270
50	48,277	46,669	45,165	43,754	42,429	41,182	38,896	36,850	35,008	33,342	31,827	30,444	29,176	26,425	24,148	22,232	20,598	19,187	17,958	15,918	14,294	12,970	10,944	9,465
55	52,923	50,996	49,205	47,536	45,976	44,515	41,856	39,496	37,388	35,494	33,782	32,228	30,811	27,758	25,257	23,168	21,399	19,881	18,564	16,392	14,675	13,284	11,166	9,631
60	57,536	55,267	53,169	51,225	49,419	47,735	44,690	42,010	39,633	37,511	35,604	33,882	32,319	28,977	26,261	24,011	22,116	20,498	19,101	16,809	15,009	13,557	11,359	9,774
65	62,118	59,481	57,059	54,826	52,761	50,847	47,406	44,401	41,755	39,406	37,307	35,421	33,716	30,095	27,176	24,774	22,762	21,052	19,581	17,180	15,303	13,796	11,526	9,898
70	66,669	63,641	60,876	58,341	56,009	53,856	50,011	46,679	43,763	41,190	38,902	36,856	35,014	31,125	28,013	25,467	23,345	21,550	20,011	17,510	15,565	14,009	11,674	10,007
75	71,189	67,747	64,622	61,773	59,165	56,767	52,512	48,850	45,666	42,871	40,399	38,196	36,221	32,075	28,781	26,100	23,876	22,002	20,400	17,807	15,799	14,198	11,805	10,103
80	75,679	71,801	68,300	65,126	62,233	59,586	54,915	50,923	47,473	44,460	41,807	39,452	37,349	32,956	29,488	26,681	24,361	22,413	20,753	18,075	16,010	14,368	11,923	10,189
85	80,138	75,803	71,912	68,401	65,217	62,317	57,226	52,904	49,190	45,962	43,132	40,631	38,403	33,775	30,142	27,214	24,805	22,788	21,074	18,319	16,201	14,522	12,028	10,266
90	84,568	79,754	75,459	71,602	68,121	64,963	59,450	54,799	50,824	47,386	44,384	41,739	39,392	34,537	30,747	27,707	25,214	23,133	21,369	18,541	16,374	14,661	12,124	10,335
95	88,968	83,656	78,942	74,732	70,947	67,528	61,591	56,614	52,380	48,736	45,566	42,784	40,321	35,249	31,310	28,164	25,591	23,450	21,639	18,744	16,532	14,787	12,210	10,398
100	93,338	87,509	82,364	77,791	73,699	70,016	63,654	58,352	53,865	50,019	46,686	43,769	41,195	35,915	31,835	28,587	25,941	23,743	21,888	18,931	16,677	14,903	12,289	10,455
105	97,680	91,314	85,727	80,784	76,380	72,431	65,644	60,020	55,284	51,240	47,748	44,701	42,020	36,540	32,325	28,982	26,265	24,014	22,119	19,103	16,811	15,010	12,361	10,507
110	101,993	95,072	89,031	83,711	78,992	74,776	67,564	61,621	56,639	52,402	48,755	45,583	42,798	37,128	32,784	29,350	26,567	24,267	22,333	19,262	16,934	15,108	12,428	10,555
115	106,277	98,784	92,278	86,576	81,537	77,053	69,418	63,159	57,936	53,511	49,713	46,419	43,535	37,681	33,214	29,695	26 849	24,502	22,532	19,410	17,048	15,199	12,489	10,600
120	110,533	102,451	95,470	89,379	84,020	79,266	71,209	64,639	59,178	54,569	50,625	47,213	44,232	38,202	33,619	30,018	27 113	24,721	22,717	19,548	17,154	15,283	12,546	10,640
125	114,761	106,073	98,608	92,124	86,441	81,417	72,940	66,062	60,369	55,580	51,494	47,968	44,894	38,695	34,000	30,321	27,360	24,927	22,890	19,676	17,253	15,361	12,599	10,678
130	118,962	109,652	101,693	94,812	88,802	83,509	74,615	67,433	61,512	56,547	52,323	48,687	45,523	39,161	34,359	30,606	27,593	25,119	23,053	19,796	17,345	15,434	12,648	10,714
135	123,135	113,188	104,727	97,444	91,107	85,545	76,236	68,754	62,609	57,472	53,115	49,372	46,121	39,603	34,699	30,876	27,811	25,300	23,205	19,908	17,431	15,502	12,694	10,746
140	127,282	116,682	107,712	100,022	93,358	87,526	77,805	70,027	63,663	58,360	53,872	50,025	46,691	40,022	35,020	31,130	28,017	25,471	23,348	20,013	17,512	15,566	12,736	10,777
145	131,401	120,134	110,647	102,549	95,555	89,454	79,325	71,256	64,678	59,211	54,596	50,649	47,234	40,421	35,325	31,370	28,212	25,632	23,484	20,112	17,588	15,626	12,776	10,806
150	135,494	123,546	113,535	105,025	97,701	91,332	80,798	72,443	65,654	60,028	55,290	51,246	47,753	40,800	35,614	31,598	28,396	25,783	23,611	20,206	17,659	15,683	12,814	10,833
155	139,561	126,918	116,376	107,451	99,798	93,162	82,227	73,589	66,594	60,813	55,956	51,817	48,248	41,161	35,889	31,815	28,571	25,927	23,731	20,294	17,727	15,736	12,849	10,858
160	143,601	130,251	119,173	109,831	101,847	94,945	83,613	74,698	67,500	61,568	56,594	52,364	48,722	41,505	36,151	32,020	28,736	26,063	23,846	20,377	17,790	15,786	12,883	10,882
165	147,616	133,546	121,925	112,164	103,850	96,684	84,959	75,770	68,374	62,294	57,207	52,888	49,176	41,834	36,400	32,215	28,394	26,193	23,954	20,456	17,850	15,833	12,914	10,904
170	151,605	136,802	124,633	114,452	105,809	98,379	86,265	76,807	69,218	62,994	57,797	53,392	49,611	42,149	36,638	32,401	29,043	26,316	24,056	20,531	17,907	15,878	12,944	10,925
175	155,569	140,022	127,300	116,697	107,725	100,033	87,534	77,811	70,033	63,668	58,364	53,875	50,028	42,449	36,865	32,579	29,186	26,433	24,154	20,602	17,961	15,920	12,972	10,945
180	159,508	143,205	129,925	118,899	109,599	101,647	88,767	78,784	70,820	64,318	58,909	54,340	50,428	42,737	37,082	32,748	29,321	26,544	24,247	20,670	18,013	15,961	12,999	10,965
185	163,422	146,351	132,510	121,061	111,432	103,223	89,967	79,728	71,581	64,945	59,435	54,787	50,813	43,013	37,289	32,910	29,451	26,650	24,336	20,734	18,061	15,999	13,024	10,983
190	167,311	149,463	135,056	123,182	113,227	104,761	91,133	80,642	72,317	65,551	59,942	55,217	51,183	43,278	37,488	33,065	29,575	26,752	24,420	20,796	18,108	16,036	13,049	11,000
195	171,176	152,540	137,563	125,264	114,984	106,263	92,268	81,529	73,030	66,136	60,431	55,632	51,539	43,532	37,679	33,213	29,693	26,848	24,501	20,854	18,152	16,070	13,072	11,016
200	175,017	155,582	140,033	127,309	116,704	107,731	93,372	82,391	73,720	66,701	60,903	56,031	51,882	43,777	37,862	33,355	29,807	26,941	24,578	20,910	18,195	16,103	13,094	11,032

1000 MCM	2 Run(s)	240 Volts	C-Value = 25,278	Magnetic Duct	Copper

Calculations:

1. $f = \dfrac{1.732 \times \text{Length} \times \text{Isca}}{\text{Runs} \times \text{C-Value} \times \text{E L-L}}$

2. $M = \dfrac{1}{1 + f}$

3. $\text{Isc} = \text{Isca} \times M$

4. Add Motor Contribution = Motor FLA x 4

* All results are given in symmetrical amperes

Short-Circuit Tables and Data: 208 Volts Three-Phase

Three-Phase Bolted Fault-Current* Table

One-Way Distance in Feet

Available Isc in Thousands (Isca)	5	10	15	20	25	30	40	50	60	70	80	90	100	125	150	175	200	225	250	300	350	400	500	600
1	903	824	757	700	651	609	539	483	438	400	369	342	318	272	237	211	189	172	157	135	118	105	85	72
3	2,271	1,827	1,528	1,313	1,151	1,025	841	712	618	546	489	443	404	332	282	245	217	194	176	148	128	112	91	76
5	3,257	2,415	1,919	1,592	1,360	1,187	947	787	674	589	523	470	427	348	293	253	223	199	180	151	130	114	92	77
10	4,830	3,184	2,375	1,894	1,574	1,347	1,046	854	722	626	552	493	446	360	302	260	228	203	183	153	132	115	93	77
15	5,757	3,562	2,579	2,021	1,662	1,411	1,084	880	740	639	562	502	453	365	305	262	230	205	185	154	132	116	93	77
20	6,368	3,787	2,695	2,092	1,709	1,445	1,103	893	749	646	567	506	456	367	307	263	231	205	185	155	133	116	93	78
25	6,801	3,936	2,769	2,136	1,739	1,466	1,116	901	755	650	571	509	459	368	308	264	231	206	185	155	133	116	93	78
30	7,124	4,042	2,821	2,167	1,759	1,480	1,124	906	759	653	573	510	460	369	308	265	232	206	186	155	133	116	93	78
35	7,375	4,121	2,860	2,190	1,774	1,491	1,130	910	762	655	574	511	461	370	309	265	232	206	186	155	133	116	93	78
40	7,574	4,183	2,889	2,207	1,785	1,499	1,135	913	764	656	576	512	462	370	309	265	232	207	186	155	133	116	93	78
45	7,737	4,232	2,913	2,221	1,794	1,505	1,138	915	765	658	576	513	462	371	309	265	232	207	186	155	133	116	93	78
50	7,872	4,272	2,932	2,232	1,801	1,510	1,141	917	767	659	577	514	463	371	310	266	232	207	186	155	133	117	93	78
55	7,986	4,306	2,947	2,241	1,807	1,514	1,144	919	768	659	578	514	463	371	310	266	233	207	186	155	133	117	93	78
60	8,084	4,334	2,961	2,248	1,812	1,518	1,146	920	769	660	578	515	464	371	310	266	233	207	186	155	133	117	93	78
65	8,169	4,358	2,972	2,255	1,816	1,521	1,147	921	769	661	579	515	464	372	310	266	233	207	186	155	133	117	93	78
70	8,243	4,379	2,982	2,260	1,820	1,523	1,149	922	770	661	579	515	464	372	310	266	233	207	186	155	133	117	93	78
75	8,308	4,398	2,990	2,265	1,823	1,526	1,150	923	771	661	579	515	464	372	310	266	233	207	186	155	133	117	93	78
80	8,366	4,414	2,998	2,270	1,826	1,527	1,151	924	771	662	580	516	464	372	310	266	233	207	186	155	133	117	93	78
85	8,418	4,428	3,004	2,273	1,828	1,529	1,152	924	772	662	580	516	465	372	310	266	233	207	186	155	133	117	93	78
90	8,464	4,441	3,010	2,277	1,831	1,531	1,153	925	772	662	580	516	465	372	310	266	233	207	186	155	133	117	93	78
95	8,507	4,453	3,016	2,280	1,833	1,532	1,154	925	772	663	580	516	465	372	310	266	233	207	186	155	133	117	93	78
100	8,545	4,463	3,020	2,282	1,834	1,533	1,154	926	773	663	581	516	465	372	310	266	233	207	187	155	133	117	93	78
105	8,580	4,473	3,025	2,285	1,836	1,534	1,155	926	773	663	581	517	465	372	311	266	233	207	187	155	133	117	93	78
110	8,612	4,481	3,029	2,287	1,837	1,535	1,156	926	773	663	581	517	465	372	311	266	233	207	187	155	133	117	93	78
115	8,641	4,489	3,032	2,289	1,839	1,536	1,156	927	773	664	581	517	465	373	311	266	233	207	187	156	133	117	93	78
120	8,668	4,497	3,036	2,291	1,840	1,537	1,157	927	774	664	581	517	465	373	311	266	233	207	187	156	133	117	93	78
125	8,693	4,503	3,039	2,293	1,841	1,538	1,157	927	774	664	581	517	465	373	311	266	233	207	187	156	133	117	93	78
130	8,717	4,510	3,042	2,295	1,842	1,539	1,157	928	774	664	581	517	465	373	311	266	233	207	187	156	133	117	93	78
135	8,738	4,515	3,044	2,296	1,843	1,539	1,158	928	774	664	581	517	466	373	311	266	233	207	187	156	133	117	93	78
140	8,759	4,521	3,047	2,297	1,844	1,540	1,158	928	774	664	582	517	466	373	311	266	233	207	187	156	133	117	93	78
145	8,778	4,526	3,049	2,299	1,845	1,541	1,159	928	774	664	582	517	466	373	311	266	233	207	187	156	133	117	93	78
150	8,795	4,530	3,051	2,300	1,846	1,541	1,159	929	775	664	582	517	466	373	311	266	233	207	187	156	133	117	93	78
155	8,812	4,535	3,053	2,301	1,846	1,542	1,159	929	775	665	582	517	466	373	311	266	233	207	187	156	133	117	93	78
160	8,828	4,539	3,055	2,302	1,847	1,542	1,159	929	775	665	582	517	466	373	311	267	233	207	187	156	133	117	93	78
165	8,842	4,543	3,057	2,303	1,848	1,543	1,160	929	775	665	582	517	466	373	311	267	233	207	187	156	133	117	93	78
170	8,856	4,547	3,058	2,304	1,848	1,543	1,160	929	775	665	582	517	466	373	311	267	233	207	187	156	133	117	93	78
175	8,870	4,550	3,060	2,305	1,849	1,543	1,160	929	775	665	582	518	466	373	311	267	233	207	187	156	133	117	93	78
180	8,882	4,553	3,061	2,306	1,849	1,544	1,160	929	775	665	582	518	466	373	311	267	233	207	187	156	133	117	93	78
185	8,894	4,557	3,063	2,307	1,850	1,544	1,161	930	775	665	582	518	466	373	311	267	233	207	187	156	133	117	93	78
190	8,905	4,559	3,064	2,307	1,850	1,545	1,161	930	775	665	582	518	466	373	311	267	233	207	187	156	133	117	93	78
195	8,916	4,562	3,065	2,308	1,851	1,545	1,161	930	776	665	582	518	466	373	311	267	233	207	187	156	133	117	93	78
200	8,926	4,565	3,067	2,309	1,851	1,545	1,161	930	776	665	582	518	466	373	311	267	233	207	187	156	133	117	93	78

14 AWG | 1 Run(s) | 208 Volts | C-Value = 389 | Magnetic Duct | Copper

Calculations:

1. $f = \dfrac{1.732 \times \text{Length} \times \text{Isca}}{\text{Runs} \times \text{C-Value} \times \text{E L-L}}$
2. $M = \dfrac{1}{1 + f}$
3. $\text{Isc} = \text{Isca} \times M$
4. Add Motor Contribution = Motor FLA x 4

* All results are given in symmetrical amperes

Three-Phase Bolted Fault-Current* Table

One-Way Distance in Feet

Available Isc in Thousands (Isca)	5	10	15	20	25	30	40	50	60	70	80	90	100	125	150	175	200	225	250	300	350	400	500	600
1	937	881	832	787	748	712	649	597	553	514	481	452	426	372	331	297	270	248	229	98	175	156	129	110
3	2,495	2,135	1,866	1,658	1,491	1,355	1,145	992	875	782	708	646	594	495	424	371	330	297	270	228	198	174	141	119
5	3,739	2,985	2,485	2,128	1,861	1,653	1,352	1,143	990	874	781	707	645	530	450	390	345	309	280	235	203	179	144	121
10	5,971	4,256	3,306	2,703	2,286	1,981	1,563	1,291	1,099	957	848	761	690	560	471	406	357	319	288	241	207	182	146	122
15	7,455	4,960	3,716	2,971	2,475	2,121	1,649	1,349	1,141	989	872	780	706	570	478	412	362	322	291	243	209	183	147	122
20	8,512	5,407	3,961	3,126	2,581	2,198	1,695	1,380	1,163	1,005	885	791	714	576	482	415	364	324	292	244	209	184	147	123
25	9,304	5,716	4,125	3,227	2,650	2,248	1,725	1,399	1,177	1,016	893	797	720	579	484	416	365	325	293	245	210	184	147	123
30	9,919	5,942	4,241	3,298	2,697	2,282	1,745	1,412	1,186	1,022	898	801	723	581	486	418	366	326	293	245	210	184	147	123
35	10,411	6,115	4,329	3,350	2,732	2,307	1,759	1,422	1,193	1,027	902	804	726	583	487	418	367	326	294	245	210	184	148	123
40	10,813	6,252	4,397	3,391	2,759	2,326	1,770	1,429	1,198	1,031	905	807	727	584	488	419	367	327	294	245	211	184	148	123
45	11,148	6,362	4,451	3,423	2,781	2,341	1,779	1,435	1,202	1,034	908	809	729	585	489	419	367	327	294	246	211	184	148	123
50	11,431	6,453	4,496	3,449	2,798	2,354	1,786	1,439	1,205	1,037	909	810	730	586	489	420	368	327	295	246	211	185	148	123
55	11,674	6,530	4,533	3,471	2,812	2,364	1,792	1,443	1,208	1,039	911	811	731	586	490	420	368	327	295	246	211	185	148	123
60	11,884	6,595	4,564	3,489	2,824	2,372	1,797	1,446	1,210	1,040	912	812	732	587	490	420	368	328	295	246	211	185	148	123
65	12,068	6,651	4,591	3,505	2,835	2,379	1,801	1,449	1,212	1,042	913	813	733	587	490	421	368	328	295	246	211	185	148	123
70	12,230	6,700	4,614	3,519	2,843	2,386	1,805	1,451	1,214	1,043	914	814	733	588	491	421	369	328	295	246	211	185	148	123
75	12,374	6,743	4,635	3,530	2,851	2,391	1,808	1,453	1,215	1,044	915	814	734	588	491	421	369	328	295	246	211	185	148	123
80	12,503	6,782	4,653	3,541	2,858	2,396	1,811	1,455	1,216	1,045	916	815	734	588	491	421	369	328	295	246	211	185	148	123
85	12,619	6,816	4,668	3,550	2,864	2,400	1,813	1,457	1,217	1,046	916	815	735	589	491	421	369	328	295	246	211	185	148	123
90	12,724	6,846	4,683	3,558	2,869	2,404	1,815	1,458	1,218	1,046	917	816	735	589	491	421	369	328	295	246	211	185	148	123
95	12,820	6,874	4,696	3,566	2,874	2,407	1,817	1,459	1,219	1,047	917	816	735	589	491	422	369	328	295	246	211	185	148	123
100	12,907	6,899	4,707	3,572	2,879	2,410	1,819	1,460	1,220	1,047	918	817	736	589	492	422	369	328	296	246	211	185	148	123
105	12,987	6,921	4,718	3,579	2,883	2,413	1,820	1,461	1,221	1,048	918	817	736	589	492	422	369	328	296	246	211	185	148	123
110	13,060	6,942	4,728	3,584	2,886	2,416	1,822	1,462	1,221	1,048	918	817	736	590	492	422	369	328	296	246	211	185	148	123
115	13,128	6,961	4,736	3,589	2,889	2,418	1,823	1,463	1,222	1,049	919	817	736	590	492	422	369	328	296	246	211	185	148	123
120	13,190	6,979	4,744	3,594	2,892	2,420	1,824	1,464	1,222	1,049	919	818	736	590	492	422	369	328	296	246	211	185	148	123
125	13,249	6,995	4,752	3,598	2,895	2,422	1,825	1,465	1,223	1,050	919	818	737	590	492	422	369	328	296	247	211	185	148	123
130	13,303	7,010	4,759	3,602	2,898	2,424	1,826	1,465	1,223	1,050	920	818	737	590	492	422	369	328	296	247	211	185	148	123
135	13,354	7,024	4,765	3,606	2,900	2,426	1,827	1,466	1,224	1,050	920	818	737	590	492	422	369	329	296	247	211	185	148	123
140	13,401	7,037	4,771	3,609	2,902	2,427	1,828	1,466	1,224	1,051	920	818	737	590	492	422	370	329	296	247	211	185	148	123
145	13,445	7,049	4,777	3,613	2,905	2,429	1,829	1,467	1,225	1,051	920	819	737	590	492	422	370	329	296	247	211	185	148	123
150	13,487	7,061	4,782	3,616	2,906	2,430	1,830	1,467	1,225	1,051	921	819	737	590	492	422	370	329	296	247	211	185	148	123
155	13,526	7,072	4,787	3,618	2,908	2,431	1,831	1,468	1,225	1,051	921	819	737	591	492	422	370	329	296	247	211	185	148	123
160	13,563	7,082	4,792	3,621	2,910	2,432	1,831	1,468	1,225	1,052	921	819	738	591	492	422	370	329	296	247	211	185	148	123
165	13,598	7,091	4,796	3,623	2,912	2,433	1,832	1,469	1,226	1,052	921	819	738	591	493	422	370	329	296	247	211	185	148	123
170	13,631	7,100	4,800	3,626	2,913	2,435	1,832	1,469	1,226	1,052	921	819	738	591	493	422	370	329	296	247	211	185	148	123
175	13,662	7,109	4,804	3,628	2,915	2,436	1,833	1,469	1,226	1,052	921	819	738	591	493	422	370	329	296	247	211	185	148	123
180	13,692	7,117	4,808	3,630	2,916	2,436	1,834	1,470	1,227	1,052	921	820	738	591	493	422	370	329	296	247	211	185	148	123
185	13,720	7,124	4,811	3,632	2,917	2,437	1,834	1,470	1,227	1,053	922	820	738	591	493	422	370	329	296	247	211	185	148	123
190	13,747	7,132	4,815	3,634	2,918	2,438	1,835	1,470	1,227	1,053	922	820	738	591	493	422	370	329	296	247	211	185	148	123
195	13,773	7,138	4,818	3,636	2,920	2,439	1,835	1,471	1,227	1,053	922	820	738	591	493	422	370	329	296	247	211	185	148	123
200	13,797	7,145	4,821	3,637	2,921	2,440	1,835	1,471	1,227	1,053	922	820	738	591	493	423	370	329	296	247	211	185	148	123

12 AWG | 1 Run(s) | 208 Volts | C-Value = 617 | Magnetic Duct | Copper

Calculations:

1. $f = \dfrac{1.732 \times \text{Length} \times \text{Isca}}{\text{Runs} \times \text{C-Value} \times E_{L-L}}$
2. $M = \dfrac{1}{1 + f}$
3. $\text{Isc} = \text{Isca} \times M$
4. Add Motor Contribution = Motor FLA x 4

* All results are given in symmetrical amperes

Three-Phase Bolted Fault-Current* Table

One-Way Distance in Feet

Available Isc in Thousands (Isca)	5	10	15	20	25	30	40	50	60	70	80	90	100	125	150	175	200	225	250	300	350	400	500	600
1	959	922	887	855	825	797	747	702	663	627	596	567	541	485	440	402	371	344	320	282	252	228	191	164
3	2,661	2,391	2,171	1,988	1,833	1,701	1,486	1,320	1,187	1,078	988	911	846	717	622	550	492	446	407	347	303	268	218	184
5	4,125	3,510	3,055	2,704	2,426	2,200	1,853	1,602	1,410	1,259	1,138	1,037	953	793	679	593	527	474	431	364	315	278	225	189
10	7,020	5,409	4,399	3,707	3,203	2,820	2,275	1,907	1,641	1,441	1,284	1,157	1,054	861	728	631	556	498	450	378	326	286	230	193
15	9,165	6,599	5,155	4,230	3,586	3,112	2,462	2,036	1,736	1,513	1,341	1,204	1,092	887	746	644	567	506	457	383	329	289	232	194
20	10,818	7,414	5,639	4,550	3,814	3,282	2,567	2,108	1,788	1,552	1,372	1,229	1,113	900	756	651	572	510	460	385	331	290	233	194
25	12,130	8,008	5,976	4,767	3,965	3,394	2,635	2,153	1,821	1,577	1,391	1,244	1,125	908	761	656	575	513	463	387	332	291	233	195
30	13,197	8,459	6,224	4,924	4,073	3,472	2,682	2,185	1,843	1,594	1,404	1,254	1,134	914	765	658	578	515	464	388	333	292	234	195
35	14,082	8,814	6,415	5,042	4,153	3,531	2,717	2,208	1,859	1,606	1,413	1,262	1,140	918	768	660	579	516	465	388	333	292	234	195
40	14,828	9,101	6,565	5,134	4,216	3,576	2,743	2,225	1,872	1,615	1,420	1,268	1,144	921	770	662	581	517	466	389	334	292	234	195
45	15,465	9,337	6,687	5,209	4,266	3,612	2,764	2,239	1,881	1,622	1,426	1,272	1,148	923	772	663	581	518	466	389	334	293	234	195
50	16,015	9,535	6,788	5,270	4,307	3,641	2,781	2,250	1,889	1,628	1,431	1,276	1,151	925	773	664	582	518	467	390	334	293	235	196
55	16,495	9,703	6,873	5,321	4,341	3,665	2,796	2,259	1,896	1,633	1,434	1,279	1,153	927	774	665	583	519	467	390	335	293	235	196
60	16,918	9,847	6,945	5,364	4,369	3,686	2,807	2,267	1,901	1,637	1,437	1,281	1,155	928	775	666	583	519	468	390	335	293	235	196
65	17,293	9,973	7,007	5,401	4,394	3,703	2,818	2,274	1,906	1,641	1,440	1,283	1,157	929	776	666	584	519	468	390	335	293	235	196
70	17,628	10,084	7,062	5,433	4,415	3,718	2,826	2,279	1,910	1,643	1,442	1,285	1,159	930	777	667	584	520	468	391	335	293	235	196
75	17,929	10,182	7,110	5,462	4,434	3,732	2,834	2,284	1,913	1,646	1,444	1,287	1,160	931	777	667	584	520	468	391	335	293	235	196
80	18,201	10,269	7,152	5,487	4,450	3,743	2,841	2,289	1,916	1,648	1,446	1,288	1,161	932	778	668	585	520	468	391	335	293	235	196
85	18,448	10,347	7,190	5,509	4,465	3,754	2,847	2,293	1,919	1,650	1,448	1,289	1,162	932	778	668	585	520	469	391	335	294	235	196
90	18,673	10,417	7,224	5,529	4,478	3,763	2,852	2,296	1,922	1,652	1,449	1,290	1,163	933	779	668	585	521	469	391	335	294	235	196
95	18,880	10,481	7,254	5,547	4,490	3,771	2,857	2,299	1,924	1,654	1,450	1,291	1,164	933	779	668	585	521	469	391	335	294	235	196
100	19,069	10,539	7,282	5,563	4,500	3,779	2,861	2,302	1,926	1,655	1,451	1,292	1,164	934	779	669	586	521	469	391	335	294	235	196
105	19,244	10,593	7,307	5,578	4,510	3,785	2,865	2,304	1,927	1,656	1,452	1,293	1,165	934	780	669	586	521	469	391	336	294	235	196
110	19,405	10,641	7,331	5,591	4,519	3,792	2,868	2,307	1,929	1,658	1,453	1,294	1,166	934	780	669	586	521	469	391	336	294	235	196
115	19,555	10,686	7,352	5,604	4,527	3,797	2,872	2,309	1,931	1,659	1,454	1,294	1,166	935	780	669	586	521	469	391	336	294	235	196
120	19,695	10,728	7,372	5,615	4,534	3,803	2,875	2,311	1,932	1,660	1,455	1,295	1,167	935	780	669	586	521	469	391	336	294	235	196
125	19,825	10,766	7,390	5,625	4,541	3,807	2,877	2,313	1,933	1,661	1,455	1,295	1,167	935	781	670	586	521	469	391	336	294	235	196
130	19,947	10,802	7,407	5,635	4,548	3,812	2,880	2,314	1,934	1,661	1,456	1,296	1,168	936	781	670	586	522	470	392	336	294	235	196
135	20,061	10,835	7,422	5,644	4,553	3,816	2,882	2,316	1,935	1,662	1,457	1,296	1,168	936	781	670	586	522	470	392	336	294	235	196
140	20,168	10,867	7,437	5,653	4,559	3,820	2,885	2,317	1,936	1,663	1,457	1,297	1,168	936	781	670	587	522	470	392	336	294	235	196
145	20,269	10,896	7,450	5,661	4,564	3,823	2,887	2,319	1,937	1,664	1,458	1,297	1,169	936	781	670	587	522	470	392	336	294	235	196
150	20,363	10,923	7,463	5,668	4,569	3,827	2,889	2,320	1,938	1,664	1,458	1,298	1,169	937	781	670	587	522	470	392	336	294	235	196
155	20,453	10,949	7,475	5,675	4,573	3,830	2,890	2,321	1,939	1,665	1,459	1,298	1,169	937	781	670	587	522	470	392	336	294	235	196
160	20,538	10,973	7,487	5,681	4,578	3,833	2,892	2,322	1,940	1,665	1,459	1,298	1,169	937	782	670	587	522	470	392	336	294	235	196
165	20,618	10,996	7,497	5,687	4,582	3,836	2,894	2,323	1,940	1,666	1,460	1,299	1,170	937	782	670	587	522	470	392	336	294	235	196
170	20,694	11,018	7,507	5,693	4,585	3,838	2,895	2,324	1,941	1,667	1,460	1,299	1,170	937	782	671	587	522	470	392	336	294	235	196
175	20,766	11,038	7,517	5,699	4,589	3,841	2,897	2,325	1,942	1,667	1,460	1,299	1,170	937	782	671	587	522	470	392	336	294	235	196
180	20,835	11,057	7,526	5,704	4,592	3,843	2,898	2,326	1,942	1,667	1,461	1,300	1,170	938	782	671	587	522	470	392	336	294	235	196
185	20,900	11,076	7,534	5,709	4,595	3,845	2,899	2,327	1,943	1,668	1,461	1,300	1,171	938	782	671	587	522	470	392	336	294	235	196
190	20,963	11,093	7,542	5,713	4,598	3,847	2,900	2,327	1,943	1,668	1,461	1,300	1,171	938	782	671	587	522	470	392	336	294	235	196
195	21,022	11,110	7,550	5,718	4,601	3,849	2,901	2,328	1,944	1,669	1,462	1,300	1,171	938	782	671	587	522	470	392	336	294	235	196
200	21,079	11,126	7,557	5,722	4,604	3,851	2,903	2,329	1,944	1,669	1,462	1,300	1,171	938	782	671	587	522	470	392	336	294	235	196

# 10 AWG	1 Run(s)	208 Volts		C-Value = 981	Magnetic Duct	Copper

Calculations:

1. $f = \dfrac{1.732 \times \text{Length} \times \text{Isca}}{\text{Runs} \times \text{C-Value} \times E_{L\text{-}L}}$

2. $M = \dfrac{1}{1 + f}$

3. $\text{Isc} = \text{Isca} \times M$

4. Add Motor Contribution = Motor FLA x 4

* All results are given in symmetrical amperes

Three-Phase Bolted Fault-Current* Table

One-Way Distance in Feet

Available Isc in Thousands (Isca)	5	10	15	20	25	30	40	50	60	70	80	90	100	125	150	175	200	225	250	300	350	400	500	600
3	2,777	2,585	2,418	2,271	2,141	2,025	1,827	1,665	1,529	1,413	1,314	1,228	1,152	998	881	788	713	651	599	5[illegible]6	453	404	333	282
5	4,410	3,945	3,569	3,258	2,997	2,774	2,416	2,139	1,920	1,741	1,593	1,468	1,361	1,151	998	880	788	713	651	5[illegible]4	483	427	348	293
7	5,896	5,093	4,483	4,003	3,616	3,297	2,803	2,437	2,156	1,933	1,752	1,602	1,476	1,232	1,058	927	825	743	676	572	496	438	355	298
10	7,890	6,515	5,549	4,832	4,279	3,840	3,186	2,722	2,376	2,108	1,894	1,720	1,575	1,301	1,108	965	855	767	696	587	507	447	360	302
15	10,706	8,323	6,808	5,759	4,991	4,403	3,564	2,993	2,580	2,267	2,022	1,825	1,663	1,360	1,151	997	880	787	712	598	516	453	365	305
20	13,031	9,664	7,679	6,371	5,444	4,752	3,789	3,151	2,696	2,356	2,093	1,882	1,710	1,392	1,173	1,014	893	798	721	604	520	457	367	307
25	14,983	10,697	8,318	6,805	5,757	4,989	3,938	3,253	2,771	2,413	2,137	1,918	1,740	1,411	1,187	1,025	901	804	726	608	523	459	368	308
30	16,646	11,519	8,806	7,128	5,987	5,161	4,044	3,325	2,823	2,453	2,168	1,943	1,760	1,425	1,197	1,032	907	809	730	611	525	460	369	308
35	18,079	12,187	9,192	7,378	6,162	5,291	4,124	3,379	2,862	2,482	2,191	1,961	1,775	1,435	1,204	1,037	911	812	732	612	526	461	370	309
40	19,327	12,742	9,504	7,578	6,301	5,393	4,185	3,420	2,891	2,504	2,208	1,975	1,786	1,442	1,209	1,041	914	814	734	614	527	462	371	309
45	20,424	13,210	9,762	7,741	6,413	5,475	4,235	3,453	2,915	2,522	2,222	1,986	1,795	1,448	1,213	1,044	916	816	736	615	528	463	371	309
50	21,395	13,609	9,978	7,876	6,506	5,542	4,275	3,479	2,934	2,536	2,233	1,995	1,802	1,452	1,216	1,046	918	817	737	616	529	463	371	310
55	22,261	13,954	10,162	7,991	6,584	5,598	4,308	3,502	2,949	2,547	2,242	2,002	1,808	1,456	1,219	1,048	919	819	738	616	529	464	371	310
60	23,038	14,256	10,321	8,089	6,650	5,646	4,337	3,520	2,963	2,557	2,250	2,008	1,813	1,459	1,221	1,050	921	820	739	617	530	464	372	310
65	23,739	14,521	10,460	8,174	6,708	5,687	4,361	3,536	2,974	2,566	2,256	2,013	1,818	1,462	1,223	1,051	922	821	739	617	530	464	372	310
70	24,375	14,757	10,581	8,248	6,757	5,723	4,382	3,550	2,984	2,573	2,262	2,018	1,821	1,465	1,225	1,052	923	821	740	618	530	464	372	310
75	24,954	14,967	10,689	8,313	6,801	5,755	4,400	3,562	2,992	2,579	2,267	2,022	1,824	1,467	1,226	1,053	923	822	741	618	530	465	372	310
80	25,484	15,156	10,785	8,371	6,840	5,782	4,417	3,573	3,000	2,585	2,271	2,025	1,827	1,468	1,227	1,054	924	822	741	618	531	465	372	310
85	25,971	15,327	10,871	8,423	6,874	5,807	4,431	3,582	3,006	2,590	2,275	2,028	1,830	1,470	1,229	1,055	925	823	741	619	531	465	372	311
90	26,419	15,482	10,949	8,469	6,905	5,829	4,444	3,590	3,012	2,594	2,278	2,031	1,832	1,471	1,230	1,056	925	823	742	619	531	465	372	311
95	26,834	15,623	11,020	8,512	6,933	5,849	4,455	3,598	3,017	2,598	2,281	2,033	1,834	1,473	1,230	1,057	926	824	742	619	531	465	373	311
100	27,218	15,753	11,084	8,550	6,959	5,867	4,466	3,605	3,022	2,602	2,284	2,035	1,836	1,474	1,231	1,057	926	824	742	619	531	465	373	311
105	27,575	15,872	11,143	8,585	6,982	5,884	4,475	3,611	3,027	2,605	2,286	2,037	1,837	1,475	1,232	1,058	927	825	743	620	532	465	373	311
110	27,909	15,982	11,197	8,617	7,003	5,899	4,484	3,617	3,031	2,608	2,289	2,039	1,839	1,476	1,233	1,058	927	825	743	620	532	465	373	311
115	28,220	16,083	11,247	8,646	7,023	5,912	4,492	3,622	3,034	2,611	2,291	2,041	1,840	1,477	1,233	1,059	927	825	743	620	532	466	373	311
120	28,511	16,178	11,293	8,673	7,041	5,925	4,499	3,627	3,038	2,613	2,293	2,042	1,841	1,477	1,234	1,059	928	825	743	620	532	466	373	311
125	28,785	16,265	11,335	8,699	7,057	5,937	4,506	3,631	3,041	2,615	2,294	2,044	1,842	1,478	1,234	1,059	928	826	743	620	532	466	373	311
130	29,042	16,347	11,375	8,722	7,072	5,948	4,512	3,635	3,043	2,617	2,296	2,045	1,843	1,479	1,235	1,060	928	826	744	620	532	466	373	311
135	29,285	16,424	11,412	8,744	7,087	5,958	4,518	3,639	3,046	2,619	2,298	2,046	1,844	1,479	1,235	1,060	928	826	744	620	532	466	373	311
140	29,513	16,495	11,446	8,764	7,100	5,967	4,524	3,642	3,049	2,621	2,299	2,047	1,845	1,480	1,236	1,060	929	826	744	621	532	466	373	311
145	29,729	16,563	11,479	8,783	7,112	5,976	4,529	3,646	3,051	2,623	2,300	2,048	1,846	1,481	1,236	1,061	929	826	744	621	532	466	373	311
150	29,934	16,626	11,509	8,801	7,124	5,984	4,533	3,649	3,053	2,624	2,301	2,049	1,847	1,481	1,236	1,061	929	826	744	621	532	466	373	311
155	30,128	16,686	11,538	8,817	7,135	5,992	4,538	3,652	3,055	2,626	2,303	2,050	1,848	1,482	1,237	1,061	929	827	744	621	532	466	373	311
160	30,312	16,742	11,565	8,833	7,145	5,999	4,542	3,654	3,057	2,627	2,304	2,051	1,848	1,482	1,237	1,061	929	827	744	621	532	466	373	311
165	30,487	16,795	11,590	8,848	7,155	6,006	4,546	3,657	3,059	2,629	2,305	2,052	1,849	1,482	1,237	1,062	930	827	745	621	533	466	373	311
170	30,654	16,846	11,614	8,862	7,164	6,012	4,549	3,659	3,060	2,630	2,306	2,053	1,849	1,483	1,237	1,062	930	827	745	621	533	466	373	311
175	30,812	16,893	11,637	8,875	7,173	6,018	4,553	3,661	3,062	2,631	2,306	2,053	1,850	1,483	1,238	1,062	930	827	745	621	533	466	373	311
180	30,964	16,939	11,658	8,888	7,181	6,024	4,556	3,664	3,063	2,632	2,307	2,054	1,851	1,484	1,238	1,062	930	827	745	621	533	466	373	311
185	31,108	16,982	11,679	8,899	7,189	6,030	4,559	3,666	3,065	2,633	2,308	2,055	1,851	1,484	1,238	1,062	930	827	745	621	533	466	373	311
190	31,247	17,023	11,698	8,911	7,196	6,035	4,562	3,667	3,066	2,634	2,309	2,055	1,852	1,484	1,238	1,063	930	827	745	621	533	466	373	311
195	31,379	17,062	11,717	8,921	7,203	6,040	4,565	3,669	3,067	2,635	2,310	2,056	1,852	1,484	1,239	1,063	930	828	745	621	533	466	373	311
200	31,506	17,100	11,734	8,932	7,210	6,044	4,568	3,671	3,069	2,636	2,310	2,056	1,853	1,485	1,239	1,063	931	828	745	621	533	466	373	311

# 8 AWG	1 Run(s)	208 Volts		C-Value = 1,557	Magnetic Duct	Copper

Calculations:

1. $f = \dfrac{1.732 \times \text{Length} \times \text{Isca}}{\text{Runs} \times \text{C-Value} \times \text{E L-L}}$

2. $M = \dfrac{1}{1 + f}$

3. $\text{Isc} = \text{Isca} \times M$

4. Add Motor Contribution = Motor FLA x 4

* All results are given in symmetrical amperes

Three-Phase Bolted Fault-Current* Table

One-Way Distance in Feet

Available Isc in Thousands (Isca)	5	10	15	20	25	30	40	50	60	70	80	90	100	125	150	175	200	225	250	300	350	400	500	600
3	2,853	2,720	2,598	2,488	2,386	2,292	2,125	1,980	1,854	1,743	1,645	1,557	1,478	1,311	1,179	1,070	980	904	839	733	651	586	488	418
5	4,605	4,267	3,976	3,722	3,498	3,300	2,964	2,690	2,463	2,271	2,107	1,964	1,840	1,589	1,398	1,249	1,128	1,028	945	813	713	636	522	442
7	6,249	5,644	5,145	4,727	4,373	4,067	3,569	3,179	2,866	2,609	2,395	2,213	2,057	1,748	1,520	1,345	1,205	1,092	999	853	744	659	538	454
10	8,535	7,444	6,600	5,929	5,381	4,926	4,213	3,681	3,268	2,938	2,669	2,445	2,255	1,890	1,626	1,427	1,271	1,146	1,043	885	768	679	550	463
15	11,928	9,901	8,462	7,389	6,557	5,893	4,902	4,195	3,667	3,257	2,929	2,662	2,439	2,017	1,719	1,498	1,327	1,192	1,081	912	788	694	561	470
20	14,888	11,857	9,852	8,426	7,361	6,535	5,338	4,511	3,906	3,444	3,080	2,785	2,542	2,087	1,770	1,536	1,357	1,216	1,101	926	799	702	566	474
25	17,492	13,452	10,928	9,202	7,946	6,992	5,639	4,724	4,065	3,567	3,178	2,865	2,608	2,131	1,802	1,560	1,376	1,231	1,113	934	805	707	569	476
30	19,801	14,777	11,787	9,803	8,391	7,334	5,859	4,878	4,178	3,654	3,246	2,921	2,655	2,162	1,823	1,577	1,389	1,241	1,121	940	810	711	571	478
35	21,863	15,896	12,488	10,283	8,740	7,600	6,027	4,993	4,263	3,718	3,297	2,962	2,689	2,184	1,839	1,589	1,398	1,248	1,127	945	813	713	573	479
40	23,714	16,853	13,071	10,675	9,022	7,812	6,159	5,084	4,328	3,768	3,337	2,994	2,715	2,202	1,852	1,598	1,405	1,254	1,132	948	815	715	574	480
45	25,386	17,680	13,563	11,001	9,254	7,985	6,267	5,157	4,381	3,808	3,368	3,019	2,735	2,215	1,861	1,605	1,410	1,258	1,136	950	817	716	575	480
50	26,904	18,403	13,985	11,277	9,448	8,129	6,355	5,217	4,424	3,841	3,393	3,039	2,752	2,226	1,869	1,611	1,415	1,262	1,138	952	818	718	576	481
55	28,288	19,040	14,350	11,513	9,613	8,251	6,429	5,267	4,460	3,868	3,414	3,056	2,766	2,235	1,875	1,615	1,419	1,265	1,141	954	820	719	576	481
60	29,555	19,606	14,668	11,718	9,755	8,356	6,493	5,309	4,490	3,891	3,432	3,070	2,777	2,243	1,881	1,619	1,422	1,267	1,143	955	821	719	577	481
65	30,719	20,112	14,950	11,896	9,879	8,446	6,547	5,345	4,516	3,910	3,447	3,082	2,787	2,249	1,885	1,623	1,424	1,269	1,144	956	822	720	577	482
70	31,792	20,566	15,199	12,054	9,987	8,525	6,595	5,377	4,539	3,927	3,460	3,093	2,796	2,255	1,889	1,625	1,426	1,271	1,146	957	822	721	578	482
75	32,784	20,977	15,423	12,194	10,083	8,595	6,636	5,405	4,559	3,942	3,472	3,102	2,803	2,260	1,893	1,628	1,428	1,272	1,147	958	823	721	578	482
80	33,705	21,350	15,623	12,319	10,168	8,657	6,673	5,429	4,576	3,955	3,482	3,110	2,810	2,264	1,895	1,630	1,430	1,274	1,148	959	824	721	578	482
85	34,562	21,691	15,805	12,432	10,245	8,712	6,706	5,451	4,592	3,966	3,491	3,117	2,816	2,268	1,898	1,632	1,432	1,275	1,149	960	824	722	578	483
90	35,361	22,003	15,970	12,533	10,314	8,762	6,736	5,470	4,605	3,977	3,499	3,124	2,821	2,271	1,900	1,634	1,433	1,276	1,150	960	824	722	579	483
95	36,107	22,290	16,120	12,626	10,377	8,807	6,762	5,488	4,618	3,986	3,506	3,129	2,826	2,274	1,903	1,635	1,434	1,277	1,151	961	825	723	579	483
100	36,807	22,554	16,258	12,710	10,434	8,849	6,787	5,504	4,629	3,994	3,512	3,134	2,830	2,277	1,905	1,637	1,435	1,278	1,151	961	825	723	579	483
105	37,463	22,799	16,385	12,788	10,486	8,886	6,809	5,518	4,639	4,002	3,518	3,139	2,834	2,279	1,906	1,638	1,436	1,279	1,152	962	826	723	579	483
110	38,081	23,026	16,502	12,859	10,533	8,920	6,829	5,532	4,649	4,009	3,524	3,143	2,837	2,281	1,908	1,639	1,437	1,279	1,153	962	826	723	579	483
115	38,663	23,238	16,611	12,925	10,578	8,952	6,847	5,544	4,657	4,015	3,529	3,147	2,840	2,284	1,909	1,640	1,438	1,280	1,153	963	826	723	580	483
120	39,212	23,435	16,711	12,985	10,618	8,981	6,864	5,555	4,665	4,021	3,533	3,151	2,843	2,285	1,911	1,641	1,439	1,281	1,154	963	826	724	580	483
125	39,732	23,620	16,805	13,042	10,656	9,008	6,880	5,565	4,672	4,026	3,537	3,154	2,846	2,287	1,912	1,642	1,439	1,281	1,154	963	827	724	580	483
130	40,223	23,792	16,892	13,094	10,691	9,033	6,894	5,575	4,679	4,031	3,541	3,157	2,848	2,289	1,913	1,643	1,440	1,282	1,155	964	827	724	580	484
135	40,690	23,955	16,974	13,144	10,724	9,056	6,908	5,584	4,685	4,036	3,545	3,160	2,851	2,290	1,914	1,644	1,441	1,282	1,155	964	827	724	580	484
140	41,132	24,108	17,050	13,189	10,754	9,078	6,921	5,592	4,691	4,040	3,548	3,163	2,853	2,292	1,915	1,645	1,441	1,282	1,155	964	827	724	580	484
145	41,553	24,252	17,122	13,232	10,783	9,098	6,933	5,600	4,697	4,044	3,551	3,165	2,855	2,293	1,916	1,645	1,442	1,283	1,156	964	827	724	580	484
150	41,954	24,388	17,190	13,273	10,809	9,117	6,944	5,607	4,702	4,048	3,554	3,167	2,857	2,294	1,917	1,646	1,442	1,283	1,156	965	827	725	580	484
155	42,336	24,516	17,254	13,311	10,835	9,135	6,954	5,614	4,706	4,052	3,557	3,170	2,859	2,295	1,917	1,646	1,443	1,284	1,156	965	828	725	580	484
160	42,701	24,638	17,314	13,347	10,858	9,152	6,964	5,620	4,711	4,055	3,559	3,172	2,860	2,296	1,918	1,647	1,443	1,284	1,156	965	828	725	580	484
165	43,049	24,753	17,371	13,380	10,881	9,168	6,973	5,626	4,715	4,058	3,562	3,174	2,862	2,297	1,919	1,648	1,443	1,284	1,157	965	828	725	580	484
170	43,382	24,863	17,425	13,412	10,902	9,183	6,982	5,632	4,719	4,061	3,564	3,175	2,863	2,298	1,920	1,648	1,444	1,285	1,157	965	828	725	580	484
175	43,700	24,967	17,476	13,443	10,922	9,197	6,990	5,637	4,723	4,064	3,566	3,177	2,865	2,299	1,920	1,648	1,444	1,285	1,157	965	828	725	581	484
180	44,005	25,067	17,525	13,471	10,941	9,211	6,998	5,642	4,726	4,066	3,568	3,179	2,866	2,300	1,921	1,649	1,444	1,285	1,157	966	828	725	581	484
185	44,298	25,162	17,571	13,499	10,959	9,223	7,005	5,647	4,730	4,069	3,570	3,180	2,867	2,301	1,921	1,649	1,445	1,285	1,158	966	828	725	581	484
190	44,579	25,252	17,615	13,525	10,976	9,236	7,012	5,651	4,733	4,071	3,572	3,182	2,868	2,302	1,922	1,650	1,445	1,286	1,158	966	828	725	581	484
195	44,849	25,338	17,657	13,549	10,992	9,247	7,019	5,656	4,736	4,073	3,574	3,183	2,869	2,302	1,922	1,650	1,445	1,286	1,158	966	829	725	581	484
200	45,108	25,421	17,697	13,573	11,008	9,258	7,025	5,660	4,739	4,076	3,575	3,184	2,870	2,303	1,923	1,650	1,446	1,286	1,158	966	829	725	581	484

 # 6 AWG | **1 Run(s)** | **208 Volts**

C-Value = 2,425 | **Magnetic Duct** | **Copper**

Calculations:

1. $f = \dfrac{1.732 \times \text{Length} \times \text{Isca}}{\text{Runs} \times \text{C-Value} \times \text{E L-L}}$
2. $M = \dfrac{1}{1 + f}$
3. $\text{Isc} = \text{Isca} \times M$
4. Add Motor Contribution = Motor FLA x 4

* All results are given in symmetrical amperes

Three-Phase Bolted Fault-Current* Table

One-Way Distance in Feet

Available Isc in Thousands (Isca)	5	10	15	20	25	30	40	50	60	70	80	90	100	125	150	175	200	225	250	300	350	400	500	600
3	2,905	2,815	2,731	2,652	2,577	2,506	2,376	2,259	2,152	2,056	1,967	1,886	1,811	1,648	1,512	1,396	1,297	1,211	1,136	1,010	910	827	701	608
5	4,741	4,507	4,295	4,102	3,926	3,765	3,478	3,232	3,019	2,832	2,666	2,519	2,388	2,112	1,893	1,716	1,568	1,445	1,339	1,168	1,035	930	773	661
7	6,502	6,070	5,692	5,359	5,062	4,796	4,341	3,964	3,648	3,378	3,146	2,943	2,765	2,402	2,123	1,902	1,723	1,575	1,450	1,251	1,101	982	809	687
10	9,014	8,205	7,529	6,956	6,464	6,037	5,333	4,776	4,324	3,950	3,636	3,368	3,137	2,678	2,335	2,071	1,860	1,688	1,546	1,322	1,155	1,025	838	708
15	12,886	11,294	10,052	9,056	8,240	7,558	6,486	5,680	5,052	4,549	4,137	3,794	3,503	2,940	2,533	2,225	1,983	1,789	1,630	1,383	1,201	1,062	862	725
20	16,410	13,912	12,075	10,666	9,551	8,648	7,272	6,274	5,517	4,922	4,444	4,050	3,720	3,091	2,644	2,310	2,051	1,844	1,675	1,416	1,226	1,081	874	734
25	19,631	16,161	13,733	11,939	10,560	9,467	7,842	6,694	5,839	5,177	4,651	4,221	3,864	3,190	2,716	2,365	2,094	1,879	1,704	1,436	1,241	1,093	882	739
30	22,587	18,112	15,117	12,972	11,360	10,104	8,275	7,006	6,075	5,362	4,799	4,343	3,966	3,259	2,766	2,403	2,124	1,903	1,723	1,450	1,251	1,101	887	743
35	25,310	19,822	16,290	13,826	12,009	10,615	8,614	7,248	6,256	5,503	4,912	4,435	4,043	3,311	2,803	2,430	2,145	1,920	1,738	1,460	1,259	1,107	891	746
40	27,825	21,332	17,296	14,544	12,548	11,033	8,888	7,441	6,399	5,613	4,999	4,506	4,102	3,350	2,831	2,452	2,162	1,933	1,748	1,468	1,265	1,111	894	748
45	30,156	22,675	18,169	15,156	13,001	11,382	9,113	7,598	6,515	5,702	5,070	4,564	4,149	3,382	2,854	2,469	2,175	1,944	1,757	1,474	1,269	1,114	896	749
50	32,321	23,879	18,933	15,685	13,388	11,677	9,301	7,728	6,611	5,775	5,127	4,610	4,188	3,407	2,872	2,482	2,185	1,952	1,764	1,479	1,273	1,117	898	750
55	34,339	24,962	19,608	16,145	13,722	11,931	9,461	7,839	6,691	5,837	5,176	4,649	4,220	3,429	2,887	2,493	2,194	1,959	1,769	1,483	1,276	1,119	899	751
60	36,224	25,944	20,208	16,550	14,013	12,150	9,599	7,933	6,760	5,889	5,217	4,682	4,247	3,447	2,900	2,503	2,202	1,965	1,774	1,486	1,278	1,121	900	752
65	37,988	26,836	20,746	16,909	14,269	12,343	9,718	8,014	6,819	5,934	5,252	4,711	4,270	3,462	2,911	2,511	2,208	1,970	1,778	1,489	1,280	1,123	901	753
70	39,643	27,652	21,230	17,229	14,497	12,512	9,823	8,086	6,870	5,972	5,282	4,735	4,291	3,475	2,920	2,518	2,213	1,974	1,782	1,491	1,282	1,124	902	754
75	41,199	28,400	21,668	17,516	14,700	12,663	9,916	8,148	6,915	6,007	5,309	4,756	4,308	3,487	2,928	2,524	2,218	1,978	1,785	1,493	1,284	1,126	903	754
80	42,664	29,088	22,066	17,776	14,882	12,798	9,999	8,204	6,956	6,037	5,333	4,775	4,324	3,497	2,935	2,529	2,222	1,981	1,787	1,495	1,285	1,127	904	755
85	44,045	29,724	22,430	18,011	15,046	12,920	10,073	8,254	6,991	6,064	5,354	4,792	4,337	3,506	2,942	2,534	2,226	1,984	1,790	1,497	1,286	1,128	904	755
90	45,351	30,313	22,764	18,226	15,196	13,030	10,139	8,299	7,023	6,088	5,372	4,807	4,350	3,514	2,947	2,538	2,229	1,987	1,792	1,498	1,287	1,128	905	755
95	46,586	30,860	23,071	18,422	15,332	13,130	10,200	8,339	7,052	6,110	5,389	4,821	4,361	3,521	2,952	2,542	2,232	1,989	1,794	1,500	1,288	1,129	905	756
100	47,757	31,369	23,355	18,602	15,457	13,221	10,255	8,376	7,079	6,129	5,405	4,833	4,371	3,528	2,957	2,545	2,234	1,991	1,795	1,501	1,289	1,130	906	756
105	48,869	31,845	23,618	18,769	15,572	13,305	10,305	8,409	7,103	6,147	5,419	4,844	4,380	3,534	2,961	2,548	2,237	1,993	1,797	1,502	1,290	1,130	906	756
110	49,925	32,290	23,861	18,922	15,677	13,382	10,351	8,440	7,124	6,164	5,431	4,854	4,388	3,539	2,965	2,551	2,239	1,995	1,798	1,503	1,291	1,131	907	757
115	50,930	32,707	24,089	19,065	15,775	13,453	10,394	8,468	7,145	6,179	5,443	4,864	4,396	3,544	2,968	2,554	2,241	1,996	1,800	1,504	1,291	1,131	907	757
120	51,887	33,100	24,301	19,197	15,866	13,519	10,433	8,494	7,163	6,193	5,454	4,872	4,403	3,548	2,972	2,556	2,243	1,998	1,801	1,504	1,292	1,132	907	757
125	52,801	33,469	24,499	19,321	15,950	13,580	10,470	8,518	7,180	6,205	5,464	4,880	4,409	3,553	2,975	2,558	2,244	1,999	1,802	1,505	1,292	1,132	908	757
130	53,673	33,817	24,685	19,437	16,029	13,637	10,504	8,541	7,196	6,217	5,473	4,888	4,415	3,557	2,977	2,560	2,246	2,000	1,803	1,506	1,293	1,133	908	757
135	54,506	34,146	24,860	19,545	16,102	13,691	10,535	8,562	7,211	6,228	5,481	4,894	4,421	3,560	2,980	2,562	2,247	2,001	1,804	1,507	1,293	1,133	908	758
140	55,303	34,457	25,025	19,646	16,171	13,740	10,565	8,581	7,225	6,239	5,489	4,901	4,426	3,563	2,982	2,564	2,249	2,002	1,805	1,507	1,294	1,133	908	758
145	56,067	34,752	25,180	19,742	16,236	13,787	10,592	8,599	7,238	6,248	5,497	4,907	4,431	3,567	2,984	2,566	2,250	2,003	1,806	1,508	1,294	1,134	908	758
150	56,799	35,032	25,327	19,832	16,297	13,831	10,618	8,616	7,250	6,257	5,504	4,912	4,436	3,570	2,986	2,567	2,251	2,004	1,806	1,508	1,295	1,134	909	758
155	57,502	35,298	25,465	19,917	16,354	13,872	10,642	8,632	7,261	6,266	5,510	4,917	4,440	3,572	2,988	2,569	2,252	2,005	1,807	1,509	1,295	1,134	909	758
160	58,176	35,551	25,597	19,997	16,408	13,911	10,665	8,647	7,272	6,274	5,516	4,922	4,444	3,575	2,990	2,570	2,253	2,006	1,808	1,509	1,295	1,135	909	758
165	58,824	35,792	25,721	20,073	16,459	13,948	10,687	8,662	7,282	6,281	5,522	4,927	4,448	3,577	2,992	2,571	2,254	2,007	1,808	1,510	1,296	1,135	909	758
170	59,448	36,022	25,840	20,145	16,508	13,983	10,707	8,675	7,291	6,288	5,528	4,931	4,451	3,580	2,993	2,572	2,255	2,007	1,809	1,510	1,296	1,135	909	758
175	60,047	36,241	25,953	20,214	16,553	14,016	10,726	8,688	7,300	6,295	5,533	4,935	4,454	3,582	2,995	2,573	2,256	2,008	1,809	1,510	1,296	1,135	909	758
180	60,625	36,451	26,060	20,279	16,597	14,047	10,745	8,700	7,309	6,301	5,538	4,939	4,458	3,584	2,996	2,574	2,257	2,009	1,810	1,511	1,297	1,135	910	759
185	61,182	36,652	26,162	20,341	16,639	14,076	10,762	8,711	7,317	6,307	5,542	4,943	4,461	3,586	2,998	2,575	2,257	2,009	1,810	1,511	1,297	1,136	910	759
190	61,719	36,844	26,260	20,400	16,678	14,105	10,779	8,722	7,324	6,313	5,547	4,946	4,463	3,588	2,999	2,573	2,258	2,010	1,811	1,511	1,297	1,136	910	759
195	62,238	37,028	26,353	20,456	16,716	14,132	10,794	8,732	7,331	6,318	5,551	4,950	4,466	3,589	3,000	2,577	2,259	2,010	1,811	1,512	1,297	1,136	910	759
200	62,738	37,205	26,443	20,510	16,752	14,157	10,809	8,742	7,338	6,323	5,555	4,953	4,469	3,591	3,001	2,573	2,260	2,011	1,812	1,512	1,297	1,136	910	759

# 4 AWG	1 Run(s)	208 Volts	C-Value = 3,806	Magnetic Duct	Copper

Calculations:

1. $f = \dfrac{1.732 \times \text{Length} \times \text{Isca}}{\text{Runs} \times \text{C-Value} \times \text{E L-L}}$

2. $M = \dfrac{1}{1 + f}$

3. Isc = Isca x M

4. Add Motor Contribution = Motor FLA x 4

* All results are given in symmetrical amperes

Three-Phase Bolted Fault-Current* Table

One-Way Distance in Feet

Available Isc in Thousands (Isca)	5	10	15	20	25	30	40	50	60	70	80	90	100	125	150	175	200	225	250	300	350	400	500	600
3	2,923	2,850	2,781	2,715	2,652	2,592	2,479	2,376	2,282	2,194	2,113	2,038	1,967	1,812	1,679	1,564	1,464	1,376	1,298	1,165	1,058	968	828	723
5	4,790	4,598	4,420	4,256	4,103	3,961	3,704	3,479	3,279	3,101	2,942	2,798	2,667	2,389	2,163	1,976	1,819	1,685	1,569	1,380	1,231	1,111	931	800
7	6,596	6,236	5,914	5,623	5,359	5,119	4,699	4,342	4,035	3,769	3,536	3,330	3,147	2,766	2,468	2,227	2,030	1,864	1,724	1,498	1,324	1,187	983	839
10	9,196	8,511	7,921	7,408	6,957	6,558	5,883	5,334	4,879	4,495	4,168	3,884	3,637	3,138	2,759	2,462	2,223	2,026	1,861	1,600	1,404	1,250	1,026	870
15	13,260	11,882	10,763	9,837	9,058	8,393	7,318	6,488	5,827	5,288	4,840	4,462	4,139	3,505	3,039	2,682	2,401	2,173	1,984	1,691	1,473	1,305	1,062	896
20	17,022	14,816	13,116	11,766	10,669	9,758	8,335	7,274	6,453	5,799	5,265	4,821	4,446	3,722	3,201	2,808	2,501	2,254	2,052	1,740	1,510	1,334	1,081	909
25	20,514	17,393	15,097	13,336	11,943	10,813	9,093	7,845	6,898	6,156	5,557	5,065	4,653	3,866	3,307	2,889	2,565	2,306	2,095	1,771	1,533	1,352	1,093	918
30	23,764	19,675	16,786	14,637	12,976	11,653	9,680	8,278	7,231	6,419	5,771	5,242	4,801	3,968	3,381	2,946	2,610	2,342	2,125	1,792	1,549	1,364	1,101	923
35	26,797	21,708	18,244	15,734	13,830	12,338	10,148	8,618	7,489	6,621	5,934	5,376	4,914	4,045	3,437	2,988	2,642	2,369	2,146	1,807	1,560	1,373	1,107	927
40	29,632	23,533	19,516	16,670	14,549	12,906	10,529	8,891	7,695	6,782	6,063	5,481	5,002	4,104	3,479	3,020	2,668	2,389	2,163	1,819	1,569	1,380	1,112	931
45	32,290	25,179	20,635	17,480	15,162	13,386	10,846	9,117	7,863	6,912	6,166	5,566	5,072	4,151	3,513	3,045	2,688	2,405	2,176	1,828	1,576	1,385	1,115	933
50	34,787	26,671	21,626	18,186	15,690	13,797	11,114	9,305	8,002	7,020	6,252	5,636	5,130	4,190	3,541	3,066	2,704	2,418	2,187	1,836	1,582	1,389	1,118	935
55	37,135	28,031	22,511	18,808	16,151	14,152	11,344	9,465	8,121	7,111	6,324	5,694	5,178	4,222	3,564	3,083	2,717	2,428	2,195	1,842	1,586	1,393	1,120	937
60	39,349	29,274	23,306	19,360	16,556	14,462	11,542	9,603	8,222	7,188	6,385	5,744	5,219	4,249	3,583	3,098	2,728	2,437	2,203	1,847	1,590	1,396	1,122	938
65	41,440	30,415	24,024	19,852	16,915	14,735	11,715	9,723	8,309	7,255	6,438	5,786	5,254	4,273	3,600	3,110	2,738	2,445	2,209	1,851	1,593	1,398	1,124	939
70	43,417	31,467	24,676	20,295	17,236	14,978	11,868	9,828	8,386	7,313	6,484	5,823	5,285	4,293	3,614	3,121	2,746	2,452	2,214	1,855	1,596	1,401	1,125	940
75	45,290	32,439	25,269	20,695	17,523	15,194	12,004	9,921	8,453	7,364	6,524	5,856	5,312	4,310	3,627	3,130	2,753	2,457	2,219	1,858	1,598	1,402	1,126	941
80	47,066	33,341	25,813	21,058	17,783	15,389	12,125	10,003	8,513	7,410	6,560	5,884	5,335	4,326	3,638	3,138	2,760	2,462	2,223	1,861	1,601	1,404	1,127	942
85	48,753	34,178	26,312	21,390	18,018	15,565	12,234	10,077	8,567	7,450	6,591	5,910	5,356	4,340	3,647	3,146	2,765	2,467	2,227	1,864	1,602	1,405	1,128	942
90	50,358	34,959	26,773	21,693	18,233	15,725	12,333	10,144	8,615	7,487	6,620	5,933	5,375	4,352	3,656	3,152	2,770	2,471	2,230	1,866	1,604	1,407	1,129	943
95	51,886	35,689	27,199	21,972	18,430	15,871	12,422	10,205	8,659	7,520	6,646	5,954	5,392	4,363	3,664	3,158	2,775	2,474	2,233	1,868	1,606	1,408	1,130	943
100	53,343	36,372	27,594	22,229	18,610	16,005	12,504	10,260	8,699	7,550	6,669	5,972	5,407	4,373	3,671	3,163	2,779	2,478	2,235	1,870	1,607	1,409	1,130	944
105	54,733	37,013	27,961	22,466	18,777	16,128	12,579	10,310	8,735	7,577	6,690	5,989	5,421	4,382	3,677	3,168	2,782	2,481	2,238	1,872	1,608	1,410	1,131	944
110	56,061	37,616	28,304	22,687	18,931	16,241	12,648	10,356	8,768	7,602	6,710	6,005	5,434	4,391	3,683	3,172	2,786	2,483	2,240	1,873	1,609	1,411	1,132	945
115	57,331	38,184	28,624	22,892	19,073	16,346	12,711	10,399	8,798	7,625	6,727	6,019	5,446	4,398	3,689	3,176	2,789	2,486	2,242	1,874	1,610	1,412	1,132	945
120	58,548	38,719	28,924	23,084	19,206	16,444	12,770	10,438	8,827	7,646	6,744	6,032	5,456	4,405	3,694	3,180	2,792	2,488	2,244	1,876	1,611	1,412	1,132	945
125	59,713	39,226	29,205	23,263	19,330	16,534	12,825	10,475	8,853	7,665	6,759	6,044	5,466	4,412	3,698	3,183	2,794	2,490	2,245	1,877	1,612	1,413	1,133	946
130	60,831	39,705	29,470	23,431	19,445	16,619	12,876	10,509	8,877	7,684	6,773	6,056	5,476	4,418	3,702	3,186	2,797	2,492	2,247	1,878	1,613	1,414	1,133	946
135	61,904	40,159	29,720	23,588	19,554	16,698	12,923	10,540	8,899	7,700	6,786	6,066	5,484	4,423	3,706	3,189	2,799	2,494	2,248	1,879	1,614	1,414	1,134	946
140	62,934	40,590	29,955	23,736	19,655	16,772	12,967	10,570	8,920	7,716	6,799	6,076	5,492	4,428	3,710	3,192	2 801	2,495	2,250	1,880	1,614	1,415	1,134	946
145	63,925	41,000	30,178	23,876	19,751	16,841	13,009	10,597	8,940	7,731	6,810	6,085	5,500	4,433	3,713	3,195	2 803	2,497	2,251	1,881	1,615	1,415	1,134	947
150	64,878	41,390	30,389	24,007	19,841	16,907	13,048	10,623	8,958	7,745	6,821	6,094	5,507	4,438	3,717	3,197	2 805	2,498	2,252	1,882	1,616	1,416	1,135	947
155	65,796	41,762	30,589	24,132	19,926	16,969	13,085	10,647	8,976	7,758	6,831	6,102	5,513	4,442	3,719	3,199	2 806	2,500	2,253	1,882	1,616	1,416	1,135	947
160	66,681	42,117	30,778	24,250	20,006	17,027	13,119	10,670	8,992	7,770	6,840	6,109	5,519	4,446	3,722	3,201	2,808	2,501	2,254	1,883	1,617	1,416	1,135	947
165	67,534	42,455	30,959	24,362	20,083	17,082	13,152	10,692	9,007	7,781	6,849	6,116	5,525	4,450	3,725	3,203	2,810	2,502	2,255	1,884	1,617	1,417	1,135	947
170	68,357	42,779	31,131	24,468	20,155	17,134	13,183	10,712	9,022	7,792	6,857	6,123	5,530	4,453	3,727	3,205	2,811	2,503	2,256	1,884	1,618	1,417	1,136	947
175	69,151	43,089	31,294	24,569	20,223	17,184	13,212	10,732	9,035	7,802	6,865	6,129	5,536	4,457	3,730	3,207	2,812	2,504	2,257	1,885	1,618	1,418	1,136	948
180	69,919	43,386	31,451	24,665	20,288	17,231	13,240	10,750	9,048	7,812	6,873	6,135	5,540	4,460	3,732	3,208	2,814	2,505	2,258	1,886	1,619	1,418	1,136	948
185	70,661	43,670	31,600	24,757	20,350	17,275	13,266	10,767	9,061	7,821	6,880	6,141	5,545	4,463	3,734	3,210	2,815	2,506	2,259	1,886	1,619	1,418	1,136	948
190	71,378	43,943	31,743	24,845	20,409	17,318	13,291	10,784	9,072	7,830	6,887	6,146	5,549	4,466	3,736	3,211	2,816	2,507	2,259	1,887	1,619	1,418	1,136	948
195	72,072	44,205	31,879	24,928	20,466	17,358	13,315	10,800	9,084	7,838	6,893	6,151	5,554	4,468	3,738	3,213	2,317	2,508	2,260	1,887	1,620	1,419	1,137	948
200	72,744	44,457	32,010	25,008	20,520	17,397	13,338	10,815	9,094	7,846	6,899	6,156	5,558	4,471	3,740	3,214	2,318	2,509	2,261	1,887	1,620	1,419	1,137	948

# 3 AWG	1 Run(s)	208 Volts	C-Value = 4,760	Magnetic Duct	Copper

Calculations:

1. $f = \dfrac{1.732 \times \text{Length} \times \text{Isca}}{\text{Runs} \times \text{C-Value} \times E_{L\text{-}L}}$
2. $M = \dfrac{1}{1 + f}$
3. $\text{Isc} = \text{Isca} \times M$
4. Add Motor Contribution = Motor FLA x 4

* All results are given in symmetrical amperes

Three-Phase Bolted Fault-Current* Table

One-Way Distance in Feet

Available Isc in Thousands (Isca)

Isca	5	10	15	20	25	30	40	50	60	70	80	90	100	125	150	175	200	225	250	300	350	400	500	600
3	2,938	2,878	2,821	2,766	2,713	2,662	2,566	2,476	2,393	2,315	2,242	2,173	2,108	1,962	1,835	1,724	1,625	1,537	1,458	1,322	1,209	1,114	963	848
5	4,830	4,671	4,522	4,382	4,251	4,127	3,900	3,697	3,514	3,348	3,197	3,059	2,933	2,658	2,430	2,238	2,075	1,933	1,810	1,605	1,442	1,309	1,105	956
7	6,671	6,371	6,097	5,846	5,615	5,401	5,019	4,687	4,397	4,140	3,912	3,707	3,523	3,134	2,822	2,567	2,354	2,174	2,019	1,757	1,572	1,415	1,180	1,011
10	9,341	8,764	8,254	7,800	7,394	7,028	6,394	5,865	5,417	5,033	4,699	4,407	4,150	3,620	3,210	2,884	2,618	2,397	2,210	1,912	1,685	1,506	1,242	1,057
15	13,566	12,381	11,388	10,541	9,812	9,177	8,126	7,291	6,611	6,047	5,572	5,166	4,816	4,117	3,595	3,191	2,868	2,605	2,386	2,042	1,785	1,586	1,296	1,096
20	17,529	15,601	14,055	12,788	11,731	10,835	9,399	8,299	7,430	6,725	6,143	5,653	5,236	4,420	3,824	3,370	3,012	2,723	2,485	2,114	1,840	1,629	1,325	1,116
25	21,254	18,485	16,354	14,663	13,289	12,151	10,374	9,050	8,026	7,210	6,545	5,992	5,525	4,625	3,976	3,488	3,106	2,799	2,548	2,160	1,875	1,656	1,342	1,129
30	24,763	21,083	18,355	16,252	14,581	13,222	11,145	9,631	8,480	7,574	6,843	6,241	5,736	4,772	4,085	3,571	3,171	2,853	2,592	2,192	1,898	1,674	1,354	1,137
35	28,073	23,435	20,113	17,615	15,669	14,111	11,769	10,094	8,837	7,858	7,074	6,432	5,898	4,883	4,166	3,632	3,220	2,892	2,624	2,215	1,916	1,688	1,363	1,143
40	31,202	25,576	21,669	18,798	16,598	14,859	12,286	10,472	9,125	8,084	7,257	6,584	6,024	4,969	4,229	3,680	3,258	2,922	2,649	2,232	1,929	1,698	1,370	1,148
45	34,163	27,532	23,057	19,833	17,400	15,499	12,720	10,785	9,362	8,270	7,407	6,706	6,127	5,039	4,279	3,718	3,287	2,946	2,669	2,246	1,939	1,706	1,375	1,152
50	36,969	29,326	24,302	20,748	18,100	16,052	13,090	11,050	9,561	8,425	7,531	6,808	6,212	5,096	4,320	3,749	3,311	2,965	2,685	2,257	1,948	1,712	1,379	1,155
55	39,633	30,978	25,426	21,561	18,716	16,535	13,409	11,277	9,730	8,556	7,635	6,893	6,282	5,143	4,354	3,775	3,332	2,981	2,698	2,267	1,954	1,718	1,383	1,157
60	42,165	32,504	26,444	22,289	19,262	16,960	13,687	11,473	9,875	8,668	7,724	6,966	6,343	5,184	4,383	3,796	3,348	2,995	2,709	2,275	1,960	1,722	1,386	1,159
65	44,575	33,917	27,372	22,945	19,750	17,336	13,931	11,644	10,002	8,766	7,802	7,029	6,395	5,219	4,408	3,815	3,363	3,006	2,718	2,281	1,965	1,726	1,388	1,161
70	46,871	35,230	28,221	23,538	20,188	17,673	14,148	11,795	10,113	8,851	7,869	7,083	6,440	5,249	4,429	3,831	3,375	3,016	2,727	2,287	1,969	1,729	1,390	1,162
75	49,061	36,453	29,001	24,078	20,584	17,976	14,341	11,929	10,212	8,926	7,929	7,131	6,480	5,275	4,448	3,845	3,386	3,025	2,734	2,292	1,973	1,732	1,392	1,164
80	51,152	37,595	29,719	24,571	20,943	18,249	14,515	12,049	10,299	8,993	7,981	7,174	6,515	5,298	4,465	3,858	3,396	3,033	2,740	2,296	1,976	1,735	1,394	1,165
85	53,151	38,664	30,383	25,023	21,271	18,497	14,671	12,157	10,378	9,053	8,028	7,212	6,546	5,319	4,479	3,868	3,404	3,040	2,745	2,300	1,979	1,737	1,395	1,166
90	55,064	39,666	30,998	25,439	21,571	18,724	14,813	12,254	10,449	9,107	8,071	7,246	6,575	5,338	4,492	3,878	3,412	3,046	2,750	2,304	1,982	1,739	1,397	1,167
95	56,896	40,608	31,571	25,823	21,846	18,931	14,943	12,342	10,513	9,156	8,109	7,277	6,600	5,354	4,504	3,887	3,419	3,051	2,755	2,307	1,984	1,741	1,398	1,168
100	58,653	41,495	32,104	26,179	22,101	19,121	15,061	12,423	10,571	9,200	8,144	7,305	6,623	5,369	4,515	3,895	3,425	3,056	2,759	2,310	1,986	1,742	1,399	1,168
105	60,338	42,332	32,603	26,510	22,336	19,297	15,170	12,497	10,625	9,241	8,176	7,331	6,644	5,383	4,525	3,902	3,430	3,060	2,762	2,312	1,988	1,744	1,400	1,169
110	61,956	43,122	33,069	26,817	22,554	19,460	15,270	12,565	10,674	9,278	8,205	7,354	6,663	5,396	4,534	3,909	3,436	3,064	2,766	2,314	1,990	1,745	1,400	1,170
115	63,511	43,870	33,507	27,105	22,757	19,611	15,363	12,628	10,719	9,312	8,231	7,375	6,681	5,407	4,542	3,915	3,440	3,068	2,769	2,317	1,991	1,746	1,401	1,170
120	65,007	44,578	33,919	27,374	22,946	19,751	15,449	12,686	10,761	9,343	8,256	7,395	6,697	5,418	4,549	3,921	3,445	3,072	2,772	2,319	1,993	1,747	1,402	1,171
125	66,447	45,251	34,307	27,626	23,123	19,882	15,529	12,740	10,800	9,373	8,279	7,413	6,712	5,428	4,556	3,926	3,448	3,075	2,774	2,320	1,994	1,748	1,403	1,171
130	67,834	45,890	34,673	27,863	23,288	20,004	15,603	12,790	10,836	9,400	8,300	7,430	6,726	5,437	4,562	3,930	3,452	3,078	2,776	2,322	1,995	1,749	1,403	1,171
135	69,171	46,498	35,019	28,085	23,444	20,119	15,673	12,836	10,869	9,425	8,319	7,446	6,739	5,445	4,568	3,935	3,456	3,080	2,779	2,324	1,997	1,750	1,404	1,172
140	70,460	47,077	35,346	28,296	23,590	20,226	15,738	12,880	10,901	9,449	8,338	7,461	6,751	5,453	4,574	3,939	3,459	3,083	2,781	2,325	1,998	1,751	1,404	1,172
145	71,705	47,629	35,657	28,494	23,728	20,328	15,800	12,921	10,930	9,471	8,355	7,474	6,762	5,460	4,579	3,943	3,462	3,085	2,783	2,326	1,999	1,752	1,405	1,173
150	72,906	48,156	35,951	28,682	23,858	20,423	15,857	12,960	10,958	9,491	8,371	7,487	6,772	5,467	4,584	3,946	3,464	3,087	2,784	2,328	1,999	1,752	1,405	1,173
155	74,068	48,660	36,232	28,860	23,981	20,513	15,911	12,996	10,983	9,511	8,386	7,499	6,782	5,474	4,588	3,950	3,467	3,089	2,786	2,329	2,000	1,753	1,406	1,173
160	75,191	49,142	36,498	29,029	24,098	20,598	15,963	13,030	11,008	9,529	8,400	7,511	6,792	5,480	4,593	3,953	3,469	3,091	2,788	2,330	2,001	1,754	1,406	1,173
165	76,277	49,604	36,752	29,190	24,208	20,679	16,011	13,062	11,031	9,546	8,414	7,521	6,800	5,485	4,597	3,956	3,472	3,093	2,789	2,331	2,002	1,754	1,406	1,174
170	77,328	50,046	36,995	29,342	24,313	20,756	16,057	13,093	11,053	9,562	8,426	7,532	6,809	5,491	4,600	3,959	3,474	3,095	2,790	2,332	2,003	1,755	1,407	1,174
175	78,346	50,471	37,226	29,488	24,413	20,828	16,100	13,122	11,073	9,578	8,438	7,541	6,816	5,496	4,604	3,961	3,476	3,097	2,792	2,333	2,003	1,755	1,407	1,174
180	79,333	50,879	37,447	29,626	24,508	20,897	16,142	13,149	11,093	9,592	8,450	7,550	6,824	5,501	4,607	3,964	3,478	3,098	2,793	2,334	2,004	1,756	1,407	1,174
185	80,289	51,270	37,659	29,759	24,598	20,963	16,181	13,175	11,111	9,606	8,460	7,559	6,831	5,505	4,611	3,966	3,480	3,099	2,794	2,334	2,005	1,756	1,408	1,175
190	81,217	51,647	37,862	29,885	24,685	21,026	16,218	13,200	11,129	9,619	8,471	7,567	6,837	5,510	4,614	3,968	3,481	3,101	2,795	2,335	2,005	1,757	1,408	1,175
195	82,117	52,009	38,056	30,006	24,767	21,086	16,254	13,223	11,145	9,632	8,480	7,575	6,844	5,514	4,616	3,970	3,483	3,102	2,796	2,336	2,006	1,757	1,408	1,175
200	82,991	52,359	38,243	30,122	24,846	21,143	16,288	13,246	11,161	9,644	8,489	7,582	6,850	5,518	4,619	3,972	3,485	3,103	2,797	2,337	2,006	1,758	1,409	1,175

# 2 AWG	1 Run(s)	208 Volts	C-Value = 5,906	Magnetic Duct	Copper

Calculations:

1. $f = \dfrac{1.732 \times \text{Length} \times \text{Isca}}{\text{Runs} \times \text{C-Value} \times \text{E L-L}}$

2. $M = \dfrac{1}{1 + f}$

3. Isc = Isca x M

4. Add Motor Contribution = Motor FLA x 4

* All results are given in symmetrical amperes

Three-Phase Bolted Fault-Current* Table

One-Way Distance in Feet

Available Isc in Thousands (Isca)

Isca	5	10	15	20	25	30	40	50	60	70	80	90	100	125	150	175	200	225	250	300	350	400	500	600
3	2,949	2,901	2,853	2,808	2,763	2,720	2,638	2,561	2,488	2,420	2,355	2,293	2,235	2,101	1,982	1,876	1,780	1,694	1,616	1,479	1,364	1,266	1,106	982
5	4,861	4,730	4,606	4,488	4,375	4,269	4,070	3,890	3,724	3,572	3,432	3,303	3,183	2,918	2,693	2,501	2,334	2,189	2,060	1,843	1,668	1,523	1,297	1,130
7	6,731	6,482	6,251	6,035	5,834	5,646	5,304	5,001	4,731	4,488	4,270	4,071	3,890	3,501	3,183	2,918	2,694	2,501	2,335	2,060	1,843	1,668	1,401	1,208
10	9,460	8,975	8,538	8,141	7,779	7,448	6,865	6,366	5,934	5,558	5,226	4,932	4,669	4,120	3,686	3,335	3,045	2,802	2,594	2,259	2,001	1,796	1,490	1,274
15	13,817	12,806	11,934	11,173	10,503	9,908	8,901	8,080	7,397	6,821	6,328	5,902	5,529	4,775	4,202	3,752	3,389	3,090	2,840	2,444	2,144	1,910	1,568	1,330
20	17,950	16,282	14,897	13,729	12,731	11,868	10,452	9,337	8,438	7,696	7,074	6,546	6,090	5,188	4,519	4,003	3,592	3,258	2,981	2,547	2,224	1,973	1,610	1,360
25	21,877	19,448	17,504	15,914	14,588	13,467	11,672	10,299	9,215	8,338	7,613	7,004	6,485	5,472	4,733	4,169	3,726	3,368	3,072	2,614	2,274	2,013	1,637	1,379
30	25,613	22,345	19,817	17,803	16,160	14,795	12,657	11,058	9,818	8,829	8,020	7,347	6,778	5,679	4,887	4,289	3,821	3,445	3,137	2,660	2,309	2,040	1,655	1,392
35	29,171	25,006	21,882	19,451	17,507	15,916	13,468	11,673	10,300	9,216	8,339	7,614	7,005	5,837	5,003	4,378	3,892	3,503	3,184	2,694	2,335	2,060	1,668	1,401
40	32,563	27,458	23,737	20,904	18,675	16,875	14,149	12,181	10,693	9,530	8,594	7,826	7,184	5,962	5,095	4,448	3,947	3,547	3,221	2,721	2,355	2,076	1,678	1,408
45	35,801	29,725	25,412	22,192	19,697	17,705	14,728	12,607	11,021	9,789	8,805	8,000	7,331	6,062	5,168	4,503	3,990	3,582	3,250	2,741	2,370	2,088	1,686	1,414
50	38,896	31,828	26,933	23,343	20,598	18,431	15,226	12,971	11,297	10,007	8,980	8,145	7,452	6,145	5,228	4,549	4,026	3,611	3,274	2,758	2,383	2,097	1,692	1,418
55	41,856	33,783	28,320	24,378	21,400	19,070	15,660	13,284	11,534	10,192	9,129	8,268	7,554	6,214	5,278	4,587	4,056	3,635	3,293	2,772	2,393	2,105	1,697	1,422
60	44,690	35,605	29,590	25,313	22,117	19,637	16,040	13,557	11,740	10,352	9,257	8,372	7,642	6,273	5,320	4,619	4 081	3,655	3,310	2,784	2,402	2,112	1,702	1,425
65	47,406	37,308	30,756	26,162	22,762	20,144	16,377	13,797	11,919	10,491	9,369	8,463	7,717	6,324	5,357	4,646	4,102	3,672	3,324	2,794	2,409	2,118	1,705	1,427
70	50,012	38,903	31,832	26,937	23,346	20,600	16,677	14,009	12,077	10,613	9,466	8,543	7,783	6,368	5,389	4,670	4,121	3,687	3,336	2,802	2,416	2,123	1,709	1,430
75	52,513	40,400	32,828	27,646	23,877	21,012	16,946	14,199	12,218	10,722	9,552	8,613	7,842	6,407	5,416	4,691	4,137	3,700	3,347	2,810	2,421	2,127	1,711	1,432
80	54,916	41,807	33,751	28,298	24,362	21,387	17,189	14,369	12,343	10,818	9,629	8,675	7,893	6,442	5,441	4,709	4,151	3,711	3,356	2,816	2,426	2,131	1,714	1,433
85	57,227	43,133	34,610	28,899	24,806	21,729	17,409	14,522	12,456	10,905	9,698	8,731	7,939	6,472	5,463	4,726	4,164	3,722	3,364	2,822	2,430	2,134	1,716	1,435
90	59,450	44,385	35,411	29,455	25,215	22,042	17,609	14,661	12,559	10,983	9,759	8,781	7,981	6,500	5,482	4,741	4,175	3,731	3,372	2,827	2,434	2,137	1,718	1,436
95	61,592	45,567	36,160	29,972	25,592	22,329	17,793	14,788	12,652	11,054	9,815	8,826	8,018	6,525	5,500	4,754	4,186	3,739	3,378	2,832	2,438	2,140	1,720	1,437
100	63,655	46,687	36,861	30,452	25,942	22,595	17,961	14,904	12,736	11,119	9,866	8,867	8,052	6,547	5,516	4,766	4,195	3,746	3,384	2,836	2,441	2,142	1,721	1,439
105	65,645	47,748	37,520	30,900	26,266	22,841	18,116	15,010	12,814	11,178	9,913	8,905	8,083	6,568	5,531	4,776	4,203	3,753	3,390	2,840	2,444	2,145	1,723	1,440
110	67,565	48,756	38,139	31,319	26,568	23,069	18,259	15,109	12,886	11,233	9,956	8,939	8,111	6,586	5,544	4,786	4,211	3,759	3,395	2,844	2,446	2,147	1,724	1,440
115	69,419	49,714	38,723	31,712	26,850	23,281	18,392	15,199	12,951	11,283	9,995	8,971	8,137	6,603	5,556	4,795	4,218	3,765	3,399	2,847	2,449	2,148	1,725	1,441
120	71,210	50,626	39,274	32,080	27,114	23,479	18,515	15,284	13,013	11,329	10,031	9,000	8,162	6,619	5,567	4,804	4,224	3,770	3,404	2,850	2,451	2,150	1,726	1,442
125	72,941	51,495	39,795	32,427	27,361	23,664	18,630	15,362	13,069	11,372	10,065	9,027	8,184	6,634	5,578	4,811	4,230	3,775	3,407	2,852	2,453	2,152	1,727	1,443
130	74,616	52,324	40,288	32,754	27,593	23,838	18,737	15,435	13,122	11,412	10,096	9,053	8,204	6,647	5,587	4,819	4,236	3,779	3,411	2,855	2,455	2,153	1,728	1,443
135	76,237	53,116	40,756	33,062	27,812	24,001	18,838	15,503	13,171	11,449	10,125	9,076	8,224	6,660	5,596	4,825	4,241	3,783	3,414	2,857	2,457	2,154	1,729	1,444
140	77,806	53,873	41,200	33,354	28,018	24,154	18,932	15,567	13,217	11,484	10,153	9,098	8,242	6,672	5,604	4,831	4,246	3,787	3,417	2,859	2,458	2,156	1,730	1,444
145	79,326	54,598	41,623	33,630	28,213	24,299	19,021	15,627	13,260	11,517	10,178	9,118	8,258	6,683	5,612	4,837	4,250	3,790	3,420	2,861	2,460	2,157	1,731	1,445
150	80,800	55,292	42,025	33,892	28,397	24,435	19,104	15,683	13,301	11,547	10,202	9,137	8,274	6,693	5,619	4,843	4,254	3,794	3,423	2,863	2,461	2,158	1,731	1,445
155	82,228	55,957	42,408	34,141	28,572	24,564	19,183	15,736	13,339	11,576	10,224	9,155	8,289	6,703	5,626	4,848	4,258	3,797	3,425	2,865	2,462	2,159	1,732	1,446
160	83,615	56,595	42,774	34,378	28,737	24,687	19,258	15,786	13,375	11,603	10,245	9,172	8,303	6,712	5,633	4,852	4,262	3,800	3,428	2,867	2,464	2,160	1,732	1,446
165	84,960	57,209	43,123	34,603	28,894	24,803	19,328	15,834	13,409	11,629	10,265	9,188	8,316	6,720	5,639	4,857	4,265	3,802	3,430	2,868	2,465	2,161	1,733	1,447
170	86,267	57,798	43,457	34,818	29,044	24,913	19,395	15,878	13,441	11,653	10,284	9,203	8,328	6,728	5,644	4,861	4,269	3,805	3,432	2,870	2,466	2,161	1,734	1,447
175	87,536	58,365	43,777	35,023	29,186	25,017	19,459	15,921	13,472	11,676	10,302	9,218	8,340	6,736	5,650	4,865	4,272	3,807	3,434	2,871	2,467	2,162	1,734	1,447
180	88,769	58,911	44,083	35,219	29,322	25,117	19,519	15,961	13,501	11,697	10,319	9,231	8,351	6,743	5,655	4,869	4,275	3,810	3,436	2,872	2,468	2,163	1,735	1,448
185	89,968	59,437	44,377	35,406	29,452	25,212	19,576	16,000	13,528	11,718	10,335	9,244	8,361	6,750	5,659	4,872	4,277	3,812	3,438	2,874	2,469	2,164	1,735	1,448
190	91,135	59,943	44,659	35,585	29,576	25,303	19,631	16,036	13,554	11,737	10,350	9,256	8,371	6,757	5,664	4,876	4,280	3,814	3,439	2,875	2,470	2,164	1,735	1,448
195	92,269	60,432	44,929	35,757	29,694	25,390	19,683	16,071	13,579	11,756	10,365	9,268	8,381	6,763	5,668	4,879	4,282	3,816	3,441	2,876	2,470	2,165	1,736	1,449
200	93,374	60,904	45,190	35,921	29,808	25,473	19,733	16,104	13,603	11,774	10,378	9,279	8,390	6,769	5,673	4,882	4,285	3,818	3,443	2,877	2,471	2,166	1,736	1,449

1 AWG | 1 Run(s) | 208 Volts | C-Value = 7,292 | Magnetic Duct | Copper

Calculations:

1. $f = \dfrac{1.732 \times \text{Length} \times \text{Isca}}{\text{Runs} \times \text{C-Value} \times E_{L-L}}$

2. $M = \dfrac{1}{1 + f}$

3. Isc = Isca x M

4. Add Motor Contribution = Motor FLA x 4

* All results are given in symmetrical amperes

Three-Phase Bolted Fault-Current* Table

One-Way Distance in Feet

Available Isc in Thousands (Isca)

Isca	5	10	15	20	25	30	40	50	60	70	80	90	100	125	150	175	200	225	250	300	350	400	500	600
3	2,959	2,918	2,879	2,841	2,804	2,768	2,398	2,632	2,569	2,508	2,451	2,396	2,344	2,222	2,113	2,014	1,923	1,841	1,765	1,631	1,515	1,415	1,250	1,120
5	4,886	4,777	4,673	4,573	4,478	4,386	4,214	4,054	3,906	3,769	3,641	3,521	3,409	3,158	2,941	2,753	2,587	2,439	2,308	2,0[illegible]4	1,899	1,744	1,500	1,316
7	6,779	6,571	6,375	6,191	6,017	5,853	5,550	5,277	5,029	4,804	4,598	4,408	4,234	3,854	3,536	3,266	3,035	2,834	2,659	2,3[illegible]5	2,130	1,938	1,641	1,423
10	9,554	9,147	8,772	8,427	8,109	7,813	7,282	6,819	6,411	6,049	5,726	5,435	5,173	4,616	4,167	3,798	3,489	3,226	3,001	2,632	2,344	2,113	1,765	1,515
15	14,019	13,158	12,397	11,719	11,112	10,564	9,616	8,824	8,153	7,577	7,076	6,638	6,251	5,455	4,840	4,349	3,948	3,615	3,334	2,845	2,543	2,273	1,875	1,596
20	18,293	16,855	15,626	14,564	13,637	12,822	11,452	10,346	9,435	8,672	8,023	7,464	6,978	6,001	5,264	4,688	4,226	3,847	3,530	3,081	2,655	2,363	1,936	1,640
25	22,389	20,271	18,520	17,047	15,791	14,707	12,933	11,540	10,418	9,495	8,722	8,066	7,501	6,384	5,557	4,919	4,413	4,001	3,659	3,126	2,728	2,420	1,974	1,667
30	26,317	23,439	21,128	19,233	17,649	16,306	14,153	12,502	11,196	10,137	9,261	8,524	7,896	6,668	5,770	5,086	4,546	4,111	3,751	3,192	2,778	2,460	2,000	1,686
35	30,087	26,384	23,492	21,172	19,268	17,679	15,176	13,293	11,826	10,651	9,688	8,885	8,205	6,887	5,933	5,212	4,647	4,193	3,819	3,241	2,816	2,489	2,020	1,699
40	33,709	29,128	25,643	22,903	20,692	18,870	16,045	13,956	12,348	11,072	10,035	9,176	8,452	7,060	6,062	5,311	4,725	4,256	3,872	3,279	2,844	2,511	2,034	1,710
45	37,192	31,693	27,610	24,459	21,954	19,914	16,794	14,519	12,786	11,424	10,323	9,416	8,656	7,202	6,166	5,390	4,788	4,307	3,914	3,310	2,867	2,529	2,046	1,718
50	40,543	34,094	29,415	25,865	23,080	20,836	17,445	15,003	13,160	11,721	10,566	9,617	8,825	7,319	6,251	5,456	4,840	4,349	3,948	3,334	2,885	2,543	2,055	1,725
55	43,769	36,347	31,077	27,142	24,091	21,657	18,016	15,423	13,483	11,976	10,772	9,789	8,969	7,417	6,323	5,510	4,883	4,384	3,977	3,354	2,901	2,555	2,063	1,730
60	46,878	38,465	32,613	28,306	25,004	22,392	18,522	15,792	13,764	12,198	10,951	9,936	9,093	7,502	6,384	5,557	4,919	4,413	4,001	3,372	2,913	2,565	2,069	1,735
65	49,875	40,460	34,036	29,372	25,832	23,053	18,972	16,119	14,011	12,391	11,107	10,064	9,200	7,575	6,437	5,597	4,950	4,438	4,022	3,386	2,924	2,573	2,075	1,738
70	52,767	42,343	35,358	30,351	26,587	23,653	19,376	16,409	14,231	12,562	11,244	10,177	9,294	7,638	6,483	5,631	4,977	4,460	4,039	3,399	2,934	2,580	2,080	1,742
75	55,559	44,122	36,590	31,255	27,277	24,198	19,741	16,670	14,426	12,715	11,366	10,276	9,377	7,694	6,523	5,662	5,001	4,479	4,055	3,410	2,942	2,587	2,084	1,745
80	58,257	45,807	37,741	32,091	27,912	24,696	20,071	16,905	14,602	12,851	11,475	10,365	9,451	7,744	6,559	5,689	5,022	4,495	4,069	3,420	2,949	2,592	2,087	1,747
85	60,864	47,403	38,818	32,866	28,496	25,152	20,371	17,118	14,760	12,973	11,572	10,445	9,517	7,788	6,591	5,712	5,041	4,510	4,081	3,428	2,956	2,597	2,091	1,749
90	63,385	48,919	39,829	33,588	29,037	25,573	20,646	17,311	14,904	13,084	11,661	10,516	9,577	7,828	6,619	5,734	5,057	4,524	4,092	3,436	2,961	2,602	2,094	1,751
95	65,825	50,359	40,779	34,260	29,539	25,961	20,899	17,488	15,035	13,185	11,741	10,581	9,631	7,864	6,645	5,753	5,072	4,536	4,102	3,443	2,966	2,606	2,096	1,753
100	68,187	51,731	41,673	34,890	30,005	26,321	21,131	17,651	15,155	13,277	11,814	10,641	9,680	7,897	6,668	5,771	5,086	4,547	4,111	3,449	2,971	2,609	2,098	1,755
105	70,476	53,037	42,517	35,479	30,440	26,655	21,346	17,800	15,265	13,362	11,881	10,695	9,724	7,926	6,690	5,787	5,098	4,556	4,119	3,455	2,975	2,613	2,101	1,756
110	72,694	54,283	43,314	36,032	30,847	26,966	21,545	17,939	15,367	13,440	11,942	10,745	9,766	7,954	6,709	5,801	5,110	4,565	4,126	3,460	2,979	2,616	2,102	1,758
115	74,844	55,474	44,068	36,553	31,228	27,257	21,730	18,067	15,460	13,511	11,999	10,791	9,803	7,979	6,727	5,814	5,120	4,574	4,133	3,465	2,983	2,618	2,104	1,759
120	76,930	56,611	44,783	37,044	31,585	27,528	21,902	18,186	15,548	13,578	12,051	10,833	9,838	8,002	6,743	5,827	5,129	4,581	4,139	3,469	2,986	2,621	2,106	1,760
125	78,955	57,700	45,462	37,507	31,921	27,783	22,063	18,297	15,629	13,639	12,100	10,872	9,871	8,023	6,758	5,838	5,138	4,588	4,145	3,473	2,989	2,623	2,107	1,761
130	80,921	58,743	46,107	37,945	32,238	28,023	22,214	18,400	15,704	13,697	12,145	10,909	9,901	8,043	6,772	5,849	5,146	4,595	4,150	3,477	2,992	2,625	2,109	1,762
135	82,830	59,743	46,721	38,359	32,536	28,248	22,356	18,497	15,775	13,751	12,187	10,943	9,929	8,062	6,786	5,858	5,154	4,601	4,155	3,480	2,994	2,627	2,110	1,763
140	84,686	60,702	47,305	38,753	32,819	28,461	22,489	18,588	15,841	13,801	12,226	10,974	9,955	8,079	6,798	5,867	5,161	4,606	4,159	3,483	2,996	2,629	2,111	1,764
145	86,490	61,624	47,863	39,126	33,086	28,662	22,614	18,674	15,903	13,848	12,263	11,004	9,979	8,095	6,809	5,876	5,168	4,612	4,164	3,486	2,999	2,631	2,112	1,764
150	88,245	62,509	48,396	39,481	33,340	28,852	22,732	18,754	15,961	13,892	12,298	11,032	10,002	8,110	6,820	5,884	5,174	4,617	4,168	3,489	3,001	2,632	2,113	1,765
155	89,952	63,361	48,905	39,819	33,581	29,032	22,844	18,830	16,016	13,934	12,331	11,058	10,024	8,124	6,830	5,891	5,179	4,621	4,171	3,492	3,003	2,634	2,114	1,766
160	91,613	64,181	49,392	40,142	33,810	29,203	22,950	18,902	16,068	13,973	12,361	11,083	10,044	8,138	6,839	5,898	5,185	4,625	4,175	3,494	3,005	2,635	2,115	1,766
165	93,231	64,971	49,858	40,449	34,028	29,366	23,050	18,970	16,117	14,010	12,390	11,106	10,063	8,150	6,848	5,905	5,190	4,629	4,178	3,497	3,006	2,636	2,116	1,767
170	94,806	65,732	50,305	40,743	34,235	29,520	23,145	19,034	16,163	14,045	12,418	11,128	10,081	8,162	6,857	5,911	5,195	4,633	4,181	3,499	3,008	2,638	2,117	1,768
175	96,341	66,466	50,734	41,024	34,433	29,667	23,235	19,095	16,207	14,078	12,444	11,149	10,099	8,173	6,864	5,917	5,199	4,637	4,184	3,501	3,009	2,639	2,117	1,768
180	97,838	67,175	51,146	41,293	34,623	29,808	23,321	19,153	16,249	14,110	12,468	11,169	10,115	8,184	6,872	5,923	5,204	4,640	4,187	3,503	3,011	2,640	2,118	1,769
185	99,296	67,859	51,542	41,550	34,804	29,942	23,403	19,209	16,289	14,140	12,492	11,188	10,130	8,194	6,879	5,928	5,208	4,644	4,190	3,505	3,012	2,641	2,119	1,769
190	100,719	68,521	51,922	41,797	34,977	30,070	23,481	19,261	16,327	14,168	12,514	11,206	10,145	8,203	6,886	5,933	5,212	4,647	4,192	3,506	3,013	2,642	2,119	1,770
195	102,107	69,160	52,289	42,034	35,143	30,192	23,556	19,311	16,363	14,196	12,535	11,223	10,159	8,213	6,892	5,938	5,215	4,650	4,195	3,508	3,015	2,643	2,120	1,770
200	103,461	69,779	52,642	42,262	35,302	30,310	23,627	19,359	16,397	14,221	12,555	11,239	10,172	8,221	6,898	5,942	5,219	4,652	4,197	3,510	3,016	2,644	2,121	1,770

1/0 AWG | 1 Run(s) | 208 Volts

C-Value = 8,924 | Magnetic Duct | Copper

Calculations:

1. $f = \dfrac{1.732 \times \text{Length} \times \text{Isca}}{\text{Runs} \times \text{C-Value} \times E_{L\text{-}L}}$

2. $M = \dfrac{1}{1 + f}$

3. Isc = Isca x M

4. Add Motor Contribution = Motor FLA x 4

* All results are given in symmetrical amperes

Three-Phase Bolted Fault-Current* Table

One-Way Distance in Feet

Available Isc in Thousands (Isca)	5	10	15	20	25	30	40	50	60	70	80	90	100	125	150	175	200	225	250	300	350	400	500	600
3	2,979	2,959	2,938	2,918	2,899	2,879	2,841	2,804	2,768	2,732	2,698	2,664	2,632	2,553	2,479	2,410	2,344	2,282	2,222	2,113	2,014	1,923	1,765	1,631
5	4,942	4,886	4,831	4,777	4,724	4,673	4,573	4,478	4,386	4,298	4,214	4,132	4,054	3,871	3,704	3,551	3,409	3,279	3,158	2,941	2,753	2,587	2,308	2,084
7	6,888	6,779	6,673	6,571	6,472	6,375	6,191	6,017	5,853	5,698	5,550	5,410	5,277	4,971	4,698	4,454	4,234	4,035	3,854	3,536	3,266	3,035	2,659	2,365
10	9,772	9,554	9,346	9,147	8,955	8,772	8,427	8,109	7,813	7,538	7,282	7,043	6,819	6,316	5,883	5,505	5,173	4,879	4,616	4,167	3,798	3,489	3,001	2,632
15	14,493	14,019	13,575	13,158	12,766	12,397	11,719	11,112	10,564	10,068	9,616	9,203	8,824	8,001	7,318	6,743	6,251	5,826	5,455	4,840	4,349	3,948	3,334	2,885
20	19,108	18,293	17,544	16,855	16,217	15,626	14,564	13,637	12,822	12,098	11,452	10,871	10,346	9,232	8,335	7,596	6,978	6,453	6,001	5,264	4,688	4,226	3,530	3,031
25	23,622	22,389	21,277	20,271	19,356	18,520	17,047	15,791	14,707	13,763	12,933	12,197	11,540	10,171	9,092	8,221	7,501	6,898	6,384	5,557	4,919	4,413	3,659	3,126
30	28,038	26,317	24,794	23,439	22,224	21,128	19,233	17,649	16,306	15,153	14,153	13,276	12,502	10,911	9,679	8,697	7,896	7,230	6,668	5,770	5,086	4,546	3,751	3,192
35	32,358	30,087	28,114	26,384	24,854	23,492	21,172	19,268	17,679	16,332	15,176	14,172	13,293	11,509	10,147	9,073	8,205	7,488	6,887	5,933	5,212	4,647	3,819	3,241
40	36,586	33,709	31,252	29,128	27,275	25,643	22,903	20,692	18,870	17,344	16,045	14,928	13,956	12,002	10,528	9,377	8,452	7,694	7,060	6,062	5,311	4,725	3,872	3,279
45	40,725	37,192	34,223	31,693	29,511	27,610	24,459	21,954	19,914	18,221	16,794	15,574	14,519	12,416	10,845	9,628	8,656	7,862	7,202	6,166	5,390	4,788	3,914	3,310
50	44,777	40,543	37,040	34,094	31,582	29,415	25,865	23,080	20,836	18,990	17,445	16,132	15,003	12,768	11,113	9,838	8,825	8,002	7,319	6,251	5,456	4,840	3,948	3,334
55	48,746	43,769	39,714	36,347	33,506	31,077	27,142	24,091	21,657	19,670	18,016	16,619	15,423	13,072	11,343	10,017	8,969	8,120	7,417	6,323	5,510	4,883	3,977	3,354
60	52,633	46,878	42,257	38,465	35,298	32,613	28,306	25,004	22,392	20,274	18,522	17,049	15,792	13,336	11,541	10,172	9,093	8,221	7,502	6,384	5,557	4,919	4,001	3,372
65	56,442	49,875	44,677	40,460	36,971	34,036	29,372	25,832	23,053	20,815	18,972	17,430	16,119	13,568	11,714	10,306	9,200	8,309	7,575	6,437	5,597	4,950	4,022	3,386
70	60,174	52,767	46,984	42,343	38,537	35,358	30,351	26,587	23,653	21,302	19,376	17,770	16,409	13,773	11,867	10,424	9,294	8,385	7,638	6,483	5,631	4,977	4,039	3,399
75	63,832	55,559	49,185	44,122	40,005	36,590	31,255	27,277	24,198	21,743	19,741	18,076	16,670	13,956	12,003	10,529	9,377	8,453	7,694	6,523	5,662	5,001	4,055	3,410
80	67,418	58,257	51,287	45,807	41,384	37,741	32,091	27,912	24,696	22,144	20,071	18,352	16,905	14,121	12,124	10,622	9,451	8,513	7,744	6,559	5,689	5,022	4,069	3,420
85	70,935	60,864	53,297	47,403	42,683	38,818	32,866	28,496	25,152	22,511	20,371	18,603	17,118	14,269	12,233	10,705	9,517	8,566	7,788	6,591	5,712	5,041	4,081	3,428
90	74,383	63,385	55,220	48,919	43,908	39,829	33,588	29,037	25,573	22,847	20,646	18,832	17,311	14,403	12,332	10,781	9,577	8,614	7,828	6,619	5,734	5,057	4,092	3,436
95	77,766	65,825	57,063	50,359	45,065	40,779	34,260	29,539	25,961	23,156	20,899	19,042	17,488	14,525	12,421	10,849	9,631	8,658	7,864	6,645	5,753	5,072	4,102	3,443
100	81,085	68,187	58,830	51,731	46,160	41,673	34,890	30,005	26,321	23,442	21,131	19,235	17,651	14,637	12,503	10,912	9,680	8,698	7,897	6,668	5,771	5,086	4,111	3,449
105	84,342	70,476	60,525	53,037	47,198	42,517	35,479	30,440	26,655	23,707	21,346	19,413	17,800	14,740	12,578	10,969	9,724	8,734	7,926	6,690	5,787	5,098	4,119	3,455
110	87,538	72,694	62,154	54,283	48,182	43,314	36,032	30,847	26,966	23,953	21,545	19,577	17,939	14,835	12,647	11,021	9,766	8,767	7,954	6,709	5,801	5,110	4,126	3,460
115	90,675	74,844	63,719	55,474	49,118	44,068	36,553	31,228	27,257	24,182	21,730	19,730	18,067	14,922	12,710	11,069	9,803	8,798	7,979	6,727	5,814	5,120	4,133	3,465
120	93,755	76,930	65,225	56,611	50,008	44,783	37,044	31,585	27,528	24,395	21,902	19,872	18,186	15,003	12,769	11,114	9,838	8,826	8,002	6,743	5,827	5,129	4,139	3,469
125	96,780	78,955	66,675	57,700	50,855	45,462	37,507	31,921	27,783	24,595	22,063	20,004	18,297	15,079	12,823	11,155	9,871	8,852	8,023	6,758	5,838	5,138	4,145	3,473
130	99,750	80,921	68,071	58,743	51,664	46,107	37,945	32,238	28,023	24,783	22,214	20,128	18,400	15,149	12,874	11,193	9,901	8,876	8,043	6,772	5,849	5,146	4,150	3,477
135	102,668	82,830	69,417	59,743	52,435	46,721	38,359	32,536	28,248	24,959	22,356	20,244	18,497	15,215	12,922	11,229	9,929	8,898	8,062	6,786	5,858	5,154	4,155	3,480
140	105,534	84,686	70,716	60,702	53,173	47,305	38,753	32,819	28,461	25,125	22,489	20,353	18,588	15,276	12,966	11,263	9,955	8,919	8,079	6,798	5,867	5,161	4,159	3,483
145	108,351	86,490	71,970	61,624	53,879	47,863	39,126	33,086	28,662	25,281	22,614	20,456	18,674	15,334	13,008	11,294	9,979	8,939	8,095	6,809	5,876	5,168	4,164	3,486
150	111,119	88,245	73,180	62,509	54,554	48,396	39,481	33,340	28,852	25,429	22,732	20,553	18,754	15,388	13,047	11,323	10,002	8,957	8,110	6,820	5,884	5,174	4,168	3,489
155	113,839	89,952	74,350	63,361	55,202	48,905	39,819	33,581	29,032	25,569	22,844	20,644	18,830	15,439	13,083	11,351	10,024	8,975	8,124	6,830	5,891	5,179	4,171	3,492
160	116,513	91,613	75,482	64,181	55,823	49,392	40,142	33,810	29,203	25,701	22,950	20,730	18,902	15,487	13,118	11,377	10,044	8,991	8,138	6,839	5,898	5,185	4,175	3,494
165	119,142	93,231	76,577	64,971	56,420	49,858	40,449	34,028	29,366	25,827	23,050	20,812	18,970	15,533	13,151	11,402	10,063	9,006	8,150	6,848	5,905	5,190	4,178	3,497
170	121,727	94,806	77,636	65,732	56,993	50,305	40,743	34,235	29,520	25,947	23,145	20,889	19,034	15,576	13,181	11,425	10,081	9,021	8,162	6,857	5,911	5,195	4,181	3,499
175	124,270	96,341	78,663	66,466	57,544	50,734	41,024	34,433	29,667	26,060	23,235	20,963	19,095	15,617	13,211	11,447	10,099	9,034	8,173	6,864	5,917	5,199	4,184	3,501
180	126,770	97,838	79,657	67,175	58,075	51,146	41,293	34,623	29,808	26,169	23,321	21,033	19,153	15,656	13,238	11,468	10,115	9,047	8,184	6,872	5,923	5,204	4,187	3,503
185	129,230	99,296	80,622	67,859	58,586	51,542	41,550	34,804	29,942	26,272	23,403	21,099	19,209	15,693	13,265	11,488	10,130	9,060	8,194	6,879	5,928	5,208	4,190	3,505
190	131,650	100,719	81,557	68,521	59,078	51,922	41,797	34,977	30,070	26,370	23,481	21,163	19,261	15,728	13,290	11,506	10,145	9,071	8,203	6,886	5,933	5,212	4,192	3,506
195	134,031	102,107	82,465	69,160	59,553	52,289	42,034	35,143	30,192	26,465	23,556	21,224	19,311	15,761	13,314	11,524	10,159	9,083	8,213	6,892	5,938	5,215	4,195	3,508
200	136,375	103,461	83,346	69,779	60,011	52,642	42,262	35,302	30,310	26,555	23,627	21,281	19,359	15,793	13,337	11,541	10,172	9,093	8,221	6,898	5,942	5,219	4,197	3,510

# 1/0 AWG	2 Run(s)	208 Volts	C-Value = 8,924	Magnetic Duct	Copper

Calculations:

1. $f = \dfrac{1.732 \times \text{Length} \times \text{Isca}}{\text{Runs} \times \text{C-Value} \times \text{E L-L}}$

2. $M = \dfrac{1}{1 + f}$

3. Isc = Isca x M

4. Add Motor Contribution = Motor FLA x 4

* All results are given in symmetrical amperes

Three-Phase Bolted Fault-Current* Table

One-Way Distance in Feet

Available Isc in Thousands (Isca)	5	10	15	20	25	30	40	50	60	70	80	90	100	125	150	175	200	225	250	300	350	400	500	600
3	2,966	2,932	2,899	2,867	2,835	2,805	2,745	2,688	2,633	2,580	2,530	2,481	2,435	2,325	2,225	2,133	2,048	1,970	1,898	1,76[illegible]	1,655	1,555	1,388	1,253
5	4,905	4,814	4,726	4,641	4,559	4,480	4,330	4,189	4,058	3,934	3,818	3,708	3,605	3,370	3,163	2,981	2,818	2,672	2,541	2,31[illegible]	2,123	1,962	1,703	1,505
7	6,815	6,640	6,474	6,315	6,165	6,021	5,753	5,508	5,282	5,075	4,883	4,705	4,540	4,173	3,861	3,593	3,359	3,154	2,973	2,66[illegible]	2,416	2,210	1,887	1,646
10	9,627	9,281	8,959	8,659	8,378	8,115	7,635	7,209	6,828	6,485	6,175	5,893	5,636	5,082	4,627	4,246	3,924	3,647	3,406	3,01[illegible]	2,696	2,441	2,053	1,771
15	14,177	13,439	12,775	12,173	11,625	11,124	10,242	9,490	8,840	8,274	7,776	7,334	6,940	6,118	5,470	4,947	4,514	4,152	3,843	3,345	2,962	2,657	2,204	1,882
20	18,563	17,318	16,230	15,271	14,418	13,656	12,350	11,272	10,368	9,597	8,933	8,356	7,848	6,813	6,019	5,391	4,882	4,460	4,106	3,548	3,115	2,780	2,288	1,943
25	22,794	20,946	19,375	18,023	16,848	15,816	14,091	12,705	11,567	10,616	9,810	9,117	8,516	7,311	6,405	5,698	5,132	4,668	4,282	3,673	3,216	2,860	2,341	1,982
30	26,878	24,345	22,248	20,484	18,979	17,680	15,351	13,880	12,533	11,425	10,496	9,707	9,029	7,686	6,690	5,923	5,314	4,818	4,407	3,765	3,286	2,915	2,378	2,009
35	30,824	27,538	24,885	22,698	20,865	19,306	16,795	14,863	13,329	12,082	11,048	10,178	9,434	7,978	6,911	6,095	5,452	4,932	4,502	3,834	3,338	2,956	2,406	2,028
40	34,637	30,541	27,312	24,701	22,545	20,735	17,867	15,696	13,995	12,627	11,502	10,562	9,763	8,212	7,085	6,231	5,560	5,020	4,575	3,887	3,379	2,988	2,426	2,043
45	38,324	33,373	29,555	26,520	24,051	22,003	18,800	16,411	14,561	13,086	11,882	10,881	10,036	8,403	7,228	6,341	5,647	5,091	4,634	3,929	3,411	3,013	2,443	2,054
50	41,892	36,046	31,632	28,181	25,409	23,134	19,620	17,032	15,048	13,478	12,204	11,151	10,264	8,563	7,346	6,431	5,719	5,149	4,683	3,964	3,437	3,033	2,456	2,064
55	45,345	38,574	33,562	29,703	26,640	24,149	20,345	17,577	15,471	13,816	12,481	11,381	10,460	8,699	7,445	6,507	5,779	5,198	4,723	3,993	3,458	3,050	2,467	2,072
60	48,691	40,968	35,360	31,103	27,760	25,067	20,992	18,058	15,843	14,112	12,722	11,581	10,628	8,815	7,530	6,572	5,830	5,239	4,757	4,0[illegible]7	3,476	3,064	2,477	2,078
65	51,932	43,240	37,040	32,395	28,785	25,899	21,573	18,485	16,171	14,372	12,933	11,756	10,775	8,915	7,603	6,628	5,874	5,275	4,786	4,038	3,492	3,076	2,484	2,084
70	55,075	45,397	38,611	33,590	29,725	26,658	22,097	18,869	16,464	14,602	13,119	11,909	10,904	9,004	7,667	6,677	5,913	5,305	4,811	4,056	3,505	3,087	2,491	2,088
75	58,124	47,448	40,085	34,700	30,591	27,352	22,572	19,214	16,726	14,808	13,285	12,046	11,018	9,082	7,724	6,719	5,946	5,332	4,833	4,072	3,517	3,096	2,497	2,093
80	61,083	49,401	41,470	35,734	31,391	27,990	23,005	19,527	16,962	14,993	13,434	12,168	11,121	9,151	7,774	6,757	5,976	5,356	4,853	4,085	3,528	3,104	2,502	2,096
85	63,955	51,263	42,775	36,698	32,133	28,578	23,400	19,811	17,177	15,160	13,568	12,278	11,212	9,213	7,819	6,791	6,002	5,377	4,870	4,098	3,537	3,111	2,507	2,099
90	66,745	53,041	44,005	37,600	32,822	29,122	23,764	20,071	17,372	15,312	13,689	12,377	11,295	9,269	7,859	6,821	6,026	5,396	4,886	4,109	3,545	3,117	2,511	2,102
95	69,456	54,738	45,167	38,445	33,465	29,627	24,099	20,309	17,550	15,450	13,800	12,468	11,370	9,319	7,895	6,848	6,047	5,413	4,900	4,119	3,552	3,123	2,515	2,105
100	72,092	56,362	46,267	39,239	34,065	30,096	24,408	20,529	17,713	15,577	13,901	12,550	11,439	9,365	7,928	6,873	6,066	5,429	4,913	4,128	3,559	3,128	2,518	2,107
105	74,655	57,917	47,310	39,986	34,626	30,533	24,695	20,732	17,864	15,694	13,993	12,625	11,501	9,407	7,958	6,896	6,084	5,443	4,924	4,136	3,565	3,133	2,521	2,109
110	77,148	59,406	48,299	40,691	35,153	30,942	24,962	20,919	18,003	15,801	14,079	12,695	11,559	9,445	7,986	6,916	6,100	5,456	4,935	4,143	3,570	3,137	2,524	2,111
115	79,575	60,835	49,239	41,356	35,649	31,326	25,211	21,094	18,132	15,900	14,157	12,759	11,612	9,481	8,011	6,935	6,115	5,467	4,944	4,150	3,576	3,141	2,526	2,113
120	81,937	62,206	50,133	41,985	36,115	31,685	25,443	21,256	18,252	15,992	14,230	12,818	11,661	9,514	8,034	6,953	6,128	5,478	4,953	4,156	3,580	3,144	2,529	2,115
125	84,238	63,523	50,985	42,581	36,555	32,023	25,661	21,408	18,364	16,078	14,298	12,873	11,706	9,544	8,056	6,969	6,141	5,488	4,961	4,162	3,584	3,148	2,531	2,116
130	86,479	64,789	51,798	43,146	36,971	32,342	25,865	21,550	18,468	16,158	14,361	12,924	11,749	9,572	8,076	6,984	6,152	5,498	4,969	4,167	3,588	3,151	2,533	2,118
135	88,664	66,008	52,574	43,683	37,365	32,643	26,057	21,683	18,566	16,233	14,420	12,972	11,788	9,598	8,094	6,998	6,163	5,506	4,976	4,172	3,592	3,154	2,535	2,119
140	90,793	67,181	53,315	44,194	37,738	32,927	26,238	21,808	18,658	16,303	14,476	13,017	11,825	9,623	8,112	7,011	6,173	5,514	4,983	4,177	3,595	3,156	2,536	2,120
145	92,870	68,311	54,025	44,680	38,092	33,196	26,409	21,926	18,744	16,368	14,527	13,059	11,860	9,645	8,128	7,023	6,183	5,522	4,989	4,181	3,599	3,159	2,538	2,121
150	94,896	69,401	54,704	45,144	38,428	33,452	26,570	22,037	18,825	16,430	14,576	13,098	11,892	9,667	8,143	7,034	6,191	5,529	4,994	4,185	3,602	3,161	2,539	2,122
155	96,873	70,452	55,355	45,586	38,748	33,694	26,723	22,142	18,901	16,489	14,622	13,135	11,922	9,687	8,157	7,045	6,200	5,535	5,000	4,189	3,604	3,163	2,541	2,123
160	98,803	71,468	55,980	46,009	39,053	33,925	26,868	22,241	18,974	16,544	14,665	13,170	11,951	9,706	8,171	7,055	6,207	5,542	5,005	4,192	3,607	3,165	2,542	2,124
165	100,687	72,448	56,580	46,414	39,344	34,144	27,005	22,335	19,042	16,596	14,706	13,203	11,978	9,724	8,184	7,065	6,215	5,547	5,010	4,196	3,610	3,167	2,543	2,125
170	102,527	73,396	57,156	46,801	39,622	34,353	27,136	22,424	19,107	16,645	14,745	13,234	12,004	9,741	8,196	7,073	6,222	5,553	5,014	4,199	3,612	3,169	2,545	2,126
175	104,325	74,313	57,711	47,172	39,888	34,553	27,260	22,509	19,169	16,691	14,781	13,263	12,028	9,757	8,207	7,082	6,228	5,558	5,018	4,202	3,614	3,170	2,546	2,126
180	106,081	75,200	58,244	47,528	40,142	34,743	27,378	22,590	19,227	16,736	14,816	13,291	12,051	9,772	8,218	7,090	6,234	5,563	5,022	4,205	3,616	3,172	2,547	2,127
185	107,798	76,058	58,758	47,869	40,386	34,925	27,491	22,667	19,283	16,778	14,849	13,318	12,073	9,786	8,228	7,097	6,240	5,568	5,026	4,207	3,618	3,174	2,548	2,128
190	109,477	76,890	59,253	48,198	40,619	35,100	27,599	22,740	19,336	16,818	14,880	13,343	12,094	9,800	8,237	7,105	6,246	5,572	5,030	4,210	3,620	3,175	2,549	2,129
195	111,119	77,697	59,731	48,513	40,843	35,267	27,703	22,810	19,386	16,856	14,910	13,367	12,114	9,813	8,246	7,111	6,251	5,576	5,033	4,212	3,622	3,176	2,549	2,129
200	112,724	78,478	60,192	48,817	41,058	35,427	27,801	22,877	19,435	16,893	14,939	13,390	12,132	9,825	8,255	7,118	6,256	5,580	5,036	4,215	3,623	3,178	2,550	2,130

# 2/0 AWG	1 Run(s)	208 Volts	C-Value = 10,755	Magnetic Duct	Copper

Calculations:

1. $f = \dfrac{1.732 \times \text{Length} \times \text{Isca}}{\text{Runs} \times \text{C-Value} \times \text{E L-L}}$

2. $M = \dfrac{1}{1 + f}$

3. Isc = Isca x M

4. Add Motor Contribution = Motor FLA x 4

* All results are given in symmetrical amperes

Three-Phase Bolted Fault-Current* Table

One-Way Distance in Feet

Available Isc in Thousands (Isca)	5	10	15	20	25	30	40	50	60	70	80	90	100	125	150	175	200	225	250	300	350	400	500	600
3	2,983	2,966	2,949	2,932	2,915	2,899	2,867	2,835	2,805	2,774	2,745	2,716	2,688	2,620	2,555	2,493	2,435	2,378	2,325	2,225	2,133	2,048	1,898	1,768
5	4,952	4,905	4,859	4,814	4,769	4,726	4,641	4,559	4,480	4,403	4,330	4,258	4,189	4,026	3,875	3,735	3,605	3,483	3,370	3,163	2,981	2,818	2,541	2,313
7	6,906	6,815	6,727	6,640	6,556	6,474	6,315	6,165	6,021	5,884	5,753	5,628	5,508	5,229	4,977	4,748	4,540	4,349	4,173	3,861	3,593	3,359	2,973	2,666
10	9,810	9,627	9,451	9,281	9,118	8,959	8,659	8,378	8,115	7,868	7,635	7,416	7,209	6,739	6,326	5,961	5,636	5,345	5,082	4,627	4,246	3,924	3,406	3,010
15	14,577	14,177	13,798	13,439	13,098	12,775	12,173	11,625	11,124	10,665	10,242	9,852	9,490	8,691	8,017	7,440	6,940	6,503	6,118	5,470	4,947	4,514	3,843	3,345
20	19,255	18,563	17,919	17,318	16,757	16,230	15,271	14,418	13,656	12,970	12,350	11,787	11,272	10,164	9,253	8,493	7,848	7,294	6,813	6,019	5,391	4,882	4,106	3,543
25	23,846	22,794	21,831	20,946	20,130	19,375	18,023	16,848	15,816	14,904	14,091	13,362	12,705	11,314	10,197	9,281	8,516	7,868	7,311	6,405	5,698	5,132	4,282	3,673
30	28,354	26,878	25,549	24,345	23,250	22,248	20,484	18,979	17,680	16,548	15,551	14,668	13,880	12,236	10,941	9,893	9,029	8,303	7,686	6,690	5,923	5,314	4,407	3,765
35	32,779	30,824	29,088	27,538	26,144	24,885	22,698	20,865	19,306	17,963	16,795	15,770	14,863	12,994	11,542	10,382	9,434	8,645	7,978	6,911	6,095	5,452	4,502	3,834
40	37,126	34,637	32,460	30,541	28,837	27,312	24,701	22,545	20,735	19,194	17,867	16,711	15,696	13,626	12,038	10,782	9,763	8,920	8,212	7,085	6,231	5,560	4,575	3,887
45	41,394	38,324	35,677	33,373	31,348	29,555	26,520	24,051	22,003	20,276	18,800	17,525	16,411	14,162	12,455	11,115	10,036	9,147	8,403	7,228	6,341	5,647	4,634	3,929
50	45,588	41,892	38,750	36,046	33,695	31,632	28,181	25,409	23,134	21,232	19,620	18,235	17,032	14,622	12,809	11,397	10,264	9,337	8,563	7,346	6,431	5,719	4,683	3,964
55	49,708	45,345	41,686	38,574	35,894	33,562	29,703	26,640	24,149	22,085	20,345	18,860	17,577	15,021	13,115	11,638	10,460	9,498	8,699	7,445	6,507	5,779	4,723	3,993
60	53,757	48,691	44,497	40,968	37,958	35,360	31,103	27,760	25,067	22,849	20,992	19,415	18,058	15,371	13,381	11,847	10,628	9,637	8,815	7,530	6,572	5,830	4,757	4,017
65	57,736	51,932	47,189	43,240	39,900	37,040	32,395	28,785	25,899	23,539	21,573	19,910	18,485	15,680	13,614	12,029	10,775	9,757	8,915	7,603	6,628	5,874	4,786	4,038
70	61,647	55,075	49,770	45,397	41,730	38,611	33,590	29,725	26,658	24,164	22,097	20,356	18,869	15,955	13,821	12,190	10,904	9,863	9,004	7,667	6,677	5,913	4,811	4,056
75	65,492	58,124	52,246	47,448	43,457	40,085	34,700	30,591	27,352	24,733	22,572	20,758	19,214	16,201	14,005	12,334	11,018	9,957	9,082	7,724	6,719	5,946	4,833	4,072
80	69,273	61,083	54,625	49,401	45,090	41,470	35,734	31,391	27,990	25,254	23,005	21,123	19,527	16,423	14,171	12,462	11,121	10,040	9,151	7,774	6,757	5,976	4,853	4,085
85	72,991	63,955	56,910	51,263	46,636	42,775	36,698	32,133	28,578	25,731	23,400	21,457	19,811	16,624	14,320	12,577	11,212	10,115	9,213	7,819	6,791	6,002	4,870	4,098
90	76,648	66,745	59,109	53,041	48,102	44,005	37,600	32,822	29,122	26,172	23,764	21,762	20,071	16,806	14,455	12,681	11,295	10,182	9,269	7,859	6,821	6,026	4,886	4,109
95	80,245	69,456	61,225	54,738	49,494	45,167	38,445	33,465	29,627	26,578	24,099	22,042	20,309	16,973	14,579	12,776	11,370	10,243	9,319	7,895	6,848	6,047	4,900	4,119
100	83,783	72,092	63,264	56,362	50,818	46,267	39,239	34,065	30,096	26,955	24,408	22,301	20,529	17,126	14,691	12,862	11,439	10,298	9,365	7,928	6,873	6,066	4,913	4,128
105	87,265	74,655	65,229	57,917	52,079	47,310	39,986	34,626	30,533	27,306	24,695	22,541	20,732	17,267	14,795	12,942	11,501	10,349	9,407	7,958	6,896	6,084	4,924	4,136
110	90,691	77,148	67,125	59,406	53,280	48,299	40,691	35,153	30,942	27,633	24,962	22,763	20,919	17,397	14,890	13,015	11,559	10,396	9,445	7,986	6,916	6,100	4,935	4,143
115	94,062	79,575	68,954	60,835	54,426	49,239	41,356	35,649	31,326	27,938	25,211	22,969	21,094	17,518	14,978	13,082	11,612	10,439	9,481	8,011	6,935	6,115	4,944	4,150
120	97,381	81,937	70,721	62,206	55,521	50,133	41,985	36,115	31,685	28,223	25,443	23,162	21,256	17,629	15,060	13,144	11,661	10,478	9,514	8,034	6,953	6,128	4,953	4,156
125	100,648	84,238	72,428	63,523	56,568	50,985	42,581	36,555	32,023	28,491	25,661	23,342	21,408	17,734	15,136	13,202	11,706	10,515	9,544	8,056	6,969	6,141	4,961	4,162
130	103,865	86,479	74,079	64,789	57,570	51,798	43,146	36,971	32,342	28,743	25,865	23,511	21,550	17,831	15,207	13,256	11,749	10,549	9,572	8,076	6,984	6,152	4,969	4,167
135	107,032	88,664	75,676	66,008	58,530	52,574	43,683	37,365	32,643	28,981	26,057	23,670	21,683	17,922	15,273	13,306	11,788	10,581	9,598	8,094	6,998	6,163	4,976	4,172
140	110,151	90,793	77,222	67,181	59,450	53,315	44,194	37,738	32,927	29,205	26,238	23,819	21,808	18,007	15,335	13,353	11,825	10,611	9,623	8,112	7,011	6,173	4,983	4,177
145	113,223	92,870	78,720	68,311	60,334	54,025	44,680	38,092	33,196	29,416	26,409	23,959	21,926	18,088	15,393	13,397	11,860	10,638	9,645	8,128	7,023	6,183	4,989	4,181
150	116,249	94,896	80,170	69,401	61,182	54,704	45,144	38,428	33,452	29,616	26,570	24,092	22,037	18,163	15,448	13,439	11,892	10,665	9,667	8,143	7,034	6,191	4,994	4,185
155	119,229	96,873	81,577	70,452	61,998	55,355	45,586	38,748	33,694	29,806	26,723	24,218	22,142	18,234	15,499	13,478	11,922	10,689	9,687	8,157	7,045	6,200	5,000	4,189
160	122,166	98,803	82,941	71,468	62,783	55,980	46,009	39,053	33,925	29,987	26,868	24,336	22,241	18,302	15,548	13,514	11,951	10,712	9,706	8,171	7,055	6,207	5,005	4,192
165	125,059	100,687	84,265	72,448	63,538	56,580	46,414	39,344	34,144	30,158	27,005	24,449	22,335	18,365	15,594	13,549	11,978	10,734	9,724	8,184	7,065	6,215	5,010	4,196
170	127,911	102,527	85,550	73,396	64,266	57,156	46,801	39,622	34,353	30,321	27,136	24,556	22,424	18,426	15,637	13,582	12,004	10,755	9,741	8,196	7,073	6,222	5,014	4,199
175	130,721	104,325	86,798	74,313	64,968	57,711	47,172	39,888	34,553	30,476	27,260	24,658	22,509	18,483	15,678	13,613	12,028	10,774	9,757	8,207	7,082	6,228	5,018	4,202
180	133,491	106,081	88,010	75,200	65,645	58,244	47,528	40,142	34,743	30,624	27,378	24,755	22,590	18,537	15,717	13,642	12,051	10,792	9,772	8,218	7,090	6,234	5,022	4,205
185	136,221	107,798	89,189	76,058	66,298	58,758	47,869	40,386	34,925	30,766	27,491	24,847	22,667	18,589	15,755	13,670	12,073	10,810	9,786	8,228	7,097	6,240	5,026	4,207
190	138,913	109,477	90,335	76,890	66,929	59,253	48,198	40,619	35,100	30,901	27,599	24,935	22,740	18,638	15,790	13,697	12,094	10,827	9,800	8,237	7,105	6,246	5,030	4,210
195	141,567	111,119	91,450	77,697	67,539	59,731	48,513	40,843	35,267	31,030	27,703	25,019	22,810	18,685	15,824	13,722	12,114	10,842	9,813	8,246	7,111	6,251	5,033	4,212
200	144,184	112,724	92,534	78,478	68,129	60,192	48,817	41,058	35,427	31,154	27,801	25,100	22,877	18,730	15,856	13,747	12,132	10,858	9,825	8,255	7,118	6,256	5,036	4,215

 # 2/0 AWG | **2 Run(s)** | **208 Volts** | **C-Value = 10,755** | **Magnetic Duct** | **Copper**

Calculations:

1. $f = \dfrac{1.732 \times \text{Length} \times \text{Isca}}{\text{Runs} \times \text{C-Value} \times E_{L\text{-}L}}$

2. $M = \dfrac{1}{1 + f}$

3. $\text{Isc} = \text{Isca} \times M$

4. Add Motor Contribution = Motor FLA x 4

* All results are given in symmetrical amperes

Three-Phase Bolted Fault-Current* Table

One-Way Distance in Feet

Available Isc in Thousands (Isca)	5	10	15	20	25	30	40	50	60	70	80	90	100	125	150	175	200	225	250	300	350	400	500	600
3	2,971	2,943	2,915	2,888	2,861	2,835	2,783	2,734	2,686	2,640	2,596	2,553	2,511	2,413	2,322	2,238	2,160	2,087	2,018	1,895	1,785	1,687	1,521	1,384
5	4,920	4,843	4,768	4,696	4,625	4,557	4,426	4,303	4,186	4,075	3,970	3,871	3,776	3,558	3,364	3,190	3,033	2,891	2,762	2,535	2,342	2,177	1,908	1,698
7	6,845	6,696	6,554	6,417	6,287	6,161	5,924	5,705	5,502	5,312	5,135	4,970	4,815	4,466	4,165	3,901	3,669	3,463	3,279	2,964	2,704	2,486	2,141	1,880
10	9,686	9,391	9,114	8,852	8,605	8,372	7,941	7,552	7,199	6,878	6,585	6,315	6,067	5,523	5,070	4,685	4,354	4,067	3,815	3,395	3,059	2,783	2,357	2,045
15	14,304	13,670	13,090	12,557	12,066	11,612	10,739	10,092	9,473	8,924	8,436	7,999	7,604	6,770	6,100	5,552	5,093	4,705	4,371	3,829	3,406	3,067	2,559	2,195
20	18,782	17,704	16,743	15,881	15,104	14,399	13,169	12,133	11,248	10,484	9,817	9,229	8,708	7,631	6,791	6,118	5,566	5,105	4,715	4,090	3,611	3,233	2,673	2,278
25	23,126	21,513	20,110	18,880	17,791	16,821	15,167	13,809	12,674	11,712	10,885	10,167	9,539	8,261	7,286	6,516	5,894	5,380	4,948	4,264	3,746	3,341	2,746	2,331
30	27,341	25,115	23,224	21,598	20,185	18,945	16,873	15,209	13,844	12,703	11,737	10,907	10,186	8,743	7,658	6,812	6,135	5,580	5,117	4,389	3,842	3,417	2,797	2,368
35	31,433	28,527	26,112	24,074	22,331	20,824	18,347	16,396	14,821	13,521	12,432	11,504	10,706	9,123	7,947	7,041	6,319	5,732	5,245	4,483	3,914	3,473	2,835	2,395
40	35,408	31,763	28,797	26,338	24,267	22,497	19,633	17,416	15,649	14,207	13,009	11,997	11,131	9,430	8,180	7,222	6,465	5,852	5,345	4,556	3,969	3,517	2,864	2,415
45	39,271	34,836	31,301	28,418	26,020	23,996	20,766	18,301	16,360	14,791	13,497	12,411	11,487	9,684	8,370	7,370	6,584	5,949	5,426	4,614	4,014	3,552	2,887	2,432
50	43,026	37,759	33,641	30,333	27,617	25,348	21,770	19,077	16,977	15,294	13,914	12,763	11,787	9,897	8,528	7,493	6,681	6,028	5,492	4,662	4,050	3,580	2,905	2,445
55	46,677	40,543	35,833	32,104	29,077	26,573	22,367	19,763	17,518	15,731	14,275	13,066	12,046	10,078	8,663	7,596	6,763	6,095	5,547	4,702	4,080	3,603	2,921	2,456
60	50,230	43,196	37,890	33,745	30,418	27,687	23,474	20,373	17,996	16,116	14,591	13,330	12,269	10,234	8,778	7,685	6,833	6,152	5,594	4,735	4,105	3,623	2,934	2,465
65	53,687	45,728	39,825	35,271	31,652	28,706	24,202	20,919	18,421	16,455	14,869	13,562	12,466	10,370	8,878	7,761	6,894	6,201	5,635	4,764	4,127	3,640	2,945	2,473
70	57,053	48,148	41,647	36,693	32,792	29,641	24,863	21,412	18,801	16,759	15,116	13,767	12,639	10,490	8,965	7,828	6,946	6,243	5,670	4,789	4,146	3,655	2,954	2,480
75	60,331	50,462	43,367	38,022	33,850	30,502	25,466	21,857	19,144	17,030	15,337	13,950	12,793	10,596	9,043	7,887	6,993	6,281	5,700	4,811	4,162	3,667	2,963	2,485
80	63,525	52,677	44,993	39,266	34,832	31,298	26,018	22,263	19,455	17,276	15,535	14,114	12,931	10,690	9,111	7,939	7,034	6,314	5,728	4,831	4,177	3,679	2,970	2,491
85	66,638	54,800	46,533	40,434	35,748	32,035	26,526	22,633	19,737	17,498	15,715	14,262	13,055	10,775	9,173	7,985	7,070	6,343	5,752	4,848	4,190	3,689	2,977	2,495
90	69,672	56,835	47,993	41,531	36,603	32,720	26,994	22,973	19,995	17,700	15,878	14,396	13,167	10,851	9,228	8,027	7,103	6,370	5,774	4,863	4,201	3,697	2,982	2,499
95	72,632	58,789	49,378	42,565	37,404	33,359	27,427	23,286	20,231	17,885	16,027	14,518	13,269	10,920	9,278	8,065	7,133	6,394	5,793	4,877	4,211	3,705	2,988	2,503
100	75,518	60,666	50,696	43,540	38,155	33,955	27,828	23,575	20,449	18,055	16,163	14,630	13,363	10,984	9,324	8,100	7,160	6,415	5,811	4,890	4,221	3,713	2,992	2,506
105	78,335	62,471	51,950	44,462	38,861	34,513	28,202	23,842	20,650	18,212	16,289	14,733	13,448	11,041	9,365	8,131	7,184	6,435	5,827	4,901	4,229	3,719	2,997	2,509
110	81,085	64,207	53,145	45,335	39,526	35,036	28,551	24,091	20,837	18,357	16,404	14,827	13,527	11,094	9,403	8,160	7,207	6,453	5,842	4,912	4,237	3,725	3,001	2,512
115	83,770	65,879	54,286	46,162	40,153	35,528	28,877	24,323	21,010	18,491	16,511	14,915	13,600	11,143	9,438	8,186	7,227	6,469	5,855	4,921	4,244	3,731	3,004	2,514
120	86,392	67,490	55,375	46,947	40,746	35,992	29,182	24,539	21,171	18,615	16,611	14,996	13,667	11,188	9,471	8,210	7,246	6,484	5,868	4,930	4,251	3,736	3,007	2,517
125	88,954	69,043	56,416	47,693	41,307	36,429	29,469	24,741	21,321	18,732	16,703	15,071	13,729	11,230	9,501	8,233	7,264	6,499	5,879	4,938	4,257	3,740	3,010	2,519
130	91,457	70,542	57,413	48,404	41,839	36,842	29,738	24,931	21,462	18,840	16,789	15,141	13,788	11,269	9,529	8,254	7,280	6,512	5,890	4,946	4,262	3,745	3,013	2,521
135	93,904	71,989	58,367	49,081	42,343	37,232	29,992	25,110	21,594	18,942	16,870	15,207	13,842	11,305	9,555	8,273	7,295	6,524	5,900	4,953	4,267	3,749	3,016	2,523
140	96,296	73,386	59,283	49,726	42,823	37,603	30,232	25,277	21,718	19,037	16,946	15,268	13,893	11,339	9,579	8,291	7,309	6,535	5,909	4,959	4,272	3,753	3,018	2,524
145	98,635	74,737	60,161	50,343	43,280	37,954	30,459	25,436	21,835	19,127	17,017	15,326	13,941	11,371	9,601	8,308	7,322	6,545	5,918	4,965	4,277	3,756	3,020	2,526
150	100,924	76,044	61,005	50,932	43,714	38,288	30,674	25,585	21,945	19,212	17,084	15,380	13,985	11,401	9,623	8,324	7,335	6,555	5,926	4,971	4,281	3,759	3,023	2,527
155	103,163	77,308	61,816	51,496	44,129	38,606	30,877	25,727	22,049	19,291	17,147	15,431	14,028	11,429	9,643	8,339	7,346	6,565	5,933	4,976	4,285	3,762	3,025	2,529
160	105,354	78,532	62,596	52,037	44,525	38,909	31,071	25,861	22,148	19,367	17,206	15,479	14,067	11,455	9,661	8,353	7,357	6,573	5,940	4,981	4,289	3,765	3,026	2,530
165	107,499	79,718	63,347	52,555	44,904	39,198	31,255	25,988	22,241	19,438	17,262	15,525	14,105	11,480	9,679	8,367	7,367	6,581	5,947	4,986	4,292	3,768	3,028	2,531
170	109,599	80,867	64,071	53,052	45,266	39,474	31,430	26,109	22,329	19,505	17,316	15,568	14,141	11,504	9,696	8,379	7,377	6,589	5,953	4,990	4,295	3,770	3,030	2,532
175	111,656	81,981	64,768	53,529	45,613	39,737	31,597	26,224	22,413	19,570	17,366	15,609	14,174	11,526	9,712	8,391	7,386	6,596	5,959	4,994	4,298	3,773	3,031	2,533
180	113,670	83,062	65,441	53,987	45,946	39,990	31,756	26,334	22,493	19,631	17,414	15,647	14,206	11,547	9,727	8,402	7,395	6,603	5,965	4,998	4,301	3,775	3,033	2,534
185	115,644	84,111	66,090	54,429	46,265	40,231	31,908	26,439	22,570	19,689	17,460	15,684	14,237	11,567	9,741	8,413	7,403	6,610	5,970	5,002	4,304	3,777	3,034	2,535
190	117,578	85,130	66,717	54,853	46,572	40,463	32,054	26,538	22,642	19,744	17,503	15,719	14,265	11,586	9,754	8,423	7,411	6,616	5,975	5,006	4,307	3,779	3,035	2,536
195	119,474	86,119	67,324	55,262	46,866	40,685	32,193	26,634	22,712	19,797	17,545	15,753	14,293	11,604	9,767	8,432	7,418	6,622	5,980	5,009	4,309	3,781	3,037	2,537
200	121,333	87,080	67,910	55,657	47,150	40,898	32,326	26,725	22,778	19,847	17,584	15,785	14,319	11,622	9,780	8,441	7,425	6,628	5,985	5,012	4,312	3,783	3,038	2,538

# 3/0 AWG	1 Run(s)	208 Volts	C-Value = 12,843	Magnetic Duct	Copper

Calculations:

1. $f = \dfrac{1.732 \times \text{Length} \times \text{Isca}}{\text{Runs} \times \text{C-Value} \times E_{L-L}}$

2. $M = \dfrac{1}{1 + f}$

3. Isc = Isca x M

4. Add Motor Contribution = Motor FLA x 4

* All results are given in symmetrical amperes

Three-Phase Bolted Fault-Current* Table

One-Way Distance in Feet

Available Isc in Thousands (Isca)	5	10	15	20	25	30	40	50	60	70	80	90	100	125	150	175	200	225	250	300	350	400	500	600
3	2,985	2,971	2,957	2,943	2,929	2,915	2,888	2,861	2,835	2,809	2,783	2,759	2,734	2,675	2,618	2,564	2,511	2,461	2,413	2,322	2,238	2,160	2,018	1,895
5	4,960	4,920	4,881	4,843	4,805	4,768	4,696	4,625	4,557	4,490	4,426	4,363	4,303	4,158	4,022	3,895	3,776	3,664	3,558	3,364	3,190	3,033	2,762	2,535
7	6,921	6,845	6,770	6,696	6,624	6,554	6,417	6,287	6,161	6,040	5,924	5,813	5,705	5,453	5,222	5,010	4,815	4,634	4,466	4,165	3,901	3,669	3,279	2,964
10	9,840	9,686	9,536	9,391	9,250	9,114	8,852	8,605	8,372	8,150	7,941	7,741	7,552	7,116	6,728	6,380	6,067	5,782	5,523	5,070	4,685	4,354	3,815	3,395
15	14,644	14,304	13,980	13,670	13,374	13,090	12,557	12,066	11,612	11,191	10,799	10,434	10,092	9,329	8,673	8,104	7,604	7,163	6,770	6,100	5,552	5,093	4,371	3,829
20	19,372	18,782	18,227	17,704	17,210	16,743	15,881	15,104	14,399	13,757	13,169	12,630	12,133	11,047	10,139	9,369	8,708	8,134	7,631	6,791	6,118	5,566	4,715	4,090
25	24,026	23,126	22,290	21,513	20,788	20,110	18,880	17,791	16,821	15,951	15,167	14,456	13,809	12,419	11,283	10,338	9,539	8,854	8,261	7,286	6,516	5,894	4,948	4,264
30	28,609	27,341	26,181	25,115	24,133	23,224	21,598	20,185	18,945	17,849	16,873	15,998	15,209	13,540	12,201	11,103	10,186	9,410	8,743	7,658	6,812	6,135	5,117	4,389
35	33,121	31,433	29,910	28,527	27,266	26,112	24,074	22,331	20,824	19,507	18,347	17,317	16,396	14,473	12,954	11,723	10,706	9,851	9,123	7,947	7,041	6,319	5,245	4,483
40	37,564	35,408	33,487	31,763	30,207	28,797	26,338	24,267	22,497	20,968	19,633	18,458	17,416	15,262	13,582	12,235	11,131	10,210	9,430	8,180	7,222	6,465	5,345	4,556
45	41,941	39,271	36,921	34,836	32,974	31,301	28,418	26,020	23,996	22,264	20,766	19,456	18,301	15,938	14,114	12,666	11,487	10,508	9,684	8,370	7,370	6,584	5,426	4,614
50	46,252	43,026	40,221	37,759	35,581	33,641	30,333	27,617	25,348	23,423	21,770	20,335	19,077	16,523	14,571	13,032	11,787	10,760	9,897	8,528	7,493	6,681	5,492	4,662
55	50,498	46,677	43,394	40,543	38,043	35,833	32,104	29,077	26,573	24,465	22,667	21,116	19,763	17,034	14,968	13,349	12,046	10,974	10,078	8,663	7,596	6,763	5,547	4,702
60	54,682	50,230	46,448	43,196	40,369	37,890	33,745	30,418	27,687	25,407	23,474	21,814	20,373	17,486	15,315	13,624	12,269	11,160	10,234	8,778	7,685	6,833	5,594	4,735
65	58,804	53,687	49,389	45,728	42,573	39,825	35,271	31,652	28,706	26,262	24,202	22,441	20,919	17,887	15,622	13,866	12 466	11,322	10,370	8,878	7,761	6,894	5,635	4,764
70	62,867	57,053	52,224	48,148	44,662	41,647	36,693	32,792	29,641	27,043	24,863	23,009	21,412	18,245	15,895	14,081	12,639	11,464	10,490	8,965	7,828	6,946	5,670	4,789
75	66,871	60,331	54,957	50,462	46,646	43,367	38,022	33,850	30,502	27,758	25,466	23,524	21,857	18,568	16,139	14,272	12,793	11,591	10,596	9,043	7,887	6,993	5,700	4,811
80	70,817	63,525	57,595	52,677	48,533	44,993	39,266	34,832	31,298	28,415	26,018	23,994	22,263	18,860	16,359	14,444	12,931	11,704	10,690	9,111	7,939	7,034	5,728	4,831
85	74,707	66,638	60,142	54,800	50,329	46,533	40,434	35,748	32,035	29,021	26,526	24,425	22,633	19,125	16,559	14,599	13,355	11,806	10,775	9,173	7,985	7,070	5,752	4,848
90	78,542	69,672	62,602	56,835	52,041	47,993	41,531	36,603	32,720	29,582	26,994	24,822	22,973	19,367	16,740	14,740	13,167	11,897	10,851	9,228	8,027	7,103	5,774	4,863
95	82,323	72,632	64,981	58,789	53,674	49,378	42,565	37,404	33,359	30,103	27,427	25,187	23,286	19,589	16,905	14,868	13,269	11,981	10,920	9,278	8,065	7,133	5,793	4,877
100	86,052	75,518	67,282	60,666	55,235	50,696	43,540	38,155	33,955	30,588	27,828	25,526	23,575	19,793	17,057	14,985	13,363	12,057	10,984	9,324	8,100	7,160	5,811	4,890
105	89,729	78,335	69,509	62,471	56,727	51,950	44,462	38,861	34,513	31,040	28,202	25,840	23,842	19,981	17,197	15,093	13,448	12,126	11,041	9,365	8,131	7,184	5,827	4,901
110	93,355	81,085	71,666	64,207	58,155	53,145	45,335	39,526	35,036	31,463	28,551	26,132	24,091	20,156	17,326	15,192	13,527	12,190	11,094	9,403	8,160	7,207	5,842	4,912
115	96,932	83,770	73,755	65,879	59,523	54,286	46,162	40,153	35,528	31,859	28,877	26,405	24,323	20,318	17,445	15,284	13,600	12,249	11,143	9,438	8,186	7,227	5,855	4,921
120	100,460	86,392	75,780	67,490	60,835	55,375	46,947	40,746	35,992	32,231	29,182	26,660	24,539	20,468	17,556	15,369	13,667	12,304	11,188	9,471	8,210	7,246	5,868	4,930
125	103,940	88,954	77,744	69,043	62,094	56,416	47,693	41,307	36,429	32,581	29,469	26,899	24,741	20,609	17,659	15,448	13,729	12,355	11,230	9,501	8,233	7,264	5,879	4,938
130	107,374	91,457	79,649	70,542	63,304	57,413	48,404	41,839	36,842	32,911	29,738	27,123	24,931	20,740	17,756	15,522	13,738	12,402	11,269	9,529	8,254	7,280	5,890	4,946
135	110,763	93,904	81,499	71,989	64,466	58,367	49,081	42,343	37,232	33,222	29,992	27,335	25,110	20,864	17,846	15,591	13,842	12,446	11,305	9,555	8,273	7,295	5,900	4,953
140	114,106	96,296	83,295	73,386	65,585	59,283	49,726	42,823	37,603	33,517	30,232	27,534	25,277	20,980	17,931	15,656	13,893	12,487	11,339	9,579	8,291	7,309	5,909	4,959
145	117,406	98,635	85,039	74,737	66,662	60,161	50,343	43,280	37,954	33,796	30,459	27,722	25,436	21,088	18,010	15,716	13,941	12,525	11,371	9,601	8,308	7,322	5,918	4,965
150	120,663	100,924	86,735	76,044	67,699	61,005	50,932	43,714	38,288	34,061	30,674	27,899	25,585	21,191	18,085	15,773	13,985	12,562	11,401	9,623	8,324	7,335	5,926	4,971
155	123,877	103,163	88,383	77,308	68,699	61,816	51,496	44,129	38,606	34,312	30,877	28,068	25,727	21,288	18,156	15,827	14,028	12,596	11,429	9,643	8,339	7,346	5,933	4,976
160	127,050	105,354	89,987	78,532	69,664	62,596	52,037	44,525	38,909	34,551	31,071	28,228	25,861	21,380	18,223	15,878	14,067	12,628	11,455	9,661	8,353	7,357	5,940	4,981
165	130,183	107,499	91,547	79,718	70,596	63,347	52,555	44,904	39,198	34,779	31,255	28,379	25,988	21,467	18,286	15,926	14,105	12,658	11,480	9,679	8,367	7,367	5,947	4,986
170	133,275	109,599	93,066	80,867	71,496	64,071	53,052	45,266	39,474	34,996	31,430	28,524	26,109	21,549	18,345	15,971	14,141	12,687	11,504	9,696	8,379	7,377	5,953	4,990
175	136,329	111,656	94,545	81,981	72,365	64,768	53,529	45,613	39,737	35,203	31,597	28,661	26,224	21,628	18,402	16,014	14,174	12,714	11,526	9,712	8,391	7,386	5,959	4,994
180	139,344	113,670	95,985	83,062	73,206	65,441	53,987	45,946	39,990	35,400	31,756	28,792	26,334	21,702	18,456	16,055	14,205	12,739	11,547	9,727	8,402	7,395	5,965	4,998
185	142,322	115,644	97,389	84,111	74,020	66,090	54,429	46,265	40,231	35,590	31,908	28,917	26,439	21,773	18,507	16,093	14,237	12,764	11,567	9,741	8,413	7,403	5,970	5,002
190	145,263	117,578	98,757	85,130	74,807	66,717	54,853	46,572	40,463	35,771	32,054	29,036	26,538	21,841	18,556	16,130	14,265	12,787	11,586	9,754	8,423	7,411	5,975	5,006
195	148,168	119,474	100,091	86,119	75,570	67,324	55,262	46,866	40,685	35,944	32,193	29,151	26,634	21,905	18,603	16,166	14,293	12,809	11,604	9,767	8,432	7,418	5,980	5,009
200	151,037	121,333	101,392	87,080	76,309	67,910	55,657	47,150	40,898	36,111	32,326	29,260	26,725	21,967	18,647	16,199	14,319	12,830	11,622	9,780	8,441	7,425	5,985	5,012

3/0 AWG | 2 Run(s) | 208 Volts | C-Value = 12,843 | Magnetic Duct | Copper

Calculations:

1. $f = \dfrac{1.732 \times \text{Length} \times \text{Isca}}{\text{Runs} \times \text{C-Value} \times E_{L\text{-}L}}$

2. $M = \dfrac{1}{1 + f}$

3. Isc = Isca x M

4. Add Motor Contribution = Motor FLA x 4

* All results are given in symmetrical amperes

Three-Phase Bolted Fault-Current* Table

One-Way Distance in Feet

Available Isc in Thousands (Isca)

	5	10	15	20	25	30	40	50	60	70	80	90	100	125	150	175	200	225	250	300	350	400	500	600
3	2,975	2,951	2,927	2,904	2,881	2,858	2,814	2,771	2,729	2,688	2,649	2,611	2,574	2,485	2,403	2,326	2,253	2,186	2,122	2,304	1,899	1,804	1,641	1,505
5	4,932	4,866	4,801	4,738	4,677	4,618	4,503	4,394	4,290	4,190	4,096	4,005	3,918	3,717	3,536	3,371	3,221	3,084	2,958	2,735	2,543	2,376	2,101	1,882
7	6,867	6,740	6,616	6,498	6,383	6,273	6,063	5,866	5,682	5,509	5,347	5,194	5,049	4,720	4,431	4,176	3,948	3,744	3,560	3,242	2,975	2,750	2,387	2,109
10	9,731	9,477	9,235	9,006	8,787	8,579	8,191	7,837	7,512	7,213	6,936	6,680	6,443	5,917	5,470	5,086	4,752	4,460	4,201	3,765	3,410	3,117	2,659	2,319
15	14,404	13,853	13,343	12,869	12,427	12,015	11,267	10,608	10,021	9,495	9,022	8,594	8,205	7,370	6,690	6,124	5,647	5,239	4,885	4,305	3,848	3,478	2,918	2,513
20	18,954	18,011	17,158	16,382	15,673	15,023	13,873	12,886	12,030	11,281	10,619	10,031	9,505	8,402	7,529	6,820	6,234	5,740	5,318	4,638	4,111	3,692	3,067	2,623
25	23,386	21,968	20,712	19,592	18,586	17,679	16,107	14,792	13,675	12,715	11,881	11,150	10,503	9,173	8,142	7,320	6,648	6,089	5,617	4,863	4,287	3,834	3,164	2,693
30	27,706	25,737	24,030	22,535	21,215	20,041	18,045	16,410	15,047	13,893	12,903	12,045	11,294	9,771	8,610	7,695	6,956	6,347	5,836	5,026	4,414	3,934	3,232	2,743
35	31,916	29,332	27,135	25,244	23,599	22,156	19,741	17,801	16,208	14,877	13,748	12,778	11,936	10,247	8,978	7,988	7,195	6,545	6,002	5,149	4,508	4,009	3,283	2,779
40	36,022	32,764	30,047	27,745	25,771	24,060	21,238	19,009	17,204	15,711	14,457	13,389	12,467	10,637	9,275	8,222	7,384	6,701	6,134	5,246	4,582	4,068	3,322	2,807
45	40,028	36,045	32,783	30,062	27,759	25,783	22,570	20,069	18,067	16,428	15,062	13,906	12,914	10,961	9,520	8,415	7,539	6,828	6,240	5,323	4,641	4,114	3,353	2,829
50	43,936	39,183	35,359	32,214	29,583	27,350	23,762	21,006	18,823	17,051	15,584	14,349	13,296	11,234	9,726	8,575	7,667	6,934	6,328	5,387	4,690	4,152	3,378	2,847
55	47,750	42,189	37,788	34,218	31,265	28,781	24,835	21,840	19,490	17,596	16,038	14,734	13,625	11,468	9,901	8,711	7,776	7,022	6,402	5,440	4,730	4,184	3,399	2,862
60	51,474	45,070	40,083	36,090	32,820	30,093	25,806	22,588	20,083	18,078	16,438	15,070	13,913	11,671	10,052	8,827	7,869	7,098	6,464	5,485	4,764	4,210	3,416	2,874
65	55,111	47,834	42,254	37,840	34,261	31,301	26,689	23,261	20,614	18,507	16,792	15,367	14,165	11,849	10,183	8,928	7,949	7,163	6,518	5,524	4,793	4,233	3,431	2,885
70	58,664	50,488	44,312	39,482	35,602	32,416	27,495	23,871	21,092	18,892	17,107	15,631	14,389	12,005	10,298	9,017	8,019	7,220	6,565	5,558	4,819	4,253	3,444	2,894
75	62,135	53,038	46,264	41,025	36,851	33,449	28,234	24,427	21,524	19,238	17,391	15,867	14,589	12,144	10,400	9,095	8,080	7,270	6,607	5,588	4,841	4,270	3,456	2,902
80	65,528	55,491	48,119	42,477	38,019	34,408	28,915	24,934	21,917	19,551	17,646	16,080	14,769	12,268	10,491	9,164	8,135	7,314	6,643	5,614	4,861	4,286	3,466	2,909
85	68,846	57,851	49,884	43,846	39,112	35,301	29,543	25,400	22,276	19,836	17,878	16,272	14,931	12,380	10,573	9,226	8,184	7,354	6,676	5,637	4,878	4,299	3,474	2,915
90	72,089	60,124	51,566	45,140	40,138	36,134	30,125	25,829	22,605	20,097	18,090	16,447	15,078	12,481	10,646	9,282	8,228	7,389	6,705	5,658	4,894	4,311	3,482	2,921
95	75,262	62,315	53,169	46,364	41,103	36,915	30,665	26,225	22,908	20,336	18,283	16,607	15,212	12,572	10,713	9,333	8,268	7,421	6,732	5,677	4,908	4,322	3,489	2,926
100	78,367	64,428	54,700	47,524	42,012	37,646	31,168	26,592	23,188	20,556	18,461	16,753	15,335	12,656	10,774	9,379	8,304	7,450	6,756	5,694	4,920	4,332	3,496	2,930
105	81,404	66,468	56,163	48,624	42,870	38,333	31,637	26,933	23,446	20,759	18,625	16,888	15,448	12,733	10,830	9,421	8,337	7,477	6,777	5,709	4,932	4,341	3,502	2,934
110	84,378	68,437	57,562	49,669	43,680	38,980	32,077	27,251	23,687	20,947	18,776	17,012	15,552	12,803	10,881	9,460	8,367	7,501	6,797	5,723	4,942	4,349	3,507	2,938
115	87,289	70,340	58,902	50,664	44,448	39,590	32,489	27,547	23,911	21,122	18,916	17,128	15,648	12,868	10,928	9,495	8,395	7,523	6,816	5,736	4,952	4,357	3,512	2,942
120	90,140	72,179	60,187	51,611	45,175	40,166	32,876	27,825	24,120	21,285	19,047	17,234	15,737	12,929	10,971	9,528	8,421	7,544	6,832	5,748	4,961	4,363	3,516	2,945
125	92,932	73,958	61,419	52,515	45,866	40,711	33,240	28,086	24,315	21,437	19,169	17,334	15,820	12,985	11,011	9,558	8,444	7,563	6,848	5,759	4,969	4,370	3,520	2,948
130	95,668	75,681	62,602	53,377	46,522	41,228	33,583	28,330	24,498	21,580	19,282	17,427	15,897	13,037	11,049	9,587	8,466	7,581	6,862	5,769	4,977	4,376	3,524	2,950
135	98,348	77,348	63,739	54,202	47,147	41,718	33,908	28,561	24,671	21,713	19,389	17,514	15,970	13,085	11,084	9,613	8,487	7,597	6,876	5,779	4,984	4,381	3,528	2,953
140	100,975	78,964	64,832	54,990	47,743	42,183	34,215	28,778	24,833	21,839	19,489	17,595	16,038	13,131	11,116	9,637	8,506	7,612	6,888	5,788	4,990	4,386	3,531	2,955
145	103,551	80,531	65,884	55,745	48,311	42,626	34,505	28,984	24,986	21,957	19,583	17,672	16,101	13,173	11,147	9,660	8,524	7,627	6,900	5,796	4,997	4,391	3,534	2,957
150	106,076	82,049	66,897	56,469	48,853	43,048	34,781	29,178	25,130	22,068	19,671	17,744	16,161	13,213	11,175	9,682	8,541	7,640	6,911	5,804	5,002	4,395	3,537	2,959
155	108,552	83,523	67,874	57,163	49,372	43,450	35,043	29,362	25,266	22,173	19,755	17,812	16,217	13,251	11,202	9,702	8,556	7,652	6,921	5,811	5,008	4,400	3,540	2,961
160	110,981	84,954	68,815	57,830	49,868	43,834	35,293	29,537	25,396	22,273	19,834	17,876	16,270	13,287	11,228	9,721	8,571	7,664	6,931	5,818	5,013	4,403	3,542	2,963
165	113,364	86,343	69,724	58,470	50,344	44,201	35,530	29,703	25,519	22,367	19,909	17,937	16,321	13,320	11,251	9,739	8,585	7,675	6,940	5,824	5,018	4,407	3,545	2,964
170	115,702	87,693	70,602	59,086	50,800	44,552	35,757	29,862	25,635	22,457	19,980	17,995	16,368	13,352	11,274	9,756	8,598	7,686	6,949	5,830	5,022	4,411	3,547	2,966
175	117,996	89,005	71,449	59,678	51,237	44,888	35,973	30,012	25,746	22,542	20,047	18,049	16,414	13,382	11,296	9,772	8,611	7,696	6,957	5,836	5,026	4,414	3,549	2,968
180	120,249	90,280	72,269	60,249	51,657	45,210	36,180	30,156	25,852	22,623	20,111	18,101	16,456	13,410	11,316	9,787	8,622	7,705	6,965	5,842	5,030	4,417	3,551	2,969
185	122,460	91,521	73,062	60,799	52,061	45,519	36,377	30,293	25,952	22,700	20,172	18,150	16,497	13,437	11,335	9,802	8,634	7,714	6,972	5,847	5,034	4,420	3,553	2,970
190	124,631	92,728	73,829	61,330	52,450	45,816	36,566	30,424	26,049	22,773	20,230	18,197	16,536	13,463	11,353	9,815	8,644	7,723	6,979	5,852	5,038	4,423	3,555	2,972
195	126,763	93,903	74,572	61,841	52,824	46,101	36,748	30,550	26,141	22,844	20,285	18,242	16,573	13,488	11,371	9,828	8,654	7,731	6,985	5,856	5,041	4,425	3,556	2,973
200	128,857	95,047	75,292	62,336	53,184	46,375	36,922	30,670	26,228	22,911	20,338	18,285	16,608	13,511	11,387	9,841	8,664	7,738	6,992	5,861	5,044	4,428	3,558	2,974

4/0 AWG | 1 Run(s) | 208 Volts

C-Value = 15,082 | Magnetic Duct | Copper

Calculations:

1. $f = \dfrac{1.732 \times \text{Length} \times \text{Isca}}{\text{Runs} \times \text{C-Value} \times \text{E L-L}}$

2. $M = \dfrac{1}{1 + f}$

3. Isc = Isca x M

4. Add Motor Contribution = Motor FLA x 4

* All results are given in symmetrical amperes

Three-Phase Bolted Fault-Current* Table

One-Way Distance in Feet

Available Isc in Thousands (Isca)

	5	10	15	20	25	30	40	50	60	70	80	90	100	125	150	175	200	225	250	300	350	400	500	600
3	2,988	2,975	2,963	2,951	2,939	2,927	2,904	2,881	2,858	2,836	2,814	2,792	2,771	2,719	2,669	2,620	2,574	2,529	2,485	2,403	2,326	2,253	2,122	2,004
5	4,966	4,932	4,899	4,866	4,833	4,801	4,738	4,677	4,618	4,559	4,503	4,448	4,394	4,264	4,142	4,027	3,918	3,815	3,717	3,536	3,371	3,221	2,958	2,735
7	6,933	6,867	6,803	6,740	6,677	6,616	6,498	6,383	6,273	6,166	6,063	5,963	5,866	5,638	5,427	5,231	5,049	4,879	4,720	4,431	4,176	3,948	3,560	3,242
10	9,864	9,731	9,602	9,477	9,354	9,235	9,006	8,787	8,579	8,381	8,191	8,010	7,837	7,435	7,072	6,743	6,443	6,169	5,917	5,470	5,086	4,752	4,201	3,765
15	14,696	14,404	14,123	13,853	13,593	13,343	12,869	12,427	12,015	11,629	11,267	10,928	10,608	9,884	9,253	8,697	8,205	7,765	7,370	6,690	6,124	5,647	4,885	4,305
20	19,463	18,954	18,470	18,011	17,574	17,158	16,382	15,673	15,023	14,425	13,873	13,361	12,886	11,833	10,940	10,172	9,505	8,920	8,402	7,529	6,820	6,234	5,318	4,638
25	24,166	23,386	22,655	21,968	21,321	20,712	19,592	18,586	17,679	16,857	16,107	15,421	14,792	13,422	12,284	11,324	10,503	9,793	9,173	8,142	7,320	6,648	5,617	4,863
30	28,807	27,706	26,685	25,737	24,854	24,030	22,535	21,215	20,041	18,991	18,045	17,189	16,410	14,741	13,379	12,248	11,294	10,477	9,771	8,610	7,695	6,956	5,836	5,026
35	33,387	31,916	30,570	29,332	28,191	27,135	25,244	23,599	22,156	20,879	19,741	18,721	17,801	15,853	14,290	13,007	11,936	11,027	10,247	8,978	7,988	7,195	6,002	5,149
40	37,907	36,022	34,316	32,764	31,347	30,047	27,745	25,771	24,060	22,561	21,238	20,062	19,009	16,805	15,058	13,641	12,467	11,479	10,637	9,275	8,222	7,384	6,134	5,246
45	42,368	40,028	37,932	36,045	34,336	32,783	30,062	27,759	25,783	24,070	22,570	21,246	20,069	17,628	15,716	14,178	12,914	11,858	10,961	9,520	8,415	7,539	6,240	5,323
50	46,772	43,936	41,424	39,183	37,173	35,359	32,214	29,583	27,350	25,430	23,762	22,299	21,006	18,346	16,284	14,639	13,296	12,178	11,234	9,726	8,575	7,667	6,328	5,387
55	51,119	47,750	44,798	42,189	39,867	37,788	34,218	31,265	28,781	26,663	24,835	23,241	21,840	18,979	16,781	15,040	13,625	12,454	11,468	9,901	8,711	7,776	6,402	5,440
60	55,411	51,474	48,060	45,070	42,430	40,083	36,090	32,820	30,093	27,785	25,806	24,090	22,588	19,541	17,219	15,390	13,913	12,694	11,671	10,052	8,827	7,869	6,464	5,485
65	59,648	55,111	51,215	47,834	44,871	42,254	37,840	34,261	31,301	28,811	26,689	24,857	23,261	20,043	17,608	15,700	14,165	12,904	11,849	10,183	8,928	7,949	6,518	5,524
70	63,833	58,664	54,270	50,488	47,199	44,312	39,482	35,602	32,416	29,753	27,495	25,555	23,871	20,495	17,955	15,976	14,389	13,089	12,005	10,298	9,017	8,019	6,565	5,558
75	67,964	62,135	57,227	53,038	49,420	46,264	41,025	36,851	33,449	30,621	28,234	26,193	24,427	20,903	18,268	16,222	14,589	13,255	12,144	10,400	9,095	8,080	6,607	5,588
80	72,045	65,528	60,093	55,491	51,543	48,119	42,477	38,019	34,408	31,423	28,915	26,777	24,934	21,273	18,550	16,445	14,769	13,403	12,268	10,491	9,164	8,135	6,643	5,614
85	76,075	68,846	62,871	57,851	53,573	49,884	43,846	39,112	35,301	32,166	29,543	27,315	25,400	21,612	18,807	16,646	14,931	13,536	12,380	10,573	9,226	8,184	6,676	5,637
90	80,055	72,089	65,565	60,124	55,517	51,566	45,140	40,138	36,134	32,857	30,125	27,812	25,829	21,921	19,041	16,829	15,078	13,657	12,481	10,646	9,282	8,228	6,705	5,658
95	83,987	75,262	68,180	62,315	57,380	53,169	46,364	41,103	36,915	33,501	30,665	28,272	26,225	22,206	19,255	16,996	15,212	13,767	12,572	10,713	9,333	8,268	6,732	5,677
100	87,871	78,367	70,717	64,428	59,167	54,700	47,524	42,012	37,646	34,102	31,168	28,699	26,592	22,468	19,452	17,150	15,335	13,867	12,656	10,774	9,379	8,304	6,756	5,694
105	91,709	81,404	73,182	66,468	60,882	56,163	48,624	42,870	38,333	34,665	31,637	29,096	26,933	22,711	19,634	17,291	15,448	13,959	12,733	10,830	9,421	8,337	6,777	5,709
110	95,500	84,378	75,576	68,437	62,530	57,562	49,669	43,680	38,980	35,193	32,077	29,467	27,251	22,937	19,802	17,421	15,552	14,044	12,803	10,881	9,460	8,367	6,797	5,723
115	99,246	87,289	77,903	70,340	64,115	58,902	50,664	44,448	39,590	35,689	32,489	29,815	27,547	23,147	19,959	17,542	15,648	14,123	12,868	10,928	9,495	8,395	6,816	5,736
120	102,948	90,140	80,166	72,179	65,640	60,187	51,611	45,175	40,166	36,157	32,876	30,140	27,825	23,343	20,104	17,654	15,737	14,195	12,929	10,971	9,528	8,421	6,832	5,748
125	106,607	92,932	82,367	73,958	67,108	61,419	52,515	45,866	40,711	36,598	33,240	30,446	28,086	23,526	20,240	17,759	15,820	14,263	12,985	11,011	9,558	8,444	6,848	5,759
130	110,222	95,668	84,508	75,681	68,523	62,602	53,377	46,522	41,228	37,015	33,583	30,734	28,330	23,697	20,366	17,857	15,897	14,326	13,037	11,049	9,587	8,466	6,862	5,769
135	113,796	98,348	86,593	77,348	69,887	63,739	54,202	47,147	41,718	37,409	33,908	31,005	28,561	23,858	20,485	17,948	15,970	14,384	13,085	11,084	9,613	8,487	6,876	5,779
140	117,328	100,975	88,624	78,964	71,204	64,832	54,990	47,743	42,183	37,783	34,215	31,262	28,778	24,010	20,597	18,033	16,038	14,439	13,131	11,116	9,637	8,506	6,888	5,788
145	120,819	103,551	90,601	80,531	72,475	65,884	55,745	48,311	42,626	38,138	34,505	31,504	28,984	24,153	20,702	18,114	16,101	14,491	13,173	11,147	9,660	8,524	6,900	5,796
150	124,271	106,076	92,528	82,049	73,703	66,897	56,469	48,853	43,048	38,476	34,781	31,734	29,178	24,287	20,801	18,190	16,161	14,539	13,213	11,175	9,682	8,541	6,911	5,804
155	127,683	108,552	94,407	83,523	74,890	67,874	57,163	49,372	43,450	38,797	35,043	31,952	29,362	24,415	20,894	18,261	16,217	14,585	13,251	11,202	9,702	8,556	6,921	5,811
160	131,057	110,981	96,239	84,954	76,038	68,815	57,830	49,868	43,834	39,102	35,293	32,160	29,537	24,536	20,983	18,329	16,270	14,628	13,287	11,228	9,721	8,571	6,931	5,818
165	134,393	113,364	98,025	86,343	77,149	69,724	58,470	50,344	44,201	39,394	35,530	32,357	29,703	24,650	21,066	18,392	16,321	14,669	13,320	11,251	9,739	8,585	6,940	5,824
170	137,691	115,702	99,769	87,693	78,224	70,602	59,086	50,800	44,552	39,673	35,757	32,544	29,862	24,759	21,146	18,453	16,368	14,707	13,352	11,274	9,756	8,598	6,949	5,830
175	140,953	117,996	101,470	89,005	79,267	71,449	59,678	51,237	44,888	39,939	35,973	32,723	30,012	24,863	21,221	18,510	16,414	14,743	13,382	11,296	9,772	8,611	6,957	5,836
180	144,179	120,249	103,131	90,280	80,277	72,269	60,249	51,657	45,210	40,194	36,180	32,894	30,156	24,961	21,293	18,565	16,456	14,778	13,410	11,316	9,787	8,622	6,965	5,842
185	147,369	122,460	104,753	91,521	81,256	73,062	60,799	52,061	45,519	40,438	36,377	33,057	30,293	25,055	21,361	18,617	16,497	14,811	13,437	11,335	9,802	8,634	6,972	5,847
190	150,525	124,631	106,338	92,728	82,206	73,829	61,330	52,450	45,816	40,672	36,566	33,214	30,424	25,145	21,426	18,666	16,536	14,842	13,463	11,353	9,815	8,644	6,979	5,852
195	153,646	126,763	107,886	93,903	83,128	74,572	61,841	52,824	46,101	40,896	36,748	33,363	30,550	25,230	21,489	18,713	16 573	14,872	13,488	11,371	9,828	8,654	6,985	5,856
200	156,733	128,857	109,399	95,047	84,024	75,292	62,336	53,184	46,375	41,112	36,922	33,507	30,670	25,312	21,548	18,758	16 608	14,900	13,511	11,387	9,841	8,664	6,992	5,861

 # 4/0 AWG | **2 Run(s)** | **208 Volts** | **C-Value = 15,082** | **Magnetic Duct** | **Copper**

Calculations:

1. $f = \dfrac{1.732 \times \text{Length} \times \text{Isca}}{\text{Runs} \times \text{C-Value} \times E_{L\text{-}L}}$
2. $M = \dfrac{1}{1 + f}$
3. $\text{Isc} = \text{Isca} \times M$
4. Add Motor Contribution = Motor FLA x 4

* All results are given in symmetrical amperes

Three-Phase Bolted Fault-Current* Table

One-Way Distance in Feet

Available Isc in Thousands (Isca)

Isca	5	10	15	20	25	30	40	50	60	70	80	90	100	125	150	175	200	225	250	300	350	400	500	600
3	2,977	2,955	2,933	2,912	2,890	2,870	2,329	2,789	2,750	2,712	2,676	2,640	2,605	2,522	2,444	2,371	2,302	2,237	2,176	2,062	1,960	1,868	1,707	1,571
5	4,938	4,877	4,817	4,760	4,703	4,648	4,541	4,439	4,342	4,249	4,159	4,074	3,992	3,800	3,626	3,467	3,322	3,188	3,065	2,845	2,654	2,487	2,209	1,988
7	6,878	6,761	6,647	6,538	6,431	6,329	6,133	5,948	5,775	5,611	5,456	5,310	5,171	4,854	4,574	4,324	4,100	3,898	3,715	3,387	3,128	2,899	2,529	2,242
10	9,754	9,519	9,296	9,082	8,879	8,684	8,319	7,983	7,674	7,388	7,122	6,874	6,644	6,129	5,689	5,308	4,974	4,680	4,419	3,975	3,613	3,310	2,836	2,481
15	14,452	13,943	13,469	13,026	12,611	12,222	11,511	10,878	10,312	9,801	9,339	8,918	8,534	7,703	7,020	6,449	5,963	5,545	5,182	4,533	4,107	3,721	3,132	2,704
20	19,038	18,165	17,368	16,638	15,967	15,348	14,244	13,287	12,452	11,715	11,060	10,475	9,948	8,838	7,951	7,225	6,621	6,110	5,672	4,951	4,409	3,967	3,305	2,832
25	23,515	22,197	21,018	19,959	19,001	18,131	16,609	15,324	14,223	13,269	12,436	11,701	11,047	9,695	8,637	7,788	7,090	6,508	6,013	5,220	4,612	4,131	3,418	2,915
30	27,887	26,052	24,443	23,022	21,757	20,623	18,677	17,067	15,712	14,557	13,560	12,690	11,926	10,365	9,165	8,214	7,442	6,803	6,265	5,409	4,759	4,248	3,497	2,972
35	32,157	29,741	27,663	25,856	24,271	22,869	20,501	18,577	16,983	15,641	14,496	13,507	12,644	10,903	9,583	8,549	7,716	7,030	6,457	5,552	4,869	4,336	3,557	3,015
40	36,329	33,276	30,696	28,487	26,575	24,903	22,120	19,897	18,080	16,567	15,287	14,191	13,242	11,345	9,923	8,818	7,934	7,212	6,610	5,664	4,955	4,404	3,602	3,048
45	40,407	36,665	33,557	30,935	28,693	26,754	23,568	21,061	19,036	17,366	15,965	14,774	13,748	11,714	10,204	9,039	8,113	7,359	6,733	5,755	5,024	4,458	3,639	3,074
50	44,393	39,917	36,261	33,219	30,647	28,445	24,871	22,095	19,876	18,063	16,552	15,275	14,181	12,027	10,441	9,225	8,262	7,481	6,835	5,829	5,081	4,503	3,668	3,095
55	48,291	43,041	38,821	35,354	32,456	29,996	25,049	23,020	20,622	18,676	17,066	15,711	14,556	12,296	10,643	9,382	8,388	7,584	6,921	5,891	5,128	4,540	3,693	3,112
60	52,103	46,044	41,247	37,355	34,134	31,425	27,119	23,852	21,287	19,220	17,519	16,094	14,884	12,529	10,817	9,517	8,496	7,673	6,995	5,945	5,168	4,572	3,714	3,127
65	55,833	48,932	43,550	39,234	35,696	32,744	28,096	24,604	21,884	19,705	17,921	16,434	15,174	12,734	10,969	9,635	8,590	7,749	7,058	5,990	5,203	4,599	3,732	3,140
70	59,483	51,713	45,738	41,001	37,154	33,966	28,991	25,288	22,423	20,142	18,281	16,736	15,431	12,914	11,103	9,738	8,671	7,815	7,113	6,030	5,233	4,622	3,747	3,151
75	63,055	54,392	47,822	42,668	38,517	35,102	29,815	25,912	22,913	20,536	18,605	17,007	15,661	13,075	11,222	9,829	8,744	7,874	7,162	6,065	5,259	4,642	3,760	3,160
80	66,552	56,974	49,806	44,241	39,794	36,159	30,574	26,484	23,359	20,893	18,898	17,251	15,868	13,219	11,328	9,910	8,808	7,926	7,205	6,096	5,282	4,660	3,772	3,168
85	69,976	59,465	51,700	45,728	40,993	37,147	31,277	27,010	23,767	21,219	19,165	17,473	16,056	13,349	11,423	9,983	8,865	7,973	7,243	6,123	5,303	4,676	3,783	3,176
90	73,330	61,870	53,508	47,137	42,122	38,071	31,930	27,495	24,142	21,517	19,408	17,675	16,226	13,466	11,509	10,048	8,917	8,014	7,278	6,148	5,321	4,691	3,792	3,182
95	76,615	64,193	55,236	48,473	43,186	38,938	32,538	27,944	24,487	21,792	19,631	17,859	16,381	13,573	11,587	10,108	8,964	8,052	7,309	6,170	5,338	4,704	3,801	3,188
100	79,835	66,437	56,890	49,742	44,190	39,753	33,105	28,361	24,807	22,045	19,836	18,029	16,524	13,671	11,658	10,162	9,006	8,086	7,337	6,190	5,353	4,715	3,808	3,194
105	82,989	68,608	58,474	50,949	45,140	40,520	33,635	28,750	25,104	22,278	20,025	18,185	16,655	13,761	11,723	10,211	9,045	8,118	7,363	6,208	5,367	4,726	3,815	3,199
110	86,082	70,708	59,993	52,098	46,040	41,243	34,132	29,112	25,380	22,495	20,200	18,329	16,776	13,843	11,783	10,257	9,080	8,146	7,386	6,225	5,379	4,736	3,821	3,203
115	89,114	72,741	61,450	53,193	46,893	41,927	34,599	29,451	25,637	22,697	20,362	18,463	16,888	13,919	11,838	10,298	9,113	8,172	7,408	6,240	5,391	4,745	3,827	3,207
120	92,087	74,710	62,849	54,239	47,703	42,573	35,038	29,769	25,877	22,885	20,514	18,587	16,992	13,990	11,889	10,337	9,143	8,197	7,428	6,254	5,401	4,753	3,833	3,211
125	95,004	76,618	64,194	55,237	48,474	43,186	35,452	30,067	26,102	23,061	20,655	18,703	17,089	14,055	11,936	10,373	9,171	8,219	7,446	6,267	5,411	4,760	3,837	3,214
130	97,864	78,467	65,488	56,193	49,208	43,768	35,843	30,348	26,314	23,226	20,787	18,812	17,179	14,116	11,980	10,406	9,197	8,240	7,463	6,280	5,420	4,767	3,842	3,217
135	100,671	80,262	66,733	57,107	49,908	44,321	36,213	30,612	26,512	23,381	20,911	18,913	17,264	14,173	12,021	10,437	9,221	8,259	7,479	6,291	5,428	4,774	3,846	3,220
140	103,426	82,003	67,932	57,983	50,575	44,846	36,563	30,862	26,700	23,526	21,027	19,008	17,343	14,227	12,060	10,466	9,244	8,278	7,494	6,301	5,436	4,780	3,850	3,223
145	106,129	83,693	69,088	58,823	51,213	45,347	36,895	31,099	26,876	23,663	21,137	19,097	17,417	14,277	12,096	10,493	9,265	8,294	7,508	6,311	5,443	4,785	3,854	3,226
150	108,783	85,335	70,203	59,629	51,824	45,825	37,211	31,323	27,043	23,793	21,240	19,182	17,487	14,324	12,129	10,513	9,285	8,310	7,521	6,320	5,450	4,791	3,857	3,228
155	111,389	86,931	71,279	60,404	52,408	46,281	37,511	31,535	27,202	23,915	21,337	19,261	17,553	14,368	12,161	10,542	9,303	8,325	7,533	6,329	5,457	4,796	3,860	3,230
160	113,948	88,481	72,318	61,148	52,967	46,717	37,797	31,737	27,352	24,031	21,430	19,336	17,615	14,410	12,191	10,564	9,321	8,339	7,545	6,337	5,463	4,800	3,863	3,232
165	116,462	89,989	73,323	61,865	53,504	47,134	38,069	31,929	27,494	24,141	21,517	19,407	17,674	14,449	12,219	10,586	9,337	8,352	7,555	6,345	5,468	4,805	3,866	3,234
170	118,931	91,456	74,294	62,555	54,019	47,533	38,329	32,112	27,629	24,245	21,600	19,475	17,730	14,486	12,246	10,606	9,353	8,365	7,566	6,352	5,474	4,809	3,869	3,236
175	121,356	92,884	75,233	63,219	54,514	47,916	38,578	32,286	27,758	24,344	21,678	19,539	17,783	14,522	12,271	10,625	9,368	8,377	7,575	6,359	5,479	4,813	3,871	3,238
180	123,740	94,274	76,142	63,860	54,990	48,283	38,816	32,452	27,881	24,439	21,753	19,599	17,834	14,555	12,295	10,643	9,382	8,388	7,584	6,365	5,483	4,816	3,874	3,240
185	126,082	95,628	77,023	64,478	55,448	48,636	39,043	32,611	27,998	24,529	21,825	19,657	17,882	14,587	12,318	10,660	9,395	8,398	7,593	6,371	5,488	4,820	3,876	3,241
190	128,385	96,946	77,876	65,075	55,889	48,975	39,261	32,763	28,110	24,615	21,892	19,712	17,927	14,618	12,340	10,676	9,407	8,408	7,601	6,377	5,492	4,823	3,878	3,243
195	130,649	98,232	78,703	65,652	56,313	49,301	39,470	32,908	28,217	24,697	21,957	19,765	17,971	14,646	12,360	10,691	9,419	8,418	7,609	6,382	5,496	4,826	3,880	3,244
200	132,874	99,484	79,506	66,209	56,723	49,614	39,671	33,048	28,320	24,775	22,019	19,815	18,012	14,674	12,380	10,706	9,431	8,427	7,616	6,388	5,500	4,829	3,882	3,246

250 MCM | 1 Run(s) | 208 Volts | C-Value = 16,483 | Magnetic Duct | Copper

Calculations:

1. $f = \dfrac{1.732 \times \text{Length} \times \text{Isca}}{\text{Runs} \times \text{C-Value} \times E_{L\text{-}L}}$

2. $M = \dfrac{1}{1 + f}$

3. Isc = Isca x M

4. Add Motor Contribution = Motor FLA x 4

* All results are given in symmetrical amperes

Three-Phase Bolted Fault-Current* Table

Available Isc in Thousands (Isca)	One-Way Distance in Feet																							
	5	10	15	20	25	30	40	50	60	70	80	90	100	125	150	175	200	225	250	300	350	400	500	600
3	2,989	2,977	2,966	2,955	2,944	2,933	2,912	2,890	2,870	2,849	2,829	2,808	2,789	2,740	2,694	2,649	2,605	2,563	2,522	2,444	2,371	2,302	2,176	2,062
5	4,969	4,938	4,907	4,877	4,847	4,817	4,760	4,703	4,648	4,594	4,541	4,490	4,439	4,318	4,204	4,095	3,992	3,894	3,800	3,626	3,467	3,322	3,065	2,845
7	6,939	6,878	6,819	6,761	6,704	6,647	6,538	6,431	6,329	6,229	6,133	6,039	5,948	5,733	5,533	5,346	5,171	5,008	4,854	4,574	4,324	4,100	3,715	3,397
10	9,875	9,754	9,635	9,519	9,406	9,296	9,082	8,879	8,684	8,498	8,319	8,148	7,983	7,600	7,252	6,935	6,644	6,376	6,129	5,689	5,308	4,974	4,419	3,975
15	14,721	14,452	14,193	13,943	13,702	13,469	13,026	12,611	12,222	11,856	11,511	11,186	10,878	10,179	9,564	9,020	8,534	8,097	7,703	7,020	6,449	5,963	5,182	4,583
20	19,507	19,038	18,591	18,165	17,757	17,368	16,638	15,967	15,348	14,775	14,244	13,749	13,287	12,259	11,378	10,615	9,948	9,360	8,838	7,951	7,225	6,621	5,672	4,961
25	24,235	23,515	22,837	22,197	21,591	21,018	19,959	19,001	18,131	17,337	16,609	15,941	15,324	13,972	12,839	11,876	11,047	10,327	9,695	8,637	7,788	7,090	6,013	5,220
30	28,905	27,887	26,938	26,052	25,222	24,443	23,022	21,757	20,623	19,602	18,677	17,836	17,067	15,407	14,041	12,897	11,926	11,091	10,365	9,165	8,214	7,442	6,265	5,409
35	33,518	32,157	30,902	29,741	28,665	27,663	25,856	24,271	22,869	21,620	20,501	19,491	18,577	16,626	15,047	13,741	12,644	11,709	10,903	9,583	8,549	7,716	6,457	5,552
40	38,076	36,329	34,736	33,276	31,934	30,696	28,487	26,575	24,903	23,429	22,120	20,950	19,897	17,676	15,901	14,450	13,242	12,220	11,345	9,923	8,818	7,934	6,610	5,664
45	42,580	40,407	38,445	36,665	35,042	33,557	30,935	28,693	26,754	25,060	23,568	22,244	21,061	18,589	16,636	15,054	13,748	12,649	11,714	10,204	9,039	8,113	6,733	5,755
50	47,030	44,393	42,036	39,917	38,001	36,261	33,219	30,647	28,445	26,538	24,871	23,401	22,095	19,390	17,275	15,575	14,181	13,015	12,027	10,441	9,225	8,262	6,835	5,829
55	51,428	48,291	45,515	43,041	40,822	38,821	35,354	32,456	29,996	27,884	26,049	24,441	23,020	20,098	17,835	16,029	14,556	13,331	12,296	10,643	9,382	8,388	6,921	5,891
60	55,774	52,103	48,887	46,044	43,513	41,247	37,355	34,134	31,425	29,114	27,119	25,381	23,852	20,729	18,330	16,428	14,884	13,605	12,529	10,817	9,517	8,496	6,995	5,945
65	60,069	55,833	52,155	48,932	46,084	43,550	39,234	35,696	32,744	30,243	28,096	26,234	24,604	21,295	18,771	16,782	15,174	13,847	12,734	10,969	9,635	8,590	7,058	5,990
70	64,314	59,483	55,326	51,713	48,543	45,738	41,001	37,154	33,966	31,282	28,991	27,013	25,288	21,806	19,166	17,097	15,431	14,061	12,914	11,103	9,738	8,671	7,113	6,030
75	68,511	63,055	58,404	54,392	50,895	47,822	42,668	38,517	35,102	32,243	29,815	27,727	25,912	22,268	19,523	17,380	15,661	14,252	13,075	11,222	9,829	8,744	7,162	6,065
80	72,659	66,552	61,392	56,974	53,150	49,806	44,241	39,794	36,159	33,133	30,574	28,382	26,484	22,689	19,846	17,636	15,868	14,423	13,219	11,328	9,910	8,808	7,205	6,096
85	76,760	69,976	64,294	59,465	55,311	51,700	45,728	40,993	37,147	33,960	31,277	28,987	27,010	23,074	20,140	17,867	16,056	14,578	13,349	11,423	9,983	8,865	7,243	6,123
90	80,814	73,330	67,114	61,870	57,386	53,508	47,137	42,122	38,071	34,731	31,930	29,547	27,495	23,427	20,408	18,078	16,226	14,718	13,466	11,509	10,048	8,917	7,278	6,148
95	84,823	76,615	69,856	64,193	59,379	55,236	48,473	43,186	38,938	35,451	32,538	30,067	27,944	23,753	20,655	18,272	16,381	14,846	13,573	11,587	10,108	8,964	7,309	6,170
100	88,787	79,835	72,522	66,437	61,294	56,890	49,742	44,190	39,753	36,125	33,105	30,550	28,361	24,054	20,882	18,449	16,524	14,963	13,671	11,658	10,162	9,006	7,337	6,190
105	92,706	82,989	75,116	68,608	63,137	58,474	50,949	45,140	40,520	36,758	33,635	31,001	28,750	24,332	21,091	18,613	16,655	15,070	13,761	11,723	10,211	9,045	7,363	6,208
110	96,582	86,082	77,641	70,708	64,911	59,993	52,098	46,040	41,243	37,352	34,132	31,423	29,112	24,591	21,286	18,764	16,776	15,169	13,843	11,783	10,257	9,080	7,386	6,225
115	100,416	89,114	80,099	72,741	66,620	61,450	53,193	46,893	41,927	37,912	34,599	31,818	29,451	24,833	21,466	18,904	16,888	15,260	13,919	11,838	10,298	9,113	7,408	6,240
120	104,207	92,087	82,493	74,710	68,268	62,849	54,239	47,703	42,573	38,440	35,038	32,189	29,769	25,058	21,635	19,034	16,992	15,345	13,990	11,889	10,337	9,143	7,428	6,254
125	107,957	95,004	84,826	76,618	69,858	64,194	55,237	48,474	43,186	38,939	35,452	32,538	30,067	25,269	21,792	19,156	17,089	15,424	14,055	11,936	10,373	9,171	7,446	6,267
130	111,666	97,864	87,099	78,467	71,392	65,488	56,193	49,208	43,768	39,411	35,843	32,867	30,348	25,467	21,939	19,269	17,179	15,498	14,116	11,980	10,406	9,197	7,463	6,280
135	115,335	100,671	89,315	80,262	72,875	66,733	57,107	49,908	44,321	39,858	36,213	33,178	30,612	25,653	22,077	19,376	17,264	15,567	14,173	12,021	10,437	9,221	7,479	6,291
140	118,965	103,426	91,477	82,003	74,307	67,932	57,983	50,575	44,846	40,283	36,563	33,472	30,862	25,829	22,207	19,476	17,343	15,631	14,227	12,060	10,466	9,244	7,494	6,301
145	122,556	106,129	93,585	83,693	75,693	69,088	58,823	51,213	45,347	40,687	36,895	33,750	31,099	25,994	22,329	19,569	17,417	15,691	14,277	12,096	10,493	9,265	7,508	6,311
150	126,109	108,783	95,643	85,335	77,033	70,203	59,629	51,824	45,825	41,071	37,211	34,014	31,323	26,150	22,444	19,658	17,487	15,748	14,324	12,129	10,518	9,285	7,521	6,320
155	129,625	111,389	97,652	86,931	78,331	71,279	60,404	52,408	46,281	41,437	37,511	34,264	31,535	26,298	22,553	19,741	17,553	15,802	14,368	12,161	10,542	9,303	7,533	6,329
160	133,103	113,948	99,613	88,481	79,588	72,318	61,148	52,967	46,717	41,786	37,797	34,503	31,737	26,438	22,656	19,820	17,615	15,852	14,410	12,191	10,564	9,321	7,545	6,337
165	136,546	116,462	101,528	89,989	80,806	73,323	61,865	53,504	47,134	42,119	38,069	34,730	31,929	26,571	22,753	19,895	17,674	15,900	14,449	12,219	10,586	9,337	7,555	6,345
170	139,952	118,931	103,400	91,456	81,986	74,294	62,555	54,019	47,533	42,438	38,329	34,946	32,112	26,698	22,846	19,966	17,730	15,945	14,486	12,246	10,606	9,353	7,566	6,352
175	143,323	121,356	105,228	92,884	83,132	75,233	63,219	54,514	47,916	42,743	38,578	35,152	32,286	26,818	22,934	20,033	17,783	15,988	14,522	12,271	10,625	9,368	7,575	6,359
180	146,660	123,740	107,016	94,274	84,244	76,142	63,860	54,990	48,283	43,035	38,816	35,350	32,452	26,933	23,018	20,097	17,834	16,029	14,555	12,295	10,643	9,382	7,584	6,365
185	149,962	126,082	108,763	95,628	85,323	77,023	64,478	55,448	48,636	43,315	39,043	35,538	32,611	27,042	23,098	20,158	17,832	16,067	14,587	12,318	10,660	9,395	7,593	6,371
190	153,231	128,385	110,472	96,946	86,371	77,876	65,075	55,889	48,975	43,583	39,261	35,719	32,763	27,147	23,174	20,216	17,927	16,104	14,618	12,340	10,676	9,407	7,601	6,377
195	156,466	130,649	112,144	98,232	87,390	78,703	65,652	56,313	49,301	43,841	39,470	35,892	32,908	27,246	23,247	20,271	17,971	16,139	14,646	12,360	10,691	9,419	7,609	6,382
200	159,669	132,874	113,780	99,484	88,380	79,506	66,209	56,723	49,614	44,089	39,671	36,058	33,048	27,342	23,316	20,324	18,012	16,173	14,674	12,380	10,706	9,431	7,616	6,388

250 MCM	2 Run(s)	208 Volts		C-Value = 16,483	Magnetic Duct	Copper

Calculations:

1. $f = \dfrac{1.732 \times \text{Length} \times \text{Isca}}{\text{Runs} \times \text{C-Value} \times E_{L-L}}$

2. $M = \dfrac{1}{1 + f}$

3. $\text{Isc} = \text{Isca} \times M$

4. Add Motor Contribution = Motor FLA x 4

* All results are given in symmetrical amperes

Three-Phase Bolted Fault-Current* Table

One-Way Distance in Feet

Available Isc in Thousands (Isca)	5	10	15	20	25	30	40	50	60	70	80	90	100	125	150	175	200	225	250	300	350	400	500	600
3	2,980	2,959	2,939	2,920	2,900	2,881	2,844	2,807	2,771	2,737	2,703	2,670	2,638	2,560	2,487	2,418	2,353	2,291	2,233	2,124	2,026	1,936	1,778	1,644
5	4,943	4,888	4,834	4,781	4,729	4,678	4,580	4,486	4,396	4,309	4,226	4,145	4,068	3,887	3,721	3,569	3,429	3,299	3,179	2,364	2,775	2,609	2,331	2,106
7	6,890	6,782	6,679	6,578	6,480	6,386	6,204	6,033	5,870	5,717	5,571	5,432	5,300	4,997	4,726	4,484	4,265	4,066	3,885	3,368	3,298	3,066	2,689	2,394
10	9,776	9,562	9,357	9,161	8,972	8,792	8,451	8,136	7,844	7,572	7,318	7,081	6,858	6,359	5,927	5,550	5,219	4,924	4,661	4,212	3,841	3,530	3,039	2,668
15	14,502	14,035	13,598	13,188	12,801	12,436	11,766	11,164	10,621	10,128	9,679	9,268	8,891	8,069	7,386	6,810	6,317	5,891	5,519	4,399	4,405	4,001	3,381	2,928
20	19,124	18,321	17,583	16,903	16,273	15,688	14,636	13,716	12,905	12,185	11,541	10,961	10,437	9,323	8,423	7,682	7,061	6,533	6,078	5,335	4,754	4,287	3,583	3,078
25	23,646	22,431	21,335	20,341	19,435	18,607	17,145	15,897	14,818	13,876	13,046	12,311	11,653	10,281	9,198	8,321	7,597	6,989	6,471	5,636	4,991	4,479	3,717	3,176
30	28,071	26,375	24,872	23,532	22,328	21,242	19,358	17,781	16,442	15,290	14,289	13,411	12,635	11,038	9,799	8,810	8,003	7,331	6,763	5,856	5,163	4,617	3,811	3,245
35	32,402	30,163	28,214	26,501	24,985	23,632	21,324	19,426	17,838	16,491	15,332	14,326	13,444	11,650	10,278	9,196	8,320	7,596	6,988	6,024	5,293	4,721	3,881	3,295
40	36,643	33,805	31,376	29,272	27,432	25,811	23,081	20,874	19,052	17,523	16,221	15,099	14,122	12,156	10,670	9,508	8,574	7,808	7,167	6,156	5,395	4,802	3,936	3,335
45	40,795	37,309	34,371	31,863	29,695	27,804	24,663	22,159	20,117	18,419	16,986	15,760	14,698	12,580	10,996	9,766	8,784	7,981	7,312	6,263	5,477	4,867	3,980	3,366
50	44,862	40,681	37,214	34,291	31,793	29,635	26,093	23,307	21,058	19,205	17,652	16,331	15,195	12,942	11,272	9,983	8,959	8,125	7,433	6,352	5,545	4,920	4,015	3,391
55	48,846	43,931	39,914	36,571	33,744	31,323	27,392	24,338	21,896	19,900	18,238	16,831	15,626	13,254	11,507	10,167	9,107	8,247	7,535	6,426	5,601	4,964	4,045	3,412
60	52,750	47,063	42,483	38,716	35,562	32,883	28,578	25,270	22,648	20,519	18,756	17,272	16,005	13,526	11,712	10,326	9,234	8,351	7,622	6,489	5,649	5,002	4,070	3,430
65	56,576	50,085	44,931	40,738	37,261	34,331	29,665	26,116	23,325	21,073	19,218	17,663	16,341	13,765	11,890	10,465	9,345	8,441	7,697	6,544	5,691	5,034	4,091	3,445
70	60,327	53,003	47,264	42,647	38,852	35,677	30,665	26,887	23,939	21,573	19,633	18,012	16,639	13,976	12,047	10,587	9,442	8,520	7,763	6,591	5,726	5,062	4,109	3,458
75	64,004	55,820	49,492	44,453	40,345	36,932	31,587	27,594	24,497	22,025	20,007	18,327	16,907	14,164	12,187	10,695	9,528	8,590	7,821	6,633	5,758	5,087	4,125	3,470
80	67,610	58,544	51,621	46,163	41,748	38,104	32,441	28,244	25,008	22,437	20,346	18,611	17,149	14,334	12,312	10,791	9,604	8,652	7,872	6,669	5,786	5,109	4,140	3,480
85	71,147	61,177	53,658	47,785	43,070	39,203	33,234	28,842	25,476	22,814	20,655	18,869	17,368	14,486	12,425	10,877	9,672	8,708	7,918	6,702	5,810	5,128	4,152	3,489
90	74,617	63,725	55,608	49,325	44,318	40,233	33,972	29,397	25,908	23,159	20,937	19,105	17,567	14,625	12,527	10,955	9,734	8,757	7,959	6,732	5,832	5,145	4,164	3,497
95	78,022	66,192	57,477	50,790	45,497	41,203	34,660	29,911	26,306	23,477	21,197	19,321	17,750	14,751	12,619	11,023	9,789	8,802	7,996	6,758	5,852	5,161	4,174	3,504
100	81,363	68,581	59,270	52,185	46,613	42,116	35,304	30,389	26,675	23,771	21,436	19,519	17,917	14,866	12,703	11,090	9,840	8,843	8,030	6,783	5,870	5,175	4,183	3,510
105	84,642	70,896	60,992	53,515	47,671	42,978	35,908	30,835	27,019	24,043	21,657	19,702	18,071	14,972	12,781	11,149	9,886	8,881	8,061	6,804	5,887	5,187	4,191	3,516
110	87,862	73,141	62,646	54,784	48,676	43,793	36,475	31,253	27,338	24,296	21,862	19,872	18,214	15,070	12,852	11,203	9,929	8,915	8,089	6,825	5,902	5,199	4,199	3,522
115	91,023	75,319	64,236	55,997	49,631	44,564	37,009	31,644	27,637	24,531	22,053	20,029	18,346	15,160	12,917	11,253	9,968	8,947	8,115	6,843	5,916	5,210	4,206	3,526
120	94,127	77,432	65,767	57,156	50,539	45,296	37,512	32,011	27,917	24,751	22,230	20,176	18,469	15,244	12,978	11,299	10,004	8,976	8,139	6,860	5,928	5,220	4,212	3,531
125	97,176	79,483	67,241	58,266	51,405	45,990	37,987	32,356	28,179	24,957	22,396	20,312	18,583	15,322	13,035	11,341	10,038	9,003	8,161	6,876	5,940	5,229	4,218	3,535
130	100,171	81,476	68,661	59,330	52,232	46,650	38,436	32,681	28,425	25,150	22,552	20,440	18,690	15,395	13,087	11,381	10,069	9,028	8,182	6,890	5,951	5,237	4,224	3,539
135	103,114	83,412	70,031	60,350	53,021	47,279	38,861	32,988	28,657	25,332	22,698	20,560	18,790	15,462	13,136	11,418	10,098	9,051	8,201	6,904	5,961	5,245	4,229	3,543
140	106,005	85,294	71,353	61,329	53,775	47,877	39,265	33,279	28,876	25,503	22,835	20,672	18,884	15,526	13,182	11,453	10,125	9,073	8,219	6,917	5,971	5,252	4,234	3,546
145	108,847	87,125	72,630	62,270	54,497	48,449	39,648	33,554	29,083	25,664	22,964	20,778	18,972	15,585	13,225	11,485	10,150	9,093	8,235	6,928	5,979	5,259	4,238	3,549
150	111,641	88,905	73,863	63,174	55,188	48,994	40,013	33,815	29,279	25,816	23,086	20,878	19,055	15,641	13,265	11,516	10,174	9,112	8,251	6,939	5,988	5,265	4,242	3,552
155	114,387	90,638	75,055	64,044	55,851	49,516	40,360	34,062	29,464	25,960	23,201	20,972	19,134	15,694	13,303	11,544	10,196	9,130	8,266	6,950	5,995	5,271	4,246	3,555
160	117,087	92,325	76,208	64,882	56,487	50,015	40,692	34,298	29,640	26,097	23,310	21,061	19,208	15,744	13,339	11,571	10,217	9,147	8,279	6,960	6,003	5,277	4,250	3,557
165	119,743	93,968	77,324	65,689	57,098	50,494	41,008	34,522	29,808	26,226	23,413	21,145	19,278	15,791	13,373	11,597	10,237	9,163	8,292	6,969	6,009	5,282	4,253	3,560
170	122,354	95,569	78,405	66,468	57,685	50,952	41,310	34,736	29,967	26,350	23,511	21,225	19,344	15,836	13,405	11,621	10,256	9,178	8,305	6,977	6,016	5,287	4,256	3,562
175	124,923	97,129	79,452	67,219	58,250	51,392	41,598	34,940	30,119	26,467	23,605	21,301	19,407	15,878	13,435	11,643	10,273	9,192	8,316	6,986	6,022	5,292	4,259	3,564
180	127,450	98,650	80,467	67,944	58,793	51,815	41,875	35,135	30,263	26,578	23,693	21,373	19,467	15,918	13,464	11,655	10,290	9,205	8,327	6,993	6,028	5,296	4,262	3,566
185	129,937	100,133	81,451	68,644	59,317	52,221	42,140	35,321	30,402	26,685	23,778	21,442	19,524	15,956	13,491	11,635	10,306	9,218	8,338	7,001	6,033	5,301	4,265	3,568
190	132,384	101,580	82,406	69,321	59,822	52,612	42,394	35,499	30,534	26,787	23,859	21,508	19,579	15,993	13,517	11,735	10,321	9,230	8,348	7,008	6,038	5,305	4,268	3,570
195	134,792	102,992	83,333	69,975	60,309	52,988	42,638	35,670	30,660	26,884	23,936	21,570	19,631	16,027	13,541	11,723	10,336	9,242	8,357	7,014	6,043	5,308	4,270	3,571
200	137,162	104,370	84,232	70,609	60,779	53,351	42,872	35,834	30,781	26,977	24,009	21,630	19,680	16,060	13,565	11,741	10,349	9,253	8,366	7,021	6,048	5,312	4,272	3,573

300 MCM | 1 Run(s) | 208 Volts | C-Value = 18,176 | Magnetic Duct | Copper

Calculations:

1. $f = \dfrac{1.732 \times \text{Length} \times \text{Isca}}{\text{Runs} \times \text{C-Value} \times \text{E L-L}}$

2. $M = \dfrac{1}{1 + f}$

3. Isc = Isca x M

4. Add Motor Contribution = Motor FLA x 4

* All results are given in symmetrical amperes

Three-Phase Bolted Fault-Current* Table

One-Way Distance in Feet

Available Isc in Thousands (Isca)	5	10	15	20	25	30	40	50	60	70	80	90	100	125	150	175	200	225	250	300	350	400	500	600
3	2,990	2,980	2,969	2,959	2,949	2,939	2,920	2,900	2,881	2,862	2,844	2,825	2,807	2,763	2,720	2,678	2,638	2,598	2,560	2,487	2,418	2,353	2,233	2,124
5	4,972	4,943	4,916	4,888	4,861	4,834	4,781	4,729	4,678	4,629	4,580	4,533	4,486	4,374	4,267	4,165	4,068	3,976	3,887	3,721	3,569	3,429	3,179	2,964
7	6,944	6,890	6,836	6,782	6,730	6,679	6,578	6,480	6,386	6,294	6,204	6,117	6,033	5,831	5,643	5,466	5,300	5,144	4,997	4,726	4,484	4,265	3,885	3,568
10	9,887	9,776	9,668	9,562	9,458	9,357	9,161	8,972	8,792	8,618	8,451	8,291	8,136	7,774	7,443	7,138	6,858	6,599	6,359	5,927	5,550	5,219	4,661	4,212
15	14,747	14,502	14,265	14,035	13,813	13,598	13,188	12,801	12,436	12,092	11,766	11,457	11,164	10,493	9,898	9,367	8,891	8,460	8,069	7,386	6,810	6,317	5,519	4,899
20	19,552	19,124	18,714	18,321	17,945	17,583	16,903	16,273	15,688	15,144	14,636	14,161	13,716	12,717	11,854	11,100	10,437	9,848	9,323	8,423	7,682	7,061	6,078	5,335
25	24,304	23,646	23,022	22,431	21,869	21,335	20,341	19,435	18,607	17,846	17,145	16,497	15,897	14,570	13,448	12,487	11,653	10,924	10,281	9,198	8,321	7,597	6,471	5,636
30	29,003	28,071	27,197	26,375	25,602	24,872	23,532	22,328	21,242	20,256	19,358	18,536	17,781	16,138	14,773	13,620	12,635	11,782	11,038	9,799	8,810	8,003	6,763	5,856
35	33,651	32,402	31,243	30,163	29,156	28,214	26,501	24,985	23,632	22,419	21,324	20,331	19,426	17,481	15,890	14,565	13,444	12,483	11,650	10,278	9,196	8,320	6,988	6,024
40	38,248	36,643	35,167	33,805	32,545	31,376	29,272	27,432	25,811	24,370	23,081	21,922	20,874	18,645	16,846	15,364	14,122	13,065	12,156	10,670	9,508	8,574	7,167	6,156
45	42,794	40,795	38,974	37,309	35,780	34,371	31,863	29,695	27,804	26,139	24,663	23,344	22,159	19,664	17,674	16,049	14,698	13,557	12,580	10,996	9,766	8,784	7,312	6,263
50	47,292	44,862	42,669	40,681	38,870	37,214	34,291	31,793	29,635	27,751	26,093	24,621	23,307	20,562	18,396	16,643	15,195	13,978	12,942	11,272	9,983	8,959	7,433	6,352
55	51,741	48,846	46,258	43,931	41,826	39,914	36,571	33,744	31,323	29,226	27,392	25,775	24,338	21,361	19,033	17,162	15,626	14,343	13,254	11,507	10,167	9,107	7,535	6,426
60	56,142	52,750	49,745	47,063	44,656	42,483	38,716	35,562	32,883	30,580	28,578	26,822	25,270	22,075	19,598	17,620	16,005	14,661	13,526	11,712	10,326	9,234	7,622	6,489
65	60,496	56,576	53,133	50,085	47,368	44,931	40,738	37,261	34,331	31,828	29,665	27,778	26,116	22,718	20,103	18,028	16,341	14,942	13,765	11,890	10,465	9,345	7,697	6,544
70	64,804	60,327	56,428	53,003	49,969	47,264	42,647	38,852	35,677	32,981	30,665	28,652	26,887	23,300	20,557	18,392	16,639	15,192	13,976	12,047	10,587	9,442	7,763	6,591
75	69,067	64,004	59,633	55,820	52,466	49,492	44,453	40,345	36,932	34,051	31,587	29,456	27,594	23,829	20,967	18,720	16,907	15,415	14,164	12,187	10,695	9,528	7,821	6,633
80	73,285	67,610	62,751	58,544	54,865	51,621	46,163	41,748	38,104	35,045	32,441	30,197	28,244	24,311	21,340	19,016	17,149	15,615	14,334	12,312	10,791	9,604	7,872	6,669
85	77,459	71,147	65,787	61,177	57,171	53,658	47,785	43,070	39,203	35,972	33,234	30,883	28,842	24,754	21,681	19,286	17,368	15,797	14,486	12,425	10,877	9,672	7,918	6,702
90	81,590	74,617	68,742	63,725	59,391	55,608	49,325	44,318	40,233	36,838	33,972	31,519	29,397	25,161	21,992	19,532	17,567	15,962	14,625	12,527	10,955	9,734	7,959	6,732
95	85,678	78,022	71,622	66,192	61,527	57,477	50,790	45,497	41,203	37,650	34,660	32,111	29,911	25,537	22,279	19,758	17,750	16,112	14,751	12,619	11,026	9,789	7,996	6,758
100	89,724	81,363	74,427	68,581	63,587	59,270	52,185	46,613	42,116	38,411	35,304	32,663	30,389	25,885	22,543	19,966	17,917	16,250	14,866	12,703	11,090	9,840	8,030	6,783
105	93,728	84,642	77,162	70,896	65,572	60,992	53,515	47,671	42,978	39,126	35,908	33,179	30,835	26,208	22,788	20,157	18,071	16,376	14,972	12,781	11,149	9,886	8,061	6,804
110	97,692	87,862	79,828	73,141	67,488	62,646	54,784	48,676	43,793	39,800	36,475	33,663	31,253	26,508	23,015	20,335	18,214	16,493	15,070	12,852	11,203	9,929	8,089	6,825
115	101,616	91,023	82,429	75,319	69,337	64,236	55,997	49,631	44,564	40,437	37,009	34,116	31,644	26,789	23,226	20,499	18,346	16,602	15,160	12,917	11,253	9,968	8,115	6,843
120	105,500	94,127	84,967	77,432	71,124	65,767	57,156	50,539	45,296	41,038	37,512	34,543	32,011	27,052	23,423	20,653	18,469	16,702	15,244	12,978	11,299	10,004	8,139	6,860
125	109,346	97,176	87,444	79,483	72,851	67,241	58,266	51,405	45,990	41,607	37,987	34,946	32,356	27,298	23,607	20,796	18,583	16,796	15,322	13,035	11,341	10,038	8,161	6,876
130	113,153	100,171	89,861	81,476	74,522	68,661	59,330	52,232	46,650	42,146	38,436	35,326	32,681	27,529	23,780	20,930	18,690	16,883	15,395	13,087	11,381	10,069	8,182	6,890
135	116,922	103,114	92,222	83,412	76,138	70,031	60,350	53,021	47,279	42,659	38,861	35,685	32,988	27,747	23,942	21,055	18,790	16,964	15,462	13,136	11,418	10,098	8,201	6,904
140	120,654	106,005	94,529	85,294	77,703	71,353	61,329	53,775	47,877	43,146	39,265	36,025	33,279	27,952	24,095	21,173	18,884	17,041	15,526	13,182	11,453	10,125	8,219	6,917
145	124,349	108,847	96,782	87,125	79,220	72,630	62,270	54,497	48,449	43,609	39,648	36,347	33,554	28,146	24,239	21,284	18,972	17,113	15,585	13,225	11,485	10,150	8,235	6,928
150	128,008	111,641	98,984	88,905	80,689	73,863	63,174	55,188	48,994	44,051	40,013	36,654	33,815	28,329	24,375	21,389	19,055	17,180	15,641	13,265	11,516	10,174	8,251	6,939
155	131,632	114,387	101,137	90,638	82,114	75,055	64,044	55,851	49,516	44,472	40,360	36,945	34,062	28,503	24,503	21,488	19,134	17,244	15,694	13,303	11,544	10,196	8,266	6,950
160	135,221	117,087	103,242	92,325	83,496	76,208	64,882	56,487	50,015	44,874	40,692	37,222	34,298	28,667	24,625	21,581	19,208	17,304	15,744	13,339	11,571	10,217	8,279	6,960
165	138,775	119,743	105,301	93,968	84,838	77,324	65,689	57,098	50,494	45,259	41,008	37,486	34,522	28,824	24,740	21,670	19,278	17,361	15,791	13,373	11,597	10,237	8,292	6,969
170	142,295	122,354	107,316	95,569	86,140	78,405	66,468	57,685	50,952	45,627	41,310	37,739	34,736	28,973	24,850	21,754	19,344	17,415	15,836	13,405	11,621	10,256	8,305	6,977
175	145,781	124,923	109,287	97,129	87,406	79,452	67,219	58,250	51,392	45,980	41,598	37,979	34,940	29,114	24,954	21,834	19,407	17,466	15,878	13,435	11,643	10,273	8,316	6,986
180	149,234	127,450	111,216	98,650	88,636	80,467	67,944	58,793	51,815	46,318	41,875	38,210	35,135	29,250	25,053	21,910	19,467	17,515	15,918	13,464	11,665	10,290	8,327	6,993
185	152,655	129,937	113,105	100,133	89,831	81,451	68,644	59,317	52,221	46,642	42,140	38,430	35,321	29,379	25,148	21,982	19,524	17,561	15,956	13,491	11,685	10,306	8,338	7,001
190	156,043	132,384	114,954	101,580	90,994	82,406	69,321	59,822	52,612	46,954	42,394	38,642	35,499	29,502	25,238	22,051	19,579	17,605	15,993	13,517	11,705	10,321	8,348	7,008
195	159,400	134,792	116,766	102,992	92,125	83,333	69,975	60,309	52,988	47,253	42,638	38,844	35,670	29,620	25,324	22,117	19,631	17,647	16,027	13,541	11,723	10,336	8,357	7,014
200	162,725	137,162	118,540	104,370	93,226	84,232	70,609	60,779	53,351	47,541	42,872	39,039	35,834	29,733	25,407	22,180	19,680	17,687	16,060	13,565	11,741	10,349	8,366	7,021

300 MCM | 2 Run(s) | 208 Volts | C-Value = 18,176 | Magnetic Duct | Copper

Calculations:

1. $f = \dfrac{1.732 \times Length \times Isca}{Runs \times C\text{-}Value \times E\ L\text{-}L}$

2. $M = \dfrac{1}{1 + f}$

3. Isc = Isca x M

4. Add Motor Contribution = Motor FLA x 4

* All results are given in symmetrical amperes

Three-Phase Bolted Fault-Current* Table

One-Way Distance in Feet

Available Isc in Thousands (Isca)

	5	10	15	20	25	30	40	50	60	70	80	90	100	125	150	175	200	225	250	300	350	400	500	600
3	2,981	2,962	2,944	2,926	2,908	2,890	2,855	2,821	2,788	2,755	2,724	2,693	2,662	2,590	2,521	2,455	2,393	2,334	2,278	2,[illegible]73	2,078	1,991	1,836	1,704
5	4,948	4,897	4,846	4,797	4,749	4,702	4,610	4,522	4,437	4,356	4,277	4,201	4,128	3,955	3,797	3,650	3,515	3,389	3,272	3,[illegible]60	2,874	2,710	2,431	2,205
7	6,898	6,799	6,703	6,609	6,518	6,429	6,259	6,098	5,945	5,799	5,660	5,528	5,402	5,110	4,848	4,612	4,398	4,203	4,024	3,709	3,439	3,206	2,824	2,523
10	9,793	9,595	9,404	9,221	9,044	8,875	8,554	8,256	7,977	7,717	7,473	7,244	7,029	6,543	6,120	5,748	5,419	5,126	4,863	4,[illegible]09	4,034	3,717	3,212	2,828
15	14,539	14,106	13,698	13,312	12,948	12,603	11,966	11,390	10,867	10,390	9,953	9,551	9,180	8,369	7,689	7,111	6,614	6,182	5,803	5,[illegible]69	4,660	4,242	3,597	3,123
20	19,189	18,441	17,750	17,108	16,511	15,954	14,947	14,059	13,270	12,565	11,932	11,359	10,839	9,725	8,819	8,067	7,434	6,892	6,424	5,[illegible]57	5,053	4,565	3,827	3,294
25	23,746	22,611	21,580	20,639	19,776	18,983	17,573	16,358	15,301	14,371	13,548	12,815	12,156	10,773	9,672	8,775	8,031	7,402	6,866	5,[illegible]96	5,321	4,784	3,979	3,406
30	28,212	26,624	25,206	23,932	22,780	21,733	19,905	18,361	17,038	15,894	14,894	14,012	13,228	11,606	10,338	9,320	8,485	7,787	7,195	6,245	5,517	4,941	4,088	3,485
35	32,590	30,490	28,644	27,010	25,551	24,242	21,989	20,120	18,543	17,195	16,030	15,013	14,118	12,285	10,874	9,753	8,842	8,087	7,450	6,437	5,666	5,060	4,169	3,544
40	36,883	34,216	31,909	29,893	28,117	26,540	23,864	21,677	19,858	18,321	17,004	15,864	14,867	12,849	11,313	10,105	9,130	8,327	7,654	6,588	5,783	5,153	4,232	3,590
45	41,093	37,809	35,012	32,600	30,499	28,653	25,558	23,066	21,017	19,303	17,847	16,595	15,508	13,324	11,680	10,397	9,368	8,524	7,820	6,711	5,878	5,228	4,282	3,626
50	45,222	41,278	37,966	35,146	32,717	30,601	27,097	24,313	22,047	20,168	18,584	17,231	16,061	13,731	11,991	10,643	9,567	8,689	7,958	6,813	5,955	5,290	4,323	3,655
55	49,273	44,627	40,781	37,546	34,786	32,404	28,501	25,437	22,968	20,936	19,234	17,788	16,544	14,083	12,259	10,853	9,737	8,828	8,075	6,898	6,020	5,341	4,357	3,680
60	53,249	47,863	43,467	39,810	36,721	34,077	29,787	26,457	23,796	21,622	19,811	18,281	16,970	14,390	12,491	11,034	9,882	8,948	8,175	6,971	6,076	5,385	4,386	3,700
65	57,150	50,992	46,032	41,951	38,535	35,634	30,970	27,385	24,545	22,238	20,328	18,719	17,347	14,660	12,694	11,193	10,009	9,052	8,262	7,034	6,124	5,422	4,411	3,718
70	60,980	54,019	48,485	43,979	40,239	37,086	32,061	28,235	25,225	22,795	20,792	19,113	17,684	14,900	12,873	11,332	10,120	9,143	8,337	7,089	6,165	5,455	4,433	3,733
75	64,740	56,949	50,832	45,902	41,843	38,444	33,071	29,015	25,846	23,301	21,212	19,467	17,987	15,115	13,033	11,456	10,219	9,223	8,404	7,137	6,202	5,483	4,451	3,747
80	68,432	59,786	53,080	47,727	43,355	39,716	34,008	29,734	26,415	23,762	21,594	19,788	18,261	15,307	13,176	11,566	10,307	9,295	8,463	7,179	6,234	5,508	4,468	3,758
85	72,057	62,535	55,236	49,463	44,782	40,911	34,880	30,399	26,938	24,185	21,942	20,080	18,509	15,482	13,305	11,665	10,385	9,359	8,516	7,218	6,262	5,531	4,483	3,769
90	75,619	65,200	57,305	51,115	46,133	42,035	35,694	31,015	27,421	24,573	22,261	20,347	18,736	15,640	13,422	11,755	10,456	9,416	8,564	7,252	6,288	5,551	4,496	3,778
95	79,118	67,785	59,292	52,690	47,412	43,094	36,455	31,588	27,868	24,932	22,555	20,592	18,944	15,784	13,528	11,836	10,521	9,468	8,607	7,283	6,311	5,569	4,508	3,786
100	82,555	70,293	61,202	54,193	48,625	44,094	37,168	32,122	28,283	25,263	22,826	20,818	19,134	15,917	13,625	11,911	10,579	9,516	8,646	7,311	6,332	5,585	4,519	3,794
105	85,933	72,727	63,039	55,629	49,778	45,040	37,838	32,621	28,669	25,571	23,077	21,026	19,310	16,038	13,714	11,979	10,633	9,559	8,682	7,336	6,352	5,600	4,528	3,801
110	89,254	75,091	64,808	57,002	50,874	45,936	38,468	33,088	29,029	25,857	23,310	21,219	19,473	16,150	13,796	12,041	10,682	9,599	8,715	7,360	6,369	5,614	4,537	3,807
115	92,518	77,388	66,512	58,316	51,918	46,785	39,062	33,527	29,366	26,124	23,526	21,399	19,624	16,254	13,872	12,099	10,727	9,635	8,745	7,381	6,385	5,626	4,545	3,813
120	95,726	79,621	68,154	59,574	52,913	47,592	39,622	33,939	29,682	26,373	23,729	21,566	19,765	16,350	13,942	12,152	10,769	9,669	8,773	7,401	6,400	5,638	4,553	3,818
125	98,882	81,791	69,738	60,781	53,863	48,359	40,153	34,328	29,978	26,607	23,918	21,722	19,896	16,440	14,007	12,201	10,808	9,700	8,799	7,419	6,414	5,648	4,560	3,823
130	101,984	83,903	71,268	61,940	54,771	49,089	40,655	34,694	30,258	26,827	24,095	21,868	20,018	16,523	14,068	12,247	10,844	9,729	8,822	7,436	6,426	5,658	4,566	3,828
135	105,036	85,958	72,745	63,052	55,639	49,786	41,131	35,040	30,521	27,034	24,262	22,005	20,133	16,602	14,124	12,290	10,878	9,756	8,845	7,452	6,438	5,667	4,572	3,832
140	108,038	87,958	74,172	64,122	56,470	50,450	41,584	35,368	30,769	27,228	24,418	22,134	20,241	16,675	14,177	12,330	10,909	9,782	8,865	7,467	6,449	5,676	4,578	3,836
145	110,992	89,906	75,552	65,151	57,267	51,085	42,014	35,679	31,004	27,412	24,566	22,256	20,342	16,744	14,227	12,368	10,938	9,805	8,885	7,480	6,459	5,684	4,583	3,839
150	113,898	91,803	76,888	66,141	58,031	51,692	42,424	35,974	31,227	27,586	24,706	22,370	20,438	16,808	14,273	12,403	10,966	9,827	8,903	7,493	6,469	5,691	4,588	3,843
155	116,758	93,652	78,180	67,096	58,764	52,273	42,815	36,255	31,438	27,751	24,838	22,478	20,528	16,869	14,317	12,436	10,992	9,848	8,920	7,505	6,478	5,698	4,592	3,846
160	119,573	95,454	79,432	68,016	59,469	52,830	43,187	36,522	31,638	27,907	24,963	22,581	20,613	16,927	14,359	12,467	11,016	9,868	8,936	7,517	6,486	5,705	4,596	3,849
165	122,343	97,212	80,646	68,904	60,146	53,364	43,544	36,776	31,829	28,055	25,081	22,677	20,694	16,981	14,398	12,497	11,039	9,886	8,951	7,527	6,494	5,711	4,600	3,852
170	125,071	98,926	81,822	69,760	60,798	53,876	43,884	37,019	32,011	28,196	25,194	22,770	20,771	17,033	14,435	12,525	11,061	9,904	8,966	7,538	6,502	5,717	4,604	3,854
175	127,756	100,598	82,963	70,588	61,426	54,369	44,210	37,250	32,184	28,330	25,301	22,857	20,844	17,082	14,470	12,551	11,082	9,920	8,979	7,547	6,509	5,722	4,608	3,857
180	130,401	102,231	84,070	71,388	62,030	54,842	44,523	37,472	32,349	28,458	25,403	22,940	20,913	17,128	14,504	12,576	11,101	9,936	8,992	7,556	6,516	5,727	4,611	3,859
185	133,005	103,825	85,145	72,161	62,614	55,297	44,822	37,684	32,507	28,580	25,500	23,020	20,979	17,172	14,535	12,600	11,120	9,951	9,004	7,565	6,522	5,732	4,614	3,861
190	135,570	105,381	86,188	72,910	63,176	55,736	45,110	37,887	32,658	28,697	25,593	23,095	21,041	17,214	14,565	12,623	11,137	9,965	9,016	7,573	6,528	5,737	4,617	3,863
195	138,096	106,901	87,203	73,634	63,720	56,158	45,386	38,082	32,802	28,809	25,682	23,167	21,101	17,254	14,594	12,644	11,154	9,978	9,027	7,581	6,534	5,741	4,620	3,865
200	140,585	108,387	88,189	74,336	64,244	56,565	45,652	38,269	32,941	28,915	25,767	23,236	21,159	17,293	14,621	12,665	11,170	9,991	9,037	7,588	6,539	5,746	4,623	3,867

350 MCM | 1 Run(s) | 208 Volts | C-Value = 19,703 | Magnetic Duct | Copper

Calculations:

1. $f = \dfrac{1.732 \times \text{Length} \times \text{Isca}}{\text{Runs} \times \text{C-Value} \times E_{L\text{-}L}}$

2. $M = \dfrac{1}{1 + f}$

3. Isc = Isca x M

4. Add Motor Contribution = Motor FLA x 4

* All results are given in symmetrical amperes

Three-Phase Bolted Fault-Current* Table

One-Way Distance in Feet

Available Isc in Thousands (Isca)	5	10	15	20	25	30	40	50	60	70	80	90	100	125	150	175	200	225	250	300	350	400	500	600
3	2,991	2,981	2,972	2,962	2,953	2,944	2,926	2,908	2,890	2,873	2,855	2,838	2,821	2,780	2,740	2,700	2,662	2,626	2,590	2,521	2,455	2,393	2,278	2,173
5	4,974	4,948	4,922	4,897	4,871	4,846	4,797	4,749	4,702	4,656	4,610	4,566	4,522	4,417	4,316	4,220	4,128	4,040	3,955	3,797	3,650	3,515	3,272	3,060
7	6,949	6,898	6,848	6,799	6,750	6,703	6,609	6,518	6,429	6,343	6,259	6,178	6,098	5,908	5,729	5,561	5,402	5,252	5,110	4,848	4,612	4,398	4,024	3,709
10	9,895	9,793	9,693	9,595	9,498	9,404	9,221	9,044	8,875	8,711	8,554	8,402	8,256	7,911	7,593	7,300	7,029	6,778	6,543	6,120	5,748	5,419	4,863	4,409
15	14,766	14,539	14,319	14,106	13,899	13,698	13,312	12,948	12,603	12,276	11,966	11,671	11,390	10,743	10,166	9,648	9,180	8,756	8,369	7,689	7,111	6,614	5,803	5,169
20	19,586	19,189	18,808	18,441	18,089	17,750	17,108	16,511	15,954	15,434	14,947	14,489	14,059	13,087	12,240	11,497	10,839	10,252	9,725	8,819	8,067	7,434	6,424	5,657
25	24,357	23,746	23,164	22,611	22,083	21,580	20,639	19,776	18,983	18,251	17,573	16,944	16,358	15,057	13,948	12,990	12,156	11,423	10,773	9,672	8,775	8,031	6,866	5,996
30	29,078	28,212	27,395	26,624	25,896	25,206	23,932	22,780	21,733	20,779	19,905	19,102	18,361	16,737	15,378	14,222	13,228	12,364	11,606	10,338	9,320	8,485	7,195	6,245
35	33,752	32,590	31,505	30,490	29,538	28,644	27,010	25,551	24,242	23,061	21,989	21,013	20,120	18,187	16,593	15,255	14,118	13,138	12,285	10,874	9,753	8,842	7,450	6,437
40	38,378	36,883	35,499	34,216	33,022	31,909	29,893	28,117	26,540	25,131	23,864	22,718	21,677	19,450	17,638	16,134	14,867	13,785	12,849	11,313	10,105	9,130	7,654	6,588
45	42,958	41,093	39,383	37,809	36,357	35,012	32,600	30,499	28,653	27,017	25,558	24,248	23,066	20,561	18,546	16,891	15,508	14,333	13,324	11,680	10,397	9,368	7,820	6,711
50	47,491	45,222	43,160	41,278	39,553	37,966	35,146	32,717	30,601	28,742	27,097	25,629	24,313	21,545	19,344	17,550	16,061	14,805	13,731	11,991	10,643	9,567	7,958	6,813
55	51,979	49,273	46,835	44,627	42,617	40,781	37,546	34,786	32,404	30,327	28,501	26,882	25,437	22,424	20,049	18,129	16,544	15,214	14,083	12,259	10,853	9,737	8,075	6,898
60	56,423	53,249	50,413	47,863	45,559	43,467	39,810	36,721	34,077	31,788	29,787	28,023	26,457	23,212	20,677	18,641	16,970	15,573	14,390	12,491	11,034	9,882	8,175	6,971
65	60,823	57,150	53,896	50,992	48,385	46,032	41,951	38,535	35,634	33,138	30,970	29,068	27,385	23,924	21,240	19,097	17,347	15,891	14,660	12,694	11,193	10,009	8,262	7,034
70	65,179	60,980	57,289	54,019	51,103	48,485	43,979	40,239	37,086	34,391	32,061	30,027	28,235	24,570	21,747	19,506	17,684	16,173	14,900	12,873	11,332	10,120	8,337	7,089
75	69,493	64,740	60,595	56,949	53,717	50,832	45,902	41,843	38,444	35,555	33,071	30,911	29,015	25,159	22,207	19,876	17,987	16,426	15,115	13,033	11,456	10,219	8,404	7,137
80	73,765	68,432	63,818	59,786	56,234	53,080	47,727	43,355	39,716	36,641	34,008	31,728	29,734	25,698	22,626	20,210	18,261	16,654	15,307	13,176	11,566	10,307	8,463	7,179
85	77,995	72,057	66,960	62,535	58,660	55,236	49,463	44,782	40,911	37,656	34,880	32,486	30,399	26,193	23,009	20,515	18,509	16,861	15,482	13,305	11,665	10,385	8,516	7,218
90	82,185	75,619	70,024	65,200	60,998	57,305	51,115	46,133	42,035	38,606	35,694	33,190	31,015	26,649	23,360	20,794	18,736	17,049	15,640	13,422	11,755	10,456	8,564	7,252
95	86,334	79,118	73,014	67,785	63,255	59,292	52,690	47,412	43,094	39,497	36,455	33,847	31,588	27,071	23,684	21,050	18,944	17,220	15,784	13,528	11,836	10,521	8,607	7,283
100	90,444	82,555	75,932	70,293	65,433	61,202	54,193	48,625	44,094	40,336	37,168	34,461	32,122	27,462	23,983	21,286	19,134	17,378	15,917	13,625	11,911	10,579	8,646	7,311
105	94,515	85,933	78,781	72,727	67,538	63,039	55,629	49,778	45,040	41,126	37,838	35,036	32,621	27,826	24,260	21,504	19,310	17,523	16,038	13,714	11,979	10,633	8,682	7,336
110	98,547	89,254	81,562	75,091	69,572	64,808	57,002	50,874	45,936	41,871	38,468	35,576	33,088	28,165	24,517	21,706	19,473	17,657	16,150	13,796	12,041	10,682	8,715	7,360
115	102,541	92,518	84,279	77,388	71,539	66,512	58,316	51,918	46,785	42,576	39,062	36,083	33,527	28,482	24,757	21,894	19,624	17,781	16,254	13,872	12,099	10,727	8,745	7,381
120	106,498	95,726	86,934	79,621	73,442	68,154	59,574	52,913	47,592	43,243	39,622	36,561	33,939	28,779	24,981	22,069	19,765	17,896	16,350	13,942	12,152	10,769	8,773	7,401
125	110,417	98,882	89,528	81,791	75,285	69,738	60,781	53,863	48,359	43,875	40,153	37,012	34,328	29,058	25,191	22,232	19,896	18,003	16,440	14,007	12,201	10,808	8,799	7,419
130	114,301	101,984	92,064	83,903	77,071	71,268	61,940	54,771	49,089	44,476	40,655	37,439	34,694	29,320	25,388	22,386	20,018	18,104	16,523	14,068	12,247	10,844	8,822	7,436
135	118,148	105,036	94,544	85,958	78,801	72,745	63,052	55,639	49,786	45,047	41,131	37,842	35,040	29,567	25,573	22,529	20,133	18,198	16,602	14,124	12,290	10,878	8,845	7,452
140	121,960	108,038	96,969	87,958	80,479	74,172	64,122	56,470	50,450	45,590	41,584	38,225	35,368	29,800	25,747	22,664	20,241	18,286	16,675	14,177	12,330	10,909	8,865	7,467
145	125,737	110,992	99,342	89,906	82,106	75,552	65,151	57,267	51,085	46,108	42,014	38,588	35,679	30,021	25,911	22,792	20,342	18,368	16,744	14,227	12,368	10,938	8,885	7,480
150	129,480	113,898	101,664	91,803	83,686	76,888	66,141	58,031	51,692	46,602	42,424	38,934	35,974	30,229	26,067	22,912	20,438	18,446	16,808	14,273	12,403	10,966	8,903	7,493
155	133,188	116,758	103,936	93,652	85,220	78,180	67,096	58,764	52,273	47,073	42,815	39,262	36,255	30,427	26,214	23,025	20,528	18,520	16,869	14,317	12,436	10,992	8,920	7,505
160	136,863	119,573	106,161	95,454	86,709	79,432	68,016	59,469	52,830	47,525	43,187	39,576	36,522	30,615	26,353	23,132	20,613	18,589	16,927	14,359	12,467	11,016	8,936	7,517
165	140,505	122,343	108,339	97,212	88,157	80,646	68,904	60,146	53,364	47,956	43,544	39,875	36,776	30,793	26,485	23,234	20,694	18,655	16,981	14,398	12,497	11,039	8,951	7,527
170	144,115	125,071	110,473	98,926	89,565	81,822	69,760	60,798	53,876	48,370	43,884	40,160	37,019	30,963	26,611	23,331	20,771	18,717	17,033	14,435	12,525	11,061	8,966	7,538
175	147,692	127,756	112,563	100,598	90,933	82,963	70,588	61,426	54,369	48,766	44,210	40,433	37,250	31,125	26,730	23,423	20,844	18,776	17,082	14,470	12,551	11,082	8,979	7,547
180	151,238	130,401	114,610	102,231	92,265	84,070	71,388	62,030	54,842	49,147	44,523	40,694	37,472	31,280	26,844	23,510	20,913	18,832	17,128	14,504	12,576	11,101	8,992	7,556
185	154,752	133,005	116,617	103,825	93,561	85,145	72,161	62,614	55,297	49,512	44,822	40,944	37,684	31,427	26,953	23,593	20,979	18,886	17,172	14,535	12,600	11,120	9,004	7,565
190	158,235	135,570	118,584	105,381	94,823	86,188	72,910	63,176	55,736	49,863	45,110	41,184	37,887	31,569	27,056	23,673	21,041	18,936	17,214	14,565	12,623	11,137	9,016	7,573
195	161,688	138,096	120,513	106,901	96,052	87,203	73,634	63,720	56,158	50,201	45,386	41,414	38,082	31,704	27,156	23,749	21,101	18,985	17,254	14,594	12,644	11,154	9,027	7,581
200	165,110	140,585	122,404	108,387	97,250	88,189	74,336	64,244	56,565	50,526	45,652	41,635	38,269	31,833	27,250	23,821	21,159	19,031	17,293	14,621	12,665	11,170	9,037	7,588

350 MCM	2 Run(s)	208 Volts	Copyright © 1994 - V.F. Christoffer - All Rights Reserved	C-Value = 19,703	Magnetic Duct	Copper

Calculations:

1. $f = \dfrac{1.732 \times \text{Length} \times \text{Isca}}{\text{Runs} \times \text{C-Value} \times \text{E L-L}}$

2. $M = \dfrac{1}{1 + f}$

3. $\text{Isc} = \text{Isca} \times M$

4. Add Motor Contribution = Motor FLA x 4

* All results are given in symmetrical amperes

Three-Phase Bolted Fault-Current* Table

One-Way Distance in Feet

Available Isc in Thousands (Isca)	5	10	15	20	25	30	40	50	60	70	80	90	100	125	150	175	200	225	250	300	350	400	500	600
3	2,982	2,964	2,946	2,929	2,912	2,895	2,861	2,828	2,796	2,765	2,734	2,704	2,675	2,605	2,538	2,474	2,414	2,356	2,301	2,199	2,105	2,019	1,866	1,735
5	4,950	4,901	4,853	4,805	4,759	4,714	4,625	4,540	4,458	4,379	4,303	4,229	4,158	3,990	3,835	3,692	3,559	3,435	3,320	3,111	2,926	2,763	2,485	2,258
7	6,902	6,807	6,715	6,624	6,537	6,451	6,287	6,131	5,983	5,841	5,706	5,577	5,454	5,169	4,912	4,679	4,468	4,274	4,097	3,783	3,514	3,281	2,896	2,592
10	9,802	9,611	9,427	9,251	9,081	8,917	8,606	8,316	8,045	7,792	7,553	7,329	7,118	6,640	6,221	5,853	5,525	5,233	4,970	4,515	4,137	3,817	3,306	2,916
15	14,558	14,141	13,748	13,375	13,023	12,688	12,068	11,506	10,994	10,525	10,095	9,699	9,332	8,527	7,849	7,271	6,773	6,338	5,956	5,315	4,799	4,374	3,716	3,230
20	19,222	18,502	17,834	17,212	16,633	16,091	15,107	14,236	13,460	12,764	12,137	11,568	11,051	9,939	9,030	8,274	7,635	7,087	6,613	5,332	5,216	4,718	3,961	3,414
25	23,796	22,702	21,704	20,791	19,951	19,176	17,795	16,599	15,553	14,632	13,814	13,082	12,424	11,036	9,927	9,020	8,266	7,628	7,081	6,193	5,503	4,951	4,124	3,534
30	28,282	26,751	25,376	24,136	23,012	21,987	20,190	18,664	17,353	16,214	15,215	14,332	13,546	11,912	10,630	9,598	8,748	8,036	7,432	6,460	5,713	5,120	4,241	3,620
35	32,684	30,656	28,864	27,271	25,844	24,559	22,338	20,485	18,916	17,570	16,403	15,382	14,480	12,629	11,197	10,057	9,128	8,356	7,704	6,665	5,872	5,248	4,329	3,683
40	37,003	34,425	32,182	30,213	28,472	26,920	24,274	22,102	20,286	18,746	17,424	16,276	15,269	13,225	11,664	10,432	9,436	8,613	7,922	6,827	5,998	5,349	4,396	3,732
45	41,243	38,064	35,341	32,981	30,917	29,096	26,029	23,547	21,498	19,776	18,310	17,046	15,946	13,730	12,054	10,743	9,690	8,824	8,101	6,959	6,100	5,429	4,451	3,771
50	45,404	41,582	38,353	35,590	33,198	31,107	27,627	24,848	22,576	20,685	19,087	17,717	16,531	14,162	12,386	11,006	9,903	9,001	8,249	7,069	6,184	5,496	4,495	3,803
55	49,489	44,982	41,228	38,052	35,330	32,972	29,088	26,023	23,543	21,494	19,773	18,307	17,044	14,536	12,671	11,231	10,084	9,150	8,375	7,161	6,254	5,551	4,532	3,830
60	53,501	48,272	43,975	40,380	37,328	34,705	30,429	27,091	24,413	22,217	20,384	18,829	17,496	14,863	12,919	11,425	10,241	9,279	8,482	7,239	6,314	5,598	4,564	3,852
65	57,441	51,457	46,602	42,584	39,204	36,322	31,665	28,066	25,202	22,869	20,930	19,295	17,897	15,152	13,137	11,595	10,377	9,391	8,575	7,307	6,365	5,639	4,591	3,871
70	61,311	54,541	49,118	44,675	40,969	37,832	32,806	28,959	25,920	23,458	21,423	19,713	18,256	15,409	13,329	11,745	10,497	9,489	8,657	7,366	6,410	5,674	4,614	3,888
75	65,113	57,529	51,528	46,660	42,633	39,246	33,864	29,781	26,576	23,994	21,869	20,090	18,579	15,638	13,501	11,878	10,603	9,575	8,729	7,418	6,449	5,705	4,634	3,902
80	68,849	60,426	53,840	48,548	44,203	40,573	34,848	30,539	27,178	24,484	22,275	20,432	18,871	15,844	13,654	11,996	10,697	9,652	8,793	7,464	6,484	5,732	4,652	3,915
85	72,520	63,236	56,059	50,345	45,688	41,820	35,764	31,240	27,732	24,933	22,646	20,744	19,137	16,031	13,793	12,103	10,782	9,721	8,850	7,505	6,515	5,756	4,668	3,926
90	76,129	65,962	58,191	52,058	47,095	42,995	36,620	31,891	28,244	25,346	22,987	21,029	19,379	16,201	13,918	12,200	10,859	9,783	8,902	7,542	6,543	5,778	4,682	3,936
95	79,676	68,609	60,241	53,693	48,428	44,104	37,421	32,497	28,718	25,727	23,300	21,291	19,601	16,356	14,033	12,287	10,928	9,840	8,948	7,576	6,568	5,797	4,695	3,945
100	83,163	71,179	62,214	55,254	49,695	45,152	38,173	33,063	29,159	26,080	23,589	21,532	19,806	16,498	14,137	12,367	10,991	9,891	8,991	7,606	6,591	5,815	4,707	3,953
105	86,592	73,676	64,113	56,747	50,900	46,145	38,880	33,592	29,570	26,408	23,857	21,755	19,994	16,629	14,233	12,440	11,049	9,938	9,029	7,634	6,612	5,831	4,717	3,961
110	89,965	76,104	65,943	58,177	52,046	47,085	39,546	34,087	29,953	26,713	24,106	21,962	20,169	16,749	14,321	12,508	11,102	9,981	9,065	7,659	6,631	5,846	4,727	3,968
115	93,282	78,464	67,708	59,546	53,140	47,978	40,174	34,553	30,312	26,998	24,338	22,155	20,331	16,861	14,403	12,570	11,151	10,020	9,097	7,682	6,648	5,860	4,736	3,974
120	96,545	80,760	69,411	60,859	54,183	48,827	40,767	34,991	30,649	27,265	24,554	22,334	20,482	16,964	14,478	12,623	11,196	10,057	9,127	7,704	6,664	5,872	4,744	3,980
125	99,755	82,994	71,055	62,119	55,179	49,635	41,329	35,404	30,965	27,515	24,757	22,501	20,622	17,061	14,548	12,681	11,238	10,090	9,155	7,724	6,679	5,884	4,752	3,985
130	102,914	85,169	72,643	63,329	56,132	50,404	41,861	35,794	31,263	27,750	24,947	22,658	20,754	17,151	14,614	12,731	11,277	10,122	9,181	7,742	6,693	5,894	4,759	3,990
135	106,023	87,287	74,178	64,493	57,045	51,139	42,366	36,163	31,544	27,971	25,126	22,805	20,878	17,235	14,675	12,777	11,314	10,151	9,205	7,759	6,706	5,904	4,765	3,994
140	109,082	89,350	75,663	65,612	57,919	51,840	42,846	36,512	31,809	28,180	25,294	22,944	20,994	17,314	14,732	12,820	11,348	10,178	9,228	7,775	6,718	5,913	4,771	3,999
145	112,094	91,361	77,100	66,690	58,757	52,511	43,303	36,843	32,061	28,377	25,452	23,074	21,103	17,388	14,786	12,861	11,379	10,204	9,249	7,790	6,729	5,922	4,777	4,003
150	115,059	93,321	78,491	67,729	59,562	53,152	43,739	37,158	32,299	28,563	25,602	23,197	21,206	17,458	14,836	12,899	11,409	10,228	9,268	7,804	6,739	5,930	4,782	4,006
155	117,978	95,232	79,839	68,730	60,334	53,767	44,154	37,457	32,524	28,740	25,744	23,314	21,303	17,524	14,884	12,935	11,437	10,251	9,287	7,817	6,749	5,938	4,787	4,010
160	120,853	97,096	81,145	69,695	61,077	54,356	44,551	37,742	32,739	28,907	25,878	23,424	21,395	17,586	14,928	12,969	11,464	10,272	9,304	7,829	6,758	5,945	4,791	4,013
165	123,684	98,915	82,412	70,628	61,792	54,921	44,930	38,014	32,943	29,066	26,006	23,528	21,482	17,645	14,971	13,001	11,489	10,292	9,321	7,841	6,767	5,952	4,796	4,016
170	126,472	100,690	83,640	71,528	62,480	55,464	45,293	38,274	33,138	29,218	26,127	23,627	21,564	17,700	15,011	13,031	11,512	10,311	9,336	7,852	6,775	5,958	4,800	4,019
175	129,219	102,424	84,833	72,399	63,143	55,986	45,640	38,521	33,324	29,362	26,242	23,721	21,643	17,753	15,049	13,059	11,535	10,329	9,351	7,862	6,783	5,964	4,804	4,022
180	131,925	104,116	85,991	73,240	63,783	56,488	45,973	38,758	33,501	29,499	26,352	23,811	21,717	17,803	15,085	13,087	11,556	10,346	9,365	7,872	6,790	5,969	4,807	4,024
185	134,591	105,770	87,115	74,055	64,399	56,972	46,293	38,985	33,670	29,631	26,456	23,897	21,788	17,851	15,119	13,112	11,576	10,362	9,378	7,882	6,797	5,975	4,811	4,027
190	137,218	107,386	88,209	74,843	64,995	57,437	46,599	39,203	33,832	29,756	26,556	23,978	21,856	17,897	15,152	13,137	11,595	10,377	9,391	7,890	6,804	5,980	4,814	4,029
195	139,807	108,965	89,271	75,607	65,570	57,886	46,894	39,411	33,987	29,876	26,652	24,056	21,921	17,940	15,183	13,160	11,613	10,392	9,402	7,899	6,810	5,985	4,817	4,031
200	142,358	110,509	90,305	76,347	66,126	58,318	47,178	39,611	34,136	29,991	26,743	24,130	21,982	17,981	15,212	13,182	11,630	10,405	9,414	7,907	6,816	5,989	4,820	4,033

400 MCM	1 Run(s)	208 Volts	C-Value = 20,565	Magnetic Duct	Copper

Calculations:

1. $f = \dfrac{1.732 \times \text{Length} \times \text{Isca}}{\text{Runs} \times \text{C-Value} \times E_{L-L}}$

2. $M = \dfrac{1}{1 + f}$

3. Isc = Isca x M

4. Add Motor Contribution = Motor FLA x 4

* All results are given in symmetrical amperes

Three-Phase Bolted Fault-Current* Table

One-Way Distance in Feet

Available Isc in Thousands (Isca)

	5	10	15	20	25	30	40	50	60	70	80	90	100	125	150	175	200	225	250	300	350	400	500	600
3	2,991	2,982	2,973	2,964	2,955	2,946	2,929	2,912	2,895	2,878	2,861	2,845	2,828	2,788	2,750	2,712	2,675	2,639	2,605	2,538	2,474	2,414	2,301	2,199
5	4,975	4,950	4,925	4,901	4,877	4,853	4,805	4,759	4,714	4,669	4,625	4,583	4,540	4,438	4,341	4,248	4,158	4,072	3,990	3,835	3,692	3,559	3,320	3,111
7	6,951	6,902	6,854	6,807	6,760	6,715	6,624	6,537	6,451	6,368	6,287	6,208	6,131	5,947	5,773	5,609	5,454	5,308	5,169	4,912	4,679	4,468	4,097	3,783
10	9,900	9,802	9,705	9,611	9,518	9,427	9,251	9,081	8,917	8,759	8,606	8,459	8,316	7,980	7,671	7,384	7,118	6,870	6,640	6,221	5,853	5,525	4,970	4,515
15	14,776	14,558	14,346	14,141	13,942	13,748	13,375	13,023	12,688	12,370	12,068	11,780	11,506	10,873	10,306	9,795	9,332	8,911	8,527	7,849	7,271	6,773	5,956	5,315
20	19,603	19,222	18,855	18,502	18,162	17,834	17,212	16,633	16,091	15,583	15,107	14,658	14,236	13,279	12,443	11,706	11,051	10,465	9,939	9,030	8,274	7,635	6,613	5,832
25	24,383	23,796	23,236	22,702	22,192	21,704	20,791	19,951	19,176	18,460	17,795	17,176	16,599	15,312	14,211	13,257	12,424	11,689	11,036	9,927	9,020	8,266	7,081	6,193
30	29,116	28,282	27,495	26,751	26,045	25,376	24,136	23,012	21,987	21,050	20,190	19,397	18,664	17,053	15,698	14,543	13,546	12,677	11,912	10,630	9,598	8,748	7,432	6,460
35	33,802	32,684	31,637	30,656	29,733	28,864	27,271	25,844	24,559	23,396	22,338	21,371	20,485	18,560	16,967	15,625	14,480	13,491	12,629	11,197	10,057	9,128	7,704	6,665
40	38,443	37,003	35,667	34,425	33,265	32,182	30,213	28,472	26,920	25,529	24,274	23,137	22,102	19,878	18,061	16,548	15,269	14,174	13,225	11,664	10,432	9,436	7,922	6,827
45	43,039	41,243	39,590	38,064	36,652	35,341	32,981	30,917	29,096	27,477	26,029	24,726	23,547	21,040	19,015	17,346	15,946	14,755	13,730	12,054	10,743	9,690	8,101	6,959
50	47,591	45,404	43,409	41,582	39,902	38,353	35,590	33,198	31,107	29,264	27,627	26,164	24,848	22,072	19,854	18,041	16,531	15,255	14,162	12,386	11,006	9,903	8,249	7,069
55	52,099	49,489	47,128	44,982	43,023	41,228	38,052	35,330	32,972	30,908	29,088	27,471	26,023	22,995	20,597	18,653	17,044	15,690	14,536	12,671	11,231	10,084	8,375	7,161
60	56,564	53,501	50,752	48,272	46,024	43,975	40,380	37,328	34,705	32,427	30,429	28,664	27,091	23,825	21,261	19,195	17,496	16,072	14,863	12,919	11,425	10,241	8,482	7,239
65	60,987	57,441	54,285	51,457	48,909	46,602	42,584	39,204	36,322	33,834	31,665	29,757	28,066	24,575	21,857	19,680	17,897	16,410	15,152	13,137	11,595	10,377	8,575	7,307
70	65,368	61,311	57,728	54,541	51,687	49,118	44,675	40,969	37,832	35,140	32,806	30,763	28,959	25,257	22,395	20,115	18,256	16,712	15,409	13,329	11,745	10,497	8,657	7,366
75	69,708	65,113	61,087	57,529	54,364	51,528	46,660	42,633	39,246	36,357	33,864	31,692	29,781	25,880	22,883	20,507	18,579	16,982	15,638	13,501	11,878	10,603	8,729	7,418
80	74,007	68,849	64,363	60,426	56,943	53,840	48,548	44,203	40,573	37,493	34,848	32,551	30,539	26,450	23,327	20,864	18,871	17,226	15,844	13,654	11,996	10,697	8,793	7,464
85	78,266	72,520	67,561	63,236	59,432	56,059	50,345	45,688	41,820	38,556	35,764	33,349	31,240	26,975	23,734	21,189	19,137	17,447	16,031	13,793	12,103	10,782	8,850	7,505
90	82,485	76,129	70,682	65,962	61,834	58,191	52,058	47,095	42,995	39,552	36,620	34,093	31,891	27,459	24,108	21,487	19,379	17,648	16,201	13,918	12,200	10,859	8,902	7,542
95	86,666	79,676	73,729	68,609	64,153	60,241	53,693	48,428	44,104	40,489	37,421	34,786	32,497	27,907	24,453	21,760	19,601	17,832	16,356	14,033	12,287	10,928	8,948	7,576
100	90,808	83,163	76,706	71,179	66,395	62,214	55,254	49,695	45,152	41,371	38,173	35,435	33,063	28,323	24,772	22,012	19,806	18,001	16,498	14,137	12,367	10,991	8,991	7,606
105	94,912	86,592	79,614	73,676	68,563	64,113	56,747	50,900	46,145	42,202	38,880	36,043	33,592	28,710	25,068	22,245	19,994	18,157	16,629	14,233	12,440	11,049	9,029	7,634
110	98,979	89,965	82,456	76,104	70,660	65,943	58,177	52,046	47,085	42,987	39,546	36,614	34,087	29,072	25,343	22,462	20,169	18,301	16,749	14,321	12,508	11,102	9,065	7,659
115	103,009	93,282	85,234	78,464	72,690	67,708	59,546	53,140	47,978	43,730	40,174	37,152	34,553	29,410	25,599	22,663	20,331	18,434	16,861	14,403	12,570	11,151	9,097	7,682
120	107,002	96,545	87,950	80,760	74,657	69,411	60,859	54,183	48,827	44,434	40,767	37,659	34,991	29,726	25,839	22,850	20,482	18,558	16,964	14,478	12,628	11,196	9,127	7,704
125	110,960	99,755	90,606	82,994	76,562	71,055	62,119	55,179	49,635	45,102	41,329	38,138	35,404	30,024	26,063	23,026	20,622	18,673	17,061	14,548	12,681	11,238	9,155	7,724
130	114,882	102,914	93,204	85,169	78,409	72,643	63,329	56,132	50,404	45,737	41,861	38,590	35,794	30,304	26,274	23,190	20,754	18,781	17,151	14,614	12,731	11,277	9,181	7,742
135	118,769	106,023	95,747	87,287	80,200	74,178	64,493	57,045	51,139	46,341	42,366	39,019	36,163	30,568	26,472	23,344	20,878	18,882	17,235	14,675	12,777	11,314	9,205	7,759
140	122,622	109,082	98,235	89,350	81,939	75,663	65,612	57,919	51,840	46,916	42,846	39,426	36,512	30,817	26,659	23,489	20,994	18,977	17,314	14,732	12,820	11,348	9,228	7,775
145	126,441	112,094	100,671	91,361	83,627	77,100	66,690	58,757	52,511	47,465	43,303	39,813	36,843	31,053	26,835	23,626	21,103	19,066	17,388	14,786	12,861	11,379	9,249	7,790
150	130,226	115,059	103,056	93,321	85,266	78,491	67,729	59,562	53,152	47,988	43,739	40,181	37,158	31,276	27,002	23,755	21,206	19,150	17,458	14,836	12,899	11,409	9,268	7,804
155	133,979	117,978	105,392	95,232	86,859	79,839	68,730	60,334	53,767	48,489	44,154	40,531	37,457	31,488	27,159	23,877	21,303	19,229	17,524	14,884	12,935	11,437	9,287	7,817
160	137,698	120,853	107,680	97,096	88,407	81,145	69,695	61,077	54,356	48,967	44,551	40,865	37,742	31,689	27,309	23,993	21,395	19,304	17,586	14,928	12,969	11,464	9,304	7,829
165	141,385	123,684	109,921	98,915	89,912	82,412	70,628	61,792	54,921	49,426	44,930	41,184	38,014	31,880	27,451	24,102	21,482	19,375	17,645	14,971	13,001	11,489	9,321	7,841
170	145,041	126,472	112,118	100,690	91,377	83,640	71,528	62,480	55,464	49,865	45,293	41,488	38,274	32,063	27,586	24,206	21,564	19,442	17,700	15,011	13,031	11,512	9,336	7,852
175	148,665	129,219	114,271	102,424	92,802	84,833	72,399	63,143	55,986	50,287	45,640	41,780	38,521	32,236	27,714	24,305	21,643	19,506	17,753	15,049	13,059	11,535	9,351	7,862
180	152,257	131,925	116,382	104,116	94,189	85,991	73,240	63,783	56,488	50,691	45,973	42,059	38,758	32,402	27,837	24,399	21,717	19,567	17,803	15,085	13,087	11,556	9,365	7,872
185	155,820	134,591	118,452	105,770	95,541	87,115	74,055	64,399	56,972	51,080	46,293	42,326	38,985	32,560	27,954	24,489	21,788	19,624	17,851	15,119	13,112	11,576	9,378	7,882
190	159,352	137,218	120,482	107,386	96,857	88,209	74,843	64,995	57,437	51,454	46,599	42,582	39,203	32,712	28,065	24,575	21,856	19,679	17,897	15,152	13,137	11,595	9,391	7,890
195	162,854	139,807	122,474	108,965	98,140	89,271	75,607	65,570	57,886	51,814	46,894	42,828	39,411	32,857	28,172	24,656	21,921	19,732	17,940	15,183	13,160	11,613	9,402	7,899
200	166,327	142,358	124,428	110,509	99,390	90,305	76,347	66,126	58,318	52,160	47,178	43,065	39,611	32,996	28,274	24,734	21,982	19,782	17,981	15,212	13,182	11,630	9,414	7,907

400 MCM	2 Run(s)	208 Volts	C-Value = 20,565	Magnetic Duct	Copper

Calculations:

1. $f = \dfrac{1.732 \times \text{Length} \times \text{Isca}}{\text{Runs} \times \text{C-Value} \times E_{L\text{-}L}}$

2. $M = \dfrac{1}{1 + f}$

3. $\text{Isc} = \text{Isca} \times M$

4. Add Motor Contribution = Motor FLA x 4

* All results are given in symmetrical amperes

Three-Phase Bolted Fault-Current* Table

One-Way Distance in Feet

Available Isc in Thousands (Isca)	5	10	15	20	25	30	40	50	60	70	80	90	100	125	150	175	200	225	250	300	350	400	500	600
3	2,983	2,967	2,950	2,934	2,918	2,902	2,871	2,840	2,810	2,781	2,752	2,724	2,696	2,630	2,567	2,506	2,449	2,394	2,341	2,242	2,152	2,068	1,919	1,790
5	4,954	4,908	4,863	4,819	4,776	4,733	4,651	4,571	4,494	4,419	4,347	4,278	4,210	4,050	3,902	3,764	3,635	3,516	3,403	3,[illegible]99	3,018	2,856	2,580	2,352
7	6,909	6,821	6,735	6,651	6,569	6,489	6,334	6,187	6,047	5,913	5,784	5,661	5,544	5,269	5,021	4,795	4,589	4,399	4,225	3,915	3,647	3,413	3,025	2,717
10	9,816	9,638	9,467	9,302	9,142	8,988	8,695	8,420	8,162	7,919	7,691	7,475	7,271	6,807	6,398	6,036	5,712	5,421	5,159	4,704	4,322	3,998	3,476	3,075
15	14,589	14,200	13,832	13,482	13,149	12,833	12,243	11,705	11,212	10,760	10,342	9,955	9,597	8,804	8,132	7,556	7,055	6,617	6,230	5,578	5,050	4,612	3,932	3,426
20	19,276	18,603	17,976	17,389	16,840	16,324	15,381	14,542	13,789	13,111	12,496	11,936	11,424	10,318	9,407	8,644	7,996	7,438	6,952	6,150	5,514	4,997	4,208	3,634
25	23,880	22,855	21,915	21,050	20,250	19,508	18,177	17,016	15,995	15,089	14,280	13,554	12,898	11,505	10,384	9,462	8,691	8,035	7,472	6,553	5,835	5,259	4,392	3,771
30	28,401	26,964	25,665	24,486	23,410	22,425	20,684	19,194	17,904	16,777	15,783	14,900	14,111	12,461	11,156	10,099	9,225	8,490	7,864	6,352	6,072	5,450	4,525	3,868
35	32,843	30,936	29,238	27,718	26,347	25,106	22,944	21,124	19,573	18,233	17,065	16,038	15,127	13,247	11,782	10,609	9,649	8,848	8,170	7,083	6,252	5,596	4,624	3,940
40	37,207	34,778	32,648	30,763	29,084	27,578	24,991	22,848	21,044	19,503	18,173	17,012	15,991	13,905	12,300	11,027	9,993	9,136	8,415	7,267	6,395	5,710	4,702	3,997
45	41,496	38,498	35,904	33,637	31,640	29,866	26,856	24,397	22,350	20,620	19,139	17,856	16,735	14,463	12,735	11,376	10,279	9,374	8,616	7,417	6,511	5,802	4,764	4,042
50	45,711	42,099	39,017	36,355	34,033	31,990	28,560	25,795	23,518	21,610	19,989	18,594	17,381	14,944	13,106	11,671	10,519	9,574	8,785	7,541	6,606	5,878	4,815	4,078
55	49,854	45,589	41,996	38,928	36,277	33,965	30,125	27,064	24,569	22,494	20,743	19,245	17,948	15,361	13,426	11,924	10,724	9,743	8,927	7,646	6,687	5,941	4,858	4,109
60	53,928	48,971	44,850	41,368	38,387	35,808	31,565	28,222	25,519	23,288	21,416	19,823	18,450	15,727	13,705	12,143	10,901	9,889	9,050	7,736	6,755	5,995	4,894	4,134
65	57,933	52,252	47,586	43,684	40,374	37,531	32,897	29,281	26,382	24,005	22,021	20,340	18,897	16,051	13,950	12,335	11,056	10,016	9,156	7,813	6,814	6,042	4,925	4,156
70	61,872	55,435	50,211	45,887	42,249	39,145	34,130	30,255	27,169	24,655	22,567	20,805	19,298	16,339	14,167	12,505	11,191	10,128	9,249	7,881	6,866	6,082	4,952	4,176
75	65,746	58,525	52,733	47,984	44,020	40,661	35,277	31,152	27,891	25,248	23,062	21,225	19,659	16,597	14,361	12,655	11,312	10,227	9,331	7,941	6,911	6,117	4,975	4,192
80	69,557	61,526	55,157	49,983	45,697	42,087	36,346	31,983	28,555	25,791	23,514	21,607	19,986	16,830	14,535	12,790	11,420	10,314	9,404	7,993	6,951	6,149	4,996	4,207
85	73,306	64,441	57,488	51,890	47,285	43,431	37,344	32,753	29,167	26,289	23,928	21,956	20,284	17,041	14,692	12,912	11,516	10,393	9,470	8,041	6,986	6,177	5,014	4,220
90	76,995	67,274	59,733	53,712	48,793	44,700	38,278	33,469	29,734	26,749	24,308	22,276	20,557	17,233	14,834	13,022	11,604	10,464	9,529	8,083	7,019	6,202	5,031	4,232
95	80,626	70,029	61,895	55,454	50,226	45,900	39,154	34,137	30,260	27,174	24,659	22,570	20,807	17,408	14,964	13,121	11,683	10,529	9,582	8,122	7,047	6,224	5,045	4,242
100	84,198	72,709	63,979	57,121	51,590	47,036	39,978	34,762	30,750	27,568	24,983	22,841	21,038	17,569	15,083	13,213	11,755	10,587	9,631	8,156	7,074	6,245	5,059	4,252
105	87,715	75,317	65,990	58,718	52,890	48,114	40,754	35,347	31,207	27,935	25,284	23,092	21,250	17,718	15,192	13,296	11,821	10,641	9,675	8,188	7,098	6,263	5,071	4,260
110	91,178	77,855	67,930	60,249	54,129	49,137	41,486	35,896	31,634	28,277	25,564	23,326	21,448	17,854	15,292	13,373	11,882	10,690	9,716	8,217	7,119	6,280	5,082	4,268
115	94,586	80,327	69,804	61,719	55,312	50,111	42,178	36,413	32,035	28,596	25,825	23,543	21,631	17,981	15,385	13,444	11,938	10,736	9,753	8,244	7,140	6,296	5,093	4,275
120	97,943	82,735	71,616	63,131	56,443	51,037	42,832	36,900	32,411	28,896	26,068	23,745	21,802	18,099	15,472	13,510	11,990	10,778	9,788	8,269	7,158	6,310	5,102	4,282
125	101,248	85,082	73,367	64,488	57,526	51,921	43,453	37,359	32,765	29,177	26,297	23,935	21,962	18,209	15,552	13,571	12,038	10,816	9,820	8,292	7,175	6,324	5,111	4,288
130	104,504	87,369	75,061	65,793	58,562	52,763	44,041	37,794	33,099	29,441	26,511	24,112	22,111	18,312	15,627	13,628	12,083	10,853	9,850	8,313	7,191	6,336	5,119	4,294
135	107,711	89,599	76,702	67,050	59,556	53,569	44,601	38,205	33,414	29,690	26,713	24,279	22,251	18,408	15,697	13,681	12,125	10,886	9,877	8,333	7,206	6,347	5,126	4,299
140	110,870	91,775	78,290	68,261	60,509	54,339	45,134	38,595	33,712	29,925	26,903	24,436	22,383	18,498	15,762	13,731	12,164	10,918	9,903	8,351	7,220	6,358	5,133	4,304
145	113,983	93,897	79,830	69,428	61,425	55,076	45,641	38,966	33,994	30,147	27,083	24,584	22,507	18,582	15,823	13,778	12,200	10,947	9,927	8,368	7,232	6,368	5,140	4,308
150	117,050	95,969	81,322	70,554	62,305	55,782	46,125	39,318	34,262	30,358	27,252	24,724	22,624	18,662	15,881	13,821	12,235	10,975	9,950	8,384	7,245	6,377	5,146	4,313
155	120,072	97,991	82,770	71,641	63,151	56,459	46,587	39,653	34,516	30,557	27,413	24,856	22,735	18,737	15,936	13,863	12,267	11,001	9,971	8,400	7,256	6,386	5,151	4,317
160	123,051	99,966	84,174	72,691	63,965	57,110	47,029	39,973	34,758	30,747	27,566	24,981	22,839	18,808	15,987	13,902	12,297	11,025	9,992	8,414	7,266	6,394	5,157	4,321
165	125,987	101,895	85,538	73,706	64,750	57,734	47,451	40,278	34,988	30,927	27,710	25,100	22,939	18,876	16,035	13,938	12,326	11,048	10,010	8,427	7,276	6,402	5,162	4,324
170	128,882	103,780	86,862	74,687	65,506	58,334	47,856	40,569	35,208	31,098	27,848	25,212	23,033	18,939	16,081	13,973	12,353	11,070	10,028	8,440	7,286	6,409	5,167	4,327
175	131,735	105,622	88,149	75,637	66,235	58,912	48,244	40,848	35,417	31,262	27,979	25,320	23,122	19,000	16,125	14,005	12,379	11,091	10,045	8,452	7,295	6,416	5,171	4,331
180	134,549	107,423	89,400	76,556	66,939	59,468	48,616	41,114	35,618	31,418	28,103	25,422	23,207	19,057	16,166	14,037	12,403	11,110	10,061	8,463	7,303	6,423	5,175	4,334
185	137,323	109,185	90,617	77,446	67,618	60,004	48,974	41,369	35,809	31,566	28,223	25,519	23,289	19,112	16,206	14,067	12,426	11,129	10,077	8,474	7,311	6,429	5,179	4,336
190	140,059	110,907	91,800	78,309	68,275	60,520	49,318	41,614	35,992	31,709	28,336	25,612	23,366	19,164	16,243	14,095	12,448	11,146	10,091	8,484	7,319	6,435	5,183	4,339
195	142,757	112,592	92,951	79,145	68,910	61,019	49,648	41,849	36,168	31,845	28,445	25,701	23,440	19,214	16,279	14,122	12,469	11,163	10,105	8,494	7,326	6,441	5,187	4,342
200	145,419	114,241	94,072	79,956	69,524	61,500	49,966	42,075	36,337	31,976	28,549	25,786	23,511	19,261	16,313	14,147	12,489	11,179	10,118	8,503	7,333	6,446	5,190	4,344

500 MCM | 1 Run(s) | 208 Volts

C-Value = 22,185 | Magnetic Duct | Copper

Calculations:

1. $f = \dfrac{1.732 \times \text{Length} \times \text{Isca}}{\text{Runs} \times \text{C-Value} \times E_{L\text{-}L}}$

2. $M = \dfrac{1}{1 + f}$

3. $\text{Isc} = \text{Isca} \times M$

4. Add Motor Contribution = Motor FLA x 4

* All results are given in symmetrical amperes

Three-Phase Bolted Fault-Current* Table

One-Way Distance in Feet

Available Isc in Thousands (Isca)	5	10	15	20	25	30	40	50	60	70	80	90	100	125	150	175	200	225	250	300	350	400	500	600
3	2,992	2,983	2,975	2,967	2,958	2,950	2,934	2,918	2,902	2,886	2,871	2,855	2,840	2,803	2,766	2,731	2,696	2,663	2,630	2,567	2,506	2,449	2,341	2,242
5	4,977	4,954	4,931	4,908	4,885	4,863	4,819	4,776	4,733	4,692	4,651	4,611	4,571	4,475	4,383	4,295	4,210	4,128	4,050	3,902	3,764	3,635	3,403	3,199
7	6,954	6,909	6,865	6,821	6,777	6,735	6,651	6,569	6,489	6,411	6,334	6,260	6,187	6,013	5,848	5,692	5,544	5,403	5,269	5,021	4,795	4,589	4,225	3,915
10	9,907	9,816	9,726	9,638	9,552	9,467	9,302	9,142	8,988	8,839	8,695	8,555	8,420	8,100	7,803	7,528	7,271	7,031	6,807	6,398	6,036	5,712	5,159	4,704
15	14,792	14,589	14,392	14,200	14,014	13,832	13,482	13,149	12,833	12,531	12,243	11,968	11,705	11,096	10,547	10,049	9,597	9,183	8,804	8,132	7,556	7,055	6,230	5,578
20	19,632	19,276	18,934	18,603	18,284	17,976	17,389	16,840	16,324	15,839	15,381	14,950	14,542	13,613	12,796	12,071	11,424	10,843	10,318	9,407	8,644	7,996	6,952	6,150
25	24,427	23,880	23,356	22,855	22,375	21,915	21,050	20,250	19,508	18,819	18,177	17,578	17,016	15,758	14,673	13,728	12,898	12,162	11,505	10,384	9,462	8,691	7,472	6,553
30	29,179	28,401	27,664	26,964	26,298	25,665	24,486	23,410	22,425	21,519	20,684	19,911	19,194	17,608	16,264	15,111	14,111	13,235	12,461	11,156	10,099	9,225	7,864	6,852
35	33,887	32,843	31,861	30,936	30,063	29,238	27,718	26,347	25,106	23,976	22,944	21,997	21,124	19,220	17,630	16,283	15,127	14,125	13,247	11,782	10,609	9,649	8,170	7,083
40	38,553	37,207	35,952	34,778	33,679	32,648	30,763	29,084	27,578	26,221	24,991	23,872	22,848	20,636	18,814	17,288	15,991	14,875	13,905	12,300	11,027	9,993	8,415	7,267
45	43,177	41,496	39,940	38,498	37,155	35,904	33,637	31,640	29,866	28,281	26,856	25,567	24,397	21,891	19,852	18,161	16,735	15,516	14,463	12,735	11,376	10,279	8,616	7,417
50	47,759	45,711	43,831	42,099	40,499	39,017	36,355	34,033	31,990	30,178	28,560	27,107	25,795	23,010	20,768	18,924	17,381	16,071	14,944	13,106	11,671	10,519	8,785	7,541
55	52,301	49,854	47,626	45,589	43,719	41,996	38,928	36,277	33,965	31,930	30,125	28,513	27,064	24,015	21,583	19,599	17,948	16,554	15,361	13,426	11,924	10,724	8,927	7,646
60	56,802	53,928	51,330	48,971	46,820	44,850	41,368	38,387	35,808	33,553	31,565	29,800	28,222	24,922	22,313	20,198	18,450	16,980	15,727	13,705	12,143	10,901	9,050	7,736
65	61,263	57,933	54,946	52,252	49,810	47,586	43,684	40,374	37,531	35,061	32,897	30,984	29,281	25,744	22,970	20,735	18,897	17,358	16,051	13,950	12,335	11,056	9,156	7,813
70	65,685	61,872	58,477	55,435	52,694	50,211	45,887	42,249	39,145	36,466	34,130	32,076	30,255	26,494	23,565	21,219	19,298	17,696	16,339	14,167	12,505	11,191	9,249	7,881
75	70,069	65,746	61,926	58,525	55,478	52,733	47,984	44,020	40,661	37,778	35,277	33,087	31,152	27,180	24,106	21,656	19,659	17,999	16,597	14,361	12,655	11,312	9,331	7,941
80	74,414	69,557	65,295	61,526	58,167	55,157	49,983	45,697	42,087	39,006	36,346	34,025	31,983	27,810	24,600	22,054	19,986	18,273	16,830	14,535	12,790	11,420	9,404	7,993
85	78,721	73,306	68,588	64,441	60,766	57,488	51,890	47,285	43,431	40,158	37,344	34,898	32,753	28,390	25,053	22,418	20,284	18,522	17,041	14,692	12,912	11,516	9,470	8,041
90	82,991	76,995	71,807	67,274	63,280	59,733	53,712	48,793	44,700	41,240	38,278	35,712	33,469	28,927	25,470	22,751	20,557	18,749	17,233	14,834	13,022	11,604	9,529	8,083
95	87,225	80,626	74,955	70,029	65,711	61,895	55,454	50,226	45,900	42,260	39,154	36,474	34,137	29,425	25,855	23,058	20,807	18,957	17,408	14,964	13,121	11,683	9,582	8,122
100	91,421	84,198	78,033	72,709	68,065	63,979	57,121	51,590	47,036	43,221	39,978	37,188	34,762	29,888	26,212	23,341	21,038	19,148	17,569	15,083	13,213	11,755	9,631	8,156
105	95,583	87,715	81,045	75,317	70,345	65,990	58,718	52,890	48,114	44,129	40,754	37,859	35,347	30,319	26,543	23,604	21,250	19,324	17,718	15,192	13,296	11,821	9,675	8,188
110	99,708	91,178	83,992	77,855	72,555	67,930	60,249	54,129	49,137	44,989	41,486	38,489	35,896	30,722	26,852	23,847	21,448	19,487	17,854	15,292	13,373	11,882	9,716	8,217
115	103,799	94,586	86,876	80,327	74,697	69,804	61,719	55,312	50,111	45,803	42,178	39,084	36,413	31,100	27,140	24,074	21,631	19,638	17,981	15,385	13,444	11,938	9,753	8,244
120	107,855	97,943	89,699	82,735	76,775	71,616	63,131	56,443	51,037	46,576	42,832	39,645	36,900	31,454	27,409	24,286	21,802	19,779	18,099	15,472	13,510	11,990	9,788	8,269
125	111,877	101,248	92,464	85,082	78,791	73,367	64,488	57,526	51,921	47,311	43,453	40,176	37,359	31,788	27,662	24,484	21,962	19,910	18,209	15,552	13,571	12,038	9,820	8,292
130	115,866	104,504	95,171	87,369	80,749	75,061	65,793	58,562	52,763	48,009	44,041	40,679	37,794	32,102	27,900	24,670	22,111	20,033	18,312	15,627	13,628	12,083	9,850	8,313
135	119,821	107,711	97,824	89,599	82,650	76,702	67,050	59,556	53,569	48,675	44,601	41,156	38,205	32,398	28,123	24,845	22,251	20,148	18,408	15,697	13,681	12,125	9,877	8,333
140	123,744	110,870	100,423	91,775	84,498	78,290	68,261	60,509	54,339	49,310	45,134	41,609	38,595	32,678	28,334	25,009	22,383	20,256	18,498	15,762	13,731	12,164	9,903	8,351
145	127,634	113,983	102,970	93,897	86,294	79,830	69,428	61,425	55,076	49,916	45,641	42,040	38,966	32,943	28,533	25,164	22,507	20,357	18,582	15,823	13,778	12,200	9,927	8,368
150	131,492	117,050	105,466	95,969	88,040	81,322	70,554	62,305	55,782	50,496	46,125	42,450	39,318	33,195	28,721	25,311	22,624	20,453	18,662	15,881	13,821	12,235	9,950	8,384
155	135,319	120,072	107,914	97,991	89,739	82,770	71,641	63,151	56,459	51,050	46,587	42,841	39,653	33,433	28,900	25,449	22,735	20,543	18,737	15,936	13,863	12,267	9,971	8,400
160	139,114	123,051	110,314	99,966	91,393	84,174	72,691	63,965	57,110	51,581	47,029	43,215	39,973	33,660	29,069	25,581	22,839	20,629	18,808	15,987	13,902	12,297	9,992	8,414
165	142,878	125,987	112,668	101,895	93,003	85,538	73,706	64,750	57,734	52,090	47,451	43,571	40,278	33,876	29,230	25,705	22,939	20,710	18,876	16,035	13,938	12,326	10,010	8,427
170	146,612	128,882	114,977	103,780	94,571	86,862	74,687	65,506	58,334	52,578	47,856	43,912	40,569	34,082	29,383	25,823	23,033	20,787	18,939	16,081	13,973	12,353	10,028	8,440
175	150,316	131,735	117,242	105,622	96,098	88,149	75,637	66,235	58,912	53,047	48,244	44,239	40,848	34,278	29,529	25,936	23,122	20,859	19,000	16,125	14,006	12,379	10,045	8,452
180	153,991	134,549	119,466	107,423	97,587	89,400	76,556	66,939	59,468	53,497	48,616	44,552	41,114	34,466	29,668	26,043	23,207	20,929	19,057	16,166	14,037	12,403	10,061	8,463
185	157,635	137,323	121,648	109,185	99,038	90,617	77,446	67,618	60,004	53,931	48,974	44,852	41,369	34,645	29,801	26,145	23,289	20,995	19,112	16,206	14,067	12,426	10,077	8,474
190	161,251	140,059	123,790	110,907	100,453	91,800	78,309	68,275	60,520	54,348	49,318	45,140	41,614	34,817	29,928	26,243	23,366	21,058	19,164	16,243	14,095	12,448	10,091	8,484
195	164,838	142,757	125,893	112,592	101,833	92,951	79,145	68,910	61,019	54,749	49,648	45,416	41,849	34,981	30,049	26,336	23,440	21,118	19,214	16,279	14,122	12,469	10,105	8,494
200	168,397	145,419	127,958	114,241	103,181	94,072	79,956	69,524	61,500	55,136	49,966	45,682	42,075	35,139	30,165	26,425	23,511	21,175	19,261	16,313	14,147	12,489	10,118	8,503

500 MCM	2 Run(s)	208 Volts	C-Value = 22,185	Magnetic Duct	Copper

Calculations:

1. $f = \dfrac{1.732 \times \text{Length} \times \text{Isca}}{\text{Runs} \times \text{C-Value} \times E_{L-L}}$

2. $M = \dfrac{1}{1 + f}$

3. Isc = Isca x M

4. Add Motor Contribution = Motor FLA x 4

* All results are given in symmetrical amperes

Three-Phase Bolted Fault-Current* Table

One-Way Distance in Feet

Available Isc in Thousands (Isca)

Isca	5	10	15	20	25	30	40	50	60	70	80	90	100	125	150	175	200	225	250	300	350	400	500	600
3	2,984	2,968	2,952	2,936	2,921	2,905	2,875	2,845	2,816	2,788	2,760	2,732	2,706	2,641	2,579	2,520	2,464	2,410	2,359	2,262	2,173	2,090	1,943	1,815
5	4,955	4,911	4,868	4,825	4,783	4,742	4,662	4,584	4,509	4,437	4,367	4,299	4,233	4,076	3,931	3,796	3,669	3,551	3,441	3,239	3,059	2,898	2,623	2,395
7	6,912	6,827	6,743	6,662	6,582	6,505	6,355	6,212	6,075	5,944	5,819	5,698	5,583	5,314	5,070	4,847	4,643	4,456	4,283	3,974	3,707	3,474	3,085	2,775
10	9,822	9,650	9,484	9,324	9,169	9,019	8,733	8,465	8,213	7,976	7,751	7,540	7,339	6,881	6,477	6,118	5,796	5,507	5,245	4,790	4,407	4,081	3,555	3,149
15	14,603	14,226	13,869	13,528	13,205	12,896	12,320	11,793	11,309	10,864	10,452	10,070	9,716	8,929	8,261	7,685	7,185	6,745	6,357	5,700	5,166	4,724	4,033	3,518
20	19,300	18,648	18,038	17,467	16,931	16,426	15,503	14,678	13,936	13,266	12,657	12,102	11,593	10,491	9,580	8,814	8,162	7,600	7,110	6,298	5,653	5,127	4,323	3,738
25	23,916	22,922	22,008	21,163	20,381	19,655	18,347	17,203	16,193	15,295	14,491	13,768	13,113	11,720	10,594	9,666	8,887	8,225	7,654	6,721	5,991	5,404	4,519	3,883
30	28,453	27,057	25,792	24,640	23,586	22,619	20,904	19,431	18,152	17,031	16,041	15,159	14,369	12,713	11,400	10,332	9,447	8,702	8,066	7,037	6,241	5,606	4,659	3,986
35	32,912	31,058	29,403	27,915	26,570	25,349	23,215	21,413	19,870	18,535	17,368	16,339	15,425	13,533	12,054	10,867	9,892	9,078	8,388	7,281	6,432	5,760	4,765	4,063
40	37,295	34,933	32,853	31,006	29,356	27,872	25,314	23,186	21,388	19,849	18,516	17,351	16,324	14,220	12,596	11,305	10,254	9,382	8,647	7,475	6,583	5,881	4,847	4,123
45	41,606	38,688	36,152	33,928	31,962	30,211	27,229	24,782	22,739	21,007	19,520	18,230	17,099	14,805	13,053	11,672	10,555	9,633	8,860	7,634	6,706	5,979	4,914	4,171
50	45,844	42,326	39,310	36,695	34,406	32,386	28,982	26,226	23,949	22,035	20,405	18,999	17,775	15,308	13,443	11,983	10,809	9,844	9,038	7,765	6,807	6,059	4,968	4,210
55	50,013	45,855	42,336	39,318	36,702	34,412	30,595	27,540	25,039	22,955	21,191	19,679	18,369	15,747	13,780	12,253	11,025	10,024	9,189	7,877	6,892	6,127	5,013	4,242
60	54,114	49,279	45,238	41,809	38,863	36,305	32,082	28,739	26,027	23,782	21,894	20,284	18,894	16,131	14,074	12,481	11,213	10,178	9,318	7,972	6,965	6,184	5,051	4,269
65	58,148	52,602	48,023	44,177	40,901	38,077	33,458	29,838	26,925	24,530	22,527	20,826	19,363	16,472	14,332	12,684	11,376	10,313	9,431	8,054	7,028	6,234	5,084	4,293
70	62,117	55,830	50,698	46,431	42,826	39,740	34,735	30,850	27,746	25,210	23,098	21,313	19,784	16,776	14,561	12,864	11,520	10,431	9,530	8,126	7,083	6,277	5,113	4,313
75	66,023	58,965	53,270	48,579	44,647	41,303	35,923	31,783	28,499	25,330	23,618	21,755	20,164	17,048	14,766	13,023	11,648	10,536	9,617	8,189	7,131	6,314	5,138	4,331
80	69,867	62,012	55,745	50,628	46,372	42,776	37,032	32,648	29,192	26,398	24,092	22,157	20,509	17,294	14,950	13,166	11,762	10,629	9,695	8,246	7,173	6,348	5,160	4,347
85	73,650	64,975	58,127	52,586	48,009	44,165	38,069	33,451	29,833	26,921	24,527	22,524	20,823	17,517	15,116	13,295	11,865	10,713	9,764	8,296	7,211	6,377	5,180	4,361
90	77,375	67,856	60,423	54,457	49,564	45,478	39,040	34,199	30,426	27,403	24,926	22,860	21,110	17,719	15,267	13,411	11,958	10,788	9,827	8,341	7,245	6,404	5,197	4,373
95	81,042	70,660	62,636	56,249	51,044	46,720	39,952	34,897	30,977	27,849	25,295	23,170	21,374	17,905	15,405	13,517	12,042	10,857	9,884	8,382	7,276	6,428	5,213	4,384
100	84,653	73,390	64,772	57,965	52,453	47,898	40,810	35,550	31,491	28,263	25,636	23,456	21,617	18,075	15,531	13,614	12,119	10,919	9,936	8,419	7,304	6,450	5,228	4,395
105	88,209	76,047	66,833	59,610	53,796	49,016	41,619	36,162	31,970	28,649	25,953	23,721	21,842	18,232	15,646	13,703	12,189	10,976	9,983	8,453	7,330	6,470	5,241	4,404
110	91,711	78,636	68,824	61,189	55,079	50,078	42,383	36,737	32,419	29,009	26,248	23,967	22,051	18,377	15,753	13,785	12,254	11,029	10,026	8,484	7,353	6,488	5,252	4,412
115	95,160	81,158	70,749	62,706	56,305	51,090	43,105	37,278	32,839	29,345	26,523	24,196	22,245	18,512	15,852	13,860	12,313	11,077	10,066	8,513	7,374	6,505	5,263	4,420
120	98,558	83,617	72,610	64,164	57,477	52,053	43,789	37,789	33,235	29,661	26,780	24,410	22,425	18,637	15,943	13,930	12,368	11,121	10,103	8,539	7,394	6,520	5,273	4,427
125	101,906	86,015	74,411	65,566	58,600	52,972	44,437	38,271	33,607	29,957	27,022	24,610	22,594	18,753	16,029	13,995	12,420	11,163	10,137	8,563	7,413	6,534	5,283	4,434
130	105,205	88,353	76,155	66,916	59,676	53,850	45,053	38,727	33,958	30,235	27,248	24,798	22,752	18,862	16,108	14,056	12,467	11,201	10,169	8,586	7,429	6,548	5,291	4,440
135	108,456	90,634	77,844	68,216	60,708	54,689	45,639	39,159	34,290	30,498	27,461	24,975	22,901	18,964	16,182	14,112	12,512	11,237	10,198	8,607	7,445	6,560	5,299	4,445
140	111,659	92,861	79,430	69,470	61,699	55,492	46,197	39,569	34,604	30,746	27,662	25,141	23,040	19,060	16,252	14,165	12,553	11,271	10,226	8,627	7,460	6,571	5,307	4,450
145	114,817	95,035	81,067	70,679	62,651	56,261	46,728	39,958	34,901	30,981	27,852	25,297	23,172	19,150	16,317	14,215	12,592	11,302	10,252	8,645	7,474	6,582	5,314	4,455
150	117,930	97,157	82,607	71,847	63,567	56,998	47,236	40,329	35,184	31,203	28,032	25,445	23,296	19,234	16,379	14,261	12,629	11,331	10,276	8,662	7,486	6,592	5,320	4,460
155	120,998	99,231	84,101	72,974	64,448	57,706	47,721	40,681	35,452	31,414	28,202	25,585	23,413	19,314	16,436	14,305	12,663	11,359	10,299	8,678	7,499	6,601	5,326	4,464
160	124,024	101,256	85,551	74,064	65,296	58,385	48,184	41,018	35,707	31,614	28,363	25,718	23,524	19,390	16,491	14,346	12,695	11,385	10,320	8,694	7,510	6,610	5,332	4,468
165	127,007	103,236	86,960	75,118	66,114	59,038	48,628	41,339	35,950	31,805	28,516	25,844	23,630	19,461	16,543	14,386	12,726	11,410	10,340	8,708	7,521	6,618	5,337	4,472
170	129,949	105,172	88,330	76,137	66,902	59,665	49,053	41,646	36,182	31,986	28,662	25,963	23,730	19,529	16,592	14,423	12,755	11,433	10,359	8,721	7,531	6,626	5,342	4,476
175	132,851	107,064	89,661	77,124	67,663	60,270	49,461	41,939	36,404	32,159	28,800	26,077	23,825	19,593	16,638	14,458	12,782	11,455	10,378	8,734	7,540	6,633	5,347	4,479
180	135,713	108,915	90,955	78,080	68,398	60,852	49,852	42,221	36,615	32,324	28,933	26,186	23,915	19,654	16,682	14,491	12,808	11,476	10,395	8,746	7,549	6,640	5,352	4,482
185	138,536	110,726	92,214	79,006	69,108	61,413	50,228	42,490	36,818	32,481	29,059	26,289	24,001	19,712	16,724	14,522	12,833	11,496	10,411	8,758	7,558	6,647	5,356	4,485
190	141,320	112,498	93,440	79,904	69,794	61,954	50,590	42,748	37,011	32,632	29,180	26,388	24,083	19,768	16,764	14,552	12,857	11,515	10,426	8,769	7,566	6,653	5,360	4,488
195	144,068	114,232	94,633	80,775	70,457	62,477	50,938	42,996	37,197	32,777	29,295	26,482	24,162	19,821	16,802	14,581	12,879	11,533	10,441	8,779	7,574	6,659	5,364	4,491
200	146,779	115,930	95,796	81,620	71,099	62,981	51,272	43,235	37,375	32,915	29,405	26,572	24,237	19,871	16,838	14,608	12,900	11,550	10,455	8,789	7,581	6,665	5,368	4,493

600 MCM	1 Run(s)	208 Volts	C-Value = 22,965	Magnetic Duct	Copper

Calculations:

1. $f = \dfrac{1.732 \times \text{Length} \times \text{Isca}}{\text{Runs} \times \text{C-Value} \times E\ L\text{-}L}$

2. $M = \dfrac{1}{1 + f}$

3. Isc = Isca x M

4. Add Motor Contribution = Motor FLA x 4

* All results are given in symmetrical amperes

Three-Phase Bolted Fault-Current* Table

One-Way Distance in Feet

Available Isc in Thousands (Isca)	5	10	15	20	25	30	40	50	60	70	80	90	100	125	150	175	200	225	250	300	350	400	500	600
3	2,992	2,984	2,976	2,968	2,960	2,952	2,936	2,921	2,905	2,890	2,875	2,860	2,845	2,809	2,774	2,739	2,706	2,673	2,641	2,579	2,520	2,464	2,359	2,262
5	4,977	4,955	4,933	4,911	4,889	4,868	4,825	4,783	4,742	4,702	4,662	4,623	4,584	4,491	4,402	4,315	4,233	4,153	4,076	3,931	3,796	3,669	3,441	3,239
7	6,956	6,912	6,869	6,827	6,785	6,743	6,662	6,582	6,505	6,429	6,355	6,282	6,212	6,042	5,881	5,728	5,583	5,445	5,314	5,070	4,847	4,643	4,283	3,974
10	9,910	9,822	9,735	9,650	9,566	9,484	9,324	9,169	9,019	8,874	8,733	8,597	8,465	8,152	7,862	7,591	7,339	7,103	6,881	6,477	6,118	5,796	5,245	4,790
15	14,799	14,603	14,412	14,226	14,045	13,869	13,528	13,205	12,896	12,601	12,320	12,051	11,793	11,195	10,654	10,163	9,716	9,306	8,929	8,261	7,685	7,185	6,357	5,700
20	19,644	19,300	18,968	18,648	18,338	18,038	17,467	16,931	16,426	15,951	15,503	15,079	14,678	13,762	12,954	12,236	11,593	11,014	10,491	9,580	8,814	8,162	7,110	6,298
25	24,446	23,916	23,409	22,922	22,456	22,008	21,163	20,381	19,655	18,979	18,347	17,757	17,203	15,959	14,882	13,942	13,113	12,378	11,720	10,594	9,666	8,887	7,654	6,721
30	29,206	28,453	27,737	27,057	26,409	25,792	24,640	23,586	22,619	21,728	20,904	20,141	19,431	17,859	16,521	15,370	14,369	13,491	12,713	11,400	10,332	9,447	8,066	7,037
35	33,924	32,912	31,958	31,058	30,208	29,403	27,915	26,570	25,349	24,235	23,215	22,278	21,413	19,519	17,932	16,584	15,425	14,417	13,533	12,054	10,867	9,892	8,388	7,281
40	38,600	37,295	36,076	34,933	33,861	32,853	31,006	29,356	27,872	26,532	25,314	24,203	23,186	20,981	19,159	17,628	16,324	15,200	14,220	12,596	11,305	10,254	8,647	7,475
45	43,236	41,606	40,094	38,688	37,377	36,152	33,928	31,962	30,211	28,643	27,229	25,948	24,782	22,280	20,236	18,536	17,099	15,870	14,805	13,053	11,672	10,555	8,860	7,634
50	47,832	45,844	44,015	42,326	40,762	39,310	36,695	34,406	32,386	30,590	28,982	27,536	26,226	23,440	21,189	19,332	17,775	16,450	15,308	13,443	11,983	10,809	9,038	7,765
55	52,388	50,013	47,844	45,855	44,025	42,336	39,318	36,702	34,412	32,391	30,595	28,987	27,540	24,484	22,038	20,037	18,369	16,957	15,747	13,780	12,250	11,025	9,189	7,877
60	56,905	54,114	51,583	49,279	47,172	45,238	41,809	38,863	36,305	34,063	32,082	30,318	28,739	25,427	22,799	20,664	18,894	17,404	16,131	14,074	12,481	11,213	9,318	7,972
65	61,383	58,148	55,236	52,602	50,208	48,023	44,177	40,901	38,077	35,618	33,458	31,544	29,838	26,284	23,486	21,226	19,363	17,801	16,472	14,332	12,684	11,376	9,431	8,054
70	65,823	62,117	58,806	55,830	53,140	50,698	46,431	42,826	39,740	37,069	34,735	32,677	30,850	27,065	24,108	21,733	19,784	18,156	16,776	14,561	12,864	11,520	9,530	8,126
75	70,226	66,023	62,295	58,965	55,973	53,270	48,579	44,647	41,303	38,426	35,923	33,727	31,783	27,781	24,674	22,193	20,164	18,476	17,048	14,766	13,023	11,648	9,617	8,189
80	74,591	69,867	65,705	62,012	58,712	55,745	50,628	46,372	42,776	39,697	37,032	34,702	32,648	28,440	25,192	22,611	20,509	18,765	17,294	14,950	13,166	11,762	9,695	8,246
85	78,919	73,650	69,041	64,975	61,361	58,127	52,586	48,009	44,165	40,891	38,069	35,611	33,451	29,047	25,668	22,993	20,823	19,027	17,517	15,116	13,295	11,865	9,764	8,296
90	83,211	77,375	72,304	67,856	63,924	60,423	54,457	49,564	45,478	42,014	39,040	36,459	34,199	29,609	26,106	23,344	21,110	19,267	17,719	15,267	13,411	11,958	9,827	8,341
95	87,468	81,042	75,496	70,660	66,407	62,636	56,249	51,044	46,720	43,072	39,952	37,254	34,897	30,131	26,511	23,667	21,374	19,486	17,905	15,405	13,517	12,042	9,884	8,382
100	91,689	84,653	78,620	73,390	68,812	64,772	57,965	52,453	47,898	44,071	40,810	37,999	35,550	30,617	26,886	23,965	21,617	19,688	18,075	15,531	13,614	12,119	9,936	8,419
105	95,875	88,209	81,678	76,047	71,143	66,833	59,610	53,796	49,016	45,016	41,619	38,699	36,162	31,070	27,234	24,242	21,842	19,875	18,232	15,646	13,703	12,189	9,983	8,453
110	100,026	91,711	84,672	78,636	73,404	68,824	61,189	55,079	50,078	45,910	42,383	39,358	36,737	31,493	27,559	24,499	22,051	20,047	18,377	15,753	13,785	12,254	10,026	8,484
115	104,144	95,160	87,603	81,158	75,597	70,749	62,706	56,305	51,090	46,759	43,105	39,980	37,278	31,890	27,863	24,739	22,245	20,207	18,512	15,852	13,860	12,313	10,066	8,513
120	108,227	98,558	90,475	83,617	77,726	72,610	64,164	57,477	52,053	47,565	43,789	40,568	37,789	32,263	28,147	24,962	22,425	20,356	18,637	15,943	13,930	12,368	10,103	8,539
125	112,278	101,906	93,288	86,015	79,793	74,411	65,566	58,600	52,972	48,331	44,437	41,124	38,271	32,614	28,414	25,172	22,594	20,495	18,753	16,029	13,995	12,420	10,137	8,563
130	116,295	105,205	96,045	88,353	81,802	76,155	66,916	59,676	53,850	49,060	45,053	41,651	38,727	32,944	28,664	25,368	22,752	20,625	18,862	16,108	14,056	12,467	10,169	8,586
135	120,281	108,456	98,747	90,634	83,753	77,844	68,216	60,708	54,689	49,756	45,639	42,151	39,159	33,256	28,900	25,553	22,901	20,747	18,964	16,182	14,112	12,512	10,198	8,607
140	124,234	111,659	101,396	92,861	85,651	79,480	69,470	61,699	55,492	50,420	46,197	42,627	39,569	33,552	29,123	25,727	23,040	20,862	19,060	16,252	14,165	12,553	10,226	8,627
145	128,155	114,817	103,993	95,035	87,497	81,067	70,679	62,651	56,261	51,054	46,728	43,079	39,958	33,831	29,333	25,891	23,172	20,970	19,150	16,317	14,215	12,592	10,252	8,645
150	132,046	117,930	106,540	97,157	89,293	82,607	71,847	63,567	56,998	51,660	47,236	43,510	40,329	34,096	29,532	26,046	23,296	21,071	19,234	16,379	14,261	12,629	10,276	8,662
155	135,905	120,998	109,039	99,231	91,041	84,101	72,974	64,448	57,706	52,240	47,721	43,921	40,681	34,348	29,721	26,193	23,413	21,167	19,314	16,436	14,305	12,663	10,299	8,678
160	139,734	124,024	111,490	101,256	92,744	85,551	74,064	65,296	58,385	52,796	48,184	44,313	41,018	34,588	29,900	26,332	23,524	21,258	19,390	16,491	14,346	12,695	10,320	8,694
165	143,532	127,007	113,895	103,236	94,402	86,960	75,118	66,114	59,038	53,330	48,628	44,688	41,339	34,816	30,071	26,464	23,630	21,344	19,461	16,543	14,386	12,726	10,340	8,708
170	147,301	129,949	116,255	105,172	96,018	88,330	76,137	66,902	59,665	53,841	49,053	45,047	41,646	35,033	30,233	26,589	23,730	21,425	19,529	16,592	14,423	12,755	10,359	8,721
175	151,040	132,851	118,572	107,064	97,593	89,661	77,124	67,663	60,270	54,333	49,461	45,391	41,939	35,241	30,387	26,709	23,825	21,503	19,593	16,638	14,458	12,782	10,378	8,734
180	154,750	135,713	120,846	108,915	99,128	90,955	78,080	68,398	60,852	54,806	49,852	45,720	42,221	35,439	30,534	26,822	23,915	21,576	19,654	16,682	14,491	12,808	10,395	8,746
185	158,431	138,536	123,079	110,726	100,626	92,214	79,006	69,108	61,413	55,260	50,228	46,036	42,490	35,629	30,675	26,931	24,001	21,646	19,712	16,724	14,522	12,833	10,411	8,758
190	162,084	141,320	125,273	112,498	102,087	93,440	79,904	69,794	61,954	55,698	50,590	46,340	42,748	35,810	30,809	27,034	24,083	21,713	19,768	16,764	14,552	12,857	10,426	8,769
195	165,709	144,068	127,427	114,232	103,513	94,633	80,775	70,457	62,477	56,120	50,938	46,631	42,996	35,984	30,938	27,133	24,162	21,777	19,821	16,802	14,581	12,879	10,441	8,779
200	169,306	146,779	129,543	115,930	104,905	95,796	81,620	71,099	62,981	56,527	51,272	46,912	43,235	36,151	31,061	27,228	24,237	21,838	19,871	16,838	14,608	12,900	10,455	8,789

600 MCM	2 Run(s)	208 Volts	C-Value = 22,965	Magnetic Duct	Copper

Calculations:

1. $f = \dfrac{1.732 \times \text{Length} \times \text{Isca}}{\text{Runs} \times \text{C-Value} \times E_{L\text{-}L}}$
2. $M = \dfrac{1}{1 + f}$
3. $\text{Isc} = \text{Isca} \times M$
4. Add Motor Contribution = Motor FLA x 4

* All results are given in symmetrical amperes

Three-Phase Bolted Fault-Current* Table

One-Way Distance in Feet

Available Isc in Thousands (Isca)

Isca	5	10	15	20	25	30	40	50	60	70	80	90	100	125	150	175	200	225	250	300	350	400	500	600
3	2,985	2,969	2,954	2,939	2,924	2,910	2,881	2,852	2,825	2,797	2,771	2,744	2,719	2,656	2,597	2,540	2,486	2,433	2,383	2,289	2,202	2,122	1,977	1,851
5	4,957	4,915	4,874	4,833	4,793	4,754	4,677	4,603	4,531	4,461	4,394	4,328	4,264	4,113	3,972	3,841	3,717	3,602	3,493	3,295	3,118	2,959	2,685	2,457
7	6,916	6,835	6,755	6,677	6,601	6,527	6,383	6,246	6,114	5,983	5,867	5,750	5,638	5,377	5,139	4,920	4,720	4,536	4,365	4,059	3,794	3,561	3,171	2,858
10	9,830	9,667	9,508	9,355	9,206	9,062	8,787	8,529	8,285	8,055	7,837	7,631	7,435	6,987	6,590	6,235	5,917	5,630	5,369	4,914	4,530	4,202	3,670	3,257
15	14,622	14,262	13,919	13,593	13,282	12,984	12,428	11,917	11,446	11,011	10,608	10,234	9,885	9,108	8,445	7,871	7,371	6,930	6,540	5,877	5,336	4,886	4,181	3,654
20	19,333	18,709	18,124	17,575	17,058	16,570	15,674	14,870	14,144	13,486	12,887	12,338	11,834	10,738	9,828	9,060	8,403	7,835	7,339	6,515	5,857	5,319	4,494	3,891
25	23,966	23,015	22,136	21,322	20,566	19,861	18,587	17,467	16,474	15,588	14,793	14,075	13,423	12,030	10,899	9,963	9,174	8,502	7,921	6,969	6,221	5,618	4,706	4,049
30	28,524	27,186	25,968	24,855	23,833	22,892	21,216	19,769	18,507	17,396	16,411	15,532	14,742	13,079	11,753	10,671	9,772	9,012	8,362	7,308	6,490	5,837	4,858	4,161
35	33,007	31,229	29,633	28,192	26,884	25,693	23,601	21,824	20,296	18,968	17,803	16,772	15,855	13,948	12,450	11,243	10,249	9,417	8,709	7,572	6,697	6,003	4,973	4,245
40	37,418	35,149	33,140	31,348	29,740	28,289	25,773	23,669	21,882	20,346	19,011	17,841	16,807	14,679	13,029	11,713	10,638	9,744	8,989	7,782	6,861	6,135	5,063	4,310
45	41,758	38,953	36,500	34,338	32,418	30,701	27,761	25,334	23,298	21,565	20,071	18,772	17,630	15,303	13,519	12,107	10,962	10,015	9,219	7,954	6,994	6,241	5,136	4,363
50	46,030	42,644	39,722	37,175	34,934	32,949	29,586	26,846	24,570	22,650	21,008	19,589	18,349	15,842	13,937	12,442	11,236	10,243	9,412	8,097	7,105	6,329	5,195	4,405
55	50,234	46,228	42,814	39,870	37,304	35,049	31,268	28,223	25,719	23,623	21,843	20,312	18,982	16,311	14,300	12,730	11,470	10,438	9,576	8,218	7,198	6,403	5,244	4,441
60	54,372	49,710	45,784	42,433	39,539	37,014	32,823	29,484	26,762	24,500	22,590	20,957	19,544	16,725	14,616	12,980	11,673	10,605	9,717	8,322	7,277	6,466	5,286	4,471
65	58,447	53,094	48,639	44,874	41,650	38,858	34,265	30,642	27,713	25,294	23,264	21,536	20,046	17,091	14,895	13,200	11,850	10,752	9,839	8,412	7,346	6,520	5,322	4,497
70	62,458	56,383	51,386	47,202	43,648	40,591	35,605	31,710	28,583	26,017	23,874	22,058	20,498	17,418	15,143	13,394	12,007	10,880	9,947	8,490	7,405	6,567	5,354	4,519
75	66,408	59,583	54,030	49,423	45,541	42,224	36,855	32,698	29,383	26,679	24,430	22,531	20,906	17,712	15,365	13,567	12,146	10,994	10,042	8,559	7,458	6,608	5,381	4,539
80	70,299	62,696	56,577	51,546	47,337	43,764	38,023	33,613	30,120	27,285	24,938	22,962	21,277	17,978	15,564	13,722	12,270	11,096	10,127	8,621	7,505	6,645	5,405	4,556
85	74,131	65,726	59,033	53,577	49,044	45,219	39,116	34,465	30,803	27,844	25,403	23,356	21,615	18,218	15,744	13,862	12,382	11,187	10,203	8,676	7,546	6,677	5,427	4,571
90	77,905	68,676	61,402	55,521	50,669	46,596	40,143	35,260	31,436	28,360	25,832	23,719	21,924	18,438	15,908	13,989	12,483	11,269	10,271	8,725	7,584	6,706	5,446	4,585
95	81,624	71,550	63,689	57,384	52,216	47,901	41,108	36,002	32,024	28,838	26,229	24,052	22,209	18,639	16,057	14,104	12,574	11,344	10,333	8,770	7,618	6,733	5,464	4,597
100	85,288	74,349	65,893	59,172	53,691	49,140	42,017	36,697	32,573	29,283	26,596	24,361	22,472	18,824	16,194	14,210	12,658	11,412	10,390	8,811	7,648	6,757	5,479	4,608
105	88,898	77,078	68,033	60,887	55,100	50,317	42,875	37,350	33,086	29,697	26,937	24,646	22,715	18,994	16,320	14,306	12,735	11,475	10,441	8,848	7,676	6,779	5,494	4,618
110	92,456	79,739	70,097	62,536	56,446	51,438	43,685	37,964	33,567	30,083	27,255	24,912	22,941	19,151	16,436	14,396	12,806	11,532	10,489	8,882	7,702	6,799	5,507	4,628
115	95,963	82,334	72,095	64,120	57,735	52,505	44,453	38,542	34,019	30,445	27,552	25,160	23,150	19,297	16,544	14,478	12,871	11,585	10,532	8,913	7,725	6,817	5,519	4,636
120	99,420	84,866	74,028	65,646	58,968	53,524	45,181	39,088	34,443	30,785	27,829	25,391	23,346	19,433	16,644	14,554	12,931	11,634	10,573	8,942	7,747	6,834	5,530	4,644
125	102,828	87,336	75,901	67,114	60,150	54,496	45,872	39,604	34,843	31,104	28,090	25,608	23,529	19,560	16,736	14,625	12,987	11,679	10,610	8,969	7,767	6,849	5,540	4,651
130	106,187	89,748	77,716	68,529	61,285	55,425	46,528	40,093	35,221	31,405	28,335	25,812	23,701	19,678	16,823	14,691	13,039	11,721	10,645	8,993	7,786	6,864	5,550	4,658
135	109,500	92,103	79,476	69,894	62,374	56,315	47,153	40,556	35,578	31,688	28,565	26,003	23,862	19,789	16,904	14,753	13,088	11,760	10,677	9,017	7,803	6,877	5,558	4,664
140	112,767	94,403	81,183	71,211	63,420	57,166	47,749	40,996	35,916	31,956	28,783	26,183	24,014	19,893	16,980	14,811	13,133	11,797	10,707	9,038	7,819	6,890	5,567	4,670
145	115,988	96,651	82,839	72,482	64,427	57,983	48,317	41,414	36,236	32,210	28,988	26,353	24,157	19,991	17,051	14,865	13,176	11,831	10,736	9,058	7,834	6,901	5,574	4,675
150	119,166	98,847	84,448	73,710	65,395	58,766	48,860	41,812	36,541	32,450	29,183	26,513	24,291	20,084	17,118	14,916	13,216	11,864	10,762	9,077	7,848	6,912	5,581	4,680
155	122,300	100,994	86,010	74,897	66,328	59,518	49,379	42,191	36,830	32,678	29,367	26,666	24,419	20,171	17,182	14,964	13,254	11,894	10,787	9,095	7,862	6,923	5,588	4,685
160	125,392	103,093	87,527	76,046	67,227	60,241	49,875	42,553	37,106	32,395	29,542	26,810	24,540	20,253	17,241	15,009	13,289	11,922	10,811	9,112	7,874	6,932	5,594	4,689
165	128,442	105,146	89,003	77,157	68,094	60,936	50,351	42,899	37,368	33,101	29,708	26,946	24,654	20,331	17,298	15,052	13,323	11,949	10,833	9,127	7,886	6,942	5,600	4,693
170	131,452	107,154	90,438	78,233	68,931	61,605	50,807	43,229	37,619	33,297	29,866	27,077	24,763	20,405	17,351	15,093	13,354	11,975	10,854	9,142	7,897	6,950	5,606	4,697
175	134,421	109,119	91,833	79,275	69,738	62,250	51,244	43,546	37,858	33,485	30,017	27,200	24,867	20,475	17,402	15,131	13,384	11,999	10,874	9,156	7,907	6,958	5,611	4,701
180	137,352	111,043	93,192	80,285	70,519	62,871	51,665	43,849	38,087	33,664	30,161	27,318	24,965	20,542	17,450	15,167	13,413	12,022	10,893	9,170	7,917	6,966	5,616	4,705
185	140,244	112,925	94,514	81,265	71,274	63,470	52,069	44,140	38,306	33,835	30,298	27,431	25,059	20,606	17,496	15,202	13,440	12,044	10,910	9,182	7,927	6,973	5,621	4,708
190	143,099	114,769	95,802	82,215	72,004	64,049	52,457	44,418	38,516	33,998	30,429	27,538	25,149	20,666	17,540	15,235	13,466	12,064	10,927	9,194	7,936	6,980	5,625	4,711
195	145,917	116,574	97,057	83,138	72,710	64,607	52,831	44,686	38,717	34,155	30,555	27,641	25,235	20,724	17,581	15,266	13,490	12,084	10,944	9,206	7,944	6,987	5,630	4,714
200	148,699	118,343	98,280	84,034	73,394	65,147	53,191	44,944	38,910	34,305	30,675	27,739	25,316	20,779	17,621	15,296	13,514	12,103	10,959	9,217	7,952	6,993	5,634	4,717

750 MCM | 1 Run(s) | 208 Volts | C-Value = 24,136 | Magnetic Duct | Copper

Calculations:

1. $f = \dfrac{1.732 \times \text{Length} \times \text{Isca}}{\text{Runs} \times \text{C-Value} \times E_{L\text{-}L}}$

2. $M = \dfrac{1}{1 + f}$

3. $\text{Isc} = \text{Isca} \times M$

4. Add Motor Contribution = Motor FLA x 4

* All results are given in symmetrical amperes

Three-Phase Bolted Fault-Current* Table

One-Way Distance in Feet

Available Isc in Thousands (Isca)	5	10	15	20	25	30	40	50	60	70	80	90	100	125	150	175	200	225	250	300	350	400	500	600
3	2,992	2,985	2,977	2,969	2,962	2,954	2,939	2,924	2,910	2,895	2,881	2,866	2,852	2,818	2,784	2,751	2,719	2,687	2,656	2,597	2,540	2,486	2,383	2,289
5	4,979	4,957	4,936	4,915	4,894	4,874	4,833	4,793	4,754	4,715	4,677	4,640	4,603	4,513	4,427	4,344	4,264	4,187	4,113	3,972	3,841	3,717	3,493	3,295
7	6,958	6,916	6,875	6,835	6,795	6,755	6,677	6,601	6,527	6,454	6,383	6,314	6,246	6,082	5,927	5,779	5,638	5,504	5,377	5,139	4,920	4,720	4,365	4,059
10	9,914	9,830	9,748	9,667	9,587	9,508	9,355	9,206	9,062	8,923	8,787	8,656	8,529	8,226	7,944	7,681	7,435	7,204	6,987	6,590	6,235	5,917	5,369	4,914
15	14,808	14,622	14,440	14,262	14,089	13,919	13,593	13,282	12,984	12,700	12,428	12,167	11,917	11,334	10,806	10,325	9,885	9,481	9,108	8,445	7,871	7,371	6,540	5,877
20	19,661	19,333	19,016	18,709	18,412	18,124	17,575	17,058	16,570	16,110	15,674	15,261	14,870	13,974	13,180	12,471	11,834	11,260	10,738	9,828	9,060	8,403	7,339	6,515
25	24,472	23,966	23,481	23,015	22,567	22,136	21,322	20,566	19,861	19,203	18,587	18,010	17,467	16,244	15,180	14,248	13,423	12,688	12,030	10,899	9,963	9,174	7,921	6,969
30	29,243	28,524	27,839	27,186	26,563	25,968	24,855	23,833	22,892	22,022	21,216	20,467	19,769	18,216	16,890	15,743	14,742	13,861	13,079	11,753	10,671	9,772	8,362	7,308
35	33,974	33,007	32,094	31,229	30,410	29,633	28,192	26,884	25,693	24,602	23,601	22,678	21,824	19,947	18,367	17,019	15,855	14,840	13,948	12,450	11,243	10,249	8,709	7,572
40	38,666	37,418	36,248	35,149	34,115	33,140	31,348	29,740	28,289	26,972	25,773	24,676	23,669	21,477	19,656	18,120	16,807	15,671	14,679	13,029	11,713	10,638	8,989	7,782
45	43,319	41,758	40,307	38,953	37,686	36,500	34,338	32,418	30,701	29,157	27,761	26,492	25,334	22,839	20,791	19,080	17,630	16,384	15,303	13,519	12,107	10,962	9,219	7,954
50	47,933	46,030	44,272	42,644	41,131	39,722	37,175	34,934	32,949	31,177	29,586	28,149	26,846	24,060	21,798	19,925	18,349	17,003	15,842	13,937	12,442	11,236	9,412	8,097
55	52,509	50,234	48,148	46,228	44,456	42,814	39,870	37,304	35,049	33,050	31,268	29,668	28,223	25,161	22,698	20,674	18,982	17,546	16,311	14,300	12,730	11,470	9,576	8,218
60	57,048	54,372	51,937	49,710	47,666	45,784	42,433	39,539	37,014	34,793	32,823	31,064	29,484	26,158	23,506	21,343	19,544	18,025	16,725	14,616	12,980	11,673	9,717	8,322
65	61,549	58,447	55,642	53,094	50,769	48,639	44,874	41,650	38,858	36,417	34,265	32,352	30,642	27,066	24,237	21,943	20,046	18,451	17,091	14,895	13,200	11,850	9,839	8,412
70	66,014	62,458	59,266	56,383	53,769	51,386	47,202	43,648	40,591	37,935	35,605	33,545	31,710	27,895	24,900	22,485	20,498	18,833	17,418	15,143	13,394	12,007	9,947	8,490
75	70,443	66,408	62,811	59,583	56,671	54,030	49,423	45,541	42,224	39,357	36,855	34,652	32,698	28,657	25,505	22,977	20,906	19,177	17,712	15,365	13,567	12,146	10,042	8,559
80	74,836	70,299	66,280	62,696	59,480	56,577	51,546	47,337	43,764	40,692	38,023	35,682	33,613	29,358	26,059	23,426	21,277	19,488	17,978	15,564	13,722	12,270	10,127	8,621
85	79,194	74,131	69,676	65,726	62,200	59,033	53,577	49,044	45,219	41,947	39,116	36,644	34,465	30,006	26,568	23,837	21,615	19,772	18,218	15,744	13,862	12,382	10,203	8,676
90	83,517	77,905	73,000	68,676	64,836	61,402	55,521	50,669	46,596	43,129	40,143	37,543	35,260	30,606	27,037	24,214	21,924	20,031	18,438	15,908	13,989	12,483	10,271	8,725
95	87,805	81,624	76,255	71,550	67,391	63,689	57,384	52,216	47,901	44,245	41,108	38,386	36,002	31,163	27,472	24,562	22,209	20,268	18,639	16,057	14,104	12,574	10,333	8,770
100	92,060	85,288	79,444	74,349	69,869	65,898	59,172	53,691	49,140	45,300	42,017	39,177	36,697	31,683	27,875	24,883	22,472	20,487	18,824	16,194	14,210	12,658	10,390	8,811
105	96,281	88,898	82,567	77,078	72,274	68,033	60,887	55,100	50,317	46,299	42,875	39,922	37,350	32,168	28,250	25,182	22,715	20,688	18,994	16,320	14,306	12,735	10,441	8,848
110	100,468	92,456	85,628	79,739	74,608	70,097	62,536	56,446	51,438	47,246	43,685	40,624	37,964	32,623	28,599	25,459	22,941	20,875	19,151	16,436	14,396	12,806	10,489	8,882
115	104,623	95,963	88,628	82,334	76,875	72,095	64,120	57,735	52,505	48,145	44,453	41,287	38,542	33,049	28,926	25,718	23,150	21,049	19,297	16,544	14,478	12,871	10,532	8,913
120	108,745	99,420	91,568	84,866	79,077	74,028	65,646	58,968	53,524	49,000	45,181	41,914	39,088	33,449	29,233	25,960	23,346	21,211	19,433	16,644	14,554	12,931	10,573	8,942
125	112,835	102,828	94,451	87,336	81,218	75,901	67,114	60,150	54,496	49,813	45,872	42,508	39,604	33,827	29,520	26,187	23,529	21,362	19,560	16,736	14,625	12,987	10,610	8,969
130	116,893	106,187	97,278	89,748	83,300	77,716	68,529	61,285	55,425	50,589	46,528	43,071	40,093	34,182	29,791	26,399	23,701	21,503	19,678	16,823	14,691	13,039	10,645	8,993
135	120,920	109,500	100,051	92,103	85,325	79,476	69,894	62,374	56,315	51,328	47,153	43,606	40,556	34,519	30,046	26,599	23,862	21,636	19,789	16,904	14,753	13,088	10,677	9,017
140	124,916	112,767	102,771	94,403	87,295	81,183	71,211	63,420	57,166	52,035	47,749	44,115	40,996	34,837	30,287	26,788	24,014	21,760	19,893	16,980	14,811	13,133	10,707	9,038
145	128,882	115,988	105,440	96,651	89,214	82,839	72,482	64,427	57,983	52,710	48,317	44,600	41,414	35,138	30,514	26,966	24,157	21,877	19,991	17,051	14,865	13,176	10,736	9,058
150	132,817	119,166	108,059	98,847	91,082	84,448	73,710	65,395	58,766	53,357	48,860	45,062	41,812	35,424	30,730	27,134	24,291	21,988	20,084	17,118	14,916	13,216	10,762	9,077
155	136,722	122,300	110,630	100,994	92,901	86,010	74,897	66,328	59,518	53,976	49,379	45,503	42,191	35,696	30,934	27,293	24,419	22,093	20,171	17,182	14,964	13,254	10,787	9,095
160	140,598	125,392	113,154	103,093	94,675	87,527	76,046	67,227	60,241	54,570	49,875	45,924	42,553	35,955	31,128	27,444	24,540	22,191	20,253	17,241	15,009	13,289	10,811	9,112
165	144,444	128,442	115,632	105,146	96,403	89,003	77,157	68,094	60,936	55,140	50,351	46,327	42,899	36,202	31,313	27,588	24,654	22,285	20,331	17,298	15,052	13,323	10,833	9,127
170	148,261	131,452	118,066	107,154	98,089	90,438	78,233	68,931	61,605	55,687	50,807	46,713	43,229	36,437	31,489	27,724	24,763	22,374	20,405	17,351	15,093	13,354	10,854	9,142
175	152,050	134,421	120,456	109,119	99,733	91,833	79,275	69,738	62,250	56,214	51,244	47,083	43,546	36,661	31,656	27,854	24,867	22,458	20,475	17,402	15,131	13,384	10,874	9,156
180	155,810	137,352	122,804	111,043	101,337	93,192	80,285	70,519	62,871	56,720	51,665	47,437	43,849	36,876	31,816	27,977	24,965	22,539	20,542	17,450	15,167	13,413	10,893	9,170
185	159,543	140,244	125,111	112,925	102,903	94,514	81,265	71,274	63,470	57,207	52,069	47,777	44,140	37,081	31,969	28,095	25,059	22,615	20,606	17,496	15,202	13,440	10,910	9,182
190	163,248	143,099	127,378	114,769	104,431	95,802	82,215	72,004	64,049	57,676	52,457	48,104	44,418	37,278	32,115	28,208	25,149	22,688	20,666	17,540	15,235	13,466	10,927	9,194
195	166,925	145,917	129,606	116,574	105,924	97,057	83,138	72,710	64,607	58,129	52,831	48,419	44,686	37,466	32,255	28,316	25,235	22,758	20,724	17,581	15,266	13,490	10,944	9,206
200	170,576	148,699	131,796	118,343	107,383	98,280	84,034	73,394	65,147	58,565	53,191	48,721	44,944	37,647	32,389	28,419	25 316	22,825	20,779	17,621	15,296	13,514	10,959	9,217

750 MCM | 2 Run(s) | 208 Volts | C-Value = 24,136 | Magnetic Duct | Copper

Calculations:

1. $f = \dfrac{1.732 \times \text{Length} \times \text{Isca}}{\text{Runs} \times \text{C-Value} \times E_{L\text{-}L}}$

2. $M = \dfrac{1}{1 + f}$

3. $\text{Isc} = \text{Isca} \times M$

4. Add Motor Contribution = Motor FLA x 4

* All results are given in symmetrical amperes

Three-Phase Bolted Fault-Current* Table

One-Way Distance in Feet

Available Isc in Thousands (Isca)

	5	10	15	20	25	30	40	50	60	70	80	90	100	125	150	175	200	225	250	300	350	400	500	600
3	2,985	2,971	2,956	2,942	2,928	2,914	2,886	2,859	2,832	2,806	2,780	2,755	2,730	2,670	2,613	2,558	2,505	2,454	2,406	2,3[illegible]4	2,229	2,150	2,008	1,883
5	4,959	4,919	4,879	4,841	4,802	4,765	4,691	4,620	4,550	4,483	4,418	4,355	4,293	4,146	4,009	3,881	3,761	3,648	3,542	3,3[illegible]6	3,172	3,014	2,742	2,515
7	6,920	6,842	6,766	6,691	6,618	6,547	6,409	6,276	6,149	6,027	5,910	5,797	5,688	5,434	5,201	4,987	4,791	4,609	4,440	4,138	3,874	3,641	3,251	2,937
10	9,838	9,681	9,529	9,382	9,239	9,101	8,836	8,586	8,350	8,126	7,914	7,713	7,522	7,083	6,693	6,343	6,028	5,743	5,484	5,030	4,645	4,315	3,778	3,360
15	14,638	14,294	13,965	13,651	13,351	13,064	12,525	12,028	11,570	11,145	10,750	10,383	10,039	9,273	8,615	8,044	7,544	7,103	6,711	6,043	5,496	5,040	4,322	3,783
20	19,362	18,764	18,201	17,672	17,172	16,699	15,829	15,044	14,334	13,688	13,097	12,555	12,057	10,968	10,059	9,290	8,629	8,057	7,556	6,719	6,050	5,502	4,658	4,038
25	24,011	23,098	22,251	21,465	20,732	20,047	18,805	17,708	16,732	15,858	15,071	14,358	13,710	12,319	11,184	10,241	9,444	8,763	8,173	7,203	6,439	5,822	4,885	4,208
30	28,587	27,302	26,127	25,049	24,057	23,140	21,501	20,079	18,833	17,733	16,754	15,878	15,089	13,421	12,085	10,991	10,079	9,307	8,644	7,567	6,728	6,057	5,049	4,329
35	33,092	31,382	29,839	28,442	27,169	26,005	23,953	22,201	20,688	19,368	18,207	17,177	16,257	14,337	12,823	11,598	10,587	9,738	9,015	7,850	6,951	6,237	5,174	4,420
40	37,528	35,343	33,399	31,657	30,088	28,668	25,194	24,113	22,339	20,808	19,473	18,299	17,259	15,111	13,439	12,100	11,003	10.089	9,315	8,076	7,128	6,379	5,271	4,491
45	41,895	39,191	36,814	34,710	32,833	31,148	23,250	25,845	23,817	22,084	20,587	19,279	18,128	15,773	13,960	12,520	11,350	10,380	9,562	8,[illegible]61	7,272	6,494	5,350	4,548
50	46,196	42,929	40,094	37,611	35,417	33,464	30,142	27,419	25,148	23,224	21,574	20,142	18,889	16,346	14,407	12,879	11,644	10,625	9,770	8,[illegible]16	7,391	6,589	5,414	4,595
55	50,431	46,564	43,247	40,371	37,854	35,633	31,889	28,858	26,353	24,248	22,454	20,908	19,561	16,847	14,794	13,188	11,896	10,834	9,947	8,547	7,492	6,669	5,468	4,633
60	54,604	50,098	46,279	43,002	40,157	37,666	33,508	30,177	27,449	25,173	23,245	21,592	20,158	17,288	15,133	13,456	12,114	11,015	10,099	8,659	7,578	6,737	5,513	4,666
65	58,714	53,537	49,198	45,511	42,337	39,577	35,013	31,392	28,450	26,012	23,959	22,206	20,693	17,680	15,433	13,693	12,305	11,173	10,231	8,756	7,652	6,796	5,553	4,694
70	62,764	56,883	52,010	47,907	44,403	41,377	36,414	32,514	29,368	26,778	24,607	22,762	21,174	18,030	15,699	13,902	12,474	11,312	10,348	8,[illegible]41	7,717	6,847	5,587	4,718
75	66,754	60,141	54,721	50,197	46,364	43,074	37,722	33,553	30,213	27,478	25,198	23,266	21,610	18,345	15,937	14,088	12,624	11,435	10,451	8,916	7,774	6,892	5,617	4,740
80	70,686	63,315	57,335	52,388	48,227	44,678	38,946	34,518	30,993	28,122	25,738	23,726	22,006	18,630	16,152	14,256	12,758	11,545	10,543	8,983	7,825	6,932	5,643	4,759
85	74,561	66,406	59,859	54,487	50,000	46,196	40,094	35,417	31,716	28,716	26,234	24,148	22,368	18,889	16,346	14,407	12,879	11,644	10,625	9,043	7,870	6,967	5,667	4,775
90	78,381	69,419	62,296	56,499	51,689	47,634	41,173	36,256	32,388	29,265	26,692	24,535	22,700	19,125	16,523	14,544	12,988	11,733	10,699	9,096	7,911	6,999	5,688	4,790
95	82,146	72,357	64,652	58,430	53,300	48,999	42,189	37,041	33,013	29,775	27,115	24,892	23,006	19,341	16,684	14,668	13,087	11,814	10,767	9,145	7,948	7,028	5,707	4,804
100	85,859	75,221	66,929	60,284	54,839	50,296	43,147	37,778	33,597	30,249	27,508	25,222	23,288	19,540	16,832	14,783	13,178	11,888	10,828	9,189	7,981	7,054	5,724	4,816
105	89,518	78,016	69,132	62,065	56,309	51,530	44,052	38,470	34,143	30,691	27,873	25,529	23,549	19,724	16,968	14,887	13,261	11,956	10,884	9,230	8,012	7,078	5,740	4,827
110	93,127	80,743	71,265	63,779	57,716	52,706	44,909	39,121	34,655	31,104	28,214	25,814	23,791	19,894	17,093	14,984	13,338	12,018	10,936	9,267	8,040	7,099	5,754	4,837
115	96,686	83,404	73,331	65,428	59,063	53,827	45,720	39,736	35,136	31,491	28,532	26,080	24,017	20,051	17,209	15,073	13,409	12,075	10,983	9,301	8,065	7,119	5,767	4,846
120	100,196	86,003	75,332	67,017	60,355	54,898	46,490	40,316	35,590	31,855	28,830	26,329	24,228	20,198	17,317	15,156	13,474	12,128	11,027	9,332	8,089	7,138	5,779	4,855
125	103,658	88,542	77,273	68,548	61,594	55,921	47,222	40,865	36,017	32,197	29,109	26,562	24,425	20,335	17,418	15,233	13,535	12,178	11,068	9,361	8,111	7,155	5,790	4,863
130	107,074	91,021	79,155	70,025	62,784	56,900	47,918	41,386	36,420	32,519	29,373	26,781	24,610	20,463	17,512	15,305	13,592	12,223	11,105	9,388	8,131	7,171	5,800	4,870
135	110,443	93,444	80,981	71,451	63,927	57,837	48,582	41,879	36,802	32,823	29,620	26,987	24,784	20,583	17,600	15,372	13,644	12,266	11,141	9,413	8,150	7,185	5,810	4,877
140	113,767	95,813	82,754	72,827	65,027	58,736	49,214	42,349	37,164	33,111	29,854	27,181	24,947	20,696	17,682	15,434	13,694	12,306	11,174	9,437	8,167	7,199	5,819	4,883
145	117,046	98,129	84,475	74,157	66,086	59,598	49,818	42,795	37,507	33,383	30,075	27,364	25,102	20,802	17,759	15,493	13,740	12,343	11,204	9,459	8,184	7,212	5,827	4,889
150	120,283	100,393	86,148	75,444	67,105	60,426	50,395	43,220	37,834	33,641	30,285	27,538	25,247	20,902	17,832	15,543	13,784	12,379	11,233	9,479	8,199	7,224	5,835	4,894
155	123,477	102,609	87,775	76,688	68,088	61,222	50,947	43,626	38,144	33,886	30,483	27,702	25,385	20,996	17,901	15,601	13,825	12,412	11,261	9,499	8,214	7,235	5,843	4,900
160	126,629	104,776	89,356	77,892	69,035	61,987	51,476	44,013	38,440	34,119	30,672	27,857	25,516	21,085	17,966	15,650	13,863	12,443	11,286	9,517	8,227	7,246	5,849	4,904
165	129,741	106,898	90,894	79,058	69,950	62,723	51,983	44,383	38,722	34,341	30,851	28,005	25,640	21,170	18,027	15,697	13,900	12,472	11,310	9,534	8,240	7,256	5,856	4,909
170	132,812	108,974	92,391	80,188	70,833	63,433	52,469	44,737	38,991	34,553	31,022	28,146	25,757	21,250	18,085	15,741	13,934	12,500	11,333	9,551	8,252	7,265	5,862	4,913
175	135,845	111,007	93,848	81,284	71,686	64,116	52,936	45,076	39,248	34,754	31,184	28,279	25,869	21,326	18,140	15,782	13,967	12,526	11,355	9,566	8,264	7,274	5,868	4,917
180	138,838	112,998	95,267	82,346	72,512	64,775	53,384	45,400	39,494	34,947	31,339	28,407	25,976	21,398	18,193	15,822	13,998	12,551	11,375	9,580	8,275	7,282	5,873	4,921
185	141,794	114,949	96,650	83,377	73,310	65,412	53,816	45,712	39,729	35,132	31,488	28,529	26,078	21,467	18,242	15,860	14,028	12,575	11,395	9,594	8,285	7,290	5,878	4,925
190	144,713	116,859	97,997	84,378	74,082	66,026	54,231	46,011	39,955	35,308	31,629	28,645	26,175	21,533	18,290	15,896	14,056	12,597	11,413	9,607	8,295	7,298	5,883	4,928
195	147,595	118,732	99,311	85,350	74,830	66,619	54,631	46,299	40,172	35,477	31,765	28,756	26,268	21,596	18,335	15,930	14,082	12,619	11,431	9,620	8,304	7,305	5,888	4,932
200	150,442	120,567	100,591	86,294	75,555	67,193	55,016	46,575	40,380	35,639	31,895	28,862	26,356	21,656	18,378	15,962	14,108	12,639	11,448	9,632	8,313	7,312	5,893	4,935

1000 MCM | 1 Run(s) | 208 Volts | C-Value = 25,278 | Magnetic Duct | Copper

Calculations:

1. $f = \frac{1.732 \times \text{Length} \times \text{Isca}}{\text{Runs} \times \text{C-Value} \times \text{E L-L}}$

2. $M = \frac{1}{1 + f}$

3. Isc = Isca x M

4. Add Motor Contribution = Motor FLA x 4

* All results are given in symmetrical amperes

Three-Phase Bolted Fault-Current* Table

One-Way Distance in Feet

Available Isc in Thousands (Isca)	5	10	15	20	25	30	40	50	60	70	80	90	100	125	150	175	200	225	250	300	350	400	500	600
3	2,993	2,985	2,978	2,971	2,963	2,956	2,942	2,928	2,914	2,900	2,886	2,872	2,859	2,825	2,793	2,761	2,730	2,700	2,670	2,613	2,558	2,505	2,406	2,314
5	4,979	4,959	4,939	4,919	4,899	4,879	4,841	4,802	4,765	4,727	4,691	4,655	4,620	4,533	4,450	4,370	4,293	4,218	4,146	4,009	3,881	3,761	3,542	3,346
7	6,960	6,920	6,881	6,842	6,804	6,766	6,691	6,618	6,547	6,477	6,409	6,342	6,276	6,118	5,968	5,825	5,688	5,558	5,434	5,201	4,987	4,791	4,440	4,138
10	9,918	9,838	9,759	9,681	9,605	9,529	9,382	9,239	9,101	8,966	8,836	8,709	8,586	8,293	8,019	7,763	7,522	7,296	7,083	6,693	6,343	6,028	5,484	5,030
15	14,817	14,638	14,464	14,294	14,127	13,965	13,651	13,351	13,064	12,788	12,525	12,271	12,028	11,461	10,944	10,472	10,039	9,641	9,273	8,615	8,044	7,544	6,711	6,043
20	19,676	19,362	19,058	18,764	18,478	18,201	17,672	17,172	16,699	16,252	15,829	15,426	15,044	14,167	13,386	12,687	12,057	11,486	10,968	10,059	9,290	8,629	7,556	6,719
25	24,496	24,011	23,546	23,098	22,667	22,251	21,465	20,732	20,047	19,406	18,805	18,240	17,708	16,505	15,455	14,530	13,710	12,977	12,319	11,184	10,241	9,444	8,173	7,203
30	29,277	28,587	27,930	27,302	26,702	26,127	25,049	24,057	23,140	22,290	21,501	20,765	20,079	18,545	17,230	16,088	15,089	14,206	13,421	12,085	10,991	10,079	8,644	7,567
35	34,019	33,092	32,214	31,382	30,591	29,839	28,442	27,169	26,005	24,937	23,953	23,044	22,201	20,342	18,770	17,423	16,257	15,237	14,337	12,823	11,598	10,587	9,015	7,850
40	38,724	37,528	36,403	35,343	34,343	33,399	31,657	30,088	28,668	27,375	26,194	25,111	24,113	21,935	20,118	18,579	17,259	16,114	15,111	13,439	12,100	11,003	9,315	8,076
45	43,392	41,895	40,498	39,191	37,965	36,814	34,710	32,833	31,148	29,628	28,250	26,994	25,845	23,359	21,309	19,590	18,128	16,869	15,773	13,960	12,520	11,350	9,562	8,261
50	48,023	46,196	44,503	42,929	41,463	40,094	37,611	35,417	33,464	31,716	30,142	28,716	27,419	24,638	22,368	20,482	18,889	17,526	16,346	14,407	12,879	11,644	9,770	8,416
55	52,617	50,431	48,420	46,564	44,844	43,247	40,371	37,854	35,633	33,657	31,889	30,298	28,858	25,793	23,317	21,274	19,561	18,103	16,847	14,794	13,188	11,896	9,947	8,547
60	57,175	54,604	52,254	50,098	48,113	46,279	43,002	40,157	37,666	35,466	33,508	31,756	30,177	26,842	24,171	21,983	20,158	18,613	17,288	15,133	13,456	12,114	10,099	8,659
65	61,697	58,714	56,006	53,537	51,276	49,198	45,511	42,337	39,577	37,155	35,013	33,104	31,392	27,799	24,943	22,620	20,693	19,068	17,680	15,433	13,693	12,305	10,231	8,756
70	66,185	62,764	59,679	56,883	54,338	52,010	47,907	44,403	41,377	38,737	36,414	34,353	32,514	28,675	25,646	23,197	21,174	19,476	18,030	15,699	13,902	12,474	10,348	8,841
75	70,637	66,754	63,275	60,141	57,303	54,721	50,197	46,364	43,074	40,221	37,722	35,515	33,553	29,480	26,289	23,721	21,610	19,844	18,345	15,937	14,088	12,624	10,451	8,916
80	75,055	70,686	66,798	63,315	60,177	57,335	52,388	48,227	44,678	41,616	38,946	36,598	34,518	30,222	26,877	24,199	22,006	20,178	18,630	16,152	14,256	12,758	10,543	8,983
85	79,439	74,561	70,248	66,406	62,963	59,859	54,487	50,000	46,196	42,929	40,094	37,610	35,417	30,909	27,419	24,638	22,368	20,482	18,889	16,346	14,407	12,879	10,625	9,043
90	83,790	78,381	73,628	69,419	65,665	62,296	56,499	51,689	47,634	44,168	41,173	38,558	36,256	31,546	27,920	25,041	22,700	20,760	19,125	16,523	14,544	12,988	10,699	9,096
95	88,107	82,146	76,941	72,357	68,287	64,652	58,430	53,300	48,999	45,340	42,189	39,448	37,041	32,139	28,383	25,413	23,006	21,015	19,341	16,684	14,668	13,087	10,767	9,145
100	92,391	85,859	80,189	75,221	70,833	66,929	60,284	54,839	50,296	46,448	43,147	40,284	37,778	32,692	28,813	25,757	23,288	21,250	19,540	16,832	14,783	13,178	10,828	9,189
105	96,643	89,518	83,372	78,016	73,306	69,132	62,065	56,309	51,530	47,499	44,052	41,072	38,470	33,209	29,214	26,077	23,549	21,467	19,724	16,968	14,887	13,261	10,884	9,230
110	100,863	93,127	86,494	80,743	75,708	71,265	63,779	57,716	52,706	48,496	44,909	41,816	39,121	33,694	29,588	26,375	23,791	21,668	19,894	17,093	14,984	13,338	10,936	9,267
115	105,051	96,686	89,556	83,404	78,044	73,331	65,428	59,063	53,827	49,443	45,720	42,518	39,736	34,148	29,939	26,653	24,017	21,856	20,051	17,209	15,073	13,409	10,983	9,301
120	109,208	100,196	92,559	86,003	80,315	75,332	67,017	60,355	54,898	50,345	46,490	43,184	40,316	34,576	30,267	26,913	24,228	22,030	20,198	17,317	15,156	13,474	11,027	9,332
125	113,333	103,658	95,505	88,542	82,524	77,273	68,548	61,594	55,921	51,205	47,222	43,814	40,865	34,979	30,575	27,156	24,425	22,193	20,335	17,418	15,233	13,535	11,068	9,361
130	117,428	107,074	98,397	91,021	84,674	79,155	70,025	62,784	56,900	52,024	47,918	44,413	41,386	35,360	30,866	27,385	24,610	22,346	20,463	17,512	15,305	13,592	11,105	9,388
135	121,493	110,443	101,235	93,444	86,767	80,981	71,451	63,927	57,837	52,807	48,582	44,982	41,879	35,720	31,140	27,601	24,784	22,489	20,583	17,600	15,372	13,644	11,141	9,413
140	125,527	113,767	104,021	95,813	88,806	82,754	72,827	65,027	58,736	53,555	49,214	45,524	42,349	36,060	31,398	27,804	24,947	22,623	20,696	17,682	15,434	13,694	11,174	9,437
145	129,532	117,046	106,756	98,129	90,792	84,475	74,157	66,086	59,598	54,271	49,818	46,040	42,795	36,384	31,643	27,995	25,102	22,750	20,802	17,759	15,493	13,740	11,204	9,459
150	133,508	120,283	109,442	100,393	92,727	86,148	75,444	67,105	60,426	54,957	50,395	46,533	43,220	36,690	31,875	28,177	25,247	22,870	20,902	17,832	15,549	13,784	11,233	9,479
155	137,454	123,477	112,080	102,609	94,614	87,775	76,688	68,088	61,222	55,614	50,947	47,003	43,626	36,982	32,095	28,348	25,385	22,983	20,996	17,901	15,601	13,825	11,261	9,499
160	141,372	126,629	114,671	104,776	96,454	89,356	77,892	69,035	61,987	56,245	51,476	47,453	44,013	37,260	32,304	28,511	25,516	23,090	21,085	17,966	15,650	13,863	11,286	9,517
165	145,261	129,741	117,217	106,898	98,248	90,894	79,058	69,950	62,723	56,850	51,983	47,883	44,383	37,525	32,503	28,666	25,640	23,191	21,170	18,027	15,697	13,900	11,310	9,534
170	149,123	132,812	119,718	108,974	100,000	92,391	80,188	70,833	63,433	57,432	52,469	48,295	44,737	37,778	32,692	28,813	25,757	23,288	21,250	18,085	15,741	13,934	11,333	9,551
175	152,956	135,845	122,176	111,007	101,709	93,848	81,284	71,686	64,116	57,992	52,936	48,690	45,076	38,019	32,873	28,954	25,869	23,379	21,326	18,140	15,782	13,967	11,355	9,566
180	156,762	138,838	124,593	112,998	103,378	95,267	82,346	72,512	64,775	58,531	53,384	49,070	45,400	38,250	33,045	29,087	25,976	23,466	21,398	18,193	15,822	13,998	11,375	9,580
185	160,541	141,794	126,968	114,949	105,008	96,650	83,377	73,310	65,412	59,050	53,816	49,434	45,712	38,471	33,210	29,215	26,078	23,549	21,467	18,242	15,860	14,028	11,395	9,594
190	164,293	144,713	129,303	116,859	106,600	97,997	84,378	74,082	66,026	59,550	54,231	49,784	46,011	38,682	33,368	29,337	26,175	23,628	21,533	18,290	15,896	14,056	11,413	9,607
195	168,018	147,595	131,600	118,732	108,156	99,311	85,350	74,830	66,619	60,032	54,631	50,121	46,299	38,885	33,519	29,453	26,258	23,704	21,596	18,335	15,930	14,082	11,431	9,620
200	171,717	150,442	133,858	120,567	109,677	100,591	86,294	75,555	67,193	60,498	55,016	50,445	46,575	39,080	33,663	29,565	26,356	23,776	21,656	18,378	15,962	14,108	11,448	9,632

1000 MCM	2 Run(s)	208 Volts	C-Value = 25,278	Magnetic Duct	Copper

Calculations:

1. $f = \dfrac{1.732 \times \text{Length} \times \text{Isca}}{\text{Runs} \times \text{C-Value} \times E_{L\text{-}L}}$
2. $M = \dfrac{1}{1 + f}$
3. $\text{Isc} = \text{Isca} \times M$
4. Add Motor Contribution = Motor FLA x 4

* All results are given in symmetrical amperes

Single-Phase L-L Fault-Current Tables

Secondary L-L Fault-Current for Single-Phase Transformers

Project: Project No.: Date: Designer: Page of

Transformer Data

XFMR KVA	% Imp.	FLA Amps @ 208 Volts	S/C Amps @ 208 Volts	208 V. w/50% Motor Load	FLA Amps @ 240 Volts	S/C Amps @ 240 Volts	240 V. w/50% Motor Load	FLA Amps @ 480 Volts	S/C Amps @ 480 Volts	480 V. w/50% Motor Load
3	8.0	14	180	209	13	156	181	6	78	91
6	3.5	29	824	882	25	714	764	13	357	382
9	3.0	43	1,442	1,529	38	1,250	1,325	19	625	663
15	2.3	72	3,135	3,280	63	2,717	2,842	31	1,359	1,421
25	2.5	120	4,808	5,048	104	4,167	4,375	52	2,083	2,188
30	3.0	144	4,808	5,096	125	4,167	4,417	63	2,083	2,208
45	3.0	216	7,212	7,644	188	6,250	6,625	94	3,125	3,313
75	3.5	361	10,302	11,023	313	8,929	9,554	156	4,464	4,777
112.5	2.0	541	27,043	28,125	469	23,438	24,375	234	11,719	12,188
150	2.0	721	36,058	37,500	625	31,250	32,500	313	15,625	16,250
225	2.0	1,082	54,087	56,250	938	46,875	48,750	469	23,438	24,375
300	4.5	1,442	32,051	34,936	1,250	27,778	30,278	625	13,889	15,139
400	4.5	1,923	42,735	46,581	1,667	37,037	40,370	833	18,519	20,185
500	4.5	2,404	53,419	58,226	2,083	46,296	50,463	1,042	23,148	25,231
750	5.75	3,606	62,709	69,921	3,125	54,348	60,598	1,563	27,174	30,299
1000	5.75	4,808	83,612	93,227	4,167	72,464	80,797	2,083	36,232	40,399

Circuit Breaker AIC Ratings

AIC in Thousands

Volts	10	14	18	22	25	35	42	50	65	100	200
120	X			X					X		
120/240	X			X			X		X		
240	X		X	X			X		X	X	X
277		X		X	X				X		
480		X		X	X	X		X	X	X	
600		X	X	X	X	X		X	X		

Note: All results are given in symmetrical amperes

Transformer Calculations

Infinite Primary Isc:

1.) ______ / ______ = ______
KVA / KV = XFMR FLA

2.) 100 / ______ x ______ = [ ] Isca
%Z, XFMR FLA

Known Primary Isc:

1.) ______ x ______ x 2 x ______ = ______
Iscp x E L-Lp, %Z, a.)

2.) 100,000 x ______ = ______
KVA, b.)

3.) f. = ______ / ______ = ______
a.) / b.)

4.) M = 1 / ______ = ______
f + 1

5.) ______ / ______ x ______ x ______ = [ ] Isca
E L-Lp / E L-Ls x M x Iscp

Legend:

Isc = Short Circuit or Fault Current
Iscp = Available Fault-Current at Primary
E L-Lp = Primary Line to Line Volts
E L-Ls = Secondary Line to Line Volts
KV = Line to Line Kilovolts
Isca = Available Fault-Current

[ ] Fuse Let-Thru

[ ] Isc Used

Single-Phase L-L Bolted Fault Calculations

Motor Contribution: (FLA) X 4 = **Proj. No.:** **By:** **Date:** **Page** of

Node	From Node	Name	Wire Size	(Isca	x	2	x	Length)	/	(Runs	x	C-Value	x	L-L Volts)	+	1	(1/x)	x	Isca	=	Isc	+	Motor Contribution	=	Isc @ Node
					x	2	x		/		x		x		+	1	(1/x)	x		=		+		=	
					x	2	x		/		x		x		+	1	(1/x)	x		=		+		=	
					x	2	x		/		x		x		+	1	(1/x)	x		=		+		=	
					x	2	x		/		x		x		+	1	(1/x)	x		=		+		=	
					x	2	x		/		x		x		+	1	(1/x)	x		=		+		=	
					x	2	x		/		x		x		+	1	(1/x)	x		=		+		=	
					x	2	x		/		x		x		+	1	(1/x)	x		=		+		=	
					x	2	x		/		x		x		+	1	(1/x)	x		=		+		=	
					x	2	x		/		x		x		+	1	(1/x)	x		=		+		=	
					x	2	x		/		x		x		+	1	(1/x)	x		=		+		=	
					x	2	x		/		x		x		+	1	(1/x)	x		=		+		=	
					x	2	x		/		x		x		+	1	(1/x)	x		=		+		=	
					x	2	x		/		x		x		+	1	(1/x)	x		=		+		=	
					x	2	x		/		x		x		+	1	(1/x)	x		=		+		=	
					x	2	x		/		x		x		+	1	(1/x)	x		=		+		=	
					x	2	x		/		x		x		+	1	(1/x)	x		=		+		=	
					x	2	x		/		x		x		+	1	(1/x)	x		=		+		=	
					x	2	x		/		x		x		+	1	(1/x)	x		=		+		=	
					x	2	x		/		x		x		+	1	(1/x)	x		=		+		=	
					x	2	x		/		x		x		+	1	(1/x)	x		=		+		=	
					x	2	x		/		x		x		+	1	(1/x)	x		=		+		=	
					x	2	x		/		x		x		+	1	(1/x)	x		=		+		=	
					x	2	x		/		x		x		+	1	(1/x)	x		=		+		=	
					x	2	x		/		x		x		+	1	(1/x)	x		=		+		=	
					x	2	x		/		x		x		+	1	(1/x)	x		=		+		=	
					x	2	x		/		x		x		+	1	(1/x)	x		=		+		=	
					x	2	x		/		x		x		+	1	(1/x)	x		=		+		=	
					x	2	x		/		x		x		+	1	(1/x)	x		=		+		=	
					x	2	x		/		x		x		+	1	(1/x)	x		=		+		=	
					x	2	x		/		x		x		+	1	(1/x)	x		=		+		=	
					x	2	x		/		x		x		+	1	(1/x)	x		=		+		=	
					x	2	x		/		x		x		+	1	(1/x)	x		=		+		=	
					x	2	x		/		x		x		+	1	(1/x)	x		=		+		=	

"C"-Values for Copper Conductors

Conductor AWG	*Three Single Conductors*						*Three-Conductor Cable*					
	Magnetic Duct			Non-Magnetic Duct			Magnetic Duct			Non-Magnetic Duct		
	600V	5KV	15KV	600V	5KV	15KV	600V	5KV	15KV	600V	5KV	15KV
# 14	**389**	389	389	**389**	389	389	**389**	389	389	**389**	389	389
# 12	**617**	617	617	**617**	617	617	**617**	617	617	**617**	617	617
# 10	**981**	981	981	**981**	981	981	**981**	981	981	**981**	981	981
# 8	**1,557**	1,551	1,557	**1,558**	1,555	1,558	**1,559**	1,557	1,559	**1,559**	1,558	1,559
# 6	**2,425**	2,406	2,389	**2,430**	2,417	2,406	**2,431**	2,424	2,414	**2,433**	2,428	2,420
# 4	**3,806**	3,750	3,695	**3,825**	3,789	3,752	**3,830**	3,811	3,778	**3,837**	3,823	3,798
# 3	**4,760**	4,760	4,760	**4,802**	4,802	4,802	**4,760**	4,790	4,760	**4,802**	4,802	4,802
# 2	**5,906**	5,736	5,574	**6,044**	5,926	5,809	**5,989**	5,929	5,827	**6,087**	6,022	5,957
# 1	**7,292**	7,029	6,758	**7,493**	7,306	7,108	**7,454**	7,364	7,188	**7,579**	7,507	7,364
# 1/0	**8,924**	8,543	7,973	**9,317**	9,033	8,590	**9,209**	9,086	8,707	**9,472**	9,372	9,052
# 2/0	**10,755**	10,061	9,389	**11,423**	10,877	10,318	**11,244**	11,045	10,500	**11,703**	11,528	11,052
# 3/0	**12,843**	11,804	11,021	**13,923**	13,048	12,360	**13,656**	13,333	12,613	**14,410**	14,118	13,461
# 4/0	**15,082**	13,605	12,542	**16,673**	15,351	14,347	**16,391**	15,890	14,813	**17,482**	17,019	16,012
# 250	**16,483**	14,924	13,643	**18,593**	17,120	15,865	**18,310**	17,850	16,465	**19,779**	19,352	18,001
# 300	**18,176**	16,292	14,768	**20,867**	18,975	17,408	**20,617**	20,051	18,318	**22,524**	21,938	20,163
# 350	**19,703**	17,385	15,678	**22,736**	20,526	18,672	**22,646**	21,914	19,821	**24,904**	24,126	21,982
# 400	**20,565**	18,235	16,365	**24,296**	21,786	19,731	**24,253**	23,371	21,042	**26,915**	26,044	23,517
# 500	**22,185**	19,172	17,492	**26,706**	23,277	21,329	**26,980**	25,449	23,125	**30,028**	28,712	25,916
# 600	**22,965**	20,567	17,962	**28,033**	25,203	22,097	**28,752**	27,974	24,896	**32,236**	31,258	27,766
# 750	**24,136**	21,386	18,888	**28,303**	25,430	22,690	**31,050**	30,024	26,932	**32,404**	31,338	28,303
# 1000	**25,278**	22,539	19,923	**31,490**	28,083	24,887	**33,864**	32,688	29,320	**37,197**	35,748	31,959

Single Phase L-L Calculations:

1. $f = \dfrac{2 \times \text{Length} \times \text{Isca}}{\text{Runs} \times \text{C-Value} \times \text{E L-L}}$

2. $M = \dfrac{1}{1 + f}$

3. Isc = Isca x M

4. Add Motor Contribution = Motor FLA x 4

"C"-Values for Aluminum Conductors

Conductor AWG	Three Single Conductors						Three-Conductor Cable					
	Magnetic Duct			Non-Magnetic Duct			Magnetic Duct			Non-Magnetic Duct		
	600V	5KV	15KV	600V	5KV	15KV	600V	5KV	15KV	600V	5KV	15KV
# 12	**375**	375	375	**375**	375	375	**375**	375	375	**375**	375	375
# 10	**598**	598	598	**598**	598	598	**598**	598	598	**598**	598	598
# 8	**951**	950	951	**951**	950	951	**951**	951	951	**951**	951	951
# 6	**1,480**	1,476	1,472	**1,481**	1,478	1,476	**1,481**	1,480	1,478	**1,482**	1,481	1,479
# 4	**2,345**	2,332	2,319	**2,350**	2,341	2,333	**2,351**	2,347	2,339	**2,353**	2,349	2,344
# 3	**2,948**	2,948	2,948	**2,958**	2,958	2,958	**2,948**	2,956	2,948	**2,958**	2,958	2,958
# 2	**3,713**	3,669	3,626	**3,729**	3,701	3,672	**3,733**	3,719	3,693	**3,739**	3,724	3,709
# 1	**4,645**	4,574	4,497	**4,678**	4,631	4,580	**4,686**	4,663	4,617	**4,699**	4,681	4,646
# 1/0	**5,777**	5,669	5,493	**5,838**	5,766	5,645	**5,852**	5,820	5,717	**5,875**	5,851	5,771
# 2/0	**7,186**	6,968	6,733	**7,301**	7,152	6,986	**7,327**	7,271	7,109	**7,372**	7,328	7,201
# 3/0	**8,826**	8,466	8,163	**9,110**	8,851	8,627	**9,077**	8,980	8,750	**9,242**	9,164	8,977
# 4/0	**10,740**	10,167	9,700	**11,174**	10,749	10,386	**11,184**	11,021	10,642	**11,4[illegible]8**	11,277	10,968
# 250	**12,122**	11,460	10,848	**12,862**	12,343	11,847	**12,796**	12,636	12,115	**13,236**	13,105	12,661
# 300	**13,909**	13,009	12,192	**14,922**	14,182	13,491	**14,916**	14,698	13,973	**15,494**	15,299	14,658
# 350	**15,484**	14,280	13,288	**16,812**	15,857	14,954	**15,413**	16,490	15,540	**17,635**	17,351	16,500
# 400	**16,670**	15,355	14,188	**18,505**	17,321	16,233	**18,461**	18,063	16,921	**19,587**	19,243	18,154
# 500	**18,755**	16,827	15,657	**21,390**	19,503	18,314	**21,394**	20,606	19,314	**22,987**	22,381	20,978
# 600	**20,093**	18,427	16,484	**23,451**	21,718	19,635	**23,633**	23,195	21,348	**25,750**	25,243	23,294
# 750	**21,766**	19,685	17,686	**23,491**	21,769	19,976	**26,431**	25,789	23,750	**25,682**	25,141	23,491
# 1000	**23,477**	21,235	19,005	**28,778**	26,109	23,482	**29,864**	29,049	26,608	**32,938**	31,919	29,135

Single Phase L-L Calculations:

1. $f = \dfrac{2 \times \text{Length} \times \text{Isca}}{\text{Runs} \times \text{C-Value} \times \text{E L-L}}$

2. $M = \dfrac{1}{1 + f}$

3. Isc = Isca x M

4. Add Motor Contribution = Motor FLA x 4

Short-Circuit Tables and Data: 600 Volts Single-Phase L-L

Single-Phase L-L Bolted Fault-Current* Table

One-Way Distance in Feet

Available Isc in Thousands (Isca)

Isca	5	10	15	20	25	30	40	50	60	70	80	90	100	125	150	175	200	225	250	300	350	400	500	600
1	959	921	886	854	824	796	745	700	660	625	593	565	539	483	438	400	368	342	318	280	250	226	189	163
3	2,658	2,387	2,165	1,981	1,826	1,694	1,479	1,313	1,180	1,072	981	905	840	712	618	546	488	442	404	344	300	266	217	183
5	4,118	3,500	3,044	2,693	2,414	2,188	1,842	1,591	1,400	1,250	1,129	1,030	946	787	673	588	523	470	427	361	313	276	223	187
10	7,001	5,385	4,376	3,685	3,182	2,801	2,259	1,892	1,628	1,429	1,273	1,148	1,045	854	722	625	551	493	446	374	323	283	228	191
15	9,131	6,564	5,123	4,201	3,560	3,089	2,442	2,020	1,722	1,500	1,329	1,193	1,083	879	740	638	562	501	453	379	326	286	230	192
20	10,771	7,370	5,601	4,517	3,785	3,257	2,546	2,090	1,773	1,539	1,360	1,218	1,103	892	749	645	567	506	456	382	328	288	231	193
25	12,071	7,956	5,933	4,731	3,934	3,366	2,613	2,135	1,805	1,563	1,378	1,233	1,115	900	755	650	570	508	458	383	329	288	231	193
30	13,127	8,402	6,178	4,885	4,039	3,443	2,659	2,166	1,827	1,579	1,391	1,243	1,123	905	758	652	572	510	460	384	330	289	232	193
35	14,002	8,752	6,365	5,001	4,119	3,501	2,693	2,188	1,843	1,591	1,400	1,250	1,129	909	761	654	574	511	461	385	330	289	232	193
40	14,740	9,034	6,513	5,092	4,180	3,545	2,719	2,205	1,855	1,600	1,407	1,256	1,134	912	763	656	575	512	461	385	331	290	232	194
45	15,369	9,267	6,633	5,165	4,229	3,580	2,740	2,219	1,864	1,608	1,413	1,260	1,138	915	765	657	576	513	462	386	331	290	232	194
50	15,912	9,462	6,732	5,225	4,269	3,609	2,757	2,230	1,872	1,613	1,417	1,264	1,140	916	766	658	577	513	462	386	331	290	232	194
55	16,386	9,627	6,816	5,275	4,303	3,633	2,771	2,239	1,879	1,618	1,421	1,267	1,143	918	767	659	577	514	463	386	331	290	232	194
60	16,803	9,770	6,887	5,318	4,331	3,653	2,782	2,247	1,884	1,622	1,424	1,269	1,145	919	768	660	578	514	463	386	332	290	232	194
65	17,173	9,894	6,948	5,354	4,355	3,670	2,792	2,253	1,888	1,625	1,427	1,271	1,146	920	769	660	578	515	463	387	332	290	233	194
70	17,504	10,002	7,002	5,386	4,376	3,685	2,801	2,259	1,892	1,628	1,429	1,273	1,148	921	769	661	579	515	464	387	332	291	233	194
75	17,800	10,099	7,049	5,414	4,394	3,698	2,808	2,264	1,896	1,631	1,431	1,275	1,149	922	770	661	579	515	464	387	332	291	233	194
80	18,069	10,184	7,090	5,438	4,411	3,710	2,815	2,268	1,899	1,633	1,433	1,276	1,150	923	771	661	579	515	464	387	332	291	233	194
85	18,312	10,261	7,128	5,460	4,425	3,720	2,821	2,272	1,901	1,635	1,434	1,277	1,151	923	771	662	580	516	464	387	332	291	233	194
90	18,534	10,330	7,161	5,480	4,438	3,729	2,826	2,275	1,904	1,637	1,435	1,278	1,152	924	771	662	580	516	464	387	332	291	233	194
95	18,737	10,393	7,191	5,497	4,449	3,737	2,831	2,278	1,906	1,638	1,437	1,279	1,153	925	772	662	580	516	465	387	332	291	233	194
100	18,923	10,450	7,218	5,513	4,460	3,744	2,835	2,281	1,908	1,640	1,438	1,280	1,154	925	772	662	580	516	465	387	332	291	233	194
105	19,095	10,503	7,243	5,528	4,469	3,751	2,839	2,283	1,910	1,641	1,439	1,281	1,154	925	772	663	580	516	465	388	332	291	233	194
110	19,255	10,551	7,266	5,541	4,478	3,757	2,842	2,286	1,911	1,642	1,440	1,282	1,155	926	773	663	580	516	465	388	332	291	233	194
115	19,402	10,595	7,287	5,553	4,486	3,763	2,845	2,288	1,913	1,643	1,440	1,282	1,155	926	773	663	581	516	465	388	332	291	233	194
120	19,540	10,636	7,306	5,564	4,493	3,768	2,848	2,289	1,914	1,644	1,441	1,283	1,156	926	773	663	581	516	465	388	333	291	233	194
125	19,668	10,674	7,324	5,575	4,500	3,773	2,851	2,291	1,915	1,645	1,442	1,283	1,156	927	773	663	581	517	465	388	333	291	233	194
130	19,787	10,709	7,341	5,584	4,506	3,777	2,853	2,293	1,916	1,646	1,443	1,284	1,157	927	773	663	581	517	465	388	333	291	233	194
135	19,900	10,741	7,356	5,593	4,512	3,781	2,856	2,294	1,917	1,647	1,443	1,284	1,157	927	774	664	581	517	465	388	333	291	233	194
140	20,005	10,772	7,370	5,602	4,517	3,785	2,858	2,296	1,918	1,648	1,444	1,285	1,157	927	774	664	581	517	465	388	333	291	233	194
145	20,104	10,801	7,384	5,609	4,522	3,788	2,860	2,297	1,919	1,648	1,444	1,285	1,158	928	774	664	581	517	465	388	333	291	233	194
150	20,197	10,828	7,396	5,617	4,527	3,792	2,862	2,298	1,920	1,649	1,445	1,286	1,158	928	774	664	581	517	465	388	333	291	233	194
155	20,285	10,853	7,408	5,623	4,532	3,795	2,864	2,299	1,921	1,649	1,445	1,286	1,158	928	774	664	581	517	465	388	333	291	233	194
160	20,369	10,877	7,419	5,630	4,536	3,798	2,865	2,300	1,922	1,650	1,446	1,286	1,159	928	774	664	581	517	465	388	333	291	233	194
165	20,448	10,899	7,430	5,636	4,540	3,800	2,867	2,301	1,922	1,650	1,446	1,287	1,159	928	774	664	581	517	465	388	333	291	233	194
170	20,522	10,920	7,440	5,641	4,543	3,803	2,868	2,302	1,923	1,651	1,446	1,287	1,159	929	774	664	582	517	466	388	333	291	233	194
175	20,593	10,940	7,449	5,647	4,547	3,805	2,870	2,303	1,924	1,651	1,447	1,287	1,159	929	775	664	582	517	466	388	333	291	233	194
180	20,661	10,959	7,458	5,652	4,550	3,808	2,871	2,304	1,924	1,652	1,447	1,287	1,159	929	775	664	582	517	466	388	333	291	233	194
185	20,725	10,978	7,466	5,657	4,553	3,810	2,872	2,305	1,925	1,652	1,447	1,288	1,160	929	775	664	582	517	466	388	333	291	233	194
190	20,787	10,995	7,474	5,661	4,556	3,812	2,873	2,306	1,925	1,653	1,448	1,288	1,160	929	775	665	582	517	466	388	333	291	233	194
195	20,845	11,011	7,482	5,665	4,559	3,814	2,874	2,306	1,926	1,653	1,448	1,288	1,160	929	775	665	582	517	466	388	333	291	233	194
200	20,901	11,027	7,489	5,670	4,562	3,816	2,876	2,307	1,926	1,653	1,448	1,288	1,160	929	775	665	582	517	466	388	333	291	233	194

# 14 AWG	1 Run(s)	600 Volts	C-Value = 389	Magnetic Duct	Copper

Calculations:

1. $f = \dfrac{\text{Length} \times 2 \times \text{Isca}}{\text{Runs} \times \text{C-Value} \times \text{E L-L}}$

2. $M = \dfrac{1}{1 + f}$

3. Isc = Isca x M

4. Add Motor Contribution = Motor FLA x 4

* All results are given in symmetrical amperes

Single-Phase L-L Bolted Fault-Current* Table

One-Way Distance in Feet

Available Isc in Thousands (Isca)	5	10	15	20	25	30	40	50	60	70	80	90	100	125	150	175	200	225	250	300	350	400	500	600
1	974	949	925	902	881	861	822	787	755	726	698	673	649	597	552	514	481	451	425	382	346	316	270	236
3	2,775	2,582	2,413	2,266	2,135	2,019	1,820	1,657	1,521	1,405	1,306	1,220	1,145	991	874	782	707	646	594	512	450	401	330	280
5	4,405	3,937	3,558	3,246	2,985	2,762	2,403	2,127	1,908	1,730	1,582	1,457	1,351	1,142	990	873	781	706	645	549	478	424	345	291
10	7,873	6,492	5,524	4,807	4,254	3,816	3,164	2,702	2,358	2,091	1,879	1,706	1,562	1,290	1,098	957	847	760	689	581	502	442	357	299
15	10,675	8,286	6,770	5,724	4,957	4,372	3,536	2,969	2,559	2,248	2,005	1,809	1,648	1,348	1,140	988	872	780	706	593	511	449	361	302
20	12,985	9,613	7,631	6,327	5,404	4,715	3,758	3,124	2,673	2,335	2,074	1,865	1,694	1,379	1,162	1,005	885	790	714	599	515	452	363	304
25	14,923	10,635	8,262	6,754	5,712	4,949	3,905	3,225	2,746	2,391	2,118	1,900	1,723	1,398	1,176	1,015	892	796	719	602	518	454	365	305
30	16,571	11,447	8,744	7,073	5,938	5,118	4,009	3,295	2,797	2,430	2,148	1,925	1,743	1,411	1,185	1,022	898	801	723	605	520	456	366	305
35	17,991	12,107	9,123	7,320	6,111	5,245	4,087	3,348	2,835	2,459	2,170	1,943	1,758	1,421	1,192	1,027	902	804	725	606	521	457	366	306
40	19,226	12,654	9,431	7,516	6,248	5,345	4,148	3,388	2,864	2,480	2,187	1,956	1,769	1,428	1,197	1,030	905	806	727	608	522	457	367	306
45	20,311	13,115	9,684	7,676	6,358	5,426	4,196	3,421	2,887	2,498	2,201	1,967	1,778	1,434	1,201	1,033	907	808	728	609	523	458	367	306
50	21,271	13,509	9,897	7,809	6,449	5,492	4,236	3,447	2,906	2,511	2,211	1,975	1,785	1,438	1,204	1,036	909	809	730	609	523	459	367	307
55	22,127	13,849	10,079	7,922	6,526	5,548	4,268	3,469	2,921	2,523	2,220	1,983	1,791	1,442	1,207	1,038	910	811	731	610	524	459	368	307
60	22,894	14,146	10,235	8,018	6,591	5,595	4,296	3,487	2,934	2,533	2,228	1,989	1,796	1,445	1,209	1,039	911	812	731	611	524	459	368	307
65	23,587	14,407	10,371	8,101	6,647	5,635	4,320	3,503	2,945	2,541	2,234	1,994	1,800	1,448	1,211	1,041	913	812	732	611	525	459	368	307
70	24,214	14,639	10,491	8,174	6,696	5,670	4,341	3,516	2,955	2,548	2,240	1,998	1,803	1,450	1,213	1,042	913	813	733	612	525	460	368	307
75	24,786	14,846	10,597	8,238	6,739	5,701	4,359	3,528	2,963	2,554	2,245	2,002	1,806	1,452	1,214	1,043	914	814	733	612	525	460	368	307
80	25,308	15,032	10,691	8,295	6,777	5,728	4,374	3,538	2,970	2,560	2,249	2,005	1,809	1,454	1,215	1,044	915	814	734	612	525	460	368	307
85	25,788	15,200	10,776	8,346	6,811	5,752	4,389	3,547	2,977	2,565	2,252	2,008	1,812	1,455	1,216	1,045	916	815	734	613	526	460	369	307
90	26,231	15,353	10,852	8,392	6,841	5,774	4,401	3,556	2,983	2,569	2,256	2,011	1,814	1,457	1,217	1,045	916	815	734	613	526	460	369	307
95	26,639	15,492	10,921	8,433	6,869	5,794	4,413	3,563	2,988	2,573	2,259	2,013	1,816	1,458	1,218	1,046	917	816	735	613	526	461	369	308
100	27,018	15,619	10,985	8,471	6,894	5,811	4,423	3,570	2,993	2,576	2,261	2,015	1,817	1,459	1,219	1,047	917	816	735	613	526	461	369	308
105	27,370	15,736	11,042	8,505	6,916	5,828	4,432	3,576	2,997	2,579	2,264	2,017	1,819	1,460	1,220	1,047	917	816	735	613	526	461	369	308
110	27,698	15,844	11,095	8,537	6,937	5,842	4,441	3,581	3,001	2,582	2,266	2,019	1,820	1,461	1,220	1,048	918	817	735	614	526	461	369	308
115	28,005	15,944	11,144	8,566	6,956	5,856	4,448	3,587	3,004	2,585	2,268	2,021	1,822	1,462	1,221	1,048	918	817	736	614	526	461	369	308
120	28,292	16,036	11,189	8,592	6,974	5,868	4,456	3,591	3,008	2,587	2,270	2,022	1,823	1,463	1,221	1,048	918	817	736	614	527	461	369	308
125	28,561	16,123	11,231	8,617	6,990	5,880	4,462	3,596	3,011	2,590	2,272	2,023	1,824	1,463	1,222	1,049	919	817	736	614	527	461	369	308
130	28,815	16,203	11,270	8,640	7,005	5,890	4,468	3,599	3,013	2,592	2,273	2,025	1,825	1,464	1,222	1,049	919	817	736	614	527	461	369	308
135	29,053	16,278	11,307	8,661	7,019	5,900	4,474	3,603	3,016	2,593	2,275	2,026	1,826	1,465	1,223	1,049	919	818	736	614	527	461	369	308
140	29,278	16,348	11,340	8,681	7,032	5,910	4,479	3,607	3,018	2,595	2,276	2,027	1,827	1,465	1,223	1,050	919	818	737	614	527	461	369	308
145	29,491	16,415	11,372	8,700	7,044	5,918	4,484	3,610	3,021	2,597	2,277	2,028	1,828	1,466	1,224	1,050	920	818	737	614	527	461	369	308
150	29,692	16,477	11,402	8,717	7,056	5,926	4,489	3,613	3,023	2,598	2,279	2,029	1,828	1,466	1,224	1,050	920	818	737	614	527	461	369	308
155	29,883	16,535	11,430	8,734	7,066	5,934	4,493	3,616	3,025	2,600	2,280	2,030	1,829	1,467	1,224	1,051	920	818	737	615	527	461	369	308
160	30,064	16,591	11,456	8,749	7,077	5,941	4,497	3,618	3,027	2,601	2,281	2,031	1,830	1,467	1,225	1,051	920	818	737	615	527	461	369	308
165	30,236	16,643	11,481	8,763	7,086	5,948	4,501	3,621	3,028	2,603	2,282	2,031	1,830	1,468	1,225	1,051	920	819	737	615	527	461	369	308
170	30,400	16,692	11,505	8,777	7,095	5,954	4,505	3,623	3,030	2,604	2,283	2,032	1,831	1,468	1,225	1,051	920	819	737	615	527	461	369	308
175	30,556	16,739	11,527	8,790	7,103	5,960	4,508	3,625	3,032	2,605	2,284	2,033	1,832	1,468	1,225	1,051	921	819	737	615	527	462	369	308
180	30,705	16,784	11,548	8,802	7,111	5,966	4,512	3,627	3,033	2,606	2,284	2,033	1,832	1,469	1,226	1,052	921	819	737	615	527	462	369	308
185	30,847	16,826	11,568	8,814	7,119	5,971	4,515	3,629	3,034	2,607	2,285	2,034	1,833	1,469	1,226	1,052	921	819	737	615	527	462	369	308
190	30,983	16,867	11,587	8,825	7,126	5,976	4,517	3,631	3,036	2,608	2,286	2,035	1,833	1,469	1,226	1,052	921	819	738	615	527	462	369	308
195	31,113	16,905	11,606	8,836	7,133	5,981	4,520	3,633	3,037	2,609	2,287	2,035	1,834	1,470	1,226	1,052	921	819	738	615	527	462	369	308
200	31,238	16,942	11,623	8,846	7,140	5,985	4,523	3,635	3,038	2,610	2,287	2,036	1,834	1,470	1,226	1,052	921	819	738	615	527	462	370	308

12 AWG | 1 Run(s) | 600 Volts | C-Value = 617 | Magnetic Duct | Copper

Calculations:

1. $f = \dfrac{\text{Length} \times 2 \times \text{Isca}}{\text{Runs} \times \text{C-Value} \times \text{E L-L}}$

2. $M = \dfrac{1}{1 + f}$

3. $\text{Isc} = \text{Isca} \times M$

4. Add Motor Contribution = Motor FLA x 4

* All results are given in symmetrical amperes

Single-Phase L-L Bolted Fault-Current* Table

One-Way Distance in Feet

Available Isc in Thousands (Isca)	5	10	15	20	25	30	40	50	60	70	80	90	100	125	150	175	200	225	250	300	350	400	500	600
1	983	967	952	936	922	907	880	855	831	808	786	766	746	702	662	627	595	567	541	495	457	424	371	329
3	2,855	2,722	2,602	2,492	2,391	2,297	2,131	1,987	1,861	1,751	1,652	1,565	1,486	1,319	1,186	1,078	987	911	845	739	657	591	492	422
5	4,609	4,274	3,985	3,732	3,509	3,312	2,977	2,703	2,476	2,284	2,119	1,977	1,853	1,601	1,409	1,258	1,137	1,037	953	820	720	641	527	447
10	8,548	7,464	6,624	5,954	5,407	4,952	4,239	3,705	3,291	2,960	2,689	2,464	2,274	1,906	1,640	1,440	1,283	1,157	1,053	893	776	685	556	468
15	11,954	9,936	8,501	7,428	6,596	5,931	4,936	4,227	3,696	3,284	2,954	2,685	2,460	2,035	1,735	1,512	1,340	1,203	1,092	921	796	701	566	475
20	14,928	11,908	9,904	8,478	7,410	6,582	5,379	4,548	3,939	3,474	3,107	2,810	2,565	2,106	1,787	1,551	1,371	1,228	1,112	935	807	710	572	479
25	17,547	13,517	10,993	9,263	8,003	7,045	5,685	4,764	4,100	3,599	3,207	2,892	2,633	2,152	1,819	1,576	1,390	1,243	1,124	944	813	715	575	481
30	19,872	14,856	11,862	9,873	8,454	7,393	5,908	4,921	4,216	3,688	3,277	2,949	2,680	2,183	1,842	1,592	1,403	1,253	1,133	950	818	718	577	483
35	21,949	15,987	12,572	10,360	8,809	7,662	6,080	5,039	4,302	3,753	3,329	2,991	2,715	2,206	1,858	1,605	1,412	1,261	1,139	954	821	721	579	484
40	23,815	16,955	13,163	10,758	9,095	7,878	6,214	5,131	4,369	3,804	3,369	3,023	2,741	2,224	1,870	1,614	1,419	1,267	1,144	958	824	722	580	485
45	25,503	17,793	13,663	11,089	9,331	8,054	6,324	5,205	4,423	3,845	3,401	3,048	2,762	2,237	1,880	1,621	1,425	1,271	1,147	960	825	724	581	485
50	27,035	18,526	14,091	11,369	9,529	8,201	6,414	5,266	4,467	3,878	3,427	3,069	2,779	2,249	1,888	1,627	1,429	1,275	1,150	962	827	725	582	486
55	28,432	19,172	14,461	11,609	9,697	8,325	6,489	5,317	4,503	3,906	3,448	3,086	2,794	2,258	1,894	1,632	1,433	1,278	1,153	964	828	726	582	486
60	29,712	19,745	14,785	11,817	9,841	8,431	6,554	5,360	4,534	3,929	3,466	3,101	2,805	2,266	1,900	1,636	1,436	1,280	1,155	965	829	727	583	487
65	30,889	20,258	15,071	11,999	9,967	8,524	6,609	5,397	4,561	3,949	3,482	3,113	2,816	2,272	1,905	1,639	1,439	1,282	1,156	966	830	728	583	487
70	31,974	20,719	15,325	12,159	10,077	8,604	6,658	5,429	4,584	3,966	3,495	3,124	2,824	2,278	1,909	1,642	1,441	1,284	1,158	967	831	728	584	487
75	32,978	21,136	15,552	12,301	10,175	8,675	6,700	5,458	4,604	3,981	3,507	3,133	2,832	2,283	1,912	1,645	1,443	1,286	1,159	968	832	729	584	487
80	33,910	21,515	15,756	12,429	10,262	8,738	6,738	5,483	4,622	3,994	3,517	3,142	2,839	2,287	1,915	1,647	1,445	1,287	1,160	969	832	729	584	488
85	34,778	21,861	15,941	12,543	10,340	8,795	6,771	5,505	4,637	4,006	3,526	3,149	2,845	2,291	1,918	1,649	1,446	1,288	1,161	970	833	729	585	488
90	35,586	22,178	16,108	12,647	10,410	8,846	6,801	5,525	4,651	4,017	3,534	3,155	2,850	2,294	1,920	1,651	1,448	1,289	1,162	970	833	730	585	488
95	36,343	22,469	16,262	12,741	10,474	8,892	6,829	5,543	4,664	4,026	3,542	3,161	2,855	2,297	1,922	1,652	1,449	1,290	1,163	971	833	730	585	488
100	37,051	22,738	16,402	12,827	10,532	8,934	6,853	5,559	4,676	4,035	3,548	3,166	2,859	2,300	1,924	1,654	1,450	1,291	1,164	971	834	730	585	488
105	37,717	22,987	16,531	12,906	10,585	8,972	6,876	5,574	4,686	4,042	3,554	3,171	2,863	2,303	1,926	1,655	1,451	1,292	1,164	972	834	731	585	488
110	38,343	23,218	16,650	12,979	10,634	9,007	6,896	5,587	4,696	4,050	3,560	3,176	2,866	2,305	1,928	1,656	1,452	1,293	1,165	972	834	731	585	488
115	38,933	23,433	16,761	13,046	10,679	9,039	6,915	5,599	4,704	4,056	3,565	3,180	2,870	2,307	1,929	1,657	1,453	1,293	1,165	973	835	731	586	488
120	39,490	23,634	16,863	13,108	10,720	9,069	6,932	5,611	4,712	4,062	3,569	3,183	2,873	2,309	1,930	1,658	1,454	1,294	1,166	973	835	731	586	489
125	40,017	23,821	16,958	13,165	10,759	9,096	6,949	5,621	4,720	4,067	3,574	3,187	2,875	2,311	1,932	1,659	1,454	1,294	1,166	973	835	731	586	489
130	40,516	23,997	17,047	13,219	10,795	9,122	6,963	5,631	4,727	4,073	3,578	3,190	2,878	2,313	1,933	1,660	1,455	1,295	1,167	974	835	732	586	489
135	40,989	24,163	17,130	13,269	10,828	9,145	6,977	5,640	4,733	4,077	3,581	3,193	2,880	2,314	1,934	1,661	1,456	1,295	1,167	974	836	732	586	489
140	41,438	24,318	17,208	13,315	10,859	9,168	6,990	5,649	4,739	4,082	3,585	3,195	2,882	2,315	1,935	1,662	1,456	1,296	1,167	974	836	732	586	489
145	41,865	24,465	17,282	13,359	10,888	9,188	7,002	5,656	4,745	4,086	3,588	3,198	2,884	2,317	1,936	1,662	1,457	1,296	1,168	974	836	732	586	489
150	42,272	24,603	17,351	13,400	10,915	9,208	7,013	5,664	4,750	4,090	3,591	3,200	2,886	2,318	1,937	1,663	1,457	1,297	1,168	975	836	732	586	489
155	42,660	24,734	17,416	13,439	10,941	9,226	7,024	5,671	4,755	4,093	3,593	3,202	2,888	2,319	1,937	1,664	1,458	1,297	1,168	975	836	732	586	489
160	43,030	24,858	17,477	13,476	10,965	9,243	7,034	5,677	4,759	4,097	3,596	3,205	2,890	2,320	1,938	1,664	1,458	1,297	1,169	975	836	732	586	489
165	43,384	24,975	17,535	13,510	10,988	9,259	7,043	5,683	4,763	4,100	3,599	3,206	2,891	2,321	1,939	1,665	1,458	1,298	1,169	975	837	732	587	489
170	43,722	25,087	17,590	13,543	11,010	9,275	7,052	5,689	4,767	4,103	3,601	3,208	2,893	2,322	1,940	1,665	1,459	1,298	1,169	975	837	733	587	489
175	44,046	25,193	17,642	13,574	11,030	9,289	7,061	5,694	4,771	4,106	3,603	3,210	2,894	2,323	1,940	1,666	1,459	1,298	1,169	976	837	733	587	489
180	44,356	25,294	17,692	13,603	11,049	9,303	7,069	5,700	4,775	4,108	3,605	3,212	2,896	2,324	1,941	1,666	1,460	1,299	1,170	976	837	733	587	489
185	44,653	25,391	17,739	13,631	11,068	9,316	7,076	5,705	4,778	4,111	3,607	3,213	2,897	2,325	1,941	1,667	1,460	1,299	1,170	976	837	733	587	489
190	44,939	25,483	17,784	13,657	11,085	9,328	7,083	5,709	4,782	4,113	3,609	3,215	2,898	2,326	1,942	1,667	1,460	1,299	1,170	976	837	733	587	489
195	45,213	25,571	17,826	13,682	11,102	9,340	7,090	5,714	4,785	4,116	3,611	3,216	2,899	2,326	1,942	1,667	1,460	1,299	1,170	976	837	733	587	489
200	45,476	25,655	17,867	13,707	11,118	9,351	7,096	5,718	4,788	4,118	3,612	3,217	2,900	2,327	1,943	1,668	1,461	1,300	1,170	976	837	733	587	489

10 AWG | 1 Run(s) | 600 Volts | C-Value = 981 | Magnetic Duct | Copper

Calculations:

1. $f = \dfrac{\text{Length} \times 2 \times \text{Isca}}{\text{Runs} \times \text{C-Value} \times \text{E L-L}}$

2. $M = \dfrac{1}{1 + f}$

3. Isc = Isca x M

4. Add Motor Contribution = Motor FLA x 4

* All results are given in symmetrical amperes

Single-Phase L-L Bolted Fault-Current* Table

One-Way Distance in Feet

Available Isc in Thousands (Isca)	5	10	15	20	25	30	40	50	60	70	80	90	100	125	150	175	200	225	250	300	350	400	500	600
3	2,907	2,819	2,736	2,659	2,585	2,515	2,337	2,271	2,166	2,070	1,982	1,901	1,827	1,664	1,528	1,412	1,313	1,227	1,151	1,025	924	841	712	618
5	4,746	4,517	4,308	4,118	3,944	3,785	3,501	3,257	3,045	2,858	2,693	2,547	2,415	2,139	1,919	1,740	1,592	1,467	1,360	1,187	1,053	947	787	674
7	6,512	6,088	5,715	5,386	5,092	4,829	4,377	4,002	3,686	3,416	3,183	2,980	2,802	2,436	2,155	1,932	1,751	1,601	1,475	1,27[illegible]	1,121	1,001	824	701
10	9,033	8,237	7,569	7,002	6,514	6,089	5,387	4,830	4,377	4,002	3,686	3,417	3,184	2,720	2,375	2,107	1,893	1,719	1,574	1,347	1,177	1,046	854	722
15	12,925	11,354	10,124	9,134	8,320	7,640	6,566	5,757	5,125	4,618	4,203	3,856	3,562	2,992	2,579	2,266	2,021	1,824	1,661	1,41[illegible]	1,226	1,083	879	740
20	16,473	14,004	12,178	10,774	9,660	8,755	7,373	6,368	5,604	5,003	4,519	4,121	3,787	3,149	2,694	2,355	2,091	1,881	1,709	1,445	1,251	1,103	893	749
25	19,722	16,284	13,867	12,075	10,693	9,595	7,960	6,801	5,936	5,267	4,733	4,298	3,936	3,251	2,769	2,412	2,136	1,917	1,738	1,465	1,267	1,116	901	755
30	22,708	18,268	15,280	13,132	11,513	10,250	8,406	7,124	6,181	5,459	4,888	4,425	4,042	3,323	2,821	2,451	2,167	1,942	1,759	1,480	1,278	1,124	906	759
35	25,461	20,008	16,479	14,008	12,181	10,776	8,756	7,374	6,368	5,604	5,004	4,520	4,121	3,376	2,860	2,480	2,189	1,960	1,774	1,491	1,286	1,130	910	762
40	28,008	21,548	17,509	14,745	12,735	11,207	9,039	7,573	6,517	5,719	5,095	4,594	4,183	3,418	2,889	2,502	2,207	1,974	1,785	1,499	1,291	1,135	913	764
45	30,371	22,920	18,404	15,375	13,202	11,568	9,272	7,736	6,637	5,811	5,168	4,653	4,232	3,450	2,912	2,520	2,220	1,984	1,794	1,505	1,296	1,138	915	765
50	32,569	24,150	19,189	15,919	13,601	11,873	9,467	7,871	6,736	5,887	5,228	4,702	4,272	3,477	2,931	2,534	2,231	1,993	1,801	1,510	1,300	1,141	917	767
55	34,619	25,259	19,883	16,394	13,946	12,135	9,632	7,986	6,820	5,951	5,278	4,742	4,305	3,499	2,947	2,546	2,240	2,000	1,807	1,514	1,303	1,143	919	768
60	36,535	26,264	20,500	16,811	14,247	12,362	9,775	8,083	6,891	6,005	5,321	4,777	4,334	3,518	2,960	2,555	2,248	2,007	1,812	1,518	1,306	1,145	920	769
65	38,330	27,179	21,054	17,182	14,512	12,561	9,899	8,168	6,952	6,052	5,358	4,806	4,358	3,534	2,972	2,564	2,254	2,012	1,816	1,521	1,308	1,147	921	769
70	40,016	28,016	21,552	17,512	14,748	12,737	10,008	8,242	7,006	6,092	5,389	4,832	4,379	3,547	2,981	2,571	2,260	2,016	1,820	1,523	1,310	1,149	922	770
75	41,601	28,784	22,004	17,809	14,958	12,893	10,104	8,307	7,053	6,128	5,417	4,854	4,397	3,559	2,990	2,577	2,265	2,020	1,823	1,525	1,311	1,150	923	771
80	43,095	29,491	22,415	18,077	15,147	13,033	10,190	8,365	7,095	6,159	5,442	4,874	4,413	3,570	2,997	2,583	2,269	2,023	1,826	1,527	1,313	1,151	923	771
85	44,506	30,145	22,791	18,321	15,317	13,159	10,267	8,417	7,132	6,187	5,463	4,891	4,428	3,579	3,004	2,588	2,273	2,027	1,828	1,529	1,314	1,152	924	771
90	45,839	30,750	23,135	18,543	15,472	13,274	10,336	8,463	7,165	6,212	5,483	4,907	4,441	3,588	3,010	2,592	2,276	2,029	1,830	1,531	1,315	1,153	925	772
95	47,102	31,314	23,453	18,746	15,613	13,377	10,399	8,506	7,195	6,235	5,501	4,921	4,452	3,595	3,015	2,596	2,279	2,032	1,832	1,532	1,316	1,154	925	772
100	48,299	31,838	23,746	18,933	15,743	13,472	10,456	8,544	7,223	6,255	5,517	4,934	4,463	3,602	3,020	2,600	2,282	2,034	1,834	1,533	1,317	1,154	926	772
105	49,436	32,328	24,017	19,105	15,862	13,559	10,509	8,579	7,248	6,274	5,531	4,946	4,472	3,608	3,024	2,603	2,285	2,036	1,836	1,534	1,318	1,155	926	773
110	50,517	32,787	24,270	19,265	15,971	13,639	10,557	8,611	7,270	6,291	5,544	4,956	4,481	3,614	3,028	2,606	2,287	2,038	1,837	1,535	1,319	1,155	926	773
115	51,546	33,218	24,505	19,413	16,073	13,713	10,601	8,640	7,291	6,307	5,557	4,966	4,489	3,619	3,032	2,609	2,289	2,039	1,839	1,536	1,319	1,156	927	773
120	52,527	33,622	24,724	19,550	16,167	13,782	10,642	8,667	7,311	6,321	5,568	4,975	4,496	3,624	3,035	2,611	2,291	2,041	1,840	1,537	1,320	1,156	927	773
125	53,464	34,004	24,930	19,678	16,254	13,845	10,680	8,692	7,329	6,335	5,578	4,983	4,503	3,628	3,038	2,613	2,293	2,042	1,841	1,538	1,320	1,157	927	774
130	54,358	34,363	25,122	19,798	16,336	13,905	10,715	8,716	7,345	6,347	5,588	4,991	4,509	3,632	3,041	2,615	2,294	2,043	1,842	1,539	1,321	1,157	928	774
135	55,213	34,703	25,303	19,910	16,413	13,960	10,748	8,737	7,361	6,359	5,597	4,998	4,515	3,636	3,044	2,617	2,296	2,045	1,843	1,539	1,322	1,158	928	774
140	56,031	35,024	25,474	20,016	16,484	14,012	10,778	8,758	7,375	6,369	5,605	5,004	4,520	3,640	3,046	2,619	2,297	2,046	1,844	1,540	1,322	1,158	928	774
145	56,815	35,329	25,635	20,115	16,551	14,060	10,807	8,777	7,388	6,379	5,613	5,011	4,525	3,643	3,049	2,621	2,298	2,047	1,845	1,540	1,322	1,158	928	774
150	57,567	35,618	25,787	20,209	16,614	14,106	10,834	8,794	7,401	6,389	5,620	5,016	4,530	3,646	3,051	2,622	2,300	2,048	1,845	1,541	1,323	1,159	928	774
155	58,289	35,893	25,930	20,297	16,674	14,149	10,859	8,811	7,413	6,397	5,627	5,022	4,534	3,649	3,053	2,624	2,301	2,049	1,846	1,542	1,323	1,159	929	775
160	58,982	36,155	26,067	20,380	16,730	14,189	10,883	8,827	7,424	6,406	5,633	5,027	4,539	3,652	3,055	2,625	2,302	2,049	1,847	1,542	1,324	1,159	929	775
165	59,648	36,404	26,196	20,459	16,783	14,227	10,906	8,841	7,434	6,413	5,639	5,032	4,542	3,654	3,056	2,627	2,303	2,050	1,847	1,542	1,324	1,160	929	775
170	60,289	36,642	26,319	20,534	16,834	14,264	10,927	8,855	7,444	6,421	5,645	5,036	4,546	3,656	3,058	2,628	2,304	2,051	1,848	1,543	1,324	1,160	929	775
175	60,906	36,869	26,436	20,605	16,882	14,298	10,947	8,869	7,453	6,428	5,650	5,041	4,550	3,659	3,060	2,629	2,305	2,052	1,849	1,543	1,324	1,160	929	775
180	61,501	37,086	26,547	20,673	16,927	14,330	10,966	8,881	7,462	6,434	5,655	5,045	4,553	3,661	3,061	2,630	2,306	2,052	1,849	1,544	1,325	1,160	929	775
185	62,074	37,294	26,654	20,737	16,970	14,361	10,984	8,893	7,471	6,441	5,660	5,048	4,556	3,663	3,062	2,631	2,306	2,053	1,850	1,544	1,325	1,160	930	775
190	62,627	37,493	26,755	20,798	17,011	14,391	11,001	8,904	7,479	6,446	5,665	5,052	4,559	3,665	3,064	2,632	2,307	2,054	1,850	1,544	1,325	1,161	930	775
195	63,161	37,683	26,852	20,857	17,050	14,419	11,018	8,915	7,486	6,452	5,669	5,055	4,562	3,667	3,065	2,633	2,308	2,054	1,851	1,545	1,325	1,161	930	775
200	63,677	37,866	26,945	20,913	17,088	14,445	11,033	8,925	7,493	6,457	5,673	5,059	4,564	3,668	3,066	2,634	2,309	2,055	1,851	1,545	1,326	1,161	930	775

# 8 AWG	1 Run(s)	600 Volts	C-Value = 1,557	Magnetic Duct	Copper

Calculations:

1. $f = \dfrac{\text{Length} \times 2 \times \text{Isca}}{\text{Runs} \times \text{C-Value} \times \text{E L-L}}$

2. $M = \dfrac{1}{1 + f}$

3. Isc = Isca x M

4. Add Motor Contribution = Motor FLA x 4

* All results are given in symmetrical amperes

Single-Phase L-L Bolted Fault-Current* Table

One-Way Distance in Feet

Available Isc in Thousands (Isca)

Isca	5	10	15	20	25	30	40	50	60	70	80	90	100	125	150	175	200	225	250	300	350	400	500	600
3	2,939	2,881	2,825	2,771	2,720	2,670	2,575	2,487	2,405	2,328	2,256	2,188	2,124	1,980	1,854	1,743	1,644	1,556	1,477	1,341	1,228	1,132	980	864
5	4,834	4,678	4,533	4,396	4,267	4,145	3,922	3,721	3,540	3,376	3,226	3,089	2,963	2,689	2,462	2,270	2,106	1,964	1,839	1,633	1,468	1,334	1,127	976
7	6,679	6,386	6,117	5,870	5,643	5,432	5,055	4,726	4,438	4,183	3,955	3,751	3,567	3,178	2,865	2,608	2,394	2,212	2,055	1,801	1,603	1,444	1,205	1,033
10	9,357	8,792	8,291	7,844	7,442	7,080	6,452	5,927	5,480	5,096	4,763	4,470	4,211	3,679	3,266	2,936	2,667	2,443	2,254	1,952	1,721	1,539	1,270	1,081
15	13,598	12,436	11,457	10,620	9,898	9,268	8,220	7,386	6,705	6,139	5,661	5,253	4,899	4,193	3,665	3,255	2,928	2,660	2,437	2,088	1,826	1,622	1,326	1,122
20	17,583	15,687	14,161	12,905	11,853	10,960	9,525	8,423	7,549	6,839	6,251	5,757	5,335	4,508	3,903	3,442	3,078	2,783	2,540	2,163	1,883	1,667	1,356	1,143
25	21,334	18,606	16,497	14,817	13,447	12,310	10,528	9,197	8,165	7,341	6,668	6,108	5,635	4,721	4,062	3,564	3,175	2,863	2,607	2,211	1,919	1,695	1,375	1,156
30	24,872	21,241	18,535	16,441	14,772	13,410	11,323	9,798	8,635	7,719	6,978	6,368	5,855	4,874	4,175	3,651	3,244	2,919	2,653	2,244	1,944	1,715	1,388	1,165
35	28,213	23,631	20,329	17,837	15,889	14,325	11,968	10,277	9,005	8,013	7,218	6,567	6,023	4,990	4,260	3,716	3,295	2,960	2,687	2,268	1,962	1,729	1,397	1,172
40	31,375	25,809	21,921	19,051	16,845	15,097	12,503	10,669	9,305	8,249	7,409	6,724	6,155	5,081	4,326	3,766	3,334	2,992	2,713	2,286	1,976	1,740	1,404	1,177
45	34,370	27,803	23,342	20,115	17,672	15,758	12,953	10,995	9,551	8,443	7,565	6,852	6,263	5,153	4,378	3,806	3,365	3,017	2,733	2,301	1,987	1,748	1,409	1,181
50	37,212	29,633	24,619	21,056	18,394	16,330	13,336	11,270	9,759	8,604	7,694	6,958	6,351	5,213	4,421	3,838	3,391	3,037	2,750	2,313	1,996	1,755	1,414	1,184
55	39,913	31,321	25,773	21,895	19,031	16,830	13,668	11,506	9,935	8,741	7,804	7,048	6,425	5,263	4,457	3,865	3,412	3,054	2,764	2,323	2,003	1,761	1,418	1,186
60	42,482	32,881	26,820	22,646	19,596	17,270	13,957	11,710	10,087	8,858	7,897	7,124	6,488	5,305	4,487	3,888	3 430	3,068	2,775	2,331	2,009	1,765	1,421	1,188
65	44,929	34,328	27,775	23,323	20,101	17,661	14,211	11,889	10,219	8,960	7,978	7,189	6,543	5,342	4,513	3,907	3 445	3,080	2,785	2,338	2,014	1,769	1,423	1,190
70	47,262	35,674	28,650	23,937	20,555	18,011	14,437	12,046	10,335	9,049	8,048	7,247	6,590	5,373	4,536	3,924	3 458	3,091	2,794	2,344	2,019	1,773	1,425	1,192
75	49,490	36,929	29,453	24,495	20,965	18,325	14,638	12,186	10,438	9,128	8,110	7,297	6,632	5,401	4,555	3,939	3,469	3,100	2,801	2,349	2,023	1,776	1,427	1,193
80	51,619	38,101	30,195	25,005	21,338	18,609	14,819	12,311	10,529	9,198	8,166	7,342	6,669	5,425	4,573	3,952	3,479	3,108	2,808	2,354	2,026	1,778	1,429	1,194
85	53,655	39,200	30,880	25,474	21,678	18,867	14,982	12,423	10,611	9,261	8,215	7,381	6,701	5,447	4,588	3,963	3,488	3,115	2,814	2,358	2,029	1,781	1,431	1,195
90	55,605	40,230	31,516	25,905	21,990	19,103	15,130	12,525	10,685	9,317	8,259	7,417	6,731	5,466	4,602	3,974	3,496	3,121	2,819	2,361	2,032	1,783	1,432	1,196
95	57,474	41,200	32,108	26,304	22,276	19,319	15,265	12,618	10,753	9,368	8,299	7,449	6,758	5,484	4,614	3,983	3,503	3,127	2,824	2,365	2,034	1,785	1,433	1,197
100	59,267	42,113	32,660	26,673	22,541	19,517	15,389	12,702	10,814	9,414	8,336	7,479	6,782	5,500	4,626	3,991	3,510	3,132	2,828	2,368	2,036	1,786	1,434	1,198
105	60,988	42,975	33,176	27,016	22,785	19,700	15,502	12,779	10,870	9,457	8,369	7,506	6,804	5,514	4,636	3,999	3,516	3,137	2,832	2,370	2,038	1,788	1,435	1,199
110	62,642	43,789	33,659	27,336	23,012	19,870	15,607	12,850	10,921	9,496	8,399	7,530	6,824	5,528	4,645	4,006	3,521	3,141	2,835	2,373	2,040	1,789	1,436	1,199
115	64,232	44,561	34,113	27,634	23,223	20,027	15,704	12,916	10,969	9,531	8,427	7,552	6,842	5,540	4,654	4,012	3,526	3,145	2,838	2,375	2,042	1,790	1,437	1,200
120	65,763	45,292	34,540	27,914	23,421	20,173	15,794	12,977	11,012	9,565	8,453	7,573	6,859	5,551	4,662	4,018	3,530	3,148	2,841	2,377	2,043	1,792	1,438	1,200
125	67,237	45,986	34,942	28,176	23,605	20,310	15,877	13,033	11,053	9,595	8,477	7,592	6,875	5,561	4,669	4,023	3,535	3,152	2,844	2,379	2,045	1,793	1,438	1,201
130	68,657	46,646	35,322	28,422	23,777	20,438	15,955	13,085	11,091	9,624	8,499	7,610	6,889	5,571	4,676	4,028	3,538	3,155	2,846	2,381	2,046	1,794	1,439	1,201
135	70,027	47,274	35,681	28,654	23,940	20,557	16,028	13,134	11,126	9,650	8,520	7,627	6,903	5,579	4,682	4,033	3,542	3,158	2,849	2,382	2,047	1,795	1,439	1,202
140	71,349	47,873	36,021	28,873	24,092	20,670	16,096	13,180	11,159	9,675	8,539	7,642	6,916	5,588	4,688	4,037	3,545	3,160	2,851	2,384	2,048	1,795	1,440	1,202
145	72,625	48,444	36,344	29,080	24,236	20,775	16,160	13,223	11,189	9,698	8,557	7,657	6,927	5,595	4,693	4,041	3,548	3,163	2,853	2,385	2,049	1,796	1,441	1,202
150	73,858	48,990	36,650	29,276	24,372	20,875	16,221	13,263	11,218	9,719	8,574	7,670	6,938	5,603	4,698	4,045	3,551	3,165	2,855	2,386	2,050	1,797	1,441	1,203
155	75,050	49,512	36,941	29,461	24,500	20,969	16,278	13,301	11,245	9,740	8,590	7,683	6,949	5,609	4,703	4,049	3,554	3,167	2,856	2,388	2,051	1,798	1,441	1,203
160	76,203	50,011	37,218	29,637	24,622	21,058	16,331	13,337	11,271	9,759	8,605	7,695	6,959	5,616	4,707	4,052	3,557	3,169	2,858	2,389	2,052	1,798	1,442	1,203
165	77,319	50,489	37,482	29,804	24,737	21,143	16,382	13,371	11,295	9,777	8,619	7,706	6,968	5,622	4,712	4,055	3,559	3,171	2,860	2,390	2,053	1,799	1,442	1,204
170	78,399	50,947	37,735	29,964	24,847	21,223	16,430	13,403	11,318	9,794	8,632	7,716	6,976	5,627	4,715	4,058	3,561	3 173	2,861	2,391	2,053	1,799	1,443	1,204
175	79,446	51,387	37,975	30,115	24,951	21,299	16,475	13,433	11,339	9,810	8,645	7,726	6,985	5,633	4,719	4,061	3,563	3,175	2,862	2,392	2,054	1,800	1,443	1,204
180	80,461	51,810	38,206	30,260	25,050	21,371	16,518	13,462	11,360	9,826	8,656	7,736	6,992	5,638	4,723	4,063	3,565	3,176	2,864	2,393	2,055	1,801	1,443	1,204
185	81,445	52,216	38,426	30,398	25,145	21,440	16,560	13,489	11,379	9,840	8,668	7,745	7,000	5,642	4,726	4,066	3,567	3,178	2,865	2,394	2,055	1,801	1,444	1,205
190	82,399	52,607	38,637	30,530	25,235	21,505	16,599	13,515	11,398	9,854	8,678	7,753	7,007	5,647	4,729	4,068	3,569	3,179	2,866	2,394	2,056	1,802	1,444	1,205
195	83,326	52,983	38,840	30,656	25,321	21,568	16,636	13,540	11,415	9,867	8,689	7,762	7,013	5,651	4,732	4,070	3,571	3,181	2,867	2,395	2,057	1,802	1,444	1,205
200	84,226	53,346	39,034	30,777	25,404	21,628	16,671	13,563	11,432	9,879	8,698	7,769	7,020	5,655	4,735	4,072	3,573	3,182	2,868	2,396	2,057	1,802	1,444	1,205

 # 6 AWG | **1 Run(s)** | **600 Volts** | **C-Value = 2,425** | **Magnetic Duct** | **Copper**

Calculations:

1. $f = \dfrac{\text{Length} \times 2 \times \text{Isca}}{\text{Runs} \times \text{C-Value} \times E_{L-L}}$

2. $M = \dfrac{1}{1 + f}$

3. $\text{Isc} = \text{Isca} \times M$

4. Add Motor Contribution = Motor FLA x 4

* All results are given in symmetrical amperes

Single-Phase L-L Bolted Fault-Current* Table

One-Way Distance in Feet

Available Isc in Thousands (Isca)	5	10	15	20	25	30	40	50	60	70	80	90	100	125	150	175	200	225	250	300	350	400	500	600
3	2,961	2,923	2,886	2,850	2,815	2,781	2,715	2,652	2,591	2,534	2,479	2,426	2,376	2,258	2,152	2,055	1,967	1,885	1,811	1,678	1,563	1,463	1,297	1,164
5	4,893	4,790	4,692	4,597	4,507	4,419	4,255	4,102	3,960	3,827	3,703	3,587	3,477	3,231	3,018	2,831	2,666	2,519	2,387	2,161	1,974	1,817	1,568	1,378
7	6,792	6,596	6,410	6,235	6,070	5,913	5,621	5,358	5,118	4,898	4,697	4,511	4,340	3,963	3,647	3,377	3,144	2,942	2,764	2,455	2,225	2,028	1,722	1,496
10	9,580	9,195	8,839	8,509	8,204	7,919	7,406	6,955	6,555	6,199	5,880	5,592	5,331	4,774	4,322	3,948	3,634	3,366	3,135	2,757	2,460	2,221	1,859	1,599
15	14,075	13,258	12,531	11,879	11,292	10,760	9,833	9,053	8,388	7,814	7,314	6,873	6,483	5,677	5,050	4,547	4,135	3,792	3,501	3,036	2,680	2,398	1,982	1,689
20	18,389	17,019	15,839	14,811	13,909	13,111	11,760	10,662	9,751	8,984	8,329	7,763	7,268	6,271	5,514	4,920	4,441	4,048	3,718	3,198	2,805	2,498	2,050	1,738
25	22,533	20,509	18,819	17,386	16,156	15,089	13,328	11,935	10,805	9,871	9,086	8,416	7,838	6,690	5,835	5,174	4,648	4,218	3,862	3,303	2,886	2,562	2,092	1,768
30	26,516	23,758	21,519	19,666	18,107	16,776	14,627	12,966	11,644	10,566	9,671	8,916	8,270	7,002	6,071	5,359	4,796	4,340	3,964	3,378	2,942	2,606	2,122	1,789
35	30,349	26,788	23,976	21,698	19,815	18,233	15,722	13,819	12,327	11,126	10,138	9,311	8,609	7,244	6,252	5,499	4,908	4,432	4,040	3,433	2,984	2,639	2,144	1,805
40	34,038	29,623	26,221	23,520	21,324	19,503	16,658	14,537	12,895	11,587	10,519	9,632	8,882	7,436	6,395	5,610	4,996	4,503	4,099	3,475	3,016	2,664	2,160	1,817
45	37,592	32,279	28,281	25,165	22,667	20,620	17,466	15,149	13,374	11,972	10,836	9,897	9,107	7,593	6,511	5,698	5,066	4,560	4,146	3,509	3,042	2,684	2,173	1,826
50	41,019	34,773	30,178	26,655	23,869	21,610	18,171	15,676	13,784	12,299	11,103	10,119	9,295	7,723	6,606	5,771	5,124	4,607	4,185	3,537	3,062	2,700	2,184	1,833
55	44,325	37,120	31,930	28,013	24,952	22,494	18,792	16,136	14,138	12,580	11,332	10,309	9,455	7,833	6,687	5,833	5,172	4,646	4,217	3,560	3,080	2,714	2,193	1,839
60	47,516	39,332	33,553	29,254	25,932	23,288	19,343	16,541	14,448	12,825	11,530	10,472	9,593	7,928	6,755	5,885	5,213	4,679	4,244	3,579	3,094	2,725	2,200	1,844
65	50,598	41,420	35,061	30,394	26,824	24,004	19,835	16,899	14,720	13,039	11,703	10,615	9,712	8,009	6,814	5,929	5,248	4,707	4,267	3,595	3,106	2,734	2,206	1,849
70	53,577	43,396	36,466	31,445	27,639	24,655	20,277	17,219	14,962	13,229	11,855	10,740	9,817	8,080	6,865	5,968	5,279	4,732	4,287	3,610	3,117	2,743	2,211	1,853
75	56,458	45,266	37,778	32,415	28,386	25,248	20,676	17,506	15,179	13,398	11,991	10,851	9,909	8,143	6,911	6,002	5,305	4,753	4,305	3,622	3,126	2,750	2,216	1,856
80	59,245	47,041	39,006	33,315	29,074	25,790	21,038	17,765	15,373	13,549	12,112	10,950	9,992	8,198	6,951	6,033	5,329	4,772	4,321	3,633	3,134	2,756	2,220	1,859
85	61,943	48,726	40,158	34,152	29,709	26,289	21,369	18,000	15,549	13,685	12,221	11,039	10,066	8,248	6,986	6,059	5,350	4,789	4,334	3,643	3,142	2,762	2,224	1,861
90	64,557	50,329	41,240	34,932	30,297	26,748	21,672	18,214	15,709	13,809	12,319	11,119	10,133	8,293	7,018	6,084	5,368	4,804	4,347	3,652	3,148	2,767	2,227	1,864
95	67,090	51,855	42,259	35,660	30,844	27,173	21,950	18,411	15,854	13,921	12,408	11,192	10,193	8,333	7,047	6,105	5,385	4,817	4,358	3,659	3,154	2,771	2,230	1,866
100	69,546	53,310	43,221	36,342	31,353	27,568	22,206	18,591	15,988	14,024	12,490	11,258	10,248	8,370	7,074	6,125	5,401	4,830	4,368	3,666	3,159	2,775	2,233	1,867
105	71,928	54,699	44,129	36,982	31,828	27,934	22,444	18,757	16,110	14,118	12,565	11,319	10,298	8,403	7,097	6,143	5,415	4,841	4,377	3,673	3,164	2,779	2,235	1,869
110	74,239	56,026	44,988	37,584	32,272	28,276	22,664	18,910	16,223	14,205	12,633	11,375	10,344	8,434	7,119	6,159	5,427	4,851	4,385	3,679	3,168	2,782	2,237	1,871
115	76,484	57,294	45,803	38,151	32,689	28,596	22,869	19,053	16,328	14,285	12,697	11,426	10,387	8,462	7,139	6,174	5,439	4,860	4,393	3,684	3,172	2,785	2,239	1,872
120	78,663	58,509	46,576	38,685	33,081	28,895	23,060	19,185	16,425	14,360	12,755	11,474	10,426	8,488	7,158	6,188	5,450	4,869	4,400	3,689	3,176	2,788	2,241	1,873
125	80,782	59,673	47,310	39,191	33,450	29,176	23,238	19,309	16,516	14,429	12,810	11,518	10,462	8,512	7,175	6,201	5,460	4,877	4,406	3,694	3,179	2,791	2,243	1,874
130	82,841	60,789	48,009	39,669	33,798	29,441	23,406	19,424	16,600	14,493	12,861	11,559	10,496	8,535	7,191	6,213	5,469	4,884	4,412	3,698	3,182	2,793	2,244	1,876
135	84,843	61,860	48,675	40,123	34,127	29,690	23,563	19,532	16,679	14,553	12,908	11,597	10,528	8,556	7,206	6,224	5,477	4,891	4,418	3,702	3,185	2,795	2,246	1,877
140	86,791	62,889	49,310	40,553	34,438	29,925	23,711	19,633	16,753	14,609	12,952	11,633	10,557	8,575	7,219	6,234	5,485	4,897	4,423	3,705	3,188	2,797	2,247	1,877
145	88,687	63,879	49,916	40,962	34,732	30,147	23,850	19,729	16,822	14,662	12,994	11,666	10,585	8,593	7,232	6,244	5,493	4,903	4,428	3,709	3,191	2,799	2,248	1,878
150	90,533	64,831	50,495	41,352	35,012	30,357	23,981	19,819	16,888	14,712	13,032	11,697	10,610	8,610	7,244	6,253	5,500	4,909	4,432	3,712	3,193	2,801	2,249	1,879
155	92,330	65,747	51,050	41,723	35,277	30,557	24,106	19,904	16,949	14,758	13,069	11,727	10,635	8,626	7,256	6,261	5,506	4,914	4,436	3,715	3,195	2,803	2,250	1,880
160	94,082	66,631	51,581	42,077	35,530	30,746	24,223	19,984	17,007	14,802	13,104	11,755	10,657	8,641	7,266	6,269	5,512	4,919	4,440	3,718	3,197	2,804	2,251	1,881
165	95,789	67,482	52,089	42,415	35,771	30,926	24,335	20,060	17,062	14,844	13,136	11,781	10,679	8,655	7,276	6,276	5,518	4,923	4,444	3,720	3,199	2,806	2,252	1,881
170	97,453	68,304	52,578	42,738	36,000	31,098	24,441	20,132	17,114	14,883	13,167	11,806	10,699	8,669	7,286	6,283	5,524	4,928	4,448	3,723	3,201	2,807	2,253	1,882
175	99,075	69,097	53,046	43,047	36,219	31,261	24,542	20,200	17,164	14,921	13,196	11,829	10,719	8,681	7,295	6,290	5,529	4,932	4,451	3,725	3,203	2,809	2,254	1,883
180	100,658	69,863	53,497	43,343	36,429	31,417	24,638	20,265	17,210	14,956	13,224	11,851	10,737	8,693	7,303	6,296	5,533	4,936	4,454	3,727	3,204	2,810	2,255	1,883
185	102,203	70,604	53,930	43,627	36,629	31,566	24,729	20,327	17,255	14,990	13,250	11,872	10,754	8,705	7,311	6,302	5,538	4,939	4,457	3,729	3,206	2,811	2,256	1,884
190	103,711	71,320	54,347	43,899	36,821	31,708	24,817	20,386	17,298	15,022	13,275	11,893	10,771	8,715	7,319	6,308	5,542	4,943	4,460	3,731	3,207	2,812	2,256	1,884
195	105,183	72,013	54,748	44,161	37,005	31,845	24,900	20,442	17,338	15,052	13,299	11,912	10,786	8,726	7,326	6,313	5,547	4,946	4,463	3,733	3,209	2,813	2,257	1,885
200	106,621	72,684	55,135	44,412	37,181	31,975	24,980	20,496	17,377	15,081	13,322	11,930	10,801	8,735	7,333	6,318	5,551	4,949	4,465	3,735	3,210	2,814	2,258	1,885

4 AWG | 1 Run(s) | 600 Volts | C-Value = 3,806 | Magnetic Duct | Copper

Calculations:

1. $f = \dfrac{\text{Length} \times 2 \times \text{Isca}}{\text{Runs} \times \text{C-Value} \times \text{E L-L}}$

2. $M = \dfrac{1}{1 + f}$

3. Isc = Isca x M

4. Add Motor Contribution = Motor FLA x 4

* All results are given in symmetrical amperes

Single-Phase L-L Bolted Fault-Current* Table

One-Way Distance in Feet

Available Isc in Thousands (Isca)	5	10	15	20	25	30	40	50	60	70	80	90	100	125	150	175	200	225	250	300	350	400	500	600
3	2,969	2,938	2,908	2,879	2,850	2,822	2,767	2,715	2,664	2,615	2,568	2,523	2,479	2,376	2,281	2,194	2,112	2,037	1,967	1,840	1,729	1,630	1,463	1,327
5	4,914	4,831	4,750	4,673	4,598	4,525	4,386	4,255	4,132	4,016	3,906	3,802	3,703	3,478	3,278	3,100	2,941	2,797	2,666	2,439	2,247	2,083	1,818	1,612
7	6,833	6,673	6,521	6,375	6,236	6,103	5,852	5,622	5,409	5,212	5,028	4,857	4,697	4,340	4,034	3,768	3,535	3,329	3,145	2,833	2,578	2,364	2,028	1,776
10	9,662	9,346	9,049	8,771	8,510	8,264	7,812	7,407	7,041	6,711	6,409	6,134	5,881	5,332	4,877	4,493	4,166	3,883	3,635	3,225	2,898	2,631	2,222	1,922
15	14,251	13,574	12,958	12,396	11,880	11,406	10,562	9,835	9,201	8,644	8,151	7,711	7,316	6,485	5,824	5,285	4,837	4,460	4,137	3,613	3,208	2,884	2,399	2,054
20	18,691	17,543	16,528	15,624	14,813	14,083	12,819	11,763	10,868	10,099	9,432	8,848	8,331	7,271	6,450	5,795	5,262	4,818	4,443	3,845	3,389	3,029	2,499	2,127
25	22,988	21,275	19,800	18,517	17,389	16,391	14,703	13,331	12,193	11,233	10,414	9,706	9,089	7,841	6,895	6,152	5,554	5,062	4,650	3,999	3,508	3,124	2,563	2,173
30	27,148	24,792	22,812	21,124	19,669	18,402	16,301	14,631	13,271	12,143	11,191	10,378	9,675	8,273	7,227	6,415	5,767	5,238	4,798	4,108	3,592	3,190	2,608	2,205
35	31,179	28,110	25,591	23,487	21,702	20,169	17,673	15,727	14,167	12,888	11,821	10,917	10,142	8,613	7,484	6,617	5,930	5,372	4,911	4,190	3,654	3,240	2,641	2,228
40	35,086	31,247	28,166	25,637	23,526	21,735	18,864	16,663	14,922	13,510	12,342	11,360	10,523	8,886	7,690	6,777	6,059	5,478	4,998	4,254	3,702	3,277	2,666	2,246
45	38,875	34,217	30,556	27,603	25,170	23,132	19,907	17,471	15,567	14,037	12,780	11,731	10,840	9,111	7,858	6,907	6,162	5,562	5,069	4,305	3,741	3,308	2,686	2,260
50	42,551	37,033	32,782	29,407	26,662	24,385	20,828	18,177	16,125	14,489	13,154	12,045	11,108	9,299	7,997	7,015	6,248	5,632	5,126	4,346	3,772	3,332	2,702	2,272
55	46,119	39,707	34,860	31,068	28,020	25,517	21,648	18,798	16,612	14,881	13,476	12,314	11,337	9,459	8,115	7,106	6,320	5,690	5,175	4,381	3,798	3,352	2,715	2,281
60	49,583	42,249	36,804	32,603	29,262	26,543	22,382	19,350	17,041	15,224	13,757	12,548	11,535	9,597	8,216	7,183	6,381	5,740	5,215	4,410	3,820	3,370	2,726	2,289
65	52,949	44,668	38,627	34,025	30,403	27,478	23,044	19,842	17,421	15,527	14,004	12,754	11,708	9,716	8,304	7,250	6,433	5,782	5,251	4,435	3,839	3,384	2,736	2,296
70	56,220	46,974	40,339	35,347	31,454	28,333	23,642	20,284	17,761	15,796	14,223	12,935	11,860	9,821	8,380	7,308	6,479	5,819	5,281	4,457	3,855	3,397	2,744	2,302
75	59,401	49,174	41,951	36,578	32,425	29,119	24,187	20,684	18,067	16,038	14,418	13,096	11,996	9,914	8,448	7,359	6,519	5,851	5,308	4,476	3,869	3,408	2,751	2,307
80	62,495	51,275	43,470	37,728	33,326	29,843	24,685	21,046	18,343	16,255	14,594	13,241	12,117	9,996	8,508	7,405	6,555	5,880	5,331	4,493	3,882	3,417	2,758	2,311
85	65,505	53,284	44,906	38,804	34,163	30,513	25,141	21,377	18,594	16,452	14,752	13,371	12,226	10,071	8,561	7,445	6,587	5,906	5,352	4,508	3,893	3,426	2,763	2,315
90	68,435	55,206	46,263	39,814	34,943	31,134	25,561	21,680	18,822	16,630	14,896	13,489	12,325	10,137	8,609	7,482	6 615	5,929	5,371	4,521	3,903	3,434	2,768	2,319
95	71,287	57,048	47,550	40,763	35,672	31,711	25,949	21,959	19,032	16,794	15,027	13,596	12,414	10,198	8,653	7,515	6,641	5,949	5,388	4,533	3,912	3,441	2,773	2,322
100	74,066	58,814	48,770	41,657	36,354	32,249	26,308	22,215	19,225	16,944	15,146	13,694	12,496	10,253	8,692	7,544	6,664	5,968	5,403	4,544	3,920	3,447	2,777	2,325
105	76,774	60,508	49,930	42,500	36,995	32,752	26,642	22,453	19,402	17,081	15,256	13,784	12,570	10,303	8,729	7,572	6,685	5,985	5,417	4,554	3,927	3,453	2,780	2,327
110	79,414	62,136	51,033	43,297	37,597	33,223	26,953	22,673	19,567	17,209	15,358	13,867	12,639	10,349	8,762	7,596	6,705	6,000	5,430	4,563	3,934	3,458	2,784	2,330
115	81,987	63,701	52,084	44,050	38,164	33,665	27,243	22,878	19,719	17,326	15,452	13,943	12,703	10,392	8,792	7,619	6,723	6,015	5,442	4,571	3,940	3,463	2,787	2,332
120	84,497	65,205	53,086	44,765	38,699	34,081	27,514	23,069	19,861	17,436	15,539	14,014	12,761	10,431	8,820	7,640	6,739	6,028	5,452	4,578	3,946	3,467	2,790	2,334
125	86,946	66,654	54,042	45,443	39,205	34,473	27,769	23,248	19,993	17,538	15,620	14,080	12,816	10,467	8,846	7,660	6,754	6,040	5,462	4,585	3,951	3,471	2,792	2,336
130	89,336	68,050	54,956	46,087	39,684	34,842	28,008	23,416	20,117	17,633	15,695	14,141	12,867	10,501	8,870	7,678	6,768	6,051	5,472	4,592	3,956	3,475	2,795	2,337
135	91,669	69,395	55,830	46,701	40,137	35,192	28,234	23,573	20,233	17,722	15,765	14,198	12,914	10,533	8,893	7,695	6,781	6,062	5,480	4,598	3,960	3,478	2,797	2,339
140	93,947	70,693	56,667	47,285	40,568	35,522	28,446	23,721	20,342	17,805	15,831	14,251	12,958	10,562	8,914	7,711	6,794	6,071	5,488	4,603	3,964	3,481	2,799	2,340
145	96,173	71,946	57,469	47,842	40,978	35,836	28,647	23,860	20,444	17,884	15,893	14,302	13,000	10,590	8,933	7,725	6,805	6,081	5,496	4,609	3,968	3,484	2,801	2,342
150	98,347	73,156	58,238	48,374	41,367	36,134	28,837	23,992	20,541	17,958	15,952	14,349	13,039	10,616	8,952	7,739	6,816	6,089	5,502	4,614	3,972	3,487	2,803	2,343
155	100,472	74,325	58,977	48,883	41,739	36,417	29,017	24,116	20,632	18,027	16,007	14,393	13,075	10,640	8,969	7,752	6,826	6,097	5,509	4,618	3,975	3,490	2,804	2,344
160	102,549	75,456	59,687	49,369	42,093	36,686	29,188	24,234	20,718	18,093	16,058	14,435	13,110	10,663	8,985	7,764	6,835	6,105	5,515	4,622	3,979	3,492	2,806	2,345
165	104,581	76,550	60,369	49,835	42,431	36,943	29,350	24,346	20,800	18,155	16,107	14,475	13,143	10,684	9,001	7,775	6,844	6,112	5,521	4,627	3,982	3,494	2,807	2,346
170	106,567	77,609	61,026	50,282	42,754	37,188	29,504	24,452	20,877	18,214	16,154	14,512	13,173	10,705	9,015	7,786	6,852	6,118	5,526	4,630	3,984	3,497	2,809	2,347
175	108,511	78,634	61,658	50,710	43,064	37,421	29,651	24,553	20,951	18,270	16,198	14,548	13,203	10,724	9,029	7,796	6,860	6,125	5,531	4,634	3,987	3,499	2,810	2,348
180	110,412	79,628	62,267	51,122	43,360	37,645	29,791	24,649	21,021	18,323	16,240	14,581	13,230	10,742	9,042	7,806	6,868	6,131	5,536	4,637	3,990	3,501	2,811	2,349
185	112,274	80,592	62,855	51,517	43,644	37,859	29,925	24,741	21,087	18,374	16,279	14,613	13,257	10,760	9,054	7,815	6,875	6,136	5,541	4,641	3,992	3,502	2,813	2,350
190	114,096	81,526	63,422	51,897	43,917	38,064	30,053	24,828	21,151	18,422	16,317	14,644	13,282	10,776	9,066	7,824	6,881	6,142	5,545	4,644	3,994	3,504	2,814	2,351
195	115,880	82,433	63,970	52,264	44,179	38,261	30,176	24,911	21,211	18,468	16,353	14,673	13,306	10,792	9,077	7,832	6,888	6,147	5,549	4,647	3,996	3,506	2,815	2,351
200	117,628	83,314	64,499	52,616	44,431	38,449	30,293	24,991	21,269	18,512	16,387	14,700	13,328	10,807	9,087	7,840	6,894	6,151	5,553	4,649	3,998	3,507	2,816	2,352

#3 AWG	1 Run(s)	600 Volts	C-Value = 4,760	Magnetic Duct	Copper

Calculations:

1. $f = \dfrac{\text{Length} \times 2 \times \text{Isca}}{\text{Runs} \times \text{C-Value} \times E_{L\text{-}L}}$

2. $M = \dfrac{1}{1 + f}$

3. $\text{Isc} = \text{Isca} \times M$

4. Add Motor Contribution = Motor FLA x 4

* All results are given in symmetrical amperes

Single-Phase L-L Bolted Fault-Current* Table

One-Way Distance in Feet

Available Isc in Thousands (Isca)	5	10	15	20	25	30	40	50	60	70	80	90	100	125	150	175	200	225	250	300	350	400	500	600
3	2,975	2,950	2,926	2,902	2,878	2,855	2,810	2,766	2,723	2,682	2,642	2,603	2,566	2,476	2,392	2,314	2,241	2,172	2,108	1,989	1,884	1,789	1,625	1,488
5	4,930	4,863	4,797	4,733	4,670	4,610	4,493	4,382	4,276	4,175	4,079	3,987	3,900	3,696	3,513	3,347	3,196	3,058	2,932	2,708	2,515	2,349	2,074	1,857
7	6,864	6,734	6,608	6,487	6,371	6,258	6,045	5,845	5,659	5,484	5,319	5,164	5,018	4,686	4,395	4,139	3,910	3,706	3,522	3,203	2,938	2,713	2,353	2,077
10	9,726	9,466	9,219	8,986	8,763	8,552	8,158	7,799	7,470	7,168	6,889	6,631	6,392	5,863	5,415	5,031	4,697	4,405	4,148	3,713	3,361	3,070	2,616	2,280
15	14,391	13,829	13,310	12,828	12,380	11,962	11,205	10,539	9,947	9,418	8,943	8,513	8,123	7,288	6,608	6,045	5,570	5,164	4,813	4,238	3,785	3,420	2,866	2,467
20	18,932	17,971	17,104	16,316	15,598	14,941	13,779	12,784	11,924	11,172	10,510	9,921	9,395	8,295	7,426	6,722	6,140	5,650	5,233	4,560	4,040	3,626	3,010	2,573
25	23,352	21,909	20,633	19,498	18,481	17,565	15,981	14,658	13,538	12,577	11,744	11,014	10,369	9,046	8,022	7,206	6,541	5,988	5,522	4,777	4,210	3,763	3,104	2,641
30	27,658	25,656	23,924	22,411	21,078	19,894	17,886	16,246	14,882	13,728	12,741	11,886	11,139	9,626	8,475	7,570	6,839	6,237	5,733	4,935	4,331	3,860	3,169	2,688
35	31,854	29,227	27,000	25,088	23,429	21,976	19,551	17,608	16,017	14,689	13,564	12,600	11,763	10,089	8,831	7,853	7,070	6,428	5,894	5,053	4,423	3,932	3,218	2,723
40	35,943	32,633	29,881	27,557	25,569	23,848	21,019	18,790	16,988	15,502	14,255	13,193	12,279	10,466	9,119	8,080	7,253	6,579	6,020	5,146	4,494	3,988	3,255	2,750
45	39,929	35,886	32,586	29,842	27,524	25,540	22,322	19,825	17,830	16,200	14,843	13,695	12,713	10,779	9,356	8,265	7,402	6,702	6,123	5,221	4,550	4,033	3,285	2,771
50	43,817	38,996	35,130	31,961	29,317	27,077	23,487	20,738	18,565	16,805	15,349	14,125	13,082	11,044	9,555	8,420	7,526	6,803	6,207	5,282	4,597	4,069	3,309	2,788
55	47,610	41,971	37,527	33,933	30,968	28,479	24,535	21,551	19,214	17,334	15,789	14,497	13,401	11,270	9,724	8,551	7,630	6,888	6,278	5,333	4,636	4,099	3,329	2,803
60	51,312	44,822	39,789	35,772	32,492	29,763	25,483	22,278	19,790	17,802	16,176	14,823	13,679	11,466	9,869	8,663	7,719	6,961	6,338	5,377	4,668	4,125	3,346	2,814
65	54,925	47,554	41,928	37,492	33,905	30,944	26,343	22,933	20,305	18,217	16,519	15,110	13,923	11,637	9,996	8,760	7,796	7,024	6,390	5,414	4,697	4,147	3,360	2,825
70	58,453	50,176	43,953	39,103	35,217	32,033	27,128	23,526	20,769	18,590	16,824	15,365	14,139	11,788	10,107	8,845	7,864	7,078	6,436	5,446	4,721	4,166	3,373	2,833
75	61,899	52,695	45,873	40,615	36,439	33,041	27,848	24,066	21,188	18,925	17,098	15,594	14,332	11,921	10,205	8,920	7,923	7,126	6,475	5,475	4,742	4,182	3,384	2,841
80	65,266	55,115	47,696	42,038	37,580	33,977	28,510	24,558	21,569	19,228	17,346	15,799	14,505	12,041	10,292	8,987	7,976	7,169	6,510	5,500	4,761	4,197	3,393	2,848
85	68,556	57,443	49,430	43,379	38,648	34,847	29,120	25,010	21,916	19,504	17,570	15,985	14,662	12,149	10,371	9,047	8,023	7,207	6,542	5,522	4,778	4,210	3,402	2,854
90	71,772	59,683	51,080	44,645	39,649	35,659	29,685	25,425	22,235	19,755	17,774	16,153	14,804	12,246	10,442	9,101	8,065	7,241	6,570	5,542	4,793	4,222	3,409	2,859
95	74,916	61,842	52,653	45,842	40,591	36,419	30,209	25,809	22,528	19,986	17,960	16,307	14,933	12,334	10,506	9,149	8,103	7,272	6,595	5,560	4,806	4,232	3,416	2,864
100	77,991	63,922	54,154	46,975	41,477	37,131	30,698	26,164	22,798	20,199	18,132	16,449	15,051	12,415	10,564	9,194	8,138	7,300	6,618	5,577	4,818	4,242	3,422	2,868
105	80,999	65,929	55,587	48,050	42,312	37,799	31,153	26,494	23,048	20,395	18,290	16,578	15,160	12,489	10,618	9,234	8,170	7,325	6,639	5,591	4,829	4,250	3,428	2,872
110	83,943	67,866	56,958	49,070	43,102	38,428	31,579	26,802	23,280	20,577	18,436	16,698	15,260	12,556	10,667	9,271	8,199	7,349	6,658	5,605	4,840	4,258	3,433	2,876
115	86,823	69,737	58,270	50,041	43,849	39,020	31,978	27,089	23,497	20,745	18,571	16,809	15,353	12,619	10,712	9,305	8,225	7,370	6,676	5,618	4,849	4,265	3,438	2,879
120	89,643	71,545	59,526	50,965	44,557	39,580	32,353	27,357	23,698	20,902	18,697	16,912	15,439	12,677	10,753	9,337	8,250	7,390	6,692	5,629	4,857	4,272	3,442	2,882
125	92,404	73,292	60,731	51,846	45,229	40,109	32,705	27,609	23,887	21,049	18,814	17,008	15,518	12,731	10,792	9,366	8,273	7,408	6,707	5,640	4,865	4,278	3,446	2,885
130	95,109	74,983	61,888	52,686	45,867	40,610	33,038	27,846	24,064	21,186	18,924	17,097	15,593	12,781	10,828	9,393	8,294	7,425	6,721	5,649	4,873	4,284	3,450	2,887
135	97,757	76,620	62,999	53,489	46,474	41,086	33,352	28,068	24,230	21,315	19,026	17,181	15,662	12,828	10,862	9,418	8,313	7,441	6,734	5,658	4,879	4,289	3,453	2,890
140	100,353	78,205	64,066	54,257	47,053	41,537	33,649	28,278	24,386	21,436	19,122	17,260	15,728	12,871	10,893	9,442	8,332	7,455	6,746	5,667	4,886	4,294	3,456	2,892
145	102,896	79,741	65,093	54,992	47,604	41,967	33,930	28,477	24,534	21,550	19,213	17,333	15,789	12,912	10,922	9,464	8,349	7,469	6,757	5,675	4,892	4,298	3,459	2,894
150	105,389	81,231	66,082	55,696	48,131	42,375	34,197	28,664	24,673	21,657	19,298	17,403	15,846	12,951	10,950	9,484	8,365	7,482	6,767	5,682	4,897	4,302	3,462	2,896
155	107,833	82,675	67,035	56,371	48,634	42,765	34,450	28,842	24,804	21,758	19,379	17,468	15,900	12,987	10,976	9,504	8,380	7,494	6,777	5,689	4,902	4,306	3,464	2,898
160	110,229	84,076	67,953	57,019	49,116	43,137	34,691	29,011	24,929	21,854	19,455	17,530	15,952	13,021	11,000	9,522	8,394	7,505	6,787	5,696	4,907	4,310	3,467	2,899
165	112,580	85,437	68,839	57,642	49,577	43,492	34,920	29,171	25,047	21,945	19,527	17,588	16,000	13,053	11,023	9,539	8,408	7,516	6,795	5,702	4,912	4,314	3,469	2,901
170	114,885	86,758	69,695	58,240	50,019	43,832	35,139	29,324	25,160	22,031	19,595	17,643	16,046	13,084	11,045	9,555	8,420	7,526	6,804	5,708	4,916	4,317	3,471	2,903
175	117,147	88,042	70,521	58,816	50,443	44,157	35,348	29,469	25,266	22,113	19,659	17,696	16,089	13,112	11,065	9,571	8,432	7,536	6,811	5,713	4,920	4,320	3,473	2,904
180	119,367	89,289	71,319	59,370	50,850	44,469	35,547	29,607	25,368	22,191	19,721	17,746	16,130	13,140	11,085	9,585	8,443	7,545	6,819	5,718	4,924	4,323	3,475	2,905
185	121,545	90,503	72,091	59,904	51,242	44,768	35,738	29,740	25,465	22,265	19,780	17,793	16,169	13,166	11,103	9,599	8,454	7,553	6,826	5,723	4,927	4,326	3,477	2,907
190	123,684	91,683	72,838	60,419	51,618	45,055	35,921	29,866	25,558	22,336	19,835	17,838	16,207	13,190	11,121	9,612	8,464	7,561	6,832	5,728	4,931	4,329	3,479	2,908
195	125,783	92,832	73,561	60,916	51,980	45,331	36,096	29,987	25,646	22,403	19,889	17,881	16,242	13,214	11,137	9,625	8,474	7,569	6,839	5,732	4,934	4,331	3,480	2,909
200	127,845	93,950	74,261	61,395	52,329	45,596	36,264	30,102	25,731	22,468	19,939	17,922	16,276	13,236	11,153	9,637	8,483	7,576	6,845	5,737	4,937	4,334	3,482	2,910

 # 2 AWG | **1 Run(s)** | **600 Volts** | **C-Value = 5,906** | **Magnetic Duct** | **Copper**

Calculations:

1. $f = \dfrac{\text{Length} \times 2 \times \text{Isca}}{\text{Runs} \times \text{C-Value} \times \text{E L-L}}$
2. $M = \dfrac{1}{1 + f}$
3. $\text{Isc} = \text{Isca} \times M$
4. Add Motor Contribution = Motor FLA x 4

* All results are given in symmetrical amperes

Single-Phase L-L Bolted Fault-Current* Table

One-Way Distance in Feet

Available Isc in Thousands (Isca)

Isca	5	10	15	20	25	30	40	50	60	70	80	90	100	125	150	175	200	225	250	300	350	400	500	600
3	2,980	2,959	2,940	2,920	2,901	2,881	2,844	2,807	2,772	2,737	2,703	2,670	2,638	2,561	2,488	2,419	2,354	2,293	2,234	2,126	2,027	1,937	1,780	1,646
5	4,944	4,888	4,834	4,781	4,730	4,679	4,581	4,487	4,397	4,310	4,227	4,147	4,070	3,889	3,723	3,571	3,431	3,302	3,182	2,966	2,778	2,612	2,333	2,108
7	6,890	6,783	6,679	6,579	6,482	6,387	6,206	6,035	5,873	5,719	5,573	5,435	5,303	5,000	4,730	4,487	4,268	4,070	3,889	3,572	3,302	3,070	2,692	2,397
10	9,777	9,563	9,358	9,162	8,974	8,794	8,454	8,140	7,848	7,576	7,322	7,085	6,863	6,364	5,932	5,556	5,224	4,930	4,667	4,217	3,846	3,535	3,044	2,672
15	14,503	14,037	13,601	13,191	12,805	12,441	11,771	11,170	10,628	10,135	9,687	9,276	8,898	8,077	7,395	6,818	6,325	5,899	5,526	4,907	4,412	4,008	3,387	2,933
20	19,126	18,325	17,588	16,908	16,279	15,695	14,645	13,726	12,915	12,195	11,551	10,972	10,448	9,334	8,434	7,693	7,071	6,542	6,087	5,344	4,762	4,295	3,590	3,084
25	23,649	22,436	21,342	20,349	19,445	18,617	17,157	15,909	14,831	13,889	13,060	12,324	11,667	10,294	9,211	8,334	7,609	7,000	6,482	5,645	5,000	4,487	3,724	3,182
30	28,075	26,382	24,882	23,543	22,341	21,255	19,373	17,797	16,458	15,306	14,306	13,427	12,651	11,053	9,813	8,824	8,016	7,343	6,774	5,866	5,173	4,626	3,818	3,251
35	32,408	30,173	28,226	26,515	25,000	23,649	21,342	19,445	17,858	16,510	15,351	14,345	13,462	11,667	10,294	9,211	8,334	7,609	7,000	6,035	5,303	4,730	3,889	3,302
40	36,649	33,817	31,390	29,289	27,451	25,831	23,103	20,896	19,074	17,544	16,242	15,119	14,142	12,174	10,687	9,524	8,589	7,822	7,180	6,168	5,406	4,811	3,944	3,341
45	40,803	37,323	34,389	31,883	29,717	27,827	24,687	22,184	20,141	18,443	17,009	15,782	14,720	12,600	11,014	9,783	8,799	7,995	7,326	6,275	5,488	4,876	3,988	3,373
50	44,872	40,698	37,234	34,314	31,819	29,662	26,120	23,334	21,085	19,231	17,677	16,356	15,218	12,963	11,291	10,000	8,975	8,140	7,447	6,364	5,556	4,930	4,023	3,398
55	48,858	43,950	39,938	36,598	33,773	31,352	27,422	24,368	21,925	19,928	18,264	16,857	15,651	13,276	11,527	10,186	9,124	8,262	7,549	6,438	5,612	4,974	4,053	3,419
60	52,764	47,086	42,511	38,746	35,594	32,916	28,611	25,302	22,679	20,549	18,784	17,299	16,031	13,549	11,732	10,345	9,251	8,367	7,637	6,502	5,661	5,012	4,078	3,437
65	56,592	50,111	44,961	40,771	37,296	34,366	29,700	26,150	23,358	21,105	19,248	17,691	16,367	13,788	11,911	10,484	9,363	8,458	7,712	6,556	5,702	5,045	4,099	3,452
70	60,345	53,031	47,298	42,684	38,890	35,715	30,703	26,924	23,973	21,606	19,664	18,042	16,667	14,001	12,069	10,606	9,460	8,537	7,778	6,604	5,738	5,073	4,118	3,465
75	64,025	55,852	49,529	44,492	40,385	36,973	31,627	27,632	24,533	22,060	20,039	18,357	16,936	14,190	12,210	10,715	9,546	8,607	7,836	6,646	5,769	5,097	4,134	3,477
80	67,633	58,578	51,661	46,206	41,792	38,148	32,483	28,284	25,046	22,473	20,379	18,642	17,179	14,360	12,335	10,811	9,622	8,669	7,888	6,683	5,797	5,119	4,148	3,487
85	71,173	61,215	53,701	47,831	43,117	39,249	33,278	28,884	25,515	22,850	20,689	18,902	17,398	14,513	12,448	10,898	9,691	8,725	7,934	6,716	5,822	5,138	4,161	3,496
90	74,645	63,766	55,655	49,374	44,367	40,282	34,018	29,440	25,948	23,197	20,973	19,138	17,598	14,652	12,550	10,976	9,753	8,775	7,975	6,745	5,844	5,156	4,172	3,504
95	78,052	66,236	57,527	50,842	45,549	41,254	34,709	29,956	26,348	23,516	21,233	19,355	17,781	14,778	12,643	11,047	9,809	8,820	8,012	6,772	5,864	5,171	4,183	3,511
100	81,396	68,628	59,323	52,240	46,668	42,170	35,355	30,436	26,718	23,810	21,473	19,554	17,949	14,894	12,728	11,112	9,860	8,861	8,046	6,796	5,883	5,185	4,192	3,518
105	84,678	70,947	61,048	53,573	47,728	43,034	35,960	30,883	27,063	24,083	21,695	19,738	18,104	15,001	12,805	11,171	9,906	8,899	8,077	6,818	5,899	5,198	4,200	3,524
110	87,900	73,195	62,705	54,845	48,735	43,851	36,529	31,302	27,384	24,337	21,901	19,908	18,247	15,099	12,877	11,225	9,949	8,933	8,106	6,839	5,914	5,210	4,208	3,529
115	91,064	75,376	64,298	56,060	49,693	44,624	37,064	31,694	27,683	24,574	22,092	20,066	18,380	15,189	12,943	11,275	9,988	8,965	8,132	6,857	5,928	5,221	4,215	3,534
120	94,171	77,492	65,832	57,222	50,604	45,358	37,568	32,062	27,964	24,794	22,270	20,213	18,503	15,273	13,004	11,321	10,024	8,994	8,156	6,874	5,941	5,231	4,221	3,538
125	97,223	79,547	67,309	58,335	51,472	46,054	38,045	32,409	28,227	25,001	22,437	20,350	18,618	15,351	13,060	11,364	10,058	9,021	8,178	6,890	5,953	5,240	4,227	3,543
130	100,221	81,543	68,733	59,401	52,300	46,716	38,495	32,735	28,474	25,195	22,593	20,478	18,725	15,424	13,113	11,404	10,089	9,046	8,199	6,905	5,964	5,248	4,233	3,547
135	103,167	83,482	70,105	60,424	53,091	47,346	38,922	33,043	28,707	25,377	22,739	20,598	18,825	15,492	13,162	11,441	10,118	9,069	8,218	6,918	5,974	5,256	4,238	3,550
140	106,062	85,367	71,430	61,405	53,848	47,947	39,327	33,334	28,927	25,548	22,877	20,711	18,920	15,556	13,208	11,476	10,145	9,091	8,236	6,931	5,983	5,263	4,243	3,553
145	108,907	87,201	72,709	62,348	54,571	48,520	39,712	33,610	29,134	25,710	23,006	20,817	19,008	15,616	13,251	11,508	10,171	9,112	8,252	6,943	5,992	5,270	4,247	3,557
150	111,703	88,985	73,945	63,255	55,265	49,067	40,078	33,872	29,331	25,863	23,129	20,917	19,092	15,672	13,292	11,539	10,195	9,131	8,268	6,954	6,000	5,277	4,251	3,559
155	114,453	90,721	75,140	64,127	55,929	49,590	40,426	34,121	29,517	26,008	23,244	21,012	19,170	15,725	13,330	11,568	10,217	9,149	8,283	6,964	6,008	5,283	4,255	3,562
160	117,156	92,411	76,296	64,967	56,567	50,091	40,758	34,357	29,694	26,145	23,354	21,101	19,245	15,775	13,366	11,595	10,238	9,166	8,297	6,974	6,015	5,288	4,259	3,565
165	119,815	94,057	77,415	65,776	57,180	50,571	41,075	34,582	29,862	26,275	23,457	21,186	19,315	15,823	13,400	11,620	10,258	9,182	8,310	6,983	6,022	5,294	4,262	3,567
170	122,430	95,661	78,498	66,557	57,769	51,031	41,378	34,797	30,021	26,399	23,556	21,266	19,382	15,867	13,432	11,644	10,277	9,197	8,322	6,992	6,029	5,299	4,265	3,569
175	125,002	97,224	79,547	67,310	58,335	51,472	41,668	35,001	30,174	26,516	23,650	21,342	19,445	15,910	13,462	11,667	10,295	9,211	8,334	7,000	6,035	5,303	4,268	3,572
180	127,532	98,748	80,565	68,036	58,880	51,896	41,946	35,197	30,319	26,628	23,739	21,415	19,505	15,950	13,491	11,689	10,311	9,224	8,345	7,008	6,041	5,308	4,271	3,574
185	130,022	100,234	81,551	68,739	59,406	52,304	42,211	35,384	30,457	26,735	23,824	21,484	19,563	15,988	13,518	11,709	10,327	9,237	8,355	7,015	6,046	5,312	4,274	3,576
190	132,472	101,684	82,508	69,417	59,912	52,696	42,466	35,563	30,590	26,837	23,905	21,550	19,617	16,025	13,544	11,729	10,343	9,249	8,365	7,022	6,051	5,316	4,277	3,577
195	134,883	103,099	83,437	70,074	60,400	53,073	42,711	35,734	30,717	26,935	23,982	21,613	19,669	16,059	13,569	11,747	10,357	9,261	8,375	7,029	6,056	5,320	4,279	3,579
200	137,257	104,480	84,340	70,709	60,872	53,437	42,946	35,899	30,838	27,028	24,056	21,673	19,719	16,093	13,593	11,765	10,371	9,272	8,384	7,035	6,061	5,323	4,282	3,581

# 1 AWG	1 Run(s)	600 Volts	C-Value = 7,292	Magnetic Duct	Copper

Calculations:

1. $f = \frac{\text{Length} \times 2 \times \text{Isca}}{\text{Runs} \times \text{C-Value} \times E_{L-L}}$
2. $M = \frac{1}{1 + f}$
3. $Isc = Isca \times M$
4. Add Motor Contribution = Motor FLA x 4

* All results are given in symmetrical amperes

Single-Phase L-L Bolted Fault-Current* Table

One-Way Distance in Feet

Available Isc in Thousands (Isca)

	5	10	15	20	25	30	40	50	60	70	80	90	100	125	150	175	200	225	250	300	350	400	500	600
3	2,983	2,967	2,950	2,934	2,918	2,902	2,871	2,841	2,811	2,782	2,753	2,725	2,698	2,631	2,568	2,508	2,451	2,396	2,343	2,245	2,155	2,071	1,923	1,794
5	4,954	4,908	4,864	4,820	4,777	4,735	4,652	4,573	4,496	4,422	4,350	4,281	4,213	4,054	3,906	3,768	3,640	3,521	3,409	3,205	3,024	2,862	2,586	2,358
7	6,910	6,822	6,736	6,652	6,571	6,491	6,337	6,191	6,051	5,917	5,789	5,667	5,549	5,276	5,028	4,803	4,596	4,407	4,233	3,923	3,655	3,422	3,034	2,725
10	9,817	9,640	9,469	9,305	9,146	8,992	8,700	8,426	8,169	7,927	7,699	7,484	7,281	6,817	6,409	6,047	5,724	5,434	5,171	4,716	4,334	4,009	3,487	3,085
15	14,591	14,204	13,837	13,489	13,157	12,842	12,254	11,717	11,226	10,774	10,357	9,972	9,614	8,822	8,150	7,574	7,074	6,635	6,248	5,595	5,066	4,628	3,946	3,439
20	19,280	18,610	17,985	17,400	16,853	16,338	15,399	14,561	13,810	13,133	12,518	11,959	11,448	10,342	9,431	8,668	8,019	7,460	6,974	6,171	5,533	5,015	4,224	3,648
25	23,885	22,865	21,928	21,066	20,268	19,529	18,201	17,043	16,023	15,118	14,310	13,584	12,928	11,535	10,414	9,491	8,718	8,062	7,497	6,576	5,857	5,280	4,410	3,786
30	28,408	26,977	25,683	24,508	23,435	22,452	20,715	19,227	17,939	16,812	15,819	14,936	14,147	12,496	11,190	10,132	9,256	8,520	7,892	6,878	6,095	5,472	4,543	3,884
35	32,853	30,953	29,262	27,745	26,379	25,140	22,982	21,165	19,614	18,275	17,108	16,080	15,169	13,287	11,820	10,645	9,683	8,880	8,200	7,111	6,277	5,619	4,644	3,957
40	37,220	34,800	32,677	30,797	29,122	27,620	25,037	22,896	21,092	19,552	18,221	17,060	16,038	13,949	12,341	11,066	10,030	9,171	8,447	7,296	6,421	5,734	4,722	4,014
45	41,511	38,525	35,939	33,678	31,685	29,915	26,908	24,451	22,405	20,674	19,192	17,908	16,786	14,511	12,779	11,417	10,317	9,410	8,650	7,447	6,538	5,826	4,785	4,059
50	45,730	42,131	39,058	36,403	34,085	32,045	28,620	25,856	23,579	21,670	20,047	18,651	17,436	14,995	13,153	11,714	10,559	9,611	8,820	7,572	6,634	5,903	4,836	4,096
55	49,877	45,627	42,044	38,983	36,337	34,028	30,191	27,131	24,635	22,559	20,806	19,305	18,007	15,415	13,475	11,969	10,766	9,782	8,964	7,678	6,715	5,967	4,879	4,127
60	53,954	49,015	44,904	41,430	38,454	35,878	31,638	28,294	25,590	23,357	21,483	19,887	18,512	15,784	13,756	12,190	10,944	9,930	9,087	7,769	6,784	6,021	4,916	4,153
65	57,963	52,302	47,647	43,754	40,449	37,608	32,975	29,359	26,458	24,078	22,091	20,407	18,962	16,109	14,003	12,384	11,100	10,058	9,194	7,847	6,844	6,068	4,947	4,175
70	61,907	55,491	50,280	45,964	42,330	39,229	34,215	30,338	27,250	24,733	22,641	20,876	19,366	16,400	14,222	12,555	11,237	10,170	9,288	7,915	6,896	6,109	4,974	4,195
75	65,785	58,587	52,809	48,068	44,108	40,751	35,368	31,241	27,976	25,329	23,140	21,299	19,729	16,660	14,417	12,706	11,359	10,269	9,371	7,975	6,941	6,145	4,998	4,211
80	69,601	61,594	55,240	50,074	45,791	42,184	36,442	32,076	28,644	25,875	23,595	21,684	20,059	16,895	14,592	12,842	11,467	10,358	9,445	8,028	6,982	6,176	5,019	4,226
85	73,355	64,516	57,579	51,988	47,387	43,534	37,445	32,851	29,260	26,377	24,012	22,035	20,359	17,107	14,751	12,965	11,565	10,438	9,511	8,076	7,018	6,204	5,037	4,239
90	77,049	67,357	59,830	53,817	48,902	44,809	38,385	33,571	29,831	26,840	24,394	22,357	20,634	17,301	14,894	13,076	11,653	10,509	9,570	8,119	7,050	6,230	5,054	4,251
95	80,685	70,119	61,999	55,565	50,341	46,015	39,266	34,244	30,360	27,268	24,747	22,653	20,886	17,477	15,025	13,176	11,733	10,574	9,624	8,158	7,079	6,252	5,069	4,262
100	84,263	72,805	64,091	57,239	51,711	47,157	40,095	34,872	30,853	27,665	25,074	22,927	21,118	17,640	15,145	13,268	11,806	10,633	9,673	8,193	7,106	6,273	5,082	4,271
105	87,785	75,420	66,108	58,843	53,017	48,240	40,875	35,461	31,313	28,034	25,377	23,180	21,333	17,789	15,255	13,353	11,872	10,688	9,718	8,225	7,130	6,292	5,095	4,280
110	91,253	77,966	68,056	60,381	54,262	49,269	41,611	36,014	31,744	28,379	25,659	23,415	21,532	17,927	15,356	13,430	11,934	10,737	9,759	8,254	7,152	6,309	5,106	4,288
115	94,668	80,445	69,937	61,858	55,452	50,248	42,307	36,534	32,147	28,701	25,922	23,633	21,716	18,055	15,450	13,502	11,990	10,783	9,797	8,281	7,172	6,325	5,116	4,295
120	98,030	82,860	71,756	63,276	56,588	51,180	42,966	37,024	32,526	29,002	26,168	23,838	21,889	18,174	15,537	13,568	12,043	10,825	9,831	8,306	7,191	6,339	5,126	4,302
125	101,342	85,213	73,514	64,639	57,676	52,068	43,590	37,487	32,882	29,285	26,398	24,029	22,050	18,285	15,618	13,630	12,091	10,864	9,864	8,329	7,208	6,353	5,134	4,308
130	104,603	87,508	75,215	65,951	58,718	52,916	44,183	37,924	33,218	29,552	26,614	24,207	22,200	18,388	15,693	13,688	12,136	10,901	9,894	8,351	7,224	6,365	5,143	4,314
135	107,816	89,745	76,862	67,214	59,717	53,725	44,746	38,338	33,536	29,803	26,817	24,376	22,341	18,485	15,764	13,741	12,178	10,935	9,922	8,371	7,239	6,377	5,150	4,319
140	110,982	91,928	78,458	68,431	60,676	54,500	45,282	38,731	33,836	30,039	27,009	24,534	22,474	18,576	15,830	13,791	12,218	10,967	9,948	8,389	7,253	6,388	5,157	4,324
145	114,101	94,057	80,004	69,604	61,597	55,242	45,793	39,104	34,120	30,263	27,190	24,683	22,599	18,661	15,892	13,833	12,255	10,996	9,972	8,407	7,266	6,398	5,164	4,329
150	117,174	96,136	81,503	70,736	62,481	55,952	46,280	39,459	34,390	30,475	27,361	24,824	22,717	18,742	15,950	13,882	12,289	11,024	9,995	8,423	7,278	6,407	5,170	4,333
155	120,203	98,166	82,957	71,828	63,332	56,634	46,745	39,796	34,646	30,676	27,523	24,957	22,829	18,817	16,005	13,924	12,322	11,050	10,017	8,438	7,289	6,416	5,176	4,337
160	123,189	100,148	84,368	72,884	64,151	57,288	47,190	40,118	34,890	30,867	27,676	25,083	22,934	18,889	16,057	13,963	12,353	11,075	10,037	8,453	7,300	6,424	5,181	4,341
165	126,132	102,084	85,738	73,904	64,940	57,916	47,615	40,426	35,122	31,049	27,822	25,203	23,035	18,957	16,106	14,000	12,382	11,098	10,056	8,466	7,310	6,432	5,186	4,345
170	129,033	103,976	87,068	74,890	65,701	58,520	48,023	40,719	35,343	31,222	27,961	25,317	23,130	19,021	16,152	14,035	12,409	11,120	10,074	8,479	7,320	6,439	5,191	4,348
175	131,893	105,825	88,361	75,845	66,435	59,102	48,414	41,000	35,555	31,386	28,093	25,425	23,220	19,082	16,196	14,068	12,435	11,141	10,091	8,491	7,329	6,446	5,195	4,351
180	134,713	107,633	89,618	76,769	67,143	59,661	48,789	41,268	35,756	31,543	28,219	25,528	23,306	19,140	16,238	14,100	12,459	11,161	10,107	8,502	7,337	6,453	5,200	4,354
185	137,494	109,401	90,841	77,664	67,826	60,201	49,149	41,525	35,949	31,694	28,339	25,626	23,388	19,195	16,278	14,130	12,483	11,180	10,123	8,513	7,345	6,459	5,204	4,357
190	140,237	111,131	92,030	78,532	68,487	60,721	49,495	41,772	36,134	31,837	28,453	25,720	23,466	19,248	16,315	14,158	12,505	11,197	10,137	8,524	7,353	6,465	5,208	4,360
195	142,942	112,823	93,187	79,373	69,126	61,222	49,828	42,009	36,311	31,974	28,563	25,810	23,540	19,298	16,351	14,185	12,526	11,214	10,151	8,533	7,360	6,471	5,211	4,362
200	145,611	114,479	94,314	80,189	69,744	61,707	50,148	42,236	36,481	32,106	28,668	25,895	23,611	19,346	16,386	14,211	12,546	11,231	10,165	8,543	7,367	6,476	5,215	4,365

\# 1/0 AWG | 1 Run(s) | 600 Volts | C-Value = 8,924 | Magnetic Duct | Copper

Calculations:

1. $f = \dfrac{\text{Length} \times 2 \times \text{Isca}}{\text{Runs} \times \text{C-Value} \times \text{E L-L}}$

2. $M = \dfrac{1}{1 + f}$

3. $\text{Isc} = \text{Isca} \times M$

4. Add Motor Contribution = Motor FLA x 4

* All results are given in symmetrical amperes

Single-Phase L-L Bolted Fault-Current* Table

One-Way Distance in Feet

Available Isc in Thousands (Isca)	5	10	15	20	25	30	40	50	60	70	80	90	100	125	150	175	200	225	250	300	350	400	500	600
3	2,992	2,983	2,975	2,967	2,959	2,950	2,934	2,918	2,902	2,887	2,871	2,856	2,841	2,804	2,767	2,732	2,698	2,664	2,631	2,568	2,508	2,451	2,343	2,245
5	4,977	4,954	4,931	4,908	4,886	4,864	4,820	4,777	4,735	4,693	4,652	4,612	4,573	4,477	4,386	4,298	4,213	4,132	4,054	3,906	3,768	3,640	3,409	3,205
7	6,955	6,910	6,865	6,822	6,778	6,736	6,652	6,571	6,491	6,413	6,337	6,263	6,191	6,017	5,852	5,697	5,549	5,409	5,276	5,028	4,803	4,596	4,233	3,923
10	9,907	9,817	9,727	9,640	9,554	9,469	9,305	9,146	8,992	8,844	8,700	8,561	8,426	8,107	7,812	7,537	7,281	7,041	6,817	6,409	6,047	5,724	5,171	4,716
15	14,793	14,591	14,395	14,204	14,018	13,837	13,489	13,157	12,842	12,541	12,254	11,980	11,717	11,110	10,562	10,065	9,614	9,201	8,822	8,150	7,574	7,074	6,248	5,595
20	19,633	19,280	18,939	18,610	18,292	17,985	17,400	16,853	16,338	15,855	15,399	14,968	14,561	13,634	12,818	12,094	11,448	10,867	10,342	9,431	8,668	8,019	6,974	6,171
25	24,430	23,885	23,364	22,865	22,387	21,928	21,066	20,268	19,529	18,842	18,201	17,603	17,043	15,786	14,703	13,758	12,928	12,192	11,535	10,414	9,491	8,718	7,497	6,576
30	29,182	28,408	27,674	26,977	26,314	25,683	24,508	23,435	22,452	21,549	20,715	19,943	19,227	17,643	16,301	15,148	14,147	13,271	12,496	11,190	10,132	9,256	7,892	6,878
35	33,892	32,853	31,875	30,953	30,084	29,262	27,745	26,379	25,140	24,013	22,982	22,036	21,165	19,262	17,672	16,325	15,169	14,166	13,287	11,820	10,645	9,683	8,200	7,111
40	38,560	37,220	35,969	34,800	33,705	32,677	30,797	29,122	27,620	26,265	25,037	23,919	22,896	20,685	18,863	17,336	16,038	14,921	13,949	12,341	11,066	10,030	8,447	7,296
45	43,185	41,511	39,962	38,525	37,187	35,939	33,678	31,685	29,915	28,332	26,908	25,621	24,451	21,945	19,906	18,213	16,786	15,566	14,511	12,779	11,417	10,317	8,650	7,447
50	47,770	45,730	43,857	42,131	40,537	39,058	36,403	34,085	32,045	30,236	28,620	27,168	25,856	23,071	20,827	18,981	17,436	16,123	14,995	13,153	11,714	10,559	8,820	7,572
55	52,313	49,877	47,657	45,627	43,762	42,044	38,983	36,337	34,028	31,995	30,191	28,579	27,131	24,081	21,647	19,660	18,007	16,610	15,415	13,475	11,969	10,766	8,964	7,678
60	56,817	53,954	51,366	49,015	46,870	44,904	41,430	38,454	35,878	33,625	31,638	29,873	28,294	24,993	22,381	20,263	18,512	17,039	15,784	13,756	12,190	10,944	9,087	7,769
65	61,280	57,963	54,987	52,302	49,866	47,647	43,754	40,449	37,608	35,140	32,975	31,062	29,359	25,820	23,042	20,804	18,962	17,420	16,109	14,003	12,384	11,100	9,194	7,847
70	65,705	61,907	58,524	55,491	52,757	50,280	45,964	42,330	39,229	36,551	34,215	32,160	30,338	26,574	23,641	21,291	19,366	17,760	16,400	14,222	12,555	11,237	9,288	7,915
75	70,091	65,785	61,978	58,587	55,548	52,809	48,068	44,108	40,751	37,869	35,368	33,176	31,241	27,264	24,185	21,731	19,729	18,065	16,660	14,417	12,706	11,359	9,371	7,975
80	74,439	69,601	65,353	61,594	58,244	55,240	50,074	45,791	42,184	39,103	36,442	34,120	32,076	27,898	24,683	22,132	23,059	18,341	16,895	14,592	12,842	11,467	9,445	8,028
85	78,749	73,355	68,652	64,516	60,850	57,579	51,988	47,387	43,534	40,261	37,445	34,998	32,851	28,482	25,139	22,498	23,359	18,592	17,107	14,751	12,965	11,565	9,511	8,076
90	83,023	77,049	71,878	67,357	63,371	59,830	53,817	48,902	44,809	41,349	38,385	35,817	33,571	29,022	25,559	22,834	20,634	18,821	17,301	14,894	13,076	11,653	9,570	8,119
95	87,259	80,685	75,031	70,119	65,810	61,999	55,565	50,341	46,015	42,373	39,266	36,583	34,244	29,523	25,947	23,143	20,886	19,030	17,477	15,025	13,176	11,733	9,624	8,158
100	91,459	84,263	78,116	72,805	68,171	64,091	57,239	51,711	47,157	43,340	40,095	37,301	34,872	29,989	26,306	23,428	21,118	19,223	17,640	15,145	13,268	11,806	9,673	8,193
105	95,624	87,785	81,134	75,420	70,458	66,108	58,843	53,017	48,240	44,253	40,875	37,976	35,461	30,424	26,640	23,693	21,333	19,400	17,789	15,255	13,353	11,872	9,718	8,225
110	99,753	91,253	84,088	77,966	72,675	68,056	60,381	54,262	49,269	45,118	41,611	38,611	36,014	30,830	26,950	23,938	21,532	19,565	17,927	15,356	13,430	11,934	9,759	8,254
115	103,848	94,668	86,979	80,445	74,824	69,937	61,858	55,452	50,248	45,937	42,307	39,209	36,534	31,210	27,241	24,167	21,716	19,717	18,055	15,450	13,502	11,990	9,797	8,281
120	107,908	98,030	89,809	82,860	76,909	71,756	63,276	56,588	51,180	46,714	42,966	39,774	37,024	31,567	27,512	24,380	21,889	19,859	18,174	15,537	13,568	12,043	9,831	8,306
125	111,934	101,342	92,580	85,213	78,932	73,514	64,639	57,676	52,068	47,453	43,590	40,309	37,487	31,903	27,767	24,580	22,050	19,991	18,285	15,618	13,630	12,091	9,864	8,329
130	115,927	104,603	95,295	87,508	80,897	75,215	65,951	58,718	52,916	48,156	44,183	40,815	37,924	32,219	28,006	24,767	22,200	20,115	18,388	15,693	13,688	12,136	9,894	8,351
135	119,887	107,816	97,954	89,745	82,806	76,862	67,214	59,717	53,725	48,826	44,746	41,295	38,338	32,517	28,231	24,943	22,341	20,231	18,485	15,764	13,741	12,178	9,922	8,371
140	123,813	110,982	100,560	91,928	84,660	78,458	68,431	60,676	54,500	49,465	45,282	41,751	38,731	32,800	28,444	25,109	22,474	20,340	18,576	15,830	13,791	12,218	9,948	8,389
145	127,708	114,101	103,114	94,057	86,463	80,004	69,604	61,597	55,242	50,075	45,793	42,185	39,104	33,067	28,644	25,265	22,599	20,442	18,661	15,892	13,838	12,255	9,972	8,407
150	131,571	117,174	105,618	96,136	88,217	81,503	70,736	62,481	55,952	50,658	46,280	42,598	39,459	33,320	28,834	25,413	22,717	20,539	18,742	15,950	13,882	12,289	9,995	8,423
155	135,402	120,203	108,073	98,166	89,923	82,957	71,828	63,332	56,634	51,216	46,745	42,992	39,796	33,561	29,014	25,553	22,829	20,630	18,817	16,005	13,924	12,322	10,017	8,438
160	139,202	123,189	110,480	100,148	91,583	84,368	72,884	64,151	57,288	51,751	47,190	43,368	40,118	33,789	29,185	25,685	22,934	20,716	18,889	16,057	13,963	12,353	10,037	8,453
165	142,971	126,132	112,841	102,084	93,200	85,738	73,904	64,940	57,916	52,263	47,615	43,727	40,426	34,007	29,347	25,810	23,035	20,798	18,957	16,106	14,000	12,382	10,056	8,466
170	146,710	129,033	115,157	103,976	94,774	87,068	74,890	65,701	58,520	52,755	48,023	44,070	40,719	34,214	29,501	25,930	23,130	20,875	19,021	16,152	14,035	12,409	10,074	8,479
175	150,419	131,893	117,430	105,825	96,308	88,361	75,845	66,435	59,102	53,226	48,414	44,399	41,000	34,412	29,648	26,043	23,220	20,949	19,082	16,196	14,068	12,435	10,091	8,491
180	154,098	134,713	119,660	107,633	97,803	89,618	76,769	67,143	59,661	53,680	48,789	44,714	41,268	34,601	29,789	26,151	23,306	21,019	19,140	16,238	14,100	12,459	10,107	8,502
185	157,748	137,494	121,850	109,401	99,261	90,841	77,664	67,826	60,201	54,116	49,149	45,017	41,525	34,782	29,922	26,254	23,388	21,085	19,195	16,278	14,130	12,483	10,123	8,513
190	161,369	140,237	123,999	111,131	100,682	92,030	78,532	68,487	60,721	54,536	49,495	45,307	41,772	34,955	30,050	26,353	23,466	21,149	19,248	16,315	14,158	12,505	10,137	8,524
195	164,962	142,942	126,109	112,823	102,069	93,187	79,373	69,126	61,222	54,940	49,828	45,585	42,009	35,120	30,173	26,447	23,540	21,209	19,298	16,351	14,185	12,526	10,151	8,533
200	168,526	145,611	128,182	114,479	103,423	94,314	80,189	69,744	61,707	55,330	50,148	45,853	42,236	35,279	30,290	26,537	23,611	21,267	19,346	16,386	14,211	12,546	10,165	8,543

1/0 AWG | 2 Run(s) | 600 Volts | C-Value = 8,924 | Magnetic Duct | Copper

Calculations:

1. $f = \dfrac{\text{Length} \times 2 \times \text{Isca}}{\text{Runs} \times \text{C-Value} \times \text{E L-L}}$

2. $M = \dfrac{1}{1 + f}$

3. $\text{Isc} = \text{Isca} \times M$

4. Add Motor Contribution = Motor FLA x 4

* All results are given in symmetrical amperes

Single-Phase L-L Bolted Fault-Current* Table

One-Way Distance in Feet

Available Isc in Thousands (Isca)	5	10	15	20	25	30	40	50	60	70	80	90	100	125	150	175	200	225	250	300	350	400	500	600
3	2,986	2,972	2,959	2,945	2,932	2,919	2,892	2,867	2,841	2,817	2,792	2,768	2,745	2,688	2,633	2,580	2,530	2,481	2,434	2,346	2,263	2,187	2,048	1,926
5	4,962	4,924	4,886	4,850	4,814	4,778	4,708	4,640	4,575	4,511	4,449	4,388	4,329	4,189	4,057	3,933	3,817	3,707	3,604	3,413	3,242	3,087	2,817	2,591
7	6,925	6,851	6,779	6,709	6,640	6,572	6,441	6,315	6,194	6,077	5,965	5,856	5,752	5,507	5,281	5,074	4,882	4,704	4,538	4,240	3,979	3,748	3,358	3,041
10	9,847	9,699	9,556	9,416	9,281	9,149	8,897	8,658	8,432	8,217	8,013	7,819	7,634	7,208	6,826	6,483	6,173	5,892	5,634	5,182	4,797	4,465	3,922	3,497
15	14,659	14,334	14,022	13,724	13,438	13,164	12,648	12,171	11,728	11,317	10,934	10,575	10,240	9,487	8,837	8,271	7,773	7,331	6,937	6,264	5,710	5,245	4,512	3,958
20	19,399	18,833	18,299	17,794	17,317	16,864	16,026	15,268	14,578	13,948	13,370	12,838	12,347	11,269	10,364	9,593	8,930	8,352	7,844	6,994	6,310	5,748	4,879	4,238
25	24,068	23,202	22,397	21,646	20,943	20,285	19,085	18,019	17,066	16,209	15,433	14,729	14,086	12,700	11,562	10,611	9,805	9,113	8,512	7,520	6,735	6,099	5,129	4,426
30	28,667	27,448	26,328	25,296	24,342	23,457	21,867	20,479	19,257	18,172	17,203	16,333	15,546	13,874	12,528	11,419	10,491	9,702	9,024	7,917	7,052	6,357	5,311	4,560
35	33,199	31,575	30,102	28,760	27,533	26,407	24,409	22,692	21,201	19,894	18,738	17,710	16,788	14,856	13,322	12,076	11,043	10,172	9,429	8,227	7,297	6,555	5,448	4,661
40	37,665	35,588	33,728	32,053	30,536	29,156	26,740	24,693	22,938	21,415	20,082	18,906	17,859	15,688	13,988	12,620	11,496	10,556	9,758	8,476	7,492	6,713	5,557	4,740
45	42,066	39,492	37,215	35,185	33,366	31,726	28,885	26,512	24,499	22,770	21,269	19,954	18,791	16,403	14,553	13,079	11,875	10,875	10,030	8,680	7,651	6,840	5,644	4,803
50	46,404	43,291	40,570	38,170	36,038	34,132	30,867	28,172	25,909	23,983	22,324	20,879	19,610	17,024	15,040	13,470	12,197	11,144	10,258	8,851	7,784	6,946	5,715	4,855
55	50,680	46,990	43,800	41,016	38,565	36,390	32,702	29,693	27,190	25,077	23,269	21,703	20,335	17,567	15,463	13,808	12,474	11,374	10,453	8,996	7,895	7,035	5,775	4,899
60	54,896	50,592	46,914	43,734	40,958	38,514	34,407	31,091	28,359	26,067	24,119	22,441	20,982	18,048	15,834	14,103	12,714	11,574	10,621	9,120	7,991	7,110	5,826	4,935
65	59,052	54,101	49,916	46,332	43,228	40,514	35,995	32,382	29,429	26,969	24,888	23,106	21,562	18,475	16,162	14,363	12,925	11,748	10,768	9,228	8,074	7,176	5,870	4,967
70	63,150	57,521	52,813	48,818	45,384	42,402	37,477	33,577	30,412	27,792	25,588	23,708	22,085	18,858	16,454	14,593	13,111	11,902	10,897	9,323	8,146	7,233	5,908	4,994
75	67,191	60,854	55,610	51,198	47,435	44,187	38,864	34,686	31,319	28,548	26,227	24,256	22,560	19,203	16,716	14,799	13,277	12,038	11,011	9,406	8,210	7,283	5,942	5,018
80	71,176	64,105	58,312	53,480	49,387	45,876	40,165	35,719	32,158	29,244	26,813	24,756	22,992	19,515	16,952	14,984	13,425	12,160	11,113	9,480	8,266	7,327	5,971	5,039
85	75,107	67,276	60,925	55,669	51,248	47,477	41,387	36,682	32,937	29,886	27,353	25,215	23,387	19,799	17,166	15,151	13,559	12,270	11,205	9,547	8,317	7,367	5,998	5,058
90	78,984	70,371	63,451	57,771	53,024	48,998	42,538	37,583	33,662	30,482	27,851	25,638	23,750	20,059	17,361	15,302	13,680	12,369	11,287	9,607	8,362	7,403	6,021	5,074
95	82,809	73,391	65,896	59,791	54,721	50,443	43,623	38,428	34,338	31,035	28,312	26,028	24,085	20,297	17,539	15,441	13,791	12,459	11,362	9,661	8,403	7,435	6,043	5,089
100	86,583	76,340	68,264	61,733	56,343	51,819	44,648	39,221	34,970	31,550	28,740	26,389	24,394	20,516	17,702	15,567	13,891	12,542	11,431	9,711	8,440	7,464	6,062	5,103
105	90,306	79,220	70,558	63,603	57,897	53,130	45,618	39,967	35,562	32,032	29,139	26,725	24,681	20,719	17,853	15,683	13,984	12,617	11,493	9,756	8,475	7,491	6,079	5,116
110	93,980	82,033	72,781	65,404	59,385	54,381	46,537	40,671	36,118	32,482	29,511	27,038	24,947	20,906	17,992	15,790	14,069	12,686	11,551	9,797	8,506	7,515	6,095	5,127
115	97,606	84,782	74,936	67,140	60,813	55,575	47,409	41,335	36,641	32,904	29,859	27,330	25,196	21,080	18,121	15,890	14,148	12,750	11,604	9,835	8,534	7,538	6,110	5,137
120	101,184	87,469	77,028	68,814	62,183	56,717	48,238	41,964	37,134	33,302	30,186	27,603	25,428	21,243	18,240	15,982	14,221	12,809	11,653	9,870	8,561	7,558	6,124	5,147
125	104,716	90,095	79,058	70,429	63,499	57,810	49,026	42,559	37,600	33,675	30,493	27,860	25,645	21,394	18,352	16,067	14,288	12,864	11,698	9,903	8,585	7,577	6,136	5,156
130	108,202	92,664	81,029	71,989	64,764	58,857	49,777	43,124	38,040	34,028	30,782	28,101	25,849	21,536	18,456	16,147	14,352	12,915	11,740	9,933	8,608	7,595	6,148	5,164
135	111,644	95,177	82,943	73,497	65,982	59,861	50,493	43,660	38,456	34,361	31,054	28,327	26,041	21,669	18,554	16,222	14,410	12,963	11,780	9,961	8,629	7,611	6,159	5,172
140	115,041	97,635	84,804	74,954	67,154	60,824	51,177	44,171	38,852	34,676	31,311	28,541	26,222	21,794	18,645	16,292	14,466	13,008	11,817	9,988	8,649	7,627	6,169	5,179
145	118,396	100,041	86,613	76,364	68,283	61,749	51,830	44,656	39,227	34,975	31,554	28,743	26,392	21,911	18,731	16,357	14,517	13,049	11,851	10,012	8,668	7,641	6,178	5,185
150	121,709	102,396	88,373	77,728	69,372	62,638	52,455	45,120	39,584	35,258	31,785	28,935	26,553	22,022	18,812	16,419	14,566	13,089	11,884	10,035	8,685	7,655	6,187	5,191
155	124,980	104,702	90,085	79,050	70,423	63,494	53,053	45,562	39,924	35,528	32,004	29,116	26,706	22,127	18,889	16,477	14,612	13,126	11,914	10,057	8,701	7,667	6,195	5,197
160	128,211	106,959	91,752	80,330	71,437	64,317	53,627	45,984	40,248	35,784	32,212	29,288	26,850	22,226	18,961	16,532	14,655	13,160	11,943	10,078	8,716	7,679	6,203	5,203
165	131,401	109,171	93,374	81,571	72,417	65,110	54,177	46,388	40,557	36,028	32,409	29,451	26,988	22,320	19,029	16,584	14,696	13,193	11,970	10,097	8,731	7,690	6,210	5,208
170	134,553	111,338	94,955	82,774	73,364	65,875	54,706	46,775	40,852	36,261	32,598	29,607	27,118	22,409	19,094	16,633	14,734	13,224	11,995	10,115	8,744	7,701	6,217	5,213
175	137,666	113,461	96,494	83,942	74,280	66,612	55,213	47,145	41,135	36,484	32,777	29,755	27,242	22,494	19,156	16,680	14,771	13,254	12,020	10,132	8,757	7,711	6,224	5,217
180	140,742	115,542	97,995	85,076	75,166	67,324	55,701	47,501	41,405	36,696	32,949	29,896	27,361	22,575	19,214	16,724	14,806	13,282	12,043	10,149	8,769	7,720	6,230	5,222
185	143,780	117,582	99,459	86,177	76,024	68,011	56,171	47,842	41,664	36,899	33,113	30,031	27,473	22,652	19,270	16,766	14,839	13,308	12,064	10,164	8,781	7,729	6,235	5,226
190	146,782	119,582	100,886	87,246	76,855	68,676	56,624	48,170	41,913	37,094	33,269	30,159	27,581	22,725	19,322	16,806	14,870	13,334	12,085	10,179	8,792	7,738	6,241	5,229
195	149,748	121,543	102,279	88,286	77,661	69,318	57,060	48,485	42,151	37,281	33,419	30,283	27,684	22,795	19,373	16,845	14,900	13,358	12,105	10,193	8,802	7,746	6,246	5,233
200	152,680	123,467	103,638	89,296	78,442	69,940	57,480	48,788	42,380	37,460	33,563	30,401	27,783	22,861	19,421	16,881	14,928	13,381	12,124	10,206	8,812	7,754	6,251	5,237

2/0 AWG | 1 Run(s) | 600 Volts | C-Value = 10,755 | Magnetic Duct | Copper

Calculations:

1. $f = \dfrac{\text{Length} \times 2 \times \text{Isca}}{\text{Runs} \times \text{C-Value} \times E_{L-L}}$
2. $M = \dfrac{1}{1 + f}$
3. $Isc = Isca \times M$
4. Add Motor Contribution = Motor FLA x 4

* All results are given in symmetrical amperes

Single-Phase L-L Bolted Fault-Current* Table

One-Way Distance in Feet

Available Isc in Thousands (Isca)	5	10	15	20	25	30	40	50	60	70	80	90	100	125	150	175	200	225	250	300	350	400	500	600
3	2,993	2,986	2,979	2,972	2,966	2,959	2,945	2,932	2,919	2,905	2,892	2,880	2,867	2,835	2,804	2,774	2,745	2,716	2,688	2,633	2,580	2,530	2,434	2,346
5	4,981	4,962	4,943	4,924	4,905	4,886	4,850	4,814	4,778	4,743	4,708	4,674	4,640	4,558	4,479	4,403	4,329	4,258	4,189	4,057	3,933	3,817	3,604	3,413
7	6,962	6,925	6,888	6,851	6,815	6,779	6,709	6,640	6,572	6,506	6,441	6,377	6,315	6,164	6,020	5,883	5,752	5,627	5,507	5,281	5,074	4,882	4,538	4,240
10	9,923	9,847	9,773	9,699	9,627	9,556	9,416	9,281	9,149	9,021	8,897	8,776	8,658	8,377	8,114	7,867	7,634	7,415	7,208	6,826	6,483	6,173	5,634	5,182
15	14,828	14,659	14,495	14,334	14,176	14,022	13,724	13,438	13,164	12,901	12,648	12,405	12,171	11,623	11,122	10,663	10,240	9,849	9,487	8,837	8,271	7,773	6,937	6,264
20	19,695	19,399	19,112	18,833	18,562	18,299	17,794	17,317	16,864	16,434	16,026	15,638	15,268	14,415	13,653	12,967	12,347	11,783	11,269	10,364	9,593	8,930	7,844	6,994
25	24,525	24,068	23,627	23,202	22,792	22,397	21,646	20,943	20,285	19,667	19,085	18,537	18,019	16,843	15,812	14,899	14,086	13,357	12,700	11,562	10,611	9,805	8,512	7,520
30	29,318	28,667	28,044	27,448	26,876	26,328	25,296	24,342	23,457	22,634	21,867	21,150	20,479	18,974	17,675	16,542	15,546	14,663	13,874	12,528	11,419	10,491	9,024	7,917
35	34,076	33,199	32,367	31,575	30,821	30,102	28,760	27,533	26,407	25,368	24,409	23,519	22,692	20,858	19,299	17,956	16,788	15,763	14,856	13,322	12,076	11,043	9,429	8,227
40	38,798	37,665	36,597	35,588	34,633	33,728	32,053	30,536	29,156	27,896	26,740	25,676	24,693	22,537	20,728	19,187	17,859	16,704	15,688	13,988	12,620	11,496	9,758	8,476
45	43,484	42,066	40,739	39,492	38,319	37,215	35,185	33,366	31,726	30,239	28,885	27,648	26,512	24,042	21,994	20,267	18,791	17,516	16,403	14,553	13,079	11,875	10,030	8,680
50	48,135	46,404	44,794	43,291	41,886	40,570	38,170	36,038	34,132	32,417	30,867	29,458	28,172	25,400	23,124	21,223	19,610	18,226	17,024	15,040	13,470	12,197	10,258	8,851
55	52,752	50,680	48,765	46,990	45,339	43,800	41,016	38,565	36,390	34,448	32,702	31,125	29,693	26,629	24,139	22,075	20,335	18,850	17,567	15,463	13,808	12,474	10,453	8,996
60	57,335	54,896	52,656	50,592	48,684	46,914	43,734	40,958	38,514	36,345	34,407	32,665	31,091	27,749	25,055	22,838	20,982	19,405	18,048	15,834	14,103	12,714	10,621	9,120
65	61,883	59,052	56,468	54,101	51,924	49,916	46,332	43,228	40,514	38,121	35,995	34,093	32,382	28,772	25,887	23,527	21,562	19,900	18,475	16,162	14,363	12,925	10,768	9,228
70	66,399	63,150	60,204	57,521	55,066	52,813	48,818	45,384	42,402	39,788	37,477	35,420	33,577	29,712	26,645	24,152	22,085	20,345	18,858	16,454	14,593	13,111	10,897	9,323
75	70,881	67,191	63,866	60,854	58,114	55,610	51,198	47,435	44,187	41,355	38,864	36,656	34,686	30,577	27,339	24,720	22,560	20,747	19,203	16,716	14,799	13,277	11,011	9,406
80	75,331	71,176	67,456	64,105	61,072	58,312	53,480	49,387	45,876	42,831	40,165	37,811	35,719	31,377	27,976	25,240	22,992	21,112	19,515	16,952	14,984	13,425	11,113	9,480
85	79,748	75,107	70,976	67,276	63,943	60,925	55,669	51,248	47,477	44,224	41,387	38,893	36,682	32,118	28,564	25,718	23,387	21,444	19,799	17,166	15,151	13,559	11,205	9,547
90	84,133	78,984	74,429	70,371	66,732	63,451	57,771	53,024	48,998	45,540	42,538	39,907	37,583	32,806	29,107	26,157	23,750	21,749	20,059	17,361	15,302	13,680	11,287	9,607
95	88,487	82,809	77,816	73,391	69,442	65,896	59,791	54,721	50,443	46,786	43,623	40,861	38,428	33,448	29,611	26,564	24,085	22,029	20,297	17,539	15,441	13,791	11,362	9,661
100	92,809	86,583	81,139	76,340	72,076	68,264	61,733	56,343	51,819	47,967	44,648	41,759	39,221	34,047	30,080	26,940	24,394	22,288	20,516	17,702	15,567	13,891	11,431	9,711
105	97,100	90,306	84,400	79,220	74,638	70,558	63,603	57,897	53,130	49,088	45,618	42,606	39,967	34,608	30,517	27,290	24,681	22,527	20,719	17,853	15,683	13,984	11,493	9,756
110	101,361	93,980	87,601	82,033	77,130	72,781	65,404	59,385	54,381	50,154	46,537	43,407	40,671	35,135	30,925	27,617	24,947	22,749	20,906	17,992	15,790	14,069	11,551	9,797
115	105,591	97,606	90,743	84,782	79,556	74,936	67,140	60,813	55,575	51,168	47,409	44,164	41,335	35,630	31,308	27,921	25,196	22,955	21,080	18,121	15,890	14,148	11,604	9,835
120	109,792	101,184	93,828	87,469	81,917	77,028	68,814	62,183	56,717	52,135	48,238	44,883	41,964	36,096	31,667	28,207	25,428	23,148	21,243	18,240	15,982	14,221	11,653	9,870
125	113,962	104,716	96,857	90,095	84,216	79,058	70,429	63,499	57,810	53,057	49,026	45,564	42,559	36,535	32,005	28,474	25,645	23,328	21,394	18,352	16,067	14,288	11,698	9,903
130	118,104	108,202	99,832	92,664	86,457	81,029	71,989	64,764	58,857	53,938	49,777	46,212	43,124	36,951	32,323	28,726	25,849	23,496	21,536	18,456	16,147	14,352	11,740	9,933
135	122,216	111,644	102,755	95,177	88,640	82,943	73,497	65,982	59,861	54,779	50,493	46,829	43,660	37,344	32,624	28,963	26,041	23,655	21,669	18,554	16,222	14,410	11,780	9,961
140	126,299	115,041	105,626	97,635	90,769	84,804	74,954	67,154	60,824	55,585	51,177	47,416	44,171	37,716	32,908	29,187	26,222	23,804	21,794	18,645	16,292	14,466	11,817	9,988
145	130,355	118,396	108,447	100,041	92,844	86,613	76,364	68,283	61,749	56,356	51,830	47,976	44,656	38,070	33,177	29,398	26,392	23,944	21,911	18,731	16,357	14,517	11,851	10,012
150	134,382	121,709	111,220	102,396	94,869	88,373	77,728	69,372	62,638	57,096	52,455	48,512	45,120	38,406	33,432	29,598	26,553	24,077	22,022	18,812	16,419	14,566	11,884	10,035
155	138,381	124,980	113,946	104,702	96,845	90,085	79,050	70,423	63,494	57,806	53,053	49,023	45,562	38,726	33,674	29,788	26,706	24,202	22,127	18,889	16,477	14,612	11,914	10,057
160	142,352	128,211	116,625	106,959	98,774	91,752	80,330	71,437	64,317	58,488	53,627	49,512	45,984	39,031	33,904	29,968	26,850	24,321	22,226	18,961	16,532	14,655	11,943	10,078
165	146,296	131,401	119,259	109,171	100,657	93,374	81,571	72,417	65,110	59,143	54,177	49,981	46,388	39,321	34,123	30,139	26,988	24,433	22,320	19,029	16,584	14,696	11,970	10,097
170	150,214	134,553	121,849	111,338	102,496	94,955	82,774	73,364	65,875	59,773	54,706	50,430	46,775	39,599	34,332	30,302	27,118	24,540	22,409	19,094	16,633	14,734	11,995	10,115
175	154,104	137,666	124,397	113,461	104,292	96,494	83,942	74,280	66,612	60,379	55,213	50,861	47,145	39,864	34,531	30,457	27,242	24,642	22,494	19,156	16,680	14,771	12,020	10,132
180	157,968	140,742	126,903	115,542	106,048	97,995	85,076	75,166	67,324	60,964	55,701	51,275	47,501	40,118	34,722	30,605	27,361	24,738	22,575	19,214	16,724	14,806	12,043	10,149
185	161,806	143,780	129,368	117,582	107,764	99,459	86,177	76,024	68,011	61,527	56,171	51,673	47,842	40,361	34,904	30,746	27,473	24,831	22,652	19,270	16,766	14,839	12,064	10,164
190	165,618	146,782	131,793	119,582	109,441	100,886	87,246	76,855	68,676	62,070	56,624	52,056	48,170	40,594	35,078	30,881	27,581	24,919	22,725	19,322	16,806	14,870	12,085	10,179
195	169,404	149,748	134,179	121,543	111,082	102,279	88,286	77,661	69,318	62,594	57,060	52,424	48,485	40,818	35,245	31,010	27,684	25,003	22,795	19,373	16,845	14,900	12,105	10,193
200	173,165	152,680	136,528	123,467	112,687	103,638	89,296	78,442	69,940	63,101	57,480	52,779	48,788	41,033	35,404	31,134	27,783	25,083	22,861	19,421	16,881	14,928	12,124	10,206

2/0 AWG | 2 Run(s) | 600 Volts | C-Value = 10,755 | Magnetic Duct | Copper

Calculations:

1. $f = \dfrac{\text{Length} \times 2 \times \text{Isca}}{\text{Runs} \times \text{C-Value} \times E_{L\text{-}L}}$
2. $M = \dfrac{1}{1 + f}$
3. $\text{Isc} = \text{Isca} \times M$
4. Add Motor Contribution = Motor FLA x 4

* All results are given in symmetrical amperes

Single-Phase L-L Bolted Fault-Current* Table

One-Way Distance in Feet

Available Isc in Thousands (Isca)

Isca	5	10	15	20	25	30	40	50	60	70	80	90	100	125	150	175	200	225	250	300	350	400	500	600
3	2,988	2,977	2,965	2,954	2,943	2,932	2,[illegible]09	2,888	2,866	2,845	2,824	2,804	2,783	2,734	2,586	2,640	2,596	2,553	2,511	2,4[illegible]2	2,358	2,288	2,159	2,045
5	4,968	4,936	4,905	4,874	4,843	4,813	4,753	4,695	4,639	4,584	4,530	4,477	4,426	4,302	4,185	4,075	3,970	3,870	3,775	3,5[illegible]9	3,438	3,291	3,032	2,811
7	6,937	6,875	6,814	6,755	6,696	6,638	6,[illegible]26	6,417	6,312	6,210	6,112	6,016	5,924	5,704	5,501	5,311	5,134	4,969	4,814	4,5[illegible]1	4,279	4,054	3,668	3,349
10	9,872	9,747	9,625	9,507	9,391	9,278	9,[illegible]59	8,851	8,653	8,463	8,281	8,106	7,939	7,550	7,198	6,877	6,583	6,313	6,065	5,6[illegible]2	5,240	4,906	4,352	3,910
15	14,714	14,438	14,172	13,916	13,670	13,431	12,979	12,556	12,160	11,788	11,438	11,108	10,797	10,090	9,470	8,922	8,433	7,996	7,602	6,9[illegible]9	6,349	5,866	5,091	4,497
20	19,494	19,013	18,555	18,119	17,703	17,305	16,561	15,879	15,250	14,670	14,132	13,632	13,166	12,130	11,245	10,480	9,813	9,225	8,704	7,8[illegible]1	7,100	6,501	5,563	4,861
25	24,214	23,477	22,783	22,128	21,511	20,926	19,848	18,876	17,994	17,192	16,457	15,783	15,162	13,804	12,669	11,707	10,880	10,163	9,534	8,4[illegible]4	7,643	6,953	5,890	5,109
30	28,876	27,833	26,863	25,958	25,112	24,319	22,875	21,593	20,447	19,417	18,485	17,639	16,867	15,203	13,838	12,698	11,731	10,901	10,181	8,9[illegible]3	8,053	7,291	6,131	5,289
35	33,479	32,085	30,803	29,619	28,522	27,504	25,672	24,068	22,653	21,395	20,270	19,257	18,340	16,390	14,814	13,515	12,425	11,498	10,700	9,3[illegible]5	8,374	7,553	6,315	5,426
40	38,026	36,238	34,610	33,123	31,758	30,501	28,263	26,332	24,647	23,165	21,851	20,679	19,625	17,409	15,642	14,200	13,002	11,991	11,125	9,7[illegible]2	8,633	7,763	6,461	5,533
45	42,517	40,294	38,292	36,479	34,830	33,324	30,671	28,410	26,459	24,758	23,263	21,939	20,757	18,293	16,352	14,784	13,490	12,404	11,480	9,991	8,845	7,934	6,579	5,620
50	46,953	44,257	41,853	39,697	37,752	35,989	32,914	30,324	28,111	26,200	24,532	23,063	21,761	19,068	16,969	15,286	13,906	12,755	11,780	10,[illegible]18	9,022	8,076	6,677	5,691
55	51,336	48,130	45,300	42,785	40,534	38,509	35,010	32,093	29,626	27,510	25,677	24,073	22,657	19,753	17,509	15,723	14,267	13,058	12,038	10,[illegible]12	9,172	8,197	6,759	5,750
60	55,666	51,915	48,639	45,751	43,187	40,895	36,971	33,734	31,018	28,707	26,716	24,984	23,463	20,363	17,986	16,106	14,582	13,322	12,262	10,[illegible]79	9,302	8,300	6,829	5,801
65	59,944	55,617	51,873	48,601	45,718	43,157	38,810	35,259	32,303	29,804	27,664	25,811	24,190	20,908	18,411	16,446	14,860	13,553	12,458	10,[illegible]24	9,414	8,389	6,889	5,844
70	64,171	59,238	55,009	51,344	48,136	45,306	40,539	36,680	33,491	30,813	28,531	26,564	24,851	21,400	18,791	16,749	15,107	13,758	12,631	10,[illegible]52	9,512	8,467	6,942	5,882
75	68,348	62,779	58,050	53,983	50,449	47,349	42,167	38,008	34,595	31,745	29,328	27,254	25,453	21,845	19,133	17,020	15,327	13,941	12,785	10,[illegible]65	9,599	8,536	6,988	5,915
80	72,476	66,245	61,001	56,526	52,663	49,294	43,703	39,251	35,622	32,607	30,063	27,887	26,005	22,250	19,443	17,265	15,526	14,105	12,922	11,[illegible]66	9,677	8,597	7,029	5,944
85	76,555	69,637	63,866	58,978	54,785	51,148	45,154	40,417	36,580	33,408	30,742	28,471	26,512	22,620	19,725	17,487	15,705	14,253	13,046	11,[illegible]57	9,746	8,652	7,065	5,970
90	80,588	72,958	66,648	61,342	56,819	52,917	46,527	41,514	37,476	34,154	31,373	29,011	26,979	22,960	19,983	17,689	15,868	14,387	13,158	11,[illegible]39	9,809	8,701	7,098	5,994
95	84,573	76,209	69,351	63,625	58,772	54,607	47,828	42,547	38,316	34,850	31,959	29,511	27,412	23,272	20,219	17,874	16,017	14,509	13,260	11,314	9,865	8,746	7,128	6,015
100	88,513	79,394	71,978	65,829	60,648	56,223	49,063	43,521	39,104	35,501	32,506	29,977	27,813	23,561	20,437	18,044	16,153	14,620	13,354	11,381	9,917	8,786	7,154	6,034
105	92,408	82,513	74,532	67,959	62,451	57,770	50,237	44,442	39,846	36,112	33,017	30,411	28,186	23,828	20,637	18,200	16,278	14,723	13,439	11,443	9,964	8,823	7,179	6,051
110	96,259	85,570	77,017	70,019	64,187	59,251	51,354	45,314	40,546	36,685	33,496	30,817	28,534	24,077	20,824	18,345	16,393	14,817	13,518	11,500	10,007	8,857	7,201	6,067
115	100,066	88,565	79,436	72,012	65,858	60,672	52,418	46,141	41,206	37,225	33,945	31,197	28,860	24,308	20,996	18,479	16,500	14,905	13,590	11,553	10,047	8,888	7,222	6,082
120	103,831	91,502	81,790	73,941	67,467	62,036	53,433	46,925	41,830	37,734	34,368	31,553	29,165	24,524	21,157	18,603	16,600	14,986	13,658	11,[illegible]01	10,083	8,917	7,241	6,095
125	107,553	94,380	84,082	75,810	69,020	63,346	54,402	47,671	42,422	38,214	34,766	31,889	29,451	24,726	21,308	18,719	16,692	15,061	13,720	11,[illegible]46	10,117	8,943	7,258	6,108
130	111,234	97,203	86,315	77,620	70,517	64,605	55,328	48,380	42,983	38,669	35,142	32,205	29,721	24,916	21,448	18,828	16,778	15,131	13,778	11,[illegible]88	10,149	8,968	7,275	6,119
135	114,875	99,971	88,491	79,376	71,963	65,817	56,214	49,057	43,516	39,100	35,498	32,503	29,974	25,094	21,580	18,929	16,859	15,196	13,832	11,727	10,178	8,991	7,290	6,130
140	118,475	102,687	90,612	81,078	73,360	66,983	57,062	49,702	44,023	39,509	35,834	32,785	30,214	25,261	21,704	19,025	16,934	15,258	13,883	11,764	10,206	9,012	7,304	6,140
145	122,036	105,352	92,681	82,730	74,710	68,106	57,876	50,318	44,505	39,897	36,153	33,052	30,440	25,420	21,821	19,114	17,005	15,315	13,931	11,798	10,232	9,032	7,317	6,149
150	125,559	107,967	94,698	84,334	76,015	69,190	58,656	50,906	44,965	40,266	36,456	33,305	30,655	25,569	21,931	19,199	17,072	15,369	13,976	11,830	10,256	9,051	7,329	6,158
155	129,043	110,533	96,667	85,892	77,278	70,235	59,406	51,470	45,404	40,618	36,744	33,545	30,858	25,710	22,035	19,278	17,135	15,420	14,018	11,860	10,278	9,069	7,341	6,166
160	132,490	113,053	98,588	87,406	78,501	71,244	60,126	52,010	45,824	40,953	37,018	33,773	31,052	25,844	22,133	19,353	17,194	15,468	14,058	11,889	10,300	9,085	7,352	6,174
165	135,900	115,526	100,464	88,877	79,686	72,218	60,818	52,527	46,225	41,273	37,280	33,991	31,235	25,972	22,226	19,425	17,250	15,514	14,095	11,916	10,320	9,101	7,362	6,181
170	139,274	117,955	102,296	90,308	80,834	73,160	61,485	53,023	46,609	41,579	37,529	34,198	31,410	26,092	22,314	19,492	17,304	15,557	14,131	11,941	10,339	9,116	7,372	6,188
175	142,612	120,341	104,086	91,700	81,948	74,071	62,127	53,500	46,977	41,872	37,767	34,396	31,577	26,207	22,398	19,556	17,354	15,598	14,164	11,965	10,357	9,130	7,381	6,194
180	145,916	122,684	105,834	93,054	83,028	74,952	62,746	53,958	47,330	42,152	37,995	34,585	31,736	26,317	22,478	19,617	17,402	15,636	14,196	11,988	10,374	9,143	7,389	6,200
185	149,184	124,987	107,543	94,373	84,076	75,805	63,342	54,399	47,669	42,420	38,213	34,765	31,888	26,421	22,554	19,675	17,448	15,673	14,226	12,009	10,390	9,156	7,398	6,206
190	152,419	127,249	109,214	95,657	85,094	76,631	63,918	54,823	47,994	42,678	38,422	34,938	32,033	26,521	22,627	19,730	17,491	15,708	14,255	12,030	10,405	9,167	7,405	6,212
195	155,620	129,472	110,843	96,908	86,082	77,432	64,475	55,232	48,307	42,925	38,622	35,103	32,172	26,616	22,696	19,783	17,532	15,742	14,283	12,049	10,420	9,179	7,413	6,217
200	158,788	131,658	112,446	98,127	87,043	78,208	65,012	55,626	48,608	43,163	38,814	35,262	32,306	26,707	22,763	19,833	17,572	15,773	14,309	12,068	10,434	9,190	7,420	6,222

 # 3/0 AWG | **1 Run(s)** | **600 Volts** | **C-Value = 12,843** | **Magnetic Duct** | **Copper**

Calculations:

1. $f = \dfrac{\text{Length} \times 2 \times \text{Isca}}{\text{Runs} \times \text{C-Value} \times \text{E L-L}}$
2. $M = \dfrac{1}{1 + f}$
3. $\text{Isc} = \text{Isca} \times M$
4. Add Motor Contribution = Motor FLA x 4

* All results are given in symmetrical amperes

Single-Phase L-L Bolted Fault-Current* Table

One-Way Distance in Feet

Available Isc in Thousands (Isca)	5	10	15	20	25	30	40	50	60	70	80	90	100	125	150	175	200	225	250	300	350	400	500	600
3	2,994	2,988	2,983	2,977	2,971	2,965	2,954	2,943	2,932	2,920	2,909	2,898	2,888	2,861	2,834	2,809	2,783	2,758	2,734	2,686	2,640	2,596	2,511	2,432
5	4,984	4,968	4,952	4,936	4,920	4,905	4,874	4,843	4,813	4,783	4,753	4,724	4,695	4,625	4,557	4,490	4,426	4,363	4,302	4,185	4,075	3,970	3,775	3,599
7	6,968	6,937	6,906	6,875	6,845	6,814	6,755	6,696	6,638	6,581	6,526	6,471	6,417	6,286	6,161	6,040	5,924	5,812	5,704	5,501	5,311	5,134	4,814	4,531
10	9,936	9,872	9,809	9,747	9,686	9,625	9,507	9,391	9,278	9,167	9,059	8,954	8,851	8,604	8,371	8,149	7,939	7,740	7,550	7,198	6,877	6,583	6,065	5,622
15	14,855	14,714	14,574	14,438	14,304	14,172	13,916	13,670	13,431	13,201	12,979	12,764	12,556	12,064	11,610	11,189	10,797	10,431	10,090	9,470	8,922	8,433	7,602	6,919
20	19,744	19,494	19,251	19,013	18,781	18,555	18,119	17,703	17,305	16,925	16,561	16,213	15,879	15,101	14,396	13,753	13,166	12,626	12,130	11,245	10,480	9,813	8,704	7,821
25	24,601	24,214	23,840	23,477	23,124	22,783	22,128	21,511	20,926	20,373	19,848	19,350	18,876	17,787	16,816	15,946	15,162	14,451	13,804	12,669	11,707	10,880	9,534	8,484
30	29,427	28,876	28,345	27,833	27,339	26,863	25,958	25,112	24,319	23,575	22,875	22,216	21,593	20,180	18,940	17,843	16,867	15,992	15,203	13,838	12,698	11,731	10,181	8,993
35	34,223	33,479	32,768	32,085	31,431	30,803	29,619	28,522	27,504	26,557	25,672	24,844	24,068	22,325	20,817	19,500	18,340	17,310	16,390	14,814	13,515	12,425	10,700	9,395
40	38,988	38,026	37,110	36,238	35,405	34,610	33,123	31,758	30,501	29,339	28,263	27,263	26,332	24,259	22,489	20,960	19,625	18,451	17,409	15,642	14,200	13,002	11,125	9,722
45	43,723	42,517	41,376	40,294	39,267	38,292	36,479	34,830	33,324	31,942	30,671	29,497	28,410	26,012	23,988	22,256	20,757	19,447	18,293	16,352	14,784	13,490	11,480	9,991
50	48,429	46,953	45,565	44,257	43,021	41,853	39,697	37,752	35,989	34,383	32,914	31,566	30,324	27,608	25,338	23,414	21,761	20,326	19,068	16,969	15,286	13,906	11,780	10,218
55	53,105	51,336	49,681	48,130	46,672	45,300	42,785	40,534	38,509	36,676	35,010	33,488	32,093	29,067	26,562	24,455	22,657	21,106	19,753	17,509	15,723	14,267	12,038	10,412
60	57,752	55,666	53,725	51,915	50,224	48,639	45,751	43,187	40,895	38,834	36,971	35,278	33,734	30,406	27,676	25,396	23,463	21,803	20,363	17,986	16,106	14,582	12,262	10,579
65	62,370	59,944	57,699	55,617	53,680	51,873	48,601	45,718	43,157	40,869	38,810	36,949	35,259	31,639	28,694	26,250	24,190	22,430	20,908	18,411	16,446	14,860	12,458	10,724
70	66,959	64,171	61,606	59,238	57,045	55,009	51,344	48,136	45,306	42,790	40,539	38,513	36,680	32,779	29,628	27,030	24,851	22,997	21,400	18,791	16,749	15,107	12,631	10,852
75	71,520	68,348	65,445	62,779	60,322	58,050	53,983	50,449	47,349	44,608	42,167	39,979	38,008	33,835	30,489	27,744	25 453	23,512	21,845	19,133	17,020	15,327	12,785	10,965
80	76,052	72,476	69,221	66,245	63,515	61,001	56,526	52,663	49,294	46,330	43,703	41,357	39,251	34,817	31,283	28,401	26 005	23,982	22,250	19,443	17,265	15,526	12,922	11,066
85	80,557	76,555	72,933	69,637	66,627	63,866	58,978	54,785	51,148	47,964	45,154	42,654	40,417	35,732	32,020	29,007	26,512	24,412	22,620	19,725	17,487	15,705	13,046	11,157
90	85,034	80,588	76,583	72,958	69,660	66,648	61,342	56,819	52,917	49,517	46,527	43,878	41,514	36,586	32,704	29,567	26,979	24,808	22,960	19,983	17,689	15,868	13,158	11,239
95	89,484	84,573	80,174	76,209	72,618	69,351	63,625	58,772	54,607	50,993	47,828	45,033	42,547	37,386	33,342	30,087	27,412	25,173	23,272	20,219	17,874	16,017	13,260	11,314
100	93,907	88,513	83,706	79,394	75,504	71,978	65,829	60,648	56,223	52,400	49,063	46,126	43,521	38,137	33,938	30,572	27,813	25,511	23,561	20,437	18,044	16,153	13,354	11,381
105	98,303	92,408	87,181	82,513	78,320	74,532	67,959	62,451	57,770	53,741	50,237	47,162	44,442	38,842	34,495	31,023	28,186	25,825	23,828	20,637	18,200	16,278	13,439	11,443
110	102,672	96,259	90,600	85,570	81,069	77,017	70,019	64,187	59,251	55,021	51,354	48,145	45,314	39,506	35,018	31,445	28,534	26,117	24,077	20,824	18,345	16,393	13,518	11,500
115	107,015	100,066	93,965	88,565	83,752	79,436	72,012	65,858	60,672	56,244	52,418	49,079	46,141	40,133	35,509	31,841	28,360	26,389	24,308	20,996	18,479	16,500	13,590	11,553
120	111,331	103,831	97,277	91,502	86,373	81,790	73,941	67,467	62,036	57,414	53,433	49,968	46,925	40,725	35,972	32,213	29,165	26,644	24,524	21,157	18,603	16,600	13,658	11,601
125	115,622	107,553	100,537	94,380	88,934	84,082	75,810	69,020	63,346	58,534	54,402	50,814	47,671	41,286	36,409	32,563	29,451	26,883	24,726	21,308	18,719	16,692	13,720	11,646
130	119,887	111,234	103,746	97,203	91,436	86,315	77,620	70,517	64,605	59,608	55,328	51,621	48,380	41,817	36,821	32,892	29,721	27,107	24,916	21,448	18,828	16,778	13,778	11,688
135	124,127	114,875	106,906	99,971	93,882	88,491	79,376	71,963	65,817	60,637	56,214	52,392	49,057	42,321	37,212	33,203	29,974	27,318	25,094	21,580	18,929	16,859	13,832	11,727
140	128,341	118,475	110,018	102,687	96,273	90,612	81,078	73,360	66,983	61,626	57,062	53,128	49,702	42,800	37,582	33,497	30,214	27,517	25,261	21,704	19,025	16,934	13,883	11,764
145	132,531	122,036	113,082	105,352	98,611	92,681	82,730	74,710	68,106	62,576	57,876	53,833	50,318	43,256	37,933	33,776	30,440	27,704	25,420	21,821	19,114	17,005	13,931	11,798
150	136,696	125,559	116,100	107,967	100,898	94,698	84,334	76,015	69,190	63,489	58,656	54,507	50,906	43,691	38,266	34,040	30,655	27,882	25,569	21,931	19,199	17,072	13,976	11,830
155	140,836	129,043	119,073	110,533	103,136	96,667	85,892	77,278	70,235	64,368	59,406	55,154	51,470	44,105	38,584	34,291	30,858	28,050	25,710	22,035	19,278	17,135	14,018	11,860
160	144,951	132,490	122,002	113,053	105,326	98,588	87,406	78,501	71,244	65,214	60,126	55,774	52,010	44,501	38,887	34,530	31,052	28,210	25,844	22,133	19,353	17,194	14,058	11,889
165	149,043	135,900	124,888	115,526	107,470	100,464	88,877	79,686	72,218	66,030	60,818	56,369	52,527	44,879	39,175	34,757	31,235	28,361	25,972	22,226	19,425	17,250	14,095	11,916
170	153,111	139,274	127,731	117,955	109,569	102,296	90,308	80,834	73,160	66,816	61,485	56,942	53,023	45,241	39,451	34,974	31,410	28,505	26,092	22,314	19,492	17,304	14,131	11,941
175	157,155	142,612	130,533	120,341	111,625	104,086	91,700	81,948	74,071	67,575	62,127	57,492	53,500	45,588	39,714	35,181	31,577	28,643	26,207	22,398	19,556	17,354	14,164	11,965
180	161,175	145,916	133,295	122,684	113,638	105,834	93,054	83,028	74,952	68,308	62,746	58,021	53,958	45,920	39,966	35,379	31,736	28,773	26,317	22,478	19,617	17,402	14,196	11,988
185	165,173	149,184	136,018	124,987	115,611	107,543	94,373	84,076	75,805	69,016	63,342	58,531	54,399	46,239	40,207	35,567	31,888	28,898	26,421	22,554	19,675	17,448	14,226	12,009
190	169,147	152,419	138,701	127,249	117,544	109,214	95,657	85,094	76,631	69,700	63,918	59,023	54,823	46,545	40,438	35,748	32,033	29,018	26,521	22,627	19,730	17,491	14,255	12,030
195	173,098	155,620	141,347	129,472	119,438	110,848	96,908	86,082	77,432	70,362	64,475	59,496	55,232	46,839	40,660	35,922	32,172	29,132	26,616	22,696	19,783	17,532	14,283	12,049
200	177,027	158,788	143,956	131,658	121,296	112,446	98,127	87,043	78,208	71,002	65,012	59,954	55,626	47,122	40,873	36,088	32,306	29,241	26,707	22,763	19,833	17,572	14,309	12,068

3/0 AWG | 2 Run(s) | 600 Volts

C-Value = 12,843 | Magnetic Duct | Copper

Calculations:

1. $f = \dfrac{\text{Length} \times 2 \times \text{Isca}}{\text{Runs} \times \text{C-Value} \times \text{E L-L}}$

2. $M = \dfrac{1}{1 + f}$

3. Isc = Isca x M

4. Add Motor Contribution = Motor FLA x 4

* All results are given in symmetrical amperes

Single-Phase L-L Bolted Fault-Current* Table

One-Way Distance in Feet

Available Isc in Thousands (Isca)

Isca	5	10	15	20	25	30	40	50	60	70	80	90	100	125	150	175	200	225	250	300	350	400	500	600
3	2,990	2,980	2,970	2,961	2,951	2,941	2,922	2,904	2,885	2,867	2,849	2,831	2,813	2,770	2,729	2,688	2,649	2,611	2,573	2,502	2,435	2,371	2,253	2,146
5	4,973	4,945	4,918	4,892	4,866	4,840	4,788	4,738	4,689	4,641	4,594	4,548	4,502	4,393	4,289	4,190	4,095	4,004	3,918	3,755	3,605	3,467	3,221	3,007
7	6,946	6,893	6,841	6,790	6,739	6,690	6,592	6,497	6,405	6,316	6,229	6,144	6,062	5,866	5,682	5,509	5,346	5,193	5,048	4,781	4,541	4,324	3,947	3,630
10	9,891	9,784	9,679	9,577	9,476	9,378	9,188	9,005	8,829	8,660	8,498	8,341	8,190	7,835	7,510	7,211	6,935	6,679	6,441	6,013	5,638	5,308	4,750	4,299
15	14,755	14,519	14,289	14,067	13,852	13,643	13,244	12,867	12,511	12,175	11,856	11,553	11,265	10,605	10,018	9,493	9,020	8,591	8,202	7,520	6,943	6,449	5,644	5,018
20	19,568	19,153	18,756	18,376	18,010	17,658	16,995	16,380	15,808	15,274	14,775	14,308	13,869	12,882	12,026	11,277	10,615	10,027	9,501	8,598	7,852	7,225	6,230	5,476
25	24,328	23,691	23,087	22,512	21,966	21,445	20,475	19,588	18,776	18,027	17,337	16,697	16,103	14,787	13,670	12,710	11,876	11,145	10,498	9,407	8,521	7,788	6,644	5,793
30	29,037	28,135	27,286	26,488	25,734	25,023	23,711	22,531	21,462	20,490	19,602	18,788	18,039	16,404	15,041	13,887	12,897	12,039	11,288	10,036	9,034	8,214	6,952	6,026
35	33,697	32,487	31,361	30,311	29,328	28,408	26,729	25,238	23,905	22,705	21,620	20,634	19,734	17,794	16,201	14,870	13,741	12,771	11,930	10,540	9,441	8,549	7,190	6,204
40	38,307	36,751	35,317	33,990	32,760	31,615	29,550	27,739	26,136	24,709	23,430	22,276	21,231	19,002	17,196	15,704	14,450	13,382	12,461	10,952	9,770	8,818	7,380	6,345
45	42,868	40,929	39,158	37,534	36,039	34,659	32,193	30,054	28,182	26,530	25,061	23,745	22,561	20,061	18,059	16,420	15,055	13,898	12,907	11,296	10,042	9,039	7,534	6,459
50	47,382	45,024	42,890	40,950	39,177	37,551	34,673	32,205	30,065	28,192	26,538	25,068	23,752	20,997	18,814	17,042	15,576	14,341	13,288	11,587	10,272	9,225	7,662	6,553
55	51,849	49,039	46,518	44,244	42,181	40,303	37,006	34,208	31,804	29,715	27,884	26,265	24,824	21,830	19,480	17,587	16,030	14,725	13,617	11,836	10,467	9,382	7,771	6,632
60	56,269	52,975	50,045	47,423	45,061	42,924	39,205	36,078	33,414	31,116	29,114	27,354	25,794	22,577	20,073	18,069	16,429	15,061	13,904	12,052	10,636	9,517	7,863	6,699
65	60,644	56,835	53,476	50,493	47,824	45,424	41,279	37,828	34,910	32,409	30,243	28,348	26,677	23,250	20,603	18,497	16,782	15,358	14,157	12,242	10,783	9,635	7,943	6,757
70	64,974	60,621	56,815	53,459	50,477	47,810	43,241	39,469	36,302	33,606	31,282	29,259	27,482	23,859	21,080	18,881	17,097	15,622	14,380	12,409	10,912	9,738	8,013	6,808
75	69,260	64,336	60,065	56,327	53,026	50,091	45,098	41,010	37,602	34,717	32,243	30,098	28,221	24,414	21,512	19,227	17,380	15,858	14,580	12,557	11,027	9,829	8,075	6,852
80	73,502	67,980	63,230	59,101	55,477	52,273	46,859	42,462	38,819	35,751	33,133	30,873	28,901	24,921	21,905	19,540	17,636	16,070	14,759	12,690	11,129	9,910	8,130	6,891
85	77,701	71,557	66,313	61,786	57,837	54,362	48,531	43,830	39,959	36,717	33,961	31,590	29,528	25,386	22,263	19,825	17,868	16,262	14,921	12,809	11,221	9,983	8,179	6,926
90	81,859	75,068	69,318	64,386	60,109	56,365	50,121	45,123	41,031	37,619	34,732	32,256	30,109	25,815	22,592	20,085	18,079	16,437	15,068	12,917	11,304	10,049	8,222	6,958
95	85,974	78,515	72,246	66,905	62,299	58,286	51,635	46,346	42,039	38,466	35,452	32,876	30,649	26,210	22,895	20,324	18,272	16,596	15,202	13,016	11,379	10,108	8,262	6,986
100	90,049	81,899	75,102	69,347	64,411	60,131	53,077	47,504	42,991	39,260	36,126	33,455	31,151	26,577	23,174	20,543	18,449	16,743	15,325	13,105	11,448	10,162	8,298	7,012
105	94,083	85,223	77,888	71,715	66,449	61,903	54,453	48,604	43,889	40,008	36,758	33,996	31,620	26,917	23,432	20,746	18,613	16,877	15,438	13,188	11,510	10,211	8,331	7,036
110	98,078	88,487	80,605	74,013	68,417	63,608	55,768	49,648	44,739	40,713	37,352	34,504	32,059	27,235	23,673	20,934	18,764	17,001	15,541	13,263	11,568	10,257	8,361	7,057
115	102,033	91,694	83,258	76,243	70,319	65,248	57,025	50,642	45,545	41,379	37,912	34,981	32,471	27,531	23,896	21,109	18,904	17,116	15,637	13,333	11,621	10,299	8,389	7,077
120	105,950	94,845	85,848	78,409	72,157	66,828	58,228	51,589	46,309	42,009	38,440	35,430	32,857	27,809	24,105	21,272	19,035	17,223	15,727	13,398	11,670	10,337	8,415	7,095
125	109,829	97,942	88,377	80,513	73,935	68,351	59,380	52,492	47,035	42,606	38,939	35,854	33,221	28,069	24,300	21,424	19,156	17,323	15,809	13,458	11,716	10,373	8,438	7,112
130	113,670	100,985	90,847	82,559	75,656	69,819	60,486	53,353	47,726	43,172	39,411	36,253	33,564	28,313	24,483	21,566	19,270	17,415	15,887	13,514	11,758	10,406	8,460	7,128
135	117,475	103,977	93,261	84,547	77,323	71,236	61,546	54,177	48,383	43,709	39,859	36,632	33,888	28,544	24,655	21,699	19,376	17,502	15,959	13,566	11,798	10,437	8,481	7,142
140	121,243	106,918	95,620	86,482	78,938	72,604	62,565	54,965	49,011	44,221	40,284	36,990	34,195	28,761	24,817	21,824	19,476	17,584	16,027	13,615	11,835	10,466	8,500	7,156
145	124,975	109,809	97,926	88,364	80,503	73,926	63,544	55,719	49,610	44,708	40,687	37,330	34,485	28,966	24,970	21,942	19,570	17,660	16,090	13,661	11,869	10,493	8,518	7,168
150	128,671	112,653	100,182	90,196	82,021	75,204	64,486	56,442	50,182	45,172	41,071	37,654	34,761	29,160	25,114	22,054	19,658	17,732	16,150	13,704	11,902	10,518	8,534	7,180
155	132,333	115,450	102,387	91,980	83,494	76,441	65,393	57,135	50,729	45,615	41,437	37,961	35,023	29,344	25,250	22,159	19,742	17,800	16,206	13,745	11,932	10,542	8,550	7,191
160	135,961	118,201	104,546	93,718	84,923	77,637	66,267	57,801	51,254	46,038	41,787	38,254	35,272	29,519	25,379	22,258	19,821	17,864	16,259	13,783	11,961	10,565	8,565	7,202
165	139,554	120,908	106,657	95,412	86,311	78,796	67,109	58,441	51,756	46,443	42,120	38,533	35,509	29,685	25,502	22,352	19,895	17,925	16,309	13,819	11,988	10,586	8,579	7,211
170	143,114	123,571	108,724	97,063	87,660	79,918	67,921	59,056	52,238	46,831	42,439	38,799	35,735	29,843	25,618	22,442	19,966	17,982	16,357	13,853	12,014	10,606	8,592	7,221
175	146,641	126,192	110,748	98,672	88,971	81,006	68,706	59,648	52,701	47,203	42,743	39,054	35,951	29,993	25,729	22,527	20,033	18,037	16,402	13,885	12,038	10,625	8,604	7,229
180	150,136	128,771	112,730	100,242	90,245	82,062	69,463	60,218	53,145	47,559	43,035	39,298	36,157	30,137	25,835	22,603	20,097	18,089	16,445	13,916	12,061	10,643	8,616	7,238
185	153,599	131,310	114,671	101,774	91,485	83,085	70,195	60,768	53,573	47,901	43,315	39,531	36,355	30,274	25,935	22,685	20,158	18,138	16,486	13,945	12,083	10,660	8,627	7,246
190	157,030	133,810	116,572	103,269	92,691	84,079	70,903	61,298	53,984	48,230	43,584	39,754	36,544	30,404	26,031	22,753	20,216	18,185	16,524	13,973	12,104	10,676	8,638	7,253
195	160,429	136,271	118,435	104,729	93,865	85,044	71,588	61,809	54,380	48,546	43,842	39,969	36,725	30,530	26,123	22,823	20,271	18,229	16,561	13,999	12,124	10,691	8,648	7,260
200	163,798	138,694	120,262	106,154	95,009	85,981	72,251	62,303	54,762	48,850	44,090	40,175	36,898	30,650	26,211	22,895	20,324	18,272	16,597	14,024	12,143	10,706	8,657	7,267

4/0 AWG | 1 Run(s) | 600 Volts | C-Value = 15,082 | Magnetic Duct | Copper

Calculations:

1. $f = \dfrac{\text{Length} \times 2 \times \text{Isca}}{\text{Runs} \times \text{C-Value} \times \text{E L-L}}$

2. $M = \dfrac{1}{1 + f}$

3. Isc = Isca x M

4. Add Motor Contribution = Motor FLA x 4

* All results are given in symmetrical amperes

Single-Phase L-L Bolted Fault-Current* Table

One-Way Distance in Feet

Available Isc in Thousands (Isca)	5	10	15	20	25	30	40	50	60	70	80	90	100	125	150	175	200	225	250	300	350	400	500	600
3	2,995	2,990	2,985	2,980	2,975	2,970	2,961	2,951	2,941	2,932	2,922	2,913	2,904	2,881	2,858	2,835	2,813	2,792	2,770	2,729	2,688	2,649	2,573	2,502
5	4,986	4,973	4,959	4,945	4,932	4,918	4,892	4,866	4,840	4,814	4,788	4,763	4,738	4,677	4,617	4,559	4,502	4,447	4,393	4,289	4,190	4,095	3,918	3,755
7	6,973	6,946	6,920	6,893	6,867	6,841	6,790	6,739	6,690	6,640	6,592	6,544	6,497	6,383	6,272	6,165	6,062	5,962	5,866	5,682	5,509	5,346	5,048	4,781
10	9,945	9,891	9,837	9,784	9,731	9,679	9,577	9,476	9,378	9,282	9,188	9,095	9,005	8,786	8,578	8,380	8,190	8,009	7,835	7,510	7,211	6,935	6,441	6,013
15	14,877	14,755	14,636	14,519	14,403	14,289	14,067	13,852	13,643	13,440	13,244	13,053	12,867	12,425	12,013	11,627	11,265	10,925	10,605	10,018	9,493	9,020	8,202	7,520
20	19,781	19,568	19,358	19,153	18,953	18,756	18,376	18,010	17,658	17,320	16,995	16,682	16,380	15,671	15,020	14,422	13,869	13,358	12,882	12,026	11,277	10,615	9,501	8,598
25	24,659	24,328	24,005	23,691	23,385	23,087	22,512	21,966	21,445	20,949	20,475	20,022	19,588	18,583	17,675	16,852	16,103	15,417	14,787	13,670	12,710	11,876	10,498	9,407
30	29,511	29,037	28,579	28,135	27,704	27,286	26,488	25,734	25,023	24,349	23,711	23,106	22,531	21,210	20,036	18,985	18,039	17,183	16,404	15,041	13,887	12,897	11,288	10,036
35	34,336	33,697	33,081	32,487	31,914	31,361	30,311	29,328	28,408	27,543	26,729	25,963	25,238	23,593	22,150	20,872	19,734	18,714	17,794	16,201	14,870	13,741	11,930	10,540
40	39,135	38,307	37,513	36,751	36,020	35,317	33,990	32,760	31,615	30,548	29,550	28,616	27,739	25,764	24,052	22,554	21,231	20,055	19,002	17,196	15,704	14,450	12,461	10,952
45	43,908	42,868	41,876	40,929	40,024	39,158	37,534	36,039	34,659	33,380	32,193	31,087	30,054	27,750	25,774	24,061	22,561	21,238	20,061	18,059	16,420	15,055	12,907	11,296
50	48,656	47,382	46,173	45,024	43,932	42,890	40,950	39,177	37,551	36,055	34,673	33,394	32,205	29,574	27,340	25,420	23,752	22,290	20,997	18,814	17,042	15,576	13,288	11,587
55	53,378	51,849	50,405	49,039	47,745	46,518	44,244	42,181	40,303	38,584	37,006	35,552	34,208	31,255	28,770	26,652	24,824	23,231	21,830	19,480	17,587	16,030	13,617	11,836
60	58,075	56,269	54,572	52,975	51,469	50,045	47,423	45,061	42,924	40,980	39,205	37,577	36,078	32,808	30,082	27,774	25,794	24,079	22,577	20,073	18,069	16,429	13,904	12,052
65	62,746	60,644	58,678	56,835	55,105	53,476	50,493	47,824	45,424	43,252	41,279	39,478	37,828	34,249	31,288	28,799	26,677	24,846	23,250	20,603	18,497	16,782	14,157	12,242
70	67,393	64,974	62,722	60,621	58,657	56,815	53,459	50,477	47,810	45,411	43,241	41,269	39,469	35,588	32,403	29,740	27,482	25,543	23,859	21,080	18,881	17,097	14,380	12,409
75	72,016	69,260	66,707	64,336	62,127	60,065	56,327	53,026	50,091	47,463	45,098	42,957	41,010	36,837	33,434	30,607	28,221	26,180	24,414	21,512	19,227	17,380	14,580	12,557
80	76,613	73,502	70,633	67,980	65,519	63,230	59,101	55,477	52,273	49,418	46,859	44,552	42,462	38,003	34,393	31,408	28,901	26,764	24,921	21,905	19,540	17,636	14,759	12,690
85	81,187	77,701	74,503	71,557	68,836	66,313	61,786	57,837	54,362	51,282	48,531	46,061	43,830	39,096	35,285	32,151	29,528	27,301	25,386	22,263	19,825	17,868	14,921	12,809
90	85,736	81,859	78,316	75,068	72,078	69,318	64,386	60,109	56,365	53,060	50,121	47,491	45,123	40,121	36,118	32,841	30,109	27,797	25,815	22,592	20,085	18,079	15,068	12,917
95	90,262	85,974	82,075	78,515	75,250	72,246	66,905	62,299	58,286	54,759	51,635	48,847	46,346	41,085	36,897	33,484	30,649	28,256	26,210	22,895	20,324	18,272	15,202	13,016
100	94,764	90,049	85,781	81,899	78,353	75,102	69,347	64,411	60,131	56,384	53,077	50,136	47,504	41,993	37,628	34,085	31,151	28,683	26,577	23,174	20,543	18,449	15,325	13,105
105	99,242	94,083	89,434	85,223	81,390	77,888	71,715	66,449	61,903	57,940	54,453	51,363	48,604	42,850	38,314	34,647	31,620	29,080	26,917	23,432	20,746	18,613	15,438	13,188
110	103,697	98,078	93,036	88,487	84,363	80,605	74,013	68,417	63,608	59,430	55,768	52,531	49,648	43,660	38,961	35,175	32,059	29,451	27,235	23,673	20,934	18,764	15,541	13,263
115	108,129	102,033	96,588	91,694	87,273	83,258	76,243	70,319	65,248	60,860	57,025	53,644	50,642	44,427	39,570	35,670	32,471	29,798	27,531	23,896	21,109	18,904	15,637	13,333
120	112,538	105,950	100,091	94,845	90,122	85,848	78,409	72,157	66,828	62,232	58,228	54,708	51,589	45,153	40,146	36,138	32,857	30,123	27,809	24,105	21,272	19,035	15,727	13,398
125	116,924	109,829	103,545	97,942	92,914	88,377	80,513	73,935	68,351	63,551	59,380	55,724	52,492	45,843	40,690	36,578	33,221	30,428	28,069	24,300	21,424	19,156	15,809	13,458
130	121,288	113,670	106,953	100,985	95,648	90,847	82,559	75,656	69,819	64,818	60,486	56,696	53,353	46,499	41,206	36,995	33,564	30,716	28,313	24,483	21,566	19,270	15,887	13,514
135	125,629	117,475	110,314	103,977	98,328	93,261	84,547	77,323	71,236	66,038	61,546	57,627	54,177	47,124	41,695	37,389	33,888	30,987	28,544	24,655	21,699	19,376	15,959	13,566
140	129,948	121,243	113,630	106,918	100,954	95,620	86,482	78,938	72,604	67,212	62,565	58,519	54,965	47,718	42,160	37,762	34,195	31,243	28,761	24,817	21,824	19,476	16,027	13,615
145	134,245	124,975	116,902	109,809	103,528	97,926	88,364	80,503	73,926	68,343	63,544	59,375	55,719	48,286	42,603	38,117	34,485	31,486	28,966	24,970	21,942	19,570	16,090	13,661
150	138,519	128,671	120,131	112,653	106,052	100,182	90,196	82,021	75,204	69,434	64,486	60,196	56,442	48,828	43,024	38,454	34,761	31,715	29,160	25,114	22,054	19,658	16,150	13,704
155	142,773	132,333	123,316	115,450	108,527	102,387	91,980	83,494	76,441	70,487	65,393	60,986	57,135	49,346	43,426	38,774	35,023	31,933	29,344	25,250	22,159	19,742	16,206	13,745
160	147,004	135,961	126,461	118,201	110,955	104,546	93,718	84,923	77,637	71,503	66,267	61,745	57,801	49,842	43,810	39,080	35,272	32,140	29,519	25,379	22,258	19,821	16,259	13,783
165	151,214	139,554	129,564	120,908	113,337	106,657	95,412	86,311	78,796	72,484	67,109	62,476	58,441	50,317	44,176	39,371	35,509	32,337	29,685	25,502	22,352	19,895	16,309	13,819
170	155,403	143,114	132,627	123,571	115,673	108,724	97,063	87,660	79,918	73,433	67,921	63,179	59,056	50,772	44,527	39,649	35,735	32,524	29,843	25,618	22,442	19,966	16,357	13,853
175	159,571	146,641	135,650	126,192	117,967	110,748	98,672	88,971	81,006	74,351	68,706	63,857	59,648	51,209	44,862	39,915	35,951	32,703	29,993	25,729	22,527	20,033	16,402	13,885
180	163,717	150,136	138,635	128,771	120,218	112,730	100,242	90,245	82,062	75,239	69,463	64,511	60,218	51,629	45,184	40,170	36,157	32,874	30,137	25,835	22,608	20,097	16,445	13,916
185	167,843	153,599	141,583	131,310	122,428	114,671	101,774	91,485	83,085	76,098	70,195	65,142	60,768	52,032	45,493	40,414	36,355	33,037	30,274	25,935	22,685	20,158	16,486	13,945
190	171,949	157,030	144,493	133,810	124,598	116,572	103,269	92,691	84,079	76,931	70,903	65,751	61,298	52,420	45,789	40,647	36,544	33,193	30,404	26,031	22,758	20,216	16,524	13,973
195	176,033	160,429	147,366	136,271	126,729	118,435	104,729	93,865	85,044	77,738	71,588	66,340	61,809	52,794	46,074	40,871	36,725	33,342	30,530	26,123	22,828	20,271	16,561	13,999
200	180,098	163,798	150,204	138,694	128,822	120,262	106,154	95,009	85,981	78,521	72,251	66,909	62,303	53,154	46,348	41,087	36,898	33,485	30,650	26,211	22,895	20,324	16,597	14,024

# 4/0 AWG	2 Run(s)	600 Volts	C-Value = 15,082	Magnetic Duct	Copper

Calculations:

1. $f = \dfrac{\text{Length} \times 2 \times \text{Isca}}{\text{Runs} \times \text{C-Value} \times \text{E L-L}}$

2. $M = \dfrac{1}{1 + f}$

3. $\text{Isc} = \text{Isca} \times M$

4. Add Motor Contribution = Motor FLA x 4

* All results are given in symmetrical amperes

Single-Phase L-L Bolted Fault-Current* Table

One-Way Distance in Feet

Available Isc in Thousands (Isca)	5	10	15	20	25	30	40	50	60	70	80	90	100	125	150	175	200	225	250	300	350	400	500	600
3	2,991	2,982	2,973	2,964	2,955	2,946	2,929	2,912	2,895	2,878	2,861	2,845	2,828	2,789	2,750	2,712	2,675	2,640	2,605	2,538	2,475	2,414	2,302	2,199
5	4,975	4,950	4,925	4,901	4,877	4,853	4,806	4,759	4,714	4,669	4,626	4,583	4,541	4,439	4,342	4,248	4,159	4,073	3,991	3,836	3,693	3,560	3,321	3,112
7	6,951	6,902	6,854	6,807	6,761	6,715	6,625	6,537	6,452	6,369	6,288	6,209	6,132	5,948	5,774	5,610	5,455	5,309	5,170	4,913	4,681	4,469	4,099	3,785
10	9,900	9,802	9,706	9,611	9,519	9,428	9,252	9,082	8,918	8,760	8,607	8,460	8,318	7,982	7,673	7,386	7,120	6,873	6,642	6,224	5,855	5,528	4,972	4,518
15	14,776	14,558	14,347	14,142	13,943	13,749	13,377	13,025	12,690	12,373	12,071	11,783	11,509	10,876	10,309	9,798	9,336	8,915	8,531	7,853	7,276	6,777	5,960	5,319
20	19,604	19,223	18,856	18,503	18,163	17,836	17,215	16,636	16,094	15,587	15,111	14,663	14,240	13,284	12,448	11,711	11,056	10,471	9,945	9,036	8,280	7,640	6,618	5,836
25	24,384	23,797	23,238	22,704	22,195	21,708	20,795	19,956	19,181	18,465	17,800	17,182	16,605	15,319	14,218	13,264	12,431	11,696	11,043	9,934	9,027	8,272	7,086	6,198
30	29,117	28,284	27,498	26,754	26,049	25,381	24,141	23,018	21,994	21,057	20,197	19,405	18,672	17,061	15,707	14,551	13,554	12,685	11,920	10,538	9,605	8,755	7,438	6,465
35	33,804	32,686	31,641	30,660	29,738	28,870	27,277	25,851	24,567	23,404	22,347	21,380	20,494	18,570	16,976	15,634	14,489	13,500	12,638	11,206	10,065	9,136	7,711	6,671
40	38,445	37,006	35,672	34,430	33,272	32,189	30,221	28,481	26,930	25,539	24,285	23,148	22,113	19,889	18,072	16,559	15,280	14,184	13,235	11,573	10,441	9,444	7,929	6,834
45	43,042	41,246	39,595	38,071	36,660	35,349	32,991	30,928	29,107	27,489	26,041	24,739	23,560	21,052	19,027	17,357	15,957	14,766	13,740	12,064	10,752	9,698	8,108	6,966
50	47,594	45,409	43,415	41,589	39,911	38,363	35,601	33,210	31,120	29,277	27,641	26,178	24,861	22,086	19,867	18,054	16,544	15,267	14,173	12,396	11,016	9,912	8,257	7,075
55	52,102	49,495	47,136	44,992	43,034	41,239	38,065	35,344	32,986	30,924	29,104	27,486	25,038	23,009	20,612	18,667	17,057	15,703	14,548	12,582	11,241	10,094	8,383	7,167
60	56,568	53,508	50,761	48,283	46,035	43,988	40,395	37,344	34,722	32,444	30,446	28,680	27,108	23,841	21,276	19,210	17,509	16,085	14,876	12,931	11,436	10,250	8,490	7,246
65	60,991	57,448	54,295	51,469	48,923	46,617	42,601	39,222	36,339	33,852	31,683	29,775	28,084	24,592	21,873	19,695	17,911	16,424	15,165	13,149	11,606	10,387	8,584	7,314
70	65,373	61,320	57,740	54,555	51,702	49,134	44,693	40,988	37,851	35,160	32,826	30,782	28,978	25,275	22,411	20,131	18,271	16,726	15,422	13,341	11,756	10,507	8,666	7,373
75	69,713	65,123	61,099	57,544	54,380	51,546	46,680	42,653	39,266	36,378	33,885	31,712	29,801	25,899	22,900	20,524	18,595	16,997	15,652	13,513	11,889	10,613	8,738	7,426
80	74,013	68,860	64,377	60,443	56,962	53,859	48,569	44,225	40,595	37,515	34,870	32,573	30,560	26,470	23,346	20,881	18,887	17,241	15,859	13,667	12,008	10,708	8,802	7,472
85	78,273	72,532	67,576	63,254	59,452	56,080	50,368	45,712	41,844	38,579	35,787	33,372	31,262	26,995	23,754	21,207	19,153	17,462	16,046	13,806	12,115	10,793	8,859	7,513
90	82,493	76,142	70,699	65,982	61,855	58,214	52,083	47,120	43,020	39,577	36,644	34,116	31,914	27,480	24,128	21,505	19,396	17,664	16,216	13,932	12,211	10,869	8,911	7,550
95	86,674	79,690	73,748	68,630	64,176	60,266	53,719	48,455	44,131	40,515	37,447	34,811	32,521	27,929	24,473	21,779	19,619	17,848	16,371	14,046	12,299	10,939	8,957	7,584
100	90,817	83,179	76,726	71,202	66,420	62,240	55,282	49,723	45,180	41,398	38,200	35,460	33,088	28,346	24,793	22,031	19,823	18,018	16,513	14,151	12,379	11,002	9,000	7,614
105	94,922	86,609	79,635	73,701	68,589	64,141	56,776	50,929	46,173	42,230	38,907	36,069	33,617	28,734	25,089	22,265	20,012	18,173	16,644	14,247	12,453	11,060	9,038	7,642
110	98,990	89,983	82,479	76,130	70,688	65,973	58,207	52,077	47,115	43,016	39,574	36,641	34,114	29,096	25,364	22,482	20,187	18,318	16,765	14,335	12,520	11,113	9,074	7,667
115	103,021	93,302	85,258	78,492	72,720	67,739	59,578	53,171	48,009	43,761	40,203	37,180	34,580	29,434	25,621	22,683	20,349	18,451	16,877	14,417	12,582	11,162	9,107	7,690
120	107,015	96,566	87,976	80,789	74,688	69,444	60,892	54,216	48,859	44,466	40,797	37,688	35,019	29,751	25,861	22,871	20,501	18,575	16,981	14,492	12,640	11,208	9,137	7,712
125	110,974	99,778	90,634	83,025	76,595	71,089	62,154	55,214	49,668	45,135	41,359	38,167	35,432	30,049	26,086	23,047	20,642	18,691	17,077	14,563	12,694	11,250	9,165	7,732
130	114,897	102,938	93,234	85,202	78,444	72,679	63,366	56,168	50,439	45,770	41,893	38,621	35,823	30,330	26,297	23,211	20,774	18,799	17,168	14,628	12,743	11,289	9,191	7,750
135	118,785	106,048	95,778	87,321	80,237	74,216	64,530	57,081	51,174	46,375	42,399	39,050	36,192	30,594	26,496	23,366	20,897	18,900	17,252	14,689	12,790	11,325	9,215	7,767
140	122,639	109,109	98,268	89,386	81,977	75,702	65,651	57,957	51,876	46,951	42,880	39,458	36,542	30,844	26,683	23,511	21,013	18,995	17,331	14,747	12,833	11,359	9,237	7,783
145	126,459	112,122	100,705	91,398	83,666	77,140	66,730	58,796	52,548	47,500	43,337	39,845	36,874	31,080	26,859	23,648	21,123	19,085	17,405	14,801	12,874	11,391	9,258	7,798
150	130,245	115,089	103,092	93,360	85,307	78,533	67,770	59,602	53,190	48,025	43,773	40,214	37,189	31,304	27,026	23,777	21,226	19,169	17,475	14,851	12,912	11,421	9,278	7,812
155	133,999	118,009	105,429	95,273	86,901	79,882	68,772	60,375	53,806	48,526	44,189	40,564	37,489	31,516	27,184	23,900	21,323	19,248	17,541	14,899	12,948	11,449	9,297	7,825
160	137,719	120,886	107,719	97,139	88,451	81,190	69,739	61,119	54,396	49,005	44,587	40,899	37,775	31,717	27,334	24,015	21,415	19,323	17,603	14,944	12,982	11,476	9,314	7,838
165	141,408	123,718	109,962	98,959	89,958	82,457	70,673	61,835	54,962	49,464	44,966	41,218	38,047	31,909	27,476	24,125	21,502	19,394	17,662	14,986	13,014	11,501	9,331	7,849
170	145,064	126,508	112,161	100,736	91,424	83,688	71,574	62,524	55,506	49,904	45,330	41,523	38,307	32,091	27,612	24,229	21,585	19,461	17,718	15,026	13,044	11,524	9,346	7,860
175	148,689	129,256	114,316	102,471	92,851	84,881	72,446	63,188	56,029	50,326	45,678	41,815	38,555	32,266	27,740	24,328	21,664	19,525	17,771	15,064	13,073	11,547	9,361	7,871
180	152,284	131,964	116,428	104,165	94,239	86,041	73,289	63,828	56,531	50,732	46,011	42,094	38,792	32,432	27,863	24,423	21,739	19,586	17,821	15,100	13,100	11,568	9,375	7,881
185	155,847	134,631	118,500	105,820	95,592	87,167	74,104	64,446	57,015	51,121	46,331	42,362	39,019	32,590	27,980	24,513	21,810	19,644	17,869	15,135	13,126	11,588	9,388	7,890
190	159,380	137,260	120,532	107,438	96,910	88,261	74,893	65,042	57,482	51,496	46,639	42,619	39,237	32,742	28,092	24,598	21,878	19,699	17,915	15,167	13,150	11,607	9,400	7,899
195	162,884	139,851	122,524	109,018	98,194	89,325	75,658	65,618	57,931	51,856	46,934	42,866	39,446	32,887	28,199	24,680	21,942	19,751	17,958	15,198	13,174	11,625	9,412	7,907
200	166,358	142,404	124,480	110,564	99,446	90,360	76,399	66,175	58,364	52,203	47,218	43,102	39,647	33,027	28,301	24,759	22,004	19,801	17,999	15,228	13,196	11,643	9,424	7,915

250 MCM	1 Run(s)	600 Volts	C-Value = 16,483	Magnetic Duct	Copper

Calculations:

1. $f = \frac{\text{Length} \times 2 \times \text{Isca}}{\text{Runs} \times \text{C-Value} \times \text{E L-L}}$

2. $M = \frac{1}{1 + f}$

3. Isc = Isca x M

4. Add Motor Contribution = Motor FLA x 4

* All results are given in symmetrical amperes

Single-Phase L-L Bolted Fault-Current* Table

One-Way Distance in Feet

Available Isc in Thousands (Isca)	5	10	15	20	25	30	40	50	60	70	80	90	100	125	150	175	200	225	250	300	350	400	500	600
3	2,995	2,991	2,986	2,982	2,977	2,973	2,964	2,955	2,946	2,938	2,929	2,920	2,912	2,890	2,869	2,849	2,828	2,808	2,789	2,750	2,712	2,675	2,605	2,538
5	4,987	4,975	4,962	4,950	4,938	4,925	4,901	4,877	4,853	4,829	4,806	4,782	4,759	4,703	4,648	4,594	4,541	4,489	4,439	4,342	4,248	4,159	3,991	3,836
7	6,975	6,951	6,926	6,902	6,878	6,854	6,807	6,761	6,715	6,670	6,625	6,581	6,537	6,431	6,328	6,229	6,132	6,038	5,948	5,774	5,610	5,455	5,170	4,913
10	9,950	9,900	9,851	9,802	9,753	9,706	9,611	9,519	9,428	9,339	9,252	9,166	9,082	8,878	8,683	8,497	8,318	8,147	7,982	7,673	7,386	7,120	6,642	6,224
15	14,887	14,776	14,666	14,558	14,452	14,347	14,142	13,943	13,749	13,560	13,377	13,198	13,025	12,609	12,220	11,854	11,509	11,184	10,876	10,309	9,798	9,336	8,531	7,853
20	19,800	19,604	19,411	19,223	19,038	18,856	18,503	18,163	17,836	17,520	17,215	16,920	16,636	15,964	15,345	14,772	14,240	13,746	13,284	12,448	11,711	11,056	9,945	9,036
25	24,688	24,384	24,087	23,797	23,514	23,238	22,704	22,195	21,708	21,241	20,795	20,366	19,956	18,997	18,127	17,333	16,605	15,936	15,319	14,218	13,264	12,431	11,043	9,934
30	29,552	29,117	28,694	28,284	27,885	27,498	26,754	26,049	25,381	24,746	24,141	23,566	23,018	21,752	20,618	19,597	18,672	17,830	17,061	15,707	14,551	13,554	11,920	10,638
35	34,391	33,804	33,236	32,686	32,155	31,641	30,660	29,738	28,870	28,051	27,277	26,545	25,851	24,266	22,863	21,614	20,494	19,485	18,570	16,976	15,634	14,489	12,638	11,206
40	39,207	38,445	37,712	37,006	36,327	35,672	34,430	33,272	32,189	31,174	30,221	29,325	28,481	26,568	24,896	23,422	22,113	20,942	19,889	18,072	16,559	15,280	13,235	11,673
45	43,999	43,042	42,125	41,246	40,404	39,595	38,071	36,660	35,349	34,129	32,991	31,926	30,928	28,685	26,746	25,052	23,560	22,236	21,052	19,027	17,357	15,957	13,740	12,064
50	48,767	47,594	46,475	45,409	44,389	43,415	41,589	39,911	38,363	36,930	35,601	34,364	33,210	30,638	28,436	26,529	24,861	23,391	22,086	19,867	18,054	16,544	14,173	12,396
55	53,512	52,102	50,765	49,495	48,287	47,136	44,992	43,034	41,239	39,589	38,065	36,654	35,344	32,445	29,986	27,873	26,038	24,430	23,009	20,612	18,667	17,057	14,548	12,682
60	58,234	56,568	54,995	53,508	52,098	50,761	48,283	46,035	43,988	42,115	40,395	38,809	37,344	34,123	31,413	29,102	27,108	25,370	23,841	21,276	19,210	17,509	14,876	12,931
65	62,932	60,991	59,167	57,448	55,827	54,295	51,469	48,923	46,617	44,518	42,601	40,841	39,222	35,684	32,731	30,230	28,084	26,222	24,592	21,873	19,695	17,911	15,165	13,149
70	67,607	65,373	63,281	61,320	59,476	57,740	54,555	51,702	49,134	46,808	44,693	42,761	40,988	37,140	33,953	31,269	28,978	27,000	25,275	22,411	20,131	18,271	15,422	13,341
75	72,260	69,713	67,340	65,123	63,047	61,099	57,544	54,380	51,546	48,992	46,680	44,576	42,653	38,502	35,087	32,229	29,801	27,713	25,899	22,900	20,524	18,595	15,652	13,513
80	76,890	74,013	71,343	68,860	66,543	64,377	60,443	56,962	53,859	51,078	48,569	46,296	44,225	39,778	36,144	33,118	30,560	28,368	26,470	23,346	20,881	18,887	15,859	13,667
85	81,498	78,273	75,293	72,532	69,966	67,576	63,254	59,452	56,080	53,071	50,368	47,927	45,712	40,977	37,131	33,945	31,262	28,973	26,995	23,754	21,207	19,153	16,046	13,806
90	86,083	82,493	79,190	76,142	73,319	70,699	65,982	61,855	58,214	54,978	52,083	49,477	47,120	42,105	38,054	34,715	31,914	29,532	27,480	24,128	21,505	19,396	16,216	13,932
95	90,646	86,674	83,036	79,690	76,604	73,748	68,630	64,176	60,266	56,804	53,719	50,951	48,455	43,167	38,920	35,434	32,521	30,051	27,929	24,473	21,779	19,619	16,371	14,046
100	95,188	90,817	86,830	83,179	79,822	76,726	71,202	66,420	62,240	58,555	55,282	52,355	49,723	44,171	39,734	36,108	33,088	30,534	28,346	24,793	22,031	19,823	16,513	14,151
105	99,707	94,922	90,575	86,609	82,976	79,635	73,701	68,589	64,141	60,234	56,776	53,694	50,929	45,120	40,501	36,739	33,617	30,984	28,734	25,089	22,265	20,012	16,644	14,247
110	104,205	98,990	94,272	89,983	86,068	82,479	76,130	70,688	65,973	61,847	58,207	54,972	52,077	46,019	41,223	37,333	34,114	31,405	29,096	25,364	22,482	20,187	16,765	14,335
115	108,681	103,021	97,920	93,302	89,099	85,258	78,492	72,720	67,739	63,397	59,578	56,193	53,171	46,871	41,906	37,892	34,580	31,800	29,434	25,621	22,683	20,349	16,877	14,417
120	113,136	107,015	101,522	96,566	92,071	87,976	80,789	74,688	69,444	64,887	60,892	57,360	54,216	47,681	42,552	38,420	35,019	32,171	29,751	25,861	22,871	20,501	16,981	14,492
125	117,570	110,974	105,078	99,778	94,986	90,634	83,025	76,595	71,089	66,322	62,154	58,479	55,214	48,451	43,165	38,918	35,432	32,520	30,049	26,086	23,047	20,642	17,077	14,563
130	121,983	114,897	108,589	102,938	97,846	93,234	85,202	78,444	72,679	67,703	63,366	59,550	56,168	49,185	43,746	39,390	35,823	32,848	30,330	26,297	23,211	20,774	17,168	14,628
135	126,375	118,785	112,056	106,048	100,652	95,778	87,321	80,237	74,216	69,035	64,530	60,578	57,081	49,884	44,298	39,837	36,192	33,159	30,594	26,496	23,366	20,897	17,252	14,689
140	130,746	122,639	115,479	109,109	103,405	98,268	89,386	81,977	75,702	70,319	65,651	61,564	57,957	50,551	44,823	40,261	36,542	33,452	30,844	26,683	23,511	21,013	17,331	14,747
145	135,096	126,459	118,860	112,122	106,107	100,705	91,398	83,666	77,140	71,559	66,730	62,512	58,796	51,188	45,323	40,664	36,874	33,730	31,080	26,859	23,648	21,123	17,405	14,801
150	139,426	130,245	122,199	115,089	108,760	103,092	93,360	85,307	78,533	72,755	67,770	63,424	59,602	51,797	45,801	41,048	37,189	33,994	31,304	27,026	23,777	21,226	17,475	14,851
155	143,736	133,999	125,497	118,009	111,365	105,429	95,273	86,901	79,882	73,912	68,772	64,301	60,375	52,381	46,256	41,414	37,489	34,244	31,516	27,184	23,900	21,323	17,541	14,899
160	148,026	137,719	128,755	120,886	113,923	107,719	97,139	88,451	81,190	75,030	69,739	65,145	61,119	52,940	46,692	41,762	37,775	34,482	31,717	27,334	24,015	21,415	17,603	14,944
165	152,296	141,408	131,973	123,718	116,435	109,962	98,959	89,958	82,457	76,112	70,673	65,959	61,835	53,476	47,108	42,095	38,047	34,709	31,909	27,476	24,125	21,502	17,662	14,986
170	156,545	145,064	135,152	126,508	118,903	112,161	100,736	91,424	83,688	77,158	71,574	66,744	62,524	53,991	47,507	42,414	38,307	34,925	32,091	27,612	24,229	21,585	17,718	15,026
175	160,775	148,689	138,293	129,256	121,328	114,316	102,471	92,851	84,881	78,172	72,446	67,501	63,188	54,485	47,889	42,718	38,555	35,131	32,266	27,740	24,328	21,664	17,771	15,064
180	164,986	152,284	141,397	131,964	123,710	116,428	104,165	94,239	86,041	79,154	73,289	68,232	63,828	54,961	48,256	43,010	38,792	35,328	32,432	27,863	24,423	21,739	17,821	15,100
185	169,177	155,847	144,464	134,631	126,052	118,500	105,820	95,592	87,167	80,106	74,104	68,938	64,446	55,418	48,608	43,289	39,019	35,516	32,590	27,980	24,513	21,810	17,869	15,135
190	173,348	159,380	147,495	137,260	128,353	120,532	107,438	96,910	88,261	81,030	74,893	69,621	65,042	55,858	48,947	43,558	39,237	35,697	32,742	28,092	24,598	21,878	17,915	15,167
195	177,501	162,884	150,491	139,851	130,615	122,524	109,018	98,194	89,325	81,926	75,658	70,282	65,618	56,283	49,272	43,815	39,446	35,869	32,887	28,199	24,680	21,942	17,958	15,198
200	181,634	166,358	153,452	142,404	132,840	124,480	110,564	99,446	90,360	82,795	76,399	70,921	66,175	56,692	49,586	44,063	39,647	36,035	33,027	28,301	24,759	22,004	17,999	15,228

250 MCM	2 Run(s)	600 Volts	C-Value = 16,483	Magnetic Duct	Copper

Calculations:

1. $f = \dfrac{\text{Length} \times 2 \times \text{Isca}}{\text{Runs} \times \text{C-Value} \times E\ L\text{-}L}$

2. $M = \dfrac{1}{1 + f}$

3. Isc = Isca x M

4. Add Motor Contribution = Motor FLA x 4

* All results are given in symmetrical amperes

Single-Phase L-L Bolted Fault-Current* Table

One-Way Distance in Feet

Available Isc in Thousands (Isca)	5	10	15	20	25	30	40	50	60	70	80	90	100	125	150	175	200	225	250	300	350	400	500	600
3	2,992	2,984	2,975	2,967	2,959	2,951	2,935	2,920	2,904	2,889	2,874	2,858	2,844	2,807	2,771	2,737	2,703	2,670	2,637	2,575	2,516	2,459	2,353	2,255
5	4,977	4,955	4,932	4,910	4,888	4,866	4,823	4,781	4,739	4,698	4,658	4,619	4,580	4,486	4,395	4,309	4,225	4,145	4,068	3,921	3,785	3,658	3,428	3,225
7	6,955	6,911	6,868	6,825	6,782	6,740	6,658	6,578	6,499	6,423	6,348	6,275	6,204	6,032	5,870	5,716	5,570	5,431	5,299	5,054	4,830	4,625	4,263	3,954
10	9,909	9,820	9,732	9,646	9,562	9,479	9,317	9,160	9,009	8,862	8,721	8,583	8,450	8,135	7,843	7,570	7,316	7,079	6,856	6,451	6,091	5,768	5,217	4,761
15	14,796	14,598	14,406	14,218	14,035	13,856	13,513	13,186	12,875	12,578	12,294	12,023	11,764	11,162	10,618	10,126	9,676	9,265	8,888	8,218	7,642	7,142	6,315	5,659
20	19,640	19,292	18,957	18,633	18,320	18,017	17,441	16,901	16,392	15,914	15,463	15,036	14,633	13,713	12,902	12,181	11,537	10,957	10,433	9,522	8,758	8,107	7,057	6,249
25	24,440	23,904	23,391	22,900	22,429	21,977	21,126	20,338	19,606	18,926	18,291	17,697	17,141	15,892	14,813	13,871	13,041	12,306	11,648	10,524	9,598	8,822	7,593	6,665
30	29,197	28,436	27,713	27,026	26,373	25,750	24,589	23,528	22,555	21,659	20,831	20,065	19,353	17,775	16,436	15,284	14,283	13,405	12,629	11,318	10,254	9,373	7,998	6,975
35	33,912	32,889	31,926	31,018	30,160	29,349	27,850	26,496	25,269	24,149	23,125	22,184	21,317	19,419	17,832	16,484	15,326	14,320	13,437	11,963	10,781	9,811	8,315	7,215
40	38,585	37,266	36,035	34,882	33,801	32,785	30,926	29,266	27,775	26,429	25,207	24,093	23,074	20,866	19,044	17,515	16,213	15,091	14,115	12,497	11,212	10,167	8,569	7,405
45	43,217	41,569	40,043	38,625	37,304	36,070	33,832	31,855	30,097	28,523	27,105	25,821	24,654	22,150	20,108	18,411	16,978	15,752	14,691	12,947	11,573	10,463	8,778	7,561
50	47,808	45,800	43,954	42,251	40,676	39,213	36,582	34,282	32,254	30,453	28,842	27,393	26,083	23,297	21,049	19,196	17,643	16,323	15,186	13,330	11,878	10,712	8,953	7,690
55	52,359	49,961	47,772	45,767	43,924	42,223	39,189	36,561	34,264	32,238	30,438	28,829	27,381	24,327	21,886	19,890	18,228	16,822	15,618	13,661	12,140	10,924	9,101	7,799
60	56,871	54,052	51,500	49,177	47,056	45,109	41,663	38,705	36,140	33,894	31,910	30,146	28,567	25,258	22,637	20,509	18,746	17,262	15,996	13,950	12,368	11,108	9,228	7,893
65	61,344	58,077	55,140	52,487	50,077	47,878	44,014	40,726	37,896	35,433	33,271	31,358	29,653	26,104	23,314	21,062	19,207	17,653	16,331	14,204	12,567	11,269	9,339	7,973
70	65,778	62,036	58,697	55,699	52,993	50,537	46,250	42,634	39,543	36,869	34,534	32,477	30,651	26,875	23,927	21,561	19,622	18,002	16,630	14,429	12,743	11,410	9,436	8,044
75	70,174	65,932	62,173	58,819	55,809	53,092	48,382	44,439	41,090	38,211	35,708	33,514	31,573	27,581	24,485	22,013	19,995	18,316	16,897	14,630	12,900	11,535	9,521	8,106
80	74,533	69,765	65,570	61,851	58,532	55,550	50,414	46,148	42,547	39,467	36,803	34,476	32,426	28,229	24,994	22,425	20,334	18,600	17,139	14,811	13,040	11,647	9,597	8,161
85	78,854	73,537	68,891	64,798	61,164	57,916	52,355	47,768	43,921	40,647	37,827	35,373	33,218	28,828	25,462	22,801	20,643	18,858	17,357	14,974	13,166	11,748	9,666	8,210
90	83,139	77,250	72,140	67,664	63,711	60,194	54,210	49,308	45,219	41,756	38,786	36,210	33,955	29,381	25,893	23,146	20,925	19,093	17,556	15,122	13,281	11,839	9,727	8,254
95	87,388	80,905	75,317	70,452	66,176	62,391	55,985	50,772	46,447	42,801	39,686	36,994	34,643	29,895	26,291	23,463	21,184	19,309	17,739	15,257	13,384	11,921	9,783	8,295
100	91,601	84,503	78,426	73,164	68,565	64,509	57,684	52,166	47,611	43,788	40,533	37,728	35,287	30,373	26,660	23,757	21,423	19,507	17,906	15,380	13,479	11,997	9,833	8,331
105	95,778	88,046	81,468	75,806	70,879	66,553	59,314	53,495	48,716	44,720	41,331	38,419	35,890	30,819	27,003	24,028	21,644	19,690	18,060	15,494	13,566	12,066	9,879	8,364
110	99,921	91,535	84,447	78,378	73,122	68,528	60,877	54,763	49,765	45,603	42,084	39,068	36,456	31,235	27,323	24,281	21,849	19,859	18,202	15,599	13,647	12,129	9,922	8,394
115	104,030	94,971	87,363	80,883	75,299	70,435	62,378	55,975	50,764	46,440	42,795	39,681	36,989	31,626	27,621	24,516	22,039	20,016	18,334	15,695	13,721	12,187	9,961	8,422
120	108,105	98,355	90,218	83,325	77,411	72,280	63,820	57,133	51,715	47,235	43,469	40,260	37,492	31,992	27,900	24,736	22,216	20,163	18,457	15,785	13,789	12,241	9,997	8,448
125	112,146	101,689	93,016	85,706	79,461	74,064	65,207	58,242	52,622	47,991	44,109	40,808	37,966	32,337	28,162	24,942	22,382	20,299	18,571	15,869	13,853	12,292	10,030	8,472
130	116,154	104,973	95,756	88,027	81,452	75,792	66,543	59,305	53,488	48,710	44,715	41,326	38,415	32,662	28,408	25,135	22,537	20,427	18,678	15,946	13,912	12,338	10,062	8,494
135	120,129	108,210	98,442	90,291	83,387	77,464	67,828	60,325	54,316	49,395	45,292	41,819	38,840	32,969	28,640	25,316	22,683	20,546	18,777	16,019	13,968	12,382	10,090	8,515
140	124,072	111,399	101,074	92,501	85,269	79,085	69,068	61,303	55,107	50,049	45,842	42,287	39,243	33,259	28,859	25,486	22,820	20,659	18,871	16,087	14,019	12,422	10,117	8,534
145	127,983	114,541	103,655	94,658	87,098	80,656	70,263	62,243	55,866	50,674	46,365	42,732	39,626	33,534	29,065	25,647	22,949	20,764	18,959	16,151	14,068	12,461	10,143	8,552
150	131,863	117,639	106,185	96,763	88,877	82,180	71,417	63,146	56,592	51,271	46,865	43,156	39,991	33,794	29,261	25,800	23,071	20,864	19,042	16,212	14,114	12,496	10,166	8,569
155	135,711	120,692	108,666	98,820	90,609	83,658	72,531	64,016	57,290	51,843	47,342	43,560	40,338	34,042	29,446	25,944	23,186	20,958	19,121	16,268	14,157	12,530	10,189	8,585
160	139,529	123,702	111,100	100,828	92,295	85,094	73,607	64,853	57,959	52,390	47,798	43,946	40,668	34,277	29,622	26,080	23,295	21,047	19,195	16,322	14,197	12,562	10,210	8,600
165	143,316	126,670	113,488	102,791	93,937	86,487	74,648	65,659	58,602	52,916	48,235	44,315	40,984	34,501	29,789	26,209	23,398	21,131	19,265	16,372	14,235	12,592	10,229	8,614
170	147,074	129,596	115,831	104,710	95,537	87,842	75,654	66,437	59,221	53,419	48,653	44,668	41,286	34,715	29,948	26,332	23,496	21,211	19,331	16,420	14,272	12,620	10,248	8,627
175	150,801	132,482	118,131	106,586	97,096	89,158	76,629	67,187	59,816	53,903	49,054	45,005	41,574	34,918	30,100	26,450	23,589	21,287	19,394	16,466	14,306	12,647	10,266	8,639
180	154,499	135,328	120,389	108,420	98,616	90,438	77,572	67,911	60,390	54,369	49,439	45,329	41,850	35,113	30,244	26,561	23,678	21,359	19,454	16,509	14,338	12,672	10,283	8,651
185	158,169	138,134	122,605	110,214	100,098	91,683	78,486	68,611	60,942	54,816	49,809	45,640	42,115	35,299	30,382	26,667	23,762	21,428	19,511	16,550	14,369	12,696	10,299	8,662
190	161,809	140,903	124,781	111,970	101,544	92,894	79,372	69,287	61,475	55,247	50,164	45,938	42,369	35,477	30,514	26,769	23,843	21,493	19,565	16,589	14,399	12,719	10,314	8,673
195	165,421	143,634	126,918	113,687	102,955	94,074	80,232	69,941	61,990	55,662	50,506	46,225	42,612	35,648	30,640	26,866	23,920	21,556	19,617	16,626	14,427	12,741	10,328	8,683
200	169,006	146,329	129,018	115,369	104,332	95,222	81,066	70,574	62,486	56,062	50,835	46,500	42,846	35,811	30,761	26,959	23,993	21,615	19,666	16,662	14,454	12,762	10,342	8,693

300 MCM	1 Run(s)	600 Volts	C-Value = 18,176	Magnetic Duct	Copper

Calculations:

1. $f = \dfrac{\text{Length} \times 2 \times \text{Isca}}{\text{Runs} \times \text{C-Value} \times \text{E L-L}}$
2. $M = \dfrac{1}{1 + f}$
3. $\text{Isc} = \text{Isca} \times M$
4. Add Motor Contribution = Motor FLA x 4

* All results are given in symmetrical amperes

Single-Phase L-L Bolted Fault-Current* Table

One-Way Distance in Feet

Available Isc in Thousands (Isca)	5	10	15	20	25	30	40	50	60	70	80	90	100	125	150	175	200	225	250	300	350	400	500	600
3	2,996	2,992	2,988	2,984	2,980	2,975	2,967	2,959	2,951	2,943	2,935	2,928	2,920	2,900	2,881	2,862	2,844	2,825	2,807	2,771	2,737	2,703	2,637	2,575
5	4,989	4,977	4,966	4,955	4,943	4,932	4,910	4,888	4,866	4,845	4,823	4,802	4,781	4,729	4,678	4,629	4,580	4,532	4,486	4,395	4,309	4,225	4,068	3,921
7	6,978	6,955	6,933	6,911	6,889	6,868	6,825	6,782	6,740	6,699	6,658	6,618	6,578	6,480	6,385	6,293	6,204	6,117	6,032	5,870	5,716	5,570	5,299	5,054
10	9,954	9,909	9,864	9,820	9,776	9,732	9,646	9,562	9,479	9,397	9,317	9,238	9,160	8,972	8,791	8,617	8,450	8,290	8,135	7,843	7,570	7,316	6,856	6,451
15	14,898	14,796	14,697	14,598	14,501	14,406	14,218	14,035	13,856	13,683	13,513	13,348	13,186	12,799	12,435	12,090	11,764	11,455	11,162	10,618	10,126	9,676	8,888	8,218
20	19,818	19,640	19,465	19,292	19,123	18,957	18,633	18,320	18,017	17,725	17,441	17,167	16,901	16,270	15,685	15,141	14,633	14,158	13,713	12,902	12,181	11,537	10,433	9,522
25	24,717	24,440	24,169	23,904	23,645	23,391	22,900	22,429	21,977	21,543	21,126	20,724	20,338	19,432	18,603	17,842	17,141	16,493	15,892	14,813	13,871	13,041	11,648	10,524
30	29,593	29,197	28,811	28,436	28,070	27,713	27,026	26,373	25,750	25,156	24,589	24,047	23,528	22,324	21,237	20,251	19,353	18,531	17,775	16,436	15,284	14,283	12,629	11,318
35	34,447	33,912	33,392	32,889	32,400	31,926	31,018	30,160	29,349	28,579	27,850	27,156	26,496	24,979	23,626	22,412	21,317	20,324	19,419	17,832	16,484	15,326	13,437	11,963
40	39,280	38,585	37,914	37,266	36,640	36,035	34,882	33,801	32,785	31,828	30,926	30,073	29,266	27,426	25,804	24,362	23,074	21,915	20,866	19,044	17,515	16,213	14,115	12,497
45	44,090	43,217	42,377	41,569	40,792	40,043	38,625	37,304	36,070	34,915	33,832	32,814	31,855	29,687	27,796	26,131	24,654	23,335	22,150	20,108	18,411	16,978	14,691	12,947
50	48,879	47,808	46,783	45,800	44,858	43,954	42,251	40,676	39,213	37,852	36,582	35,395	34,282	31,784	29,626	27,742	26,083	24,611	23,297	21,049	19,196	17,643	15,186	13,330
55	53,647	52,359	51,132	49,961	48,842	47,772	45,767	43,924	42,223	40,650	39,189	37,829	36,561	33,734	31,312	29,215	27,381	25,764	24,327	21,886	19,890	18,228	15,618	13,661
60	58,394	56,871	55,426	54,052	52,745	51,500	49,177	47,056	45,109	43,317	41,663	40,130	38,705	35,551	32,872	30,568	28,567	26,811	25,258	22,637	20,509	18,746	15,996	13,950
65	63,119	61,344	59,666	58,077	56,571	55,140	52,487	50,077	47,878	45,865	44,014	42,306	40,726	37,249	34,318	31,815	29,653	27,765	26,104	23,314	21,062	19,207	16,331	14,204
70	67,823	65,778	63,852	62,036	60,320	58,697	55,699	52,993	50,537	48,299	46,250	44,369	42,634	38,838	35,663	32,968	30,651	28,639	26,875	23,927	21,561	19,622	16,630	14,429
75	72,507	70,174	67,987	65,932	63,997	62,173	58,819	55,809	53,092	50,628	48,382	46,326	44,439	40,330	36,917	34,037	31,573	29,442	27,581	24,485	22,013	19,995	16,897	14,630
80	77,170	74,533	72,070	69,765	67,602	65,570	61,851	58,532	55,550	52,858	50,414	48,187	46,148	41,733	38,089	35,030	32,426	30,183	28,229	24,994	22,425	20,334	17,139	14,811
85	81,812	78,854	76,103	73,537	71,138	68,891	64,798	61,164	57,916	54,995	52,355	49,957	47,768	43,054	39,186	35,956	33,218	30,868	28,828	25,462	22,801	20,643	17,357	14,974
90	86,433	83,139	80,086	77,250	74,607	72,140	67,664	63,711	60,194	57,046	54,210	51,643	49,308	44,300	40,216	36,822	33,955	31,503	29,381	25,893	23,146	20,925	17,556	15,122
95	91,035	87,388	84,021	80,905	78,011	75,317	70,452	66,176	62,391	59,014	55,985	53,251	50,772	45,479	41,185	37,632	34,643	32,095	29,895	26,291	23,463	21,184	17,739	15,257
100	95,616	91,601	87,909	84,503	81,351	78,426	73,164	68,565	64,509	60,906	57,684	54,787	52,166	46,594	42,097	38,392	35,287	32,646	30,373	26,660	23,757	21,423	17,906	15,380
105	100,177	95,778	91,749	88,046	84,630	81,468	75,806	70,879	66,553	62,725	59,314	56,254	53,495	47,651	42,959	39,107	35,890	33,162	30,819	27,003	24,028	21,644	18,060	15,494
110	104,719	99,921	95,544	91,535	87,848	84,447	78,378	73,122	68,528	64,476	60,877	57,658	54,763	48,655	43,773	39,781	36,456	33,645	31,235	27,323	24,281	21,849	18,202	15,599
115	109,240	104,030	99,294	94,971	91,008	87,363	80,883	75,299	70,435	66,162	62,378	59,003	55,975	49,609	44,543	40,416	36,989	34,098	31,626	27,621	24,516	22,039	18,334	15,695
120	113,742	108,105	103,000	98,355	94,111	90,218	83,325	77,411	72,280	67,787	63,820	60,292	57,133	50,517	45,274	41,017	37,492	34,525	31,992	27,900	24,736	22,216	18,457	15,785
125	118,225	112,146	106,662	101,689	97,159	93,016	85,706	79,461	74,064	69,354	65,207	61,528	58,242	51,382	45,968	41,586	37,966	34,926	32,337	28,162	24,942	22,382	18,571	15,869
130	122,688	116,154	110,281	104,973	100,153	95,756	88,027	81,452	75,792	70,867	66,543	62,716	59,305	52,208	46,627	42,125	38,415	35,306	32,662	28,408	25,135	22,537	18,678	15,946
135	127,131	120,129	113,858	108,210	103,095	98,442	90,291	83,387	77,464	72,327	67,828	63,857	60,325	52,996	47,255	42,636	38,840	35,665	32,969	28,640	25,316	22,683	18,777	16,019
140	131,556	124,072	117,394	111,399	105,985	101,074	92,501	85,269	79,085	73,738	69,068	64,954	61,303	53,749	47,853	43,123	39,243	36,004	33,259	28,859	25,486	22,820	18,871	16,087
145	135,961	127,983	120,890	114,541	108,826	103,655	94,658	87,098	80,656	75,102	70,263	66,010	62,243	54,471	48,424	43,586	39,626	36,326	33,534	29,065	25,647	22,949	18,959	16,151
150	140,348	131,863	124,346	117,639	111,619	106,185	96,763	88,877	82,180	76,421	71,417	67,027	63,146	55,161	48,969	44,027	39,991	36,632	33,794	29,261	25,800	23,071	19,042	16,212
155	144,716	135,711	127,762	120,692	114,364	108,666	98,820	90,609	83,658	77,698	72,531	68,008	64,016	55,823	49,490	44,448	40,338	36,923	34,042	29,446	25,944	23,186	19,121	16,268
160	149,065	139,529	131,140	123,702	117,063	111,100	100,828	92,295	85,094	78,935	73,607	68,953	64,853	56,459	49,989	44,849	40,668	37,200	34,277	29,622	26,080	23,295	19,195	16,322
165	153,396	143,316	134,480	126,670	119,717	113,488	102,791	93,937	86,487	80,133	74,648	69,865	65,659	57,069	50,467	45,234	40,984	37,464	34,501	29,789	26,209	23,398	19,265	16,372
170	157,708	147,074	137,783	129,596	122,328	115,831	104,710	95,537	87,842	81,294	75,654	70,746	66,437	57,656	50,925	45,601	41,286	37,716	34,715	29,948	26,332	23,496	19,331	16,420
175	162,002	150,801	141,049	132,482	124,896	118,131	106,586	97,096	89,158	82,420	76,629	71,598	67,187	58,220	51,365	45,954	41,574	37,957	34,918	30,100	26,450	23,589	19,394	16,466
180	166,278	154,499	144,279	135,328	127,422	120,389	108,420	98,616	90,438	83,512	77,572	72,421	67,911	58,763	51,787	46,291	41,850	38,187	35,113	30,244	26,561	23,678	19,454	16,509
185	170,535	158,169	147,474	138,134	129,907	122,605	110,214	100,098	91,683	84,573	78,486	73,217	68,611	59,286	52,193	46,615	42,115	38,407	35,299	30,382	26,667	23,762	19,511	16,550
190	174,775	161,809	150,634	140,903	132,353	124,781	111,970	101,544	92,894	85,603	79,372	73,987	69,287	59,790	52,583	46,926	42,369	38,618	35,477	30,514	26,769	23,843	19,565	16,589
195	178,997	165,421	153,760	143,634	134,760	126,918	113,687	102,955	94,074	86,603	80,232	74,734	69,941	60,277	52,959	47,225	42,612	38,820	35,648	30,640	26,866	23,920	19,617	16,626
200	183,201	169,006	156,852	146,329	137,129	129,018	115,369	104,332	95,222	87,575	81,066	75,457	70,574	60,746	53,321	47,513	42,846	39,014	35,811	30,761	26,959	23,993	19,666	16,662

300 MCM	2 Run(s)	600 Volts	C-Value = 18,176	Magnetic Duct	Copper

Calculations:

1. $f = \dfrac{\text{Length} \times 2 \times \text{Isca}}{\text{Runs} \times \text{C-Value} \times \text{E L-L}}$
2. $M = \dfrac{1}{1 + f}$
3. $\text{Isc} = \text{Isca} \times M$
4. Add Motor Contribution = Motor FLA x 4

* All results are given in symmetrical amperes

Single-Phase L-L Bolted Fault-Current* Table

One-Way Distance in Feet

Available Isc in Thousands (Isca)

Isca	5	10	15	20	25	30	40	50	60	70	80	90	100	125	150	175	200	225	250	300	350	400	500	600
3	2,992	2,985	2,977	2,970	2,962	2,955	2,940	2,926	2,911	2,897	2,883	2,869	2,855	2,821	2,788	2,755	2,724	2,693	2,662	2,604	2,547	2,494	2,393	2,300
5	4,979	4,958	4,937	4,917	4,896	4,876	4,836	4,797	4,758	4,720	4,683	4,646	4,610	4,522	4,437	4,355	4,277	4,201	4,127	3,988	3,858	3,736	3,514	3,317
7	6,959	6,918	6,878	6,838	6,799	6,760	6,683	6,609	6,536	6,464	6,394	6,326	6,259	6,097	5,944	5,798	5,660	5,527	5,401	5,165	4,949	4,750	4,397	4,092
10	9,916	9,834	9,753	9,673	9,594	9,517	9,366	9,220	9,078	8,941	8,808	8,679	8,553	8,254	7,976	7,716	7,472	7,243	7,028	6,633	6,281	5,964	5,417	4,963
15	14,812	14,629	14,450	14,275	14,105	13,939	13,618	13,311	13,018	12,737	12,469	12,211	11,964	11,388	10,864	10,387	9,950	9,548	9,178	8,516	7,944	7,444	6,611	5,946
20	19,667	19,345	19,034	18,732	18,440	18,157	17,616	17,106	16,625	16,170	15,740	15,331	14,944	14,055	13,267	12,562	11,928	11,355	10,835	9,925	9,156	8,498	7,430	6,600
25	24,482	23,986	23,509	23,050	22,609	22,185	21,383	20,636	19,940	19,289	18,680	18,107	17,569	16,354	15,296	14,367	13,544	12,810	12,151	11,019	10,079	9,288	8,026	7,067
30	29,258	28,551	27,878	27,235	26,622	26,036	24,937	23,928	22,997	22,136	21,337	20,593	19,900	18,355	17,033	15,888	14,888	14,006	13,223	11,892	10,805	9,901	8,480	7,416
35	33,994	33,043	32,145	31,294	30,487	29,721	28,298	27,005	25,825	24,744	23,750	22,832	21,983	20,113	18,536	17,189	16,024	15,007	14,111	12,606	11,392	10,390	8,837	7,688
40	38,691	37,465	36,314	35,232	34,212	33,250	31,479	29,887	28,449	27,143	25,951	24,859	23,856	21,670	19,850	18,313	16,996	15,857	14,860	13,201	11,875	10,791	9,125	7,905
45	43,350	41,816	40,388	39,054	37,805	36,633	34,495	32,593	30,890	29,356	27,967	26,703	25,549	23,058	21,009	19,294	17,839	16,587	15,500	13,703	12,280	11,124	9,362	8,082
50	47,971	46,100	44,370	42,765	41,272	39,880	37,359	35,138	33,167	31,405	29,820	28,388	27,087	24,303	22,038	20,159	18,575	17,222	16,053	14,134	12,624	11,406	9,561	8,230
55	52,555	50,318	48,264	46,371	44,620	42,997	40,082	37,536	35,295	33,306	31,530	29,933	28,490	25,426	22,958	20,926	19,224	17,779	16,535	14,506	12,921	11,648	9,730	8,355
60	57,102	54,471	52,072	49,875	47,856	45,994	42,673	39,800	37,289	35,076	33,112	31,355	29,776	26,445	23,785	21,611	19,801	18,271	16,960	14,832	13,179	11,857	9,876	8,462
65	61,612	58,560	55,796	53,282	50,984	48,876	45,143	41,940	39,161	36,728	34,579	32,668	30,957	27,373	24,533	22,227	20,317	18,709	17,337	15,120	13,405	12,040	10,003	8,555
70	66,087	62,588	59,441	56,595	54,010	51,650	47,499	43,966	40,922	38,273	35,945	33,885	32,048	28,222	25,213	22,783	20,781	19,102	17,674	15,375	13,606	12,201	10,114	8,636
75	70,526	66,555	63,008	59,820	56,939	54,322	49,750	45,888	42,582	39,721	37,219	35,015	33,057	29,002	25,833	23,288	21,200	19,456	17,977	15,604	13,784	12,345	10,212	8,708
80	74,929	70,463	66,500	62,958	59,775	56,898	51,902	47,712	44,149	41,080	38,411	36,067	33,993	29,720	26,401	23,749	21,582	19,776	18,250	15,809	13,945	12,473	10,300	8,771
85	79,298	74,314	69,918	66,014	62,523	59,382	53,961	49,447	45,630	42,360	39,527	37,050	34,864	30,384	26,924	24,172	21,930	20,068	18,498	15,995	14,089	12,589	10,378	8,828
90	83,633	78,107	73,267	68,991	65,187	61,780	55,934	51,098	47,033	43,566	40,575	37,969	35,677	31,000	27,406	24,560	22,248	20,335	18,725	16,164	14,220	12,693	10,449	8,880
95	87,934	81,846	76,546	71,891	67,770	64,096	57,825	52,672	48,363	44,705	41,562	38,831	36,438	31,572	27,853	24,917	22,542	20,580	18,932	16,319	14,339	12,788	10,514	8,926
100	92,201	85,530	79,760	74,718	70,277	66,333	59,640	54,174	49,626	45,782	42,491	39,641	37,150	32,105	28,267	25,248	22,812	20,805	19,122	16,460	14,448	12,875	10,572	8,968
105	96,435	89,162	82,908	77,475	72,710	68,497	61,384	55,609	50,827	46,803	43,369	40,404	37,819	32,604	28,653	25,556	23,063	21,013	19,298	16,590	14,548	12,954	10,625	9,006
110	100,636	92,741	85,995	80,164	75,073	70,590	63,059	56,981	51,971	47,770	44,198	41,123	38,449	33,071	29,013	25,842	23,296	21,206	19,461	16,710	14,641	13,027	10,675	9,042
115	104,805	96,270	89,021	82,787	77,369	72,616	64,671	58,293	53,061	48,690	44,984	41,803	39,042	33,509	29,349	26,108	23,512	21,385	19,612	16,821	14,726	13,095	10,720	9,074
120	108,942	99,749	91,988	85,347	79,600	74,578	66,223	59,551	54,101	49,564	45,730	42,446	39,602	33,920	29,665	26,358	23,714	21,552	19,752	16,924	14,805	13,157	10,762	9,104
125	113,047	103,180	94,898	87,846	81,770	76,480	67,718	60,757	55,094	50,397	46,438	43,055	40,132	34,308	29,961	26,591	23,903	21,708	19,883	17,020	14,878	13,215	10,800	9,132
130	117,121	106,563	97,752	90,286	83,880	78,323	69,159	61,915	56,044	51,191	47,111	43,633	40,634	34,674	30,240	26,811	24,080	21,854	20,005	17,110	14,947	13,269	10,836	9,158
135	121,164	109,900	100,552	92,670	85,934	80,110	70,549	63,026	56,954	51,948	47,752	44,182	41,109	35,020	30,502	27,017	24,246	21,991	20,120	17,194	15,010	13,319	10,870	9,181
140	125,176	113,191	103,300	94,999	87,933	81,845	71,891	64,095	57,825	52,672	48,363	44,705	41,561	35,348	30,751	27,211	24,403	22,120	20,228	17,272	15,070	13,366	10,901	9,204
145	129,158	116,437	105,997	97,275	89,879	83,529	73,187	65,123	58,660	53,364	48,946	45,203	41,991	35,658	30,985	27,395	24,551	22,241	20,329	17,346	15,126	13,411	10,931	9,225
150	133,110	119,639	108,644	99,500	91,776	85,164	74,439	66,113	59,462	54,027	49,503	45,677	42,401	35,953	31,208	27,569	24,690	22,355	20,424	17,415	15,179	13,452	10,958	9,244
155	137,033	122,799	111,243	101,676	93,623	86,753	75,650	67,067	60,232	54,662	50,035	46,130	42,791	36,233	31,418	27,733	24,822	22,463	20,514	17,481	15,229	13,491	10,984	9,263
160	140,927	125,916	113,796	103,804	95,425	88,297	76,822	67,986	60,973	55,271	50,545	46,563	43,163	36,500	31,619	27,889	24,946	22,566	20,600	17,543	15,276	13,528	11,008	9,280
165	144,791	128,992	116,302	105,885	97,181	89,799	77,956	68,873	61,685	55,856	51,034	46,978	43,519	36,754	31,809	28,037	25,065	22,662	20,680	17,601	15,320	13,563	11,031	9,296
170	148,627	132,028	118,764	107,922	98,894	91,260	79,054	69,729	62,371	56,418	51,502	47,374	43,859	36,996	31,991	28,178	25,177	22,754	20,757	17,657	15,362	13,595	11,053	9,312
175	152,435	135,024	121,183	109,916	100,566	92,681	80,119	70,556	63,032	56,958	51,952	47,755	44,185	37,228	32,164	28,312	25,284	22,842	20,829	17,709	15,402	13,627	11,074	9,326
180	156,215	137,982	123,560	111,868	102,197	94,065	81,151	71,355	63,669	57,478	52,384	48,119	44,497	37,449	32,329	28,440	25,386	22,925	20,899	17,759	15,440	13,656	11,093	9,340
185	159,967	140,901	125,896	113,779	103,790	95,413	82,152	72,127	64,283	57,978	52,799	48,470	44,796	37,661	32,486	28,562	25,483	23,004	20,964	17,807	15,476	13,684	11,112	9,353
190	163,691	143,783	128,191	115,651	105,345	96,726	83,123	72,875	64,877	58,460	53,199	48,806	45,084	37,864	32,637	28,678	25,576	23,080	21,027	17,852	15,510	13,711	11,129	9,366
195	167,389	146,628	130,448	117,484	106,864	98,005	84,066	73,599	65,450	58,925	53,583	49,130	45,359	38,058	32,781	28,790	25,665	23,152	21,087	17,895	15,542	13,736	11,146	9,378
200	171,060	149,437	132,667	119,281	108,349	99,252	84,982	74,300	66,003	59,374	53,954	49,441	45,625	38,245	32,920	28,896	25,749	23,221	21,144	17,936	15,573	13,761	11,162	9,389

350 MCM | 1 Run(s) | 600 Volts | C-Value = 19,703 | Magnetic Duct | Copper

Calculations:

1. $f = \dfrac{\text{Length} \times 2 \times \text{Isca}}{\text{Runs} \times \text{C-Value} \times \text{E L-L}}$

2. $M = \dfrac{1}{1 + f}$

3. Isc = Isca x M

4. Add Motor Contribution = Motor FLA x 4

* All results are given in symmetrical amperes

Single-Phase L-L Bolted Fault-Current* Table

One-Way Distance in Feet

Available Isc in Thousands (Isca)	5	10	15	20	25	30	40	50	60	70	80	90	100	125	150	175	200	225	250	300	350	400	500	600
3	2,996	2,992	2,989	2,985	2,981	2,977	2,970	2,962	2,955	2,948	2,940	2,933	2,926	2,908	2,890	2,872	2,855	2,838	2,821	2,788	2,755	2,724	2,662	2,604
5	4,989	4,979	4,968	4,958	4,948	4,937	4,917	4,896	4,876	4,856	4,836	4,817	4,797	4,749	4,702	4,655	4,610	4,566	4,522	4,437	4,355	4,277	4,127	3,988
7	6,979	6,959	6,938	6,918	6,898	6,878	6,838	6,799	6,760	6,721	6,683	6,646	6,609	6,518	6,429	6,343	6,259	6,177	6,097	5,944	5,798	5,660	5,401	5,165
10	9,958	9,916	9,875	9,834	9,793	9,753	9,673	9,594	9,517	9,441	9,366	9,293	9,220	9,044	8,874	8,711	8,553	8,401	8,254	7,976	7,716	7,472	7,028	6,633
15	14,905	14,812	14,720	14,629	14,539	14,450	14,275	14,105	13,939	13,776	13,618	13,463	13,311	12,947	12,602	12,274	11,964	11,669	11,388	10,864	10,387	9,950	9,178	8,516
20	19,832	19,667	19,505	19,345	19,188	19,034	18,732	18,440	18,157	17,882	17,616	17,357	17,106	16,509	15,952	15,431	14,944	14,486	14,055	13,267	12,562	11,928	10,835	9,925
25	24,738	24,482	24,231	23,986	23,745	23,509	23,050	22,609	22,185	21,776	21,383	21,003	20,636	19,773	18,980	18,247	17,569	16,940	16,354	15,296	14,367	13,544	12,151	11,019
30	29,624	29,258	28,900	28,551	28,210	27,878	27,235	26,622	26,036	25,475	24,937	24,422	23,928	22,775	21,729	20,774	19,900	19,096	18,355	17,033	15,888	14,888	13,223	11,892
35	34,489	33,994	33,512	33,043	32,588	32,145	31,294	30,487	29,721	28,992	28,298	27,636	27,005	25,546	24,237	23,055	21,983	21,007	20,113	18,536	17,189	16,024	14,111	12,606
40	39,335	38,691	38,068	37,465	36,880	36,314	35,232	34,212	33,250	32,340	31,479	30,663	29,887	28,111	26,533	25,124	23,856	22,710	21,670	19,850	18,313	16,996	14,860	13,201
45	44,160	43,350	42,569	41,816	41,090	40,388	39,054	37,805	36,633	35,532	34,495	33,517	32,593	30,492	28,645	27,008	25,549	24,240	23,058	21,009	19,294	17,839	15,500	13,703
50	48,965	47,971	47,017	46,100	45,219	44,370	42,765	41,272	39,880	38,578	37,359	36,215	35,138	32,708	30,592	28,733	27,087	25,620	24,303	22,038	20,159	18,575	16,053	14,134
55	53,750	52,555	51,412	50,318	49,269	48,264	46,371	44,620	42,997	41,488	40,082	38,767	37,536	34,776	32,394	30,317	28,490	26,871	25,426	22,958	20,926	19,224	16,535	14,506
60	58,515	57,102	55,755	54,471	53,244	52,072	49,875	47,856	45,994	44,271	42,673	41,187	39,800	36,710	34,066	31,776	29,776	28,012	26,445	23,785	21,611	19,801	16,960	14,832
65	63,261	61,612	60,048	58,560	57,145	55,796	53,282	50,984	48,876	46,935	45,143	43,483	41,940	38,523	35,621	33,126	30,957	29,055	27,373	24,533	22,227	20,317	17,337	15,120
70	67,987	66,087	64,290	62,588	60,974	59,441	56,595	54,010	51,650	49,488	47,499	45,665	43,966	40,226	37,073	34,377	32,048	30,013	28,222	25,213	22,783	20,781	17,674	15,375
75	72,694	70,526	68,483	66,555	64,733	63,008	59,820	56,939	54,322	51,936	49,750	47,741	45,888	41,829	38,429	35,541	33,057	30,897	29,002	25,833	23,288	21,200	17,977	15,604
80	77,382	74,929	72,628	70,463	68,424	66,500	62,958	59,775	56,898	54,285	51,902	49,719	47,712	43,339	39,701	36,626	33,993	31,713	29,720	26,401	23,749	21,582	18,250	15,809
85	82,050	79,298	76,725	74,314	72,049	69,918	66,014	62,523	59,382	56,542	53,961	51,606	49,447	44,766	40,895	37,639	34,864	32,470	30,384	26,924	24,172	21,930	18,498	15,995
90	86,700	83,633	80,776	78,107	75,610	73,267	68,991	65,187	61,780	58,712	55,934	53,407	51,098	46,115	42,018	38,589	35,677	33,174	31,000	27,406	24,560	22,248	18,725	16,164
95	91,330	87,934	84,781	81,846	79,107	76,546	71,891	67,770	64,096	60,799	57,825	55,129	52,672	47,393	43,076	39,480	36,438	33,831	31,572	27,853	24,917	22,542	18,932	16,319
100	95,942	92,201	88,740	85,530	82,544	79,760	74,718	70,277	66,333	62,809	59,640	56,776	54,174	48,606	44,075	40,317	37,150	34,444	32,105	28,267	25,248	22,812	19,122	16,460
105	100,535	96,435	92,656	89,162	85,921	82,908	77,475	72,710	68,497	64,746	61,384	58,354	55,609	49,757	45,020	41,107	37,819	35,018	32,604	28,653	25,556	23,063	19,298	16,590
110	105,110	100,636	96,527	92,741	89,241	85,995	80,164	75,073	70,590	66,613	63,059	59,866	56,981	50,853	45,915	41,851	38,449	35,557	33,071	29,013	25,842	23,296	19,461	16,710
115	109,666	104,805	100,356	96,270	92,504	89,021	82,787	77,369	72,616	68,414	64,671	61,317	58,293	51,896	46,764	42,555	39,042	36,064	33,509	29,349	26,108	23,512	19,612	16,821
120	114,204	108,942	104,143	99,749	95,711	91,988	85,347	79,600	74,578	70,153	66,223	62,710	59,551	52,890	47,570	43,222	39,602	36,542	33,920	29,665	26,358	23,714	19,752	16,924
125	118,723	113,047	107,888	103,180	98,866	94,898	87,846	81,770	76,480	71,833	67,718	64,049	60,757	53,840	48,336	43,854	40,132	36,992	34,308	29,961	26,591	23,903	19,883	17,020
130	123,225	117,121	111,593	106,563	101,968	97,752	90,286	83,880	78,323	73,456	69,159	65,337	61,915	54,747	49,066	44,453	40,634	37,418	34,674	30,240	26,811	24,080	20,005	17,110
135	127,708	121,164	115,257	109,900	105,018	100,552	92,670	85,934	80,110	75,026	70,549	66,576	63,026	55,614	49,762	45,024	41,109	37,821	35,020	30,502	27,017	24,246	20,120	17,194
140	132,174	125,176	118,882	113,191	108,019	103,300	94,999	87,933	81,845	76,546	71,891	67,769	64,095	56,444	50,425	45,566	41,561	38,204	35,348	30,751	27,211	24,403	20,228	17,272
145	136,621	129,158	122,468	116,437	110,972	105,997	97,275	89,879	83,529	78,016	73,187	68,920	65,123	57,240	51,060	46,084	41,991	38,567	35,658	30,985	27,395	24,551	20,329	17,346
150	141,051	133,110	126,016	119,639	113,877	108,644	99,500	91,776	85,164	79,441	74,439	70,029	66,113	58,003	51,666	46,577	42,401	38,912	35,953	31,208	27,569	24,690	20,424	17,415
155	145,464	137,033	129,526	122,799	116,736	111,243	101,676	93,623	86,753	80,822	75,650	71,100	67,067	58,736	52,247	47,048	42,791	39,240	36,233	31,418	27,733	24,822	20,514	17,481
160	149,859	140,927	132,999	125,916	119,549	113,796	103,804	95,425	88,297	82,161	76,822	72,134	67,986	59,440	52,803	47,499	43,163	39,553	36,500	31,619	27,889	24,946	20,600	17,543
165	154,236	144,791	136,436	128,992	122,319	116,302	105,885	97,181	89,799	83,459	77,956	73,133	68,873	60,117	53,336	47,930	43,519	39,851	36,754	31,809	28,037	25,065	20,680	17,601
170	158,597	148,627	139,837	132,028	125,045	118,764	107,922	98,894	91,260	84,720	79,054	74,099	69,729	60,768	53,848	48,343	43,859	40,136	36,996	31,991	28,178	25,177	20,757	17,657
175	162,940	152,435	143,202	135,024	127,730	121,183	109,916	100,566	92,681	85,944	80,119	75,034	70,556	61,395	54,340	48,739	44,185	40,409	37,228	32,164	28,312	25,284	20,829	17,709
180	167,266	156,215	146,533	137,982	130,373	123,560	111,868	102,197	94,065	87,132	81,151	75,938	71,355	61,999	54,813	49,119	44,497	40,670	37,449	32,329	28,440	25,386	20,899	17,759
185	171,575	159,967	149,830	140,901	132,976	125,896	113,779	103,790	95,413	88,287	82,152	76,814	72,127	62,582	55,267	49,484	44,796	40,920	37,661	32,486	28,562	25,483	20,964	17,807
190	175,867	163,691	153,092	143,783	135,540	128,191	115,651	105,345	96,726	89,410	83,123	77,663	72,875	63,144	55,705	49,835	45,084	41,159	37,864	32,637	28,678	25,576	21,027	17,852
195	180,143	167,389	156,322	146,628	138,065	130,448	117,484	106,864	98,005	90,502	84,066	78,485	73,599	63,687	56,127	50,172	45,359	41,389	38,058	32,781	28,790	25,665	21,087	17,895
200	184,402	171,060	159,519	149,437	140,553	132,667	119,281	108,349	99,252	91,564	84,982	79,283	74,300	64,211	56,534	50,497	45,625	41,610	38,245	32,920	28,896	25,749	21,144	17,936

350 MCM | 2 Run(s) | 600 Volts

C-Value = 19,703 | Magnetic Duct | Copper

Calculations:

1. $f = \dfrac{\text{Length} \times 2 \times \text{Isca}}{\text{Runs} \times \text{C-Value} \times \text{E L-L}}$

2. $M = \dfrac{1}{1 + f}$

3. $\text{Isc} = \text{Isca} \times M$

4. Add Motor Contribution = Motor FLA x 4

* All results are given in symmetrical amperes

Single-Phase L-L Bolted Fault-Current* Table

One-Way Distance in Feet

Available Isc in Thousands (Isca)

	5	10	15	20	25	30	40	50	60	70	80	90	100	125	150	175	200	225	250	300	350	400	500	600
3	2,993	2,985	2,978	2,971	2,964	2,957	2,943	2,929	2,915	2,901	2,888	2,874	2,861	2,828	2,796	2,765	2,734	2,704	2,675	2,618	2,564	2,512	2,413	2,322
5	4,980	4,960	4,940	4,920	4,901	4,881	4,843	4,805	4,768	4,732	4,696	4,660	4,625	4,540	4,458	4,379	4,303	4,229	4,158	4,022	3,895	3,776	3,558	3,364
7	6,961	6,921	6,883	6,845	6,807	6,770	6,696	6,624	6,554	6,485	6,417	6,351	6,287	6,131	5,982	5,840	5,705	5,576	5,453	5,222	5,010	4,815	4,466	4,165
10	9,920	9,840	9,763	9,686	9,611	9,536	9,391	9,250	9,114	8,981	8,852	8,727	8,605	8,315	8,044	7,790	7,552	7,328	7,116	6,728	6,380	6,067	5,524	5,070
15	14,820	14,644	14,472	14,304	14,140	13,980	13,671	13,374	13,090	12,818	12,558	12,307	12,066	11,504	10,991	10,523	10,092	9,696	9,329	8,674	8,104	7,604	6,770	6,101
20	19,681	19,372	19,073	18,782	18,501	18,227	17,704	17,210	16,743	16,301	15,881	15,483	15,104	14,233	13,457	12,761	12,133	11,565	11,047	10,139	9,369	8,708	7,631	6,791
25	24,504	24,026	23,567	23,126	22,700	22,290	21,513	20,788	20,111	19,476	18,880	18,319	17,791	16,594	15,549	14,627	13,809	13,077	12,419	11,283	10,338	9,539	8,261	7,286
30	29,288	28,609	27,961	27,341	26,748	26,181	25,115	24,133	23,224	22,382	21,598	20,868	20,185	18,659	17,347	16,208	15,209	14,326	13,540	12,201	11,103	10,187	8,743	7,658
35	34,035	33,121	32,255	31,434	30,653	29,910	28,527	27,266	26,112	25,052	24,074	23,170	22,331	20,478	18,909	17,563	16,396	15,375	14,473	12,954	11,723	10,706	9,123	7,948
40	38,744	37,565	36,455	35,409	34,421	33,487	31,763	30,207	28,797	27,513	26,339	25,260	24,267	22,094	20,279	18,739	17,416	16,268	15,262	13,582	12,235	11,132	9,430	8,180
45	43,417	41,941	40,562	39,271	38,060	36,921	34,836	32,974	31,301	29,790	28,418	27,166	26,021	23,539	21,489	19,763	18,302	17,038	15,938	14,115	12,666	11,487	9,684	8,370
50	48,053	46,252	44,581	43,026	41,576	40,221	37,759	35,582	33,641	31,902	30,333	28,912	27,618	24,838	22,567	20,673	19,078	17,709	16,523	14,572	13,033	11,788	9,897	8,529
55	52,653	50,498	48,513	46,678	44,976	43,394	40,543	38,043	35,833	33,866	32,104	30,516	29,078	26,013	23,532	21,484	19,763	18,298	17,035	14,968	13,349	12,046	10,078	8,663
60	57,218	54,682	52,362	50,230	48,265	46,448	43,196	40,370	37,890	35,698	33,745	31,995	30,418	27,080	24,402	22,206	20,373	18,819	17,486	15,316	13,624	12,270	10,234	8,778
65	61,747	58,805	56,130	53,687	51,449	49,389	45,729	42,573	39,825	37,410	35,271	33,364	31,652	28,054	25,190	22,857	20,920	19,285	17,887	15,622	13,867	12,466	10,370	8,878
70	66,242	62,867	59,819	57,053	54,532	52,224	48,148	44,663	41,648	39,014	36,694	34,634	32,793	28,946	25,907	23,446	21,412	19,702	18,246	15,895	14,081	12,639	10,490	8,966
75	70,702	66,871	63,433	60,332	57,519	54,957	50,462	46,647	43,368	40,520	38,022	35,815	33,850	29,767	26,563	23,982	21,858	20,079	18,568	16,140	14,273	12,793	10,596	9,043
80	75,129	70,817	66,973	63,525	60,415	57,595	52,677	48,533	44,994	41,936	39,266	36,917	34,833	30,524	27,164	24,471	22,263	20,421	18,860	16,360	14,444	12,931	10,690	9,111
85	79,522	74,707	70,442	66,638	63,224	60,142	54,800	50,329	46,533	43,270	40,434	37,947	35,748	31,225	27,718	24,919	22,634	20,732	19,125	16,559	14,600	13,055	10,775	9,173
90	83,882	78,542	73,842	69,673	65,949	62,603	56,836	52,041	47,993	44,529	41,531	38,912	36,603	31,875	28,229	25,332	22,973	21,017	19,367	16,740	14,740	13,167	10,851	9,228
95	88,209	82,324	77,175	72,632	68,594	64,982	58,790	53,675	49,379	45,720	42,565	39,818	37,404	32,481	28,703	25,712	23,286	21,278	19,589	16,905	14,868	13,269	10,921	9,278
100	92,503	86,052	80,442	75,519	71,163	67,283	60,667	55,235	50,696	46,847	43,541	40,670	38,155	33,046	29,143	26,065	23,575	21,519	19,793	7,057	14,986	13,363	10,984	9,324
105	96,766	89,729	83,646	78,336	73,659	69,510	62,471	56,727	51,951	47,916	44,463	41,474	38,861	33,574	29,553	26,393	23,843	21,742	19,982	7,197	15,093	13,448	11,041	9,365
110	100,996	93,355	86,789	81,085	76,086	71,666	64,208	58,155	53,146	48,931	45,335	42,232	39,526	34,069	29,936	26,698	24,091	21,949	20,156	7,326	15,193	13,527	11,094	9,403
115	105,196	96,932	89,872	83,770	78,445	73,756	65,880	59,524	54,286	49,896	46,162	42,949	40,154	34,534	30,295	26,983	24,323	22,141	20,318	7,445	15,284	13,600	11,143	9,439
120	109,364	100,460	92,897	86,392	80,739	75,781	67,491	60,836	55,375	50,814	46,948	43,628	40,746	34,972	30,631	27,249	24,539	22,320	20,469	7,556	15,369	13,667	11,189	9,471
125	113,502	103,941	95,865	88,954	82,972	77,745	69,044	62,095	56,417	51,690	47,694	44,272	41,307	35,385	30,947	27,499	24,742	22,487	20,609	17,660	15,449	13,730	11,230	9,501
130	117,609	107,375	98,779	91,457	85,146	79,650	70,543	63,304	57,413	52,525	48,404	44,883	41,839	35,774	31,245	27,733	24,932	22,644	20,741	17,756	15,522	13,788	11,269	9,529
135	121,686	110,763	101,639	93,904	87,263	81,499	71,989	64,467	58,368	53,323	49,081	45,464	42,344	36,142	31,525	27,954	25,110	22,791	20,864	17,846	15,591	13,842	11,306	9,555
140	125,734	114,107	104,448	96,296	89,325	83,295	73,387	65,586	59,283	54,086	49,727	46,018	42,824	36,491	31,790	28,163	25,278	22,929	20,980	17,931	15,656	13,893	11,340	9,579
145	129,752	117,406	107,206	98,636	91,335	85,040	74,738	66,663	60,162	54,816	50,343	46,545	43,280	36,822	32,041	28,359	25,436	23,059	21,089	18,011	15,717	13,941	11,371	9,602
150	133,742	120,663	109,914	100,924	93,294	86,736	76,045	67,700	61,006	55,516	50,933	47,049	43,715	37,137	32,279	28,545	25,586	23,182	21,192	18,085	15,774	13,986	11,401	9,623
155	137,702	123,878	112,575	103,163	95,204	88,384	77,309	68,700	61,817	56,187	51,497	47,530	44,130	37,436	32,505	28,722	25,727	23,298	21,289	18,156	15,827	14,028	11,429	9,643
160	141,634	127,051	115,190	105,355	97,067	89,988	78,533	69,665	62,597	56,831	52,037	47,989	44,526	37,720	32,719	28,839	25,861	23,408	21,380	18,223	15,878	14,068	11,456	9,662
165	145,538	130,183	117,759	107,500	98,885	91,548	79,719	70,597	63,348	57,449	52,555	48,430	44,905	37,992	32,923	29,048	25,989	23,513	21,467	18,286	15,926	14,105	11,480	9,679
170	149,414	133,276	120,284	109,600	100,659	93,067	80,868	71,496	64,071	58,043	53,052	48,851	45,267	38,251	33,118	29,199	26,110	23,612	21,550	18,346	15,971	14,141	11,504	9,696
175	153,263	136,330	122,766	111,656	102,391	94,546	81,982	72,366	64,769	58,615	53,529	49,256	45,614	38,498	33,303	29,343	26,225	23,706	21,628	18,402	16,014	14,174	11,526	9,712
180	157,085	139,345	125,205	113,671	104,083	95,986	83,063	73,207	65,442	59,166	53,988	49,644	45,947	38,735	33,480	29,480	26,334	23,795	21,703	18,456	16,055	14,206	11,547	9,727
185	160,879	142,323	127,604	115,645	105,735	97,390	84,112	74,020	66,091	59,696	54,429	50,017	46,266	38,961	33,649	29,611	26,439	23,881	21,774	18,508	16,094	14,237	11,567	9,741
190	164,647	145,264	129,963	117,579	107,350	98,758	85,131	74,808	66,718	60,207	54,854	50,375	46,572	39,179	33,811	29,737	26,539	23,962	21,841	18,557	16,131	14,266	11,587	9,755
195	168,389	148,168	132,284	119,475	108,928	100,092	86,120	75,571	67,324	60,700	55,263	50,720	46,867	39,387	33,966	29,856	26,634	24,040	21,906	18,603	16,166	14,293	11,605	9,767
200	172,104	151,037	134,566	121,333	110,470	101,393	87,081	76,310	67,911	61,177	55,658	51,052	47,150	39,587	34,114	29,971	26,725	24,114	21,967	18,648	16,199	14,319	11,622	9,780

400 MCM | 1 Run(s) | 600 Volts | | C-Value = 20,565 | Magnetic Duct | Copper

Calculations:

1. $f = \dfrac{\text{Length} \times 2 \times \text{Isca}}{\text{Runs} \times \text{C-Value} \times \text{E L-L}}$

2. $M = \dfrac{1}{1 + f}$

3. Isc = Isca x M

4. Add Motor Contribution = Motor FLA x 4

* All results are given in symmetrical amperes

Single-Phase L-L Bolted Fault-Current* Table

One-Way Distance in Feet

Available Isc in Thousands (Isca)	5	10	15	20	25	30	40	50	60	70	80	90	100	125	150	175	200	225	250	300	350	400	500	600
3	2,996	2,993	2,989	2,985	2,982	2,978	2,971	2,964	2,957	2,950	2,943	2,936	2,929	2,912	2,894	2,878	2,861	2,844	2,828	2,796	2,765	2,734	2,675	2,618
5	4,990	4,980	4,970	4,960	4,950	4,940	4,920	4,901	4,881	4,862	4,843	4,824	4,805	4,759	4,713	4,669	4,625	4,582	4,540	4,458	4,379	4,303	4,158	4,022
7	6,980	6,961	6,941	6,921	6,902	6,883	6,845	6,807	6,770	6,733	6,696	6,660	6,624	6,536	6,451	6,368	6,287	6,208	6,131	5,982	5,840	5,705	5,453	5,222
10	9,960	9,920	9,880	9,840	9,801	9,763	9,686	9,611	9,536	9,463	9,391	9,320	9,250	9,080	8,916	8,758	8,605	8,458	8,315	8,044	7,790	7,552	7,116	6,728
15	14,909	14,820	14,731	14,644	14,558	14,472	14,304	14,140	13,980	13,824	13,671	13,521	13,374	13,021	12,687	12,369	12,066	11,778	11,504	10,991	10,523	10,092	9,329	8,674
20	19,839	19,681	19,525	19,372	19,221	19,073	18,782	18,501	18,227	17,962	17,704	17,454	17,210	16,631	16,088	15,581	15,104	14,655	14,233	13,457	12,761	12,133	11,047	10,139
25	24,749	24,504	24,263	24,026	23,795	23,567	23,126	22,700	22,290	21,895	21,513	21,144	20,788	19,948	19,173	18,456	17,791	17,172	16,594	15,549	14,627	13,809	12,419	11,283
30	29,640	29,288	28,944	28,609	28,281	27,961	27,341	26,748	26,181	25,637	25,115	24,614	24,133	23,008	21,983	21,046	20,185	19,392	18,659	17,347	16,208	15,209	13,540	12,201
35	34,511	34,035	33,572	33,121	32,682	32,255	31,434	30,653	29,910	29,202	28,527	27,882	27,266	25,839	24,553	23,390	22,331	21,365	20,478	18,909	17,563	16,396	14,473	12,954
40	39,362	38,744	38,145	37,565	37,001	36,455	35,409	34,421	33,487	32,602	31,763	30,966	30,207	28,465	26,913	25,521	24,267	23,129	22,094	20,279	18,739	17,416	15,262	13,582
45	44,194	43,417	42,666	41,941	41,240	40,562	39,271	38,060	36,921	35,848	34,836	33,880	32,974	30,909	29,088	27,469	26,021	24,718	23,539	21,489	19,768	18,302	15,938	14,115
50	49,007	48,053	47,135	46,252	45,401	44,581	43,026	41,576	40,221	38,951	37,759	36,638	35,582	33,189	31,098	29,255	27,618	26,154	24,838	22,567	20,676	19,078	16,523	14,572
55	53,801	52,653	51,553	50,498	49,486	48,513	46,678	44,976	43,394	41,920	40,543	39,253	38,043	35,320	32,962	30,898	29,078	27,460	26,013	23,532	21,484	19,763	17,035	14,968
60	58,576	57,218	55,921	54,682	53,497	52,362	50,230	48,265	46,448	44,763	43,196	41,735	40,370	37,317	34,694	32,416	30,418	28,652	27,080	24,402	22,206	20,373	17,486	15,316
65	63,332	61,747	60,240	58,805	57,436	56,130	53,687	51,449	49,389	47,489	45,729	44,095	42,573	39,192	36,309	33,821	31,652	29,745	28,054	25,190	22,857	20,920	17,887	15,622
70	68,069	66,242	64,510	62,867	61,305	59,819	57,053	54,532	52,224	50,103	48,148	46,340	44,663	40,956	37,818	35,127	32,793	30,750	28,946	25,907	23,446	21,412	18,246	15,895
75	72,788	70,702	68,733	66,871	65,107	63,433	60,332	57,519	54,957	52,614	50,462	48,479	46,647	42,619	39,231	36,342	33,850	31,677	29,767	26,563	23,982	21,858	18,568	16,140
80	77,488	75,129	72,909	70,817	68,842	66,973	63,525	60,415	57,595	55,026	52,677	50,520	48,533	44,188	40,557	37,478	34,833	32,536	30,524	27,164	24,471	22,263	18,860	16,360
85	82,170	79,522	77,039	74,707	72,512	70,442	66,638	63,224	60,142	57,347	54,800	52,470	50,329	45,672	41,804	38,540	35,748	33,334	31,225	27,718	24,919	22,634	19,125	16,559
90	86,833	83,882	81,124	78,542	76,120	73,842	69,673	65,949	62,603	59,580	56,836	54,333	52,041	47,077	42,978	39,535	36,603	34,076	31,875	28,229	25,332	22,973	19,367	16,740
95	91,478	88,209	85,165	82,324	79,666	77,175	72,632	68,594	64,982	61,731	58,790	56,116	53,675	48,410	44,086	40,471	37,404	34,769	32,481	28,703	25,712	23,286	19,589	16,905
100	96,106	92,503	89,161	86,052	83,153	80,442	75,519	71,163	67,283	63,804	60,667	57,824	55,235	49,676	45,133	41,352	38,155	35,417	33,046	29,143	26,065	23,575	19,793	17,057
105	100,715	96,766	93,114	89,729	86,581	83,646	78,336	73,659	69,510	65,803	62,471	59,461	56,727	50,879	46,125	42,183	38,861	36,025	33,574	29,553	26,393	23,843	19,982	17,197
110	105,306	100,996	97,026	93,355	89,952	86,789	81,085	76,086	71,666	67,732	64,208	61,032	58,155	52,025	47,064	42,967	39,526	36,595	34,069	29,936	26,698	24,091	20,156	17,326
115	109,880	105,196	100,895	96,932	93,268	89,872	83,770	78,445	73,756	69,596	65,880	62,541	59,524	53,118	47,957	43,709	40,154	37,133	34,534	30,295	26,983	24,323	20,318	17,445
120	114,435	109,364	104,723	100,460	96,530	92,897	86,392	80,739	75,781	71,396	67,491	63,991	60,836	54,160	48,805	44,413	40,746	37,639	34,972	30,631	27,249	24,539	20,469	17,556
125	118,974	113,502	108,511	103,941	99,740	95,865	88,954	82,972	77,745	73,136	69,044	65,385	62,095	55,156	49,612	45,080	41,307	38,117	35,385	30,947	27,499	24,742	20,609	17,660
130	123,494	117,609	112,259	107,375	102,898	98,779	91,457	85,146	79,650	74,820	70,543	66,728	63,304	56,108	50,381	45,714	41,839	38,570	35,774	31,245	27,733	24,932	20,741	17,756
135	127,998	121,686	115,968	110,763	106,005	101,639	93,904	87,263	81,499	76,450	71,989	68,021	64,467	57,019	51,114	46,318	42,344	38,998	36,142	31,525	27,954	25,110	20,864	17,846
140	132,484	125,734	119,638	114,107	109,064	104,448	96,296	89,325	83,295	78,028	73,387	69,267	65,586	57,893	51,815	46,892	42,824	39,405	36,491	31,790	28,163	25,278	20,980	17,931
145	136,953	129,752	123,271	117,406	112,074	107,206	98,636	91,335	85,040	79,557	74,738	70,470	66,663	58,730	52,485	47,440	43,280	39,791	36,822	32,041	28,359	25,436	21,089	18,011
150	141,405	133,742	126,866	120,663	115,038	109,914	100,924	93,294	86,736	81,039	76,045	71,630	67,700	59,534	53,126	47,963	43,715	40,158	37,137	32,279	28,545	25,586	21,192	18,085
155	145,840	137,702	130,425	123,878	117,956	112,575	103,163	95,204	88,384	82,476	77,309	72,751	68,700	60,306	53,740	48,463	44,130	40,508	37,436	32,505	28,722	25,727	21,289	18,156
160	150,258	141,634	133,947	127,051	120,830	115,190	105,355	97,067	89,988	83,871	78,533	73,834	69,665	61,048	54,328	48,941	44,526	40,842	37,720	32,719	28,889	25,861	21,380	18,223
165	154,659	145,538	137,433	130,183	123,660	117,759	107,500	98,885	91,548	85,225	79,719	74,881	70,597	61,762	54,893	49,399	44,905	41,160	37,992	32,923	29,048	25,989	21,467	18,286
170	159,044	149,414	140,885	133,276	126,447	120,284	109,600	100,659	93,067	86,539	80,868	75,894	71,496	62,450	55,436	49,838	45,267	41,464	38,251	33,118	29,199	26,110	21,550	18,346
175	163,412	153,263	144,301	136,330	129,193	122,766	111,656	102,391	94,546	87,817	81,982	76,874	72,366	63,112	55,957	50,259	45,614	41,755	38,498	33,303	29,343	26,225	21,628	18,402
180	167,763	157,085	147,684	139,345	131,897	125,205	113,671	104,083	95,986	89,058	83,063	77,824	73,207	63,751	56,458	50,663	45,947	42,034	38,735	33,480	29,480	26,334	21,703	18,456
185	172,099	160,879	151,033	142,323	134,562	127,604	115,645	105,735	97,390	90,265	84,112	78,744	74,020	64,367	56,941	51,051	46,266	42,301	38,961	33,649	29,611	26,439	21,774	18,508
190	176,417	164,647	154,349	145,264	137,188	129,963	117,579	107,350	98,758	91,439	85,131	79,636	74,808	64,962	57,406	51,425	46,572	42,557	39,179	33,811	29,737	26,539	21,841	18,557
195	180,720	168,389	157,633	148,168	139,776	132,284	119,475	108,928	100,092	92,582	86,120	80,501	75,571	65,536	57,854	51,784	46,867	42,803	39,387	33,966	29,856	26,634	21,906	18,603
200	185,006	172,104	160,884	151,037	142,327	134,566	121,333	110,470	101,393	93,694	87,081	81,341	76,310	66,092	58,287	52,130	47,150	43,039	39,587	34,114	29,971	26,725	21,967	18,648

400 MCM	2 Run(s)	600 Volts		C-Value = 20,565	Magnetic Duct	Copper

Calculations:

1. $f = \dfrac{\text{Length} \times 2 \times \text{Isca}}{\text{Runs} \times \text{C-Value} \times \text{E L-L}}$

2. $M = \dfrac{1}{1 + f}$

3. Isc = Isca x M

4. Add Motor Contribution = Motor FLA x 4

* All results are given in symmetrical amperes

Single-Phase L-L Bolted Fault-Current* Table

One-Way Distance in Feet

Available Isc in Thousands (Isca)	5	10	15	20	25	30	40	50	60	70	80	90	100	125	150	175	200	225	250	300	350	400	500	600
3	2,993	2,987	2,980	2,973	2,967	2,960	2,947	2,934	2,921	2,908	2,896	2,883	2,871	2,840	2,810	2,781	2,752	2,724	2,696	2,643	2,591	2,542	2,448	2,361
5	4,981	4,963	4,944	4,926	4,908	4,890	4,854	4,819	4,784	4,750	4,717	4,683	4,651	4,571	4,494	4,419	4,347	4,277	4,209	4,080	3,959	3,845	3,635	3,446
7	6,963	6,927	6,891	6,856	6,821	6,786	6,717	6,650	6,584	6,520	6,457	6,395	6,334	6,187	6,046	5,912	5,783	5,660	5,543	5,321	5,117	4,927	4,588	4,292
10	9,925	9,852	9,780	9,708	9,638	9,569	9,433	9,301	9,173	9,048	8,927	8,809	8,694	8,419	8,161	7,918	7,689	7,473	7,269	6,893	6,554	6,246	5,710	5,259
15	14,833	14,669	14,509	14,353	14,200	14,050	13,760	13,481	13,213	12,956	12,709	12,470	12,241	11,703	11,210	10,757	10,339	9,953	9,594	8,949	8,385	7,888	7,053	6,377
20	19,704	19,417	19,137	18,866	18,602	18,346	17,854	17,387	16,945	16,524	16,124	15,742	15,379	14,539	13,786	13,107	12,492	11,932	11,420	10,518	9,748	9,083	7,992	7,135
25	24,539	24,095	23,667	23,253	22,854	22,468	21,734	21,047	20,402	19,795	19,223	18,684	18,174	17,012	15,990	15,084	14,275	13,549	12,893	11,754	10,801	9,990	8,686	7,683
30	29,339	28,706	28,100	27,519	26,962	26,426	25,417	24,482	23,614	22,805	22,049	21,342	20,679	19,188	17,898	16,771	15,777	14,894	14,105	12,754	11,639	10,703	9,220	8,098
35	34,103	33,251	32,441	31,669	30,933	30,231	28,917	27,713	26,605	25,583	24,636	23,756	22,938	21,118	19,566	18,223	17,059	16,031	15,121	13,578	12,321	11,278	9,643	8,423
40	38,833	37,732	36,692	35,708	34,775	33,890	32,248	30,757	29,399	28,155	27,012	25,959	24,984	22,841	21,036	19,495	18,165	17,005	15,984	14,270	12,889	11,751	9,987	8,684
45	43,528	42,150	40,856	39,640	38,493	37,411	35,420	33,631	32,013	30,544	29,204	27,976	26,848	24,388	22,341	20,612	19,130	17,848	16,727	14,859	13,367	12,147	10,272	8,899
50	48,190	46,506	44,936	43,469	42,094	40,804	38,447	36,347	34,465	32,768	31,230	29,831	28,551	25,786	23,509	21,601	19,980	18,585	17,372	15,367	13,776	12,484	10,512	9,078
55	52,818	50,802	48,934	47,199	45,583	44,074	41,336	38,919	36,769	34,844	33,110	31,541	30,114	27,054	24,558	22,484	20,733	19,235	17,939	15,808	14,130	12,774	10,717	9,231
60	57,412	55,038	52,853	50,834	48,965	47,227	44,098	41,358	38,938	36,786	34,859	33,124	31,554	28,210	25,507	23,277	21,405	19,812	18,440	16,196	14,439	13,026	10,894	9,362
65	61,974	59,217	56,695	54,378	52,244	50,271	46,741	43,673	40,984	38,607	36,490	34,593	32,884	29,269	26,370	23,993	22,009	20,329	18,887	16,540	14,712	13,248	11,048	9,475
70	66,503	63,338	60,461	57,834	55,426	53,211	49,271	45,875	42,917	40,317	38,014	35,960	34,117	30,241	27,157	24,643	22,555	20,793	19,287	16,846	14,954	13,443	11,184	9,575
75	71,000	67,404	64,156	61,206	58,515	56,051	51,697	47,971	44,746	41,927	39,442	37,236	35,263	31,138	27,878	25,235	23,050	21,213	19,648	17,121	15,170	13,618	11,305	9,663
80	75,465	71,416	67,779	64,495	61,515	58,797	54,025	49,969	46,479	43,445	40,783	38,428	36,330	31,968	28,541	25,777	23,502	21,595	19,975	17,368	15,364	13,774	11,412	9,742
85	79,898	75,374	71,334	67,706	64,429	61,454	56,260	51,875	48,124	44,879	42,044	39,545	37,328	32,737	29,152	26,275	23,915	21,944	20,273	7,593	15,539	13,915	11,509	9,812
90	84,300	79,279	74,823	70,841	67,261	64,026	58,407	53,695	49,686	46,235	43,232	40,595	38,261	33,453	29,719	26,734	24,295	22,263	20,545	7,798	15,699	14,043	11,596	9,875
95	88,672	83,134	78,247	73,902	70,015	66,516	60,473	55,436	51,173	47,520	44,353	41,582	39,137	34,121	30,244	27,159	24,645	22,557	20,795	7,985	15,844	14,159	11,675	9,933
100	93,012	86,937	81,608	76,893	72,694	68,930	62,461	57,102	52,590	48,739	45,413	42,512	39,960	34,745	30,734	27,553	24,969	22,828	21,025	8,157	15,977	14,265	11,747	9,985
105	97,323	90,692	84,907	79,816	75,301	71,269	64,375	58,698	53,941	49,897	46,417	43,391	40,735	35,329	31,190	27,919	25,269	23,078	21,237	18,315	16,100	14,363	11,813	10,033
110	101,604	94,398	88,147	82,672	77,838	73,538	66,221	60,228	55,230	50,998	47,369	44,221	41,466	35,878	31,617	28,261	25,548	23,311	21,434	18,462	16,213	14,453	11,874	10,076
115	105,855	98,057	91,329	85,465	80,309	75,739	68,001	61,697	56,463	52,047	48,272	45,008	42,157	36,394	32,017	28,580	25,809	23,528	21,618	18,597	16,318	14,536	11,930	10,117
120	110,076	101,669	94,454	88,196	82,716	77,876	69,718	63,108	57,642	53,048	49,132	45,754	42,811	36,880	32,393	28,879	26,053	23,730	21,788	18,723	16,415	14,613	11,982	10,154
125	114,269	105,235	97,525	90,868	85,061	79,952	71,377	64,464	58,771	54,003	49,950	46,463	43,431	37,339	32,746	29,160	26,281	23,920	21,948	18,841	16,505	14,684	12,030	10,188
130	118,433	108,757	100,542	93,481	87,347	81,968	72,980	65,768	59,854	54,915	50,729	47,137	44,019	37,773	33,080	29,424	26,495	24,097	22,097	18,951	16,589	14,751	12,075	10,220
135	122,569	112,234	103,507	96,039	89,576	83,928	74,530	67,024	60,892	55,788	51,473	47,778	44,578	38,184	33,394	29,672	26,697	24,264	22,237	19,054	16,668	14,813	12,116	10,250
140	126,677	115,669	106,421	98,543	91,750	85,834	76,029	68,234	61,889	56,624	52,184	48,390	45,110	38,574	33,692	29,907	26,887	24,420	22,368	19,150	16,742	14,871	12,155	10,278
145	130,756	119,061	109,286	100,994	93,872	87,688	77,480	69,400	62,847	57,425	52,863	48,974	45,617	38,944	33,974	30,129	27,066	24,568	22,492	19,241	16,811	14,926	12,192	10,304
150	134,809	122,411	112,102	103,394	95,942	89,492	78,885	70,526	63,768	58,193	53,514	49,531	46,100	39,296	34,241	30,339	27,235	24,708	22,609	19,327	16,876	14,977	12,226	10,329
155	138,834	125,721	114,871	105,746	97,963	91,248	80,246	71,612	64,655	58,930	54,137	50,064	46,562	39,631	34,495	30,538	27,396	24,840	22,720	19,407	16,938	15,026	12,258	10,352
160	142,831	128,990	117,595	108,049	99,937	92,958	81,566	72,661	65,509	59,639	54,734	50,575	47,003	39,950	34,737	30,728	27,548	24,965	22,824	19,483	16,996	15,071	12,289	10,373
165	146,803	132,221	120,274	110,307	101,865	94,624	82,845	73,675	66,332	60,320	55,307	51,064	47,425	40,254	34,967	30,907	27,692	25,083	22,923	19,556	17,051	15,115	12,317	10,394
170	150,747	135,412	122,909	112,519	103,749	96,247	84,087	74,655	67,126	60,976	55,858	51,533	47,830	40,545	35,186	31,079	27,830	25,196	23,017	19,624	17,103	15,155	12,344	10,413
175	154,666	138,566	125,501	114,688	105,590	97,830	85,293	75,604	67,891	61,607	56,388	51,983	48,217	40,823	35,396	31,242	27,961	25,303	23,107	19,689	17,152	15,194	12,370	10,431
180	158,559	141,682	128,052	116,814	107,390	99,373	86,463	76,522	68,631	62,215	56,897	52,416	48,589	41,090	35,596	31,398	28,085	25,405	23,192	19,751	17,199	15,231	12,394	10,449
185	162,426	144,761	130,562	118,900	109,150	100,878	87,600	77,411	69,346	62,802	57,387	52,832	48,946	41,345	35,787	31,546	28,204	25,502	23,273	19,809	17,243	15,266	12,418	10,465
190	166,267	147,805	133,033	120,945	110,872	102,347	88,706	78,273	70,037	63,368	57,859	53,232	49,289	41,589	35,970	31,688	28,318	25,595	23,350	19,865	17,286	15,299	12,440	10,481
195	170,084	150,813	135,465	122,952	112,556	103,780	89,781	79,109	70,705	63,915	58,315	53,617	49,619	41,824	36,146	31,825	28,426	25,684	23,424	19,919	17,326	15,331	12,460	10,495
200	173,875	153,787	137,859	124,921	114,204	105,180	90,826	79,920	71,352	64,443	58,754	53,988	49,937	42,050	36,314	31,955	28,530	25,769	23,495	19,970	17,365	15,361	12,480	10,510

500 MCM	1 Run(s)	600 Volts	C-Value = 22,185	Magnetic Duct	Copper

Calculations:

1. $f = \dfrac{\text{Length} \times 2 \times \text{Isca}}{\text{Runs} \times \text{C-Value} \times \text{E L-L}}$

2. $M = \dfrac{1}{1 + f}$

3. Isc = Isca x M

4. Add Motor Contribution = Motor FLA x 4

* All results are given in symmetrical amperes

Single-Phase L-L Bolted Fault-Current* Table

One-Way Distance in Feet

Available Isc in Thousands (Isca)	5	10	15	20	25	30	40	50	60	70	80	90	100	125	150	175	200	225	250	300	350	400	500	600
3	2,997	2,993	2,990	2,987	2,983	2,980	2,973	2,967	2,960	2,953	2,947	2,940	2,934	2,918	2,902	2,886	2,871	2,855	2,840	2,810	2,781	2,752	2,696	2,643
5	4,991	4,981	4,972	4,963	4,953	4,944	4,926	4,908	4,890	4,872	4,854	4,836	4,819	4,776	4,733	4,692	4,651	4,610	4,571	4,494	4,419	4,347	4,209	4,080
7	6,982	6,963	6,945	6,927	6,909	6,891	6,856	6,821	6,786	6,751	6,717	6,684	6,650	6,568	6,488	6,410	6,334	6,259	6,187	6,046	5,912	5,783	5,543	5,321
10	9,963	9,925	9,889	9,852	9,816	9,780	9,708	9,638	9,569	9,500	9,433	9,367	9,301	9,142	8,987	8,838	8,694	8,554	8,419	8,161	7,918	7,689	7,269	6,893
15	14,916	14,833	14,751	14,669	14,589	14,509	14,353	14,200	14,050	13,903	13,760	13,619	13,481	13,148	12,831	12,529	12,241	11,966	11,703	11,210	10,757	10,339	9,594	8,949
20	19,851	19,704	19,559	19,417	19,276	19,137	18,866	18,602	18,346	18,097	17,854	17,618	17,387	16,838	16,322	15,836	15,379	14,947	14,539	13,786	13,107	12,492	11,420	10,518
25	24,767	24,539	24,315	24,095	23,879	23,667	23,253	22,854	22,468	22,095	21,734	21,385	21,047	20,247	19,505	18,816	18,174	17,574	17,012	15,990	15,084	14,275	12,893	11,754
30	29,666	29,339	29,019	28,706	28,400	28,100	27,519	26,962	26,426	25,912	25,417	24,941	24,482	23,406	22,420	21,514	20,679	19,906	19,188	17,898	16,771	15,777	14,105	12,754
35	34,546	34,103	33,672	33,251	32,841	32,441	31,669	30,933	30,231	29,559	28,917	28,302	27,713	26,342	25,100	23,970	22,938	21,990	21,118	19,566	18,226	17,059	15,121	13,578
40	39,408	38,833	38,275	37,732	37,205	36,692	35,708	34,775	33,890	33,048	32,248	31,485	30,757	29,078	27,572	26,214	24,984	23,864	22,841	21,036	19,495	18,165	15,984	14,270
45	44,252	43,528	42,828	42,150	41,493	40,856	39,640	38,493	37,411	36,389	35,420	34,502	33,631	31,633	29,859	28,273	26,848	25,559	24,388	22,341	20,612	19,130	16,727	14,859
50	49,078	48,190	47,333	46,506	45,708	44,936	43,469	42,094	40,804	39,590	38,447	37,367	36,347	34,024	31,981	30,169	28,551	27,098	25,786	23,509	21,601	19,980	17,372	15,367
55	53,887	52,818	51,790	50,802	49,851	48,934	47,199	45,583	44,074	42,661	41,336	40,091	38,919	36,268	33,955	31,919	30,114	28,502	27,054	24,558	22,484	20,733	17,939	15,808
60	58,678	57,412	56,200	55,038	53,923	52,853	50,834	48,965	47,227	45,609	44,098	42,684	41,358	38,377	35,797	33,542	31,554	29,789	28,210	25,507	23,277	21,405	18,440	16,196
65	63,451	61,974	60,564	59,217	57,928	56,695	54,378	52,244	50,271	48,442	46,741	45,155	43,673	40,363	37,519	35,049	32,884	30,971	29,269	26,370	23,993	22,009	18,887	16,540
70	68,207	66,503	64,882	63,338	61,866	60,461	57,834	55,426	53,211	51,165	49,271	47,513	45,875	42,236	39,132	36,453	34,117	32,063	30,241	27,157	24,643	22,555	19,287	16,846
75	72,945	71,000	69,155	67,404	65,740	64,156	61,206	58,515	56,051	53,786	51,697	49,764	47,971	44,006	40,647	37,764	35,263	33,072	31,138	27,878	25,235	23,050	19,648	17,121
80	77,666	75,465	73,384	71,416	69,550	67,779	64,495	61,515	58,797	56,310	54,025	51,918	49,969	45,681	42,072	38,991	36,330	34,010	31,968	28,541	25,777	23,502	19,975	17,368
85	82,370	79,898	77,570	75,374	73,298	71,334	67,706	64,429	61,454	58,742	56,260	53,978	51,875	47,269	43,415	40,142	37,328	34,882	32,737	29,152	26,275	23,915	20,273	17,593
90	87,057	84,300	81,713	79,279	76,987	74,823	70,841	67,261	64,026	61,088	58,407	55,952	53,695	48,776	44,683	41,223	38,261	35,696	33,453	29,719	26,734	24,295	20,545	17,798
95	91,727	88,672	85,813	83,134	80,616	78,247	73,902	70,015	66,516	63,351	60,473	57,845	55,436	50,208	45,882	42,242	39,137	36,457	34,121	30,244	27,159	24,645	20,795	17,985
100	96,380	93,012	89,872	86,937	84,188	81,608	76,893	72,694	68,930	65,536	62,461	59,661	57,102	51,571	47,017	43,202	39,960	37,170	34,745	30,734	27,553	24,969	21,025	18,157
105	101,016	97,323	93,891	90,692	87,704	84,907	79,816	75,301	71,269	67,647	64,375	61,406	58,698	52,869	48,094	44,110	40,735	37,840	35,329	31,190	27,919	25,269	21,237	18,315
110	105,635	101,604	97,868	94,398	91,166	88,147	82,672	77,838	73,538	69,688	66,221	63,083	60,228	54,108	49,116	44,968	41,466	38,470	35,878	31,617	28,261	25,548	21,434	18,462
115	110,238	105,855	101,807	98,057	94,573	91,329	85,465	80,309	75,739	71,662	68,001	64,696	61,697	55,290	50,089	45,782	42,157	39,064	36,394	32,017	28,580	25,809	21,618	18,597
120	114,824	110,076	105,706	101,669	97,929	94,454	88,196	82,716	77,876	73,572	69,718	66,249	63,108	56,420	51,015	46,554	42,811	39,625	36,880	32,393	28,879	26,053	21,788	18,723
125	119,394	114,269	109,566	105,235	101,234	97,525	90,868	85,061	79,952	75,422	71,377	67,745	64,464	57,502	51,897	47,288	43,431	40,155	37,339	32,746	29,160	26,281	21,948	18,841
130	123,947	118,433	113,389	108,757	104,488	100,542	93,481	87,347	81,968	77,213	72,980	69,187	65,768	58,538	52,739	47,986	44,019	40,658	37,773	33,080	29,424	26,495	22,097	18,951
135	128,485	122,569	117,174	112,234	107,694	103,507	96,039	89,576	83,928	78,950	74,530	70,578	67,024	59,530	53,544	48,651	44,578	41,134	38,184	33,394	29,672	26,697	22,237	19,054
140	133,005	126,677	120,923	115,669	110,852	106,421	98,543	91,750	85,834	80,634	76,029	71,921	68,234	60,483	54,313	49,286	45,110	41,587	38,574	33,692	29,907	26,887	22,368	19,150
145	137,510	130,756	124,635	119,061	113,964	109,286	100,994	93,872	87,688	82,268	77,480	73,218	69,400	61,398	55,050	49,891	45,617	42,017	38,944	33,974	30,129	27,066	22,492	19,241
150	141,999	134,809	128,311	122,411	117,030	112,102	103,394	95,942	89,492	83,854	78,885	74,471	70,526	62,277	55,755	50,470	46,100	42,427	39,296	34,241	30,339	27,235	22,609	19,327
155	146,472	138,834	131,952	125,721	120,051	114,871	105,746	97,963	91,248	85,394	80,246	75,683	71,612	63,122	56,432	51,024	46,562	42,818	39,631	34,495	30,538	27,396	22,720	19,407
160	150,929	142,831	135,559	128,990	123,029	117,595	108,049	99,937	92,958	86,890	81,566	76,856	72,661	63,936	57,081	51,554	47,003	43,190	39,950	34,737	30,728	27,548	22,824	19,483
165	155,370	146,803	139,131	132,221	125,964	120,274	110,307	101,865	94,624	88,344	82,845	77,991	73,675	64,719	57,705	52,063	47,425	43,547	40,254	34,967	30,907	27,692	22,923	19,556
170	159,796	150,747	142,669	135,412	128,858	122,909	112,519	103,749	96,247	89,757	84,087	79,091	74,655	65,475	58,305	52,550	47,830	43,887	40,545	35,186	31,079	27,830	23,017	19,624
175	164,206	154,666	146,174	138,566	131,710	125,501	114,688	105,590	97,830	91,132	85,293	80,156	75,604	66,203	58,882	53,019	48,217	44,213	40,823	35,396	31,242	27,961	23,107	19,689
180	168,600	158,559	149,646	141,682	134,522	128,052	116,814	107,390	99,373	92,470	86,463	81,189	76,522	66,906	59,437	53,469	48,589	44,526	41,090	35,596	31,398	28,085	23,192	19,751
185	172,979	162,426	153,086	144,761	137,296	130,562	118,900	109,150	100,878	93,772	87,600	82,191	77,411	67,585	59,973	53,901	48,946	44,826	41,345	35,787	31,546	28,204	23,273	19,809
190	177,343	166,267	156,493	147,805	140,030	133,033	120,945	110,872	102,347	95,039	88,706	83,164	78,273	68,241	60,489	54,318	49,289	45,113	41,589	35,970	31,688	28,318	23,350	19,865
195	181,692	170,084	159,870	150,813	142,728	135,465	122,952	112,556	103,780	96,274	89,781	84,108	79,109	68,876	60,986	54,719	49,619	45,390	41,824	36,146	31,825	28,426	23,424	19,919
200	186,025	173,875	163,215	153,787	145,388	137,859	124,921	114,204	105,180	97,477	90,826	85,024	79,920	69,489	61,467	55,105	49,937	45,655	42,050	36,314	31,955	28,530	23,495	19,970

500 MCM	2 Run(s)	600 Volts	C-Value = 22,185	Magnetic Duct	Copper

Calculations:

1. $f = \dfrac{\text{Length} \times 2 \times \text{Isca}}{\text{Runs} \times \text{C-Value} \times \text{E L-L}}$
2. $M = \dfrac{1}{1 + f}$
3. Isc = Isca x M
4. Add Motor Contribution = Motor FLA x 4

* All results are given in symmetrical amperes

Single-Phase L-L Bolted Fault-Current* Table

One-Way Distance in Feet

Available Isc in Thousands (Isca)	5	10	15	20	25	30	40	50	60	70	80	90	100	125	150	175	200	225	250	300	350	400	500	600
3	2,993	2,987	2,981	2,974	2,968	2,961	2,949	2,936	2,924	2,911	2,899	2,887	2,875	2,845	2,816	2,788	2,760	2,732	2,705	2,653	2,603	2,555	2,464	2,379
5	4,982	4,964	4,946	4,928	4,911	4,893	4,859	4,825	4,791	4,758	4,726	4,693	4,662	4,584	4,509	4,437	4,366	4,298	4,232	4,106	3,987	3,875	3,669	3,483
7	6,965	6,930	6,895	6,861	6,827	6,793	6,727	6,662	6,598	6,535	6,474	6,414	6,354	6,211	6,074	5,943	5,818	5,698	5,582	5,365	5,164	4,977	4,642	4,349
10	9,923	9,857	9,787	9,718	9,650	9,583	9,451	9,323	9,199	9,078	8,960	8,845	8,732	8,464	8,212	7,974	7,750	7,538	7,337	6,966	6,631	6,327	5,795	5,345
15	14,838	14,680	14,526	14,374	14,226	14,080	13,798	13,527	13,267	13,016	12,775	12,542	12,318	11,791	11,307	10,862	10,450	10,068	9,713	[illegible],073	8,513	8,018	7,182	6,504
20	19,7[illegible]4	19,436	19,165	18,903	18,647	18,398	17,919	17,465	17,033	16,622	16,231	15,857	15,500	14,675	13,933	13,262	12,653	12,098	11,589	1[illegible],690	9,920	9,254	8,158	7,295
25	24,554	24,125	23,709	23,308	22,921	22,546	21,831	21,161	20,530	19,936	19,375	18,845	18,344	17,199	16,188	15,290	14,487	13,763	13,108	1[illegible],970	11,013	10,198	8,883	7,868
30	29,361	28,748	28,161	27,597	27,055	26,534	25,550	24,636	23,786	22,992	22,249	21,553	20,899	19,426	18,147	17,026	16,035	15,153	14,364	1[illegible],008	11,886	10,942	9,442	8,304
35	34,133	33,308	32,522	31,772	31,056	30,371	29,089	27,910	26,824	25,819	24,886	24,018	23,209	21,406	19,863	18,528	17,361	16,332	15,418	1[illegible],867	12,599	11,543	9,887	8,646
40	38,872	37,805	36,796	35,838	34,930	34,066	32,461	31,001	29,666	28,441	27,314	26,272	25,307	23,178	21,380	19,84[illegible]	18,508	17,344	16,317	1[illegible],589	13,192	12,040	10,249	8,921
45	43,577	42,241	40,985	39,801	38,683	37,627	35,678	33,922	32,330	30,881	29,556	28,340	27,220	24,773	22,730	20,998	19,511	18,221	17,091	1[illegible],205	13,694	12,456	10,549	9,148
50	48,249	46,617	45,091	43,662	42,321	41,060	38,751	36,687	34,832	33,156	31,634	30,245	28,973	26,217	23,939	22,026	20,396	18,990	17,766	1[illegible],737	14,124	12,811	10,802	9,338
55	52,839	50,934	49,118	47,428	45,849	44,373	41,688	39,309	37,188	35,283	33,564	32,005	30,584	27,529	25,029	22,945	21,181	19,669	18,359	1[illegible],201	14,496	13,116	11,019	9,499
60	57,436	55,193	53,068	51,100	49,272	47,571	44,499	41,799	39,408	37,276	35,363	33,636	32,070	28,727	26,015	23,771	21,884	20,274	18,884	1[illegible],608	14,822	13,382	11,206	9,638
65	62,072	59,396	56,942	54,682	52,595	50,661	47,191	44,166	41,505	39,147	37,042	35,152	33,445	29,826	26,913	24,513	22,515	20,815	19,353	1[illegible],970	15,109	13,616	11,369	9,759
70	66,616	63,544	60,742	58,178	55,821	53,648	49,772	46,418	43,488	40,906	38,614	36,564	34,722	30,836	27,733	25,197	23,086	21,302	19,773	[illegible],292	15,364	13,823	11,513	9,864
75	71,128	67,637	64,472	61,590	58,955	56,536	52,249	48,565	45,367	42,565	40,088	37,884	35,909	31,769	28,485	25,817	23,605	21,743	20,153	[illegible],582	15,592	14,007	11,640	9,958
80	75,610	71,677	68,133	64,923	62,001	59,332	54,627	50,614	47,150	44,130	41,473	39,118	37,017	32,633	29,178	26,385	24,079	22,144	20,497	[illegible],843	15,797	14,172	11,754	10,041
85	80,061	75,665	71,726	68,177	64,963	62,038	56,913	52,570	48,843	45,610	42,778	40,277	38,052	33,436	29,818	26,907	24,513	22,511	20,811	[illegible],080	15,983	14,322	11,857	10,116
90	84,482	79,601	75,254	71,357	67,843	64,660	59,112	54,441	50,454	47,011	44,008	41,366	39,023	34,183	30,411	27,388	24,912	22,847	21,098	[illegible]8,296	16,152	14,457	11,950	10,183
95	88,873	83,488	78,718	74,464	70,646	67,201	61,229	56,231	51,988	48,340	45,171	42,391	39,934	34,880	30,961	27,834	25,281	23,156	21,361	[illegible]8,494	16,306	14,580	12,034	10,244
100	93,234	87,325	82,121	77,502	73,375	69,665	63,267	57,946	53,450	49,602	46,271	43,359	40,792	35,532	31,474	28,248	25,622	23,442	21,604	[illegible]8,676	16,447	14,693	12,110	10,300
105	97,565	91,114	85,463	80,471	76,031	72,055	65,233	59,590	54,846	50,802	47,313	44,273	41,600	36,144	31,953	28,633	25,938	23,707	21,829	[illegible]8,844	16,577	14,797	12,181	10,351
110	101,368	94,855	88,746	83,376	78,619	74,375	67,128	61,168	56,180	51,945	48,303	45,138	42,363	36,718	32,401	28,992	26,233	23,952	22,037	[illegible]8,999	16,696	14,892	12,245	10,397
115	106,141	98,550	91,972	86,217	81,140	76,628	68,958	62,684	57,456	53,033	49,243	45,958	43,084	37,259	32,821	29,328	26,507	24,181	22,231	[illegible]9,142	16,807	14,980	12,305	10,440
120	110,387	102,199	95,142	88,997	83,598	78,816	70,725	64,141	58,678	54,072	50,137	46,736	43,767	37,769	33,216	29,643	26,764	24,395	22,411	[illegible]9,276	16,910	15,062	12,360	10,480
125	114,303	105,803	98,259	91,718	85,994	80,942	72,433	65,542	59,848	55,065	50,990	47,476	44,415	38,250	33,588	29,939	27,005	24,595	22,580	[illegible]9,401	17,006	15,138	12,411	10,516
130	118,792	109,364	101,322	94,382	88,331	83,010	74,084	66,891	60,971	56,014	51,802	48,180	45,031	38,706	33,939	30,218	27,232	24,783	22,738	[illegible]9,517	17,096	15,209	12,458	10,551
135	122,954	112,881	104,334	96,990	90,612	85,021	75,681	68,190	62,049	56,922	52,578	48,850	45,616	39,137	34,270	30,480	27,445	24,959	22,886	[illegible]9,626	17,179	15,275	12,503	10,582
140	127,087	116,356	107,295	99,544	92,837	86,977	77,227	69,443	63,084	57,793	53,320	49,490	46,173	39,547	34,584	30,728	27,645	25,125	23,026	[illegible]9,729	17,258	15,337	12,544	10,612
145	131,194	119,789	110,208	102,046	95,010	88,881	78,725	70,652	64,080	58,627	54,029	50,100	46,704	39,936	34,881	30,962	27,835	25,281	23,157	[illegible]9,825	17,331	15,395	12,583	10,640
150	135,274	123,181	113,072	104,497	97,131	90,735	80,176	71,818	65,038	59,428	54,709	50,684	47,211	40,306	35,163	31,184	28,014	25,429	23,281	[illegible]9,916	17,401	15,450	12,620	10,666
155	139,327	126,533	115,890	106,899	99,203	92,541	81,582	72,945	65,961	60,197	55,360	51,243	47,695	40,658	35,431	31,395	28,184	25,569	23,398	[illegible]0,002	17,466	15,501	12,654	10,691
160	143,354	129,845	118,663	109,254	101,228	94,300	82,947	74,033	66,850	60,937	55,985	51,778	48,158	40,994	35,686	31,595	28,345	25,701	23,509	[illegible]0,083	17,528	15,550	12,686	10,714
165	147,355	133,119	121,391	111,563	103,206	96,015	84,271	75,086	67,707	61,648	56,585	52,290	48,602	41,315	35,929	31,785	28,498	25,827	23,614	[illegible]0,159	17,586	15,596	12,717	10,735
170	151,330	136,354	124,076	113,826	105,141	97,687	85,556	76,105	68,534	62,333	57,162	52,782	49,026	41,622	36,160	31,966	28,643	25,947	23,714	20,232	17,642	15,639	12,746	10,756
175	155,279	139,552	126,718	116,046	107,032	99,317	86,804	77,091	69,333	62,993	57,716	53,255	49,434	41,915	36,381	32,139	28,782	26,060	23,809	20,301	17,694	15,680	12,773	10,775
180	159,203	142,714	129,320	118,224	108,882	100,908	88,017	78,046	70,104	63,630	58,250	53,709	49,825	42,196	36,593	32,303	28,914	26,168	23,899	20,367	17,744	15,720	12,799	10,794
185	163,102	145,839	131,880	120,360	110,692	102,461	89,195	78,971	70,850	64,243	58,764	54,145	50,200	42,465	36,795	32,461	29,040	26,272	23,985	20,429	17,791	15,757	12,824	10,811
190	166,976	148,928	134,402	122,457	112,462	103,976	90,342	79,869	71,571	64,836	59,259	54,566	50,561	42,723	36,989	32,611	29,161	26,370	24,067	20,489	17,836	15,792	12,847	10,828
195	170,825	151,983	136,884	124,515	114,195	105,456	91,457	80,739	72,269	65,408	59,737	54,971	50,909	42,971	37,174	32,756	29,276	26,464	24,146	20,545	17,879	15,826	12,870	10,844
200	174,650	155,003	139,330	126,535	115,892	106,901	92,542	81,583	72,945	65,961	60,198	55,361	51,243	43,209	37,352	32,834	29,386	26,555	24,221	20,600	17,921	15,858	12,891	10,859

600 MCM | 1 Run(s) | 600 Volts

C-Value = 22,965 | Magnetic Duct | Copper

Calculations:

1. $f = \dfrac{\text{Length} \times 2 \times \text{Isca}}{\text{Runs} \times \text{C-Value} \times E_{L\text{-}L}}$

2. $M = \dfrac{1}{1 + f}$

3. Isc = Isca x M

4. Add Motor Contribution = Motor FLA x 4

* All results are given in symmetrical amperes

Single-Phase L-L Bolted Fault-Current* Table

One-Way Distance in Feet

Available Isc in Thousands (Isca)	5	10	15	20	25	30	40	50	60	70	80	90	100	125	150	175	200	225	250	300	350	400	500	600
3	2,997	2,993	2,990	2,987	2,984	2,981	2,974	2,968	2,961	2,955	2,949	2,942	2,936	2,921	2,905	2,890	2,875	2,860	2,845	2,816	2,788	2,760	2,705	2,653
5	4,991	4,982	4,973	4,964	4,955	4,946	4,928	4,911	4,893	4,876	4,859	4,842	4,825	4,783	4,742	4,701	4,662	4,623	4,584	4,509	4,437	4,366	4,232	4,106
7	6,982	6,965	6,947	6,930	6,912	6,895	6,861	6,827	6,793	6,760	6,727	6,694	6,662	6,582	6,504	6,428	6,354	6,282	6,211	6,074	5,943	5,818	5,582	5,365
10	9,964	9,928	9,892	9,857	9,822	9,787	9,718	9,650	9,583	9,517	9,451	9,387	9,323	9,168	9,018	8,873	8,732	8,596	8,464	8,212	7,974	7,750	7,337	6,966
15	14,919	14,838	14,759	14,680	14,603	14,526	14,374	14,226	14,080	13,938	13,798	13,662	13,527	13,203	12,894	12,600	12,318	12,049	11,791	11,307	10,862	10,450	9,713	9,073
20	19,856	19,714	19,574	19,436	19,300	19,165	18,903	18,647	18,398	18,155	17,919	17,689	17,465	16,929	16,424	15,949	15,500	15,076	14,675	13,933	13,262	12,653	11,589	10,690
25	24,775	24,554	24,338	24,125	23,915	23,709	23,308	22,921	22,546	22,183	21,831	21,491	21,161	20,378	19,652	18,975	18,344	17,753	17,199	16,188	15,290	14,487	13,108	11,970
30	29,677	29,361	29,051	28,748	28,451	28,161	27,597	27,055	26,534	26,032	25,550	25,085	24,636	23,582	22,614	21,723	20,899	20,136	19,426	18,147	17,026	16,035	14,364	13,008
35	34,561	34,133	33,715	33,308	32,910	32,522	31,772	31,056	30,371	29,716	29,089	28,488	27,910	26,565	25,344	24,230	23,209	22,271	21,406	19,863	18,528	17,361	15,418	13,867
40	39,428	38,872	38,331	37,805	37,293	36,796	35,838	34,930	34,066	33,244	32,461	31,714	31,001	29,350	27,866	26,525	25,307	24,196	23,178	21,380	19,841	18,508	16,317	14,589
45	44,277	43,577	42,899	42,241	41,603	40,985	39,801	38,683	37,627	36,627	35,678	34,778	33,922	31,955	30,204	28,635	27,220	25,939	24,773	22,730	20,998	19,511	17,091	15,205
50	49,109	48,249	47,419	46,617	45,841	45,091	43,662	42,321	41,060	39,872	38,751	37,691	36,687	34,398	32,377	30,581	28,973	27,526	26,217	23,939	22,026	20,396	17,766	15,737
55	53,924	52,889	51,893	50,934	50,010	49,118	47,428	45,849	44,373	42,989	41,688	40,464	39,309	36,692	34,402	32,381	30,584	28,976	27,529	25,029	22,945	21,181	18,359	16,201
60	58,722	57,496	56,321	55,193	54,110	53,068	51,100	49,272	47,571	45,984	44,499	43,107	41,799	38,852	36,294	34,052	32,070	30,307	28,727	26,015	23,771	21,884	18,884	16,608
65	63,502	62,072	60,705	59,396	58,143	56,942	54,682	52,595	50,661	48,864	47,191	45,628	44,166	40,889	38,065	35,606	33,445	31,532	29,826	26,913	24,518	22,515	19,353	16,970
70	68,266	66,616	65,043	63,544	62,112	60,742	58,178	55,821	53,648	51,637	49,772	48,037	46,418	42,813	39,727	37,056	34,722	32,664	30,836	27,733	25,197	23,086	19,773	17,292
75	73,013	71,128	69,339	67,637	66,017	64,472	61,590	58,955	56,536	54,308	52,249	50,340	48,565	44,633	41,289	38,412	35,909	33,713	31,769	28,485	25,817	23,605	20,153	17,582
80	77,743	75,610	73,591	71,677	69,860	68,133	64,923	62,001	59,332	56,882	54,627	52,544	50,614	46,357	42,760	39,682	37,017	34,687	32,633	29,178	26,385	24,079	20,497	17,843
85	82,457	80,061	77,801	75,665	73,643	71,726	68,177	64,963	62,038	59,365	56,913	54,656	52,570	47,993	44,148	40,874	38,052	35,595	33,436	29,818	26,907	24,513	20,811	18,080
90	87,154	84,482	81,969	79,601	77,367	75,254	71,357	67,843	64,660	61,762	59,112	56,680	54,441	49,547	45,460	41,996	39,023	36,443	34,183	30,411	27,388	24,912	21,098	18,296
95	91,834	88,873	86,096	83,488	81,033	78,718	74,464	70,646	67,201	64,076	61,229	58,624	56,231	51,025	46,702	43,054	39,934	37,236	34,880	30,961	27,834	25,281	21,361	18,494
100	96,498	93,234	90,183	87,325	84,643	82,121	77,502	73,375	69,665	66,312	63,267	60,490	57,946	52,434	47,879	44,052	40,792	37,981	35,532	31,474	28,248	25,622	21,604	18,676
105	101,146	97,565	94,229	91,114	88,198	85,463	80,471	76,031	72,055	68,474	65,233	62,284	59,590	53,776	48,996	44,996	41,600	38,680	36,144	31,953	28,633	25,938	21,829	18,844
110	105,778	101,868	98,236	94,855	91,699	88,746	83,376	78,619	74,375	70,566	67,128	64,010	61,168	55,058	50,057	45,890	42,363	39,339	36,718	32,401	28,992	26,233	22,037	18,999
115	110,393	106,141	102,205	98,550	95,147	91,972	86,217	81,140	76,628	72,591	68,958	65,671	62,684	56,283	51,068	46,737	43,084	39,960	37,259	32,821	29,328	26,507	22,231	19,142
120	114,993	110,387	106,135	102,199	98,545	95,142	88,997	83,598	78,816	74,552	70,725	67,272	64,141	57,454	52,031	47,542	43,767	40,547	37,769	33,216	29,643	26,764	22,411	19,276
125	119,576	114,603	110,028	105,803	101,892	98,259	91,718	85,994	80,942	76,451	72,433	68,815	65,542	58,576	52,949	48,308	44,415	41,103	38,250	33,588	29,939	27,005	22,580	19,401
130	124,144	118,792	113,883	109,364	105,189	101,322	94,382	88,331	83,010	78,293	74,084	70,304	66,891	59,651	53,826	49,037	45,031	41,629	38,706	33,939	30,218	27,232	22,738	19,517
135	128,696	122,954	117,702	112,881	108,439	104,334	96,990	90,612	85,021	80,079	75,681	71,741	68,190	60,683	54,664	49,732	45,616	42,129	39,137	34,270	30,480	27,445	22,886	19,626
140	133,232	127,087	121,485	116,356	111,642	107,295	99,544	92,837	86,977	81,813	77,227	73,129	69,443	61,673	55,466	50,395	46,173	42,604	39,547	34,584	30,728	27,645	23,026	19,729
145	137,752	131,194	125,232	119,789	114,799	110,208	102,046	95,010	88,881	83,495	78,725	74,470	70,652	62,624	56,234	51,028	46,704	43,056	39,936	34,881	30,962	27,835	23,157	19,825
150	142,257	135,274	128,944	123,181	117,910	113,072	104,497	97,131	90,735	85,129	80,176	75,767	71,818	63,539	56,971	51,634	47 211	43,486	40,306	35,163	31,184	28,014	23,281	19,916
155	146,746	139,327	132,622	126,533	120,978	115,890	106,899	99,203	92,541	86,717	81,582	77,022	72,945	64,419	57,678	52,214	47 695	43,897	40,658	35,431	31,395	28,184	23,398	20,002
160	151,220	143,354	136,266	129,845	124,002	118,663	109,254	101,228	94,300	88,260	82,947	78,237	74,033	65,267	58,356	52,769	48 158	44,289	40,994	35,686	31,595	28,345	23,509	20,083
165	155,679	147,355	139,875	133,119	126,985	121,391	111,563	103,206	96,015	89,760	84,271	79,414	75,086	66,083	59,008	53,302	48 602	44,663	41,315	35,929	31,785	28,498	23,614	20,159
170	160,122	151,330	143,452	136,354	129,926	124,076	113,826	105,141	97,687	91,220	85,556	80,554	76,105	66,871	59,636	53,813	49,026	45,022	41,622	36,160	31,966	28,643	23,714	20,232
175	164,551	155,279	146,996	139,552	132,826	126,718	116,046	107,032	99,317	92,640	86,804	81,660	77,091	67,631	60,239	54,304	49,434	45,365	41,915	36,381	32,139	28,782	23,809	20,301
180	168,964	159,203	150,508	142,714	135,687	129,320	118,224	108,882	100,908	94,023	88,017	82,732	78,046	68,365	60,821	54,776	49,825	45,694	42,196	36,593	32,303	28,914	23,899	20,367
185	173,362	163,102	153,988	145,839	138,509	131,880	120,360	110,692	102,461	95,369	89,195	83,773	78,971	69,074	61,382	55,231	50,200	46,010	42,465	36,795	32,461	29,040	23,985	20,429
190	177,745	166,976	157,436	148,928	141,293	134,402	122,457	112,462	103,976	96,680	90,342	84,783	79,869	69,760	61,922	55,668	50,561	46,313	42,723	36,989	32,611	29,161	24,067	20,489
195	182,114	170,825	160,854	151,983	144,039	136,884	124,515	114,195	105,456	97,959	91,457	85,764	80,739	70,423	62,444	56,089	50,909	46,604	42,971	37,174	32,756	29,276	24,146	20,545
200	186,467	174,650	164,241	155,003	146,749	139,330	126,535	115,892	106,901	99,204	92,542	86,718	81,583	71,064	62,948	56,496	51,243	46,884	43,209	37,352	32,894	29,386	24,221	20,600

600 MCM | 2 Run(s) | 600 Volts

C-Value = 22,965 | Magnetic Duct | Copper

Calculations:

1. $f = \dfrac{\text{Length} \times 2 \times \text{Isca}}{\text{Runs} \times \text{C-Value} \times E\ \text{L-L}}$

2. $M = \dfrac{1}{1 + f}$

3. Isc = Isca x M

4. Add Motor Contribution = Motor FLA x 4

* All results are given in symmetrical amperes

Single-Phase L-L Bolted Fault-Current* Table

One-Way Distance in Feet

Available Isc in Thousands (Isca)	5	10	15	20	25	30	40	50	60	70	80	90	100	125	150	175	200	225	250	300	350	400	500	600
3	2,994	2,988	2,981	2,975	2,969	2,963	2,951	2,939	2,927	2,915	2,904	2,892	2,881	2,852	2,824	2,797	2,770	2,744	2,718	2,668	2,620	2,574	2,485	2,403
5	4,983	4,966	4,949	4,932	4,915	4,899	4,866	4,833	4,801	4,769	4,738	4,707	4,677	4,603	4,531	4,461	4,393	4,328	4,264	4,142	4,027	3,918	3,717	3,535
7	6,966	6,933	6,900	6,867	6,835	6,803	6,739	6,677	6,616	6,556	6,497	6,440	6,383	6,245	6,113	5,987	5,866	5,749	5,637	5,426	5,230	5,048	4,719	4,430
10	9,931	9,864	9,797	9,731	9,666	9,602	9,476	9,354	9,235	9,118	9,005	8,894	8,787	8,528	8,284	8,054	7,836	7,629	7,433	7,071	6,741	6,442	5,915	5,469
15	14,846	14,696	14,548	14,403	14,261	14,122	13,852	13,592	13,342	13,100	12,868	12,643	12,426	11,915	11,444	11,009	10,606	10,231	9,882	9,251	8,695	8,203	7,368	6,688
20	19,728	19,462	19,204	18,953	18,708	18,470	18,010	17,573	17,157	16,760	16,380	16,018	15,671	14,867	14,141	13,483	12,883	12,334	11,831	10,937	10,169	9,502	8,400	7,527
25	24,576	24,166	23,769	23,385	23,014	22,654	21,966	21,320	20,710	20,134	19,589	19,073	18,584	17,463	16,470	15,584	14,788	14,070	13,418	12,280	11,320	10,500	9,170	8,139
30	29,391	28,806	28,245	27,704	27,184	26,683	25,735	24,852	24,027	23,255	22,532	21,852	21,212	19,764	18,502	17,391	16,406	15,526	14,736	13,375	12,244	11,290	9,767	8,606
35	34,174	33,386	32,634	31,915	31,226	30,567	29,329	28,187	27,131	26,151	25,240	24,390	23,595	21,818	20,289	18,961	17,796	16,766	15,848	14,285	13,002	11,931	10,243	8,974
40	38,925	37,906	36,939	36,020	35,146	34,313	32,761	31,343	30,042	28,846	27,740	26,717	25,766	23,661	21,874	20,338	19,004	17,834	16,799	15,053	13,636	12,462	10,632	9,271
45	43,644	42,367	41,163	40,025	38,949	37,928	36,041	34,332	32,778	31,358	30,056	28,859	27,752	25,326	23,289	21,556	20,063	18,763	17,622	15,710	14,172	12,909	10,956	9,516
50	48,331	46,770	45,307	43,933	42,639	41,420	39,178	37,167	35,353	33,707	32,208	30,836	29,576	26,836	24,560	22,641	20,999	19,579	18,340	16,278	14,633	13,290	11,229	9,722
55	52,988	51,117	49,374	47,746	46,223	44,793	42,183	39,861	37,781	35,908	34,211	32,668	31,257	28,213	25,708	23,613	21,833	20,302	18,972	16,775	15,033	13,619	11,463	9,897
60	57,613	55,409	53,367	51,470	49,703	48,054	45,063	42,423	40,075	37,974	36,081	34,369	32,811	29,472	26,750	24,489	22,580	20,947	19,534	17,212	15,384	13,906	11,666	10,047
65	62,208	59,646	57,286	55,106	53,086	51,209	47,827	44,863	42,246	39,917	37,831	35,953	34,252	30,630	27,700	25,282	23,253	21,525	20,036	17,601	15,693	14,159	11,843	10,178
70	66,772	63,829	61,135	58,658	56,375	54,263	50,480	47,190	44,302	41,748	39,472	37,432	35,592	31,697	28,570	26,005	23,862	22,046	20,487	17,948	15,969	14,383	11,999	10,293
75	71,307	67,961	64,914	62,129	59,573	57,220	53,029	49,410	46,254	43,477	41,014	38,816	36,841	32,683	29,369	26,665	24,417	22,519	20,894	18,260	16,215	14,582	12,138	10,395
80	75,812	72,041	68,627	65,522	62,685	60,085	55,481	51,532	48,108	45,111	42,466	40,113	38,007	33,598	30,106	27,271	24,924	22,950	21,265	18,542	16,437	14,762	12,262	10,486
85	80,288	76,070	72,274	68,838	65,714	62,862	57,840	53,562	49,873	46,659	43,834	41,332	39,100	34,450	30,788	27,829	25,390	23,343	21,602	18,798	16,638	14,924	12,374	10,568
90	84,734	80,050	75,857	72,081	68,664	65,555	60,113	55,505	51,553	48,127	45,127	42,480	40,126	35,243	31,420	28,345	25,818	23,705	21,912	19,032	16,821	15,071	12,474	10,641
95	89,152	83,982	79,378	75,253	71,536	68,169	62,303	57,367	53,156	49,520	46,350	43,562	41,090	35,985	32,008	28,823	26,214	24,038	22,196	19,246	16,988	15,205	12,566	10,708
100	93,541	87,865	82,839	78,357	74,335	70,705	64,415	59,153	54,686	50,845	47,509	44,584	41,998	36,679	32,556	29,267	26,581	24,346	22,458	19,443	17,142	15,327	12,650	10,768
105	97,902	91,702	86,241	81,394	77,063	73,169	66,454	60,868	56,148	52,107	48,609	45,551	42,855	37,331	33,069	29,680	26,921	24,632	22,701	19,625	17,283	15,440	12,726	10,824
110	102,234	95,493	89,586	84,367	79,722	75,562	68,422	62,515	57,546	53,310	49,654	46,467	43,665	37,945	33,549	30,067	27,239	24,897	22,927	19,793	17,413	15,544	12,797	10,875
115	106,540	99,239	92,874	87,277	82,316	77,889	70,324	64,099	58,886	54,457	50,648	47,337	44,432	38,522	34,000	30,428	27,535	25,145	23,136	19,949	17,534	15,640	12,862	10,922
120	110,817	102,940	96,108	90,127	84,846	80,151	72,163	65,623	60,170	55,553	51,595	48,163	45,159	39,068	34,424	30,767	27,813	25,376	23,332	20,094	17,646	15,729	12,922	10,965
125	115,068	106,598	99,289	92,918	87,316	82,351	73,941	67,090	61,401	56,601	52,498	48,949	45,849	39,583	34,824	31,086	28,073	25,593	23,515	20,230	17,750	15,812	12,978	11,005
130	119,291	110,213	102,418	95,653	89,727	84,492	75,663	68,504	62,583	57,605	53,359	49,697	46,505	40,071	35,201	31,386	28,318	25,796	23,686	20,357	17,848	15,889	13,030	11,043
135	123,488	113,785	105,496	98,333	92,080	86,576	77,330	69,868	63,719	58,566	54,183	50,411	47,130	40,534	35,558	31,670	28,548	25,987	23,847	20,475	17,939	15,962	13,079	11,078
140	127,659	117,317	108,525	100,959	94,380	88,605	78,945	71,184	64,812	59,487	54,971	51,092	47,725	40,973	35,895	31,937	28,765	26,167	23,998	20,587	18,024	16,029	13,124	11,110
145	131,803	120,808	111,506	103,534	96,626	90,582	80,510	72,454	65,863	60,372	55,726	51,743	48,292	41,391	36,215	32,190	28,971	26,336	24,141	20,692	18,105	16,093	13,167	11,141
150	135,921	124,259	114,439	106,058	98,821	92,508	82,028	73,681	66,876	61,222	56,449	52,366	48,835	41,789	36,520	32,430	29,165	26,497	24,276	20,791	18,181	16,153	13,207	11,169
155	140,014	127,670	117,327	108,534	100,967	94,386	83,501	74,868	67,852	62,038	57,142	52,963	49,353	42,168	36,809	32,653	29,349	26,649	24,403	20,884	18,252	16,209	13,244	11,196
160	144,081	131,043	120,169	110,962	103,065	96,217	84,931	76,015	68,793	62,824	57,808	53,534	49,849	42,529	37,084	32,875	29,524	26,792	24,524	20,972	18,319	16,262	13,280	11,222
165	148,123	134,378	122,968	113,344	105,116	98,003	86,320	77,125	69,701	63,581	58,448	54,083	50,324	42,874	37,346	33,081	29,690	26,929	24,638	21,056	18,383	16,312	13,313	11,246
170	152,140	137,676	125,724	115,681	107,124	99,745	87,669	78,200	70,578	64,310	59,064	54,609	50,780	43,205	37,596	33,277	29,848	27,059	24,747	21,135	18,444	16,360	13,345	11,268
175	156,132	140,937	128,438	117,974	109,088	101,446	88,980	79,242	71,425	65,012	59,656	55,115	51,217	43,521	37,835	33,464	29,998	27,183	24,850	21,211	18,501	16,405	13,375	11,289
180	160,100	144,162	131,111	120,226	111,010	103,106	90,254	80,251	72,244	65,690	60,226	55,602	51,636	43,823	38,064	33,643	30,142	27,300	24,949	21,282	18,555	16,448	13,403	11,310
185	164,044	147,352	133,743	122,436	112,892	104,727	91,494	81,230	73,037	66,344	60,776	56,070	52,040	44,114	38,283	33,813	30,279	27,413	25,043	21,351	18,607	16,489	13,430	11,329
190	167,963	150,507	136,337	124,606	114,734	106,311	92,701	82,180	73,803	66,977	61,306	56,520	52,428	44,392	38,492	33,977	30,410	27,520	25,132	21,416	18,657	16,527	13,456	11,347
195	171,859	153,627	138,893	126,737	116,538	107,859	93,875	83,101	74,546	67,587	61,817	56,955	52,802	44,660	38,693	34,133	30,535	27,623	25,218	21,478	18,704	16,564	13,480	11,365
200	175,731	156,714	141,411	128,831	118,306	109,371	95,019	83,996	75,265	68,178	62,311	57,374	53,161	44,917	38,886	34,283	30,655	27,721	25,299	21,537	18,749	16,600	13,504	11,381

750 MCM	1 Run(s)	600 Volts	C-Value = 24,136	Magnetic Duct	Copper

Calculations:

1. $f = \dfrac{\text{Length} \times 2 \times \text{Isca}}{\text{Runs} \times \text{C-Value} \times E\ \text{L-L}}$

2. $M = \dfrac{1}{1 + f}$

3. $\text{Isc} = \text{Isca} \times M$

4. Add Motor Contribution = Motor FLA x 4

* All results are given in symmetrical amperes

Single-Phase L-L Bolted Fault-Current* Table

One-Way Distance in Feet

Available Isc in Thousands (Isca)	5	10	15	20	25	30	40	50	60	70	80	90	100	125	150	175	200	225	250	300	350	400	500	600
3	2,997	2,994	2,991	2,988	2,985	2,981	2,975	2,969	2,963	2,957	2,951	2,945	2,939	2,924	2,910	2,895	2,881	2,866	2,852	2,824	2,797	2,770	2,718	2,668
5	4,991	4,983	4,974	4,966	4,957	4,949	4,932	4,915	4,899	4,882	4,866	4,849	4,833	4,793	4,754	4,715	4,677	4,640	4,603	4,531	4,461	4,393	4,264	4,142
7	6,983	6,966	6,950	6,933	6,916	6,900	6,867	6,835	6,803	6,771	6,739	6,708	6,677	6,601	6,527	6,454	6,383	6,313	6,245	6,113	5,987	5,866	5,637	5,426
10	9,966	9,931	9,897	9,864	9,830	9,797	9,731	9,666	9,602	9,539	9,476	9,415	9,354	9,205	9,061	8,922	8,787	8,655	8,528	8,284	8,054	7,836	7,433	7,071
15	14,923	14,846	14,771	14,696	14,621	14,548	14,403	14,261	14,122	13,986	13,852	13,721	13,592	13,281	12,983	12,698	12,426	12,165	11,915	11,444	11,009	10,606	9,882	9,251
20	19,863	19,728	19,594	19,462	19,333	19,204	18,953	18,708	18,470	18,237	18,010	17,789	17,573	17,056	16,568	16,107	15,671	15,259	14,867	14,141	13,483	12,883	11,831	10,937
25	24,786	24,576	24,369	24,166	23,966	23,769	23,385	23,014	22,654	22,305	21,966	21,638	21,320	20,563	19,858	19,200	18,584	18,006	17,463	16,470	15,584	14,788	13,418	12,280
30	29,692	29,391	29,096	28,806	28,523	28,245	27,704	27,184	26,683	26,201	25,735	25,286	24,852	23,829	22,888	22,018	21,212	20,462	19,764	18,502	17,391	16,406	14,736	13,375
35	34,582	34,174	33,776	33,386	33,006	32,634	31,915	31,226	30,567	29,936	29,329	28,747	28,187	26,880	25,688	24,597	23,595	22,671	21,818	20,289	18,961	17,796	15,848	14,285
40	39,455	38,925	38,409	37,906	37,416	36,939	36,020	35,146	34,313	33,519	32,761	32,036	31,343	29,734	28,282	26,966	25,766	24,669	23,661	21,874	20,338	19,004	16,799	15,053
45	44,312	43,644	42,996	42,367	41,756	41,163	40,025	38,949	37,928	36,960	36,041	35,165	34,332	32,411	30,693	29,149	27,752	26,484	25,326	23,289	21,556	20,063	17,622	15,710
50	49,151	48,331	47,538	46,770	46,027	45,307	43,933	42,639	41,420	40,268	39,178	38,146	37,167	34,926	32,940	31,168	29,576	28,140	26,836	24,560	22,641	20,999	18,340	16,278
55	53,975	52,988	52,036	51,117	50,231	49,374	47,746	46,223	44,793	43,449	42,183	40,989	39,861	37,295	35,039	33,040	31,257	29,657	28,213	25,708	23,613	21,833	18,972	16,775
60	58,782	57,613	56,489	55,409	54,369	53,367	51,470	49,703	48,054	46,511	45,063	43,704	42,423	39,528	37,003	34,781	32,811	31,052	29,472	26,750	24,489	22,580	19,534	17,212
65	63,573	62,208	60,900	59,646	58,442	57,286	55,106	53,086	51,209	49,460	47,827	46,298	44,863	41,638	38,846	36,405	34,252	32,340	30,630	27,700	25,282	23,253	20,036	17,601
70	68,348	66,772	65,268	63,829	62,453	61,135	58,658	56,375	54,263	52,303	50,480	48,779	47,190	43,635	40,578	37,922	35,592	33,532	31,697	28,570	26,005	23,862	20,487	17,948
75	73,107	71,307	69,594	67,961	66,403	64,914	62,129	59,573	57,220	55,045	53,029	51,156	49,410	45,527	42,210	39,343	36,841	34,638	32,683	29,369	26,665	24,417	20,894	18,260
80	77,850	75,812	73,878	72,041	70,292	68,627	65,522	62,685	60,085	57,691	55,481	53,434	51,532	47,322	43,748	40,676	38,007	35,667	33,598	30,106	27,271	24,924	21,265	18,542
85	82,577	80,288	78,122	76,070	74,123	72,274	68,838	65,714	62,862	60,247	57,840	55,619	53,562	49,028	45,202	41,930	39,100	36,628	34,450	30,788	27,829	25,390	21,602	18,798
90	87,288	84,734	82,325	80,050	77,897	75,857	72,081	68,664	65,555	62,716	60,113	57,717	55,505	50,651	46,579	43,112	40,126	37,526	35,243	31,420	28,345	25,818	21,912	19,032
95	91,983	89,152	86,489	83,982	81,615	79,378	75,253	71,536	68,169	65,104	62,303	59,733	57,367	52,198	47,883	44,227	41,090	38,368	35,985	32,008	28,823	26,214	22,196	19,246
100	96,663	93,541	90,614	87,865	85,278	82,839	78,357	74,335	70,705	67,414	64,415	61,672	59,153	53,672	49,121	45,281	41,998	39,159	36,679	32,556	29,267	26,581	22,458	19,443
105	101,327	97,902	94,700	91,702	88,888	86,241	81,394	77,063	73,169	69,650	66,454	63,538	60,868	55,080	50,297	46,279	42,855	39,903	37,331	33,069	29,680	26,921	22,701	19,625
110	105,975	102,234	98,749	95,493	92,445	89,586	84,367	79,722	75,562	71,815	68,422	65,335	62,515	56,425	51,417	47,225	43,665	40,604	37,945	33,549	30,067	27,239	22,927	19,793
115	110,608	106,540	102,760	99,239	95,951	92,874	87,277	82,316	77,889	73,913	70,324	67,067	64,099	57,712	52,483	48,123	44,432	41,267	38,522	34,000	30,428	27,535	23,136	19,949
120	115,226	110,817	106,733	102,940	99,407	96,108	90,127	84,846	80,151	75,947	72,163	68,737	65,623	58,945	53,501	48,977	45,159	41,893	39,068	34,424	30,767	27,813	23,332	20,094
125	119,828	115,068	110,671	106,598	102,814	99,289	92,918	87,316	82,351	77,920	73,941	70,349	67,090	60,126	54,472	49,790	45,849	42,486	39,583	34,824	31,086	28,073	23,515	20,230
130	124,416	119,291	114,572	110,213	106,172	102,418	95,653	89,727	84,492	79,834	75,663	71,906	68,504	61,260	55,401	50,565	46,505	43,049	40,071	35,201	31,386	28,318	23,686	20,357
135	128,988	123,488	118,438	113,785	109,484	105,496	98,333	92,080	86,576	81,692	77,330	73,410	69,868	62,348	56,289	51,304	47,130	43,584	40,534	35,558	31,670	28,548	23,847	20,475
140	133,545	127,659	122,270	117,317	112,750	108,525	100,959	94,380	88,605	83,496	78,945	74,864	71,184	63,393	57,140	52,010	47,725	44,092	40,973	35,895	31,937	28,765	23,998	20,587
145	138,087	131,803	126,066	120,808	115,971	111,506	103,534	96,626	90,582	85,250	80,510	76,270	72,454	64,399	57,956	52,685	48,292	44,576	41,391	36,215	32,190	28,971	24,141	20,692
150	142,614	135,921	129,829	124,259	119,147	114,439	106,058	98,821	92,508	86,954	82,028	77,631	73,681	65,367	58,738	53,331	48,835	45,038	41,789	36,520	32,430	29,165	24,276	20,791
155	147,126	140,014	133,558	127,670	122,280	117,327	108,534	100,967	94,386	88,611	83,501	78,949	74,868	66,299	59,490	53,949	49,353	45,478	42,168	36,809	32,658	29,349	24,403	20,884
160	151,624	144,081	137,253	131,043	125,371	120,169	110,962	103,065	96,217	90,222	84,931	80,226	76,015	67,197	60,212	54,543	49,849	45,899	42,529	37,084	32,875	29,524	24,524	20,972
165	156,107	148,123	140,916	134,378	128,420	122,968	113,344	105,116	98,003	91,791	86,320	81,464	77,125	68,063	60,907	55,112	50,324	46,302	42,874	37,346	33,081	29,690	24,638	21,056
170	160,575	152,140	144,547	137,676	131,429	125,724	115,681	107,124	99,745	93,318	87,669	82,664	78,200	68,899	61,575	55,659	50,780	46,687	43,205	37,596	33,277	29,848	24,747	21,135
175	165,029	156,132	148,146	140,937	134,398	128,438	117,974	109,088	101,446	94,805	88,980	83,829	79,242	69,706	62,219	56,184	51,217	47,056	43,521	37,835	33,464	29,998	24,850	21,211
180	169,468	160,100	151,714	144,162	137,327	131,111	120,226	111,010	103,106	96,253	90,254	84,959	80,251	70,486	62,840	56,690	51,636	47,410	43,823	38,064	33,643	30,142	24,949	21,282
185	173,893	164,044	155,251	147,352	140,218	133,743	122,436	112,892	104,727	97,665	91,494	86,057	81,230	71,240	63,438	57,176	52,040	47,750	44,114	38,283	33,813	30,279	25,043	21,351
190	178,303	167,963	158,756	150,507	143,072	136,337	124,606	114,734	106,311	99,041	92,701	87,124	82,180	71,969	64,016	57,645	52,428	48,077	44,392	38,492	33,977	30,410	25,132	21,416
195	182,699	171,859	162,232	153,627	145,889	138,893	126,737	116,538	107,859	100,382	93,875	88,160	83,101	72,675	64,574	58,097	52,802	48,391	44,660	38,693	34,133	30,535	25,218	21,478
200	187,081	175,731	165,678	156,714	148,670	141,411	128,831	118,306	109,371	101,691	95,019	89,168	83,996	73,359	65,113	58,533	53,161	43,693	44,917	38,886	34,283	30,655	25,299	21,537

750 MCM | **2 Run(s)** | **600 Volts** | **C-Value = 24,136** | **Magnetic Duct** | **Copper**

Calculations:

1. $f = \dfrac{\text{Length} \times 2 \times \text{Isca}}{\text{Runs} \times \text{C-Value} \times \text{E L-L}}$
2. $M = \dfrac{1}{1 + f}$
3. $\text{Isc} = \text{Isca} \times M$
4. Add Motor Contribution = Motor FLA x 4

* All results are given in symmetrical amperes

Single-Phase L-L Bolted Fault-Current* Table

One-Way Distance in Feet

Available Isc in Thousands (Isca)

Isca	5	10	15	20	25	30	40	50	60	70	80	90	100	125	150	175	200	225	250	300	350	400	500	600
3	2,994	2,988	2,982	2,976	2,971	2,965	2,953	2,942	2,930	2,919	2,908	2,897	2,886	2,859	2,832	2,806	2,780	2,755	2,730	2,68[illegible]	2,635	2,590	2,505	2,425
5	4,984	4,967	4,951	4,935	4,919	4,903	4,872	4,840	4,810	4,779	4,749	4,720	4,691	4,619	4,550	4,483	4,417	4,354	4,292	4,17[illegible]	4,063	3,957	3,760	3,583
7	6,968	6,936	6,904	6,873	6,842	6,811	6,751	6,691	6,633	6,575	6,519	6,463	6,408	6,276	6,149	6,026	5,909	5,796	5,688	5,48[illegible]	5,291	5,112	4,789	4,505
10	9,934	9,870	9,806	9,743	9,681	9,619	9,439	9,381	9,267	9,155	9,046	8,939	8,835	8,585	8,349	8,125	7,913	7,712	7,521	7,16[illegible]	6,842	6,547	6,027	5,583
15	14,853	14,709	14,568	14,429	14,293	14,160	13,900	13,650	13,409	13,176	12,951	12,733	12,523	12,026	11,568	11,143	10,748	10,380	10,037	9,41[illegible]	8,864	8,374	7,541	6,859
20	19,740	19,486	19,239	18,998	18,763	18,534	18,091	17,670	17,268	16,883	16,515	16,163	15,826	15,041	14,331	13,684	13,094	12,552	12,053	11,166	10,400	9,733	8,626	7,745
25	24,595	24,202	23,822	23,454	23,096	22,750	22,087	21,462	20,872	20,313	19,783	19,280	18,802	17,704	16,728	15,854	15,066	14,353	13,705	12,5[illegible]9	11,607	10,782	9,440	8,395
30	29,418	28,858	28,320	27,800	27,300	26,817	25,901	25,046	24,245	23,494	22,788	22,123	21,496	20,074	18,828	17,727	16,749	15,872	15,083	13,7[illegible]9	12,581	11,617	10,074	8,893
35	34,211	33,456	32,734	32,042	31,379	30,743	29,545	28,438	27,410	26,454	25,562	24,728	23,947	22,195	20,682	19,362	18,200	17,170	16,250	14,6[illegible]7	13,382	12,297	10,581	9,286
40	38,972	37,996	37,067	36,183	35,340	34,535	33,031	31,652	30,384	29,214	28,130	27,124	26,187	24,106	22,331	20,800	19,465	18,292	17,251	15,4[illegible]9	14,054	12,862	10,997	9,604
45	43,703	42,479	41,322	40,226	39,187	38,200	36,368	34,703	33,185	31,794	30,514	29,334	28,241	25,836	23,808	22,076	20,578	19,271	18,120	16,136	14,625	13,339	11,344	9,868
50	48,404	46,907	45,500	44,175	42,925	41,743	39,565	37,603	35,827	34,211	32,734	31,379	30,133	27,410	25,138	23,214	21,564	20,133	18,880	16,730	15,116	13,746	11,637	10,089
55	53,075	51,281	49,604	48,033	46,558	45,172	42,632	40,363	38,323	36,480	34,805	33,278	31,879	28,847	26,342	24,237	22,444	20,898	19,551	17,3[illegible]8	15,544	14,099	11,888	10,277
60	57,717	55,601	53,635	51,803	50,092	48,490	45,576	42,992	40,686	38,614	36,743	35,045	33,497	30,166	27,437	25,161	23,234	21,581	20,148	17,735	15,918	14,406	12,107	10,440
65	62,329	59,868	57,595	55,488	53,529	51,705	48,404	45,500	42,925	40,625	38,559	36,694	35,000	31,379	28,438	26,000	23,948	22,195	20,682	18,200	16,250	14,678	12,297	10,581
70	66,912	64,085	61,487	59,091	56,875	54,819	51,124	47,895	45,050	42,524	40,266	38,235	36,400	32,500	29,355	26,765	24,595	22,750	21,163	18,572	16,546	14,918	12,466	10,706
75	71,466	68,250	65,311	62,615	60,132	57,839	53,740	50,184	47,069	44,318	41,871	39,680	37,707	33,538	30,199	27,465	25,185	23,254	21,598	18,[illegible]06	16,810	15,133	12,616	10,816
80	75,992	72,366	69,070	66,062	63,304	60,768	56,260	52,374	48,991	46,018	43,385	41,037	38,931	34,503	30,979	28,108	25,725	23,714	21,994	19,209	17,049	15,326	12,750	10,915
85	80,489	76,433	72,766	69,435	66,395	63,610	58,688	54,472	50,821	47,629	44,815	42,314	40,078	35,401	31,701	28,701	26,220	24,134	22,356	19,[illegible]84	17,266	15,501	12,870	11,003
90	84,959	80,452	76,399	72,735	69,407	66,370	61,028	56,483	52,568	49,160	46,167	43,518	41,156	36,239	32,372	29,250	26,678	24,521	22,687	19,[illegible]35	17,463	15,660	12,979	11,083
95	89,400	84,424	79,972	75,967	72,343	69,050	63,287	58,412	54,235	50,615	47,448	44,654	42,171	37,024	32,996	29,759	27,100	24,878	22,992	19,[illegible]65	17,643	15,805	13,079	11,155
100	93,814	88,350	83,486	79,131	75,207	71,654	65,468	60,265	55,828	52,000	48,663	45,729	43,128	37,760	33,580	30,233	27,493	25,208	23,274	20,[illegible]78	17,808	15,937	13,169	11,221
105	98,201	92,230	86,943	82,229	78,000	74,185	67,574	62,046	57,353	53,321	49,818	46,747	44,032	38,451	34,125	30,674	27,857	25,514	23,535	20,373	17,961	16,059	13,253	11,281
110	102,562	96,065	90,343	85,264	80,726	76,646	69,611	63,758	58,813	54,580	50,916	47,712	44,888	39,102	34,637	31,087	28,197	25,799	23,777	20,555	18,101	16,171	13,329	11,336
115	106,895	99,857	93,689	88,238	83,387	79,041	71,580	65,406	60,213	55,784	51,962	48,629	45,699	39,716	35,118	31,474	28,515	26,065	24,002	20,723	18,232	16,275	13,400	11,387
120	111,202	103,605	96,981	91,152	85,984	81,371	73,486	66,994	61,556	56,935	52,959	49,502	46,468	40,295	35,570	31,837	28,813	26,313	24,213	20,[illegible]80	18,353	16,372	13,465	11,435
125	115,482	107,311	100,220	94,008	88,522	83,640	75,331	68,524	62,846	58,036	53,910	50,332	47,199	40,844	35,997	32,178	29,092	26,546	24,410	21,026	18,466	16,462	13,526	11,478
130	119,737	110,976	103,409	96,809	91,000	85,849	77,119	70,000	64,085	59,091	54,820	51,124	47,895	41,364	36,400	32,500	29,355	26,765	24,595	21,163	18,572	16,546	13,582	11,519
135	123,966	114,599	106,548	99,555	93,422	88,002	78,851	71,425	65,277	60,103	55,689	51,880	48,558	41,857	36,782	32,804	29,603	26,971	24,768	21,291	18,670	16,624	13,635	11,557
140	128,169	118,182	109,639	102,247	95,790	90,099	80,531	72,800	66,424	61,074	56,522	52,601	49,189	42,326	37,143	33,091	29,836	27,164	24,932	21,412	18,763	16,697	13,684	11,592
145	132,347	121,725	112,682	104,889	98,104	92,144	82,161	74,130	67,528	62,007	57,320	53,292	49,793	42,772	37,486	33,363	30,057	27,347	25,086	21,525	18,850	16,766	13,731	11,626
150	136,500	125,230	115,678	107,481	100,368	94,138	83,743	75,415	68,593	62,904	58,085	53,953	50,369	43,196	37,812	33,621	30,266	27,520	25,231	21,632	18,932	16,831	13,774	11,657
155	140,628	128,695	118,629	110,024	102,582	96,083	85,278	76,658	69,620	63,766	58,820	54,586	50,921	43,602	38,122	33,866	30,465	27,684	25,369	21,734	19,010	16,892	13,815	11,686
160	144,732	132,124	121,536	112,520	104,749	97,981	86,770	77,861	70,611	64,597	59,526	55,194	51,449	43,988	38,417	34,099	30,653	27,840	25,499	2[illegible],829	19,083	16,950	13,854	11,714
165	148,811	135,515	124,400	114,970	106,869	99,834	88,220	79,027	71,569	65,397	60,205	55,777	51,955	44,358	38,699	34,320	30,832	27,987	25,623	2[illegible],920	19,152	17,005	13,890	11,740
170	152,866	138,869	127,221	117,375	108,944	101,643	89,630	80,156	72,493	66,168	60,858	56,337	52,441	44,711	38,968	34,531	31,002	28,127	25,741	2[illegible],006	19,218	17,056	13,925	11,764
175	156,897	142,188	130,000	119,737	110,976	103,409	91,000	81,250	73,387	66,912	61,487	56,875	52,907	45,050	39,224	34,733	31,165	28,261	25,852	2[illegible],088	19,280	17,105	13,957	11,788
180	160,904	145,471	132,739	122,057	112,966	105,135	92,334	82,312	74,252	67,630	62,093	57,393	53,355	45,374	39,470	34,926	31,320	28,388	25,959	2[illegible],165	19,339	17,152	13,988	11,810
185	164,888	148,719	135,439	124,336	114,915	106,821	93,632	83,342	75,090	68,324	62,677	57,892	53,786	45,686	39,705	35,110	31,468	28,510	26,061	2[illegible],239	19,395	17,196	14,018	11,831
190	168,848	151,933	138,099	126,574	116,825	108,470	94,896	84,342	75,900	68,995	63,241	58,373	54,201	45,984	39,931	35,286	31,609	28,626	26,158	2[illegible],310	19,449	17,238	14,046	11,851
195	172,785	155,114	140,722	128,774	118,696	110,081	96,127	85,313	76,686	69,643	63,785	58,837	54,600	46,271	40,147	35,455	31,744	28,737	26,250	22,377	19,500	17,279	14,072	11,870
200	176,699	158,261	143,307	130,936	120,530	111,657	97,327	86,256	77,447	70,271	64,311	59,284	54,985	46,548	40,355	35,617	31,874	28,843	26,339	22,442	19,549	17,317	14,098	11,888

1000 MCM | 1 Run(s) | 600 Volts | C-Value = 25,278 | Magnetic Duct | Copper

Calculations:

1. $f = \dfrac{\text{Length} \times 2 \times \text{Isca}}{\text{Runs} \times \text{C-Value} \times E\ \text{L-L}}$

2. $M = \dfrac{1}{1 + f}$

3. $\text{Isc} = \text{Isca} \times M$

4. Add Motor Contribution = Motor FLA x 4

* All results are given in symmetrical amperes

Single-Phase L-L Bolted Fault-Current* Table

One-Way Distance in Feet

Available Isc in Thousands (Isca)	5	10	15	20	25	30	40	50	60	70	80	90	100	125	150	175	200	225	250	300	350	400	500	600
3	2,997	2,994	2,991	2,988	2,985	2,982	2,976	2,971	2,965	2,959	2,953	2,948	2,942	2,928	2,914	2,900	2,886	2,872	2,859	2,832	2,806	2,780	2,730	2,682
5	4,992	4,984	4,975	4,967	4,959	4,951	4,935	4,919	4,903	4,887	4,872	4,856	4,840	4,802	4,764	4,727	4,691	4,655	4,619	4,550	4,483	4,417	4,292	4,174
7	6,984	6,968	6,952	6,936	6,920	6,904	6,873	6,842	6,811	6,781	6,751	6,721	6,691	6,618	6,547	6,477	6,408	6,341	6,276	6,149	6,026	5,909	5,688	5,482
10	9,967	9,934	9,902	9,870	9,838	9,806	9,743	9,681	9,619	9,559	9,499	9,440	9,381	9,239	9,100	8,966	8,835	8,708	8,585	8,349	8,125	7,913	7,521	7,165
15	14,926	14,853	14,781	14,709	14,638	14,568	14,429	14,293	14,160	14,029	13,900	13,774	13,650	13,350	13,062	12,787	12,523	12,270	12,026	11,568	11,143	10,748	10,037	9,414
20	19,869	19,740	19,612	19,486	19,362	19,239	18,998	18,763	18,534	18,310	18,091	17,878	17,670	17,170	16,697	16,250	15,826	15,424	15,041	14,331	13,684	13,094	12,053	11,166
25	24,796	24,595	24,397	24,202	24,011	23,822	23,454	23,096	22,750	22,414	22,087	21,770	21,462	20,729	20,044	19,403	18,802	18,237	17,704	16,728	15,854	15,066	13,705	12,569
30	29,706	29,418	29,136	28,858	28,586	28,320	27,800	27,300	26,817	26,351	25,901	25,466	25,046	24,053	23,136	22,286	21,496	20,761	20,074	18,828	17,727	16,749	15,083	13,719
35	34,601	34,211	33,829	33,456	33,091	32,734	32,042	31,379	30,743	30,132	29,545	28,981	28,438	27,164	26,000	24,932	23,947	23,038	22,195	20,682	19,362	18,200	16,250	14,677
40	39,479	38,972	38,478	37,996	37,526	37,067	36,183	35,340	34,535	33,766	33,031	32,327	31,652	30,083	28,661	27,368	26,187	25,104	24,106	22,331	20,800	19,465	17,251	15,489
45	44,342	43,703	43,083	42,479	41,893	41,322	40,226	39,187	38,200	37,261	36,368	35,516	34,703	32,826	31,141	29,620	28,241	26,985	25,836	23,808	22,076	20,578	18,120	16,186
50	49,189	48,404	47,644	46,907	46,193	45,500	44,175	42,925	41,743	40,625	39,565	38,559	37,603	35,409	33,456	31,707	30,133	28,707	27,410	25,138	23,214	21,564	18,880	16,790
55	54,021	53,075	52,163	51,281	50,428	49,604	48,033	46,558	45,172	43,865	42,632	41,467	40,363	37,845	35,623	33,647	31,879	30,288	28,847	26,342	24,237	22,444	19,551	17,318
60	58,836	57,717	56,639	55,601	54,600	53,635	51,803	50,092	48,490	46,988	45,576	44,246	42,992	40,147	37,655	35,455	33,497	31,744	30,166	27,437	25,161	23,234	20,148	17,785
65	63,636	62,329	61,074	59,868	58,710	57,595	55,488	53,529	51,705	50,000	48,404	46,907	45,500	42,326	39,565	37,143	35,000	33,091	31,379	28,438	26,000	23,948	20,682	18,200
70	68,421	66,912	65,468	64,085	62,759	61,487	59,091	56,875	54,819	52,907	51,124	49,457	47,895	44,390	41,364	38,724	36,400	34,340	32,500	29,355	26,765	24,595	21,163	18,572
75	73,190	71,466	69,821	68,250	66,748	65,311	62,615	60,132	57,839	55,714	53,740	51,901	50,184	46,350	43,060	40,206	37,707	35,501	33,538	30,199	27,465	25,185	21,598	18,906
80	77,944	75,992	74,134	72,366	70,680	69,070	66,062	63,304	60,768	58,427	56,260	54,248	52,374	48,212	44,663	41,600	38,931	36,583	34,503	30,979	28,108	25,725	21,994	19,209
85	82,683	80,489	78,409	76,433	74,554	72,766	69,435	66,395	63,610	61,050	58,688	56,501	54,472	49,984	46,179	42,913	40,378	37,594	35,401	31,701	28,701	26,220	22,356	19,484
90	87,407	84,959	82,644	80,452	78,373	76,399	72,735	69,407	66,370	63,587	61,028	58,668	56,483	51,672	47,616	44,151	41,156	38,541	36,239	32,372	29,250	26,678	22,687	19,735
95	92,115	89,400	86,841	84,424	82,138	79,972	75,967	72,343	69,050	66,043	63,287	60,752	58,412	53,282	48,980	45,321	42,171	39,430	37,024	32,996	29,759	27,100	22,992	19,965
100	96,809	93,814	91,000	88,350	85,849	83,486	79,131	75,207	71,654	68,421	65,468	62,759	60,265	54,819	50,276	46,429	43,128	40,266	37,760	33,580	30,233	27,493	23,274	20,178
105	101,487	98,201	95,122	92,230	89,508	86,943	82,229	78,000	74,185	70,726	67,574	64,692	62,046	56,289	51,510	47,478	44,032	41,053	38,451	34,125	30,674	27,857	23,535	20,373
110	106,151	102,562	99,207	96,065	93,116	90,343	85,264	80,726	76,646	72,959	69,611	66,556	63,758	57,695	52,684	48,475	44,888	41,796	39,102	34,637	31,087	28,197	23,777	20,555
115	110,799	106,895	103,256	99,857	96,675	93,689	88,238	83,387	79,041	75,126	71,580	68,354	65,406	59,041	53,805	49,422	45,699	42,498	39,716	35,118	31,474	28,515	24,002	20,723
120	115,433	111,202	107,269	103,605	100,184	96,981	91,152	85,984	81,371	77,228	73,486	70,090	66,994	60,332	54,875	50,323	46,468	43,162	40,295	35,570	31,837	28,813	24,213	20,880
125	120,053	115,482	111,247	107,311	103,645	100,220	94,008	88,522	83,640	79,269	75,331	71,767	68,524	61,570	55,897	51,181	47,199	43,792	40,844	35,997	32,178	29,092	24,410	21,026
130	124,658	119,737	115,190	110,976	107,059	103,409	96,809	91,000	85,849	81,250	77,119	73,387	70,000	62,759	56,875	52,000	47,895	44,391	41,364	36,400	32,500	29,355	24,595	21,163
135	129,248	123,966	119,099	114,599	110,427	106,548	99,555	93,422	88,002	83,176	78,851	74,955	71,425	63,901	57,812	52,782	48,558	44,959	41,857	36,782	32,804	29,603	24,768	21,291
140	133,824	128,169	122,973	118,182	113,750	109,639	102,247	95,790	90,099	85,047	80,531	76,471	72,800	65,000	58,710	53,530	49,139	45,500	42,326	37,143	33,091	29,836	24,932	21,412
145	138,385	132,347	126,814	121,725	117,029	112,682	104,889	98,104	92,144	86,867	82,161	77,939	74,130	66,058	59,571	54,245	49,793	46,016	42,772	37,486	33,363	30,057	25,086	21,525
150	142,932	136,500	130,622	125,230	120,265	115,678	107,481	100,368	94,138	88,637	83,743	79,361	75,415	67,076	60,399	54,930	50,369	46,508	43,196	37,812	33,621	30,266	25,231	21,632
155	147,465	140,628	134,397	128,695	123,458	118,629	110,024	102,582	96,083	90,359	85,278	80,739	76,658	68,058	61,193	55,587	50,921	46,978	43,602	38,122	33,866	30,465	25,369	21,734
160	151,983	144,732	138,141	132,124	126,609	121,536	112,520	104,749	97,981	92,036	86,770	82,075	77,861	69,005	61,958	56,217	51,449	47,427	43,988	38,417	34,099	30,653	25,499	21,829
165	156,488	148,811	141,852	135,515	129,719	124,400	114,970	106,869	99,834	93,668	88,220	83,371	79,027	69,919	62,693	56,822	51,955	47,857	44,358	38,699	34,320	30,832	25,623	21,920
170	160,978	152,866	145,532	138,869	132,790	127,221	117,375	108,944	101,643	95,259	89,630	84,628	80,156	70,801	63,402	57,403	52,441	48,269	44,711	38,968	34,531	31,002	25,741	22,006
175	165,455	156,897	149,181	142,188	135,821	130,000	119,737	110,976	103,409	96,809	91,000	85,849	81,250	71,654	64,085	57,962	52,907	48,563	45,050	39,224	34,733	31,165	25,852	22,088
180	169,917	160,904	152,799	145,471	138,814	132,739	122,057	112,966	105,135	98,320	92,334	87,035	82,312	72,478	64,743	58,500	53,355	49,042	45,374	39,470	34,926	31,320	25,959	22,165
185	174,366	164,888	156,387	148,719	141,769	135,439	124,336	114,915	106,821	99,793	93,632	88,188	83,342	73,276	65,379	59,019	53,786	49,406	45,686	39,705	35,110	31,468	26,061	22,239
190	178,801	168,848	159,945	151,933	144,686	138,099	126,574	116,825	108,470	101,230	94,896	89,308	84,342	74,048	65,993	59,518	54,201	49,756	45,984	39,931	35,286	31,609	26,158	22,310
195	183,222	172,785	163,473	155,114	147,568	140,722	128,774	118,696	110,081	102,632	96,127	90,398	85,313	74,795	66,586	60,000	54,600	50,092	46,271	40,147	35,455	31,744	26,250	22,377
200	187,629	176,699	166,973	158,261	150,414	143,307	130,936	120,530	111,657	104,000	97,327	91,458	86,256	75,519	67,159	60,465	54,985	50,416	46,548	40,355	35,617	31,874	26,339	22,442

1000 MCM	2 Run(s)	600 Volts	C-Value = 25,278	Magnetic Duct	Copper

Calculations:

1. $f = \dfrac{\text{Length} \times 2 \times \text{Isca}}{\text{Runs} \times \text{C-Value} \times E_{L-L}}$
2. $M = \dfrac{1}{1 + f}$
3. $\text{Isc} = \text{Isca} \times M$
4. Add Motor Contribution = Motor FLA x 4

* All results are given in symmetrical amperes

Short-Circuit Tables and Data: 480 Volts Single-Phase L-L

Single-Phase L-L Bolted Fault-Current* Table

One-Way Distance in Feet

Available Isc in Thousands (Isca)	5	10	15	20	25	30	40	50	60	70	80	90	100	125	150	175	200	225	250	300	350	400	500	600
1	949	903	862	824	789	757	700	651	609	571	539	509	483	428	384	348	318	293	272	237	211	189	157	135
3	2,585	2,270	2,024	1,826	1,664	1,527	1,313	1,151	1,025	923	840	771	712	598	515	453	404	365	332	282	245	217	176	148
5	3,944	3,256	2,773	2,414	2,138	1,918	1,591	1,360	1,187	1,053	946	859	787	650	554	482	427	383	347	293	253	223	180	151
10	6,512	4,828	3,836	3,182	2,719	2,373	1,892	1,573	1,346	1,177	1,045	940	854	695	586	506	446	398	360	302	260	228	183	153
15	8,318	5,754	4,399	3,560	2,990	2,577	2,020	1,661	1,410	1,225	1,083	970	879	711	598	515	453	404	364	305	262	230	184	154
20	9,657	6,365	4,747	3,785	3,147	2,693	2,090	1,708	1,444	1,250	1,103	986	892	720	604	520	456	406	367	306	263	231	185	154
25	10,689	6,798	4,983	3,934	3,249	2,768	2,135	1,737	1,465	1,266	1,115	996	900	725	607	522	458	408	368	307	264	231	185	155
30	11,509	7,120	5,155	4,039	3,321	2,820	2,166	1,758	1,479	1,277	1,123	1,003	905	729	610	524	460	409	369	308	264	232	186	155
35	12,176	7,370	5,284	4,119	3,374	2,858	2,188	1,773	1,490	1,285	1,129	1,007	909	731	612	525	461	410	369	308	265	232	186	155
40	12,730	7,569	5,386	4,180	3,416	2,887	2,205	1,784	1,498	1,291	1,134	1,011	912	733	613	526	461	411	370	309	265	232	186	155
45	13,196	7,732	5,468	4,229	3,448	2,911	2,219	1,793	1,504	1,295	1,138	1,014	915	735	614	527	462	411	370	309	265	232	186	155
50	13,595	7,867	5,535	4,269	3,475	2,930	2,230	1,800	1,509	1,299	1,140	1,016	916	736	615	528	462	412	371	309	265	232	186	155
55	13,940	7,981	5,591	4,303	3,497	2,945	2,239	1,806	1,513	1,302	1,143	1,018	918	737	615	528	463	412	371	309	265	232	186	155
60	14,240	8,079	5,639	4,331	3,516	2,959	2,247	1,811	1,517	1,305	1,145	1,020	919	738	616	529	463	412	371	310	266	232	186	155
65	14,505	8,163	5,680	4,355	3,532	2,970	2,253	1,815	1,520	1,307	1,146	1,021	920	738	616	529	463	412	371	310	266	233	186	155
70	14,740	8,237	5,716	4,376	3,545	2,980	2,259	1,819	1,522	1,309	1,148	1,022	921	739	617	529	464	412	371	310	266	233	186	155
75	14,950	8,303	5,747	4,394	3,557	2,988	2,264	1,822	1,524	1,310	1,149	1,023	922	740	617	530	464	413	372	310	266	233	186	155
80	15,139	8,360	5,775	4,411	3,568	2,995	2,268	1,825	1,526	1,312	1,150	1,024	923	740	618	530	464	413	372	310	266	233	186	155
85	15,309	8,412	5,799	4,425	3,577	3,002	2,272	1,827	1,528	1,313	1,151	1,025	923	740	618	530	464	413	372	310	266	233	186	155
90	15,464	8,459	5,821	4,438	3,586	3,008	2,275	1,829	1,530	1,314	1,152	1,026	924	741	618	530	464	413	372	310	266	233	186	155
95	15,605	8,501	5,841	4,449	3,593	3,013	2,278	1,831	1,531	1,315	1,153	1,026	925	741	618	531	465	413	372	310	266	233	186	155
100	15,734	8,539	5,859	4,460	3,600	3,018	2,281	1,833	1,532	1,316	1,154	1,027	925	741	619	531	465	413	372	310	266	233	186	155
105	15,853	8,574	5,876	4,469	3,606	3,022	2,283	1,835	1,533	1,317	1,154	1,027	925	742	619	531	465	413	372	310	266	233	186	155
110	15,962	8,606	5,891	4,478	3,612	3,026	2,286	1,836	1,534	1,318	1,155	1,028	926	742	619	531	465	413	372	310	266	233	186	155
115	16,064	8,635	5,904	4,486	3,617	3,030	2,288	1,837	1,535	1,318	1,155	1,028	926	742	619	531	465	413	372	310	266	233	186	155
120	16,158	8,662	5,917	4,493	3,622	3,033	2,289	1,839	1,536	1,319	1,156	1,028	926	742	619	531	465	414	372	310	266	233	186	155
125	16,245	8,687	5,929	4,500	3,626	3,036	2,291	1,840	1,537	1,320	1,156	1,029	927	742	619	531	465	414	372	310	266	233	186	155
130	16,327	8,710	5,940	4,506	3,630	3,039	2,293	1,841	1,538	1,320	1,157	1,029	927	743	619	531	465	414	372	310	266	233	186	155
135	16,403	8,732	5,950	4,512	3,634	3,042	2,294	1,842	1,538	1,321	1,157	1,029	927	743	620	531	465	414	372	310	266	233	186	155
140	16,475	8,752	5,959	4,517	3,637	3,044	2,296	1,843	1,539	1,321	1,157	1,030	927	743	620	531	465	414	372	311	266	233	186	155
145	16,542	8,771	5,968	4,522	3,641	3,047	2,297	1,843	1,539	1,322	1,158	1,030	928	743	620	532	465	414	372	311	266	233	186	155
150	16,605	8,789	5,976	4,527	3,644	3,049	2,298	1,844	1,540	1,322	1,158	1,030	928	743	620	532	465	414	373	311	266	233	186	155
155	16,665	8,806	5,984	4,532	3,647	3,051	2,299	1,845	1,541	1,322	1,158	1,030	928	743	620	532	465	414	373	311	266	233	186	155
160	16,721	8,821	5,991	4,536	3,649	3,053	2,300	1,846	1,541	1,323	1,159	1,031	928	743	620	532	465	414	373	311	266	233	187	155
165	16,774	8,836	5,998	4,540	3,652	3,054	2,301	1,846	1,541	1,323	1,159	1,031	928	744	620	532	465	414	373	311	266	233	187	155
170	16,824	8,850	6,004	4,543	3,654	3,056	2,302	1,847	1,542	1,323	1,159	1,031	929	744	620	532	466	414	373	311	266	233	187	155
175	16,872	8,863	6,010	4,547	3,656	3,058	2,303	1,847	1,542	1,324	1,159	1,031	929	744	620	532	466	414	373	311	266	233	187	155
180	16,917	8,876	6,016	4,550	3,658	3,059	2,304	1,848	1,543	1,324	1,159	1,031	929	744	620	532	466	414	373	311	266	233	187	155
185	16,960	8,887	6,021	4,553	3,661	3,061	2,305	1,849	1,543	1,324	1,160	1,032	929	744	620	532	466	414	373	311	266	233	187	155
190	17,001	8,899	6,027	4,556	3,662	3,062	2,306	1,849	1,543	1,324	1,160	1,032	929	744	620	532	456	414	373	311	266	233	187	155
195	17,040	8,909	6,031	4,559	3,664	3,063	2,306	1,849	1,544	1,325	1,160	1,032	929	744	620	532	456	414	373	311	266	233	187	155
200	17,078	8,920	6,036	4,562	3,666	3,064	2,307	1,850	1,544	1,325	1,160	1,032	929	744	620	532	456	414	373	311	266	233	187	155

14 AWG | 1 Run(s) | 480 Volts | C-Value = 389 | Magnetic Duct | Copper

Calculations:

1. $f = \dfrac{\text{Length} \times 2 \times \text{Isca}}{\text{Runs} \times \text{C-Value} \times \text{E L-L}}$

2. $M = \dfrac{1}{1 + f}$

3. Isc = Isca x M

4. Add Motor Contribution = Motor FLA x 4

* All results are given in symmetrical amperes

Single-Phase L-L Bolted Fault-Current* Table

One-Way Distance in Feet

Available Isc in Thousands (Isca)

Isca	5	10	15	20	25	30	40	50	60	70	80	90	100	125	150	175	200	225	250	300	350	400	500	600
1	967	937	908	881	856	832	787	748	712	679	649	622	597	542	497	458	425	397	372	330	297	270	228	198
3	2,724	2,495	2,301	2,135	1,991	1,866	1,657	1,490	1,354	1,241	1,145	1,063	991	849	743	660	594	540	495	424	371	330	270	228
5	4,278	3,738	3,319	2,985	2,711	2,484	2,127	1,860	1,652	1,487	1,351	1,238	1,142	958	824	724	645	582	530	449	390	345	280	235
10	7,476	5,969	4,968	4,254	3,720	3,305	2,702	2,285	1,979	1,746	1,562	1,413	1,290	1,059	899	780	689	617	559	470	406	357	288	241
15	9,957	7,452	5,954	4,957	4,246	3,714	2,969	2,473	2,119	1,854	1,648	1,483	1,348	1,098	926	801	706	630	570	478	411	361	290	243
20	11,938	8,508	6,610	5,404	4,570	3,959	3,124	2,580	2,197	1,913	1,694	1,520	1,379	1,118	94[illegible]	812	714	637	575	482	414	363	292	244
25	13,556	9,300	7,077	5,712	4,789	4,122	3,225	2,648	2,246	1,950	1,723	1,544	1,398	1,131	950	818	719	641	579	484	416	365	293	244
30	14,903	9,914	7,428	5,938	4,947	4,239	3,295	2,696	2,280	1,976	1,743	1,560	1,411	1,140	956	823	723	644	581	486	417	366	293	245
35	16,042	10,406	7,700	6,111	5,066	4,326	3,348	2,731	2,305	1,995	1,753	1,571	1,421	1,146	960	826	725	646	582	487	418	366	294	245
40	17,017	10,807	7,918	6,248	5,159	4,394	3,388	2,757	2,325	2,009	1,769	1,580	1,428	1,151	963	829	727	647	584	488	419	367	294	245
45	17,861	11,142	8,096	6,358	5,234	4,448	3,421	2,779	2,340	2,020	1,778	1,587	1,434	1,154	966	831	728	649	585	488	419	367	294	245
50	18,599	11,425	8,244	6,449	5,296	4,493	3,447	2,796	2,352	2,030	1,785	1,593	1,438	1,157	968	832	730	650	585	489	420	367	294	246
55	19,250	11,667	8,370	6,526	5,347	4,529	3,459	2,810	2,362	2,037	1,791	1,598	1,442	1,160	970	833	731	650	586	489	420	368	295	246
60	19,829	11,877	8,477	6,591	5,391	4,561	3,487	2,822	2,370	2,043	1,796	1,601	1,445	1,162	971	834	731	651	587	490	420	368	295	246
65	20,346	12,060	8,570	6,647	5,429	4,588	3,503	2,833	2,378	2,049	1,800	1,605	1,448	1,163	972	835	732	652	587	490	420	368	295	246
70	20,811	12,222	8,652	6,696	5,461	4,611	3,516	2,841	2,384	2,053	1,803	1,608	1,450	1,165	973	836	733	652	587	490	421	368	295	246
75	21,232	12,366	8,724	6,739	5,490	4,631	3,528	2,849	2,389	2,057	1,806	1,610	1,452	1,166	974	837	733	652	588	490	421	368	295	246
80	21,614	12,495	8,788	6,777	5,515	4,649	3,538	2,856	2,394	2,061	1,809	1,612	1,454	1,167	975	837	734	653	588	49[illegible]	421	368	295	246
85	21,963	12,611	8,845	6,811	5,537	4,665	3,547	2,862	2,398	2,064	1,812	1,614	1,455	1,168	976	838	734	653	588	49[illegible]	421	369	295	246
90	22,283	12,716	8,896	6,841	5,557	4,679	3,556	2,867	2,402	2,067	1,814	1,616	1,457	1,169	976	838	734	653	588	49[illegible]	421	369	295	246
95	22,578	12,811	8,943	6,869	5,576	4,692	3,563	2,872	2,406	2,069	1,816	1,617	1,458	1,170	977	839	735	654	589	49[illegible]	421	369	295	246
100	22,849	12,898	8,985	6,894	5,592	4,704	3,570	2,876	2,409	2,072	1,817	1,619	1,459	1,171	978	839	735	654	589	49[illegible]	421	369	295	246
105	23,100	12,978	9,024	6,916	5,607	4,714	3,576	2,880	2,411	2,074	1,819	1,620	1,460	1,171	978	839	735	654	589	49[illegible]	421	369	295	246
110	23,334	13,051	9,059	6,937	5,621	4,724	3,581	2,884	2,414	2,076	1,820	1,621	1,461	1,172	978	840	735	654	589	491	421	369	295	246
115	23,551	13,119	9,092	6,956	5,633	4,733	3,587	2,887	2,416	2,077	1,822	1,622	1,462	1,173	979	840	736	654	589	491	422	369	295	246
120	23,754	13,181	9,122	6,974	5,645	4,741	3,591	2,890	2,418	2,079	1,823	1,623	1,463	1,173	979	840	736	655	589	492	422	369	295	246
125	23,943	13,240	9,149	6,990	5,655	4,748	3,596	2,893	2,420	2,080	1,824	1,624	1,463	1,174	979	840	736	655	590	492	422	369	295	246
130	24,121	13,294	9,175	7,005	5,665	4,755	3,599	2,896	2,422	2,082	1,825	1,625	1,464	1,174	980	841	736	655	590	492	422	369	295	246
135	24,288	13,344	9,199	7,019	5,674	4,762	3,603	2,898	2,424	2,083	1,826	1,626	1,465	1,174	980	841	736	655	590	492	422	369	296	246
140	24,445	13,392	9,222	7,032	5,683	4,768	3,607	2,900	2,425	2,084	1,827	1,626	1,465	1,175	980	841	737	655	590	492	422	369	296	246
145	24,593	13,436	9,243	7,044	5,691	4,774	3,610	2,902	2,427	2,085	1,828	1,627	1,466	1,175	981	841	737	655	590	492	422	369	296	246
150	24,733	13,478	9,262	7,056	5,698	4,779	3,613	2,904	2,428	2,086	1,828	1,627	1,466	1,175	981	841	737	655	590	492	422	369	296	246
155	24,865	13,517	9,281	7,066	5,705	4,784	3,616	2,906	2,429	2,087	1,829	1,628	1,467	1,176	981	842	737	655	590	492	422	369	296	246
160	24,990	13,554	9,298	7,077	5,712	4,788	3,618	2,908	2,431	2,088	1,830	1,629	1,467	1,176	981	842	737	655	590	492	422	369	296	246
165	25,109	13,588	9,315	7,086	5,718	4,793	3,621	2,909	2,432	2,089	1,830	1,629	1,468	1,176	981	842	737	656	590	492	422	369	296	246
170	25,222	13,621	9,330	7,095	5,724	4,797	3,623	2,911	2,433	2,089	1,831	1,630	1,468	1,176	982	842	737	656	590	492	422	369	296	246
175	25,329	13,653	9,345	7,103	5,729	4,801	3,625	2,912	2,434	2,090	1,832	1,630	1,468	1,177	982	842	737	656	590	[illegible]92	422	369	296	246
180	25,432	13,682	9,359	7,111	5,734	4,804	3,627	2,914	2,435	2,091	1,832	1,630	1,469	1,177	982	842	737	656	590	[illegible]92	422	369	296	246
185	25,529	13,711	9,372	7,119	5,739	4,808	3,629	2,915	2,436	2,092	1,833	1,631	1,469	1,177	982	842	737	656	590	492	422	369	296	246
190	25,622	13,737	9,384	7,126	5,744	4,811	3,631	2,916	2,436	2,092	1,833	1,631	1,469	1,177	982	842	738	656	590	492	422	369	296	246
195	25,711	13,763	9,396	7,133	5,749	4,814	3,633	2,917	2,437	2,093	1,834	1,632	1,470	1,177	982	843	738	656	591	492	422	369	296	246
200	25,796	13,787	9,408	7,140	5,753	4,817	3,635	2,918	2,438	2,093	1,834	1,632	1,470	1,178	982	843	738	656	591	492	422	370	296	246

# 12 AWG	1 Run(s)	480 Volts	C-Value = 617	Magnetic Duct	Copper

Calculations:

1. $f = \dfrac{\text{Length} \times 2 \times \text{Isca}}{\text{Runs} \times \text{C-Value} \times \text{E L-L}}$

2. $M = \dfrac{1}{1 + f}$

3. Isc = Isca x M

4. Add Motor Contribution = Motor FLA x 4

* All results are given in symmetrical amperes

Single-Phase L-L Bolted Fault-Current* Table

One-Way Distance in Feet

Available Isc in Thousands (Isca)	5	10	15	20	25	30	40	50	60	70	80	90	100	125	150	175	200	225	250	300	350	400	500	600
1	979	959	940	922	904	887	855	825	797	771	746	723	702	653	611	574	541	511	485	440	402	371	320	282
3	2,820	2,661	2,519	2,391	2,275	2,170	1,987	1,833	1,700	1,586	1,486	1,397	1,319	1,157	1,030	929	845	776	717	622	549	492	407	347
5	4,520	4,124	3,792	3,509	3,266	3,054	2,703	2,425	2,199	2,011	1,853	1,717	1,601	1,368	1,195	1,060	953	865	792	678	593	527	430	364
10	8,248	7,019	6,108	5,407	4,850	4,397	3,705	3,201	2,818	2,517	2,274	2,074	1,906	1,585	1,357	1,186	1,053	947	861	728	630	556	450	378
15	11,376	9,163	7,670	6,596	5,785	5,152	4,227	3,584	3,110	2,747	2,460	2,228	2,035	1,673	1,421	1,235	1,092	978	886	746	644	566	457	382
20	14,038	10,814	8,794	7,410	6,403	5,636	4,548	3,811	3,280	2,879	2,565	2,313	2,106	1,721	1,455	1,261	1,112	994	899	755	651	572	460	385
25	16,330	12,125	9,642	8,003	6,841	5,973	4,764	3,962	3,392	2,965	2,633	2,368	2,152	1,752	1,477	1,277	1,124	1,004	908	761	655	575	462	386
30	18,325	13,191	10,305	8,454	7,168	6,221	4,921	4,070	3,470	3,024	2,680	2,406	2,183	1,772	1,492	1,288	1,133	1,011	913	765	658	577	464	387
35	20,077	14,076	10,836	8,809	7,421	6,411	5,039	4,150	3,528	3,069	2,715	2,434	2,206	1,787	1,502	1,296	1,139	1,016	917	768	660	579	465	388
40	21,628	14,821	11,273	9,095	7,623	6,561	5,131	4,213	3,573	3,103	2,741	2,455	2,224	1,799	1,510	1,302	1,144	1,020	920	770	662	580	465	389
45	23,010	15,457	11,637	9,331	7,788	6,683	5,205	4,263	3,609	3,130	2,762	2,472	2,237	1,808	1,517	1,306	1,147	1,023	922	771	663	581	466	389
50	24,250	16,007	11,946	9,529	7,925	6,783	5,266	4,304	3,638	3,151	2,779	2,486	2,249	1,815	1,522	1,310	1,150	1,025	924	773	664	582	466	389
55	25,369	16,487	12,211	9,697	8,041	6,868	5,317	4,337	3,663	3,170	2,794	2,497	2,258	1,821	1,526	1,313	1,153	1,027	926	774	665	582	467	390
60	26,383	16,909	12,441	9,841	8,140	6,940	5,360	4,366	3,683	3,185	2,805	2,507	2,266	1,826	1,530	1,316	1,155	1,028	927	775	665	583	467	390
65	27,306	17,284	12,643	9,967	8,226	7,003	5,397	4,391	3,701	3,198	2,816	2,515	2,272	1,830	1,533	1,318	1,156	1,030	928	775	666	583	467	390
70	28,151	17,618	12,821	10,077	8,301	7,057	5,429	4,412	3,716	3,209	2,824	2,522	2,278	1,834	1,535	1,320	1,158	1,031	929	776	666	584	468	390
75	28,927	17,919	12,980	10,175	8,367	7,105	5,458	4,431	3,729	3,219	2,832	2,528	2,283	1,837	1,537	1,322	1,159	1,032	930	777	667	584	468	390
80	29,641	18,191	13,122	10,262	8,426	7,147	5,483	4,447	3,741	3,228	2,839	2,533	2,287	1,840	1,539	1,323	1,160	1,033	931	777	667	584	468	390
85	30,302	18,437	13,249	10,340	8,478	7,185	5,505	4,462	3,751	3,235	2,845	2,538	2,291	1,843	1,541	1,324	1,161	1,034	931	778	667	585	468	391
90	30,914	18,662	13,365	10,410	8,525	7,219	5,525	4,475	3,760	3,242	2,850	2,542	2,294	1,845	1,543	1,326	1,162	1,034	932	778	668	585	468	391
95	31,483	18,868	13,470	10,474	8,568	7,249	5,543	4,486	3,768	3,248	2,855	2,546	2,297	1,847	1,544	1,327	1,163	1,035	933	778	668	585	469	391
100	32,013	19,057	13,567	10,532	8,607	7,277	5,559	4,497	3,776	3,254	2,859	2,549	2,300	1,849	1,545	1,328	1,164	1,036	933	779	668	585	469	391
105	32,509	19,232	13,655	10,585	8,642	7,302	5,574	4,507	3,783	3,259	2,863	2,552	2,303	1,850	1,546	1,328	1,164	1,036	933	779	668	585	469	391
110	32,973	19,393	13,736	10,634	8,675	7,325	5,587	4,516	3,789	3,264	2,866	2,555	2,305	1,852	1,548	1,329	1,165	1,037	934	779	669	585	469	391
115	33,409	19,543	13,811	10,679	8,705	7,347	5,599	4,524	3,795	3,268	2,870	2,558	2,307	1,853	1,548	1,330	1,165	1,037	934	779	669	586	469	391
120	33,818	19,682	13,880	10,720	8,732	7,366	5,611	4,531	3,800	3,272	2,873	2,560	2,309	1,854	1,549	1,330	1,166	1,037	934	780	669	586	469	391
125	34,203	19,812	13,945	10,759	8,758	7,384	5,621	4,538	3,805	3,275	2,875	2,562	2,311	1,856	1,550	1,331	1,166	1,038	935	780	669	586	469	391
130	34,567	19,934	14,005	10,795	8,781	7,401	5,631	4,544	3,809	3,279	2,878	2,564	2,313	1,857	1,551	1,332	1,167	1,038	935	780	669	586	469	391
135	34,911	20,048	14,061	10,828	8,803	7,417	5,640	4,550	3,813	3,282	2,880	2,566	2,314	1,858	1,552	1,332	1,167	1,038	935	780	669	586	469	391
140	35,236	20,155	14,114	10,859	8,824	7,431	5,649	4,556	3,817	3,285	2,882	2,568	2,315	1,859	1,552	1,333	1,167	1,039	935	780	669	586	469	391
145	35,545	20,255	14,163	10,888	8,843	7,445	5,656	4,561	3,821	3,287	2,884	2,570	2,317	1,859	1,553	1,333	1,168	1,039	936	781	670	586	469	391
150	35,838	20,350	14,209	10,915	8,861	7,458	5,664	4,565	3,824	3,290	2,886	2,571	2,318	1,860	1,553	1,333	1,168	1,039	936	781	670	586	469	391
155	36,116	20,439	14,253	10,941	8,878	7,470	5,671	4,570	3,827	3,292	2,888	2,573	2,319	1,861	1,554	1,334	1,163	1,039	936	781	670	586	469	391
160	36,381	20,524	14,294	10,965	8,894	7,481	5,677	4,574	3,830	3,294	2,890	2,574	2,320	1,862	1,554	1,334	1,169	1,040	936	781	670	586	469	391
165	36,633	20,604	14,333	10,988	8,909	7,492	5,683	4,578	3,833	3,296	2,891	2,575	2,321	1,862	1,555	1,334	1,169	1,040	936	781	670	587	470	391
170	36,874	20,680	14,369	11,010	8,923	7,502	5,689	4,582	3,835	3,298	2,893	2,576	2,322	1,863	1,555	1,335	1,169	1,040	937	781	670	587	470	391
175	37,104	20,752	14,404	11,030	8,937	7,511	5,694	4,585	3,838	3,300	2,894	2,577	2,323	1,863	1,556	1,335	1,169	1,040	937	781	670	587	470	392
180	37,324	20,821	14,437	11,049	8,949	7,520	5,700	4,589	3,840	3,302	2,896	2,579	2,324	1,864	1,556	1,335	1,170	1,040	937	781	670	587	470	392
185	37,534	20,886	14,468	11,068	8,961	7,529	5,705	4,592	3,842	3,303	2,897	2,580	2,325	1,865	1,556	1,336	1,170	1,041	937	781	670	587	470	392
190	37,736	20,948	14,498	11,085	8,973	7,537	5,709	4,595	3,845	3,305	2,898	2,580	2,326	1,865	1,557	1,336	1,170	1,041	937	782	670	587	470	392
195	37,929	21,008	14,527	11,102	8,984	7,544	5,714	4,598	3,847	3,306	2,899	2,581	2,326	1,866	1,557	1,336	1,170	1,041	937	782	670	587	470	392
200	38,114	21,064	14,554	11,118	8,994	7,552	5,718	4,600	3,848	3,308	2,900	2,582	2,327	1,866	1,557	1,336	1,170	1,041	937	782	670	587	470	392

# 10 AWG	1 Run(s)	480 Volts	C-Value = 981	Magnetic Duct	Copper

Calculations:

1. $f = \dfrac{\text{Length} \times 2 \times \text{Isca}}{\text{Runs} \times \text{C-Value} \times \text{E L-L}}$

2. $M = \dfrac{1}{1 + f}$

3. Isc = Isca x M

4. Add Motor Contribution = Motor FLA x 4

* All results are given in symmetrical amperes

Single-Phase L-L Bolted Fault-Current* Table

One-Way Distance in Feet

Available Isc in Thousands (Isca)

	5	10	15	20	25	30	40	50	60	70	80	90	100	125	150	175	200	225	250	300	350	400	500	600
3	2,884	2,777	2,678	2,585	2,499	2,418	2,271	2,141	2,025	1,921	1,827	1,742	1,664	1,497	1,361	1,247	1,151	1,069	998	[illegible]	787	712	598	516
5	4,686	4,410	4,164	3,944	3,747	3,568	3,257	2,996	2,773	2,582	2,415	2,268	2,139	1,871	1,663	1,496	1,360	1,247	1,151	[illegible]	880	787	650	554
7	6,401	5,896	5,465	5,092	4,767	4,481	4,002	3,615	3,296	3,029	2,802	2,606	2,436	2,095	1,837	1,636	1,475	1,342	1,232	1,057	926	824	675	572
10	8,820	7,889	7,136	6,514	5,992	5,547	4,830	4,277	3,838	3,480	3,184	2,934	2,720	2,301	1,994	1,760	1,574	1,424	1,300	1,108	965	854	695	586
15	12,493	10,703	9,363	8,320	7,487	6,805	5,757	4,988	4,401	3,937	3,562	3,252	2,992	2,493	2,136	1,869	1,661	1,495	1,359	1,150	997	879	712	598
20	15,778	13,027	11,094	9,660	8,554	7,676	6,368	5,441	4,749	4,214	3,787	3,438	3,149	2,601	2,215	1,929	1,709	1,533	1,391	1,173	1,014	893	720	604
25	18,733	14,979	12,478	10,693	9,354	8,314	6,801	5,754	4,986	4,399	3,936	3,561	3,251	2,670	2,265	1,967	1,738	1,557	1,410	1,186	1,024	901	726	608
30	21,407	16,641	13,610	11,513	9,977	8,802	7,124	5,983	5,157	4,532	4,042	3,647	3,323	2,719	2,300	1,993	1,759	1,574	1,424	1,196	1,031	906	729	610
35	23,837	18,073	14,553	12,181	10,474	9,187	7,374	6,159	5,287	4,632	4,121	3,712	3,376	2,754	2,326	2,013	1,774	1,586	1,434	1,203	1,036	910	732	612
40	26,055	19,320	15,351	12,735	10,881	9,498	7,573	6,297	5,389	4,710	4,183	3,762	3,418	2,782	2,345	2,027	1,785	1,595	1,441	1,208	1,040	913	734	613
45	28,088	20,415	16,035	13,202	11,220	9,756	7,736	6,409	5,471	4,772	4,232	3,801	3,450	2,803	2,361	2,039	1,794	1,602	1,447	1,212	1,043	915	735	614
50	29,958	21,385	16,628	13,601	11,507	9,972	7,871	6,502	5,538	4,823	4,272	3,834	3,477	2,821	2,373	2,048	1,801	1,607	1,451	1,215	1,045	917	736	615
55	31,683	22,251	17,146	13,946	11,753	10,156	7,986	6,580	5,594	4,866	4,305	3,861	3,499	2,835	2,383	2,056	1,807	1,612	1,455	1,218	1,047	919	737	616
60	33,281	23,027	17,603	14,247	11,966	10,315	8,083	6,646	5,642	4,902	4,334	3,883	3,518	2,848	2,392	2,062	1,812	1,616	1,458	1,220	1,049	920	738	616
65	34,764	23,727	18,010	14,512	12,153	10,453	8,168	6,703	5,683	4,933	4,358	3,903	3,534	2,858	2,399	2,067	1,816	1,619	1,461	1,222	1,050	921	739	617
70	36,145	24,363	18,373	14,748	12,317	10,574	8,242	6,753	5,719	4,960	4,379	3,920	3,547	2,867	2,406	2,072	1,820	1,622	1,463	1,224	1,052	922	739	617
75	37,434	24,941	18,700	14,958	12,463	10,682	8,307	6,796	5,750	4,984	4,397	3,934	3,559	2,875	2,411	2,076	1,823	1,625	1,466	1,225	1,053	923	740	618
80	38,639	25,471	18,996	15,147	12,594	10,778	8,365	6,835	5,778	5,004	4,413	3,947	3,570	2,882	2,416	2,080	1,826	1,627	1,467	1,227	1,054	923	740	618
85	39,769	25,957	19,266	15,317	12,712	10,864	8,417	6,870	5,803	5,023	4,428	3,959	3,579	2,888	2,420	2,083	1,828	1,629	1,469	1,228	1,054	924	741	618
90	40,830	26,405	19,511	15,472	12,818	10,942	8,463	6,901	5,825	5,039	4,441	3,969	3,588	2,893	2,424	2,086	1,830	1,631	1,470	1,229	1,055	925	741	619
95	41,829	26,819	19,736	15,613	12,915	11,012	8,506	6,929	5,845	5,054	4,452	3,978	3,595	2,898	2,428	2,088	1,832	1,632	1,472	1,229	1,056	925	742	619
100	42,771	27,203	19,944	15,743	13,004	11,076	8,544	6,954	5,863	5,068	4,463	3,986	3,602	2,903	2,431	2,091	1,834	1,634	1,473	1,230	1,056	926	742	619
105	43,660	27,560	20,135	15,862	13,085	11,135	8,579	6,977	5,879	5,080	4,472	3,994	3,608	2,907	2,433	2,093	1,836	1,635	1,474	1,231	1,057	926	742	619
110	44,501	27,893	20,312	15,971	13,159	11,189	8,611	6,998	5,894	5,091	4,481	4,001	3,614	2,910	2,436	2,095	1,837	1,636	1,475	1,232	1,057	926	742	619
115	45,298	28,204	20,476	16,073	13,228	11,239	8,640	7,018	5,908	5,101	4,489	4,007	3,619	2,914	2,438	2,096	1,839	1,637	1,476	1,232	1,058	927	743	619
120	46,054	28,495	20,629	16,167	13,292	11,285	8,667	7,035	5,921	5,111	4,496	4,013	3,624	2,917	2,441	2,098	1,840	1,638	1,476	1,233	1,058	927	743	620
125	46,772	28,768	20,772	16,254	13,351	11,327	8,692	7,052	5,932	5,120	4,503	4,019	3,628	2,920	2,443	2,099	1,841	1,639	1,477	1,233	1,059	927	743	620
130	47,455	29,025	20,906	16,336	13,406	11,367	8,716	7,067	5,943	5,128	4,509	4,023	3,632	2,922	2,444	2,101	1,842	1,640	1,478	1,234	1,059	928	743	620
135	48,105	29,267	21,031	16,413	13,457	11,404	8,737	7,082	5,953	5,135	4,515	4,028	3,636	2,925	2,446	2,102	1,843	1,641	1,478	1,234	1,059	928	743	620
140	48,725	29,495	21,149	16,484	13,505	11,438	8,758	7,095	5,963	5,142	4,520	4,032	3,640	2,927	2,448	2,103	1,844	1,641	1,479	1,235	1,060	928	743	620
145	49,317	29,711	21,259	16,551	13,550	11,471	8,777	7,107	5,972	5,149	4,525	4,036	3,643	2,929	2,449	2,104	1,845	1,642	1,479	1,235	1,060	928	744	620
150	49,883	29,915	21,364	16,614	13,593	11,501	8,794	7,119	5,980	5,155	4,530	4,040	3,646	2,931	2,451	2,105	1,845	1,643	1,480	1,235	1,060	928	744	620
155	50,423	30,109	21,462	16,674	13,633	11,529	8,811	7,130	5,987	5,161	4,534	4,044	3,649	2,933	2,452	2,105	1,846	1,643	1,480	1,236	1,060	929	744	620
160	50,941	30,293	21,556	16,730	13,670	11,556	8,827	7,140	5,995	5,166	4,539	4,047	3,652	2,935	2,453	2,107	1,847	1,644	1,481	1,236	1,061	929	744	620
165	51,438	30,468	21,644	16,783	13,706	11,582	8,841	7,150	6,001	5,171	4,542	4,050	3,654	2,936	2,454	2,108	1,847	1,644	1,481	1,236	1,061	929	744	620
170	51,914	30,634	21,728	16,834	13,739	11,606	8,855	7,159	6,008	5,176	4,546	4,053	3,656	2,938	2,455	2,109	1,848	1,645	1,482	1,237	1,061	929	744	621
175	52,371	30,793	21,808	16,882	13,771	11,628	8,869	7,168	6,014	5,180	4,550	4,056	3,659	2,939	2,456	2,110	1,849	1,645	1,482	1,237	1,061	929	744	621
180	52,809	30,944	21,883	16,927	13,801	11,650	8,881	7,176	6,020	5,185	4,553	4,058	3,661	2,941	2,457	2,110	1,849	1,646	1,482	1,237	1,061	929	744	621
185	53,232	31,088	21,955	16,970	13,830	11,670	8,893	7,183	6,025	5,189	4,556	4,061	3,663	2,942	2,458	2,111	1,850	1,646	1,483	1,237	1,062	930	744	621
190	53,638	31,227	22,024	17,011	13,857	11,690	8,904	7,191	6,030	5,192	4,559	4,063	3,665	2,943	2,459	2,112	1,850	1,646	1,483	1,237	1,062	930	744	621
195	54,029	31,359	22,090	17,050	13,883	11,708	8,915	7,198	6,035	5,196	4,562	4,065	3,667	2,944	2,460	2,112	1,851	1,647	1,483	1,238	1,062	930	745	621
200	54,406	31,485	22,153	17,088	13,908	11,726	8,925	7,204	6,040	5,200	4,564	4,068	3,668	2,945	2,461	2,113	1,851	1,647	1,484	1,238	1,062	930	745	621

# 8 AWG	1 Run(s)	480 Volts	C-Value = 1,557	Magnetic Duct	Copper

Calculations:

1. $f = \dfrac{\text{Length} \times 2 \times \text{Isca}}{\text{Runs} \times \text{C-Value} \times \text{E L-L}}$

2. $M = \dfrac{1}{1 + f}$

3. Isc = Isca x M

4. Add Motor Contribution = Motor FLA x 4

* All results are given in symmetrical amperes

Single-Phase L-L Bolted Fault-Current* Table

One-Way Distance in Feet

Available Isc in Thousands (Isca)	5	10	15	20	25	30	40	50	60	70	80	90	100	125	150	175	200	225	250	300	350	400	500	600
3	2,925	2,853	2,785	2,720	2,658	2,598	2,487	2,385	2,291	2,205	2,124	2,049	1,980	1,824	1,692	1,577	1,477	1,389	1,311	1,178	1,070	980	839	733
5	4,794	4,604	4,429	4,267	4,116	3,975	3,721	3,498	3,299	3,122	2,963	2,820	2,689	2,411	2,185	1,997	1,839	1,705	1,588	1,398	1,248	1,127	944	812
7	6,603	6,248	5,930	5,643	5,382	5,144	4,726	4,371	4,066	3,800	3,567	3,361	3,178	2,796	2,496	2,255	2,055	1,889	1,747	1,519	1,344	1,205	998	852
10	9,209	8,534	7,951	7,442	6,995	6,599	5,927	5,379	4,924	4,540	4,211	3,927	3,679	3,177	2,795	2,496	2,254	2,055	1,888	1,625	1,426	1,270	1,043	884
15	13,288	11,926	10,818	9,898	9,122	8,459	7,386	6,554	5,891	5,349	4,899	4,519	4,193	3,553	3,083	2,722	2,437	2,206	2,015	1,718	1,497	1,326	1,080	911
20	17,067	14,885	13,197	11,853	10,758	9,848	8,423	7,358	6,532	5,873	5,335	4,887	4,508	3,777	3,250	2,852	2,540	2,290	2,085	1,768	1,535	1,356	1,100	925
25	20,580	17,488	15,204	13,447	12,055	10,923	9,197	7,942	6,988	6,239	5,635	5,138	4,721	3,925	3,359	2,935	2,607	2,344	2,130	1,800	1,559	1,375	1,112	934
30	23,852	19,796	16,919	14,772	13,108	11,781	9,798	8,386	7,330	6,510	5,855	5,320	4,874	4,030	3,436	2,994	2,653	2,381	2,160	1,822	1,576	1,388	1,121	940
35	26,909	21,856	18,401	15,889	13,981	12,482	10,277	8,735	7,595	6,718	6,023	5,458	4,990	4,109	3,493	3,037	2,687	2,409	2,183	1,838	1,587	1,397	1,127	944
40	29,770	23,707	19,695	16,845	14,716	13,064	10,669	9,016	7,807	6,884	6,155	5,567	5,081	4,171	3,537	3,070	2,713	2,430	2,200	1,850	1,596	1,404	1,131	947
45	32,454	25,378	20,835	17,672	15,343	13,556	10,995	9,248	7,980	7,018	6,263	5,654	5,153	4,219	3,572	3,097	2,733	2,446	2,213	1,860	1,604	1,409	1,135	950
50	34,976	26,895	21,847	18,394	15,884	13,977	11,270	9,442	8,124	7,129	6,351	5,726	5,213	4,259	3,601	3,118	2,750	2,459	2,224	1,868	1,609	1,414	1,138	952
55	37,351	28,277	22,751	19,031	16,357	14,341	11,506	9,607	8,246	7,222	6,425	5,786	5,263	4,293	3,624	3,136	2,764	2,470	2,233	1,874	1,614	1,418	1,140	953
60	39,592	29,543	23,563	19,596	16,772	14,660	11,710	9,749	8,350	7,302	6,488	5,838	5,305	4,321	3,644	3,151	2,775	2,480	2,241	1,879	1,618	1,421	1,142	955
65	41,709	30,706	24,297	20,101	17,141	14,941	11,889	9,872	8,440	7,371	6,543	5,882	5,342	4,345	3,661	3,164	2,785	2,488	2,248	1,884	1,621	1,423	1,144	956
70	43,712	31,778	24,963	20,555	17,470	15,190	12,046	9,980	8,519	7,432	6,590	5,920	5,373	4,366	3,676	3,175	2,794	2,494	2,253	1,888	1,624	1,425	1,145	957
75	45,611	32,770	25,571	20,965	17,766	15,413	12,186	10,076	8,589	7,485	6,632	5,953	5,401	4,384	3,689	3,185	2,801	2,500	2,258	1,891	1,627	1,427	1,146	958
80	47,413	33,690	26,128	21,338	18,033	15,614	12,311	10,162	8,651	7,532	6,669	5,983	5,425	4,400	3,701	3,193	2 808	2,506	2,262	1,894	1,629	1,429	1,147	958
85	49,126	34,546	26,640	21,678	18,275	15,795	12,423	10,238	8,706	7,573	6,701	6,009	5,447	4,414	3,711	3,200	2 814	2,510	2,266	1,897	1,631	1,431	1,148	959
90	50,756	35,344	27,112	21,990	18,496	15,960	12,525	10,307	8,756	7,611	6,731	6,033	5,466	4,427	3,720	3,207	2 819	2,514	2,269	1,899	1,633	1,432	1,149	960
95	52,308	36,090	27,549	22,276	18,698	16,110	12,618	10,369	8,801	7,645	6,758	6,055	5,484	4,438	3,728	3,213	2,824	2,518	2,272	1,901	1,634	1,433	1,150	960
100	53,789	36,789	27,954	22,541	18,884	16,248	12,702	10,426	8,842	7,676	6,782	6,074	5,500	4,449	3,735	3,219	2,828	2,521	2,275	1,903	1,636	1,434	1,151	961
105	55,203	37,445	28,331	22,785	19,055	16,375	12,779	10,478	8,880	7,704	6,804	6,092	5,514	4,458	3,742	3,224	2,832	2,524	2,278	1,905	1,637	1,435	1,151	961
110	56,555	38,062	28,683	23,012	19,214	16,491	12,850	10,526	8,914	7,730	6,824	6,108	5,528	4,467	3,748	3,228	2,335	2,527	2,280	1,906	1,638	1,436	1,152	962
115	57,848	38,643	29,012	23,223	19,361	16,600	12,916	10,570	8,945	7,754	6,842	6,122	5,540	4,475	3,753	3,232	2,338	2,530	2,282	1,908	1,639	1,437	1,152	962
120	59,086	39,192	29,320	23,421	19,497	16,700	12,977	10,611	8,975	7,776	6,859	6,136	5,551	4,482	3,758	3,236	2,341	2,532	2,284	1,909	1,640	1,438	1,153	962
125	60,273	39,711	29,609	23,605	19,625	16,794	13,033	10,648	9,001	7,796	6,875	6,149	5,561	4,489	3,763	3,240	2,344	2,534	2,285	1,910	1,641	1,438	1,153	963
130	61,412	40,202	29,882	23,777	19,744	16,881	13,085	10,683	9,026	7,815	6,889	6,160	5,571	4,495	3,768	3,243	2,846	2,536	2,287	1,911	1,642	1,439	1,154	963
135	62,506	40,668	30,138	23,940	19,856	16,962	13,134	10,716	9,050	7,832	6,903	6,171	5,579	4,501	3,772	3,246	2,849	2,538	2,289	1,913	1,643	1,439	1,154	963
140	63,557	41,110	30,380	24,092	19,961	17,039	13,180	10,747	9,071	7,848	6,916	6,181	5,588	4,506	3,775	3,249	2,851	2,540	2,290	1,913	1,643	1,440	1,154	963
145	64,568	41,531	30,609	24,236	20,059	17,111	13,223	10,775	9,092	7,863	6,927	6,191	5,595	4,511	3,779	3,251	2,853	2,541	2,291	1,914	1,644	1,441	1,155	964
150	65,541	41,931	30,826	24,372	20,152	17,178	13,263	10,802	9,111	7,878	6,938	6,199	5,603	4,516	3,782	3,254	2,855	2,543	2,292	1,915	1,645	1,441	1,155	964
155	66,478	42,312	31,032	24,500	20,240	17,242	13,301	10,827	9,129	7,891	6,949	6,208	5,609	4,520	3,785	3,256	2,856	2,544	2,294	1,916	1,645	1,441	1,155	964
160	67,381	42,676	31,227	24,622	20,323	17,302	13,337	10,851	9,146	7,904	6,959	6,215	5,616	4,524	3,788	3,258	2,858	2,546	2,295	1,917	1,646	1,442	1,156	964
165	68,252	43,024	31,413	24,737	20,402	17,359	13,371	10,873	9,161	7,915	6,968	6,223	5,622	4,528	3,791	3,260	2,860	2,547	2,296	1,917	1,646	1,442	1,156	964
170	69,092	43,357	31,590	24,847	20,476	17,413	13,403	10,894	9,176	7,927	6,976	6,230	5,627	4,532	3,793	3,262	2,861	2,548	2,297	1,918	1,647	1,443	1,156	964
175	69,904	43,675	31,759	24,951	20,547	17,464	13,433	10,914	9,191	7,937	6,985	6,236	5,633	4,535	3,796	3,264	2,862	2,549	2,297	1,919	1,647	1,443	1,156	965
180	70,688	43,980	31,920	25,050	20,614	17,513	13,462	10,933	9,204	7,947	6,992	6,242	5,638	4,539	3,798	3,265	2,834	2,550	2,298	1,919	1,648	1,443	1,157	965
185	71,447	44,272	32,073	25,145	20,678	17,559	13,489	10,951	9,217	7,957	7,000	6,248	5,642	4,542	3,800	3,267	2,835	2,551	2,299	1,920	1,648	1,444	1,157	965
190	72,180	44,553	32,220	25,235	20,739	17,603	13,515	10,968	9,229	7,966	7,007	6,254	5,647	4,545	3,802	3,269	2,866	2,552	2,300	1,920	1,648	1,444	1,157	965
195	72,890	44,822	32,361	25,321	20,797	17,645	13,540	10,984	9,240	7,974	7,013	6,259	5,651	4,547	3,804	3,270	2,867	2,553	2,301	1,921	1,649	1,444	1,157	965
200	73,578	45,081	32,496	25,404	20,853	17,685	13,563	11,000	9,251	7,982	7,020	6,264	5,655	4,550	3,806	3,271	2,868	2 554	2,301	1,921	1,649	1,444	1,157	965

# 6 AWG	1 Run(s)	480 Volts	C-Value = 2,425	Magnetic Duct	Copper

Calculations:

1. $f = \dfrac{\text{Length} \times 2 \times \text{Isca}}{\text{Runs} \times \text{C-Value} \times E_{L\text{-}L}}$

2. $M = \dfrac{1}{1 + f}$

3. Isc = Isca x M

4. Add Motor Contribution = Motor FLA x 4

* All results are given in symmetrical amperes

Single-Phase L-L Bolted Fault-Current* Table

One-Way Distance in Feet

Available Isc in Thousands (Isca)

Isca	5	10	15	20	25	30	40	50	60	70	80	90	100	125	150	175	200	225	250	300	350	400	500	600
3	2,952	2,905	2,859	2,815	2,772	2,731	2,652	2,577	2,506	2,439	2,376	2,316	2,258	2,127	2,010	1,905	1,811	1,725	1,647	1,51[illegible]	1,396	1,297	1,135	1,010
5	4,867	4,741	4,621	4,507	4,398	4,295	4,102	3,926	3,764	3,615	3,477	3,350	3,231	2,969	2,746	2,554	2,387	2,241	2,111	1,89[illegible]	1,715	1,568	1,338	1,167
7	6,742	6,502	6,278	6,070	5,875	5,692	5,358	5,061	4,795	4,556	4,340	4,143	3,963	3,575	3,257	2,990	2,764	2,570	2,401	2,12[illegible]	1,901	1,722	1,449	1,250
10	9,481	9,013	8,589	8,204	7,851	7,528	6,955	6,463	6,036	5,661	5,331	5,037	4,774	4,222	3,785	3,430	3,135	2,887	2,676	2,334	2,070	1,859	1,545	1,321
15	13,862	12,884	12,035	11,292	10,634	10,049	9,053	8,237	7,556	6,978	6,483	6,053	5,677	4,914	4,331	3,872	3,501	3,195	2,938	2,53[illegible]	2,223	1,982	1,629	1,382
20	18,027	16,408	15,055	13,909	12,925	12,071	10,362	9,548	8,644	7,897	7,268	6,733	6,271	5,352	4,668	4,139	3,718	3,375	3,089	2,643	2,309	2,050	1,674	1,415
25	21,991	19,628	17,724	16,156	14,844	13,728	11,935	10,555	9,462	8,574	7,838	7,219	6,690	5,655	4,897	4,318	3,862	3,493	3,188	2,714	2,363	2,092	1,702	1,435
30	25,768	22,583	20,099	18,107	16,474	15,111	12,966	11,354	10,099	9,094	8,270	7,584	7,002	5,876	5,062	4,446	3,964	3,576	3,257	2,764	2,401	2,122	1,722	1,449
35	29,373	25,304	22,226	19,815	17,876	16,283	13,819	12,003	10,609	9,505	8,609	7,868	7,244	6,045	5,187	4,542	4,040	3,638	3,308	2,801	2,429	2,144	1,736	1,459
40	32,815	27,818	24,142	21,324	19,095	17,288	14,537	12,541	11,027	9,839	8,882	8,095	7,436	6,179	5,285	4,617	4,099	3,686	3,348	2,829	2,450	2,160	1,747	1,467
45	36,106	30,148	25,877	22,667	20,165	18,160	15,149	12,994	11,376	10,116	9,107	8,282	7,593	6,287	5,364	4,677	4,146	3,724	3,379	2,852	2,467	2,173	1,756	1,473
50	39,256	32,313	27,456	23,869	21,111	18,924	15,676	13,380	11,671	10,348	9,295	8,437	7,723	6,376	5,428	4,726	4,185	3,755	3,405	2,870	2,480	2,184	1,762	1,477
55	42,273	34,330	28,899	24,952	21,953	19,598	16,136	13,714	11,924	10,547	9,455	8,568	7,833	6,450	5,483	4,767	4,217	3,781	3,426	2,885	2,492	2,193	1,768	1,481
60	45,166	36,213	30,222	25,932	22,709	20,198	16,541	14,005	12,143	10,718	9,593	8,681	7,928	6,514	5,528	4,802	4,244	3,802	3,444	2,898	2,501	2,200	1,773	1,485
65	47,942	37,976	31,441	26,824	23,390	20,735	16,899	14,261	12,335	10,867	9,712	8,779	8,009	6,569	5,568	4,832	4,267	3,821	3,459	2,909	2,509	2,206	1,777	1,488
70	50,608	39,630	32,566	27,639	24,007	21,219	17,219	14,488	12,504	10,999	9,817	8,864	8,080	6,617	5,602	4,857	4,287	3,837	3,473	2,918	2,516	2,211	1,780	1,490
75	53,171	41,185	33,608	28,386	24,569	21,656	17,506	14,690	12,655	11,115	9,909	8,940	8,143	6,659	5,632	4,880	4,305	3,851	3,484	2,926	2,522	2,216	1,783	1,492
80	55,636	42,648	34,576	29,074	25,082	22,054	17,765	14,873	12,790	11,219	9,992	9,007	8,198	6,696	5,659	4,900	4,321	3,864	3,494	2,933	2,527	2,220	1,786	1,494
85	58,010	44,029	35,478	29,709	25,553	22,418	18,000	15,037	12,911	11,312	10,066	9,067	8,248	6,729	5,682	4,918	4,334	3,875	3,503	2,940	2,532	2,224	1,788	1,496
90	60,296	45,334	36,321	30,297	25,987	22,751	18,214	15,186	13,021	11,397	10,133	9,121	8,293	6,759	5,704	4,934	4,347	3,885	3,511	2,945	2,536	2,227	1,791	1,497
95	62,499	46,568	37,109	30,844	26,388	23,058	18,411	15,322	13,121	11,473	10,193	9,170	8,333	6,786	5,723	4,948	4,358	3,893	3,518	2,950	2,540	2,230	1,792	1,498
100	64,625	47,738	37,848	31,353	26,760	23,341	18,591	15,447	13,213	11,543	10,248	9,214	8,370	6,810	5,740	4,961	4,368	3,901	3,525	2,955	2,543	2,233	1,794	1,500
105	66,677	48,849	38,543	31,828	27,106	23,603	18,757	15,561	13,296	11,607	10,298	9,255	8,403	6,832	5,756	4,972	4,377	3,909	3,531	2,959	2,547	2,235	1,796	1,501
110	68,659	49,904	39,197	32,272	27,427	23,847	18,910	15,667	13,373	11,665	10,344	9,292	8,434	6,852	5,770	4,983	4,385	3,915	3,536	2,963	2,549	2,237	1,797	1,502
115	70,574	50,908	39,814	32,689	27,728	24,074	19,053	15,764	13,444	11,719	10,387	9,326	8,462	6,871	5,783	4,993	4,393	3,921	3,541	2,966	2,552	2,239	1,798	1,503
120	72,426	51,865	40,396	33,081	28,009	24,286	19,185	15,855	13,510	11,769	10,426	9,358	8,488	6,888	5,795	5,002	4,400	3,927	3,546	2,969	2,554	2,241	1,799	1,503
125	74,218	52,777	40,948	33,450	28,273	24,484	19,309	15,939	13,571	11,816	10,462	9,387	8,512	6,904	5,807	5,010	4,406	3,932	3,550	2,972	2,556	2,243	1,801	1,504
130	75,953	53,648	41,470	33,798	28,521	24,670	19,424	16,018	13,628	11,859	10,496	9,414	8,535	6,919	5,817	5,018	4,412	3,937	3,554	2,975	2,558	2,244	1,802	1,505
135	77,632	54,481	41,966	34,127	28,755	24,845	19,532	16,091	13,681	11,899	10,528	9,440	8,556	6,932	5,827	5,025	4,418	3,941	3,557	2,978	2,560	2,246	1,802	1,505
140	79,260	55,278	42,437	34,438	28,975	25,009	19,633	16,160	13,731	11,937	10,557	9,463	8,575	6,945	5,836	5,032	4,423	3,945	3,561	2,980	2,562	2,247	1,803	1,506
145	80,838	56,041	42,885	34,732	29,184	25,164	19,729	16,225	13,777	11,972	10,585	9,485	8,593	6,957	5,844	5,038	4,428	3,949	3,564	2,982	2,564	2,248	1,804	1,507
150	82,369	56,772	43,312	35,012	29,381	25,310	19,819	16,285	13,821	12,005	10,610	9,506	8,610	6,968	5,852	5,044	4,432	3,953	3,567	2,984	2,565	2,249	1,805	1,507
155	83,854	57,474	43,720	35,277	29,568	25,449	19,904	16,343	13,862	12,036	10,635	9,526	8,626	6,979	5,859	5,050	4,436	3,956	3,570	2,986	2,567	2,250	1,806	1,508
160	85,296	58,148	44,108	35,530	29,745	25,580	19,984	16,397	13,901	12,065	10,657	9,544	8,641	6,988	5,866	5,055	4,440	3,959	3,572	2,988	2,568	2,251	1,806	1,508
165	86,697	58,795	44,480	35,771	29,914	25,705	20,060	16,448	13,938	12,093	10,679	9,561	8,655	6,998	5,873	5,060	4,444	3,962	3,575	2,990	2,569	2,252	1,807	1,508
170	88,058	59,418	44,835	36,000	30,074	25,823	20,132	16,496	13,973	12,119	10,699	9,578	8,669	7,006	5,879	5,064	4,448	3,965	3,577	2,991	2,570	2,253	1,807	1,509
175	89,381	60,017	45,176	36,219	30,227	25,936	20,200	16,542	14,006	12,144	10,719	9,593	8,681	7,015	5,885	5,068	4,451	3,968	3,579	2,993	2,571	2,254	1,808	1,509
180	90,667	60,594	45,502	36,429	30,372	26,043	20,265	16,585	14,037	12,167	10,737	9,608	8,693	7,022	5,890	5,073	4,454	3,970	3,581	2,994	2,573	2,255	1,809	1,510
185	91,918	61,151	45,815	36,629	30,512	26,145	20,327	16,627	14,066	12,189	10,754	9,621	8,705	7,030	5,896	5,073	4,457	3,973	3,583	2,995	2,574	2,256	1,809	1,510
190	93,136	61,687	46,116	36,821	30,645	26,243	20,386	16,666	14,095	12,211	10,771	9,635	8,715	7,037	5,900	5,080	4,460	3,975	3,585	2,997	2,574	2,256	1,809	1,510
195	94,322	62,205	46,404	37,005	30,772	26,336	20,442	16,704	14,122	12,231	10,786	9,647	8,726	7,044	5,905	5,084	4,463	3,977	3,587	2,998	2,575	2,257	1,810	1,511
200	95,476	62,705	46,682	37,181	30,894	26,425	20,496	16,740	14,147	12,250	10,801	9,659	8,735	7,050	5,910	5,087	4,465	3,979	3,588	2,999	2,576	2,258	1,810	1,511

4 AWG | 1 Run(s) | 480 Volts | C-Value = 3,806 | Magnetic Duct | Copper

Calculations:

1. $f = \dfrac{\text{Length} \times 2 \times \text{Isca}}{\text{Runs} \times \text{C-Value} \times \text{E L-L}}$

2. $M = \dfrac{1}{1 + f}$

3. Isc = Isca x M

4. Add Motor Contribution = Motor FLA x 4

* All results are given in symmetrical amperes

Single-Phase L-L Bolted Fault-Current* Table

One-Way Distance in Feet

Available Isc in Thousands (Isca)	5	10	15	20	25	30	40	50	60	70	80	90	100	125	150	175	200	225	250	300	350	400	500	600
3	2,961	2,923	2,886	2,850	2,815	2,781	2,715	2,652	2,592	2,534	2,479	2,427	2,376	2,259	2,152	2,055	1,967	1,886	1,811	1,678	1,563	1,463	1,297	1,165
5	4,893	4,790	4,692	4,598	4,507	4,420	4,255	4,102	3,960	3,827	3,703	3,587	3,478	3,232	3,018	2,831	2,666	2,519	2,388	2,162	1,975	1,818	1,568	1,379
7	6,792	6,596	6,411	6,236	6,070	5,913	5,622	5,358	5,118	4,899	4,697	4,512	4,340	3,964	3,648	3,378	3,145	2,943	2,765	2,466	2,226	2,028	1,723	1,497
10	9,581	9,195	8,839	8,510	8,205	7,920	7,407	6,956	6,556	6,201	5,881	5,593	5,332	4,775	4,323	3,950	3,635	3,368	3,136	2,758	2,461	2,222	1,860	1,599
15	14,076	13,259	12,532	11,880	11,293	10,761	9,835	9,055	8,390	7,816	7,316	6,875	6,485	5,679	5,051	4,548	4,137	3,793	3,503	3,037	2,681	2,399	1,983	1,690
20	18,390	17,020	15,840	14,813	13,911	13,113	11,763	10,665	9,754	8,987	8,331	7,765	7,271	6,273	5,516	4,922	4,443	4,049	3,720	3,199	2,806	2,499	2,051	1,738
25	22,534	20,511	18,822	17,389	16,159	15,092	13,331	11,938	10,808	9,874	9,089	8,419	7,841	6,693	5,838	5,176	4,650	4,220	3,863	3,305	2,887	2,563	2,093	1,769
30	26,518	23,760	21,522	19,669	18,110	16,780	14,631	12,970	11,648	10,570	9,675	8,919	8,273	7,005	6,074	5,361	4,798	4,342	3,966	3,379	2,944	2,608	2,123	1,790
35	30,351	26,792	23,980	21,702	19,820	18,238	15,727	13,824	12,332	11,130	10,142	9,315	8,613	7,247	6,255	5,502	4,911	4,434	4,042	3,434	2,986	2,641	2,145	1,806
40	34,041	29,627	26,226	23,526	21,329	19,508	16,663	14,542	12,900	11,591	10,523	9,636	8,886	7,439	6,398	5,612	4,998	4,505	4,101	3,477	3,018	2,666	2,161	1,817
45	37,595	32,283	28,287	25,170	22,673	20,626	17,471	15,154	13,379	11,977	10,840	9,901	9,111	7,596	6,514	5,701	5,069	4,563	4,148	3,511	3,043	2,686	2,174	1,827
50	41,023	34,778	30,184	26,662	23,876	21,617	18,177	15,682	13,789	12,304	11,108	10,123	9,299	7,727	6,609	5,774	5,126	4,609	4,187	3,539	3,064	2,702	2,185	1,834
55	44,329	37,126	31,937	28,020	24,959	22,501	18,798	16,142	14,144	12,586	11,337	10,313	9,459	7,837	6,690	5,835	5,175	4,648	4,219	3,561	3,081	2,715	2,194	1,840
60	47,521	39,339	33,561	29,262	25,940	23,295	19,350	16,547	14,453	12,830	11,535	10,477	9,597	7,931	6,758	5,887	5,215	4,681	4,246	3,581	3,096	2,726	2,201	1,845
65	50,604	41,428	35,069	30,403	26,832	24,012	19,842	16,906	14,726	13,045	11,708	10,620	9,716	8,013	6,817	5,932	5,251	4,709	4,269	3,597	3,108	2,736	2,207	1,850
70	53,583	43,404	36,475	31,454	27,648	24,663	20,284	17,226	14,969	13,234	11,860	10,745	9,821	8,084	6,869	5,971	5,281	4,734	4,290	3,612	3,119	2,744	2,213	1,854
75	56,465	45,276	37,788	32,425	28,395	25,256	20,684	17,513	15,185	13,403	11,996	10,856	9,914	8,147	6,914	6,005	5,308	4,755	4,307	3,624	3,128	2,751	2,217	1,857
80	59,253	47,051	39,016	33,326	29,084	25,799	21,046	17,772	15,380	13,555	12,117	10,955	9,996	8,202	6,954	6,036	5,331	4,774	4,323	3,635	3,136	2,758	2,221	1,860
85	61,952	48,737	40,169	34,163	29,719	26,298	21,377	18,008	15,556	13,691	12,226	11,044	10,071	8,252	6,990	6,062	5,352	4,791	4,336	3,645	3,143	2,763	2,225	1,862
90	64,567	50,341	41,252	34,943	30,308	26,758	21,680	18,222	15,715	13,815	12,325	11,124	10,137	8,297	7,022	6,087	5,371	4,806	4,349	3,653	3,150	2,768	2,228	1,865
95	67,100	51,868	42,272	35,672	30,855	27,184	21,959	18,418	15,861	13,927	12,414	11,197	10,198	8,337	7,051	6,108	5,388	4,820	4,360	3,661	3,156	2,773	2,231	1,867
100	69,557	53,323	43,233	36,354	31,364	27,578	22,215	18,599	15,995	14,030	12,496	11,264	10,253	8,374	7,077	6,128	5,403	4,832	4,370	3,668	3,161	2,777	2,234	1,868
105	71,940	54,713	44,142	36,995	31,839	27,945	22,453	18,765	16,117	14,125	12,570	11,324	10,303	8,407	7,101	6,146	5,417	4,843	4,379	3,675	3,166	2,780	2,236	1,870
110	74,252	56,040	45,002	37,597	32,284	28,287	22,673	18,918	16,231	14,212	12,639	11,380	10,349	8,438	7,123	6,162	5,430	4,853	4,387	3,681	3,170	2,784	2,238	1,872
115	76,497	57,309	45,817	38,164	32,702	28,607	22,878	19,061	16,335	14,292	12,703	11,432	10,392	8,466	7,143	6,177	5,442	4,863	4,395	3,686	3,174	2,787	2,240	1,873
120	78,678	58,525	46,591	38,699	33,094	28,907	23,069	19,194	16,433	14,366	12,761	11,479	10,431	8,492	7,161	6,191	5,452	4,871	4,402	3,691	3,178	2,790	2,242	1,874
125	80,797	59,689	47,326	39,205	33,463	29,188	23,248	19,317	16,523	14,435	12,816	11,523	10,467	8,517	7,179	6,204	5,462	4,879	4,408	3,695	3,181	2,792	2,244	1,875
130	82,857	60,806	48,025	39,684	33,811	29,453	23,416	19,433	16,608	14,500	12,867	11,564	10,501	8,539	7,195	6,216	5,472	4,886	4,414	3,700	3,184	2,795	2,245	1,877
135	84,860	61,878	48,691	40,137	34,140	29,702	23,573	19,541	16,687	14,560	12,914	11,602	10,533	8,560	7,209	6,227	5,480	4,893	4,420	3,704	3,187	2,797	2,247	1,878
140	86,809	62,907	49,326	40,568	34,451	29,937	23,721	19,642	16,761	14,616	12,958	11,638	10,562	8,579	7,223	6,237	5,488	4,900	4,425	3,707	3,190	2,799	2,248	1,878
145	88,705	63,898	49,933	40,978	34,746	30,159	23,860	19,738	16,830	14,669	13,000	11,672	10,590	8,597	7,236	6,247	5,496	4,906	4,430	3,711	3,192	2,801	2,249	1,879
150	90,552	64,850	50,513	41,367	35,026	30,370	23,992	19,828	16,895	14,719	13,039	11,703	10,616	8,614	7,248	6,256	5,502	4,911	4,435	3,714	3,194	2,803	2,251	1,880
155	92,350	65,767	51,068	41,739	35,292	30,570	24,116	19,913	16,957	14,765	13,075	11,733	10,640	8,630	7,259	6,264	5,509	4,916	4,439	3,717	3,197	2,804	2,252	1,881
160	94,102	66,651	51,599	42,093	35,544	30,759	24,234	19,993	17,015	14,809	13,110	11,760	10,663	8,645	7,270	6,272	5,515	4,921	4,443	3,719	3,199	2,806	2,253	1,882
165	95,810	67,503	52,108	42,431	35,785	30,940	24,346	20,069	17,070	14,851	13,143	11,787	10,684	8,660	7,280	6,280	5,521	4,926	4,446	3,722	3,201	2,807	2,254	1,882
170	97,474	68,325	52,597	42,754	36,015	31,111	24,452	20,141	17,122	14,891	13,173	11,811	10,705	8,673	7,289	6,287	5,526	4,930	4,450	3,725	3,203	2,809	2,254	1,883
175	99,098	69,119	53,066	43,064	36,234	31,275	24,553	20,209	17,172	14,928	13,203	11,835	10,724	8,686	7,298	6,293	5,531	4,934	4,453	3,727	3,204	2,810	2,255	1,884
180	100,682	69,886	53,517	43,360	36,444	31,431	24,649	20,274	17,219	14,963	13,230	11,857	10,742	8,698	7,307	6,300	5,536	4,938	4,456	3,729	3,206	2,811	2,256	1,884
185	102,227	70,627	53,950	43,644	36,645	31,580	24,741	20,336	17,263	14,997	13,257	11,878	10,760	8,709	7,315	6,306	5,541	4,942	4,459	3,731	3,207	2,813	2,257	1,885
190	103,735	71,344	54,367	43,917	36,837	31,722	24,828	20,395	17,306	15,029	13,282	11,898	10,776	8,720	7,322	6,311	5,545	4,945	4,462	3,733	3,209	2,814	2,258	1,885
195	105,208	72,037	54,769	44,179	37,021	31,859	24,911	20,452	17,346	15,060	13,306	11,918	10,792	8,730	7,330	6,317	5,549	4,948	4,465	3,735	3,210	2,815	2,258	1,886
200	106,647	72,709	55,156	44,431	37,197	31,989	24,991	20,505	17,385	15,089	13,328	11,936	10,807	8,740	7,337	6,322	5,553	4,952	4,468	3,737	3,212	2,816	2,259	1,886

3 AWG | 1 Run(s) | 480 Volts

C-Value = 4,760 | Magnetic Duct | Copper

Calculations:

1. $f = \dfrac{\text{Length} \times 2 \times \text{Isca}}{\text{Runs} \times \text{C-Value} \times \text{E L-L}}$

2. $M = \dfrac{1}{1 + f}$

3. Isc = Isca x M

4. Add Motor Contribution = Motor FLA x 4

* All results are given in symmetrical amperes

Single-Phase L-L Bolted Fault-Current* Table

One-Way Distance in Feet

Available Isc in Thousands (Isca)

Isca	5	10	15	20	25	30	40	50	60	70	80	90	100	125	150	175	200	225	250	300	350	400	500	600
3	2,969	2,938	2,908	2,878	2,849	2,821	2,766	2,713	2,662	2,613	2,566	2,520	2,476	2,372	2,277	2,189	2,108	2,032	1,962	1,835	1,723	1,625	1,458	1,322
5	4,913	4,830	4,749	4,670	4,595	4,522	4,382	4,250	4,127	4,010	3,900	3,795	3,696	3,470	3,270	3,092	2,932	2,788	2,657	2,429	2,238	2,074	1,809	1,604
7	6,831	6,671	6,517	6,371	6,231	6,097	5,845	5,614	5,400	5,202	5,018	4,846	4,686	4,328	4,021	3,755	3,522	3,316	3,133	2,821	2,566	2,353	2,018	1,766
10	9,659	9,341	9,043	8,763	8,501	8,253	7,799	7,392	7,026	6,694	6,392	6,116	5,863	5,314	4,859	4,475	4,148	3,865	3,618	3,209	2,882	2,616	2,209	1,911
15	14,246	13,565	12,945	12,380	11,862	11,385	10,539	9,810	9,175	8,617	8,123	7,683	7,288	6,458	5,797	5,260	4,813	4,436	4,115	3,593	3,189	2,866	2,384	2,041
20	18,682	17,527	16,506	15,598	14,785	14,052	12,784	11,727	10,831	10,062	9,395	8,811	8,295	7,237	6,417	5,765	5,233	4,791	4,417	3,822	3,368	3,010	2,483	2,113
25	22,974	21,252	19,770	18,481	17,350	16,349	14,658	13,285	12,146	11,188	10,369	9,662	9,046	7,801	6,858	6,118	5,522	5,032	4,622	3,974	3,485	3,104	2,546	2,158
30	27,129	24,760	22,771	21,078	19,619	18,349	16,246	14,576	13,216	12,089	11,139	10,328	9,626	8,229	7,186	6,378	5,733	5,206	4,769	4,082	3,568	3,169	2,590	2,190
35	31,154	28,069	25,540	23,429	21,641	20,106	17,608	15,663	14,104	12,828	11,763	10,862	10,089	8,565	7,441	6,578	5,894	5,339	4,879	4,163	3,630	3,218	2,622	2,213
40	35,054	31,196	28,104	25,569	23,454	21,661	18,790	16,591	14,852	13,444	12,279	11,300	10,466	8,835	7,644	6,736	6,020	5,443	4,966	4,226	3,677	3,255	2,647	2,231
45	38,835	34,156	30,483	27,524	25,088	23,048	19,825	17,392	15,491	13,965	12,713	11,666	10,779	9,057	7,810	6,864	6,123	5,526	5,035	4,276	3,715	3,285	2,667	2,245
50	42,503	36,962	32,698	29,317	26,569	24,293	20,733	18,091	16,044	14,412	13,082	11,977	11,044	9,243	7,948	6,970	6,207	5,595	5,092	4,317	3,746	3,309	2,683	2,256
55	46,063	39,625	34,765	30,968	27,918	25,415	21,551	18,707	16,526	14,800	13,401	12,243	11,270	9,401	8,064	7,060	6,278	5,652	5,140	4,351	3,772	3,329	2,696	2,265
60	49,519	42,156	36,698	32,492	29,151	26,433	22,278	19,252	16,950	15,140	13,679	12,475	11,466	9,537	8,164	7,136	6,338	5,701	5,180	4,380	3,794	3,346	2,707	2,273
65	52,876	44,564	38,510	33,905	30,283	27,360	22,933	19,740	17,327	15,439	13,923	12,678	11,637	9,655	8,250	7,202	6,390	5,743	5,215	4,405	3,812	3,360	2,716	2,280
70	56,138	46,859	40,212	35,217	31,325	28,208	23,526	20,177	17,663	15,706	14,139	12,857	11,788	9,759	8,326	7,260	6,436	5,780	5,245	4,426	3,828	3,373	2,725	2,285
75	59,309	49,048	41,813	36,439	32,289	28,987	24,066	20,573	17,965	15,944	14,332	13,016	11,921	9,850	8,392	7,310	6,475	5,812	5,271	4,445	3,842	3,384	2,732	2,290
80	62,393	51,138	43,323	37,580	33,181	29,705	24,558	20,932	18,238	16,159	14,505	13,159	12,041	9,932	8,451	7,355	6,510	5,840	5,295	4,461	3,855	3,393	2,738	2,295
85	65,393	53,136	44,748	38,648	34,011	30,368	25,010	21,259	18,486	16,353	14,662	13,287	12,149	10,005	8,504	7,395	6,542	5,865	5,315	4,476	3,866	3,402	2,743	2,299
90	68,313	55,048	46,097	39,649	34,784	30,983	25,425	21,558	18,712	16,530	14,804	13,404	12,246	10,071	8,552	7,431	6,570	5,888	5,334	4,489	3,875	3,409	2,748	2,302
95	71,155	56,879	47,374	40,591	35,507	31,554	25,809	21,833	18,919	16,691	14,933	13,510	12,334	10,130	8,595	7,463	6,595	5,908	5,350	4,50[illegible]	3,884	3,416	2,753	2,305
100	73,924	58,634	48,585	41,477	36,183	32,087	26,[illegible]64	22,087	19,110	16,839	15,051	13,606	12,415	10,185	8,634	7,493	6,618	5,926	5,366	4,51[illegible]	3,892	3,422	2,757	2,308
105	76,621	60,318	49,736	42,312	36,817	32,585	26,494	22,322	19,285	16,975	15,160	13,695	12,489	10,234	8,669	7,520	6,639	5,943	5,379	4,52	3,899	3,428	2,760	2,310
110	79,249	61,935	50,830	43,102	37,413	33,051	26,802	22,540	19,447	17,101	15,260	13,777	12,556	10,280	8,702	7,544	6,658	5,958	5,392	4,53[illegible]	3,906	3,433	2,764	2,313
115	81,812	63,490	51,872	43,849	37,975	33,489	27,089	22,743	19,598	17,217	15,353	13,852	12,619	10,322	8,732	7,567	6,676	5,973	5,403	4,538	3,912	3,438	2,767	2,315
120	84,311	64,984	52,866	44,557	38,505	33,900	27,357	22,931	19,738	17,326	15,439	13,922	12,677	10,360	8,760	7,588	6,692	5,986	5,414	4,54[illegible]	3,918	3,442	2,769	2,317
125	86,749	66,423	53,814	45,229	39,005	34,288	27,609	23,108	19,869	17,426	15,518	13,987	12,731	10,396	8,785	7,607	6,707	5,997	5,424	4,553	3,923	3,446	2,772	2,319
130	89,128	67,809	54,720	45,867	39,479	34,653	27,846	23,274	19,991	17,520	15,593	14,047	12,781	10,430	8,809	7,625	6,721	6,009	5,433	4,559	3,927	3,450	2,774	2,320
135	91,450	69,145	55,587	46,474	39,928	34,999	28,068	23,429	20,106	17,608	15,662	14,104	12,828	10,461	8,831	7,641	6,734	6,019	5,441	4,565	3,932	3,453	2,777	2,322
140	93,718	70,433	56,416	47,053	40,355	35,326	28,278	23,575	20,213	17,690	15,728	14,157	12,871	10,490	8,852	7,657	6,746	6,028	5,449	4,571	3,936	3,456	2,779	2,323
145	95,932	71,677	57,211	47,604	40,760	35,636	28,477	23,713	20,314	17,768	15,789	14,206	12,912	10,517	8,871	7,671	6,757	6,037	5,456	4,576	3,940	3,459	2,781	2,325
150	98,095	72,878	57,974	48,131	41,145	35,930	28,664	23,843	20,410	17,841	15,846	14,253	12,951	10,543	8,890	7,685	6,767	6,046	5,463	4,581	3,943	3,462	2,782	2,326
155	100,209	74,038	58,706	48,634	41,513	36,210	28,842	23,966	20,500	17,909	15,900	14,297	12,987	10,566	8,907	7,697	6,777	6,054	5,470	4,585	3,947	3,464	2,784	2,327
160	102,276	75,160	59,409	49,116	41,863	36,476	29,011	24,082	20,585	17,974	15,952	14,338	13,021	10,589	8,923	7,709	6,787	6,061	5,476	4,589	3,950	3,467	2,786	2,328
165	104,296	76,245	60,085	49,577	42,198	36,730	29,171	24,192	20,665	18,036	16,000	14,377	13,053	10,610	8,938	7,721	6,795	6,068	5,481	4,593	3,953	3,469	2,787	2,329
170	106,272	77,296	60,736	50,019	42,517	36,972	29,324	24,297	20,742	18,094	16,046	14,414	13,084	10,630	8,952	7,731	6,804	6,075	5,487	4,597	3,956	3,471	2,788	2,330
175	108,204	78,313	61,362	50,443	42,823	37,203	29,469	24,397	20,814	18,149	16,089	14,449	13,112	10,649	8,965	7,741	6,811	6,081	5,492	4,601	3,958	3,473	2,790	2,331
180	110,095	79,299	61,965	50,850	43,116	37,424	29,607	24,492	20,883	18,202	16,130	14,482	13,140	10,667	8,978	7,751	6,819	6,087	5,497	4,604	3,961	3,475	2,791	2,332
185	111,946	80,254	62,547	51,242	43,397	37,636	29,740	24,582	20,949	18,251	16,169	14,514	13,166	10,685	8,990	7,760	6,826	6,092	5,501	4,607	3,963	3,477	2,792	2,333
190	113,757	81,181	63,109	51,618	43,667	37,839	29,866	24,668	21,011	18,299	16,207	14,544	13,190	10,701	9,002	7,768	6,832	6,098	5,505	4,610	3,965	3,479	2,793	2,333
195	115,531	82,080	63,651	51,980	43,926	38,033	29,987	24,751	21,071	18,344	16,242	14,572	13,214	10,716	9,013	7,777	6,839	6,103	5,510	4,613	3,967	3,480	2,794	2,334
200	117,268	82,953	64,175	52,329	44,175	38,219	30,102	24,829	21,128	18,387	16,276	14,600	13,236	10,731	9,023	7,784	6,845	6,107	5,513	4,616	3,969	3,482	2,795	2,335

# 2 AWG	1 Run(s)	480 Volts	C-Value = 5,906	Magnetic Duct	Copper

Calculations:

1. $f = \frac{\text{Length} \times 2 \times \text{Isca}}{\text{Runs} \times \text{C-Value} \times \text{E L-L}}$

2. $M = \frac{1}{1 + f}$

3. Isc = Isca x M

4. Add Motor Contribution = Motor FLA x 4

* All results are given in symmetrical amperes

Single-Phase L-L Bolted Fault-Current* Table

One-Way Distance in Feet

Available Isc in Thousands (Isca)	5	10	15	20	25	30	40	50	60	70	80	90	100	125	150	175	200	225	250	300	350	400	500	600
3	2,975	2,949	2,925	2,901	2,877	2,853	2,807	2,763	2,720	2,679	2,638	2,599	2,561	2,471	2,386	2,308	2,234	2,165	2,100	1,981	1,875	1,780	1,615	1,479
5	4,930	4,861	4,795	4,730	4,667	4,605	4,487	4,375	4,268	4,167	4,070	3,977	3,889	3,684	3,500	3,333	3,182	3,044	2,917	2,692	2,500	2,333	2,059	1,842
7	6,863	6,731	6,604	6,482	6,364	6,250	6,035	5,833	5,645	5,469	5,303	5,147	5,000	4,667	4,375	4,118	3,889	3,684	3,500	3,182	2,917	2,692	2,333	2,059
10	9,722	9,459	9,211	8,974	8,750	8,537	8,140	7,778	7,447	7,143	6,863	6,604	6,364	5,833	5,385	5,000	4,667	4,375	4,118	3,684	3,333	3,044	2,593	2,258
15	14,384	13,816	13,291	12,805	12,353	11,932	11,170	10,500	9,906	9,375	8,898	8,468	8,077	7,242	6,563	6,000	5,526	5,122	4,773	4,200	3,750	3,387	2,838	2,442
20	18,919	17,949	17,073	16,279	15,556	14,894	13,726	12,727	11,865	11,111	10,448	9,859	9,334	8,236	7,369	6,667	6,087	5,600	5,185	4,516	4,000	3,590	2,979	2,546
25	23,333	21,875	20,588	19,445	18,421	17,500	15,909	14,584	13,462	12,500	11,667	10,938	10,294	8,975	7,955	7,143	6,482	5,932	5,469	4,730	4,167	3,724	3,070	2,612
30	27,632	25,610	23,864	22,341	21,000	19,812	17,797	16,154	14,789	13,637	12,651	11,798	11,053	9,546	8,400	7,500	6,774	6,177	5,676	4,884	4,286	3,818	3,134	2,658
35	31,818	29,167	26,923	25,000	23,334	21,875	19,445	17,500	15,909	14,584	13,462	12,500	11,667	10,000	8,750	7,778	7,000	6,364	5,834	5,000	4,375	3,889	3,182	2,692
40	35,898	32,558	29,788	27,451	25,455	23,729	20,896	18,667	16,868	15,385	14,142	13,085	12,174	10,371	9,033	8,000	7,180	6,512	5,958	5,091	4,445	3,944	3,219	2,719
45	39,874	35,796	32,475	29,717	27,392	25,404	22,184	19,688	17,697	16,072	14,720	13,578	12,600	10,678	9,265	8,182	7,326	6,632	6,058	5,164	4,500	3,988	3,248	2,739
50	43,750	38,889	35,000	31,819	29,167	26,924	23,334	20,589	18,422	16,667	15,218	14,000	12,963	10,938	9,460	8,334	7,447	6,731	6,141	5,224	4,546	4,023	3,271	2,756
55	47,531	41,848	37,379	33,773	30,801	28,309	24,368	21,389	19,060	17,188	15,651	14,366	13,276	11,160	9,625	8,462	7,549	6,814	6,210	5,274	4,584	4,053	3,291	2,770
60	51,220	44,681	39,623	35,594	32,308	29,578	25,302	22,106	19,627	17,648	16,031	14,686	13,549	11,352	9,768	8,572	7,637	6,886	6,269	5,317	4,616	4,078	3,307	2,782
65	54,820	47,396	41,744	37,296	33,704	30,744	26,150	22,751	20,133	18,056	16,367	14,968	13,788	11,519	9,892	8,667	7,712	6,947	6,320	5,353	4,643	4,099	3,321	2,792
70	58,334	50,001	43,751	38,890	35,001	31,819	26,924	23,334	20,589	18,422	16,667	15,218	14,001	11,667	10,000	8,750	7,778	7,000	6,364	5,385	4,667	4,118	3,333	2,800
75	61,765	52,501	45,653	40,385	36,208	32,813	27,632	23,864	21,001	18,751	16,936	15,442	14,190	11,798	10,097	8,824	7,836	7,047	6,403	5,413	4,688	4,134	3,344	2,808
80	65,117	54,903	47,459	41,792	37,334	33,736	28,284	24,349	21,375	19,048	17,179	15,643	14,360	11,915	10,182	8,889	7,888	7,089	6,437	5,437	4,706	4,148	3,353	2,814
85	68,391	57,212	49,175	43,117	38,388	34,594	28,884	24,792	21,716	19,319	17,398	15,825	14,513	12,021	10,259	8,948	7,934	7,126	6,468	5,459	4,722	4,161	3,362	2,820
90	71,592	59,435	50,807	44,367	39,376	35,394	29,440	25,201	22,029	19,566	17,598	15,990	14,652	12,116	10,328	9,000	7,975	7,159	6,495	5,478	4,737	4,172	3,369	2,825
95	74,720	61,575	52,363	45,549	40,304	36,142	29,956	25,578	22,316	19,792	17,781	16,141	14,778	12,202	10,391	9,048	8,012	7,189	6,520	5,496	4,750	4,183	3,376	2,830
100	77,779	63,637	53,847	46,668	41,178	36,843	30,436	25,927	22,581	20,001	17,949	16,280	14,894	12,281	10,448	9,091	8,046	7,217	6,542	5,512	4,762	4,192	3,382	2,834
105	80,770	65,626	55,264	47,728	42,001	37,501	30,883	26,251	22,827	20,193	18,104	16,407	15,001	12,353	10,500	9,131	8,077	7,242	6,563	5,527	4,773	4,200	3,387	2,838
110	83,697	67,545	56,619	48,735	42,779	38,120	31,302	26,553	23,055	20,371	18,247	16,524	15,099	12,420	10,548	9,167	8,106	7,264	6,581	5,540	4,783	4,208	3,392	2,841
115	86,560	69,398	57,915	49,693	43,515	38,703	31,694	26,834	23,267	20,536	18,380	16,633	15,189	12,481	10,593	9,200	8,132	7,285	6,599	5,552	4,792	4,215	3,397	2,845
120	89,363	71,188	59,156	50,604	44,212	39,254	32,062	27,098	23,465	20,690	18,503	16,734	15,273	12,538	10,633	9,231	8,156	7,305	6,614	5,563	4,800	4,221	3,401	2,848
125	92,106	72,918	60,346	51,472	44,873	39,774	32,409	27,345	23,650	20,834	18,618	16,828	15,351	12,590	10,671	9,260	8,178	7,322	6,629	5,573	4,808	4,227	3,405	2,850
130	94,793	74,592	61,488	52,300	45,501	40,267	32,735	27,577	23,823	20,969	18,725	16,915	15,424	12,639	10,706	9,286	8,199	7,339	6,643	5,583	4,815	4,233	3,408	2,853
135	97,424	76,211	62,584	53,091	46,099	40,734	33,043	27,795	23,986	21,095	18,825	16,997	15,492	12,685	10,739	9,311	8,218	7,354	6,655	5,592	4,822	4,238	3,412	2,855
140	100,001	77,779	63,638	53,848	46,668	41,178	33,334	28,001	24,139	21,213	18,920	17,074	15,556	12,728	10,770	9,334	8,236	7,369	6,667	5,600	4,828	4,243	3,415	2,857
145	102,527	79,299	64,651	54,571	47,211	41,600	33,610	28,195	24,283	21,324	19,008	17,146	15,616	12,768	10,798	9,355	8,252	7,382	6,678	5,608	4,834	4,247	3,418	2,859
150	105,001	80,771	65,627	55,265	47,729	42,001	33,872	28,379	24,420	21,429	19,092	17,214	15,672	12,805	10,825	9,375	8,268	7,395	6,688	5,615	4,839	4,251	3,420	2,861
155	107,427	82,199	66,566	55,929	48,224	42,384	34,121	28,554	24,548	21,529	19,170	17,278	15,725	12,841	10,850	9,394	8,283	7,406	6,698	5,622	4,844	4,255	3,423	2,863
160	109,805	83,584	67,472	56,567	48,697	42,750	34,357	28,719	24,671	21,622	19,245	17,338	15,775	12,874	10,874	9,412	8,297	7,413	6,707	5,628	4,849	4,259	3,425	2,865
165	112,138	84,928	68,345	57,180	49,151	43,098	34,582	28,876	24,786	21,711	19,315	17,395	15,823	12,906	10,897	9,429	8,310	7,428	6,715	5,634	4,853	4,262	3,427	2,866
170	114,425	86,234	69,188	57,769	49,585	43,432	34,797	29,025	24,896	21,796	19,382	17,449	15,867	12,935	10,918	9,445	8,322	7,438	6,723	5,640	4,857	4,265	3,430	2,868
175	116,668	87,502	70,002	58,335	50,002	43,751	35,001	29,168	25,001	21,876	19,445	17,501	15,910	12,964	10,938	9,460	8,334	7,447	6,731	5,645	4,861	4,268	3,432	2,869
180	118,870	88,734	70,788	58,880	50,402	44,057	35,197	29,303	25,101	21,952	19,505	17,549	15,950	12,990	10,957	9,474	8,345	7,456	6,738	5,650	4,865	4,271	3,433	2,870
185	121,030	89,933	71,549	59,406	50,786	44,351	35,384	29,433	25,196	22,025	19,563	17,596	15,988	13,016	10,975	9,488	8,355	7,464	6,745	5,655	4,869	4,274	3,435	2,872
190	123,150	91,098	72,285	59,912	51,156	44,632	35,563	29,557	25,286	22,094	19,617	17,640	16,025	13,040	10,992	9,500	8,365	7,472	6,752	5,660	4,872	4,277	3,437	2,873
195	125,231	92,232	72,997	60,400	51,511	44,903	35,734	29,675	25,373	22,160	19,669	17,682	16,059	13,063	11,009	9,513	8,375	7,480	6,758	5,664	4,875	4,279	3,438	2,874
200	127,275	93,336	73,686	60,872	51,854	45,163	35,899	29,788	25,456	22,223	19,719	17,722	16,093	13,085	11,024	9,524	8,384	7,487	6,764	5,668	4,878	4,282	3,440	2,875

# 1 AWG	1 Run(s)	480 Volts	C-Value = 7,292	Magnetic Duct	Copper

Calculations:

1. $f = \dfrac{\text{Length} \times 2 \times \text{Isca}}{\text{Runs} \times \text{C-Value} \times \text{E L-L}}$

2. $M = \dfrac{1}{1 + f}$

3. Isc = Isca x M

4. Add Motor Contribution = Motor FLA x 4

* All results are given in symmetrical amperes

Single-Phase L-L Bolted Fault-Current* Table

One-Way Distance in Feet

Available Isc in Thousands (Isca)	5	10	15	20	25	30	40	50	60	70	80	90	100	125	150	175	200	225	250	300	350	400	500	600
3	2,979	2,959	2,938	2,918	2,899	2,879	2,841	2,804	2,767	2,732	2,698	2,664	2,631	2,553	2,479	2,409	2,343	2,281	2,222	2,112	2,013	1,923	1,764	1,630
5	4,942	4,886	4,831	4,777	4,724	4,673	4,573	4,477	4,386	4,298	4,213	4,132	4,054	3,871	3,703	3,550	3,409	3,278	3,157	2,941	2,752	2,586	2,307	2,083
7	6,887	6,778	6,673	6,571	6,471	6,375	6,191	6,017	5,852	5,697	5,549	5,409	5,276	4,970	4,697	4,453	4,233	4,034	3,852	3,534	3,265	3,034	2,657	2,364
10	9,772	9,554	9,345	9,146	8,955	8,771	8,426	8,107	7,812	7,537	7,281	7,041	6,817	6,315	5,881	5,503	5,171	4,877	4,614	4,165	3,796	3,487	2,999	2,631
15	14,493	14,018	13,574	13,157	12,765	12,396	11,717	11,110	10,562	10,065	9,614	9,201	8,822	7,998	7,315	6,740	6,248	5,823	5,453	4,837	4,346	3,946	3,332	2,883
20	19,108	18,292	17,543	16,853	16,215	15,623	14,561	13,634	12,818	12,094	11,448	10,867	10,342	9,228	8,331	7,593	6,974	6,449	5,998	5,261	4,686	4,224	3,528	3,029
25	23,621	22,387	21,275	20,268	19,353	18,516	17,043	15,786	14,703	13,758	12,928	12,192	11,535	10,166	9,088	8,216	7,497	6,894	6,381	5,553	4,916	4,410	3,657	3,124
30	28,036	26,314	24,791	23,435	22,219	21,124	19,227	17,643	16,301	15,148	14,147	13,271	12,496	10,906	9,674	8,692	7,892	7,226	6,664	5,767	5,083	4,543	3,748	3,190
35	32,356	30,084	28,110	26,379	24,848	23,486	21,165	19,262	17,672	16,325	15,169	14,166	13,287	11,503	10,141	9,068	8,200	7,484	6,882	5,930	5,209	4,644	3,816	3,239
40	36,584	33,705	31,246	29,122	27,268	25,636	22,896	20,685	18,863	17,336	16,038	14,921	13,949	11,996	10,522	9,371	8,447	7,689	7,056	6,058	5,307	4,722	3,869	3,277
45	40,722	37,187	34,216	31,685	29,503	27,602	24,451	21,945	19,906	18,213	16,786	15,566	14,511	12,409	10,839	9,622	8,650	7,857	7,197	6,162	5,387	4,785	3,911	3,307
50	44,774	40,537	37,032	34,085	31,573	29,406	25,856	23,071	20,827	18,981	17,436	16,123	14,995	12,761	11,107	9,832	8,820	7,997	7,314	6,247	5,452	4,836	3,946	3,332
55	48,742	43,762	39,706	36,337	33,496	31,067	27,131	24,081	21,647	19,660	18,007	16,610	15,415	13,064	11,336	10,011	8,964	8,115	7,412	6,319	5,507	4,879	3,974	3,352
60	52,628	46,870	42,247	38,454	35,287	32,601	28,294	24,993	22,381	20,263	18,512	17,039	15,784	13,328	11,534	10,165	9,087	8,216	7,497	6,380	5,553	4,916	3,998	3,369
65	56,436	49,866	44,666	40,449	36,959	34,023	29,359	25,820	23,042	20,804	18,962	17,420	16,109	13,560	11,707	10,299	9,194	8,303	7,569	6,433	5,593	4,947	4,019	3,384
70	60,168	52,757	46,972	42,330	38,523	35,345	30,338	26,574	23,641	21,291	19,366	17,760	16,400	13,765	11,859	10,417	9,288	8,379	7,633	6,478	5,627	4,974	4,037	3,396
75	63,825	55,548	49,172	44,108	39,990	36,576	31,241	27,264	24,185	21,731	19,729	18,065	16,660	13,948	11,995	10,522	9,371	8,447	7,689	6,519	5,658	4,998	4,052	3,407
80	67,410	58,244	51,273	45,791	41,369	37,726	32,076	27,898	24,683	22,132	20,059	18,341	16,895	14,112	12,116	10,615	9,445	8,507	7,738	6,554	5,684	5,019	4,066	3,417
85	70,926	60,850	53,281	47,387	42,667	38,802	32,851	28,482	25,139	22,498	20,359	18,592	17,107	14,260	12,225	10,698	9,511	8,560	7,783	6,586	5,708	5,037	4,078	3,426
90	74,374	63,371	55,204	48,902	43,891	39,812	33,571	29,022	25,559	22,834	20,634	18,821	17,301	14,394	12,323	10,774	9,570	8,608	7,822	6,615	5,730	5,054	4,089	3,433
95	77,755	65,810	57,045	50,341	45,047	40,761	34,244	29,523	25,947	23,143	20,886	19,030	17,477	14,516	12,413	10,842	9,624	8,652	7,858	6,640	5,749	5,069	4,099	3,440
100	81,073	68,171	58,811	51,711	46,141	41,654	34,872	29,989	26,306	23,428	21,118	19,223	17,640	14,628	12,494	10,904	9,673	8,692	7,891	6,663	5,766	5,082	4,108	3,447
105	84,329	70,458	60,506	53,017	47,178	42,497	35,461	30,424	26,640	23,693	21,333	19,400	17,789	14,730	12,569	10,961	9,718	8,728	7,921	6,685	5,782	5,095	4,116	3,452
110	87,524	72,675	62,133	54,262	48,161	43,294	36,014	30,830	26,950	23,938	21,532	19,565	17,927	14,825	12,638	11,013	9,759	8,761	7,948	6,704	5,797	5,106	4,123	3,457
115	90,660	74,824	63,697	55,452	49,096	44,047	36,534	31,210	27,241	24,167	21,716	19,717	18,055	14,912	12,701	11,061	9,797	8,791	7,973	6,722	5,810	5,116	4,130	3,462
120	93,739	76,909	65,202	56,588	49,985	44,762	37,024	31,567	27,512	24,380	21,889	19,859	18,174	14,993	12,760	11,103	9,831	8,819	7,996	6,738	5,822	5,126	4,136	3,466
125	96,763	78,932	66,651	57,676	50,832	45,440	37,487	31,903	27,767	24,580	22,050	19,991	18,285	15,069	12,815	11,147	9,864	8,845	8,018	6,753	5,834	5,134	4,142	3,470
130	99,732	80,897	68,046	58,718	51,640	46,084	37,924	32,219	28,006	24,767	22,200	20,115	18,388	15,139	12,865	11,186	9,894	8,869	8,037	6,768	5,844	5,143	4,147	3,474
135	102,649	82,806	69,391	59,717	52,411	46,697	38,338	32,517	28,231	24,943	22,341	20,231	18,485	15,204	12,913	11,221	9,922	8,892	8,056	6,781	5,854	5,150	4,152	3,478
140	105,514	84,660	70,689	60,676	53,148	47,281	38,731	32,800	28,444	25,109	22,474	20,340	18,576	15,266	12,957	11,255	9,948	8,913	8,073	6,793	5,863	5,157	4,156	3,481
145	108,330	86,463	71,942	61,597	53,853	47,838	39,104	33,067	28,644	25,265	22,599	20,442	18,661	15,323	12,998	11,286	9,972	8,933	8,089	6,804	5,872	5,164	4,161	3,484
150	111,096	88,217	73,152	62,481	54,528	48,370	39,459	33,320	28,834	25,413	22,717	20,539	18,742	15,378	13,037	11,315	9,995	8,951	8,104	6,815	5,879	5,170	4,165	3,487
155	113,816	89,923	74,321	63,332	55,175	48,879	39,796	33,561	29,014	25,553	22,829	20,630	18,817	15,429	13,074	11,343	10,017	8,968	8,118	6,825	5,887	5,176	4,168	3,489
160	116,489	91,583	75,451	64,151	55,795	49,365	40,118	33,789	29,185	25,685	22,934	20,716	18,889	15,477	13,109	11,369	10,037	8,984	8,132	6,834	5,894	5,181	4,172	3,492
165	119,117	93,200	76,545	64,940	56,391	49,831	40,426	34,007	29,347	25,810	23,035	20,798	18,957	15,522	13,141	11,394	10,056	9,000	8,144	6,843	5,900	5,186	4,175	3,494
170	121,701	94,774	77,604	65,701	56,964	50,278	40,719	34,214	29,501	25,930	23,130	20,875	19,021	15,565	13,172	11,417	10,074	9,014	8,156	6,851	5,907	5,191	4,178	3,496
175	124,242	96,308	78,630	66,435	57,514	50,706	41,000	34,412	29,648	26,043	23,220	20,949	19,082	15,606	13,201	11,439	10,091	9,028	8,167	6,859	5,913	5,195	4,181	3,498
180	126,741	97,803	79,623	67,143	58,044	51,118	41,268	34,601	29,789	26,151	23,306	21,019	19,140	15,645	13,229	11,459	10,107	9,041	8,178	6,867	5,918	5,200	4,184	3,500
185	129,200	99,261	80,587	67,826	58,555	51,513	41,525	34,782	29,922	26,254	23,388	21,085	19,195	15,682	13,255	11,479	10,123	9,053	8,188	6,874	5,923	5,204	4,187	3,502
190	131,619	100,682	81,521	68,487	59,047	51,893	41,772	34,955	30,050	26,353	23,466	21,149	19,248	15,717	13,280	11,498	10,137	9,065	8,197	6,881	5,928	5,208	4,189	3,504
195	133,999	102,069	82,428	69,126	59,521	52,259	42,009	35,120	30,173	26,447	23,540	21,209	19,298	15,750	13,304	11,516	10,151	9,076	8,206	6,887	5,933	5,211	4,191	3,505
200	136,341	103,423	83,308	69,744	59,978	52,612	42,236	35,279	30,290	26,537	23,611	21,267	19,346	15,782	13,327	11,533	10,165	9,086	8,215	6,893	5,938	5,215	4,194	3,507

1/0 AWG | 1 Run(s) | 480 Volts | C-Value = 8,924 | Magnetic Duct | Copper

Calculations:

1. $f = \dfrac{\text{Length} \times 2 \times \text{Isca}}{\text{Runs} \times \text{C-Value} \times \text{E L-L}}$

2. $M = \dfrac{1}{1 + f}$

3. Isc = Isca x M

4. Add Motor Contribution = Motor FLA x 4

* All results are given in symmetrical amperes

Single-Phase L-L Bolted Fault-Current* Table

One-Way Distance in Feet

Available Isc in Thousands (Isca)

Isca	5	10	15	20	25	30	40	50	60	70	80	90	100	125	150	175	200	225	250	300	350	400	500	600
3	2,990	2,979	2,969	2,959	2,948	2,938	2,918	2,899	2,879	2,860	2,841	2,822	2,804	2,759	2,715	2,672	2,631	2,592	2,553	2,479	2,409	2,343	2,222	2,112
5	4,971	4,942	4,914	4,886	4,858	4,831	4,777	4,724	4,673	4,622	4,573	4,525	4,477	4,363	4,255	4,152	4,054	3,960	3,871	3,703	3,550	3,409	3,157	2,941
7	6,943	6,887	6,833	6,778	6,725	6,673	6,571	6,471	6,375	6,281	6,191	6,102	6,017	5,813	5,622	5,443	5,276	5,118	4,970	4,697	4,453	4,233	3,852	3,534
10	9,885	9,772	9,662	9,554	9,449	9,345	9,146	8,955	8,771	8,595	8,426	8,264	8,107	7,741	7,406	7,100	6,817	6,556	6,315	5,881	5,503	5,171	4,614	4,165
15	14,742	14,493	14,251	14,018	13,793	13,574	13,157	12,765	12,396	12,047	11,717	11,405	11,110	10,433	9,834	9,301	8,822	8,390	7,998	7,315	6,740	6,248	5,453	4,837
20	19,544	19,108	18,691	18,292	17,909	17,543	16,853	16,215	15,623	15,073	14,561	14,082	13,634	12,629	11,762	11,007	10,342	9,754	9,228	8,331	7,593	6,974	5,998	5,261
25	24,291	23,621	22,988	22,387	21,817	21,275	20,268	19,353	18,516	17,749	17,043	16,391	15,786	14,455	13,330	12,368	11,535	10,808	10,166	9,088	8,216	7,497	6,381	5,553
30	28,985	28,036	27,148	26,314	25,530	24,791	23,435	22,219	21,124	20,131	19,227	18,401	17,643	15,996	14,630	13,479	12,496	11,647	10,906	9,674	8,692	7,892	6,664	5,767
35	33,626	32,356	31,179	30,084	29,063	28,110	26,379	24,848	23,486	22,265	21,165	20,169	19,262	17,315	15,726	14,404	13,287	12,331	11,503	10,141	9,068	8,200	6,882	5,930
40	38,216	36,584	35,086	33,705	32,429	31,246	29,122	27,268	25,636	24,189	22,896	21,734	20,685	18,456	16,662	15,185	13,949	12,899	11,996	10,522	9,371	8,447	7,056	6,058
45	42,754	40,722	38,874	37,187	35,640	34,216	31,685	29,503	27,602	25,931	24,451	23,130	21,945	19,454	17,470	15,854	14,511	13,378	12,409	10,839	9,622	8,650	7,197	6,162
50	47,243	44,774	42,550	40,537	38,705	37,032	34,085	31,573	29,406	27,517	25,856	24,384	23,071	20,333	18,176	16,433	14,995	13,788	12,761	11,107	9,832	8,820	7,314	6,247
55	51,682	48,742	46,118	43,762	41,635	39,706	36,337	33,496	31,067	28,966	27,131	25,515	24,081	21,113	18,797	16,939	15,415	14,143	13,064	11,336	10,011	8,964	7,412	6,319
60	56,073	52,628	49,582	46,870	44,439	42,247	38,454	35,287	32,601	30,295	28,294	26,541	24,993	21,811	19,348	17,385	15,784	14,452	13,328	11,534	10,165	9,087	7,497	6,380
65	60,416	56,436	52,948	49,866	47,123	44,666	40,449	36,959	34,023	31,520	29,359	27,476	25,820	22,439	19,840	17,781	16,109	14,725	13,560	11,707	10,299	9,194	7,569	6,433
70	64,712	60,168	56,219	52,757	49,697	46,972	42,330	38,523	35,345	32,650	30,338	28,331	26,574	23,006	20,282	18,136	16,400	14,967	13,765	11,859	10,417	9,288	7,633	6,478
75	68,963	63,825	59,400	55,548	52,166	49,172	44,108	39,990	36,576	33,698	31,241	29,117	27,264	23,521	20,682	18,454	16,660	15,184	13,948	11,995	10,522	9,371	7,689	6,519
80	73,168	67,410	62,493	58,244	54,537	51,273	45,791	41,369	37,726	34,672	32,076	29,841	27,898	23,991	21,045	18,743	16,895	15,378	14,112	12,116	10,615	9,445	7,738	6,554
85	77,328	70,926	65,503	60,850	56,815	53,281	47,387	42,667	38,802	35,579	32,851	30,511	28,482	24,422	21,375	19,005	17,107	15,554	14,260	12,225	10,698	9,511	7,783	6,586
90	81,444	74,374	68,433	63,371	59,006	55,204	48,902	43,891	39,812	36,426	33,571	31,131	29,022	24,818	21,678	19,244	17,301	15,714	14,394	12,323	10,774	9,570	7,822	6,615
95	85,517	77,755	71,285	65,810	61,115	57,045	50,341	45,047	40,761	37,219	34,244	31,709	29,523	25,184	21,957	19,463	17,477	15,860	14,516	12,413	10,842	9,624	7,858	6,640
100	89,547	81,073	74,064	68,171	63,146	58,811	51,711	46,141	41,654	37,963	34,872	32,247	29,989	25,522	22,213	19,664	17,640	15,993	14,628	12,494	10,904	9,673	7,891	6,663
105	93,536	84,329	76,772	70,458	65,104	60,506	53,017	47,178	42,497	38,662	35,461	32,750	30,424	25,836	22,451	19,850	17,789	16,116	14,730	12,569	10,961	9,718	7,921	6,685
110	97,483	87,524	79,411	72,675	66,992	62,133	54,262	48,161	43,294	39,320	36,014	33,221	30,830	26,128	22,671	20,022	17,927	16,229	14,825	12,638	11,013	9,759	7,948	6,704
115	101,390	90,660	81,984	74,824	68,814	63,697	55,452	49,096	44,047	39,940	36,534	33,663	31,210	26,401	22,876	20,182	18,055	16,334	14,912	12,701	11,061	9,797	7,973	6,722
120	105,257	93,739	84,494	76,909	70,573	65,202	56,588	49,985	44,762	40,527	37,024	34,078	31,567	26,656	23,067	20,330	18,174	16,431	14,993	12,760	11,106	9,831	7,996	6,738
125	109,084	96,763	86,943	78,932	72,274	66,651	57,676	50,832	45,440	41,082	37,487	34,470	31,903	26,895	23,246	20,469	18,285	16,522	15,069	12,815	11,147	9,864	8,018	6,753
130	112,872	99,732	89,333	80,897	73,917	68,046	58,718	51,640	46,084	41,608	37,924	34,839	32,219	27,119	23,414	20,599	18,388	16,606	15,139	12,865	11,186	9,894	8,037	6,768
135	116,623	102,649	91,666	82,806	75,507	69,391	59,717	52,411	46,697	42,107	38,338	35,189	32,517	27,331	23,571	20,720	18,485	16,685	15,204	12,913	11,221	9,922	8,056	6,781
140	120,335	105,514	93,944	84,660	77,046	70,689	60,676	53,148	47,281	42,581	38,731	35,519	32,800	27,530	23,719	20,835	18,576	16,759	15,266	12,957	11,255	9,948	8,073	6,793
145	124,011	108,330	96,169	86,463	78,537	71,942	61,597	53,853	47,838	43,032	39,104	35,833	33,067	27,718	23,858	20,942	18,661	16,828	15,323	12,998	11,286	9,972	8,089	6,804
150	127,650	111,096	98,343	88,217	79,981	73,152	62,481	54,528	48,370	43,462	39,459	36,131	33,320	27,895	23,990	21,043	18,742	16,894	15,378	13,037	11,315	9,995	8,104	6,815
155	131,253	113,816	100,468	89,923	81,381	74,321	63,332	55,175	48,879	43,873	39,796	36,413	33,561	28,064	24,114	21,139	18,817	16,955	15,429	13,074	11,343	10,017	8,118	6,825
160	134,821	116,489	102,545	91,583	82,738	75,451	64,151	55,795	49,365	44,264	40,118	36,683	33,789	28,223	24,232	21,230	18,889	17,013	15,477	13,109	11,369	10,037	8,132	6,834
165	138,353	119,117	104,576	93,200	84,055	76,545	64,940	56,391	49,831	44,638	40,426	36,939	34,007	28,375	24,344	21,315	18,957	17,068	15,522	13,141	11,394	10,056	8,144	6,843
170	141,852	121,701	106,563	94,774	85,334	77,604	65,701	56,964	50,278	44,996	40,719	37,184	34,214	28,519	24,450	21,397	19,021	17,121	15,565	13,172	11,417	10,074	8,156	6,851
175	145,316	124,242	108,506	96,308	86,576	78,630	66,435	57,514	50,706	45,339	41,000	37,418	34,412	28,657	24,551	21,474	19,082	17,170	15,606	13,201	11,439	10,091	8,167	6,859
180	148,747	126,741	110,408	97,803	87,782	79,623	67,143	58,044	51,118	45,668	41,268	37,642	34,601	28,788	24,647	21,547	19,140	17,217	15,645	13,229	11,459	10,107	8,178	6,867
185	152,145	129,200	112,269	99,261	88,954	80,587	67,826	58,555	51,513	45,983	41,525	37,856	34,782	28,913	24,738	21,617	19,195	17,262	15,682	13,255	11,479	10,123	8,188	6,874
190	155,511	131,619	114,091	100,682	90,094	81,521	68,487	59,047	51,893	46,286	41,772	38,061	34,955	29,032	24,826	21,684	19,248	17,304	15,717	13,280	11,498	10,137	8,197	6,881
195	158,844	133,999	115,875	102,069	91,203	82,428	69,126	59,521	52,259	46,577	42,009	38,257	35,120	29,146	24,909	21,747	19,298	17,345	15,750	13,304	11,516	10,151	8,206	6,887
200	162,146	136,341	117,622	103,423	92,282	83,308	69,744	59,978	52,612	46,857	42,236	38,446	35,279	29,255	24,989	21,808	19,346	17,383	15,782	13,327	11,533	10,165	8,215	6,893

# 1/0 AWG	2 Run(s)	480 Volts	C-Value = 8,924	Magnetic Duct	Copper

Calculations:

1. $f = \dfrac{\text{Length} \times 2 \times \text{Isca}}{\text{Runs} \times \text{C-Value} \times E_{L-L}}$

2. $M = \dfrac{1}{1 + f}$

3. Isc = Isca x M

4. Add Motor Contribution = Motor FLA x 4

* All results are given in symmetrical amperes

Single-Phase L-L Bolted Fault-Current* Table

One-Way Distance in Feet

Available Isc in Thousands (Isca)	5	10	15	20	25	30	40	50	60	70	80	90	100	125	150	175	200	225	250	300	350	400	500	600
3	2,983	2,966	2,949	2,932	2,915	2,899	2,867	2,835	2,804	2,774	2,745	2,716	2,688	2,619	2,555	2,493	2,434	2,378	2,325	2,224	2,133	2,048	1,897	1,767
5	4,952	4,905	4,859	4,814	4,769	4,725	4,640	4,558	4,479	4,403	4,329	4,258	4,189	4,025	3,874	3,734	3,604	3,482	3,369	3,162	2,980	2,817	2,540	2,312
7	6,906	6,815	6,726	6,640	6,556	6,473	6,315	6,164	6,020	5,883	5,752	5,627	5,507	5,228	4,976	4,747	4,538	4,347	4,172	3,860	3,591	3,358	2,971	2,664
10	9,810	9,627	9,451	9,281	9,117	8,959	8,658	8,377	8,114	7,867	7,634	7,415	7,208	6,737	6,325	5,960	5,634	5,343	5,080	4,625	4,245	3,922	3,405	3,008
15	14,576	14,176	13,797	13,438	13,097	12,773	12,171	11,623	11,122	10,663	10,240	9,849	9,487	8,689	8,014	7,437	6,937	6,500	6,115	5,468	4,944	4,512	3,841	3,343
20	19,254	18,562	17,918	17,317	16,755	16,228	15,268	14,415	13,653	12,967	12,347	11,783	11,269	10,160	9,250	8,489	7,844	7,290	6,809	6,016	5,388	4,879	4,103	3,540
25	23,845	22,792	21,829	20,943	20,127	19,371	18,019	16,843	15,812	14,899	14,086	13,357	12,700	11,309	10,192	9,277	8,512	7,864	7,307	6,401	5,695	5,129	4,279	3,670
30	28,352	26,876	25,546	24,342	23,246	22,244	20,479	18,974	17,675	16,542	15,546	14,663	13,874	12,231	10,935	9,888	9,024	8,299	7,681	6,686	5,920	5,311	4,404	3,762
35	32,778	30,821	29,084	27,533	26,139	24,879	22,692	20,858	19,299	17,956	16,788	15,763	14,856	12,987	11,536	10,377	9,429	8,640	7,973	6,906	6,091	5,448	4,499	3,831
40	37,124	34,633	32,456	30,536	28,831	27,306	24,693	22,537	20,728	19,187	17,859	16,704	15,688	13,619	12,032	10,776	9,758	8,915	8,207	7,081	6,227	5,557	4,572	3,884
45	41,392	38,319	35,672	33,366	31,340	29,547	26,512	24,042	21,994	20,267	18,791	17,516	16,403	14,154	12,448	11,109	10,030	9,142	8,398	7,223	6,336	5,644	4,631	3,927
50	45,585	41,886	38,743	36,038	33,687	31,623	28,172	25,400	23,124	21,223	19,610	18,226	17,024	14,614	12,802	11,390	10,258	9,331	8,558	7,341	6,427	5,715	4,679	3,961
55	49,705	45,339	41,679	38,565	35,884	33,552	29,693	26,629	24,139	22,075	20,335	18,850	17,567	15,013	13,107	11,631	10,453	9,492	8,693	7,440	6,503	5,775	4,719	3,990
60	53,753	48,684	44,488	40,958	37,948	35,349	31,091	27,749	25,055	22,838	20,982	19,405	18,048	15,362	13,373	11,839	10,621	9,631	8,809	7,525	6,568	5,826	4,753	4,014
65	57,731	51,924	47,179	43,228	39,888	37,027	32,382	28,772	25,887	23,527	21,562	19,900	18,475	15,671	13,606	12,022	10,768	9,751	8,910	7,598	6,623	5,870	4,783	4,035
70	61,642	55,066	49,759	45,384	41,717	38,598	33,577	29,712	26,645	24,152	22,085	20,345	18,858	15,946	13,812	12,183	10,897	9,857	8,998	7,662	6,672	5,908	4,808	4,053
75	65,486	58,114	52,234	47,435	43,443	40,071	34,686	30,577	27,339	24,720	22,560	20,747	19,203	16,192	13,997	12,325	11,011	9,950	9,075	7,719	6,715	5,942	4,830	4,069
80	69,266	61,072	54,611	49,387	45,075	41,455	35,719	31,377	27,976	25,240	22,992	21,112	19,515	16,413	14,162	12,454	11,113	10,033	9,145	7,768	6,752	5,971	4,849	4,082
85	72,983	63,943	56,896	51,248	46,620	42,758	36,682	32,118	28,564	25,718	23,387	21,444	19,799	16,614	14,311	12,569	11,205	10,108	9,207	7,813	6,786	5,998	4,867	4,095
90	76,639	66,732	59,093	53,024	48,085	43,988	37,583	32,806	29,107	26,157	23,750	21,749	20,059	16,796	14,446	12,673	11,287	10,175	9,262	7,853	6,816	6,021	4,882	4,106
95	80,235	69,442	61,209	54,721	49,476	45,149	38,428	33,448	29,611	26,564	24,085	22,029	20,297	16,963	14,569	12,767	11,362	10,236	9,313	7,889	6,844	6,043	4,896	4,116
100	83,773	72,076	63,246	56,343	50,799	46,248	39,221	34,047	30,080	26,940	24,394	22,288	20,516	17,115	14,682	12,854	11,431	10,291	9,359	7,922	6,868	6,062	4,909	4,125
105	87,253	74,638	65,210	57,897	52,058	47,290	39,967	34,608	30,517	27,290	24,681	22,527	20,719	17,256	14,785	12,933	11,493	10,342	9,400	7,952	6,891	6,079	4,920	4,133
110	90,678	77,130	67,104	59,385	53,259	48,278	40,671	35,135	30,925	27,617	24,947	22,749	20,906	17,386	14,880	13,006	11,551	10,389	9,439	7,980	6,911	6,095	4,931	4,140
115	94,049	79,556	68,933	60,813	54,404	49,217	41,335	35,630	31,308	27,921	25,196	22,955	21,080	17,506	14,968	13,073	11,604	10,431	9,474	8,005	6,930	6,110	4,941	4,147
120	97,367	81,917	70,698	62,183	55,498	50,111	41,964	36,096	31,667	28,207	25,428	23,148	21,243	17,618	15,050	13,135	11,653	10,471	9,507	8,028	6,948	6,124	4,949	4,153
125	100,633	84,216	72,405	63,499	56,544	50,962	42,559	36,535	32,005	28,474	25,645	23,328	21,394	17,722	15,126	13,193	11,698	10,508	9,537	8,050	6,964	6,136	4,958	4,159
130	103,849	86,457	74,055	64,764	57,545	51,774	43,124	36,951	32,323	28,726	25,849	23,496	21,536	17,819	15,196	13,247	11,740	10,542	9,565	8,070	6,979	6,148	4,965	4,164
135	107,015	88,640	75,651	65,982	58,504	52,549	43,660	37,344	32,624	28,963	26,041	23,655	21,669	17,910	15,263	13,297	11,780	10,573	9,591	8,088	6,993	6,159	4,972	4,169
140	110,133	90,769	77,196	67,154	59,424	53,290	44,171	37,716	32,908	29,187	26,222	23,804	21,794	17,995	15,324	13,344	11,817	10,603	9,616	8,106	7,006	6,169	4,979	4,174
145	113,204	92,844	78,692	68,283	60,306	53,998	44,656	38,070	33,177	29,398	26,392	23,944	21,911	18,075	15,382	13,338	11,851	10,631	9,638	8,122	7,018	6,178	4,985	4,178
150	116,228	94,869	80,142	69,372	61,154	54,677	45,120	38,406	33,432	29,598	26,553	24,077	22,022	18,151	15,437	13,429	11,884	10,657	9,660	8,137	7,029	6,187	4,991	4,182
155	119,208	96,845	81,547	70,423	61,969	55,328	45,562	38,726	33,674	29,788	26,706	24,202	22,127	18,222	15,488	13,458	11,914	10,681	9,680	8,152	7,040	6,195	4,996	4,186
160	122,144	98,774	82,910	71,437	62,753	55,952	45,984	39,031	33,904	29,968	26,850	24,321	22,226	18,289	15,537	13,505	11,943	10,704	9,699	8,165	7,050	6,203	5,001	4,189
165	125,036	100,657	84,233	72,417	63,508	56,551	46,388	39,321	34,123	30,139	26,988	24,433	22,320	18,353	15,583	13,539	11,970	10,726	9,717	8,178	7,059	6,210	5,006	4,193
170	127,886	102,496	85,517	73,364	64,235	57,127	46,775	39,599	34,332	30,302	27,118	24,540	22,409	18,413	15,626	13,572	11,995	10,747	9,734	8,190	7,068	6,217	5,010	4,196
175	130,696	104,292	86,764	74,280	64,936	57,681	47,145	39,864	34,531	30,457	27,242	24,642	22,494	18,470	15,667	13,603	12,020	10,766	9,750	8,201	7,077	6,224	5,014	4,199
180	133,464	106,048	87,975	75,166	65,613	58,214	47,501	40,118	34,722	30,605	27,361	24,738	22,575	18,524	15,706	13,633	12,043	10,785	9,765	8,211	7,085	6,230	5,018	4,202
185	136,194	107,764	89,153	76,024	66,265	58,727	47,842	40,361	34,904	30,746	27,473	24,831	22,652	18,576	15,744	13,661	12,064	10,802	9,779	8,222	7,092	6,235	5,022	4,204
190	138,884	109,441	90,298	76,855	66,896	59,222	48,170	40,594	35,078	30,881	27,581	24,919	22,725	18,625	15,779	13,687	12,085	10,819	9,793	8,231	7,099	6,241	5,026	4,207
195	141,537	111,082	91,412	77,661	67,505	59,699	48,485	40,818	35,245	31,010	27,684	25,003	22,795	18,672	15,813	13,713	12,105	10,835	9,806	8,240	7,106	6,246	5,029	4,209
200	144,153	112,687	92,496	78,442	68,095	60,159	48,788	41,033	35,404	31,134	27,783	25,083	22,861	18,717	15,845	13,737	12,124	10,850	9,818	8,249	7,113	6,251	5,033	4,211

2/0 AWG | 1 Run(s) | 480 Volts | C-Value = 10,755 | Magnetic Duct | Copper

Calculations:

1. $f = \dfrac{\text{Length} \times 2 \times \text{Isca}}{\text{Runs} \times \text{C-Value} \times \text{E L-L}}$

2. $M = \dfrac{1}{1 + f}$

3. Isc = Isca x M

4. Add Motor Contribution = Motor FLA x 4

* All results are given in symmetrical amperes

Single-Phase L-L Bolted Fault-Current* Table

One-Way Distance in Feet

Available Isc in Thousands (Isca)	5	10	15	20	25	30	40	50	60	70	80	90	100	125	150	175	200	225	250	300	350	400	500	600
3	2,991	2,983	2,974	2,966	2,957	2,949	2,932	2,915	2,899	2,883	2,867	2,851	2,835	2,797	2,759	2,723	2,688	2,653	2,619	2,555	2,493	2,434	2,325	2,224
5	4,976	4,952	4,928	4,905	4,882	4,859	4,814	4,769	4,725	4,683	4,640	4,599	4,558	4,460	4,366	4,275	4,189	4,105	4,025	3,874	3,734	3,604	3,369	3,162
7	6,953	6,906	6,860	6,815	6,770	6,726	6,640	6,556	6,473	6,393	6,315	6,239	6,164	5,985	5,817	5,658	5,507	5,364	5,228	4,976	4,747	4,538	4,172	3,860
10	9,904	9,810	9,718	9,627	9,538	9,451	9,281	9,117	8,959	8,806	8,658	8,515	8,377	8,051	7,749	7,468	7,208	6,965	6,737	6,325	5,960	5,634	5,080	4,625
15	14,785	14,576	14,374	14,176	13,984	13,797	13,438	13,097	12,773	12,465	12,171	11,891	11,623	11,003	10,447	9,944	9,487	9,070	8,689	8,014	7,437	6,937	6,115	5,468
20	19,620	19,254	18,902	18,562	18,234	17,918	17,317	16,755	16,228	15,733	15,268	14,829	14,415	13,475	12,649	11,919	11,269	10,686	10,160	9,250	8,489	7,844	6,809	6,016
25	24,409	23,845	23,307	22,792	22,300	21,829	20,943	20,127	19,371	18,671	18,019	17,411	16,843	15,573	14,481	13,532	12,700	11,964	11,309	10,192	9,277	8,512	7,307	6,401
30	29,153	28,352	27,595	26,876	26,194	25,546	24,342	23,246	22,244	21,325	20,479	19,698	18,974	17,377	16,028	14,874	13,874	13,001	12,231	10,935	9,888	9,024	7,681	6,686
35	33,852	32,778	31,769	30,821	29,927	29,084	27,533	26,139	24,879	23,735	22,692	21,737	20,858	18,945	17,353	16,008	14,856	13,859	12,987	11,536	10,377	9,429	7,973	6,906
40	38,508	37,124	35,835	34,633	33,509	32,456	30,536	28,831	27,306	25,934	24,693	23,566	22,537	20,320	18,499	16,978	15,688	14,581	13,619	12,032	10,776	9,758	8,207	7,081
45	43,121	41,392	39,796	38,319	36,948	35,672	33,366	31,340	29,547	27,947	26,512	25,217	24,042	21,535	19,501	17,819	16,403	15,196	14,154	12,448	11,109	10,030	8,398	7,223
50	47,690	45,585	43,657	41,886	40,253	38,743	36,038	33,687	31,623	29,798	28,172	26,714	25,400	22,618	20,385	18,553	17,024	15,727	14,614	12,802	11,390	10,258	8,558	7,341
55	52,218	49,705	47,422	45,339	43,432	41,679	38,565	35,884	33,552	31,505	29,693	28,078	26,629	23,587	21,169	19,201	17,567	16,190	15,013	13,107	11,631	10,453	8,693	7,440
60	56,705	53,753	51,093	48,684	46,491	44,488	40,958	37,948	35,349	33,084	31,091	29,325	27,749	24,462	21,871	19,776	18,048	16,597	15,362	13,373	11,839	10,621	8,809	7,525
65	61,150	57,731	54,674	51,924	49,438	47,179	43,228	39,888	37,027	34,549	32,382	30,471	28,772	25,254	22,502	20,291	18,475	16,958	15,671	13,606	12,022	10,768	8,910	7,598
70	65,555	61,642	58,169	55,066	52,278	49,759	45,384	41,717	38,598	35,913	33,577	31,526	29,712	25,975	23,072	20,753	18,858	17,280	15,946	13,812	12,183	10,897	8,998	7,662
75	69,921	65,486	61,580	58,114	55,017	52,234	47,435	43,443	40,071	37,185	34,686	32,502	30,577	26,633	23,591	21,172	19,203	17,569	16,192	13,997	12,326	11,011	9,075	7,719
80	74,247	69,266	64,911	61,072	57,661	54,611	49,387	45,075	41,455	38,374	35,719	33,407	31,377	27,238	24,064	21,552	19,515	17,830	16,413	14,162	12,454	11,113	9,145	7,768
85	78,535	72,983	68,165	63,943	60,214	56,896	51,248	46,620	42,758	39,488	36,682	34,248	32,118	27,795	24,497	21,899	19,799	18,067	16,614	14,311	12,569	11,205	9,207	7,813
90	82,784	76,639	71,343	66,732	62,681	59,093	53,024	48,085	43,988	40,534	37,583	35,033	32,806	28,309	24,896	22,217	20,059	18,283	16,796	14,446	12,673	11,287	9,262	7,853
95	86,995	80,235	74,449	69,442	65,066	61,209	54,721	49,476	45,149	41,518	38,428	35,765	33,448	28,785	25,264	22,510	20,297	18,481	16,963	14,569	12,767	11,362	9,313	7,889
100	91,170	83,773	77,486	72,076	67,373	63,246	56,343	50,799	46,248	42,446	39,221	36,451	34,047	29,228	25,604	22,780	20,516	18,662	17,115	14,682	12,854	11,431	9,359	7,922
105	95,308	87,253	80,454	74,638	69,606	65,210	57,897	52,058	47,290	43,321	39,967	37,095	34,608	29,641	25,920	23,029	20,719	18,829	17,256	14,785	12,933	11,493	9,400	7,952
110	99,409	90,678	83,357	77,130	71,769	67,104	59,385	53,259	48,278	44,149	40,671	37,701	35,135	30,026	26,214	23,261	20,906	18,984	17,386	14,880	13,006	11,551	9,439	7,980
115	103,475	94,049	86,197	79,556	73,864	68,933	60,813	54,404	49,217	44,933	41,335	38,271	35,630	30,387	26,489	23,477	21,080	19,128	17,506	14,968	13,073	11,604	9,474	8,005
120	107,505	97,367	88,976	81,917	75,895	70,698	62,183	55,498	50,111	45,677	41,964	38,809	36,096	30,725	26,745	23,679	21,243	19,261	17,618	15,050	13,135	11,653	9,507	8,028
125	111,501	100,633	91,696	84,216	77,865	72,405	63,499	56,544	50,962	46,383	42,559	39,318	36,535	31,043	26,986	23,867	21,394	19,386	17,722	15,126	13,193	11,698	9,537	8,050
130	115,462	103,849	94,358	86,457	79,777	74,055	64,764	57,545	51,774	47,055	43,124	39,799	36,951	31,342	27,212	24,044	21,536	19,502	17,819	15,196	13,247	11,740	9,565	8,070
135	119,389	107,015	96,965	88,640	81,632	75,651	65,982	58,504	52,549	47,694	43,660	40,256	37,344	31,625	27,425	24,209	21,669	19,611	17,910	15,263	13,297	11,780	9,591	8,088
140	123,283	110,133	99,518	90,769	83,434	77,196	67,154	59,424	53,290	48,303	44,171	40,689	37,716	31,891	27,625	24,365	21,794	19,713	17,995	15,324	13,344	11,817	9,616	8,106
145	127,144	113,204	102,018	92,844	85,184	78,692	68,283	60,306	53,998	48,885	44,656	41,101	38,070	32,144	27,814	24,512	21,911	19,809	18,075	15,382	13,388	11,851	9,638	8,122
150	130,972	116,228	104,468	94,869	86,886	80,142	69,372	61,154	54,677	49,441	45,120	41,493	38,406	32,383	27,993	24,651	22,022	19,900	18,151	15,437	13,429	11,884	9,660	8,137
155	134,768	119,208	106,869	96,845	88,540	81,547	70,423	61,969	55,328	49,972	45,562	41,867	38,726	32,610	28,163	24,783	22,127	19,986	18,222	15,488	13,468	11,914	9,680	8,152
160	138,532	122,144	109,222	98,774	90,149	82,910	71,437	62,753	55,952	50,481	45,984	42,223	39,031	32,826	28,324	24,907	22,226	20,066	18,289	15,537	13,505	11,943	9,699	8,165
165	142,265	125,036	111,530	100,657	91,715	84,233	72,417	63,508	56,551	50,968	46,388	42,563	39,321	33,031	28,476	25,025	22,320	20,143	18,353	15,583	13,539	11,970	9,717	8,178
170	145,966	127,886	113,792	102,496	93,240	85,517	73,364	64,235	57,127	51,435	46,775	42,889	39,599	33,227	28,622	25,137	22,409	20,216	18,413	15,626	13,572	11,995	9,734	8,190
175	149,637	130,696	116,010	104,292	94,724	86,764	74,280	64,936	57,681	51,884	47,145	43,200	39,864	33,414	28,760	25,244	22,494	20,285	18,470	15,667	13,603	12,020	9,750	8,201
180	153,278	133,464	118,187	106,048	96,170	87,975	75,166	65,613	58,214	52,315	47,501	43,498	40,118	33,592	28,892	25,346	22,575	20,350	18,524	15,706	13,633	12,043	9,765	8,211
185	156,889	136,194	120,322	107,764	97,579	89,153	76,024	66,265	58,727	52,729	47,842	43,784	40,361	33,762	29,018	25,442	22,652	20,412	18,576	15,744	13,661	12,064	9,779	8,222
190	160,470	138,884	122,417	109,441	98,952	90,298	76,855	66,896	59,222	53,127	48,170	44,059	40,594	33,925	29,138	25,535	22,725	20,472	18,625	15,779	13,687	12,085	9,793	8,231
195	164,022	141,537	124,474	111,082	100,292	91,412	77,661	67,505	59,699	53,511	48,485	44,322	40,818	34,081	29,253	25,623	22,795	20,529	18,672	15,813	13,713	12,105	9,806	8,240
200	167,545	144,153	126,492	112,687	101,598	92,496	78,442	68,095	60,159	53,881	48,788	44,576	41,033	34,231	29,363	25,708	22,861	20,583	18,717	15,845	13,737	12,124	9,818	8,249

# 2/0 AWG	2 Run(s)	480 Volts	C-Value = 10,755	Magnetic Duct	Copper

Calculations:

1. $f = \dfrac{\text{Length} \times 2 \times \text{Isca}}{\text{Runs} \times \text{C-Value} \times \text{E L-L}}$

2. $M = \dfrac{1}{1 + f}$

3. Isc = Isca x M

4. Add Motor Contribution = Motor FLA x 4

* All results are given in symmetrical amperes

Single-Phase L-L Bolted Fault-Current* Table

One-Way Distance in Feet

Available Isc in Thousands (Isca)

	5	10	15	20	25	30	40	50	60	70	80	90	100	125	150	175	200	225	250	300	350	400	500	600
3	2,985	2,971	2,957	2,943	2,929	2,915	2,888	2,861	2,834	2,809	2,783	2,758	2,734	2,675	2,618	2,563	2,511	2,461	2,413	2,322	2,238	2,159	2,018	1,894
5	4,960	4,920	4,881	4,843	4,805	4,768	4,695	4,625	4,557	4,490	4,426	4,363	4,302	4,157	4,021	3,894	3,775	3,663	3,557	3,363	3,189	3,032	2,761	2,534
7	6,921	6,845	6,769	6,696	6,624	6,554	6,417	6,286	6,161	6,040	5,924	5,812	5,704	5,452	5,221	5,009	4,814	4,633	4,465	4,163	3,900	3,668	3,278	2,963
10	9,840	9,686	9,536	9,391	9,250	9,113	8,351	8,604	8,371	8,149	7,939	7,740	7,550	7,115	6,727	6,379	6,065	5,780	5,522	5,068	4,683	4,352	3,814	3,394
15	14,644	14,304	13,980	13,670	13,373	13,089	12,556	12,064	11,610	11,189	10,797	10,431	10,090	9,327	8,671	8,101	7,602	7,160	6,767	6,098	5,549	5,091	4,369	3,827
20	19,372	18,781	18,226	17,703	17,209	16,741	15,879	15,101	14,396	13,753	13,166	12,626	12,130	11,043	10,135	9,365	8,704	8,130	7,627	6,738	6,114	5,563	4,712	4,087
25	24,026	23,124	22,288	21,511	20,785	20,107	18,876	17,787	16,816	15,946	15,162	14,451	13,804	12,414	11,278	10,333	9,534	8,850	8,257	7,232	6,513	5,890	4,945	4,262
30	28,608	27,339	26,178	25,112	24,129	23,220	21,593	20,180	18,940	17,843	16,867	15,992	15,203	13,534	12,195	11,098	10,181	9,405	8,738	7,633	6,808	6,131	5,114	4,386
35	33,120	31,431	29,906	28,522	27,261	26,107	24,068	22,325	20,817	19,500	18,340	17,310	16,390	14,466	12,947	11,717	10,700	9,846	9,118	7,943	7,036	6,315	5,241	4,480
40	37,563	35,405	33,482	31,758	30,202	28,791	26,332	24,259	22,489	20,960	19,625	18,451	17,409	15,255	13,575	12,229	11,125	10,204	9,424	8,175	7,218	6,461	5,341	4,553
45	41,939	39,267	36,916	34,830	32,967	31,294	28,410	26,012	23,988	22,256	20,757	19,447	18,293	15,930	14,107	12,659	11,480	10,502	9,678	8,355	7,365	6,579	5,422	4,611
50	46,249	43,021	40,215	37,752	35,574	33,633	30,324	27,608	25,338	23,414	21,761	20,326	19,068	16,514	14,564	13,025	11,780	10,753	9,890	8,523	7,488	6,677	5,488	4,659
55	50,495	46,672	43,387	40,534	38,034	35,823	32,093	29,067	26,562	24,455	22,657	21,106	19,753	17,025	14,960	13,341	12,038	10,967	10,072	8,657	7,591	6,759	5,543	4,698
60	54,678	50,224	46,440	43,187	40,359	37,879	33,734	30,406	27,676	25,396	23,463	21,803	20,363	17,476	15,307	13,616	12,262	11,153	10,228	8,772	7,679	6,829	5,590	4,732
65	58,800	53,680	49,380	45,718	42,562	39,813	35,259	31,639	28,694	26,250	24,190	22,430	20,908	17,877	15,613	13,858	12,458	11,315	10,364	8,872	7,756	6,889	5,631	4,761
70	62,862	57,045	52,213	48,136	44,650	41,634	36,680	32,779	29,628	27,030	24,851	22,997	21,400	18,235	15,886	14,072	12,631	11,457	10,483	8,959	7,822	6,942	5,666	4,786
75	66,865	60,322	54,946	50,449	46,633	43,353	33,008	33,835	30,489	27,744	25,453	23,512	21,845	18,557	16,130	14,264	12,785	11,583	10,589	9,036	7,881	6,988	5,696	4,808
80	70,811	63,515	57,582	52,663	48,518	44,978	39,251	34,817	31,283	28,401	26,005	23,982	22,250	18,849	16,349	14,435	12,922	11,696	10,683	9,105	7,933	7,029	5,724	4,827
85	74,700	66,627	60,128	54,785	50,313	46,517	40,417	35,732	32,020	29,007	26,512	24,412	22,620	19,114	16,548	14,590	13,046	11,798	10,767	9,166	7,980	7,065	5,748	4,844
90	78,534	69,660	62,588	56,819	52,024	47,975	41,514	36,586	32,704	29,567	26,979	24,808	22,960	19,355	16,729	14,730	13,158	11,889	10,844	9,222	8,022	7,098	5,769	4,860
95	82,315	72,618	64,966	58,772	53,656	49,360	42,547	37,386	33,342	30,087	27,412	25,173	23,272	19,577	16,894	14,858	13,260	11,973	10,913	9,272	8,060	7,128	5,789	4,874
100	86,043	75,504	67,265	60,648	55,216	50,677	43,521	38,137	33,938	30,572	27,813	25,511	23,561	19,781	17,046	14,976	13,354	12,049	10,976	9,317	8,094	7,154	5,807	4,886
105	89,719	78,320	69,491	62,451	56,707	51,930	44,442	38,842	34,495	31,023	28,186	25,825	23,828	19,969	17,186	15,083	13,439	12,118	11,034	9,359	8,125	7,179	5,823	4,898
110	93,344	81,069	71,647	64,187	58,134	53,124	45,314	39,506	35,018	31,445	28,534	26,117	24,077	20,143	17,314	15,182	13,518	12,182	11,087	9,397	8,154	7,201	5,837	4,908
115	96,920	83,752	73,735	65,858	59,501	54,264	46,141	40,133	35,509	31,841	28,860	26,389	24,308	20,305	17,434	15,274	13,590	12,241	11,135	9,432	8,180	7,222	5,851	4,918
120	100,447	86,373	75,759	67,467	60,812	55,352	46,925	40,725	35,972	32,213	29,165	26,644	24,524	20,455	17,544	15,359	13,658	12,296	11,181	9,464	8,205	7,241	5,863	4,926
125	103,927	88,934	77,721	69,020	62,070	56,392	47,671	41,286	36,409	32,563	29,451	26,883	24,726	20,596	17,648	15,438	13,720	12,346	11,222	9,494	8,227	7,258	5,875	4,934
130	107,360	91,436	79,626	70,517	63,279	57,388	48,380	41,817	36,821	32,892	29,721	27,107	24,916	20,727	17,744	15,512	13,778	12,393	11,261	9,522	8,248	7,275	5,886	4,942
135	110,747	93,882	81,474	71,963	64,441	58,342	49,057	42,321	37,212	33,203	29,974	27,318	25,094	20,850	17,834	15,580	13,832	12,437	11,298	9,548	8,267	7,290	5,895	4,949
140	114,090	96,273	83,269	73,360	65,558	59,257	49,702	42,800	37,582	33,497	30,214	27,517	25,261	20,966	17,919	15,645	13,883	12,478	11,331	9,572	8,285	7,304	5,905	4,955
145	117,389	98,611	85,012	74,710	66,634	60,134	50,318	43,256	37,933	33,776	30,440	27,704	25,420	21,075	17,998	15,705	13,931	12,517	11,363	9,595	8,302	7,317	5,913	4,961
150	120,644	100,898	86,707	76,015	67,671	60,977	50,906	43,691	38,266	34,040	30,655	27,882	25,569	21,177	18,073	15,762	13,976	12,553	11,393	9,616	8,318	7,329	5,921	4,967
155	123,858	103,136	88,354	77,278	68,670	61,787	51,470	44,105	38,584	34,291	30,858	28,050	25,710	21,274	18,143	15,816	14,018	12,587	11,421	9,636	8,333	7,341	5,929	4,972
160	127,030	105,326	89,957	78,501	69,634	62,567	52,010	44,501	38,887	34,530	31,052	28,210	25,844	21,366	18,210	15,867	14,058	12,619	11,447	9,654	8,347	7,352	5,936	4,977
165	130,162	107,470	91,516	79,686	70,565	63,317	52,527	44,879	39,175	34,757	31,235	28,361	25,972	21,453	18,273	15,914	14,095	12,649	11,472	9,672	8,360	7,362	5,943	4,982
170	133,253	109,569	93,033	80,834	71,464	64,040	53,023	45,241	39,451	34,974	31,410	28,505	26,092	21,535	18,333	15,960	14,131	12,678	11,496	9,689	8,373	7,372	5,949	4,987
175	136,306	111,625	94,511	81,948	72,332	64,737	53,500	45,588	39,714	35,181	31,577	28,643	26,207	21,613	18,389	16,003	14,164	12,705	11,518	9,705	8,385	7,381	5,955	4,991
180	139,320	113,638	95,951	83,028	73,173	65,409	53,958	45,920	39,966	35,379	31,736	28,773	26,317	21,688	18,443	16,043	14,196	12,730	11,539	9,720	8,396	7,389	5,961	4,995
185	142,297	115,611	97,353	84,076	73,985	66,057	54,399	46,239	40,207	35,567	31,888	28,898	26,421	21,758	18,495	16,082	14,226	12,755	11,559	9,734	8,406	7,398	5,966	4,998
190	145,237	117,544	98,720	85,094	74,772	66,684	54,823	46,545	40,438	35,748	32,033	29,018	26,521	21,826	18,543	16,119	14,255	12,778	11,578	9,747	8,417	7,405	5,971	5,002
195	148,140	119,438	100,053	86,082	75,535	67,290	55,232	46,839	40,660	35,922	32,172	29,132	26,616	21,890	18,590	16,154	14,283	12,800	11,596	9,760	8,426	7,413	5,976	5,005
200	151,008	121,296	101,353	87,043	76,273	67,875	55,626	47,122	40,873	36,088	32,306	29,241	26,707	21,952	18,634	16,188	14,309	12,821	11,613	9,772	8,435	7,420	5,980	5,009

# 3/0 AWG	1 Run(s)	480 Volts	C-Value = 12,843	Magnetic Duct	Copper

Calculations:

1. $f = \dfrac{\text{Length} \times 2 \times \text{Isca}}{\text{Runs} \times \text{C-Value} \times \text{E L-L}}$

2. $M = \dfrac{1}{1 + f}$

3. Isc = Isca x M

4. Add Motor Contribution = Motor FLA x 4

* All results are given in symmetrical amperes

Single-Phase L-L Bolted Fault-Current* Table

One-Way Distance in Feet

Available Isc in Thousands (Isca)	5	10	15	20	25	30	40	50	60	70	80	90	100	125	150	175	200	225	250	300	350	400	500	600
3	2,993	2,985	2,978	2,971	2,964	2,957	2,943	2,929	2,915	2,901	2,888	2,874	2,861	2,828	2,796	2,765	2,734	2,704	2,675	2,618	2,563	2,511	2,413	2,322
5	4,980	4,960	4,940	4,920	4,901	4,881	4,843	4,805	4,768	4,731	4,695	4,660	4,625	4,540	4,458	4,379	4,302	4,228	4,157	4,021	3,894	3,775	3,557	3,363
7	6,960	6,921	6,883	6,845	6,807	6,769	6,696	6,624	6,554	6,485	6,417	6,351	6,286	6,130	5,981	5,840	5,704	5,576	5,452	5,221	5,009	4,814	4,465	4,163
10	9,920	9,840	9,762	9,686	9,610	9,536	9,391	9,250	9,113	8,980	8,851	8,726	8,604	8,314	8,043	7,789	7,550	7,326	7,115	6,727	6,379	6,065	5,522	5,068
15	14,820	14,644	14,472	14,304	14,140	13,980	13,670	13,373	13,089	12,817	12,556	12,305	12,064	11,502	10,989	10,520	10,090	9,693	9,327	8,671	8,101	7,602	6,767	6,098
20	19,681	19,372	19,072	18,781	18,500	18,226	17,703	17,209	16,741	16,299	15,879	15,480	15,101	14,229	13,453	12,757	12,130	11,561	11,043	10,135	9,365	8,704	7,627	6,788
25	24,503	24,026	23,566	23,124	22,699	22,288	21,511	20,785	20,107	19,472	18,876	18,315	17,787	16,590	15,544	14,623	13,804	13,072	12,414	11,278	10,333	9,534	8,257	7,282
30	29,287	28,608	27,959	27,339	26,746	26,178	25,112	24,129	23,220	22,377	21,593	20,863	20,180	18,653	17,341	16,202	15,203	14,320	13,534	12,195	11,098	10,181	8,738	7,653
35	34,034	33,120	32,253	31,431	30,650	29,906	28,522	27,261	26,107	25,046	24,068	23,164	22,325	20,472	18,902	17,556	16,390	15,368	14,466	12,947	11,717	10,700	9,118	7,943
40	38,743	37,563	36,452	35,405	34,417	33,482	31,758	30,202	28,791	27,506	26,332	25,253	24,259	22,086	20,271	18,731	17,409	16,261	15,255	13,575	12,229	11,125	9,424	8,175
45	43,415	41,939	40,559	39,267	38,055	36,916	34,830	32,967	31,294	29,782	28,410	27,158	26,012	23,530	21,480	19,759	18,293	17,030	15,930	14,107	12,659	11,480	9,678	8,365
50	48,051	46,249	44,577	43,021	41,571	40,215	37,752	35,574	33,633	31,893	30,324	28,902	27,608	24,828	22,557	20,666	19,068	17,700	16,514	14,564	13,025	11,780	9,890	8,523
55	52,651	50,495	48,508	46,672	44,970	43,387	40,534	38,034	35,823	33,856	32,093	30,505	29,067	26,002	23,522	21,473	19,753	18,288	17,025	14,960	13,341	12,038	10,072	8,657
60	57,216	54,678	52,356	50,224	48,258	46,440	43,187	40,359	37,879	35,687	33,734	31,984	30,406	27,068	24,391	22,195	20,363	18,809	17,476	15,307	13,616	12,262	10,228	8,772
65	61,745	58,800	56,124	53,680	51,440	49,380	45,718	42,562	39,813	37,398	35,259	33,351	31,639	28,041	25,178	22,845	20,908	19,274	17,877	15,613	13,858	12,458	10,364	8,872
70	66,239	62,862	59,812	57,045	54,522	52,213	48,136	44,650	41,634	39,000	36,680	34,620	32,779	28,933	25,895	23,434	21,400	19,691	18,235	15,886	14,072	12,631	10,483	8,959
75	70,699	66,865	63,425	60,322	57,509	54,946	50,449	46,633	43,353	40,505	38,008	35,800	33,835	29,753	26,549	23,969	21,845	20,068	18,557	16,130	14,264	12,785	10,589	9,036
80	75,125	70,811	66,965	63,515	60,403	57,582	52,663	48,518	44,978	41,920	39,251	36,901	34,817	30,509	27,150	24,457	22,250	20,409	18,849	16,349	14,435	12,922	10,683	9,105
85	79,518	74,700	70,433	66,627	63,211	60,128	54,785	50,313	46,517	43,253	40,417	37,930	35,732	31,209	27,703	24,905	22,620	20,720	19,114	16,548	14,590	13,046	10,767	9,166
90	83,877	78,534	73,832	69,660	65,935	62,588	56,819	52,024	47,975	44,511	41,514	38,895	36,586	31,859	28,214	25,317	22,960	21,004	19,355	16,729	14,730	13,158	10,844	9,222
95	88,204	82,315	77,163	72,618	68,579	64,966	58,772	53,656	49,360	45,701	42,547	39,800	37,386	32,464	28,687	25,698	23,272	21,265	19,577	16,894	14,858	13,260	10,913	9,272
100	92,498	86,043	80,430	75,504	71,147	67,265	60,648	55,216	50,677	46,827	43,521	40,651	38,137	33,028	29,127	26,050	23,561	21,506	19,781	17,046	14,976	13,354	10,976	9,317
105	96,760	89,719	83,633	78,320	73,642	69,491	62,451	56,707	51,930	47,895	44,442	41,454	38,842	33,556	29,537	26,377	23,828	21,729	19,969	17,186	15,083	13,439	11,034	9,359
110	100,990	93,344	86,774	81,069	76,067	71,647	64,187	58,134	53,124	48,909	45,314	42,211	39,506	34,051	29,919	26,682	24,077	21,935	20,143	17,314	15,182	13,518	11,087	9,397
115	105,189	96,920	89,856	83,752	78,425	73,735	65,858	59,501	54,264	49,873	46,141	42,928	40,133	34,515	30,277	26,966	24,308	22,127	20,305	17,434	15,274	13,590	11,135	9,432
120	109,356	100,447	92,880	86,373	80,719	75,759	67,467	60,812	55,352	50,791	46,925	43,606	40,725	34,952	30,613	27,232	24,524	22,306	20,455	17,544	15,359	13,658	11,181	9,464
125	113,493	103,927	95,848	88,934	82,950	77,721	69,020	62,070	56,392	51,666	47,671	44,249	41,286	35,365	30,929	27,482	24,726	22,473	20,596	17,648	15,438	13,720	11,222	9,494
130	117,600	107,360	98,760	91,436	85,123	79,626	70,517	63,279	57,388	52,501	48,380	44,860	41,817	35,754	31,226	27,716	24,916	22,629	20,727	17,744	15,512	13,778	11,261	9,522
135	121,677	110,747	101,619	93,882	87,239	81,474	71,963	64,441	58,342	53,298	49,057	45,441	42,321	36,122	31,506	27,937	25,094	22,776	20,850	17,834	15,580	13,832	11,298	9,548
140	125,724	114,090	104,427	96,273	89,300	83,269	73,360	65,558	59,257	54,060	49,702	45,993	42,800	36,470	31,771	28,145	25,261	22,914	20,966	17,919	15,645	13,883	11,331	9,572
145	129,742	117,389	107,184	98,611	91,308	85,012	74,710	66,634	60,134	54,790	50,318	46,520	43,256	36,801	32,022	28,341	25,420	23,044	21,075	17,998	15,705	13,931	11,363	9,595
150	133,730	120,644	109,891	100,898	93,266	86,707	76,015	67,671	60,977	55,489	50,906	47,023	43,691	37,115	32,259	28,527	25,569	23,167	21,177	18,073	15,762	13,976	11,393	9,616
155	137,690	123,858	112,551	103,136	95,175	88,354	77,278	68,670	61,787	56,159	51,470	47,504	44,105	37,413	32,484	28,703	25,710	23,283	21,274	18,143	15,816	14,018	11,421	9,636
160	141,621	127,030	115,164	105,326	97,037	89,957	78,501	69,634	62,567	56,802	52,010	47,963	44,501	37,698	32,699	28,870	25,844	23,393	21,366	18,210	15,867	14,058	11,447	9,654
165	145,525	130,162	117,732	107,470	98,853	91,516	79,686	70,565	63,317	57,419	52,527	48,403	44,879	37,969	32,902	29,029	25,972	23,497	21,453	18,273	15,914	14,095	11,472	9,672
170	149,400	133,253	120,256	109,569	100,627	93,033	80,834	71,464	64,040	58,013	53,023	48,824	45,241	38,227	33,097	29,180	26,092	23,596	21,535	18,333	15,960	14,131	11,496	9,689
175	153,248	136,306	122,737	111,625	102,358	94,511	81,948	72,332	64,737	58,584	53,500	49,228	45,588	38,475	33,282	29,324	26,207	23,690	21,613	18,389	16,003	14,164	11,518	9,705
180	157,069	139,320	125,175	113,638	104,048	95,951	83,028	73,173	65,409	59,134	53,958	49,616	45,920	38,711	33,458	29,461	26,317	23,779	21,688	18,443	16,043	14,196	11,539	9,720
185	160,863	142,297	127,573	115,611	105,699	97,353	84,076	73,985	66,057	59,664	54,399	49,988	46,239	38,937	33,627	29,592	26,421	23,864	21,758	18,495	16,082	14,226	11,559	9,734
190	164,630	145,237	129,931	117,544	107,313	98,720	85,094	74,772	66,684	60,175	54,823	50,346	46,545	39,154	33,789	29,717	26,521	23,945	21,826	18,543	16,119	14,255	11,578	9,747
195	168,371	148,140	132,250	119,438	108,890	100,053	86,082	75,535	67,290	60,668	55,232	50,690	46,839	39,362	33,944	29,837	26,616	24,023	21,890	18,590	16,154	14,283	11,596	9,760
200	172,085	151,008	134,531	121,296	110,432	101,353	87,043	76,273	67,875	61,143	55,626	51,022	47,122	39,562	34,092	29,951	26,707	24,097	21,952	18,634	16,188	14,309	11,613	9,772

# 3/0 AWG	2 Run(s)	480 Volts		C-Value = 12,843	Magnetic Duct	Copper

Calculations:

1. $f = \dfrac{\text{Length} \times 2 \times \text{Isca}}{\text{Runs} \times \text{C-Value} \times \text{E L-L}}$

2. $M = \dfrac{1}{1 + f}$

3. Isc = Isca x M

4. Add Motor Contribution = Motor FLA x 4

* All results are given in symmetrical amperes

Single-Phase L-L Bolted Fault-Current* Table

One-Way Distance in Feet

Available Isc in Thousands (Isca)

	5	10	15	20	25	30	40	50	60	70	80	90	100	125	150	175	200	225	250	300	350	400	500	600
3	2,988	2,975	2,963	2,951	2,939	2,927	2,904	2,881	2,858	2,835	2,813	2,792	2,770	2,718	2,668	2,620	2,573	2,528	2,485	2,403	2,325	2,253	2,121	2,004
5	4,966	4,932	4,899	4,866	4,833	4,801	4,738	4,677	4,617	4,559	4,502	4,447	4,393	4,264	4,142	4,027	3,918	3,814	3,717	3,535	3,370	3,221	2,957	2,734
7	6,933	6,867	6,803	6,739	6,677	6,616	6,497	6,383	6,272	6,165	6,062	5,962	5,866	5,637	5,426	5,230	5,048	4,878	4,719	4,430	4,174	3,947	3,559	3,240
10	9,864	9,731	9,602	9,476	9,354	9,235	9,005	8,786	8,578	8,380	8,190	8,009	7,835	7,433	7,070	6,741	6,441	6,167	5,915	5,[illegible]68	5,084	4,750	4,199	3,763
15	14,696	14,403	14,122	13,852	13,592	13,341	12,867	12,425	12,013	11,627	11,265	10,925	10,605	9,881	9,250	8,695	8,202	7,762	7,367	6,[illegible]87	6,121	5,644	4,883	4,302
20	19,462	18,953	18,469	18,010	17,573	17,156	16,380	15,671	15,020	14,422	13,869	13,358	12,882	11,830	10,936	10,168	9,501	8,916	8,399	7,526	6,817	6,230	5,315	4,635
25	24,165	23,385	22,653	21,966	21,319	20,709	19,588	18,583	17,675	16,852	16,103	15,417	14,787	13,417	12,279	11,319	10,498	9,789	9,169	8,[illegible]38	7,316	6,644	5,614	4,860
30	28,806	27,704	26,683	25,734	24,851	24,026	22,531	21,210	20,036	18,985	18,039	17,183	16,404	14,735	13,374	12,243	11,288	10,472	9,766	8,[illegible]05	7,691	6,952	5,832	5,023
35	33,386	31,914	30,567	29,328	28,186	27,130	25,238	23,593	22,150	20,872	19,734	18,714	17,794	15,847	14,283	13,001	11,930	11,022	10,242	8,972	7,983	7,190	5,999	5,146
40	37,906	36,020	34,312	32,760	31,341	30,041	27,739	25,764	24,052	22,554	21,231	20,055	19,002	16,797	15,051	13,634	12,461	11,473	10,631	9,270	8,217	7,380	6,130	5,242
45	42,366	40,024	37,927	36,039	34,330	32,776	30,054	27,750	25,774	24,061	22,561	21,238	20,061	17,619	15,708	14,171	12,907	11,851	10,954	9,515	8,409	7,534	6,236	5,320
50	46,770	43,932	41,418	39,177	37,165	35,351	32,205	29,574	27,340	25,420	23,752	22,290	20,997	18,337	16,276	14,631	13,288	12,171	11,228	9,720	8,569	7,662	6,324	5,383
55	51,116	47,745	44,791	42,181	39,859	37,779	34,208	31,255	28,770	26,652	24,824	23,231	21,830	18,970	16,772	15,031	13,617	12,447	11,461	9,895	8,705	7,771	6,397	5,436
60	55,408	51,469	48,052	45,061	42,421	40,073	36,078	32,808	30,082	27,774	25,794	24,079	22,577	19,531	17,210	15,381	13,904	12,686	11,664	10,046	8,821	7,863	6,460	5,482
65	59,645	55,105	51,207	47,824	44,861	42,243	37,828	34,249	31,288	28,799	26,677	24,846	23,250	20,033	17,598	15,691	14,157	12,896	11,841	10,177	8,922	7,943	6,514	5,520
70	63,828	58,657	54,260	50,477	47,187	44,299	39,469	35,588	32,403	29,740	27,482	25,543	23,859	20,484	17,945	15,966	14,380	13,081	11,997	10,292	9,011	8,013	6,561	5,554
75	67,959	62,127	57,217	53,026	49,407	46,251	41,010	36,837	33,434	30,607	28,221	26,180	24,414	20,891	18,257	16,213	14,580	13,246	12,136	10,394	9,089	8,075	6,602	5,584
80	72,039	65,519	60,082	55,477	51,529	48,105	42,462	38,003	34,393	31,408	28,901	26,764	24,921	21,261	18,539	16,435	14,759	13,394	12,260	10,484	9,158	8,130	6,639	5,610
85	76,069	68,836	62,859	57,837	53,558	49,869	43,830	39,096	35,285	32,151	29,528	27,301	25,386	21,599	18,795	16,636	14,921	13,527	12,371	10,566	9,220	8,179	6,671	5,633
90	80,048	72,078	65,552	60,109	55,501	51,549	45,123	40,121	36,118	32,841	30,109	27,797	25,815	21,908	19,029	16,819	15,068	13,648	12,472	10,639	9,276	8,222	6,700	5,654
95	83,980	75,250	68,165	62,299	57,362	53,151	46,346	41,085	36,897	33,484	30,649	28,256	26,210	22,193	19,243	16,986	15,202	13,758	12,564	10,706	9,327	8,262	6,727	5,673
100	87,863	78,353	70,701	64,411	59,148	54,681	47,504	41,993	37,628	34,085	31,151	28,683	26,577	22,455	19,440	17,139	15,325	13,858	12,648	10,767	9,373	8,298	6,751	5,690
105	91,700	81,390	73,165	66,449	60,862	56,142	48,604	42,850	38,314	34,647	31,620	29,080	26,917	22,698	19,622	17,280	15,438	13,950	12,724	10,822	9,415	8,331	6,772	5,705
110	95,490	84,363	75,558	68,417	62,509	57,541	49,648	43,660	38,961	35,175	32,059	29,451	27,235	22,923	19,790	17,410	15,541	14,035	12,795	10,873	9,453	8,361	6,792	5,719
115	99,236	87,273	77,884	70,319	64,093	58,880	50,642	44,427	39,570	35,670	32,471	29,798	27,531	23,133	19,946	17,531	15,637	14,113	12,860	10,920	9,489	8,389	6,811	5,732
120	102,937	90,122	80,145	72,157	65,617	60,164	51,589	45,153	40,146	36,138	32,857	30,123	27,809	23,328	20,091	17,643	15,727	14,186	12,920	10,963	9,521	8,415	6,827	5,744
125	106,595	92,914	82,345	73,935	67,084	61,395	52,492	45,843	40,690	36,578	33,221	30,428	28,069	23,511	20,226	17,747	15,809	14,253	12,976	11,003	9,552	8,438	6,843	5,755
130	110,209	95,648	84,486	75,656	68,498	62,577	53,353	46,499	41,206	36,995	33,564	30,716	28,313	23,682	20,353	17,845	15,887	14,316	13,028	11,041	9,580	8,460	6,857	5,765
135	113,782	98,328	86,569	77,323	69,861	63,713	54,177	47,124	41,695	37,389	33,888	30,987	28,544	23,843	20,472	17,936	15,959	14,375	13,076	11,076	9,606	8,481	6,871	5,775
140	117,313	100,954	88,599	78,938	71,177	64,805	54,965	47,718	42,160	37,762	34,195	31,243	28,761	23,994	20,583	18,021	16,027	14,429	13,122	11,108	9,631	8,500	6,883	5,784
145	120,804	103,528	90,575	80,503	72,447	65,856	55,719	48,286	42,603	38,117	34,485	31,486	28,966	24,137	20,688	18,102	16,090	14,481	13,164	11,139	9,653	8,518	6,895	5,792
150	124,254	106,052	92,501	82,021	73,674	66,869	56,442	48,828	43,024	38,454	34,761	31,715	29,160	24,272	20,787	18,177	16,150	14,529	13,204	11,167	9,675	8,534	6,906	5,800
155	127,666	108,527	94,379	83,494	74,860	67,844	57,135	49,346	43,426	38,774	35,023	31,933	29,344	24,399	20,880	18,249	16,206	14,575	13,242	11,194	9,695	8,550	6,916	5,807
160	131,039	110,955	96,209	84,923	76,007	68,785	57,801	49,842	43,810	39,080	35,272	32,140	29,519	24,520	20,969	18,316	16,259	14,618	13,277	11,220	9,714	8,565	6,926	5,814
165	134,373	113,337	97,995	86,311	77,117	69,693	58,441	50,317	44,176	39,371	35,509	32,337	29,685	24,634	21,052	18,380	16,309	14,658	13,311	11,243	9,732	8,579	6,935	5,820
170	137,671	115,673	99,737	87,660	78,192	70,570	59,056	50,772	44,527	39,649	35,735	32,524	29,843	24,743	21,132	18,440	16,357	14,697	13,342	11,266	9,749	8,592	6,944	5,826
175	140,932	117,967	101,437	88,971	79,233	71,417	59,648	51,209	44,862	39,915	35,951	32,703	29,993	24,846	21,207	18,458	16,402	14,733	13,372	11,287	9,765	8,604	6,952	5,832
180	144,157	120,218	103,097	90,245	80,242	72,236	60,218	51,629	45,184	40,170	36,157	32,874	30,137	24,945	21,279	18,552	16,445	14,768	13,401	11,308	9,780	8,616	6,959	5,837
185	147,346	122,428	104,718	91,485	81,221	73,028	60,768	52,032	45,493	40,414	36,355	33,037	30,274	25,038	21,347	18,604	16,486	14,800	13,428	11,327	9,794	8,627	6,967	5,842
190	150,501	124,598	106,302	92,691	82,170	73,794	61,298	52,420	45,789	40,647	36,544	33,193	30,404	25,128	21,412	18,653	16,524	14,832	13,454	[illegible]1,345	9,808	8,638	6,974	5,847
195	153,621	126,729	107,849	93,865	83,092	74,537	61,809	52,794	46,074	40,871	36,725	33,342	30,530	25,213	21,474	18,700	16,561	14,861	13,478	[illegible]1,363	9,821	8,648	6,980	5,852
200	156,707	128,822	109,361	95,009	83,986	75,256	62,303	53,154	46,348	41,087	36,898	33,485	30,650	25,295	21,533	18,745	16,597	14,890	13,501	[illegible]1,379	9,833	8,657	6,986	5,856

# 4/0 AWG	1 Run(s)	480 Volts		C-Value = 15,082	Magnetic Duct	Copper

Calculations:

1. $f = \dfrac{\text{Length} \times 2 \times \text{Isca}}{\text{Runs} \times \text{C-Value} \times \text{E L-L}}$

2. $M = \dfrac{1}{1 + f}$

3. Isc = Isca x M

4. Add Motor Contribution = Motor FLA x 4

* All results are given in symmetrical amperes

Single-Phase L-L Bolted Fault-Current* Table

One-Way Distance in Feet

Available Isc in Thousands (Isca)	5	10	15	20	25	30	40	50	60	70	80	90	100	125	150	175	200	225	250	300	350	400	500	600
3	2,994	2,988	2,981	2,975	2,969	2,963	2,951	2,939	2,927	2,915	2,904	2,892	2,881	2,852	2,824	2,797	2,770	2,744	2,718	2,668	2,620	2,573	2,485	2,403
5	4,983	4,966	4,949	4,932	4,915	4,899	4,866	4,833	4,801	4,769	4,738	4,707	4,677	4,603	4,531	4,461	4,393	4,328	4,264	4,142	4,027	3,918	3,717	3,535
7	6,966	6,933	6,900	6,867	6,835	6,803	6,739	6,677	6,616	6,556	6,497	6,440	6,383	6,245	6,113	5,987	5,866	5,749	5,637	5,426	5,230	5,048	4,719	4,430
10	9,931	9,864	9,797	9,731	9,666	9,602	9,476	9,354	9,235	9,118	9,005	8,894	8,786	8,528	8,284	8,053	7,835	7,629	7,433	7,070	6,741	6,441	5,915	5,468
15	14,846	14,696	14,548	14,403	14,261	14,122	13,852	13,592	13,341	13,100	12,867	12,642	12,425	11,914	11,443	11,008	10,605	10,231	9,881	9,250	8,695	8,202	7,367	6,687
20	19,727	19,462	19,204	18,953	18,708	18,469	18,010	17,573	17,156	16,759	16,380	16,017	15,671	14,866	14,140	13,482	12,882	12,333	11,830	10,936	10,168	9,501	8,399	7,526
25	24,576	24,165	23,769	23,385	23,013	22,653	21,966	21,319	20,709	20,133	19,588	19,072	18,583	17,462	16,469	15,583	14,787	14,069	13,417	12,279	11,319	10,498	9,169	8,138
30	29,391	28,806	28,244	27,704	27,184	26,683	25,734	24,851	24,026	23,254	22,531	21,851	21,210	19,763	18,500	17,389	16,404	15,525	14,735	13,374	12,243	11,288	9,766	8,605
35	34,174	33,386	32,633	31,914	31,226	30,567	29,328	28,186	27,130	26,150	25,238	24,388	23,593	21,816	20,287	18,959	17,794	16,764	15,847	14,283	13,001	11,930	10,242	8,972
40	38,925	37,906	36,939	36,020	35,145	34,312	32,760	31,341	30,041	28,844	27,739	26,715	25,764	23,659	21,872	20,336	19,002	17,832	16,797	15,051	13,634	12,461	10,631	9,270
45	43,644	42,366	41,162	40,024	38,948	37,927	36,039	34,330	32,776	31,356	30,054	28,856	27,750	25,324	23,287	21,554	20,061	18,761	17,619	15,708	14,171	12,907	10,954	9,515
50	48,331	46,770	45,306	43,932	42,638	41,418	39,177	37,165	35,351	33,705	32,205	30,834	29,574	26,834	24,558	22,638	20,997	19,577	18,337	16,276	14,631	13,288	11,228	9,720
55	52,987	51,116	49,373	47,745	46,221	44,791	42,181	39,859	37,779	35,905	34,208	32,665	31,255	28,210	25,706	23,610	21,830	20,300	18,970	16,772	15,031	13,617	11,461	9,895
60	57,613	55,408	53,366	51,469	49,702	48,052	45,061	42,421	40,073	37,971	36,078	34,366	32,808	29,470	26,747	24,486	22,577	20,944	19,531	17,210	15,381	13,904	11,664	10,046
65	62,207	59,645	57,285	55,105	53,084	51,207	47,824	44,861	42,243	39,914	37,828	35,950	34,249	30,627	27,697	25,279	23,250	21,522	20,033	17,598	15,691	14,157	11,841	10,177
70	66,772	63,828	61,133	58,657	56,373	54,260	50,477	47,187	44,299	41,745	39,469	37,428	35,588	31,693	28,567	26,002	23,859	22,043	20,484	17,945	15,966	14,380	11,997	10,292
75	71,306	67,959	64,913	62,127	59,571	57,217	53,026	49,407	46,251	43,473	41,010	38,812	36,837	32,680	29,366	26,662	24,414	22,516	20,891	18,257	16,213	14,580	12,136	10,394
80	75,811	72,039	68,625	65,519	62,683	60,082	55,477	51,529	48,105	45,107	42,462	40,109	38,003	33,595	30,102	27,268	24,921	22,946	21,261	18,539	16,435	14,759	12,260	10,484
85	80,287	76,069	72,272	68,836	65,711	62,859	57,837	53,558	49,869	46,655	43,830	41,328	39,096	34,445	30,784	27,826	25,386	23,340	21,599	18,795	16,636	14,921	12,371	10,566
90	84,733	80,048	75,855	72,078	68,660	65,552	60,109	55,501	51,549	48,122	45,123	42,475	40,121	35,239	31,416	28,341	25,815	23,702	21,908	19,029	16,819	15,068	12,472	10,639
95	89,151	83,980	79,376	75,250	71,532	68,165	62,299	57,362	53,151	49,516	46,346	43,557	41,085	35,980	32,004	28,819	26,210	24,035	22,193	19,243	16,986	15,202	12,564	10,706
100	93,540	87,863	82,836	78,353	74,331	70,701	64,411	59,148	54,681	50,840	47,504	44,579	41,993	36,675	32,552	29,263	26,577	24,343	22,455	19,440	17,139	15,325	12,648	10,767
105	97,900	91,700	86,238	81,390	77,059	73,165	66,449	60,862	56,142	52,102	48,604	45,546	42,850	37,327	33,065	29,676	26,917	24,628	22,698	19,622	17,280	15,438	12,724	10,822
110	102,233	95,490	89,582	84,363	79,718	75,558	68,417	62,509	57,541	53,304	49,648	46,462	43,660	37,940	33,545	30,062	27,235	24,894	22,923	19,790	17,410	15,541	12,795	10,873
115	106,538	99,236	92,871	87,273	82,311	77,884	70,319	64,093	58,880	54,451	50,642	47,331	44,427	38,517	33,995	30,424	27,531	25,141	23,133	19,946	17,531	15,637	12,860	10,920
120	110,816	102,937	96,105	90,122	84,842	80,145	72,157	65,617	60,164	55,547	51,589	48,157	45,153	39,062	34,419	30,763	27,809	25,372	23,328	20,091	17,643	15,727	12,920	10,963
125	115,066	106,595	99,285	92,914	87,311	82,345	73,935	67,084	61,395	56,595	52,492	48,943	45,843	39,578	34,819	31,082	28,069	25,588	23,511	20,226	17,747	15,809	12,976	11,003
130	119,289	110,209	102,414	95,648	89,721	84,486	75,656	68,498	62,577	57,598	53,353	49,691	46,499	40,066	35,196	31,382	28,313	25,792	23,682	20,353	17,845	15,887	13,028	11,041
135	123,486	113,782	105,492	98,328	92,075	86,569	77,323	69,861	63,713	58,559	54,177	50,405	47,124	40,528	35,552	31,665	28,544	25,982	23,843	20,472	17,936	15,959	13,076	11,076
140	127,656	117,313	108,520	100,954	94,373	88,599	78,938	71,177	64,805	59,481	54,965	51,086	47,718	40,968	35,890	31,932	28,761	26,162	23,994	20,583	18,021	16,027	13,122	11,108
145	131,801	120,804	111,501	103,528	96,619	90,575	80,503	72,447	65,856	60,365	55,719	51,737	48,286	41,385	36,210	32,185	28,966	26,332	24,137	20,688	18,102	16,090	13,164	11,139
150	135,919	124,254	114,434	106,052	98,814	92,501	82,021	73,674	66,869	61,214	56,442	52,360	48,828	41,783	36,514	32,425	29,160	26,492	24,272	20,787	18,177	16,150	13,204	11,167
155	140,011	127,666	117,321	108,527	100,960	94,379	83,494	74,860	67,844	62,031	57,135	52,956	49,346	42,161	36,803	32,653	29,344	26,644	24,399	20,880	18,249	16,206	13,242	11,194
160	144,078	131,039	120,163	110,955	103,057	96,209	84,923	76,007	68,785	62,817	57,801	53,527	49,842	42,523	37,078	32,869	29,519	26,788	24,520	20,969	18,316	16,259	13,277	11,220
165	148,120	134,373	122,962	113,337	105,109	97,995	86,311	77,117	69,693	63,573	58,441	54,076	50,317	42,868	37,340	33,075	29,685	26,925	24,634	21,052	18,380	16,309	13,311	11,243
170	152,137	137,671	125,717	115,673	107,116	99,737	87,660	78,192	70,570	64,302	59,056	54,602	50,772	43,198	37,591	33,271	29,843	27,054	24,743	21,132	18,440	16,357	13,342	11,266
175	156,129	140,932	128,431	117,967	109,079	101,437	88,971	79,233	71,417	65,004	59,648	55,108	51,209	43,514	37,830	33,459	29,993	27,178	24,846	21,207	18,498	16,402	13,372	11,287
180	160,097	144,157	131,103	120,218	111,001	103,097	90,245	80,242	72,236	65,682	60,218	55,594	51,629	43,817	38,058	33,637	30,137	27,296	24,945	21,279	18,552	16,445	13,401	11,308
185	164,040	147,346	133,736	122,428	112,883	104,718	91,485	81,221	73,028	66,336	60,768	56,062	52,032	44,107	38,277	33,808	30,274	27,408	25,038	21,347	18,604	16,486	13,428	11,327
190	167,959	150,501	136,330	124,598	114,725	106,302	92,691	82,170	73,794	66,968	61,298	56,513	52,420	44,386	38,486	33,971	30,404	27,515	25,128	21,412	18,653	16,524	13,454	11,345
195	171,855	153,621	138,885	126,729	116,529	107,849	93,865	83,092	74,537	67,579	61,809	56,947	52,794	44,653	38,687	34,128	30,530	27,618	25,213	21,474	18,700	16,561	13,478	11,363
200	175,726	156,707	141,403	128,822	118,296	109,361	95,009	83,986	75,256	68,169	62,303	57,366	53,154	44,910	38,880	34,278	30,650	27,716	25,295	21,533	18,745	16,597	13,501	11,379

# 4/0 AWG	2 Run(s)	480 Volts	C-Value = 15,082	Magnetic Duct	Copper

Calculations:

1. $f = \dfrac{\text{Length} \times 2 \times \text{Isca}}{\text{Runs} \times \text{C-Value} \times \text{E L-L}}$
2. $M = \dfrac{1}{1 + f}$
3. Isc = Isca x M
4. Add Motor Contribution = Motor FLA x 4

* All results are given in symmetrical amperes

Single-Phase L-L Bolted Fault-Current* Table

One-Way Distance in Feet

Available Isc in Thousands (Isca)

Isca	5	10	15	20	25	30	40	50	60	70	80	90	100	125	150	175	200	225	250	300	350	400	500	600
3	2,989	2,977	2,966	2,955	2,944	2,933	2,912	2,890	2,869	2,849	2,828	2,808	2,789	2,740	2,694	2,649	2,605	2,563	2,522	2,444	2,371	2,302	2,175	2,062
5	4,969	4,938	4,907	4,877	4,847	4,817	4,759	4,703	4,648	4,594	4,541	4,489	4,439	4,318	4,203	4,094	3,991	3,893	3,799	3,625	3,467	3,321	3,064	2,844
7	6,939	6,878	6,819	6,761	6,703	6,647	6,537	6,431	6,328	6,229	6,132	6,038	5,948	5,732	5,532	5,345	5,170	5,007	4,853	4,573	4,323	4,099	3,714	3,395
10	9,875	9,753	9,635	9,519	9,406	9,295	9,082	8,878	8,683	8,497	8,318	8,147	7,982	7,599	7,251	6,933	6,642	6,374	6,128	5,687	5,306	4,972	4,417	3,973
15	14,721	14,452	14,193	13,943	13,701	13,468	13,025	12,609	12,220	11,854	11,509	11,184	10,876	10,177	9,562	9,017	8,531	8,094	7,700	7,0[illegible]7	6,446	5,960	5,180	4,580
20	19,507	19,038	18,590	18,163	17,756	17,366	16,636	15,964	15,345	14,772	14,240	13,746	13,284	12,255	11,374	10,611	9,945	9,357	8,834	7,947	7,222	6,618	5,669	4,959
25	24,234	23,514	22,835	22,195	21,589	21,016	19,956	18,997	18,127	17,333	16,605	15,936	15,319	13,967	12,834	11,871	11,043	10,322	9,690	8,633	7,784	7,086	6,010	5,217
30	28,904	27,885	26,936	26,049	25,219	24,440	23,018	21,752	20,618	19,597	18,672	17,830	17,061	15,401	14,035	12,891	11,920	11,085	10,359	9,150	8,210	7,438	6,261	5,405
35	33,517	32,155	30,899	29,738	28,661	27,659	25,851	24,266	22,863	21,614	20,494	19,485	18,570	16,620	15,040	13,735	12,638	11,703	10,897	9,578	8,544	7,711	6,453	5,548
40	38,075	36,327	34,732	33,272	31,929	30,690	28,481	26,568	24,896	23,422	22,113	20,942	19,889	17,668	15,894	14,443	13,235	12,213	11,338	9,917	8,813	7,929	6,605	5,660
45	42,578	40,404	38,441	36,660	35,036	33,551	30,928	28,685	26,746	25,052	23,560	22,236	21,052	18,580	16,628	15,047	13,740	12,642	11,707	10,[illegible]98	9,034	8,108	6,729	5,751
50	47,028	44,389	42,031	39,911	37,994	36,253	33,210	30,638	28,436	26,529	24,861	23,391	22,086	19,381	17,266	15,567	14,173	13,008	12,020	10,[illegible]35	9,219	8,257	6,831	5,825
55	51,425	48,287	45,509	43,034	40,814	38,812	35,344	32,445	29,986	27,873	26,038	24,430	23,009	20,088	17,825	16,021	14,548	13,323	12,288	10,[illegible]36	9,376	8,383	6,917	5,887
60	55,771	52,098	48,880	46,035	43,504	41,237	37,344	34,123	31,413	29,102	27,108	25,370	23,841	20,719	18,320	16,419	14,876	13,597	12,521	10,[illegible]11	9,511	8,490	6,990	5,940
65	60,065	55,827	52,147	48,923	46,074	43,538	39,222	35,684	32,731	30,230	28,084	26,222	24,592	21,284	18,761	16,772	15,165	13,839	12,726	10,[illegible]62	9,628	8,584	7,053	5,986
70	64,310	59,476	55,317	51,702	48,531	45,726	40,988	37,140	33,953	31,269	28,978	27,000	25,275	21,794	19,156	17,087	15,422	14,052	12,906	11,[illegible]96	9,731	8,666	7,108	6,026
75	68,506	63,047	58,394	54,380	50,883	47,808	42,653	38,502	35,087	32,229	29,801	27,713	25,899	22,256	19,512	17,370	15,652	14,243	13,067	11,215	9,822	8,738	7,157	6,060
80	72,654	66,543	61,381	56,962	53,136	49,792	44,225	39,778	36,144	33,118	30,560	28,368	26,470	22,677	19,834	17,625	15,859	14,414	13,211	11,320	9,903	8,802	7,200	6,091
85	76,754	69,966	64,282	59,452	55,296	51,684	45,712	40,977	37,131	33,945	31,262	28,973	26,995	23,061	20,128	17,856	16,046	14,568	13,340	11,415	9,976	8,859	7,238	6,119
90	80,808	73,319	67,101	61,855	57,370	53,491	47,120	42,105	38,054	34,715	31,914	29,532	27,480	23,414	20,396	18,067	16,216	14,709	13,458	11,501	10,042	8,911	7,273	6,143
95	84,816	76,604	69,842	64,176	59,361	55,218	48,455	43,167	38,920	35,434	32,521	30,051	27,929	23,739	20,642	18,260	16,371	14,836	13,564	11,579	10,101	8,957	7,304	6,165
100	88,779	79,822	72,507	66,420	61,276	56,871	49,723	44,171	39,734	36,108	33,088	30,534	28,346	24,039	20,869	18,437	16,513	14,953	13,662	11,650	10,155	9,000	7,332	6,185
105	92,698	82,976	75,100	68,589	63,118	58,454	50,929	45,120	40,501	36,739	33,617	30,984	28,734	24,318	21,079	18,601	16,644	15,060	13,751	11,715	10,204	9,038	7,357	6,204
110	96,573	86,068	77,624	70,688	64,891	59,972	52,077	46,019	41,223	37,333	34,114	31,405	29,096	24,577	21,273	18,752	16,765	15,159	13,834	11,775	10,249	9,074	7,381	6,220
115	100,406	89,099	80,080	72,720	66,599	61,428	53,171	46,871	41,906	37,892	34,580	31,800	29,434	24,818	21,453	18,892	16,877	15,250	13,910	11,830	10,291	9,107	7,403	6,236
120	104,196	92,071	82,473	74,688	68,246	62,826	54,216	47,681	42,552	38,420	35,019	32,171	29,751	25,043	21,621	19,022	16,981	15,335	13,980	11,881	10,330	9,137	7,422	6,250
125	107,946	94,986	84,805	76,595	69,834	64,170	55,214	48,451	43,165	38,918	35,432	32,520	30,049	25,254	21,778	19,143	17,077	15,414	14,046	11,928	10,365	9,165	7,441	6,263
130	111,654	97,846	87,077	78,444	71,368	65,463	56,168	49,185	43,746	39,390	35,823	32,848	30,330	25,451	21,925	19,257	17,168	15,487	14,107	11,972	10,399	9,191	7,458	6,275
135	115,322	100,652	89,292	80,237	72,849	66,707	57,081	49,884	44,298	39,837	36,192	33,159	30,594	25,637	22,063	19,363	17,252	15,556	14,164	12,013	10,429	9,215	7,474	6,286
140	118,952	103,405	91,452	81,977	74,280	67,905	57,957	50,551	44,823	40,261	36,542	33,452	30,844	25,812	22,192	19,463	17,331	15,620	14,217	12,051	10,458	9,237	7,489	6,297
145	122,542	106,107	93,560	83,666	75,665	69,060	58,796	51,188	45,323	40,664	36,874	33,730	31,080	25,978	22,314	19,556	17,405	15,681	14,267	12,087	10,485	9,258	7,502	6,306
150	126,094	108,760	95,616	85,307	77,004	70,174	59,602	51,797	45,801	41,048	37,189	33,994	31,304	26,134	22,429	19,645	17,475	15,737	14,314	12,121	10,511	9,278	7,515	6,316
155	129,609	111,365	97,624	86,901	78,301	71,250	60,375	52,381	46,256	41,414	37,489	34,244	31,516	26,281	22,538	19,728	17,541	15,791	14,358	12,153	10,534	9,297	7,528	6,324
160	133,086	113,923	99,584	88,451	79,557	72,288	61,119	52,940	46,692	41,762	37,775	34,482	31,717	26,421	22,641	19,807	17,603	15,841	14,400	12,182	10,557	9,314	7,539	6,332
165	136,527	116,435	101,498	89,958	80,774	73,291	61,835	53,476	47,108	42,095	38,047	34,709	31,909	26,554	22,738	19,881	17,662	15,889	14,439	12,211	10,578	9,331	7,550	6,340
170	139,933	118,903	103,368	91,424	81,954	74,262	62,524	53,991	47,507	42,414	38,307	34,925	32,091	26,680	22,831	19,952	17,718	15,934	14,476	12,237	10,598	9,346	7,560	6,347
175	143,303	121,328	105,196	92,851	83,098	75,200	63,188	54,485	47,889	42,718	38,555	35,131	32,266	26,801	22,919	20,019	17,771	15,977	14,512	12,262	10,617	9,361	7,570	6,354
180	146,639	123,710	106,982	94,239	84,209	76,109	63,828	54,961	48,256	43,010	38,792	35,328	32,432	26,915	23,003	20,083	17,821	16,017	14,545	12,286	10,635	9,375	7,579	6,360
185	149,940	126,052	108,729	95,592	85,287	76,988	64,446	55,418	48,608	43,289	39,019	35,516	32,590	27,024	23,082	20,144	17,869	16,056	14,577	12,309	10,652	9,388	7,587	6,366
190	153,208	128,353	110,437	96,910	86,335	77,841	65,042	55,858	48,947	43,558	39,237	35,697	32,742	27,129	23,158	20,202	17,915	16,093	14,607	12,331	10,668	9,400	7,596	6,372
195	156,442	130,615	112,108	98,194	87,353	78,667	65,618	56,283	49,272	43,815	39,446	35,869	32,887	27,228	23,231	20,257	17,958	16,128	14,636	12,351	10,683	9,412	7,603	6,378
200	159,644	132,840	113,743	99,446	88,342	79,469	66,175	56,692	49,586	44,063	39,647	36,035	33,027	27,324	23,300	20,310	17,999	16,161	14,664	12,371	10,698	9,424	7,611	6,383

250 MCM | 1 Run(s) | 480 Volts

C-Value = 16,483 | Magnetic Duct | Copper

Calculations:

1. $f = \dfrac{\text{Length} \times 2 \times \text{Isca}}{\text{Runs} \times \text{C-Value} \times \text{E L-L}}$

2. $M = \dfrac{1}{1 + f}$

3. Isc = Isca x M

4. Add Motor Contribution = Motor FLA x 4

* All results are given in symmetrical amperes

Single-Phase L-L Bolted Fault-Current* Table

One-Way Distance in Feet

Available Isc in Thousands (Isca)

Isca	5	10	15	20	25	30	40	50	60	70	80	90	100	125	150	175	200	225	250	300	350	400	500	600
3	2,994	2,989	2,983	2,977	2,972	2,966	2,955	2,944	2,933	2,922	2,912	2,901	2,890	2,864	2,839	2,813	2,789	2,764	2,740	2,694	2,649	2,605	2,522	2,444
5	4,984	4,969	4,953	4,938	4,922	4,907	4,877	4,847	4,817	4,788	4,759	4,731	4,703	4,634	4,567	4,502	4,439	4,378	4,318	4,203	4,094	3,991	3,799	3,625
7	6,969	6,939	6,908	6,878	6,849	6,819	6,761	6,703	6,647	6,592	6,537	6,484	6,431	6,303	6,180	6,061	5,948	5,838	5,732	5,532	5,345	5,170	4,853	4,573
10	9,937	9,875	9,814	9,753	9,694	9,635	9,519	9,406	9,295	9,187	9,082	8,979	8,878	8,636	8,406	8,189	7,982	7,786	7,599	7,251	6,933	6,642	6,128	5,687
15	14,859	14,721	14,585	14,452	14,321	14,193	13,943	13,701	13,468	13,243	13,025	12,814	12,609	12,126	11,679	11,263	10,876	10,515	10,177	9,562	9,017	8,531	7,700	7,017
20	19,750	19,507	19,269	19,038	18,811	18,590	18,163	17,756	17,366	16,993	16,636	16,293	15,964	15,198	14,501	13,866	13,284	12,749	12,255	11,374	10,611	9,945	8,834	7,947
25	24,611	24,234	23,869	23,514	23,170	22,835	22,195	21,589	21,016	20,472	19,956	19,465	18,997	17,921	16,961	16,098	15,319	14,612	13,967	12,834	11,871	11,043	9,690	8,633
30	29,442	28,904	28,386	27,885	27,402	26,936	26,049	25,219	24,440	23,707	23,018	22,367	21,752	20,353	19,123	18,034	17,061	16,189	15,401	14,035	12,891	11,920	10,359	9,160
35	34,243	33,517	32,822	32,155	31,515	30,899	29,738	28,661	27,659	26,724	25,851	25,033	24,266	22,537	21,039	19,728	18,570	17,541	16,620	15,040	13,735	12,638	10,897	9,578
40	39,014	38,075	37,180	36,327	35,512	34,732	33,272	31,929	30,690	29,544	28,481	27,491	26,568	24,510	22,749	21,223	19,889	18,713	17,668	15,894	14,443	13,235	11,338	9,917
45	43,756	42,578	41,463	40,404	39,398	38,441	36,660	35,036	33,551	32,186	30,928	29,764	28,685	26,301	24,283	22,553	21,052	19,739	18,580	16,628	15,047	13,740	11,707	10,198
50	48,468	47,028	45,671	44,389	43,178	42,031	39,911	37,994	36,253	34,665	33,210	31,872	30,638	27,934	25,668	23,742	22,086	20,645	19,381	17,266	15,567	14,173	12,020	10,435
55	53,153	51,425	49,806	48,287	46,857	45,509	43,034	40,814	38,812	36,997	35,344	33,833	32,445	29,428	26,925	24,814	23,009	21,450	20,088	17,825	16,021	14,548	12,288	10,636
60	57,808	55,771	53,872	52,098	50,438	48,880	46,035	43,504	41,237	39,194	37,344	35,661	34,123	30,802	28,070	25,783	23,841	22,170	20,719	18,320	16,419	14,876	12,521	10,811
65	62,435	60,065	57,869	55,827	53,925	52,147	48,923	46,074	43,538	41,268	39,222	37,369	35,684	32,068	29,118	26,664	24,592	22,819	21,284	18,761	16,772	15,165	12,726	10,962
70	67,035	64,310	61,799	59,476	57,321	55,317	51,702	48,531	45,726	43,228	40,988	38,970	37,140	33,239	30,080	27,469	25,275	23,406	21,794	19,156	17,087	15,422	12,906	11,096
75	71,606	68,506	65,663	63,047	60,631	58,394	54,380	50,883	47,808	45,084	42,653	40,472	38,502	34,326	30,967	28,207	25,899	23,940	22,256	19,512	17,370	15,652	13,067	11,215
80	76,150	72,654	69,464	66,543	63,858	61,381	56,962	53,136	49,792	46,844	44,225	41,884	39,778	35,337	31,787	28,886	26,470	24,427	22,677	19,834	17,625	15,859	13,211	11,320
85	80,667	76,754	73,203	69,966	67,004	64,282	59,452	55,296	51,684	48,515	45,712	43,215	40,977	36,279	32,548	29,513	26,995	24,874	23,061	20,128	17,856	16,046	13,340	11,415
90	85,157	80,808	76,882	73,319	70,073	67,101	61,855	57,370	53,491	50,104	47,120	44,471	42,105	37,161	33,256	30,093	27,480	25,285	23,414	20,396	18,067	16,216	13,458	11,501
95	89,620	84,816	80,501	76,604	73,067	69,842	64,176	59,361	55,218	51,616	48,455	45,659	43,167	37,986	33,915	30,633	27,929	25,664	23,739	20,642	18,260	16,371	13,564	11,579
100	94,056	88,779	84,063	79,822	75,989	72,507	66,420	61,276	56,871	53,057	49,723	46,783	44,171	38,761	34,532	31,134	28,346	26,016	24,039	20,869	18,437	16,513	13,662	11,650
105	98,466	92,698	87,568	82,976	78,842	75,100	68,589	63,118	58,454	54,433	50,929	47,849	45,120	39,490	35,109	31,603	28,734	26,342	24,318	21,079	18,601	16,644	13,751	11,715
110	102,850	96,573	91,018	86,068	81,628	77,624	70,688	64,891	59,972	55,746	52,077	48,861	46,019	40,177	35,651	32,041	29,096	26,646	24,577	21,273	18,752	16,765	13,834	11,775
115	107,209	100,406	94,415	89,099	84,349	80,080	72,720	66,599	61,428	57,002	53,171	49,823	46,871	40,825	36,160	32,452	29,434	26,929	24,818	21,453	18,892	16,877	13,910	11,830
120	111,541	104,196	97,759	92,071	87,008	82,473	74,688	68,246	62,826	58,204	54,216	50,739	47,681	41,438	36,640	32,838	29,751	27,195	25,043	21,621	19,022	16,981	13,980	11,881
125	115,848	107,946	101,052	94,986	89,607	84,805	76,595	69,834	64,170	59,356	55,214	51,612	48,451	42,018	37,093	33,202	30,049	27,444	25,254	21,778	19,143	17,077	14,046	11,928
130	120,131	111,654	104,295	97,846	92,148	87,077	78,444	71,368	65,463	60,460	56,168	52,445	49,185	42,569	37,522	33,545	30,330	27,677	25,451	21,925	19,257	17,168	14,107	11,972
135	124,388	115,322	107,489	100,652	94,632	89,292	80,237	72,849	66,707	61,520	57,081	53,240	49,884	43,091	37,927	33,868	30,594	27,897	25,637	22,063	19,363	17,252	14,164	12,013
140	128,620	118,952	110,635	103,405	97,062	91,452	81,977	74,280	67,905	62,538	57,957	54,001	50,551	43,588	38,312	34,174	30,844	28,105	25,812	22,192	19,463	17,331	14,217	12,051
145	132,828	122,542	113,734	106,107	99,439	93,560	83,666	75,665	69,060	63,516	58,796	54,729	51,188	44,061	38,677	34,465	31,080	28,301	25,978	22,314	19,556	17,405	14,267	12,087
150	137,012	126,094	116,788	108,760	101,766	95,616	85,307	77,004	70,174	64,457	59,602	55,426	51,797	44,512	39,023	34,740	31,304	28,486	26,134	22,429	19,645	17,475	14,314	12,121
155	141,172	129,609	119,796	111,365	104,043	97,624	86,901	78,301	71,250	65,363	60,375	56,095	52,381	44,942	39,354	35,001	31,516	28,662	26,281	22,538	19,728	17,541	14,358	12,153
160	145,307	133,086	122,761	113,923	106,272	99,584	88,451	79,557	72,288	66,236	61,119	56,736	52,940	45,353	39,668	35,250	31,717	28,828	26,421	22,641	19,807	17,603	14,400	12,182
165	149,419	136,527	125,683	116,435	108,455	101,498	89,958	80,774	73,291	67,078	61,835	57,353	53,476	45,746	39,969	35,487	31,909	28,986	26,554	22,738	19,881	17,662	14,439	12,211
170	153,508	139,933	128,564	118,903	110,593	103,368	91,424	81,954	74,262	67,889	62,524	57,945	53,991	46,122	40,256	35,713	32,091	29,137	26,680	22,831	19,952	17,718	14,476	12,237
175	157,573	143,303	131,403	121,328	112,687	105,196	92,851	83,098	75,200	68,673	63,188	58,515	54,485	46,483	40,530	35,929	32,266	29,280	26,801	22,919	20,019	17,771	14,512	12,262
180	161,616	146,639	134,202	123,710	114,740	106,982	94,239	84,209	76,109	69,430	63,828	59,064	54,961	46,828	40,792	36,135	32,432	29,417	26,915	23,003	20,083	17,821	14,545	12,286
185	165,635	149,940	136,962	126,052	116,751	108,729	95,592	85,287	76,988	70,161	64,446	59,592	55,418	47,160	41,044	36,332	32,590	29,548	27,024	23,082	20,144	17,869	14,577	12,309
190	169,632	153,208	139,683	128,353	118,723	110,437	96,910	86,335	77,841	70,868	65,042	60,102	55,858	47,478	41,285	36,520	32,742	29,672	27,129	23,158	20,202	17,915	14,607	12,331
195	173,606	156,442	142,367	130,615	120,656	112,108	98,194	87,353	78,667	71,553	65,618	60,593	56,283	47,784	41,516	36,701	32,887	29,792	27,228	23,231	20,257	17,958	14,636	12,351
200	177,558	159,644	145,014	132,840	122,552	113,743	99,446	88,342	79,469	72,215	66,175	61,067	56,692	48,079	41,738	36,875	33,027	29,906	27,324	23,300	20,310	17,999	14,664	12,371

250 MCM | 2 Run(s) | 480 Volts | C-Value = 16,483 | Magnetic Duct | Copper

Calculations:

1. $f = \dfrac{\text{Length} \times 2 \times \text{Isca}}{\text{Runs} \times \text{C-Value} \times E_{L-L}}$
2. $M = \dfrac{1}{1 + f}$
3. $\text{Isc} = \text{Isca} \times M$
4. Add Motor Contribution = Motor FLA x 4

* All results are given in symmetrical amperes

Single-Phase L-L Bolted Fault-Current* Table

One-Way Distance in Feet

Available Isc in Thousands (Isca)	5	10	15	20	25	30	40	50	60	70	80	90	100	125	150	175	200	225	250	300	350	400	500	600
3	2,990	2,980	2,969	2,959	2,949	2,939	2,920	2,900	2,881	2,862	2,844	2,825	2,807	2,763	2,719	2,678	2,637	2,598	2,560	2,487	2,418	2,353	2,232	2,124
5	4,972	4,943	4,915	4,888	4,861	4,834	4,781	4,729	4,678	4,629	4,580	4,532	4,486	4,373	4,266	4,165	4,068	3,975	3,886	3,721	3,568	3,428	3,178	2,963
7	6,944	6,889	6,835	6,782	6,730	6,678	6,578	6,480	6,385	6,293	6,204	6,117	6,032	5,830	5,642	5,465	5,299	5,143	4,996	4,725	4,482	4,263	3,884	3,566
10	9,887	9,776	9,668	9,562	9,458	9,357	9,160	8,972	8,791	8,617	8,450	8,290	8,135	7,773	7,441	7,137	6,856	6,597	6,357	5,925	5,548	5,217	4,659	4,210
15	14,746	14,501	14,264	14,035	13,813	13,597	13,186	12,799	12,435	12,090	11,764	11,455	11,162	10,491	9,896	9,365	8,888	8,457	8,066	7,383	6,807	6,315	5,516	4,897
20	19,552	19,123	18,713	18,320	17,943	17,582	16,901	16,270	15,685	15,141	14,633	14,158	13,713	12,714	11,850	11,097	10,433	9,845	9,319	8,419	7,678	7,057	6,075	5,332
25	24,304	23,645	23,021	22,429	21,867	21,332	20,338	19,432	18,603	17,842	17,141	16,493	15,892	14,566	13,443	12,482	11,648	10,920	10,276	9,154	8,317	7,593	6,467	5,632
30	29,003	28,070	27,195	26,373	25,599	24,869	23,528	22,324	21,237	20,251	19,353	18,531	17,775	16,132	14,767	13,615	12,629	11,777	11,032	9,754	8,805	7,998	6,759	5,852
35	33,650	32,400	31,240	30,160	29,152	28,210	26,496	24,979	23,626	22,412	21,317	20,324	19,419	17,474	15,884	14,558	13,437	12,477	11,644	10,273	9,191	8,315	6,984	6,020
40	38,246	36,640	35,163	33,801	32,540	31,370	29,266	27,426	25,804	24,362	23,074	21,915	20,866	18,638	16,839	15,357	14,115	13,058	12,149	10,654	9,503	8,569	7,162	6,152
45	42,793	40,792	38,970	37,304	35,774	34,365	31,855	29,687	27,796	26,131	24,654	23,335	22,150	19,655	17,665	16,041	14,691	13,550	12,574	10,990	9,760	8,778	7,308	6,259
50	47,290	44,858	42,665	40,676	38,864	37,206	34,282	31,784	29,626	27,742	26,083	24,611	23,297	20,553	18,387	16,634	15,186	13,971	12,935	11,235	9,977	8,953	7,428	6,347
55	51,738	48,842	46,253	43,924	41,819	39,906	36,561	33,734	31,312	29,215	27,381	25,764	24,327	21,351	19,023	17,153	15,618	14,335	13,246	11,500	10,161	9,101	7,530	6,422
60	56,139	52,745	49,738	47,056	44,648	42,474	38,705	35,551	32,872	30,568	28,567	26,811	25,258	22,065	19,588	17,611	15,996	14,653	13,518	11,704	10,320	9,228	7,617	6,485
65	60,493	56,571	53,126	50,077	47,358	44,920	40,726	37,249	34,318	31,815	29,653	27,765	26,104	22,707	20,092	18,017	16,331	14,933	13,756	11,883	10,458	9,339	7,692	6,539
70	64,801	60,320	56,420	52,993	49,958	47,252	42,634	38,838	35,663	32,968	30,651	28,639	26,875	23,288	20,546	18,381	16,630	15,183	13,967	12,040	10,580	9,436	7,758	6,586
75	69,063	63,997	59,623	55,809	52,454	49,479	44,439	40,330	36,917	34,037	31,573	29,442	27,581	23,816	20,956	18,709	16,897	15,405	14,156	12,179	10,687	9,521	7,815	6,628
80	73,280	67,602	62,741	58,532	54,852	51,607	46,148	41,733	38,089	35,030	32,426	30,183	28,229	24,298	21,328	19,005	17,139	15,606	14,325	12,304	10,784	9,597	7,867	6,665
85	77,454	71,138	65,775	61,164	57,157	53,643	47,768	43,054	39,186	35,956	33,218	30,868	28,828	24,740	21,668	19,275	17,357	15,787	14,477	12,417	10,870	9,666	7,912	6,698
90	81,584	74,607	68,730	63,711	59,375	55,592	49,308	44,300	40,216	36,822	33,955	31,503	29,381	25,147	21,979	19,521	17,556	15,951	14,615	12,518	10,947	9,727	7,953	6,727
95	85,671	78,011	71,608	66,176	61,511	57,460	50,772	45,479	41,185	37,632	34,643	32,095	29,895	25,522	22,266	19,746	17,739	16,102	14,741	12,611	11,018	9,783	7,991	6,754
100	89,717	81,351	74,413	68,565	63,569	59,252	52,166	46,594	42,097	38,392	35,287	32,646	30,373	25,870	22,530	19,953	17,906	16,239	14,857	12,695	11,082	9,833	8,024	6,778
105	93,721	84,630	77,146	70,879	65,553	60,972	53,495	47,651	42,959	39,107	35,890	33,162	30,819	26,193	22,774	20,145	18,060	16,366	14,962	12,772	11,141	9,879	8,055	6,800
110	97,684	87,848	79,812	73,122	67,468	62,625	54,763	48,655	43,773	39,781	36,456	33,645	31,235	26,493	23,001	20,322	18,202	16,483	15,060	12,843	11,195	9,922	8,083	6,820
115	101,607	91,008	82,411	75,299	69,316	64,214	55,975	49,609	44,543	40,416	36,989	34,098	31,626	26,773	23,212	20,486	18,334	16,591	15,150	12,909	11,245	9,961	8,109	6,838
120	105,490	94,111	84,948	77,411	71,102	65,744	57,133	50,517	45,274	41,017	37,492	34,525	31,992	27,036	23,409	20,640	18,457	16,691	15,234	12,969	11,291	9,997	8,133	6,855
125	109,335	97,159	87,423	79,461	72,828	67,217	58,242	51,382	45,968	41,586	37,966	34,926	32,337	27,281	23,593	20,783	18,571	16,784	15,312	13,026	11,333	10,030	8,155	6,871
130	113,141	100,153	89,840	81,452	74,497	68,636	59,305	52,208	46,627	42,125	38,415	35,306	32,662	27,512	23,765	20,916	18,678	16,872	15,384	13,078	11,373	10,062	8,176	6,885
135	116,910	103,095	92,200	83,387	76,113	70,005	60,325	52,996	47,255	42,636	38,840	35,665	32,969	27,730	23,927	21,042	18,777	16,953	15,452	13,127	11,410	10,090	8,195	6,899
140	120,641	105,985	94,505	85,269	77,677	71,326	61,303	53,749	47,853	43,123	39,243	36,004	33,259	27,935	24,080	21,160	18,871	17,029	15,515	13,173	11,445	10,117	8,213	6,911
145	124,336	108,826	96,757	87,098	79,192	72,602	62,243	54,471	48,424	43,586	39,626	36,326	33,534	28,128	24,223	21,270	18,959	17,101	15,575	13,216	11,477	10,143	8,229	6,923
150	127,994	111,619	98,958	88,877	80,660	73,834	63,146	55,161	48,969	44,027	39,991	36,632	33,794	28,311	24,359	21,375	19,042	17,169	15,631	13,256	11,507	10,166	8,245	6,934
155	131,617	114,364	101,110	90,609	82,084	75,025	64,016	55,823	49,490	44,448	40,338	36,923	34,042	28,485	24,487	21,474	19,121	17,232	15,683	13,294	11,536	10,189	8,260	6,945
160	135,205	117,063	103,214	92,295	83,465	76,178	64,853	56,459	49,989	44,849	40,668	37,200	34,277	28,649	24,609	21,567	19,195	17,292	15,733	13,329	11,563	10,210	8,273	6,954
165	138,758	119,717	105,272	93,937	84,806	77,293	65,659	57,069	50,467	45,234	40,984	37,464	34,501	28,805	24,724	21,656	19,265	17,349	15,780	13,363	11,588	10,229	8,286	6,964
170	142,277	122,328	107,285	95,537	86,108	78,373	66,437	57,656	50,925	45,601	41,286	37,716	34,715	28,954	24,833	21,739	19,331	17,403	15,825	13,395	11,612	10,248	8,299	6,972
175	145,762	124,896	109,255	97,096	87,372	79,419	67,187	58,220	51,365	45,954	41,574	37,957	34,918	29,096	24,937	21,819	19,394	17,454	15,867	13,425	11,635	10,266	8,310	6,980
180	149,215	127,422	111,183	98,616	88,601	80,433	67,911	58,763	51,787	46,291	41,850	38,187	35,113	29,231	25,037	21,895	19,454	17,503	15,907	13,454	11,656	10,283	8,321	6,988
185	152,634	129,907	113,071	100,098	89,796	81,416	68,611	59,286	52,193	46,615	42,115	38,407	35,299	29,360	25,131	21,967	19,511	17,549	15,945	13,481	11,677	10,299	8,332	6,995
190	156,022	132,353	114,919	101,544	90,957	82,370	69,287	59,790	52,583	46,926	42,369	38,618	35,477	29,483	25,221	22,036	19,565	17,593	15,981	13,507	11,696	10,314	8,341	7,002
195	159,378	134,760	116,730	102,955	92,088	83,296	69,941	60,277	52,959	47,225	42,612	38,820	35,648	29,601	25,307	22,102	19,617	17,634	16,016	13,532	11,715	10,328	8,351	7,009
200	162,702	137,129	118,503	104,332	93,188	84,195	70,574	60,746	53,321	47,513	42,846	39,014	35,811	29,713	25,390	22,165	19,666	17,674	16,049	13,555	11,732	10,342	8,360	7,015

300 MCM	1 Run(s)	480 Volts	C-Value = 18,176	Magnetic Duct	Copper

Calculations:

1. $f = \dfrac{\text{Length} \times 2 \times \text{Isca}}{\text{Runs} \times \text{C-Value} \times \text{E L-L}}$
2. $M = \dfrac{1}{1 + f}$
3. $\text{Isc} = \text{Isca} \times M$
4. Add Motor Contribution = Motor FLA x 4

* All results are given in symmetrical amperes

Single-Phase L-L Bolted Fault-Current* Table

One-Way Distance in Feet

Available Isc in Thousands (Isca)	5	10	15	20	25	30	40	50	60	70	80	90	100	125	150	175	200	225	250	300	350	400	500	600
3	2,995	2,990	2,985	2,980	2,974	2,969	2,959	2,949	2,939	2,929	2,920	2,910	2,900	2,876	2,853	2,830	2,807	2,785	2,763	2,719	2,678	2,637	2,560	2,487
5	4,986	4,972	4,957	4,943	4,929	4,915	4,888	4,861	4,834	4,807	4,781	4,755	4,729	4,666	4,604	4,544	4,486	4,429	4,373	4,266	4,165	4,068	3,886	3,721
7	6,972	6,944	6,917	6,889	6,862	6,835	6,782	6,730	6,678	6,628	6,578	6,529	6,480	6,362	6,248	6,138	6,032	5,930	5,830	5,642	5,465	5,299	4,996	4,725
10	9,943	9,887	9,831	9,776	9,721	9,668	9,562	9,458	9,357	9,257	9,160	9,065	8,972	8,747	8,533	8,329	8,135	7,950	7,773	7,441	7,137	6,856	6,357	5,925
15	14,872	14,746	14,623	14,501	14,382	14,264	14,035	13,813	13,597	13,389	13,186	12,990	12,799	12,347	11,925	11,531	11,162	10,816	10,491	9,896	9,365	8,888	8,066	7,383
20	19,773	19,552	19,335	19,123	18,916	18,713	18,320	17,943	17,582	17,234	16,901	16,579	16,270	15,545	14,883	14,274	13,713	13,194	12,714	11,850	11,097	10,433	9,319	8,419
25	24,647	24,304	23,970	23,645	23,329	23,021	22,429	21,867	21,332	20,823	20,338	19,874	19,432	18,407	17,485	16,650	15,892	15,200	14,566	13,443	12,482	11,648	10,276	9,194
30	29,493	29,003	28,529	28,070	27,625	27,195	26,373	25,599	24,869	24,180	23,528	22,910	22,324	20,982	19,792	18,729	17,775	16,914	16,132	14,767	13,615	12,629	11,032	9,794
35	34,312	33,650	33,013	32,400	31,810	31,240	30,160	29,152	28,210	27,326	26,496	25,715	24,979	23,311	21,851	20,563	19,419	18,396	17,474	15,884	14,558	13,437	11,644	10,273
40	39,104	38,246	37,426	36,640	35,887	35,163	33,801	32,540	31,370	30,282	29,266	28,316	27,426	25,427	23,701	22,193	20,866	19,689	18,638	16,839	15,357	14,115	12,149	10,664
45	43,869	42,793	41,768	40,792	39,860	38,970	37,304	35,774	34,365	33,063	31,855	30,733	29,687	27,360	25,371	23,651	22,150	20,828	19,655	17,665	16,041	14,691	12,574	10,990
50	48,607	47,290	46,042	44,858	43,734	42,665	40,676	38,864	37,206	35,684	34,282	32,986	31,784	29,131	26,887	24,963	23,297	21,839	20,553	18,387	16,634	15,186	12,935	11,265
55	53,319	51,738	50,248	48,842	47,512	46,253	43,924	41,819	39,906	38,160	36,561	35,091	33,734	30,760	28,269	26,150	24,327	22,742	21,351	19,023	17,153	15,618	13,246	11,500
60	58,005	56,139	54,389	52,745	51,198	49,738	47,056	44,648	42,474	40,502	38,705	37,061	35,551	32,264	29,534	27,229	25,258	23,554	22,065	19,588	17,611	15,996	13,518	11,704
65	62,666	60,493	58,466	56,571	54,794	53,126	50,077	47,358	44,920	42,720	40,726	38,910	37,249	33,656	30,696	28,214	26,104	24,287	22,707	20,092	18,017	16,331	13,756	11,883
70	67,300	64,801	62,480	60,320	58,305	56,420	52,993	49,958	47,252	44,825	42,634	40,648	38,838	34,949	31,767	29,117	26,875	24,953	23,288	20,546	18,381	16,630	13,967	12,040
75	71,909	69,063	66,434	63,997	61,733	59,623	55,809	52,454	49,479	46,824	44,439	42,285	40,330	36,152	32,759	29,947	27,581	25,561	23,816	20,956	18,709	16,897	14,156	12,179
80	76,493	73,280	70,327	67,602	65,081	62,741	58,532	54,852	51,607	48,725	46,148	43,829	41,733	37,275	33,678	30,714	28,229	26,117	24,298	21,328	19,005	17,139	14,325	12,304
85	81,052	77,454	74,162	71,138	68,352	65,775	61,164	57,157	53,643	50,535	47,768	45,289	43,054	38,326	34,533	31,424	28,828	26,628	24,740	21,668	19,275	17,357	14,477	12,417
90	85,586	81,584	77,940	74,607	71,548	68,730	63,711	59,375	55,592	52,262	49,308	46,670	44,300	39,310	35,331	32,083	29,381	27,100	25,147	21,979	19,521	17,556	14,615	12,518
95	90,095	85,671	81,662	78,011	74,672	71,608	66,176	61,511	57,460	53,909	50,772	47,980	45,479	40,235	36,076	32,696	29,895	27,536	25,522	22,266	19,746	17,739	14,741	12,611
100	94,580	89,717	85,329	81,351	77,727	74,413	68,565	63,569	59,252	55,483	52,166	49,223	46,594	41,106	36,774	33,268	30,373	27,941	25,870	22,530	19,953	17,906	14,857	12,695
105	99,040	93,721	88,943	84,630	80,715	77,146	70,879	65,553	60,972	56,989	53,495	50,404	47,651	41,926	37,430	33,804	30,319	28,318	26,193	22,774	20,145	18,060	14,962	12,772
110	103,477	97,684	92,505	87,848	83,637	79,812	73,122	67,468	62,625	58,431	54,763	51,529	48,655	42,701	38,046	34,306	31,235	28,669	26,493	23,001	20,322	18,202	15,060	12,843
115	107,889	101,607	96,016	91,008	86,497	82,411	75,299	69,316	64,214	59,812	55,975	52,600	49,609	43,435	38,627	34,778	31,626	28,998	26,773	23,212	20,486	18,334	15,150	12,909
120	112,278	105,490	99,476	94,111	89,295	84,948	77,411	71,102	65,744	61,137	57,133	53,622	50,517	44,129	39,175	35,221	31,992	29,306	27,036	23,409	20,640	18,457	15,234	12,969
125	116,644	109,335	102,888	97,159	92,034	87,423	79,461	72,828	67,217	62,409	58,242	54,598	51,382	44,788	39,694	35,640	32,337	29,595	27,281	23,593	20,783	18,571	15,312	13,026
130	120,986	113,141	106,252	100,153	94,717	89,840	81,452	74,497	68,636	63,631	59,305	55,531	52,208	45,414	40,184	36,035	32,662	29,867	27,512	23,765	20,916	18,678	15,384	13,078
135	125,305	116,910	109,569	103,095	97,343	92,200	83,387	76,113	70,005	64,805	60,325	56,423	52,996	46,009	40,650	36,409	32,969	30,123	27,730	23,927	21,042	18,777	15,452	13,127
140	129,602	120,641	112,839	105,985	99,916	94,505	85,269	77,677	71,326	65,936	61,303	57,278	53,749	46,576	41,092	36,763	33,259	30,365	27,935	24,080	21,160	18,871	15,515	13,173
145	133,875	124,336	116,065	108,826	102,437	96,757	87,098	79,192	72,602	67,024	62,243	58,098	54,471	47,116	41,512	37,099	33,534	30,594	28,128	24,223	21,270	18,959	15,575	13,216
150	138,126	127,994	119,247	111,619	104,908	98,958	88,877	80,660	73,834	68,073	63,146	58,884	55,161	47,632	41,912	37,418	33,794	30,811	28,311	24,359	21,375	19,042	15,631	13,256
155	142,355	131,617	122,385	114,364	107,329	101,110	90,609	82,084	75,025	69,085	64,016	59,639	55,823	48,125	42,293	37,721	34,042	31,016	28,485	24,487	21,474	19,121	15,683	13,294
160	146,561	135,205	125,482	117,063	109,703	103,214	92,295	83,465	76,178	70,060	64,853	60,365	56,459	48,597	42,657	38,011	34,277	31,211	28,649	24,609	21,567	19,195	15,733	13,329
165	150,745	138,758	128,536	119,717	112,031	105,272	93,937	84,806	77,293	71,003	65,659	61,063	57,069	49,048	43,004	38,286	34,501	31,397	28,805	24,724	21,656	19,265	15,780	13,363
170	154,908	142,277	131,550	122,328	114,314	107,285	95,537	86,108	78,373	71,913	66,437	61,735	57,656	49,481	43,336	38,549	34,715	31,574	28,954	24,833	21,739	19,331	15,825	13,395
175	159,049	145,762	134,525	124,896	116,553	109,255	97,096	87,372	79,419	72,792	67,187	62,383	58,220	49,896	43,654	38,801	34,918	31,742	29,096	24,937	21,819	19,394	15,867	13,425
180	163,168	149,215	137,460	127,422	118,750	111,183	98,616	88,601	80,433	73,643	67,911	63,007	58,763	50,294	43,959	39,041	35,113	31,903	29,231	25,037	21,895	19,454	15,907	13,454
185	167,266	152,634	140,357	129,907	120,906	113,071	100,098	89,796	81,416	74,467	68,611	63,608	59,286	50,677	44,251	39,271	35,299	32,057	29,360	25,131	21,967	19,511	15,945	13,481
190	171,343	156,022	143,216	132,353	123,022	114,919	101,544	90,957	82,370	75,264	69,287	64,189	59,790	51,045	44,531	39,492	35,477	32,203	29,483	25,221	22,036	19,565	15,981	13,507
195	175,398	159,378	146,039	134,760	125,098	116,730	102,955	92,088	83,296	76,036	69,941	64,750	60,277	51,399	44,800	39,703	35,648	32,344	29,601	25,307	22,102	19,617	16,016	13,532
200	179,433	162,702	148,825	137,129	127,137	118,503	104,332	93,188	84,195	76,785	70,574	65,292	60,746	51,740	45,059	39,907	35,811	32,479	29,713	25,390	22,165	19,666	16,049	13,555

300 MCM | 2 Run(s) | 480 Volts

C-Value = 18,176 | Magnetic Duct | Copper

Calculations:

1. $f = \dfrac{\text{Length} \times 2 \times \text{Isca}}{\text{Runs} \times \text{C-Value} \times E_{L-L}}$

2. $M = \dfrac{1}{1 + f}$

3. Isc = Isca x M

4. Add Motor Contribution = Motor FLA x 4

* All results are given in symmetrical amperes

Single-Phase L-L Bolted Fault-Current* Table

One-Way Distance in Feet

Available Isc in Thousands (Isca)

	5	10	15	20	25	30	40	50	60	70	80	90	100	125	150	175	200	225	250	300	350	400	500	600
3	2,991	2,981	2,972	2,962	2,953	2,944	2,926	2,908	2,890	2,872	2,855	2,838	2,821	2,780	2,739	2,700	2,662	2,625	2,589	2,520	2,455	2,393	2,278	2,173
5	4,974	4,948	4,922	4,896	4,871	4,846	4,797	4,749	4,702	4,655	4,610	4,566	4,522	4,416	4,316	4,219	4,127	4,039	3,955	3,75[illegible]	3,649	3,514	3,271	3,059
7	6,949	6,898	6,848	6,799	6,750	6,702	6,609	6,518	6,429	6,343	6,259	6,177	6,097	5,907	5,728	5,560	5,401	5,251	5,109	4,847	4,611	4,397	4,023	3,707
10	9,895	9,793	9,693	9,594	9,498	9,403	9,220	9,044	8,874	8,711	8,553	8,401	8,254	7,909	7,592	7,299	7,028	6,776	6,542	6,1[illegible]8	5,747	5,417	4,861	4,408
15	14,766	14,539	14,319	14,105	13,898	13,697	13,311	12,947	12,602	12,274	11,964	11,669	11,388	10,741	10,164	9,646	9,178	8,753	8,366	7,686	7,108	6,611	5,800	5,167
20	19,586	19,188	18,807	18,440	18,087	17,748	17,106	16,509	15,952	15,431	14,944	14,486	14,055	13,083	12,237	11,493	10,835	10,248	9,721	8,8[illegible]5	8,063	7,430	6,421	5,653
25	24,356	23,745	23,163	22,609	22,081	21,578	20,636	19,773	18,980	18,247	17,569	16,940	16,354	15,052	13,943	12,986	12,151	11,418	10,768	9,6[illegible]7	8,771	8,026	6,862	5,992
30	29,078	28,210	27,393	26,622	25,893	25,203	23,928	22,775	21,729	20,774	19,900	19,096	18,355	16,731	15,372	14,216	13,223	12,359	11,601	10,333	9,315	8,480	7,191	6,242
35	33,751	32,588	31,502	30,487	29,535	28,640	27,005	25,546	24,237	23,055	21,983	21,007	20,113	18,180	16,586	15,249	14,111	13,131	12,279	10,658	9,748	8,837	7,446	6,433
40	38,377	36,880	35,496	34,212	33,018	31,904	29,887	28,111	26,533	25,124	23,856	22,710	21,670	19,442	17,630	16,127	14,860	13,778	12,842	11,[illegible]07	10,099	9,125	7,649	6,584
45	42,956	41,090	39,379	37,805	36,352	35,006	32,593	30,492	28,645	27,008	25,549	24,240	23,058	20,552	18,538	16,883	15,500	14,326	13,317	11,673	10,391	9,362	7,815	6,707
50	47,489	45,219	43,155	41,272	39,546	37,959	35,138	32,708	30,592	28,733	27,087	25,620	24,303	21,536	19,334	17,541	16,053	14,797	13,723	11,984	10,637	9,561	7,953	6,808
55	51,977	49,269	46,830	44,620	42,610	40,773	37,536	34,776	32,394	30,317	28,490	26,871	25,426	22,413	20,039	18,119	16,535	15,206	14,075	12,251	10,846	9,730	8,070	6,893
60	56,421	53,244	50,406	47,856	45,551	43,458	39,800	36,710	34,066	31,776	29,776	28,012	26,445	23,201	20,666	18,631	16,960	15,565	14,381	12,[illegible]83	11,027	9,876	8,170	6,966
65	60,820	57,145	53,889	50,984	48,376	46,022	41,940	38,523	35,621	33,126	30,957	29,055	27,373	23,913	21,229	19,087	17,337	15,882	14,651	12,[illegible]86	11,186	10,003	8,256	7,029
70	65,176	60,974	57,281	54,010	51,092	48,473	43,966	40,226	37,073	34,377	32,048	30,013	28,222	24,558	21,736	19,496	17,674	16,164	14,891	12,[illegible]65	11,325	10,114	8,332	7,084
75	69,489	64,733	60,586	56,939	53,705	50,819	45,888	41,829	38,429	35,541	33,057	30,897	29,002	25,146	22,195	19,864	17,977	16,416	15,105	13,[illegible]25	11,448	10,212	8,398	7,132
80	73,761	68,424	63,808	59,775	56,221	53,067	47,712	43,339	39,701	36,626	33,993	31,713	29,720	25,684	22,614	20,199	18,250	16,644	15,298	13,[illegible]68	11,559	10,300	8,458	7,174
85	77,991	72,049	66,949	62,523	58,646	55,221	49,447	44,766	40,895	37,639	34,864	32,470	30,384	26,179	22,996	20,503	18,498	16,850	15,472	13,[illegible]97	11,658	10,378	8,511	7,212
90	82,180	75,610	70,012	65,187	60,983	57,289	51,098	46,115	42,018	38,589	35,677	33,174	31,000	26,634	23,347	20,782	18,725	17,038	15,630	13,413	11,747	10,449	8,558	7,247
95	86,328	79,107	73,001	67,770	63,238	59,275	52,672	47,393	43,076	39,480	36,438	33,831	31,572	27,056	23,670	21,037	18,932	17,209	15,774	13,519	11,828	10,514	8,601	7,277
100	90,437	82,544	75,918	70,277	65,416	61,184	54,174	48,606	44,075	40,317	37,150	34,444	32,105	27,447	23,969	21,273	19,122	17,367	15,906	13,[illegible]16	11,903	10,572	8,640	7,305
105	94,507	85,921	78,765	72,710	67,519	63,020	55,609	49,757	45,020	41,107	37,819	35,018	32,604	27,810	24,245	21,491	19,298	17,511	16,028	13,705	11,970	10,625	8,676	7,331
110	98,539	89,241	81,546	75,073	69,552	64,787	56,981	50,853	45,915	41,851	38,449	35,557	33,071	28,149	24,503	21,693	19,461	17,645	16,140	13,787	12,033	10,675	8,709	7,354
115	102,532	92,504	84,262	77,369	71,518	66,490	58,293	51,896	46,764	42,555	39,042	36,064	33,509	28,466	24,742	21,880	19,612	17,769	16,243	13,862	12,090	10,720	8,739	7,376
120	106,488	95,711	86,915	79,600	73,420	68,131	59,551	52,890	47,570	43,222	39,602	36,542	33,920	28,762	24,966	22,055	19,752	17,884	16,339	13,932	12,143	10,762	8,767	7,395
125	110,407	98,866	89,509	81,770	75,262	69,715	60,757	53,840	48,336	43,854	40,132	36,992	34,308	29,041	25,176	22,218	19,883	17,992	16,429	13,997	12,193	10,800	8,792	7,414
130	114,290	101,968	92,044	83,880	77,047	71,243	61,915	54,747	49,066	44,453	40,634	37,418	34,674	29,303	25,372	22,371	20,005	18,092	16,512	14,058	12,239	10,836	8,816	7,431
135	118,137	105,018	94,522	85,934	78,776	72,719	63,026	55,614	49,762	45,024	41,109	37,821	35,020	29,549	25,557	22,515	20,120	18,185	16,590	14,114	12,282	10,870	8,838	7,446
140	121,948	108,019	96,947	87,933	80,452	74,145	64,095	56,444	50,425	45,566	41,561	38,204	35,348	29,782	25,731	22,650	20,228	18,273	16,664	14,167	12,322	10,901	8,859	7,461
145	125,724	110,972	99,318	89,879	82,079	75,524	65,123	57,240	51,060	46,084	41,991	38,567	35,658	30,002	25,895	22,777	20,329	18,356	16,732	14,217	12,359	10,931	8,878	7,475
150	129,466	113,877	101,639	91,776	83,657	76,859	66,113	58,003	51,666	46,577	42,401	38,912	35,953	30,211	26,050	22,897	20,424	18,434	16,797	14,264	12,394	10,958	8,897	7,488
155	133,174	116,736	103,910	93,623	85,190	78,150	67,067	58,736	52,247	47,048	42,791	39,240	36,233	30,408	26,197	23,010	20,514	18,507	16,858	14,307	12,427	10,984	8,914	7,500
160	136,848	119,549	106,133	95,425	86,679	79,402	67,986	59,440	52,803	47,499	43,163	39,553	36,500	30,596	26,336	23,117	20,600	18,576	16,915	14,349	12,459	11,008	8,930	7,511
165	140,489	122,319	108,311	97,181	88,126	80,614	68,873	60,117	53,336	47,930	43,519	39,851	36,754	30,774	26,468	23,219	20,680	18,642	16,970	1[illegible],388	12,488	11,031	8,945	7,522
170	144,098	125,045	110,443	98,894	89,532	81,789	69,729	60,768	53,848	48,343	43,859	40,136	36,996	30,944	26,593	23,315	20,757	18,704	17,021	1[illegible],425	12,516	11,053	8,959	7,532
175	147,674	127,730	112,532	100,566	90,900	82,929	70,556	61,395	54,340	48,739	44,185	40,409	37,228	31,106	26,713	23,407	20,829	18,763	17,070	1[illegible],460	12,542	11,074	8,973	7,542
180	151,219	130,373	114,578	102,197	92,231	84,035	71,355	61,999	54,813	49,119	44,497	40,670	37,449	31,260	26,826	23,494	20,899	18,819	17,116	1[illegible],493	12,567	11,093	8,985	7,551
185	154,732	132,976	116,584	103,790	93,526	85,109	72,127	62,582	55,267	49,484	44,796	40,920	37,661	31,407	26,935	23,578	20,964	18,873	17,160	1[illegible],525	12,591	11,112	8,997	7,559
190	158,215	135,540	118,550	105,345	94,787	86,152	72,875	63,144	55,705	49,835	45,084	41,159	37,864	31,548	27,039	23,657	21,027	18,923	17,202	1[illegible],555	12,614	11,129	9,009	7,567
195	161,666	138,065	120,477	106,864	96,015	87,166	73,599	63,687	56,127	50,172	45,359	41,389	38,058	31,683	27,138	23,733	21,087	18,972	17,242	14,584	12,635	11,146	9,020	7,575
200	165,088	140,553	122,367	108,349	97,212	88,151	74,300	64,211	56,534	50,497	45,625	41,610	38,245	31,812	27,232	23,805	21,144	19,018	17,281	14,611	12,656	11,162	9,030	7,582

350 MCM | **1 Run(s)** | **480 Volts** | **C-Value = 19,703** | **Magnetic Duct** | **Copper**

Calculations:

1. $f = \dfrac{\text{Length} \times 2 \times \text{Isca}}{\text{Runs} \times \text{C-Value} \times E_{L-L}}$
2. $M = \dfrac{1}{1 + f}$
3. $\text{Isc} = \text{Isca} \times M$
4. Add Motor Contribution = Motor FLA x 4

* All results are given in symmetrical amperes

Single-Phase L-L Bolted Fault-Current* Table

One-Way Distance in Feet

Available Isc in Thousands (Isca)	5	10	15	20	25	30	40	50	60	70	80	90	100	125	150	175	200	225	250	300	350	400	500	600
3	2,995	2,991	2,986	2,981	2,976	2,972	2,962	2,953	2,944	2,935	2,926	2,917	2,908	2,886	2,864	2,842	2,821	2,800	2,780	2,739	2,700	2,662	2,589	2,520
5	4,987	4,974	4,961	4,948	4,935	4,922	4,896	4,871	4,846	4,822	4,797	4,773	4,749	4,690	4,633	4,577	4,522	4,468	4,416	4,316	4,219	4,127	3,955	3,796
7	6,974	6,949	6,923	6,898	6,873	6,848	6,799	6,750	6,702	6,655	6,609	6,563	6,518	6,407	6,300	6,197	6,097	6,001	5,907	5,728	5,560	5,401	5,109	4,847
10	9,947	9,895	9,844	9,793	9,742	9,693	9,594	9,498	9,403	9,311	9,220	9,131	9,044	8,833	8,631	8,439	8,254	8,078	7,909	7,592	7,299	7,028	6,542	6,118
15	14,882	14,766	14,651	14,539	14,428	14,319	14,105	13,898	13,697	13,501	13,311	13,126	12,947	12,518	12,117	11,741	11,388	11,055	10,741	10,164	9,646	9,178	8,366	7,686
20	19,791	19,586	19,385	19,188	18,996	18,807	18,440	18,087	17,748	17,421	17,106	16,802	16,509	15,819	15,184	14,598	14,055	13,552	13,083	12,237	11,493	10,835	9,721	8,815
25	24,674	24,356	24,047	23,745	23,450	23,163	22,609	22,081	21,578	21,096	20,636	20,195	19,773	18,791	17,902	17,093	16,354	15,676	15,052	13,943	12,986	12,151	10,768	9,667
30	29,532	29,078	28,637	28,210	27,796	27,393	26,622	25,893	25,203	24,549	23,928	23,337	22,775	21,482	20,328	19,291	18,355	17,506	16,731	15,372	14,216	13,223	11,601	10,333
35	34,364	33,751	33,159	32,588	32,036	31,502	30,487	29,535	28,640	27,799	27,005	26,255	25,546	23,930	22,506	21,243	20,113	19,098	18,180	16,586	15,249	14,111	12,279	10,868
40	39,172	38,377	37,614	36,880	36,175	35,496	34,212	33,018	31,904	30,863	29,887	28,972	28,111	26,166	24,473	22,986	21,670	20,496	19,442	17,630	16,127	14,860	12,842	11,307
45	43,954	42,956	42,002	41,090	40,216	39,379	37,805	36,352	35,006	33,757	32,593	31,507	30,492	28,217	26,259	24,554	23,058	21,733	20,552	18,538	16,883	15,500	13,317	11,673
50	48,712	47,489	46,326	45,219	44,163	43,155	41,272	39,546	37,959	36,494	35,138	33,880	32,708	30,105	27,886	25,971	24,303	22,836	21,536	19,334	17,541	16,053	13,723	11,984
55	53,446	51,977	50,587	49,269	48,019	46,830	44,620	42,610	40,773	39,088	37,536	36,104	34,776	31,848	29,375	27,259	25,426	23,825	22,413	20,039	18,119	16,535	14,075	12,251
60	58,155	56,421	54,786	53,244	51,786	50,406	47,856	45,551	43,458	41,549	39,800	38,193	36,710	33,463	30,744	28,433	26,445	24,717	23,201	20,666	18,631	16,960	14,381	12,483
65	62,841	60,820	58,925	57,145	55,469	53,889	50,984	48,376	46,022	43,886	41,940	40,159	38,523	34,963	32,005	29,508	27,373	25,526	23,913	21,229	19,087	17,337	14,651	12,686
70	67,502	65,176	63,005	60,974	59,070	57,281	54,010	51,092	48,473	46,110	43,966	42,013	40,226	36,360	33,172	30,497	28,222	26,263	24,558	21,736	19,496	17,674	14,891	12,865
75	72,140	69,489	67,027	64,733	62,591	60,586	56,939	53,705	50,819	48,228	45,888	43,764	41,829	37,664	34,254	31,410	29,002	26,937	25,146	22,195	19,864	17,977	15,105	13,025
80	76,754	73,761	70,992	68,424	66,035	63,808	59,775	56,221	53,067	50,247	47,712	45,421	43,339	38,885	35,260	32,254	29,720	27,555	25,684	22,614	20,199	18,250	15,298	13,168
85	81,345	77,991	74,902	72,049	69,405	66,949	62,523	58,646	55,221	52,175	49,447	46,990	44,766	40,029	36,199	33,037	30 384	28,125	26,179	22,996	20,503	18,498	15,472	13,297
90	85,912	82,180	78,758	75,610	72,703	70,012	65,187	60,983	57,289	54,017	51,098	48,479	46,115	41,105	37,076	33,767	31,000	28,652	26,634	23,347	20,782	18,725	15,630	13,413
95	90,457	86,328	82,560	79,107	75,932	73,001	67,770	63,238	59,275	55,779	52,672	49,894	47,393	42,117	37,898	34,447	31,572	29,140	27,056	23,670	21,037	18,932	15,774	13,519
100	94,979	90,437	86,311	82,544	79,093	75,918	70,277	65,416	61,184	57,466	54,174	51,239	48,606	43,072	38,669	35,083	32,105	29,594	27,447	23,969	21,273	19,122	15,906	13,616
105	99,478	94,507	90,010	85,921	82,188	78,765	72,710	67,519	63,020	59,083	55,609	52,521	49,757	43,974	39,394	35,679	32,504	30,017	27,810	24,245	21,491	19,298	16,028	13,705
110	103,954	98,539	93,660	89,241	85,220	81,546	75,073	69,552	64,787	60,634	56,981	53,743	50,853	44,827	40,078	36,239	33,071	30,412	28,149	24,503	21,693	19,461	16,140	13,787
115	108,409	102,532	97,260	92,504	88,191	84,262	77,369	71,518	66,490	62,122	58,293	54,909	51,896	45,636	40,723	36,765	33,509	30,782	28,466	24,742	21,880	19,612	16,243	13,862
120	112,841	106,488	100,813	95,711	91,102	86,915	79,600	73,420	68,131	63,553	59,551	56,024	52,890	46,403	41,333	37,262	33,920	31,129	28,762	24,966	22,055	19,752	16,339	13,932
125	117,251	110,407	104,318	98,866	93,955	89,509	81,770	75,262	69,715	64,928	60,757	57,090	53,840	47,132	41,910	37,730	34,308	31,456	29,041	25,176	22,218	19,883	16,429	13,997
130	121,640	114,290	107,778	101,968	96,752	92,044	83,880	77,047	71,243	66,252	61,915	58,110	54,747	47,825	42,458	38,173	34,674	31,763	29,303	25,372	22,371	20,005	16,512	14,058
135	126,007	118,137	111,192	105,018	99,494	94,522	85,934	78,776	72,719	67,527	63,026	59,089	55,614	48,486	42,978	38,593	35,020	32,053	29,549	25,557	22,515	20,120	16,590	14,114
140	130,352	121,948	114,562	108,019	102,184	96,947	87,933	80,452	74,145	68,755	64,095	60,027	56,444	49,116	43,472	38,991	35,348	32,327	29,782	25,731	22,650	20,228	16,664	14,167
145	134,676	125,724	117,888	110,972	104,822	99,318	89,879	82,079	75,524	69,939	65,123	60,928	57,240	49,717	43,942	39,369	35,658	32,587	30,002	25,895	22,777	20,329	16,732	14,217
150	138,979	129,466	121,172	113,877	107,410	101,639	91,776	83,657	76,859	71,082	66,113	61,793	58,003	50,292	44,391	39,729	35,953	32,833	30,211	26,050	22,897	20,424	16,797	14,264
155	143,260	133,174	124,414	116,736	109,950	103,910	93,623	85,190	78,150	72,186	67,067	62,626	58,736	50,842	44,819	40,071	36,233	33,066	30,408	26,197	23,010	20,514	16,858	14,307
160	147,521	136,848	127,615	119,549	112,443	106,133	95,425	86,679	79,402	73,252	67,986	63,426	59,440	51,369	45,227	40,398	36,500	33,288	30,596	26,336	23,117	20,600	16,915	14,349
165	151,761	140,489	130,776	122,319	114,889	108,311	97,181	88,126	80,614	74,282	68,873	64,198	60,117	51,873	45,618	40,709	36,754	33,499	30,774	26,468	23,219	20,680	16,970	14,388
170	155,981	144,098	133,897	125,045	117,291	110,443	98,894	89,532	81,789	75,279	69,729	64,941	60,768	52,358	45,992	41,007	36,996	33,701	30,944	26,593	23,315	20,757	17,021	14,425
175	160,180	147,674	136,980	127,730	119,650	112,532	100,566	90,900	82,929	76,244	70,556	65,657	61,395	52,822	46,350	41,291	37,228	33,892	31,106	26,713	23,407	20,829	17,070	14,460
180	164,359	151,219	140,024	130,373	121,966	114,578	102,197	92,231	84,035	77,178	71,355	66,349	61,999	53,269	46,694	41,564	37,449	34,076	31,260	26,826	23,494	20,899	17,116	14,493
185	168,518	154,732	143,032	132,976	124,242	116,584	103,790	93,526	85,109	78,082	72,127	67,016	62,582	53,698	47,024	41,825	37,661	34,251	31,407	26,935	23,578	20,964	17,160	14,525
190	172,657	158,215	146,002	135,540	126,477	118,550	105,345	94,787	86,152	78,959	72,875	67,661	63,144	54,112	47,340	42,075	37,864	34,419	31,548	27,039	23,657	21,027	17,202	14,555
195	176,776	161,666	148,937	138,065	128,673	120,477	106,864	96,015	87,166	79,810	73,599	68,285	63,687	54,510	47,645	42,315	38,058	34 579	31,683	27,138	23,733	21,087	17,242	14,584
200	180,875	165,088	151,836	140,553	130,831	122,367	108,349	97,212	88,151	80,635	74,300	68,888	64,211	54,893	47,937	42,546	38,245	34,733	31,812	27,232	23,805	21,144	17,281	14,611

350 MCM	2 Run(s)	480 Volts	C-Value = 19,703	Magnetic Duct	Copper

Calculations:

1. $f = \frac{\text{Length} \times 2 \times \text{Isca}}{\text{Runs} \times \text{C-Value} \times \text{E L-L}}$

2. $M = \frac{1}{1 + f}$

3. Isc = Isca x M

4. Add Motor Contribution = Motor FLA x 4

* All results are given in symmetrical amperes

Single-Phase L-L Bolted Fault-Current* Table

One-Way Distance in Feet

Available Isc in Thousands (Isca)

	5	10	15	20	25	30	40	50	60	70	80	90	100	125	150	175	200	225	250	300	350	400	500	600
3	2,991	2,982	2,973	2,964	2,955	2,946	2,929	2,912	2,894	2,878	2,861	2,844	2,828	2,788	2,749	2,712	2,675	2,639	2,604	2,537	2,474	2,413	2,301	2,198
5	4,975	4,950	4,925	4,901	4,876	4,853	4,805	4,759	4,713	4,669	4,625	4,582	4,540	4,438	4,340	4,247	4,158	4,072	3,990	3,835	3,691	3,558	3,319	3,110
7	6,951	6,902	6,854	6,807	6,760	6,714	6,624	6,536	6,451	6,368	6,287	6,208	6,131	5,946	5,772	5,608	5,453	5,307	5,168	4,911	4,678	4,466	4,096	3,782
10	9,900	9,801	9,705	9,611	9,518	9,427	9,250	9,080	8,916	8,758	8,605	8,458	8,315	7,979	7,669	7,382	7,116	6,869	6,638	6,220	5,851	5,524	4,968	4,513
15	14,775	14,558	14,346	14,140	13,941	13,747	13,374	13,021	12,687	12,369	12,066	11,778	11,504	10,870	10,303	9,792	9,329	8,908	8,524	7,846	7,268	6,770	5,953	5,313
20	19,603	19,221	18,854	18,501	18,160	17,832	17,210	16,631	16,088	15,581	15,104	14,655	14,233	13,276	12,439	11,702	11,047	10,462	9,935	9,027	8,270	7,631	6,609	5,829
25	24,382	23,795	23,235	22,700	22,190	21,702	20,788	19,948	19,173	18,456	17,791	17,172	16,594	15,308	14,206	13,253	12,419	11,684	11,031	9,822	9,016	8,261	7,077	6,189
30	29,115	28,281	27,493	26,748	26,043	25,373	24,133	23,008	21,983	21,046	20,185	19,392	18,659	17,048	15,692	14,537	13,540	12,671	11,907	10,525	9,593	8,743	7,427	6,456
35	33,802	32,682	31,635	30,653	29,729	28,860	27,266	25,839	24,553	23,390	22,331	21,365	20,478	18,554	16,960	15,618	14,473	13,485	12,622	11,191	10,052	9,123	7,700	6,661
40	38,442	37,001	35,664	34,421	33,261	32,177	30,207	28,465	26,913	25,521	24,267	23,129	22,094	19,870	18,053	16,541	15,262	14,167	13,218	11,657	10,426	9,430	7,917	6,823
45	43,038	41,240	39,586	38,060	36,647	35,335	32,974	30,909	29,088	27,469	26,021	24,718	23,539	21,031	19,006	17,337	15,938	14,747	13,722	12,047	10,737	9,684	8,095	6,955
50	47,589	45,401	43,404	41,576	39,896	38,346	35,582	33,189	31,098	29,255	27,618	26,154	24,838	22,062	19,845	18,032	16,523	15,247	14,154	12,379	10,999	9,897	8,244	7,064
55	52,097	49,486	47,123	44,976	43,016	41,220	38,043	35,320	32,962	30,898	29,078	27,460	26,013	22,984	20,587	18,643	17,035	15,682	14,528	12,664	11,224	10,078	8,369	7,156
60	56,562	53,497	50,746	48,265	46,015	43,966	40,370	37,317	34,694	32,416	30,418	28,652	27,080	23,814	21,250	19,185	17,486	16,063	14,855	12,912	11,418	10,234	8,477	7,234
65	60,984	57,436	54,278	51,449	48,900	46,592	42,573	39,192	36,309	33,821	31,652	29,745	28,054	24,563	21,845	19,669	17,887	16,401	15,143	13,129	11,588	10,370	8,570	7,302
70	65,365	61,305	57,721	54,532	51,677	49,106	44,663	40,956	37,818	35,127	32,793	30,750	28,946	25,245	22,383	20,104	18,246	16,702	15,399	13,321	11,737	10,490	8,651	7,361
75	69,704	65,107	61,078	57,519	54,352	51,516	46,647	42,619	39,231	36,342	33,850	31,677	29,767	25,867	22,870	20,496	18,568	16,972	15,628	13,492	11,870	10,596	8,723	7,413
80	74,003	68,842	64,354	60,415	56,931	53,826	48,533	44,188	40,557	37,478	34,833	32,536	30,524	26,437	23,315	20,852	18,860	17,216	15,835	13,646	11,988	10,690	8,787	7,459
85	78,261	72,512	67,550	63,224	59,418	56,044	50,329	45,672	41,804	38,540	35,748	33,334	31,225	26,961	23,721	21,177	19,125	17,436	16,021	13,784	12,095	10,775	8,844	7,500
90	82,480	76,120	70,670	65,949	61,819	58,175	52,041	47,077	42,978	39,535	36,603	34,076	31,875	27,444	24,095	21,474	19,367	17,637	16,191	13,909	12,191	10,851	8,896	7,537
95	86,660	79,666	73,717	68,594	64,137	60,224	53,675	48,410	44,086	40,471	37,404	34,769	32,481	27,892	24,439	21,747	19,589	17,821	16,346	14,023	12,279	10,921	8,942	7,570
100	90,801	83,153	76,692	71,163	66,378	62,196	55,235	49,676	45,133	41,352	38,155	35,417	33,046	28,308	24,758	21,999	19,793	17,990	16,487	14,128	12,359	10,984	8,984	7,601
105	94,905	86,581	79,599	73,659	68,545	64,094	56,727	50,879	46,125	42,183	38,861	36,025	33,574	28,694	25,053	22,232	19,982	18,145	16,618	14,223	12,432	11,041	9,023	7,628
110	98,971	89,952	82,440	76,086	70,641	65,923	58,155	52,025	47,064	42,967	39,526	36,595	34,069	29,055	25,328	22,448	20,156	18,289	16,738	14,312	12,499	11,094	9,058	7,654
115	103,000	93,268	85,217	78,445	72,670	67,687	59,524	53,118	47,957	43,709	40,154	37,133	34,534	29,393	25,584	22,649	20,318	18,422	16,850	14,393	12,561	11,143	9,091	7,677
120	106,993	96,530	87,932	80,739	74,635	69,388	60,836	54,160	48,805	44,413	40,746	37,639	34,972	29,709	25,823	22,836	20,469	18,546	16,953	14,468	12,619	11,189	9,121	7,698
125	110,950	99,740	90,587	82,972	76,539	71,031	62,095	55,156	49,612	45,080	41,307	38,117	35,385	30,006	26,047	23,011	20,609	18,661	17,050	14,539	12,672	11,230	9,149	7,718
130	114,872	102,898	93,184	85,146	78,385	72,618	63,304	56,108	50,381	45,714	41,839	38,570	35,774	30,286	26,258	23,176	20,741	18,769	17,140	14,604	12,722	11,269	9,175	7,736
135	118,758	106,005	95,725	87,263	80,175	74,153	64,467	57,019	51,114	46,318	42,344	38,998	36,142	30,550	26,456	23,330	20,864	18,870	17,224	14,665	12,768	11,306	9,199	7,754
140	122,611	109,064	98,213	89,325	81,913	75,636	65,586	57,893	51,815	46,892	42,824	39,405	36,491	30,799	26,642	23,474	20,980	18,965	17,302	14,722	12,811	11,340	9,221	7,769
145	126,429	112,074	100,647	91,335	83,600	77,072	66,663	58,730	52,485	47,440	43,280	39,791	36,822	31,034	26,818	23,611	21,089	19,054	17,377	14,776	12,852	11,371	9,242	7,784
150	130,213	115,038	103,031	93,294	85,238	78,462	67,700	59,534	53,126	47,963	43,715	40,158	37,137	31,257	26,985	23,740	21,192	19,137	17,446	14,826	12,890	11,401	9,262	7,798
155	133,965	117,956	105,366	95,204	86,829	79,809	68,700	60,306	53,740	48,463	44,130	40,508	37,436	31,468	27,142	23,862	21,289	19,216	17,512	14,873	12,926	11,429	9,280	7,811
160	137,683	120,830	107,653	97,067	88,376	81,114	69,665	61,048	54,328	48,941	44,526	40,842	37,720	31,669	27,292	23,977	21,380	19,291	17,574	14,918	12,960	11,456	9,298	7,824
165	141,370	123,660	109,893	98,885	89,881	82,380	70,597	61,762	54,893	49,399	44,905	41,160	37,992	31,861	27,433	24,086	21,467	19,362	17,633	14,960	12,991	11,480	9,314	7,835
170	145,024	126,447	112,089	100,659	91,344	83,608	71,496	62,450	55,436	49,838	45,267	41,464	38,251	32,042	27,568	24,190	21,550	19,429	17,688	15,000	13,022	11,504	9,329	7,846
175	148,647	129,193	114,241	102,391	92,768	84,799	72,366	63,112	55,957	50,259	45,614	41,755	38,498	32,216	27,696	24,289	21,628	19,493	17,741	15,038	13,050	11,526	9,344	7,857
180	152,239	131,897	116,351	104,083	94,155	85,956	73,207	63,751	56,458	50,663	45,947	42,034	38,735	32,382	27,819	24,383	21,703	19,553	17,791	15,074	13,077	11,547	9,358	7,867
185	155,801	134,562	118,420	105,735	95,505	87,080	74,020	64,367	56,941	51,051	46,266	42,301	38,961	32,540	27,935	24,473	21,774	19,611	17,839	15,108	13,103	11,567	9,371	7,876
190	159,332	137,188	120,449	107,350	96,820	88,172	74,808	64,962	57,406	51,425	46,572	42,557	39,179	32,691	28,047	24,558	21,841	19,666	17,884	15,141	13,127	11,587	9,384	7,885
195	162,833	139,776	122,439	108,928	98,102	89,234	75,571	65,536	57,854	51,784	46,867	42,803	39,387	32,836	28,153	24,640	21,906	19,718	17,927	15,172	13,151	11,605	9,396	7,893
200	166,305	142,327	124,391	110,470	99,352	90,267	76,310	66,092	58,287	52,130	47,150	43,039	39,587	32,975	28,255	24,718	21,967	19,768	17,969	15,202	13,173	11,622	9,407	7,901

400 MCM	1 Run(s)	480 Volts	C-Value = 20,565	Magnetic Duct	Copper

Calculations:

1. $f = \frac{\text{Length} \times 2 \times \text{Isca}}{\text{Runs} \times \text{C-Value} \times \text{E L-L}}$

2. $M = \frac{1}{1 + f}$

3. Isc = Isca x M

4. Add Motor Contribution = Motor FLA x 4

* All results are given in symmetrical amperes

Single-Phase L-L Bolted Fault-Current* Table

One-Way Distance in Feet

Available Isc in Thousands (Isca)	5	10	15	20	25	30	40	50	60	70	80	90	100	125	150	175	200	225	250	300	350	400	500	600
3	2,995	2,991	2,986	2,982	2,977	2,973	2,964	2,955	2,946	2,938	2,929	2,920	2,912	2,890	2,869	2,849	2,828	2,808	2,788	2,749	2,712	2,675	2,604	2,537
5	4,987	4,975	4,962	4,950	4,937	4,925	4,901	4,876	4,853	4,829	4,805	4,782	4,759	4,702	4,647	4,593	4,540	4,488	4,438	4,340	4,247	4,158	3,990	3,835
7	6,975	6,951	6,926	6,902	6,878	6,854	6,807	6,760	6,714	6,669	6,624	6,580	6,536	6,430	6,327	6,227	6,131	6,037	5,946	5,772	5,608	5,453	5,168	4,911
10	9,950	9,900	9,850	9,801	9,753	9,705	9,611	9,518	9,427	9,338	9,250	9,164	9,080	8,876	8,681	8,494	8,315	8,144	7,979	7,669	7,382	7,116	6,638	6,220
15	14,887	14,775	14,666	14,558	14,451	14,346	14,140	13,941	13,747	13,558	13,374	13,195	13,021	12,606	12,216	11,849	11,504	11,178	10,870	10,303	9,792	9,329	8,524	7,846
20	19,799	19,603	19,410	19,221	19,036	18,854	18,501	18,160	17,832	17,516	17,210	16,915	16,631	15,958	15,338	14,765	14,233	13,737	13,276	12,439	11,702	11,047	9,935	9,027
25	24,687	24,382	24,085	23,795	23,511	23,235	22,700	22,190	21,702	21,235	20,788	20,359	19,948	18,989	18,117	17,323	16,594	15,925	15,308	14,206	13,253	12,419	11,031	9,922
30	29,551	29,115	28,692	28,281	27,882	27,493	26,748	26,043	25,373	24,737	24,133	23,557	23,008	21,741	20,606	19,584	18,659	17,817	17,048	15,692	14,537	13,540	11,907	10,625
35	34,390	33,802	33,233	32,682	32,150	31,635	30,653	29,729	28,860	28,040	27,266	26,533	25,839	24,252	22,848	21,598	20,478	19,469	18,554	16,960	15,618	14,473	12,622	11,191
40	39,206	38,442	37,708	37,001	36,321	35,664	34,421	33,261	32,177	31,161	30,207	29,311	28,465	26,551	24,878	23,404	22,094	20,923	19,870	18,053	16,541	15,262	13,218	11,657
45	43,997	43,038	42,120	41,240	40,396	39,586	38,060	36,647	35,335	34,114	32,974	31,908	30,909	28,665	26,725	25,031	23,539	22,214	21,031	19,006	17,337	15,938	13,722	12,047
50	48,765	47,589	46,469	45,401	44,380	43,404	41,576	39,896	38,346	36,912	35,582	34,344	33,189	30,616	28,413	26,505	24,838	23,368	22,062	19,845	18,032	16,523	14,154	12,379
55	53,509	52,097	50,758	49,486	48,276	47,123	44,976	43,016	41,220	39,568	38,043	36,631	35,320	32,420	29,960	27,847	26,013	24,405	22,984	20,587	18,643	17,035	14,528	12,664
60	58,230	56,562	54,987	53,497	52,085	50,746	48,265	46,015	43,966	42,091	40,370	38,784	37,317	34,095	31,385	29,074	27,080	25,342	23,814	21,250	19,185	17,486	14,855	12,912
65	62,928	60,984	59,157	57,436	55,812	54,278	51,449	48,900	46,592	44,492	42,573	40,813	39,192	35,654	32,701	30,200	28,054	26,193	24,563	21,845	19,669	17,887	15,143	13,129
70	67,603	65,365	63,270	61,305	59,459	57,721	54,532	51,677	49,106	46,779	44,663	42,729	40,956	37,107	33,920	31,236	28,946	26,969	25,245	22,383	20,104	18,246	15,399	13,321
75	72,255	69,704	67,327	65,107	63,028	61,078	57,519	54,352	51,516	48,960	46,647	44,542	42,619	38,467	35,052	32,194	29,767	27,680	25,867	22,870	20,496	18,568	15,628	13,492
80	76,884	74,003	71,329	68,842	66,522	64,354	60,415	56,931	53,826	51,043	48,533	46,259	44,188	39,741	36,107	33,082	30,524	28,334	26,437	23,315	20,852	18,860	15,835	13,646
85	81,491	78,261	75,277	72,512	69,943	67,550	63,224	59,418	56,044	53,033	50,329	47,888	45,672	40,937	37,091	33,906	31,225	28,937	26,961	23,721	21,177	19,125	16,021	13,784
90	86,076	82,480	79,172	76,120	73,294	70,670	65,949	61,819	58,175	54,938	52,041	49,435	47,077	42,062	38,013	34,675	31,875	29,494	27,444	24,095	21,474	19,367	16,191	13,909
95	90,638	86,660	83,016	79,666	76,576	73,717	68,594	64,137	60,224	56,761	53,675	50,907	48,410	43,123	38,877	35,392	32,481	30,012	27,892	24,439	21,747	19,589	16,346	14,023
100	95,179	90,801	86,809	83,153	79,792	76,692	71,163	66,378	62,196	58,509	55,235	52,308	49,676	44,125	39,689	36,064	33,046	30,494	28,308	24,758	21,999	19,793	16,487	14,128
105	99,698	94,905	90,552	86,581	82,943	79,599	73,659	68,545	64,094	60,186	56,727	53,644	50,879	45,072	40,454	36,694	33,574	30,943	28,694	25,053	22,232	19,982	16,618	14,223
110	104,195	98,971	94,246	89,952	86,032	82,440	76,086	70,641	65,923	61,796	58,155	54,920	52,025	45,969	41,175	37,287	34,069	31,363	29,055	25,328	22,448	20,156	16,738	14,312
115	108,670	103,000	97,893	93,268	89,061	85,217	78,445	72,670	67,687	63,343	59,524	56,139	53,118	46,819	41,856	37,844	34,534	31,757	29,393	25,584	22,649	20,318	16,850	14,393
120	113,124	106,993	101,493	96,530	92,031	87,932	80,739	74,635	69,388	64,831	60,836	57,304	54,160	47,627	42,501	38,371	34,972	32,127	29,709	25,823	22,836	20,469	16,953	14,468
125	117,557	110,950	105,047	99,740	94,943	90,587	82,972	76,539	71,031	66,263	62,095	58,420	55,156	48,395	43,111	38,868	35,385	32,474	30,006	26,047	23,011	20,609	17,050	14,539
130	121,969	114,872	108,555	102,898	97,800	93,184	85,146	78,385	72,618	67,642	63,304	59,489	56,108	49,127	43,691	39,338	35,774	32,802	30,286	26,258	23,176	20,741	17,140	14,604
135	126,359	118,758	112,020	106,005	100,603	95,725	87,263	80,175	74,153	68,971	64,467	60,515	57,019	49,824	44,242	39,784	36,142	33,111	30,550	26,456	23,330	20,864	17,224	14,665
140	130,730	122,611	115,441	109,064	103,354	98,213	89,325	81,913	75,636	70,253	65,586	61,500	57,893	50,490	44,766	40,207	36,491	33,404	30,799	26,642	23,474	20,980	17,302	14,722
145	135,079	126,429	118,820	112,074	106,054	100,647	91,335	83,600	77,072	71,490	66,663	62,445	58,730	51,126	45,265	40,609	36,822	33,681	31,034	26,818	23,611	21,089	17,377	14,776
150	139,408	130,213	122,156	115,038	108,704	103,031	93,294	85,238	78,462	72,685	67,700	63,355	59,534	51,734	45,741	40,992	37,137	33,944	31,257	26,985	23,740	21,192	17,446	14,826
155	143,717	133,965	125,452	117,956	111,306	105,366	95,204	86,829	79,809	73,839	68,700	64,230	60,306	52,316	46,195	41,357	37,436	34,194	31,468	27,142	23,862	21,289	17,512	14,873
160	148,005	137,683	128,707	120,830	113,861	107,653	97,067	88,376	81,114	74,955	69,665	65,073	61,048	52,873	46,629	41,704	37,720	34,431	31,669	27,292	23,977	21,380	17,574	14,918
165	152,274	141,370	131,923	123,660	116,371	109,893	98,885	89,881	82,380	76,034	70,597	65,885	61,762	53,408	47,045	42,036	37,992	34,657	31,861	27,433	24,086	21,467	17,633	14,960
170	156,522	145,024	135,100	126,447	118,836	112,089	100,659	91,344	83,608	77,079	71,496	66,668	62,450	53,922	47,443	42,354	38 251	34,872	32,042	27,568	24,190	21,550	17,688	15,000
175	160,751	148,647	138,239	129,193	121,258	114,241	102,391	92,768	84,799	78,091	72,366	67,423	63,112	54,415	47,824	42,657	38 498	35,078	32,216	27,696	24,289	21,628	17,741	15,038
180	164,960	152,239	141,340	131,897	123,637	116,351	104,083	94,155	85,956	79,071	73,207	68,152	63,751	54,889	48,190	42,948	38,735	35,274	32,382	27,819	24,383	21,703	17,791	15,074
185	169,150	155,801	144,405	134,562	125,976	118,420	105,735	95,505	87,080	80,021	74,020	68,857	64,367	55,345	48,541	43,227	38,961	35,462	32,540	27,935	24,473	21,774	17,839	15,108
190	173,320	159,332	147,433	137,188	128,275	120,449	107,350	96,820	88,172	80,942	74,808	69,538	64,962	55,784	48,879	43,494	39,179	35,642	32,691	28,047	24,558	21,841	17,884	15,141
195	177,471	162,833	150,426	139,776	130,534	122,439	108,928	98,102	89,234	81,836	75,571	70,197	65,536	56,207	49,203	43,751	39,387	35,814	32,836	28,153	24,640	21,906	17,927	15,172
200	181,603	166,305	153,384	142,327	132,756	124,391	110,470	99,352	90,267	82,704	76,310	70,834	66,092	56,615	49,515	43,998	39,587	35,980	32,975	28,255	24,718	21,967	17,969	15,202

400 MCM	2 Run(s)	480 Volts		C-Value = 20,565	Magnetic Duct	Copper

Calculations:

1. $f = \dfrac{\text{Length} \times 2 \times \text{Isca}}{\text{Runs} \times \text{C-Value} \times E\ L\text{-}L}$

2. $M = \dfrac{1}{1 + f}$

3. Isc = Isca x M

4. Add Motor Contribution = Motor FLA x 4

* All results are given in symmetrical amperes

Single-Phase L-L Bolted Fault-Current* Table

One-Way Distance in Feet

Available Isc in Thousands (Isca)	5	10	15	20	25	30	40	50	60	70	80	90	100	125	150	175	200	225	250	300	350	400	500	600
3	2,992	2,983	2,975	2,967	2,958	2,950	2,934	2,918	2,902	2,886	2,871	2,855	2,840	2,803	2,766	2,731	2,696	2,662	2,630	2,566	2,506	2,448	2,341	2,242
5	4,977	4,953	4,931	4,908	4,885	4,863	4,819	4,776	4,733	4,692	4,651	4,610	4,571	4,475	4,383	4,294	4,209	4,128	4,049	3,901	3,763	3,635	3,402	3,198
7	6,954	6,909	6,865	6,821	6,777	6,734	6,650	6,568	6,488	6,410	6,334	6,259	6,187	6,012	5,847	5,691	5,543	5,402	5,268	5,020	4,794	4,588	4,224	3,913
10	9,907	9,816	9,726	9,638	9,552	9,467	9,301	9,142	8,987	8,838	8,694	8,554	8,419	8,099	7,802	7,526	7,269	7,029	6,805	6,396	6,034	5,710	5,157	4,702
15	14,792	14,589	14,392	14,200	14,013	13,831	13,481	13,148	12,831	12,529	12,241	11,966	11,703	11,093	10,544	10,047	9,594	9,181	8,801	8,129	7,553	7,053	6,228	5,576
20	19,631	19,276	18,933	18,602	18,283	17,974	17,387	16,838	16,322	15,836	15,379	14,947	14,539	13,610	12,792	12,067	11,420	10,839	10,314	9,403	8,640	7,992	6,949	6,147
25	24,427	23,879	23,355	22,854	22,374	21,913	21,047	20,247	19,505	18,816	18,174	17,574	17,012	15,754	14,669	13,724	12,893	12,157	11,500	10,379	9,458	8,686	7,468	6,549
30	29,178	28,400	27,662	26,962	26,296	25,662	24,482	23,406	22,420	21,514	20,679	19,906	19,188	17,602	16,259	15,106	14,105	13,229	12,455	11,151	10,094	9,220	7,859	6,848
35	33,886	32,841	31,859	30,933	30,060	29,235	27,713	26,342	25,100	23,970	22,938	21,990	21,118	19,213	17,623	16,276	15,121	14,118	13,241	11,776	10,604	9,643	8,165	7,079
40	38,552	37,205	35,949	34,775	33,675	32,643	30,757	29,078	27,572	26,214	24,984	23,864	22,841	20,628	18,807	17,281	15,984	14,868	13,898	12,293	11,021	9,987	8,410	7,263
45	43,175	41,493	39,937	38,493	37,150	35,898	33,631	31,633	29,859	28,273	26,848	25,559	24,388	21,882	19,843	18,152	16,727	15,509	14,456	12,728	11,369	10,272	8,611	7,412
50	47,758	45,708	43,827	42,094	40,493	39,010	36,347	34,024	31,981	30,169	28,551	27,098	25,786	23,001	20,759	18,915	17,372	16,062	14,936	13,099	11,664	10,512	8,779	7,536
55	52,299	49,851	47,621	45,583	43,712	41,988	38,919	36,268	33,955	31,919	30,114	28,502	27,054	24,005	21,573	19,589	17,939	16,545	15,353	13,418	11,917	10,717	8,921	7,641
60	56,800	53,923	51,324	48,965	46,812	44,841	41,358	38,377	35,797	33,542	31,554	29,789	28,210	24,911	22,302	20,188	18,440	16,971	15,718	13,697	12,136	10,894	9,044	7,731
65	61,261	57,928	54,940	52,244	49,801	47,576	43,673	40,363	37,519	35,049	32,884	30,971	29,269	25,732	22,959	20,724	18,887	17,348	16,042	13,941	12,327	11,048	9,150	7,808
70	65,682	61,866	58,470	55,426	52,684	50,200	45,875	42,236	39,132	36,453	34,117	32,063	30,241	26,481	23,553	21,207	19,287	17,685	16,329	14,158	12,497	11,184	9,243	7,876
75	70,065	65,740	61,917	58,515	55,467	52,721	47,971	44,006	40,647	37,764	35,263	33,072	31,138	27,166	24,093	21,645	19,648	17,988	16,587	14,352	12,647	11,305	9,325	7,935
80	74,410	69,550	65,286	61,515	58,155	55,144	49,969	45,681	42,072	38,991	36,330	34,010	31,968	27,796	24,587	22,042	19,975	18,262	16,820	14,526	12,782	11,412	9,398	7,988
85	78,717	73,298	68,578	64,429	60,753	57,474	51,875	47,269	43,415	40,142	37,328	34,882	32,737	28,376	25,040	22,405	20,273	18,511	17,030	14,682	12,903	11,509	9,463	8,035
90	82,986	76,987	71,796	67,261	63,265	59,717	53,695	48,776	44,683	41,223	38,261	35,696	33,453	28,912	25,456	22,733	20,545	18,737	17,222	14,825	13,013	11,596	9,522	8,078
95	87,219	80,616	74,943	70,015	65,696	61,878	55,436	50,208	45,882	42,242	39,137	36,457	34,121	29,409	25,841	23,045	20,795	18,945	17,397	14,954	13,113	11,675	9,575	8,116
100	91,415	84,188	78,020	72,694	68,049	63,961	57,102	51,571	47,017	43,202	39,960	37,170	34,745	29,871	26,197	23,328	21,025	19,136	17,558	15,073	13,204	11,747	9,624	8,151
105	95,576	87,704	81,031	75,301	70,328	65,971	58,698	52,869	48,094	44,110	40,735	37,840	35,329	30,302	26,528	23,590	21,237	19,312	17,706	15,182	13,287	11,813	9,668	8,182
110	99,701	91,166	83,976	77,838	72,536	67,910	60,228	54,108	49,116	44,968	41,466	38,470	35,878	30,705	26,836	23,833	21,434	19,475	17,843	15,282	13,364	11,874	9,709	8,212
115	103,791	94,573	86,859	80,309	74,677	69,783	61,697	55,290	50,089	45,782	42,157	39,064	36,394	31,082	27,124	24,060	21,618	19,626	17,970	15,375	13,435	11,930	9,746	8,238
120	107,847	97,929	89,682	82,716	76,754	71,593	63,108	56,420	51,015	46,554	42,811	39,625	36,880	31,436	27,393	24,271	21,788	19,766	18,087	15,461	13,501	11,982	9,781	8,263
125	111,868	101,234	92,445	85,061	78,769	73,344	64,464	57,502	51,897	47,288	43,431	40,155	37,339	31,769	27,646	24,469	21,948	19,897	18,197	15,541	13,562	12,030	9,813	8,286
130	115,856	104,488	95,152	87,347	80,725	75,037	65,768	58,538	52,739	47,986	44,019	40,658	37,773	32,083	27,883	24,655	22,097	20,020	18,300	15,616	13,619	12,075	9,843	8,307
135	119,811	107,694	97,803	89,576	82,626	76,676	67,024	59,530	53,544	48,651	44,578	41,134	38,184	32,379	28,106	24,829	22,237	20,135	18,396	15,686	13,672	12,116	9,870	8,327
140	123,733	110,852	100,401	91,750	84,472	78,264	68,234	60,483	54,313	49,286	45,110	41,587	38,574	32,659	28,317	24,953	22,368	20,242	18,485	15,751	13,722	12,155	9,896	8,345
145	127,622	113,964	102,947	93,872	86,267	79,802	69,400	61,398	55,050	49,891	45,617	42,017	38,944	32,924	28,515	25,148	22,492	20,344	18,570	15,813	13,768	12,192	9,920	8,362
150	131,480	117,030	105,442	95,942	88,012	81,294	70,526	62,277	55,755	50,470	46,100	42,427	39,296	33,175	28,704	25,295	22,609	20,439	18,650	15,870	13,812	12,226	9,943	8,378
155	135,305	120,051	107,888	97,963	89,710	82,740	71,612	63,122	56,432	51,024	46,562	42,818	39,631	33,413	28,882	25,433	22,720	20,530	18,725	15,925	13,853	12,258	9,964	8,393
160	139,100	123,029	110,287	99,937	91,363	84,144	72,661	63,936	57,081	51,554	47,003	43,190	39,950	33,640	29,051	25,564	22,824	20,615	18,796	15,976	13,892	12,289	9,984	8,408
165	142,864	125,964	112,640	101,865	92,972	85,506	73,675	64,719	57,705	52,063	47,425	43,547	40,254	33,855	29,212	25,688	22,923	20,696	18,863	16,024	13,928	12,317	10,003	8,421
170	146,597	128,858	114,948	103,749	94,538	86,830	74,655	65,475	58,305	52,550	47,830	43,887	40,545	34,061	29,365	25,807	23,017	20,772	18,926	16,070	13,963	12,344	10,021	8,434
175	150,300	131,710	117,213	105,590	96,065	88,116	75,604	66,203	58,882	53,019	48,217	44,213	40,823	34,257	29,510	25,919	23,107	20,845	18,987	16,114	13,996	12,370	10,038	8,446
180	153,973	134,522	119,435	107,390	97,552	89,366	76,522	66,906	59,437	53,469	48,589	44,526	41,090	34,444	29,649	26,026	23,192	20,914	19,044	16,155	14,027	12,394	10,054	8,457
185	157,617	137,296	121,616	109,150	99,002	90,581	77,411	67,585	59,973	53,901	48,946	44,826	41,345	34,623	29,782	26,128	23,273	20,980	19,099	16,194	14,057	12,418	10,069	8,468
190	161,232	140,030	123,757	110,872	100,417	91,763	78,273	68,241	60,489	54,318	49,289	45,113	41,589	34,795	29,908	26,226	23,350	21,043	19,151	16,232	14,085	12,440	10,084	8,478
195	164,819	142,728	125,859	112,556	101,796	92,914	79,109	68,876	60,986	54,719	49,619	45,390	41,824	34,959	30,030	26,319	23,424	21,103	19,201	16,267	14,112	12,460	10,097	8,488
200	168,376	145,388	127,923	114,204	103,142	94,034	79,920	69,489	61,467	55,105	49,937	45,655	42,050	35,116	30,146	26,408	23,495	21,160	19,248	16,301	14,137	12,480	10,110	8,497

500 MCM	1 Run(s)	480 Volts	C-Value = 22,185	Magnetic Duct	Copper

Calculations:

1. $f = \dfrac{\text{Length} \times 2 \times \text{Isca}}{\text{Runs} \times \text{C-Value} \times E_{L\text{-}L}}$
2. $M = \dfrac{1}{1 + f}$
3. $\text{Isc} = \text{Isca} \times M$
4. Add Motor Contribution = Motor FLA x 4

* All results are given in symmetrical amperes

Single-Phase L-L Bolted Fault-Current* Table

One-Way Distance in Feet

Available Isc in Thousands (Isca)	5	10	15	20	25	30	40	50	60	70	80	90	100	125	150	175	200	225	250	300	350	400	500	600
3	2,996	2,992	2,987	2,983	2,979	2,975	2,967	2,958	2,950	2,942	2,934	2,926	2,918	2,898	2,878	2,859	2,840	2,821	2,803	2,766	2,731	2,696	2,630	2,566
5	4,988	4,977	4,965	4,953	4,942	4,931	4,908	4,885	4,863	4,841	4,819	4,797	4,776	4,723	4,671	4,620	4,571	4,522	4,475	4,383	4,294	4,209	4,049	3,901
7	6,977	6,954	6,932	6,909	6,887	6,865	6,821	6,777	6,734	6,692	6,650	6,609	6,568	6,468	6,372	6,278	6,187	6,098	6,012	5,847	5,691	5,543	5,268	5,020
10	9,953	9,907	9,861	9,816	9,771	9,726	9,638	9,552	9,467	9,383	9,301	9,221	9,142	8,949	8,765	8,589	8,419	8,256	8,099	7,802	7,526	7,269	6,805	6,396
15	14,895	14,792	14,690	14,589	14,490	14,392	14,200	14,013	13,831	13,654	13,481	13,312	13,148	12,754	12,383	12,034	11,703	11,390	11,093	10,544	10,047	9,594	8,801	8,129
20	19,814	19,631	19,452	19,276	19,103	18,933	18,602	18,283	17,974	17,676	17,387	17,108	16,838	16,197	15,604	15,053	14,539	14,059	13,610	12,792	12,067	11,420	10,314	9,403
25	24,710	24,427	24,150	23,879	23,614	23,355	22,854	22,374	21,913	21,471	21,047	20,639	20,247	19,328	18,489	17,720	17,012	16,359	15,754	14,669	13,724	12,893	11,500	10,379
30	29,583	29,178	28,784	28,400	28,026	27,662	26,962	26,296	25,662	25,058	24,482	23,932	23,406	22,187	21,088	20,094	19,188	18,361	17,602	16,259	15,106	14,105	12,455	11,151
35	34,434	33,886	33,356	32,841	32,342	31,859	30,933	30,060	29,235	28,454	27,713	27,010	26,342	24,808	23,443	22,220	21,118	20,120	19,213	17,623	16,276	15,121	13,241	11,776
40	39,263	38,552	37,866	37,205	36,566	35,949	34,775	33,675	32,643	31,672	30,757	29,894	29,078	27,219	25,585	24,135	22,841	21,678	20,628	18,807	17,281	15,984	13,898	12,293
45	44,069	43,175	42,318	41,493	40,700	39,937	38,493	37,150	35,898	34,727	33,631	32,601	31,633	29,446	27,542	25,869	24,388	23,067	21,882	19,843	18,152	16,727	14,456	12,728
50	48,853	47,758	46,710	45,708	44,747	43,827	42,094	40,493	39,010	37,631	36,347	35,147	34,024	31,508	29,337	27,447	25,786	24,314	23,001	20,759	18,915	17,372	14,936	13,099
55	53,615	52,299	51,045	49,851	48,710	47,621	45,583	43,712	41,988	40,395	38,919	37,547	36,268	33,422	30,991	28,889	27,054	25,438	24,005	21,573	19,589	17,939	15,353	13,418
60	58,356	56,800	55,324	53,923	52,592	51,324	48,965	46,812	44,841	43,029	41,358	39,812	38,377	35,205	32,517	30,211	28,210	26,458	24,911	22,302	20,188	18,440	15,718	13,697
65	63,075	61,261	59,548	57,928	56,394	54,940	52,244	49,801	47,576	45,541	43,673	41,953	40,363	36,869	33,932	31,428	29,269	27,387	25,732	22,959	20,724	18,887	16,042	13,941
70	67,772	65,682	63,717	61,866	60,120	58,470	55,426	52,684	50,200	47,940	45,875	43,980	42,236	38,426	35,246	32,553	30,241	28,237	26,481	23,553	21,207	19,287	16,329	14,158
75	72,449	70,065	67,834	65,740	63,771	61,917	58,515	55,467	52,721	50,234	47,971	45,903	44,006	39,886	36,470	33,594	31,138	29,017	27,166	24,093	21,645	19,648	16,587	14,352
80	77,104	74,410	71,898	69,550	67,351	65,286	61,515	58,155	55,144	52,429	49,969	47,729	45,681	41,257	37,614	34,562	31,968	29,736	27,796	24,587	22,042	19,975	16,820	14,526
85	81,738	78,717	75,911	73,298	70,860	68,578	64,429	60,753	57,474	54,531	51,875	49,465	47,269	42,548	38,684	35,463	32,737	30,401	28,376	25,040	22,405	20,273	17,030	14,682
90	86,351	82,986	79,874	76,987	74,301	71,796	67,261	63,265	59,717	56,546	53,695	51,118	48,776	43,765	39,687	36,304	33,453	31,017	28,912	25,456	22,738	20,545	17,222	14,825
95	90,943	87,219	83,788	80,616	77,676	74,943	70,015	65,696	61,878	58,480	55,436	52,693	50,208	44,914	40,630	37,092	34,121	31,590	29,409	25,841	23,045	20,795	17,397	14,954
100	95,515	91,415	87,653	84,188	80,987	78,020	72,694	68,049	63,961	60,337	57,102	54,196	51,571	46,002	41,518	37,830	34,745	32,124	29,871	26,197	23,328	21,025	17,558	15,073
105	100,067	95,576	91,471	87,704	84,235	81,031	75,301	70,328	65,971	62,122	58,698	55,631	52,869	47,032	42,355	38,524	35,329	32,623	30,302	26,528	23,590	21,237	17,706	15,182
110	104,598	99,701	95,242	91,166	87,423	83,976	77,838	72,536	67,910	63,839	60,228	57,004	54,108	48,009	43,146	39,178	35,878	33,091	30,705	26,836	23,833	21,434	17,843	15,282
115	109,109	103,791	98,968	94,573	90,552	86,859	80,309	74,677	69,783	65,491	61,697	58,318	55,290	48,938	43,895	39,794	36,394	33,529	31,082	27,124	24,060	21,618	17,970	15,375
120	113,599	107,847	102,649	97,929	93,624	89,682	82,716	76,754	71,593	67,083	63,108	59,577	56,420	49,821	44,604	40,376	36,880	33,941	31,436	27,393	24,271	21,788	18,087	15,461
125	118,070	111,868	106,286	101,234	96,640	92,445	85,061	78,769	73,344	68,618	64,464	60,784	57,502	50,663	45,277	40,927	37,339	34,330	31,769	27,646	24,469	21,948	18,197	15,541
130	122,521	115,856	109,879	104,488	99,602	95,152	87,347	80,725	75,037	70,098	65,768	61,943	58,538	51,465	45,917	41,449	37,773	34,696	32,083	27,883	24,655	22,097	18,300	15,616
135	126,953	119,811	113,430	107,694	102,511	97,803	89,576	82,626	76,676	71,526	67,024	63,055	59,530	52,231	46,526	41,944	38,184	35,043	32,379	28,106	24,829	22,237	18,396	15,686
140	131,365	123,733	116,939	110,852	105,368	100,401	91,750	84,472	78,264	72,906	68,234	64,125	60,483	52,963	47,105	42,415	38,574	35,371	32,659	28,317	24,993	22,368	18,485	15,751
145	135,757	127,622	120,407	113,964	108,176	102,947	93,872	86,267	79,802	74,239	69,400	65,154	61,398	53,663	47,658	42,863	38,944	35,682	32,924	28,515	25,148	22,492	18,570	15,813
150	140,131	131,480	123,835	117,030	110,934	105,442	95,942	88,012	81,294	75,528	70,526	66,145	62,277	54,333	48,186	43,289	39,296	35,977	33,175	28,704	25,295	22,609	18,650	15,870
155	144,485	135,305	127,223	120,051	113,645	107,888	97,963	89,710	82,740	76,775	71,612	67,099	63,122	54,975	48,691	43,696	39,631	36,257	33,413	28,882	25,433	22,720	18,725	15,925
160	148,820	139,100	130,572	123,029	116,310	110,287	99,937	91,363	84,144	77,982	72,661	68,020	63,936	55,591	49,174	44,084	39,950	36,524	33,640	29,051	25,564	22,824	18,796	15,976
165	153,136	142,864	133,883	125,964	118,930	112,640	101,865	92,972	85,506	79,151	73,675	68,907	64,719	56,183	49,636	44,456	40,254	36,779	33,855	29,212	25,688	22,923	18,863	16,024
170	157,433	146,597	137,156	128,858	121,506	114,948	103,749	94,538	86,830	80,283	74,655	69,764	65,475	56,751	50,079	44,811	40,545	37,021	34,061	29,365	25,807	23,017	18,926	16,070
175	161,712	150,300	140,392	131,710	124,039	117,213	105,590	96,065	88,116	81,382	75,604	70,592	66,203	57,298	50,504	45,151	40,823	37,253	34,257	29,510	25,919	23,107	18,987	16,114
180	165,973	153,973	143,592	134,522	126,530	119,435	107,390	97,552	89,366	82,447	76,522	71,392	66,906	57,824	50,912	45,477	41,090	37,475	34,444	29,649	26,026	23,192	19,044	16,155
185	170,214	157,617	146,756	137,296	128,981	121,616	109,150	99,002	90,581	83,480	77,411	72,165	67,585	58,330	51,304	45,789	41,345	37,687	34,623	29,782	26,128	23,273	19,099	16,194
190	174,438	161,232	149,885	140,030	131,392	123,757	110,872	100,417	91,763	84,483	78,273	72,914	68,241	58,818	51,682	46,089	41,589	37,890	34,795	29,908	26,226	23,350	19,151	16,232
195	178,643	164,819	152,980	142,728	133,763	125,859	112,556	101,796	92,914	85,458	79,109	73,638	68,876	59,289	52,045	46,378	41,824	38,085	34,959	30,030	26,319	23,424	19,201	16,267
200	182,831	168,376	156,040	145,388	136,097	127,923	114,204	103,142	94,034	86,404	79,920	74,340	69,489	59,743	52,394	46,655	42,050	38,271	35,116	30,146	26,408	23,495	19,248	16,301

500 MCM	2 Run(s)	480 Volts	C-Value = 22,185	Magnetic Duct	Copper

Calculations:

1. $f = \dfrac{\text{Length} \times 2 \times \text{Isca}}{\text{Runs} \times \text{C-Value} \times \text{E L-L}}$

2. $M = \dfrac{1}{1 + f}$

3. Isc = Isca x M

4. Add Motor Contribution = Motor FLA x 4

* All results are given in symmetrical amperes

Single-Phase L-L Bolted Fault-Current* Table

One-Way Distance in Feet

Available Isc in Thousands (Isca)	5	10	15	20	25	30	40	50	60	70	80	90	100	125	150	175	200	225	250	300	350	400	500	600
3	2,992	2,984	2,976	2,968	2,960	2,952	2,936	2,921	2,905	2,890	2,875	2,860	2,845	2,809	2,774	2,739	2,705	2,673	2,641	2,57[illegible]	2,520	2,464	2,358	2,261
5	4,977	4,955	4,933	4,911	4,889	4,868	4,825	4,783	4,742	4,701	4,662	4,623	4,584	4,491	4,401	4,315	4,232	4,152	4,076	3,93[illegible]	3,795	3,669	3,440	3,238
7	6,956	6,912	6,869	6,827	6,785	6,743	6,662	6,582	6,504	6,428	6,354	6,282	6,211	6,041	5,880	5,727	5,582	5,444	5,313	5,06[illegible]	4,846	4,642	4,281	3,973
10	9,910	9,822	9,735	9,650	9,566	9,484	9,323	9,168	9,018	8,873	8,732	8,596	8,464	8,151	7,861	7,590	7,337	7,101	6,880	6,475	6,116	5,795	5,243	4,788
15	14,799	14,603	14,412	14,226	14,044	13,868	13,527	13,203	12,894	12,600	12,318	12,049	11,791	11,192	10,652	10,161	9,713	9,303	8,927	8,253	7,682	7,182	6,354	5,697
20	19,644	19,300	18,968	18,647	18,337	18,037	17,465	16,929	16,424	15,949	15,500	15,076	14,675	13,759	12,951	12,232	11,589	11,010	10,487	9,575	8,810	8,158	7,106	6,295
25	24,446	23,915	23,407	22,921	22,454	22,006	21,161	20,378	19,652	18,975	18,344	17,753	17,199	15,954	14,878	13,937	13,108	12,373	11,715	10,5[illegible]0	9,662	8,883	7,650	6,718
30	29,205	28,451	27,736	27,055	26,407	25,789	24,536	23,582	22,614	21,723	20,899	20,136	19,426	17,853	16,516	15,365	14,364	13,485	12,708	11,3[illegible]4	10,327	9,442	8,061	7,033
35	33,923	32,910	31,956	31,056	30,205	29,399	27,910	26,565	25,344	24,230	23,209	22,271	21,406	19,512	17,925	16,578	15,418	14,410	13,526	12,0[illegible]8	10,861	9,887	8,383	7,276
40	38,599	37,293	36,073	34,930	33,857	32,848	31,001	29,350	27,866	26,525	25,307	24,196	23,178	20,973	19,151	17,621	16,317	15,192	14,213	12,530	11,299	10,249	8,642	7,470
45	43,235	41,603	40,090	38,683	37,372	36,146	33,922	31,955	30,204	28,635	27,220	25,939	24,773	22,271	20,228	18,528	17,091	15,862	14,797	13,0[illegible]6	11,665	10,549	8,854	7,629
50	47,830	45,841	44,011	42,321	40,757	39,303	36,687	34,398	32,377	30,581	28,973	27,526	26,217	23,430	21,180	19,323	17,766	16,441	15,300	13,435	11,976	10,802	9,032	7,760
55	52,386	50,010	47,839	45,849	44,019	42,328	39,309	36,692	34,402	32,381	30,584	28,976	27,529	24,473	22,028	20,027	18,359	16,948	15,738	13,772	12,242	11,019	9,183	7,871
60	56,903	54,110	51,578	49,272	47,164	45,229	41,799	38,852	36,294	34,052	32,070	30,307	28,727	25,415	22,788	20,654	18,884	17,394	16,122	14,035	12,474	11,206	9,312	7,966
65	61,381	58,143	55,230	52,595	50,200	48,013	44,166	40,889	38,065	35,606	33,445	31,532	29,826	26,272	23,474	21,215	19,353	17,791	16,463	14,224	12,676	11,369	9,425	8,049
70	65,820	62,112	58,798	55,821	53,130	50,687	46,418	42,813	39,727	37,056	34,722	32,664	30,836	27,053	24,096	21,722	19,773	18,146	16,766	14,553	12,855	11,513	9,523	8,120
75	70,222	66,017	62,286	58,955	55,962	53,258	48,565	44,633	41,289	38,412	35,909	33,713	31,769	27,768	24,662	22,181	20,153	18,465	17,038	14,757	13,015	11,640	9,611	8,184
80	74,587	69,860	65,696	62,001	58,700	55,732	50,614	46,357	42,760	39,682	37,017	34,687	32,633	28,426	25,179	22,598	20,497	18,754	17,283	14,941	13,157	11,754	9,688	8,240
85	78,915	73,643	69,031	64,963	61,347	58,113	52,570	47,993	44,148	40,874	38,052	35,595	33,436	29,033	25,654	22,980	20,811	19,016	17,506	15,[illegible]07	13,286	11,857	9,758	8,290
90	83,207	77,367	72,293	67,843	63,910	60,408	54,441	49,547	45,460	41,996	39,023	36,443	34,183	29,594	26,092	23,331	21,098	19,255	17,709	15,[illegible]57	13,402	11,950	9,820	8,335
95	87,462	81,033	75,484	70,646	66,391	62,620	56,231	51,025	46,702	43,054	39,934	37,236	34,880	30,115	26,496	23,653	21,361	19,474	17,894	15,395	13,508	12,034	9,877	8,376
100	91,683	84,643	78,607	73,375	68,795	64,754	57,946	52,434	47,879	44,052	40,792	37,981	35,532	30,600	26,871	23,951	21,604	19,676	18,064	15,521	13,605	12,110	9,929	8,413
105	95,868	88,198	81,664	76,031	71,125	66,814	59,590	53,776	48,996	44,996	41,600	38,680	36,144	31,053	27,219	24,228	21,829	19,862	18,221	15,636	13,694	12,181	9,976	8,447
110	100,019	91,699	84,657	78,619	73,385	68,804	61,168	55,058	50,057	45,890	42,363	39,339	36,718	31,476	27,543	24,485	22,037	20,034	18,366	15,743	13,775	12,245	10,019	8,478
115	104,136	95,147	87,587	81,140	75,577	70,728	62,684	56,283	51,068	46,737	43,084	39,960	37,259	31,872	27,847	24,724	22,231	20,194	18,500	15,341	13,851	12,305	10,059	8,507
120	108,219	98,545	90,458	83,598	77,705	72,588	64,141	57,454	52,031	47,542	43,767	40,547	37,769	32,245	28,130	24,947	22,411	20,343	18,625	15,933	13,921	12,360	10,096	8,533
125	112,269	101,892	93,270	85,994	79,771	74,388	65,542	58,576	52,949	48,308	44,415	41,103	38,250	32,595	28,397	25,156	22,580	20,482	18,741	16,018	13,986	12,411	10,130	8,557
130	116,286	105,189	96,026	88,331	81,778	76,130	66,891	59,651	53,826	49,037	45,031	41,629	38,706	32,925	28,647	25,353	22,738	20,612	18,850	16,097	14,046	12,458	10,162	8,580
135	120,271	108,439	98,727	90,612	83,729	77,818	68,190	60,683	54,664	49,732	45,616	42,129	39,137	33,237	28,883	25,537	22,886	20,734	18,951	16,171	14,102	12,503	10,191	8,601
140	124,223	111,642	101,375	92,837	85,626	79,454	69,443	61,673	55,466	50,395	46,173	42,604	39,547	33,532	29,105	25,711	23,026	20,848	19,047	16,241	14,155	12,544	10,219	8,620
145	128,144	114,799	103,971	95,010	87,470	81,040	70,652	62,624	56,234	51,028	46,704	43,056	39,936	33,811	29,315	25,875	23,157	20,956	19,137	16,306	14,205	12,583	10,244	8,639
150	132,033	117,910	106,517	97,131	89,265	82,578	71,818	63,539	56,971	51,634	47,211	43,486	40,306	34,076	29,514	26,030	23,281	21,057	19,221	16,367	14,251	12,620	10,269	8,656
155	135,892	120,978	109,014	99,203	91,013	84,071	72,945	64,419	57,678	52,214	47,695	43,897	40,658	34,328	29,703	26,176	23,398	21,153	19,301	16,425	14,295	12,654	10,291	8,672
160	139,720	124,002	111,464	101,228	92,714	85,521	74,033	65,267	58,356	52,769	48,158	44,289	40,994	34,567	29,882	26,315	23,509	21,244	19,377	16,480	14,336	12,686	10,313	8,687
165	143,518	126,985	113,867	103,206	94,371	86,929	75,086	66,083	59,008	53,302	48,602	44,663	41,315	34,795	30,052	26,447	23,614	21,329	19,448	16,531	14,375	12,717	10,333	8,702
170	147,286	129,926	116,227	105,141	95,985	88,297	76,105	66,871	59,636	53,813	49,026	45,022	41,622	35,012	30,214	26,572	23,714	21,411	19,516	16,580	14,412	12,746	10,352	8,715
175	151,024	132,826	118,542	107,032	97,559	89,627	77,091	67,631	60,239	54,304	49,434	45,365	41,915	35,219	30,368	26,691	23,809	21,488	19,580	16,627	14,447	12,773	10,370	8,728
180	154,733	135,687	120,815	108,882	99,094	90,920	78,046	68,365	60,821	54,776	49,825	45,694	42,196	35,417	30,515	26,805	23,899	21,562	19,641	16,670	14,481	12,799	10,387	8,740
185	158,414	138,509	123,048	110,692	100,591	92,179	78,971	69,074	61,382	55,231	50,200	46,010	42,465	35,606	30,655	26,913	23,985	21,632	19,699	16,712	14,512	12,824	10,403	8,751
190	162,066	141,293	125,240	112,462	102,051	93,404	79,869	69,760	61,922	55,668	50,561	46,313	42,723	35,788	30,790	27,017	24,067	21,698	19,754	16,752	14,542	12,847	10,419	8,762
195	165,690	144,039	127,393	114,195	103,476	94,596	80,739	70,423	62,444	56,089	50,909	46,604	42,971	35,961	30,918	27,115	24,146	21,762	19,807	16,790	14,571	12,870	10,433	8,773
200	169,286	146,749	129,508	115,892	104,867	95,757	81,583	71,064	62,948	56,496	51,243	46,884	43,209	36,128	31,041	27,210	24,221	21,823	19,857	16,826	14,598	12,891	10,447	8,783

600 MCM	1 Run(s)	480 Volts	C-Value = 22,965	Magnetic Duct	Copper

Calculations:

1. $f = \dfrac{\text{Length} \times 2 \times \text{Isca}}{\text{Runs} \times \text{C-Value} \times \text{E L-L}}$

2. $M = \dfrac{1}{1 + f}$

3. Isc = Isca x M

4. Add Motor Contribution = Motor FLA x 4

* All results are given in symmetrical amperes

Single-Phase L-L Bolted Fault-Current* Table

One-Way Distance in Feet

Available Isc in Thousands (Isca)	5	10	15	20	25	30	40	50	60	70	80	90	100	125	150	175	200	225	250	300	350	400	500	600
3	2,996	2,992	2,988	2,984	2,980	2,976	2,968	2,960	2,952	2,944	2,936	2,928	2,921	2,901	2,882	2,864	2,845	2,827	2,809	2,774	2,739	2,705	2,641	2,579
5	4,989	4,977	4,966	4,955	4,944	4,933	4,911	4,889	4,868	4,846	4,825	4,804	4,783	4,732	4,681	4,632	4,584	4,537	4,491	4,401	4,315	4,232	4,076	3,930
7	6,978	6,956	6,934	6,912	6,891	6,869	6,827	6,785	6,743	6,702	6,662	6,622	6,582	6,485	6,391	6,300	6,211	6,125	6,041	5,880	5,727	5,582	5,313	5,069
10	9,955	9,910	9,866	9,822	9,778	9,735	9,650	9,566	9,484	9,403	9,323	9,245	9,168	8,982	8,802	8,630	8,464	8,305	8,151	7,861	7,590	7,337	6,880	6,475
15	14,899	14,799	14,700	14,603	14,507	14,412	14,226	14,044	13,868	13,695	13,527	13,363	13,203	12,819	12,457	12,115	11,791	11,484	11,192	10,652	10,161	9,713	8,927	8,258
20	19,820	19,644	19,470	19,300	19,132	18,968	18,647	18,337	18,037	17,746	17,465	17,193	16,929	16,303	15,721	15,180	14,675	14,202	13,759	12,951	12,232	11,589	10,487	9,576
25	24,720	24,446	24,178	23,915	23,659	23,407	22,921	22,454	22,006	21,575	21,161	20,762	20,378	19,478	18,654	17,897	17,199	16,553	15,954	14,878	13,937	13,108	11,715	10,590
30	29,597	29,205	28,823	28,451	28,089	27,736	27,055	26,407	25,789	25,199	24,636	24,098	23,582	22,385	21,303	20,322	19,426	18,606	17,853	16,516	15,365	14,364	12,708	11,394
35	34,453	33,923	33,409	32,910	32,426	31,956	31,056	30,205	29,399	28,636	27,910	27,221	26,565	25,056	23,708	22,499	21,406	20,415	19,512	17,925	16,578	15,418	13,526	12,048
40	39,287	38,599	37,935	37,293	36,673	36,073	34,930	33,857	32,848	31,898	31,001	30,153	29,350	27,518	25,902	24,464	23,178	22,021	20,973	19,151	17,621	16,317	14,213	12,590
45	44,100	43,235	42,403	41,603	40,833	40,090	38,683	37,372	36,146	34,999	33,922	32,909	31,955	29,796	27,910	26,248	24,773	23,456	22,271	20,228	18,528	17,091	14,797	13,046
50	48,891	47,830	46,815	45,841	44,908	44,011	42,321	40,757	39,303	37,950	36,687	35,506	34,398	31,908	29,755	27,874	26,217	24,745	23,430	21,180	19,323	17,766	15,300	13,435
55	53,661	52,386	51,170	50,010	48,900	47,839	45,849	44,019	42,328	40,763	39,309	37,956	36,692	33,874	31,457	29,362	27,529	25,911	24,473	22,028	20,027	18,359	15,738	13,772
60	58,410	56,903	55,471	54,110	52,813	51,578	49,272	47,164	45,229	43,446	41,799	40,272	38,852	35,706	33,031	30,729	28,727	26,970	25,415	22,788	20,654	18,884	16,122	14,065
65	63,138	61,381	59,718	58,143	56,649	55,230	52,595	50,200	48,013	46,009	44,166	42,464	40,889	37,419	34,492	31,990	29,826	27,936	26,272	23,474	21,215	19,353	16,463	14,324
70	67,846	65,820	63,912	62,112	60,410	58,798	55,821	53,130	50,687	48,459	46,418	44,543	42,813	39,024	35,851	33,155	30,836	28,821	27,053	24,096	21,722	19,773	16,766	14,553
75	72,533	70,222	68,055	66,017	64,097	62,286	58,955	55,962	53,258	50,804	48,565	46,516	44,633	40,530	37,118	34,236	31,769	29,634	27,768	24,662	22,181	20,153	17,038	14,757
80	77,199	74,587	72,146	69,860	67,714	65,696	62,001	58,700	55,732	53,050	50,614	48,392	46,357	41,947	38,303	35,242	32,633	30,385	28,426	25,179	22,598	20,497	17,283	14,941
85	81,844	78,915	76,188	73,643	71,262	69,031	64,963	61,347	58,113	55,203	52,570	50,177	47,993	43,282	39,413	36,179	33,436	31,079	29,033	25,654	22,980	20,811	17,506	15,107
90	86,470	83,207	80,180	77,367	74,744	72,293	67,843	63,910	60,408	57,269	54,441	51,879	49,547	44,542	40,455	37,055	34,183	31,723	29,594	26,092	23,331	21,098	17,709	15,257
95	91,075	87,462	84,125	81,033	78,160	75,484	70,646	66,391	62,620	59,254	56,231	53,502	51,025	45,733	41,435	37,876	34,380	32,323	30,115	26,496	23,653	21,361	17,894	15,395
100	95,661	91,683	88,022	84,643	81,513	78,607	73,375	68,795	64,754	61,161	57,946	55,052	52,434	46,861	42,359	38,646	35,332	32,882	30,600	26,871	23,951	21,604	18,064	15,521
105	100,227	95,868	91,873	88,198	84,805	81,664	76,031	71,125	66,814	62,996	59,590	56,534	53,776	47,931	43,231	39,371	36,144	33,405	31,053	27,219	24,228	21,829	18,221	15,636
110	104,772	100,019	95,678	91,699	88,037	84,657	78,619	73,385	68,804	64,762	61,168	57,953	55,058	48,946	44,056	40,054	36,718	33,896	31,476	27,543	24,485	22,037	18,366	15,743
115	109,299	104,136	99,439	95,147	91,211	87,587	81,140	75,577	70,728	66,463	62,684	59,311	56,283	49,912	44,836	40,698	37,259	34,356	31,872	27,847	24,724	22,231	18,500	15,841
120	113,805	108,219	103,156	98,545	94,328	90,458	83,598	77,705	72,588	68,103	64,141	60,614	57,454	50,831	45,577	41,307	37,769	34,789	32,245	28,130	24,947	22,411	18,625	15,933
125	118,293	112,269	106,829	101,892	97,390	93,270	85,994	79,771	74,388	69,685	65,542	61,864	58,576	51,707	46,280	41,884	38,250	35,197	32,595	28,397	25,156	22,580	18,741	16,018
130	122,761	116,286	110,460	105,189	100,399	96,026	88,331	81,778	76,130	71,212	66,891	63,064	59,651	52,543	46,948	42,431	38,706	35,582	32,925	28,647	25,353	22,738	18,850	16,097
135	127,210	120,271	114,049	108,439	103,355	98,727	90,612	83,729	77,818	72,687	68,190	64,218	60,683	53,342	47,585	42,950	39,137	35,947	33,237	28,883	25,537	22,886	18,951	16,171
140	131,641	124,223	117,597	111,642	106,261	101,375	92,837	85,626	79,454	74,112	69,443	65,328	61,673	54,105	48,192	43,443	39,547	36,292	33,532	29,105	25,711	23,026	19,047	16,241
145	136,052	128,144	121,105	114,799	109,117	103,971	95,010	87,470	81,040	75,490	70,652	66,396	62,624	54,836	48,770	43,913	39,936	36,619	33,811	29,315	25,875	23,157	19,137	16,306
150	140,444	132,033	124,573	117,910	111,924	106,517	97,131	89,265	82,578	76,823	71,818	67,425	63,539	55,536	49,323	44,361	40,306	36,930	34,076	29,514	26,030	23,281	19,221	16,367
155	144,818	135,892	128,002	120,978	114,685	109,014	99,203	91,013	84,071	78,114	72,945	68,417	64,419	56,207	49,852	44,788	40,658	37,226	34,328	29,703	26,176	23,398	19,301	16,425
160	149,174	139,720	131,393	124,002	117,399	111,464	101,228	92,714	85,521	79,364	74,033	69,374	65,267	56,851	50,358	45,196	40,994	37,507	34,567	29,882	26,315	23,509	19,377	16,480
165	153,511	143,518	134,746	126,985	120,069	113,867	103,206	94,371	86,929	80,575	75,086	70,298	66,083	57,470	50,843	45,587	41,315	37,776	34,795	30,052	26,447	23,614	19,448	16,531
170	157,830	147,286	138,062	129,926	122,695	116,227	105,141	95,985	88,297	81,749	76,105	71,190	66,871	58,065	51,308	45,960	41,622	38,032	35,012	30,214	26,572	23,714	19,516	16,580
175	162,130	151,024	141,342	132,826	125,278	118,542	107,032	97,559	89,627	82,888	77,091	72,052	67,631	58,637	51,755	46,318	41,915	38,276	35,219	30,368	26,691	23,809	19,580	16,627
180	166,413	154,733	144,586	135,687	127,820	120,815	108,882	99,094	90,920	83,993	78,046	72,885	68,365	59,188	52,183	46,661	42,196	38,510	35,417	30,515	26,805	23,899	19,641	16,670
185	170,678	158,414	147,794	138,509	130,321	123,048	110,692	100,591	92,179	85,065	78,971	73,692	69,074	59,719	52,595	46,990	42,465	38,734	35,606	30,655	26,913	23,985	19,699	16,712
190	174,925	162,066	150,968	141,293	132,783	125,240	112,462	102,051	93,404	86,107	79,869	74,473	69,760	60,231	52,992	47,306	42,723	38,949	35,788	30,790	27,017	24,067	19,754	16,752
195	179,154	165,690	154,108	144,039	135,205	127,393	114,195	103,476	94,596	87,120	80,739	75,229	70,423	60,724	53,374	47,610	42,971	39,155	35,961	30,918	27,115	24,146	19,807	16,790
200	183,365	169,286	157,214	146,749	137,590	129,508	115,892	104,867	95,757	88,104	81,583	75,961	71,064	61,201	53,741	47,903	43,209	39,352	36,128	31,041	27,210	24,221	19,857	16,826

600 MCM | 2 Run(s) | 480 Volts

C-Value = 22,965 | Magnetic Duct | Copper

Calculations:

1. $f = \dfrac{\text{Length} \times 2 \times \text{Isca}}{\text{Runs} \times \text{C-Value} \times \text{E L-L}}$

2. $M = \dfrac{1}{1 + f}$

3. Isc = Isca x M

4. Add Motor Contribution = Motor FLA x 4

* All results are given in symmetrical amperes

Single-Phase L-L Bolted Fault-Current* Table

One-Way Distance in Feet

Available Isc in Thousands (Isca)

Isca	5	10	15	20	25	30	40	50	60	70	80	90	100	125	150	175	200	225	250	300	350	400	500	600
3	2,992	2,985	2,977	2,969	2,962	2,954	2,939	2,924	2,910	2,895	2,881	2,866	2,852	2,818	2,784	2,751	2,718	2,687	2,656	2,597	2,540	2,485	2,383	2,289
5	4,979	4,957	4,936	4,915	4,894	4,874	4,833	4,793	4,754	4,715	4,677	4,640	4,603	4,513	4,427	4,344	4,264	4,187	4,113	3,972	3,840	3,717	3,493	3,294
7	6,958	6,916	6,875	6,835	6,795	6,755	6,677	6,601	6,527	6,454	6,383	6,313	6,245	6,081	5,926	5,778	5,637	5,504	5,376	5,138	4,919	4,719	4,364	4,058
10	9,914	9,830	9,748	9,666	9,586	9,508	9,354	9,205	9,061	8,922	8,787	8,655	8,528	8,225	7,943	7,680	7,433	7,202	6,985	6,588	6,234	5,915	5,367	4,912
15	14,808	14,621	14,439	14,261	14,088	13,919	13,592	13,281	12,983	12,698	12,426	12,165	11,915	11,332	10,804	10,322	9,882	9,478	9,105	8,442	7,869	7,368	6,537	5,874
20	19,661	19,333	19,015	18,708	18,411	18,123	17,573	17,056	16,568	16,107	15,671	15,259	14,867	13,971	13,176	12,467	11,831	11,256	10,734	9,824	9,056	8,400	7,336	6,511
25	24,472	23,966	23,480	23,014	22,565	22,134	21,320	20,563	19,858	19,200	18,584	18,006	17,463	16,239	15,176	14,243	13,418	12,684	12,025	10,894	9,958	9,170	7,917	6,965
30	29,243	28,523	27,837	27,184	26,561	25,966	24,852	23,829	22,888	22,018	21,212	20,462	19,764	18,211	16,884	15,737	14,736	13,855	13,073	11,748	10,666	9,767	8,358	7,304
35	33,974	33,006	32,091	31,226	30,407	29,629	28,187	26,880	25,688	24,597	23,595	22,671	21,818	19,940	18,360	17,012	15,848	14,834	13,941	12,444	11,237	10,243	8,704	7,567
40	38,665	37,416	36,246	35,146	34,111	33,136	31,343	29,734	28,282	26,966	25,766	24,669	23,661	21,469	19,648	18,112	16,799	15,664	14,672	13,023	11,707	10,632	8,983	7,777
45	43,317	41,756	40,304	38,949	37,682	36,495	34,332	32,411	30,693	29,149	27,752	26,484	25,326	22,830	20,783	19,072	17,622	16,376	15,295	13,511	12,100	10,956	9,213	7,949
50	47,931	46,027	44,268	42,639	41,125	39,716	37,167	34,926	32,940	31,168	29,576	28,140	26,836	24,051	21,789	19,916	18,340	16,995	15,833	13,930	12,434	11,229	9,406	8,092
55	52,507	50,231	48,143	46,223	44,449	42,807	39,861	37,295	35,039	33,040	31,257	29,657	28,213	25,150	22,688	20,664	18,972	17,536	16,303	14,291	12,722	11,463	9,570	8,213
60	57,046	54,369	51,931	49,703	47,659	45,776	42,423	39,528	37,003	34,781	32,811	31,052	29,472	26,147	23,495	21,332	19,534	18,015	16,715	14,608	12,972	11,666	9,710	8,316
65	61,547	58,442	55,636	53,086	50,760	48,630	44,863	41,638	38,846	36,405	34,252	32,340	30,630	27,054	24,225	21,932	20,036	18,441	17,082	14,887	13,192	11,843	9,833	8,406
70	66,011	62,453	59,259	56,375	53,759	51,375	47,190	43,635	40,578	37,922	35,592	33,532	31,697	27,882	24,888	22,474	20,487	18,822	17,408	15,134	13,386	11,999	9,940	8,484
75	70,440	66,403	62,803	59,573	56,660	54,018	49,410	45,527	42,210	39,343	36,841	34,638	32,683	28,643	25,492	22,965	20,894	19,166	17,702	15,356	13,558	12,138	10,035	8,553
80	74,833	70,292	66,271	62,685	59,468	56,564	51,532	47,322	43,748	40,676	38,007	35,667	33,598	29,343	26,045	23,413	21,265	19,477	17,967	15,555	13,713	12,262	10,120	8,615
85	79,190	74,123	69,666	65,714	62,187	59,019	53,562	49,028	45,202	41,930	39,100	36,628	34,450	29,991	26,554	23,823	21,602	19,760	18,207	15,735	13,853	12,374	10,196	8,670
90	83,512	77,897	72,989	68,664	64,822	61,387	55,505	50,651	46,579	43,112	40,126	37,526	35,243	30,590	27,023	24,208	21,912	20,019	18,427	15,898	13,980	12,474	10,264	8,719
95	87,800	81,615	76,244	71,536	67,376	63,673	57,367	52,198	47,883	44,227	41,090	38,368	35,985	31,147	27,457	24,548	22,196	20,256	18,627	16,047	14,095	12,566	10,326	8,764
100	92,054	85,278	79,431	74,335	69,853	65,881	59,153	53,672	49,121	45,281	41,998	39,159	36,679	31,667	27,859	24,869	22,458	20,474	18,812	16,184	14,200	12,650	10,382	8,804
105	96,274	88,888	82,554	77,063	72,256	68,014	60,868	55,080	50,297	46,279	42,855	39,903	37,331	32,151	28,234	25,167	22,701	20,676	18,982	16,310	14,297	12,726	10,434	8,841
110	100,461	92,445	85,613	79,722	74,589	70,078	62,515	56,425	51,417	47,225	43,665	40,604	37,945	32,605	28,583	25,444	22,927	20,862	19,139	16,426	14,386	12,797	10,481	8,875
115	104,615	95,951	88,612	82,316	76,855	72,074	64,099	57,712	52,483	48,123	44,432	41,267	38,522	33,031	28,910	25,703	23,136	21,036	19,285	16,533	14,468	12,862	10,525	8,907
120	108,737	99,407	91,551	84,846	79,057	74,007	65,623	58,945	53,501	48,977	45,159	41,893	39,068	33,431	29,216	25,944	23,332	21,197	19,421	16,633	14,544	12,922	10,565	8,936
125	112,827	102,814	94,433	87,316	81,196	75,878	67,090	60,126	54,472	49,790	45,849	42,486	39,583	33,808	29,503	26,171	23,515	21,348	19,547	16,725	14,615	12,978	10,603	8,962
130	116,884	106,172	97,259	89,727	83,277	77,692	68,504	61,260	55,401	50,565	46,505	43,049	40,071	34,163	29,773	26,383	23,686	21,489	19,665	16,812	14,681	13,030	10,637	8,987
135	120,911	109,484	100,031	92,080	85,301	79,451	69,868	62,348	56,289	51,304	47,130	43,584	40,534	34,499	30,028	26,583	23,847	21,622	19,776	16,893	14,743	13,079	10,670	9,010
140	124,906	112,750	102,750	94,380	87,270	81,157	71,184	63,393	57,140	52,010	47,725	44,092	40,973	34,817	30,268	26,771	23,998	21,746	19,880	16,969	14,801	13,124	10,700	9,032
145	128,871	115,971	105,418	96,626	89,187	82,812	72,454	64,399	57,956	52,685	48,292	44,576	41,391	35,118	30,496	26,949	24,141	21,863	19,978	17,040	14,855	13,167	10,728	9,052
150	132,805	119,147	108,036	98,821	91,054	84,419	73,681	65,367	58,738	53,331	48,835	45,038	41,789	35,404	30,711	27,117	24,276	21,974	20,070	17,107	14,906	13,207	10,755	9,071
155	136,710	122,280	110,606	100,967	92,873	85,980	74,868	66,299	59,490	53,949	49,353	45,478	42,168	35,675	30,915	27,275	24,403	22,078	20,157	17,170	14,954	13,244	10,780	9,088
160	140,584	125,371	113,129	103,065	94,645	87,497	76,015	67,197	60,212	54,543	49,849	45,899	42,529	35,934	31,109	27,427	24,524	22,177	20,240	17,230	14,999	13,280	10,803	9,105
165	144,430	128,420	115,606	105,116	96,372	88,971	77,125	68,063	60,907	55,112	50,324	46,302	42,874	36,180	31,293	27,570	24,638	22,270	20,317	17,286	15,042	13,313	10,825	9,121
170	148,247	131,429	118,038	107,124	98,057	90,405	78,200	68,899	61,575	55,659	50,780	46,687	43,205	36,415	31,469	27,706	24,747	22,359	20,391	17,339	15,082	13,345	10,846	9,136
175	152,035	134,398	120,427	109,088	99,700	91,800	79,242	69,706	62,219	56,184	51,217	47,056	43,521	36,639	31,636	27,836	24,850	22,443	20,461	17,390	15,120	13,375	10,866	9,150
180	155,794	137,327	122,774	111,010	101,303	93,157	80,251	70,486	62,840	56,690	51,636	47,410	43,823	36,853	31,796	27,959	24,949	22,524	20,528	17,438	15,157	13,403	10,885	9,163
185	159,526	140,218	125,080	112,892	102,868	94,479	81,230	71,240	63,438	57,176	52,040	47,750	44,114	37,058	31,949	28,077	25,043	22,600	20,592	17,484	15,191	13,430	10,903	9,176
190	163,230	143,072	127,346	114,734	104,395	95,766	82,180	71,969	64,016	57,645	52,428	48,077	44,392	37,255	32,094	28,190	25,132	22,673	20,652	17,528	15,224	13,456	10,919	9,188
195	166,907	145,889	129,572	116,538	105,887	97,020	83,101	72,675	64,574	58,097	52,802	48,391	44,660	37,443	32,234	28,297	25,218	22,742	20,710	17,569	15,256	13,480	10,936	9,199
200	170,556	148,670	131,761	118,306	107,344	98,242	83,996	73,359	65,113	58,533	53,161	48,693	44,917	37,624	32,368	28,400	25,299	22,809	20,765	17,609	15,285	13,504	10,951	9,210

750 MCM | **1 Run(s)** | **480 Volts** | **C-Value = 24,136** | **Magnetic Duct** | **Copper**

Calculations:

1. $f = \dfrac{\text{Length} \times 2 \times \text{Isca}}{\text{Runs} \times \text{C-Value} \times \text{E L-L}}$

2. $M = \dfrac{1}{1 + f}$

3. Isc = Isca x M

4. Add Motor Contribution = Motor FLA x 4

* All results are given in symmetrical amperes

Single-Phase L-L Bolted Fault-Current* Table

One-Way Distance in Feet

Available Isc in Thousands (Isca)

Isca	5	10	15	20	25	30	40	50	60	70	80	90	100	125	150	175	200	225	250	300	350	400	500	600
3	2,996	2,992	2,988	2,985	2,981	2,977	2,969	2,962	2,954	2,947	2,939	2,932	2,924	2,906	2,888	2,870	2,852	2,835	2,818	2,784	2,751	2,718	2,656	2,597
5	4,989	4,979	4,968	4,957	4,947	4,936	4,915	4,894	4,874	4,853	4,833	4,813	4,793	4,744	4,696	4,649	4,603	4,557	4,513	4,427	4,344	4,264	4,113	3,972
7	6,979	6,958	6,937	6,916	6,896	6,875	6,835	6,795	6,755	6,716	6,677	6,639	6,601	6,508	6,418	6,331	6,245	6,162	6,081	5,926	5,778	5,637	5,376	5,138
10	9,957	9,914	9,872	9,830	9,789	9,748	9,666	9,586	9,508	9,430	9,354	9,279	9,205	9,026	8,854	8,688	8,528	8,374	8,225	7,943	7,680	7,433	6,985	6,588
15	14,904	14,808	14,714	14,621	14,530	14,439	14,261	14,088	13,919	13,753	13,592	13,435	13,281	12,911	12,561	12,229	11,915	11,616	11,332	10,804	10,322	9,882	9,105	8,442
20	19,829	19,661	19,495	19,333	19,173	19,015	18,708	18,411	18,123	17,844	17,573	17,310	17,056	16,450	15,886	15,360	14,867	14,405	13,971	13,176	12,467	11,831	10,734	9,824
25	24,733	24,472	24,216	23,966	23,720	23,480	23,014	22,565	22,134	21,719	21,320	20,934	20,563	19,689	18,887	18,147	17,463	16,829	16,239	15,176	14,243	13,418	12,025	10,894
30	29,617	29,243	28,878	28,523	28,176	27,837	27,184	26,561	25,966	25,397	24,852	24,330	23,829	22,664	21,607	20,645	19,764	18,956	18,211	16,884	15,737	14,736	13,073	11,748
35	34,479	33,974	33,483	33,006	32,542	32,091	31,226	30,407	29,629	28,890	28,187	27,518	26,880	25,406	24,085	22,895	21,818	20,837	19,940	18,360	17,012	15,848	13,941	12,444
40	39,321	38,665	38,030	37,416	36,822	36,246	35,146	34,111	33,136	32,214	31,343	30,517	29,734	27,941	26,352	24,934	23,661	22,512	21,469	19,648	18,112	16,799	14,672	13,023
45	44,143	43,317	42,522	41,756	41,017	40,304	38,949	37,682	36,495	35,380	34,332	33,344	32,411	30,292	28,434	26,790	25,326	24,013	22,830	20,783	19,072	17,622	15,295	13,511
50	48,944	47,931	46,960	46,027	45,131	44,268	42,639	41,125	39,716	38,399	37,167	36,012	34,926	32,479	30,351	28,486	26,836	25,367	24,051	21,789	19,916	18,340	15,833	13,930
55	53,725	52,507	51,344	50,231	49,165	48,143	46,223	44,449	42,807	41,281	39,861	38,535	37,295	34,517	32,124	30,042	28,213	26,594	25,150	22,688	20,664	18,972	16,303	14,291
60	58,486	57,046	55,675	54,369	53,122	51,931	49,703	47,659	45,776	44,036	42,423	40,925	39,528	36,422	33,768	31,474	29,472	27,710	26,147	23,495	21,332	19,534	16,715	14,608
65	63,226	61,547	59,954	58,442	57,004	55,636	53,086	50,760	48,630	46,671	44,863	43,191	41,638	38,206	35,296	32,798	30,630	28,731	27,054	24,225	21,932	20,036	17,082	14,887
70	67,947	66,011	64,183	62,453	60,814	59,259	56,375	53,759	51,375	49,194	47,190	45,343	43,635	39,880	36,720	34,024	31,697	29,668	27,882	24,888	22,474	20,487	17,408	15,134
75	72,648	70,440	68,362	66,403	64,553	62,803	59,573	56,660	54,018	51,612	49,410	47,389	45,527	41,454	38,051	35,163	32,683	30,530	28,643	25,492	22,965	20,894	17,702	15,356
80	77,330	74,833	72,491	70,292	68,223	66,271	62,685	59,468	56,564	53,931	51,532	49,338	47,322	42,938	39,297	36,225	33,598	31,327	29,343	26,045	23,413	21,265	17,967	15,555
85	81,992	79,190	76,573	74,123	71,826	69,666	65,714	62,187	59,019	56,158	53,562	51,195	49,028	44,338	40,466	37,216	34,450	32,066	29,991	26,554	23,823	21,602	18,207	15,735
90	86,635	83,512	80,607	77,897	75,364	72,989	68,664	64,822	61,387	58,298	55,505	52,967	50,651	45,661	41,565	38,144	35,243	32,752	30,590	27,023	24,200	21,912	18,427	15,898
95	91,258	87,800	84,595	81,615	78,838	76,244	71,536	67,376	63,673	60,356	57,367	54,660	52,198	46,913	42,601	39,014	35,985	33,392	31,147	27,457	24,548	22,196	18,627	16,047
100	95,863	92,054	88,537	85,278	82,251	79,431	74,335	69,853	65,881	62,336	59,153	56,279	53,672	48,101	43,578	39,832	36,679	33,989	31,667	27,859	24,869	22,458	18,812	16,184
105	100,448	96,274	92,434	88,888	85,604	82,554	77,063	72,256	68,014	64,243	60,868	57,829	55,080	49,229	44,501	40,602	37,331	34,548	32,151	28,234	25,167	22,701	18,982	16,310
110	105,015	100,461	96,287	92,445	88,898	85,613	79,722	74,589	70,078	66,080	62,515	59,314	56,425	50,301	45,375	41,329	37,945	35,073	32,605	28,583	25,444	22,927	19,139	16,426
115	109,562	104,615	100,096	95,951	92,136	88,612	82,316	76,855	72,074	67,853	64,099	60,738	57,712	51,321	46,204	42,015	38,522	35,566	33,031	28,910	25,703	23,136	19,285	16,533
120	114,091	108,737	103,863	99,407	95,318	91,551	84,846	79,057	74,007	69,563	65,623	62,105	58,945	52,293	46,991	42,664	39,068	36,030	33,431	29,216	25,944	23,332	19,421	16,633
125	118,602	112,827	107,588	102,814	98,445	94,433	87,316	81,196	75,878	71,214	67,090	63,418	60,126	53,221	47,738	43,280	39,583	36,468	33,808	29,503	26,171	23,515	19,547	16,725
130	123,094	116,884	111,271	106,172	101,521	97,259	89,727	83,277	77,692	72,810	68,504	64,680	61,260	54,107	48,450	43,864	40,071	36,882	34,163	29,773	26,383	23,686	19,665	16,812
135	127,567	120,911	114,914	109,484	104,544	100,031	92,080	85,301	79,451	74,352	69,868	65,894	62,348	54,954	49,128	44,419	40,534	37,274	34,499	30,028	26,583	23,847	19,776	16,893
140	132,023	124,906	118,517	112,750	107,518	102,750	94,380	87,270	81,157	75,844	71,184	67,063	63,393	55,765	49,775	44,947	40,973	37,645	34,817	30,268	26,771	23,998	19,880	16,969
145	136,460	128,871	122,081	115,971	110,443	105,418	96,626	89,187	82,812	77,288	72,454	68,189	64,399	56,542	50,393	45,451	41,391	37,997	35,118	30,496	26,949	24,141	19,978	17,040
150	140,880	132,805	125,606	119,147	113,320	108,036	98,821	91,054	84,419	78,686	73,681	69,275	65,367	57,286	50,984	45,930	41,789	38,332	35,404	30,711	27,117	24,276	20,070	17,107
155	145,281	136,710	129,093	122,280	116,150	110,606	100,967	92,873	85,980	80,040	74,868	70,323	66,299	58,001	51,549	46,389	42,168	38,651	35,675	30,915	27,276	24,403	20,157	17,170
160	149,665	140,584	132,543	125,371	118,936	113,129	103,065	94,645	87,497	81,353	76,015	71,334	67,197	58,687	52,090	46,827	42,529	38,954	35,934	31,109	27,427	24,524	20,240	17,230
165	154,031	144,430	135,955	128,420	121,676	115,606	105,116	96,372	88,971	82,626	77,125	72,311	68,063	59,347	52,609	47,246	42,874	39,244	36,180	31,293	27,570	24,638	20,317	17,286
170	158,380	148,247	139,332	131,429	124,374	118,038	107,124	98,057	90,405	83,861	78,200	73,256	68,899	59,981	53,107	47,647	43,205	39,520	36,415	31,469	27,706	24,747	20,391	17,339
175	162,711	152,035	142,673	134,398	127,029	120,427	109,088	99,700	91,800	85,060	79,242	74,169	69,706	60,592	53,586	48,032	43,521	39,784	36,639	31,636	27,836	24,850	20,461	17,390
180	167,025	155,794	145,979	137,327	129,643	122,774	111,010	101,303	93,157	86,224	80,251	75,052	70,486	61,180	54,045	48,401	43,823	40,037	36,853	31,796	27,959	24,949	20,528	17,438
185	171,321	159,526	149,250	140,218	132,217	125,080	112,892	102,868	94,479	87,355	81,230	75,908	71,240	61,748	54,487	48,755	44,114	40,279	37,058	31,949	28,077	25,043	20,592	17,484
190	175,601	163,230	152,488	143,072	134,751	127,346	114,734	104,395	95,766	88,454	82,180	76,736	71,969	62,295	54,913	49,095	44,392	40,511	37,255	32,094	28,190	25,132	20,652	17,528
195	179,863	166,907	155,692	145,889	137,247	129,572	116,538	105,887	97,020	89,523	83,101	77,539	72,675	62,823	55,323	49,423	44,560	40,734	37,443	32,234	28,297	25,218	20,710	17,569
200	184,108	170,556	158,863	148,670	139,706	131,761	118,306	107,344	98,242	90,562	83,996	78,318	73,359	63,333	55,718	49,738	44,317	40,948	37,624	32,368	28,400	25,299	20,765	17,609

750 MCM	2 Run(s)	480 Volts	C-Value = 24,136	Magnetic Duct	Copper

Calculations:

1. $f = \dfrac{\text{Length} \times 2 \times \text{Isca}}{\text{Runs} \times \text{C-Value} \times \text{E L-L}}$

2. $M = \dfrac{1}{1 + f}$

3. Isc = Isca x M

4. Add Motor Contribution = Motor FLA x 4

* All results are given in symmetrical amperes

Single-Phase L-L Bolted Fault-Current* Table

One-Way Distance in Feet

Available Isc in Thousands (Isca)

	5	10	15	20	25	30	40	50	60	70	80	90	100	125	150	175	200	225	250	300	350	400	500	600
3	2,993	2,985	2,978	2,971	2,963	2,956	2,942	2,928	2,914	2,900	2,886	2,872	2,859	2,825	2,793	2,761	2,730	2,700	2,670	2,612	2,557	2,505	2,405	2,314
5	4,979	4,959	4,939	4,919	4,899	4,879	4,840	4,802	4,764	4,727	4,691	4,655	4,619	4,533	4,450	4,370	4,292	4,218	4,146	4,009	3,881	3,760	3,541	3,346
7	6,960	6,920	6,881	6,842	6,804	6,766	6,691	6,618	6,547	6,477	6,408	6,341	6,276	6,118	5,967	5,824	5,688	5,557	5,433	5,200	4,986	4,789	4,439	4,136
10	9,918	9,838	9,759	9,681	9,604	9,529	9,381	9,239	9,100	8,966	8,835	8,708	8,585	8,292	8,018	7,761	7,521	7,295	7,082	6,691	6,341	6,027	5,482	5,028
15	14,817	14,638	14,464	14,293	14,127	13,964	13,650	13,350	13,062	12,787	12,523	12,270	12,026	11,459	10,942	10,470	10,037	9,638	9,270	8,412	8,041	7,541	6,708	6,040
20	19,676	19,362	19,058	18,763	18,477	18,200	17,670	17,170	16,697	16,250	15,826	15,424	15,041	14,163	13,382	12,683	12,053	11,483	10,964	10,355	9,286	8,626	7,552	6,716
25	24,495	24,011	23,545	23,096	22,665	22,249	21,462	20,729	20,044	19,403	18,802	18,237	17,704	16,501	15,450	14,525	13,705	12,972	12,314	11,179	10,236	9,440	8,169	7,199
30	29,276	28,586	27,928	27,300	26,699	26,124	25,046	24,053	23,136	22,286	21,496	20,761	20,074	18,540	17,224	16,083	15,083	14,200	13,415	12,080	10,986	10,074	8,639	7,562
35	34,019	33,091	32,212	31,379	30,588	29,836	28,438	27,164	26,000	24,932	23,947	23,038	22,195	20,335	18,763	17,416	16,250	15,230	14,331	12,817	11,592	10,581	9,010	7,845
40	38,723	37,526	36,400	35,340	34,340	33,395	31,652	30,083	28,661	27,368	26,187	25,104	24,106	21,928	20,111	18,572	17,251	16,106	15,104	13,432	12,093	10,997	9,310	8,071
45	43,391	41,893	40,494	39,187	37,961	36,809	34,703	32,826	31,141	29,620	28,241	26,985	25,836	23,350	21,300	19,582	18,120	16,861	15,765	13,952	12,513	11,344	9,557	8,256
50	48,021	46,193	44,499	42,925	41,458	40,088	37,603	35,409	33,456	31,707	30,133	28,707	27,410	24,628	22,359	20,473	18,880	17,517	16,338	14,399	12,871	11,637	9,764	8,410
55	52,615	50,428	48,416	46,558	44,838	43,240	40,363	37,845	35,623	33,647	31,879	30,288	28,847	25,782	23,306	21,264	19,551	18,093	16,838	14,786	13,180	11,888	9,940	8,541
60	57,173	54,600	52,249	50,092	48,106	46,271	42,992	40,147	37,655	35,455	33,497	31,744	30,166	26,831	24,159	21,972	20,148	18,603	17,279	15,125	13,448	12,107	10,092	8,653
65	61,695	58,710	56,000	53,529	51,268	49,189	45,500	42,326	39,565	37,143	35,000	33,091	31,379	27,786	24,932	22,609	20,682	19,058	17,670	15,424	13,684	12,297	10,225	8,750
70	66,182	62,759	59,672	56,875	54,328	52,000	47,895	44,390	41,364	38,724	36,400	34,340	32,500	28,662	25,634	23,185	21,163	19,465	18,020	15,690	13,893	12,466	10,341	8,835
75	70,634	66,748	63,268	60,132	57,293	54,710	50,184	46,350	43,060	40,206	37,707	35,501	33,538	29,466	26,275	23,708	21,598	19,833	18,335	15,928	14,080	12,616	10,444	8,910
80	75,052	70,680	66,789	63,304	60,165	57,323	52,374	48,212	44,663	41,600	38,931	36,583	34,503	30,208	26,864	24,186	21,994	20,166	18,619	16,142	14,247	12,750	10,536	8,977
85	79,435	74,554	70,238	66,395	62,950	59,845	54,472	49,984	46,179	42,913	40,078	37,594	35,401	30,894	27,405	24,624	22,356	20,470	18,877	16,336	14,397	12,870	10,618	9,036
90	83,785	78,373	73,618	69,407	65,651	62,282	56,483	51,672	47,616	44,151	41,156	38,541	36,239	31,530	27,905	25,027	22,687	20,747	19,113	16,512	14,534	12,979	10,692	9,090
95	88,102	82,138	76,930	72,343	68,273	64,636	58,412	53,282	48,980	45,321	42,171	39,430	37,024	32,123	28,368	25,399	22,992	21,002	19,329	16,673	14,659	13,079	10,759	9,139
100	92,386	85,849	80,176	75,207	70,817	66,912	60,265	54,819	50,276	46,429	43,128	40,266	37,760	32,675	28,798	25,743	23,274	21,237	19,528	16,821	14,773	13,169	10,821	9,183
105	96,637	89,508	83,359	78,000	73,289	69,114	62,046	56,289	51,510	47,478	44,032	41,053	38,451	33,192	29,198	26,062	23,535	21,454	19,711	16,957	14,877	13,253	10,877	9,223
110	100,856	93,116	86,480	80,726	75,690	71,246	63,758	57,695	52,684	48,475	44,888	41,796	39,102	33,676	29,572	26,360	23,777	21,655	19,881	17,082	14,974	13,329	10,928	9,260
115	105,044	96,675	89,540	83,387	78,024	73,310	65,406	59,041	53,805	49,422	45,699	42,498	39,716	34,130	29,922	26,637	24,002	21,842	20,038	17,198	15,063	13,400	10,975	9,294
120	109,200	100,184	92,543	85,984	80,294	75,311	66,994	60,332	54,875	50,323	46,468	43,162	40,295	34,557	30,250	26,897	24,213	22,016	20,185	17,306	15,146	13,465	11,019	9,325
125	113,325	103,645	95,488	88,522	82,503	77,250	68,524	61,570	55,897	51,181	47,199	43,792	40,844	34,960	30,558	27,140	24,410	22,179	20,322	17,406	15,223	13,526	11,060	9,355
130	117,419	107,059	98,379	91,000	84,651	79,131	70,000	62,759	56,875	52,000	47,895	44,391	41,364	35,340	30,848	27,369	24,595	22,331	20,450	17,500	15,294	13,582	11,098	9,382
135	121,483	110,427	101,215	93,422	86,743	80,956	71,425	63,901	57,812	52,782	48,558	44,959	41,857	35,699	31,121	27,584	24,768	22,474	20,569	17,588	15,361	13,635	11,133	9,407
140	125,517	113,750	104,000	95,790	88,781	82,728	72,800	65,000	58,710	53,530	49,189	45,500	42,326	36,040	31,380	27,786	24,932	22,609	20,682	17,670	15,424	13,684	11,166	9,430
145	129,522	117,029	106,734	98,104	90,766	84,448	74,130	66,058	59,571	54,245	49,793	46,016	42,772	36,363	31,624	27,978	25,086	22,735	20,788	17,747	15,483	13,731	11,197	9,452
150	133,496	120,265	109,419	100,368	92,700	86,120	75,415	67,076	60,399	54,930	50,369	46,508	43,196	36,669	31,856	28,159	25,231	22,855	20,888	17,820	15,538	13,774	11,225	9,473
155	137,442	123,458	112,056	102,582	94,585	87,745	76,658	68,058	61,193	55,587	50,921	46,978	43,602	36,961	32,075	28,331	25,369	22,968	20,982	17,889	15,590	13,815	11,253	9,492
160	141,359	126,609	114,646	104,749	96,424	89,326	77,861	69,005	61,958	56,217	51,449	47,427	43,988	37,238	32,284	28,493	25,499	23,075	21,071	17,953	15,639	13,854	11,278	9,510
165	145,248	129,719	117,191	106,869	98,218	90,863	79,027	69,919	62,693	56,822	51,955	47,857	44,358	37,503	32,483	28,648	25,623	23,176	21,155	18,015	15,686	13,890	11,302	9,527
170	149,109	132,790	119,691	108,944	99,968	92,359	80,156	70,801	63,402	57,403	52,441	48,269	44,711	37,755	32,672	28,795	25,741	23,272	21,236	18,073	15,730	13,925	11,325	9,544
175	152,941	135,821	122,148	110,976	101,676	93,815	81,250	71,654	64,085	57,962	52,907	48,663	45,050	37,996	32,852	28,935	25,852	23,363	21,312	18,128	15,771	13,957	11,347	9,559
180	156,747	138,814	124,563	112,966	103,344	95,233	82,312	72,478	64,743	58,500	53,355	49,042	45,374	38,227	33,024	29,069	25,959	23,450	21,384	18,180	15,811	13,988	11,367	9,573
185	160,525	141,769	126,937	114,915	104,973	96,614	83,342	73,276	65,379	59,019	53,786	49,406	45,686	38,447	33,189	29,195	26,061	23,533	21,453	18,230	15,849	14,018	11,387	9,587
190	164,276	144,686	129,271	116,825	106,564	97,961	84,342	74,048	65,993	59,518	54,201	49,756	45,984	38,659	33,346	29,313	26,158	23,612	21,519	18,277	15,884	14,046	11,405	9,600
195	168,000	147,568	131,567	118,696	108,119	99,273	85,313	74,795	66,586	60,000	54,600	50,092	46,271	38,861	33,497	29,434	26,250	23,688	21,581	18,322	15,918	14,072	11,423	9,613
200	171,698	150,414	133,824	120,530	109,639	100,553	86,256	75,519	67,159	60,465	54,985	50,416	46,548	39,056	33,642	29,546	26,339	23,760	21,641	18,365	15,951	14,098	11,439	9,625

1000 MCM	1 Run(s)	480 Volts	C-Value = 25,278	Magnetic Duct	Copper

Calculations:

1. $f = \dfrac{\text{Length} \times 2 \times \text{Isca}}{\text{Runs} \times \text{C-Value} \times \text{E L-L}}$
2. $M = \dfrac{1}{1 + f}$
3. Isc = Isca x M
4. Add Motor Contribution = Motor FLA x 4

* All results are given in symmetrical amperes

Single-Phase L-L Bolted Fault-Current* Table

One-Way Distance in Feet

Available Isc in Thousands (Isca)	5	10	15	20	25	30	40	50	60	70	80	90	100	125	150	175	200	225	250	300	350	400	500	600
3	2,996	2,993	2,989	2,985	2,982	2,978	2,971	2,963	2,956	2,949	2,942	2,935	2,928	2,910	2,893	2,876	2,859	2,842	2,825	2,793	2,761	2,730	2,670	2,612
5	4,990	4,979	4,969	4,959	4,949	4,939	4,919	4,899	4,879	4,860	4,840	4,821	4,802	4,755	4,709	4,664	4,619	4,576	4,533	4,450	4,370	4,292	4,146	4,009
7	6,980	6,960	6,940	6,920	6,900	6,881	6,842	6,804	6,766	6,728	6,691	6,654	6,618	6,529	6,442	6,358	6,276	6,196	6,118	5,967	5,824	5,688	5,433	5,200
10	9,959	9,918	9,878	9,838	9,798	9,759	9,681	9,604	9,529	9,455	9,381	9,309	9,239	9,066	8,900	8,740	8,585	8,436	8,292	8,018	7,761	7,521	7,082	6,691
15	14,908	14,817	14,727	14,638	14,550	14,464	14,293	14,127	13,964	13,805	13,650	13,498	13,350	12,992	12,654	12,332	12,026	11,736	11,459	10,942	10,470	10,037	9,270	8,612
20	19,837	19,676	19,517	19,362	19,208	19,058	18,763	18,477	18,200	17,931	17,670	17,416	17,170	16,583	16,035	15,522	15,041	14,589	14,163	13,382	12,683	12,053	10,964	10,055
25	24,745	24,495	24,251	24,011	23,775	23,545	23,096	22,665	22,249	21,849	21,462	21,089	20,729	19,880	19,098	18,375	17,704	17,081	16,501	15,450	14,525	13,705	12,314	11,179
30	29,634	29,276	28,927	28,586	28,254	27,928	27,300	26,699	26,124	25,574	25,046	24,539	24,053	22,917	21,884	20,940	23,074	19,276	18,540	17,224	16,083	15,083	13,415	12,080
35	34,502	34,019	33,548	33,091	32,646	32,212	31,379	30,588	29,836	29,120	28,438	27,786	27,164	25,724	24,430	23,259	22,195	21,225	20,335	18,763	17,416	16,250	14,331	12,817
40	39,351	38,723	38,115	37,526	36,954	36,400	35,340	34,340	33,395	32,500	31,652	30,848	30,083	28,327	26,765	25,366	24,106	22,965	21,928	20,111	18,572	17,251	15,104	13,432
45	44,181	43,391	42,629	41,893	41,182	40,494	39,187	37,961	36,809	35,725	34,703	33,738	32,826	30,746	28,914	27,289	25,836	24,530	23,350	21,300	19,582	18,120	15,765	13,952
50	48,991	48,021	47,089	46,193	45,330	44,499	42,925	41,458	40,088	38,806	37,603	36,473	35,409	33,001	30,900	29,050	27,410	25,945	24,628	22,359	20,473	18,880	16,338	14,399
55	53,781	52,615	51,498	50,428	49,402	48,416	46,558	44,838	43,240	41,752	40,363	39,064	37,845	35,108	32,739	30,670	28,847	27,229	25,782	23,306	21,264	19,551	16,838	14,786
60	58,552	57,173	55,857	54,600	53,399	52,249	50,092	48,106	46,271	44,572	42,992	41,521	40,147	37,080	34,448	32,165	30,166	28,401	26,831	24,159	21,972	20,148	17,279	15,125
65	63,304	61,695	60,165	58,710	57,323	56,000	53,529	51,268	49,189	47,273	45,500	43,856	42,326	38,931	36,040	33,549	31,379	29,474	27,786	24,932	22,609	20,682	17,670	15,424
70	68,037	66,182	64,425	62,759	61,177	59,672	56,875	54,328	52,000	49,863	47,895	46,076	44,390	40,671	37,526	34,833	32,500	30,460	28,662	25,634	23,185	21,163	18,020	15,690
75	72,752	70,634	68,636	66,748	64,961	63,268	60,132	57,293	54,710	52,349	50,184	48,191	46,350	42,309	38,917	36,028	33,538	31,370	29,466	26,275	23,708	21,598	18,335	15,928
80	77,447	75,052	72,800	70,680	68,679	66,789	63,304	60,165	57,323	54,737	52,374	50,207	48,212	43,856	40,221	37,143	34,503	32,213	30,208	26,864	24,186	21,994	18,619	16,142
85	82,123	79,435	76,917	74,554	72,332	70,238	66,395	62,950	59,845	57,032	54,472	52,132	49,984	45,317	41,447	38,186	35,401	32,994	30,894	27,405	24,624	22,356	18,877	16,336
90	86,781	83,785	80,989	78,373	75,921	73,618	69,407	65,651	62,282	59,241	56,483	53,971	51,672	46,700	42,601	39,163	36,239	33,721	31,530	27,905	25,027	22,687	19,113	16,512
95	91,421	88,102	85,015	82,138	79,449	76,930	72,343	68,273	64,636	61,367	58,412	55,729	53,282	48,011	43,689	40,081	37,024	34,400	32,123	28,368	25,399	22,992	19,329	16,673
100	96,042	92,386	88,998	85,849	82,916	80,176	75,207	70,817	66,912	63,415	60,265	57,413	54,819	49,256	44,718	40,945	37,760	35,034	32,675	28,798	25,743	23,274	19,528	16,821
105	100,645	96,637	92,936	89,508	86,324	83,359	78,000	73,289	69,114	65,389	62,046	59,027	56,289	50,439	45,691	41,759	38,451	35,628	33,192	29,198	26,062	23,535	19,711	16,957
110	105,230	100,856	96,832	93,116	89,675	86,480	80,726	75,690	71,246	67,294	63,758	60,575	57,695	51,565	46,613	42,528	39,102	36,186	33,676	29,572	26,360	23,777	19,881	17,082
115	109,797	105,044	100,686	96,675	92,971	89,540	83,387	78,024	73,310	69,133	65,406	62,061	59,041	52,638	47,487	43,255	39,716	36,711	34,130	29,922	26,637	24,002	20,038	17,198
120	114,346	109,200	104,498	100,184	96,212	92,543	85,984	80,294	75,311	70,909	66,994	63,489	60,332	53,661	48,319	43,944	40,295	37,206	34,557	30,250	26,897	24,213	20,185	17,306
125	118,877	113,325	108,269	103,645	99,399	95,488	88,522	82,503	77,250	72,626	68,524	64,861	61,570	54,639	49,110	44,597	40,844	37,674	34,960	30,558	27,140	24,410	20,322	17,406
130	123,390	117,419	112,000	107,059	102,535	98,379	91,000	84,651	79,131	74,286	70,000	66,182	62,759	55,573	49,863	45,218	41 364	38,115	35,340	30,848	27,369	24,595	20,450	17,500
135	127,886	121,483	115,692	110,427	105,621	101,215	93,422	86,743	80,956	75,892	71,425	67,454	63,901	56,467	50,582	45,808	41 857	38,534	35,699	31,121	27,584	24,768	20,569	17,588
140	132,364	125,517	119,344	113,750	108,657	104,000	95,790	88,781	82,728	77,447	72,800	68,680	65,000	57,323	51,268	46,370	42 326	38,931	36,040	31,380	27,786	24,932	20,682	17,670
145	136,824	129,522	122,959	117,029	111,645	106,734	98,104	90,766	84,448	78,953	74,130	69,861	66,058	58,144	51,924	46,905	42 772	39,308	36,363	31,624	27,978	25,086	20,788	17,747
150	141,268	133,496	126,536	120,265	114,586	109,419	100,368	92,700	86,120	80,413	75,415	71,002	67,076	58,932	52,551	47,417	43,196	39,666	36,669	31,856	28,159	25,231	20,888	17,820
155	145,694	137,442	130,075	123,458	117,481	112,056	102,582	94,585	87,745	81,828	76,658	72,103	68,058	59,688	53,152	47,905	43,602	40,007	36,961	32,075	28,331	25,369	20,982	17,889
160	150,103	141,359	133,578	126,609	120,331	114,646	104,749	96,424	89,326	83,200	77,861	73,166	69,005	60,415	53,727	48,372	43,988	40,333	37,238	32,284	28,493	25,499	21,071	17,953
165	154,495	145,248	137,045	129,719	123,137	117,191	106,869	98,218	90,863	84,532	79,027	74,194	69,919	61,115	54,280	48,820	44,358	40,643	37,503	32,483	28,648	25,623	21,155	18,015
170	158,870	149,109	140,477	132,790	125,901	119,691	108,944	99,968	92,359	85,826	80,156	75,189	70,801	61,788	54,810	49,248	44,711	40,940	37,755	32,672	28,795	25,741	21,236	18,073
175	163,229	152,941	143,874	135,821	128,622	122,148	110,976	101,676	93,815	87,082	81,250	76,151	71,654	62,436	55,319	49,659	45,350	41,223	37,996	32,852	28,935	25,852	21,312	18,128
180	167,570	156,747	147,236	138,814	131,303	124,563	112,966	103,344	95,233	88,302	82,312	77,083	72,478	63,061	55,810	50,054	45,374	41,495	38,227	33,024	29,069	25,959	21,384	18,180
185	171,895	160,525	150,565	141,769	133,944	126,937	114,915	104,973	96,614	89,489	83,342	77,985	73,276	63,664	56,281	50,433	45,386	41,755	38,447	33,189	29,196	26,061	21,453	18,230
190	176,204	164,276	153,860	144,686	136,545	129,271	116,825	106,564	97,961	90,643	84,342	78,860	74,048	64,246	56,735	50,797	45,984	42,005	38,659	33,346	29,318	26,158	21,519	18,277
195	180,496	168,000	157,123	147,568	139,109	131,567	118,696	108,119	99,273	91,765	85,313	79,708	74,795	64,808	57,173	51,148	46,271	42,244	38,861	33,497	29,434	26,250	21,581	18,322
200	184,772	171,698	160,353	150,414	141,635	133,824	120,530	109,639	100,553	92,858	86,256	80,531	75,519	65,350	57,595	51,485	46,548	42,474	39,056	33,642	29,546	26,339	21,641	18,365

1000 MCM	2 Run(s)	480 Volts	C-Value = 25,278	Magnetic Duct	Copper

Calculations:

1. $f = \dfrac{\text{Length} \times 2 \times \text{Isca}}{\text{Runs} \times \text{C-Value} \times \text{E L-L}}$

2. $M = \dfrac{1}{1 + f}$

3. Isc = Isca x M

4. Add Motor Contribution = Motor FLA x 4

* All results are given in symmetrical amperes

Short-Circuit Tables and Data: 240 Volts Single-Phase L-L

Single-Phase L-L Bolted Fault-Current* Table

One-Way Distance in Feet

Available Isc in Thousands (Isca)	5	10	15	20	25	30	40	50	60	70	80	90	100	125	150	175	200	225	250	300	350	400	500	600
1	903	824	757	700	651	609	539	483	438	400	368	342	318	272	237	211	189	172	157	135	118	105	85	72
3	2,270	1,826	1,527	1,313	1,151	1,025	840	712	618	546	488	442	404	332	282	245	217	194	176	148	128	112	91	76
5	3,256	2,414	1,918	1,591	1,360	1,187	946	787	673	588	523	470	427	347	293	253	223	199	180	151	130	114	92	77
10	4,828	3,182	2,373	1,892	1,573	1,346	1,045	854	722	625	551	493	446	360	302	260	228	203	183	153	132	115	92	77
15	5,754	3,560	2,577	2,020	1,661	1,410	1,083	879	740	638	562	501	453	364	305	262	230	205	184	154	132	116	93	77
20	6,365	3,785	2,693	2,090	1,708	1,444	1,103	892	749	645	567	506	456	367	306	263	231	205	185	154	132	116	93	77
25	6,798	3,934	2,768	2,135	1,737	1,465	1,115	900	755	650	570	508	458	368	307	264	231	206	185	155	133	116	93	78
30	7,120	4,039	2,820	2,166	1,758	1,479	1,123	905	758	652	572	510	460	369	308	264	232	206	186	155	133	116	93	78
35	7,370	4,119	2,858	2,188	1,773	1,490	1,129	909	761	654	574	511	461	369	308	265	232	206	186	155	133	116	93	78
40	7,569	4,180	2,887	2,205	1,784	1,498	1,134	912	763	656	575	512	461	370	309	265	232	206	186	155	133	116	93	78
45	7,732	4,229	2,911	2,219	1,793	1,504	1,138	915	765	657	576	513	462	370	309	265	232	207	186	155	133	116	93	78
50	7,867	4,269	2,930	2,230	1,800	1,509	1,140	916	766	658	577	513	462	371	309	265	232	207	186	155	133	116	93	78
55	7,981	4,303	2,945	2,239	1,806	1,513	1,143	918	767	659	577	514	463	371	309	265	232	207	186	155	133	116	93	78
60	8,079	4,331	2,959	2,247	1,811	1,517	1,145	919	768	660	578	514	463	371	310	266	232	207	186	155	133	116	93	78
65	8,163	4,355	2,970	2,253	1,815	1,520	1,146	920	769	660	578	515	463	371	310	266	233	207	186	155	133	116	93	78
70	8,237	4,376	2,980	2,259	1,819	1,522	1,148	921	769	661	579	515	464	371	310	266	233	207	186	155	133	117	93	78
75	8,303	4,394	2,988	2,264	1,822	1,524	1,149	922	770	661	579	515	464	372	310	266	233	207	186	155	133	117	93	78
80	8,360	4,411	2,995	2,268	1,825	1,526	1,150	923	771	661	579	515	464	372	310	266	233	207	186	155	133	117	93	78
85	8,412	4,425	3,002	2,272	1,827	1,528	1,151	923	771	662	580	516	464	372	310	266	233	207	186	155	133	117	93	78
90	8,459	4,438	3,008	2,275	1,829	1,530	1,152	924	771	662	580	516	464	372	310	266	233	207	186	155	133	117	93	78
95	8,501	4,449	3,013	2,278	1,831	1,531	1,153	925	772	662	580	516	465	372	310	266	233	207	186	155	133	117	93	78
100	8,539	4,460	3,018	2,281	1,833	1,532	1,154	925	772	662	580	516	465	372	310	266	233	207	186	155	133	117	93	78
105	8,574	4,469	3,022	2,283	1,835	1,533	1,154	925	772	663	580	516	465	372	310	266	233	207	186	155	133	117	93	78
110	8,606	4,478	3,026	2,286	1,836	1,534	1,155	926	773	663	580	516	465	372	310	266	233	207	186	155	133	117	93	78
115	8,635	4,486	3,030	2,288	1,837	1,535	1,155	926	773	663	581	516	465	372	310	266	233	207	186	155	133	117	93	78
120	8,662	4,493	3,033	2,289	1,839	1,536	1,156	926	773	663	581	516	465	372	310	266	233	207	186	155	133	117	93	78
125	8,687	4,500	3,036	2,291	1,840	1,537	1,156	927	773	663	581	517	465	372	310	266	233	207	186	155	133	117	93	78
130	8,710	4,506	3,039	2,293	1,841	1,538	1,157	927	773	663	581	517	465	372	310	266	233	207	186	155	133	117	93	78
135	8,732	4,512	3,042	2,294	1,842	1,538	1,157	927	774	664	581	517	465	372	310	266	233	207	186	155	133	117	93	78
140	8,752	4,517	3,044	2,296	1,843	1,539	1,157	927	774	664	581	517	465	372	311	266	233	207	186	155	133	117	93	78
145	8,771	4,522	3,047	2,297	1,843	1,539	1,158	928	774	664	581	517	465	372	311	266	233	207	186	155	133	117	93	78
150	8,789	4,527	3,049	2,298	1,844	1,540	1,158	928	774	664	581	517	465	373	311	266	233	207	186	155	133	117	93	78
155	8,806	4,532	3,051	2,299	1,845	1,541	1,158	928	774	664	581	517	465	373	311	266	233	207	186	155	133	117	93	78
160	8,821	4,536	3,053	2,300	1,846	1,541	1,159	928	774	664	581	517	465	373	311	266	233	207	187	155	133	117	93	78
165	8,836	4,540	3,054	2,301	1,846	1,541	1,159	928	774	664	581	517	465	373	311	266	233	207	187	155	133	117	93	78
170	8,850	4,543	3,056	2,302	1,847	1,542	1,159	929	774	664	582	517	466	373	311	266	233	207	187	155	133	117	93	78
175	8,863	4,547	3,058	2,303	1,847	1,542	1,159	929	775	664	582	517	466	373	311	266	233	207	187	155	133	117	93	78
180	8,876	4,550	3,059	2,304	1,848	1,543	1,159	929	775	664	582	517	466	373	311	266	233	207	187	155	133	117	93	78
185	8,887	4,553	3,061	2,305	1,849	1,543	1,160	929	775	664	582	517	466	373	311	266	233	207	187	155	133	117	93	78
190	8,899	4,556	3,062	2,306	1,849	1,543	1,160	929	775	665	582	517	466	373	311	266	233	207	187	155	133	117	93	78
195	8,909	4,559	3,063	2,306	1,849	1,544	1,160	929	775	665	582	517	466	373	311	266	233	207	187	155	133	117	93	78
200	8,920	4,562	3,064	2,307	1,850	1,544	1,160	929	775	665	582	517	466	373	311	266	233	207	187	155	133	117	93	78

# 14 AWG	1 Run(s)	240 Volts	C-Value = 389	Magnetic Duct	Copper

Calculations:

1. $f = \frac{\text{Length} \times 2 \times \text{Isca}}{\text{Runs} \times \text{C-Value} \times E_{L-L}}$

2. $M = \frac{1}{1 + f}$

3. $Isc = Isca \times M$

4. Add Motor Contribution = Motor FLA x 4

* All results are given in symmetrical amperes

Single-Phase L-L Bolted Fault-Current* Table

One-Way Distance in Feet

Available Isc in Thousands (Isca)

Isca	5	10	15	20	25	30	40	50	60	70	80	90	100	125	150	175	200	225	250	300	350	400	500	600
1	937	881	832	787	748	712	649	597	552	514	481	451	425	372	330	297	270	248	228	194	175	156	129	110
3	2,495	2,135	1,866	1,657	1,490	1,354	1,145	991	874	782	707	646	594	495	424	371	330	297	270	223	198	174	141	119
5	3,738	2,985	2,484	2,127	1,860	1,652	1,351	1,142	990	873	781	706	645	530	449	390	345	309	280	235	203	178	144	120
10	5,969	4,254	3,305	2,702	2,285	1,979	1,562	1,290	1,098	957	847	760	689	559	470	406	357	319	288	24[illegible]	207	182	146	122
15	7,452	4,957	3,714	2,969	2,473	2,119	1,648	1,348	1,140	988	872	780	706	570	478	411	361	322	290	243	209	183	147	122
20	8,508	5,404	3,959	3,124	2,580	2,197	1,694	1,379	1,162	1,005	885	790	714	575	482	414	363	324	292	244	209	183	147	123
25	9,300	5,712	4,122	3,225	2,648	2,246	1,723	1,398	1,176	1,015	892	796	719	579	484	416	365	325	293	244	210	184	147	123
30	9,914	5,938	4,239	3,295	2,696	2,280	1,743	1,411	1,185	1,022	898	801	723	581	486	417	366	325	293	245	210	184	147	123
35	10,406	6,111	4,326	3,348	2,731	2,305	1,758	1,421	1,192	1,027	902	804	725	582	487	418	366	326	294	245	210	184	147	123
40	10,807	6,248	4,394	3,388	2,757	2,325	1,769	1,428	1,197	1,030	905	806	727	584	488	419	367	326	294	245	210	184	148	123
45	11,142	6,358	4,448	3,421	2,779	2,340	1,778	1,434	1,201	1,033	907	808	728	585	488	419	367	327	294	245	211	184	148	123
50	11,425	6,449	4,493	3,447	2,796	2,352	1,785	1,438	1,204	1,036	909	809	730	585	489	420	367	327	294	246	211	184	148	123
55	11,667	6,526	4,529	3,469	2,810	2,362	1,791	1,442	1,207	1,038	910	811	731	586	489	420	368	327	295	246	211	184	148	123
60	11,877	6,591	4,561	3,487	2,822	2,370	1,796	1,445	1,209	1,039	911	812	731	587	490	420	368	327	295	246	211	185	148	123
65	12,060	6,647	4,588	3,503	2,833	2,378	1,800	1,448	1,211	1,041	913	812	732	587	490	420	368	327	295	246	211	185	148	123
70	12,222	6,696	4,611	3,516	2,841	2,384	1,803	1,450	1,213	1,042	913	813	733	587	490	421	368	328	295	246	211	185	148	123
75	12,366	6,739	4,631	3,528	2,849	2,389	1,806	1,452	1,214	1,043	914	814	733	588	490	421	368	328	295	246	211	185	148	123
80	12,495	6,777	4,649	3,538	2,856	2,394	1,809	1,454	1,215	1,044	915	814	734	588	491	421	368	328	295	246	211	185	148	123
85	12,611	6,811	4,665	3,547	2,862	2,398	1,812	1,455	1,216	1,045	916	815	734	588	491	421	369	328	295	246	211	185	148	123
90	12,716	6,841	4,679	3,556	2,867	2,402	1,814	1,457	1,217	1,045	916	815	734	588	491	421	369	328	295	246	211	185	148	123
95	12,811	6,869	4,692	3,563	2,872	2,406	1,816	1,458	1,218	1,046	917	816	735	589	491	421	369	328	295	246	211	185	148	123
100	12,898	6,894	4,704	3,570	2,876	2,409	1,817	1,459	1,219	1,047	917	816	735	589	491	421	369	328	295	246	211	185	148	123
105	12,978	6,916	4,714	3,576	2,880	2,411	1,819	1,460	1,220	1,047	917	816	735	589	491	421	369	328	295	246	211	185	148	123
110	13,051	6,937	4,724	3,581	2,884	2,414	1,820	1,461	1,220	1,048	918	817	735	589	491	421	369	328	295	246	211	185	148	123
115	13,119	6,956	4,733	3,587	2,887	2,416	1,822	1,462	1,221	1,048	918	817	736	589	491	422	369	328	295	246	211	185	148	123
120	13,181	6,974	4,741	3,591	2,890	2,418	1,823	1,463	1,221	1,048	918	817	736	589	492	422	369	328	295	246	211	185	148	123
125	13,240	6,990	4,748	3,596	2,893	2,420	1,824	1,463	1,222	1,049	919	817	736	590	492	422	369	328	295	246	211	185	148	123
130	13,294	7,005	4,755	3,599	2,896	2,422	1,825	1,464	1,222	1,049	919	817	736	590	492	422	369	328	295	246	211	185	148	123
135	13,344	7,019	4,762	3,603	2,898	2,424	1,826	1,465	1,223	1,049	919	818	736	590	492	422	369	328	296	246	211	185	148	123
140	13,392	7,032	4,768	3,607	2,900	2,425	1,827	1,465	1,223	1,050	919	818	737	590	492	422	369	328	296	246	211	185	148	123
145	13,436	7,044	4,774	3,610	2,902	2,427	1,828	1,466	1,224	1,050	920	818	737	590	492	422	369	328	296	246	211	185	148	123
150	13,478	7,056	4,779	3,613	2,904	2,428	1,828	1,466	1,224	1,050	920	818	737	590	492	422	369	328	296	246	211	185	148	123
155	13,517	7,066	4,784	3,616	2,906	2,429	1,829	1,467	1,224	1,051	920	818	737	590	492	422	369	328	296	246	211	185	148	123
160	13,554	7,077	4,788	3,618	2,908	2,431	1,830	1,467	1,225	1,051	920	818	737	590	492	422	369	328	296	246	211	185	148	123
165	13,588	7,086	4,793	3,621	2,909	2,432	1,830	1,468	1,225	1,051	920	819	737	590	492	422	369	328	296	246	211	185	148	123
170	13,621	7,095	4,797	3,623	2,911	2,433	1,831	1,468	1,225	1,051	920	819	737	590	492	422	369	328	296	246	211	185	148	123
175	13,653	7,103	4,801	3,625	2,912	2,434	1,832	1,468	1,225	1,051	921	819	737	590	492	422	369	328	296	246	211	185	148	123
180	13,682	7,111	4,804	3,627	2,914	2,435	1,832	1,469	1,226	1,052	921	819	737	590	492	422	369	328	296	246	211	185	148	123
185	13,711	7,119	4,808	3,629	2,915	2,436	1,833	1,469	1,226	1,052	921	819	737	590	492	422	369	328	296	246	211	185	148	123
190	13,737	7,126	4,811	3,631	2,916	2,436	1,833	1,469	1,226	1,052	921	819	738	590	492	422	369	328	296	246	211	185	148	123
195	13,763	7,133	4,814	3,633	2,917	2,437	1,834	1,470	1,226	1,052	921	819	738	591	492	422	369	329	296	246	211	185	148	123
200	13,787	7,140	4,817	3,635	2,918	2,438	1,834	1,470	1,226	1,052	921	819	738	591	492	422	370	329	296	246	211	185	148	123

12 AWG | 1 Run(s) | 240 Volts | C-Value = 617 | Magnetic Duct | Copper

Calculations:

1. $f = \dfrac{\text{Length} \times 2 \times \text{Isca}}{\text{Runs} \times \text{C-Value} \times E_{L\text{-}L}}$

2. $M = \dfrac{1}{1 + f}$

3. Isc = Isca x M

4. Add Motor Contribution = Motor FLA x 4

* All results are given in symmetrical amperes

Single-Phase L-L Bolted Fault-Current* Table

One-Way Distance in Feet

Available Isc in Thousands (Isca)	5	10	15	20	25	30	40	50	60	70	80	90	100	125	150	175	200	225	250	300	350	400	500	600
1	959	922	887	855	825	797	746	702	662	627	595	567	541	485	440	402	371	343	320	282	252	227	191	164
3	2,661	2,391	2,170	1,987	1,833	1,700	1,486	1,319	1,186	1,078	987	911	845	717	622	549	492	446	407	347	302	268	218	184
5	4,124	3,509	3,054	2,703	2,425	2,199	1,853	1,601	1,409	1,258	1,137	1,037	953	792	678	593	527	474	430	364	315	278	225	189
10	7,019	5,407	4,397	3,705	3,201	2,818	2,274	1,906	1,640	1,440	1,283	1,157	1,053	861	728	630	556	497	450	378	325	286	230	192
15	9,163	6,596	5,152	4,227	3,584	3,110	2,460	2,035	1,735	1,512	1,340	1,203	1,092	886	746	644	566	506	457	382	329	289	232	194
20	10,814	7,410	5,636	4,548	3,811	3,280	2,565	2,106	1,787	1,551	1,371	1,228	1,112	899	755	651	572	510	460	385	331	290	233	194
25	12,125	8,003	5,973	4,764	3,962	3,392	2,633	2,152	1,819	1,576	1,390	1,243	1,124	908	761	655	575	512	462	386	332	291	233	195
30	13,191	8,454	6,221	4,921	4,070	3,470	2,680	2,183	1,842	1,592	1,403	1,253	1,133	913	765	658	577	514	464	387	333	291	234	195
35	14,076	8,809	6,411	5,039	4,150	3,528	2,715	2,206	1,858	1,605	1,412	1,261	1,139	917	768	660	579	515	465	388	333	292	234	195
40	14,821	9,095	6,561	5,131	4,213	3,573	2,741	2,224	1,870	1,614	1,419	1,267	1,144	920	770	662	580	516	465	389	334	292	234	195
45	15,457	9,331	6,683	5,205	4,263	3,609	2,762	2,237	1,880	1,621	1,425	1,271	1,147	922	771	663	581	517	466	389	334	292	234	195
50	16,007	9,529	6,783	5,266	4,304	3,638	2,779	2,249	1,888	1,627	1,429	1,275	1,150	924	773	664	582	518	466	389	334	293	234	195
55	16,487	9,697	6,868	5,317	4,337	3,663	2,794	2,258	1,894	1,632	1,433	1,278	1,153	926	774	665	582	518	467	390	334	293	234	196
60	16,909	9,841	6,940	5,360	4,366	3,683	2,805	2,266	1,900	1,636	1,436	1,280	1,155	927	775	665	583	519	467	390	334	293	235	196
65	17,284	9,967	7,003	5,397	4,391	3,701	2,816	2,272	1,905	1,639	1,439	1,282	1,156	928	775	666	583	519	467	390	335	293	235	196
70	17,618	10,077	7,057	5,429	4,412	3,716	2,824	2,278	1,909	1,642	1,441	1,284	1,158	929	776	666	584	519	468	390	335	293	235	196
75	17,919	10,175	7,105	5,458	4,431	3,729	2,832	2,283	1,912	1,645	1,443	1,286	1,159	930	777	667	584	520	468	390	335	293	235	196
80	18,191	10,262	7,147	5,483	4,447	3,741	2,839	2,287	1,915	1,647	1,445	1,287	1,160	931	777	667	584	520	468	390	335	293	235	196
85	18,437	10,340	7,185	5,505	4,462	3,751	2,845	2,291	1,918	1,649	1,446	1,288	1,161	931	778	667	585	520	468	391	335	293	235	196
90	18,662	10,410	7,219	5,525	4,475	3,760	2,850	2,294	1,920	1,651	1,448	1,289	1,162	932	778	668	585	520	468	391	335	293	235	196
95	18,868	10,474	7,249	5,543	4,486	3,768	2,855	2,297	1,922	1,652	1,449	1,290	1,163	933	778	668	585	520	469	391	335	293	235	196
100	19,057	10,532	7,277	5,559	4,497	3,776	2,859	2,300	1,924	1,654	1,450	1,291	1,164	933	779	668	585	520	469	391	335	293	235	196
105	19,232	10,585	7,302	5,574	4,507	3,783	2,863	2,303	1,926	1,655	1,451	1,292	1,164	933	779	668	585	521	469	391	335	293	235	196
110	19,393	10,634	7,325	5,587	4,516	3,789	2,866	2,305	1,928	1,656	1,452	1,293	1,165	934	779	669	585	521	469	391	335	294	235	196
115	19,543	10,679	7,347	5,599	4,524	3,795	2,870	2,307	1,929	1,657	1,453	1,293	1,165	934	779	669	586	521	469	391	335	294	235	196
120	19,682	10,720	7,366	5,611	4,531	3,800	2,873	2,309	1,930	1,658	1,454	1,294	1,166	934	780	669	586	521	469	391	335	294	235	196
125	19,812	10,759	7,384	5,621	4,538	3,805	2,875	2,311	1,932	1,659	1,454	1,294	1,166	935	780	669	586	521	469	391	335	294	235	196
130	19,934	10,795	7,401	5,631	4,544	3,809	2,878	2,313	1,933	1,660	1,455	1,295	1,167	935	780	669	586	521	469	391	335	294	235	196
135	20,048	10,828	7,417	5,640	4,550	3,813	2,880	2,314	1,934	1,661	1,456	1,295	1,167	935	780	669	586	521	469	391	336	294	235	196
140	20,155	10,859	7,431	5,649	4,556	3,817	2,882	2,315	1,935	1,662	1,456	1,296	1,167	935	780	669	586	521	469	391	336	294	235	196
145	20,255	10,888	7,445	5,656	4,561	3,821	2,884	2,317	1,936	1,662	1,457	1,296	1,168	936	781	670	586	521	469	391	336	294	235	196
150	20,350	10,915	7,458	5,664	4,565	3,824	2,886	2,318	1,937	1,663	1,457	1,297	1,168	936	781	670	586	521	469	391	336	294	235	196
155	20,439	10,941	7,470	5,671	4,570	3,827	2,888	2,319	1,937	1,664	1,458	1,297	1,168	936	781	670	586	521	469	391	336	294	235	196
160	20,524	10,965	7,481	5,677	4,574	3,830	2,890	2,320	1,938	1,664	1,458	1,297	1,169	936	781	670	586	521	469	391	336	294	235	196
165	20,604	10,988	7,492	5,683	4,578	3,833	2,891	2,321	1,939	1,665	1,458	1,298	1,169	936	781	670	587	522	470	391	336	294	235	196
170	20,680	11,010	7,502	5,689	4,582	3,835	2,893	2,322	1,940	1,665	1,459	1,298	1,169	937	781	670	587	522	470	391	336	294	235	196
175	20,752	11,030	7,511	5,694	4,585	3,838	2,894	2,323	1,940	1,666	1,459	1,298	1,169	937	781	670	587	522	470	392	336	294	235	196
180	20,821	11,049	7,520	5,700	4,589	3,840	2,896	2,324	1,941	1,666	1,460	1,299	1,170	937	781	670	587	522	470	392	336	294	235	196
185	20,886	11,068	7,529	5,705	4,592	3,842	2,897	2,325	1,941	1,667	1,460	1,299	1,170	937	781	670	587	522	470	392	336	294	235	196
190	20,948	11,085	7,537	5,709	4,595	3,845	2,898	2,326	1,942	1,667	1,460	1,299	1,170	937	782	670	587	522	470	392	336	294	235	196
195	21,008	11,102	7,544	5,714	4,598	3,847	2,899	2,326	1,942	1,667	1,460	1,299	1,170	937	782	670	587	522	470	392	336	294	235	196
200	21,064	11,118	7,552	5,718	4,600	3,848	2,900	2,327	1,943	1,668	1,461	1,300	1,170	937	782	670	587	522	470	392	336	294	235	196

10 AWG | 1 Run(s) | 240 Volts | C-Value = 981 | Magnetic Duct | Copper

Calculations:

1. $f = \frac{\text{Length} \times 2 \times \text{Isca}}{\text{Runs} \times \text{C-Value} \times \text{E L-L}}$

2. $M = \frac{1}{1 + f}$

3. Isc = Isca x M

4. Add Motor Contribution = Motor FLA x 4

* All results are given in symmetrical amperes

Single-Phase L-L Bolted Fault-Current* Table

One-Way Distance in Feet

Available Isc in Thousands (Isca)	5	10	15	20	25	30	40	50	60	70	80	90	100	125	150	175	200	225	250	300	350	400	500	600
3	2,777	2,585	2,418	2,271	2,141	2,025	1,327	1,664	1,528	1,412	1,313	1,227	1,151	998	880	787	712	650	598	513	453	404	332	282
5	4,410	3,944	3,568	3,257	2,996	2,773	2,415	2,139	1,919	1,740	1,592	1,467	1,360	1,151	997	880	787	712	650	554	482	427	348	293
7	5,896	5,092	4,481	4,002	3,615	3,296	2,802	2,436	2,155	1,932	1,751	1,601	1,475	1,232	1,057	926	824	742	675	572	496	438	355	298
10	7,889	6,514	5,547	4,830	4,277	3,838	3,184	2,720	2,375	2,107	1,893	1,719	1,574	1,300	1,108	965	854	767	695	585	507	446	360	302
15	10,703	8,320	6,805	5,757	4,988	4,401	3,562	2,992	2,579	2,266	2,021	1,824	1,661	1,359	1,150	997	879	787	712	598	515	453	365	305
20	13,027	9,660	7,676	6,368	5,441	4,749	3,787	3,149	2,694	2,355	2,091	1,881	1,709	1,391	1,173	1,014	893	797	720	604	520	456	367	307
25	14,979	10,693	8,314	6,801	5,754	4,986	3,936	3,251	2,769	2,412	2,136	1,917	1,738	1,410	1,186	1,024	901	804	726	608	523	459	368	308
30	16,641	11,513	8,802	7,124	5,983	5,157	4,042	3,323	2,821	2,451	2,167	1,942	1,759	1,424	1,196	1,031	906	808	729	610	524	460	369	308
35	18,073	12,181	9,187	7,374	6,159	5,287	4,121	3,376	2,860	2,480	2,189	1,960	1,774	1,434	1,203	1,036	910	811	732	[illegible]	526	461	370	309
40	19,320	12,735	9,498	7,573	6,297	5,389	4,183	3,418	2,889	2,502	2,207	1,974	1,785	1,441	1,208	1,040	913	814	734	[illegible]	527	462	370	309
45	20,415	13,202	9,756	7,736	6,409	5,471	4,232	3,450	2,912	2,520	2,220	1,984	1,794	1,447	1,212	1,043	915	815	735	[illegible]	528	462	371	309
50	21,385	13,601	9,972	7,871	6,502	5,538	4,272	3,477	2,931	2,534	2,231	1,993	1,801	1,451	1,215	1,045	917	817	736	615	528	463	371	309
55	22,251	13,946	10,156	7,986	6,580	5,594	4,305	3,499	2,947	2,546	2,240	2,000	1,807	1,455	1,218	1,047	919	818	737	616	529	463	371	310
60	23,027	14,247	10,315	8,083	6,646	5,642	4,334	3,518	2,960	2,555	2,248	2,007	1,812	1,458	1,220	1,049	920	819	738	616	529	463	371	310
65	23,727	14,512	10,453	8,168	6,703	5,683	4,358	3,534	2,972	2,564	2,254	2,012	1,816	1,461	1,222	1,050	921	820	739	617	529	464	372	310
70	24,363	14,748	10,574	8,242	6,753	5,719	4,379	3,547	2,981	2,571	2,260	2,016	1,820	1,463	1,224	1,052	922	821	739	617	530	464	372	310
75	24,941	14,958	10,682	8,307	6,796	5,750	4,397	3,559	2,990	2,577	2,265	2,020	1,823	1,466	1,225	1,053	923	821	740	618	530	464	372	310
80	25,471	15,147	10,778	8,365	6,835	5,778	4,413	3,570	2,997	2,583	2,269	2,023	1,826	1,467	1,227	1,054	923	822	740	618	530	464	372	310
85	25,957	15,317	10,864	8,417	6,870	5,803	4,428	3,579	3,004	2,588	2,273	2,027	1,828	1,469	1,228	1,054	924	822	741	618	530	465	372	310
90	26,405	15,472	10,942	8,463	6,901	5,825	4,441	3,588	3,010	2,592	2,276	2,029	1,830	1,470	1,229	1,055	925	823	741	619	531	465	372	310
95	26,819	15,613	11,012	8,506	6,929	5,845	4,452	3,595	3,015	2,596	2,279	2,032	1,832	1,472	1,229	1,056	925	823	742	619	531	465	372	310
100	27,203	15,743	11,076	8,544	6,954	5,863	4,463	3,602	3,020	2,600	2,282	2,034	1,834	1,473	1,230	1,056	926	824	742	619	531	465	372	310
105	27,560	15,862	11,135	8,579	6,977	5,879	4,472	3,608	3,024	2,603	2,285	2,036	1,836	1,474	1,231	1,057	926	824	742	619	531	465	372	310
110	27,893	15,971	11,189	8,611	6,998	5,894	4,481	3,614	3,028	2,606	2,287	2,038	1,837	1,475	1,232	1,057	926	824	742	619	531	465	372	311
115	28,204	16,073	11,239	8,640	7,018	5,908	4,489	3,619	3,032	2,609	2,289	2,039	1,839	1,476	1,232	1,058	927	824	743	619	531	465	372	311
120	28,495	16,167	11,285	8,667	7,035	5,921	4,496	3,624	3,035	2,611	2,291	2,041	1,840	1,476	1,233	1,058	927	825	743	620	531	465	373	311
125	28,768	16,254	11,327	8,692	7,052	5,932	4,503	3,628	3,038	2,613	2,293	2,042	1,841	1,477	1,233	1,059	927	825	743	620	532	465	373	311
130	29,025	16,336	11,367	8,716	7,067	5,943	4,509	3,632	3,041	2,615	2,294	2,043	1,842	1,478	1,234	1,059	928	825	743	620	532	465	373	311
135	29,267	16,413	11,404	8,737	7,082	5,953	4,515	3,636	3,044	2,617	2,296	2,045	1,843	1,478	1,234	1,059	928	825	743	620	532	465	373	311
140	29,495	16,484	11,438	8,758	7,095	5,963	4,520	3,640	3,046	2,619	2,297	2,046	1,844	1,479	1,235	1,060	928	826	743	620	532	466	373	311
145	29,711	16,551	11,471	8,777	7,107	5,972	4,525	3,643	3,049	2,621	2,298	2,047	1,845	1,479	1,235	1,060	928	826	744	620	532	466	373	311
150	29,915	16,614	11,501	8,794	7,119	5,980	4,530	3,646	3,051	2,622	2,300	2,048	1,845	1,480	1,235	1,060	928	826	744	620	532	466	373	311
155	30,109	16,674	11,529	8,811	7,130	5,987	4,534	3,649	3,053	2,624	2,301	2,049	1,846	1,480	1,236	1,060	929	826	744	620	532	466	373	311
160	30,293	16,730	11,556	8,827	7,140	5,995	4,539	3,652	3,055	2,625	2,302	2,049	1,847	1,481	1,236	1,061	929	826	744	620	532	466	373	311
165	30,468	16,783	11,582	8,841	7,150	6,001	4,542	3,654	3,056	2,627	2,303	2,050	1,847	1,481	1,236	1,061	929	826	744	620	532	466	373	311
170	30,634	16,834	11,606	8,855	7,159	6,008	4,546	3,656	3,058	2,628	2,304	2,051	1,848	1,482	1,237	1,061	929	826	744	621	532	466	373	311
175	30,793	16,882	11,628	8,869	7,168	6,014	4,550	3,659	3,060	2,629	2,305	2,052	1,849	1,482	1,237	1,061	929	826	744	621	532	466	373	311
180	30,944	16,927	11,650	8,881	7,176	6,020	4,553	3,661	3,061	2,630	2,306	2,052	1,849	1,482	1,237	1,061	929	827	744	621	532	466	373	311
185	31,088	16,970	11,670	8,893	7,183	6,025	4,556	3,663	3,062	2,631	2,306	2,053	1,850	1,483	1,237	1,062	930	827	744	621	532	466	373	311
190	31,227	17,011	11,690	8,904	7,191	6,030	4,559	3,665	3,064	2,632	2,307	2,054	1,850	1,483	1,237	1,062	930	827	744	621	532	466	373	311
195	31,359	17,050	11,708	8,915	7,198	6,035	4,562	3,667	3,065	2,633	2,308	2,054	1,851	1,483	1,238	1,062	930	827	745	621	532	466	373	311
200	31,485	17,088	11,726	8,925	7,204	6,040	4,564	3,668	3,066	2,634	2,309	2,055	1,851	1,484	1,238	1,062	930	827	745	621	532	466	373	311

# 8 AWG	1 Run(s)	240 Volts	C-Value = 1,557	Magnetic Duct	Copper

Calculations:

1. $f = \dfrac{\text{Length} \times 2 \times \text{Isca}}{\text{Runs} \times \text{C-Value} \times \text{E L-L}}$

2. $M = \dfrac{1}{1 + f}$

3. $\text{Isc} = \text{Isca} \times M$

4. Add Motor Contribution = Motor FLA x 4

* All results are given in symmetrical amperes

Single-Phase L-L Bolted Fault-Current* Table

One-Way Distance in Feet

Available Isc in Thousands (Isca)	5	10	15	20	25	30	40	50	60	70	80	90	100	125	150	175	200	225	250	300	350	400	500	600
3	2,853	2,720	2,598	2,487	2,385	2,291	2,124	1,980	1,854	1,743	1,644	1,556	1,477	1,311	1,178	1,070	980	904	839	733	651	586	487	418
5	4,604	4,267	3,975	3,721	3,498	3,299	2,963	2,689	2,462	2,270	2,106	1,964	1,839	1,588	1,398	1,248	1,127	1,028	944	812	713	635	521	442
7	6,248	5,643	5,144	4,726	4,371	4,066	3,567	3,178	2,865	2,608	2,394	2,212	2,055	1,747	1,519	1,344	1,205	1,092	998	852	743	659	537	454
10	8,534	7,442	6,599	5,927	5,379	4,924	4,211	3,679	3,266	2,936	2,667	2,443	2,254	1,888	1,625	1,426	1,270	1,145	1,043	884	768	678	550	463
15	11,926	9,898	8,459	7,386	6,554	5,891	4,899	4,193	3,665	3,255	2,928	2,660	2,437	2,015	1,718	1,497	1,326	1,191	1,080	911	788	694	560	470
20	14,885	11,853	9,848	8,423	7,358	6,532	5,335	4,508	3,903	3,442	3,078	2,783	2,540	2,085	1,768	1,535	1,356	1,215	1,100	925	798	702	566	474
25	17,488	13,447	10,923	9,197	7,942	6,988	5,635	4,721	4,062	3,564	3,175	2,863	2,607	2,130	1,800	1,559	1,375	1,230	1,112	934	805	707	569	476
30	19,796	14,772	11,781	9,798	8,386	7,330	5,855	4,874	4,175	3,651	3,244	2,919	2,653	2,160	1,822	1,576	1,388	1,240	1,121	940	809	710	571	477
35	21,856	15,889	12,482	10,277	8,735	7,595	6,023	4,990	4,260	3,716	3,295	2,960	2,687	2,183	1,838	1,587	1,397	1,247	1,127	944	812	713	572	478
40	23,707	16,845	13,064	10,669	9,016	7,807	6,155	5,081	4,326	3,766	3,334	2,992	2,713	2,200	1,850	1,596	1,404	1,253	1,131	947	814	715	574	479
45	25,378	17,672	13,556	10,995	9,248	7,980	6,263	5,153	4,378	3,806	3,365	3,017	2,733	2,213	1,860	1,604	1,409	1,257	1,135	950	816	716	575	480
50	26,895	18,394	13,977	11,270	9,442	8,124	6,351	5,213	4,421	3,838	3,391	3,037	2,750	2,224	1,868	1,609	1,414	1,261	1,138	952	818	717	575	480
55	28,277	19,031	14,341	11,506	9,607	8,246	6,425	5,263	4,457	3,865	3,412	3,054	2,764	2,233	1,874	1,614	1,418	1,264	1,140	953	819	718	576	481
60	29,543	19,596	14,660	11,710	9,749	8,350	6,488	5,305	4,487	3,888	3,430	3,068	2,775	2,241	1,879	1,618	1,421	1,266	1,142	955	820	719	576	481
65	30,706	20,101	14,941	11,889	9,872	8,440	6,543	5,342	4,513	3,907	3,445	3,080	2,785	2,248	1,884	1,621	1,423	1,268	1,144	956	821	719	577	481
70	31,778	20,555	15,190	12,046	9,980	8,519	6,590	5,373	4,536	3,924	3,458	3,091	2,794	2,253	1,888	1,624	1,425	1,270	1,145	957	822	720	577	482
75	32,770	20,965	15,413	12,186	10,076	8,589	6,632	5,401	4,555	3,939	3,469	3,100	2,801	2,258	1,891	1,627	1,427	1,271	1,146	958	822	721	578	482
80	33,690	21,338	15,614	12,311	10,162	8,651	6,669	5,425	4,573	3,952	3,479	3,108	2,808	2,262	1,894	1,629	1,429	1,273	1,147	958	823	721	578	482
85	34,546	21,678	15,795	12,423	10,238	8,706	6,701	5,447	4,588	3,963	3,488	3,115	2,814	2,266	1,897	1,631	1,431	1,274	1,148	959	823	721	578	482
90	35,344	21,990	15,960	12,525	10,307	8,756	6,731	5,466	4,602	3,974	3,496	3,121	2,819	2,269	1,899	1,633	1,432	1,275	1,149	960	824	722	578	482
95	36,090	22,276	16,110	12,618	10,369	8,801	6,758	5,484	4,614	3,983	3,503	3,127	2,824	2,272	1,901	1,634	1 433	1,276	1,150	960	824	722	578	483
100	36,789	22,541	16,248	12,702	10,426	8,842	6,782	5,500	4,626	3,991	3,510	3,132	2,828	2,275	1,903	1,636	1 434	1,277	1,151	961	825	722	579	483
105	37,445	22,785	16,375	12,779	10,478	8,880	6,804	5,514	4,636	3,999	3,516	3,137	2,832	2,278	1,905	1,637	1 435	1,278	1,151	961	825	722	579	483
110	38,062	23,012	16,491	12,850	10,526	8,914	6,824	5,528	4,645	4,006	3,521	3,141	2,835	2,280	1,906	1,638	1,436	1,278	1,152	962	825	723	579	483
115	38,643	23,223	16,600	12,916	10,570	8,945	6,842	5,540	4,654	4,012	3,526	3,145	2,838	2,282	1,908	1,639	1,437	1,279	1,152	962	825	723	579	483
120	39,192	23,421	16,700	12,977	10,611	8,975	6,859	5,551	4,662	4,018	3,530	3,148	2,841	2,284	1,909	1,640	1,438	1,280	1,153	962	826	723	579	483
125	39,711	23,605	16,794	13,033	10,648	9,001	6,875	5,561	4,669	4,023	3,535	3,152	2,844	2,285	1,910	1,641	1,438	1,280	1,153	963	826	723	579	483
130	40,202	23,777	16,881	13,085	10,683	9,026	6,889	5,571	4,676	4,028	3,538	3,155	2,846	2,287	1,911	1,642	1,439	1,281	1,154	963	826	723	579	483
135	40,668	23,940	16,962	13,134	10,716	9,050	6,903	5,579	4,682	4,033	3,542	3,158	2,849	2,289	1,913	1,643	1,439	1,281	1,154	963	826	724	580	483
140	41,110	24,092	17,039	13,180	10,747	9,071	6,916	5,588	4,688	4,037	3,545	3,160	2,851	2,290	1,913	1,643	1,440	1,281	1,154	963	827	724	580	483
145	41,531	24,236	17,111	13,223	10,775	9,092	6,927	5,595	4,693	4,041	3,548	3,163	2,853	2,291	1,914	1,644	1,441	1,282	1,155	964	827	724	580	483
150	41,931	24,372	17,178	13,263	10,802	9,111	6,938	5,603	4,698	4,045	3,551	3,165	2,855	2,292	1,915	1,645	1,441	1,282	1,155	964	827	724	580	483
155	42,312	24,500	17,242	13,301	10,827	9,129	6,949	5,609	4,703	4,049	3,554	3,167	2,856	2,294	1,916	1,645	1,441	1,283	1,155	964	827	724	580	483
160	42,676	24,622	17,302	13,337	10,851	9,146	6,959	5,616	4,707	4,052	3,557	3,169	2,858	2,295	1,917	1,646	1,442	1,283	1,156	964	827	724	580	484
165	43,024	24,737	17,359	13,371	10,873	9,161	6,968	5,622	4,712	4,055	3,559	3,171	2,860	2,296	1,917	1,646	1,442	1,283	1,156	964	827	724	580	484
170	43,357	24,847	17,413	13,403	10,894	9,176	6,976	5,627	4,715	4,058	3,561	3,173	2,861	2,297	1,918	1,647	1,443	1,284	1,156	964	827	724	580	484
175	43,675	24,951	17,464	13,433	10,914	9,191	6,985	5,633	4,719	4,061	3,563	3,175	2,862	2,297	1,919	1,647	1,443	1,284	1,156	965	827	724	580	484
180	43,980	25,050	17,513	13,462	10,933	9,204	6,992	5,638	4,723	4,063	3,565	3,176	2,864	2,298	1,919	1,648	1,443	1,284	1,157	965	828	725	580	484
185	44,272	25,145	17,559	13,489	10,951	9,217	7,000	5,642	4,726	4,066	3,567	3,178	2,865	2,299	1,920	1,648	1,444	1,284	1,157	965	828	725	580	484
190	44,553	25,235	17,603	13,515	10,968	9,229	7,007	5,647	4,729	4,068	3,569	3,179	2,866	2,300	1,920	1,648	1,444	1,285	1,157	965	828	725	580	484
195	44,822	25,321	17,645	13,540	10,984	9,240	7,013	5,651	4,732	4,070	3,571	3,181	2,867	2,301	1,921	1,649	1,444	1,285	1,157	965	828	725	580	484
200	45,081	25,404	17,685	13,563	11,000	9,251	7,020	5,655	4,735	4,072	3,573	3,182	2,868	2,301	1,921	1,649	1,444	1,285	1,157	965	828	725	580	484

# 6 AWG	1 Run(s)	240 Volts	C-Value = 2,425	Magnetic Duct	Copper

Calculations:

1. $f = \dfrac{\text{Length} \times 2 \times \text{Isca}}{\text{Runs} \times \text{C-Value} \times E_{L-L}}$

2. $M = \dfrac{1}{1 + f}$

3. $\text{Isc} = \text{Isca} \times M$

4. Add Motor Contribution = Motor FLA x 4

* All results are given in symmetrical amperes

Single-Phase L-L Bolted Fault-Current* Table

One-Way Distance in Feet

Available Isc in Thousands (Isca)	5	10	15	20	25	30	40	50	60	70	80	90	100	125	150	175	200	225	250	300	350	400	500	600
3	2,905	2,815	2,731	2,652	2,577	2,506	2,376	2,258	2,152	2,055	1,967	1,885	1,811	1,647	1,511	1,396	1,297	1,211	1,135	1,010	909	827	700	607
5	4,741	4,507	4,295	4,102	3,926	3,764	3,477	3,231	3,018	2,831	2,666	2,519	2,387	2,111	1,892	1,715	1,568	1,444	1,338	1,157	1,035	930	772	661
7	6,502	6,070	5,692	5,358	5,061	4,795	4,340	3,963	3,647	3,377	3,144	2,942	2,764	2,401	2,122	1,901	1,722	1,574	1,449	1,250	1,100	982	808	687
10	9,013	8,204	7,528	6,955	6,463	6,036	5,331	4,774	4,322	3,948	3,634	3,366	3,135	2,676	2,334	2,070	1,859	1,687	1,545	1,321	1,154	1,025	837	707
15	12,884	11,292	10,049	9,053	8,237	7,556	6,483	5,677	5,050	4,547	4,135	3,792	3,501	2,938	2,531	2,223	1,982	1,788	1,629	1,382	1,200	1,061	861	724
20	16,408	13,909	12,071	10,662	9,548	8,644	7,268	6,271	5,514	4,920	4,441	4,048	3,718	3,089	2,643	2,309	2,050	1,843	1,674	1,415	1,225	1,080	874	733
25	19,628	16,156	13,728	11,935	10,555	9,462	7,838	6,690	5,835	5,174	4,648	4,218	3,862	3,188	2,714	2,363	2,092	1,877	1,702	1,435	1,240	1,092	881	739
30	22,583	18,107	15,111	12,966	11,354	10,099	8,270	7,002	6,071	5,359	4,796	4,340	3,964	3,257	2,764	2,401	2,122	1,901	1,722	1,449	1,251	1,100	886	742
35	25,304	19,815	16,283	13,819	12,003	10,609	8,609	7,244	6,252	5,499	4,908	4,432	4,040	3,308	2,801	2,429	2,144	1,919	1,736	1,459	1,258	1,106	890	745
40	27,818	21,324	17,288	14,537	12,541	11,027	8,882	7,436	6,395	5,610	4,996	4,503	4,099	3,348	2,829	2,450	2,160	1,932	1,747	1,467	1,264	1,110	893	747
45	30,148	22,667	18,160	15,149	12,994	11,376	9,107	7,593	6,511	5,698	5,066	4,560	4,146	3,379	2,852	2,467	2,173	1,942	1,756	1,473	1,268	1,114	895	749
50	32,313	23,869	18,924	15,676	13,380	11,671	9,295	7,723	6,606	5,771	5,124	4,607	4,185	3,405	2,870	2,480	2,184	1,951	1,762	1,477	1,272	1,116	897	750
55	34,330	24,952	19,598	16,136	13,714	11,924	9,455	7,833	6,687	5,833	5,172	4,646	4,217	3,426	2,885	2,492	2,193	1,958	1,768	1,481	1,275	1,119	899	751
60	36,213	25,932	20,198	16,541	14,005	12,143	9,593	7,928	6,755	5,885	5,213	4,679	4,244	3,444	2,898	2,501	2,200	1,963	1,773	1,485	1,277	1,120	900	752
65	37,976	26,824	20,735	16,899	14,261	12,335	9,712	8,009	6,814	5,929	5,248	4,707	4,267	3,459	2,909	2,509	2,206	1,968	1,777	1,488	1,279	1,122	901	752
70	39,630	27,639	21,219	17,219	14,488	12,504	9,817	8,080	6,865	5,968	5,279	4,732	4,287	3,473	2,918	2,516	2,211	1,973	1,780	1,490	1,281	1,123	902	753
75	41,185	28,386	21,656	17,506	14,690	12,655	9,909	8,143	6,911	6,002	5,305	4,753	4,305	3,484	2,926	2,522	2,216	1,976	1,783	1,492	1,283	1,125	902	754
80	42,648	29,074	22,054	17,765	14,873	12,790	9,992	8,198	6,951	6,033	5,329	4,772	4,321	3,494	2,933	2,527	2,220	1,980	1,786	1,494	1,284	1,126	903	754
85	44,029	29,709	22,418	18,000	15,037	12,911	10,066	8,248	6,986	6,059	5,350	4,789	4,334	3,503	2,940	2,532	2,224	1,983	1,788	1,496	1,285	1,127	904	754
90	45,334	30,297	22,751	18,214	15,186	13,021	10,133	8,293	7,018	6,084	5,368	4,804	4,347	3,511	2,945	2,536	2,227	1,985	1,791	1,497	1,286	1,127	904	755
95	46,568	30,844	23,058	18,411	15,322	13,121	10,193	8,333	7,047	6,105	5,385	4,817	4,358	3,518	2,950	2,540	2,230	1,987	1,792	1,498	1,287	1,128	905	755
100	47,738	31,353	23,341	18,591	15,447	13,213	10,248	8,370	7,074	6,125	5,401	4,830	4,368	3,525	2,955	2,543	2,233	1,989	1,794	1,500	1,288	1,129	905	755
105	48,849	31,828	23,603	18,757	15,561	13,296	10,298	8,403	7,097	6,143	5,415	4,841	4,377	3,531	2,959	2,547	2,235	1,991	1,796	1,501	1,289	1,130	906	756
110	49,904	32,272	23,847	18,910	15,667	13,373	10,344	8,434	7,119	6,159	5,427	4,851	4,385	3,536	2,963	2,549	2,237	1,993	1,797	1,502	1,290	1,130	906	756
115	50,908	32,689	24,074	19,053	15,764	13,444	10,387	8,462	7,139	6,174	5,439	4,860	4,393	3,541	2,966	2,552	2,239	1,995	1,798	1,503	1,290	1,131	906	756
120	51,865	33,081	24,286	19,185	15,855	13,510	10,426	8,488	7,158	6,188	5,450	4,869	4,400	3,546	2,969	2,554	2,241	1,996	1,799	1,503	1,291	1,131	907	756
125	52,777	33,450	24,484	19,309	15,939	13,571	10,462	8,512	7,175	6,201	5,460	4,877	4,406	3,550	2,972	2,556	2,243	1,997	1,801	1,504	1,291	1,131	907	757
130	53,648	33,798	24,670	19,424	16,018	13,628	10,496	8,535	7,191	6,213	5,469	4,884	4,412	3,554	2,975	2,558	2,244	1,999	1,802	1,505	1,292	1,132	907	757
135	54,481	34,127	24,845	19,532	16,091	13,681	10,528	8,556	7,206	6,224	5,477	4,891	4,418	3,557	2,978	2,560	2,246	2,000	1,802	1,505	1,292	1,132	907	757
140	55,278	34,438	25,009	19,633	16,160	13,731	10,557	8,575	7,219	6,234	5,485	4,897	4,423	3,561	2,980	2,562	2,247	2,001	1,803	1,506	1,293	1,133	908	757
145	56,041	34,732	25,164	19,729	16,225	13,777	10,585	8,593	7,232	6,244	5,493	4,903	4,428	3,564	2,982	2,564	2,248	2,002	1,804	1,507	1,293	1,133	908	757
150	56,772	35,012	25,310	19,819	16,285	13,821	10,610	8,610	7,244	6,253	5,500	4,909	4,432	3,567	2,984	2,565	2,249	2,003	1,805	1,507	1,294	1,133	908	757
155	57,474	35,277	25,449	19,904	16,343	13,862	10,635	8,626	7,256	6,261	5,506	4,914	4,436	3,570	2,986	2,567	2,250	2,004	1,806	1,508	1,294	1,133	908	757
160	58,148	35,530	25,580	19,984	16,397	13,901	10,657	8,641	7,266	6,269	5,512	4,919	4,440	3,572	2,988	2,568	2,251	2,004	1,806	1,508	1,294	1,134	908	758
165	58,795	35,771	25,705	20,060	16,448	13,938	10,679	8,655	7,276	6,276	5,518	4,923	4,444	3,575	2,990	2,569	2,252	2,005	1,807	1,508	1,295	1,134	908	758
170	59,418	36,000	25,823	20,132	16,496	13,973	10,699	8,669	7,286	6,283	5,524	4,928	4,448	3,577	2,991	2,570	2,253	2,006	1,807	1,509	1,295	1,134	909	758
175	60,017	36,219	25,936	20,200	16,542	14,006	10,719	8,681	7,295	6,290	5,529	4,932	4,451	3,579	2,993	2,571	2,254	2,007	1,808	1,509	1,295	1,134	909	758
180	60,594	36,429	26,043	20,265	16,585	14,037	10,737	8,693	7,303	6,296	5,533	4,936	4,454	3,581	2,994	2,573	2,255	2,007	1,809	1,510	1,296	1,135	909	758
185	61,151	36,629	26,145	20,327	16,627	14,066	10,754	8,705	7,311	6,302	5,538	4,939	4,457	3,583	2,995	2,574	2,256	2,008	1,809	1,510	1,296	1,135	909	758
190	61,687	36,821	26,243	20,386	16,666	14,095	10,771	8,715	7,319	6,308	5,542	4,943	4,460	3,585	2,997	2,574	2,256	2,008	1,809	1,510	1,296	1,135	909	758
195	62,205	37,005	26,336	20,442	16,704	14,122	10,786	8,726	7,326	6,313	5,547	4,946	4,463	3,587	2,998	2,575	2,257	2,009	1,810	1,511	1,296	1,135	909	758
200	62,705	37,181	26,425	20,496	16,740	14,147	10,801	8,735	7,333	6,318	5,551	4,949	4,465	3,588	2,999	2,576	2,258	2,009	1,810	1,511	1,296	1,135	909	758

#4 AWG	1 Run(s)	240 Volts	C-Value = 3,806	Magnetic Duct	Copper

Calculations:

1. $f = \dfrac{\text{Length} \times 2 \times \text{Isca}}{\text{Runs} \times \text{C-Value} \times E_{L-L}}$

2. $M = \dfrac{1}{1 + f}$

3. $Isc = Isca \times M$

4. Add Motor Contribution = Motor FLA x 4

* All results are given in symmetrical amperes

Single-Phase L-L Bolted Fault-Current* Table

One-Way Distance in Feet

Available Isc in Thousands (Isca)	5	10	15	20	25	30	40	50	60	70	80	90	100	125	150	175	200	225	250	300	350	400	500	600
3	2,923	2,850	2,781	2,715	2,652	2,592	2,479	2,376	2,281	2,194	2,112	2,037	1,967	1,811	1,678	1,563	1,463	1,375	1,297	1,165	1,057	967	827	723
5	4,790	4,598	4,420	4,255	4,102	3,960	3,703	3,478	3,278	3,100	2,941	2,797	2,666	2,388	2,162	1,975	1,818	1,684	1,568	1,379	1,230	1,111	930	800
7	6,596	6,236	5,913	5,622	5,358	5,118	4,697	4,340	4,034	3,768	3,535	3,329	3,145	2,765	2,466	2,226	2,028	1,863	1,723	1,497	1,323	1,186	982	838
10	9,195	8,510	7,920	7,407	6,956	6,556	5,881	5,332	4,877	4,493	4,166	3,883	3,635	3,136	2,758	2,461	2,222	2,025	1,860	1,599	1,403	1,250	1,025	869
15	13,259	11,880	10,761	9,835	9,055	8,390	7,316	6,485	5,824	5,285	4,837	4,460	4,137	3,503	3,037	2,681	2,399	2,171	1,983	1,690	1,472	1,304	1,062	895
20	17,020	14,813	13,113	11,763	10,665	9,754	8,331	7,271	6,450	5,795	5,262	4,818	4,443	3,720	3,199	2,806	2,499	2,253	2,051	1,738	1,509	1,333	1,081	909
25	20,511	17,389	15,092	13,331	11,938	10,808	9,089	7,841	6,895	6,152	5,554	5,062	4,650	3,863	3,305	2,887	2,563	2,305	2,093	1,769	1,532	1,351	1,092	917
30	23,760	19,669	16,780	14,631	12,970	11,648	9,675	8,273	7,227	6,415	5,767	5,238	4,798	3,966	3,379	2,944	2,608	2,341	2,123	1,790	1,548	1,363	1,100	923
35	26,792	21,702	18,238	15,727	13,824	12,332	10,142	8,613	7,484	6,617	5,930	5,372	4,911	4,042	3,434	2,986	2,641	2,367	2,145	1,806	1,559	1,372	1,106	927
40	29,627	23,526	19,508	16,663	14,542	12,900	10,523	8,886	7,690	6,777	6,059	5,478	4,998	4,101	3,477	3,018	2,666	2,387	2,161	1,817	1,568	1,379	1,111	930
45	32,283	25,170	20,626	17,471	15,154	13,379	10,840	9,111	7,858	6,907	6,162	5,562	5,069	4,148	3,511	3,043	2,686	2,403	2,174	1,827	1,575	1,384	1,114	932
50	34,778	26,662	21,617	18,177	15,682	13,789	11,108	9,299	7,997	7,015	6,248	5,632	5,126	4,187	3,539	3,064	2 702	2,416	2,185	1,834	1,580	1,388	1,117	934
55	37,126	28,020	22,501	18,798	16,142	14,144	11,337	9,459	8,115	7,106	6,320	5,690	5,175	4,219	3,561	3,081	2 715	2,427	2,194	1,840	1,585	1,392	1,119	936
60	39,339	29,262	23,295	19,350	16,547	14,453	11,535	9,597	8,216	7,183	6,381	5,740	5,215	4,246	3,581	3,096	2,726	2,436	2,201	1,845	1,589	1,395	1,121	937
65	41,428	30,403	24,012	19,842	16,906	14,726	11,708	9,716	8,304	7,250	6,433	5,782	5,251	4,269	3,597	3,108	2,736	2,443	2,207	1,850	1,592	1,397	1,123	938
70	43,404	31,454	24,663	20,284	17,226	14,969	11,860	9,821	8,380	7,308	6,479	5,819	5,281	4,290	3,612	3,119	2,744	2,450	2,213	1,854	1,595	1,399	1,124	939
75	45,276	32,425	25,256	20,684	17,513	15,185	11,996	9,914	8,448	7,359	6,519	5,851	5,308	4,307	3,624	3,128	2,751	2,456	2,217	1,857	1,597	1,401	1,125	940
80	47,051	33,326	25,799	21,046	17,772	15,380	12,117	9,996	8,508	7,405	6,555	5,880	5,331	4,323	3,635	3,136	2,758	2,461	2,221	1,860	1,599	1,403	1,126	941
85	48,737	34,163	26,298	21,377	18,008	15,556	12,226	10,071	8,561	7,445	6,587	5,906	5,352	4,336	3,645	3,143	2,763	2,465	2,225	1,862	1,601	1,404	1,127	941
90	50,341	34,943	26,758	21,680	18,222	15,715	12,325	10,137	8,609	7,482	6,615	5,929	5,371	4,349	3,653	3,150	2,768	2,469	2,228	1,865	1,603	1,406	1,128	942
95	51,868	35,672	27,184	21,959	18,418	15,861	12,414	10,198	8,653	7,515	6,641	5,949	5,388	4,360	3,661	3,156	2,773	2,473	2,231	1,867	1,604	1,407	1,129	943
100	53,323	36,354	27,578	22,215	18,599	15,995	12,496	10,253	8,692	7,544	6,664	5,968	5,403	4,370	3,668	3,161	2,777	2,476	2,234	1,868	1,606	1,408	1,129	943
105	54,713	36,995	27,945	22,453	18,765	16,117	12,570	10,303	8,729	7,572	6,685	5,985	5,417	4,379	3,675	3,166	2,780	2,479	2,236	1,870	1,607	1,409	1,130	943
110	56,040	37,597	28,287	22,673	18,918	16,231	12,639	10,349	8,762	7,596	6,705	6,000	5,430	4,387	3,681	3,170	2,784	2,481	2,238	1,872	1,608	1,410	1,131	944
115	57,309	38,164	28,607	22,878	19,061	16,335	12,703	10,392	8,792	7,619	6,723	6,015	5,442	4,395	3,686	3,174	2,787	2,484	2,240	1,873	1,609	1,410	1,131	944
120	58,525	38,699	28,907	23,069	19,194	16,433	12,761	10,431	8,820	7,640	6,739	6,028	5,452	4,402	3,691	3,178	2,790	2,486	2,242	1,874	1,610	1,411	1,132	945
125	59,689	39,205	29,188	23,248	19,317	16,523	12,816	10,467	8,846	7,660	6,754	6,040	5,462	4,408	3,695	3,181	2,792	2,488	2,244	1,875	1,611	1,412	1,132	945
130	60,806	39,684	29,453	23,416	19,433	16,608	12,867	10,501	8,870	7,678	6,768	6,051	5,472	4,414	3,700	3,184	2,795	2,490	2,245	1,877	1,612	1,412	1,132	945
135	61,878	40,137	29,702	23,573	19,541	16,687	12,914	10,533	8,893	7,695	6,781	6,062	5,480	4,420	3,704	3,187	2,797	2,492	2,247	1,878	1,613	1,413	1,133	945
140	62,907	40,568	29,937	23,721	19,642	16,761	12,958	10,562	8,914	7,711	6,794	6,071	5,488	4,425	3,707	3,190	2,799	2,493	2,248	1,878	1,613	1,414	1,133	946
145	63,898	40,978	30,159	23,860	19,738	16,830	13,000	10,590	8,933	7,725	6,805	6,081	5,496	4,430	3,711	3,192	2,801	2,495	2,249	1,879	1,614	1,414	1,133	946
150	64,850	41,367	30,370	23,992	19,828	16,895	13,039	10,616	8,952	7,739	6,816	6,089	5,502	4,435	3,714	3,194	2,803	2,496	2,251	1,880	1,614	1,415	1,134	946
155	65,767	41,739	30,570	24,116	19,913	16,957	13,075	10,640	8,969	7,752	6,826	6,097	5,509	4,439	3,717	3,197	2,804	2,498	2,252	1,881	1,615	1,415	1,134	946
160	66,651	42,093	30,759	24,234	19,993	17,015	13,110	10,663	8,985	7,764	6,835	6,105	5,515	4,443	3,719	3,199	2,806	2,499	2,253	1,882	1,616	1,415	1,134	946
165	67,503	42,431	30,940	24,346	20,069	17,070	13,143	10,684	9,001	7,775	6,844	6,112	5,521	4,446	3,722	3,201	2,807	2 500	2,254	1,882	1,616	1,416	1,135	947
170	68,325	42,754	31,111	24,452	20,141	17,122	13,173	10,705	9,015	7,786	6,852	6,118	5,526	4,450	3,725	3,203	2,809	2 501	2,254	1,883	1,616	1,416	1,135	947
175	69,119	43,064	31,275	24,553	20,209	17,172	13,203	10,724	9,029	7,796	6,860	6,125	5,531	4,453	3,727	3,204	2,810	2,502	2,255	1,884	1,617	1,416	1,135	947
180	69,886	43,360	31,431	24,649	20,274	17,219	13,230	10,742	9,042	7,806	6,868	6,131	5,536	4,456	3,729	3,206	2,811	2,503	2,256	1,884	1,617	1,417	1,135	947
185	70,627	43,644	31,580	24,741	20,336	17,263	13,257	10,760	9,054	7,815	6,875	6,136	5,541	4,459	3,731	3,207	2,813	2,504	2,257	1,885	1,618	1,417	1,135	947
190	71,344	43,917	31,722	24,828	20,395	17,306	13,282	10,776	9,066	7,824	6,881	6,142	5,545	4,462	3,733	3,209	2,814	2,505	2,258	1,885	1,618	1,417	1,136	947
195	72,037	44,179	31,859	24,911	20,452	17,346	13,306	10,792	9,077	7,832	6,888	6,147	5,549	4,465	3,735	3,210	2,815	2,506	2,258	1,886	1,618	1,418	1,136	947
200	72,709	44,431	31,989	24,991	20,505	17,385	13,328	10,807	9,087	7,840	6,894	6,151	5,553	4,468	3,737	3,212	2,816	2,507	2,259	1,886	1,619	1,418	1,136	947

# 3 AWG	1 Run(s)	240 Volts	C-Value = 4,760	Magnetic Duct	Copper

Calculations:

1. $f = \dfrac{\text{Length} \times 2 \times \text{Isca}}{\text{Runs} \times \text{C-Value} \times \text{E L-L}}$
2. $M = \dfrac{1}{1 + f}$
3. $\text{Isc} = \text{Isca} \times M$
4. Add Motor Contribution = Motor FLA x 4

* All results are given in symmetrical amperes

Single-Phase L-L Bolted Fault-Current* Table

One-Way Distance in Feet

Available Isc in Thousands (Isca)

Isca	5	10	15	20	25	30	40	50	60	70	80	90	100	125	150	175	200	225	250	300	350	400	500	600
3	2,938	2,878	2,821	2,766	2,713	2,662	2 566	2,476	2,392	2,314	2,241	2,172	2,108	1,962	1,835	1,723	1,625	1,537	1,458	1,322	1,209	1,114	963	848
5	4,830	4,670	4,522	4,382	4,250	4,127	3 900	3,696	3,513	3,347	3,196	3,058	2,932	2,657	2,429	2,238	2,074	1,932	1,809	1,604	1,441	1,308	1,104	955
7	6,671	6,371	6,097	5,845	5,614	5,400	5 018	4,686	4,395	4,139	3,910	3,706	3,522	3,133	2,821	2,566	2,353	2,172	2,018	1,766	1,571	1,414	1,179	1,011
10	9,341	8,763	8,253	7,799	7,392	7,026	6,392	5,863	5,415	5,031	4,697	4,405	4,148	3,618	3,209	2,882	2,616	2,395	2,209	1,911	1,684	1,505	1,241	1,056
15	13,565	12,380	11,385	10,539	9,810	9,175	8,123	7,288	6,608	6,045	5,570	5,164	4,813	4,115	3,593	3,189	2,866	2,603	2,384	2,041	1,784	1,585	1,295	1,095
20	17,527	15,598	14,052	12,784	11,727	10,831	9,395	8,295	7,426	6,722	6,140	5,650	5,233	4,417	3,822	3,368	3,010	2,721	2,483	2,113	1,839	1,628	1,324	1,115
25	21,252	18,481	16,349	14,658	13,285	12,146	10,369	9,046	8,022	7,206	6,541	5,988	5,522	4,622	3,974	3,485	3,104	2,797	2,546	2,158	1,873	1,655	1,341	1,128
30	24,760	21,078	18,349	16,246	14,576	13,216	11,139	9,626	8,475	7,570	6,839	6,237	5,733	4,769	4,082	3,568	3,169	2,851	2,590	2,190	1,897	1,673	1,353	1,136
35	28,069	23,429	20,106	17,608	15,663	14,104	11,763	10,089	8,831	7,853	7,070	6,428	5,894	4,879	4,163	3,630	3,218	2,890	2,622	2,213	1,914	1,686	1,362	1,143
40	31,196	25,569	21,661	18,790	16,591	14,852	12,279	10,466	9,119	8,080	7,253	6,579	6,020	4,966	4,226	3,677	3,255	2,920	2,647	2,231	1,927	1,697	1,369	1,147
45	34,156	27,524	23,048	19,825	17,392	15,491	12,713	10,779	9,356	8,265	7,402	6,702	6,123	5,035	4,276	3,715	3,285	2,944	2,667	2,245	1,938	1,705	1,374	1,151
50	36,962	29,317	24,293	20,738	18,091	16,044	13,082	11,044	9,555	8,420	7,526	6,803	6,207	5,092	4,317	3,746	3,309	2,963	2,683	2,256	1,946	1,711	1,378	1,154
55	39,625	30,968	25,415	21,551	18,707	16,526	13,401	11,270	9,724	8,551	7,630	6,888	6,278	5,140	4,351	3,772	3,329	2,979	2,696	2,265	1,953	1,717	1,382	1,156
60	42,156	32,492	26,433	22,278	19,252	16,950	13,679	11,466	9,869	8,663	7,719	6,961	6,338	5,180	4,380	3,794	3,346	2,993	2,707	2,273	1,959	1,721	1,385	1,158
65	44,564	33,905	27,360	22,933	19,740	17,327	13,923	11,637	9,996	8,760	7,796	7,024	6,390	5,215	4,405	3,812	3,360	3,004	2,716	2,280	1,964	1,725	1,387	1,160
70	46,859	35,217	28,208	23,526	20,177	17,663	14,139	11,788	10,107	8,845	7,864	7,078	6,436	5,245	4,426	3,828	3,373	3,014	2,725	2,285	1,968	1,728	1,389	1,162
75	49,048	36,439	28,987	24,066	20,573	17,965	14,332	11,921	10,205	8,920	7,923	7,126	6,475	5,271	4,445	3,842	3,384	3,023	2,732	2,290	1,972	1,731	1,391	1,163
80	51,138	37,580	29,705	24,558	20,932	18,238	14,505	12,041	10,292	8,987	7,976	7,169	6,510	5,295	4,461	3,855	3,393	3,031	2,738	2,295	1,975	1,733	1,393	1,164
85	53,136	38,648	30,368	25,010	21,259	18,486	14,662	12,149	10,371	9,047	8,023	7,207	6,542	5,315	4,476	3,866	3,402	3,037	2,743	2,299	1,978	1,736	1,394	1,165
90	55,048	39,649	30,983	25,425	21,558	18,712	14,804	12,246	10,442	9,101	8,065	7,241	6,570	5,334	4,489	3,875	3,409	3,043	2,748	2,302	1,980	1,738	1,395	1,166
95	56,879	40,591	31,554	25,809	21,833	18,919	14,933	12,334	10,506	9,149	8,103	7,272	6,595	5,350	4,501	3,884	3,416	3,049	2,753	2,305	1,983	1,739	1,397	1,167
100	58,634	41,477	32,087	26,164	22,087	19,110	15,051	12,415	10,564	9,194	8,138	7,300	6,618	5,366	4,512	3,892	3,422	3,054	2,757	2,308	1,985	1,741	1,398	1,167
105	60,318	42,312	32,585	26,494	22,322	19,285	15,160	12,489	10,618	9,234	8,170	7,325	6,639	5,379	4,521	3,899	3,428	3,058	2,760	2,310	1,987	1,742	1,399	1,168
110	61,935	43,102	33,051	26,802	22,540	19,447	15,260	12,556	10,667	9,271	8,199	7,349	6,658	5,392	4,530	3,906	3,433	3,062	2,764	2,313	1,988	1,744	1,399	1,169
115	63,490	43,849	33,489	27,089	22,743	19,598	15,353	12,619	10,712	9,305	8,225	7,370	6,676	5,403	4,538	3,912	3,438	3,066	2,767	2,315	1,990	1,745	1,400	1,169
120	64,984	44,557	33,900	27,357	22,931	19,738	15,439	12,677	10,753	9,337	8,250	7,390	6,692	5,414	4,546	3,918	3,442	3,069	2,769	2,317	1,991	1,746	1,401	1,170
125	66,423	45,229	34,288	27,609	23,108	19,869	15,518	12,731	10,792	9,366	8,273	7,408	6,707	5,424	4,553	3,923	3,446	3,072	2,772	2,319	1,993	1,747	1,402	1,170
130	67,809	45,867	34,653	27,846	23,274	19,991	15,593	12,781	10,828	9,393	8,294	7,425	6,721	5,433	4,559	3,927	3,450	3,075	2,774	2,320	1,994	1,748	1,402	1,171
135	69,145	46,474	34,999	28,068	23,429	20,106	15,662	12,828	10,862	9,418	8,313	7,441	6,734	5,441	4,565	3,932	3,453	3,078	2,777	2,322	1,995	1,749	1,403	1,171
140	70,433	47,053	35,326	28,278	23,575	20,213	15,728	12,871	10,893	9,442	8,332	7,455	6,746	5,449	4,571	3,936	3,456	3,081	2,779	2,323	1,996	1,750	1,403	1,171
145	71,677	47,604	35,636	28,477	23,713	20,314	15,789	12,912	10,922	9,464	8,349	7,469	6,757	5,456	4,576	3,940	3,459	3,083	2,781	2,325	1,997	1,750	1,404	1,172
150	72,878	48,131	35,930	28,664	23,843	20,410	15,846	12,951	10,950	9,484	8,365	7,482	6,767	5,463	4,581	3,943	3,462	3,085	2,782	2,326	1,998	1,751	1,404	1,172
155	74,038	48,634	36,210	28,842	23,966	20,500	15,900	12,987	10,976	9,504	8,380	7,494	6,777	5,470	4,585	3,947	3,464	3,087	2,784	2,327	1,999	1,752	1,405	1,172
160	75,160	49,116	36,476	29,011	24,082	20,585	15,952	13,021	11,000	9,522	8,394	7,505	6,787	5,476	4,589	3,950	3,467	3,089	2,786	2,328	2,000	1,752	1,405	1,173
165	76,245	49,577	36,730	29,171	24,192	20,665	16,000	13,053	11,023	9,539	8,408	7,516	6,795	5,481	4,593	3,953	3,469	3,091	2,787	2,329	2,000	1,753	1,405	1,173
170	77,296	50,019	36,972	29,324	24,297	20,742	16,046	13,084	11,045	9,555	8,420	7,526	6,804	5,487	4,597	3,956	3,471	3,093	2,788	2,330	2,001	1,754	1,406	1,173
175	78,313	50,443	37,203	29,469	24,397	20,814	16,089	13,112	11,065	9,571	8,432	7,536	6,811	5,492	4,601	3,958	3,473	3,094	2,790	2,331	2,002	1,754	1,406	1,173
180	79,299	50,850	37,424	29,607	24,492	20,883	16,130	13,140	11,085	9,585	8,443	7,545	6,819	5,497	4,604	3,961	3,475	3,096	2,791	2,332	2,002	1,755	1,406	1,173
185	80,254	51,242	37,636	29,740	24,582	20,949	16,169	13,166	11,103	9,599	8,454	7,553	6,826	5,501	4,607	3,963	3,477	3,097	2,792	2,333	2,003	1,755	1,407	1,174
190	81,181	51,618	37,839	29,866	24,668	21,011	16,207	13,190	11,121	9,612	8,464	7,561	6,832	5,505	4,610	3,965	3,479	3,098	2,793	2,333	2,004	1,755	1,407	1,174
195	82,080	51,980	38,033	29,987	24,751	21,071	16,242	13,214	11,137	9,625	8,474	7,569	6,839	5,510	4,613	3,967	3,480	3,100	2,794	2,334	2,004	1,756	1,407	1,174
200	82,953	52,329	38,219	30,102	24,829	21,128	16,276	13,236	11,153	9,637	8,483	7,576	6,845	5,513	4,616	3,969	3,482	3,101	2,795	2,335	2,005	1,756	1,407	1,174

# 2 AWG	1 Run(s)	240 Volts	C-Value = 5,906	Magnetic Duct	Copper

Calculations:

1. $f = \frac{\text{Length} \times 2 \times \text{Isca}}{\text{Runs} \times \text{C-Value} \times \text{E L-L}}$
2. $M = \frac{1}{1 + f}$
3. Isc = Isca x M
4. Add Motor Contribution = Motor FLA x 4

* All results are given in symmetrical amperes

Single-Phase L-L Bolted Fault-Current* Table

One-Way Distance in Feet

Available Isc in Thousands (Isca)	5	10	15	20	25	30	40	50	60	70	80	90	100	125	150	175	200	225	250	300	350	400	500	600
3	2,949	2,901	2,853	2,807	2,763	2,720	2,638	2,561	2,488	2,419	2,354	2,293	2,234	2,100	1,981	1,875	1,780	1,694	1,615	1,479	1,364	1,265	1,105	981
5	4,861	4,730	4,605	4,487	4,375	4,268	4,070	3,889	3,723	3,571	3,431	3,302	3,182	2,917	2,692	2,500	2,333	2,188	2,059	1,842	1,667	1,522	1,296	1,129
7	6,731	6,482	6,250	6,035	5,833	5,645	5,303	5,000	4,730	4,487	4,268	4,070	3,889	3,500	3,182	2,917	2,692	2,500	2,333	2,059	1,842	1,667	1,400	1,207
10	9,459	8,974	8,537	8,140	7,778	7,447	6,863	6,364	5,932	5,556	5,224	4,930	4,667	4,118	3,684	3,333	3,044	2,800	2,593	2,258	2,000	1,795	1,489	1,273
15	13,816	12,805	11,932	11,170	10,500	9,906	8,898	8,077	7,395	6,818	6,325	5,899	5,526	4,773	4,200	3,750	3,387	3,088	2,838	2,442	2,143	1,909	1,567	1,329
20	17,949	16,279	14,894	13,726	12,727	11,865	10,448	9,334	8,434	7,693	7,071	6,542	6,087	5,185	4,516	4,000	3,590	3,256	2,979	2,546	2,222	1,972	1,609	1,359
25	21,875	19,445	17,500	15,909	14,584	13,462	11,667	10,294	9,211	8,334	7,609	7,000	6,482	5,469	4,730	4,167	3,724	3,366	3,070	2,612	2,273	2,012	1,636	1,378
30	25,610	22,341	19,812	17,797	16,154	14,789	12,651	11,053	9,813	8,824	8,016	7,343	6,774	5,676	4,884	4,286	3,818	3,443	3,134	2,658	2,308	2,039	1,654	1,391
35	29,167	25,000	21,875	19,445	17,500	15,909	13,462	11,667	10,294	9,211	8,334	7,609	7,000	5,834	5,000	4,375	3,889	3,500	3,182	2,692	2,333	2,059	1,667	1,400
40	32,558	27,451	23,729	20,896	18,667	16,868	14,142	12,174	10,687	9,524	8,589	7,822	7,180	5,958	5,091	4,445	3,944	3,544	3,219	2,719	2,353	2,074	1,677	1,407
45	35,796	29,717	25,404	22,184	19,688	17,697	14,720	12,600	11,014	9,783	8,799	7,995	7,326	6,058	5,164	4,500	3,988	3,580	3,248	2,739	2,369	2,086	1,685	1,413
50	38,889	31,819	26,924	23,334	20,589	18,422	15,218	12,963	11,291	10,000	8,975	8,140	7,447	6,141	5,224	4,546	4,023	3,608	3,271	2,756	2,381	2,096	1,691	1,417
55	41,848	33,773	28,309	24,368	21,389	19,060	15,651	13,276	11,527	10,186	9,124	8,262	7,549	6,210	5,274	4,584	4,053	3,632	3,291	2,770	2,391	2,104	1,696	1,421
60	44,681	35,594	29,578	25,302	22,106	19,627	16,031	13,549	11,732	10,345	9,251	8,367	7,637	6,269	5,317	4,616	4,078	3,652	3,307	2,782	2,400	2,111	1,700	1,424
65	47,396	37,296	30,744	26,150	22,751	20,133	16,367	13,788	11,911	10,484	9,363	8,458	7,712	6,320	5,353	4,643	4,099	3,670	3,321	2,792	2,408	2,116	1,704	1,426
70	50,001	38,890	31,819	26,924	23,334	20,589	16,667	14,001	12,069	10,606	9,460	8,537	7,778	6,364	5,385	4,667	4,118	3,684	3,333	2,800	2,414	2,121	1,707	1,429
75	52,501	40,385	32,813	27,632	23,864	21,001	16,936	14,190	12,210	10,715	9,546	8,607	7,836	6,403	5,413	4,688	4,134	3,697	3,344	2,808	2,419	2,126	1,710	1,431
80	54,903	41,792	33,736	28,284	24,349	21,375	17,179	14,360	12,335	10,811	9,622	8,669	7,888	6,437	5,437	4,706	4,148	3,709	3,353	2,814	2,424	2,129	1,713	1,432
85	57,212	43,117	34,594	28,884	24,792	21,716	17,398	14,513	12,448	10,898	9,691	8,725	7,934	6,468	5,459	4,722	4,161	3,719	3,362	2,820	2,429	2,133	1,715	1,434
90	59,435	44,367	35,394	29,440	25,201	22,029	17,598	14,652	12,550	10,976	9,753	8,775	7,975	6,495	5,478	4,737	4,172	3,728	3,369	2,825	2,433	2,136	1,717	1,435
95	61,575	45,549	36,142	29,956	25,578	22,316	17,781	14,778	12,643	11,047	9,809	8,820	8,012	6,520	5,496	4,750	4,183	3,736	3,376	2,830	2,436	2,138	1,718	1,436
100	63,637	46,668	36,843	30,436	25,927	22,581	17,949	14,894	12,728	11,112	9,860	8,861	8,046	6,542	5,512	4,762	4,192	3,743	3,382	2,834	2,439	2,141	1,720	1,437
105	65,626	47,728	37,501	30,883	26,251	22,827	18,104	15,001	12,805	11,171	9,906	8,899	8,077	6,563	5,527	4,773	4,200	3,750	3,387	2,838	2,442	2,143	1,721	1,438
110	67,545	48,735	38,120	31,302	26,553	23,055	18,247	15,099	12,877	11,225	9,949	8,933	8,106	6,581	5,540	4,783	4,208	3,756	3,392	2,841	2,445	2,145	1,723	1,439
115	69,398	49,693	38,703	31,694	26,834	23,267	18,380	15,189	12,943	11,275	9,988	8,965	8,132	6,599	5,552	4,792	4,215	3,762	3,397	2,845	2,447	2,147	1,724	1,440
120	71,188	50,604	39,254	32,062	27,098	23,465	18,503	15,273	13,004	11,321	10,024	8,994	8,156	6,614	5,563	4,800	4,221	3,767	3,401	2,848	2,449	2,148	1,725	1,441
125	72,918	51,472	39,774	32,409	27,345	23,650	18,618	15,351	13,060	11,364	10,058	9,021	8,178	6,629	5,573	4,808	4,227	3,772	3,405	2,850	2,451	2,150	1,726	1,442
130	74,592	52,300	40,267	32,735	27,577	23,823	18,725	15,424	13,113	11,404	10,089	9,046	8,199	6,643	5,583	4,815	4,233	3,776	3,408	2,853	2,453	2,151	1,727	1,442
135	76,211	53,091	40,734	33,043	27,795	23,986	18,825	15,492	13,162	11,441	10,118	9,069	8,218	6,655	5,592	4,822	4,238	3,780	3,412	2,855	2,455	2,153	1,728	1,443
140	77,779	53,848	41,178	33,334	28,001	24,139	18,920	15,556	13,208	11,476	10,145	9,091	8,236	6,667	5,600	4,828	4,243	3,784	3,415	2,857	2,456	2,154	1,728	1,443
145	79,299	54,571	41,600	33,610	28,195	24,283	19,008	15,616	13,251	11,508	10,171	9,112	8,252	6,678	5,608	4,834	4,247	3,787	3,418	2,859	2,458	2,155	1,729	1,444
150	80,771	55,265	42,001	33,872	28,379	24,420	19,092	15,672	13,292	11,539	10,195	9,131	8,268	6,688	5,615	4,839	4,251	3,791	3,420	2,861	2,459	2,156	1,730	1,444
155	82,199	55,929	42,384	34,121	28,554	24,548	19,170	15,725	13,330	11,568	10,217	9,149	8,283	6,698	5,622	4,844	4,255	3,794	3,423	2,863	2,460	2,157	1,731	1,445
160	83,584	56,567	42,750	34,357	28,719	24,671	19,245	15,775	13,366	11,595	10,238	9,166	8,297	6,707	5,628	4,849	4,259	3,797	3,425	2,865	2,462	2,158	1,731	1,445
165	84,928	57,180	43,098	34,582	28,876	24,786	19,315	15,823	13,400	11,620	10,258	9,182	8,310	6,715	5,634	4,853	4,262	3,800	3,427	2,866	2,463	2,159	1,732	1,446
170	86,234	57,769	43,432	34,797	29,025	24,896	19,382	15,867	13,432	11,644	10,277	9,197	8,322	6,723	5,640	4,857	4,265	3,802	3,430	2,868	2,464	2,160	1,732	1,446
175	87,502	58,335	43,751	35,001	29,168	25,001	19,445	15,910	13,462	11,667	10,295	9,211	8,334	6,731	5,645	4,861	4,268	3,805	3,432	2,869	2,465	2,161	1,733	1,446
180	88,734	58,880	44,057	35,197	29,303	25,101	19,505	15,950	13,491	11,689	10,311	9,224	8,345	6,738	5,650	4,865	4,271	3,807	3,433	2,870	2,466	2,161	1,733	1,447
185	89,933	59,406	44,351	35,384	29,433	25,196	19,563	15,988	13,518	11,709	10,327	9,237	8,355	6,745	5,655	4,869	4,274	3,809	3,435	2,872	2,467	2,162	1,734	1,447
190	91,098	59,912	44,632	35,563	29,557	25,286	19,617	16,025	13,544	11,729	10,343	9,249	8,365	6,752	5,660	4,872	4,277	3,811	3,437	2,873	2,468	2,163	1,734	1,447
195	92,232	60,400	44,903	35,734	29,675	25,373	19,669	16,059	13,569	11,747	10,357	9,261	8,375	6,758	5,664	4,875	4,279	3,813	3,438	2,874	2,468	2,163	1,735	1,448
200	93,336	60,872	45,163	35,899	29,788	25,456	19,719	16,093	13,593	11,765	10,371	9,272	8,384	6,764	5,668	4,878	4,282	3,815	3,440	2,875	2,469	2,164	1,735	1,448

# 1 AWG	1 Run(s)	240 Volts	C-Value = 7,292	Magnetic Duct	Copper

Calculations:

1. $f = \dfrac{\text{Length} \times 2 \times \text{Isca}}{\text{Runs} \times \text{C-Value} \times E_{L\text{-}L}}$
2. $M = \dfrac{1}{1 + f}$
3. $\text{Isc} = \text{Isca} \times M$
4. Add Motor Contribution = Motor FLA x 4

* All results are given in symmetrical amperes

Single-Phase L-L Bolted Fault-Current* Table

One-Way Distance in Feet

Available Isc in Thousands (Isca)

Isca	5	10	15	20	25	30	40	50	60	70	80	90	100	125	150	175	200	225	250	300	350	400	500	600
3	2,959	2,918	2,879	2,841	2,804	2,767	2,698	2,631	2,568	2,508	2,451	2,396	2,343	2,222	2,112	2,013	1,923	1,840	1,764	1,630	1,515	1,415	1,250	1,119
5	4,886	4,777	4,673	4,573	4,477	4,386	4,213	4,054	3,906	3,768	3,640	3,521	3,409	3,157	2,941	2,752	2,586	2,438	2,307	2,083	1,898	1,744	1,499	1,315
7	6,778	6,571	6,375	6,191	6,017	5,852	5,549	5,276	5,028	4,803	4,596	4,407	4,233	3,852	3,534	3,265	3,034	2,833	2,657	2,364	2,129	1,937	1,640	1,422
10	9,554	9,146	8,771	8,426	8,107	7,812	7,281	6,817	6,409	6,047	5,724	5,434	5,171	4,614	4,165	3,796	3,487	3,225	2,999	2,631	2,343	2,112	1,764	1,514
15	14,018	13,157	12,396	11,717	11,110	10,562	9,614	8,822	8,150	7,574	7,074	6,635	6,248	5,453	4,837	4,346	3,946	3,613	3,332	2,883	2,541	2,272	1,874	1,595
20	18,292	16,853	15,623	14,561	13,634	12,818	11,448	10,342	9,431	8,668	8,019	7,460	6,974	5,998	5,261	4,686	4,224	3,845	3,528	3,029	2,654	2,361	1,935	1,639
25	22,387	20,268	18,516	17,043	15,786	14,703	12,928	11,535	10,414	9,491	8,718	8,062	7,497	6,381	5,553	4,916	4,410	3,998	3,657	3,124	2,726	2,418	1,973	1,666
30	26,314	23,435	21,124	19,227	17,643	16,301	14,147	12,496	11,190	10,132	9,256	8,520	7,892	6,664	5,767	5,083	4,543	4,108	3,748	3,190	2,776	2,458	1,999	1,685
35	30,084	26,379	23,486	21,165	19,262	17,672	15,169	13,287	11,820	10,645	9,683	8,880	8,200	6,882	5,930	5,209	4,644	4,190	3,816	3,239	2,814	2,487	2,018	1,698
40	33,705	29,122	25,636	22,896	20,685	18,863	16,038	13,949	12,341	11,066	10,030	9,171	8,447	7,056	6,058	5,307	4,722	4,253	3,869	3,277	2,842	2,509	2,033	1,709
45	37,187	31,685	27,602	24,451	21,945	19,906	16,786	14,511	12,779	11,417	10,317	9,410	8,650	7,197	6,162	5,387	4,785	4,304	3,911	3,307	2,865	2,527	2,044	1,717
50	40,537	34,085	29,406	25,856	23,071	20,827	17,436	14,995	13,153	11,714	10,559	9,611	8,820	7,314	6,247	5,452	4,836	4,346	3,946	3,332	2,883	2,541	2,054	1,723
55	43,762	36,337	31,067	27,131	24,081	21,647	18,007	15,415	13,475	11,969	10,766	9,782	8,964	7,412	6,319	5,507	4,879	4,380	3,974	3,352	2,898	2,553	2,061	1,729
60	46,870	38,454	32,601	28,294	24,993	22,381	18,512	15,784	13,756	12,190	10,944	9,930	9,087	7,497	6,380	5,553	4,916	4,410	3,998	3,369	2,911	2,563	2,068	1,733
65	49,866	40,449	34,023	29,359	25,820	23,042	18,962	16,109	14,003	12,384	11,100	10,058	9,194	7,569	6,433	5,593	4,947	4,435	4,019	3,384	2,922	2,571	2,073	1,737
70	52,757	42,330	35,345	30,338	26,574	23,641	19,366	16,400	14,222	12,555	11,237	10,170	9,288	7,633	6,478	5,627	4,974	4,456	4,037	3,396	2,932	2,579	2,078	1,740
75	55,548	44,108	36,576	31,241	27,264	24,185	19,729	16,660	14,417	12,706	11,359	10,269	9,371	7,689	6,519	5,658	4,998	4,475	4,052	3,407	2,940	2,585	2,082	1,743
80	58,244	45,791	37,726	32,076	27,898	24,683	20,059	16,895	14,592	12,842	11,467	10,358	9,445	7,738	6,554	5,684	5,019	4,492	4,066	3,417	2,947	2,591	2,086	1,746
85	60,850	47,387	38,802	32,851	28,482	25,139	20,359	17,107	14,751	12,955	11,565	10,438	9,511	7,783	6,586	5,708	5,037	4,507	4,078	3,426	2,953	2,595	2,089	1,748
90	63,371	48,902	39,812	33,571	29,022	25,559	20,634	17,301	14,894	13,076	11,653	10,509	9,570	7,822	6,615	5,730	5,054	4,520	4,089	3,433	2,959	2,600	2,092	1,750
95	65,810	50,341	40,761	34,244	29,523	25,947	20,886	17,477	15,025	13,176	11,733	10,574	9,624	7,858	6,640	5,749	5,069	4,532	4,099	3,440	2,964	2,604	2,095	1,752
100	68,171	51,711	41,654	34,872	29,989	26,306	21,118	17,640	15,145	13,268	11,806	10,633	9,673	7,891	6,663	5,766	5,082	4,543	4,108	3,447	2,969	2,607	2,097	1,754
105	70,458	53,017	42,497	35,461	30,424	26,640	21,333	17,789	15,255	13,353	11,872	10,688	9,718	7,921	6,685	5,782	5,095	4,553	4,116	3,452	2,973	2,611	2,099	1,755
110	72,675	54,262	43,294	36,014	30,830	26,950	21,532	17,927	15,356	13,430	11,934	10,737	9,759	7,948	6,704	5,797	5,106	4,562	4,123	3,457	2,977	2,614	2,101	1,756
115	74,824	55,452	44,047	36,534	31,210	27,241	21,716	18,055	15,450	13,502	11,990	10,783	9,797	7,973	6,722	5,810	5,116	4,570	4,130	3,462	2,980	2,616	2,103	1,758
120	76,909	56,588	44,762	37,024	31,567	27,512	21,889	18,174	15,537	13,568	12,043	10,825	9,831	7,996	6,738	5,822	5,126	4,578	4,136	3,466	2,984	2,619	2,104	1,759
125	78,932	57,676	45,440	37,487	31,903	27,767	22,050	18,285	15,618	13,630	12,091	10,864	9,864	8,018	6,753	5,834	5,134	4,585	4,142	3,470	2,987	2,621	2,106	1,760
130	80,897	58,718	46,084	37,924	32,219	28,006	22,200	18,388	15,693	13,688	12,136	10,901	9,894	8,037	6,768	5,844	5,143	4,591	4,147	3,474	2,989	2,623	2,107	1,761
135	82,806	59,717	46,697	38,338	32,517	28,231	22,341	18,485	15,764	13,741	12,178	10,935	9,922	8,056	6,781	5,854	5,150	4,597	4,152	3,478	2,992	2,625	2,108	1,762
140	84,660	60,676	47,281	38,731	32,800	28,444	22,474	18,576	15,830	13,791	12,218	10,967	9,948	8,073	6,793	5,863	5,157	4,603	4,156	3,481	2,994	2,627	2,109	1,762
145	86,463	61,597	47,838	39,104	33,067	28,644	22,599	18,661	15,892	13,838	12,255	10,996	9,972	8,089	6,804	5,872	5,164	4,608	4,161	3,484	2,996	2,629	2,111	1,763
150	88,217	62,481	48,370	39,459	33,320	28,834	22,717	18,742	15,950	13,882	12,289	11,024	9,995	8,104	6,815	5,879	5,170	4,613	4,165	3,487	2,998	2,630	2,112	1,764
155	89,923	63,332	48,879	39,796	33,561	29,014	22,829	18,817	16,005	13,924	12,322	11,050	10,017	8,118	6,825	5,887	5,176	4,618	4,168	3,489	3,000	2,632	2,113	1,764
160	91,583	64,151	49,365	40,118	33,789	29,185	22,934	18,889	16,057	13,963	12,353	11,075	10,037	8,132	6,834	5,894	5,181	4,622	4,172	3,492	3,002	2,633	2,113	1,765
165	93,200	64,940	49,831	40,426	34,007	29,347	23,035	18,957	16,106	14,000	12,382	11,098	10,056	8,144	6,843	5,900	5,186	4,626	4,175	3,494	3,004	2,634	2,114	1,766
170	94,774	65,701	50,278	40,719	34,214	29,501	23,130	19,021	16,152	14,035	12,409	11,120	10,074	8,156	6,851	5,907	5,191	4,630	4,178	3,496	3,006	2,636	2,115	1,766
175	96,308	66,435	50,706	41,000	34,412	29,648	23,220	19,082	16,196	14,068	12,435	11,141	10,091	8,167	6,859	5,913	5,195	4,633	4,181	3,498	3,007	2,637	2,116	1,767
180	97,803	67,143	51,118	41,268	34,601	29,789	23,306	19,140	16,238	14,100	12,459	11,161	10,107	8,178	6,867	5,918	5,200	4,637	4,184	3,500	3,009	2,638	2,117	1,767
185	99,261	67,826	51,513	41,525	34,782	29,922	23,388	19,195	16,278	14,130	12,483	11,180	10,123	8,188	6,874	5,923	5,204	4,640	4,187	3,502	3,010	2,639	2,117	1,768
190	100,682	68,487	51,893	41,772	34,955	30,050	23,466	19,248	16,315	14,158	12,505	11,197	10,137	8,197	6,881	5,928	5,208	4,643	4,189	3,504	3,011	2,640	2,118	1,768
195	102,069	69,126	52,259	42,009	35,120	30,173	23,540	19,298	16,351	14,185	12,526	11,214	10,151	8,206	6,887	5,933	5,211	4,646	4,191	3,505	3,012	2,641	2,118	1,769
200	103,423	69,744	52,612	42,236	35,279	30,290	23,611	19,346	16,386	14,211	12,546	11,231	10,165	8,215	6,893	5,938	5,215	4,649	4,194	3,507	3,014	2,642	2,119	1,769

# 1/0 AWG	1 Run(s)	240 Volts	C-Value = 8,924	Magnetic Duct	Copper

Calculations:

1. $f = \dfrac{\text{Length} \times 2 \times \text{Isca}}{\text{Runs} \times \text{C-Value} \times E\ \text{L-L}}$

2. $M = \dfrac{1}{1 + f}$

3. Isc = Isca x M

4. Add Motor Contribution = Motor FLA x 4

* All results are given in symmetrical amperes

Single-Phase L-L Bolted Fault-Current* Table

One-Way Distance in Feet

Available Isc in Thousands (Isca)	5	10	15	20	25	30	40	50	60	70	80	90	100	125	150	175	200	225	250	300	350	400	500	600
3	2,979	2,959	2,938	2,918	2,899	2,879	2,841	2,804	2,767	2,732	2,698	2,664	2,631	2,553	2,479	2,409	2,343	2,281	2,222	2,112	2,013	1,923	1,764	1,630
5	4,942	4,886	4,831	4,777	4,724	4,673	4,573	4,477	4,386	4,298	4,213	4,132	4,054	3,871	3,703	3,550	3,409	3,278	3,157	2,941	2,752	2,586	2,307	2,083
7	6,887	6,778	6,673	6,571	6,471	6,375	6,191	6,017	5,852	5,697	5,549	5,409	5,276	4,970	4,697	4,453	4,233	4,034	3,852	3,534	3,265	3,034	2,657	2,364
10	9,772	9,554	9,345	9,146	8,955	8,771	8,426	8,107	7,812	7,537	7,281	7,041	6,817	6,315	5,881	5,503	5,171	4,877	4,614	4,165	3,796	3,487	2,999	2,631
15	14,493	14,018	13,574	13,157	12,765	12,396	11,717	11,110	10,562	10,065	9,614	9,201	8,822	7,998	7,315	6,740	6,248	5,823	5,453	4,837	4,346	3,946	3,332	2,883
20	19,108	18,292	17,543	16,853	16,215	15,623	14,561	13,634	12,818	12,094	11,448	10,867	10,342	9,228	8,331	7,593	6,974	6,449	5,998	5,261	4,686	4,224	3,528	3,029
25	23,621	22,387	21,275	20,268	19,353	18,516	17,043	15,786	14,703	13,758	12,928	12,192	11,535	10,166	9,088	8,216	7,497	6,894	6,381	5,553	4,916	4,410	3,657	3,124
30	28,036	26,314	24,791	23,435	22,219	21,124	19,227	17,643	16,301	15,148	14,147	13,271	12,496	10,906	9,674	8,692	7,892	7,226	6,664	5,767	5,083	4,543	3,748	3,190
35	32,356	30,084	28,110	26,379	24,848	23,486	21,165	19,262	17,672	16,325	15,169	14,166	13,287	11,503	10,141	9,068	8,200	7,484	6,882	5,930	5,209	4,644	3,816	3,239
40	36,584	33,705	31,246	29,122	27,268	25,636	22,896	20,685	18,863	17,336	16,038	14,921	13,949	11,996	10,522	9,371	8,447	7,689	7,056	6,058	5,307	4,722	3,869	3,277
45	40,722	37,187	34,216	31,685	29,503	27,602	24,451	21,945	19,906	18,213	16,786	15,566	14,511	12,409	10,839	9,622	8,650	7,857	7,197	6,162	5,387	4,785	3,911	3,307
50	44,774	40,537	37,032	34,085	31,573	29,406	25,856	23,071	20,827	18,981	17,436	16,123	14,995	12,761	11,107	9,832	8,820	7,997	7,314	6,247	5,452	4,836	3,946	3,332
55	48,742	43,762	39,706	36,337	33,496	31,067	27,131	24,081	21,647	19,660	18,007	16,610	15,415	13,064	11,336	10,011	8,964	8,115	7,412	6,319	5,507	4,879	3,974	3,352
60	52,628	46,870	42,247	38,454	35,287	32,601	28,294	24,993	22,381	20,263	18,512	17,039	15,784	13,328	11,534	10,165	9,087	8,216	7,497	6,380	5,553	4,916	3,998	3,369
65	56,436	49,866	44,666	40,449	36,959	34,023	29,359	25,820	23,042	20,804	18,962	17,420	16,109	13,560	11,707	10,299	9,194	8,303	7,569	6,433	5,593	4,947	4,019	3,384
70	60,168	52,757	46,972	42,330	38,523	35,345	30,338	26,574	23,641	21,291	19,366	17,760	16,400	13,765	11,859	10,417	9,288	8,379	7,633	6,478	5,627	4,974	4,037	3,396
75	63,825	55,548	49,172	44,108	39,990	36,576	31,241	27,264	24,185	21,731	19,729	18,065	16,660	13,948	11,995	10,522	9,371	8,447	7,689	6,519	5,658	4,998	4,052	3,407
80	67,410	58,244	51,273	45,791	41,369	37,726	32,076	27,898	24,683	22,132	20,059	18,341	16,895	14,112	12,116	10,615	9,445	8,507	7,738	6,554	5,684	5,019	4,066	3,417
85	70,926	60,850	53,281	47,387	42,667	38,802	32,851	28,482	25,139	22,498	20,359	18,592	17,107	14,260	12,225	10,698	9,511	8,560	7,783	6,586	5,708	5,037	4,078	3,426
90	74,374	63,371	55,204	48,902	43,891	39,812	33,571	29,022	25,559	22,834	20,634	18,821	17,301	14,394	12,323	10,774	9,570	8,608	7,822	6,615	5,730	5,054	4,089	3,433
95	77,755	65,810	57,045	50,341	45,047	40,761	34,244	29,523	25,947	23,143	20,886	19,030	17,477	14,516	12,413	10,842	9,624	8,652	7,858	6,640	5,749	5,069	4,099	3,440
100	81,073	68,171	58,811	51,711	46,141	41,654	34,872	29,989	26,306	23,428	21,118	19,223	17,640	14,628	12,494	10,904	9,673	8,692	7,891	6,663	5,766	5,082	4,108	3,447
105	84,329	70,458	60,506	53,017	47,178	42,497	35,461	30,424	26,640	23,693	21,333	19,400	17,789	14,730	12,569	10,961	9,718	8,728	7,921	6,685	5,782	5,095	4,116	3,452
110	87,524	72,675	62,133	54,262	48,161	43,294	36,014	30,830	26,950	23,938	21,532	19,565	17,927	14,825	12,638	11,013	9,759	8,761	7,948	6,704	5,797	5,106	4,123	3,457
115	90,660	74,824	63,697	55,452	49,096	44,047	36,534	31,210	27,241	24,167	21,716	19,717	18,055	14,912	12,701	11,061	9,797	8,791	7,973	6,722	5,810	5,116	4,130	3,462
120	93,739	76,909	65,202	56,588	49,985	44,762	37,024	31,567	27,512	24,380	21,889	19,859	18,174	14,993	12,760	11,106	9,831	8,819	7,996	6,738	5,822	5,126	4,136	3,466
125	96,763	78,932	66,651	57,676	50,832	45,440	37,487	31,903	27,767	24,580	22,050	19,991	18,285	15,069	12,815	11,147	9,864	8,845	8,018	6,753	5,834	5,134	4,142	3,470
130	99,732	80,897	68,046	58,718	51,640	46,084	37,924	32,219	28,006	24,767	22,200	20,115	18,388	15,139	12,865	11,186	9,894	8,869	8,037	6,768	5,844	5,143	4,147	3,474
135	102,649	82,806	69,391	59,717	52,411	46,697	38,338	32,517	28,231	24,943	22,341	20,231	18,485	15,204	12,913	11,221	9,922	8,892	8,056	6,781	5,854	5,150	4,152	3,478
140	105,514	84,660	70,689	60,676	53,148	47,281	38,731	32,800	28,444	25,109	22,474	20,340	18,576	15,266	12,957	11,255	9,948	8,913	8,073	6,793	5,863	5,157	4,156	3,481
145	108,330	86,463	71,942	61,597	53,853	47,838	39,104	33,067	28,644	25,265	22,599	20,442	18,661	15,323	12,998	11,286	9,972	8,933	8,089	6,804	5,872	5,164	4,161	3,484
150	111,096	88,217	73,152	62,481	54,528	48,370	39,459	33,320	28,834	25,413	22,717	20,539	18,742	15,378	13,037	11,315	9,995	8,951	8,104	6,815	5,879	5,170	4,165	3,487
155	113,816	89,923	74,321	63,332	55,175	48,879	39,796	33,561	29,014	25,553	22,829	20,630	18,817	15,429	13,074	11,343	10,017	8,968	8,118	6,825	5,887	5,176	4,168	3,489
160	116,489	91,583	75,451	64,151	55,795	49,365	40,118	33,789	29,185	25,685	22,934	20,716	18,889	15,477	13,109	11,369	10,037	8,984	8,132	6,834	5,894	5,181	4,172	3,492
165	119,117	93,200	76,545	64,940	56,391	49,831	40,426	34,007	29,347	25,810	23,035	20,798	18,957	15,522	13,141	11,394	10,056	9,000	8,144	6,843	5,900	5,186	4,175	3,494
170	121,701	94,774	77,604	65,701	56,964	50,278	40,719	34,214	29,501	25,930	23,130	20,875	19,021	15,565	13,172	11,417	10,074	9,014	8,156	6,851	5,907	5,191	4,178	3,496
175	124,242	96,308	78,630	66,435	57,514	50,706	41,000	34,412	29,648	26,043	23,220	20,949	19,082	15,606	13,201	11,439	10,091	9,028	8,167	6,859	5,913	5,195	4,181	3,498
180	126,741	97,803	79,623	67,143	58,044	51,118	41,268	34,601	29,789	26,151	23,306	21,019	19,140	15,645	13,229	11,459	10,107	9,041	8,178	6,867	5,918	5,200	4,184	3,500
185	129,200	99,261	80,587	67,826	58,555	51,513	41,525	34,782	29,922	26,254	23,388	21,085	19,195	15,682	13,255	11,479	10,123	9,053	8,188	6,874	5,923	5,204	4,187	3,502
190	131,619	100,682	81,521	68,487	59,047	51,893	41,772	34,955	30,050	26,353	23,466	21,149	19,248	15,717	13,280	11,498	10,137	9,065	8,197	6,881	5,928	5,208	4,189	3,504
195	133,999	102,069	82,428	69,126	59,521	52,259	42,009	35,120	30,173	26,447	23,540	21,209	19,298	15,750	13,304	11,516	10,151	9,076	8,206	6,887	5,933	5,211	4,191	3,505
200	136,341	103,423	83,308	69,744	59,978	52,612	42,236	35,279	30,290	26,537	23,611	21,267	19,346	15,782	13,327	11,533	10,165	9,086	8,215	6,893	5,938	5,215	4,194	3,507

# 1/0 AWG	2 Run(s)	240 Volts	C-Value = 8,924	Magnetic Duct	Copper

Calculations:

1. $f = \dfrac{\text{Length} \times 2 \times \text{Isca}}{\text{Runs} \times \text{C-Value} \times \text{E L-L}}$

2. $M = \dfrac{1}{1 + f}$

3. $\text{Isc} = \text{Isca} \times M$

4. Add Motor Contribution = Motor FLA x 4

* All results are given in symmetrical amperes

Single-Phase L-L Bolted Fault-Current* Table

One-Way Distance in Feet

Available Isc in Thousands (Isca)

	5	10	15	20	25	30	40	50	60	70	80	90	100	125	150	175	200	225	250	300	350	400	500	600
3	2,966	2,932	2,899	2,867	2,835	2,804	2,745	2,688	2,633	2,580	2,530	2,481	2,434	2,325	2,224	2,133	2,048	1,970	1,897	1,767	1,654	1,555	1,387	1,253
5	4,905	4,814	4,725	4,640	4,558	4,479	4,329	4,189	4,057	3,933	3,817	3,707	3,604	3,369	3,162	2,980	2,817	2,671	2,540	2,312	2,122	1,961	1,702	1,504
7	6,815	6,640	6,473	6,315	6,164	6,020	5,752	5,507	5,281	5,074	4,882	4,704	4,538	4,172	3,860	3,591	3,358	3,153	2,971	2,664	2,415	2,209	1,886	1,645
10	9,627	9,281	8,959	8,658	8,377	8,114	7,634	7,208	6,826	6,483	6,173	5,892	5,634	5,080	4,625	4,245	3,922	3,645	3,405	3,008	2,694	2,439	2,052	1,770
15	14,176	13,438	12,773	12,171	11,623	11,122	10,240	9,487	8,837	8,271	7,773	7,331	6,937	6,115	5,468	4,944	4,512	4,149	3,841	3,343	2,960	2,655	2,202	1,881
20	18,562	17,317	16,228	15,268	14,415	13,653	12,347	11,269	10,364	9,593	8,930	8,352	7,844	6,809	6,016	5,388	4,879	4,458	4,103	3,540	3,113	2,778	2,286	1,942
25	22,792	20,943	19,371	18,019	16,843	15,812	14,086	12,700	11,562	10,611	9,805	9,113	8,512	7,307	6,401	5,695	5,129	4,666	4,279	3,670	3,213	2,858	2,340	1,981
30	26,876	24,342	22,244	20,479	18,974	17,675	15,546	13,874	12,528	11,419	10,491	9,702	9,024	7,681	6,686	5,920	5,311	4,815	4,404	3,762	3,284	2,913	2,377	2,007
35	30,821	27,533	24,879	22,692	20,858	19,299	16,788	14,856	13,322	12,076	11,043	10,172	9,429	7,973	6,906	6,091	5,448	4,928	4,499	3,831	3,336	2,954	2,404	2,026
40	34,633	30,536	27,306	24,693	22,537	20,728	17,859	15,688	13,988	12,620	11,496	10,556	9,758	8,207	7,081	6,227	5,557	5,017	4,572	3,884	3,376	2,986	2,425	2,041
45	38,319	33,366	29,547	26,512	24,042	21,994	18,791	16,403	14,553	13,079	11,875	10,875	10,030	8,398	7,223	6,336	5,644	5,088	4,631	3,927	3,408	3,011	2,441	2,053
50	41,886	36,038	31,623	28,172	25,400	23,124	19,610	17,024	15,040	13,470	12,197	11,144	10,258	8,558	7,341	6,427	5,715	5,146	4,679	3,961	3,434	3,031	2,454	2,062
55	45,339	38,565	33,552	29,693	26,629	24,139	20,335	17,567	15,463	13,808	12,474	11,374	10,453	8,693	7,440	6,503	5,775	5,194	4,719	3,990	3,456	3,048	2,465	2,070
60	48,684	40,958	35,349	31,091	27,749	25,055	20,982	18,048	15,834	14,103	12,714	11,574	10,621	8,809	7,525	6,568	5,826	5,235	4,753	4,014	3,474	3,062	2,475	2,077
65	51,924	43,228	37,027	32,382	28,772	25,887	21,562	18,475	16,162	14,363	12,925	11,748	10,768	8,910	7,598	6,623	5,870	5,271	4,783	4,035	3,489	3,074	2,483	2,082
70	55,066	45,384	38,598	33,577	29,712	26,645	22,085	18,858	16,454	14,593	13,111	11,902	10,897	8,998	7,662	6,672	5,908	5,302	4,808	4,053	3,503	3,084	2,489	2,087
75	58,114	47,435	40,071	34,686	30,577	27,339	22,560	19,203	16,716	14,799	13,277	12,038	11,011	9,075	7,719	6,715	5,942	5,328	4,830	4,069	3,515	3,093	2,495	2,091
80	61,072	49,387	41,455	35,719	31,377	27,976	22,992	19,515	16,952	14,984	13,425	12,160	11,113	9,145	7,768	6,752	5,971	5,352	4,849	4,082	3,525	3,101	2,501	2,095
85	63,943	51,248	42,758	36,682	32,118	28,564	23,387	19,799	17,166	15,151	13,559	12,270	11,205	9,207	7,813	6,786	5,998	5,373	4,867	4,095	3,534	3,109	2,505	2,098
90	66,732	53,024	43,988	37,583	32,806	29,107	23,750	20,059	17,361	15,302	13,680	12,369	11,287	9,262	7,853	6,816	6,021	5,392	4,882	4,106	3,542	3,115	2,509	2,101
95	69,442	54,721	45,149	38,428	33,448	29,611	24,085	20,297	17,539	15,441	13,791	12,459	11,362	9,313	7,889	6,844	6,043	5,409	4,896	4,116	3,550	3,121	2,513	2,103
100	72,076	56,343	46,248	39,221	34,047	30,080	24,394	20,516	17,702	15,567	13,891	12,542	11,431	9,359	7,922	6,868	6,062	5,425	4,909	4,125	3,556	3,126	2,516	2,106
105	74,638	57,897	47,290	39,967	34,608	30,517	24,681	20,719	17,853	15,683	13,984	12,617	11,493	9,400	7,952	6,891	6,079	5,439	4,920	4,133	3,562	3,130	2,519	2,108
110	77,130	59,385	48,278	40,671	35,135	30,925	24,947	20,906	17,992	15,790	14,069	12,686	11,551	9,439	7,980	6,911	6,095	5,452	4,931	4,140	3,568	3,135	2,522	2,110
115	79,556	60,813	49,217	41,335	35,630	31,308	25,196	21,080	18,121	15,890	14,148	12,750	11,604	9,474	8,005	6,930	6,110	5,463	4,941	4,147	3,573	3,138	2,525	2,112
120	81,917	62,183	50,111	41,964	36,096	31,667	25,428	21,243	18,240	15,982	14,221	12,809	11,653	9,507	8,028	6,943	6,124	5,474	4,949	4,153	3,577	3,142	2,527	2,113
125	84,216	63,499	50,962	42,559	36,535	32,005	25,645	21,394	18,352	16,067	14,288	12,864	11,698	9,537	8,050	6,964	6,136	5,484	4,958	4,159	3,582	3,145	2,529	2,115
130	86,457	64,764	51,774	43,124	36,951	32,323	25,849	21,536	18,456	16,147	14,352	12,915	11,740	9,565	8,070	6,973	6,148	5,494	4,965	4,164	3,586	3,148	2,531	2,116
135	88,640	65,982	52,549	43,660	37,344	32,624	26,041	21,669	18,554	16,222	14,410	12,963	11,780	9,591	8,088	6,993	6,159	5,502	4,972	4,169	3,589	3,151	2,533	2,117
140	90,769	67,154	53,290	44,171	37,716	32,908	26,222	21,794	18,645	16,292	14,466	13,008	11,817	9,616	8,106	7,005	6,169	5,510	4,979	4,174	3,593	3,154	2,534	2,118
145	92,844	68,283	53,998	44,656	38,070	33,177	26,392	21,911	18,731	16,357	14,517	13,049	11,851	9,638	8,122	7,013	6,178	5,518	4,985	4,178	3,596	3,156	2,536	2,120
150	94,869	69,372	54,677	45,120	38,406	33,432	26,553	22,022	18,812	16,419	14,566	13,089	11,884	9,660	8,137	7,029	6,187	5,525	4,991	4,182	3,599	3,159	2,538	2,121
155	96,845	70,423	55,328	45,562	38,726	33,674	26,706	22,127	18,889	16,477	14,612	13,126	11,914	9,680	8,152	7,040	6,195	5,531	4,996	4,186	3,602	3,161	2,539	2,122
160	98,774	71,437	55,952	45,984	39,031	33,904	26,850	22,226	18,961	16,532	14,655	13,160	11,943	9,699	8,165	7,050	6,203	5,537	5,001	4,189	3,604	3,163	2,540	2,122
165	100,657	72,417	56,551	46,388	39,321	34,123	26,988	22,320	19,029	16,584	14,696	13,193	11,970	9,717	8,178	7,059	6,210	5,543	5,006	4,193	3,607	3,165	2,541	2,123
170	102,496	73,364	57,127	46,775	39,599	34,332	27,118	22,409	19,094	16,633	14,734	13,224	11,995	9,734	8,190	7,068	6,217	5,549	5,010	4,196	3,609	3,166	2,543	2,124
175	104,292	74,280	57,681	47,145	39,864	34,531	27,242	22,494	19,156	16,680	14,771	13,254	12,020	9,750	8,201	7,077	6,224	5,554	5,014	4,199	3,611	3,168	2,544	2,125
180	106,048	75,166	58,214	47,501	40,118	34,722	27,361	22,575	19,214	16,724	14,806	13,282	12,043	9,765	8,211	7,085	6,230	5,559	5,018	4,202	3,613	3,170	2,545	2,126
185	107,764	76,024	58,727	47,842	40,361	34,904	27,473	22,652	19,270	16,766	14,839	13,308	12,064	9,779	8,222	7,092	6,235	5,564	5,022	4,204	3,615	3,171	2,546	2,126
190	109,441	76,855	59,222	48,170	40,594	35,078	27,581	22,725	19,322	16,806	14,870	13,334	12,085	9,793	8,231	7,099	6,241	5,568	5,026	4,207	3,617	3,173	2,547	2,127
195	111,082	77,661	59,699	48,485	40,818	35,245	27,684	22,795	19,373	16,845	14,900	13,358	12,105	9,806	8,240	7,106	6,246	5,572	5,029	4,209	3,619	3,174	2,547	2,128
200	112,687	78,442	60,159	48,788	41,033	35,404	27,783	22,861	19,421	16,881	14,928	13,381	12,124	9,818	8,249	7,113	6,251	5,576	5,033	4,211	3,621	3,175	2,548	2,128

# 2/0 AWG	1 Run(s)	240 Volts

C-Value = 10,755	Magnetic Duct	Copper

Calculations:

1. $f = \dfrac{\text{Length} \times 2 \times \text{Isca}}{\text{Runs} \times \text{C-Value} \times E_{L\text{-}L}}$

2. $M = \dfrac{1}{1 + f}$

3. Isc = Isca x M

4. Add Motor Contribution = Motor FLA x 4

* All results are given in symmetrical amperes

Single-Phase L-L Bolted Fault-Current* Table

One-Way Distance in Feet

Available Isc in Thousands (Isca)	5	10	15	20	25	30	40	50	60	70	80	90	100	125	150	175	200	225	250	300	350	400	500	600
3	2,983	2,966	2,949	2,932	2,915	2,899	2,867	2,835	2,804	2,774	2,745	2,716	2,688	2,619	2,555	2,493	2,434	2,378	2,325	2,224	2,133	2,048	1,897	1,767
5	4,952	4,905	4,859	4,814	4,769	4,725	4,640	4,558	4,479	4,403	4,329	4,258	4,189	4,025	3,874	3,734	3,604	3,482	3,369	3,162	2,980	2,817	2,540	2,312
7	6,906	6,815	6,726	6,640	6,556	6,473	6,315	6,164	6,020	5,883	5,752	5,627	5,507	5,228	4,976	4,747	4,538	4,347	4,172	3,860	3,591	3,358	2,971	2,664
10	9,810	9,627	9,451	9,281	9,117	8,959	8,658	8,377	8,114	7,867	7,634	7,415	7,208	6,737	6,325	5,960	5,634	5,343	5,080	4,625	4,245	3,922	3,405	3,008
15	14,576	14,176	13,797	13,438	13,097	12,773	12,171	11,623	11,122	10,663	10,240	9,849	9,487	8,689	8,014	7,437	6,937	6,500	6,115	5,468	4,944	4,512	3,841	3,343
20	19,254	18,562	17,918	17,317	16,755	16,228	15,268	14,415	13,653	12,967	12,347	11,783	11,269	10,160	9,250	8,489	7,844	7,290	6,809	6,016	5,388	4,879	4,103	3,540
25	23,845	22,792	21,829	20,943	20,127	19,371	18,019	16,843	15,812	14,899	14,086	13,357	12,700	11,309	10,192	9,277	8,512	7,864	7,307	6,401	5,695	5,129	4,279	3,670
30	28,352	26,876	25,546	24,342	23,246	22,244	20,479	18,974	17,675	16,542	15,546	14,663	13,874	12,231	10,935	9,888	9,024	8,299	7,681	6,686	5,920	5,311	4,404	3,762
35	32,778	30,821	29,084	27,533	26,139	24,879	22,692	20,858	19,299	17,956	16,788	15,763	14,856	12,987	11,536	10,377	9,429	8,640	7,973	6,906	6,091	5,448	4,499	3,831
40	37,124	34,633	32,456	30,536	28,831	27,306	24,693	22,537	20,728	19,187	17,859	16,704	15,688	13,619	12,032	10,776	9,758	8,915	8,207	7,081	6,227	5,557	4,572	3,884
45	41,392	38,319	35,672	33,366	31,340	29,547	26,512	24,042	21,994	20,267	18,791	17,516	16,403	14,154	12,448	11,109	10,030	9,142	8,398	7,223	6,336	5,644	4,631	3,927
50	45,585	41,886	38,743	36,038	33,687	31,623	28,172	25,400	23,124	21,223	19,610	18,226	17,024	14,614	12,802	11,390	10,258	9,331	8,558	7,341	6,427	5,715	4,679	3,961
55	49,705	45,339	41,679	38,565	35,884	33,552	29,693	26,629	24,139	22,075	20,335	18,850	17,567	15,013	13,107	11,631	10,453	9,492	8,693	7,440	6,503	5,775	4,719	3,990
60	53,753	48,684	44,488	40,958	37,948	35,349	31,091	27,749	25,055	22,838	20,982	19,405	18,048	15,362	13,373	11,839	10,621	9,631	8,809	7,525	6,568	5,826	4,753	4,014
65	57,731	51,924	47,179	43,228	39,888	37,027	32,382	28,772	25,887	23,527	21,562	19,900	18,475	15,671	13,606	12,022	10,768	9,751	8,910	7,598	6,623	5,870	4,783	4,035
70	61,642	55,066	49,759	45,384	41,717	38,598	33,577	29,712	26,645	24,152	22,085	20,345	18,858	15,946	13,812	12,183	10,897	9,857	8,998	7,662	6,672	5,908	4,808	4,053
75	65,486	58,114	52,234	47,435	43,443	40,071	34,686	30,577	27,339	24,720	22,560	20,747	19,203	16,192	13,997	12,326	11,011	9,950	9,075	7,719	6,715	5,942	4,830	4,069
80	69,266	61,072	54,611	49,387	45,075	41,455	35,719	31,377	27,976	25,240	22,992	21,112	19,515	16,413	14,162	12,454	11,113	10,033	9,145	7,768	6,752	5,971	4,849	4,082
85	72,983	63,943	56,896	51,248	46,620	42,758	36,682	32,118	28,564	25,718	23,387	21,444	19,799	16,614	14,311	12,569	11,205	10,108	9,207	7,813	6,786	5,998	4,867	4,095
90	76,639	66,732	59,093	53,024	48,085	43,988	37,583	32,806	29,107	26,157	23,750	21,749	20,059	16,796	14,446	12,673	11,287	10,175	9,262	7,853	6,816	6,021	4,882	4,106
95	80,235	69,442	61,209	54,721	49,476	45,149	38,428	33,448	29,611	26,564	24,085	22,029	20,297	16,963	14,569	12,767	11,362	10,236	9,313	7,889	6,844	6,043	4,896	4,116
100	83,773	72,076	63,246	56,343	50,799	46,248	39,221	34,047	30,080	26,940	24,394	22,288	20,516	17,115	14,682	12,854	11,431	10,291	9,359	7,922	6,868	6,062	4,909	4,125
105	87,253	74,638	65,210	57,897	52,058	47,290	39,967	34,608	30,517	27,290	24,681	22,527	20,719	17,256	14,785	12,933	11,493	10,342	9,400	7,952	6,891	6,079	4,920	4,133
110	90,678	77,130	67,104	59,385	53,259	48,278	40,671	35,135	30,925	27,617	24,947	22,749	20,906	17,386	14,880	13,006	11,551	10,389	9,439	7,980	6,911	6,095	4,931	4,140
115	94,049	79,556	68,933	60,813	54,404	49,217	41,335	35,630	31,308	27,921	25,196	22,955	21,080	17,506	14,968	13,073	11,604	10,431	9,474	8,005	6,930	6,110	4,941	4,147
120	97,367	81,917	70,698	62,183	55,498	50,111	41,964	36,096	31,667	28,207	25,428	23,148	21,243	17,618	15,050	13,135	11,653	10,471	9,507	8,028	6,948	6,124	4,949	4,153
125	100,633	84,216	72,405	63,499	56,544	50,962	42,559	36,535	32,005	28,474	25,645	23,328	21,394	17,722	15,126	13,193	11,698	10,508	9,537	8,050	6,964	6,136	4,958	4,159
130	103,849	86,457	74,055	64,764	57,545	51,774	43,124	36,951	32,323	28,726	25,849	23,496	21,536	17,819	15,196	13,247	11,740	10,542	9,565	8,070	6,979	6,148	4,965	4,164
135	107,015	88,640	75,651	65,982	58,504	52,549	43,660	37,344	32,624	28,963	26,041	23,655	21,669	17,910	15,263	13,297	11,780	10,573	9,591	8,088	6,993	6,159	4,972	4,169
140	110,133	90,769	77,196	67,154	59,424	53,290	44,171	37,716	32,908	29,187	26,222	23,804	21,794	17,995	15,324	13,344	11,817	10,603	9,616	8,106	7,006	6,169	4,979	4,174
145	113,204	92,844	78,692	68,283	60,306	53,998	44,656	38,070	33,177	29,398	26,392	23,944	21,911	18,075	15,382	13,388	11,851	10,631	9,638	8,122	7,018	6,178	4,985	4,178
150	116,228	94,869	80,142	69,372	61,154	54,677	45,120	38,406	33,432	29,598	26,553	24,077	22,022	18,151	15,437	13,429	11,884	10,657	9,660	8,137	7,029	6,187	4,991	4,182
155	119,208	96,845	81,547	70,423	61,969	55,328	45,562	38,726	33,674	29,788	26,706	24,202	22,127	18,222	15,488	13,468	11,914	10,681	9,680	8,152	7,040	6,195	4,996	4,186
160	122,144	98,774	82,910	71,437	62,753	55,952	45,984	39,031	33,904	29,968	26,850	24,321	22,226	18,289	15,537	13,505	11,943	10,704	9,699	8,165	7,050	6,203	5,001	4,189
165	125,036	100,657	84,233	72,417	63,508	56,551	46,388	39,321	34,123	30,139	26,988	24,433	22,320	18,353	15,583	13,539	11,970	10,726	9,717	8,178	7,059	6,210	5,006	4,193
170	127,886	102,496	85,517	73,364	64,235	57,127	46,775	39,599	34,332	30,302	27,118	24,540	22,409	18,413	15,626	13,572	11,995	10,747	9,734	8,190	7,068	6,217	5,010	4,196
175	130,696	104,292	86,764	74,280	64,936	57,681	47,145	39,864	34,531	30,457	27,242	24,642	22,494	18,470	15,667	13,603	12,020	10,766	9,750	8,201	7,077	6,224	5,014	4,199
180	133,464	106,048	87,975	75,166	65,613	58,214	47,501	40,118	34,722	30,605	27,361	24,738	22,575	18,524	15,706	13,633	12,043	10,785	9,765	8,211	7,085	6,230	5,018	4,202
185	136,194	107,764	89,153	76,024	66,265	58,727	47,842	40,361	34,904	30,746	27,473	24,831	22,652	18,576	15,744	13,661	12,064	10,802	9,779	8,222	7,092	6,235	5,022	4,204
190	138,884	109,441	90,298	76,855	66,896	59,222	48,170	40,594	35,078	30,881	27,581	24,919	22,725	18,625	15,779	13,687	12,085	10,819	9,793	8,231	7,099	6,241	5,026	4,207
195	141,537	111,082	91,412	77,661	67,505	59,699	48,485	40,818	35,245	31,010	27,684	25,003	22,795	18,672	15,813	13,713	12,105	10,835	9,806	8,240	7,106	6,246	5,029	4,209
200	144,153	112,687	92,496	78,442	68,095	60,159	48,788	41,033	35,404	31,134	27,783	25,083	22,861	18,717	15,845	13,737	12,124	10,850	9,818	8,249	7,113	6,251	5,033	4,211

# 2/0 AWG	2 Run(s)	240 Volts	C-Value = 10,755	Magnetic Duct	Copper

Calculations:

1. $f = \dfrac{\text{Length} \times 2 \times \text{Isca}}{\text{Runs} \times \text{C-Value} \times E_{L\text{-}L}}$

2. $M = \dfrac{1}{1 + f}$

3. $\text{Isc} = \text{Isca} \times M$

4. Add Motor Contribution = Motor FLA x 4

* All results are given in symmetrical amperes

Single-Phase L-L Bolted Fault-Current* Table

One-Way Distance in Feet

Available Isc in Thousands (Isca)	5	10	15	20	25	30	40	50	60	70	80	90	100	125	150	175	200	225	250	300	350	400	500	600
3	2,971	2,943	2,915	2,888	2,861	2,834	2,783	2,734	2,686	2,640	2,596	2,553	2,511	2,413	2,322	2,238	2,159	2,086	2,018	1,894	1,784	1,687	1,520	1,384
5	4,920	4,843	4,768	4,695	4,625	4,557	4,426	4,302	4,185	4,075	3,970	3,870	3,775	3,557	3,363	3,189	3,032	2,890	2,761	2,534	2,341	2,176	1,907	1,697
7	6,845	6,696	6,554	6,417	6,286	6,161	5,924	5,704	5,501	5,311	5,134	4,969	4,814	4,465	4,163	3,900	3,668	3,462	3,278	2,963	2,703	2,485	2,140	1,879
10	9,686	9,391	9,113	8,851	8,604	8,371	7,939	7,550	7,198	6,877	6,583	6,313	6,065	5,522	5,068	4,683	4,352	4,065	3,814	3,394	3,057	2,781	2,356	2,044
15	14,304	13,670	13,089	12,556	12,064	11,610	10,797	10,090	9,470	8,922	8,433	7,996	7,602	6,767	6,098	5,549	5,091	4,702	4,369	3,827	3,404	3,065	2,557	2,193
20	18,781	17,703	16,741	15,879	15,101	14,396	13,166	12,130	11,245	10,480	9,813	9,225	8,704	7,627	6,788	6,114	5,563	5,102	4,712	4,087	3,609	3,231	2,671	2,276
25	23,124	21,511	20,107	18,876	17,787	16,816	15,162	13,804	12,669	11,707	10,880	10,163	9,534	8,257	7,282	6,513	5,890	5,377	4,945	4,262	3,744	3,338	2,744	2,329
30	27,339	25,112	23,220	21,593	20,180	18,940	16,867	15,203	13,838	12,698	11,731	10,901	10,181	8,738	7,653	6,808	6,131	5,576	5,114	4,386	3,840	3,414	2,795	2,366
35	31,431	28,522	26,107	24,068	22,325	20,817	18,340	16,390	14,814	13,515	12,425	11,498	10,700	9,118	7,943	7,036	6,315	5,729	5,241	4,480	3,911	3,471	2,833	2,393
40	35,405	31,758	28,791	26,332	24,259	22,489	19,625	17,409	15,642	14,200	13,002	11,991	11,125	9,424	8,175	7,218	6,461	5,848	5,341	4,553	3,967	3,514	2,862	2,414
45	39,267	34,830	31,294	28,410	26,012	23,988	20,757	18,293	16,352	14,784	13,490	12,404	11,480	9,678	8,365	7,365	6,579	5,945	5,422	4,611	4,011	3,549	2,885	2,430
50	43,021	37,752	33,633	30,324	27,608	25,338	21,761	19,068	16,969	15,286	13,906	12,755	11,780	9,890	8,523	7,488	6,677	6,024	5,488	4,659	4,047	3,577	2,903	2,443
55	46,672	40,534	35,823	32,093	29,067	26,562	22,657	19,753	17,509	15,723	14,267	13,058	12,038	10,072	8,657	7,591	6,759	6,091	5,543	4,698	4,077	3,601	2,919	2,454
60	50,224	43,187	37,879	33,734	30,406	27,676	23,463	20,363	17,986	16,106	14,582	13,322	12,262	10,228	8,772	7,679	6,829	6,148	5,590	4,732	4,102	3,620	2,932	2,463
65	53,680	45,718	39,813	35,259	31,639	28,694	24,190	20,908	18,411	16,446	14,860	13,553	12,458	10,364	8,872	7,756	6,889	6,197	5,631	4,761	4,124	3,637	2,943	2,471
70	57,045	48,136	41,634	36,680	32,779	29,628	24,851	21,400	18,791	16,749	15,107	13,758	12,631	10,483	8,959	7,822	6,942	6,239	5,666	4,786	4,143	3,652	2,952	2,478
75	60,322	50,449	43,353	38,008	33,835	30,489	25,453	21,845	19,133	17,020	15,327	13,941	12,785	10,589	9,036	7,881	6,988	6,276	5,696	4,808	4,159	3,665	2,961	2,484
80	63,515	52,663	44,978	39,251	34,817	31,283	26,005	22,250	19,443	17,265	15,526	14,105	12,922	10,683	9,105	7,933	7,029	6,309	5,724	4,827	4,174	3,676	2,968	2,489
85	66,627	54,785	46,517	40,417	35,732	32,020	26,512	22,620	19,725	17,487	15,705	14,253	13,046	10,767	9,166	7,980	7,065	6,339	5,748	4,844	4,186	3,686	2,974	2,493
90	69,660	56,819	47,975	41,514	36,586	32,704	26,979	22,960	19,983	17,689	15,868	14,387	13,158	10,844	9,222	8,022	7,098	6,365	5,769	4,860	4,198	3,695	2,980	2,497
95	72,618	58,772	49,360	42,547	37,386	33,342	27,412	23,272	20,219	17,874	16,017	14,509	13,260	10,913	9,272	8,060	7,128	6,389	5,789	4,874	4,208	3,703	2,985	2,501
100	75,504	60,648	50,677	43,521	38,137	33,938	27,813	23,561	20,437	18,044	16,153	14,620	13,354	10,976	9,317	8,094	7,154	6,411	5,807	4,886	4,218	3,710	2,990	2,504
105	78,320	62,451	51,930	44,442	38,842	34,495	28,186	23,828	20,637	18,200	16,278	14,723	13,439	11,034	9,359	8,125	7,179	6,430	5,823	4,898	4,226	3,717	2,994	2,507
110	81,069	64,187	53,124	45,314	39,506	35,018	28,534	24,077	20,824	18,345	16,393	14,817	13,518	11,087	9,397	8,154	7,201	6,448	5,837	4,908	4,234	3,723	2,998	2,510
115	83,752	65,858	54,264	46,141	40,133	35,509	28,860	24,308	20,996	18,479	16,500	14,905	13,590	11,135	9,432	8,180	7,222	6,465	5,851	4,918	4,241	3,728	3,002	2,512
120	86,373	67,467	55,352	46,925	40,725	35,972	29,165	24,524	21,157	18,603	16,600	14,986	13,658	11,181	9,464	8,205	7,241	6,480	5,863	4,926	4,247	3,733	3,005	2,515
125	88,934	69,020	56,392	47,671	41,286	36,409	29,451	24,726	21,308	18,719	16,692	15,061	13,720	11,222	9,494	8,227	7,258	6,494	5,875	4,934	4,253	3,738	3,008	2,517
130	91,436	70,517	57,388	48,380	41,817	36,821	29,721	24,916	21,448	18,828	16,778	15,131	13,778	11,261	9,522	8,248	7,275	6,507	5,886	4,942	4,259	3,742	3,011	2,519
135	93,882	71,963	58,342	49,057	42,321	37,212	29,974	25,094	21,580	18,929	16,859	15,196	13,832	11,298	9,548	8,267	7,290	6,519	5,895	4,949	4,264	3,746	3,014	2,521
140	96,273	73,360	59,257	49,702	42,800	37,582	30,214	25,261	21,704	19,025	16,934	15,258	13,883	11,331	9,572	8,285	7,304	6,530	5,905	4,955	4,269	3,750	3,016	2,522
145	98,611	74,710	60,134	50,318	43,256	37,933	30,440	25,420	21,821	19,114	17,005	15,315	13,931	11,363	9,595	8,302	7,317	6,541	5,913	4,961	4,274	3,753	3,018	2,524
150	100,898	76,015	60,977	50,906	43,691	38,266	30,655	25,569	21,931	19,199	17,072	15,369	13,976	11,393	9,616	8,318	7,329	6,550	5,921	4,967	4,278	3,756	3,020	2,525
155	103,136	77,278	61,787	51,470	44,105	38,584	30,858	25,710	22,035	19,278	17,135	15,420	14,018	11,421	9,636	8,333	7,341	6,560	5,929	4,972	4,282	3,759	3,022	2,527
160	105,326	78,501	62,567	52,010	44,501	38,887	31,052	25,844	22,133	19,353	17,194	15,468	14,058	11,447	9,654	8,347	7,352	6,568	5,936	4,977	4,285	3,762	3,024	2,528
165	107,470	79,686	63,317	52,527	44,879	39,175	31,235	25,972	22,226	19,425	17,250	15,514	14,095	11,472	9,672	8,360	7,362	6,577	5,943	4,982	4,289	3,765	3,026	2,529
170	109,569	80,834	64,040	53,023	45,241	39,451	31,410	26,092	22,314	19,492	17,304	15,557	14,131	11,496	9,689	8,373	7,372	6,584	5,949	4,987	4,292	3,768	3,027	2,530
175	111,625	81,948	64,737	53,500	45,588	39,714	31,577	26,207	22,398	19,556	17,354	15,598	14,164	11,518	9,705	8,385	7,381	6,592	5,955	4,991	4,295	3,770	3,029	2,531
180	113,638	83,028	65,409	53,958	45,920	39,966	31,736	26,317	22,478	19,617	17,402	15,636	14,196	11,539	9,720	8,396	7,389	6,599	5,961	4,995	4,298	3,772	3,030	2,532
185	115,611	84,076	66,057	54,399	46,239	40,207	31,888	26,421	22,554	19,675	17,448	15,673	14,226	11,559	9,734	8,405	7,398	6,605	5,966	4,998	4,301	3,774	3,032	2,533
190	117,544	85,094	66,684	54,823	46,545	40,438	32,033	26,521	22,627	19,730	17,491	15,708	14,255	11,578	9,747	8,417	7,405	6,611	5,971	5,002	4,304	3,776	3,033	2,534
195	119,438	86,082	67,290	55,232	46,839	40,660	32,172	26,616	22,696	19,783	17,532	15,742	14,283	11,596	9,760	8,423	7,413	6,617	5,976	5,005	4,306	3,778	3,034	2,535
200	121,296	87,043	67,875	55,626	47,122	40,873	32,306	26,707	22,763	19,833	17,572	15,773	14,309	11,613	9,772	8,435	7,420	6,623	5,980	5,009	4,308	3,780	3,036	2,536

# 3/0 AWG	1 Run(s)	240 Volts		C-Value = 12,843	Magnetic Duct	Copper

Calculations:

1. $f = \dfrac{\text{Length} \times 2 \times \text{Isca}}{\text{Runs} \times \text{C-Value} \times \text{E L-L}}$

2. $M = \dfrac{1}{1 + f}$

3. $\text{Isc} = \text{Isca} \times M$

4. Add Motor Contribution = Motor FLA x 4

* All results are given in symmetrical amperes

Single-Phase L-L Bolted Fault-Current* Table

One-Way Distance in Feet

Available Isc in Thousands (Isca)	5	10	15	20	25	30	40	50	60	70	80	90	100	125	150	175	200	225	250	300	350	400	500	600
3	2,985	2,971	2,957	2,943	2,929	2,915	2,888	2,861	2,834	2,809	2,783	2,758	2,734	2,675	2,618	2,563	2,511	2,461	2,413	2,322	2,238	2,159	2,018	1,894
5	4,960	4,920	4,881	4,843	4,805	4,768	4,695	4,625	4,557	4,490	4,426	4,363	4,302	4,157	4,021	3,894	3,775	3,663	3,557	3,363	3,189	3,032	2,761	2,534
7	6,921	6,845	6,769	6,696	6,624	6,554	6,417	6,286	6,161	6,040	5,924	5,812	5,704	5,452	5,221	5,009	4,814	4,633	4,465	4,163	3,900	3,668	3,278	2,963
10	9,840	9,686	9,536	9,391	9,250	9,113	8,851	8,604	8,371	8,149	7,939	7,740	7,550	7,115	6,727	6,379	6,065	5,780	5,522	5,068	4,683	4,352	3,814	3,394
15	14,644	14,304	13,980	13,670	13,373	13,089	12,556	12,064	11,610	11,189	10,797	10,431	10,090	9,327	8,671	8,101	7,602	7,160	6,767	6,098	5,549	5,091	4,369	3,827
20	19,372	18,781	18,226	17,703	17,209	16,741	15,879	15,101	14,396	13,753	13,166	12,626	12,130	11,043	10,135	9,365	8,704	8,130	7,627	6,788	6,114	5,563	4,712	4,087
25	24,026	23,124	22,288	21,511	20,785	20,107	18,876	17,787	16,816	15,946	15,162	14,451	13,804	12,414	11,278	10,333	9,534	8,850	8,257	7,282	6,513	5,890	4,945	4,262
30	28,608	27,339	26,178	25,112	24,129	23,220	21,593	20,180	18,940	17,843	16,867	15,992	15,203	13,534	12,195	11,098	10,181	9,405	8,738	7,653	6,808	6,131	5,114	4,386
35	33,120	31,431	29,906	28,522	27,261	26,107	24,068	22,325	20,817	19,500	18,340	17,310	16,390	14,466	12,947	11,717	10,700	9,846	9,118	7,943	7,036	6,315	5,241	4,480
40	37,563	35,405	33,482	31,758	30,202	28,791	26,332	24,259	22,489	20,960	19,625	18,451	17,409	15,255	13,575	12,229	11,125	10,204	9,424	8,175	7,218	6,461	5,341	4,553
45	41,939	39,267	36,916	34,830	32,967	31,294	28,410	26,012	23,988	22,256	20,757	19,447	18,293	15,930	14,107	12,659	11,480	10,502	9,678	8,365	7,365	6,579	5,422	4,611
50	46,249	43,021	40,215	37,752	35,574	33,633	30,324	27,608	25,338	23,414	21,761	20,326	19,068	16,514	14,564	13,025	11,780	10,753	9,890	8,523	7,488	6,677	5,488	4,659
55	50,495	46,672	43,387	40,534	38,034	35,823	32,093	29,067	26,562	24,455	22,657	21,106	19,753	17,025	14,960	13,341	12,038	10,967	10,072	8,657	7,591	6,759	5,543	4,698
60	54,678	50,224	46,440	43,187	40,359	37,879	33,734	30,406	27,676	25,396	23,463	21,803	20,363	17,476	15,307	13,616	12,262	11,153	10,228	8,772	7,679	6,829	5,590	4,732
65	58,800	53,680	49,380	45,718	42,562	39,813	35,259	31,639	28,694	26,250	24,190	22,430	20,908	17,877	15,613	13,858	12,458	11,315	10,364	8,872	7,756	6,889	5,631	4,761
70	62,862	57,045	52,213	48,136	44,650	41,634	36,680	32,779	29,628	27,030	24,851	22,997	21,400	18,235	15,886	14,072	12,631	11,457	10,483	8,959	7,822	6,942	5,666	4,786
75	66,865	60,322	54,946	50,449	46,633	43,353	38,008	33,835	30,489	27,744	25,453	23,512	21,845	18,557	16,130	14,264	12,785	11,583	10,589	9,036	7,881	6,988	5,696	4,808
80	70,811	63,515	57,582	52,663	48,518	44,978	39,251	34,817	31,283	28,401	26,005	23,982	22,250	18,849	16,349	14,435	12,922	11,696	10,683	9,105	7,933	7,029	5,724	4,827
85	74,700	66,627	60,128	54,785	50,313	46,517	40,417	35,732	32,020	29,007	26,512	24,412	22,620	19,114	16,548	14,590	13,046	11,798	10,767	9,166	7,980	7,065	5,748	4,844
90	78,534	69,660	62,588	56,819	52,024	47,975	41,514	36,586	32,704	29,567	26,979	24,808	22,960	19,355	16,729	14,730	13,158	11,889	10,844	9,222	8,022	7,098	5,769	4,860
95	82,315	72,618	64,966	58,772	53,656	49,360	42,547	37,386	33,342	30,087	27,412	25,173	23,272	19,577	16,894	14,858	13,260	11,973	10,913	9,272	8,060	7,128	5,789	4,874
100	86,043	75,504	67,265	60,648	55,216	50,677	43,521	38,137	33,938	30,572	27,813	25,511	23,561	19,781	17,046	14,976	13,354	12,049	10,976	9,317	8,094	7,154	5,807	4,886
105	89,719	78,320	69,491	62,451	56,707	51,930	44,442	38,842	34,495	31,023	28,186	25,825	23,828	19,969	17,186	15,083	13,439	12,118	11,034	9,359	8,125	7,179	5,823	4,898
110	93,344	81,069	71,647	64,187	58,134	53,124	45,314	39,506	35,018	31,445	28,534	26,117	24,077	20,143	17,314	15,182	13,518	12,182	11,087	9,397	8,154	7,201	5,837	4,908
115	96,920	83,752	73,735	65,858	59,501	54,264	46,141	40,133	35,509	31,841	28,860	26,389	24,308	20,305	17,434	15,274	13,590	12,241	11,135	9,432	8,180	7,222	5,851	4,918
120	100,447	86,373	75,759	67,467	60,812	55,352	46,925	40,725	35,972	32,213	29,165	26,644	24,524	20,455	17,544	15,359	13,658	12,296	11,181	9,464	8,205	7,241	5,863	4,926
125	103,927	88,934	77,721	69,020	62,070	56,392	47,671	41,286	36,409	32,563	29,451	26,883	24,726	20,596	17,648	15,438	13,720	12,346	11,222	9,494	8,227	7,258	5,875	4,934
130	107,360	91,436	79,626	70,517	63,279	57,388	48,380	41,817	36,821	32,892	29,721	27,107	24,916	20,727	17,744	15,512	13,778	12,393	11,261	9,522	8,248	7,275	5,886	4,942
135	110,747	93,882	81,474	71,963	64,441	58,342	49,057	42,321	37,212	33,203	29,974	27,318	25,094	20,850	17,834	15,580	13,832	12,437	11,298	9,548	8,267	7,290	5,895	4,949
140	114,090	96,273	83,269	73,360	65,558	59,257	49,702	42,800	37,582	33,497	30,214	27,517	25,261	20,966	17,919	15,645	13,883	12,478	11,331	9,572	8,285	7,304	5,905	4,955
145	117,389	98,611	85,012	74,710	66,634	60,134	50,318	43,256	37,933	33,776	30,440	27,704	25,420	21,075	17,998	15,705	13,931	12,517	11,363	9,595	8,302	7,317	5,913	4,961
150	120,644	100,898	86,707	76,015	67,671	60,977	50,906	43,691	38,266	34,040	30,655	27,882	25,569	21,177	18,073	15,762	13,976	12,553	11,393	9,616	8,318	7,329	5,921	4,967
155	123,858	103,136	88,354	77,278	68,670	61,787	51,470	44,105	38,584	34,291	30,858	28,050	25,710	21,274	18,143	15,816	14,018	12,587	11,421	9,636	8,333	7,341	5,929	4,972
160	127,030	105,326	89,957	78,501	69,634	62,567	52,010	44,501	38,887	34,530	31,052	28,210	25,844	21,366	18,210	15,867	14,058	12,619	11,447	9,654	8,347	7,352	5,936	4,977
165	130,162	107,470	91,516	79,686	70,565	63,317	52,527	44,879	39,175	34,757	31,235	28,361	25,972	21,453	18,273	15,914	14,095	12,649	11,472	9,672	8,360	7,362	5,943	4,982
170	133,253	109,569	93,033	80,834	71,464	64,040	53,023	45,241	39,451	34,974	31,410	28,505	26,092	21,535	18,333	15,960	14,131	12,678	11,496	9,689	8,373	7,372	5,949	4,987
175	136,306	111,625	94,511	81,948	72,332	64,737	53,500	45,588	39,714	35,181	31,577	28,643	26,207	21,613	18,389	16,003	14,164	12,705	11,518	9,705	8,385	7,381	5,955	4,991
180	139,320	113,638	95,951	83,028	73,173	65,409	53,958	45,920	39,966	35,379	31,736	28,773	26,317	21,688	18,443	16,043	14,196	12,730	11,539	9,720	8,396	7,389	5,961	4,995
185	142,297	115,611	97,353	84,076	73,985	66,057	54,399	46,239	40,207	35,567	31,888	28,898	26,421	21,758	18,495	16,082	14,226	12,755	11,559	9,734	8,406	7,398	5,966	4,998
190	145,237	117,544	98,720	85,094	74,772	66,684	54,823	46,545	40,438	35,748	32,033	29,018	26,521	21,826	18,543	16,119	14,255	12,778	11,578	9,747	8,417	7,405	5,971	5,002
195	148,140	119,438	100,053	86,082	75,535	67,290	55,232	46,839	40,660	35,922	32,172	29,132	26,616	21,890	18,590	16,154	14,283	12,800	11,596	9,760	8,426	7,413	5,976	5,005
200	151,008	121,296	101,353	87,043	76,273	67,875	55,626	47,122	40,873	36,088	32,306	29,241	26,707	21,952	18,634	16,188	14,309	12,821	11,613	9,772	8,435	7,420	5,980	5,009

3/0 AWG | 2 Run(s) | 240 Volts | C-Value = 12,843 | Magnetic Duct | Copper

Calculations:

1. $f = \dfrac{\text{Length} \times 2 \times \text{Isca}}{\text{Runs} \times \text{C-Value} \times E_{L-L}}$
2. $M = \dfrac{1}{1 + f}$
3. $\text{Isc} = \text{Isca} \times M$
4. Add Motor Contribution = Motor FLA x 4

* All results are given in symmetrical amperes

Single-Phase L-L Bolted Fault-Current* Table

One-Way Distance in Feet

Available Isc in Thousands (Isca)

Isca	5	10	15	20	25	30	40	50	60	70	80	90	100	125	150	175	200	225	250	300	350	400	500	600
3	2,975	2,951	2,927	2,904	2,881	2,858	2,813	2,770	2,729	2,688	2,649	2,611	2,573	2,485	2,403	2,325	2,253	2,185	2,121	2,004	1,899	1,804	1,640	1,504
5	4,932	4,866	4,801	4,738	4,677	4,617	4,502	4,393	4,289	4,190	4,095	4,004	3,918	3,717	3,535	3,370	3,221	3,083	2,957	2,734	2,542	2,375	2,100	1,881
7	6,867	6,739	6,616	6,497	6,383	6,272	6,062	5,866	5,682	5,509	5,346	5,193	5,048	4,719	4,430	4,174	3,947	3,743	3,559	3,240	2,974	2,748	2,386	2,108
10	9,731	9,476	9,235	9,005	8,786	8,578	8,190	7,835	7,510	7,211	6,935	6,679	6,441	5,915	5,468	5,084	4,750	4,458	4,199	3,763	3,408	3,115	2,658	2,317
15	14,403	13,852	13,341	12,867	12,425	12,013	11,265	10,605	10,018	9,493	9,020	8,591	8,202	7,367	6,687	6,121	5,644	5,236	4,883	4,302	3,845	3,476	2,916	2,511
20	18,953	18,010	17,156	16,380	15,671	15,020	13,869	12,882	12,026	11,277	10,615	10,027	9,501	8,399	7,526	6,817	6,230	5,737	5,315	4,635	4,109	3,690	3,065	2,621
25	23,385	21,966	20,709	19,588	18,583	17,675	16,103	14,787	13,670	12,710	11,876	11,145	10,498	9,169	8,138	7,316	6,644	6,086	5,614	4,860	4,285	3,831	3,162	2,692
30	27,704	25,734	24,026	22,531	21,210	20,036	18,039	16,404	15,041	13,887	12,897	12,039	11,288	9,766	8,605	7,691	6,952	6,343	5,832	5,023	4,411	3,932	3,230	2,741
35	31,914	29,328	27,130	25,238	23,593	22,150	19,734	17,794	16,201	14,870	13,741	12,771	11,930	10,242	8,972	7,983	7,190	6,541	5,999	5,146	4,505	4,007	3,280	2,777
40	36,020	32,760	30,041	27,739	25,764	24,052	21,231	19,002	17,196	15,704	14,450	13,382	12,461	10,631	9,270	8,217	7,380	6,697	6,130	5,242	4,579	4,065	3,319	2,805
45	40,024	36,039	32,776	30,054	27,750	25,774	22,561	20,061	18,059	16,420	15,055	13,898	12,907	10,954	9,515	8,409	7,534	6,824	6,236	5,320	4,638	4,111	3,350	2,827
50	43,932	39,177	35,351	32,205	29,574	27,340	23,752	20,997	18,814	17,042	15,576	14,341	13,288	11,228	9,720	8,569	7,662	6,929	6,324	5,383	4,686	4,149	3,375	2,845
55	47,745	42,181	37,779	34,208	31,255	28,770	24,824	21,830	19,480	17,587	16,030	14,725	13,617	11,461	9,895	8,705	7,771	7,017	6,397	5,436	4,727	4,181	3,396	2,860
60	51,469	45,061	40,073	36,078	32,808	30,082	25,794	22,577	20,073	18,069	16,429	15,061	13,904	11,664	10,046	8,821	7,863	7,093	6,460	5,482	4,761	4,207	3,414	2,872
65	55,105	47,824	42,243	37,828	34,249	31,288	26,677	23,250	20,603	18,497	16,782	15,358	14,157	11,841	10,177	8,922	7,943	7,158	6,514	5,520	4,790	4,230	3,429	2,883
70	58,657	50,477	44,299	39,469	35,588	32,403	27,482	23,859	21,080	18,881	17,097	15,622	14,380	11,997	10,292	9,011	8,013	7,215	6,561	5,554	4,815	4,250	3,442	2,892
75	62,127	53,026	46,251	41,010	36,837	33,434	28,221	24,414	21,512	19,227	17,380	15,858	14,580	12,136	10,394	9,089	8,075	7,265	6,602	5,584	4,837	4,267	3,453	2,900
80	65,519	55,477	48,105	42,462	38,003	34,393	28,901	24,921	21,905	19,540	17,636	16,070	14,759	12,260	10,484	9,158	8,130	7,309	6,639	5,610	4,857	4,282	3,463	2,907
85	68,836	57,837	49,869	43,830	39,096	35,285	29,528	25,386	22,263	19,825	17,868	16,262	14,921	12,371	10,566	9,220	8,179	7,348	6,671	5,633	4,874	4,296	3,472	2,913
90	72,078	60,109	51,549	45,123	40,121	36,118	30,109	25,815	22,592	20,085	18,079	16,437	15,068	12,472	10,639	9,276	8,222	7,384	6,700	5,654	4,890	4,308	3,480	2,919
95	75,250	62,299	53,151	46,346	41,085	36,897	30,649	26,210	22,895	20,324	18,272	16,596	15,202	12,564	10,706	9,327	8,262	7,416	6,727	5,673	4,904	4,319	3,487	2,924
100	78,353	64,411	54,681	47,504	41,993	37,628	31,151	26,577	23,174	20,543	18,449	16,743	15,325	12,648	10,767	9,373	8,298	7,445	6,751	5,690	4,917	4,329	3,493	2,928
105	81,390	66,449	56,142	48,604	42,850	38,314	31,620	26,917	23,432	20,746	18,613	16,877	15,438	12,724	10,822	9,415	8,331	7,471	6,772	5,705	4,928	4,338	3,499	2,932
110	84,363	68,417	57,541	49,648	43,660	38,961	32,059	27,235	23,673	20,934	18,764	17,001	15,541	12,795	10,873	9,453	8,361	7,496	6,792	5,719	4,939	4,346	3,504	2,936
115	87,273	70,319	58,880	50,642	44,427	39,570	32,471	27,531	23,896	21,109	18,904	17,116	15,637	12,860	10,920	9,489	8,389	7,518	6,811	5,732	4,948	4,353	3,509	2,939
120	90,122	72,157	60,164	51,589	45,153	40,146	32,857	27,809	24,105	21,272	19,035	17,223	15,727	12,920	10,963	9,521	8,415	7,538	6,827	5,744	4,957	4,360	3,514	2,942
125	92,914	73,935	61,395	52,492	45,843	40,690	33,221	28,069	24,300	21,424	19,156	17,323	15,809	12,976	11,003	9,552	8,438	7,557	6,843	5,755	4,966	4,367	3,518	2,945
130	95,648	75,656	62,577	53,353	46,499	41,206	33,564	28,313	24,483	21,566	19,270	17,415	15,887	13,028	11,041	9,580	8,460	7,575	6,857	5,765	4,973	4,372	3,522	2,948
135	98,328	77,323	63,713	54,177	47,124	41,695	33,888	28,544	24,655	21,699	19,376	17,502	15,959	13,076	11,076	9,606	8,481	7,591	6,871	5,775	4,980	4,378	3,525	2,950
140	100,954	78,938	64,805	54,965	47,718	42,160	34,195	28,761	24,817	21,824	19,476	17,584	16,027	13,122	11,108	9,631	8,500	7,607	6,883	5,784	4,987	4,383	3,528	2,953
145	103,528	80,503	65,856	55,719	48,286	42,603	34,485	28,966	24,970	21,942	19,570	17,660	16,090	13,164	11,139	9,653	8,518	7,621	6,895	5,792	4,993	4,388	3,532	2,955
150	106,052	82,021	66,869	56,442	48,828	43,024	34,761	29,160	25,114	22,054	19,658	17,732	16,150	13,204	11,167	9,675	8,534	7,634	6,906	5,800	4,999	4,392	3,534	2,957
155	108,527	83,494	67,844	57,135	49,346	43,426	35,023	29,344	25,250	22,159	19,742	17,800	16,206	13,242	11,194	9,695	8,550	7,647	6,916	5,807	5,004	4,396	3,537	2,959
160	110,955	84,923	68,785	57,801	49,842	43,810	35,272	29,519	25,379	22,258	19,821	17,864	16,259	13,277	11,220	9,714	8,565	7,659	6,926	5,814	5,009	4,400	3,540	2,961
165	113,337	86,311	69,693	58,441	50,317	44,176	35,509	29,685	25,502	22,352	19,895	17,925	16,309	13,311	11,243	9,732	8,579	7,670	6,935	5,820	5,014	4,404	3,542	2,962
170	115,673	87,660	70,570	59,056	50,772	44,527	35,735	29,843	25,618	22,442	19,966	17,982	16,357	13,342	11,266	9,749	8,592	7,680	6,944	5,826	5,018	4,407	3,544	2,964
175	117,967	88,971	71,417	59,648	51,209	44,862	35,951	29,993	25,729	22,527	20,033	18,037	16,402	13,372	11,287	9,765	8,604	7,690	6,952	5,832	5,023	4,411	3,546	2,965
180	120,218	90,245	72,236	60,218	51,629	45,184	36,157	30,137	25,835	22,608	20,097	18,089	16,445	13,401	11,308	9,780	8,616	7,700	6,959	5,837	5,027	4,414	3,548	2,967
185	122,428	91,485	73,028	60,768	52,032	45,493	36,355	30,274	25,935	22,685	20,158	18,138	16,486	13,428	11,327	9,794	8,627	7,709	6,967	5,842	5,030	4,417	3,550	2,968
190	124,598	92,691	73,794	61,298	52,420	45,789	36,544	30,404	26,031	22,758	20,216	18,185	16,524	13,454	11,345	9,808	8,638	7,717	6,974	5,847	5,034	4,419	3,552	2,969
195	126,729	93,865	74,537	61,809	52,794	46,074	36,725	30,530	26,123	22,828	20,271	18,229	16,561	13,478	11,363	9,821	8,648	7,725	6,980	5,852	5,037	4,422	3,554	2,970
200	128,822	95,009	75,256	62,303	53,154	46,348	36,898	30,650	26,211	22,895	20,324	18,272	16,597	13,501	11,379	9,833	8,657	7,733	6,986	5,856	5,041	4,425	3,555	2,972

4/0 AWG | 1 Run(s) | 240 Volts

C-Value = 15,082 | Magnetic Duct | Copper

Calculations:

1. $f = \frac{\text{Length} \times 2 \times \text{Isca}}{\text{Runs} \times \text{C-Value} \times \text{E L-L}}$

2. $M = \frac{1}{1 + f}$

3. Isc = Isca x M

4. Add Motor Contribution = Motor FLA x 4

* All results are given in symmetrical amperes

Single-Phase L-L Bolted Fault-Current* Table

One-Way Distance in Feet

Available Isc in Thousands (Isca)	5	10	15	20	25	30	40	50	60	70	80	90	100	125	150	175	200	225	250	300	350	400	500	600
3	2,988	2,975	2,963	2,951	2,939	2,927	2,904	2,881	2,858	2,835	2,813	2,792	2,770	2,718	2,668	2,620	2,573	2,528	2,485	2,403	2,325	2,253	2,121	2,004
5	4,966	4,932	4,899	4,866	4,833	4,801	4,738	4,677	4,617	4,559	4,502	4,447	4,393	4,264	4,142	4,027	3,918	3,814	3,717	3,535	3,370	3,221	2,957	2,734
7	6,933	6,867	6,803	6,739	6,677	6,616	6,497	6,383	6,272	6,165	6,062	5,962	5,866	5,637	5,426	5,230	5,048	4,878	4,719	4,430	4,174	3,947	3,559	3,240
10	9,864	9,731	9,602	9,476	9,354	9,235	9,005	8,786	8,578	8,380	8,190	8,009	7,835	7,433	7,070	6,741	6,441	6,167	5,915	5,468	5,084	4,750	4,199	3,763
15	14,696	14,403	14,122	13,852	13,592	13,341	12,867	12,425	12,013	11,627	11,265	10,925	10,605	9,881	9,250	8,695	8,202	7,762	7,367	6,687	6,121	5,644	4,883	4,302
20	19,462	18,953	18,469	18,010	17,573	17,156	16,380	15,671	15,020	14,422	13,869	13,358	12,882	11,830	10,936	10,168	9,501	8,916	8,399	7,526	6,817	6,230	5,315	4,635
25	24,165	23,385	22,653	21,966	21,319	20,709	19,588	18,583	17,675	16,852	16,103	15,417	14,787	13,417	12,279	11,319	10,498	9,789	9,169	8,138	7,316	6,644	5,614	4,860
30	28,806	27,704	26,683	25,734	24,851	24,026	22,531	21,210	20,036	18,985	18,039	17,183	16,404	14,735	13,374	12,243	11,288	10,472	9,766	8,605	7,691	6,952	5,832	5,023
35	33,386	31,914	30,567	29,328	28,186	27,130	25,238	23,593	22,150	20,872	19,734	18,714	17,794	15,847	14,283	13,001	11,930	11,022	10,242	8,972	7,983	7,190	5,999	5,146
40	37,906	36,020	34,312	32,760	31,341	30,041	27,739	25,764	24,052	22,554	21,231	20,055	19,002	16,797	15,051	13,634	12,461	11,473	10,631	9,270	8,217	7,380	6,130	5,242
45	42,366	40,024	37,927	36,039	34,330	32,776	30,054	27,750	25,774	24,061	22,561	21,238	20,061	17,619	15,708	14,171	12,907	11,851	10,954	9,515	8,409	7,534	6,236	5,320
50	46,770	43,932	41,418	39,177	37,165	35,351	32,205	29,574	27,340	25,420	23,752	22,290	20,997	18,337	16,276	14,631	13,288	12,171	11,228	9,720	8,569	7,662	6,324	5,383
55	51,116	47,745	44,791	42,181	39,859	37,779	34,208	31,255	28,770	26,652	24,824	23,231	21,830	18,970	16,772	15,031	13,617	12,447	11,461	9,895	8,705	7,771	6,397	5,436
60	55,408	51,469	48,052	45,061	42,421	40,073	36,078	32,808	30,082	27,774	25,794	24,079	22,577	19,531	17,210	15,381	13,904	12,686	11,664	10,046	8,821	7,863	6,460	5,482
65	59,645	55,105	51,207	47,824	44,861	42,243	37,828	34,249	31,288	28,799	26,677	24,846	23,250	20,033	17,598	15,691	14,157	12,896	11,841	10,177	8,922	7,943	6,514	5,520
70	63,828	58,657	54,260	50,477	47,187	44,299	39,469	35,588	32,403	29,740	27,482	25,543	23,859	20,484	17,945	15,966	14,380	13,081	11,997	10,292	9,011	8,013	6,561	5,554
75	67,959	62,127	57,217	53,026	49,407	46,251	41,010	36,837	33,434	30,607	28,221	26,180	24,414	20,891	18,257	16,213	14,580	13,246	12,136	10,394	9,089	8,075	6,602	5,584
80	72,039	65,519	60,082	55,477	51,529	48,105	42,462	38,003	34,393	31,408	28,901	26,764	24,921	21,261	18,539	16,435	14,759	13,394	12,260	10,484	9,158	8,130	6,639	5,610
85	76,069	68,836	62,859	57,837	53,558	49,869	43,830	39,096	35,285	32,151	29,528	27,301	25,386	21,599	18,795	16,636	14,921	13,527	12,371	10,566	9,220	8,179	6,671	5,633
90	80,048	72,078	65,552	60,109	55,501	51,549	45,123	40,121	36,118	32,841	30,109	27,797	25,815	21,908	19,029	16,819	15,068	13,648	12,472	10,639	9,276	8,222	6,700	5,654
95	83,980	75,250	68,165	62,299	57,362	53,151	46,346	41,085	36,897	33,484	30,649	28,256	26,210	22,193	19,243	16,986	15,202	13,758	12,564	10,706	9,327	8,262	6,727	5,673
100	87,863	78,353	70,701	64,411	59,148	54,681	47,504	41,993	37,628	34,085	31,151	28,683	26,577	22,455	19,440	17,139	15,325	13,858	12,648	10,767	9,373	8,298	6,751	5,690
105	91,700	81,390	73,165	66,449	60,862	56,142	48,604	42,850	38,314	34,647	31,620	29,080	26,917	22,698	19,622	17,280	15,438	13,950	12,724	10,822	9,415	8,331	6,772	5,705
110	95,490	84,363	75,558	68,417	62,509	57,541	49,648	43,660	38,961	35,175	32,059	29,451	27,235	22,923	19,790	17,410	15,541	14,035	12,795	10,873	9,453	8,361	6,792	5,719
115	99,236	87,273	77,884	70,319	64,093	58,880	50,642	44,427	39,570	35,670	32,471	29,798	27,531	23,133	19,946	17,531	15,637	14,113	12,860	10,920	9,489	8,389	6,811	5,732
120	102,937	90,122	80,145	72,157	65,617	60,164	51,589	45,153	40,146	36,138	32,857	30,123	27,809	23,328	20,091	17,643	15,727	14,186	12,920	10,963	9,521	8,415	6,827	5,744
125	106,595	92,914	82,345	73,935	67,084	61,395	52,492	45,843	40,690	36,578	33,221	30,428	28,069	23,511	20,226	17,747	15,809	14,253	12,976	11,003	9,552	8,438	6,843	5,755
130	110,209	95,648	84,486	75,656	68,498	62,577	53,353	46,499	41,206	36,995	33,564	30,716	28,313	23,682	20,353	17,845	15,887	14,316	13,028	11,041	9,580	8,460	6,857	5,765
135	113,782	98,328	86,569	77,323	69,861	63,713	54,177	47,124	41,695	37,389	33,888	30,987	28,544	23,843	20,472	17,936	15,959	14,375	13,076	11,076	9,606	8,481	6,871	5,775
140	117,313	100,954	88,599	78,938	71,177	64,805	54,965	47,718	42,160	37,762	34,195	31,243	28,761	23,994	20,583	18,021	16,027	14,429	13,122	11,108	9,631	8,500	6,883	5,784
145	120,804	103,528	90,575	80,503	72,447	65,856	55,719	48,286	42,603	38,117	34,485	31,486	28,966	24,137	20,688	18,102	16,090	14,481	13,164	11,139	9,653	8,518	6,895	5,792
150	124,254	106,052	92,501	82,021	73,674	66,869	56,442	48,828	43,024	38,454	34,761	31,715	29,160	24,272	20,787	18,177	16,150	14,529	13,204	11,167	9,675	8,534	6,906	5,800
155	127,666	108,527	94,379	83,494	74,860	67,844	57,135	49,346	43,426	38,774	35,023	31,933	29,344	24,399	20,880	18,249	16,206	14,575	13,242	11,194	9,695	8,550	6,916	5,807
160	131,039	110,955	96,209	84,923	76,007	68,785	57,801	49,842	43,810	39,080	35,272	32,140	29,519	24,520	20,969	18,316	16,259	14,618	13,277	11,220	9,714	8,565	6,926	5,814
165	134,373	113,337	97,995	86,311	77,117	69,693	58,441	50,317	44,176	39,371	35,509	32,337	29,685	24,634	21,052	18,380	16,309	14,658	13,311	11,243	9,732	8,579	6,935	5,820
170	137,671	115,673	99,737	87,660	78,192	70,570	59,056	50,772	44,527	39,649	35,735	32,524	29,843	24,743	21,132	18,440	16,357	14,697	13,342	11,266	9,749	8,592	6,944	5,826
175	140,932	117,967	101,437	88,971	79,233	71,417	59,648	51,209	44,862	39,915	35,951	32,703	29,993	24,846	21,207	18,498	16,402	14,733	13,372	11,287	9,765	8,604	6,952	5,832
180	144,157	120,218	103,097	90,245	80,242	72,236	60,218	51,629	45,184	40,170	36,157	32,874	30,137	24,945	21,279	18,552	16,445	14,768	13,401	11,308	9,780	8,616	6,959	5,837
185	147,346	122,428	104,718	91,485	81,221	73,028	60,768	52,032	45,493	40,414	36,355	33,037	30,274	25,038	21,347	18,604	16,486	14,800	13,428	11,327	9,794	8,627	6,967	5,842
190	150,501	124,598	106,302	92,691	82,170	73,794	61,298	52,420	45,789	40,647	36,544	33,193	30,404	25,128	21,412	18,653	16,524	14,832	13,454	11,345	9,808	8,638	6,974	5,847
195	153,621	126,729	107,849	93,865	83,092	74,537	61,809	52,794	46,074	40,871	36,725	33,342	30,530	25,213	21,474	18,700	16,561	14,861	13,478	11,363	9,821	8,648	6,980	5,852
200	156,707	128,822	109,361	95,009	83,986	75,256	62,303	53,154	46,348	41,087	36,898	33,485	30,650	25,295	21,533	18,745	16,597	14,890	13,501	11,379	9,833	8,657	6,986	5,856

# 4/0 AWG	2 Run(s)	240 Volts

C-Value =	15,082	Magnetic Duct	Copper

Calculations:

1. $f = \dfrac{\text{Length} \times 2 \times \text{Isca}}{\text{Runs} \times \text{C-Value} \times \text{E L-L}}$

2. $M = \dfrac{1}{1 + f}$

3. Isc = Isca x M

4. Add Motor Contribution = Motor FLA x 4

* All results are given in symmetrical amperes

Single-Phase L-L Bolted Fault-Current* Table

One-Way Distance in Feet

Available Isc in Thousands (Isca)	5	10	15	20	25	30	40	50	60	70	80	90	100	125	150	175	200	225	250	300	350	400	500	600
3	2,977	2,955	2,933	2,912	2,890	2,869	2,828	2,789	2,750	2,712	2,675	2,640	2,605	2,522	2,444	2,371	2,302	2,237	2,175	2,062	1,960	1,867	1,706	1,571
5	4,938	4,877	4,817	4,759	4,703	4,648	4,541	4,439	4,342	4,248	4,159	4,073	3,991	3,799	3,625	3,467	3,321	3,187	3,064	2,844	2,653	2,486	2,209	1,987
7	6,878	6,761	6,647	6,537	6,431	6,328	6,132	5,948	5,774	5,610	5,455	5,309	5,170	4,853	4,573	4,323	4,099	3,897	3,714	3,395	3,127	2,898	2,528	2,241
10	9,753	9,519	9,295	9,082	8,878	8,683	8,318	7,982	7,673	7,386	7,120	6,873	6,642	6,128	5,687	5,306	4,972	4,678	4,417	3,973	3,611	3,309	2,835	2,479
15	14,452	13,943	13,468	13,025	12,609	12,220	11,509	10,876	10,309	9,798	9,336	8,915	8,531	7,700	7,017	6,446	5,960	5,543	5,180	4,580	4,105	3,719	3,130	2,703
20	19,038	18,163	17,366	16,636	15,964	15,345	14,240	13,284	12,448	11,711	11,056	10,471	9,945	8,834	7,947	7,222	6,618	6,107	5,669	4,959	4,406	3,965	3,303	2,830
25	23,514	22,195	21,016	19,956	18,997	18,127	16,605	15,319	14,218	13,264	12,431	11,696	11,043	9,690	8,633	7,784	7,086	6,504	6,010	5,217	4,609	4,128	3,415	2,913
30	27,885	26,049	24,440	23,018	21,752	20,618	18,672	17,061	15,707	14,551	13,554	12,685	11,920	10,359	9,160	8,210	7,438	6,799	6,261	5,405	4,755	4,245	3,495	2,970
35	32,155	29,738	27,659	25,851	24,266	22,863	20,494	18,570	16,976	15,634	14,489	13,500	12,638	10,897	9,578	8,544	7,711	7,026	6,453	5,548	4,866	4,333	3,554	3,013
40	36,327	33,272	30,690	28,481	26,568	24,896	22,113	19,889	18,072	16,559	15,280	14,184	13,235	11,338	9,917	8,813	7,929	7,207	6,605	5,660	4,952	4,401	3,600	3,046
45	40,404	36,660	33,551	30,928	28,685	26,746	23,560	21,052	19,027	17,357	15,957	14,766	13,740	11,707	10,198	9,034	8,108	7,354	6,729	5,751	5,021	4,455	3,636	3,072
50	44,389	39,911	36,253	33,210	30,638	28,436	24,861	22,086	19,867	18,054	16,544	15,267	14,173	12,020	10,435	9,219	8,257	7,476	6,831	5,825	5,077	4,500	3,666	3,093
55	48,287	43,034	38,812	35,344	32,445	29,986	26,038	23,009	20,612	18,667	17,057	15,703	14,548	12,288	10,636	9,376	8,383	7,579	6,917	5,887	5,125	4,537	3,690	3,110
60	52,098	46,035	41,237	37,344	34,123	31,413	27,108	23,841	21,276	19,210	17,509	16,085	14,876	12,521	10,811	9,511	8,490	7,668	6,990	5,940	5,165	4,568	3,711	3,125
65	55,827	48,923	43,538	39,222	35,684	32,731	28,084	24,592	21,873	19,695	17,911	16,424	15,165	12,726	10,962	9,628	8,584	7,744	7,053	5,986	5,199	4,595	3,729	3,137
70	59,476	51,702	45,726	40,988	37,140	33,953	28,978	25,275	22,411	20,131	18,271	16,726	15,422	12,906	11,096	9,731	8,666	7,810	7,108	6,026	5,229	4,619	3,744	3,148
75	63,047	54,380	47,808	42,653	38,502	35,087	29,801	25,899	22,900	20,524	18,595	16,997	15,652	13,067	11,215	9,822	8,738	7,869	7,157	6,060	5,255	4,639	3,758	3,158
80	66,543	56,962	49,792	44,225	39,778	36,144	30,560	26,470	23,346	20,881	18,887	17,241	15,859	13,211	11,320	9,903	8,802	7,921	7,200	6,091	5,278	4,657	3,770	3,166
85	69,966	59,452	51,684	45,712	40,977	37,131	31,262	26,995	23,754	21,207	19,153	17,462	16,046	13,340	11,415	9,976	8,859	7,967	7,238	6,119	5,299	4,673	3,780	3,174
90	73,319	61,855	53,49[illegible]	47,120	42,105	38,054	31,914	27,480	24,128	21,505	19,396	17,664	16,216	13,458	11,501	10,042	8,911	8,009	7,273	6,143	5,317	4,687	3,789	3,180
95	76,604	64,176	55,218	48,455	43,167	38,920	32,521	27,929	24,473	21,779	19,619	17,848	16,371	13,564	11,579	10,10[illegible]	8,957	8,046	7,304	6,165	5,334	4,700	3,798	3,186
100	79,822	66,420	56,87[illegible]	49,723	44,171	39,734	33,088	28,346	24,793	22,031	19,823	18,018	16,513	13,662	11,650	10,155	9,000	8,081	7,332	6,185	5,349	4,712	3,805	3,191
105	82,976	68,589	58,454	50,929	45,120	40,501	33,617	28,734	25,089	22,265	20,012	18,173	16,644	13,751	11,715	10,204	9,038	8,112	7,357	6,204	5,363	4,722	3,812	3,196
110	86,068	70,688	59,972	52,077	46,019	41,223	34,114	29,096	25,364	22,482	20,187	18,318	16,765	13,834	11,775	10,249	9,074	8,140	7,381	6,220	5,375	4,732	3,819	3,201
115	89,099	72,720	61,428	53,171	46,871	41,906	34,580	29,434	25,621	22,683	20,349	18,451	16,877	13,910	11,830	10,29[illegible]	9,107	8,167	7,403	6,236	5,387	4,741	3,824	3,205
120	92,071	74,688	62,826	54,216	47,681	42,552	35,019	29,751	25,861	22,871	20,501	18,575	16,981	13,980	11,881	10,330	9,137	8,191	7,422	6,250	5,397	4,749	3,830	3,208
125	94,986	76,595	64,170	55,214	48,451	43,165	35,432	30,049	26,086	23,047	20,642	18,691	17,077	14,046	11,928	10,365	9,165	8,213	7,441	6,263	5,407	4,757	3,835	3,212
130	97,846	78,444	65,463	56,168	49,185	43,746	35,823	30,330	26,297	23,211	20,774	18,799	17,168	14,107	11,972	10,399	9,191	8,234	7,458	6,275	5,416	4,764	3,839	3,215
135	100,652	80,237	66,707	57,081	49,884	44,298	36,192	30,594	26,496	23,366	20,897	18,900	17,252	14,164	12,013	10,429	9,215	8,253	7,474	6,286	5,424	4,770	3,843	3,218
140	103,405	81,977	67,905	57,957	50,551	44,823	36,542	30,844	26,683	23,511	21,013	18,995	17,331	14,217	12,051	10,458	9,237	8,272	7,489	6,297	5,432	4,776	3,847	3,221
145	106,107	83,666	69,060	58,796	51,188	45,323	36,874	31,080	26,859	23,648	21,123	19,085	17,405	14,267	12,087	10,485	9,258	8,288	7,502	6,306	5,439	4,782	3,851	3,223
150	108,760	85,307	70,174	59,602	51,797	45,801	37,189	31,304	27,026	23,777	21,226	19,169	17,475	14,314	12,121	10,51[illegible]	9,278	8,304	7,515	6,316	5,446	4,787	3,854	3,226
155	111,365	86,901	71,250	60,375	52,381	46,256	37,489	31,516	27,184	23,900	21,323	19,248	17,541	14,358	12,153	10,534	9,297	8,319	7,528	6,324	5,453	4,792	3,857	3,228
160	113,923	88,451	72,288	61,119	52,940	46,692	37,775	31,717	27,334	24,015	21,415	19,323	17,603	14,400	12,182	10,557	9,314	8,333	7,539	6,332	5,459	4,797	3,860	3,230
165	116,435	89,958	73,291	61,835	53,476	47,108	38,047	31,909	27,476	24,125	21,502	19,394	17,662	14,439	12,211	10,578	9,331	8,346	7,550	6,340	5,464	4,801	3,863	3,232
170	118,903	91,424	74,262	62,524	53,991	47,507	38,307	32,091	27,612	24,229	21,585	19,461	17,718	14,476	12,237	10,598	9,346	8,359	7,560	6,347	5,469	4,805	3,866	3,234
175	121,328	92,851	75,200	63,188	54,485	47,889	38,555	32,266	27,740	24,328	21,664	19,525	17,771	14,512	12,262	10,617	9,361	8,370	7,570	6,354	5,475	4,809	3,868	3,236
180	123,710	94,239	76,109	63,828	54,961	48,256	38,792	32,432	27,863	24,423	21,739	19,586	17,821	14,545	12,286	10,635	9,375	8,382	7,579	6,360	5,479	4,813	3,871	3,237
185	126,052	95,592	76,988	64,446	55,418	48,608	39,019	32,590	27,980	24,513	21,810	19,644	17,869	14,577	12,309	10,652	9,388	8,392	7,587	6,366	5,484	4,816	3,873	3,239
190	128,353	96,910	77,841	65,042	55,858	48,947	39,237	32,742	28,092	24,598	21,878	19,699	17,915	14,607	12,331	10,668	9,400	8,402	7,596	6,372	5,488	4,819	3,875	3,240
195	130,615	98,194	78,667	65,618	56,283	49,272	39,446	32,887	28,199	24,680	21,942	19,751	17,958	14,636	12,351	10,683	9,412	8,412	7,603	6,378	5,492	4,823	3,877	3,242
200	132,840	99,446	79,469	66,175	56,692	49,586	39,647	33,027	28,301	24,759	22,004	19,801	17,999	14,664	12,371	10,698	9,424	8,421	7,611	6,383	5,496	4,826	3,879	3,243

250 MCM	1 Run(s)	240 Volts	C-Value = 16,483	Magnetic Duct	Copper

Calculations:

1. $f = \dfrac{\text{Length} \times 2 \times \text{Isca}}{\text{Runs} \times \text{C-Value} \times \text{E L-L}}$

2. $M = \dfrac{1}{1 + f}$

3. Isc = Isca x M

4. Add Motor Contribution = Motor FLA x 4

* All results are given in symmetrical amperes

Single-Phase L-L Bolted Fault-Current* Table

One-Way Distance in Feet

Available Isc in Thousands (Isca)	5	10	15	20	25	30	40	50	60	70	80	90	100	125	150	175	200	225	250	300	350	400	500	600
3	2,989	2,977	2,966	2,955	2,944	2,933	2,912	2,890	2,869	2,849	2,828	2,808	2,789	2,740	2,694	2,649	2,605	2,563	2,522	2,444	2,371	2,302	2,175	2,062
5	4,969	4,938	4,907	4,877	4,847	4,817	4,759	4,703	4,648	4,594	4,541	4,489	4,439	4,318	4,203	4,094	3,991	3,893	3,799	3,625	3,467	3,321	3,064	2,844
7	6,939	6,878	6,819	6,761	6,703	6,647	6,537	6,431	6,328	6,229	6,132	6,038	5,948	5,732	5,532	5,345	5,170	5,007	4,853	4,573	4,323	4,099	3,714	3,395
10	9,875	9,753	9,635	9,519	9,406	9,295	9,082	8,878	8,683	8,497	8,318	8,147	7,982	7,599	7,251	6,933	6,642	6,374	6,128	5,687	5,306	4,972	4,417	3,973
15	14,721	14,452	14,193	13,943	13,701	13,468	13,025	12,609	12,220	11,854	11,509	11,184	10,876	10,177	9,562	9,017	8,531	8,094	7,700	7,017	6,446	5,960	5,180	4,580
20	19,507	19,038	18,590	18,163	17,756	17,366	16,636	15,964	15,345	14,772	14,240	13,746	13,284	12,255	11,374	10,611	9,945	9,357	8,834	7,947	7,222	6,618	5,669	4,959
25	24,234	23,514	22,835	22,195	21,589	21,016	19,956	18,997	18,127	17,333	16,605	15,936	15,319	13,967	12,834	11,871	11,043	10,322	9,690	8,633	7,784	7,086	6,010	5,217
30	28,904	27,885	26,936	26,049	25,219	24,440	23,018	21,752	20,618	19,597	18,672	17,830	17,061	15,401	14,035	12,891	11,920	11,085	10,359	9,160	8,210	7,438	6,261	5,405
35	33,517	32,155	30,899	29,738	28,661	27,659	25,851	24,266	22,863	21,614	20,494	19,485	18,570	16,620	15,040	13,735	12,638	11,703	10,897	9,578	8,544	7,711	6,453	5,548
40	38,075	36,327	34,732	33,272	31,929	30,690	28,481	26,568	24,896	23,422	22,113	20,942	19,889	17,668	15,894	14,443	13,235	12,213	11,338	9,917	8,813	7,929	6,605	5,660
45	42,578	40,404	38,441	36,660	35,036	33,551	30,928	28,685	26,746	25,052	23,560	22,236	21,052	18,580	16,628	15,047	13,740	12,642	11,707	10,198	9,034	8,108	6,729	5,751
50	47,028	44,389	42,031	39,911	37,994	36,253	33,210	30,638	28,436	26,529	24,861	23,391	22,086	19,381	17,266	15,567	14,173	13,008	12,020	10,435	9,219	8,257	6,831	5,825
55	51,425	48,287	45,509	43,034	40,814	38,812	35,344	32,445	29,986	27,873	26,038	24,430	23,009	20,088	17,825	16,021	14,548	13,323	12,288	10,636	9,376	8,383	6,917	5,887
60	55,771	52,098	48,880	46,035	43,504	41,237	37,344	34,123	31,413	29,102	27,108	25,370	23,841	20,719	18,320	16,419	14,876	13,597	12,521	10,811	9,511	8,490	6,990	5,940
65	60,065	55,827	52,147	48,923	46,074	43,538	39,222	35,684	32,731	30,230	28,084	26,222	24,592	21,284	18,761	16,772	15,165	13,839	12,726	10,962	9,628	8,584	7,053	5,986
70	64,310	59,476	55,317	51,702	48,531	45,726	40,988	37,140	33,953	31,269	28,978	27,000	25,275	21,794	19,156	17,087	15,422	14,052	12,906	11,096	9,731	8,666	7,108	6,026
75	68,506	63,047	58,394	54,380	50,883	47,808	42,653	38,502	35,087	32,229	29,801	27,713	25,899	22,256	19,512	17,370	15,652	14,243	13,067	11,215	9,822	8,738	7,157	6,060
80	72,654	66,543	61,381	56,962	53,136	49,792	44,225	39,778	36,144	33,118	30,560	28,368	26,470	22,677	19,834	17,625	15,859	14,414	13,211	11,320	9,903	8,802	7,200	6,091
85	76,754	69,966	64,282	59,452	55,296	51,684	45,712	40,977	37,131	33,945	31,262	28,973	26,995	23,061	20,128	17,856	16,046	14,568	13,340	11,415	9,976	8,859	7,238	6,119
90	80,808	73,319	67,101	61,855	57,370	53,491	47,120	42,105	38,054	34,715	31,914	29,532	27,480	23,414	20,396	18,067	16,216	14,709	13,458	11,501	10,042	8,911	7,273	6,143
95	84,816	76,604	69,842	64,176	59,361	55,218	48,455	43,167	38,920	35,434	32,521	30,051	27,929	23,739	20,642	18,260	16,371	14,836	13,564	11,579	10,101	8,957	7,304	6,165
100	88,779	79,822	72,507	66,420	61,276	56,871	49,723	44,171	39,734	36,108	33,088	30,534	28,346	24,039	20,869	18,437	16,513	14,953	13,662	11,650	10,155	9,000	7,332	6,185
105	92,698	82,976	75,100	68,589	63,118	58,454	50,929	45,120	40,501	36,739	33,617	30,984	28,734	24,318	21,079	18,601	16,644	15,060	13,751	11,715	10,204	9,038	7,357	6,204
110	96,573	86,068	77,624	70,688	64,891	59,972	52,077	46,019	41,223	37,333	34,114	31,405	29,096	24,577	21,273	18,752	16,765	15,159	13,834	11,775	10,249	9,074	7,381	6,220
115	100,406	89,099	80,080	72,720	66,599	61,428	53,171	46,871	41,906	37,892	34,580	31,800	29,434	24,818	21,453	18,892	16,877	15,250	13,910	11,830	10,291	9,107	7,403	6,236
120	104,196	92,071	82,473	74,688	68,246	62,826	54,216	47,681	42,552	38,420	35,019	32,171	29,751	25,043	21,621	19,022	16,981	15,335	13,980	11,881	10,330	9,137	7,422	6,250
125	107,946	94,986	84,805	76,595	69,834	64,170	55,214	48,451	43,165	38,918	35,432	32,520	30,049	25,254	21,778	19,143	17,077	15,414	14,046	11,928	10,365	9,165	7,441	6,263
130	111,654	97,846	87,077	78,444	71,368	65,463	56,168	49,185	43,746	39,390	35,823	32,848	30,330	25,451	21,925	19,257	17,168	15,487	14,107	11,972	10,399	9,191	7,458	6,275
135	115,322	100,652	89,292	80,237	72,849	66,707	57,081	49,884	44,298	39,837	36,192	33,159	30,594	25,637	22,063	19,363	17,252	15,556	14,164	12,013	10,429	9,215	7,474	6,286
140	118,952	103,405	91,452	81,977	74,280	67,905	57,957	50,551	44,823	40,261	36,542	33,452	30,844	25,812	22,192	19,463	17,331	15,620	14,217	12,051	10,458	9,237	7,489	6,297
145	122,542	106,107	93,560	83,666	75,665	69,060	58,796	51,188	45,323	40,664	36,874	33,730	31,080	25,978	22,314	19,556	17,405	15,681	14,267	12,087	10,485	9,258	7,502	6,306
150	126,094	108,760	95,616	85,307	77,004	70,174	59,602	51,797	45,801	41,048	37,189	33,994	31,304	26,134	22,429	19,645	17,475	15,737	14,314	12,121	10,511	9,278	7,515	6,316
155	129,609	111,365	97,624	86,901	78,301	71,250	60,375	52,381	46,256	41,414	37,489	34,244	31,516	26,281	22,538	19,728	17,541	15,791	14,358	12,153	10,534	9,297	7,528	6,324
160	133,086	113,923	99,584	88,451	79,557	72,288	61,119	52,940	46,692	41,762	37,775	34,482	31,717	26,421	22,641	19,807	17,603	15,841	14,400	12,182	10,557	9,314	7,539	6,332
165	136,527	116,435	101,498	89,958	80,774	73,291	61,835	53,476	47,108	42,095	38,047	34,709	31,909	26,554	22,738	19,881	17,662	15,889	14,439	12,211	10,578	9,331	7,550	6,340
170	139,933	118,903	103,368	91,424	81,954	74,262	62,524	53,991	47,507	42,414	38,307	34,925	32,091	26,680	22,831	19,952	17,718	15,934	14,476	12,237	10,598	9,346	7,560	6,347
175	143,303	121,328	105,196	92,851	83,098	75,200	63,188	54,485	47,889	42,718	38,555	35,131	32,266	26,801	22,919	20,019	17,771	15,977	14,512	12,262	10,617	9,361	7,570	6,354
180	146,639	123,710	106,982	94,239	84,209	76,109	63,828	54,961	48,256	43,010	38,792	35,328	32,432	26,915	23,003	20,083	17,821	16,017	14,545	12,286	10,635	9,375	7,579	6,360
185	149,940	126,052	108,729	95,592	85,287	76,988	64,446	55,418	48,608	43,289	39,019	35,516	32,590	27,024	23,082	20,144	17,869	16,056	14,577	12,309	10,652	9,388	7,587	6,366
190	153,208	128,353	110,437	96,910	86,335	77,841	65,042	55,858	48,947	43,558	39,237	35,697	32,742	27,129	23,158	20,202	17,915	16,093	14,607	12,331	10,668	9,400	7,596	6,372
195	156,442	130,615	112,108	98,194	87,353	78,667	65,618	56,283	49,272	43,815	39,446	35,869	32,887	27,228	23,231	20,257	17,958	16,128	14,636	12,351	10,683	9,412	7,603	6,378
200	159,644	132,840	113,743	99,446	88,342	79,469	66,175	56,692	49,586	44,063	39,647	36,035	33,027	27,324	23,300	20,310	17,999	16,161	14,664	12,371	10,698	9,424	7,611	6,383

250 MCM	2 Run(s)	240 Volts	C-Value = 16,483	Magnetic Duct	Copper

Calculations:

1. $f = \frac{\text{Length} \times 2 \times \text{Isca}}{\text{Runs} \times \text{C-Value} \times \text{E L-L}}$

2. $M = \frac{1}{1 + f}$

3. Isc = Isca x M

4. Add Motor Contribution = Motor FLA x 4

* All results are given in symmetrical amperes

Single-Phase L-L Bolted Fault-Current* Table

One-Way Distance in Feet

Available Isc in Thousands (Isca)	5	10	15	20	25	30	40	50	60	70	80	90	100	125	150	175	200	225	250	300	350	400	500	600
3	2,980	2,959	2,939	2,920	2,900	2,881	2,844	2,807	2,771	2,737	2,703	2,670	2,637	2,560	2,487	2,418	2,353	2,291	2,232	2,[illegible]24	2,025	1,935	1,778	1,644
5	4,943	4,888	4,834	4,781	4,729	4,678	4,580	4,486	4,395	4,309	4,225	4,145	4,068	3,886	3,721	3,568	3,428	3,299	3,178	2,963	2,774	2,608	2,330	2,105
7	6,889	6,782	6,678	6,578	6,480	6,385	6,204	6,032	5,870	5,716	5,570	5,431	5,299	4,996	4,725	4,482	4,263	4,065	3,884	3,566	3,297	3,065	2,687	2,393
10	9,776	9,562	9,357	9,160	8,972	8,791	8,450	8,135	7,843	7,570	7,316	7,079	6,856	6,357	5,925	5,548	5,217	4,922	4,659	4,[illegible]10	3,839	3,529	3,037	2,666
15	14,501	14,035	13,597	13,186	12,799	12,435	11,764	11,162	10,618	10,126	9,676	9,265	8,888	8,066	7,383	6,807	6,315	5,888	5,516	4,897	4,403	3,999	3,379	2,926
20	19,123	18,320	17,582	16,901	16,270	15,685	14,633	13,713	12,902	12,181	11,537	10,957	10,433	9,319	8,419	7,678	7,057	6,529	6,075	5,[illegible]32	4,751	4,285	3,581	3,076
25	23,645	22,429	21,332	20,338	19,432	18,603	17,141	15,892	14,813	13,871	13,041	12,306	11,648	10,276	9,194	8,317	7,593	6,985	6,467	5,[illegible]32	4,988	4,476	3,714	3,174
30	28,070	26,373	24,869	23,528	22,324	21,237	19,353	17,775	16,436	15,284	14,283	13,405	12,629	11,032	9,794	8,805	7,998	7,326	6,759	5,[illegible]52	5,160	4,614	3,808	3,242
35	32,400	30,160	28,210	26,496	24,979	23,626	21,317	19,419	17,832	16,484	15,326	14,320	13,437	11,644	10,273	9,191	8,315	7,591	6,984	6,[illegible]20	5,290	4,718	3,879	3,293
40	36,640	33,801	31,370	29,266	27,426	25,804	23,074	20,866	19,044	17,515	16,213	15,091	14,115	12,149	10,664	9,503	8,569	7,803	7,162	6,[illegible]52	5,392	4,799	3,933	3,332
45	40,792	37,304	34,365	31,855	29,687	27,796	24,654	22,150	20,108	18,411	16,978	15,752	14,691	12,574	10,990	9,760	8,778	7,976	7,308	6,[illegible]59	5,474	4,863	3,977	3,363
50	44,858	40,676	37,206	34,282	31,784	29,626	25,083	23,297	21,049	19,196	17,643	16,323	15,186	12,935	11,265	9,977	8,953	8,120	7,428	6,347	5,541	4,917	4,012	3,389
55	48,842	43,924	39,906	36,561	33,734	31,312	27,381	24,327	21,886	19,890	18,228	16,822	15,618	13,246	11,500	10,161	9,101	8,241	7,530	6,[illegible]22	5,598	4,961	4,042	3,410
60	52,745	47,056	42,474	38,705	35,551	32,872	28,567	25,258	22,637	20,509	18,746	17,262	15,996	13,518	11,704	10,320	9,228	8,346	7,617	6,[illegible]85	5,645	4,999	4,067	3,428
65	56,571	50,077	44,920	40,726	37,249	34,318	29,653	26,104	23,314	21,062	19,207	17,653	16,331	13,756	11,883	10,458	9,339	8,436	7,692	6,539	5,687	5,031	4,088	3,443
70	60,320	52,993	47,252	42,634	38,838	35,663	30,651	26,875	23,927	21,561	19,622	18,002	16,630	13,967	12,040	10,580	9,436	8,515	7,758	6,586	5,722	5,059	4,106	3,456
75	63,997	55,809	49,479	44,439	40,330	36,917	31,573	27,581	24,485	22,013	19,995	18,316	16,897	14,156	12,179	10,687	9,521	8,584	7,815	6,628	5,754	5,083	4,122	3,467
80	67,602	58,532	51,607	46,148	41,733	38,089	32,426	28,229	24,994	22,425	20,334	18,600	17,139	14,325	12,304	10,784	9,597	8,646	7,867	6,665	5,781	5,105	4,137	3,477
85	71,133	61,164	53,643	47,768	43,054	39,186	33,218	28,828	25,462	22,801	20,643	18,858	17,357	14,477	12,417	10,870	9,666	8,701	7,912	6,698	5,806	5,124	4,149	3,486
90	74,607	63,711	55,592	49,308	44,300	40,216	33,955	29,381	25,893	23,146	20,925	19,093	17,556	14,615	12,518	10,947	9,727	8,751	7,953	6,727	5,828	5,141	4,161	3,494
95	78,011	66,176	57,460	50,772	45,479	41,185	34,643	29,895	26,291	23,463	21,184	19,309	17,739	14,741	12,611	11,018	9,783	8,796	7,991	6,754	5,848	5,157	4,171	3,501
100	81,351	68,565	59,252	52,166	46,594	42,097	35,287	30,373	26,660	23,757	21,423	19,507	17,906	14,857	12,695	11,082	9,833	8,837	8,024	6,778	5,866	5,171	4,180	3,508
105	84,630	70,879	60,972	53,495	47,651	42,959	35,890	30,819	27,003	24,028	21,644	19,690	18,060	14,962	12,772	11,141	9,879	8,875	8,055	6,800	5,883	5,184	4,188	3,514
110	87,848	73,122	62,625	54,763	48,655	43,773	36,456	31,235	27,323	24,281	21,849	19,859	18,202	15,060	12,843	11,195	9,922	8,909	8,083	6,820	5,898	5,195	4,196	3,519
115	91,008	75,299	64,214	55,975	49,609	44,543	36,989	31,626	27,621	24,516	22,039	20,016	18,334	15,150	12,909	11,245	9,961	8,940	8,109	6,838	5,911	5,206	4,203	3,524
120	94,111	77,411	65,744	57,133	50,517	45,274	37,492	31,992	27,900	24,736	22,216	20,163	18,457	15,234	12,969	11,291	9,997	8,969	8,133	6,855	5,924	5,216	4,209	3,528
125	97,159	79,461	67,217	58,242	51,382	45,968	37,966	32,337	28,162	24,942	22,382	20,299	18,571	15,312	13,026	11,333	10,030	8,996	8,155	6,871	5,936	5,225	4,215	3,532
130	100,153	81,452	68,636	59,305	52,208	46,627	38,415	32,662	28,408	25,135	22,537	20,427	18,678	15,384	13,078	11,373	10,062	9,021	8,176	6,885	5,947	5,233	4,221	3,536
135	103,095	83,387	70,005	60,325	52,996	47,255	38,840	32,969	28,640	25,316	22,683	20,546	18,777	15,452	13,127	11,410	10,090	9,044	8,195	6,899	5,957	5,241	4,226	3,540
140	105,985	85,269	71,326	61,303	53,749	47,853	39,243	33,259	28,859	25,486	22,820	20,659	18,871	15,515	13,173	11,445	10,117	9,066	8,213	6,911	5,966	5,248	4,230	3,543
145	108,826	87,098	72,602	62,243	54,471	48,424	39,626	33,534	29,065	25,647	22,949	20,764	18,959	15,575	13,216	11,477	10,143	9,086	8,229	6,923	5,975	5,255	4,235	3,546
150	111,619	88,877	73,834	63,146	55,161	48,969	39,991	33,794	29,261	25,800	23,071	20,864	19,042	15,631	13,256	11,507	10,166	9,105	8,245	6,934	5,983	5,262	4,239	3,549
155	114,354	90,609	75,025	64,016	55,823	49,490	40,338	34,042	29,446	25,944	23,186	20,958	19,121	15,683	13,294	11,536	10,189	9,123	8,260	6,945	5,991	5,267	4,243	3,552
160	117,063	92,295	76,178	64,853	56,459	49,989	40,668	34,277	29,622	26,080	23,295	21,047	19,195	15,733	13,329	11,563	10,210	9,140	8,273	6,954	5,998	5,273	4,246	3,554
165	119,717	93,937	77,293	65,659	57,069	50,467	40,984	34,501	29,789	26,209	23,398	21,131	19,265	15,780	13,363	11,588	10,229	9,156	8,286	6,964	6,005	5,278	4,250	3,557
170	122,328	95,537	78,373	66,437	57,656	50,925	41,286	34,715	29,948	26,332	23,496	21,211	19,331	15,825	13,395	11,612	10,248	9,171	8,299	6,972	6,011	5,283	4,253	3,559
175	124,896	97,096	79,419	67,187	58,220	51,365	41,574	34,918	30,100	26,450	23,589	21,287	19,394	15,867	13,425	11,635	10,266	9,185	8,310	6,980	6,017	5,288	4,256	3,561
180	127,422	98,616	80,433	67,911	58,763	51,787	41,850	35,113	30,244	26,561	23,678	21,359	19,454	15,907	13,454	11,656	10,283	9,198	8,321	6,988	6,023	5,292	4,259	3,563
185	129,907	100,098	81,416	68,611	59,286	52,193	42,115	35,299	30,382	26,667	23,762	21,428	19,511	15,945	13,481	11,677	10,299	9,211	8,332	6,995	6,029	5,297	4,262	3,565
190	132,353	101,544	82,370	69,287	59,790	52,583	42,369	35,477	30,514	26,769	23,843	21,493	19,565	15,981	13,507	11,696	10,314	9,223	8,341	7,002	6,034	5,301	4,264	3,567
195	134,760	102,955	83,296	69,941	60,277	52,959	42,612	35,648	30,640	26,866	23,920	21,556	19,617	16,016	13,532	11,715	10,328	9,235	8,351	7,009	6,039	5,304	4,267	3,569
200	137,129	104,332	84,195	70,574	60,746	53,321	42,846	35,811	30,761	26,959	23,993	21,615	19,666	16,049	13,555	11,732	10,342	9,246	8,360	7,015	6,043	5,308	4,269	3,570

300 MCM	1 Run(s)	240 Volts	C-Value = 18,176	Magnetic Duct	Copper

Calculations:

1. $f = \dfrac{\text{Length} \times 2 \times \text{Isca}}{\text{Runs} \times \text{C-Value} \times \text{E L-L}}$
2. $M = \dfrac{1}{1 + f}$
3. Isc = Isca x M
4. Add Motor Contribution = Motor FLA x 4

* All results are given in symmetrical amperes

Single-Phase L-L Bolted Fault-Current* Table

One-Way Distance in Feet

Available Isc in Thousands (Isca)

	5	10	15	20	25	30	40	50	60	70	80	90	100	125	150	175	200	225	250	300	350	400	500	600
3	2,990	2,980	2,969	2,959	2,949	2,939	2,920	2,900	2,881	2,862	2,844	2,825	2,807	2,763	2,719	2,678	2,637	2,598	2,560	2,487	2,418	2,353	2,232	2,124
5	4,972	4,943	4,915	4,888	4,861	4,834	4,781	4,729	4,678	4,629	4,580	4,532	4,486	4,373	4,266	4,165	4,068	3,975	3,886	3,721	3,568	3,428	3,178	2,963
7	6,944	6,889	6,835	6,782	6,730	6,678	6,578	6,480	6,385	6,293	6,204	6,117	6,032	5,830	5,642	5,465	5,299	5,143	4,996	4,725	4,482	4,263	3,884	3,566
10	9,887	9,776	9,668	9,562	9,458	9,357	9,160	8,972	8,791	8,617	8,450	8,290	8,135	7,773	7,441	7,137	6,856	6,597	6,357	5,925	5,548	5,217	4,659	4,210
15	14,746	14,501	14,264	14,035	13,813	13,597	13,186	12,799	12,435	12,090	11,764	11,455	11,162	10,491	9,896	9,365	8,888	8,457	8,066	7,383	6,807	6,315	5,516	4,897
20	19,552	19,123	18,713	18,320	17,943	17,582	16,901	16,270	15,685	15,141	14,633	14,158	13,713	12,714	11,850	11,097	10,433	9,845	9,319	8,419	7,678	7,057	6,075	5,332
25	24,304	23,645	23,021	22,429	21,867	21,332	20,338	19,432	18,603	17,842	17,141	16,493	15,892	14,566	13,443	12,482	11,648	10,920	10,276	9,194	8,317	7,593	6,467	5,632
30	29,003	28,070	27,195	26,373	25,599	24,869	23,528	22,324	21,237	20,251	19,353	18,531	17,775	16,132	14,767	13,615	12,629	11,777	11,032	9,794	8,805	7,998	6,759	5,852
35	33,650	32,400	31,240	30,160	29,152	28,210	26,496	24,979	23,626	22,412	21,317	20,324	19,419	17,474	15,884	14,558	13,437	12,477	11,644	10,273	9,191	8,315	6,984	6,020
40	38,246	36,640	35,163	33,801	32,540	31,370	29,266	27,426	25,804	24,362	23,074	21,915	20,866	18,638	16,839	15,357	14,115	13,058	12,149	10,664	9,503	8,569	7,162	6,152
45	42,793	40,792	38,970	37,304	35,774	34,365	31,855	29,687	27,796	26,131	24,654	23,335	22,150	19,655	17,665	16,041	14,691	13,550	12,574	10,990	9,760	8,778	7,308	6,259
50	47,290	44,858	42,665	40,676	38,864	37,206	34,282	31,784	29,626	27,742	26,083	24,611	23,297	20,553	18,387	16,634	15,186	13,971	12,935	11,265	9,977	8,953	7,428	6,347
55	51,738	48,842	46,253	43,924	41,819	39,906	36,561	33,734	31,312	29,215	27,381	25,764	24,327	21,351	19,023	17,153	15,618	14,335	13,246	11,500	10,161	9,101	7,530	6,422
60	56,139	52,745	49,738	47,056	44,648	42,474	38,705	35,551	32,872	30,568	28,567	26,811	25,258	22,065	19,588	17,611	15,996	14,653	13,518	11,704	10,320	9,228	7,617	6,485
65	60,493	56,571	53,126	50,077	47,358	44,920	40,726	37,249	34,318	31,815	29,653	27,765	26,104	22,707	20,092	18,017	16,331	14,933	13,756	11,883	10,458	9,339	7,692	6,539
70	64,801	60,320	56,420	52,993	49,958	47,252	42,634	38,838	35,663	32,968	30,651	28,639	26,875	23,288	20,546	18,381	16,630	15,183	13,967	12,040	10,580	9,436	7,758	6,586
75	69,063	63,997	59,623	55,809	52,454	49,479	44,439	40,330	36,917	34,037	31,573	29,442	27,581	23,816	20,956	18,709	16,897	15,405	14,156	12,179	10,687	9,521	7,815	6,628
80	73,280	67,602	62,741	58,532	54,852	51,607	46,148	41,733	38,089	35,030	32,426	30,183	28,229	24,298	21,328	19,005	17,139	15,606	14,325	12,304	10,784	9,597	7,867	6,665
85	77,454	71,138	65,775	61,164	57,157	53,643	47,768	43,054	39,186	35,956	33,218	30,868	28,828	24,740	21,668	19,275	17,357	15,787	14,477	12,417	10,870	9,666	7,912	6,698
90	81,584	74,607	68,730	63,711	59,375	55,592	49,308	44,300	40,216	36,822	33,955	31,503	29,381	25,147	21,979	19,521	17,556	15,951	14,615	12,518	10,947	9,727	7,953	6,727
95	85,671	78,011	71,608	66,176	61,511	57,460	50,772	45,479	41,185	37,632	34,643	32,095	29,895	25,522	22,266	19,746	17,739	16,102	14,741	12,611	11,018	9,783	7,991	6,754
100	89,717	81,351	74,413	68,565	63,569	59,252	52,166	46,594	42,097	38,392	35,287	32,646	30,373	25,870	22,530	19,953	17,906	16,239	14,857	12,695	11,082	9,833	8,024	6,778
105	93,721	84,630	77,146	70,879	65,553	60,972	53,495	47,651	42,959	39,107	35,890	33,162	30,819	26,193	22,774	20,145	18,060	16,366	14,962	12,772	11,141	9,879	8,055	6,800
110	97,684	87,848	79,812	73,122	67,468	62,625	54,763	48,655	43,773	39,781	36,456	33,645	31,235	26,493	23,001	20,322	18,202	16,483	15,060	12,843	11,195	9,922	8,083	6,820
115	101,607	91,008	82,411	75,299	69,316	64,214	55,975	49,609	44,543	40,416	36,989	34,098	31,626	26,773	23,212	20,486	18,334	16,591	15,150	12,909	11,245	9,961	8,109	6,838
120	105,490	94,111	84,948	77,411	71,102	65,744	57,133	50,517	45,274	41,017	37,492	34,525	31,992	27,036	23,409	20,640	18,457	16,691	15,234	12,969	11,291	9,997	8,133	6,855
125	109,335	97,159	87,423	79,461	72,828	67,217	58,242	51,382	45,968	41,586	37,966	34,926	32,337	27,281	23,593	20,783	18,571	16,784	15,312	13,026	11,333	10,030	8,155	6,871
130	113,141	100,153	89,840	81,452	74,497	68,636	59,305	52,208	46,627	42,125	38,415	35,306	32,662	27,512	23,765	20,916	18,578	16,872	15,384	13,078	11,373	10,062	8,176	6,885
135	116,910	103,095	92,200	83,387	76,113	70,005	60,325	52,996	47,255	42,636	38,840	35,665	32,969	27,730	23,927	21,042	18,777	16,953	15,452	13,127	11,410	10,090	8,195	6,899
140	120,641	105,985	94,505	85,269	77,677	71,326	61,303	53,749	47,853	43,123	39,243	36,004	33,259	27,935	24,080	21,160	18,371	17,029	15,515	13,173	11,445	10,117	8,213	6,911
145	124,336	108,826	96,757	87,098	79,192	72,602	62,243	54,471	48,424	43,586	39,626	36,326	33,534	28,128	24,223	21,270	18,959	17,101	15,575	13,216	11,477	10,143	8,229	6,923
150	127,994	111,619	98,958	88,877	80,660	73,834	63,146	55,161	48,969	44,027	39,991	36,632	33,794	28,311	24,359	21,375	19,042	17,169	15,631	13,256	11,507	10,166	8,245	6,934
155	131,617	114,364	101,110	90,609	82,084	75,025	64,016	55,823	49,490	44,448	40,338	36,923	34,042	28,485	24,487	21,474	19,121	17,232	15,683	13,294	11,536	10,189	8,260	6,945
160	135,205	117,063	103,214	92,295	83,465	76,178	64,853	56,459	49,989	44,849	40,668	37,200	34,277	28,649	24,609	21,567	19,195	17,292	15,733	13,329	11,563	10,210	8,273	6,954
165	138,758	119,717	105,272	93,937	84,806	77,293	65,659	57,069	50,467	45,234	40,984	37,464	34,501	28,805	24,724	21,656	19,265	17,349	15,780	13,363	11,588	10,229	8,286	6,964
170	142,277	122,328	107,285	95,537	86,108	78,373	66,437	57,656	50,925	45,601	41,286	37,716	34,715	28,954	24,833	21,739	19,331	17,403	15,825	13,395	11,612	10,248	8,299	6,972
175	145,762	124,896	109,255	97,096	87,372	79,419	67,187	58,220	51,365	45,954	41,574	37,957	34,918	29,096	24,937	21,819	19,394	17,454	15,867	13,425	11,635	10,266	8,310	6,980
180	149,215	127,422	111,183	98,616	88,601	80,433	67,911	58,763	51,787	46,291	41,850	38,187	35,113	29,231	25,037	21,895	19,454	17,503	15,907	13,454	11,656	10,283	8,321	6,988
185	152,634	129,907	113,071	100,098	89,796	81,416	68,611	59,286	52,193	46,615	42,115	38,407	35,299	29,360	25,131	21,967	19,511	17,549	15,945	13,481	11,677	10,299	8,332	6,995
190	156,022	132,353	114,919	101,544	90,957	82,370	69,287	59,790	52,583	46,926	42,369	38,618	35,477	29,483	25,221	22,036	19,565	17,593	15,981	13,507	11,696	10,314	8,341	7,002
195	159,378	134,760	116,730	102,955	92,088	83,296	69,941	60,277	52,959	47,225	42,612	38,820	35,648	29,601	25,307	22,102	19,617	17,634	16,016	13,532	11,715	10,328	8,351	7,009
200	162,702	137,129	118,503	104,332	93,188	84,195	70,574	60,746	53,321	47,513	42,846	39,014	35,811	29,713	25,390	22,165	19,666	17,674	16,049	13,555	11,732	10,342	8,360	7,015

300 MCM | 2 Run(s) | 240 Volts

C-Value = 18,176 | Magnetic Duct | Copper

Calculations:

1. $f = \dfrac{\text{Length} \times 2 \times \text{Isca}}{\text{Runs} \times \text{C-Value} \times \text{E L-L}}$

2. $M = \dfrac{1}{1 + f}$

3. Isc = Isca x M

4. Add Motor Contribution = Motor FLA x 4

* All results are given in symmetrical amperes

Single-Phase L-L Bolted Fault-Current* Table

One-Way Distance in Feet

Available Isc in Thousands (Isca)	5	10	15	20	25	30	40	50	60	70	80	90	100	125	150	175	200	225	250	300	350	400	500	600
3	2,981	2,962	2,944	2,926	2,908	2,890	2,855	2,821	2,788	2,755	2,724	2,693	2,662	2,589	2,520	2,455	2,393	2,334	2,278	2,173	2,077	1,990	1,836	1,703
5	4,948	4,896	4,846	4,797	4,749	4,702	4,610	4,522	4,437	4,355	4,277	4,201	4,127	3,955	3,796	3,649	3,514	3,388	3,271	3,059	2,873	2,709	2,430	2,204
7	6,898	6,799	6,702	6,609	6,518	6,429	6,259	6,097	5,944	5,798	5,660	5,527	5,401	5,109	4,847	4,611	4,397	4,201	4,023	3,707	3,438	3,205	2,822	2,521
10	9,793	9,594	9,403	9,220	9,044	8,874	8,553	8,254	7,976	7,716	7,472	7,243	7,028	6,542	6,118	5,747	5,417	5,124	4,861	4,408	4,032	3,715	3,211	2,827
15	14,539	14,105	13,697	13,311	12,947	12,602	11,964	11,388	10,864	10,387	9,950	9,548	9,178	8,366	7,686	7,108	6,611	6,179	5,800	5,167	4,658	4,240	3,595	3,121
20	19,188	18,440	17,748	17,106	16,509	15,952	14,944	14,055	13,267	12,562	11,928	11,355	10,835	9,721	8,815	8,063	7,430	6,889	6,421	5,653	5,050	4,562	3,824	3,292
25	23,745	22,609	21,578	20,636	19,773	18,980	17,569	16,354	15,296	14,367	13,544	12,810	12,151	10,768	9,667	8,771	8,026	7,398	6,862	5,992	5,318	4,781	3,977	3,404
30	28,210	26,622	25,203	23,928	22,775	21,729	19,900	18,355	17,033	15,888	14,888	14,006	13,223	11,601	10,333	9,315	8,480	7,782	7,191	6,242	5,514	4,938	4,085	3,483
35	32,588	30,487	28,640	27,005	25,546	24,237	21,983	20,113	18,536	17,189	16,024	15,007	14,111	12,279	10,868	9,748	8,837	8,082	7,446	6,433	5,662	5,057	4,166	3,542
40	36,880	34,212	31,904	29,887	28,111	26,533	23,856	21,670	19,850	18,313	16,996	15,857	14,860	12,842	11,307	10,099	9,125	8,322	7,649	6,584	5,779	5,150	4,229	3,587
45	41,090	37,805	35,006	32,593	30,492	28,645	25,549	23,058	21,009	19,294	17,839	16,587	15,500	13,317	11,673	10,391	9,362	8,519	7,815	6,707	5,874	5,225	4,279	3,623
50	45,219	41,272	37,959	35,138	32,708	30,592	27,087	24,303	22,038	20,159	18,575	17,222	16,053	13,723	11,984	10,637	9,561	8,683	7,953	6,808	5,951	5,286	4,320	3,653
55	49,269	44,620	40,773	37,536	34,776	32,394	28,490	25,426	22,958	20,926	19,224	17,779	16,535	14,075	12,251	10,846	9,730	8,823	8,070	6,893	6,016	5,337	4,354	3,677
60	53,244	47,856	43,458	39,800	36,710	34,066	29,776	26,445	23,785	21,611	19,801	18,271	16,960	14,381	12,483	11,027	9,876	8,942	8,170	6,966	6,072	5,381	4,383	3,698
65	57,145	50,984	46,022	41,940	38,523	35,621	30,957	27,373	24,533	22,227	20,317	18,709	17,337	14,651	12,686	11,186	10,003	9,046	8,256	7,029	6,119	5,418	4,408	3,715
70	60,974	54,010	48,473	43,966	40,226	37,073	32,048	28,222	25,213	22,783	20,781	19,102	17,674	14,891	12,865	11,325	10,114	9,137	8,332	7,084	6,161	5,451	4,429	3,731
75	64,733	56,939	50,819	45,888	41,829	38,429	33,057	29,002	25,833	23,288	21,200	19,456	17,977	15,105	13,025	11,443	10,212	9,217	8,398	7,132	6,197	5,479	4,448	3,744
80	68,424	59,775	53,067	47,712	43,339	39,701	33,993	29,720	26,401	23,749	21,582	19,776	18,250	15,298	13,168	11,559	10,300	9,288	8,458	7,174	6,229	5,504	4,465	3,756
85	72,049	62,523	55,221	49,447	44,766	40,895	34,864	30,384	26,924	24,172	21,930	20,068	18,498	15,472	13,297	11,653	10,378	9,352	8,511	7,212	6,258	5,527	4,480	3,766
90	75,610	65,187	57,289	51,098	46,115	42,018	35,677	31,000	27,406	24,560	22,248	20,335	18,725	15,630	13,413	11,747	10,449	9,410	8,558	7,247	6,284	5,547	4,493	3,775
95	79,107	67,770	59,275	52,672	47,393	43,076	36,438	31,572	27,853	24,917	22,542	20,580	18,932	15,774	13,519	11,823	10,514	9,462	8,601	7,277	6,307	5,565	4,505	3,784
100	82,544	70,277	61,184	54,174	48,606	44,075	37,150	32,105	28,267	25,248	22,812	20,805	19,122	15,906	13,616	11,903	10,572	9,509	8,640	7,305	6,328	5,581	4,515	3,791
105	85,921	72,710	63,020	55,609	49,757	45,020	37,819	32,604	28,653	25,556	23,063	21,013	19,298	16,028	13,705	11,973	10,625	9,552	8,676	7,331	6,347	5,596	4,525	3,798
110	89,241	75,073	64,787	56,981	50,853	45,915	38,449	33,071	29,013	25,842	23,296	21,206	19,461	16,140	13,787	12,033	10,675	9,592	8,709	7,354	6,364	5,609	4,534	3,804
115	92,504	77,369	66,490	58,293	51,896	46,764	39,042	33,509	29,349	26,108	23,512	21,385	19,612	16,243	13,862	12,090	10,720	9,628	8,739	7,376	6,381	5,622	4,542	3,810
120	95,711	79,600	68,131	59,551	52,890	47,570	39,602	33,920	29,665	26,358	23,714	21,552	19,752	16,339	13,932	12,143	10,762	9,662	8,767	7,395	6,395	5,633	4,549	3,815
125	98,866	81,770	69,715	60,757	53,840	48,336	40,132	34,308	29,961	26,591	23,903	21,708	19,883	16,429	13,997	12,193	10,800	9,693	8,792	7,414	6,409	5,644	4,556	3,820
130	101,968	83,880	71,243	61,915	54,747	49,066	40,634	34,674	30,240	26,811	24,080	21,854	20,005	16,512	14,058	12,239	10,836	9,722	8,816	7,431	6,422	5,654	4,563	3,825
135	105,018	85,934	72,719	63,026	55,614	49,762	41,109	35,020	30,502	27,017	24,246	21,991	20,120	16,590	14,114	12,282	10,870	9,749	8,838	7,446	6,433	5,663	4,569	3,829
140	108,019	87,933	74,145	64,095	56,444	50,425	41,561	35,348	30,751	27,211	24,403	22,120	20,228	16,664	14,167	12,322	10,901	9,775	8,859	7,461	6,444	5,671	4,574	3,833
145	110,972	89,879	75,524	65,123	57,240	51,060	41,991	35,658	30,985	27,395	24,551	22,241	20,329	16,732	14,217	12,359	10,931	9,798	8,878	7,475	6,455	5,679	4,579	3,836
150	113,877	91,776	76,859	66,113	58,003	51,666	42,401	35,953	31,208	27,569	24,690	22,355	20,424	16,797	14,264	12,394	10,958	9,820	8,897	7,488	6,464	5,687	4,584	3,840
155	116,736	93,623	78,150	67,067	58,736	52,247	42,791	36,233	31,418	27,733	24,822	22,463	20,514	16,858	14,307	12,427	10,984	9,841	8,914	7,500	6,473	5,694	4,589	3,843
160	119,549	95,425	79,402	67,986	59,440	52,803	43,163	36,500	31,619	27,889	24,946	22,566	20,600	16,915	14,349	12,459	11,008	9,861	8,930	7,511	6,482	5,700	4,593	3,846
165	122,319	97,181	80,614	68,873	60,117	53,336	43,519	36,754	31,809	28,037	25,065	22,662	20,680	16,970	14,388	12,488	11,031	9,879	8,945	7,522	6,490	5,706	4,597	3,849
170	125,045	98,894	81,789	69,729	60,768	53,848	43,859	36,996	31,991	28,178	25,177	22,754	20,757	17,021	14,425	12,516	11,053	9,897	8,959	7,532	6,497	5,712	4,601	3,851
175	127,730	100,566	82,929	70,556	61,395	54,340	44,185	37,228	32,164	28,312	25,284	22,842	20,829	17,070	14,460	12,542	11,074	9,913	8,973	7,542	6,504	5,718	4,604	3,854
180	130,373	102,197	84,035	71,355	61,999	54,813	44,497	37,449	32,329	28,440	25,386	22,925	20,899	17,116	14,493	12,567	11,093	9,929	8,985	7,551	6,511	5,723	4,608	3,856
185	132,976	103,790	85,109	72,127	62,582	55,267	44,796	37,661	32,486	28,562	25,483	23,004	20,964	17,160	14,525	12,591	11,112	9,943	8,997	7,559	6,517	5,728	4,611	3,858
190	135,540	105,345	86,152	72,875	63,144	55,705	45,084	37,864	32,637	28,678	25,576	23,080	21,027	17,202	14,555	12,614	11,129	9,958	9,009	7,567	6,523	5,733	4,614	3,861
195	138,065	106,864	87,166	73,599	63,687	56,127	45,359	38,058	32,781	28,790	25,665	23,152	21,087	17,242	14,584	12,635	11,146	9,971	9,020	7,575	6,529	5,737	4,617	3,863
200	140,553	108,349	88,151	74,300	64,211	56,534	45,625	38,245	32,920	28,896	25,749	23,221	21,144	17,281	14,611	12,656	11,162	9,984	9,030	7,582	6,535	5,741	4,619	3,864

350 MCM | 1 Run(s) | 240 Volts | C-Value = 19,703 | Magnetic Duct | Copper

Calculations:

1. $f = \dfrac{\text{Length} \times 2 \times \text{Isca}}{\text{Runs} \times \text{C-Value} \times \text{E L-L}}$
2. $M = \dfrac{1}{1 + f}$
3. Isc = Isca x M
4. Add Motor Contribution = Motor FLA x 4

* All results are given in symmetrical amperes

Single-Phase L-L Bolted Fault-Current* Table

One-Way Distance in Feet

Available Isc in Thousands (Isca)	5	10	15	20	25	30	40	50	60	70	80	90	100	125	150	175	200	225	250	300	350	400	500	600
3	2,991	2,981	2,972	2,962	2,953	2,944	2,926	2,908	2,890	2,872	2,855	2,838	2,821	2,780	2,739	2,700	2,662	2,625	2,589	2,520	2,455	2,393	2,278	2,173
5	4,974	4,948	4,922	4,896	4,871	4,846	4,797	4,749	4,702	4,655	4,610	4,566	4,522	4,416	4,316	4,219	4,127	4,039	3,955	3,796	3,649	3,514	3,271	3,059
7	6,949	6,898	6,848	6,799	6,750	6,702	6,609	6,518	6,429	6,343	6,259	6,177	6,097	5,907	5,728	5,560	5,401	5,251	5,109	4,847	4,611	4,397	4,023	3,707
10	9,895	9,793	9,693	9,594	9,498	9,403	9,220	9,044	8,874	8,711	8,553	8,401	8,254	7,909	7,592	7,299	7,028	6,776	6,542	6,118	5,747	5,417	4,861	4,408
15	14,766	14,539	14,319	14,105	13,898	13,697	13,311	12,947	12,602	12,274	11,964	11,669	11,388	10,741	10,164	9,646	9,178	8,753	8,366	7,686	7,108	6,611	5,800	5,167
20	19,586	19,188	18,807	18,440	18,087	17,748	17,106	16,509	15,952	15,431	14,944	14,486	14,055	13,083	12,237	11,493	10,835	10,248	9,721	8,815	8,063	7,430	6,421	5,653
25	24,356	23,745	23,163	22,609	22,081	21,578	20,636	19,773	18,980	18,247	17,569	16,940	16,354	15,052	13,943	12,986	12,151	11,418	10,768	9,667	8,771	8,026	6,862	5,992
30	29,078	28,210	27,393	26,622	25,893	25,203	23,928	22,775	21,729	20,774	19,900	19,096	18,355	16,731	15,372	14,216	13,223	12,359	11,601	10,333	9,315	8,480	7,191	6,242
35	33,751	32,588	31,502	30,487	29,535	28,640	27,005	25,546	24,237	23,055	21,983	21,007	20,113	18,180	16,586	15,249	14,111	13,131	12,279	10,868	9,748	8,837	7,446	6,433
40	38,377	36,880	35,496	34,212	33,018	31,904	29,887	28,111	26,533	25,124	23,856	22,710	21,670	19,442	17,630	16,127	14,860	13,778	12,842	11,307	10,099	9,125	7,649	6,584
45	42,956	41,090	39,379	37,805	36,352	35,006	32,593	30,492	28,645	27,008	25,549	24,240	23,058	20,552	18,538	16,883	15,500	14,326	13,317	11,673	10,391	9,362	7,815	6,707
50	47,489	45,219	43,155	41,272	39,546	37,959	35,138	32,708	30,592	28,733	27,087	25,620	24,303	21,536	19,334	17,541	16,053	14,797	13,723	11,984	10,637	9,561	7,953	6,808
55	51,977	49,269	46,830	44,620	42,610	40,773	37,536	34,776	32,394	30,317	28,490	26,871	25,426	22,413	20,039	18,119	16,535	15,206	14,075	12,251	10,846	9,730	8,070	6,893
60	56,421	53,244	50,406	47,856	45,551	43,458	39,800	36,710	34,066	31,776	29,776	28,012	26,445	23,201	20,666	18,631	16,960	15,565	14,381	12,483	11,027	9,876	8,170	6,966
65	60,820	57,145	53,889	50,984	48,376	46,022	41,940	38,523	35,621	33,126	30,957	29,055	27,373	23,913	21,229	19,087	17,337	15,882	14,651	12,686	11,186	10,003	8,256	7,029
70	65,176	60,974	57,281	54,010	51,092	48,473	43,966	40,226	37,073	34,377	32,048	30,013	28,222	24,558	21,736	19,496	17,674	16,164	14,891	12,865	11,325	10,114	8,332	7,084
75	69,489	64,733	60,586	56,939	53,705	50,819	45,888	41,829	38,429	35,541	33,057	30,897	29,002	25,146	22,195	19,864	17,977	16,416	15,105	13,025	11,448	10,212	8,398	7,132
80	73,761	68,424	63,808	59,775	56,221	53,067	47,712	43,339	39,701	36,626	33,993	31,713	29,720	25,684	22,614	20,199	18,250	16,644	15,298	13,168	11,559	10,300	8,458	7,174
85	77,991	72,049	66,949	62,523	58,646	55,221	49,447	44,766	40,895	37,639	34,864	32,470	30,384	26,179	22,996	20,503	18,498	16,850	15,472	13,297	11,658	10,378	8,511	7,212
90	82,180	75,610	70,012	65,187	60,983	57,289	51,098	46,115	42,018	38,589	35,677	33,174	31,000	26,634	23,347	20,782	18,725	17,038	15,630	13,413	11,747	10,449	8,558	7,247
95	86,328	79,107	73,001	67,770	63,238	59,275	52,672	47,393	43,076	39,480	36,438	33,831	31,572	27,056	23,670	21,037	18,932	17,209	15,774	13,519	11,828	10,514	8,601	7,277
100	90,437	82,544	75,918	70,277	65,416	61,184	54,174	48,606	44,075	40,317	37,150	34,444	32,105	27,447	23,969	21,273	19,122	17,367	15,906	13,616	11,903	10,572	8,640	7,305
105	94,507	85,921	78,765	72,710	67,519	63,020	55,609	49,757	45,020	41,107	37,819	35,018	32,604	27,810	24,245	21,491	19,298	17,511	16,028	13,705	11,970	10,625	8,676	7,331
110	98,539	89,241	81,546	75,073	69,552	64,787	56,981	50,853	45,915	41,851	38,449	35,557	33,071	28,149	24,503	21,693	19,461	17,645	16,140	13,787	12,033	10,675	8,709	7,354
115	102,532	92,504	84,262	77,369	71,518	66,490	58,293	51,896	46,764	42,555	39,042	36,064	33,509	28,466	24,742	21,880	19,612	17,769	16,243	13,862	12,090	10,720	8,739	7,376
120	106,488	95,711	86,915	79,600	73,420	68,131	59,551	52,890	47,570	43,222	39,602	36,542	33,920	28,762	24,966	22,055	19,752	17,884	16,339	13,932	12,143	10,762	8,767	7,395
125	110,407	98,866	89,509	81,770	75,262	69,715	60,757	53,840	48,336	43,854	40,132	36,992	34,308	29,041	25,176	22,218	19,883	17,992	16,429	13,997	12,193	10,800	8,792	7,414
130	114,290	101,968	92,044	83,880	77,047	71,243	61,915	54,747	49,066	44,453	40,634	37,418	34,674	29,303	25,372	22,371	20,005	18,092	16,512	14,058	12,239	10,836	8,816	7,431
135	118,137	105,018	94,522	85,934	78,776	72,719	63,026	55,614	49,762	45,024	41,109	37,821	35,020	29,549	25,557	22,515	20,120	18,185	16,590	14,114	12,282	10,870	8,838	7,446
140	121,948	108,019	96,947	87,933	80,452	74,145	64,095	56,444	50,425	45,566	41,561	38,204	35,348	29,782	25,731	22,650	20,228	18,273	16,664	14,167	12,322	10,901	8,859	7,461
145	125,724	110,972	99,318	89,879	82,079	75,524	65,123	57,240	51,060	46,084	41,991	38,567	35,658	30,002	25,895	22,777	20,329	18,356	16,732	14,217	12,359	10,931	8,878	7,475
150	129,466	113,877	101,639	91,776	83,657	76,859	66,113	58,003	51,666	46,577	42,401	38,912	35,953	30,211	26,050	22,897	20,424	18,434	16,797	14,264	12,394	10,958	8,897	7,488
155	133,174	116,736	103,910	93,623	85,190	78,150	67,067	58,736	52,247	47,048	42,791	39,240	36,233	30,408	26,197	23,010	20,514	18,507	16,858	14,307	12,427	10,984	8,914	7,500
160	136,848	119,549	106,133	95,425	86,679	79,402	67,986	59,440	52,803	47,499	43,163	39,553	36,500	30,596	26,336	23,117	20,600	18,576	16,915	14,349	12,459	11,008	8,930	7,511
165	140,489	122,319	108,311	97,181	88,126	80,614	68,873	60,117	53,336	47,930	43,519	39,851	36,754	30,774	26,468	23,219	20,680	18,642	16,970	14,388	12,488	11,031	8,945	7,522
170	144,098	125,045	110,443	98,894	89,532	81,789	69,729	60,768	53,848	48,343	43,859	40,136	36,996	30,944	26,593	23,315	20,757	18,704	17,021	14,425	12,516	11,053	8,959	7,532
175	147,674	127,730	112,532	100,566	90,900	82,929	70,556	61,395	54,340	48,739	44,185	40,409	37,228	31,106	26,713	23,407	20,829	18,763	17,070	14,460	12,542	11,074	8,973	7,542
180	151,219	130,373	114,578	102,197	92,231	84,035	71,355	61,999	54,813	49,119	44,497	40,670	37,449	31,260	26,826	23,494	20,899	18,819	17,116	14,493	12,567	11,093	8,985	7,551
185	154,732	132,976	116,584	103,790	93,526	85,109	72,127	62,582	55,267	49,484	44,796	40,920	37,661	31,407	26,935	23,578	20,964	18,873	17,160	14,525	12,591	11,112	8,997	7,559
190	158,215	135,540	118,550	105,345	94,787	86,152	72,875	63,144	55,705	49,835	45,084	41,159	37,864	31,548	27,039	23,657	21,027	18,923	17,202	14,555	12,614	11,129	9,009	7,567
195	161,666	138,065	120,477	106,864	96,015	87,166	73,599	63,687	56,127	50,172	45,359	41,389	38,058	31,683	27,138	23,733	21,087	18,972	17,242	14,584	12,635	11,146	9,020	7,575
200	165,088	140,553	122,367	108,349	97,212	88,151	74,300	64,211	56,534	50,497	45,625	41,610	38,245	31,812	27,232	23,805	21,144	19,018	17,281	14,611	12,656	11,162	9,030	7,582

350 MCM | 2 Run(s) | 240 Volts

C-Value = 19,703 | Magnetic Duct | Copper

Calculations:

1. $f = \dfrac{\text{Length} \times 2 \times \text{Isca}}{\text{Runs} \times \text{C-Value} \times \text{E L-L}}$

2. $M = \dfrac{1}{1 + f}$

3. Isc = Isca x M

4. Add Motor Contribution = Motor FLA x 4

* All results are given in symmetrical amperes

Single-Phase L-L Bolted Fault-Current* Table

One-Way Distance in Feet

Available Isc in Thousands (Isca)	5	10	15	20	25	30	40	50	60	70	80	90	100	125	150	175	200	225	250	300	350	400	500	600
3	2,982	2,964	2,946	2,929	2,912	2,894	2,861	2,828	2,796	2,765	2,734	2,704	2,675	2,604	2,537	2,474	2,413	2,356	2,301	2,198	2,105	2,018	1,866	1,735
5	4,950	4,901	4,853	4,805	4,759	4,713	4,625	4,540	4,458	4,379	4,303	4,229	4,158	3,990	3,835	3,691	3,558	3,434	3,319	3,110	2,925	2,762	2,484	2,257
7	6,902	6,807	6,714	6,624	6,536	6,451	6,287	6,131	5,982	5,840	5,705	5,576	5,453	5,168	4,911	4,678	4,466	4,273	4,096	3,782	3,513	3,279	2,895	2,591
10	9,801	9,611	9,427	9,250	9,080	8,916	8,605	8,315	8,044	7,790	7,552	7,328	7,116	6,638	6,220	5,851	5,524	5,231	4,968	4,513	4,135	3,816	3,305	2,914
15	14,558	14,140	13,747	13,374	13,021	12,687	12,066	11,504	10,991	10,523	10,092	9,696	9,329	8,524	7,846	7,268	6,770	6,335	5,953	5,313	4,796	4,372	3,714	3,228
20	19,221	18,501	17,832	17,210	16,631	16,088	15,104	14,233	13,457	12,761	12,133	11,565	11,047	9,935	9,027	8,270	7,631	7,083	6,609	5,829	5,213	4,715	3,959	3,411
25	23,795	22,700	21,702	20,788	19,948	19,173	17,791	16,594	15,549	14,627	13,809	13,077	12,419	11,031	9,922	9,016	8,261	7,623	7,077	6,189	5,500	4,948	4,122	3,532
30	28,281	26,748	25,373	24,133	23,008	21,983	20,185	18,659	17,347	16,208	15,209	14,326	13,540	11,907	10,625	9,593	8,743	8,032	7,427	6,456	5,709	5,117	4,238	3,617
35	32,682	30,653	28,860	27,266	25,839	24,553	22,331	20,478	18,909	17,563	16,396	15,375	14,473	12,622	11,191	10,052	9,123	8,351	7,700	6,661	5,869	5,245	4,326	3,680
40	37,001	34,421	32,177	30,207	28,465	26,913	24,267	22,094	20,279	18,739	17,416	16,268	15,262	13,218	11,657	10,426	9,430	8,608	7,917	6,823	5,994	5,345	4,393	3,730
45	41,240	38,060	35,335	32,974	30,909	29,088	26,021	23,539	21,489	19,768	18,302	17,038	15,938	13,722	12,047	10,737	9,684	8,819	8,095	6,955	6,096	5,426	4,448	3,769
50	45,401	41,576	38,346	35,582	33,189	31,098	27,618	24,838	22,567	20,676	19,078	17,709	16,523	14,154	12,379	10,999	9,897	8,995	8,244	7,064	6,179	5,492	4,492	3,800
55	49,436	44,976	41,220	38,043	35,320	32,962	29,078	26,013	23,532	21,484	19,763	18,298	17,035	14,528	12,664	11,224	10,078	9,144	8,369	7,156	6,250	5,547	4,529	3,827
60	53,497	48,265	43,966	40,370	37,317	34,694	30,418	27,080	24,402	22,206	20,373	18,819	17,486	14,855	12,912	11,413	10,234	9,273	8,477	7,234	6,309	5,594	4,560	3,849
65	57,436	51,449	46,592	42,573	39,192	36,309	31,652	28,054	25,190	22,857	20,920	19,285	17,887	15,143	13,129	11,583	10,370	9,384	8,570	7,302	6,361	5,635	4,587	3,868
70	61,305	54,532	49,106	44,663	40,956	37,818	32,793	28,946	25,907	23,446	21,412	19,702	18,246	15,399	13,321	11,737	10,490	9,482	8,651	7,361	6,406	5,670	4,611	3,885
75	65,107	57,519	51,516	46,647	42,619	39,231	33,850	29,767	26,563	23,982	21,858	20,079	18,568	15,628	13,492	11,873	10,596	9,569	8,723	7,413	6,445	5,701	4,631	3,899
80	68,842	60,415	53,826	48,533	44,188	40,557	34,833	30,524	27,164	24,471	22,263	20,421	18,860	15,835	13,646	11,988	10,690	9,646	8,787	7,459	6,480	5,728	4,649	3,912
85	72,512	63,224	56,044	50,329	45,672	41,804	35,748	31,225	27,718	24,919	22,634	20,732	19,125	16,021	13,784	12,095	10,775	9,714	8,844	7,500	6,511	5,752	4,665	3,923
90	76,120	65,949	58,175	52,041	47,077	42,978	36,603	31,875	28,229	25,332	22,973	21,017	19,367	16,191	13,909	12,191	10,851	9,777	8,896	7,537	6,539	5,774	4,679	3,933
95	79,666	68,594	60,224	53,675	48,410	44,086	37,404	32,481	28,703	25,712	23,286	21,278	19,589	16,346	14,023	12,279	10,921	9,833	8,942	7,570	6,564	5,793	4,692	3,942
100	83,153	71,163	62,196	55,235	49,676	45,133	38,155	33,046	29,143	26,065	23,575	21,519	19,793	16,487	14,128	12,359	10,984	9,884	8,984	7,601	6,586	5,811	4,703	3,951
105	86,581	73,659	64,094	56,727	50,879	46,125	38,861	33,574	29,553	26,393	23,843	21,742	19,982	16,618	14,223	12,432	11,041	9,931	9,023	7,628	6,607	5,827	4,714	3,958
110	89,952	76,086	65,923	58,155	52,025	47,064	39,526	34,069	29,936	26,698	24,091	21,949	20,156	16,738	14,312	12,499	11,094	9,974	9,058	7,654	6,626	5,842	4,724	3,965
115	93,268	78,445	67,687	59,524	53,118	47,957	40,154	34,534	30,295	26,983	24,323	22,141	20,318	16,850	14,393	12,561	11,143	10,013	9,091	7,677	6,644	5,855	4,732	3,971
120	96,530	80,739	69,388	60,836	54,160	48,805	40,746	34,972	30,631	27,249	24,539	22,320	20,469	16,953	14,468	12,619	11,189	10,049	9,121	7,698	6,660	5,868	4,741	3,977
125	99,740	82,972	71,031	62,095	55,156	49,612	41,307	35,385	30,947	27,499	24,742	22,487	20,609	17,050	14,539	12,672	11,230	10,083	9,149	7,718	6,674	5,879	4,748	3,982
130	102,898	85,146	72,618	63,304	56,108	50,381	41,839	35,774	31,245	27,733	24,932	22,644	20,741	17,140	14,604	12,722	11,269	10,115	9,175	7,736	6,688	5,890	4,755	3,987
135	106,005	87,263	74,153	64,467	57,019	51,114	42,344	36,142	31,525	27,954	25,110	22,791	20,864	17,224	14,665	12,768	11,306	10,144	9,199	7,754	6,701	5,900	4,762	3,991
140	109,064	89,325	75,636	65,586	57,893	51,815	42,824	36,491	31,790	28,163	25,278	22,929	20,980	17,302	14,722	12,811	11,340	10,171	9,221	7,769	6,713	5,909	4,768	3,996
145	112,074	91,335	77,072	66,663	58,730	52,485	43,280	36,822	32,041	28,359	25,436	23,059	21,089	17,377	14,776	12,852	11,371	10,197	9,242	7,784	6,724	5,918	4,773	4,000
150	115,038	93,294	78,462	67,700	59,534	53,126	43,715	37,137	32,279	28,545	25,586	23,182	21,192	17,446	14,826	12,890	11,401	10,221	9,262	7,798	6,734	5,926	4,778	4,003
155	117,956	95,204	79,809	68,700	60,306	53,740	44,130	37,436	32,505	28,722	25,727	23,298	21,289	17,512	14,873	12,926	11,429	10,243	9,280	7,811	6,744	5,933	4,783	4,007
160	120,830	97,067	81,114	69,665	61,048	54,328	44,526	37,720	32,719	28,889	25,861	23,408	21,380	17,574	14,918	12,960	11,456	10,264	9,298	7,824	6,753	5,940	4,788	4,010
165	123,660	98,885	82,380	70,597	61,762	54,893	44,905	37,992	32,923	29,048	25,989	23,513	21,467	17,633	14,960	12,991	11,480	10,284	9,314	7,835	6,762	5,947	4,792	4,013
170	126,447	100,659	83,608	71,496	62,450	55,436	45,267	38,251	33,118	29,199	26,110	23,612	21,550	17,688	15,000	13,022	11,504	10,303	9,329	7,846	6,770	5,953	4,796	4,016
175	129,193	102,391	84,799	72,366	63,112	55,957	45,614	38,498	33,303	29,343	26,225	23,706	21,628	17,741	15,038	13,050	11,526	10,321	9,344	7,857	6,778	5,959	4,800	4,019
180	131,897	104,083	85,956	73,207	63,751	56,458	45,947	38,735	33,480	29,480	26,334	23,795	21,703	17,791	15,074	13,077	11,547	10,338	9,358	7,867	6,785	5,965	4,804	4,021
185	134,562	105,735	87,080	74,020	64,367	56,941	46,266	38,961	33,649	29,611	26,439	23,881	21,774	17,839	15,108	13,103	11,567	10,354	9,371	7,876	6,792	5,970	4,807	4,024
190	137,188	107,350	88,172	74,808	64,962	57,406	46,572	39,179	33,811	29,737	26,539	23,962	21,841	17,884	15,141	13,127	11,587	10,369	9,384	7,885	6,799	5,975	4,811	4,026
195	139,776	108,928	89,234	75,571	65,536	57,854	46,867	39,387	33,966	29,856	26,634	24,040	21,906	17,927	15,172	13,151	11,605	10,384	9,396	7,893	6,805	5,980	4,814	4,028
200	142,327	110,470	90,267	76,310	66,092	58,287	47,150	39,587	34,114	29,971	26,725	24,114	21,967	17,969	15,202	13,173	11,622	10,398	9,407	7,901	6,811	5,985	4,817	4,030

400 MCM | 1 Run(s) | 240 Volts | | C-Value = 20,565 | Magnetic Duct | Copper

Calculations:

1. $f = \dfrac{\text{Length} \times 2 \times \text{Isca}}{\text{Runs} \times \text{C-Value} \times \text{E L-L}}$
2. $M = \dfrac{1}{1 + f}$
3. Isc = Isca x M
4. Add Motor Contribution = Motor FLA x 4

* All results are given in symmetrical amperes

Single-Phase L-L Bolted Fault-Current* Table

One-Way Distance in Feet

Available Isc in Thousands (Isca)	5	10	15	20	25	30	40	50	60	70	80	90	100	125	150	175	200	225	250	300	350	400	500	600
3	2,991	2,982	2,973	2,964	2,955	2,946	2,929	2,912	2,894	2,878	2,861	2,844	2,828	2,788	2,749	2,712	2,675	2,639	2,604	2,537	2,474	2,413	2,301	2,198
5	4,975	4,950	4,925	4,901	4,876	4,853	4,805	4,759	4,713	4,669	4,625	4,582	4,540	4,438	4,340	4,247	4,158	4,072	3,990	3,835	3,691	3,558	3,319	3,110
7	6,951	6,902	6,854	6,807	6,760	6,714	6,624	6,536	6,451	6,368	6,287	6,208	6,131	5,946	5,772	5,608	5,453	5,307	5,168	4,911	4,678	4,466	4,096	3,782
10	9,900	9,801	9,705	9,611	9,518	9,427	9,250	9,080	8,916	8,758	8,605	8,458	8,315	7,979	7,669	7,382	7,116	6,869	6,638	6,220	5,851	5,524	4,968	4,513
15	14,775	14,558	14,346	14,140	13,941	13,747	13,374	13,021	12,687	12,369	12,066	11,778	11,504	10,870	10,303	9,792	9,329	8,908	8,524	7,846	7,268	6,770	5,953	5,313
20	19,603	19,221	18,854	18,501	18,160	17,832	17,210	16,631	16,088	15,581	15,104	14,655	14,233	13,276	12,439	11,702	11,047	10,462	9,935	9,027	8,270	7,631	6,609	5,829
25	24,382	23,795	23,235	22,700	22,190	21,702	20,788	19,948	19,173	18,456	17,791	17,172	16,594	15,308	14,206	13,253	12,419	11,684	11,031	9,922	9,016	8,261	7,077	6,189
30	29,115	28,281	27,493	26,748	26,043	25,373	24,133	23,008	21,983	21,046	20,185	19,392	18,659	17,048	15,692	14,537	13,540	12,671	11,907	10,625	9,593	8,743	7,427	6,456
35	33,802	32,682	31,635	30,653	29,729	28,860	27,266	25,839	24,553	23,390	22,331	21,365	20,478	18,554	16,960	15,618	14,473	13,485	12,622	11,191	10,052	9,123	7,700	6,661
40	38,442	37,001	35,664	34,421	33,261	32,177	30,207	28,465	26,913	25,521	24,267	23,129	22,094	19,870	18,053	16,541	15,262	14,167	13,218	11,657	10,426	9,430	7,917	6,823
45	43,038	41,240	39,586	38,060	36,647	35,335	32,974	30,909	29,088	27,469	26,021	24,718	23,539	21,031	19,006	17,337	15,938	14,747	13,722	12,047	10,737	9,684	8,095	6,955
50	47,589	45,401	43,404	41,576	39,896	38,346	35,582	33,189	31,098	29,255	27,618	26,154	24,838	22,062	19,845	18,032	16,523	15,247	14,154	12,379	10,999	9,897	8,244	7,064
55	52,097	49,486	47,123	44,976	43,016	41,220	38,043	35,320	32,962	30,898	29,078	27,460	26,013	22,984	20,587	18,643	17,035	15,682	14,528	12,664	11,224	10,078	8,369	7,156
60	56,562	53,497	50,746	48,265	46,015	43,966	40,370	37,317	34,694	32,416	30,418	28,652	27,080	23,814	21,250	19,185	17,486	16,063	14,855	12,912	11,418	10,234	8,477	7,234
65	60,984	57,436	54,278	51,449	48,900	46,592	42,573	39,192	36,309	33,821	31,652	29,745	28,054	24,563	21,845	19,669	17,887	16,401	15,143	13,129	11,588	10,370	8,570	7,302
70	65,365	61,305	57,721	54,532	51,677	49,106	44,663	40,956	37,818	35,127	32,793	30,750	28,946	25,245	22,383	20,104	18,246	16,702	15,399	13,321	11,737	10,490	8,651	7,361
75	69,704	65,107	61,078	57,519	54,352	51,516	46,647	42,619	39,231	36,342	33,850	31,677	29,767	25,867	22,870	20,496	18,568	16,972	15,628	13,492	11,870	10,596	8,723	7,413
80	74,003	68,842	64,354	60,415	56,931	53,826	48,533	44,188	40,557	37,478	34,833	32,536	30,524	26,437	23,315	20,852	18,860	17,216	15,835	13,646	11,988	10,690	8,787	7,459
85	78,261	72,512	67,550	63,224	59,418	56,044	50,329	45,672	41,804	38,540	35,748	33,334	31,225	26,961	23,721	21,177	19,125	17,436	16,021	13,784	12,095	10,775	8,844	7,500
90	82,480	76,120	70,670	65,949	61,819	58,175	52,041	47,077	42,978	39,535	36,603	34,076	31,875	27,444	24,095	21,474	19,367	17,637	16,191	13,909	12,191	10,851	8,896	7,537
95	86,660	79,666	73,717	68,594	64,137	60,224	53,675	48,410	44,086	40,471	37,404	34,769	32,481	27,892	24,439	21,747	19,589	17,821	16,346	14,023	12,279	10,921	8,942	7,570
100	90,801	83,153	76,692	71,163	66,378	62,196	55,235	49,676	45,133	41,352	38,155	35,417	33,046	28,308	24,758	21,999	19,793	17,990	16,487	14,128	12,359	10,984	8,984	7,601
105	94,905	86,581	79,599	73,659	68,545	64,094	56,727	50,879	46,125	42,183	38,861	36,025	33,574	28,694	25,053	22,232	19,982	18,145	16,618	14,223	12,432	11,041	9,023	7,628
110	98,971	89,952	82,440	76,086	70,641	65,923	58,155	52,025	47,064	42,967	39,526	36,595	34,069	29,055	25,328	22,448	20,156	18,289	16,738	14,312	12,499	11,094	9,058	7,654
115	103,000	93,268	85,217	78,445	72,670	67,687	59,524	53,118	47,957	43,709	40,154	37,133	34,534	29,393	25,584	22,649	20,318	18,422	16,850	14,393	12,561	11,143	9,091	7,677
120	106,993	96,530	87,932	80,739	74,635	69,388	60,836	54,160	48,805	44,413	40,746	37,639	34,972	29,709	25,823	22,836	20,469	18,546	16,953	14,468	12,619	11,189	9,121	7,698
125	110,950	99,740	90,587	82,972	76,539	71,031	62,095	55,156	49,612	45,080	41,307	38,117	35,385	30,006	26,047	23,011	20,609	18,661	17,050	14,539	12,672	11,230	9,149	7,718
130	114,872	102,898	93,184	85,146	78,385	72,618	63,304	56,108	50,381	45,714	41,839	38,570	35,774	30,286	26,258	23,176	20,741	18,769	17,140	14,604	12,722	11,269	9,175	7,736
135	118,758	106,005	95,725	87,263	80,175	74,153	64,467	57,019	51,114	46,318	42,344	38,998	36,142	30,550	26,456	23,330	20,864	18,870	17,224	14,665	12,768	11,306	9,199	7,754
140	122,611	109,064	98,213	89,325	81,913	75,636	65,586	57,893	51,815	46,892	42,824	39,405	36,491	30,799	26,642	23,474	20,980	18,965	17,302	14,722	12,811	11,340	9,221	7,769
145	126,429	112,074	100,647	91,335	83,600	77,072	66,663	58,730	52,485	47,440	43,280	39,791	36,822	31,034	26,818	23,611	21,089	19,054	17,377	14,776	12,852	11,371	9,242	7,784
150	130,213	115,038	103,031	93,294	85,238	78,462	67,700	59,534	53,126	47,963	43,715	40,158	37,137	31,257	26,985	23,740	21,192	19,137	17,446	14,826	12,890	11,401	9,262	7,798
155	133,965	117,956	105,366	95,204	86,829	79,809	68,700	60,306	53,740	48,463	44,130	40,508	37,436	31,468	27,142	23,862	21,289	19,216	17,512	14,873	12,926	11,429	9,280	7,811
160	137,683	120,830	107,653	97,067	88,376	81,114	69,665	61,048	54,328	48,941	44,526	40,842	37,720	31,669	27,292	23,977	21,380	19,291	17,574	14,918	12,960	11,456	9,298	7,824
165	141,370	123,660	109,893	98,885	89,881	82,380	70,597	61,762	54,893	49,399	44,905	41,160	37,992	31,861	27,433	24,086	21,467	19,362	17,633	14,960	12,991	11,480	9,314	7,835
170	145,024	126,447	112,089	100,659	91,344	83,608	71,496	62,450	55,436	49,838	45,267	41,464	38,251	32,042	27,568	24,190	21,550	19,429	17,688	15,000	13,022	11,504	9,329	7,846
175	148,647	129,193	114,241	102,391	92,768	84,799	72,366	63,112	55,957	50,259	45,614	41,755	38,498	32,216	27,696	24,289	21,628	19,493	17,741	15,038	13,050	11,526	9,344	7,857
180	152,239	131,897	116,351	104,083	94,155	85,956	73,207	63,751	56,458	50,663	45,947	42,034	38,735	32,382	27,819	24,383	21,703	19,553	17,791	15,074	13,077	11,547	9,358	7,867
185	155,801	134,562	118,420	105,735	95,505	87,080	74,020	64,367	56,941	51,051	46,266	42,301	38,961	32,540	27,935	24,473	21,774	19,611	17,839	15,108	13,103	11,567	9,371	7,876
190	159,332	137,188	120,449	107,350	96,820	88,172	74,808	64,962	57,406	51,425	46,572	42,557	39,179	32,691	28,047	24,558	21,841	19,666	17,884	15,141	13,127	11,587	9,384	7,885
195	162,833	139,776	122,439	108,928	98,102	89,234	75,571	65,536	57,854	51,784	46,867	42,803	39,387	32,836	28,153	24,640	21,906	19,718	17,927	15,172	13,151	11,605	9,396	7,893
200	166,305	142,327	124,391	110,470	99,352	90,267	76,310	66,092	58,287	52,130	47,150	43,039	39,587	32,975	28,255	24,718	21,967	19,768	17,969	15,202	13,173	11,622	9,407	7,901

400 MCM | **2 Run(s)** | **240 Volts** | **C-Value = 20,565** | **Magnetic Duct** | **Copper**

Calculations:

1. $f = \dfrac{\text{Length} \times 2 \times \text{Isca}}{\text{Runs} \times \text{C-Value} \times \text{E L-L}}$
2. $M = \dfrac{1}{1 + f}$
3. $\text{Isc} = \text{Isca} \times M$
4. Add Motor Contribution = Motor FLA x 4

* All results are given in symmetrical amperes

Single-Phase L-L Bolted Fault-Current* Table

One-Way Distance in Feet

Available Isc in Thousands (Isca)	5	10	15	20	25	30	40	50	60	70	80	90	100	125	150	175	200	225	250	300	350	400	500	600
3	2,983	2,967	2,950	2,934	2,918	2,902	2,871	2,840	2,810	2,781	2,752	2,724	2,696	2,630	2,566	2,506	2,448	2,393	2,341	2,242	2,151	2,068	1,919	1,790
5	4,953	4,908	4,863	4,819	4,776	4,733	4,651	4,571	4,494	4,419	4,347	4,277	4,209	4,049	3,901	3,763	3,635	3,515	3,402	3,198	3,017	2,855	2,579	2,351
7	6,909	6,821	6,734	6,650	6,568	6,488	6,334	6,187	6,046	5,912	5,783	5,660	5,543	5,268	5,020	4,794	4,588	4,398	4,224	3,913	3,645	3,412	3,024	2,716
10	9,816	9,638	9,467	9,301	9,142	8,987	8,694	8,419	8,161	7,918	7,689	7,473	7,269	6,805	6,396	6,034	5,710	5,420	5,157	4,702	4,320	3,996	3,474	3,073
15	14,589	14,200	13,831	13,481	13,148	12,831	12,241	11,703	11,210	10,757	10,339	9,953	9,594	8,801	8,129	7,553	7,053	6,614	6,228	5,576	5,047	4,610	3,930	3,424
20	19,276	18,602	17,974	17,387	16,838	16,322	15,379	14,539	13,786	13,107	12,492	11,932	11,420	10,314	9,403	8,640	7,992	7,434	6,949	6,147	5,511	4,994	4,205	3,631
25	23,879	22,854	21,913	21,047	20,247	19,505	18,174	17,012	15,990	15,084	14,275	13,549	12,893	11,500	10,379	9,458	8,686	8,031	7,468	6,549	5,832	5,256	4,390	3,768
30	28,400	26,962	25,662	24,482	23,406	22,420	20,679	19,188	17,898	16,771	15,777	14,894	14,105	12,455	11,151	10,094	9,220	8,485	7,859	6,848	6,068	5,447	4,522	3,865
35	32,841	30,933	29,235	27,713	26,342	25,100	22,938	21,118	19,566	18,226	17,059	16,031	15,121	13,241	11,776	10,604	9,643	8,843	8,165	7,079	6,248	5,592	4,621	3,938
40	37,205	34,775	32,643	30,757	29,078	27,572	24,984	22,841	21,036	19,495	18,165	17,005	15,984	13,898	12,293	11,027	9,987	9,131	8,410	7,263	6,391	5,706	4,699	3,994
45	41,493	38,493	35,898	33,631	31,633	29,859	26,848	24,388	22,341	20,612	19,130	17,848	16,727	14,456	12,728	11,369	10,272	9,369	8,611	7,412	6,507	5,798	4,761	4,039
50	45,708	42,094	39,010	36,347	34,024	31,981	28,551	25,786	23,509	21,601	19,980	18,585	17,372	14,936	13,099	11,664	10,512	9,568	8,779	7,536	6,602	5,874	4,812	4,075
55	49,851	45,583	41,988	38,919	36,268	33,955	30,114	27,054	24,558	22,484	20,733	19,235	17,939	15,353	13,418	11,917	10,717	9,737	8,921	7,641	6,682	5,937	4,854	4,106
60	53,923	48,965	44,841	41,358	38,377	35,797	31,554	28,210	25,507	23,277	21,405	19,812	18,440	15,718	13,697	12,136	10,894	9,883	9,044	7,731	6,751	5,991	4,890	4,131
65	57,928	52,244	47,576	43,673	40,363	37,519	32,884	29,269	26,370	23,993	22,009	20,329	18,887	16,042	13,941	12,327	11,048	10,010	9,150	7,808	6,809	6,037	4,921	4,153
70	61,866	55,426	50,200	45,875	42,236	39,132	34,117	30,241	27,157	24,643	22,555	20,793	19,287	16,329	14,158	12,497	11,184	10,121	9,243	7,876	6,861	6,078	4,948	4,173
75	65,740	58,515	52,721	47,971	44,006	40,647	35,263	31,138	27,878	25,235	23,050	21,213	19,648	16,587	14,352	12,647	11,305	10,220	9,325	7,935	6,906	6,113	4,971	4,189
80	69,550	61,515	55,144	49,969	45,681	42,072	36,330	31,968	28,541	25,777	23,502	21,595	19,975	16,820	14,526	12,782	11,412	10,308	9,398	7,988	6,946	6,144	4,992	4,204
85	73,298	64,429	57,474	51,875	47,269	43,415	37,328	32,737	29,152	26,275	23,915	21,944	20,273	17,030	14,682	12,903	11,509	10,386	9,463	8,035	6,982	6,172	5,011	4,217
90	76,987	67,261	59,717	53,695	48,776	44,683	38,261	33,453	29,719	26,734	24,295	22,263	20,545	17,222	14,825	13,013	11,596	10,457	9,522	8,078	7,014	6,197	5,027	4,229
95	80,616	70,015	61,878	55,436	50,208	45,882	39,137	34,121	30,244	27,159	24,645	22,557	20,795	17,397	14,954	13,113	11,675	10,522	9,575	8,116	7,042	6,220	5,042	4,239
100	84,188	72,694	63,961	57,102	51,571	47,017	39,960	34,745	30,734	27,553	24,969	22,828	21,025	17,558	15,073	13,204	11,747	10,580	9,624	8,151	7,069	6,240	5,055	4,248
105	87,704	75,301	65,971	58,698	52,869	48,094	40,735	35,329	31,190	27,919	25,269	23,078	21,237	17,706	15,182	13,287	11,813	10,634	9,668	8,182	7,092	6,259	5,067	4,257
110	91,166	77,838	67,910	60,228	54,108	49,116	41,466	35,878	31,617	28,261	25,548	23,311	21,434	17,843	15,282	13,364	11,874	10,683	9,709	8,212	7,114	6,276	5,079	4,265
115	94,573	80,309	69,783	61,697	55,290	50,089	42,157	36,394	32,017	28,580	25,809	23,528	21,618	17,970	15,375	13,435	11,930	10,728	9,746	8,238	7,134	6,291	5,089	4,272
120	97,929	82,716	71,593	63,108	56,420	51,015	42,811	36,880	32,393	28,879	26,053	23,730	21,788	18,087	15,461	13,501	11,982	10,770	9,781	8,263	7,153	6,306	5,098	4,279
125	101,234	85,061	73,344	64,464	57,502	51,897	43,431	37,339	32,746	29,160	26,281	23,920	21,948	18,197	15,541	13,562	12,030	10,809	9,813	8,286	7,170	6,319	5,107	4,285
130	104,488	87,347	75,037	65,768	58,538	52,739	44,019	37,773	33,080	29,424	26,495	24,097	22,097	18,300	15,616	13,619	12,075	10,845	9,843	8,307	7,186	6,331	5,115	4,291
135	107,694	89,576	76,676	67,024	59,530	53,544	44,578	38,184	33,394	29,672	26,697	24,264	22,237	18,396	15,686	13,672	12,116	10,879	9,870	8,327	7,201	6,343	5,122	4,296
140	110,852	91,750	78,264	68,234	60,483	54,313	45,110	38,574	33,692	29,907	26,887	24,420	22,368	18,485	15,751	13,722	12,155	10,910	9,896	8,345	7,214	6,353	5,129	4,301
145	113,964	93,872	79,802	69,400	61,398	55,050	45,617	38,944	33,974	30,129	27,066	24,568	22,492	18,570	15,813	13,768	12,192	10,939	9,920	8,362	7,227	6,363	5,136	4,305
150	117,030	95,942	81,294	70,526	62,277	55,755	46,100	39,296	34,241	30,339	27,235	24,708	22,609	18,650	15,870	13,812	12,226	10,967	9,943	8,378	7,239	6,373	5,142	4,310
155	120,051	97,963	82,740	71,612	63,122	56,432	46,562	39,631	34,495	30,538	27,396	24,840	22,720	18,725	15,925	13,853	12,258	10,993	9,964	8,393	7,250	6,381	5,148	4,314
160	123,029	99,937	84,144	72,661	63,936	57,081	47,003	39,950	34,737	30,728	27,548	24,965	22,824	18,796	15,976	13,892	12,289	11,017	9,984	8,408	7,261	6,390	5,153	4,317
165	125,964	101,865	85,506	73,675	64,719	57,705	47,425	40,254	34,967	30,907	27,692	25,083	22,923	18,863	16,024	13,928	12,317	11,040	10,003	8,421	7,271	6,397	5,158	4,321
170	128,858	103,749	86,830	74,655	65,475	58,305	47,830	40,545	35,186	31,079	27,830	25,196	23,017	18,926	16,070	13,963	12,344	11,062	10,021	8,434	7,281	6,405	5,163	4,324
175	131,710	105,590	88,116	75,604	66,203	58,882	48,217	40,823	35,396	31,242	27,961	25,303	23,107	18,987	16,114	13,996	12,370	11,083	10,038	8,446	7,289	6,412	5,167	4,327
180	134,522	107,390	89,366	76,522	66,906	59,437	48,589	41,090	35,596	31,398	28,085	25,405	23,192	19,044	16,155	14,027	12,394	11,102	10,054	8,457	7,298	6,418	5,171	4,330
185	137,296	109,150	90,581	77,411	67,585	59,973	48,946	41,345	35,787	31,546	28,204	25,502	23,273	19,099	16,194	14,057	12,418	11,121	10,069	8,468	7,306	6,424	5,175	4,333
190	140,030	110,872	91,763	78,273	68,241	60,489	49,289	41,589	35,970	31,688	28,318	25,595	23,350	19,151	16,232	14,085	12,440	11,138	10,084	8,478	7,314	6,430	5,179	4,336
195	142,728	112,556	92,914	79,109	68,876	60,986	49,619	41,824	36,146	31,825	28,426	25,684	23,424	19,201	16,267	14,112	12,460	11,155	10,097	8,488	7,321	6,436	5,183	4,338
200	145,388	114,204	94,034	79,920	69,489	61,467	49,937	42,050	36,314	31,955	28,530	25,769	23,495	19,248	16,301	14,137	12,480	11,171	10,110	8,497	7,328	6,441	5,186	4,341

500 MCM | 1 Run(s) | 240 Volts

C-Value = 22,185 | Magnetic Duct | Copper

Calculations:

1. $f = \dfrac{\text{Length} \times 2 \times \text{Isca}}{\text{Runs} \times \text{C-Value} \times E_{L-L}}$

2. $M = \dfrac{1}{1 + f}$

3. $Isc = Isca \times M$

4. Add Motor Contribution = Motor FLA x 4

* All results are given in symmetrical amperes

Single-Phase L-L Bolted Fault-Current* Table

One-Way Distance in Feet

Available Isc in Thousands (Isca)	5	10	15	20	25	30	40	50	60	70	80	90	100	125	150	175	200	225	250	300	350	400	500	600
3	2,992	2,983	2,975	2,967	2,958	2,950	2,934	2,918	2,902	2,886	2,871	2,855	2,840	2,803	2,766	2,731	2,696	2,662	2,630	2,566	2,506	2,448	2,341	2,242
5	4,977	4,953	4,931	4,908	4,885	4,863	4,819	4,776	4,733	4,692	4,651	4,610	4,571	4,475	4,383	4,294	4,209	4,128	4,049	3,901	3,763	3,635	3,402	3,198
7	6,954	6,909	6,865	6,821	6,777	6,734	6,650	6,568	6,488	6,410	6,334	6,259	6,187	6,012	5,847	5,691	5,543	5,402	5,268	5,020	4,794	4,588	4,224	3,913
10	9,907	9,816	9,726	9,638	9,552	9,467	9,301	9,142	8,987	8,838	8,694	8,554	8,419	8,099	7,802	7,526	7,269	7,029	6,805	6,396	6,034	5,710	5,157	4,702
15	14,792	14,589	14,392	14,200	14,013	13,831	13,481	13,148	12,831	12,529	12,241	11,966	11,703	11,093	10,544	10,047	9,594	9,181	8,801	8,129	7,553	7,053	6,228	5,576
20	19,631	19,276	18,933	18,602	18,283	17,974	17,387	16,838	16,322	15,836	15,379	14,947	14,539	13,610	12,792	12,067	11,420	10,839	10,314	9,403	8,640	7,992	6,949	6,147
25	24,427	23,879	23,355	22,854	22,374	21,913	21,047	20,247	19,505	18,816	18,174	17,574	17,012	15,754	14,669	13,724	12,893	12,157	11,500	10,379	9,458	8,686	7,468	6,549
30	29,178	28,400	27,662	26,962	26,296	25,662	24,482	23,406	22,420	21,514	20,679	19,906	19,188	17,602	16,259	15,106	14,105	13,229	12,455	11,151	10,094	9,220	7,859	6,848
35	33,886	32,841	31,859	30,933	30,060	29,235	27,713	26,342	25,100	23,970	22,938	21,990	21,118	19,213	17,623	16,276	15,121	14,118	13,241	11,776	10,604	9,643	8,165	7,079
40	38,552	37,205	35,949	34,775	33,675	32,643	30,757	29,078	27,572	26,214	24,984	23,864	22,841	20,628	18,807	17,281	15,984	14,868	13,898	12,293	11,021	9,987	8,410	7,263
45	43,175	41,493	39,937	38,493	37,150	35,898	33,631	31,633	29,859	28,273	26,848	25,559	24,388	21,882	19,843	18,152	16,727	15,509	14,456	12,728	11,369	10,272	8,611	7,412
50	47,758	45,708	43,827	42,094	40,493	39,010	36,347	34,024	31,981	30,169	28,551	27,098	25,786	23,001	20,759	18,915	17,372	16,062	14,936	13,099	11,664	10,512	8,779	7,536
55	52,299	49,851	47,621	45,583	43,712	41,988	38,919	36,268	33,955	31,919	30,114	28,502	27,054	24,005	21,573	19,589	17,939	16,545	15,353	13,418	11,917	10,717	8,921	7,641
60	56,800	53,923	51,324	48,965	46,812	44,841	41,358	38,377	35,797	33,542	31,554	29,789	28,210	24,911	22,302	20,188	18,440	16,971	15,718	13,697	12,136	10,894	9,044	7,731
65	61,261	57,928	54,940	52,244	49,801	47,576	43,673	40,363	37,519	35,049	32,884	30,971	29,269	25,732	22,959	20,724	18,887	17,348	16,042	13,941	12,327	11,048	9,150	7,808
70	65,682	61,866	58,470	55,426	52,684	50,200	45,875	42,236	39,132	36,453	34,117	32,063	30,241	26,481	23,553	21,207	19,287	17,685	16,329	14,158	12,497	11,184	9,243	7,876
75	70,065	65,740	61,917	58,515	55,467	52,721	47,971	44,006	40,647	37,764	35,263	33,072	31,138	27,166	24,093	21,645	19,648	17,988	16,587	14,352	12,647	11,305	9,325	7,935
80	74,410	69,550	65,286	61,515	58,155	55,144	49,969	45,681	42,072	38,991	36,330	34,010	31,968	27,796	24,587	22,042	19,975	18,262	16,820	14,526	12,782	11,412	9,398	7,988
85	78,717	73,298	68,578	64,429	60,753	57,474	51,875	47,269	43,415	40,142	37,328	34,882	32,737	28,376	25,040	22,405	20,273	18,511	17,030	14,682	12,903	11,509	9,463	8,035
90	82,986	76,987	71,796	67,261	63,265	59,717	53,695	48,776	44,683	41,223	38,261	35,696	33,453	28,912	25,456	22,738	20,545	18,737	17,222	14,825	13,013	11,596	9,522	8,078
95	87,219	80,616	74,943	70,015	65,696	61,878	55,436	50,208	45,882	42,242	39,137	36,457	34,121	29,409	25,841	23,045	20,795	18,945	17,397	14,954	13,113	11,675	9,575	8,116
100	91,415	84,188	78,020	72,694	68,049	63,961	57,102	51,571	47,017	43,202	39,960	37,170	34,745	29,871	26,197	23,328	21,025	19,136	17,558	15,073	13,204	11,747	9,624	8,151
105	95,576	87,704	81,031	75,301	70,328	65,971	58,698	52,869	48,094	44,110	40,735	37,840	35,329	30,302	26,528	23,590	21,237	19,312	17,706	15,182	13,287	11,813	9,668	8,182
110	99,701	91,166	83,976	77,838	72,536	67,910	60,228	54,108	49,116	44,968	41,466	38,470	35,878	30,705	26,836	23,833	21,434	19,475	17,843	15,282	13,364	11,874	9,709	8,212
115	103,791	94,573	86,859	80,309	74,677	69,783	61,697	55,290	50,089	45,782	42,157	39,064	36,394	31,082	27,124	24,060	21,618	19,626	17,970	15,375	13,435	11,930	9,746	8,238
120	107,847	97,929	89,682	82,716	76,754	71,593	63,108	56,420	51,015	46,554	42,811	39,625	36,880	31,436	27,393	24,271	21,788	19,766	18,087	15,461	13,501	11,982	9,781	8,263
125	111,868	101,234	92,445	85,061	78,769	73,344	64,464	57,502	51,897	47,288	43,431	40,155	37,339	31,769	27,646	24,469	21,948	19,897	18,197	15,541	13,562	12,030	9,813	8,286
130	115,856	104,488	95,152	87,347	80,725	75,037	65,768	58,538	52,739	47,986	44,019	40,658	37,773	32,083	27,883	24,655	22,097	20,020	18,300	15,616	13,619	12,075	9,843	8,307
135	119,811	107,694	97,803	89,576	82,626	76,676	67,024	59,530	53,544	48,651	44,578	41,134	38,184	32,379	28,106	24,829	22,237	20,135	18,396	15,686	13,672	12,116	9,870	8,327
140	123,733	110,852	100,401	91,750	84,472	78,264	68,234	60,483	54,313	49,286	45,110	41,587	38,574	32,659	28,317	24,993	22,368	20,242	18,485	15,751	13,722	12,155	9,896	8,345
145	127,622	113,964	102,947	93,872	86,267	79,802	69,400	61,398	55,050	49,891	45,617	42,017	38,944	32,924	28,515	25,148	22,492	20,344	18,570	15,813	13,768	12,192	9,920	8,362
150	131,480	117,030	105,442	95,942	88,012	81,294	70,526	62,277	55,755	50,470	46,100	42,427	39,296	33,175	28,704	25,295	22,609	20,439	18,650	15,870	13,812	12,226	9,943	8,378
155	135,305	120,051	107,888	97,963	89,710	82,740	71,612	63,122	56,432	51,024	46,562	42,818	39,631	33,413	28,882	25,433	22,720	20,530	18,725	15,925	13,853	12,258	9,964	8,393
160	139,100	123,029	110,287	99,937	91,363	84,144	72,661	63,936	57,081	51,554	47,003	43,190	39,950	33,640	29,051	25,564	22,824	20,615	18,796	15,976	13,892	12,289	9,984	8,408
165	142,864	125,964	112,640	101,865	92,972	85,506	73,675	64,719	57,705	52,063	47,425	43,547	40,254	33,855	29,212	25,688	22,923	20,696	18,863	16,024	13,928	12,317	10,003	8,421
170	146,597	128,858	114,948	103,749	94,538	86,830	74,655	65,475	58,305	52,550	47,830	43,887	40,545	34,061	29,365	25,807	23,017	20,772	18,926	16,070	13,963	12,344	10,021	8,434
175	150,300	131,710	117,213	105,590	96,065	88,116	75,604	66,203	58,882	53,019	48,217	44,213	40,823	34,257	29,510	25,919	23,107	20,845	18,987	16,114	13,996	12,370	10,038	8,446
180	153,973	134,522	119,435	107,390	97,552	89,366	76,522	66,906	59,437	53,469	48,589	44,526	41,090	34,444	29,649	26,026	23,192	20,914	19,044	16,155	14,027	12,394	10,054	8,457
185	157,617	137,296	121,616	109,150	99,002	90,581	77,411	67,585	59,973	53,901	48,946	44,826	41,345	34,623	29,782	26,128	23,273	20,980	19,099	16,194	14,057	12,418	10,069	8,468
190	161,232	140,030	123,757	110,872	100,417	91,763	78,273	68,241	60,489	54,318	49,289	45,113	41,589	34,795	29,908	26,226	23,350	21,043	19,151	16,232	14,085	12,440	10,084	8,478
195	164,819	142,728	125,859	112,556	101,796	92,914	79,109	68,876	60,986	54,719	49,619	45,390	41,824	34,959	30,030	26,319	23,424	21,103	19,201	16,267	14,112	12,460	10,097	8,488
200	168,376	145,388	127,923	114,204	103,142	94,034	79,920	69,489	61,467	55,105	49,937	45,655	42,050	35,116	30,146	26,408	23,495	21,160	19,248	16,301	14,137	12,480	10,110	8,497

500 MCM	2 Run(s)	240 Volts	C-Value = 22,185	Magnetic Duct	Copper

Calculations:

1. $f = \dfrac{\text{Length} \times 2 \times \text{Isca}}{\text{Runs} \times \text{C-Value} \times \text{E L-L}}$
2. $M = \dfrac{1}{1 + f}$
3. Isc = Isca x M
4. Add Motor Contribution = Motor FLA x 4

* All results are given in symmetrical amperes

Single-Phase L-L Bolted Fault-Current* Table

One-Way Distance in Feet

Available Isc in Thousands (Isca)	5	10	15	20	25	30	40	50	60	70	80	90	100	125	150	175	200	225	250	300	350	400	500	600
3	2,984	2,968	2,952	2,936	2,921	2,905	2,875	2,845	2,816	2,788	2,760	2,732	2,705	2,641	2,579	2,523	2,464	2,410	2,358	2,261	2,172	2,090	1,943	1,815
5	4,955	4,911	4,868	4,825	4,783	4,742	4,662	4,584	4,509	4,437	4,366	4,298	4,232	4,076	3,930	3,795	3,669	3,551	3,440	3,238	3,058	2,897	2,622	2,394
7	6,912	6,827	6,743	6,662	6,582	6,504	6,354	6,211	6,074	5,943	5,818	5,698	5,582	5,313	5,069	4,845	4,642	4,454	4,281	3,973	3,706	3,472	3,084	2,773
10	9,822	9,650	9,484	9,323	9,168	9,018	8,732	8,464	8,212	7,974	7,750	7,538	7,337	6,880	6,475	6,113	5,795	5,505	5,243	4,788	4,405	4,079	3,553	3,147
15	14,603	14,226	13,868	13,527	13,203	12,894	12,318	11,791	11,307	10,862	10,450	10,068	9,713	8,927	8,258	7,682	7,182	6,743	6,354	5,697	5,163	4,721	4,031	3,516
20	19,300	18,647	18,037	17,465	16,929	16,424	15,500	14,675	13,933	13,262	12,653	12,098	11,589	10,487	9,576	8,813	8,158	7,596	7,106	6,295	5,650	5,124	4,321	3,735
25	23,915	22,921	22,006	21,161	20,378	19,652	18,344	17,199	16,188	15,290	14,487	13,763	13,108	11,715	10,590	9,662	8,883	8,221	7,650	6,718	5,988	5,401	4,516	3,880
30	28,451	27,055	25,789	24,636	23,582	22,614	20,899	19,426	18,147	17,026	16,035	15,153	14,364	12,708	11,394	10,327	9,442	8,697	8,061	7,033	6,237	5,603	4,656	3,983
35	32,910	31,056	29,399	27,910	26,565	25,344	23,209	21,406	19,863	18,528	17,361	16,332	15,418	13,526	12,048	10,861	9,887	9,073	8,383	7,276	6,428	5,756	4,762	4,060
40	37,293	34,930	32,848	31,001	29,350	27,866	25,307	23,178	21,380	19,841	18,508	17,344	16,317	14,213	12,590	11,299	10,249	9,377	8,642	7,470	6,579	5,877	4,844	4,120
45	41,603	38,683	36,146	33,922	31,955	30,204	27,220	24,773	22,730	20,998	19,511	18,221	17,091	14,797	13,046	11,665	10,549	9,628	8,854	7,629	6,701	5,975	4,910	4,168
50	45,841	42,321	39,303	36,687	34,398	32,377	28,973	26,217	23,939	22,026	20,396	18,990	17,766	15,300	13,435	11,976	10,802	9,838	9,032	7,760	6,802	6,055	4,964	4,207
55	50,010	45,849	42,328	39,309	36,692	34,402	30,584	27,529	25,029	22,945	21,181	19,669	18,359	15,738	13,772	12,242	11,019	10,017	9,183	7,871	6,888	6,123	5,010	4,239
60	54,110	49,272	45,229	41,799	38,852	36,294	32,070	28,727	26,015	23,771	21,884	20,274	18,884	16,122	14,065	12,474	11,206	10,172	9,312	7,966	6,960	6,180	5,048	4,266
65	58,143	52,595	48,013	44,166	40,889	38,065	33,445	29,826	26,913	24,518	22,515	20,815	19,353	16,463	14,324	12,676	11,369	10,306	9,425	8,049	7,023	6,229	5,081	4,290
70	62,112	55,821	50,687	46,418	42,813	39,727	34,722	30,836	27,733	25,197	23,086	21,302	19,773	16,766	14,553	12,855	11,513	10,424	9,523	8,120	7,078	6,272	5,109	4,310
75	66,017	58,955	53,258	48,565	44,633	41,289	35,909	31,769	28,485	25,817	23,605	21,743	20,153	17,038	14,757	13,075	11,640	10,529	9,611	8,184	7,126	6,310	5,134	4,328
80	69,860	62,001	55,732	50,614	46,357	42,760	37,017	32,633	29,178	26,385	24,079	22,144	20,497	17,283	14,941	13,157	11,754	10,622	9,688	8,240	7,168	6,343	5,156	4,344
85	73,643	64,963	58,113	52,570	47,993	44,148	38,052	33,436	29,818	26,907	24,513	22,511	20,811	17,506	15,107	13,286	11,857	10,705	9,758	8,290	7,206	6,373	5,176	4,358
90	77,367	67,843	60,408	54,441	49,547	45,460	39,023	34,183	30,411	27,388	24,912	22,847	21,098	17,709	15,257	13,402	11,950	10,781	9,820	8,335	7,240	6,400	5,194	4,370
95	81,033	70,646	62,620	56,231	51,025	46,702	39,934	34,880	30,961	27,834	25,281	23,156	21,361	17,894	15,395	13,508	12,034	10,849	9,877	8,376	7,271	6,424	5,209	4,381
100	84,643	73,375	64,754	57,946	52,434	47,879	40,792	35,532	31,474	28,248	25,622	23,442	21,604	18,064	15,521	13,605	12,110	10,912	9,929	8,413	7,299	6,445	5,224	4,391
105	88,198	76,031	66,814	59,590	53,776	48,996	41,600	36,144	31,953	28,633	25,938	23,707	21,829	18,221	15,636	13,694	12,181	10,969	9,976	8,447	7,324	6,465	5,237	4,401
110	91,699	78,619	68,804	61,168	55,058	50,057	42,363	36,718	32,401	28,992	26,233	23,952	22,037	18,366	15,743	13,775	12,245	11,021	10,019	8,478	7,348	6,483	5,249	4,409
115	95,147	81,140	70,728	62,684	56,283	51,068	43,084	37,259	32,821	29,328	26,507	24,181	22,231	18,500	15,841	13,851	12,305	11,069	10,059	8,507	7,369	6,500	5,260	4,417
120	98,545	83,598	72,588	64,141	57,454	52,031	43,767	37,769	33,216	29,643	26,764	24,395	22,411	18,625	15,933	13,921	12,360	11,114	10,096	8,533	7,389	6,515	5,270	4,424
125	101,892	85,994	74,388	65,542	58,576	52,949	44,415	38,250	33,588	29,939	27,005	24,595	22,580	18,741	16,018	13,986	12,411	11,155	10,130	8,557	7,407	6,530	5,279	4,430
130	105,189	88,331	76,130	66,891	59,651	53,826	45,031	38,706	33,939	30,218	27,232	24,783	22,738	18,850	16,097	14,046	12,458	11,193	10,162	8,580	7,424	6,543	5,287	4,436
135	108,439	90,612	77,818	68,190	60,683	54,664	45,616	39,137	34,270	30,480	27,445	24,959	22,886	18,951	16,171	14,102	12,503	11,229	10,191	8,601	7,440	6,555	5,295	4,442
140	111,642	92,837	79,454	69,443	61,673	55,466	46,173	39,547	34,584	30,728	27,645	25,125	23,026	19,047	16,241	14,155	12,544	11,263	10,219	8,620	7,454	6,566	5,303	4,447
145	114,799	95,010	81,040	70,652	62,624	56,234	46,704	39,936	34,881	30,962	27,835	25,281	23,157	19,137	16,306	14,205	12,583	11,294	10,244	8,639	7,468	6,577	5,310	4,452
150	117,910	97,131	82,578	71,818	63,539	56,971	47,211	40,306	35,163	31,184	28,014	25,429	23,281	19,221	16,367	14,251	12,620	11,323	10,269	8,656	7,481	6,587	5,316	4,457
155	120,978	99,203	84,071	72,945	64,419	57,678	47,695	40,658	35,431	31,395	28,184	25,569	23,398	19,301	16,425	14,295	12,654	11,351	10,291	8,672	7,493	6,596	5,322	4,461
160	124,002	101,228	85,521	74,033	65,267	58,356	48,158	40,994	35,686	31,595	28,345	25,701	23,509	19,377	16,480	14,336	12,686	11,377	10,313	8,687	7,504	6,605	5,328	4,465
165	126,985	103,206	86,929	75,086	66,083	59,008	48,602	41,315	35,929	31,785	28,498	25,827	23,614	19,448	16,531	14,375	12,717	11,402	10,333	8,702	7,515	6,613	5,333	4,469
170	129,926	105,141	88,297	76,105	66,871	59,636	49,026	41,622	36,160	31,966	28,643	25,947	23,714	19,516	16,580	14,412	12,746	11,425	10,352	8,715	7,525	6,621	5,339	4,472
175	132,826	107,032	89,627	77,091	67,631	60,239	49,434	41,915	36,381	32,139	28,782	26,060	23,809	19,580	16,627	14,447	12,773	11,447	10,370	8,728	7,535	6,629	5,343	4,476
180	135,687	108,882	90,920	78,046	68,365	60,821	49,825	42,196	36,593	32,303	28,914	26,168	23,899	19,641	16,670	14,481	12,799	11,468	10,387	8,740	7,544	6,636	5,348	4,479
185	138,509	110,692	92,179	78,971	69,074	61,382	50,200	42,465	36,795	32,461	29,040	26,272	23,985	19,699	16,712	14,512	12,824	11,487	10,403	8,751	7,552	6,642	5,352	4,482
190	141,293	112,462	93,404	79,869	69,760	61,922	50,561	42,723	36,989	32,611	29,161	26,370	24,067	19,754	16,752	14,542	12,847	11,506	10,419	8,762	7,560	6,648	5,356	4,485
195	144,039	114,195	94,596	80,739	70,423	62,444	50,909	42,971	37,174	32,756	29,276	26,464	24,146	19,807	16,790	14,571	12,870	11,524	10,433	8,773	7,568	6,654	5,360	4,487
200	146,749	115,892	95,757	81,583	71,064	62,948	51,243	43,209	37,352	32,894	29,386	26,555	24,221	19,857	16,826	14,598	12,891	11,541	10,447	8,783	7,575	6,660	5,364	4,490

600 MCM	1 Run(s)	240 Volts	C-Value = 22,965	Magnetic Duct	Copper

Calculations:

1. $f = \dfrac{\text{Length} \times 2 \times \text{Isca}}{\text{Runs} \times \text{C-Value} \times \text{E L-L}}$

2. $M = \dfrac{1}{1 + f}$

3. Isc = Isca x M

4. Add Motor Contribution = Motor FLA x 4

* All results are given in symmetrical amperes

Single-Phase L-L Bolted Fault-Current* Table

One-Way Distance in Feet

Available Isc in Thousands (Isca)

	5	10	15	20	25	30	40	50	60	70	80	90	100	125	150	175	200	225	250	300	350	400	500	600
3	2,992	2,984	2,976	2,968	2,960	2,952	2,936	2,921	2,905	2,890	2,875	2,860	2,845	2,809	2,774	2,739	2,705	2,673	2,641	2,579	2,520	2,464	2,358	2,261
5	4,977	4,955	4,933	4,911	4,889	4,868	4,825	4,783	4,742	4,701	4,662	4,623	4,584	4,491	4,401	4,315	4,232	4,152	4,076	3,930	3,795	3,669	3,440	3,238
7	6,956	6,912	6,869	6,827	6,785	6,743	6,662	6,582	6,504	6,428	6,354	6,282	6,211	6,041	5,880	5,727	5,582	5,444	5,313	5,069	4,846	4,642	4,281	3,973
10	9,910	9,822	9,735	9,650	9,566	9,484	9,323	9,168	9,018	8,873	8,732	8,596	8,464	8,151	7,861	7,590	7,337	7,101	6,880	6,475	6,116	5,795	5,243	4,788
15	14,799	14,603	14,412	14,226	14,044	13,868	13,527	13,203	12,894	12,600	12,318	12,049	11,791	11,192	10,652	10,161	9,713	9,303	8,927	8,258	7,682	7,182	6,354	5,697
20	19,644	19,300	18,968	18,647	18,337	18,037	17,465	16,929	16,424	15,949	15,500	15,076	14,675	13,759	12,951	12,232	11,589	11,010	10,487	9,576	8,810	8,158	7,106	6,295
25	24,446	23,915	23,407	22,921	22,454	22,006	21,161	20,378	19,652	18,975	18,344	17,753	17,199	15,954	14,878	13,937	13,108	12,373	11,715	10,590	9,662	8,883	7,650	6,718
30	29,205	28,451	27,736	27,055	26,407	25,789	24,636	23,582	22,614	21,723	20,899	20,136	19,426	17,853	16,516	15,365	14,364	13,485	12,708	11,394	10,327	9,442	8,061	7,033
35	33,923	32,910	31,956	31,056	30,205	29,399	27,910	26,565	25,344	24,230	23,209	22,271	21,406	19,512	17,925	16,578	15,418	14,410	13,526	12,048	10,861	9,887	8,383	7,276
40	38,599	37,293	36,073	34,930	33,857	32,848	31,001	29,350	27,866	26,525	25,307	24,196	23,178	20,973	19,151	17,621	16,317	15,192	14,213	12,590	11,299	10,249	8,642	7,470
45	43,235	41,603	40,090	38,683	37,372	36,146	33,922	31,955	30,204	28,635	27,220	25,939	24,773	22,271	20,228	18,528	17,091	15,862	14,797	13,046	11,665	10,549	8,854	7,629
50	47,830	45,841	44,011	42,321	40,757	39,303	36,687	34,398	32,377	30,581	28,973	27,526	26,217	23,430	21,180	19,323	17,766	16,441	15,300	13,435	11,976	10,802	9,032	7,760
55	52,386	50,010	47,839	45,849	44,019	42,328	39,309	36,692	34,402	32,381	30,584	28,976	27,529	24,473	22,028	20,027	18,359	16,948	15,738	13,772	12,242	11,019	9,183	7,871
60	56,903	54,110	51,578	49,272	47,164	45,229	41,799	38,852	36,294	34,052	32,070	30,307	28,727	25,415	22,788	20,654	18,884	17,394	16,122	14,065	12,474	11,206	9,312	7,966
65	61,381	58,143	55,230	52,595	50,200	48,013	44,166	40,889	38,065	35,606	33,445	31,532	29,826	26,272	23,474	21,215	19,353	17,791	16,463	14,324	12,676	11,369	9,425	8,049
70	65,820	62,112	58,798	55,821	53,130	50,687	46,418	42,813	39,727	37,056	34,722	32,664	30,836	27,053	24,096	21,722	19,773	18,146	16,766	14,553	12,855	11,513	9,523	8,120
75	70,222	66,017	62,286	58,955	55,962	53,258	48,565	44,633	41,289	38,412	35,909	33,713	31,769	27,768	24,662	22,181	20,153	18,465	17,038	14,757	13,015	11,640	9,611	8,184
80	74,587	69,860	65,696	62,001	58,700	55,732	50,614	46,357	42,760	39,682	37,017	34,687	32,633	28,426	25,179	22,598	20,497	18,754	17,283	14,941	13,157	11,754	9,688	8,240
85	78,915	73,643	69,031	64,963	61,347	58,113	52,570	47,993	44,148	40,874	38,052	35,595	33,436	29,033	25,654	22,980	20,811	19,016	17,506	15,107	13,286	11,857	9,758	8,290
90	83,207	77,367	72,293	67,843	63,910	60,408	54,441	49,547	45,460	41,996	39,023	36,443	34,183	29,594	26,092	23,331	21,098	19,255	17,709	15,257	13,402	11,950	9,820	8,335
95	87,462	81,033	75,484	70,646	66,391	62,620	56,231	51,025	46,702	43,054	39,934	37,236	34,880	30,115	26,496	23,653	21,361	19,474	17,894	15,395	13,508	12,034	9,877	8,376
100	91,683	84,643	78,607	73,375	68,795	64,754	57,946	52,434	47,879	44,052	40,792	37,981	35,532	30,600	26,871	23,951	21,604	19,676	18,064	15,521	13,605	12,110	9,929	8,413
105	95,868	88,198	81,664	76,031	71,125	66,814	59,590	53,776	48,996	44,996	41,600	38,680	36,144	31,053	27,219	24,228	21,829	19,862	18,221	15,636	13,694	12,181	9,976	8,447
110	100,019	91,699	84,657	78,619	73,385	68,804	61,168	55,058	50,057	45,890	42,363	39,339	36,718	31,476	27,543	24,485	22,037	20,034	18,366	15,743	13,775	12,245	10,019	8,478
115	104,136	95,147	87,587	81,140	75,577	70,728	62,684	56,283	51,068	46,737	43,084	39,960	37,259	31,872	27,847	24,724	22,231	20,194	18,500	15,841	13,851	12,305	10,059	8,507
120	108,219	98,545	90,458	83,598	77,705	72,588	64,141	57,454	52,031	47,542	43,767	40,547	37,769	32,245	28,130	24,947	22,411	20,343	18,625	15,933	13,921	12,360	10,096	8,533
125	112,269	101,892	93,270	85,994	79,771	74,388	65,542	58,576	52,949	48,308	44,415	41,103	38,250	32,595	28,397	25,156	22,580	20,482	18,741	16,018	13,986	12,411	10,130	8,557
130	116,286	105,189	96,026	88,331	81,778	76,130	66,891	59,651	53,826	49,037	45,031	41,629	38,706	32,925	28,647	25,353	22,738	20,612	18,850	16,097	14,046	12,458	10,162	8,580
135	120,271	108,439	98,727	90,612	83,729	77,818	68,190	60,683	54,664	49,732	45,616	42,129	39,137	33,237	28,883	25,537	22,886	20,734	18,951	16,171	14,102	12,503	10,191	8,601
140	124,223	111,642	101,375	92,837	85,626	79,454	69,443	61,673	55,466	50,395	46,173	42,604	39,547	33,532	29,105	25,711	23,026	20,848	19,047	16,241	14,155	12,544	10,219	8,620
145	128,144	114,799	103,971	95,010	87,470	81,040	70,652	62,624	56,234	51,028	46,704	43,056	39,936	33,811	29,315	25,875	23,157	20,956	19,137	16,306	14,205	12,583	10,244	8,639
150	132,033	117,910	106,517	97,131	89,265	82,578	71,818	63,539	56,971	51,634	47,211	43,486	40,306	34,076	29,514	26,030	23,281	21,057	19,221	16,367	14,251	12,620	10,269	8,656
155	135,892	120,978	109,014	99,203	91,013	84,071	72,945	64,419	57,678	52,214	47,695	43,897	40,658	34,328	29,703	26,176	23,398	21,153	19,301	16,425	14,295	12,654	10,291	8,672
160	139,720	124,002	111,464	101,228	92,714	85,521	74,033	65,267	58,356	52,769	48,158	44,289	40,994	34,567	29,882	26,315	23,509	21,244	19,377	16,480	14,336	12,686	10,313	8,687
165	143,518	126,985	113,867	103,206	94,371	86,929	75,086	66,083	59,008	53,302	48,602	44,663	41,315	34,795	30,052	26,447	23,614	21,329	19,448	16,531	14,375	12,717	10,333	8,702
170	147,286	129,926	116,227	105,141	95,985	88,297	76,105	66,871	59,636	53,813	49,026	45,022	41,622	35,012	30,214	26,572	23,714	21,411	19,516	16,580	14,412	12,746	10,352	8,715
175	151,024	132,826	118,542	107,032	97,559	89,627	77,091	67,631	60,239	54,304	49,434	45,365	41,915	35,219	30,368	26,691	23,809	21,488	19,580	16,627	14,447	12,773	10,370	8,728
180	154,733	135,687	120,815	108,882	99,094	90,920	78,046	68,365	60,821	54,776	49,825	45,694	42,196	35,417	30,515	26,805	23,899	21,562	19,641	16,670	14,481	12,799	10,387	8,740
185	158,414	138,509	123,048	110,692	100,591	92,179	78,971	69,074	61,382	55,231	50,200	46,010	42,465	35,606	30,655	26,913	23,985	21,632	19,699	16,712	14,512	12,824	10,403	8,751
190	162,066	141,293	125,240	112,462	102,051	93,404	79,869	69,760	61,922	55,668	50,561	46,313	42,723	35,788	30,790	27,017	24,067	21,698	19,754	16,752	14,542	12,847	10,419	8,762
195	165,690	144,039	127,393	114,195	103,476	94,596	80,739	70,423	62,444	56,089	50,909	46,604	42,971	35,961	30,918	27,115	24,146	21,762	19,807	16,790	14,571	12,870	10,433	8,773
200	169,286	146,749	129,508	115,892	104,867	95,757	81,583	71,064	62,948	56,496	51,243	46,884	43,209	36,128	31,041	27,210	24,221	21,823	19,857	16,826	14,598	12,891	10,447	8,783

600 MCM | **2 Run(s)** | **240 Volts** | **C-Value = 22,965** | **Magnetic Duct** | **Copper**

Calculations:

1. $f = \dfrac{\text{Length} \times 2 \times \text{Isca}}{\text{Runs} \times \text{C-Value} \times \text{E L-L}}$
2. $M = \dfrac{1}{1 + f}$
3. $\text{Isc} = \text{Isca} \times M$
4. Add Motor Contribution = Motor FLA x 4

* All results are given in symmetrical amperes

Single-Phase L-L Bolted Fault-Current* Table

One-Way Distance in Feet

Available Isc in Thousands (Isca)

Isca	5	10	15	20	25	30	40	50	60	70	80	90	100	125	150	175	200	225	250	300	350	400	500	600
3	2,985	2,969	2,954	2,939	2,924	2,910	2,881	2,852	2,824	2,797	2,770	2,744	2,718	2,656	2,597	2,540	2,485	2,433	2,383	2,289	2,202	2,121	1,976	1,850
5	4,957	4,915	4,874	4,833	4,793	4,754	4,677	4,603	4,531	4,461	4,393	4,328	4,264	4,113	3,972	3,840	3,717	3,601	3,493	3,294	3,117	2,958	2,684	2,456
7	6,916	6,835	6,755	6,677	6,601	6,527	6,383	6,245	6,113	5,987	5,866	5,749	5,637	5,376	5,138	4,9[illegible]9	4,719	4,534	4,364	4,058	3,792	3,559	3,170	2,857
10	9,830	9,666	9,508	9,354	9,205	9,061	8,787	8,528	8,284	8,054	7,836	7,629	7,433	6,985	6,588	6,234	5,915	5,628	5,367	4,912	4,528	4,200	3,668	3,256
15	14,621	14,261	13,919	13,592	13,281	12,983	12,426	11,915	11,444	11,009	10,606	10,231	9,882	9,105	8,442	7,869	7,368	6,928	6,537	5,874	5,333	4,883	4,179	3,652
20	19,333	18,708	18,123	17,573	17,056	16,568	15,671	14,867	14,141	13,483	12,883	12,334	11,831	10,734	9,824	9,056	8,400	7,832	7,336	6,511	5,853	5,316	4,492	3,889
25	23,966	23,014	22,134	21,320	20,563	19,858	18,584	17,463	16,470	15,584	14,788	14,070	13,418	12,025	10,894	9,958	9,170	8,497	7,917	6,965	6,217	5,615	4,703	4,046
30	28,523	27,184	25,966	24,852	23,829	22,888	21,212	19,764	18,502	17,391	16,406	15,526	14,736	13,073	11,748	10,656	9,767	9,008	8,358	7,304	6,486	5,833	4,855	4,158
35	33,006	31,226	29,629	28,187	26,880	25,688	23,595	21,818	20,289	18,961	17,796	16,766	15,848	13,941	12,444	11,237	10,243	9,411	8,704	7,567	6,693	6,000	4,970	4,242
40	37,416	35,146	33,136	31,343	29,734	28,282	25,766	23,661	21,874	20,338	19,004	17,834	16,799	14,672	13,023	11,707	10,632	9,739	8,983	7,777	6,857	6,131	5,060	4,307
45	41,756	38,949	36,495	34,332	32,411	30,693	27,752	25,326	23,289	21,556	20,063	18,763	17,622	15,295	13,511	12,100	10,956	10,009	9,213	7,949	6,990	6,237	5,132	4,360
50	46,027	42,639	39,716	37,167	34,926	32,940	29,576	26,836	24,560	22,641	20,999	19,579	18,340	15,833	13,930	12,434	11,229	10,237	9,406	8,092	7,100	6,325	5,191	4,402
55	50,231	46,223	42,807	39,861	37,295	35,039	31,257	28,213	25,708	23,613	21,833	20,302	18,972	16,303	14,291	12,722	11,463	10,431	9,570	8,213	7,193	6,398	5,241	4,438
60	54,369	49,703	45,776	42,423	39,528	37,003	32,811	29,472	26,750	24,489	22,580	20,947	19,534	16,715	14,608	12,972	11,666	10,599	9,710	8,316	7,272	6,461	5,283	4,468
65	58,442	53,086	48,630	44,863	41,638	38,846	34,252	30,630	27,700	25,282	23,253	21,525	20,036	17,082	14,887	13,192	11,843	10,745	9,833	8,406	7,341	6,515	5,319	4,493
70	62,453	56,375	51,375	47,190	43,635	40,578	35,592	31,697	28,570	26,005	23,862	22,046	20,487	17,408	15,134	13,386	11,999	10,873	9,940	8,484	7,400	6,562	5,350	4,516
75	66,403	59,573	54,018	49,410	45,527	42,210	36,841	32,683	29,369	26,665	24,417	22,519	20,894	17,702	15,356	13,558	12,138	10,987	10,035	8,553	7,453	6,603	5,377	4,535
80	70,292	62,685	56,564	51,532	47,322	43,748	38,007	33,598	30,106	27,271	24,924	22,950	21,265	17,967	15,555	13,713	12,262	11,088	10,120	8,615	7,499	6,640	5,402	4,553
85	74,123	65,714	59,019	53,562	49,028	45,202	39,100	34,450	30,788	27,829	25,390	23,343	21,602	18,207	15,735	13,853	12,374	11,179	10,196	8,670	7,541	6,672	5,423	4,568
90	77,897	68,664	61,387	55,505	50,651	46,579	40,126	35,243	31,420	28,345	25,818	23,705	21,912	18,427	15,898	13,980	12,474	11,262	10,264	8,719	7,578	6,702	5,442	4,581
95	81,615	71,536	63,673	57,367	52,198	47,883	41,090	35,985	32,008	28,823	26,214	24,038	22,196	18,627	16,047	14,095	12,566	11,336	10,326	8,764	7,612	6,728	5,460	4,594
100	85,278	74,335	65,881	59,153	53,672	49,121	41,998	36,679	32,556	29,267	26,581	24,346	22,458	18,812	16,184	14,200	12,650	11,404	10,382	8,804	7,643	6,752	5,475	4,605
105	88,888	77,063	68,014	60,868	55,080	50,297	42,855	37,331	33,069	29,680	26,921	24,632	22,701	18,982	16,310	14,297	12,726	11,467	10,434	8,841	7,671	6,774	5,490	4,615
110	92,445	79,722	70,078	62,515	56,425	51,417	43,665	37,945	33,549	30,067	27,239	24,897	22,927	19,139	16,426	14,386	12,797	11,524	10,481	8,875	7,696	6,794	5,503	4,624
115	95,951	82,316	72,074	64,099	57,712	52,483	44,432	38,522	34,000	30,428	27,535	25,145	23,136	19,285	16,533	14,468	12,862	11,577	10,525	8,907	7,720	6,812	5,515	4,633
120	99,407	84,846	74,007	65,623	58,945	53,501	45,159	39,068	34,424	30,767	27,813	25,376	23,332	19,421	16,633	14,544	12,922	11,625	10,565	8,936	7,741	6,829	5,526	4,641
125	102,814	87,316	75,878	67,090	60,126	54,472	45,849	39,583	34,824	31,086	28,073	25,593	23,515	19,547	16,725	14,615	12,978	11,671	10,603	8,962	7,761	6,844	5,536	4,648
130	106,172	89,727	77,692	68,504	61,260	55,401	46,505	40,071	35,201	31,386	28,318	25,796	23,686	19,665	16,812	14,681	13,030	11,713	10,637	8,987	7,780	6,859	5,546	4,654
135	109,484	92,080	79,451	69,868	62,348	56,289	47,130	40,534	35,558	31,670	28,548	25,987	23,847	19,776	16,893	14,743	13,079	11,752	10,670	9,010	7,797	6,872	5,554	4,661
140	112,750	94,380	81,157	71,184	63,393	57,140	47,725	40,973	35,895	31,937	28,765	26,167	23,998	19,880	16,969	14,801	13,124	11,789	10,700	9,032	7,813	6,885	5,562	4,666
145	115,971	96,626	82,812	72,454	64,399	57,956	48,292	41,391	36,215	32,190	28,971	26,336	24,141	19,978	17,040	14,855	13,167	11,823	10,728	9,052	7,828	6,896	5,570	4,672
150	119,147	98,821	84,419	73,681	65,367	58,738	48,835	41,789	36,520	32,430	29,165	26,497	24,276	20,070	17,107	14,906	13,207	11,855	10,755	9,071	7,843	6,907	5,577	4,677
155	122,280	100,967	85,980	74,868	66,299	59,490	49,353	42,168	36,809	32,658	29,349	26,649	24,403	20,157	17,170	14,954	13,244	11,885	10,780	9,088	7,856	6,918	5,584	4,681
160	125,371	103,065	87,497	76,015	67,197	60,212	49,849	42,529	37,084	32,875	29,524	26,792	24,524	20,240	17,230	14,999	13,280	11,914	10,803	9,105	7,868	6,927	5,590	4,686
165	128,420	105,116	88,971	77,125	68,063	60,907	50,324	42,874	37,346	33,081	29,690	26,929	24,638	20,317	17,286	15,042	13,313	11,941	10,825	9,121	7,880	6,936	5,596	4,690
170	131,429	107,124	90,405	78,200	68,899	61,575	50,780	43,205	37,596	33,277	29,848	27,059	24,747	20,391	17,339	15,082	13,345	11,966	10,846	9,136	7,891	6,945	5,602	4,694
175	134,398	109,088	91,800	79,242	69,706	62,219	51,217	43,521	37,835	33,464	29,998	27,183	24,850	20,461	17,390	15,120	13,375	11,991	10,866	9,150	7,902	6,953	5,607	4,698
180	137,327	111,010	93,157	80,251	70,486	62,840	51,636	43,823	38,064	33,643	30,142	27,300	24,949	20,528	17,438	15,157	13,403	12,013	10,885	9,163	7,911	6,961	5,612	4,701
185	140,218	112,892	94,479	81,230	71,240	63,438	52,040	44,114	38,283	33,813	30,279	27,413	25,043	20,592	17,484	15,191	13,430	12,035	10,903	9,176	7,921	6,968	5,617	4,704
190	143,072	114,734	95,766	82,180	71,969	64,016	52,428	44,392	38,492	33,977	30,410	27,520	25,132	20,652	17,528	15,224	13,456	12,056	10,919	9,188	7,930	6,975	5,621	4,708
195	145,889	116,538	97,020	83,101	72,675	64,574	52,802	44,660	38,693	34,133	30,535	27,623	25,218	20,710	17,569	15,256	13,480	12,075	10,936	9,199	7,938	6,982	5,626	4,711
200	148,670	118,306	98,242	83,996	73,359	65,113	53,161	44,917	38,886	34,283	30,655	27,721	25,299	20,765	17,609	15,285	13,504	12,094	10,951	9,210	7,946	6,988	5,630	4,713

750 MCM | 1 Run(s) | 240 Volts | C-Value = 24,136 | Magnetic Duct | Copper

Calculations:

1. $f = \dfrac{\text{Length} \times 2 \times \text{Isca}}{\text{Runs} \times \text{C-Value} \times E_{L\text{-}L}}$

2. $M = \dfrac{1}{1 + f}$

3. Isc = Isca x M

4. Add Motor Contribution = Motor FLA x 4

* All results are given in symmetrical amperes

Single-Phase L-L Bolted Fault-Current* Table

Available Isc in Thousands (Isca)	One-Way Distance in Feet																						
	5	10	15	20	25	30	40	50	60	70	80	90	100	125	150	175	200	225	250	300	350	400	500	600
3	2,992	2,985	2,977	2,969	2,962	2,954	2,939	2,924	2,910	2,895	2,881	2,866	2,852	2,818	2,784	2,751	2,718	2,687	2,656	2,597	2,540	2,485	2,383	2,289
5	4,979	4,957	4,936	4,915	4,894	4,874	4,833	4,793	4,754	4,715	4,677	4,640	4,603	4,513	4,427	4,344	4,264	4,187	4,113	3,972	3,840	3,717	3,493	3,294
7	6,958	6,916	6,875	6,835	6,795	6,755	6,677	6,601	6,527	6,454	6,383	6,313	6,245	6,081	5,926	5,778	5,637	5,504	5,376	5,138	4,919	4,719	4,364	4,058
10	9,914	9,830	9,748	9,666	9,586	9,508	9,354	9,205	9,061	8,922	8,787	8,655	8,528	8,225	7,943	7,680	7,433	7,202	6,985	6,588	6,234	5,915	5,367	4,912
15	14,808	14,621	14,439	14,261	14,088	13,919	13,592	13,281	12,983	12,698	12,426	12,165	11,915	11,332	10,804	10,322	9,882	9,478	9,105	8,442	7,869	7,368	6,537	5,874
20	19,661	19,333	19,015	18,708	18,411	18,123	17,573	17,056	16,568	16,107	15,671	15,259	14,867	13,971	13,176	12,467	11,831	11,256	10,734	9,824	9,056	8,400	7,336	6,511
25	24,472	23,966	23,480	23,014	22,565	22,134	21,320	20,563	19,858	19,200	18,584	18,006	17,463	16,239	15,176	14,243	13,418	12,684	12,025	10,894	9,958	9,170	7,917	6,965
30	29,243	28,523	27,837	27,184	26,561	25,966	24,852	23,829	22,888	22,018	21,212	20,462	19,764	18,211	16,884	15,737	14,736	13,855	13,073	11,748	10,666	9,767	8,358	7,304
35	33,974	33,006	32,091	31,226	30,407	29,629	28,187	26,880	25,688	24,597	23,595	22,671	21,818	19,940	18,360	17,012	15,848	14,834	13,941	12,444	11,237	10,243	8,704	7,567
40	38,665	37,416	36,246	35,146	34,111	33,136	31,343	29,734	28,282	26,966	25,766	24,669	23,661	21,469	19,648	18,112	16,799	15,664	14,672	13,023	11,707	10,632	8,983	7,777
45	43,317	41,756	40,304	38,949	37,682	36,495	34,332	32,411	30,693	29,149	27,752	26,484	25,326	22,830	20,783	19,072	17,622	16,376	15,295	13,511	12,100	10,956	9,213	7,949
50	47,931	46,027	44,268	42,639	41,125	39,716	37,167	34,926	32,940	31,168	29,576	28,140	26,836	24,051	21,789	19,916	18,340	16,995	15,833	13,930	12,434	11,229	9,406	8,092
55	52,507	50,231	48,143	46,223	44,449	42,807	39,861	37,295	35,039	33,040	31,257	29,657	28,213	25,150	22,688	20,664	18,972	17,536	16,303	14,291	12,722	11,463	9,570	8,213
60	57,046	54,369	51,931	49,703	47,659	45,776	42,423	39,528	37,003	34,781	32,811	31,052	29,472	26,147	23,495	21,332	19,534	18,015	16,715	14,608	12,972	11,666	9,710	8,316
65	61,547	58,442	55,636	53,086	50,760	48,630	44,863	41,638	38,846	36,405	34,252	32,340	30,630	27,054	24,225	21,932	20,036	18,441	17,082	14,887	13,192	11,843	9,833	8,406
70	66,011	62,453	59,259	56,375	53,759	51,375	47,190	43,635	40,578	37,922	35,592	33,532	31,697	27,882	24,888	22,474	20,487	18,822	17,408	15,134	13,386	11,999	9,940	8,484
75	70,440	66,403	62,803	59,573	56,660	54,018	49,410	45,527	42,210	39,343	36,841	34,638	32,683	28,643	25,492	22,965	20,894	19,166	17,702	15,356	13,558	12,138	10,035	8,553
80	74,833	70,292	66,271	62,685	59,468	56,564	51,532	47,322	43,748	40,676	38,007	35,667	33,598	29,343	26,045	23,413	21,265	19,477	17,967	15,555	13,713	12,262	10,120	8,615
85	79,190	74,123	69,666	65,714	62,187	59,019	53,562	49,028	45,202	41,930	39,100	36,628	34,450	29,991	26,554	23,823	21,602	19,760	18,207	15,735	13,853	12,374	10,196	8,670
90	83,512	77,897	72,989	68,664	64,822	61,387	55,505	50,651	46,579	43,112	40,126	37,526	35,243	30,590	27,023	24,200	21,912	20,019	18,427	15,898	13,980	12,474	10,264	8,719
95	87,800	81,615	76,244	71,536	67,376	63,673	57,367	52,198	47,883	44,227	41,090	38,368	35,985	31,147	27,457	24,548	22,196	20,256	18,627	16,047	14,095	12,566	10,326	8,764
100	92,054	85,278	79,431	74,335	69,853	65,881	59,153	53,672	49,121	45,281	41,998	39,159	36,679	31,667	27,859	24,869	22,458	20,474	18,812	16,184	14,200	12,650	10,382	8,804
105	96,274	88,888	82,554	77,063	72,256	68,014	60,868	55,080	50,297	46,279	42,855	39,903	37,331	32,151	28,234	25,167	22,701	20,676	18,982	16,310	14,297	12,726	10,434	8,841
110	100,461	92,445	85,613	79,722	74,589	70,078	62,515	56,425	51,417	47,225	43,665	40,604	37,945	32,605	28,583	25,444	22,927	20,862	19,139	16,426	14,386	12,797	10,481	8,875
115	104,615	95,951	88,612	82,316	76,855	72,074	64,099	57,712	52,483	48,123	44,432	41,267	38,522	33,031	28,910	25,703	23,136	21,036	19,285	16,533	14,468	12,862	10,525	8,907
120	108,737	99,407	91,551	84,846	79,057	74,007	65,623	58,945	53,501	48,977	45,159	41,893	39,068	33,431	29,216	25,944	23,332	21,197	19,421	16,633	14,544	12,922	10,565	8,936
125	112,827	102,814	94,433	87,316	81,196	75,878	67,090	60,126	54,472	49,790	45,849	42,486	39,583	33,808	29,503	26,171	23,515	21,348	19,547	16,725	14,615	12,978	10,603	8,962
130	116,884	106,172	97,259	89,727	83,277	77,692	68,504	61,260	55,401	50,565	46,505	43,049	40,071	34,163	29,773	26,383	23,686	21,489	19,665	16,812	14,681	13,030	10,637	8,987
135	120,911	109,484	100,031	92,080	85,301	79,451	69,868	62,348	56,289	51,304	47,130	43,584	40,534	34,499	30,028	26,583	23,847	21,622	19,776	16,893	14,743	13,079	10,670	9,010
140	124,906	112,750	102,750	94,380	87,270	81,157	71,184	63,393	57,140	52,010	47,725	44,092	40,973	34,817	30,268	26,771	23,998	21,746	19,880	16,969	14,801	13,124	10,700	9,032
145	128,871	115,971	105,418	96,626	89,187	82,812	72,454	64,399	57,956	52,685	48,292	44,576	41,391	35,118	30,496	26,949	24,141	21,863	19,978	17,040	14,855	13,167	10,728	9,052
150	132,805	119,147	108,036	98,821	91,054	84,419	73,681	65,367	58,738	53,331	48,835	45,038	41,789	35,404	30,711	27,117	24,276	21,974	20,070	17,107	14,906	13,207	10,755	9,071
155	136,710	122,280	110,606	100,967	92,873	85,980	74,868	66,299	59,490	53,949	49,353	45,478	42,168	35,675	30,915	27,276	24,403	22,078	20,157	17,170	14,954	13,244	10,780	9,088
160	140,584	125,371	113,129	103,065	94,645	87,497	76,015	67,197	60,212	54,543	49,849	45,899	42,529	35,934	31,109	27,427	24,524	22,177	20,240	17,230	14,999	13,280	10,803	9,105
165	144,430	128,420	115,606	105,116	96,372	88,971	77,125	68,063	60,907	55,112	50,324	46,302	42,874	36,180	31,293	27,570	24,638	22,270	20,317	17,286	15,042	13,313	10,825	9,121
170	148,247	131,429	118,038	107,124	98,057	90,405	78,200	68,899	61,575	55,659	50,780	46,687	43,205	36,415	31,469	27,706	24,747	22,359	20,391	17,339	15,082	13,345	10,846	9,136
175	152,035	134,398	120,427	109,088	99,700	91,800	79,242	69,706	62,219	56,184	51,217	47,056	43,521	36,639	31,636	27,836	24,850	22,443	20,461	17,390	15,120	13,375	10,866	9,150
180	155,794	137,327	122,774	111,010	101,303	93,157	80,251	70,486	62,840	56,690	51,636	47,410	43,823	36,853	31,796	27,959	24,949	22,524	20,528	17,438	15,157	13,403	10,885	9,163
185	159,526	140,218	125,080	112,892	102,868	94,479	81,230	71,240	63,438	57,176	52,040	47,750	44,114	37,058	31,949	28,077	25,043	22,600	20,592	17,484	15,191	13,430	10,903	9,176
190	163,230	143,072	127,346	114,734	104,395	95,766	82,180	71,969	64,016	57,645	52,428	48,077	44,392	37,255	32,094	28,190	25,132	22,673	20,652	17,528	15,224	13,456	10,919	9,188
195	166,907	145,889	129,572	116,538	105,887	97,020	83,101	72,675	64,574	58,097	52,802	48,391	44,660	37,443	32,234	28,297	25,218	22,742	20,710	17,569	15,256	13,480	10,936	9,199
200	170,556	148,670	131,761	118,306	107,344	98,242	83,996	73,359	65,113	58,533	53,161	48,693	44,917	37,624	32,368	28,400	25,299	22,809	20,765	17,609	15,285	13,504	10,951	9,210

750 MCM	2 Run(s)	240 Volts	C-Value = 24,136	Magnetic Duct	Copper

Calculations:

1. $f = \dfrac{\text{Length} \times 2 \times \text{Isca}}{\text{Runs} \times \text{C-Value} \times \text{E L-L}}$

2. $M = \dfrac{1}{1 + f}$

3. Isc = Isca x M

4. Add Motor Contribution = Motor FLA x 4

* All results are given in symmetrical amperes

Single-Phase L-L Bolted Fault-Current* Table

One-Way Distance in Feet

Available Isc in Thousands (Isca)

	5	10	15	20	25	30	40	50	60	70	80	90	100	125	150	175	200	225	250	300	350	400	500	600
3	2,985	2,971	2,956	2,942	2,928	2,914	2,886	2,859	2,832	2,806	2,780	2,755	2,730	2,670	2,612	2,557	2,505	2,454	2,405	2,314	2,229	2,150	2,007	1,883
5	4,959	4,919	4,879	4,840	4,802	4,764	4,691	4,619	4,550	4,483	4,417	4,354	4,292	4,146	4,009	3,881	3,760	3,647	3,541	3,346	3,171	3,013	2,741	2,514
7	6,920	6,842	6,766	6,691	6,618	6,547	6,408	6,276	6,149	6,026	5,909	5,796	5,688	5,433	5,200	4,986	4,789	4,608	4,439	4,136	3,872	3,640	3,250	2,935
10	9,838	9,681	9,529	9,381	9,239	9,100	8,835	8,585	8,349	8,125	7,913	7,712	7,521	7,082	6,691	6,341	6,027	5,741	5,482	5,028	4,643	4,313	3,776	3,358
15	14,638	14,293	13,964	13,650	13,350	13,062	12,523	12,026	11,568	11,143	10,748	10,380	10,037	9,270	8,612	8,041	7,541	7,100	6,708	6,040	5,493	5,037	4,320	3,781
20	19,362	18,763	18,200	17,670	17,170	16,697	15,826	15,041	14,331	13,684	13,094	12,552	12,053	10,964	10,055	9,286	8,626	8,053	7,552	6,[illegible]16	6,047	5,499	4,655	4,036
25	24,011	23,096	22,249	21,462	20,729	20,044	18,802	17,704	16,728	15,854	15,066	14,353	13,705	12,314	11,179	10,236	9,440	8,758	8,169	7,[illegible]99	6,436	5,818	4,882	4,205
30	28,586	27,300	26,124	25,046	24,053	23,136	21,496	20,074	18,828	17,727	16,749	15,872	15,083	13,415	12,080	10,986	10,074	9,302	8,639	7,562	6,724	6,053	5,046	4,326
35	33,091	31,379	29,836	28,438	27,164	26,000	23,947	22,195	20,682	19,362	18,200	17,170	16,250	14,331	12,817	11,592	10,581	9,733	9,010	7,[illegible]45	6,947	6,233	5,170	4,418
40	37,526	35,340	33,395	31,652	30,083	28,661	26,187	24,106	22,331	20,800	19,465	18,292	17,251	15,104	13,432	12,093	10,997	10,083	9,310	8,[illegible]71	7,123	6,375	5,268	4,488
45	41,893	39,187	36,809	34,703	32,826	31,141	28,241	25,836	23,808	22,076	20,578	19,271	18,120	15,765	13,952	12,513	11,344	10,374	9,557	8,[illegible]56	7,267	6,490	5,346	4,545
50	46,193	42,925	40,088	37,603	35,409	33,456	30,133	27,410	25,138	23,214	21,564	20,133	18,880	16,338	14,399	12,871	11,637	10,619	9,764	8,[illegible]10	7,386	6,585	5,410	4,591
55	50,428	46,558	43,240	40,363	37,845	35,623	31,879	28,847	26,342	24,237	22,444	20,898	19,551	16,838	14,786	13,180	11,888	10,828	9,940	8,541	7,487	6,664	5,464	4,630
60	54,600	50,092	46,271	42,992	40,147	37,655	33,497	30,166	27,437	25,161	23,234	21,581	20,148	17,279	15,125	13,448	12,107	11,008	10,092	8,653	7,573	6,732	5,510	4,663
65	58,710	53,529	49,189	45,500	42,326	39,565	35,000	31,379	28,438	26,000	23,948	22,195	20,682	17,670	15,424	13,684	12,297	11,166	10,225	8,[illegible]50	7,647	6,791	5,549	4,691
70	62,759	56,875	52,000	47,895	44,390	41,364	36,400	32,500	29,355	26,765	24,595	22,750	21,163	18,020	15,690	13,893	12,466	11,304	10,341	8,[illegible]35	7,712	6,842	5,583	4,715
75	66,748	60,132	54,710	50,184	46,350	43,060	37,707	33,538	30,199	27,465	25,185	23,254	21,598	18,335	15,928	14,080	12,616	11,427	10,444	8,[illegible]10	7,769	6,887	5,613	4,736
80	70,680	63,304	57,323	52,374	48,212	44,663	38,931	34,503	30,979	28,108	25,725	23,714	21,994	18,619	16,142	14,247	12,750	11,537	10,536	8,[illegible]77	7,820	6,927	5,639	4,755
85	74,554	66,395	59,845	54,472	49,984	46,179	40,078	35,401	31,701	28,701	26,220	24,134	22,356	18,877	16,336	14,397	12,870	11,636	10,618	9,[illegible]36	7,865	6,962	5,663	4,772
90	78,373	69,407	62,282	56,483	51,672	47,616	41,156	36,239	32,372	29,250	26,678	24,521	22,687	19,113	16,512	14,534	12,979	11,725	10,692	9,[illegible]90	7,905	6,994	5,684	4,787
95	82,138	72,343	64,636	58,412	53,282	48,980	42,171	37,024	32,996	29,759	27,100	24,878	22,992	19,329	16,673	14,659	13,079	11,806	10,759	9,[illegible]39	7,942	7,023	5,703	4,800
100	85,849	75,207	66,912	60,265	54,819	50,276	43,128	37,760	33,580	30,233	27,493	25,208	23,274	19,528	16,821	14,773	13,169	11,880	10,821	9,[illegible]83	7,976	7,049	5,720	4,812
105	89,508	78,000	69,114	62,046	56,289	51,510	44,032	38,451	34,125	30,674	27,857	25,514	23,535	19,711	16,957	14,877	13,253	11,948	10,877	9,223	8,006	7,073	5,735	4,823
110	93,116	80,726	71,246	63,758	57,695	52,684	44,888	39,102	34,637	31,087	28,197	25,799	23,777	19,881	17,082	14,974	13,329	12,010	10,928	9,260	8,034	7,094	5,750	4,833
115	96,675	83,387	73,310	65,406	59,041	53,805	45,699	39,716	35,118	31,474	28,515	26,065	24,002	20,038	17,198	15,063	13,400	12,067	10,975	9,294	8,059	7,114	5,763	4,843
120	100,184	85,984	75,311	66,994	60,332	54,875	46,468	40,295	35,570	31,837	28,813	26,313	24,213	20,185	17,306	15,146	13,465	12,120	11,019	9,325	8,083	7,133	5,775	4,851
125	103,645	88,522	77,250	68,524	61,570	55,897	47,199	40,844	35,997	32,178	29,092	26,546	24,410	20,322	17,406	15,223	13,526	12,169	11,060	9,355	8,105	7,150	5,786	4,859
130	107,059	91,000	79,131	70,000	62,759	56,875	47,895	41,364	36,400	32,500	29,355	26,765	24,595	20,450	17,500	15,294	13,582	12,215	11,098	9,382	8,125	7,165	5,796	4,866
135	110,427	93,422	80,956	71,425	63,901	57,812	48,558	41,857	36,782	32,804	29,603	26,971	24,768	20,569	17,588	15,361	13,635	12,258	11,133	9,407	8,144	7,180	5,806	4,873
140	113,750	95,790	82,728	72,800	65,000	58,710	49,189	42,326	37,143	33,091	29,836	27,164	24,932	20,632	17,670	15,424	13,684	12,297	11,166	9,430	8,162	7,194	5,815	4,879
145	117,029	98,104	84,448	74,130	66,058	59,571	49,793	42,772	37,486	33,363	30,057	27,347	25,086	20,788	17,747	15,483	13,731	12,335	11,197	9,452	8,178	7,207	5,823	4,885
150	120,265	100,368	86,120	75,415	67,076	60,399	50,369	43,196	37,812	33,621	30,266	27,520	25,231	20,888	17,820	15,538	13,774	12,370	11,225	9,473	8,193	7,218	5,831	4,891
155	123,458	102,582	87,745	76,658	68,058	61,193	50,921	43,602	38,122	33,866	30,465	27,684	25,369	20,982	17,889	15,590	13,815	12,403	11,253	9,492	8,208	7,230	5,838	4,896
160	126,609	104,749	89,326	77,861	69,005	61,958	51,449	43,988	38,417	34,099	30,653	27,840	25,499	21,071	17,953	15,639	13,854	12,434	11,278	9,510	8,221	7,240	5,845	4,901
165	129,719	106,869	90,863	79,027	69,919	62,693	51,955	44,358	38,699	34,320	30,832	27,987	25,623	21,155	18,015	15,686	13,890	12,463	11,302	9,527	8,234	7,250	5,852	4,905
170	132,790	108,944	92,359	80,156	70,801	63,402	52,441	44,711	38,968	34,531	31,002	28,127	25,741	21,236	18,073	15,730	13,925	12,491	11,325	9,544	8,246	7,260	5,858	4,910
175	135,821	110,976	93,815	81,250	71,654	64,085	52,907	45,050	39,224	34,733	31,165	28,261	25,852	21,312	18,128	15,771	13,957	12,517	11,347	9,559	8,258	7,268	5,863	4,914
180	138,814	112,966	95,233	82,312	72,478	64,743	53,355	45,374	39,470	34,926	31,320	28,388	25,959	21,384	18,180	15,811	13,988	12,542	11,367	9,573	8,269	7,277	5,869	4,917
185	141,769	114,915	96,614	83,342	73,276	65,379	53,786	45,686	39,705	35,110	31,468	28,510	26,061	21,453	18,230	15,849	14,018	12,566	11,387	9,587	8,279	7,285	5,874	4,921
190	144,686	116,825	97,961	84,342	74,048	65,993	54,201	45,984	39,931	35,286	31,609	28,626	26,158	21,519	18,277	15,884	14,046	12,588	11,405	9,600	8,289	7,292	5,879	4,925
195	147,568	118,696	99,273	85,313	74,795	66,586	54,600	46,271	40,147	35,455	31,744	28,737	26,250	21,581	18,322	15,918	14,072	12,610	11,423	9,613	8,298	7,300	5,884	4,928
200	150,414	120,530	100,553	86,256	75,519	67,159	54,985	46,548	40,355	35,617	31,874	28,843	26,339	21,641	18,365	15,951	14,098	12,630	11,439	9,625	8,307	7,306	5,888	4,931

1000 MCM	1 Run(s)	240 Volts	C-Value = 25,278	Magnetic Duct	Copper

Calculations:

1. $f = \frac{\text{Length} \times 2 \times \text{Isca}}{\text{Runs} \times \text{C-Value} \times \text{E L-L}}$
2. $M = \frac{1}{1 + f}$
3. $\text{Isc} = \text{Isca} \times M$
4. Add Motor Contribution = Motor FLA x 4

* All results are given in symmetrical amperes

Single-Phase L-L Bolted Fault-Current* Table

One-Way Distance in Feet

Available Isc in Thousands (Isca)	5	10	15	20	25	30	40	50	60	70	80	90	100	125	150	175	200	225	250	300	350	400	500	600
3	2,993	2,985	2,978	2,971	2,963	2,956	2,942	2,928	2,914	2,900	2,886	2,872	2,859	2,825	2,793	2,761	2,730	2,700	2,670	2,612	2,557	2,505	2,405	2,314
5	4,979	4,959	4,939	4,919	4,899	4,879	4,840	4,802	4,764	4,727	4,691	4,655	4,619	4,533	4,450	4,370	4,292	4,218	4,146	4,009	3,881	3,760	3,541	3,346
7	6,960	6,920	6,881	6,842	6,804	6,766	6,691	6,618	6,547	6,477	6,408	6,341	6,276	6,118	5,967	5,824	5,688	5,557	5,433	5,200	4,986	4,789	4,439	4,136
10	9,918	9,838	9,759	9,681	9,604	9,529	9,381	9,239	9,100	8,966	8,835	8,708	8,585	8,292	8,018	7,761	7,521	7,295	7,082	6,691	6,341	6,027	5,482	5,028
15	14,817	14,638	14,464	14,293	14,127	13,964	13,650	13,350	13,062	12,787	12,523	12,270	12,026	11,459	10,942	10,470	10,037	9,638	9,270	8,612	8,041	7,541	6,708	6,040
20	19,676	19,362	19,058	18,763	18,477	18,200	17,670	17,170	16,697	16,250	15,826	15,424	15,041	14,163	13,382	12,683	12,053	11,483	10,964	10,055	9,286	8,626	7,552	6,716
25	24,495	24,011	23,545	23,096	22,665	22,249	21,462	20,729	20,044	19,403	18,802	18,237	17,704	16,501	15,450	14,525	13,705	12,972	12,314	11,179	10,236	9,440	8,169	7,199
30	29,276	28,586	27,928	27,300	26,699	26,124	25,046	24,053	23,136	22,286	21,496	20,761	20,074	18,540	17,224	16,083	15,083	14,200	13,415	12,080	10,986	10,074	8,639	7,562
35	34,019	33,091	32,212	31,379	30,588	29,836	28,438	27,164	26,000	24,932	23,947	23,038	22,195	20,335	18,763	17,416	16,250	15,230	14,331	12,817	11,592	10,581	9,010	7,845
40	38,723	37,526	36,400	35,340	34,340	33,395	31,652	30,083	28,661	27,368	26,187	25,104	24,106	21,928	20,111	18,572	17,251	16,106	15,104	13,432	12,093	10,997	9,310	8,071
45	43,391	41,893	40,494	39,187	37,961	36,809	34,703	32,826	31,141	29,620	28,241	26,985	25,836	23,350	21,300	19,582	18,120	16,861	15,765	13,952	12,513	11,344	9,557	8,256
50	48,021	46,193	44,499	42,925	41,458	40,088	37,603	35,409	33,456	31,707	30,133	28,707	27,410	24,628	22,359	20,473	18,880	17,517	16,338	14,399	12,871	11,637	9,764	8,410
55	52,615	50,428	48,416	46,558	44,838	43,240	40,363	37,845	35,623	33,647	31,879	30,288	28,847	25,782	23,306	21,264	19,551	18,093	16,838	14,786	13,180	11,888	9,940	8,541
60	57,173	54,600	52,249	50,092	48,106	46,271	42,992	40,147	37,655	35,455	33,497	31,744	30,166	26,831	24,159	21,972	20,148	18,603	17,279	15,125	13,448	12,107	10,092	8,653
65	61,695	58,710	56,000	53,529	51,268	49,189	45,500	42,326	39,565	37,143	35,000	33,091	31,379	27,786	24,932	22,609	20,682	19,058	17,670	15,424	13,684	12,297	10,225	8,750
70	66,182	62,759	59,672	56,875	54,328	52,000	47,895	44,390	41,364	38,724	36,400	34,340	32,500	28,662	25,634	23,185	21,163	19,465	18,020	15,690	13,893	12,466	10,341	8,835
75	70,634	66,748	63,268	60,132	57,293	54,710	50,184	46,350	43,060	40,206	37,707	35,501	33,538	29,466	26,275	23,708	21,598	19,833	18,335	15,928	14,080	12,616	10,444	8,910
80	75,052	70,680	66,789	63,304	60,165	57,323	52,374	48,212	44,663	41,600	38,931	36,583	34,503	30,208	26,864	24,186	21,994	20,166	18,619	16,142	14,247	12,750	10,536	8,977
85	79,435	74,554	70,238	66,395	62,950	59,845	54,472	49,984	46,179	42,913	40,078	37,594	35,401	30,894	27,405	24,624	22,356	20,470	18,877	16,336	14,397	12,870	10,618	9,036
90	83,785	78,373	73,618	69,407	65,651	62,282	56,483	51,672	47,616	44,151	41,156	38,541	36,239	31,530	27,905	25,027	22,687	20,747	19,113	16,512	14,534	12,979	10,692	9,090
95	88,102	82,138	76,930	72,343	68,273	64,636	58,412	53,282	48,980	45,321	42,171	39,430	37,024	32,123	28,368	25,399	22,992	21,002	19,329	16,673	14,659	13,079	10,759	9,139
100	92,386	85,849	80,176	75,207	70,817	66,912	60,265	54,819	50,276	46,429	43,128	40,266	37,760	32,675	28,798	25,743	23,274	21,237	19,528	16,821	14,773	13,169	10,821	9,183
105	96,637	89,508	83,359	78,000	73,289	69,114	62,046	56,289	51,510	47,478	44,032	41,053	38,451	33,192	29,198	26,062	23,535	21,454	19,711	16,957	14,877	13,253	10,877	9,223
110	100,856	93,116	86,480	80,726	75,690	71,246	63,758	57,695	52,684	48,475	44,888	41,796	39,102	33,676	29,572	26,360	23,777	21,655	19,881	17,082	14,974	13,329	10,928	9,260
115	105,044	96,675	89,540	83,387	78,024	73,310	65,406	59,041	53,805	49,422	45,699	42,498	39,716	34,130	29,922	26,637	24,002	21,842	20,038	17,198	15,063	13,400	10,975	9,294
120	109,200	100,184	92,543	85,984	80,294	75,311	66,994	60,332	54,875	50,323	46,468	43,162	40,295	34,557	30,250	26,897	24,213	22,016	20,185	17,306	15,146	13,465	11,019	9,325
125	113,325	103,645	95,488	88,522	82,503	77,250	68,524	61,570	55,897	51,181	47,199	43,792	40,844	34,960	30,558	27,140	24,410	22,179	20,322	17,406	15,223	13,526	11,060	9,355
130	117,419	107,059	98,379	91,000	84,651	79,131	70,000	62,759	56,875	52,000	47,895	44,391	41,364	35,340	30,848	27,369	24,595	22,331	20,450	17,500	15,294	13,582	11,098	9,382
135	121,483	110,427	101,215	93,422	86,743	80,956	71,425	63,901	57,812	52,782	48,558	44,959	41,857	35,699	31,121	27,584	24,768	22,474	20,569	17,588	15,361	13,635	11,133	9,407
140	125,517	113,750	104,000	95,790	88,781	82,728	72,800	65,000	58,710	53,530	49,189	45,500	42,326	36,040	31,380	27,786	24,932	22,609	20,682	17,670	15,424	13,684	11,166	9,430
145	129,522	117,029	106,734	98,104	90,766	84,448	74,130	66,058	59,571	54,245	49,793	46,016	42,772	36,363	31,624	27,978	25,086	22,735	20,788	17,747	15,483	13,731	11,197	9,452
150	133,496	120,265	109,419	100,368	92,700	86,120	75,415	67,076	60,399	54,930	50,369	46,508	43,196	36,669	31,856	28,159	25,231	22,855	20,888	17,820	15,538	13,774	11,225	9,473
155	137,442	123,458	112,056	102,582	94,585	87,745	76,658	68,058	61,193	55,587	50,921	46,978	43,602	36,961	32,075	28,331	25,369	22,968	20,982	17,889	15,590	13,815	11,253	9,492
160	141,359	126,609	114,646	104,749	96,424	89,326	77,861	69,005	61,958	56,217	51,449	47,427	43,988	37,238	32,284	28,493	25,499	23,075	21,071	17,953	15,639	13,854	11,278	9,510
165	145,248	129,719	117,191	106,869	98,218	90,863	79,027	69,919	62,693	56,822	51,955	47,857	44,358	37,503	32,483	28,648	25,623	23,176	21,155	18,015	15,686	13,890	11,302	9,527
170	149,109	132,790	119,691	108,944	99,968	92,359	80,156	70,801	63,402	57,403	52,441	48,269	44,711	37,755	32,672	28,795	25,741	23,272	21,236	18,073	15,730	13,925	11,325	9,544
175	152,941	135,821	122,148	110,976	101,676	93,815	81,250	71,654	64,085	57,962	52,907	48,663	45,050	37,996	32,852	28,935	25,852	23,363	21,312	18,128	15,771	13,957	11,347	9,559
180	156,747	138,814	124,563	112,966	103,344	95,233	82,312	72,478	64,743	58,500	53,355	49,042	45,374	38,227	33,024	29,069	25,959	23,450	21,384	18,180	15,811	13,988	11,367	9,573
185	160,525	141,769	126,937	114,915	104,973	96,614	83,342	73,276	65,379	59,019	53,786	49,406	45,686	38,447	33,189	29,196	26,061	23,533	21,453	18,230	15,849	14,018	11,387	9,587
190	164,276	144,686	129,271	116,825	106,564	97,961	84,342	74,048	65,993	59,518	54,201	49,756	45,984	38,659	33,346	29,318	26,158	23,612	21,519	18,277	15,884	14,046	11,405	9,600
195	168,000	147,568	131,567	118,696	108,119	99,273	85,313	74,795	66,586	60,000	54,600	50,092	46,271	38,861	33,497	29,434	26,250	23,688	21,581	18,322	15,918	14,072	11,423	9,613
200	171,698	150,414	133,824	120,530	109,639	100,553	86,256	75,519	67,159	60,465	54,985	50,416	46,548	39,056	33,642	29,546	26,339	23,760	21,641	18,365	15,951	14,098	11,439	9,625

1000 MCM | 2 Run(s) | 240 Volts

C-Value = 25,278 | Magnetic Duct | Copper

Calculations:

1. $f = \dfrac{\text{Length} \times 2 \times \text{Isca}}{\text{Runs} \times \text{C-Value} \times \text{E L-L}}$

2. $M = \dfrac{1}{1 + f}$

3. $\text{Isc} = \text{Isca} \times M$

4. Add Motor Contribution = Motor FLA x 4

* All results are given in symmetrical amperes

Short-Circuit Tables and Data: 208 Volts Single-Phase L-L

Single-Phase L-L Bolted Fault-Current* Table

One-Way Distance in Feet

Available Isc in Thousands (Isca)	5	10	15	20	25	30	40	50	60	70	80	90	100	125	150	175	200	225	250	300	350	400	500	600
1	890	802	730	669	618	574	503	447	403	366	336	310	288	245	212	188	168	152	139	119	104	92	75	63
3	2,189	1,723	1,420	1,208	1,051	930	756	637	551	485	433	391	356	292	247	215	190	170	154	129	111	98	79	66
5	3,090	2,236	1,752	1,440	1,223	1,062	841	696	594	518	459	412	374	304	256	221	194	174	157	131	113	99	80	67
10	4,472	2,880	2,124	1,682	1,393	1,188	919	749	632	546	481	430	389	314	263	226	198	177	159	133	114	100	80	67
15	5,256	3,186	2,286	1,782	1,461	1,237	948	768	645	557	489	436	394	317	265	228	200	178	160	134	115	100	80	67
20	5,761	3,365	2,377	1,837	1,497	1,263	963	778	652	562	493	440	397	318	266	229	200	178	161	134	115	101	81	67
25	6,113	3,482	2,434	1,871	1,520	1,280	972	784	657	565	496	442	398	320	267	229	201	179	161	134	115	101	81	67
30	6,372	3,565	2,475	1,895	1,535	1,291	978	788	659	567	497	443	399	320	267	229	201	179	161	134	115	101	81	67
35	6,572	3,626	2,504	1,912	1,547	1,299	983	791	662	569	498	444	400	321	268	230	201	179	161	134	115	101	81	67
40	6,730	3,674	2,527	1,925	1,555	1,305	986	793	663	570	499	445	401	321	268	230	201	179	161	134	115	101	81	67
45	6,858	3,712	2,545	1,936	1,562	1,309	989	795	664	571	500	445	401	321	268	230	201	179	161	134	115	101	81	67
50	6,964	3,743	2,559	1,944	1,568	1,313	991	796	665	571	501	446	401	322	268	230	201	179	161	134	115	101	81	67
55	7,054	3,768	2,571	1,951	1,572	1,316	993	797	666	572	501	446	402	322	268	230	202	179	161	135	115	101	81	67
60	7,130	3,790	2,581	1,957	1,576	1,319	995	798	667	572	501	446	402	322	268	230	202	179	161	135	115	101	81	67
65	7,196	3,809	2,590	1,962	1,579	1,321	996	799	667	573	502	446	402	322	269	230	202	179	161	135	115	101	81	67
70	7,253	3,825	2,597	1,966	1,582	1,323	997	800	668	573	502	447	402	322	269	230	202	179	161	135	115	101	81	67
75	7,303	3,839	2,603	1,970	1,584	1,325	998	800	668	574	502	447	402	322	269	230	202	179	161	135	115	101	81	67
80	7,348	3,851	2,609	1,973	1,586	1,326	999	801	669	574	503	447	403	322	269	231	202	179	161	135	115	101	81	67
85	7,388	3,862	2,614	1,976	1,588	1,327	1,000	801	669	574	503	447	403	322	269	231	202	179	162	135	115	101	81	67
90	7,424	3,872	2,619	1,978	1,590	1,329	1,000	802	669	574	503	447	403	322	269	231	202	179	162	135	115	101	81	67
95	7,456	3,880	2,623	1,981	1,591	1,330	1,001	802	670	574	503	447	403	323	269	231	202	179	162	135	115	101	81	67
100	7,486	3,888	2,626	1,983	1,592	1,331	1,001	803	670	575	503	447	403	323	269	231	202	179	162	135	115	101	81	67
105	7,512	3,896	2,630	1,985	1,594	1,331	1,002	803	670	575	503	448	403	323	269	231	202	179	162	135	115	101	81	67
110	7,537	3,902	2,633	1,986	1,595	1,332	1,002	803	670	575	503	448	403	323	269	231	202	180	162	135	115	101	81	67
115	7,559	3,908	2,635	1,988	1,596	1,333	1,003	803	670	575	503	448	403	323	269	231	202	180	162	135	115	101	81	67
120	7,580	3,914	2,638	1,989	1,597	1,334	1,003	804	670	575	504	448	403	323	269	231	202	180	162	135	115	101	81	67
125	7,599	3,919	2,640	1,991	1,598	1,334	1,003	804	671	575	504	448	403	323	269	231	202	180	162	135	115	101	81	67
130	7,617	3,924	2,642	1,992	1,598	1,335	1,004	804	671	575	504	448	403	323	269	231	202	180	162	135	115	101	81	67
135	7,634	3,928	2,644	1,993	1,599	1,335	1,004	804	671	575	504	448	403	323	269	231	202	180	162	135	115	101	81	67
140	7,649	3,932	2,646	1,994	1,600	1,336	1,004	804	671	576	504	448	403	323	269	231	202	180	162	135	115	101	81	67
145	7,664	3,936	2,648	1,995	1,600	1,336	1,004	805	671	576	504	448	403	323	269	231	202	180	162	135	115	101	81	67
150	7,677	3,939	2,649	1,996	1,601	1,337	1,005	805	671	576	504	448	403	323	269	231	202	180	162	135	115	101	81	67
155	7,690	3,943	2,651	1,997	1,602	1,337	1,005	805	671	576	504	448	404	323	269	231	202	180	162	135	116	101	81	67
160	7,702	3,946	2,652	1,998	1,602	1,337	1,005	805	671	576	504	448	404	323	269	231	202	180	162	135	116	101	81	67
165	7,713	3,949	2,654	1,998	1,603	1,338	1,005	805	672	576	504	448	404	323	269	231	202	180	162	135	116	101	81	67
170	7,724	3,952	2,655	1,999	1,603	1,338	1,005	805	672	576	504	448	404	323	269	231	202	180	162	135	116	101	81	67
175	7,734	3,954	2,656	2,000	1,603	1,338	1,006	805	672	576	504	448	404	323	269	231	202	180	162	135	116	101	81	67
180	7,743	3,957	2,657	2,000	1,604	1,339	1,006	805	672	576	504	448	404	323	269	231	202	180	162	135	116	101	81	67
185	7,752	3,959	2,658	2,001	1,604	1,339	1,006	806	672	576	504	448	404	323	269	231	202	180	162	135	116	101	81	67
190	7,761	3,961	2,659	2,001	1,605	1,339	1,006	806	672	576	504	448	404	323	269	231	202	180	162	135	116	101	81	67
195	7,769	3,963	2,660	2,002	1,605	1,339	1,006	806	672	576	504	448	404	323	269	231	202	180	162	135	116	101	81	67
200	7,777	3,965	2,661	2,003	1,605	1,340	1,006	806	672	576	504	449	404	323	269	231	202	180	162	135	116	101	81	67

 # 14 AWG | **1 Run(s)** | **208 Volts** | **C-Value = 389** | **Magnetic Duct** | **Copper**

Calculations:

1. $f = \dfrac{\text{Length} \times 2 \times \text{Isca}}{\text{Runs} \times \text{C-Value} \times E_{L-L}}$
2. $M = \dfrac{1}{1 + f}$
3. $Isc = Isca \times M$
4. Add Motor Contribution = Motor FLA x 4

* All results are given in symmetrical amperes

Single-Phase L-L Bolted Fault-Current* Table

One-Way Distance in Feet

Available Isc in Thousands (Isca)	5	10	15	20	25	30	40	50	60	70	80	90	100	125	150	175	200	225	250	300	350	400	500	600
1	928	865	811	762	720	681	616	562	517	478	445	416	391	339	300	268	243	222	204	176	155	138	114	97
3	2,432	2,044	1,763	1,550	1,383	1,249	1,045	899	788	702	633	576	529	438	374	327	290	260	236	200	173	152	123	103
5	3,598	2,810	2,305	1,954	1,696	1,498	1,215	1,021	881	775	691	624	569	466	394	342	301	270	244	205	177	155	125	105
10	5,620	3,909	2,996	2,429	2,042	1,762	1,382	1,137	966	840	743	666	603	488	410	354	311	277	250	209	180	158	127	106
15	6,916	4,494	3,329	2,643	2,192	1,872	1,449	1,182	998	864	761	681	615	496	416	358	314	280	252	211	181	159	127	106
20	7,817	4,858	3,524	2,765	2,275	1,932	1,485	1,206	1,015	877	771	688	622	500	419	360	316	281	253	212	182	159	128	106
25	8,480	5,106	3,653	2,843	2,328	1,970	1,507	1,221	1,026	884	777	693	626	503	421	361	317	282	254	212	182	159	128	106
30	8,988	5,286	3,744	2,898	2,364	1,997	1,523	1,231	1,033	890	781	696	628	505	422	362	317	283	254	212	182	160	128	107
35	9,390	5,423	3,812	2,939	2,391	2,016	1,534	1,238	1,038	893	784	699	630	506	423	363	318	283	255	213	182	160	128	107
40	9,716	5,530	3,865	2,970	2,412	2,030	1,542	1,243	1,042	896	786	700	632	507	423	363	318	283	255	213	183	160	128	107
45	9,986	5,616	3,907	2,995	2,428	2,042	1,549	1,248	1,045	898	788	702	633	508	424	364	319	283	255	213	183	160	128	107
50	10,212	5,687	3,941	3,015	2,441	2,051	1,554	1,251	1,047	900	789	703	634	508	424	364	319	284	255	213	183	160	128	107
55	10,406	5,746	3,969	3,032	2,452	2,059	1,559	1,254	1,049	902	791	704	634	509	424	364	319	284	255	213	183	160	128	107
60	10,572	5,797	3,993	3,046	2,461	2,065	1,562	1,256	1,051	903	792	705	635	509	425	364	319	284	256	213	183	160	128	107
65	10,718	5,840	4,014	3,057	2,469	2,071	1,566	1,259	1,052	904	792	705	635	509	425	365	319	284	256	213	183	160	128	107
70	10,845	5,878	4,031	3,068	2,476	2,076	1,568	1,260	1,053	905	793	706	636	510	425	365	319	284	256	213	183	160	128	107
75	10,958	5,911	4,047	3,077	2,482	2,080	1,571	1,262	1,054	906	794	706	636	510	425	365	319	284	256	213	183	160	128	107
80	11,059	5,940	4,061	3,085	2,487	2,083	1,573	1,263	1,055	906	794	707	637	510	426	365	320	284	256	213	183	160	128	107
85	11,150	5,966	4,073	3,092	2,491	2,086	1,574	1,264	1,056	907	795	707	637	510	426	365	320	284	256	213	183	160	128	107
90	11,232	5,990	4,084	3,098	2,496	2,089	1,576	1,265	1,057	907	795	707	637	510	426	365	320	284	256	213	183	160	128	107
95	11,306	6,011	4,094	3,104	2,499	2,092	1,578	1,266	1,058	908	795	708	637	511	426	365	320	284	256	213	183	160	128	107
100	11,374	6,030	4,102	3,109	2,502	2,094	1,579	1,267	1,058	908	796	708	638	511	426	365	320	284	256	213	183	160	128	107
105	11,436	6,047	4,110	3,113	2,505	2,096	1,580	1,268	1,059	909	796	708	638	511	426	365	320	284	256	213	183	160	128	107
110	11,493	6,063	4,118	3,117	2,508	2,098	1,581	1,269	1,059	909	796	708	638	511	426	365	320	284	256	213	183	160	128	107
115	11,545	6,078	4,124	3,121	2,511	2,100	1,582	1,269	1,060	909	797	709	638	511	426	366	320	284	256	213	183	160	128	107
120	11,594	6,091	4,131	3,125	2,513	2,101	1,583	1,270	1,060	910	797	709	638	511	426	366	320	285	256	214	183	160	128	107
125	11,639	6,103	4,136	3,128	2,515	2,103	1,584	1,270	1,060	910	797	709	638	511	426	366	320	285	256	214	183	160	128	107
130	11,681	6,115	4,142	3,131	2,517	2,104	1,585	1,271	1,061	910	797	709	639	511	426	366	320	285	256	214	183	160	128	107
135	11,720	6,126	4,146	3,134	2,519	2,106	1,585	1,271	1,061	911	797	709	639	511	426	366	320	285	256	214	183	160	128	107
140	11,756	6,136	4,151	3,137	2,521	2,107	1,586	1,272	1,061	911	798	709	639	511	426	366	320	285	256	214	183	160	128	107
145	11,790	6,145	4,155	3,139	2,522	2,108	1,587	1,272	1,062	911	798	709	639	512	427	366	320	285	256	214	183	160	128	107
150	11,822	6,154	4,159	3,141	2,524	2,109	1,587	1,272	1,062	911	798	710	639	512	427	366	320	285	256	214	183	160	128	107
155	11,852	6,162	4,163	3,143	2,525	2,110	1,588	1,273	1,062	911	798	710	639	512	427	366	320	285	256	214	183	160	128	107
160	11,881	6,169	4,166	3,145	2,526	2,111	1,588	1,273	1,062	911	798	710	639	512	427	366	320	285	256	214	183	160	128	107
165	11,907	6,177	4,170	3,147	2,527	2,112	1,589	1,273	1,063	912	798	710	639	512	427	366	320	285	256	214	183	160	128	107
170	11,933	6,183	4,173	3,149	2,529	2,112	1,589	1,274	1,063	912	798	710	639	512	427	366	320	285	256	214	183	160	128	107
175	11,957	6,190	4,176	3,151	2,530	2,113	1,590	1,274	1,063	912	798	710	639	512	427	366	320	285	256	214	183	160	128	107
180	11,979	6,196	4,179	3,152	2,531	2,114	1,590	1,274	1,063	912	799	710	639	512	427	366	320	285	256	214	183	160	128	107
185	12,001	6,202	4,181	3,154	2,532	2,114	1,590	1,275	1,063	912	799	710	639	512	427	366	320	285	256	214	183	160	128	107
190	12,022	6,207	4,184	3,155	2,533	2,115	1,591	1,275	1,063	912	799	710	640	512	427	366	320	285	256	214	183	160	128	107
195	12,041	6,212	4,186	3,156	2,533	2,116	1,591	1,275	1,064	912	799	710	640	512	427	366	320	285	256	214	183	160	128	107
200	12,060	6,217	4,188	3,158	2,534	2,116	1,591	1,275	1,064	913	799	710	640	512	427	366	320	285	256	214	183	160	128	107

 # 12 AWG | **1 Run(s)** | **208 Volts** | **C-Value = 617** | **Magnetic Duct** | **Copper**

Calculations:

1. $f = \dfrac{\text{Length} \times 2 \times \text{Isca}}{\text{Runs} \times \text{C-Value} \times \text{E L-L}}$

2. $M = \dfrac{1}{1 + f}$

3. $\text{Isc} = \text{Isca} \times M$

4. Add Motor Contribution = Motor FLA x 4

* All results are given in symmetrical amperes

Single-Phase L-L Bolted Fault-Current* Table

One-Way Distance in Feet

Available Isc in Thousands (Isca)

Isca	5	10	15	20	25	30	40	50	60	70	80	90	100	125	150	175	200	225	250	300	350	400	500	600
1	953	911	872	836	803	773	718	671	630	593	560	531	505	449	405	368	338	312	290	254	226	203	169	145
3	2,615	2,318	2,082	1,889	1,729	1,594	1,379	1,214	1,085	981	895	823	761	642	554	488	436	394	359	305	266	235	191	161
5	4,016	3,356	2,882	2,525	2,247	2,024	1,689	1,449	1,269	1,129	1,016	924	847	702	599	522	463	416	377	318	275	243	196	164
10	6,711	5,050	4,048	3,378	2,898	2,538	2,032	1,695	1,453	1,272	1,131	1,018	926	755	637	551	485	434	392	329	283	249	200	167
15	8,645	6,072	4,680	3,807	3,208	2,772	2,180	1,796	1,527	1,328	1,175	1,054	955	774	651	561	493	440	397	333	286	251	201	168
20	10,100	6,756	5,076	4,065	3,389	2,907	2,262	1,852	1,567	1,358	1,199	1,073	971	784	658	566	497	443	400	334	287	252	202	169
25	11,235	7,246	5,347	4,237	3,508	2,994	2,314	1,887	1,592	1,377	1,213	1,084	980	790	662	570	500	445	402	336	288	252	202	169
30	12,145	7,613	5,545	4,360	3,592	3,055	2,351	1,911	1,609	1,390	1,223	1,092	987	795	665	572	502	447	403	336	289	253	203	169
35	12,890	7,900	5,695	4,452	3,655	3,100	2,377	1,928	1,622	1,399	1,230	1,098	991	798	667	573	503	448	403	337	289	253	203	169
40	13,512	8,129	5,813	4,524	3,703	3,134	2,398	1,941	1,631	1,406	1,236	1,102	995	800	669	575	504	448	404	337	289	253	203	169
45	14,039	8,317	5,909	4,582	3,742	3,162	2,414	1,952	1,638	1,412	1,240	1,106	998	802	670	576	504	449	404	338	290	254	203	169
50	14,491	8,473	5,987	4,629	3,773	3,184	2,427	1,960	1,644	1,416	1,244	1,108	1,000	803	671	576	505	449	405	338	290	254	203	169
55	14,883	8,606	6,053	4,668	3,799	3,203	2,438	1,967	1,649	1,420	1,246	1,111	1,002	804	672	577	505	450	405	338	290	254	203	170
60	15,227	8,720	6,109	4,701	3,821	3,218	2,447	1,973	1,654	1,423	1,249	1,113	1,003	805	673	577	506	450	405	338	290	254	203	170
65	15,530	8,818	6,157	4,730	3,840	3,232	2,454	1,978	1,657	1,426	1,251	1,114	1,004	806	673	578	506	450	406	338	290	254	203	170
70	15,799	8,905	6,199	4,755	3,856	3,243	2,461	1,983	1,660	1,428	1,252	1,116	1,006	807	674	578	506	451	406	338	290	254	203	170
75	16,041	8,981	6,236	4,776	3,870	3,253	2,467	1,986	1,663	1,430	1,254	1,117	1,007	807	674	578	507	451	406	339	290	254	203	170
80	16,258	9,048	6,269	4,795	3,883	3,262	2,472	1,990	1,665	1,431	1,255	1,118	1,007	808	674	579	507	451	406	339	290	254	204	170
85	16,455	9,109	6,298	4,812	3,894	3,270	2,476	1,993	1,667	1,433	1,256	1,119	1,008	808	675	579	507	451	406	339	291	254	204	170
90	16,634	9,164	6,324	4,828	3,904	3,277	2,480	1,995	1,669	1,434	1,257	1,119	1,009	809	675	579	507	451	406	339	291	254	204	170
95	16,797	9,213	6,347	4,841	3,913	3,283	2,484	1,998	1,670	1,435	1,258	1,120	1,009	809	675	579	507	451	406	339	291	254	204	170
100	16,947	9,258	6,368	4,854	3,921	3,289	2,487	2,000	1,672	1,437	1,259	1,121	1,010	810	676	580	508	451	406	339	291	254	204	170
105	17,085	9,299	6,388	4,865	3,928	3,294	2,490	2,002	1,673	1,438	1,260	1,121	1,010	810	676	580	508	451	407	339	291	254	204	170
110	17,212	9,336	6,406	4,875	3,935	3,299	2,493	2,003	1,675	1,438	1,261	1,122	1,011	810	676	580	508	452	407	339	291	254	204	170
115	17,330	9,371	6,422	4,885	3,941	3,303	2,495	2,005	1,676	1,439	1,261	1,123	1,011	810	676	580	508	452	407	339	291	254	204	170
120	17,439	9,403	6,437	4,893	3,947	3,307	2,498	2,006	1,677	1,440	1,262	1,123	1,012	811	676	580	508	452	407	339	291	255	204	170
125	17,541	9,433	6,451	4,901	3,952	3,311	2,500	2,008	1,678	1,441	1,262	1,123	1,012	811	676	580	508	452	407	339	291	255	204	170
130	17,637	9,460	6,463	4,909	3,957	3,314	2,502	2,009	1,678	1,441	1,263	1,124	1,012	811	677	580	508	452	407	339	291	255	204	170
135	17,726	9,486	6,475	4,915	3,961	3,317	2,503	2,010	1,679	1,442	1,263	1,124	1,013	811	677	580	508	452	407	339	291	255	204	170
140	17,809	9,509	6,486	4,922	3,965	3,320	2,505	2,011	1,680	1,442	1,264	1,124	1,013	811	677	581	508	452	407	339	291	255	204	170
145	17,888	9,532	6,497	4,928	3,969	3,323	2,507	2,012	1,681	1,443	1,264	1,125	1,013	812	677	581	508	452	407	339	291	255	204	170
150	17,961	9,553	6,507	4,933	3,973	3,325	2,508	2,013	1,681	1,443	1,265	1,125	1,013	812	677	581	508	452	407	339	291	255	204	170
155	18,031	9,572	6,516	4,939	3,976	3,328	2,509	2,014	1,682	1,444	1,265	1,125	1,014	812	677	581	508	452	407	339	291	255	204	170
160	18,097	9,591	6,524	4,944	3,979	3,330	2,511	2,015	1,683	1,444	1,265	1,126	1,014	812	677	581	508	452	407	339	291	255	204	170
165	18,159	9,608	6,532	4,948	3,982	3,332	2,512	2,016	1,683	1,445	1,266	1,126	1,014	812	677	581	509	452	407	339	291	255	204	170
170	18,218	9,625	6,540	4,953	3,985	3,334	2,513	2,016	1,684	1,445	1,266	1,126	1,014	812	677	581	509	452	407	339	291	255	204	170
175	18,274	9,640	6,547	4,957	3,988	3,336	2,514	2,017	1,684	1,445	1,266	1,126	1,014	812	678	581	509	452	407	339	291	255	204	170
180	18,327	9,655	6,554	4,961	3,990	3,338	2,515	2,018	1,684	1,446	1,266	1,127	1,014	813	678	581	509	452	407	339	291	255	204	170
185	18,378	9,669	6,560	4,964	3,993	3,339	2,516	2,018	1,685	1,446	1,267	1,127	1,015	813	678	581	509	452	407	339	291	255	204	170
190	18,426	9,682	6,567	4,968	3,995	3,341	2,517	2,019	1,685	1,446	1,267	1,127	1,015	813	678	581	509	452	407	339	291	255	204	170
195	18,472	9,695	6,572	4,971	3,997	3,343	2,518	2,019	1,686	1,447	1,267	1,127	1,015	813	678	581	509	452	407	339	291	255	204	170
200	18,516	9,707	6,578	4,974	3,999	3,344	2,518	2,020	1,686	1,447	1,267	1,127	1,015	813	678	581	509	452	407	340	291	255	204	170

# 10 AWG	1 Run(s)	208 Volts	C-Value = 981	Magnetic Duct	Copper

Calculations:

1. $f = \dfrac{\text{Length} \times 2 \times \text{Isca}}{\text{Runs} \times \text{C-Value} \times \text{E L-L}}$

2. $M = \dfrac{1}{1 + f}$

3. $\text{Isc} = \text{Isca} \times M$

4. Add Motor Contribution = Motor FLA x 4

* All results are given in symmetrical amperes

Single-Phase L-L Bolted Fault-Current* Table

One-Way Distance in Feet

Available Isc in Thousands (Isca)	5	10	15	20	25	30	40	50	60	70	80	90	100	125	150	175	200	225	250	300	350	400	500	600
3	2,746	2,531	2,348	2,189	2,050	1,928	1,723	1,557	1,421	1,306	1,209	1,125	1,052	905	794	707	638	580	533	45[illegible]	401	357	292	248
5	4,331	3,820	3,417	3,091	2,822	2,596	2,237	1,965	1,753	1,582	1,441	1,323	1,223	1,029	838	781	697	629	573	48[illegible]	423	374	304	256
7	5,756	4,887	4,246	3,754	3,364	3,048	2,565	2,214	1,948	1,739	1,570	1,431	1,315	1,093	935	817	726	653	593	50[illegible]	434	383	310	260
10	7,641	6,182	5,191	4,474	3,931	3,505	2,882	2,446	2,125	1,879	1,683	1,525	1,394	1,147	974	847	749	671	608	51[illegible]	442	389	314	263
15	10,252	7,787	6,277	5,258	4,524	3,969	3,788	2,663	2,287	2,004	1,783	1,607	1,462	1,192	1,007	872	768	687	621	52[illegible]	449	394	317	265
20	12,364	8,948	7,011	5,763	4,893	4,250	3,367	2,787	2,378	2,073	1,838	1,651	1,498	1,217	1,024	884	778	695	627	52[illegible]	452	397	319	266
25	14,109	9,827	7,540	6,116	5,144	4,439	3,484	2,867	2,436	2,117	1,872	1,678	1,521	1,232	1,035	892	784	700	631	52[illegible]	454	398	320	267
30	15,574	10,516	7,939	6,376	5,327	4,575	3,567	2,923	2,476	2,148	1,896	1,697	1,536	1,242	1,042	898	788	703	634	53[illegible]	456	399	320	267
35	16,821	11,071	8,250	6,575	5,466	4,676	3,629	2,964	2,506	2,170	1,913	1,711	1,548	1,249	1,047	901	791	705	636	53[illegible]	457	400	321	268
40	17,896	11,527	8,501	6,733	5,574	4,756	3,676	2,996	2,528	2,187	1,927	1,722	1,556	1,255	1,051	904	794	707	637	53[illegible]	457	401	321	268
45	18,832	11,908	8,707	6,862	5,662	4,820	3,714	3,021	2,546	2,200	1,937	1,730	1,563	1,259	1,054	907	795	708	639	53[illegible]	458	401	322	268
50	19,655	12,232	8,878	6,968	5,734	4,872	3,745	3,042	2,561	2,211	1,945	1,737	1,568	1,263	1,057	908	797	709	639	53[illegible]	458	402	322	268
55	20,383	12,510	9,024	7,057	5,795	4,915	3,771	3,058	2,573	2,220	1,952	1,742	1,573	1,266	1,059	910	798	710	640	535	459	402	322	269
60	21,033	12,751	9,149	7,134	5,846	4,952	3,792	3,073	2,583	2,227	1,958	1,747	1,577	1,268	1,060	911	799	711	641	535	459	402	322	269
65	21,616	12,963	9,258	7,200	5,890	4,984	3,811	3,085	2,591	2,234	1,963	1,751	1,580	1,270	1,062	912	800	712	641	535	459	402	322	269
70	22,142	13,151	9,353	7,257	5,929	5,011	3,827	3,095	2,599	2,239	1,967	1,754	1,583	1,272	1,063	913	800	712	642	535	460	402	322	269
75	22,619	13,317	9,437	7,308	5,962	5,035	3,841	3,105	2,605	2,244	1,971	1,757	1,585	1,273	1,064	914	801	713	642	535	460	403	322	269
80	23,053	13,467	9,512	7,352	5,992	5,056	3,853	3,113	2,611	2,248	1,974	1,760	1,587	1,275	1,065	915	802	713	643	535	460	403	323	269
85	23,451	13,602	9,579	7,392	6,019	5,075	3,864	3,120	2,616	2,252	1,977	1,762	1,589	1,276	1,066	915	802	714	643	536	460	403	323	269
90	23,816	13,724	9,639	7,428	6,042	5,092	3,874	3,126	2,620	2,255	1,980	1,764	1,591	1,277	1,067	916	802	714	643	537	460	403	323	269
95	24,152	13,835	9,694	7,461	6,064	5,107	3,883	3,132	2,624	2,258	1,982	1,766	1,592	1,278	1,067	916	803	714	643	537	460	403	323	269
100	24,463	13,936	9,743	7,490	6,083	5,121	3,891	3,137	2,628	2,261	1,984	1,767	1,593	1,279	1,068	917	803	715	644	537	461	403	323	269
105	24,751	14,029	9,789	7,517	6,101	5,134	3,898	3,142	2,631	2,263	1,986	1,769	1,595	1,280	1,069	917	803	715	644	537	461	403	323	269
110	25,019	14,115	9,830	7,541	6,117	5,145	3,905	3,146	2,634	2,266	1,988	1,770	1,596	1,280	1,069	918	804	715	644	537	461	403	323	269
115	25,269	14,194	9,869	7,564	6,132	5,156	3,911	3,150	2,637	2,268	1,989	1,771	1,597	1,281	1,069	918	804	715	644	537	461	403	323	269
120	25,503	14,268	9,904	7,585	6,145	5,165	3,916	3,153	2,639	2,270	1,991	1,773	1,598	1,282	1,070	918	804	715	644	537	461	403	323	269
125	25,722	14,336	9,937	7,604	6,158	5,174	3,921	3,157	2,642	2,271	1,992	1,774	1,599	1,282	1,070	919	804	716	644	537	461	404	323	269
130	25,927	14,399	9,967	7,622	6,170	5,182	3,926	3,160	2,644	2,273	1,993	1,775	1,599	1,283	1,071	919	805	716	645	538	461	404	323	269
135	26,120	14,459	9,996	7,638	6,181	5,190	3,930	3,163	2,646	2,274	1,994	1,776	1,600	1,283	1,071	919	805	716	645	538	461	404	323	269
140	26,301	14,514	10,022	7,654	6,191	5,197	3,934	3,165	2,648	2,276	1,995	1,776	1,601	1,284	1,071	919	805	716	645	538	461	404	323	269
145	26,473	14,566	10,047	7,668	6,200	5,204	3,938	3,168	2,649	2,277	1,996	1,777	1,601	1,284	1,072	919	805	716	645	538	461	404	323	269
150	26,635	14,615	10,070	7,682	6,209	5,210	3,942	3,170	2,651	2,278	1,997	1,778	1,602	1,284	1,072	920	805	716	645	538	461	404	323	269
155	26,788	14,661	10,092	7,694	6,217	5,216	3,945	3,172	2,653	2,279	1,998	1,779	1,603	1,285	1,072	920	805	716	645	538	461	404	323	269
160	26,934	14,705	10,113	7,706	6,225	5,221	3,948	3,174	2,654	2,280	1,999	1,779	1,603	1,285	1,072	920	806	716	645	538	461	404	323	269
165	27,072	14,746	10,132	7,718	6,232	5,227	3,951	3,176	2,655	2,281	2,000	1,780	1,604	1,285	1,073	920	806	717	645	538	461	404	323	269
170	27,203	14,785	10,151	7,728	6,239	5,231	3,954	3,178	2,657	2,282	2,000	1,780	1,604	1,286	1,073	920	806	717	645	538	461	404	323	269
175	27,328	14,821	10,168	7,738	6,246	5,236	3,957	3,180	2,658	2,283	2,001	1,781	1,604	1,286	1,073	920	806	717	645	538	461	404	323	269
180	27,447	14,856	10,184	7,748	6,252	5,240	3,959	3,181	2,659	2,284	2,002	1,781	1,605	1,286	1,073	921	806	717	645	538	461	404	323	269
185	27,561	14,890	10,200	7,757	6,258	5,245	3,962	3,183	2,660	2,285	2,002	1,782	1,605	1,286	1,073	921	806	717	645	538	461	404	323	269
190	27,669	14,921	10,215	7,765	6,264	5,248	3,964	3,184	2,661	2,285	2,003	1,782	1,606	1,287	1,073	921	806	717	646	538	462	404	323	269
195	27,773	14,951	10,229	7,774	6,269	5,252	3,966	3,186	2,662	2,286	2,003	1,783	1,606	1,287	1,074	921	806	717	646	538	462	404	323	270
200	27,872	14,980	10,242	7,781	6,274	5,256	3,968	3,187	2,663	2,287	2,004	1,783	1,606	1,287	1,074	921	806	717	646	538	462	404	323	270

# 8 AWG	1 Run(s)	208 Volts

C-Value = 1,557	Magnetic Duct	Copper

Calculations:

1. $f = \dfrac{\text{Length} \times 2 \times \text{Isca}}{\text{Runs} \times \text{C-Value} \times \text{E L-L}}$

2. $M = \dfrac{1}{1 + f}$

3. Isc = Isca x M

4. Add Motor Contribution = Motor FLA x 4

* All results are given in symmetrical amperes

Single-Phase L-L Bolted Fault-Current* Table

One-Way Distance in Feet

Available Isc in Thousands (Isca)	5	10	15	20	25	30	40	50	60	70	80	90	100	125	150	175	200	225	250	300	350	400	500	600
3	2,832	2,681	2,546	2,423	2,312	2,211	2,033	1,881	1,751	1,637	1,537	1,449	1,370	1,206	1,077	973	888	816	755	657	581	521	432	369
5	4,549	4,173	3,854	3,580	3,343	3,135	2,789	2,511	2,284	2,094	1,933	1,796	1,676	1,438	1,258	1,119	1,007	916	839	720	630	560	458	388
7	6,147	5,479	4,942	4,501	4,132	3,820	3,317	2,932	2,626	2,379	2,174	2,001	1,854	1,566	1,356	1,195	1,069	966	882	751	653	578	470	397
10	8,345	7,161	6,271	5,577	5,022	4,567	3,867	3,353	2,959	2,649	2,397	2,189	2,014	1,679	1,439	1,260	1,120	1,008	916	775	672	593	480	403
15	11,562	9,406	7,927	6,851	6,032	5,387	4,439	3,775	3,283	2,905	2,605	2,361	2,159	1,778	1,512	1,315	1,163	1,043	945	796	688	605	488	409
20	14,321	11,154	9,134	7,734	6,706	5,919	4,794	4,028	3,473	3,053	2,723	2,458	2,240	1,833	1,551	1,344	1,186	1,061	960	807	696	611	492	412
25	16,715	12,555	10,053	8,382	7,188	6,291	5,035	4,197	3,598	3,149	2,799	2,520	2,291	1,867	1,575	1,363	1,200	1,073	970	813	700	615	494	413
30	18,812	13,702	10,775	8,878	7,549	6,567	5,210	4,318	3,687	3,217	2,853	2,563	2,326	1,890	1,592	1,375	1,210	1,081	976	818	704	618	496	415
35	20,662	14,658	11,357	9,270	7,831	6,779	5,343	4,409	3,753	3,267	2,892	2,594	2,352	1,908	1,604	1,384	1,217	1,086	981	821	706	619	497	415
40	22,309	15,468	11,838	9,588	8,056	6,947	5,446	4,479	3,804	3,305	2,922	2,619	2,372	1,921	1,614	1,391	1,222	1,090	984	823	708	621	498	416
45	23,782	16,162	12,240	9,850	8,241	7,083	5,530	4,536	3,844	3,336	2,946	2,638	2,388	1,931	1,621	1,396	1,227	1,094	987	825	709	622	499	416
50	25,110	16,764	12,582	10,070	8,394	7,197	5,599	4,582	3,877	3,361	2,966	2,654	2,401	1,939	1,627	1,401	1,230	1,096	989	827	710	623	499	417
55	26,311	17,291	12,877	10,258	8,524	7,292	5,657	4,620	3,905	3,381	2,982	2,666	2,411	1,946	1,631	1,404	1,233	1,099	991	828	711	623	500	417
60	27,403	17,756	13,133	10,420	8,636	7,374	5,705	4,653	3,928	3,399	2,995	2,677	2,420	1,952	1,636	1,407	1,235	1,100	992	829	712	624	500	417
65	28,401	18,170	13,358	10,561	8,733	7,444	5,747	4,681	3,948	3,414	3,007	2,686	2,428	1,957	1,639	1,410	1,237	1,102	993	830	713	624	501	418
70	29,316	18,540	13,557	10,685	8,817	7,505	5,784	4,705	3,965	3,426	3,017	2,694	2,434	1,961	1,642	1,412	1,239	1,103	994	831	713	625	501	418
75	30,158	18,873	13,734	10,795	8,892	7,559	5,816	4,726	3,980	3,438	3,025	2,701	2,440	1,965	1,644	1,414	1,240	1,104	995	831	714	625	501	418
80	30,935	19,175	13,893	10,893	8,958	7,607	5,844	4,745	3,994	3,448	3,033	2,707	2,445	1,968	1,647	1,416	1,241	1,105	996	832	714	626	501	418
85	31,655	19,449	14,037	10,981	9,018	7,650	5,870	4,761	4,005	3,456	3,040	2,713	2,449	1,971	1,649	1,417	1,243	1,106	997	832	715	626	501	418
90	32,324	19,700	14,167	11,060	9,071	7,689	5,892	4,776	4,016	3,464	3,046	2,718	2,453	1,973	1,650	1,418	1,244	1,107	998	833	715	626	502	418
95	32,947	19,929	14,285	11,132	9,120	7,723	5,913	4,790	4,025	3,471	3,051	2,722	2,457	1,976	1,652	1,420	1,244	1,108	998	833	715	626	502	418
100	33,528	20,141	14,393	11,198	9,164	7,755	5,931	4,802	4,034	3,478	3,056	2,726	2,460	1,978	1,654	1,421	1,245	1,108	999	834	715	627	502	419
105	34,072	20,336	14,493	11,258	9,204	7,783	5,948	4,813	4,042	3,483	3,061	2,729	2,463	1,980	1,655	1,422	1,246	1,109	999	834	716	627	502	419
110	34,582	20,516	14,584	11,313	9,241	7,810	5,963	4,823	4,049	3,489	3,065	2,733	2,465	1,981	1,656	1,423	1,247	1,110	1,000	834	716	627	502	419
115	35,062	20,684	14,669	11,364	9,274	7,834	5,977	4,832	4,055	3,493	3,068	2,736	2,468	1,983	1,657	1,423	1,247	1,110	1,000	835	716	627	502	419
120	35,513	20,840	14,747	11,411	9,306	7,856	5,990	4,841	4,061	3,498	3,072	2,738	2,470	1,984	1,658	1,424	1,248	1,111	1,000	835	716	627	502	419
125	35,938	20,986	14,820	11,454	9,335	7,877	6,002	4,848	4,067	3,502	3,075	2,741	2,472	1,986	1,659	1,425	1,248	1,111	1,001	835	716	627	502	419
130	36,340	21,122	14,888	11,495	9,362	7,896	6,013	4,856	4,072	3,506	3,078	2,743	2,474	1,987	1,660	1,425	1,249	1,111	1,001	835	717	627	502	419
135	36,720	21,250	14,951	11,533	9,387	7,914	6,024	4,862	4,076	3,509	3,081	2,745	2,476	1,988	1,661	1,426	1,249	1,112	1,001	835	717	628	503	419
140	37,080	21,370	15,011	11,568	9,410	7,930	6,033	4,869	4,081	3,512	3,083	2,747	2,477	1,989	1,661	1,426	1,250	1,112	1,002	836	717	628	503	419
145	37,422	21,483	15,066	11,601	9,432	7,946	6,042	4,874	4,085	3,516	3,085	2,749	2,479	1,990	1,662	1,427	1,250	1,112	1,002	836	717	628	503	419
150	37,747	21,590	15,119	11,632	9,452	7,961	6,051	4,880	4,089	3,518	3,088	2,751	2,480	1,991	1,663	1,427	1,250	1,113	1,002	836	717	628	503	419
155	38,056	21,691	15,168	11,661	9,472	7,974	6,059	4,885	4,092	3,521	3,090	2,752	2,482	1,992	1,663	1,428	1,251	1,113	1,002	836	717	628	503	419
160	38,350	21,786	15,215	11,689	9,490	7,987	6,066	4,890	4,096	3,524	3,092	2,754	2,483	1,992	1,664	1,428	1,251	1,113	1,002	836	717	628	503	419
165	38,631	21,876	15,259	11,715	9,507	7,999	6,073	4,894	4,099	3,526	3,093	2,755	2,484	1,993	1,664	1,429	1,251	1,113	1,003	836	717	628	503	419
170	38,899	21,962	15,300	11,739	9,523	8,011	6,080	4,899	4,102	3,528	3,095	2,757	2,485	1,994	1,665	1,429	1,252	1,114	1,003	837	718	628	503	419
175	39,155	22,043	15,340	11,762	9,538	8,021	6,086	4,903	4,105	3,530	3,097	2,758	2,486	1,995	1,665	1,429	1,252	1,114	1,003	837	718	628	503	419
180	39,399	22,121	15,377	11,784	9,553	8,032	6,092	4,907	4,107	3,532	3,098	2,759	2,487	1,995	1,666	1,430	1,252	1,114	1,003	837	718	628	503	419
185	39,634	22,194	15,413	11,805	9,566	8,041	6,097	4,910	4,110	3,534	3,100	2,760	2,488	1,996	1,666	1,430	1,252	1,114	1,003	837	718	628	503	419
190	39,859	22,265	15,446	11,825	9,579	8,050	6,102	4,914	4,112	3,536	3,101	2,761	2,489	1,996	1,667	1,430	1,253	1,114	1,003	837	718	628	503	419
195	40,074	22,332	15,479	11,844	9,592	8,059	6,108	4,917	4,115	3,537	3,102	2,763	2,490	1,997	1,667	1,431	1,253	1,114	1,004	837	718	628	503	419
200	40,281	22,396	15,510	11,862	9,604	8,068	6,112	4,920	4,117	3,539	3,104	2,764	2,491	1,997	1,667	1,431	1,253	1,115	1,004	837	718	629	503	419

# 6 AWG	1 Run(s)	208 Volts	C-Value = 2,425	Magnetic Duct	Copper

Calculations:

1. $f = \dfrac{\text{Length} \times 2 \times \text{Isca}}{\text{Runs} \times \text{C-Value} \times E_{L-L}}$
2. $M = \dfrac{1}{1 + f}$
3. Isc = Isca x M
4. Add Motor Contribution = Motor FLA x 4

* All results are given in symmetrical amperes

Single-Phase L-L Bolted Fault-Current* Table

One-Way Distance in Feet

Available Isc in Thousands (Isca)	5	10	15	20	25	30	40	50	60	70	80	90	100	125	150	175	200	225	250	300	350	400	500	600
3	2,890	2,789	2,694	2,605	2,522	2,444	2,302	2,176	2,062	1,960	1,868	1,783	1,707	1,541	1,404	1,290	1,192	1,109	1,036	916	821	744	626	541
5	4,703	4,439	4,204	3,992	3,800	3,626	3,322	3,064	2,844	2,654	2,487	2,340	2,209	1,939	1,727	1,557	1,418	1,301	1,203	1,044	922	826	683	583
7	6,431	5,948	5,532	5,171	4,854	4,574	4,100	3,715	3,396	3,128	2,899	2,701	2,528	2,180	1,916	1,709	1,543	1,406	1,291	1,110	974	867	711	603
10	8,878	7,983	7,252	6,643	6,129	5,689	4,974	4,419	3,975	3,612	3,310	3,055	2,836	2,405	2,088	1,845	1,652	1,496	1,367	1,166	1,016	900	734	619
15	12,611	10,878	9,564	8,533	7,703	7,020	5,962	5,182	4,582	4,107	3,721	3,401	3,132	2,615	2,244	1,965	1,748	1,575	1,432	1,213	1,052	928	752	632
20	15,966	13,287	11,377	9,948	8,837	7,950	6,620	5,672	4,961	4,408	3,967	3,605	3,304	2,734	2,331	2,032	1,801	1,617	1,467	1,238	1,070	943	762	639
25	19,000	15,322	12,838	11,046	9,694	8,636	7,089	6,013	5,220	4,612	4,130	3,740	3,417	2,811	2,387	2,074	1,834	1,644	1,489	1,253	1,082	952	767	643
30	21,756	17,066	14,039	11,925	10,363	9,164	7,441	6,264	5,408	4,758	4,247	3,836	3,497	2,864	2,425	2,103	1,857	1,662	1,504	1,264	1,090	958	771	646
35	24,270	18,575	15,045	12,642	10,901	9,582	7,714	6,456	5,551	4,868	4,335	3,907	3,556	2,904	2,454	2,125	1,873	1,675	1,515	1,271	1,096	962	774	648
40	26,573	19,895	15,899	13,240	11,343	9,921	7,933	6,609	5,663	4,954	4,403	3,962	3,602	2,934	2,476	2,141	1,886	1,685	1,523	1,277	1,100	966	776	649
45	28,691	21,059	16,634	13,746	11,712	10,203	8,112	6,732	5,754	5,023	4,458	4,006	3,638	2,958	2,493	2,154	1,896	1,693	1,529	1,282	1,103	968	778	650
50	30,645	22,093	17,272	14,179	12,025	10,439	8,261	6,834	5,828	5,080	4,502	4,042	3,668	2,978	2,507	2,164	1,904	1,699	1,535	1,285	1,106	970	779	651
55	32,453	23,017	17,832	14,554	12,294	10,641	8,387	6,920	5,891	5,127	4,539	4,072	3,692	2,994	2,518	2,173	1,910	1,705	1,539	1,289	1,108	972	780	652
60	34,131	23,849	18,328	14,882	12,527	10,816	8,495	6,994	5,944	5,168	4,571	4,098	3,713	3,008	2,528	2,180	1,916	1,709	1,543	1,291	1,110	974	781	653
65	35,693	24,601	18,769	15,172	12,732	10,968	8,588	7,057	5,989	5,202	4,598	4,119	3,731	3,019	2,536	2,186	1,921	1,713	1,546	1,293	1,112	975	782	653
70	37,150	25,285	19,164	15,429	12,912	11,102	8,670	7,112	6,029	5,232	4,621	4,138	3,746	3,030	2,543	2,191	1,925	1,716	1,548	1,295	1,113	976	783	654
75	38,513	25,909	19,520	15,659	13,073	11,220	8,742	7,161	6,064	5,258	4,642	4,154	3,760	3,038	2,549	2,196	1,928	1,719	1,551	1,297	1,114	977	783	654
80	39,790	26,480	19,843	15,866	13,217	11,326	8,806	7,204	6,094	5,281	4,660	4,169	3,772	3,046	2,555	2,200	1,931	1,721	1,553	1,298	1,115	977	784	654
85	40,989	27,006	20,137	16,053	13,347	11,421	8,864	7,242	6,122	5,302	4,676	4,182	3,782	3,053	2,559	2,203	1,934	1,724	1,554	1,299	1,116	978	784	655
90	42,118	27,492	20,405	16,224	13,464	11,507	8,915	7,276	6,147	5,320	4,690	4,193	3,791	3,059	2,564	2,206	1,937	1,725	1,556	1,300	1,117	979	785	655
95	43,181	27,941	20,652	16,379	13,571	11,585	8,962	7,308	6,169	5,337	4,703	4,203	3,800	3,064	2,568	2,209	1,939	1,727	1,557	1,301	1,118	979	785	655
100	44,185	28,358	20,879	16,521	13,669	11,656	9,005	7,336	6,189	5,352	4,715	4,213	3,808	3,069	2,571	2,212	1,941	1,729	1,559	1,302	1,118	980	785	655
105	45,135	28,746	21,088	16,652	13,758	11,721	9,043	7,361	6,207	5,366	4,725	4,221	3,814	3,074	2,574	2,214	1,943	1,730	1,560	1,303	1,119	980	786	656
110	46,035	29,108	21,283	16,773	13,841	11,781	9,079	7,385	6,224	5,378	4,735	4,229	3,821	3,078	2,577	2,216	1,944	1,732	1,561	1,304	1,119	981	786	656
115	46,888	29,447	21,463	16,885	13,917	11,836	9,112	7,407	6,239	5,390	4,744	4,236	3,827	3,082	2,580	2,218	1,946	1,733	1,562	1,304	1,120	981	786	656
120	47,698	29,764	21,631	16,989	13,987	11,887	9,142	7,427	6,253	5,400	4,752	4,243	3,832	3,085	2,582	2,220	1,947	1,734	1,563	1,305	1,120	981	786	656
125	48,469	30,063	21,789	17,086	14,053	11,934	9,170	7,445	6,266	5,410	4,759	4,249	3,837	3,088	2,584	2,222	1,948	1,735	1,563	1,306	1,121	982	787	656
130	49,202	30,343	21,936	17,176	14,114	11,978	9,196	7,462	6,278	5,419	4,766	4,254	3,841	3,091	2,586	2,223	1,949	1,736	1,564	1,306	1,121	982	787	656
135	49,902	30,608	22,074	17,261	14,171	12,019	9,220	7,478	6,290	5,427	4,773	4,259	3,845	3,094	2,588	2,225	1,951	1,737	1,565	1,307	1,122	982	787	656
140	50,570	30,858	22,203	17,340	14,224	12,058	9,242	7,493	6,300	5,435	4,779	4,264	3,849	3,097	2,590	2,226	1,952	1,737	1,566	1,307	1,122	983	787	657
145	51,207	31,094	22,325	17,414	14,274	12,094	9,263	7,507	6,310	5,442	4,785	4,269	3,853	3,099	2,592	2,227	1,952	1,738	1,566	1,308	1,122	983	787	657
150	51,817	31,318	22,440	17,484	14,321	12,127	9,283	7,520	6,319	5,449	4,790	4,273	3,856	3,101	2,593	2,228	1,953	1,739	1,567	1,308	1,122	983	787	657
155	52,401	31,530	22,549	17,550	14,366	12,159	9,302	7,532	6,328	5,456	4,795	4,277	3,860	3,103	2,595	2,229	1,954	1,739	1,567	1,308	1,123	983	788	657
160	52,961	31,732	22,652	17,613	14,407	12,189	9,319	7,543	6,336	5,462	4,799	4,280	3,863	3,105	2,596	2,230	1,955	1,740	1,568	1,309	1,123	983	788	657
165	53,497	31,924	22,750	17,672	14,447	12,217	9,336	7,554	6,343	5,467	4,804	4,284	3,866	3,107	2,597	2,231	1,956	1,741	1,568	1,309	1,123	984	788	657
170	54,013	32,107	22,843	17,727	14,484	12,244	9,351	7,564	6,351	5,473	4,808	4,287	3,868	3,109	2,598	2,232	1,956	1,741	1,569	1,309	1,123	984	788	657
175	54,507	32,281	22,931	17,780	14,519	12,269	9,366	7,574	6,357	5,478	4,812	4,290	3,871	3,110	2,600	2,233	1,957	1,742	1,569	1,310	1,124	984	788	657
180	54,983	32,447	23,014	17,831	14,553	12,293	9,380	7,583	6,364	5,482	4,815	4,293	3,873	3,112	2,601	2,234	1,958	1,742	1,569	1,310	1,124	984	788	657
185	55,441	32,606	23,094	17,879	14,585	12,316	9,393	7,592	6,370	5,487	4,819	4,296	3,875	3,113	2,602	2,235	1,958	1,743	1,570	1,310	1,124	984	788	657
190	55,881	32,758	23,170	17,924	14,615	12,337	9,406	7,600	6,376	5,491	4,822	4,299	3,877	3,115	2,603	2,235	1,959	1,743	1,570	1,310	1,124	984	788	657
195	56,306	32,903	23,243	17,968	14,644	12,358	9,418	7,608	6,381	5,495	4,825	4,301	3,879	3,116	2,604	2,236	1,959	1,743	1,571	1,311	1,124	985	788	657
200	56,715	33,043	23,312	18,009	14,671	12,378	9,429	7,615	6,386	5,499	4,828	4,303	3,881	3,117	2,604	2,237	1,960	1,744	1,571	1,311	1,125	985	789	658

# 4 AWG	1 Run(s)	208 Volts	C-Value = 3,806	Magnetic Duct	Copper

Calculations:

1. $f = \dfrac{\text{Length} \times 2 \times \text{Isca}}{\text{Runs} \times \text{C-Value} \times \text{E L-L}}$
2. $M = \dfrac{1}{1 + f}$
3. $\text{Isc} = \text{Isca} \times M$
4. Add Motor Contribution = Motor FLA x 4

* All results are given in symmetrical amperes

Single-Phase L-L Bolted Fault-Current* Table

One-Way Distance in Feet

Available Isc in Thousands (Isca)	5	10	15	20	25	30	40	50	60	70	80	90	100	125	150	175	200	225	250	300	350	400	500	600
3	2,912	2,829	2,750	2,676	2,605	2,538	2,415	2,302	2,200	2,106	2,020	1,941	1,868	1,707	1,571	1,456	1,356	1,269	1,193	1,065	961	876	744	647
5	4,760	4,541	4,342	4,160	3,992	3,837	3,561	3,322	3,113	2,929	2,765	2,619	2,488	2,210	1,988	1,807	1,656	1,528	1,418	1,241	1,103	992	826	708
7	6,538	6,133	5,775	5,457	5,172	4,915	4,471	4,101	3,787	3,518	3,285	3,080	2,900	2,529	2,243	2,015	1,829	1,674	1,544	1,335	1,177	1,052	867	738
10	9,083	8,319	7,675	7,122	6,644	6,227	5,531	4,975	4,521	4,142	3,823	3,549	3,311	2,837	2,481	2,205	1,984	1,803	1,653	1,416	1,239	1,101	901	762
15	13,026	11,512	10,313	9,340	8,535	7,857	6,781	5,964	5,323	4,806	4,381	4,025	3,722	3,133	2,705	2,380	2,125	1,919	1,749	1,487	1,293	1,143	929	782
20	16,639	14,245	12,453	11,062	9,950	9,041	7,645	6,622	5,841	5,225	4,726	4,314	3,968	3,306	2,833	2,478	2,203	1,982	1,802	1,524	1,321	1,165	943	792
25	19,960	16,611	14,225	12,438	11,050	9,940	8,278	7,092	6,203	5,513	4,960	4,508	4,132	3,419	2,915	2,541	2,252	2,022	1,835	1,548	1,339	1,179	952	799
30	23,024	18,680	15,715	13,562	11,928	10,646	8,762	7,444	6,471	5,723	5,130	4,648	4,249	3,498	2,973	2,585	2,287	2,050	1,858	1,564	1,351	1,189	958	803
35	25,859	20,504	16,986	14,499	12,647	11,214	9,143	7,718	6,677	5,883	5,258	4,753	4,337	3,558	3,016	2,617	2,312	2,070	1,874	1,576	1,359	1,195	963	806
40	28,490	22,124	18,083	15,290	13,245	11,682	9,452	7,936	6,840	6,010	5,359	4,836	4,405	3,604	3,049	2,642	2,331	2,085	1,887	1,585	1,366	1,200	966	808
45	30,938	23,572	19,039	15,969	13,751	12,074	9,706	8,115	6,972	6,112	5,440	4,901	4,460	3,640	3,075	2,661	2,346	2,098	1,897	1,592	1,371	1,204	969	810
50	33,222	24,875	19,880	16,556	14,184	12,407	9,920	8,264	7,082	6,196	5,507	4,955	4,504	3,670	3,096	2,677	2,358	2,107	1,905	1,597	1,375	1,208	971	812
55	35,358	26,054	20,626	17,070	14,560	12,693	10,103	8,390	7,174	6,266	5,562	5,000	4,542	3,694	3,113	2,690	2,369	2,116	1,911	1,602	1,379	1,210	973	813
60	37,360	27,124	21,291	17,523	14,888	12,942	10,260	8,498	7,253	6,326	5,609	5,039	4,573	3,715	3,128	2,701	2,377	2,122	1,917	1,606	1,382	1,213	974	814
65	39,239	28,102	21,889	17,926	15,178	13,160	10,397	8,592	7,321	6,378	5,650	5,071	4,600	3,733	3,141	2,711	2,384	2,128	1,922	1,609	1,384	1,214	975	815
70	41,007	28,997	22,428	18,286	15,435	13,353	10,517	8,674	7,381	6,423	5,685	5,100	4,623	3,748	3,152	2,719	2,391	2,133	1,926	1,612	1,386	1,216	976	815
75	42,674	29,821	22,918	18,610	15,666	13,525	10,623	8,746	7,433	6,463	5,716	5,125	4,644	3,762	3,161	2,726	2,396	2,137	1,929	1,615	1,388	1,218	977	816
80	44,247	30,581	23,364	18,903	15,873	13,680	10,718	8,810	7,479	6,498	5,744	5,147	4,662	3,774	3,170	2,732	2,401	2,141	1,932	1,617	1,390	1,219	978	817
85	45,735	31,284	23,773	19,170	16,060	13,819	10,803	8,868	7,521	6,529	5,768	5,166	4,678	3,784	3,177	2,738	2,405	2,145	1,935	1,619	1,391	1,220	979	817
90	47,145	31,937	24,148	19,413	16,231	13,945	10,880	8,920	7,558	6,557	5,790	5,184	4,692	3,793	3,184	2,743	2,409	2,148	1,938	1,620	1,393	1,221	979	818
95	48,481	32,545	24,494	19,636	16,386	14,059	10,950	8,966	7,591	6,582	5,810	5,199	4,705	3,802	3,189	2,747	2,412	2,150	1,940	1,622	1,394	1,222	980	818
100	49,751	33,112	24,814	19,841	16,529	14,164	11,013	9,009	7,622	6,605	5,827	5,214	4,717	3,809	3,195	2,751	2,415	2,153	1,942	1,623	1,395	1,222	980	818
105	50,958	33,643	25,110	20,030	16,660	14,260	11,071	9,048	7,650	6,626	5,844	5,227	4,728	3,816	3,200	2,755	2,418	2,155	1,944	1,625	1,396	1,223	981	819
110	52,107	34,140	25,386	20,205	16,781	14,349	11,124	9,083	7,675	6,645	5,858	5,238	4,737	3,823	3,204	2,758	2,421	2,157	1,945	1,626	1,396	1,224	981	819
115	53,203	34,607	25,644	20,368	16,893	14,431	11,174	9,116	7,698	6,662	5,872	5,249	4,746	3,828	3,208	2,761	2,423	2,159	1,947	1,627	1,397	1,224	982	819
120	54,249	35,046	25,884	20,520	16,997	14,507	11,219	9,146	7,720	6,678	5,885	5,259	4,754	3,834	3,212	2,764	2,425	2,161	1,948	1,628	1,398	1,225	982	819
125	55,248	35,461	26,109	20,661	17,094	14,577	11,261	9,174	7,740	6,693	5,896	5,269	4,762	3,839	3,215	2,766	2,427	2,162	1,949	1,629	1,399	1,225	982	820
130	56,203	35,852	26,321	20,793	17,184	14,643	11,300	9,200	7,758	6,707	5,907	5,277	4,769	3,843	3,219	2,769	2,429	2,164	1,950	1,629	1,399	1,226	983	820
135	57,118	36,222	26,520	20,917	17,269	14,704	11,337	9,224	7,775	6,720	5,917	5,285	4,775	3,847	3,222	2,771	2,431	2,165	1,952	1,630	1,400	1,226	983	820
140	57,994	36,572	26,707	21,033	17,348	14,761	11,371	9,247	7,791	6,732	5,926	5,293	4,781	3,851	3,224	2,773	2,432	2,166	1,953	1,631	1,400	1,227	983	820
145	58,835	36,905	26,884	21,143	17,422	14,815	11,403	9,268	7,806	6,743	5,935	5,299	4,787	3,855	3,227	2,775	2,434	2,167	1,953	1,632	1,401	1,227	983	820
150	59,641	37,220	27,051	21,246	17,492	14,866	11,433	9,288	7,821	6,754	5,943	5,306	4,792	3,858	3,229	2,776	2,435	2,168	1,954	1,632	1,401	1,227	984	821
155	60,416	37,521	27,209	21,344	17,558	14,914	11,461	9,306	7,834	6,763	5,950	5,312	4,797	3,862	3,231	2,778	2,436	2,169	1,955	1,633	1,402	1,228	984	821
160	61,161	37,807	27,359	21,436	17,621	14,959	11,487	9,324	7,846	6,773	5,958	5,318	4,802	3,865	3,234	2,780	2,437	2,170	1,956	1,633	1,402	1,228	984	821
165	61,878	38,079	27,502	21,523	17,680	15,001	11,512	9,340	7,858	6,781	5,964	5,323	4,806	3,867	3,236	2,781	2,439	2,171	1,957	1,634	1,402	1,228	984	821
170	62,568	38,340	27,637	21,606	17,736	15,041	11,536	9,356	7,869	6,790	5,971	5,328	4,810	3,870	3,237	2,782	2,440	2,172	1,957	1,634	1,403	1,229	984	821
175	63,233	38,588	27,766	21,685	17,789	15,079	11,559	9,371	7,879	6,797	5,977	5,333	4,814	3,873	3,239	2,784	2,441	2,173	1,958	1,635	1,403	1,229	985	821
180	63,874	38,826	27,889	21,760	17,839	15,116	11,580	9,385	7,889	6,805	5,982	5,337	4,818	3,875	3,241	2,785	2,442	2,174	1,959	1,635	1,403	1,229	985	821
185	64,493	39,054	28,007	21,831	17,887	15,150	11,600	9,398	7,898	6,812	5,988	5,342	4,821	3,877	3,242	2,786	2,443	2,174	1,959	1,636	1,404	1,229	985	821
190	65,090	39,272	28,119	21,899	17,933	15,183	11,619	9,410	7,907	6,818	5,993	5,346	4,825	3,879	3,244	2,787	2,443	2,175	1,960	1,636	1,404	1,230	985	821
195	65,667	39,481	28,226	21,964	17,976	15,214	11,637	9,422	7,916	6,824	5,998	5,350	4,828	3,881	3,245	2,788	2,444	2,176	1,960	1,636	1,404	1,230	985	822
200	66,224	39,682	28,328	22,026	18,018	15,244	11,655	9,434	7,924	6,830	6,002	5,353	4,831	3,883	3,247	2,789	2,445	2,176	1,961	1,637	1,404	1,230	985	822

3 AWG | 1 Run(s) | 208 Volts | C-Value = 4,760 | Magnetic Duct | Copper

Calculations:

1. $f = \dfrac{\text{Length} \times 2 \times \text{Isca}}{\text{Runs} \times \text{C-Value} \times \text{E L-L}}$

2. $M = \dfrac{1}{1 + f}$

3. Isc = Isca x M

4. Add Motor Contribution = Motor FLA x 4

* All results are given in symmetrical amperes

Single-Phase L-L Bolted Fault-Current* Table

One-Way Distance in Feet

Available Isc in Thousands (Isca)

Isca	5	10	15	20	25	30	40	50	60	70	80	90	100	125	150	175	200	225	250	300	350	400	500	600
3	2,928	2,860	2,795	2,733	2,674	2,617	2,510	2,411	2,320	2,236	2,157	2,084	2,016	1,863	1,731	1,617	1,518	1,429	1,351	1,217	1,107	1,016	872	763
5	4,804	4,624	4,456	4,300	4,155	4,019	3,772	3,554	3,359	3,185	3,028	2,886	2,756	2,478	2,251	2,062	1,903	1,766	1,647	1,453	1,299	1,175	986	850
7	6,623	6,284	5,978	5,701	5,448	5,217	4,808	4,459	4,157	3,894	3,662	3,456	3,272	2,887	2,584	2,338	2,135	1,964	1,819	1,584	1,403	1,259	1,045	893
10	9,247	8,600	8,037	7,544	7,107	6,719	6,056	5,513	5,059	4,674	4,343	4,056	3,805	3,295	2,905	2,598	2,350	2,144	1,972	1,699	1,493	1,331	1,094	929
15	13,368	12,056	10,978	10,078	9,314	8,657	7,588	6,754	6,085	5,536	5,078	4,691	4,358	3,701	3,217	2,844	2,549	2,310	2,111	1,802	1,571	1,393	1,135	958
20	17,200	15,087	13,437	12,112	11,025	10,117	8,686	7,610	6,771	6,099	5,548	5,088	4,699	3,945	3,399	2,986	2,662	2,402	2,188	1,857	1,613	1,426	1,157	974
25	20,773	17,768	15,523	13,781	12,391	11,256	9,513	8,237	7,263	6,495	5,874	5,361	4,931	4,107	3,519	3,078	2,735	2,461	2,237	1,892	1,640	1,447	1,171	983
30	24,112	20,156	17,315	15,176	13,507	12,169	10,157	8,716	7,633	6,789	6,113	5,560	5,098	4,222	3,603	3,142	2,786	2,502	2,271	1,917	1,658	1,461	1,180	990
35	27,239	22,295	18,871	16,358	14,436	12,918	10,673	9,093	7,920	7,016	6,297	5,711	5,225	4,309	3,666	3,190	2,823	2,532	2,296	1,934	1,671	1,471	1,187	995
40	30,175	24,224	20,234	17,373	15,220	13,542	11,096	9,398	8,151	7,196	6,441	5,830	5,325	4,376	3,715	3,227	2,852	2,555	2,315	1,948	1,681	1,479	1,192	998
45	32,935	25,972	21,439	18,254	15,892	14,072	11,449	9,650	8,340	7,343	6,559	5,926	5,405	4,430	3,753	3,256	2,875	2,574	2,330	1,958	1,689	1,485	1,196	1,001
50	35,536	27,563	22,512	19,025	16,474	14,526	11,748	9,862	8,497	7,465	6,656	6,005	5,470	4,474	3,785	3,280	2,893	2,589	2,342	1,967	1,695	1,490	1,199	1,003
55	37,991	29,017	23,473	19,707	16,983	14,920	12,004	10,042	8,631	7,567	6,737	6,071	5,525	4,511	3,811	3,299	2,909	2,601	2,352	1,974	1,701	1,494	1,202	1,005
60	40,311	30,351	24,338	20,314	17,431	15,265	12,227	10,197	8,745	7,655	6,807	6,128	5,572	4,542	3,833	3,316	2,922	2,611	2,360	1,980	1,705	1,497	1,204	1,007
65	42,508	31,580	25,122	20,857	17,830	15,570	12,421	10,332	8,844	7,731	6,867	6,176	5,612	4,568	3,852	3,330	2,933	2,620	2,367	1,985	1,709	1,500	1,206	1,008
70	44,591	32,716	25,835	21,346	18,186	15,841	12,593	10,450	8,931	7,797	6,919	6,218	5,647	4,591	3,869	3,342	2,942	2,627	2,374	1,989	1,712	1,503	1,207	1,009
75	46,569	33,768	26,487	21,789	18,506	16,084	12,746	10,556	9,008	7,856	6,965	6,255	5,677	4,612	3,883	3,353	2,950	2,634	2,379	1,993	1,715	1,505	1,209	1,010
80	48,449	34,746	27,085	22,192	18,796	16,302	12,883	10,649	9,076	7,907	7,005	6,288	5,704	4,629	3,895	3,362	2,958	2,640	2,384	1,996	1,717	1,507	1,210	1,011
85	50,238	35,656	27,635	22,560	19,060	16,500	13,006	10,733	9,137	7,954	7,042	6,317	5,728	4,645	3,907	3,371	2,964	2,645	2,388	1,999	1,719	1,508	1,211	1,012
90	51,944	36,507	28,144	22,898	19,300	16,680	13,118	10,809	9,192	7,995	7,074	6,344	5,750	4,659	3,917	3,378	2,970	2,650	2,392	2,002	1,721	1,510	1,212	1,012
95	53,571	37,304	28,614	23,208	19,521	16,844	13,219	10,878	9,241	8,033	7,104	6,367	5,769	4,672	3,926	3,385	2,975	2,654	2,395	2,004	1,723	1,511	1,213	1,013
100	55,126	38,051	29,052	23,495	19,723	16,995	13,312	10,940	9,286	8,067	7,130	6,389	5,787	4,684	3,934	3,391	2,980	2,657	2,398	2,006	1,725	1,512	1,214	1,013
105	56,612	38,753	29,460	23,761	19,910	17,133	13,396	10,998	9,328	8,098	7,155	6,408	5,803	4,694	3,941	3,396	2,984	2,661	2,401	2,008	1,726	1,513	1,214	1,014
110	58,034	39,414	29,840	24,008	20,083	17,261	13,475	11,050	9,365	8,126	7,177	6,426	5,817	4,704	3,948	3,401	2,988	2,664	2,403	2,010	1,727	1,514	1,215	1,014
115	59,397	40,038	30,196	24,238	20,244	17,380	13,547	11,099	9,400	8,153	7,197	6,442	5,831	4,712	3,954	3,406	2,991	2,667	2,406	2,012	1,729	1,515	1,215	1,015
120	60,703	40,627	30,530	24,453	20,394	17,490	13,614	11,144	9,432	8,177	7,216	6,457	5,843	4,720	3,960	3,410	2,994	2,669	2,408	2,013	1,730	1,516	1,216	1,015
125	61,957	41,185	30,844	24,654	20,533	17,593	13,676	11,185	9,462	8,199	7,234	6,471	5,855	4,728	3,965	3,414	2,997	2,672	2,410	2,014	1,731	1,517	1,216	1,015
130	63,161	41,714	31,140	24,842	20,664	17,688	13,733	11,224	9,490	8,220	7,250	6,484	5,865	4,735	3,970	3,418	3,000	2,674	2,411	2,016	1,732	1,518	1,217	1,016
135	64,318	42,215	31,418	25,020	20,786	17,778	13,787	11,260	9,516	8,239	7,265	6,496	5,875	4,741	3,974	3,421	3,003	2,676	2,413	2,017	1,732	1,518	1,217	1,016
140	65,431	42,692	31,682	25,186	20,901	17,862	13,838	11,294	9,540	8,257	7,279	6,507	5,884	4,747	3,978	3,424	3,005	2,678	2,415	2,018	1,733	1,519	1,218	1,016
145	66,503	43,146	31,931	25,343	21,009	17,941	13,885	11,325	9,562	8,274	7,292	6,518	5,893	4,753	3,982	3,427	3,007	2,679	2,416	2,019	1,734	1,519	1,218	1,017
150	67,536	43,578	32,167	25,492	21,111	18,015	13,930	11,355	9,583	8,290	7,304	6,528	5,901	4,758	3,986	3,430	3,010	2,681	2,417	2,020	1,735	1,520	1,218	1,017
155	68,531	43,990	32,391	25,632	21,207	18,085	13,971	11,382	9,603	8,305	7,315	6,537	5,908	4,763	3,989	3,432	3,011	2,683	2,419	2,021	1,735	1,520	1,219	1,017
160	69,491	44,384	32,604	25,766	21,298	18,151	14,011	11,409	9,621	8,318	7,326	6,546	5,915	4,767	3,993	3,435	3,013	2,684	2,420	2,022	1,736	1,521	1,219	1,017
165	70,418	44,760	32,807	25,892	21,385	18,214	14,048	11,433	9,639	8,332	7,336	6,554	5,922	4,772	3,996	3,437	3,015	2,685	2,421	2,022	1,736	1,521	1,219	1,017
170	71,313	45,120	33,000	26,012	21,467	18,273	14,083	11,457	9,656	8,344	7,346	6,561	5,928	4,776	3,999	3,439	3,017	2,687	2,422	2,023	1,737	1,522	1,220	1,018
175	72,178	45,465	33,184	26,126	21,544	18,330	14,117	11,479	9,671	8,356	7,355	6,569	5,934	4,780	4,001	3,441	3,018	2,688	2,423	2,024	1,738	1,522	1,220	1,018
180	73,015	45,795	33,359	26,235	21,618	18,383	14,149	11,500	9,686	8,367	7,364	6,575	5,940	4,783	4,004	3,443	3,020	2,689	2,424	2,024	1,738	1,523	1,220	1,018
185	73,824	46,112	33,527	26,339	21,689	18,434	14,179	11,520	9,700	8,377	7,372	6,582	5,945	4,787	4,006	3,445	3,021	2,690	2,425	2,025	1,738	1,523	1,220	1,018
190	74,607	46,417	33,688	26,438	21,756	18,482	14,207	11,538	9,714	8,387	7,380	6,588	5,950	4,790	4,008	3,446	3,022	2,691	2,426	2,026	1,739	1,523	1,221	1,018
195	75,366	46,710	33,842	26,533	21,820	18,529	14,235	11,556	9,726	8,397	7,387	6,594	5,955	4,793	4,011	3,448	3,024	2,692	2,426	2,026	1,739	1,524	1,221	1,018
200	76,101	46,991	33,989	26,623	21,881	18,573	14,261	11,574	9,739	8,406	7,394	6,600	5,959	4,796	4,013	3,449	3,025	2,693	2,427	2,027	1,740	1,524	1,221	1,018

# 2 AWG	1 Run(s)	208 Volts	C-Value = 5,906	Magnetic Duct	Copper

Calculations:

1. $f = \dfrac{\text{Length} \times 2 \times \text{Isca}}{\text{Runs} \times \text{C-Value} \times \text{E L-L}}$

2. $M = \dfrac{1}{1 + f}$

3. Isc = Isca x M

4. Add Motor Contribution = Motor FLA x 4

* All results are given in symmetrical amperes

Single-Phase L-L Bolted Fault-Current* Table

One-Way Distance in Feet

Available Isc in Thousands (Isca)	5	10	15	20	25	30	40	50	60	70	80	90	100	125	150	175	200	225	250	300	350	400	500	600
3	2,942	2,886	2,832	2,780	2,730	2,682	2,590	2,505	2,425	2,349	2,279	2,212	2,150	2,007	1,883	1,773	1,675	1,587	1,508	1,372	1,258	1,162	1,007	889
5	4,840	4,691	4,550	4,417	4,292	4,174	3,957	3,760	3,583	3,421	3,273	3,138	3,013	2,741	2,514	2,321	2,156	2,013	1,888	1,679	1,512	1,375	1,164	1,009
7	6,691	6,408	6,149	5,909	5,688	5,482	5,112	4,790	4,505	4,252	4,027	3,824	3,640	3,250	2,936	2,677	2,460	2,275	2,116	1,857	1,655	1,492	1,247	1,071
10	9,381	8,835	8,349	7,913	7,521	7,165	6,547	6,027	5,583	5,200	4,866	4,573	4,313	3,776	3,358	3,023	2,749	2,521	2,327	2,018	1,781	1,594	1,317	1,122
15	13,650	12,523	11,568	10,748	10,037	9,414	8,374	7,542	6,859	6,290	5,809	5,395	5,037	4,320	3,781	3,362	3,027	2,752	2,523	2,163	1,893	1,683	1,377	1,166
20	17,670	15,826	14,331	13,094	12,053	11,166	9,733	8,626	7,745	7,027	6,431	5,929	5,499	4,655	4,036	3,562	3,188	2,884	2,634	2,244	1,955	1,732	1,410	1,189
25	21,462	18,802	16,728	15,066	13,705	12,569	10,782	9,440	8,395	7,558	6,873	6,302	5,819	4,882	4,205	3,693	3,292	2,970	2,705	2,296	1,994	1,762	1,430	1,203
30	25,046	21,496	18,828	16,749	15,083	13,719	11,617	10,074	8,893	7,959	7,203	6,579	6,053	5,046	4,327	3,787	3,366	3,030	2,755	2,331	2,021	1,783	1,444	1,213
35	28,438	23,948	20,682	18,200	16,250	14,678	12,298	10,582	9,286	8,273	7,459	6,791	6,233	5,171	4,418	3,856	3,421	3,074	2,792	2,358	2,040	1,798	1,454	1,220
40	31,652	26,187	22,332	19,466	17,252	15,490	12,863	10,997	9,605	8,525	7,663	6,960	6,375	5,268	4,488	3,910	3,464	3,109	2,820	2,378	2,055	1,810	1,461	1,225
45	34,704	28,242	23,809	20,578	18,120	16,186	13,339	11,344	9,868	8,732	7,830	7,097	6,490	5,346	4,545	3,953	3,497	3,136	2,842	2,393	2,067	1,819	1,467	1,229
50	37,604	30,133	25,139	21,565	18,880	16,790	13,747	11,637	10,089	8,904	7,969	7,211	6,585	5,410	4,592	3,988	3,525	3,158	2,860	2,406	2,077	1,827	1,472	1,233
55	40,363	31,880	26,343	22,445	19,551	17,319	14,099	11,889	10,278	9,051	8,086	7,307	6,665	5,464	4,630	4,017	3,547	3,176	2,875	2,417	2,085	1,833	1,476	1,236
60	42,993	33,498	27,438	23,235	20,148	17,786	14,407	12,107	10,440	9,177	8,186	7,389	6,733	5,510	4,663	4,042	3,566	3,191	2,887	2,426	2,091	1,838	1,479	1,238
65	45,501	35,001	28,438	23,948	20,682	18,201	14,678	12,298	10,582	9,286	8,273	7,459	6,791	5,549	4,691	4,063	3,583	3,204	2,898	2,433	2,097	1,842	1,482	1,240
70	47,895	36,401	29,356	24,595	21,163	18,572	14,919	12,466	10,706	9,382	8,349	7,521	6,842	5,583	4,715	4,081	3,597	3,216	2,907	2,440	2,102	1,846	1,485	1,242
75	50,185	37,708	30,200	25,185	21,599	18,906	15,134	12,616	10,817	9,466	8,416	7,575	6,887	5,613	4,736	4,097	3,609	3,226	2,916	2,445	2,106	1,849	1,487	1,243
80	52,375	38,931	30,980	25,725	21,995	19,209	15,327	12,750	10,915	9,542	8,475	7,623	6,927	5,639	4,755	4,111	3,620	3,234	2,923	2,450	2,110	1,852	1,489	1,244
85	54,473	40,079	31,702	26,221	22,356	19,484	15,502	12,871	11,003	9,609	8,528	7,666	6,962	5,663	4,772	4,123	3,630	3,242	2,929	2,455	2,113	1,855	1,490	1,245
90	56,484	41,157	32,372	26,678	22,688	19,736	15,660	12,980	11,083	9,670	8,576	7,705	6,994	5,684	4,787	4,134	3,639	3,249	2,935	2,459	2,116	1,857	1,492	1,246
95	58,413	42,172	32,997	27,101	22,993	19,966	15,805	13,079	11,155	9,725	8,620	7,740	7,023	5,703	4,800	4,144	3,646	3,255	2,940	2,462	2,118	1,859	1,493	1,247
100	60,266	43,129	33,580	27,493	23,274	20,178	15,938	13,170	11,221	9,775	8,659	7,771	7,049	5,720	4,812	4,154	3,653	3,261	2,944	2,466	2,121	1,861	1,494	1,248
105	62,047	44,033	34,126	27,858	23,535	20,374	16,059	13,253	11,281	9,821	8,695	7,800	7,073	5,736	4,824	4,162	3,660	3,266	2,948	2,468	2,123	1,862	1,495	1,249
110	63,759	44,889	34,638	28,198	23,778	20,555	16,172	13,329	11,337	9,862	8,727	7,827	7,095	5,750	4,834	4,169	3,665	3,270	2,952	2,471	2,125	1,864	1,496	1,250
115	65,408	45,700	35,119	28,516	24,003	20,724	16,276	13,400	11,388	9,901	8,758	7,851	7,115	5,763	4,843	4,176	3,671	3,275	2,956	2,474	2,127	1,865	1,497	1,250
120	66,995	46,469	35,571	28,814	24,214	20,880	16,372	13,465	11,435	9,937	8,786	7,873	7,133	5,775	4,851	4,182	3,676	3,278	2,959	2,476	2,128	1,866	1,498	1,251
125	68,526	47,201	35,998	29,093	24,411	21,027	16,462	13,526	11,479	9,970	8,811	7,894	7,150	5,786	4,859	4,188	3,680	3,282	2,962	2,478	2,130	1,868	1,499	1,251
130	70,001	47,896	36,401	29,356	24,596	21,164	16,546	13,583	11,519	10,000	8,835	7,913	7,166	5,796	4,867	4,194	3,684	3,285	2,964	2,480	2,131	1,869	1,499	1,252
135	71,426	48,559	36,783	29,603	24,769	21,292	16,624	13,635	11,557	10,029	8,858	7,931	7,180	5,806	4,873	4,199	3,688	3,288	2,967	2,481	2,133	1,870	1,500	1,252
140	72,802	49,191	37,144	29,837	24,932	21,413	16,698	13,685	11,593	10,056	8,878	7,948	7,194	5,815	4,880	4,203	3,692	3,291	2,969	2,483	2,134	1,871	1,500	1,253
145	74,131	49,794	37,487	30,058	25,086	21,526	16,767	13,731	11,626	10,081	8,898	7,964	7,207	5,823	4,885	4,208	3,695	3,294	2,971	2,485	2,135	1,871	1,501	1,253
150	75,416	50,371	37,813	30,267	25,232	21,633	16,832	13,775	11,657	10,104	8,916	7,978	7,219	5,831	4,891	4,212	3,698	3,296	2,973	2,486	2,136	1,872	1,502	1,253
155	76,659	50,922	38,123	30,465	25,370	21,734	16,893	13,815	11,686	10,126	8,933	7,992	7,230	5,838	4,896	4,216	3,701	3,299	2,975	2,487	2,137	1,873	1,502	1,254
160	77,863	51,450	38,418	30,654	25,500	21,830	16,951	13,854	11,714	10,147	8,949	8,005	7,240	5,845	4,901	4,219	3,704	3,301	2,977	2,489	2,138	1,874	1,502	1,254
165	79,028	51,957	38,700	30,833	25,624	21,921	17,005	13,890	11,740	10,166	8,965	8,017	7,250	5,852	4,905	4,223	3,707	3,303	2,979	2,490	2,139	1,874	1,503	1,254
170	80,157	52,442	38,969	31,003	25,741	22,007	17,057	13,925	11,765	10,185	8,979	8,028	7,260	5,858	4,910	4,226	3,709	3,305	2,980	2,491	2,139	1,875	1,503	1,255
175	81,252	52,909	39,226	31,166	25,853	22,088	17,106	13,958	11,788	10,202	8,992	8,039	7,269	5,864	4,914	4,229	3,711	3,307	2,982	2,492	2,140	1,876	1,504	1,255
180	82,314	53,357	39,471	31,320	25,960	22,166	17,153	13,989	11,810	10,219	9,005	8,049	7,277	5,869	4,918	4,232	3,714	3,309	2,983	2,493	2,141	1,876	1,504	1,255
185	83,344	53,788	39,707	31,468	26,061	22,240	17,197	14,018	11,831	10,234	9,018	8,059	7,285	5,874	4,921	4,234	3,716	3,310	2,985	2,494	2,142	1,877	1,504	1,255
190	84,344	54,202	39,932	31,610	26,158	22,311	17,239	14,046	11,851	10,249	9,029	8,068	7,293	5,879	4,925	4,237	3,718	3,312	2,986	2,495	2,142	1,877	1,505	1,256
195	85,315	54,602	40,149	31,745	26,251	22,378	17,279	14,073	11,870	10,264	9,040	8,077	7,300	5,884	4,928	4,239	3,720	3,313	2,987	2,496	2,143	1,878	1,505	1,256
200	86,258	54,987	40,356	31,875	26,340	22,442	17,318	14,098	11,888	10,277	9,051	8,086	7,307	5,888	4,931	4,242	3,721	3,315	2,988	2,496	2,144	1,878	1,505	1,256

# 1 AWG	1 Run(s)	208 Volts	C-Value = 7,292	Magnetic Duct	Copper

Calculations:

1. $f = \dfrac{\text{Length} \times 2 \times \text{Isca}}{\text{Runs} \times \text{C-Value} \times E_{L\text{-}L}}$

2. $M = \dfrac{1}{1 + f}$

3. $\text{Isc} = \text{Isca} \times M$

4. Add Motor Contribution = Motor FLA x 4

* All results are given in symmetrical amperes

Single-Phase L-L Bolted Fault-Current* Table

One-Way Distance in Feet

Available Isc in Thousands (Isca)

Isca	5	10	15	20	25	30	40	50	60	70	80	90	100	125	150	175	200	225	250	300	350	400	500	600
3	2,952	2,906	2,861	2,818	2,776	2,735	2,657	2,583	2,513	2,446	2,384	2,324	2,267	2,137	2,020	1,916	1,822	1,737	1,659	1,523	1,408	1,308	1,147	1,021
5	4,869	4,744	4,626	4,514	4,407	4,304	4,114	3,939	3,779	3,631	3,494	3,367	3,249	2,988	2,765	2,574	2,407	2,260	2,131	1,971	1,733	1,585	1,354	1,181
7	6,746	6,509	6,289	6,082	5,889	5,708	5,378	5,083	4,819	4,581	4,366	4,170	3,990	3,603	3,284	3,017	2,791	2,595	2,426	2,145	1,923	1,743	1,467	1,267
10	9,489	9,027	8,609	8,227	7,878	7,557	6,988	6,499	6,074	5,700	5,371	5,077	4,814	4,261	3,822	3,466	3,170	2,920	2,707	2,363	2,096	1,883	1,566	1,340
15	13,878	12,913	12,073	11,336	10,683	10,102	9,110	8,296	7,615	7,038	6,542	6,111	5,733	4,966	4,380	3,918	3,544	3,235	2,976	2,565	2,253	2,009	1,652	1,402
20	18,055	16,454	15,114	13,976	12,998	12,147	10,741	9,627	8,722	7,973	7,342	6,804	6,339	5,415	4,725	4,192	3,767	3,420	3,131	2,679	2,341	2,079	1,699	1,436
25	22,033	19,695	17,806	16,247	14,939	13,827	12,034	10,653	9,556	8,664	7,924	7,301	6,768	5,725	4,960	4,375	3,914	3,541	3,232	2,753	2,397	2,123	1,728	1,457
30	25,826	22,672	20,204	18,221	16,592	15,231	13,083	11,467	10,206	9,195	8,366	7,674	7,088	5,952	5,129	4,507	4,019	3,626	3,304	2,804	2,436	2,154	1,748	1,471
35	29,447	25,415	22,355	19,952	18,015	16,422	13,953	12,129	10,727	9,616	8,713	7,965	7,336	6,125	5,258	4,606	4,097	3,690	3,356	2,842	2,465	2,176	1,763	1,481
40	32,908	27,953	24,294	21,483	19,254	17,445	14,685	12,678	11,155	9,958	8,993	8,199	7,533	6,262	5,358	4,683	4,158	3,739	3,397	2,872	2,487	2,193	1,774	1,489
45	36,219	30,306	26,052	22,846	20,342	18,333	15,309	13,141	11,511	10,241	9,223	8,390	7,694	6,373	5,439	4,744	4,207	3,779	3,429	2,895	2,504	2,206	1,783	1,495
50	39,390	32,494	27,653	24,068	21,305	19,112	15,848	13,537	11,814	10,480	9,416	8,549	7,823	6,465	5,506	4,795	4,246	3,811	3,456	2,913	2,518	2,217	1,790	1,500
55	42,428	34,534	29,117	25,169	22,164	19,800	16,318	13,878	12,073	10,683	9,580	8,684	7,941	6,542	5,562	4,837	4,279	3,837	3,478	2,929	2,530	2,226	1,796	1,505
60	45,343	36,441	30,461	26,167	22,934	20,412	16,732	14,176	12,298	10,859	9,722	8,800	8,038	6,607	5,609	4,873	4,307	3,860	3,496	2,942	2,539	2,234	1,800	1,508
65	48,142	38,227	31,699	27,075	23,629	20,960	17,099	14,439	12,495	11,012	9,844	8,900	8,121	6,664	5,650	4,903	4,331	3,879	3,512	2,953	2,548	2,240	1,805	1,511
70	50,831	39,903	32,843	27,906	24,259	21,455	17,426	14,671	12,669	11,147	9,952	8,988	8,194	6,713	5,685	4,930	4,352	3,895	3,525	2,963	2,555	2,246	1,808	1,513
75	53,417	41,480	33,904	28,667	24,832	21,902	17,720	14,879	12,824	11,267	10,047	9,066	8,259	6,756	5,716	4,953	4,370	3,910	3,537	2,971	2,561	2,251	1,811	1,516
80	55,905	42,965	34,889	29,369	25,357	22,309	17,986	15,066	12,962	11,374	10,132	9,135	8,316	6,794	5,743	4,974	4,386	3,923	3,548	2,978	2,567	2,255	1,814	1,517
85	58,302	44,367	35,808	30,017	25,839	22,681	18,227	15,235	13,087	11,469	10,208	9,196	8,367	6,828	5,767	4,992	4,400	3,934	3,557	2,985	2,571	2,259	1,817	1,519
90	60,612	45,692	36,666	30,618	26,283	23,023	18,447	15,388	13,200	11,556	10,277	9,252	8,413	6,859	5,789	5,008	4,413	3,944	3,565	2,991	2,576	2,262	1,819	1,521
95	62,839	46,946	37,469	31,176	26,693	23,337	18,648	15,528	13,302	11,635	10,339	9,302	8,455	6,887	5,809	5,023	4,424	3,953	3,573	2,996	2,580	2,265	1,821	1,522
100	64,988	48,135	38,223	31,696	27,073	23,627	18,833	15,656	13,396	11,706	10,395	9,348	8,493	6,912	5,827	5,036	4,435	3,961	3,579	3,001	2,583	2,268	1,822	1,523
105	67,064	49,265	38,932	32,182	27,427	23,896	19,003	15,773	13,482	11,772	10,447	9,390	8,527	6,934	5,843	5,048	4,444	3,969	3,586	3,005	2,586	2,270	1,824	1,524
110	69,069	50,338	39,599	32,637	27,756	24,146	19,161	15,882	13,561	11,832	10,494	9,428	8,559	6,955	5,858	5,059	4,453	3,976	3,591	3,009	2,589	2,272	1,825	1,525
115	71,007	51,360	40,229	33,063	28,064	24,378	19,307	15,982	13,634	11,888	10,538	9,464	8,588	6,974	5,871	5,070	4,460	3,982	3,596	3,013	2,592	2,274	1,827	1,526
120	72,883	52,334	40,824	33,464	28,353	24,596	19,443	16,075	13,702	11,939	10,579	9,496	8,615	6,992	5,884	5,079	4,468	3,988	3,601	3,016	2,594	2,276	1,828	1,527
125	74,697	53,263	41,387	33,842	28,623	24,799	19,570	16,162	13,765	11,987	10,616	9,526	8,639	7,008	5,895	5,088	4,474	3,993	3,605	3,019	2,597	2,278	1,829	1,528
130	76,454	54,150	41,921	34,198	28,877	24,990	19,688	16,243	13,823	12,031	10,651	9,554	8,663	7,024	5,906	5,096	4,481	3,998	3,609	3,022	2,599	2,280	1,830	1,529
135	78,157	54,999	42,428	34,534	29,117	25,169	19,799	16,318	13,878	12,073	10,683	9,580	8,684	7,038	5,916	5,103	4,486	4,003	3,613	3,024	2,601	2,281	1,831	1,529
140	79,807	55,811	42,909	34,852	29,343	25,338	19,904	16,389	13,929	12,112	10,713	9,605	8,704	7,051	5,925	5,110	4,492	4,007	3,616	3,027	2,602	2,282	1,832	1,530
145	81,407	56,589	43,368	35,154	29,557	25,497	20,002	16,455	13,977	12,148	10,742	9,627	8,723	7,063	5,934	5,116	4,497	4,011	3,620	3,029	2,604	2,284	1,833	1,530
150	82,960	57,335	43,804	35,441	29,759	25,647	20,094	16,518	14,022	12,182	10,768	9,649	8,740	7,075	5,942	5,122	4,501	4,014	3,623	3,031	2,606	2,285	1,834	1,531
155	84,467	58,051	44,221	35,713	29,950	25,789	20,181	16,577	14,065	12,214	10,793	9,669	8,757	7,085	5,950	5,128	4,506	4,018	3,626	3,033	2,607	2,286	1,834	1,532
160	85,930	58,738	44,619	35,972	30,132	25,924	20,264	16,632	14,105	12,244	10,817	9,688	8,772	7,096	5,957	5,133	4,510	4,021	3,628	3,035	2,608	2,287	1,835	1,532
165	87,352	59,399	44,999	36,219	30,305	26,052	20,342	16,685	14,142	12,272	10,839	9,706	8,787	7,105	5,964	5,138	4,514	4,024	3,631	3,037	2,610	2,288	1,836	1,532
170	88,733	60,034	45,363	36,454	30,470	26,173	20,416	16,735	14,178	12,299	10,860	9,722	8,801	7,114	5,970	5,143	4,517	4,027	3,633	3,038	2,611	2,289	1,836	1,533
175	90,077	60,646	45,711	36,679	30,627	26,289	20,486	16,782	14,212	12,325	10,880	9,738	8,814	7,123	5,976	5,147	4,521	4,030	3,635	3,040	2,612	2,290	1,837	1,533
180	91,383	61,236	46,045	36,893	30,776	26,399	20,553	16,827	14,244	12,349	10,899	9,753	8,826	7,131	5,982	5,152	4,524	4,032	3,637	3,041	2,613	2,291	1,837	1,534
185	92,655	61,804	46,366	37,099	30,919	26,504	20,617	16,869	14,275	12,372	10,917	9,768	8,838	7,138	5,987	5,156	4,527	4,035	3,639	3,043	2,614	2,292	1,838	1,534
190	93,892	62,352	46,674	37,296	31,056	26,605	20,677	16,910	14,304	12,394	10,934	9,781	8,849	7,146	5,992	5,159	4,530	4,037	3,641	3,044	2,615	2,292	1,838	1,534
195	95,097	62,881	46,970	37,484	31,187	26,701	20,735	16,949	14,331	12,414	10,950	9,794	8,859	7,152	5,997	5,163	4,533	4,039	3,643	3,045	2,616	2,293	1,839	1,535
200	96,271	63,392	47,254	37,666	31,312	26,792	20,790	16,986	14,358	12,434	10,965	9,807	8,869	7,159	6,002	5,166	4,535	4,042	3,645	3,047	2,617	2,294	1,839	1,535

1/0 AWG | 1 Run(s) | 208 Volts | C-Value = 8,924 | Magnetic Duct | Copper

Calculations:

1. $f = \dfrac{\text{Length} \times 2 \times \text{Isca}}{\text{Runs} \times \text{C-Value} \times E_{L-L}}$

2. $M = \dfrac{1}{1 + f}$

3. $\text{Isc} = \text{Isca} \times M$

4. Add Motor Contribution = Motor FLA x 4

* All results are given in symmetrical amperes

Single-Phase L-L Bolted Fault-Current* Table

One-Way Distance in Feet

Available Isc in Thousands (Isca)	5	10	15	20	25	30	40	50	60	70	80	90	100	125	150	175	200	225	250	300	350	400	500	600
3	2,976	2,952	2,929	2,906	2,883	2,861	2,818	2,776	2,735	2,695	2,657	2,619	2,583	2,496	2,415	2,339	2,267	2,200	2,137	2,020	1,916	1,822	1,659	1,523
5	4,934	4,869	4,806	4,744	4,685	4,626	4,514	4,407	4,304	4,207	4,114	4,024	3,939	3,741	3,561	3,398	3,249	3,113	2,988	2,765	2,574	2,407	2,131	1,911
7	6,870	6,746	6,625	6,509	6,397	6,289	6,082	5,889	5,708	5,538	5,378	5,226	5,083	4,757	4,471	4,217	3,990	3,787	3,603	3,284	3,017	2,791	2,426	2,145
10	9,738	9,489	9,252	9,027	8,813	8,609	8,227	7,878	7,557	7,262	6,988	6,735	6,499	5,976	5,531	5,147	4,814	4,520	4,261	3,822	3,466	3,170	2,707	2,363
15	14,417	13,878	13,378	12,913	12,479	12,073	11,336	10,683	10,102	9,581	9,110	8,684	8,296	7,462	6,781	6,213	5,733	5,322	4,966	4,380	3,918	3,544	2,976	2,565
20	18,978	18,055	17,217	16,454	15,756	15,114	13,976	12,998	12,147	11,401	10,741	10,154	9,627	8,522	7,645	6,931	6,339	5,841	5,415	4,725	4,192	3,767	3,131	2,679
25	23,423	22,033	20,798	19,695	18,703	17,806	16,247	14,939	13,827	12,868	12,034	11,301	10,653	9,316	8,277	7,447	6,768	6,203	5,725	4,960	4,375	3,914	3,232	2,753
30	27,757	25,826	24,146	22,672	21,367	20,204	18,221	16,592	15,231	14,076	13,083	12,222	11,467	9,933	8,761	7,836	7,088	6,470	5,952	5,129	4,507	4,019	3,304	2,804
35	31,985	29,447	27,283	25,415	23,787	22,355	19,952	18,015	16,422	15,087	13,953	12,977	12,129	10,426	9,142	8,140	7,336	6,676	6,125	5,258	4,606	4,097	3,356	2,842
40	36,109	32,908	30,229	27,953	25,995	24,294	21,483	19,254	17,445	15,946	14,685	13,608	12,678	10,829	9,451	8,384	7,533	6,839	6,262	5,358	4,683	4,158	3,397	2,872
45	40,135	36,219	33,000	30,306	28,019	26,052	22,846	20,342	18,333	16,685	15,309	14,143	13,141	11,165	9,706	8,584	7,694	6,972	6,373	5,439	4,744	4,207	3,429	2,895
50	44,065	39,390	35,611	32,494	29,879	27,653	24,068	21,305	19,112	17,328	15,848	14,601	13,537	11,449	9,920	8,751	7,828	7,081	6,465	5,506	4,795	4,246	3,456	2,913
55	47,903	42,428	38,077	34,534	31,595	29,117	25,169	22,164	19,800	17,891	16,318	15,000	13,878	11,693	10,102	8,892	7,941	7,174	6,542	5,562	4,837	4,279	3,478	2,929
60	51,652	45,343	40,408	36,441	33,184	30,461	26,167	22,934	20,412	18,390	16,732	15,348	14,176	11,904	10,259	9,013	8,038	7,253	6,607	5,609	4,873	4,307	3,496	2,942
65	55,315	48,142	42,615	38,227	34,658	31,699	27,075	23,629	20,960	18,834	17,099	15,657	14,439	12,088	10,396	9,119	8,121	7,321	6,664	5,650	4,903	4,331	3,512	2,953
70	58,895	50,831	44,709	39,903	36,031	32,843	27,906	24,259	21,455	19,232	17,426	15,931	14,671	12,251	10,516	9,211	8,194	7,380	6,713	5,685	4,930	4,352	3,525	2,963
75	62,395	53,417	46,698	41,480	37,311	33,904	28,667	24,832	21,902	19,591	17,720	16,176	14,879	12,395	10,622	9,293	8,259	7,432	6,756	5,716	4,953	4,370	3,537	2,971
80	65,817	55,905	48,588	42,965	38,508	34,889	29,369	25,357	22,309	19,916	17,986	16,397	15,066	12,525	10,717	9,365	8,316	7,479	6,794	5,743	4,974	4,386	3,548	2,978
85	69,164	58,302	50,389	44,367	39,630	35,808	30,017	25,839	22,681	20,212	18,227	16,597	15,235	12,641	10,802	9,430	8,367	7,520	6,828	5,767	4,992	4,400	3,557	2,985
90	72,439	60,612	52,105	45,692	40,684	36,666	30,618	26,283	23,023	20,482	18,447	16,779	15,388	12,746	10,879	9,489	8,413	7,557	6,859	5,789	5,008	4,413	3,565	2,991
95	75,643	62,839	53,742	46,946	41,676	37,469	31,176	26,693	23,337	20,731	18,648	16,946	15,528	12,842	10,948	9,541	8,455	7,591	6,887	5,809	5,023	4,424	3,573	2,996
100	78,779	64,988	55,306	48,135	42,610	38,223	31,696	27,073	23,627	20,959	18,833	17,098	15,656	12,930	11,012	9,590	8,493	7,621	6,912	5,827	5,036	4,435	3,579	3,001
105	81,850	67,064	56,802	49,265	43,493	38,932	32,182	27,427	23,896	21,171	19,003	17,238	15,773	13,010	11,070	9,634	8,527	7,649	6,934	5,843	5,048	4,444	3,586	3,005
110	84,857	69,069	58,234	50,338	44,328	39,599	32,637	27,756	24,146	21,366	19,161	17,368	15,882	13,083	11,123	9,674	8,559	7,674	6,955	5,858	5,059	4,453	3,591	3,009
115	87,801	71,007	59,606	51,360	45,118	40,229	33,063	28,064	24,378	21,548	19,307	17,488	15,982	13,151	11,172	9,711	8,588	7,698	6,974	5,871	5,070	4,460	3,596	3,013
120	90,686	72,883	60,922	52,334	45,868	40,824	33,464	28,353	24,596	21,718	19,443	17,600	16,075	13,214	11,218	9,745	8,615	7,719	6,992	5,884	5,079	4,468	3,601	3,016
125	93,513	74,697	62,185	53,263	46,580	41,387	33,842	28,623	24,799	21,876	19,570	17,703	16,162	13,273	11,260	9,777	8,639	7,739	7,008	5,895	5,088	4,474	3,605	3,019
130	96,284	76,454	63,398	54,150	47,257	41,921	34,198	28,877	24,990	22,025	19,688	17,800	16,243	13,327	11,299	9,807	8,663	7,757	7,024	5,906	5,096	4,481	3,609	3,022
135	98,999	78,157	64,564	54,999	47,902	42,428	34,534	29,117	25,169	22,164	19,799	17,891	16,318	13,378	11,336	9,834	8,634	7,775	7,038	5,916	5,103	4,486	3,613	3,024
140	101,662	79,807	65,686	55,811	48,517	42,909	34,852	29,343	25,338	22,294	19,904	17,976	16,389	13,426	11,370	9,860	8,704	7,791	7,051	5,925	5,110	4,492	3,616	3,027
145	104,273	81,407	66,766	56,589	49,104	43,368	35,154	29,557	25,497	22,417	20,002	18,056	16,455	13,470	11,402	9,884	8,723	7,806	7,063	5,934	5,116	4,497	3,620	3,029
150	106,834	82,960	67,807	57,335	49,665	43,804	35,441	29,759	25,647	22,534	20,094	18,131	16,518	13,512	11,432	9,906	8,740	7,820	7,075	5,942	5,122	4,501	3,623	3,031
155	109,346	84,467	68,810	58,051	50,201	44,221	35,713	29,950	25,789	22,643	20,181	18,202	16,577	13,551	11,460	9,927	8,757	7,833	7,085	5,950	5,128	4,506	3,626	3,033
160	111,811	85,930	69,779	58,738	50,714	44,619	35,972	30,132	25,924	22,747	20,264	18,269	16,632	13,588	11,486	9,947	8,772	7,845	7,096	5,957	5,133	4,510	3,628	3,035
165	114,230	87,352	70,713	59,399	51,206	44,999	36,219	30,305	26,052	22,846	20,342	18,333	16,685	13,623	11,511	9,966	8,787	7,857	7,105	5,964	5,138	4,514	3,631	3,037
170	116,604	88,733	71,616	60,034	51,677	45,363	36,454	30,470	26,173	22,939	20,416	18,393	16,735	13,657	11,535	9,984	8,801	7,868	7,114	5,970	5,143	4,517	3,633	3,038
175	118,935	90,077	72,488	60,646	52,130	45,711	36,679	30,627	26,289	23,028	20,486	18,450	16,782	13,688	11,557	10,001	8,814	7,878	7,123	5,976	5,147	4,521	3,635	3,040
180	121,223	91,383	73,332	61,236	52,565	46,045	36,893	30,776	26,399	23,112	20,553	18,504	16,827	13,718	11,579	10,017	8,826	7,888	7,131	5,982	5,152	4,524	3,637	3,041
185	123,471	92,655	74,148	61,804	52,983	46,366	37,099	30,919	26,504	23,193	20,617	18,556	16,869	13,746	11,599	10,032	8,838	7,898	7,138	5,987	5,156	4,527	3,639	3,043
190	125,678	93,892	74,939	62,352	53,386	46,674	37,296	31,056	26,605	23,269	20,677	18,605	16,910	13,773	11,618	10,046	8,849	7,906	7,146	5,992	5,159	4,530	3,641	3,044
195	127,846	95,097	75,704	62,881	53,773	46,970	37,484	31,187	26,701	23,343	20,735	18,652	16,949	13,799	11,636	10,060	8,859	7,915	7,152	5,997	5,163	4,533	3,643	3,045
200	129,977	96,271	76,446	63,392	54,146	47,254	37,666	31,312	26,792	23,413	20,790	18,696	16,986	13,823	11,654	10,073	8,869	7,923	7,159	6,002	5,166	4,535	3,645	3,047

# 1/0 AWG	2 Run(s)	208 Volts	C-Value = 8,924	Magnetic Duct	Copper

Calculations:

1. $f = \dfrac{\text{Length} \times 2 \times \text{Isca}}{\text{Runs} \times \text{C-Value} \times \text{E L-L}}$
2. $M = \dfrac{1}{1 + f}$
3. $\text{Isc} = \text{Isca} \times M$
4. Add Motor Contribution = Motor FLA x 4

* All results are given in symmetrical amperes

Single-Phase L-L Bolted Fault-Current* Table

One-Way Distance in Feet

Available Isc in Thousands (Isca)

	5	10	15	20	25	30	40	50	60	70	80	90	100	125	150	175	200	225	250	300	350	400	500	600
3	2,960	2,922	2,884	2,847	2,811	2,777	2,709	2,645	2,584	2,526	2,470	2,417	2,366	2,247	2,139	2,042	1,953	1,871	1,796	1,662	1,547	1,447	1,281	1,150
5	4,891	4,786	4,686	4,590	4,497	4,409	4,242	4,087	3,943	3,808	3,683	3,566	3,455	3,208	2,993	2,805	2,640	2,493	2,361	2,136	1,950	1,793	1,546	1,358
7	6,788	6,588	6,399	6,221	6,053	5,894	5,599	5,332	5,089	4,868	4,665	4,478	4,306	3,928	3,611	3,341	3,109	2,907	2,730	2,433	2,194	1,998	1,695	1,472
10	9,572	9,179	8,818	8,483	8,173	7,885	7,366	6,911	6,509	6,151	5,830	5,541	5,280	4,722	4,272	3,899	3,587	3,321	3,091	2,716	2,422	2,185	1,828	1,571
15	14,057	13,226	12,488	11,828	11,234	10,697	9,763	8,979	8,312	7,737	7,236	6,797	6,407	5,605	4,981	4,482	4,074	3,734	3,446	2,986	2,634	2,357	1,947	1,658
20	18,359	16,966	15,770	14,732	13,822	13,017	11,660	10,559	9,649	8,882	8,229	7,665	7,173	6,182	5,432	4,844	4,370	3,982	3,656	3,143	2,755	2,453	2,012	1,705
25	22,487	20,433	18,723	17,277	16,038	14,965	13,199	11,806	10,679	9,748	8,967	8,301	7,728	6,590	5,744	5,090	4,570	4,147	3,795	3,245	2,834	2,515	2,053	1,735
30	26,453	23,655	21,393	19,526	17,958	16,624	14,473	12,815	11,497	10,426	9,537	8,788	8,147	6,892	5,972	5,269	4,714	4,265	3,893	3,316	2,888	2,558	2,082	1,755
35	30,265	26,658	23,820	21,528	19,638	18,053	15,544	13,648	12,163	10,970	9,991	9,171	8,476	7,126	6,147	5,405	4,822	4,353	3,967	3,369	2,928	2,589	2,103	1,770
40	33,933	29,463	26,034	23,320	21,119	19,297	16,458	14,347	12,716	11,418	10,360	9,482	8,741	7,312	6,285	5,511	4,907	4,422	4,024	3,411	2,959	2,614	2,119	1,781
45	37,464	32,090	28,064	24,936	22,435	20,390	17,246	14,942	13,181	11,792	10,667	9,738	8,958	7,464	6,397	5,597	4,974	4,477	4,069	3,443	2,984	2,633	2,131	1,790
50	40,866	34,554	29,931	26,399	23,612	21,358	17,934	15,455	13,579	12,109	10,926	9,954	9,140	7,590	6,489	5,667	5,030	4,522	4,107	3,470	3,004	2,648	2,141	1,797
55	44,146	36,870	31,653	27,730	24,671	22,221	18,538	15,902	13,923	12,382	11,148	10,137	9,295	7,696	6,567	5,726	5,076	4,559	4,138	3,492	3,020	2,661	2,150	1,803
60	47,311	39,052	33,248	28,946	25,629	22,995	19,074	16,295	14,223	12,618	11,339	10,295	9,428	7,787	6,633	5,776	5,116	4,591	4,164	3,510	3,034	2,672	2,157	1,808
65	50,366	41,110	34,728	30,061	26,500	23,693	19,552	16,643	14,487	12,826	11,506	10,433	9,543	7,865	6,689	5,819	5,150	4,618	4,186	3,526	3,046	2,681	2,163	1,812
70	53,317	43,055	36,106	31,088	27,295	24,327	19,931	16,953	14,721	13,009	11,654	10,554	9,644	7,934	6,739	5,857	5,179	4,642	4,205	3,540	3,056	2,689	2,168	1,816
75	56,169	44,896	37,392	32,037	28,024	24,904	20,359	17,231	14,931	13,172	11,785	10,661	9,734	7,994	6,782	5,890	5,205	4,662	4,222	3,552	3,065	2,696	2,172	1,819
80	58,927	46,641	38,594	32,916	28,694	25,432	20,720	17,482	15,119	13,319	11,901	10,757	9,813	8,048	6,821	5,919	5,227	4,680	4,237	3,562	3,073	2,702	2,176	1,822
85	61,596	48,297	39,721	33,732	29,312	25,916	21,041	17,710	15,289	13,450	12,007	10,843	9,884	8,096	6,855	5,945	5,247	4,697	4,250	3,572	3,080	2,707	2,180	1,824
90	64,179	49,872	40,780	34,492	29,885	26,363	21,334	17,917	15,443	13,570	12,102	10,920	9,949	8,139	6,886	5,968	5,265	4,711	4,262	3,580	3,086	2,712	2,183	1,826
95	66,682	51,370	41,777	35,202	30,416	26,776	21,604	18,107	15,584	13,678	12,188	10,990	10,007	8,178	6,914	5,989	5,282	4,724	4,273	3,588	3,092	2,716	2,186	1,828
100	69,108	52,797	42,716	35,867	30,911	27,158	21,852	18,281	15,713	13,777	12,266	11,054	10,060	8,213	6,939	6,008	5,296	4,736	4,282	3,59[illegible]	3,097	2,720	2,188	1,830
105	71,459	54,159	43,603	36,490	31,373	27,514	22,082	18,441	15,831	13,868	12,339	11,113	10,108	8,245	6,962	6,025	5,310	4,746	4,291	3,60[illegible]	3,101	2,724	2,190	1,832
110	73,740	55,459	44,442	37,076	31,805	27,846	22,295	18,590	15,941	13,952	12,405	11,166	10,153	8,275	6,983	6,041	5,322	4,756	4,299	3,606	3,106	2,727	2,192	1,833
115	75,954	56,702	45,236	37,627	32,210	28,156	22,494	18,727	16,042	14,030	12,466	11,216	10,194	8,302	7,003	6,055	5,333	4,765	4,307	3,611	3,109	2,730	2,194	1,834
120	78,103	57,891	45,990	38,147	32,590	28,446	22,678	18,855	16,135	14,101	12,522	11,262	10,232	8,327	7,021	6,068	5,344	4,773	4,313	3,615	3,113	2,733	2,196	1,836
125	80,191	59,031	46,706	38,639	32,948	28,718	22,851	18,975	16,223	14,168	12,575	11,304	10,267	8,350	7,037	6,081	5,353	4,781	4,319	3,620	3,116	2,735	2,198	1,837
130	82,220	60,123	47,387	39,104	33,285	28,974	23,013	19,086	16,304	14,230	12,624	11,344	10,299	8,372	7,052	6,092	5,362	4,788	4,325	3,624	3,119	2,737	2,199	1,838
135	84,192	61,170	48,035	39,544	33,604	29,215	23,165	19,190	16,380	14,288	12,669	11,380	10,329	8,392	7,066	6,103	5,370	4,795	4,331	3,628	3,122	2,740	2,201	1,839
140	86,110	62,177	48,654	39,962	33,905	29,443	23,308	19,288	16,451	14,342	12,712	11,415	10,358	8,411	7,080	6,112	5,378	4,801	4,336	3,632	3,124	2,742	2,202	1,840
145	87,976	63,144	49,244	40,359	34,191	29,658	23,442	19,380	16,518	14,393	12,752	11,447	10,384	8,428	7,092	6,122	5,385	4,806	4,340	3,635	3,127	2,743	2,203	1,841
150	89,792	64,074	49,808	40,737	34,462	29,862	23,569	19,467	16,581	14,441	12,789	11,477	10,409	8,444	7,104	6,130	5,392	4,812	4,344	3,638	3,129	2,745	2,204	1,841
155	91,560	64,969	50,347	41,097	34,719	30,055	23,689	19,549	16,641	14,486	12,825	11,505	10,432	8,460	7,115	6,138	5,398	4,817	4,349	3,641	3,131	2,747	2,205	1,842
160	93,282	65,831	50,863	41,441	34,964	30,238	23,803	19,626	16,697	14,528	12,858	11,532	10,454	8,474	7,125	6,146	5,404	4,821	4,352	3,643	3,133	2,748	2,206	1,843
165	94,960	66,662	51,358	41,769	35,197	30,412	23,911	19,700	16,750	14,568	12,889	11,557	10,475	8,488	7,134	6,153	5,409	4,826	4,356	3,646	3,135	2,750	2,207	1,843
170	96,595	67,464	51,832	42,082	35,419	30,578	24,013	19,769	16,800	14,606	12,919	11,581	10,495	8,501	7,143	6,160	5,414	4,830	4,359	3,648	3,137	2,751	2,208	1,844
175	98,189	68,238	52,288	42,382	35,631	30,736	24,110	19,835	16,847	14,642	12,947	11,604	10,513	8,513	7,152	6,166	5,419	4,834	4,363	3,651	3,138	2,752	2,209	1,845
180	99,743	68,985	52,726	42,669	35,834	30,886	24,203	19,898	16,893	14,676	12,974	11,625	10,531	8,524	7,160	6,172	5,424	4,838	4,366	3,653	3,140	2,754	2,210	1,845
185	101,260	69,707	53,146	42,944	36,028	31,030	24,291	19,957	16,935	14,708	12,999	11,646	10,547	8,535	7,168	6,178	5,428	4,841	4,368	3,655	3,142	2,755	2,210	1,846
190	102,740	70,405	53,551	43,208	36,213	31,168	24,376	20,014	16,976	14,739	13,023	11,665	10,563	8,546	7,175	6,184	5,433	4,844	4,371	3,657	3,143	2,756	2,211	1,846
195	104,184	71,080	53,941	43,461	36,391	31,300	24,456	20,068	17,015	14,769	13,046	11,683	10,578	8,556	7,182	6,189	5,437	4,848	4,374	3,658	3,144	2,757	2,212	1,847
200	105,594	71,734	54,317	43,705	36,562	31,426	24,533	20,120	17,053	14,797	13,068	11,701	10,593	8,565	7,189	6,194	5,440	4,851	4,376	3,660	3,146	2,758	2,212	1,847

\# 2/0 AWG | 1 Run(s) | 208 Volts | C-Value = 10,755 | Magnetic Duct | Copper

Calculations:

1. $f = \dfrac{\text{Length} \times 2 \times \text{Isca}}{\text{Runs} \times \text{C-Value} \times E_{L-L}}$
2. $M = \dfrac{1}{1 + f}$
3. $\text{Isc} = \text{Isca} \times M$
4. Add Motor Contribution = Motor FLA x 4

* All results are given in symmetrical amperes

Single-Phase L-L Bolted Fault-Current* Table

One-Way Distance in Feet

Available Isc in Thousands (Isca)

Isca	5	10	15	20	25	30	40	50	60	70	80	90	100	125	150	175	200	225	250	300	350	400	500	600
3	2,980	2,960	2,941	2,922	2,903	2,884	2,847	2,811	2,777	2,743	2,709	2,677	2,645	2,569	2,498	2,430	2 366	2,305	2,247	2,139	2,042	1,953	1,796	1,662
5	4,945	4,891	4,838	4,786	4,735	4,686	4,590	4,497	4,409	4,324	4,242	4,163	4,087	3,908	3,745	3,594	3,455	3,327	3,208	2,993	2,805	2,640	2,361	2,136
7	6,892	6,788	6,686	6,588	6,492	6,399	6,221	6,053	5,894	5,742	5,599	5,462	5,332	5,032	4,764	4,523	4,306	4,108	3,928	3,611	3,341	3,109	2,730	2,433
10	9,781	9,572	9,372	9,179	8,995	8,818	8,483	8,173	7,885	7,617	7,366	7,131	6,911	6,415	5,986	5,611	5,280	4,986	4,722	4,272	3,899	3,587	3,091	2,716
15	14,513	14,057	13,629	13,226	12,847	12,488	11,828	11,234	10,697	10,208	9,763	9,355	8,979	8,160	7,478	6,902	6,407	5,979	5,605	4,981	4,482	4,074	3,446	2,986
20	19,144	18,359	17,635	16,966	16,346	15,770	14,732	13,822	13,017	12,301	11,660	11,083	10,559	9,445	8,543	7,799	7,173	6,641	6,182	5,432	4,844	4,370	3,656	3,143
25	23,677	22,487	21,411	20,433	19,541	18,723	17,277	16,038	14,965	14,027	13,199	12,464	11,806	10,430	9,341	8,458	7,728	7,113	6,590	5,744	5,090	4,570	3,795	3,245
30	28,115	26,453	24,976	23,655	22,467	21,393	19,526	17,958	16,624	15,474	14,473	13,593	12,815	11,209	9,962	8,964	8,147	7,468	6,892	5,972	5,269	4,714	3,893	3,316
35	32,461	30,265	28,347	26,658	25,159	23,820	21,528	19,638	18,053	16,705	15,544	14,534	13,648	11,841	10,458	9,363	8,476	7,743	7,126	6,147	5,405	4,822	3,967	3,369
40	36,717	33,933	31,540	29,463	27,643	26,034	23,320	21,119	19,297	17,765	16,458	15,330	14,347	12,364	10,863	9,687	8,741	7,963	7,312	6,285	5,511	4,907	4,024	3,411
45	40,888	37,464	34,569	32,090	29,942	28,064	24,936	22,435	20,390	18,687	17,246	16,012	14,942	12,804	11,201	9,955	8,958	8,143	7,464	6,397	5,597	4,974	4,069	3,443
50	44,974	40,866	37,446	34,554	32,076	29,931	26,399	23,612	21,358	19,496	17,934	16,603	15,455	13,179	11,487	10,180	9,140	8,293	7,590	6,489	5,667	5,030	4,107	3,470
55	48,979	44,146	40,181	36,870	34,063	31,653	27,730	24,671	22,221	20,213	18,538	17,119	15,902	13,503	11,732	10,372	9,295	8,420	7,696	6,567	5,726	5,076	4,138	3,492
60	52,905	47,311	42,786	39,052	35,917	33,248	28,946	25,629	22,995	20,852	19,074	17,575	16,295	13,785	11,945	10,538	9,428	8,529	7,787	6,633	5,776	5,116	4,164	3,510
65	56,755	50,366	45,270	41,110	37,650	34,728	30,061	26,500	23,693	21,424	19,552	17,980	16,643	14,033	12,130	10,682	9,543	8,623	7,865	6,689	5,819	5,150	4,186	3,526
70	60,530	53,317	47,639	43,055	39,275	36,106	31,088	27,295	24,327	21,941	19,981	18,343	16,953	14,253	12,294	10,809	9,644	8 706	7,934	6,739	5,857	5,179	4,205	3,540
75	64,233	56,169	49,904	44,896	40,802	37,392	32,037	28,024	24,904	22,409	20,369	18,669	17,231	14,449	12,440	10,922	9,734	8,779	7,994	6,782	5,890	5,205	4,222	3,552
80	67,865	58,927	52,069	46,641	42,238	38,594	32,916	28,694	25,432	22,836	20,720	18,964	17,482	14,625	12,570	11,022	9,813	8,843	8,048	6,821	5,919	5,227	4,237	3,562
85	71,430	61,596	54,142	48,297	43,592	39,721	33,732	29,312	25,916	23,226	21,041	19,232	17,710	14,784	12,688	11,112	9,884	8,901	8,096	6,855	5,945	5,247	4,250	3,572
90	74,928	64,179	56,128	49,872	44,870	40,780	34,492	29,885	26,363	23,584	21,334	19,477	17,917	14,928	12,794	11,193	9,949	8,953	8,139	6,886	5,968	5,265	4,262	3,580
95	78,361	66,682	58,033	51,370	46,079	41,777	35,202	30,416	26,776	23,913	21,604	19,701	18,107	15,059	12,890	11,267	10,007	9,000	8,178	6,914	5,989	5,282	4,273	3,588
100	81,732	69,108	59,861	52,797	47,224	42,716	35,867	30,911	27,158	24,218	21,852	19,908	18,281	15,180	12,978	11,334	10,060	9,043	8,213	6,939	6,008	5,296	4,282	3,594
105	85,042	71,459	61,618	54,159	48,311	43,603	36,490	31,373	27,514	24,501	22,082	20,098	18,441	15,290	13,059	11,396	10,108	9,082	8,245	6,962	6,025	5,310	4,291	3,601
110	88,292	73,740	63,306	55,459	49,343	44,442	37,076	31,805	27,846	24,763	22,295	20,275	18,590	15,392	13,133	11,452	10,153	9,118	8,275	6,983	6,041	5,322	4,299	3,606
115	91,485	75,954	64,931	56,702	50,324	45,236	37,627	32,210	28,156	25,008	22,494	20,438	18,727	15,486	13,202	11,504	10,194	9,151	8,302	7,003	6,055	5,333	4,307	3,611
120	94,621	78,103	66,495	57,891	51,259	45,990	38,147	32,590	28,446	25,237	22,678	20,591	18,855	15,574	13,265	11,552	10,232	9,182	8,327	7,021	6,068	5,344	4,313	3,616
125	97,703	80,191	68,003	59,031	52,150	46,706	38,639	32,948	28,718	25,451	22,851	20,733	18,975	15,655	13,324	11,597	10,267	9,210	8,350	7,037	6,081	5,353	4,319	3,620
130	100,731	82,220	69,456	60,123	53,000	47,387	39,104	33,285	28,974	25,652	23,013	20,866	19,086	15,731	13,379	11,639	10,299	9,236	8,372	7,052	6,092	5,362	4,325	3,624
135	103,708	84,192	70,858	61,170	53,813	48,035	39,544	33,604	29,215	25,841	23,165	20,991	19,190	15,802	13,430	11,677	10,329	9,260	8,392	7,066	6,103	5,370	4,331	3,628
140	106,633	86,110	72,212	62,177	54,590	48,654	39,962	33,905	29,443	26,018	23,308	21,108	19,288	15,868	13,478	11,714	10,358	9,283	8,411	7,080	6,112	5,378	4,336	3,632
145	109,509	87,976	73,519	63,144	55,334	49,244	40,359	34,191	29,658	26,186	23,442	21,219	19,380	15,930	13,523	11,747	10,384	9,304	8,428	7,092	6,122	5,385	4,340	3,635
150	112,337	89,792	74,783	64,074	56,047	49,808	40,737	34,462	29,862	26,345	23,569	21,323	19,467	15,989	13,565	11,779	10,409	9,324	8,444	7,104	6,130	5,392	4,344	3,638
155	115,118	91,560	76,006	64,969	56,731	50,347	41,097	34,719	30,055	26,495	23,689	21,421	19,549	16,044	13,605	11,809	10,432	9,343	8,460	7,115	6,138	5,398	4,349	3,641
160	117,854	93,282	77,189	65,831	57,387	50,863	41,441	34,964	30,238	26,637	23,803	21,514	19,626	16,096	13,642	11,837	10,454	9,361	8,474	7,125	6,146	5,404	4,352	3,643
165	120,544	94,960	78,334	66,662	58,018	51,358	41,769	35,197	30,412	26,772	23,911	21,602	19,700	16,145	13,677	11,864	10,475	9,377	8,488	7,134	6,153	5,409	4,356	3,646
170	123,191	96,595	79,443	67,464	58,624	51,832	42,082	35,419	30,578	26,901	24,013	21,685	19,769	16,192	13,711	11,889	10,495	9,393	8,501	7,143	6,160	5,414	4,359	3,648
175	125,796	98,189	80,518	68,238	59,207	52,288	42,382	35,631	30,736	27,023	24,110	21,765	19,835	16,236	13,742	11,913	10,513	9,408	8,513	7,152	6,166	5,419	4,363	3,651
180	128,359	99,743	81,560	68,985	59,769	52,726	42,669	35,834	30,886	27,139	24,203	21,840	19,898	16,278	13,773	11,935	10,531	9,422	8,524	7,160	6,172	5,424	4,366	3,653
185	130,881	101,260	82,572	69,707	60,310	53,146	42,944	36,028	31,030	27,250	24,291	21,912	19,957	16,318	13,801	11,957	10,547	9,435	8,535	7,168	6,178	5,428	4,368	3,655
190	133,364	102,740	83,553	70,405	60,832	53,551	43,208	36,213	31,168	27,356	24,376	21,980	20,014	16,356	13,828	11,977	10,563	9,448	8,546	7,175	6,184	5,433	4,371	3,657
195	135,809	104,184	84,506	71,080	61,336	53,941	43,461	36,391	31,300	27,458	24,456	22,046	20,068	16,392	13,854	11,997	10,578	9,460	8,556	7,182	6,189	5,437	4,374	3,658
200	138,215	105,594	85,431	71,734	61,822	54,317	43,705	36,562	31,426	27,555	24,533	22,108	20,120	16,426	13,879	12,015	10,593	9,472	8,565	7,189	6,194	5,440	4,376	3,660

# 2/0 AWG	2 Run(s)	208 Volts		C-Value = 10,755	Magnetic Duct	Copper

Calculations:

1. $f = \frac{\text{Length} \times 2 \times \text{Isca}}{\text{Runs} \times \text{C-Value} \times \text{E L-L}}$

2. $M = \frac{1}{1 + f}$

3. $\text{Isc} = \text{Isca} \times M$

4. Add Motor Contribution = Motor FLA x 4

* All results are given in symmetrical amperes

Single-Phase L-L Bolted Fault-Current* Table

One-Way Distance in Feet

Available Isc in Thousands (Isca)	5	10	15	20	25	30	40	50	60	70	80	90	100	125	150	175	200	225	250	300	350	400	500	600
3	2,967	2,934	2,902	2,871	2,841	2,811	2,753	2,697	2,644	2,592	2,543	2,496	2,450	2,342	2,244	2,154	2,070	1,993	1,921	1,792	1,680	1,580	1,413	1,278
5	4,908	4,820	4,734	4,652	4,572	4,495	4,349	4,212	4,083	3,962	3,848	3,740	3,638	3,406	3,202	3,021	2,859	2,714	2,583	2,355	2,164	2,002	1,741	1,540
7	6,821	6,651	6,490	6,336	6,189	6,049	5,787	5,547	5,325	5,121	4,932	4,756	4,593	4,229	3,919	3,651	3,418	3,212	3,030	2,72[illegible]	2,470	2,261	1,933	1,689
10	9,639	9,303	8,990	8,698	8,423	8,166	7,695	7,276	6,900	6,561	6,254	5,974	5,719	5,166	4,710	4,329	4,004	3,725	3,482	3,08[illegible]	2,762	2,503	2,108	1,821
15	14,203	13,486	12,837	12,249	11,712	11,220	10,350	9,606	8,962	8,398	7,901	7,460	7,065	6,240	5,588	5,059	4,621	4,253	3,940	3,433	3,042	2,731	2,268	1,938
20	18,607	17,395	16,332	15,391	14,552	13,801	12,508	11,437	10,535	9,765	9,100	8,519	8,008	6,964	6,161	5,524	5,007	4,578	4,216	3,642	3,205	2,861	2,357	2,003
25	22,861	21,058	19,520	18,191	17,031	16,010	14,296	12,914	11,776	10,822	10,011	9,313	8,706	7,486	6,566	5,847	5,270	4,797	4,402	3,779	3,311	2,946	2,413	2,044
30	26,971	24,498	22,440	20,701	19,212	17,923	15,303	14,131	12,779	11,663	10,726	9,929	9,242	7,879	6,866	6,084	5,462	4,956	4,535	3,877	3,386	3,005	2,453	2,072
35	30,946	27,733	25,125	22,965	21,147	19,596	17,388	15,150	13,607	12,349	11,304	10,422	9,667	8,186	7,099	6,266	5,608	5,075	4,635	3,950	3,441	3,048	2,482	2,093
40	34,791	30,782	27,601	25,016	22,874	21,070	18,199	16,017	14,302	12,919	11,779	10,825	10,013	8,433	7,283	6,409	5,723	5,169	4,713	4,006	3,484	3,082	2,504	2,109
45	38,512	33,660	29,893	26,885	24,426	22,380	19,168	16,763	14,893	13,399	12,178	11,160	10,300	8,635	7,434	6,526	5,815	5,244	4,776	4,051	3,518	3,109	2,522	2,121
50	42,117	36,381	32,020	28,593	25,828	23,551	20,021	17,411	15,403	13,811	12,516	11,444	10,541	8,804	7,558	6,622	5,891	5,306	4,827	4,088	3,546	3,130	2,536	2,131
55	45,610	38,958	34,000	30,161	27,101	24,605	20,777	17,980	15,847	14,166	12,808	11,687	10,747	8,947	7,564	6,702	5,955	5,358	4,870	4,1[illegible]9	3,569	3,148	2,548	2,140
60	48,995	41,402	35,846	31,605	28,261	25,558	21,453	18,484	16,237	14,477	13,061	11,898	10,925	9,070	7,754	6,771	6,009	5,402	4,906	4,1[illegible]5	3,588	3,163	2,557	2,146
65	52,279	43,723	37,573	32,940	29,324	26,423	22,059	18,933	16,582	14,751	13,284	12,082	11,080	9,177	7,832	6,830	6,056	5,440	4,937	4,1[illegible]7	3,605	3,176	2,566	2,152
70	55,466	45,929	39,191	34,177	30,300	27,214	22,607	19,335	16,890	14,994	13,481	12,245	11,216	9,270	7,900	6,882	6,097	5,472	4,964	4,1[illegible]6	3,619	3,187	2,573	2,158
75	58,559	48,030	40,711	35,327	31,201	27,938	23,105	19,698	17,166	15,211	13,656	12,389	11,338	9,353	7,959	6,927	6,132	5,501	4,987	4,2[illegible]3	3,631	3,197	2,579	2,162
80	61,563	50,033	42,140	36,398	32,034	28,604	23,559	20,026	17,415	15,406	13,813	12,518	11,446	9,426	8,013	6,968	6,164	5,526	5,008	4,2[illegible]8	3,642	3,205	2,585	2,166
85	64,482	51,944	43,488	37,399	32,806	29,218	23,974	20,326	17,641	15,583	13,955	12,635	11,543	9,492	8,060	7,004	6,192	5,549	5,027	4,231	3,652	3,213	2,590	2,169
90	67,319	53,769	44,760	38,336	33,525	29,787	24,355	20,599	17,847	15,743	14,083	12,740	11,631	9,551	8,103	7,036	6,217	5,569	5,043	4,2[illegible]2	3,661	3,220	2,594	2,172
95	70,078	55,515	45,963	39,216	34,196	30,315	24,707	20,850	18,035	15,890	14,200	12,836	11,710	9,605	8,141	7,065	6,240	5,587	5,058	4,2[illegible]3	3,669	3,226	2,598	2,175
100	72,762	57,186	47,102	40,042	34,822	30,807	25,033	21,082	18,208	16,024	14,307	12,923	11,783	9,654	8,176	7,091	6,260	5,604	5,072	4,2[illegible]2	3,676	3,231	2,602	2,178
105	75,374	58,787	48,183	40,820	35,410	31,265	25,335	21,296	18,367	16,147	14,405	13,003	11,849	9,698	8,208	7,115	6,279	5,619	5,084	4,2[illegible]1	3,682	3,236	2,605	2,180
110	77,916	60,322	49,210	41,555	35,961	31,694	25,616	21,494	18,514	16,260	14,496	13,077	11,910	9,739	8,238	7,137	6,296	5,632	5,095	4,2[illegible]9	3,688	3,241	2,608	2,182
115	80,392	61,795	50,186	42,249	36,479	32,096	25,878	21,678	18,651	16,366	14,579	13,144	11,967	9,777	8,265	7,157	6,312	5,645	5,105	4,2[illegible]6	3,694	3,245	2,611	2,184
120	82,804	63,210	51,115	42,905	36,968	32,474	26,123	21,849	18,778	16,463	14,657	13,207	12,019	9,812	8,289	7,176	6,326	5,656	5,115	4,2[illegible]3	3,699	3,249	2,613	2,186
125	85,154	64,571	52,001	43,528	37,429	32,829	26,352	22,010	18,896	16,554	14,729	13,266	12,067	9,844	8,312	7,193	6,340	5,667	5,124	4,2[illegible]9	3,703	3,252	2,615	2,187
130	87,445	65,880	52,847	44,119	37,865	33,164	26,568	22,160	19,007	16,639	14,796	13,320	12,112	9,874	8,334	7,209	6,352	5,677	5,132	4,305	3,707	3,256	2,618	2,189
135	89,679	67,140	53,655	44,680	38,278	33,481	26,770	22,301	19,110	16,718	14,858	13,371	12,154	9,902	8,353	7,224	6,364	5,686	5,139	4,310	3,711	3,259	2,620	2,190
140	91,859	68,354	54,427	45,215	38,670	33,780	26,961	22,433	19,207	16,792	14,917	13,418	12,193	9,928	8,372	7,238	6,374	5,695	5,146	4,315	3,715	3,261	2,621	2,191
145	93,985	69,524	55,167	45,724	39,042	34,063	27,141	22,558	19,298	16,862	14,972	13,463	12,230	9,952	8,389	7,251	6,384	5,703	5,153	4,320	3,718	3,264	2,623	2,192
150	96,061	70,654	55,875	46,210	39,395	34,332	27,312	22,675	19,384	16,928	15,024	13,505	12,265	9,975	8,406	7,263	6,394	5,710	5,159	4,324	3,722	3,266	2,625	2,194
155	98,087	71,744	56,555	46,674	39,732	34,587	27,473	22,786	19,466	16,990	15,072	13,544	12,297	9,996	8,421	7,274	6,403	5,717	5,165	4,328	3,725	3,269	2,626	2,195
160	100,066	72,797	57,207	47,117	40,053	34,830	27,626	22,891	19,542	17,048	15,118	13,581	12,328	10,016	8,435	7,285	6,411	5,724	5,170	4,332	3,727	3,271	2,627	2,196
165	101,999	73,814	57,834	47,541	40,359	35,062	27,772	22,991	19,615	17,103	15,162	13,616	12,356	10,035	8,449	7,295	6,419	5,730	5,175	4,335	3,730	3,273	2,629	2,196
170	103,888	74,799	58,436	47,948	40,651	35,282	27,910	23,086	19,684	17,155	15,203	13,649	12,384	10,053	8,461	7,304	6,426	5,736	5,180	4,339	3,732	3,275	2,630	2,197
175	105,734	75,751	59,016	48,337	40,931	35,493	28,041	23,176	19,749	17,205	15,242	13,681	12,410	10,070	8,473	7,313	6,433	5,742	5,184	4,342	3,735	3,277	2,631	2,198
180	107,539	76,673	59,574	48,711	41,199	35,694	28,167	23,261	19,811	17,252	15,279	13,710	12,434	10,087	8,485	7,322	6,439	5,747	5,189	4,345	3,737	3,278	2,632	2,199
185	109,303	77,566	60,112	49,070	41,455	35,886	28,286	23,343	19,870	17,297	15,314	13,739	12,457	10,102	8,496	7,330	6,446	5,752	5,193	4,348	3,739	3,280	2,633	2,200
190	111,030	78,431	60,630	49,415	41,701	36,070	28,401	23,421	19,927	17,340	15,347	13,766	12,479	10,116	8,506	7,338	6,452	5,756	5,197	4,350	3,741	3,282	2,634	2,200
195	112,719	79,270	61,130	49,746	41,937	36,247	28,510	23,495	19,980	17,380	15,379	13,791	12,500	10,130	8,516	7,345	6,457	5,761	5,200	4,353	3,743	3,283	2,635	2,201
200	114,372	80,084	61,613	50,066	42,164	36,416	28,614	23,566	20,032	17,419	15,410	13,816	12,521	10,143	8,525	7,352	6,463	5,765	5,204	4,355	3,745	3,284	2,636	2,202

# 3/0 AWG	1 Run(s)	208 Volts	C-Value = 12,843	Magnetic Duct	Copper

Calculations:

1. $f = \dfrac{\text{Length} \times 2 \times \text{Isca}}{\text{Runs} \times \text{C-Value} \times \text{E L-L}}$

2. $M = \dfrac{1}{1 + f}$

3. $\text{Isc} = \text{Isca} \times M$

4. Add Motor Contribution = Motor FLA x 4

* All results are given in symmetrical amperes

Single-Phase L-L Bolted Fault-Current* Table

One-Way Distance in Feet

Available Isc in Thousands (Isca)	5	10	15	20	25	30	40	50	60	70	80	90	100	125	150	175	200	225	250	300	350	400	500	600
3	2,983	2,967	2,950	2,934	2,918	2,902	2,871	2,841	2,811	2,781	2,753	2,725	2,697	2,631	2,567	2,507	2,450	2,395	2,342	2,244	2,154	2,070	1,921	1,792
5	4,954	4,908	4,863	4,820	4,776	4,734	4,652	4,572	4,495	4,421	4,349	4,279	4,212	4,052	3,904	3,766	3,638	3,518	3,406	3,202	3,021	2,859	2,583	2,355
7	6,909	6,821	6,735	6,651	6,570	6,490	6,336	6,189	6,049	5,915	5,787	5,664	5,547	5,273	5,025	4,799	4,593	4,404	4,229	3,919	3,651	3,418	3,030	2,721
10	9,816	9,639	9,468	9,303	9,144	8,990	8,698	8,423	8,166	7,924	7,695	7,480	7,276	6,812	6,404	6,042	5,719	5,428	5,166	4,710	4,329	4,004	3,482	3,081
15	14,590	14,203	13,835	13,486	13,154	12,837	12,249	11,712	11,220	10,768	10,350	9,964	9,606	8,814	8,142	7,566	7,065	6,627	6,240	5,588	5,059	4,621	3,940	3,433
20	19,278	18,607	17,981	17,395	16,847	16,332	15,391	14,552	13,801	13,123	12,508	11,949	11,437	10,331	9,420	8,657	8,008	7,450	6,964	6,161	5,524	5,007	4,216	3,642
25	23,882	22,861	21,923	21,058	20,260	19,520	18,191	17,031	16,010	15,105	14,296	13,570	12,914	11,522	10,400	9,478	8,706	8,050	7,486	6,566	5,847	5,270	4,402	3,779
30	28,405	26,971	25,675	24,498	23,424	22,440	20,701	19,212	17,923	16,796	15,803	14,920	14,131	12,480	11,175	10,117	9,242	8,506	7,879	6,866	6,084	5,462	4,535	3,877
35	32,848	30,946	29,251	27,733	26,364	25,125	22,965	21,147	19,596	18,256	17,088	16,061	15,150	13,269	11,803	10,629	9,667	8,865	8,186	7,099	6,266	5,608	4,635	3,950
40	37,214	34,791	32,664	30,782	29,105	27,601	25,016	22,874	21,070	19,530	18,199	17,038	16,017	13,929	12,323	11,048	10,013	9,155	8,433	7,283	6,409	5,723	4,713	4,006
45	41,504	38,512	35,923	33,660	31,665	29,893	26,885	24,426	22,380	20,650	19,168	17,885	16,763	14,490	12,759	11,398	10,300	9,394	8,635	7,434	6,526	5,815	4,776	4,051
50	45,721	42,117	39,039	36,381	34,062	32,020	28,593	25,828	23,551	21,643	20,021	18,625	17,411	14,972	13,132	11,695	10,541	9,594	8,804	7,558	6,622	5,891	4,827	4,088
55	49,867	45,610	42,022	38,958	36,310	34,000	30,161	27,101	24,605	22,530	20,777	19,278	17,980	15,391	13,453	11,949	10,747	9,765	8,947	7,664	6,702	5,955	4,870	4,119
60	53,942	48,995	44,880	41,402	38,424	35,846	31,605	28,261	25,558	23,326	21,453	19,858	18,484	15,758	13,733	12,169	10,925	9,911	9,070	7,754	6,771	6,009	4,906	4,145
65	57,950	52,279	47,620	43,723	40,415	37,573	32,940	29,324	26,423	24,045	22,059	20,377	18,933	16,083	13,979	12,362	11,080	10,039	9,177	7,832	6,830	6,056	4,937	4,167
70	61,891	55,466	50,249	45,929	42,293	39,191	34,177	30,300	27,214	24,698	22,607	20,843	19,335	16,372	14,197	12,532	11,216	10,151	9,270	7,900	6,882	6,097	4,964	4,186
75	65,768	58,559	52,775	48,030	44,069	40,711	35,327	31,201	27,938	25,293	23,105	21,266	19,698	16,632	14,392	12,683	11,338	10,250	9,353	7,959	6,927	6,132	4,987	4,203
80	69,581	61,563	55,202	50,033	45,749	42,140	36,398	32,034	28,604	25,837	23,559	21,649	20,026	16,865	14,566	12,819	11,446	10,338	9,426	8,013	6,968	6,164	5,008	4,218
85	73,333	64,482	57,538	51,944	47,341	43,488	37,399	32,806	29,218	26,337	23,974	21,999	20,326	17,077	14,724	12,941	11,543	10,418	9,492	8,060	7,004	6,192	5,027	4,231
90	77,025	67,319	59,786	53,769	48,853	44,760	38,336	33,525	29,787	26,799	24,355	22,320	20,599	17,270	14,867	13,051	11,631	10,489	9,551	8,103	7,036	6,217	5,043	4,242
95	80,658	70,078	61,952	55,515	50,289	45,963	39,216	34,196	30,315	27,225	24,707	22,616	20,850	17,446	14,997	13,152	11,710	10,554	9,605	8,141	7,065	6,240	5,058	4,253
100	84,234	72,762	64,040	57,186	51,657	47,102	40,042	34,822	30,807	27,621	25,033	22,888	21,082	17,608	15,117	13,243	11,783	10,613	9,654	8,176	7,091	6,260	5,072	4,262
105	87,754	75,374	66,055	58,787	52,959	48,183	40,820	35,410	31,265	27,989	25,335	23,140	21,296	17,757	15,226	13,327	11,849	10,667	9,698	8,208	7,115	6,279	5,084	4,271
110	91,219	77,916	67,999	60,322	54,202	49,210	41,555	35,961	31,694	28,333	25,616	23,374	21,494	17,894	15,327	13,405	11,910	10,716	9,739	8,238	7,137	6,296	5,095	4,279
115	94,631	80,392	69,877	61,795	55,389	50,186	42,249	36,479	32,096	28,654	25,878	23,592	21,678	18,022	15,421	13,476	11,967	10,762	9,777	8,265	7,157	6,312	5,105	4,286
120	97,991	82,804	71,692	63,210	56,523	51,115	42,905	36,968	32,474	28,954	26,123	23,796	21,849	18,140	15,508	13,542	12,019	10,804	9,812	8,289	7,176	6,326	5,115	4,293
125	101,300	85,154	73,448	64,571	57,608	52,001	43,528	37,429	32,829	29,236	26,352	23,986	22,010	18,251	15,588	13,604	12,067	10,843	9,844	8,312	7,193	6,340	5,124	4,299
130	104,558	87,445	75,146	65,880	58,648	52,847	44,119	37,865	33,164	29,502	26,568	24,164	22,160	18,354	15,663	13,661	12,112	10,879	9,874	8,334	7,209	6,352	5,132	4,305
135	107,769	89,679	76,790	67,140	59,645	53,655	44,680	38,278	33,481	29,752	26,770	24,332	22,301	18,450	15,733	13,714	12,154	10,913	9,902	8,353	7,224	6,364	5,139	4,310
140	110,931	91,859	78,382	68,354	60,601	54,427	45,215	38,670	33,780	29,988	26,961	24,490	22,433	18,541	15,799	13,764	12,193	10,944	9,928	8,372	7,238	6,374	5,146	4,315
145	114,048	93,985	79,925	69,524	61,519	55,167	45,724	39,042	34,063	30,211	27,141	24,638	22,558	18,626	15,861	13,811	12,230	10,974	9,952	8,389	7,251	6,384	5,153	4,320
150	117,118	96,061	81,421	70,654	62,402	55,875	46,210	39,395	34,332	30,422	27,312	24,778	22,675	18,706	15,919	13,855	12,265	11,002	9,975	8,406	7,263	6,394	5,159	4,324
155	120,144	98,087	82,872	71,744	63,250	56,555	46,674	39,732	34,587	30,623	27,473	24,911	22,786	18,781	15,974	13,896	12,297	11,028	9,996	8,421	7,274	6,403	5,165	4,328
160	123,127	100,066	84,280	72,797	64,067	57,207	47,117	40,053	34,830	30,813	27,626	25,037	22,891	18,853	16,025	13,935	12,328	11,053	10,016	8,435	7,285	6,411	5,170	4,332
165	126,066	101,999	85,648	73,814	64,854	57,834	47,541	40,359	35,062	30,994	27,772	25,156	22,991	18,920	16,074	13,972	12,356	11,076	10,035	8,449	7,295	6,419	5,175	4,335
170	128,965	103,888	86,975	74,799	65,613	58,436	47,948	40,651	35,282	31,166	27,910	25,270	23,086	18,984	16,120	14,007	12,334	11,098	10,053	8,461	7,304	6,426	5,180	4,339
175	131,822	105,734	88,266	75,751	66,344	59,016	48,337	40,931	35,493	31,330	28,041	25,377	23,176	19,045	16,164	14,040	12,410	11,118	10,070	8,473	7,313	6,433	5,184	4,342
180	134,639	107,539	89,520	76,673	67,050	59,574	48,711	41,199	35,694	31,487	28,167	25,480	23,261	19,103	16,206	14,071	12,434	11,138	10,087	8,485	7,322	6,439	5,189	4,345
185	137,417	109,303	90,740	77,566	67,732	60,112	49,070	41,455	35,886	31,636	28,286	25,578	23,343	19,158	16,245	14,101	12,457	11,157	10,102	8,496	7,330	6,446	5,193	4,348
190	140,157	111,030	91,926	78,431	68,391	60,630	49,415	41,701	36,070	31,779	28,401	25,671	23,421	19,210	16,283	14,130	12,479	11,174	10,116	8,506	7,338	6,452	5,197	4,350
195	142,859	112,719	93,081	79,270	69,028	61,130	49,746	41,937	36,247	31,916	28,510	25,761	23,495	19,260	16,319	14,157	12,500	11,191	10,130	8,516	7,345	6,457	5,200	4,353
200	145,524	114,372	94,205	80,084	69,645	61,613	50,066	42,164	36,416	32,047	28,614	25,846	23,566	19,308	16,353	14,182	12,521	11,207	10,143	8,525	7,352	6,463	5,204	4,355

# 3/0 AWG	2 Run(s)	208 Volts	C-Value = 12,843	Magnetic Duct	Copper

Calculations:

1. $f = \dfrac{\text{Length} \times 2 \times \text{Isca}}{\text{Runs} \times \text{C-Value} \times \text{E L-L}}$
2. $M = \dfrac{1}{1 + f}$
3. Isc = Isca x M
4. Add Motor Contribution = Motor FLA x 4

* All results are given in symmetrical amperes

Single-Phase L-L Bolted Fault-Current* Table

One-Way Distance in Feet

Available Isc in Thousands (Isca)

Isca	5	10	15	20	25	30	40	50	60	70	80	90	100	125	150	175	200	225	250	300	350	400	500	600
3	2,972	2,944	2,916	2,889	2,863	2,837	2,767	2,738	2,691	2,646	2,602	2,559	2,518	2,421	2,331	2,248	2,170	2,097	2,030	1,906	1,797	1,700	1,533	1,397
5	4,922	4,846	4,772	4,700	4,631	4,564	4,435	4,313	4,197	4,088	3,984	3,885	3,791	3,575	3,383	3,210	3,053	2,912	2,783	2,556	2,363	2,198	1,928	1,717
7	6,847	6,701	6,561	6,426	6,297	6,173	5,940	5,723	5,522	5,334	5,158	4,994	4,840	4,493	4,193	3,930	3,699	3,493	3,309	2,993	2,732	2,513	2,166	1,903
10	9,691	9,401	9,127	8,869	8,625	8,394	7,968	7,583	7,233	6,914	6,622	6,354	6,107	5,565	5,112	4,727	4,395	4,108	3,855	3,433	3,095	2,817	2,388	2,072
15	14,315	13,691	13,118	12,592	12,106	11,656	10,850	10,148	9,531	8,985	8,498	8,062	7,667	6,833	6,162	5,611	5,150	4,759	4,424	3,877	3,451	3,109	2,594	2,226
20	18,801	17,738	16,789	15,936	15,166	14,466	13,245	12,213	11,331	10,568	9,901	9,313	8,791	7,711	6,857	6,189	5,634	5,169	4,776	4,145	3,661	3,279	2,712	2,312
25	23,155	21,563	20,176	18,957	17,877	16,913	15,267	13,913	12,779	11,816	10,989	10,269	9,638	8,355	7,373	6,598	5,970	5,451	5,015	4,324	3,800	3,390	2,787	2,367
30	27,381	25,183	23,312	21,699	20,296	19,062	16,997	15,335	13,969	12,827	11,857	11,024	10,300	8,848	7,754	6,901	6,217	5,657	5,189	4,452	3,899	3,468	2,840	2,405
35	31,487	28,615	26,223	24,200	22,467	20,965	18,494	16,543	14,965	13,661	12,567	11,635	10,831	9,237	8,051	7,136	6,407	5,813	5,320	4,549	3,973	3,526	2,879	2,433
40	35,476	31,872	28,933	26,489	24,427	22,662	19,801	17,582	15,810	14,362	13,157	12,139	11,267	9,552	8,290	7,322	6,557	5,937	5,423	4,621	4,030	3,571	2,909	2,454
45	39,355	34,968	31,461	28,593	26,205	24,185	20,954	18,485	16,536	14,959	13,656	12,563	11,631	9,812	8,485	7,474	6,679	6,036	5,506	4,684	4,076	3,607	2,933	2,471
50	43,126	37,914	33,826	30,534	27,825	25,558	21,977	19,276	17,167	15,473	14,084	12,923	11,940	10,031	8,648	7,601	6,779	6,118	5,575	4,733	4,113	3,636	2,952	2,484
55	46,796	40,721	36,043	32,328	29,308	26,804	22,892	19,976	17,720	15,921	14,454	13,234	12,205	10,217	8,786	7,707	6,864	6,187	5,632	4,775	4,144	3,660	2,968	2,496
60	50,367	43,399	38,125	33,993	30,670	27,939	23,715	20,600	18,209	16,315	14,778	13,505	12,435	10,378	8,905	7,798	6,936	6,246	5,680	4,809	4,170	3,681	2,981	2,505
65	53,844	45,956	40,084	35,542	31,925	28,976	24,458	21,159	18,644	16,663	15,063	13,743	12,636	10,518	9,008	7,877	6,998	6,296	5,722	4,839	4,192	3,698	2,993	2,513
70	57,230	48,400	41,931	36,987	33,086	29,929	25,134	21,662	19,034	16,974	15,317	13,954	12,314	10,641	9,098	7,946	7,052	6,340	5,758	4,865	4,212	3,713	3,002	2,520
75	60,529	50,739	43,675	38,337	34,163	30,808	25,750	22,119	19,385	17,253	15,543	14,142	12,972	10,750	9,177	8,006	7,100	6,378	5,790	4,883	4,229	3,726	3,011	2,526
80	63,744	52,979	45,325	39,603	35,164	31,619	26,315	22,534	19,703	17,505	15,747	14,311	13,114	10,847	9,248	8,060	7,142	6,412	5,818	4,903	4,244	3,738	3,019	2,531
85	66,879	55,126	46,887	40,790	36,097	32,372	26,834	22,914	19,993	17,733	15,932	14,463	13,242	10,934	9,311	8,108	7,180	6,443	5,843	4,925	4,257	3,748	3,025	2,536
90	69,936	57,187	48,369	41,908	36,969	33,072	27,313	23,262	20,258	17,941	16,099	14,601	13,357	11,013	9,368	8,151	7,214	6,470	5,865	4,941	4,269	3,758	3,031	2,540
95	72,918	59,166	49,777	42,961	37,786	33,724	27,756	23,583	20,501	18,131	16,252	14,726	13,463	11,084	9,420	8,190	7,245	6,495	5,885	4,956	4,280	3,766	3,037	2,544
100	75,828	61,067	51,117	43,954	38,553	34,333	28,168	23,879	20,724	18,306	16,393	14,841	13,559	11,149	9,467	8,226	7,272	6,517	5,904	4,969	4,289	3,773	3,042	2,548
105	78,669	62,896	52,392	44,894	39,274	34,904	28,551	24,154	20,931	18,467	16,522	14,947	13,647	11,209	9,510	8,258	7,298	6,537	5,920	4,980	4,298	3,780	3,046	2,551
110	81,442	64,657	53,608	45,784	39,953	35,439	28,908	24,409	21,122	18,615	16,641	15,044	13,728	11,263	9,549	8,288	7,321	6,556	5,936	4,991	4,306	3,786	3,050	2,554
115	84,151	66,352	54,768	46,628	40,594	35,943	29,242	24,647	21,300	18,753	16,751	15,134	13,803	11,314	9,585	8,315	7,342	6,573	5,950	5,001	4,313	3,792	3,054	2,556
120	86,798	67,987	55,877	47,429	41,200	36,417	29,555	24,869	21,466	18,882	16,853	15,218	13,872	11,360	9,619	8,340	7,362	6,588	5,962	5,010	4,320	3,797	3,057	2,558
125	89,384	69,563	56,938	48,191	41,774	36,865	29,849	25,077	21,620	19,001	16,948	15,296	13,936	11,403	9,650	8,363	7,380	6,603	5,974	5,019	4,326	3,802	3,060	2,561
130	91,912	71,085	57,953	48,916	42,318	37,288	30,126	25,272	21,765	19,113	17,037	15,368	13,997	11,444	9,678	8,385	7,396	6,616	5,985	5,026	4,332	3,807	3,063	2,563
135	94,383	72,554	58,926	49,608	42,834	37,688	30,387	25,455	21,901	19,218	17,120	15,435	14,053	11,481	9,705	8,405	7,412	6,629	5,995	5,033	4,338	3,811	3,066	2,565
140	96,800	73,974	59,859	50,267	43,325	38,068	30,633	25,628	22,029	19,316	17,198	15,499	14,105	11,516	9,730	8,424	7,427	6,641	6,005	5,040	4,343	3,814	3,068	2,566
145	99,165	75,347	60,755	50,897	43,792	38,428	30,866	25,791	22,149	19,408	17,271	15,558	14,154	11,549	9,753	8,441	7,440	6,651	6,014	5,046	4,347	3,818	3,071	2,568
150	101,478	76,675	61,615	51,500	44,238	38,770	31,087	25,945	22,262	19,495	17,340	15,614	14,200	11,580	9,775	8,458	7,453	6,662	6,022	5,052	4,352	3,821	3,073	2,569
155	103,742	77,960	62,443	52,077	44,663	39,096	31,296	26,090	22,369	19,577	17,405	15,667	14,244	11,608	9,796	8,473	7,465	6,671	6,030	5,058	4,356	3,825	3,075	2,571
160	105,958	79,205	63,239	52,629	45,068	39,407	31,494	26,228	22,471	19,655	17,466	15,716	14,285	11,636	9,815	8,488	7,476	6,680	6,037	5,063	4,359	3,828	3,077	2,572
165	108,128	80,412	64,005	53,159	45,456	39,703	31,683	26,359	22,567	19,728	17,524	15,763	14,324	11,661	9,834	8,501	7,487	6,689	6,044	5,068	4,363	3,830	3,079	2,573
170	110,253	81,581	64,744	53,668	45,828	39,986	31,863	26,483	22,658	19,798	17,579	15,808	14,360	11,686	9,851	8,514	7,497	6,697	6,051	5,072	4,366	3,833	3,080	2,575
175	112,334	82,715	65,456	54,156	46,183	40,257	32,035	26,602	22,744	19,864	17,631	15,850	14,395	11,709	9,867	8,526	7,506	6,704	6,057	5,077	4,370	3,835	3,082	2,576
180	114,374	83,816	66,143	54,626	46,524	40,516	32,199	26,715	22,827	19,927	17,681	15,890	14,428	11,730	9,883	8,538	7,515	6,711	6,063	5,081	4,373	3,838	3,083	2,577
185	116,372	84,884	66,807	55,078	46,852	40,764	32,355	26,822	22,905	19,987	17,728	15,928	14,459	11,751	9,897	8,549	7,524	6,718	6,068	5,085	4,376	3,840	3,085	2,578
190	118,331	85,921	67,448	55,512	47,166	41,001	32,505	26,925	22,980	20,044	17,773	15,964	14,489	11,771	9,911	8,559	7,532	6,725	6,074	5,088	4,378	3,842	3,086	2,579
195	120,251	86,929	68,067	55,931	47,468	41,230	32,648	27,023	23,052	20,098	17,815	15,998	14,518	11,790	9,925	8,569	7,539	6,731	6,079	5,092	4,381	3,844	3,087	2,580
200	122,134	87,909	68,667	56,335	47,759	41,449	32,785	27,117	23,120	20,150	17,856	16,031	14,545	11,807	9,937	8,579	7,547	6,736	6,083	5,095	4,383	3,846	3,089	2,580

# 4/0 AWG	1 Run(s)	208 Volts	C-Value = 15,082	Magnetic Duct	Copper

Calculations:

1. $f = \dfrac{\text{Length} \times 2 \times \text{Isca}}{\text{Runs} \times \text{C-Value} \times \text{E L-L}}$

2. $M = \dfrac{1}{1 + f}$

3. Isc = Isca x M

4. Add Motor Contribution = Motor FLA x 4

* All results are given in symmetrical amperes

Single-Phase L-L Bolted Fault-Current* Table

One-Way Distance in Feet

Available Isc in Thousands (Isca)	5	10	15	20	25	30	40	50	60	70	80	90	100	125	150	175	200	225	250	300	350	400	500	600
3	2,986	2,972	2,958	2,944	2,930	2,916	2,889	2,863	2,837	2,812	2,787	2,762	2,738	2,680	2,624	2,570	2,518	2,469	2,421	2,331	2,248	2,170	2,030	1,906
5	4,960	4,922	4,883	4,846	4,808	4,772	4,700	4,631	4,564	4,498	4,435	4,373	4,313	4,169	4,035	3,910	3,791	3,680	3,575	3,383	3,210	3,053	2,783	2,556
7	6,923	6,847	6,773	6,701	6,630	6,561	6,426	6,297	6,173	6,054	5,940	5,829	5,723	5,473	5,245	5,034	4,840	4,660	4,493	4,193	3,930	3,699	3,309	2,993
10	9,843	9,691	9,544	9,401	9,262	9,127	8,869	8,625	8,394	8,176	7,968	7,771	7,583	7,151	6,765	6,419	6,107	5,823	5,565	5,112	4,727	4,395	3,855	3,433
15	14,650	14,315	13,996	13,691	13,398	13,118	12,592	12,106	11,656	11,238	10,850	10,487	10,148	9,389	8,735	8,167	7,667	7,226	6,833	6,162	5,611	5,150	4,424	3,877
20	19,382	18,801	18,254	17,738	17,251	16,789	15,936	15,166	14,466	13,829	13,245	12,708	12,213	11,130	10,223	9,453	8,791	8,215	7,711	6,867	6,189	5,634	4,776	4,145
25	24,042	23,155	22,331	21,563	20,847	20,176	18,957	17,877	16,913	16,048	15,267	14,558	13,913	12,524	11,388	10,440	9,638	8,951	8,355	7,373	6,598	5,970	5,015	4,324
30	28,631	27,381	26,236	25,183	24,212	23,312	21,699	20,296	19,062	17,970	16,997	16,123	15,335	13,665	12,323	11,221	10,300	9,519	8,848	7,754	6,901	6,217	5,189	4,452
35	33,151	31,487	29,982	28,615	27,367	26,223	24,200	22,467	20,965	19,652	18,494	17,464	16,543	14,616	13,091	11,854	10,831	9,971	9,237	8,051	7,136	6,407	5,320	4,549
40	37,603	35,476	33,578	31,872	30,331	28,933	26,489	24,427	22,662	21,135	19,801	18,626	17,582	15,421	13,733	12,379	11,267	10,339	9,552	8,290	7,322	6,557	5,423	4,624
45	41,988	39,355	37,032	34,968	33,122	31,461	28,593	26,205	24,185	22,454	20,954	19,642	18,485	16,111	14,278	12,819	11,631	10,644	9,812	8,485	7,474	6,679	5,506	4,684
50	46,309	43,126	40,353	37,914	35,754	33,826	30,534	27,825	25,558	23,633	21,977	20,538	19,276	16,709	14,746	13,195	11,940	10,902	10,031	8,648	7,601	6,779	5,575	4,733
55	50,567	46,796	43,548	40,721	38,239	36,043	32,328	29,308	26,804	24,694	22,892	21,335	19,976	17,233	15,152	13,520	12,205	11,123	10,217	8,786	7,707	6,864	5,632	4,775
60	54,763	50,367	46,624	43,399	40,591	38,125	33,993	30,670	27,939	25,654	23,715	22,048	20,600	17,695	15,508	13,802	12,435	11,314	10,378	8,905	7,798	6,936	5,680	4,809
65	58,898	53,844	49,588	45,956	42,819	40,084	35,542	31,925	28,976	26,526	24,458	22,689	21,159	18,106	15,823	14,051	12,636	11,480	10,518	9,008	7,877	6,998	5,722	4,839
70	62,974	57,230	52,446	48,400	44,934	41,931	36,987	33,086	29,929	27,323	25,134	23,269	21,662	18,473	16,103	14,271	12,814	11,627	10,641	9,098	7,946	7,052	5,758	4,865
75	66,992	60,529	55,203	50,739	46,943	43,675	38,337	34,163	30,808	28,053	25,750	23,797	22,119	18,804	16,354	14,468	12,972	11,757	10,750	9,177	8,006	7,100	5,790	4,888
80	70,953	63,744	57,865	52,979	48,854	45,325	39,603	35,164	31,619	28,724	26,315	24,278	22,534	19,104	16,579	14,645	13,114	11,873	10,847	9,248	8,060	7,142	5,818	4,908
85	74,858	66,879	60,437	55,126	50,674	46,887	40,790	36,097	32,372	29,344	26,834	24,719	22,914	19,376	16,784	14,804	13,242	11,978	10,934	9,311	8,108	7,180	5,843	4,925
90	78,709	69,936	62,922	57,187	52,410	48,369	41,908	36,969	33,072	29,918	27,313	25,125	23,262	19,624	16,970	14,949	13,357	12,072	11,013	9,368	8,151	7,214	5,865	4,941
95	82,507	72,918	65,326	59,166	54,067	49,777	42,961	37,786	33,724	30,450	27,756	25,500	23,583	19,852	17,140	15,080	13,453	12,158	11,084	9,420	8,190	7,245	5,885	4,956
100	86,253	75,828	67,652	61,067	55,651	51,117	43,954	38,553	34,333	30,946	28,168	25,847	23,879	20,062	17,296	15,201	13,559	12,236	11,149	9,467	8,226	7,272	5,904	4,969
105	89,947	78,669	69,904	62,896	57,166	52,392	44,894	39,274	34,904	31,409	28,551	26,169	24,154	20,255	17,440	15,312	13,647	12,308	11,209	9,510	8,258	7,298	5,920	4,980
110	93,591	81,442	72,085	64,657	58,616	53,608	45,784	39,953	35,439	31,842	28,908	26,469	24,409	20,434	17,573	15,414	13,728	12,374	11,263	9,549	8,288	7,321	5,936	4,991
115	97,186	84,151	74,199	66,352	60,006	54,768	46,628	40,594	35,943	32,248	29,242	26,749	24,647	20,601	17,696	15,509	13,803	12,435	11,314	9,585	8,315	7,342	5,950	5,001
120	100,733	86,798	76,249	67,987	61,340	55,877	47,429	41,200	36,417	32,629	29,555	27,010	24,869	20,756	17,810	15,596	13,872	12,491	11,360	9,619	8,340	7,362	5,962	5,010
125	104,233	89,384	78,238	69,563	62,620	56,938	48,191	41,774	36,865	32,988	29,849	27,256	25,077	20,900	17,916	15,678	13,936	12,543	11,403	9,650	8,363	7,380	5,974	5,019
130	107,687	91,912	80,168	71,085	63,851	57,953	48,916	42,318	37,288	33,326	30,126	27,486	25,272	21,036	18,015	15,754	13,997	12,592	11,444	9,678	8,385	7,396	5,985	5,026
135	111,096	94,383	82,041	72,554	65,034	58,926	49,608	42,834	37,688	33,646	30,387	27,703	25,455	21,162	18,108	15,825	14,053	12,637	11,481	9,705	8,405	7,412	5,995	5,033
140	114,460	96,800	83,862	73,974	66,172	59,859	50,267	43,325	38,068	33,948	30,633	27,908	25,628	21,282	18,196	15,891	14,105	12,680	11,516	9,730	8,424	7,427	6,005	5,040
145	117,780	99,165	85,630	75,347	67,268	60,755	50,897	43,792	38,428	34,234	30,866	28,101	25,791	21,394	18,277	15,954	14,154	12,719	11,549	9,753	8,441	7,440	6,014	5,046
150	121,058	101,478	87,350	76,675	68,325	61,615	51,500	44,238	38,770	34,506	31,087	28,284	25,945	21,499	18,355	16,012	14,200	12,757	11,580	9,775	8,458	7,453	6,022	5,052
155	124,294	103,742	89,022	77,960	69,344	62,443	52,077	44,663	39,096	34,764	31,296	28,457	26,090	21,599	18,427	16,068	14,244	12,792	11,608	9,796	8,473	7,465	6,030	5,058
160	127,488	105,958	90,649	79,205	70,327	63,239	52,629	45,068	39,407	35,009	31,494	28,621	26,228	21,694	18,496	16,120	14,285	12,825	11,636	9,815	8,488	7,476	6,037	5,063
165	130,643	108,128	92,233	80,412	71,277	64,005	53,159	45,456	39,703	35,243	31,683	28,777	26,359	21,783	18,561	16,169	14,324	12,856	11,661	9,834	8,501	7,487	6,044	5,068
170	133,758	110,253	93,774	81,581	72,194	64,744	53,668	45,828	39,986	35,466	31,863	28,925	26,483	21,868	18,623	16,216	14,360	12,886	11,686	9,851	8,514	7,497	6,051	5,072
175	136,834	112,334	95,276	82,715	73,080	65,456	54,156	46,183	40,257	35,678	32,035	29,067	26,602	21,949	18,681	16,260	14,395	12,914	11,709	9,867	8,526	7,506	6,057	5,077
180	139,872	114,374	96,739	83,816	73,938	66,143	54,626	46,524	40,516	35,882	32,199	29,201	26,715	22,026	18,737	16,302	14,428	12,940	11,730	9,883	8,538	7,515	6,063	5,081
185	142,872	116,372	98,165	84,884	74,768	66,807	55,078	46,852	40,764	36,076	32,355	29,330	26,822	22,099	18,790	16,342	14,459	12,965	11,751	9,897	8,549	7,524	6,068	5,085
190	145,836	118,331	99,555	85,921	75,572	67,448	55,512	47,166	41,001	36,262	32,505	29,453	26,925	22,168	18,840	16,381	14,489	12,989	11,771	9,911	8,559	7,532	6,074	5,088
195	148,764	120,251	100,911	86,929	76,351	68,067	55,931	47,468	41,230	36,440	32,648	29,570	27,023	22,235	18,888	16,417	14,518	13,012	11,790	9,925	8,569	7,539	6,079	5,092
200	151,656	122,134	102,233	87,909	77,105	68,667	56,335	47,759	41,449	36,611	32,785	29,683	27,117	22,298	18,934	16,451	14,545	13,034	11,807	9,937	8,579	7,547	6,083	5,095

# 4/0 AWG	2 Run(s)	208 Volts

C-Value =	15,082	Magnetic Duct	Copper

Calculations:

1. $f = \dfrac{\text{Length} \times 2 \times \text{Isca}}{\text{Runs} \times \text{C-Value} \times \text{E L-L}}$

2. $M = \dfrac{1}{1 + f}$

3. Isc = Isca x M

4. Add Motor Contribution = Motor FLA x 4

* All results are given in symmetrical amperes

Single-Phase L-L Bolted Fault-Current* Table

One-Way Distance in Feet

Available Isc in Thousands (Isca)	5	10	15	20	25	30	40	50	60	70	80	90	100	125	150	175	200	225	250	300	350	400	500	600
3	2,974	2,948	2,923	2,899	2,874	2,850	2,804	2,759	2,715	2,673	2,632	2,592	2,553	2,462	2,376	2,297	2,222	2,152	2,087	1,967	1,860	1,765	1,600	1,463
5	4,928	4,858	4,790	4,724	4,660	4,598	4,473	4,364	4,255	4,152	4,054	3,960	3,871	3,664	3,478	3,310	3,158	3,019	2,892	2,667	2,474	2,308	2,034	1,818
7	6,860	6,725	6,596	6,471	6,352	6,236	6,017	5,813	5,622	5,444	5,276	5,119	4,970	4,634	4,341	4,083	3,853	3,648	3,464	3,146	2,882	2,658	2,301	2,029
10	9,717	9,449	9,195	8,955	8,727	8,511	8,108	7,742	7,407	7,101	6,818	6,557	6,316	5,783	5,333	4,948	4,615	4,324	4,068	3,636	3,288	3,000	2,553	2,222
15	14,371	13,793	13,260	12,766	12,308	11,881	11,111	10,435	9,836	9,302	8,823	8,391	8,000	7,164	6,486	5,926	5,454	5,053	4,706	4,138	3,692	3,333	2,791	2,400
20	18,898	17,910	17,021	16,216	15,484	14,815	13,636	12,631	11,765	11,009	10,345	9,756	9,231	8,135	7,273	6,575	6,000	5,517	5,106	4,444	3,934	3,529	2,927	2,500
25	23,301	21,818	20,513	19,355	18,320	17,391	15,739	14,458	13,333	12,371	11,538	10,811	10,159	8,856	7,843	7,038	6,383	5,839	5,381	4,651	4,095	3,658	3,015	2,564
30	27,586	25,532	23,762	22,222	20,869	19,672	17,647	16,000	14,634	13,483	12,500	11,650	10,909	9,412	8,276	7,384	6,667	6,076	5,581	4,800	4,210	3,750	3,077	2,609
35	31,758	29,066	26,794	24,852	23,172	21,705	19,256	17,319	15,730	14,408	13,291	12,335	11,507	9,853	8,615	7,654	6,885	6,257	5,734	4,912	4,297	3,818	3,123	2,641
40	35,821	32,432	29,629	27,272	25,263	23,529	20,689	18,461	16,666	15,190	13,953	12,903	12,000	10,213	8,889	7,869	7,059	6,400	5,854	5,000	4,364	3,871	3,158	2,667
45	39,779	35,643	32,287	29,508	27,169	25,174	21,951	19,459	17,475	15,859	14,516	13,383	12,414	10,511	9,114	8,044	7,200	6,516	5,950	5,070	4,417	3,913	3,186	2,686
50	43,636	38,709	34,782	31,579	28,915	26,666	23,077	20,339	18,181	16,438	15,000	13,793	12,766	10,762	9,332	8,191	7,317	6,611	6,030	5,128	4,461	3,947	3,208	2,703
55	47,397	41,640	37,130	33,502	30,520	28,025	24,087	21,120	18,803	16,944	15,420	14,148	13,069	10,977	9,462	8,315	7,416	6,692	6,097	5,176	4,497	3,976	3,227	2,716
60	51,064	44,444	39,344	35,294	32,000	29,268	25,000	21,818	19,354	17,391	15,789	14,457	13,333	11,163	9,600	8,421	7,500	6,760	6,154	5,217	4,528	4,000	3,243	2,727
65	54,641	47,130	41,434	36,966	33,368	30,409	25,827	22,446	19,847	17,788	16,115	14,731	13,565	11,325	9,719	8,513	7,573	6,819	6,203	5,252	4,555	4,021	3,257	2,737
70	58,131	49,704	43,410	38,532	34,639	31,460	26,582	23,013	20,289	18,142	16,406	14,973	13,770	11,467	9,824	8,593	7,636	6,871	6,245	5,284	4,578	4,038	3,268	2,745
75	61,538	52,173	45,282	39,999	35,820	32,432	27,272	23,529	20,689	18,461	16,666	15,189	13,953	11,594	9,917	8,664	7,692	6,916	6,283	5,310	4,598	4,054	3,279	2,752
80	64,864	54,545	47,058	41,379	36,922	33,333	27,906	23,999	21,052	18,750	16,901	15,384	14,117	11,707	10,000	8,727	7,742	6,956	6,316	5,338	4,615	4,068	3,288	2,759
85	68,113	56,824	48,745	42,677	37,953	34,170	28,491	24,431	21,383	19,012	17,114	15,560	14,265	11,809	10,074	8,783	7,786	6,992	6,345	5,354	4,631	4,080	3,296	2,764
90	71,287	59,016	50,349	43,902	38,918	34,951	29,032	24,827	21,686	19,251	17,307	15,720	14,400	11,901	10,141	8,834	7,826	7,024	6,371	5,373	4,645	4,091	3,303	2,769
95	74,388	61,125	51,876	45,059	39,825	35,680	29,533	25,193	21,965	19,470	17,484	15,866	14,522	11,984	10,201	8,880	7,862	7,053	6,395	5,390	4,658	4,101	3,309	2,774
100	77,419	63,157	53,333	46,153	40,677	36,363	29,999	25,531	22,222	19,672	17,647	16,000	14,634	12,060	10,256	8,922	7,895	7,079	6,417	5,405	4,669	4,109	3,315	2,778
105	80,382	65,116	54,722	47,190	41,481	37,004	30,434	25,846	22,459	19,858	17,796	16,122	14,736	12,130	10,306	8,960	7,924	7,103	6,437	5,419	4,680	4,118	3,320	2,781
110	83,280	67,004	56,050	48,174	42,239	37,606	30,840	26,138	22,680	20,030	17,934	16,236	14,831	12,194	10,353	8,995	7,952	7,125	6,455	5,432	4,689	4,125	3,325	2,785
115	86,115	68,827	57,320	49,109	42,956	38,173	31,221	26,411	22,885	20,190	18,062	16,341	14,919	12,253	10,395	9,027	7,977	7,145	6,471	5,444	4,698	4,132	3,329	2,788
120	88,888	70,587	58,536	49,999	43,635	38,709	31,578	26,666	23,076	20,338	18,181	16,438	15,000	12,307	10,434	9,056	8,000	7,164	6,486	5,454	4,706	4,138	3,333	2,791
125	91,602	72,288	59,701	50,847	44,280	39,215	31,914	26,905	23,255	20,477	18,292	16,528	15,075	12,358	10,471	9,084	8,021	7,181	6,500	5,464	4,713	4,144	3,337	2,793
130	94,259	73,933	60,818	51,655	44,891	39,694	32,231	27,130	23,423	20,607	18,396	16,613	15,145	12,405	10,505	9,109	8,041	7,197	6,513	5,474	4,720	4,149	3,340	2,796
135	96,860	75,523	61,890	52,426	45,473	40,148	32,529	27,341	23,580	20,729	18,493	16,692	15,211	12,449	10,536	9,133	8,059	7,212	6,525	5,482	4,726	4,154	3,344	2,798
140	99,407	77,063	62,920	53,164	46,026	40,579	32,812	27,540	23,728	20,843	18,584	16,766	15,272	12,490	10,566	9,155	8,077	7,226	6,537	5,490	4,732	4,158	3,347	2,800
145	101,902	78,554	63,911	53,869	46,554	40,988	33,079	27,728	23,868	20,951	18,669	16,836	15,330	12,529	10,593	9,176	8,093	7,238	6,547	5,497	4,738	4,163	3,349	2,802
150	104,347	79,999	64,864	54,544	47,058	41,378	33,333	27,906	23,999	21,052	18,749	16,901	15,384	12,565	10,619	9,195	8,108	7,251	6,557	5,504	4,743	4,167	3,352	2,804
155	106,742	81,399	65,781	55,192	47,539	41,750	33,573	28,075	24,124	21,148	18,825	16,963	15,435	12,599	10,643	9,213	8,122	7,262	6,566	5,511	4,748	4,170	3,354	2,805
160	109,090	82,757	66,665	55,813	47,999	42,104	33,802	28,235	24,242	21,238	18,897	17,021	15,483	12,631	10,666	9,230	8,135	7,273	6,575	5,517	4,752	4,174	3,357	2,807
165	111,391	84,075	67,518	56,409	48,439	42,443	34,020	28,386	24,354	21,324	18,965	17,076	15,529	12,662	10,688	9,247	8,148	7,283	6,583	5,523	4,757	4,177	3,359	2,808
170	113,648	85,354	68,340	56,982	48,861	42,766	34,227	28,531	24,460	21,405	19,029	17,128	15,572	12,690	10,708	9,262	8,160	7,292	6,591	5,528	4,761	4,180	3,361	2,810
175	115,861	86,597	69,134	57,533	49,266	43,076	34,425	28,668	24,561	21,483	19,090	17,177	15,613	12,717	10,728	9,276	8,171	7,301	6,598	5,533	4,764	4,183	3,363	2,811
180	118,032	87,803	69,902	58,063	49,654	43,372	34,615	28,799	24,657	21,556	19,148	17,224	15,652	12,743	10,746	9,290	8,182	7,309	6,605	5,538	4,768	4,186	3,364	2,812
185	120,161	88,977	70,643	58,574	50,027	43,657	34,795	28,924	24,748	21,626	19,204	17,269	15,689	12,767	10,763	9,303	8,192	7,317	6,612	5,543	4,771	4,189	3,366	2,814
190	122,251	90,117	71,360	59,066	50,386	43,930	34,968	29,044	24,836	21,693	19,256	17,312	15,724	12,791	10,780	9,315	8,201	7,325	6,618	5,547	4,775	4,191	3,368	2,815
195	124,301	91,227	72,054	59,541	50,731	44,192	35,134	29,158	24,919	21,757	19,306	17,352	15,757	12,813	10,796	9,327	8,210	7,332	6,624	5,551	4,778	4,193	3,369	2,816
200	126,314	92,306	72,726	59,999	51,063	44,443	35,293	29,268	24,999	21,818	19,354	17,391	15,789	12,834	10,810	9,338	8,219	7,339	6,630	5,555	4,781	4,196	3,371	2,817

250 MCM | **1 Run(s)** | **208 Volts** | **C-Value = 16,483** | **Magnetic Duct** | **Copper**

Calculations:

1. $f = \dfrac{\text{Length} \times 2 \times \text{Isca}}{\text{Runs} \times \text{C-Value} \times E_{L-L}}$

2. $M = \dfrac{1}{1 + f}$

3. $\text{Isc} = \text{Isca} \times M$

4. Add Motor Contribution = Motor FLA x 4

* All results are given in symmetrical amperes

Single-Phase L-L Bolted Fault-Current* Table

One-Way Distance in Feet

Available Isc in Thousands (Isca)	5	10	15	20	25	30	40	50	60	70	80	90	100	125	150	175	200	225	250	300	350	400	500	600
3	2,987	2,974	2,961	2,948	2,936	2,923	2,899	2,874	2,850	2,827	2,804	2,781	2,759	2,704	2,652	2,602	2,553	2,507	2,462	2,376	2,297	2,222	2,087	1,967
5	4,964	4,928	4,893	4,858	4,824	4,790	4,724	4,660	4,598	4,537	4,478	4,420	4,364	4,229	4,103	3,983	3,871	3,765	3,664	3,478	3,310	3,158	2,892	2,667
7	6,929	6,860	6,792	6,725	6,660	6,596	6,471	6,352	6,236	6,125	6,017	5,913	5,813	5,577	5,359	5,157	4,970	4,797	4,634	4,341	4,083	3,853	3,464	3,146
10	9,856	9,717	9,581	9,449	9,320	9,195	8,955	8,727	8,511	8,304	8,108	7,921	7,742	7,328	6,956	6,621	6,316	6,038	5,783	5,333	4,948	4,615	4,068	3,636
15	14,679	14,371	14,076	13,793	13,521	13,260	12,766	12,308	11,881	11,483	11,111	10,762	10,435	9,697	9,056	8,495	8,000	7,559	7,164	6,486	5,926	5,454	4,706	4,138
20	19,433	18,898	18,391	17,910	17,454	17,021	16,216	15,484	14,815	14,201	13,636	13,115	12,631	11,566	10,667	9,897	9,231	8,648	8,135	7,273	6,575	6,000	5,106	4,444
25	24,121	23,301	22,535	21,818	21,145	20,513	19,355	18,320	17,391	16,552	15,789	15,094	14,458	13,079	11,940	10,984	10,169	9,467	8,856	7,843	7,038	6,383	5,381	4,651
30	28,742	27,586	26,519	25,532	24,615	23,762	22,222	20,869	19,672	18,604	17,647	16,783	16,000	14,328	12,973	11,852	10,909	10,105	9,412	8,276	7,384	6,667	5,581	4,800
35	33,300	31,758	30,352	29,066	27,884	26,794	24,852	23,172	21,705	20,413	19,266	18,241	17,319	15,377	13,827	12,560	11,507	10,616	9,853	8,615	7,654	6,885	5,734	4,912
40	37,795	35,821	34,042	32,432	30,968	29,629	27,272	25,263	23,529	22,018	20,689	19,512	18,461	16,271	14,545	13,150	12,000	11,034	10,213	8,889	7,869	7,059	5,854	5,000
45	42,229	39,779	37,598	35,643	33,882	32,287	29,508	27,169	25,174	23,452	21,951	20,630	19,459	17,041	15,158	13,649	12,414	11,383	10,511	9,114	8,044	7,200	5,950	5,070
50	46,602	43,636	41,025	38,709	36,641	34,782	31,579	28,915	26,666	24,742	23,077	21,621	20,339	17,712	15,686	14,076	12,766	11,679	10,762	9,302	8,191	7,317	6,030	5,128
55	50,916	47,397	44,332	41,640	39,256	37,130	33,502	30,520	28,025	25,907	24,087	22,506	21,120	18,301	16,146	14,446	13,069	11,932	10,977	9,462	8,315	7,416	6,097	5,176
60	55,172	51,064	47,524	44,444	41,739	39,344	35,294	32,000	29,268	26,966	25,000	23,301	21,818	18,823	16,551	14,769	13,333	12,152	11,163	9,600	8,421	7,500	6,154	5,217
65	59,372	54,641	50,608	47,130	44,098	41,434	36,966	33,368	30,409	27,931	25,827	24,018	22,446	19,289	16,910	15,054	13,565	12,344	11,325	9,719	8,513	7,573	6,203	5,252
70	63,516	58,131	53,588	49,704	46,344	43,410	38,532	34,639	31,460	28,816	26,582	24,669	23,013	19,706	17,230	15,307	13,770	12,514	11,467	9,824	8,593	7,636	6,245	5,283
75	67,605	61,538	56,470	52,173	48,484	45,282	39,999	35,820	32,432	29,629	27,272	25,263	23,529	20,083	17,518	15,534	13,953	12,665	11,594	9,917	8,664	7,692	6,283	5,310
80	71,642	64,864	59,259	54,545	50,526	47,058	41,379	36,922	33,333	30,379	27,906	25,806	23,999	20,425	17,777	15,737	14,117	12,800	11,707	10,000	8,727	7,742	6,316	5,333
85	75,625	68,113	61,958	56,824	52,475	48,745	42,677	37,953	34,170	31,073	28,491	26,305	24,431	20,736	18,013	15,922	14,265	12,921	11,809	10,074	8,783	7,786	6,345	5,354
90	79,558	71,287	64,573	59,016	54,339	50,349	43,902	38,918	34,951	31,717	29,032	26,765	24,827	21,021	18,227	16,089	14,400	13,031	11,901	10,141	8,834	7,826	6,371	5,373
95	83,440	74,388	67,108	61,125	56,122	51,876	45,059	39,825	35,680	32,317	29,533	27,191	25,193	21,283	18,424	16,242	14,522	13,131	11,984	10,201	8,880	7,862	6,395	5,390
100	87,272	77,419	69,565	63,157	57,831	53,333	46,153	40,677	36,363	32,876	29,999	27,586	25,531	21,524	18,604	16,382	14,634	13,223	12,060	10,256	8,922	7,895	6,417	5,405
105	91,057	80,382	71,948	65,116	59,468	54,722	47,190	41,481	37,004	33,399	30,434	27,953	25,846	21,747	18,770	16,511	14,736	13,307	12,130	10,306	8,960	7,924	6,437	5,419
110	94,793	83,280	74,261	67,004	61,040	56,050	48,174	42,239	37,606	33,889	30,840	28,295	26,138	21,954	18,924	16,629	14,831	13,384	12,194	10,353	8,995	7,952	6,455	5,432
115	98,483	86,115	76,506	68,827	62,549	57,320	49,109	42,956	38,173	34,349	31,221	28,615	26,411	22,146	19,067	16,740	14,919	13,455	12,253	10,395	9,027	7,977	6,471	5,444
120	102,127	88,888	78,688	70,587	63,999	58,536	49,999	43,635	38,709	34,782	31,578	28,915	26,666	22,325	19,199	16,842	15,000	13,521	12,307	10,434	9,056	8,000	6,486	5,454
125	105,726	91,602	80,807	72,288	65,394	59,701	50,847	44,280	39,215	35,190	31,914	29,196	26,905	22,492	19,323	16,937	15,075	13,582	12,358	10,471	9,084	8,021	6,500	5,464
130	109,281	94,259	82,868	73,933	66,737	60,818	51,655	44,891	39,694	35,575	32,231	29,461	27,130	22,649	19,439	17,025	15,145	13 639	12,405	10,505	9,109	8,041	6,513	5,474
135	112,793	96,860	84,871	75,523	68,030	61,890	52,426	45,473	40,148	35,939	32,529	29,710	27,341	22,796	19,547	17,108	15,211	13,692	12,449	10,536	9,133	8,059	6,525	5,482
140	116,262	99,407	86,821	77,063	69,277	62,920	53,164	46,026	40,579	36,284	32,812	29,946	27,540	22,935	19,649	17,186	15,272	13,742	12,490	10,566	9,155	8,077	6,537	5,490
145	119,690	101,902	88,718	78,554	70,480	63,911	53,869	46,554	40,988	36,611	33,079	30,168	27,728	23,065	19,744	17,259	15,330	13,789	12,529	10,593	9,176	8,093	6,547	5,497
150	123,076	104,347	90,565	79,999	71,641	64,864	54,544	47,058	41,378	36,922	33,333	30,379	27,906	23,188	19,834	17,328	15,384	13,832	12,565	10,619	9,195	8,108	6,557	5,504
155	126,422	106,742	92,364	81,399	72,762	65,781	55,192	47,539	41,750	37,218	33,573	30,579	28,075	23,304	19,919	17,393	15,435	13,874	12,599	10,643	9,213	8,122	6,566	5,511
160	129,729	109,090	94,116	82,757	73,845	66,665	55,813	47,999	42,104	37,499	33,802	30,768	28,235	23,414	19,999	17,454	15,483	13,913	12,631	10,666	9,230	8,135	6,575	5,517
165	132,997	111,391	95,825	84,075	74,892	67,518	56,409	48,439	42,443	37,767	34,020	30,949	28,386	23,518	20,075	17,512	15,529	13,949	12,662	10,688	9,247	8,148	6,583	5,523
170	136,226	113,648	97,490	85,354	75,906	68,340	56,982	48,861	42,766	38,023	34,227	31,120	28,531	23,617	20,148	17,567	15,572	13,984	12,690	10,708	9,262	8,160	6,591	5,528
175	139,418	115,861	99,114	86,597	76,887	69,134	57,533	49,266	43,076	38,268	34,425	31,284	28,668	23,711	20,216	17,619	15,613	14,017	12,717	10,728	9,276	8,171	6,598	5,533
180	142,573	118,032	100,698	87,803	77,836	69,902	58,063	49,654	43,372	38,502	34,615	31,440	28,799	23,801	20,281	17,668	15,652	14,048	12,743	10,746	9,290	8,182	6,605	5,538
185	145,692	120,161	102,244	88,977	78,757	70,643	58,574	50,027	43,657	38,726	34,795	31,589	28,924	23,886	20,343	17,715	15,689	14,078	12,767	10,763	9,303	8,192	6,612	5,543
190	148,775	122,251	103,753	90,117	79,649	71,360	59,066	50,386	43,930	38,940	34,968	31,732	29,044	23,968	20,402	17,760	15,724	14,106	12,791	10,780	9,315	8,201	6,618	5,547
195	151,824	124,301	105,226	91,227	80,515	72,054	59,541	50,731	44,192	39,146	35,134	31,868	29,158	24,046	20,458	17,803	15,757	14,133	12,813	10,796	9,327	8,210	6,624	5,551
200	154,838	126,314	106,665	92,306	81,354	72,726	59,999	51,063	44,443	39,343	35,293	31,999	29,268	24,120	20,512	17,843	15,789	14,159	12,834	10,810	9,338	8,219	6,630	5,555

250 MCM | 2 Run(s) | 208 Volts

C-Value = 16,483 | Magnetic Duct | Copper

Calculations:

1. $f = \dfrac{\text{Length} \times 2 \times \text{Isca}}{\text{Runs} \times \text{C-Value} \times \text{E L-L}}$

2. $M = \dfrac{1}{1 + f}$

3. Isc = Isca x M

4. Add Motor Contribution = Motor FLA x 4

* All results are given in symmetrical amperes

Single-Phase L-L Bolted Fault-Current* Table

One-Way Distance in Feet

Available Isc in Thousands (Isca)	5	10	15	20	25	30	40	50	60	70	80	90	100	125	150	175	200	225	250	300	350	400	500	600
3	2,976	2,953	2,930	2,908	2,886	2,864	2,821	2,779	2,739	2,700	2,662	2,625	2,589	2,503	2,423	2,348	2,277	2,211	2,148	2,032	1,929	1,835	1,673	1,537
5	4,935	4,871	4,809	4,749	4,690	4,632	4,522	4,416	4,315	4,219	4,127	4,039	3,954	3,758	3,580	3,418	3,270	3,135	3,010	2,788	2,596	2,430	2,153	1,933
7	6,873	6,750	6,632	6,517	6,407	6,300	6,097	5,906	5,727	5,559	5,400	5,250	5,108	4,785	4,500	4,247	4,022	3,818	3,635	3,346	3,049	2,821	2,455	2,173
10	9,742	9,498	9,265	9,043	8,832	8,630	8,254	7,908	7,591	7,298	7,026	6,775	6,540	6,019	5,576	5,193	4,859	4,566	4,306	3,845	3,507	3,209	2,743	2,396
15	14,428	13,897	13,404	12,945	12,517	12,116	11,386	10,739	10,162	9,643	9,175	8,751	8,363	7,530	6,848	6,280	5,798	5,385	5,027	4,427	3,971	3,594	3,020	2,604
20	18,995	18,086	17,261	16,507	15,816	15,181	14,053	13,080	12,234	11,490	10,832	10,245	9,718	8,611	7,731	7,014	6,418	5,916	5,487	4,751	4,253	3,823	3,180	2,722
25	23,449	22,080	20,861	19,771	18,788	17,899	16,350	15,049	13,939	12,982	12,148	11,414	10,764	9,423	8,379	7,543	6,859	6,288	5,805	5,033	4,441	3,974	3,284	2,798
30	27,794	25,891	24,232	22,772	21,478	20,324	18,351	16,727	15,367	14,212	13,218	12,354	11,596	10,054	8,874	7,942	7,187	6,563	6,039	5,217	4,577	4,083	3,357	2,851
35	32,034	29,532	27,392	25,542	23,925	22,501	20,108	18,174	16,580	15,243	14,106	13,126	12,274	10,560	9,266	8,254	7,442	6,775	6,218	5,340	4,679	4,164	3,412	2,890
40	36,173	33,014	30,363	28,105	26,161	24,468	21,663	19,436	17,624	16,121	14,854	13,772	12,837	10,974	9,583	8,505	7,645	6,943	6,359	5,444	4,758	4,226	3,454	2,920
45	40,213	36,347	33,159	30,485	28,211	26,252	23,051	20,545	18,531	16,877	15,493	14,320	13,311	11,319	9,845	8,711	7,811	7,080	6,473	5,527	4,822	4,277	3,488	2,944
50	44,160	39,541	35,797	32,701	30,097	27,878	24,295	21,528	19,327	17,534	16,046	14,790	13,717	11,611	10,065	8,883	7,949	7,193	6,568	5,596	4,874	4,318	3,515	2,964
55	48,015	42,604	38,289	34,768	31,840	29,367	25,418	22,405	20,031	18,112	16,528	15,199	14,068	11,861	10,253	9,029	8,065	7,288	6,647	5,653	4,918	4,352	3,537	2,980
60	51,782	45,544	40,647	36,701	33,454	30,734	26,436	23,192	20,658	18,623	16,953	15,557	14,374	12,078	10,415	9,154	8,165	7,369	6,715	5,702	4,955	4,381	3,557	2,993
65	55,464	48,368	42,882	38,513	34,953	31,995	27,363	23,903	21,220	19,078	17,329	15,874	14,644	12,268	10,556	9,262	8,252	7,440	6,773	5,744	4,987	4,405	3,573	3,005
70	59,064	51,083	45,003	40,216	36,349	33,161	28,212	24,548	21,727	19,487	17,666	16,156	14,884	12,436	10,679	9,358	8,327	7,501	6,824	5,781	5,014	4,427	3,587	3,015
75	62,584	53,696	47,018	41,817	37,652	34,242	28,991	25,136	22,186	19,855	17,968	16,408	15,098	12,585	10,789	9,442	8,394	7,555	6,869	5,813	5,038	4,446	3,599	3,023
80	66,028	56,211	48,935	43,327	38,872	35,248	29,708	25,673	22,604	20,189	18,241	16,636	15,290	12,718	10,887	9,517	8,453	7,603	6,908	5,841	5,059	4,462	3,610	3,031
85	69,397	58,634	50,762	44,753	40,016	36,186	30,372	26,167	22,986	20,494	18,489	16,842	15,464	12,838	10,975	9,584	8,506	7,646	6,944	5,866	5,078	4,477	3,620	3,038
90	72,695	60,971	52,504	46,101	41,091	37,062	30,987	26,623	23,336	20,772	18,715	17,029	15,622	12,947	11,054	9,644	8,553	7,684	6,975	5,889	5,095	4,490	3,628	3,044
95	75,922	63,225	54,167	47,378	42,102	37,883	31,559	27,044	23,659	21,027	18,922	17,201	15,766	13,046	11,126	9,699	8,596	7,719	7,004	5,909	5,110	4,502	3,636	3,049
100	79,082	65,402	55,756	48,590	43,056	38,654	32,092	27,434	23,957	21,263	19,113	17,358	15,898	13,136	11,192	9,749	8,635	7,750	7,030	5,928	5,124	4,513	3,643	3,054
105	82,177	67,504	57,277	49,741	43,958	39,379	32,590	27,797	24,234	21,480	19,288	17,502	16,019	13,219	11,252	9,794	8,671	7,779	7,053	5,944	5,137	4,522	3,649	3,059
110	85,208	69,536	58,733	50,836	44,810	40,062	33,056	28,136	24,491	21,682	19,451	17,636	16,131	13,295	11,307	9,836	8,704	7,805	7,075	5,960	5,148	4,531	3,655	3,063
115	88,178	71,501	60,129	51,878	45,618	40,706	33,494	28,452	24,730	21,869	19,601	17,760	16,235	13,365	11,357	9,874	8,734	7,829	7,095	5,974	5,159	4,539	3,660	3,066
120	91,088	73,403	61,468	52,872	46,385	41,316	33,905	28,749	24,954	22,044	19,742	17,875	16,331	13,430	11,404	9,910	8,761	7,852	7,113	5,987	5,168	4,547	3,665	3,070
125	93,940	75,244	62,754	53,820	47,113	41,893	34,293	29,027	25,163	22,207	19,872	17,982	16,420	13,490	11,448	9,943	8,787	7,872	7,130	5,999	5,177	4,554	3,670	3,073
130	96,736	77,027	63,990	54,727	47,806	42,440	34,659	29,289	25,359	22,360	19,995	18,082	16,503	13,547	11,488	9,973	8,811	7,891	7,146	6,010	5,185	4,560	3,674	3,076
135	99,478	78,755	65,178	55,593	48,467	42,959	35,004	29,535	25,544	22,503	20,109	18,176	16,581	13,599	11,526	10,001	8,833	7,909	7,160	6,020	5,193	4,566	3,678	3,079
140	102,167	80,431	66,321	56,423	49,096	43,453	35,331	29,768	25,718	22,638	20,217	18,263	16,654	13,648	11,561	10,028	8,854	7,926	7,174	6,030	5,200	4,571	3,681	3,081
145	104,804	82,057	67,423	57,219	49,697	43,923	35,642	29,987	25,882	22,765	20,318	18,346	16,723	13,694	11,594	10,053	8,873	7,941	7,186	6,039	5,207	4,577	3,685	3,084
150	107,391	83,634	68,484	57,981	50,271	44,371	35,936	30,196	26,037	22,884	20,413	18,424	16,787	13,737	11,625	10,076	8,891	7,956	7,198	6,047	5,213	4,581	3,688	3,086
155	109,930	85,166	69,508	58,713	50,821	44,799	36,216	30,393	26,183	22,998	20,503	18,497	16,848	13,778	11,654	10,098	8,908	7,969	7,210	6,055	5,219	4,586	3,691	3,088
160	112,422	86,654	70,496	59,417	51,347	45,207	36,482	30,580	26,322	23,105	20,588	18,566	16,906	13,817	11,682	10,119	8,924	7,982	7,220	6,062	5,225	4,590	3,693	3,090
165	114,867	88,100	71,450	60,093	51,851	45,597	36,736	30,758	26,454	23,206	20,669	18,632	16,960	13,853	11,708	10,138	8,939	7,994	7,230	6,069	5,230	4,594	3,696	3,091
170	117,269	89,505	72,372	60,744	52,335	45,971	36,978	30,928	26,579	23,303	20,745	18,694	17,011	13,887	11,732	10,156	8,954	8,006	7,239	6,076	5,235	4,598	3,698	3,093
175	119,626	90,872	73,263	61,370	52,799	46,329	37,209	31,090	26,699	23,394	20,818	18,753	17,060	13,920	11,755	10,174	8,967	8,016	7,248	6,082	5,239	4,601	3,701	3,095
180	121,942	92,202	74,125	61,974	53,245	46,672	37,431	31,244	26,812	23,482	20,887	18,809	17,107	13,950	11,777	10,190	8,980	8,027	7,256	6,088	5,244	4,605	3,703	3,096
185	124,216	93,497	74,959	62,556	53,675	47,002	37,642	31,391	26,921	23,565	20,953	18,862	17,151	13,980	11,798	10,206	8,992	8,036	7,264	6,093	5,248	4,608	3,705	3,098
190	126,450	94,757	75,767	63,117	54,088	47,318	37,845	31,532	27,024	23,644	21,015	18,913	17,193	14,008	11,818	10,221	9,004	8,046	7,272	6,099	5,252	4,611	3,707	3,099
195	128,646	95,984	76,550	63,660	54,485	47,622	38,039	31,667	27,123	23,720	21,075	18,961	17,233	14,034	11,837	10,235	9,015	8,054	7,279	6,104	5,255	4,614	3,709	3,100
200	130,803	97,180	77,308	64,184	54,869	47,915	38,225	31,796	27,218	23,792	21,132	19,007	17,271	14,059	11,855	10,248	9,025	8,063	7,286	6,109	5,259	4,617	3,710	3,102

300 MCM	1 Run(s)	208 Volts	C-Value = 18,176	Magnetic Duct	Copper

Calculations:

1. $f = \dfrac{\text{Length} \times 2 \times \text{Isca}}{\text{Runs} \times \text{C-Value} \times E_{L-L}}$

2. $M = \dfrac{1}{1 + f}$

3. Isc = Isca x M

4. Add Motor Contribution = Motor FLA x 4

* All results are given in symmetrical amperes

Single-Phase L-L Bolted Fault-Current* Table

One-Way Distance in Feet

Available Isc in Thousands (Isca)	5	10	15	20	25	30	40	50	60	70	80	90	100	125	150	175	200	225	250	300	350	400	500	600
3	2,988	2,976	2,965	2,953	2,942	2,930	2,908	2,886	2,864	2,842	2,821	2,800	2,779	2,729	2,681	2,634	2,589	2,546	2,503	2,423	2,348	2,277	2,148	2,032
5	4,967	4,935	4,903	4,871	4,840	4,809	4,749	4,690	4,632	4,576	4,522	4,468	4,416	4,291	4,172	4,060	3,954	3,853	3,758	3,580	3,418	3,270	3,010	2,788
7	6,936	6,873	6,811	6,750	6,690	6,632	6,517	6,407	6,300	6,197	6,097	6,000	5,906	5,684	5,478	5,287	5,108	4,941	4,785	4,500	4,247	4,022	3,635	3,316
10	9,869	9,742	9,618	9,498	9,380	9,265	9,043	8,832	8,630	8,438	8,254	8,077	7,908	7,515	7,159	6,836	6,540	6,269	6,019	5,576	5,193	4,859	4,306	3,865
15	14,708	14,428	14,157	13,897	13,646	13,404	12,945	12,517	12,116	11,740	11,386	11,053	10,739	10,027	9,404	8,853	8,363	7,925	7,530	6,848	6,280	5,798	5,027	4,437
20	19,485	18,995	18,530	18,086	17,664	17,261	16,507	15,816	15,181	14,595	14,053	13,549	13,080	12,039	11,151	10,385	9,718	9,131	8,611	7,731	7,014	6,418	5,487	4,791
25	24,200	23,449	22,744	22,080	21,453	20,861	19,771	18,788	17,899	17,089	16,350	15,673	15,049	13,687	12,551	11,589	10,764	10,049	9,423	8,379	7,543	6,859	5,805	5,033
30	28,855	27,794	26,809	25,891	25,034	24,232	22,772	21,478	20,324	19,287	18,351	17,501	16,727	15,061	13,697	12,559	11,596	10,770	10,054	8,874	7,942	7,187	6,039	5,207
35	33,452	32,034	30,732	29,532	28,422	27,392	25,542	23,925	22,501	21,237	20,108	19,092	18,174	16,225	14,653	13,358	12,274	11,353	10,560	9,266	8,254	7,442	6,218	5,340
40	37,990	36,173	34,521	33,014	31,633	30,363	28,105	26,161	24,468	22,980	21,663	20,489	19,436	17,223	15,462	14,027	12,837	11,832	10,974	9,583	8,505	7,645	6,359	5,444
45	42,472	40,213	38,183	36,347	34,680	33,159	30,485	28,211	26,252	24,547	23,051	21,726	20,545	18,088	16,156	14,596	13,311	12,234	11,319	9,845	8,711	7,811	6,473	5,527
50	46,899	44,160	41,723	39,541	37,576	35,797	32,701	30,097	27,878	25,964	24,295	22,828	21,528	18,845	16,757	15,085	13,717	12,576	11,611	10,065	8,883	7,949	6,568	5,596
55	51,271	48,015	45,148	42,604	40,332	38,289	34,768	31,840	29,367	27,250	25,418	23,817	22,405	19,514	17,284	15,511	14,068	12,871	11,861	10,253	9,029	8,065	6,647	5,653
60	55,589	51,782	48,463	45,544	42,957	40,647	36,701	33,454	30,734	28,423	26,436	24,708	23,192	20,109	17,748	15,884	14,374	13,127	12,078	10,415	9,154	8,165	6,715	5,702
65	59,855	55,464	51,674	48,368	45,460	42,882	38,513	34,953	31,995	29,498	27,363	25,517	23,903	20,641	18,162	16,214	14,644	13,351	12,268	10,556	9,262	8,252	6,773	5,744
70	64,069	59,064	54,785	51,083	47,851	45,003	40,216	36,349	33,161	30,487	28,212	26,253	24,548	21,120	18,532	16,509	14,884	13,550	12,436	10,679	9,358	8,327	6,824	5,781
75	68,232	62,584	57,800	53,696	50,135	47,018	41,817	37,652	34,242	31,398	28,991	26,926	25,136	21,553	18,865	16,772	15,098	13,727	12,585	10,789	9,442	8,394	6,869	5,813
80	72,346	66,028	60,725	56,211	52,321	48,935	43,327	38,872	35,248	32,242	29,708	27,544	25,673	21,947	19,166	17,010	15,290	13,886	12,718	10,887	9,517	8,453	6,908	5,841
85	76,410	69,397	63,563	58,634	54,415	50,762	44,753	40,016	36,186	33,025	30,372	28,113	26,167	22,307	19,440	17,225	15,464	14,029	12,838	10,975	9,584	8,506	6,944	5,866
90	80,427	72,695	66,319	60,971	56,421	52,504	46,101	41,091	37,062	33,753	30,987	28,640	26,623	22,637	19,690	17,422	15,622	14,159	12,947	11,054	9,644	8,553	6,975	5,889
95	84,396	75,922	68,994	63,225	58,346	54,167	47,378	42,102	37,883	34,433	31,559	29,127	27,044	22,941	19,919	17,601	15,766	14,277	13,046	11,126	9,699	8,596	7,004	5,909
100	88,319	79,082	71,594	65,402	60,195	55,756	48,590	43,056	38,654	35,069	32,092	29,581	27,434	23,222	20,130	17,766	15,898	14,386	13,136	11,192	9,749	8,635	7,030	5,928
105	92,197	82,177	74,121	67,504	61,971	57,277	49,741	43,958	39,379	35,664	32,590	30,003	27,797	23,481	20,325	17,917	16,019	14,485	13,219	11,252	9,794	8,671	7,053	5,944
110	96,030	85,208	76,578	69,536	63,680	58,733	50,836	44,810	40,062	36,223	33,056	30,398	28,136	23,722	20,506	18,057	16,131	14,576	13,295	11,307	9,836	8,704	7,075	5,960
115	99,818	88,178	78,969	71,501	65,324	60,129	51,878	45,618	40,706	36,750	33,494	30,768	28,452	23,947	20,673	18,187	16,235	14,661	13,365	11,357	9,874	8,734	7,095	5,974
120	103,564	91,088	81,295	73,403	66,907	61,468	52,872	46,385	41,316	37,246	33,905	31,115	28,749	24,156	20,829	18,308	16,331	14,739	13,430	11,404	9,910	8,761	7,113	5,987
125	107,267	93,940	83,559	75,244	68,434	62,754	53,820	47,113	41,893	37,714	34,293	31,441	29,027	24,353	20,975	18,420	16,420	14,812	13,490	11,448	9,943	8,787	7,130	5,999
130	110,928	96,736	85,764	77,027	69,906	63,990	54,727	47,806	42,440	38,157	34,659	31,748	29,289	24,536	21,111	18,525	16,503	14,880	13,547	11,488	9,973	8,811	7,146	6,010
135	114,548	99,478	87,912	78,755	71,326	65,178	55,593	48,467	42,959	38,576	35,004	32,038	29,535	24,709	21,239	18,623	16,581	14,943	13,599	11,526	10,001	8,833	7,160	6,020
140	118,128	102,167	90,005	80,431	72,698	66,321	56,423	49,096	43,453	38,974	35,331	32,312	29,768	24,872	21,359	18,715	16,654	15,002	13,648	11,561	10,028	8,854	7,174	6,030
145	121,668	104,804	92,046	82,057	74,023	67,423	57,219	49,697	43,923	39,351	35,642	32,571	29,987	25,025	21,472	18,802	16,723	15,058	13,694	11,594	10,053	8,873	7,186	6,039
150	125,169	107,391	94,035	83,634	75,305	68,484	57,981	50,271	44,371	39,711	35,936	32,817	30,196	25,170	21,578	18,884	16,787	15,110	13,737	11,625	10,076	8,891	7,198	6,047
155	128,631	109,930	95,976	85,166	76,544	69,508	58,713	50,821	44,799	40,053	36,216	33,050	30,393	25,307	21,679	18,961	16,848	15,159	13,778	11,654	10,098	8,908	7,210	6,055
160	132,056	112,422	97,870	86,654	77,744	70,496	59,417	51,347	45,207	40,379	36,482	33,272	30,580	25,437	21,774	19,034	16,906	15,206	13,817	11,682	10,119	8,924	7,220	6,062
165	135,444	114,867	99,719	88,100	78,906	71,450	60,093	51,851	45,597	40,690	36,736	33,483	30,758	25,560	21,864	19,102	16,960	15,250	13,853	11,708	10,138	8,939	7,230	6,069
170	138,795	117,269	101,523	89,505	80,032	72,372	60,744	52,335	45,971	40,987	36,978	33,684	30,928	25,677	21,950	19,168	17,011	15,291	13,887	11,732	10,156	8,954	7,239	6,076
175	142,110	119,626	103,285	90,872	81,123	73,263	61,370	52,799	46,329	41,271	37,209	33,875	31,090	25,788	22,031	19,230	17,060	15,331	13,920	11,755	10,174	8,967	7,248	6,082
180	145,389	121,942	105,007	92,202	82,181	74,125	61,974	53,245	46,672	41,544	37,431	34,058	31,244	25,894	22,108	19,288	17,107	15,368	13,950	11,777	10,190	8,980	7,256	6,088
185	148,634	124,216	106,689	93,497	83,208	74,959	62,556	53,675	47,002	41,804	37,642	34,234	31,391	25,995	22,182	19,345	17,151	15,404	13,980	11,798	10,206	8,992	7,264	6,093
190	151,844	126,450	108,333	94,757	84,204	75,767	63,117	54,088	47,318	42,054	37,845	34,401	31,532	26,092	22,252	19,398	17,193	15,437	14,008	11,818	10,221	9,004	7,272	6,099
195	155,021	128,646	109,941	95,984	85,172	76,550	63,660	54,485	47,622	42,294	38,039	34,562	31,667	26,184	22,319	19,449	17,233	15,470	14,034	11,837	10,235	9,015	7,279	6,104
200	158,164	130,803	111,512	97,180	86,113	77,308	64,184	54,869	47,915	42,525	38,225	34,715	31,796	26,272	22,383	19,497	17,271	15,500	14,059	11,855	10,248	9,025	7,286	6,109

300 MCM | 2 Run(s) | 208 Volts

C-Value = 18,176 | Magnetic Duct | Copper

Calculations:

1. $f = \frac{\text{Length} \times 2 \times \text{Isca}}{\text{Runs} \times \text{C-Value} \times E_{L-L}}$

2. $M = \frac{1}{1 + f}$

3. $\text{Isc} = \text{Isca} \times M$

4. Add Motor Contribution = Motor FLA x 4

* All results are given in symmetrical amperes

Single-Phase L-L Bolted Fault-Current* Table

One-Way Distance in Feet

Available Isc in Thousands (Isca)

Isca	5	10	15	20	25	30	40	50	60	70	80	90	100	125	150	175	200	225	250	300	350	400	500	600
3	2,978	2,957	2,936	2,915	2,894	2,874	2,834	2,795	2,758	2,721	2,685	2,651	2,617	2,536	2,460	2,388	2,321	2,257	2,196	2,084	1,984	1,892	1,732	1,597
5	4,940	4,881	4,823	4,767	4,713	4,659	4,555	4,456	4,361	4,271	4,183	4,100	4,019	3,831	3,660	3,504	3,360	3,228	3,106	2,887	2,697	2,530	2,252	2,029
7	6,882	6,769	6,659	6,552	6,449	6,349	6,158	5,979	5,809	5,649	5,498	5,354	5,218	4,905	4,628	4,381	4,159	3,958	3,776	3,457	3,188	2,958	2,585	2,295
10	9,762	9,535	9,318	9,111	8,913	8,723	8,367	8,039	7,735	7,454	7,192	6,948	6,720	6,211	5,774	5,394	5,061	4,766	4,504	4,053	3,693	3,387	2,907	2,546
15	14,470	13,977	13,516	13,084	12,680	12,299	11,503	10,981	10,422	9,918	9,460	9,043	8,660	7,833	7,150	6,576	6,088	5,667	5,300	4,653	4,211	3,819	3,219	2,782
20	19,069	18,222	17,446	16,734	16,077	15,470	14,384	13,441	12,613	11,882	11,231	10,647	10,121	9,009	8,117	7,385	6,775	6,258	5,814	5,052	4,529	4,078	3,401	2,917
25	23,563	22,282	21,133	20,096	19,157	18,301	16,801	15,528	14,434	13,484	12,652	11,916	11,261	9,901	8,834	7,974	7,267	6,675	6,173	5,365	4,744	4,252	3,521	3,005
30	27,954	26,169	24,598	23,205	21,962	20,845	18,920	17,321	15,971	14,816	13,817	12,944	12,175	10,600	9,387	8,422	7,637	6,986	6,438	5,564	4,899	4,376	3,606	3,066
35	32,246	29,894	27,862	26,088	24,527	23,142	20,793	18,878	17,285	15,941	14,790	13,794	12,924	11,164	9,826	8,774	7,926	7,227	6,641	5,715	5,016	4,469	3,669	3,112
40	36,443	33,467	30,940	28,768	26,881	25,227	22,461	20,243	18,423	16,903	15,615	14,509	13,550	11,628	10,183	9,058	8,156	7,418	6,803	5,834	5,107	4,541	3,717	3,147
45	40,548	36,897	33,850	31,267	29,051	27,128	23,956	21,449	19,416	17,736	16,323	15,119	14,080	12,016	10,479	9,292	8,345	7,574	6,934	5,930	5,181	4,599	3,756	3,174
50	44,563	40,193	36,603	33,602	31,056	28,868	25,303	22,522	20,292	18,463	16,937	15,644	14,535	12,345	10,729	9,487	8,503	7,704	7,042	6,049	5,241	4,647	3,788	3,197
55	48,492	43,361	39,213	35,788	32,914	30,467	26,523	23,484	21,069	19,105	17,475	16,102	14,929	12,629	10,943	9,654	8,637	7,813	7,133	6,076	5,291	4,686	3,814	3,216
60	52,338	46,411	41,689	37,840	34,642	31,942	27,634	24,350	21,764	19,674	17,951	16,505	15,275	12,875	11,127	9,797	8,751	7,907	7,211	6,132	5,334	4,720	3,836	3,231
65	56,102	49,347	44,043	39,769	36,252	33,305	28,649	25,135	22,389	20,183	18,374	16,862	15,580	13,091	11,288	9,922	8,851	7,988	7,279	6,181	5,371	4,749	3,855	3,245
70	59,788	52,176	46,284	41,587	37,756	34,571	29,580	25,849	22,953	20,641	18,752	17,180	15,851	13,282	11,430	10,031	8,937	8,059	7,337	6,223	5,403	4,773	3,872	3,256
75	63,398	54,904	48,418	43,302	39,164	35,748	30,438	26,501	23,466	21,055	19,093	17,466	16,094	13,453	11,556	10,128	9,014	8,121	7,389	6,260	5,431	4,795	3,886	3,266
80	66,934	57,537	50,453	44,923	40,485	36,845	31,230	27,100	23,934	21,431	19,402	17,724	16,313	13,605	11,668	10,214	9,082	8,176	7,435	6,293	5,455	4,814	3,899	3,275
85	70,399	60,079	52,397	46,458	41,727	37,871	31,964	27,651	24,363	21,774	19,683	17,958	16,511	13,743	11,769	10,292	9,143	8,226	7,476	6,322	5,477	4,832	3,910	3,283
90	73,794	62,534	54,255	47,912	42,897	38,833	32,646	28,160	24,757	22,089	19,939	18,171	16,691	13,867	11,860	10,361	9,198	8,270	7,512	6,349	5,497	4,847	3,920	3,290
95	77,122	64,908	56,033	49,294	44,001	39,735	33,281	28,631	25,121	22,378	20,174	18,366	16,855	13,980	11,943	10,424	9,248	8,310	7,545	6,372	5,515	4,861	3,929	3,297
100	80,385	67,204	57,736	50,606	45,044	40,584	33,875	29,069	25,458	22,644	20,391	18,545	17,006	14,084	12,019	10,482	9,293	8,347	7,576	6,394	5,531	4,873	3,937	3,302
105	83,585	69,425	59,368	51,856	46,032	41,383	34,430	29,477	25,770	22,891	20,591	18,711	17,145	14,179	12,088	10,534	9,335	8,380	7,603	6,413	5,545	4,884	3,944	3,308
110	86,723	71,576	60,934	53,047	46,968	42,138	34,951	29,858	26,061	23,120	20,776	18,864	17,273	14,267	12,152	10,583	9,373	8,411	7,628	6,431	5,559	4,895	3,951	3,312
115	89,801	73,660	62,438	54,183	47,856	42,852	35,441	30,215	26,332	23,334	20,948	19,005	17,392	14,348	12,210	10,627	9,407	8,439	7,651	6,447	5,571	4,904	3,957	3,317
120	92,821	75,680	63,883	55,268	48,700	43,528	35,902	30,549	26,586	23,532	21,108	19,137	17,502	14,423	12,265	10,668	9,440	8,465	7,672	6,463	5,582	4,913	3,963	3,321
125	95,785	77,639	65,273	56,305	49,504	44,169	36,336	30,863	26,823	23,719	21,258	19,260	17,605	14,492	12,315	10,706	9,469	8,489	7,692	6,476	5,593	4,921	3,968	3,324
130	98,693	79,539	66,611	57,298	50,270	44,777	36,747	31,159	27,047	23,893	21,398	19,375	17,701	14,557	12,362	10,742	9,497	8,511	7,710	6,489	5,602	4,929	3,973	3,328
135	101,549	81,383	67,899	58,249	51,000	45,356	37,136	31,438	27,257	24,057	21,529	19,482	17,791	14,618	12,405	10,775	9,523	8,532	7,727	6,501	5,611	4,935	3,977	3,331
140	104,352	83,174	69,141	59,160	51,698	45,907	37,504	31,702	27,455	24,211	21,652	19,583	17,875	14,675	12,446	10,805	9,547	8,551	7,743	6,513	5,620	4,942	3,982	3,334
145	107,105	84,913	70,339	60,035	52,364	46,432	37,854	31,952	27,641	24,356	21,769	19,678	17,954	14,728	12,485	10,834	9,569	8,569	7,758	6,523	5,627	4,948	3,986	3,337
150	109,809	86,604	71,495	60,875	53,002	46,933	38,186	32,188	27,818	24,493	21,878	19,767	18,028	14,778	12,520	10,861	9,590	8,586	7,772	6,533	5,635	4,954	3,989	3,339
155	112,465	88,247	72,612	61,683	53,613	47,411	38,503	32,412	27,986	24,623	21,981	19,852	18,098	14,825	12,554	10,887	9,610	8,602	7,785	6,542	5,642	4,959	3,993	3,342
160	115,074	89,846	73,691	62,460	54,199	47,869	38,804	32,626	28,144	24,746	22,079	19,932	18,165	14,869	12,586	10,911	9,629	8,617	7,797	6,551	5,648	4,964	3,996	3,344
165	117,638	91,401	74,734	63,207	54,761	48,307	39,091	32,828	28,295	24,862	22,172	20,007	18,227	14,911	12,616	10,933	9,647	8,631	7,809	6,559	5,654	4,969	3,999	3,346
170	120,157	92,915	75,743	63,928	55,301	48,726	39,365	33,022	28,439	24,973	22,260	20,079	18,287	14,951	12,645	10,955	9,663	8,644	7,819	6,567	5,660	4,973	4,002	3,348
175	122,634	94,389	76,719	64,622	55,820	49,129	39,628	33,206	28,575	25,078	22,344	20,147	18,343	14,989	12,672	10,975	9,679	8,657	7,830	6,574	5,665	4,977	4,004	3,350
180	125,068	95,825	77,665	65,292	56,319	49,515	39,878	33,382	28,705	25,178	22,423	20,211	18,397	15,025	12,697	10,994	9,694	8,669	7,839	6,581	5,670	4,981	4,007	3,352
185	127,462	97,224	78,581	65,938	56,799	49,886	40,119	33,550	28,830	25,274	22,499	20,273	18,448	15,059	12,721	11,012	9,708	8,680	7,849	6,587	5,675	4,985	4,009	3,353
190	129,816	98,587	79,470	66,562	57,262	50,242	40,349	33,711	28,948	25,365	22,571	20,332	18,496	15,091	12,744	11,029	9,721	8,691	7,857	6,593	5,680	4,988	4,012	3,355
195	132,130	99,916	80,331	67,166	57,708	50,585	40,570	33,865	29,062	25,452	22,640	20,387	18,543	15,122	12,766	11,046	9,734	8,701	7,866	6,599	5,684	4,992	4,014	3,356
200	134,407	101,213	81,167	67,749	58,138	50,915	40,782	34,013	29,171	25,536	22,706	20,441	18,587	15,151	12,787	11,062	9,746	8,711	7,874	6,605	5,688	4,995	4,016	3,358

350 MCM	1 Run(s)	208 Volts	C-Value = 19,703	Magnetic Duct	Copper

Calculations:

1. $f = \dfrac{\text{Length} \times 2 \times \text{Isca}}{\text{Runs} \times \text{C-Value} \times \text{E L-L}}$

2. $M = \dfrac{1}{1 + f}$

3. $\text{Isc} = \text{Isca} \times M$

4. Add Motor Contribution = Motor FLA x 4

* All results are given in symmetrical amperes

Single-Phase L-L Bolted Fault-Current* Table

One-Way Distance in Feet

Available Isc in Thousands (Isca)	5	10	15	20	25	30	40	50	60	70	80	90	100	125	150	175	200	225	250	300	350	400	500	600
3	2,989	2,978	2,967	2,957	2,946	2,936	2,915	2,894	2,874	2,854	2,834	2,815	2,795	2,749	2,703	2,659	2,617	2,576	2,536	2,460	2,388	2,321	2,196	2,084
5	4,970	4,940	4,910	4,881	4,852	4,823	4,767	4,713	4,659	4,607	4,555	4,505	4,456	4,338	4,227	4,120	4,019	3,923	3,831	3,660	3,504	3,360	3,106	2,887
7	6,941	6,882	6,825	6,769	6,713	6,659	6,552	6,449	6,349	6,252	6,158	6,067	5,979	5,768	5,572	5,389	5,218	5,057	4,905	4,628	4,381	4,159	3,776	3,457
10	9,879	9,762	9,647	9,535	9,425	9,318	9,111	8,913	8,723	8,541	8,367	8,199	8,039	7,663	7,321	7,008	6,720	6,456	6,211	5,774	5,394	5,061	4,504	4,058
15	14,730	14,470	14,219	13,977	13,743	13,516	13,084	12,680	12,299	11,941	11,603	11,283	10,981	10,291	9,684	9,143	8,660	8,226	7,833	7,150	6,576	6,088	5,300	4,693
20	19,524	19,069	18,636	18,222	17,825	17,446	16,734	16,077	15,470	14,907	14,384	13,896	13,441	12,422	11,547	10,787	10,121	9,533	9,009	8,117	7,385	6,775	5,814	5,092
25	24,260	23,563	22,904	22,282	21,692	21,133	20,096	19,157	18,301	17,519	16,801	16,139	15,528	14,184	13,055	12,092	11,261	10,537	9,901	8,834	7,974	7,267	6,173	5,365
30	28,941	27,954	27,032	26,169	25,359	24,598	23,205	21,962	20,845	19,836	18,920	18,085	17,321	15,666	14,299	13,152	12,175	11,333	10,600	9,387	8,422	7,637	6,438	5,564
35	33,567	32,246	31,026	29,894	28,842	27,862	26,088	24,527	23,142	21,905	20,793	19,789	18,878	16,928	15,344	14,031	12,924	11,980	11,164	9,826	8,774	7,926	6,641	5,715
40	38,139	36,443	34,892	33,467	32,154	30,940	28,768	26,881	25,227	23,764	22,461	21,294	20,243	18,018	16,233	14,771	13,550	12,515	11,628	10,183	9,058	8,156	6,803	5,834
45	42,658	40,548	38,636	36,897	35,308	33,850	31,267	29,051	27,128	25,443	23,956	22,633	21,449	18,967	17,000	15,403	14,080	12,966	12,016	10,479	9,292	8,345	6,934	5,930
50	47,125	44,563	42,265	40,193	38,314	36,603	33,602	31,056	28,868	26,968	25,303	23,832	22,522	19,802	17,667	15,949	14,535	13,351	12,345	10,729	9,487	8,503	7,042	6,009
55	51,541	48,492	45,783	43,361	41,183	39,213	35,788	32,914	30,467	28,359	26,523	24,911	23,484	20,541	18,254	16,425	14,929	13,683	12,629	10,943	9,654	8,637	7,133	6,076
60	55,907	52,338	49,196	46,411	43,923	41,689	37,840	34,642	31,942	29,632	27,634	25,888	24,350	21,201	18,773	16,844	15,275	13,973	12,875	11,127	9,797	8,751	7,211	6,132
65	60,224	56,102	52,508	49,347	46,544	44,043	39,769	36,252	33,305	30,802	28,649	26,777	25,135	21,793	19,236	17,216	15,580	14,227	13,091	11,288	9,922	8,851	7,279	6,181
70	64,492	59,788	55,723	52,176	49,053	46,284	41,587	37,756	34,571	31,881	29,580	27,589	25,849	22,328	19,651	17,548	15,851	14,453	13,282	11,430	10,031	8,937	7,337	6,223
75	68,713	63,398	58,846	54,904	51,457	48,418	43,302	39,164	35,748	32,880	30,438	28,333	26,501	22,813	20,026	17,846	16,094	14,655	13,453	11,556	10,128	9,014	7,389	6,260
80	72,886	66,934	61,881	57,537	53,763	50,453	44,923	40,485	36,845	33,806	31,230	29,019	27,100	23,255	20,366	18,115	16,313	14,836	13,605	11,668	10,214	9,082	7,435	6,293
85	77,013	70,399	64,831	60,079	55,976	52,397	46,458	41,727	37,871	34,668	31,964	29,651	27,651	23,660	20,676	18,360	16,511	15,000	13,743	11,769	10,292	9,143	7,476	6,322
90	81,095	73,794	67,699	62,534	58,101	54,255	47,912	42,897	38,833	35,471	32,646	30,237	28,160	24,031	20,959	18,583	16,691	15,149	13,867	11,860	10,361	9,198	7,512	6,349
95	85,133	77,122	70,490	64,908	60,145	56,033	49,294	44,001	39,735	36,223	33,281	30,781	28,631	24,374	21,219	18,787	16,855	15,284	13,980	11,943	10,424	9,248	7,545	6,372
100	89,126	80,385	73,206	67,204	62,111	57,736	50,606	45,044	40,584	36,927	33,875	31,288	29,069	24,691	21,459	18,975	17,006	15,408	14,084	12,019	10,482	9,293	7,576	6,394
105	93,077	83,585	75,850	69,425	64,004	59,368	51,856	46,032	41,383	37,588	34,430	31,762	29,477	24,984	21,680	19,148	17,145	15,522	14,179	12,088	10,534	9,335	7,603	6,413
110	96,984	86,723	78,425	71,576	65,828	60,934	53,047	46,968	42,138	38,210	34,951	32,204	29,858	25,258	21,886	19,308	17,273	15,627	14,267	12,152	10,583	9,373	7,628	6,431
115	100,850	89,801	80,934	73,660	67,586	62,438	54,183	47,856	42,852	38,795	35,441	32,620	30,215	25,512	22,077	19,456	17,392	15,724	14,348	12,210	10,627	9,407	7,651	6,447
120	104,675	92,821	83,379	75,680	69,283	63,883	55,268	48,700	43,528	39,349	35,902	33,010	30,549	25,750	22,255	19,594	17,502	15,814	14,423	12,265	10,668	9,440	7,672	6,463
125	108,459	95,785	85,762	77,639	70,921	65,273	56,305	49,504	44,169	39,872	36,336	33,377	30,863	25,973	22,421	19,723	17,605	15,898	14,492	12,315	10,706	9,469	7,692	6,476
130	112,204	98,693	88,087	79,539	72,503	66,611	57,298	50,270	44,777	40,367	36,747	33,723	31,159	26,183	22,577	19,844	17,701	15,976	14,557	12,362	10,742	9,497	7,710	6,489
135	115,909	101,549	90,354	81,383	74,032	67,899	58,249	51,000	45,356	40,836	37,136	34,051	31,438	26,379	22,723	19,957	17,791	16,049	14,618	12,405	10,775	9,523	7,727	6,501
140	119,576	104,352	92,567	83,174	75,511	69,141	59,160	51,698	45,907	41,282	37,504	34,360	31,702	26,565	22,860	20,062	17,875	16,117	14,675	12,446	10,805	9,547	7,743	6,513
145	123,204	107,105	94,727	84,913	76,942	70,339	60,035	52,364	46,432	41,706	37,854	34,653	31,952	26,740	22,990	20,162	17,954	16,182	14,728	12,485	10,834	9,569	7,758	6,523
150	126,796	109,809	96,836	86,604	78,328	71,495	60,875	53,002	46,933	42,110	38,186	34,932	32,188	26,905	23,112	20,256	18,028	16,242	14,778	12,520	10,861	9,590	7,772	6,533
155	130,350	112,465	98,895	88,247	79,670	72,612	61,683	53,613	47,411	42,495	38,503	35,196	32,412	27,062	23,227	20,345	18,098	16,299	14,825	12,554	10,887	9,610	7,785	6,542
160	133,868	115,074	100,907	89,846	80,970	73,691	62,460	54,199	47,869	42,862	38,804	35,448	32,626	27,210	23,337	20,428	18,165	16,353	14,869	12,586	10,911	9,629	7,797	6,551
165	137,350	117,638	102,873	91,401	82,231	74,734	63,207	54,761	48,307	43,213	39,091	35,687	32,828	27,351	23,440	20,508	18,227	16,404	14,911	12,616	10,933	9,647	7,809	6,559
170	140,798	120,157	104,795	92,915	83,455	75,743	63,928	55,301	48,726	43,548	39,365	35,916	33,022	27,485	23,539	20,583	18,237	16,452	14,951	12,645	10,955	9,663	7,819	6,567
175	144,210	122,634	106,673	94,389	84,642	76,719	64,622	55,820	49,129	43,870	39,628	36,134	33,206	27,613	23,632	20,654	18,343	16,497	14,989	12,672	10,975	9,679	7,830	6,574
180	147,588	125,068	108,511	95,825	85,794	77,665	65,292	56,319	49,515	44,177	39,878	36,342	33,382	27,734	23,721	20,722	18,397	16,541	15,025	12,697	10,994	9,694	7,839	6,581
185	150,933	127,462	110,308	97,224	86,914	78,581	65,938	56,799	49,886	44,472	40,119	36,542	33,550	27,850	23,806	20,787	18,448	16,582	15,059	12,721	11,012	9,708	7,849	6,587
190	154,245	129,816	112,066	98,587	88,002	79,470	66,562	57,262	50,242	44,755	40,349	36,732	33,711	27,961	23,887	20,849	18,496	16,621	15,091	12,744	11,029	9,721	7,857	6,593
195	157,524	132,130	113,787	99,916	89,060	80,331	67,166	57,708	50,585	45,027	40,570	36,915	33,865	28,067	23,964	20,908	18,543	16,658	15,122	12,766	11,046	9,734	7,866	6,599
200	160,771	134,407	115,472	101,213	90,088	81,167	67,749	58,138	50,915	45,289	40,782	37,091	34,013	28,168	24,038	20,964	18,587	16,694	15,151	12,787	11,062	9,746	7,874	6,605

350 MCM | 2 Run(s) | 208 Volts

C-Value = 19,703 | Magnetic Duct | Copper

Calculations:

1. $f = \frac{\text{Length} \times 2 \times \text{Isca}}{\text{Runs} \times \text{C-Value} \times E_{L-L}}$

2. $M = \frac{1}{1 + f}$

3. Isc = Isca x M

4. Add Motor Contribution = Motor FLA x 4

* All results are given in symmetrical amperes

Single-Phase L-L Bolted Fault-Current* Table

One-Way Distance in Feet

Available Isc in Thousands (Isca)	5	10	15	20	25	30	40	50	60	70	80	90	100	125	150	175	200	225	250	300	350	400	500	600
3	2,979	2,959	2,938	2,918	2,898	2,879	2,84[illegible]	2,803	2,767	2,732	2,697	2,664	2,631	2,552	2,479	2,409	2 343	2,280	2,221	2,111	2,012	1,922	1,763	1,629
5	4,942	4,886	4,831	4,777	4,724	4,672	4,572	4,477	4,385	4,297	4,212	4,131	4,053	3,869	3,702	3,548	3,407	3,277	3,156	2,939	2,750	2,584	2,305	2,081
7	6,887	6,778	6,672	6,570	6,471	6,374	6,190	6,016	5,851	5,695	5,547	5,407	5,274	4,968	4,695	4,451	4,231	4,031	3,850	3,532	3,263	3,031	2,655	2,362
10	9,772	9,553	9,345	9,145	8,953	8,770	8,424	8,105	7,809	7,534	7,278	7,038	6,814	6,311	5,878	5,500	5,168	4,873	4,611	4,162	3,793	3,484	2,996	2,628
15	14,492	14,017	13,572	13,155	12,762	12,393	11,7[illegible]4	11,106	10,557	10,061	9,609	9,196	8,817	7,993	7,310	6,734	6,243	5,818	5,448	4,832	4,342	3,942	3,328	2,880
20	19,107	18,290	17,540	16,849	16,210	15,618	14,556	13,628	12,812	12,088	11,441	10,860	10,335	9,221	8,324	7,586	6,968	6,443	5,992	5,256	4,681	4,219	3,524	3,025
25	23,620	22,384	21,271	20,263	19,346	18,509	17,035	15,778	14,694	13,750	12,919	12,183	11,527	10,158	9,080	8,209	7,490	6,887	6,374	5,547	4,910	4,405	3,653	3,120
30	28,034	26,310	24,785	23,428	22,211	21,115	19,218	17,633	16,290	15,137	14,137	13,260	12,486	10,896	9,665	8,684	7,884	7,218	6,657	5,760	5,077	4,538	3,744	3,186
35	32,353	30,078	28,102	26,369	24,838	23,475	21,153	19,249	17,660	16,313	15,157	14,154	13,275	11,492	10,131	9,058	8,191	7,475	6,875	5,923	5,202	4,638	3,812	3,235
40	36,579	33,698	31,237	29,111	27,256	25,623	22,832	20,671	18,849	17,322	16,024	14,907	13,936	11,984	10,511	9,361	8,438	7,680	7,048	6,051	5,301	4,716	3,864	3,273
45	40,717	37,178	34,205	31,672	29,489	27,587	24,435	21,930	19,890	18,198	16,771	15,551	14,497	12,397	10,828	9,611	8,640	7,848	7,188	6,154	5,380	4,779	3,906	3,303
50	44,767	40,526	37,019	34,070	31,557	29,389	25,838	23,053	20,810	18,965	17,420	16,108	14,980	12,748	11,095	9,821	8,810	7,987	7,305	6,240	5,445	4,830	3,940	3,327
55	48,734	43,749	39,690	36,320	33,477	31,048	27,112	24,062	21,628	19,642	17,990	16,594	15,399	13,050	11,323	10,000	8,953	8,105	7,403	6,31[illegible]	5,500	4,873	3,969	3,348
60	52,619	46,855	42,230	38,435	35,266	32,580	28,273	24,972	22,361	20,245	18,494	17,022	15,767	13,313	11,521	10,153	9,076	8,206	7,487	6,37[illegible]	5,546	4,909	3,993	3,365
65	56,426	49,850	44,647	40,427	36,936	34,000	29,337	25,798	23,021	20,784	18,943	17,402	16,093	13,545	11,693	10,287	9,183	8,293	7,560	6,42[illegible]	5,586	4,940	4,013	3,379
70	60,156	52,739	46,950	42,307	38,499	35,320	30,314	26,551	23,619	21,270	19,346	17,741	16,382	13,749	11,846	10,405	9,277	8,369	7,623	6,47[illegible]	5,620	4,967	4,031	3,392
75	63,812	55,528	49,148	44,083	39,964	36,550	31,215	27,240	24,162	21,710	19,709	18,046	16,642	13,932	11,981	10,509	9,359	8,436	7,679	6,51[illegible]	5,650	4,991	4,047	3,403
80	67,395	58,222	51,247	45,764	41,341	37,698	32,049	27,872	24,659	22,110	20,038	18,322	16,876	14,095	12,102	10,602	9,433	8,496	7,729	6,546	5,677	5,012	4,060	3,413
85	70,909	60,826	53,253	47,358	42,637	38,772	32,322	28,455	25,114	22,475	20,338	18,572	17,088	14,243	12,210	10,685	9,499	8,550	7,773	6,573	5,701	5,030	4,073	3,421
90	74,355	63,344	55,174	48,870	43,859	39,781	33,542	28,995	25,533	22,810	20,612	18,800	17,281	14,377	12,308	10,760	9,558	8,598	7,812	6,60[illegible]	5,722	5,047	4,083	3,429
95	77,736	65,781	57,013	50,308	45,014	40,728	34,213	29,495	25,920	23,118	20,863	19,009	17,457	14,499	12 398	10,828	9,612	8,641	7,848	6,632	5,741	5,062	4,093	3,436
100	81,052	68,140	58,777	51,676	46,106	41,620	34,840	29,960	26,279	23,403	21,095	19,201	17,619	14,610	12,479	10,891	9,661	8,680	7,881	6,655	5,759	5,076	4,102	3,442
105	84,306	70,425	60,470	52,980	47,141	42,462	35,428	30,393	26,612	23,667	21,309	19,378	17,768	14,713	12,554	10,947	9,705	8,716	7,911	6,676	5,775	5,088	4,110	3,448
110	87,499	72,640	62,095	54,224	48,123	43,257	35,980	30,799	26,922	23,912	21,507	19,542	17,906	14,807	12,622	10,999	9,746	8,750	7,938	6,695	5,789	5,099	4,117	3,453
115	90,633	74,787	63,658	55,411	49,056	44,009	36,499	31,178	27,211	24,140	21,692	19,694	18,034	14,894	12,686	11,047	9,784	8,730	7,963	6,7[illegible]3	5,802	5,109	4,124	3,457
120	93,711	76,870	65,160	56,547	49,944	44,722	36,988	31,534	27,482	24,353	21,864	19,836	18,152	14,975	12,744	11,092	9,819	8,808	7,986	6,729	5,815	5,119	4,130	3,462
125	96,732	78,892	66,607	57,633	50,790	45,399	37,450	31,869	27,736	24,552	22,024	19,968	18,263	15,050	12,799	11,133	9,851	8,834	8,007	6,745	5,826	5,128	4,136	3,466
130	99,700	80,854	68,001	58,673	51,596	46,042	37,886	32,185	27,975	24,739	22,174	20,091	18,366	15,120	12,849	11,171	9,881	8,858	8,027	6,759	5,836	5,136	4,141	3,469
135	102,615	82,761	69,344	59,671	52,366	46,654	38,300	32,483	28,200	24,915	22,315	20,207	18,463	15,185	12,896	11,207	9,909	8,880	8,045	6,772	5,846	5,143	4,146	3,473
140	105,478	84,613	70,640	60,628	53,101	47,237	38,692	32,764	28,412	25,080	22,448	20,316	18,553	15,247	12,940	11,240	9,935	8,901	8,062	6,784	5,855	5,150	4,151	3,476
145	108,291	86,414	71,891	61,547	53,805	47,793	39,064	33,031	28,612	25,236	22,573	20,418	18,638	15,304	12,982	11,271	9,959	8,921	8,078	6,795	5,864	5,157	4,155	3,479
150	111,056	88,166	73,099	62,430	54,479	48,324	39,418	33,284	28,802	25,383	22,690	20,514	18,719	15,358	13,021	11,301	9,982	8,939	8,093	6,806	5,872	5,163	4,159	3,482
155	113,773	89,870	74,266	63,280	55,125	48,832	39,755	33,524	28,981	25,523	22,802	20,605	18,794	15,409	13,057	11,328	10,004	8,956	8,108	6,816	5,879	5,169	4,163	3,484
160	116,444	91,528	75,395	64,098	55,744	49,317	40,076	33,752	29,151	25,655	22,907	20,691	18,866	15,457	13,092	11,354	10,024	8,973	8,121	6,825	5,886	5,174	4,166	3,487
165	119,070	93,143	76,488	64,885	56,339	49,782	40,383	33,969	29,313	25,780	23,007	20,772	18,933	15,503	13,124	11,379	10,043	8 988	8,133	6,834	5,893	5,179	4,169	3,489
170	121,652	94,715	77,545	65,645	56,911	50,228	40,676	34,176	29,467	25,899	23,102	20,849	18,998	15,545	13,155	11,402	10,061	9,002	8,145	6,842	5,899	5,184	4,173	3,491
175	124,191	96,247	78,569	66,377	57,460	50,656	40,956	34,373	29,614	26,012	23,192	20,923	19,058	15,586	13,184	11,424	10,078	9,016	8,156	6,850	5,905	5,188	4,175	3,493
180	126,689	97,741	79,561	67,084	57,989	51,066	41,224	34,562	29,754	26,120	23,277	20,993	19,116	15,625	13,212	11,444	10,094	9,029	8,167	6,858	5,910	5,193	4,178	3,495
185	129,145	99,196	80,523	67,766	58,499	51,461	41,480	34,742	29,887	26,223	23,359	21,059	19,171	15,662	13,238	11,464	10,109	9,041	8,177	6,865	5,915	5,197	4,181	3,497
190	131,562	100,616	81,456	68,426	58,989	51,840	41,727	34,915	30,015	26,321	23,437	21,122	19,224	15,697	13,263	11,483	10,124	9 053	8,186	6,871	5,920	5,201	4,183	3,499
195	133,940	102,001	82,361	69,064	59,463	52,206	41,963	35,080	30,137	26,415	23,511	21,183	19,274	15,730	13,287	11,501	10,138	9,064	8,195	6,878	5,925	5,204	4,186	3,501
200	136,281	103,353	83,240	69,681	59,920	52,557	42,190	35,239	30,254	26,505	23,582	21,240	19,321	15,762	13,310	11,518	10,151	9,074	8,204	6,884	5,930	5,208	4,188	3,502

400 MCM | 1 Run(s) | 208 Volts

C-Value = 20,565 | Magnetic Duct | Copper

Calculations:

1. $f = \dfrac{\text{Length} \times 2 \times \text{Isca}}{\text{Runs} \times \text{C-Value} \times \text{E L-L}}$

2. $M = \dfrac{1}{1 + f}$

3. Isc = Isca x M

4. Add Motor Contribution = Motor FLA x 4

* All results are given in symmetrical amperes

Single-Phase L-L Bolted Fault-Current* Table

One-Way Distance in Feet

Available Isc in Thousands (Isca)	5	10	15	20	25	30	40	50	60	70	80	90	100	125	150	175	200	225	250	300	350	400	500	600
3	2,990	2,979	2,969	2,959	2,948	2,938	2,918	2,898	2,879	2,860	2,841	2,822	2,803	2,758	2,714	2,672	2,631	2,591	2,552	2,479	2,409	2,343	2,221	2,111
5	4,971	4,942	4,914	4,886	4,858	4,831	4,777	4,724	4,672	4,622	4,572	4,524	4,477	4,363	4,254	4,151	4,053	3,959	3,869	3,702	3,548	3,407	3,156	2,939
7	6,943	6,887	6,832	6,778	6,725	6,672	6,570	6,471	6,374	6,281	6,190	6,101	6,016	5,811	5,620	5,442	5,274	5,116	4,968	4,695	4,451	4,231	3,850	3,532
10	9,884	9,772	9,661	9,553	9,448	9,345	9,145	8,953	8,770	8,594	8,424	8,262	8,105	7,739	7,404	7,097	6,814	6,553	6,311	5,878	5,500	5,168	4,611	4,162
15	14,742	14,492	14,250	14,017	13,791	13,572	13,155	12,762	12,393	12,044	11,714	11,402	11,106	10,429	9,830	9,296	8,817	8,385	7,993	7,310	6,734	6,243	5,448	4,832
20	19,543	19,107	18,689	18,290	17,907	17,540	16,849	16,210	15,618	15,068	14,556	14,077	13,628	12,623	11,755	11,000	10,335	9,747	9,221	8,324	7,586	6,968	5,992	5,256
25	24,290	23,620	22,985	22,384	21,813	21,271	20,263	19,346	18,509	17,742	17,035	16,383	15,778	14,446	13,321	12,359	11,527	10,799	10,158	9,080	8,209	7,490	6,374	5,547
30	28,984	28,034	27,144	26,310	25,525	24,785	23,428	22,211	21,115	20,122	19,218	18,391	17,633	15,986	14,620	13,469	12,486	11,637	10,896	9,665	8,684	7,884	6,657	5,760
35	33,624	32,353	31,174	30,078	29,056	28,102	26,369	24,838	23,475	22,254	21,153	20,157	19,249	17,303	15,714	14,392	13,275	12,320	11,492	10,131	9,058	8,191	6,875	5,923
40	38,213	36,579	35,079	33,698	32,421	31,237	29,111	27,256	25,623	24,175	22,882	21,720	20,671	18,443	16,648	15,172	13,936	12,887	11,984	10,511	9,361	8,438	7,048	6,051
45	42,751	40,717	38,867	37,178	35,629	34,205	31,672	29,489	27,587	25,916	24,435	23,115	21,930	19,438	17,455	15,839	14,497	13,365	12,397	10,828	9,611	8,640	7,188	6,154
50	47,239	44,767	42,541	40,526	38,693	37,019	34,070	31,557	29,389	27,499	25,838	24,366	23,053	20,316	18,160	16,417	14,380	13,774	12,748	11,095	9,821	8,810	7,305	6,240
55	51,678	48,734	46,107	43,749	41,621	39,690	36,320	33,477	31,048	28,947	27,112	25,496	24,062	21,095	18,780	16,922	15,399	14,128	13,050	11,323	10,000	8,953	7,403	6,311
60	56,068	52,619	49,570	46,855	44,422	42,230	38,435	35,266	32,580	30,274	28,273	26,520	24,972	21,792	19,330	17,368	15,767	14,437	13,313	11,521	10,153	9,076	7,487	6,372
65	60,410	56,426	52,934	49,850	47,105	44,647	40,427	36,936	34,000	31,497	29,337	27,454	25,798	22,418	19,821	17,763	16,093	14,709	13,545	11,693	10,287	9,183	7,560	6,425
70	64,706	60,156	56,204	52,739	49,677	46,950	42,307	38,499	35,320	32,626	30,314	28,308	26,551	22,984	20,262	18,117	16,382	14,951	13,749	11,846	10,405	9,277	7,623	6,470
75	68,955	63,812	59,382	55,528	52,144	49,148	44,083	39,964	36,550	33,672	31,215	29,092	27,240	23,499	20,661	18,435	16,642	15,167	13,932	11,981	10,509	9,359	7,679	6,510
80	73,159	67,395	62,474	58,222	54,512	51,247	45,764	41,341	37,698	34,644	32,049	29,815	27,872	23,968	21,023	18,723	16,876	15,361	14,095	12,102	10,602	9,433	7,729	6,546
85	77,318	70,909	65,482	60,826	56,788	53,253	47,358	42,637	38,772	35,550	32,822	30,483	28,455	24,398	21,353	18,984	17,088	15,536	14,243	12,210	10,685	9,499	7,773	6,578
90	81,433	74,355	68,410	63,344	58,977	55,174	48,870	43,859	39,781	36,396	33,542	31,103	28,995	24,793	21,655	19,222	17,281	15,696	14,377	12,308	10,760	9,558	7,812	6,606
95	85,505	77,736	71,260	65,781	61,084	57,013	50,308	45,014	40,728	37,187	34,213	31,679	29,495	25,158	21,933	19,441	17,457	15,841	14,499	12,398	10,828	9,612	7,848	6,632
100	89,534	81,052	74,037	68,140	63,113	58,777	51,676	46,106	41,620	37,930	34,840	32,216	29,960	25,496	22,189	19,642	17,619	15,974	14,610	12,479	10,891	9,661	7,881	6,655
105	93,522	84,306	76,743	70,425	65,069	60,470	52,980	47,141	42,462	38,627	35,428	32,718	30,393	25,809	22,426	19,827	17,768	16,097	14,713	12,554	10,947	9,705	7,911	6,676
110	97,468	87,499	79,380	72,640	66,955	62,095	54,224	48,123	43,257	39,284	35,980	33,188	30,799	26,100	22,646	19,999	17,906	16,210	14,807	12,622	10,999	9,746	7,938	6,695
115	101,373	90,633	81,951	74,787	68,775	63,658	55,411	49,056	44,009	39,904	36,499	33,629	31,178	26,373	22,851	20,158	18,034	16,314	14,894	12,686	11,047	9,784	7,963	6,713
120	105,238	93,711	84,459	76,870	70,533	65,160	56,547	49,944	44,722	40,489	36,988	34,044	31,534	26,627	23,041	20,307	18,152	16,411	14,975	12,744	11,092	9,819	7,986	6,729
125	109,064	96,732	86,906	78,892	72,231	66,607	57,633	50,790	45,399	41,043	37,450	34,435	31,869	26,865	23,220	20,445	18,263	16,501	15,050	12,799	11,133	9,851	8,007	6,745
130	112,851	99,700	89,294	80,854	73,873	68,001	58,673	51,596	46,042	41,568	37,886	34,804	32,185	27,089	23,387	20,574	18,366	16,586	15,120	12,849	11,171	9,881	8,027	6,759
135	116,600	102,615	91,624	82,761	75,461	69,344	59,671	52,366	46,654	42,066	38,300	35,152	32,483	27,300	23,544	20,696	18,463	16,664	15,185	12,896	11,207	9,909	8,045	6,772
140	120,311	105,478	93,901	84,613	76,998	70,640	60,628	53,101	47,237	42,540	38,692	35,482	32,764	27,499	23,691	20,810	18,553	16,738	15,247	12,940	11,240	9,935	8,062	6,784
145	123,986	108,291	96,124	86,414	78,486	71,891	61,547	53,805	47,793	42,990	39,064	35,795	33,031	27,686	23,830	20,917	18,638	16,808	15,304	12,982	11,271	9,959	8,078	6,795
150	127,623	111,056	98,296	88,166	79,929	73,099	62,430	54,479	48,324	43,419	39,418	36,092	33,284	27,864	23,961	21,018	18,719	16,873	15,358	13,021	11,301	9,982	8,093	6,806
155	131,225	113,773	100,419	89,870	81,326	74,266	63,280	55,125	48,832	43,828	39,755	36,374	33,524	28,031	24,086	21,113	18,794	16,934	15,409	13,057	11,328	10,004	8,108	6,816
160	134,791	116,444	102,494	91,528	82,682	75,395	64,098	55,744	49,317	44,219	40,076	36,643	33,752	28,191	24,203	21,204	18,866	16,992	15,457	13,092	11,354	10,024	8,121	6,825
165	138,322	119,070	104,523	93,143	83,998	76,488	64,885	56,339	49,782	44,593	40,383	36,899	33,969	28,342	24,315	21,289	18,933	17,047	15,503	13,124	11,379	10,043	8,133	6,834
170	141,819	121,652	106,507	94,715	85,274	77,545	65,645	56,911	50,228	44,950	40,676	37,144	34,176	28,486	24,420	21,370	18,998	17,099	15,545	13,155	11,402	10,061	8,145	6,842
175	145,282	124,191	108,448	96,247	86,514	78,569	66,377	57,460	50,656	45,292	40,956	37,377	34,373	28,623	24,521	21,447	19,058	17,148	15,586	13,184	11,424	10,078	8,156	6,850
180	148,711	126,689	110,348	97,741	87,719	79,561	67,084	57,989	51,066	45,620	41,224	37,600	34,562	28,754	24,617	21,521	19,116	17,195	15,625	13,212	11,444	10,094	8,167	6,858
185	152,107	129,145	112,207	99,196	88,890	80,523	67,766	58,499	51,461	45,935	41,480	37,813	34,742	28,878	24,708	21,590	19,171	17,240	15,662	13,238	11,464	10,109	8,177	6,865
190	155,471	131,562	114,027	100,616	90,028	81,456	68,426	58,989	51,840	46,237	41,727	38,018	34,915	28,998	24,795	21,657	19,224	17,282	15,697	13,263	11,483	10,124	8,186	6,871
195	158,803	133,940	115,809	102,001	91,135	82,361	69,064	59,463	52,206	46,527	41,963	38,214	35,080	29,111	24,879	21,720	19,274	17,322	15,730	13,287	11,501	10,138	8,195	6,878
200	162,103	136,281	117,554	103,353	92,213	83,240	69,681	59,920	52,557	46,806	42,190	38,402	35,239	29,221	24,958	21,781	19,321	17,361	15,762	13,310	11,518	10,151	8,204	6,884

400 MCM	2 Run(s)	208 Volts	C-Value = 20,565	Magnetic Duct	Copper

Calculations:

1. $f = \dfrac{\text{Length} \times 2 \times \text{Isca}}{\text{Runs} \times \text{C-Value} \times \text{E L-L}}$

2. $M = \dfrac{1}{1 + f}$

3. Isc = Isca x M

4. Add Motor Contribution = Motor FLA x 4

* All results are given in symmetrical amperes

Single-Phase L-L Bolted Fault-Current* Table

One-Way Distance in Feet

Available Isc in Thousands (Isca)	5	10	15	20	25	30	40	50	60	70	80	90	100	125	150	175	200	225	250	300	350	400	500	600
3	2,981	2,961	2,943	2,924	2,906	2,887	2,852	2,817	2,783	2,750	2,717	2,686	2,655	2,581	2,510	2,444	2,381	2,321	2,264	2,158	2,062	1,974	1,818	1,685
5	4,946	4,894	4,843	4,792	4,743	4,695	4,601	4,511	4,425	4,341	4,261	4,184	4,109	3,934	3,773	3,625	3,488	3,361	3,243	3,030	2,843	2,678	2,400	2,174
7	6,895	6,794	6,695	6,600	6,506	6,416	6,242	6,078	5,922	5,774	5,633	5,499	5,371	5,075	4,811	4,572	4,357	4,160	3,981	3,665	3,395	3,162	2,781	2,482
10	9,788	9,585	9,390	9,202	9,022	8,849	8,522	8,219	7,936	7,672	7,425	7,194	6,976	6,486	6,060	5,687	5,357	5,063	4,800	4,347	3,973	3,658	3,157	2,777
15	14,528	14,084	13,667	13,274	12,903	12,552	11,904	11,320	10,791	10,309	9,868	9,463	9,090	8,275	7,594	7,017	6,521	6,091	5,714	5,084	4,580	4,166	3,529	3,061
20	19,169	18,405	17,699	17,045	16,438	15,872	14,851	13,953	13,157	12,447	11,810	11,235	10,713	9,599	8,695	7,946	7,316	6,779	6,315	5,555	4,958	4,477	3,749	3,225
25	23,715	22,556	21,505	20,547	19,671	18,867	17,441	16,215	15,150	14,217	13,392	12,657	11,999	10,618	9,523	8,632	7,894	7,272	6,741	5,88[illegible]	5,217	4,687	3,895	3,333
30	28,169	26,548	25,104	23,809	22,640	21,582	19,736	18,180	16,852	15,705	14,704	13,823	13,042	11,427	10,168	9,159	8,332	7,642	7,058	6,12[illegible]	5,405	4,838	3,999	3,408
35	32,532	30,390	28,512	26,853	25,376	24,054	21,783	19,904	18,323	16,975	15,812	14,797	13,906	12,085	10,686	9,577	8,676	7,931	7,303	6,30[illegible]	5,547	4,952	4,077	3,465
40	36,809	34,090	31,745	29,701	27,905	26,314	23,620	21,427	19,606	18,070	16,758	15,623	14,632	12,630	11,110	9,916	8,954	8,162	7,499	6,45[illegible]	5,659	5,041	4,137	3,508
45	41,002	37,656	34,815	32,372	30,250	28,389	25,279	22,783	20,735	19,025	17,576	16,332	15,252	13,089	11,463	10,197	9,182	8,351	7,658	6,568	5,750	5,113	4,185	3,543
50	45,112	41,094	37,734	34,882	32,430	30,301	26,783	23,998	21,737	19,865	18,290	16,947	15,787	13,481	11,763	10,433	9,374	8,509	7,791	6,665	5,824	5,172	4,225	3,571
55	49,143	44,413	40,514	37,244	34,462	32,067	28,154	25,092	22,631	20,610	18,920	17,486	16,254	13,820	12,020	10,635	9,536	8,643	7,903	6,747	5,887	5,221	4,257	3,594
60	53,096	47,617	43,163	39,471	36,361	33,705	29,409	26,084	23,435	21,274	19,478	17,962	16,664	14,116	12,243	10,809	9,676	8,758	7,999	6,817	5,940	5,262	4,285	3,614
65	56,975	50,713	45,692	41,575	38,139	35,227	30,561	26,987	24,161	21,870	19,977	18,385	17,028	14,376	12,438	10,961	9,797	8,857	8,082	6,877	5,985	5,298	4,309	3,631
70	60,780	53,706	48,107	43,565	39,807	36,646	31,623	27,811	24,820	22,409	20,425	18,764	17,353	14,606	12,611	11,095	9,904	8,944	8,154	6,929	6,025	5,329	4,329	3,645
75	64,514	56,601	50,417	45,451	41,376	37,971	32,605	28,568	25,420	22,898	20,830	19,105	17,644	14,812	12,764	11,213	9,998	9,021	8,218	6,976	6,060	5,356	4,347	3,658
80	68,180	59,403	52,628	47,240	42,853	39,212	33,516	29,265	25,971	23,343	21,198	19,415	17,908	14,998	12,901	11,319	10,082	9,089	8,274	7,0[illegible]6	6,090	5,380	4,363	3,669
85	71,778	62,116	54,747	48,940	44,247	40,376	34,363	29,908	26,476	23,751	21,534	19,696	18,147	15,165	13,025	11,414	10,158	9,150	8,325	7,053	6,118	5,402	4,377	3,679
90	75,311	64,745	56,778	50,557	45,565	41,470	35,152	30,505	26,942	24,125	21,841	19,953	18,364	15,317	13,136	11,500	10,225	9,206	8,371	7,0[illegible]5	6,142	5,421	4,389	3,688
95	78,781	67,292	58,728	52,098	46,813	42,501	35,890	31,059	27,374	24,470	22,124	20,188	18,564	15,455	13,238	11,577	10,287	9,255	8,412	7,1[illegible]5	6,164	5,438	4,401	3,696
100	82,189	69,763	60,601	53,567	47,995	43,473	36,581	31,575	27,774	24,790	22,385	20,405	18,747	15,582	13,331	11,648	10,343	9,301	8,449	7,1[illegible]2	6,184	5,454	4,411	3,703
105	85,537	72,161	62,402	54,969	49,118	44,392	37,229	32,057	28,146	25,086	22,626	20,605	18,916	15,698	13,416	11,713	10,394	9,342	8,483	7,136	6,203	5,468	4,420	3,710
110	88,826	74,487	64,135	56,309	50,185	45,262	37,839	32,508	28,493	25,361	22,850	20,791	19,072	15,806	13,495	11,773	10,441	9,380	8,515	7,1[illegible]8	6,219	5,481	4,429	3,716
115	92,058	76,747	65,803	57,590	51,200	46,087	38,414	32,931	28,818	25,618	23,058	20,963	19,217	15,905	13,567	11,828	10,484	9,415	8,543	7,209	6,235	5,493	4,436	3,721
120	95,234	78,942	67,410	58,818	52,168	46,869	38,956	33,329	29,122	25,858	23,252	21,123	19,352	15,997	13,634	11,879	10,524	9,447	8,570	7,228	6,249	5,504	4,444	3,726
125	98,357	81,075	68,959	59,994	53,091	47,613	39,468	33,703	29,407	26,083	23,434	21,273	19,477	16,083	13,696	11,926	10,561	9,477	8,594	7,245	6,262	5,514	4,450	3,731
130	101,426	83,150	70,454	61,122	53,973	48,321	39,954	34,056	29,676	26,294	23,604	21,413	19,595	16,163	13,754	11,970	10,596	9,505	8,617	7,261	6,274	5,523	4,456	3,735
135	104,444	85,167	71,898	62,205	54,816	48,996	40,414	34,390	29,929	26,492	23,764	21,545	19,705	16,238	13,808	12,011	10,628	9,530	8,638	7,276	6,285	5,532	4,462	3,739
140	107,412	87,130	73,292	63,246	55,623	49,639	40,850	34,706	30,168	26,679	23,914	21,668	19,808	16,308	13,859	12,049	10,658	9,555	8,658	7,290	6,296	5,540	4,467	3,743
145	110,331	89,041	74,639	64,247	56,395	50,254	41,266	35,005	30,394	26,856	24,056	21,785	19,905	16,374	13,906	12,085	10,686	9,577	8,677	7,303	6,305	5,547	4,472	3,746
150	113,202	90,902	75,942	65,210	57,136	50,841	41,661	35,289	30,607	27,023	24,190	21,894	19,997	16,435	13,951	12,119	10,712	9,598	8,694	7,316	6,315	5,555	4,477	3,749
155	116,027	92,715	77,203	66,138	57,847	51,403	42,037	35,559	30,810	27,181	24,316	21,998	20,083	16,494	13,993	12,151	10,737	9,618	8,710	7,327	6,323	5,561	4,481	3,752
160	118,806	94,481	78,424	67,031	58,529	51,941	42,397	35,815	31,003	27,330	24,436	22,096	20,165	16,549	14,033	12,181	10,760	9,637	8,726	7,338	6,331	5,567	4,485	3,755
165	121,541	96,202	79,606	67,893	59,185	52,457	42,740	36,060	31,186	27,473	24,549	22,189	20,242	16,601	14,070	12,209	10,782	9,654	8,740	7,348	6,339	5,573	4,489	3,758
170	124,232	97,881	80,752	68,725	59,816	52,952	43,068	36,293	31,360	27,608	24,657	22,277	20,315	16,650	14,105	12,235	10,803	9,671	8,754	7,358	6,346	5,579	4,493	3,760
175	126,881	99,518	81,863	69,528	60,424	53,428	43,382	36,516	31,526	27,737	24,760	22,360	20,385	16,697	14,139	12,261	10,823	9,687	8,767	7,367	6,353	5,584	4,496	3,763
180	129,489	101,115	82,941	70,304	61,009	53,885	43,683	36,729	31,685	27,859	24,858	22,440	20,451	16,741	14,171	12,284	10,841	9,702	8,779	7,376	6,359	5,589	4,499	3,765
185	132,057	102,674	83,986	71,054	61,573	54,324	43,971	36,933	31,836	27,976	24,951	22,516	20,514	16,783	14,201	12,307	10,859	9,716	8,790	7,384	6,365	5,594	4,502	3,767
190	134,585	104,196	85,002	71,780	62,117	54,747	44,248	37,128	31,981	28,088	25,040	22,588	20,574	16,824	14,230	12,329	10,876	9,729	8,801	7,392	6,371	5,598	4,505	3,769
195	137,075	105,682	85,988	72,482	62,642	55,155	44,514	37,315	32,120	28,195	25,125	22,657	20,631	16,862	14,257	12,349	10,892	9,742	8,812	7,399	6,377	5,602	4,508	3,771
200	139,527	107,133	86,947	73,162	63,149	55,548	44,769	37,494	32,253	28,297	25,206	22,723	20,686	16,898	14,283	12,369	10,907	9,754	8,822	7,406	6,382	5,606	4,510	3,773

500 MCM | 1 Run(s) | 208 Volts | C-Value = 22,185 | Magnetic Duct | Copper

Calculations:

1. $f = \frac{\text{Length} \times 2 \times \text{Isca}}{\text{Runs} \times \text{C-Value} \times E_{L-L}}$

2. $M = \frac{1}{1 + f}$

3. Isc = Isca x M

4. Add Motor Contribution = Motor FLA x 4

* All results are given in symmetrical amperes

Single-Phase L-L Bolted Fault-Current* Table

One-Way Distance in Feet

Available Isc in Thousands (Isca)	5	10	15	20	25	30	40	50	60	70	80	90	100	125	150	175	200	225	250	300	350	400	500	600
3	2,990	2,981	2,971	2,961	2,952	2,943	2,924	2,906	2,887	2,869	2,852	2,834	2,817	2,775	2,733	2,694	2,655	2,617	2,581	2,510	2,444	2,381	2,264	2,158
5	4,973	4,946	4,920	4,894	4,868	4,843	4,792	4,743	4,695	4,647	4,601	4,556	4,511	4,404	4,301	4,203	4,109	4,020	3,934	3,773	3,625	3,488	3,243	3,030
7	6,947	6,895	6,844	6,794	6,744	6,695	6,600	6,506	6,416	6,328	6,242	6,159	6,078	5,884	5,702	5,532	5 371	5,219	5,075	4,811	4,572	4,357	3,981	3,665
10	9,893	9,788	9,685	9,585	9,486	9,390	9,202	9,022	8,849	8,683	8,522	8,368	8,219	7,869	7,547	7,250	6,976	6,722	6,486	6,060	5,687	5,357	4,800	4,347
15	14,760	14,528	14,303	14,084	13,873	13,667	13,274	12,903	12,552	12,220	11,904	11,605	11,320	10,666	10,083	9,561	9,090	8,664	8,275	7,594	7,017	6,521	5,714	5,084
20	19,576	19,169	18,779	18,405	18,045	17,699	17,045	16,438	15,872	15,345	14,851	14,388	13,953	12,972	12,120	11,373	10,713	10,126	9,599	8,695	7,946	7,316	6,315	5,555
25	24,341	23,715	23,121	22,556	22,018	21,505	20,547	19,671	18,867	18,126	17,441	16,806	16,215	14,906	13,792	12,833	11,999	11,266	10,618	9,523	8,632	7,894	6,741	5,881
30	29,056	28,169	27,334	26,548	25,806	25,104	23,809	22,640	21,582	20,617	19,736	18,926	18,180	16,550	15,188	14,034	13 042	12,181	11,427	10,168	9,159	8,332	7,058	6,121
35	33,721	32,532	31,425	30,390	29,421	28,512	26,853	25,376	24,054	22,862	21,783	20,801	19,904	17,966	16,373	15,039	13,906	12,931	12,085	10,686	9,577	8,676	7,303	6,305
40	38,338	36,809	35,397	34,090	32,876	31,745	29,701	27,905	26,314	24,894	23,620	22,470	21,427	19,198	17,389	15,892	14,632	13,558	12,630	11,110	9,916	8,954	7,499	6,451
45	42,908	41,002	39,257	37,656	36,180	34,815	32,372	30,250	28,389	26,744	25,279	23,966	22,783	20,280	18,272	16,626	15,252	14,088	13,089	11,463	10,197	9,182	7,658	6,568
50	47,430	45,112	43,010	41,094	39,343	37,734	34,882	32,430	30,301	28,434	26,783	25,314	23,998	21,237	19,045	17,264	15,787	14,543	13,481	11,763	10,433	9,374	7,791	6,666
55	51,907	49,143	46,658	44,413	42,374	40,514	37,244	34,462	32,067	29,984	28,154	26,535	25,092	22,089	19,728	17,823	16,254	14,938	13,820	12,020	10,635	9,536	7,903	6,747
60	56,337	53,096	50,208	47,617	45,281	43,163	39,471	36,361	33,705	31,411	29,409	27,647	26,084	22,854	20,336	18,318	16,664	15,284	14,116	12,243	10,809	9,676	7,999	6,817
65	60,723	56,975	53,662	50,713	48,072	45,692	41,575	38,139	35,227	32,729	30,561	28,663	26,987	23,544	20,881	18,759	17,028	15,590	14,376	12,438	10,961	9,797	8,082	6,877
70	65,065	60,780	57,024	53,706	50,753	48,107	43,565	39,807	36,646	33,950	31,623	29,595	27,811	24,170	21,371	19,153	17,353	15,862	14,606	12,611	11,095	9,904	8,154	6,929
75	69,363	64,514	60,299	56,601	53,330	50,417	45,451	41,376	37,971	35,084	32,605	30,453	28,568	24,739	21,815	19,509	17,644	16,105	14,812	12,764	11,213	9,998	8,218	6,976
80	73,618	68,180	63,489	59,403	55,811	52,628	47,240	42,853	39,212	36,141	33,516	31,246	29,265	25,260	22,219	19,832	17,908	16,324	14,998	12,901	11,319	10,082	8,274	7,016
85	77,832	71,778	66,599	62,116	58,199	54,747	48,940	44,247	40,376	37,127	34,363	31,981	29,908	25,738	22,588	20,125	18,147	16,522	15,165	13,025	11,414	10,158	8,325	7,053
90	82,003	75,311	69,629	64,745	60,500	56,778	50,557	45,565	41,470	38,051	35,152	32,664	30,505	26,178	22,927	20,394	18,364	16,703	15,317	13,136	11,500	10,225	8,371	7,085
95	86,134	78,781	72,585	67,292	62,719	58,728	52,098	46,813	42,501	38,917	35,890	33,300	31,059	26,585	23,238	20,640	18,564	16,867	15,455	13,238	11,577	10,287	8,412	7,115
100	90,224	82,189	75,468	69,763	64,860	60,601	53,567	47,995	43,473	39,730	36,581	33,894	31,575	26,962	23,526	20,866	18,747	17,019	15,582	13,331	11,648	10,343	8,449	7,142
105	94,274	85,537	78,281	72,161	66,928	62,402	54,969	49,118	44,392	40,497	37,229	34,450	32,057	27,313	23,792	21,076	18,916	17,158	15,698	13,416	11,713	10,394	8,483	7,166
110	98,285	88,826	81,027	74,487	68,924	64,135	56,309	50,185	45,262	41,219	37,839	34,971	32,508	27,640	24,040	21,270	19,072	17,286	15,806	13,495	11,773	10,441	8,515	7,188
115	102,258	92,058	83,708	76,747	70,855	65,803	57,590	51,200	46,087	41,902	38,414	35,462	32,931	27,945	24,271	21,450	19,217	17 405	15,905	13,567	11,828	10,484	8,543	7,209
120	106,192	95,234	86,326	78,942	72,722	67,410	58,818	52,168	46,869	42,548	38,956	35,923	33,329	28,231	24,486	21,618	19,352	17 515	15,997	13,634	11,879	10,524	8,570	7,228
125	110,089	98,357	88,884	81,075	74,528	68,959	59,994	53,091	47,613	43,160	39,468	36,359	33,703	28,499	24,687	21,775	19,477	17,618	16,083	13,696	11,926	10,561	8,594	7,245
130	113,949	101,426	91,383	83,150	76,277	70,454	61,122	53,973	48,321	43,741	39,954	36,770	34,056	28,751	24,876	21,922	19,595	17,714	16,163	13,754	11,970	10,596	8,617	7,261
135	117,772	104,444	93,826	85,167	77,972	71,898	62,205	54,816	48,996	44,293	40,414	37,159	34,390	28,989	25,054	22,060	19,705	17,804	16,238	13,808	12,011	10,628	8,638	7,276
140	121,560	107,412	96,214	87,130	79,614	73,292	63,246	55,623	49,639	44,818	40,850	37,528	34,706	29,213	25,221	22,189	19,808	17,388	16,308	13,859	12,049	10,658	8,658	7,290
145	125,312	110,331	98,549	89,041	81,207	74,639	64,247	56,395	50,254	45,318	41,266	37,878	35,005	29,425	25,379	22,311	19,905	17,967	16,374	13,906	12,085	10,686	8,677	7,303
150	129,029	113,202	100,834	90,902	82,751	75,942	65,210	57,136	50,841	45,795	41,661	38,211	35,289	29,625	25,528	22,426	19,997	18,042	16,435	13,951	12,119	10,712	8,694	7,316
155	132,711	116,027	103,069	92,715	84,251	77,203	66,138	57,847	51,403	46,251	42,037	38,528	35,559	29,815	25,669	22,535	20,083	18,112	16,494	13,993	12,151	10,737	8,710	7,327
160	136,360	118,806	105,256	94,481	85,706	78,424	67,031	58,529	51,941	46,686	42,397	38,829	35,815	29,995	25,802	22,638	20,165	18,179	16,549	14,033	12,181	10,760	8,726	7,338
165	139,975	121,541	107,397	96,202	87,121	79,606	67,893	59,185	52,457	47,103	42,740	39,117	36,060	30,167	25,929	22,735	20,242	18,241	16,601	14,070	12,209	10,782	8,740	7,348
170	143,556	124,232	109,493	97,881	88,495	80,752	68,725	59,816	52,952	47,501	43,068	39,392	36,293	30,330	26,049	22,828	20,315	18,301	16,650	14,105	12,235	10,803	8,754	7,358
175	147,106	126,881	111,546	99,518	89,831	81,863	69,528	60,424	53,428	47,884	43,382	39,654	36,516	30,485	26,164	22,916	20,385	18,357	16,697	14,139	12,261	10,823	8,767	7,367
180	150,623	129,489	113,556	101,115	91,130	82,941	70,304	61,009	53,885	48,250	43,683	39,905	36,729	30,633	26,273	22,999	20,451	18,411	16,741	14,171	12,284	10,841	8,779	7,376
185	154,108	132,057	115,526	102,674	92,395	83,986	71,054	61,573	54,324	48,603	43,971	40,146	36,933	30,775	26,377	23,079	20,514	18,462	16,783	14,201	12,307	10,859	8,790	7,384
190	157,562	134,585	117,456	104,196	93,625	85,002	71,780	62,117	54,747	48,941	44,248	40,376	37,128	30,910	26,476	23,155	20,574	18,511	16,824	14,230	12,329	10,876	8,801	7,392
195	160,985	137,075	119,348	105,682	94,823	85,988	72,482	62,642	55,155	49,266	44,514	40,598	37,315	31,040	26,571	23,228	20,631	18,557	16,862	14,257	12,349	10,892	8,812	7,399
200	164,378	139,527	121,203	107,133	95,990	86,947	73,162	63,149	55,548	49,579	44,769	40,810	37,494	31,164	26,662	23,297	20,686	18,601	16,898	14,283	12,369	10,907	8,822	7,406

500 MCM	2 Run(s)	208 Volts	C-Value = 22,185	Magnetic Duct	Copper

Calculations:

1. $f = \dfrac{\text{Length} \times 2 \times \text{Isca}}{\text{Runs} \times \text{C-Value} \times \text{E L-L}}$
2. $M = \dfrac{1}{1 + f}$
3. $\text{Isc} = \text{Isca} \times M$
4. Add Motor Contribution = Motor FLA x 4

* All results are given in symmetrical amperes

Single-Phase L-L Bolted Fault-Current* Table

One-Way Distance in Feet

Available Isc in Thousands (Isca)	5	10	15	20	25	30	40	50	60	70	80	90	100	125	150	175	200	225	250	300	350	400	500	600
3	2,981	2,963	2,945	2,926	2,909	2,891	2,856	2,823	2,790	2,758	2,726	2,695	2,665	2,593	2,524	2,459	2,398	2,339	2,283	2,179	2,084	1,997	1,843	1,711
5	4,948	4,897	4,848	4,799	4,751	4,705	4,614	4,526	4,442	4,361	4,283	4,207	4,134	3,963	3,805	3,659	3,524	3,399	3,282	3,071	2,886	2,721	2,443	2,216
7	6,899	6,801	6,705	6,612	6,522	6,434	6,265	6,105	5,953	5,808	5,670	5,539	5,413	5,123	4,862	4,627	4,413	4,218	4,040	3,725	3,455	3,222	2,839	2,538
10	9,795	9,598	9,409	9,227	9,052	8,884	8,565	8,269	7,992	7,733	7,491	7,263	7,049	6,564	6,142	5,771	5,442	5,149	4,886	4,432	4,056	3,739	3,233	2,847
15	14,543	14,114	13,709	13,326	12,964	12,622	11,983	11,415	10,895	10,419	9,984	9,583	9,213	8,403	7,724	7,146	6,649	6,216	5,836	5,201	4,690	4,271	3,623	3,146
20	19,196	18,455	17,768	17,131	16,538	15,984	14,982	14,097	13,312	12,609	11,977	11,405	10,885	9,772	8,865	8,112	7,477	6,934	6,465	5,694	5,088	4,598	3,856	3,320
25	23,757	22,631	21,607	20,672	19,815	19,026	17,622	16,411	15,356	14,428	13,606	12,873	12,215	10,830	9,727	8,828	8,081	7,451	6,912	6,038	5,361	4,820	4,010	3,434
30	28,227	26,652	25,244	23,977	22,831	21,789	19,968	18,427	17,107	15,964	14,964	14,081	13,297	11,673	10,402	9,380	8,542	7,841	7,246	6,292	5,559	4,980	4,121	3,514
35	32,611	30,527	28,693	27,067	25,616	24,312	22,066	20,199	18,624	17,277	16,112	15,093	14,196	12,360	10,944	9,819	8,904	8,145	7,505	6,486	5,711	5,101	4,203	3,574
40	36,909	34,262	31,969	29,963	28,195	26,623	23,953	21,770	19,951	18,413	17,095	15,953	14,954	12,930	11,389	10,176	9,196	8,389	7,712	6,640	5,829	5,195	4,267	3,620
45	41,126	37,866	35,084	32,684	30,591	28,750	25,661	23,171	21,122	19,406	17,948	16,693	15,603	13,412	11,761	10,472	9,437	8,589	7,880	6,764	5,925	5,271	4,318	3,657
50	45,262	41,345	38,051	35,244	32,822	30,712	27,212	24,429	22,162	20,280	18,693	17,336	16,163	13,824	12,077	10,721	9,640	8,756	8,021	6,868	6,004	5,334	4,360	3,687
55	49,321	44,705	40,879	37,657	34,905	32,528	28,629	25,565	23,093	21,057	19,351	17,900	16,652	14,181	12,348	10,934	9,811	8,898	8,140	6,955	6,071	5,386	4,395	3,712
60	53,304	47,953	43,578	39,935	36,854	34,214	29,927	26,595	23,930	21,751	19,935	18,399	17,083	14,492	12,583	11,119	9,960	9,019	8,241	7,029	6,127	5,430	4,424	3,733
65	57,214	51,094	46,157	42,090	38,682	35,784	31,121	27,533	24,687	22,375	20,458	18,844	17,466	14,766	12,789	11,279	10,088	9,125	8,329	7,093	6,176	5,469	4,450	3,751
70	61,053	54,134	48,624	44,131	40,399	37,249	32,223	28,392	25,376	22,939	20,929	19,242	17,808	15,010	12,972	11,421	10,201	9,217	8,406	7,148	6,218	5,502	4,472	3,766
75	64,822	57,077	50,985	46,067	42,015	38,619	33,243	29,182	26,004	23,451	21,354	19,602	18,115	15,228	13,134	11,547	10,302	9,299	8,474	7,197	6,255	5,531	4,491	3,780
80	68,524	59,927	53,247	47,907	43,540	39,903	34,190	29,909	26,580	23,918	21,741	19,927	18,393	15,423	13,279	11,659	10,391	9,371	8,534	7,24[illegible]	6,288	5,556	4,508	3,792
85	72,159	62,689	55,417	49,656	44,980	41,109	35,072	30,581	27,110	24,347	22,094	20,223	18,645	15,600	13,410	11,760	10,471	9,436	8,588	7,279	6,317	5,579	4,523	3,803
90	75,731	65,368	57,499	51,321	46,342	42,244	35,895	31,205	27,599	24,740	22,418	20,494	18,875	15,761	13,529	11,851	10,543	9,495	8,637	7,31[illegible]	6,343	5,599	4,536	3,812
95	79,241	67,966	59,500	52,909	47,633	43,314	36,665	31,785	28,052	25,103	22,716	20,743	19,085	15,907	13,637	11,933	10,608	9,548	8,681	7,34[illegible]	6,367	5,618	4,548	3,821
100	82,689	70,487	61,423	54,425	48,858	44,324	37,386	32,326	28,472	25,440	22,991	20,972	19,279	16,042	13,735	12,009	10,668	9,596	8,720	7,37[illegible]	6,388	5,634	4,559	3,828
105	86,079	72,935	63,274	55,873	50,022	45,280	38,064	32,831	28,864	25,752	23,245	21,183	19,458	16,165	13,826	12,078	10,722	9,640	8,757	7,400	6,407	5,650	4,569	3,835
110	89,410	75,313	65,056	57,258	51,129	46,185	38,701	33,305	29,229	26,042	23,482	21,380	19,623	16,279	13,909	12,141	10,772	9,681	8,790	7,424	6,425	5,663	4,578	3,842
115	92,686	77,624	66,773	58,584	52,184	47,044	39,303	33,749	29,570	26,313	23,701	21,562	19,776	16,385	13,986	12,200	10,818	9,718	8,821	7,446	6,442	5,676	4,586	3,847
120	95,907	79,870	68,429	59,854	53,189	47,860	39,370	34,167	29,891	26,566	23,907	21,732	19,919	16,482	14,057	12,254	10,861	9,752	8,849	7,465	6,457	5,688	4,594	3,853
125	99,074	82,055	70,026	61,073	54,149	48,636	40,407	34,560	30,192	26,803	24,099	21,890	20,052	16,574	14,123	12,304	10,900	9,784	8,875	7,485	6,471	5,699	4,601	3,858
130	102,189	84,180	71,568	62,242	55,067	49,375	40,916	34,932	30,475	27,026	24,279	22,039	20,177	16,658	14,185	12,351	10,937	9,814	8,899	7,502	6,484	5,709	4,607	3,862
135	105,253	86,249	73,057	63,366	55,945	50,079	41,399	35,283	30,742	27,236	24,448	22,178	20,293	16,738	14,243	12,395	10,971	9,841	8,922	7,5[illegible]8	6,496	5,718	4,613	3,867
140	108,268	88,263	74,497	64,446	56,785	50,752	41,857	35,615	30,994	27,434	24,607	22,309	20,403	16,812	14,296	12,436	11,003	9,867	8,943	7,533	6,507	5,727	4,619	3,871
145	111,234	90,224	75,890	65,486	57,590	51,394	42,293	35,931	31,232	27,620	24,757	22,432	20,506	16,882	14,347	12,474	11,033	9,891	8,963	7,547	6,517	5,735	4,624	3,874
150	114,153	92,135	77,237	66,487	58,363	52,009	42,708	36,230	31,458	27,797	24,899	22,548	20,603	16,948	14,394	12,510	11,061	9,913	8,981	7,560	6,527	5,742	4,629	3,878
155	117,026	93,997	78,542	67,451	59,105	52,597	43,104	36,514	31,672	27,964	25,033	22,658	20,695	17,010	14,439	12,543	11,088	9,935	8,999	7,572	6,536	5,749	4,634	3,881
160	119,854	95,813	79,806	68,381	59,818	53,161	43,482	36,785	31,876	28,122	25,160	22,762	20,781	17,069	14,481	12,575	11,112	9,955	9,015	7,584	6,545	5,756	4,638	3,884
165	122,638	97,584	81,030	69,278	60,503	53,701	43,843	37,043	32,069	28,273	25,280	22,861	20,864	17,124	14,521	12,605	11,136	9,973	9,031	7,595	6,553	5,762	4,642	3,887
170	125,379	99,312	82,218	70,144	61,163	54,220	44,189	37,289	32,254	28,416	25,395	22,954	20,941	17,176	14,559	12,634	11,158	9,991	9,045	7,605	6,561	5,768	4,646	3,890
175	128,078	100,997	83,370	70,981	61,798	54,719	44,519	37,525	32,429	28,553	25,504	23,043	21,015	17,226	14,595	12,660	11,179	10,008	9,059	7,615	6,568	5,774	4,650	3,892
180	130,735	102,643	84,488	71,790	62,410	55,198	44,836	37,749	32,597	28,683	25,607	23,128	21,086	17,273	14,628	12,686	11,199	10,024	9,072	7,624	6,575	5,779	4,653	3,894
185	133,353	104,249	85,573	72,572	63,001	55,660	45,140	37,965	32,758	28,807	25,706	23,208	21,153	17,318	14,661	12,710	11,218	10,039	9,084	7,633	6,581	5,784	4,656	3,897
190	135,932	105,819	86,628	73,329	63,570	56,104	45,432	38,171	32,911	28,925	25,800	23,285	21,217	17,361	14,691	12,733	11,236	10,053	9,096	7,641	6,587	5,789	4,660	3,899
195	138,472	107,352	87,653	74,062	64,120	56,532	45,712	38,368	33,058	29,039	25,891	23,359	21,278	17,402	14,720	12,755	11,253	10,067	9,107	7,649	6,593	5,794	4,663	3,901
200	140,974	108,850	88,649	74,772	64,652	56,945	45,981	38,558	33,199	29,147	25,977	23,429	21,336	17,441	14,748	12,776	11,269	10,080	9,118	7,656	6,599	5,798	4,665	3,903

600 MCM	1 Run(s)	208 Volts	C-Value = 22,965	Magnetic Duct	Copper

Calculations:

1. $f = \dfrac{\text{Length} \times 2 \times \text{Isca}}{\text{Runs} \times \text{C-Value} \times \text{E L-L}}$

2. $M = \dfrac{1}{1 + f}$

3. Isc = Isca x M

4. Add Motor Contribution = Motor FLA x 4

* All results are given in symmetrical amperes

Single-Phase L-L Bolted Fault-Current* Table

One-Way Distance in Feet

Available Isc in Thousands (Isca)	5	10	15	20	25	30	40	50	60	70	80	90	100	125	150	175	200	225	250	300	350	400	500	600
3	2,991	2,981	2,972	2,963	2,954	2,945	2,926	2,909	2,891	2,874	2,856	2,839	2,823	2,782	2,742	2,703	2,665	2,629	2,593	2,524	2,459	2,398	2,283	2,179
5	4,974	4,948	4,923	4,897	4,872	4,848	4,799	4,751	4,705	4,659	4,614	4,570	4,526	4,421	4,321	4,226	4 134	4,047	3,963	3,805	3,659	3,524	3,282	3,071
7	6,949	6,899	6,849	6,801	6,753	6,705	6,612	6,522	6,434	6,349	6,265	6,184	6,105	5,916	5,739	5,571	5,413	5,264	5,123	4,862	4,627	4,413	4,040	3,725
10	9,896	9,795	9,696	9,598	9,503	9,409	9,227	9,052	8,884	8,722	8,565	8,415	8,269	7,926	7,610	7,319	7,049	6,798	6,564	6,142	5,771	5,442	4,886	4,432
15	14,768	14,543	14,325	14,114	13,908	13,709	13,326	12,964	12,622	12,297	11,988	11,695	11,415	10,772	10,197	9,680	9,213	8,790	8,403	7,724	7,146	6,649	5,836	5,201
20	19,590	19,196	18,818	18,455	18,105	17,768	17,131	16,538	15,984	15,467	14,982	14,526	14,097	13,129	12,285	11,543	10,885	10,298	9,772	8,865	8,112	7,477	6,465	5,694
25	24,362	23,757	23,180	22,631	22,107	21,607	20,672	19,815	19,026	18,297	17,622	16,995	16,411	15,113	14,005	13,049	12 215	11,481	10,830	9,727	8,828	8,081	6,912	6,038
30	29,087	28,227	27,417	26,652	25,929	25,244	23,977	22,831	21,789	20,839	19,968	19,166	18,427	16,806	15,447	14,292	13,297	12,432	11,673	10,402	9,380	8,542	7,246	6,292
35	33,763	32,611	31,534	30,527	29,581	28,693	27,067	25,616	24,312	23,134	22,066	21,091	20,199	18,268	16,674	15,336	14,196	13,214	12,360	10,944	9,819	8,904	7,505	6,486
40	38,393	36,909	35,536	34,262	33,076	31,969	29,963	28,195	26,623	25,218	23,953	22,810	21,770	19,543	17,730	16,224	14,954	13,869	12,930	11,389	10,176	9,196	7,712	6,640
45	42,976	41,126	39,428	37,866	36,422	35,084	32,684	30,591	28,750	27,117	25,661	24,352	23,171	20,665	18,648	16,990	15,303	14,425	13,412	11,761	10,472	9,437	7,880	6,764
50	47,513	45,262	43,215	41,345	39,630	38,051	35,244	32,822	30,712	28,856	27,212	25,746	24,429	21,660	19,454	17,657	16,163	14,902	13,824	12,077	10,721	9,640	8,021	6,868
55	52,006	49,321	46,900	44,705	42,707	40,879	37,657	34,905	32,528	30,454	28,629	27,010	25,565	22,548	20,168	18,242	16,652	15,317	14,181	12,348	10,934	9,811	8,140	6,955
60	56,454	53,304	50,487	47,953	45,661	43,578	39,935	36,854	34,214	31,927	29,927	28,163	26,595	23,345	20,803	18,761	17,083	15,681	14,492	12,583	11,119	9,960	8,241	7,029
65	60,859	57,214	53,982	51,094	48,501	46,157	42,090	38,682	35,784	33,290	31,121	29,218	27,533	24,066	21,373	19,223	17,466	16,003	14,766	12,789	11,279	10,088	8,329	7,093
70	65,221	61,053	57,386	54,134	51,231	48,624	44,131	40,399	37,249	34,554	32,223	30,187	28,392	24,719	21,888	19,638	17,808	16,290	15,010	12,972	11,421	10,201	8,406	7,148
75	69,541	64,822	60,703	57,077	53,859	50,985	46,067	42,015	38,619	35,730	33,243	31,080	29,182	25,315	22,354	20,012	18,115	16,546	15,228	13,134	11,547	10,302	8,474	7,197
80	73,818	68,524	63,938	59,927	56,390	53,247	47,907	43,540	39,903	36,826	34,190	31,907	29,909	25,861	22,778	20,352	18,393	16,778	15,423	13,279	11,659	10,391	8,534	7,241
85	78,055	72,159	67,092	62,689	58,829	55,417	49,656	44,980	41,109	37,851	35,072	32,673	30,581	26,362	23,166	20,661	18,645	16,987	15,600	13,410	11,760	10,471	8,588	7,279
90	82,251	75,731	70,169	65,368	61,181	57,499	51,321	46,342	42,244	38,812	35,895	33,386	31,205	26,824	23,522	20,944	18,875	17,178	15,761	13,529	11,851	10,543	8,637	7,314
95	86,408	79,241	73,171	67,966	63,452	59,500	52,909	47,633	43,314	39,713	36,665	34,051	31,785	27,252	23,850	21,203	19,035	17,352	15,907	13,637	11,933	10,608	8,681	7,346
100	90,524	82,689	76,102	70,487	65,644	61,423	54,425	48,858	44,324	40,561	37,386	34,672	32,326	27,648	24,153	21,443	19,279	17,512	16,042	13,735	12,009	10,668	8,720	7,374
105	94,602	86,079	78,964	72,935	67,762	63,274	55,873	50,022	45,280	41,360	38,064	35,254	32,831	28,017	24,434	21,664	19,458	17,659	16,165	13,826	12,078	10,722	8,757	7,400
110	98,642	89,410	81,759	75,313	69,810	65,056	57,258	51,129	46,185	42,114	38,701	35,801	33,305	28,361	24,695	21,869	19,623	17,795	16,279	13,909	12,141	10,772	8,790	7,424
115	102,644	92,686	84,489	77,624	71,791	66,773	58,584	52,184	47,044	42,826	39,303	36,315	33,749	28,683	24,939	22,060	19,776	17,921	16,385	13,986	12,200	10,818	8,821	7,446
120	106,609	95,907	87,157	79,870	73,708	68,429	59,854	53,189	47,860	43,501	39,870	36,799	34,167	28,984	25,166	22,237	19,919	18,039	16,482	14,057	12,254	10,861	8,849	7,466
125	110,537	99,074	89,765	82,055	75,565	70,026	61,073	54,149	48,636	44,142	40,407	37,256	34,560	29,267	25,379	22,403	20,052	18,148	16,574	14,123	12,304	10,900	8,875	7,485
130	114,429	102,189	92,314	84,180	77,363	71,568	62,242	55,067	49,375	44,749	40,916	37,688	34,932	29,533	25,579	22,559	20,177	18,250	16,658	14,185	12,351	10,937	8,899	7,502
135	118,285	105,253	94,808	86,249	79,107	73,057	63,366	55,945	50,079	45,327	41,399	38,097	35,283	29,783	25,767	22,705	20,293	18,345	16,738	14,243	12,395	10,971	8,922	7,518
140	122,106	108,268	97,247	88,263	80,798	74,497	64,446	56,785	50,752	45,877	41,857	38,485	35,615	30,020	25,944	22,842	20,403	18,434	16,812	14,296	12,436	11,003	8,943	7,533
145	125,892	111,234	99,634	90,224	82,438	75,890	65,486	57,590	51,394	46,402	42,293	38,853	35,931	30,243	26,110	22,971	20,506	18,519	16,882	14,347	12,474	11,033	8,963	7,547
150	129,644	114,153	101,969	92,135	84,031	77,237	66,487	58,363	52,009	46,902	42,708	39,203	36,230	30,455	26,268	23,093	20,603	18,598	16,948	14,394	12,510	11,061	8,981	7,560
155	133,363	117,026	104,255	93,997	85,577	78,542	67,451	59,105	52,597	47,380	43,104	39,537	36,514	30,656	26,417	23,209	20,695	18,672	17,010	14,439	12,543	11,088	8,999	7,572
160	137,047	119,854	106,494	95,813	87,080	79,806	68,381	59,818	53,161	47,837	43,482	39,854	36,785	30,847	26,559	23,318	20,781	18,743	17,069	14,481	12,575	11,112	9,015	7,584
165	140,699	122,638	108,686	97,584	88,540	81,030	69,278	60,503	53,701	48,274	43,843	40,157	37,043	31,028	26,693	23,421	20,864	18,810	17,124	14,521	12,605	11,136	9,031	7,595
170	144,319	125,379	110,833	99,312	89,960	82,218	70,144	61,163	54,220	48,693	44,189	40,447	37,289	31,200	26,821	23,519	20,941	18,873	17,176	14,559	12,634	11,158	9,045	7,605
175	147,906	128,078	112,937	100,997	91,341	83,370	70,981	61,798	54,719	49,095	44,519	40,724	37,525	31,365	26,942	23,613	21,015	18,933	17,226	14,595	12,660	11,179	9,059	7,615
180	151,462	130,735	114,998	102,643	92,685	84,488	71,790	62,410	55,198	49,481	44,836	40,989	37,749	31,522	27,058	23,701	21,086	18,990	17,273	14,628	12,686	11,199	9,072	7,624
185	154,987	133,353	117,019	104,249	93,993	85,573	72,572	63,001	55,660	49,851	45,140	41,243	37,965	31,672	27,168	23,786	21,153	19,044	17,318	14,661	12,710	11,218	9,084	7,633
190	158,481	135,932	119,000	105,819	95,266	86,628	73,329	63,570	56,104	50,207	45,432	41,486	38,171	31,815	27,274	23,867	21,217	19,096	17,361	14,691	12,733	11,236	9,096	7,641
195	161,945	138,472	120,942	107,352	96,507	87,653	74,062	64,120	56,532	50,549	45,712	41,720	38,368	31,952	27,374	23,944	21,278	19,145	17,402	14,720	12,755	11,253	9,107	7,649
200	165,378	140,974	122,847	108,850	97,716	88,649	74,772	64,652	56,945	50,879	45,981	41,944	38,558	32,084	27,471	24,018	21,336	19,193	17,441	14,748	12,776	11,269	9,118	7,656

600 MCM	2 Run(s)	208 Volts	C-Value = 22,965	Magnetic Duct	Copper

Calculations:

1. $f = \dfrac{\text{Length} \times 2 \times \text{Isca}}{\text{Runs} \times \text{C-Value} \times \text{E L-L}}$

2. $M = \dfrac{1}{1 + f}$

3. $\text{Isc} = \text{Isca} \times M$

4. Add Motor Contribution = Motor FLA x 4

* All results are given in symmetrical amperes

Single-Phase L-L Bolted Fault-Current* Table

One-Way Distance in Feet

Available Isc in Thousands (Isca)	5	10	15	20	25	30	40	50	60	70	80	90	100	125	150	175	200	225	250	300	350	400	500	600
3	2,982	2,965	2,947	2,930	2,913	2,896	2,863	2,831	2,799	2,768	2,738	2,709	2,680	2,610	2,544	2,481	2,421	2,364	2,310	2,208	2,115	2,030	1,878	1,747
5	4,951	4,902	4,855	4,808	4,763	4,718	4,631	4,547	4,466	4,388	4,313	4,240	4,169	4,003	3,850	3,708	3,576	3,453	3,338	3,130	2,946	2,783	2,505	2,278
7	6,904	6,810	6,719	6,630	6,544	6,460	6,298	6,143	5,997	5,857	5,723	5,596	5,474	5,191	4,935	4,704	4,494	4,301	4,125	3,811	3,542	3,309	2,924	2,619
10	9,805	9,617	9,436	9,262	9,094	8,932	8,625	8,339	8,071	7,819	7,583	7,361	7,151	6,676	6,259	5,892	5,566	5,273	5,010	4,555	4,177	3,856	3,342	2,950
15	14,565	14,154	13,766	13,399	13,050	12,720	12,106	11,549	11,041	10,576	10,148	9,754	9,389	8,586	7,910	7,332	6,833	6,398	6,015	5,371	4,852	4,424	3,761	3,271
20	19,234	18,524	17,865	17,251	16,678	16,142	15,166	14,302	13,531	12,839	12,214	11,648	11,131	10,020	9,111	8,353	7,711	7,161	6,685	5,899	5,279	4,777	4,013	3,460
25	23,814	22,736	21,751	20,847	20,016	19,249	17,878	16,689	15,649	14,730	13,914	13,183	12,525	11,136	10,024	9,114	8,356	7,714	7,164	6,269	5,573	5,016	4,181	3,584
30	28,308	26,797	25,439	24,212	23,098	22,082	20,297	18,778	17,471	16,334	15,336	14,453	13,666	12,029	10,742	9,704	8,849	8,132	7,523	6,542	5,788	5,190	4,301	3,672
35	32,719	30,717	28,946	27,368	25,953	24,677	22,468	20,623	19,057	17,712	16,545	15,522	14,618	12,760	11,321	10,174	9,238	8,460	7,802	6,753	5,952	5,321	4,391	3,737
40	37,048	34,502	32,283	30,333	28,604	27,062	24,429	22,262	20,449	18,908	17,584	16,433	15,423	13,369	11,798	10,558	9,553	8,723	8,026	6,920	6,081	5,424	4,460	3,787
45	41,298	38,159	35,464	33,124	31,073	29,262	26,207	23,730	21,680	19,956	18,487	17,219	16,113	13,885	12,198	10,877	9,814	8,940	8,209	7,055	6,186	5,507	4,516	3,828
50	45,471	41,695	38,497	35,756	33,378	31,297	27,828	25,051	22,777	20,883	19,279	17,904	16,712	14,327	12,538	11,146	10,032	9,121	8,361	7,168	6,272	5,576	4,562	3,861
55	49,569	45,115	41,395	38,242	35,535	33,186	29,311	26,246	23,762	21,707	19,979	18,506	17,235	14,710	12,830	11,377	10,219	9,275	8,491	7,262	6,345	5,633	4,600	3,888
60	53,595	48,425	44,165	40,594	37,557	34,943	30,673	27,333	24,649	22,445	20,603	19,040	17,698	15,046	13,085	11,576	10,380	9,407	8,601	7,343	6,406	5,681	4,633	3,911
65	57,549	51,630	46,816	42,822	39,457	36,582	31,929	28,326	25,453	23,110	21,162	19,516	18,108	15,342	13,308	11,751	10,520	9,522	8,697	7,413	6,459	5,723	4,660	3,931
70	61,434	54,736	49,355	44,937	41,245	38,114	33,090	29,236	26,186	23,712	21,665	19,944	18,476	15,605	13,506	11,904	10,643	9,623	8,781	7,474	6,505	5,759	4,684	3,948
75	65,252	57,746	51,789	46,946	42,931	39,549	34,166	30,073	26,855	24,260	22,122	20,330	18,807	15,840	13,682	12,041	10,752	9,712	8,855	7,52[illegible]	6,546	5,791	4,705	3,963
80	69,004	60,665	54,125	48,858	44,524	40,897	35,168	30,846	27,470	24,761	22,537	20,681	19,106	16,052	13,839	12,163	10,849	9,791	8,921	7,575	6,582	5,819	4,724	3,976
85	72,692	63,498	56,368	50,678	46,031	42,165	36,101	31,562	28,037	25,220	22,917	21,000	19,379	16,244	13,982	12,273	10,936	9,862	8,980	7,61[illegible]	6,614	5,844	4,740	3,987
90	76,318	66,247	58,524	52,414	47,459	43,360	36,973	32,227	28,560	25,642	23,266	21,292	19,627	16,418	14,111	12,372	11,015	9,926	9,033	7,655	6,643	5,866	4,755	3,998
95	79,883	68,917	60,598	54,072	48,814	44,488	37,791	32,846	29,045	26,033	23,587	21,561	19,855	16,577	14,228	12,462	11,086	9,984	9,081	7,690	6,668	5,887	4,768	4,007
100	83,389	71,511	62,595	55,656	50,101	45,555	38,557	33,423	29,496	26,394	23,883	21,808	20,065	16,723	14,335	12,544	11,151	10,037	9,124	7,72[illegible]	6,692	5,905	4,780	4,016
105	86,838	74,032	64,518	57,171	51,326	46,565	39,279	33,964	29,916	26,730	24,158	22,037	20,258	16,857	14,434	12,620	11,211	10,085	9,164	7,750	6,713	5,921	4,791	4,023
110	90,230	76,483	66,372	58,622	52,492	47,523	39,958	34,471	30,309	27,043	24,413	22,249	20,438	16,981	14,525	12,689	11,265	10,129	9,201	7,775	6,733	5,937	4,801	4,030
115	93,567	78,868	68,160	60,012	53,604	48,433	40,599	34,947	30,676	27,335	24,651	22,447	20,604	17,096	14,609	12,753	11,316	10,170	9,234	7,800	6,751	5,951	4,810	4,037
120	96,850	81,187	69,886	61,346	54,666	49,298	41,205	35,395	31,021	27,609	24,873	22,631	20,759	17,202	14,686	12,812	11,362	10,207	9,265	7,822	6,767	5,964	4,819	4,043
125	100,081	83,446	71,552	62,627	55,681	50,122	41,779	35,818	31,345	27,865	25,081	22,803	20,904	17,302	14,759	12,867	11,406	10,242	9,294	7,842	6,783	5,975	4,826	4,048
130	103,261	85,645	73,163	63,857	56,651	50,907	42,323	36,217	31,650	28,106	25,276	22,964	21,039	17,394	14,826	12,918	11,446	10,274	9,321	7,861	6,797	5,986	4,834	4,053
135	106,391	87,787	74,721	65,040	57,580	51,656	42,840	36,594	31,938	28,333	25,459	23,115	21,166	17,481	14,889	12,966	11,483	10,305	9,346	7,8[illegible]9	6,810	5,997	4,840	4,058
140	109,472	89,874	76,228	66,179	58,471	52,371	43,331	36,952	32,210	28,547	25,632	23,257	21,285	17,562	14,948	13,011	11,518	10,333	9,369	7,8[illegible]5	6,822	6,006	4,846	4,062
145	112,505	91,908	77,686	67,276	59,326	53,056	43,798	37,292	32,468	28,749	25,795	23,391	21,397	17,638	15,003	13,053	11,551	10,359	9,390	7,9[illegible]1	6,834	6,015	4,852	4,066
150	115,492	93,892	79,099	68,332	60,146	53,711	44,244	37,614	32,712	28,941	25,949	23,518	21,503	17,710	15,055	13,092	11,582	10,384	9,411	7,925	6,845	6,023	4,858	4,070
155	118,434	95,827	80,468	69,352	60,934	54,339	44,669	37,921	32,944	29,122	26,094	23,637	21,603	17,778	15,104	13,129	11,611	10,407	9,430	7,939	6,855	6,031	4,863	4,074
160	121,331	97,715	81,795	70,335	61,692	54,941	45,075	38,213	33,164	29,294	26,232	23,750	21,697	17,842	15,150	13,164	11,638	10,429	9,448	7,951	6,864	6,039	4,868	4,077
165	124,185	99,558	83,082	71,285	62,421	55,518	45,463	38,491	33,374	29,457	26,363	23,858	21,787	17,902	15,193	13,196	11,664	10,450	9,465	7,963	6,873	6,045	4,872	4,080
170	126,996	101,356	84,331	72,202	63,124	56,073	45,834	38,757	33,574	29,613	26,488	23,960	21,872	17,960	15,235	13,228	11,688	10,469	9,481	7,975	6,882	6,052	4,876	4,083
175	129,766	103,113	85,543	73,089	63,800	56,607	46,190	39,012	33,764	29,761	26,606	24,057	21,953	18,014	15,274	13,257	11,711	10,488	9,496	7,985	6,889	6,058	4,880	4,086
180	132,495	104,828	86,720	73,947	64,453	57,120	46,531	39,255	33,946	29,902	26,719	24,149	22,029	18,066	15,311	13,285	11,733	10,505	9,510	7,995	6,897	6,064	4,884	4,089
185	135,184	106,505	87,865	74,777	65,083	57,614	46,859	39,487	34,120	30,037	26,827	24,237	22,102	18,115	15,346	13,312	11,753	10,522	9,524	8,005	6,904	6,069	4,888	4,091
190	137,835	108,143	88,977	75,581	65,691	58,090	47,173	39,710	34,286	30,166	26,930	24,320	22,172	18,162	15,380	13,337	11,773	10,537	9,537	8,014	6,911	6,075	4,891	4,093
195	140,447	109,745	90,058	76,360	66,279	58,549	47,475	39,924	34,446	30,289	27,028	24,401	22,239	18,206	15,412	13,361	11,792	10,552	9,549	8,023	6,917	6,080	4,894	4,096
200	143,022	111,311	91,110	77,115	66,847	58,992	47,766	40,130	34,598	30,407	27,122	24,477	22,302	18,249	15,442	13,384	11,810	10,567	9,561	8,031	6,924	6,084	4,897	4,098

750 MCM | 1 Run(s) | 208 Volts

C-Value = 24,136 | Magnetic Duct | Copper

Calculations:

1. $f = \dfrac{\text{Length} \times 2 \times \text{Isca}}{\text{Runs} \times \text{C-Value} \times \text{E L-L}}$

2. $M = \dfrac{1}{1 + f}$

3. Isc = Isca x M

4. Add Motor Contribution = Motor FLA x 4

* All results are given in symmetrical amperes

Single-Phase L-L Bolted Fault-Current* Table

One-Way Distance in Feet

Available Isc in Thousands (Isca)	5	10	15	20	25	30	40	50	60	70	80	90	100	125	150	175	200	225	250	300	350	400	500	600
3	2,991	2,982	2,973	2,965	2,956	2,947	2,930	2,913	2,896	2,880	2,863	2,847	2,831	2,791	2,753	2,716	2,680	2,644	2,610	2,544	2,481	2,421	2,310	2,208
5	4,975	4,951	4,926	4,902	4,879	4,855	4,808	4,763	4,718	4,674	4,631	4,589	4,547	4,446	4,350	4,258	4,169	4,085	4,003	3,850	3,708	3,576	3,338	3,130
7	6,952	6,904	6,857	6,810	6,764	6,719	6,630	6,544	6,460	6,378	6,298	6,220	6,143	5,961	5,789	5,627	5,474	5,328	5,191	4,935	4,704	4,494	4,125	3,811
10	9,901	9,805	9,710	9,617	9,526	9,436	9,262	9,094	8,932	8,776	8,625	8,480	8,339	8,006	7,699	7,415	7,151	6,905	6,676	6,259	5,892	5,566	5,010	4,555
15	14,779	14,565	14,357	14,154	13,957	13,766	13,399	13,050	12,720	12,405	12,106	11,821	11,549	10,921	10,358	9,850	9,389	8,970	8,586	7,910	7,332	6,833	6,015	5,371
20	19,609	19,234	18,872	18,524	18,189	17,865	17,251	16,678	16,142	15,639	15,166	14,722	14,302	13,351	12,519	11,784	11,131	10,547	10,020	9,111	8,353	7,711	6,685	5,899
25	24,393	23,814	23,262	22,736	22,232	21,751	20,847	20,016	19,249	18,538	17,878	17,263	16,689	15,409	14,310	13,359	12,525	11,790	11,136	10,024	9,114	8,356	7,164	6,269
30	29,130	28,308	27,532	26,797	26,101	25,439	24,212	23,098	22,082	21,152	20,297	19,508	18,778	17,173	15,820	14,665	13,666	12,796	12,029	10,742	9,704	8,849	7,523	6,542
35	33,821	32,719	31,686	30,717	29,805	28,946	27,368	25,953	24,677	23,521	22,468	21,506	20,623	18,702	17,109	15,765	14,618	13,626	12,760	11,321	10,174	9,238	7,802	6,753
40	38,468	37,048	35,730	34,502	33,356	32,283	30,333	28,604	27,062	25,678	24,429	23,295	22,262	20,040	18,222	16,706	15,423	14,323	13,369	11,798	10,558	9,553	8,026	6,920
45	43,070	41,298	39,667	38,159	36,762	35,464	33,124	31,073	29,262	27,651	26,207	24,907	23,730	21,222	19,193	17,519	16,113	14,916	13,885	12,198	10,877	9,814	8,209	7,055
50	47,628	45,471	43,501	41,695	40,032	38,497	35,756	33,378	31,297	29,461	27,828	26,366	25,051	22,272	20,049	18,229	16,712	15,428	14,327	12,538	11,146	10,032	8,361	7,168
55	52,144	49,569	47,237	45,115	43,175	41,395	38,242	35,535	33,186	31,128	29,311	27,694	26,246	23,212	20,807	18,854	17,235	15,873	14,710	12,830	11,377	10,219	8,491	7,262
60	56,617	53,595	50,879	48,425	46,197	44,165	40,594	37,557	34,943	32,669	30,673	28,907	27,333	24,058	21,484	19,408	17,698	16 264	15,046	13,085	11,576	10,380	8,601	7,343
65	61,048	57,549	54,429	51,630	49,105	46,816	42,822	39,457	36,582	34,097	31,929	30,019	28,326	24,824	22,093	19,903	18,108	16,611	15,342	13,308	11,751	10,520	8,697	7,413
70	65,438	61,434	57,892	54,736	51,906	49,355	44,937	41,245	38,114	35,424	33,090	31,043	29,236	25,520	22,643	20,348	18,476	16,919	15,605	13,506	11,904	10,643	8,781	7,474
75	69,787	65,252	61,270	57,746	54,606	51,789	46,946	42,931	39,549	36,661	34,166	31,989	30,073	26,156	23,142	20,750	18,807	17,196	15,840	13,682	12,041	10,752	8,855	7,527
80	74,096	69,004	64,567	60,665	57,209	54,125	48,858	44,524	40,897	37,817	35,168	32,865	30,846	26,739	23,597	21,116	19,106	17,446	16,052	13,839	12,163	10,849	8,921	7,575
85	78,366	72,692	67,785	63,498	59,721	56,368	50,678	46,031	42,165	38,898	36,101	33,679	31,562	27,275	24,013	21,449	19,379	17,673	16,244	13,982	12,273	10,936	8,980	7,617
90	82,596	76,318	70,927	66,247	62,147	58,524	52,414	47,459	43,360	39,913	36,973	34,437	32,227	27,770	24,396	21,753	19,627	17,880	16,418	14,111	12,372	11,015	9,033	7,655
95	86,788	79,883	73,996	68,917	64,491	60,598	54,072	48,814	44,488	40,867	37,791	35,145	32,846	28,228	24,749	22,034	19,855	18,069	16,577	14,228	12,462	11,086	9,081	7,690
100	90,943	83,389	76,995	71,511	66,757	62,595	55,656	50,101	45,555	41,765	38,557	35,807	33,423	28,654	25,076	22,292	20,065	18,242	16,723	14,335	12,544	11,151	9,124	7,721
105	95,059	86,838	79,925	74,032	68,948	64,518	57,171	51,326	46,565	42,613	39,279	36,428	33,964	29,051	25,379	22,531	20,258	18,402	16,857	14,434	12,620	11,211	9,164	7,750
110	99,139	90,230	82,790	76,483	71,070	66,372	58,622	52,492	47,523	43,413	39,958	37,012	34,471	29,421	25,661	22,753	20,438	18,550	16,981	14,525	12,689	11,265	9,201	7,776
115	103,182	93,567	85,591	78,868	73,124	68,160	60,012	53,604	48,433	44,171	40,599	37,562	34,947	29,767	25,924	22,960	20,604	18,687	17,096	14,609	12,753	11,316	9,234	7,800
120	107,189	96,850	88,330	81,187	75,114	69,886	61,346	54,666	49,298	44,890	41,205	38,080	35,395	30,091	26,170	23,152	20,759	18,814	17,202	14,686	12,812	11,362	9,265	7,822
125	111,161	100,081	91,009	83,446	77,043	71,552	62,627	55,681	50,122	45,572	41,779	38,569	35,818	30,396	26,400	23,333	20,904	18,933	17,302	14,759	12,867	11,406	9,294	7,842
130	115,098	103,261	93,631	85,645	78,914	73,163	63,857	56,651	50,907	46,220	42,323	39,033	36,217	30,683	26,616	23,501	21,039	19,044	17,394	14,826	12,918	11,446	9,321	7,861
135	119,000	106,391	96,197	87,787	80,728	74,721	65,040	57,580	51,656	46,837	42,840	39,472	36,594	30,954	26,820	23,660	21,166	19,148	17,481	14,889	12,966	11,483	9,346	7,879
140	122,868	109,472	98,710	89,874	82,490	76,228	66,179	58,471	52,371	47,424	43,331	39,888	36,952	31,209	27,011	23,809	21,285	19,245	17,562	14,948	13,011	11,518	9,369	7,895
145	126,702	112,505	101,169	91,908	84,201	77,686	67,276	59,326	53,056	47,985	43,798	40,284	37,292	31,451	27,192	23,949	21,397	19,337	17,638	15,003	13,053	11,551	9,390	7,911
150	130,504	115,492	103,578	93,892	85,863	79,099	68,332	60,146	53,711	48,520	44,244	40,660	37,614	31,680	27,363	24,082	21,503	19,423	17,710	15,055	13,092	11,582	9,411	7,925
155	134,272	118,434	105,938	95,827	87,478	80,468	69,352	60,934	54,339	49,032	44,669	41,019	37,921	31,897	27,525	24,207	21,603	19,505	17,778	15,104	13,129	11,611	9,430	7,939
160	138,008	121,331	108,250	97,715	89,049	81,795	70,335	61,692	54,941	49,521	45,075	41,361	38,213	32,104	27,679	24,326	21,697	19,582	17,842	15,150	13,164	11,638	9,448	7,951
165	141,712	124,185	110,516	99,558	90,576	83,082	71,285	62,421	55,518	49,990	45,463	41,688	38,491	32,300	27,825	24,438	21,787	19,655	17,902	15,193	13,196	11,664	9,465	7,963
170	145,385	126,996	112,737	101,356	92,063	84,331	72,202	63,124	56,073	50,439	45,834	42,000	38,757	32,487	27,963	24,545	21,872	19,724	17,960	15,235	13,228	11,688	9,481	7,975
175	149,026	129,766	114,914	103,113	93,510	85,543	73,089	63,800	56,607	50,871	46,190	42,298	39,012	32,666	28,095	24,647	21,953	19,789	18,014	15,274	13,257	11,711	9,496	7,985
180	152,636	132,495	117,049	104,828	94,919	86,720	73,947	64,453	57,120	51,285	46,531	42,584	39,255	32,836	28,221	24,744	22,029	19,852	18,066	15,311	13,285	11,733	9,510	7,995
185	156,217	135,184	119,143	106,505	96,291	87,865	74,777	65,083	57,614	51,683	46,859	42,858	39,487	32,999	28,341	24,836	22,102	19,911	18,115	15,346	13,312	11,753	9,524	8,005
190	159,767	137,835	121,197	108,143	97,628	88,977	75,581	65,691	58,090	52,065	47,173	43,121	39,710	33,154	28,456	24,924	22,172	19,968	18,162	15,380	13,337	11,773	9,537	8,014
195	163,288	140,447	123,212	109,745	98,932	90,058	76,360	66,279	58,549	52,434	47,475	43,374	39,924	33,303	28,566	25,008	22,239	20,021	18,206	15,412	13,361	11,792	9,549	8,023
200	166,779	143,022	125,190	111,311	100,202	91,110	77,115	66,847	58,992	52,789	47,766	43,616	40,130	33,446	28,671	25,089	22,302	20,073	18,249	15,442	13,384	11,810	9,561	8,031

750 MCM | **2 Run(s)** | **208 Volts**

C-Value = 24,136 | **Magnetic Duct** | **Copper**

Calculations:

1. $f = \dfrac{\text{Length} \times 2 \times \text{Isca}}{\text{Runs} \times \text{C-Value} \times \text{E L-L}}$

2. $M = \dfrac{1}{1 + f}$

3. Isc = Isca x M

4. Add Motor Contribution = Motor FLA x 4

* All results are given in symmetrical amperes

Single-Phase L-L Bolted Fault-Current* Table

One-Way Distance in Feet

Available Isc in Thousands (Isca)

	5	10	15	20	25	30	40	50	60	70	80	90	100	125	150	175	200	225	250	300	350	400	500	600
3	2,983	2,966	2,950	2,933	2,917	2,901	2,869	2,838	2,808	2,778	2,749	2,721	2,693	2,625	2,562	2,501	2,443	2,387	2,334	2,235	2,144	2,060	1,910	1,781
5	4,953	4,907	4,861	4,817	4,773	4,730	4,647	4,566	4,488	4,413	4,340	4,269	4,201	4,040	3,890	3,751	3,622	3,502	3,389	3,184	3,002	2,840	2,563	2,335
7	6,908	6,818	6,731	6,646	6,563	6,482	6,326	6,178	6,036	5,900	5,771	5,647	5,528	5,252	5,002	4,775	4,568	4,377	4,203	3,891	3,623	3,390	3,003	2,695
10	9,813	9,634	9,460	9,293	9,132	8,976	8,679	8,402	8,142	7,897	7,667	7,450	7,244	6,777	6,367	6,004	5,679	5,388	5,126	4,670	4,289	3,966	3,446	3,047
15	14,584	14,190	13,817	13,464	13,127	12,808	12,2[illegible]3	11,671	11,174	10,719	10,299	9,911	9,551	8,755	8,082	7,506	7,006	6,568	6,182	5,532	5,005	4,570	3,893	3,391
20	19,267	18,586	17,951	17,359	16,804	16,284	15,334	14,489	13,732	13,050	12,433	11,872	11,359	10,251	9,341	8,579	7,932	7,375	6,892	6,093	5,460	4,947	4,163	3,594
25	23,865	22,829	21,879	21,005	20,198	19,451	18,111	16,944	15,918	15,009	14,198	13,471	12,814	11,422	10,303	9,384	8,615	7,963	7,402	6,489	5,776	5,204	4,344	3,728
30	28,381	26,927	25,615	24,425	23,341	22,349	20,598	19,101	17,807	16,678	15,683	14,800	14,011	12,364	11,053	10,010	9,140	8,409	7,786	6,782	6,007	5,391	4,474	3,823
35	32,816	30,888	29,174	27,640	26,260	25,011	22,838	21,013	19,457	18,116	16,949	15,922	15,013	13,137	11,678	10,511	9,556	8,760	8,086	7,008	6,184	5,533	4,571	3,894
40	37,172	34,718	32,567	30,668	28,977	27,464	24,856	22,717	20,910	19,370	18,040	16,882	15,863	13,784	12,187	10,921	9,893	9,043	8,327	7,188	6,324	5,645	4,647	3,949
45	41,452	38,423	35,806	33,523	31,514	29,732	26,711	24,247	22,200	20,471	18,992	17,713	16,595	14,333	12,614	11,263	10,173	9,276	8,524	7,335	6,437	5,735	4,708	3,993
50	45,658	42,010	38,902	36,222	33,887	31,835	28,397	25,628	23,352	21,447	19,829	18,438	17,230	14,804	12,977	11,552	10,408	9,471	8,688	7,456	6,530	5,809	4,758	4,029
55	49,792	45,484	41,863	38,775	36,112	33,791	29,543	26,881	24,387	22,317	20,571	19,078	17,787	15,214	13,291	11,800	10,609	9,637	8,828	7,559	6,609	5,871	4,799	4,058
60	53,854	48,851	44,698	41,196	38,202	35,615	31,366	28,022	25,323	23,098	21,233	19,646	18,280	15,573	13,564	12,014	10,782	9,780	8,948	7,646	6,675	5,923	4,834	4,083
65	57,848	52,115	47,415	43,493	40,170	37,319	32,680	29,066	26,173	23,803	21,827	20,153	18,718	15,890	13,804	12,202	10,934	9,904	9,051	7,72[illegible]	6,733	5,969	4,864	4,105
70	61,776	55,280	50,021	45,676	42,025	38,915	33,897	30,026	26,948	24,442	22,363	20,610	19,112	16,172	14,017	12,368	11,066	10,013	9,142	7,78[illegible]	6,783	6,008	4,890	4,123
75	65,637	58,353	52,523	47,753	43,777	40,412	35,028	30,909	27,658	25,025	22,850	21,023	19,466	16,425	14,206	12,516	11,184	10,109	9,223	7,84[illegible]	6,827	6,043	4,913	4,140
80	69,435	61,335	54,928	49,732	45,435	41,821	36,081	31,727	28,310	25,558	23,293	21,397	19,787	16,653	14,377	12,647	11,290	10,195	9,294	7,898	6,866	6,073	4,934	4,154
85	73,171	64,232	57,239	51,620	47,005	43,148	37,064	32,484	28,912	26,047	23,699	21,739	20,079	16,860	14,530	12,766	11,384	10,272	9,358	7,944	6,901	6,101	4,952	4,167
90	76,846	67,047	59,464	53,422	48,495	44,400	37,384	33,189	29,469	26,498	24,072	22,053	20,346	17,048	14,669	12,874	11,469	10,341	9,416	7,985	6,933	6,125	4,968	4,178
95	80,462	69,783	61,606	55,145	49,910	45,583	38,347	33,846	29,986	26,916	24,416	22,341	20,591	17,219	14,796	12,971	11,547	10,404	9,468	8,023	6,961	6,147	4,982	4,188
100	84,020	72,444	63,671	56,793	51,257	46,704	39,658	34,460	30,466	27,302	24,734	22,607	20,817	17,377	14,913	13,060	11,617	10,462	9,515	8,057	6,986	6,167	4,995	4,198
105	87,522	75,032	65,662	58,372	52,539	47,766	40,422	35,035	30,915	27,662	25,028	22,853	21,025	17,522	15,019	13,142	11,682	10,514	9,558	8,088	7,010	6,185	5,007	4,206
110	90,968	77,551	67,583	59,885	53,762	48,775	41,142	35,574	31,334	27,997	25,303	23,081	21,218	17,656	15,117	13,217	11,741	10,562	9,598	8,1[illegible]6	7,031	6,202	5,018	4,214
115	94,361	80,003	69,438	61,337	54,929	49,733	41,822	36,082	31,727	28,310	25,558	23,294	21,398	17,780	15,208	13,287	11,796	10,606	9,635	8,1[illegible]3	7,051	6,217	5,028	4,221
120	97,701	82,391	71,230	62,731	56,044	50,646	42,465	36,560	32,096	28,604	25,797	23,492	21,565	17,895	15,293	13,351	11,847	10,647	9,668	8,1[illegible]7	7,069	6,231	5,037	4,227
125	100,990	84,718	72,962	64,071	57,111	51,516	43,075	37,011	32,443	28,879	26,021	23,677	21,721	18,002	15,371	13,411	11,894	10,685	9,700	8,1[illegible]9	7,085	6,244	5,046	4,233
130	104,229	86,986	74,638	65,359	58,133	52,345	43,653	37,437	32,770	29,138	26,231	23,851	21,867	18,103	15,444	13,466	11,938	10,721	9,729	8,2[illegible]0	7,101	6,256	5,053	4,239
135	107,419	89,196	76,259	66,600	59,112	53,138	44,203	37,841	33,079	29,382	26,428	24,014	22,004	18,197	15,512	13,518	11,978	10,753	9,756	8,229	7,115	6,267	5,061	4,244
140	110,561	91,352	77,829	67,794	60,051	53,896	44,726	38,223	33,371	29,612	26,614	24,168	22,133	18,285	15,576	13,567	12,016	10,784	9,781	8,247	7,129	6,278	5,068	4,249
145	113,656	93,454	79,350	68,945	60,953	54,621	45,224	38,586	33,648	29,830	26,790	24,312	22,254	18,367	15,636	13,612	12,052	10,813	9,805	8,254	7,141	6,287	5,074	4,253
150	116,705	95,506	80,825	70,056	61,819	55,315	45,699	38,932	33,910	30,036	26,956	24,449	22,369	18,445	15,693	13,655	12,086	10,840	9,827	8,279	7,153	6,296	5,080	4,257
155	119,710	97,509	82,254	71,127	62,652	55,981	46,153	39,261	34,159	30,231	27,113	24,578	22,477	18,519	15,746	13,695	12,117	10,865	9,848	8,294	7,164	6,305	5,085	4,261
160	122,670	99,464	83,642	72,162	63,453	56,620	46,587	39,574	34,396	30,416	27,262	24,701	22,579	18,588	15,796	13,733	12,147	10,889	9,867	8,308	7,174	6,313	5,091	4,265
165	125,588	101,374	84,988	73,162	64,225	57,234	47,001	39,873	34,622	30,593	27,404	24,817	22,676	18,654	15,843	13,769	12,175	10,911	9,886	8,321	7,184	6,321	5,095	4,268
170	128,464	103,240	86,295	74,129	64,969	57,824	47,398	40,158	34,837	30,760	27,538	24,927	22,768	18,716	15,888	13,803	12,201	10,933	9,903	8,333	7,193	6,328	5,100	4,271
175	131,299	105,063	87,565	75,064	65,686	58,391	47,779	40,431	35,042	30,920	27,666	25,032	22,856	18,775	15,931	13,835	12,226	10,953	9,920	8,345	7,202	6,334	5,104	4,274
180	134,093	106,844	88,799	75,969	66,378	58,938	48,144	40,692	35,238	31,073	27,788	25,132	22,939	18,831	15,971	13,865	12,250	10,972	9,935	8,356	7,210	6,341	5,109	4,277
185	136,849	108,586	89,999	76,846	67,046	59,464	48,495	40,942	35,425	31,218	27,905	25,227	23,018	18,884	16,009	13,894	12,273	10,990	9,950	8,367	7,218	6,347	5,113	4,280
190	139,566	110,290	91,167	77,695	67,692	59,971	48,832	41,182	35,605	31,358	28,016	25,318	23,094	18,935	16,046	13,922	12,294	11,007	9,964	8,377	7,226	6,353	5,116	4,283
195	142,245	111,956	92,302	78,518	68,316	60,460	49,155	41,412	35,776	31,491	28,122	25,405	23,166	18,984	16,081	13,948	12,314	11,024	9,978	8,386	7,233	6,358	5,120	4,285
200	144,887	113,587	93,407	79,317	68,920	60,933	49,467	41,633	35,941	31,619	28,224	25,488	23,235	19,030	16,114	13,973	12,334	11,039	9,990	8,395	7,239	6,363	5,123	4,288

1000 MCM | 1 Run(s) | 208 Volts | C-Value = 25,278 | Magnetic Duct | Copper

Calculations:

1. $f = \dfrac{\text{Length} \times 2 \times \text{Isca}}{\text{Runs} \times \text{C-Value} \times \text{E L-L}}$
2. $M = \dfrac{1}{1 + f}$
3. $\text{Isc} = \text{Isca} \times M$
4. Add Motor Contribution = Motor FLA x 4

* All results are given in symmetrical amperes

Single-Phase L-L Bolted Fault-Current* Table

One-Way Distance in Feet

Available Isc in Thousands (Isca)

	5	10	15	20	25	30	40	50	60	70	80	90	100	125	150	175	200	225	250	300	350	400	500	600
3	2,991	2,983	2,975	2,966	2,958	2,950	2,933	2,917	2,901	2,885	2,869	2,853	2,838	2,800	2,763	2,728	2,693	2,659	2,625	2,562	2,501	2,443	2,334	2,235
5	4,976	4,953	4,930	4,907	4,884	4,861	4,817	4,773	4,730	4,688	4,647	4,606	4,566	4,469	4,376	4,287	4,201	4,119	4,040	3,890	3,751	3,622	3,389	3,184
7	6,954	6,908	6,863	6,818	6,775	6,731	6,646	6,563	6,482	6,403	6,326	6,251	6,178	6,001	5,835	5,677	5,528	5,386	5,252	5,002	4,775	4,568	4,203	3,891
10	9,906	9,813	9,723	9,634	9,546	9,460	9,293	9,132	8,976	8,825	8,679	8,538	8,402	8,079	7,780	7,503	7,244	7,003	6,777	6,367	6,004	5,679	5,126	4,670
15	14,789	14,584	14,384	14,190	14,001	13,817	13,464	13,127	12,808	12,503	12,213	11,935	11,671	11,057	10,505	10,005	9,551	9,136	8,755	8,082	7,506	7,006	6,182	5,532
20	19,627	19,267	18,920	18,586	18,263	17,951	17,359	16,804	16,284	15,794	15,334	14,899	14,489	13,555	12,734	12,007	11,359	10,777	10,251	9,341	8,579	7,932	6,892	6,093
25	24,419	23,865	23,336	22,829	22,344	21,879	21,005	20,198	19,451	18,757	18,111	17,508	16,944	15,680	14,592	13,646	12,814	12,078	11,422	10,303	9,384	8,615	7,402	6,489
30	29,168	28,381	27,635	26,927	26,255	25,615	24,425	23,341	22,349	21,438	20,598	19,821	19,101	17,511	16,165	15,011	14,011	13,136	12,364	11,063	10,010	9,140	7,786	6,782
35	33,873	32,816	31,822	30,888	30,006	29,174	27,640	26,260	25,011	23,875	22,838	21,887	21,013	19,104	17,513	16,167	15,013	14,013	13,137	11,678	10,511	9,556	8,086	7,008
40	38,534	37,172	35,903	34,718	33,608	32,567	30,668	28,977	27,464	26,100	24,866	23,743	22,717	20,503	18,681	17,157	15,863	14,751	13,784	12,187	10,921	9,893	8,327	7,188
45	43,153	41,452	39,880	38,423	37,069	35,806	33,523	31,514	29,732	28,141	26,711	25,420	24,247	21,741	19,704	18,016	16,595	15,381	14,333	12,614	11,263	10,173	8,524	7,335
50	47,731	45,658	43,758	42,010	40,396	38,902	36,222	33,887	31,835	30,018	28,397	26,942	25,628	22,845	20,606	18,767	17,230	15,925	14,804	12,977	11,552	10,408	8,688	7,456
55	52,266	49,792	47,540	45,484	43,598	41,863	38,775	36,112	33,791	31,751	29,943	28,329	26,881	23,835	21,408	19,430	17,787	16,400	15,214	13,291	11,800	10,609	8,828	7,559
60	56,761	53,854	51,231	48,851	46,682	44,698	41,196	38,202	35,615	33,355	31,366	29,600	28,022	24,728	22,126	20,020	18,280	16,818	15,573	13,564	12,014	10,782	8,948	7,646
65	61,216	57,848	54,832	52,115	49,654	47,415	43,493	40,170	37,319	34,846	32,680	30,767	29,066	25,537	22,772	20,547	18,718	17,189	15,890	13,804	12,202	10,934	9,051	7,722
70	65,631	61,776	58,348	55,280	52,520	50,021	45,676	42,025	38,915	36,233	33,897	31,844	30,026	26,274	23,357	21,022	19,112	17,520	16,172	14,017	12,368	11,066	9,142	7,788
75	70,007	65,637	61,781	58,353	55,285	52,523	47,753	43,777	40,412	37,528	35,028	32,840	30,909	26,949	23,888	21,451	19,466	17,817	16,425	14,206	12,516	11,184	9,223	7,846
80	74,344	69,435	65,134	61,335	57,955	54,928	49,732	45,435	41,821	38,739	36,081	33,764	31,727	27,568	24,373	21,842	19,787	18 085	16,653	14,377	12,647	11,290	9,294	7,898
85	78,643	73,171	68,411	64,232	60,534	57,239	51,620	47,005	43,148	39,875	37,064	34,624	32,484	28,138	24,818	22,198	20,079	18,329	16,860	14,530	12,766	11,384	9,358	7,944
90	82,904	76,846	71,613	67,047	63,028	59,464	53,422	48,495	44,400	40,942	37,984	35,425	33,189	28,665	25,227	22,525	20,346	18,551	17,048	14,669	12,874	11,469	9,416	7,986
95	87,129	80,462	74,743	69,783	65,440	61,606	55,145	49,910	45,583	41,947	38,847	36,175	33,846	29,154	25,605	22,826	20,591	18,755	17,219	14,796	12,971	11,547	9,468	8,023
100	91,316	84,020	77,804	72,444	67,774	63,671	56,793	51,257	46,704	42,894	39,658	36,877	34,460	29,608	25,955	23,103	20,817	18,942	17,377	14,913	13,060	11,617	9,515	8,057
105	95,467	87,522	80,797	75,032	70,035	65,662	58,372	52,539	47,766	43,788	40,422	37,536	35,035	30,032	26,279	23,360	21,025	19,114	17,522	15,019	13,142	11,682	9,558	8,088
110	99,583	90,968	83,725	77,551	72,224	67,583	59,885	53,762	48,775	44,634	41,142	38,156	35,574	30,428	26,582	23,599	21,218	19,274	17,656	15,117	13,217	11,741	9,598	8,116
115	103,663	94,361	86,591	80,003	74,347	69,438	61,337	54,929	49,733	45,436	41,822	38,740	36,082	30,798	26,864	23,821	21,398	19,422	17,780	15,208	13,287	11,796	9,635	8,143
120	107,709	97,701	89,396	82,391	76,405	71,230	62,731	56,044	50,646	46,196	42,465	39,292	36,560	31,145	27,128	24,029	21,565	19,559	17,895	15,293	13,351	11,847	9,668	8,167
125	111,720	100,990	92,141	84,718	78,402	72,962	64,071	57,111	51,516	46,919	43,075	39,813	37,011	31,472	27,376	24,223	21,721	19,688	18,002	15,371	13,411	11,894	9,700	8,189
130	115,697	104,229	94,830	86,986	80,340	74,638	65,359	58,133	52,345	47,606	43,653	40,307	37,437	31,780	27,608	24,405	21,867	19,808	18,103	15,444	13,466	11,938	9,729	8,210
135	119,641	107,419	97,463	89,196	82,222	76,259	66,600	59,112	53,138	48,260	44,203	40,775	37,841	32,070	27,827	24,575	22,004	19,920	18,197	15,512	13,518	11,978	9,756	8,229
140	123,551	110,561	100,043	91,352	84,050	77,829	67,794	60,051	53,896	48,885	44,726	41,220	38,223	32,345	28,033	24,736	22,133	20,026	18,285	15,576	13,567	12,016	9,781	8,247
145	127,429	113,656	102,570	93,454	85,827	79,350	68,945	60,953	54,621	49,480	45,224	41,643	38,586	32,604	28,228	24,888	22,254	20,125	18,367	15,636	13,612	12,052	9,805	8,264
150	131,274	116,705	105,047	95,506	87,554	80,825	70,056	61,819	55,315	50,050	45,699	42,045	38,932	32,851	28,413	25,031	22,369	20,218	18,445	15,693	13,655	12,086	9,827	8,279
155	135,088	119,710	107,475	97,509	89,234	82,254	71,127	62,652	55,981	50,594	46,153	42,429	39,261	33,084	28,587	25,167	22,477	20,307	18,519	15,746	13,695	12,117	9,848	8,294
160	138,870	122,670	109,855	99,464	90,869	83,642	72,162	63,453	56,620	51,116	46,587	42,795	39,574	33,307	28,753	25,295	22,579	20,390	18,588	15,796	13,733	12,147	9,867	8,308
165	142,621	125,588	112,189	101,374	92,461	84,988	73,162	64,225	57,234	51,615	47,001	43,144	39,873	33,518	28,910	25,417	22,676	20,469	18,654	15,843	13,769	12,175	9,886	8,321
170	146,342	128,464	114,479	103,240	94,010	86,295	74,129	64,969	57,824	52,095	47,398	43,479	40,158	33,719	29,060	25,532	22,768	20,544	18,716	15,888	13,803	12,201	9,903	8,333
175	150,032	131,299	116,725	105,063	95,519	87,565	75,064	65,686	58,391	52,555	47,779	43,799	40,431	33,912	29,203	25,642	22,856	20,615	18,775	15,931	13,835	12,226	9,920	8,345
180	153,692	134,093	118,928	106,844	96,990	88,799	75,969	66,378	58,938	52,997	48,144	44,106	40,692	34,095	29,339	25,747	22,939	20,683	18,831	15,971	13,865	12,250	9,935	8,356
185	157,323	136,849	121,090	108,586	98,423	89,999	76,846	67,046	59,464	53,422	48,495	44,400	40,942	34,271	29,469	25,847	23,018	20,747	18,884	16,009	13,894	12,273	9,950	8,367
190	160,924	139,566	123,213	110,290	99,821	91,167	77,695	67,692	59,971	53,831	48,832	44,682	41,182	34,439	29,593	25,942	23,094	20,809	18,935	16,046	13,922	12,294	9,964	8,377
195	164,496	142,245	125,296	111,956	101,184	92,302	78,518	68,316	60,460	54,225	49,155	44,953	41,412	34,599	29,711	26,034	23,166	20,867	18,984	16,081	13,948	12,314	9,978	8,386
200	168,040	144,887	127,342	113,587	102,513	93,407	79,317	68,920	60,933	54,605	49,467	45,213	41,633	34,753	29,825	26,121	23,235	20,923	19,030	16,114	13,973	12,334	9,990	8,395

1000 MCM | **2 Run(s)** | **208 Volts** | **C-Value = 25,278** | **Magnetic Duct** | **Copper**

Calculations:

1. $f = \dfrac{\text{Length} \times 2 \times \text{Isca}}{\text{Runs} \times \text{C-Value} \times \text{E L-L}}$

2. $M = \dfrac{1}{1 + f}$

3. Isc = Isca x M

4. Add Motor Contribution = Motor FLA x 4

* All results are given in symmetrical amperes

Single-Phase L-N Fault-Current Tables

Single-Phase L-N Bolted Fault Calculations

Motor Contribution: (FLA) X 4 = Proj. No.: By: Date: Page of

Node	From Node	Name	Wire Size	(Isca x	1.5	x	2	x	Length)	/	(Runs	x	C-Value	x	E L-N)	+	1	(1/x)	x	Isca	x	1.5	=	Isc	+	Motor Contribution	=	Isc @ Node
					1.5	x	2	x		/		x		x		+	1	(1/x)	x		x	1.5	=		+		=	
					1.5	x	2	x		/		x		x		+	1	(1/x)	x		x	1.5	=		+		=	
					1.5	x	2	x		/		x		x		+	1	(1/x)	x		x	1.5	=		+		=	
					1.5	x	2	x		/		x		x		+	1	(1/x)	x		x	1.5	=		+		=	
					1.5	x	2	x		/		x		x		+	1	(1/x)	x		x	1.5	=		+		=	
					1.5	x	2	x		/		x		x		+	1	(1/x)	x		x	1.5	=		+		=	
					1.5	x	2	x		/		x		x		+	1	(1/x)	x		x	1.5	=		+		=	
					1.5	x	2	x		/		x		x		+	1	(1/x)	x		x	1.5	=		+		=	
					1.5	x	2	x		/		x		x		+	1	(1/x)	x		x	1.5	=		+		=	
					1.5	x	2	x		/		x		x		+	1	(1/x)	x		x	1.5	=		+		=	
					1.5	x	2	x		/		x		x		+	1	(1/x)	x		x	1.5	=		+		=	
					1.5	x	2	x		/		x		x		+	1	(1/x)	x		x	1.5	=		+		=	
					1.5	x	2	x		/		x		x		+	1	(1/x)	x		x	1.5	=		+		=	
					1.5	x	2	x		/		x		x		+	1	(1/x)	x		x	1.5	=		+		=	
					1.5	x	2	x		/		x		x		+	1	(1/x)	x		x	1.5	=		+		=	
					1.5	x	2	x		/		x		x		+	1	(1/x)	x		x	1.5	=		+		=	
					1.5	x	2	x		/		x		x		+	1	(1/x)	x		x	1.5	=		+		=	
					1.5	x	2	x		/		x		x		+	1	(1/x)	x		x	1.5	=		+		=	
					1.5	x	2	x		/		x		x		+	1	(1/x)	x		x	1.5	=		+		=	
					1.5	x	2	x		/		x		x		+	1	(1/x)	x		x	1.5	=		+		=	
					1.5	x	2	x		/		x		x		+	1	(1/x)	x		x	1.5	=		+		=	
					1.5	x	2	x		/		x		x		+	1	(1/x)	x		x	1.5	=		+		=	
					1.5	x	2	x		/		x		x		+	1	(1/x)	x		x	1.5	=		+		=	
					1.5	x	2	x		/		x		x		+	1	(1/x)	x		x	1.5	=		+		=	
					1.5	x	2	x		/		x		x		+	1	(1/x)	x		x	1.5	=		+		=	
					1.5	x	2	x		/		x		x		+	1	(1/x)	x		x	1.5	=		+		=	
					1.5	x	2	x		/		x		x		+	1	(1/x)	x		x	1.5	=		+		=	
					1.5	x	2	x		/		x		x		+	1	(1/x)	x		x	1.5	=		+		=	
					1.5	x	2	x		/		x		x		+	1	(1/x)	x		x	1.5	=		+		=	
					1.5	x	2	x		/		x		x		+	1	(1/x)	x		x	1.5	=		+		=	
					1.5	x	2	x		/		x		x		+	1	(1/x)	x		x	1.5	=		+		=	
					1.5	x	2	x		/		x		x		+	1	(1/x)	x		x	1.5	=		+		=	
					1.5	x	2	x		/		x		x		+	1	(1/x)	x		x	1.5	=		+		=	

"C"-Values for Copper Conductors

Conductor AWG	Three Single Conductors						Three-Conductor Cable					
	Magnetic Duct			Non-Magnetic Duct			Magnetic Duct			Non-Magnetic Duct		
	600V	5KV	15KV	600V	5KV	15KV	600V	5KV	15KV	600V	5KV	15KV
# 14	**389**	389	389	**389**	389	389	**389**	389	389	**389**	389	389
# 12	**617**	617	617	**617**	617	617	**617**	617	617	**617**	617	617
# 10	**981**	981	981	**981**	981	981	**981**	981	981	**981**	981	981
# 8	**1,557**	1,551	1,557	**1,558**	1,555	1,558	**1,559**	1,557	1,559	**1,559**	1,558	1,559
# 6	**2,425**	2,406	2,389	**2,430**	2,417	2,406	**2,431**	2,424	2,414	**2,433**	2,428	2,420
# 4	**3,806**	3,750	3,695	**3,825**	3,789	3,752	**3,830**	3,811	3,778	**3,837**	3,823	3,798
# 3	**4,760**	4,760	4,760	**4,802**	4,802	4,802	**4,760**	4,790	4,760	**4,802**	4,802	4,802
# 2	**5,906**	5,736	5,574	**6,044**	5,926	5,809	**5,989**	5,929	5,827	**6,087**	6,022	5,957
# 1	**7,292**	7,029	6,758	**7,493**	7,306	7,108	**7,454**	7,364	7,188	**7,579**	7,507	7,364
# 1/0	**8,924**	8,543	7,973	**9,317**	9,033	8,590	**9,209**	9,086	8,707	**9,472**	9,372	9,052
# 2/0	**10,755**	10,061	9,389	**11,423**	10,877	10,318	**11,244**	11,045	10,500	**11,703**	11,528	11,052
# 3/0	**12,843**	11,804	11,021	**13,923**	13,048	12,360	**13,656**	13,333	12,613	**14,410**	14,118	13,461
# 4/0	**15,082**	13,605	12,542	**16,673**	15,351	14,347	**16,391**	15,890	14,813	**17,482**	17,019	16,012
# 250	**16,483**	14,924	13,643	**18,593**	17,120	15,865	**18,310**	17,850	16,465	**19,779**	19,352	18,001
# 300	**18,176**	16,292	14,768	**20,867**	18,975	17,408	**20,617**	20,051	18,318	**22,524**	21,938	20,163
# 350	**19,703**	17,385	15,678	**22,736**	20,526	18,672	**22,646**	21,914	19,821	**24,904**	24,126	21,982
# 400	**20,565**	18,235	16,365	**24,296**	21,786	19,731	**24,253**	23,371	21,042	**26,915**	26,044	23,517
# 500	**22,185**	19,172	17,492	**26,706**	23,277	21,329	**26,980**	25,449	23,125	**30,028**	28,712	25,916
# 600	**22,965**	20,567	17,962	**28,033**	25,203	22,097	**28,752**	27,974	24,896	**32,236**	31,258	27,766
# 750	**24,136**	21,386	18,888	**28,303**	25,430	22,690	**31,050**	30,024	26,932	**32,404**	31,338	28,303
# 1000	**25,278**	22,539	19,923	**31,490**	28,083	24,887	**33,864**	32,688	29,320	**37,197**	35,748	31,959

Single Phase L-N Calculations:

1. $f = \dfrac{2 \times \text{Length} \times \text{Isca} \times 1.5}{\text{Runs} \times \text{C-Value} \times \text{E L-N}}$

2. $M = \dfrac{1}{1 + f}$

3. Isc = Isca x 1.5 x M

4. Add Motor Contribution = Motor FLA x 4

"C"-Values for Aluminum Conductors

Conductor AWG	Three Single Conductors						Three-Conductor Cable					
	Magnetic Duct			Non-Magnetic Duct			Magnetic Duct			Non-Magnetic Duct		
	600V	5KV	15KV	**600V**	5KV	15KV	**600V**	5KV	15KV	**600V**	5KV	15KV
# 12	**375**	375	375	**375**	375	375	**375**	375	375	**375**	375	375
# 10	**598**	598	598	**598**	598	598	**598**	598	598	**598**	598	598
# 8	**951**	950	951	**951**	950	951	**951**	951	951	**951**	951	951
# 6	**1,480**	1,476	1,472	**1,481**	1,478	1,476	**1,481**	1,480	1,478	**1,482**	1,481	1,479
# 4	**2,345**	2,332	2,319	**2,350**	2,341	2,333	**2,351**	2,347	2,339	**2,353**	2,349	2,344
# 3	**2,948**	2,948	2,948	**2,958**	2,958	2,958	**2,948**	2,956	2,948	**2,958**	2,958	2,958
# 2	**3,713**	3,669	3,626	**3,729**	3,701	3,672	**3,733**	3,719	3,693	**3,739**	3,724	3,709
# 1	**4,645**	4,574	4,497	**4,678**	4,631	4,580	**4,686**	4,663	4,617	**4,699**	4,681	4,646
# 1/0	**5,777**	5,669	5,493	**5,838**	5,766	5,645	**5,852**	5,820	5,717	**5,875**	5,851	5,771
# 2/0	**7,186**	6,968	6,733	**7,301**	7,152	6,986	**7,327**	7,271	7,109	**7,372**	7,328	7,201
# 3/0	**8,826**	8,466	8,163	**9,110**	8,851	8,627	**9,077**	8,980	8,750	**9,242**	9,164	8,977
# 4/0	**10,740**	10,167	9,700	**11,174**	10,749	10,386	**11,184**	11,021	10,642	**11,408**	11,277	10,968
# 250	**12,122**	11,460	10,848	**12,862**	12,343	11,847	**12,796**	12,636	12,115	**13,236**	13,105	12,661
# 300	**13,909**	13,009	12,192	**14,922**	14,182	13,491	**14,916**	14,698	13,973	**15,494**	15,299	14,658
# 350	**15,484**	14,280	13,288	**16,812**	15,857	14,954	**15,413**	16,490	15,540	**17,635**	17,351	16,500
# 400	**16,670**	15,355	14,188	**18,505**	17,321	16,233	**18,461**	18,063	16,921	**19,587**	19,243	18,154
# 500	**18,755**	16,827	15,657	**21,390**	19,503	18,314	**21,394**	20,606	19,314	**22,987**	22,381	20,978
# 600	**20,093**	18,427	16,484	**23,451**	21,718	19,635	**23,633**	23,195	21,348	**25,750**	25,243	23,294
# 750	**21,766**	19,685	17,686	**23,491**	21,769	19,976	**26,431**	25,789	23,750	**25,682**	25,141	23,491
# 1000	**23,477**	21,235	19,005	**28,778**	26,109	23,482	**29,864**	29,049	26,608	**32,938**	31,919	29,135

Single Phase L-N Calculations:

1. $f = \dfrac{2 \times \text{Length} \times \text{Isca} \times 1.5}{\text{Runs} \times \text{C-Value} \times \text{E L-N}}$

2. $M = \dfrac{1}{1 + f}$

3. $\text{Isc} = \text{Isca} \times 1.5 \times M$

4. Add Motor Contribution = Motor FLA x 4

Short-Circuit Tables and Data: 277 Volts Single-Phase L-N

Single-Phase L-N Bolted Fault-Current* Table

One-Way Distance in Feet

Available Isc in Thousands (Isca)	5	10	15	20	25	30	40	50	60	70	80	90	100	125	150	175	200	225	250	300	350	400	500	600
1	1,317	1,173	1,058	963	884	817	710	627	562	509	465	428	396	335	290	255	228	206	188	160	140	124	101	85
3	3,174	2,452	1,997	1,685	1,457	1,284	1,037	869	749	657	586	528	481	393	333	288	254	227	206	173	149	131	105	88
5	4,422	3,135	2,429	1,982	1,674	1,449	1,142	942	802	698	618	554	503	408	343	296	260	232	209	175	151	132	106	89
10	6,271	3,964	2,898	2,284	1,884	1,604	1,236	1,005	847	732	645	576	520	419	351	302	265	236	212	177	152	133	107	89
15	7,286	4,347	3,097	2,406	1,967	1,663	1,271	1,028	863	744	654	583	526	423	354	304	266	237	213	178	153	134	107	89
20	7,928	4,567	3,208	2,472	2,011	1,694	1,289	1,040	872	750	659	587	529	425	355	305	267	238	214	179	153	134	107	90
25	8,370	4,711	3,278	2,513	2,038	1,714	1,300	1,047	877	754	662	589	531	426	356	305	267	238	214	179	153	134	107	90
30	8,694	4,812	3,326	2,542	2,057	1,727	1,308	1,052	880	757	664	591	532	427	356	306	268	238	214	179	153	134	107	90
35	8,940	4,886	3,362	2,562	2,070	1,736	1,313	1,056	883	759	665	592	533	428	357	306	268	238	215	179	153	134	108	90
40	9,135	4,944	3,389	2,578	2,080	1,744	1,317	1,059	885	760	666	593	534	428	357	306	268	238	215	179	154	134	108	90
45	9,292	4,989	3,410	2,590	2,088	1,749	1,321	1,061	886	761	667	593	534	428	357	306	268	239	215	179	154	134	108	90
50	9,422	5,027	3,428	2,600	2,095	1,754	1,323	1,062	887	762	667	594	535	429	357	307	268	239	215	179	154	134	108	90
55	9,531	5,057	3,442	2,609	2,100	1,758	1,325	1,064	888	763	668	594	535	429	358	307	269	239	215	179	154	134	108	90
60	9,623	5,083	3,454	2,616	2,105	1,761	1,327	1,065	889	763	668	595	536	429	358	307	269	239	215	179	154	134	108	90
65	9,703	5,106	3,464	2,621	2,108	1,763	1,329	1,066	890	764	669	595	536	429	358	307	269	239	215	179	154	135	108	90
70	9,772	5,125	3,473	2,626	2,112	1,766	1,330	1,067	890	764	669	595	536	429	358	307	269	239	215	179	154	135	108	90
75	9,833	5,141	3,481	2,631	2,115	1,768	1,331	1,067	891	764	669	595	536	429	358	307	269	239	215	179	154	135	108	90
80	9,887	5,156	3,487	2,635	2,117	1,769	1,332	1,068	891	765	670	596	536	429	358	307	269	239	215	179	154	135	108	90
85	9,936	5,169	3,493	2,638	2,119	1,771	1,333	1,068	892	765	670	596	536	430	358	307	269	239	215	179	154	135	108	90
90	9,979	5,181	3,499	2,641	2,121	1,772	1,334	1,069	892	765	670	596	537	430	358	307	269	239	215	179	154	135	108	90
95	10,018	5,191	3,503	2,644	2,123	1,774	1,334	1,069	892	766	670	596	537	430	358	307	269	239	215	179	154	135	108	90
100	10,053	5,201	3,508	2,646	2,125	1,775	1,335	1,070	893	766	670	596	537	430	358	307	269	239	215	179	154	135	108	90
105	10,085	5,209	3,512	2,649	2,126	1,776	1,335	1,070	893	766	671	596	537	430	358	307	269	239	215	179	154	135	108	90
110	10,115	5,217	3,515	2,651	2,127	1,777	1,336	1,071	893	766	671	596	537	430	358	307	269	239	215	179	154	135	108	90
115	10,142	5,224	3,519	2,652	2,128	1,777	1,336	1,071	893	766	671	597	537	430	358	307	269	239	215	179	154	135	108	90
120	10,167	5,231	3,521	2,654	2,130	1,778	1,337	1,071	893	766	671	597	537	430	358	307	269	239	215	179	154	135	108	90
125	10,190	5,237	3,524	2,656	2,131	1,779	1,337	1,071	894	767	671	597	537	430	358	307	269	239	215	179	154	135	108	90
130	10,211	5,243	3,527	2,657	2,132	1,779	1,338	1,072	894	767	671	597	537	430	359	307	269	239	215	179	154	135	108	90
135	10,231	5,248	3,529	2,658	2,132	1,780	1,338	1,072	894	767	671	597	537	430	359	307	269	239	215	179	154	135	108	90
140	10,249	5,253	3,531	2,660	2,133	1,781	1,338	1,072	894	767	671	597	537	430	359	307	269	239	215	179	154	135	108	90
145	10,267	5,257	3,533	2,661	2,134	1,781	1,339	1,072	894	767	671	597	537	430	359	307	269	239	215	179	154	135	108	90
150	10,283	5,262	3,535	2,662	2,135	1,782	1,339	1,072	894	767	671	597	537	430	359	307	269	239	215	179	154	135	108	90
155	10,298	5,266	3,537	2,663	2,135	1,782	1,339	1,073	894	767	672	597	538	430	359	307	269	239	215	179	154	135	108	90
160	10,312	5,269	3,539	2,664	2,136	1,783	1,339	1,073	895	767	672	597	538	430	359	307	269	239	215	179	154	135	108	90
165	10,326	5,273	3,540	2,665	2,136	1,783	1,340	1,073	895	767	672	597	538	430	359	307	269	239	215	179	154	135	108	90
170	10,338	5,276	3,542	2,666	2,137	1,783	1,340	1,073	895	767	672	597	538	430	359	307	269	239	215	179	154	135	108	90
175	10,350	5,279	3,543	2,666	2,138	1,784	1,340	1,073	895	767	672	597	538	430	359	308	269	239	215	179	154	135	108	90
180	10,362	5,282	3,545	2,667	2,138	1,784	1,340	1,073	895	767	672	597	538	430	359	308	269	239	215	179	154	135	108	90
185	10,373	5,285	3,546	2,668	2,138	1,784	1,340	1,073	895	768	672	597	538	430	359	308	269	239	215	179	154	135	108	90
190	10,383	5,288	3,547	2,669	2,139	1,785	1,341	1,073	895	768	672	597	538	430	359	308	269	239	215	179	154	135	108	90
195	10,392	5,290	3,548	2,669	2,139	1,785	1,341	1,074	895	768	672	597	538	430	359	308	269	239	215	179	154	135	108	90
200	10,402	5,293	3,549	2,670	2,140	1,785	1,341	1,074	895	768	672	597	538	430	359	308	269	239	215	179	154	135	108	90

# 14 AWG	1 Run(s)	277 Volts	C-Value = 389	Magnetic Duct	Copper

Calculations:

1. $f = \dfrac{2 \times \text{Length} \times \text{Isca} \times 1.5}{\text{Runs} \times \text{C-Value} \times \text{E L-N}}$

2. $M = \dfrac{1}{1 + f}$

3. $\text{Isc} = \text{Isca} \times 1.5 \times M$

4. Add Motor Contribution = Motor FLA x 4

* All results are given in symmetrical amperes

Single-Phase L-N Bolted Fault-Current* Table

One-Way Distance in Feet

Available Isc in Thousands (Isca)	5	10	15	20	25	30	40	50	60	70	80	90	100	125	150	175	200	225	250	300	350	400	500	600
1	1,379	1,276	1,187	1,110	1,043	983	881	799	731	673	624	581	544	470	413	368	333	303	278	239	210	187	153	130
3	3,562	2,948	2,514	2,192	1,943	1,744	1,449	1,239	1,082	960	863	784	718	593	506	441	390	350	318	268	232	204	165	138
5	5,213	3,994	3,238	2,722	2,348	2,064	1,663	1,392	1,197	1,050	935	843	767	627	529	458	404	361	327	274	236	208	167	140
10	7,989	5,444	4,129	3,325	2,784	2,394	1,870	1,534	1,301	1,129	997	893	808	654	549	473	415	370	334	280	240	211	169	141
15	9,713	6,193	4,546	3,591	2,967	2,528	1,951	1,588	1,339	1,158	1,020	911	823	663	556	478	419	373	337	281	242	212	170	142
20	10,888	6,651	4,788	3,740	3,069	2,601	1,994	1,617	1,360	1,173	1,031	920	831	668	559	480	421	375	338	282	242	212	170	142
25	11,740	6,960	4,946	3,836	3,133	2,647	2,021	1,635	1,372	1,182	1,039	926	836	671	561	482	422	376	339	283	243	212	170	142
30	12,387	7,182	5,057	3,902	3,177	2,679	2,040	1,647	1,381	1,189	1,043	930	839	673	563	483	423	377	339	283	243	213	170	142
35	12,894	7,349	5,139	3,951	3,209	2,702	2,053	1,655	1,387	1,193	1,047	933	841	675	564	484	424	377	340	283	243	213	170	142
40	13,302	7,480	5,203	3,989	3,234	2,719	2,063	1,662	1,391	1,196	1,049	935	843	676	564	484	424	377	340	284	243	213	170	142
45	13,638	7,585	5,254	4,018	3,253	2,733	2,071	1,667	1,395	1,199	1,052	936	844	677	565	485	425	378	340	284	243	213	170	142
50	13,919	7,671	5,295	4,042	3,269	2,744	2,077	1,671	1,398	1,201	1,053	938	845	677	565	485	425	378	340	284	243	213	171	142
55	14,158	7,743	5,329	4,062	3,282	2,753	2,082	1,674	1,400	1,203	1,055	939	846	678	566	485	425	378	340	284	243	213	171	142
60	14,363	7,804	5,358	4,079	3,293	2,761	2,087	1,677	1,402	1,204	1,056	940	847	678	566	486	425	378	341	284	243	213	171	142
65	14,542	7,857	5,382	4,093	3,302	2,768	2,091	1,680	1,404	1,206	1,057	940	847	679	566	486	425	378	341	284	244	213	171	142
70	14,698	7,902	5,404	4,106	3,310	2,773	2,094	1,682	1,405	1,207	1,057	941	848	679	567	486	426	378	341	284	244	213	171	142
75	14,837	7,942	5,422	4,116	3,317	2,778	2,097	1,684	1,406	1,208	1,058	942	848	680	567	486	426	379	341	284	244	213	171	142
80	14,960	7,977	5,439	4,126	3,324	2,782	2,099	1,685	1,408	1,208	1,059	942	849	680	567	486	426	379	341	284	244	213	171	142
85	15,071	8,009	5,453	4,134	3,329	2,786	2,101	1,686	1,409	1,209	1,059	942	849	680	567	486	426	379	341	284	244	213	171	142
90	15,170	8,037	5,466	4,142	3,334	2,790	2,103	1,688	1,409	1,210	1,060	943	849	680	567	487	426	379	341	284	244	213	171	142
95	15,261	8,062	5,478	4,148	3,338	2,793	2,105	1,689	1,410	1,210	1,060	943	849	680	567	487	426	379	341	284	244	213	171	142
100	15,343	8,085	5,489	4,154	3,342	2,795	2,106	1,690	1,411	1,211	1,061	944	850	681	568	487	426	379	341	284	244	213	171	142
105	15,418	8,106	5,498	4,160	3,346	2,798	2,108	1,691	1,411	1,211	1,061	944	850	681	568	487	426	379	341	284	244	213	171	142
110	15,487	8,125	5,507	4,165	3,349	2,800	2,109	1,692	1,412	1,212	1,061	944	850	681	568	487	426	379	341	284	244	213	171	142
115	15,550	8,142	5,515	4,169	3,352	2,802	2,110	1,692	1,413	1,212	1,062	944	850	681	568	487	426	379	341	284	244	213	171	142
120	15,609	8,158	5,522	4,174	3,354	2,804	2,111	1,693	1,413	1,213	1,062	945	851	681	568	487	426	379	341	284	244	213	171	142
125	15,663	8,173	5,529	4,178	3,357	2,806	2,112	1,694	1,414	1,213	1,062	945	851	681	568	487	426	379	341	284	244	213	171	142
130	15,714	8,187	5,535	4,181	3,359	2,807	2,113	1,694	1,414	1,213	1,062	945	851	681	568	487	426	379	341	284	244	213	171	142
135	15,761	8,199	5,541	4,184	3,361	2,809	2,114	1,695	1,414	1,213	1,063	945	851	681	568	487	426	379	341	284	244	213	171	142
140	15,805	8,211	5,546	4,188	3,363	2,810	2,115	1,695	1,415	1,214	1,063	945	851	681	568	487	426	379	341	284	244	213	171	142
145	15,846	8,222	5,552	4,190	3,365	2,812	2,116	1,696	1,415	1,214	1,063	945	851	681	568	487	426	379	341	284	244	213	171	142
150	15,884	8,233	5,556	4,193	3,367	2,813	2,116	1,696	1,415	1,214	1,063	946	851	682	568	487	426	379	341	284	244	213	171	142
155	15,921	8,243	5,561	4,196	3,369	2,814	2,117	1,697	1,416	1,214	1,063	946	851	682	568	487	426	379	341	284	244	213	171	142
160	15,955	8,252	5,565	4,198	3,370	2,815	2,118	1,697	1,416	1,215	1,063	946	852	682	568	487	427	379	341	285	244	213	171	142
165	15,987	8,260	5,569	4,200	3,372	2,816	2,118	1,697	1,416	1,215	1,064	946	852	682	568	487	427	379	341	285	244	213	171	142
170	16,017	8,268	5,572	4,202	3,373	2,817	2,119	1,698	1,416	1,215	1,064	946	852	682	568	487	427	379	341	285	244	213	171	142
175	16,046	8,276	5,576	4,204	3,374	2,818	2,119	1,698	1,417	1,215	1,064	946	852	682	568	487	427	379	341	285	244	213	171	142
180	16,073	8,283	5,579	4,206	3,375	2,819	2,120	1,698	1,417	1,215	1,064	946	852	682	568	487	427	379	341	285	244	213	171	142
185	16,099	8,290	5,582	4,208	3,377	2,820	2,120	1,699	1,417	1,215	1,064	946	852	682	569	487	427	379	341	285	244	213	171	142
190	16,124	8,297	5,585	4,210	3,378	2,820	2,120	1,699	1,417	1,216	1,064	946	852	682	569	487	427	379	341	285	244	213	171	142
195	16,147	8,303	5,588	4,211	3,379	2,821	2,121	1,699	1,417	1,216	1,064	946	852	682	569	487	427	379	341	285	244	213	171	142
200	16,170	8,309	5,591	4,213	3,380	2,822	2,121	1,699	1,418	1,216	1,064	946	852	682	569	488	427	379	341	285	244	213	171	142

12 AWG | 1 Run(s) | 277 Volts | | C-Value = 617 | Magnetic Duct | Copper

Calculations:

1. $f = \dfrac{2 \times \text{Length} \times \text{Isca} \times 1.5}{\text{Runs} \times \text{C-Value} \times \text{E L-N}}$

2. $M = \dfrac{1}{1 + f}$

3. $\text{Isc} = \text{Isca} \times 1.5 \times M$

4. Add Motor Contribution = Motor FLA x 4

* All results are given in symmetrical amperes

Single-Phase L-N Bolted Fault-Current* Table

One-Way Distance in Feet

Available Isc in Thousands (Isca)	5	10	15	20	25	30	40	50	60	70	80	90	100	125	150	175	200	225	250	300	350	400	500	600
1	1,422	1,351	1,287	1,229	1,176	1,127	1,041	966	902	846	797	752	713	630	565	512	468	431	399	348	308	277	230	197
3	3,861	3,380	3,006	2,707	2,462	2,257	1,936	1,694	1,506	1,356	1,233	1,130	1,044	875	754	662	590	532	485	411	357	316	256	216
5	5,878	4,832	4,103	3,565	3,151	2,824	2,338	1,995	1,739	1,542	1,385	1,257	1,150	949	808	704	623	559	507	427	369	325	262	220
10	9,665	7,129	5,648	4,676	3,989	3,479	2,770	2,301	1,967	1,719	1,526	1,372	1,246	1,014	854	738	650	580	524	440	378	332	267	223
15	12,308	8,471	6,458	5,218	4,377	3,770	2,951	2,425	2,057	1,787	1,579	1,415	1,281	1,037	871	750	659	588	531	444	382	335	268	224
20	14,258	9,352	6,957	5,539	4,601	3,935	3,051	2,492	2,106	1,823	1,607	1,437	1,300	1,049	879	757	664	592	534	446	383	336	269	225
25	15,756	9,973	7,296	5,751	4,747	4,041	3,115	2,534	2,136	1,845	1,625	1,451	1,311	1,056	884	761	667	594	536	447	384	337	270	225
30	16,943	10,436	7,540	5,902	4,849	4,115	3,158	2,563	2,156	1,861	1,637	1,461	1,319	1,061	888	763	669	596	537	448	385	337	270	225
35	17,906	10,794	7,725	6,015	4,925	4,169	3,190	2,584	2,171	1,872	1,645	1,467	1,324	1,065	890	765	671	597	538	449	385	337	270	225
40	18,703	11,078	7,870	6,102	4,983	4,211	3,215	2,600	2,182	1,880	1,652	1,473	1,329	1,068	892	766	672	598	539	450	386	338	271	226
45	19,374	11,310	7,986	6,172	5,030	4,244	3,234	2,612	2,191	1,887	1,657	1,477	1,332	1,070	894	768	673	599	539	450	386	338	271	226
50	19,947	11,503	8,082	6,229	5,068	4,271	3,250	2,622	2,198	1,892	1,661	1,480	1,335	1,071	895	768	673	599	540	450	386	338	271	226
55	20,441	11,666	8,162	6,277	5,099	4,293	3,262	2,631	2,204	1,896	1,664	1,483	1,337	1,073	896	769	674	599	540	450	386	338	271	226
60	20,872	11,805	8,230	6,317	5,125	4,312	3,273	2,638	2,209	1,900	1,667	1,485	1,338	1,074	897	770	674	600	540	451	387	338	271	226
65	21,251	11,925	8,288	6,351	5,148	4,328	3,282	2,644	2,213	1,903	1,669	1,487	1,340	1,075	897	770	675	600	540	451	387	338	271	226
70	21,587	12,030	8,339	6,381	5,167	4,342	3,290	2,649	2,217	1,906	1,671	1,488	1,341	1,076	898	771	675	600	541	451	387	339	271	226
75	21,887	12,123	8,383	6,407	5,184	4,354	3,297	2,653	2,220	1,908	1,673	1,490	1,342	1,077	899	771	675	601	541	451	387	339	271	226
80	22,156	12,205	8,422	6,429	5,199	4,364	3,303	2,657	2,223	1,910	1,675	1,491	1,343	1,077	899	771	676	601	541	451	387	339	271	226
85	22,400	12,278	8,457	6,450	5,213	4,374	3,309	2,661	2,225	1,912	1,676	1,492	1,344	1,078	899	772	676	601	541	451	387	339	271	226
90	22,620	12,344	8,488	6,468	5,224	4,382	3,313	2,664	2,227	1,913	1,677	1,493	1,345	1,078	900	772	676	601	541	451	387	339	271	226
95	22,822	12,404	8,517	6,484	5,235	4,389	3,318	2,667	2,229	1,915	1,678	1,494	1,346	1,079	900	772	676	601	541	451	387	339	271	226
100	23,006	12,458	8,542	6,499	5,245	4,396	3,321	2,669	2,231	1,916	1,679	1,495	1,346	1,079	900	772	676	601	542	452	387	339	271	226
105	23,175	12,508	8,565	6,513	5,253	4,402	3,325	2,671	2,232	1,917	1,680	1,495	1,347	1,079	901	773	676	602	542	452	387	339	271	226
110	23,331	12,553	8,587	6,525	5,261	4,408	3,328	2,673	2,234	1,918	1,681	1,496	1,348	1,080	901	773	677	602	542	452	387	339	271	226
115	23,476	12,595	8,606	6,536	5,269	4,413	3,331	2,675	2,235	1,919	1,682	1,497	1,348	1,080	901	773	677	602	542	452	387	339	271	226
120	23,609	12,633	8,624	6,546	5,275	4,418	3,334	2,677	2,236	1,920	1,682	1,497	1,349	1,080	901	773	677	602	542	452	387	339	271	226
125	23,734	12,669	8,640	6,556	5,282	4,422	3,336	2,679	2,237	1,921	1,683	1,498	1,349	1,081	901	773	677	602	542	452	387	339	271	226
130	23,850	12,702	8,656	6,565	5,287	4,426	3,339	2,680	2,238	1,922	1,684	1,498	1,349	1,081	902	773	677	602	542	452	387	339	271	226
135	23,959	12,733	8,670	6,573	5,293	4,430	3,341	2,681	2,239	1,923	1,684	1,498	1,350	1,081	902	773	677	602	542	452	387	339	271	226
140	24,060	12,761	8,683	6,581	5,298	4,433	3,343	2,683	2,240	1,923	1,685	1,499	1,350	1,081	902	774	677	602	542	452	387	339	271	226
145	24,156	12,788	8,696	6,588	5,302	4,437	3,344	2,684	2,241	1,924	1,685	1,499	1,350	1,082	902	774	677	602	542	452	388	339	271	226
150	24,246	12,813	8,707	6,594	5,307	4,440	3,346	2,685	2,242	1,924	1,686	1,500	1,351	1,082	902	774	677	602	542	452	388	339	271	226
155	24,330	12,837	8,718	6,601	5,311	4,442	3,348	2,686	2,243	1,925	1,686	1,500	1,351	1,082	902	774	677	602	542	452	388	339	271	226
160	24,410	12,859	8,728	6,606	5,314	4,445	3,349	2,687	2,243	1,925	1,686	1,500	1,351	1,082	902	774	677	602	542	452	388	339	271	226
165	24,485	12,880	8,738	6,612	5,318	4,448	3,351	2,688	2,244	1,926	1,687	1,500	1,351	1,082	902	774	677	602	542	452	388	339	271	226
170	24,557	12,900	8,747	6,617	5,321	4,450	3,352	2,689	2,245	1,926	1,687	1,501	1,351	1,082	903	774	678	602	542	452	388	339	271	226
175	24,625	12,918	8,756	6,622	5,325	4,452	3,353	2,690	2,245	1,927	1,687	1,501	1,352	1,082	903	774	678	602	542	452	388	339	271	226
180	24,689	12,936	8,764	6,627	5,328	4,454	3,355	2,690	2,246	1,927	1,688	1,501	1,352	1,083	903	774	678	603	542	452	388	339	271	226
185	24,750	12,953	8,772	6,631	5,330	4,456	3,356	2,691	2,246	1,927	1,688	1,501	1,352	1,083	903	774	678	603	542	452	388	339	271	226
190	24,808	12,969	8,779	6,635	5,333	4,458	3,357	2,692	2,247	1,928	1,688	1,502	1,352	1,083	903	774	678	603	542	452	388	339	271	226
195	24,864	12,984	8,786	6,639	5,336	4,460	3,358	2,692	2,247	1,928	1,689	1,502	1,352	1,083	903	774	678	603	542	452	388	339	271	226
200	24,917	12,998	8,792	6,643	5,338	4,462	3,359	2,693	2,248	1,929	1,689	1,502	1,353	1,083	903	774	678	603	542	452	388	339	271	226

# 10 AWG	1 Run(s)	277 Volts	C-Value = 981	Magnetic Duct	Copper

Calculations:

1. $f = \dfrac{2 \times \text{Length} \times \text{Isca} \times 1.5}{\text{Runs} \times \text{C-Value} \times E_{L\text{-}N}}$

2. $M = \dfrac{1}{1 + f}$

3. $\text{Isc} = \text{Isca} \times 1.5 \times M$

4. Add Motor Contribution = Motor FLA x 4

* All results are given in symmetrical amperes

Single-Phase L-N Bolted Fault-Current* Table

One-Way Distance in Feet

Available Isc in Thousands (Isca)	5	10	15	20	25	30	40	50	60	70	80	90	100	125	150	175	200	225	250	300	350	400	500	600
3	4,075	3,723	3,427	3,175	2,957	2,767	2,453	2,202	1,998	1,829	1,686	1,564	1,458	1,247	1,090	967	870	790	724	620	542	481	394	333
5	6,389	5,565	4,929	4,423	4,012	3,670	3,137	2,738	2,430	2,184	1,983	1,816	1,675	1,403	1,206	1,058	943	850	774	656	569	503	408	343
7	8,444	7,062	6,068	5,320	4,736	4,267	3,562	3,057	2,678	2,382	2,145	1,951	1,789	1,482	1,264	1,103	978	878	797	673	582	513	414	348
10	11,129	8,846	7,341	6,273	5,477	4,859	3,966	3,350	2,899	2,556	2,285	2,066	1,885	1,547	1,312	1,139	1,006	901	816	686	592	520	419	351
15	14,786	11,011	8,772	7,289	6,235	5,448	4,349	3,619	3,099	2,710	2,407	2,165	1,968	1,602	1,351	1,168	1,029	919	831	697	600	526	423	354
20	17,693	12,546	9,719	7,932	6,699	5,799	4,570	3,771	3,210	2,794	2,473	2,219	2,012	1,631	1,372	1,184	1,041	929	838	702	604	530	425	355
25	20,059	13,691	10,392	8,374	7,013	6,032	4,713	3,868	3,280	2,847	2,515	2,252	2,039	1,649	1,385	1,193	1,048	935	843	705	606	531	426	356
30	22,022	14,578	10,895	8,698	7,238	6,198	4,814	3,936	3,328	2,883	2,543	2,275	2,058	1,661	1,393	1,199	1,053	938	846	708	608	533	427	357
35	23,678	15,286	11,286	8,945	7,409	6,322	4,889	3,985	3,364	2,910	2,564	2,291	2,071	1,670	1,399	1,204	1,057	941	849	709	609	534	428	357
40	25,092	15,863	11,597	9,140	7,542	6,419	4,947	4,024	3,391	2,930	2,580	2,304	2,082	1,677	1,404	1,207	1,059	943	850	710	610	534	428	357
45	26,315	16,343	11,852	9,297	7,648	6,496	4,992	4,054	3,412	2,946	2,592	2,314	2,090	1,682	1,408	1,210	1,061	945	852	711	611	535	429	358
50	27,383	16,749	12,064	9,427	7,736	6,559	5,030	4,078	3,430	2,959	2,602	2,322	2,096	1,686	1,411	1,212	1,063	946	853	712	611	535	429	358
55	28,323	17,096	12,243	9,536	7,809	6,612	5,060	4,099	3,444	2,970	2,610	2,328	2,102	1,690	1,413	1,214	1,064	947	854	713	612	536	429	358
60	29,157	17,396	12,396	9,629	7,871	6,657	5,086	4,116	3,456	2,979	2,617	2,334	2,106	1,693	1,415	1,216	1,065	948	854	713	612	536	429	358
65	29,902	17,659	12,529	9,709	7,925	6,695	5,109	4,130	3,466	2,986	2,623	2,339	2,110	1,695	1,417	1,217	1,066	949	855	714	612	536	429	358
70	30,572	17,890	12,645	9,778	7,971	6,728	5,128	4,143	3,475	2,993	2,628	2,343	2,113	1,697	1,418	1,218	1,067	950	856	714	613	536	430	358
75	31,177	18,096	12,747	9,839	8,012	6,756	5,145	4,154	3,483	2,999	2,632	2,346	2,116	1,699	1,419	1,219	1,068	950	856	714	613	537	430	358
80	31,726	18,280	12,838	9,893	8,047	6,782	5,159	4,163	3,490	3,004	2,636	2,349	2,118	1,701	1,421	1,220	1,069	951	856	715	613	537	430	358
85	32,227	18,445	12,920	9,942	8,079	6,805	5,172	4,172	3,496	3,008	2,640	2,352	2,121	1,702	1,422	1,220	1,069	951	857	715	613	537	430	358
90	32,686	18,594	12,993	9,985	8,108	6,825	5,184	4,179	3,501	3,012	2,643	2,354	2,123	1,703	1,422	1,221	1,070	952	857	715	613	537	430	358
95	33,108	18,730	13,059	10,024	8,133	6,843	5,195	4,186	3,506	3,015	2,646	2,356	2,124	1,705	1,423	1,222	1,070	952	857	715	613	537	430	359
100	33,497	18,854	13,119	10,059	8,157	6,859	5,204	4,192	3,510	3,019	2,648	2,358	2,126	1,706	1,424	1,222	1,071	952	858	715	614	537	430	359
105	33,858	18,967	13,174	10,091	8,178	6,874	5,213	4,198	3,514	3,022	2,650	2,360	2,127	1,706	1,425	1,223	1,071	953	858	716	614	537	430	359
110	34,192	19,072	13,224	10,121	8,197	6,888	5,221	4,203	3,517	3,024	2,652	2,362	2,129	1,707	1,425	1,223	1,071	953	858	716	614	537	430	359
115	34,502	19,168	13,270	10,148	8,215	6,901	5,228	4,208	3,521	3,027	2,654	2,363	2,130	1,708	1,426	1,224	1,072	953	858	716	614	537	430	359
120	34,792	19,257	13,313	10,173	8,231	6,912	5,234	4,212	3,524	3,029	2,656	2,365	2,131	1,709	1,426	1,224	1,072	953	858	716	614	538	430	359
125	35,064	19,340	13,353	10,196	8,246	6,923	5,240	4,216	3,526	3,031	2,657	2,366	2,132	1,709	1,427	1,224	1,072	954	859	716	614	538	430	359
130	35,318	19,417	13,389	10,217	8,260	6,933	5,246	4,220	3,529	3,033	2,659	2,367	2,133	1,710	1,427	1,225	1,072	954	859	716	614	538	430	359
135	35,556	19,489	13,423	10,237	8,273	6,942	5,251	4,223	3,531	3,034	2,660	2,368	2,134	1,711	1,427	1,225	1,073	954	859	716	614	538	430	359
140	35,780	19,556	13,455	10,256	8,285	6,950	5,256	4,226	3,534	3,036	2,661	2,369	2,135	1,711	1,428	1,225	1,073	954	859	716	614	538	430	359
145	35,992	19,619	13,485	10,273	8,297	6,958	5,261	4,229	3,536	3,038	2,663	2,370	2,135	1,712	1,428	1,225	1,073	954	859	716	614	538	430	359
150	36,192	19,678	13,513	10,289	8,307	6,966	5,265	4,232	3,538	3,039	2,664	2,371	2,136	1,712	1,429	1,226	1,073	954	859	717	614	538	430	359
155	36,380	19,734	13,539	10,304	8,317	6,973	5,269	4,234	3,539	3,040	2,665	2,372	2,137	1,712	1,429	1,226	1,073	954	859	717	614	538	430	359
160	36,559	19,787	13,564	10,319	8,327	6,979	5,273	4,237	3,541	3,042	2,666	2,372	2,137	1,713	1,429	1,226	1,073	955	859	717	615	538	431	359
165	36,729	19,836	13,587	10,332	8,335	6,985	5,276	4,239	3,543	3,043	2,667	2,373	2,138	1,713	1,429	1,226	1,074	955	860	717	615	538	431	359
170	36,890	19,883	13,609	10,345	8,344	6,991	5,279	4,241	3,544	3,044	2,667	2,374	2,138	1,714	1,430	1,226	1,074	955	860	717	615	538	431	359
175	37,043	19,927	13,630	10,357	8,351	6,997	5,283	4,243	3,546	3,045	2,668	2,374	2,139	1,714	1,430	1,226	1,074	955	860	717	615	538	431	359
180	37,189	19,970	13,650	10,368	8,359	7,002	5,286	4,245	3,547	3,046	2,669	2,375	2,139	1,714	1,430	1,227	1,074	955	860	717	615	538	431	359
185	37,327	20,010	13,668	10,379	8,366	7,007	5,288	4,247	3,548	3,047	2,670	2,376	2,140	1,714	1,430	1,227	1,074	955	860	717	615	538	431	359
190	37,460	20,048	13,686	10,389	8,372	7,011	5,291	4,249	3,549	3,048	2,670	2,376	2,140	1,715	1,430	1,227	1,074	955	860	717	615	538	431	359
195	37,587	20,084	13,703	10,399	8,379	7,016	5,294	4,250	3,550	3,049	2,671	2,377	2,141	1,715	1,431	1,227	1,074	955	860	717	615	538	431	359
200	37,708	20,118	13,719	10,408	8,385	7,020	5,296	4,252	3,552	3,049	2,672	2,377	2,141	1,715	1,431	1,227	1,074	955	860	717	615	538	431	359

# 8 AWG	1 Run(s)	277 Volts	C-Value = 1,557	Magnetic Duct	Copper

Calculations:

1. $f = \dfrac{2 \times \text{Length} \times \text{Isca} \times 1.5}{\text{Runs} \times \text{C-Value} \times \text{E L-N}}$

2. $M = \dfrac{1}{1 + f}$

3. $\text{Isc} = \text{Isca} \times 1.5 \times M$

4. Add Motor Contribution = Motor FLA x 4

* All results are given in symmetrical amperes

Single-Phase L-N Bolted Fault-Current* Table

One-Way Distance in Feet

Available Isc in Thousands (Isca)	5	10	15	20	25	30	40	50	60	70	80	90	100	125	150	175	200	225	250	300	350	400	500	600
3	4,217	3,968	3,747	3,549	3,371	3,210	2,930	2,695	2,495	2,322	2,172	2,040	1,923	1,682	1,495	1,345	1,223	1,121	1,035	897	791	708	584	498
5	6,747	6,131	5,618	5,185	4,813	4,491	3,961	3,544	3,205	2,926	2,692	2,492	2,320	1,978	1,724	1,528	1,372	1,245	1,139	974	851	755	617	521
7	9,081	7,999	7,148	6,461	5,894	5,418	4,666	4,097	3,651	3,293	2,999	2,753	2,545	2,139	1,846	1,623	1,448	1,307	1,191	1,012	879	777	631	531
10	12,262	10,369	8,982	7,923	7,087	6,411	5,383	4,640	4,076	3,635	3,280	2,988	2,744	2,279	1,948	1,702	1,510	1,358	1,233	1,042	902	795	643	540
15	16,854	13,474	11,223	9,616	8,412	7,476	6,115	5,173	4,483	3,955	3,538	3,201	2,922	2,400	2,036	1,768	1,563	1,400	1,268	1,066	920	809	652	546
20	20,738	15,846	12,821	10,766	9,279	8,153	6,560	5,488	4,717	4,136	3,683	3,319	3,020	2,466	2,084	1,804	1,590	1,422	1,286	1,079	930	817	657	550
25	24,065	17,718	14,020	11,599	9,891	8,622	6,860	5,697	4,871	4,254	3,776	3,394	3,083	2,507	2,113	1,826	1,607	1,436	1,297	1,087	936	821	660	552
30	26,947	19,232	14,951	12,229	10,346	8,965	7,076	5,845	4,978	4,336	3,840	3,446	3,125	2,536	2,133	1,841	1,619	1,445	1,305	1,092	940	824	662	553
35	29,468	20,483	15,696	12,723	10,697	9,228	7,239	5,955	5,058	4,396	3,887	3,484	3,157	2,556	2,147	1,852	1,627	1,451	1,310	1,096	942	826	663	554
40	31,692	21,533	16,306	13,121	10,977	9,435	7,366	6,041	5,120	4,443	3,924	3,513	3,181	2,572	2,159	1,860	1,634	1,456	1,314	1,099	945	828	664	555
45	33,668	22,427	16,814	13,448	11,204	9,603	7,468	6,109	5,169	4,480	3,952	3,536	3,199	2,584	2,167	1,866	1,639	1,460	1,317	1,101	946	829	665	555
50	35,435	23,198	17,243	13,721	11,394	9,741	7,551	6,165	5,209	4,510	3,976	3,555	3,215	2,594	2,174	1,871	1,643	1,464	1,320	1,103	947	830	666	556
55	37,026	23,869	17,611	13,953	11,553	9,858	7,621	6,212	5,242	4,534	3,995	3,570	3,227	2,602	2,180	1,876	1,646	1,466	1,322	1,105	949	831	666	556
60	38,464	24,459	17,930	14,152	11,690	9,957	7,680	6,251	5,270	4,555	4,011	3,583	3,238	2,609	2,185	1,879	1,649	1,468	1,324	1,106	949	832	667	556
65	39,772	24,981	18,209	14,326	11,808	10,042	7,731	6,284	5,294	4,573	4,025	3,594	3,247	2,615	2,189	1,882	1,651	1,470	1,325	1,107	950	832	667	557
70	40,965	25,447	18,455	14,478	11,911	10,117	7,775	6,313	5,314	4,588	4,037	3,604	3,255	2,620	2,192	1,885	1,653	1,472	1,326	1,108	951	833	667	557
75	42,059	25,865	18,674	14,612	12,001	10,182	7,813	6,339	5,332	4,602	4,047	3,612	3,261	2,624	2,195	1,887	1,655	1,473	1,328	1,109	951	833	668	557
80	43,066	26,242	18,870	14,732	12,082	10,240	7,847	6,361	5,348	4,614	4,056	3,619	3,267	2,628	2,198	1,889	1,656	1,474	1,329	1,109	952	834	668	557
85	43,994	26,584	19,046	14,839	12,154	10,292	7,878	6,381	5,362	4,624	4,064	3,626	3,272	2,631	2,200	1,891	1,657	1,475	1,329	1,110	952	834	668	557
90	44,854	26,895	19,205	14,935	12,219	10,338	7,905	6,399	5,375	4,633	4,072	3,631	3,277	2,634	2,203	1,892	1,659	1,476	1,330	1,110	953	834	668	557
95	45,653	27,180	19,350	15,023	12,277	10,380	7,929	6,415	5,386	4,642	4,078	3,637	3,281	2,637	2,204	1,894	1,660	1,477	1,331	1,111	953	835	669	558
100	46,396	27,442	19,483	15,102	12,330	10,418	7,951	6,429	5,396	4,649	4,084	3,641	3,285	2,640	2,206	1,895	1,661	1,478	1,332	1,111	954	835	669	558
105	47,089	27,683	19,604	15,175	12,379	10,452	7,972	6,442	5,406	4,656	4,089	3,645	3,288	2,642	2,208	1,896	1,662	1,479	1,332	1,112	954	835	669	558
110	47,738	27,906	19,715	15,242	12,423	10,484	7,990	6,454	5,414	4,662	4,094	3,649	3,292	2,644	2,209	1,897	1,662	1,479	1,333	1,112	954	835	669	558
115	48,346	28,113	19,818	15,303	12,464	10,513	8,007	6,465	5,422	4,668	4,099	3,653	3,294	2,646	2,210	1,898	1,663	1,480	1,333	1,112	954	836	669	558
120	48,917	28,305	19,914	15,360	12,501	10,540	8,022	6,476	5,429	4,673	4,103	3,656	3,297	2,647	2,212	1,899	1,664	1,480	1,333	1,113	955	836	669	558
125	49,455	28,484	20,002	15,413	12,536	10,565	8,037	6,485	5,435	4,678	4,106	3,659	3,300	2,649	2,213	1,900	1,664	1,481	1,334	1,113	955	836	669	558
130	49,962	28,651	20,085	15,462	12,569	10,588	8,050	6,494	5,442	4,683	4,110	3,662	3,302	2,650	2,214	1,901	1,665	1,481	1,334	1,113	955	836	669	558
135	50,441	28,808	20,162	15,507	12,599	10,609	8,062	6,502	5,447	4,687	4,113	3,664	3,304	2,652	2,215	1,901	1,666	1,482	1,335	1,113	955	836	670	558
140	50,893	28,955	20,233	15,550	12,627	10,629	8,074	6,509	5,452	4,691	4,116	3,667	3,306	2,653	2,215	1,902	1,666	1,482	1,335	1,114	955	836	670	558
145	51,322	29,094	20,301	15,589	12,653	10,647	8,084	6,516	5,457	4,694	4,119	3,669	3,308	2,654	2,216	1,902	1,666	1,483	1,335	1,114	955	836	670	558
150	51,729	29,224	20,364	15,627	12,678	10,665	8,094	6,523	5,462	4,698	4,121	3,671	3,309	2,655	2,217	1,903	1,667	1,483	1,335	1,114	956	837	670	558
155	52,116	29,347	20,424	15,662	12,701	10,681	8,104	6,529	5,466	4,701	4,124	3,673	3,311	2,656	2,218	1,904	1,667	1,483	1,336	1,114	956	837	670	558
160	52,483	29,463	20,480	15,695	12,722	10,696	8,113	6,534	5,470	4,704	4,126	3,675	3,312	2,657	2,218	1,904	1,668	1,483	1,336	1,114	956	837	670	558
165	52,833	29,573	20,533	15,726	12,743	10,711	8,121	6,540	5,474	4,707	4,128	3,676	3,314	2,658	2,219	1,904	1,668	1,484	1,336	1,115	956	837	670	559
170	53,167	29,677	20,583	15,756	12,762	10,725	8,129	6,545	5,477	4,709	4,130	3,678	3,315	2,659	2,220	1,905	1,668	1,484	1,336	1,115	956	837	670	559
175	53,486	29,776	20,631	15,783	12,780	10,737	8,136	6,550	5,481	4,712	4,132	3,679	3,316	2,660	2,220	1,905	1,669	1,484	1,337	1,115	956	837	670	559
180	53,790	29,871	20,676	15,810	12,798	10,750	8,143	6,554	5,484	4,714	4,134	3,681	3,317	2,660	2,221	1,906	1,669	1,485	1,337	1,115	956	837	670	559
185	54,081	29,960	20,719	15,835	12,814	10,761	8,150	6,558	5,487	4,716	4,136	3,682	3,318	2,661	2,221	1,906	1,669	1,485	1,337	1,115	956	837	670	559
190	54,360	30,045	20,760	15,859	12,830	10,772	8,156	6,563	5,490	4,719	4,137	3,684	3,320	2,662	2,222	1,906	1,669	1,485	1,337	1,115	956	837	670	559
195	54,627	30,127	20,799	15,881	12,845	10,783	8,162	6,566	5,493	4,721	4,139	3,685	3,320	2,662	2,222	1,907	1,670	1,485	1,337	1,115	956	837	670	559
200	54,884	30,205	20,836	15,903	12,859	10,793	8,168	6,570	5,495	4,723	4,140	3,686	3,321	2,663	2,222	1,907	1,670	1,485	1,337	1,115	957	837	670	559

# 6 AWG	1 Run(s)	277 Volts	C-Value = 2,425	Magnetic Duct	Copper

Calculations:

1. $f = \dfrac{2 \times \text{Length} \times \text{Isca} \times 1.5}{\text{Runs} \times \text{C-Value} \times E_{L\text{-}N}}$
2. $M = \dfrac{1}{1 + f}$
3. $\text{Isc} = \text{Isca} \times 1.5 \times M$
4. Add Motor Contribution = Motor FLA x 4

* All results are given in symmetrical amperes

Single-Phase L-N Bolted Fault-Current* Table

One-Way Distance in Feet

Available Isc in Thousands (Isca)

Isca	5	10	15	20	25	30	40	50	60	70	80	90	100	125	150	175	200	225	250	300	350	400	500	600
3	4,316	4,146	3,989	3,844	3,709	3,583	3,355	3,154	2,976	2,817	2,674	2,545	2,428	2,177	1,973	1,804	1,662	1,541	1,436	1,264	1,128	1,019	854	735
5	7,002	6,566	6,181	5,839	5,532	5,256	4,780	4,382	4,046	3,758	3,508	3,289	3,096	2,699	2,393	2,149	1,950	1,785	1,646	1,424	1,254	1,121	924	786
7	9,549	8,756	8,084	7,509	7,009	6,572	5,844	5,261	4,783	4,385	4,049	3,760	3,509	3,009	2,633	2,341	2,107	1,915	1,756	1,505	1,317	1,171	958	811
10	13,132	11,677	10,513	9,560	8,765	8,092	7,015	6,191	5,540	5,014	4,578	4,212	3,901	3,292	2,847	2,508	2,242	2,026	1,849	1,573	1,369	1,211	985	830
15	18,543	15,769	13,717	12,138	10,885	9,866	8,311	7,179	6,318	5,642	5,097	4,647	4,271	3,551	3,039	2,657	2,359	2,122	1,928	1,630	1,412	1,245	1,007	846
20	23,354	19,119	16,134	14,030	12,382	11,081	9,156	7,801	6,795	6,019	5,403	4,900	4,484	3,697	3,146	2,737	2,423	2,173	1,970	1,660	1,434	1,262	1,018	854
25	27,661	21,912	18,141	15,478	13,497	11,965	9,751	8,229	7,118	6,271	5,604	5,066	4,622	3,791	3,213	2,788	2,463	2,205	1,996	1,678	1,448	1,273	1,025	858
30	31,538	24,276	19,732	16,621	14,358	12,637	10,193	8,542	7,350	6,451	5,748	5,182	4,719	3,856	3,260	2,823	2,490	2,227	2,014	1,691	1,457	1,280	1,030	862
35	35,047	26,303	21,051	17,547	15,043	13,165	10,534	8,780	7,526	6,586	5,854	5,269	4,790	3,904	3,294	2,849	2,510	2,243	2,027	1,700	1,464	1,286	1,034	864
40	38,238	28,061	22,162	18,312	15,602	13,591	10,805	8,967	7,663	6,691	5,937	5,336	4,846	3,940	3,320	2,868	2,525	2,255	2,037	1,707	1,469	1,290	1,036	866
45	41,152	29,599	23,110	18,955	16,066	13,942	11,026	9,118	7,774	6,775	6,003	5,389	4,889	3,969	3,340	2,884	2,537	2,264	2,045	1,713	1,473	1,293	1,038	867
50	43,824	30,956	23,930	19,503	16,458	14,236	11,209	9,243	7,864	6,843	6,057	5,433	4,925	3,993	3,357	2,895	2,546	2,272	2,051	1,717	1,476	1,295	1,040	868
55	46,282	32,163	24,644	19,975	16,793	14,486	11,363	9,348	7,940	6,901	6,102	5,469	4,955	4,012	3,371	2,905	2,554	2,278	2,056	1,720	1,479	1,297	1,041	869
60	48,552	33,243	25,274	20,386	17,083	14,701	11,495	9,437	8,004	6,949	6,140	5,499	4,980	4,028	3,382	2,915	2,561	2,283	2,060	1,723	1,481	1,299	1,042	870
65	50,654	34,215	25,832	20,748	17,336	14,888	11,609	9,514	8,059	6,991	6,172	5,525	5,001	4,042	3,392	2,922	2,566	2,288	2,064	1,726	1,483	1,300	1,043	871
70	52,606	35,095	26,330	21,068	17,559	15,052	11,709	9,581	8,107	7,027	6,200	5,548	5,019	4,054	3,400	2,923	2,571	2,292	2,067	1,728	1,485	1,301	1,044	871
75	54,424	35,894	26,777	21,354	17,757	15,197	11,796	9,639	8,149	7,058	6,225	5,567	5,035	4,065	3,408	2,934	2,575	2,295	2,070	1,730	1,486	1,303	1,044	872
80	56,121	36,625	27,182	21,610	17,934	15,327	11,874	9,691	8,186	7,086	6,246	5,584	5,049	4,074	3,414	2,933	2,579	2,298	2,072	1,732	1,487	1,304	1,045	872
85	57,709	37,294	27,549	21,842	18,093	15,443	11,944	9,737	8,219	7,110	6,265	5,600	5,062	4,082	3,420	2,943	2,582	2,301	2,074	1,733	1,489	1,304	1,046	873
90	59,197	37,910	27,884	22,051	18,237	15,547	12,006	9,779	8,249	7,133	6,282	5,613	5,073	4,089	3,425	2,945	2,585	2,303	2,076	1,735	1,489	1,305	1,046	873
95	60,596	38,479	28,190	22,243	18,367	15,642	12,063	9,816	8,275	7,152	6,298	5,626	5,083	4,096	3,430	2,950	2,588	2,305	2,078	1,736	1,490	1,306	1,047	873
100	61,912	39,006	28,472	22,418	18,487	15,729	12,114	9,850	8,299	7,170	6,312	5,637	5,092	4,102	3,434	2,953	2,590	2,307	2,079	1,737	1,491	1,306	1,047	873
105	63,153	39,495	28,731	22,578	18,596	15,808	12,161	9,881	8,321	7,187	6,325	5,647	5,101	4,107	3,438	2,956	2,592	2,308	2,081	1,738	1,492	1,307	1,047	874
110	64,326	39,950	28,972	22,726	18,696	15,880	12,204	9,909	8,341	7,202	6,336	5,656	5,108	4,112	3,441	2,958	2,594	2,310	2,082	1,739	1,492	1,307	1,048	874
115	65,435	40,375	29,195	22,863	18,789	15,947	12,243	9,935	8,360	7,215	6,347	5,665	5,115	4,116	3,444	2,960	2,596	2,311	2,083	1,739	1,493	1,308	1,048	874
120	66,486	40,773	29,402	22,990	18,874	16,008	12,279	9,959	8,377	7,228	6,356	5,672	5,121	4,121	3,447	2,963	2,598	2,313	2,084	1,740	1,494	1,308	1,048	874
125	67,483	41,146	29,595	23,108	18,954	16,066	12,313	9,981	8,392	7,240	6,365	5,680	5,127	4,124	3,450	2,965	2,599	2,314	2,085	,741	1,494	1,309	1,048	874
130	68,430	41,496	29,776	23,218	19,028	16,119	12,344	10,002	8,407	7,250	6,374	5,686	5,133	4,128	3,452	2,966	2,601	2,315	2,086	,741	1,495	1,309	1,049	875
135	69,331	41,825	29,945	23,321	19,097	16,168	12,373	10,021	8,420	7,260	6,381	5,692	5,138	4,131	3,454	2,968	2,602	2,316	2,087	,742	1,495	1,309	1,049	875
140	70,189	42,136	30,104	23,417	19,161	16,214	12,400	10,039	8,433	7,270	6,389	5,698	5,142	4,134	3,456	2,970	2,603	2,317	2,088	,743	1,495	1,310	1,049	875
145	71,008	42,430	30,254	23,508	19,222	16,258	12,425	10,055	8,444	7,278	6,395	5,703	5,147	4,137	3,458	2,971	2,604	2,318	2,088	,743	1,496	1,310	1,049	875
150	71,789	42,708	30,395	23,593	19,279	16,298	12,449	10,071	8,455	7,287	6,402	5,708	5,151	4,139	3,460	2,972	2,605	2,319	2,089	,743	1,496	1,310	1,049	875
155	72,535	42,971	30,528	23,673	19,332	16,336	12,471	10,085	8,466	7,294	6,408	5,713	5,154	4,142	3,462	2,974	2,606	2,319	2,090	,744	1,496	1,310	1,050	875
160	73,249	43,220	30,654	23,749	19,382	16,372	12,492	10,099	8,475	7,301	6,413	5,717	5,158	4,144	3,463	2,975	2,607	2,320	2,090	,744	1,497	1,311	1,050	875
165	73,933	43,457	30,773	23,820	19,430	16,406	12,512	10,112	8,484	7,308	6,418	5,722	5,161	4,146	3,465	2,976	2,608	2,321	2,091	,745	1,497	1,311	1,050	875
170	74,589	43,683	30,886	23,888	19,475	16,438	12,531	10,124	8,493	7,314	6,423	5,726	5,165	4,148	3,466	2,977	2,609	2,321	2,091	,745	1,497	1,311	1,050	876
175	75,217	43,898	30,993	23,952	19,518	16,469	12,548	10,136	8,501	7,320	6,428	5,729	5,168	4,150	3,468	2,978	2,609	2,322	2,092	,745	1,497	1,311	1,050	876
180	75,821	44,103	31,095	24,013	19,558	16,497	12,565	10,146	8,509	7,326	6,432	5,733	5,170	4,152	3,469	2,979	2,610	2,323	2,092	,746	1,498	1,311	1,050	876
185	76,401	44,298	31,192	24,070	19,596	16,525	12,581	10,157	8,516	7,331	6,436	5,736	5,173	4,154	3,470	2,980	2,611	2,323	2,093	,746	1,498	1,312	1,050	876
190	76,958	44,485	31,285	24,125	19,633	16,551	12,596	10,167	8,523	7,337	6,440	5,739	5,176	4,156	3,471	2,981	2,612	2,324	2,093	,746	1,498	1,312	1,050	876
195	77,495	44,664	31,373	24,178	19,667	16,575	12,610	10,176	8,529	7,341	6,444	5,742	5,178	4,157	3,472	2,981	2,612	2,324	2,093	,747	1,498	1,312	1,050	876
200	78,011	44,835	31,457	24,228	19,701	16,599	12,624	10,185	8,536	7,346	6,448	5,745	5,180	4,159	3,474	2,982	2,613	2,325	2,094	,747	1,499	1,312	1,051	876

# 4 AWG	1 Run(s)	277 Volts	C-Value = 3,806	Magnetic Duct	Copper

Calculations:

1. $f = \dfrac{2 \times \text{Length} \times \text{Isca} \times 1.5}{\text{Runs} \times \text{C-Value} \times \text{E L-N}}$

2. $M = \dfrac{1}{1 + f}$

3. Isc = Isca x 1.5 x M

4. Add Motor Contribution = Motor FLA x 4

* All results are given in symmetrical amperes

Single-Phase L-N Bolted Fault-Current* Table

One-Way Distance in Feet

Available Isc in Thousands (Isca)	5	10	15	20	25	30	40	50	60	70	80	90	100	125	150	175	200	225	250	300	350	400	500	600
3	4,351	4,212	4,082	3,959	3,844	3,735	3,535	3,355	3,193	3,045	2,911	2,788	2,674	2,428	2,223	2,051	1,903	1,775	1,663	1,476	1,328	1,206	1,020	883
5	7,096	6,734	6,407	6,110	5,839	5,592	5,154	4,781	4,457	4,175	3,926	3,706	3,509	3,097	2,771	2,508	2,290	2,107	1,951	1,700	1,505	1,351	1,121	958
7	9,726	9,057	8,475	7,963	7,510	7,105	6,414	5,845	5,369	4,965	4,617	4,315	4,050	3,511	3,098	2,772	2,509	2,291	2,108	1,817	1,597	1,425	1,171	995
10	13,468	12,220	11,183	10,309	9,561	8,915	7,853	7,017	6,342	5,785	5,319	4,922	4,580	3,902	3,399	3,011	2,702	2,451	2,243	1,917	1,673	1,485	1,212	1,024
15	19,220	16,775	14,882	13,372	12,141	11,117	9,513	8,313	7,382	6,639	6,032	5,526	5,099	4,273	3,677	3,227	2,875	2,592	2,360	2,002	1,738	1,536	1,246	1,048
20	24,439	20,618	17,830	15,706	14,034	12,684	10,637	9,160	8,042	7,168	6,465	5,888	5,405	4,486	3,833	3,347	2,970	2,669	2,424	2,048	1,772	1,562	1,263	1,060
25	29,196	23,903	20,235	17,543	15,483	13,856	11,449	9,755	8,498	7,527	6,756	6,128	5,607	4,624	3,934	3,423	3,030	2,718	2,464	2,076	1,794	1,579	1,274	1,067
30	33,550	26,745	22,235	19,026	16,627	14,765	12,063	10,197	8,831	7,788	6,965	6,300	5,750	4,721	4,004	3,476	3,071	2,751	2,491	2,095	1,808	1,590	1,281	1,073
35	37,549	29,226	23,923	20,249	17,553	15,491	12,544	10,538	9,086	7,985	7,123	6,428	5,857	4,793	4,056	3,515	3,102	2,775	2,511	2,109	1,818	1,598	1,286	1,076
40	41,236	31,412	25,368	21,275	18,319	16,084	12,930	10,810	9,287	8,140	7,246	6,528	5,940	4,848	4,095	3,545	3,125	2,794	2,526	2,120	1,826	1,604	1,290	1,079
45	44,645	33,352	26,619	22,147	18,962	16,578	13,247	11,031	9,449	8,265	7,344	6,608	6,006	4,892	4,126	3,568	3,143	2,808	2,538	2,128	1,832	1,609	1,293	1,081
50	47,807	35,085	27,711	22,899	19,510	16,996	13,512	11,214	9,584	8,367	7,425	6,673	6,060	4,928	4,152	3,587	3,158	2,820	2,547	2,135	1,837	1,613	1,296	1,083
55	50,747	36,644	28,675	23,553	19,983	17,353	13,737	11,368	9,696	8,453	7,492	6,728	6,105	4,957	4,173	3,603	3,170	2,830	2,555	2,141	1,842	1,616	1,298	1,084
60	53,489	38,052	29,530	24,127	20,395	17,663	13,930	11,500	9,792	8,526	7,549	6,774	6,143	4,982	4,190	3,616	3,180	2,838	2,562	2,145	1,845	1,619	1,299	1,086
65	56,052	39,331	30,295	24,635	20,756	17,933	14,098	11,615	9,875	8,588	7,599	6,813	6,175	5,003	4,205	3,627	3,189	2,845	2,568	2,149	1,848	1,621	1,301	1,087
70	58,452	40,498	30,982	25,087	21,077	18,172	14,245	11,714	9,947	8,643	7,641	6,847	6,203	5,022	4,218	3,637	3,196	2,851	2,572	2,152	1,850	1,623	1,302	1,087
75	60,705	41,567	31,604	25,493	21,363	18,384	14,375	11,802	10,010	8,690	7,678	6,877	6,228	5,038	4,230	3,645	3,202	2,856	2,577	2,155	1,853	1,624	1,303	1,088
80	62,824	42,550	32,169	25,860	21,619	18,574	14,491	11,880	10,066	8,733	7,711	6,904	6,249	5,052	4,240	3,653	3,208	2,860	2,580	2,158	1,854	1,626	1,304	1,089
85	64,820	43,456	32,684	26,192	21,851	18,745	14,595	11,949	10,116	8,770	7,740	6,927	6,268	5,065	4,249	3,659	3,213	2,864	2,584	2,160	1,856	1,627	1,305	1,089
90	66,704	44,295	33,156	26,494	22,061	18,899	14,688	12,012	10,161	8,804	7,767	6,948	6,286	5,076	4,256	3,665	3,218	2,868	2,587	2,162	1,858	1,628	1,306	1,090
95	68,485	45,073	33,590	26,770	22,252	19,039	14,773	12,069	10,201	8,834	7,790	6,967	6,301	5,086	4,264	3,670	3,222	2,871	2,589	2,164	1,859	1,629	1,306	1,090
100	70,171	45,798	33,991	27,024	22,428	19,167	14,850	12,120	10,238	8,862	7,812	6,984	6,315	5,095	4,270	3,675	3,225	2,874	2,591	2,166	1,860	1,630	1,307	1,091
105	71,770	46,473	34,362	27,258	22,588	19,285	14,920	12,167	10,271	8,887	7,831	7,000	6,328	5,103	4,276	3,679	3,229	2,877	2,594	2,167	1,861	1,631	1,308	1,091
110	73,288	47,105	34,706	27,474	22,737	19,393	14,985	12,210	10,302	8,909	7,849	7,014	6,339	5,111	4,281	3,683	3,232	2,879	2,596	2,169	1,862	1,632	1,308	1,091
115	74,731	47,697	35,026	27,675	22,874	19,492	15,044	12,249	10,330	8,930	7,865	7,027	6,350	5,118	4,286	3,687	3,234	2,881	2,597	2,170	1,863	1,633	1,309	1,092
120	76,105	48,253	35,325	27,861	23,001	19,584	15,099	12,285	10,356	8,950	7,880	7,039	6,360	5,124	4,290	3,690	3,237	2,883	2,599	2,171	1,864	1,633	1,309	1,092
125	77,414	48,776	35,605	28,034	23,119	19,670	15,150	12,319	10,379	8,968	7,894	7,050	6,369	5,130	4,294	3,693	3,239	2,885	2,600	2,172	1,865	1,634	1,309	1,092
130	78,663	49,269	35,867	28,197	23,229	19,750	15,197	12,350	10,402	8,984	7,907	7,060	6,377	5,135	4,298	3,696	3,242	2,887	2,602	2,173	1,866	1,634	1,310	1,093
135	79,856	49,734	36,113	28,348	23,332	19,824	15,241	12,379	10,422	8,999	7,919	7,069	6,385	5,140	4,302	3,698	3,244	2,888	2,603	2,174	1,866	1,635	1,310	1,093
140	80,997	50,175	36,344	28,491	23,428	19,894	15,282	12,406	10,441	9,014	7,930	7,078	6,392	5,145	4,305	3,701	3,245	2,890	2,604	2,175	1,867	1,635	1,310	1,093
145	82,089	50,591	36,562	28,625	23,519	19,959	15,321	12,432	10,459	9,027	7,940	7,086	6,399	5,149	4,308	3,703	3,247	2,891	2,605	2,176	1,867	1,636	1,311	1,093
150	83,134	50,987	36,768	28,751	23,604	20,020	15,357	12,455	10,476	9,040	7,950	7,094	6,405	5,153	4,311	3,705	3,249	2,892	2,606	2,176	1,868	1,636	1,311	1,093
155	84,137	51,362	36,963	28,870	23,684	20,078	15,390	12,478	10,492	9,051	7,959	7,101	6,411	5,157	4,314	3,707	3,250	2,894	2,607	2,177	1,868	1,637	1,311	1,094
160	85,100	51,719	37,148	28,982	23,760	20,132	15,422	12,499	10,507	9,062	7,967	7,108	6,416	5,161	4,316	3,709	3,252	2,895	2,608	2,178	1,869	1,637	1,311	1,094
165	86,024	52,059	37,323	29,089	23,831	20,183	15,452	12,518	10,521	9,073	7,975	7,115	6,422	5,164	4,318	3,711	3,253	2,896	2,609	2,178	1,869	1,637	1,312	1,094
170	86,912	52,383	37,489	29,190	23,899	20,232	15,481	12,537	10,534	9,083	7,983	7,121	6,426	5,167	4,321	3,712	3,254	2,897	2,610	2,179	1,870	1,638	1,312	1,094
175	87,767	52,692	37,647	29,286	23,963	20,278	15,508	12,555	10,546	9,092	7,990	7,126	6,431	5,170	4,323	3,714	3,255	2,898	2,611	2,179	1,870	1,638	1,312	1,094
180	88,590	52,988	37,798	29,377	24,024	20,321	15,533	12,571	10,558	9,101	7,997	7,132	6,435	5,173	4,325	3,715	3,257	2,899	2,612	2,180	1,871	1,638	1,312	1,094
185	89,383	53,270	37,941	29,463	24,082	20,363	15,557	12,587	10,569	9,109	8,003	7,137	6,440	5,176	4,327	3,717	3,258	2,899	2,612	2,180	1,871	1,638	1,312	1,094
190	90,147	53,541	38,078	29,546	24,137	20,402	15,580	12,602	10,580	9,117	8,009	7,142	6,444	5,178	4,328	3,718	3,259	2,900	2,613	2,181	1,871	1,639	1,312	1,095
195	90,884	53,800	38,209	29,624	24,190	20,440	15,602	12,616	10,590	9,124	8,015	7,146	6,447	5,181	4,330	3,719	3,260	2,901	2,613	2,181	1,872	1,639	1,313	1,095
200	91,595	54,049	38,335	29,700	24,240	20,475	15,623	12,630	10,599	9,131	8,020	7,151	6,451	5,183	4,332	3,720	3,260	2,902	2,614	2,182	1,872	1,639	1,313	1,095

\# 3 AWG | 1 Run(s) | 277 Volts | C-Value = 4,760 | Magnetic Duct | Copper

Calculations:

1. $f = \dfrac{2 \times \text{Length} \times \text{Isca} \times 1.5}{\text{Runs} \times \text{C-Value} \times \text{E L-N}}$

2. $M = \dfrac{1}{1 + f}$

3. $\text{Isc} = \text{Isca} \times 1.5 \times M$

4. Add Motor Contribution = Motor FLA x 4

* All results are given in symmetrical amperes

Single-Phase L-N Bolted Fault-Current* Table

One-Way Distance in Feet

Available Isc in Thousands (Isca)

Isca	5	10	15	20	25	30	40	50	60	70	80	90	100	125	150	175	200	225	250	300	350	400	500	600
3	4,380	4,265	4,157	4,054	3,956	3,863	3,688	3,529	3,383	3,249	3,125	3,010	2,903	2,666	2,465	2,293	2,143	2,011	1,894	1,698	1,538	1,406	1,200	1,046
5	7,171	6,870	6,593	6,338	6,101	5,882	5,487	5,142	4,838	4,568	4,326	4,109	3,913	3,495	3,157	2,880	2,647	2,449	2,278	2,000	1,782	1,607	1,343	1,154
7	9,867	9,306	8,805	8,355	7,949	7,581	6,938	6,395	5,932	5,531	5,180	4,872	4,598	4,031	3,589	3,234	2,943	2,700	2,495	2,165	1,912	1,712	1,415	1,207
10	13,740	12,676	11,764	10,975	10,285	9,677	8,653	7,825	7,142	6,568	6,080	5,660	5,293	4,556	3,999	3,564	3,214	2,926	2,686	2,307	2,022	1,800	1,475	1,250
15	19,780	17,646	15,928	14,515	13,332	12,327	10,713	9,472	8,489	7,691	7,030	6,474	5,999	5,069	4,389	3,870	3,461	3,130	2,857	2,432	2,117	1,875	1,525	1,285
20	25,351	21,950	19,353	17,306	15,650	14,284	12,160	10,587	9,373	8,410	7,626	6,975	6,427	5,372	4,614	4,044	3,599	3,243	2,950	2,499	2,168	1,914	1,551	1,304
25	30,507	25,712	22,220	19,563	17,473	15,787	13,233	11,390	9,998	8,909	8,034	7,316	6,715	5,572	4,761	4,156	3,688	3,314	3,009	2,542	2,200	1,939	1,568	1,315
30	35,292	29,030	24,655	21,426	18,945	16,979	14,060	11,998	10,463	9,277	8,332	7,561	6,922	5,713	4,864	4,234	3,749	3,364	3,050	2,571	2,222	1,956	1,579	1,323
35	39,745	31,977	26,748	22,990	20,157	17,946	14,717	12,473	10,823	9,558	8,558	7,747	7,077	5,819	4,940	4,292	3,794	3,400	3,080	2,592	2,237	1,968	1,587	1,329
40	43,900	34,612	28,568	24,321	21,173	18,747	15,251	12,855	11,109	9,781	8,736	7,893	7,198	5,900	4,999	4,336	3,829	3,428	3,103	2,608	2,249	1,978	1,593	1,333
45	47,784	36,982	30,164	25,468	22,037	19,421	15,695	13,168	11,342	9,961	8,880	8,010	7,296	5,966	5,046	4,371	3,856	3,450	3,121	2,621	2,259	1,985	1,597	1,336
50	51,425	39,126	31,574	26,466	22,781	19,996	16,068	13,430	11,536	10,110	8,998	8,106	7,375	6,019	5,084	4,400	3,878	3,467	3,135	2,631	2,266	1,991	1,601	1,339
55	54,843	41,074	32,831	27,344	23,428	20,493	16,388	13,652	11,700	10,236	9,097	8,187	7,442	6,063	5,115	4,424	3,897	3,482	3,147	2,639	2,273	1,995	1,604	1,341
60	58,059	42,852	33,957	28,120	23,996	20,926	16,663	13,843	11,840	10,343	9,182	8,255	7,498	6,100	5,142	4,443	3,912	3,494	3,157	2,646	2,278	2,000	1,607	1,343
65	61,091	44,481	34,972	28,813	24,498	21,307	16,904	14,009	11,961	10,435	9,254	8,314	7,547	6,132	5,164	4,460	3,925	3,505	3,166	2,652	2,282	2,003	1,609	1,345
70	63,953	45,979	35,892	29,434	24,946	21,645	17,116	14,154	12,066	10,515	9,317	8,365	7,589	6,160	5,184	4,475	3,937	3,514	3,173	2,658	2,286	2,006	1,611	1,346
75	66,660	47,362	36,729	29,995	25,347	21,947	17,304	14,283	12,159	10,586	9,373	8,409	7,625	6,184	5,201	4,488	3,946	3,522	3,179	2,662	2,290	2,008	1,613	1,347
80	69,224	48,642	37,494	30,503	25,709	22,218	17,472	14,397	12,242	10,649	9,422	8,449	7,658	6,205	5,216	4,499	3,955	3,529	3,185	2,666	2,292	2,011	1,614	1,348
85	71,655	49,830	38,196	30,966	26,037	22,462	17,623	14,499	12,316	10,704	9,466	8,484	7,687	6,224	5,230	4,509	3,963	3,535	3,190	2,670	2,295	2,013	1,615	1,349
90	73,964	50,936	38,842	31,389	26,336	22,684	17,759	14,591	12,383	10,755	9,505	8,515	7,713	6,241	5,241	4,518	3,970	3,540	3,195	2,673	2,297	2,014	1,616	1,350
95	76,161	51,968	39,439	31,778	26,609	22,887	17,883	14,675	12,443	10,800	9,540	8,544	7,736	6,257	5,252	4,526	3,976	3,545	3,198	2,675	2,299	2,016	1,617	1,350
100	78,252	52,933	39,993	32,137	26,860	23,072	17,996	14,751	12,497	10,841	9,572	8,569	7,757	6,270	5,262	4,533	3,981	3,549	3,202	2,678	2,301	2,017	1,618	1,351
105	80,245	53,837	40,507	32,468	27,091	23,242	18,100	14,820	12,547	10,878	9,601	8,593	7,776	6,283	5,271	4,539	3,986	3,553	3,205	2,680	2,303	2,019	1,619	1,352
110	82,148	54,687	40,986	32,775	27,305	23,399	18,195	14,884	12,593	10,913	9,628	8,614	7,793	6,294	5,279	4,545	3,991	3,557	3,208	2,682	2,304	2,020	1,620	1,352
115	83,965	55,487	41,434	33,061	27,503	23,544	18,282	14,942	12,634	10,944	9,653	8,634	7,809	6,305	5,286	4,551	3,995	3,560	3,211	2,684	2,306	2,021	1,621	1,353
120	85,703	56,241	41,853	33,327	27,687	23,679	18,363	14,997	12,673	10,973	9,675	8,652	7,824	6,314	5,293	4,556	3,999	3,563	3,214	2,686	2,307	2,022	1,621	1,353
125	87,367	56,952	42,245	33,575	27,858	23,804	18,439	15,047	12,709	11,000	9,696	8,668	7,838	6,323	5,299	4,560	4,003	3,566	3,216	2,688	2,308	2,023	1,622	1,353
130	88,962	57,626	42,615	33,808	28,018	23,921	18,509	15,093	12,742	11,025	9,715	8,684	7,850	6,331	5,305	4,565	4,006	3,569	3,218	2,689	2,309	2,024	1,622	1,354
135	90,491	58,263	42,963	34,027	28,168	24,030	18,574	15,137	12,773	11,048	9,733	8,698	7,862	6,339	5,310	4,569	4,009	3,571	3,220	2,690	2,310	2,025	1,623	1,354
140	91,958	58,868	43,291	34,232	28,309	24,133	18,635	15,177	12,802	11,069	9,750	8,712	7,873	6,346	5,315	4,572	4,012	3,574	3,222	2,692	2,311	2,025	1,623	1,355
145	93,368	59,443	43,600	34,426	28,441	24,229	18,692	15,215	12,829	11,090	9,766	8,724	7,883	6,353	5,320	4,576	4,014	3,576	3,223	2,693	2,312	2,026	1,624	1,355
150	94,723	59,989	43,894	34,608	28,565	24,319	18,746	15,251	12,854	11,109	9,780	8,736	7,893	6,359	5,324	4,579	4,017	3,578	3,225	2,694	2,313	2,027	1,624	1,355
155	96,027	60,510	44,172	34,781	28,683	24,404	18,796	15,284	12,878	11,126	9,794	8,747	7,902	6,365	5,328	4,582	4,019	3,580	3,227	2,695	2,314	2,027	1,625	1,355
160	97,283	61,006	44,436	34,944	28,794	24,484	18,844	15,316	12,900	11,143	9,807	8,757	7,910	6,370	5,332	4,585	4,021	3,581	3,228	2,696	2,315	2,028	1,625	1,356
165	98,493	61,479	44,686	35,099	28,899	24,560	18,889	15,345	12,921	11,159	9,819	8,767	7,918	6,375	5,336	4,588	4,023	3,583	3,229	2,697	2,315	2,028	1,625	1,356
170	99,659	61,932	44,925	35,246	28,998	24,632	18,931	15,373	12,941	11,173	9,831	8,776	7,926	6,380	5,339	4,590	4,025	3,584	3,230	2,698	2,316	2,029	1,626	1,356
175	100,785	62,365	45,152	35,386	29,093	24,700	18,972	15,400	12,960	11,187	9,841	8,785	7,933	6,385	5,342	4,592	4,027	3,586	3,232	2,699	2,316	2,029	1,626	1,356
180	101,871	62,779	45,369	35,519	29,183	24,765	19,010	15,425	12,978	11,201	9,852	8,793	7,939	6,389	5,345	4,595	4,029	3,587	3,233	2,699	2,317	2,030	1,626	1,356
185	102,921	63,176	45,576	35,645	29,268	24,827	19,046	15,449	12,995	11,213	9,861	8,800	7,946	6,393	5,348	4,597	4,031	3,588	3,234	2,700	2,318	2,030	1,626	1,357
190	103,935	63,557	45,774	35,766	29,350	24,885	19,080	15,472	13,011	11,225	9,871	8,808	7,952	6,397	5,351	4,599	4,032	3,590	3,235	2,701	2,318	2,030	1,627	1,357
195	104,916	63,922	45,963	35,882	29,427	24,941	19,113	15,493	13,026	11,237	9,879	8,815	7,957	6,401	5,353	4,601	4,034	3,591	3,236	2,701	2,319	2,031	1,627	1,357
200	105,866	64,273	46,144	35,992	29,502	24,994	19,145	15,514	13,040	11,247	9,888	8,821	7,963	6,404	5,356	4,602	4,035	3,592	3,237	2,702	2,319	2,031	1,627	1,357

# 2 AWG	1 Run(s)	277 Volts	C-Value = 5,906	Magnetic Duct	Copper

Calculations:

1. $f = \dfrac{2 \times \text{Length} \times \text{Isca} \times 1.5}{\text{Runs} \times \text{C-Value} \times E_{L-N}}$

2. $M = \dfrac{1}{1 + f}$

3. Isc = Isca x 1.5 x M

4. Add Motor Contribution = Motor FLA x 4

* All results are given in symmetrical amperes

Single-Phase L-N Bolted Fault-Current* Table

One-Way Distance in Feet

Available Isc in Thousands (Isca)

Isca	5	10	15	20	25	30	40	50	60	70	80	90	100	125	150	175	200	225	250	300	350	400	500	600
3	4,402	4,308	4,218	4,132	4,049	3,969	3,819	3,680	3,551	3,430	3,317	3,212	3,113	2,890	2,697	2,528	2,380	2,247	2,129	1,926	1,758	1,617	1,394	1,225
5	7,231	6,982	6,748	6,530	6,326	6,134	5,782	5,469	5,188	4,935	4,705	4,495	4,304	3,889	3,548	3,261	3,018	2,808	2,626	2,324	2,084	1,889	1,591	1,375
7	9,981	9,511	9,083	8,693	8,334	8,004	7,416	6,909	6,466	6,077	5,732	5,424	5,148	4,566	4,102	3,724	3,410	3,144	2,917	2,549	2,264	2,035	1,694	1,451
10	13,963	13,060	12,267	11,565	10,938	10,377	9,410	8,608	7,932	7,354	6,855	6,419	6,036	5,251	4,647	4,168	3,778	3,455	3,183	2,749	2,420	2,161	1,780	1,513
15	20,245	18,401	16,864	15,565	14,451	13,486	11,898	10,644	9,629	8,791	8,087	7,487	6,971	5,945	5,182	4,593	4,124	3,742	3,425	2,928	2,558	2,270	1,853	1,566
20	26,120	23,129	20,753	18,819	17,215	15,863	13,710	12,071	10,783	9,742	8,885	8,167	7,556	6,365	5,499	4,840	4,322	3,904	3,560	3,027	2,632	2,329	1,892	1,594
25	31,628	27,346	24,085	21,519	19,447	17,740	15,089	13,128	11,618	10,419	9,445	8,637	7,957	6,647	5,708	5,001	4,450	4,009	3,647	3,089	2,679	2,366	1,917	1,611
30	36,801	31,130	26,973	23,795	21,287	19,258	16,174	13,941	12,250	10,925	9,859	8,982	8,248	6,850	5,857	5,115	4,540	4,082	3,707	3,132	2,712	2,391	1,933	1,623
35	41,669	34,543	29,499	25,740	22,830	20,512	17,049	14,587	12,746	11,318	10,177	9,245	8,470	7,002	5,968	5,200	4,607	4,135	3,751	3,164	2,735	2,409	1,945	1,631
40	46,259	37,639	31,727	27,420	24,143	21,565	17,771	15,112	13,145	11,631	10,430	9,454	8,644	7,121	6,054	5,265	4,658	4,176	3,785	3,188	2,753	2,423	1,954	1,637
45	50,593	40,459	33,707	28,887	25,272	22,462	18,375	15,547	13,473	11,887	10,635	9,622	8,785	7,216	6,122	5,317	4,698	4,209	3,812	3,207	2,767	2,434	1,961	1,642
50	54,692	43,039	35,479	30,178	26,256	23,235	18,889	15,913	13,747	12,100	10,805	9,761	8,901	7,294	6,178	5,359	4,731	4,235	3,833	3,222	2,779	2,443	1,967	1,646
55	58,575	45,408	37,073	31,324	27,119	23,909	19,332	16,226	13,980	12,280	10,949	9,878	8,998	7,359	6,225	5,394	4,758	4,257	3,851	3,234	2,788	2,450	1,972	1,650
60	62,259	47,590	38,516	32,348	27,882	24,500	19,717	16,497	14,180	12,434	11,071	9,978	9,080	7,414	6,264	5,423	4,781	4,275	3,866	3,245	2,796	2,456	1,976	1,652
65	65,758	49,608	39,827	33,267	28,563	25,024	20,055	16,732	14,354	12,568	11,177	10,063	9,151	7,461	6,298	5,449	4,801	4,291	3,879	3,254	2,803	2,461	1,979	1,655
70	69,087	51,479	41,024	34,098	29,173	25,492	20,354	16,940	14,507	12,685	11,269	10,138	9,213	7,502	6,327	5,470	4,818	4,305	3,890	3,262	2,808	2,466	1,982	1,657
75	72,256	53,219	42,121	34,853	29,724	25,911	20,621	17,124	14,642	12,788	11,351	10,204	9,267	7,538	6,353	5,489	4,333	4,316	3,900	3,269	2,813	2,469	1,984	1,658
80	75,278	54,840	43,130	35,541	30,223	26,289	20,860	17,289	14,762	12,879	11,423	10,262	9,315	7,570	6,375	5,506	4,346	4,327	3,908	3,275	2,818	2,473	1,986	1,660
85	78,162	56,355	44,062	36,171	30,678	26,633	21,075	17,437	14,869	12,961	11,487	10,314	9,358	7,598	6,395	5,521	4,357	4,336	3,916	3,280	2,822	2,476	1,988	1,661
90	80,918	57,774	44,924	36,750	31,093	26,945	21,270	17,570	14,966	13,035	11,545	10,360	9,396	7,623	6,413	5,535	4,368	4,344	3,922	3,285	2,825	2,479	1,990	1,663
95	83,554	59,105	45,725	37,285	31,475	27,231	21,448	17,691	15,054	13,101	11,597	10,402	9,431	7,646	6,429	5,546	4,877	4,352	3,928	3,289	2,828	2,481	1,992	1,664
100	86,077	60,356	46,471	37,779	31,826	27,494	21,611	17,802	15,134	13,162	11,644	10,441	9,462	7,667	6,444	5,557	4,885	4,358	3,934	3,293	2,831	2,483	1,993	1,665
105	88,496	61,536	47,166	38,238	32,151	27,736	21,760	17,903	15,207	13,217	11,687	10,475	9,491	7,685	6,457	5,567	4,893	4,364	3,939	3,296	2,834	2,485	1,994	1,665
110	90,815	62,648	47,817	38,664	32,452	27,960	21,898	17,996	15,274	13,268	11,727	10,507	9,517	7,702	6,469	5,576	4,900	4,370	3,943	3,299	2,836	2,487	1,995	1,666
115	93,042	63,700	48,427	39,062	32,732	28,168	22,025	18,082	15,336	13,314	11,763	10,536	9,541	7,718	6,480	5,584	4,906	4,375	3,947	3,302	2,838	2,488	1,997	1,667
120	95,181	64,695	49,001	39,434	32,993	28,361	22,143	18,161	15,393	13,357	11,797	10,563	9,563	7,732	6,490	5,592	4,912	4,379	3,951	3,305	2,840	2,490	1,997	1,668
125	97,237	65,639	49,540	39,783	33,237	28,540	22,252	18,234	15,446	13,397	11,828	10,588	9,583	7,746	6,500	5,599	4 917	4,384	3,955	3,307	2,842	2,491	1,998	1,668
130	99,216	66,535	50,049	40,110	33,465	28,709	22,354	18,303	15,495	13,434	11,857	10,611	9,602	7,758	6,508	5,605	4 922	4,388	3,958	3,309	2,843	2,493	1,999	1,669
135	101,122	67,386	50,529	40,418	33,679	28,866	22,449	18,367	15,541	13,468	11,883	10,632	9,620	7,770	6,516	5,611	4 927	4,391	3,961	3,311	2,845	2,494	2,000	1,669
140	102,958	68,197	50,983	40,708	33,880	29,014	22,539	18,426	15,583	13,500	11,908	10,652	9,636	7,780	6,524	5,617	4,931	4,395	3,964	3,313	2,846	2,495	2,001	1,670
145	104,729	68,969	51,414	40,982	34,070	29,152	22,622	18,482	15,623	13,530	11,932	10,671	9,651	7,790	6,531	5,622	4,935	4,398	3,966	3,315	2,848	2,496	2,001	1,670
150	106,437	69,706	51,822	41,241	34,249	29,283	22,701	18,535	15,661	13,558	11,954	10,689	9,666	7,799	6,537	5,627	4,939	4,401	3,969	3,317	2,849	2,497	2,002	1,671
155	108,086	70,409	52,210	41,487	34,418	29,407	22,775	18,584	15,696	13,585	11,974	10,705	9,679	7,808	6,543	5,631	4,942	4,404	3,971	3,318	2,850	2,498	2,002	1,671
160	109,680	71,082	52,579	41,719	34,577	29,523	22,845	18,631	15,729	13,610	11,993	10,720	9,692	7,816	6,549	5,636	4,946	4,406	3,973	3,320	2,851	2,499	2,003	1,672
165	111,220	71,726	52,930	41,940	34,729	29,634	22,911	18,675	15,761	13,633	12,012	10,735	9,703	7,824	6,555	5,640	4,949	4,409	3,975	3,321	2,852	2,499	2,004	1,672
170	112,710	72,343	53,265	42,150	34,873	29,739	22,974	18,716	15,790	13,655	12,029	10,749	9,715	7,831	6,560	5,643	4,952	4,411	3,977	3,323	2,853	2,500	2,004	1,672
175	114,151	72,934	53,585	42,350	35,010	29,838	23,033	18,756	15,818	13,676	12,045	10,762	9,725	7,838	6,565	5,647	4,954	4,413	3,979	3,324	2,854	2,501	2,004	1,673
180	115,547	73,501	53,891	42,541	35,140	29,933	23,089	18,793	15,845	13,696	12,060	10,774	9,735	7,845	6,569	5,650	4,957	4,415	3,980	3,325	2,855	2,501	2,005	1,673
185	116,899	74,046	54,183	42,723	35,264	30,023	23,143	18,828	15,870	13,715	12,075	10,785	9,745	7,851	6,573	5,654	4,959	4,417	3,982	3,326	2,856	2,502	2,005	1,673
190	118,210	74,569	54,463	42,897	35,382	30,108	23,194	18,862	15,894	13,733	12,089	10,796	9,754	7,857	6,578	5,657	4,962	4,419	3,983	3,327	2,857	2,503	2,006	1,673
195	119,480	75,073	54,731	43,063	35,495	30,190	23,242	18,894	15,916	13,750	12,102	10,807	9,762	7,862	6,581	5,659	4,964	4,421	3,985	3,328	2,857	2,503	2,006	1,674
200	120,713	75,558	54,988	43,222	35,603	30,268	23,289	18,925	15,938	13,766	12,114	10,817	9,770	7,868	6,585	5,662	4,966	4,422	3,986	3,329	2,858	2,504	2,006	1,674

# 1 AWG	1 Run(s)	277 Volts	C-Value = 7,292	Magnetic Duct	Copper

Calculations:

1. $f = \dfrac{2 \times \text{Length} \times \text{Isca} \times 1.5}{\text{Runs} \times \text{C-Value} \times \text{E L-N}}$

2. $M = \dfrac{1}{1 + f}$

3. $\text{Isc} = \text{Isca} \times 1.5 \times M$

4. Add Motor Contribution = Motor FLA x 4

* All results are given in symmetrical amperes

Single-Phase L-N Bolted Fault-Current* Table

One-Way Distance in Feet

Available Isc in Thousands (Isca)

	5	10	15	20	25	30	40	50	60	70	80	90	100	125	150	175	200	225	250	300	350	400	500	600
3	4,420	4,342	4,267	4,195	4,125	4,057	3,928	3,807	3,693	3,586	3,485	3,389	3,299	3,093	2,910	2,749	2,604	2,474	2,356	2,151	1,979	1,832	1,596	1,413
5	7,279	7,071	6,874	6,688	6,512	6,345	6,035	5,754	5,498	5,264	5,049	4,851	4,668	4,265	3,926	3,637	3,388	3,171	2,980	2,659	2,401	2,188	1,859	1,616
7	10,072	9,678	9,313	8,975	8,661	8,367	7,837	7,370	6,955	6,584	6,251	5,950	5,677	5,092	4,617	4,222	3,890	3,606	3,361	2,959	2,643	2,387	2,001	1,722
10	14,142	13,377	12,690	12,070	11,508	10,996	10,098	9,335	8,680	8,110	7,611	7,169	6,776	5,959	5,318	4,802	4,377	4,021	3,718	3,232	2,858	2,562	2,122	1,811
15	20,623	19,035	17,674	16,495	15,463	14,552	13,020	11,779	10,754	9,893	9,160	8,528	7,978	6,869	6,031	5,375	4,848	4,415	4,053	3,482	3,052	2,717	2,227	1,887
20	26,753	24,141	21,993	20,196	18,671	17,359	15,222	13,552	12,213	11,115	10,198	9,421	8,753	7,437	6,464	5,717	5,124	4,643	4,244	3,622	3,159	2,801	2,284	1,928
25	32,560	28,771	25,771	23,338	21,325	19,631	16,941	14,899	13,296	12,004	10,942	10,052	9,296	7,825	6,755	5,943	5,306	4,791	4,368	3,712	3,227	2,855	2,319	1,953
30	38,070	32,989	29,105	26,039	23,558	21,508	18,320	15,955	14,131	12,681	11,501	10,522	9,696	8,107	6,965	6,105	5,434	4,896	4,455	3,774	3,274	2,891	2,343	1,970
35	43,303	36,848	32,068	28,386	25,462	23,084	19,451	16,806	14,795	13,213	11,937	10,886	10,004	8,321	7,122	6,225	5,529	4,973	4,518	3,820	3,309	2,918	2,361	1,982
40	48,281	40,392	34,719	30,443	27,105	24,427	20,396	17,507	15,335	13,642	12,286	11,175	10,249	8,489	7,245	6,319	5,603	5,032	4,568	3,855	3,335	2,939	2,374	1,992
45	53,022	43,657	37,104	32,262	28,537	25,584	21,196	18,093	15,783	13,996	12,572	11,411	10,447	8,624	7,343	6,394	5,662	5,080	4,607	3,883	3,356	2,955	2,385	1,999
50	57,542	46,676	39,263	33,881	29,797	26,592	21,884	18,592	16,161	14,292	12,811	11,608	10,611	8,736	7,424	6,455	5,709	5,118	4,638	3,905	3,373	2,968	2,393	2,005
55	61,856	49,476	41,225	35,332	30,914	27,477	22,480	19,020	16,484	14,544	13,013	11,773	10,749	8,830	7,492	6,506	5,749	5,150	4,664	3,924	3,386	2,978	2,400	2,010
60	65,978	52,078	43,016	36,640	31,910	28,262	23,002	19,393	16,763	14,761	13,186	11,915	10,867	8,909	7,549	6,549	5,783	5,177	4,686	3,940	3,398	2,987	2,406	2,014
65	69,921	54,504	44,658	37,824	32,805	28,961	23,463	19,720	17,006	14,949	13,336	12,038	10,969	8,977	7,598	6,586	5,812	5,200	4,705	3,953	3,408	2,995	2,411	2,017
70	73,696	56,771	46,168	38,902	33,613	29,589	23,874	20,009	17,221	15,115	13,468	12,145	11,058	9,037	7,640	6,618	5,836	5,220	4,722	3,964	3,416	3,002	2,415	2,020
75	77,314	58,894	47,562	39,888	34,346	30,156	24,241	20,266	17,411	15,261	13,584	12,239	11,136	9,089	7,678	6,646	5,858	5,237	4,736	3,974	3,424	3,007	2,419	2,023
80	80,784	60,886	48,853	40,792	35,014	30,669	24,572	20,497	17,581	15,392	13,687	12,323	11,206	9,135	7,710	6,670	5,877	5,253	4,748	3,983	3,430	3,012	2,422	2,025
85	84,115	62,759	50,052	41,624	35,625	31,138	24,872	20,705	17,734	15,509	13,780	12,398	11,267	9,176	7,740	6,692	5,894	5,266	4,759	3,991	3,436	3,017	2,425	2,027
90	87,315	64,524	51,168	42,393	36,187	31,566	25,144	20,894	17,872	15,615	13,863	12,465	11,323	9,213	7,766	6,712	5,909	5,278	4,769	3,998	3,441	3,021	2,427	2,029
95	90,392	66,189	52,209	43,105	36,705	31,959	25,393	21,065	17,998	15,710	13,938	12,526	11,373	9,246	7,789	6,723	5,923	5,289	4,778	4,004	3,446	3,024	2,430	2,031
100	93,353	67,762	53,183	43,767	37,184	32,322	25,621	21,222	18,112	15,797	14,007	12,581	11,419	9,276	7,811	6,743	5,935	5,299	4,786	4,010	3,450	3,028	2,432	2,032
105	96,204	69,252	54,097	44,384	37,628	32,657	25,832	21,366	18,217	15,877	14,070	12,632	11,460	9,304	7,830	6,760	5,947	5,308	4,793	4,015	3,454	3,030	2,434	2,033
110	98,951	70,664	54,955	44,960	38,041	32,967	26,026	21,499	18,313	15,950	14,127	12,678	11,498	9,329	7,848	6,773	5,957	5,316	4,800	4,020	3,457	3,033	2,435	2,035
115	101,600	72,005	55,762	45,499	38,426	33,256	26,205	21,621	18,402	16,017	14,180	12,720	11,533	9,352	7,864	6,785	5,966	5,324	4,806	4,024	3,461	3,036	2,437	2,036
120	104,156	73,280	56,524	46,004	38,786	33,526	26,372	21,735	18,484	16,079	14,228	12,760	11,566	9,373	7,879	6,796	5,975	5,331	4,812	4,028	3,463	3,038	2,438	2,037
125	106,624	74,493	57,243	46,479	39,123	33,777	26,528	21,840	18,560	16,137	14,274	12,796	11,595	9,392	7,893	6,806	5,983	5,337	4,817	4,031	3,466	3,040	2,440	2,038
130	109,008	75,649	57,923	46,927	39,440	34,013	26,673	21,938	18,631	16,191	14,315	12,830	11,623	9,411	7,906	6,816	5,990	5,343	4,822	4,035	3,469	3,042	2,441	2,038
135	111,313	76,752	58,567	47,349	39,737	34,234	26,809	22,030	18,698	16,241	14,355	12,861	11,649	9,427	7,918	6,825	5,997	5,348	4,826	4,038	3,471	3,043	2,442	2,039
140	113,542	77,805	59,178	47,748	40,018	34,442	26,936	22,116	18,759	16,287	14,391	12,890	11,673	9,443	7,929	6,833	6,003	5,353	4,830	4,041	3,473	3,045	2,443	2,040
145	115,699	78,812	59,759	48,125	40,283	34,638	27,056	22,197	18,817	16,331	14,425	12,917	11,695	9,458	7,939	6,84[illegible]	6,009	5,358	4,834	4,043	3,475	3,047	2,444	2,041
150	117,788	79,775	60,311	48,482	40,533	34,823	27,168	22,273	18,872	16,372	14,457	12,943	11,716	9,472	7,949	6,848	6,015	5,362	4,838	4,046	3,477	3,048	2,445	2,041
155	119,811	80,698	60,837	48,822	40,770	34,998	27,275	22,344	18,923	16,411	14,487	12,967	11,736	9,484	7,958	6,854	6,020	5,366	4,841	[illegible],048	3,479	3,049	2,446	2,042
160	121,772	81,583	61,339	49,144	40,994	35,163	27,375	22,411	18,971	16,447	14,515	12,990	11,754	9,497	7,966	6,861	6,025	5,370	4,844	[illegible],050	3,480	3,051	2,447	2,042
165	123,674	82,432	61,818	49,451	41,208	35,320	27,470	22,475	19,017	16,481	14,542	13,011	11,772	9,508	7,974	6,867	6,029	5,374	4,847	[illegible],052	3,482	3,052	2,448	2,043
170	125,518	83,248	62,275	49,743	41,410	35,469	27,560	22,535	19,060	16,513	14,567	13,031	11,788	9,519	7,982	6,872	6,034	5,377	4,850	[illegible],054	3,483	3,053	2,448	2,043
175	127,309	84,031	62,713	50,022	41,603	35,610	27,645	22,592	19,101	16,544	14,591	13,050	11,804	9,529	7,989	6,878	6,038	5,381	4,853	[illegible],056	3,484	3,054	2,449	2,044
180	129,047	84,785	63,132	50,288	41,787	35,745	27,726	22,646	19,139	16,573	14,613	13,068	11,819	9,538	7,996	6,883	6,042	5,384	4,855	[illegible],058	3,486	3,055	2,450	2,044
185	130,736	85,511	63,533	50,543	41,963	35,873	27,803	22,698	19,176	16,601	14,635	13,085	11,833	9,548	8,002	6,887	6,045	5,387	4,857	[illegible],060	3,487	3,056	2,450	2,045
190	132,377	86,210	63,918	50,786	42,131	35,996	27,877	22,747	19,211	16,627	14,655	13,102	11,846	9,556	8,008	6,892	6,049	5,389	4,860	[illegible],061	3,488	3,057	2,451	2,045
195	133,973	86,884	64,288	51,019	42,291	36,113	27,947	22,793	19,244	16,652	14,675	13,117	11,859	9,564	8,014	6,896	6,052	5,392	4,862	[illegible],063	3,489	3,058	2,451	2,046
200	135,525	87,534	64,643	51,243	42,444	36,224	28,014	22,838	19,276	16,675	14,693	13,132	11,871	9,572	8,020	6,900	6,055	5,394	4,864	[illegible],064	3,490	3,058	2,452	2,046

# 1/0 AWG	1 Run(s)	277 Volts	C-Value = 8,924	Magnetic Duct	Copper

Calculations:

1. $f = \dfrac{2 \times \text{Length} \times \text{Isca} \times 1.5}{\text{Runs} \times \text{C-Value} \times \text{E L-N}}$

2. $M = \dfrac{1}{1 + f}$

3. Isc = Isca x 1.5 x M

4. Add Motor Contribution = Motor FLA x 4

* All results are given in symmetrical amperes

Single-Phase L-N Bolted Fault-Current* Table

One-Way Distance in Feet

Available Isc in Thousands (Isca)	5	10	15	20	25	30	40	50	60	70	80	90	100	125	150	175	200	225	250	300	350	400	500	600
3	4,433	4,368	4,305	4,244	4,184	4,126	4,015	3,909	3,809	3,714	3,624	3,538	3,456	3,266	3,097	2,944	2,805	2,679	2,564	2,361	2,187	2,038	1,792	1,600
5	7,316	7,140	6,973	6,814	6,661	6,516	6,243	5,992	5,760	5,545	5,346	5,161	4,988	4,603	4,273	3,987	3,737	3,516	3,320	2,987	2,715	2,488	2,132	1,865
7	10,143	9,809	9,496	9,203	8,927	8,667	8,191	7,764	7,379	7,031	6,714	6,424	6,159	5,582	5,104	4,701	4,357	4,060	3,801	3,371	3,028	2,749	2,321	2,008
10	14,281	13,628	13,032	12,485	11,983	11,520	10,693	9,977	9,350	8,798	8,307	7,869	7,474	6,641	5,975	5,430	4,977	4,593	4,264	3,730	3,315	2,983	2,485	2,130
15	20,920	19,547	18,344	17,280	16,332	15,484	14,026	12,819	11,803	10,936	10,188	9,536	8,962	7,791	6,890	6,176	5,596	5,115	4,711	4,068	3,579	3,195	2,631	2,236
20	27,255	24,971	23,040	21,386	19,953	18,701	16,615	14,948	13,584	12,449	11,489	10,666	9,954	8,529	7,461	6,631	5,967	5,423	4,971	4,260	3,727	3,313	2,710	2,293
25	33,307	29,958	27,221	24,942	23,015	21,364	18,685	16,602	14,937	13,576	12,442	11,483	10,661	9,043	7,851	6,937	6,214	5,627	5,141	4,385	3,822	3,388	2,760	2,328
30	39,095	34,560	30,967	28,051	25,637	23,606	20,377	17,925	15,999	14,448	13,170	12,100	11,191	9,422	8,135	7,158	6,390	5,771	5,262	4,472	3,888	3,439	2,794	2,353
35	44,634	38,818	34,343	30,794	27,909	25,518	21,786	19,006	16,856	15,142	13,745	12,584	11,603	9,712	8,351	7,324	6,523	5,879	5,351	4,536	3,937	3,477	2,819	2,371
40	49,942	42,772	37,402	33,230	29,895	27,169	22,978	19,907	17,560	15,708	14,210	12,972	11,933	9,942	8,520	7,454	6,625	5,962	5,420	4,586	3,974	3,506	2,838	2,384
45	55,031	46,451	40,185	35,409	31,647	28,608	23,999	20,669	18,150	16,179	14,594	13,292	12,203	10,128	8,657	7,559	6,708	6,029	5,475	4,625	4,003	3,529	2,853	2,395
50	59,916	49,884	42,729	37,369	33,204	29,874	24,884	21,322	18,652	16,576	14,916	13,559	12,427	10,283	8,769	7,644	6,775	6,083	5,520	4,657	4,027	3,548	2,865	2,403
55	64,608	53,094	45,063	39,142	34,597	30,997	25,658	21,888	19,083	16,916	15,191	13,785	12,618	10,413	8,864	7,716	6,831	6,129	5,557	4,683	4,047	3,563	2,875	2,410
60	69,119	56,103	47,212	40,753	35,849	31,999	26,340	22,382	19,459	17,210	15,428	13,980	12,780	10,523	8,944	7,776	6,879	6,167	5,588	4,706	4,064	3,576	2,884	2,416
65	73,459	58,928	49,197	42,224	36,983	32,899	26,947	22,819	19,788	17,467	15,634	14,149	12,922	10,619	9,013	7,828	6,919	6,199	5,615	4,725	4,078	3,587	2,891	2,421
70	77,637	61,587	51,036	43,572	38,012	33,711	27,490	23,207	20,079	17,694	15,815	14,297	13,045	10,702	9,072	7,874	6,955	6,228	5,638	4,741	4,090	3,596	2,897	2,425
75	81,662	64,093	52,746	44,812	38,953	34,448	27,978	23,554	20,338	17,895	15,976	14,428	13,154	10,775	9,125	7,913	6,985	6,252	5,659	4,755	4,101	3,605	2,902	2,429
80	85,543	66,460	54,338	45,956	39,814	35,120	28,420	23,866	20,570	18,074	16,119	14,545	13,251	10,840	9,171	7,948	7,013	6,274	5,676	4,768	4,110	3,612	2,907	2,432
85	89,287	68,698	55,825	47,015	40,607	35,736	28,821	24,149	20,780	18,236	16,247	14,649	13,337	10,898	9,213	7,979	7,037	6,294	5,692	4,779	4,118	3,618	2,911	2,435
90	92,902	70,818	57,217	47,998	41,338	36,301	29,188	24,406	20,970	18,382	16,363	14,743	13,415	10,950	9,250	8,007	7,058	6,311	5,706	4,789	4,126	3,624	2,915	2,438
95	96,393	72,828	58,522	48,913	42,015	36,822	29,524	24,640	21,143	18,515	16,468	14,828	13,486	10,997	9,284	8,032	7,078	6,326	5,719	4,798	4,132	3,629	2,918	2,440
100	99,767	74,738	59,749	49,768	42,644	37,304	29,833	24,855	21,301	18,636	16,564	14,906	13,550	11,040	9,314	8,055	7,096	6,340	5,731	4,806	4,138	3,634	2,921	2,442
105	103,030	76,555	60,904	50,567	43,229	37,751	30,118	25,053	21,446	18,747	16,651	14,977	13,609	11,078	9,341	8,075	7,112	6,353	5,741	4,813	4,144	3,638	2,924	2,444
110	106,188	78,284	61,994	51,315	43,775	38,167	30,382	25,235	21,579	18,849	16,732	15,042	13,662	11,114	9,367	8,094	7,126	6,365	5,751	4,820	4,149	3,642	2,926	2,446
115	109,244	79,933	63,023	52,019	44,286	38,555	30,627	25,404	21,703	18,943	16,806	15,102	13,712	11,147	9,390	8,112	7,140	6,376	5,759	4,826	4,153	3,645	2,929	2,447
120	112,205	81,507	63,998	52,681	44,765	38,917	30,856	25,561	21,817	19,030	16,874	15,157	13,757	11,177	9,411	8,127	7,152	6,385	5,767	4,832	4,158	3,648	2,931	2,449
125	115,075	83,010	64,921	53,305	45,215	39,257	31,069	25,707	21,923	19,111	16,938	15,208	13,799	11,204	9,431	8,142	7,163	6,395	5,775	4,837	4,161	3,651	2,933	2,450
130	117,857	84,448	65,797	53,894	45,638	39,575	31,268	25,843	22,022	19,186	16,997	15,256	13,839	11,230	9,449	8,156	7,174	6,403	5,782	4,842	4,165	3,654	2,934	2,451
135	120,555	85,825	66,630	54,451	46,037	39,875	31,455	25,971	22,115	19,256	17,052	15,300	13,875	11,254	9,466	8,168	7,184	6,411	5,788	4,846	4,168	3,657	2,936	2,453
140	123,174	87,144	67,422	54,979	46,414	40,157	31,630	26,090	22,201	19,322	17,103	15,342	13,909	11,277	9,482	8,180	7,193	6,418	5,794	4,851	4,171	3,659	2,937	2,454
145	125,717	88,409	68,177	55,480	46,770	40,424	31,795	26,202	22,283	19,383	17,151	15,380	13,941	11,298	9,497	8,191	7,201	6,425	5,799	4,854	4,174	3,661	2,939	2,455
150	128,187	89,623	68,897	55,956	47,108	40,676	31,951	26,308	22,359	19,441	17,197	15,417	13,971	11,317	9,511	8,202	7,209	6,431	5,805	4,858	4,177	3,663	2,940	2,456
155	130,587	90,790	69,584	56,409	47,428	40,915	32,098	26,408	22,431	19,495	17,239	15,451	13,999	11,336	9,524	8,211	7,217	6,437	5,809	4,861	4,179	3,665	2,941	2,456
160	132,920	91,912	70,241	56,840	47,733	41,141	32,237	26,502	22,499	19,546	17,279	15,483	14,025	11,353	9,536	8,220	7,224	6,443	5,814	4,865	4,182	3,667	2,943	2,457
165	135,188	92,991	70,870	57,250	48,022	41,356	32,369	26,591	22,563	19,595	17,317	15,513	14,050	11,369	9,547	8,229	7,230	6,448	5,818	4,868	4,184	3,669	2,944	2,458
170	137,396	94,030	71,471	57,643	48,298	41,560	32,494	26,675	22,624	19,641	17,353	15,542	14,074	11,385	9,558	8,237	7,236	6,453	5,822	4,870	4,186	3,670	2,945	2,459
175	139,544	95,031	72,048	58,017	48,560	41,754	32,613	26,755	22,681	19,684	17,386	15,569	14,096	11,399	9,568	8,244	7,242	6,457	5,826	4,873	4,188	3,672	2,946	2,459
180	141,635	95,996	72,602	58,376	48,811	41,940	32,726	26,831	22,736	19,725	17,418	15,595	14,117	11,413	9,578	8,252	7,248	6,462	5,830	4,876	4,190	3,673	2,947	2,460
185	143,672	96,928	73,133	58,719	49,051	42,116	32,833	26,903	22,787	19,764	17,449	15,619	14,137	11,426	9,587	8,258	7,253	6,466	5,833	4,878	4,192	3,675	2,947	2,461
190	145,657	97,827	73,644	59,048	49,280	42,285	32,936	26,972	22,837	19,801	17,478	15,642	14,156	11,438	9,596	8,265	7,258	6,470	5,836	4,880	4,193	3,676	2,948	2,461
195	147,591	98,696	74,135	59,363	49,500	42,447	33,034	27,038	22,884	19,836	17,505	15,664	14,174	11,450	9,604	8,271	7,263	6,474	5,839	4,882	4,195	3,677	2,949	2,462
200	149,477	99,535	74,608	59,666	49,710	42,601	33,127	27,100	22,929	19,870	17,531	15,685	14,191	11,461	9,612	8,277	7,267	6,477	5,842	4,884	4,196	3,678	2,950	2,462

2/0 AWG | 1 Run(s) | 277 Volts | C-Value = 10,755 | Magnetic Duct | Copper

Calculations:

1. $f = \dfrac{2 \times \text{Length} \times \text{Isca} \times 1.5}{\text{Runs} \times \text{C-Value} \times \text{E L-N}}$

2. $M = \dfrac{1}{1 + f}$

3. $\text{Isc} = \text{Isca} \times 1.5 \times M$

4. Add Motor Contribution = Motor FLA x 4

* All results are given in symmetrical amperes

Single-Phase L-N Bolted Fault-Current* Table

One-Way Distance in Feet

Available Isc in Thousands (Isca)	5	10	15	20	25	30	40	50	60	70	80	90	100	125	150	175	200	225	250	300	350	400	500	600
3	4,444	4,389	4,335	4,283	4,232	4,183	4,086	3,995	3,907	3,823	3,743	3,665	3,591	3,419	3,262	3,119	2,988	2,868	2,757	2,558	2,387	2,237	1,987	1,787
5	7,345	7,197	7,054	6,917	6,785	6,658	6,418	6,194	5,986	5,791	5,608	5,437	5,276	4,911	4,594	4,316	4,069	3,849	3,651	3,311	3,029	2,792	2,413	2,125
7	10,199	9,915	9,646	9,391	9,150	8,920	8,494	8,107	7,754	7,430	7,132	6,857	6,603	6,042	5,569	5,165	4,815	4,510	4,241	3,789	3,425	3,124	2,657	2,312
10	14,393	13,833	13,316	12,835	12,388	11,971	11,217	10,551	9,960	9,432	8,957	8,528	8,138	7,302	6,623	6,059	5,583	5,177	4,826	4,249	3,796	3,430	2,876	2,475
15	21,162	19,973	18,912	17,957	17,094	16,311	14,941	13,783	12,792	11,933	11,183	10,522	9,934	8,717	7,766	7,001	6,374	5,850	5,406	4,693	4,146	3,713	3,072	2,619
20	27,667	25,670	23,943	22,433	21,102	19,921	17,914	16,275	14,911	13,758	12,770	11,915	11,167	9,652	8,499	7,592	6,860	6,257	5,751	4,951	4,346	3,873	3,180	2,698
25	33,924	30,971	28,490	26,378	24,557	22,971	20,344	18,256	16,557	15,147	13,958	12,943	12,065	10,316	9,009	7,997	7,189	6,529	5,980	5,120	4,476	3,975	3,249	2,747
30	39,947	35,914	32,621	29,881	27,566	25,583	22,366	19,868	17,872	16,240	14,882	13,733	12,748	10,811	9,385	8,291	7,426	6,724	6,144	5,239	4,566	4,047	3,297	2,781
35	45,749	40,536	36,389	33,013	30,209	27,845	24,076	21,206	18,947	17,123	15,619	14,359	13,286	11,196	9,673	8,516	7,605	6,871	6,266	5,328	4,634	4,100	3,332	2,806
40	51,341	44,866	39,841	35,829	32,551	29,822	25,540	22,333	19,842	17,851	16,223	14,867	13,720	11,502	9,901	8,692	7,746	6,985	6,361	5,396	4,685	4,140	3,358	2,825
45	56,735	48,932	43,015	38,375	34,639	31,565	26,808	23,297	20,599	18,461	16,725	15,288	14,078	11,752	10,086	8,834	7,858	7,077	6,437	5,450	4,726	4,172	3,379	2,840
50	61,941	52,756	45,943	40,688	36,512	33,114	27,917	24,130	21,247	18,980	17,150	15,642	14,378	11,961	10,239	8,951	7,951	7,152	6,499	5,495	4,760	4,198	3,396	2,852
55	66,970	56,360	48,652	42,799	38,203	34,498	28,894	24,857	21,809	19,427	17,514	15,944	14,633	12,137	10,368	9,049	8,028	7,214	6,550	5,532	4,787	4,219	3,410	2,862
60	71,828	59,762	51,167	44,733	39,736	35,744	29,763	25,497	22,300	19,816	17,830	16,205	14,852	12,287	10,478	9,133	8,094	7,267	6,594	5,563	4,811	4,238	3,422	2,870
65	76,527	62,979	53,507	46,511	41,133	36,870	30,540	26,065	22,734	20,157	18,106	16,433	15,043	12,418	10,572	9,205	8,150	7,313	6,631	5,589	4,830	4,253	3,432	2,877
70	81,072	66,025	55,690	48,152	42,411	37,894	31,239	26,572	23,119	20,459	18,349	16,633	15,211	12,532	10,655	9,267	8,199	7,352	6,663	5,612	4,848	4,266	3,441	2,883
75	85,471	68,914	57,731	49,670	43,585	38,828	31,871	27,028	23,463	20,729	18,565	16,811	15,359	12,632	10,728	9,322	8,242	7,387	6,692	5,632	4,862	4,278	3,448	2,888
80	89,732	71,658	59,644	51,080	44,667	39,684	32,445	27,440	23,773	20,970	18,759	16,969	15,491	12,721	10,792	9,371	8,280	7,417	6,717	5,650	4,876	4,288	3,455	2,893
85	93,861	74,266	61,440	52,392	45,666	40,471	32,970	27,814	24,053	21,188	18,933	17,111	15,610	12,801	10,849	9,414	8,314	7,444	6,739	5,666	4,887	4,297	3,461	2,897
90	97,863	76,750	63,130	53,616	46,594	41,198	33,450	28,156	24,308	21,385	19,090	17,240	15,717	12,873	10,901	9,453	8,344	7,468	6,759	5,680	4,898	4,305	3,466	2,901
95	101,745	79,117	64,723	54,760	47,456	41,870	33,892	28,468	24,540	21,565	19,233	17,357	15,814	12,938	10,947	9,488	8,371	7,490	6,777	5,692	4,907	4,312	3,471	2,904
100	105,512	81,376	66,227	55,833	48,259	42,495	34,300	28,755	24,754	21,730	19,364	17,463	15,902	12,997	10,990	9,519	8,396	7,510	6,793	5,704	4,916	4,319	3,475	2,907
105	109,168	83,534	67,650	56,841	49,010	43,076	34,678	29,020	24,950	21,881	19,484	17,560	15,983	13,051	11,028	9,548	8,418	7,528	6,807	5,714	4,923	4,325	3,479	2,910
110	112,720	85,598	68,997	57,789	49,713	43,618	35,028	29,265	25,131	22,020	19,594	17,650	16,057	13,100	11,063	9,575	8,439	7,544	6,821	5,724	4,930	4,330	3,482	2,912
115	116,170	87,573	70,274	58,682	50,373	44,125	35,355	29,493	25,298	22,148	19,696	17,732	16,125	13,146	11,096	9,599	8,458	7,559	6,833	5,732	4,937	4,335	3,486	2,915
120	119,524	89,466	71,488	59,526	50,994	44,600	35,659	29,704	25,454	22,267	19,790	17,809	16,188	13,187	11,125	9,621	8,475	7,573	6,844	5,740	4,943	4,340	3,489	2,917
125	122,785	91,281	72,642	60,324	51,578	45,047	35,944	29,902	25,599	22,378	19,877	17,879	16,246	13,226	11,153	9,642	8,491	7,586	6,855	5,747	4,948	4,344	3,491	2,918
130	125,958	93,022	73,741	61,080	52,130	45,467	36,211	30,086	25,734	22,481	19,959	17,945	16,301	13,262	11,179	9,661	8,506	7,598	6,865	5,754	4,953	4,348	3,494	2,920
135	129,045	94,695	74,783	61,797	52,651	45,863	36,462	30,259	25,860	22,578	20,035	18,007	16,351	13,296	11,202	9,679	8,520	7,609	6,874	5,761	4,958	4,351	3,496	2,922
140	132,051	96,304	75,788	62,478	53,144	46,237	36,698	30,422	25,979	22,668	20,106	18,064	16,399	13,327	11,225	9,695	8,532	7,619	6,882	5,766	4,962	4,355	3,498	2,923
145	134,977	97,851	76,743	63,125	53,612	46,591	36,920	30,574	26,090	22,753	20,172	18,118	16,443	13,356	11,245	9,711	8,544	7,628	6,890	5,772	4,966	4,358	3,500	2,925
150	137,828	99,341	77,656	63,742	54,056	46,926	37,130	30,718	26,195	22,832	20,235	18,168	16,484	13,384	11,265	9,725	8,556	7,637	6,897	5,777	4,970	4,361	3,502	2,926
155	140,607	100,776	78,530	64,330	54,479	47,244	37,329	30,854	26,293	22,907	20,294	18,216	16,523	13,409	11,283	9,739	8,566	7,646	6,904	5,782	4,973	4,363	3,504	2,927
160	143,315	102,160	79,368	64,891	54,880	47,546	37,517	30,983	26,387	22,978	20,349	18,260	16,560	13,434	11,300	9,751	8,576	7,653	6,910	5,786	4,977	4,366	3,506	2,928
165	145,956	103,495	80,171	65,427	55,263	47,833	37,696	31,104	26,475	23,045	20,402	18,302	16,595	13,456	11,316	9,763	8,585	7,661	6,916	5,790	4,980	4,368	3,507	2,930
170	148,533	104,784	80,943	65,940	55,629	48,106	37,866	31,220	26,558	23,108	20,451	18,342	16,628	13,478	11,331	9,775	8,594	7,668	6,922	5,794	4,983	4,371	3,509	2,931
175	151,046	106,028	81,683	66,430	55,978	48,367	38,027	31,329	26,638	23,168	20,498	18,380	16,659	13,498	11,346	9,785	8,602	7,674	6,927	5,798	4,986	4,373	3,510	2,931
180	153,500	107,232	82,396	66,901	56,311	48,616	38,181	31,433	26,713	23,225	20,543	18,416	16,688	13,518	11,359	9,796	8,610	7,681	6,932	5,802	4,988	4,375	3,511	2,932
185	155,895	108,395	83,081	67,352	56,630	48,854	38,327	31,533	26,784	23,279	20,585	18,450	16,716	13,536	11,372	9,805	8,618	7,687	6,937	5,805	4,991	4,377	3,512	2,933
190	158,235	109,521	83,741	67,785	56,936	49,081	38,467	31,627	26,853	23,331	20,625	18,482	16,743	13,553	11,385	9,814	8,625	7,692	6,942	5,808	4,993	4,379	3,514	2,934
195	160,520	110,611	84,376	68,201	57,229	49,299	38,600	31,717	26,918	23,380	20,664	18,513	16,768	13,570	11,396	9,823	8,631	7,698	6,946	5,811	4,995	4,380	3,515	2,935
200	162,753	111,666	84,989	68,601	57,511	49,507	38,728	31,804	26,980	23,427	20,700	18,542	16,792	13,586	11,407	9,831	8,638	7,703	6,950	5,814	4,997	4,382	3,516	2,936

3/0 AWG | 1 Run(s) | 277 Volts

C-Value = 12,843 | Magnetic Duct | Copper

Calculations:

1. $f = \dfrac{2 \times \text{Length} \times \text{Isca} \times 1.5}{\text{Runs} \times \text{C-Value} \times E_{L\text{-}N}}$

2. $M = \dfrac{1}{1 + f}$

3. $\text{Isc} = \text{Isca} \times 1.5 \times M$

4. Add Motor Contribution = Motor FLA x 4

* All results are given in symmetrical amperes

Single-Phase L-N Bolted Fault-Current* Table

One-Way Distance in Feet

Available Isc in Thousands (Isca)	5	10	15	20	25	30	40	50	60	70	80	90	100	125	150	175	200	225	250	300	350	400	500	600
3	4,452	4,405	4,359	4,314	4,270	4,227	4,143	4,062	3,985	3,910	3,838	3,769	3,702	3,545	3,401	3,268	3,145	3,031	2,925	2,733	2,566	2,417	2,166	1,963
5	7,368	7,240	7,117	6,998	6,882	6,771	6,558	6,358	6,171	5,994	5,826	5,668	5,519	5,177	4,875	4,606	4,365	4,149	3,952	3,611	3,323	3,079	2,683	2,378
7	10,243	9,997	9,764	9,541	9,328	9,124	8,742	8,391	8,067	7,767	7,489	7,229	6,988	6,448	5,986	5,586	5,236	4,927	4,653	4,187	3,805	3,488	2,989	2,615
10	14,480	13,995	13,541	13,116	12,717	12,341	11,653	11,037	10,483	9,982	9,527	9,111	8,731	7,905	7,221	6,647	6,157	5,735	5,366	4,755	4,269	3,874	3,268	2,826
15	21,350	20,312	19,370	18,512	17,727	17,005	15,725	14,624	13,667	12,828	12,086	11,425	10,832	9,589	8,602	7,799	7,133	6,572	6,093	5,317	4,717	4,238	3,523	3,015
20	27,990	26,233	24,683	23,306	22,074	20,966	19,054	17,461	16,114	14,960	13,960	13,086	12,314	10,733	9,511	8,539	7,747	7,090	6,535	5,651	4,978	4,448	3,667	3,119
25	34,411	31,792	29,544	27,593	25,883	24,373	21,826	19,762	18,054	16,617	15,393	14,336	13,416	11,560	10,155	9,054	8,169	7,442	6,833	5,872	5,149	4,584	3,759	3,186
30	40,624	37,024	34,010	31,450	29,248	27,334	24,171	21,664	19,629	17,943	16,523	15,312	14,266	12,186	10,635	9,434	8,477	7,696	7,047	6,030	5,269	4,679	3,823	3,231
35	46,639	41,955	38,126	34,938	32,242	29,932	26,180	23,264	20,933	19,026	17,438	16,094	14,943	12,676	11,006	9,725	8,711	7,889	7,208	6,148	5,359	4,750	3,870	3,265
40	52,465	46,611	41,933	38,108	34,922	32,228	27,921	24,629	22,031	19,929	18,193	16,736	15,494	13,071	11,302	9,956	8,896	8,040	7,334	6,239	5,428	4,804	3,906	3,291
45	58,111	51,015	45,463	41,001	37,337	34,274	29,443	25,806	22,968	20,693	18,828	17,271	15,952	13,395	11,544	10,143	9,045	8,161	7,435	6,312	5,483	4,847	3,934	3,311
50	63,585	55,186	48,746	43,653	39,523	36,107	30,786	26,831	23,777	21,347	19,368	17,724	16,338	13,666	11,745	10,297	9,168	8,261	7,518	6,371	5,528	4,882	3,957	3,327
55	68,895	59,142	51,808	46,092	41,512	37,760	31,979	27,733	24,483	21,914	19,834	18,114	16,668	13,896	11,915	10,428	9,271	8,345	7,587	6,421	5,566	4,911	3,976	3,340
60	74,048	62,899	54,668	48,343	43,329	39,257	33,047	28,533	25,104	22,410	20,239	18,451	16,954	14,094	12,060	10,539	9,358	8,416	7,646	6,463	5,597	4,936	3,992	3,352
65	79,051	66,473	57,348	50,426	44,995	40,620	34,007	29,246	25,654	22,848	20,595	18,747	17,203	14,266	12,185	10,634	9,434	8,477	7,696	6,499	5,624	4,957	4,006	3,361
70	83,910	69,876	59,863	52,360	46,529	41,866	34,876	29,886	26,145	23,237	20,911	19,008	17,423	14,416	12,295	10,718	9,499	8,530	7,740	6,530	5,647	4,975	4,018	3,370
75	88,633	73,120	62,228	54,161	47,945	43,009	35,666	30,464	26,587	23,585	21,192	19,240	17,617	14,550	12,392	10,791	9,557	8,576	7,778	6,557	5,668	4,990	4,028	3,377
80	93,223	76,216	64,457	55,841	49,257	44,062	36,387	30,989	26,985	23,898	21,445	19,448	17,792	14,668	12,478	10,856	9,608	8,617	7,812	6,581	5,685	5,004	4,037	3,383
85	97,687	79,174	66,560	57,413	50,476	45,035	37,048	31,467	27,347	24,181	21,672	19,635	17,948	14,774	12,554	10,915	9,654	8,654	7,842	6,602	5,701	5,017	4,045	3,389
90	102,030	82,003	68,548	58,886	51,611	45,936	37,655	31,904	27,677	24,439	21,879	19,805	18,090	14,870	12,624	10,967	9,694	8,686	7,868	6,621	5,715	5,028	4,052	3,394
95	106,256	84,711	70,430	60,269	52,671	46,774	38,216	32,306	27,979	24,674	22,067	19,959	18,218	14,957	12,686	11,014	9,731	8,716	7,893	6,638	5,728	5,038	4,059	3,398
100	110,371	87,306	72,215	61,571	53,663	47,554	38,736	32,676	28,256	24,889	22,239	20,100	18,335	15,036	12,743	11,056	9,764	8,743	7,915	6,654	5,740	5,046	4,065	3,402
105	114,379	89,795	73,909	62,799	54,593	48,283	39,218	33,019	28,512	25,088	22,398	20,229	18,443	15,108	12,794	11,095	9,795	8,767	7,935	6,668	5,750	5,055	4,070	3,406
110	118,284	92,184	75,520	63,958	55,467	48,966	39,667	33,337	28,748	25,271	22,543	20,347	18,541	15,174	12,842	11,131	9,823	8,789	7,953	6,681	5,760	5,062	4,075	3,409
115	122,089	94,479	77,053	65,055	56,289	49,606	40,086	33,632	28,968	25,440	22,678	20,457	18,632	15,235	12,885	11,164	9,848	8,810	7,969	6,693	5,769	5,069	4,079	3,413
120	125,799	96,685	78,514	66,093	57,065	50,207	40,478	33,907	29,172	25,597	22,803	20,559	18,717	15,291	12,926	11,194	9,872	8,828	7,985	6,704	5,777	5,075	4,083	3,415
125	129,417	98,808	79,909	67,078	57,798	50,774	40,845	34,165	29,362	25,744	22,919	20,653	18,795	15,343	12,963	11,222	9,893	8,846	7,999	6,714	5,784	5,081	4,087	3,418
130	132,946	100,852	81,240	68,014	58,492	51,308	41,191	34,406	29,540	25,880	23,027	20,741	18,867	15,392	12,998	11,248	9,913	8,862	8,012	6,723	5,791	5,086	4,090	3,420
135	136,390	102,822	82,513	68,904	59,149	51,813	41,515	34,632	29,707	26,008	23,128	20,823	18,935	15,437	13,030	11,272	9,932	8,877	8,024	6,731	5,797	5,091	4,093	3,423
140	139,751	104,721	83,732	69,752	59,772	52,291	41,822	34,845	29,863	26,128	23,223	20,900	18,999	15,479	13,060	11,294	9,949	8,891	8,036	6,739	5,803	5,095	4,096	3,425
145	143,034	106,553	84,899	70,560	60,365	52,744	42,111	35,046	30,011	26,241	23,312	20,972	19,058	15,519	13,088	11,315	9,966	8,904	8,046	6,747	5,809	5,100	4,099	3,427
150	146,239	108,322	86,019	71,331	60,928	53,173	42,384	35,235	30,149	26,347	23,396	21,039	19,114	15,556	13,114	11,335	9,981	8,916	8,056	6,754	5,814	5,104	4,102	3,428
155	149,371	110,031	87,093	72,068	61,465	53,582	42,643	35,414	30,280	26,446	23,474	21,103	19,167	15,590	13,139	11,353	9,995	8,927	8,066	6,760	5,819	5,107	4,104	3,430
160	152,432	111,682	88,124	72,773	61,977	53,971	42,889	35,583	30,404	26,541	23,549	21,163	19,216	15,623	13,162	11,371	10,009	8,938	8,074	6,767	5,823	5,111	4,106	3,432
165	155,423	113,280	89,116	73,448	62,466	54,341	43,123	35,744	30,521	26,630	23,619	21,220	19,263	15,654	13,184	11,387	10,021	8,948	8,083	6,772	5,828	5,114	4,108	3,433
170	158,348	114,825	90,070	74,095	62,933	54,694	43,345	35,896	30,632	26,715	23,685	21,273	19,307	15,683	13,205	11,403	10,033	8,958	8,090	6,778	5,832	5,117	4,110	3,435
175	161,208	116,322	90,988	74,715	63,380	55,031	43,556	36,041	30,738	26,795	23,748	21,324	19,349	15,711	13,224	11,417	10,045	8,967	8,098	6,783	5,835	5,120	4,112	3,436
180	164,005	117,772	91,872	75,311	63,808	55,354	43,758	36,179	30,838	26,871	23,808	21,372	19,389	15,737	13,243	11,431	10,055	8,975	8,105	6,788	5,839	5,123	4,114	3,437
185	166,743	119,177	92,725	75,883	64,218	55,662	43,951	36,311	30,933	26,943	23,865	21,418	19,426	15,762	13,260	11,444	10,065	8,983	8,111	6,792	5,843	5,126	4,116	3,438
190	169,422	120,539	93,548	76,433	64,612	55,958	44,135	36,436	31,024	27,012	23,919	21,462	19,462	15,785	13,277	11,457	10,075	8,991	8,117	6,797	5,846	5,128	4,117	3,439
195	172,044	121,860	94,342	76,962	64,990	56,241	44,310	36,556	31,111	27,078	23,971	21,503	19,496	15,808	13,293	11,468	10,084	8,998	8,123	6,801	5,849	5,131	4,119	3,440
200	174,612	123,143	95,109	77,472	65,353	56,512	44,479	36,671	31,194	27,141	24,020	21,543	19,529	15,829	13,308	11,480	10,093	9,005	8,129	6,805	5,852	5,133	4,120	3,441

4/0 AWG | 1 Run(s) | 277 Volts

C-Value = 15,082 | Magnetic Duct | Copper

Calculations:

1. $f = \frac{2 \times \text{Length} \times \text{Isca} \times 1.5}{\text{Runs} \times \text{C-Value} \times \text{E L-N}}$

2. $M = \frac{1}{1 + f}$

3. $\text{Isc} = \text{Isca} \times 1.5 \times M$

4. Add Motor Contribution = Motor FLA x 4

* All results are given in symmetrical amperes

Single-Phase L-N Bolted Fault-Current* Table

One-Way Distance in Feet

Available Isc in Thousands (Isca)

	5	10	15	20	25	30	40	50	60	70	80	90	100	125	150	175	200	225	250	300	350	400	500	600
3	4,456	4,413	4,371	4,329	4,289	4,249	4,171	4,096	4,024	3,954	3,887	3,822	3,759	3,610	3,473	3,346	3,228	3,117	3,014	2,828	2,663	2,516	2,266	2,062
5	7,379	7,261	7,148	7,038	6,931	6,827	6,629	6,442	6,265	6,098	5,939	5,788	5,645	5,317	5,024	4,762	4,526	4,312	4,118	3,777	3,489	3,241	2,838	2,524
7	10,264	10,038	9,822	9,615	9,417	9,227	8,868	8,537	8,229	7,943	7,676	7,426	7,192	6,667	6,213	5,878	5,469	5,160	4,884	4,412	4,023	3,697	3,182	2,793
10	14,523	14,075	13,654	13,258	12,884	12,530	11,878	11,291	10,759	10,274	9,832	9,426	9,052	8,236	7,554	6,977	6,482	6,052	5,676	5,048	4,546	4,134	3,500	3,035
15	21,443	20,481	19,602	18,795	18,052	17,365	16,138	15,072	14,139	13,314	12,581	11,923	11,332	10,081	9,078	8,258	7,573	6,993	6,495	5,686	5,057	4,552	3,796	3,254
20	28,150	26,516	25,060	23,756	22,581	21,517	19,664	18,104	16,774	15,626	14,625	13,744	12,964	11,352	10,097	9,092	8,268	7,582	7,001	6,070	5,358	4,795	3,963	3,377
25	34,654	32,209	30,087	28,227	26,583	25,121	22,630	20,589	18,886	17,443	16,205	15,131	14,190	12,282	10,826	9,678	8,751	7,986	7,343	6,326	5,556	4,953	4,070	3,454
30	40,963	37,590	34,731	32,276	30,145	28,278	25,161	22,663	20,617	18,909	17,463	16,222	15,145	12,991	11,373	10,113	9,105	8,279	7,591	6,509	5,697	5,065	4,145	3,508
35	47,086	42,684	39,035	35,960	33,335	31,067	27,345	24,420	22,060	20,117	18,487	17,102	15,910	13,550	11,799	10,449	9,376	8,503	7,779	6,646	5,802	5,148	4,200	3,548
40	53,031	47,513	43,034	39,328	36,209	33,548	29,250	25,928	23,283	21,128	19,339	17,828	16,537	14,001	12,140	10,715	9,590	8,679	7,925	6,753	5,883	5,212	4,243	3,578
45	58,806	52,096	46,761	42,417	38,811	35,770	30,925	27,235	24,333	21,989	20,057	18,437	17,059	14,374	12,419	10,932	9,763	8,820	8,043	6,839	5,948	5,262	4,277	3,602
50	64,418	56,453	50,241	45,261	41,179	37,772	32,410	28,381	25,242	22,729	20,671	18,955	17,502	14,687	12,652	11,112	9,907	8,937	8,140	6,909	6,001	5,304	4,304	3,621
55	69,874	60,600	53,499	47,888	43,342	39,584	33,735	29,392	26,039	23,373	21,202	19,401	17,881	14,953	12,849	11,264	10,027	9,035	8,222	6,967	6,045	5,338	4,326	3,637
60	75,181	64,551	56,556	50,322	45,327	41,233	34,925	30,291	26,743	23,938	21,666	19,788	18,210	15,182	13,018	11,394	10,130	9,118	8,290	7,016	6,082	5,367	4,345	3,650
65	80,343	68,321	59,428	52,584	47,153	42,739	36,000	31,096	27,368	24,438	22,075	20,129	18,498	15,382	13,164	11,506	10,218	9,190	8,350	7,059	6,114	5,392	4,362	3,662
70	85,368	71,921	62,133	54,691	48,840	44,121	36,975	31,821	27,928	24,884	22,438	20,430	18,752	15,557	13,293	11,604	10,295	9,252	8,401	7,095	6,141	5,413	4,376	3,672
75	90,260	75,362	64,635	56,658	50,403	45,392	37,864	32,477	28,432	25,283	22,762	20,699	18,978	15,712	13,406	11,690	10,363	9,307	8,446	7,128	6,165	5,432	4,388	3,680
80	95,025	78,655	67,096	58,500	51,856	46,567	38,677	33,074	28,889	25,644	23,054	20,939	19,180	15,851	13,506	11,766	10,423	9,355	8,486	7,156	6,186	5,448	4,398	3,688
85	99,668	81,809	69,378	60,227	53,208	47,655	39,425	33,619	29,304	25,970	23,317	21,157	19,362	15,975	13,596	11,834	10,477	9,398	8,521	7,181	6,205	5,463	4,408	3,695
90	104,193	84,833	71,541	61,850	54,471	48,665	40,114	34,119	29,683	26,267	23,557	21,353	19,527	16,087	13,677	11,896	10,525	9,437	8,553	7,204	6,222	5,476	4,416	3,701
95	108,604	87,735	73,594	63,378	55,653	49,606	40,751	34,579	30,030	26,539	23,775	21,533	19,677	16,188	13,751	11,951	10,568	9,472	8,582	7,224	6,237	5,487	4,424	3,706
100	112,907	90,522	75,544	64,819	56,761	50,485	41,342	35,003	30,350	26,788	23,975	21,697	19,813	16,281	13,817	12,001	10,607	9,503	8,608	7,242	6,251	5,498	4,431	3,711
105	117,104	93,200	77,401	66,181	57,803	51,307	41,892	35,397	30,645	27,018	24,159	21,847	19,939	16,365	13,878	12,047	10,643	9,532	8,631	7,259	6,263	5,508	4,437	3,715
110	121,200	95,776	79,169	67,470	58,783	52,078	42,405	35,762	30,919	27,231	24,329	21,986	20,054	16,443	13,934	12,089	10,676	9,558	8,653	7,274	6,275	5,516	4,443	3,719
115	125,199	98,256	80,856	68,691	59,708	52,803	42,884	36,102	31,173	27,427	24,486	22,114	20,161	16,515	13,985	12,128	10,706	9,583	8,672	7,288	6,285	5,524	4,448	3,723
120	129,103	100,645	82,466	69,850	60,582	53,485	43,333	36,420	31,409	27,610	24,631	22,233	20,259	16,581	14,033	12,164	10,734	9,605	8,691	7,301	6,294	5,532	4,453	3,726
125	132,916	102,947	84,006	70,951	61,409	54,129	43,754	36,717	31,630	27,781	24,767	22,343	20,351	16,642	14,077	12,197	10,759	9,625	8,708	7,313	6,303	5,539	4,457	3,729
130	136,642	105,168	85,479	71,999	62,192	54,736	44,150	36,996	31,836	27,940	24,893	22,446	20,436	16,699	14,117	12,227	10,783	9,644	8,723	7,324	6,311	5,545	4,461	3,732
135	140,283	107,311	86,889	72,998	62,936	55,311	44,524	37,257	32,030	28,089	25,012	22,542	20,516	16,752	14,155	12,256	10,805	9,662	8,738	7,334	6,319	5,551	4,465	3,735
140	143,841	109,382	88,242	73,950	63,642	55,856	44,876	37,504	32,212	28,229	25,122	22,632	20,591	16,802	14,191	12,282	10,826	9,679	8,751	7,344	6,326	5,556	4,469	3,737
145	147,321	111,382	89,539	74,859	64,314	56,373	45,209	37,736	32,383	28,360	25,226	22,716	20,660	16,848	14,224	12,307	10,845	9,694	8,764	7,352	6,333	5,561	4,472	3,739
150	150,724	113,316	90,785	75,727	64,954	56,865	45,525	37,956	32,545	28,484	25,324	22,796	20,726	16,892	14,255	12,330	10,863	9,708	8,775	7,361	6,339	5,566	4,475	3,742
155	154,053	115,188	91,982	76,559	65,565	57,332	45,824	38,163	32,697	28,601	25,417	22,870	20,788	16,933	14,284	12,352	10,880	9,722	8,786	7,368	6,345	5,570	4,478	3,744
160	157,310	116,999	93,134	77,355	66,148	57,777	46,108	38,360	32,842	28,711	25,504	22,941	20,846	16,972	14,312	12,373	10,896	9,735	8,797	7,376	6,350	5,575	4,481	3,745
165	160,498	118,753	94,242	78,118	66,705	58,202	46,378	38,547	32,978	28,816	25,586	23,008	20,901	17,008	14,338	12,392	10,911	9,747	8,807	7,383	6,355	5,579	4,483	3,747
170	163,619	120,453	95,309	78,850	67,238	58,607	46,635	38,724	33,108	28,915	25,664	23,071	20,953	17,043	14,362	12,410	10,925	9,758	8,816	7,389	6,360	5,582	4,485	3,749
175	166,674	122,101	96,338	79,552	67,748	58,995	46,880	38,893	33,231	29,009	25,738	23,130	21,002	17,075	14,385	12,426	10,939	9,769	8,825	7,395	6,364	5,586	4,488	3,750
180	169,667	123,700	97,330	80,228	68,237	59,365	47,114	39,054	33,349	29,098	25,809	23,187	21,049	17,106	14,407	12,444	10,951	9,779	8,833	7,401	6,369	5,589	4,490	3,752
185	172,598	125,250	98,288	80,877	68,707	59,720	47,337	39,207	33,460	29,183	25,875	23,241	21,094	17,135	14,428	12,459	10,964	9,788	8,841	7,407	6,373	5,592	4,492	3,753
190	175,470	126,756	99,213	81,502	69,157	60,060	47,550	39,353	33,567	29,264	25,939	23,292	21,136	17,163	14,448	12,474	10,975	9,797	8,848	7,412	6,377	5,595	4,494	3,755
195	178,285	128,218	100,106	82,104	69,590	60,386	47,755	39,493	33,669	29,341	26,000	23,341	21,176	17,190	14,467	12,483	10,986	9,806	8,855	7,417	6,380	5,598	4,496	3,756
200	181,044	129,639	100,970	82,685	70,007	60,700	47,950	39,627	33,766	29,415	26,058	23,388	21,215	17,215	14,484	12,502	10,996	9,814	8,862	7,421	6,384	5,601	4,497	3,757

250 MCM | 1 Run(s) | 277 Volts

C-Value = 16,483 | Magnetic Duct | Copper

Calculations:

1. $f = \dfrac{2 \times \text{Length} \times \text{Isca} \times 1.5}{\text{Runs} \times \text{C-Value} \times \text{E L-N}}$

2. $M = \dfrac{1}{1 + f}$

3. Isc = Isca x 1.5 x M

4. Add Motor Contribution = Motor FLA x 4

* All results are given in symmetrical amperes

Single-Phase L-N Bolted Fault-Current* Table

One-Way Distance in Feet

Available Isc in Thousands (Isca)	5	10	15	20	25	30	40	50	60	70	80	90	100	125	150	175	200	225	250	300	350	400	500	600
3	4,460	4,421	4,382	4,345	4,308	4,271	4,200	4,131	4,064	4,000	3,937	3,876	3,818	3,678	3,549	3,428	3,315	3,209	3,110	2,929	2,768	2,624	2,376	2,171
5	7,390	7,283	7,179	7,078	6,980	6,885	6,701	6,528	6,363	6,206	6,056	5,914	5,778	5,465	5,184	4,930	4,700	4,490	4,298	3,960	3,672	3,422	3,012	2,691
7	10,285	10,080	9,882	9,692	9,508	9,332	8,999	8,688	8,398	8,127	7,873	7,634	7,409	6,902	6,459	6,070	5,725	5,417	5,140	4,664	4,269	3,935	3,403	2,998
10	14,566	14,156	13,769	13,403	13,055	12,725	12,113	11,557	11,050	10,585	10,158	9,764	9,399	8,597	7,921	7,343	6,844	6,408	6,025	5,381	4,861	4,433	3,770	3,279
15	21,538	20,654	19,840	19,088	18,391	17,743	16,574	15,551	14,646	13,841	13,119	12,469	11,881	10,627	9,613	8,775	8,072	7,473	6,956	6,112	5,450	4,918	4,114	3,536
20	28,313	26,806	25,451	24,226	23,114	22,099	20,316	18,799	17,492	16,356	15,358	14,475	13,688	12,050	10,762	9,723	8,867	8,149	7,539	6,557	5,802	5,202	4,311	3,681
25	34,901	32,638	30,651	28,892	27,324	25,918	23,498	21,492	19,802	18,358	17,110	16,021	15,062	13,102	11,594	10,397	9,424	8,617	7,938	6,857	6,035	5,389	4,439	3,773
30	41,308	38,176	35,485	33,149	31,101	29,292	26,239	23,762	21,712	19,988	18,518	17,249	16,143	13,913	12,224	10,900	9,836	8,960	8,228	7,072	6,201	5,521	4,528	3,838
35	47,542	43,440	39,990	37,047	34,508	32,295	28,623	25,701	23,320	21,343	19,675	18,248	17,015	14,556	12,717	11,291	10,153	9,223	8,449	7,235	6,326	5,620	4,594	3,885
40	53,611	48,452	44,198	40,631	37,597	34,985	30,716	27,376	24,691	22,485	20,642	19,077	17,733	15,078	13,114	11,603	10,404	9,430	8,622	7,362	6,423	5,696	4,645	3,921
45	59,520	53,228	48,138	43,938	40,411	37,408	32,569	28,838	25,874	23,462	21,462	19,776	18,336	15,511	13,441	11,858	10,609	9,598	8,762	7,463	6,500	5,757	4,685	3,950
50	65,276	57,784	51,835	46,997	42,984	39,603	34,220	30,125	26,905	24,307	22,167	20,373	18,848	15,876	13,714	12,070	10,778	9,736	8,878	7,547	6,563	5,806	4,718	3,973
55	70,885	62,136	55,310	49,836	45,347	41,600	35,701	31,266	27,812	25,045	22,779	20,889	19,288	16,187	13,946	12,249	10,921	9,852	8,974	7,617	6,616	5,847	4,745	3,993
60	76,352	66,298	58,583	52,477	47,524	43,425	37,036	32,286	28,616	25,695	23,315	21,339	19,671	16,457	14,145	12,403	11,043	9,951	9,056	7,676	6,660	5,882	4,768	4,009
65	81,682	70,280	61,671	54,941	49,536	45,099	38,247	33,202	29,333	26,272	23,789	21,735	20,008	16,691	14,318	12,536	11,148	10,037	9,127	7,726	6,698	5,912	4,788	4,023
70	86,881	74,095	64,589	57,246	51,401	46,640	39,349	34,030	29,978	26,788	24,211	22,087	20,306	16,898	14,470	12,652	11,240	10,111	9,188	7,770	6,731	5,938	4,804	4,034
75	91,953	77,753	67,352	59,405	53,135	48,063	40,358	34,782	30,559	27,251	24,589	22,401	20,571	17,081	14,604	12,754	11,320	10,176	9,242	7,809	6,760	5,960	4,819	4,045
80	96,904	81,263	69,970	61,432	54,752	49,381	41,283	35,467	31,087	27,670	24,930	22,684	20,809	17,245	14,723	12,845	11,392	10,234	9,290	7,843	6,786	5,980	4,832	4,054
85	101,736	84,634	72,455	63,340	56,262	50,606	42,136	36,094	31,568	28,051	25,238	22,939	21,023	17,392	14,830	12,927	11,456	10,286	9,332	7,873	6,808	5,997	4,843	4,062
90	106,455	87,875	74,817	65,137	57,676	51,748	42,924	36,671	32,008	28,398	25,519	23,170	21,217	17,525	14,927	13,000	11,513	10,332	9,371	7,900	6,829	6,013	4,854	4,069
95	111,065	90,992	77,065	66,835	59,002	52,813	43,655	37,203	32,413	28,716	25,775	23,381	21,394	17,645	15,014	13,066	11,565	10,374	9,405	7,925	6,847	6,027	4,863	4,076
100	115,569	93,993	79,206	68,439	60,250	53,810	44,334	37,695	32,786	29,008	26,011	23,575	21,556	17,755	15,094	13,126	11,612	10,412	9,436	7,947	6,863	6,040	4,871	4,081
105	119,970	96,884	81,249	69,959	61,424	54,745	44,967	38,152	33,131	29,277	26,227	23,753	21,705	17,856	15,166	13,181	11,655	10,446	9,464	7,967	6,878	6,052	4,879	4,087
110	124,273	99,671	83,200	71,401	62,533	55,624	45,558	38,576	33,450	29,527	26,427	23,917	21,841	17,948	15,233	13,231	11,695	10,478	9,490	7,985	6,892	6,062	4,886	4,092
115	128,480	102,359	85,065	72,770	63,581	56,452	46,111	38,973	33,748	29,759	26,613	24,068	21,968	18,034	15,295	13,278	11,731	10,507	9,514	8,002	6,905	6,072	4,892	4,096
120	132,595	104,954	86,850	74,072	64,572	57,232	46,631	39,343	34,025	29,974	26,785	24,209	22,085	18,113	15,351	13,320	11,764	10,534	9,536	8,017	6,916	6,081	4,898	4,100
125	136,621	107,461	88,559	75,312	65,512	57,969	47,119	39,690	34,285	30,175	26,945	24,340	22,194	18,186	15,404	13,360	11,795	10,558	9,556	8,032	6,927	6,089	4,903	4,104
130	140,560	109,883	90,197	76,494	66,405	58,667	47,579	40,016	34,527	30,363	27,095	24,462	22,295	18,254	15,453	13,397	11,824	10,581	9,575	8,045	6,937	6,097	4,908	4,107
135	144,415	112,225	91,770	77,621	67,253	59,328	48,013	40,322	34,755	30,539	27,235	24,576	22,390	18,317	15,498	13,431	11,850	10,603	9,593	8,057	6,946	6,104	4,913	4,110
140	148,190	114,491	93,279	78,699	68,060	59,955	48,423	40,611	34,970	30,704	27,366	24,683	22,479	18,377	15,541	13,463	11,875	10,622	9,609	8,069	6,954	6,110	4,917	4,113
145	151,886	116,685	94,730	79,729	68,829	60,551	48,811	40,884	35,172	30,860	27,490	24,784	22,562	18,432	15,580	13,493	11,898	10,641	9,624	8,080	6,962	6,116	4,921	4,116
150	155,506	118,810	96,126	80,715	69,563	61,119	49,179	41,141	35,362	31,007	27,606	24,878	22,641	18,485	15,618	13,521	11,920	10,658	9,638	8,090	6,970	6,122	4,925	4,119
155	159,052	120,868	97,469	81,660	70,264	61,659	49,528	41,386	35,542	31,145	27,716	24,967	22,714	18,534	15,653	13,547	11,940	10,675	9,652	8,099	6,977	6,128	4,928	4,121
160	162,526	122,864	98,763	82,567	70,934	62,174	49,860	41,617	35,713	31,276	27,820	25,051	22,784	18,580	15,686	13,572	11,960	10,690	9,664	8,108	6,983	6,133	4,931	4,124
165	165,931	124,800	100,010	83,436	71,575	62,666	50,176	41,837	35,875	31,400	27,918	25,131	22,850	18,624	15,717	13,595	11,978	10,704	9,676	8,116	6,989	6,137	4,934	4,126
170	169,269	126,679	101,213	84,272	72,189	63,136	50,477	42,046	36,028	31,518	28,011	25,206	22,912	18,665	15,746	13,617	11,995	10,718	9,687	8,124	6,995	6,142	4,937	4,128
175	172,541	128,503	102,374	85,075	72,778	63,586	50,764	42,245	36,174	31,629	28,099	25,277	22,971	18,704	15,774	13,638	12,011	10,731	9,698	8,131	7,001	6,146	4,940	4,130
180	175,750	130,275	103,495	85,848	73,342	64,017	51,038	42,435	36,313	31,736	28,183	25,345	23,027	18,741	15,800	13,657	12,026	10,743	9,707	8,138	7,006	6,150	4,943	4,131
185	178,897	131,996	104,579	86,592	73,885	64,430	51,300	42,616	36,446	31,837	28,262	25,410	23,080	18,776	15,825	13,676	12,041	10,755	9,717	8,145	7,011	6,154	4,945	4,133
190	181,985	133,669	105,626	87,309	74,406	64,826	51,551	42,789	36,572	31,933	28,338	25,471	23,131	18,810	15,849	13,694	12,054	10,766	9,726	8,151	7,015	6,157	4,947	4,135
195	185,014	135,296	106,639	88,000	74,908	65,206	51,791	42,954	36,693	32,025	28,411	25,530	23,179	18,842	15,872	13,711	12,068	10,776	9,734	8,157	7,020	6,161	4,950	4,136
200	187,987	136,879	107,620	88,667	75,390	65,572	52,021	43,112	36,808	32,113	28,480	25,585	23,225	18,872	15,893	13,727	12,080	10,786	9,742	8,163	7,024	6,164	4,952	4,138

300 MCM	1 Run(s)	277 Volts	C-Value = 18,176	Magnetic Duct	Copper

Calculations:

1. $f = \dfrac{2 \times \text{Length} \times \text{Isca} \times 1.5}{\text{Runs} \times \text{C-Value} \times \text{E L-N}}$

2. $M = \dfrac{1}{1 + f}$

3. $\text{Isc} = \text{Isca} \times 1.5 \times M$

4. Add Motor Contribution = Motor FLA x 4

* All results are given in symmetrical amperes

Single-Phase L-N Bolted Fault-Current* Table

One-Way Distance in Feet

Available Isc in Thousands (Isca)

Isca	5	10	15	20	25	30	40	50	60	70	80	90	100	125	150	175	200	225	250	300	350	400	500	600
3	4,463	4,427	4,391	4,356	4,322	4,288	4,222	4,157	4,095	4,034	3,976	3,918	3,863	3,731	3,608	3,492	3,384	3,282	3,186	3,011	2,853	2,711	2,466	2,262
5	7,398	7,299	7,203	7,109	7,018	6,929	6,757	6,594	6,438	6,290	6,148	6,013	5,883	5,582	5,311	5,064	4,840	4,634	4,446	4,111	3,823	3,573	3,159	2,831
7	10,302	10,111	9,927	9,750	9,579	9,413	9,099	8,806	8,531	8,272	8,029	7,799	7,582	7,090	6,658	6,275	5,934	5,628	5,352	4,874	4,474	4,135	3,591	3,173
10	14,599	14,218	13,857	13,514	13,188	12,877	12,296	11,766	11,280	10,832	10,419	10,035	9,679	8,891	8,221	7,646	7,145	6,706	6,318	5,662	5,130	4,689	4,002	3,490
15	21,609	20,786	20,024	19,315	18,655	18,038	16,920	15,932	15,053	14,266	13,557	12,916	12,332	11,080	10,059	9,210	8,494	7,880	7,350	6,478	5,790	5,235	4,392	3,783
20	28,437	27,029	25,753	24,593	23,532	22,560	20,837	19,359	18,076	16,953	15,962	15,080	14,290	12,636	11,325	10,260	9,379	8,637	8,003	6,980	6,188	5,558	4,618	3,949
25	35,089	32,969	31,091	29,415	27,911	26,553	24,199	22,228	20,553	19,114	17,863	16,765	15,795	13,798	12,250	11,014	10,004	9,164	8,455	7,321	6,455	5,772	4,764	4,056
30	41,572	38,630	36,076	33,839	31,864	30,106	27,115	24,664	22,620	20,888	19,403	18,115	16,987	14,700	12,955	11,581	10,470	9,553	8,785	7,567	6,645	5,924	4,867	4,131
35	47,893	44,029	40,742	37,912	35,450	33,288	29,669	26,759	24,370	22,372	20,677	19,220	17,956	15,419	13,511	12,023	10,830	9,852	9,037	7,753	6,789	6,038	4,944	4,186
40	54,057	49,186	45,119	41,674	38,718	36,153	31,924	28,580	25,871	23,630	21,747	20,142	18,758	16,007	13,960	12,377	11,116	10,089	9,235	7,899	6,900	6,126	5,003	4,228
45	60,071	54,114	49,233	45,159	41,708	38,747	33,929	30,177	27,172	24,712	22,660	20,922	19,433	16,496	14,330	12,667	11,350	10,281	9,396	8,016	6,989	6,196	5,049	4,261
50	65,939	58,831	53,106	48,397	44,455	41,107	35,725	31,590	28,312	25,651	23,447	21,592	20,009	16,909	14,641	12,909	11,544	10,440	9,529	8,112	7,063	6,253	5,088	4,288
55	71,667	63,348	56,760	51,413	46,987	43,262	37,342	32,847	29,318	26,474	24,133	22,172	20,506	17,263	14,906	13,115	11,708	10,574	9,640	8,193	7,124	6,301	5,119	4,310
60	77,260	67,679	60,212	54,229	49,328	45,239	38,806	33,975	30,213	27,201	24,736	22,680	20,940	17,569	15,133	13,291	11,848	10,688	9,735	8,261	7,175	6,341	5,146	4,329
65	82,722	71,834	63,479	56,865	51,499	47,059	40,137	34,991	31,014	27,849	25,270	23,128	21,321	17,837	15,332	13,443	11,969	10,787	9,816	8,320	7,219	6,376	5,168	4,345
70	88,059	75,825	66,575	59,337	53,519	48,739	41,353	35,911	31,735	28,429	25,747	23,527	21,660	18,073	15,506	13,577	12,075	10,872	9,888	8,371	7,258	6,406	5,188	4,359
75	93,274	79,660	69,514	61,660	55,401	50,296	42,468	36,749	32,388	28,951	26,175	23,884	21,962	18,283	15,660	13,695	12,168	10,948	9,950	8,416	7,291	6,432	5,205	4,371
80	98,371	83,348	72,306	63,847	57,160	51,741	43,494	37,515	32,981	29,425	26,561	24,205	22,233	18,471	15,797	13,800	12,251	11,015	10,005	8,455	7,321	6,455	5,220	4,382
85	103,355	86,899	74,963	65,910	58,808	53,088	44,442	38,218	33,523	29,855	26,911	24,495	22,478	18,639	15,921	13,894	12,325	11,075	10,055	8,490	7,347	6,476	5,234	4,391
90	108,229	90,318	77,494	67,859	60,355	54,345	45,320	38,865	34,020	30,249	27,230	24,760	22,700	18,792	16,032	13,979	12,392	11,129	10,099	8,522	7,371	6,494	5,246	4,400
95	112,997	93,615	79,908	69,703	61,809	55,521	46,135	39,463	34,477	30,610	27,523	25,001	22,903	18,931	16,133	14,055	12,452	11,177	10,139	8,550	7,392	6,510	5,256	4,407
100	117,662	96,794	82,213	71,450	63,179	56,624	46,894	40,017	34,899	30,942	27,791	25,222	23,088	19,057	16,225	14,125	12,507	11,221	10,175	8,576	7,412	6,525	5,266	4,414
105	122,227	99,863	84,417	73,109	64,472	57,661	47,602	40,532	35,290	31,249	28,038	25,426	23,259	19,173	16,309	14,189	12,557	11,261	10,208	8,600	7,429	6,539	5,275	4,420
110	126,697	102,826	86,525	74,685	65,695	58,637	48,266	41,012	35,653	31,534	28,267	25,614	23,416	19,280	16,386	14,247	12,602	11,298	10,238	8,621	7,445	6,551	5,283	4,426
115	131,073	105,690	88,543	76,184	66,852	59,557	48,887	41,460	35,992	31,798	28,479	25,788	23,561	19,378	16,457	14,301	12,644	11,332	10,266	8,641	7,460	6,563	5,290	4,431
120	135,358	108,459	90,479	77,612	67,949	60,426	49,471	41,879	36,307	32,044	28,676	25,950	23,696	19,470	16,523	14,350	12,683	11,363	10,291	8,659	7,473	6,573	5,297	4,436
125	139,556	111,138	92,335	78,974	68,991	61,249	50,021	42,273	36,603	32,274	28,860	26,100	23,822	19,554	16,583	14,396	12,719	11,391	10,315	8,675	7,485	6,583	5,303	4,440
130	143,668	113,730	94,118	80,274	69,981	62,028	50,540	42,642	36,879	32,489	29,032	26,241	23,939	19,633	16,640	14,439	12,752	11,418	10,337	8,691	7,497	6,592	5,309	4,444
135	147,699	116,241	95,831	81,517	70,924	62,767	51,030	42,991	37,140	32,690	29,193	26,372	24,048	19,706	16,693	14,479	12,783	11,443	10,357	8,705	7,508	6,600	5,314	4,448
140	151,649	118,674	97,478	82,706	71,822	63,470	51,493	43,319	37,384	32,880	29,344	26,495	24,150	19,775	16,742	14,516	12,812	11,466	10,376	8,719	7,518	6,608	5,319	4,452
145	155,522	121,033	99,064	83,845	72,680	64,138	51,932	43,629	37,615	33,059	29,486	26,611	24,247	19,840	16,788	14,550	12,839	11,488	10,394	8,731	7,527	6,615	5,324	4,455
150	159,319	123,320	100,591	84,937	73,498	64,775	52,349	43,923	37,833	33,227	29,620	26,720	24,337	19,900	16,832	14,583	12,864	11,508	10,410	8,743	7,536	6,621	5,328	4,458
155	163,043	125,540	102,063	85,984	74,281	65,382	52,745	44,201	38,040	33,386	29,747	26,823	24,422	19,957	16,872	14,613	12,888	11,527	10,426	8,754	7,544	6,628	5,333	4,461
160	166,696	127,694	103,483	86,989	75,030	65,962	53,121	44,466	38,235	33,536	29,866	26,920	24,503	20,011	16,911	14,642	12,910	11,545	10,441	8,764	7,551	6,634	5,336	4,464
165	170,280	129,787	104,853	87,955	75,748	66,516	53,480	44,717	38,421	33,679	29,979	27,012	24,579	20,061	16,947	14,669	12,931	11,562	10,454	8,774	7,559	6,639	5,340	4,466
170	173,797	131,820	106,176	88,884	76,436	67,046	53,822	44,956	38,597	33,814	30,086	27,099	24,651	20,109	16,981	14,695	12,951	11,578	10,467	8,783	7,565	6,644	5,343	4,468
175	177,249	133,796	107,454	89,778	77,096	67,553	54,149	45,183	38,765	33,943	30,188	27,181	24,719	20,155	17,013	14,719	12,970	11,593	10,480	8,792	7,572	6,649	5,347	4,471
180	180,637	135,718	108,690	90,639	77,730	68,040	54,461	45,400	38,924	34,065	30,285	27,260	24,784	20,198	17,044	14,742	12,988	11,607	10,491	8,800	7,578	6,654	5,350	4,473
185	183,963	137,587	109,885	91,469	78,340	68,506	54,759	45,607	39,077	34,182	30,377	27,334	24,845	20,239	17,073	14,764	13,005	11,620	10,502	8,808	7,584	6,658	5,352	4,475
190	187,230	139,406	111,042	92,269	78,926	68,954	55,045	45,806	39,222	34,293	30,465	27,405	24,904	20,278	17,101	14,785	13,021	11,633	10,513	8,815	7,589	6,663	5,355	4,477
195	190,437	141,177	112,163	93,042	79,490	69,385	55,319	45,995	39,361	34,399	30,548	27,473	24,960	20,315	17,127	14,804	13,036	11,645	10,523	8,822	7,594	6,667	5,358	4,478
200	193,588	142,901	113,249	93,788	80,034	69,799	55,582	46,177	39,494	34,501	30,628	27,538	25,013	20,350	17,152	14,823	13,051	11,657	10,532	8,829	7,599	6,670	5,360	4,480

350 MCM | 1 Run(s) | 277 Volts | C-Value = 19,703 | Magnetic Duct | Copper

Calculations:

1. $f = \dfrac{2 \times \text{Length} \times \text{Isca} \times 1.5}{\text{Runs} \times \text{C-Value} \times \text{E L-N}}$

2. $M = \dfrac{1}{1 + f}$

3. $\text{Isc} = \text{Isca} \times 1.5 \times M$

4. Add Motor Contribution = Motor FLA x 4

* All results are given in symmetrical amperes

Single-Phase L-N Bolted Fault-Current* Table

One-Way Distance in Feet

Available Isc in Thousands (Isca)	5	10	15	20	25	30	40	50	60	70	80	90	100	125	150	175	200	225	250	300	350	400	500	600
3	4,465	4,430	4,396	4,362	4,329	4,296	4,233	4,171	4,110	4,052	3,995	3,940	3,886	3,758	3,638	3,525	3 419	3,320	3,226	3,053	2,898	2,757	2,514	2,310
5	7,403	7,308	7,215	7,125	7,037	6,951	6,785	6,627	6,477	6,333	6,195	6,063	5,937	5,643	5,376	5,134	4 913	4,710	4,523	4,190	3,903	3,653	3,238	2,907
7	10,310	10,127	9,950	9,779	9,614	9,454	9,151	8,866	8,598	8,346	8,109	7,884	7,672	7,188	6,761	6,382	6 044	5,739	5,464	4,986	4,585	4,243	3,693	3,269
10	14,615	14,250	13,902	13,571	13,255	12,953	12,390	11,873	11,398	10,960	10,554	10,177	9,826	9,045	8,380	7,806	7,305	6,865	6,475	5,814	5,276	4,829	4,129	3,606
15	21,645	20,853	20,116	19,430	18,789	18,189	17,097	16,129	15,265	14,488	13,787	13,150	12,570	11,321	10,298	9,444	8,721	8,101	7,563	6,677	5,976	5,409	4,546	3,920
20	28,499	27,141	25,907	24,780	23,747	22,797	21,107	19,651	18,383	17,268	16,281	15,401	14,611	12,950	11,628	10,551	9,657	8,902	8,257	7,212	6,401	5,755	4,787	4,099
25	35,184	33,137	31,316	29,684	28,214	26,882	24,564	22,614	20,950	19,515	18,263	17,163	16,188	14,174	12,605	11,350	10,321	9,464	8,738	7,576	6,687	5,984	4,945	4,214
30	41,705	38,860	36,379	34,195	32,259	30,530	27,574	25,140	23,101	21,368	19,877	18,580	17,442	15,127	13,354	11,953	10,818	9,880	9,091	7,840	6,892	6,148	5,056	4,294
35	48,070	44,329	41,129	38,359	35,939	33,806	30,219	27,321	24,929	22,923	21,216	19,745	18,465	15,890	13,945	12,424	11,202	10,200	9,361	8,040	7,046	6,270	5,139	4,353
40	54,283	49,560	45,593	42,215	39,302	36,765	32,562	29,222	26,503	24,246	22,344	20,719	19,314	16,514	14,424	12,803	11,509	10,453	9,575	8,197	7,166	6,365	5,203	4,399
45	60,349	54,568	49,798	45,795	42,387	39,451	34,652	30,893	27,870	25,386	23,309	21,546	20,030	17,035	14,819	13,114	11,760	10,660	9,748	8,323	7,262	6,441	5,253	4,435
50	66,274	59,367	53,764	49,128	45,227	41,900	36,527	32,375	29,071	26,378	24,143	22,256	20,643	17,476	15,152	13,374	11,969	10,831	9,891	8,427	7,341	6,503	5,294	4,465
55	72,063	63,971	57,512	52,238	47,850	44,142	38,219	33,697	30,132	27,250	24,870	22,873	21,173	17,855	15,436	13,594	12,145	10,975	10,011	8,514	7,407	6,555	5,329	4,489
60	77,721	68,390	61,059	55,148	50,281	46,202	39,754	34,885	31,078	28,021	25,511	23,414	21,635	18,183	15,680	13,783	12,296	11,098	10,113	8,588	7,463	6,599	5,357	4,509
65	83,251	72,636	64,421	57,876	52,538	48,102	41,152	35,957	31,926	28,709	26,080	23,892	22,043	18,470	15,893	13,947	12,426	11,204	10,201	8,652	7,511	6,636	5,382	4,527
70	88,658	76,718	67,612	60,439	54,641	49,859	42,431	36,930	32,691	29,325	26,588	24,318	22,405	18,723	16,080	14,091	12,540	11,297	10,278	8,707	7,553	6,668	5,403	4,542
75	93,947	80,646	70,645	62,851	56,605	51,489	43,606	37,816	33,384	29,882	27,044	24,699	22,728	18,948	16,246	14,219	12,641	11,379	10,345	8,755	7,589	6,697	5,422	4,555
80	99,120	84,429	73,531	65,125	58,443	53,005	44,689	38,628	34,015	30,386	27,457	25,043	23,019	19,150	16,394	14,332	12,730	11,451	10,405	8,798	7,621	6,722	5,438	4,566
85	104,182	88,074	76,280	67,272	60,167	54,419	45,690	39,374	34,592	30,846	27,831	25,354	23,282	19,331	16,527	14,433	12,810	11,516	10,458	8,836	7,650	6,744	5,453	4,577
90	109,136	91,589	78,903	69,304	61,787	55,741	46,618	40,061	35,121	31,266	28,173	25,637	23,520	19,495	16,647	14,525	12,882	11,574	10,506	8,870	7,675	6,764	5,466	4,586
95	113,986	94,981	81,407	71,228	63,312	56,979	47,481	40,696	35,609	31,652	28,486	25,896	23,738	19,645	16,756	14,607	12,947	11,626	10,550	8,901	7,698	6,782	5,478	4,594
100	118,735	98,255	83,801	73,054	64,750	58,141	48,285	41,286	36,059	32,007	28,774	26,134	23,937	19,781	16,855	14,683	13,006	11,674	10,589	8,929	7,719	6,798	5,488	4,601
105	123,386	101,419	86,091	74,788	66,109	59,235	49,037	41,834	36,477	32,336	29,039	26,352	24,121	19,906	16,945	14,751	13,060	11,717	10,624	8,954	7,738	6,813	5,498	4,608
110	127,942	104,476	88,285	76,438	67,395	60,265	49,741	42,346	36,865	32,640	29,284	26,554	24,290	20,021	17,029	14,814	13,110	11,757	10,657	8,978	7,755	6,826	5,506	4,614
115	132,405	107,434	90,387	78,010	68,613	61,237	50,401	42,823	37,226	32,923	29,512	26,741	24,446	20,127	17,105	14,872	13,155	11,793	10,687	8,999	7,771	6,838	5,514	4,620
120	136,780	110,296	92,405	79,508	69,770	62,157	51,022	43,271	37,564	33,187	29,724	26,915	24,591	20,226	17,176	14,926	13,197	11,827	10,715	9,018	7,786	6,850	5,522	4,625
125	141,068	113,068	94,342	80,938	70,868	63,027	51,607	43,691	37,880	33,434	29,922	27,077	24,726	20,317	17,242	14,976	13,236	11,858	10,740	9,037	7,799	6,860	5,529	4,630
130	145,271	115,752	96,204	82,304	71,914	63,853	52,160	44,086	38,177	33,665	30,106	27,228	24,852	20,402	17,303	15,022	13,272	11,887	10,764	9,053	7,812	6,870	5,535	4,634
135	149,393	118,354	97,994	83,611	72,910	64,637	52,682	44,458	38,456	33,881	30,279	27,370	24,970	20,481	17,360	15,065	13,306	11,914	10,786	9,069	7,823	6,879	5,541	4,638
140	153,436	120,878	99,718	84,863	73,860	65,382	53,176	44,810	38,718	34,085	30,442	27,503	25,081	20,556	17,414	15,105	13,337	11,939	10,807	9,084	7,834	6,887	5,546	4,642
145	157,402	123,325	101,378	86,062	74,766	66,092	53,644	45,142	38,966	34,277	30,595	27,627	25,185	20,625	17,464	15,143	13,366	11,963	10,826	9,097	7,844	6,895	5,551	4,646
150	161,293	125,701	102,978	87,212	75,633	66,768	54,089	45,456	39,200	34,458	30,739	27,745	25,282	20,691	17,511	15,178	13,394	11,985	10,844	9,110	7,854	6,902	5,556	4,649
155	165,111	128,008	104,521	88,316	76,462	67,413	54,511	45,755	39,422	34,629	30,875	27,856	25,374	20,752	17,555	15,211	13,419	12,005	10,861	9,122	7,863	6,909	5,560	4,652
160	168,858	130,249	106,010	89,377	77,256	68,030	54,914	46,038	39,632	34,791	31,004	27,960	25,461	20,810	17,596	15,242	13,444	12,025	10,877	9,133	7,871	6,915	5,564	4,655
165	172,537	132,427	107,448	90,397	78,017	68,619	55,297	46,307	39,831	34,944	31,126	28,059	25,543	20,865	17,635	15,271	13,466	12,043	10,892	9,143	7,879	6,921	5,568	4,658
170	176,148	134,544	108,838	91,379	78,747	69,183	55,663	46,563	40,021	35,090	31,241	28,153	25,621	20,917	17,672	15,299	13,488	12,060	10,906	9,153	7,886	6,927	5,572	4,660
175	179,695	136,604	110,182	92,324	79,448	69,724	56,012	46,807	40,201	35,229	31,351	28,242	25,695	20,966	17,707	15,326	13,508	12,077	10,919	9,163	7,893	6,933	5,576	4,663
180	183,178	138,607	111,482	93,235	80,122	70,242	56,346	47,040	40,373	35,360	31,455	28,327	25,765	21,013	17,741	15,350	13,528	12,092	10,932	9,172	7,900	6,938	5,579	4,665
185	186,600	140,557	112,740	94,114	80,769	70,739	56,666	47,263	40,536	35,486	31,555	28,408	25,831	21,057	17,772	15,374	13,546	12,107	10,944	9,180	7,906	6,942	5,582	4,667
190	189,961	142,456	113,958	94,961	81,393	71,217	56,972	47,476	40,693	35,606	31,649	28,484	25,895	21,099	17,802	15,396	13,564	12,121	10,955	9,188	7,912	6,947	5,585	4,669
195	193,264	144,306	115,139	95,779	81,993	71,676	57,266	47,679	40,842	35,720	31,740	28,557	25,955	21,139	17,831	15,418	13,580	12,134	10,966	9,196	7,918	6,951	5,588	4,671
200	196,510	146,108	116,283	96,570	82,572	72,118	57,547	47,874	40,985	35,830	31,826	28,627	26,013	21,178	17,858	15,438	13,596	12,146	10,976	9,203	7,923	6,956	5,590	4,673

400 MCM | 1 Run(s) | 277 Volts

C-Value = 20,565 | Magnetic Duct | Copper

Calculations:

1. $f = \dfrac{2 \times \text{Length} \times \text{Isca} \times 1.5}{\text{Runs} \times \text{C-Value} \times \text{E L-N}}$

2. $M = \dfrac{1}{1 + f}$

3. Isc = Isca x 1.5 x M

4. Add Motor Contribution = Motor FLA x 4

* All results are given in symmetrical amperes

Single-Phase L-N Bolted Fault-Current* Table

One-Way Distance in Feet

Available Isc in Thousands (Isca)	5	10	15	20	25	30	40	50	60	70	80	90	100	125	150	175	200	225	250	300	350	400	500	600
3	4,467	4,435	4,403	4,372	4,341	4,311	4,251	4,193	4,137	4,082	4,028	3,976	3,925	3,804	3,689	3,582	3,481	3,385	3,294	3,126	2,975	2,838	2,598	2,395
5	7,410	7,321	7,235	7,151	7,069	6,988	6,833	6,684	6,542	6,406	6,275	6,149	6,028	5,747	5,490	5,255	5,040	4,841	4,658	4,330	4,045	3,795	3,378	3,043
7	10,324	10,153	9,988	9,828	9,674	9,524	9,237	8,968	8,713	8,473	8,246	8,030	7,826	7,357	6,942	6,571	6,237	5,936	5,662	5,185	4,781	4,436	3,876	3,442
10	14,643	14,302	13,977	13,666	13,368	13,084	12,549	12,057	11,602	11,180	10,787	10,421	10,079	9,315	8,659	8,089	7,590	7,148	6,755	6,086	5,538	5,080	4,359	3,818
15	21,705	20,965	20,273	19,626	19,018	18,447	17,403	16,470	15,632	14,875	14,188	13,562	12,989	11,747	10,722	9,862	9,129	8,498	7,949	7,038	6,315	5,727	4,827	4,172
20	28,604	27,331	26,168	25,099	24,114	23,203	21,574	20,159	18,918	17,820	16,844	15,968	15,179	13,511	12,173	11,076	10,160	9,384	8,719	7,635	6,792	6,116	5,100	4,374
25	35,343	33,421	31,697	30,142	28,733	27,450	25,199	23,289	21,648	20,223	18,974	17,871	16,888	14,848	13,248	11,959	10,898	10,011	9,257	8,045	7,114	6,376	5,280	4,506
30	41,930	39,251	36,895	34,805	32,940	31,264	28,377	25,977	23,952	22,220	20,722	19,413	18,259	15,897	14,076	12,630	11,453	10,477	9,654	8,343	7,346	6,561	5,407	4,598
35	48,368	44,839	41,790	39,129	36,786	34,709	31,186	28,312	25,924	23,907	22,181	20,687	19,382	16,742	14,735	13,158	11,885	10,837	9,959	8,570	7,521	6,701	5,501	4,666
40	54,663	50,198	46,407	43,149	40,318	37,835	33,687	30,359	27,629	25,350	23,418	21,759	20,320	17,437	15,271	13,583	12,231	11,124	10,201	8,749	7,658	6,810	5,574	4,718
45	60,820	55,342	50,770	46,896	43,571	40,686	35,929	32,167	29,119	26,598	24,479	22,673	21,115	18,019	15,715	13,934	12,515	11,358	10,397	8,893	7,769	6,897	5,632	4,760
50	66,842	60,285	54,899	50,397	46,577	43,296	37,948	33,777	30,432	27,689	25,400	23,461	21,797	18,513	16,090	14,227	12,751	11,553	10,560	9,011	7,859	6,968	5,680	4,794
55	72,735	65,037	58,813	53,676	49,364	45,694	39,778	35,219	31,597	28,651	26,207	24,148	22,388	18,938	16,410	14,477	12,951	11,717	10,697	9,111	7,935	7,027	5,719	4,822
60	78,503	69,610	62,528	56,753	51,955	47,905	41,443	36,518	32,639	29,505	26,920	24,751	22,906	19,308	16,686	14,692	13,123	11,857	10,814	9,196	7,999	7,077	5,752	4,845
65	84,149	74,014	66,058	59,646	54,369	49,950	42,965	37,694	33,575	30,268	27,554	25,286	23,363	19,632	16,928	14,879	13,272	11,978	10,915	9,268	8,054	7,121	5,781	4,865
70	89,677	78,257	69,417	62,372	56,625	51,847	44,362	38,765	34,422	30,954	28,121	25,763	23,770	19,918	17,140	15,042	13,402	12,084	11,003	9,332	8,102	7,158	5,805	4,883
75	95,092	82,349	72,618	64,944	58,736	53,612	45,647	39,743	35,191	31,575	28,633	26,192	24,135	20,173	17,329	15,188	13,517	12,178	11,080	9,387	8,143	7,191	5,827	4,898
80	100,395	86,297	75,671	67,374	60,718	55,258	46,835	40,640	35,893	32,139	29,095	26,579	24,463	20,402	17,497	15,317	13,619	12,261	11,149	9,437	8,180	7,219	5,846	4,911
85	105,592	90,109	78,586	69,676	62,580	56,796	47,936	41,466	36,536	32,653	29,516	26,929	24,759	20,608	17,649	15,433	13,711	12,335	11,210	9,481	8,213	7,245	5,863	4,923
90	110,685	93,791	81,372	71,857	64,334	58,238	48,958	42,229	37,127	33,124	29,901	27,249	25,029	20,795	17,785	15,537	13,793	12,402	11,265	9,520	8,243	7,268	5,878	4,934
95	115,676	97,351	84,038	73,928	65,990	59,591	49,911	42,936	37,672	33,558	30,254	27,542	25,276	20,965	17,910	15,632	13,868	12,462	11,315	9,555	8,269	7,289	5,891	4,943
100	120,570	100,794	86,591	75,897	67,554	60,863	50,800	43,593	38,177	33,958	30,578	27,811	25,502	21,120	18,023	15,718	13,936	12,517	11,360	9,587	8,294	7,307	5,903	4,952
105	125,369	104,126	89,039	77,771	69,034	62,062	51,633	44,205	38,645	34,328	30,878	28,058	25,710	21,263	18,127	15,797	13,998	12,567	11,401	9,617	8,315	7,324	5,914	4,960
110	130,075	107,352	91,387	79,556	70,438	63,194	52,414	44,776	39,081	34,671	31,156	28,287	25,903	21,394	18,222	15,869	14,055	12,612	11,438	9,643	8,335	7,340	5,925	4,967
115	134,691	110,477	93,642	81,260	71,770	64,264	53,148	45,311	39,488	34,991	31,413	28,500	26,081	21,515	18,310	15,936	14,107	12,654	11,473	9,668	8,354	7,354	5,934	4,973
120	139,221	113,506	95,809	82,887	73,036	65,278	53,839	45,812	39,868	35,289	31,654	28,697	26,246	21,628	18,391	15,997	14,155	12,693	11,505	9,691	8,371	7,367	5,942	4,979
125	143,666	116,443	97,894	84,442	74,241	66,238	54,491	46,283	40,224	35,568	31,878	28,881	26,400	21,732	18,467	16,054	14,200	12,729	11,534	9,712	8,386	7,379	5,950	4,985
130	148,028	119,293	99,900	85,931	75,389	67,151	55,107	46,727	40,559	35,829	32,088	29,054	26,544	21,829	18,537	16,108	14,241	12,762	11,562	9,731	8,401	7,390	5,957	4,990
135	152,310	122,058	101,832	87,356	76,484	68,018	55,690	47,145	40,874	36,075	32,284	29,215	26,678	21,920	18,602	16,157	14,280	12,793	11,587	9,749	8,414	7,401	5,964	4,995
140	156,515	124,743	103,694	88,723	77,530	68,844	56,243	47,541	41,171	36,306	32,469	29,366	26,804	22,005	18,664	16,203	14,316	12,822	11,611	9,766	8,427	7,410	5,971	4,999
145	160,643	127,352	105,491	90,035	78,529	69,631	56,767	47,915	41,451	36,524	32,643	29,508	26,923	22,085	18,721	16,246	14,350	12,849	11,633	9,781	8,438	7,420	5,976	5,003
150	164,698	129,887	107,224	91,295	79,486	70,382	57,265	48,269	41,716	36,729	32,807	29,642	27,034	22,160	18,775	16,287	14,381	12,875	11,654	9,796	8,449	7,428	5,982	5,007
155	168,681	132,352	108,898	92,505	80,402	71,100	57,739	48,605	41,967	36,924	32,963	29,769	27,140	22,231	18,826	16,325	14,411	12,898	11,673	9,810	8,460	7,436	5,987	5,011
160	172,594	134,749	110,516	93,670	81,281	71,786	58,191	48,925	42,205	37,108	33,109	29,889	27,239	22,297	18,873	16,361	14,439	12,921	11,692	9,823	8,469	7,443	5,992	5,014
165	176,439	137,081	112,080	94,791	82,124	72,443	58,621	49,229	42,431	37,282	33,248	30,002	27,333	22,360	18,918	16,395	14,465	12,942	11,709	9,835	8,478	7,450	5,996	5,017
170	180,218	139,351	113,593	95,871	82,933	73,072	59,033	49,519	42,646	37,448	33,380	30,109	27,422	22,420	18,961	16,427	14,490	12,962	11,725	9,847	8,487	7,457	6,001	5,020
175	183,932	141,561	115,057	96,912	83,711	73,675	59,426	49,795	42,851	37,606	33,505	30,211	27,507	22,476	19,001	16,457	14,514	12,981	11,741	9,857	8,495	7,463	6,005	5,023
180	187,583	143,714	116,475	97,916	84,459	74,254	59,802	50,059	43,046	37,756	33,625	30,308	27,587	22,530	19,040	16,486	14,536	12,999	11,755	9,868	8,502	7,469	6,008	5,026
185	191,172	145,812	117,849	98,886	85,179	74,810	60,162	50,311	43,232	37,900	33,738	30,400	27,663	22,581	19,076	16,513	14,557	13,016	11,769	9,878	8,510	7,475	6,012	5,028
190	194,702	147,856	119,181	99,822	85,873	75,344	60,507	50,552	43,410	38,036	33,846	30,488	27,736	22,629	19,111	16,539	14,577	13,032	11,782	9,887	8,517	7,480	6,016	5,031
195	198,174	149,850	120,473	100,726	86,541	75,858	60,838	50,783	43,580	38,167	33,950	30,572	27,805	22,675	19,144	16,564	14,596	13,047	11,795	9,896	8,523	7,485	6,019	5,033
200	201,588	151,794	121,726	101,601	87,186	76,353	61,156	51,005	43,743	38,292	34,049	30,652	27,872	22,719	19,175	16,587	14,615	13,062	11,807	9,904	8,529	7,490	6,022	5,035

500 MCM | 1 Run(s) | 277 Volts | C-Value = 22,185 | Magnetic Duct | Copper

Calculations:

1. $f = \frac{2 \times \text{Length} \times \text{Isca} \times 1.5}{\text{Runs} \times \text{C-Value} \times E\ L\text{-}N}$

2. $M = \frac{1}{1 + f}$

3. Isc = Isca x 1.5 x M

4. Add Motor Contribution = Motor FLA x 4

* All results are given in symmetrical amperes

Single-Phase L-N Bolted Fault-Current* Table

One-Way Distance in Feet

Available Isc in Thousands (Isca)	5	10	15	20	25	30	40	50	60	70	80	90	100	125	150	175	200	225	250	300	350	400	500	600
3	4,468	4,437	4,406	4,376	4,346	4,317	4,259	4,203	4,148	4,094	4,042	3,992	3,942	3,824	3,712	3,607	3,508	3,413	3,324	3,159	3,010	2,874	2,636	2,434
5	7,413	7,327	7,244	7,162	7,082	7,005	6,854	6,709	6,570	6,437	6,310	6,187	6,069	5,793	5,540	5,309	5,096	4,900	4,718	4,393	4,109	3,860	3,442	3,106
7	10,330	10,164	10,005	9,850	9,699	9,554	9,275	9,012	8,764	8,529	8,306	8,095	7,894	7,433	7,023	6,655	6,324	6,025	5,752	5,275	4,871	4,525	3,961	3,523
10	14,654	14,324	14,009	13,707	13,418	13,141	12,619	12,138	11,692	11,277	10,891	10,530	10,193	9,437	8,785	8,218	7,719	7,278	6,884	6,212	5,659	5,197	4,467	3,917
15	21,731	21,014	20,342	19,711	19,119	18,561	17,538	16,621	15,796	15,048	14,369	13,747	13,178	11,941	10,916	10,054	9,318	8,682	8,127	7,206	6,473	5,875	4,959	4,290
20	28,649	27,414	26,282	25,239	24,276	23,383	21,782	20,386	19,158	18,070	17,098	16,226	15,438	13,768	12,423	11,318	10,394	9,609	8,934	7,834	6,975	6,286	5,248	4,505
25	35,412	33,545	31,865	30,345	28,963	27,702	25,482	23,592	21,963	20,545	19,298	18,194	17,210	15,159	13,545	12,242	11,167	10,266	9,500	8,265	7,315	6,561	5,439	4,645
30	42,027	39,422	37,122	35,075	33,242	31,591	28,737	26,356	24,339	22,609	21,108	19,795	18,635	16,254	14,413	12,946	11,751	10,757	9,918	8,581	7,561	6,758	5,573	4,742
35	48,497	45,062	42,081	39,470	37,164	35,113	31,622	28,762	26,377	24,357	22,625	21,122	19,807	17,139	15,104	13,501	12,206	11,137	10,241	8,821	7,747	6,906	5,674	4,815
40	54,829	50,478	46,767	43,564	40,772	38,316	34,197	30,877	28,145	25,857	23,913	22,241	20,787	17,868	15,667	13,950	12,571	11,441	10,497	9,010	7,892	7,021	5,752	4,871
45	61,025	55,683	51,201	47,387	44,102	41,242	36,509	32,749	29,692	27,157	25,021	23,196	21,619	18,479	16,136	14,319	12,871	11,688	10,705	9,163	8,009	7,114	5,813	4,915
50	67,090	60,689	55,404	50,965	47,185	43,926	38,596	34,419	31,058	28,295	25,984	24,021	22,335	18,999	16,531	14,630	13,121	11,894	10,877	9,289	8,105	7,189	5,864	4,951
55	73,029	65,508	59,392	54,321	50,047	46,397	40,490	35,918	32,273	29,300	26,829	24,742	22,956	19,447	16,869	14,894	13,333	12,068	11,023	9,395	8,186	7,253	5,906	4,981
60	78,845	70,150	63,183	57,474	52,712	48,678	42,217	37,270	33,361	30,194	27,576	25,376	23,501	19,837	17,161	15,121	13,515	12,217	11,147	9,485	8,254	7,306	5,941	5,006
65	84,542	74,625	66,789	60,443	55,199	50,791	43,797	38,496	34,340	30,994	28,242	25,939	23,983	20,179	17,417	15,319	13,673	12,346	11,254	9,562	8,313	7,352	5,972	5,028
70	90,124	78,940	70,226	63,244	57,525	52,754	45,249	39,614	35,226	31,714	28,838	26,441	24,412	20,482	17,642	15,493	13,811	12,459	11,348	9,630	8,364	7,392	5,998	5,046
75	95,594	83,106	73,503	65,890	59,705	54,582	46,588	40,636	36,032	32,366	29,376	26,893	24,796	20,752	17,842	15,647	13,934	12,558	11,430	9,689	8,408	7,427	6,021	5,063
80	100,956	87,128	76,632	68,393	61,754	56,289	47,825	41,574	36,768	32,958	29,864	27,300	25,142	20,994	18,020	15,784	14,042	12,646	11,503	9,741	8,448	7,457	6,041	5,077
85	106,212	91,015	79,623	70,766	63,681	57,887	48,974	42,439	37,443	33,499	30,307	27,671	25,456	21,212	18,181	15,908	14,140	12,725	11,568	9,788	8,483	7,485	6,059	5,089
90	111,366	94,774	82,485	73,017	65,499	59,384	50,041	43,239	38,064	33,996	30,713	28,008	25,742	21,410	18,326	16,019	14,227	12,796	11,627	9,830	8,514	7,509	6,075	5,101
95	116,421	98,410	85,226	75,157	67,215	60,792	51,037	43,980	38,637	34,452	31,085	28,318	26,003	21,590	18,458	16,119	14,307	12,860	11,680	9,868	8,543	7,531	6,089	5,111
100	121,379	101,930	87,853	77,192	68,839	62,117	51,968	44,669	39,168	34,874	31,428	28,602	26,242	21,755	18,578	16,211	14,379	12,919	11,728	9,902	8,568	7,551	6,103	5,120
105	126,243	105,338	90,373	79,131	70,377	63,366	52,839	45,312	39,662	35,264	31,745	28,864	26,463	21,906	18,688	16,295	14,445	12,972	11,772	9,933	8,592	7,569	6,114	5,128
110	131,017	108,641	92,793	80,981	71,836	64,547	53,658	45,912	40,121	35,627	32,038	29,106	26,666	22,046	18,790	16,372	14,505	13,021	11,812	9,962	8,613	7,586	6,125	5,136
115	135,702	111,843	95,119	82,746	73,222	65,664	54,427	46,475	40,550	35,965	32,311	29,331	26,855	22,174	18,883	16,443	14,561	13,066	11,849	9,988	8,633	7,601	6,135	5,143
120	140,300	114,948	97,356	84,434	74,540	66,722	55,152	47,002	40,951	36,280	32,565	29,541	27,030	22,294	18,970	16,508	14,612	13,107	11,883	10,012	8,651	7,615	6,144	5,149
125	144,816	117,961	99,509	86,049	75,796	67,726	55,837	47,498	41,327	36,575	32,803	29,736	27,194	22,405	19,050	16,569	14,660	13,145	11,914	10,035	8,667	7,628	6,153	5,155
130	149,249	120,887	101,582	87,595	76,993	68,680	56,484	47,966	41,680	36,851	33,025	29,918	27,346	22,508	19,125	16,626	14,704	13,181	11,943	10,055	8,683	7,640	6,160	5,161
135	153,603	123,727	103,581	89,077	78,135	69,588	57,096	48,407	42,013	37,111	33,233	30,089	27,489	22,605	19,194	16,678	14,745	13,214	11,971	10,075	8,697	7,651	6,168	5,166
140	157,880	126,488	105,508	90,498	79,227	70,453	57,677	48,823	42,326	37,355	33,429	30,250	27,623	22,695	19,260	16,727	14,784	13,245	11,996	10,093	8,711	7,662	6,174	5,171
145	162,082	129,170	107,369	91,863	80,271	71,277	58,228	49,218	42,623	37,586	33,614	30,401	27,749	22,780	19,321	16,773	14,820	13,274	12,020	10,109	8,723	7,671	6,181	5,175
150	166,211	131,779	109,165	93,175	81,271	72,064	58,753	49,592	42,903	37,804	33,788	30,543	27,867	22,860	19,378	16,817	14,853	13,301	12,042	10,125	8,735	7,680	6,186	5,179
155	170,268	134,317	110,901	94,437	82,229	72,817	59,252	49,947	43,168	38,010	33,952	30,678	27,979	22,935	19,432	16,857	14,885	13,326	12,063	10,140	8,746	7,689	6,192	5,183
160	174,256	136,786	112,579	95,651	83,148	73,536	59,728	50,285	43,420	38,205	34,108	30,805	28,085	23,006	19,483	16,896	14,915	13,350	12,082	10,154	8,756	7,697	6,197	5,187
165	178,177	139,190	114,202	96,820	84,031	74,226	60,181	50,606	43,660	38,390	34,255	30,925	28,185	23,073	19,531	16,932	14,943	13,372	12,101	10,167	8,766	7,704	6,202	5,190
170	182,031	141,531	115,773	97,947	84,878	74,886	60,615	50,912	43,887	38,566	34,395	31,039	28,279	23,137	19,576	16,966	14,970	13,394	12,118	10,179	8,775	7,711	6,206	5,193
175	185,821	143,812	117,295	99,034	85,693	75,520	61,029	51,204	44,104	38,733	34,528	31,147	28,369	23,197	19,620	16,998	14,995	13,414	12,134	10,191	8,783	7,718	6,211	5,196
180	189,548	146,034	118,769	100,083	86,477	76,128	61,426	51,483	44,311	38,893	34,655	31,250	28,455	23,254	19,660	17,029	15,019	13,433	12,150	10,202	8,792	7,724	6,215	5,199
185	193,214	148,201	120,198	101,096	87,232	76,713	61,806	51,750	44,508	39,045	34,776	31,348	28,536	23,308	19,699	17,058	15,041	13,451	12,165	10,212	8,799	7,730	6,219	5,202
190	196,820	150,313	121,584	102,074	87,960	77,275	62,170	52,005	44,697	39,190	34,891	31,442	28,613	23,360	19,736	17,086	15,063	13,468	12,179	10,222	8,807	7,736	6,222	5,204
195	200,368	152,374	122,928	103,020	88,662	77,816	62,520	52,250	44,878	39,328	35,001	31,531	28,687	23,409	19,771	17,112	15,083	13,485	12,192	10,231	8,814	7,741	6,226	5,207
200	203,860	154,384	124,234	103,936	89,339	78,337	62,856	52,484	45,050	39,461	35,106	31,616	28,758	23,456	19,805	17,137	15,103	13,500	12,205	10,240	8,820	7,746	6,229	5,209

600 MCM | 1 Run(s) | 277 Volts

C-Value = 22,965 | Magnetic Duct | Copper

Calculations:

1. $f = \dfrac{2 \times \text{Length} \times \text{Isca} \times 1.5}{\text{Runs} \times \text{C-Value} \times \text{E L-N}}$

2. $M = \dfrac{1}{1 + f}$

3. Isc = Isca x 1.5 x M

4. Add Motor Contribution = Motor FLA x 4

* All results are given in symmetrical amperes

Single-Phase L-N Bolted Fault-Current* Table

One-Way Distance in Feet

Available Isc in Thousands (Isca)	5	10	15	20	25	30	40	50	60	70	80	90	100	125	150	175	200	225	250	300	350	400	500	600
3	4,470	4,440	4,411	4,382	4,353	4,325	4,270	4,216	4,164	4,112	4,062	4,014	3,966	3,852	3,744	3,642	3,545	3,454	3,367	3,205	3,059	2,925	2,690	2,489
5	7,417	7,335	7,256	7,178	7,102	7,027	6,882	6,744	6,610	6,482	6,359	6,240	6,126	5,857	5,612	5,385	5,177	4,984	4,805	4,483	4,201	3,953	3,535	3,197
7	10,338	10,180	10,028	9,879	9,736	9,596	9,328	9,075	8,835	8,607	8,391	8,186	7,990	7,540	7,137	6,776	6,449	6,152	5,881	5,406	5,002	4,653	4,085	3,640
10	14,671	14,356	14,054	13,765	13,487	13,220	12,717	12,251	11,818	11,415	11,038	10,685	10,354	9,610	8,965	8,402	7,905	7,464	7,069	6,393	5,835	5,367	4,624	4,062
15	21,767	21,081	20,437	19,830	19,259	18,720	17,727	16,835	16,027	15,294	14,625	14,012	13,448	12,219	11,196	10,331	9,590	8,948	8,387	7,452	6,705	6,094	5,154	4,466
20	28,712	27,529	26,441	25,435	24,503	23,636	22,075	20,708	19,500	18,425	17,463	16,596	15,811	14,139	12,787	11,671	10,734	9,936	9,249	8,125	7,245	6,536	5,467	4,699
25	35,508	33,718	32,099	30,628	29,287	28,058	25,885	24,025	22,414	21,005	19,763	18,660	17,674	15,610	13,978	12,655	11,561	10,641	9,857	8,590	7,612	6,834	5,674	4,851
30	42,162	39,661	37,440	35,454	33,669	32,055	29,250	26,896	24,894	23,168	21,667	20,348	19,180	16,774	14,904	13,410	12,187	11,169	10,308	8,931	7,879	7,048	5,821	4,958
35	48,678	45,374	42,490	39,951	37,698	35,686	32,244	29,407	27,030	25,008	23,267	21,753	20,424	17,718	15,645	14,006	12,678	11,580	10,657	9,192	8,081	7,209	5,930	5,037
40	55,059	50,870	47,273	44,151	41,416	39,000	34,925	31,622	28,889	26,591	24,631	22,941	21,468	18,498	16,250	14,489	13,073	11,908	10,935	9,398	8,239	7,335	6,015	5,098
45	61,310	56,160	51,808	48,082	44,856	42,036	37,340	33,588	30,522	27,968	25,809	23,959	22,357	19,154	16,754	14,889	13,397	12,177	11,161	9,564	8,367	7,436	6,083	5,147
50	67,435	61,256	56,115	51,770	48,049	44,827	39,527	35,347	31,967	29,177	26,835	24,841	23,122	19,713	17,181	15,224	13,668	12,401	11,348	9,701	8,472	7,519	6,138	5,186
55	73,438	66,170	60,210	55,236	51,021	47,403	41,516	36,930	33,256	30,247	27,737	25,612	23,789	20,196	17,546	15,511	13,898	12,590	11,506	9,817	8,560	7,588	6,184	5,219
60	79,322	70,909	64,109	58,500	53,793	49,787	43,333	38,361	34,412	31,200	28,536	26,292	24,375	20,617	17,863	15,757	14,096	12,752	11,642	9,915	8,635	7,647	6,223	5,247
65	85,091	75,484	67,826	61,579	56,385	52,000	45,000	39,661	35,454	32,055	29,250	26,896	24,893	20,986	18,139	15,973	14,268	12,892	11,759	10,000	8,699	7,697	6,257	5,270
70	90,748	79,902	71,372	64,488	58,815	54,059	46,534	40,848	36,400	32,826	29,890	27,437	25,356	21,314	18,384	16,162	14,419	13,015	11,861	10,074	8,755	7,741	6,285	5,291
75	96,296	84,173	74,760	67,241	61,096	55,981	47,951	41,935	37,261	33,524	30,469	27,923	25,771	21,607	18,601	16,329	14,552	13,124	11,951	10,139	8,804	7,779	6,311	5,308
80	101,739	88,302	78,000	69,851	63,243	57,778	49,263	42,936	38,049	34,160	30,993	28,363	26,145	21,869	18,795	16,479	14,671	13,220	12,031	10,196	8,847	7,813	6,333	5,324
85	107,079	92,297	81,101	72,327	65,266	59,462	50,482	43,859	38,772	34,742	31,471	28,763	26,485	22,106	18,970	16,613	14,777	13,306	12,102	10,247	8,885	7,843	6,353	5,338
90	112,320	96,164	84,072	74,681	67,177	61,043	51,617	44,713	39,438	35,276	31,909	29,128	26,794	22,321	19,128	16,734	14,873	13,384	12,166	10,293	8,920	7,870	6,370	5,351
95	117,464	99,910	86,921	76,920	68,983	62,531	52,678	45,506	40,054	35,768	32,311	29,463	27,077	22,517	19,272	16,844	14,960	13,454	12,224	10,335	8,951	7,894	6,386	5,362
100	122,513	103,540	89,655	79,054	70,695	63,934	53,670	46,245	40,625	36,223	32,681	29,771	27,336	22,696	19,403	16,944	15,038	13,518	12,277	10,372	8,979	7,916	6,400	5,372
105	127,471	107,059	92,281	81,089	72,318	65,259	54,600	46,934	41,156	36,644	33,024	30,055	27,576	22,861	19,523	17,036	15,111	13,576	12,325	10,407	9,005	7,936	6,413	5,381
110	132,339	110,472	94,806	83,032	73,859	66,511	55,474	47,578	41,650	37,036	33,342	30,318	27,797	23,013	19,634	17,120	15,177	13,630	12,369	10,438	9,028	7,954	6,425	5,389
115	137,121	113,784	97,236	84,889	75,325	67,698	56,297	48,182	42,112	37,401	33,637	30,562	28,002	23,153	19,736	17,138	15,238	13,679	12,409	10,467	9,050	7,971	6,436	5,397
120	141,818	117,000	99,574	86,666	76,721	68,823	57,073	48,750	42,545	37,742	33,913	30,789	28,193	23,283	19,830	17,239	15,294	13,724	12,447	10,493	9,070	7,986	6,446	5,404
125	146,433	120,123	101,827	88,368	78,052	69,892	57,806	49,284	42,951	38,061	34,170	31,001	28,370	23,405	19,918	17,336	15,346	13,766	12,481	10,518	9,088	8,000	6,455	5,411
130	150,968	123,158	104,000	90,000	79,322	70,909	58,500	49,787	43,333	38,360	34,412	31,200	28,536	23,517	20,000	17,338	15,395	13,805	12,513	10,540	9,105	8,014	6,464	5,417
135	155,424	126,107	106,095	91,565	80,535	71,877	59,157	50,262	43,693	38,642	34,638	31,386	28,692	23,623	20,076	17,455	15,440	13,842	12,543	10,562	9,121	8,026	6,472	5,422
140	159,805	128,976	108,118	93,068	81,695	72,800	59,781	50,712	44,032	38,907	34,851	31,561	28,838	23,722	20,147	17,509	15,482	13,875	12,571	10,581	9,135	8,037	6,479	5,427
145	164,111	131,767	110,073	94,512	82,806	73,680	60,373	51,138	44,353	39,157	35,051	31,725	28,975	23,815	20,214	17,560	15,521	13,907	12,597	10,600	9,149	8,048	6,486	5,432
150	168,345	134,482	111,961	95,901	83,871	74,522	60,937	51,542	44,656	39,394	35,241	31,880	29,104	23,902	20,277	17,607	15,558	13,937	12,621	10,617	9,162	8,058	6,493	5,437
155	172,509	137,126	113,788	97,238	84,891	75,327	61,474	51,925	44,944	39,617	35,420	32,026	29,226	23,984	20,336	17,652	15,593	13,965	12,644	10,633	9,174	8,067	6,499	5,441
160	176,603	139,701	115,555	98,526	85,871	76,097	61,986	52,290	45,217	39,830	35,589	32,165	29,342	24,062	20,392	17,694	15,626	13,991	12,666	10,648	9,185	8,076	6,504	5,445
165	180,631	142,210	117,266	99,767	86,812	76,835	62,475	52,638	45,477	40,031	35,750	32,296	29,451	24,135	20,445	17,733	15,657	14,016	12,686	10,663	9,196	8,084	6,510	5,449
170	184,594	144,654	118,923	100,964	87,717	77,544	62,943	52,969	45,724	40,222	35,902	32,420	29,554	24,204	20,494	17,771	15,686	14,039	12,705	0,676	9,206	8,092	6,515	5,452
175	188,492	147,037	120,529	102,119	88,588	78,223	63,390	53,285	45,959	40,404	36,047	32,539	29,652	24,270	20,542	17,806	15,714	14,061	12,723	0,689	9,216	8,099	6,520	5,456
180	192,328	149,361	122,087	103,235	89,426	78,876	63,818	53,588	46,184	40,578	36,185	32,651	29,746	24,333	20,586	17,840	15,740	14,082	12,740	0,701	9,225	8,106	6,524	5,459
185	196,104	151,628	123,597	104,313	90,234	79,504	64,228	53,877	46,398	40,743	36,317	32,758	29,834	24,392	20,629	17,872	15,765	14,102	12,757	0,713	9,233	8,113	6,528	5,462
190	199,820	153,840	125,063	105,355	91,013	80,108	64,622	54,153	46,604	40,901	36,442	32,860	29,919	24,449	20,669	17,902	15,788	14,121	12,772	0,724	9,241	8,119	6,532	5,465
195	203,478	156,000	126,486	106,363	91,764	80,689	65,000	54,418	46,800	41,052	36,562	32,958	30,000	24,502	20,708	17,931	15,811	14,139	12,787	0,734	9,249	8,125	6,536	5,467
200	207,079	158,108	127,868	107,339	92,490	81,250	65,363	54,673	46,988	41,197	36,677	33,051	30,077	24,554	20,745	17,958	15,832	14,156	12,801	0,744	9,256	8,131	6,540	5,470

750 MCM | 1 Run(s) | 277 Volts | C-Value = 24,136 | Magnetic Duct | Copper

Calculations:

1. $f = \dfrac{2 \times \text{Length} \times \text{Isca} \times 1.5}{\text{Runs} \times \text{C-Value} \times \text{E L-N}}$

2. $M = \dfrac{1}{1 + f}$

3. $\text{Isc} = \text{Isca} \times 1.5 \times M$

4. Add Motor Contribution = Motor FLA x 4

* All results are given in symmetrical amperes

374

Single-Phase L-N Bolted Fault-Current* Table

One-Way Distance in Feet

Available Isc in Thousands (Isca)	5	10	15	20	25	30	40	50	60	70	80	90	100	125	150	175	200	225	250	300	350	400	500	600
3	4,471	4,443	4,415	4,387	4,360	4,333	4,280	4,228	4,178	4,129	4,080	4,033	3,987	3,877	3,773	3,674	3,580	3,491	3,406	3,248	3,104	2,972	2,739	2,541
5	7,421	7,343	7,267	7,192	7,119	7,047	6,908	6,774	6,646	6,522	6,403	6,288	6,177	5,916	5,676	5,455	5,250	5,061	4,884	4,566	4,286	4,039	3,621	3,282
7	10,345	10,194	10,048	9,906	9,768	9,633	9,375	9,131	8,899	8,678	8,468	8,268	8,077	7,637	7,242	6,886	6,563	6,269	6,001	5,527	5,123	4,773	4,201	3,751
10	14,685	14,384	14,094	13,816	13,549	13,292	12,805	12,354	11,933	11,539	11,171	10,826	10,501	9,768	9,131	8,572	8,078	7,637	7,242	6,564	6,001	5,527	4,774	4,201
15	21,800	21,141	20,522	19,937	19,385	18,863	17,899	17,028	16,238	15,519	14,860	14,255	13,697	12,477	11,456	10,590	9,845	9,199	8,632	7,684	6,924	6,301	5,340	4,633
20	28,767	27,632	26,583	25,611	24,707	23,865	22,342	21,002	19,813	18,752	17,799	16,938	16,156	14,485	13,127	12,002	11,055	10,246	9,547	8,402	7,502	6,776	5,677	4,885
25	35,594	33,872	32,309	30,884	29,579	28,380	26,252	24,421	22,829	21,431	20,195	19,094	18,106	16,033	14,386	13,046	11,934	10,997	10,196	8,900	7,897	7,096	5,900	5,049
30	42,283	39,875	37,726	35,798	34,056	32,477	29,720	27,394	25,406	23,687	22,186	20,864	19,691	17,263	15,369	13,849	12,603	11,562	10,680	9,267	8,184	7,327	6,059	5,165
35	48,838	45,654	42,859	40,387	38,185	36,210	32,816	30,004	27,635	25,614	23,867	22,344	21,004	18,264	16,157	14,486	13,128	12,003	11,055	9,548	8,402	7,502	6,178	5,251
40	55,264	51,222	47,730	44,684	42,004	39,626	35,597	32,312	29,582	27,277	25,305	23,600	22,109	19,095	16,803	15,003	13,551	12,356	11,354	9,770	8,574	7,638	6,270	5,318
45	61,565	56,589	52,358	48,715	45,546	42,765	38,110	34,368	31,296	28,728	26,550	24,678	23,053	19,795	17,343	15,432	13,900	12,645	11,598	9,950	8,712	7,748	6,344	5,371
50	67,744	61,768	56,761	52,505	48,842	45,657	40,390	36,212	32,818	30,005	27,637	25,615	23,868	20,393	17,800	15,793	14,192	12,886	11,801	10,099	8,826	7,838	6,404	5,414
55	73,804	66,767	60,954	56,073	51,916	48,332	42,469	37,875	34,177	31,138	28,594	26,435	24,579	20,909	18,193	16,101	14,441	13,091	11,972	10,224	8,921	7,913	6,454	5,450
60	79,749	71,595	64,954	59,440	54,789	50,813	44,373	39,381	35,399	32,149	29,445	27,161	25,205	21,361	18,534	16,367	14,655	13,266	12,118	10,330	9,002	7,977	6,497	5,480
65	85,583	76,262	68,772	62,621	57,480	53,120	46,122	40,753	36,504	33,057	30,205	27,806	25,760	21,758	18,832	16,600	14,841	13,419	12,245	10,423	9,072	8,032	6,533	5,506
70	91,308	80,775	72,420	65,632	60,007	55,271	47,735	42,007	37,507	33,878	30,889	28,384	26,256	22,110	19,095	16,804	15,004	13,552	12,356	10,503	9,133	8,079	6,564	5,528
75	96,927	85,141	75,911	68,486	62,384	57,281	49,227	43,158	38,422	34,622	31,506	28,905	26,701	22,425	19,330	16,985	15,148	13,669	12,454	10,573	9,186	8,121	6,592	5,547
80	102,443	89,368	79,253	71,195	64,624	59,164	50,611	44,219	39,260	35,301	32,068	29,377	27,103	22,708	19,540	17,147	15,277	13,774	12,541	10,636	9,233	8,158	6,616	5,564
85	107,860	93,463	82,456	73,769	66,738	60,931	51,898	45,198	40,030	35,923	32,580	29,806	27,468	22,964	19,729	17,292	15,392	13,868	12,618	10,691	9,275	8,190	6,637	5,580
90	113,179	97,430	85,529	76,219	68,737	62,592	53,099	46,106	40,741	36,494	33,049	30,198	27,800	23,196	19,900	17,424	15,496	13,952	12,688	10,741	9,313	8,220	6,657	5,593
95	118,403	101,277	88,480	78,553	70,630	64,158	54,222	46,950	41,398	37,021	33,480	30,558	28,105	23,407	20,055	17,543	15,590	14,028	12,751	10,787	9,347	8,246	6,674	5,605
100	123,536	105,009	91,315	80,780	72,425	65,636	55,273	47,737	42,009	37,508	33,878	30,889	28,385	23,601	20,197	17,652	15,676	14,098	12,808	10,828	9,378	8,270	6,690	5,617
105	128,578	108,630	94,041	82,906	74,129	67,032	56,260	48,471	42,576	37,960	34,247	31,195	28,643	23,779	20,328	17,751	15,754	14,161	12,861	10,865	9,406	8,292	6,704	5,627
110	133,533	112,146	96,664	84,938	75,749	68,355	57,189	49,159	43,106	38,381	34,589	31,479	28,882	23,944	20,448	17,842	15,826	14,219	12,908	10,899	9,431	8,312	6,717	5,636
115	138,403	115,561	99,191	86,883	77,292	69,608	58,064	49,804	43,601	38,773	34,907	31,742	29,103	24,096	20,558	17,927	15,892	14,273	12,952	10,931	9,455	8,330	6,729	5,644
120	143,190	118,879	101,626	88,745	78,763	70,799	58,890	50,410	44,065	39,139	35,204	31,987	29,309	24,237	20,661	18,005	15,954	14,322	12,993	10,959	9,476	8,347	6,740	5,652
125	147,896	122,105	103,974	90,531	80,166	71,931	59,671	50,982	44,501	39,483	35,481	32,216	29,502	24,368	20,756	18,077	16,010	14,368	13,031	10,986	9,496	8,362	6,750	5,659
130	152,523	125,242	106,240	92,244	81,506	73,008	60,410	51,520	44,911	39,805	35,741	32,431	29,681	24,490	20,845	18,144	16,063	14,410	13,066	11,011	9,515	8,377	6,759	5,665
135	157,074	128,294	108,427	93,889	82,788	74,034	61,111	52,029	45,298	40,108	35,986	32,632	29,849	24,605	20,928	18,207	16,112	14,450	13,098	11,034	9,532	8,390	6,768	5,672
140	161,549	131,264	110,541	95,469	84,014	75,014	61,777	52,511	45,662	40,394	36,215	32,820	30,007	24,712	21,005	18,266	16,158	14,487	13,129	11,056	9,548	8,402	6,776	5,677
145	165,951	134,156	112,585	96,990	85,190	75,949	62,410	52,968	46,007	40,664	36,432	32,998	30,156	24,813	21,078	18,321	16,201	14,521	13,157	11,076	9,563	8,414	6,784	5,683
150	170,282	136,972	114,562	98,453	86,317	76,844	63,013	53,401	46,334	40,919	36,637	33,166	30,296	24,908	21,146	18,372	16,241	14,554	13,183	11,095	9,577	8,425	6,791	5,688
155	174,543	139,716	116,475	99,863	87,398	77,700	63,587	53,814	46,644	41,160	36,830	33,324	30,428	24,997	21,211	18,421	16,279	14,584	13,208	11,112	9,590	8,435	6,797	5,692
160	178,736	142,389	118,327	101,222	88,437	78,520	64,136	54,206	46,938	41,389	37,013	33,474	30,553	25,081	21,271	18,466	16,315	14,613	13,232	11,129	9,603	8,445	6,804	5,697
165	182,863	144,996	120,122	102,532	89,436	79,306	64,659	54,579	47,218	41,607	37,187	33,616	30,671	25,161	21,329	18,510	16,349	14,640	13,254	11,145	9,614	8,454	6,809	5,701
170	186,925	147,538	121,861	103,797	90,396	80,061	65,160	54,935	47,484	41,813	37,352	33,751	30,784	25,236	21,383	18,550	16,381	14,665	13,275	11,159	9,625	8,462	6,815	5,704
175	190,924	150,018	123,548	105,018	91,321	80,785	65,639	55,276	47,738	42,010	37,509	33,879	30,890	25,308	21,434	18,589	16,411	14,689	13,295	11,173	9,636	8,470	6,820	5,708
180	194,861	152,438	125,185	106,198	92,213	81,482	66,098	55,601	47,981	42,198	37,659	34,001	30,991	25,376	21,483	18,626	16,439	14,712	13,313	11,187	9,646	8,478	6,825	5,712
185	198,737	154,800	126,773	107,339	93,072	82,152	66,538	55,912	48,212	42,377	37,801	34,117	31,088	25,440	21,529	18,660	16,466	14,734	13,331	11,199	9,655	8,485	6,830	5,715
190	202,555	157,107	128,316	108,443	93,900	82,797	66,961	56,210	48,434	42,548	37,937	34,228	31,180	25,502	21,573	18,694	16,492	14,754	13,348	11,211	9,664	8,492	6,834	5,718
195	206,315	159,359	129,815	109,512	94,700	83,418	67,367	56,496	48,646	42,711	38,067	34,334	31,268	25,561	21,615	18,725	16,517	14,774	13,364	11,222	9,672	8,498	6,838	5,721
200	210,018	161,560	131,271	110,546	95,473	84,017	67,757	56,770	48,849	42,868	38,191	34,435	31,351	25,616	21,655	18,755	16,540	14,793	13,379	11,233	9,680	8,504	6,842	5,724

1000 MCM	1 Run(s)	277 Volts	C-Value = 25,278	Magnetic Duct	Copper

Calculations:

1. $f = \dfrac{2 \times \text{Length} \times \text{Isca} \times 1.5}{\text{Runs} \times \text{C-Value} \times \text{E L-N}}$
2. $M = \dfrac{1}{1 + f}$
3. Isc = Isca x 1.5 x M
4. Add Motor Contribution = Motor FLA x 4

* All results are given in symmetrical amperes

Short-Circuit Tables and Data: 120 Volts Single-Phase L-N

Single-Phase L-N Bolted Fault-Current* Table

One-Way Distance in Feet

Available Isc in Thousands (Isca)

Isca	5	10	15	20	25	30	40	50	60	70	80	90	100	125	150	175	200	225	250	300	350	400	500	600
1	1,135	913	764	656	575	512	420	356	309	273	244	221	202	166	141	122	108	97	88	74	64	56	45	38
3	2,291	1,537	1,156	927	773	663	517	423	358	310	274	245	222	179	150	130	114	101	91	76	66	58	46	39
5	2,877	1,780	1,289	1,010	830	705	541	439	370	319	281	251	226	182	152	131	115	102	92	77	66	58	46	39
10	3,560	2,020	1,410	1,083	879	740	562	453	379	326	286	255	230	184	154	132	116	103	93	77	66	58	47	39
15	3,866	2,115	1,455	1,109	896	752	569	457	382	329	288	256	231	185	155	133	116	103	93	78	66	58	47	39
20	4,039	2,166	1,479	1,123	905	758	572	460	384	330	289	257	232	186	155	133	116	103	93	78	67	58	47	39
25	4,151	2,197	1,494	1,132	911	762	575	461	385	330	289	258	232	186	155	133	116	103	93	78	67	58	47	39
30	4,229	2,219	1,504	1,138	915	765	576	462	386	331	290	258	232	186	155	133	116	103	93	78	67	58	47	39
35	4,287	2,235	1,511	1,142	917	767	577	463	386	331	290	258	232	186	155	133	116	104	93	78	67	58	47	39
40	4,331	2,247	1,517	1,145	919	768	578	463	386	332	290	258	232	186	155	133	116	104	93	78	67	58	47	39
45	4,366	2,256	1,521	1,147	921	769	578	464	387	332	290	258	233	186	155	133	116	104	93	78	67	58	47	39
50	4,394	2,264	1,524	1,149	922	770	579	464	387	332	291	258	233	186	155	133	116	104	93	78	67	58	47	39
55	4,418	2,270	1,527	1,151	923	771	579	464	387	332	291	259	233	186	155	133	117	104	93	78	67	58	47	39
60	4,438	2,275	1,530	1,152	924	771	580	464	387	332	291	259	233	186	155	133	117	104	93	78	67	58	47	39
65	4,455	2,279	1,532	1,153	925	772	580	465	387	332	291	259	233	186	155	133	117	104	93	78	67	58	47	39
70	4,469	2,283	1,533	1,154	925	772	580	465	388	332	291	259	233	186	155	133	117	104	93	78	67	58	47	39
75	4,482	2,287	1,535	1,155	926	773	580	465	388	332	291	259	233	186	155	133	117	104	93	78	67	58	47	39
80	4,493	2,289	1,536	1,156	926	773	581	465	388	333	291	259	233	186	155	133	117	104	93	78	67	58	47	39
85	4,503	2,292	1,537	1,156	927	773	581	465	388	333	291	259	233	186	155	133	117	104	93	78	67	58	47	39
90	4,512	2,294	1,538	1,157	927	774	581	465	388	333	291	259	233	186	155	133	117	104	93	78	67	58	47	39
95	4,520	2,296	1,539	1,158	928	774	581	465	388	333	291	259	233	186	155	133	117	104	93	78	67	58	47	39
100	4,527	2,298	1,540	1,158	928	774	581	465	388	333	291	259	233	186	155	133	117	104	93	78	67	58	47	39
105	4,534	2,300	1,541	1,158	928	774	581	465	388	333	291	259	233	186	155	133	117	104	93	78	67	58	47	39
110	4,540	2,301	1,541	1,159	928	774	581	465	388	333	291	259	233	187	155	133	117	104	93	78	67	58	47	39
115	4,545	2,303	1,542	1,159	929	775	582	466	388	333	291	259	233	187	155	133	117	104	93	78	67	58	47	39
120	4,550	2,304	1,543	1,159	929	775	582	466	388	333	291	259	233	187	155	133	117	104	93	78	67	58	47	39
125	4,555	2,305	1,543	1,160	929	775	582	466	388	333	291	259	233	187	155	133	117	104	93	78	67	58	47	39
130	4,559	2,306	1,544	1,160	929	775	582	466	388	333	291	259	233	187	155	133	117	104	93	78	67	58	47	39
135	4,563	2,307	1,544	1,160	929	775	582	466	388	333	291	259	233	187	155	133	117	104	93	78	67	58	47	39
140	4,566	2,308	1,545	1,161	929	775	582	466	388	333	291	259	233	187	155	133	117	104	93	78	67	58	47	39
145	4,570	2,309	1,545	1,161	930	775	582	466	388	333	291	259	233	187	155	133	117	104	93	78	67	58	47	39
150	4,573	2,310	1,545	1,161	930	775	582	466	388	333	291	259	233	187	155	133	117	104	93	78	67	58	47	39
155	4,576	2,311	1,546	1,161	930	775	582	466	388	333	291	259	233	187	155	133	117	104	93	78	67	58	47	39
160	4,579	2,312	1,546	1,161	930	775	582	466	388	333	291	259	233	187	155	133	117	104	93	78	67	58	47	39
165	4,582	2,312	1,546	1,162	930	776	582	466	388	333	291	259	233	187	156	133	117	104	93	78	67	58	47	39
170	4,584	2,313	1,547	1,162	930	776	582	466	388	333	291	259	233	187	156	133	117	104	93	78	67	58	47	39
175	4,586	2,313	1,547	1,162	930	776	582	466	388	333	291	259	233	187	156	133	117	104	93	78	67	58	47	39
180	4,589	2,314	1,547	1,162	930	776	582	466	388	333	291	259	233	187	156	133	117	104	93	78	67	58	47	39
185	4,591	2,315	1,547	1,162	930	776	582	466	388	333	291	259	233	187	156	133	117	104	93	78	67	58	47	39
190	4,593	2,315	1,548	1,162	931	776	582	466	388	333	291	259	233	187	156	133	117	104	93	78	67	58	47	39
195	4,595	2,316	1,548	1,162	931	776	582	466	388	333	291	259	233	187	156	133	117	104	93	78	67	58	47	39
200	4,596	2,316	1,548	1,162	931	776	582	466	388	333	291	259	233	187	156	133	117	104	93	78	67	58	47	39

# 14 AWG	1 Run(s)	120 Volts	C-Value = 389	Magnetic Duct	Copper

Calculations:

1. $f = \frac{2 \times \text{Length} \times \text{Isca} \times 1.5}{\text{Runs} \times \text{C-Value} \times \text{E L-N}}$

2. $M = \frac{1}{1 + f}$

3. $\text{Isc} = \text{Isca} \times 1.5 \times M$

4. Add Motor Contribution = Motor FLA x 4

* All results are given in symmetrical amperes

Single-Phase L-N Bolted Fault-Current* Table

One-Way Distance in Feet

Available Isc in Thousands (Isca)	5	10	15	20	25	30	40	50	60	70	80	90	100	125	150	175	200	225	250	300	350	400	500	600
1	1,247	1,067	933	829	745	677	572	496	437	391	354	323	297	247	212	185	165	148	135	114	99	87	71	59
3	2,799	2,031	1,594	1,312	1,114	968	768	636	543	473	420	377	342	278	234	202	178	159	143	120	103	91	73	61
5	3,726	2,479	1,857	1,485	1,237	1,060	824	674	570	494	436	390	353	285	239	206	181	161	145	121	104	91	73	61
10	4,957	2,969	2,119	1,648	1,348	1,140	872	706	593	511	449	400	361	290	243	209	183	163	147	122	105	92	74	61
15	5,571	3,179	2,224	1,710	1,389	1,170	889	717	601	517	453	404	364	292	244	210	184	163	147	123	105	92	74	62
20	5,938	3,295	2,280	1,743	1,411	1,185	898	723	605	520	456	406	366	293	245	210	184	164	147	123	105	92	74	62
25	6,183	3,369	2,316	1,764	1,425	1,195	903	726	607	522	457	407	367	294	245	210	184	164	147	123	105	92	74	62
30	6,358	3,421	2,340	1,778	1,434	1,201	907	728	609	523	458	408	367	294	245	211	184	164	148	123	106	92	74	62
35	6,489	3,458	2,357	1,788	1,440	1,206	909	730	610	524	459	408	368	294	246	211	184	164	148	123	106	92	74	62
40	6,591	3,487	2,370	1,796	1,445	1,209	911	731	611	524	459	409	368	295	246	211	185	164	148	123	106	92	74	62
45	6,672	3,510	2,381	1,802	1,449	1,212	913	732	611	525	460	409	368	295	246	211	185	164	148	123	106	92	74	62
50	6,739	3,528	2,389	1,806	1,452	1,214	914	733	612	525	460	409	368	295	246	211	185	164	148	123	106	92	74	62
55	6,794	3,543	2,396	1,810	1,455	1,216	915	734	612	525	460	409	369	295	246	211	185	164	148	123	106	92	74	62
60	6,841	3,556	2,402	1,814	1,457	1,217	916	734	613	526	460	409	369	295	246	211	185	164	148	123	106	92	74	62
65	6,881	3,567	2,407	1,817	1,459	1,219	917	735	613	526	461	410	369	295	246	211	185	164	148	123	106	92	74	62
70	6,916	3,576	2,411	1,819	1,460	1,220	917	735	613	526	461	410	369	295	246	211	185	164	148	123	106	92	74	62
75	6,947	3,584	2,415	1,821	1,462	1,221	918	736	614	526	461	410	369	295	246	211	185	164	148	123	106	92	74	62
80	6,974	3,591	2,418	1,823	1,463	1,221	918	736	614	527	461	410	369	295	246	211	185	164	148	123	106	92	74	62
85	6,998	3,598	2,421	1,825	1,464	1,222	919	736	614	527	461	410	369	295	246	211	185	164	148	123	106	92	74	62
90	7,019	3,603	2,424	1,826	1,465	1,223	919	736	614	527	461	410	369	296	246	211	185	164	148	123	106	92	74	62
95	7,038	3,608	2,426	1,827	1,466	1,223	920	737	614	527	461	410	369	296	246	211	185	164	148	123	106	92	74	62
100	7,056	3,613	2,428	1,828	1,466	1,224	920	737	614	527	461	410	369	296	246	211	185	164	148	123	106	92	74	62
105	7,072	3,617	2,430	1,829	1,467	1,224	920	737	615	527	461	410	369	296	246	211	185	164	148	123	106	92	74	62
110	7,086	3,621	2,432	1,830	1,468	1,225	920	737	615	527	461	410	369	296	246	211	185	164	148	123	106	92	74	62
115	7,099	3,624	2,433	1,831	1,468	1,225	921	737	615	527	462	410	369	296	246	211	185	164	148	123	106	93	74	62
120	7,111	3,627	2,435	1,832	1,469	1,226	921	737	615	527	462	410	369	296	246	211	185	164	148	123	106	93	74	62
125	7,123	3,630	2,436	1,833	1,469	1,226	921	737	615	527	462	410	369	296	246	211	185	164	148	123	106	93	74	62
130	7,133	3,633	2,437	1,834	1,470	1,226	921	738	615	527	462	410	369	296	246	211	185	164	148	123	106	93	74	62
135	7,143	3,636	2,438	1,834	1,470	1,227	921	738	615	527	462	410	370	296	246	211	185	164	148	123	106	93	74	62
140	7,152	3,638	2,439	1,835	1,470	1,227	921	738	615	528	462	411	370	296	247	211	185	164	148	123	106	93	74	62
145	7,160	3,640	2,440	1,835	1,471	1,227	922	738	615	528	462	411	370	296	247	211	185	164	148	123	106	93	74	62
150	7,168	3,642	2,441	1,836	1,471	1,227	922	738	615	528	462	411	370	296	247	211	185	164	148	123	106	93	74	62
155	7,175	3,644	2,442	1,836	1,471	1,227	922	738	615	528	462	411	370	296	247	211	185	164	148	123	106	93	74	62
160	7,182	3,646	2,443	1,837	1,472	1,228	922	738	615	528	462	411	370	296	247	211	185	164	148	123	106	93	74	62
165	7,189	3,647	2,444	1,837	1,472	1,228	922	738	615	528	462	411	370	296	247	211	185	164	148	123	106	93	74	62
170	7,195	3,649	2,444	1,838	1,472	1,228	922	738	616	528	462	411	370	296	247	211	185	164	148	123	106	93	74	62
175	7,201	3,651	2,445	1,838	1,472	1,228	922	738	616	528	462	411	370	296	247	211	185	164	148	123	106	93	74	62
180	7,206	3,652	2,446	1,838	1,473	1,228	922	738	616	528	462	411	370	296	247	211	185	164	148	123	106	93	74	62
185	7,212	3,653	2,446	1,839	1,473	1,229	922	738	616	528	462	411	370	296	247	211	185	164	148	123	106	93	74	62
190	7,217	3,655	2,447	1,839	1,473	1,229	923	738	616	528	462	411	370	296	247	211	185	164	148	123	106	93	74	62
195	7,221	3,656	2,447	1,839	1,473	1,229	923	739	616	528	462	411	370	296	247	211	185	164	148	123	106	93	74	62
200	7,226	3,657	2,448	1,840	1,474	1,229	923	739	616	528	462	411	370	296	247	211	185	164	148	123	106	93	74	62

# 12 AWG	1 Run(s)	120 Volts	C-Value = 617	Magnetic Duct	Copper

Calculations:

1. $f = \dfrac{2 \times \text{Length} \times \text{Isca} \times 1.5}{\text{Runs} \times \text{C-Value} \times \text{E L-N}}$

2. $M = \dfrac{1}{1 + f}$

3. $\text{Isc} = \text{Isca} \times 1.5 \times M$

4. Add Motor Contribution = Motor FLA x 4

* All results are given in symmetrical amperes

Single-Phase L-N Bolted Fault-Current* Table

One-Way Distance in Feet

Available Isc in Thousands (Isca)	5	10	15	20	25	30	40	50	60	70	80	90	100	125	150	175	200	225	250	300	350	400	500	600
1	1,330	1,195	1,085	994	916	850	743	660	593	539	494	455	423	358	311	275	246	223	203	174	151	134	109	92
3	3,256	2,550	2,096	1,779	1,546	1,366	1,109	933	805	708	632	571	521	426	361	313	276	247	224	188	162	142	115	96
5	4,581	3,298	2,576	2,114	1,792	1,555	1,230	1,017	868	756	670	602	546	443	373	322	283	253	228	191	164	144	116	97
10	6,596	4,227	3,110	2,460	2,035	1,735	1,340	1,092	921	796	701	627	566	457	382	329	289	257	232	194	166	146	117	97
15	7,728	4,666	3,341	2,603	2,131	1,805	1,381	1,119	940	811	712	636	574	461	386	331	291	259	233	195	167	146	117	98
20	8,454	4,921	3,470	2,680	2,183	1,842	1,403	1,133	950	818	718	640	577	464	387	333	291	259	234	195	167	146	117	98
25	8,959	5,087	3,552	2,729	2,215	1,864	1,416	1,141	956	822	722	643	580	465	388	333	292	260	234	195	167	147	117	98
30	9,331	5,205	3,609	2,762	2,237	1,880	1,425	1,147	960	825	724	645	581	466	389	334	292	260	234	195	168	147	117	98
35	9,616	5,293	3,651	2,787	2,253	1,891	1,431	1,151	963	828	726	646	582	467	389	334	293	260	234	195	168	147	117	98
40	9,841	5,360	3,683	2,805	2,266	1,900	1,436	1,155	965	829	727	647	583	467	390	334	293	260	235	196	168	147	117	98
45	10,024	5,414	3,708	2,820	2,275	1,907	1,440	1,157	967	831	728	648	584	468	390	335	293	261	235	196	168	147	118	98
50	10,175	5,458	3,729	2,832	2,283	1,912	1,443	1,159	968	832	729	648	584	468	390	335	293	261	235	196	168	147	118	98
55	10,302	5,494	3,746	2,842	2,289	1,916	1,446	1,161	969	832	729	649	584	468	391	335	293	261	235	196	168	147	118	98
60	10,410	5,525	3,760	2,850	2,294	1,920	1,448	1,162	970	833	730	649	585	468	391	335	293	261	235	196	168	147	118	98
65	10,504	5,551	3,772	2,857	2,299	1,923	1,450	1,163	971	834	730	650	585	469	391	335	293	261	235	196	168	147	118	98
70	10,585	5,574	3,783	2,863	2,303	1,926	1,451	1,164	972	834	731	650	585	469	391	335	293	261	235	196	168	147	118	98
75	10,657	5,593	3,792	2,868	2,306	1,928	1,453	1,165	973	835	731	650	586	469	391	335	294	261	235	196	168	147	118	98
80	10,720	5,611	3,800	2,873	2,309	1,930	1,454	1,166	973	835	731	650	586	469	391	335	294	261	235	196	168	147	118	98
85	10,777	5,626	3,807	2,877	2,312	1,932	1,455	1,166	974	835	732	651	586	469	391	335	294	261	235	196	168	147	118	98
90	10,828	5,640	3,813	2,880	2,314	1,934	1,456	1,167	974	836	732	651	586	469	391	336	294	261	235	196	168	147	118	98
95	10,874	5,653	3,819	2,883	2,316	1,935	1,456	1,168	974	836	732	651	586	469	391	336	294	261	235	196	168	147	118	98
100	10,915	5,664	3,824	2,886	2,318	1,937	1,457	1,168	975	836	732	651	586	469	391	336	294	261	235	196	168	147	118	98
105	10,953	5,674	3,829	2,889	2,320	1,938	1,458	1,168	975	836	732	651	586	469	391	336	294	261	235	196	168	147	118	98
110	10,988	5,683	3,833	2,891	2,321	1,939	1,458	1,169	975	837	732	651	587	470	391	336	294	261	235	196	168	147	118	98
115	11,020	5,692	3,837	2,894	2,323	1,940	1,459	1,169	975	837	733	652	587	470	392	336	294	261	235	196	168	147	118	98
120	11,049	5,700	3,840	2,896	2,324	1,941	1,460	1,170	976	837	733	652	587	470	392	336	294	261	235	196	168	147	118	98
125	11,077	5,707	3,844	2,898	2,325	1,942	1,460	1,170	976	837	733	652	587	470	392	336	294	261	235	196	168	147	118	98
130	11,102	5,714	3,847	2,899	2,326	1,942	1,460	1,170	976	837	733	652	587	470	392	336	294	261	235	196	168	147	118	98
135	11,125	5,720	3,849	2,901	2,327	1,943	1,461	1,170	976	837	733	652	587	470	392	336	294	261	235	196	168	147	118	98
140	11,147	5,726	3,852	2,902	2,328	1,944	1,461	1,171	976	838	733	652	587	470	392	336	294	261	235	196	168	147	118	98
145	11,168	5,731	3,854	2,904	2,329	1,944	1,462	1,171	977	838	733	652	587	470	392	336	294	261	235	196	168	147	118	98
150	11,187	5,736	3,857	2,905	2,330	1,945	1,462	1,171	977	838	733	652	587	470	392	336	294	261	235	196	168	147	118	98
155	11,205	5,741	3,859	2,906	2,331	1,946	1,462	1,171	977	838	733	652	587	470	392	336	294	261	235	196	168	147	118	98
160	11,222	5,745	3,861	2,907	2,332	1,946	1,463	1,171	977	838	734	652	587	470	392	336	294	261	235	196	168	147	118	98
165	11,238	5,749	3,863	2,908	2,332	1,947	1,463	1,172	977	838	734	652	587	470	392	336	294	261	235	196	168	147	118	98
170	11,253	5,753	3,865	2,909	2,333	1,947	1,463	1,172	977	838	734	652	587	470	392	336	294	261	235	196	168	147	118	98
175	11,267	5,757	3,866	2,910	2,333	1,947	1,463	1,172	977	838	734	652	587	470	392	336	294	261	235	196	168	147	118	98
180	11,280	5,760	3,868	2,911	2,334	1,948	1,464	1,172	977	838	734	652	587	470	392	336	294	261	235	196	168	147	118	98
185	11,293	5,764	3,869	2,912	2,335	1,948	1,464	1,172	978	838	734	652	587	470	392	336	294	261	235	196	168	147	118	98
190	11,305	5,767	3,871	2,913	2,335	1,949	1,464	1,172	978	838	734	653	587	470	392	336	294	261	235	196	168	147	118	98
195	11,317	5,770	3,872	2,914	2,336	1,949	1,464	1,172	978	838	734	653	587	470	392	336	294	261	235	196	168	147	118	98
200	11,328	5,773	3,873	2,914	2,336	1,949	1,464	1,173	978	839	734	653	587	470	392	336	294	261	235	196	168	147	118	98

# 10 AWG	1 Run(s)	120 Volts	C-Value = 981	Magnetic Duct	Copper

Calculations:

1. $f = \dfrac{2 \times \text{Length} \times \text{Isca} \times 1.5}{\text{Runs} \times \text{C-Value} \times \text{E L-N}}$

2. $M = \dfrac{1}{1 + f}$

3. $\text{Isc} = \text{Isca} \times 1.5 \times M$

4. Add Motor Contribution = Motor FLA x 4

* All results are given in symmetrical amperes

Single-Phase L-N Bolted Fault-Current* Table

One-Way Distance in Feet

Available Isc in Thousands (Isca)	5	10	15	20	25	30	40	50	60	70	80	90	100	125	150	175	200	225	250	300	350	400	500	600
3	3,627	3,037	2,612	2,292	2,042	1,840	1,538	1,320	1,157	1,029	927	843	774	641	547	477	423	380	345	291	252	222	179	150
5	5,352	4,160	3,403	2,878	2,494	2,200	1,781	1,496	1,289	1,133	1,010	912	831	680	575	498	440	393	356	299	258	226	182	153
7	6,722	4,944	3,909	3,233	2,756	2,402	1,911	1,586	1,356	1,184	1,051	945	858	698	588	508	447	399	361	302	260	228	184	153
10	8,320	5,757	4,401	3,562	2,992	2,579	2,021	1,661	1,411	1,226	1,083	971	879	712	598	515	453	404	365	305	262	230	185	154
15	10,208	6,601	4,878	3,868	3,205	2,735	2,116	1,725	1,456	1,260	1,110	992	897	723	606	521	458	408	368	307	264	231	185	155
20	11,513	7,124	5,157	4,042	3,323	2,821	2,167	1,759	1,480	1,278	1,124	1,003	906	729	610	524	460	410	369	308	265	232	186	155
25	12,471	7,479	5,341	4,154	3,398	2,875	2,199	1,780	1,495	1,289	1,132	1,010	911	733	613	526	461	411	370	309	265	232	186	155
30	13,202	7,736	5,471	4,232	3,450	2,912	2,220	1,794	1,505	1,296	1,138	1,015	915	735	614	528	462	411	371	309	265	232	186	155
35	13,780	7,931	5,568	4,289	3,488	2,940	2,236	1,804	1,512	1,301	1,142	1,018	918	737	615	528	463	412	371	310	266	233	186	155
40	14,247	8,083	5,642	4,334	3,518	2,960	2,248	1,812	1,518	1,306	1,145	1,020	920	738	616	529	463	412	371	310	266	233	186	155
45	14,633	8,206	5,702	4,369	3,541	2,977	2,257	1,818	1,522	1,309	1,148	1,022	921	739	617	530	464	413	372	310	266	233	186	155
50	14,958	8,307	5,750	4,397	3,559	2,990	2,265	1,823	1,525	1,311	1,150	1,024	923	740	618	530	464	413	372	310	266	233	186	155
55	15,234	8,392	5,791	4,421	3,575	3,001	2,271	1,827	1,528	1,313	1,151	1,025	924	741	618	530	464	413	372	310	266	233	186	155
60	15,472	8,463	5,825	4,441	3,588	3,010	2,276	1,830	1,531	1,315	1,153	1,026	925	741	619	531	465	413	372	310	266	233	186	155
65	15,679	8,525	5,854	4,457	3,599	3,018	2,281	1,833	1,533	1,317	1,154	1,027	925	742	619	531	465	413	372	310	266	233	186	155
70	15,862	8,579	5,879	4,472	3,608	3,024	2,285	1,836	1,534	1,318	1,155	1,028	926	742	619	531	465	414	372	310	266	233	187	155
75	16,023	8,626	5,901	4,485	3,617	3,030	2,288	1,838	1,536	1,319	1,156	1,029	927	742	619	531	465	414	372	311	266	233	187	155
80	16,167	8,667	5,921	4,496	3,624	3,035	2,291	1,840	1,537	1,320	1,156	1,029	927	743	620	531	465	414	373	311	266	233	187	155
85	16,296	8,704	5,938	4,506	3,630	3,040	2,293	1,841	1,538	1,321	1,157	1,030	927	743	620	532	465	414	373	311	266	233	187	156
90	16,413	8,737	5,953	4,515	3,636	3,044	2,296	1,843	1,539	1,322	1,158	1,030	928	743	620	532	465	414	373	311	266	233	187	156
95	16,518	8,767	5,967	4,523	3,641	3,047	2,298	1,844	1,540	1,322	1,158	1,030	928	743	620	532	466	414	373	311	266	233	187	156
100	16,614	8,794	5,980	4,530	3,646	3,051	2,300	1,845	1,541	1,323	1,159	1,031	928	744	620	532	466	414	373	311	266	233	187	156
105	16,703	8,819	5,991	4,536	3,650	3,054	2,301	1,846	1,542	1,323	1,159	1,031	929	744	620	532	466	414	373	311	266	233	187	156
110	16,783	8,841	6,001	4,542	3,654	3,056	2,303	1,847	1,542	1,324	1,160	1,032	929	744	620	532	466	414	373	311	266	233	187	156
115	16,858	8,862	6,011	4,548	3,658	3,059	2,304	1,848	1,543	1,324	1,160	1,032	929	744	621	532	466	414	373	311	267	233	187	156
120	16,927	8,881	6,020	4,553	3,661	3,061	2,306	1,849	1,544	1,325	1,160	1,032	929	744	621	532	466	414	373	311	267	233	187	156
125	16,991	8,899	6,028	4,557	3,664	3,063	2,307	1,850	1,544	1,325	1,161	1,032	930	744	621	532	466	414	373	311	267	233	187	156
130	17,050	8,915	6,035	4,562	3,667	3,065	2,308	1,851	1,545	1,325	1,161	1,033	930	745	621	532	466	414	373	311	267	233	187	156
135	17,106	8,930	6,042	4,566	3,669	3,067	2,309	1,851	1,545	1,326	1,161	1,033	930	745	621	532	466	414	373	311	267	233	187	156
140	17,157	8,944	6,049	4,569	3,671	3,068	2,310	1,852	1,546	1,326	1,161	1,033	930	745	621	532	466	414	373	311	267	233	187	156
145	17,206	8,957	6,055	4,573	3,674	3,070	2,311	1,852	1,546	1,326	1,162	1,033	930	745	621	533	466	414	373	311	267	233	187	156
150	17,251	8,970	6,060	4,576	3,676	3,071	2,312	1,853	1,546	1,327	1,162	1,033	930	745	621	533	466	414	373	311	267	233	187	156
155	17,294	8,981	6,066	4,579	3,678	3,073	2,312	1,854	1,547	1,327	1,162	1,033	930	745	621	533	466	414	373	311	267	233	187	156
160	17,335	8,992	6,070	4,582	3,680	3,074	2,313	1,854	1,547	1,327	1,162	1,034	931	745	621	533	466	414	373	311	267	233	187	156
165	17,373	9,002	6,075	4,584	3,681	3,075	2,314	1,854	1,547	1,327	1,162	1,034	931	745	621	533	466	415	373	311	267	233	187	156
170	17,408	9,012	6,080	4,587	3,683	3,076	2,314	1,855	1,548	1,328	1,162	1,034	931	745	621	533	466	415	373	311	267	233	187	156
175	17,442	9,021	6,084	4,589	3,684	3,077	2,315	1,855	1,548	1,328	1,163	1,034	931	745	621	533	466	415	373	311	267	233	187	156
180	17,475	9,030	6,088	4,592	3,686	3,078	2,315	1,856	1,548	1,328	1,163	1,034	931	745	621	533	466	415	373	311	267	233	187	156
185	17,505	9,038	6,091	4,594	3,687	3,079	2,316	1,856	1,548	1,328	1,163	1,034	931	745	621	533	466	415	373	311	267	233	187	156
190	17,534	9,045	6,095	4,596	3,688	3,080	2,317	1,856	1,549	1,328	1,163	1,034	931	745	621	533	466	415	373	311	267	233	187	156
195	17,562	9,053	6,098	4,598	3,690	3,081	2,317	1,857	1,549	1,329	1,163	1,034	931	745	621	533	466	415	373	311	267	233	187	156
200	17,589	9,060	6,101	4,599	3,691	3,082	2,317	1,857	1,549	1,329	1,163	1,034	931	746	622	533	466	415	373	311	267	233	187	156

8 AWG | 1 Run(s) | 120 Volts | | C-Value = 1,557 | Magnetic Duct | Copper

Calculations:

1. $f = \frac{2 \times \text{Length} \times \text{Isca} \times 1.5}{\text{Runs} \times \text{C-Value} \times \text{E L-N}}$

2. $M = \frac{1}{1 + f}$

3. Isc = Isca x 1.5 x M

4. Add Motor Contribution = Motor FLA x 4

* All results are given in symmetrical amperes

Single-Phase L-N Bolted Fault-Current* Table

One-Way Distance in Feet

Available Isc in Thousands (Isca)	5	10	15	20	25	30	40	50	60	70	80	90	100	125	150	175	200	225	250	300	350	400	500	600
3	3,897	3,437	3,074	2,780	2,538	2,334	2,012	1,767	1,576	1,422	1,295	1,189	1,099	925	798	702	626	565	515	438	381	337	273	230
5	5,963	4,949	4,230	3,693	3,277	2,945	2,449	2,097	1,832	1,628	1,464	1,330	1,219	1,008	859	748	663	595	540	456	394	347	280	235
7	7,716	6,099	5,042	4,297	3,744	3,318	2,702	2,279	1,970	1,735	1,550	1,401	1,278	1,048	888	770	680	609	551	464	400	352	283	237
10	9,898	7,386	5,891	4,899	4,193	3,665	2,928	2,437	2,088	1,826	1,622	1,459	1,326	1,080	911	788	694	620	560	470	405	355	285	239
15	12,689	8,836	6,778	5,497	4,624	3,990	3,131	2,577	2,189	1,903	1,683	1,508	1,367	1,107	930	802	705	629	567	475	408	358	287	240
20	14,772	9,798	7,330	5,855	4,874	4,175	3,244	2,653	2,244	1,944	1,715	1,534	1,388	1,121	940	809	710	633	571	477	410	359	288	241
25	16,385	10,483	7,707	6,093	5,038	4,295	3,316	2,700	2,278	1,969	1,735	1,550	1,401	1,129	946	813	714	636	573	479	411	360	289	241
30	17,672	10,995	7,980	6,263	5,153	4,378	3,365	2,733	2,301	1,987	1,748	1,561	1,409	1,135	950	816	716	638	575	480	412	361	289	241
35	18,722	11,393	8,187	6,390	5,239	4,440	3,402	2,757	2,318	1,999	1,758	1,568	1,416	1,139	952	818	718	639	576	481	412	361	289	241
40	19,596	11,710	8,350	6,488	5,305	4,487	3,430	2,775	2,331	2,009	1,765	1,574	1,421	1,142	955	820	719	640	576	481	413	362	290	242
45	20,334	11,970	8,481	6,567	5,358	4,525	3,452	2,790	2,341	2,016	1,771	1,579	1,424	1,144	956	821	720	641	577	482	413	362	290	242
50	20,965	12,186	8,589	6,632	5,401	4,555	3,469	2,801	2,349	2,023	1,776	1,583	1,427	1,146	958	822	721	641	578	482	413	362	290	242
55	21,512	12,369	8,680	6,685	5,436	4,581	3,484	2,811	2,356	2,027	1,780	1,586	1,430	1,148	959	823	721	642	578	482	414	362	290	242
60	21,990	12,525	8,756	6,731	5,466	4,602	3,496	2,819	2,361	2,032	1,783	1,588	1,432	1,149	960	824	722	642	578	482	414	362	290	242
65	22,411	12,661	8,822	6,770	5,492	4,620	3,507	2,826	2,366	2,035	1,785	1,590	1,434	1,150	960	824	722	642	579	483	414	362	290	242
70	22,785	12,779	8,880	6,804	5,514	4,636	3,516	2,832	2,370	2,038	1,788	1,592	1,435	1,151	961	825	722	643	579	483	414	362	290	242
75	23,120	12,884	8,930	6,833	5,534	4,650	3,524	2,837	2,374	2,041	1,790	1,594	1,436	1,152	962	825	723	643	579	483	414	363	290	242
80	23,421	12,977	8,975	6,859	5,551	4,662	3,530	2,841	2,377	2,043	1,792	1,595	1,438	1,153	962	826	723	643	579	483	414	363	290	242
85	23,693	13,060	9,014	6,882	5,566	4,672	3,537	2,845	2,380	2,045	1,793	1,596	1,439	1,153	963	826	723	643	579	483	414	363	290	242
90	23,940	13,134	9,050	6,903	5,579	4,682	3,542	2,849	2,382	2,047	1,795	1,598	1,439	1,154	963	826	724	644	580	483	414	363	290	242
95	24,165	13,202	9,082	6,922	5,592	4,690	3,547	2,852	2,384	2,049	1,796	1,599	1,440	1,155	963	827	724	644	580	483	415	363	290	242
100	24,372	13,263	9,111	6,938	5,603	4,698	3,551	2,855	2,386	2,050	1,797	1,599	1,441	1,155	964	827	724	644	580	483	415	363	290	242
105	24,562	13,320	9,137	6,954	5,613	4,705	3,555	2,857	2,388	2,051	1,798	1,600	1,442	1,155	964	827	724	644	580	484	415	363	290	242
110	24,737	13,371	9,161	6,968	5,622	4,712	3,559	2,860	2,390	2,053	1,799	1,601	1,442	1,156	964	827	724	644	580	484	415	363	290	242
115	24,900	13,418	9,184	6,981	5,630	4,717	3,562	2,862	2,391	2,054	1,800	1,602	1,443	1,156	965	827	724	644	580	484	415	363	291	242
120	25,050	13,462	9,204	6,992	5,638	4,723	3,565	2,864	2,393	2,055	1,801	1,602	1,443	1,157	965	828	725	644	580	484	415	363	291	242
125	25,190	13,502	9,223	7,003	5,645	4,728	3,568	2,866	2,394	2,056	1,801	1,603	1,444	1,157	965	828	725	644	580	484	415	363	291	242
130	25,321	13,540	9,240	7,013	5,651	4,732	3,571	2,867	2,395	2,057	1,802	1,603	1,444	1,157	965	828	725	645	580	484	415	363	291	242
135	25,444	13,575	9,257	7,023	5,657	4,737	3,573	2,869	2,396	2,057	1,803	1,604	1,445	1,157	965	828	725	645	580	484	415	363	291	242
140	25,558	13,607	9,272	7,031	5,663	4,741	3,576	2,870	2,397	2,058	1,803	1,604	1,445	1,158	966	828	725	645	580	484	415	363	291	242
145	25,666	13,638	9,286	7,040	5,668	4,744	3,578	2,872	2,398	2,059	1,804	1,605	1,445	1,158	966	828	725	645	580	484	415	363	291	242
150	25,767	13,666	9,299	7,047	5,673	4,748	3,580	2,873	2,399	2,060	1,804	1,605	1,446	1,158	966	828	725	645	580	484	415	363	291	242
155	25,863	13,693	9,312	7,054	5,678	4,751	3,581	2,874	2,400	2,060	1,805	1,606	1,446	1,158	966	828	725	645	581	484	415	363	291	242
160	25,953	13,718	9,323	7,061	5,682	4,754	3,583	2,875	2,401	2,061	1,805	1,606	1,446	1,158	966	829	725	645	581	484	415	363	291	242
165	26,039	13,742	9,334	7,067	5,686	4,757	3,585	2,876	2,401	2,061	1,805	1,606	1,446	1,159	966	829	725	645	581	484	415	363	291	242
170	26,119	13,765	9,345	7,073	5,690	4,759	3,586	2,877	2,402	2,062	1,806	1,606	1,447	1,159	966	829	725	645	581	484	415	363	291	242
175	26,196	13,786	9,354	7,079	5,694	4,762	3,588	2,878	2,403	2,062	1,806	1,607	1,447	1,159	966	829	725	645	581	484	415	363	291	242
180	26,269	13,806	9,364	7,084	5,697	4,764	3,589	2,879	2,403	2,063	1,807	1,607	1,447	1,159	967	829	726	645	581	484	415	363	291	242
185	26,338	13,825	9,372	7,089	5,700	4,767	3,590	2,880	2,404	2,063	1,807	1,607	1,447	1,159	967	829	726	645	581	484	415	363	291	242
190	26,404	13,843	9,381	7,094	5,704	4,769	3,592	2,881	2,405	2,064	1,807	1,608	1,448	1,159	967	829	726	645	581	484	415	363	291	242
195	26,467	13,861	9,389	7,098	5,706	4,771	3,593	2,881	2,405	2,064	1,808	1,608	1,448	1,159	967	829	726	645	581	484	415	363	291	242
200	26,527	13,877	9,396	7,103	5,709	4,773	3,594	2,882	2,406	2,064	1,808	1,608	1,448	1,160	967	829	726	645	581	484	415	363	291	242

# 6 AWG	1 Run(s)	120 Volts	C-Value = 2,425	Magnetic Duct	Copper

Calculations:

1. $f = \dfrac{2 \times \text{Length} \times \text{Isca} \times 1.5}{\text{Runs} \times \text{C-Value} \times \text{E L-N}}$

2. $M = \dfrac{1}{1 + f}$

3. $\text{Isc} = \text{Isca} \times 1.5 \times M$

4. Add Motor Contribution = Motor FLA x 4

* All results are given in symmetrical amperes

Single-Phase L-N Bolted Fault-Current* Table

One-Way Distance in Feet

Available Isc in Thousands (Isca)	5	10	15	20	25	30	40	50	60	70	80	90	100	125	150	175	200	225	250	300	350	400	500	600
3	4,096	3,759	3,473	3,228	3,015	2,828	2,516	2,267	2,062	1,891	1,747	1,622	1,515	1,299	1,138	1,012	911	828	759	651	570	507	415	351
5	6,442	5,646	5,025	4,527	4,118	3,778	3,242	2,839	2,525	2,273	2,068	1,896	1,751	1,469	1,266	1,112	991	894	814	691	600	531	431	362
7	8,537	7,193	6,214	5,470	4,885	4,413	3,698	3,183	2,793	2,489	2,244	2,044	1,876	1,556	1,330	1,16[illegible]	1,030	925	840	710	614	541	438	367
10	11,292	9,053	7,556	6,483	5,677	5,050	4,135	3,501	3,036	2,680	2,398	2,170	1,982	1,629	1,382	1,200	1,061	951	861	724	625	550	443	371
15	15,074	11,333	9,080	7,574	6,497	5,688	4,554	3,797	3,255	2,849	2,533	2,280	2,073	1,690	1,426	1,233	1,087	971	878	736	634	557	448	374
20	18,107	12,966	10,099	8,270	7,002	6,071	4,796	3,964	3,378	2,942	2,606	2,339	2,122	1,722	1,449	1,251	1,100	982	886	742	639	560	450	376
25	20,592	14,193	10,828	8,753	7,345	6,328	4,955	4,071	3,455	3,001	2,653	2,377	2,153	1,742	1,463	1,261	1,108	988	892	746	641	562	451	377
30	22,667	15,149	11,376	9,107	7,593	6,511	5,066	4,146	3,509	3,042	2,684	2,402	2,173	1,756	1,473	1,268	1,114	993	895	749	643	564	452	377
35	24,424	15,914	11,802	9,378	7,781	6,648	5,149	4,202	3,549	3,071	2,707	2,420	2,188	1,765	1,479	1,273	1,117	996	898	750	644	565	453	378
40	25,932	16,541	12,143	9,593	7,928	6,755	5,213	4,244	3,579	3,094	2,725	2,434	2,200	1,773	1,485	1,277	1,120	998	900	752	645	566	453	378
45	27,240	17,063	12,422	9,766	8,046	6,841	5,264	4,278	3,603	3,112	2,739	2,445	2,209	1,779	1,489	1,280	1,123	1,000	901	753	646	566	454	378
50	28,386	17,506	12,655	9,909	8,143	6,911	5,305	4,305	3,622	3,126	2,750	2,454	2,216	1,783	1,492	1,283	1,125	1,001	902	754	647	567	454	379
55	29,398	17,885	12,852	10,030	8,224	6,969	5,340	4,328	3,638	3,138	2,759	2,462	2,222	1,787	1,495	1,285	1,126	1,003	903	754	647	567	454	379
60	30,297	18,214	13,021	10,133	8,293	7,018	5,368	4,347	3,652	3,148	2,767	2,468	2,227	1,791	1,497	1,286	1,127	1,004	904	755	648	567	454	379
65	31,103	18,502	13,168	10,221	8,352	7,061	5,393	4,363	3,663	3,157	2,773	2,473	2,231	1,793	1,499	1,288	1,129	1,004	905	755	648	568	455	379
70	31,828	18,757	13,296	10,298	8,403	7,097	5,415	4,377	3,673	3,164	2,779	2,477	2,235	1,796	1,501	1,289	1,130	1,005	906	756	648	568	455	379
75	32,484	18,983	13,409	10,366	8,448	7,130	5,433	4,389	3,681	3,170	2,784	2,481	2,238	1,798	1,502	1,290	1,130	1,006	906	756	649	568	455	379
80	33,081	19,185	13,510	10,426	8,488	7,158	5,450	4,400	3,689	3,176	2,788	2,485	2,241	1,799	1,503	1,291	1,131	1,006	907	756	649	568	455	379
85	33,627	19,367	13,600	10,480	8,524	7,183	5,464	4,409	3,696	3,181	2,792	2,488	2,243	1,801	1,504	1,292	1,132	1,007	907	757	649	568	455	379
90	34,127	19,532	13,681	10,528	8,556	7,206	5,477	4,418	3,702	3,185	2,795	2,491	2,246	1,802	1,505	1,292	1,132	1,007	907	757	649	568	455	380
95	34,587	19,682	13,755	10,571	8,584	7,226	5,489	4,425	3,707	3,189	2,798	2,493	2,248	1,804	1,506	1,293	1,133	1,008	908	757	649	569	455	380
100	35,012	19,819	13,821	10,610	8,610	7,244	5,500	4,432	3,712	3,193	2,801	2,495	2,249	1,805	1,507	1,294	1,133	1,008	908	757	650	569	455	380
105	35,405	19,944	13,882	10,646	8,634	7,261	5,509	4,438	3,716	3,196	2,804	2,497	2,251	1,806	1,508	1,294	1,134	1,008	908	758	650	569	455	380
110	35,771	20,060	13,938	10,679	8,655	7,276	5,518	4,444	3,720	3,199	2,806	2,499	2,252	1,807	1,508	1,295	1,134	1,009	908	758	650	569	455	380
115	36,111	20,166	13,989	10,709	8,675	7,290	5,526	4,449	3,724	3,202	2,808	2,501	2,254	1,808	1,509	1,295	1,134	1,009	909	758	650	569	456	380
120	36,429	20,265	14,037	10,737	8,693	7,303	5,533	4,454	3,727	3,204	2,810	2,502	2,255	1,809	1,510	1,296	1,135	1,009	909	758	650	569	456	380
125	36,726	20,357	14,081	10,763	8,710	7,315	5,540	4,459	3,730	3,206	2,812	2,503	2,256	1,809	1,510	1,296	1,135	1,009	909	758	650	569	456	380
130	37,005	20,442	14,122	10,786	8,726	7,326	5,547	4,463	3,733	3,209	2,813	2,505	2,257	1,810	1,511	1,296	1,135	1,010	909	758	650	569	456	380
135	37,267	20,522	14,159	10,809	8,740	7,336	5,552	4,466	3,736	3,211	2,815	2,506	2,258	1,811	1,511	1,297	1,135	1,010	909	758	650	569	456	380
140	37,513	20,596	14,195	10,829	8,754	7,346	5,558	4,470	3,738	3,212	2,816	2,507	2,259	1,811	1,511	1,297	1,136	1,010	909	758	650	569	456	380
145	37,746	20,666	14,228	10,848	8,766	7,355	5,563	4,473	3,741	3,214	2,818	2,508	2,260	1,812	1,512	1,297	1,136	1,010	910	759	651	569	456	380
150	37,966	20,732	14,259	10,867	8,778	7,363	5,568	4,476	3,743	3,216	2,819	2,509	2,261	1,812	1,512	1,297	1,136	1,010	910	759	651	569	456	380
155	38,173	20,794	14,288	10,884	8,789	7,371	5,572	4,479	3,745	3,217	2,820	2,510	2,261	1,813	1,512	1,298	1,136	1,011	910	759	651	570	456	380
160	38,370	20,852	14,316	10,899	8,799	7,378	5,576	4,482	3,747	3,219	2,821	2,511	2,262	1,813	1,513	1,298	1,136	1,011	910	759	651	570	456	380
165	38,557	20,907	14,342	10,914	8,809	7,385	5,580	4,484	3,748	3,220	2,822	2,512	2,263	1,813	1,513	1,298	1,137	1,011	910	759	651	570	456	380
170	38,734	20,959	14,366	10,929	8,819	7,391	5,584	4,487	3,750	3,221	2,823	2,512	2,263	1,814	1,513	1,298	1,137	1,011	910	759	651	570	456	380
175	38,903	21,008	14,389	10,942	8,827	7,397	5,587	4,489	3,752	3,222	2,824	2,513	2,264	1,814	1,514	1,298	1,137	1,011	910	759	651	570	456	380
180	39,064	21,055	14,411	10,955	8,835	7,403	5,591	4,491	3,753	3,223	2,825	2,514	2,264	1,815	1,514	1,299	1,137	1,011	910	759	651	570	456	380
185	39,217	21,100	14,432	10,967	8,843	7,409	5,594	4,493	3,755	3,224	2,825	2,514	2,265	1,815	1,514	1,299	1,137	1,011	910	759	651	570	456	380
190	39,364	21,142	14,452	10,978	8,851	7,414	5,597	4,495	3,756	3,225	2,826	2,515	2,265	1,815	1,514	1,299	1,137	1,011	911	759	651	570	456	380
195	39,504	21,182	14,471	10,989	8,858	7,419	5,600	4,497	3,757	3,226	2,827	2,516	2,266	1,816	1,515	1,299	1,137	1,011	911	759	651	570	456	380
200	39,638	21,221	14,489	10,999	8,864	7,424	5,602	4,499	3,758	3,227	2,828	2,516	2,266	1,816	1,515	1,299	1,137	1,012	911	759	651	570	456	380

# 4 AWG	1 Run(s)	120 Volts	C-Value = 3,806	Magnetic Duct	Copper

Calculations:

1. $f = \dfrac{2 \times \text{Length} \times \text{Isca} \times 1.5}{\text{Runs} \times \text{C-Value} \times \text{E L-N}}$

2. $M = \dfrac{1}{1 + f}$

3. $\text{Isc} = \text{Isca} \times 1.5 \times M$

4. Add Motor Contribution = Motor FLA x 4

* All results are given in symmetrical amperes

Single-Phase L-N Bolted Fault-Current* Table

One-Way Distance in Feet

Available Isc in Thousands (Isca)

Isca	5	10	15	20	25	30	40	50	60	70	80	90	100	125	150	175	200	225	250	300	350	400	500	600
3	4,171	3,887	3,640	3,422	3,228	3,056	2,760	2,517	2,313	2,140	1,991	1,861	1,747	1,515	1,338	1,198	1,084	990	911	786	691	616	507	430
5	6,630	5,940	5,381	4,917	4,528	4,195	3,658	3,243	2,912	2,642	2,419	2,230	2,068	1,751	1,519	1,340	1,200	1,086	991	845	736	652	531	448
7	8,870	7,677	6,768	6,051	5,471	4,993	4,250	3,699	3,275	2,938	2,664	2,437	2,245	1,876	1,612	1,412	1,257	1,132	1,030	873	757	669	542	455
10	11,880	9,835	8,390	7,316	6,485	5,824	4,837	4,137	3,613	3,208	2,884	2,619	2,399	1,983	1,690	1,472	1,304	1,170	1,062	895	774	682	550	461
15	16,142	12,585	10,313	8,736	7,577	6,690	5,420	4,556	3,929	3,454	3,081	2,781	2,534	2,074	1,755	1,522	1,343	1,202	1,087	913	787	692	557	466
20	19,669	14,631	11,648	9,675	8,273	7,227	5,767	4,798	4,108	3,592	3,190	2,870	2,608	2,123	1,790	1,548	1,363	1,218	1,100	923	794	697	561	469
25	22,638	16,213	12,628	10,342	8,756	7,593	5,998	4,957	4,224	3,680	3,260	2,926	2,654	2,154	1,812	1,564	1,376	1,228	1,109	928	799	701	563	470
30	25,170	17,471	13,379	10,840	9,111	7,858	6,162	5,069	4,305	3,741	3,308	2,964	2,686	2,174	1,827	1,575	1,384	1,235	1,114	932	801	703	564	471
35	27,356	18,497	13,973	11,226	9,382	8,059	6,285	5,152	4,364	3,786	3,343	2,992	2,709	2,190	1,837	1,583	1,390	1,239	1,118	935	804	704	565	472
40	29,262	19,350	14,453	11,535	9,597	8,216	6,381	5,215	4,410	3,820	3,370	3,014	2,726	2,201	1,845	1,589	1,395	1,243	1,121	937	805	706	566	472
45	30,939	20,069	14,851	11,787	9,770	8,343	6,457	5,266	4,446	3,847	3,391	3,031	2,740	2,210	1,852	1,593	1,398	1,246	1,123	939	806	707	566	473
50	32,425	20,684	15,185	11,996	9,914	8,448	6,519	5,308	4,476	3,869	3,408	3,045	2,751	2,217	1,857	1,597	1,401	1,248	1,125	940	807	707	567	473
55	33,752	21,216	15,470	12,173	10,034	8,535	6,571	5,342	4,500	3,888	3,422	3,056	2,760	2,223	1,861	1,600	1,404	1,250	1,127	941	808	708	567	473
60	34,943	21,680	15,715	12,325	10,137	8,609	6,615	5,371	4,521	3,903	3,434	3,065	2,768	2,228	1,865	1,603	1,406	1,252	1,128	942	809	708	568	473
65	36,019	22,089	15,929	12,456	10,226	8,673	6,653	5,396	4,538	3,916	3,444	3,073	2,775	2,232	1,868	1,605	1,407	1,253	1,129	943	809	709	568	474
70	36,995	22,453	16,117	12,570	10,303	8,729	6,685	5,417	4,554	3,927	3,453	3,080	2,780	2,236	1,870	1,607	1,409	1,254	1,130	943	810	709	568	474
75	37,885	22,778	16,284	12,672	10,371	8,777	6,714	5,436	4,567	3,937	3,460	3,086	2,785	2,239	1,872	1,609	1,410	1,255	1,131	944	810	709	568	474
80	38,699	23,069	16,433	12,761	10,431	8,820	6,739	5,452	4,578	3,946	3,467	3,092	2,790	2,242	1,874	1,610	1,411	1,256	1,132	945	810	710	568	474
85	39,448	23,333	16,566	12,842	10,485	8,859	6,761	5,467	4,589	3,953	3,473	3,096	2,793	2,245	1,876	1,611	1,412	1,257	1,132	945	811	710	569	474
90	40,137	23,573	16,687	12,914	10,533	8,893	6,781	5,480	4,598	3,960	3,478	3,100	2,797	2,247	1,878	1,613	1,413	1,258	1,133	945	811	710	569	474
95	40,775	23,792	16,796	12,979	10,576	8,924	6,799	5,492	4,606	3,966	3,483	3,104	2,800	2,249	1,879	1,614	1,414	1,258	1,133	946	811	710	569	474
100	41,367	23,992	16,895	13,039	10,616	8,952	6,816	5,502	4,614	3,972	3,487	3,108	2,803	2,251	1,880	1,614	1,415	1,259	1,134	946	812	711	569	474
105	41,918	24,176	16,987	13,093	10,651	8,977	6,830	5,512	4,620	3,977	3,491	3,111	2,805	2,252	1,881	1,615	1,415	1,259	1,134	946	812	711	569	475
110	42,431	24,346	17,070	13,143	10,684	9,001	6,844	5,521	4,627	3,982	3,494	3,113	2,807	2,254	1,882	1,616	1,416	1,260	1,135	947	812	711	569	475
115	42,911	24,503	17,147	13,188	10,714	9,022	6,856	5,529	4,632	3,986	3,498	3,116	2,809	2,255	1,883	1,617	1,416	1,260	1,135	947	812	711	569	475
120	43,360	24,649	17,219	13,230	10,742	9,042	6,868	5,536	4,637	3,990	3,501	3,118	2,811	2,256	1,884	1,617	1,417	1,260	1,135	947	812	711	569	475
125	43,782	24,785	17,285	13,269	10,768	9,060	6,878	5,543	4,642	3,993	3,503	3,121	2,813	2,257	1,885	1,618	1,417	1,261	1,135	947	812	711	569	475
130	44,179	24,911	17,346	13,306	10,792	9,077	6,888	5,549	4,647	3,996	3,506	3,123	2,815	2,258	1,886	1,618	1,418	1,261	1,136	947	813	711	570	475
135	44,553	25,030	17,404	13,339	10,814	9,093	6,897	5,555	4,651	3,999	3,508	3,124	2,816	2,259	1,886	1,619	1,418	1,261	1,136	948	813	711	570	475
140	44,906	25,141	17,457	13,371	10,835	9,107	6,905	5,561	4,654	4,002	3,510	3,126	2,818	2,260	1,887	1,619	1,418	1,262	1,136	948	813	712	570	475
145	45,239	25,245	17,507	13,400	10,854	9,121	6,913	5,566	4,658	4,005	3,512	3,128	2,819	2,261	1,887	1,620	1,419	1,262	1,136	948	813	712	570	475
150	45,555	25,343	17,554	13,428	10,872	9,134	6,920	5,571	4,661	4,007	3,514	3,129	2,820	2,262	1,888	1,620	1,419	1,262	1,137	948	813	712	570	475
155	45,855	25,436	17,599	13,454	10,889	9,146	6,927	5,575	4,665	4,010	3,516	3,131	2,821	2,263	1,889	1,621	1,419	1,262	1,137	948	813	712	570	475
160	46,139	25,523	17,641	13,478	10,905	9,157	6,934	5,579	4,667	4,012	3,518	3,132	2,822	2,263	1,889	1,621	1,420	1,263	1,137	948	813	712	570	475
165	46,409	25,605	17,680	13,501	10,920	9,167	6,940	5,583	4,670	4,014	3,519	3,133	2,823	2,264	1,889	1,621	1,420	1,263	1,137	948	813	712	570	475
170	46,667	25,683	17,717	13,523	10,934	9,177	6,946	5,587	4,673	4,016	3,521	3,134	2,824	2,265	1,890	1,622	1,420	1,263	1,137	948	813	712	570	475
175	46,912	25,758	17,752	13,543	10,948	9,187	6,951	5,590	4,675	4,018	3,522	3,135	2,825	2,265	1,890	1,622	1,420	1,263	1,137	949	813	712	570	475
180	47,146	25,828	17,786	13,563	10,960	9,196	6,956	5,594	4,678	4,019	3,523	3,136	2,826	2,266	1,891	1,622	1,420	1,263	1,138	949	814	712	570	475
185	47,370	25,895	17,817	13,581	10,972	9,204	6,961	5,597	4,680	4,021	3,525	3,137	2,827	2,266	1,891	1,622	1,421	1,264	1,138	949	814	712	570	475
190	47,583	25,959	17,848	13,599	10,984	9,212	6,965	5,600	4,682	4,022	3,526	3,138	2,828	2,267	1,891	1,623	1,421	1,264	1,138	949	814	712	570	475
195	47,788	26,019	17,876	13,615	10,995	9,220	6,970	5,603	4,684	4,024	3,527	3,139	2,828	2,267	1,892	1,623	1,421	1,264	1,138	949	814	712	570	475
200	47,984	26,077	17,904	13,631	11,005	9,227	6,974	5,605	4,686	4,025	3,528	3,140	2,829	2,268	1,892	1,623	1,421	1,264	1,138	949	814	712	570	475

3 AWG | 1 Run(s) | 120 Volts | C-Value = 4,760 | Magnetic Duct | Copper

Calculations:

1. $f = \dfrac{2 \times \text{Length} \times \text{Isca} \times 1.5}{\text{Runs} \times \text{C-Value} \times \text{E L-N}}$

2. $M = \dfrac{1}{1 + f}$

3. $\text{Isc} = \text{Isca} \times 1.5 \times M$

4. Add Motor Contribution = Motor FLA x 4

* All results are given in symmetrical amperes

Single-Phase L-N Bolted Fault-Current* Table

One-Way Distance in Feet

Available Isc in Thousands (Isca)

Isca	5	10	15	20	25	30	40	50	60	70	80	90	100	125	150	175	200	225	250	300	350	400	500	600
3	4,231	3,993	3,780	3,589	3,416	3,259	2,984	2,752	2,554	2,382	2,232	2,100	1,982	1,739	1,549	1,397	1,271	1,167	1,078	936	827	740	612	522
5	6,782	6,190	5,693	5,269	4,905	4,587	4,062	3,644	3,304	3,022	2,785	2,582	2,407	2,057	1,797	1,594	1,433	1,302	1,192	1,020	892	792	648	547
7	9,145	8,100	7,269	6,593	6,032	5,559	4,805	4,231	3,780	3,416	3,115	2,864	2,649	2,232	1,929	1,693	1,516	1,370	1,249	1,062	923	817	664	559
10	12,380	10,539	9,175	8,123	7,288	6,608	5,570	4,813	4,238	3,785	3,420	3,119	2,866	2,384	2,041	1,784	1,585	1,425	1,295	1,095	948	836	677	568
15	17,078	13,762	11,524	9,912	8,696	7,746	6,356	5,390	4,678	4,133	3,701	3,351	3,061	2,518	2,138	1,858	1,642	1,472	1,333	1,122	969	852	687	575
20	21,078	16,246	13,216	11,139	9,626	8,475	6,839	5,733	4,935	4,331	3,860	3,481	3,169	2,590	2,190	1,897	1,673	1,496	1,353	1,136	979	860	692	579
25	24,524	18,219	14,493	12,033	10,286	8,983	7,166	5,961	5,102	4,460	3,962	3,563	3,238	2,636	2,222	1,921	1,692	1,511	1,366	1,145	986	865	696	581
30	27,524	19,825	15,491	12,713	10,779	9,356	7,402	6,123	5,221	4,550	4,033	3,621	3,285	2,667	2,245	1,938	1,705	1,522	1,374	1,151	990	869	698	583
35	30,159	21,156	16,293	13,247	11,161	9,643	7,580	6,244	5,309	4,617	4,085	3,663	3,320	2,690	2,261	1,950	1,714	1,529	1,380	1,155	993	871	699	584
40	32,492	22,278	16,950	13,679	11,466	9,869	7,719	6,338	5,377	4,668	4,125	3,695	3,346	2,707	2,273	1,959	1,721	1,535	1,385	1,158	996	873	700	585
45	34,572	23,237	17,499	14,034	11,714	10,053	7,831	6,414	5,431	4,709	4,157	3,720	3,367	2,721	2,283	1,966	1,726	1,539	1,388	1,161	997	874	701	585
50	36,439	24,066	17,965	14,332	11,921	10,205	7,923	6,475	5,475	4,742	4,182	3,741	3,384	2,732	2,290	1,972	1,731	1,543	1,391	1,163	999	876	702	586
55	38,123	24,789	18,365	14,586	12,096	10,333	8,000	6,527	5,511	4,770	4,204	3,758	3,398	2,741	2,297	1,976	1,735	1,545	1,393	1,165	1,000	876	703	586
60	39,649	25,425	18,712	14,804	12,246	10,442	8,065	6,570	5,542	4,793	4,222	3,772	3,409	2,748	2,302	1,980	1,738	1,548	1,395	1,166	1,001	877	703	587
65	41,040	25,990	19,016	14,993	12,375	10,536	8,121	6,607	5,569	4,812	4,237	3,785	3,419	2,755	2,307	1,984	1,740	1,550	1,397	1,167	1,002	878	704	587
70	42,312	26,494	19,285	15,160	12,489	10,618	8,170	6,639	5,591	4,829	4,250	3,795	3,428	2,760	2,310	1,987	1,742	1,552	1,399	1,168	1,003	878	704	587
75	43,480	26,948	19,524	15,307	12,588	10,690	8,212	6,667	5,611	4,844	4,262	3,804	3,435	2,765	2,314	1,989	1,744	1,553	1,400	1,169	1,003	879	704	588
80	44,557	27,357	19,738	15,439	12,677	10,753	8,250	6,692	5,629	4,857	4,272	3,812	3,442	2,769	2,317	1,991	1,746	1,555	1,401	1,170	1,004	879	705	588
85	45,552	27,729	19,931	15,556	12,756	10,810	8,283	6,714	5,645	4,869	4,281	3,819	3,448	2,773	2,319	1,993	1,748	1,556	1,402	1,170	1,004	880	705	588
90	46,474	28,068	20,106	15,662	12,828	10,862	8,313	6,734	5,658	4,879	4,289	3,826	3,453	2,777	2,322	1,995	1,749	1,557	1,403	1,171	1,005	880	705	588
95	47,332	28,379	20,265	15,759	12,892	10,908	8,340	6,751	5,671	4,889	4,296	3,831	3,458	2,780	2,324	1,997	1,750	1,558	1,403	1,171	1,005	880	705	588
100	48,131	28,664	20,410	15,846	12,951	10,950	8,365	6,767	5,682	4,897	4,302	3,837	3,462	2,782	2,326	1,998	1,751	1,559	1,404	1,172	1,006	881	705	588
105	48,878	28,928	20,543	15,926	13,004	10,988	8,387	6,782	5,693	4,905	4,308	3,841	3,466	2,785	2,327	1,999	1,752	1,559	1,405	1,172	1,006	881	706	588
110	49,577	29,171	20,665	16,000	13,053	11,023	8,408	6,795	5,702	4,912	4,314	3,846	3,469	2,787	2,329	2,000	1,753	1,560	1,405	1,173	1,006	881	706	588
115	50,233	29,397	20,778	16,068	13,098	11,055	8,426	6,808	5,710	4,918	4,319	3,849	3,472	2,789	2,330	2,001	1,754	1,561	1,406	1,173	1,007	881	706	589
120	50,850	29,607	20,883	16,130	13,140	11,085	8,443	6,819	5,718	4,924	4,323	3,853	3,475	2,791	2,332	2,002	1,755	1,561	1,406	1,173	1,007	882	706	589
125	51,432	29,803	20,981	16,188	13,178	11,112	8,459	6,829	5,726	4,929	4,327	3,856	3,478	2,793	2,333	2,003	1,755	1,562	1,407	1,174	1,007	882	706	589
130	51,980	29,987	21,071	16,242	13,214	11,137	8,474	6,839	5,732	4,934	4,331	3,859	3,480	2,794	2,334	2,004	1,756	1,562	1,407	1,174	1,007	882	706	589
135	52,498	30,158	21,156	16,292	13,247	11,161	8,488	6,848	5,739	4,939	4,335	3,862	3,483	2,796	2,335	2,005	1,756	1,563	1,408	1,174	1,007	882	706	589
140	52,989	30,320	21,235	16,339	13,278	11,183	8,500	6,856	5,744	4,943	4,338	3,865	3,485	2,797	2,336	2,006	1,757	1,563	1,408	1,175	1,008	882	706	589
145	53,454	30,471	21,309	16,383	13,307	11,204	8,512	6,864	5,750	4,947	4,341	3,867	3,487	2,798	2,337	2,006	1,757	1,564	1,408	1,175	1,008	882	706	589
150	53,896	30,614	21,379	16,425	13,334	11,223	8,523	6,871	5,755	4,951	4,344	3,870	3,489	2,800	2,338	2,007	1,758	1,564	1,409	1,175	1,008	882	706	589
155	54,315	30,749	21,445	16,463	13,360	11,241	8,534	6,878	5,760	4,954	4,347	3,872	3,490	2,801	2,339	2,007	1,758	1,564	1,409	1,175	1,008	883	707	589
160	54,715	30,877	21,507	16,500	13,384	11,258	8,544	6,884	5,764	4,958	4,349	3,874	3,492	2,802	2,339	2,008	1,759	1,565	1,409	1,175	1,008	883	707	589
165	55,095	30,998	21,566	16,534	13,407	11,274	8,553	6,890	5,768	4,961	4,352	3,876	3,494	2,803	2,340	2,008	1,759	1,565	1,409	1,176	1,008	883	707	589
170	55,458	31,112	21,621	16,567	13,428	11,289	8,562	6,896	5,772	4,964	4,354	3,877	3,495	2,804	2,341	2,009	1,760	1,565	1,410	1,176	1,008	883	707	589
175	55,805	31,221	21,673	16,598	13,448	11,303	8,570	6,901	5,776	4,967	4,356	3,879	3,496	2,805	2,341	2,009	1,760	1,566	1,410	1,176	1,009	883	707	589
180	56,137	31,325	21,723	16,627	13,467	11,317	8,578	6,906	5,780	4,969	4,358	3,881	3,498	2,805	2,342	2,010	1,760	1,566	1,410	1,176	1,009	883	707	589
185	56,454	31,423	21,771	16,655	13,486	11,330	8,585	6,911	5,783	4,972	4,360	3,882	3,499	2,806	2,342	2,010	1,761	1,566	1,410	1,176	1,009	883	707	589
190	56,758	31,517	21,816	16,681	13,503	11,342	8,592	6,915	5,786	4,974	4,362	3,884	3,500	2,807	2,343	2,011	1,761	1,566	1,410	1,176	1,009	883	707	589
195	57,049	31,607	21,859	16,706	13,519	11,354	8,599	6,920	5,789	4,976	4,363	3,885	3,501	2,808	2,343	2,011	1,761	1,566	1,411	1,176	1,009	883	707	589
200	57,329	31,692	21,899	16,730	13,535	11,365	8,605	6,924	5,792	4,978	4,365	3,886	3,502	2,808	2,344	2,011	1,761	1,567	1,411	1,177	1,009	883	707	589

# 2 AWG	1 Run(s)	120 Volts	C-Value = 5,906	Magnetic Duct	Copper

Calculations:

1. $f = \dfrac{2 \times \text{Length} \times \text{Isca} \times 1.5}{\text{Runs} \times \text{C-Value} \times E_{L\text{-}N}}$

2. $M = \dfrac{1}{1 + f}$

3. Isc = Isca x 1.5 x M

4. Add Motor Contribution = Motor FLA x 4

* All results are given in symmetrical amperes

Single-Phase L-N Bolted Fault-Current* Table

One-Way Distance in Feet

Available Isc in Thousands (Isca)	5	10	15	20	25	30	40	50	60	70	80	90	100	125	150	175	200	225	250	300	350	400	500	600
3	4,280	4,080	3,899	3,732	3,580	3,439	3,188	2,972	2,783	2,616	2,469	2,337	2,218	1,969	1,770	1,607	1,472	1,358	1,260	1,101	978	880	733	628
5	6,908	6,402	5,966	5,585	5,250	4,953	4,449	4,039	3,697	3,409	3,163	2,950	2,763	2,386	2,100	1,875	1,694	1,544	1,419	1,221	1,071	955	784	665
7	9,375	8,468	7,721	7,095	6,563	6,105	5,357	4,773	4,303	3,918	3,596	3,323	3,088	2,625	2,283	2,019	1,810	1,641	1,500	1,281	1,117	991	808	682
10	12,805	11,170	9,906	8,898	8,077	7,395	6,325	5,526	4,907	4,412	4,008	3,671	3,387	2,838	2,442	2,143	1,909	1,721	1,567	1,329	1,154	1,019	827	695
15	17,898	14,859	12,702	11,092	9,844	8,849	7,360	6,300	5,507	4,891	4,400	3,998	3,663	3,029	2,582	2,250	1,994	1,790	1,624	1,370	1,184	1,043	842	706
20	22,341	17,797	14,789	12,651	11,053	9,813	8,016	6,774	5,866	5,173	4,626	4,183	3,818	3,134	2,658	2,308	2,039	1,826	1,654	1,391	1,200	1,055	850	712
25	26,250	20,193	16,407	13,816	11,932	10,500	8,468	7,095	6,105	5,357	4,773	4,303	3,918	3,201	2,706	2,344	2,067	1,849	1,672	1,404	1,210	1,063	855	715
30	29,717	22,184	17,697	14,720	12,600	11,014	8,799	7,326	6,275	5,488	4,876	4,387	3,988	3,248	2,739	2,369	2,086	1,864	1,685	1,413	1,216	1,068	858	718
35	32,813	23,864	18,751	15,442	13,125	11,413	9,052	7,500	6,403	5,585	4,953	4,449	4,039	3,281	2,763	2,386	2,100	1,875	1,694	1,419	1,221	1,071	861	719
40	35,594	25,302	19,627	16,031	13,549	11,732	9,251	7,637	6,502	5,661	5,012	4,497	4,078	3,307	2,782	2,400	2,111	1,883	1,700	1,424	1,225	1,074	862	720
45	38,106	26,546	20,367	16,522	13,898	11,993	9,413	7,746	6,581	5,721	5,059	4,535	4,109	3,328	2,796	2,411	2,119	1,890	1,706	1,428	1,227	1,076	864	721
50	40,385	27,632	21,001	16,936	14,190	12,210	9,546	7,836	6,646	5,769	5,097	4,565	4,134	3,344	2,808	2,419	2,126	1,895	1,710	1,431	1,230	1,078	865	722
55	42,464	28,590	21,549	17,291	14,438	12,393	9,658	7,911	6,700	5,810	5,129	4,591	4,155	3,358	2,817	2,427	2,131	1,900	1,714	1,433	1,231	1,079	866	723
60	44,367	29,440	22,029	17,598	14,652	12,550	9,753	7,975	6,745	5,844	5,156	4,612	4,172	3,369	2,825	2,433	2,136	1,903	1,717	1,435	1,233	1,081	867	723
65	46,116	30,200	22,451	17,867	14,838	12,686	9,835	8,030	6,785	5,874	5,179	4,630	4,187	3,379	2,832	2,438	2,140	1,907	1,719	1,437	1,234	1,082	867	724
70	47,728	30,883	22,827	18,104	15,001	12,805	9,906	8,077	6,818	5,899	5,198	4,646	4,200	3,387	2,838	2,442	2,143	1,909	1,721	1,438	1,235	1,083	868	724
75	49,220	31,501	23,163	18,315	15,145	12,910	9,969	8,119	6,848	5,921	5,215	4,660	4,211	3,395	2,843	2,446	2,146	1,911	1,723	1,440	1,236	1,083	868	725
80	50,604	32,062	23,465	18,503	15,273	13,004	10,024	8,156	6,874	5,941	5,231	4,672	4,221	3,401	2,848	2,449	2,148	1,914	1,725	1,441	1,237	1,084	869	725
85	51,891	32,574	23,738	18,672	15,389	13,087	10,074	8,188	6,898	5,958	5,244	4,683	4,230	3,407	2,852	2,452	2,151	1,915	1,726	1,442	1,238	1,084	869	725
90	53,091	33,043	23,986	18,825	15,492	13,162	10,118	8,218	6,918	5,974	5,256	4,692	4,238	3,412	2,855	2,455	2,153	1,917	1,728	1,443	1,239	1,085	869	725
95	54,213	33,474	24,212	18,965	15,587	13,230	10,158	8,244	6,937	5,988	5,267	4,701	4,245	3,416	2,858	2,457	2,155	1,918	1,729	1,444	1,239	1,085	870	725
100	55,265	33,872	24,420	19,092	15,672	13,292	10,195	8,268	6,954	6,000	5,277	4,709	4,251	3,420	2,861	2,459	2,156	1,920	1,730	1,444	1,240	1,086	870	726
105	56,252	34,240	24,610	19,208	15,751	13,348	10,228	8,290	6,969	6,012	5,285	4,716	4,257	3,424	2,864	2,461	2,158	1,921	1,731	1,445	1,240	1,086	870	726
110	57,180	34,582	24,786	19,315	15,823	13,400	10,258	8,310	6,983	6,022	5,294	4,722	4,262	3,427	2,866	2,463	2,159	1,922	1,732	1,446	1,241	1,087	870	726
115	58,055	34,900	24,949	19,414	15,889	13,447	10,286	8,328	6,996	6,032	5,301	4,728	4,267	3,431	2,868	2,464	2,160	1,923	1,733	1,446	1,241	1,087	871	726
120	58,880	35,197	25,101	19,505	15,950	13,491	10,311	8,345	7,008	6,041	5,308	4,733	4,271	3,433	2,870	2,466	2,161	1,924	1,733	1,447	1,241	1,087	871	726
125	59,661	35,474	25,241	19,590	16,007	13,532	10,335	8,360	7,019	6,049	5,314	4,738	4,275	3,436	2,872	2,467	2,162	1,925	1,734	1,447	1,242	1,087	871	726
130	60,400	35,734	25,373	19,669	16,059	13,569	10,357	8,375	7,029	6,056	5,320	4,743	4,279	3,438	2,874	2,468	2,163	1,925	1,735	1,448	1,242	1,088	871	726
135	61,101	35,979	25,496	19,743	16,109	13,604	10,377	8,388	7,039	6,063	5,325	4,747	4,283	3,441	2,875	2,470	2,164	1,926	1,735	1,448	1,242	1,088	871	727
140	61,767	36,208	25,611	19,812	16,155	13,637	10,396	8,400	7,047	6,070	5,330	4,751	4,286	3,443	2,877	2,471	2,165	1,927	1,736	1,448	1,243	1,088	871	727
145	62,400	36,425	25,719	19,877	16,197	13,668	10,414	8,412	7,055	6,076	5,335	4,755	4,289	3,445	2,878	2,472	2,166	1,927	1,736	1,449	1,243	1,088	872	727
150	63,002	36,629	25,821	19,938	16,238	13,696	10,431	8,423	7,063	6,081	5,339	4,759	4,292	3,447	2,879	2,473	2,167	1,928	1,737	1,449	1,243	1,089	872	727
155	63,576	36,823	25,917	19,995	16,276	13,723	10,447	8,433	7,070	6,087	5,343	4,762	4,294	3,448	2,881	2,474	2,167	1,928	1,737	1,449	1,243	1,089	872	727
160	64,124	37,006	26,007	20,049	16,311	13,749	10,461	8,443	7,077	6,092	5,347	4,765	4,297	3,450	2,882	2,474	2,168	1,929	1,737	1,450	1,244	1,089	872	727
165	64,648	37,180	26,093	20,099	16,345	13,772	10,475	8,452	7,083	6,096	5,351	4,768	4,299	3,451	2,883	2,475	2,168	1,929	1,738	1,450	1,244	1,089	872	727
170	65,148	37,345	26,174	20,148	16,377	13,795	10,488	8,460	7,089	6,101	5,354	4,770	4,301	3,453	2,884	2,476	2,169	1,930	1,738	1,450	1,244	1,089	872	727
175	65,627	37,501	26,251	20,193	16,407	13,816	10,500	8,468	7,095	6,105	5,357	4,773	4,303	3,454	2,885	2,477	2,170	1,930	1,738	1,450	1,244	1,089	872	727
180	66,086	37,651	26,324	20,236	16,435	13,837	10,512	8,476	7,100	6,109	5,360	4,775	4,305	3,455	2,886	2,477	2,170	1,931	1,739	1,451	1,244	1,089	872	727
185	66,526	37,793	26,394	20,277	16,463	13,856	10,523	8,483	7,105	6,113	5,363	4,778	4,307	3,457	2,886	2,478	2,170	1,931	1,739	1,451	1,244	1,090	872	727
190	66,949	37,929	26,460	20,317	16,488	13,874	10,534	8,490	7,110	6,116	5,366	4,780	4,309	3,458	2,887	2,478	2,171	1,931	1,739	1,451	1,245	1,090	872	727
195	67,354	38,059	26,523	20,354	16,513	13,891	10,544	8,496	7,115	6,120	5,369	4,782	4,311	3,459	2,888	2,479	2,171	1,932	1,740	1,451	1,245	1,090	872	727
200	67,744	38,183	26,583	20,389	16,536	13,908	10,553	8,502	7,119	6,123	5,371	4,784	4,312	3,460	2,889	2,479	2,172	1,932	1,740	1,451	1,245	1,090	872	727

1 AWG | 1 Run(s) | 120 Volts | C-Value = 7,292 | Magnetic Duct | Copper

Calculations:

1. $f = \dfrac{2 \times \text{Length} \times \text{Isca} \times 1.5}{\text{Runs} \times \text{C-Value} \times E_{L\text{-}N}}$
2. $M = \dfrac{1}{1 + f}$
3. $\text{Isc} = \text{Isca} \times 1.5 \times M$
4. Add Motor Contribution = Motor FLA x 4

* All results are given in symmetrical amperes

Single-Phase L-N Bolted Fault-Current* Table

One-Way Distance in Feet

Available Isc in Thousands (Isca)	5	10	15	20	25	30	40	50	60	70	80	90	100	125	150	175	200	225	250	300	350	400	500	600
3	4,319	4,151	3,996	3,852	3,719	3,594	3,368	3,169	2,992	2,833	2,691	2,562	2,445	2,195	1,991	1,821	1,679	1,557	1,451	1,278	1,142	1,032	865	745
5	7,009	6,579	6,198	5,859	5,555	5,281	4,807	4,411	4,075	3,787	3,537	3,318	3,124	2,726	2,419	2,173	1,973	1,807	1,666	1,442	1,271	1,136	937	798
7	9,562	8,779	8,113	7,542	7,046	6,611	5,884	5,302	4,824	4,425	4,088	3,798	3,546	3,042	2,664	2,369	2,133	1,940	1,779	1,525	1,335	1,187	972	822
10	13,157	11,717	10,562	9,614	8,822	8,150	7,074	6,248	5,595	5,066	4,628	4,260	3,946	3,332	2,883	2,541	2,272	2,054	1,874	1,595	1,388	1,229	1,000	842
15	18,593	15,843	13,801	12,225	10,973	9,953	8,393	7,256	6,390	5,708	5,159	4,705	4,325	3,598	3,081	2,693	2,393	2,152	1,956	1,654	1,432	1,263	1,022	858
20	23,435	19,227	16,301	14,147	12,496	11,190	9,256	7,892	6,878	6,095	5,472	4,965	4,543	3,748	3,190	2,776	2,458	2,205	1,999	1,685	1,456	1,281	1,034	867
25	27,774	22,054	18,288	15,620	13,632	12,093	9,865	8,330	7,209	6,353	5,679	5,135	4,685	3,844	3,259	2,829	2,499	2,238	2,026	1,704	1,470	1,292	1,041	872
30	31,685	24,451	19,906	16,786	14,511	12,779	10,317	8,650	7,447	6,538	5,826	5,255	4,785	3,911	3,307	2,865	2,527	2,260	2,044	1,717	1,480	1,300	1,046	875
35	35,229	26,508	21,249	17,730	15,212	13,320	10,666	8,895	7,627	6,676	5,936	5,344	4,859	3,960	3,342	2,891	2,547	2,277	2,058	1,726	1,487	1,305	1,049	877
40	38,454	28,294	22,381	18,512	15,784	13,756	10,944	9,087	7,769	6,784	6,021	5,413	4,916	3,998	3,369	2,911	2,563	2,289	2,068	1,733	1,492	1,309	1,052	879
45	41,403	29,859	23,349	19,169	16,259	14,116	11,171	9,242	7,882	6,871	6,089	5,467	4,961	4,028	3,390	2,927	2,575	2,299	2,076	1,739	1,496	1,313	1,054	881
50	44,108	31,241	24,185	19,729	16,660	14,417	11,359	9,371	7,975	6,941	6,145	5,512	4,998	4,052	3,407	2,940	2,585	2,307	2,082	1,743	1,499	1,315	1,056	882
55	46,600	32,470	24,916	20,213	17,003	14,674	11,517	9,478	8,053	7,000	6,191	5,549	5,028	4,072	3,422	2,950	2,593	2,313	2,088	1,747	1,502	1,317	1,057	883
60	48,902	33,571	25,559	20,634	17,301	14,894	11,653	9,570	8,119	7,050	6,230	5,580	5,054	4,089	3,433	2,959	2,600	2,318	2,092	1,750	1,504	1,319	1,058	884
65	51,035	34,563	26,130	21,004	17,560	15,086	11,770	9,649	8,176	7,093	6,263	5,607	5,076	4,103	3,444	2,967	2,606	2,323	2,096	1,753	1,506	1,320	1,059	884
70	53,017	35,461	26,640	21,333	17,789	15,255	11,872	9,718	8,225	7,130	6,292	5,630	5,095	4,116	3,452	2,973	2,611	2,327	2,099	1,755	1,508	1,322	1,060	885
75	54,864	36,278	27,098	21,626	17,992	15,404	11,963	9,778	8,268	7,162	6,317	5,651	5,111	4,126	3,460	2,979	2,615	2,330	2,102	1,757	1,509	1,323	1,061	885
80	56,588	37,024	27,512	21,889	18,174	15,537	12,043	9,831	8,306	7,191	6,339	5,668	5,126	4,136	3,466	2,984	2,619	2,333	2,104	1,759	1,511	1,324	1,061	886
85	58,203	37,708	27,888	22,126	18,337	15,656	12,114	9,879	8,340	7,216	6,359	5,684	5,139	4,144	3,472	2,988	2,622	2,336	2,106	1,760	1,512	1,325	1,062	886
90	59,717	38,338	28,231	22,341	18,485	15,764	12,178	9,922	8,371	7,239	6,377	5,698	5,150	4,152	3,478	2,992	2,625	2,339	2,108	1,762	1,513	1,325	1,062	887
95	61,141	38,920	28,545	22,538	18,619	15,861	12,237	9,960	8,398	7,259	6,393	5,711	5,160	4,159	3,482	2,995	2,628	2,341	2,110	1,763	1,514	1,326	1,063	887
100	62,481	39,459	28,834	22,717	18,742	15,950	12,289	9,995	8,423	7,278	6,407	5,722	5,170	4,165	3,487	2,998	2,630	2,343	2,112	1,764	1,514	1,327	1,063	887
105	63,746	39,959	29,101	22,882	18,854	16,031	12,337	10,027	8,445	7,295	6,420	5,733	5,178	4,170	3,490	3,001	2,632	2,344	2,113	1,765	1,515	1,327	1,064	887
110	64,940	40,426	29,347	23,035	18,957	16,106	12,382	10,056	8,466	7,310	6,432	5,742	5,186	4,175	3,494	3,004	2,634	2,346	2,114	1,766	1,516	1,328	1,064	888
115	66,071	40,861	29,576	23,175	19,052	16,174	12,422	10,083	8,485	7,324	6,443	5,751	5,193	4,180	3,497	3,006	2,636	2,347	2,115	1,767	1,516	1,328	1,064	888
120	67,143	41,268	29,789	23,306	19,140	16,238	12,459	10,107	8,502	7,337	6,453	5,759	5,200	4,184	3,500	3,009	2,638	2,349	2,117	1,767	1,517	1,329	1,065	888
125	68,160	41,650	29,987	23,427	19,222	16,297	12,494	10,130	8,519	7,349	6,462	5,766	5,206	4,188	3,503	3,011	2,640	2,350	2,118	1,768	1,517	1,329	1,065	888
130	69,126	42,009	30,173	23,540	19,298	16,351	12,526	10,151	8,533	7,360	6,471	5,773	5,211	4,191	3,505	3,012	2,641	2,351	2,118	1,769	1,518	1,329	1,065	888
135	70,046	42,347	30,347	23,646	19,369	16,402	12,556	10,171	8,547	7,371	6,479	5,780	5,216	4,195	3,508	3,014	2,642	2,352	2,119	1,769	1,518	1,330	1,065	888
140	70,922	42,666	30,510	23,745	19,435	16,450	12,584	10,189	8,560	7,380	6,486	5,785	5,221	4,198	3,510	3,016	2,643	2,353	2,120	1,770	1,519	1,330	1,065	889
145	71,758	42,967	30,664	23,838	19,498	16,494	12,610	10,206	8,572	7,389	6,493	5,791	5,226	4,201	3,512	3,017	2,645	2,354	2,121	1,770	1,519	1,330	1,066	889
150	72,555	43,251	30,808	23,925	19,556	16,536	12,634	10,222	8,584	7,398	6,500	5,796	5,230	4,203	3,514	3,019	2,646	2,355	2,122	1,771	1,519	1,331	1,066	889
155	73,318	43,521	30,945	24,008	19,611	16,576	12,657	10,237	8,594	7,406	6,506	5,801	5,234	4,206	3,516	3,020	2,647	2,356	2,122	1,771	1,520	1,331	1,066	889
160	74,048	43,777	31,074	24,085	19,663	16,613	12,679	10,251	8,604	7,413	6,511	5,805	5,238	4,208	3,517	3,021	2,648	2,356	2,123	1,772	1,520	1,331	1,066	889
165	74,747	44,021	31,197	24,159	19,712	16,647	12,699	10,265	8,613	7,420	6,517	5,810	5,241	4,211	3,519	3,022	2,649	2,357	2,123	1,772	1,520	1,331	1,066	889
170	75,417	44,252	31,313	24,228	19,758	16,680	12,718	10,277	8,622	7,426	6,522	5,814	5,244	4,213	3,520	3,023	2,649	2,358	2,124	1,772	1,521	1,332	1,066	889
175	76,059	44,473	31,423	24,294	19,802	16,712	12,737	10,289	8,631	7,433	6,527	5,817	5,247	4,215	3,522	3,024	2,650	2,358	2,124	1,773	1,521	1,332	1,067	889
180	76,676	44,683	31,528	24,357	19,844	16,741	12,754	10,300	8,638	7,438	6,531	5,821	5,250	4,217	3,523	3,025	2,651	2,359	2,125	1,773	1,521	1,332	1,067	889
185	77,269	44,884	31,628	24,416	19,883	16,769	12,770	10,311	8,646	7,444	6,535	5,824	5,253	4,218	3,524	3,026	2,652	2,359	2,125	1,773	1,521	1,332	1,067	890
190	77,840	45,075	31,723	24,473	19,921	16,796	12,785	10,321	8,653	7,449	6,539	5,828	5,256	4,220	3,525	3,027	2,652	2,360	2,126	1,774	1,522	1,332	1,067	890
195	78,389	45,259	31,814	24,527	19,956	16,822	12,800	10,331	8,660	7,454	6,543	5,831	5,258	4,222	3,527	3,028	2,653	2,361	2,126	1,774	1,522	1,333	1,067	890
200	78,918	45,435	31,900	24,579	19,990	16,846	12,814	10,340	8,666	7,459	6,547	5,834	5,261	4,223	3,528	3,029	2,654	2,361	2,127	1,774	1,522	1,333	1,067	890

1/0 AWG | 1 Run(s) | 120 Volts

C-Value = 8,924 | Magnetic Duct | Copper

Calculations:

1. $f = \dfrac{2 \times \text{Length} \times \text{Isca} \times 1.5}{\text{Runs} \times \text{C-Value} \times E_{L\text{-}N}}$

2. $M = \dfrac{1}{1 + f}$

3. $\text{Isc} = \text{Isca} \times 1.5 \times M$

4. Add Motor Contribution = Motor FLA x 4

* All results are given in symmetrical amperes

Single-Phase L-N Bolted Fault-Current* Table

One-Way Distance in Feet

Available Isc in Thousands (Isca)	5	10	15	20	25	30	40	50	60	70	80	90	100	125	150	175	200	225	250	300	350	400	500	600
3	4,348	4,207	4,074	3,949	3,832	3,721	3,519	3,337	3,173	3,024	2,889	2,765	2,651	2,404	2,199	2,027	1,879	1,752	1,640	1,455	1,308	1,188	1,003	868
5	7,088	6,719	6,387	6,085	5,811	5,561	5,120	4,743	4,419	4,135	3,886	3,666	3,469	3,058	2,734	2,472	2,256	2,075	1,920	1,672	1,480	1,328	1,101	941
7	9,710	9,031	8,440	7,922	7,464	7,056	6,360	5,790	5,313	4,909	4,562	4,261	3,997	3,461	3,052	2,729	2,468	2,253	2,072	1,785	1,568	1,398	1,149	976
10	13,438	12,171	11,122	10,240	9,487	8,837	7,773	6,937	6,264	5,710	5,245	4,851	4,512	3,841	3,343	2,960	2,655	2,408	2,202	1,881	1,642	1,457	1,188	1,004
15	19,160	16,683	14,773	13,256	12,021	10,997	9,396	8,202	7,277	6,539	5,938	5,437	5,015	4,199	3,611	3,168	2,822	2,544	2,316	1,963	1,704	1,505	1,221	1,026
20	24,342	20,479	17,675	15,546	13,874	12,528	10,491	9,024	7,917	7,052	6,357	5,787	5,311	4,404	3,762	3,284	2,913	2,618	2,377	2,007	1,737	1,531	1,237	1,038
25	29,057	23,717	20,035	17,343	15,289	13,669	11,280	9,602	8,358	7,400	6,638	6,019	5,506	4,538	3,859	3,357	2,971	2,664	2,415	2,034	1,757	1,547	1,248	1,046
30	33,366	26,512	21,994	18,791	16,403	14,553	11,875	10,030	8,680	7,651	6,840	6,185	5,644	4,631	3,927	3,408	3,011	2,696	2,441	2,053	1,771	1,557	1,255	1,050
35	37,319	28,948	23,645	19,984	17,304	15,258	12,340	10,359	8,926	7,842	6,992	6,308	5,747	4,700	3,976	3,445	3,040	2,719	2,460	2,066	1,781	1,565	1,260	1,054
40	40,958	31,091	25,055	20,982	18,048	15,834	12,714	10,621	9,120	7,991	7,110	6,405	5,826	4,753	4,014	3,474	3,062	2,737	2,475	2,077	1,789	1,571	1,263	1,057
45	44,320	32,991	26,274	21,830	18,672	16,312	13,021	10,834	9,277	8,111	7,205	6,482	5,890	4,796	4,044	3,496	3,079	2,751	2,486	2,085	1,795	1,576	1,266	1,059
50	47,435	34,686	27,339	22,560	19,203	16,716	13,277	11,011	9,406	8,210	7,283	6,544	5,942	4,830	4,069	3,515	3,093	2,762	2,495	2,091	1,799	1,579	1,269	1,060
55	50,328	36,208	28,276	23,194	19,661	17,062	13,494	11,160	9,515	8,292	7,348	6,597	5,985	4,858	4,089	3,530	3,105	2,772	2,503	2,096	1,803	1,582	1,271	1,062
60	53,024	37,583	29,107	23,750	20,059	17,361	13,680	11,287	9,607	8,362	7,403	6,641	6,021	4,882	4,106	3,542	3,115	2,779	2,509	2,101	1,807	1,585	1,272	1,063
65	55,541	38,830	29,849	24,243	20,409	17,622	13,842	11,397	9,687	8,422	7,450	6,679	6,052	4,903	4,120	3,553	3,123	2,786	2,515	2,105	1,809	1,587	1,274	1,064
70	57,897	39,967	30,517	24,681	20,719	17,853	13,984	11,493	9,756	8,475	7,491	6,712	6,079	4,920	4,133	3,562	3,130	2,792	2,519	2,108	1,812	1,589	1,275	1,065
75	60,106	41,008	31,120	25,074	20,995	18,057	14,109	11,578	9,817	8,520	7,527	6,740	6,103	4,936	4,144	3,570	3,137	2,797	2,523	2,111	1,814	1,590	1,276	1,065
80	62,183	41,964	31,667	25,428	21,243	18,240	14,221	11,653	9,870	8,561	7,558	6,766	6,124	4,949	4,153	3,577	3,142	2,801	2,527	2,113	1,816	1,592	1,277	1,066
85	64,138	42,845	32,167	25,749	21,466	18,405	14,321	11,720	9,918	8,597	7,586	6,788	6,142	4,962	4,162	3,584	3,147	2,805	2,530	2,115	1,817	1,593	1,278	1,067
90	65,982	43,660	32,624	26,041	21,669	18,554	14,410	11,780	9,961	8,629	7,611	6,808	6,159	4,972	4,169	3,589	3,151	2,808	2,533	2,117	1,819	1,594	1,278	1,067
95	67,724	44,416	33,044	26,308	21,854	18,689	14,492	11,834	10,000	8,658	7,634	6,827	6,173	4,982	4,176	3,594	3,155	2,811	2,535	2,119	1,820	1,595	1,279	1,067
100	69,372	45,120	33,432	26,553	22,022	18,812	14,566	11,884	10,035	8,685	7,655	6,843	6,187	4,991	4,182	3,599	3,159	2,814	2,538	2,121	1,821	1,596	1,280	1,068
105	70,934	45,775	33,790	26,779	22,177	18,925	14,634	11,929	10,068	8,709	7,673	6,858	6,199	4,999	4,188	3,603	3,162	2,817	2,540	2,122	1,822	1,597	1,280	1,068
110	72,417	46,388	34,123	26,988	22,320	19,029	14,696	11,970	10,097	8,731	7,690	6,871	6,210	5,006	4,193	3,607	3,165	2,819	2,541	2,123	1,823	1,598	1,281	1,069
115	73,826	46,962	34,433	27,181	22,452	19,125	14,753	12,008	10,124	8,751	7,706	6,884	6,220	5,012	4,197	3,610	3,167	2,821	2,543	2,125	1,824	1,598	1,281	1,069
120	75,166	47,501	34,722	27,361	22,575	19,214	14,806	12,043	10,149	8,769	7,720	6,895	6,230	5,018	4,202	3,613	3,170	2,823	2,545	2,126	1,825	1,599	1,281	1,069
125	76,443	48,008	34,992	27,528	22,689	19,296	14,854	12,075	10,172	8,787	7,734	6,906	6,238	5,024	4,206	3,616	3,172	2,825	2,546	2,127	1,826	1,599	1,282	1,069
130	77,661	48,485	35,245	27,684	22,795	19,373	14,900	12,105	10,193	8,802	7,746	6,916	6,246	5,029	4,209	3,619	3 174	2,826	2,547	2,128	1,826	1,600	1,282	1,070
135	78,823	48,936	35,482	27,831	22,894	19,445	14,942	12,133	10,213	8,817	7,757	6,925	6,254	5,034	4,213	3,621	3 176	2,828	2,549	2,128	1,827	1,600	1,282	1,070
140	79,935	49,362	35,705	27,968	22,987	19,511	14,982	12,159	10,231	8,831	7,768	6,933	6,261	5,039	4,216	3,624	3 178	2,829	2,550	2,129	1,828	1,601	1,283	1,070
145	80,998	49,765	35,916	28,097	23,074	19,574	15,019	12,183	10,248	8,844	7,778	6,941	6,267	5,043	4,219	3,626	3 179	2,831	2,551	2,130	1,828	1,601	1,283	1,070
150	82,016	50,148	36,115	28,218	23,156	19,633	15,053	12,206	10,264	8,856	7,787	6,949	6,273	5,047	4,221	3,628	3,181	2,832	2,552	2,131	1,829	1,602	1,283	1,070
155	82,992	50,511	36,303	28,333	23,233	19,688	15,086	12,227	10,279	8,867	7,796	6,956	6,279	5,050	4,224	3,630	3,182	2,833	2,553	2,131	1,829	1,602	1,283	1,071
160	83,928	50,856	36,481	28,441	23,305	19,741	15,116	12,247	10,294	8,878	7,804	6,962	6,284	5,054	4,226	3,632	3,184	2,834	2,554	2,132	1,830	1,602	1,284	1,071
165	84,827	51,185	36,650	28,544	23,374	19,790	15,145	12,266	10,307	8,888	7,812	6,968	6,289	5,057	4,229	3,633	3,185	2,835	2,555	2,132	1,830	1,603	1,284	1,071
170	85,691	51,498	36,810	28,641	23,439	19,837	15,173	12,284	10,320	8,897	7,819	6,974	6,294	5,060	4,231	3,635	3,186	2,836	2,555	2,133	1,830	1,603	1,284	1,071
175	86,521	51,797	36,962	28,733	23,501	19,881	15,198	12,301	10,332	8,906	7,826	6,979	6,298	5,063	4,233	3,636	3,187	2,837	2,556	2,134	1,831	1,603	1,284	1,071
180	87,321	52,082	37,108	28,821	23,560	19,923	15,223	12,317	10,343	8,914	7,832	6,985	6,302	5,066	4,235	3,638	3,188	2,838	2,557	2,134	1,831	1,604	1,284	1,071
185	88,091	52,355	37,246	28,904	23,615	19,963	15,246	12,332	10,354	8,922	7,838	6,989	6,306	5,068	4,236	3,639	3,189	2,839	2,557	2,134	1,832	1,604	1,285	1,071
190	88,833	52,617	37,378	28,984	23,668	20,000	15,268	12,347	10,364	8,930	7,844	6,994	6,310	5,071	4,238	3,640	3,190	2,839	2,558	2,135	1,832	1,604	1,285	1,071
195	89,548	52,867	37,504	29,060	23,719	20,037	15,289	12,361	10,374	8,937	7,850	6,998	6,314	5,073	4,240	3,642	3,191	2,840	2,559	2,135	1,832	1,604	1,285	1,072
200	90,239	53,107	37,625	29,132	23,767	20,071	15,309	12,374	10,383	8,944	7,855	7,003	6,317	5,075	4,241	3,643	3,192	2,841	2,559	2,136	1,832	1,605	1,285	1,072

2/0 AWG | 1 Run(s) | 120 Volts

C-Value = 10,755 | Magnetic Duct | Copper

Calculations:

1. $f = \dfrac{2 \times \text{Length} \times \text{Isca} \times 1.5}{\text{Runs} \times \text{C-Value} \times E\ \text{L-N}}$

2. $M = \dfrac{1}{1 + f}$

3. $\text{Isc} = \text{Isca} \times 1.5 \times M$

4. Add Motor Contribution = Motor FLA x 4

* All results are given in symmetrical amperes

Single-Phase L-N Bolted Fault-Current* Table

One-Way Distance in Feet

Available Isc in Thousands (Isca)	5	10	15	20	25	30	40	50	60	70	80	90	100	125	150	175	200	225	250	300	350	400	500	600
3	4,372	4,252	4,138	4,029	3,927	3,829	3,648	3,483	3,332	3,194	3,067	2,950	2,841	2,601	2,399	2,226	2,076	1,945	1,829	1,535	1,478	1,349	1,148	999
5	7,152	6,835	6,545	6,278	6,032	5,805	5,398	5,045	4,735	4,461	4,217	3,998	3,801	3,384	3,049	2,774	2,545	2,351	2,185	1,913	1,702	1,533	1,278	1,097
7	9,830	9,241	8,718	8,251	7,832	7,453	6,796	6,245	5,777	5,374	5,024	4,716	4,444	3,884	3,450	3,102	2,819	2,582	2,383	2,064	1,820	1,628	1,344	1,144
10	13,670	12,556	11,610	10,797	10,090	9,470	8,433	7,602	6,919	6,349	5,866	5,451	5,091	4,369	3,827	3,404	3,065	2,788	2,557	2,193	1,920	1,707	1,398	1,183
15	19,634	17,415	15,647	14,205	13,006	11,994	10,378	9,147	8,176	7,392	6,745	6,202	5,740	4,839	4,182	3,683	3,290	2,972	2,711	2,305	2,005	1,775	1,442	1,215
20	25,112	21,593	18,940	16,867	15,203	13,838	11,731	10,181	8,993	8,053	7,291	6,661	6,131	5,114	4,386	3,840	3,414	3,074	2,795	2,366	2,051	1,810	1,466	1,232
25	30,161	25,225	21,677	19,004	16,918	15,244	12,727	10,923	9,567	8,510	7,664	6,970	6,392	5,294	4,518	3,941	3,494	3,138	2,848	2,404	2,080	1,832	1,480	1,242
30	34,830	28,410	23,988	20,757	18,293	16,352	13,490	11,480	9,991	8,845	7,934	7,193	6,579	5,422	4,611	4,011	3,549	3,183	2,885	2,430	2,099	1,847	1,490	1,249
35	39,160	31,226	25,965	22,221	19,421	17,248	14,093	11,914	10,319	9,100	8,139	7,361	6,720	5,517	4,679	4,063	3,589	3,215	2,911	2,449	2,113	1,858	1,497	1,254
40	43,187	33,734	27,676	23,463	20,363	17,986	14,582	12,262	10,579	9,302	8,300	7,493	6,829	5,590	4,732	4,102	3,620	3,240	2,932	2,463	2,124	1,867	1,503	1,257
45	46,941	35,982	29,171	24,528	21,160	18,606	14,987	12,547	10,790	9,465	8,429	7,598	6,916	5,649	4,774	4,134	3,645	3,259	2,948	2,474	2,132	1,873	1,507	1,260
50	50,449	38,008	30,489	25,453	21,845	19,133	15,327	12,785	10,965	9,599	8,536	7,685	6,988	5,696	4,808	4,159	3,665	3,275	2,961	2,484	2,139	1,878	1,510	1,263
55	53,735	39,843	31,659	26,263	22,439	19,588	15,618	12,986	11,113	9,712	8,625	7,757	7,048	5,736	4,836	4,180	3,681	3,288	2,971	2,491	2,144	1,882	1,513	1,265
60	56,819	41,514	32,704	26,979	22,960	19,983	15,868	13,158	11,239	9,809	8,701	7,818	7,098	5,769	4,860	4,198	3,695	3,299	2,980	2,497	2,149	1,886	1,515	1,266
65	59,719	43,041	33,645	27,616	23,419	20,330	16,086	13,308	11,348	9,891	8,766	7,871	7,141	5,798	4,880	4,213	3,706	3,309	2,988	2,503	2,153	1,889	1,517	1,268
70	62,451	44,442	34,495	28,186	23,828	20,637	16,278	13,439	11,443	9,964	8,823	7,916	7,179	5,823	4,898	4,226	3,717	3,317	2,994	2,507	2,156	1,892	1,519	1,269
75	65,030	45,733	35,267	28,700	24,194	20,911	16,448	13,555	11,527	10,027	8,873	7,956	7,212	5,844	4,913	4,237	3,725	3,324	3,000	2,511	2,159	1,894	1,520	1,270
80	67,467	46,925	35,972	29,165	24,524	21,157	16,600	13,658	11,601	10,083	8,917	7,992	7,241	5,863	4,926	4,247	3,733	3,330	3,005	2,515	2,162	1,896	1,522	1,271
85	69,775	48,030	36,618	29,588	24,822	21,379	16,736	13,750	11,668	10,133	8,956	8,023	7,267	5,880	4,938	4,256	3,740	3,335	3,010	2,518	2,164	1,898	1,523	1,271
90	71,963	49,057	37,212	29,974	25,094	21,580	16,859	13,832	11,727	10,178	8,991	8,051	7,290	5,895	4,949	4,264	3,746	3,340	3,014	2,521	2,166	1,899	1,524	1,272
95	74,040	50,013	37,760	30,329	25,342	21,763	16,970	13,907	11,781	10,219	9,022	8,077	7,310	5,909	4,958	4,271	3,751	3,344	3,017	2,523	2,168	1,901	1,525	1,273
100	76,015	50,906	38,266	30,655	25,569	21,931	17,072	13,976	11,830	10,256	9,051	8,100	7,329	5,921	4,967	4,278	3,756	3,348	3,020	2,525	2,170	1,902	1,525	1,273
105	77,895	51,743	38,737	30,956	25,778	22,084	17,165	14,038	11,875	10,289	9,077	8,121	7,346	5,932	4,975	4,284	3,761	3,352	3,023	2,527	2,171	1,903	1,526	1,274
110	79,686	52,527	39,175	31,235	25,972	22,226	17,250	14,095	11,916	10,320	9,101	8,140	7,362	5,943	4,982	4,289	3,765	3,355	3,026	2,529	2,173	1,904	1,527	1,274
115	81,395	53,264	39,584	31,494	26,150	22,357	17,329	14,148	11,953	10,348	9,123	8,157	7,376	5,952	4,989	4,294	3,769	3,358	3,028	2,531	2,174	1,905	1,528	1,275
120	83,028	53,958	39,966	31,736	26,317	22,478	17,402	14,196	11,988	10,374	9,143	8,173	7,389	5,961	4,995	4,298	3,772	3,361	3,030	2,532	2,175	1,906	1,528	1,275
125	84,588	54,613	40,324	31,961	26,472	22,591	17,470	14,241	12,020	10,398	9,162	8,188	7,402	5,968	5,000	4,302	3,775	3,363	3,032	2,534	2,176	1,907	1,529	1,276
130	86,082	55,232	40,660	32,172	26,616	22,696	17,532	14,283	12,049	10,420	9,179	8,202	7,413	5,976	5,005	4,306	3,778	3,366	3,034	2,535	2,177	1,908	1,529	1,276
135	87,513	55,818	40,977	32,370	26,751	22,795	17,591	14,322	12,077	10,441	9,195	8,215	7,423	5,983	5,010	4,310	3,781	3,368	3,036	2,536	2,178	1,908	1,530	1,276
140	88,885	56,373	41,275	32,556	26,878	22,887	17,646	14,358	12,103	10,460	9,210	8,227	7,433	5,989	5,015	4,313	3,783	3,370	3,038	2,538	2,179	1,909	1,530	1,276
145	90,201	56,899	41,557	32,731	26,997	22,973	17,697	14,392	12,127	10,478	9,224	8,238	7,442	5,995	5,019	4,316	3,786	3,372	3,039	2,539	2,180	1,910	1,530	1,277
150	91,466	57,400	41,823	32,896	27,109	23,054	17,745	14,424	12,150	10,495	9,237	8,248	7,451	6,000	5,023	4,319	3,788	3,373	3,041	2,540	2,180	1,910	1,531	1,277
155	92,681	57,876	42,075	33,052	27,215	23,131	17,790	14,454	12,171	10,511	9,249	8,258	7,459	6,005	5,026	4,321	3,790	3,375	3,042	2,541	2,181	1,911	1,531	1,277
160	93,850	58,330	42,315	33,199	27,315	23,203	17,833	14,482	12,191	10,526	9,261	8,267	7,466	6,010	5,030	4,324	3,792	3,377	3,043	2,541	2,182	1,911	1,531	1,277
165	94,976	58,763	42,542	33,339	27,410	23,271	17,873	14,508	12,209	10,540	9,271	8,276	7,473	6,015	5,033	4,326	3,794	3,378	3,044	2,542	2,182	1,912	1,532	1,278
170	96,060	59,176	42,758	33,472	27,499	23,335	17,911	14,533	12,227	10,553	9,282	8,284	7,480	6,019	5,036	4,329	3,796	3,379	3,046	2,543	2,183	1,912	1,532	1,278
175	97,105	59,571	42,964	33,598	27,584	23,397	17,947	14,557	12,244	10,565	9,291	8,292	7,486	6,023	5,039	4,331	3,797	3,381	3,047	2,544	2,183	1,912	1,532	1,278
180	98,113	59,949	43,160	33,718	27,665	23,455	17,982	14,579	12,260	10,577	9,300	8,299	7,492	6,027	5,041	4,333	3,799	3,382	3,048	2,544	2,184	1,913	1,532	1,278
185	99,086	60,311	43,347	33,832	27,742	23,510	18,014	14,601	12,275	10,588	9,309	8,306	7,498	6,031	5,044	4,335	3,800	3,383	3,048	2,545	2,184	1,913	1,533	1,278
190	100,026	60,657	43,526	33,941	27,815	23,562	18,045	14,621	12,289	10,599	9,317	8,312	7,503	6,034	5,046	4,336	3,802	3,384	3,049	2,546	2,185	1,914	1,533	1,279
195	100,934	60,990	43,697	34,045	27,885	23,612	18,074	14,640	12,303	10,609	9,325	8,319	7,508	6,037	5,049	4,338	3,803	3,385	3,050	2,546	2,185	1,914	1,533	1,279
200	101,813	61,310	43,861	34,144	27,951	23,660	18,102	14,659	12,316	10,619	9,333	8,324	7,513	6,041	5,051	4,340	3,804	3,386	3,051	2,547	2,186	1,914	1,533	1,279

3/0 AWG | 1 Run(s) | 120 Volts | C-Value = 12,843 | Magnetic Duct | Copper

Calculations:

1. $f = \dfrac{2 \times \text{Length} \times \text{Isca} \times 1.5}{\text{Runs} \times \text{C-Value} \times \text{E L-N}}$

2. $M = \dfrac{1}{1 + f}$

3. Isc = Isca x 1.5 x M

4. Add Motor Contribution = Motor FLA x 4

* All results are given in symmetrical amperes

Single-Phase L-N Bolted Fault-Current* Table

One-Way Distance in Feet

Available Isc in Thousands (Isca)	5	10	15	20	25	30	40	50	60	70	80	90	100	125	150	175	200	225	250	300	350	400	500	600
3	4,391	4,287	4,188	4,093	4,002	3,916	3,753	3,604	3,466	3,338	3,219	3,109	3,005	2,775	2,577	2,406	2,256	2,124	2,006	1,806	1,642	1,505	1,291	1,130
5	7,202	6,926	6,671	6,434	6,213	6,007	5,633	5,303	5,009	4,746	4,510	4,296	4,101	3,684	3,343	3,061	2,822	2,618	2,441	2,151	1,923	1,738	1,458	1,256
7	9,924	9,408	8,943	8,522	8,139	7,789	7,171	6,645	6,190	5,794	5,445	5,136	4,860	4,285	3,831	3,465	3,162	2,908	2,692	2,343	2,075	1,861	1,544	1,319
10	13,852	12,867	12,013	11,265	10,605	10,018	9,020	8,202	7,520	6,943	6,449	6,020	5,644	4,883	4,302	3,845	3,476	3,172	2,916	2,511	2,205	1,966	1,615	1,370
15	20,012	18,020	16,388	15,027	13,875	12,887	11,281	10,030	9,029	8,210	7,527	6,949	6,454	5,477	4,757	4,205	3,767	3,412	3,118	2,660	2,319	2,056	1,675	1,413
20	25,734	22,531	20,036	18,039	16,404	15,041	12,897	11,288	10,036	9,034	8,214	7,531	6,952	5,832	5,023	4,411	3,932	3,546	3,230	2,741	2,380	2,104	1,707	1,436
25	31,064	26,513	23,125	20,505	18,418	16,717	14,110	12,207	10,756	9,613	8,690	7,929	7,290	6,068	5,197	4,544	4,037	3,632	3,301	2,792	2,419	2,134	1,727	1,450
30	36,039	30,054	25,774	22,561	20,061	18,059	15,055	12,907	11,296	10,042	9,039	8,218	7,534	6,236	5,320	4,638	4,111	3,692	3,350	2,827	2,445	2,154	1,740	1,459
35	40,695	33,224	28,071	24,302	21,425	19,157	15,810	13,459	11,716	10,373	9,306	8,439	7,719	6,362	5,411	4,707	4,166	3,736	3,386	2,853	2,464	2,169	1,750	1,466
40	45,061	36,078	30,082	25,794	22,577	20,073	16,429	13,904	12,052	10,636	9,517	8,612	7,863	6,460	5,482	4,761	4,207	3,769	3,414	2,872	2,479	2,180	1,757	1,471
45	49,164	38,662	31,856	27,088	23,562	20,848	16,944	14,272	12,328	10,850	9,688	8,751	7,979	6,538	5,538	4,803	4,240	3,796	3,435	2,887	2,490	2,189	1,763	1,475
50	53,026	41,010	33,434	28,221	24,414	21,512	17,380	14,580	12,557	11,027	9,829	8,866	8,075	6,602	5,584	4,837	4,267	3,817	3,453	2,900	2,499	2,196	1,767	1,478
55	56,668	43,156	34,847	29,220	25,159	22,088	17,754	14,842	12,751	11,176	9,948	8,962	8,155	6,655	5,622	4,866	4,289	3,835	3,468	2,910	2,507	2,202	1,771	1,481
60	60,109	45,123	36,118	30,109	25,815	22,592	18,079	15,068	12,917	11,304	10,049	9,044	8,222	6,700	5,654	4,890	4,308	3,850	3,480	2,919	2,513	2,207	1,774	1,483
65	63,364	46,933	37,268	30,904	26,397	23,037	18,362	15,265	13,062	11,414	10,136	9,115	8,281	6,739	5,681	4,911	4,324	3,863	3,490	2,926	2,519	2,211	1,777	1,485
70	66,449	48,604	38,314	31,620	26,917	23,432	18,613	15,438	13,188	11,510	10,211	9,176	8,331	6,772	5,705	4,928	4,338	3,873	3,499	2,932	2,523	2,215	1,779	1,487
75	69,376	50,151	39,270	32,268	27,386	23,786	18,835	15,590	13,299	11,595	10,278	9,230	8,375	6,802	5,726	4,944	4,350	3,883	3,507	2,938	2,527	2,218	1,781	1,488
80	72,157	51,589	40,146	32,857	27,809	24,105	19,035	15,727	13,398	11,670	10,337	9,277	8,415	6,827	5,744	4,957	4,360	3,891	3,514	2,942	2,531	2,220	1,783	1,489
85	74,803	52,927	40,951	33,395	28,193	24,393	19,214	15,849	13,487	11,737	10,390	9,320	8,450	6,850	5,760	4,969	4,370	3,899	3,520	2,947	2,534	2,223	1,785	1,491
90	77,323	54,177	41,695	33,888	28,544	24,655	19,376	15,959	13,566	11,798	10,437	9,358	8,481	6,871	5,775	4,980	4,378	3,906	3,525	2,950	2,537	2,225	1,786	1,492
95	79,726	55,346	42,384	34,342	28,865	24,894	19,523	16,059	13,639	11,852	10,480	9,392	8,509	6,889	5,788	4,990	4,385	3,911	3,530	2,954	2,539	2,227	1,787	1,492
100	82,021	56,442	43,024	34,761	29,160	25,114	19,658	16,150	13,704	11,902	10,518	9,423	8,534	6,906	5,800	4,999	4,392	3,917	3,534	2,957	2,542	2,229	1,788	1,493
105	84,214	57,472	43,620	35,149	29,433	25,316	19,782	16,233	13,764	11,947	10,554	9,451	8,558	6,921	5,810	5,007	4,398	3,922	3,538	2,960	2,544	2,230	1,789	1,494
110	86,311	58,441	44,176	35,509	29,685	25,502	19,895	16,309	13,819	11,988	10,586	9,477	8,579	6,935	5,820	5,014	4,404	3,926	3,542	2,962	2,546	2,232	1,790	1,495
115	88,320	59,355	44,696	35,844	29,919	25,674	20,000	16,380	13,869	12,026	10,615	9,501	8,598	6,948	5,829	5,020	4,409	3,930	3,545	2,965	2,547	2,233	1,791	1,495
120	90,245	60,218	45,184	36,157	30,137	25,835	20,097	16,445	13,916	12,061	10,643	9,523	8,616	6,959	5,837	5,027	4,414	3,934	3,548	2,967	2,549	2,234	1,792	1,496
125	92,092	61,035	45,643	36,450	30,340	25,984	20,187	16,505	13,959	12,094	10,668	9,543	8,633	6,970	5,845	5,032	4,418	3,937	3,551	2,969	2,550	2,235	1,793	1,496
130	93,865	61,809	46,074	36,725	30,530	26,123	20,271	16,561	13,999	12,124	10,691	9,562	8,648	6,980	5,852	5,037	4,422	3,941	3,554	2,970	2,552	2,236	1,793	1,497
135	95,569	62,543	46,481	36,983	30,708	26,253	20,350	16,614	14,037	12,152	10,713	9,579	8,662	6,989	5,858	5,042	4,426	3,944	3,556	2,972	2,553	2,237	1,794	1,497
140	97,208	63,241	46,865	37,226	30,875	26,375	20,423	16,662	14,071	12,178	10,733	9,595	8,675	6,998	5,864	5,047	4,429	3,946	3,558	2,974	2,554	2,238	1,794	1,497
145	98,784	63,904	47,228	37,454	31,032	26,490	20,492	16,708	14,104	12,202	10,752	9,610	8,688	7,006	5,870	5,051	4,432	3,949	3,560	2,975	2,555	2,239	1,795	1,498
150	100,303	64,536	47,573	37,671	31,181	26,598	20,556	16,751	14,135	12,225	10,770	9,625	8,699	7,014	5,875	5,055	4,435	3,951	3,562	2,976	2,556	2,240	1,795	1,498
155	101,766	65,139	47,899	37,875	31,321	26,700	20,617	16,791	14,163	12,247	10,787	9,638	8,710	7,021	5,880	5,058	4,438	3,953	3,564	2,978	2,557	2,240	1,796	1,498
160	103,178	65,714	48,210	38,069	31,453	26,796	20,674	16,829	14,190	12,267	10,802	9,650	8,720	7,027	5,885	5,062	4,441	3,956	3,566	2,979	2,558	2,241	1,796	1,499
165	104,540	66,264	48,505	38,253	31,578	26,887	20,728	16,865	14,216	12,286	10,817	9,662	8,730	7,034	5,889	5,065	4,443	3,958	3,568	2,980	2,559	2,242	1,797	1,499
170	105,855	66,790	48,786	38,428	31,697	26,973	20,779	16,899	14,240	12,304	10,831	9,673	8,739	7,040	5,893	5,068	4,446	3,959	3,569	2,981	2,560	2,242	1,797	1,499
175	107,125	67,294	49,054	38,594	31,810	27,055	20,828	16,931	14,263	12,321	10,844	9,684	8,748	7,045	5,897	5,071	4,448	3,961	3,570	2,982	2,560	2,243	1,797	1,500
180	108,353	67,776	49,310	38,752	31,918	27,133	20,874	16,961	14,284	12,337	10,857	9,694	8,756	7,050	5,901	5,074	4,450	3,963	3,572	2,983	2,561	2,244	1,798	1,500
185	109,542	68,239	49,555	38,903	32,020	27,207	20,918	16,990	14,305	12,352	10,868	9,703	8,763	7,055	5,904	5,076	4,452	3,964	3,573	2,984	2,562	2,244	1,798	1,500
190	110,691	68,684	49,789	39,047	32,118	27,277	20,959	17,018	14,324	12,366	10,880	9,712	8,771	7,060	5,908	5,079	4,454	3,966	3,574	2,985	2,562	2,244	1,798	1,500
195	111,805	69,111	50,013	39,185	32,211	27,344	20,999	17,044	14,342	12,380	10,890	9,721	8,778	7,065	5,911	5,081	4,456	3,967	3,575	2,986	2,563	2,245	1,799	1,500
200	112,884	69,522	50,228	39,316	32,300	27,408	21,037	17,069	14,360	12,393	10,900	9,729	8,784	7,069	5,914	5,083	4,457	3,969	3,577	2,986	2,563	2,245	1,799	1,501

4/0 AWG | 1 Run(s) | 120 Volts | C-Value = 15,082 | Magnetic Duct | Copper

Calculations:

1. $f = \dfrac{2 \times \text{Length} \times \text{Isca} \times 1.5}{\text{Runs} \times \text{C-Value} \times \text{E L-N}}$

2. $M = \dfrac{1}{1 + f}$

3. $\text{Isc} = \text{Isca} \times 1.5 \times M$

4. Add Motor Contribution = Motor FLA x 4

* All results are given in symmetrical amperes

Single-Phase L-N Bolted Fault-Current* Table

One-Way Distance in Feet

Available Isc in Thousands (Isca)	5	10	15	20	25	30	40	50	60	70	80	90	100	125	150	175	200	225	250	300	350	400	500	600
3	4,400	4,304	4,212	4,125	4,040	3,960	3,807	3,666	3,535	3,413	3,299	3,193	3,093	2,868	2,675	2,505	2,356	2,224	2,105	1,903	1,736	1,596	1,374	1,206
5	7,226	6,971	6,734	6,512	6,305	6,110	5,754	5,438	5,155	4,899	4,668	4,458	4,265	3,850	3,509	3,223	2,980	2,771	2,590	2,290	2,052	1,859	1,565	1,351
7	9,971	9,492	9,058	8,661	8,298	7,964	7,370	6,859	6,414	6,023	5,678	5,369	5,093	4,512	4,050	3,674	3,362	3,098	2,873	2,509	2,226	2,001	1,664	1,425
10	13,943	13,025	12,220	11,509	10,876	10,309	9,336	8,531	7,853	7,276	6,777	6,342	5,960	5,180	4,580	4,105	3,719	3,399	3,130	2,703	2,378	2,123	1,748	1,485
15	20,202	18,330	16,775	15,464	14,342	13,373	11,780	10,526	9,514	8,679	7,979	7,383	6,870	5,854	5,099	4,517	4,054	3,677	3,364	2,875	2,510	2,228	1,818	1,536
20	26,049	23,018	20,618	18,672	17,061	15,707	13,554	11,920	10,638	9,605	8,755	8,043	7,438	6,261	5,405	4,755	4,245	3,834	3,495	2,970	2,582	2,284	1,856	1,562
25	31,523	27,190	23,904	21,327	19,251	17,544	14,900	12,949	11,450	10,262	9,297	8,498	7,826	6,533	5,607	4,911	4,369	3,934	3,578	3,030	2,628	2,320	1,879	1,579
30	36,660	30,928	26,746	23,560	21,052	19,027	15,957	13,740	12,064	10,752	9,698	8,832	8,108	6,729	5,751	5,021	4,455	4,004	3,636	3,072	2,659	2,344	1,895	1,590
35	41,488	34,295	29,227	25,464	22,560	20,250	16,809	14,367	12,545	11,132	10,006	9,087	8,322	6,876	5,858	5,102	4,519	4,056	3,679	3,102	2,681	2,361	1,906	1,598
40	46,035	37,344	31,413	27,108	23,841	21,276	17,509	14,876	12,931	11,436	10,250	9,288	8,490	6,990	5,940	5,165	4,568	4,095	3,711	3,125	2,699	2,375	1,915	1,604
45	50,326	40,118	33,353	28,541	24,942	22,149	18,096	15,297	13,248	11,683	10,449	9,450	8,626	7,082	6,007	5,215	4,607	4,127	3,737	3,143	2,712	2,385	1,922	1,609
50	54,380	42,653	35,087	29,801	25,899	22,900	18,595	15,652	13,513	11,889	10,613	9,584	8,738	7,157	6,060	5,255	4,639	4,152	3,758	3,158	2,723	2,394	1,927	1,613
55	58,218	44,979	36,646	30,918	26,738	23,554	19,023	15,954	13,738	12,063	10,751	9,697	8,831	7,219	6,105	5,289	4,665	4,173	3,775	3,170	2,732	2,401	1,932	1,616
60	61,855	47,120	38,054	31,914	27,480	24,128	19,396	16,216	13,932	12,211	10,869	9,793	8,911	7,273	6,143	5,317	4,687	4,191	3,789	3,180	2,740	2,406	1,935	1,619
65	65,308	49,097	39,334	32,809	28,141	24,636	19,723	16,444	14,099	12,340	10,971	9,876	8,979	7,318	6,176	5,342	4,706	4,206	3,802	3,189	2,746	2,411	1,939	1,621
70	68,589	50,929	40,501	33,617	28,734	25,089	20,012	16,644	14,247	12,453	11,060	9,948	9,038	7,357	6,204	5,363	4,722	4,219	3,812	3,196	2,752	2,416	1,941	1,623
75	71,712	52,631	41,570	34,350	29,268	25,495	20,270	16,822	14,377	12,552	11,138	10,011	9,091	7,392	6,228	5,381	4,737	4,230	3,822	3,203	2,756	2,419	1,944	1,624
80	74,688	54,216	42,552	35,019	29,751	25,861	20,501	16,981	14,492	12,640	11,208	10,067	9,137	7,422	6,250	5,397	4,749	4,240	3,830	3,208	2,761	2,423	1,946	1,626
85	77,526	55,696	43,459	35,630	30,192	26,193	20,709	17,123	14,596	12,719	11,270	10,117	9,178	7,450	6,269	5,411	4,760	4,249	3,837	3,214	2,764	2,425	1,948	1,627
90	80,237	57,081	44,298	36,192	30,594	26,496	20,897	17,252	14,689	12,790	11,325	10,162	9,215	7,474	6,286	5,424	4,770	4,257	3,843	3,218	2,768	2,428	1,949	1,628
95	82,828	58,381	45,076	36,710	30,963	26,772	21,069	17,369	14,774	12,854	11,375	10,202	9,248	7,496	6,302	5,436	4,779	4,264	3,849	3,222	2,771	2,430	1,951	1,629
100	85,307	59,602	45,801	37,189	31,304	27,026	21,226	17,475	14,851	12,912	11,421	10,239	9,278	7,515	6,316	5,446	4,787	4,270	3,854	3,226	2,773	2,432	1,952	1,630
105	87,681	60,751	46,476	37,634	31,618	27,260	21,370	17,573	14,921	12,965	11,463	10,272	9,305	7,533	6,328	5,456	4,794	4,276	3,859	3,229	2,776	2,434	1,953	1,631
110	89,958	61,835	47,108	38,047	31,909	27,476	21,502	17,662	14,986	13,014	11,501	10,303	9,331	7,550	6,340	5,464	4,801	4,281	3,863	3,232	2,778	2,436	1,955	1,632
115	92,142	62,859	47,700	38,432	32,180	27,677	21,625	17,745	15,045	13,059	11,536	10,331	9,354	7,565	6,350	5,472	4,807	4,286	3,867	3,235	2,780	2,438	1,956	1,633
120	94,239	63,828	48,256	38,792	32,432	27,863	21,739	17,821	15,100	13,100	11,568	10,356	9,375	7,579	6,360	5,479	4,813	4,291	3,871	3,237	2,782	2,439	1,956	1,633
125	96,255	64,747	48,779	39,129	32,667	28,037	21,844	17,892	15,151	13,138	11,598	10,380	9,394	7,592	6,369	5,486	4,818	4,295	3,874	3,240	2,784	2,440	1,957	1,634
130	98,194	65,618	49,272	39,446	32,887	28,199	21,942	17,958	15,198	13,174	11,625	10,402	9,412	7,603	6,378	5,492	4,823	4,299	3,877	3,242	2,785	2,441	1,958	1,634
135	100,060	66,447	49,738	39,744	33,094	28,351	22,034	18,020	15,242	13,207	11,651	10,423	9,429	7,614	6,385	5,498	4,827	4,302	3,880	3,244	2,787	2,443	1,959	1,635
140	101,858	67,234	50,178	40,024	33,288	28,493	22,120	18,077	15,283	13,238	11,675	10,442	9,445	7,625	6,392	5,503	4,831	4,305	3,883	3,246	2,788	2,444	1,960	1,635
145	103,590	67,985	50,595	40,289	33,471	28,627	22,201	18,131	15,322	13,267	11,697	10,460	9,460	7,634	6,399	5,508	4,835	4,308	3,885	3,247	2,789	2,445	1,960	1,636
150	105,261	68,701	50,990	40,539	33,644	28,753	22,277	18,181	15,358	13,294	11,718	10,477	9,473	7,643	6,405	5,513	4,839	4,311	3,888	3,249	2,791	2,446	1,961	1,636
155	106,874	69,384	51,366	40,776	33,807	28,872	22,348	18,229	15,392	13,319	11,738	10,493	9,486	7,651	6,411	5,517	4,842	4,314	3,890	3,251	2,792	2,446	1,961	1,637
160	108,432	70,037	51,723	41,001	33,961	28,985	22,415	18,274	15,424	13,343	11,757	10,508	9,498	7,659	6,417	5,521	4,845	4,316	3,892	3,252	2,793	2,447	1,962	1,637
165	109,937	70,662	52,063	41,215	34,108	29,091	22,479	18,316	15,454	13,365	11,774	10,522	9,510	7,667	6,422	5,525	4,848	4,319	3,894	3,253	2,794	2,448	1,962	1,637
170	111,392	71,261	52,387	41,417	34,246	29,192	22,539	18,356	15,482	13,387	11,791	10,535	9,521	7,674	6,427	5,529	4,851	4,321	3,895	3,255	2,795	2,449	1,963	1,638
175	112,800	71,834	52,696	41,611	34,378	29,288	22,596	18,394	15,509	13,407	11,806	10,547	9,531	7,680	6,432	5,532	4,853	4,323	3,897	3,256	2,796	2,449	1,963	1,638
180	114,163	72,384	52,992	41,795	34,504	29,379	22,650	18,429	15,535	13,426	11,821	10,559	9,540	7,687	6,436	5,535	4,856	4,325	3,899	3,257	2,796	2,450	1,964	1,638
185	115,483	72,913	53,274	41,970	34,623	29,466	22,702	18,464	15,559	13,444	11,835	10,570	9,549	7,693	6,440	5,539	4,858	4,327	3,900	3,258	2,797	2,451	1,964	1,639
190	116,761	73,420	53,545	42,138	34,737	29,548	22,751	18,496	15,582	13,461	11,848	10,581	9,558	7,698	6,444	5,541	4,861	4,329	3,902	3,259	2,798	2,451	1,964	1,639
195	118,001	73,909	53,804	42,298	34,846	29,627	22,797	18,527	15,604	13,477	11,861	10,591	9,566	7,703	6,448	5,544	4,863	4,330	3,903	3,260	2,799	2,452	1,965	1,639
200	119,203	74,378	54,053	42,452	34,950	29,702	22,842	18,556	15,625	13,493	11,873	10,600	9,574	7,709	6,451	5,547	4,865	4,332	3,904	3,261	2,799	2,452	1,965	1,639

250 MCM | 1 Run(s) | 120 Volts

C-Value = 16,483 | Magnetic Duct | Copper

Calculations:

1. $f = \frac{2 \times \text{Length} \times \text{Isca} \times 1.5}{\text{Runs} \times \text{C-Value} \times \text{E L-N}}$

2. $M = \frac{1}{1 + f}$

3. $\text{Isc} = \text{Isca} \times 1.5 \times M$

4. Add Motor Contribution = Motor FLA x 4

* All results are given in symmetrical amperes

Single-Phase L-N Bolted Fault-Current* Table

One-Way Distance in Feet

Available Isc in Thousands (Isca)

Isca	5	10	15	20	25	30	40	50	60	70	80	90	100	125	150	175	200	225	250	300	350	400	500	600
3	4,409	4,322	4,238	4,157	4,079	4,004	3,862	3,730	3,607	3,492	3,383	3,281	3,186	2,969	2,780	2,613	2,465	2,334	2,215	2,011	1,841	1,698	1,469	1,295
5	7,251	7,017	6,799	6,593	6,400	6,217	5,882	5,581	5,309	5,063	4,838	4,633	4,444	4,033	3,692	3,404	3,157	2,944	2,758	2,448	2,201	2,000	1,690	1,463
7	10,018	9,578	9,175	8,805	8,463	8,147	7,581	7,088	6,655	6,273	5,931	5,625	5,349	4,765	4,296	3,911	3,589	3,316	3,082	2,700	2,403	2,164	1,806	1,549
10	14,035	13,186	12,435	11,764	11,162	10,618	9,676	8,888	8,218	7,642	7,142	6,703	6,315	5,516	4,897	4,403	3,999	3,663	3,379	2,926	2,580	2,307	1,904	1,621
15	20,396	18,652	17,182	15,928	14,844	13,898	12,327	11,075	10,054	9,205	8,489	7,876	7,345	6,287	5,495	4,880	4,389	3,988	3,654	3,130	2,737	2,432	1,988	1,682
20	26,373	23,528	21,237	19,353	17,775	16,436	14,283	12,629	11,318	10,254	9,373	8,631	7,998	6,759	5,852	5,160	4,614	4,173	3,808	3,242	2,823	2,499	2,033	1,714
25	31,998	27,905	24,740	22,219	20,165	18,459	15,787	13,790	12,242	11,007	9,998	9,158	8,449	7,078	6,090	5,344	4,761	4,292	3,908	3,314	2,877	2,542	2,061	1,734
30	37,304	31,855	27,796	24,654	22,150	20,108	16,978	14,691	12,947	11,573	10,463	9,547	8,778	7,308	6,259	5,474	4,863	4,376	3,977	3,363	2,914	2,571	2,080	1,747
35	42,315	35,439	30,486	26,747	23,826	21,479	17,945	15,409	13,502	12,014	10,822	9,845	9,030	7,481	6,386	5,571	4,940	4,437	4,028	3,400	2,941	2,592	2,094	1,757
40	47,056	38,705	32,872	28,567	25,258	22,637	18,746	15,996	13,950	12,368	11,108	10,081	9,228	7,617	6,485	5,645	4,999	4,485	4,067	3,428	2,962	2,608	2,105	1,764
45	51,547	41,694	35,003	30,162	26,498	23,627	19,420	16,485	14,320	12,658	11,342	10,273	9,389	7,726	6,563	5,705	5,045	4,522	4,097	3,449	2,978	2,621	2,113	1,770
50	55,809	44,439	36,917	31,573	27,581	24,485	19,995	16,897	14,630	12,900	11,535	10,432	9,521	7,815	6,628	5,754	5,083	4,553	4,122	3,467	2,992	2,631	2,119	1,775
55	59,859	46,969	38,646	32,829	28,535	25,233	20,492	17,251	14,895	13,105	11,699	10,566	9,632	7,890	6,682	5,794	5,115	4,578	4,143	3,482	3,002	2,639	2,125	1,778
60	63,711	49,308	40,216	33,955	29,381	25,893	20,925	17,556	15,122	13,281	11,839	10,679	9,727	7,953	6,727	5,828	5,141	4,599	4,161	3,494	3,012	2,646	2,130	1,782
65	67,380	51,477	41,648	34,970	30,138	26,479	21,306	17,824	15,320	13,433	11,960	10,778	9,808	8,008	6,766	5,857	5,164	4,617	4,175	3,505	3,019	2,652	2,133	1,784
70	70,879	53,495	42,959	35,890	30,819	27,003	21,644	18,060	15,494	13,566	12,066	10,864	9,879	8,055	6,800	5,883	5,184	4,633	4,188	3,514	3,026	2,657	2,137	1,787
75	74,219	55,376	44,163	36,727	31,434	27,474	21,946	18,269	15,648	13,684	12,159	10,939	9,942	8,097	6,829	5,905	5,201	4,647	4,199	3,521	3,032	2,662	2,140	1,789
80	77,411	57,133	45,274	37,492	31,992	27,900	22,216	18,457	15,785	13,789	12,241	11,006	9,997	8,133	6,855	5,924	5,216	4,659	4,209	3,528	3,037	2,666	2,142	1,790
85	80,464	58,779	46,302	38,194	32,502	28,287	22,461	18,625	15,908	13,883	12,315	11,066	10,046	8,166	6,878	5,941	5,229	4,669	4,218	3,534	3,042	2,669	2,144	1,792
90	83,387	60,325	47,255	38,840	32,969	28,640	22,683	18,777	16,019	13,968	12,382	11,119	10,090	8,195	6,899	5,957	5,241	4,679	4,226	3,540	3,046	2,672	2,146	1,793
95	86,189	61,777	48,142	39,437	33,398	28,963	22,885	18,916	16,120	14,044	12,442	11,168	10,130	8,221	6,917	5,971	5,252	4,687	4,233	3,545	3,049	2,675	2,148	1,795
100	88,877	63,146	48,969	39,991	33,794	29,261	23,071	19,042	16,212	14,114	12,496	11,212	10,166	8,245	6,934	5,983	5,262	4,695	4,239	3,549	3,052	2,678	2,150	1,796
105	91,458	64,438	49,742	40,505	34,161	29,535	23,241	19,158	16,295	14,177	12,546	11,252	10,199	8,267	6,950	5,995	5,270	4,702	4,245	3,553	3,055	2,680	2,151	1,797
110	93,937	65,659	50,467	40,984	34,501	29,789	23,398	19,265	16,372	14,235	12,592	11,288	10,229	8,286	6,964	6,005	5,278	4,709	4,250	3,557	3,058	2,682	2,153	1,798
115	96,321	66,815	51,147	41,431	34,818	30,025	23,543	19,363	16,443	14,289	12,634	11,322	10,257	8,304	6,976	6,014	5,286	4,714	4,255	3,560	3,061	2,684	2,154	1,799
120	98,616	67,911	51,787	41,850	35,113	30,244	23,678	19,454	16,509	14,338	12,672	11,353	10,283	8,321	6,988	6,023	5,292	4,720	4,259	3,563	3,063	2,686	2,155	1,799
125	100,825	68,952	52,390	42,243	35,389	30,449	23,803	19,538	16,570	14,384	12,708	11,382	10,306	8,337	6,999	6,031	5,299	4,725	4,263	3,566	3,065	2,687	2,156	1,800
130	102,955	69,941	52,959	42,612	35,648	30,640	23,920	19,617	16,626	14,427	12,741	11,408	10,328	8,351	7,009	6,039	5,304	4,729	4,267	3,569	3,067	2,689	2,157	1,801
135	105,008	70,882	53,497	42,960	35,891	30,819	24,029	19,690	16,679	14,466	12,772	11,433	10,348	8,364	7,018	6,046	5,310	4,734	4,270	3,571	3,069	2,690	2,158	1,801
140	106,990	71,780	54,006	43,288	36,119	30,988	24,131	19,759	16,728	14,503	12,801	11,456	10,367	8,376	7,027	6,052	5,315	4,738	4,273	3,573	3,070	2,691	2,159	1,802
145	108,903	72,636	54,490	43,598	36,335	31,146	24,227	19,823	16,774	14,538	12,828	11,478	10,385	8,388	7,035	6,058	5,319	4,741	4,276	3,575	3,072	2,693	2,159	1,803
150	110,751	73,454	54,949	43,891	36,538	31,296	24,317	19,884	16,817	14,571	12,853	11,498	10,401	8,399	7,043	6,064	5,324	4,745	4,279	3,577	3,073	2,694	2,160	1,803
155	112,538	74,235	55,385	44,169	36,731	31,437	24,402	19,941	16,858	14,601	12,877	11,517	10,417	8,409	7,050	6,069	5,328	4,748	4,282	3,579	3,075	2,695	2,161	1,804
160	114,267	74,983	55,800	44,433	36,913	31,570	24,483	19,994	16,896	14,630	12,899	11,535	10,432	8,418	7,057	6,074	5,332	4,751	4,284	3,581	3,076	2,696	2,161	1,804
165	115,939	75,700	56,196	44,684	37,086	31,697	24,559	20,045	16,933	14,657	12,920	11,552	10,445	8,427	7,063	6,079	5,335	4,754	4,287	3,583	3,077	2,697	2,162	1,804
170	117,559	76,387	56,574	44,922	37,250	31,816	24,631	20,093	16,967	14,682	12,940	11,568	10,458	8,436	7,069	6,083	5,339	4,757	4,289	3,584	3,078	2,698	2,163	1,805
175	119,128	77,047	56,935	45,149	37,406	31,930	24,699	20,138	16,999	14,707	12,959	11,583	10,471	8,444	7,074	6,087	5,342	4,759	4,291	3,586	3,079	2,698	2,163	1,805
180	120,649	77,680	57,280	45,366	37,555	32,038	24,763	20,181	17,030	14,730	12,977	11,597	10,482	8,451	7,080	6,091	5,345	4,761	4,293	3,587	3,080	2,699	2,164	1,805
185	122,124	78,289	57,610	45,573	37,697	32,142	24,825	20,222	17,059	14,751	12,994	11,610	10,493	8,459	7,085	6,095	5,348	4,764	4,295	3,588	3,081	2,700	2,164	1,806
190	123,555	78,874	57,927	45,771	37,832	32,240	24,884	20,261	17,086	14,772	13,010	11,623	10,504	8,465	7,090	6,098	5,350	4,766	4,296	3,589	3,082	2,701	2,165	1,806
195	124,944	79,438	58,230	45,960	37,961	32,334	24,939	20,298	17,113	14,792	13,025	11,635	10,514	8,472	7,094	6,102	5,353	4,768	4,298	3,591	3,083	2,701	2,165	1,806
200	126,292	79,981	58,522	46,141	38,085	32,423	24,993	20,333	17,138	14,810	13,039	11,647	10,523	8,478	7,098	6,105	5,355	4,770	4,300	3,592	3,084	2,702	2,165	1,807

300 MCM | 1 Run(s) | 120 Volts | C-Value = 18,176 | Magnetic Duct | Copper

Calculations:

1. $f = \dfrac{2 \times \text{Length} \times \text{Isca} \times 1.5}{\text{Runs} \times \text{C-Value} \times \text{E L-N}}$

2. $M = \dfrac{1}{1 + f}$

3. $\text{Isc} = \text{Isca} \times 1.5 \times M$

4. Add Motor Contribution = Motor FLA x 4

* All results are given in symmetrical amperes

Single-Phase L-N Bolted Fault-Current* Table

One-Way Distance in Feet

Available Isc in Thousands (Isca)	5	10	15	20	25	30	40	50	60	70	80	90	100	125	150	175	200	225	250	300	350	400	500	600
3	4,416	4,335	4,257	4,182	4,109	4,039	3,905	3,780	3,663	3,553	3,450	3,352	3,259	3,049	2,864	2,701	2,555	2,424	2,306	2,101	1,929	1,784	1,550	1,370
5	7,269	7,053	6,848	6,656	6,473	6,301	5,982	5,694	5,432	5,194	4,975	4,774	4,589	4,183	3,843	3,554	3,306	3,090	2,900	2,583	2,329	2,120	1,798	1,560
7	10,054	9,643	9,266	8,916	8,592	8,291	7,748	7,271	6,850	6,475	6,138	5,835	5,561	4,976	4,502	4,111	3,782	3,502	3,260	2,865	2,556	2,306	1,930	1,659
10	14,105	13,311	12,602	11,964	11,388	10,864	9,950	9,178	8,516	7,944	7,444	7,003	6,611	5,800	5,167	4,658	4,240	3,891	3,595	3,121	2,757	2,469	2,042	1,742
15	20,545	18,902	17,503	16,297	15,246	14,322	12,775	11,529	10,504	9,647	8,919	8,294	7,750	6,659	5,837	5,195	4,681	4,259	3,907	3,353	2,937	2,612	2,140	1,812
20	26,622	23,928	21,729	19,900	18,355	17,033	14,888	13,223	11,892	10,805	9,901	9,135	8,480	7,191	6,242	5,514	4,938	4,471	4,085	3,483	3,036	2,690	2,192	1,849
25	32,367	28,469	25,410	22,944	20,914	19,215	16,528	14,501	12,916	11,644	10,600	9,728	8,988	7,553	6,512	5,724	5,106	4,608	4,199	3,566	3,099	2,740	2,224	1,872
30	37,805	32,593	28,645	25,549	23,058	21,009	17,839	15,500	13,703	12,280	11,124	10,167	9,362	7,815	6,707	5,874	5,225	4,705	4,279	3,623	3,142	2,773	2,246	1,888
35	42,961	36,355	31,510	27,804	24,879	22,510	18,910	16,302	14,326	12,778	11,531	10,507	9,649	8,014	6,853	5,985	5,313	4,776	4,338	3,665	3,173	2,798	2,262	1,899
40	47,856	39,800	34,066	29,776	26,445	23,785	19,801	16,960	14,832	13,179	11,857	10,776	9,876	8,170	6,966	6,072	5,381	4,831	4,383	3,698	3,198	2,817	2,275	1,908
45	52,509	42,967	36,359	31,513	27,807	24,881	20,555	17,510	15,251	13,509	12,123	10,996	10,060	8,295	7,057	6,141	5,435	4,875	4,419	3,723	3,217	2,831	2,284	1,914
50	56,939	45,888	38,429	33,057	29,002	25,833	21,200	17,977	15,604	13,784	12,345	11,178	10,212	8,398	7,132	6,197	5,479	4,910	4,448	3,744	3,232	2,843	2,292	1,920
55	61,160	48,590	40,307	34,436	30,058	26,668	21,759	18,377	15,905	14,019	12,532	11,331	10,340	8,485	7,194	6,244	5,516	4,940	4,472	3,761	3,245	2,853	2,298	1,924
60	65,187	51,098	42,018	35,677	31,000	27,406	22,248	18,725	16,164	14,220	12,693	11,462	10,449	8,558	7,247	6,284	5,547	4,964	4,493	3,775	3,255	2,861	2,304	1,928
65	69,033	53,432	43,583	36,799	31,843	28,064	22,680	19,029	16,391	14,395	12,832	11,576	10,543	8,621	7,292	6,318	5,573	4,985	4,510	3,788	3,265	2,868	2,308	1,931
70	72,710	55,609	45,020	37,819	32,604	28,653	23,063	19,298	16,590	14,548	12,954	11,675	10,625	8,676	7,331	6,347	5,596	5,004	4,525	3,798	3,272	2,875	2,312	1,934
75	76,229	57,644	46,345	38,749	33,293	29,184	23,406	19,537	16,767	14,684	13,062	11,762	10,698	8,724	7,365	6,373	5,616	5,020	4,538	3,807	3,279	2,880	2,316	1,936
80	79,600	59,551	47,570	39,602	33,920	29,665	23,714	19,752	16,924	14,805	13,157	11,839	10,762	8,767	7,395	6,395	5,633	5,034	4,549	3,815	3,285	2,884	2,319	1,938
85	82,832	61,342	48,706	40,386	34,494	30,102	23,993	19,945	17,066	14,913	13,242	11,908	10,819	8,804	7,422	6,415	5,649	5,046	4,560	3,822	3,290	2,888	2,321	1,940
90	85,934	63,026	49,762	41,109	35,020	30,502	24,246	20,120	17,194	15,010	13,319	11,971	10,870	8,838	7,446	6,433	5,663	5,057	4,569	3,829	3,295	2,892	2,324	1,942
95	88,912	64,614	50,746	41,779	35,505	30,870	24,478	20,279	17,310	15,099	13,389	12,027	10,916	8,869	7,468	6,450	5,675	5,067	4,577	3,835	3,299	2,895	2,326	1,943
100	91,776	66,113	51,666	42,401	35,953	31,208	24,690	20,424	17,415	15,179	13,452	12,078	10,958	8,897	7,488	6,464	5,687	5,076	4,584	3,840	3,303	2,898	2,328	1,945
105	94,530	67,530	52,528	42,979	36,368	31,520	24,885	20,558	17,512	15,253	13,510	12,124	10,996	8,922	7,506	6,477	5,697	5,085	4,591	3,844	3,307	2,901	2,329	1,946
110	97,181	68,873	53,336	43,519	36,754	31,809	25,065	20,680	17,601	15,320	13,563	12,167	11,031	8,945	7,522	6,490	5,706	5,092	4,597	3,849	3,310	2,903	2,331	1,947
115	99,735	70,146	54,096	44,024	37,113	32,078	25,232	20,794	17,683	15,382	13,611	12,206	11,064	8,966	7,537	6,501	5,715	5,099	4,603	3,853	3,313	2,906	2,332	1,948
120	102,197	71,355	54,813	44,497	37,449	32,329	25,386	20,899	17,759	15,440	13,656	12,242	11,093	8,985	7,551	6,511	5,723	5,105	4,608	3,856	3,315	2,908	2,334	1,949
125	104,572	72,504	55,488	44,941	37,763	32,562	25,530	20,996	17,829	15,493	13,698	12,275	11,121	9,003	7,563	6,520	5,730	5,111	4,612	3,859	3,318	2,910	2,335	1,950
130	106,864	73,599	56,127	45,359	38,058	32,781	25,665	21,087	17,895	15,542	13,736	12,306	11,146	9,020	7,575	6,529	5,737	5,116	4,617	3,863	3,320	2,911	2,336	1,951
135	109,078	74,642	56,732	45,754	38,335	32,987	25,790	21,172	17,956	15,588	13,772	12,335	11,170	9,035	7,586	6,537	5,743	5,121	4,621	3,865	3,322	2,913	2,337	1,951
140	111,218	75,638	57,306	46,126	38,596	33,180	25,908	21,251	18,013	15,631	13,806	12,362	11,192	9,050	7,596	6,545	5,749	5,126	4,625	3,868	3,324	2,914	2,338	1,952
145	113,287	76,589	57,850	46,478	38,842	33,362	26,019	21,325	18,066	15,671	13,837	12,387	11,212	9,063	7,606	6,552	5,755	5,130	4,628	3,870	3,326	2,916	2,339	1,953
150	115,288	77,499	58,367	46,811	39,075	33,533	26,123	21,395	18,117	15,709	13,867	12,411	11,232	9,076	7,614	6,558	5,760	5,134	4,631	3,873	3,328	2,917	2,340	1,953
155	117,226	78,370	58,860	47,128	39,295	33,695	26,221	21,461	18,164	15,745	13,894	12,433	11,250	9,088	7,623	6,565	5,764	5,138	4,634	3,875	3,329	2,918	2,341	1,954
160	119,102	79,204	59,329	47,428	39,504	33,848	26,314	21,523	18,208	15,778	13,920	12,454	11,267	9,099	7,631	6,570	5,769	5,142	4,637	3,877	3,331	2,919	2,341	1,954
165	120,921	80,004	59,777	47,714	39,702	33,994	26,402	21,582	18,250	15,810	13,945	12,473	11,283	9,109	7,638	6,576	5,773	5,145	4,640	3,879	3,332	2,921	2,342	1,955
170	122,684	80,772	60,205	47,986	39,890	34,132	26,485	21,637	18,290	15,839	13,968	12,492	11,298	9,119	7,645	6,581	5,777	5,148	4,643	3,881	3,334	2,922	2,343	1,955
175	124,394	81,510	60,614	48,245	40,069	34,263	26,564	21,690	18,327	15,867	13,990	12,509	11,312	9,129	7,651	6,586	5,781	5,151	4,645	3,882	3,335	2,923	2,343	1,956
180	126,053	82,219	61,005	48,493	40,240	34,387	26,639	21,740	18,363	15,894	14,010	12,526	11,326	9,137	7,658	6,590	5,784	5,154	4,647	3,884	3,336	2,923	2,344	1,956
185	127,664	82,901	61,380	48,729	40,402	34,506	26,710	21,787	18,397	15,919	14,030	12,542	11,339	9,146	7,664	6,595	5,788	5,157	4,649	3,885	3,337	2,924	2,344	1,956
190	129,228	83,558	61,739	48,956	40,558	34,619	26,778	21,832	18,429	15,944	14,049	12,557	11,351	9,154	7,669	6,599	5,791	5,159	4,652	3,887	3,338	2,925	2,345	1,957
195	130,748	84,191	62,084	49,172	40,706	34,727	26,842	21,875	18,460	15,966	14,067	12,571	11,363	9,161	7,674	6,603	5,794	5,161	4,653	3,888	3,339	2,926	2,345	1,957
200	132,226	84,801	62,415	49,380	40,848	34,831	26,904	21,916	18,489	15,988	14,084	12,584	11,374	9,168	7,679	6,607	5,797	5,164	4,655	3,890	3,340	2,927	2,346	1,957

350 MCM | 1 Run(s) | 120 Volts | C-Value = 19,703 | Magnetic Duct | Copper

Calculations:

1. $f = \dfrac{2 \times \text{Length} \times \text{Isca} \times 1.5}{\text{Runs} \times \text{C-Value} \times \text{E L-N}}$

2. $M = \dfrac{1}{1 + f}$

3. Isc = Isca x 1.5 x M

4. Add Motor Contribution = Motor FLA x 4

* All results are given in symmetrical amperes

Single-Phase L-N Bolted Fault-Current* Table

One-Way Distance in Feet

Available Isc in Thousands (Isca)	5	10	15	20	25	30	40	50	60	70	80	90	100	125	150	175	200	225	250	300	350	400	500	600
3	4,419	4,342	4,267	4,194	4,124	4,056	3,927	3,806	3,692	3,585	3,484	3,388	3,297	3,091	2,909	2,747	2,602	2,472	2,354	2,149	1,977	1,830	1,594	1,411
5	7,279	7,070	6,873	6,687	6,511	6,343	6,033	5,752	5,496	5,261	5,046	4,848	4,665	4,262	3,923	3,634	3,385	3,168	2,977	2,656	2,398	2,186	1,857	1,614
7	10,071	9,677	9,311	8,973	8,658	8,365	7,834	7,366	6,951	6,580	6,247	5,946	5,673	5,088	4,612	4,218	3,886	3,602	3,357	2,955	2,639	2,384	1,998	1,720
10	14,140	13,374	12,687	12,066	11,504	10,991	10,092	9,329	8,674	8,104	7,604	7,163	6,770	5,953	5,313	4,796	4,372	4,016	3,714	3,228	2,855	2,559	2,119	1,809
15	20,620	19,030	17,668	16,487	15,455	14,544	13,010	11,769	10,745	9,884	9,151	8,519	7,969	6,861	6,024	5,369	4,842	4,409	4,048	3,477	3,048	2,713	2,224	1,884
20	26,748	24,133	21,983	20,185	18,659	17,347	15,209	13,540	12,201	11,103	10,187	9,410	8,743	7,427	6,456	5,709	5,117	4,636	4,238	3,617	3,155	2,797	2,280	1,925
25	32,553	28,760	25,758	23,323	21,309	19,616	16,925	14,883	13,281	11,991	10,929	10,040	9,284	7,814	6,746	5,935	5,298	4,784	4,362	3,706	3,222	2,850	2,315	1,950
30	38,060	32,974	29,088	26,021	23,539	21,489	18,302	15,938	14,115	12,666	11,487	10,508	9,684	8,095	6,955	6,096	5,426	4,888	4,448	3,769	3,269	2,887	2,340	1,967
35	43,290	36,830	32,047	28,364	25,440	23,062	19,431	16,787	14,777	13,196	11,921	10,871	9,991	8,309	7,112	6,216	5,521	4,965	4,511	3,814	3,304	2,914	2,357	1,979
40	48,265	40,370	34,694	30,418	27,080	24,402	20,373	17,486	15,316	13,624	12,270	11,160	10,234	8,477	7,234	6,309	5,594	5,025	4,560	3,849	3,330	2,934	2,370	1,988
45	53,003	43,632	37,076	32,234	28,510	25,557	21,172	18,071	15,763	13,977	12,555	11,395	10,432	8,612	7,332	6,384	5,653	5,072	4,599	3,877	3,350	2,950	2,381	1,996
50	57,519	46,647	39,231	33,850	29,767	26,563	21,858	18,568	16,140	14,273	12,793	11,591	10,596	8,723	7,413	6,445	5,701	5,110	4,631	3,899	3,367	2,963	2,389	2,002
55	61,830	49,442	41,190	35,298	30,881	27,447	22,452	18,996	16,462	14,524	12,994	11,756	10,734	8,816	7,480	6,496	5,740	5,142	4,657	3,918	3,381	2,974	2,396	2,006
60	65,949	52,041	42,978	36,603	31,875	28,229	22,973	19,367	16,740	14,740	13,167	11,898	10,851	8,896	7,537	6,539	5,774	5,169	4,679	3,933	3,393	2,983	2,402	2,011
65	69,888	54,464	44,617	37,785	32,768	28,927	23,434	19,693	16,983	14,928	13,317	12,020	10,953	8,964	7,586	6,575	5,802	5,192	4,698	3,947	3,402	2,990	2,407	2,014
70	73,659	56,727	46,125	38,861	33,574	29,553	23,843	19,982	17,197	15,093	13,448	12,127	11,041	9,023	7,628	6,607	5,827	5,212	4,714	3,958	3,411	2,997	2,411	2,017
75	77,273	58,847	47,516	39,844	34,305	30,119	24,209	20,238	17,387	15,239	13,564	12,221	11,119	9,075	7,665	6,635	5,849	5,229	4,728	3,968	3,418	3,002	2,415	2,020
80	80,739	60,836	48,805	40,746	34,972	30,631	24,539	20,469	17,556	15,369	13,667	12,304	11,189	9,121	7,698	6,660	5,868	5,244	4,741	3,977	3,425	3,007	2,418	2,022
85	84,067	62,706	50,001	41,577	35,582	31,098	24,838	20,676	17,709	15,486	13,759	12,379	11,250	9,162	7,727	6,681	5,885	5,258	4,752	3,984	3,431	3,012	2,421	2,024
90	87,263	64,467	51,114	42,344	36,142	31,525	25,110	20,864	17,846	15,591	13,842	12,446	11,306	9,199	7,754	6,701	5,900	5,270	4,762	3,991	3,436	3,016	2,423	2,026
95	90,336	66,129	52,154	43,055	36,659	31,918	25,358	21,035	17,971	15,687	13,917	12,507	11,356	9,232	7,777	6,718	5,913	5,281	4,770	3,998	3,440	3,019	2,426	2,027
100	93,294	67,700	53,126	43,715	37,137	32,279	25,586	21,192	18,085	15,774	13,986	12,562	11,401	9,262	7,798	6,734	5,926	5,291	4,778	4,003	3,444	3,023	2,428	2,029
105	96,141	69,187	54,037	44,330	37,580	32,613	25,795	21,335	18,190	15,853	14,048	12,612	11,443	9,289	7,818	6,749	5,937	5,299	4,786	4,008	3,448	3,025	2,430	2,030
110	98,885	70,597	54,893	44,905	37,992	32,923	25,989	21,467	18,286	15,926	14,105	12,658	11,480	9,314	7,835	6,762	5,947	5,308	4,792	4,013	3,452	3,028	2,431	2,031
115	101,530	71,935	55,699	45,442	38,376	33,211	26,168	21,589	18,374	15,993	14,158	12,701	11,515	9,337	7,852	6,774	5,956	5,315	4,798	4,017	3,455	3,031	2,433	2,032
120	104,083	73,207	56,458	45,947	38,735	33,480	26,334	21,703	18,456	16,055	14,206	12,740	11,547	9,358	7,867	6,785	5,965	5,322	4,804	4,021	3,458	3,033	2,434	2,033
125	106,547	74,417	57,176	46,421	39,071	33,731	26,489	21,808	18,532	16,112	14,251	12,776	11,577	9,378	7,880	6,795	5,973	5,328	4,809	4,025	3,460	3,035	2,436	2,034
130	108,928	75,571	57,854	46,867	39,387	33,966	26,634	21,906	18,603	16,166	14,293	12,809	11,605	9,396	7,893	6,805	5,980	5,334	4,814	4,028	3,463	3,037	2,437	2,035
135	111,229	76,671	58,497	47,288	39,684	34,186	26,770	21,997	18,669	16,216	14,332	12,841	11,630	9,412	7,905	6,814	5,987	5,339	4,818	4,031	3,465	3,038	2,438	2,036
140	113,455	77,722	59,107	47,686	39,963	34,394	26,897	22,083	18,731	16,262	14,368	12,870	11,654	9,428	7,916	6,822	5,993	5,344	4,822	4,034	3,467	3,040	2,439	2,037
145	115,608	78,727	59,686	48,062	40,227	34,589	27,016	22,163	18,789	16,306	14,402	12,897	11,677	9,443	7,926	6,829	5,999	5,349	4,826	4,037	3,469	3,042	2,440	2,037
150	117,694	79,689	60,237	48,419	40,477	34,773	27,128	22,239	18,843	16,347	14,434	12,923	11,698	9,456	7,936	6,837	6,005	5,354	4,830	4,039	3,471	3,043	2,441	2,038
155	119,714	80,610	60,762	48,757	40,713	34,948	27,234	22,310	18,894	16,385	14,464	12,947	11,717	9,469	7,945	6,843	6,010	5,358	4,833	4,042	3,473	3,044	2,442	2,038
160	121,671	81,493	61,262	49,079	40,937	35,113	27,334	22,377	18,942	16,421	14,492	12,969	11,736	9,481	7,953	6,850	6,015	5,361	4,836	4,044	3,474	3,046	2,443	2,039
165	123,570	82,340	61,740	49,385	41,150	35,269	27,429	22,440	18,987	16,455	14,519	12,990	11,753	9,493	7,961	6,856	6,019	5,365	4,839	4,046	3,476	3,047	2,443	2,040
170	125,411	83,153	62,196	49,676	41,352	35,417	27,519	22,500	19,030	16,487	14,544	13,010	11,769	9,503	7,969	6,861	6,024	5,369	4,842	4,048	3,477	3,048	2,444	2,040
175	127,199	83,936	62,633	49,954	41,545	35,558	27,604	22,557	19,071	16,518	14,568	13,029	11,785	9,513	7,976	6,866	6,028	5,372	4,845	4,050	3,479	3,049	2,445	2,041
180	128,934	84,688	63,051	50,220	41,728	35,693	27,685	22,611	19,109	16,547	14,590	13,047	11,800	9,523	7,983	6,871	6,032	5,375	4,847	4,051	3,480	3,050	2,445	2,041
185	130,620	85,412	63,451	50,474	41,903	35,821	27,761	22,663	19,146	16,574	14,612	13,065	11,814	9,532	7,989	6,876	6,035	5,378	4,849	4,053	3,481	3,051	2,446	2,041
190	132,258	86,109	63,835	50,716	42,070	35,943	27,835	22,711	19,181	16,600	14,632	13,081	11,827	9,541	7,995	6,881	6,039	5,380	4,852	4,054	3,482	3,052	2,447	2,042
195	133,851	86,782	64,204	50,949	42,230	36,059	27,905	22,758	19,214	16,625	14,651	13,096	11,840	9,549	8,001	6,885	6,042	5,383	4,854	4,056	3,483	3,053	2,447	2,042
200	135,400	87,430	64,558	51,172	42,383	36,171	27,971	22,802	19,246	16,649	14,670	13,111	11,852	9,557	8,006	6,889	6,045	5,386	4,856	4,057	3,484	3,053	2,448	2,042

400 MCM	1 Run(s)	120 Volts	C-Value = 20,565	Magnetic Duct	Copper

Calculations:

1. $f = \dfrac{2 \times \text{Length} \times \text{Isca} \times 1.5}{\text{Runs} \times \text{C-Value} \times E_{L\text{-}N}}$

2. $M = \dfrac{1}{1 + f}$

3. $\text{Isc} = \text{Isca} \times 1.5 \times M$

4. Add Motor Contribution = Motor FLA x 4

* All results are given in symmetrical amperes

Single-Phase L-N Bolted Fault-Current* Table

One-Way Distance in Feet

Available Isc in Thousands (Isca)

Isca	5	10	15	20	25	30	40	50	60	70	80	90	100	125	150	175	200	225	250	300	350	400	500	600
3	4,425	4,353	4,283	4,215	4,149	4,086	3,964	3,849	3,741	3,639	3,542	3,450	3,363	3,163	2,986	2,827	2,685	2,556	2,439	2,234	2,061	1,913	1,673	1,486
5	7,294	7,100	6,916	6,740	6,574	6,416	6,121	5,852	5,605	5,379	5,170	4,976	4,797	4,401	4,065	3,776	3,526	3,307	3,114	2,788	2,524	2,305	1,965	1,712
7	10,102	9,732	9,389	9,069	8,770	8,491	7,982	7,530	7,127	6,765	6,438	6,141	5,870	5,287	4,809	4,411	4,073	3,784	3,533	3,[illegible]19	2,792	2,527	2,124	1,832
10	14,200	13,481	12,831	12,241	11,703	11,210	10,339	9,594	8,949	8,385	7,888	7,447	7,053	6,228	5,576	5,047	4,610	4,243	3,930	3,424	3,034	2,724	2,261	1,933
15	20,747	19,247	17,949	16,815	15,816	14,929	13,424	12,194	11,171	10,306	9,565	8,924	8,363	7,228	6,364	5,685	5,136	4,684	4,306	3,706	3,253	2,899	2,381	2,019
20	26,962	24,482	22,420	20,679	19,188	17,898	15,777	14,105	12,754	11,639	10,703	9,906	9,220	7,859	6,848	6,068	5,447	4,942	4,522	3,865	3,375	2,995	2,445	2,066
25	32,870	29,258	26,361	23,986	22,003	20,323	17,631	15,569	13,939	12,618	11,525	10,607	9,824	8,294	7,176	6,324	5,652	5,110	4,662	3,968	3,453	3,057	2,486	2,095
30	38,493	33,631	29,859	26,848	24,388	22,341	19,130	16,727	14,859	13,367	12,147	11,131	10,272	8,611	7,412	6,507	5,798	5,229	4,761	4,039	3,507	3,099	2,514	2,114
35	43,852	37,650	32,985	29,349	26,435	24,047	20,367	17,665	15,595	13,960	12,635	11,539	10,619	8,853	7,591	6,644	5,907	5,317	4,834	4,091	3,546	3,129	2,534	2,129
40	48,965	41,358	35,797	31,554	28,210	25,507	21,405	18,440	16,196	14,439	13,026	11,865	10,894	9,044	7,731	6,751	5,991	5,385	4,890	4,131	3,576	3,153	2,549	2,139
45	53,847	44,788	38,338	33,512	29,765	26,772	22,289	19,092	16,697	14,836	13,348	12,132	11,118	9,198	7,843	6,836	6,058	5,439	4,935	4,163	3,600	3,171	2,561	2,148
50	58,515	47,971	40,647	35,263	31,138	27,878	23,050	19,648	17,121	15,170	13,618	12,354	11,305	9,325	7,935	6,906	6,113	5,483	4,971	4,189	3,620	3,186	2,571	2,155
55	62,982	50,933	42,753	36,837	32,360	28,853	23,713	20,127	17,484	15,454	13,846	12,542	11,462	9,431	8,012	6,964	6,159	5,520	5,002	4,211	3,636	3,199	2,579	2,160
60	67,261	53,695	44,683	38,261	33,453	29,719	24,295	20,545	17,798	15,699	14,043	12,703	11,596	9,522	8,078	7,014	6,197	5,551	5,027	4,229	3,649	3,209	2,586	2,165
65	71,364	56,278	46,457	39,554	34,438	30,493	24,810	20,912	18,073	15,912	14,213	12,842	11,712	9,600	8,134	7,056	6,230	5,578	5,049	4,244	3,660	3,218	2,591	2,169
70	75,301	58,698	48,094	40,735	35,329	31,190	25,269	21,237	18,315	16,100	14,363	12,964	11,813	9,668	8,182	7,092	6,259	5,600	5,067	4,257	3,670	3,226	2,596	2,173
75	79,082	60,970	49,609	41,816	36,140	31,820	25,681	21,528	18,531	16,266	14,495	13,072	11,903	9,728	8,225	7,125	6,284	5,620	5,084	4,269	3,679	3,232	2,601	2,176
80	82,716	63,108	51,015	42,811	36,880	32,393	26,053	21,788	18,723	16,415	14,613	13,167	11,982	9,781	8,263	7,153	6,306	5,638	5,098	4,279	3,686	3,238	2,604	2,178
85	86,211	65,122	52,323	43,729	37,559	32,915	26,390	22,023	18,897	16,548	14,718	13,253	12,053	9,828	8,297	7,178	6,325	5,654	5,111	4,288	3,693	3,243	2,608	2,181
90	89,576	67,024	53,544	44,578	38,184	33,394	26,697	22,237	19,054	16,668	14,813	13,330	12,116	9,870	8,327	7,201	6,343	5,668	5,122	4,296	3,699	3,248	2,611	2,183
95	92,817	68,823	54,685	45,366	38,761	33,835	26,978	22,431	19,196	16,777	14,899	13,399	12,174	9,908	8,354	7,221	6,359	5,680	5,133	4,303	3,704	3,252	2,613	2,184
100	95,942	70,526	55,755	46,100	39,296	34,241	27,235	22,609	19,327	16,876	14,977	13,463	12,226	9,943	8,378	7,239	6,373	5,692	5,142	4,310	3,709	3,256	2,616	2,186
105	98,956	72,141	56,760	46,785	39,792	34,618	27,473	22,773	19,446	16,967	15,049	13,520	12,274	9,974	8,401	7,256	6,386	5,702	5,150	4,315	3,713	3,259	2,618	2,188
110	101,865	73,675	57,705	47,425	40,254	34,967	27,692	22,923	19,556	17,051	15,115	13,573	12,317	10,003	8,421	7,271	6,397	5,711	5,158	4,321	3,717	3,262	2,620	2,189
115	104,675	75,133	58,596	48,026	40,686	35,292	27,896	23,063	19,657	17,128	15,175	13,622	12,357	10,030	8,440	7,285	6,408	5,720	5,165	4,326	3,721	3,265	2,622	2,190
120	107,390	76,522	59,437	48,589	41,090	35,596	28,085	23,192	19,751	17,199	15,231	13,667	12,394	10,054	8,457	7,298	6,418	5,728	5,171	4,330	3,724	3,267	2,623	2,191
125	110,016	77,846	60,233	49,120	41,468	35,879	28,262	23,312	19,838	17,265	15,283	13,709	12,429	10,077	8,473	7,310	6,427	5,735	5,177	4,334	3,728	3,270	2,625	2,193
130	112,556	79,109	60,986	49,619	41,824	36,146	28,426	23,424	19,919	17,326	15,331	13,747	12,460	10,097	8,488	7,321	6,436	5,742	5,183	4,338	3,730	3,272	2,626	2,194
135	115,014	80,316	61,701	50,092	42,159	36,395	28,581	23,529	19,994	17,383	15,375	13,783	12,490	10,117	8,501	7,331	6,444	5,748	5,188	4,342	3,733	3,274	2,628	2,194
140	117,396	81,470	62,380	50,538	42,475	36,630	28,726	23,627	20,065	17,437	15,417	13,817	12,518	10,135	8,514	7,340	6,451	5,754	5,193	4,345	3,735	3,276	2,629	2,195
145	119,703	82,574	63,026	50,961	42,773	36,852	28,862	23,719	20,132	17,487	15,456	13,848	12,543	10,152	8,526	7,349	6,458	5,759	5,197	4,348	3,738	3,278	2,630	2,196
150	121,940	83,633	63,640	51,362	43,055	37,061	28,990	23,805	20,194	17,534	15,493	13,878	12,568	10,168	8,537	7,358	6,464	5,764	5,201	4,351	3,740	3,279	2,631	2,197
155	124,110	84,648	64,226	51,743	43,323	37,259	29,111	23,887	20,253	17,578	15,528	13,905	12,590	10,182	8,548	7,365	6,470	5,769	5,205	4,354	3,742	3,281	2,632	2,198
160	126,215	85,622	64,786	52,105	43,577	37,447	29,225	23,964	20,308	17,620	15,560	13,931	12,612	10,196	8,558	7,373	6,476	5,774	5,209	4,356	3,744	3,282	2,633	2,198
165	128,259	86,558	65,320	52,451	43,818	37,625	29,333	24,037	20,360	17,659	15,591	13,956	12,632	10,210	8,567	7,379	6,481	5,778	5,212	4,359	3,746	3,284	2,634	2,199
170	130,245	87,457	65,831	52,780	44,047	37,794	29,436	24,105	20,409	17,696	15,620	13,979	12,651	10,222	8,576	7,386	6,486	5,782	5,216	4,361	3,747	3,285	2,635	2,199
175	132,173	88,323	66,320	53,094	44,265	37,955	29,533	24,171	20,456	17,731	15,647	14,001	12,669	10,234	8,584	7,392	6,491	5,786	5,219	4,363	3,749	3,286	2,635	2,200
180	134,048	89,156	66,789	53,394	44,474	38,108	29,626	24,233	20,501	17,765	15,673	14,022	12,686	10,245	8,592	7,398	6,495	5,789	5,221	4,365	3,750	3,287	2,636	2,200
185	135,871	89,959	67,238	53,680	44,673	38,254	29,714	24,292	20,543	17,796	15,698	14,042	12,702	10,255	8,599	7,403	6,500	5,793	5,224	4,367	3,752	3,288	2,637	2,201
190	137,645	90,733	67,670	53,955	44,863	38,393	29,798	24,348	20,583	17,826	15,721	14,060	12,717	10,265	8,606	7,409	6,504	5,796	5,227	4,369	3,753	3,289	2,638	2,201
195	139,371	91,480	68,084	54,218	45,045	38,526	29,878	24,401	20,621	17,855	15,743	14,078	12,732	10,275	8,613	7,414	6,507	5,799	5,229	4,371	3,754	3,290	2,638	2,202
200	141,051	92,201	68,483	54,471	45,219	38,653	29,955	24,452	20,657	17,882	15,764	14,095	12,745	10,284	8,619	7,418	6,511	5,802	5,232	4,372	3,756	3,291	2,639	2,202

500 MCM | 1 Run(s) | 120 Volts

C-Value = 22,185 | Magnetic Duct | Copper

Calculations:

1. $f = \dfrac{2 \times \text{Length} \times \text{Isca} \times 1.5}{\text{Runs} \times \text{C-Value} \times \text{E L-N}}$

2. $M = \dfrac{1}{1 + f}$

3. $\text{Isc} = \text{Isca} \times 1.5 \times M$

4. Add Motor Contribution = Motor FLA x 4

* All results are given in symmetrical amperes

Single-Phase L-N Bolted Fault-Current* Table

One-Way Distance in Feet

Available Isc in Thousands (Isca)	5	10	15	20	25	30	40	50	60	70	80	90	100	125	150	175	200	225	250	300	350	400	500	600
3	4,428	4,358	4,290	4,224	4,160	4,098	3,980	3,868	3,763	3,663	3,568	3,478	3,392	3,196	3,020	2,863	2,722	2,594	2,477	2,273	2,100	1,951	1,709	1,521
5	7,301	7,113	6,934	6,764	6,602	6,447	6,159	5,896	5,654	5,431	5,225	5,034	4,857	4,463	4,129	3,841	3,591	3,371	3,177	2,849	2,582	2,361	2,015	1,758
7	10,115	9,757	9,423	9,111	8,820	8,546	8,047	7,603	7,206	6,847	6,523	6,228	5,959	5,378	4,900	4,500	4,160	3,868	3,614	3,195	2,863	2,594	2,183	1,884
10	14,226	13,527	12,894	12,318	11,791	11,307	10,450	9,713	9,073	8,513	8,018	7,577	7,182	6,354	5,697	5,163	4,721	4,349	4,031	3,516	3,118	2,801	2,328	1,992
15	20,802	19,342	18,073	16,961	15,978	15,102	13,610	12,387	11,365	10,499	9,756	9,111	8,546	7,399	6,523	5,833	5,274	4,814	4,427	3,814	3,351	2,987	2,455	2,084
20	27,055	24,636	22,614	20,899	19,426	18,147	16,035	14,364	13,008	11,886	10,942	10,137	9,442	8,061	7,033	6,237	5,603	5,086	4,656	3,983	3,480	3,090	2,524	2,133
25	33,008	29,478	26,629	24,283	22,316	20,645	17,954	15,885	14,243	12,908	11,803	10,872	10,076	8,519	7,379	6,507	5,820	5,264	4,805	4,092	3,563	3,155	2,567	2,164
30	38,683	33,922	30,204	27,220	24,773	22,730	19,511	17,091	15,205	13,694	12,456	11,423	10,549	8,854	7,629	6,701	5,975	5,390	4,910	4,168	3,620	3,200	2,597	2,185
35	44,099	38,016	33,407	29,795	26,888	24,498	20,800	18,072	15,976	14,316	12,969	11,853	10,914	9,110	7,818	6,847	6,090	5,484	4,988	4,224	3,662	3,233	2,618	2,200
40	49,272	41,799	36,294	32,070	28,727	26,015	21,884	18,884	16,608	14,822	13,382	12,198	11,206	9,312	7,966	6,960	6,180	5,557	5,048	4,266	3,694	3,258	2,635	2,212
45	54,220	45,306	38,909	34,095	30,341	27,332	22,808	19,569	17,135	15,240	13,722	12,479	11,443	9,476	8,086	7,051	6,251	5,615	5,096	4,300	3,720	3,277	2,648	2,221
50	58,955	48,565	41,289	35,909	31,769	28,485	23,605	20,153	17,582	15,592	14,007	12,715	11,640	9,611	8,184	7,126	6,310	5,662	5,134	4,328	3,741	3,293	2,658	2,228
55	63,492	51,603	43,464	37,543	33,042	29,504	24,301	20,658	17,964	15,892	14,249	12,914	11,807	9,724	8,266	7,188	6,359	5,701	5,166	4,351	3,758	3,307	2,667	2,234
60	67,843	54,441	45,460	39,023	34,183	30,411	24,912	21,098	18,296	16,152	14,457	13,084	11,950	9,820	8,335	7,240	6,400	5,734	5,194	4,370	3,772	3,318	2,674	2,239
65	72,020	57,098	47,298	40,369	35,211	31,222	25,454	21,485	18,587	16,378	14,638	13,232	12,073	9,904	8,395	7,285	6,435	5,762	5,217	4,386	3,784	3,327	2,680	2,244
70	76,031	59,590	48,996	41,600	36,144	31,953	25,938	21,829	18,844	16,577	14,797	13,362	12,181	9,976	8,447	7,324	6,465	5,787	5,237	4,401	3,795	3,335	2,685	2,247
75	79,888	61,934	50,569	42,728	36,993	32,615	26,372	22,136	19,072	16,753	14,937	13,476	12,275	10,039	8,493	7,359	6,492	5,808	5,254	4,413	3,804	3,342	2,690	2,251
80	83,598	64,141	52,031	43,767	37,769	33,216	26,764	22,411	19,276	16,910	15,062	13,578	12,360	10,096	8,533	7,389	6,515	5,827	5,270	4,424	3,812	3,349	2,694	2,253
85	87,170	66,223	53,392	44,727	38,481	33,766	27,120	22,660	19,460	17,052	15,174	13,669	12,435	10,146	8,569	7,416	6,536	5,843	5,283	4,433	3,819	3,354	2,697	2,256
90	90,612	68,190	54,664	45,616	39,137	34,270	27,445	22,886	19,626	17,179	15,275	13,751	12,503	10,191	8,601	7,440	6,555	5,358	5,295	4,442	3,825	3,359	2,701	2,258
95	93,930	70,053	55,854	46,442	39,744	34,735	27,741	23,092	19,778	17,295	15,366	13,825	12,564	10,232	8,630	7,461	6,572	5,372	5,306	4,450	3,831	3,363	2,704	2,260
100	97,131	71,818	56,971	47,211	40,306	35,163	28,014	23,281	19,916	17,401	15,450	13,892	12,620	10,269	8,656	7,481	6,587	5,384	5,316	4,457	3,836	3,367	2,706	2,262
105	100,221	73,494	58,020	47,929	40,828	35,560	28,265	23,454	20,043	17,497	15,526	13,954	12,671	10,302	8,680	7,499	6,601	5,395	5,325	4,463	3,841	3,371	2,708	2,263
110	103,206	75,086	59,008	48,602	41,315	35,929	28,498	23,614	20,159	17,586	15,596	14,010	12,717	10,333	8,702	7,515	6,613	5,905	5,333	4,469	3,845	3,374	2,711	2,265
115	106,092	76,602	59,940	49,232	41,770	36,272	28,714	23,762	20,267	17,668	15,660	14,062	12,760	10,361	8,722	7,530	6,625	5,914	5,341	4,474	3,849	3,377	2,712	2,266
120	108,882	78,046	60,821	49,825	42,196	36,593	28,914	23,899	20,367	17,744	15,720	14,110	12,799	10,387	8,740	7,544	6,636	5,923	5,348	4,479	3,853	3,380	2,714	2,268
125	111,582	79,423	61,654	50,382	42,595	36,893	29,101	24,027	20,459	17,814	15,775	14,154	12,836	10,411	8,757	7,556	6,645	5,930	5,354	4,483	3,856	3,383	2,716	2,269
130	114,195	80,739	62,444	50,909	42,971	37,174	29,276	24,146	20,545	17,879	15,826	14,195	12,870	10,433	8,773	7,568	6,654	5,938	5,360	4,487	3,859	3,385	2,717	2,270
135	116,727	81,996	63,194	51,406	43,324	37,438	29,440	24,257	20,626	17,940	15,874	14,234	12,901	10,454	8,787	7,579	6,663	5,944	5,366	4,491	3,862	3,387	2,719	2,271
140	119,181	83,199	63,906	51,876	43,658	37,687	29,593	24,361	20,701	17,997	15,918	14,270	12,931	10,473	8,801	7,589	6,671	5,950	5,371	4,495	3,864	3,389	2,720	2,272
145	121,560	84,352	64,583	52,322	43,973	37,922	29,738	24,459	20,772	18,051	15,960	14,303	12,958	10,491	8,814	7,599	6,678	5,956	5,375	4,498	3,867	3,391	2,721	2,273
150	123,867	85,456	65,229	52,745	44,271	38,144	29,874	24,551	20,838	18,101	15,999	14,335	12,984	10,508	8,826	7,607	6,685	5,962	5,380	4,501	3,869	3,393	2,722	2,273
155	126,107	86,516	65,845	53,146	44,554	38,353	30,002	24,638	20,901	18,148	16,036	14,364	13,008	10,524	8,837	7,616	6,691	5,967	5,384	4,504	3,871	3,394	2,724	2,274
160	128,281	87,534	66,433	53,529	44,823	38,552	30,124	24,720	20,959	18,192	16,070	14,392	13,031	10,539	8,847	7,624	6,697	5,972	5,388	4,507	3,873	3,396	2,725	2,275
165	130,393	88,513	66,995	53,893	45,078	38,741	30,239	24,797	21,015	18,234	16,103	14,418	13,052	10,553	8,857	7,631	6,703	5,976	5,392	4,509	3,875	3,397	2,725	2,275
170	132,445	89,454	67,532	54,240	45,320	38,920	30,348	24,870	21,068	18,274	16,134	14,443	13,073	10,566	8,867	7,638	6,708	5,980	5,395	4,512	3,877	3,399	2,726	2,276
175	134,441	90,359	68,047	54,572	45,552	39,090	30,451	24,940	21,118	18,311	16,163	14,466	13,092	10,579	8,875	7,644	6,713	5,984	5,398	4,514	3,879	3,400	2,727	2,277
180	136,381	91,232	68,541	54,889	45,772	39,253	30,550	25,006	21,165	18,347	16,191	14,488	13,110	10,591	8,884	7,651	6,718	5,988	5,401	4,516	3,880	3,401	2,728	2,277
185	138,268	92,072	69,014	55,192	45,983	39,408	30,644	25,068	21,210	18,380	16,217	14,509	13,127	10,602	8,892	7,656	6,723	5,992	5,404	4,518	3,882	3,403	2,729	2,278
190	140,105	92,883	69,469	55,483	46,184	39,555	30,733	25,128	21,252	18,413	16,242	14,529	13,144	10,613	8,899	7,662	6,727	5,995	5,407	4,520	3,883	3,404	2,729	2,278
195	141,894	93,666	69,906	55,761	46,377	39,697	30,818	25,185	21,293	18,443	16,266	14,549	13,159	10,623	8,906	7,667	6,731	5,998	5,410	4,522	3,885	3,405	2,730	2,279
200	143,636	94,422	70,326	56,028	46,562	39,832	30,899	25,239	21,332	18,472	16,289	14,567	13,174	10,633	8,913	7,672	6,735	6,001	5,412	4,524	3,886	3,406	2,731	2,279

600 MCM | 1 Run(s) | 120 Volts

C-Value = 22,965 | Magnetic Duct | Copper

Calculations:

1. $f = \dfrac{2 \times \text{Length} \times \text{Isca} \times 1.5}{\text{Runs} \times \text{C-Value} \times \text{E L-N}}$

2. $M = \dfrac{1}{1 + f}$

3. Isc = Isca x 1.5 x M

4. Add Motor Contribution = Motor FLA x 4

* All results are given in symmetrical amperes

Single-Phase L-N Bolted Fault-Current* Table

One-Way Distance in Feet

Available Isc in Thousands (Isca)	5	10	15	20	25	30	40	50	60	70	80	90	100	125	150	175	200	225	250	300	350	400	500	600
3	4,431	4,364	4,300	4,237	4,176	4,116	4,003	3,895	3,793	3,696	3,604	3,517	3,433	3,241	3,069	2,915	2,775	2,648	2,533	2,329	2,156	2,006	1,762	1,571
5	7,311	7,131	6,959	6,796	6,640	6,491	6,213	5,957	5,722	5,504	5,303	5,116	4,941	4,553	4,221	3,934	3,684	3,464	3,268	2,937	2,667	2,442	2,089	1,826
7	10,133	9,790	9,470	9,170	8,889	8,624	8,139	7,706	7,317	6,965	6,645	6,354	6,087	5,508	5,030	4,628	4,286	3,990	3,733	3,307	2,968	2,692	2,270	1,962
10	14,261	13,592	12,983	12,426	11,915	11,444	10,606	9,882	9,251	8,695	8,203	7,763	7,368	6,537	5,874	5,333	4,883	4,504	4,179	3,652	3,243	2,916	2,428	2,079
15	20,878	19,474	18,247	17,166	16,205	15,347	13,876	12,663	11,645	10,778	10,031	9,382	8,811	7,648	6,756	6,050	5,478	5,005	4,607	3,975	3,495	3,119	2,566	2,180
20	27,184	24,852	22,888	21,212	19,764	18,502	16,406	14,736	13,375	12,244	11,290	10,473	9,767	8,358	7,304	6,486	5,833	5,299	4,855	4,158	3,636	3,231	2,641	2,234
25	33,201	29,787	27,009	24,705	22,764	21,105	18,420	16,342	14,685	13,333	12,209	11,259	10,447	8,851	7,678	6,779	6,069	5,493	5,018	4,277	3,726	3,302	2,689	2,268
30	38,949	34,332	30,693	27,752	25,326	23,289	20,063	17,622	15,710	14,172	12,909	11,853	10,956	9,213	7,949	6,990	6,237	5,631	5,132	4,360	3,789	3,351	2,721	2,291
35	44,444	38,531	34,007	30,434	27,540	25,149	21,428	18,666	16,535	14,840	13,461	12,316	11,351	9,491	8,155	7,148	6,363	5,733	5,217	4,421	3,835	3,387	2,745	2,308
40	49,703	42,423	37,003	32,811	29,472	26,750	22,580	19,534	17,212	15,384	13,906	12,688	11,666	9,710	8,316	7,272	6,461	5,813	5,283	4,468	3,871	3,414	2,763	2,320
45	54,742	46,040	39,725	34,934	31,174	28,145	23,565	20,267	17,779	15,835	14,274	12,993	11,924	9,888	8,446	7,371	6,539	5,876	5,335	4,505	3,899	3,436	2,777	2,330
50	59,573	49,410	42,210	36,841	32,683	29,369	24,417	20,894	18,260	16,215	14,582	13,248	12,138	10,035	8,553	7,453	6,603	5,928	5,377	4,535	3,921	3,454	2,789	2,338
55	64,210	52,558	44,486	38,563	34,032	30,453	25,162	21,437	18,673	16,540	14,845	13,465	12,319	10,159	8,643	7,521	6,657	5,970	5,413	4,560	3,940	3,468	2,798	2,345
60	68,664	55,505	46,579	40,126	35,243	31,420	25,818	21,912	19,032	16,821	15,071	13,650	12,474	10,264	8,719	7,578	6,702	6,007	5,442	4,581	3,956	3,480	2,806	2,351
65	72,944	58,269	48,510	41,551	36,338	32,287	26,401	22,330	19,347	17,067	15,267	13,811	12,609	10,355	8,785	7,628	6,740	6,038	5,468	4,599	3,969	3,491	2,813	2,355
70	77,063	60,868	50,297	42,855	37,331	33,069	26,921	22,701	19,625	17,283	15,440	13,953	12,726	10,434	8,841	7,671	6,774	6,065	5,490	4,615	3,981	3,500	2,819	2,359
75	81,027	63,314	51,957	44,054	38,238	33,778	27,390	23,033	19,873	17,475	15,593	14,077	12,830	10,504	8,891	7,708	6,803	6,088	5,509	4,629	3,991	3,508	2,824	2,363
80	84,846	65,623	53,501	45,159	39,068	34,424	27,813	23,332	20,094	17,646	15,729	14,188	12,922	10,565	8,936	7,741	6,829	6,109	5,526	4,641	4,000	3,514	2,828	2,366
85	88,529	67,804	54,942	46,181	39,830	35,015	28,197	23,602	20,294	17,800	15,851	14,288	13,005	10,620	8,975	7,771	6,852	6,127	5,541	4,651	4,008	3,520	2,832	2,369
90	92,080	69,868	56,289	47,130	40,534	35,558	28,548	23,847	20,475	17,939	15,962	14,377	13,079	10,670	9,010	7,797	6,872	6,143	5,554	4,661	4,015	3,526	2,835	2,371
95	95,509	71,824	57,552	48,012	41,185	36,057	28,869	24,071	20,640	18,065	16,062	14,458	13,146	10,714	9,042	7,821	6,891	6,158	5,566	4,669	4,021	3,531	2,839	2,373
100	98,821	73,681	58,738	48,835	41,789	36,520	29,165	24,276	20,791	18,181	16,153	14,532	13,207	10,755	9,071	7,843	6,907	6,171	5,577	4,677	4,027	3,535	2,841	2,375
105	102,021	75,446	59,855	49,604	42,350	36,948	29,437	24,464	20,929	18,286	16,236	14,599	13,262	10,791	9,097	7,862	6,923	6,184	5,587	4,684	4,032	3,539	2,844	2,377
110	105,116	77,125	60,907	50,324	42,874	37,346	29,690	24,638	21,056	18,383	16,312	14,661	13,313	10,825	9,121	7,880	6,936	6,195	5,596	4,690	4,036	3,543	2,846	2,379
115	108,111	78,725	61,900	51,000	43,364	37,717	29,924	24,799	21,173	18,473	16,383	14,718	13,360	10,856	9,143	7,896	6,949	6,205	5,604	4,696	4,041	3,546	2,848	2,380
120	111,010	80,251	62,840	51,636	43,823	38,064	30,142	24,949	21,282	18,555	16,448	14,770	13,403	10,885	9,163	7,911	6,961	6,214	5,612	4,701	4,045	3,549	2,850	2,382
125	113,818	81,708	63,730	52,236	44,254	38,389	30,345	25,088	21,383	18,632	16,508	14,819	13,443	10,911	9,182	7,925	6,972	6,223	5,619	4,706	4,048	3,552	2,852	2,383
130	116,538	83,101	64,574	52,802	44,660	38,693	30,535	25,218	21,478	18,704	16,564	14,864	13,480	10,936	9,199	7,938	6,982	6,231	5,626	4,711	4,052	3,554	2,854	2,384
135	119,176	84,434	65,376	53,336	45,042	38,980	30,713	25,339	21,566	18,770	16,617	14,906	13,515	10,958	9,215	7,950	6,991	6,238	5,632	4,715	4,055	3,557	2,855	2,385
140	121,735	85,710	66,138	53,843	45,403	39,250	30,880	25,453	21,648	18,833	16,665	14,946	13,547	10,980	9,230	7,961	6,999	6,245	5,637	4,719	4,058	3,559	2,857	2,386
145	124,218	86,934	66,864	54,323	45,744	39,504	31,038	25,560	21,725	18,891	16,711	14,982	13,578	10,999	9,244	7,972	7,008	6,251	5,642	4,722	4,060	3,561	2,858	2,387
150	126,629	88,108	67,557	54,779	46,067	39,745	31,186	25,660	21,798	18,946	16,754	15,017	13,606	11,018	9,257	7,982	7,015	6,257	5,647	4,726	4,063	3,563	2,860	2,388
155	128,970	89,235	68,217	55,213	46,373	39,973	31,326	25,755	21,866	18,998	16,794	15,049	13,632	11,035	9,269	7,991	7,022	6,263	5,652	4,729	4,065	3,565	2,861	2,389
160	131,245	90,318	68,849	55,626	46,664	40,189	31,458	25,844	21,931	19,046	16,832	15,080	13,658	11,052	9,281	7,999	7,029	6,268	5,656	4,732	4,067	3,567	2,862	2,390
165	133,457	91,360	69,452	56,019	46,940	40,394	31,584	25,929	21,991	19,092	16,868	15,108	13,681	11,067	9,292	8,007	7,035	6,273	5,660	4,735	4,070	3,568	2,863	2,390
170	135,607	92,363	70,030	56,395	47,204	40,589	31,703	26,009	22,049	19,136	16,902	15,136	13,703	11,082	9,302	8,015	7,041	6,278	5,664	4,738	4,072	3,570	2,864	2,391
175	137,700	93,329	70,584	56,753	47,455	40,774	31,816	26,085	22,104	19,177	16,934	15,161	13,724	11,096	9,312	8,022	7,046	6,282	5,668	4,740	4,073	3,571	2,865	2,392
180	139,736	94,259	71,115	57,096	47,694	40,951	31,923	26,157	22,155	19,216	16,965	15,186	13,744	11,109	9,321	8,029	7,052	6,286	5,671	4,742	4,075	3,572	2,866	2,392
185	141,718	95,157	71,625	57,424	47,923	41,119	32,026	26,226	22,205	19,253	16,993	15,209	13,763	11,121	9,330	8,036	7,057	6,290	5,674	4,745	4,077	3,574	2,866	2,393
190	143,649	96,024	72,115	57,739	48,142	41,280	32,123	26,291	22,252	19,288	17,021	15,231	13,781	11,133	9,338	8,042	7,061	6,294	5,677	4,747	4,078	3,575	2,867	2,393
195	145,529	96,861	72,586	58,040	48,351	41,434	32,216	26,354	22,296	19,321	17,047	15,252	13,798	11,144	9,346	8,048	7,066	6,298	5,680	4,749	4,080	3,576	2,868	2,394
200	147,362	97,669	73,039	58,330	48,552	41,581	32,305	26,413	22,339	19,353	17,072	15,272	13,815	11,155	9,353	8,053	7,070	6,301	5,683	4,751	4,081	3,577	2,869	2,394

750 MCM	1 Run(s)	120 Volts	C-Value = 24,136	Magnetic Duct	Copper

Calculations:

1. $f = \dfrac{2 \times \text{Length} \times \text{Isca} \times 1.5}{\text{Runs} \times \text{C-Value} \times E_{L\text{-}N}}$

2. $M = \dfrac{1}{1 + f}$

3. Isc = Isca x 1.5 x M

4. Add Motor Contribution = Motor FLA x 4

* All results are given in symmetrical amperes

Single-Phase L-N Bolted Fault-Current* Table

One-Way Distance in Feet

Available Isc in Thousands (Isca)	5	10	15	20	25	30	40	50	60	70	80	90	100	125	150	175	200	225	250	300	350	400	500	600
3	4,434	4,370	4,308	4,248	4,189	4,132	4,023	3,919	3,820	3,726	3,637	3,552	3,470	3,283	3,114	2,962	2,824	2,699	2,584	2,381	2,208	2,058	1,812	1,619
5	7,319	7,147	6,982	6,825	6,675	6,531	6,261	6,013	5,784	5,571	5,374	5,190	5,018	4,635	4,306	4,021	3,771	3,550	3,354	3,020	2,746	2,518	2,160	1,891
7	10,149	9,820	9,512	9,223	8,951	8,694	8,223	7,800	7,418	7,073	6,757	6,469	6,205	5,629	5,151	4,748	4,403	4,105	3,845	3,413	3,067	2,786	2,353	2,037
10	14,293	13,650	13,062	12,523	12,026	11,568	10,748	10,037	9,414	8,864	8,374	7,936	7,541	6,708	6,040	5,493	5,037	4,651	4,320	3,781	3,362	3,027	2,523	2,163
15	20,946	19,593	18,405	17,352	16,413	15,570	14,121	12,918	11,904	11,038	10,289	9,635	9,060	7,883	6,976	6,257	5,672	5,187	4,778	4,128	3,634	3,245	2,673	2,272
20	27,300	25,046	23,136	21,496	20,074	18,828	16,749	15,083	13,719	12,581	11,617	10,791	10,074	8,639	7,562	6,724	6,053	5,504	5,046	4,326	3,786	3,366	2,755	2,331
25	33,374	30,066	27,355	25,092	23,175	21,530	18,854	16,769	15,100	13,732	12,592	11,627	10,799	9,167	7,964	7,040	6,308	5,714	5,222	4,455	3,884	3,444	2,806	2,368
30	39,187	34,703	31,141	28,241	25,836	23,808	20,578	18,120	16,186	14,625	13,339	12,261	11,344	9,557	8,256	7,267	6,490	5,863	5,346	4,545	3,953	3,497	2,842	2,393
35	44,754	39,000	34,557	31,023	28,144	25,755	22,016	19,225	17,063	15,337	13,929	12,757	11,767	9,856	8,478	7,439	6,626	5,974	5,438	4,612	4,003	3,536	2,868	2,412
40	50,092	42,992	37,655	33,497	30,166	27,437	23,234	20,148	17,785	15,918	14,406	13,157	12,107	10,092	8,653	7,573	6,732	6,060	5,510	4,663	4,041	3,566	2,887	2,426
45	55,214	46,711	40,478	35,712	31,951	28,906	24,279	20,929	18,391	16,402	14,801	13,485	12,384	10,285	8,794	7,681	6,817	6,129	5,566	4,703	4,072	3,590	2,903	2,437
50	60,132	50,184	43,060	37,707	33,538	30,199	25,185	21,598	18,906	16,810	15,133	13,760	12,616	10,444	8,910	7,769	6,887	6,185	5,613	4,736	4,097	3,609	2,915	2,445
55	64,860	53,434	45,431	39,513	34,959	31,347	25,978	22,179	19,349	17,160	15,416	13,994	12,812	10,578	9,007	7,843	6,945	6,232	5,651	4,764	4,117	3,625	2,926	2,453
60	69,407	56,483	47,616	41,156	36,239	32,372	26,678	22,687	19,735	17,463	15,660	14,194	12,979	10,692	9,090	7,905	6,994	6,271	5,684	4,787	4,134	3,638	2,934	2,459
65	73,784	59,348	49,637	42,656	37,397	33,293	27,300	23,136	20,074	17,727	15,872	14,369	13,125	10,791	9,161	7,959	7,036	6,305	5,711	4,806	4,149	3,650	2,942	2,464
70	78,000	62,046	51,510	44,032	38,451	34,125	27,857	23,535	20,373	17,961	16,059	14,521	13,253	10,877	9,223	8,006	7,073	6,334	5,735	4,823	4,162	3,660	2,948	2,468
75	82,064	64,590	53,251	45,299	39,413	34,881	28,359	23,892	20,640	18,168	16,224	14,657	13,365	10,952	9,277	8,047	7,104	6,360	5,756	4,838	4,173	3,668	2,954	2,472
80	85,984	66,994	54,875	46,468	40,295	35,570	28,813	24,213	20,880	18,353	16,372	14,777	13,465	11,019	9,325	8,083	7,133	6,382	5,775	4,851	4,182	3,676	2,959	2,476
85	89,768	69,269	56,392	47,551	41,107	36,201	29,226	24,504	21,096	18,520	16,504	14,885	13,554	11,079	9,368	8,115	7,158	6,402	5,791	4,863	4,191	3,682	2,963	2,479
90	93,422	71,425	57,812	48,558	41,857	36,782	29,603	24,768	21,291	18,670	16,624	14,982	13,635	11,133	9,407	8,144	7,180	6,420	5,806	4,873	4,199	3,688	2,967	2,481
95	96,954	73,471	59,145	49,495	42,552	37,317	29,948	25,010	21,470	18,807	16,732	15,070	13,708	11,181	9,441	8,170	7,200	6,436	5,819	4,882	4,205	3,693	2,970	2,484
100	100,368	75,415	60,399	50,369	43,196	37,812	30,266	25,231	21,632	18,932	16,831	15,150	13,774	11,225	9,473	8,193	7,218	6,451	5,831	4,891	4,212	3,698	2,973	2,486
105	103,671	77,264	61,579	51,188	43,797	38,271	30,560	25,435	21,782	19,047	16,922	15,223	13,835	11,266	9,501	8,215	7,235	6,464	5,842	4,898	4,217	3,703	2,976	2,488
110	106,869	79,027	62,693	51,955	44,358	38,699	30,832	25,623	21,920	19,152	17,005	15,290	13,890	11,302	9,527	8,234	7,250	6,476	5,852	4,905	4,222	3,707	2,979	2,490
115	109,965	80,707	63,747	52,676	44,882	39,097	31,084	25,797	22,047	19,249	17,081	15,352	13,941	11,336	9,551	8,252	7,264	6,487	5,861	4,912	4,227	3,710	2,981	2,491
120	112,966	82,312	64,743	53,355	45,374	39,470	31,320	25,959	22,165	19,339	17,152	15,409	13,988	11,367	9,573	8,269	7,277	6,497	5,869	4,917	4,232	3,713	2,983	2,493
125	115,875	83,846	65,689	53,996	45,836	39,819	31,539	26,110	22,275	19,422	17,218	15,462	14,032	11,396	9,594	8,284	7,289	6,507	5,877	4,923	4,235	3,717	2,985	2,494
130	118,696	85,313	66,586	54,600	46,271	40,147	31,744	26,250	22,377	19,500	17,279	15,511	14,072	11,423	9,613	8,298	7,300	6,516	5,884	4,928	4,239	3,719	2,987	2,495
135	121,434	86,718	67,439	55,173	46,682	40,456	31,937	26,382	22,473	19,573	17,336	15,557	14,110	11,448	9,630	8,311	7,310	6,524	5,890	4,932	4,243	3,722	2,989	2,497
140	124,091	88,065	68,250	55,715	47,069	40,747	32,118	26,505	22,562	19,640	17,389	15,600	14,145	11,471	9,647	8,323	7,319	6,531	5,896	4,937	4,246	3,724	2,990	2,498
145	126,672	89,357	69,024	56,229	47,436	41,021	32,288	26,621	22,646	19,704	17,438	15,640	14,178	11,492	9,662	8,335	7,328	6,538	5,902	4,941	4,249	3,727	2,992	2,499
150	129,180	90,598	69,762	56,718	47,783	41,281	32,449	26,730	22,725	19,764	17,485	15,678	14,209	11,513	9,676	8,345	7,336	6,545	5,907	4,945	4,251	3,729	2,993	2,500
155	131,618	91,790	70,467	57,183	48,113	41,526	32,600	26,833	22,799	19,820	17,529	15,713	14,238	11,532	9,690	8,355	7,344	6,551	5,912	4,948	4,254	3,731	2,994	2,501
160	133,988	92,937	71,141	57,626	48,426	41,759	32,744	26,930	22,869	19,873	17,571	15,746	14,265	11,550	9,702	8,365	7,351	6,557	5,917	4,951	4,257	3,733	2,995	2,501
165	136,294	94,040	71,785	58,048	48,724	41,981	32,880	27,022	22,936	19,923	17,610	15,778	14,291	11,566	9,714	8,374	7,358	6,562	5,922	4,954	4,259	3,734	2,997	2,502
170	138,538	95,103	72,403	58,451	49,008	42,191	33,009	27,109	22,998	19,970	17,647	15,807	14,315	11,582	9,726	8,382	7,364	6,567	5,926	4,957	4,261	3,736	2,998	2,503
175	140,722	96,127	72,995	58,837	49,278	42,392	33,131	27,191	23,058	20,015	17,681	15,835	14,338	11,597	9,736	8,390	7,370	6,572	5,930	4,960	4,263	3,738	2,999	2,504
180	142,849	97,115	73,563	59,205	49,537	42,583	33,248	27,270	23,114	20,057	17,715	15,862	14,360	11,612	9,746	8,397	7,376	6,577	5,933	4,963	4,265	3,739	3,000	2,504
185	144,922	98,069	74,109	59,558	49,784	42,765	33,359	27,345	23,168	20,098	17,746	15,887	14,381	11,625	9,756	8,404	7,382	6,581	5,937	4,965	4,267	3,741	3,001	2,505
190	146,941	98,989	74,634	59,896	50,020	42,939	33,465	27,416	23,219	20,136	17,776	15,911	14,400	11,638	9,765	8,411	7,387	6,585	5,940	4,967	4,268	3,742	3,001	2,506
195	148,910	99,879	75,138	60,221	50,246	43,106	33,566	27,483	23,267	20,173	17,804	15,934	14,419	11,650	9,773	8,417	7,392	6,589	5,943	4,970	4,270	3,743	3,002	2,506
200	150,829	100,739	75,624	60,533	50,462	43,265	33,662	27,548	23,314	20,207	17,832	15,956	14,437	11,662	9,782	8,423	7,396	6,593	5,946	4,972	4,272	3,744	3,003	2,507

1000 MCM	1 Run(s)	120 Volts

C-Value = 25,278	Magnetic Duct	Copper

Calculations:

1. $f = \dfrac{2 \times \text{Length} \times \text{Isca} \times 1.5}{\text{Runs} \times \text{C-Value} \times \text{E L-N}}$

2. $M = \dfrac{1}{1 + f}$

3. $\text{Isc} = \text{Isca} \times 1.5 \times M$

4. Add Motor Contribution = Motor FLA x 4

* All results are given in symmetrical amperes

ABOUT THE AUTHOR

V. F. Christoffer is project manager with ARJO Engineers, a mechanical/electrical engineering consulting firm in Dallas, Texas. He previously worked as a facilities project electrical engineer with Vought Aircraft, Hercules, Texas Instruments, IBM, and other companies.

SOFTWARE AND INFORMATION LICENSE

The software and information on this diskette (collectively referred to as the "Product") are the property of The McGraw-Hill Companies, Inc. ("McGraw-Hill") and are protected by both United States copyright law and international copyright treaty provision. You must treat this Product just like a book, except that you may copy it into a computer to be used and you may make archival copies of the Products for the sole purpose of backing up our software and protecting your investment from loss.

By saying "just like a book," McGraw-Hill means, for example, that the Product may be used by any number of people and may be freely moved from one computer location to another, so long as there is no possibility of the Product (or any part of the Product) being used at one location or on one computer while it is being used at another. Just as a book cannot be read by two different people in two different places at the same time, neither can the Product be used by two different people in two different places at the same time (unless, of course, McGraw-Hill's rights are being violated).

McGraw-Hill reserves the right to alter or modify the contents of the Product at any time.

This agreement is effective until terminated. The Agreement will terminate automatically without notice if you fail to comply with any provisions of this Agreement. In the event of termination by reason of your breach, you will destroy or erase all copies of the Product installed on any computer system or made for backup purposes and shall expunge the Product from your data storage facilities.

LIMITED WARRANTY

McGraw-Hill warrants the physical diskette(s) enclosed herein to be free of defects in materials and workmanship for a period of sixty days from the purchase date. If McGraw-Hill receives written notification within the warranty period of defects in materials or workmanship, and such notification is determined by McGraw-Hill to be correct, McGraw-Hill will replace the defective diskette(s). Send request to:

Customer Service
McGraw-Hill
Gahanna Industrial Park
860 Taylor Station Road
Blacklick, OH 43004-9615

The entire and exclusive liability and remedy for breach of this Limited Warranty shall be limited to replacement of defective diskette(s) and shall not include or extend to any claim for or right to cover any other damages, including but not limited to, loss of profit, data, or use of the software, or special, incidental, or consequential damages or other similar claims, even if McGraw-Hill has been specifically advised as to the possibility of such damages. In no event will McGraw-Hill's liability for any damages to you or any other person ever exceed the lower of suggested list price or actual price paid for the license to use the Product, regardless of any form of the claim.

THE McGRAW-HILL COMPANIES, INC. SPECIFICALLY DISCLAIMS ALL OTHER WARRANTIES, EXPRESS OR IMPLIED, INCLUDING BUT NOT LIMITED TO, ANY IMPLIED WARRANTY OF MERCHANTABILITY OR FITNESS FOR A PARTICULAR PURPOSE. Specifically, McGraw-Hill makes no representation or warranty that the Product is fit for any particular purpose and any implied warranty of merchantability is limited to the sixty day duration of the Limited Warranty covering the physical diskette(s) only (and not the software or in-formation) and is otherwise expressly and specifically disclaimed.

This Limited Warranty gives you specific legal rights; you may have others which may vary from state to state. Some states do not allow the exclusion of incidental or consequential damages, or the limitation on how long an implied warranty lasts, so some of the above may not apply to you.

This Agreement constitutes the entire agreement between the parties relating to use of the Product. The terms of any purchase order shall have no effect on the terms of this Agreement. Failure of McGraw-Hill to insist at any time on strict compliance with this Agreement shall not constitute a waiver of any rights under this Agreement. This Agreement shall be construed and governed in accordance with the laws of New York. If any provision of this Agreement is held to be contrary to law, that provision will be enforced to the maximum extent permissible and the remaining provisions will remain in force and effect.